CW00435460

THERAPEUTIC MONOCLONAL ANTIBODIES

THERAPEUTIC MONOCLONAL ANTIBODIES

FROM BENCH TO CLINIC

Edited by

Zhiqiang An

A JOHN WILEY & SONS, INC., PUBLICATION

Published by John Wiley & Sons, Inc., Hoboken, New Jersey
Published simultaneously in Canada

For general information on our other products and services or for technical support, please contact our Customer Care Department within the United States at (800) 762-2974, outside the United States at (317) 572-3993 or fax (317) 572-4002.

Wiley also publishes its books in a variety of electronic formats. Some content that appears in print may not be available in electronic formats. For more information about Wiley products, visit our web site at www.wiley.com.

Library of Congress Cataloging-in-Publication Data:

Therapeutic monoclonal antibodies: from the bench to the clinic / [edited by] Zhiqiang An.
p.; cm.
Includes bibliographical references and index.
ISBN 978-0-470-11791-0 (cloth)
1. Monoclonal antibodies—Therapeutic use. I. An, Zhiqiang, Dr.
[DNLM: 1. Antibodies, Monoclonal—therapeutic use. 2. Drug Discovery.
QW 575.5.A6 T398 2009]

RM282.M65T495 2009
616.07′98—dc22 2008053435

Printed in the United States of America

10 9 8 7 6 5 4 3 2 1

CONTENTS

FOREWORD

In the latter part of the 19th century Emil von Behring and Shibasaburo Kitasato showed that serum from human patients (or animals, typically horses) who had recovered from an infectious disease (typhus, diphtheria, etc) could be used to prevent or treat the same disease in other humans (indeed hyperimmune horse serum is still used to treat diphtheria today). Hyperimmune globulins obtained from human donors are used to treat a variety of infectious diseases today. However its use is restricted by availability and limited potency. For more than a century the widespread use of antibodies for treatment of a variety of diseases has awaited a practical method for production of specific antibodies, as well as the identification of the specific targets associated with a particular disease. Today many of those limitations have been resolved, and antibody therapy is the most active field in therapeutics.

In Köhler and Milstein's classic 1975 Nature paper, the authors state that "the manufacture of predefined specific antibodies by means of permanent tissue culture cell lines is of general interest", and after demonstrating convincingly that this could be accomplished by fusing (mouse) spleen cells with multiple myeloma cells, they conclude "such cells can be grown *in vitro* in massive cultures to provide specific antibody" and "such cultures could be valuable for medical and industrial use". Today, we can attest to the validity of those prescient remarks. It unleashed an avalanche of scientific and commercial interest. In my view it represented, in both practical and heuristic terms, the most significant methodological advance toward the treatment (not to mention diagnosis and prevention) of human disease in the past century. It initiated the era of biological therapeutics.

The first antibody for clinical use (OKT3), for tissue rejection, was approved 11 years after the Köhler and Milstein paper. Initially the field developed slowly–partially because of the time required to develop methods for "humanizing" the mouse monoclonals and to develop manufacturing processes and capability, but also significantly because of the widely held view by many pharmaceutical scientists/executives that antibodies were "transitional therapeutic products" and would be ultimately replaced by small molecules. With time it became clear that there were advantages to these macromolecules: the size could confer longer half lives, the problems of antigenicity could be limited by more effective humanization procedures. The inherent great specificity for the epitopes could be used to limit side effects, and provide progressively greater potency, and finally manufacturing processes were improving rapidly. Today, antibodies are always considered as a possible if not preferred therapeutic modality when feasible (extracellular, or cell surface targets, but also some modes for addressing intracellular targets).

In 2000, nine of the top 10 molecules were small molecule drugs and one (Epogen/Procrit) was a recombinant protein product. By 2008, the situation was dramatically different: three monoclonal antibody products (Rituxan, Remicade, and Avastin) were in the top 10, along with two other recombinant protein products (Enbrel and Epogen/Procrit). By 2014, it is predicted that five of the top 10 products will be antibodies (Avastin, Humira, Rituxan, Herceptin, and Remicade), along with two other recombinant protein products (Enbrel and Lantus). Furthermore about half of new therapeutic products under development are antibodies. It is clear that most of the high value products will be antibodies . . . but the largest share of the therapeutic market will be small molecules. The game is not yet over!

The opportunity for innovation exists at every level. The challenge is to develop ever increasing potency while decreasing the cost of development and production. This book comprehensively addresses the technology and the development of antibody therapeutics. It provides both basic and sophisticated information. It should be of great interest to scientists and executives in the biotech and pharmaceutical industry, and academic scientists who are interested in meeting that challenge for the benefit of healthcare, worldwide.

William J. Rutter

PREFACE

The study of antibodies has been a focal point in modern biology and medicine since the early 1900s. However, the ability to use antibodies as weapons against diseases or as tools to study disease state was mostly confined to crude, undefined preparations until César Milstein and Georges Köhler developed methods for the isolation of monoclonal antibodies from hybridoma cells in 1975. Since then, antibodies have not only been used as subjects and tools for breakthrough basic research, but have also been used as clinical diagnostics, reagents for high throughput drug screening, and most importantly, as life-saving medicines.

Progress in the therapeutic antibody field was initially slow and intermittent. The first therapeutic antibody, murine-derived Murononab OKT3 for acute organ rejection, was approved by the FDA in 1986, more than a decade after the discovery of the hybridoma technology. As a result of technological breakthroughs in the 1980s and 1990s, progress in the therapeutic antibody field accelerated dramatically (Chapter 1). This book provides readers with a comprehensive review of the history and tools of discovery, development, characterization, and clinical application of therapeutic antibodies.

An antibody contains two light chains and two heavy chains, which are linked by multiple disulphide bonds (Chapter 2). The antigen-binding complementarity-determining regions (CDRs) are short, hypervariable amino acid sequences found in the variable domains of both light and heavy chains. The binding affinity and specificity of an antibody to its antigen can be readily manipulated by *in vitro* genetic engineering approaches (Chapter 13). Powerful bioinformatics tools are being developed to annotate the genetic diversity of antibodies (Chapter 4).

After binding to a target, the fragment crystallizable region (Fc region) of an antibody can recruit effector cells such as natural killer cells, macrophages, or neutrophils, and/or activate the complement system to destroy the target-associated cells. These properties, referred to as "antibody-dependent cell cytotoxicity" (ADCC) and "complement-dependent cytotoxicity" (CDC), respectively, are fundamental aspects of natural antibody biology that are being manipulated to create therapeutics with more potent biological activities (Chapter 16). In addition to ADCC and CDC activities, the Fc region of an antibody is also responsible for the long half-life of the molecule through its interaction with the neonatal receptor FcRn (Chapter 19). Finally, the Fc domain has interactions with certain bacterial proteins such as Protein A/G, which demonstrate the power of evolution in the interaction between microorganisms and the molecules made by the body to defend against them (Chapter 17).

Like many other mammalian proteins, antibodies are glycoproteins. Glycosylation plays an important role in the biological activities of antibodies and manipulation of the glycosylation pattern of an antibody has been applied to the improvement of pharmaceutical properties of the molecule (Chapters 3 and 26). Genetic manipulation of the Fc region of an antibody has also been utilized to improve the serum half-life, ADCC, and CDC activities of the molecule (Chapters 14 and 15).

One of the major sources of therapeutic antibodies is monoclonal antibodies isolated from immunized animals using hybridoma technology (Chapters 5 and 6). Monoclonal antibodies isolated from wild-type animals, such as murine species, induce immunological responses in humans. To reduce this response, monoclonal antibodies are commonly modified and produced as murine/human chimeric antibodies or humanized antibodies for therapeutic applications (Chapters 1, 13, and 31). In addition, fully human monoclonal antibodies can be generated in transgenic mice to circumvent the immunogenicity issue of murine sequence (Chapter 5). Phage-displayed antibody libraries represent another source of fully human antibodies (Chapters 7 and 8). In addition to phage, antibody fragments or, in some cases, full IgG molecules can also be displayed on yeast (Chapter 9), bacteria (Chapter 11), mammalian cells (Chapter 12), and on other *in vitro* systems such as ribosomes (Chapter 10).

Most therapeutic antibodies are full length IgG molecules (Chapter 31). In addition to IgGs, antibody fragments have also been developed as therapeutics (Chapters 27, 31, and 33) and as imaging reagents (Chapter 20). Monoclonal antibodies have been used as tissue targeting reagents as well; there are many examples of antibodies used as targeting agents for small molecule toxins (Chapter 35) or radiolabeled isotopes (Chapter 34).

The history of therapeutic antibody development parallels the desire of the industry to reduce the potential immunogenicity of the drugs. *In silico* tools have been developed; to analyze antibody sequences to be humanized (Chapter 4) and to predict the immunogenicity potential of antibodies before they are tested in the clinic (Chapter 18).

The manufacturing of therapeutic monoclonal antibodies has been, to date, an expensive proposition. A large scale facility can take multiple years to build, at a cost of several hundreds of million dollars. Mammalian cell culture (Chinese hamster ovary cells; CHO) is the dominant production cell platform for antibody therapeutics (Chapter 25). Other exploratory methods of antibody production include the use of plants (Chapter 29), transgenic animals (milk), eggs (Chapter 28), and yeast (Chapter 26). An antibody fragment made in a bacterial cell line was approved for clinical use in 2008 (Chapter 27).

Antibodies can engage a wide range of extracellular drug targets such as membrane bound proteins or circulating ligands and cytokines, but they do not readily cross cell membranes or the brain blood barrier (BBB). Efforts are being made to facilitate the transfer of antibodies across cell membranes and the BBB (Chapter 21). Unlike small-molecule drugs, monoclonal antibodies are large, complex molecules that are not easily formulated and delivered (Chapter 30). Additionally, antibody therapeutics are produced as heterogeneous mixtures of molecules including different glycoforms that can vary slightly in molecular structure (Chapter 24). Complex analytical tools have been developed and optimized for the molecular and functional characterization of antibody therapeutics (Chapters 22, 23, and 24).

The complex nature of antibodies, as mentioned above, has contributed to the lack of a consensus regarding the definition of generic biopharmaceuticals. Multiple terms are used to describe generic biopharmaceuticals, such as biogenerics, biosimilars, and follow-on biologies (Chapter 32). The development of follow-on (or biosimilar) antibody therapeutics will be expensive as compared with small molecule generics, as it is highly likely that regulatory authorities will require that clinical trials be run to provide comparability data. Experts predict that it will take at least a decade before technology is advanced to a stage whereby the safety and bioequivalence of a biosimilar can be verified without clinical testing. Despite the regulatory and technological barriers to the development of generic biopharmaceuticals, it is certain that therapeutic antibodies will eventually face "generic" competition.

It seems fitting that this book should be written a hundred years after Paul Ehrlich received the Nobel Prize in 1908 for his studies in medicine, hematology, immunology, and chemotherapy. It was Ehrlich who popularized the term "magic bullet" which was an apt and prescient description for many of the therapeutic monoclonal antibodies on the market or in development today.

In closing, I would like to express my gratitude to Ms. Anita Lekhwani for the opportunity to edit this book and I am indebted to the expert authors who contributed to this endeavor. I want to thank Dr. William R. Strohl for his input on the project. I also want to thank Mr. Nick Barber for his assistance during the production stage of the project and Ms. Michelle Snider for typing the index words. Finally, I want to thank my family for their patience and support throughout this complex undertaking.

ZHIQIANG AN

CONTRIBUTORS

Jodie Abrahams, National Institute for Bioprocessing Research and Training, Belfield, Dublin, Ireland

Stephen C. Alley, Seattle Genetics, Inc., Bothell, WA

Juan C. Almagro, Centocor R&D, Inc., Radnor, PA

Zhiqiang An, Epitomics, Burlingame, CA

Dennis Benjamin, Seattle Genetics, Inc., Bothell, WA

Alan J. Bitonti, Syntonix Pharmaceuticals, Waltham, MA

Leigh Bowering, UCB-Celltech, Slough, Berkshire, UK

Lorenzo Chen, Merck Research Laboratories, West Point, PA

Hung-Wei Chih, Genentech, Inc., South San Francisco, CA

Chen-Ni Chin, Merck Research Laboratories, West Point, PA

Jennifer R. Cochran, Stanford University, Stanford, CA

Neal Connors, Merck Research Laboratories, Rahway, NJ

Leslie Cope, Merck Research Laboratories, West Point, PA

Kevin M. Cox, Biolex Therapeutics, Pittsboro, NC

William F. Dall'Acqua, MedImmune, Inc., Gaithersburg, MD

Ann L. Daugherty, Genentech, Inc., South San Francisco, CA

Anne S. De Groot, Brown University and EpiVax, Inc., Providence, RI

John R. Desjarlais, Xencor, Inc., Monrovia, CA

Lynn F. Dickey, Biolex Therapeutics, Pittsboro, NC

Chaitanya R. Divgi, University of Pennsylvania, Philadelphia, PA

Stefan Dübel, Technical University of Braunschweig, Braunschweig, Germany

Jennifer A. Dumont, Syntonix Pharmaceuticals, Waltham, MA

Robin E. Ernst, Merck Research Laboratories, West Point, PA

Robert J. Etches, Origen Therapeutics, Burlingame, CA

Christopher R. Gibson, Merck Research Laboratories, West Point, PA

Tom R. Glass, Sapidyne Instruments Inc., Boise, Idaho

Si-Han Hai, Brown University, Providence, RI

William D. Hanley, Merck Research Laboratories, West Point, PA

Barrett R. Harvey, The University of Texas Health Sciences Center, Houston, TX

Katrina N. High, Merck Research Laboratories, West Point, PA

David P. Humphreys, UCB-Celltech, Slough, Berkshire, UK

Michael Hust, Technical University of Braunschweig, Braunschweig, Germany

Roy Jefferis, University of Birmingham, Edgbaston, Birmingham, UK

Angela R. Jones, University of Wisconsin, Madison, WI

Paul M. Knopf, Brown University and EpiVax, Inc., Providence, RI

Jennifer L. Lahti, Stanford University, Stanford, CA

Greg A. Lazar, Xencor, Inc., Monrovia, CA

Che-Leung Law, Seattle Genetics, Inc., Bothell, WA

Marie-Paule Lefranc, Institut de Génétique Humaine, Montpellier, France

Nils Lonberg, Medarex, Milpitas, California

Susan C. Low, Syntonix Pharmaceuticals, Waltham, MA

David Lowe, MedImmune Limited, Granta Park, Cambridge, UK

Ping Lu, Merck Research Laboratories, West Point, PA

Henryk Mach, Merck Research Laboratories, West Point, PA

William Martin, EpiVax, Inc., Providence, RI

Julie A. McMurry, EpiVax, Inc., Providence, RI

Tessie McNeely, Merck Research Laboratories, West Point, PA

Kileen L. Mershon, University of California, Los Angeles, CA

Yusuke Mimura, NHO Yamaguchi-Ube Medical Center, Ube, Japan

Yuka Mimura-Kimura, National Institute for Bioprocessing Research and Training, Belfield, Dublin, Ireland

Sherie L. Morrison, University of California, Los Angeles, CA

Randall J. Mrsny, Genentech, Inc., South San Francisco, CA

Juergen H. Nett, GlycoFi, Inc., Lebanon, NH

Tove Olafsen, University of California, Los Angeles, CA

Neeta Pandit-Taskar, Memorial Sloan-Kettering Cancer Center, New York, NY

Robert T. Peters, Syntonix Pharmaceuticals, Waltham, MA

Jeffrey T. Regan, Biolex Therapeutics, Pittsboro, NC

Pauline M. Rudd, National Institute for Bioprocessing Research and Training and Conway Institute, University College Dublin, Belfield, Dublin, Ireland

Punam Sandhu, Merck Research Laboratories, West Point, PA

Thomas Schirrmann, Technical University of Braunschweig, Braunschweig, Germany

Vikas K. Sharma, Genentech, Inc., South San Francisco, CA

Eric V. Shusta, University of Wisconsin, Madison, WI

Ernest S. Smith, Vaccinex, Inc., Rochester, NY

Terrance A. Stadheim, GlycoFi, Inc., Lebanon, NH

Robyn L. Stanfield, The Scripps Research Institute, La Jolla, CA

Jason D. Sterling, Biolex Therapeutics, Pittsboro, NC

William R. Strohl, Centocor R&D, Inc., Radnor, PA

Holger Thie, Technical University of Braunschweig, Braunschweig, Germany

George Thom, MedImmune Limited, Granta Park, Cambridge, UK

Thomas J. Van Blarcom, University of Texas at Austin, Austin, Texas

Tristan J. Vaughan, MedImmune Limited, Granta Park, Cambridge, UK

Fubao Wang, Merck Research Laboratories, West Point, PA

Yang Wang, Merck Research Laboratories, West Point, PA

Michael W. Washabaugh, Merck Research Laboratories, West Point, PA

Vincent P. M. Wingate, Biolex Therapeutics, Pittsboro, NC

Brent R. Williams, Merck Research Laboratories, West Point, PA

Ian A. Wilson, The Scripps Research Institute, La Jolla, CA

Anna M. Wu, University of California, Los Angeles, CA

Herren Wu, MedImmune, Inc., Gaithersburg, MD

Guo-Liang Yu, Epitomics, Burlingame, CA

Maurice Zauderer, Vaccinex, Inc., Rochester, NY

Ningyan Zhang, Merck Research Laboratories, West Point, PA

Qinjian Zhao, Merck Research Laboratories, West Point, PA

Lei Zhu, Origen Therapeutics, Burlingame, CA

Weimin Zhu, Epitomics, Burlingame, CA

PART I

ANTIBODY BASICS

CHAPTER 1

Therapeutic Monoclonal Antibodies: Past, Present, and Future

WILLIAM R. STROHL

ABSTRACT

In this chapter, an overview of the therapeutic antibody industry today, including the many commercial antibodies and Fc fusions and the rich clinical pipeline, is presented and analyzed. The long history of antibodies is given to bring context to the therapeutic antibody industry. This history includes serum therapy, the use of IVIG, and the evolution of those therapies into the development of the monoclonal

Therapeutic Monoclonal Antibodies: From Bench to Clinic. Edited by Zhiqiang An

antibody business as we know it today. The history of technologies that fostered the revolution of therapeutic antibody development in the 1990s is also described. Finally, the future of the therapeutic monoclonal antibody and Fc fusion business is presented along with opportunities and challenges facing the business and those who work in it.

1.1 INTRODUCTION

Protein therapeutics in general, and more specifically, therapeutic monoclonal antibodies (Mabs; Fig. 1.1) and Fc fusion proteins, have become a significant addition to the pharmaceutical repertoire over the past 20 years, and promise to play an even more significant role in the future of pharmaceutical intervention in diseases (Carter 2006; Riley 2006; Dimitrov and Marks 2008; Leader, Baca, and Golan 2008). In total, protein therapeutics produced by the BioPharm industry had over $55 billion in sales in 2005 (Table 1.1), approximately 20 percent of the roughly $280 billion 2005 pharmaceutical market. Based on the increase in value of protein therapeutics already on the market, therapeutic proteins are projected to reach about $94 billion by 2010 (Table 1.1), which calculates to an approximately 12 percent compound annual growth rate over that period. The 27 currently marketed monoclonal antibodies and Fc fusion proteins (see Table 1.2 for a complete listing) combine to make up 35 percent of the market value of all therapeutic proteins (based on 2006 data; Table 1.1), but are projected to increase in proportion by 2010, especially now that the market for epoetins has weakened based on safety concerns raised in mid-2007. Sales in 2006 for therapeutic Mabs and Fc fusion proteins topped $23 billion (Table 1.1), led by Enbrel®, Rituxan®, Remicade®, and Herceptin®, all of which were approved in the 1997–1998 time frame (Fig. 1.2). Of the six Mabs and Fc fusions brought to market in 1997–1998,

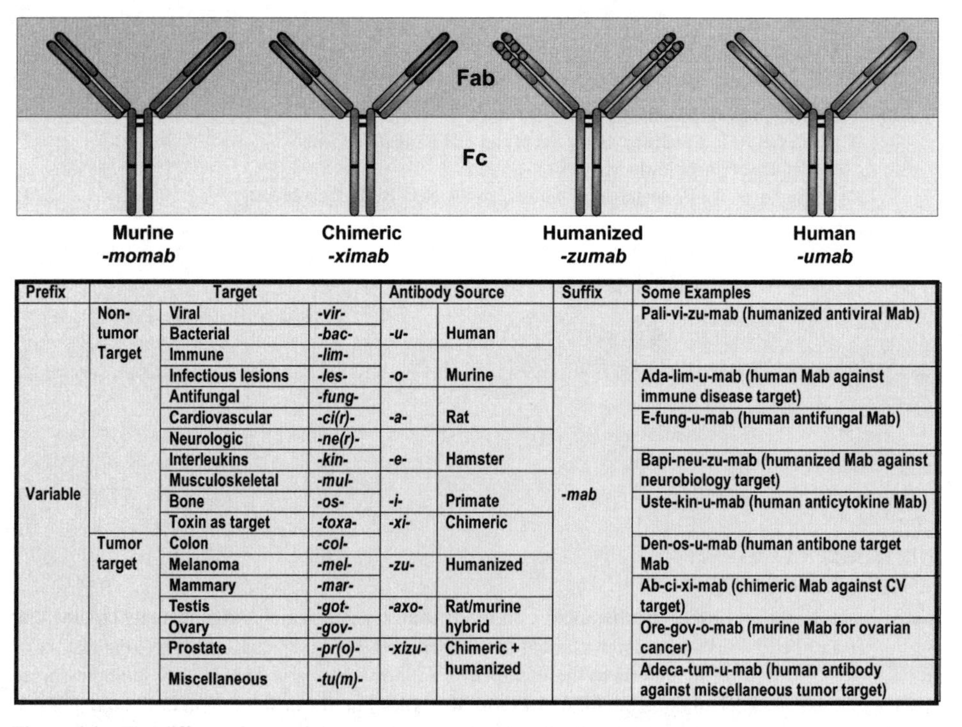

Prefix	Target			Antibody Source		Suffix	Some Examples
Variable	Non-tumor Target	Viral	*-vir-*			-mab	Pali-vi-zu-mab (humanized antiviral Mab)
		Bacterial	*-bac-*	*-u-*	Human		
		Immune	*-lim-*				
		Infectious lesions	*-les-*	*-o-*	Murine		Ada-lim-u-mab (human Mab against immune disease target)
		Antifungal	*-fung-*				
		Cardiovascular	*-ci(r)-*	*-a-*	Rat		E-fung-u-mab (human antifungal Mab)
		Neurologic	*-ne(r)-*				
		Interleukins	*-kin-*	*-e-*	Hamster		Bapi-neu-zu-mab (humanized Mab against neurobiology target)
		Musculoskeletal	*-mul-*				
		Bone	*-os-*	*-i-*	Primate		Uste-kin-u-mab (human anticytokine Mab)
		Toxin as target	*-toxa-*	*-xi-*	Chimeric		
	Tumor target	Colon	*-col-*				Den-os-u-mab (human antibone target Mab)
		Melanoma	*-mel-*	*-zu-*	Humanized		
		Mammary	*-mar-*				Ab-ci-xi-mab (chimeric Mab against CV target)
		Testis	*-got-*	*-axo-*	Rat/murine hybrid		
		Ovary	*-gov-*				Ore-gov-o-mab (murine Mab for ovarian cancer)
		Prostate	*-pr(o)-*	*-xizu-*	Chimeric + humanized		
		Miscellaneous	*-tu(m)-*				Adeca-tum-u-mab (human antibody against miscellaneous tumor target)

Figure 1.1 The different forms of therapeutic monoclonal antibodies that have been approved for marketing, including murine, chimeric, humanized, and fully human antibodies, as well as how the generic names are applied to each of them based on structure, source, and target. (See color insert.)

TABLE 1.1 Breakdown of Estimated Market for Protein Therapeutics Based on the Sales of the Top-Selling Biologics Drugs*

Category	2005 (in $B)	2006 (in $B)	2010 (Projected) (in $B)
Monoclonal antibodies and Fc fusions	17.3	23.1	41.2
Epoetins	11.2	12.0	12.8**
Insulin-related	7.6	9.0	13.0
Interferons	6.4	6.8	7.7
Antifibrinolytics	4.1	4.5	6.3
Immunostimulatory (xCSF)	3.9	4.3	5.5
Growth hormones	2.2	2.4	2.5
Other (mixed mechanisms)	2.9	3.2	4.7
Totals	55.6	65.3	93.7

*Based on published 2005 and 2006 sales, and 2010 projected sales of the top 70 biologics currently on the market, excluding vaccines (multiple sources).
**Projections were made prior to published safety concerns in mid-2007, which have depressed overall sales of epoetins.

four (Rituxan®, Remicade®, Enbrel®, and Herceptin®) are significant blockbusters, each with markets great than $3 billion (Fig. 1.2). Of the more recently approved Mabs, Humira®, Erbitux®, and Avastin®, all approved in the 2003–2004 time frame (Table 1.2, Fig. 1.2), had not yet hit their peak sales by 2006, the time at which these numbers were generated. Thus, it is expected that these successful new entries also will hit blockbuster status like some of their predecessors. The key inflection points for success in bringing these Mab-based biologics to market appear to be the late 1990s, 2003–2004, and the period 2007–2012 (Fig. 1.3), the latter being the period in which we are currently working. This near-term future inflection point is likely to be a direct result of the success of monoclonal antibodies marketed in the late 1990s, as well as a maturation of antibody engineering technologies and strategies to make more commercially successful biologics molecules.

At the time of this writing, Biologic License Application (BLAs) for four additional antibodies had been submitted for regulatory approval in the United States (Table 1.3), and an additional 30 Mabs and Fc fusion proteins are in advanced clinical trials (defined here as Phase III or entering into Phase III based on successful completion of Phase II clinical trials; Table 1.3). Between 2007 and 2012, the growth of marketed monoclonal antibodies promises to be extraordinary (Fig. 1.3). Although not probable, if all current Phase III candidates listed in Table 1.3 were to achieve registration, this would translate to over 60 monoclonal antibodies and Fc fusion proteins on the market by the 2012–2013 time frame (Fig. 1.3). Even if only 50 percent are successful in being marketed, that number still reaches 46, a 50 percent increase in numbers over the currently marketed antibodies and Fc fusion proteins (Fig. 1.3). With the current rate of success for monoclonal antibodies transitioning from Phase III to the market at 75 percent (KMR Group, Inc. 2007), this would translate into about 53 to 54 Mabs and Fc fusion proteins on the market by the 2012–2013 time frame. These additional marketed biologics should have a substantial impact on the pharmaceutical industry over the next five years. It has been projected that 60 percent of the total growth in the pharmaceutical industry between 2004 ($271 billion total) and 2010 ($317 billion) will be driven by biologics (Riley 2006), and the data presented herein (Table 1.1, Fig. 1.2) suggest that a significant fraction of that growth will be accounted for by Mab and Fc fusion proteins.

There are currently over 140 additional publicly stated, commercially funded monoclonal antibodies in early clinical trials (defined here as Phase I and Phase II candidates combined), many of which are listed in Table 1.4. With the probability of success (POS) for antibodies transitioning between Phase II and Phase III currently at about 62 percent (KMR Group, Inc. 2007), approximately 50 of the 84 Phase II candidates shown in Table 1.4 should result in Phase III candidates. Of the 58 known Phase I candidates listed in Table 1.4, 35 should transition to Phase II (based on 61 percent POS; KMR Group, Inc. 2007), and of those, 22 would be predicted to make it to Phase III based on the 62 percent POS for that transition. Thus, of the 142 early phase candidates listed in Table 1.4

TABLE 1.2 Marketed Monoclonal Antibodies and Fusion Proteins*

U.S. Trade Name (Generic Name)	Company	Approval Date (U.S.)	Molecular Target	Major Indication	Protein Format	Route & Form	Antibody Source	Production Cell Line
Orthoclone OKT3® (Muromonab-CD3)	Ortho Biotech (J&J)	06/19/1986	CD3 on T-cells	OTR	Murine IgG2a	IV, Liquid	Hybridoma	Hybridoma
ReoPro® (Abciximab)	Centocor (now J&J)/Lilly	12/22/1994	gPIIb/IIIa on platelets	CVD	Chimeric Fab	IV, Liquid	Hybridoma	*E. coli*
Rituxan® (Rituximab)	Biogen/Idec/ Genentech	11/26/1997	CD20 on B-cells	NHL, RA added 2/8/06	IgG1κ, Chimeric	IV, Liquid	Hybridoma	CHO
Zenapax® (Daclizumab)	PDL/Roche	12/10/1997	IL-2Rα (CD25; tac)	OTR	IgG1, Humanized	IV, Liquid	Hybridoma	NS0
Synagis® (Palivizumab)	MedImmune	06/19/1998	A-antigenic site of RSV F-protein	RSV (infant)	IgG1κ, Chimeric	IM, Lyo	Hybridoma	NS0
Remicade® (Infliximab)	Centocor (now J&J)	08/24/1998	TNF-α	RA	IgG1κ, Chimeric	IV, Lyo	Hybridoma	NS0
Herceptin® (Trastuzumab)	Genentech	09/25/1998	HER2/Neu	Breast cancer	IgG1κ, Humanized	IV, Lyo	Hybridoma	CHO
Enbrel® (Etanercept)	Immunex (now Amgen)	11/02/1998	TNF-α	RA	IgG1-Fc conjugated to p75exodomain of TNFR	SC, Lyo	Recombinant Fc fusion	CHO
Simulect® (Basiliximab)	Novartis	12/05/1998	IL-2Rα (CD25; tac)	OTR	IgG1κ, Chimeric	IV, Lyo	Hybridoma	SPII/0
Mylotarg® (Gemtuzumab ozogamicin)	Wyeth	05/17/2000	CD33	Leukemia	Humanized IgG4κ-Ozogamicin conjugate**	IV, Lyo	Hybridoma	NS0
Campath-1H® (Alemtuzumab)	ILEX/ Millenium	05/07/2001	CD52 on B- and T-cells	Leukemia	IgG1κ, Humanized	IV, Liquid	Hybridoma	CHO
Zevalin® (Ibritumomab tiuxetan)	Biogen/Idec	02/19/2002	CD20 on B-cells	NHL	Murine IgG1κ conjugate, Y-90 or In-111	IV, Liquid	Hybridoma	CHO

Humira® (Adalimumab)	CAT, Abbott	12/31/2002	TNF-α; Blocks interaction with p55 and p75	RA, Crohn disease	IgG1κ, Human	SC, Liquid	Phage display	HEK293
Amevive® (Alefacept)	Biogen	01/30/2003	CD2—inhibits CD2-LFA-3 interaction on activated T-cells	Psoriasis	CD2-binding domain of LFA-3::Fc fusion protein	IM/IV, Form not known	Recombinant Fc fusion	CHO
Xolair® (Omalizumab)	Genentech	06/20/2003	IgE	Asthma	IgG1κ, Humanized	SC, Lyo	Hybridoma	CHO
Raptiva® (Efalizumab)	Genentech	10/27/2003	CD11a, α-subunit of LFA-1; inhibits binding to ICAM-1	Psoriasis	IgG1κ, Humanized	SC, Lyo	Hybridoma	CHO
Bexxar® (Tositumomab-I131)	Corixa	06/27/2003	CD20 on B cells	NHL	Murine IgG2a/λ-I-131	IV, Liquid	Hybridoma	Mammalian
Erbitux® (Cetuximab)	ImClone/ BMS	02/12/2004	Binds EGF-R (HER1, c-ErbB-1)	Colorectal cancer	IgG1κ, Chimeric	IV, Liquid	Hybridoma	SPII/0
Avastin® (Bevacizumab)	Genentech	02/26/2004	VEGF (ligand)	Colorectal cancer	IgG1, Humanized	IV, Liquid	Hybridoma	CHO
Tysabri® (Natalizumab)	Biogen/Elan	11/23/2004***	α4 subunit of α4β1 or α4β7	Multiple sclerosis	IgG4 k, Humanized	IV, Liquid	Hybridoma	Murine myeloma
Orencia® (Abatacept)	BMS	12/23/2005	CD80/CD86 – T-cell ostimulatory	RA	CTLA4-Fc fusion protein	IV, Lyo	Recombinant Fc fusion	Mammalian
Lucentis® (Ranibizumab)	Genentech/ Novartis	06/30/2006	VEGF-A	Wet AMD	Humanized IgG1 k Fab fragment	Intravitreal injection	Hybridoma	*E. coli*
Vectibix® (Panitumumab)	Amgen	09/27/2006	EGFR	Colorectal cancer	Human IgG2 k	IV infusion, Lyo	Transgenic humanized mouse	CHO
Soliris® (Ecolizumab)	Alexion Pharma	03/16/2007	Complement C5	PNH (reduce hemolysis)	Humanized IgG2/4 hybrid	IV, Liquid	Hybridoma	Murine myeloma

(*Continued*)

TABLE 1.2 *Continued*

U.S. Trade Name (Generic Name)	Company	Approval Date (U.S.)	Molecular Target	Major Indication	Protein Format	Route & Form	Antibody Source	Production Cell Line
Arcalyst® (Rilonacept)	Regeneron	02/27/2008	IL-1	CAPS	Dimeric Fc fusion protein with IL-1R & IL-1 accessory protein in-line	Lyo	Recombinant Fc fusion	CHO
Cimzia® (Certolizumab pegol)	UCB/ Schwartz	04/22/2008	TNF-α	RA	PEGylated humanized Fab	SC	Hybridoma	*E. coli*
Nplate® (Romiplostim, AMG-531)	Amgen	08/22/08	TPO-R	Thrombocytopenia	Fc-peptide fusion (peptibody)	SC	Not applicable	*E. coli*

Abbreviations: OTR, organ transplantation rejection; CV, cardiovascular disease; RA, rheumatoid arthritis; RSV, respiratory syncytial virus; NHL, non-Hodgkin's lymphoma; TNF, tissue necrosis factor; PNH, paroxysmal nocturnal hemoglobinuria; AMD, age-related macular degeneration; CAPS, cropyrin-associated periodic syndrome; LYO, lyophilized; IV, intravenous; SC, subcutaneous; ND, not disclosed.

*Data obtained from prescribing information released by the manufacturers, company websites, Prous Science Integrity.

**Conjugate is ozogamicin, a calecheamycin (natural product cytotoxin).

***Suspended 2/28/05; reinstated under specified conditions.

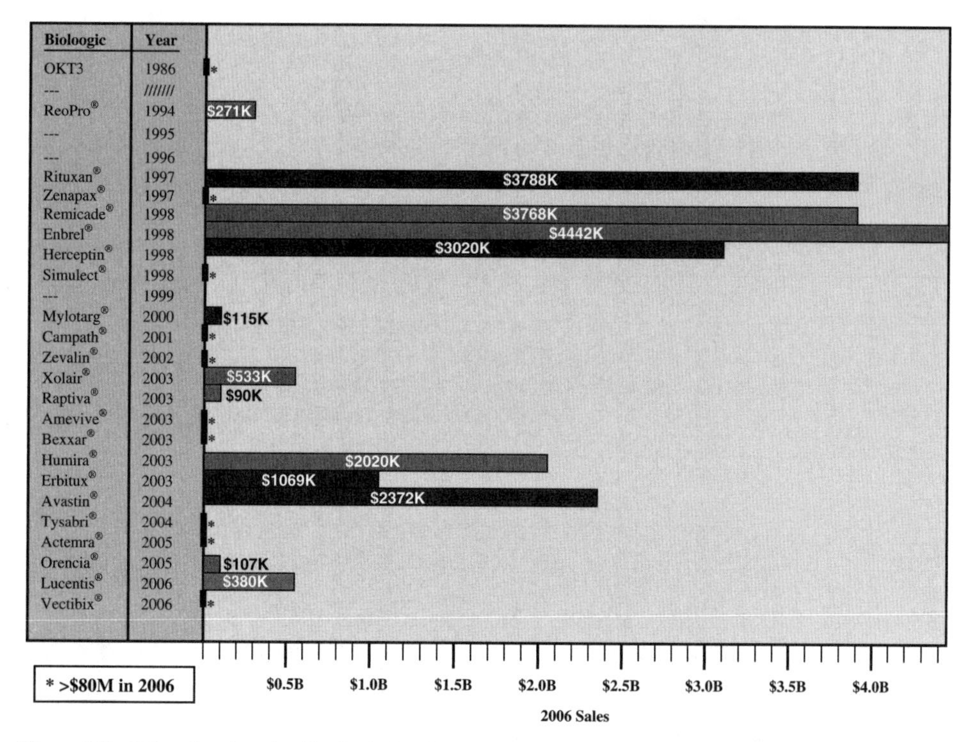

Figure 1.2 Sales of marketed antibodies in 2006 as a function of the year in which they reached the market. Two important features can be observed: (1) only 9 of 23 of the marketed antibodies and Fc fusion proteins have achieved substantial sales; (2) several of the Mabs and Fc fusion proteins on the market from the 1997–1998 period have been major blockbusters, which has driven broad interest by the pharmaceutical interest in biologics.

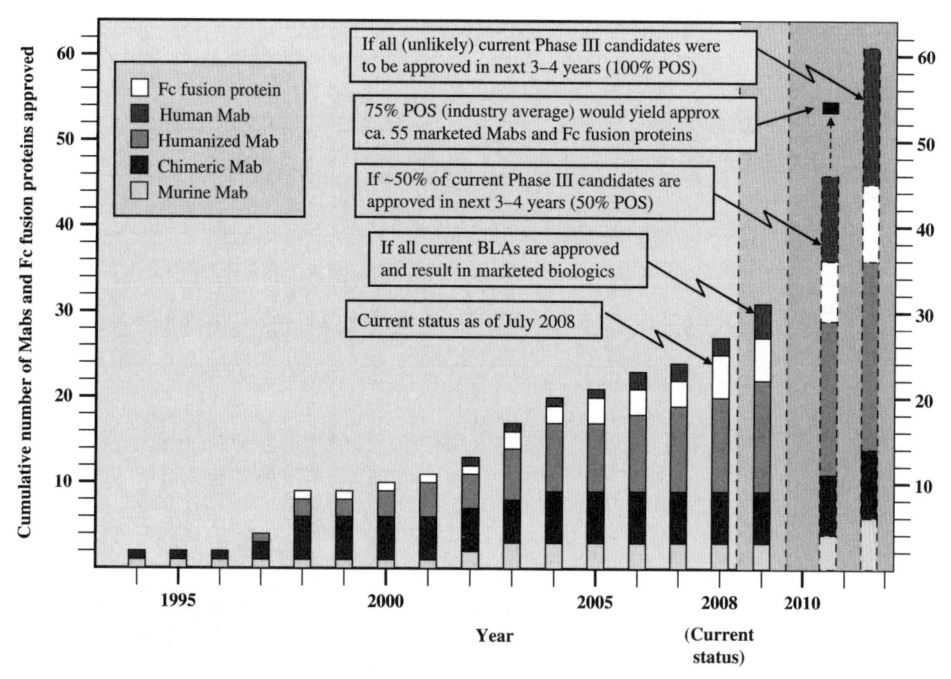

Figure 1.3 Accumulated Mabs and Fc fusion proteins on the market in the United States up to 2008, with projections for near-term (2012–2013 timeframe) future numbers based on current Phase III clinical candidates.

TABLE 1.3 Examples of Important Monoclonal Antibodies and Fc Fusion Proteins in Advanced Clinical Trials*

U.S. Trade Name (Generic Name)	Company	Current Status (U.S.)	Molecular Target	Major Indication	Protein Format	Antibody Source
Golimumab (CNTO-148)	Centocor	BLA filed June 2008	TNF-α	Psoriasis	Human IgG1	Transgenic humanized mouse
Ustikinumab (CNTO-1275)	Centocor	BLA filed November 2007	P40 subunit of IL-12 & IL-23	Psoriasis	Human IgG1	Transgenic humanized mouse
ABT-874	Abbott	Phase III initiated	P40 subunit of IL-12 & IL-23	Psoriasis, Crohn disease	Human IgG	Phage displayed human antibody library
Belatacept (LEA-29Y)	Bristol-Myers Squibb	Phase II/III	CD80/CD86	Renal transplantation	CTLA-4 Fc fusion LEA (higher affinity than abatacept)	Not applicable; Fc fusion
Lymphostat B® (Belimumab)	Glaxo Smith-Kline	Phase III	BLyS	Lupus	Human IgG	Phage displayed human antibody library
Atacicept (TACI-Ig)	Merck-Serono/ Zymogenetics	Phase III	BlyS, April antagonist	SLE (PII for MS, RA)	Fc fusion protein	Not applicable; Fc fusion
Ocrelizumab (2nd gen. anti-CD20)	Genentech	Phase III	CD20	RA, lupus, relapsing MS	Humanized IgG1	Hybridoma
ACZ-885	Novartis	Phase III	IL-1β	Muckle Wells syndrome	Human IgG1 k	Transgenic humanized mouse
Actemra® (Tocilizumab; Atlizumab)	Roche/Chugai	BLA filed Nov 2007	IL-6R	Castlemans disease	IgG1, humanized	Hybridoma
Bosatria® (Mepolizumab)	Glaxo Smith-Kline	Phase III	IL-5	Hyper eosinophilic syndrome	Humanized IgG	Hybridoma
Omnitarg® (Pertuzumab)	Genentech	Phase III	Her2	Oncology	Humanized IgG1	Hybridoma
Ofatumumab (Humax CD20)	GenMab/Glaxo Smith-Kline	Phase III	N-terminal epitope of CD20	B-cell chronic leukemia; NHL, RA	Humanized IgG1	Hybridoma
Tremelimumab (CP-675,206; Ticilimumab)	Pfizer	Phase III	CTLA4	Oncology	IgG2 human antibody	Transgenic humanized mice

Ipilimumab (MDX-010)	Medarex/Bristol-Myers Squibb	Phase III	CTLA4	Melanoma, other cancers	Human IgG1	Transgenic humanized mouse
Galiximab (IDEC-114)	Biogen/IDEC	Phase III	CD80 (B7-1)	NHL	Primatized IgG1λ	Hybridoma
Inotuzumab ozogamycin	Wyeth	Phase III	CD22	NHL	Humanized IgG4, Mab-conjugate	Hybridoma
Zalutumumab (HuMax EGFR)	GenMab/Medarex	Phase III	EGFR	Head and neck cancer	Human IgG1	Transgenic humanized mice
Aflibercept	Sanofi-Aventis/Regeneron	Phase III	VEGF	NSCLC	Fc fusion	Exodomain 1 of human VEGFR1 and 2 of VEGFR2 – Fc fusion
VEGF Trap-Eye	Bayer-Schering Pharma/Regeneron	Phase III	VEGF	Wet age-related macular degeneration	Fc fusion	Exodomain 1 of human VEGFR1 and 2 of VEGFR2 – Fc fusion
Zanolimumab (HuMax-CD4)	GenMab/Medarex	Phase III	CD4	Cutaneous T-cell lymphoma	Human IgG1	Transgenic humanized mice
Teplizumab; HOKT3γ1(Ala-Ala)	Macrogenics/Eli Lilly	Phase III	CD3	Diabetes	Humanized modified Fc	Hybridoma
Otelixizumab (ChAglyCD3; TRX4)	GSK/Tolerx	Phase III	CD3	Type 1 diabetes	Aglycosylated, humanized IgG	Hybridoma
Removab® (Catumaxomab)	Fresenius/Trion	Phase II/III	EpCAM and CD3	Malignant ascites; cancer	Rat-murine hybrid	Hybridomas; modified Fc; trifunctional bispecific
IGN101	Aphton	Phase II/III	EpCAM (CD326)	Oncology, specifically non-small-cell lung cancer	Murine Mab 17A-1 absorbed on aluminum hydroxide to provoke immune response on cells containing EpCAM	Hybridoma
Adecatumumab (MT-201)	Micromet/Merck-Serono	Phase II completed	EpCAM (CD326)	Oncology	Human IgG1	Phage displayed human antibody library

(Continued)

TABLE 1.3 *Continued*

U.S. Trade Name (Generic Name)	Company	Current Status (U.S.)	Molecular Target	Major Indication	Protein Format	Antibody Source
OvaRex® (Oregovomab)	ViRess/United Therapeutics	Phase III	CA125 tumor antigen	Ovarian cancer	Murine IgG	Hybridoma
CH-14.18	NCI	Phase III	GD2 ganglioside	Neuroblastoma	Chimeric IgG1 k	Hybridoma
Rencarex® (WX-G250)	Wilex AG	Phase III	Carbonic anhydrase IX	Nonmetastatic renal cell cancer	Chimeric IgG1	Hybridoma
Denosumab (AMG-162)	Amgen	Phase III	RANK-ligand	Osteoporosis	Human IgG2	Transgenic humanized mouse
Bapineuzumab (AAB-001)	Wyeth	Phase II completed	Amyloid beta	Alzheimer disease	Humanized	Hybridoma
Numax® (Motavizumab; MEDI-524)	Astra-Zeneca	BLA filed January 2008	Respiratory syncytial virus	Respiratory infection	Humanized IgG1; affinity optimized	Hybridoma
Mycograb® (efungumab)	Novartis	Phase III	Fungal HSP90	Fungal diseases	Human scFv	Phage displayed human antibody library
Aurograb®	Novartis/Neutec Pharma (now part of Novartis)	Phase III	Staph ABC transporter GrfA	MRSA, to be used with vancomycin	Human scFv	Phage displayed human antibody library
Abthrax® (Raxibacumab)	Human Genome Sciences	Phase III	B. anthracis PA toxin	Anthrax biodefense	Human IgG	Phage displayed human antibody library

Abbreviations: MS, multiple sclerosis; NHL, non-Hodgkin's lymphoma; MRSA, methicillin-resistant *Staphylococcus aureus*; NSCLC, non-small cell lung cancer; RA, rheumatoid arthritis; SLE, systemic lupus erythematosus.

*Data as of August 2008; data obtained from company websites, Prous Science Integrity, and www.clinicaltrials.gov.

TABLE 1.4 Key Phase I and Phase II Clinical Candidates by Indication and Target*

U.S. Trade Name (Generic Name)	Company	Current Status (U.S.)	Molecular Target	Major Indication	Protein Format
			Mostly Inflammatory Diseases		
Anrukinzumab (IMA-638)	Wyeth	Phase II	IL-13	Asthma	Humanized IgG
CAT-354	Astra-Zeneca	Phase II	IL-13	Asthma	Human IgG4
QAX-576	Novartis	Phase II	IL-13	Asthma	IgG
Anti-IL-13 Ab	Genentech	Phase I	IL-13	Asthma	IgG
AMG-317	Amgen	Phase II	IL-4 R	Asthma	Human IgG
MEDI-528	Astra-Zeneca	Phase II	IL-9	Asthma	Humanized IgG
GSK-679586A	GSK	Phase II	Not disclosed	Asthma	IgG
AMG-714 (HuMax-IL-15)	Genmab/Amgen	Phase II	IL-15	RA	Human IgG
CNTO-328	Johnson & Johnson	Phase II	IL-6	Multiple myeloma	IgG
CNTO-136	Johnson & Johnson	Phase II	IL-6	RA	IgG
REGN-88	Sanofi-Aventis	Phase I	IL-6	RA	IL-6R Fc-fusion trap
AMG-220	Amgen	Phase I	IL-6	Crohn disease	IgG
Baminercept alpha (LTβR-Ig)	Biogen/IDEC	Phase II	LTβR-Ig	RA	Lymphotoxin-βR – Fc fusion protein
IL-1βAb (Hu-007)	Eli Lilly	Phase II	IL-1β	RA	Humanized IgG
AMG-108	Amgen	Phase II	IL-1	Osteoarthritis	Human IgG
Xoma 052	Xoma	Phase I	IL-1β	Type 2 diabetes	Humanized IgG2
AMG-827	Amgen	Phase I	IL-17	RA	Human IgG
AIN-457	Norvartis	Phase I	IL-17A	Psoriasis	IgG
HuMax IL-8 (MDX-018)	Genmab/Medarex	Phase I/II	IL-8	Palmoplantar pustulosis	Human IgG
MEDI-563 (formerly BIW-8405)	Astra-Zeneca	Phase II	IL-5R	SLE	Afucosylated IgG based on BioWa's Potelligent technol.
MEDI-545	Astra-Zeneca/ Medarex	Phase II	IFNα	SLE	IgG
Fontolizumab (HuZAF)	Biogen IDEC/PDL	Phase II	IFNγ	IBD, inflammatory disorders	Humanized IgG1
AMG-811	Amgen	Phase I	IFN-γ	SLE	Human IgG
Anti-IFNα	Genentech	Phase I	IFN-R	SLE	IgG
MEDI-502	Astra-Zeneca	Phase II	CD2	Psoriasis	Humanized IgG

(*Continued*)

TABLE 1.4 *Continued*

U.S. Trade Name (Generic Name)	Company	Current Status (U.S.)	Molecular Target	Major Indication	Protein Format
Nuvion® (Visilizumab)	PDL	Phase II	CD3	GVHD; ulcerative colitis	Humanized IgG with modified Fc for decreased FcγR binding
NI-0401	NovImmune	Phase I/II	CD3	Crohn disease; renal transplantation	Human IgG
MT-110	Micromet	Phase I	CD3, EpCAM (CD326)	Gastrointestinal cancer	CD3, EpCAM bispecific scFv-based BiTE
TRX1	Genentech	Phase I	CD4	RA	IgG
MLN-0002	Millenium	Phase II	$\alpha 4\beta 7$ on T-cells	Crohn disease	Humanized IgG
Anti-beta7	Genentech	Phase I	Beta7	Ulcerative colitis	IgG
MDX-1100	Medarex	Phase II	CXCL10 (IP-10)	Ulcerative colitis, RA	Human IgG
MLN-1202	Millenium	Phase II	CCR2	Scleroderma	Humanized IgG
Anti-OX40L	Genentech	Phase I	OX40-L	Asthma	IgG
PD-360324	Pfizer	Phase I	GM-CSF	RA	IgG
CAM-3001	Zenith	Phase I	GM-CSF R	RA	Human IgG (CAT Library)
Anti-VAP1 MAb	BioTie	Phase I	VAP1	Inflammatory disease	Human IgG
AMG-557	Amgen	Phase I	B7RP-1	SLE	Human IgG
FG-3019	Fibrogen	Phase I	CTGF	IPF, diabetic nephropathy	Human IgG
GC-1008	Genzyme	Phase I/II	TGF-β	IPF	Human IgG (CAT)
			Mostly Oncology		
Milatuzumab	Immunomedics	Phase I	CD74	MM, other tumors	IgG
MDX-1411	Medarex	Phase I	CD70 (ligand for CD27)	RCC	Human IgG
SGN-70	Seattle Genetics	Phase I	CD70 (ligand for CD27)	Autoimmune diseases	Human IgG
IMGN901 (formerly huN901-DM1)	Immunogen	Phase II	CD56	SCLC, MM	Humanized IgG-DM1 cytotoxin conjugate
CP-870893	Pfizer	Phase II	CD40 agonist	Oncology	IgG
Dacetuzumab (SGN-40)	Genentech	Phase II	CD40	Oncology	Humanized IgG
HCD122 (formerly CHIR-12.12)	Xoma/Novartis	Phase I	CD40	B-cell malignancies	IgG
HuMax CD-38	GenMab	Phase II	CD38	Multiple myeloma	IgG
Lintuzumab (SGN-33)	Seattle Genetics	Phase II	CD33	AML	IgG
AVE-9633 (huMy9-6)	Sanofi-Aventis	Phase I	CD33	AML	Humanized IgG-DM4 cytotoxin conjugate

MDX-060	Medarex	Phase II	CD30	Hodgkin lymphoma	Human IgG
MDX-1401	Medarex	Phase I	CD30 (backup to MDX-060)	Oncology	Human IgG
SGN-35	Seattle Genetics	Phase I	CD30 auristatin conjugate	Hodgkin lymphoma	IgG conjugated NP-toxin
Lumiliximab (IDEC-152)	Biogen/IDEC	Phase II	CD23	CLL	Primatized IgG1 k
Epratuzumab	Immunomedics/UCB	Phase II	CD22	NHL, SLE	Humanized IgG
Epratuzumab tetraxetan	Immunomedics	Phase I/II	CD22	NHL, SLE	Humanized IgG-Y90 radio-conjugate
Anti-CD22-PE (CAT-8015)	Astra-Zeneca	Phase I	CD22	Oncology; CLL	anti-CD22-pseudomonas exotoxin fusion protein
Veltuzumab (IMMU-106)	Immunomedics	Phase I/II	CD20	NHL, autoimmune diseases	Humanized IgG
PRO-131921	Genentech	Phase I/II	CD20 (third generation)	Oncology	IgG with modified Fc for increased ADCC
AME-133v (LY2469298)	Eli Lilly	Phase I/II	CD20	NHL	IgG with modified Fc for increased ADCC
TRU-015	Wyeth	Phase II	CD20	RA	Small modular immunopharmaceutical product (SMIP)
R-7159	Roche	Phase II	CD20	NHL	IgG
Blinatumomab (MT103/MEDI-538)	Astra Zeneca/Micromet	Phase II	CD19, CD3	ALL, NHL, CLL	CD19, CD3 bispecific scFv-based BiTE
MDX-1342	Medarex	Phase I	CD19	CLL, RA	Human IgG
SAR-3419	Sanofi-Aventis/Immunogen	Phase I	CD19	NHL	IgG
AMG-102	Amgen	Phase II	HGF/SF (cmet-L)	RCC	Human IgG
MetMAb	Roche/Genentech	Phase I/II	c-met	Oncology	IgG
AMG-386	Amgen	Phase II	Angiopoietin-related target	RCC	Fc-peptide "peptibody"
Mapatumumab (HGS-ETR1)	HGS	Phase II	TRAIL-R1	Oncology	Agonist MAb
AMG-951; RhApo2L/TRAIL	Amgen; Genentech	Phase II	rhApo2L/TRAIL	Oncology	Therapeutic protein
AMG-655	Amgen	Phase I/II	DR5 (TRAIL-2)	Pancreatic cancer	Human IgG
Apomab	Genentech	Phase II	DR5 (TRAIL-R2) agonist	Oncology	Human IgG
Lexatumumab (HGS-ETR2)	GSK	Phase I	TRAIL-2	Solid tumors	Human IgG (agonist)

(*Continued*)

TABLE 1.4 *Continued*

U.S. Trade Name (Generic Name)	Company	Current Status (U.S.)	Molecular Target	Major Indication	Protein Format
LBY-135	Norvartis	Phase I	DR5	Solid tumors	Chimeric IgG
CT011	CureTech	Phase II	PD-1	Oncology	Humanized IgG
MDX-1106 (ONO-4538)	Ono Pharma/ Medarex	Phase I	PD-1	Oncology; infectious diseases	Human IgG
MDX-1105	Medarex	Phase I	PD-L1	Oncology	Human IgG
Farletuzumab (MORAb-003)	Morphotek	Phase II	Folate Receptor Alpha	Oncology	Humanized IgG
Trastuzumab-DM1	Genentech	Phase II	HER2	Oncology	Humanized IgG- DM1 cytotoxin conjugate
Ertumaxomab (Rexomun)	Fresenius/Trion	Phase II/III	Her2/neu and CD3	Malignant ascites; Cancer	Rat-murine hybrid; modified Fc; trifunctional bispecific
CP-751871	Pfizer	Phase II	IGF-1R	NSCLC	Human IgG2
MK-0646 (H7C10)	Merck/Pierre-Fabre	Phase II	IGF-1R	Oncology	Humanized IgG
AMG-479	Amgen	Phase II	IGF-1R	Ewings sarcoma	Human IgG
IMC-A12	ImClone	Phase II	IGF-1R	Oncology (multiple)	Human IgG
R-1507 (formerly Roche 1)	GenMab	Phase II	IGF-1R	Solid tumors	Human IgG
AVE-1642	Sanofi-Aventis/ Immunogen	Phase I	IGF-1R	Solid tumors	Human IgG
Anti-IGF-1R Mab	Biogen/IDEC	Phase I	IGF-1R	Solid tumors	IgG
Nimotuzumab (DE-766)	Oncoscience/YM Biosciences	Phase II (launched in India)	EGFR	Oncology, several indications	Humanized IgG1
IMC-11F8	ImClone, Dyax	Phase II	EGFR	Solid tumors	Human IgG (CAT Library)
IMC-1121B	Imclone, Dyax	Phase II	VEGF-B	Solid tumors	Human IgG (CAT Library)
Alacizumab pegol (CDP-791)	UCB-Celltech/ ImClone	Phase II	VEGF-R2	Lung cancer	Di-Fab-PEG conjugate
Angiocept (CT-322)	BMS	Phase II	VEGF-R2 (FLK-1/ KDR)	Glioblastoma	PEGylated adnectin
IMC-18F1	ImClone	Phase I	VEGF-R1	Oncology	Human IgG
IMC-3G3	ImClone	Phase I	PDGFRα	Oncology	Human IgG
CVX-045	Pfizer	Phase I	Angiogenesis inhibitor	Oncology	Thrombospondin-1 mimetic-IgG conjugate
Volociximab (M200)	Biogen Idec/PDL	Phase II	$\alpha5\beta1$ integrin	Solid tumors	Chimeric MAb
CNTO-95	Johnson & Johnson	Phase II	αV integrins	Melanoma	IgG

MEDI-522	Astra-Zeneca	Phase II	$\alpha V\beta 3$ integrin	Systemic psoriasis	Humanized IgG
IMGN388	Immunogen	Phase I	αV integrins	Oncology	Humanized IgG-DM4 cytotoxin conjugate
CNTO-888	J&J	Phase I	MCP-1	Oncology	Human IgG1
CME-548	Wyeth	Phase II	5T4	Solid tumors	IgG-calicheamicin conjugate
MK-4721	Merck/Astellas	Phase II	PSCA	Oncology	Human IgG
Apolizumab (Hu1D10)	NCI/PDL	Phase II	HLA-DR beta-chain epitope	NHL, other oncology	Humanized IgG1
3F8	NCI	Phase II	GD2 ganglioside	Oncology	Murine IgG
Tucotuzumab celmo-leukin (EMD 273066)	Merck-Serono	Phase II	GD-2 tumor antigen	SCLC	Humanized IgG linked to cytokine IL-2
IMGN242	Immunogen	Phase II	CanAg	Oncology	Humanized IgG- DM4 cytotoxin conjugate
BrevaRex(R) AR20.5	AltaRex	Phase I/II	MUC1	Multiple myeloma	IgG
Y90-hPAM4	Immunomedics	Phase I	MUC1	Oncology	IgG-Y90 radioconjugate
MORAb-009	Morphotek	Phase II	Mesothelin on ovarian cancer	Ovarian cancer	Humanized IgG
CDX-1307	Celldex/Medarex	Phase I	Mannose receptor; hCGβ	Oncology	Human IgG
Elotuzumab (HuLuc 63)	PDL/BMS	Phase I	CS1 surface antigen	Myeloma	IgG
BMS-663513	BMS	Phase I/II	CD137 agonist	Oncology	Human IgG
anti-ALK1 Ab	Pfizer	Phase I	ALK1	Solid tumors	IgG
anti-P-cadherin Ab	Pfizer	Phase I	P-cadherin	Solid tumors	IgG
MT-293	Micromet/Tracon Pharma	Phase I	Cleaved collagen	Oncology	IgG
BIIB-015	Biogen/IDEC	Phase I	Cripto	Solid tumors	IgG
		Mostly Nonimmunology/Nononcology Mabs			
Stamulumab (MYO-029)	Wyeth	Phase II	Myostatin (GDF8)	Muscular dystrophy; sarcopenia	IgG
AMG-745	Amgen	Phase I	Myostatin	Muscle loss	IgG
CAL	Chugai/Roche	Phase II	PTHrP	Hypercalcemia, bone metastases	IgG
NPVBHQ-880	Norvartis	Phase I	DKK-1	Osteoporosis	IgG
AMG-785	Amgen/UCB	Phase I	Sclerostin	Osteoporosis	IgG
LY-2062430	Eli Lilly	Phase II	Amyloid-β	Alzheimer disease	IgG

(*Continued*)

TABLE 1.4 *Continued*

U.S. Trade Name (Generic Name)	Company	Current Status (U.S.)	Molecular Target	Major Indication	Protein Format
PF-4260365	Pfizer	Phase II	Amyloid-β	Alzheimer disease	IgG
Ganteneruma (R-1450)	Roche/MorphoSys	Phase I	Amyloid-β	Alzheimer disease	Human IgG
TTP-4000	Pfizer	Phase I	Amyloid-β	Alzheimer disease	RAGE Fc fusion protein
anti-Nogo Ab	Norvartis	Phase I	NOGO	Spinal cord injury	IgG
Tanezumab (RI-624)	Pfizer	Phase II	NGF	Pain	IgG
AMG-403	Johnson & Johnson/ Amgen	Phase I	NGF	Pain	IgG
Bertilimumab (iCo-008, CAT-213)	iCo Therpaeutics	Phase II	Eotaxin (CCL11)	Vernal keratoconjunctivitis	Human IgG4 (CAT Library)
CVX-096	Pfizer	Phase I	GLP-1R	Diabetes	Peptide mimetic-IgG conjugate
AMG-477	Amgen	Phase I	Glucagon receptor	Type 2 diabetes	IgG
TGFβ Ab	Eli Lilly	Phase I	TGFβ Ab	Diabetic nephropathy	IgG
R-7025	Roche	Phase I	HCV	Antiviral therapy	Therapeutic protein
Ibalizumab (TNX-355; Hu5A)	Genentech (Tanox)	Phase II	CD4	HIV	Humanized IgG1
Pagibaximab (BSYX-A110)	Biosynexis/ Medimmune	Phase II	Staphylococcus lipoteichoic acid	Staphylococcus infections	Chimeric IgG1
Aurexis (Tefibazumab)	Inhibitex	Phase II	Staphylococcus clumping factor A	Staphylococcus infections	Humanized IgG1
MDX-066 (CDA-1)+ MDX-1388 (CDA-2)	Medarex/MBL	Phase II	*Clostridium difficile* toxins A and B	*C. difficile*-associated diarrhea (CDAD)	Human IgGs
MEDI-557	Astra-Zeneca	Phase I	F-protein on RSV	RSV	YTE mutant—longer half-life Mab
Valortim™ (MDX-1303)	Pharm Athene/ Medarex	Phase I	*Bacillus anthracis* PA toxin	Anthrax—biodefense	Human IgG
CytoFab™	Astra-Zeneca	Phase II	TNF-α	Severe sepsis	IgG

Abbreviations: ADCC, antibody-dependent cellular cytotoxicity; ALL, acute lymphoblastic leukemia; AML, acute myelogenous leukemia; CLL, chronic lymphocytic leukemia; GVHD, graft-versus-host disease; IBD, irritable bowel disease; IPF, idiopathic pulmonary fibrosis; MM, multiple myeloma; NHL, non-Hodgkin's lymphoma; NSCLC, non-small cell lung cancer; RA, rheumatoid arthritis; RCC, renal cell carcinoma; RSV, respiratory syncytial virus; SCLC, small cell lung cancer; SLE, systemic lupus erythematosus.

*Data as of August 2008; data obtained from company websites, Prous Science Integrity, and www.clinicaltrials.gov.

(these represent the majority, but likely not all, early phase commercially funded clinical candidates), 72 (50 from current Phase II candidates and another 22 from the current Phase I candidates) should eventually reach Phase III. Based on the 75 percent POS for transition from Phase III to marketing approval (KMR Group, Inc. 2007), this could result approximately in an additional 54 Mabs and Fc fusion proteins on the market between 2013 and 2018 (Fig. 1.3).

Taking into account the currently marketed Mabs and Fc fusion proteins (Table 1.2, Fig. 1.3), as well as the POS-adjusted clinical candidates listed in Table 1.3 (late clinical phase) and Table 1.4 (early clinical phase), there could potentially be a total of 135 Mabs and Fc fusion proteins on the U.S. market by the 2018 time frame, a decade from now. This number is approximately five times the current 27 Mabs and Fc fusion products on the market today, most of which have been marketed in the 10 year period of 1997–2007. These calculations, while necessarily forward projecting, suggest that there will be significant expansion of Mab and Fc fusion products reaching the market over the next decade, which could have a profound and lasting impact on the pharmaceutical industry in general. This could be especially noticeable if the overall POS for Mabs remains in the 18 to 20 percent range, as compared with the historical POS for small molecules, at about 7 to 8 percent. Thus, by 2018, a decade from now, therapeutic proteins in general and, more specifically, Mabs and Fc fusion proteins, should comprise a significant fraction of worldwide pharmaceutical revenues, considerably higher than the 20 percent fraction that biologics make up today.

In a sampling of the more than 1500 clinical trials testing Mabs today (www.clinicaltrials.gov), approximately 45 percent of these represent oncology studies using "naked" antibodies (i.e., no toxin- or radioconjugate attached). Another 31 percent, many funded by the National Cancer Institue (NCI) and/or academic groups, are being conducted using radioconjugated Mabs to kill tumors, and 2 percent represent toxin-conjugates for oncology. Thus, 78 percent of all current clinical trials on Mabs are focused on the therapeutic area of oncology. Fourteen percent are focused on immunology-related indications, and the final eight percent on non-oncology, non-immunology-related indications. These clinical trials data are somewhat skewed, however, by the large number of academic- and government-funded Phase I clinical trials focused on testing radioconjugated monoclonal antibodies for oncology indications. In another view of the therapeutic area breakdown, of the more than 200 combined clinical candidate Mabs and Fc fusion proteins listed in Tables 1.3 and 1.4, approximately 50 percent are either used for, or are being tested primarily for, oncology indications, 32 percent for immunology-related indications, and about 18 percent for non-oncology, non-immunology indications. This final category includes a wide range of indications, including, for examples, atherosclerosis, diabetes, infectious diseases, bone loss, muscle wasting and dystrophy, and other assorted indications.

1.2 HISTORICAL ASPECTS

It now has been approximately a third of a century since Köhler and Milstein described methods for producing murine hybridomas, in what is accepted by most as the dawn of the era of therapeutic monoclonal antibodies (Köhler and Milstein 1975). After Mab hybridoma technology was first described in 1975, it took 11 years, until 1986, before the first commercial therapeutic antibody, Orthoclone OKT3® (muronomab-CD3), was licensed by Ortho Biotech, a subsidiary of Johnson & Johnson, for inhibition of transplanted organ rejection. It was yet another eight years before the second antibody, the chimeric Fab antibody, ReoPro® (abciximab), was developed by Centocor (now a subsidiary of Johnson & Johnson), and marketed by Eli Lilly to inhibit platelet aggregation post-cardiovascular surgery. Thus, even two decades after the seminal paper was published on monoclonal antibodies, only two monoclonal antibody products had been brought to the market. This changed dramatically in 1997–1998, when a total of five monoclonal antibody drugs were introduced to the market (Table 1.2, Fig. 1.3), generating considerable interest in the field of therapeutic monoclonal antibodies. How did we get from discovery to market, and why did it take so long? In the next section, the history leading up to the current status of the monoclonal antibody field will be addressed.

1.2.1 Historical Aspects: Origins of Serum Therapy, Forerunner to the Monoclonal Antibody Business

The first concept of using antibodies as therapeutics came long before the generation of hybridomas, as shown in Figure 1.4. It started when Robert Koch, discoverer of the tubercle bacillus and 1905 Nobel Laureate, was named director of the Institute of Hygiene in Berlin in 1885. There he assembled a team of the brightest minds in the newly forming field of immunotherapeutics, including Paul Ehrlich (known for the "magic bullet" hypothesis; 1908 Nobel Laureate), Emil von Behring (father of immunotherapy; 1901 Nobel Laureate), Erich Wernicke, and Shibasaburo Kitasato (eventual founder of Japan's famed Kitasato Institute), all of whom would have a significant impact on the beginnings of antibody-based therapy (Winau, Westphal, and Winau 2004). Working initially on iodoform chemotherapeutics, Behring made several key observations that led to the concept of *Blutserumtherapie*, or serum therapy. He noticed that the blood of those rats resistant to anthrax was able to kill the anthrax bacterium (Chung n.d.), and together with his friend Wernicke, he developed the first working serum therapy for diphtheria. Behring and Kitasato, a student of Koch's who had isolated the tetanus-forming bacillus and had determined that its pathogenesis lay in the activity of its toxin, together demonstrated that the transfer of serum from a guinea pig immunized with diphtheria toxin to another guinea pig offered protection from the toxin (Behring and Kitasato 1890). Behring and Kitasato also obtained anti-sera against tetanus toxin, demonstrating the breadth of the principle.

Behring's diphtheria serum therapy was first tested clinically in 1891 at Charite' Hospital in Berlin. A year later, Behring began working with the pharmaceutical manufacturer Faberwerke Hoechst to develop the diphtheria serum treatment. In 1894, Hoechst launched the first immunobiological therapeutic, dispatching the first 25,000 doses of anti-diphtheria serum to fight the diphtheria epidemic that was claiming the lives of 50,000 children annually in Germany alone (Fig. 1.4). The serum therapy was

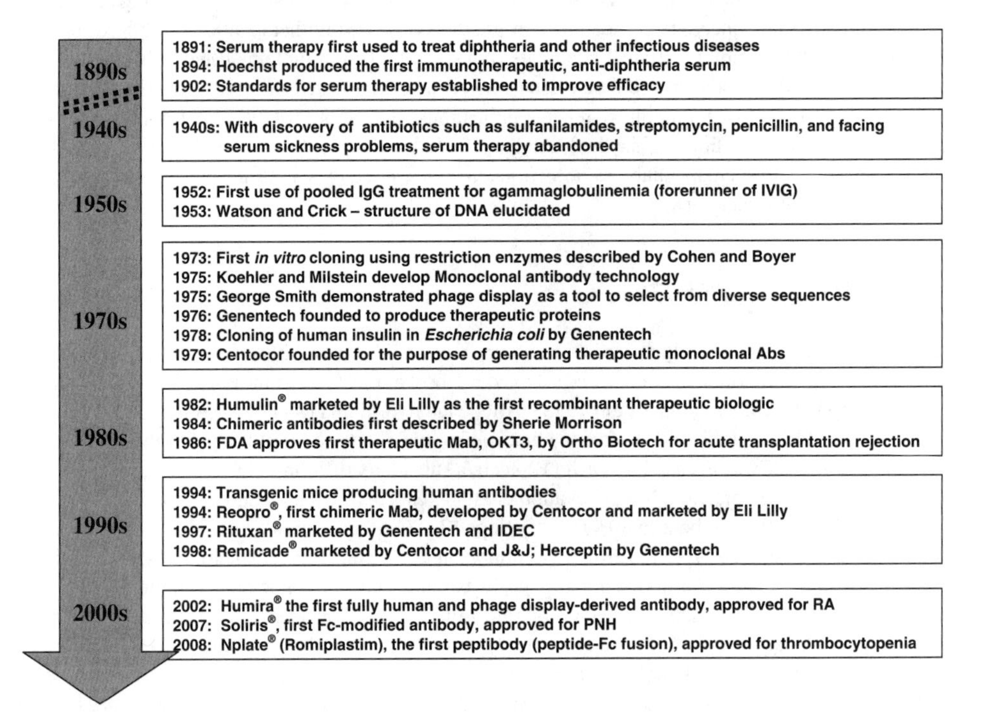

Figure 1.4 Timeline for important discoveries leading to therapeutic Mabs and Fc fusion proteins.

not without detractors—several scientists and politicians across Europe criticized, and some even poked fun at, the serum-based therapeutic approach (Winau and Winau 2002). However, with introduction of the serum therapy, mortality in Paris dropped from 52 percent to 25 percent (Llewelyn, Hawkins, and Russell 1992), silencing most doubters. The discovery of serum therapy led to the awarding of the first Nobel Prize in Medicine and Physiology to Behring in 1901. In what was likely the first example of venture capital funding in biotechnology, Behring used the funds from his Nobel Prize to seed a new company in 1904, which still exists today as Novartis (Chiron) Behring, a vaccine manufacturer, located on Emil Von Behring Strasse in Marburg (Winau and Winau 2002).

The serum used for serum therapy was crude and from immunized, nonhuman (heterologous) sources (e.g., rabbits, horses), containing many foreign proteins as well as the antibodies, and gave rise to a phenomenon that has been generally called "serum sickness" (Gronski, Seiler, and Schwick 1991; Lawley et al. 1984). The natural onset of serum sickness in virtually 100 percent of patients treated with serum therapy led to a variety of efforts to improve on the therapeutic approach. Behring even came to realize that the toxic side effects of serum therapy were interrelated with the efficacy of the preparation (Gronski, Seiler, and Schwick 1991). He tried many methods of purifying the serum, without substantial success. At about the same time, Paul Ehrlich recognized the need for standardization of serum therapies, which led to the development of methods for quantitation of the serum therapeutic effect, including the concept of LD50 (dilution of serum preventing death of 50 percent of animals treated), still used today. In 1908, Ehrlich was awarded the Nobel Prize (shared with Ilya Metchnikoff, who discovered the basis of phagocytosis), for his characterization and standardization of the anti-serum therapies. Others worked on the concept of protease-treated sera, with the notion of deriving a formulation that would remain efficacious but would lose the side effects caused by the serum itself. The use of proteases to purify heterologous (nonhuman origin) immune sera (dubbed *fermo-sera*) was not perfected until the late 1930s (Weil, Pafentjiev, and Bowman 1938), but these preparations still resulted in serum sickness and also were prone to sensitizing and anaphylactic reactions.

Besides the obvious issues with serum sickness, other significant problems besetting serum therapy included lack of batch-to-batch consistency, difficult administration, and variable pharmacokinetics (Casadevall 1996). Nevertheless, heterologous serum therapy was used widely until approximately the onset of World War II for a variety of diseases, including the bacterial diseases: diphtheria, streptococcal pneumonia, meningitis, tularemia, shigella dysentery, brucellosis, gas gangrene, tetanus, botulism, anthrax, whooping cough; and the viral diseases: measles, poliomyelitis, mumps, and chickenpox (Casadevall and Scharff 1995). Some of these treatments, for example, for diphtheria, meningitis, and pneumonia, proved to be fairly successful, whereas others, for example, anthrax, whooping cough, and shigella dysentery, were apparently less so (Casadevall and Scharff 1995). An example for how widespread serum therapy was used is that 86 percent of patients diagnosed with type I streptococcal pneumonia in the late 1930s at Boston City Hospital were treated with a type-specific serum therapy (Casadevall and Scharff 1995).

With the discovery of sulfonamides in the mid-1930s, and later penicillin, streptomycin and other natural product antibiotics (many of which were broad spectrum), the practice of passive immunization using heterologous serum declined precipitously. The combination of serum sickness, lack of consistency, narrow spectrum of use, unknown pharmacokinetics, and intravenous administration made heterologous serum therapy largely noncompetitive with the, then, newly found chemotherapeutics (Casadevall 1996). The exception to this paradigm is the third world, in which health care is substantially different from that in Western nations or nations with large, robust economies. In many countries in which antimicrobial chemotherapeutics are not readily available, serum-based therapy still plays an important role in overall health care (Wilde et al. 1996).

1.2.2 IVIG Therapeutics and Prophylactics

Later, the concept of serum therapy was modified by isolating natural antibodies in either vaccinated (or convalescing, called "specific immunoglobulins") or "naïve" humans, followed by isolation of the IgG fraction from the pool sera for therapeutic use. The first reported use of fractionated IgG as a therapeutic agent was in 1952, in which a patient with primary immunodeficiency was treated with intramuscular (IM) injections of purified human IgG (Bruton 1952; Fig. 1.4). IM delivery resulted in limited dosing regimens, marginal IgG replenishment, and thus marginal clinical benefit. Early attempts at intravenous delivery of human IgG fractions (IVIG or, in the United States, IGIV), however, resulted in immunological reactions thought at the time to be due to activation of complement (Weiler 2004). Later methods for producing IVIG were able to remove the cause for this immune reaction, allowing for transfusions to take place. Thus, widespread use of IVIG to treat primary antibody deficiencies did not occur until the early 1980s (Mouthon and Lortholary 2003).

There are two types of IVIG, specific immunoglobulins and normal immunoglobulins. Specific immunoglobulins are obtained from convalescent donors or from healthy volunteers specifically vaccinated to provide the antibodies (as in the case of anti-rhesus D antigen) (Llewelyn, Hawkins, and Russell 1992). Another example of a specific immunoglobulin is the passive administration of human anti-rabies IgG to patients not previously vaccinated against rabies. The huIgG, generated by hyperimmunized human donors, provides virus-neutralizing antibodies immediately to bridge the gap until the patient produces his or her own antibodies in response to concomitant vaccine administration. Two anti-rabies IgG formulations are licensed for use in the United States: Imogam® Rabies-HT (Sanofi-Pasteur) and HyperRab™ S/D (Talecris Biotherapeutics).

The term *normal immunoglobulin* has been used for IgG pools obtained from a large number of random donors. These antibodies generally provide four to six weeks of protection against pathogens that are relatively widespread in populations, including hepatitis A, measles, mumps, and other viral diseases (Llewelyn, Hawkins, and Russell 1992). This approach, generally known as IVIG or gamma globulin treatment, was initially used as replacement therapy for patients unable to generate their own immunoglobulins (Orange et al. 2006). As of 2006, the U.S. Food and Drug Administration (FDA) had approved 11 products for primary immunodeficiency or humoral immunodeficiency, another 5 for idiopathic thrombocytopenic purpurea, 3 for Kawasaki syndrome, 2 for B-cell chronic lymphocytic leukemia, and one each for HIV infection and bone marrow transplantation (Weiler 2004; Orange et al. 2006). Doses typically are in the range of 300 to 600 mg/kg on a monthly or biweekly basis (Orange et al. 2006). It is noteworthy as well as ironic that the approved use for IVIG in the United States for various infectious diseases is severely limited, even though the conceptual origins of IVIG use sprang directly from Berhing's work on serum therapy for infectious diseases. Several new uses for IVIG have been proposed recently, including the expansion of use for infectious diseases (Wallington 2004; Casadevall and Scharff 1995) and protection from potential biological warfare agents (Casadevall 2002).

Significantly, the IVIG approach led directly to the development of one of the early significant licensed monoclonal antibodies, Synagis®. Medimmune first developed an IVIG prophylactic, Respigam®, which was licensed in 1996 to protect infants from respiratory syncytial virus (RSV). While pushing forward with their development of Respigam®, MedImmune already had begun clinical trial development of Synagis® as early as 1994, understanding that a monoclonal antibody would be both preferable over IVIG and ultimately more profitable. Synagis®, a human-mouse chimeric IgG1 targeting a key epitope of the A-antigenic site of RSV F-protein, was then licensed in 1998, essentially replacing Respigam® as the primary anti-RSV prophylactic for premature infants.

The concept of IVIG also has led to an approach similar to IVIG, but yet significantly more refined and sophisticated, as developed by the biotech company, Symphogen. The scientists at Symphogen isolate multiple antibodies directed against a single target or target entity (a virus in the case of antiviral, or a cell for antibacterial or antitumor) and then produce the multiple Mabs in a single pot cell culture based on a mixed inoculum from individual master cell banks (Rasmussen et al., 2007). This concept apparently is meant to simulate the natural mechanism the body uses to defeat a foreign antigen or invader, while lacking the huge volume of nonspecific antibodies that would be present in an IVIG type of preparation. Symphogen is currently in Phase I with an anti-RhD product candidate containing 25 different Mabs (Wilberg et al. 2006).

1.3 TECHNOLOGIES LEADING TO THE CURRENT MONOCLONAL ANTIBODY ENGINEERING ENVIRONMENT

1.3.1 Fundamental Breakthroughs Allowing for Recombinant Monoclonal Antibodies

The 1970s and 1980s proved to be an incubator period that spawned the dawn of the biologics revolution (Fig. 1.4). A series of technologies were developed in this time period that ultimately converged to provide all of the technological and fundamental bases for development of the therapeutic monoclonal antibody industry. These technologies include the use of restriction enzymes to clone a gene into a plasmid (Cohen et al. 1973), development of hybridoma technology by Köhler and Milstein (1975), site-directed mutagenesis as a tool for protein engineering (Hutchinson et al. 1978; Zoller and Smith 1982; Dalbadie-McFarland et al. 1982), and development of an understanding of the genetics of antibody expression [Hozumi and Tonegawa 1976; Early et al. 1980; Gough and Bernard 1981; Tonegawa 1983; also, the debate concerning germline versus somatic mutation as the basis for diversity was laid out nicely by Silverstein (2003)]. Additional technologies and scientific knowledge leading to the development of recombinant antibodies were added in the 1980s, including phage display technology (Smith 1985), polymerase chain reaction (PCR; Mullis et al. 1986; Saiki et al. 1988), sequencing and characterization of human germline antibody genes (Kabat et al. 1987), and expression of antibody genes in cell cultures (Neuberger 1983; Neuberger and Williams 1986) and in *Escherichia coli* (Better et al. 1988; Skerra and Plückthun 1988). In the next few sections, the fundamental breakthroughs in antibody engineering, built on the shoulders of the technologies mentioned above, are described.

The commercial path for therapeutic monoclonal antibodies was paved by Genentech and Eli Lilly, who teamed up to produce the first recombinant human protein, the human insulin product Humulin®, approved on October 30, 1982 for marketing in the United States. Leading up to this achievement, Genentech had produced somatostatin, the first recombinant human protein from a chemically synthesized gene, in *E. coli* (Itakura et al. 1977). Shortly thereafter, scientists at Genentech used the same approach to clone out the human insulin gene for expression in *E. coli* (Goeddel et al. 1979b). Genentech then licensed the recombinant human insulin to Eli Lilly, who developed it clinically and obtained marketing approval for the first recombinant human protein, Humulin®, in 1982 (Fig. 1.4). Prior to this seminal achievement, diabetic patients had been limited to taking Iletin®, a heterologous insulin product purified from the pancreas of animals (mostly pigs and cows), since 1923 when Eli Lilly had developed the first commercial process for its production (Shook 2007). Genentech scientists then went on to clone and express human growth hormone in *E. coli* (Goeddel et al. 1979a), which in 1985 became their first internally marketed product, Protropin®.

1.3.2 Hybridoma Technology

In the early 1970s, Georges Köhler was having difficulty finding a way to obtain antibodies from mortal B-cells in culture. Caesar Milstein and his colleagues, on the other hand, had worked out how to transform myeloma cell lines and generate myeloma-myeloma fusions to secrete antibodies (Milstein 1985). These myeloma fusions, however, produced antibodies lacking specificity (Alkan 2004). Another key piece to the puzzle was a critically important hemolytic plaque assay developed by Jerne, which allows direct visualization of antibody-producing B-cells (Jerne and Nordin 1963). Köhler joined Milstein's lab as a postdoctoral fellow in 1973, where the two joined forces to generate B-cell-myeloma fusions that secreted single (i.e., monoclonal) antibodies that recognized a specific antigen (Köhler and Milstein 1975), as visualized using Jerne's plaque assay (Alkan 2004). This discovery led to the awarding of the Nobel Prize in Physiology or Medicine in 1984 to Milstein, Köhler, and Jerne. As has been discussed on many occasions, Köhler and Milstein did not patent their discovery, which opened up the use of their hybridoma technology to academics and industry alike for generation of future potential therapeutic monoclonal antibodies.

1.3.3 Transfectomas and Chimeric Antibodies

The leap from the use of purely murine antibodies from hybridomas, as originally described by Köhler and Milstein (1975) and developed into an early industrial process by Ortho Biotech for licensure of Orthoclone OKT3® in 1984 (Table 1.2), to recombinant antibodies with human Fc domains that could be developed more fully and used more widely came with three significant developments, as described below.

The first development was the ability to clone out, using PCR methodology, the murine VH and VL genes for recombinant expression (Orlandi et al. 1989). The second requirement was to express both heavy and light chain antibody genes in stable human cell lines after transfection (originally called *transfectomas*; Neuberger 1983; Neuberger and Williams 1986; Beidler et al. 1988). Coupled with that was the third development, which was the method of making chimeric antibodies possessing murine VH and VL chains fused with human constant regions (Morrison et al. 1984; Boulianne, Hozumi, and Shulman 1984). Chimeric antibodies possess about one-third murine sequences (2VH and 2VL subunits) and two-thirds human sequences, including a human Fc. The first descriptions for the construction of chimeric antibodies occurred in 1984 (Fig. 1.4).

Vectors such as pSV2 and murine myeloma cell lines such as SP2/0 were popular early on (Shin and Morrison 1989). CHO-dg44-DHFR was the expression system of choice for several years, but in more recent years, Chinese hamster ovary (CHO) cell lines, and in particular the glutamine synthetase (GS) expression system coupled with the cell line CHO-K1SV (de la Cruz Edmonds et al. 2006), as widely licensed by Lonza, has seen more widespread use. Using the GS-CHO expression system, current levels of production routinely hit 1 g/L, with ranges of 3 to 5 g/L antibody production often achieved after cell cloning and process optimization (Kalwy, Rance, and Young 2006; personal communications with several bioprocess colleagues in the industry).

The first chimeric antibody to be marketed was ReoPro®, which was made chimeric and then cleaved to a Fab and purified to make the drug. The first chimeric IgG was Rituxan®, which is still a strong product with worldwide sales of more than $3 billion annually (Fig. 1.2). There are a total of six chimeric antibodies on the market, with an additional two in Phase III clinical trials (Fig. 1.5). With the development of humanization technologies and the concomitant reduction in immunogenicity that humanization brings, it is likely that there will be very few additional chimeric antibodies brought through clinical trials these days, as indicated by the trends shown in Figure 1.5.

1.3.4 Humanization Technology

Chimeric antibodies, as described earlier in this chapter, still retain 30 to 35 percent murine sequence, which may lead to enhanced immunogenicity (Pendley, Schantz, and Wagner 2003; Hwang and Foote 2005; Almagro and Strohl, Chapter 13 in this volume). Humanization, the idea of making the V-chains from a murine or other mammalian antibody "more human," was first described in 1986 by Winter and colleagues (Jones et al. 1986). They grafted the complementary determining regions (CDRs) from a murine antibody into the most closely related human framework, followed by making amino acid changes required to stabilize the engineered constructs. Queen and colleagues (Queen et al. 1989; Co and Queen 1991; Ostberg and Queen 1995) at Protein Design Labs (now PDL BioPharma) developed a detailed process for humanizing antibodies via CDR grafting, which has been the basis for humanization of many of the antibodies currently on the market or in advanced clinical trials. Most of the humanized antibodies on the market, or currently in development, have been humanized by some form of CDR grafting. Other forms of humanization that do not include CDR grafting also have been developed, however, most notably resurfacing (also called veneering) of antibodies to remove B-cell epitopes (Roguska et al. 1994, 1996; Staelens et al. 2006).

The first humanized antibody to reach the marketplace was Zenapax®, an anti-CD25 (IL-2 alpha subunit) Mab which was humanized and developed at Protein Design Labs and licensed by Roche, to combat transplant rejection. As can be seen in Figure 1.5 and in Tables 1.3 and 1.4, humanization

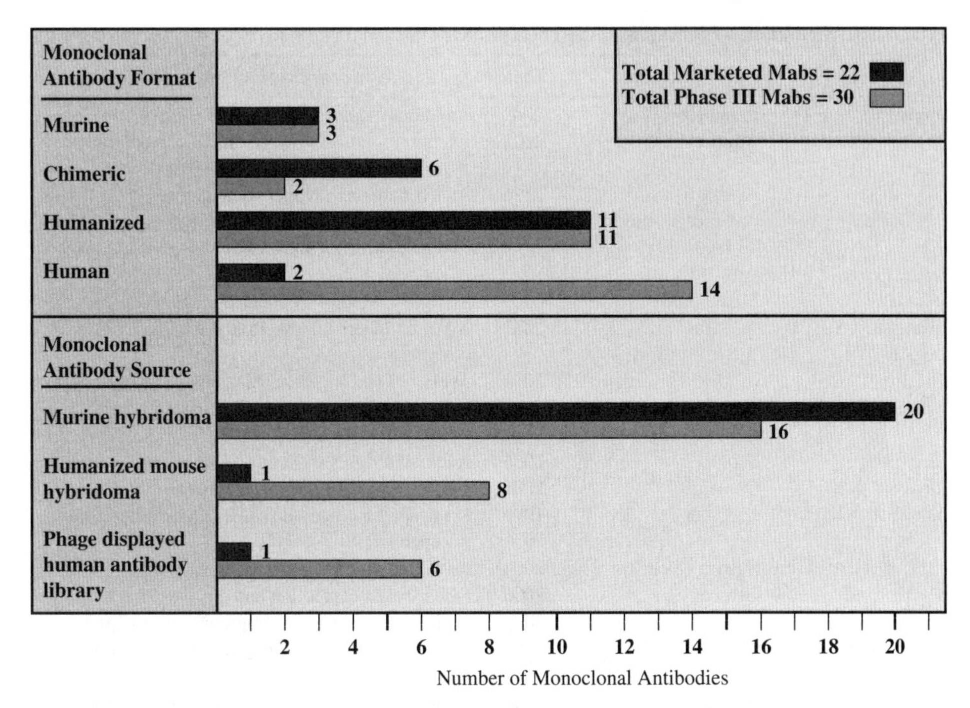

Figure 1.5 Format and primary source of current commercial Mabs and Phase III clinical candidates.

of murine hybridoma-derived monoclonal antibodies remains a major source of therapeutic candidates in the biopharm industry pipeline.

1.3.5 Humanized Mice

It became apparent after Orthoclone OKT3® was marketed that murine antibodies were not going to be acceptable as mainstream therapeutics, especially for chronic indications. As mentioned previously, *in vitro* manipulations were being used to generate chimeric and humanized monoclonal antibodies, and even libraries of phage-displayed human antibodies. Alt, Blackwell, and Yancopoulos (1985) suggested a different possibility altogether, that is, the generation of transgenic mice producing human antibodies ("humanized mice"). Subsequently, two groups separately and independently were successful in developing approaches to generate functional human antibodies directly from transgenic mice. Scientists at Cell Genesys (which later spun off Abgenix, which was acquired by Amgen in 2007; Table 1.5) and GenPharm (acquired in 1997 by Medarex) each engineered mice by disabling the ability of the mice to produce their own murine antibodies, and replacing that function with human antibody genes (Lonberg et al. 1994; Green et al. 1994). Thus, with both systems, immunization of the resultant transgenic "humanized" mice would result in the generation of fully human antibodies by those mice against the antigen (reviewed by Green 1999; Lonberg 2005). The first fully human antibody to be developed and marketed from one of these humanized mouse systems was Vectibix®, a human IgG2 antibody discovered using the Abgenix XenoMouse™ technology, in 2006, 12 years after the publication of the key paper demonstrating the construction of the mice (Fig. 1.6). Fully human antibodies derived from humanized mice make up 27 percent of the current Phase III candidates (Table 1.3, Fig. 1.5), and are expected to continue to feed the pipeline. Amgen, for example, has more than a dozen fully human antibodies in clinical trials (Tables 1.2 and 1.3) derived from the Abgenix mouse platform, which likely led to their acquisition of Abgenix in 2005 (Table 1.5).

TABLE 1.5 Significant Acquisitions in the Therapeutic Monoclonal Antibody Space*

Acquirer	Acquired	Date	Apparent Driver(s) for Acquisition
Amgen	Immunex	2002	Additional rights to Enbrel
	Abgenix	2005	Vectibix® (anti-EGFR); transgenic humanized mouse technology
	Avidia	2006	Anti-IL6 avimer; avimer technology
Astra-Zeneca	Cambridge Antibody Technology	2006	CAT354 (anti-IL-13) and CAT-3888 (anti-CD22-Ps. immunotoxin fusion); phage displayed human antibody libraries
	Medimmune	2007	Synagis®, Numax®, extended pipeline
Bristol-Meyers Squibb	Adnexus	2007	Adnectin (alternative scaffold) technology; discovery engine
Eli Lilly	Applied Molecular Evolution (AME)	2004	AME-133 (second generation anti-CD20), AME-527 (second generation anti-TNF-α), and discovery engine
Genentech	Tanox	2006	Additional rights to Xolair
Glaxo Smith-Kline	Domantis	2006	Domain antibodies; discovery engine
Johnson & Johnson	Centocor	1999	Remicade® antibody discovery and development capabilities
	Egea	2004	Advanced protein optimization technology
Merck	GlycoFi	2006	*Pichia*-based expression and glycosylation technology
	Abmaxis	2006	Antibody structure-based optimization technology
Novartis	Chiron	2006	Vaccines, but also antibody discovery and early development experience
	Neutec	2006	Aurograb® (antistaphylococcal Mab) and Mycograb® (antifungal Mab)
Pfizer	Bioren	2005	Antibody optimization technology
	Rinat	2006	RN624 (anti-NGF Mab), RN219 (anti-Aβ mab), pipeline, discovery engine
	Biorexis	2007	GLP-1 lead; transferring-fusion protein technology
	Coley	2007	Vaccines; TLR technology
	CovX	2007	CVX-045 (thrombospondin 1 mimetic); CVX-60 (angiopoietin-2 binder); CVX-096 (GLP-1 mimetic); peptide-Mab conjugation technology
Roche	Genentech	1990	Biologics capabilities
	Chugai	2001	Expansion of biologics capabilities and market in Japan
	Therapeutic Human Polyclonals	2007	Polyclonal antibody technology
	Glycart	2007	Afucosyl glycosylation and cell culture technology
Schering-Plough	DNAX	1982	Fundamental biology expertise, especially in cytokines
	Canji	1996	Gene therapy
	Organon	2007	Biologics manufacturing
Wyeth	Haptogen	2007	Shark antibodies; discovery engine

*Data as of August 2008; data obtained from company websites and news releases.

1.3.6 Phage Display Technology

George Smith (1985) demonstrated that peptides could be displayed as fusions of P3 on the tail fibers of the *Escherichia coli* filamentous phage M13. Shortly afterwards, it was realized that this method of display could be used more broadly, including the display of proteins (Markland et al. 1991). It became apparent that phage display would make a great tool for selection of mutants for protein engineering (Bass, Greene, and Wells 1990; Lowman et al. 1991; Markland et al. 1991). Additionally, a phagemid system was described which allowed for monovalent display, which helped considerably in selections,

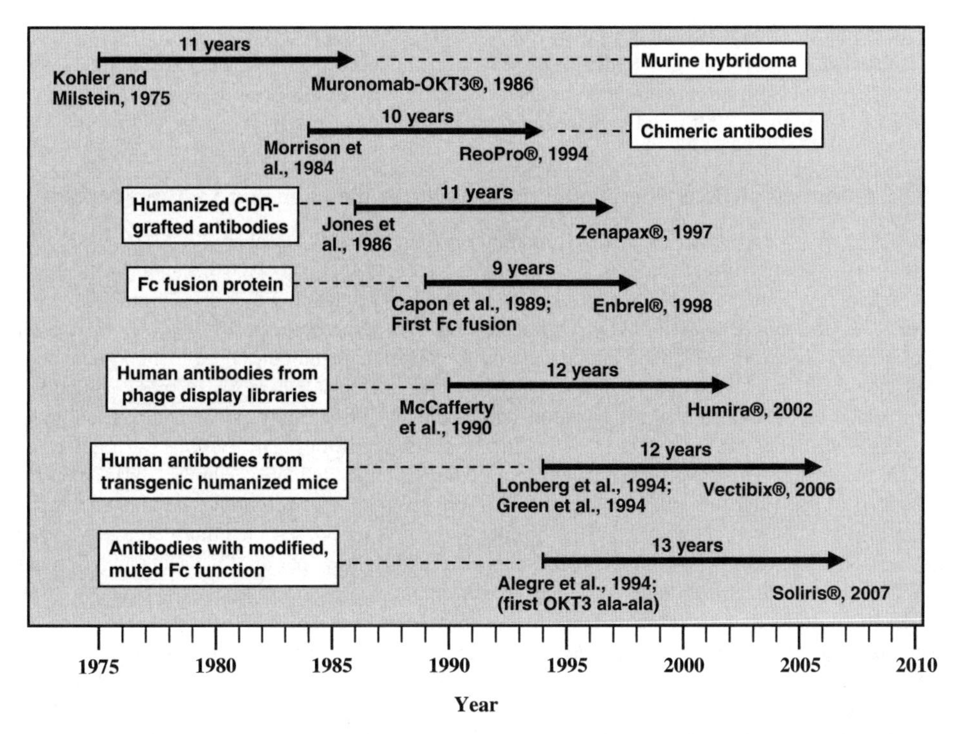

Figure 1.6 Timelines for maturation of technologies to reach the market in the form of a therapeutic Mab or Fc fusion protein. Given is the year of discovery or first publication and the year the technology resulted in the first therapeutic using that technology. Adapted, updated, and modified from Lonberg (2005).

especially for high affinity binders rather than high avidity binders (Barbas et al. 1991). Since antibodies function via binding ligands, M13 P3-based phage display technology became an optimal methodology for selecting modified antibody fragments capable of increased binding capabilities (Gram et al. 1992; Hawkins, Russell, and Winter 1992; Marks et al. 1992a,b).

1.3.7 Human Antibody Libraries

McCafferty et al. (1990) first demonstrated the potential for bypassing immunization altogether by building a library of antibody genes, displayed on the P3 protein of M13 phage, using PCR methodology to recover the human genes from either B-cells or hybridomas. This was followed by the construction of large, human libraries from either synthetic repertoires (Barbas et al. 1992; Griffiths et al. 1994) or from multiple "naïve" human donors (Marks et al. 1991; Vaughan et al. 1996). The latter library from Cambridge Antibody Technology (CaT, now part of MedImmune, a wholly owned subsidiary of Astra-Zeneca; Table 1.5), constructed using single chain Fv constructs fused with His and myc tags, produced sub-nM binders and became the prototype for many later libraries. Since then, there have been many fully human antibody libraries displayed on M13 P3 using either Fab or scFv formats, including (among others), the following examples: a synthetic library used by Crucell (de Kruif et al., 1995), synthetic libraries by MorphoSys (Knappik et al. 2000; Rothe et al. 2008), a large Fab-based library built by scientists at Dyax (Hoet et al. 2005), and a minimalist library generated at Genentech based on a single, well-characterized framework (Fellouse et al. 2007). A wide variety of strategies for building either large libraries or more focused libraries have been recently published and have been reviewed by Hoogenboom (2005), Sidhu and Fellouse (2006), and Mondon et al. (2008). The first antibody from a phage-displayed human antibody library to be approved for therapeutic use is Humira®, which was approved in 2002 (Table 1.2). Interestingly enough, Humira® was not isolated *de novo* from the CaT human antibody library, but instead was isolated via a "guided selection" strategy, using a murine antibody as the primary binder, as has been described by Osbourn, Groves, and Vaughan (2005), and reviewed in Chapter 13 in this book by Almagro and Strohl.

1.3.8 Summary of Core Therapeutic Mab Technologies Leading to Therapeutics

In his 2005 review on human antibodies from transgenic animals, Nils Lonberg (2005) showed timelines from first reports on technologies, such as chimerization, to approval of the first antibody to utilize that technology. Figure 1.6 shows a similar set of timelines, adapted and updated from Lonberg's paper. The first monoclonal antibody, Orthoclone OKT3® was approved in 1986, 11 years after Köhler and Milstein first described the generation of monoclonal antibodies. The first chimeric antibody, ReoPro®, was approved in 1994, 10 years after the first papers on chimeric antibodies (Morrison et al. 1984; Boulianne, Hozumi, and Shulman 1984). The first humanized antibody, Zenapax®, was approved in 1997, 11 years after the first report on CDR grafting by Jones et al. (1986), and the first antibody from a phage-displayed human antibody library, Humira®, was approved in 2002, 12 years after the initial paper by McCafferty et al. (1990) describing the construction of phage-displayed human antibody libraries. Similarly, the first antibody from a transgenic humanized mouse, Vectibix®, was approved in 2006, 12 years after the first descriptions of the generation of functional transgenic humanized mice producing human antibodies (Lonberg et al. 1994; Green et al. 1994). In an antibody-related technology development, Capon et al. (1989) described the first Fc fusion proteins (then termed *immunoadhesin*) using the Fc portion of IgG attached to the exodomain of CD4. Enbrel®, the first Fc fusion protein to be marketed, was approved in 1998 (Table 1.2), nine years later (Fig. 1.6). Finally, in 2007, Soliris® became the first antibody with a modified, or nonnatural IgG (the Fc of Soliris® is an IgG2-4 chimera), to be approved for marketing in the United States (Rother et al. 2007). This antibody, which has significantly decreased ability to bind Fcγ receptors, and thus substantially diminished immunological activity, was approved 13 years after Alegre et al. (1994) described the generation and characterization of the ala-ala mutations of a humanized OKT3, the first detailed description of an Fc-muted antibody.

In summary, it has taken approximately 9 to 13 years for each of these key technologies to result in an FDA-approved therapeutic monoclonal antibody or Fc fusion protein. Considering that discovery, preclinical development, preclinical toxicology studies, clinical trials, and registration of a new molecular entity often take from 8 to 10 years, it is apparent that these technologies were taken up rapidly by the industry and incorporated into the pipeline without significant delay.

1.4 FROM BIOTECHNOLOGY TO BIOPHARMA

1.4.1 From OKT3 to Remicade: Early Successes and Disappointments

The first therapeutic antibody, Orthoclone OKT3® (muronomab CD3), from Ortho Biotech, a subsidiary of Johnson & Johnson, was approved by the U.S. FDA for use in transplantation in June 1986. The original OKT3 monoclonal antibody was isolated in 1979 as a T-cell recognizing antibody (Kung et al. 1979). It took a total of nine years from the discovery of hybridoma technology to put Orthoclone OKT3® on the market, which is remarkably short, considering the time required for development of the technology at Ortho Biotech, use of the technology to target T-cell binding antibodies, understanding of the biology enough to determine a therapeutic indication, and then adding the time required for preclinical toxicology, clinical development, and registration (Fig. 1.6).

While the marketing of Orthoclone OKT3® was a huge breakthrough for the biotechnology industry as a whole, it also came with a serious reality check for two significant issues that have laid the foundation for modern therapeutic monoclonal antibodies. Muronomab-CD3®, a fully murine antibody of the IgG2a murine isotype, was found to be highly immunogenic in people (Kimball et al. 1995), which, given its murine nature, is hardly surprising. Muronomab-CD3® also generated a cytokine storm known as systemic cytokine release syndrome (Chatenoud et al. 1990), shown to be via the interaction of its murine IgG2a Fc with FcγRs on human immune effector cells (Tax et al. 1984; Lobo and Patel 1997). This led to several important studies concerning the nature of Fc-based functionality in therapeutic monoclonal antibodies (as exemplified by Brüggeman et al. 1987; Xu et al. 2000). Ultimately, this has led to the generation of three different humanized versions of anti-CD3 (Teplizumab [hOKT3γ1ala-ala], Visilizumab, and Otelixizumab [ChAglyCD3; TRX4]), all of which lack significant Fc functionality; these antibodies are all currently in clinical trials.

In part because Orthoclone OKT3® caused these significant adverse events, in part because the biotechnology field was rapidly changing with new technologies being proposed and developed, in part because the discovery and development timelines for drugs are so long (averaging approximately 8 to 10 years from discovery to market), and in part because of a devastating clinical failure, the next therapeutic Mab to be approved in the United States was not until 1994, eight years later. This next market entry was ReoPro®, an anti-gpIIb/IIIa murine-human chimeric Fab developed by Centocor and marketed by Eli Lilly. The situation surrounding the development of ReoPro® by Centocor and its licensing to Eli Lilly have been described in detail by Shook (2007). ReoPro® was not intended, initially, to be the first therapeutic monoclonal antibody from Centocor. Instead, in the early 1990s, most of the resources were focused on the development of a sepsis monoclonal antibody, targeting bacterial lipopolysaccharide, named Centoxin. With Centoxin having looked positive in late stage clinical trials, and even achieving licensure in several European countries, Centocor invested heavily in development and manufacturing capabilities as it ran Phase III clinical trials to demonstrate unequivocal efficacy. Unfortunately, those trials were not unequivocal and the FDA ordered additional trials to be run. At that time, Centocor was operating with substantial and rising losses, and no drug on the market to buffer the costs with incoming revenue, so they made a licensing deal with Eli Lilly to share the risk, as well as give rights to Lilly for another Centocor drug in development, ReoPro®. The final clinical trials on Centoxin ultimately demonstrated not only lack of efficacy, but also higher death rates in one group of patients, resulting in a complete cessation in the development of that candidate. At this point, Centocor licensed the rights to Panorex® (murine monoclonal IgG2a antibody known originally as 171A, which recognizes a 37 to 40 kDa cell surface glycoprotein expressed on malignant and normal epithelial cells; approved in 1995 in Europe for colorectal cancer therapy) to Wellcome, reduced its workforce and overhead costs, and focused all remaining resources on development of ReoPro®, which eventually was approved in 1994, effectively keeping Centocor afloat as an independent company (Shook 2007). By 1997, Centocor became Pennsylvania's first biotech company to make a profit (Shook 2007).

Shortly thereafter, Centocor initiated collaborative studies with Jan Vilcek, New York University School of Medicine, on a murine anti-TNF-α antibody named mA2. The antibody was converted to a murine-human chimeric antibody (cA2), and developed both for Crohn disease (FDA approval granted in 1998) and rheumatoid arthritis (FDA approval granted in 1999), marketed by Centocor in the United States and Schering-Plough overseas. The successes of ReoPro and Remicade ultimately led to the acquisition of Centocor by Johnson & Johnson in 1999 (Table 1.6). Notably, through its ownership of Ortho Biotech and its acquisition of Centocor in 1999, J&J had a stake in four (Orthoclone OKT3®, ReoPro®, and Remicade® worldwide; Panorex® in Europe) of the first seven monoclonal antibodies to be approved.

1.4.2 Examples of Other Early Mabs

In the meantime, Rituxan®, a mouse-human chimeric anti-CD20 monoclonal antibody (originally known as C2B8), was being developed by IDEC for non-Hodgkin's lymphoma. IDEC, which was founded in 1985, had begun work on C2B8 in the early 1990s (Reff et al. 1994), and entered Phase I clinical trials with it in 1993. Significantly, it was recognized from the very beginning that C2B8 possessed the ability to kill target B-cells via antibody-dependent cellular cytotoxicity (ADCC) and complement-mediated cytotoxicity (CDC), which laid the groundwork for literally thousands of

TABLE 1.6 Comparison of the Top Biopharmaceutical Companies with Respect to Biopharma Pipelines and Future Potential*

Company	Estimated Sales of Top Biologics (2006) in millions	Public Late Pipeline (Phase II or III)*	Key Biologics Pipeline Candidates	Key Technologies	Potential Future Ranking
Amgen	12,140	$N = 10$	Romiplostim; Denosumab	Peptibody; Abgenix mice; Avidia avimers	2
Roche (not including Genentech)	6,895	$N = 4$	Omnitarg; Ocrelizumab	Glycart afucosyl glycosylation; THP polyclonals; alternative scaffolds (MP)	1**
Genentech	6,765	$N = 7$	Omnitarg; Ocrelizumab	Fc engineering; Afucosyl Mabs; antibody libraries, humanization; protein engineering; toxin conjugates	See Roche**
Novo Nordisk	5,930	$N = 11$	NovoSeven; Liraglutide	Focus on diabetes and hormone replacement therapies; no Mabs	3
Johnson & Johnson	5,590	$N = 5$	Ustikinumab; Golimumab	Library; Medarex mice; humanization	4
Eli Lilly	3,635	$N = 5$	Teplizumab	Focus on diabetes and neurobiology	8
Merck-Serono	2,535	$N = 1$	Cetuximab	—	—
Sanofi-Aventis	2,485	$N = 3$	Aflibercept (VEGF trap)	Regeneron license—humanized mice	9
Schering Plough	2,305	$N = 2$	Golimumab; Acadesine	—	10
Wyeth	2,265	$N = 7$	Bapineuumab (anti-Aβ)	Trubion, alternative scaffolds, shark domain Abs (small binders)	5
Abbott	2,190	$N = 1$	ABT874 (anti-IL12/23)	—	—
Bayer AG	2,145	$N = ?$	—	—	—
Biogen Idec	1,775	$N = 4$	Galixumab (anti-CD80); Lumilixumab (anti-CD23)	Bispecifics; protein engineering	—
Baxter	1,700	$N = ?$	—	—	—
Astra-Zeneca (including Medimmune and CaT)	910	$N = 5$	Motavizumab; anti-IL-5R, IL-9, IL-13 MAbs	CaT Library; half-life extension; BiTE technology (Micromet)	7
Pfizer	880	$N = 10$	Ticilizumab (anti-CTLA4)	Medarex license; Biorexus transferrin fusions; CovX Ab-peptide fusions	6
Chugai	855	$N = 1$	Tocilizumab (Actemra)	—	—
Bristol-Myers Squibb	771	$N = 3$	Belatacept; Ipilizumab	Adnectin alternative scaffold	11
Kirin	515	$N = 1$	TPO mimetic	Humanized mice	—
Glaxo-Smith Kline	—	$N = 6$	Belimumab; Mepolizumab; Ofatumumab	Domantis domain antibodies; Afucosyl Mabs (via BioWa license)	12

*Based on publicly stated, novel biologics (not market extension or expansion of existing biologics) in Phase II or III; data as of August 2008; data obtained from company websites, Prous Science Integrity, and www.clinicaltrials.gov.

**Roche and Genentech combined, based on the 7/21/08 offer by Roche to acquire remaining shares of Genentech they currently do not own.

studies since then on these important mechanisms of action for oncology indications. In 1995, after strong Phase II clinical results were reported, IDEC signed an agreement with Genentech to collaborate on the Phase III clinical development and marketing of C2B8. Simultaneously, Genentech also was conducting Phase III clinical trials on Herceptin® for breast cancer and Actimmune® (interferon-γ1b; eventually discontinued in 1996 by Genentech) for renal cell carcinoma, as well as preclinical studies on an anti-VEGF monoclonal antibody (which eventually was to be developed into bevacizimab, approved as Avastin® in 2004). In 1997, Rituxan® (C2B8) was approved as the third monoclonal antibody to reach the marketplace in the United States, and in 1998, Herceptin® was approved as the sixth Mab to be marketed in the United States.

1.4.3 Evolution of the Biotechnology Industry to the New BioPharma Industry

The period of technology development in the 1970s and 1980s quickly gave way to an era of biotechnology start-up companies, many of which were in the San Francisco Bay area, that effectively drove the biotechnology boom. Cetus was the first major biotechnology company, formed in 1971 in the Berkeley area. Genentech was founded in South San Francisco in 1976 as a shortened name for "genetic engineering technology," Amgen was formed in 1980 as AMGen (for Applied Molecular Genetics), Xoma was started in 1980, Chiron was founded in 1981 to find a vaccine for hepatitis B, and SCIOS was formed in 1981 as Cal Bio (for California Biotechnology). In all, 112 biotechnology companies were started in the Bay Area by 1987. In Europe, Biogen was formed in 1979, and Celltech, the first biotechnology company founded in the United Kingdom, was started in 1980. Toward the end of this period, Medimmune was started in Maryland in 1987 as Molecular Vaccines, Centocor was founded in Philadelphia in 1989 to take advantage of the new antibody revolution, and Cambridge Antibody Technology was opened in Cambridge, UK, in 1989.

Since then, there have been many changes in the biopharma industry, with a significant number of acquisitions, mergers, and licensing deals, driven by the need of large pharmaceutical companies to build their pipelines. As mentioned earlier in this chapter, Mabs and Fc fusions proteins have been playing an increasing role in pharma pipeline portfolios, so the intense competition for the most important technologies, drug candidates, and intellectual property has driven and will continue to drive such mergers and acquisitions (Table 1.5). Notably, of the companies mentioned above, Roche now owns a majority stake in Genentech and has recently (July 21, 2008) tendered an offer for all remaining shares it does not own. This would make Roche/Genentech the largest biopharma player with the strongest pipeline in the business (Table 1.6), taking over from Amgen, who had the strongest combined sales and portfolio in 2006 when the marketing data used in this analysis were gathered. Additionally, of those biotechnology companies listed above, Biogen has since merged with IDEC, Celltech has become a part of UCB, MedImmune and Cambridge Antibody Technology were acquired by Astra-Zeneca in 2006 and 2007, respectively, and, as mentioned previously in this chapter, Centocor was acquired in 1999 by Johnson & Johnson (Table 1.5). Other major acquisitions relevant to the monoclonal antibody field are also listed in Table 1.5.

Table 1.6 shows the top 20 major biopharma, their public pipelines, important advanced clinical candidates, key technologies, a current ranking (based on 2006 market information), and a projected future ranking, taking the factors mentioned into consideration. Other than Novo Nordisk, whose pipeline is focused almost entirely on its strong diabetes franchise, and Bayer-Schering and Baxter, which have diversified biologics pipelines, the top biopharma companies have placed significant efforts and resources into the discovery, development, and commercialization of monoclonal antibodies and Fc fusion proteins (Table 1.6).

Significantly, virtually all of the major large pharma companies have joined in the search for monoclonal antibody and Fc fusion protein drugs. Even companies not invested, or not significantly so, in the late 1990s, such as Merck, Sanofi-Aventis, and Astra-Zeneca, are now players, along with other large pharma that have been in the field longer, for example, Johnson & Johnson (the first company to bring any monoclonal antibody to the market), Roche, Novartis, and Wyeth. With the greater number of players, and the increased significance placed on biologics approaches by all of these companies, the competition for new biologics, well validated biologics-friendly targets, and biologics markets has become incredibly intense.

1.5 CHALLENGES AND OPPORTUNITIES FOR MONOCLONAL ANTIBODIES AND Fc FUSION PROTEINS

1.5.1 SWOT Analysis

Figure 1.7 shows a SWOT (strengths, weaknesses, opportunities, threats) diagram for the Mab and Fc fusion market. The key threats include safety concerns in the post-Tegenero TGN-1412 debacle (Haller, Cosenza, and Sullivan 2008), the high cost-of-goods (including production, purification, formulation, packaging, and delivery) of Mabs, which lead to significant market pressures from third-part payors, and small molecules that could enter markets currently dominated by biologics (Ziegelbauer and Light 2008). Additionally, the impending follow-on biologics revolution is eventually expected to impact several Mabs coming off patent, although the regulatory environment supporting this still remains unclear today (see Williams and Strohl, Chapter 32 in this volume). The most significant of these threats to innovator biologics is probably cost and impending follow-on biologics. A final threat, or perhaps weakness, in the field is that there is the perception that there is a limited number of targets, and thus, many companies compete for market share on the same "hot" targets.

1.5.2 Competition on "Hot" Targets

As pointed out earlier in this chapter, there are 27 marketed therapeutic antibodies and Fc fusion proteins, and another 170+ in various stages of clinical trials. While many of these clinical candidate biologics target novel mechanisms, there are also several that target the same molecule. Some of these already face significant competition on the market. For example, Remicade®, Enbrel®, Humira®, and Cimzia® are now all marketed as anti-TNF-α therapeutics for the treatment of rheumatoid arthritis (Table 1.2). A BLA was recently submitted for golimumab in the same field and CytoFab is still in

Major Strengths:	**Major Weaknesses:**
• Targets that cannot be addressed with small molecules (e.g., protein–protein interactions) • Half-life leads to less frequent dosing • Efficacy • ADCC/CDC (i.e., immune system functionality)	• Parenteral delivery (IV, SC) • Limitation to extracellular and cell-surface targets • Cost, and cost of goods driving the pricing • Immunogenicity and injection site reactions
Major Opportunities:	**Major Threats:**
• Delivery improvements (e.g., transdermal, oral, intranasal) • Modified Fc; fine-tuning immune system functionality • Extended and/or tunable T1/2 • Multispecificity (e.g., ability to engage multiple targets while retaining long T1/2 • Novel scaffolds, approaches • Tissue targeting, e.g., ability to cross BBB	• Safety concerns in a post-Tegenero TGN-1412 era • Small molecules functioning in same pathways as biologic • Third party payer restrictions on reimbursement • Follow-on-biologics • Perception of a limited number of high quality targets leading to intense competition on certain "hot" targets (e.g., TNF-α, CD20)

Figure 1.7 SWOT diagram for therapeutic Mabs and Fc fusion proteins.

clinical trials for sepsis (Tables 1.2 and 1.3), bringing the total to six competitive molecules directed against TNF-α. Those anti-TNF-α antibodies able to out-compete on the basis of potency, dosing frequency, route of administration, and safety (particularly low infection rates and injection site reactions, and lack of immunogenicity) should ultimately garner the most significant shares of the steady-state market. It should be noted that there are public reports on efforts to make follow-on anti-TNF-α biologics as well, although how these will be developed and commercialized is still unknown due to a lack of clarity concerning U.S. policy in this area (see Chapter 32).

The most competitive target, based on the number of known molecules under development, is CD20 (Table 1.7). To date, there are at least 11 known anti-CD20 Mabs in development for oncology and/or rheumatoid arthritis indications (Table 1.7). These known candidates roughly fall into two categories known as Type I and Type II anti-CD20 Mabs. The type I anti-CD20 antibodies, such as Rituxan®, Ofatumumab (Humax CD20; Hagenbook et al. 2005), Ocrelizumab (Genentech's second generation anti-CD20), and Veltuzumab, are thought to function primarily by antibody-dependent cellular cytotoxicity (ADCC) and complement-dependent cytotoxicity (CDC), and promote translocation into lipid rafts, but probably do not induce apoptosis to a substantial degree (data summarized from Teeling et al. 2004; Bello and Sotomayor 2007; Maloney 2007; Glennie et al. 2007; Leonard et al. 2008; Beers et al. 2008). Type II anti-CD20 antibodies, such as Tositumumab, AME-133v (Weiner et al. 2005), and GA-101 (Cragg and Glennie 2004; Umana et al. 2006), are generally thought to function through ADCC and apoptosis, but do not significantly function via CDC nor allow translocation into lipid rafts (Table 1.7).

The competition and efforts to sort out the biology of CD20 and the Mabs attacking it will likely lead to clear winners and losers in the marketplace based on efficacy measured by patient survival, as well as use in patients who are heterozygous FcγRIIIa 158V/F or homozygous FcγRIIIa 158F/F, polymorphs of FcγRIIIa for which IgGs have lower affinity (Cartron et al. 2002; Weng and Levy 2003; Ghielmini et al. 2005). These patients have been shown clinically to respond more poorly to Rituxan® than patients who are homozygous FcγRIIIa 158V/V, the higher affinity form of the receptor. It is believed that anti-CD20 antibodies with modified Fc functionality to impart tighter binding to FcγRIIIa will help to overcome the problems observed with the V/F and F/F polymorphisms (Bello and Sotomayer, 2007). The intense competition also should lead to a significantly greater understanding of tumor biology and disease mechanisms, as well as tumor killing mechanisms, and ultimately should help to improve the efficacy of future antibodies targeting CD20, as well as other oncology targets.

Another hot target for oncology is IGF-1R, for which there are at least seven anti-IGF-1R clinical candidates in Phase I or II, as noted in Table 1.4. Similarly, Feng and Dimitrov (2008) described eight different antibodies in clinical programs targeting that receptor, making it perhaps the single most competitive Mab target for which there is not yet a marketed drug. Other targets on which there is considerable competition include: IL-6/IL-6R (five known competitors in the clinic), VEGF/VEGFR (6 known competitors in the clinic, although two receptor types covered), IL-13 (four known competitors in the clinic), and amyloid-beta (four known competitors in the clinic). Targets for which there are at least three known competitor Mabs/Fc fusion proteins in clinical trial include: CD19, CD22, CD30, CD40, IFN-γ/IFN-R, IL-1, EpCAM, and alpha-V integrin (Tables 1.2 to 1.4).

1.5.3 Targets

Perhaps the two most important opportunities in the field of Mabs and Fc fusion proteins are (1) the increase in basic scientific knowledge around many novel targets that may provide new opportunities for therapeutic interventions using biologics; and (2) the burgeoning understanding of antibody biology and how to engineer antibodies to function optimally for a desired target.

Figure 1.8 shows the general classes of targets for all of the marketed therapeutic Mabs and Fc fusion proteins, as well as those in Phase III clinical trials. Interestingly, nearly two-thirds of these targets are receptor or cell-surface targets, whereas only approximately 30 percent fall into the cytokine or soluble protein group. Five belong to infectious disease entities. Of the receptor-based targets, there are no marketed Mabs or Phase III candidates for proteases, G-protein coupled receptors (GPCRs), or ion

TABLE 1.7 Intense Competition: Well-Known Antibodies Targeting CD20

Name of Molecule	Company	Date Approved or Current Phase	Type*	Primary Indication	Comments
Rituxan® (Rituximab)	Biogen/Idec/ Genentech	11/26/1997	I	NHL, RA added 2/8/06	IgG1κ, chimeric
Zevalin® (Ibritumomab tiuxetan)	Biogen/Idec	02/19/2002	UNK	NHL	Murine IgG1κ conjugate, Y-90 or In-111
Bexxar® (Tositumomab-I131; B-1 Mab)	Corixa	06/27/2003	II	NHL	Murine IgG2a/ λ-I-131
Ofatumumab (Humax CD20)	GenMab/Glaxo Smith-Kline	Phase III	I	B-cell chronic leukemia; NHL, RA	Humanized IgG1; Targets N-terminus of CD20; strong CDC activity
Ocrelizumab (Second gen. anti-CD20; 2H7; PRO-70769; R-1594)	Genentech	Phase III	I	RA, Lupus, relapsing MS	Humanized IgG1
Veltuzumab (Ha-20; IMMU-106)	Immunomedics	Phase I/II	I	NHL, autoimmune diseases	Humanized IgG
PRO-131921 (Third generation ant-CD20)	Genentech	Phase I/II	I	Oncology	IgG with modified Fc for increased ADCC
AME-133v (LY2469298)	Eli Lilly	Phase I/II	II	NHL	IgG with modified Fc for increased ADCC
TRU-015	Wyeth	Phase II	I	RA	Small modular immunopharmaceutical product (SMIP)
R-7159	Roche	Phase II	II	NHL	IgG; glycoengineered, afucosylated Mab; strong apoptotic efect
GA-101	Roche	Preclinical	II	Oncology	Afucosylated form generated using Glycart technology

Abbreviations: MS, multiple sclerosis; NHL, non-Hodgkin's lymphoma; RA, rheumatoid arthritis; UNK, unknown.

*Type I anti-CD20 antibodies are thought to possess the following general characteristics: function by antibody-dependent cellular cytotoxicity (ADCC) and complement-dependent cytotoxicity (CDC), and allow translocation by lipid rafts, but probably do not induce apoptosis to a substantial degree. Type II anti-CD20 antibodies function through ADCC and apoptosis but do not function via CDC, nor promote translocation into lipid rafts. Data accumulated and summarized from Cragg and Glennie (2004); Teeling et al. (2004); Hagenbook et al. (2005); Weiner et al. (2005); Umana et al. (2006); Bello and Sotomayor (2007); Maloney (2007); Glennie et al. (2007); Leonard et al. (2008); Beers et al. (2008).

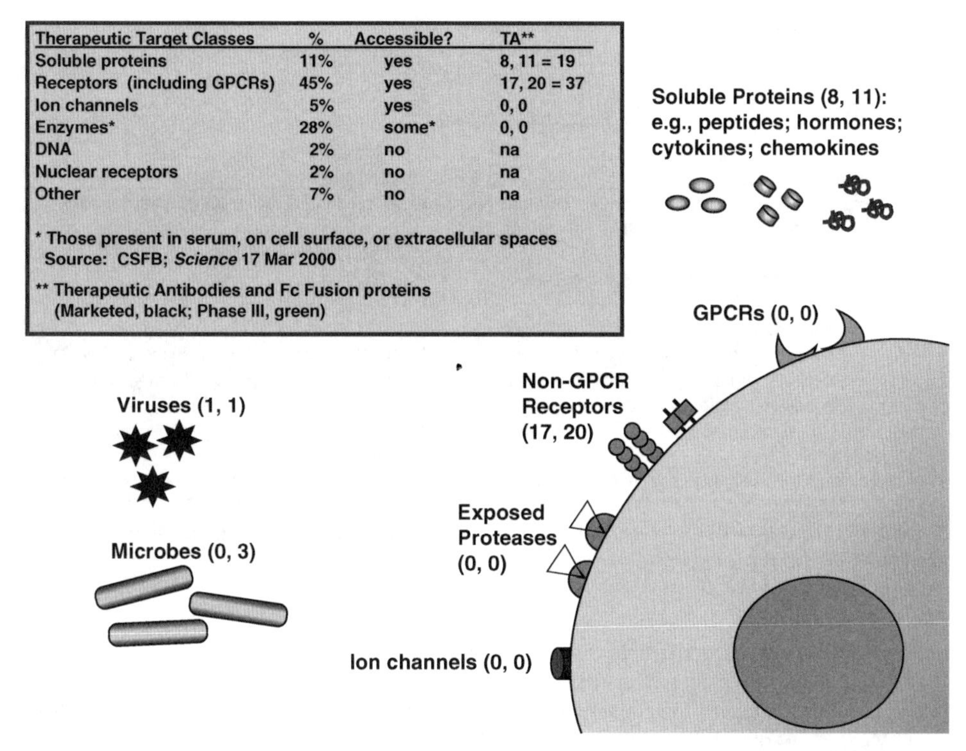

Figure 1.8 Cartoon exemplifying the types of targets for currently marketed and Phase III candidate Mabs and Fc fusion proteins. The first number in each pair indicates the current marketed number, while the second in each pair indicates the number in Phase III clinical trials (based on data from Tables 1.1 and 1.2).

channels. The overall trend also was observed with the earlier clinical candidates, in which 78 percent of the candidates targeted receptors, 37 percent targeted soluble ligands, and 5 percent targeted infectious agents. Figure 1.9 shows a comparison of a profile on what might be considered a perfect target for a therapeutic Mab or Fc fusion protein versus a profile that might be considered more challenging. This comparison, which reflects the author's bias, is interesting in that the majority of clinically validated targets are more closely aligned with what may be considered the more challenging type of targets. This type of comparison, however, can be skewed significantly by the needs of a therapeutic area. The significant number of Mabs against oncology cell-surface targets is a prime example of this kind of skewing.

Obtaining intellectual property rights on important targets appropriate for biologics approaches will be a critical issue that cannot ignored. There are only a limited number of new targets coming available each year for which the biology is compelling; that is, the target is appropriate for a biologic, and the human or rodent genetics data are strongly supportive. A few recent new targets (i.e., not yet validated with a marketed drug, but possessing strong preclinical proof-of-pharmacology) that fall into this class are IGF-1R, IL-13, TRAIL-R1, IL-1β, IL-6, and myostatin, all of which have attracted significant competition (Tables 1.3 and 1.4). It is clear that as more companies focus on biologics approaches, the competition on each new target will become only greater, and will require companies to figure out strategies to differentiate their biologics molecules from the entries of their competitors.

1.5.4 Differentiation and Fit-for-Purpose Biologics

The first antibodies to be marketed were largely based on a standard IgG1 platform. The concept of making antibodies that go beyond what nature gives us, that is, to design antibodies that fit the

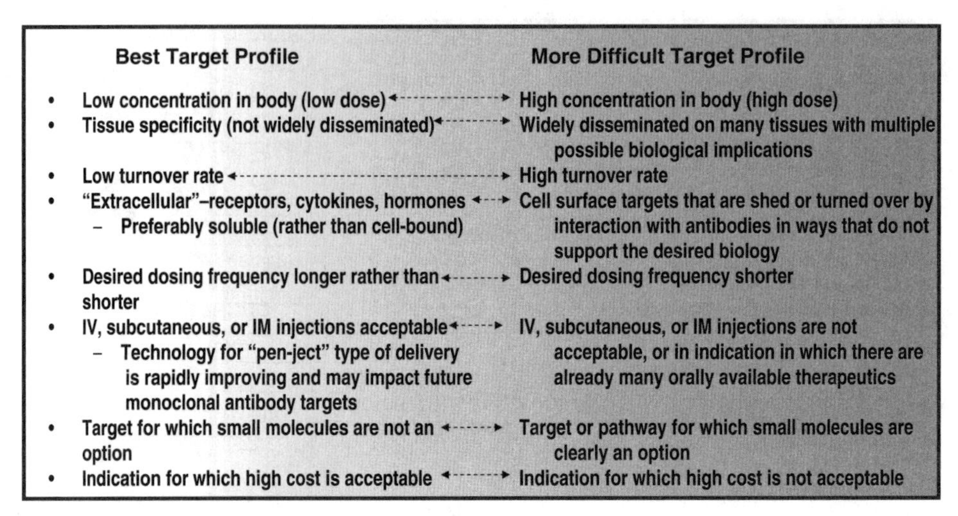

Figure 1.9 Characteristics of an optimal target for a therapeutic Mab as compared with the characteristics of a significantly more challenging target.

needs rather than just accept the biology dictated by natural IgG isotypes (e.g., IgG1, 2, or 4), has been around since the early 1980s (see Neuberger, Williams, and Fox 1984), so it is nothing new to speak of "fit-for-purpose" biologics. In another striking example of this, a search of "monoclonal + ADCC" in PubMed reveals over 1700 references; another for "monoclonal + engineered" yielded more than 1100 references. Moreover, dozens of recent reviews have, in fact, highlighted the need for making antibodies that are more tuned to the biology required of them; a few are listed here for reference (Stockwin and Holmes 2003; Chowdhury and Wu 2005; Laffly and Sodoyer 2005; Carter 2006; Presta 2006, 2008; Jefferis 2007; Dimitrov and Marks 2008).

Several companies already have made significant efforts in modifying Mabs or Fc fusion proteins to make them more fit-for-purpose, represented by the presence of marketed products or advanced clinical candidates. As shown in Table 1.8, several antibodies and Fc fusion proteins already on the market have incorporated significant modifications or alternative formats to a standard human IgG1 to be considered engineered for fit-for-purpose. Certainly, the efforts to generate humanized, and then fully human, Mabs to reduce immunogenicity were the first step in this direction (review by Almagro and Strohl, Chapter 13). Antibodies and Fc fusion proteins concentrated enough for dosing subcutaneously, such as Enbrel®, Xolair®, and Humira®, also fall into this category, as they provide the patient and health care workers with a preferred form of dosing (Table 1.8). Important firsts also include the first alternative isotypes (IgG4 [Bexxar®], IgG2 [Vectibix®]), conjugation with either cytotoxic chemicals or radionuclides for targeting cancer tissue, the peptibody (NPlate®) from Amgen, recently approved for marketing by the FDA, and alternative formats such as a site-specifically PEGylated Fab (Cimzia®) (Table 1.8).

Table 1.9 shows an impressive list of fit-for-purpose engineered molecules currently in clinical trials. These molecules generally fall into three categories: (1) modification of Fc functionality relating to either FcγR, complement, or FcRn interactions; (2) half-life extension of peptides, scFvs, or Fabs using either PEGylation or fusion to Ig domains; and (3) multispecificity. It is clear that several companies are working hard to identify strategies to develop molecules that can be differentiated from the competition (Carter 2006; Dimitrov and Marks 2008).

From a therapeutic standpoint, the most critical feature of a product is that it must have demonstrated clinical efficacy, which relies on a combination of many factors, including target biology, potency, safety, proper dosing, selection of patient population, and so forth. These are all fundamental issues that must be considered when starting a therapeutic biologics project. Nevertheless, there are several

TABLE 1.8 Table of Significant "Marketplace Firsts" in Therapeutic Antibody and Fc-Fusion Protein Drug Discovery

Technology or Process "First"	Product	Company or Inventor	Effective Date	Notes/Comments
IgG marketed	Orthoclone OKT3® (Muromonab-CD3)	Ortho Biotech (Johnson & Johnson)	06/19/1986	Murine IgG
Fab marketed	ReoPro® (Abciximab)	Centocor (now Johnson & Johnson)/Lilly	12/22/1994	Also first chimeric antibody of any kind and first recombinant antibody produced in *E. coli*
Recombinant antibody produced by CHO or NS0	Rituxan® (Rituximab)	Biogen/Idec/Genentech	11/26/1997	Also first chimeric Mab (IgG) marketed
Humanized Mab marketed	Zenapax® (Daclizumab)	PDL/Roche	12/10/1997	Now, 11 humanized antibodies are on the market
Subcutaneous formulation and administration	Enbrel® (Etanercept)	Immunex (now Amgen)	11/02/1998	Now also available with a "SureClick®" Autoinjector device
Mab-conjugate marketed	Mylotarg® (Gemtuzumab ozogamicin)	Wyeth	05/17/2000	Conjugated to the cytotoxic drug, ozogamicin, a calecheamycin (natural product cytotoxin)
Phage displayed human antibody marketed	Humira® (Adalimumab)	Cambridge Antibody Technology/Abbott	12/31/2002	Known before marketing as "D2E7," the indicators for ELISA plate and wells from which it was isolated
Mab radionuclide conj	Bexxar® (Tositumomab-I131)	Corixa	06/27/2003	Reconstitution with radiolabel takes place at clinical site
First IgG4 isotype (or any other than IgG1) antibody to be approved	Mylotarg® (Gemtuzumab ozogamicin)	Wyeth	05/17/2000	Tysabri (Natalizumab) is second example of IgG4 to be marketed; Vectibix is first IgG2 to be marketed
Intravitreal injection	Lucentis® (Ranibizumab)	Genentech/Novartis	06/30/2006	Different V chains than used in Avastin
Mab from transgenic humanized mouse source	Vectibix® (Panitumumab)	Amgen	09/27/2006	Also first IgG2 isotype antibody to be approved
Modified Fc functionality	Soliris® (Ecolizumab)	Alexion Pharma	03/16/2007	Humanized IgG2/4 hybrid to significantly reduce Fc functionality while retaining FcRn-binding mediated half-life
PEGylated antibody marketed	Cimzia® (Certolizumab pegol)	UCB/Schwartz	04/22/2008	Site-specific PEGylation
Peptide-Fc construct	Nplate® (AMG531; Romiplostim)	Amgen	08/22/2008	"Peptibody" construct recently approved for marketing

TABLE 1.9 Examples of Mab and Fc Fusion Protein Engineering to Generate Greater Molecule Fitness for Therapeutic Purpose

Property	Effect	Example(s)
Fc mutations resulting in increased binding to FcγRIIIa	Increased antibody-dependent cellular cytotoxicity (ADCC) as measured preclinically in antibody-dependent cell-killing assays. While widely hypothesized, clinical efficacy improvements attributed to ADCC are still to be determined.	PRO-131921; AME-133v (LY2469298)
Afucosylated antibody	Increased ADCC as measured preclinically in antibody-dependent cell-killing assays. Multiple approaches to achieve afucosylated Mabs have been published, as described in the text.	Anti-CD20 GA-101
Fc mutations resulting in increased binding to FcγRIIa	Increased opsonophagocytosis or ADCP (antibody-dependent phagocytic cytotoxicity), as measured preclinically, in antibody-dependent bacterial or tumor cell-killing assays. Stills need clinical proof-of-concept.	Richards et al., 2008
Fc mutations resulting in decreased binding to FcγRIIa and FcγRIIIa for safety, or to provide a greater efficacy to toxicity window	Murine Orthoclone OKT3® elicited both strong anti-antibody responses because it was a murine antibody, as well as systemic cytokine release syndrome (SCRS), as described in the text. Teplizumab, humanized OKT3γ1ala-ala, yields lowered immunogenicity as well as significantly decreased incidence of SCRS.	Teplizumab, humanized OKT3γ1(ala-ala)
Aglycosylated Mab	Descreased binding to FcγRs resulting in no ADCC; complement activation relatively unaffected; Otelixizumab (ChAglyCD3) to reduce FcγR interaction.	Aglycosylated anti-CD3
Modified pH-dependent binding to FcRn	Longer or shorter half-life as compared with wild-type Fc.	MEDI-557 (Numax-YTE; MEDI-524-YTE)
Tissue distribution	> In making the Numax® FcRn mutant of MED-524 (yielding MEDI-557), it was observed that the change in FcRn binding significantly altered tissue distribution into the lung (Dall'Acqua, Kiener, and Wu 2006). > Also, it has long been known that the size of an antibody construct can have an impact on biodistribution and tissue penetration (Colcher et al. 1998).	MEDI-557 (Numax-YTE)
Affinity—K_{on} vs K_{off}	> In the affinity maturation of Numax®, differences in K_{on} vs K_{off} resulted in significant differences in potency, indicating the importance of both parameters contributing to K_D (Wu et al. 2005). > Additionally, affinity has been shown to be an important factor in both tumor penetration and strength of ADCC response, as described in the text (Adams et al. 2001; Tang et al. 2007).	MEDI-557 (Numax-YTE)

(*Continued*)

TABLE 1.9 *Continued*

Property	Effect	Example(s)
Peptibody or mimetibody approach	Peptide agonists fused with Fc to impart substantially longer half-life; replacement of PEGylation as a method for improving serum half-life of peptides.	Romiplostim
Antibody-peptide fusions or conjugates	An agonist or tissue-targeting peptide is conjugated with a biologically active antibody possessing antagonist or agonist activity. CovX (now Pfizer) utilizes a semisynthetic linker-based method to achieve peptide stabilization.	Pfizer/CovX CVX-045 (thrombospondin 1 mimetic); CVX-60 (angiopoietin-2 binder); CVX-096 (GLP-1 mimetic)
Antibody conjugates	Toxin-conjugates, radioconjugates, siRNA conjugates, and perhaps small molecule conjugates, all fundamentally tissue targeting modalities, are likely to become more sophisticated approaches to delivery of small molecule entities to specific tissues, targets, and compartments.	Toxin conjugates (Mylotarg®) and radio-conjugate (Bexxar®, Zevalin®)
Antibody fragments such as Fabs, scFvs, domain antibodies, etc.	Small, short half-life molecules that may possess increased ability to penetrate tissues; or retain other desired properties inherent with their smaller size, lack of Fc effector activity, and/or short serum half-life.	ReoPro and Lucentis are Fab constructs; Cimzia is a site-specifically PEGylated Fab construct
Bispecific antibodies lacking Fc domains	Bispecificity coupled with short serum half-life may be perfect fit for some targets and indications.	BiTEs (Micromet MT103/MEDI-538)
Bispecific approaches using IgG scaffolds	Ability to engage two targets simultaneously while retaining long half-life of typically IgGs and effector functionality (Ridgeway, Presta, and Carter 1996; Wu et al. 2007b). One clinical example is the Trion Triomab® technology, using a bispecific CD3 and either CD20 or Her2/neu built into IgG with modified Fc for improved FcγR binding (Shen et al. 2006).	Catumaxomab; Ertumaxomab

additional factors that can help, in some cases, to differentiate molecules against the same target from one another. So what are the technologies of the future that will help to differentiate a clinical candidate from the competition? A few of the high-level considerations are listed below, followed by a more in-depth treatise on a few of them:

- Molecules that provide a greater margin of safety, or larger efficacy to toxicity window
- Delivery—route and/or ease of administration (subcutaneous route being more preferred in most cases over IV)
- Tissue distribution and penetration
- Tuning Fc functionality to desired biology—interaction of the Fc with FcγRs and complement and the biology they confer
- Affinity, including differentiation of K_{on} and K_{off} where appropriate
- Epitope—it is clear that for many targets, especially cell surface targets, epitope differences can lead to significant differences in biology, and therefore efficacy
- Multispecificity—in a molecular format that can be stabilized and manufactured

- Size, shape, and flexibility—affects biodistribution and tissue penetration
- Behavior of the molecule—stability and efficient folding (Honegger 2008), solubility (especially at high concentrations), aggregation characteristics, degradation and amino acid reactions, spurious glycosylation sites, and so forth
- Mixed modality—using antibodies as carriers for peptides, siRNA, toxins, small molecules either for half-life extension and/or for tissue targeting

Included below are several examples highlighting how fit-for-purpose Mabs and Fc fusion proteins are being pursued. Soliris® is the first example of a marketed Mab in which the Fc domain has been mutated away from that of a natural Fc (Rother et al. 2007). Examples of other Mabs in late clinical trials having modified Fc domains include Teplizumab (humanized OKT3-γ1-ala-ala, an anti-CD3 Mab with substantially reduced FcγR binding to down-modulate mitogenic response; H. Li et al. 2006), Visilizumab (anti-CD3 Mab with substantially reduced FcγR binding), AME-133v (anti-CD20 with increased affinity to FcγRIIIa; Weiner et al. 2005), and rhuMab V114 (anti-CD20 with increased affinity to FcγRIIIa). As mentioned previously, the field of Fc engineering is incredibly competitive, with significant activity in both the research and development phases; a few reviews and key papers are cited for reference (Shields et al. 2001; Lazar et al. 2006; Presta 2006, 2008; Richards et al. 2008).

There are two important findings with respect to how glycosylation can affect the functionality of IgG Fc. In the first example, now quite well known, it was determined that lack of a fucose residue in the glycoside that binds to residue ser297 in the CH2 domain of IgG results in a tighter binding of the antibody to FcγRIIIa, and with that, higher ADCC (Shields et al. 2002; Niwa et al. 2004; Masuda et al. 2007). Interestingly, the second example for glycosylation effects on biological function is the opposite in nature. Ravetch and colleagues (Kaneko, Nimmerjahn, and Ravetch 2006; Nimmerjahn and Ravetch 2007; Anthony et al. 2008) have shown that sialylated antibodies or Fc domains can have an immunosuppressive effect, and have proposed that this might explain at least part of the mechanism for why IVIG treatment has immunosuppressive properties. It is likely that as we understand more about the relationship between Mab glycoform and immune-related functionality, additional unique glycoform-specific activities may be found. The greatest challenge to these experiments is that normal CHO or other mammalian cell systems produce heterogeneous *N*-glycans. It seems likely that GlycoFi, now a wholly owned subsidiary of Merck (Table 1.5), is the company that should have the greatest opportunity in the near future to address these kinds of questions, as they have shown that they can produce Mabs that possess a single major species of glycoside in recombinant humanized *Pichia pastoris* (see H. Li et al. 2006).

Another example of Fc engineering to improve biologic molecule fitness would be the modification of Fc sequence to potentially improve half-life, and therefore reduce frequency of administration (Petkova et al. 2006). Perhaps the best example of this modification is the YTE mutant Fc from Medimmune (now part of Astra-Zeneca), in which residues M252Y, S254T, and T256E were modified to increase the binding of the IgG1 Fc to human FcRn specifically at pH 6.0, while at neutral pH there was little binding (Dall'Acqua, Kiener, and Wu 2006). This pH-dependent increase in binding to FcRn has been hypothesized to be a way to increase the half-life of antibodies through an improvement in the recycling mechanism (Dall'Acqua, Kiener, and Wu 2006). Indeed, the YTE Fc mutant of an IgG1 possessed approximately four times the half-life in nonhuman primates as compared to the wild-type version of the same antibody (Dall'Acqua, Kiener, and Wu 2006). Another benefit of engineering FcRn was that the biodistribution of the resultant antibody into the lungs was significantly increased (Dall'Acqua, Kiener, and Wu 2006), which is perhaps not too surprising since it had been determined previously that antibodies are transcytosed across lung epithelium via FcRn (Bitonti et al. 2004). The lead molecule, MEDI-557 (also known as MEDI-524-YTE and Numax-YTE), entered Phase I clinical trials in December 2007, so clinical validation (or refutation) of this mechanism should be publicly available soon. If these results hold true in human trials, with little or no mutant-associated immunogenicity, this type of approach could result in many enhanced half-life antibodies in the future, as well as antibodies that distribute better into the lungs. Such a biodistribution pattern could be quite attractive with many of the antibody candidates targeting asthma (e.g., anti-IL-13, anti-IL-9, anti-IL-4R, and anti-IL-13R).

Another area of Mab engineering that has been of significant interest for both research and development concerns antibody fragments such as Fabs, scFvs, and domain antibody fragments. Thus far, there are three marketed Fab-based products, ReoPro®, Lucentis®, and Cimzia® (Table 1.2). Additionally, both Aurograb® and Mycograb® are in Phase III clinical trials (Table 1.3), and Alacizumuab pegol, a di-Fab-PEGylated construct, is in Phase II trials (Table 1.4). BiTEs, as described below, are constructs consisting of two scFvs linked together (Baeuerle, Reinhardt, and Kufer 2008). The constructs that have recently made the most visible splash are domain antibodies (Dumoulin et al. 2002; Harmsen and De Haard 2007), as also evidenced by the recent acquisition of Domantis by Glaxo-Smith Kline (Table 1.5).

The idea of making a bispecific antibody has been around for over 20 years (Paulus 1985; Brennan, Davison, and Paulus 1985). The concepts of using a bispecific antibody to either bind and neutralize two targets simultaneously, to carry a molecule, for example, toxin, to a specific targeted site, or, alternatively, to bring two targets together, were already being discussed by the mid-1980s (Paulus 1985). The first bispecific construct to bring together an effector cell (e.g., T-cell) and its target cell was published in 1986 (Staerz and Bevan 1986), which was essentially the forerunner to what is now known as the BiTE technology from Micromet (Baeuerle, Reinhardt, and Kufer 2008). There are currently four bispecific constructs in clinical trials, two from Micromet and two using the Trion Triomab® technology. MT-103 and MT-110 are bispecific T-cell engagers (BiTE) scFv constructs from Micromet targeting CD3 on T-cells and CD19 and EpCAM, respectively. Blinatumomab (MT-103) is in Phase II clinical trials for non-Hodgkin's lymphoma and is being tested for several other B-cell malignancies. Catumaxomab is a trifunctional, bispecific hybrid mouse-rat monoclonal antibody (Triomab® technology) against human EpCAM and human CD3. The tri-specificity comes from the fact that the hybrid murine IgG2a/rat IgG2a Fc also binds to FcγRs I and III to trigger ADCC (Zeidler et al. 1999; Shen and Zhu 2008). Catumaxomab is currently in the preregistration phase, so it is expected to reach the marketplace very shortly. A second Triomab® is Ertumaxomab, a trifunctional, bispecific hybrid Mab consisting of a dimer comprised of the subunits anti-HER2/neu mouse IgG2a and anti-CD3 rat IgG2a. This antibody also is functional on FcγRs I and III, giving it its reported trifunctionality. There are many reports in the literature highlighting other strategies to make bispecific antibodies containing functional Fc domains (e.g., Ridgeway, Presta, and Carter 1996; Coloma and Morrison 1997; Shen et al. 2006; Wu et al. 2007b), but none of these approaches has yet been incorporated into Mabs that made it into the clinic. A new approach just recently published is the construction of Surrobodies, which are antibody-pre-BCR subunit chimeras that lend themselves to multispecificity due to an extra fragment hanging off the surrogate light chain components (Xu et al. 2008).

Affinity has always been an issue of discussion. There are now several examples of antibodies being affinity matured to K_D values of 1 to 10 pM or even sub-pM (reviewed in Chapter 13). The question of how tight is tight enough will be debated for some time to come, and is likely to have target-specific answers. The most highly successful antibodies on the market, for the most part, are not particularly high affinity antibodies, many of them having K_D values in the 0.1 to 3 nM range (Carter 2006). There are a few examples, however, of cases in which affinity, and the type of affinity, matter. In a classical study, Wu et al. (2005) showed that affinity matured mutants of an anti-RSV antibody possessed very different characteristics based on whether the maturation improved K_{on} or K_{off}, the two components that make up K_D. This study exemplifies the importance of understanding the details of the biology of a system and how the antibody will interact with that system. In another interesting study of affinity versus functionality, Adams et al. (2001) demonstrated that affinity of an scFv (monovalent) antibody had a significant impact on the ability of that antibody to penetrate tumors; the higher the affinity, the poorer the tissue penetration. The same group, however, later showed that the higher the affinity of an antitumor IgG, the stronger the ADCC (Tang et al. 2007). Taken together, these two studies suggest that there might be a delicate balance in affinity when targeting solid tumors with antibodies.

A novel approach to the stabilization of peptides is the fusion of those peptides to Ig domains, followed by engineering to stabilize the fusion construct (Kuter 2007). The most advanced molecule in this class is Romiplostim, a "peptibody" from Amgen comprised of an Fc derived from IgG fused to a TPO peptide mimetic for specific binding to TPO receptor (Kuter 2007). Romiplostim (Nplate®) was approved by the FDA for marketing in August 2008, making it the first of this type of construct to reach the market. Scientists at Centocor also have reported the construction of EPO- (Bugelski et al. 2008) and GLP-1 (Picha et al. 2008) "mimetibody constructs" which should extend the half-life of the biologically active peptide mimetics. An alternative approach for using the IgG scaffold to stabilize and extend the half-life of peptides or small molecules is the CovX body™, as recently described by Doppalapudi et al. (2007). They used an aldolase antibody engineered so that it possessed a highly reactive lysine in the V-chain, which allowed for highly specific placement of a chemical linker. This linker can then be used to attach a pharmacophore of interest, such as a biologically active peptide, a small molecule, or any other molecule for which a longer half-life may be desired (Doppalapudi et al. 2007).

1.6 SUMMARY, AND "WHERE DO WE GO FROM HERE"?

The examples mentioned above are a few of the many ways in which scientists are looking to engineer Mabs and Fc fusion proteins to impact biology conferred by them. It is probably fair to say that most of the conceptual modifications that could be made to alter the activity of an antibody have been proposed, and most of them are either being made now or have already been made and tested (see Tables 1.8 and 1.9). Variable chain maturation and humanization are well worked out, and now there are several approaches to apply information from *in silico* and *in vitro* deimmunization to improve, at least in theory, the humanness of the molecules made (Abhinanden and Martin 2007; see also Chapter 13). In the near future (e.g., the next decade), the major advancements that will come to the fore will be from the clinical validation (or lack thereof) of constructs that have already been conceptualized. Additionally, recent advances have been made in generation of domain antibodies, nonantibody binding scaffolds, bispecific and multispecific antibodies, and generation of Fc mutants that result in differential pharmacology, that is, increased or decreased ADCC, CMC, and half-life. Many of these types of molecules will make their way into clinical validation. In some cases, this has already begun. Phase I trials are being conducted on MedImmune's YTE half-life extension mutant, MEDI-557. The BiTE technology (CD3/CD19) is already in advanced clinical trials and holds significant promise as a cancer therapeutic (Baeuerle, Reinhardt, and Kufer 2008). As stated previously, Fc modified versions of IgGs already are in the clinic that have improved binding to FcγRs, with the concept of improving ADCC functionality (a hypothesis that still requires clinical validation). By the 2010–2015 time frame, it is expected that many antibodies entering the clinic will be modified from the natural IgG isotypes, IgG1, IgG2, or IgG4, either to increase effector function or, like with Soliris®, to down-modulate it.

In summary, Dimitrov and Marks (2008) recently proposed that there are two major eras in antibody discovery, the original serum therapy period in the early 1900s and, today, in which significant changes are impacting the way we design and make therapeutic proteins. I propose that the "antibody era" be considered in four phases: For the first phase, I concur with Dimitrov and Marks that the seminal work done by von Behring and his colleagues in the early 1900s set the stage for treatment of infectious diseases and the field of immunology. For the second phase, I propose that the era of IVIG therapy (starting in 1952), which led to both the concept and practice of monoclonal therapy, was a crucial step to get to where we are today. The third phase is the 1990s, which is the decade of the "first generation antibody therapeutics," exemplified by therapeutic chimeric and humanized Mabs based largely on standard IgG1 scaffolds.

Finally, I propose that in the 2006–2008 period, we entered the fourth phase, the expansion decade, an era in which many engineered antibodies and antibody-like constructs will be developed and commercialized. Additionally, over the next 10 years, nonantibody scaffolds (not covered here) will likely be validated and made commercially successful for certain applications, and a wide variety of new IgG isoytpes, modified Fc constructs, peptibodies, and similar second-generation biologics will be built and tested clinically for validation. Collectively, these "fit-for-purpose" molecules will revolutionize how we view biologics. This fourth era also will likely usher in follow-on biologics to those marketed in the first era, which should add pressure on the innovators to continue to innovate.

REFERENCES

Abhinanden, K.R., and A.C. Martin. 2007. Analyzing the "degree of humanness" of antibody sequences. *J. Mol. Biol.* 369:852–862.

Adams, G.P., R. Schier, A.M. McCall, H.H. Simmons, E.M. Horak, R.K. Alpaugh, J.D. Marks, and L.M. Weiner. 2001. High affinity restricts the localization and tumor penetration of single-chain Fv antibody molecules. *Cancer Res.* 61:4750–4755.

Alkan, S.S. 2004. Monoclonal antibodies: The story of a discovery that revolutionized science and medicine. *Nature Rev. Immunol.* 4:153–156.

Alegre, M.L., L.J. Peterson, D. Xu, H.A. Sattar, D.R. Jeyarajah, K. Kowalkowski, J.R. Thistlewaite, R.A. Zivin, L. Jolliffe, and J.A. Bluestone. 1994. A non-activating "humanized" anti-CD3 monoclonal antibody retains innunosuppressive properties in vivo. *Transplantation* 57:1537–1543.

Alt, F.W., T.K. Blackwell, and G.D. Yancopoulos. 1985. Immunoglobulin genes in transgenic mice. *Trends Genet.* 1:231–236.

Anthony, R.M., F. Nimmerjahn, D.J. Ashline, V.N. Reinhold, J.C. Paulson, and J.V. Ravetch. 2008. Recapitulation of IVIG anti-inflammatory activity with a recombinant IgG Fc. *Science* 320:373–376.

Baeuerle, P.A., C. Reinhardt, and P. Kufer. 2008. BiTE: A new class of antibodies that recruit T-cells. *Drugs Future* 33:137–147.

Barbas, C.F. III, J.D. Bain, D.M. Hoekstra, and R.A. Lerner. 1992. Semisynthetic combinatorial antibody libraries: A chemical solution to the diversity problem. *Proc. Natl. Acad. Sci. USA* 89:4457–4461.

Barbas, C.F. III, A.S. Kang, R.A. Lerner, and S.J. Benkovic. 1991. Assembly of combinatorial antibody libraries on phage surfaces: The gene III site. *Proc. Natl. Acad. Sci. USA* 88:7978–7982.

Bass, S., R. Greene, and J.A. Wells. 1990. Hormone phage: An enrichment method for variant proteins with altered binding properties. *Proteins* 8:309–314.

Beers, S.A., C.H. Chan, S. James, R.R. French, K.E. Attfield, C.M. Brennan, A. Ahuja, M.J. Shlomchik, M.S. Cragg, and M.J. Glennie. 2008. Type II (tositumomab) anti-CD20 monoclonal antibody out performs Type I (rituximab-like) reagents in B-cell depletion regardless of complement activation. *Blood* 112:4170–4177.

Behring, E.A., and S. Kitasato. 1890. Ueber das Zustandekommen der Diphtherie-Immunitaet und der Tetanus-Immunitaet bei Thieren. *Deutsch. Med. Wochenschr.* 49:1113–1114.

Beidler, C.B., J.R. Ludwig, J. Cardenal, J. Phelps, C.G. Papworth, E. Melcher, M. Sierzega, L.J. Myers, B.W. Unger, M. Fisher, et al. 1988. Cloning and high level expression of a chimeric antibody with specificity for human carcinoembryonic antigen. *J. Immunol.* 141:4053–4060.

Bello, C., and E.M. Sotomayor. 2007. Monoclonal antibodies for B-cell lymphomas: Rituximab and beyond. *Hematology Am. Soc. Hematol. Educ. Program* 2007:233–242.

Better, M., C.P. Chang, R.R. Robinson, and A.H. Horwitz. 1988. *Escherichia coli* secretion of an active chimeric antibody fragment. *Science* 240:1041–1043.

Bitonti, A.J., J.A. Dumont, S.C. Low, R.T. Peters, K.E. Kropp, V.J. Palombella, J.M. Stattel, Y. Lu, C.A. Tan, J.J. Song, A.M. Garcia, N.E. Simister, G.M. Spiekermann, W.I. Lencer, and R.S. Blumberg. 2004. Pulmonary delivery of an erythropoietin Fc fusion protein in non-human primates through an immunoglobulin transport pathway. *Proc. Natl. Acad. Sci. USA* 101:9763–9768.

Boulianne, G.L., N. Hozumi, and M.L. Shulman. 1984. Production of functional chimaeric mouse/human antibody. *Nature* 312:643–646.

Brennan, M., P.F. Davison, and H. Paulus. 1985. Preparation of bispecific antibodies by chemical recombination of monoclonal immunoglobulin G1 fragments. *Science* 229:81–83.

Brüggemann, M., G.T. Williams, C.I. Bindon, M.R. Clark, M.R. Walker, R. Jefferis, H. Waldmann, and M.S. Neuberger. 1987. Comparison of the effector functions of human immunoglobulins using a matched set of chimeric antibodies. *J. Exp. Med.* 66:1351–1361.

Bruton, O.C. 1952. Agammaglobulinemia. *Pediatrics* 9:722–727.

Bugelski, P.J., R.J. Capocasale, D. Makropoulos, D. Marshall, P.W. Fisher, J. Lu, R. Achuthanandam, T. Spinka-Doms, D. Kwok, D. Graden, A. Voklk, T. Nesspor, I.E. James, and C. Haung. 2008. CNTO 530: Molecular pharmacology in human UT-7EPO cells and pharmacokinetics in mice. *J. Biotechnol.* 134:171–180.

Capon, D.J., S.M. Chamow, J. Mordenti, S.A. Marsters, T. Gregory, H. Mitsuya, R.A. Byrn, C. Lucas, F.M. Wurm, J.E. Groopman, et al. 1989. Designing CD4 immunoadhesins for AIDS therapy. *Nature* 337:525–531.

Carter, P.J. 2006. Potent antibody therapeutics by design. *Nature Rev. Immunol.* 6:343–357.

Cartron, G., L. Dacheux, G. Salles, P. Solal-Celigny, P. Bordos, P. Colombat, and H. Watier. 2002. Therapeutic activity of humanized anti-CD20 monoclonal antibody and polymorphism in igG Fc receptor FcgammaRilla gene. *Blood* 99:754–758.

Casadevall, A. 1996. Antibody-based therapies for emerging infectious diseases. *Emerg. Infect. Dis.* 2:200–208.

Casadevall, A. 2002. Passive antibody administration (immediate immunity) as a specific defense against biological weapons. *Emerg. Infect. Dis.* 8:833–841.

Casadevall, A., and M.D. Scharff. 1995. Return to the past: The case for antibody-based therapies in infectious diseases. *Clin. Infect. Dis.* 21:150–161.

Chatenoud, L., C. Ferran, C. Legendre, I. Thouard, S. Merite, A. Reuter, Y. Gevaert, H. Kreis, P. Franchimont, and J.F. Bach. 1990. In vivo cell activation following OKT3 administration. Systemic cytokine release and modulation by corticosteroids. *Transplantation* 49:697–702.

Chowdhury, P.S., and H. Wu. 2005. Tailor-made antibody therapeutics. *Methods* 36:11–24.

Chung, K.T. n.d. Emil Von Behring (1854–1917). Pioneer of serology. http://www.mhhe.com/biosci/cellmicro/nester/graphics/nester3ehp/common/vonbehr.html

Co, M.S., and C. Queen. 1991. Humanized antibodies for therapy. *Nature* 351:501–502.

Cohen, S.N., A.C. Chang, H.W. Boyer, and R.B. Helling. 1973. Construction of biologically functional bacterial plasmids in vitro. *Proc. Natl. Acad. Sci. USA* 70:3240–3244.

Colcher, D., G. Pavlinkova, G. Bresford, B.J. Booth, A. Choudhury, and S.K. Batra. 1998. Pharmacokinetics and biodistribution of genetically-engineered antibodies. *Quarterly J. Nucl. Med.* 42:225–241.

Coloma, M.J., and S.L. Morrison. 1997. Design and production of novel tetravalent bispecific antibodies. *Nature Biotechnol.* 15:159–163.

Cragg, M.S., and M.J. Glennie. 2004. Antibody specificity controls in vivo effector mechanisms of anti CD20 reagents. *Blood* 103:2738–2743.

Dalbadie-McFarland, G., L.W. Cohen, A.D. Riggs, C. Morin, K. Itakura, and J.H. Richards. 1982. Oligonucleotide-directed mutagenesis as a general and powerful method for studies of protein function. *Proc. Natl. Acad. Sci. USA* 79:6409–6413.

Dall'Acqua, W.F., P.A. Kiener, and H. Wu. 2006. Properties of human IgG1 engineered for enhanced binding to the neonatal Fc receptor (FcRn). *J. Biol. Chem.* 281:23514–23524.

De Kruif, J., L. Terstappen, E. Boel, and T. Logtenberg. 1995. Rapid selection of cell subpopulation-specific human monoclonal antibodies from a synthetic phage display library. *Proc. Natl. Acad. Sci. USA* 92:3938–3942.

De la Cruz Edmonds, M.C., M. Tellers, C. Chan, P. Salmon, D.K. Robinson, and J. Markussen. 2006. Development of transfection and high-producer screening protocols for the CHOK1SV cell system. *Mol. Biotechnol.* 34:179–190.

Dimitrov, D.S., and J.D. Marks. 2008. Therapeutic antibodies: Current state and future trends—is a paradigm change coming soon? *Methods Mol. Biol.* 525:1–27.

Doppalapudi, V.R., N. Tryder, L. Li, D. Griffith, F.F. Liao, G. Roxas, M.P. Ramprasad, C. Bradshaw, and C.F. Barbas, III. 2007. Chemically programmed antibodies: Endothelin receptor targeting CovX-bodies. *Bioorg. Med. Chem. Lett.* 17:501–506.

Dumoulin, M., K. Conrath, A. Van Meiraeghe, E. Meersman, K. Meersman, L.G. Frenken, S. Muyldermans, L. Wyns, and A. Matagne. 2002. Single-domain antibody fragments with high conformational stability. *Protein Sci.* 11:500–515.

Early, P., H. Huang, M. Davis, K. Calame, and L. Hood. 1980. An immunoglobulin chain variable region gene is generated from three segments of DNAP: V_H, D and J_H. *Cell* 19:981–992.

Fellouse, F.A., K. Esaki, S. Birtalan, D. Raptis, V.J. Cancasci, A. Koide, P. Jhurani, M. Vasser, C. Weismann, A.A. Kossiakoff, S. Koide, and S.S. Sidhu. 2007. High-throughput generation of synthetic antibodies from highly functional minimalist phage-displayed libraries. *J. Mol. Biol.* 373:924–940.

Feng, Y., and D.S. Dimitrov. 2008. Monoclonal antibodies against components of the IGF system for cancer treatment. *Curr. Opin. Drug Disc. Develop.* 11:178–185.

Ghielmini, M., K. Rufibach, G. Salles, L. Leoncini-Franscini, C. Léger-Falandry, S. Cogliatti, M. Fey, G. Martinelli, R. Stahel, A. Lohri, N. Ketterer, M. Wernli, T. Cerny, and S.F. Schmitz. 2005. Single agent rituximab in patients with follicular or mantle cell lymphoma: Clinical and biological factors that are predictive of response and event-free survival as well as the effect of rituximab on the immune system: A study of the Swiss Group for Clinical Cancer Research (SAKK). *Ann. Oncol.* 16:1675–1682.

Glennie, M.J., R.R. French, M.S. Cragg, and R.P. Taylor. 2007. Mechanisms of killing by anti-CD20 monoclonal antibodies. *Mol. Immunol.* 44:3823–3837.

Goeddel, D.V., H.L. Heyneker, T. Hozume, R. Arentzen, K. Itakura, D.G. Yansura, M.J. Ross, G. Miozzari, R. Crea, and P.H. Seeburg. 1979a. Direct expression in *Escherichia coli* of a DNA sequence coding for human growth hormone. *Nature* 281:544–548.

Goeddel, D.V., D.G. Kleid, F. Bolivar, G.L. Heyneker, D.G. Yansura, R. Crea, T. Hirose, A. Kraszewski, K. Itakura, and A.D. Riggs. 1979b. Expression in *Escherichia coli* of chemically synthesized genes for human insulin. *Proc. Natl. Acad. Sci. USA* 76:106–110.

Gough, N.M., and O. Bernard. 1981. Sequences of the joining region genes for immunoglobulin heavy chains and their sole in generation of antibody diversity. *Proc. Natl. Acad. Sci. USA* 78:509–513.

Gram, H., L.A. Marconi, C.F. Barbas III, T.A. Collet, R.A. Lerner, and A.S. Kang. 1992. In vitro selection and affinity maturation of antibodies from a naïve combinatorial immunoglobulin library. *Proc. Natl. Acad. Sci. USA* 89:3576–3580.

Green, L.L. 1999. Antobody engineering via genetic engineering of the mouse: XenoMouse strains are a vehicle for the facile generation of therapeutic human monoclonal antibodies. *J. Immunol. Methods* 231:11–23.

Green, L.L., M.C. Hardy, C.E. Maynard-Currie, H. Tsuda, D.M. Tsuda, D.M. Louie, M.J. Mendez, H. Abderrahim, M. Noguchi, D.H. Smith, Y. Zeng, et al. 1994. Antigen-specific human monoclonal antibodies from mice engineered with human Ig heavy and light chain YACs. *Nature Genet.* 7:13–21.

Griffiths, A.D., S.C. Williams, O. Hartley, I.M. Tomlinson, P. Waterhouse, W.L. Crosby, R.E. Kontermann, P.T. Jones, N.M. Low, T.J. Allison, et al. 1994. Isolation of high affinity human antibodies directly from large synthetic repertoires. *EMBO J.* 13:3245–3260.

Gronski, P., F.R. Seiler, and H.G. Schwick. 1991. Discovery of antitoxins and development of antibody preparations for clinical uses from 1890 to 1990. *Mol. Immunol.* 28:1321–1332.

Hagenbook, A., T. Plesner, P. Johnson, et al. 2005. HuMax-Cd20, a novel fully human anti-CD20 monoclonal antibody: Results of a phase I/II trial in relapsed or refractory follicular non-Hodgkins's lymphoma. *Blood* 106:4760a.

Haller, C.A., M.E. Cosenza, and J.T. Sullivan. 2008. Safety issues specific to clinical development of protein therapeutics. *Clin. Pharm. Therap.* 84:624–627.

Harmsen, M.M., and H. De Haard. 2007. Properties, production, and applications of camelid single-domain antibody fragments. *Appl. Microbiol. Biotechnol.* 77:13–22.

Hawkins, R.E., S.J. Russel, and G. Winter. 1992. Selection of phage antibodies by binding affinity. Mimicking affinity maturation. *J. Mol. Biol.* 226:889–896.

He, M., and F. Khan. 2005. Ribosome display: Next generation display technologies for production of antibodies in vitro. *Expert Rev. Proteom.* 2:421–430.

Hoet, R.M., E.H. Cohen, R.B. Kent, K. Rookey, S. Schoonbroodt, S. Hogan, L. Rem, N. Frans, M. Daukandt, H. Pieters, R. van Hegelsom, N.C. Neer, H.G. Nastri, I.J. Rondon, J.A. Leeds, S.E. Hufton, L. Huang, I. Kashin, M. Devlin, G. Kuang, M. Steukers, M. Viswanathan, A.E. Nixon, D.J. Sexton, G.R. Hoogenboom, and R.C. Charles Ladner. 2005. Generation of high-affinity human antibodies by combining donor-derived and synthetic complementarity-determining-region diversity. *Nature Biotechnol.* 23:344–348.

Honegger, A. 2008. Engineering antibodies for stability and efficient folding. *Handbook Exp. Pharmacol.* 181:47–68.

Hoogenboom, H.R. 2005. Selecting and screening recombinant antibody libraries. *Nature Biotechnol.* 23:1105–1116.

Hozumi, N., and S. Tonegawa. 1976. Evidence for somatic rearrangement of immunoglobulin genes coding for variable and constant regions. *Proc. Natl. Acad. Sci. USA* 73:3628–3632.

Hutchinson, C.A., S. Philips, M.H. Edgell, S. Gillam, P. Jahnke, and M. Smith. 1978. Mutagenesis at a specific position in a DNA sequence. *J. Biol. Chem.* 253:6551–6560.

Hwang, W.Y., and J. Foote. 2005. Immunogenicity of engineered antibodies. *Methods* 36:3–10.

Itakura, K., T. Hirose, R. Crea, A.D. Riggs, H.L. Heyneker, F. Bolivar, and H.W. Boyer. 1977. Expression in *Escherichia coli* of a chemically synthesized gene for the hormone somatostatin. *Science* 198:1056–1063.

Jefferis, R. 2007. Antibody therapeutics: Isotype and glycoform selection. *Expert Opin. Ther.* 7:1401–1413.

Jerne, N.K., and A.A. Nordin. 1963. Plaque formation in agar by single antibody-producing cells. *Science* 140:405.

Jones, P.T., P.H. Dear, J. Foote, M.S. Neuberger, and G. Winter. 1986. Replacing the complementarity-determining regions in a human antibody with those from a mouse. *Nature* 321:522–525.

Kabat, E.A., T.T. Wu, M. Reid-Miller, H.M. Perry, and K.S. Gottesman. 1987. *Sequences of proteins of immunological interest.* Washington, D.C.: U.S. Department of Health and Human Services. Publ. No. 91-3242.

Kalwy, S., J. Rance, and R. Young. 2006. Toward more efficient protein expression: Keep the message simple. *Mol. Biotechnol.* 34:151–156.

Kaneko, Y., F. Nimmerjahn, and J.V. Ravetch. 2006. Anti-inflammatory activity of immunoglobulin G resulting from Fc sialylation. *Science* 313:670–673.

Kimball, J.A., D.J. Norman, C.F. Shield, T.J. Schroeder, P. Lisi, M. Garovoy, J.B. O'Connell, F. Stuart, S.V. McDiarmid, and W. Wall. 1995. The OKT3 antibody response study: A multicentre study of human anti-mouse antibody (HAMA) production following OKT3 use in solid organ transplantation. *Transpl. Immunol.* 3(3):212–221.

KMR Group, Inc. 2007. Benchmarking biopharmaceutical industry, Chicago.

Knappik, A., L. Ge, A. Honegger, P. Pack, M. Fischer, G. Wellnhofer, A. Hoess, J. Wölle, A. Plückthun, and B. Virnekäs. 2000. Fully synthetic human combinatorial antibody libraries (HuCAL) based on modular consensus frameworks and CDRs randomized with trinucleotides. *J. Mol. Biol.* 296:57–86.

Köhler, G., and C. Milstein. 1975. Continuous cultures of fused cells secreting antibody of predefined specificity. *Nature* 256:495–497.

Kung, P., G. Goldstein, E.L. Reinherz, and S.F. Schlossman. 1979. Monoclonal antibodies defining distinctive human T cell surface antigens. *Science* 206:347–349.

Kuter, D.J. 2007. New thrombopoietic growth factors. *Blood* 109:4607–4616.

Laffly, E., and R. Sodoyer. 2005. Monoclonal and recombinant antibodies, 30 years after... *Human Antibodies* 14:33–55.

Lawley, T.J., L. Bielory, P. Gascon, K.B. Yancey, N.S. Yound, and M.M. Frank. 1984. A prospective clinical and immunologic analysis of patients with serum sickness. *New Engl. J. Med.* 311:1407–1413.

Lazar, G.A., W. Dang, S. Karki, O. Vafa, J.S. Peng, L. Hyun, C. Chan, H.S. Chung, A. Eivazi, S.C. Yoder, J. Veilmetter, D.F. Carmichael, R.J. Hayes, and B.I. Dahiyat. 2006. Engineered antibody Fc variants with enhanced effector function. *Proc. Natl. Acad. Sci. USA* 103:4005–4010.

Leader, B., Q.J. Baca, and D.E. Golan. 2008. Protein therapeutics: A summary and pharmacological classification. *Nature Rev. Drug Discov.* 7:21–39.

Leonard, J.P., P. Martin, J. Ruan, R. Elstom, J. Barrientos, M. Coleman, and R.R. Furman. 2008. New monoclonal antibodies for non-Hodgkin's lymphoma. *Annals Oncol.* 19:iv60–iv62.

Li, H., N. Sethuraman, T.A. Stadheim, D. Zha, B. Prinz, N. Ballew, P. Bobrowicz, B.K. Choi, W.J. Cook, M. Cukan, N.R. Houston-Cummings, R. Davidson, B. Gong, S.R. Hamilton, J.P. Hoopes, Y. Jiang, N. Kim, R. Mansfield, J.H. Nett, S. Rios, R. Strawbridge, S. Wildt, and T.U. Gerngross. 2006. Optimization of humanized IgGs in glycoengineered *Pichia pastoris*. *Nature Biotechnol.* 24:210–215.

Llewelyn, M.B., R.E. Hawkins, and S.J. Russell. 1992. Discovery of antibodies. *Br. Med. J.* 305:1269–1272.

Lobo, P.I., and H.C. Patel. 1997. Murine monoclonal IgG antibodies: Differences in their IgG isotypes can affect the antibody effector activity when using human cells. *Immunol. Cell Biol.* 75:267–274.

Lonberg, N. 2005. Human antibodies from transgenic animals. *Nature Biotechnol.* 23:1117–1125.

Lonberg, N., L.D. Taylor, F.A. Harding, M. Trounstine, K.M. Higgins, S.R. Schramm, C.C. Kuo, R. Mashayekh, K. Wymore, J.G. McCabe, et al. 1994. Antigen-specific human antibodies from mice comprising four distinct genetic modifications. *Nature* 368:856–859.

Lowman, H.B., S.H. Bass, N. Simpson, and J.A. Wells. 1991. Selecting high-affinity binding proteins by monovalent phage display. *Biochemistry.* 30:10832–10838.

Maloney, D.G. 2007. Follicular NHL: From antibodies and vaccines to graft-versus-lymphoma effects. *Hematology Am. Soc. Hematol. Educ. Program* 2007:226–232.

Markland, W., B.L. Roberts, M.J. Saxena, S.K. Guterman and R.C. Ladner. 1991. Design, construction and function of a multicopy display vector using fusions to the major coat protein of bacteriophage M13. *Gene* 109:13–19.

Marks, J.D., H.R. Hoogenboom, T.P. Bonnert, J. McCafferty, A.D. Griffiths, and G. Winter. 1991. By-passing immunization. Human antibodies from V-gene libraries displayed on phage. *J. Mol. Biol.* 222:581–597.

Marks, J.D., A.D. Griffiths, M. Malmgvist, T.P. Clackson, J.M. Bye, and G. Winter. 1992. By-passing immunization: Building high affinity human antibodies by chain shuffling. *Biotechnology (NY)* 10:779–783.

Marks, J.D., H.R. Hoogenboom, A.D. Griffiths, and G. Winter. 1992. Molecular evolution of proteins on filamentous phage. Mimicking the strategy of the immune system. *J. Biol. Chem.* 267:16007–16010.

Masuda, K., T. Kubota, E. Kaneko, S. Iida, M. Wakitani, Y. Kobayashi-Natsume, A. Kubota, K. Shitara, and K. Nakamura. 2007. Enhanced binding affinity for FcgammaRIIIa of fucose-negative antibody is sufficient to induce maximal antibody-dependent cellular cytotoxicity. *Mol. Immunol.* 44:3122–3131.

McCafferty, J., A.D. Griffiths, G. Winter, and D.J. Chiswell. 1990. Phage antibodies: Filamentous phage displaying antibody variable domains. *Nature* 348:552–554.

Milstein, C. 1985. From the structure of antibodies to the diversification of the immune response. Nobel lecture, 8 December 1984. *Biosci. Rep.* 5:275–297.

Mondon, P., O. Dubreuli, K. Bouayadi, and H. Kharrat. 2008. Human antibody libraries: A race to engineer and explore a larger diversity. *Front. Biosci.* 13:1117–1129.

Morrison, S.L., M.J. Johnson, L.A. Herzenberg, and V.T. Oi. 1984. Chimeric human antibody molecules: Mouse antigen-binding domains with human constant region domains. *Proc. Natl. Acad. Sci. USA* 81:6851–6855.

Mouthon, L., and O. Lortholary. 2003. Intravenous immunoglobulins in infectious diseases: Where do we stand? *Clin. Microbiol. Infect.* 9:333–338.

Mullis, K., F. Faloona, S. Scharf, R. Saiki, G. Horn, and H. Erlich. 1986. Specific enzymatic amplification of DNA in vitro: The polymerase chain reaction. *Cold Spring Harbor Symp. Quant. Biol.* 51(pt. 1):263–273.

Neuberger, M.S. 1983. Expression and regulation of immunoglobulin heave chain gene transfected into lymphoid cells. *EMBO J.* 1373–1378.

Neuberger, M.S., and G.T. Williams. 1986. Construction of novel antibodies by use of DNA transfection: Design of plasmid vectors. *Philos. Trans. R. Soc. London A* 317:425–432.

Neuberger, M.S., G.T. Williams, and R.O. Fox. 1984. Recombinant antibodies possessing novel effector functions. *Nature* 312:604–608.

Nimmerjahn, F., and J.V. Ravetch. 2007. The anti-inflammatory activity of IgG: The intravenous IgG paradox. *J. Exp. Med.* 204:11–15.

Niwa, R., S. Hatanaka, E. Shoji-Hosaka, M. Sakurada, Y. Kobayashi, A. Uehara, H. Yodoi, K. Nakamura, and K. Shitara. 2004. Enhancement of the antibody-dependent cellular cytotoxicity of low-fucose IgG1 is independent of FcgammaRIIIa functional polymorphism. *Clin. Cancer Res.* 10:6248–6255.

Orange, J.S., E.M. Hossny, C.R. Weiler, M. Ballow, M. Berger, F.A. Bonilla, R. Buckley, J. Chinen, Y. El-Gamal, B.D. Mazer, R.P. Nelson, Jr., D.D. Patel, E. Secord, R.U. Sorensen, R.L. Wasserman, and C. Cunningham-Rundles; Primary Immunodeficiency Committee of the American Academy of Allergy, Asthma and Immunology. 2006. Use of intravenous immunoglobulin in human disease: A review of evidence by members of the Primary Immunodeficiency Committee of the American Academy of Allergy, Asthma and Immunology. *J. Allergy Clin. Immunol.* 117:S525–S553.

Orlandi, R., D.H. Güssow, P.T. Jones, and G. Winter. 1989. Cloning immunoglobulin variable domains for expression by the polymerase chain reaction. *Proc. Natl. Acad. Sci. USA* 86:3833–3837.

Osbourn, J., M. Groves, and T. Vaughan. 2005. From rodent reagents to human therapeutics using antibody guided selection. *Methods* 36:61–68.

Ostberg, L., and C. Queen. 1995. Human and humanized monoclonal antibodies: Preclinical studies and clinical experience. *Biochem. Soc. Trans.* 23:1038–1043.

Paulus, H. 1985. Preparation and biomedical applications of bispecific antibodies. *Behring Inst. Mitt.* 78:118–132.

Pendley, C., A. Schantz, and C. Wagner. 2003. Immunogenicity of therapeutic monoclonal antibodies. *Curr. Opin. Mol. Therap.* 5:172–179.

Petkova, S.B., S. Akilesh, T.J. Sproule, G.J. Christianson, H. Al Khabbaz, A.C. Brown, L.G. Presta, Y.G. Meng, and D.C. Roopenian. 2006. Enhanced half-life of genetically engineered human IgG1 antibodies in a humanized FcRn mouse model: Potential application in humorally mediated autoimmune disease. *Int. Immunol.* 18:1759–1769.

Picha, K.M., M.R. Cunningham, D.J. Drucker, A. Mathur, T. Ort, M. Scully, A. Soderman, T. Spinka-Doms, V. Stojanovic-Susulic, B.A. Thomas, and K.T. O'Neil. 2008. Protein engineering strategies for sustained glucagon-like peptide-1 receptor-dependent control of glucose homeostasis. *Diabetes* 57:1926–1934.

Presta, L.G. 2006. Engineering of therapeutic antibodies to minimize immunogenicity and optimize function. *Adv. Drug Deliv. Rev.* 58:640–656.

Presta, L.G. 2008. Molecular engineering and design of therapeutic antibodies. *Curr. Opin. Immunol.* 20:460–470.

Queen, C., W.P. Schneider, H.E. Selick, P.W. Payne, N.F. Landolfi, J.F. Duncan, N.M. Avdalovic, M. Levitt, R.P. Junghans, and T.A. Waldmann. 1989. A humanized antibody that binds to the interleukin 2 receptor. *Proc. Natl. Acad. Sci. USA* 86:10029–10033.

Rasmussen, S.K., L.K. Rasmussen, D. Weilguny, and A.B. Tolstrup. 2007. Manufacture of recombinant polyclonal antibodies. *Biotechnol. Lett.* 29:845–852.

Reff, M.E., K. Carner, K.S. Chambers, P.C. Chinn, J.E. Loenard, R. Raab, R.A. Newman, N. Hanna, and D.R. Anderson. 1994. Depletion of B cells in vivo by a chimeric mouse human monoclonal antibody to CD20. *Blood* 83:435–445.

Richards, J.O., S. Karki, G.A. Lazar, H. Chen, W. Dang, and J.R. Desjarlais. 2008. Optimization of antibody binding to FcγRIIa enhances macrophage phagocytosis of tumor cells. *Mol. Cancer Therap.* 7:2517–2527.

Ridgeway, J.B., L.G. Presta, and P. Carter. 1996. "Knobs-into-holes" engineering of antibody CH2 domains for heavy chain heterodimerization. *Protein Eng.* 9:617–621.

Riley, S. 2006. The future of monoclonal antibodies therapeutics: Innovation in antibody engineering, key growth strategies and forecasts to 2011. Datamonitor Report. London: Business Insights, Inc.

Roguska, M.A., J.T. Pedersen, A.H. Henry, S.M.J. Searle, C.M. Roja, B. Avery, M. Hoffee, S. Cook, J.M. Lambert, W.A. Blaettler, A.R. Rees, and B.C. Guild. 1996. A comparison of two murine monoclonal antibodies humanized by CDR-grafting and variable domain resurfacing. *Protein Eng.* 9:895–904.

Roguska, M.A., J.T. Pedersen, C.A. Keddy, A.H. Henry, S.J. Searle, J.M. Lambert, V.S. Goldmacher, W.A. Blaettler, A.R. Rees, and B.C. Guild. 1994. Humanization of murine monoclonal antibodies through variable domain resurfacing. *Proc. Natl. Acad. Sci. USA* 91:969–973.

Rothe, C., S. Urlinger, C. Löhning, J. Prassier, Y. Stark, U. Jäger, B. Hubner, M. Bardroff, I. Pradel, M. Boss, R. Bittlingmaier, T. Bataa, C. Frisch, B. Brocks, A. Honegger, and M. Urban. 2008. The human combinatorial antibody library HuCAC GOLD combines diversification of all six CDRs according to the natural immune system with a novel display method for efficient selection of high-affinity antibodies. *J. Mol. Biol.* 376:1182–1200.

Rother, R.P., S.A. Rollins, C.J. Mojcik, R.A. Brodsky, and L. Bell. 2007. Discovery and development of the complement inhibitor eculizumab for the treatment of paroxysmal nocturnal hemoglobinuria. *Nature Biotechnol.* 25:1256–1264.

Saiki, R.K., D.H. Gelfand, S. Stoffel, S.J. Scharf, R. Higuchi, G.T. Horn, K.B. Mullis, and H.A. Erlich. 1988. Primer-directed enzymatic amplification of DNA with a thermostable DNA polymerase. *Science* 239:487–491.

Shen, J., M.D. Vil, X. Jimenez, M. Iacolina, H. Zhang, and Z. Zhu. 2006. Single variable domain-IgG fusion. A novel recombinant approach to Fc domain-containing bispecific antibodies. *J. Biol. Chem.* 281:10706–10714.

Shen, J. and Z. Zhu. 2008. Catumaxomab, a rat murine hybrid trifunctional bispecific monoclonal antibody for the treatment of cancer. *Curr. Opin. Mol. Ther.* 10:273–284.

Shields, R.L., J. Lai, R. Keck, L.Y. O'Connell, K. Hong, Y.G. Meng, S.H. Weikert, and L.G. Presta. 2002. Lack of fructose on human IgG1 N-linked oligosaccharide improves binding to human Fcgamma RIII and antibody-dependent cellular toxicity. *J. Biol. Chem.* 277:26733–26740.

Shields, R.L., A.K. Namenuk, K. Hog, Y.G. Meng, J. Rae, J. Briggs, D. Xie, J. Lai, A. Stadlen, B. Li, J.A. Fox, and L.G. Presta. 2001. High resolution mapping of the binding site on human IgG1 for Fc gamma RI, Fc gamma RII, Fc gamma RIII, and FcRn and design of IgG1 variants with improved binding to the Fc gamma R. *J. Biol. Chem.* 276:6591–6604.

Shin, S.U., and S.L. Morrison. 1989. Production and properties of chimeric antibody molecules. *Methods Enzymol.* 178:459–476.

Shook, R.L. 2007. *Miracle medicines. Seven lifesaving drugs and the people who created them.* New York: Penguin Books.

Sidhu, S.S., and F.A. Fellouse. 2006. Synthetic therapeutic antibodies. *Nature Chem. Biol.* 2:682–688.

Silverstein, A.M. 2003. Splitting the difference: The germline-somatic mutation debate on generating antibody diversity. *Nature Immunol.* 4:829–833.

Skerra, A., and A. Plückthun. 1988. Assembly of a functional immunoglobulin Fv fragment in *Escherichia coli. Science* 240:1038–1040.

Smith, G.P. 1985. Filamentous fusion phage: Novel expression vectors that display cloned antigens on the virion surface. *Science* 228:1315–1317.

Staelens, S., J. Desmet, T.H. Ngo, S. Vauterin, I. Pareyn, P. Barbeaux, I. Van Rompaey, J.-M. Stassen, H. Deckmyn, and K. Vanhoorelbeke. 2006. Humanization by variable domain resurfacing and grafting on a human IgG4, using a new approach for determination of non-human like surface accessible framework residues based on homology modeling of variable domains. *Mol. Immunol.* 43:1243–1257.

Staerz, U.D., and M.J. Bevan. 1986. Hybrid hybridoma producing a bispecific monoclonal antibody that can focus effector T-cell activity. *Proc. Natl. Acad. Sci. USA* 83:1453–1457.

Stockwin, L.H. and S. Holmes. 2003. Antibodies as therapeutic agents: vive la renaissance! *Expert Opin. Biol. Ther.* 3:1133–1152.

Tang, Y., J. Lou, R.K. Alpaugh, M.K. Robinson, J.D. Marks, and L.M. Weiner. 2007. Regulation of antibody-dependent cellular cytotoxicity by IgG intrinsic and apparent affinity for target antigen. *J. Immunol.* 179:2815–2823.

Tax, W.F., F.F. Hermes, R.W. Willems, P.J. Capel, and R.A. Koene. 1984. Fc receptors for mouse IgG1 on human monocytes: Polymorphism and role in antibody-induced T cell proliferation. *J. Immunol.* 133:1185–1189.

Teeling, J.L., R.R. French, M.S. Cragg, J. van den Brakel, M. Pluyter, H. Huang, C. Chan, P.W. Parren, C.E. Hack, M. Dechant, T. Valerius, J.G. van de Winkel, and M.J. Glennie. 2004. Characterization of new human CD20 monoclonal antibodies with potent cytolytic activity against non-Hodgkin lymphomas. *Blood* 104:1793–1800.

Tonegawa, S. 1983. Somatic generation of antibody diversity. *Nature* 302:575–581.

Umana, P., E. Moessner, P. Bruenker, et al. 2006. Novel 3rd generation humanized type II CD20 antibody with glycoengineered Fc and modified elbow hinge for enhanced ADCC and superior apoptosis induction. *Blood* 108:229a.

Vaughan, T.J., A.J. Williams, K. Pritchard, J.K. Osbourn, A.R. Pope, J.C. Earnshaw, J. McCafferty, R.A. Hodits, F. Wilton, and K.S. Johnson. 1996. Human antibodies with sub-nanomolar affinities isolated from a large non-immunized phage display library. *Nature Biotechnol.* 14:309–314.

Wallington, T. 2004. New uses for IVIgG immunoglobulin therapies. *Vox Sanguinis* 87:S155–S157.

Weil, A.J., J.A. Pafentjiev, and K.L. Bowman. 1938. Antigenic qualities of antitoxins. *J. Immunol.* 35:399–413.

Weiler, C.R. 2004. Immunoglobulin therapy: History, indications, and routes of administration. *Int. J. Dermatol.* 43:163–166.

Weiner, G.J., J.A. Bowles, B.K. Link, M.A. Campbell, J.E. Wooldridge, J.B. Breitmeyer. 2005. Anti-CD20 monoclonal antibody (mAb) with enhanced affinity for CD16 activates NK cells at lower concentrations and more effectively than rituximab (R). *Blood* 106:348a.

Weng, W.K., and R. Levy. 2003. Two immunoglobulin G fragment C receptor polymorphisms independently predict response to rituximab in patients with follicular lymphoma. *J. Clin. Oncol.* 21:3940–3947.

Wilberg, F.C., S.K. Rasmussen, T.P. Frandsen, L.K. Rasmussen, K. Tengbjerg, V.W. Coljee, J. Sharon, C.Y. Yang, S. Bregenholt, L.S. Nielsen, J.S. Harum, and A.B. Tolstrup. 2006. Production of target-specific recombinant human polyclonal antibodies in mammalian cells. *Biotechnol. Bioeng.* 94:396–405.

Wilde, H., P. Thipkong, V. Sitprija, and N. Chaiyabutr. 1996. Heterologous antisera and antivenins are essential biologics: Perspectives on a worldwide crisis. *Ann. Intern. Med.* 125:233–236.

Winau, F., and R. Winau. 2002. Emil von Behring and serum therapy. *Microbes Infect.* 4:185–188.

Winau, F., O. Westphal, and R. Winau. 2004. Paul Ehrlich—in search of the magic bullet. *Microbes Infect.* 6:786–789.

Wu, H., D.S. Pfarr, Y. Tang, L.L. An, N.K. Patel, J.D. Watkins, W.D. Huse, P.A. Kiener, and J.F. Young. 2005. Ultra-potent antibodies against respiratory syncytial virus: Effects of binding kinetics and binding valence on viral neutralization. *J. Mol. Biol.* 350:126–144.

Wu, H., D.S. Pfarr, S. Johnson, Y.A. Brewah, R.M. Woods, N.K. Patel, W.I. White, J.F. Young, and P.A. Kiener. 2007a. Development of motavizumab, an ultra-potent antibody for the prevention of respiratory syncytial virus infection in the upper and lower respiratory tract. *J. Mol. Biol.* 368:652–665.

Wu, C., H. Ting, C. Grinnell, S. Bryant, R. Miller, A. Clabbers, S. Bose, D. McCarthy, R.R. Zhu, L. Santora, R. Davis-Taber, Y. Kunes, E. Fung, A. Schwartz, P. Sakorafas, J. Gu, E. Tarcsa, A. Murtaza, and T. Ghayur. 2007b. Simultaneous targeting of multiple disease mediators by a dual-variable-domain immunoglobulin. *Nature Biotechnol.* 25:1290–1297.

Xu, D., M.L. Alegre, S.S. Varga, A.L. Rothermel, A.M. Collins, V.L. Pulito, L.S. Hanna, K.P. Dolan, P.W. Parren, J.A. Bluestone, L.K. Jolliffe, and R.A. Zivin. 2000. In vitro characterization of five humanized OKT3 effector function variant antibodies. *Cell Immunol.* 200:16–26.

Xu, L., H. Yee, C. Chan, A.K. Kashyap, L. Horowitz, M. Horowitz, R.R. Bhatt, and R.A. Lerner. 2008. Combinatorial surrobody libraries. *Proc. Natl. Acad. Sci. USA* 105:10762–10767.

Zeidler, R., G. Reisbach, B. Wollenberg, S. Lang, S. Chaubel, B. Schmitt, and H. Lindhofer. 1999. Simultaneous activation of T cells and accessory cells by a new class of intact bispecific antibody results in efficient tumor cell killing. *J. Immunol.* 163:1246–1252.

Ziegelbauer, K., and D.R. Light. 2008. Monoclonal antibody therapeutics: Leading companies to maximize sales and market share. *J. Commercial Biotechnol.* 14:65–72.

Zoller, M.J., and M. Smith. 1982. Oligonucleotide-directed mutagenesis using M13-derived vectors: An efficient and general procedure for the production of point mutations in any fragment of DNA. *Nucleic Acids Res.* 10:6487–6500.

CHAPTER 2

Antibody Molecular Structure

ROBYN L. STANFIELD and IAN A. WILSON

ABSTRACT

The structural features of antibodies have been studied extensively over the years by many techniques, including electron microscopy, x-ray crystallography, and NMR. Consequently, a wealth of structural information is available for antibodies, alone and in complex with antigens ranging in size from small haptens to whole viruses. The knowledge gained from these studies has greatly facilitated the engineering of antibodies for use as human therapeutics.

2.1 INTRODUCTION

Antibodies are the key component of the humoral adaptive immune response against foreign pathogens. Their enormous sequence and structural diversity allows for recognition of any foreign antigen imaginable, with high affinity and specificity. While antibodies have been best studied in mice and humans, they are also found in animals as evolutionarily distant as the cartilaginous fish, such as sharks, skates, and rays. An abundance of structural information has accumulated for antibodies, including over 800 crystal structures that are predominantly Fab, Fab′, Fv, or occasionally V_H fragments, many in complex with antigens ranging in size from small haptens, to peptide and DNA fragments, to proteins. In addition, combined cryo-electron microscope and crystallographic studies have revealed how Fab and IgG molecules interact with intact viruses. This structural information

Therapeutic Monoclonal Antibodies: From Bench to Clinic. Edited by Zhiqiang An

has been of great value for our understanding of the antibody-antigen recognition process, and also for development of antibodies as human therapeutics.

2.2 GENERAL STRUCTURAL FEATURES

Mammalian antibodies are constructed from two types of protein sequences, called the heavy and light chains. The five classes of antibodies in humans and other placental mammals differ in their heavy chain sequences, with heavy chain types μ, δ, γ, ε, and α found in IgM, IgD, IgG, IgE, and IgA antibodies, respectively. Each heavy chain can pair with one of two types of light chain, called λ or κ. Most antibodies are made up of four chains: two copies of the heavy chain and two copies of the light chain (Fig. 2.1). However, the IgM class can also exist as a pentamer of these four chains, with a resultant 10 heavy and 10 light chains, and IgA can exist in a form called secretory IgA, where an additional J chain stabilizes the dimerization of two antibodies, to give a total of four heavy and four light chains.

The majority of antibodies found in the serum belong to the IgG class, and most structural information has been derived for this class of antibody; thus, most of our discussion will deal with IgG antibodies. An intact IgG molecule has two heavy (~55,000 Da each) and two light chains (~24,000 Da each) that fold into three large domains: two Fab fragments (one light and the N-terminal half of a heavy chain) and one Fc fragment (two C-terminal heavy chain halves) (Fig. 2.1). Fab and Fc are abbreviations for *fragment-antibody binding* and *fragment-crystallizable*, so called because the Fab fragment binds antigen, and the first Fc fragments studied were easy to crystallize, although that clearly now is somewhat of a misnomer. The overall shape of the IgG can be described as a Y, with the Fc fragment forming the base of the Y, and the two Fab fragments forming the two arms (Fig. 2.2).

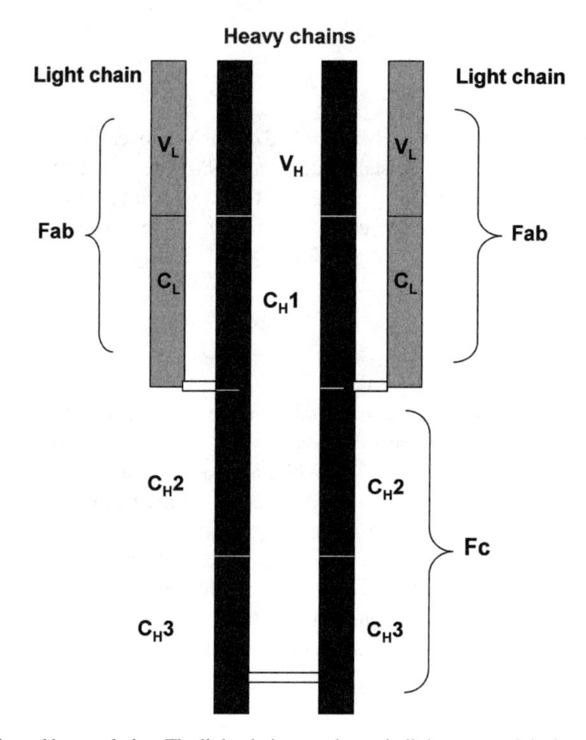

Figure 2.1 IgG light and heavy chains. The light chains are shown in light gray, and the heavy chains in dark gray, with the different Ig domains indicated.

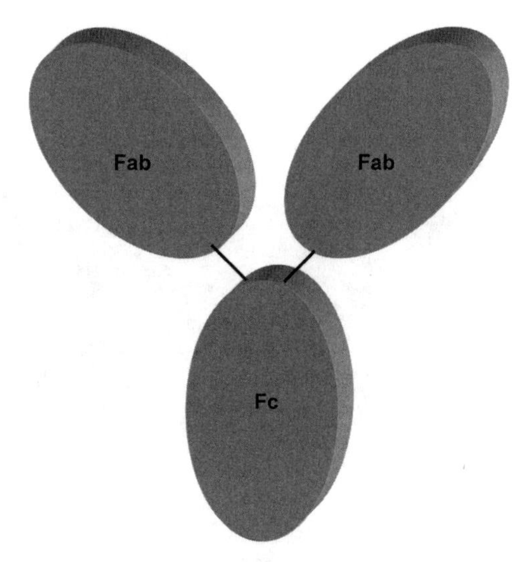

Figure 2.2 IgG domain organization. The Fc and two Fab domains form an overall Y shape.

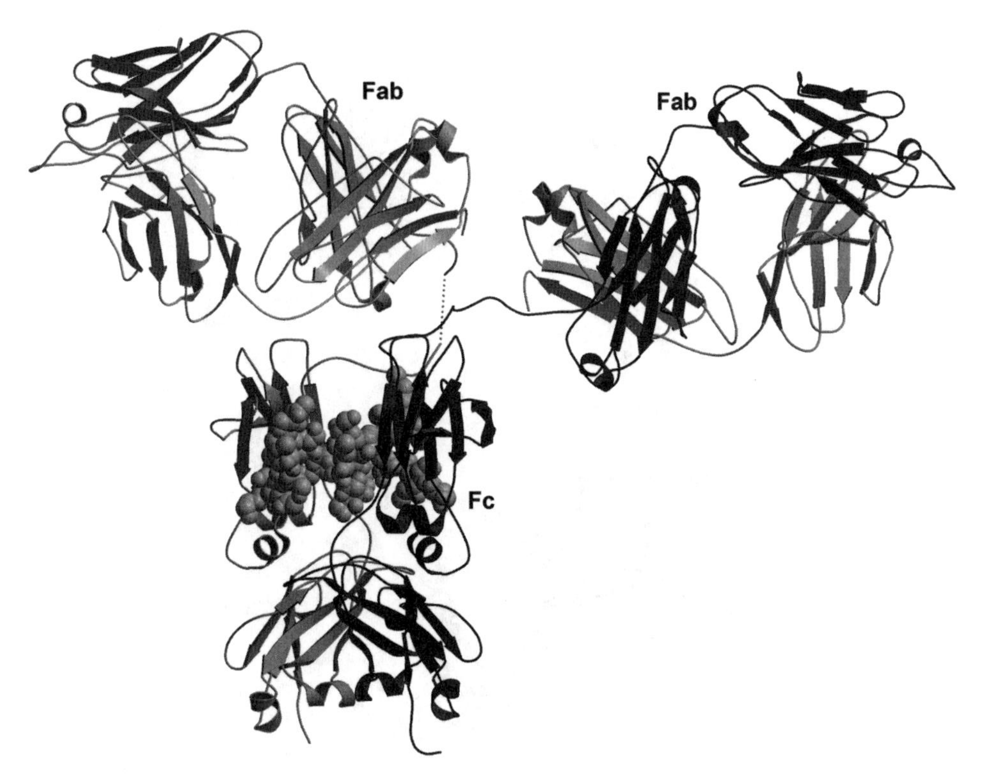

Figure 2.3 Structure of IgG b12 determined by x-ray crystallography. Residues in the linker region of the heavy chain that have no visible electron density are shown as a dotted line. Carbohydrate within the Fc domain is shown in a gray CPK representation. Rather than a rigid Y shape, as depicted in Figure 2.2, in a real IgG, the Fab arms are very dynamic with respect to the Fc domain, allowing for greater flexibility in binding antigen. (See color insert.)

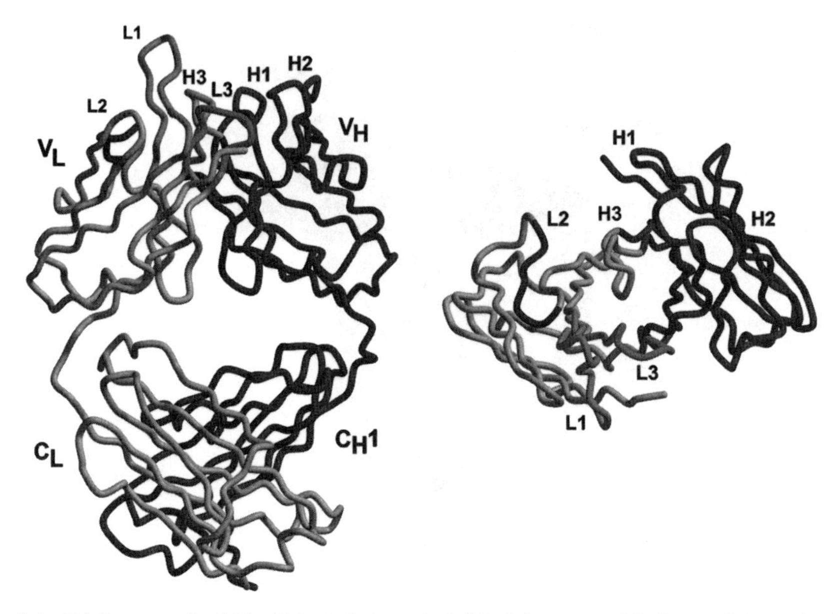

Figure 2.4 Fab fragment. (Left) The light chain is on the left in light gray, and the heavy chain on the right in darker gray. The CDR loops that contact antigen are labeled. (Right) Looking down onto the Fab antigen-binding site. This view is the same as on the left, but rotated 90 degrees about a horizontal axis.

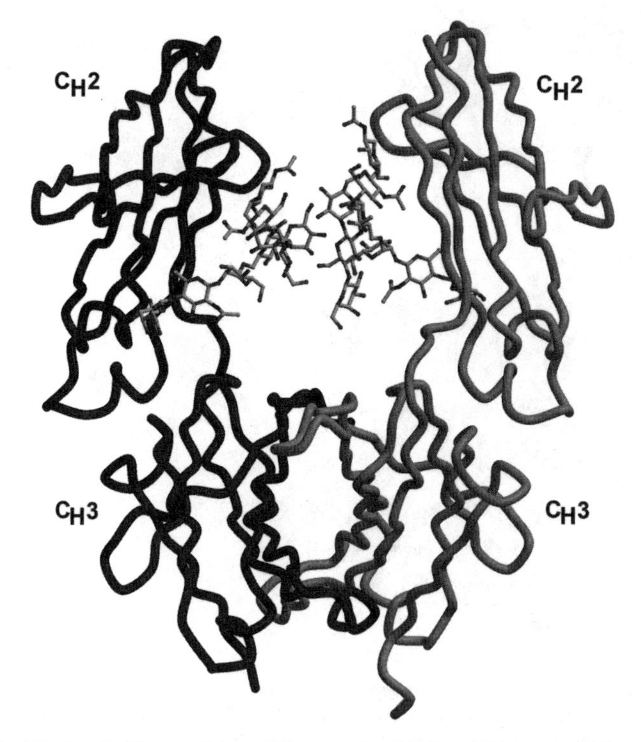

Figure 2.5 An IgG Fc domain. The two heavy chains are colored dark and light gray, and the carbohydrate found at the center of the Fc in an IgG is shown in a ball-and-stick representation.

However, highly flexible hinge regions join the Fc to the two Fabs, and electron microscopy (Roux, Strelets, and Michaelsen 1997) and crystal structures of IgGs with intact hinge regions (Harris et al. 1992, 1997; Saphire et al. 2001) show that the Fabs have a wide range of orientations with respect to the Fc (Fig. 2.3). The Fab fragment can be further subdivided into smaller domains called the variable domain (the N-terminal half of the Fab; V_L and V_H) and constant domain (the C-terminal half of the Fab; C_L and C_H1) (Fig. 2.4). As suggested by their names, the variable domain contains regions with high sequence diversity, while the constant domain sequences are highly conserved. The Fc fragment also has two domains, formed by dimerization of C_H2-C_H2 and C_H3-C_H3 regions (Fig. 2.5).

The smallest unit of the domain architecture is the immunoglobulin or Ig domain. Each individual V_L, V_H, C_L, C_H1, C_H2, or C_H3 domain is an Ig domain, consisting of a characteristic Ig fold with either seven or nine β strands, in two β sheets, that form a Greek-key β-barrel, with a highly conserved disulfide connecting the two sheets at the core of the barrel (Fig. 2.6). A constant-type Ig domain has seven strands, three in one sheet and four in the other, while a variable-type Ig domain adds two extra strands (C′ and C″) to the three-stranded sheet, so that the barrel is composed of five- and four-stranded β-sheets (Fig. 2.6). While first identified in immunoglobulins, the Ig fold is found in many other types of molecules and, in addition to the variable- and constant-type Ig domain, other different but related types of Ig domains exist, some without the canonical disulfide bridge (Bork, Holm, and Sander 1994; Halaby, Poupon, and Mornon 1999).

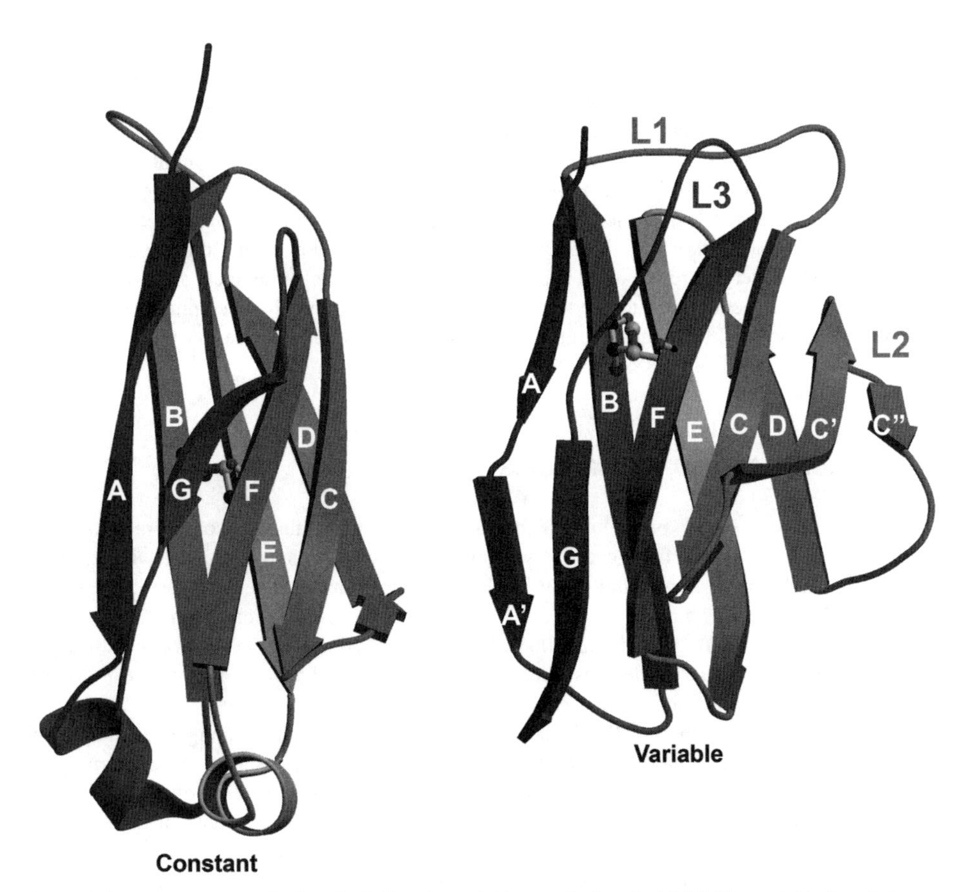

Figure 2.6 A constant-type Ig domain (left) and a variable-type Ig domain (right). The variable domain has two extra strands (C′ and C″) that form the CDR2 loop. This domain is from a light chain; however, a heavy chain variable domain has the same domain architecture.

The Fab fragment recognizes antigen at a site formed by six linear polypeptide segments called complementarity determining region (CDR) loops. Three CDRs are contributed by the V_L (L1, L2, and L3) and three by the V_H domain (H1, H2, and H3) (Fig. 2.4). These loops are hypervariable in sequence, and were predicted to be the interaction site for antigen recognition long before any structural information existed for the Fab fragment (Wu and Kabat 1970; Kabat and Wu 1971). However, the structure of the first Fab-hapten complex (McPC603-phosphocholine) determined by x-ray crystallography in 1973 (Padlan et al. 1973; Segal et al. 1974) verified that the antigen interacts mainly with the CDR loop regions of the Fab. While recognition of protein antigens by Fab generally, but not always, utilizes all six of the CDR loops, recognition of smaller antigens may use a more limited set of these CDRs; in fact, some isolated V_H fragments containing CDRs H1-3 are sufficient for recognition of their antigen (Davies and Riechmann 1996). Typical V_H-V_L antibody-antigen interface sizes (the combined buried molecular surface area for both the antigen and antibody) range from around 300–600 Å^2 for small haptens, to 800–900 Å^2 for peptides, to 1500–1800 Å^2 for proteins (Wilson et al. 1991).

2.3 CANONICAL CONFORMATIONS

While the CDR loops are the most variable part of the Fab (in both their sequence and structure), five of the six loops have been shown to have a limited number of conformations, termed canonical structures (Chothia and Lesk 1987). Based on their sequence, and a set of rules derived from known structures, the conformations of these loops can usually be predicted with reasonable accuracy. However, CDR H3 has been far more difficult to classify. H3 is the most variable of the CDR loops in sequence and length, and structures of many Fab fragments have shown that H3 is also the most structurally variable and flexible of the CDRs. The base of CDR H3 can usually be found in one of two different conformations that can often be predicted from the sequence (Shirai, Kidera, and Nakamura 1996; Al-Lazikani, Lesk, and Chothia 1997; Koliasnikov et al. 2006), but the remainder of the loop has no conserved structure, is frequently seen to undergo large changes in conformation upon binding of antigen, and is often poorly ordered in crystal structures of unliganded Fabs. More recent studies have emphasized important functional differences in the lengths of the CDR H3s in humans versus mice that can be exploited for antigen recognition (Collis, Brouwer, and Martin 2003).

2.4 Fab CONFORMATIONAL CHANGES

In addition to hinge flexibility between the Fab and Fc fragments, flexibility also occurs between the domains that make up the Fab fragment. The variable and constant domains can move with respect to each other around what is termed the elbow angle. This angle has been seen to vary between 117 degrees and 227 degrees (Stanfield et al. 2006), with larger values seen more often with antibodies that contain λ light chains (Stanfield et al. 2006). The V_L and V_H domains can also rotate with respect to each other by as much as 16 degrees (Stanfield et al. 1993), and these movements can serve to dramatically reconfigure the antigen binding site.

Currently structures for about 100 Fabs have been determined in both the unliganded and ligand-bound form. Comparisons of these liganded and ligand-free structures have shown that conformational changes often occur in the antigen-binding site to enable greater complementarity of fit to the antigen (Stanfield and Wilson 1994; Wilson and Stanfield 1994). These changes usually take place in the CDR loops and, as mentioned earlier, the largest changes are often seen in CDR H3. While often termed induced-fit binding, studies of antibodies in solution indicate that the conformational changes are probably not completely induced by binding of the antigen, but rather the antigen may in some cases bind to one of several preexisting antibody conformations found in solution, where small structural changes then allow the antibody to adapt and custom-fit its binding site to the antigen in question (Foote and Milstein 1994; James, Roversi, and Tawfik 2003).

Conformational changes in antibodies include changes in side-chain rotamers, movements of the CDR loops as either a rigid unit or by more extensive structural rearrangements, and through changes in the relative disposition of their V_H/V_L domains. An excellent illustration of a key side-chain rearrangement is found in the anti-progesterone antibody DB3 (Arevalo et al. 1993; Arevalo, Taussig, and Wilson 1993; Arevalo et al. 1994). When the unliganded and steroid-bound structures

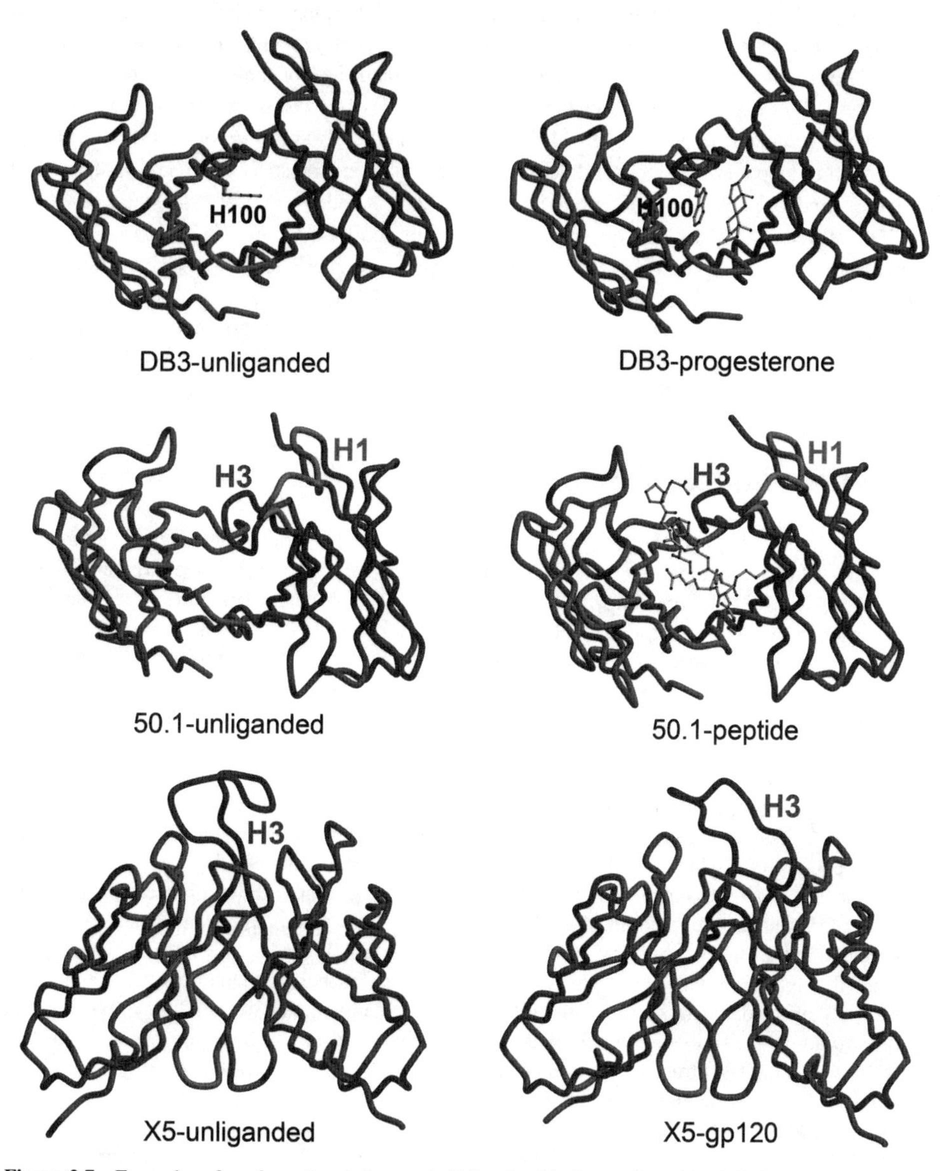

Figure 2.7 Examples of conformational changes in Fabs after binding antigen. Top left: The anti-progesterone Fab DB3 in its unliganded form, with Trp^{H100} filling its antigen-binding pocket. Top right: DB3 with progesterone (ball-and-stick) bound, and Trp^{H100} moved away from its position in the unliganded Fab. (Middle left) The anti-HIV-1 Fab 50.1 in its unliganded form. Middle right: 50.1 with bound peptide (ball-and-stick). The H3 and H1 CDRs are labeled. H3 undergoes a structural rearrangement, while H1 moves away from the binding site while maintaining its overall shape. Bottom left: The unliganded anti-HIV-1 Fab X5, with its long CDR H3 labeled. Bottom right: Fab X5 in its gp120-bound conformation, with H3 labeled. A large conformational change occurs in the CDR H3 between the unliganded and liganded forms. (See color insert.)

for DB3 are compared, a substantial conformational change in the position of the Trp^{H100} side chain is observed (Fig. 2.7). Interestingly, this side chain fills the antigen-binding site in the unliganded Fab structure, and then moves out of the way to allow binding of steroid (Fig. 2.7). Thus, Trp^{H100} acts as a surrogate ligand for the antibody in the absence of steroid. One of the most extensive conformational changes has been seen for antibody 50.1, an anti-HIV-1 neutralizing antibody that recognizes the HIV-1 gp120 V3 loop (Rini et al. 1993; Stanfield et al. 1993). A structural comparison of the unliganded and V3 peptide-bound Fab showed that the H3 CDR loop undergoes a large structural rearrangement, with the side chain of Tyr^{H97} moving by about 6 Å. The CDR H1 also moves, in a rigid body or segmental fashion, with a root-mean square deviation for the H1 main-chain atoms of 1.1 to 1.4 Å (for multiple copies of both the native and peptide-bound Fab in the crystals). In addition, the V_L and V_H domains of 50.1 move with respect to each other by about 16 degrees (Fig. 2.7). The combination of loop rearrangement and domain rotation result in a drastically differently shaped and configured binding pocket for the peptide ligand (Fig. 2.7). Another example of an extremely large H3 CDR movement is seen in crystal structures of the anti-HIV-1 antibody X5 that recognizes the recessed CD4-binding site of gp120. The long X5 CDR loop undergoes an extensive rearrangement and moves by as much as 17 Å (Darbha et al. 2004; Huang et al. 2005) between the unbound and gp120 bound states (Fig. 2.7). Unusually long H3 CDR loops are not uncommon in human antiviral antibodies, and it has been proposed that long CDR loops may help antibodies target recessed clefts in viral antigens (Burton et al. 2005).

2.5 HUMAN ANTI-HIV-1 ANTIBODIES

While most of the early crystal structures for antibodies were obtained for human myeloma antibodies, the advent of the monoclonal antibody technology made it very easy to produce large amounts of mouse monoclonal antibodies to known antigens. Thus, many more crystal structures have been determined for antibody fragments from mice than from humans. However, subsequent developments, such as Epstein-Barr virus immortalized human B-cells (Steinitz et al. 1977) and human-mice heterohybridomas (Cole et al. 1984), production of human antibodies and antibody fragments in bacteria (Skerra and Pluckthun 1988), and phage-display of antibody fragments (Huse et al. 1989), have made it possible to produce human monoclonal antibodies against known antigens. In addition, techniques, such as humanization of mouse or rat antibodies (Jones et al. 1986) or the production of human/mouse chimeric antibodies (Boulianne, Hozumi, and Shulman 1984; Morrison et al. 1984; Neuberger et al. 1985; Better et al. 1988), have been made possible by now standard molecular biology techniques. Currently about 40 fully human Fab or Fv fragments have crystal structures deposited in the Protein Data Bank (PDB), although many of these are for the same fragment with multiple, related antigens.

Many of the human antibodies that have been studied by x-ray crystallography are against viral antigens, including the HIV-1 virus. The HIV-1 virus is able to rapidly evolve to evade the host immune system, but only a handful of antibodies have been discovered that are able to effectively neutralize a wide variety of the different strains of the virus (Burton, Stanfield, and Wilson 2005). A successful vaccine would induce similar antibodies that are able to neutralize any strain of the virus that an individual might encounter. Structures of these rare antibodies have shown that the antibodies outwit the virus by using novel structural features in their mechanisms of antigen recognition (Burton, Stanfield, and Wilson 2005).

One of the most interesting of the anti-HIV-1 antibodies is 2G12. 2G12 was isolated from an HIV-1 infected patient, and is one of the most potent, broadly neutralizing antibodies known. While initial studies indicated that 2G12 recognized a carbohydrate epitope on the HIV-1 surface (Sanders et al. 2002; Scanlan et al. 2002), it was difficult to understand how the antibody might recognize carbohydrate with high affinity. The carbohydrate on the HIV-1 viral surface is transferred onto the viral coat proteins by the human host cellular machinery. Thus, these sugars should appear as self to our immune system and, hence, give rise to tolerance and not lead to a strong antigenic response. Such anti-carbohydrate antibodies should be exceedingly rare and, as for other anti-carbohydrate antibodies against non-self sugars, be of low affinity. The crystal structure of the 2G12 Fab fragment was, therefore, a huge surprise (Calarese et al. 2003), and revealed the Fab had dimerized via a domain swap of its V_H domains (Fig. 2.8). This domain-swapped dimer has two closely spaced antigen-binding sites and a potential third binding region at the interface of the two newly associated V_H domains (V_H-V_H'). These intertwined Fab regions give rise to the unusual linear shape of the intact IgG that differs from the more typical Y configuration seen for other IgGs. Electron microscope (EM) studies have also clearly shown this unusual linear structure in the intact 2G12 IgG (Roux et al. 2004), and analytical ultracentrifugation also confirmed that the Fab is an obligate dimer in solution (Calarese et al. 2003). Crystal structures have been determined for the dimeric 2G12 Fab in complex with carbohydrates (Calarese et al. 2003, 2005) ranging in size from Manα1-2Man to $Man_9GlcNac_2$, and these structures indicate that the Fab can bind to the terminal Manα1-2Man in either the D1 or D3 arms of $Man_9GlcNac_2$. Mutagenesis and modeling studies of 2G12 have pointed to a conserved cluster of high mannose moieties, especially those linked to gp120 residues 332, 339, and 392, as being the likely epitope for this antibody (Sanders et al. 2002; Scanlan et al. 2002; Calarese et al. 2003). In addition, glycan array and solution-phase ELISA analyses have helped to define the carbohydrate specificity of 2G12 (Blixt et al. 2004; Bryan et al. 2004; Calarese et al. 2005). Modeling studies show that the distances between these groups on gp120 carbohydrates are compatible with the distance between the two antigen-binding sites on the dimeric Fab (Calarese et al. 2003) and that a third, novel antigen-binding site may exist at the V_H-V_H' interface. The inherent high avidity (nM) of the dimeric 2G12 has thus given the antibody high affinity for carbohydrate on the virus due to the multivalency. Analysis of the 2G12 sequence uncovers several unusual residues that may favor the domain swap event. One rare residue is Pro^{H113}, located in the linker between V_H and CH1. Other unusual residues include the hydrophobic Ile^{H19} and Phe^{H77} that may help to stabilize the novel V_H-V_H' interface,

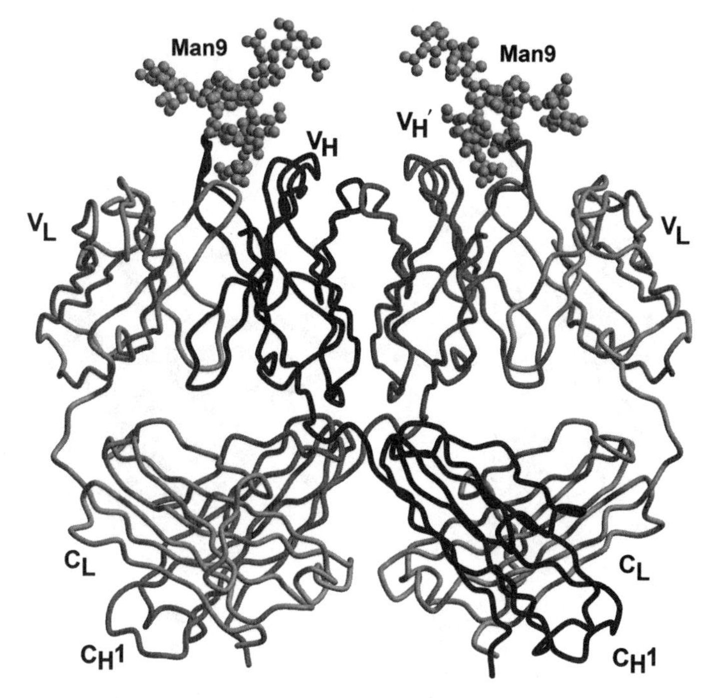

Figure 2.8 The broadly neutralizing, anti-HIV-1 Fab 2G12. 2G12 forms an unusual domain swapped dimer, with the V_H domains from each heavy chain (labeled V_H-V_H') swapping to the other Fab, resulting in a tightly linked dimer of Fabs. Bound $Man_9GlcNac_2$ is shown in a CPK representation.

whereas an Arg at position H39 disrupts a highly conserved Gln^{L38}-Gln^{H39} interaction that is found in all antibodies, as well as in T cell receptors.

Two other unusual anti-HIV-1 antibodies that recognize neighboring epitopes are 2F5 and 4E10. Both of these antibodies recognize epitopes in the membrane proximal external region (MPER) of gp41. The MPER region, as its name implies, immediately precedes the membrane spanning region of gp41, and is thought to be only transiently exposed during the viral-cell fusion process. The 2F5 and 4E10 Fab structures include complexes with gp41 peptides (Ofek et al. 2004; Cardoso et al. 2005; Cardoso et al. 2007), with the 2F5 peptide forming a β-turn, and the 4E10 peptide forming an α-helix. Estimations of the distances of these gp41 epitopes to the membrane spanning regions of gp41 indicate that the Fabs must come very close to or even contact the host cell membrane in order to bind these epitopes. Interestingly, the Fab CDR residues surrounding the peptide-binding site include a large number of hydrophobic residues, including a number of Trp and Gly residues in 4E10 that may facilitate interaction with the membrane.

An interesting class of anti-HIV-1 antibodies has been found that only bind to gp120 after it binds to its primary receptor, CD4. Thus, these antibodies are called CD4-induced, or CD4i. The binding of CD4 to gp120 induces conformational changes in the gp120 that are necessary for binding to the co-receptor (usually CCR5 or CXCR4) and, hence, required for viral entry into the cell. The CCR5 co-receptor has sulfated tyrosine residues in its N terminus, and this N-terminal region is proposed to interact with gp120 during the binding process. Surprisingly, some of the CD4i antibodies were also found to have acquired sulfated tyrosine residues in their highly acidic CDR H3 loops through posttranslational modification, suggesting that they may be structural mimics of the co-receptor (Choe et al. 2003; Huang et al. 2004, 2007). A recent structure of gp120 in complex with CD4 and sulfated antibody 412D (Fig. 2.9) shows how these two sulfated tyrosine residues in the CDR

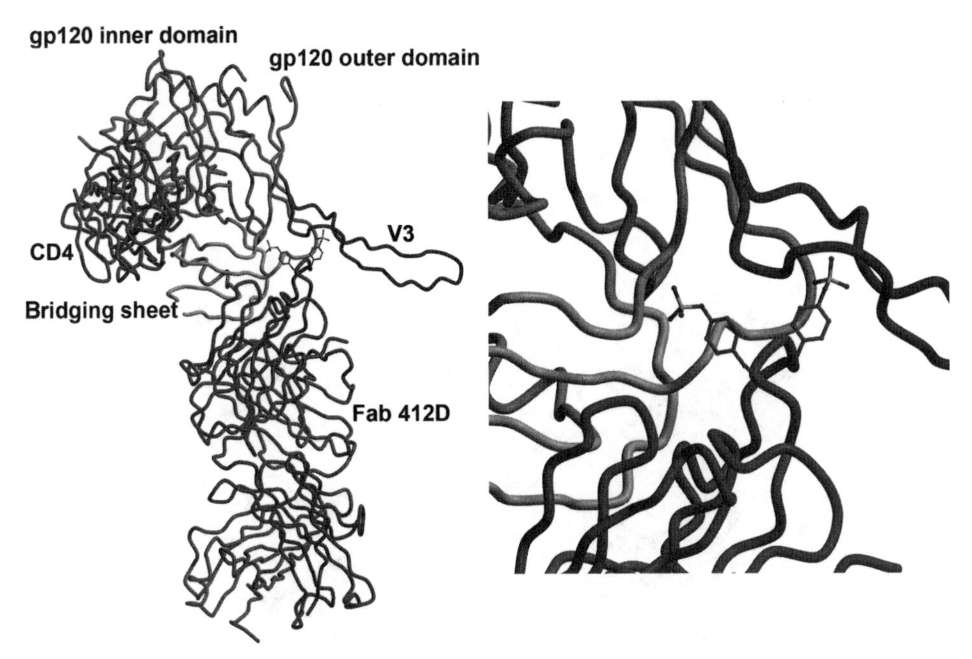

Figure 2.9 The anti-HIV-1 Fab 412D has sulfated tyrosine residues in CDR H3. Left: The complex of Fab 412d with CD4 and gp120. Right: enlargement of the sulfated tyrosine residues and their interaction with gp120. It is thought that the gp120 co-receptor (CCR5) binds to the bridging sheet and V3 regions of the gp120 molecule. CCR5 has four sulfated tyrosine residues in its N-terminal region, and at least two of these are thought to take part in the interaction with gp120. Several antibodies that recognize the same region of gp120 have also evolved to have sulfated tyrosine residues. (See color insert.)

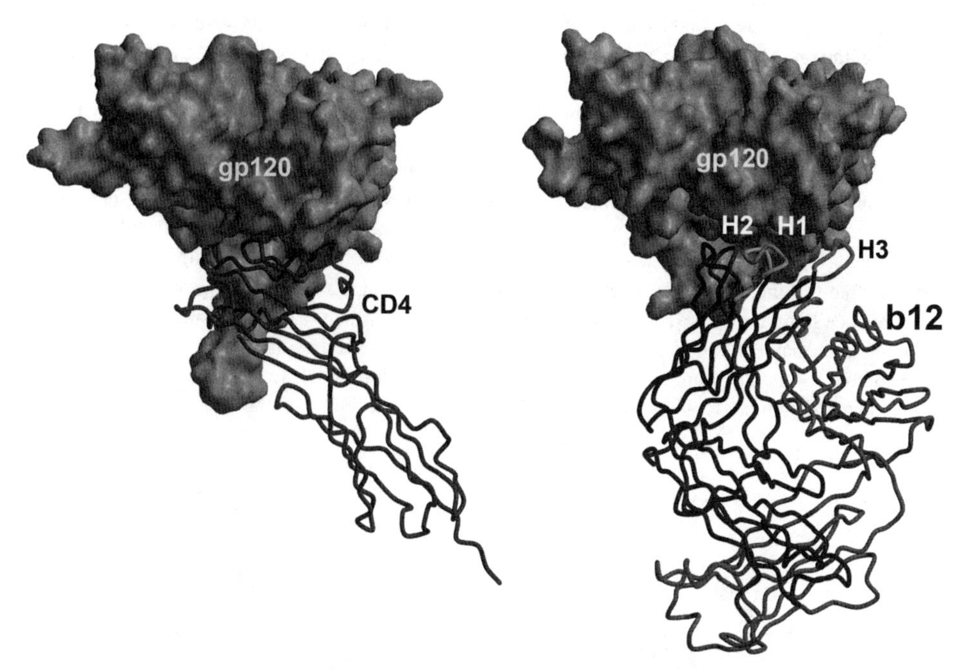

Figure 2.10 Broadly neutralizing anti-HIV-1 antibody b12 interactions with gp120. Left: The CD4 (tubes) binding site on gp120 (solid surface) is shown. Right: b12 binds to the CD4-binding site, accessing a deep cleft with its long CDR loop (H2) and clasping the CD4-binding loop between CDRs H2 and H1 on one side and H3 on the other side. (See color insert.)

H3 region interact with the gp120 bridging sheet/V3 regions, and mimic the interaction with CCR5 (Choe et al. 2003; Huang et al. 2004, 2007).

Antibody b12 was discovered by phage-display technology using a library developed from blood marrow taken from an HIV-1 infected, long-term nonprogressing patient. Crystal structures of the intact b12 IgG (Saphire et al. 2001) and the recently determined structure of b12 Fab in complex with gp120 (Zhou et al. 2007) revealed that the Fab recognizes the highly conserved, but deeply recessed CD4-binding site (Fig. 2.10). The Fab CDR H2 loop accesses the CD4-binding site, with Tyr^{H53} binding in the same pocket as Phe43 of CD4. The long CDR H3 unexpectedly was found to bind on the outside of the binding site, so that the CD4-binding loop is sandwiched between the H3 CDR on one side, and H1 and H2 on the other side. Interestingly, the antibody light chain makes no contact with the gp120 monomer, but may possibly contact part of the intact gp120 trimer on the viral surface. Thus, this antibody has managed to find a site of vulnerability on the virus, its Achilles' heel, and interacts primarily with the structurally conserved outer domain that is the primary site of attachment for CD4. But unlike CD4, it can remain bound to the outer domain by itself with high affinity, whereas CD4 reorganizes the highly flexible inner domain of gp120 to decrease its off-rate and assemble the bridging sheet that constitutes a major portion of the CCR5 co-receptor binding site (Zhou et al. 2007).

2.6 SHARK AND CAMEL ANTIBODIES

Crystal structures have also been determined for antibodies from rats, hamsters, camels, and sharks. While the rodent and human antibodies are very similar, camels and sharks both have, in addition to a conventional antibody repertoire, unusual antibodies that exist as dimers of heavy chains with no associated light chains. In the case of the camel (and llama), these antibodies are the result of a gene deletion of the IgG C_H1 domain, resulting in a heavy chain with V_H, C_H2, and C_H3. Light chains do not associate with the C_H1-free heavy chain; however, the free heavy chains still pair up through their two associated constant domains. The camel heavy chain antibodies are closely related in sequence to their normal IgG antibodies. Sharks and other cartilaginous fish, such as skates and rays, also have heavy chain antibodies, called new antigen receptors or IgNAR, that contain two heavy chains and no light chains. However, unlike the camel heavy chain dimers, the IgNAR antibodies are not closely related by sequence to the other, more typical antibodies found in the shark (IgM and IgX). The five constant domains on each heavy chain dimerize to form a long stalk, leaving the two variable domains free to bind antigen independently. Although the shark and camel V_H domains have very low sequence homology, they have co-evolved many unique structural features, including mutations in the interface that would normally be involved in the V_L-V_H dimerization, to make their V_H domains more soluble. Both camel and shark V_H domains also have long CDR3 regions that usually contain noncanonical disulfides to tether their long CDR H3s to the body of the V_H domain. While the

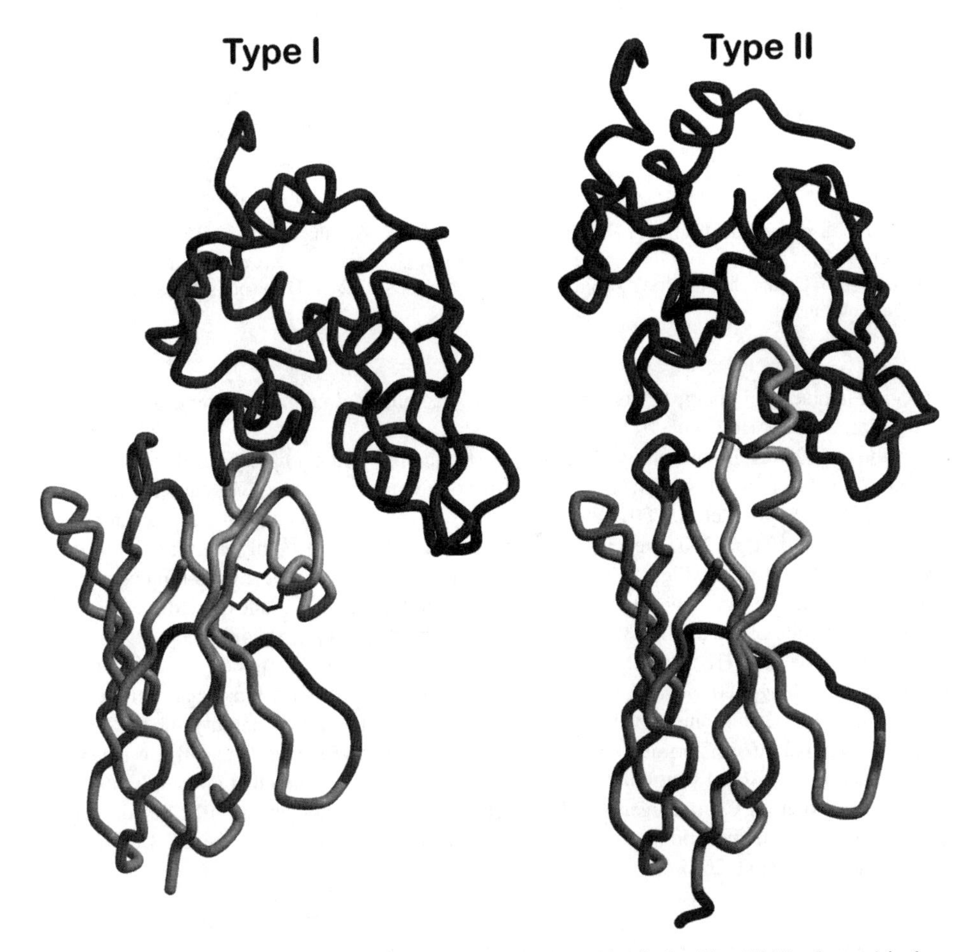

Figure 2.11 Shark IgNAR single domain antibodies bound to lysozyme. Left: a Type I IgNar (bottom) is shown bound to lysozyme (top). The Type I IgNAR is characterized by a long CDR3 region stabilized by two noncanonical disulfide bonds tethering the CDR to the IgNAr framework. Other regions that have a large number of somatic mutations and may interact with antigen are HV2 and HV4. Right: a Type II IgNAR is characterized by a long CDR3 region with one noncanonical disulfide that tethers it to the CDR1 loop.

camel V_H domain has the three CDR regions found in a typical V_H domain (Desmyter et al. 1996; Spinelli et al. 1996), the shark IgNAR domains have only CDR1 and 3, and are missing the C′ and C″ strands that make up a typical CDR2 (Stanfield et al. 2004) (Fig. 2.11). Despite having only three (camel) or two (shark) CDR loops, these small, independent V_H domains are able to bind antigen with very high affinity. The production of camel and shark V_H domains has been adapted to phage-display technology (Arbabi Ghahroudi et al. 1997; Dooley, Flajnik, and Porter 2003) and the resulting molecules show great promise as small antigen-binding fragments for many different uses. Several studies have shown that the shark and camel single-domain antibodies are able to access deeply recessed clefts better than their Fab counterparts (Stanfield et al. 2007). In addition, the solubility-enhancing mutations in these single-domain fragments have been adapted into engineered, human single-domain V_H fragments to enhance their solubility (Davies and Riechmann 1996).

2.7 SUMMARY

An enormous wealth of structural information is available for antibodies and their fragments, making the antibody molecule one of the most highly studied in the entire protein universe. This information has proven to be invaluable for scientists working towards the development and production of antibodies and antibody fragments for human therapeutic products. Nevertheless, although we know a great deal about antibody structure, recent structures of several human antibodies have revealed surprisingly novel and diverse structural features, such as Fab domain swapping, long but relatively rigid CDR H3s, and posttranslational modifications, that were totally unexpected from the study of an enormous number of mouse antibodies. Given the tremendous diversity of the antibody immune response, we expect to continue to uncover further novel aspects of antibody-antigen recognition that will provide new ideas and inspiration for the development of antibody-based therapeutics.

ACKNOWLEDGMENTS

This is manuscript number MB-19299 from the Scripps Research Institute. The authors acknowledge support from National Institutes of Health grants GM-46192 and the Neutralizing Antibody Consortium of the International Aids Vaccine Initiative.

REFERENCES

Al-Lazikani, B., A.M. Lesk, and C. Chothia. 1997. Standard conformations for the canonical structures of immunoglobulins. *J. Mol. Biol.* 273:927–948.

Arbabi Ghahroudi, M., A. Desmyter, L. Wyns, R. Hamers, and S. Muyldermans. 1997. Selection and identification of single domain antibody fragments from camel heavy-chain antibodies. *FEBS Lett.* 414:521–526.

Arevalo, J.H., C.A. Hassig, E.A. Stura, M.J. Sims, M.J. Taussig, and I.A. Wilson. 1994. Structural analysis of antibody specificity. Detailed comparison of five Fab'-steroid complexes. *J. Mol. Biol.* 241:663–690.

Arevalo, J.H., E.A. Stura, M.J. Taussig, and I.A. Wilson. 1993. Three-dimensional structure of an anti-steroid Fab' and progesterone-Fab' complex. *J. Mol. Biol.* 231:103–118.

Arevalo, J.H., M.J. Taussig, and I.A. Wilson. 1993. Molecular basis of crossreactivity and the limits of antibody-antigen complementarity. *Nature* 365:859–863.

Better, M., C.P. Chang, R.R. Robinson, and A.H. Horwitz. 1988. *Escherichia coli* secretion of an active chimeric antibody fragment. *Science* 240:1041–1043.

Blixt, O., S. Head, T. Mondala, C. Scanlan, M.E. Huflejt, R. Alvarez, M.C. Bryan, F. Fazio, D. Calarese, J. Stevens, N. Razi, D.J. Stevens, J.J. Skehel, I. van Die, D.R. Burton, I.A. Wilson, R. Cummings, N. Bovin, C.H. Wong, and J.C. Paulson. 2004. Printed covalent glycan array for ligand profiling of diverse glycan binding proteins. *Proc. Natl. Acad. Sci. USA* 101:17033–17038.

Bork, P., L. Holm, and C. Sander. 1994. The immunoglobulin fold. Structural classification, sequence patterns and common core. *J. Mol. Biol.* 242:309–320.

Boulianne, G.L., N. Hozumi, and M.J. Shulman. 1984. Production of functional chimaeric mouse/human antibody. *Nature* 312:643–646.

Bryan, M.C., F. Fazio, H.K. Lee, C.Y. Huang, A. Chang, M.D. Best, D.A. Calarese, O. Blixt, J.C. Paulson, D. Burton, I.A. Wilson, and C.H. Wong. 2004. Covalent display of oligosaccharide arrays in microtiter plates. *J. Am. Chem. Soc.* 126:8640–8641.

Burton, D.R., R.L. Stanfield, and I.A. Wilson. 2005. Antibody vs. HIV in a clash of evolutionary titans. *Proc. Natl. Acad. Sci. USA* 102:14943–14948.

Calarese, D.A., H.K. Lee, C.Y. Huang, M.D. Best, R.D. Astronomo, R.L. Stanfield, H. Katinger, D.R. Burton, C.H. Wong, and I.A. Wilson. 2005. Dissection of the carbohydrate specificity of the broadly neutralizing anti-HIV-1 antibody 2G12. *Proc. Natl. Acad. Sci. USA* 102:13372–13377.

Calarese, D.A., C.N. Scanlan, M.B. Zwick, S. Deechongkit, Y. Mimura, R. Kunert, P. Zhu, M.R. Wormald, R.L. Stanfield, K.H. Roux, J.W. Kelly, P.M. Rudd, R.A. Dwek, H. Katinger, D.R. Burton, and I.A. Wilson. 2003. Antibody domain exchange is an immunological solution to carbohydrate cluster recognition. *Science* 300:2065–2071.

Cardoso, R.M., F.M. Brunel, S. Ferguson, M. Zwick, D.R. Burton, P.E. Dawson, and I.A. Wilson. 2007. Structural basis of enhanced binding of extended and helically constrained peptide epitopes of the broadly neutralizing HIV-1 antibody 4E10. *J. Mol. Biol.* 365:1533–1544.

Cardoso, R.M., M.B. Zwick, R.L. Stanfield, R. Kunert, J.M. Binley, H. Katinger, D.R. Burton, and I.A. Wilson. 2005. Broadly neutralizing anti-HIV antibody 4E10 recognizes a helical conformation of a highly conserved fusion-associated motif in gp41. *Immunity* 22:163–173.

Choe, H., W. Li, P.L. Wright, N. Vasilieva, M. Venturi, C.C. Huang, C. Grundner, T. Dorfman, M.B. Zwick, L. Wang, E.S. Rosenberg, P.D. Kwong, D.R. Burton, J.E. Robinson, J.G. Sodroski, and M. Farzan. 2003. Tyrosine sulfation of human antibodies contributes to recognition of the CCR5 binding region of HIV-1 gp120. *Cell* 114:161–170.

Chothia, C., and A.M. Lesk. 1987. Canonical structures for the hypervariable regions of immunoglobulins. *J. Mol. Biol.* 196:901–917.

Cole, S.P., B.G. Campling, T. Atlaw, D. Kozbor, and J.C. Roder. 1984. Human monoclonal antibodies. *Mol. Cell Biochem.* 62:109–120.

Collis, A.V., A.P. Brouwer, and A.C. Martin. 2003. Analysis of the antigen combining site: Correlations between length and sequence composition of the hypervariable loops and the nature of the antigen. *J. Mol. Biol.* 325:337–354.

Darbha, R., S. Phogat, A.F. Labrijn, Y. Shu, Y. Gu, M. Andrykovitch, M.Y. Zhang, R. Pantophlet, L. Martin, C. Vita, D.R. Burton, D.S. Dimitrov, and X. Ji. 2004. Crystal structure of the broadly cross-reactive HIV-1-neutralizing Fab X5 and fine mapping of its epitope. *Biochemistry* 43:1410–1417.

Davies, J., and L. Riechmann. 1996. Single antibody domains as small recognition units: Design and in vitro antigen selection of camelized, human VH domains with improved protein stability. *Protein Eng.* 9:531–537.

Desmyter, A., T.R. Transue, M.A. Ghahroudi, M.H. Thi, F. Poortmans, R. Hamers, S. Muyldermans, and L. Wyns. 1996. Crystal structure of a camel single-domain VH antibody fragment in complex with lysozyme. *Nat. Struct. Biol.* 3:803–811.

Dooley, H., M.F. Flajnik, and A.J. Porter. 2003. Selection and characterization of naturally occurring single-domain (IgNAR) antibody fragments from immunized sharks by phage display. *Mol. Immunol.* 40:25–33.

Foote, J., and C. Milstein. 1994. Conformational isomerism and the diversity of antibodies. *Proc. Natl. Acad. Sci. USA* 91:10370–10374.

Halaby, D.M., A. Poupon, and J. Mornon. 1999. The immunoglobulin fold family: Sequence analysis and 3D structure comparisons. *Protein Eng.* 12:563–571.

Harris, L.J., S.B. Larson, K.W. Hasel, J. Day, A. Greenwood, and A. McPherson. 1992. The three-dimensional structure of an intact monoclonal antibody for canine lymphoma. *Nature* 360:369–372.

Harris, L.J., S.B. Larson, K.W. Hasel, and A. McPherson. 1997. Refined structure of an intact IgG2a monoclonal antibody. *Biochemistry* 36:1581–1597.

Huang, C.C., S.N. Lam, P. Acharya, M. Tang, S.H. Xiang, S.S. Hussan, R.L. Stanfield, J. Robinson, J. Sodroski, I.A. Wilson, R. Wyatt, C.A. Bewley, and P.D. Kwong. 2007. Structures of the CCR5 N terminus and of a tyrosine-sulfated antibody with HIV-1 gp120 and CD4. *Science* 317:1930–1934.

Huang, C.C., M. Tang, M.Y. Zhang, S. Majeed, E. Montabana, R.L. Stanfield, D.S. Dimitrov, B. Korber, J. Sodroski, I.A. Wilson, R. Wyatt, and P.D. Kwong. 2005. Structure of a V3-containing HIV-1 gp120 core. *Science* 310:1025–1028.

Huang, C.C., M. Venturi, S. Majeed, M.J. Moore, S. Phogat, M.Y. Zhang, D.S. Dimitrov, W.A. Hendrickson, J. Robinson, J. Sodroski, R. Wyatt, H. Choe, M. Farzan, and P.D. Kwong. 2004. Structural basis of tyrosine sulfation and VH-gene usage in antibodies that recognize the HIV type 1 coreceptor-binding site on gp120. *Proc. Natl. Acad. Sci. USA* 101:2706–2711.

Huse, W.D., L. Sastry, S.A. Iverson, A.S. Kang, M. Alting-Mees, D.R. Burton, S.J. Benkovic, and R.A. Lerner. 1989. Generation of a large combinatorial library of the immunoglobulin repertoire in phage lambda. *Science* 246:1275–1281.

James, L.C., P. Roversi, and D.S. Tawfik. 2003. Antibody multispecificity mediated by conformational diversity. *Science* 299:1362–1367.

Jones, P.T., P.H. Dear, J. Foote, M.S. Neuberger, and G. Winter. 1986. Replacing the complementarity-determining regions in a human antibody with those from a mouse. *Nature* 321:522–525.

Kabat, E.A., and T.T. Wu. 1971. Attempts to locate complementarity-determining residues in the variable positions of light and heavy chains. *Ann. NY Acad. Sci.* 190:382–393.

Koliasnikov, O.V., M.O. Kiral, V.G. Grigorenko, and A.M. Egorov. 2006. Antibody CDR H3 modeling rules: Extension for the case of absence of Arg H94 and Asp H101. *J. Bioinform. Comput. Biol.* 4:415–424.

Morrison, S.L., M.J. Johnson, L.A. Herzenberg, and V.T. Oi. 1984. Chimeric human antibody molecules: Mouse antigen-binding domains with human constant region domains. *Proc. Natl. Acad. Sci. USA* 81:6851–6855.

Neuberger, M.S., G.T. Williams, E.B. Mitchell, S.S. Jouhal, J.G. Flanagan, and T.H. Rabbitts. 1985. A hapten-specific chimaeric IgE antibody with human physiological effector function. *Nature* 314:268–270.

Ofek, G., M. Tang, A. Sambor, H. Katinger, J.R. Mascola, R. Wyatt, and P.D. Kwong. 2004. Structure and mechanistic analysis of the anti-human immunodeficiency virus type 1 antibody 2F5 in complex with its gp41 epitope. *J. Virol.* 78:10724–10737.

Padlan, E.A., D.M. Segal, T.F. Spande, D.R. Davies, S. Rudikoff, and M. Potter. 1973. Structure at 4.5 Å resolution of a phosphorylcholine-binding Fab. *Nat. New Biol.* 245:165–167.

Rini, J.M., R.L. Stanfield, E.A. Stura, P.A. Salinas, A.T. Profy, and I.A. Wilson. 1993. Crystal structure of a human immunodeficiency virus type 1 neutralizing antibody, 50.1, in complex with its V3 loop peptide antigen. *Proc. Natl. Acad. Sci. USA* 90:6325–6329.

Roux, K.H., L. Strelets, and T.E. Michaelsen. 1997. Flexibility of human IgG subclasses. *J. Immunol.* 159: 3372–3382.

Roux, K.H., P. Zhu, M. Seavy, H. Katinger, R. Kunert, and V. Seamon. 2004. Electron microscopic and immunochemical analysis of the broadly neutralizing HIV-1-specific, anti-carbohydrate antibody, 2G12. *Mol. Immunol.* 41:1001–1011.

Sanders, R.W., M. Venturi, L. Schiffner, R. Kalyanaraman, H. Katinger, K.O. Lloyd, P.D. Kwong, and J.P. Moore. 2002. The mannose-dependent epitope for neutralizing antibody 2G12 on human immunodeficiency virus type 1 glycoprotein gp120. *J. Virol.* 76:7293–7305.

Saphire, E.O., P.W. Parren, R. Pantophlet, M.B. Zwick, G.M. Morris, P.M. Rudd, R.A. Dwek, R.L. Stanfield, D.R. Burton, and I.A. Wilson. 2001. Crystal structure of a neutralizing human IgG against HIV-1: A template for vaccine design. *Science* 293:1155–1159.

Scanlan, C.N., R. Pantophlet, M.R. Wormald, E. Ollmann Saphire, R. Stanfield, I.A. Wilson, H. Katinger, R.A. Dwek, P.M. Rudd, and D.R. Burton. 2002. The broadly neutralizing anti-human immunodeficiency virus type 1 antibody 2G12 recognizes a cluster of alpha1-->2 mannose residues on the outer face of gp120. *J. Virol.* 76:7306–7321.

Segal, D.M., E.A. Padlan, G.H. Cohen, S. Rudikoff, M. Potter, and D.R. Davies. 1974. The three-dimensional structure of a phosphorylcholine-binding mouse immunoglobulin Fab and the nature of the antigen binding site. *Proc. Natl. Acad. Sci. USA* 71:4298–4302.

Shirai, H., A. Kidera, and H. Nakamura. 1996. Structural classification of CDR-H3 in antibodies. *FEBS Lett.* 399:1–8.

Skerra, A., and A. Pluckthun. 1988. Assembly of a functional immunoglobulin Fv fragment in *Escherichia coli. Science* 240:1038–1041.

Spinelli, S., L. Frenken, D. Bourgeois, L. de Ron, W. Bos, T. Verrips, C. Anguille, C. Cambillau, and M. Tegoni. 1996. The crystal structure of a llama heavy chain variable domain. *Nat. Struct. Biol.* 3:752–757.

Stanfield, R.L., H. Dooley, M.F. Flajnik, and I.A. Wilson. 2004. Crystal structure of a shark single-domain antibody V region in complex with lysozyme. *Science* 305:1770–1773.

Stanfield, R.L., H. Dooley, P. Verdino, M.F. Flajnik, and I.A. Wilson. 2007. Maturation of shark single-domain (IgNAR) antibodies: Evidence for induced-fit binding. *J. Mol. Biol.* 367:358–372.

Stanfield, R.L., M. Takimoto-Kamimura, J.M. Rini, A.T. Profy, and I.A. Wilson. 1993. Major antigen-induced domain rearrangements in an antibody. *Structure* 1:83–93.

Stanfield, R.L., and I.A. Wilson. 1994. Antigen-induced conformational changes in antibodies: A problem for structural prediction and design. *Trends Biotechnol.* 12:275–279.

Stanfield, R.L., A. Zemla, I.A. Wilson, and B. Rupp. 2006. Antibody elbow angles are influenced by their light chain class. *J. Mol. Biol.* 357:1566–1574.

Steinitz, M., G. Klein, S. Koskimies, and O. Makel. 1977. EB virus-induced B lymphocyte cell lines producing specific antibody. *Nature* 269:420–422.

Wilson, I.A., and R.L. Stanfield. 1994. Antibody-antigen interactions: New structures and new conformational changes. *Curr. Opin. Struct. Biol.* 4:857–867.

Wilson, I.A., R.L. Stanfield, J.M. Rini, J.H. Arevalo, U. Schulze-Gahmen, D.H. Fremont, and E.A. Stura. 1991. Structural aspects of antibodies and antibody-antigen complexes. *Ciba Found. Symp.* 159:13–28; discussion 28–39.

Wu, T.T., and E.A. Kabat. 1970. An analysis of the sequences of the variable regions of Bence Jones proteins and myeloma light chains and their implications for antibody complementarity. *J. Exp. Med.* 132:211–250.

Zhou, T., L. Xu, B. Dey, A.J. Hessell, D. Van Ryk, S.H. Xiang, X. Yang, M.Y. Zhang, M.B. Zwick, J. Arthos, D.R. Burton, D.S. Dimitrov, J. Sodroski, R. Wyatt, G.J. Nabel, and P.D. Kwong. 2007. Structural definition of a conserved neutralization epitope on HIV-1 gp120. *Nature* 445:732–737.

CHAPTER 3

Glycosylation of Therapeutic IgGs

YUSUKE MIMURA, ROY JEFFERIS, YUKA MIMURA-KIMURA,
JODIE ABRAHAMS, and PAULINE M. RUDD

ABSTRACT

Glycosylation has become a focus of interest for the biopharmaceutical industry as a means to control the efficacy and safety of biological pharmaceuticals. IgG has a conserved glycosylation site at Asn297 on each C_H2 domain of the Fc, and in healthy human serum IgG these sites contain a family of diantennary complex-type oligosaccharides. The Fc glycans are highly heterogeneous, due to

Therapeutic Monoclonal Antibodies: From Bench to Clinic. Edited by Zhiqiang An

variable processing of outer-arm monosaccharide residues onto the core heptasaccharide. Glycosylation at Asn297 is essential for stabilization of the C_H2 domains and to achieve optimal antibody effector functions, including Fcγ receptor and complement activation. Furthermore, individual IgG glycoforms may play a role in modulating antibody effector functions. Recently, IgG-Fc bearing nonfucosylated oligosaccharides has been shown to enhance antibody-dependent cellular cytotoxicity, while IgG-Fc bearing sialylated oligosaccharides may modulate antibody-induced inflammation. These properties may be exploited when developing antibody therapeutics to target tumors or to modulate inflammation in autoimmune diseases.

ABBREVIATIONS

2-AB	2-aminobenzamide
ADCC	antibody-dependent cellular cytotoxicity
GU	glucose unit
G0	nongalactosylated glycoforms
G1	monogalactosylated glycoforms
G2	digalactosylated glycoforms
Fuc	fucose
Gal	galactose
Glc	glucose
GlcNAc	*N*-acetylglucosamine
GalNAc	*N*-acetylgalactosamine
Man	mannose
NeuAc	*N*-acetylneuraminic acid
NeuGc	*N*-glycolylneuraminic acid
FcγR	receptor for Fc portion of IgG
FcRn	neonatal Fc receptor
HPLC	high performance liquid chromatography
MBL	mannose-binding lectin
PNGase F	peptide-*N*-glycosidase F
RA	rheumatoid arthritis

3.1 INTRODUCTION

Glycosylation of proteins represents an extensive posttranslational modification that can influence biological activity, protein conformation, stability, solubility, secretion, pharmacokinetics, and antigenicity (Dwek 1998). Glycosylation of IgG-Fc is highly heterogeneous and plays an essential role in the modulation of the biological activities (Arnold et al. 2006; Jefferis 2005, 2007). It has been demonstrated that individual outer-arm sugar residues contribute to the stability and functionality of IgG-Fc using homogeneously glycosylated species (glycoforms) generated by enzymatic remodeling or produced in glycosyltransferase-deficient or engineered cell lines. Quantitative glycan analysis has become important in the endeavor to control the efficacy and safety of recombinant antibody therapeutics (Sheridan 2007). This chapter focuses on our glycan analysis method based on high performance liquid chromatography (HPLC) and the influence of glycosylation on the structure and functions of IgG.

3.2 OLIGOSACCHARIDE STRUCTURE AND HETEROGENEITY

The IgG antibody molecule has a conserved glycosylation site at Asn297 on each of the C_H2 domains. Oligosaccharides released from human serum IgG are highly heterogeneous, due to variable

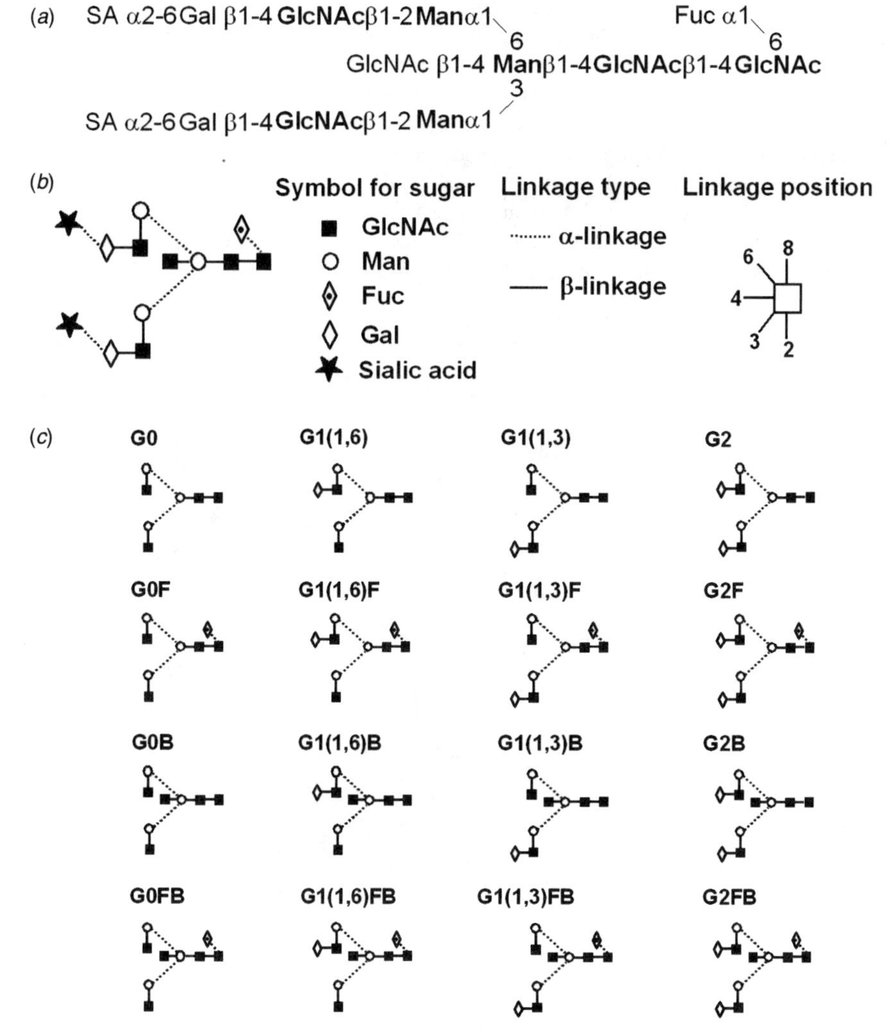

Figure 3.1 Structures of neutral glycans derived from human IgG antibody molecules. (a) The sugars in bold belong to the core heptasaccharide which is found in all naturally occurring IgG glycoforms. (b) Symbols of monosaccharides and lines for showing glycosidic linkages. (c) Schematic representation of diantennary complex-type glycans found in IgG after desialylation. Glycans are designated by G0, G1, and G2 according to the numbers of terminal galactose followed by the letter F and/or B indicating the presence of fucose and/or bisecting GlcNAc, respectively. (1,6) and (1,3) in the G1 glycans indicate the attachment of galactose on the α(1-6)- and α(1-3)-arm, respectively.

attachments of outer-arm sugar residues to a core complex diantennary heptasaccharide ($GlcNAc_2Man_3GlcNAc_2$) (Fig. 3.1) (Jefferis et al. 1990; Routier et al. 1998; Rudd and Dwek 1997; Takahashi et al. 1995). Each Asn297 site can contain one of a family of 36 glycans, depending on the presence or absence of galactose, fucose, bisecting GlcNAc, and NeuAc (Figs. 3.1, 3.3, and 3.4). x-ray crystallography of the human IgG1-Fc fragment reveals the oligosaccharide moiety as integral to the glycoprotein structure, sequestered within the interstitial space enclosed by the two C_H2 domains (see Chapter 2, Fig. 2.6). The glycosylation sequon 297Asn-Ser-Thr299 is proximal to the N-terminal region of the C_H2 domains and the glycan runs toward the C_H2/C_H3 domain interface.

The electron density map provides coherent diffraction for the oligosaccharide and allows the possibility of 85 contacts with 14 amino acid residues of the C_H2 domains (Deisenhofer 1981; Padlan 1990). The interactions between the glycan and C_H2 cover approximately 522 Å^2 of surface area where both hydrophobic and polar residues are involved (Fig. 3.2); within the C_H3 domains analogous residues are involved in the C_H3-C_H3 contacts (1090 Å^2) (Deisenhofer 1981). The branching mannose residues on the α(1-3)-arm of opposed heavy chains make weak lateral contacts for all IgG-Fc glycoform structures in space group $P2_12_12_1$ (Krapp et al. 2003). Although x-ray crystallography indicates that the two Asn297 residues of IgG-Fc bear identical oligosaccharides, electrospray ionization mass spectrometry of intact monoclonal IgG and IgG-Fc fragments have revealed six to seven pairs of Fc glycoforms with both symmetrical and asymmetrical glycoform pairs including G0F + G0F,

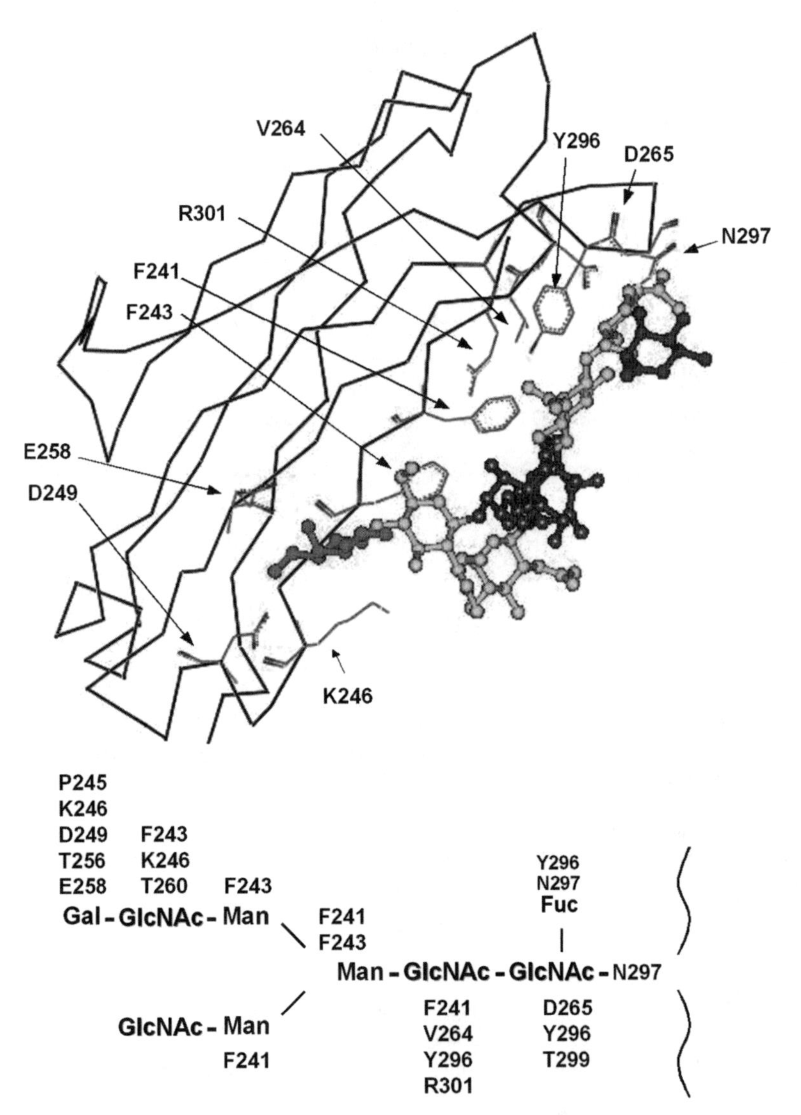

Figure 3.2 Contacts between the C_H2 domain and the oligosaccharide. The C^{α}-atoms of the polypeptide chain of the C_H2 domain is drawn with a line while the oligosaccharide chain is shown in colored ball-and-stick representation (PDB code 1Fc). Residues that make contacts with the oligosaccharide are labeled with the Eu numbering system. (See color insert.)

G0F + G1F, G1F + G1F, G0F + G2F, G1F + G2F (Masuda et al. 2000; Mimura et al. 2007). Interestingly, in fetal human IgG1 fully galactosylated glycoform pairings (G2F + G2F and G2F + G2FS1) including a monosialylated glycoform pairing (G2FS1 + G2FS1) were mainly observed (Masuda et al. 2000).

The heavy and/or light chain variable regions of 15 to 20 percent of normal polyclonal IgG bear N-linked complex diantennary oligosaccharides. In contrast to IgG-Fc, glycan profiles of IgG-Fab are highly galactosylated and sialylated (Holland et al. 2006; Youings et al. 1996), which is presumed to reflect increased exposure and accessibility of glycans to glycosyltransferases. Conserved processing of complex diantennary oligosaccharides on human serum IgG is somewhat surprising as site-related differences might be anticipated for Fab glycosylation. For mouse anti-dextran monoclonal antibodies (Mabs) the oligosaccharides were of the complex or high mannose type, depending on the location of the glycosylation site in the V_H-region (Gala and Morrison 2004; Wright et al. 1991). Detailed analysis of two recombinant antibodies produced in Sp2/0 cells bearing V-region N-linked oligosaccharides reported extreme heterogeneity including the addition of high mannose-type and tri- and tetra-antennary complex-type oligosaccharides (Huang et al. 2006; Qian et al. 2007). In humans follicular lymphoma cells have been reported to express B-cell receptors that often bear high mannose-type glycans in the Fab, which is described in Section 3.7.3.

Although normal human IgG antibodies are not O-glycosylated there has been a single report of O-mannosylation in the V_L region of a recombinant IgG2 antibody produced in Chinese hamster ovary (CHO) cells (Martinez et al. 2007). In addition, it has been reported that approximately 40 percent of murine IgG2b has an O-glycosylation site at Thr221A in the hinge region (Kim et al. 1994). The *O*-glycans are mainly tetrasaccharides composed of GalNAcGlcNAc(NeuGc)$_2$. Human serum IgA1 and IgD each bear multiple O-linked oligosaccharides within their hinge regions (Arnold et al. 2004). Antibodies bearing *O*-glycans on the hinge are more resistant to proteolysis.

3.3 ASSEMBLY AND PROCESSING OF N-LINKED OLIGOSACCHARIDES ON IgG

The biosynthesis pathway of *N*-glycans is reviewed in detail elsewhere (Kornfeld and Kornfeld 1985). N-glycosylation is initiated cotranslationally by the transfer of the preformed dolicol-linked oligosaccharide onto Asn residues of a nascent polypeptide chain in the lumen of endoplasmic reticulum (ER), an event catalyzed by oligosaccharyltransferase. The membrane-bound precursor consisting of 14 monosaccharide residues ($Glc_3Man_9GlcNAc_2$) linked to dolicol phosphate is synthesized first on the cytosolic and then on the luminal side of the ER, with the proximal core portion synthesized in the cytosol and the outer-arm mannose and glucose residues added in the lumen. Oligosaccharyltransferase transfers the oligosaccharide to the N-glycosylation sequon Asn-Xaa-Thr/Ser (Xaa: any amino acid except Pro). N-glycosylation does not occur at every potential glycosylation site, and this variation is one of the causes of microheterogeneity in glycoproteins (Jones, Krag, and Betenbaugh 2005). After the precursor oligosaccharide is added to a glycosylation site, terminal glucose and mannose residues undergo trimming, resulting in the formation of high mannose-type structures ($Man_{8-9}GlcNAc_2$). Nascent heavy chains are associated with heavy chain-binding protein (BiP) until they are polymerized with light chains in the ER (Bole, Hendershot, and Kearney 1986; Hendershot et al. 1987). As only assembled immunoglobulin molecules can be transported to the Golgi apparatus, the complex glycans on the proteins are therefore processed on intact IgG molecules.

In the *cis*-Golgi cisternae α1,2-mannosidase-I truncates the high mannose glycans to $Man_5GlcNAc_2$ (Man5). In the *medial* Golgi β1,2-*N*-acetylglucosaminyltransferase-I transfers a GlcNAc residue to form a hybrid oligosaccharide ($GlcNAc_1Man_5GlcNAc_2$). α1,2-Mannosidase-II trims up to two mannose residues on the α(1-6)-arm, which is available for core fucosylation (FA1 in Fig. 3.4) by α1,6-fucosyltransferase and extension by *N*-acetylglucosaminyltransferase-II to generate the diantennary heptasaccharide core structure. In the *trans*-Golgi, additions of galactose to a terminal GlcNAc and bisecting GlcNAc to the β-mannose are catalyzed by β1,4-galactosyltransferase and *N*-acetylglucosaminyltransferase-III, respectively. Fucose can be added at any time after $Man_5GlcNAc_2$ is synthesized but not after the addition of galactose or bisecting GlcNAc (Kornfeld and Kornfeld 1985; Longmore and Schachter 1982; Restelli and Butler 2002). Extension of the oligosaccharide chain can terminate upon addition of sialic acid to a terminal galactose by α2,6-sialyltransferase which preferentially sialylates the galactose residue on the α(1-3)-arm (Grey et al. 1982; van den Eijnden et al. 1980). In CHO cells only α2,3-sialyltransferase is active and therefore sialic acids are added in an α(2-3)-linkage. The α(1-6)-arm branch of the diantennary glycan on IgG-Fc associates with the hydrophobic surface of the C_H2 domains and the terminal sugar residue on the arm is exposed at the C_H2/C_H3 domain interface (Fig. 3.2).

3.4 GLYCAN ANALYSIS OF IgG BY HIGH PERFORMANCE LIQUID CHROMATOGRAPHY (HPLC)

3.4.1 Method

HPLC of fluorescently labeled glycans provides robust, sensitive, quantitative, and reproducible analysis. IgG samples were purified using a protein G affinity column from cell culture supernatants or serum. Purified IgG was digested with papain and the Fc was purified on a diethylaminoethyl-cellulose anion exchange column (Fig. 3.3, inset) (Radcliffe et al. 2007). *N*-glycans of Fc were released by in-gel digestion with peptide-*N*-glycosidase F (PNGase F), which has been previously described in

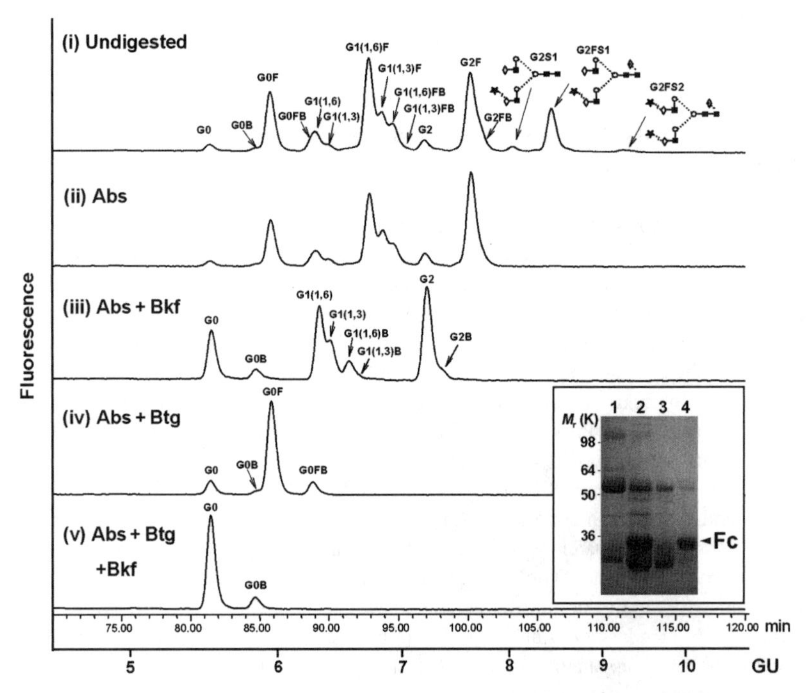

Figure 3.3 NP-HPLC analysis of 2-AB labeled total glycan pool from healthy human serum IgG-Fc. The top profile shows undigested whole pool glycans, followed by a series of exoglycosidase digestions. Abs: *Arthrobacter ureafaciens* sialidase; Bkf: bovine kidney fucosidase; Btg: bovine testis galactosidase. Human serum IgG purified on protein G column was digested with papain. Fab and Fc fractions were obtained after DEAE chromatography of papain-digested IgG (inset). Lane 1, purified IgG; lane 2, papain digest; lane 3, flow-through fraction (Fab and intact IgG); lane 4, eluate (Fc). The Fc band was cut and used to release *N*-glycans by PNGase F. The structures of the neutral glycans are shown in Figure 3.1.

detail (Royle et al. 2006). Approximately, 10 to 50 μg of purified IgG or Fc was reduced with SDS sample buffer containing 50 mM dithiothreitol, alkylated with 10 mM iodoacetamide after heating at 70°C for 10 min, and separated by SDS-PAGE. Coomassie Blue-stained gel bands were excised, cut into 1 mm^3, frozen for 2 h at −20°C, washed alternately with acetonitrile and 20 mM $NaHCO_3$, pH 7. The glycans were released from the protein in the gel bands by PNGase F digestion at 37°C for 18 h. The glycans were extracted from the gel pieces by sonication alternately with water and acetonitrile. Samples were dried under vacuum and fluorescently labeled with 2-aminobenzamide (2-AB) dye. Unlabeled glycans can be also used for mass spectrometry after removal of Coomassie Blue dye with MicroPure-EZ filter (Amicon). 2-AB derivatives were prepared by reductive amination (Bigge et al. 1995). Glycans were dissolved in the mixture of dimethylsulfoxide, acetic acid, an excess of 2-AB and sodium cyanoborohydride (LudgerTag™ 2-AB Glycan Labeling Kit, Ludger, UK) and heated for 2 h at 65°C. The solution was applied 1 cm from the end of a 10 cm strip of Whatman 3MM Chr paper, allowed to dry and the excess reagents were removed by ascending chromatography in acetonitrile leaving the derivatized glycans at the origin. These glycans were extracted with water and ready for HPLC analysis following concentration under vacuum. HPLC was by normal phase (NP) on a 4.6 × 250 mm TSK amide-80 column (Anachem, Luton, U.K.). The elution times of glycans are expressed in glucose units (GU) by reference to a dextran ladder (Guile et al. 1996). Each individual glycan has a GU value that is directly related to the number and linkage of its constituent monosaccharides. The use of arrays of exoglycosidases in combination with NP-HPLC profiling enables the identification of individual monosaccharides and linkages through enzyme specificity (Royle et al. 2003). Peaks from glycan profiles can be assigned via web-based software that accesses our database (GlycoBase, v2.0) of over 350 *N*-glycan structures (http://glycobase.nibrt.ie) (Royle et al. 2008).

3.4.2 Glycan Profile of Human Serum IgG

The glycans of polyclonal IgG-Fc from a healthy human control are highly heterogeneous but can be classified into three sets (i.e., G0, G1, and G2), depending on the number of galactose residues in the outer arms of diantennary glycans. Within each of these sets are four species that result from the presence or absence of core fucose and bisecting GlcNAc, namely, 16 neutral complex-type structures after desialylation [Figs. 3.1 and 3.3 (i) and (ii)]. The G1 and G2 glycans may be sialylated (in humans with NeuAc only), and the sialylated glycans account for 10 to 20 percent in human polyclonal IgG, with a preference for sialylation on G2 [G2FS1, Fig. 3.3 (i)]. Arm-specific galactosylation is clearly resolved by NP-HPLC, showing a preference of galactosylation on the α(1-6)-arm over the α(1-3)-arm for polyclonal IgG; however, analysis of IgG2 myeloma proteins shows the opposite preference (Farooq et al. 1997). Identification of the elution positions of nonfucosylated or bisected glycans are clearly determined following digestion with sialidase and fucosidase [Fig. 3.3 (iii)]. Similarly, digestion with sialidase and galactosidase reveals the overall fucosylation level [about 90 percent, Fig. 3.3 (iv), Table 3.1], while digestion with sialidase, fucosidase, and galacotosidase shows the abundance of bisected glycans [10 to 20 percent, Fig. 3.3 (v), Table 3.1]. Analysis of IgG myeloma proteins indicates that each protein exhibits a unique glycoform profile (Farooq et al. 1997; Jefferis et al. 1990). It is evident, therefore, that the glycoform profile of polyclonal IgG is the sum of the contributions from individually unique clones of plasma cells.

3.4.3 Glycan Profiles of Recombinant IgG Produced in Rodent Cell Lines

Currently approved therapeutic IgG antibodies are produced in CHO and murine myeloma cells (NS0 and Sp2/0). Here we analyzed glycans of the two anti-TNFα antibodies produced in NS0 (infliximab, Remicade®, Fig. 3.4a) and CHO cells (adalimumab, Humira®, Fig. 3.4b), which act as blocking antibodies for the treatments of rheumatoid arthritis, psoriatic arthritis, Crohn's disease, and ankylosing spondylitis. Glycan profiles of the IgG produced in the NS0 and CHO expression systems differ in terms of the galactosylation level (Fig. 3.4). G0F predominated for adalimumab (Fig. 3.4b,

TABLE 3.1 Analysis of the Key Features of the *N*-Glycans Released from the IgGs*

	Complex-Type Glycans								High Mannose-Type Glycans
	Sialylation (%)		Term.	Term.	Bisecting	Core	Predominant	α(1-3)Gal	
IgG	S1	S2	Gal (%)	GlcNAc (%)	GlcNAc (%)	Fucose (%)	Glycan(s)	Epitope (%)	$Man_{5-7}GlcNAc_2$
Serum IgG	13	1	70	16	12	88	G1F	0	0
Infliximab	4	2.5	49	36	0	88	G0F, G1F	1.8	6.7
Adalimumab	0	0	23	70	0	95	G0F	0	6.4

*Glycans were quantitated by measuring peak areas in the HPLC profiles (Figs. 3.3 and 3.4).

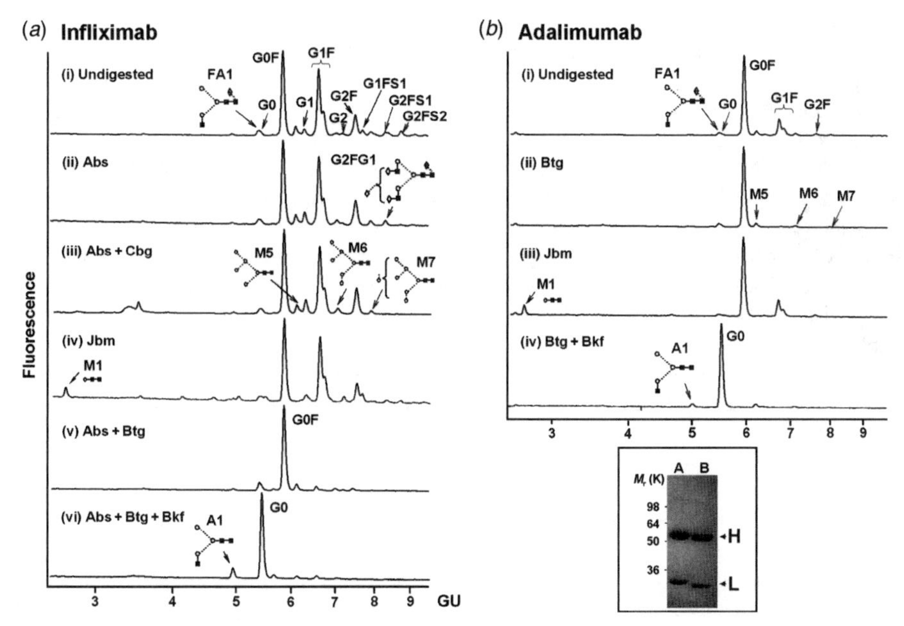

Figure 3.4 NP-HPLC profiles of recombinant therapeutic IgGs. Infliximab (anti-TNFα mouse/human chimeric IgG1) derived from NS0 cell line (a) and adalimumab (anti-TNFα humanized IgG1) from CHO cell line (b) were purchased from Schering Plough Corporation and Abbott Laboratories, respectively. The top chromatograms show *N*-glycans released from the heavy chain bands of the antibodies (inset), followed by a series of exoglycosidase digestions. Cbg, coffee bean α-galactosidase; Jbm, Jack bean α-mannosidase. The structures of the neutral glycans are shown in Figure 3.1. FA1 denotes the monoantennary (A1), fucosylated *N*-glycan.

CHO-derived) while the G1F glycoforms predominated for infliximab (Fig. 3.4a, NS0-derived) as observed for human polyclonal IgG (Fig. 3.3, Table 3.1). In contrast to human serum IgG, recombinant antibodies produced in rodent cells do not contain bisecting GlcNAc. Sialylated glycans were absent or present at a very low level.

Sialic acid linkage can be analyzed by digestion with *Streptococcus pneumoniae* sialidase and *Arthrobacter ureafaciens* sialidase, which are specific for α(2-3,8)-linked and α(2-3,6,8)-linked sialic acids, respectively. The glycans released from infliximab were not sensitive to the former sialidase, indicating that the NS0-derived antibody is sialylated in an α(2-6)-linkage (data not shown). On the other hand, CHO cells only produce α(2-3)-linked sialyl oligosaccharide structures (Takeuchi et al. 1988). The extent to which the oligosaccharides are sialylated with either NeuAc or NeuGc residues is species specific, with mouse-derived IgG containing only NeuGc (Raju et al. 2000). Analysis of sialic acid variants associated with CHO-derived recombinant proteins has shown that NeuAc is prevalent on the *N*-glycans (Baker et al. 2001; Hokke et al. 1995).

Another major difference between NS0-derived and CHO-derived recombinant IgG is the presence of terminal α(1-3)-linked galactose on the NS0-derived oligosaccharides [1.8 percent, G2FG1, Fig. 3.4(a-ii)], which was identified following digestion with sialidase and coffee bean α-galactosidase [Fig. 3.4(a-iii)]. This difference is presumably due to the presence of α1,3-galactosyltransferase in the NS0 host cell (Sheeley, Merrill, and Taylor 1997). As up to 1 percent of all circulating human IgG is directed against the Galα(1-3)Gal motif (Galili et al. 1993), this epitope is presumed to be immunogenic in humans. NeuGc is present in glycoproteins derived from nonhuman mammalian cells and might also be immunogenic (Nguyen, Tangvoranuntakul, and Varki 2005; Tangvoranuntakul et al. 2003). In fact, SP2/0-derived cetuximab (Erbitux) bears glycans containing both α(1-3)-linked Gal (30 percent) and NeuGc (12 percent) on the Fab portion (Qian et al. 2007), and a high prevalence of IgE-dependent hypersensitivity reactions to the Galα(1-3)Gal epitope on cetuximab has been reported in some areas of the United States (Chung et al. 2008). Although severe events occur only occasionally, premedication is required to prevent infusion reactions with all monoclonal antibodies (Chung 2008).

Interestingly, both infliximab and adalimumab contained high mannose-type glycans ($Man_{5-7}GlcNAc_2$) [Figs. 3.4(a-iv) and 3.4(b-iii)] as reported for a humanized anti-CD18 antibody produced in NSO cells (Ip et al. 1994). High mannose-type glycans are not usually present in normal human and mouse serum IgG, and rapid clearance of IgGs bearing high mannose-type glycans has been demonstrated in mice (Kanda et al. 2007; Wright and Morrison 1998). On the other hand, the presence or absence of galactose on the IgG-Fc glycans does not influence the clearance rate *in vivo* (Huang et al. 2006). The neonatal Fc receptor (FcRn) protects IgG from degradation, improving its serum half-life (Ghetie and Ward 2000; Roopenian and Akilesh 2007). As human FcRn exhibits comparable affinity to mouse/human chimeric anti-CD20 IgG1 bearing complex-type, hybrid-type, or high mannose-type glycans (Kanda et al. 2007), rapid clearance of IgG glycoforms with high mannose-type glycans is likely to be mediated by the mannose receptor (McGreal, Miller, and Gordon 2005).

3.5 PREPARATION OF HOMOGENEOUS Fc GLYCOFORMS

Separating various glycoforms to investigate the biological relevance of glycosylation is a real challenge with glycoproteins. Jefferis's laboratory remodeled glycoforms of IgG-Fc *in vitro* to investigate thermal stability, complement activation, FcγRIIb binding, and crystal structures (Ghirlando et al. 1999; Krapp et al. 2003; Mimura et al. 2000; Mimura et al. 2001). Homogeneous glycoforms G2F, G0F, M3N2F, MN2F, and deglycosylated Fc were prepared as below: The *G0F glycoform* of IgG1-Fc was generated by exposure of the Fc protein (1 mg) in 50 mM acetate buffer (pH 5.0) to sialidase (100 mU; *Arthrobacter ureafaciens*, Roche) at 37°C for 24 h. The buffer was replaced by dialysis against 50 mM sodium phosphate (pH 5.0) and β(1-3,4,6) galactosidase (0.5 U; bovine testis, Prozyme) added and the solution incubated at 37°C for 24 h. The *G2F glycoform* was prepared by exposure of IgG-Fc (1 mg) to sialidase as described above and subsequently with galactosyltransferase (2 U, bovine milk, Calbiochem) in 20 mM Hepes, 0.15 M NaCl, 5 mM $MnCl_2$, 1 mM UDP-Gal (pH 7.0) at 37°C for 24 h. *M3N2F glycoform*: G0F glycoform (1 mg), in 50 mM sodium phosphate (pH 5.0) was exposed to β-*N*-acetylhexosaminidase (1.6 U, *Streptococcus pneumoniae*, Prozyme) at 37°C for 24 h. *MN2F glycoform*: M3N2F glycoform (1 mg) in 50 mM acetate buffer containing 2 mM zinc acetate (pH 5.0) was exposed to α-mannosidase (3 U, Jack bean, Prozyme) at 37°C for 48 h. *Deglycosylated Fc*: native Fc (1 mg) was exposed to PNGase F (5 U, Roche) in 40 mM phosphate buffer containing 1 mM EDTA, pH 7.0, for 72 h. Following exposure to glycosidases all proteins were affinity purified using a Streptococcal protein G Sepharose 4B (Amersham), eluted with 0.1 M glycine-HCl buffer (pH 2.7). Eluates were immediately neutralized by the addition of 1 M Tris/HCl (pH 9.0), and dialyzed extensively against phosphate-buffered saline. Glycan profiles of the homogeneous Fc glycoforms can be confirmed by electrospray ionization mass spectrometry (Mimura et al. 2001) or NP-HPLC (Fig. 3.5). Although there is no enzyme available that can defucosylate IgG-Fc, the *FUT8*$^{-/-}$ CHO cell line has been established to produce completely nonfucosylated antibodies by targeted disruption of α1,6-fucosyltransferase-encoding gene *FUT8* (Yamane-Ohnuki et al. 2004). The cell line has been shown to be applicable for the large-scale production of nonfucosylated IgG glycoforms. The influence of the individual outer-arm sugar residues on the structure and functions of IgG-Fc is discussed below.

3.6 INFLUENCE OF Fc GLYCOSYLATION ON BIOLOGICAL ACTIVITIES OF IgG

Although the carbohydrate moiety of the IgG molecule accounts for only 2 to 3 percent of its mass, it exerts a profound influence on IgG effector mechanisms that lead to the clearance and destruction of

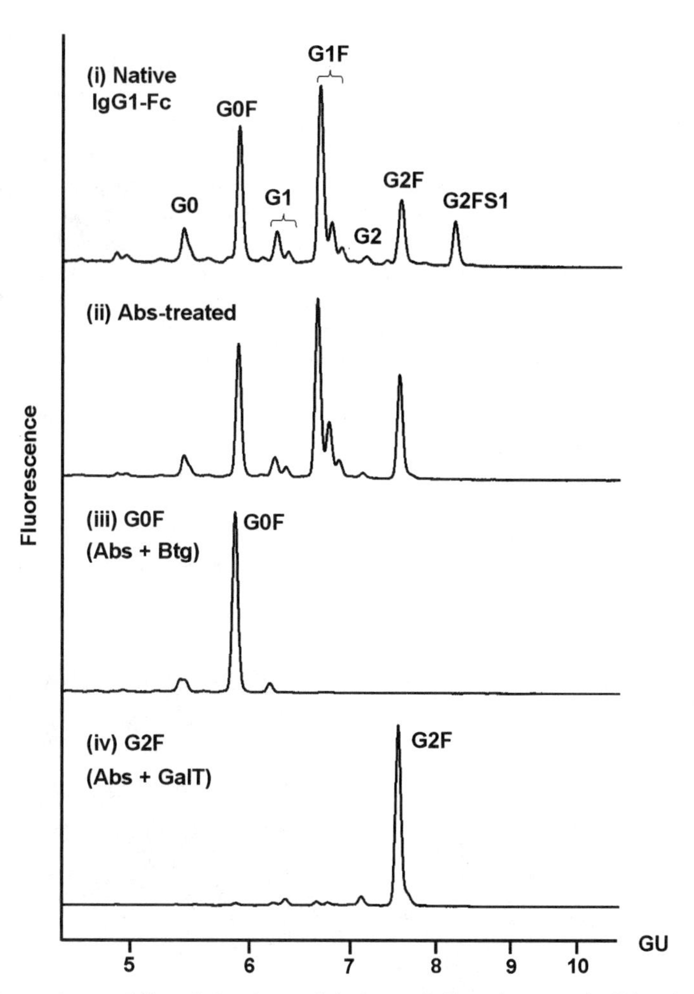

Figure 3.5 Enzymatic remodeling of glycoforms of the human IgG myeloma protein Cri (native, Abs-treated, G0F, G2F). *N*-glycans on the IgG-Fc protein (i) were truncated with galactosidase (iii) or galactosylated with galactosyltransferase (iv) after desialylation (ii). These homogeneous Fc glycoforms were crystallizable (see Fig. 3.6).

bound antigens and pathogens. Elimination mechanisms mediated through Fcγ receptors and C1q have been shown to be abrogated or severely compromised for aglycosylated or deglycosylated forms of IgG (Nose and Wigzell 1983; Pound, Lund, and Jefferis 1993; Sarmay et al. 1992; Tao and Morrison 1989). Principal ligands for the activation of clearance mechanisms are the three types of cellular receptors (FcγR) expressed constitutively on various hematopoietic cell types, including macrophages, eosinophils, neutrophils, natural killer cells, and lymphocytes, and FcγRs link cellular and humoral immunity by serving as a bridge between antibody specificity and effector cell functions. Stimulation of cells through FcγRs may result in the activation of one or more of a variety of effector functions, including antibody-dependent cellular cytotoxicity (ADCC), phagocytosis, oxidative burst, and release of inflammatory mediators (Burton and Woof 1992; Woof and Burton 2004). FcγR interaction sites in the IgG-Fc were investigated by mutagenesis studies before the structures of antibody-FcγR complexes were solved, and the sequence 234Leu-Leu-Gly-Gly237 in the hinge proximal region of the C_H2 domain was identified as a crucial region for the interaction (Canfield and Morrison 1991; Chappel et al. 1991; Duncan et al. 1988; Lund et al. 1991; Woof et al. 1986). Replacement of residues with Ala, including the ones that make contacts with the oligosaccharides (Asp265, Tyr296, and

Arg301), has also been demonstrated to be influential (Shields et al. 2001). x-ray crystallographic analysis of a complex between IgG-Fc and a soluble form of FcγRIIIa (sFcγRIII) has demonstrated that the FcγR binds to the lower hinge and the hinge proximal regions of the two C_H2 domains asymmetrically with a 1 : 1 stoichiometry (Radaev et al. 2001; Sondermann et al. 2000). Importantly, the carbohydrate moieties are not directly associated with sFcγRIII except the primary GlcNAc of one oligosaccharide although deglycosylation of Fc abrogates sFcγRIII binding. Interestingly, the profound influence of Fc glycosylation on FcγR binding has not been paralleled by gross conformational differences between glycosylated and aglycosylated Fc forms. NMR studies using His residues as reporter groups have revealed a subtle difference of the environment of His-268 in the vicinity of the lower hinge region (Lund et al. 1990). Differential scanning calorimetry of homogeneous Fc glycoforms (G0, M3N2, MN2, and deglycosylated Fc) has provided evidence for destabilization of the C_H2 domains by truncation of terminal sugar residues as indicated by progressively lowered thermal unfolding temperature of the C_H2 domains. Importantly, deglycosylation resulted in the lowest unfolding temperature and the loss of cooperativity of the C_H2 domains, indicating the presence of at least one intermediate between the folded and unfolded state during thermal unfolding of the domain structure (Mimura et al. 2000, 2001). Although crystallization of deglycosylated Fc has not been successful, probably due to instability of the C_H2 domains, crystal structures of homogeneous Fc glycoforms (G2F, G0F, M3N2F, and MN2F) have provided a rationale for the influence of individual sugar residues of the Fc glycan on the structure and function of IgG-Fc (Figs. 3.6 and 3.7).

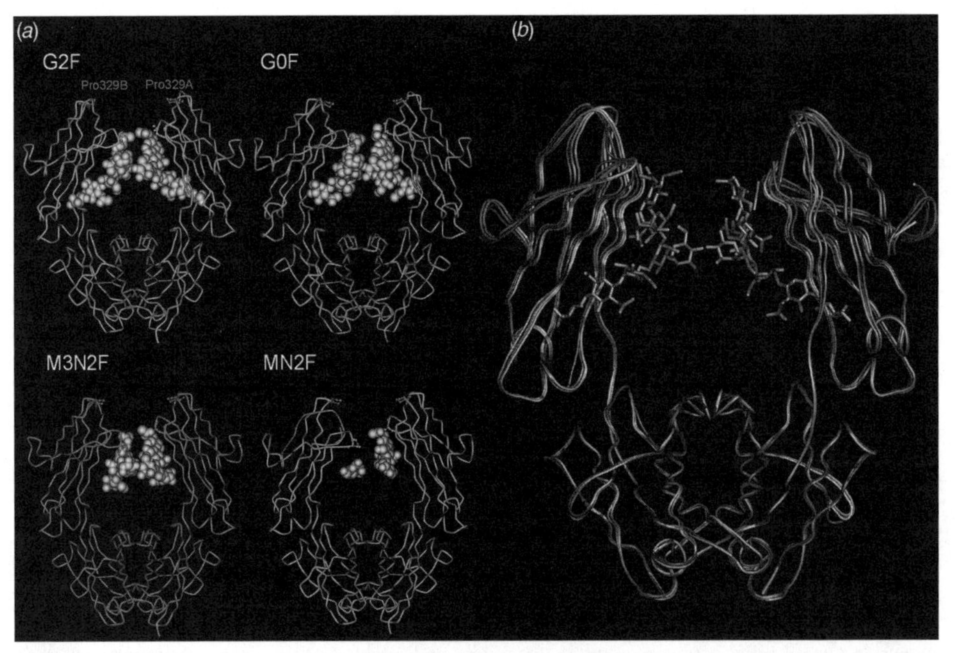

Figure 3.6 Crystal structures of homogeneous glycoforms of IgG-Fc Cri (a) and superimposed image (b). Structures of homogeneous Fc glycoforms G2F, G0F, M3N2F, and MN2F in space group $P2_12_12_1$ are shown (PDB codes 1H3V, 1H3X, 1H3U, and 1H3T, respectively). Pro329 residues are shown on the tips of the C_H2 domains. Note that the tetrasaccharide on the chain-B (left) of the MN2F glycoform is not completely defined but present, due to less/weaker crystal contacts. The C_H2 domains of the MN2F glycoform exhibit a higher B-factor and "softer" structure than in the other glycoforms. The superimposed image was created using PYMOL (www.pymol.org). (Courtesy of Dr. Peter Sondermann). Only the G2F glycan is shown in the superimposed image. Note that the conformational difference between the glycoforms is prominent in the hinge proximal region of the C_H2 domain in contrast to the C_H3 domain and C_H2/C_H3 domain interface. (See color insert.)

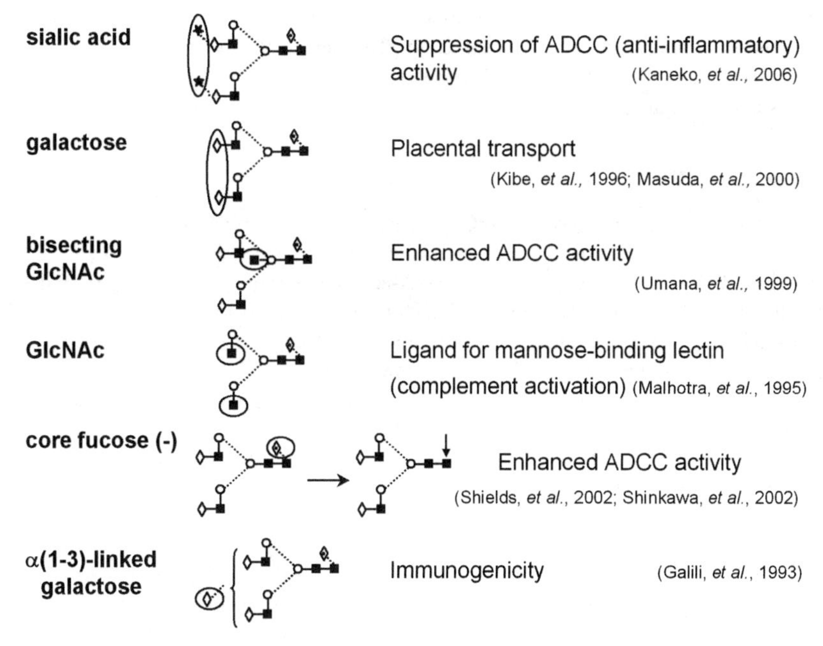

Figure 3.7 Biological activities of individual glycoforms of IgG-Fc.

3.6.1 Core Fucose Residue

Depletion of core fucose from IgG-Fc has been demonstrated to dramatically enhance ADCC mediated through the activation of FcγRIIIa on natural killer cells (Ferrara et al. 2006b; Kanda et al. 2007; Okazaki et al. 2004; Shields et al. 2002; Shinkawa et al. 2003) although natural IgG molecules are mainly fucosylated. Absence of fucose on the oligosaccharides of IgG-Fc does not affect binding to human FcγRI, C1q and FcRn. For FcγRII, differential effects are evident between allotypes in that FcγRIIa-Arg131 polymorphic form and FcγRIIb show a slight improvement in binding for nonfucosylated IgG, while the FcγRIIa-His131 allotype does not (Shields et al. 2002). The binding affinity of nonfucosylated IgG to FcγRIIIa is higher for the FcγRIIIa-Val158 allotype than the Phe158 allotype (Ferrara et al. 2006b; Shields et al. 2002) as demonstrated with native IgG (Cartron et al. 2002; Koene et al. 1997). Shinkawa et al. (2003) expressed anti-CD20 IgG1 with a low fucosylation level (9 percent) from rat hybridoma YB2/0 cells and IgG1 with a high content of bisecting GlcNAc (74 percent) from CHO-LEC10 cells and isolated nonfucosylated IgG and fucosylated IgG with varying bisecting GlcNAc contents by serial lectin affinity chromatography using *Phaseolus vulgaris E4* (bisecting GlcNAc binding) and *Lens culinaris agglutinin* (fucose binding). Using these reagents they demonstrated the enhanced ability of YB2/0-derived, nonfucosylated IgG to mediate ADCC up to 50-fold. On the other hand, the Lec10-produced IgG containing glycans with bisecting GlcNAc shows two- to threefold greater ADCC than control CHO-derived IgG. It has been noted that affinity of IgG to glycosylated (CHO-produced) FcγRIIIa is lower by about threefold than bacterially expressed aglycosylated FcγRIII, possibly due to steric hindrance (Ferrara et al. 2006b; Ghirlando et al. 1995; Maenaka et al. 2001; Sondermann et al. 2000). Although the presence or absence of core fucose in IgG-Fc does not affect its binding to aglycosylated FcγRIIIa, fucose-deficient IgG-Fc shows over 13 times higher affinity to FcγRIIIa glycosylated at Asn162 than to aglycosylated FcγRIIIa (Ferrara et al. 2006b). The oligosaccharide at Asn162 on FcγRIIIa might participate in the complex formation with fucose-deficient IgG-Fc. The presence of core fucose does not influence the conformation of the C_H2 domain as the thermal stability is almost the same between fucosylated and nonfucosylated C_H2 domains (Mimura et al. 2000, 2001). Furthermore, crystal structures of fucosylated and nonfucosylated

Fc fragments are similar, and stable-isotope-assisted NMR analyses confirmed the similarity of the overall structures in solution (Matsumiya et al. 2007). To elucidate the structural basis for the enhanced affinity of fucose-deficient IgG-Fc to glycosylated FcγRIIIa, a crystal complex between these components may be required.

3.6.2 Bisecting GlcNAc Residue

The presence of bisecting GlcNAc on IgG glycans has been implicated in enhanced ADCC of therapeutic antibodies including Campath-1H (Lifely et al. 1995). Two independent laboratories have generated CHO mutants that over-express *N*-acetylglucosaminyltransferase III (GnTIII) and secrete anti-neuroblastoma IgG (Umana et al. 1999) or anti-CD20 IgG (Davies et al. 2001) with improved ADCC up to 10- to 20-fold. However, nonfucosylation has been found to exert this effect because bisected Fc glycans are often deficient in core fucose. Addition of bisecting GlcNAc residue is a relatively early event in glycoprotein processing in CHO cells and its presence inhibits the later addition of fucose to the primary GlcNAc residue (Ferrara et al. 2006a). The reevaluated increase of ADCC mediated by bisecting GlcNAc was two- to threefold by using fucosylated and bisected IgG as described above (Shinkawa et al. 2003).

3.6.3 Sialic Acid Residues

IgG-Fc glycans are hyposialylated (10 to 20 percent). Sialic acid is added preferentially on the α(1-3)-arm to digalactosylated (G2) glycoforms in healthy human individuals (Grey et al. 1982; Holland et al. 2002, 2006; van den Eijnden et al. 1980). Disialylated glycoforms are either totally absent or present at a very low level in CHO- and NS0-derived recombinant IgG (Fig. 3.4). Although there are several strategies to improve sialylation of recombinant proteins, including over-expression of sialyltransferase or CMP-sialic acid transporter and supplementation with Mn^{2+} ion or *N*-acetylmannosamine (Wong, Yap, and Wang 2006), sialylation enhancement of IgG-Fc has not been efficiently achieved, probably due to poor accessibility of sialyltransferases to the Fc glycans which are interposed between the two C_H2 domains.

Although the effect of Fc sialylation on antibody function has received relatively little attention, two groups recently suggested that the addition of sialic acid to the oligosaccharides on human IgG molecules enhances their anti-inflammatory properties (Kaneko, Nimmerjahn, and Ravetch 2006; Scallon et al. 2007). A sialylated IgG-enriched preparation isolated on the lectins *Sambucus nigra* agglutinin or wheat germ agglutinin shows reduced ADCC activity. Ravetch and colleagues hypothesized that sialylated IgG might be involved in the effects of a polyclonal IgG preparation known as IVIG (intravenous Ig) that is widely used to treat inflammatory diseases such as immune-mediated thrombocytopenia and Kawasaki disease. The protective effect of IVIG was associated with the induced expression of an inhibitory Fc receptor FcγRIIb (Bruhns et al. 2003) although a correlation between the sialylation level and the anti-inflammatory activity of IgG was not provided. The molecular basis for Fcγ receptor discrimination of sialylated IgG glycoforms presents an intriguing problem because the sialic acid residues are distant from Fcγ receptor binding sites on IgG-Fc (i.e., lower hinge) (Burton and Dwek 2006; Jefferis 2006).

3.6.4 Galactose Residues

Monogalactosylated (G1F) glycoforms predominate in serum IgG from healthy individuals (Fig. 3.3). Heterogeneity of IgG-Fc glycosylation mainly results from variable addition of galactose residue, and galactosylation level varies with age (Parekh et al. 1988; Yamada et al. 1997), pregnancy (Kibe et al. 1996), and inflammatory conditions (Holland et al. 2002; Itoh, Takahashi, and Hirayama 1993; Wormald et al. 1997). Galactosylated IgG glycoforms predominate in pregnant women and galactosylated IgG is found to be preferentially transported to the fetus (Kibe et al. 1996; Masuda et al. 2000). Patients with rheumatoid arthritis (RA) are known to have high levels of nongalactosylated glycoforms, but during pregnancy galactosylation is activated by an unknown mechanism and the disease severity is reduced. Galactosylation is preferred on the α(1-6)-arm of human IgG1 while galactosylation on the α(1-3)-arm predominates in human IgG2 (Jefferis et al. 1990) and bovine IgG (Fujii et al. 1990). In the IgG-G2F glycoform, the α(1-3)-arm galactose residue is mobile as revealed by the crystal structures with two space groups (Krapp et al. 2003) and NMR (Yamaguchi et al. 1998). The influence of the galactose residue on FcγR and complement activation has been investigated using the G0F glycoform; however, no significant difference was observed for superoxide generation by monocyte mediated via FcγRI, binding to FcγRIIb and complement activation (Mimura et al. 2000, 2001). Furthermore, no correlation is observed between ADCC and the galactose content (Shinkawa et al. 2003). Nonetheless, the presence of galactose appears to have a positive influence on the stability of the C_H2 domain structure as shown by differential scanning microcalorimetry of the G0F and G2F glycoforms of IgG4-Fc (Ghirlando et al. 1999).

3.6.5 GlcNAc Residues

The absence of terminal galactose residues exposes GlcNAc residues on the Fc glycans. The G0(F) glycan has been shown to serve as a ligand for mannose-binding lectin (MBL) that activates the complement system *in vitro* (Malhotra et al. 1995). On the basis of increased binding of MBL to multiply presented IgG-G0(F) glycoforms, it has been suggested that the MBL pathway of complement activation may be involved in the pathogenesis of RA. As the G0F glycoform predominates in CHO-derived IgG (Fig. 3.4b), concern has been expressed at the low level of galactosylation and its possible impact on activation of the complement pathway. On the other hand, Ravetch's laboratory has recently reported the dependence of activating Fcγ receptor(s) on IgG autoantibody-mediated inflammation in mice and lack of involvement of the MBL pathway of complement activation mediated by the G0(F) glycoforms (Nimmerjahn, Anthony, and Ravetch 2007). However, there is a lack of correspondence between mouse and human IgG subclasses and Fc receptors, and further investigation in humans may be required.

The terminal GlcNAc residues are found to play an important role in stabilization of the C_H2 domains by microcalorimetry (Mimura et al. 2000, 2001). This is evidenced by two crystal structures of the G2F glycoform with different space groups in which the GlcNAc residue on the α(1-6)-arm maintains a contact with Phe243 of the C_H2 domain in contrast to the terminal galactose and branching mannose residues (Krapp et al. 2003). Furthermore, conformational changes are demonstrated in the C′E-loop after trimming of the terminal GlcNAc residues by crystallography of homogeneous Fc glycoforms (Fig. 3.6b). When the amino acid residues Phe243 and Phe241, which make contacts with GlcNAc on the α(1-6)-arm, are replaced with Ala, the oligosaccharides attached on the mutants are highly galactosylated and sialylated (Lund et al. 1996). This indicates that loss of the contacts between the GlcNAc residue and the protein moiety results in increased conformational freedom of the oligosaccharide, giving the glycosyltransferases higher accessibility.

3.6.6 Mannose Residues

The α(1-3)- and α(1-6)-arm mannose residues are among key sugar residues that influence FcγR binding. Removal of the branching mannose residues leaves the linear tetrasaccharide core *N*-glycan (MN2F). Crystal structures of homogenous Fc glycoforms have revealed an increased B-factor (a measure for vibration/mobility of atoms/amino acid residues) in the C_H2 domain upon removal of the branching mannose residues. The distance between the two C_H2 domains (Pro329A–Pro329B) is maximal (26.6 Å) for the G2F glycoform but minimum (21.9 Å) for the MN2F glycoform (Fig. 3.6a). The crystal structure of the Fc:FcγRIII complex shows that the horseshoe-shaped Fc opens up on FcγR binding (Pro329 distance: 30.3 Å). The affinity of the MN2F glycoform for FcγRIIb is three times lower than that of native Fc by isothermal titration calorimetry, which is correlated with the structural changes following truncation of the mannose residues.

3.6.7 Chitobiose Core

The replacement of Asp265, which makes contacts with the primary GlcNAc, with Ala abrogates FcγRI binding and the complement activation (Lund et al. 1995, 1996). The glycan profile of the mutant IgG shows a high level of digalactosylated oligosaccharides, indicating that the glycans are more mobile and acquire higher accessibility to glycosyltransferases. This result may explain why IgG-Fc glycans do not protrude from the tip of the C_H2 domains but run towards the C_H2/C_H3 interface. Even the trisaccharide (MN2) is associated with the C_H2 domains and provides certain stability with the protein (Fig. 3.6). The interactions of the core GlcNAc residues with Asp265, Val264, and R301 in the hinge-proximal region of the C_H2 domain may control the orientation of the glycan, resulting in the unique reciprocal influence on conformations of both the oligosaccharide and protein.

3.7 IgG GLYCOSYLATION IN DISEASES

3.7.1 Rheumatoid Arthritis (RA)

Alterations in the glycosylation of human IgG have been shown to occur in RA (Parekh et al. 1985). In RA there is a marked increase in proportion of serum IgG glycans lacking sialic acid and galactose. The level of IgG-G0(F) glycoforms is more than 2 standard deviations above those of age-matched healthy controls, they correlate with disease activity and they are directly associated with pathogenesis in a mouse model (Arnold et al. 2006). IgG-G0(F) levels reflect the severity of the disease because they fall during remission of the disease and also during pregnancy. Interestingly, the increase in the G0(F) level is restricted to IgG isotypes and is only observed in the Fc glycans but not in the Fab glycans (Youings et al. 1996). Although the reduction in galactosylation of IgG-Fc glycans has been reported to be associated with reduced galactosyltransferase (GalT) activity in B cells (Axford et al. 1987), there is no evidence for reduced transcription or translation of GalTs in B-cells of patients with RA (Alavi, Pool, and Axford 2005). It is known that β1,4-galactosyltransferase-I (β4GalT-I) is responsible for galactosylation of serum glycoproteins including immunoglobulins. β4GalT-I-deficient mice have been created by multiple laboratories (Asano et al. 1997; Nishie et al. 2007) and shown complete lack of galactose on N- and O-linked glycans of immunoglobulins. Unexpectedly, β4GalT-I-deficient mice develop IgA nephropathy-like but not RA-like symptoms (Nishie et al. 2007).

3.7.2 Congenital Disorders of Glycosylation (CDGs)

CDGs are a group of genetic disorders characterized by low efficiency of one of the steps in the glycan processing pathway. These disorders affect the glycosylation of all glycoproteins produced by the patients and have severe pathological consequences. CDGs can arise from genetic mutations leading to molecular defects in glycan processing enzymes, sugar nucleotides donors, or protein trafficking machinery within a cell (Jaeken and Matthijs 2007). Aberrant IgG glycoforms are diagnostic for CDG. In CDG type IIa caused by mutations in the *Mgat2* gene encoding *N*-acetylglucosaminyltransferase-II, the heavy chain glycan pool consists of only mono-antennary structures that were predominantly fucosylated and sialylated on the α(1-3)-arm (Butler et al. 2003). This deficiency would be expected to reduce the stability of IgG-Fc by the lack of the terminal GlcNAc residue on the α(1-6)-arm and be susceptible to proteolysis. The patient VT with CDG-IIa suffered from recurrent infections and died from a severe pulmonary infection. In CDG type IId caused by mutations in *B4GalT*1 encoding β4GalT-I, the patient SM showed psychomotor retardation, Dandy-Walker malformation, severe coagulation abnormalities, a transient cholestatic syndrome, hydrocephalus, and myopathy (Hansske et al. 2002). Analysis of oligosaccharides from serum transferrin and IgG revealed the loss of galactose and sialic acid residues. The IgG of the patient SM contains only G0 structures, namely, G0 (5 percent), G0F (82 percent), and G0B (13 percent) (Critchley 2006). This indicates that β4GalT-I is the main functional galactosyltransferase in B-cells.

3.7.3 Follicular Lymphoma (FL)

Follicular lymphoma is a B-cell malignancy that represents about 40 percent of all non-Hodgkin lymphomas. A significant increase in potential N-glycosylation sites in variable regions of the heavy chains of B-cell receptors (BCR) is observed by cDNA sequencing for FL cells (79 percent) compared with normal B-cells (9 percent). Analysis of glycans that are located in the antigen-binding sites of the variable regions has shown that the glycans are mainly of high mannose type (Radcliffe et al. 2007), which might provide an alternative stimulating pathway by engaging with soluble or cell surface mannose-specific lectins. This hypothesis has been supported by direct binding of MBL to FL cells expressing BCR bearing high mannose-type glycans in the Fab. Functional importance of the high mannose-type glycans in the V_H region on survival of germline-associated lymphoma may be questionable because of the presence of natural N-glycosylation sites in certain germline GLV region genes. However, by using single-chain Fv-Cκ expressed in 293F cells, the high mannose-type glycans have been shown to be added only to the glycosylation motifs from tumor-derived V-region sequences but not to that of $V_{4\text{-}34}$ gene which is used by 10 percent of normal B-cells (McCann et al. 2008).

3.8 CONCLUSION

The IgG glycoform can profoundly influence Fc effector functions. Glycoengineering of IgG provides possibilities for the production of therapeutic antibodies with tailored effector functions. The biopharmaceutical industry endeavors to produce fucose-deficient IgG for anti-tumor antibodies. Sialylation of IgG-Fc may provide another possibility to reduce unnecessary immune responses and/or enhance the anti-inflammatory properties of IVIG. Production of selective IgG glycoforms may be inevitable in order to reduce the dose and cost of antibody drugs and make their administration safer. The development has been underway of humanized yeast expression systems (*Pichia pastoris*, Chapter 26) that can manipulate the glycosylation machinery to obtain human-like homogeneous IgG glycoforms (Hamilton et al. 2003; Li et al. 2006). The establishment of further optimal production vehicles to enhance the productivity as well as the manipulation of glycosylation of therapeutic antibodies is still a challenging issue.

ACKNOWLEDGMENTS

The authors thank Peter Sondermann (GLYCART Biotechnology) for generating Figure 3.6b, Oliver FitzGerald (St Vincent's Hospital, University College Dublin) for providing infliximab and adalimumab, and Raymond A. Dwek (Glycobiology Institute, University of Oxford) for comments and discussion.

REFERENCES

Alavi, A., A.J. Pool, and J.S. Axford. 2005. New insights into rheumatoid arthritis associated glycosylation changes. *Adv. Exp. Med. Biol.* 564:129–138.

Arnold, J.N., C.M. Radcliffe, M.R. Wormald, L. Royle, D.J. Harvey, M. Crispin, R.A. Dwek, R.B. Sim, and P.M. Rudd. 2004. The glycosylation of human serum IgD and IgE and the accessibility of identified oligomannose structures for interaction with mannan-binding lectin. *J. Immunol.* 173(11):6831–6840.

Arnold, J.N., M.R. Wormald, R.B. Sim, P.M. Rudd, and R.A. Dwek. 2006. The impact of glycosylation on the biological function and structure of human immunoglobulins. *Annu. Rev. Immunol.* 25:21–50.

Asano, M., K. Furukawa, M. Kido, S. Matsumoto, Y. Umesaki, N. Kochibe, and Y. Iwakura. 1997. Growth retardation and early death of beta-1,4-galactosyltransferase knockout mice with augmented proliferation and abnormal differentiation of epithelial cells. *EMBO J.* 16(8):1850–1857.

Axford, J.S., L. Mackenzie, P.M. Lydyard, F.C. Hay, D.A. Isenberg, and I.M. Roitt. 1987. Reduced B-cell galactosyltransferase activity in rheumatoid arthritis. *Lancet* 330(8574):1486–1488.

Baker, K.N., M.H. Rendall, A.E. Hills, M. Hoare, R.B. Freedman, and D.C. James. 2001. Metabolic control of recombinant protein N-glycan processing in NS0 and CHO cells. *Biotechnol. Bioeng.* 73(3):188–202.

Bigge, J.C., T.P. Patel, J.A. Bruce, P.N. Goulding, S.M. Charles, and R.B. Parekh. 1995. Nonselective and efficient fluorescent labeling of glycans using 2-amino benzamide and anthranilic acid. *Anal. Biochem.* 230(2):229–238.

Bole, D.G., L.M. Hendershot, and J.F. Kearney. 1986. Posttranslational association of immunoglobulin heavy chain binding protein with nascent heavy chains in nonsecreting and secreting hybridomas. *J. Cell. Biol.* 102(5):1558–1566.

Bruhns, P., A. Samuelsson, J.W. Pollard, and J.V. Ravetch. 2003. Colony-stimulating factor-1-dependent macrophages are responsible for IVIG protection in antibody-induced autoimmune disease. *Immunity* 18(4):573–581.

Burton, D.R., and R.A. Dwek. 2006. Immunology. Sugar determines antibody activity. *Science* 313(5787): 627–628.

Burton, D.R., and J.M. Woof. 1992. Human antibody effector function. *Adv. Immunol.* 51:1–84.

Butler, M., D. Quelhas, A.J. Critchley, H. Carchon, H.F. Hebestreit, R.G. Hibbert, L. Vilarinho, E. Teles, G. Matthijs, E. Schollen, and others. 2003. Detailed glycan analysis of serum glycoproteins of patients with congenital disorders of glycosylation indicates the specific defective glycan processing step and provides an insight into pathogenesis. *Glycobiology* 13(9):601–622.

Canfield, S.M., and S.L. Morrison. 1991. The binding affinity of human IgG for its high affinity Fc receptor is determined by multiple amino acids in the CH2 domain and is modulated by the hinge region. *J. Exp. Med.* 173(6):1483–1491.

Cartron, G., L. Dacheux, G. Salles, P. Solal-Celigny, P. Bardos, P. Colombat, and H. Watier. 2002. Therapeutic activity of humanized anti-CD20 monoclonal antibody and polymorphism in IgG Fc receptor FcgammaRIIIa gene. *Blood* 99(3):754–758.

Chappel, M.S., D.E. Isenman, M. Everett, Y.Y. Xu, K.J. Dorrington, and M.H. Klein. 1991. Identification of the Fc gamma receptor class I binding site in human IgG through the use of recombinant IgG1/IgG2 hybrid and point-mutated antibodies. *Proc. Natl. Acad. Sci. U.S.A* 88(20):9036–9040.

Chung, C.H. 2008. Managing premedications and the risk for reactions to infusional monoclonal antibody therapy. *Oncologist* 13(6):725–732.

Chung, C.H., B. Mirakhur, E. Chan, Q.T. Le, J. Berlin, M. Morse, B.A. Murphy, S.M. Satinover, J. Hosen, D. Mauro, and others. 2008. Cetuximab-induced anaphylaxis and IgE specific for galactose-alpha-1,3-galactose. *N. Engl. J. Med.* 358(11):1109–1117.

Critchley, A.J. 2006. Congenital disorders of glycosylation and disease pathogenesis. D. Phil thesis. Oxford: University of Oxford.

Davies, J., L. Jiang, L.Z. Pan, M.J. LaBarre, D. Anderson, and M. Reff. 2001. Expression of GnTIII in a recombinant anti-CD20 CHO production cell line: Expression of antibodies with altered glycoforms leads to an increase in ADCC through higher affinity for FC gamma RIII. *Biotechnol. Bioeng.* 74(4):288–294.

Deisenhofer, J. 1981. Crystallographic refinement and atomic models of a human Fc fragment and its complex with fragment B of protein A from *Staphylococcus aureus* at 2.9- and 2.8-A resolution. *Biochemistry* 20(9):2361–2370.

Duncan, A.R., J.M. Woof, L.J. Partridge, D.R. Burton, and G. Winter. 1988. Localization of the binding site for the human high-affinity Fc receptor on IgG. *Nature* 332(6164):563–564.

Dwek, R.A. 1998. Biological importance of glycosylation. *Dev. Biol. Stand.* 96:43–47.

Farooq, M., N. Takahashi, H. Arrol, M. Drayson, and R. Jefferis. 1997. Glycosylation of polyclonal and paraprotein IgG in multiple myeloma. *Glycoconj. J.* 14(4):489–492.

Ferrara, C., P. Brunker, T. Suter, S. Moser, U. Puntener, and R. Umana. 2006a. Modulation of therapeutic antibody effector functions by glycosylation engineering: Influence of Golgi enzyme localization domain and co-expression of heterologous beta1, 4-N-acetylglucosaminyltransferase III and Golgi alpha-mannosidase II. *Biotechnol. Bioeng.* 93(5):851–861.

Ferrara, C., F. Stuart, P. Sondermann, P. Brunker, and P. Umana. 2006b. The carbohydrate at FcgammaRIIIa Asn-162. An element required for high affinity binding to non-fucosylated IgG glycoforms. *J. Biol. Chem.* 281(8):5032–5036.

Fujii, S., T. Nishiura, A. Nishikawa, R. Miura, and N. Taniguchi. 1990. Structural heterogeneity of sugar chains in immunoglobulin G. Conformation of immunoglobulin G molecule and substrate specificities of glycosyltransferases. *J. Biol. Chem.* 265(11):6009–6018.

Gala, F.A., and S.L. Morrison. 2004. V region carbohydrate and antibody expression. *J. Immunol.* 172(9): 5489–5494.

Galili, U., F. Anaraki, A. Thall, C. Hill-Black, and M. Radic. 1993. One percent of human circulating B lymphocytes are capable of producing the natural anti-Gal antibody. *Blood* 82(8):2485–2493.

Ghetie, V., and E.S. Ward. 2000. Multiple roles for the major histocompatibility complex class I-related receptor FcRn. *Annu. Rev. Immunol.* 18:739–766.

Ghirlando, R., M.B. Keown, G.A. Mackay, M.S. Lewis, J.C. Unkeless, and H.J. Gould. 1995. Stoichiometry and thermodynamics of the interaction between the Fc fragment of human IgG1 and its low-affinity receptor Fc gamma RIII. *Biochemistry* 34(41):13320–13327.

Ghirlando, R., J. Lund, M. Goodall, and R. Jefferis. 1999. Glycosylation of human IgG-Fc: Influences on structure revealed by differential scanning micro-calorimetry. *Immunol. Lett.* 68(1):47–52.

Grey, A.A., S. Narasimhan, J.R. Brisson, H. Schachter, and J.P. Carver. 1982. Structure of the glycopeptides of a human gamma 1-immunoglobulin G (Tem) myeloma protein as determined by 360-megahertz nuclear magnetic resonance spectroscopy. *Can. J. Biochem.* 60(12):1123–1131.

Guile, G.R., P.M. Rudd, D.R. Wing, S.B. Prime, and R.A. Dwek. 1996. A rapid high-resolution high-performance liquid chromatographic method for separating glycan mixtures and analyzing oligosaccharide profiles. *Anal. Biochem.* 240(2):210–226.

Hamilton, S.R., P. Bobrowicz, B. Bobrowicz, R.C. Davidson, H. Li, T. Mitchell, J.H. Nett, S. Rausch, T.A. Stadheim, H. Wischnewski, and others. 2003. Production of complex human glycoproteins in yeast. *Science* 301(5637):1244–1246.

Hansske, B., C. Thiel, T. Lubke, M. Hasilik, S. Honing, V. Peters, P.H. Heidemann, G.F. Hoffmann, E.G. Berger, K. von Figura, and others. 2002. Deficiency of UDP-galactose: N-acetylglucosamine beta-1,4-galactosyltransferase I causes the congenital disorder of glycosylation type IId. *J. Clin. Invest.* 109(6):725–733.

Hendershot, L., D. Bole, G. Kohler, and J.F. Kearney. 1987. Assembly and secretion of heavy chains that do not associate posttranslationally with immunoglobulin heavy chain-binding protein. *J. Cell. Biol.* 104(3):761–767.

Hokke, C.H., A.A. Bergwerff, G.W. van Dedem, J.P. Kamerling, and J.F. Vliegenthart. 1995. Structural analysis of the sialylated N- and O-linked carbohydrate chains of recombinant human erythropoietin expressed in Chinese hamster ovary cells. Sialylation patterns and branch location of dimeric N-acetyllactosamine units. *Eur. J. Biochem.* 228(3):981–1008.

Holland, M., K. Takada, T. Okumoto, N. Takahashi, K. Kato, D. Adu, A. Ben-Smith, L. Harper, C.O. Savage, and R. Jefferis. 2002. Hypogalactosylation of serum IgG in patients with ANCA-associated systemic vasculitis. *Clin. Exp. Immunol.* 129(1):183–190.

Holland, M., H. Yagi, N. Takahashi, K. Kato, C.O. Savage, D.M. Goodall, and R. Jefferis. 2006. Differential glycosylation of polyclonal IgG, IgG-Fc and IgG-Fab isolated from the sera of patients with ANCA-associated systemic vasculitis. *Biochim. Biophys. Acta* 1760(4):669–677.

Huang, L., S. Biolsi, K.R. Bales, and U. Kuchibhotla. 2006. Impact of variable domain glycosylation on antibody clearance: An LC/MS characterization. *Anal. Biochem.* 349(2):197–207.

Ip, C.C., W.J. Miller, M. Silberklang, G.E. Mark, R.W. Ellis, L. Huang, J. Glushka, H. van Halbeek, J. Zhu, and J.A. Alhadeff. 1994. Structural characterization of the N-glycans of a humanized anti-CD18 murine immunoglobulin G. *Arch. Biochem. Biophys.* 308(2):387–399.

Itoh, K., N. Takahashi, and M. Hirayama. 1993. Abnormalities in the oligosaccharide moieties of immunoglobulin G in patients with mytonic dystrophy. *J. Clin. Biochem. Nutr.* 14:61–69.

Jaeken, J., and G. Matthijs. 2007. Congenital disorders of glycosylation: A rapidly expanding disease family. *Annu. Rev. Genomics Hum. Genet.* 8:261–278.

Jefferis, R. 2005. Glycosylation of recombinant antibody therapeutics. *Biotechnol. Prog.* 21(1):11–16.

Jefferis, R. 2006. A sugar switch for anti-inflammatory antibodies. *Nature Biotechnol.* 24(10):1230–1231.

Jefferis, R. 2007. Antibody therapeutics: Isotype and glycoform selection. *Expert Opin. Biol. Ther.* 7(9): 1401–1413.

Jefferis, R., J. Lund, H. Mizutani, H. Nakagawa, Y. Kawazoe, Y. Arata, and N. Takahashi. 1990. A comparative study of the N-linked oligosaccharide structures of human IgG subclass proteins. *Biochem. J.* 268(3):529–537.

Jones, J., S.S. Krag, and M.J. Betenbaugh. 2005. Controlling N-linked glycan site occupancy. *Biochim. Biophys. Acta* 1726(2):121–137.

Kanda, Y., T. Yamada, K. Mori, A. Okazaki, M. Inoue, K. Kitajima-Miyama, R. Kuni-Kamochi, R. Nakano, K. Yano, S. Kakita, and others. 2007. Comparison of biological activity among nonfucosylated therapeutic IgG1 antibodies with three different N-linked Fc oligosaccharides: The high-mannose, hybrid, and complex types. *Glycobiology* 17(1):104–118.

Kaneko, Y., F. Nimmerjahn, and J.V. Ravetch. 2006. Anti-inflammatory activity of immunoglobulin G resulting from Fc sialylation. *Science* 313(5787):670–673.

Kibe, T., S. Fujimoto, C. Ishida, H. Togari, S. Okada, H. Nakagawa, Y. Tsukamoto, and N. Takahashi. 1996. Glycosylation and placental transport of IgG. *J. Clin. Biochem. Nutr.* 21:57–63.

Kim, H., Y. Yamaguchi, K. Masuda, C. Matsunaga, K. Yamamoto, T. Irimura, N. Takahashi, K. Kato, and Y. Arata. 1994. O-glycosylation in hinge region of mouse immunoglobulin G2b. *J. Biol. Chem.* 269(16): 12345–12350.

Koene, H.R., M. Kleijer, J. Algra, D. Roos, A.E. von dem Borne and M. de Haas. 1997. Fc gammaRIIIa-158V/F polymorphism influences the binding of IgG by natural killer cell Fc gammaRIIIa, independently of the Fc gammaRIIIa-48L/R/H phenotype. *Blood* 90(3):1109–1114.

Kornfeld, R., and M. Kornfeld. 1985. Assembly of asparagine-linked oligosaccharides. *Annu. Rev. Biochem.* 54:631–664.

Krapp, S., Y. Mimura, R. Jefferis, R. Huber, and P. Sondermann. 2003. Structural analysis of human IgG-Fc glycoforms reveals a correlation between glycosylation and structural integrity. *J. Mol. Biol.* 325(5):979–989.

Li, H., N. Sethuraman, T.A. Stadheim, D. Zha, B. Prinz, N. Ballew, P. Bobrowicz, B.K. Choi, W.J. Cook, M. Cukan, and others. 2006. Optimization of humanized IgGs in glycoengineered Pichia pastoris. *Nature Biotechnol.* 24(2):210–215.

Lifely, M.R., C. Hale, S. Boyce, M.J. Keen, and J. Phillips. 1995. Glycosylation and biological activity of CAMPATH-1H expressed in different cell lines and grown under different culture conditions. *Glycobiology* 5(8):813–822.

Longmore, G.D., and H. Schachter. 1982. Product-identification and substrate-specificity studies of the GDP-L-fucose: 2-acetamido-2-deoxy-beta-D-glucoside (FUC goes to Asn-linked GlcNAc) 6-alpha-L-fucosyltransferase in a Golgi-rich fraction from porcine liver. *Carbohydr. Res.* 100:365–392.

Lund, J., N. Takahashi, J.D. Pound, M. Goodall, and R. Jefferis. 1996. Multiple interactions of IgG with its core oligosaccharide can modulate recognition by complement and human Fc gamma receptor I and influence the synthesis of its oligosaccharide chains. *J. Immunol.* 157(11):4963–4969.

Lund, J., N. Takahashi, J.D. Pound, M. Goodall, H. Nakagawa, and R. Jefferis. 1995. Oligosaccharide-protein interactions in IgG can modulate recognition by Fc gamma receptors. *FASEB J.* 9(1):115–119.

Lund, J., T. Tanaka, N. Takahashi, G. Sarmay, Y. Arata, and R. Jefferis. 1990. A protein structural change in aglycosylated IgG3 correlates with loss of huFc gamma R1 and huFc gamma R111 binding and/or activation. *Mol. Immunol.* 27(11):1145–1153.

Lund, J., G. Winter, P.T. Jones, J.D. Pound, T. Tanaka, M.R. Walker, P.J. Artymiuk, Y. Arata, D.R. Burton, R. Jefferis, and others. 1991. Human Fc gamma RI and Fc gamma RII interact with distinct but overlapping sites on human IgG. *J. Immunol.* 147(8):2657–2662.

Maenaka, K., P.A. van der Merwe, D.I. Stuart, E.Y. Jones, and P. Sondermann. 2001. The human low affinity Fcgamma receptors IIa, IIb, and III bind IgG with fast kinetics and distinct thermodynamic properties. *J. Biol. Chem.* 276(48):44898–44904.

Malhotra, R., M.R. Wormald, P.M. Rudd, P.B. Fischer, R.A. Dwek, and R.B. Sim. 1995. Glycosylation changes of IgG associated with rheumatoid arthritis can activate complement via the mannose-binding protein. *Nature Med.* 1(3):237–243.

Martinez, T., D. Pace, L. Brady, M. Gerhart, and A. Balland. 2007. Characterization of a novel modification on IgG2 light chain. Evidence for the presence of O-linked mannosylation. *J. Chromatogr. A* 1156(1-2):183–187.

Masuda, K., Y. Yamaguchi, K. Kato, N. Takahashi, I. Shimada, and Y. Arata. 2000. Pairing of oligosaccharides in the Fc region of immunoglobulin G. *FEBS Lett.* 473(3):349–357.

Matsumiya, S., Y. Yamaguchi, J. Saito, M. Nagano, H. Sasakawa, S. Otaki, M. Satoh, K. Shitara, and K. Kato. 2007. Structural comparison of fucosylated and nonfucosylated Fc fragments of human immunoglobulin G1. *J. Mol. Biol.* 368(3):767–779.

McCann, K.J., C.H. Ottensmeier, A. Callard, C.M. Radcliffe, D.J. Harvey, R.A. Dwek, P.M. Rudd, B.J. Sutton, P. Hobby, and F.K. Stevenson. 2008. Remarkable selective glycosylation of the immunoglobulin variable region in follicular lymphoma. *Mol. Immunol.* 45(6):1567–1572.

McGreal, E.P., J.L. Miller, and S. Gordon. 2005. Ligand recognition by antigen-presenting cell C-type lectin receptors. *Curr. Opin. Immunol.* 17(1):18–24.

Mimura, Y., P.R. Ashton, N. Takahashi, D.J. Harvey, and R. Jefferis. 2007. Contrasting glycosylation profiles between Fab and Fc of a human IgG protein studied by electrospray ionization mass spectrometry. *J. Immunol. Methods* 326(1-2):116–126.

Mimura, Y., S. Church, R. Ghirlando, P.R. Ashton, S. Dong, M. Goodall, J. Lund, and R. Jefferis. 2000. The influence of glycosylation on the thermal stability and effector function expression of human IgG1-Fc: Properties of a series of truncated glycoforms. *Mol. Immunol.* 37(12-13):697–706.

Mimura, Y., P. Sondermann, R. Ghirlando, J. Lund, S.P. Young, M. Goodall, and R. Jefferis. 2001. Role of oligosaccharide residues of IgG1-Fc in Fc gamma RIIb binding. *J. Biol. Chem.* 276(49):45539–45547.

Nguyen, D.H., P. Tangvoranuntakul, and A. Varki. 2005. Effects of natural human antibodies against a nonhuman sialic acid that metabolically incorporates into activated and malignant immune cells. *J. Immunol.* 175(1):228–236.

Nimmerjahn, F., R.M. Anthony, and J.V. Ravetch. 2007. Agalactosylated IgG antibodies depend on cellular Fc receptors for in vivo activity. *Proc. Natl. Acad. Sci. USA* 104(20):8433–8437.

Nishie, T., O. Miyaishi, H. Azuma, A. Kameyama, C. Naruse, N. Hashimoto, H. Yokoyama, H. Narimatsu, T. Wada, and M. Asano. 2007. Development of immunoglobulin A nephropathy-like disease in beta-1,4-galactosyltransferase-I-deficient mice. *Am. J. Pathol.* 170(2):447–456.

Nose, M., and H. Wigzell. 1983. Biological significance of carbohydrate chains on monoclonal antibodies. *Proc. Natl. Acad. Sci. USA* 80(21):6632–6636.

Okazaki, A., E. Shoji-Hosaka, K. Nakamura, M. Wakitani, K. Uchida, S. Kakita, K. Tsumoto, I. Kumagai, and K. Shitara. 2004. Fucose depletion from human IgG1 oligosaccharide enhances binding enthalpy and association rate between IgG1 and FcgammaRIIIa. *J. Mol. Biol.* 336(5):1239–1249.

Padlan, E.A. 1990. X-ray diffraction studies of antibody constant regions. In *Fc receptors and the action of antibodies*, ed. Metzger, H. 12–30. Washington, D.C.: American Society for Microbiology.

Parekh, R.B., R.A. Dwek, B.J. Sutton, D.L. Fernandes, A. Leung, D. Stanworth, T.W. Rademacher, T. Mizuochi, T. Taniguchi, K. Matsuta, and others. 1985. Association of rheumatoid arthritis and primary osteoarthritis with changes in the glycosylation pattern of total serum IgG. *Nature* 316(6027):452–457.

Parekh, R., I. Roitt, D. Isenberg, R. Dwek, and T. Rademacher. 1988. Age-related galactosylation of the N-linked oligosaccharides of human serum IgG. *J. Exp. Med.* 167(5):1731–1736.

Pound, J.D., J. Lund, and R. Jefferis. 1993. Aglycosylated chimaeric human IgG3 can trigger the human phagocyte respiratory burst. *Mol. Immunol.* 30(3):233–241.

Qian, J., T. Liu, L. Yang, A. Daus, R. Crowley, and Q. Zhou. 2007. Structural characterization of N-linked oligosaccharides on monoclonal antibody cetuximab by the combination of orthogonal matrix-assisted laser desorption/ionization hybrid quadrupole-quadrupole time-of-flight tandem mass spectrometry and sequential enzymatic digestion. *Anal. Biochem.* 364(1):8–18.

Radaev, S., S. Motyka, W.H. Fridman, C. Sautes-Fridman, and P.D. Sun. 2001. The structure of a human type III Fcgamma receptor in complex with Fc. *J. Biol. Chem.* 276(19):16469–16477.

Radcliffe, C.M., J.N. Arnold, D.M. Suter, M.R. Wormald, D.J. Harvey, L. Royle, Y. Mimura, Y. Kimura, R.B. Sim, S. Inoges, and others. 2007. Human follicular lymphoma cells contain oligomannose glycans in the antigen-binding site of the B-cell receptor. *J. Biol. Chem.* 282(10):7405–7415.

Raju, T.S., J.B. Briggs, S.M. Borge, and A.J. Jones. 2000. Species-specific variation in glycosylation of IgG: Evidence for the species-specific sialylation and branch-specific galactosylation and importance for engineering recombinant glycoprotein therapeutics. *Glycobiology* 10(5):477–486.

Restelli, V., and M. Butler. 2002. The effect of cell culture parameters on protein glycosylation. In *Cell engineering*, ed. Al-Rubeai, M. 61–92. Dordrecht: Kluwer.

Roopenian, D.C., and S. Akilesh. 2007. FcRn: The neonatal Fc receptor comes of age. *Nature Rev. Immunol.* 7(9):715–725.

Routier, F.H., E.F. Hounsell, P.M. Rudd, N. Takahashi, A. Bond, F.C. Hay, A. Alavi, J.S. Axford, and R. Jefferis. 1998. Quantitation of the oligosaccharides of human serum IgG from patients with rheumatoid arthritis: A critical evaluation of different methods. *J. Immunol. Methods* 213(2):113–130.

Royle, L., M.P. Campbell, C.M. Radcliffe, D.M. White, D.J. Harvey, J.L. Abrahams, Y.G. Kim, G.W. Henry, N.A. Shadick, M.E. Weinblatt, and others. 2008. HPLC-based analysis of serum N-glycans on a 96-well plate platform with dedicated database software. *Anal. Biochem.* 376(1):1–12.

Royle, L., C.M. Radcliffe, R.A. Dwek, and P.M. Rudd. 2006. Detailed structural analysis of N-glycans released from glycoproteins in SDS-PAGE gel bands using HPLC combined with exoglycosidase array digestions. *Methods Mol. Biol.* 347:125–143.

Royle, L., A. Roos, D.J. Harvey, M.R. Wormald, D. van Gijlswijk-Janssen, R.M. Redwan el, I.A. Wilson, M.R. Daha, R.A. Dwek, and P.M. Rudd. 2003. Secretory IgA N- and O-glycans provide a link between the innate and adaptive immune systems. *J. Biol. Chem.* 278(22):20140–20153.

Rudd, P.M., and R.A. Dwek. 1997. Glycosylation: Heterogeneity and the 3D structure of proteins. *Crit. Rev. Biochem. Mol. Biol.* 32(1):1–100.

Sarmay, G., J. Lund, Z. Rozsnyay, J. Gergely, and R. Jefferis. 1992. Mapping and comparison of the interaction sites on the Fc region of IgG responsible for triggering antibody dependent cellular cytotoxicity (ADCC) through different types of human Fc gamma receptor. *Mol. Immunol.* 29(5):633–639.

Scallon, B.J., S.H. Tam, S.G. McCarthy, A.N. Cai, and T.S. Raju. 2007. Higher levels of sialylated Fc glycans in immunoglobulin G molecules can adversely impact functionality. *Mol. Immunol.* 44(7):1524–1534.

Sheeley, D.M., B.M. Merrill, and L.C. Taylor. 1997. Characterization of monoclonal antibody glycosylation: Comparison of expression systems and identification of terminal alpha-linked galactose. *Anal. Biochem.* 247(1):102–110.

Sheridan, C. 2007. Commercial interest grows in glycan analysis. *Nature Biotechnol.* 25(2):145–146.

Shields, R.L., J. Lai, R. Keck, L.Y. O'Connell, K. Hong, Y.G. Meng, S.H. Weikert, and L.G. Presta. 2002. Lack of fucose on human IgG1 N-linked oligosaccharide improves binding to human Fcgamma RIII and antibody-dependent cellular toxicity. *J. Biol. Chem.* 277(30):26733–26740.

Shields, R.L., A.K. Namenuk, K. Hong, Y.G. Meng, J. Rae, J. Briggs, D. Xie, J. Lai, A. Stadlen, B. Li, and others. 2001. High resolution mapping of the binding site on human IgG1 for Fc gamma RI, Fc gamma RII, Fc gamma RIII, and FcRn and design of IgG1 variants with improved binding to the Fc gamma R. *J. Biol. Chem.* 276(9):6591–6604.

Shinkawa, T., K. Nakamura, N. Yamane, E. Shoji-Hosaka, Y. Kanda, M. Sakurada, K. Uchida, H. Anazawa, M. Satoh, M. Yamasaki, and others. 2003. The absence of fucose but not the presence of galactose or bisecting N-acetylglucosamine of human IgG1 complex-type oligosaccharides shows the critical role of enhancing antibody-dependent cellular cytotoxicity. *J. Biol. Chem.* 278(5):3466–3473.

Sondermann, P., R. Huber, V. Oosthuizen, and U. Jacob. 2000. The 3.2-A crystal structure of the human IgG1 Fc fragment-Fc gammaRIII complex. *Nature* 406(6793):267–273.

Takahashi, N., H. Nakagawa, K. Fujikawa, Y. Kawamura, and N. Tomiya. 1995. Three-dimensional elution mapping of pyridylaminated N-linked neutral and sialyl oligosaccharides. *Anal. Biochem.* 226(1):139–146.

Takeuchi, M., S. Takasaki, H. Miyazaki, T. Kato, S. Hoshi, N. Kochibe, and A. Kobata. 1988. Comparative study of the asparagine-linked sugar chains of human erythropoietins purified from urine and the culture medium of recombinant Chinese hamster ovary cells. *J. Biol. Chem.* 263(8):3657–3663.

Tangvoranuntakul, P., P. Gagneux, S. Diaz, M. Bardor, N. Varki, A. Varki, and E. Muchmore. 2003. Human uptake and incorporation of an immunogenic nonhuman dietary sialic acid. *Proc. Natl. Acad. Sci. USA* 100(21):12045–12050.

Tao, M.H., and S.L. Morrison. 1989. Studies of aglycosylated chimeric mouse-human IgG. Role of carbohydrate in the structure and effector functions mediated by the human IgG constant region. *J. Immunol.* 143(8):2595–2601.

Umana, P., J. Jean-Mairet, R. Moudry, H. Amstutz, and J.E. Bailey. 1999. Engineered glycoforms of an antineuroblastoma IgG1 with optimized antibody-dependent cellular cytotoxic activity. *Nature Biotechnol.* 17(2):176–180.

van den Eijnden, D.H., D.H. Joziasse, L. Dorland, H. van Halbeek, J.F. Vliegenthart, and K. Schmid. 1980. Specificity in the enzymic transfer of sialic acid to the oligosaccharide branches of b1- and triantennary glycopeptides of alpha 1-acid glycoprotein. *Biochem. Biophys. Res. Commun.* 92(3):839–845.

Wong, N.S., M.G. Yap, and D.I. Wang. 2006. Enhancing recombinant glycoprotein sialylation through CMP-sialic acid transporter over expression in Chinese hamster ovary cells. *Biotechnol. Bioeng.* 93(5):1005–1016.

Woof, J.M., and D.R. Burton. 2004. Human antibody-Fc receptor interactions illuminated by crystal structures. *Nature Rev. Immunol.* 4(2):89–99.

Woof, J.M., L.J. Partridge, R. Jefferis, and D.R. Burton. 1986. Localisation of the monocyte-binding region on human immunoglobulin G. *Mol. Immunol.* 23(3):319–330.

Wormald, M.R., P.M. Rudd, D.J. Harvey, S.C. Chang, I.G. Scragg, and R.A. Dwek. 1997. Variations in oligosaccharide-protein interactions in immunoglobulin G determine the site-specific glycosylation profiles and modulate the dynamic motion of the Fc oligosaccharides. *Biochemistry* 36(6):1370–1380.

Wright, A., and S.L. Morrison. 1998. Effect of C2-associated carbohydrate structure on Ig effector function: Studies with chimeric mouse-human IgG1 antibodies in glycosylation mutants of Chinese hamster ovary cells. *J. Immunol.* 160(7):3393–3402.

Wright, A., M.H. Tao, E.A. Kabat, and S.L. Morrison. 1991. Antibody variable region glycosylation: Position effects on antigen binding and carbohydrate structure. *EMBO J.* 10(10):2717–2723.

Yamada, E., Y. Tsukamoto, R. Sasaki, K. Yagyu, and N. Takahashi. 1997. Structural changes of immunoglobulin G oligosaccharides with age in healthy human serum. *Glycoconj. J.* 14(3):401–405.

Yamaguchi, Y., K. Kato, M. Shindo, S. Aoki, K. Furusho, K. Koga, N. Takahashi, Y. Arata, and I. Shimada. 1998. Dynamics of the carbohydrate chains attached to the Fc portion of immunoglobulin G as studied by NMR spectroscopy assisted by selective 13C labeling of the glycans. *J. Biomol. NMR* 12(3):385–394.

Yamane-Ohnuki, N., S. Kinoshita, M. Inoue-Urakubo, M. Kusunoki, S. Iida, R. Nakano, M. Wakitani, R. Niwa, M. Sakurada, K. Uchida, and others. 2004. Establishment of FUT8 knockout Chinese hamster ovary cells: An ideal host cell line for producing completely defucosylated antibodies with enhanced antibody-dependent cellular cytotoxicity. *Biotechnol. Bioeng.* 87(5):614–622.

Youings, A., S.C. Chang, R.A. Dwek, and I.G. Scragg. 1996. Site-specific glycosylation of human immunoglobulin G is altered in four rheumatoid arthritis patients. *Biochem. J.* 314 (Pt 2):621–630.

CHAPTER 4

Antibody Databases and Tools: The IMGT® Experience

MARIE-PAULE LEFRANC

Therapeutic Monoclonal Antibodies: From Bench to Clinic. Edited by Zhiqiang An

ABSTRACT

To solve the complexity of immunogenetics knowledge, a conceptualization has been developed in IMGT-ONTOLOGY, in an unprecedented approach. To allow data comparison on antibodies, IMGT® databases and tools are highly standardized and combine genomic, genetic, and structural approaches, providing an immunoinformatics framework essential for a systemic immunotherapy.

4.1 INTRODUCTION

The number of potential protein forms of the antigen receptors, immunoglobulins (IG) and T cell receptors (TR), is almost unlimited. The potential repertoire of each individual is estimated to comprise about 10^{12} different IG (or antibodies) and TR, and the limiting factor is only the number of B and T cells that an organism is genetically programmed to produce. This huge diversity is inherent to the particularly complex and unique molecular synthesis and genetics of the antigen receptor chains. This includes biological mechanisms such as DNA molecular rearrangements in multiple loci (three for IG and four for TR in humans) located on different chromosomes (four in humans), nucleotide deletions and insertions at the rearrangement junctions (or N-diversity), and somatic hypermutations in the IG loci (for review see Lefranc and Lefranc 2001a, 2001b). IMGT®, the international ImMunoGeneTics information system® (http://www.imgt.org) (Lefranc et al. 2009), was created in 1989, by the Laboratoire d'ImmunoGénétique Moléculaire (LIGM) (Université Montpellier 2 and CNRS) at Montpellier, France, in order to standardize and manage the complexity of the immunogenetics data. IMGT® is the international reference in immunogenetics and immunoinformatics. IMGT® is acknowledged as providing the molecular standards for antibodies and T cell receptors by the World Health Organization-International Union of Immunological Societies (WHO-IUIS) Nomenclature Subcommittee for immunoglobulins and T cell receptors (Lefranc 2007, 2008), the Human Genome Organisation (HUGO) Nomenclature Committee (HGNC) (Wain et al. 2002), and the Antibody Society (http://www.antibodysociety.org). IMGT® is a high quality integrated knowledge resource, specialized in (1) the IG, TR, major histocompatibility complex (MHC) of human and other vertebrates; (2) proteins that belong to the immunoglobulin superfamily (IgSF) and to the MHC superfamily (MhcSF); and (3) related proteins of the immune systems (RPI) of any species. IMGT® provides a common access to standardized data from genome, proteome, genetics, and three-dimensional (3D) structures for the IG, TR, MHC, IgSF, MhcSF, and RPI (Lefranc et al. 2005a, 2009).

The IMGT® information system consists of databases, tools, and Web resources (Lefranc et al. 2005a, 2009) (Fig. 4.1). IMGT® databases include one genome database, three sequence databases, and one 3D structure database. IMGT® interactive online tools are provided for genome, sequence, and 3D structure analysis. IMGT® Web resources comprise more than 10,000 HTML pages of synthesis and knowledge (IMGT Repertoire, IMGT Index, IMGT Scientific chart, IMGT Education, IMGT Medical page, IMGT Veterinary page, IMGT Biotechnology page) and external links (IMGT Bloc-notes, IMGT Immunoinformatics page). Other IMGT accesses include Sequence Retrieval System (SRS) at the European Bioinformatics Institute (EBI), DNA Data Bank of Japan (DDBJ), Institut Pasteur Paris, Deutschen Krebsforschungszentrum (DKFZ) Heidelberg, Columbia University New York and Indiana University (United States), and FTP IMGT/LIGM-DB flat file downloading at CINES (ftp://ftp.cines.fr/IMGT/) and its FTP copy mirror at EBI (ftp://ftp.ebi.ac.uk/pub/databases/imgt) (Lefranc et al. 2005a, 2009). Despite the heterogeneity of these different components, all data in the IMGT® information system are expertly annotated. The accuracy, consistency, and integration of the IMGT® data, as well as the coherence between the different IMGT® components (databases, tools, and Web resources) are based on IMGT-ONTOLOGY (Giudicelli and Lefranc 1999), the first ontology in immunogenetics and immunoinformatics. IMGT-ONTOLOGY provides a semantic

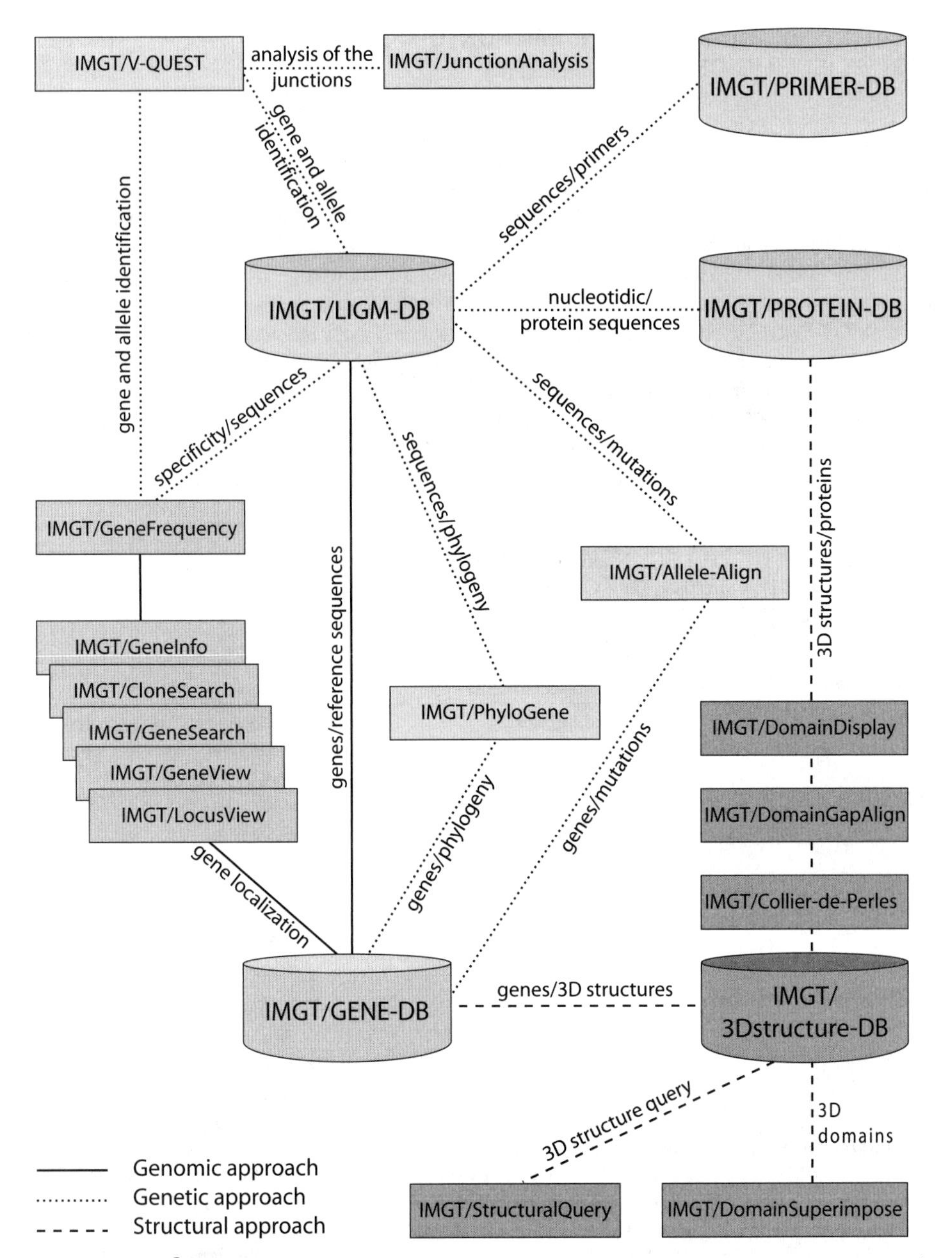

Figure 4.1 IMGT® databases and tools with their interactions according to the genomic, genetic, and/or structural approaches.

specification of the terms to be used in the domain and, thus, allows the management of immunogenetics knowledge for all vertebrate species. In this chapter, we will review the IMGT® standardization and the IMGT® genomic, genetic, and structural approaches that are the foundations of the IMGT® information system. The three following sections will describe examples of IMGT® databases and tools for antibody genomics, antibody genetics, and antibody structural analysis.

4.2 IMGT® STANDARDIZATION: IMGT-ONTOLOGY AND IMGT SCIENTIFIC CHART

The IMGT-ONTOLOGY axioms and the derived concepts (Giudicelli and Lefranc 1999; Lefranc et al. 2005a; Duroux et al. 2008) are available, for biologists and IMGT® users, in the IMGT Scientific chart (Lefranc et al. 2005a, 2009), and have been formalized, for computing scientists, in IMGT-ML (Chaume, Giudicelli, and Lefranc 2001; Chaume et al. 2003). The IMGT Scientific chart comprises the controlled vocabulary and the annotation rules necessary for the immunogenetics data identification, description, classification, and numbering, and for knowledge management in the IMGT® information system (Lefranc et al. 2005a, 2009). All IMGT® data are expertly annotated according to the IMGT Scientific chart rules. Standardized keywords, labels and annotation rules, standardized IG and TR gene nomenclature, the IMGT unique numbering, localization, orientation and standardized origin/methodology were defined, respectively, based on the seven main axioms of IMGT-ONTOLOGY (Duroux et al. 2008) (Table 4.1). These axioms, "IDENTIFICATION," "CLASSIFICATION," "DESCRIPTION," "LOCALIZATION," "NUMEROTATION," "ORIENTATION," and "OBTENTION" postulate that objects, processes, and relations have to be identified,

TABLE 4.1 IMGT-ONTOLOGY Main Concepts, IMGT Scientific Chart Rules, and Examples of IMGT® Expertised Data Concepts

IMGT-ONTOLOGY Axioms (Duroux et al. 2008)	IMGT Scientific Chart Rules (Lefranc et al. 2005a, 2009)	Examples of IMGT® Expertised Data Concepts
IDENTIFICATION	Standardized **keywords**	Species, molecule type, receptor type, chain type, gene type, structure, functionality, specificity
DESCRIPTION	Standardized **labels** and annotations	Core (V-, D-, J-, C-REGION) Prototypes Labels for sequences Labels for 2D and 3D structures
CLASSIFICATION	Standardized IG and TR gene **nomenclature** (group, subgroup, gene, allele) and reference sequences	Nomenclature of the human IG and TR genes (entry in 1999 in GDB), HGNC, and LocusLink and Entrez Gene at NCBI (Lefranc 2000a, 2000b; Lefranc and Lefranc 2001a, 2001b) Alignment of alleles Nomenclature of the IG and TR genes of all vertebrate species
NUMEROTATION	**IMGT unique numbering** for: V- and V-LIKE-DOMAIN (Lefranc 1997, 1999; Lefranc et al. 2003) C- and C-LIKE-DOMAIN (Lefranc et al. 2005c) G- and G-LIKE-DOMAIN (Lefranc et al. 2005b)	Protein displays IMGT Colliers de PerlesFR-IMGT and CDR-IMGT delimitations Structural loops and beta strand delimitations (Ruiz and Lefranc 2002)
LOCALIZATION	**Localization** of instances in space and time	Gene positions in loci
ORIENTATION	**Orientation** of instances relative to each other	Examples for genomic instances: Chromosome orientation Locus orientation Gene orientation DNA strand orientation
OBTENTION	Standardized **origin** and **methodology**	

described, classified, numerotated, localized, and orientated, and that the way they are obtained has to be determined. The axioms constitute the Formal IMGT-ONTOLOGY, also designated as IMGT-Kaleidoscope (Duroux et al. 2008). Each axiom gives rise to a set of concepts and to standardized rules defined in the IMGT Scientific chart, available as a section of the IMGT® Web resources. Examples of IMGT® expertised data concepts derived from the IMGT Scientific chart rules are shown in Table 4.1.

4.2.1 IDENTIFICATION Axiom: Standardized Keywords

IMGT® standardized keywords for IG and TR include the following:

(1) **General keywords:** indispensable for the sequence assignments, they are described in an exhaustive and nonredundant list, and are organized in a tree structure.
(2) **Specific keywords:** they are more specifically associated to particularities of the sequences (orphon, transgene, etc.). The list is not definitive and new specific keywords can easily be added if needed. IMGT/LIGM-DB standardized keywords have been assigned to all entries.

4.2.2 DESCRIPTION Axiom: Standardized Labels and Annotation

Two hundred twenty-one feature labels are necessary to describe all structural and functional subregions that compose IG and TR sequences, whereas only seven of them are available in EMBL, GenBank, or DDBJ (Benson et al. 2007; Cochrane et al. 2006; Okubo et al. 2006). Levels of annotation have been defined, which allow the users to query sequences in IMGT/LIGM-DB even though they are not fully annotated. Prototypes represent the organizational relationship between labels and give information on the order and expected length (in number of nucleotides) of the labels. This provides rules to verify the manual annotation, and to design automatic annotation tool. Two hundred eighty-five feature labels have been defined for the 3D structures. Annotation of sequences and 3D structures with these labels (in capital letters) constitutes the main part of the expertise. Interestingly, 64 IMGT® specific labels have been entered in the newly created Sequence Ontology (Eilbeck et al. 2005).

4.2.3 CLASSIFICATION Axiom: Standardized Gene Nomenclature

The objective is to provide immunologists and geneticists with a standardized nomenclature per locus and per species that allows extraction and comparison of data for the complex B and T cell antigen receptor molecules. The concepts of classification have been used to set up a unique nomenclature of human IG and TR genes. The IMGT® gene nomenclature was approved at the international level by the HGNC in 1999 (Wain et al. 2002). All the IMGT human IG and TR gene names (Lefranc 2000a, 2000b; Lefranc and Lefranc 2001a, 2001b) were entered in Genome Database (GDB) (Letovsky et al. 1998) and in LocusLink at the National Center for Biotechnology Information (NCBI) in 1999–2000, in Entrez Gene (NCBI) (Maglott et al. 2007) when this gene database superseded LocusLink, and in IMGT/GENE-DB (Giudicelli, Chaume, and Lefranc 2005). The IMGT® IG and TR gene names are the official references for the WHO-IUIS (Lefranc 2007, 2008) and for the genome projects and, as such, have been integrated in the MapViewer at NCBI and, since 2006, in the Ensembl server at EBI, and in the Vega database of the Wellcome Trust Sanger Institute.

IMGT reference sequences have been defined for each allele of each gene based on one or, whenever possible, several of the following criteria: germline sequence, first sequence published, longest sequence, mapped sequence. They are listed in the germline gene tables of the IMGT Repertoire (http://www.imgt.org). The IMGT Protein displays show the translated sequences of the alleles *01 of the functional or ORF genes (Lefranc and Lefranc 2001a, 2001b).

4.2.4 NUMEROTATION Axiom: The IMGT Unique Numbering

A uniform numbering system for IG and TR sequences of all species has been established to facilitate sequence comparison and cross-referencing between experiments from different laboratories whatever the antigen receptor (IG or TR), the chain type, or the species (Lefranc 1997, 1999; Lefranc et al. 2003, 2005c).

This numbering results from the analysis of more than 5000 IG and TR variable region sequences of vertebrate species from fish to human. It takes into account and combines the definition of the framework (FR) and complementarity determining region (CDR) (Kabat et al. 1991), structural data from x-ray diffraction studies (Satow et al. 1986), and the characterization of the hypervariable loops (Chothia and Lesk 1987). In the IMGT unique numbering, conserved amino acids from frameworks always have the same number whatever the IG or TR variable sequence, and whatever the species they come from. As examples: cysteine 23 (in FR1-IMGT), tryptophan 41 (in FR2-IMGT), leucine (or other hydrophobic amino acid) 89, and cysteine 104 (in FR3-IMGT). Tables and 2D graphical representations designated as IMGT Colliers de Perles (Lefranc and Lefranc 2001a, 2001b, Ruiz and Lefranc, 2002) are available on the IMGT® Web site at http://www.imgt.org. The IMGT Colliers de Perles of the variable domains (or V-DOMAIN) of the humanized antibody alemtuzumab are shown, as examples, in Figure 4.2. The same IMGT unique numbering is used for the VH, the variable domain of the IG heavy chain that corresponds to the V-D-J-REGION, and for V-KAPPA, the variable domain of the IG light chain kappa that corresponds to the V-J-REGION.

This IMGT unique numbering has several advantages:

(1) It has allowed the redefinition of the limits of the FR and CDR of the IG and TR variable domains. The FR-IMGT and CDR-IMGT lengths become in themselves crucial information that characterizes variable regions belonging to a group, a subgroup, and/or a gene. For that reason, information on antibody CDR-IMGT lengths is required by the WHO-International Nonproprietary Names (INN) program on its application form. The CDR-IMGT lengths are shown between brackets and separated with dots. Thus, the CDR-IMGT lengths of the alemtuzumab VH are [8.10.12] and those of the V-KAPPA are [6.3.9] (Fig. 4.2).

(2) Framework amino acids (and codons) located at the same position in different sequences can be compared without requiring sequence alignments. This also holds for amino acids belonging to CDR-IMGT of the same length.

(3) The unique numbering is used as the output of the IMGT/V-QUEST alignment tool. The aligned sequences are displayed according to the IMGT unique numbering and with the FR-IMGT and CDR-IMGT delimitations (Lefranc 2004).

(4) The unique numbering has allowed a standardization of the description of mutations and the description of IG and TR allele polymorphisms. The mutations and allelic polymorphisms of each gene are described by comparison to the IMGT reference sequences of the allele *01 (Lefranc and Lefranc 2001a, 2001b).

(5) The unique numbering allows the description and comparison of somatic hypermutations of the IG variable domains.

(6) The unique numbering allows calculation of the percentage of identity of the FR-IMGT whatever the chain type and whatever the species. This is particularly useful to compare humanized antibodies with human sequences for evaluating potential immunogenicity (Fig. 4.3).

By facilitating the comparison between sequences and by allowing the description of alleles and mutations, the IMGT unique numbering represents a big step forward in the analysis of the IG and TR sequences of all vertebrate species. Moreover, it gives insight into the structural configuration of the domains. Hydrogen bonds from experimental crystallographic data can be displayed in a standardized way in IMGT Colliers de Perles on two layers (Fig. 4.2). Structural and functional domains of the IG and TR chains comprise the V-DOMAIN (9-strand beta-sandwich) which corresponds to the

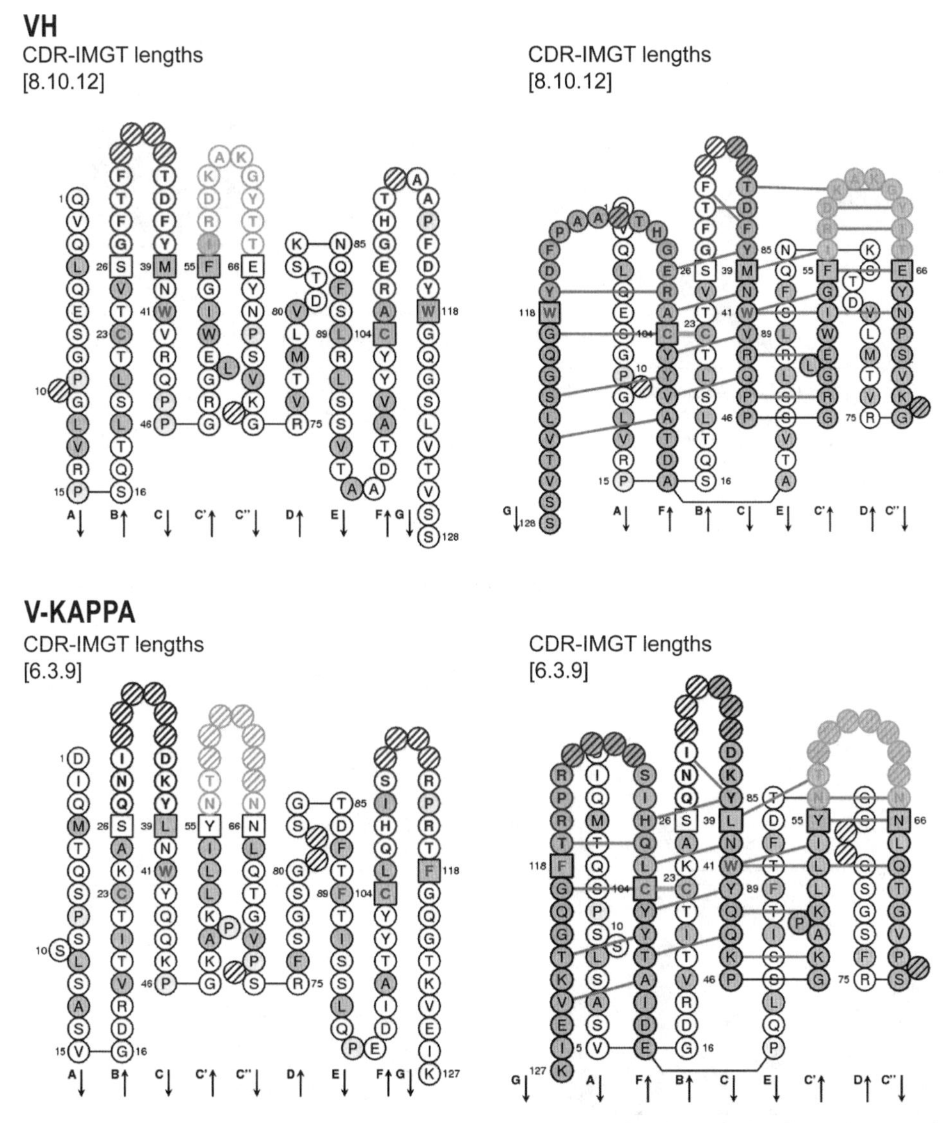

Figure 4.2 IMGT Collier de Perles of the VH and V-KAPPA domains of the humanized antibody alemtuzumab (Magdelaine-Beuzelin et al. 2007). The IMGT Colliers de Perles shown on one layer (left) and two layers (right) are based on the IMGT unique numbering for V-DOMAIN and V-LIKE-DOMAIN (Lefranc et al. 2003). Hydrogen bonds of the [GFCC′C″] sheet in the two layers IMGT Colliers de Perles are shown with lines (green in IMGT® (http://www.imgt.org) and in the E-book colored figure). Amino acids are shown in the one-letter abbreviation. The CDR-IMGT are limited by amino acids shown in squares, which belong to the neighboring FR-IMGT. The CDR3-IMGT extend from position 105 to position 117. Hatched circles correspond to missing positions according to the IMGT unique numbering for V -DOMAIN and V-LIKE-DOMAIN (Lefranc et al. 2003). Arrows indicate the direction of the nine beta strands that form the two beta sheets of the immunoglobulin fold (Lefranc 2001a, 2001b; Kaas, Ehrenmann, and Lefranc 2007). In IMGT® and in the E-book colored figure: (1) positions at which hydrophobic amino acids (hydropathy index with positive value: I, V, L, F, C, M, A) and tryptophan (W) are found in more than 50 percent of analyzed IG and TR sequences are shown in blue; (2) all proline (P) are shown in yellow; (3) CDR-IMGT are colored as follows: VH CDR1-IMGT (red), CDR2-IMGT (orange), CDR3-IMGT (purple); V-KAPPA (also valid for a V-LAMBDA) CDR1-IMGT (blue), CDR2-IMGT (bright green), CDR3-IMGT (dark green).

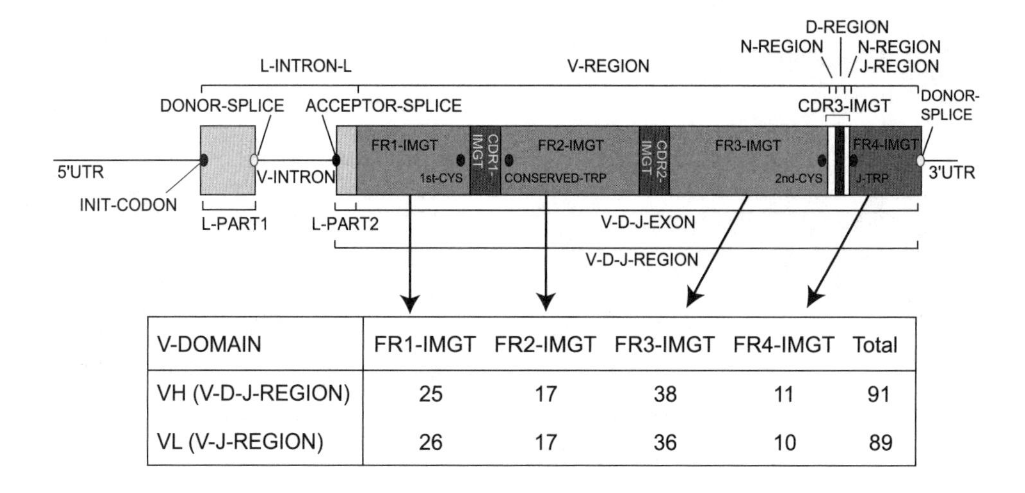

V-DOMAIN	FR1-IMGT	FR2-IMGT	FR3-IMGT	FR4-IMGT	Total
VH (V-D-J-REGION)	25	17	38	11	91
VL (V-J-REGION)	26	17	36	10	89

Humanized antibodies used in oncology

	Humanized antibody	CDR-IMGT antibody	Closest *Homo sapiens* gene and allele	FR-IMGT identity	AA with IMGT class change
VH	alemtuzumab	[8.10.12]	IGHV4-59*01	84.61% (77/91)	2
VH	bevacizumab	[8.8.16]	IGHV7-4-1*02	74.72% (68/91)	1
VH	trastuzumab	[8.8.13]	IGHV3-66*01	90.10% (82/91)	0
VL	alemtuzumab	[6.3.9]	IGKV1-33*01	97.75% (87/89)	1
VL	bevacizumab	[6.3.9]	IGKV1-33*01	92.13% (82/89)	2
VL	trastuzumab	[6.3.9]	IGKV1-39*01	93.25% (83/89)	1

Figure 4.3 FR-IMGT labels representation, lengths, and percentage of identity of three humanized antibodies compared to human. The prototype of a rearranged V-D-J gene is shown with the IMGT® labels. Lengths in number of amino acids (or codons) of the four FR-IMGT of a VH (V-D-J-REGION) and VL (V-J-REGION), here V-KAPPA, are shown [note that for a V-LAMBDA, FR1-IMGT is 25 amino acids (AA) long and the total is 88]. The percentage of identity of the 91 VH and 89 V-KAPPA FR-IMGT positions of three humanized antibodies alemtuzumab, bevacizumab, and trastuzumab obtained by comparison with the FR-IMGT of the closest *Homo sapiens* gene and allele, and the number of amino acids (AA) with IMGT class change (Pommié et al. 2004) are indicated (Magdelaine-Beuzelin et al. 2007). CDR-IMGT lengths are shown between brackets.

V-J-REGION or V-D-J-REGION and is encoded by two or three genes (Lefranc and Lefranc 2001a, 2001b), and the constant domain or C-DOMAIN (7-strand beta-sandwich). The IMGT unique numbering initially defined for the V-DOMAIN (Lefranc et al. 2003) has been extended to the C-DOMAIN of the IG and TR (Lefranc et al. 2005c). The IMGT unique numbering also opens interesting views on the evolution of all domain sequences that belong to the V-set and C-set of the IgSF as it has been extended to the V-LIKE-DOMAIN and to the C-LIKE-DOMAIN of IgSF proteins other than IG and TR (Lefranc et al. 2003, 2005c; Duprat et al. 2004; Bertrand et al. 2004; Garapati and Lefranc 2007). More recently, the IMGT unique numbering has also been defined for the groove domain or G-DOMAIN (4 beta-strand and one alpha-helix) of the MHC class I and II chains (Lefranc et al. 2005b; Kaas and Lefranc 2005, 2007; Kaas, Ehrenmann, and Lefranc 2007), and for the G-LIKE-DOMAIN of MhcSF proteins other than MHC (Frigoul and Lefranc 2005; Lefranc et al. 2005b; Duprat, Lefranc, and Gascuel 2006).

4.2.5 LOCALIZATION Axiom: Localization of Instances

The LOCALIZATION axiom and the concepts of localization allow specification of the position of a gene, sequence, or structural feature in a coordinate system (space or time).

4.2.6 ORIENTATION Axiom: Orientation of Instances Relative to Each Other

The ORIENTATION axiom and the concepts of orientation allow setting up of the orientation of instances relative to each other, for example, genomic orientation (for chromosome, locus, and gene) and DNA strand orientation. The genomic orientation is particularly useful in large genomic projects to localize a gene in a locus and/or a sequence (or a clone) in a contig or on a chromosome. The forward (FWD or "Watson") orientation from pter (short arm telomeric end) to qter (long arm telomeric end) of the human IGL locus (22q11.2) and the reverse (REV) orientation from qter to pter of the human IGH locus (14q32.33) and of the human IGK locus (2p11.2) are shown in IMGT Repertoire "Chromosomal localizations" (http://www.imgt.org). Opposite orientation of transcription of genes, such as that of the *Homo sapiens* IGKV5-2 and IGKV4-1, and that of the 36 IGKV genes of the distal cluster in the human IGK locus, are indicated with arrows in IMGT Repertoire "Locus representations" (http://www.imgt.org). The concepts of orientation also include the structural orientation (chain relative to the membrane, strands in a domain, etc.).

4.2.7 OBTENTION Axiom: Controlled Vocabulary for Biological Origin and Experimental Methodology

The OBTENTION axiom and the concepts of obtention, which are still in development, will be particularly useful for clinical data integration. This will help us to compare the repertoires of the IG antibody recognition sites and of the TR recognition sites in normal and pathological situations (autoimmune diseases, infectious diseases, leukemias, lymphomas, and myelomas).

4.3 IMGT® GENOMIC, GENETIC, AND STRUCTURAL APPROACHES

In order to extract knowledge from IMGT® standardized immunogenetics data, three main IMGT® biological approaches have been developed: genomic, genetic, and structural approaches (Table 4.2). The IMGT® genomic approach is gene-centered and mainly oriented towards the study of the genes within their loci and on the chromosomes. The IMGT® genetic approach refers to the study of the genes in relation with their sequence polymorphisms and mutations, their expression, their specificity, and their evolution.

The genetic approach relies heavily on the DESCRIPTION axiom (and particularly on the V-, D-, J-, and C-REGION core concepts for the IG and TR), on the CLASSIFICATION axiom (IMGT® gene and allele names) and on the NUMEROTATION concept [IMGT unique numbering (Lefranc 1997, 1999; Lefranc et al. 2003, 2005b, 2005c)]. The IMGT® structural approach refers to the study of the 2D and 3D structures of the IG, TR, MHC, and RPI, and to the antigen- or ligand-binding characteristics in relationship with the protein functions, polymorphisms, and evolution. The structural approach relies on the CLASSIFICATION axiom (IMGT® gene and allele names), DESCRIPTION axiom (receptor and chain description, domain delimitations), and NUMEROTATION axiom

TABLE 4.2 IMGT® Databases, Tools, and Web Resources for Antibody Genomic, Genetic, and Structural Approaches

Approaches	Databases[a]	Tools[a]	Web Resources[b]
Genomic	IMGT/GENE-DB (Giudicelli, Chaume, and Lefranc 2005)	IMGT/LocusView IMGT/GeneView IMGT/GeneSearch IMGT/CloneSearch IMGT/GeneInfo (Baum et al. 2006) IMGT/GeneFrequency	IMGT Repertoire "Locus and genes" section: Chromosomal localizations Locus description Gene tables, and so on Potential germline repertoires Lists of genes Correspondence between nomenclatures (Lefranc 2000a, 2000b, 2001a, 2001b)
Genetic	IMGT/LIGM-DB (Giudicelli et al. 2006) IMGT/MHC-DB (Robinson et al. 2003) IMGT/PRIMER-DB (Folch et al. 2004)	IMGT/V-QUEST (Giudicelli, Chaume, and Lefranc 2004, Lefranc 2004) IMGT/JunctionAnalysis (Yousfi Monod et al. 2004) IMGT/Allele-Align IMGT/PhyloGene (Elemento and Lefranc 2003) IMGT/DomainDisplay	IMGT Repertoire "Proteins and alleles" section: Alignments of alleles Protein displays Tables of alleles, and so on
Structural	IMGT/3Dstructure-DB (Kaas, Ruiz, and Lefranc 2004)	IMGT/DomainGapAlign IMGT/Collier-de-Perles IMGT/DomainSuperimpose IMGT/StructuralQuery (Kaas, Ruiz, and Lefranc 2004)	IMGT Repertoire "2D and 3D structures" section: IMGT Colliers de Perles (2D representations on one layer or two layers) IMGT® classes for amino acid characteristics (Pommié et al. 2004) IMGT Colliers de Perles reference profiles (Pommié et al. 2004) 3D representations

[a]All IMGT® databases and tools manage antibody data, except IMGT/MHC-DB and IMGT/GeneInfo.
[b]Only Web resources examples from the IMGT Repertoire section are shown.

(amino acid positions according to the IMGT unique numbering (Lefranc 1997, 1999; Lefranc et al. 2003, 2005b, 2005c).

For each approach, IMGT® provides databases [one genome database (IMGT/GENE-DB), three sequence databases (IMGT/LIGM-DB, IMGT/MHC-DB, IMGT/PRIMER-DB), one 3D structure database (IMGT/3Dstructure-DB)], interactive tools [15 online tools for genome, sequence, and 3D structure analysis], and IMGT Repertoire Web resources (providing an easy-to-use interface to carefully and expertly annotated data on the genome, proteome, polymorphism, and structural data of the IG and TR, MHC and RPI) (Table 4.2). Examples of databases, tools, and Web resources fo antibody genomics, genetics, and structural analysis are detailed in the following Sections 4.4 to 4.6.

Other IMGT® Web resources providing information on antibodies include:

- IMGT Bloc-notes (Interesting links, etc.): provides numerous hyperlinks to the Web servers specializing in immunology, genetics, molecular biology, and bioinformatics (associations, collections, companies, databases, immunology themes, journals, molecular biology servers, resources, societies, tools, etc.) (Lefranc 2006).

- IMGT Lexique.
- The IMGT Immunoinformatics page.
- The IMGT Medical page.
- The IMGT Veterinary page (domestic species, model species, and wild-life species).
- The IMGT Biotechnology page: provides information on combinatorial libraries and phage displays, reagents monoclonal antibodies for TR variable domains, antibody humanization, antibody camelization, characteristics of the Camelidae (camel, llama antibody synthesis), monoclonal antibodies with clinical indications (murine, rat, chimeric, humanized, human antibodies, immunotoxins, immunoadhesins, bispecific antibodies).
- IMGT Education (Aide-mémoire, Tutorials, Questions and answers, etc.): provides useful biological resources for students and includes figures and tutorials (in English and/or in French) in immunogenetics.
- IMGT Aide-mémoire: provides an easy access to information such as genetic code, splicing sites, amino acid structures, restriction enzyme sites, and so on.
- IMGT Index: is a fast way to access data when information has to be retrieved from different parts of the IMGT® site. For example, "allele" provides links to the IMGT Scientific chart rules for the allele description, and to the IMGT Repertoire "Alignments of alleles" and "Tables of alleles" (http://www.imgt.org).

4.4 IMGT® DATABASES, TOOLS, AND WEB RESOURCES FOR ANTIBODY GENOMICS

4.4.1 IMGT/GENE-DB Database

Antibody genomic data are managed in IMGT/GENE-DB (Giudicelli, Chaume, and Lefranc 2005), which is the comprehensive IMGT® genome database. In March 2009, IMGT/GENE-DB contained 1998 IG and TR genes and 3024 alleles [672 genes and 1242 alleles from human, 832 genes and 1264 alleles from mouse, 397 IGH genes and 400 alleles from rat, and 97 genes and 118 alleles from diverse other species (*Rattus norvegicus*)]. Based on the IMGT® CLASSIFICATION concept, all the human IMGT® gene names (Lefranc 2000a, 2000b; Lefranc and Lefranc 2001a, 2001b) approved by the HGNC in 1999 are available in IMGT/GENE-DB (France) and in Entrez Gene at NCBI (United States). All the mouse IMGT® gene and allele names and the corresponding IMGT reference sequences were provided to the Mouse Genome Informatics (MGI) Mouse Genome Database (MGD) in July 2002 and were presented by IMGT® at the 19th International Mouse Genome Conference IMGC 2005, in Strasbourg, France. IMGT/GENE-DB allows a query per gene and allele name. IMGT/GENE-DB interacts dynamically with IMGT/LIGM-DB (Giudicelli et al. 2006) to download and display human and mouse gene-related sequence data. Thus, all IMGT/LIGM-DB accession numbers of cDNA that express a given gene are displayed in the IMGT/GENE-DB entry in the "Known IMGT/LIGM-DB cDNA sequences" section, with direct links to the corresponding IMGT/LIGM-DB entries.

4.4.2 IMGT® Genome Analysis Tools for IG Genes

The IMGT® genome analysis tools manage the locus organization and gene location and provide the display of physical maps for the human and mouse IG loci. They allow viewing of IG genes in a locus (IMGT/GeneView, IMGT/LocusView), searching for clones (IMGT/CloneSearch), searching for IG genes in a locus (IMGT/GeneSearch) based on IMGT® gene names, functionality, or localization on the chromosome, and provide information on the clones that were used to build the locus contigs (accession numbers are from IMGT/LIGM-DB, gene names from IMGT/GENE-DB).

TABLE 4.3 Number of Functional Human Immunoglobulin Genes per Haploid Genome

Locus	Chromosomal Localization	Locus Size (kb)	V	D	J	C	Number of Functional Genes	Combinatorial (Range per Locus)[a]
IGH	14q32.33	1250	38–46	23	6	9[b]	76–84	$38 \times 23 \times 6 = 5244$ (m)
								$46 \times 23 \times 6 = 6348$ (M)
IGK	2p11.2	1820	34–38	0	5	1	40–44	$34 \times 5 = 170$ (m)
								$38 \times 5 = 190$ (M)
		500[c]	17–19[c]	0	5	1	23–25[c]	$17 \times 5 = 85$ (m)[c]
								$19 \times 5 = 95$ (M)[c]
IGL	22q11.2	1050	29–33	0	4–5	4–5	37–43	$29 \times 4 = 116$ (m)
								$33 \times 5 = 165$ (M)

Source: Lefranc and Lefranc (2001a).

[a]The range of the theoretical combinatorial diversity indicated takes into account the minimum (m) and the maximum (M) number of functional V, D, and J genes in each of the major IGH, IGK and IGL loci.

[b]In haplotypes with multigene deletion, the number of functional IGHC genes is five (deletions I, III, and V), six (deletions IV and VI), or eight (deletion II), per haploid genome. In haplotypes with multigene duplication or triplication, the exact number of functional IGHC genes per haploid genome is not known.

[c]In the rare IGKV haplotype without the distal V-CLUSTER.

IMGT/GeneFrequency provides IG gene histograms proportional to the number of IMGT/LIGM-DB rearranged sequences and localized at the V, D, and J gene positions along the loci.

4.4.3 IMGT Repertoire for IG Genome Data

The IMGT Repertoire genome data include chromosomal localizations, locus representations, locus description, germline gene tables, potential germline repertoires, number of functional IG genes in the loci (Table 4.3), lists of IG genes, and links between IMGT, HGNC, Entrez Gene, and OMIM, correspondence between nomenclatures (Lefranc and Lefranc 2001a).

4.5 IMGT® DATABASES, TOOLS, AND WEB RESOURCES FOR ANTIBODY GENETICS

4.5.1 IMGT/LIGM-DB Database

IMGT/LIGM-DB (Giudicelli et al. 2006) is the comprehensive IMGT® database of IG nucleotide sequences from human and other vertebrate species, with translation for fully annotated sequences. Created in 1989 by LIGM, Montpellier, France, on the Web since 1995, IMGT/LIGM-DB is the first and the largest IMGT® database. The first demonstration on the Internet took place at the nineth International Congress of Immunology in San Francisco (July 1995). In March 2009, IMGT/LIGM-DB contained 136,033 nucleotide sequences of IG and TR from 235 species. The unique source of data for IMGT/LIGM-DB is EMBL, which shares data with the other two generalist databases GenBank and DDBJ. IMGT/LIGM-DB sequence data are identified with the GEDI (for GenBank/EMBL/DDBJ and IMGT/LIGM-DB) accession number. Based on expert analysis, specific detailed annotations are added to IMGT flat files.

Since August 1996, the IMGT/LIGM-DB content closely follows the EMBL one for the IG and TR, with the following advantages: IMGT/LIGM-DB does not contain sequences that have previously been wrongly assigned to IG and TR; conversely, IMGT/LIGM-DB contains IG and TR entries that have disappeared from the generalist databases; thus, in 1999, IMGT/LIGM-DB detected the disappearance of 20 IG sequences that inadvertently had been lost by GenBank, and allowed the recuperation of these sequences in the generalist databases.

The IMGT/LIGM-DB annotations (gene and allele name assignment, labels) allow data retrieval not only from IMGT/LIGM-DB, but also from other IMGT® databases. As an example, the IMGT/GENE-DB entries provide the IMGT/LIGM-DB accession numbers of the IG and TR cDNA sequences that contain a given V, D, J, or C gene. The automatic annotation of rearranged human and mouse cDNA sequences in IMGT/LIGM-DB is performed by IMGT/Automat (Giudicelli et al. 2005), an internal Java tool that implements IMGT/V-QUEST and IMGT/JunctionAnalysis.

4.5.2 IMGT/PRIMER-DB Database

Standardized information on oligonucleotides (or Primers) and combinations of primers (Sets, Couples) for IG (and TR) are managed in IMGT/PRIMER-DB (Folch et al. 2004), the IMGT® oligonucleotide database on the Web since February 2002.

4.5.3 IMGT® Analysis Tools for IG Sequences

4.5.3.1 IMGT/V-QUEST The IMGT® tools for antibody genetics comprise IMGT/V-QUEST (Giudicelli, Chaume, and Lefranc 2004; Lefranc, 2004), for the identification of the V, D, and J genes and of their mutations, IMGT/JunctionAnalysis (Yousfi Monod et al. 2004; Lefranc, 2004) for the analysis of the V-J and V-D-J junctions which confer the antigen receptor specificity, IMGT/Allele-Align for the detection of polymorphisms, and IMGT/Phylogene (Elemento and Lefranc 2003) for gene evolution analyses. IMGT/V-QUEST (V-QUEry and STandardization) (http://www.imgt.org) is an integrated software for IG and TR (Giudicelli, Chaume, and Lefranc 2004; Lefranc, 2004). This tool, easy to use, analyzes an input IG (or TR) germline or rearranged variable nucleotide sequence (Fig. 4.4). IMGT/V-QUEST results comprise the identification of the V, D, and J genes and alleles and the nucleotide alignment by comparison with sequences from the IMGT reference directory, the delimitations of the FR-IMGT and CDR-IMGT based on the IMGT unique numbering, the protein translation of the input sequence, the identification of the JUNCTION, the description of the mutations and amino acid changes of the V-REGION and the two-dimensional (2D) IMGT Collier de Perles representation of the V-REGION or V-DOMAIN. The set of sequences from the IMGT reference directory, used for IMGT/V-QUEST can be downloaded in FASTA format from the IMGT® site.

4.5.3.2 IMGT/JunctionAnalysis IMGT/JunctionAnalysis (Yousfi Monod et al. 2004; Lefranc, 2004) is a tool developed by LIGM, complementary to IMGT/V-QUEST, which provides a thorough analysis of the V-J and V-D-J junction of IG and TR rearranged genes (Fig. 4.5). The JUNCTION extends from 2nd-CYS 104 to J-PHE or J-TRP 118 inclusive. J-PHE or J-TRP is easily identified for in-frame rearranged sequences when the conserved Phe/Trp-Gly-X-Gly motif of the J-REGION is present. The length of the CDR3-IMGT of rearranged V-J-GENE or V-D-J-GENE is a crucial piece of information. It is the number of amino acids or codons from position 105 to 117 (J-PHE or J-TRP noninclusive). CDR3-IMGT amino acid and codon numbers are according to the IMGT unique numbering for V-DOMAIN (Lefranc et al. 2003). IMGT/JunctionAnalysis identifies the D-GENE and allele involved in the IGH V-D-J rearrangements by comparison with the IMGT reference directory, and delimits precisely the P, N, and D regions (Lefranc and Lefranc 2001a). Results from IMGT/JunctionAnalysis are more accurate than those given by IMGT/V-QUEST regarding the D-GENE identification. Indeed, IMGT/JunctionAnalysis works on shorter sequences (JUNCTION), and with a higher constraint since the identification of the V-GENE and J-GENE and alleles is a prerequisite to perform the analysis. Several hundreds of junction sequences can be analyzed simultaneously.

4.5.3.3 IMGT/Allele-Align and IMGT/PhyloGene Other IMGT® Tools for antibody nucleotide sequence analysis comprise IMGT/Allele-Align that allows the comparison of two alleles highlighting the nucleotide (and amino acid) differences and IMGT/PhyloGene (Elemento and (*a*) **Detailed results for the IMGT/V-QUEST analysed sequences**

Number of analysed sequences: 4

user_seq_80 user_seq_42 user_seq_33 user_seq_22

This release of IMGT/V-QUEST uses IMGT/JunctionAnalysis for the analysis of the JUNCTION

Hyphens (-) show nucleotide identity, dots (.) represent gaps

---

Sequence number 1: user_seq_80

Sequence compared with the human IG set from the IMGT reference directory

```
>user_seq_80
gaggtgcagctggtggagtctgggcctgaggtgaagaagcctgggacctcagtgaaggtc
tcctgcaaggcttctggattcacctttactagctctgctgtgcagtgggtgcgacaggct
cgtggacaacgccttgagtggataggacggatcgtcgttggcagtggtaacacaaactac
gcacagaagttccaggaaagagtcaccattaccagggacatgtccacaagtacagcctac
atggagctgagcagcctgagatccgaggacacggccgtgtattactgtgcggcagattct
ggattattactatggttcggggagagcccttactactactacggtatggacgtctggggc
caagggaccacggtcaccgtctcgagt
```

Result summary:	**Productive IGH rearranged sequence** (no stop codon and in frame junction)		
V-GENE and allele	IGHV1-58*01	score = 1381	identity = **97,92%** (282/288 nt)
J-GENE and allele	IGHJ6*02	score = 229	identity = 85,48% (53/62 nt)
D-GENE and allele by IMGT/JunctionAnalysis	IGHD3-10*01	D-REGION is in reading frame 1	
[CDR1-IMGT.CDR2-IMGT.CDR3-IMGT] lengths and AA JUNCTION	[8.8.22]	CAADSGLLLWFGESPYYYYGMDVW	

---

1. Alignment for V-GENE and allele identification

Closest V-REGIONs (evaluated from the V-REGION first nucleotide to the 2nd-CYS codon)

```
                           Score       Identity
M29809 IGHV1-58*01         1381        97,92% (282/288 nt)
AB019438 IGHV1-58*02       1372        97,57% (281/288 nt)
X62109 IGHV1-3*01          1102        87,15% (251/288 nt)
X62107 IGHV1-3*02          1093        86,81% (250/288 nt)
M99637 IGHV1-8*01          1075        86,11% (248/288 nt)
```

Alignment with FR-IMGT and CDR-IMGT delimitations

```
                      <---------------------------------- FR1-IMGT --------------
user_seq_80           gaggtgcagctggtggagtctgggcct...gaggtgaagaagcctgggacctcagtgaag
M29809 IGHV1-58*01    c-aa-----------c-----------...------------------------------
AB019438 IGHV1-58*02  c-aa-----------c-----------...------------------------------
X62109 IGHV1-3*01     c----c-----t---c--------g--...------------------g-----------
X62107 IGHV1-3*02     c----t---------c--------g--...------------------g-----------
M99637 IGHV1-8*01     c--------------c--------g--...------------------g-----------
```

Figure 4.4 IMGT/V-QUEST tool for the analysis of rearranged IG and TR sequences. Two types of displays are available (only the top of the page is shown). (a) Detailed results for each sequence. (b) Synthesis for the analyzed sequences. In that display, the user sequences that express the same gene and allele are aligned with the germline sequence.

Lefranc 2003), an easy to use tool for phylogenetic analysis of IMGT standardized reference sequences.

4.5.3.4 IMGT Repertoire for IG Genetics Data The IMGT Repertoire polymorphism data are represented by "Alignments of alleles," "Tables of alleles," "Allotypes," "Protein displays," particularities in protein designations, IMGT reference directory in FASTA format, correspondence between chain and receptor IMGT designations.

(*b*) Synthesis for the IMGT/V-QUEST analysed sequences

Number of analysed sequences: 4

Summary table:

Sequence	V-GENE and allele	Functionality	V Score	V Identity	J-GENE and allele	D-GENE and allele	D reading frame	CDR-IMGT lengths	AA JUNCTION	JUNCTION frame
user_seq_80	IGHV1-58*01	Productive	1381	97,92% (282/288 nt)	IGHJ6*02	IGHD3-10*01	1	[12, 10, 22]	CAADSGLLLWFGESPYYYYGMDVW	in frame
user_seq_42	IGHV1-58*01	Productive	1435	100,00% (288/288 nt)	IGHJ4*02	IGHD1-26*01	1	[12, 10, 13]	CAAPPLVGATTIGYW	in frame
user_seq_33	IGHV1-58*01	Productive	1255	93,06% (268/288 nt)	IGHJ5*02	IGHD1-26*01	3	[12, 10, 14]	CAAERYSGSCCWFDPW	in frame
user_seq_22	IGHV1-58*01	Productive	1435	100,00% (288/288 nt)	IGHJ4*02	IGHD6-19*01	3	[12, 10, 13]	CAAEGEQWLANFDYW	in frame

Results of IMGT/JunctionAnalysis for : IGH junctions

Alignment with the closest alleles:

The analysed sequences are aligned with the closest allele (with number of aligned sequences in parentheses):

IGHV1-58*01(4)

Sequences aligned with IGHV1-58*01

1. Alignment for V-GENE

```
                            <---------------------------------- FR1-IMGT --------------
M29809 IGHV1-58*01          caaatgcagctggtgcagtctgggcct...gaggtgaagaagcctgggacctcagtgaag
user_seq_80                 g-gg-----------g-----------...------------------------------
user_seq_42                 ---------------------------...------------------------------
user_seq_33                 ---------g-----------------...-------g-g-------a------------
user_seq_22                 ---------------------------...------------------------------
```

Figure 4.4 (*Continued*).

Translation of the JUNCTIONs

Click on mutated (underlined) amino acid to see the original one:

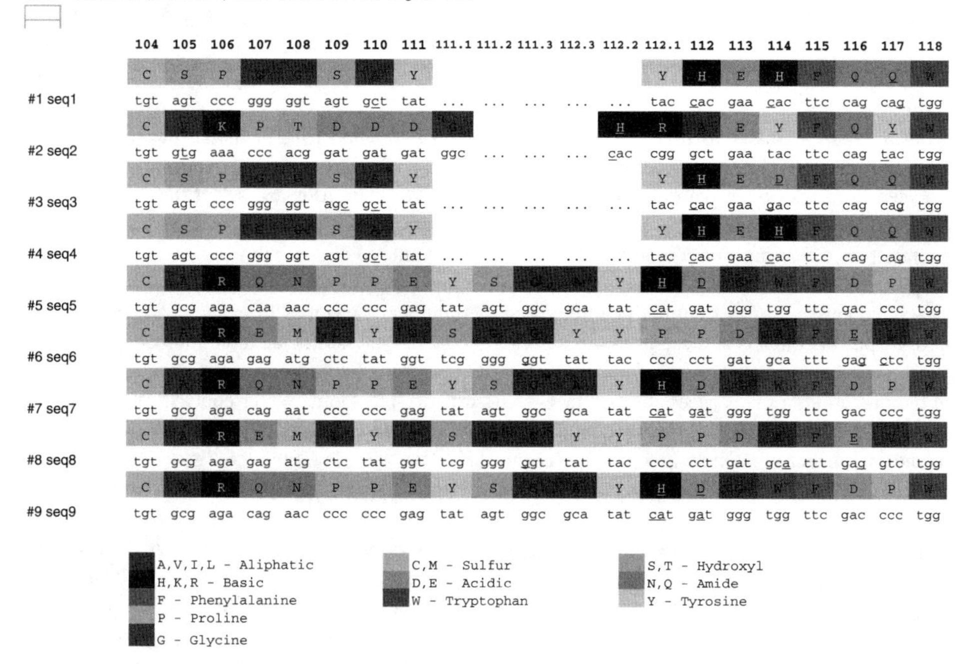

Figure 4.5 IMGT/JunctionAnalysis for the analysis of the V-J and V-D-J junctions. The translation of the junctions is shown with amino acids colored according to the IMGT physicochemical characteristics classes (Pommié et al. 2004). (See color insert.)

4.6 IMGT® DATABASES, TOOLS, AND WEB RESOURCES FOR ANTIBODY STRUCTURAL ANALYSIS

4.6.1 IMGT/3Dstructure-DB Database

4.6.1.1 IMGT/3Dstructure-DB Cards Antibody structural data are compiled and annotated in IMGT/3Dstructure-DB (Kaas, Ruiz, and Lefranc 2004), the IMGT® 3D structure database, created by LIGM, on the Web since November 2001. IMGT/3Dstructure-DB comprises IG, TR, MHC, and RPI with known 3D structures. In March 2009, IMGT/3Dstructure-DB contained 1570 atomic coordinate files, of which 1034 are from antibodies (622 from mouse, 267 from human, 23 humanized, 30 chimeric, 34 from camel, 10 from llama, 11 from rat, and 37 from diverse other sources). These coordinate files, extracted from the Protein Data Bank (PDB) (Berman et al. 2000), are renumbered according to the standardized IMGT unique numbering (Lefranc et al. 2003, 2005b, 2005c). The IMGT/3Dstructure-DB cards provide IMGT® annotations that comprise assignment of IMGT® genes and alleles, IMGT® chain and domain labels, IMGT Colliers de Perles on one layer and two layers, and for complexes, paratope/epitope analysis (for IG/antigen and TR/pMHC interactions) and "IMGT pMHC contact sites" (for MHC interface with peptide). IMGT/3Dstructure-DB downloadable flat files renumbered with the IMGT unique numbering, vizualization tools, and external links are available.

4.6.1.2 Contact Definition The contacts are provided by a local program. Atoms are considered to be in contact when no water molecule can exist between them. The contacts are classified as backbone-backbone, side chain-side chain, and backbone-side chain contacts, and are identified as covalent or noncovalent, as polar or nonpolar, or as hydrogen bond. A contact is a covalent link if the distance between the two atom centers is less than the sum of the Van der Waals radius of each atom. A noncovalent contact is either a polar contact, if the two atoms are oxygen and/or nitrogen, or nonpolar. A polar contact is a hydrogen bond if the distance between the atom centers is less than 3.5 Å and the hydrogen bond angles are correct. A hydrogen bond contact counts also for a polar contact and a noncovalent contact. The contact types and categories between domains (e.g., a VH domain with a V-KAPPA domain) are provided in "IMGT/3Dstructure-DB Domain contacts." The atom contacts are described at the residue and position level in "IMGT/3Dstructure-DB Residue@Position contacts."

4.6.1.3 IMGT/3Dstructure-DB Domain Contacts The "IMGT/3Dstructure-DB Domain contacts (overview)" provides information for a given IMGT/3Dstructure-DB entry on the contacts between domains or between domain and ligand (Fig. 4.6). The user can select the types of atom contacts (noncovalent, polar, hydrogen bond, nonpolar, covalent, and disulfide) and the atom contact pair categories (backbone-backbone, side chain-side chain, backbone-side chain, and side chain-backbone contacts). The resulting table provides (1) the number of residue pair contacts, the total number of residues from both partners, and the number of residues from each partner (involved in the types of contacts and categories as selected by the user); and (2) the total number of atom pair contacts and the detailed description of the contacts (as selected by the user). Clicking on "Details" in the first column gives access to the "IMGT/3Dstructure-DB Domain pair contacts" that gives the list of the Residue@Position pair contacts between one given domain and another domain (or ligand) in the entry. Clicking on any Residue@Position (R@P) (Fig. 4.6) gives access to the "IMGT/3Dstructure-DB Residue@Position contacts" described in the next paragraph.

4.6.1.4 IMGT/3Dstructure-DB Residue@Position Contacts The "IMGT/3Dstructure-DB Residue@Position contacts" gives the list of the Residue@Position that are in contact with a given Residue@Position. This page is created from the IMGT/3Dstructure-DB card by clicking on any amino acid either in a sequence or in an IMGT Collier de Perles, provided that this residue has 3D coordinates (amino acids without 3D coordinates are in italic in a sequence). The header of the "IMGT/3Dstructure-DB Residue@Position contacts" shows the complete identification of Residue@Position that comprises the position number, the residue name (with three letters and

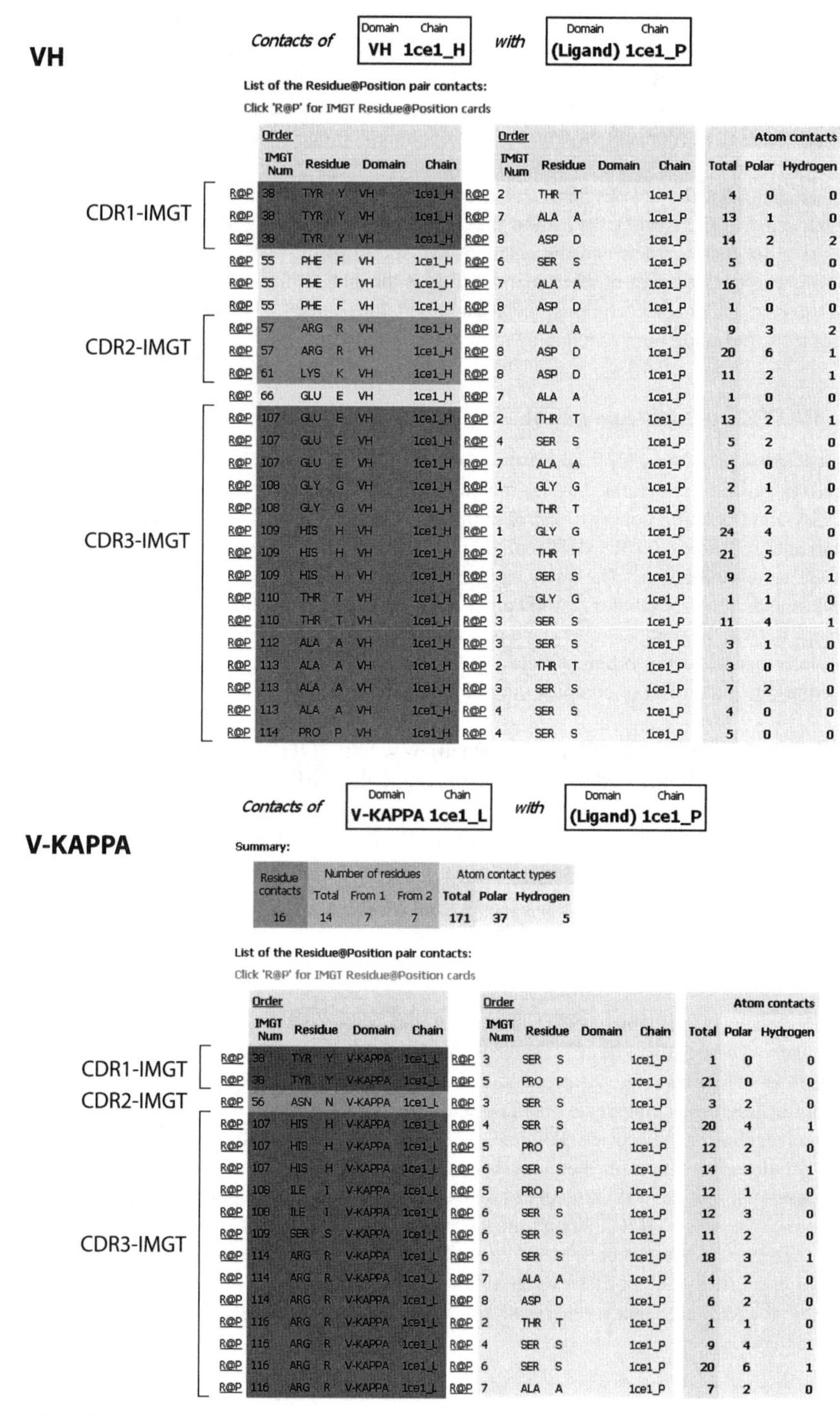

VH

Contacts of Domain VH Chain 1ce1_H *with* Domain (Ligand) Chain 1ce1_P

List of the Residue@Position pair contacts:
Click 'R@P' for IMGT Residue@Position cards

		Order IMGT Num	Residue		Domain	Chain		Order IMGT Num	Residue		Domain	Chain	Atom contacts Total	Polar	Hydrogen
CDR1-IMGT	R@P	38	TYR	Y	VH	1ce1_H	R@P	2	THR	T		1ce1_P	4	0	0
CDR1-IMGT	R@P	38	TYR	Y	VH	1ce1_H	R@P	7	ALA	A		1ce1_P	13	1	0
CDR1-IMGT	R@P	38	TYR	Y	VH	1ce1_H	R@P	8	ASP	D		1ce1_P	14	2	2
	R@P	55	PHE	F	VH	1ce1_H	R@P	6	SER	S		1ce1_P	5	0	0
	R@P	55	PHE	F	VH	1ce1_H	R@P	7	ALA	A		1ce1_P	16	0	0
	R@P	55	PHE	F	VH	1ce1_H	R@P	8	ASP	D		1ce1_P	1	0	0
CDR2-IMGT	R@P	57	ARG	R	VH	1ce1_H	R@P	7	ALA	A		1ce1_P	9	3	2
CDR2-IMGT	R@P	57	ARG	R	VH	1ce1_H	R@P	8	ASP	D		1ce1_P	20	6	1
CDR2-IMGT	R@P	61	LYS	K	VH	1ce1_H	R@P	8	ASP	D		1ce1_P	11	2	1
	R@P	66	GLU	E	VH	1ce1_H	R@P	7	ALA	A		1ce1_P	1	0	0
CDR3-IMGT	R@P	107	GLU	E	VH	1ce1_H	R@P	2	THR	T		1ce1_P	13	2	1
CDR3-IMGT	R@P	107	GLU	E	VH	1ce1_H	R@P	4	SER	S		1ce1_P	5	2	0
CDR3-IMGT	R@P	107	GLU	E	VH	1ce1_H	R@P	7	ALA	A		1ce1_P	5	0	0
CDR3-IMGT	R@P	108	GLY	G	VH	1ce1_H	R@P	1	GLY	G		1ce1_P	2	1	0
CDR3-IMGT	R@P	108	GLY	G	VH	1ce1_H	R@P	2	THR	T		1ce1_P	9	2	0
CDR3-IMGT	R@P	109	HIS	H	VH	1ce1_H	R@P	1	GLY	G		1ce1_P	24	4	0
CDR3-IMGT	R@P	109	HIS	H	VH	1ce1_H	R@P	2	THR	T		1ce1_P	21	5	0
CDR3-IMGT	R@P	109	HIS	H	VH	1ce1_H	R@P	3	SER	S		1ce1_P	9	2	1
CDR3-IMGT	R@P	110	THR	T	VH	1ce1_H	R@P	1	GLY	G		1ce1_P	1	1	0
CDR3-IMGT	R@P	110	THR	T	VH	1ce1_H	R@P	3	SER	S		1ce1_P	11	4	1
CDR3-IMGT	R@P	112	ALA	A	VH	1ce1_H	R@P	3	SER	S		1ce1_P	3	1	0
CDR3-IMGT	R@P	113	ALA	A	VH	1ce1_H	R@P	2	THR	T		1ce1_P	3	0	0
CDR3-IMGT	R@P	113	ALA	A	VH	1ce1_H	R@P	3	SER	S		1ce1_P	7	2	0
CDR3-IMGT	R@P	113	ALA	A	VH	1ce1_H	R@P	4	SER	S		1ce1_P	4	0	0
CDR3-IMGT	R@P	114	PRO	P	VH	1ce1_H	R@P	4	SER	S		1ce1_P	5	0	0

V-KAPPA

Contacts of Domain V-KAPPA Chain 1ce1_L *with* Domain (Ligand) Chain 1ce1_P

Summary:

Residue contacts	Number of residues Total	From 1	From 2	Atom contact types Total	Polar	Hydrogen
16	14	7	7	171	37	5

List of the Residue@Position pair contacts:
Click 'R@P' for IMGT Residue@Position cards

		Order IMGT Num	Residue		Domain	Chain		Order IMGT Num	Residue		Domain	Chain	Atom contacts Total	Polar	Hydrogen
CDR1-IMGT	R@P	38	TYR	Y	V-KAPPA	1ce1_L	R@P	3	SER	S		1ce1_P	1	0	0
CDR1-IMGT	R@P	38	TYR	Y	V-KAPPA	1ce1_L	R@P	5	PRO	P		1ce1_P	21	0	0
CDR2-IMGT	R@P	56	ASN	N	V-KAPPA	1ce1_L	R@P	3	SER	S		1ce1_P	3	2	0
CDR3-IMGT	R@P	107	HIS	H	V-KAPPA	1ce1_L	R@P	4	SER	S		1ce1_P	20	4	1
CDR3-IMGT	R@P	107	HIS	H	V-KAPPA	1ce1_L	R@P	5	PRO	P		1ce1_P	12	2	0
CDR3-IMGT	R@P	107	HIS	H	V-KAPPA	1ce1_L	R@P	6	SER	S		1ce1_P	14	3	1
CDR3-IMGT	R@P	108	ILE	I	V-KAPPA	1ce1_L	R@P	5	PRO	P		1ce1_P	12	1	0
CDR3-IMGT	R@P	108	ILE	I	V-KAPPA	1ce1_L	R@P	6	SER	S		1ce1_P	12	3	0
CDR3-IMGT	R@P	109	SER	S	V-KAPPA	1ce1_L	R@P	6	SER	S		1ce1_P	11	2	0
CDR3-IMGT	R@P	114	ARG	R	V-KAPPA	1ce1_L	R@P	6	SER	S		1ce1_P	18	3	1
CDR3-IMGT	R@P	114	ARG	R	V-KAPPA	1ce1_L	R@P	7	ALA	A		1ce1_P	4	2	0
CDR3-IMGT	R@P	114	ARG	R	V-KAPPA	1ce1_L	R@P	8	ASP	D		1ce1_P	6	2	0
CDR3-IMGT	R@P	116	ARG	R	V-KAPPA	1ce1_L	R@P	2	THR	T		1ce1_P	1	1	0
CDR3-IMGT	R@P	116	ARG	R	V-KAPPA	1ce1_L	R@P	4	SER	S		1ce1_P	9	4	1
CDR3-IMGT	R@P	116	ARG	R	V-KAPPA	1ce1_L	R@P	6	SER	S		1ce1_P	20	6	1
CDR3-IMGT	R@P	116	ARG	R	V-KAPPA	1ce1_L	R@P	7	ALA	A		1ce1_P	7	2	0

Figure 4.6 List of IMGT Residue@Position pair contacts between the VH and V-KAPPA domains of alemtuzumab with its ligand, a CD52 mimotope. The list shows that for VH and V-KAPPA the contacts involve positions in the three CDR-IMGT. The only FR-IMGT positions with contacts are the two anchor positions 55 and 66 in VH, but they do not have hydrogen bonds (IMGT/3Dstructure-DB entry: 1ce1) (Kaas et al. 2004).

eventually one letter abbreviation), the domain label, and the IMGT/3Dstructure-DB chain ID. Under the header are provided the original numbering in the PDB file, the residue full name, the secondary structure computed by Stride (Frishman and Argos 1995) (turn, coil, extended, 3-10 helix, etc.), the phi and psi angles (in degree), and the accessible surface area (ASA in square angstroms).

The user can select, for result display, the type of contacts (noncovalent, polar, hydrogen bond, nonpolar, covalent bond, or disulfide bond) and the atom contact pair category (backbone-backbone, side chain-side chain, backbone-side chain, and side chain-backbone atoms). The results are shown as a table with a list of the Residue@Position that are in contact. For each Residue@Position, the total number of atom pair contacts and the detailed description of the contacts, as selected by the user, are shown. The total number of atom pair contacts is the sum of noncovalent contacts and covalent bonds. Noncovalent contacts comprise polar and nonpolar contacts. Hydrogen bonds are included in polar contacts. Disulfide bonds are included in covalent bonds.

4.6.2 IMGT/StructuralQuery Tool

The IMGT/StructuralQuery tools allows the retrieval of IMGT/3Dstructure-DB entries using domain type and/or position structural criteria, including distance, phi and psi angles, accessible surface area (ASA), and position contacts, and amino acid sequence criteria, which are amino acids in the one letter code, positions, CDR-IMGT lengths, CDR-IMGT patterns (as a regular expression), species, and IMGT genes and groups. The query can be performed with a form or by typing a complex query that understands logical operators (AND and OR) and parentheses (that can be nested). A dynamic form helps to type the complex query and a quick description of each mnemonic is provided.

The information in the different fields and the complex query are combined to mine IMGT/3Dstructure-DB. Examples of complex queries are shown below.

group = IGHV AND length(CDR3) = 13: retrieves, from IMGT/3Dstructure-DB, structures that have a V-DOMAIN encoded by a gene belonging to the IGHV group with a CDR3-IMGT of 13 amino acids.

length(CDR1) = 8 AND aa(29V) = V: retrieves structures that have a V-DOMAIN with a CDR1-IMGT of 8 amino acids and a valine at position 29.

pattern(CDR3) = W AND length(CDR3) = 13 AND species = Human: retrieves structures that have a human V-DOMAIN with a CDR3-IMGT of 13 amino acids that contains a tryptophan (W).

In order to analyze the amino acid resemblances and differences between proteins appropriately, 11 IMGT® classes were defined for the chemical characteristics amino acid properties and used to set up IMGT Colliers de Perles reference profiles (Pommié et al. 2004). The IMGT Colliers de Perles reference profiles allow easy comparison of amino acid properties at each position whatever the domain, the chain, the receptor, or the species (Pommié et al. 2004). The IG variable and constant domains represent a privileged situation for the analysis of amino acid properties in relation with 3D structures, by the conservation of their 3D structure despite divergent amino acid sequences, and by the considerable amount of genomic (IMGT Repertoire), structural (IMGT/3Dstructure-DB), and functional data available. These data are not only useful to study mutations and allele polymorphisms, but are also needed to establish correlations between amino acids in the protein sequences and 3D structures, to analyze the domain interactions, and to determine amino acids potentially involved in the immunogenicity.

4.6.3 IMGT/DomainGapAlign

Antibody amino acid sequences can be analyzed per domain using the IMGT/DomainGapAlign tool. Several sequences of the same domain type (V or C) may be analyzed simultaneously (Fig. 4.7).

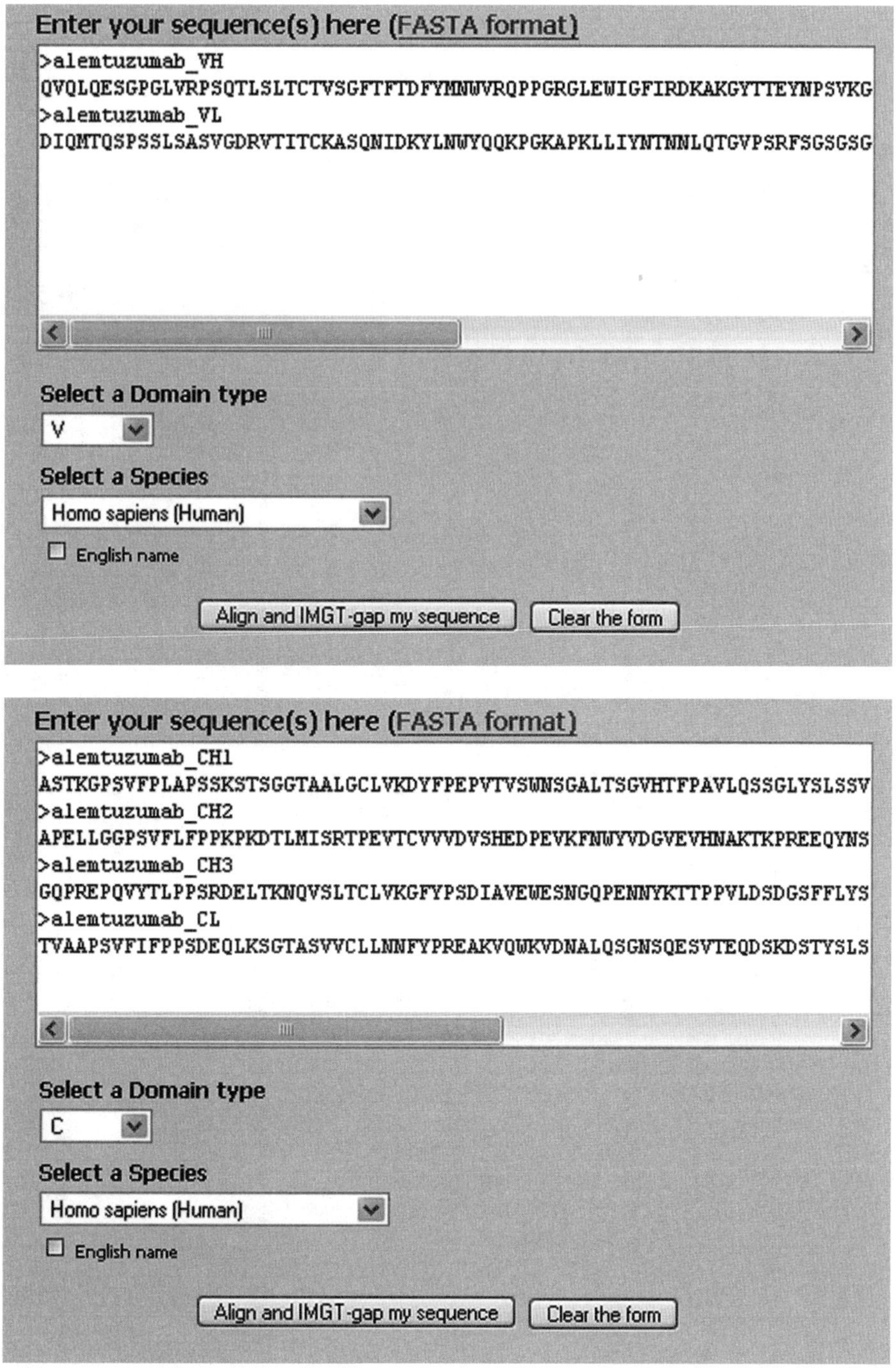

Enter your sequence(s) here (FASTA format)

```
>alemtuzumab_VH
QVQLQESGPGLVRPSQTLSLTCTVSGFTFTDFYMNWVRQPPGRGLEWIGFIRDKAKGYTTEYNPSVKG
>alemtuzumab_VL
DIQMTQSPSSLSASVGDRVTITCKASQNIDKYLNWYQQKPGKAPKLLIYNTNNLQTGVPSRFSGSGSG
```

Select a Domain type

V

Select a Species

Homo sapiens (Human)

☐ English name

Align and IMGT-gap my sequence | Clear the form

Enter your sequence(s) here (FASTA format)

```
>alemtuzumab_CH1
ASTKGPSVFPLAPSSKSTSGGTAALGCLVKDYFPEPVTVSWNSGALTSGVHTFPAVLQSSGLYSLSSV
>alemtuzumab_CH2
APELLGGPSVFLFPPKPKDTLMISRTPEVTCVVVDVSHEDPEVKFNWYVDGVEVHNAKTKPREEQYNS
>alemtuzumab_CH3
GQPREPQVYTLPPSRDELTKNQVSLTCLVKGFYPSDIAVEWESNGQPENNYKTTPPVLDSDGSFFLYS
>alemtuzumab_CL
TVAAPSVFIFPPSDEQLKSGTASVVCLLNNFYPREAKVQWKVDNALQSGNSQESVTEQDSKDSTYSLS
```

Select a Domain type

C

Select a Species

Homo sapiens (Human)

☐ English name

Align and IMGT-gap my sequence | Clear the form

Figure 4.7 IMGT/DomainGapAlign tool for the analysis of domain sequences. Several antibody domain sequences can be analyzed simultaneously per domain type (V or C). As examples, VH and VL of alemtuzumab are displayed for a V domain analysis (top), whereas CH1, CH2, CH3 of the IG heavy chain and CL of IG light chain are displayed for a C domain analysis (bottom).

V-DOMAIN

Gene and allele	Species	Domain	Smith-Waterman	Identity percentage	Overlap
IGHV4-59*01	*Homo sapiens*	1	494	73.0	100
IGKV1-33*01	*Homo sapiens*	1	551	86.3	95

```
                  FR1-IMGT                 CDR1-IMGT     FR2-IMGT          CDR2-IMGT                 FR3-IMGT
                   (1-26)                   (27-38)       (39-55)           (56-65)                  (66-104)

                  1         10        20        30         40        50        60          70        80        90        100
                  |.........|.........|......  ...|........ .|.........|.....  ....|.....  ....|.........|.........|.........|....  ....
alemtuzumab_VH    QVQLQESGP.GLVRPSQTLSLTCTVS GFTF....TDFY MNWVRQPPGRGLEWIGF IRDKAKGYTT EYNPSVK.GRVTMLVDTSKNQFSLRLSSVTAADTAVYYC ARE
IGHV4-59*01       QVQLQESGP.GLVKPSETLSLTCTVS GGSI....SSYY WSWIRQPPGKGLEWIGY IYYS...GST NYNPSLK.SRVTISVDTSKNQFSLKLSSVTAADTAVYYC AR.
(Homo sapiens)               R  Q            FTF     TDF  MN V     R     F  RDK    YT  E    V  G  ML            R

alemtuzumab_VL    DIQMTQSPSSLSASVGDRVTITCKAS QNI......DKY LNWYQQKPGKAPKLLIY NT.......N NLQTGVP.SRFSGSG..SGTDFTFTISSLQPEDIATYYC LQHISRPR
IGKV1-33*01       DIQMTQSPSSLSASVGDRVTITCQAS QDI......SNY LNWYQQKPGKAPKLLIY DA.......S NLETGVP.SRFSGSG..SGTDFTFTISSLQPEDIATYYC QQYDNLP.
(Homo sapiens)                           K    N        DK                     NT        N   Q                                         L HISR
```

C-DOMAIN

Gene and allele	Species	Domain	Smith-Waterman	Identity percentage	Overlap
IGHG1*01 or *02	*Homo sapiens*	1	634	100	98
IGHG1*01 or *02 or *03	*Homo sapiens*	2	742	100	110
IGHG1*01 or *02	*Homo sapiens*	3	738	100	107

```
                          A          AB       B          BC        C        CD     D                DE               E      EF    F            FG            G
                       (1-15)                (16-26)    (27-38)   (39-45)          (77-84)                           (86-96)     (97-104)    (105-118)     (119-125)

                     1        10         20        30        40                80                        90        100        110        120
                  87654321|........|.....123....|......  ...|......  .|.....1234567...|....12345677654321.....|......12...|....  .....|.......  ..|.....
alemtuzumab_CH1   ....ASTKGPSVFPLAPSSKSTS...GGTAALGCLVK DYFP..EPVT VSWNSGALTS....GVHTFPAVLQSS......GLYSLSSVVTVPSSSL...GTQTYIC NVNHKP..SNTKV DKKV
IGHG1*01          ....ASTKGPSVFPLAPSSKSTS...GGTAALGCLVK DYFP..EPVT VSWNSGALTS....GVHTFPAVLQSS......GLYSLSSVVTVPSSSL...GTQTYIC NVNHKP..SNTKV DKKV
(Homo sapiens)

alemtuzumab_CH2   ..APELLGGPSVFLFPPKPKDTLMI.SRTPEVTCVVV DVSHEDPEVK FNWYVDGVEVH...NAKTKPREEQYN......STYRVVSVLTVLHQDW..LNGKEYKC KVSNKA..LPAPI EKTISKAK
IGHG1*01          ..APELLGGPSVFLFPPKPKDTLMI.SRTPEVTCVVV DVSHEDPEVK FNWYVDGVEVH...NAKTKPREEQYN......STYRVVSVLTVLHQDW..LNGKEYKC KVSNKA..LPAPI EKTISKAK
(Homo sapiens)

alemtuzumab_CH3   ....GQPREPQVYTLPPSRDELT...KNQVSLTCLVK GFYP..SDIA VEWESNGQPEN...NYKTTPPVLDSD......GSFFLYSKLTVDKSRW..QQGNVFSC SVMHEA.LHNHYT QKSLSLSPGK
IGHG1*01          ....GQPREPQVYTLPPSRDELT...KNQVSLTCLVK GFYP..SDIA VEWESNGQPEN...NYKTTPPVLDSD......GSFFLYSKLTVDKSRW..QQGNVFSC SVMHEA.LHNHYT QKSLSLSPGK
(Homo sapiens)
```

Gene and allele	Species	Domain	Smith-Waterman	Identity percentage	Overlap
IGKC*01	*Homo sapiens*	1	659	100.0	102

```
                          A          AB       B          BC        C        CD     D                DE               E      EF    F            FG            G
                       (1-15)                (16-26)    (27-38)   (39-45)          (77-84)                           (86-96)     (97-104)    (105-118)     (119-125)

                     1        10         20        30        40                80                        90        100        110        120
                  87654321|........|.....123....|......  ...|......  .|.....1234567...|....12345677654321.....|......12...|....  .....|.......  ..|......
alemtuzumab_CL    ....RTVAAPSVFIFPPSDEQLK...SGTASVVCLLN NFYP..REAK VQWKVDNALQSG..NSQESVTEQDSKD.....STYSLSSTLTLSKADY..EKHKVYAC EVTHQG..LSSPV TKSFNRGEC
IGKC*01           ....RTVAAPSVFIFPPSDEQLK...SGTASVVCLLN NFYP..REAK VQWKVDNALQSG..NSQESVTEQDSKD.....STYSLSSTLTLSKADY..EKHKVYAC EVTHQG..LSSPV TKSFNRGEC
(Homo sapiens)
```

Figure 4.8 Results of IMGT/DomainGapAlign tool for the alemtuzumab domain sequences. The variable region of the VH and VL alemtuzumab domains are identified as being closest of *Homo sapiens* IGHV4-59*01 and IGKV1-33*01, respectively (Magdelaine-Beuzelin et al. 2007). The constant regions are 100 percent identical to those of the *Homo sapiens* IGHG1*01 and IGKC1*01. Dots indicate gaps according to the IMGT unique numbering for V-DOMAIN (Lefranc et al. 2003) and to the IMGT unique numbering for C-DOMAIN (Lefranc et al. 2005c).

IMGT/DomainGapAlign creates gaps in the user amino acid sequences, according to the IMGT unique numbering for V-REGION (or C-DOMAIN), identifies for each user sequence the closest germline V-REGION (or the closest C-DOMAIN) by comparison with the IMGT domain directory (Fig. 4.8), and provides the IMGT Collier de Perles of the V-REGION (or C-DOMAIN). For amino acid sequences that are already gapped according to the IMGT unique numbering, IMGT Colliers de Perles can also be obtained using the IMGT/Collier-de-Perles tool. Several displays are available, on one layer or two layers, and with colored positions by comparison with the conserved hydrophobic positions found in more than 50 percent of the analyzed sequences, and by comparison with the IMGT reference profiles of rearranged *Homo sapiens* IGHV, IGKV, and IGLV sequences (threshold of 80 percent for the hydropathy classes, the volume classes, or the physicochemical characteristic classes (Pommié et al. 2004).

4.6.4 IMGT Repertoire for Antibody Structural Data

The IMGT Repertoire structural data comprise IMGT Colliers de Perles (Lefranc and Lefranc 2001a; Ruiz and Lefranc 2002; Lefranc et al. 2003, 2005c), FR-IMGT, and CDR-IMGT lengths, and 3D representations of variable domains. This visualization permits rapid correlation between protein sequences and 3D data retrieved from the PDB.

4.7 CITING IMGT®

Authors who make use of the information provided by IMGT® should cite Lefranc et al. (2009) as a general reference for the access to and content of IMGT®, and quote the IMGT® home page URL, http://www.imgt.org.

4.8 CONCLUSION

Since July 1995, IMGT® has been available on the Web at http://www.imgt.org. IMGT® has an exceptional response with more than 150,000 requests a month. The information is of much value to clinicians and biological scientists in general. IMGT® databases, tools, and Web resources are queried extensively and used by scientists from both academic and industrial laboratories, who are equally distributed between the United States, Europe, and the remaining world. IMGT® is used in very diverse domains, and particularly in the field of antibody in: (1) fundamental and medical research (repertoire analysis of the IG antibody recognition sites in normal and pathological situations such as autoimmune diseases, infectious diseases, AIDS, leukemias, lymphomas, myelomas); (2) veterinary research (IG repertoires in domestic and wild-life species); (3) genome diversity and genome evolution studies of the adaptive immune responses; (4) structural evolution of the IgSF proteins which interact with antibodies such as Fc receptors; (5) biotechnology related to antibody engineering [single chain Fragment variable (scFv), phage displays, combinatorial libraries, chimeric, humanized, and human antibodies]; (6) diagnostics (clonalities, detection and follow-up of residual diseases); and (7) therapeutical approaches (immunotherapy).

ACKNOWLEDGMENTS

I thank Chantal Ginestoux for the figures and Gérard Lefranc for helpful discussion. I am deeply grateful to the IMGT® team for its expertise and constant motivation. IMGT® is currently supported by the CNRS, the Ministère de l'Enseignement Supérieur et de la Recherche, the Région Languedoc-Roussillon, Agence Nationale de la Recherche (BIOSYS, ANR-06-BYOS-005-01), and the EU ImmunoGrid (IST-028069) program.

REFERENCES

Baum, T.P., V. Hierle, N. Pascal, F. Bellahcene, D. Chaume, M.-P., Lefranc, E. Jouvin-Marche, P.N. Marche, and J. Demongeot. 2006. IMGT/GeneInfo: T cell receptor gamma TRG and delta TRD genes in database give access to all TR potential V(D)J recombinations. *BMC Bioinformatics* 7:224.

Benson, D.A., I. Karsch-Mizrachi, D.J. Lipman, J. Ostell, and D.L. Wheeler. 2007. GenBank. *Nucleic Acids Res.* 35:D21–D25.

Berman, H.M., J. Westbrook, Z. Feng, G. Gilliland, T.N. Bhat, H. Weissig, I.N. Shindyalov, and P.E. Bourne. 2000. The Protein Data Bank. *Nucleic Acids Res.* 28:235–242.

Bertrand, G., E. Duprat, M.-P. Lefranc, J. Marti, and J. Coste. 2004. Characterization of human FCGR3B*02 (HNA-1b, NA2) cDNAs and IMGT standardized description of FCGR3B alleles. *Tissue Antigens* 64:119–131.

Chaume, D., V. Giudicelli, and M.-P. Lefranc. 2001. IMGT-ML: A language for IMGT-ONTOLOGY and IMGT/LIGM-DB data. In: *CORBA and XML: Towards a bioinformatics integrated network environment.* Proceedings of NETTAB 2001, Network tools and applications in biology, 71–75, Genova, 2001, May 17–18.

Chaume, D., V. Giudicelli, K. Combres, and M.-P. Lefranc. 2003. IMGT-ONTOLOGY and IMGT-ML for immunogenetics and immunoinformatics. In Abstract book of the Sequence databases and Ontologies satellite event, European Congress in Computational Biology ECCB'2003, September 27–30, Paris, France, pp. 22–23.

Chothia, C., and A.M. Lesk. 1987. Canonical structures for the hypervariable regions of immunoglobulins. *J. Mol. Biol.* 196:901–917.

Cochrane, G., P. Aldebert, N. Althorpe, M. Andersson, W. Baker, A. Baldwin, K. Bates, S. Bhattacharyya, P. Browne, A. van den Broek, M. Castro, K. Duggan, R. Eberhardt, N. Faruque, J. Gamble, C. Kanz, T. Kulikova, C. Lee, R. Leinonen, Q. Lin, V. Lombard, R. Lopez, M. McHale, H. McWilliam, G. Mukherjee, F. Nardone, M.P. Garcia Pastor, S. Sobhany, P. Stoehr, K. Tzouvara, R. Vaughan, D. Wu, W. Zhu, and R. Apweiler. 2006. EMBL Nucleotide Sequence Database: Developments in 2005. *Nucleic Acids Res.* 34:D10–D15.

Duprat, E., Q. Kaas, V. Garelle, G. Lefranc, and M.-.P. Lefranc. 2004. IMGT standardization for alleles and mutations of the V-LIKE-DOMAINs and C-LIKE-DOMAINs of the immunoglobulin superfamily. In: S.G. Pandalai (ed.), *Recent research developments in human genetics*, vol. 2, 111–136. Trivandrum, Kerala, India: Research Signpost.

Duprat, E., M.-P. Lefranc, and O. Gascuel. 2006. A simple method to predict protein binding from aligned sequences: Application to MHC superfamily and beta2-microglobulin. *Bioinformatics* 22:453–459.

Duroux, P., Q. Kaas, X. Brochet, J. Lane, C. Ginestoux, M.-P. Lefranc, and V. Giudicelli. 2008. IMGT-Kaleidoscope, the Formal IMGT-ONTOLOGY paradigm. *Biochimie*, 90:570–583.

Eilbeck, K., S.E. Lewis, C.J. Mungall, M. Yandell, L. Stein, R. Durbin, and M. Ashburner. 2005. The Sequence Ontology: A tool for the unification of genome annotations. *Genome Biol.* 6(5):R44. Epub April 29, 2005. http://genomebiology.com/2005/6/5/R44.

Elemento, O., and M.-P. Lefranc. 2003. IMGT/PhyloGene: An on-line tool for comparative analysis of immunoglobulin and T cell receptor genes. *Dev. Comp. Immunol.* 27:763–779.

Folch, G., J. Bertrand, M. Lemaitre, and M.-P. Lefranc. 2004. IMGT/PRIMER-DB. In: M.Y. Galperin (ed.), *Database listing*. The Molecular Biology Database Collection: 2004 update. *Nucleic Acids Res.* 32:D3–D22.

Frigoul, A., and M.-P. Lefranc. 2005. MICA: Standardized IMGT allele nomenclature, polymorphisms and diseases. In: S.G. Pandalai (ed.), *Recent research developments in human genetics*, vol. 3, 95–145. Trivandrum, Kerala, India: Research Signpost.

Frishman, D., and P. Argos. 1995. Knowledge-based secondary structure assignment. *Proteins* 23:566–579.

Garapati, V.P., and M.-P. Lefranc. 2007. IMGT Colliers de Perles and IgSF domain standardization for T cell costimulatory activatory (CD28, ICOS) and inhibitory (CTLA4, PDCD1 and BTLA) receptors. *Dev. Comp. Immunol.* 31:1050–1072.

Giudicelli, V., D. Chaume, J. Jabado-Michaloud, and M.-P. Lefranc. 2005. Immunogenetics sequence annotation: The strategy of IMGT based on IMGT-ONTOLOGY. *Stud. Health Technol. Inform.* 116:3–8.

Giudicelli, V., D. Chaume, and M.-P. Lefranc. 2004. IMGT/V-QUEST, an integrated software program for immunoglobulin and T cell receptor V-J and V-D-J rearrangement analysis. *Nucleic Acids Res.* 32:W435–W440.

Giudicelli, V., D. Chaume, and M.-P. Lefranc. 2005. IMGT/GENE-DB: A comprehensive database for human and mouse immunoglobulin and T cell receptor genes. *Nucleic Acids Res.* 33:D256–D261.

Giudicelli, V., P. Duroux, C. Ginestoux, G. Folch, J. Jabado-Michaloud, D. Chaume, and M.-P. Lefranc. 2006. IMGT/LIGM-DB, the IMGT® comprehensive database of immunoglobulin and T cell receptor nucleotide sequences. *Nucleic Acids Res.* 34:D781–D784.

Giudicelli, V., and M.-P. Lefranc. 1999. Ontology for Immunogenetics: The IMGT-ONTOLOGY. *Bioinformatics* 12:1047–1054.

Kaas, Q., F. Ehrenmann, and M.-P. Lefranc. 2007. IG, TR and IgSf, MHC and MhcSF: What do we learn from the IMGT Colliers de Perles? *Brief. Funct. Genomic Proteomic*, 6:253–264.

Kaas, Q., and M.-P. Lefranc. 2005. T cell receptor/peptide/MHC molecular characterization and standardized pMHC contact sites in IMGT/3Dstructure-DB. Epub 5 0046, October 20, 2005. http://www.bioinfo.de/isb/2005/05/0046/. *In Silico Biol.* 5:505–528.

Kaas, Q., and M.-P. Lefranc. 2007. IMGT Colliers de Perles: Standardized sequence-structure representations of the IgSF and MhcSF superfamily domains. *Current Bioinformatics* 2:21–30.

Kaas, Q., M. Ruiz, and M.-P. Lefranc. 2004. IMGT/3Dstructure-DB and IMGT/StructuralQuery, a database and a tool for immunoglobulin, T cell receptor and MHC structural data. *Nucleic Acids Res.* 32:D208–D210.

Kabat, E.A., T.T. Wu, H.M. Perry, K.S. Gottesman, and C. Foeller. 1991. *Sequences of proteins of immunological interest.* Publication no. 91-3242. Washington, D.C.: National Institutes of Health.

Lefranc, M.-P. 1997. Unique database numbering system for immunogenetic analysis. *Immunol. Today* 18:509.

Lefranc, M.-P. 1999. The IMGT unique numbering for Immunoglobulins, T cell receptors and Ig-like domains. *Immunologist* 7:132–136.

Lefranc, M.-P. 2000a. Nomenclature of the human immunoglobulin genes. In: J.E. Coligan, B.E. Bierer, D.E. Margulies, E.M. Shevach, and W. Strober (eds), *Current protocols in immunology*, A.1P.1–A.1P.37. Hoboken, NJ: Wiley.

Lefranc, M.-P. 2000b. Nomenclature of the human T cell receptor genes. In: J.E. Coligan, B.E. Bierer, D.E. Margulies, E.M. Shevach, and W. Strober (eds), *Current protocols in immunology*, A.1O.1–A.1O.23. Hoboken, NJ: Wiley.

Lefranc, M.-P. 2004. IMGT, The International ImMunoGeneTics Information System®. *Antibody engineering: Methods and Protocols*, (ed.) B.K.C. Lo. Totowa, NJ: Humana. *Methods Mol. Biol.* 248:27–49.

Lefranc, M.-P. 2006. Web sites of interest to immunologists. In: J.E. Coligan, B.E. Bierer, D.E. Margulies, E.M. Shevach, and W. Strober (eds), *Current Protocols in Immunology*, A.1J.1–A.1J.74. Hoboken, NJ: Wiley.

Lefranc, M.-P. 2007. WHO-IUIS Nomenclature Subcommittee for Immunoglobulins and T cell receptors report. *Immunogenetics* 59:899–902.

Lefranc, M.-P. 2008. WHO-IUIS Nomenclature Subcommittee for Immunoglobulins and T cell receptors report. *Dev. Comp. Immunol.* 32:461–463.

Lefranc, M.-P., O. Clément, Q. Kaas, E. Duprat, P. Chastellan, I. Coelho, K. Combres, C. Ginestoux, V. Giudicelli, D. Chaume, and G. Lefranc. 2005a. IMGT-Choreography for Immunogenetics and Immunoinformatics. E pub 5 0006, December 24, 2004. http://www.bioinfo.de/isb/2004/05/0006/. *In Silico Biol.* 5:45–60.

Lefranc, M.-P., E. Duprat, Q. Kaas, M. Tranne, A. Thiriot, and G. Lefranc. 2005b. IMGT unique numbering for MHC groove G-DOMAIN and MHC superfamily (MhcSF) G-LIKE-DOMAIN. *Dev. Comp. Immunol.* 29:917–938.

Lefranc, M.-P., V. Giudicelli, C. Ginestoux, J. Jabado-Michaloud, G. Folch, F. Bellahcene, Y. Wu, E. Gemrot, X. Brochet, J. Lane, L. Regnier, F. Ehrenmann, G. Lefranc, and P. Duroux. 2009. IMGT®, the international ImMunoGeneTics information system®. *Nucleic Acids Res.* 37:D1006–D1012.

Lefranc, M.-P., and G. Lefranc. 2001a. *The Immunoglobulin FactsBook.* London: Academic Press, 1–458.

Lefranc, M.-P., and G. Lefranc. 2001b. *The T cell receptor FactsBook.* London: Academic Press, 1–398.

Lefranc, M.-P., C. Pommié, Q. Kaas, E. Duprat, N. Bosc, D. Guiraudou, C. Jean, M. Ruiz, I. Da Piedade, M. Rouard, E. Foulquier, V. Thouvenin, and G. Lefranc. 2005c. IMGT unique numbering for immunoglobulin and T cell receptor constant domains and Ig superfamily C-like domains. *Dev. Comp. Immunol.* 29:185–203.

Lefranc, M.-P., C. Pommié, M. Ruiz, V. Giudicelli, E. Foulquier, L. Truong, V. Thouvenin-Contet, and G. Lefranc. 2003. IMGT unique numbering for immunoglobulin and T cell receptor variable domains and Ig superfamily V-like domains. *Dev. Comp. Immunol.* 27:55–77.

Letovsky, S.I., R.W. Cottingham, C.J. Porter, and P.W. Li. 1998. GDB: The Human Genome Database. *Nucleic Acids Res.* 26:94–99.

Magdelaine-Beuzelin, C., Q. Kaas, V. Wehbi, M. Ohresser, R. Jefferis, M.-P. Lefranc, and H. Watier. 2007. Structure-function relationships of the variable domains of monoclonal antibodies approved for cancer treatment. *Crit. Rev. Oncol. Hematol.* 64:210–225.

Maglott, D., J. Ostell, K.D. Pruitt, and T. Tatusova. 2007. Entrez Gene: Gene-centered information at NCBI. *Nucleic Acids Res.* 35:D26–D31.

Okubo, K., H. Sugawara, T. Gojobori, and Y. Tateno. 2006. DDBJ in preparation for overview of research activities behind data submissions. *Nucleic Acids Res.* 34:D6–D9.

Pommié, C., S. Sabatier, G. Lefranc, and M.-P. Lefranc. 2004. IMGT standardized criteria for statistical analysis of immunoglobulin V-REGION amino acid properties. *J. Mol. Recognit.* 17:17–32.

Robinson, J., M.J. Waller, P. Parham, N. de Groot, R. Bontrop, L.J. Kennedy, P. Stoehr, and S.G. Marsh. 2003. IMGT/HLA and IMGT/MHC sequence databases for the study of the major histocompatibility complex. *Nucleic Acids Res.* 31:311–314.

Ruiz, M., and M.-P. Lefranc. 2002. IMGT gene identification and Colliers de Perles of human immunoglobulins with known 3D structures. *Immunogenetics* 53:857–883.

Satow, Y., G.H. Cohen, E.A. Padlan, and D.R. Davies. 1986. Phosphocholine binding immunoglobulin Fab McPC603. *J. Mol. Biol.* 190:593–604.

Wain, H.M., E.A. Bruford, R.C. Lovering, M.J. Lush, M.W. Wright, and S. Povey. 2002. Guidelines for human gene nomenclature. *Genomics* 79:464–470.

Yousfi Monod, M., V. Giudicelli, D. Chaume, and M.-P. Lefranc. 2004. IMGT/JunctionAnalysis: The first tool for the analysis of the immunoglobulin and T cell receptor complex V-J and V-D-J JUNCTIONs. *Bioinformatics* 20:i379–i385.

PART II

ANTIBODY SOURCES

CHAPTER 5

Human Antibodies from Transgenic Mice

NILS LONBERG

ABSTRACT

Transgenic mice comprising germline configuration human immunoglobulin gene loci, and inactivated endogenous mouse loci, can generate diverse human sequence antibody repertoires. The human transgenes functionally replace the inactivated mouse loci, so that the mice can be used as a drug discovery tool for generating high affinity human monoclonal antibodies. Because these antibodies are derived from human gene sequences, they offer an alternative to the humanization and phage display methods that have been employed for reducing the immunogenicity of rodent antibodies. Two of these transgenic mouse antibodies, panitumumab and ustekinumab, have now been approved for human therapeutic use, and over 40 additional such antibodies are now in human clinical testing. Reported data from these human clinical tests supports the hypothesis that transgenic mouse-derived antibodies can be well tolerated and clinically active.

Therapeutic Monoclonal Antibodies: From Bench to Clinic. Edited by Zhiqiang An

5.1 INTRODUCTION

As a class of therapeutic agents, monoclonal antibodies (MAbs) promise broad applicability across a variety of different disease indications. Potential attributes of this drug class include diversity, tolerability, potency, and slow *in vivo* clearance. Diversity provides access to multiple targets. Tolerability is evident from the fact that natural antibodies do not generally provoke immune responses, despite their sequence diversity. Potency is augmented by Fc-mediated complement killing and recruitment of cellular effector functions. In addition, binding to the neonatal Fc receptor (FcRn) provides for very long *in vivo* residence time, which is of critical importance for a biological drug, which must usually be delivered parenterally. Despite these potential attributes, the first MAb therapeutics tested in human clinical trails did not show great promise. Early antibody-based drugs were derived from rodents, and patient immune responses to these foreign protein sequences made it readily apparent that these agents would only be used in niche indications (Goldstein et al. 1985; Pendley, Schantz, and Wagner 2003; Kuus-Reichel et al. 1994; Baert et al. 2003). For MAbs to become a significant component of the therapeutic arsenal it would be necessary to obtain molecules that more closely resemble authentic human antibodies.

5.2 SOLUTIONS TO THE PROBLEM OF IMMUNOGENICITY

The first set of solutions to the problem of MAb immunogenicity came from protein engineering. Advances in molecular biology, involving the manipulation of gene sequences *in vitro*, and the expression of these manipulated sequences in bacterial, fungal, and mammalian cell culture systems, provided methods for reengineering rodent antibodies to partially replace the rodent antibody sequences with functionally equivalent human amino acid sequences, thus reducing the overall immunogenicity without destroying the recognition properties of the original antibody (Morrison et al. 1984; Jones et al. 1986). As of March 2009, 17 of these reengineered rodent MAbs have been approved for therapeutic use in the United States. Molecular biology tools also allow for the generation and screening of very large libraries of diverse sequences, thus providing a laboratory analog of the natural *in vivo* assembly of human antibody repertoires (McCafferty et al. 1990). While the synthetic MAbs isolated from these libraries are typically of low to moderate affinity, additional techniques have been developed that mirror the *in vivo* process of affinity maturation, allowing for the generation of completely synthetic high affinity antibodies derived *in vitro* from human sequences (Barbas and Burton 1996; Razai et al. 2005). These synthetic MAbs have been commonly referred to as "fully human sequence" antibodies, despite the presence of laboratory-introduced amino acid residues necessary for high affinity. One of these "fully human sequence" antibodies, adalimumab, has now been approved in the United States (van de Putte et al. 2004). With the 2006 approval of the transgenic mouse-derived panitumumab, an alternative source of "fully human sequence" antibodies has arrived (Jakobovits et al. 2007; Gibson, Ranganathan, and Grothey 2006). Just as advances in protein engineering led to methods for reducing the immunogenicity of preexisting rodent antibodies, advances in techniques for the manipulation of the germline of mammalian embryos led to the creation of genetically engineered strains of mice that directly generate high affinity human sequence antibodies *de novo* in response to immunization. These advances are outlined in the next section.

5.3 GENETICALLY ENGINEERED MICE

Basic research in embryology and molecular biology led to the development in the early 1980s of a set of tools for the manipulation of the mouse genome (Nagy et al. 2003). The generation of genetically engineered mice by direct microinjection of cloned DNA sequences into the pronuclei of single-cell half-day embryos was reported by several groups in 1981 (Gordon and Ruddle 1981; Costantini and Lacy 1981; Brinster et al. 1981; Harbers, Jahner, and Jaenisch 1981; E. Wagner, Stewart, and

Mintz 1981; T. Wagner et al. 1981). The microinjected DNA constructs, which insert into mouse chromosomes and are propagated through the germline, could include transcriptional regulatory sequences to direct expression to restricted differentiated cell types, including B-cell expression of antibody genes (Brinster et al. 1983). This first report of an expressed immunoglobulin gene in transgenic mice involved a very small transgene; however, despite the fact that very fine glass needles are employed for pronuclear microinjection, the sheer forces experienced by the injected DNA do not prevent the use of this technique for introducing much larger (>100 kb) transgenes into the mouse germline (Costantini and Lacy 1981; Taylor et al. 1992; Schedl et al. 1993; Lonberg and Huszar 1995; Fishwild et al. 1996).

Because microinjected transgenes integrate relatively randomly over a large number of potential sites within the mouse genome, it does not provide for easy manipulation of specific endogenous mouse genes. Microinjection could generate mice expressing human genes, but the mouse ortholog was typically still active. This technical hurdle was overcome with the development of positive-negative selection vectors that allow for the selection and screening of specifically targeted homologous recombination events in cultured cells, and with the parallel development of embryonic stem (ES) cell lines that could be cultured and manipulated *in vitro* and reintroduced into 3.5-day-old blastocyst-stage embryos to populate the germline of the resulting chimeric mice. The combination of these two technologies led to the generation of strains of engineered mice comprising specifically targeted modifications of their germlines (Mansour, Thomas, and Capecchi 1988; Zijlstra et al. 1989; Schwartzberg, Goff, and Robertson 1989). The most commonly introduced specific modification leads to the inactivation of an endogenous gene and the creation of what are commonly referred to as gene knockout mouse strains. Gene knockout technology has proved to be of enormous value for basic research, and applied to the endogenous mouse immunoglobulin loci, important for the development of transgenic mouse platforms for human antibody drug discovery.

In addition to applications for modifying endogenous mouse genes, ES cells have also proved useful as an alternative to pronuclear microinjection for the introduction of large DNA clones such as YAC clones (Strauss et al. 1993; Choi et al. 1993; Jakobovits et al. 1993; Davies et al. 1993). Very large human chromosome fragments have also been introduced into the mouse germline using ES cell technology. In this approach, called microcell-mediated chromosome transfer (MMCT), human fibroblast-derived microcells are fused with mouse ES cells resulting in pluripotent cell lines having a single human chromosome or chromosome fragment, including a centromere and both telomeres, that replicates and assorts during cell division without insertion into an endogenous mouse chromosome (Tomizuka et al. 1997).

It was recognized by the mid-1980s that these advances in germline manipulation could be exploited to produce a genetically engineered mouse platform for discovering human antibodies (Alt, Blackwell, and Yancopoulos 1985; Buttin 1987). Creation of a transgenic mouse that uses introduced transgene sequences as a substrate for the generation of an antibody repertoire asks more of the transgene than, for example, the creation of a mouse that expresses human albumin in its liver. Unlike the albumin-secreting liver cell, the B-cell that expresses the immunoglobulin transgene is itself dependent on those transgene sequences to encode components of membrane-bound receptor complexes that are necessary for the development and survival of the cell. The structure of these receptor complexes, and their role in B-cell development, is discussed in the following sections.

5.4 THE ROLE OF IMMUNOGLOBULIN GENES IN B-CELL DEVELOPMENT

5.4.1 Antibody Structure

Most marketed and development-stage antibody drugs are intact, IgG class, antibodies. IgG antibodies are large (MW ~150 kilodalton) tetrameric proteins comprised of two heavy chains and two light chains. The light chains each comprise two immunoglobulin superfamily (Ig) domains, and the heavy chains each comprise four. The C-terminal half of the two heavy chains combine to form the

Fc region, which mediates effector functions through interaction with lymphocyte Fc receptors and serum complement proteins. This part of the molecule is also responsible for the long *in vivo* half-life of IgG antibodies, mediated through interaction with the neonatal Fc receptor, FcRn. FcRn is widely expressed throughout the body and prevents pinocytosis-mediated IgG catabolism by recycling internalized antibody molecules back to the surface. The N-terminal light and heavy chain Ig domains combine to form the variable region, which includes six different loop structures, three from each chain, which together form the antigen combining site. These loops, designated heavy chain complementarity determining regions (CDRs) 1 through 3 and light chain CDR1 through 3, comprise most of the antigen contact residues and include the greatest sequence variability between individual antibodies.

5.4.2 B-Cell Development

The mature antibody repertoire is not directly encoded in the germline. Instead, the DNA sequences of the immunoglobulin heavy and light chain loci undergo a variety of rearrangement and modification steps during B-cell development, so that individual mature B-cells can express individually tailored antibodies having specific binding properties that are selected-for during adaptive immune responses (Fig. 5.1). This process comprises editing checkpoints where B-cells with rearranged and modified immunoglobulin genes encoding nonfunctional, unstable, or autoreactive antibodies are deleted.

Early in B-cell development, proB-cells first rearrange the heavy chain locus to join a single diversity (D) segment to a single joining (J) segment. The proB-cell then undergoes a second rearrangement step that joins a single heavy chain V gene segment to the already joined D and J segments. This second rearrangement marks the transition of the proB-cell to the preB-cell stage, and the resulting heavy chain variable region is expressed as a component of the preB-cell receptor complex (preBCR, Fig. 5.2). In addition to the recombined heavy chain protein, the preBCR includes a surrogate light chain comprised of two different germline encoded proteins, λ5 and VpreB, as well as membrane-associated signaling molecules Igα and Igβ (Melchers 2005).

The expressed preB-cell repertoire draws on the combinatorial diversity that resides within the library of individual germline V, D, and J gene segments. In the human heavy chain locus there are approximately 50 V gene segments, 25 D gene segments, and 6 J gene segments that contribute to this diversity. Because the process of D to J and V to DJ joining is imprecise, there is also a second layer of diversity added on top of the combinatorial diversity encoded in the germline. Nucleotide

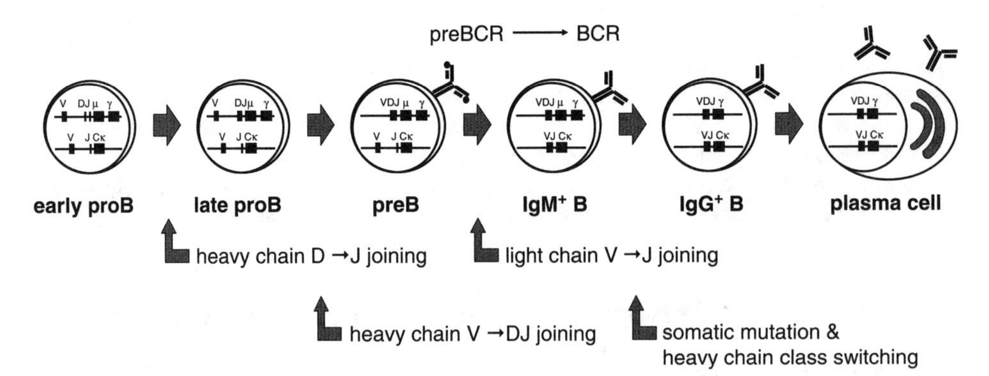

Figure 5.1 Immunoglobulin locus rearrangement and modification events in B-cell development. During the course of B-cell ontogeny and maturation, germline configuration immunoglobulin heavy and light chain gene loci undergo a series of rearrangement events to assemble a repertoire of preB-cell receptors (preBCR), B-cell receptors, and secreted antibodies. T-cell-mediated adaptive immune responses are associated with further modification of the immunoglobulin genes through somatic hypermutation and heavy chain class switching.

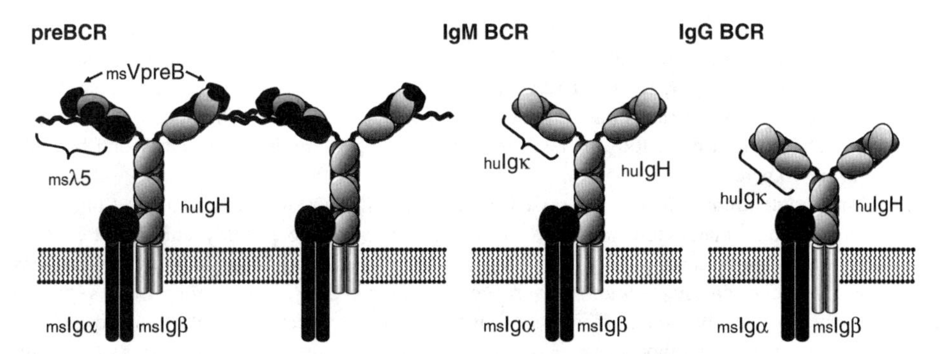

Figure 5.2 Hybrid PreB- and B-cell receptor complexes in human immunoglobulin transgenic mice. The preB-cell receptor (preBCR) and B-cell receptor (BCR) complexes in human immunoglobulin transgenic mice are creoss-species hybrids that comprise both human and mouse encoded protein chains. In the preBCR, the surrogate light chains, VpreB and λ5, and the signaling molecules Igα and Igβ are encoded by the endogenous mouse genes. In the BCR, Igα and Igβ are encoded by the endogenous mouse genes. The preBCR complexes are depicted as associating through interactions with λ5, according to the model of Bankovich et al. (2007).

deletion during joint formation provides variability at the 5′ and 3′ ends of the D segment, and the 5′ end of the J segment. Furthermore, short templated (P) and nontemplated (N) sequences can be added at each of the joints (Benedict et al. 2000). The V to D and D to J joints fall within the heavy chain CDR3 loop, which is the structural feature that comprises the greatest diversity within the antibody repertoire. However, the preB-cell repertoire is not simply a random output of this combinatorial and junctional diversity. It is instead shaped and edited by signaling through the preBCR. Correct preBCR signaling is important for maturation and survival of preB-cells, as evidenced by the blockade in B-cell development seen in mouse and human genetic deficiencies in λ5 or VpreB (Shimizu et al. 2002). And, because the preBCR distinguishes between different recombined heavy chain variable region structures, the heavy chain repertoire is partially shaped by editing even before an actual light chain is formed. This editing process results in a preB repertoire with V, D, and J segment usage similar to that of the B-cell repertoire (Milili et al. 1996; Meffre et al. 2001). Structural studies suggest the λ5/VpreB surrogate light chain directly contacts the hypervariable antigen-combining portion of the rearranged heavy chain variable region to form a complex that can then form oligomers with adjacent preBCR complexes (Bankovich et al. 2007). In this model, the oligomerized complexes transmit an intracellular signal that is necessary for cell survival and maturation. The model is consistent with the observed cell autonomy of preBCR signaling, and with data suggesting that the site of activity of the preBCR complex may be within intracellular membranes such as the endoplasmic reticulum and the trans-Golgi network, rather than the cell surface (Guloglu and Roman 2006). Because the direct contacts between the surrogate light chains and the heavy chain CDR3 residues may be critical for selecting a subset of rearranged heavy chain structures for survival, sequence differences between the mouse and human λ5 and VpreB proteins could account for observed species differences in CDR3 repertoires. Nevertheless, a human λ5 transgene has been shown to functionally rescue λ5-deficient mice (Donohoe et al. 2000).

The preB to immature B-cell transition is marked by light chain V to J joining, and the formation of a functional cell surface B-cell receptor (BCR). At this stage, the V(D)J recombination machinery is shut down, contributing to allelic exclusion, and providing for BCR monospecificity of individual B-cells. The functional rearranged heavy and light chain loci appear to be marked by chromatin accessibility and demethylation directed by cis-acting enhancer sequences (Inlay et al. 2002), and this marked state appears to be maintained in later stages of B-cell development, leading to preferential somatic mutation of active alleles (Fraenkel et al. 2007). The maintenance of the immature B-cell phenotype, characterized by surface expression of certain B-cell markers such as CD20 and CD23, and by silencing of genes associated with V(D)J recombination, such as Rag-1, Rag-2, and TdT, requires signaling through a functional BCR comprising IgM heavy and κ or λ light chains (Tze et al. 2005). In humans, the light chain combinatorial diversity comes from two separate loci, λ and κ, comprising about 31 and 40 V gene segments, 4 and 5 J gene segments, and 4 and 1 C gene segments, respectively (Williams et al. 1996; Tomlinson et al. 1995; Cook and Tomlinson 1995). While the λ locus contributes to only about 1 to 5 percent of the adult mouse B-cell repertoire, in humans the contribution is approximately 40 percent.

The immunoglobulin molecules expressed by immature B-cells, which have not yet undergone T-cell-driven affinity maturation and class switching, comprise the primary antibody repertoire. Primary repertoire antibodies typically have low to moderate binding affinities for individual antigens, with dissociation constants in the micromolar range. Heavy chain CDR3 lengths for primary repertoire antibodies are also typically longer than those found in secondary repertoire antibodies (Rosner et al. 2001). Primary repertoire antibodies may play a role in immediate, innate, immune responses to pathogens; however, the role of the primary repertoire in adaptive immune responses is primarily through its contribution to the BCR repertoire. The relatively low affinity V regions of this repertoire are sufficient for recognition and internalization of antigens. The internalized antigens are then proteolytically processed and presented on the B-cell surface by class II major histocompatibility complex (MHC) molecules. In an adaptive immune response, activated helper T-cells having T-cell receptors that recognize B-cell antigen/MHC complexes can initiate a program of B-cell maturation that leads to the generation of high affinity derivatives of primary repertoire antibodies. Two different types of gene modification events take place during this process: class switching and somatic mutation. Class switching involves the deletion of heavy chain Cμ and Cδ constant region gene segments through recombination between so-called switch sequences, located upstream of the μ gene and upstream of each of the Cγ, Cε, and Cα genes, leading to the expression of IgG, IgE, and IgA antibodies. Somatic mutation involves the introduction of single nucleotide substitutions into the recombined heavy and light chain V gene segments. Many of these substitutions encode deleterious amino acid replacements that either destabilize the overall structure of the BCR or destroy its ability to bind and internalize antigen, either of which can result in growth arrest or cell death. A fraction of the introduced mutations will provide for higher affinity interactions between the BCR and antigen, and these mutant B-cells will be selected for in an evolutionary process that generates the high affinity secondary antibody repertoire. This process of affinity maturation can improve the binding affinity of a V region by several orders of magnitude, resulting in antibodies with nanomolar to picomolar dissociation constants.

Class switching and somatic mutation are not mechanistically linked, and each can take place independently of the other; however, they are temporally associated within the same cell type, and they are both dependent on the activity of the AID gene (Muramatsu et al. 2000; Barreto et al. 2003). This correlation makes class switching, and the expression of IgG isotype antibodies in particular, a useful surrogate marker for identifying B-cells that may express those high affinity secondary repertoire antibodies which are the best candidates for therapeutic MAbs.

In addition to providing for expression of secreted IgG antibodies, class switching also changes the structure of the BCR for mature B-cells. One of these changes involves the replacement of the relatively short, three amino acid, cytoplasmic domain of the IgM BCR with the extended IgG gene encoded cytoplasmic domain of the IgG BCR (Fig. 5.2). This extended cytoplasmic domain appears to alter BCR signaling responses to antigen so as to promote differentiation to high IgG-secreting plasma cells (Horikawa et al. 2007). Functional signaling through the class switched BCR has also been shown to be just as critical for survival of mature B-cells as IgM BCR signaling is for immature B-cells (Kraus et al. 2004).

In addition to the role of immunoglobulin-encoded BCR components in directing and maintaining B-cell development and survival, secreted IgM also appears to have a feedback effect. Analysis of knockout mice, in which the sequences encoding the μ secretory tailpiece and polyadenylation site have been deleted, show reduced survival of plasma cells (Kumazaki et al. 2007), reduced immune responses to foreign antigens, and reduced tolerance, as manifested by an increased propensity for autoantibodies and even autoimmune disease (Ehrenstein, Cook, and Neuberger 2000).

5.5 HUMAN IMMUNOGLOBULIN TRANSGENIC MICE

For a transgenic mouse to express a human antibody repertoire, the introduced human transgene sequences must functionally replace the role of endogenous mouse immunoglobulin genes in B-cell development as outlined above. Given the complexity of immunoglobulin gene rearrangement and modification required for the expression of an antibody repertoire, and given the critical roles in B-cell development and survival described above for the proteins encoded by those genes, it is perhaps surprising that it is possible to generate such mice. Human immunoglobulin gene sequences must direct chromatin accessibility and marking of the transgene locus, and proper V(D)J joining, somatic mutation, and class switching. The encoded proteins must also form functional preBCR and BCR complexes. Finally, the secreted human immunoglobulins may have feedback effects on downstream B-cell responses. However, it turns out that even relatively small transgenes comprising limited combinatorial repertoires can functionally replace the endogenous mouse gene loci to generate high affinity antibodies (Xu and Davis 2000). Despite the fact that the preBCR and BCR complexes in these animals is a cross-species hybrid (Fig. 5.2), different B-cell compartments are populated, heavy and light chain repertoires are generated, and class switching and affinity maturation takes place. In some of these systems the overall B-cell numbers are reduced; nevertheless, the resulting antibodies have qualities of affinity, stability, and low immunogenicity that are unlikely to be improved upon. As discussed above, lower secreted IgM levels might even have a beneficial effect, reducing B-cell tolerance, and making it easier to generate antibodies to targets that are highly conserved between mice and humans (Ehrenstein, Cook, and Neuberger 2000). Some of the key steps toward the creation of transgenic mice with human immunoglobulin repertoires are outlined below.

5.5.1 Early Demonstrations of Immunoglobulin Transgene Rearrangement

In 1986, Yamamura et al. reported the cell type-specific expression of a human immunoglobulin gamma heavy chain transgene. This was followed by reports of expression and rearrangement of germline configuration (unrearranged) chicken and rabbit light chain transgenes in transgenic mice (Bucchini et al. 1987; Goodhardt et al. 1987). In 1989, Bruggemann et al. reported the expression of a repertoire of human IgM heavy chains from a hybrid, germline configuration, mini-locus transgene comprising one human and one mouse V gene segment, one mouse and one human D segment, two synthetic D segments, the six human J segments, mouse and human μ enchancer sequences, and a hybrid human/mouse μ constant region. Three years later, my laboratory (Taylor et al. 1992) reported mice comprising germline configuration human heavy and κ light chain transgenes that produced a repertoire of human IgM and IgG antibodies. We showed in a later paper (Taylor et al. 1994) that the IgG antibodies were a product of class switching, and that they comprised somatic mutations consistent with functional affinity maturation. Transcripts were also detected that included human V regions and mouse $C\gamma$ regions. These transcripts, which represent a potential useful source of human heavy chain V regions, are produced by trans-switching of the rearranged human transgene VDJ into the endogenous mouse constant region. Trans-switching is consistent with the recent finding that 17 percent of the IgG3 transcripts in normal mice are produced by intrachromosomal class switch recombination (Reynaud et al. 2005).

5.5.2 Immunoglobulin Knockout Mice with Human Heavy and Light Chain Repertoires

Initial experiments with germline configuration transgenes, outlined in the previous section, demonstrated that human gene sequences can direct cell type-specific expression of human immunoglobulins in mice, and that those exogenous gene sequences can undergo the normal rearrangements and modifications required for generating primary and secondary antibody repertoires. However, human immunoglobulin transgenic mice with intact functional endogenous immunoglobulin loci also express mouse antibodies and chimeric mouse-human antibodies. To create a more useful platform for human antibody drug discovery, human immunoglobulin transgenic mice were combined with mice having disrupted endogenous immunoglobulin loci.

In 1994, two papers, one from my laboratory (Lonberg et al. 1994) and the other from Green et al. (1994), reported the generation of mice with four different germline modifications: two targeted disruptions (the endogenous mouse heavy and κ light chain genes) and two introduced human transgenes (encoding the heavy chain and κ light chain). Although both papers reported the use of homologous recombination in mouse ES cells to engineer similar disruptions of the endogenous mouse loci, different technologies were used to construct and deliver the human sequence transgenes. Lonberg et al. (1994) used pronuclear microinjection to introduce reconstructed minilocus transgenes, with the heavy chain minilocus comprising 3 heavy chain variable (V_H), 16 diversity (D), and all 6 heavy chain joining (J_H) regions, together with μ and $\gamma 1$ constant-region gene segments. This construct underwent VDJ joining, together with somatic mutation and correlated class switching (Taylor et al. 1994). The light chain transgene included four Vκ, all five Jκ, and the κ constant region (Cκ). In contrast, Green et al. (1994) used fusion of yeast protoplasts to deliver yeast artificial chromosome (YAC)-based minilocus transgenes. In this case, the heavy chain included 5 V_H, all 25 D, and all 6 J_H gene segments together with μ and δ constant region gene segments. This construct underwent VDJ joining and expressed both IgM and IgD. The light chain YAC construct included two functional Vκ and all five Jκ segments, together with Cκ. Neither group inactivated the endogenous λ light chain locus, which in typical laboratory mouse strains contributes to only ~5 percent of the B-cell repertoire. Functional λ light chain expression leads to a subpopulation of B-cells producing hybrid B-cell receptors and secreted antibodies that have human heavy and mouse λ light chains. These hybrid molecules could be avoided by inactivation of the mouse λ locus (Zou et al. 2003). However, the presence of the λ expressing subpopulation did not appear to be a significant issue. Hybridoma cell lines were isolated that secreted fully human monoclonal IgMκ (Green et al. 1994) and IgGκ (Lonberg et al. 1994) MAbs specifically recognizing the target antigens against which the mice had been immunized.

The ability of these initial engineered mouse strains, each comprising only a fraction of the natural human primary V gene segment repertoire, to generate antibodies to a variety of targets may reflect the relative importance of combinatorial diversity (encoded in the germline library of V, D, and J gene segments) and junctional and somatic diversity (a product of the assembly and maturation of antibody genes). Although naive B-cell CDR1 and CDR2 sequences are completely encoded by the germline, junctional diversity, which is intact in minilocus transgenes, creates much of the heavy chain CDR3 repertoire. CDR3 sequences appear to be critical for antigen recognition by unmutated B-cell receptors, and may be largely responsible for the primary repertoire (Ignatovitch et al. 1997; Davis 2004; Tomlinson et al. 1996). Primary repertoire B-cells having low affinity for the immunogen can then enter into the T-cell-mediated process of affinity maturation, which has been shown to generate high affinity antibodies from a very limited V gene repertoire. An extreme example of this is offered by a report of an engineered mouse strain having only a single functional human V_H gene and three mouse Vλ genes (Xu and Davis 2000). These animals demonstrated specific antibody responses to a variety of T-dependent antigens. High affinity, somatically mutated MAbs were characterized, including a very high affinity (25 pM) MAb against hen egg-white lysozyme. However, the minimal V-repertoire mice did not respond to the T-independent antigen, dextran B512, and the authors suggested that responses to carbohydrate antigens might drive evolutionary selection for large primary repertoires. Germline encoded recognition of such antigens may be important for developing a rapid primary protective response to pathogens, a feature that would be selected for in the wild, but less important for isolating high affinity antibodies from laboratory mice using hyperimmunization protocols that trigger T-cell-dependent affinity maturation. The differential impact of primary repertoire size on T-dependent and T-independent responses was also seen in knockout mice in which all but one germline D segment is deleted (Schelonka et al. 2005). The reduced repertoire appeared to only affect T-independent responses.

In addition to affecting the response to T-independent antigens and the kinetics of overall immune reactions, repertoire size may have an impact on B-cell development and the size of different B-cell compartments. Fishwild et al. (1996) compared mice having different numbers of light chain

V gene segments and found that the introduction of larger repertoires encoded by a κ light chain YAC clone comprising approximately half the Vκ repertoire led to increased population of the peripheral and bone marrow B-cell compartments relative to transgenic strains comprising only four Vκ genes. The relative number of mature and immature B-cells in these compartments also appeared more normal in mice with larger V gene repertoires. Mendez et al. (1997) generated transgenic mice having nearly complete heavy chain V repertoires and approximately half the κ light chain V repertoire, and compared them with the minilocus mice of Green et al. (1994). This paper, and a later analysis of the same mouse strains by Green and Jakobovits (1998), showed that V region repertoire size had a profound effect on multiple checkpoints in B-cell development, with larger repertoires capable of restoring B-cell compartments to near normal levels. However, as noted above, restoration of normal B-cell and normal secreted immunoglobulin levels could also contribute to B-cell tolerance, which might be an undesirable feature for an MAb drug discovery platform (Ehrenstein, Cook, and Neuberger 2000).

Despite the fact that human immunoglobulin transgenic mice express B-cell receptors that are essentially hybrids of mouse and human components (e.g., human immunoglobulin, mouse Igα, Igβ, and other signaling molecules, Fig. 5.2), their B-cells develop and mature into what appear to be all of the normal B-cell subtypes. Furthermore, the immunoglobulin transgenes undergo V(D)J joining, random nucleotide (N region) addition, class switching, and somatic mutation to generate high affinity MAbs to a variety of different antigens. The process of affinity maturation in these animals even recapitulates the normal pattern of somatic mutation hotspots observed in authentic human secondary repertoire antibodies (Harding and Lonberg 1995).

5.5.3 Different Strains of Human Immunoglobulin Transgenic Mice

There have now been multiple reports in the literature of transgenic mice having immunoglobulin repertoires comprising human heavy and light chain sequences in the background of disrupted endogenous heavy and κ light chain loci (Fig. 5.3). Several different technologies, including pronuclear microinjection and yeast protoplast fusion with ES cells, have been employed for engineering these mouse strains. The introduction of the largest fraction of the human germline repertoire has been facilitated by microcell-mediated chromosome transfer. Using this technique, Tomizuka et al. (1997)

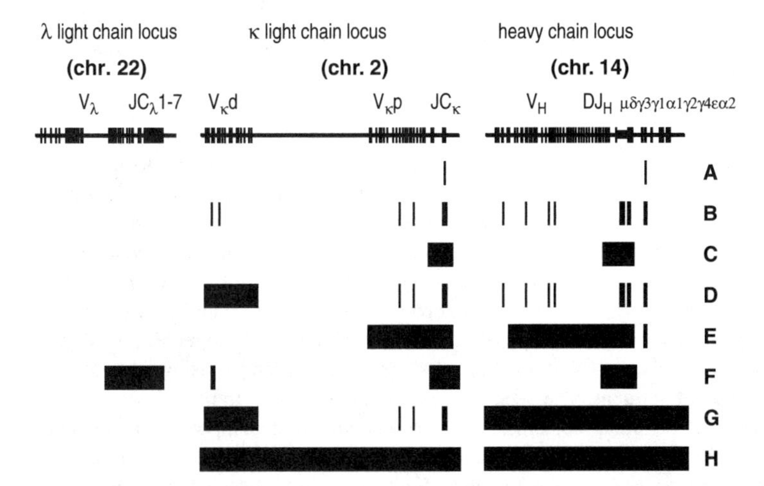

Figure 5.3 Human immunoglobulin sequences introduced in the germline of mice comprising endogenous Ig heavy chain and κ light chain gene inactivations. The germline configuration of the human immunoglobulin λ light chain, κ light chain and heavy chain is depicted above bars representing those sequences used to assemble transgenes introduced into strains of mice used for generation and isolation of human sequence MAbs. [From Zou et al. 1994 (A); Lonberg et al. 1994 (B); Green et al. 1994 (C); Fishwild et al. 1996 (D); Mendez et al. 1997 (E); Nicholson et al. 1999 (F); Ishida et al. 2002 (G); Tomizuka et al. 2000 (H).]

generated ES cell lines and chimeric mice containing fragments of human chromosomes 2 and 14, including the human κ light chain and heavy chain loci, respectively. In addition, they generated chimeric mice that incorporated an apparently intact human chromosome 22, comprising the λ light chain locus. Germline transmission was obtained with the human κ light chain ES cell lines. In a subsequent report, germline transmission was obtained with a human heavy chain ES cell line, and mice were created that expressed complete human heavy and light chain repertoires in a genetic background that included disruptions of the mouse heavy and κ light chain loci (Tomizuka et al. 2000). Completely human, high affinity (<nanomolar) MAbs were isolated from the animals. Although both chromosome fragments could be transmitted through the germline, the κ light chain-containing chromosome 2 fragment was found to be less mitotically stable. The observed stability of the heavy chain-containing fragment may derive from the fortuitous location of the immunoglobulin heavy chain locus (*IgH*) at the very telomere of the long arm of human chromosome 14. Because of the structure of chromosome 14, a random deletion between *IgH* and the centromere removed most of the non-immunoglobulin genes leaving *IgH*, the centromere, and both telomeres functionally intact. The resulting 10 to 20 Mb fragment minimizes cross-species trisomy, which would presumably be selected against during cell division. The mitotic stability of this fragment has now been exploited to create artificially constructed human chromosome fragments that include the entire human heavy chain locus together with the entire human λ light chain locus (Kuroiwa et al. 2000).

Bruggemann and colleagues (Popov et al. 1999) have also generated human λ light chain locus transgenes, using them to create transgenic mice that express partial repertoires of all three human immunoglobulin loci in the background of disrupted endogenous κ light chain and heavy chain loci (Nicholson et al. 1999).

Another transgenic mouse platform, which generates chimeric antibodies rather than fully human sequence antibodies, was developed by Rajewsky and colleagues in 1994 (Zou et al. 1994). These mice comprise relatively precise replacements of the mouse κ and γ1 constant region gene segments with the corresponding human gene sequences. The κ constant region gene segment was replaced using homologous recombination in mouse ES cells. For the γ1 gene, only the secreted exons were replaced, and the engineering was accomplished in two steps using the Cre-*loxP* recombination system, also in mouse ES cells.

5.6 HUMAN THERAPEUTIC APPLICATIONS OF TRANSGENIC MOUSE DERIVED MAbs

It is clear, from the published literature discussed in the previous section, that human sequence germline configuration transgenes can function in developing mouse B-cells to undergo V(D)J joining, heavy chain class switching, and somatic hypermutation, to generate a secondary immune response comprised of human sequence antibodies. However, the goal for creating these mice was not to recapitulate the normal humoral immune compartment in mice with inactivated endogenous immunoglobulin loci. Instead, the goal was to create a drug discovery tool for identifying lead compounds for human therapy that are active and well tolerated. Therefore, the validation of these platforms lies not in an *a priori* estimate of utility based on the functionality of the reconstituted immune systems, but in an *a posteriori* analysis of the drugs that have been generated using these mice.

A diverse set of antigens have been successfully targeted with transgenic derived MAbs (Lonberg 2005). These include small molecules, pathogen-encoded proteins, polysaccharide antigens, human-secreted proteins, and human tumor-associated glycosylation variants (Table 5.1). Most of the transgenic mouse-derived MAbs have binding affinities in the 0.1 to 10 nM range, the same affinity range typically seen for MAbs derived from wild-type mice (Ball et al. 1999; Yang et al. 1999a, 1999b; Keler et al. 2003; B.D. Cohen et al. 2005; Burgess et al. 2006). This range is probably a function of the natural constraints on affinity maturation operating *in vivo* (Foote and Eisen 1995; Roost et al. 1995). However, there are also examples of transgenic-derived human MAbs with picomolar and even subpicomolar affinities (Wang et al. 2005; Rathanaswami et al. 2005).

TABLE 5.1 Development Status of Transgenic Mouse-Derived Human Antibodies

Target	Drug	Indication	Company (Developer)	Company (Technology)	Highest Development Stage	References
EGFR	Panitumumab	Colorectal cancer and non-small cell lung cancer, renal cell carcinoma	Amgen	Abgenix	Launched	Jakobovits et al. 2007; Gibson, Ranganathan, and Grothey 2006; Mendez et al. 1997; Yang et al. 1999b, 2001; Rowinsky et al. 2004; Foon et al. 2004; Cohenuram and Saif 2007; Calvo and Rowinsky 2005
IL-12/IL-23 p40	Ustekinumab	Psoriasis and multiple sclerosis	Johnson & Johnson	Medarex	Launched	Kasper et al. 2006; Kauffman et al. 2004; Toichi et al. 2006; Krueger et al. 2007; McClung et al. 2006
CD20	Ofatumumab	Non-Hodgkin lymphoma	Genmab	Medarex	Phase III	Teeling et al. 2004; Coiffier et al. 1998, 2008; Glennie et al. 2007; Ishida et al. 2002; Bleeker et al. 2008
CD4	Zanolimumab	Lymphoma	Genmab	Medarex	Phase III	Fishwild et al. 1996, 1999; Rider et al. 2007; Villadsen et al. 2007; Skov et al. 2003; Kim et al. 2007
CTLA-4	Ipilimumab	Melanoma and various other cancers	Medarex	Medarex	Phase III	Keler et al. 2003; Korman, Peggs, and Allison 2006; Phan et al. 2003; Hodi et al. 2003; Ribas et al. 2004 Attia et al. 2005; Blansfield et al. 2005; Maker et al. 2005, 2006; Maker, Attia, and Rosenberg 2005; Sanderson et al. 2005; Beck et al. 2006; Thompson, Allison, and Kwon 2006; Weber 2007; Small et al. 2007; O'Mahony et al. 2007; Downey et al. 2007; Jaber et al. 2006; Zang and Allison 2007
CTLA-4	Tremelimumab	Melanoma	Pfizer	Abgenix	Phase III	Ribas et al. 2004, 2005, 2007; Reuben et al. 2006; Tarhini and Kirkwood 2007
EGFR	Zalutumumab	Head and neck cancer	Genmab	Medarex	Phase III	Bleeker et al. 2004; Lammerts van Bueren et al. 2006; Bastholt et al. 2007

(*Continued*)

TABLE 5.1 *Continued*

Target	Drug	Indication	Company (Developer)	Company (Technology)	Highest Development Stage	References
IL-1beta	ACZ885	Muckle–Wells syndrome	Novartis	Medarex	Phase III	Company information
RANKL	Denosumab	Osteoporosis and treatment-induced bone loss	Amgen	Abgenix	Phase III	Schwarz and Ritchlin 2007; Bekker et al. 2004; Body et al. 2006; Lipton et al. 2007; Lewiecki et al. 2007
TNFα	Golimumab	Inflammatory disease	Johnson & Johnson	Medarex	Phase III	Martin, Oneda, and Treacy 2007; Zhou et al. 2007
αv Integrins	CNTO 95	Solid tumors	Johnson & Johnson	Medarex	Phase II	Martin et al. 2005; Trikha et al. 2004
CCR5	CCR5 mAb	HIV infection	Human Genome Sciences	Abgenix	Phase II	Jakobovits et al. 2007
CD30	MDX-060, MDX-1401[a]	Lymphoma	Medarex	Medarex	Phase II	Borchmann et al. 2003, 2004; Borchmann, Schnell, and Engert 2005; Heuck et al. 2004; Boll et al. 2005; Klimm et al. 2005
Clostridium difficile toxins A and B	MDX-066/ MDX-1388[b]	Hospital acquired *C. difficile* associated diarrhea	Medarex/ Massachusetts Biologic Laboratories	Medarex	Phase II	Babcock et al. 2006
IGF-1R	CP-751,871	Cancer	Pfizer	Abgenix	Phase II	B.D. Cohen et al. 2005
IGF-1R	R1507	Solid tumors	Roche	Medarex	Phase II	Company information
IL-15	AMG 714	Rheumatoid arthritis/ psoriasis	Amgen/Genmab	Medarex	Phase II	Villadsen et al. 2003; Baslund et al. 2005
PSMA	MDX-070	Prostate cancer	Medarex	Medarex	Phase II	Holmes 2001
Alpha Interferons	MDX-1103/ MEDI-545	Lupus	Medimmune/ Medarex	Medarex	Phase I	Company information, meeting abstracts
Anthrax protective antigen	MDX-1303/ Valortim™	*B. Anthracis* infection	Pharmathene/ Medarex	Medarex	Phase I	Vitale et al. 2006
CD19	MDX-1342	Lymphoma/RA	Medarex	Medarex	Phase I	Company information
CD38	HuMax-CD38	Multiple myeloma	Genmab	Medarex	Phase I	Company information
CD3ε	NI-0401	Autoimmune disease	NovImmune	Medarex	Phase I	Company information
CD40 (agonist)	CP-870,893	Cancer	Pfizer	Abgenix	Phase I	Vonderheide 2007; Vonderheide et al. 2007
CD40 (antagonist)	CHIR-12.12/ HCD122	Chronic lymphocytic leukemia	Novartis/Xoma	Abgenix	Phase I	Tai et al. 2005

CD70	MDX-1411	Renal cell carcinoma	Medarex	Medarex	Phase I	Company information, meeting abstracts
CD89	MDX-214[c]	Solid tumors	Medarex	Medarex	Phase I	Company information
CDw137	BMS-66513	Cancer	Bristol-Myers Squibb	Medarex	Phase I	Company information
CTGF	FG-3019	Diabetic nephropathy/pulmonary fibrosis/pancreatic cancer	Fibrogen	Medarex	Phase I	Dornhofer et al. 2006; Aikawa et al. 2006
CXCL10	MDX-1100	Ulcerative colitis	Medarex	Medarex	Phase I	Company information
Dendritic cell mannose receptor	MDX-1307[d]	Human gonadotropin-positive cancers	Celldex	Medarex	Phase I	Ramakrishna et al. 2004
HGF/SF	AMG 102	Solid tumors	Amgen	Abgenix	Phase I	Burgess et al. 2006
IL-8	ABX-IL8	Psoriasis	Abgenix	Abgenix	Phase I	Huang et al. 2002; Melnikova and Bar-Eli 2006; Mori et al. 2004; Mendez et al. 1997; Yang et al. 1999a; Rathanaswami et al. 2005
IL-8	HuMax-Inflam	Palmoplantar pustulosis	Genmab	Medarex	Phase I	Company information
Melanoma antigen glycoprotein NMB	CR011-vcMMAE[e]	Melanoma	Curagen	Abgenix	Phase I	Ribas et al. 2007
Muc18	ABX-MA1	Melanoma	Abgenix	Abgenix	Phase I	Melnikova and Bar-Eli 2006
OX40L		Asthma	Roche	Medarex	Phase I	Seshasayee et al. 2007
Parathyroid hormone	ABX-PTH	Hyperparathyroidism	Amgen	Abgenix	Phase I	Company information
PD-1	MDX-1106/ONO-4538	Cancer/HCV	Ono Pharmaceuticals/Medarex	Medarex	Phase I	Wong et al. 2007
PDGF-D	CR002	Inflammatory kidney disease	Curagen	Abgenix	Phase I	Ostendorf et al. 2003
PDGFRα	IMC-3G3	Cancer	ImClone	Medarex	Phase I	Dolloff et al. 2007
PSCA	AGS-PSCA/MK-4721	Prostate cancer	Agensys/Merck	Abgenix	Phase I	Jakobovits et al. 2007
PSMA		Prostate cancer	Progenics	Abgenix	Phase I	Ma et al. 2006
TRAIL-R2	AMG 655	Solid tumors	Amgen	Abgenix	Phase I	Jakobovits et al. 2007
TRAIL-R2	HGS-TR2J	Solid tumors	Human Genome Sciences	Kirin	Phase I	Mori et al. 2004
VAP1		Inflammatory disease	BioTie Therapeutics	Medarex	Phase I	Company information
VEGFR-1	IMC-18F1	Cancer	ImClone	Medarex	Phase I	Weinblatt et al. 2003
CD154	ABI793	Inflammatory disease	Novartis	Medarex	Preclinical	Schuler et al. 2004; Kanmaz et al. 2004
CD32		Autoimmune/inflammatory disease		Medarex	Preclinical	van Royen-Kerkhof et al. 2005
CEA		Cancer		Abgenix	Preclinical	Garambois et al. 2004

(*Continued*)

TABLE 5.1 *Continued*

Target	Drug	Indication	Company (Developer)	Company (Technology)	Highest Development Stage	References
CEA		Cancer		Kirin	Preclinical	Imakiire et al. 2004
Cryptococcus neoformans capsular polysaccharide glucuronoxylomannan (GXM)		Cryptococcus infection		Abgenix	Preclinical	Maitta et al. 2004
Digoxin		Digoxin overdose		Medarex	Preclinical	Ball et al. 1999; Farr et al. 2002
EpCAM		Cancer		Kirin	Preclinical	Kuroki et al. 2005
HIV gp120		HIV infection		Abgenix	Preclinical	He et al. 2002
IFNAR	MDX-1333	Lupus	Medimmune/ Medarex	Medarex	Preclinical	Company information
IGF-1R	19D12	Cancer	Schering Plough	Medarex	Preclinical	Wang et al. 2005
Lymphoma Ig-id determinants		Lymphoma		Abgenix	Preclinical	Suarez et al. 2004
Mindin/RG-1		Cancer	Schering AG	Medarex	Preclinical	Parry et al. 2005
Mullerian duct carcinoma specific O-glycan		Cancer		Kirin	Preclinical	Nozawa et al. 2004
Ovarian carcinoma specific glycoprotein	HMOCC-1	Cancer		Kirin	Preclinical	Suzuki et al. 2004
Pneumococcal capsular polysaccharide		Pneumococcal infection		Abgenix	Preclinical	Chang et al. 2002
Rabies virus	17c7	Rabies	Massachusetts Biologic Laboratories	Medarex	Preclinical	Sloan et al. 2007
SARS Virus S1 Protein		SARS		Abgenix	Preclinical	Coughlin et al. 2006
SARS Virus S1 Protein		SARS		Medarex	Preclinical	Greenough et al. 2005
STX1		*E. coli* 0157 : H7 infection		Medarex	Preclinical	Mukherjee et al. 2002b; Tzipori et al. 2004
STX2		*E. coli* 0157 : H7 infection		Medarex	Preclinical	Mukherjee et al. 2002a; Sheoran et al. 2005; Tzipori et al. 2004; Akiyoshi et al. 2005
TRAIL-R1		Cancer		Kirin	Preclinical	Mori et al. 2004

[a]Nonfucosylated glycovariant of MDX-060.
[b]Combination of two different monoclonal antibodies directed against each of two toxins.
[c]Human antigen-binding fragment (Fab) fused to epidermal growth factor.
[d]Human Fab fused to βhCG.
[e]Antibody-drug conjugate with small molecule microtubule inhibitor MMAE.

Over 40 different transgenic-derived human MAbs have been tested in human patients (Table 5.1). These MAbs are being tested in a variety of different therapeutic indications ranging from cancer and autoimmune diseases to defense against bio-warfare pathogens. Oncology indications represent the largest subset of these applications, followed by inflammatory and autoimmune diseases. Because most of these drugs have entered clinical development in the last few years, published clinical data is not available for all of them. However, the existing clinical literature is very encouraging, and supports the original hypothesis that led to the creation of human immunoglobulin transgenic mice: the hypothesis that MAbs derived from such animals would be better tolerated than MAbs derived from nongenetically modified mice.

5.6.1 Oncology Applications

5.6.1.1 Epidermal Growth Factor Receptor Targeted MAbs Panitumumab, which in 2006 obtained regulatory approval in the United States, is the first marketed product derived from immunoglobulin transgenic mouse drug discovery platforms. Panitumumab binds to the epidermal growth factor receptor (EGFR) with very high affinity ($K_d = 5 \times 10^{-11}\,M^{-1}$) and blocks ligand binding (Rowinski et al. 2004; Yang et al. 1999a, 2001; Foon et al. 2004). In preclinical mouse xenograft models, it was found to be more potent than the lower affinity mouse antibody m225 (Yang et al. 1999a), the parent of the already marketed mouse/human IgG1 chimeric anti-EGFR antibody, cetuximab (Cunningham et al. 2004).

There has been no direct comparison of the safety and efficacy of cetuximab and panitumumab in a side-by-side clinical study. In addition, the fact that cetuximab is an IgG1 antibody and panitumumab an IgG2 antibody further complicates any attempt to compare the two drugs. However, an initial survey of the available literature suggests that the fact that panitumumab is a fully human antibody derived from a transgenic mouse may differentiate it from the chimeric cetuximab (Cohenuram and Saif 2007). In early phase I and phase II trials, panitumumab was associated with a higher frequency of skin rashes than cetuximab; however, skin rashes (which are related to the mechanism of action of EGFR-targeted drugs, including small molecules, and in this case are not a product of drug immunogenicity) have been positively correlated with activity for cetuximab (Calvo and Rowinsky 2005), and in a renal cell carcinoma trial, involving a relatively small number of patients, skin rashes correlated with longer cancer progression-free survival for panitumumab (Rowinski et al. 2004). Later trials appear to indicate that the two molecules have similar clinical activity. In a randomized, two-arm (231 patients in the treatment arm) phase III trial in second line, chemotherapy refractory, $EGFR^+$, metastatic colorectal carcinoma patients (Van Cutsem et al. 2007), there was a 10 percent objective response rate, with 20 percent of the patients having stable disease (compared to a 0 percent response rate and 10 percent stable disease in the control cohort). This response rate is comparable to that seen for cetuximab monotherapy in a 346 patient phase II trial in a similar set of refractory, $EGFR^+$, metastatic colorectal carcinoma patients (Lenz et al. 2006). Approximately 12 percent of the cetuximab-treated patients were classified as objective responders, and 32 percent as having stable disease. Panitumumab was dosed at 6 mg/kg every two weeks while the chimeric cetuximab was first given at a (roughly) 50 percent higher loading dose (400 mg/m^2), followed by a similar weekly dose of 250 mg/m^2. The lower dosing schedule selected for panitumumab was a reflection of the longer clearance time for the fully human antibody; however, the terminal half-life (7.5 days) is still shorter than is typically found for human IgG2 molecules. This is presumably due to the large antigen sink provided by normal tissue expression of EGFR, which is consistent with the observed dose dependency of the pharmacokinetics (Rowinski et al. 2004). None of the patients in the panitumumab phase III had detectable levels of anti-drug antibodies after treatment, and there was a low rate of infusion reactions, which have correlated with immunogenicity for other MAbs (Baert et al. 2003). This may be the most significant point of differentiation between the two drugs: only 5 percent of the panitumumab-treated patients experienced adverse events that could be classified as infusion reactions, and no patients experienced a grade 3 or 4 reaction (Gibson, Ranganathan, and Grothey 2006). In contrast, 7.5 percent of the cetuximab patients experienced hypersensitivity reactions, with 1.7 percent having grade 3 or 4 reactions, despite the fact that most of those patients had been pretreated with antihistamines to prevent infusion reactions. Lenz et al. (2006) also reported that over 4 percent of the cetuximab-treated patients developed human anti-chimeric antibodies.

Because panitumumab is a human IgG2 antibody, and because IgG2 antibodies are poor mediators of Fc-dependent cell killing, the activity of the drug has been ascribed to non-Fc-mediated mechanisms (Yang et al. 2001). These could involve blockade of ligand-induced receptor signaling and/or altered signaling directed by MAb binding. This is consistent with the observation that the MAb is active in mouse xenograft models while a sibling human IgG2 antibody that does not block ligand binding has no activity (Yang et al. 2001). However, although IgG2 molecules do not show significant binding to human FcγRIII (CD16), they do bind to the common H131 variant of FcγRIIa (CD32A, Parren et al. 1992). This variant is also associated with clinical responses to rituximab (Weng and Levy 2003). It remains a formal possibility that in human patients some of the activity of the panitumumab is mediated through FcγRIIa in H131 individuals. Van Cutsem et al. (2007) did not report any data on the FcγRIIa allotype of the patients who responded to panitumumab; however, if a positive correlation between the H131 allotype and clinical response is found, it might indicate that some of the activity of the MAb is Fc mediated.

Because IgG1 is a more potent mediator of Fc-dependent activity, a human IgG1 variant of panitumumab might have improved activity. This theory could be tested in the near future as late stage clinical data becomes available for zalutumumab, a second EGFR binding MAb derived from transgenic mice. Preclinical studies show that like panitumumab, zalutumumab is also more potent than m225 in mouse xenograft models (Bleeker et al. 2004; Lammerts van Bueren et al. 2006). However, unlike panitumumab, zalutumumab is an IgG1 antibody and may function by eliciting Fc-mediated effector cell activity in addition to blocking ligand binding and normal receptor functioning. In a 28 patient, dose escalation, safety trial, zalutumumab was found to be safely administrable at the highest dose of 8 mg/kg (Bastholt et al. 2007). This initial study suggests that the rate of infusion reactions for the IgG1 MAb may be higher than observed for the IgG2 panitumumab; however, with the exception of a single patient developing grade 3 dyspnea, all of the adverse events that could be classified as possibly infusion reactions were grade 1 or 2, and the patient having grade 3 dyspnea was reported to have entered the study with severe lung disease, which could have been a contributing factor. The relatively small size of the phase I trial makes it impossible to draw conclusions about the relative activities of zalutumumab and panitumumab. Zalutumumab is now in phase III testing for treatment of EGFR-positive squamous cell cancer of the head and neck, and when the results of larger trials become available it may then be possible to draw conclusions about the role of Fc receptor interaction on the efficacy and safety of these drugs.

5.6.1.2 CD20 Targeted MAbs Ofatumumab (Teeling et al. 2004), is a transgenic mouse-derived human IgG1 antibody that binds to the B-cell surface antigen CD20. While this antibody shares the same target as the mouse-human IgG1 chimeric MAb rituximab, which is currently approved for treatment of non-Hodgkin's lymphoma (NHL) and rheumatoid arthritis (Coiffier et al. 1998; S.B. Cohen et al. 2006), it recognizes a distinct epitope and may, as a result, have a different mechanism of action (Teeling et al. 2006; Glennie et al. 2007). While rituximab appears to recognize only one of the two extracellular loops of CD20, the ofatumumab epitope comprises residues from both loops. The human antibody is also a more potent mediator of complement-dependent cytotoxicity *in vitro* than rituximab. This difference in potency is more pronounced at lower antigen density, and could theoretically translate into greater activity in low CD20 expressing lymphomas such as chronic lymphocytic leukemia (CLL). Results of an initial safety/efficacy trial of ofatumumab in 33 patients with relapsed or refractory CLL showed tolerability and promising clinical activity. The trial included 3 patients each in low dose cohorts, with an expansion cohort of 27 patients treated with three weekly doses of 2 g following an initial 500 mg dose. The objective response rate in the expansion cohort was 50 percent. This patient population is relatively resistant to rituximab treatment, with standard therapy of four weekly doses of 325 mg producing objective response rates of 25 to 35 percent in small trials (Huhn et al. 2001; Itala et al. 2002). However, increased response rates have been observed at higher doses of rituximab (O'Brien et al. 2001). Thus, given the relatively high doses of ofatumumab used in the reported trial, it is impossible to use this data to compare the efficacy of the human antibody to the efficacy of the chimeric rituximab antibody. It is also difficult to compare the safety profiles of the two antibodies without a side-by-side test. There were five serious adverse events that were considered treatment related, including two infectious events, two neutropenias, and one cytolytic hepatitis in a patient with preexisting liver enzyme elevations. Most of the non-serious drug-related adverse events occurred on the day of infusion. These infusion-related adverse events were most common on first infusion, and decreased in frequency and intensity with subsequent infusions. Most of these events appeared to be consistent with cytokine release syndrome, which has previously been shown to be associated with rituximab treatment in CLL patients (Winkler et al. 1999). No anti-ofatumumab antibodies were detected in treated patients.

5.6.1.3 CD4 Targeted MAbs Zanolimumab is a transgenic mouse-derived human IgG1 antibody directed against the T-cell antigen CD4. In a preclinical study, the antibody was found to be nonimmunogenic in chimpanzees; however, it did induce a blocking antibody response in a majority of the dosed cynomolgus monkeys, suggesting that monkey models may, in some cases, overestimate immunogenicity of human antibodies (Fishwild et al. 1999). The antibody potentially affects targeted T-cells through three separate mechanisms of action: binding to CD4 rapidly inhibits T-cell receptor signal transduction, binding mediates T-cell depletion through antibody-dependent cell-mediated cytotoxicity, and binding induces down-modulation of CD4 from the cell surface (Rider et al. 2007; Villadsen et al. 2007). The *in vitro* studies found that primary $CD4^+CD45RO^+$ T-cells are more sensitive to zanolimumab/NK cell-mediated destruction than naïve $CD4^+CD45RA^+$ T-cells, consistent with the results of the first clinical study. In an 85 patient, placebo-controlled, phase II trial in psoriasis, there was an observed dose-dependent decrease in circulating $CD4^+$ cells, particularly in the $CD45RO^+$ memory T-cell population (Skov et al. 2003). The drug was well tolerated, with one likely drug-related serious adverse event, a rash appearing after the second dose at 160 mg. No patients in this study developed anti-drug antibodies. In phase II trials, in cutaneous T-cell lymphoma patients (mycosis fungoides and Sezary syndrome), only one of 47 patients had measurable anti-drug antibodies. The anti-zanolimumab titer was very low in this patient, and did not appear to be neutralizing, as evidenced by continued clinical activity of the drug. These phase II trials comprised patient cohorts treated with up to 17 weekly infusions of either 280, 560, or 980 mg. Zanolimumab was judged to be well tolerated, with no dose-dependent toxicity, other than the mechanism-related depletion of T-cells. T-cell counts were found to recover at a median rate of 137 cells/microliter/year after treatment was stopped. Several, possibly drug-related, serious adverse events were observed, including infections that could have been associated with this T-cell depletion. However, infections would not be uncommon in this patient population. Low grade cytokine release syndrome was seen in a few patients, and premedication with paracetamol was introduced into the protocol, after which no further such cases were observed. Dose-dependent clinical activity of zanolimumab was observed within the mycosis fungoides patient cohorts, with three out of four patients treated at 980 mg experiencing an objective clinical response. Objective responses were also reported for Sezary syndrome patients, and for mycosis fungoides patients in the lower dose cohorts. The overall objective response rate was 32 percent.

5.6.1.4 CD30 Targeted MAbs A transgenic mouse-derived anti-CD30 MAb, MDX-060 (Borchmann et al. 2003; Heuck et al. 2004; Boll et al. 2005), has been tested in Hodgkin's lymphoma and anaplastic large cell lymphoma patients (Borchmann et al. 2004; Borchmann, Schnell, and Engert 2005; Klimm et al. 2005). Fifty-six patients were reported to have been treated with up to 15 mg/kg every week for four weeks without significant infusion reactions. The preliminary results were interpreted to indicate that the drug was well tolerated and had clinical activity.

5.6.1.5 Cancer Immunotherapy As seen with other transgenic mouse-derived human antibodies, the two human MAbs directed against CTLA-4, ipilimumab and tremelimumab, also do not appear to elicit strong patient anti-drug antibody responses. It is notable that neither drug is particularly immunogenic in light of the observation that their mechanisms of action appear to result in very potent up-modulation of immune responses to other antigens. CTLA-4 is a negative T-cell signaling molecule that binds to the two ligands CD80 and CD86, both of which are also recognized by the positive T-cell signaling molecule CD28 (Korman, Peggs, and Allison 2006). Ipilimumab (Keler et al. 2003) is a human IgG1 antibody, while tremelimumab (Ribas et al. 2005) is an IgG2 antibody. Both molecules bind to human CTLA-4 so as to block ligand binding and antagonize CTLA-4 signaling, resulting in the activation of certain T-cell responses. Laboratory experiments with surrogate hamster MAbs that block mouse CTLA-4 show that the resulting enhanced immune responses can mediate tumor rejection in syngeneic mouse tumor models (Leach, Krummel, and Allison 1996). Preclinical experiments in cynomolgus monkey models demonstrated that ipilimumab could stimulate humoral immune responses to co-administered vaccines (Keler et al. 2003). Clinical data in cancer patients has been reported for both ipilimumab (Phan et al. 2003; Hodi et al. 2003; Ribas et al. 2004; Attia et al. 2005; Blansfield et al. 2005; Maker et al. 2005; Maker, Attia, and Rosenberg 2005; Sanderson et al. 2005; Beck et al. 2006; Maker et al. 2006; Thompson, Allison, and Kwon 2006; Weber 2007; Small et al. 2007; O'Mahony et al. 2007; Downey et al. 2007) and CP-675,206 (Ribas et al. 2004, 2005; Reuben et al. 2006; Tarhini and Kirkwood 2007; Ribas et al. 2007). Objective and durable antitumor responses were observed for both drugs.

Rosenberg and colleagues conducted a trial in patients with metastatic melanoma who were treated with ipilimumab at 3 mg/kg every three weeks for up to six cycles or were given a loading dose of ipilimumab at 3 mg/kg followed by 1 mg/kg every three weeks for up to six cycles. All patients were administered a subcutaneous gp100 peptide vaccine (Attia et al. 2005). The overall objective response rate for the 56 patients in the combined cohorts was 13 percent, with ongoing complete and partial responses reported at 25, 26, 30, 31, and 34 months. A follow-up paper by this group included additional metastatic melanoma patients treated with and without the vaccine, some receiving ipilimumab doses as high as 9 mg/kg, together with 61 renal cell carcinoma patients treated with ipilimumab at up to 3 mg/kg (Beck et al. 2006). The overall objective response rate for the 198 patients in this report was 14 percent. An additional follow-up paper evaluated clinical outcomes for 139 metastatic melanoma patients from these trials (Downey et al. 2007). The overall median survival was 15.7 months, with a median duration of response at 30.6 months.

In a phase I single-dose, monotherapy, dose escalation trial of tremelimumab in metastatic melanoma, with patients receiving doses as high as 15 mg/kg, the authors reported a 10 percent objective response rate (Ribas et al. 2005), although one of the four responders had also received ipilimumab (Ribas et al. 2004).

The serious adverse events reported for both ipilimumab and tremelimumab comprise a spectrum of immune-related inflammatory responses, including rash, enterocolitis, and hypophysitis (Jaber et al. 2006; Blansfield et al. 2005; Ribas et al. 2005; Beck et al. 2006). However, because the mechanism of action of CTLA-4 blocking MAbs involves the activation of immune responses, these have been considered target-related toxicities, and have in fact correlated with clinical responses (Beck et al. 2006; Reuben et al. 2006). Beck et al. (2006) reported 36 percent and 35 percent objective response rates for melanoma and renal cell cancer patients having enterocolitis, with response rates of only 11 percent and 2 percent for patients without enterocolitis. In a later analysis of the melanoma patients in these trials, Downey et al. (2007) found that 28 percent of the patients who experienced grade 3 or 4 inflammatory adverse events were objective responders, with a median duration of response of 35 months, while only a single patient (2 percent) showed an objective response without also experiencing an inflammatory adverse event. This patient's response was reported as ongoing at 18 months. Twenty-two percent of patients with only grade 1 or 2 inflammatory adverse events had objective responses, with a medion duration of 11 months. The inflammatory adverse events have been reported to respond to medical management, which may include corticosteroids. Interestingly, corticosteroid treatment does not appear to abrogate objective tumor responses (Attia et al. 2005; Beck et al. 2006).

Despite the observed upregulation of immune responses in patients treated with these two MAbs, the drugs themselves do not appear to be readily recognized and cleared by the human immune system. A terminal half-life of 22 days was reported for tremelimumab (Ribas et al. 2005), and one month post-dosing serum trough levels of 10 μg/mL ipilimumab were reported after five months of repeated monthly dosing at 3 mg/kg (Sanderson et al. 2005). Sanderson et al. (2005) also reported that these repeatedly dosed patients did not develop a measurable antibody response to ipilimumab. This data is consistent with data from preclinical studies that showed no evidence of monkey anti-human antibody formation in cynomolgus macaques dosed five times over 140 days (Keler et al. 2003), despite the fact that the MAb upregulated the monkey humoral immune responses to co-administered vaccines. There was no sign of immune clearance by monkey anti-human antibodies, with drug titers never falling below 20 μg/mL over the course of the five month study.

In addition to CTLA-4, three other lymphocyte regulatory molecules are being explored in clinical studies as immunotherapy targets for transgenic mouse-derived human antibodies. Agonist MAbs directed against CDw137 and CD40, and a PD-1 blocking MAb are all in clinical development (Table 5.1). Reuben et al. (2006) identified the T-cell signaling molecule PD-1 as a possible biomarker for clinical unresponsiveness to CTLA-4 blockade in a study of patients treated with tremelimumab. The study looked at surface expression of PD-1 protein on peripheral $CD4^+$, $CD25^+$ T-cells from patients who experienced inflammatory adverse events. Lower levels of PD-1 correlated with objective antitumor responses. PD-1 is structurally homologous to CTLA-4, and is also a negative regulator of T-cell activity (Zang and Allison 2007). A human IgG4 αPD-1 antibody, derived from transgenic mice, has been shown to block the PD-1 pathway *in vitro* (Wong et al. 2007). The MAb is now in a phase I trial in cancer patients; however, human clinical data has not yet been published.

CD40 is a cell surface protein of the tumor necrosis factor receptor superfamily. It is expressed on a variety of lymphocytes and epithelial cells, as well as some tumor cells. Agonist antibodies to CD40 can activate immune responses to tumor antigens through engagement with CD40 expressing antigen-presenting cells, and can also directly mediate tumor cell killing for $CD40^+$ cancers (Vonderheide 2007). An initial clinical study of an agonist αCD40 human IgG2 MAb derived from transgenic mice showed tolerability and evidence of biological activity (Vonderheide et al. 2007). The single dose, dose escalation study included 29 patients with advanced solid tumors. Consistent with the mechanism of action of the drug, and the widespread normal expression of the target, dose-limiting toxicities were observed at 0.3 mg/kg. These included one patient with a thromboembolism, and a patient with cytokine release syndrome associated severe headache that lasted for eight days. The maximum tolerated dose was set at 0.2 mg/kg. At this dose, grade 1 and 2 cytokine release syndrome was commonly observed; however, it was considered manageable. Dose-related hepatotoxicity was also observed, with one dose-limiting toxicity at 0.2 mg/kg. The study included 15 melanoma patients, and 4 of these experienced an objective partial clinical response. The overall objective response rate for the trial was 14 percent (all partial responses), with a 24 percent stable disease rate. These results are consistent with direct MAb-mediated tumor cell killing and cannot yet be used to evaluate the possibility that the drug also has immune-mediated antitumor effects.

5.6.2 Inflammatory and Autoimmune Disease Applications

5.6.2.1 IL-12/IL-23 Targeted MAbs Ustekinumab, which has received regulatory approval in Europe and Canada, is the second marketed transgenic-derived MAb after panitumumab. Ustekinumab is a human IgG1κ MAb directed against the common p40 subunit shared by IL-12 and IL-23. Results have been reported from a phase I trial in multiple sclerosis (Kasper et al. 2006) and from phase I and phase II trials in psoriasis (Kauffman et al. 2004; Toichi et al. 2006; Krueger et al. 2007). In the phase I psoriasis trial, the drug showed sustained activity over 16 weeks of follow-up with a single intravenous administration, with 67 percent of the patients achieving at least a 75 percent improvement (assessed by the Psoriasis Area and Severity Index). There were no treatment-related serious adverse events, and no infusion reactions. Anti-drug antibodies were detected in one of 18 patients; however, presence of drug in the serum because of the very long terminal half-life, 19 to 27 days, precluded accurate assessment in most of the patients. A similar 20 to 31 day terminal half-life was observed in the multiple sclerosis trial where the drug was given by subcutaneous administration. One of the 16 treated patients developed a detectable anti-drug response; however, as with the psoriasis trial, the persistence of the drug in the serum made it difficult to accurately measure anti-drug antibodies. In the phase II psoriasis trial, 237 patients received the drug for up to four weekly 90 mg subcutaneous doses. During the 52 week monitoring period, anti-drug antibodies were detected in 12 (4 percent) of the treated patients. However, the measured antibody response did not correlate with injection site reactions, which occurred at the same 2 percent frequency in both placebo- and drug-treated cohorts. Patients given only a single subcutaneous dose, at either 45 or 90 mg, showed sustained disease-modifying responses for over six months following treatment. Together with the observed sustained clinical benefit, the approximately 20 to 30 day terminal half-life of ustekinumab appears to indicate that it does not elicit a strong drug-clearing antibody response. As further clinical data is reported, it will be interesting to compare the immunogenicity, pharmacokinetics, safety, and efficacy of the transgenic mouse-derived ustekinumab to the phage display-derived ABT-874, which is also directed against the common p40 subunit of IL12 and IL-23 (Mannon et al. 2004; Fuss et al. 2006). The phage display antibody also showed some signs of immunogenicity, with anti-drug antibodies detectable in 3 of 63 patients, and 2 of those patients showing evidence of early clearance of the drug from the serum (Mannon et al. 2004); however, because of differences in dosing, and inherent difference in measurement of anti-drug antibodies, it is difficult to compare the data to that reported for ustekinumab. The terminal half-life of ABT-874 was not reported.

5.6.2.2 Tumor Necrosis Factor-α Targeted MAbs Golimumab is a transgenic mouse-derived human IgG1κ antibody that binds to, and neutralizes, tumor necrosis factor-α (TNFα; Martin, Oneda, and Treacy 2007). It is currently being tested in a variety of phase III trials in rheumatoid arthritis and other indications (Johnson & Johnson, New Brunswick, NJ; company information). Two different dosing regimens are being explored in these trials: a 50 mg dose given subcutaneously once a month, and a higher dose intravenous infusion given once every three months. Published studies of clinical trials involving these dosing schedules are not yet available; however, a single-dose, placebo-controlled, dose-escalating, safety and pharmacokinetic study has been reported (Zhou et al. 2007). The drug was found to be tolerable, with no significant differences in infusion reactions or other adverse events between the drug-treated and the placebo groups. The clearance rate of the MAb was found to be comparable to that expected for a human IgG1 antibody, with a terminal half-life estimated at 11 to 19 days for the 3 and 10 mg/kg cohorts. Three out of the 26 patients treated showed detectable, but low, titers of anti-golimumab antibodies; however, the presence of these antibodies did not correlate with increased clearance of the drug. A larger trial with repeat dosing was deemed to be necessary for an evaluation of the immunogenicity of the antibody. When the results of the phase III studies become available it should be possible to compare safety and efficacy of the human antibody to the approved chimeric antibody, infliximab (Maini et al. 1998), and the approved phage display-derived antibody, adalimumab (Weinblatt et al. 2003). Like golimumab, adalimumab is a human sequence IgG1 MAb formulated for subcutaneous administration. It is dosed once every two weeks, compared to the subcutaneous once a month dosing being tested for golimumab. A comparison of these two drugs may be of particular interest because adalimumab has been reported to elicit anti-drug antibodies, despite the fact that it was genetically engineered from a lead molecule originally isolated from a phage display library constructed from human immunoglobulin sequences. It is not clear if this immunogenicity is a specific property of the molecule, if it is an inherent property for TNFα targeting MAbs, or a general property of protein-based drugs administered to this patient population. For adalimumab, the formation of these anti-drug antibodies correlated with adverse events and reduced efficacy in a study of 15 rheumatoid arthritis patients (Bender et al. 2007). The chimeric mouse-human antibody, infliximab, also elicits a strong anti-drug antibody response which also correlates with infusion reactions and reduced efficacy (Baert et al. 2003).

5.6.2.3 Interleukin-15 Targeted MAbs AMG-714 is a human IgG1κ MAb that binds to the proinflammatory cytokine interleukin-15 (IL-15; Villadsen et al. 2003). IL-15 signals through a hetrotrimeric receptor complex comprising a high affinity α chain, that binds the cytokine and then forms a signaling complex with the IL-15 receptor β chain and the cytokine receptor common γ chain. AMG-714 binds to an epitope that does not block α chain binding, but prevents formation of the signaling complex. For this reason, the MAb binds to receptor expressing cells in the presence of ligand, but blocks IL-15 induced activation *in vitro* (e.g., IFNγ or CD69 induction). The drug has been tested in a relatively small (30 rheumatoid arthritis patients) phase I and phase II, dose-escalating, randomized, placebo-controlled, clinical trial. Patients received up to five intravenous doses, with the highest dose cohort receiving an initial dose of 8 mg/kg followed four weeks later by four weekly doses of 4 mg/kg each. The MAb was well tolerated, with no evidence of dose-limiting toxicity, and no serious adverse events that were considered to be drug related. Other adverse events were low grade, including minor injection-site reactions seen at the 8 mg/kg dose. No anti-drug antibodies were detected. The treated patient groups included responders (by the criteria of the American College of Rheumatology); however, there was no clear evidence of dose-related activity, and a larger trial will be necessary to assess efficacy.

5.6.3 Other Applications

Only a small handful of transgenic mouse-derived MAbs that target indications outside of cancer, inflammatory, and autoimmune diseases have entered clinical development. The largest subset of such indications is infectious diseases. Preclinical descriptions of MAbs directed at anthrax (Vitale et al. 2006) and *Clostridium difficile* (Babcock et al. 2006) infections have been reported; however, human clinical data is not yet available. The most advanced drug in the noncancer, noninflammatory/autoimmune disease category is directed at modulation of bone metabolism. The clinical development of this molecule is discussed below.

5.6.3.1 RANKL Targeted MAbs Denosumab is a transgenic mouse-derived, human IgG2, antibody directed against RANKL, a TNF family member that stimulates the maturation and activation of osteoclasts, which mediate bone resorption. A recent review listed 21 different clinical trials for the drug, including 10 ongoing phase III studies for treatment of bone loss in postmenopausal women, and in cancer patients with treatment-induced bone loss or skeletal disease caused by bone metastases (Schwarz and Ritchlin 2007). Results from some of these studies have appeared in the published literature. A single subcutaneous administration, dose-escalation, phase I study in osteoporotic patients showed dose-dependent and sustained activity (up to 6 months) in blocking bone resorption, with no reported serious drug-related adverse events (Bekker et al. 2004). Denosumab was found to have dose-dependent pharmacokinetics, with a terminal half-life of 32 days at the highest 3 mg/kg dose. A second trial in patients with multiple myeloma or bone metastases from breast cancer showed decreased bone metabolism that persisted for the 84 day study follow-up period after a single 3 mg/kg dose (Body et al. 2006). These studies measured bone metabolism using urine concentrations of peptide products of collagen catabolism (N-telopeptide) as an indirect measure.

In a randomized, multidose, dose ranging trial in 255 metastatic breast cancer patients with bone disease, denosumab was tested directly against the bisphosphonate drugs currently approved for bone metastases (Lipton et al. 2007). Patients in the denosumab cohorts received monthly subcutaneous doses of 30, 120, or 180 mg, or three monthly subcutaneous doses of 60 or 180 mg. Patients in the bisphosphonate cohort received monthly intravenous doses. The human MAb was found to have biological activity (reduction in levels of the bone turnover marker, urine N-telopeptide) comparable to the standard bisphosphonate drugs. Even the three monthly 60 mg subcutaneous dosing regimen showed activity comparable to conventional monthly intravenous bisphosphonate therapy. The study was not powered to compare actual clinical efficacies of the two treatments; however; on-study skeletal-related events (fracture, surgery or radiation to bone, or spinal chord compression) were lower in the MAb-treated patient groups (9 percent for denosomab-treated and 16 percent for bisphosphonate-treated patients). There were no serious or severe adverse events attributed to denosumab, and the overall rate of treatment-related adverse events was lower in the denosumab-treated patients (19 percent) compared to the bisphosphonate-treated patients (30 percent). There was a low

(3 percent) occurrence of injection site reactions; however, no dose-dependent increase in overall adverse events was observed for the denosumab cohorts. No anti-denosumab antibodies were detected.

In the largest published clinical study of denosumab, 412 postmenopausal women with low bone mineral density were randomly assigned to placebo, denosumab, or bisphosphonate arms, and efficacy and safety evaluated for two years of treatment (McClung et al. 2006; Lewiecki et al. 2007). In the MAb treatment arms, patients were given subcutaneous denosumab at 6, 14, or 30 mg every three months; or 14, 60, 100, or 210 mg every six months. Overall adverse event profiles were similar between the placebo, denosumab, or bisphosphonate arms. Two of the 314 MAb-treated patients showed transient levels of anti-denosumab antibodies in single blood samples in the first 12 months; however, these measurements were not confirmed in later blood samples, and no neutralizing anti-denosumab antibodies were detected over the two year study. Both clinical activity (bone mineral density increases) and biomarker activity (serum C-telopeptide and urine N-telopeptide level decreases) were observed in all denosumab treatment cohorts, with sustained two year clinical benefit comparable to bisphosphonate therapy seen even in the group treated with only 60 mg of MAb every six months. The low incidence of measurable anti-drug antibodies, the safety profile, and the very long half-life and sustained drug activity are all consistent with an antibody that is relatively nonimmunogenic. Because infrequent dosing may be very important for patient compliance for a parenterally delivered protein based therapeutic that is directed at chronic indications such as osteoporosis, low immunogenicity could be a critical feature for the success of this product.

5.7 SUMMARY

Despite the complexity of immunoglobulin gene rearrangement and modification events associated with the assembly of natural antibody repertoires, and despite the critical role of the protein products of these genes in the function, differentiation, and survival of the B-cells that express these antibody repertoires, human sequence immunoglobulin transgenes are surprisingly capable of replacing endogenous mouse gene sequences and directing the expression of human antibody repertoires. The resulting transgenic mouse platforms lend themselves to a drug discovery process that has certain advantages over other methods for MAb drug discovery. Unlike antibody engineering technologies for making low immunogenicity MAbs, where an early lead candidate is then modified or optimized *in vitro* to reduce immunogenicity, with the transgenic mouse platforms, the process of lead optimization is bypassed, making it possible to test each potential lead candidate in a series of increasingly sophisticated *in vitro* and *in vivo* assays in essentially the same molecular form as it will eventually be used in humans. Resources that would otherwise be devoted to optimization of a small number of lead hits can be devoted to better characterization of a larger number of lead candidates comprising a wider variety of functional properties. However, a more objective assessment of the value of transgenic mice for antibody drug discovery comes from an evaluation of the actual discovered drugs themselves. The reported properties of these drugs in human clinical trials is encouraging, with many showing promising signs of biological and clinical activity, as well as relatively low immunogenicity, and relatively slow *in vivo* clearance. Now that the EGFR targeted panitumumab antibody has obtained regulatory approval, and over 40 additional transgenic mouse-derived MAbs have entered human clinical testing, it is becoming clear that these genetically engineered mice are a significant addition to the drug discovery toolbox employed by the pharmaceutical industry.

REFERENCES

Aikawa, T., J. Gunn, S.M. Spong, S.J. Klaus, and M. Korc. 2006. Connective tissue growth factor-specific antibody attenuates tumor growth, metastasis, and angiogenesis in an orthotopic mouse model of pancreatic cancer. *Mol. Cancer Ther.* 5(5):1108–1116.

Akiyoshi, D.E., C.M. Rich, S. O'Sullivan-Murphy, L. Richard, J. Dilo, A. Donohue-Rolfe, A.S. Sheoran, S. Chapman-Bonofiglio, and S. Tzipori. 2005. Characterization of a human monoclonal antibody against Shiga toxin 2 expressed in Chinese hamster ovary cells. *Infect. Immun.* 73(7):4054–4061.

Alt, F.W., T.K. Blackwell, and G.D. Yancopoulos. 1985. Immunoglobulin genes in transgenic mice. *Trends Genet.* 1, 231–236.

Attia, P., G.Q. Phan, A.V. Maker, M.R. Robinson, M.M. Quezado, J.C. Yang, R.M. Sherry, S.L. Topalian, U.S. Kammula, R.E. Royal, N.P. Restifo, L.R. Haworth, C. Levy, S.A. Mavroukakis, G. Nichol, M.J. Yellin, and S.A. Rosenberg. 2005. Autoimmunity correlates with tumor regression in patients with metastatic melanoma treated with anti-cytotoxic T-lymphocyte antigen-4. *J. Clin. Oncol.* 23(25):6043–6053.

Babcock, G.J., T.J. Broering, H.J. Hernandez, R.B. Mandell, K. Donahue, N. Boatright, A.M. Stack, I. Lowy, R. Graziano, D. Molrine, D.M. Ambrosino, and W.D. Thomas, Jr. 2006. Human monoclonal antibodies directed against toxins A and B prevent *Clostridium difficile*-induced mortality in hamsters. *Infect. Immun.* 74(11):6339–6347.

Baert, F., M. Noman, S. Vermeire, G. Van Assche, A. Carbonez, and P. Rutgeerts. 2003. Influence of immunogenicity on the long-term efficacy of infliximab in Crohn's disease. *N. Engl. J. Med.* 348(7):601–608.

Ball, W.J., Jr., R. Kasturi, P. Dey, M.Tabet, S. O'Donnell, D. Hudson, and D. Fishwild. 1999. Isolation and characterization of human monoclonal antibodies to digoxin. *J. Immunol.* 163(4):2291–2298.

Bankovich, A.J., S. Raunser, Z.S. Juo, T. Walz, M.M. Davis, and K.C. Garcia. 2007. Structural insight into pre-B cell receptor function. *Science* 316(5822):291–294.

Barbas, C.F. III, and D.R. Burton. 1996. Selection and evolution of high-affinity human anti-viral antibodies. *Trends Biotechnol.* 14(7):230–234.

Barreto, V., B. Reina-San-Martin, A.R. Ramiro, K.M. McBride, and M.C. Nussenzweig. 2003. C-terminal deletion of AID uncouples class switch recombination from somatic hypermutation and gene conversion. *Mol. Cell* 12(2):501–508.

Baslund, B., N. Tvede, B. Danneskiold-Samsoe, P. Larsson, G. Panayi, J. Petersen, L.J. Petersen, F. J. Beurskens, J. Schuurman, J.G. van de Winkel, P.W. Parren, J.A. Gracie, S. Jongbloed, F.Y. Liew, and I.B. McInnes. 2005. Targeting interleukin-15 in patients with rheumatoid arthritis: a proof-of-concept study. *Arthritis Rheum.* 52(9):2686–2692.

Bastholt, L., L. Specht, K. Jensen, E. Brun, A. Loft, J. Petersen, H. Kastberg, and J.G. Eriksen. 2007. Phase I/II clinical and pharmacokinetic study evaluating a fully human monoclonal antibody against EGFr (HuMax-EGFr) in patients with advanced squamous cell carcinoma of the head and neck. *Radiother. Oncol.* 85(1):24–28.

Beck, K.E., J.A. Blansfield, K.Q. Tran, A.L. Feldman, M.S. Hughes, R.E. Royal, U.S. Kammula, S.L. Topalian, R.M. Sherry, D. Kleiner, M. Quezado, I. Lowy, M. Yellin, S.A. Rosenberg, and J.C. Yang. 2006. Enterocolitis in patients with cancer after antibody blockade of cytotoxic T-lymphocyte-associated antigen 4. *J. Clin. Oncol.* 24(15):2283–2289.

Bekker, P.J., D.L. Holloway, A.S. Rasmussen, R. Murphy, S.W. Martin, P.T. Leese, G.B. Holmes, C.R. Dunstan, and A.M. DePaoli. 2004. A single-dose placebo-controlled study of AMG 162, a fully human monoclonal antibody to RANKL, in postmenopausal women. *J. Bone Miner. Res.* 19(7):1059–1066.

Bender, N.K., C.E. Heilig, B. Droll, J. Wohlgemuth, F.P. Armbruster, and B. Heilig. 2007. Immunogenicity, efficacy and adverse events of adalimumab in RA patients. *Rheumatol Int.* 27(3):269–274.

Benedict, C.L., S. Gilfillan, T.H. Thai, and J.F. Kearney. 2000. Terminal deoxynucleotidyl transferase and repertoire development. *Immunol. Rev.* 175:150–157.

Blansfield, J.A., K.E. Beck, K. Tran, J.C. Yang, M.S. Hughes, U.S. Kammula, R.E. Royal, S.L. Topalian, L.R. Haworth, C. Levy, S.A. Rosenberg, and R.M. Sherry. 2005. Cytotoxic T-lymphocyte-associated antigen-4 blockage can induce autoimmune hypophysitis in patients with metastatic melanoma and renal cancer. *J. Immunother. (1997)* 28(6):593–598.

Bleeker, W.K., J.J. Lammerts van Bueren, H.H. van Ojik, A.F. Gerritsen, M. Pluyter, M. Houtkamp, E. Halk, J. Goldstein, J. Schuurman, M.A. van Dijk, J.G. van de Winkel, and P.W. Parren. 2004. Dual mode of action of a human anti-epidermal growth factor receptor monoclonal antibody for cancer therapy. *J. Immunol.* 173(7):4699–4707.

Bleeker, W.K., M.E. Munk, W.J. Mackus, J.H. van den Brakel, M. Pluyter, M.J. Glennie, J.G. van de Winkel, and P.W. Parren. 2008. Estimation of dose requirements for sustained in vivo activity of a therapeutic human anti-CD20 antibody. *Br. J. Haematol.* 140(3):303–312.

Body, J.J., T. Facon, R.E. Coleman, A. Lipton, F. Geurs, M. Fan, D. Holloway, M.C. Peterson, and P.J. Bekker. 2006. A study of the biological receptor activator of nuclear factor-kappaB ligand inhibitor, denosumab, in patients with multiple myeloma or bone metastases from breast cancer. *Clin. Cancer Res.* 12(4):1221–1228.

Boll, B., H. Hansen, F. Heuck, K. Reiners, P. Borchmann, A. Rothe, A. Engert, and E. Pogge von Strandmann. 2005. The fully human anti-CD30 antibody 5F11 activates NF-{kappa}B and sensitizes lymphoma cells to bortezomib-induced apoptosis. *Blood* 106(5):1839–1842.

Borchmann, P., R. Schnell, and A. Engert. 2005. Immunotherapy of Hodgkin's lymphoma. *Eur. J. Haematol. Suppl.* 75(66):159–165.

Borchmann, P., R. Schnell, H. Schulz, and A. Engert. 2004. Monoclonal antibody-based immunotherapy of Hodgkin's lymphoma. *Curr. Opin. Investig. Drugs* 5(12):1262–1267.

Borchmann, P., J.F. Treml, H. Hansen, C. Gottstein, R. Schnell, O. Staak, H.F. Zhang, T. Davis, T. Keler, V. Diehl, R.F. Graziano, and A. Engert. 2003. The human anti-CD30 antibody 5F11 shows in vitro and in vivo activity against malignant lymphoma. *Blood* 102(10):3737–3742.

Brinster, R.L., H.Y. Chen, M. Trumbauer, A.W. Senear, R. Warren, and R.D. Palmiter. 1981. Somatic expression of herpes thymidine kinase in mice following injection of a fusion gene into eggs. *Cell* 27(1, Pt 2):223–231.

Brinster, R.L., K.A. Ritchie, R.E. Hammer, R.L. O'Brien, B. Arp, and U. Storb. 1983. Expression of a microinjected immunoglobulin gene in the spleen of transgenic mice. *Nature* 306(5941):332–336.

Bruggemann, M., H.M. Caskey, C. Teale, H. Waldmann, G.T. Williams, M.A. Surani, and M.S. Neuberger. 1989. A repertoire of monoclonal antibodies with human heavy chains from transgenic mice. *Proc. Natl. Acad. Sci. USA* 86(17):6709–6713.

Bucchini, D., C.A. Reynaud, M.A. Ripoche, H. Grimal, J. Jami, and J.C. Weill. 1987. Rearrangement of a chicken immunoglobulin gene occurs in the lymphoid lineage of transgenic mice. *Nature* 326(6111):409–411.

Burgess, T., A. Coxon, S. Meyer, J. Sun, K. Rex, T. Tsuruda, Q. Chen, S.Y. Ho, L. Li, S. Kaufman, K. McDorman, R.C. Cattley, G. Elliott, K. Zhang, X. Feng, X.C. Jia, L. Green, R. Radinsky, and R. Kendall. 2006. Fully human monoclonal antibodies to hepatocyte growth factor with therapeutic potential against hepatocyte growth factor/c-Met-dependent human tumors. *Cancer Res.* 66(3):1721–1729.

Buttin, G. 1987. Exogenous Ig gene rearrangement in transgenic mice: A new strategy for human monoclonal antibody production? *Trends Genet.* 3:205–206.

Calvo, E., and E.K. Rowinsky. 2005. Clinical experience with monoclonal antibodies to epidermal growth factor receptor. *Curr. Oncol. Rep.* 7(2):96–103.

Chang, Q., Z. Zhong, A. Lees, M. Pekna, and L. Pirofski. 2002. Structure-function relationships for human antibodies to pneumococcal capsular polysaccharide from transgenic mice with human immunoglobulin Loci. *Infect. Immun.* 70:4977–4986.

Choi, T.K., P.W. Hollenbach, B.E. Pearson, R.M. Ueda, G.N. Weddell, C.G. Kurahara, C.S. Woodhouse, R.M. Kay, and J.F. Loring. 1993. Transgenic mice containing a human heavy chain immunoglobulin gene fragment cloned in a yeast artificial chromosome. *Nature Genet.* 4(2):117–123.

Cohen, B.D., D.A. Baker, C. Soderstrom, G. Tkalcevic, A.M. Rossi, P.E. Miller, M.W. Tengowski, F. Wang, A. Gualberto, J.S. Beebe, and J.D. Moyer. 2005. Combination therapy enhances the inhibition of tumor growth with the fully human anti-type 1 insulin-like growth factor receptor monoclonal antibody CP-751,871. *Clin. Cancer Res.* 11(5):2063–2073.

Cohen, S.B.; P. Emery, M.W. Greenwald, M. Dougados, R.A. Furie, M.C. Genovese, E.C. Keystone, J.E. Loveless, G.R. Burmester, M.W. Cravets, E.W. Hessey, T. Shaw, and M.C. Totoritis. 2006. Rituximab for rheumatoid arthritis refractory to anti-tumor necrosis factor therapy: Results of a multicenter, randomized, double-blind, placebo-controlled, phase III trial evaluating primary efficacy and safety at twenty-four weeks. *Arthritis Rheum.* 54(9):2793–2806.

Cohenuram, M., and M.W. Saif. 2007. Panitumumab the first fully human monoclonal antibody: From the bench to the clini. *Anticancer Drugs* 18(1):7–15.

Coiffier, B., C. Haioun, N. Ketterer, A. Engert, H. Tilly, D. Ma, P. Johnson, A. Lister, M. Feuring-Buske, J.A. Radford, R. Capdeville, V. Diehl, and F. Reyes. 1998. Rituximab (anti-CD20 monoclonal antibody) for the treatment of patients with relapsing or refractory aggressive lymphoma: A multicenter phase II study. *Blood* 92(6):1927–1932.

Coiffier, B., S. Lepretre, L.M. Pedersen, O. Gadeberg, H. Fredriksen, M.H. van Oers, J. Wooldridge, J. Kloczko, J. Holowiecki, A. Hellmann, J. Walewski, M. Flensburg, J. Petersen, and T. Robak. 2008. Safety and efficacy of ofatumumab, a fully human monoclonal anti-CD20 antibody, in patients with relapsed or refractory B-cell chronic lymphocytic leukemia. A phase I-II study. *Blood* 111(3):1094–1100.

Cook, G.P., and I.M. Tomlinson. 1995. The human immunoglobulin VH repertoire. *Immunol. Today* 16(5):237–242.

Costantini, F., and E. Lacy. 1981. Introduction of a rabbit beta-globin gene into the mouse germ line. *Nature* 294(5836):92–94.

Coughlin, M., G. Lou, O. Martinez, S.K. Masterman, O.A. Olsen, A.A. Moksa, M. Farzan, J.S. Babcook, and B.S. Prabhakar. 2006. Generation and characterization of human monoclonal neutralizing antibodies with distinct binding and sequence features against SARS coronavirus using XenoMouse((R)). *Virology* 361(1):93–102.

Cunningham, D., Y. Humblet, S. Siena, D. Khayat, H. Bleiberg, A. Santoro, D. Bets, M. Mueser, A. Harstrick, C. Verslype, I. Chau, and E. Van Cutsem. 2004. Cetuximab monotherapy and cetuximab plus irinotecan in irinotecan-refractory metastatic colorectal cancer. *N. Engl. J. Med.* 351(4):337–345.

Davis, M.M. 2004. The evolutionary and structural "logic" of antigen receptor diversity. *Semin. Immunol.* 16(4):239–243.

Davies, N.P., I.R. Rosewell, J.C. Richardson, G.P. Cook, M.S. Neuberger, B.H. Brownstein, M.L. Norris, and M. Bruggemann. 1993. Creation of mice expressing human antibody light chains by introduction of a yeast artificial chromosome containing the core region of the human immunoglobulin kappa locus. *Biotechnology (NY)* 11(8):911–914.

Dolloff, N.G., M.R. Russell, N. Loizos, and A. Fatatis. 2007. Human bone marrow activates the Akt pathway in metastatic prostate cells through transactivation of the alpha-platelet-derived growth factor receptor. *Cancer Res.* 67(2):555–562.

Donohoe, M.E., G.B. Beck-Engeser, N. Lonberg, H. Karasuyama, R.L. Riley, H.M. Jack, and B.B. Blomberg. 2000. Transgenic human lambda 5 rescues the murine lambda 5 nullizygous phenotype. *J. Immunol.* 164(10):5269–5276.

Dornhofer, N., S. Spong, K. Bennewith, A. Salim, S. Klaus, N. Kambham, C. Wong, F. Kaper, P. Sutphin, R. Nacamuli, M. Hockel, Q. Le, M. Longaker, G. Yang, A. Koong, and A. Giaccia. 2006. Connective tissue growth factor-specific monoclonal antibody therapy inhibits pancreatic tumor growth and metastasis. *Cancer Res.* 66(11):5816–5827.

Downey, S.G., J.A. Klapper, F.O. Smith, J.C. Yang, R.M. Sherry, R.E. Royal, U.S. Kammula, M.S. Hughes, T.E. Allen, C.L. Levy, M. Yellin, G. Nichol, D.E. White, S.M. Steinberg, and S.A. Rosenberg. 2007. Prognostic factors related to clinical response in patients with metastatic melanoma treated by CTL-associated antigen-4 blockade. *Clin. Cancer Res.* 13(22):6681–6688.

Ehrenstein, M.R., H.T. Cook, and M.S. Neuberger. 2000. Deficiency in serum immunoglobulin (Ig)M predisposes to development of IgG autoantibodies. *J. Exp. Med.* 191(7):1253–1258.

Farr, C.D., M.R. Tabet, W.J. Ball, D.M. Fishwild, X. Wang, A.C. Nair, and W.J. Welsh. 2002. Three-dimensional quantitative structure-activity relationship analysis of ligand binding to human sequence antidigoxin monoclonal antibodies using comparative molecular field analysis. *J. Med. Chem.* 45(15):3257–3270.

Fishwild, D.M., D.V. Hudson, U. Deshpande, and A.H. Kung. 1999. Differential effects of administration of a human anti-CD4 monoclonal antibody, HM6G, in nonhuman primates. *Clin. Immunol.* 92(2):138–152.

Fishwild, D.M., S.L. O'Donnell, T. Bengoechea, D.V. Hudson, F. Harding, S.L. Bernhard, D. Jones, R.M. Kay, K.M. Higgins, S.R. Schramm, and N. Lonberg. 1996. High-avidity human IgG kappa monoclonal antibodies from a novel strain of minilocus transgenic mice. *Nature Biotechnol.* 14(7):845–851.

Foon, K.A., X.D. Yang, L.M. Weiner, A.S. Belldegrun, R.A. Figlin, J. Crawford, E.K. Rowinsky, J.P. Dutcher, N.J. Vogelzang, J. Gollub, J.A. Thompson, G. Schwartz, R.M. Bukowski, L.K. Roskos, and G.M. Schwab. 2004. Preclinical and clinical evaluations of ABX-EGF, a fully human anti-epidermal growth factor receptor antibody. *Int. J. Radiat. Oncol. Biol. Phys.* 58(3):984–990.

Foote, J., and H.N. Eisen. 1995. Kinetic and affinity limits on antibodies produced during immune responses. *Proc. Natl. Acad. Sci. USA* 92(5):1254–1256.

Fraenkel, S., R. Mostoslavsky, T.I. Novobrantseva, R. Pelanda, J. Chaudhuri, G. Esposito, S. Jung, F.W. Alt, K. Rajewsky, H. Cedar, and Y. Bergman. 2007. Allelic "choice" governs somatic hypermutation in vivo at the immunoglobulin kappa-chain locus. *Nature Immunol.* 8(7):715–722.

Fuss, I.J., C. Becker, Z. Yang, C. Groden, R.L. Hornung, F. Heller, M.F. Neurath, W. Strober, and P.J. Mannon. 2006. Both IL-12p70 and IL-23 are synthesized during active Crohn's disease and are down-regulated by treatment with anti-IL-12 p40 monoclonal antibody. *Inflamm. Bowel Dis.* 12(1):9–15.

Garambois, V., F. Glaussel, E. Foulquier, M. Ychou, M. Pugniere, R.X. Luo, B. Bezabeh, and A. Pelegrin. 2004. Fully human IgG and IgM antibodies directed against the carcinoembryonic antigen (CEA) Gold 4 epitope and designed for radioimmunotherapy (RIT) of colorectal cancers. *BMC Cancer* 4:75.

Gibson, T.B., A. Ranganathan, and A. Grothey. 2006. Randomized phase III trial results of panitumumab, a fully human anti-epidermal growth factor receptor monoclonal antibody, in metastatic colorectal cancer. *Clin. Colorectal Cancer* 6(1):29–31.

Glennie, M. J., R.R. French, M.S. Cragg, and R.P. Taylor. 2007. Mechanisms of killing by anti-CD20 monoclonal antibodies. *Mol. Immunol.* 44(16):3823–3837.

Goldstein, G. et al. 1985. A randomized clinical trial of OKT3 monoclonal antibody for acute rejection of cadaveric renal transplants. Ortho Multicenter Transplant Study Group. *N. Engl. J. Med.* 313(6):337–342.

Goodhardt, M., P. Cavelier, M.A. Akimenko, G. Lutfalla, C. Babinet, and F. Rougeon. 1987. Rearrangement and expression of rabbit immunoglobulin kappa light chain gene in transgenic mice. *Proc. Natl. Acad. Sci. USA* 84(12):4229–4233.

Gordon, J.W., and F.H. Ruddle. 1981. Integration and stable germ line transmission of genes injected into mouse pronuclei. *Science* 214(4526):1244–1246.

Green, L.L., M.C. Hardy, C.E. Maynard-Currie, H. Tsuda, D.M. Louie, M.J. Mendez, H. Abderrahim, M. Noguchi, D.H. Smith, Y. Zeng, et al. 1994. Antigen-specific human monoclonal antibodies from mice engineered with human Ig heavy and light chain YACs. *Nature Genet.* 7(1):13–21.

Green, L.L. and A. Jakobovits. 1998. Regulation of B cell development by variable gene complexity in mice reconstituted with human immunoglobulin yeast artificial chromosomes. *J. Exp. Med.* 188(3):483–495.

Greenough, T.C., G.J. Babcock, A. Roberts, H.J. Hernandez, W.D. Thomas, J.A. Coccia, Jr, R.F. Graziano, M. Srinivasan, I. Lowy, R.W. Finberg, K. Subbarao, L. Vogel, M. Somasundaran, K. Luzuriaga, J.L. Sullivan, and D.M. Ambrosino. 2005. Development and characterization of a severe acute respiratory syndrome-associated coronavirus-neutralizing human monoclonal antibody that provides effective immunoprophylaxis in mice. *J. Infect. Dis.* 191(4):507–514.

Guloglu, F.B., and C.A. Roman. 2006. Precursor B cell receptor signaling activity can be uncoupled from surface expression. *J. Immunol.* 176(11):6862–6872.

Harbers, K., D. Jahner, and R. Jaenisch. 1981. Microinjection of cloned retroviral genomes into mouse zygotes: Integration and expression in the animal. *Nature* 293(5833):540–542.

Harding, F.A., and N. Lonberg. 1995. Class switching in human immunoglobulin transgenic mice. *Ann. NY Acad. Sci.* 764:536–546.

He, Y., W.J. Honnen, C.P. Krachmarov, M. Burkhart, S.C. Kayman, J. Corvalan, and A. Pinter. 2002. Efficient isolation of novel human monoclonal antibodies with neutralizing activity against HIV-1 from transgenic mice expressing human Ig loci. *J. Immunol.* 169(1):595–605.

Heuck, F., J. Ellermann, P. Borchmann, A. Rothe, H. Hansen, A. Engert, and E.P. von Strandmann. 2004. Combination of the human anti-CD30 antibody 5F11 with cytostatic drugs enhances its antitumor activity against Hodgkin and anaplastic large cell lymphoma cell lines. *J. Immunother. (1997)* 27(5):347–353.

Hodi, F.S., M.C. Mihm, R.J. Soiffer, F.G. Haluska, M. Butler, M.V. Seiden, T. Davis, R. Henry-Spires, S. MacRae, A. Willman, R. Padera, M.T. Jaklitsch, S. Shankar, T.C. Chen, A. Korman, J.P. Allison, and G. Dranoff. 2003. Biologic activity of cytotoxic T lymphocyte-associated antigen 4 antibody blockade in previously vaccinated metastatic melanoma and ovarian carcinoma patients. *Proc. Natl. Acad. Sci. USA* 100(8):4712–4717.

Holmes, E.H. 2001. PSMA specific antibodies and their diagnostic and therapeutic use. *Expert Opin. Investig. Drugs* 10(3):511–519.

Horikawa, K., S.W. Martin, S.L. Pogue, K. Silver, K. Peng, K. Takatsu, and C.C. Goodnow. 2007. Enhancement and suppression of signaling by the conserved tail of IgG memory-type B cell antigen receptors. *J. Exp. Med.* 204(4):759–769.

Huang, S., L. Mills, B. Mian, C. Tellez, M. McCarty, X.D. Yang, J.M. Gudas, and M. Bar-Eli. 2002. Fully humanized neutralizing antibodies to interleukin-8 (ABX-IL8) inhibit angiogenesis, tumor growth, and metastasis of human melanoma. *Am. J. Pathol.* 161(1):125–134.

Huhn, D., C. von Schilling, M. Wilhelm, A.D. Ho, M. Hallek, R. Kuse, W. Knauf, U. Riedel, A. Hinke, S. Srock, S. Serke, C. Peschel, and B. Emmerich. 2001. Rituximab therapy of patients with B-cell chronic lymphocytic leukemia. *Blood* 98(5):1326–1331.

Ignatovich, O., I.M. Tomlinson, P.T. Jones, and G. Winter. 1997. The creation of diversity in the human immunoglobulin V(lambda) repertoire. *J. Mol. Biol.* 268(1):69–77.

Imakiire, T., M. Kuroki, H. Shibaguchi, H. Abe, Y. Yamauchi, A. Ueno, Y. Hirose, H. Yamada, Y. Yamashita, T. Shirakusa, and I. Ishida. 2004. Generation, immunologic characterization and antitumor effects of human monoclonal antibodies for carcinoembryonic antigen. *Int. J. Cancer* 108(8):564–570.

Inlay, M., F.W. Alt, D. Baltimore, and Y. Xu. 2002. Essential roles of the kappa light chain intronic enhancer and 3′ enhancer in kappa rearrangement and demethylation. *Nature Immunol.* 3(5):463–468.

Ishida, I., K. Tomizuka, H. Yoshida, T. Tahara, N. Takahashi, A. Ohguma, S. Tanaka, M. Umehashi, H. Maeda, C. Nozaki, E. Halk, and N. Lonberg. 2002. Production of human monoclonal and polyclonal antibodies in TransChromo animals. *Cloning Stem Cells* 4(1):91–102.

Itala, M., C.H. Geisler, E. Kimby, E. Juvonen, G. Tjonnfjord, K. Karlsson, and K. Remes. 2002. Standard-dose anti-CD20 antibody rituximab has efficacy in chronic lymphocytic leukaemia: Results from a Nordic multicentre study. *Eur. J. Haematol.* 69(3):129–134.

Jaber, S.H., E.W. Cowen, L.R. Haworth, S.L. Booher, D.M. Berman, S.A. Rosenberg, and S.T. Hwang. 2006. Skin reactions in a subset of patients with stage IV melanoma treated with anti-cytotoxic T-lymphocyte antigen 4 monoclonal antibody as a single agent. *Arch. Dermatol.* 142(2):166–172.

Jakobovits, A., R.G. Amado, X. Yang, L. Roskos, and G. Schwab. 2007. From XenoMouse technology to panitumumab, the first fully human antibody product from transgenic mice. *Nature Biotechnol.* 25(10):1134–1143.

Jakobovits, A., A.L. Moore, L.L. Green, G.J. Vergara, C.E. Maynard-Currie, H.A. Austin, and S. Klapholz. 1993. Germ-line transmission and expression of a human-derived yeast artificial chromosome. *Nature* 362(6417):255–258.

Jones, P.T., P.H. Dear, J. Foote, M.S. Neuberger, and G. Winter. 1986. Replacing the complementarity-determining regions in a human antibody with those from a mouse. *Nature* 321(6069):522–525.

Kanmaz, T., J.J. Fechner, J. Torrealba, Jr, H.T. Kim, Y. Dong, T.D. Oberley, J.M. Schultz, D.D. Bloom, M. Katayama, W. Dar, J. Markovits, W. Schuler, H. Hu, M.M. Hamawy, and S.J. Knechtle. 2004. Monotherapy with the novel human anti-CD154 monoclonal antibody ABI793 in rhesus monkey renal transplantation model. *Transplantation* 77(6):914–920.

Kasper, L.H., D. Everitt, T.P. Leist, K.A. Ryan, M.A. Mascelli, K. Johnson, A. Raychaudhuri, and T. Vollmer. 2006. A phase I trial of an interleukin-12/23 monoclonal antibody in relapsing multiple sclerosis. *Curr. Med. Res. Opin.* 22(9):1671–1678.

Kauffman, C.L., N. Aria, E. Toichi, T.S. McCormick, K.D. Cooper, A.B. Gottlieb, D.E. Everitt, B. Frederick, Y. Zhu, M.A. Graham, C.E. Pendley, and M.A. Mascelli. 2004. A phase I study evaluating the safety, pharmacokinetics, and clinical response of a human IL-12 p40 antibody in subjects with plaque psoriasis. *J. Invest. Dermatol.* 123(6):1037–1044.

Keler, T., E. Halk, L. Vitale, T. O'Neill, D. Blanset, S. Lee, M. Srinivasan, R.F. Graziano, T. Davis, N. Lonberg, and A. Korman. 2003. Activity and safety of CTLA-4 blockade combined with vaccines in cynomolgus macaques. *J. Immunol.* 171(11):6251–6259.

Kim, Y.H., M. Duvic, E. Obitz, R. Gniadecki, L. Iversen, A. Osterborg, S. Whittaker, T.M. Illidge, T. Schwarz, R. Kaufmann, K. Cooper, K.M. Knudsen, S. Lisby, O. Baadsgaard, and S.J. Knox. 2007. Clinical efficacy of zanolimumab (HuMax-CD4): two phase 2 studies in refractory cutaneous T-cell lymphoma. *Blood* 109(11): 4655–4662.

Klimm, B., R. Schnell, V. Diehl, and A. Engert. 2005. Current treatment and immunotherapy of Hodgkin's lymphoma. *Haematologica* 90(12):1680–1692.

Korman, A.J., K. S. Peggs, and J.P. Allison. 2006. Checkpoint blockade in cancer immunotherapy. *Adv. Immunol.* 90:297–339.

Kraus, M., M.B. Alimzhanov, N. Rajewsky, and K. Rajewsky. 2004. Survival of resting mature B lymphocytes depends on BCR signaling via the Igalpha/beta heterodimer. *Cell* 117(6):787–800.

Krueger, G.G., R.G. Langley, C. Leonardi, N. Yeilding, C. Guzzo, Y. Wang, L.T. Dooley, and M. Lebwohl. 2007. A human interleukin-12/23 monoclonal antibody for the treatment of psoriasis. *N. Engl. J. Med.* 356(6):580–592.

Kumazaki, K., B. Tirosh, R. Maehr, M. Boes, T. Honjo, and H.L. Ploegh. 2007. AID−/−mus−/− mice are agammaglobulinemic and fail to maintain B220-CD138+ plasma cells. *J. Immunol.* 178(4):2192–2203.

Kuroiwa, Y., K. Tomizuka, T. Shinohara, Y. Kazuki, H. Yoshida, A. Ohguma, T. Yamamoto, S. Tanaka, M. Oshimura, and I. Ishida. 2000. Manipulation of human minichromosomes to carry greater than megabase-sized chromosome inserts. *Nature Biotechnol.* 18(10):1086–1090.

Kuroki, M., H. Yamada, H. Shibaguchi, K. Hachimine, Y. Hirose, T. Kinugasa, and I. Ishida. 2005. Preparation of human IgG and IgM monoclonal antibodies for MK-1/Ep-CAM by using human immunoglobulin gene-transferred mouse and gene cloning of their variable regions. *Anticancer Res.* 25(6A):3733–3739.

Kuus-Reichel, K., L.S. Grauer, L.M. Karavodin, C. Knott, M. Krusemeier, and N.E. Kay. 1994. Will immunogenicity limit the use, efficacy, and future development of therapeutic monoclonal antibodies? *Clin. Diagn. Lab. Immunol.* 1(4):365–372.

Lammerts van Bueren J.J., W.K. Bleeker, H.O. Bogh, M. Houtkamp, J. Schuurman, J.G. van de Winkel, and P.W. Parren. 2006. Effect of target dynamics on pharmacokinetics of a novel therapeutic antibody against the epidermal growth factor receptor: Implications for the mechanisms of action. *Cancer Res.* 66(15):7630–7638.

Leach, D.R., M.F. Krummel, and J.P. Allison. 1996. Enhancement of antitumor immunity by CTLA-4 blockade. *Science* 271(5256):1734–1736.

Lenz, H.J., E. Van Cutsem, S. Khambata-Ford, R.J. Mayer, P. Gold, P. Stella, B. Mirtsching, A.L. Cohn, A.W. Pippas, N. Azarnia, Z. Tsuchihashi, D.J. Mauro, and E.K. Rowinsky. 2006. Multicenter phase II and translational study of cetuximab in metastatic colorectal carcinoma refractory to irinotecan, oxaliplatin, and fluoropyrimidines. *J. Clin. Oncol.* 24(30):4914–4921.

Lewiecki, E.M., P.D. Miller, M.R. McClung, S.B. Cohen, M.A. Bolognese, Y. Liu, A. Wang, S. Siddhanti, and L.A. Fitzpatrick. 2007. Two-year treatment with denosumab (AMG 162) in a randomized phase 2 study of postmenopausal women with low bone mineral density. *J. Bone Miner. Res.* 22(12):1832–1841.

Lipton, A., G.G. Steger, J. Figueroa, C. Alvarado, P. Solal-Celigny, J.J. Body, R. de Boer, R. Berardi, P. Gascon, K.S. Tonkin, R. Coleman, A.H. Paterson, M.C. Peterson, M. Fan, A. Kinsey, and S. Jun. 2007. Randomized active-controlled phase II study of denosumab efficacy and safety in patients with breast cancer-related bone metastases. *J. Clin. Oncol.* 25(28):4431–4437.

Lonberg, N. 2005. Human antibodies from transgenic animals. *Nature Biotechnol.* 23(9):1117–1125.

Lonberg, N., and D. Huszar. 1995. Human antibodies from transgenic mice. *Int. Rev. Immunol.* 13(1):65–93.

Lonberg, N., L.D. Taylor, F.A. Harding, M. Trounstine, K.M. Higgins, S.R. Schramm, C.C. Kuo, R. Mashayekh, K. Wymore, J.G. McCabe, et al. 1994. Antigen-specific human antibodies from mice comprising four distinct genetic modifications. *Nature* 368(6474):856–859.

Ma, D., C.E. Hopf, A.D. Malewicz, G.P. Donovan, P.D. Senter, W.F. Goeckeler, P.J. Maddon, and W.C. Olson. 2006. Potent antitumor activity of an auristatin-conjugated, fully human monoclonal antibody to prostate-specific membrane antigen. *Clin. Cancer Res.* 12(8):2591–2596.

Maini, R.N., F.C. Breedveld, J.R. Kalden, J.S. Smolen, D. Davis, J.D. Macfarlane, C. Antoni, B. Leeb, M.J. Elliott, J.N. Woody, T.F. Schaible, and M. Feldmann. 1998. Therapeutic efficacy of multiple intravenous infusions of anti-tumor necrosis factor alpha monoclonal antibody combined with low-dose weekly methotrexate in rheumatoid arthritis. *Arthritis Rheum.* 41(9):1552–1563.

Maitta, R.W., K. Datta, Q. Chang, R.X. Luo, B. Witover, K. Subramaniam, and L.A. Pirofski. 2004. Protective and nonprotective human immunoglobulin M monoclonal antibodies to Cryptococcus neoformans glucuronoxylomannan manifest different specificities and gene use profiles. *Infect Immun.* 72(8):4810–4818.

Maker, A.V., P. Attia, and S.A. Rosenberg. 2005. Analysis of the cellular mechanism of antitumor responses and autoimmunity in patients treated with CTLA-4 blockade. *J. Immunol.* 175(11):7746–7754.

Maker, A.V., G.Q. Phan, P. Attia, J.C. Yang, R.M. Sherry, S.L. Topalian, U.S. Kammula, R.E. Royal, L.R. Haworth, C. Levy, D. Kleiner, S.A. Mavroukakis, M. Yellin, and S.A. Rosenberg. 2005. Tumor regression and autoimmunity in patients treated with cytotoxic T lymphocyte-associated antigen 4 blockade and interleukin 2: A phase I/II study. *Ann. Surg. Oncol.* 12(12):1005–1016.

Maker, A.V., J.C. Yang, R.M. Sherry, S.L. Topalian, U.S. Kammula, R.E. Royal, M. Hughes, M.J. Yellin, L.R. Haworth, C. Levy, T. Allen, S.A. Mavroukakis, P. Attia, and S.A. Rosenberg. 2006. Intrapatient dose escalation of anti-CTLA-4 antibody in patients with metastatic melanoma. *J. Immunother. (1997)* 29(4):455–463.

Mannon, P.J., I.J. Fuss, L. Mayer, C.O. Elson, W.J. Sandborn, D. Present, B. Dolin, N. Goodman, C. Groden, R.L. Hornung, M. Quezado, Z. Yang, M.F. Neurath, J. Salfeld, G.M. Veldman, U. Schwertschlag, and W. Strober. 2004. Anti-interleukin-12 antibody for active Crohn's disease. *N. Engl. J. Med.* 351(20):2069–2079.

Mansour, S.L., K.R. Thomas, and M.R. Capecchi. 1988. Disruption of the proto-oncogene int-2 in mouse embryo-derived stem cells: A general strategy for targeting mutations to non-selectable genes. *Nature* 336(6197):348–352.

Martin, P.L., Q. Jiao, J. Cornacoff, W. Hall, B. Saville, J.A. Nemeth, A. Schantz, M. Mata, H. Jang, A.A. Fasanmade, L. Anderson, M.A. Graham, H.M. Davis, and G. Treacy. 2005. Absence of adverse effects in cynomolgus macaques treated with CNTO 95, a fully human anti-alphav integrin monoclonal antibody, despite widespread tissue binding. *Clin. Cancer Res.* 11(19):6959–6965.

Martin, P.L., S. Oneda, and G. Treacy. 2007. Effects of an anti-TNF-alpha monoclonal antibody, administered throughout pregnancy and lactation, on the development of the macaque immune system. *Am. J. Reprod. Immunol.* 58(2):138–149.

McCafferty, J., A.D. Griffiths, G. Winter, and D.J. Chiswell. 1990. Phage antibodies: Filamentous phage displaying antibody variable domains. *Nature* 348(6301):552–554.

McClung, M.R., E.M. Lewiecki, S.B. Cohen, M.A. Bolognese, G.C. Woodson, A.H. Moffett, M. Peacock, P.D. Miller, S.N. Lederman, C.H. Chesnut, D. Lain, A.J. Kivitz, D.L. Holloway, C. Zhang, M.C. Peterson, and P.J. Bekker. 2006. Denosumab in postmenopausal women with low bone mineral density. *N. Engl. J. Med.* 354(8):821–831.

Meffre, E., M. Milili, C. Blanco-Betancourt, H. Antunes, M.C. Nussenzweig, and C. Schiff. 2001. Immunoglobulin heavy chain expression shapes the B cell receptor repertoire in human B cell development. *J. Clin. Invest.* 108(6):879–886.

Melchers, F. 2005. The pre-B-cell receptor: Selector of fitting immunoglobulin heavy chains for the B-cell repertoire. *Nature Rev. Immunol.* 5(7):578–584.

Mendez, M.J., L.L. Green, J.R. Corvalan, X.C. Jia, C.E. Maynard-Currie, X.D. Yang, M.L. Gallo, D.M. Louie, D.V. Lee, K.L. Erickson, J. Luna, C.M. Roy, H. Abderrahim, F. Kirschenbaum, M. Noguchi, D.H. Smith, A. Fukushima, J.F. Hales, S. Klapholz, M.H. Finer, C.G. Davis, K.M. Zsebo, and A. Jakobovits. 1997. Functional transplant of megabase human immunoglobulin loci recapitulates human antibody response in mice. *Nature Genet.* 15(2):146–156.

Melnikova, V.O., and M. Bar-Eli. 2006. Bioimmunotherapy for melanoma using fully human antibodies targeting MCAM/MUC18 and IL-8. *Pigment Cell Res.* 19(5):395–405.

Milili, M., C. Schiff, M. Fougereau, and C. Tonnelle. 1996. The VDJ repertoire expressed in human preB cells reflects the selection of bona fide heavy chains. *Eur. J. Immunol.* 26(1):63–69.

Morrison, S.L., M.J. Johnson, L.A. Herzenberg, and V.T. Oi. 1984. Chimeric human antibody molecules: Mouse antigen-binding domains with human constant region domains. *Proc. Natl. Acad. Sci. USA* 81(21):6851–6855.

Mori, E., M. Thomas, K. Motoki, K. Nakazawa, T. Tahara, K. Tomizuka, I. Ishida, and S. Kataoka. 2004. Human normal hepatocytes are susceptible to apoptosis signal mediated by both TRAIL-R1 and TRAIL-R2. *Cell Death Differ.* 11(2):203–207.

Mukherjee, J., K. Chios, D. Fishwild, D. Hudson, S. O'Donnell, S.M. Rich, A. Donohue-Rolfe, and S. Tzipori. 2002a. Human Stx2-specific monoclonal antibodies prevent systemic complications of Escherichia coli O157:H7 infection. *Infect. Immun.* 70(2):612–619.

Mukherjee, J., K. Chios, D. Fishwild, D. Hudson, S. O'Donnell, S.M. Rich, A. Donohue-Rolfe, and S. Tzipori. 2002b. Production and characterization of protective human antibodies against Shiga toxin 1. *Infect. Immun.* 70(10):5896–5899.

Muramatsu, M., K. Kinoshita, S. Fagarasan, S. Yamada, Y. Shinkai, and T. Honjo. 2000. Class switch recombination and hypermutation require activation-induced cytidine deaminase (AID), a potential RNA editing enzyme. *Cell* 102(5):553–563.

Nagy, A.G.M., K. Vinterstein, et al. 2003. *Manipulating the mouse embryo: A laboratory manual*, 3rd edition. Cold Spring Harbor, NY: Cold Spring Harbor Laboratory Press.

Nicholson, I.C., X. Zou, A.V. Popov, G.P. Cook, E.M. Corps, S. Humphries, C. Ayling, B. Goyenechea, J. Xian, M.J. Taussig, M.S. Neuberger, and M. Bruggemann. 1999. Antibody repertoires of four- and five-feature translocus mice carrying human immunoglobulin heavy chain and kappa and lambda light chain yeast artificial chromosomes. *J. Immunol.* 163(12):6898–6906.

Nozawa, S., D. Aoki, K. Tsukazaki, N. Susumu, M. Sakayori, N. Suzuki, A. Suzuki, R. Wakita, M. Mukai, Y. Egami, K. Kojima-Aikawa, I. Ishida, F. Belot, O. Hindsgaul, M. Fukuda, and M.N. Fukuda. 2004.

HMMC-1: A humanized monoclonal antibody with therapeutic potential against Mullerian duct-related carcinomas. *Clin. Cancer Res.* 10(20):7071–7078.

O'Brien, S.M., H. Kantarjian, D.A. Thomas, F.J. Giles, E.J. Freireich, J. Cortes, S. Lerner, and M.J. Keating. 2001. Rituximab dose-escalation trial in chronic lymphocytic leukemia. *J. Clin. Oncol.* 19(8):2165–2170.

O'Mahony, D., J.C. Morris, C. Quinn, W. Gao, W.H. Wilson, B. Gause, S. Pittaluga, S. Neelapu, M. Brown, T.A. Fleisher, J.L. Gulley, J. Schlom, R. Nussenblatt, P. Albert, T.A. Davis, I. Lowy, M. Petrus, T.A. Waldmann, and J.E. Janik. 2007. A pilot study of CTLA-4 blockade after cancer vaccine failure in patients with advanced malignancy. *Clin. Cancer Res.* 13(3):958–964.

Ostendorf, T., C.R.C. van Roeyen, J.D. Peterson, U. Kunter, R. Eitner, A.J. Hamad, G. Chan, X.C. Jia, J. Macaluso, G. Gazit-Bornstein, B.A. Keyt, H.S. Lichenstein, W.J. LaRochelle, and J. Floege. 2003. A fully human monoclonal antibody (CR002) identifies PDGF-D as a novel mediator of mesangioproliferative glomerulonephritis. *J. Am. Soc. Nephrol.* 14(9):2237–2247.

Parren, P.W., P.A. Warmerdam, L.C. Boeije, J. Arts, N.A. Westerdaal, A. Vlug, P.J. Capel, L.A. Aarden, and J.G. van de Winkel. 1992. On the interaction of IgG subclasses with the low affinity Fc gamma RIIa (CD32) on human monocytes, neutrophils, and platelets. Analysis of a functional polymorphism to human IgG2. *J. Clin. Invest.* 90(4):1537–1546.

Parry, R., D. Schneider, D. Hudson, D. Parkes, J.A. Xuan, A. Newton, P. Toy, R. Lin, R. Harkins, B. Alicke, S. Biroc, P.J. Kretschmer, M. Halks-Miller, H. Klocker, Y. Zhu, B. Larsen, R.R. Cobb, P. Bringmann, G. Roth, J.S. Lewis, H. Dinter, and G. Parry. 2005. Identification of a novel prostate tumor target, mindin/RG-1, for antibody-based radiotherapy of prostate cancer. *Cancer Res.* 65(18):8397–8405.

Pendley, C., A. Schantz, and C. Wagner. 2003. Immunogenicity of therapeutic monoclonal antibodies. *Curr. Opin. Mol. Ther.* 5(2):172–179.

Phan, G.Q., J.C. Yang, R.M. Sherry, P. Hwu, S.L. Topalian, D.J. Schwartzentruber, N.P. Restifo, L.R. Haworth, C.A. Seipp, L.J. Freezer, K.E. Morton, S.A. Mavroukakis, P.H. Duray, S.M. Steinberg, J.P. Allison, T.A. Davis, and S.A. Rosenberg. 2003. Cancer regression and autoimmunity induced by cytotoxic T lymphocyte-associated antigen 4 blockade in patients with metastatic melanoma. *Proc. Natl. Acad. Sci. USA* 100(14):8372–8377.

Popov, A.V., X. Zou, J. Xian, I.C. Nicholson, and M. Bruggemann. 1999. A human immunoglobulin lambda locus is similarly well expressed in mice and humans. *J. Exp. Med.* 189(10):1611–1620.

Ramakrishna, V., J.F. Treml, L. Vitale, J.E. Connolly, T. O'Neill, P.A. Smith, C.L. Jones, L.Z. He, J. Goldstein, P.K. Wallace, T. Keler, and M.J. Endres. 2004. Mannose receptor targeting of tumor antigen pmel17 to human dendritic cells directs anti-melanoma T cell responses via multiple HLA molecules. *J. Immunol.* 172(5):2845–2852.

Rathanaswami, P., S. Roalstad, L. Roskos, Q.J. Su, S. Lackie, and J. Babcook. 2005. Demonstration of an in vivo generated sub-picomolar affinity fully human monoclonal antibody to interleukin-8. *Biochem. Biophys. Res. Commun.* 334(4):1004–1013.

Razai, A., C. Garcia-Rodriguez, J. Lou, I.N. Geren, C.M. Forsyth, Y. Robles, R. Tsai, T.J. Smith, L.A. Smith, R.W. Siegel, M. Feldhaus, and J.D. Marks. 2005. Molecular evolution of antibody affinity for sensitive detection of botulinum neurotoxin type A. *J. Mol. Biol.* 351(1):158–169.

Reuben, J.M., B.N. Lee, C. Li, J. Gomez-Navarro, V.A. Bozon, C.A. Parker, I.M. Hernandez, C. Gutierrez, G. Lopez-Berestein, and L.H. Camacho. 2006. Biologic and immunomodulatory events after CTLA-4 blockade with ticilimumab in patients with advanced malignant melanoma. *Cancer* 106(11):2437–2444.

Reynaud, S., L. Delpy, L. Fleury, H.L. Dougier, C. Sirac, and M. Cogne. 2005. Interallelic class switch recombination contributes significantly to class switching in mouse B cells. *J. Immunol.* 174(10):6176–6183.

Ribas, A., L.H. Camacho, G. Lopez-Berestein, D. Pavlov, C.A. Bulanhagui, R. Millham, B. Comin-Anduix, J.M. Reuben, E. Seja, C.A. Parker, A. Sharma, J.A. Glaspy, and J. Gomez-Navarro. 2005. Antitumor activity in melanoma and anti-self responses in a phase I trial with the anti-cytotoxic T lymphocyte-associated antigen 4 monoclonal antibody CP-675,206. *J. Clin. Oncol.* 23(35):8968–8977.

Ribas, A., J.A. Glaspy, Y. Lee, V.B. Dissette, E. Seja, H.T. Vu, N.S. Tchekmedyian, D. Oseguera, B. Comin-Anduix, J.A. Wargo, S.N. Amarnani, W.H. McBride, J.S. Economou, and L.H. Butterfield. 2004. Role of dendritic cell phenotype, determinant spreading, and negative costimulatory blockade in dendritic cell-based melanoma immunotherapy. *J. Immunother. (1997)* 27(5):354–367.

Ribas, A., D.C. Hanson, D.A. Noe, R. Millham, D.J. Guyot, S.H. Bernstein, P.C. Canniff, A. Sharma, and J. Gomez-Navarro. 2007. Tremelimumab (CP-675,206), a cytotoxic T lymphocyte associated antigen 4 blocking monoclonal antibody in clinical development for patients with cancer. *Oncologist* 12(7):873–883.

Rider, D.A., C.E. Havenith, R. de Ridder, J. Schuurman, C. Favre, J.C. Cooper, S. Walker, O. Baadsgaard, S. Marschner, J.G. vandeWinkel, J. Cambier, P.W. Parren, D.R. Alexander. 2007. A human CD4 monoclonal antibody for the treatment of T-cell lymphoma combines inhibition of T-cell signaling by a dual mechanism with potent Fc-dependent effector activity. *Cancer Res.* 67(20):9945–9953.

Roost, H.P., M.F. Bachmann, A. Haag, U. Kalinke, V. Pliska, H. Hengartner, and R.M. Zinkernagel. 1995. Early high-affinity neutralizing anti-viral IgG responses without further overall improvements of affinity. *Proc. Natl. Acad. Sci. USA* 92(5):1257–1261.

Rosner, K., D.B. Winter, R.E. Tarone, G.L. Skovgaard, V.A. Bohr, and P.J. Gearhart. 2001. Third complementarity-determining region of mutated VH immunoglobulin genes contains shorter V, D, J, P, and N components than non-mutated genes. *Immunology* 103(2):179–187.

Rowinsky, E.K., G.H. Schwartz, J.A. Gollob, J.A. Thompson, N.J. Vogelzang, R. Figlin, R. Bukowski, N. Haas, P. Lockbaum, Y.P. Li, R. Arends, K.A. Foon, G. Schwab, and J. Dutcher. 2004. Safety, pharmacokinetics, and activity of ABX-EGF, a fully human anti-epidermal growth factor receptor monoclonal antibody in patients with metastatic renal cell cancer. *J. Clin. Oncol.* 22(15):3003–3015.

Sanderson, K., R. Scotland, P. Lee, D. Liu, S. Groshen, J. Snively, S. Sian, G. Nichol, T. Davis, T. Keler, M. Yellin, and J. Weber. 2005. Autoimmunity in a phase I trial of a fully human anti-cytotoxic T-lymphocyte antigen-4 monoclonal antibody with multiple melanoma peptides and Montanide ISA 51 for patients with resected stages III and IV melanoma. *J. Clin. Oncol.* 23(4):741–750.

Schedl, A., Z. Larin, L. Montoliu, E. Thies, G. Kelsey, H. Lehrach, and G. Schutz. 1993. A method for the generation of YAC transgenic mice by pronuclear microinjection. *Nucleic Acids Res.* 21(20):4783–4787.

Schelonka, R.L., I.I. Ivanov, D.H. Jung, G.C. Ippolito, L. Nitschke, Y. Zhuang, G.L. Gartland, J. Pelkonen, F.W. Alt, K. Rajewsky, and H.W. Schroeder, Jr. 2005. A single DH gene segment creates its own unique CDR-H3 repertoire and is sufficient for B cell development and immune function. *J. Immunol.* 175(10):6624–6632.

Schuler, W., M. Bigaud, V. Brinkmann, F. Di Padova, S. Geisse, H. Gram, V. Hungerford, B. Kleuser, C. Kristofic, K. Menninger, R. Tees, G. Wieczorek, C. Wilt, C. Wioland, and M. Zurini. 2004. Efficacy and safety of ABI793, a novel human anti-human CD154 monoclonal antibody, in cynomolgus monkey renal allotransplantation. *Transplantation* 77(5):717–726.

Schwartzberg, P.L., S.P. Goff, and E.J. Robertson. 1989. Germ-line transmission of a c-abl mutation produced by targeted gene disruption in ES cells. *Science* 246(4931):799–803.

Schwarz, E.M. and C.T. Ritchlin. 2007. Clinical development of anti-RANKL therapy. *Arthritis Res. Ther.* 9 (Suppl 1):S7.

Seshasayee, D., W.P. Lee, M. Zhou, J. Shu, E. Suto, J. Zhang, L. Diehl, C.D. Austin, Y.G. Meng, M. Tan, S.L. Bullens, S. Seeber, M.E. Fuentes, A.F. Labrijn, Y.M. Graus, L.A. Miller, E.S. Schelegle, D.M. Hyde, L.C. Wu, S.G. Hymowitz, and F. Martin. 2007. In vivo blockade of OX40 ligand inhibits thymic stromal lymphopoietin driven atopic inflammation. *J. Clin. Invest.* 117(12):3868–3878.

Sheoran, A.S., S. Chapman-Bonofiglio, B.R. Harvey, J. Mukherjee, G. Georgiou, A. Donohue-Rolfe, and S. Tzipori. 2005. Human antibody against shiga toxin 2 administered to piglets after the onset of diarrhea due to Escherichia coli O157:H7 prevents fatal systemic complications. *Infect. Immun.* 73(8):4607–4613.

Shimizu, T., C. Mundt, S. Licence, F. Melchers, and I.L. Martensson. 2002. VpreB1/VpreB2/lambda 5 triple-deficient mice show impaired B cell development but functional allelic exclusion of the IgH locus. *J. Immunol.* 168(12):6286–6293.

Skov, L., K. Kragballe, C. Zachariae, E.R. Obitz, E.A. Holm, G.B. Jemec, H. Solvsten, H.H. Ibsen, L. Knudsen, P. Jensen, J.H. Petersen, T. Menne, and O. Baadsgaard. 2003. HuMax-CD4: A fully human monoclonal anti-CD4 antibody for the treatment of psoriasis vulgaris. *Arch. Dermatol.* 139(11):1433–1439.

Sloan, S.E., C. Hanlon, W. Weldon, M. Niezgoda, J. Blanton, J. Self, K.J. Rowley, R.B. Mandell, G.J. Babcock, W.D. Thomas, C.D. Rupprecht, Jr, and D.M. Ambrosino. 2007. Identification and characterization of a human monoclonal antibody that potently neutralizes a broad panel of rabies virus isolates. *Vaccine* 25(15):2800–2810.

Small, E.J., N.S. Tchekmedyian, B.I. Rini, L. Fong, I. Lowy, and J.P. Allison. 2007. A pilot trial of CTLA-4 blockade with human anti-CTLA-4 in patients with hormone-refractory prostate cancer. *Clin. Cancer Res.* 13(6):1810–1815.

Strauss, W.M., J. Dausman, C. Beard, C. Johnson, J.B. Lawrence, and R. Jaenisch. 1993. Germ line transmission of a yeast artificial chromosome spanning the murine alpha 1(I) collagen locus. *Science* 259(5103):1904–1907.

Suarez, E., R. Yáñez, Y. Barrios, and R. Díaz-Espada. 2004. Human monoclonal antibodies produced in transgenic BABkappa,lambda mice recognising idiotypic immunoglobulins of human lymphoma cells. *Mol. Immunol.* 41(5):519–526.

Suzuki, N., D. Aoki, Y. Tamada, N. Susumu, K. Orikawa, K. Tsukazaki, M. Sakayori, A. Suzuki, T. Fukuchi, M. Mukai, K. Kojima-Aikawa, I. Ishida, and S. Nozawa. 2004. HMOCC-1, a human monoclonal antibody that inhibits adhesion of ovarian cancer cells to human mesothelial cells. *Gynecol. Oncol.* 95(2):290–298.

Tai, Y.T., X. Li, X. Tong, D. Santos, T. Otsuki, L. Catley, O. Tournilhac, K. Podar, T. Hideshima, R. Schlossman, P. Richardson, N.C. Munshi, M. Luqman, and K.C. Anderson. 2005. Human anti-CD40 antagonist antibody triggers significant antitumor activity against human multiple myeloma. *Cancer Res.* 65(13):5898–5906.

Tarhini, A.A., and J.M. Kirkwood. 2007. Tremelimumab, a fully human monoclonal IgG2 antibody against CTLA4 for the potential treatment of cancer. *Curr. Opin. Mol. Ther.* 9(5):505–514.

Taylor, L.D., C.E. Carmack, D. Huszar, K.M. Higgins, R. Mashayekh, G. Sequar, S.R. Schramm, C.C. Kuo, S.L. O'Donnell, R.M. Kay, et al. 1994. Human immunoglobulin transgenes undergo rearrangement, somatic mutation and class switching in mice that lack endogenous IgM. *Int. Immunol.* 6(4):579–591.

Taylor, L.D., C.E. Carmack, S.R. Schramm, R. Mashayekh, K.M. Higgins, C.C. Kuo, C. Woodhouse, R.M. Kay, and N. Lonberg. 1992. A transgenic mouse that expresses a diversity of human sequence heavy and light chain immunoglobulins. *Nucleic Acids Res.* 20(23):6287–6295.

Teeling, J.L., R.R. French, M.S. Cragg, J. van den Brakel, M. Pluyter, H. Huang, C. Chan, P.W. Parren, C.E. Hack, M. Dechant, T. Valerius, J.G. van de Winkel, and M.J. Glennie. 2004. Characterization of new human CD20 monoclonal antibodies with potent cytolytic activity against non-Hodgkin lymphomas. *Blood* 104(6):1793–1800.

Teeling, J.L., W.J. Mackus, L.J. Wiegman, J.H. van den Brakel, S.A. Beers, R.R. French, T. van Meerten, S. Ebeling, T. Vink, J.W. Slootstra, P.W. Parren, M.J. Glennie, and J.G. van de Winkel. 2006. The biological activity of human CD20 monoclonal antibodies is linked to unique epitopes on CD20. *J. Immunol.* 177(1):362–371.

Thompson, R.H., J.P. Allison, and E.D. Kwon. 2006. Anti-cytotoxic T lymphocyte antigen-4 (CTLA-4) immunotherapy for the treatment of prostate cancer. *Urol. Oncol.* 24(5):442–447.

Toichi, E., G. Torres, T.S. McCormick, T. Chang, M.A. Mascelli, C.L. Kauffman, N. Aria, A.B. Gottlieb, D.E. Everitt, B. Frederick, C.E. Pendley, and K.D. Cooper. 2006. An anti-IL-12p40 antibody down-regulates type 1 cytokines, chemokines, and IL-12/IL-23 in psoriasis. *J. Immunol.* 177(7):4917–4926.

Tomizuka, K., T. Shinohara, H. Yoshida, H. Uejima, A. Ohguma, S. Tanaka, K. Sato, M. Oshimura, and I. Ishida. 2000. Double trans-chromosomic mice: Maintenance of two individual human chromosome fragments containing Ig heavy and kappa loci and expression of fully human antibodies. *Proc. Natl. Acad. Sci. USA* 97(2):722–727.

Tomizuka, K., H. Yoshida, H. Uejima, H. Kugoh, K. Sato, A. Ohguma, M. Hayasaka, K. Hanaoka, M. Oshimura, and I. Ishida. 1997. Functional expression and germline transmission of a human chromosome fragment in chimaeric mice. *Nature Genet.* 16(2):133–143.

Tomlinson, I.M., J.P. Cox, E. Gherardi, A.M. Lesk, and C. Chothia. 1995. The structural repertoire of the human V kappa domain. *EMBO J.* 14(18):4628–4638.

Tomlinson, I.M., G. Walter, P.T. Jones, P.H. Dear, E.L. Sonnhammer, and G. Winter. 1996. The imprint of somatic hypermutation on the repertoire of human germline V genes. *J. Mol. Biol.* 256(5):813–817.

Trikha, M., Z. Zhou, J.A. Nemeth, Q. Chen, C. Sharp, E. Emmell, J. Giles-Komar, and M.T. Nakada. 2004. CNTO 95, a fully human monoclonal antibody that inhibits alpha v integrins, has antitumor and antiangiogenic activity in vivo. *Int. J. Cancer* 110(3):326–335.

Tze, L.E., B.R. Schram, K.P. Lam, K.A. Hogquist, K.L. Hippen, J. Liu, S.A. Shinton, K.L. Otipoby, P.R. Rodine, A.L. Vegoe, M. Kraus, R.R. Hardy, M.S. Schlissel, K. Rajewsky, and T.W. Behrens. 2005. Basal immunoglobulin signaling actively maintains developmental stage in immature B cells. *PLoS Biol.* 3(3):e82.

Tzipori, S., A. Sheoran, D. Akiyoshi, A. Donohue-Rolfe, and H. Trachtman. 2004. Antibody therapy in the management of shiga toxin-induced hemolytic uremic syndrome. *Clin. Microbiol. Rev.* 17(4):926–941.

Van Cutsem, E., M. Peeters, S. Siena, Y. Humblet, A. Hendlisz, B. Neyns, J.L. Canon, J.L. Van Laethem, J. Maurel, G. Richardson, M. Wolf, and R.G. Amado. 2007. Open-label phase III trial of panitumumab plus best supportive care compared with best supportive care alone in patients with chemotherapy-refractory metastatic colorectal cancer. *J. Clin. Oncol.* 25(13):1658–1664.

van de Putte, L.B., C. Atkins, M. Malaise, J. Sany, A.S. Russell, P.L. van Riel, L. Settas, J.W. Bijlsma, S. Todesco, M. Dougados, P. Nash, P. Emery, N. Walter, M. Kaul, S. Fischkoff, and H. Kupper. 2004. Efficacy and safety of adalimumab as monotherapy in patients with rheumatoid arthritis for whom previous disease modifying antirheumatic drug treatment has failed. *Ann. Rheum. Dis.* 63(5):508–516.

van Royen-Kerkhof, A., E.A. Sanders, V. Walraven, M. Voorhorst-Ogink, E. Saeland, J.L. Teeling, A. Gerritsen, M.A. van Dijk, W. Kuis, G.T. Rijkers, L. Vitale, T. Keler, S.E. McKenzie, J.H. Leusen, and J.G. van de Winkel. 2005. A novel human CD32 mAb blocks experimental immune haemolytic anaemia in FcgammaRIIA transgenic mice. *Br. J. Haematol.* 130(1):130–137.

Villadsen, L.S., J. Schuurman, F. Beurskens, T.N. Dam, F. Dagnaes-Hansen, L. Skov, J. Rygaard, M.M. Voorhorst-Ogink, A.F. Gerritsen, M.A. van Dijk, P.W. Parren, O. Baadsgaard, and J.G. van de Winkel. 2003. Resolution of psoriasis upon blockade of IL-15 biological activity in a xenograft mouse model. *J. Clin. Invest.* 112(10):1571–1580.

Villadsen, L.S., L. Skov, T.N. Dam, F. Dagnaes-Hansen, J. Rygaard, J. Schuurman, P.W. Parren, J.G. van de Winkel, and O. Baadsgaard. 2007. In situ depletion of CD4+ T cells in human skin by Zanolimumab. *Arch. Dermatol. Res.* 298(9):449–455.

Vitale, L., D. Blanset, I. Lowy, T. O'Neill, J. Goldstein, S.F. Little, G.P. Andrews, G. Dorough, R.K. Taylor, and T. Keler. 2006. Prophylaxis and therapy of inhalational anthrax by a novel monoclonal antibody to protective antigen that mimics vaccine-induced immunity. *Infect. Immun.* 74(10):5840–5847.

Vonderheide, R.H. 2007. Prospect of targeting the CD40 pathway for cancer therapy. *Clin. Cancer Res.* 13(4):1083–1088.

Vonderheide, R.H., K.T. Flaherty, M. Khalil, M.S. Stumacher, D.L. Bajor, N.A. Hutnick, P. Sullivan, J.J. Mahany, M. Gallagher, A. Kramer, S.J. Green, P.J. O'Dwyer, K.L. Running, R.D. Huhn, and S.J. Antonia. 2007. Clinical activity and immune modulation in cancer patients treated with CP-870,893, a novel CD40 agonist monoclonal antibody. *J. Clin. Oncol.* 25(7):876–883.

Wagner, E.F., T.A. Stewart, and B. Mintz. 1981. The human beta-globin gene and a functional viral thymidine kinase gene in developing mice. *Proc. Natl. Acad. Sci. USA* 78(8):5016–5020.

Wagner, T.E., P.C. Hoppe, J.D. Jollick, D.R. Scholl, R.L. Hodinka, and J.B. Gault. 1981. Microinjection of a rabbit beta-globin gene into zygotes and its subsequent expression in adult mice and their offspring. *Proc. Natl. Acad. Sci. USA* 78(10):6376–6380.

Wang, Y., J. Hailey, D. Williams, P. Lipari, M. Malkowski, X. Wang, L. Xie, G. Li, D. Saha, W.L. Ling, S. Cannon-Carlson, R. Greenberg, R.A. Ramos, R. Shields, L. Presta, P. Brams, W.R. Bishop, and J.A. Pachter. 2005. Inhibition of insulin-like growth factor-I receptor (IGF-IR) signaling and tumor cell growth by a fully human neutralizing anti-IGF-IR antibody. *Mol. Cancer Ther.* 4(8):1214–1221.

Weber, J. 2007. Review: Anti-CTLA-4 antibody ipilimumab—case studies of clinical response and immune-related adverse events. *Oncologist* 12(7):864–872.

Weinblatt, M.E., E.C. Keystone, D.E. Furst, L.W. Moreland, M.H. Weisman, C.A. Birbara, L.A. Teoh, S.A. Fischkoff, and E.K. Chartash. 2003. Adalimumab, a fully human anti-tumor necrosis factor alpha monoclonal antibody, for the treatment of rheumatoid arthritis in patients taking concomitant methotrexate: The ARMADA trial. *Arthritis Rheum.* 48(1):35–45.

Weng, W.K. and R. Levy. 2003. Two immunoglobulin G fragment C receptor polymorphisms independently predict response to rituximab in patients with follicular lymphoma. *J. Clin. Oncol.* 21(21):3940–3947.

Williams, S.C., J.P. Frippiat, I.M. Tomlinson, O. Ignatovich, M.P. Lefranc, and G. Winter. 1996. Sequence and evolution of the human germline V lambda repertoire. *J. Mol. Biol.* 264(2):220–232.

Winkler, U., M. Jensen, O. Manzke, H. Schulz, V. Diehl, and A. Engert. 1999. Cytokine-release syndrome in patients with B-cell chronic lymphocytic leukemia and high lymphocyte counts after treatment with an anti-CD20 monoclonal antibody (rituximab, IDEC-C2B8). *Blood* 94(7):2217–2224.

Wong, R.M., R.R. Scotland, R.L. Lau, C. Wang, A.J. Korman, W.M. Kast, and J.S. Weber. 2007. Programmed death-1 blockade enhances expansion and functional capacity of human melanoma antigen-specific CTLs. *Int. Immunol.* 19(10):1223–1234.

Xu, J.L., and M.M. Davis. 2000. Diversity in the CDR3 region of V(H) is sufficient for most antibody specificities. *Immunity* 13(1):37–45.

Yamamura, K., A. Kudo, T. Ebihara, K. Kamino, K. Araki, Y. Kumahara, and T. Watanabe. 1986. Cell-type-specific and regulated expression of a human gamma 1 heavy-chain immunoglobulin gene in transgenic mice. *Proc. Natl. Acad. Sci. USA* 83(7):2152–2156.

Yang, X.D., J.R. Corvalan, P. Wang, C.M. Roy, and C.G. Davis. 1999. Fully human anti-interleukin-8 monoclonal antibodies: Potential therapeutics for the treatment of inflammatory disease states. *J. Leukoc. Biol.* 66(3):401–410.

Yang, X.D., X.C. Jia, J.R. Corvalan, P. Wang, and C.G. Davis. 2001. Development of ABX-EGF, a fully human anti-EGF receptor monoclonal antibody, for cancer therapy. *Crit. Rev. Oncol. Hematol.* 38(1):17–23.

Yang, X.D., X.C. Jia, J.R. Corvalan, P. Wang, C.G. Davis, and A. Jakobovits. 1999. Eradication of established tumors by a fully human monoclonal antibody to the epidermal growth factor receptor without concomitant chemotherapy. *Cancer Res.* 59(6):1236–1243.

Zang, X., and J.P. Allison. 2007. The b7 family and cancer therapy: Costimulation and coinhibition. *Clin. Cancer Res.* 13(18, Pt 1):5271–5279.

Zhou, H., H. Jang, R.M. Fleischmann, E. Bouman-Thio, Z. Xu, J.C. Marini, C. Pendley, Q. Jiao, G. Shankar, S.J. Marciniak, S.B. Cohen, M.U. Rahman, D. Baker, M.A. Mascelli, H.M. Davis, and D.E. Everitt. 2007. Pharmacokinetics and safety of golimumab, a fully human anti-TNF-alpha monoclonal antibody, in subjects with rheumatoid arthritis. *J. Clin. Pharmacol.* 47(3):383–396.

Zijlstra, M., E. Li, F. Sajjadi, S. Subramani, and R. Jaenisch. 1989. Germ-line transmission of a disrupted beta 2-microglobulin gene produced by homologous recombination in embryonic stem cells. *Nature* 342(6248):435–438.

Zou, Y.R., W. Muller, H. Gu, and K. Rajewsky. 1994. Cre-loxP-mediated gene replacement: A mouse strain producing humanized antibodies. *Curr. Biol.* 4(12):1099–1103.

Zou, X., T.A. Piper, J.A. Smith, N.D. Allen, J. Xian, and M. Bruggemann. 2003. Block in development at the pre-B-II to immature B cell stage in mice without Ig kappa and Ig lambda light chain. *J. Immunol.* 170(3):1354–1361.

CHAPTER 6

Rabbit Hybridoma

WEIMIN ZHU and GUO-LIANG YU

ABSTRACT

Rabbit monoclonal antibody (RabMAb) is a novel class of monoclonal antibodies that offer a number of benefits, including ultra-high affinity, diversity in epitope recognition, and ease of molecular engineering. Hundreds of RabMAbs are used as reagents by researchers to study biological functions of various proteins and are used as biomarkers to evaluate clinical candidates and their performance in clinical trials. RabMAbs are also being evaluated for the potential to become therapeutic antibodies.

Therapeutic Monoclonal Antibodies: From Bench to Clinic. Edited by Zhiqiang An

6.1 IMMUNOLOGY OF THE RABBIT

Rabbits are known to mount a strong immune response against foreign antigens. Rabbit polyclonal antibodies have been used widely in many immunological assays. Rabbit antibodies are especially useful when the antigens are not immunogenic in mice (Krause 1970a, 1970b; Bystryn et al. 1982; Weller, Meek, and Adamson 1987; Raybould and Takahashi 1988). A novel fusion partner cell, 240E-1 was developed by Katherine Knight in the 1990s using a transgenic rabbit carrying the oncogenes c-myc and V-abl (Spieker-Polet et al. 1995). Subsequent improvements of the rabbit monoclonal antibody (RabMAb) technology by Epitomics led to the successful development of thousands of high affinity antibodies that are being used for research, diagnostics, and therapeutic development.

The antibody response in rabbits is driven by a series of interactions among immunogens, antigen-presenting cells (APC), T-cells, and B-cells. After the primary immunization, naïve B-cells are stimulated to differentiate into antibody-secreting plasma cells. For most protein antigens, the first specific antibody begins to appear in the serum within five days of immunization and the majority of the antibody type is IgM. After this initial response, under the influence of cytokines secreted by T-cells, the B-cells will switch to the production of IgG. The antibody concentration (titer) will continue to rise and will peak around two weeks, and then decreases slowly. The immune response and average affinity of an antibody for an antigen drastically increase with repeated injections of the immunogen. In a rabbit, antibody affinity maturation is achieved through clonal selection, hyper-somatic mutations, and gene conversion. The New Zealand White laboratory rabbit (*Oryctolagus cuniculus*) is the species that is widely used because of its size and physiological attributes.

Rabbit antibody is somewhat simplified compared with mouse and human which are known to have five classes of antibody, defined by the type of heavy chain: $C\gamma$ for IgG; $C\mu$ for IgM; $C\alpha$ for IgA; $C\varepsilon$ for IgE, and $C\delta$ for IgD. No rabbit IgD has been found so far (Mage, Lanning, and Knight 2006). The abundant class in rabbit serum is IgG, with serum concentration of 5 to 20 mg/ml, followed by IgA, with serum concentration of 3 to 4 mg/ml. Unlike IgG from other animal species, the rabbit IgG has no subclass. Rabbit IgG consists of two light and two heavy chains. The light chain is composed of one variable domain and one constant domain, and the heavy chain is composed of one variable domain and three constant domains. Two light chains and two heavy chains assemble into Y-shaped molecules exhibiting two antigen-binding sites (Fig. 6.1). The antigen-binding sites are embedded into the variable heavy and light chain domains (VH and VL). The domains have similar three-dimensional structures, but in each variable domain the hypervariable regions form three loops that protrude from one end of the molecule, forming the antigen-binding site. The hypervariable regions are also known as the complementary determining regions (CDRs). Each binding site is therefore made up of six CDRs, three from the light chain and three from the heavy chain. Rabbit IgG is also structurally different from that of mouse and human. Compared with mouse and human IgGs, rabbit IgG tends to have fewer amino acids at the N terminus and in the D-E loop, and have extra disulfide bonds in the variable region of the heavy chain. For the light chain, in several different allotypes of rabbits there is a disulfide bond between $V\kappa$ and $C\kappa$. Table 6.1 lists the IgG subclasses from various species (Howard and Kaser 2007).

The organization of rabbit IgG genes is similar to that of other mammals, except that rabbit has one $C\gamma$ gene and 13 $C\alpha$ genes. This is compared with four $C\gamma$ genes and only one or two $C\alpha$ genes in human, mouse, rat, and cow. IgG from 70 to 90 percent of B-cells carries the V_Ha+ allotypes. The three alleles of VH1 in laboratory strains, VH1a1, VH1a2, and VH1a3, encode allotypes a1, a2, and a3, respectively. The other 10 to 30 percent of IgG in serum and mucosa are called VHa-negative (V_Ha^-) because they do not react with anti-V_Ha allotype antisera. V_Ha+ IgGs are encoded predominately by the V_H1 gene segment, whereas the V_Ha^- IgG molecules are encoded by multiple V_Hx, V_Hy and V_Hz gene segments (Mage, Lanning, and Knight 2006; Knight and Crane 1994).

The majority (90 to 95 percent) of the light chains is derived from $C_\kappa1$ (isotype $\kappa1$). Five allelic $C_\kappa1$ are known as C_κ allotypes b4, b4v, b5, b6, and b9. The $C_\kappa2$ gene encoding the isotype $\kappa2$ is rarely expressed, except in some b9 rabbits. Only 5 to 10 percent of total IgG light chains are isotype λ. An unusual feature of rabbit $\kappa1$ light chain is that there is an extra disulfide bond between cysteine residues at $V\kappa$ position 80 and $C\kappa$ position 171. This extra disulfide bond is not found in $\kappa2$ and λ

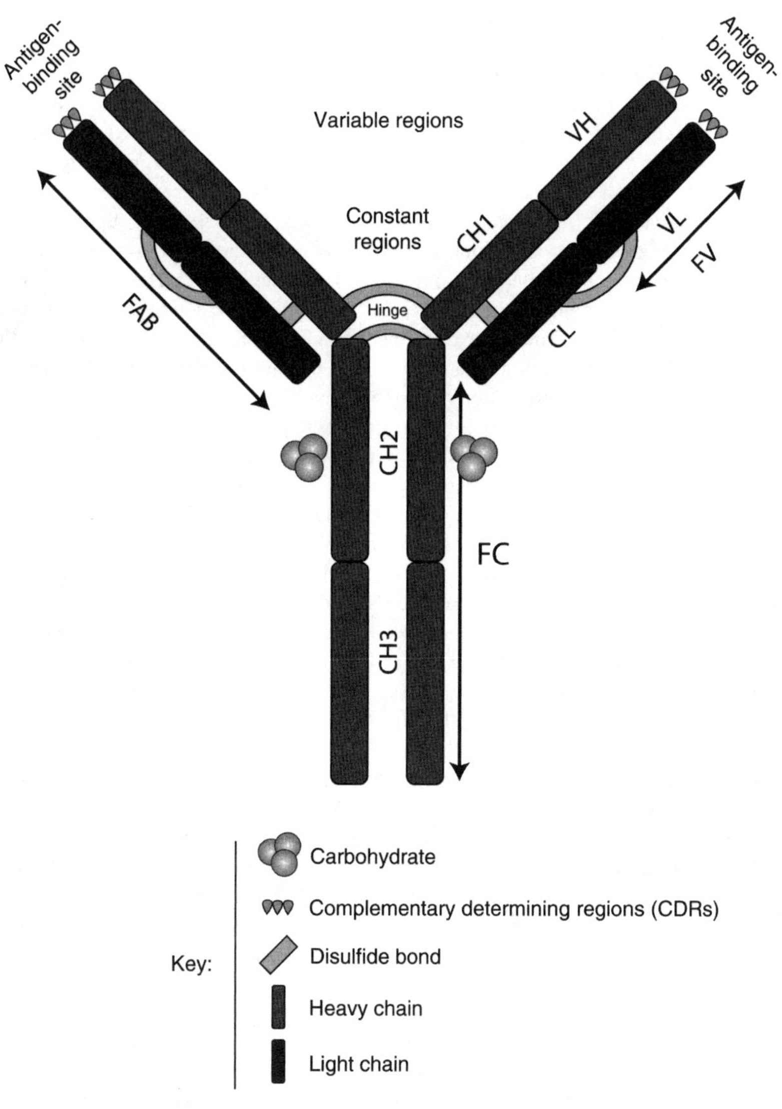

Figure 6.1 Structure of rabbit IgG.

TABLE 6.1 Immunoglobulin (Ig) Subclasses of Different Species

Species	IgG Subclasses
Human	IgG1, IgG2, IgG3, IgG4
Mouse	IgG1, IgG2a, IgG2b, IgG2c, IgG3
Rat	IgG1, IgG2a, IgG2b, IgG2c
Guinea pig	IgG1, IgG2
Rabbit	Only one IgG isotype
Sheep	IgG1, IgG2, IgG3
Chicken	IgY (comparable to IgG)
Pig	IgG1, IgG2a, IgG2b, IgG3, IgG4
Horse	IgGa, IgGb, IgGc, IgG(B), IgG(Ta), IgG(Tb)

light chains. The great stability and long shelf-life of rabbit antibodies may result in part from stabilization of the κ1 light chain structure by the unusual interdomain disulfide bond (Mage, Lanning, and Knight 2006; Knight and Crane 1994).

Rabbit has a different mechanism for the generation of its antibody repertoire. Most mammals create their primary antibody repertoire through combinational joining of multiple VH, D, and JH gene segments for heavy chains, and multiple Vκ (λ) and Jκ (λ) gene segments for light chains, and association of H and L chains. The resulting VJ and VDJ gene rearrangements can then be further diversified by somatic mutations that develop the secondary antibody repertoire. The rabbit VH chromosomal region contains an estimated 200 VH genes that are generally more than 80 percent similar in sequences. The high similarity among rabbit VH genes enhances their ability in gene conversion by codon insertions and deletions, as well as clusters of nucleotide changes from upstream VH genes. Although they have multiple germline VH genes, like human and mouse, the rabbit rearranges mostly VH1. Thus, the combinational joining of multiple VH, D, and JH gene segments contributes relatively little to the generation of antibody diversity. The repertoire is also limited by the use of a small number of JH and D gene segments in the VDJ rearrangements. JH4 is found in approximately 80 percent of the VDJ gene rearrangements and JH2 in the other 20 percent; the other three functional JH gene segments are rarely used. Also, from the total of the 12 D gene segments, most VDJ gene rearrangements use D2a (D9), the rest are less frequently or rarely utilized. The limited use of VH, D, and JH gene segments result in a limited VDJ repertoire; therefore, the repertoire that is generated in the primary sites of B-cell development is insufficient. The limited development of lymphoid tissue at birth and the delayed onset of Ig production may explain why new-born rabbits are relatively immuno-incompetent and do not develop good response within the first one to two weeks of age (Knight and Crane 1994). Newly developed B-cells from the bone marrow, fetal liver, and omentum seed the gut-associated lymphoid tissues (GALT), including the appendix and sacculus rotundus, where B-cell follicles form and extensive B-cell expansion occurs in response to intestinal microflora. By six weeks of age, the appendix becomes vastly cellular and the primary high copy number repertoire is developed by somatic diversification of Ig genes through gene conversion and somatic hypermutation. Appendix follicle development is arrested in the absence of intestinal microflora, indicating the importance of intestinal bacteria in rabbit GALT development. Similarly, rabbits raised in a germfree environment are found to have underdeveloped GALT. Moreover, germfree rabbits responded poorly to immunization with several test antigens if at all. These results suggest that in addition to GALT development intestinal bacteria have an important impact on antibody repertoire development. The mechanism of bacterially induced GALT development and primary antibody repertoire diversification in the rabbit explains in part the nature of rabbit antibody development, but somatic gene conversion is a major mechanism for rabbit VDJ gene diversification and primary antibody repertoire development. The sequences of rearranged heavy and light chain genes from the developing appendix at four to five weeks of age are still mostly undiversified. Interestingly, at seven to eight weeks of age, the sequences are to a great extent diversified largely through a gene conversion mechanism (Mage, Lanning, and Knight 2006; Knight and Crane 1994).

B-cells in the rabbit GALT are selected by both positive and negative selective events, and the cells that survive the selection exit to the periphery to participate in further engagement with foreign antigens. Gene conversion and somatic hypermutation occur not only in the germinal centers of the young rabbit appendix, but also in secondary lymphoid tissues, such as spleen, lymph nodes, and Peyer's patches, for diversification during immune responses. Analysis of lineage trees to quantify the differences between primary and secondary diversification in rabbits indicates that primary diversification appears to occur at a constant rate in the appendix and at a higher rate during immune responses in splenic germinal centers. This diversification presumably occurs in the germinal centers of secondary lymphoid tissues and results in formation of the secondary antibody repertoire (Mage, Lanning, and Knight 2006; Schiaffella et al. 1999).

6.2 DEVELOPMENT OF RABBIT MONOCLONAL ANTIBODY

The rabbit is well known to make higher affinity and more diverse antibodies to many molecules, including phospho-peptides, carbohydrates, and immunogens that are not immunogenic in mouse. However, until recently rabbit polyclonal antibodies were the only type of antibodies generated by rabbits. It is desirable to develop monoclonal antibodies for many applications for consistency and sustainable supply. Rabbit monoclonal antibody (RabMAb) combines the best properties of monoclonal antibodies with the most desirable attributes of rabbit polyclonal antibodies (Krause 1970a, 1970b; Bystryn et al. 1982; Weller, Meek, and Adamson 1987; Raybould and Takahashi 1988). Since the development of mouse hybridoma technology several efforts have been made to generate RabMAbs (Collins et al. 1974; Kuo et al. 1985; Verbanac et al. 1993a, 1993b). No myeloma-like cell line was developed in rabbits through chemically induced tumor sources. In addition, it is not feasible to transform rabbit B cells *in vivo* or *in vitro* with viruses (Collins et al. 1974). Extensive experimentation was carried out on mouse-rabbit heterohybridomas to make RabMAbs (Raybould and Takahashi 1988; Kuo et al. 1985; Verbanac et al. 1993a, 1993b). However, these heterohybridomas were difficult to clone, and the clones were generally unstable and did not secrete antibody over a prolonged period of time.

Katherine Knight and colleagues at Loyola University of Chicago achieved a major breakthrough in 1995. They generated a double transgenic rabbit over-expressing the oncogenes v-abl and c-myc under the control of the immunoglobulin heavy and light chain enhancers. The rabbit developed a myeloma-like tumor, thus allowing the isolation of a plasmacytoma cell line, termed 240E-1. Fusion of 240E-1 cells with rabbit lymphocytes produced hybridomas that secreted rabbit monoclonal antibodies (Spieker-Polet et al. 1995). For the first time, it was possible to make a rabbit–rabbit hybridoma using a rabbit myeloma-like cell line as a fusion partner cell.

However, like the early mouse myeloma lines developed in the 1970s, the stability of 240E-1-derived hybridomas was a concern. The resulting hybridomas were not genetically stable and the antibody activity was often lost after multiple passages of the cells. Indeed, it took many years of dedicated effort to develop stable mouse fusion partner cell lines, such as Sp2/0, that could be used as a reliable fusion partner cell line for mouse hybridomas (Kearney et al. 1979; Shulman, Wilde, and Kohler 1978). A number of laboratories that had received the 240E-1 cell line from Dr. Knight's laboratory reported stability problems of the rabbit hybridomas obtained by fusion with 240E-1 (Michel et al. 2001). Furthermore, after subculturing the 240E-1 cells, it was observed that fusion efficiency decreased significantly, apparently due to the formation of cell clumps. Moreover, Rief et al. (1998) reported that the majority of the rabbit hybridoma clones lost the ability to produce IgG even after multiple hybridoma subcloning.

In 1996, Weimin Zhu and Robert Pytela, then at the University of California San Francisco (UCSF), obtained 240E-1 from Dr. Knight's laboratory and attempted to make rabbit hybridomas using the procedures provided (Spieker-Polet et al. 1995). They too observed the instability problem in rabbit hybridomas. They decided to improve 240E-1 by repeated subcloning and medium optimization. 240E-1 progeny subclones were selected for high fusion efficiency and robust growth. Morphological characteristics such as a bright appearance under phase contrast microscopy and formation of a uniform adherent layer were taken into consideration during the selection. Selected subclones were further tested for their ability to produce stable hybridomas and RabMAb secretion. After 11 rounds of subcloning and selection processes, a new cell line named 240E-W was identified to be a better fusion partner cell line. 240E-W performed better as a fusion partner, with increased fusion efficiency, and produced more stable hybridomas. A number of characteristics that distinguish 240E-W from its parent 240E-1 were observed. 240E-W appears to be more uniform, shiny, and round. 240E-W cells adhered to a plastic surface better and form a uniform layer. Karyotyping analysis showed that 240E-W has a higher average chromosome number and less variation in chromosome number, ranging from 79 to 90 (average of 84), compared with 240E-1, which ranges from 44 to 70 (average of 60). AFLP (DNA fingerprinting) and gene expression pattern analysis by gene chip technique also revealed significant differences between the two cell lines (data not shown). Fusion of 240E-W cells with splenocytes of immunized rabbits yielded greater numbers of stable hybridomas as compared to that of the 240E-1 cell line. Hybridoma cell lines were robust, more stable, tolerant to high density growth, and could be maintained in confluent culture for several days without loss of viability. Thus, this new fusion partner cell line has made rabbit hybridoma a routine technique that has enabled large-scale development of rabbit monoclonal antibodies. Several hundred hybridomas lines were developed using this fusion partner cell line.

Similar to the original mouse fusion partner cell line, the 240E-W cell line inherited the undesirable traits of expression of endogenous IgG chains (Spieker-Polet et al. 1995). While the expression of antigen-specific antibodies can still be obtained using this cell line, the expression of endogenous heavy and light chains interferes with assembly of IgG proteins when the splenocyte-derived heavy and light chains form hetero-tetramers with the endogenous chains. The expression of endogenous heavy chain is readily detected by SDS polyacrylamide gel electrophoresis (SDS-PAGE), appearing as a doublet, with the endogenous heavy chain (endogenous H) migrating slightly faster than the splenocyte-derived heavy chain. The N-terminal protein sequence, obtained by MS–MS sequencing, of the fast migrating band on the SDS-PAGE is identical to the deduced amino acid sequence of the endogenous IgG gene. A rearranged, potentially active gene encoding an endogenous kappa light chain (endogenous L) is also present in the 240E-W cell line and its parental line 240E-1 although the product of this gene is seldom detected at the protein level. The presence of multiple bands or a smear near the 150 kDa position on a nonreducing SDS-PAGE indicated the formation of hetero-tetramers. Although not all hybridomas express the endogenous IgG chains, an estimate of 10 to 20 percent of the hybridomas appear to have reduced or lost specific activity over time. Antibodies from these hybridomas have reduced affinity and specificity. This problem escalated when the hybridoma cells were cultured for a long time or under unhealthy conditions such as higher cell density.

To mitigate the endogenous IgG problem of 240E-W, Yaohuang Ke, Robert Pytela, and Weimin Zhu at Epitomics developed a new derivative line of 240E-W, termed 240E-W2, which carries no endogenous heavy chain gene. A strategy was set to take advantage of an observation that a subset of hybridoma clones do not secrete detectable amounts of IgG after fusion of 240E-W cells with rabbit splenocytes. Candidate clones that showed absence of intracellular IgG by immunocytochemistry were cultured in duplicate 96-well plates. One set of the plates was stained with polyclonal goat-anti-rabbit IgG antibodies using immunocytochemistry. Several clones were identified that contained undetectable levels of IgG. One of these hybridoma clones was selected for subcloning and further testing. The cells were subcloned by the limiting-dilution method. Single-cell subclones were expanded and again tested for intracellular IgG expression by immunocytochemistry. The subcloning process was repeated twice, until all progenies for one of the clones were IgG-negative. This candidate line was again subcloned and it was verified that all the subclones remained 100 percent IgG-negative. The endogenous heavy chain protein, mRNA and DNA were all undetectable by Western blot, reverse transcriptase polymerase chain reaction (RT-PCR) and genomic PCR. These clones were cultured in the presence of azaguanine at consecutively decreasing concentrations for several months to select revertants that had regained sensitivity to HAT medium selection. One of these subclones was selected as the new fusion partner cell line, named 240E-W2.

To further qualify 240E-W2 as the new generation of fusion partner, a side-by-side comparison using 240E-W and 240E-W2 as fusion partner cell lines was conducted. Fifty-three immunogens were targeted in these hybridoma projects. The results indicated that the overall performance of 240E-W2 improved significantly over that of 240E-W. The average fusion efficiency, measured by the percentage of hybridoma growth in wells, increased to 73 percent using 240E-W2, compared with 55 percent using 240E-W. The average number of hybridomas producing immunogen-binding antibodies, measured by ELISA, increased 40 percent using 240E-W2 compared to that using 240E-W. Finally, the Western blot positive rate, measured by the ability of resulting antibodies to detect a specific band on Western blots, increased 24 percent using 240E-W2 compared with that using 240E-W. In addition, SDS-PAGE analysis confirmed that hybridoma clones derived from

TABLE 6.2 Fusion Partner Cell Lines for Rabbit Hybridoma Development

Feature	240E-1	240E-W	240E-W2	240E-W3
Endogenous H	+	+	−	−
Endogenous L	+	+	+	−
Average chromosome no.	60	84	89	ND
Fusion efficiency	low	high	high	high
Hybridoma stability	poor	intermediate	good	good

240E-W2 do not secrete any detectable endogenous IgG heavy chain, whereas endogenous light chain was detected in only one of seven hybridoma clones. The removal of endogenous heavy chain has resulted in improved antibody activity and specificity. This new fusion partner 240E-W2 has replaced 240E-W for generating rabbit hybridomas since 2006 and was used to produce RabMAbs against more than 1000 antigens.

Although there is no detectable endogenous light chain either in the fusion partner cell line 240E-W2 or in the majority of RabMAbs produced by hybridomas derived from fusion with 240E-W2, endogenous light chain was found in a subset of hybridoma clones. In most cases hybridomas were generally stable and there was no compromised sensitivity or specificity observed from the secreted antibodies in those hybridomas. Nonetheless, the presence of endogenous light chain is still a concern for instability of hybridomas and reduced antibody activities. Currently, Epitomics is in the process of developing a fourth generation of the rabbit fusion partner line, to be called 240E-W3, in which both endogenous heavy chain and light chain genes will be removed or made inactive. Currently a few candidates for 240E-W3 are under evaluation. Table 6.2 summarizes the features of different rabbit fusion partner cell lines.

6.3 ADVANTAGES OF RABBIT MONOCLONAL ANTIBODY

6.3.1 Ultra-High Binding Affinity

The dissociation constant (Kd) is commonly used to represent the affinity between an antibody and its antigen. A nanomolar Kd is considered to be high affinity and is used as a cut off for therapeutic antibody candidates. Drugs can produce harmful side effects through interactions with proteins with which they were not designed to interact. Therefore a great deal of pharmaceutical research is aimed at designing drugs that bind specifically to their target proteins with high affinity (typically 0.1 to 10 nM). Most therapeutic antibodies are nanomolar or subnanomolar in Kd value (Kd $= 10^{-9}$ to 10^{-10} M). RabMAbs have very high affinities, typically in the picomolar level. Table 6.3 lists a number of RabMAbs generated against various targets for which their affinities are measured by the BIACore machines. The high affinity of RabMAbs is attributed to the rabbit's unique immune system, described in previous sections of this chapter. The ultra-high binding affinity feature of RabMAbs results in more sensitive and more specific detection of targets in diagnostic tests and is hypothesized to lead to more efficacious therapeutics.

6.3.2 High Specificity

RabMAbs are highly specific to their targets. Analysis of approximate 1000 RabMAbs by Western blot using whole cell lysates indicate that RabMAbs mostly only recognize the intended targets (www.epitomics.com) visualized as a single band on the blots. The high specificity of RabMAbs is further demonstrated in the application of these RabMAbs in immunohistochemistry. In comparison to mouse monoclonal antibodies available to the same targets, RabMAbs detect target proteins with higher sensitivity and specificity, giving cleaner staining patterns with lower background on paraffin-embedded tissue. Rossi et al. (2005) compared RabMAbs and mouse mAbs against estrogen receptor,

TABLE 6.3 Dissociate Constant (Kd) of Selected RabMAbs to Human Targets

Antigen	Immunogen	Kd
ERα	Peptide	4.57×10^{-10} M
ERα	Peptide	2.19×10^{-10} M
ERα	Peptide	1.28×10^{-12} M
ERα	Peptide	5.63×10^{-10} M
Id1	Protein	2.82×10^{-12} M
Id3	Protein	2.42×10^{-12} M
Id3	Protein	5.30×10^{-12} M
Id3	Protein	1.26×10^{-12} M
TNFα	Protein	5.43×10^{-11} M
TNFα	Protein	1.25×10^{-11} M
IL-1β	Protein	1.99×10^{-10} M
IL-1β	Protein	2.41×10^{-10} M
VEGF	Protein	1.30×10^{-13} M
VEGF	Protein	6.99×10^{-13} M
VEGF Receptor-2	Protein fragment	1.30×10^{-11} M
VEGF Receptor-2	Protein fragment	1.40×10^{-11} M

ERα: estrogen receptor alpha; Id1, Id3: inhibitor of differentiation/DNA binding protein 1 and 3; TNFα: tumor necrosis factor alpha; IL-1β: interleukin 1 beta; VEGF: vascular endothelial growth factor.

progesterone receptor, Ki-67, cyclinD1, CD3, CD5, CD23, and synaptophysin on several tumor types as well as normal tissues. The results indicated that RabMAbs appear to offer increased sensitivity with no apparent loss of specificity. Routine use of RabMAbs permits higher working dilutions, allowing significant improvement in terms of laboratory efficiency. The robustness of RabMAbs is further demonstrated by the fact that in some instances optimal staining can be obtained even without antigen retrieval (Z. Huang et al. 2005, 2006).

6.3.3 Novel Epitope Recognition

Rabbits were shown to recognize epitopes that were not recognized in mouse and to recognize small differences on epitopes, in a parallel experiment where the same protein immunogen, human fibronectin, was used to immunize three rabbits and five mice. The ability of epitope recognition was evaluated by counting bands on Western blots containing protease-digested fibronectin. Rabbit polyclonal antibodies were shown to detect more fibronectin fragments on the Western blots than the fragments detected by mouse polyclonal antibodies, which suggested that rabbits developed antibodies that recognized more epitopes (data not shown). It has been demonstrated that for certain antigens, such as phosphorylated proteins and other post-translationally modified proteins, the success rate to make monoclonal antibody in rabbit is much higher than in mouse. The ability to develop monoclonal antibodies against novel epitopes may allow one to identify important bioactive sites on a protein target that may lead to novel mechanisms for new drug development. Another observation of the rabbit immune response is that the immuno-dominance is significantly reduced in rabbits compared with that in mice. For example, analysis of polyclonal antibodies from postimmunization with a keyhole limpet hemocyanin (KLH) conjugated peptide showed more than 90 percent of mouse mAbs were directed to KLH (<10 percent to the peptide) while 70 percent of RabMAbs were directed to KLH (30 percent to the peptide) (data not shown). The significantly reduced immuno-dominance may allow the identification of a low abundant antibody that may be valuable.

6.3.4 Larger Number of Independent Hybridomas

Generally, it is desirable to screen a large compound library to identify sufficient drug leads. This principle also applies to screening hybridomas for therapeutic antibody leads. The rabbit has an advantage over the mouse because each immunized rabbit spleen contains as much as 50 times more lymphocytes than a mouse spleen. From each rabbit spleen, thousands of hybridomas can be generated providing a much greater number of independent MAbs that recognize more epitopes. In addition, rabbit hybridomas are relatively tolerant to high density cell culture, and thus are suited for large scale antibody generation because the batch maintenance during the fusion and expansion process can replace the traditional individual maintenance method used for mouse hybridomas. The robust process offers a higher success rate and makes it possible to rapidly produce large numbers of antibodies and screen for the most desirable ones in a shorter period of time.

6.3.5 Monoclonal Antibodies against Rodent Proteins

Mouse and rat are the most commonly used animals for scientific research. Scientists use rodents as human disease models to study the underlying mechanisms of human diseases. Rodent disease models are further used to test new therapeutic agents for their effectiveness. One of the major challenges for mouse mAbs is that mouse-derived antibodies generally do not recognize mouse proteins. In order to analyze rodent proteins for their response to a therapeutic treatment, a surrogate antibody is necessary from a different source. Rabbits provide an ideal resource for making anti-mouse protein antibodies. In addition, certain epitopes are shared between human proteins and their mouse protein equivalents. RabMAbs that cross-react with human protein and its rodent counterpart can be identified in the screening, providing a convenient tool for animal model studies.

Taken together, these advantageous properties of RabMAbs have led to new research discoveries as well as increased opportunities for the development of better antibody therapeutics. The potential for more sensitive and reliable diagnostic tests and highly potent antibody drugs using RabMAbs is certain to be exploited further.

6.4 GENERATION OF RABBIT HYBRIDOMAS

6.4.1 Immunization

The process of rabbit hybridoma is illustrated in Figure 6.2. Immunization of rabbits to generate polyclonal antibodies has been practiced for decades. The most commonly used rabbit is the New Zealand White. The sex of the rabbit does not appear to be important, although female rabbits are used more frequently for immunization due to convenience of handling them. It is usually advisable to immunize three- to four-month-old young adult animals. Using rabbits that are more than 18 months old may reduce the yield of desirable hybridomas. However, rabbits that are less than six weeks old should not be considered for immunization when the animals are still developing their primary immune repertoire, and thus are relatively immuno-incompetent.

Many types of antigen preparations have been used successfully for raising rabbit monoclonal antibodies, including DNA, peptide, protein, whole cells, and tissues. Among them, peptides and proteins are the most commonly used forms of immunogens. The nature of the antigen preparation is dependent on several factors, such as the intended application of the desired antibodies, the ease and cost in preparing the antigen, and the screening assays to be used. Peptides are particularly useful for raising rabbit antibodies to specific regions of a protein, such as a specific domain or active site. Due to the challenge to produce purified proteins, especially in their native forms, using synthetic peptide as immunogens has become a routine approach. Peptides are usually coupled to KLH, which has been found to be an ideal carrier protein because it strongly stimulates the humoral immune response. In spite of the strong immunogenicity of KLH itself, rabbits also produce large numbers of high affinity antibodies to the conjugated peptide. To couple peptides to KLH, in most cases, a cysteine residue are added to the N or C terminus of the peptide and sulfhydryl-specific cross-linkers are used. The length of the peptide can vary from a few residues to a few dozen, with an average of 15 to 20 residues.

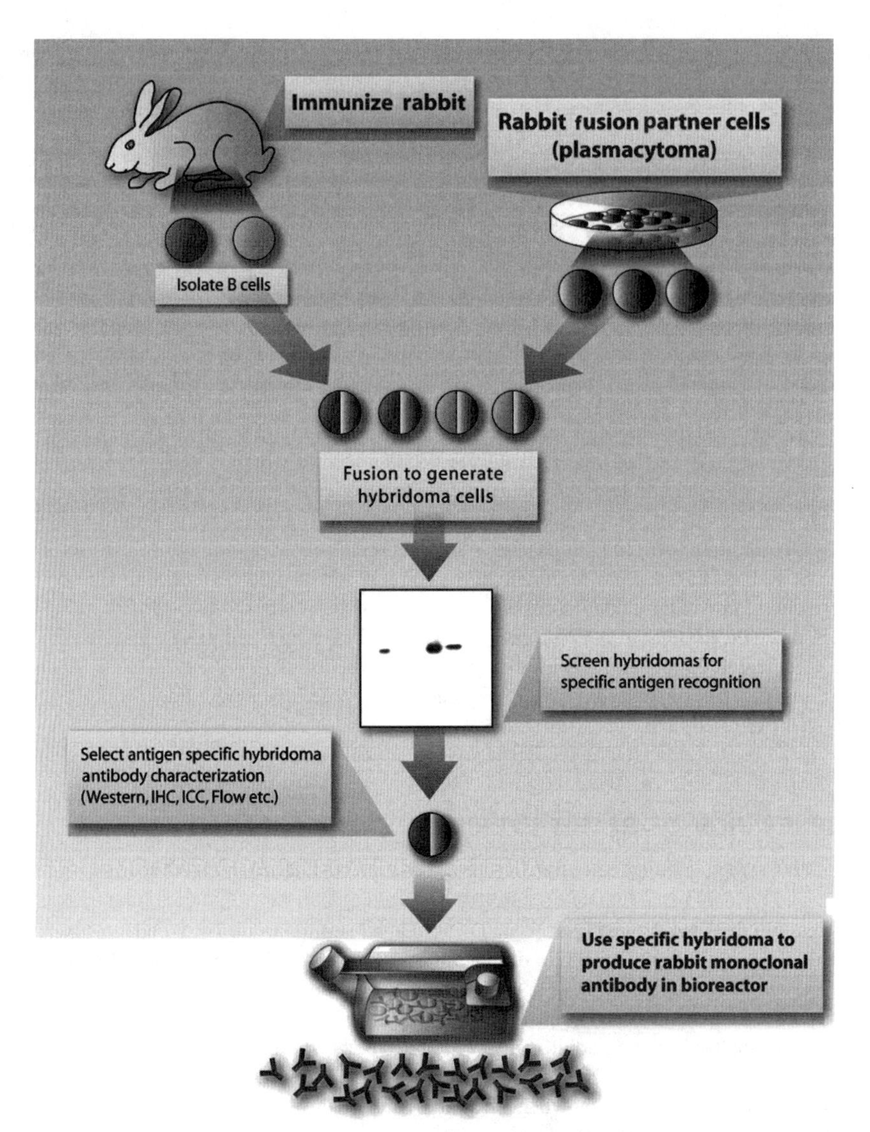

Figure 6.2 Process of rabbit hybridoma generation. B-cells are isolated from a rabbit that is immunized with an immunogen and are fused with rabbit fusion partner cell line to generate hybridomas. The antibodies from hybridomas are screened for the antigen recognition by ELISA, Western blot, immunohistochemistry, and FACS. The selected hybridoma clones are subcloned to ensure that they are monoclonal. The final clone is chosen to produce a larger quantity of RabMAbs in a CELLine Integra flask.

There are a number of immunization protocols used to produce rabbit polyclonal antibodies (Howard and Kaser 2007; Harlow and Lane 1988; Coligan et al. 1994). These protocols are all applicable to make RabMAbs with the addition of a hyperimmunization step a few days before splenectomy. Although it is widely believed that a larger amount of immunogen is needed for larger animals, there has not been systematic study to demonstrate this speculation. The minimum amount of immunogen required to elicit a response depends on the immunogenecity of the antigen. Multiple subcutaneous injection sites may also enhance the immune response. In a typical immunization, two New Zealand White rabbits are immunized with 0.1 to 0.2 mg per injection for proteins and 0.2 to 0.4 mg per injection for peptides mixed with adjuvant. Freund's adjuvant is still the most commonly used adjuvant for rabbit immunizations. To reduce the risk of inflammation and eczema formation at the injection site, TiterMax™ Gold adjuvant (Sigma-Aldrich Co.) can be used as an alternative. For peptide immunization, four to six immunizations are sufficient to induce a strong antiserum titer to the peptide. For protein immunization fewer injections are needed to achieve high titers. For booster immunizations, less immunogen can be used. Booster immunizations are usually given at 2 to 4 week intervals, and antibody titers are determined 10 to 14 days after each boost. Four days before harvesting the lymphocytes, a hyperimmunization with a high dose of intravenously injected antigen is required. For toxic or large particle immunogens, intraperitoneal injection is substituted and hyperimmunization is performed seven days before the spleen is harvested.

6.4.2 Rabbit Fusion Partner Cell Line 240E-W2 Culture and Rabbit Lymphocyte Isolation

The rabbit fusion partner cell line 240E-W2 is different from mouse fusion partner cell lines in terms of morphology, growth rate, and culture medium. A 2 L spinner flask is used to produce large quantity of 240E-W2 cells. Before being transferred to the spinner, cells are grown in six 150 mm petri dishes to 70–80 percent confluency and are combined to seed the spinner flask containing 500 mL medium.

To isolate splenocytes, rabbit spleen is harvested by opening the abdominal wall and taking out the spleen, which is connected by the mesogastrium to the left side of the dorsal surface of the stomach. After excess fat is trimmed from the spleen, the spleen is washed extensively with sterile RPMI medium and is punctured at several different sites with a needle attached to a 3 mL syringe containing RPMI. The medium is injected into the spleen, forming a balloon that releases the lymphocytes. The spleen is then crushed using the end of a 20 cc syringe. The smashed tissue is added to 10–15 mL RPMI medium and pipetted in and out a few times to loosen the lymphocytes from the red blood cells. The splenocytes are then filtered through two 100 μm cell strainers to a clean 100 mm dish. Five milliliters of Red Cell Lysis Buffer is added to deplete red blood cells. The splenocytes are counted and the viability is checked with a standard procedure. It is preferred to use freshly isolated lymphocytes for fusion immediately. However, the lymphocytes can be frozen in LN2 without significant loss of viability or fusion efficiency within one year.

6.4.3 Cell Fusion

Performing a fusion between rabbit splenocytes and 240E-W2 is an experience-dependent process in which many factors determine the fusion efficiency and the success of generating RabMAb producing hybridomas. The growth condition of fusion partner 240E-W2 cells is critical and should be inspected under a microscope before use. 240E-W2 cells are harvested and concentrated in a 500 mL V-bottle. The cell pellet is carefully resuspended in 20 mL of prewarmed RPMI. A typical fusion procedure involves mixing 2×10^8 lymphocytes with 1×10^8 240E-W2 by a sterile transfer pipette in a 50 mL tube. The tube is then centrifuged at 1600 rpm for five minutes, followed by careful aspiration of supernatant, and then resuspended with 45 mL of fresh RPMI. This washing step is repeated and the pellet is loosened by gently tapping the bottom of the tube at a 45 degree angle in a circular motion. One milliliter of PEG solution is carefully added along the side of the tube while keeping the tube in slow rotation to gently mix the PEG solution with the cell pellet. To the mixture, 25 mL fresh prewarmed RPMI medium is added slowly, then 25 mL prewarmed RPMI with 10 percent FBS is added to bring the total volume to 50 mL with gentle mixing. After being centrifuged and gently resuspended in RPMI growth medium, the fused cells are plated into 20 96-well plates at 100 μL/well. Cells are incubated in CO_2 incubators for one to two days and 100 μL 2× HAT/20 percent FBS/240E medium is added for hybridoma selection. Hybridoma colonies typically appear 10 to 14 days after fusion, and are ready for screening in 3 to 4 weeks. Positive colonies are identified by antigen-binding assays and expanded into 24-well plates for further screening. The confirmed positive clones are then subcloned by limited dilution.

6.5 RabMAb PRODUCTION

Supernatant from a rabbit hybridoma culture usually contains a low concentration of antibody, typically 0.1 to 5 μg/mL. This is approximately 10 percent of the average yield for a mouse hybridoma. Inoculation of nude mice with rabbit hybridomas to produce antibodies has not been successful. The most commonly used approach is to use an *in vitro* bioreactor such as a CELLine Integra flask. This is a relatively efficient RabMAb production system, with an average yield of 5 mg per liter in three to four weeks of incubation time. To initiate the production, 2×10^7 cells in 15 mL BD medium are carefully added to the cell compartment, and 1 L of BD medium with 2 percent FBS is added into the nutrient compartment. Hybridomas are allowed to grow for about two weeks and the cell suspension in the lower compartment is carefully removed and centrifuged briefly to pellet hybridoma cells. Supernatant containing IgG is transferred to a new container and centrifuged for 10 minutes at 3000 rpm to remove cell debris. Tris buffer (pH 8.0) is added to adjust the pH. The antibody is stable at 4°C for weeks. For longer term storage, 30 percent glycerol can be added to the antibody supernatant and stored at −20°C in small aliquots.

6.6 BIOMARKER DEVELOPMENT USING RabMAbs

With high binding affinity and specificity, RabMAbs have been widely used in various applications in life science research. RabMAbs have been recognized as superior reagents in immunohistochemistry (IHC) and in detection of posttranslational modification of proteins such as phosphorylation. Approximately 100 RabMAbs have been developed to date for IHC clinical and research uses in cancer classifications. Among them are IHC tests for Her2 and c-Kit, used as companion diagnostics for cancer drugs Herceptin (Genentech) and Gleevec (Novartis), respectively. Figure 6.3 shows examples of IHC images of human tumor tissues stained by RabMAbs to Her2, c-kit, CD34, and PR. RabMAbs have been used to facilitate scientific research (X. Huang et al. 1998; Xue et al. 2001; Atabai et al. 2005; Tao et al. 2006) and a number of RabMAbs are also under development against biomarkers that can specify the state or the progression of diseases after therapeutic treatment. Protein modification, such as phosphorylation, of a signaling protein in a specific pathway may indicate the activation of a disease related pathway. RabMAbs that specifically distinguish the phosphorylated and nonphosphorylated state of proteins are extremely useful research tools. More than 200 such phospho-specific RabMAbs have been developed.

6.7 THERAPEUTIC RabMAbs DEVELOPMENT

6.7.1 High Affinity to Increase Potency and Reduce Side Effects

Most therapeutic antibodies in clinical use have affinity in the nanomolar range (10^{-9} M). It is anticipated that higher affinity leads to high potency. Rabbits generate monoclonal antibodies with ultra-high affinity up to 10^{-11} to 10^{-13} M without *in vitro* affinity maturation. High potency not only reduces the cost of antibody production but also potentially reduces side effects caused by the high amount of antibody administrated. *In vitro* affinity maturation is commonly used to improve the affinity of antibody drug leads when the initial antibodies have lower affinity; the process typically takes 8 to 12 months. RabMAbs on the other hand naturally have ultra-high affinity to start with. The advantages of the ultra-high affinity remain to be shown in animal and clinical studies.

6.7.2 Large Antibody Repertoire and High Specificity to Increase the Success Rate of Therapeutic Antibody Development

Like other drug discovery methods, a large number of bioactive compounds with significant diversity lead to higher possibilities to find a molecule that has the most desirable activities and

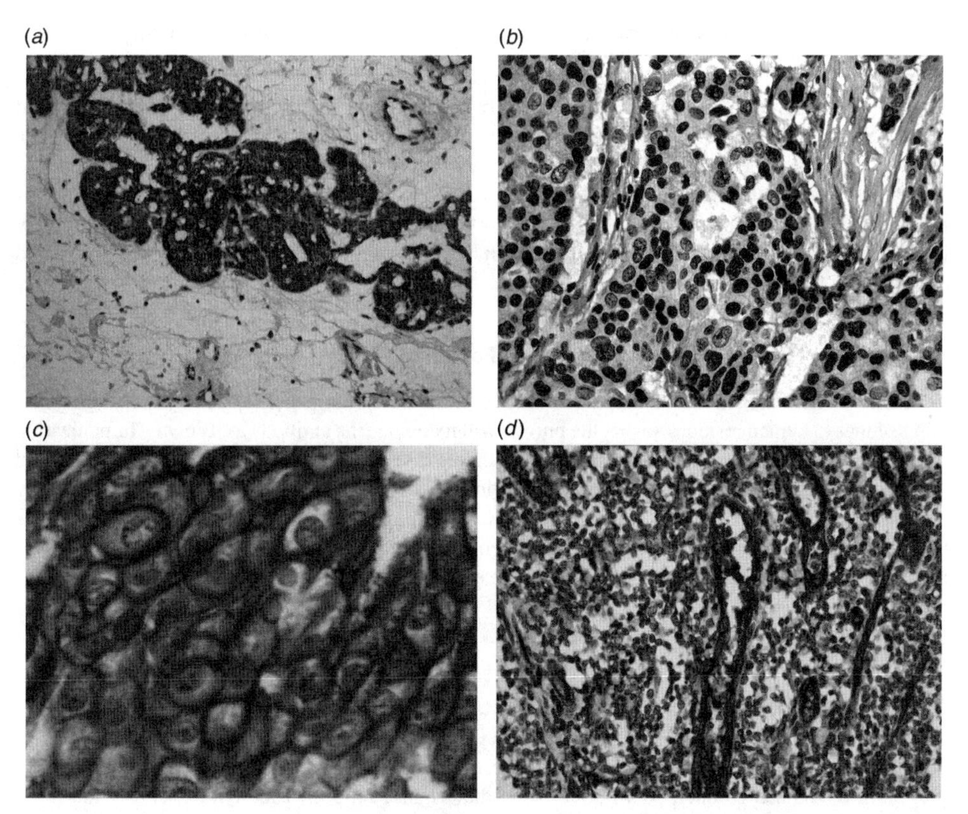

Figure 6.3 Immunohistochemistry staining of paraffin embedded tissues by RabMAbs. Human paraffin embedded tissues were deparaffinized by heating followed by a series of washes, with two xylene washes and two 100 percent ethanol rinses, followed by 95 percent ethanol, 70 percent ethanol, 50 percent ethanol, 30 percent ethanol, followed by H_2O and a TBST wash for 5 min on a shaker. Antigen retrieval was done using a rice cooker to cook in antigen retrieval solution for about 20 to 30 min. After slides were washed with TBST for 5 min on a shaker, endogenous peroxidase was inactivated by covering tissue with 3 percent hydrogen peroxide for 10 min. Slides were washed three times with TBST and blocked with blocking solution for one hour. Diluted primary antibody in the blocking buffer was applied to each section and incubated overnight in a humidified chamber (4°C). After washing slides three times with TBST, secondary HRP-conjugated anti-rabbit antibody diluted in the blocking solution was applied to incubate for one hour at room temperature. Slides were washed three times with TBST and freshly prepared DAB substrate was added to the tissue sections. The tissue sections with the substrate were incubated at room temperature until suitable staining developed, followed by rinsing with water and counterstaining with Hematoxylin. Slides were rinsed with water again and dehydrated using two rinses with 100 percent ethanol followed by two rinses with xylene. Slides were mounted with coverslips using permount medium and visualized under a microscope. (a) Ovarian adenocarcinoma tissue stained with anti-phospho-AKT RabMAb. (b) Breast ductal infiltrating carcinoma tissue stained with anti-PR RabMAb. (c) Breast ductal infiltrating carcinoma tissue stained with anti-Her2 RabMAb. (d) Vascular endothelial cells stained with anti-CD34 RabMAb. (See color insert.)

features. The RabMAb technology generates larger numbers of antibodies targeting more diverse epitopes due to the rabbit's immune system and the larger number of splenocytes available from each rabbit. In addition, RabMAbs recognize certain antigens that are not immunogenic in mice or cannot be identified by other antibody technologies. The availability of a large antibody collection and a broad diversity should help to increase the chance of identifying the desired therapeutic antibodies.

6.7.3 Cross-Reactivity to Animal Targets to Eliminate Surrogate Antibody

When a mouse antibody is used as a therapeutic lead, this antibody often does not recognize the mouse ortholog of the human target. A surrogate antibody is normally necessary to study a disease model in mouse. Sometimes, the surrogate is difficult to develop and the result can be misleading because a different antibody other than the therapeutic antibody candidate is used. When screened for bioactive antibodies, in addition to RabMAbs that are specific to human targets, many RabMAbs are capable of recognizing both human proteins and their mouse orthologs. These RabMAbs can then be used directly in preclinical studies in mouse disease models, thus increasing the efficiency of drug discovery and development.

6.7.4 Novel Humanization Approach to Facilitate Protein Engineering

To reduce immunogenicity in humans, antibodies of animal origins are humanized by exchanging certain residues to human residues so that the human will recognize the antibody as its own. Humanization of RabMAbs is relatively straightforward due to the unique characteristics of RabMAbs and a novel humanization technology. As described in Section 6.1, rabbit antibodies are mainly encoded by one heavy chain and one light chain gene, and the predominant use of a single V_H gene simplifies the cloning process. Analysis of RabMAb IgG sequences indicates that RAbMAbs share more homology (60 to 76 percent identity to human) than mouse mAbs (57 to 72 percent identity to human) to human IgG sequences.

6.8 MUTATIONAL-LINEAGE GUIDED (MLG) HUMANIZATION

Antibody humanization replaces the nonhuman residues in a parent antibody with the residues that are present in the human counterpart without losing the original antibody activity. To achieve this goal, several antibody humanization methods have been developed. The most widely used one is complementarity determining region (CDR) grafting (Jones et al. 1986; Kettleborough et al. 1991). In this method, the parental antibody framework regions are substituted by those of the human counterparts, and then some structurally critical residues in the framework regions are mutated back to the parental in order to reconstitute the original antigen-binding affinity and specificity. Another method was derived from analysis of antibody structures (Padlan 1991, 1994). The nonhuman residues in the framework regions are humanized except for the important residues that are potentially important to forming the antigen-binding pocket with the CDR residues. Both methods involve the identification of the residues that potentially affect antigen-binding activity. These residues, likely disturbing the CDR conformation, are usually identified by structural modeling analysis. However, the uncertainties of the structures derived from homology modeling and the cut-off distance under which a residue will be in contact with another may result in inaccurate prediction (Roguska et al. 1996). Moreover, the unchanged nonhuman CDRs have been found to be immunogenic in human (Ritter et al. 2001; Welt et al. 2003).

Other than the humanization approaches described, mutational lineage-guided (MLG) humanization is a novel method of antibody engineering (Couter et al. 2004, 2006), based on gene sequences of each IgG that have the same biological functions and are from B-cells deriving from the same parental B-cell during the *in vivo* antibody affinity maturation. In order to apply the MLG humanization method, a large number of bioactive (agonistic or antagonistic) antibodies are necessary. Typically 30 to 50 bioactive antibodies are identified from each immunized rabbit by either a cell-based assay or a biochemical assay. A large number of bioactive antibodies are often difficult to obtain by other technology platforms. Amino acid sequences of the variable regions of the heavy and light chains (VH and VL) from this collection of IgG sequence are aligned to form a phylogenetic tree. An example of such a lineage tree is shown in Figure 6.4. Related antibodies are grouped according to their sequence similarity to each other. The antibodies in each lineage likely share a common B-cell ancestor. Conserved

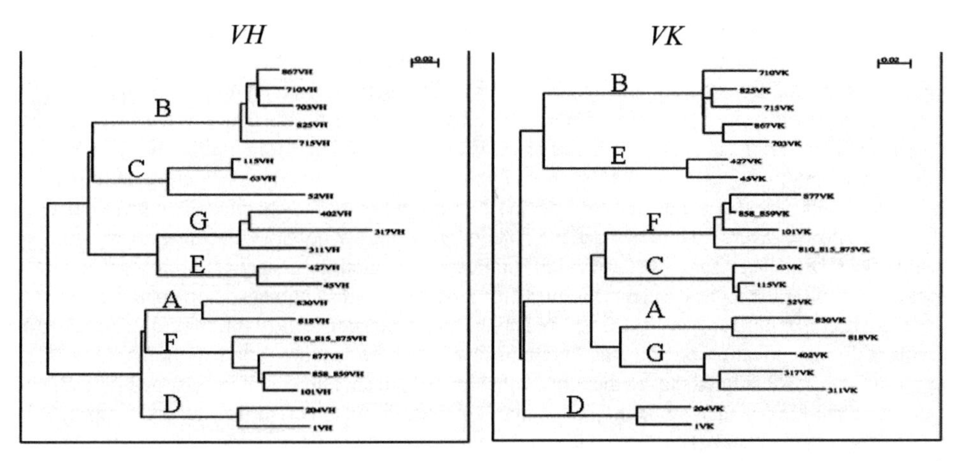

Figure 6.4 Phylogenetic tree of neutralizing RabMAbs to TNF-α. VH and VK lineage trees of related groups of rabbit anti-human TNF-α antibodies obtained by sequence alignment and phylogenetic analysis. The sequence alignment and phylogenetic analysis were performed by using ClustalX software. In a lineage group (any group from A to F), the VH sequences are related and so are the respective VK sequences.

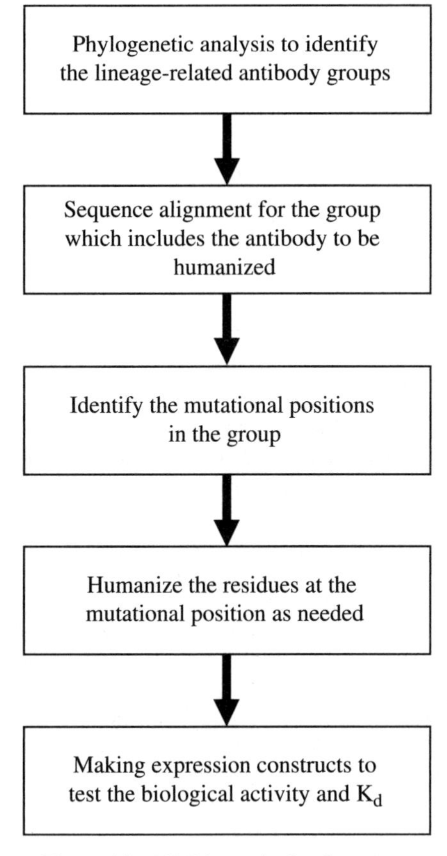

Figure 6.5 MLG humanization flow chart.

sequences in a lineage-related group represent critical residues that are important for structure and function of IgG. Conversely, the residues that are not conserved thus have no effects on the biological activities are not critical. Since these mutational positions were obtained from a group of antibodies (from one parental B-cell), they must have been originally produced and effectively tested by an animal immune system. Thus, substitution of amino acids at these positions from one antibody to another in the same group should be well tolerated without sacrificing antibody specificity and affinity. If the mutations occurred at the positions to humanize, substitution of human residues at these positions should be allowed. More importantly, such mutations are found not only in the framework regions, but also in the CDRs. Therefore, MLG humanization can be applied to humanization of the framework regions as well as the CDRs. Figure 6.5 illustrates the flow of MLG humanization approach.

The MLG method has been successfully applied to humanization of two RabMAbs. The process involves the identification of potential "humanizable" residues that are found in the MLG lineages because they are substituted during the affinity maturation and are known not to be structurally critical based on three-dimensional modeling. The parental RabMAb and a most closely related human germline sequence is aligned. All humanizable residuals are changed to human residues. To examine effects of MLG humanization on changes of antibody property, the humanized RabMAbs are tested for antigen-binding affinity and neutralizing activity. The humanized antibodies showed no significant change in either the affinity or activity, compared with the parental unhumanized antibody (data not shown). The MLG humanization is a novel technology and is still in early stages of application. Whether humanized RabMAbs have reduced immunogenicity in humans remains to be demonstrated in clinical studies.

6.9 SUMMARY

RabMAb has been demonstrated to be a superior reagent for studying various biological functions and can be employed in many applications, including Western blots, immunohistochemistry, flow cytometry, and other immunoassays. Due to its high affinity and specificity, ability to recognize novel epitopes, and ease for engineering such as humanization, RabMAb is becoming an attractive technology for therapeutic antibody development in addition to biomarker discovery.

REFERENCES

Atabai, K., X. Huang, I. Ueki, A. Kline, Y. Li, S. Sadatmansoori, W. Zhu, R. Pytela, Z. Werb, and D. Sheppard. 2005. MFG-8 is critical for mammary gland remodeling during involution. *Mol. Biol. Cell* 16:5528–5535.

Bystryn, J.C., J.S. Jacobsen, P. Liu, et al. 1982. Comparison of cell surface human melanoma-associated antigens identified by rabbit and murine antibodies. *Hybridoma* 1:465–472.

Coligan, J.E., A.M. Kruisbeek, et al. eds. 1994. *Current protocols in immunology.* New York: Greene and Wiley.

Collins, J.J., P.H. Black, A.D. Strosberg, et al. 1974. Transformation by simian virus 40 of spleen cells from a hyperimmune rabbit: Evidence for synthesis of immunoglobulin by the transformed cells. *Proc. Natl. Acad. Sci. USA* 71:260–262.

Couto, J., K. Hendricks, S.E. Wallace, H. Au, S. Li, R. Pytela, J. Villanueva, H. Wen, L. Xie, G.L. Yu, A. Zhang, D.X. Zhang, and W. Zhu. 2004. Isolation and characterization of fifty neutralizing anti-TNFa rabbit monoclonal antibodies. Poster presentation at Antibody Engineering, San Diego.

Couto, J., K. Hendricks, S.E. Wallace, and G.L. Yu. 2006. Methods for antibody engineering. International patent application No: PCT/US2005/039930, Publication Number: WO/2006/050491.

Harlow, E., and D. Lane. 1988. *Antibodies: A laboratory manual.* Cold Spring Harbor, NY: Cold Spring Harbor Laboratory.

Howard, G.C., and M.R. Kaser, eds. 2007. *Making and using antibodies: A practical handbook.* Boca Raton, FL: CRC Press.

Huang, X., J. Wu, W. Zhu, R. Pytela, and D. Sheppard. 1998. Expression of the human integrin brta6 subunit in alveolar type II cells and bronchiolar epithelial cells reverses lung inflammation in brta6 knockout mice. *Am. J. Resp. Cell Mol. Biol.* 19:636–642.

Huang, Z., W. Zhu, Y. Meng, and H. Xia. 2006. Development of new rabbit monoclonal antibody to progesterone receptor (Clone SP2): No heat pretreatment but effective for paraffin section immunohistochemistry. *Appl. Immunohistochem. Mol. Morphol.* 14:229–233.

Huang, Z., W. Zhu, G. Szekeres, and H. Xia. 2005. Development of new rabbit monoclonal antibody to estrogen receptor: Immunohistochemical assessment on formalin-fixed, paraffin-embedded tissue sections. *Appl. Immunohistochem. Mol. Morphol.* 13:91–95.

Jones, P.T., P.T. Dear, J. Foote, M.S. Neuberger, and G. Winter. 1986. Replacing the complementarity-determining regions in a human antibody with those from a mouse. *Nature* 321:522–525.

Kearney, J.F., A. Radbruch, B. Liesegand, and K. Rajewsky. 1979. A new mouse myeloma cell line that has lost immunoglobulin expression but permits the construction of antibody-secreting hybrid cell lines. *J. Immunol.* 123:1548–1554.

Kettleborough, C.A., J. Saldanha, V.J. Heath, C.J. Morrison, and M.M. Bendig. 1991. Humanization of a mouse monoclonal antibody by CDR-grafting: The importance of framework residues on loop confirmation. *Protein Eng.* 4:773–783.

Knight, K.L., and M.A. Crane. 1994. Generating the antibody repertoire in rabbit. *Adv. Immunol.* 56:179–218.

Krause, R.M. 1970a. The search for antibodies with molecular uniformity. *Adv. Immunol.* 12:1–56.

Krause, R.M. 1970b. Experimental approaches to homogenous antibody populations: Factors controlling the occurrence of antibodies with uniform properties. *Fed. Proc.* 29:59–65.

Kuo, M.C., J.A. Sogn, E.E. Max, et al. 1985. Rabbit-mouse hybridomas secreting intact rabbit immunoglobulin. *Mol. Immunol.* 22:351–359.

Mage, R.G., D. Lanning, and K.L. Knight. 2006. B cell and antibody repertoire development in rabbits: The requirement of gut-associated lymphoid tissues. *Dev. Comp. Immunol.* 30:137–153.

Michel, J., J. Liguori, J.A. Hoff-velk, and D.H. Ostrow. 2001. Recombinant human interleukin-6 enhances the immunoglobulin secretion of a rabbit-rabbit hybridoma. *Hybridoma* 20:189–198.

Padlan, E.A. 1991. A possible procedure for reducing the immunogenicity of antibody variable domains while preserving their ligand-binding properties. *Mol. Immunol.* 28:489–498.

Padlan, E.A. 1994. Anatomy of the antibody molecule. *Mol. Immunol.* 31:161–217.

Raybould, T.J., and M. Takahashi. 1988. Production of stable rabbit-mouse hybridomas that secrete rabbit mAb of defined specificity. *Science* 240:1788–1790.

Rief, N., C. Waschow, W. Nastainczyk, M. Montenarh, and C. Gotz. 1998. Production and characterization of a rabbit monoclonal antibody against human CDC25C phosphatase. *Hybridoma* 17:389–394.

Ritter, G., L.S. Cohen, C. Williams, Jr., E.C. Richards, L.J. Old, and S. Welt. 2001. Serological analysis of human anti-human antibody responses in colon cancer patients treated with repeated doses of humanized monoclonal antibody A 33. *Cancer Res.* 61:6851–6859.

Roguska, M.A., J.T. Pedersen, A.H. Henry, S.M. Searle, C.M. Roja, B. Avery, M. Hoffee, S. Cook, J.M. Lambert, W.A. Blättler, A.R. Rees, and B.C. Guild. 1996. A comparison of two murine monoclonal antibodies humanized by CDR-grafting and variable domain resurfacing. *Protein Eng.* 9:895–904.

Rossi, S., L. Laurino, A. Furlanetto, S. Chinellato, E. Orvieto, F. Canal, F. Facchetti, and A.P. Dei Tos. 2005. A comparative study between a novel category of immunoreagents and the corresponding mouse monoclonal antibodies. *Am. J. Clin. Pathol.* 124:295–302.

Schiaffella, E., D. Sehgal, A.O. Anderson, and R.G. Mage. 1999. Gene conversion and hypermutation during diversification of Vh sequences in developing splenic germinal centers of immunized rabbits. *J. Immunol.* 162:3984–3995.

Shulman, M., C.D. Wilde, and G. Kohler. 1978. A better cell line for making hybridomas secreting specific antibodies. *Nature* 276:269–270.

Spieker-Polet, H., P. Sethupathi, P.C. Yam, et al. 1995. Rabbit monoclonal antibodies: Generating a fusion partner to produce rabbit-rabbit hybridomas. *Proc. Natl. Acad. Sci. USA* 92:9348–9352.

Tao, G.Z., I. Nakamichia, N.O. Kua, J. Wang, M. Frolkis, X. Gong, W. Zhu, R. Pytela, M.B. Omary. 2006. Bispecific and human disease-related anti-keratin rabbit monoclonal antibodies. *Exp. Cell Res.* 312:411–422.

Verbanac, K.M., U.M. Gross, L.M. Rebellato, et al. 1993a. Production of stable rabbit-mouse heterohybridomas: Characterization of a rabbit monoclonal antibody recognizing a 180 kDa human lymphocyte membrane antigen. *Hybridoma* 12:285–295.

Verbanac, K.M., U. Gross, L.M. Rebellato, et al. 1993b. Generation of rabbit anti-lymphocyte monoclonal antibodies. *Transplant Proc.* 25(1, pt 1):837–838.

Weller, A., J. Meek, and E.D. Adamson. 1987. Preparation and properties of monoclonal and polyclonal antibodies to mouse epidermal growth factor (EGF) receptors: Evidence for cryptic EGF receptors in embryonal carcinoma cells. *Development* 100:351–363.

Welt, S., G. Ritter, C. Williams, Jr., L.S. Cohen, M. John, A. Junbluth, E.C. Richards, L.J. Old, and N.E. Kemeny. 2003. Phase I study of anti-colon cancer humanized antibody A33. *Clin. Cancer Res.* 9:1338–1346.

Xue, H., A. Atakilit, W. Zhu, X. Li, D.M. Ramos, and R. Pytela. 2001. Role of the alpha-v beta-6 integrin in human oral squamous cell carcinoma growth in vivo and in vitro. *Biochem. Biophys. Res. Commun.* 288:610–618.

CHAPTER 7

Human Antibody Repertoire Libraries

DAVID LOWE and TRISTAN J. VAUGHAN

ABSTRACT

Human antibody repertoire libraries are used extensively for the generation of monoclonal antibodies for research and therapeutic purposes. We describe a method for the construction of a large human scFv phagemid library and discuss methods for assessing the size, diversity, and functionality of such a repertoire. Alternative strategies, including semisynthetic antibody fragment libraries, and different display formats are also discussed.

7.1 INTRODUCTION

The construction of human antibody libraries from natural variable (V) gene sources is now a well-established technology for the generation of individual antibodies with defined specificities to any conceivable antigen. Large ($>10^{10}$) libraries have been reported from many commercial and academic

Therapeutic Monoclonal Antibodies: From Bench to Clinic. Edited by Zhiqiang An

laboratories, along with examples of isolated antibodies that are in clinical development for the treatment of severe and debilitating diseases. A wide variety of V gene sources, as well as antibody fragment formats, vectors, and display systems now exist to serve the rapidly increasing demand for human antibodies.

7.2 HISTORICAL PERSPECTIVE

The first attempts to mimic the natural immune repertoire in the test tube were kick-started by the development of the polymerase chain reaction (PCR) (Saiki et al. 1985), which facilitated the amplification of V genes from lymphocytes of immunized animals (Orlandi et al. 1989; Sastry et al. 1989). Expression of these genes in *Escherichia coli* produced sufficiently large repertoires such that binding fragments with specificities to either a hapten (Huse et al. 1989) or to lysozyme (Ward et al. 1989) could be identified by screening a few thousand clones. Subsequently, the development of phage display of antibody fragments (McCafferty et al. 1990) allowed for the creation of antibody fragment libraries that could be iteratively selected against a given antigen, to enrich for binders to the chosen antigen (Clackson et al. 1991). The random recombination of different human V_H and V_L genes, producing *de novo* repertoires of V_H/V_L combined fragments, was subsequently shown to produce antibody fragments that specifically recognize protein antigens and haptens (Marks et al. 1991a) as well as anti-self antigens (Griffiths et al. 1993).

The first such libraries of human antibody fragments were of 10^7 to 10^8 in size and generally produced a small panel of antibodies specific to the target but of only modest affinity, typically micromolar to hundreds of nanomolar. These were found to be impractical for most research, and certainly all therapeutic applications. Library repertoire sizes were subsequently further developed either by optimizing the transformation efficiencies and increasing the numbers of transformations (Vaughan et al. 1996), by employing more efficient cloning techniques (de Haard et al. 1999), or by exploiting *in vivo* bacterial recombination of the V genes (Griffiths et al. 1994; Sblattero and Bradbury 2000).

These approaches have had two major benefits. The first is to increase the number of isolated antibodies to any given target, with over 1000 different antibody fragments now being reported as possible (Edwards et al. 2003). The second is the practical demonstration that greater antibody population diversity leads to the isolation of antibodies with greater affinity, as predicted by Perelson and Oster (1979). Antibody fragments with subnanomolar affinities to a given antigen are now routinely isolated (Vaughan et al. 1996).

7.3 CONSTRUCTION OF AN scFv LIBRARY IN A PHAGEMID VECTOR

Our laboratory has constructed a natural library of human scFv antibody fragments containing $>10^{11}$ transformants, that has been used to successfully isolate antibodies suitable for clinical development for more than 10 years (Fig. 7.1). To maximize diversity and affinity, we employed three main strategies:

(1) Use of multiple tissue sources of B-cells.
(2) A large number of donors (>100), to eradicate any potential bias that might be seen in a small population.
(3) Primers designed to cover all Ig classes, including IgG, IgM, IgA, and IgD.

The source tissues of the B-cell-derived antibody V genes include spleen, bone marrow, tonsil, and peripheral blood mononuclear cells (PBMCs). These tissue sources have been chosen to represent as wide a diversity of V genes as possible, and to represent both the primary and secondary immune response. Sourcing tissues from multiple human donors ensures that V gene diversity is maximized and minimizes the effects of biases in any given individual's immune repertoire. For example, the

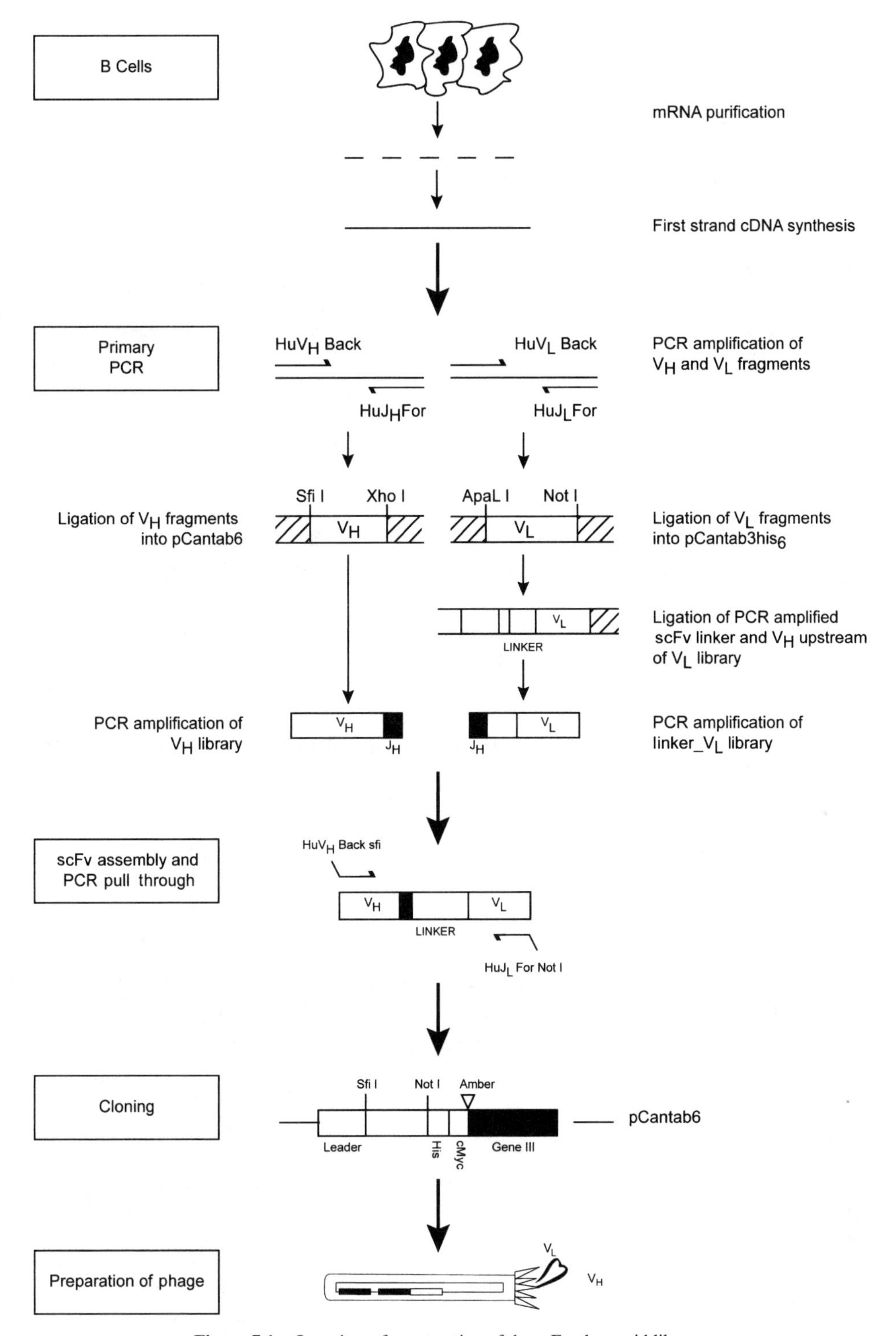

Figure 7.1 Overview of construction of the scFv phagemid library.

libraries that have been generated in our laboratory have been derived from B-lymphocytes from over 100 healthy human donors.

7.3.1 Materials

(1) Source of lymphoid tissue from which mRNA can be isolated, or commercial cDNA source (e.g., Clontech Quick-Clone cDNA).
(2) mRNA preparation kit [e.g., an oligo(dT)-purification system, such as the Illustra Quickprep Micro mRNA Purification kit; GE Healthcare].
(3) cDNA synthesis kit (e.g., SuperScript Choice System; Invitrogen).
(4) PCR reagents: Taq DNA polymerase with 10X Taq DNA polymerase buffer (Roche), a stock of deoxyribonucleoside triphosphates (dNTPs) (5 mM each) (GE Healthcare), PCR-grade water (Sigma-Aldrich).
(5) Oligonucleotide primers at 10 μM. Tables 7.1 to 7.3 show primer sets currently used in our laboratory for the construction of human scFv libraries.
(6) PCR purification kit (e.g., Qiagen minElute PCR purification kit; Qiagen).
(7) Buffer-saturated phenol (e.g., Invitrogen UltraPure).
(8) Chloroform.
(9) Absolute ethanol (−20°C); 70 percent (v/v) ethanol (−20°C).
(10) Kit for the isolation and purification of DNA from gels (e.g., GeneClean; QBiogene).
(11) PCR-grade water (Sigma-Aldrich).
(12) Phage display vectors and appropriate restriction enzymes for cloning. In the example described, the vectors pCANTAB6 and pCANTAB3his_6 are used with the enzymes *Sfi*I, *Not*I, *Xho*I, *Hin*dIII, and *Apa*LI, although the principles of library construction are not limited to these vectors or restriction enzymes.
(13) Kits for medium- and large-scale isolation of plasmid DNA (e.g., Qiagen Midiprep and Maxiprep kits).
(14) Ligation kit (e.g., DNA Ligation Kit; GE Healthcare).
(15) Electrocompetent *Escherichia coli* TG1.
(16) Bio-Rad Gene Pulser and suitable electroporation cuvettes.
(17) 2TY as liquid and solid (agar) medium [see Sambrook, Fritsch, and Maniatis (1989) for recipe].
(18) 20 percent (w/v) glucose solution (sterile filtered).
(19) Ampicillin (100 mg/mL stock) and kanamycin (50 mg/mL), both filter sterilized.
(20) 20 percent (w/v) polyethylene glycol (PEG) 8000, 2.5 M NaCl solution.
(21) TE buffer (10 mM Tris-HCl, 1 mM ethylene diamine tetraacetic acid, pH 8.0).
(22) Ultrapure caesium chloride (CsCl).

7.3.2 Preparation of cDNA Template

B-lymphocyte cDNA template can be purchased from manufacturers, for example, Clontech Quick-Clone, or can be prepared directly from appropriate tissue. For PBMCs, 50 mL of whole blood yields approximately 10^7 cells, following Ficol gradient centrifugation as detailed by Marks et al. (1991a). The mRNA is immediately isolated using an oligo(dT)-purification system, such as the Illustra Quickprep Micro mRNA Purification kit (GE Healthcare). First-strand cDNA from the mRNA template is synthesized using a kit such as the SuperScript Choice System (Invitrogen), using random hexamer primers, according to the manufacturer's instructions. In order to amplify

TABLE 7.1 Oligonucleotide Primers for Amplification of Human V_H Sequences

Primer	Sequence
Human VH BackSfiI Primers	
VH1b/7a Back*Sfi*I	5′ – GTC CTC GCA ACT GCG GCC CAG CCG GCC ATG GCC CAG (AG)TG CAG CTG GTG CA(AG) TCT GG – 3′
VH1c Back*Sfi*I	5′ – GTC CTC GCA ACT GCG GCC CAG CCG GCC ATG GCC (GC)AG GTC CAG CTG GT(AG) CAG TCT GG – 3′
VH2b Back*Sfi*I	5′ – GTC CTC GCA ACT GCG GCC CAG CCG GCC ATG GCC CAG (AG)TC ACC TTG AAG GAG TCT GG – 3′
VH3b Back*Sfi*I	5′ – GTC CTC GCA ACT GCG GCC CAG CCG GCC ATG GCC (GC)AG GTG CAG CTG GTG GAG TCT GG – 3′
VH3c Back*Sfi*I	5′ – GTC CTC GCA ACT GCG GCC CAG CCG GCC ATG GCC GAG GTG CAG CTG GTG GAG (AT)C(TC) GG – 3′
VH4b Back*Sfi*I	5′ – GTC CTC GCA ACT GCG GCC CAG CCG GCC ATG GCC CAG GTG CAG CTA CAG CAG TGG GG – 3′
VH4c Back*Sfi*I	5′ – GTC CTC GCA ACT GCG GCC CAG CCG GCC ATG GCC CAG (GC)TG CAG CTG CAG GAG TC(GC) GG – 3′
VH5b Back*Sfi*I	5′ – GTC CTC GCA ACT GCG GCC CAG CCG GCC ATG GCC GA(AG) GTG CAG CTG GTG CAG TCT GG – 3′
VH6a Back*Sfi*I	5′ – GTC CTC GCA ACT GCG GCC CAG CCG GCC ATG GCC CAG GTA CAG CTG CAG CAG TCA GG – 3′
Human VH ForXhoI Primers	
HuJHFor1-2*Xho*I	5′ – ACC GCC TCC ACC ACT CGA GAC GGT GAC CAG GGT GCC (TC)(TC)(GT) GCC CCA – 3′
HuJHFor3*Xho*I	5′ – ACC GCC TCC ACC ACT CGA GAC GGT GAC CAT TGT CCC (TC)(TC)(GT) GCC CCA – 3′
HuJHFor4-5*Xho*I	5′ – ACC GCC TCC ACC ACT CGA GAC GGT GAC CAG GGT TCC (TC)(TC)(TG) GCC CCA – 3′
HuJHFor6*Xho*I	5′ – ACC GCC TCC ACC ACT CGA GAC GGT GAC CGT GGT CCC (TC)(TC)(TG) CCC CCA – 3′

the different V_H and V_L repertoires, 0.5 ng of cDNA is required for each initial reaction, giving a total requirement of approximately 15 ng.

7.3.3 Construction of a Naïve V_H Repertoire

The primers used to amplify the V_H genes are based on those originally described by Marks et al. (1991b). The 5′ and 3′ primers are designed to contain *Sfi*I and *Xho*I restriction sites, respectively (Table 7.1). The reaction conditions for the amplification are detailed below.

TABLE 7.2 Oligonucleotide Primers for Amplification of Human V_L Sequences

Human Vλ BackApaLI Primers

Huλ1a Back*Apa*LI

5′ – ACC GCC TCC ACC AGT GCA CAG TCT GTG CTG ACT CAG CCA CC – 3′

Huλ1b Back*Apa*LI

5′ – ACC GCC TCC ACC AGT GCA CAG TCT GTG (TC)TG ACG CAG CCG CC – 3′

Huλ1c Back*Apa*LI

5′ – ACC GCC TCC ACC AGT GCA CAG TCT GTC GTG ACG CAG CCG CC – 3′

Huλ2 Back*Apa*LI

5′ – ACC GCC TCC ACC AGT GCA CA(AG) TCT GCC CTG ACT CAG CCT – 3′

Huλ3a Back*Apa*LI

5′ – ACC GCC TCC ACC AGT GCA CTT TCC TAT G(AT)G CTG ACT CAG CCA CC – 3′

Huλ3b Back*Apa*LI

5′ – ACC GCC TCC ACC AGT GCA CTT TCT TCT GAG CTG ACT CAG GAC CC – 3′

Huλ4 Back*Apa*LI

5′ – ACC GCC TCC ACC AGT GCA CAC GTT ATA CTG ACT CAA CCG CC – 3′

Huλ5 Back*Apa*LI

5′ – ACC GCC TCC ACC AGT GCA CAG GCT GTG CTG ACT CAG CCG TC – 3′

Huλ6 Back*Apa*LI

5′ – ACC GCC TCC ACC AGT GCA CTT AAT TTT ATG CTG ACT CAG CCC CA – 3′

Huλ7/8 Back*Apa*LI

5′ – ACC GCC TCC ACC AGT GCA CAG (AG)CT GTG GTG AC(TC) CAG GAG CC – 3′

Huλ9 Back*Apa*LI

5′ – ACC GCC TCC ACC AGT GCA C(AT)G CCT GTG CTG ACT CAG CC(AC) CC – 3′

Human Vκ BackApaLI Primers

Huκ1b Back*Apa*LI

5′ – ACC GCC TCC ACC AGT GCA CTT GAC ATC CAG (AT)TG ACC CAG TCT CC – 3′

Huκ2 Back*Apa*LI

5′ – ACC GCC TCC ACC AGT GCA CTT GAT GTT GTG ATG ACT CAG TCT CC – 3′

Huκ3b Back*Apa*LI

5′ – ACC GCC TCC ACC AGT GCA CTT GAA ATT GTG (AT)TG AC(AG) CAG TCT CC – 3′

Huκ4b Back*Apa*LI

5′ – ACC GCC TCC ACC AGT GCA CTT GAT ATT GTG ATG ACC CAC ACT CC – 3′

Huκ5 Back*Apa*LI

5′ – ACC GCC TCC ACC AGT GCA CTT GAA ACG ACA CTC ACG CAG TCT CC – 3′

Huκ6 Back*Apa*LI

5′ – ACC GCC TCC ACC AGT GCA CTT GAA ATT GTG CTG ACT CAG TCT CC – 3′

TABLE 7.2 *Continued*

*Human Vλ For*Not*I Primers*

HuJλ1 For*Not*I

5′ – GAG TCA TTC TCG ACT TGC GGC CGC ACC TAG GAC GGT GAC CTT GGT CCC – 3′

HuJλ2-3 For*Not*I

5′ – GAG TCA TTC TCG ACT TGC GGC CGC ACC TAG GAC GGT CAG CTT GGT CCC – 3′

HuJλ4-5 For*Not*I

5′ – GAG TCA TTC TCG ACT TGC GGC CGC ACT TAA AAC GGT GAG CTG GGT CCC – 3′

*Human Vκ For*Not*I Primers*

HuJκ1 For*Not*I

5′ – GAG TCA TTC TCG ACT TGC GGC CGC ACG TTT GAT TTC CAC CTT GGT CCC – 3′

HuJκ2 For*Not*I

5′ – GAG TCA TTC TCG ACT TGC GGC CGC ACG TTT GAT CTC CAG CTT GGT CCC – 3′

HuJκ3 For*Not*I

5′ – GAG TCA TTC TCG ACT TGC GGC CGC ACG TTT GAT ATC CAC TTT GGT CCC – 3′

HuJκ4 For*Not*I

5′ – GAG TCA TTC TCG ACT TGC GGC CGC ACG TTT GAT CTC CAC CTT GGT CCC – 3′

HuJκ5 For*Not*I

5′ – GAG TCA TTC TCG ACT TGC GGC CGC ACG TTT AAT CTC CAG TCG TGT CCC – 3′

(1) Prepare separate 50 μL PCR reactions, for each of the V_H Back primers as follows, remembering to also prepare a negative control with no template for each reaction:

5.0 μL	10× *Taq* buffer
2.5 μL	5 mM dNTPs
2.5 μL	10 μM of an individual V_H Back*Sfi* primer
2.5 μL	10 μM J_H1-6For*Xho* primer mix
5.0 μL	Approximately 0.5 ng of first strand cDNA
31.5 μL	PCR-grade water
1.0 μL	*Taq* polymerase (5 U)

(2) Place in a PCR thermocycler and carry out the following reaction: 94°C for 1 minute, 55°C for 1 minute, 72°C for 2 minutes, for 30 cycles, followed by a final extension at 72°C for 10 minutes.

(3) Run a 5 μL sample from each reaction on a 1 percent (w/v) TAE agarose gel, to check that a PCR product of approximately 400 bp has been produced from each reaction mix.

(4) Purify each product separately, either by phenol/chloroform extraction followed by ethanol precipitation, or by using a kit, such as the Qiagen MinElute PCR purification kit (Qiagen). Resuspend the purified products in PCR-grade water and quantify by measurement of A_{260} nm using a spectrophotometer.

(5) Pool the different products in a normalized fashion, based on relative size of the different V_H families. This can be calculated based on the guide below (based on the number of different V_H

TABLE 7.3 Oligonucleotide Primers for Pull-Through PCR

pUC19 Rev

5′ – AGC GGA TAA CAA TTT CAC ACA GG – 3′

fdtetseq

5′ – GTC GTC TTT CCA GAC GTT AGT – 3′

JH Forward Primers

HuJH1-2For

5′ – TGA GGA GAC GGT GAC CAG GGT GCC – 3′

HuJH3For

5′ – TGA AGA GAC GGT GAC CAT TGT CCC – 3′

HuJH4-5For

5′ – TGA GGA GAC GGT GAC CAG GGT TCC – 3′

HuJH6For

5′ – TGA GGA GAC GGT GAC CGT GGT CCC – 3′

RevJH Primers

RHuJH1-2

5′ – GCA CCC TGG TCA CCG TCT CCT CAG GTG G – 3′

RHuJH3

5′ – GGA CAA TGG TCA CCG TCT CTT CAG GTG G – 3′

RHuJH4-5

5′ – GAA CCC TGG TCA CCG TCT CCT CAG GTG G – 3′

RHuJH6

5′ – GGA CCA CGG TCA CCG TCT CCT CAG GTG C – 3′

exons in the V-BASE directory, MRC Centre for Protein Engineering, Cambridge, UK; http://vbase.mrc-cpe.cam.ac.uk):

Family		Relative % Proportion of V_H Repertoire
$V_H1/7$	12 members	23
V_H2	3 members	6
V_H3	22 members	43
V_H4	11 members	22
V_H5	2 members	4
V_H6	1 member	2

(6) Digest the pooled products with *Sfi*I and *Xho*I (New England Biolabs) according to the manufacturer's instructions. The *Sfi*I digest is carried out at 37°C and the *Xho*I digest at 50°C.

(7) Run the digested repertoire on a 1 percent (w/v) TAE low melting point agarose (Invitrogen) gel and purify using GeneClean (Qbiogene), following the manufacturer's instructions. The VH repertoire should run at approximately 350 bp, following digestion.

The phagemid vector pCANTAB6 (McCafferty et al. 1994) should be prepared in parallel, using the Qiagen Maxiprep kit (Qiagen) following the manufacturer's instructions.

(1) Digest the vector with *Sfi*I and *Xho*I restriction enzymes as per the manufacturer's instructions.
(2) Concentrate the cut vector by phenol/chloroform extraction, followed by ethanol precipitation, as described in Sambrook, Fritsch, and Maniatis (1989).
(3) Estimate the concentration of recovered DNA by measurement of A_{260} nm using a spectrophotometer.

The digested V_H gene repertoire is ligated into the digested pCANTAB6 phagemid vector, using the DNA Ligation Kit (GE Healthcare), following the manufacturer's instructions. Set up a series of trial ligations covering a range of molar ratios of insert to vector, such as 1 : 1, 2 : 1, 4 : 1, 1 : 2, and so on, using 50 ng cut vector as a baseline amount. This will help to ascertain the optimal ratio of insert to vector. Include a vector-only control ligation to quantify nonrecombinant background levels. Prepare electrocompetent *E. coli* TG1 cells and subsequent electroporations as described in Sambrook, Fritsch, and Maniatis (1989).

(1) Plate aliquots from the transformations for each ligation ratio on to 2TYG plates containing ampicillin (100 μg/mL; 2TYAG) and incubate overnight at 30°C. Confirm that background levels of vector self-ligation are minimal and determine which ligation ratio yields the highest number of transformants.
(2) Set up a larger scale set of ligations based on the determined optimal insert to vector ratio as before. As a guide, in order to generate a 10^8 library size, a total of approximately 1.5 μg of digested vector will be required, with an appropriate amount of vector.
(3) Perform at least 50 electroporations (2.5 V, 200 Ω, 25 μF) into electrocompetent *E. coli* TG1. Add 2TY media containing 2 percent (w/v) glucose and allow the cells to recover for 1 hour with shaking at 200 rpm in a 37°C incubator.
(4) Pool all of the electroporations, taking a small aliquot for plating at 10-fold serial dilutions onto 2TYAG plates to determine the size of the library. Centrifuge the remaining cells at 1200 *g* for 10 minutes, remove the supernatant, and resuspend in 1 to 2 mL 2TY medium. Plate out the total library onto four 243 × 243 mm 2TYAG agar plates.
(5) Incubate the plates overnight at 30°C. Determine the size of the library from the titer plates, which should be between 10^7 and 10^8 in total. Scrape the large plates, using 5 mL of 2TY per plate and pool the cells in a 50 mL Falcon tube. Add 0.5 volume of 50 percent (v/v) glycerol and ensure homogeneous resuspension of the cells by mixing on a rotating wheel for 30 minutes. Determine the cell density by optical density measurement at 600 nm (which will require serial dilution, 10^{-1}, 10^{-2}) and store in aliquots at −70°C.

7.3.4 Construction of a Naïve V_L Repertoire

Human V_L kappa (κ) and V_L lambda (λ) gene fragments are amplified separately using primers based on those published by Marks et al. (1991b). The 5′ back primers contain an *Apa*L1 site and the 3′ forward primers contain a *Not*I site, to facilitate cloning (Table 7.2). The same cycling parameters are employed for the PCR reactions as outlined for the amplification of the V_H fragments.

(1) Perform 50 μL PCR reactions using 0.5 ng of first strand cDNA as template using the following primer combinations: VκBack*Apa*LI + Jκ1-5For*Not*I primer mix and VλBack*Apa*LI + Jλ1-5 For*Not*I primer mix.
(2) Purify the PCR products as described for the V_H repertoire. Pool the different products in a normalized fashion, based on relative size of the different V_L families. This can be calculated based

on the guide below (based on the number of different Vλ and Vκ exons in the V-BASE directory, MRC Centre for Protein Engineering, Cambridge, UK; http://vbase.mrc-cpe.cam.ac.uk). At this stage the Vλ and Vκ pools can either be kept separate to generate two separate λ and κ libraries, or can be pooled together.

Family		Relative % Proportion of V_L Repertoire
Vκ1	17 members	48
Vκ2	7 members	20
Vκ3	7 members	20
Vκ4	1 member	3
Vκ5	1 member	3
Vκ6	2 members	6
Vλ1	5 members	17
Vλ2	5 members	17
Vλ3	9 members	30
Vλ4	3 members	10
Vλ5	3 members	10
Vλ6	1 member	3
Vλ7/8	3 members	10
Vλ9	1 member	3

(3) Digest the pooled Vκ and Vλ repertoires with *Apa*LI followed by *Not*I (New England Biolabs) following the manufacturer's instructions.

(4) Gel purify the V_L fragments as described previously for the V_H repertoire. The size of the fragments should be approximately 350 bp.

Purified pCANTAB3*his*$_6$ vector (McCafferty et al. 1994) is prepared in parallel, using the Qiagen Maxiprep kit following the manufacturer's instructions. This vector is sequentially digested with *Apa*LI and *Not*I restriction enzymes, followed by concentration by phenol/chloroform extraction and ethanol precipitation. The concentration of the digested vector is calculated by measuring absorption at A_{260} nm using a spectrophotometer. Perform trial ligation reactions and subsequent electroporations as described for the V_H repertoire, to generate final Vλ and Vκ repertoires of 10^5 to 10^6.

The next stage is to introduce a $(Gly_4Ser)_3$ scFv-linker into the V_L repertoire, by amplifying the linker from an existing scFv, together with an irrelevant dummy V_H and introducing this fragment into the V_L repertoire upstream of the V_L gene segments as a *Hin*dIII-*Apa*LI digested fragment:

(1) Pick a single colony of an irrelevant clone that possesses the required scFv linker into a 50 μL PCR reaction using the primers pUC19rev and fdtetseq, following the reaction conditions described previously for the V gene repertoire amplifications.

(2) Assuming that agarose gel analysis shows a single product, isolate the PCR fragment using the Qiagen QIAquick PCR Purification Kit (Qiagen). Digest with *Apa*LI, followed by *Hin*dIII (New England Biolabs) as per the manufacturer's instructions. Heat inactivate the restriction enzyme by heating to 65°C for 20 minutes.

(3) Purify the DNA by running a 1 percent TAE low melting point agarose gel and using the GeneClean kit (Qbiogene). Calculate the concentration of purified DNA by measuring absorption at A_{260} nm using a spectrophotometer.

(4) Prepare a plasmid DNA midi-prep of the pCANTAB3*his*$_6$ vector containing the VL repertoire using a Qiagen Plasmid Midi Kit, following the manufacturer's instructions.

(5) Sequentially digest 10 μg of purified plasmid DNA with *Apa*LI and *Hind*III restriction enzymes as previously described. Gel-purify the large DNA fragment as previously described and then concentrate the cut vector by phenol/chloroform extraction followed by ethanol precipitation. Quantify the final concentration of DNA by measuring absorbance at A_{260} nm using a spectrophotometer.

(6) Perform sufficient reactions to ligate approximately 0.4 μg dummy V_H-linker DNA fragment into a 1 μg pool of Vκ and Vλ libraries (or into two separate Vκ and Vλ repertoires if preferred) using a DNA ligation kit (GE Healthcare).

(7) Electroporate into *E. coli* TG1 cells and plate out as described previously. This should generate a V_L repertoire with an upstream scFv linker of between 10^6 and 10^7 recombinants.

7.3.5 Construction of the scFv Library

The V_H and linker-V_L DNA fragments are amplified separately from each of the cloned repertoires (V_H in pCANTAB6 and linker-V_L in pCANTAB3*his*$_6$), assembled on the J_H region and amplified by pull-through PCR (see Table 7.3 for pull-through PCR oligonucleotide primers). The resulting assembled scFv constructs (V_H-linker-V_L) are then sequentially digested with *Sfi*I and *Not*I and ligated into *Sfi*I/*Not*I digested pCANTAB6.

(1) Perform 50 μL PCR reactions using the cycling parameters described previously, amplifying the V_H repertoire with pUC19rev and J_HFor primers and the V_L repertoire with reverse J_H and fdtetseq primers. Purify the products from 1 percent TAE low melting point agarose gels and quantify the DNA concentrations.

(2) Mix equal amounts of the V_H and V_L PCR products (10 to 20 ng of each) and make the volume up to 100 μL with PCR-grade water, followed by recovery of the DNA by ethanol precipitation. Resuspend the DNA pellet in 25 μL of PCR-grade water.

(3) For the assembly reaction, add the following reagents to the pooled V_H and linker-V_L products, and perform 25 cycles of 94°C for 1 minute, followed by 65°C for 4 minutes: 3.0 μL 10X *Taq* buffer, 1.5 μL 5 mM dNTP stock, and 0.5 μL *Taq* polymerase (2.5 U).

(4) Once the assembly has been performed, prepare at least two 50 μL pull-through PCR reactions per VH Back*Sfi* primer, combined with either the Jκ1-5For*Not* primer mix or the Jλ1-5For*Not* primer mix. Use 5.0 μL of assembly DNA per reaction and amplify using the PCR cycling parameters described previously. The correct size of the assembled construct is approximately 700 bp.

(5) Pool and concentrate the PCR products by phenol/chloroform extraction, followed by ethanol precipitation. Sequentially digest with *Sfi*I and *Not*I restriction endonucleases as previously described.

(6) Gel-purify the digested scFv assembly construct and ligate with *Sfi*I/*Not*I digested pCANTAB6 after determining the optimum insert to vector ratio as described previously. Perform at least 100 electroporations, divided into batches, with each batch ultimately spread onto 243 × 243 mm 2TYAG agar plates. To calculate the total size of the library, aliquots from each batch should be plated out in serial dilutions and the titers calculated as described previously. The size of the final library will be in the region of 10^9. In order to generate larger libraries this entire procedure should be scaled up as required.

(7) Scrape the large plates, using 5 mL 2TY medium per plate, and pool the cells in 50 mL Falcon tubes. Add 0.5 volume of 50 percent (v/v) glycerol to each tube, and ensure homogeneous resuspension of the cells by mixing on a rotating wheel for 30 minutes. Determine the cell density by optical density measurement at 600 nm. Store the library in aliquots at −70°C.

7.3.6 Preparation of Library Phage

To generate the phage library for use in selections, the phagemid particles are rescued with the helper phage M13K07 (New England Biolabs). The resultant library phage is PEG precipitated and caesium banded, allowing storage at 4°C for long periods.

(1) Inoculate 500 mL of 2TYAG medium (100 μg/mL ampicillin, 2 percent w/v glucose) with 10^{10} cells from the library glycerol stock and incubate at 37°C with shaking at 250 rpm until the optical density at 600 nm reaches 0.5 to 1.0.
(2) Add M13K07 helper phage to a final concentration of 5×10^9 pfu/mL.
(3) Incubate the cells for 30 minutes at 37°C without shaking, then for 30 minutes with gentle shaking (200 rpm) to allow infection.
(4) Centrifuge cultures at 2200 *g* for 15 minutes to recover the infected cells and resuspend the pellet in the same volume of 2TY medium containing 100 μg/mL ampicillin and 50 μg/mL kanamycin. Incubate overnight at 30°C with shaking at 300 rpm.
(5) Pellet the cells by centrifugation at 7000 *g* for 15 minutes at 4°C and recover the supernatant containing the phage into prechilled 1 L bottles.
(6) Add 0.3 volume of PEG/NaCl (20 percent w/v polyethylene glycol 8000, 2.5 M NaCl) and mix by swirling. Allow the phage to precipitate for at least 1 hour on ice.
(7) Concentrate the phage pellet by twice centrifuging at 7000 *g* for 15 minutes in the same bottle at 4°C. Remove as much of the supernatant as possible and resuspend the pellet in 8 mL TE buffer.
(8) Recentrifuge the phage in smaller tubes at 12,000 *g* for 10 minutes and recover the supernatant, which will now contain the phage. Ensure that any bacterial pellet that appears is left undisturbed.
(9) Add 3.6 g of caesium chloride to the phage suspension and raise the total volume to 9 mL with TE buffer. Using an ultracentrifuge, spin the samples at 110,000 *g* at 23°C, for at least 24 hours.
(10) After ultracentrifugation the phage should be visible as a tight band, which can be recovered by puncturing the tube with a 19-guage needle plus syringe and careful extraction.
(11) Dialyze the phage against two changes of 1 L TE buffer at 4°C for 24 hours. The purified phage stocks can be stored in aliquots at 4°C for long periods.
(12) Calculate the library size by infecting TG1 cells with serial dilutions of the phage stocks, plating on to 2TYAG agar, incubation overnight at 30°C, and enumeration of the numbers of ampicillin-resistant colonies that appear.

7.4 LIBRARY QUALITY CONTROL

In order to assess the quality and hence likely performance of a given antibody library, three major criteria are considered, namely size, diversity, and functionality.

7.4.1 Assessing Library Size

The sizes of antibody fragment libraries that are generally quoted refer to the total number of transformants produced, following cloning of the V genes into the vector of interest and will typically range from 10^7 to 10^{10}. By comparison, a theoretical antibody repertoire of 10^7 different molecules has been estimated to recognize 99 percent of epitopes, with an average affinity in the micromolar range (Perelson and Oster 1979). The correlation of increased antibody repertoire sizes with the average affinity of isolated members has been demonstrated practically by a number of groups (Vaughan et al. 1996; de Haard et al. 1999; Söderlind et al. 2000). The repertoire sizes of phage/phagemid display libraries have to date reached a ceiling of $\sim 10^{11}$ transformants, given the practical limitations of bacterial transformation.

7.4.2 Assessing Diversity in Human Antibody Libraries

Sequence analysis of individual antibodies from the library can be carried out in high throughput (potentially thousands of clones), both pre- and post-selection on antigen, to provide diversity data for the V gene starting repertoire and/or any bias in usage as a consequence of antigen recognition. Bioinformatic techniques can be employed to analyze diversity on this scale, allowing comparison between different libraries or selections, as well as with the *in vivo* humoral response. For a large library, at least 100 individual transformants should be sequenced and should all be unique from each other by at least one amino acid.

In our laboratory, a sequence analysis package called Blaze2 has been developed for antibody sequence analysis. This software first imports data directly from the automated capillary sequencers and compares it to an internal database of known antibody sequences. The Kabat numbering scheme (Kabat et al. 1991) is then applied to the sequence, allowing automatic positioning of the CDR and framework regions. The annotated sequences are then stored in an Oracle database and BLAST (Altschul et al. 1990) is used to assign the closest germline match from the internal database.

Sequence data for approximately 500 random scFv antibody clones taken directly from a large library in our laboratory were used to give an indicator of diversity. Virtually the entire repertoire of functional V_H germline gene segments is captured in this sample, with 44/49 of the total V_H gene segments observed (Fig. 7.2). V_L usage in the sample was also extensive, with a total of 22/30 $V\lambda$ and 16/35 $V\kappa$ germline segments observed (Figs. 7.3 and 7.4). This extensive sampling of the possible V gene repertoire by such libraries is advantageous over libraries designed around single synthetic frameworks, as the range of different antibodies will more closely match that of the natural immune response and represents a more diverse coverage of possible epitopes. For example, we have previously reported the isolation of over 1000 different antibodies to a single protein antigen (Edwards et al. 2003) from naïve human scFv phagemid libraries in our laboratory. This diverse panel was made up of 42/49 different V_H genes, coupled with light chains representing 19/30 and 13/35 of the possible $V\lambda$ and $V\kappa$ genes and represents both biologically functional as well as nonfunctional epitopes.

7.4.3 Assessing Functionality in Human Antibody Libraries

During library construction, it is possible to generate out-of-frame antibody sequences or those containing stop codons due to naturally occurring errors in the PCR process. These are likely to be more readily expressed by the bacterial host, as they will present less of a burden to the protein expression machinery and so will confer a growth advantage within the population. In addition, some full-length antibody sequences may be preferentially expressed in *E. coli* relative to others, due to differential toxicity to the host cell. It is most important, therefore, to be able to determine and monitor the percentage of functional antibodies within an antibody library, in order to maximize the possible paratope coverage of the encoded repertoire. The functional expression of antibody fragments is typically assayed using high throughput protein analysis techniques, such as Western blotting using anti-tag antibodies to detect full-length antibody fragments (Løset et al. 2005).

A number of different strategies have been used to maintain a high level of functionality within antibody repertoires, including:

(1) *Maintaining tight control of antibody fragment expression in the host cell.* The phagemid or phage vector systems used in the development of large antibody library repertoires can be engineered to more tightly control expression of the antibody fragment, such as by reducing copy number or by more tightly controlling induction of gene expression by use of alternative promoters (Beekwilder et al. 1999).

(2) *Enrichment for library phage expressing full-length functional antibody fragments via protease treatment.* Helper phage constructs have been engineered that contain a protease (e.g., trypsin) cleavage site within the pIII phage protein that renders them uninfective, following

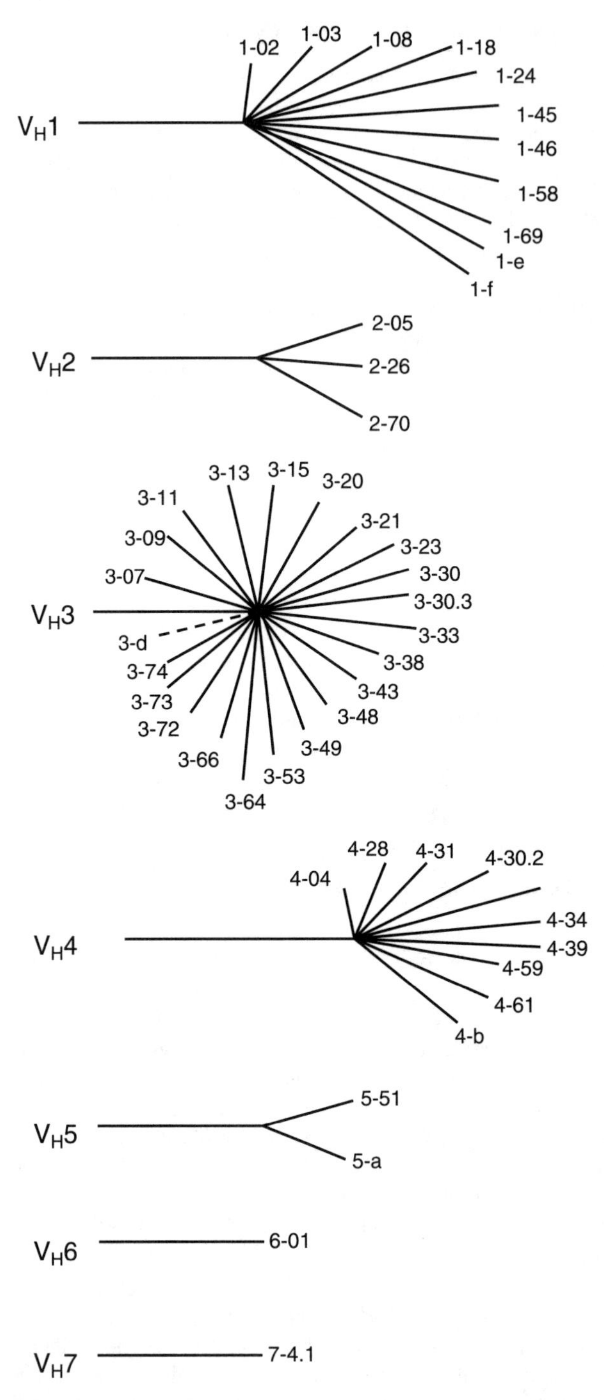

Figure 7.2 Human heavy chain V gene germline usage in an unselected naïve phagemid library: 500 randomly chosen clones were sequenced and the V genes aligned to a database of human germline V_H sequences. Solid lines represent those genes that were present at least once in the sample. Broken lines indicate those genes that were not present in the sample.

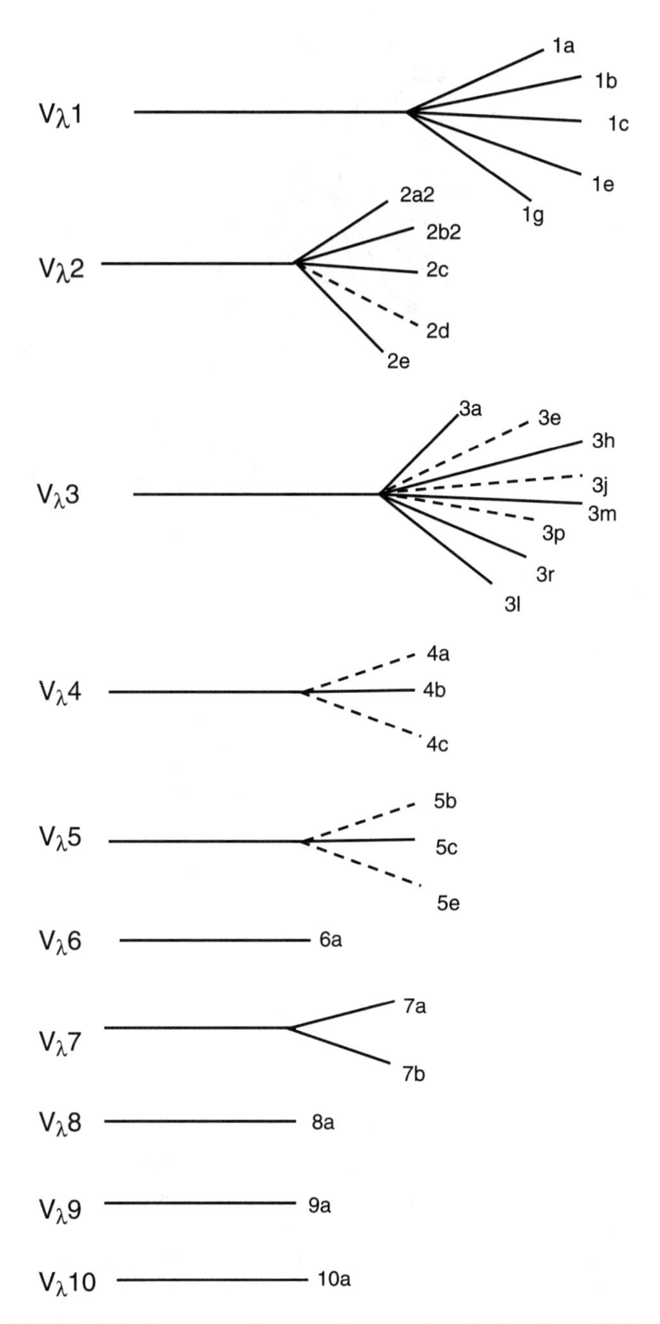

Figure 7.3 Lambda light chain V gene germline usage in an unselected naïve phagemid library: 500 randomly chosen clones were sequenced and the V genes aligned to a database of human germline V_{λ} sequences. Solid lines represent those genes that were present at least once in the sample. Broken lines indicate those genes that were not present in the sample.

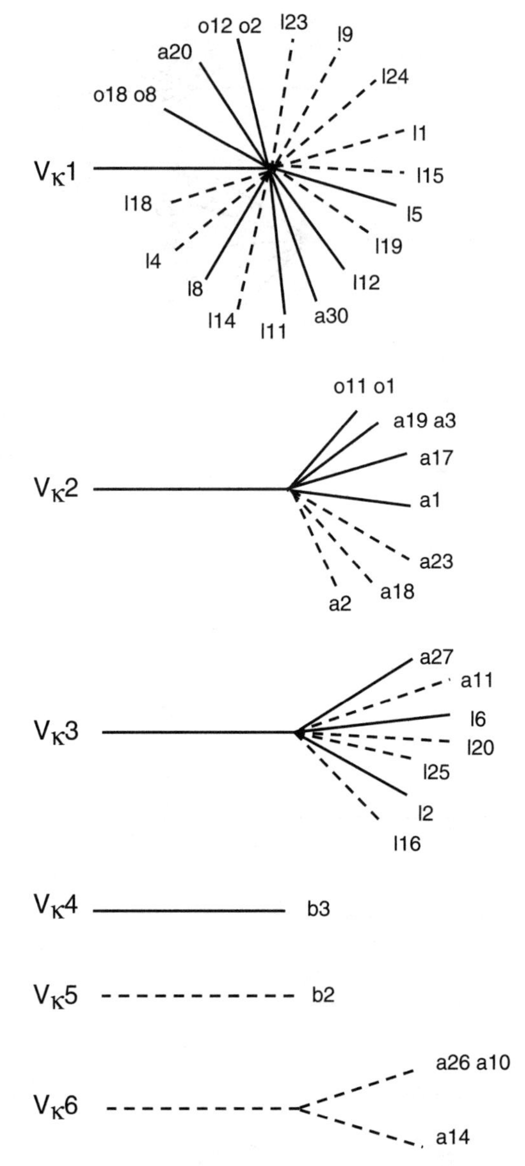

Figure 7.4 Kappa light chain V gene germline usage in an unselected naïve phagemid library: 500 randomly chosen clones were sequenced and the V genes aligned to a database of human germline V_κ sequences. Solid lines represent those genes that were present at least once in the sample. Broken lines indicate those genes that were not present in the sample.

protease treatment (Goletz et al. 2002). Use of this type of helper phage during rescue of antibody fragment phagemid libraries reduces the background of phages that carry no fragment and has been reported to increase the efficiency of library performance.

(3) *Purification of full-length antibody fragments via chromatography with the bacterial superantigens Protein A and L.* Protein A is a *Staphylococcus* B-cell superantigen that can bind with high avidity to both full-length immunoglobulin, but also V_H3 family heavy chains in a pH-dependent manner. Similarly, Protein L, derived from *Peptostreptococcus magnus*, is

capable of binding mammalian Vκ light chains. Restricting an antibody library to a heavy chain repertoire based on the V_H3 family and/or a light chain repertoire based on Vκ chains allows the implementation of a functional purification step to ensure correctly folded antibodies (Løset et al. 2005), but does so at the risk of limiting the diversity of the repertoire.

7.5 SEMI-SYNTHETIC HUMAN ANTIBODY REPERTOIRES

One of the earliest human antibody phage libraries constructed employed a strategy of cloning a synthetic repertoire of V_H genes and pairing them with a single V_L gene amplified from a human donor (Hoogenboom and Winter 1992; Nissim et al. 1994). Several semisynthetic libraries have subsequently been generated, including one whereby synthetic V_H genes were coupled with seven cloned light chains (de Kruif, Boel, and Logtenberg 1995) and one that involved the isolation of heavy and light chain CDRs from human lymphoid tissue, and subsequent transplantation into synthetically constructed V_H (3-23) and V_L (1-g) frameworks chosen for improved bacterial expression and folding (Söderlind et al. 2000). A continuation of this theme employed a single synthetic V_H framework (3-23), containing targeted mutations to the antigen-binding CDR1 and 2 regions, being paired with a repertoire of V_H CDR3 sequences and entire V_L genes isolated from human donors (Hoet et al. 2005).

While we are not describing these repertoires in any detail here, it is important to point out their relative advantages and disadvantages compared to libraries derived from natural V gene sources. Several attractive features of these semisynthetic repertoires include:

(1) *Modular Design.* Incorporation of specific unique restriction enzyme recognition sites into the chosen framework sequence facilitates further cloning/engineering of antibody fragments through CDR loop or framework shuffling.

(2) *Improved Bacterial Expression.* The bacterial expression and folding properties of the different V_H and V_L families have been studied as both individual domains and as paired Fab fragments (Ewert et al. 2003). Designing a library around a framework that is known to express and fold efficiently in *E. coli* may improve both selection efficiency and the protein yield of selected individual clones.

(3) *Facilitated Purification.* Building a semisynthetic antibody repertoire around particular frameworks, such as V_H3 and κ chains, allows a more generic and reproducible purification via the B-cell superantigens protein A or protein L, respectively.

However, as mentioned previously, there are also some potential disadvantages and these include:

(1) *Decreased Diversity and Paratope Coverage.* The major potential disadvantage of semisynthetic repertoires is a loss of diversity and therefore paratope coverage, compared to the natural immune response. By restricting a repertoire to a particular set of V genes, such as those giving optimal expression and folding in bacteria, antibody paratopes formed by other V genes will be missed, leading to smaller panels of antibodies to a given target antigen, with reduced epitope coverage and potentially lower affinity.

(2) *Immunogenicity.* The causal factors for immunogenicity for an antibody are multifactorial and include non-sequence-dependent factors such as patient genetics, product formulation, and site of inoculation. Care should be taken, however, when constructing synthetic CDR sequences, to ensure that potential T-cell epitopes are not introduced into the final antibody sequence.

(3) *Prokaryotic Codon Usage.* Semisynthetic libraries are often constructed using codons optimized for bacterial expression to increase the functional expression of the antibody fragments. By contrast, repertoires derived from human lymphoid tissue will contain a natural mammalian codon bias. This can result in antibody fragments with reduced expression in bacteria, making

downstream screening of the purified fragments more difficult. However, when generating an antibody for therapeutic use, mammalian cell lines such as NS0 or Chinese hamster ovary cells (CHO) are used for expression. An antibody fragment with a prokaryotic codon bias to its sequence may be difficult to generate as a full-length immunoglobulin for functional testing and will require codon optimization prior to any manufacturing cell line being developed.

7.6 HUMAN ANTIBODY REPERTOIRES IN OTHER DISPLAY FORMATS

Over the last 10 years a number of alternative protein and peptide display formats have been developed, such as ribosome display of scFv fragments (Hanes et al. 1998) and bacterial and yeast surface display of scFvs and Fab fragments (Chen et al. 2001; Feldhaus et al. 2003). Libraries of antibody fragments and even whole immunoglobulin molecules (Mazor et al. 2007) have been successfully displayed and enriched for a given antigen, and highly improved affinity variants of clones isolated. For example, high affinity antibodies to human insulin have been isolated from a phage displayed scFv population, following conversion and selection in ribosome display format (Groves et al. 2006). These newer display technologies have so far tended to use either antibody repertoires derived from immunized animals or subjects, or have been demonstrated using mutagenized variants of a common precursor clone. To date there is no report of a working naïve human antibody repertoire in any of these alternate formats, indicating that there remain significant technical challenges to overcome regarding their generation, utility, and/or performance. By contrast, the well-validated phage/phagemid antibody libraries are being used by researchers and biopharmaceutical companies around the world.

7.7 SUMMARY

The development of naïve repertoires of human antibodies allows the rapid isolation of antibodies to all conceivable antigens, including toxic moieties and those that are highly conserved across mammalian species, which have traditionally been difficult to isolate via conventional immunization. Moreover, such libraries typically yield large (hundreds to thousands) panels of different, specific antibodies covering a wide range of epitopes. These diverse repertoires, when combined with automated antibody selection, screening, and analysis have led to the effective industrialization of the humoral antibody response. Thus, more and more frequently, designer mAbs with exquisite specificity, functionality, and picomolar affinities are entering the clinic.

REFERENCES

Altschul, S.F., W. Gish, W. Miller, E.W. Myers, and D.J. Lipman. 1990. Basic local alignment search tool. *J. Mol. Biol.* 215:403–410.

Baum, P.D., and J.M. McCune. 2006. Direct measurement of T-cell receptor repertoire diversity with AmpliCot. *Nature Methods* 3:895–901.

Beekwilder, J., J. Rakonjac, M. Jongsma, and D. Bosch. 1999. A phagemid vector using the *E. coli* phage shock promoter facilitates phage display of toxic proteins. *Gene* 228:23–31.

Chen, G., A. Hayhurst, J.G. Thomas, B.R. Harvey, B.L. Iverson, and G. Georgiou. 2001. Isolation of high-affinity ligand-binding proteins by periplasmic expression with cytometric screening (PECS). *Nature Biotechnol.* 19:537–542.

Clackson, T., H.R. Hoogenboom, A.D. Griffiths, and G. Winter. 1991. Making antibody fragments using phage display libraries. *Nature* 352:624–628.

de Haard, H.J., N. van Neer, A. Reurs, S.E. Hufton, R.C. Roovers, P. Henderikx, A.P. de Bruine, J.W. Arends, and H.R. Hoogenboom. 1999. A large non-immunized human Fab fragment phage library that permits rapid isolation and kinetic analysis of high affinity antibodies. *J. Biol. Chem.* 274:18218–18230.

de Kruif, J., E. Boel, and T. Logtenberg. 1995. Selection and application of human single chain Fv antibody fragments from a semi-synthetic phage antibody display library with designed CDR3 regions. *J. Mol. Biol.* 248:97–105.

Edwards, B.M., S.C. Barash, S.H. Main, G.H. Choi, R. Minter, S. Ullrich, E. Williams, L. Du Fou, J. Wilton, V.R. Albert, S.M. Ruben, and T.J. Vaughan. 2003. The remarkable flexibility of the human antibody repertoire: Isolation of over one thousand different antibodies to a single protein, BLys. *J. Mol. Biol.* 334:103–118.

Ewert, S., T. Huber, A. Honegger, and A. Pluckthun. 2003. Biophysical properties of human antibody variable domains. *J. Mol. Biol.* 325:531–553.

Feldhaus, M.J., R.W. Siegel, L.K. Opresko, J.R. Coleman, J.M. Feldhaus, Y.A. Yeung, J.R. Cochran, P. Heinzelman, D. Colby, J. Swers, C. Graff, H.S. Wiley, and K.D. Wittrup. 2003. Flow-cytometric isolation of human antibodies from a nonimmune *Saccharomyces cerevisiae* surface display library. *Nature Biotechnol.* 21:163–170.

Goletz, S., P.A. Christensen, P. Kristensen, D. Blohm, I. Tomlinson, G. Winter, and U. Karsten. 2002. Selection of large diversities of antiidiotypic antibody fragments by phage display. *J. Mol. Biol.* 315:1087–1097.

Griffiths, A.D., M. Malmqvist, J.D. Marks, J.M. Bye, M.J. Embleton, J. McCafferty, M. Baier, K.P. Holliger, B.D. Gorick, N.C. Hughes-Jones, H.R. Hoogenboom, and G. Winter. 1993. Human anti-self antibodies with high specificity from phage display libraries. *EMBO J.* 12:725–734.

Griffiths, A.D., S.C. Williams, O. Hartley, I.M. Tomlinson, P. Waterhouse, W.L. Crosby, R.E. Kontermann, P.T. Jones, N.M. Low, T.J. Allison, T.D. Prospero, H.R. Hoogenboom, A. Nissim, J.P.L. Cox, J.L. Harrison, M. Zaccolo, E. Gherardi, and G. Winter. 1994. Isolation of high affinity human antibodies directly from large synthetic repertoires. *EMBO J.* 13:3245–3260.

Groves, M., S. Lane, J. Douthwaite, D. Lowne, D.G. Rees, B. Edwards, and R.H. Jackson. 2006. Affinity maturation of phage display antibody populations using ribosome display. *J. Immunol. Meth.* 313:129–139.

Hanes, J., L. Jermutus, S. Weber-Bornhauser, H.R. Bosshard, and A. Pluckthun. 1998. Ribosome display efficiently selects and evolves high-affinity antibodies in vitro from immune libraries. *Proc. Natl. Acad. Sci. USA* 95:14130–14135.

Hoet, R.M., E.H. Cohen, R.B. Kent, K. Rookey, S. Schoonbroodt, S. Hogan, L. Rem, N. Frans, M. Daukandt, H. Pieters, R. van Hegelsom, N.C. Neer, H.G. Nastri, I.J. Rondon, J.A. Leeds, S.E. Hufton, L. Huang, I. Kashin, M. Devlin, G. Kuang, M. Steukers, M. Viswanathan, A.E. Nixon, D.J. Sexton, H.R. Hoogenboom, and R.C. Ladner. 2005. Generation of high-affinity human antibodies by combining donor-derived and synthetic complementarity-determining-region diversity. *Nature Biotechnol.* 23:344–348.

Hoogenboom, H.R., and G. Winter. 1992. By-passing immunization: Human antibodies from synthetic repertoires of germline VH gene segments rearranged in vitro. *J. Mol. Biol.* 227:381–388.

Huse, W.D., L. Sastry, S.A. Iverson, A.S. Kang, M. Alting-Mees, D.R. Burton, S.J. Benkovic, and R.A. Lerner. 1989. Generation of a large combinatorial library of the immunoglobulin repertoire in phage lambda. *Science* 246:1275–1281.

Kabat, E.A., T.T. Wu, H.M. Perry, K.S. Gottesman, and C. Foeller. 1991. *Sequences of proteins of immunological interest*, 5th edition. Washington, D.C.: Public Health Service, National Institutes of Health.

Løset, G.A., I. Løbersli, A. Kavlie, J.E. Stacy, T. Borgen, L. Kausmally, E. Hvattum, B. Simonsen, M. Hovda, and O. Brekke. 2005. Construction, evaluation and refinement of a large human antibody phage library based on the IgD and IgM variable gene repertoire. *J. Immunol. Meth.* 299:47–62.

Marks, J.D., H.R. Hoogenboom, T.P. Bonnert, J. McCafferty, A.D. Griffiths, and G. Winter. 1991. By-passing immunization: Human antibodies from V-gene libraries displayed on phage. *J. Mol. Biol.* 222:581–597.

Marks, J.D., M. Tristem, A. Karpas, and G. Winter. 1991. Oligonucleotide primers for polymerase chain reaction amplification of human immunoglobulin variable genes and design of family-specific oligonucleotide probes. *Eur. J. Immunol.* 21:985–991.

Mazor, Y., T. van Blarcom, R. Mabry, B.L. Iverson, and G. Georgiou. 2007. Isolation of engineered, full-length antibodies from libraries expressed in *Escherichia coli*. *Nature Biotechnol.* 25:563–565.

McCafferty, J., K.J. Fitzgerald, J. Earnshaw, D.J. Chiswell, J. Link, R. Smith, and J. Kenten. 1994. Selection and rapid purification of murine antibody fragments that bind a transition-state analog by phage display. *Appl. Biochem. Biotech.* 47:157–173.

McCafferty, J., A.D. Griffiths, G. Winter, and D.J. Chiswell. 1990. Phage antibodies: Filamentous phage displaying antibody variable domains. *Nature* 348:552–554.

Nissim, A., H.R. Hoogenboom, I.M. Tomlinson, G. Flynn, C. Midgley, D. Lane, and G. Winter. 1994. Antibody fragments from a "single pot" phage display library as immunochemical reagents. *EMBO J.* 13:692–698.

Orlandi, R., D.H. Gussow, P.T. Jones, and G. Winter. 1989. Cloning immunoglobulin variable domains for expression by the polymerase chain reaction. *Proc. Natl. Acad. Sci. USA* 86:3833–3837.

Perelson, A.S., and G.F. Oster. 1979. Theoretical studies of clonal selection: Minimal antibody repertoire size and reliability of self-non-self discrimination. *J. Theor. Biol.* 81:645–670.

Saiki, R.K., S. Scharf, F. Faloona, K.B. Mullis, G.T. Horn, H.A. Erlich, and N. Arnheim. 1985. Enzymatic amplification of beta-globin genomic sequences and restriction site analysis for diagnosis of sickle cell anemia. *Science* 230:1350–1354.

Sambrook, J., E.F. Fritsch, and T. Maniatis. 1989. *Molecular cloning: A laboratory handbook.* Cold Spring Harbor, NY: Cold Spring Harbor Laboratory.

Sastry, L., M. Alting-Mees, W.D. Huse, J.M. Short, J.A. Sorge, B.N. Hay, K.D. Janda, S.J. Benkovic, and R.A. Lerner. 1989. Cloning of the immunological repertoire in *Escherichia coli* for generation of monoclonal catalytic antibodies: Construction of a heavy chain variable region-specific cDNA library. *Proc. Natl. Acad. Sci. USA* 86:5728–5732.

Sblattero, D., and A. Bradbury. 2000. Exploiting recombination in single bacteria to make large phage antibody libraries. *Nature Biotechnol.* 18:75–80.

Söderlind, E., L. Strandberg, P. Jirholt, N. Kobayashi, V. Alexeiva, A.M. Aberg, A. Nilsson, B. Jansson, M. Ohlin, C. Wingren, L. Danielsson, R. Carlsson, and C.A. Borrebaeck. 2000. Recombining germline-derived CDR sequences for creating diverse single-framework antibody libraries. *Nature Biotechnol.* 18:852–856.

Vaughan, T.J., A.J. Williams, K. Pritchard, J.K. Osbourn, A.R. Pope, J.C. Earnshaw, J. McCafferty, R.A. Hodits, J. Wilton, and K.S. Johnson. 1996. Human antibodies with sub-nanomolar affinities isolated from a large non-immunized phage display library. *Nature Biotechnol.* 14:309–314.

Ward, E.S., D. Gussow, A.D. Griffiths, P.T. Jones, and G. Winter. 1989. Binding activities of a repertoire of single immunoglobulin variable domains secreted from *Escherichia coli. Nature* 341:544–546.

Waterhouse, P., A.D. Griffiths, K.S. Johnson, and G. Winter. 1993. Combinatorial infection and in vivo recombination: A strategy for making large phage antibody repertoires. *Nucleic Acids Res.* 21:2265–2266.

PART III

IN VITRO DISPLAY TECHNOLOGY

CHAPTER 8

Antibody Phage Display

MICHAEL HUST, HOLGER THIE, THOMAS SCHIRRMANN, and STEFAN DÜBEL

8.1 INTRODUCTION

The production of antisera, that is, polyclonal antibodies, by immunization of animals is a method established for more than a century. The first serum was directed against diphtheria and produced in horses (von Behring and Kitasato 1890), yielding the first passive vaccination method as well as the first Nobel prize in medicine. Next came hybridoma technology, allowing the production of monoclonal antibodies by fusion of an immortal myeloma cell with an antibody-producing spleen cell (Köhler and Milstein 1975). However, hybridomas have some limitations, such as the possible instability of the aneuploid cell lines, but most of all its inability to produce human antibodies and to provide antibodies against toxic or highly conserved antigens (Winter and Milstein 1991). To overcome these limitations, antibodies or antibody fragments can be generated by recombinant gene technologies.

The most commonly used recombinant antibody fragments are the Fragment antigen binding (Fab) and the single chain Fragment variable (scFv). The Fab fragment consists of the fd fragment of the heavy chain and the light chain linked by a disulfide bond. In the scFv the variable region of the heavy chain (V_H) and the variable region of the light chain (V_L) are connected by a short peptide linker. A major breakthrough in the field of antibody engineering was the generation of antibody fragments in the periplasmatic space of *Escherichia coli* (Better et al. 1988; Huston et al. 1988; Skerra and

Therapeutic Monoclonal Antibodies: From Bench to Clinic. Edited by Zhiqiang An

TABLE 8.1 Comparison of Recombinant Antibody Selection Systems

Selection System			Advantages	Disadvantages
Transgenic mice			Somatic hypermutation	Immunization required, not freely available
Cellular display	Bacteria		N- and C-terminal and sandwich fusion	Not matured, requires individual sorting
	Yeast		Display of larger proteins, N- and C-terminal and sandwich fusion	Requires individual sorting
Intracellular display	Yeast two hybrid		Screening library versus library possible	Cytoplasm not optimal for antibody folding
Molecular display	Puromycin/ribosomal/CIS/covalent		Largest achievable library size *in vitro*	Tricky method, reducing milieu not optimal for antibodies
Phage display	Filamentous	Genomic	*In vitro*, robust, most widely used, multivalent display	Prone to mutation, only C-terminal fusion
		Phagemid	*In vitro*, robust, most widely used	Only C-terminal fusion
	T7		Well suited for peptide display	No display of antibody fragments
Arrays	Gridded clones		Robust, simple	Small library sizes

Source: Modified from Hust, Toleikis, and Dübel (2005).

Plückthun 1988). To circumvent the instability of hybridoma cell lines, the genes encoding V_H and V_L of a monoclonal antibody can be cloned into an *E. coli* expression vector in order to produce antibody fragments in the periplasmatic space of *E. coli* that preserve the binding specificity of the parental hybridoma antibody (Toleikis, Broders, and Dubel 2004).

The production of mouse-derived monoclonal antibody fragments in *E. coli* did not remove the major barrier for the broad application of antibodies in therapy as repeated administration of mouse-derived antibodies causes a human anti-mouse antibody (HAMA) response (Courtenay-Luck et al. 1986). This problem can be overcome by two approaches: humanization of mouse antibodies (Studnicka et al. 1994) or employing repertoires of human antibody genes. The second approach was achieved in two ways. First, human antibody gene repertoires were inserted into the genomes of IgG-knockout animals, allowing the generation of hybridoma cell lines that produce human immunoglobulins (Jakobovits 1995; Lonberg und Huszar 1995; Fishwild et al. 1996). However, this method still requires immunization and has limitations in respect of toxic and conserved antigens.

These restrictions do not apply for the second approach: the complete *in vitro* generation of specific antibodies from human antibody gene repertoires. There, despite the constant suggestion of novel methods like bacterial surface display (Fuchs et al. 1991; for review see Jostock and Dübel 2005), ribosomal display (Hanes and Plückthun 1997; He and Taussig 1997), puromycin display (Roberts and Szostak 1997), covalent display (Reiersen et al. 2005), CIS-display (Odegrip et al. 2004) or yeast surface display (Boder and Wittrup et al. 1997) (for comparision see Table 8.1), phage display has become the most widely used and most robust selection method. It is based on the groundbreaking work of Smith (1985).

8.2 HOW PHAGE DISPLAY WORKS

The currently dominating antibody phage display systems are based on the groundbreaking work of Smith (1985) on peptide display on the surface of filamentous phage. Historically, the first antibody gene repertoires in phage were generated and screened by using the lytic phage Lambda (Huse et al. 1989; Persson, Caothien, and Burton 1991), but with quite limited success. This changed when antibody fragments were presented on the surface of filametous phage of the M13/fd family, fused to pIII (McCafferty et al. 1990; Barbas et al. 1991; Breitling et al. 1991; Clackson et al. 1991; Hogenboom et al. 1991; Marks et al. 1991). Here, genotype and phenotype of an antibody fragment were linked by fusing its encoding gene fragment to the minor coat protein III gene of the filamentous bacteriophage M13. This resulted in the expression of a fusion protein on the surface of phage, allowing affinity purification of the antibody gene by antibody fragment binding to its antigen (Fig. 8.1). By uncoupling antibody gene replication and expression from the phage life cycle by locating them on a separate plasmid (phagemid)—containing a phage packaging sequence—genetic stability, propagation, and screening of antibody libraries was greatly facilitated (Barbas et al. 1991; Breitling et al. 1991; Hoogenboom et al. 1991; Marks et al. 1991). To date, single-pot (see below) antibody libraries with a theoretical diversity of up to 10^{11} independent clones have been assembled (Sblattero and Bradbury 2000) to serve as a molecular repertoire for phage display selections.

Due to its robustness and straightforwardness, phage display has been the selection method most widely used in the past decade. Display systems employing insertion of antibody genes into the phage genome have been developed for phage T7 (Danner and Balesco 2001), phage Lambda (Huse et al. 1989; Mullinax et al. 1990; Kang, Jones, and Burton 1991) and the Ff class (genus *Inovirus*) of the filamentous phage f1, fd, and M13 (McCafferty et al. 1990). Being well established for peptide display, the phage T7 is not suited for antibody phage display because it is assembled in the reducing enviroment of the cytoplasm, thus leaving most antibodies unfolded (Danner and Balesco 2001). In contrast, the oxidizing milieu of the bacterial periplasm allows antibody fragments to be folded and assembled properly (Skerra and Plückthun 1988). The Ff class of nonlytic bacteriophage is assembled in this cell compartment and allows the production of phage without killing the host cell. This is a major advantage compared to the lytic phage Lambda. In addition, filamentous phage allow the production of soluble proteins by introducing an amber stop codon between the antibody gene and gene III when using phagemid vectors. In an *E. coli supE* suppressor strain, the fusion proteins will be produced, whereas soluble antibodies are made in a nonsuppressor strain (Marks et al. 1992a; Griffiths et al. 1994), but expression in suppressor strains is also possible (Kirsch et al. 2005). Therefore, the members of the Ff class were found to be the phage of choice for antibody phage display.

To achieve surface display, five of the M13 coat proteins have been used in fusion to foreign proteins, protein fragments, or peptides. First, the antibody fragment was coupled to the amino terminus or second domain of the minor coat protein pIII (Barbas et al. 1991; Breitling et al. 1991; Hoogenboom et al. 1991). The function of the 3–5 copies of pIII, in particular their N-terminal domain, is to provide interaction of the phage with *E. coli* F pili (Crissman and Smith 1984). The major coat protein pVIII has been considered as an alternative fusion partner, with only very few success reports in the past decade (Kang et al. 1991b). pVIII fusions are obviously more useful for the display of short peptides (Cwirla et al. 1990; Felici et al. 1991). Fusions to pVI have also been tried, but not yet with antibody fragments (Jespers et al. 1995). pVII and pIX were used in combination, by fusing the V_L domain to pIX and the V_H domain to pVII, allowing the presentation of a Fv fragment on the phage surface. Thus, this format offers the potential for heterodimeric display (Gao et al. 1999). However, the fusion to pIII remains the only system of practical relevance.

Two different paradigms have been developed for the expression of the antibody::pIII fusion proteins. First, the fusion gene was be inserted directly into the phage genome substituting the wild-type (wt) pIII (McCafferty et al. 1990). Second, the fusion gene encoding the antibody fusion protein can be provided on a separate plasmid with an autonomous replication signal, a promoter, a resistance marker, and a phage morphogenetic signal, allowing this phagemid to be packaged into assembled phage particles. Here, a helper phage, usually M13K07, is necessary for the production of the antibody phage to complement the phage genes not encoded on the plasmid. Due to mutations, the M13K07 helper phage genome is not efficiently replicated and packaged during antibody phage assembly when compared to the phagemid (Vieira and Messing 1987), thereby limiting the unwanted packaging of helper plasmid into antibody phage particles to a few percent or less.

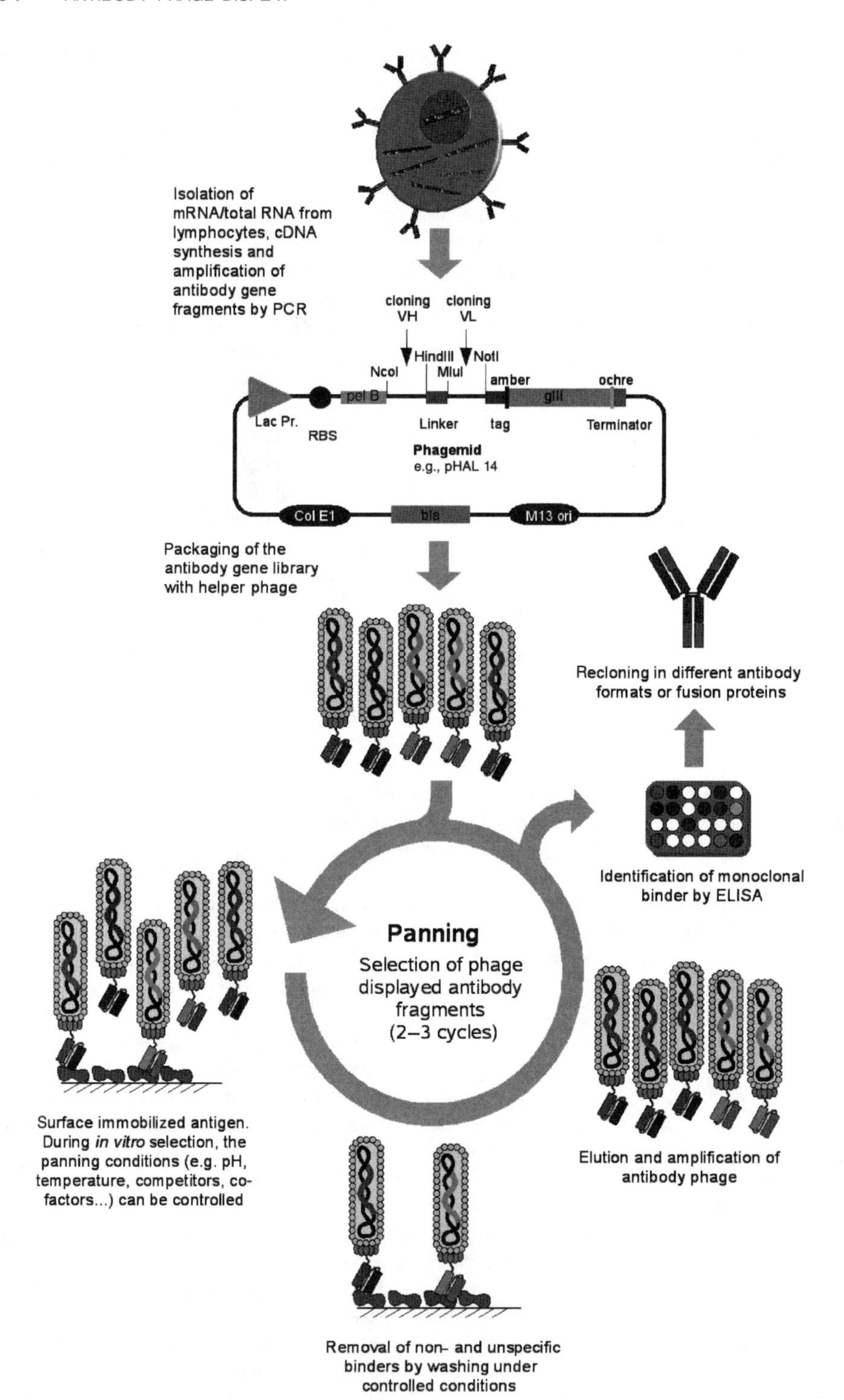

Figure 8.1 Construction of antibody gene libraries in the scFv format and selection of antibodies from antibody libraries ("panning") by phage display. (See color insert.)

In the system using direct insertion into the phage genome, every pIII protein on a phage is fused to an antibody fragment. The resulting oligovalency is of particular advantage in the first round of panning, where the desired binder is diluted in millions to billions of phage with unwanted specificity. The oligovalency of these phage improves the chances of a specific binder to be enriched due to the improvement of apparent affinity by the avidity effect. This advantage, however, has to be weighted to a number of disadvantages. The transformation efficiency of phagemids is two to three orders of magnitude better than the respective efficiency obtained with phage vectors, thus direct insertion into the phage genome results in limitations in respect of the ability to generate large libraries. Second, the additional protein domains fused to pIII may reduce the function of pIII during reinfection. In a phagemid system, the vast majority of the pIII assembled into phage are wt proteins, thus providing normal pilus interaction. Finally, the constant production of the pIII which destabilizes the *E. coli* membrane imposes a constant selection pressure against the expression of functional antibody fusion protein, which impacts the amplification and propagation of these libraries. This may explain why only two single-pot antibody libraries (Griffiths et al. 1994; O'Connel et al. 2002) have been made using a phage vector.

In contrast, in phagemid systems (e.g., Breitling et al. 1991) both replication and expression of foreign fusion proteins are regulated independently of the phage genome. Negative selection pressure can be avoided as propagation of the phagemid occurs in the absence of helper phage. The fusion proteins can be produced in adjustable quantities and the amber/suppressor system can be used for switching to soluble expression of antibody fragments without a pIII domain. Finally, despite usually not being derived from highest copy plasmids, the dsDNA of phagemids is easier to handle than phage DNA, facilitating cloning and analysis. Therefore, most currently succesful pIII display systems use the phagemid approach. There is, however, one disadvantage, which originates from the two independent sources for the pIII during phage packaging. During assembly, the wt pIII of the phage are inserted into the phage particles at a much higher rate than the antibody-pIII fusion protein. As a result, the vast majority (usually 95 percent or more) of the resulting phage particles carry no antibody fragments at all. The few antibody phage in these mixtures are mainly monovalent, with phage carrying two or more antibody fragments being extremely rare. This allows selection for antibodies with a high monovalent affinity, since avidity effects decreasing the dissociation rate from the panning antigen can be avoided. In the first panning round, however, when a few binders have to be fished out of a huge excess of unwanted antibody phage, the fact that only a few percent of the phage carry antibodies hampers the efficiency of the system. This problem can be overcome by the genetically engineered helper phage, Hyperphage. Hyperphage provides all functions of M13KO7, but does not carry a functional pIII gene. Therefore, the phagemid encoded pIII antibody fusion is the sole source of pIII in phage assembly. This enforces not only multivalent display for phagemid vectors, but improved antibody display per phage by two orders of magnitude and thus vastly improved panning efficiency (Rondot et al. 2001). Besides Hyperphage, two other helper phage offering multivalent display are avaible: Ex-phage (Baek et al. 2002) and Phaberge (Soltes et al. 2003). Due to its origin in a plasmid free *E. coli* line, Hyperhage seems to be the most efficient of those (Soltes et al. 2007).

8.3 SELECTION (PANNING) OF BINDERS

The novel procedure for isolating antibody fragments by their binding activity *in vitro* was called *panning*, referring to the gold washers' tool (Parmley and Smith 1988). The antigen is immobilized to a solid surface, such as nitrocellulose (Hawlisch et al. 2001), magnetic beads (Moghaddam 2003), a column matrix (Breitling et al. 1991) or, most widely used, plastic surfaces like polystyrol tubes (Hust et al. 2002) or 96-well microtiter plates (Barbas et al. 1991). The antibody phage are incubated with the surface-bound antigen, followed by thorough washing to remove the vast excess of nonspecific binding antibody phage. The bound antibody phage can subsequently be eluted and reamplified by infection of *E. coli*. This amplification allows detection of a single molecular interaction during panning, as a single antibody phage, by its resistance marker, can give rise to a bacterial colony after elution. This illustrates the tremendous sensitivity of the method, but also explains the need for thorough suppression of unspecific binding. The selection cycle can be repeated by infection of the phagemid bearing *E. coli* colonies from the former panning round with a helper phage to produce new antibody phage, which can be used for further rounds of panning until a significant enrichment is achieved. The number of antigen-specific antibody clones should increase with every panning round. Usually two to five panning rounds are necessary to select specifically binding antibody fragments (Fig. 8.1). After more than three to four rounds, however, increased numbers of clones with mutations are usually observed in the eluted phage. This limits the number of panning rounds. A specific enrichment of 100- to 10,000-fold per panning round can be achieved depending on careful adjustment of the binding conditions, so the lack of a specific enrichment after three to four rounds usually indicates that no specific binder was captured in the very first round. For an overview of available stratagies and protocols, refer to McCafferty, Hoogenboom, and Chiswell (1996) and Kontermann and Dübel (2001). High throughput methods using microtiter plates and robotics can facilitate and enhance the panning procedure (for review see Konthur, Hust, and Dübel 2005).

8.4 EVALUATION OF BINDERS

In most cases, the first step in the evaluation process of potential binders is an ELISA with polyclonal phage preparations from each panning round on coated target antigen and a control protein, for example, bovine serum albumin (BSA). The next step is the production of soluble monoclonal antibody fragments in microtiter plates from the panning rounds showing a significant enrichment of specific binders in polyclonal phage ELISA. After production, an ELISA on coated antigen and in parallel on control protein is performed. Soluble Fab fragments can be detected by their constant domains, whereas soluble scFvs can be detected by their engineered tags, for example, a penta-his- or c-myc-tag. The clones producing specific antibody fragments will be further analyzed by sequencing. Here, specific binders in duplicate can be rejected. A subcloning into *E. coli* expression vectors, such as pOPE101 (Schmiedl, Breitling, and Dübel 2000) offers better production rates for further analysis, for example, analysis by flow cytometry. Another important feature of an antibody is its affinity, which is analyzed by surface plasmon resonance (BIAcore) (Lauer et al. 2005). After analysis of specificity and affinity the selected antibody fragment can be subcloned into other formats like IgG or scFv::Fc-fusion in order to achieve avidity and immunological effector functions (Jostock et al. 2004).

Due to the fact that antibody phage display can yield antibodies to almost any target, it is strongly recommended to check the specificity of the selected antibody fragments as soon as possible with an assay not depending on the panning antigen. There are many cases where antibodies were selected to very minute impurities or denatured forms of the panning antigen. So whenever possible, after the initial clone picking ELISA, a binding test on naive tissues or cells or other materials not identical to the material used for phage enrichment is recommended.

8.5 PHAGE DISPLAY VECTORS

A large number of different filamentous phage display vectors have been constructed. Table 8.2 lists a selection, without pretending to be complete. Some of them have not been used for the construction of a library up to now, but have been included because they offer ideas and alternatives, for example, a system that allows control of the success of antibody gene cloning by green fluorescent protein (GFP) expression (Paschke et al. 2001). In the following, genetic elements critical to vector function are discussed in detail.

8.5.1 Promoters

A variety of different promoters have been employed for the expression of antibody fragments on the surface of phage. Widely used is the lacZ promoter (lacZ) derived from the lactose operon (Jacob

TABLE 8.2 Antibody Phage Display Vectors in Alphabetical Order

Phage Display Vector	Promoter	Secretion	Antibody Form at Used by Reference	C-Domains in Vector	Sites Heavy Chain	Sites Light Chain	Tags	gIII	Expression of Soluble Ab	References
pAALFab	1x lacZ, 2x RBS	2x pelB	Fab	no	*EcoR*I–*BstP*I	*Spe*I–*Xno*I	–	truncated	subcloning	Iba et al. (1997)
pAALFv	1x lacZ, 2x RBS	2x pelB	Fv	no	*EcoR*I–*BstP*I	*Spe*I–*Xno*I	–	truncated	subcloning	Iba et al. (1997)
pAALSC	1x lacZ, 1x RBS	1x pelB	scFv	no	*EcoR*I–*BstP*I	*Spe*I–*Xno*I	–	truncated	subcloning	Iba et al. (1997)
pAK100	1x lacZ, 1x RBS	1x pelB	scFv	no	*Sfi*I (tet resistance will be removed)		FLAG, myc	truncated	amber, supE strain	Krebber et al. (1997)
$pAPIII_6$ scFv	1x phoA, 1x RBS	1x OmpA	scFv	no	*Hind*III–*Sal*I		FLAG, His	truncated	*Sal*I–KI digest, deletion of gIII	Haidaris et al. (2001)
pCANTAB3his	1x lacZ, 1x RBS	1x g3p	scFv	no	*Nco*I/*Sfi*I–*Not*I		His, myc	full	amber, supE strain	McCafferty et al. (1994)
pCANTAB5his/ pCANTAB 6	1x lacZ, 1x RBS	1x cat	scFv	no	*Nco*I/*Sfi*I–*Not*I		His, myc	full	amber, supE strain	
pCANTAB 5 E	1x lacZ, 1x RBS	1x g3p	scFv	no	*Sfi*I–*Not*I		E tag	full	amber, supE strain	www.amershambiosciences.com
pCES1	1x lacZ, 2x RBS	1x gIII (L) 1x pelB (H)	Fab	yes	*Sfi*I–*Pst*I/*BstE*II (VH)	*ApaL*1–*Asc*I (L chain). *ApaL*1–*Xho*I (VL)	His, myc	full	amber, supE strain	Haardt et al. (1999)
pComb3	2x lacZ, 2x RBS	2x pelB	Fab	no	*Xho*I–*Spe*I	*Sac*I–*Xba*I	–	truncated	*Nhe*I–*Spe*I digest, deletion of gIII	Barbas et al. (1991)
pComb3H	1x lacZ, 2x RBS	ompA (LC) pelB (HC)	Fab, scFv	yes	*Xho*I–*Spe*I	*Sac*I–*Xba*I	–	truncated	*Nhe*I–*Spe*I digest, deletion of gIII	Barbas et al. (2001)
pComb3X	1x lacZ, 2x RBS	ompA (LC) pelB (HC)	Fab, scFv	yes	*Xho*I–*Spe*I	*Sac*I–*Xba*I	His, HA	truncated	amber, supE strain	Barbas et al. (2001)
PCW93/H, PCW99/L[1]	1x lacZ, 1x RBS	1x pelB	scFv	no	*Nco*I–*Nhe*I	*Sac*I–*Bgl*II	myc	truncated	amber, supE strain	Tsurushita et al. (1996)
pDAN5[1]	1x lacZ, 1x RBS	undiscribed leader	scFv	no	*Xho*I–*Nhe*I	*BssH*II–*Sal*I	SV5, his	full	amber, supE strain	Sblattero and Bradbury (2000)
pDH188	2x phoA, 2x RBS	2x stll	Fab	no	n.d.	n.d.	–	truncated	subcloning	Garrard et al. (1991)
PDN322	1x lacZ, 1x RBS	1x pelB	scFv	no	*Nco*I–*Not*I		FLAG, His	full	amber, supE strain	Pini et al. (1998)

(Continued)

TABLE 8.2 *Continued*

Phage Display Vector	Promoter	Secretion	Antibody Form at Used by Reference	C-Domains in Vector	Sites Heavy Chain	Sites Light Chain	Tags	gIII	Expression of Soluble Ab	References
pDNEK	1x lacZ, 1x RBS	1x pelB	scFv	no	*Nco*I–*Not*I		FLAG, His	full	amber, supE strain	Viti et al. (2000)
pEXmide3	1x lacZ, 2x RBS	2x pelB	Fab	yes	*Sfi*I/*Nco*I–*Kpn*I/*Apa*I	*Eag*I/*Not*I–*Nhe*I/*Spe*I	–	full	amber, supE strain	Söderlind et al. (1993)
pEXmide4	1x lacZ, 1x RBS	1x pelB	scFv	CH1	*Nco*I–*Sal*I		–	full	amber, supE strain	Kobayashi et al. (1997)
pEXmide5	1x lacZ, 1x RBS	1x pelB	scFv	?	*Nco*I–*Sal*I		–	full	amber, supE strain	Jirholt et al. (1998)
pFAB4	2x lacZ, 2x RBS	2x pelB	Fab	no	*Sfi*II–*Not*I		–	truncated	amber, supE strain	Ørum et al. (1993)
pFAB4H	1x lacZ, 2x RBS	2x pelB	Fab	CH1	*Sfi*II–*Not*I		–	truncated	subcloning	Dziegel et al. (1995)
pFAB5c	2x lacZ, 2x RBS	2x pelB	Fab	no	*Sfi*I–*Not*I		–	truncated	amber, supE strain	Ørum et al. (1993)
pFAB5c-His	1x lacZ, 2x RBS	2x pelB	scFv	no	*Sfi*I–*Not*I		his	truncated	amber, supE strain	Söderlind et al. (2000)
pFAB60	1x lacZ, 1x RBS	2x pelB	Fab	CH1	*Sfi*I–*Spe*I (VH) *Sfi*I–*Not*I (Fd)	*Nhe*I–*Asc*I (L chain)	his	truncated	*Eag*I digest, deletion of gIII	Johansen et al. (1995)
pFAB73H	1x lacZ, 1x RBS	2x pelB	Fab	CH1	*Nhe*I–*Apa*I (VH)	*Sfi*I–*Asc*I (L chain)	his	truncated	*Eag*I digest, deletion of gIII	Engberg et al. (1996)
pGP-F100	1x tet$^{o/p}$, 1x RBS	1x pelB	scFv	no	*Sfi*I (GFP uv will be removed)		myc	truncated	TEV protease site	Paschke et al. (2001)
pGZ1	1x tet$^{o/p}$, 1x RBS	1x pelB	scFv	no	*Sfi*I–*Not*I		myc	full	amber, supE strain	Zahn et al. (1999)
pHAL14	1x lacZ, 1x RBS	1x pelB	scFv	no	*Nco*I–*Hind*III	*Mlu*I–*Not*I	Yol1/34, his, myc	full	amber, supE strain	Pelat et al. (2007)
pHEN1	1x lacZ, 1x RBS	1x pelB	scFv, Fab, Fd, LC	no	*Sfi*I–*Not*I		myc	full	amber, supE strain	Hoogenboom et al. (1991)
pHEN1-Vλ3	1x lacZ, 1x RBS	1x pelB	scFv	no	*Nco*I–*Xho*I	Vλ 3anti-BSA Ab chain	myc	full	amber, supE strain	Hoogenboom and Winter (1992)

pHEN2	1x lacZ, 1x RBS	1x pelB	scFv	no	*Nco*I–*Xho*I	*ApaL*I–*Not*I	his, myc	full	amber, supE strain	http://www.mrccpe.cam.ac.uk
pHENIX	1x lacZ, 1x RBS	1x pelB	scFv	no	*Sfi*I/*Nco*I–*Sal*I/*Xho*I	*ApaL*I–*Not*I	myc	full	amber, supE strain	Finnern et al. (1997)
pHG-1 m/A27Jκ1	1x lacZ, 1x RBS	1x pelB	scFv	no	*ApaL*I–*Sfi*I	A27Jκ1 (VL)	his, myc	full	amber, supE strain	Rojas et al. (2002)
phh3mu-γ[a]	2x lacZ, 2x RBS	2x pelB	Fab bidirectional	yes	*Xho*I–*EcoR*I	*Sac*I–*Hind*III	–	truncated	subcloning	Den et al. (1999)
pIG10	1x lacZ, 1x RBS	1x OmpA	scFv	no	*EcoR*V–*EcoR*I		myc	full	amber, supE strain	Ge et al. (1995)
pIGT2 (vector)	1x lacZ, 1x RBS	1x g3p	scFv	no	*Sfi*I–*Not*I		myc	full	amber, supE strain	Back et al. (2002)
pIGT3 (vector)	1x lacZ, 1x RBS	1x g3p	scFv	no	*Sfi*I–*sfi*I		myc	full	subcloning	Back et al. (2002)
pIT2	1x lacZ, 1x RBS	1x pelB	scFv	no	SfiI/*Nco*I–*Xho*I	*Sal*LI–*Not*I	His, myc	full	amber, supE strain	Goletz et al. (2002)
pLG18	1x phoA, 2x RBS	2x stll	Fab	yes	*BssH*II–*Nco*I (CDR2-3)	*BstE*II–*Asp*718 (CDR1-3)	–	truncated	subcloning	Garrard and Henner (1993)
pM834, pM827[b]	2x lacZ, 2x RBS	2x pelB	Fab	no	*Xho*I–*Spe*I	*Sac*I–*Xba*I	–	full	amber, supE strain	Geoffroy et al. (1994)
pMorph series[c]	1x lacZ?	1x phoA	scFv	no	*Xba*I–*EcoR*I		FLAG?	?	subcloning	Knappik et al. (2000)
pScUAGΔcp3	1x lacZ, 1x RBS	1x pelB	scFv with Cκ	Cκ	*Xho*I–*Nhe*I	*Sst*I–*Bgl*II	–	truncated	amber, supE strain	Akamatsu et al. (1993)
pSEX	1x PA1/04/03, 1x RBS	1x pelB	scFv	no	–		Yol1/34	full	subcloning	Breitling et al. (1991)
pSEX20	1x PA1/04/03, 1x RBS	1x pelB	scFv	no	–		Yol1/34	full	subcloning	Dübel et al. (1993)
pSEX81	1x lacZ, 1x RBS	1x pelB	scFv	no	*Nco*I–*Hind*III	*Mlu*I–*Not*I	Yol1/34	full	subcloning	Welschof et al. (1997)

Source: Modified from Hust and Dübel (2005).
[a]Cre/lox recombination
[b]λ recombination
[c]Construction of the HuCAL library is described, but the pMorph vectorsystem is unpublished.

and Monod 1961). The gIII promoter (gIII) from the bacteriophage M13 (Smith 1985), the tetracycline promoter (1x $tet^{o/p}$) (Zahn, Skerra, and Höhne 1999) and the phoA promotor of the *E. coli* alkaline phosphatase (Garrard et al. 1991) were also successfully used. It seems that very strong promoters, for example, the synthetic promoter PA1/04/03 (Bujard et al. 1987), are rather a disadvantage (Dübel, personnal communication), since only a small amount of antibody pIII fusion proteins are necessary to provide presentation on the quite limited number of phage particles produced per cell. Overexpression—due to the toxicicity of both pIII and larger amounts of most antibody fragments to *E. coli*—strongly disturbs phage production. So, for every system, a delicate balance has to be kept to obtain material with optimal antibody presentation but still sufficient phage titers. To our knowledge, a systematic comparison of the different promotors has not been done.

8.5.2 Signal Peptides

The targeting of the antibody fragments to the periplasmatic space of *E. coli* requires the use of signal peptides. The pelB leader of the pectate lyase gene of *Erwinia caratovora* (Lei et al. 1987) is commonly used. The gIII leader (Smith 1985), the phoA leader of the *E. coli* alkaline phosphatase, and the ompA leader of *E. coli* outer membrane protein OmpA have also been used, being common to many protein expression vectors (Skerra, Pfitzinger, and Plückthun 1993; Skerra and Schmidt 1999). Further examples are the heat-stable enterotoxin II (stII) signal sequence (Garrard et al. 1991) and the bacterial chloramphenicol acetyltransferase (cat) leader (McCafferty et al. 1994). The use of various SRP leaders was also succesful (Thie et al. 2008b).

8.5.3 Antibody Format

Due to the very poor capability of *E. coli* to assemble complete IgG (only very few examples are described; Simmons et al. 2002), smaller antibody fragments are used for phage display. In particular, Fabs and scFvs have been shown to be the antibody formats of choice. As mentioned, in Fabs, the fd fragment and light chain are connected by a disulfide bond. In scFvs, the V_H and V_L are connected by a 15- to 25-amino-acid linker (Bird et al. 1988; Bird and Walker 1991; Huston et al. 1988). Some soluble scFvs tend to form dimers, in particular when the peptide linker is reduced to 3 to 12 amino acid residues. Diabodies or tetrabodies are produced if the linker between V_H and V_L is reduced to a few amino acids (Kortt et al. 1997; Arndt, Müller, and Plückthun 1998; Le Gall et al. 1999). The dimerization aggravates the determination of the affinity, due to the possible avidity effect of the antibody complex (Marks et al. 1992b). A few scFvs showed a reduced affinity up to one order of magnitude compared to the corresponding Fabs (Bird and Walker 1991). ScFvs with a higher affinity than the corresponding Fabs were rarely found (Iliades et al. 1998). On the other hand, small antibody fragments like Fv and scFv can more easily be produced in *E. coli*. The yield of functional Fvs expressed in *E. coli* is higher than the yield of the corresponding Fabs, due to a lower folding rate of the Fabs (Plückthun 1990, 1991). In one example, the stability in long term storage was much higher for Fabs than for scFvs. After six months the functionality of scFvs stored at 4°C was reduced by 50 percent; Fabs, however, showed no significant loss of functionality after one year (Kramer et al. 2002). The overall yield of antibody fragments expressed in *E. coli* varies from low μg/L culture to 1.2 g/L of scFvs, up to 1–2 g/L culture for Fabs and 2.5 g/L functional $F(ab')_2$ have been reported (Schirrmann et al. 2008). Recently, scFab fragments, consisting of the fd fragment and light chain joined by a polypeptide linker, were shown to combine benefits of Fab and scFv fragments (Jordan et al. 2007).

8.5.4 Vector Design

For the expression of Fabs in *E. coli* two polypeptide chains have to be assembled. In the monocistronic systems, for example, pComb3, the antibody genes are under control of two promoters and each has its own leader peptide (Barbas et al. 1991), whereas in plasmids like pCES1 with a bicistronic Fab operon, both chains are under control of a single promoter, leading to an mRNA with two ribosomal binding sites (De Haardt et al. 1999). The bicistronic system is more efficient for the expression of Fabs (Kirsch et al. 2005). An alternative to Fab fragments is the recently desribed scFabs (Hust et al. 2007; Jordan et al. 2007). This novel format, however, has to be further validated to determine its value for the construction of large libraries.

8.5.5 pIII Fusion Protein

Two variants of the antibody::pIII-fusion have been made. Either full size pIII or truncated versions of pIII were used. The truncated version was made by deleting the pIII N-terminal domain. This domain mediates the interaction with the F pili of *E. coli*. Infection is provided by wt pIII, as only a small percentage of phage in phagemid-based systems are carrying an antibody. These truncated vectors are therefore not compatible with the use of polyvalent display, for example, by using Hyperphage or Ex-phage, as the fullsize pIII is necessary for infection (Rondot et al. 2001; Baek et al. 2002).

In many panning protocols, elution of bound antibody phage from antigen is achieved by treatment with low or high pH. However, this method is not optimal as it may leave very strong binders still attached to the antigen and may damage phage integrity. Some phagemids, for example, pSEX81 (Welschof et al. 1997) or pHAL14 (Pelat et al. 2007) allow the elution of antibody phage during panning by protease digestion instead of pH shift. This is possible due to a protease cleavage site between pIII and the antibody fragment. Here, complete recovery of antigen bound antibody phage is possible, independent of type and affinity of antigen binding. Further advantages of protease elution in contrast to acid or alkaline elution are less chemical stress to the phage particles and a reconstitution of wild-type infectivity by the removal of the fusion portion from pIII.

Most of the phagemids described have an amber stop codon between the antibody gene and gIII. This allows the production of soluble antibody fragments after transformation of the phagemid to a nonsuppressor bacterial strain like HB2151 (Griffiths et al. 1994). Alternatively, recloning is necessary (Kirsch et al. 2005). In some phagemids, it is possible to delete the gIII by digestion and religation of the vector before tranformation into *E. coli* (Barbas et al. 1991). Other plasmids require subcloning of the antibody genes insert into a separate *E. coli* expression vector, for example, into the pOPE series (Breitling et al. 1991; Dübel et al. 1993; Schmiedl, Breitling, and Dübel 2000).

8.6 PHAGE DISPLAY LIBRARIES

Various types of phage display libraries have been constructed in respect of the source of the antibody genes. Immune libraries are generated by amplification of V genes isolated from IgG secreting plasma cells of immunized donors (Clackson et al. 1991). From immune libraries, antibody fragments with monovalent dissociation constants in the nanomolar range can be isolated. Immune libraries are typically created and used in medical research to select an antibody fragment against one particular antigen, typically an infectious pathogen, and therefore would not be the source of choice for the selection of a large number of different specificities. Naive, semisynthetic, and synthetic libraries have been subsumed as single-pot or universal libraries, as they are designed to isolate antibody fragments binding to every possible antigen. A clear correlation is seen between the size of the repertoire and the affinities of the isolated antibodies. Antibody fragments with a micromolar affinity have been isolated from a single-pot library consisting of approximately 10^7 clones, whereas antibody fragments with nanomolar affinities were obtained from a library consisting of 10^9 independent clones (Hoogenboom 1997). It is evident that the chance to isolate an antibody with a high affinity for a particular antigen increases almost linearly to the size of the library. According to the source of antibody genes, single-pot libraries (Table 8.3) can be naive libraries, semisynthetic libraries, or fully synthetic libraries. Naive libraries are constructed from rearranged V genes from B cells (IgM) of nonimmunized donors. An example of this library type is the naive human Fab library constructed by De Haardt et al. (1999), yielding antibodies with affinities up to 2.7×10^{-9} M. Semisynthetic libraries are derived from unrearranged V genes from pre-B-cells (germline cells) or from one antibody framework with genetically randomized

TABLE 8.3 Human Single-Pot Phage Display Libraries

Library Vector	Library Type	Antibody Type	Library Cloning Strategy	Library Size	References
DY3F63	Synthetic and naive repertoire	Fab	ONCL, 4 step cloning with integration of naive CDRH3 in synthetic HC	3.5×10^{10}	Hoet et al. (2005)
fdDOG-2lox, pUC19-2lox	Semisynthetic	Fab	PCR with random CDR3 primers, Cre-lox	6.5×10^{10}	Griffiths et al. (1994)
fdTet	Naïve	scFv	Recloning of a naive library[a]	5×10^{8}	O'Connel et al. (2002)
pAALFab	Semisynthetic (anti-hen egg white lysozyme Ab framework)	Fab	PCR with random CDR primers, assembly PCR	2×10^{8}	Iba, Ito, and Kurosawa (1997)
pAP-III6 scFv	Naïve	scFv	Assembly PCR	n.d.	Haidaris et al. (2001)
pCANTAB 6	Naïve	scFv	Assembly PCR	1.4×10^{10}	Vaughan et al. (1996)
pCES1	Naïve	Fab	3 step cloning (L chain, VH)	3.7×10^{10}	De Haardt et al. (1999)
pComb3	Semisynthetic (anti-tetanus Ab framework)[b]	Fab	PCR with random CDR H3 primers	5×10^{7}	Barbas et al. (1992)
pComb3	Semisynthetic (anti-tetanus Ab framework)[b]	Fab	PCR with random CDR H3 primers	$>10^{8}$	Barbas et al. (1993)
pDAN5	Naïve	scFv	Cre-lox	3×10^{11}	Sblattero and Bradbury (2000)
pDN322	Semisynthetic (VH DP47 and VL DPK22 V genes)	scFv	Random CDR3 primer, assembly PCR	3×10^{8}	Pini et al. (1998)
pDN322	Semisynthetic (anti-AMCV CP ab framework)	scFv	Random CDR3 primer, assembly PCR	3.75×10^{7}	Desiderio et al. (2001)
pDNEK (ETH2 library)	Semisynthetic (VH DP47, Vλ DPL16 and Vκ DPK 22 V genes)	scFv	Random CDR3 primer, assembly PCR	5×10^{8}	Viti et al. (2000)
pEXmide5	Semisynthetic (germline VH-DP47 and VL-DPL3 framework)	scFv	Assembly PCR, CDR shuffling	9×10^{6}	Söderlind et al. (1993)
pFAB5c-His (n-CoDeR library)	Semisynthetic (germline VH-DP47 and VL-DPL3 framework)	scFv	Assembly PCR, CDR shuffling	2×10^{9}	Söderlind et al. (2000)
pHAL14 (HAL4)	Naive, kappa	scFv	2 step cloning	2.2×10^{9}	Hust et al. unpublished
pHAL14 (HAL7)	Naive, lambda	scFv	2 step cloning	2.8×10^{9}	Hust et al. unpublished
pHEN1	Naïve	scFv	Assembly PCR	$10^{7}-10^{8}$	Marks et al. (1991)

TABLE 8.3 *Continued*

Library Vector	Library Type	Antibody Type	Library Cloning Strategy	Library Size	References
pHEN1	Naïve	scFv	Assembly PCR	2×10^5/ 2×10^6	Marks et al. (1992a)
pHEN1	Naïve	scFv	Assembly PCR	6.7×10^9	Sheets et al. (1998)
pHEN1-Vλ3	Semisynthetic (Vλ3 anti-BSA Ab light chain)	scFv	PCR with random CDR H3 primers	10^7	Hoogenboom and Winter (1992)
pHEN1-Vλ3	Semisynthetic (Vλ3 anti-BSA Ab light chain)	scFV	PCR with random CDR H3 primers	$>10^8$	Nissim et al. (1994)
pHEN1-Vκ3	Semisynthetic (VH)/naive (VL)	scFv	3 step cloning, PCR with random CDR H3 primers	3.6×10^8	de Kruif, Boel, and Logtenberg (1995)
pHEN2 (Griffin 1.library)	Semisynthetic	scFv	Recloning of the lox library in scFv format[c]	1.2×10^9	www.mrccpe.cam.ac.uk
pIT2 (Tom I/J library)	Semisynthetic (3x VH and 4x Vκ V genes)	scFv	PCR with random CDR2 and CDR3 primers	1.47×10^8/ 1.37×10^8	Goletz et al. (2002)
pLG18	Semisynthetic (anti-HER2 Ab framework)	Fab	PCR with random CDR primers, 2 step cloning	$2-3\times10^8$	Garrard and Henner (1993)
pMorph series (HuCAL library)	Synthetic	scFv	2 step cloning, CDR3 replacement	2×10^9	Knappik et al. (2000)
pMorph series (HuCAL GOLD library)	Synthetic	Fab	2 step cloning, all CDR replacement	1.6×10^{10}	www.morphosys.com
pMID21	Synthetic and naïve repertoire	Fab	ONCL, 4 step cloning with integration of naïve CDRH3 in synthetic HC	1×10^{10}	Hoet et al. (2005)
pScUAGDcp3	Semisynthetic	scFv connected to Ck	3 step cloning with random CDR3 primers	1.7×10^7	Akamatsu et al. (1993)
pSEX81	Naïve	scFv (with N terminus of CH1 and CL)	2 step cloning	4×10^7	–
pSEX81	Naïve	scFv (with N terminus of CH1 and CL)	2 step cloning	4×10^9	Little et al. (1999)
pSEX81	Naive	scFv (with N terminus of CH1 and CL)	4 step cloning	1.6×10^7/ 1.8×10^7/ 4×10^7	Schmiedl, Breitling, and Dübel (2000)
pSEX81	Naive	scFv (with N terminus of CH1 and CL)	2 step cloning	6.4×10^9	Løset et al. (2005)

Source: Modified from Hust and Dübel (2004).
[a]Sheets et al. (1998).
[b]Persson, Caothien, and Burton (1991).
[c]Griffiths et al. (1994).

complementary determining region (CDR) 3 regions, as described by Pini et al. (1998). Antibody fragments obtained from the latter library showed affinities between 10^{-8} and 10^{-9} M, with one scFv having a dissociation constant of 5×10^{-11} M. A combination of naive and synthetic repertoire was used by Hoet et al. (2005). They combined light chains from autoimmune patients with an fd fragment containing synthetic CDR1 and CDR2 in the human V_H3-23 framework and naive, origined from autoimmune patients, CDR3 regions. The fully synthetic libraries have a human framework with randomly integrated CDR cassettes (Hayashi et al. 1994). Antibody fragments selected from fully synthetic libraries exhibit affinities between 10^{-6} and 10^{-11} M (Knappik et al. 2000). All library types—immune, naive, synthetic, and their intermediates—have been used succesfully as a source for the selection of antibodies for diagnostic and therapeutic purposes.

8.7 GENERATION OF PHAGE DISPLAY LIBRARIES

Various methods have been employed to clone the genetic diversity of antibody repertoires. As single-pot or universal antibody libraries are required to be among the largest technically feasible human gene collections, every step of their construction has to be very carefully designed, controlled, and optimized to obtain as much complexity as possible. There is no assay to determine library complexity, and the transfection into *E. coli* is the experimental bottleneck limiting complexity. Therefore, it is commonly agreed that the number of independent clones after primary transfection into *E. coli* gives a good measure of the maximal number of different antibody genes. However, this may not reflect true complexities, as various factors, like different folding rates of different antibody fragments or the fraction of junk sequences in libraries with randomized parts can significantly decrease the real functional complexity.

After the isolation of mRNA from the desired cell type and the preparation of cDNA, the construction of immune libraries is usually done by a two step cloning or assembly polymerase chain reaction (PCR) (see below). Naive libraries are most commonly constructed by two or three cloning steps. In the two step cloning strategy, the amplified repertoire of light chain genes is cloned into the phage display vector first, as the heavy chain contributes more to diversity, due to its highly variable CDRH3. In the second step the heavy chain gene repertoire is cloned into the phagemids containing the light chain gene repertoire (Johansen et al. 1995; Welschof et al. 1997; Little et al. 1999). In the three step cloning strategy, separate heavy and light chain libraries are engineered. The V_H gene repertoire has then to be excised and cloned into the phage display vector containing the repertoire of V_L genes (De Haardt et al. 1999). Another common method used for the cloning of naive (McCafferty et al. 1994; Vaughan et al. 1996), immune (Clackson et al. 1991), or hybridoma (Krebber et al. 1997) scFv phage display libraries is the assembly PCR. The V_H and V_L genes are amplified separately and connected by a subsequent PCR, before the scFv encoding gene fragments are cloned into the vector. The assembly PCR is usually combined with a randomization of the CDR3 regions, leading to semisynthetic libraries. To achieve this, oligonucleotide primers encoding various CDR3 and J gene segments were used for the amplication of the V gene segments of human germlines (Akamatsu et al. 1993). The CDRH3 is considered as a major source of sequence variety (Shirai, Kidera, and Nakamura 1999), despite accumulating evidence that other CDRs can contribute as much as CDRH3 to structural complexity. Hoogenboom and Winter (1992) and Nissim et al. (1994) used degenerated CDRH3 oligonucleotide primers to produce a semisynthetic heavy chain repertoire derived from human V gene germline segments and combined this repertoire with an anti-BSA light chain. In some cases a framework of a well-known antibody was used as scaffold for the integration of randomly created CDRH3 and CDRL3 (Barbas et al. 1992; Desiderio et al. 2001). Jirholt et al. (1998) and Söderlind et al. (2000) amplified all CDR regions derived from B-cells before shuffling them into one antibody framework in an assembly PCR reaction. As an example for an entirely synthetic library, Knappik et al. (2000) utilized seven different V_H and V_L germline master frameworks combined with six synthetically created CDR cassettes. The construction of large naive and semisynthetic libraries (Hoet et al. 2005; Løset et al. 2005; Little et al. 1999; Sheets et al. 1998; Vaughan et al. 1996) requires significant effort to tunnel the genetic diversity through the bottleneck of *E. coli* transformation, for example, 600 transformations were necessary for the generation of a 3.5×10^{10} phage library (Hoet et al. 2005), 69 VH transformations for HAL4 kappa library containing 2.2×10^9 clones, or 86 VH transformation for HAL7 lambda library resulting in 2.8×10^9 independent clones (Hust et al. unpublished data).

To move the diversity potentiating step of random V_H/V_L combination behind the bottleneck of transformation, the Cre-lox or lamda phage recombination system has been employed (Waterhouse et al. 1993; Griffiths et al. 1994; Geoffroy, Sodoyer, and Aujame 1994). However, libraries with more than 10^{10} independent clones have now been accomplished by conventional transformation, rendering most of these complicated methods unnecessary in particular as they may result in decreased genetic stability. An interesting exception is the use of a genomically integrated CRE recombinase gene (Sblattero and Bradbury 2000) which is expected to solve the instability issue and allows the generation of libraries with complexities above the limit achievable by conventional cloning.

In summary, antibodies with nanomolar affinities can be selected from either type of library, naive or synthetic. If the assembly by cloning or PCR and preservation of molecular complexity is carefully controlled at every step of its construction, libraries of more than 10^{10} independent clones can be generated.

Antibody phage display has matured into a robust and reliable, albeit not yet kit level, technology to select antibodies to all kinds of targets. By combining the largest possible human gene libraries with a selection paradigm based on single molecule interactions, it has allowed antibodies to be generated that could not be obtained from animals. It is currently being adapted for the high throughput generation of binders to the human proteome (Taussig et al. 2007) and in particular is delivering and will continue to deliver high affinity human antibodies for medical use (Thie et al. 2008a).

ACKNOWLEDGMENTS

HT and TS were supported by the German Ministry of Reserach and Technology, through the NGFN2 grant "SMP Antibody Factory." This chapter is revised and updated from an article printed first in S. Dübel, *Handbook of therapeutic antibodies* (New York: Wiley-VCH, 2007). We are grateful to Ulrike Dübel for reading the manuscript.

REFERENCES

Akamatsu, Y., M.S. Cole, J.Y. Tso, and N. Tsurushita. 1993. Construction of a human Ig combinatorial library from genomic V segments and synthetic CDR3 fragments. *J. Immunol.* 151:4651–4659.

Arndt, K.M., K.M. Müller, and A. Plückthun. 1998. Factors influencing the dimer to monomer transition of an antibody single-chain Fv fragment. *Biochemistry* 37:12918–12926.

Baek, H., K.H. Suk, Y.H. Kim, and S. Cha, 2002. An improved helper phage system for efficient isolation of specific antibody molecules in phage display. *Nucleic Acids Res.* 30:e18.

Barbas, C.F. III, W. Amberg, A. Simoncsitis, T.M. Jones, and R.A. Lerner. 1993. Selection of human anti-hapten antibodies from semisynthetic libraries. *Gene* 137:57–62.

Barbas, C.F. III, J.D. Bain, M. Hoekstra, and R.A. Lerner. 1992. Semisynthetic combinatorial antibody libraries: A chemical solution to the diversity problems. *Proc. Natl. Acad. Sci. USA* 89:4457–4461.

Barbas, C.F. III, D.R. Burton, J.K. Scott, and G.J. Silverman. 2001. *Phage display: A laboratory manual.* Cold Spring Harbor, NY: Cold Spring Harbor Laboratory Press.

Barbas, C.F. III, A.S. Kang, R.A. Lerner, and S.J. Benkovic. 1991. Assembly of combinatorial antibody libraries on phages surfaces: Yhe gene III site. *Proc. Natl. Acad. Sci. USA* 88:7987–7982.

Better, M., C.P. Chang, R.R. Robinson, and A.H. Horwitz. 1988. *Escherichia coli* secretion of an active chimeric antibody fragment. *Science* 240:1041–1043.

Bird, R.E., K.D. Hardman, J.W. Jacobsen, S. Johnson, B.M. Kaufman, S.M. Lee, T. Lee, S.H. Pope, G.S. Riordan, and M. Whitlow. 1988. Single-chain antigen-binding proteins. *Science* 242:423–426.

Bird, R.E. and B.W. Walker. 1991. Single chain variable regions. *Trends Biotech.* 9:132–137.

Boder, E.T., and K.D. Wittrup. 1997. Yeast surface display for screening combinatorial polypeptide libraries. *Nature Biotechnol.* 15:553–558.

Breitling, F., S. Dübel, T. Seehaus, I. Kleewinghaus, and M. Little. 1991. A surface expression vector for antibody screening. *Gene* 104:1047–1153.

Bujard, H., R. Gentz, M. Lanzer, D. Stueber, M. Mueller, I. Ibrahimi, M.T. Haeuptle, and B. Dobberstein. 1987. A T5 promoter-based transcription-translation system for the analysis of proteins in vitro and in vivo. *Methods Enzymol.* 155:416–433.

Carter, P., R.F. Kelley, M.L. Rodrigues, B. Snedecor, M. Covarrubias, M.D. Velligan, W.L. Wong, A.M. Rowland, C.E. Kotts, M.E. Carver, et al. 1992. High level *Escherichia coli* expression and production of a bivalent humanized antibody fragment. *Biotechnology* 10:163–167.

Clackson, T., H.R. Hoogenboom, A.D. Griffiths, and G. Winter. 1991. Making antibody fragments using phage display libraries. *Nature* 352:624–628.

Courtenay-Luck, N.S., A.A. Epenetos, R. Moore, M. Larche, D. Pectasides, B. Dhokia, and M.A. Ritter. 1986. Development of primary and secondary immune responses to mouse monoclonal antibodies used in the diagnosis and therapy of malignant neoplasms. *Cancer Res.* 46:6489–6493.

Crissman, J.W., and G.P. Smith. 1984. Gene 3 protein of filamentous phages: Evidences for a carboxyl-terminal domain with a role in morphogenesis. *Virology* 132:445–455.

Cwirla, S.E., E.A. Peters, R.W. Barrett, and W.J. Dower. 1990. Peptides on phage: A vast library of peptides for identifying ligands. *Proc. Natl. Acad. Sci. USA* 87:6378–6382.

Danner, S., and J.G. Belasco. 2001. T7 phage display: A novel genetic selection system for cloning RNA-binding protein from cDNA libraries. *Proc. Natl. Acad. Sci. USA* 98:12954–12959.

De Haardt, H.J., N. van Neer, A. Reurst, S.E. Hufton, R.C. Roovers, P. Henderikx, A.P. de Bruine, J.-W. Arends, and H.R. Hoogenboom. 1999. A large non-immunized human Fab fragment phage library that permits rapid isolation and kinetic analysis of high affinity antibodies. *J. Biol. Chem.* 274:18218–18230.

De Kruif, J., E. Boel, and T. Logtenberg. 1995. Selection and application of human single chain Fv antibody fragments from a semi-synthetic phage antibody display library with designed CDR3 regions. *J. Mol. Biol.* 248:97–105.

Den, W., S.R. Sompuram, S. Sarantopoulos, and J. Sharon. 1999. A bidirectional phage display vector for the selection and mass transfer of polyclonal antibody libraries. *J. Immunol. Meth.* 222:45–57.

Desiderio, A., R. Franconi, M. Lopez, A.E. Villani, F. Viti, R. Chiaraluce, V. Consalvi, D. Neri, and E. Benvenuto. 2001. A semi-synthetic repertoire of intrinsically stable antibody fragments derived from a single-framework scaffold. *J. Mol. Biol.* 310:603–615.

Dübel, S., F. Breitling, P. Fuchs, M. Braunagel, I. Klewinghaus, and M. Little. 1993. A family of vectors for surface display and production of antibodies. *Gene* 128:97–101.

Dziegiel, M., L.K. Nielsen, P.S. Andersen, A. Blancher, E. Dickmeiss, and J. Engberg. 1995. Phage display used for gene cloning of human recombinant antibody against the erythrocyte surface antigen, rhesus D. *J. Immunol. Meth.* 182:7–19.

Engberg, J., P.S. Andersen, L.K. Nielsen, M. Dziegiel, L.K. Johansen, and B. Albrechtsen. 1996. Phage-display libraries of murine and human Fab fragments. *Mol. Biotechnol.* 6:287–310.

Felici, F., L. Castagnoli, A. Musacchio, R. Jappelli, and G. Cesareni. 1991. Selection of antibody ligands from a large library of oligopeptides expressed on a multivalent exposition vector. *J. Mol. Biol.* 222:301–310.

Finnern, R., E. Pedrollo, I. Fisch, J. Wieslander, J.D. Marks, C.M. Lockwood, and W.H. Ouwehand. 1997. Human autoimmune anti-proteinase 3 scFv from a phage display library. *Clin. Exp. Immunol.* 107:269–281.

Fishwild, D.M., S.L. O'Donnel, T. Bengoechea, D.V. Hudson, F. Harding, S.L. Bernhar, D. Jones, R.M. Kay, K.M. Higgins, S.R. Schramm, and N. Lonberg. 1996. High-avidity human IgG kappa monoclonal antibodies from a novel strain of minilocus transgenic mice. *Nature Biotechnol.* 14:845–851.

Fuchs, P., F. Breitling, S. Dübel, T. Seehaus, and M. Little. 1991. Targeting recombinant antibodies to the surface of *E. coli*: Fusion to a peptidoglycan associated lipoprotein. *Bio/Technology* 9:1369–1372.

Gao, C., S. Mao, C.H. Lo, P. Wirsching, R.A. Lerner, and K.D. Janda. 1999. Making artificial antibodies: A format for phage display of combinatorial heterodimeric arrays. *Proc. Natl. Acad. Sci. USA* 96:6025–6030.

Garrard, L.J., and D.J. Henner. 1993. Selection of an anti-IGF-1 Fab from a Fab phage library created by mutagenesis of multiple CDR loops. *Gene* 128:103–109.

Garrard, L.J., M. Yang, M.P. O'Connel, R. Kelley, and D.J. Henner. 1991. Fab assembly and enrichment in a monovalent phage display system. *Bio/Technology* 9:1373–1377.

Ge, L., A. Knappik, P. Pack, C. Freund, and A. Plückthun. 1995. Expressing antibodies in *Escherichia coli*. In *Antibody engineering*, ed. C.A.K. Borrebaeck. London: Oxford University Press.

Geoffroy, F., R. Sodoyer, and L. Aujame. 1994. A new phage display system to construct multicombinatorial libraries of very large antibody repertoires. *Gene* 151:109–113.

Goletz, A., P.A. Cristensen, P. Kristensen, D. Blohm, I. Tomlinson, G. Winter, and U. Karsten. 2002. Selection of large diversities of antiidiotypic antibody fragments by phage display. *J. Mol. Biol.* 315:1087–1097.

Griffiths, A.D., S.C. Williams, O. Hartley, I.M. Tomlinson, P. Waterhouse, W. Crosby, R.E. Kontermann, P.T. Jones, N.M. Low, T.J. Allison, T.D. Prospero, H.R. Hoogenboom, A. Nissim, J.P.L. Cox, J.L. Harrison, M. Zaccolo, E. Gherardi, and G. Winter. 1994. Isolation of high affinity human antibodies directly from large synthetic repertoires. *EMBO J.* 13:3245–3260.

Haidaris, C.G., J. Malone, L.A. Sherrill, J.M. Bliss, A.A. Gaspari, R.A. Insel, and M.A. Sullivan. 2001. Recombinant human antibody single chain variable fragments reactive with *Candida albicans* surface antigens. *J. Immunol. Meth.* 257:185–202.

Hanes, J., and A. Plückthun. 1997. In vitro selection and evolution of functional proteins by using ribosome display. *Proc. Natl. Acad. Sci. USA* 94:4937–4942.

Hawlisch, H., M. Müller, R. Frank, W. Bautsch, A. Klos, and J. Köhl. 2001. Site specific anti-C3a receptor single-chain antibodies selected by differential panning on cellulose sheets. *Anal. Biochem.* 293:142–145.

Hayashi, N., M. Welschoff, M. Zewe, M. Braunagel, S. Dübel, F. Breitling, and M. Little. 1994. Simultaneous mutagenesis of antibody CDR regions by overlap extension and PCR. *Biotechniques* 17:310–316.

He, M., and M.J. Taussig. 1997. Antibody-ribosome-mRNA (ARM) complexes as efficient selection particles for in vitro display and evolution of antibody combining sites. *Nucleic Acids Res.* 25:5132–5134.

Hoet, R.M., E.H. Cohen, R.B. Kent, K. Rookey, S. Schoonbroodt, S. Hogan, L. Rem, N. Frans, M. Daukandt, H. Pieters, R. van Hegelsom, N.C. Neer, H.G. Nastri, I.J. Rondon, J.A. Leeds, S.E. Hufton, L. Huang, I. Kashin, M. Devlin, G. Kuang, M. Steukers, M. Viswanathan, A.E. Nixon, D.J. Sexton, H.R. Hoogenboom, and R.C. Ladner. 2005. Generation of high-affinity human antibodies by combining donor-derived and synthetic complementarity-determining-region diversity. *Nature Biotechnol.* 23:344–348.

Hoogenboom, H.R. 1997. Designing and optimizing library selection strategies for generating high-affinity antibodies. *Trends Biotechnol.* 15:62–70.

Hoogenboom, H.R., A.D. Griffiths, K.S. Johnson, D.J. Chiswell, P. Hudson, and G. Winter. 1991. Multi-subunit proteins on the surface of filamentous phage: Methodologies for displaying antibody (Fab) heavy and light chains. *Nucleic Acids Res.* 19:4133–4137.

Hoogenboom, H.R., and G. Winter. 1992. By-passing immunisation: Human antibodies from synthetic repertoires of germline V_H gene segments rearranged *in vitro*. *J. Mol. Biol.* 227:381–388.

Huse, W.D., L. Sastry, S.A. Iverson, A.S. Kang, M. Alting-Mees, D.R. Burton, S.J. Benkovic, and R. Lerner. 1989. Generation of a large combinatorial library of the immunoglobulin repertoire in phage lambda. *Science* 246:1275–1281.

Hust, M., and S. Dübel. 2004. Mating antibody phage display with proteomics. *Trends Biotechnol.* 22:8–14.

Hust, M., and S. Dübel. 2005. Phage display vectors for the in vitro generation of human antibody fragments. In *Immunochemical protocols*, 3rd ed., ed. R. Burns. *Meth. Mol. Biol.* 295:71–95.

Hust, M., T. Jostock, C. Menzel, B. Voedisch, A. Mohr, M. Brenneis, M.I. Kirsch, D. Meier, and S. Dübel. 2007. Single chain Fab (scFab) fragment. *BMC Biotechnol.* 7:14.

Hust, M., E. Maiss, H.-J. Jacobsen, and T. Reinard. 2002. The production of a genus specific recombinant antibody (scFv) using a recombinant Potyvirus protease. *J. Virol. Meth.* 106:225–233.

Hust, M., L. Toleikis, and S. Dübel. 2005. Antibody phage display. *Mod. Asp. Immunobiol.* 15:47–49.

Huston, J.S., D. Levinson, H.M. Mudgett, M.S. Tai, J. Novotny, M.N. Margolies, R.J. Ridge, R.E. Bruccoloreri, E. Haber, R. Crea, and H. Oppermann. 1988. Protein engineering of antibody binding sites: Recovery of specific activity in an anti-digosin single-chain Fv analogue produced in *Escherichia coli*. *Proc. Natl. Acad. Sci. USA* 85:5879–5883.

Iba, Y., W. Ito, and Y. Kurosawa. 1997. Expression vectors for the introduction of higly diverged sequences into the six complementarity-determining regions of an antibody. *Gene* 194:35–46.

Iliades, P., D.A. Dougan, G.W. Oddie, D.W. Metzger, P.J. Hudson, and A.A. Kortt. 1998. Single-chain Fv of anti-idiotype 11-1G10 antibody interacts with antibody NC41 single-chain Fv with a higher affinity than the affinity for the interaction of the parent Fab fragments. *J. Protein Chem.* 17:245–254.

Jacob, F., and J. Monod. 1961. Genetic regulatory mechanism in the synthesis of proteins. *J. Mol. Biol.* 3:318–356.

Jakobovits, A. 1995. Production of fully human antibodies by transgenic mice. *Curr. Opin. Biotechnol.* 6:561–566.

Jespers, L.S., J.H. Messens, A. de Keyser, D. Eeckhout, I. van den Brande, Y.G. Gansemans, M.J. Lauwerey, G.P. Vlasuk, and P.E. Stanssens. 1995. Surface expression and ligand based selection of cDNAs fused to filamentous phage gene VI. *Bio/Technology* 13:378–381.

Jirholt, P., M. Ohlin, C.A.K. Borrebaeck, and E. Söderlind. 1998. Exploiting sequences space: Shuffling *in vivo* formed complementarity determining regions into a master framework. *Gene* 215:471–476.

Johansen, L.K., B. Albrechtsen, H.W. Andersen, and J. Engberg. 1995. pFab60: A new, efficient vector for expression of antibody Fab fragments displayed on phage. *Protein Eng.* 8:1063–1067.

Jordan, E., L. Al-Halabi, T. Schirrmann, M. Hust, and S. Dübel. 2007. Production of single chain Fab (scFab) fragments in *Bacillus megaterium. Microb. Cell Fact.* 6:38.

Jostock, T., and S. Dübel. 2005. Screening of molecular repertoires by microbial surface display. *Comb. Chem. High Throughput Screen* 8:127–133.

Jostock, T., M. Vanhove, E. Brepoels, R. Van Gool, M. Daukandt, A. Wehnert, R. Van Hegelsom, D. Dransfield, D. Sexton, M. Devlin, A. Ley, H. Hoogenboom, and J. Mullberg. 2004. Rapid generation of functional human IgG antibodies derived from Fab-on-phage display libraries. *J. Immunol. Methods* 289:65–80.

Kang, A.S., C.F. Barbas, K.D. Janda, S.J. Bencovic, and R.A. Lerner. 1991. Linkage of recognition and replication functions by assembling combinatorial antibody Fab libraries along phage surfaces. *Proc. Natl. Acad. Sci. USA* 88:4363–4366.

Kang, A.S., T.M. Jones, and D.R. Burton. 1991. Antibody redesign by chain shuffling from random combinatorial immunoglobulin libraries. *Proc. Natl. Acad. Sci. USA* 88:11120–11123.

Kirsch, M., M. Zaman, D. Meier, S. Dübel, and M. Hust, 2005. Parameters affecting the display of antibodies on phage. *J. Immunol. Meth.* 301:173–185.

Knappik, A., L. Ge, A. Honegger, P. Pack, M. Fischer, G. Wellnhofer, A. Hoess, J. Wölle, A. Plückthun, and B. Virnekäs. 2000. Fully synthetic human combinatorial antibody libraries (HuCAL) based on modular consensus framework and CDRs randomized with trinucleotides. *J. Mol. Biol.* 296:57–86.

Kobayashi, N., E. Söderlind, and C.A.K. Borrebaeck. 1997. Analysis of assembly of synthetic antibody fragments: Expression of functional scFv with predifined specificity. *BioTechniques* 23:500–503.

Köhler, G., and C. Milstein. 1975. Continuous cultures of fused cells secreting antibody of predefined specificity. *Nature* 256:495–497.

Kontermann, R., and S. Dübel, eds. 2001. *Antibody engineering*. New York: Springer-Verlag.

Konthur, Z., M. Hust, and S. Dübel. 2005. Perspectives for systematic in vitro antibody generation. *Gene* 364:19–29.

Kortt, A.A., M. Lah, G.W. Oddie, L.C. Gruen, J.E. Burns, L.A. Pearce, J.L. Atwell, A.J. McCoy, G.J. Howlett, D.W. Metzger, R.G. Webster, and P.J. Hudson. 1997. Single chain Fv fragments of anti-neurominidase antibody NC10 containing five and ten residue linkers form dimers and with zero residue linker a trimer. *Protein Eng.* 10:423–428.

Kramer, K., M. Fiedler, A. Skerra, and B. Hock. 2002. A generic strategy for subcloning antibody variable regions from the scFv phage display vector pCANTAB 5 E into pASK85 permits the economical production of Fab fragments and leads to improved recombinant immunoglobulin stability. *Biosensors & Bioelectronics* 17:305–313.

Krebber, A., S. Bornhauser, J. Burmester, A. Honegger, J. Willuda, H.R. Bosshard, and A. Plückthun. 1997. Reliable cloning of functional antibody variable domains from hybridomas and spleen cell repertoires employing a reengineered phage display system. *J. Immunol. Meth.* 201:35–55.

Lauer, B., I. Ottleben, H.J. Jacobsen, and T. Reinard. 2005. Production of a single-chain variable fragment antibody against fumonisin B1. *J. Agric. Food Chem.* 53:899–904.

Le Gall, F., S.M. Kipriyanov, G. Moldenhauer, and M. Little. 1999. Di-, tri- and tetrameric single chain Fv antibody fragments against human CD19: Effect of valency on cell binding. *FEBS Lett.* 453:164–168.

Lei, S.-P., H.-C. Lin, S.-S. Wang, J. Callaway, and G. Wilcox. 1987. Characterization of the *Erwinia caratovora pelB* gene and its product pectate lyase. *J. Bacteriol.* 169:4379–4383.

Little, M., M. Welschof, M. Braunagel, I. Hermes, C. Christ, A. Keller, P. Rohrbach, T. Kürschner, S. Schmidt, C. Kleist, and P. Terness. 1999. Generation of a large complex antibody library from multiple donors. *J. Immunol. Meth.* 231:3–9.

Lonberg, N., and D. Huszar. 1995. Human antibodies from transgenic mice. *Int. Rev. Immunol.* 13:65–93.

Løset, G.Å., I. Løbersli, A. Kavlie, J.E. Stacy, T. Borgen, L. Kausmally, E. Hvattum, B. Simonsen, M.B. Hovda, and O.H. Brekke. 2005. Construction, evaluation and refinement of a large human antibody phage library based on the IgD and IgM variable gene repertoire. *J. Immunol. Meth.* 299:47–62.

Lowman, H.B., S.H. Bass, N. Simpson, and J.A. Wells. 1991. Selecting high-affinity binding proteins by monovalent phage display. *Biochemistry* 30:10832–10838.

Marks, J.D., A.D. Griffiths, M. Malmqvist, T.P. Clackson, J.M. Bye, and G. Winter. 1992a. By-passing immunization: Building high affinity human antibodies by chain shuffling. *Bio/Technology* 10:779–783.

Marks, J.D., H.R. Hoogenboom, T.P. Bonnert, J. McCafferty, A.D. Griffiths, and G. Winter. 1991. By-passing immunization: Human antibodies from V-gene libraries diplayed on phage. *J. Mol. Biol.* 222:581–597.

Marks, J.D., H.R. Hoogenboom, A.D. Griffiths, and G. Winter. 1992b. Molecular evolution of proteins on filamentous phage. *J. Biol. Chem.* 267:16007–16010.

McCafferty, J., K.J. Fitzgerald, J. Earnshaw, D.J. Chiswell, J. Link, R. Smith, and J. Kenten. 1994. Selection and rapid purification of murine antibody fragments that bind a transition-state analog by phage-display. *Appl. Biochem. Biotechnol.* 47:157–173.

McCafferty, J., A.D. Griffiths, G. Winter, and D.J. Chiswell. 1990. Phage antibodies: Filamentous phage displaying antibody variable domain. *Nature* 348:552–554.

McCafferty, J., H.R. Hoogenboom, and D.J. Chiswell, eds. 1996. *Antibody engineering*. Oxford: IRL Press.

Moghaddam, A., T. Borgen, J. Stacy, L. Kausmally, B. Simonsen, O.J. Marvik, O.H. Brekke, and M. Braunagel. 2003. Identification of scFv antibody fragments that specifically recognise the heroin metabolite 6-monoacetylmorphine but not morphine. *J. Immunol. Meth.* 280:139–155.

Mullinax, R.L, E.A. Gross, J.R. Amberg, B.N. Hay, H.H. Hogreffe, M.M. Kubitz, A. Greener, M. Alting-Mees, D. Ardourel, J.M. Short, J.A. Sorge, and B. Shopes. 1990. Identification of human antibody fragment clones specific for tetanus toxoid in a bacteriophage λ immunoexpression library. *Proc. Natl. Acad. Sci. USA* 87:8095–8099.

Nissim, A., H.R. Hoogenboom, I.M. Tomlinson, G. Flynn, C. Midgley, D. Lane, and G. Winter. 1994. Antibody fragments from a "single pot" phage display library as immunochemical reagents. *EMBO J.* 13:692–698.

O'Connel, D., B. Becerril, A. Roy-Burman, M. Daws, and J.D. Marks. 2002. Phage versus phagemid libraries for generation of human monoclonal antibodies. *J. Mol. Biol.* 321:49–56.

Odegrip, R., D. Coomber, B. Eldridge, R. Hederer, P.A. Kuhlman, C. Ullman, K. FitzGerald, and D. McGregor. 2004. CIS display: In vitro selection of peptides from libraries of protein-DNA complexes. *PNAS* 101:2806–2810.

Ørum, H., P.S. Andersen, A. Øster, L.K. Johansen, E. Riise, M. Bjørnevad, I. Svendsen, and J. Engberg. 1993. Efficient method for constructing comprehensive murine Fab antibody libraries displayed on phage. *Nucleic Acids Res.* 21:4491–4498.

Parmley, S.F., and G.P. Smith. 1988. Antibody selectable filamentous fd phage vectors: Affinity purification of target genes. *Gene* 73:305–318.

Paschke, M., G. Zahn, A. Warsinke, and W. Höhne. 2001. New series of vectors for phage display and prokaryotic expression of proteins. *BioTechniques* 30:720–726.

Pelat, T., M. Hust, E. Laffly, F. Condemine, C. Bottex, D. Vidal, M.-P. Lefranc, S. Dübel, and P. Thullier. 2007. High-affinity, human antibody-like antibody fragment (single-chain variable fragment) neutralizing the lethal factor (LF) of *Bacillus anthracis* by inhibiting protective antigen-LF complex formation. *Antimicrob. Agents Chemotherap.* 51:2758–2764.

Persson, M.A.A., R.H. Caothien, and D.R. Burton. 1991. Generation of diverse high-affinity human monoclonal antibodies by repertoire cloning. *Proc. Natl. Acad. Sci. USA* 88:2432–2436.

Pini, A., F. Viti, A. Santucci, B. Carnemolla, L. Zardi, P. Neri, and D. Neri. 1998. Design and use of a phage display library. *J. Biol. Chem.* 273:21769–21776.

Plückthun, A. 1990. Antibodies from *Escherichia coli. Nature* 347:497–498.

Plückthun, A. 1991. Antibody engineering: Advances from the use of *Escherichia coli* expression systems. *Bio/Technology* 9:545–551.

Reiersen, H., I. Løbersli, G.A. Løset, E. Hvattum, B. Simonsen, J.E. Stacy, D. McGregor, K. Fitzgerald, M. Welschof, O.H. Brekke, and O.J. Marvik. 2005. Covalent antibody display: An in vitro antibody-DNA library selection system. *Nucleic Acids Res.* 33:e10.

Roberts, R.W., and J.W. Szostak. 1997. RNA-peptide fusions for the in vitro selection of peptides and proteins. *Proc. Natl. Acad. Sci. USA* 94:12297–12302.

Rojas, G., J.C. Almagro, B. Acevedo, and J.V. Gavilondo. 2002. Phage antibody fragments library combining a single human light chain variable region with immune mouse heavy chain variable regions. *J. Biotechnol.* 94:287–298.

Rondot, S., J. Koch, F. Breitling, and S. Dübel. 2001. A helperphage to improve single chain antibody presentation in phage display. *Nature Biotechnol.* 19:75–78.

Sblattero, D., and A. Bradbury. 2000. Exploiting recombination in single bacteria to make large phage antibody libraries. *Nature Biotechnol.* 18:75–80.

Schirrmann, T., L. Al-Halabi, S. Dübel, and M. Hust. 2008. Production systems for recombinant antibodies. *Frontiers Biosci.* 13:4576–4594.

Schmiedl, A., F. Breitling, and S. Dübel. 2000. Expression of a bispecific dsFv-dsFv′ antibody fragment in Escherichia coli. *Protein Eng.* 13:725–734.

Sheets, M.D., P. Amersdorfer, R. Finnern, P. Sargent, E. Lindqvist, R. Schier, G. Hemingsen, C. Wong, J.C. Gerhart, and J.D. Marks. 1998. Efficient construction of a large nonimmune phage antibody library: The production of high-affinity human single-chain antibodies to protein antigens. *Proc. Natl. Acad. Sci. USA* 95:6157–6162.

Shirai, H., A. Kidera, and H. Nakamura. 1999. H3-rules: Identification of CDR3-H3 structures in antibodies. *FEBS Lett.* 455:188–197.

Simmons, L.C., D. Reilly, L. Klimowski, T.S. Raju, G. Meng, P. Sims, K. Hong, R.L. Shields, L.A. Damico, P. Rancatore, and D.G. Yansura. 2002. Expression of full-length immunoglobulins in *Escherichia coli*: Rapid and efficient production of aglycosylated antibodies. *J. Immunol. Meth.* 263:133–147.

Skerra, A., I. Pfitzinger, and A. Plückthun. 1993. The functional expression of antibody Fv fragments in *Escherichia coli*: Improved vectors and a generally applicable purification technique. *Bio/Technology* 9:273–278.

Skerra, A., and A. Plückthun. 1988. Assembly of a functional immunoglobulin Fv fragment in *Escherichia coli. Science* 240:1038–1041.

Skerra, A., and T.G.M. Schmidt. 1999. Applications of a peptide ligand for streptavidin: The Strep-tag. *Biomol. Eng.* 16:79–86.

Smith, G.P. 1985. Filamentous fusion phage: Novel expression vectors that display cloned antigens on the virion surface. *Science* 228:1315–1317.

Söderlind, E., A.C.S. Lagerkvist, M. Dueñas, A.-C. Malmborg, M. Ayala, L. Danielsson, and C.A.K. Borrebaeck. 1993. Chaperonin assisted phage display of antibody fragments of filamentous bacteriophages. *Bio/Technology* 11:503–507.

Söderlind, E., L. Strandberg, P. Jirholt, N. Kobayashi, V. Alexeiva, A.-M. Aberg, A. Nilsson, B. Jansson, M. Ohlin, C. Wingren, L. Danielsson, R. Carlsson, and C.A.K. Borrebaeck. 2000. Recombining germline-derived CDR sequences for creating diverse single-framework antibody libraries. *Nature Biotechnol.* 18:852–856.

Soltes, G., H. Barker, K. Marmai, E. Pun, A. Yuen, and E.J. Wiersma. 2003. A new helper phage and phagemid vector system improves viral display of antibody Fab fragments and avoids propagation of insert-less virions. *J. Immunol. Meth.* 274:233–244.

Soltes, G., M. Hust, K.K. Ng, A. Bansal, J. Field, D.I. Stewart, S. Dübel, S. Cha, and E.J. Wiersma. 2007. On the influence of vector design on antibody phage display. *J. Biotechnol.* 127:626–637.

Studnicka, G.M., S. Soares, M. Better, R.E. Williams, R. Nadell, and A.H. Horwitz. 1994. Human-engineered monoclonal antibodies retain full specific binding activity by preserving non-CDR complementarity-modulating residues. *Protein Eng.* 6:805–814.

Taussig, M.J., O. Stoevesandt, C. Borrebaeck, A. Bradbury, S. Dübel, R. Frank, T. Gibson, L. Gold, F. Herberg, H. Hermjakob, J. Hoheisel, T. Joos, Z. Konthur, U. Landegren, A. Plückthun, M. Ueffing, and M. Uhlén. 2007.

ProteomeBinders: Planning a European resource of affinity reagents for analysis of the human proteome, *Nature Meth.* 4:13–17.

Thie, H., T. Meyer, T. Schirrmann, M. Hust, and S. Dübel. 2008a. Phage display derived therapeutic antibodies. *Curr. Pharm. Biotech.* 9:439–446.

Thie, H., T. Schirrmann, M. Paschke, S. Dübel, and M. Hust. 2008b. Leader peptides using the Sec pathway are superior to those using the SRP pathway for antibody phage display and antibody fragment production in *E. coli. New Biotechnol.* 5:49–54.

Toleikis, L., O. Broders, and S. Dübel. 2004. Cloning single-chain antibody fragments (scFv) from hybridoma cells. *Meth. Mol. Med.* 94:447–458.

Tsurushita, N., H. Fu, and C. Warren. 1996. Phage display vectors for *in vivo* recombination of immunoglobulin heavy and light chain genes to make large combinatoríal libraries. *Gene* 172:59–63.

Vaughan, T.J., A.J. Williams, K. Pritchard, J.K. Osbourn, A.R. Pope, J.C. Earnshaw, J. McCafferty, R.A. Hodits, J. Wilton, and K.S. Johnson. 1996. Human antibodies with sub-nanomolar affinities isolated from a large non-immunized phage display library. *Nature Biotechnol.* 14:309–314.

Vieira, J., and J. Messing. 1987. Production of single-stranded plasmid DNA. *Methods Enzymol.* 153:3–11.

Viti, F., V. Nilsson, S. Demartis, A. Huber, and D. Neri. 2000. Design and use of phage display libraries for the selection of antibodies and enzymes. *Meth. Enzymol.* 326:480–497.

Von Behring, E., and S. Kitasato. 1890. Über das Zustandekommen der Diphtherie-Immunität und der Tetanus-Immunität bei Thieren. *Deutsche Medizinische Wochenzeitschrift* 16:1113–1114.

Waterhouse, P., A.D. Griffiths, K.S. Johnson, and G. Winter. 1993. Combinatorial infection and *in vivo* recombination: A strategy for making large phage antibody repertoires. *Nucleic Acids Res.* 21:2265–2266.

Welschof, M., P. Terness, S. Kipriyanov, D. Stanescu, F. Breitling, H. Dörsam, S. Dübel, M. Little, and G. Opelz. 1997. The antigen binding domain of a human IgG-anti-F(ab′)2 autoantibody. *Proc. Natl. Acad. Sci. USA* 94:1902–1907.

Winter, G., and C. Milstein. 1991. Man-made antibodies. *Nature* 349:293–299.

Zahn, G., A. Skerra, and W. Höhne. 1999. Investigation of a tetracycline-regulated phage display system. *Protein Eng.* 12:1031–1034.

CHAPTER 9

Yeast Surface Display

JENNIFER L. LAHTI and JENNIFER R. COCHRAN

ABSTRACT

Yeast display has accelerated the identification and engineering of antigen-specific antibodies and antibody-like proteins with high affinity, stability, and soluble expression yields. First, the use of a eukaryotic host offers the important advantage of post-translational processing and modification of mammalian proteins. Second, yeast-displayed libraries can be quantitatively screened by fluorescence

Therapeutic Monoclonal Antibodies: From Bench to Clinic. Edited by Zhiqiang An

activated cell sorting, providing unparalleled quantitative discrimination between clones and eliminating artifacts due to host expression biases. Third, antigen-binding properties, and epitope mapping and characterization of individual clones, can be rapidly performed by flow cytometry while the antibodies are still tethered to the yeast surface, eliminating the need for laborious soluble production and purification steps. These characteristics make yeast surface display an ideal platform for the directed evolution of antibodies and other proteins. Antibodies engineered through yeast surface display are being further developed for a myriad of clinical and biotechnology applications.

Yeast surface display has been used for antibody discovery, modification, epitope mapping, and biophysical characterization (Chao et al. 2006). When combined with directed evolution, yeast surface display is a powerful method for engineering antibodies for increased antigen-binding affinity and specificity (Boder, Midelfort, and Wittrup 2000; Boder and Wittrup 1997), and enhanced stability and soluble expression (Shusta et al. 1999, 2000). With this technology, combinatorial antibody libraries are expressed as fusion proteins on the cell wall of *Saccharomyces cerevisiae* and are screened to identify mutants with desired properties. Several other display platforms exist for antibody engineering, including phage, bacterial, ribosome, and mRNA display. Advantages and disadvantages of yeast display compared to these other methods have been reviewed (Hoogenboom 2005). However, all of these *in vitro* evolution methods avoid self-tolerance of the immune system, which directs the natural antibody maturation process *in vivo*, thus resulting in unique antibodies that do not exist in the natural repertoire.

9.1 YEAST SURFACE DISPLAY CONSTRUCT

In the yeast surface display system, the yeast mating adhesion receptor *a*-agglutinin (Aga) serves as an anchor for tethering recombinant proteins to the cell surface (Boder and Wittrup 1997). Aga is a glycoprotein composed of two subunits: Aga1p, a 725-residue anchoring subunit tethered to the

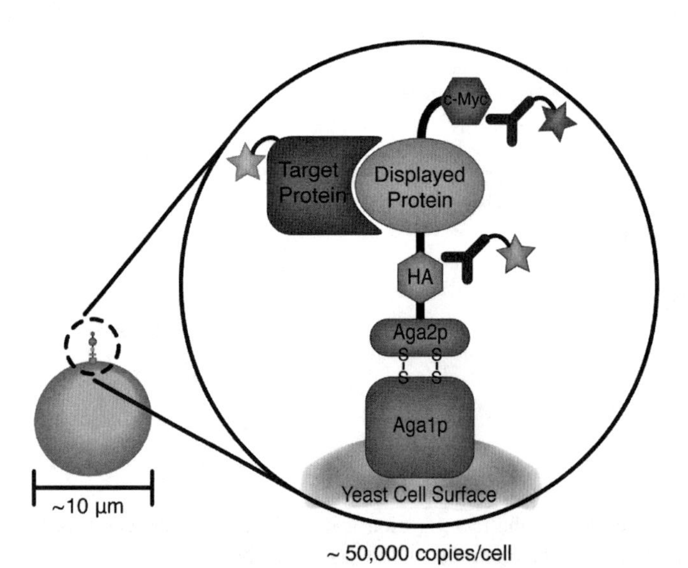

Figure 9.1 Yeast surface display construct. The protein of interest is displayed as a fusion to Aga2p and is anchored to the yeast cell wall via two disulfide bonds with Aga1p. Protein expression can be detected by flow cytometry, using antibodies that bind either the N-terminal hemagglutinin (HA) epitope tag (YPYDVPDYA) or the C-terminal c-Myc epitope tag (EQKLISEEDL). Binding of the displayed protein to a soluble, fluorescently labeled target protein can also be measured.

yeast cell wall via a ß-glucan covalent linkage at its carboxy terminus, and Aga2p, a 69-residue adhesion subunit covalently linked to Aga1p by two disulfide bonds (Cappellaro et al. 1994; Fig. 9.1). The protein of interest is expressed as a fusion to the Aga2p subunit under the control of a galactose-inducible promoter (GAL1) on the yeast display plasmid, an *S. cerevisiae*/*E. coli* shuttle plasmid maintained episomally with an essential amino acid nutritional marker. The yeast display plasmid also encodes a signal peptide (AGA2) to ensure secretory processing of the Aga2p protein fusion. The Aga1p construct is simultaneously expressed from a chromosomally integrated galactose-inducible expression cassette (Chao et al. 2006; Boder and Wittrup 1997; Feldhaus and Siegel 2004a). Upon induction of expression, the Aga1p protein and the Aga2p fusion protein associate in the Golgi along the yeast secretory pathway, and are exported to the cell surface (Chao et al. 2006; Boder and Wittrup 1997; Feldhaus and Siegel 2004a). The displayed protein of interest is flanked by N- and C-terminal peptide epitope tags that are used to detect and quantify protein expression on the yeast cell surface (Fig. 9.1). Proteins of sizes ranging from less than 1 kDa to over 100 kDa have been displayed on the yeast cell surface as Aga fusions.

9.2 ANTIBODY FRAGMENTS ENGINEERED WITH YEAST SURFACE DISPLAY

Yeast surface display has been used extensively to engineer antibody fragments and antibody-like proteins for increased affinity, specificity, stability, and expression levels. These molecular scaffolds and their particular orientation on the yeast cell wall are described in detail below.

9.2.1 scFv

The yeast surface display platform was originally developed and validated using single-chain variable antibody fragments (scFvs) (Boder and Wittrup 1997). In this system, an scFv constructed in the configuration variable heavy chain (V_H)-linker-variable light chain (V_L) is displayed on the surface of yeast via a tether between the amino terminus of the V_H and the carboxy terminus of the Aga2p anchoring protein (Fig. 9.2). The V_H and V_L domains are separated by a flexible Gly_4Ser (gly-gly-gly-gly-ser) linker and are flanked by N- and C-terminal epitope tags. These epitope tags allow for detection of the scFv on the yeast cell surface independent of function, and thus can be used to normalize the antigen-binding signal against protein display levels.

Yeast surface display has been used extensively to evolve scFvs that possess high affinity for soluble antigen targets (Boder, Midelfort, and Wittrup 2000; Boder and Wittrup 1997; Colby et al. 2004c; Feldhaus and Siegel 2004b; Feldhaus et al. 2003; Graff et al. 2004; Griffin et al. 2001; Kieke et al. 1997). Notably, scFvs with femtomolar antigen-binding affinity (Boder, Midelfort, and Wittrup 2000) and four-day dissociation half-life (Graff et al. 2004) have been engineered with this technology. In addition, yeast surface display has been used to isolate scFv antibody fragments from nonimmune (Feldhaus et al. 2003), pseudo-immune (Lee et al. 2006), and immune (Bowley et al. 2007) libraries. A nonimmune library of 10^9 human scFvs was constructed from a pool of adult human spleen and lymph node mRNA and displayed on the surface of yeast (Feldhaus et al. 2003). From this library, scFvs with nanomolar binding affinities were isolated against numerous protein, peptide, and hapten antigens. A yeast-displayed pseudo-immune scFv library was constructed from adult human lymphocyte total cellular RNA (Lee et al. 2006). This library was of considerably smaller diversity (10^6), but had significant bias for binding death receptor 5, the target antigen. Such bias facilitated the rapid enrichment of target-specific scFvs and ultimately led to the isolation of clones with higher affinity and specificity for the target antigen than those obtained from the nonimmune scFv library (Feldhaus et al. 2003; Lee et al. 2006). An immune scFv antibody library of 10^7 clones was constructed from RNA isolated from the bone marrow of an HIV-1 seropositive individual with broad HIV-1 neutralizing antibody titers (Bowley et al. 2007). Yeast surface display was used to isolate 18 unique scFvs from this immune library against the HIV-1 antigen gp120. Collectively, these studies demonstrate that yeast surface

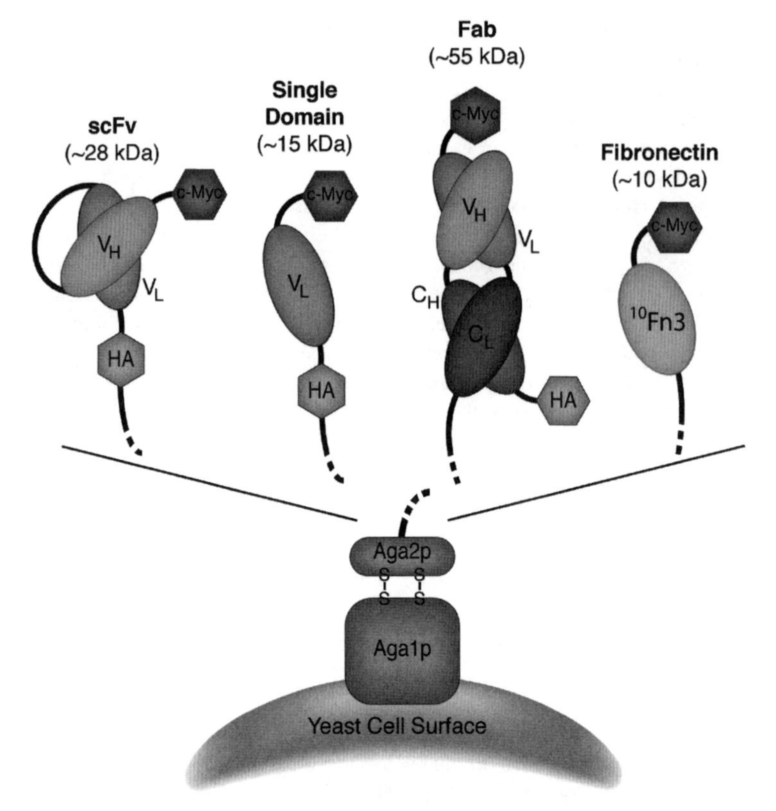

Figure 9.2 Yeast surface display of antibody fragments. A variety of antibody fragments and antibody-like proteins have been displayed on the surface of yeast as Aga2p fusions. HA and c-Myc epitope tags (hexagons) are included in these constructs for detection and quantification of cell surface protein expression levels.

display of naïve or biased scFv libraries is useful for the identification and affinity maturation of antibody fragments against a variety of antigen targets.

9.2.2 Fab

The yeast surface display platform has also been used to isolate and affinity mature Fab antibody fragments (Blaise et al. 2004; Lin et al. 2003; van den Beucken et al. 2003; Weaver-Feldhaus et al. 2004). In this system (van den Beucken et al. 2003), antibody heavy chain (C_{H1} and V_H) and light chain (C_L and V_L) domains exist as distinct polypeptides that associate along the yeast secretory pathway to form an assembled heterodimeric Fab on the yeast cell surface. Both the heavy and light chains are encoded in a single yeast display vector containing two identical protein expression cassettes driven by galactose-inducible promoters. The heavy chain is fused to the Aga2p yeast cell surface anchoring protein via an amino-terminal linker, while the light chain is produced as a soluble fragment. Inclusion of different epitope tags at the termini of the light chain and Aga2p-heavy chain polypeptides allows for the simultaneous and independent detection of each of the domains on the yeast cell surface (Fig. 9.2). The system was validated against three previously identified Fab antibodies. Displayed Fab antibodies were accessible to large soluble protein antigens, and quantitative selection of a yeast-displayed Fab library using flow cytometric sorting yielded mutants with low nanomolar affinity against the streptavidin target (van den Beucken et al. 2003).

Hetero-oligomeric catalytic Fab antibodies have been functionally expressed on the surface of yeast by combining yeast surface display and extracellular secretion systems (Lin et al. 2003). The light chain of an antibody (6D9) capable of catalyzing the synthesis of choloamphenicol was fused to Aga2p, while the heavy chain was solubly expressed. Cotransformation of yeast with both the display and secretion plasmids resulted in the production of separate antibody components that, upon association along the yeast secretory pathway, formed an intermolecular disulfide linkage, ultimately allowing for the display of the hetero-oligomeric 6D9 Fab on the yeast surface. Importantly, the yeast-displayed 6D9 Fab antibody fragment retained the catalytic properties of the original antibody, as evidenced by its ability to hydrolyze a chloramphenicol monoester derivative and stably bind a transition-state analog (Lin et al. 2003).

There are several advantages of using the Fab fragment over other antibody formats. Stable pairing of antibody variable region heavy and light chains is essential for effective, high-affinity antigen binding. Fab V_H and V_L domains more strongly associate than those of an scFv fragment due to an additional interface resulting from the C_{H1}-C_L domain association. Therefore, mutations selected through affinity maturation of a Fab fragment are more likely to be concentrated at the antigen-binding site (van den Beucken et al. 2003), and Fab libraries are proposed to provide, on average, higher affinity clones than a similarly sized scFv library (Weaver-Feldhaus et al. 2004). In comparison, yeast-displayed scFv clones screened for enhanced antigen-binding affinity have been found to contain mutations along the V_H-V_L domain interface, suggesting that improvements in protein stability or orientation are needed prior to changes directly involved in antigen binding (Boder, Midelfort, and Wittrup 2000). Furthermore, scFv antibodies have been shown to dimerize through mutations in the linker region, which can complicate the affinity maturation process.

9.2.3 Camelid Fragments

Unlike other mammalian antibodies, camelid antibodies lack a light chain and consist only of two identical heavy-chain molecules, each with a single N-terminal variable domain that is individually responsible for antigen binding. Fragments derived from these N-terminal heavy-chain variable domains (V_{HH}) can functionally serve as single domain antibodies (Harmsen and De Haard 2007). Similar to antibody V_H domains, V_{HH} fragments contain three complimentarity determining regions (CDRs) that are involved in antigen binding. However, to compensate for the lack of light chain binding interactions, the third CDR of the V_{HH} fragment is expanded, thereby creating an interface for antigen binding comparable to that of conventional antibodies (Harmsen and De Haard 2007; van der Linden et al. 2000). V_{HH} fragments may be preferable to intact antibodies, scFv, and Fab fragments for biotechnology applications because they consist of a single domain, are well expressed with high solubility, and are easily produced in multivalent forms (Harmsen and De Haard 2007). However, potential immunogenicity concerns preclude their use as human therapeutics.

Directed evolution has been used to improve stability and soluble production levels of llama V_{HH} in *S. cerevisiae* while retaining functional characteristics (Harmsen and De Haard 2007; van der Linden et al. 2000). Toward this end, DNA shuffling was used to generate a library of V_{HH} clones from three parental llama V_{HH} fragments with specificity for a common antigen. Rather than being tethered to the surface of yeast, the V_{HH} clones were produced solubly with C-terminal c-Myc and hexahistidine epitope tags and analyzed using a colony-blot assay (van der Linden et al. 2000). The yeast-secreted V_{HH} library was screened for mutants with increased stability and secretion levels. Notably, the best engineered V_{HH} clones showed threefold higher secretion levels and twofold higher binding affinity for azo-dye reactive red-6 antigen, even when binding was performed at 90°C (van der Linden et al. 2000).

9.2.4 ^{10}Fn3 Antibody Mimics

ScFv and Fab fragments have been shown to have reduced affinities and solubilities compared to their intact IgG counterparts (Harmsen and De Haard 2007). Antibody-like alternatives have been developed to address these shortcomings. The 10th human fibronectin type III domain (^{10}Fn3) is one of several alternative protein scaffolds developed as an antibody mimic (Koide et al. 1998; Lipovsek et al. 2007b; Xu et al. 2002). Despite the absence of sequence homology, ^{10}Fn3 is structurally homologous to antibodies in that it has an immunoglobulin-like fold and three solvent-accessible loops similar to the CDRs of antibodies. The ^{10}Fn3 scaffold is small in size and offers the practical benefits of high structural stability without metal ions or disulfide bonds, soluble expression at high yields in *E. coli*, and compatibility with intracellular applications (Koide et al. 1998, 2002; Koide and Koide 2007; Skerra 2007). ^{10}Fn3 variants have been engineered to bind targets with high affinity and specificity, and have been used as functional replacements for antibodies in traditional roles (Lipovsek et al. 2007b), such as Western blot detection (Karatan et al. 2004), affinity purification (Huang et al. 2006), interference of receptor-ligand interactions (Parker et al. 2005), and conformation-specific recognition of molecular targets (Koide et al. 2002). Yeast surface display has been used to engineer ^{10}Fn3 variants with subnanomolar antigen binding affinities (Lipovsek et al. 2007b). ^{10}Fn3 clones were displayed on the surface of yeast as N-terminal fusions to the Aga2p yeast mating protein with C-terminal c-Myc epitope tags (Fig. 9.2). Antibody engineering by CDR walking (fixing seven residues of one loop while randomizing the seven residues on the adjacent loop) was used to isolate ^{10}Fn3 clones with 10 nM to 350 pM affinities for hen egg white lysozyme, a model antigen. The best clones showed convergent evolution of cysteine residue pairs at adjacent locations sufficiently close to allow for the formation of a disulfide bond between the two randomized loops (Lipovsek et al. 2007b). The introduced disulfide bond was shown to be important for high-affinity antigen interaction, as its disruption resulted in the loss of binding (Lipovsek et al. 2007b). Interestingly, this interloop disulfide bond is structurally homologous to the disulfide bond naturally present in some antibody structures, namely antibody counterparts of cartilaginous fish and camelid heavy chain variable domains.

9.3 ENGINEERING ANTIBODIES FOR AFFINITY, SPECIFICITY, STABILITY, AND EXPRESSION

Yeast surface display can be combined with directed evolution to engineer antibodies with increased antigen-binding affinity and specificity, and increased protein stability and expression levels. Improvements in antibody function are often attributable to mutations with long-range or complementary effects, which are difficult to predict *a priori*. Consequently, directed evolution is generally performed using several rounds of random mutagenesis and shuffling of genetic material, rather than by a more rational approach. Individual favorable mutations often have only a small effect on properties such as antigen-binding affinity (2- to 10-fold), yet multiple combined mutations can yield large improvements in affinity due to additive or synergistic effects (Boder and Wittrup 1998; Midelfort et al. 2004; Midelfort and Wittrup 2006). Therefore, the ability to isolate and identify all library clones with even marginal improvements is essential for obtaining the most desirable mutant (VanAntwerp and Wittrup 2000).

To take full advantage of the potential offered by yeast display to quantitatively isolate mutant antibodies with improved characteristics, it is necessary to optimize experimental screening conditions. The idiom "you get what you screen for" is simplistic, yet holds true for directed evolution using the yeast display platform, making the choice of experimental parameters essential to the success of the engineering process. Using tunable screening conditions, mutants with specified improvements in target properties (e.g., affinity, specificity, stability, and expression) can be differentially labeled, allowing for their isolation from the background population (Boder and Wittrup 1998; Razai et al. 2005).

9.3.1 Engineering Affinity

High affinity antigen recognition (low nM) is critical for therapeutic and biotechnology applications of engineered antibodies; therefore, extensive efforts have been made to generate ultra-high affinity antibodies using directed evolution. Yeast surface display is particularly well suited for the affinity maturation of antibodies. Yeast cells possess eukaryotic protein processing machinery and are sufficiently large to permit combinatorial library sorting by flow cytometry, which is critical for fine affinity

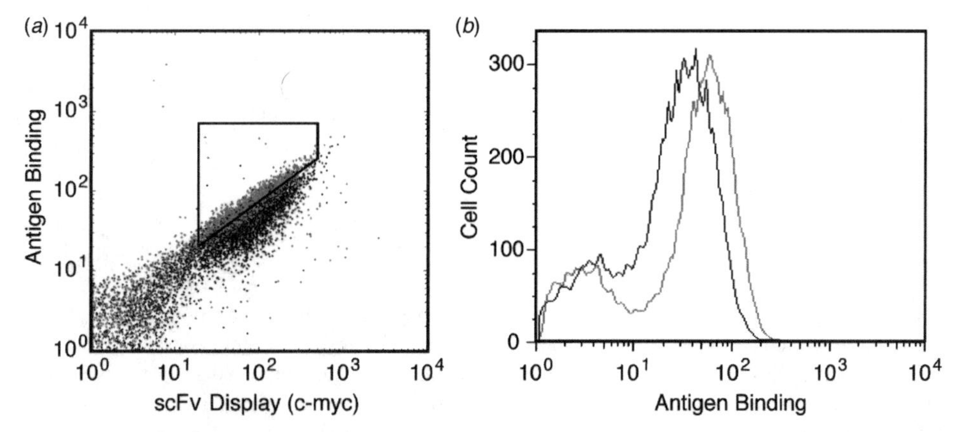

Figure 9.3 Multiparameter flow cytometry allows for fine affinity discrimination between scFv clones. (a) Overlaid flow cytometry dot plots for two yeast-displayed scFvs with a twofold difference in antigen-binding affinity. Antigen binding (y-axis) can be normalized with scFv display levels (x-axis) to allow isolation of clones with only slight affinity improvements (light color) over wild-type (dark color) using a diagonal sort window. (b) Overlaid flow cytometry histograms of antigen-binding fluorescence of these same two clones highlights the usefulness of multicolor FACS in discriminating clones with similar affinities. There is always a non-expressing, non-binding population in *S. cerevisiae* yeast surface display; this population serves as a convenient internal negative control. From Chao et al. (2006). Reprinted by permission from Macmillan Publishers, Ltd. © 2006.

discrimination between yeast-displayed antibodies with similar binding capacities (Fig. 9.3; Chao et al. 2006; VanAntwerp and Wittrup 2000).

In vivo, antibodies have an empirical affinity limit of approximately 0.1 nM due to the lack of selective pressure for dissociation rates slower than intrinsic B-cell internalization rates. This affinity limit is attributed to the somatic hypermutation process rather than an inherent restraint of the antibody scaffold architecture (Boder, Midelfort, and Wittrup 2000). In general, affinity maturation using yeast surface display results in engineered antibodies with improved antigen affinities in the low nanomolar to picomolar range, and even femtomolar affinities have been achieved (Boder, Midelfort, and Wittrup 2000; Razai et al. 2005). Antibody library screening strategies and experimental conditions are influenced by characteristics of the antigen-antibody pair of interest, and typically employ equilibrium binding and/or kinetic components to allow for targeted improvements of specific binding parameters in addition to overall affinity. Approaches for engineering affinity are described below for flow cytometry using fluorescently labeled antigens; however, these principles are also applicable to other library screening methods.

9.3.1.1 Equilibrium Binding Screen Equilibrium binding screens provide a way to improve the overall affinity of low to modest antibody-antigen binding interactions [equilibrium dissociation constant (K_d) > 10 nM]. In this approach, yeast cells displaying a mutant antibody library are incubated with fluorescently labeled antigen at a concentration below the equilibrium dissociation constant for the wild-type antibody-antigen interaction, and cells with improved binding/expression ratios are selected. Optimal conditions for engineering antibody-antigen binding properties in this manner have been described with mathematical equations by Boder and Wittrup (1998) using expected cell fluorescence intensities.

The equilibrium dissociation constant for the reversible binding of a monovalent antigen (L) to an antibody (R) displayed on the surface of yeast resulting in bound complex (C) is:

$$K_d = \frac{[R][L]}{[C]} = \frac{k_{off}}{k_{on}},$$

where k_{off} and k_{on} are the dissociation and association rate constants, respectively. When considering a yeast-displayed library, the antigen concentration ($[L]_{opt}$) that will yield the most robust signal (maximum ratio of mutant to wild-type fluorescence when analyzed by flow cytometry, where the output signal is amplified logarithmically) for a specified target affinity (K_d^{mut}) is (Boder and Wittrup 1998):

$$[L]_{opt} = \frac{K_d^{wt}}{\left[\left(\frac{F_{max}}{F_{background}}\right)\left(\frac{K_d^{wt}}{K_d^{mut}}\right)\right]^{1/2}},$$

where F_{max} is the maximal fluorescence when all yeast-displayed constructs are bound, $F_{background}$ is the background fluorescence, and K_d^{wt} is the wild-type equilibrium dissociation constant. Yeast cells displaying antibodies with improved equilibrium binding constants compared to that of wild-type will more fully saturate at the optimal antigen concentration, allowing for the discrimination and selection of mutants with enhanced antigen-binding properties from a background of lower-affinity binders.

9.3.1.2 Kinetic Binding Screen: Dissociation Rate

Ultra-high affinity antibody-antigen interactions are most easily obtained by screening for decreased antigen dissociation rates (k_{off}) using a kinetic screen. Differences in the antigen-binding affinities of antibodies are typically dominated by dissociation rates. Furthermore, reaction volumes required to perform equilibrium binding screens at very low antigen concentrations without incurring antigen depletion upon antibody binding are prohibitively large. Therefore, improvement of high-affinity antibody-antigen interactions ($K_d <$ 10 nM) is most easily accomplished using a kinetic screen to select for decreased dissociation rate (Boder and Wittrup 1998). In this method, yeast-displayed antibody libraries are incubated with fluorescently labeled antigen at concentrations sufficiently high to bind to saturation, followed by the addition of excess competing unlabeled antigen. In the presence of a large excess of unlabeled antigen, the rate of antibody-fluorescent antigen complex formation is effectively zero and the mean cellular fluorescence intensity (F) of a yeast-displayed antibody library decays exponentially over the time course (t) of the competition reaction, as described mathematically by Boder and Wittrup (1998):

$$F = (F_o - F_\infty)e^{-k_{off}t} + F_\infty$$

where F_o is the fluorescence intensity at saturation with labeled antigen and F_∞ is the background fluorescence. Yeast cells displaying mutant antibodies with decreased dissociation rates will retain greater fluorescence over time and can be isolated from a background of lower-affinity antibodies. The duration of the competition reaction will determine the magnitude of comparable difference in fluorescence between library clones, making it necessary to optimize the incubation time to fit the desired range of kinetic improvements. The optimal reaction time (t_{opt}) for the kinetic screen is that at which the fluorescence ratio between mutants of a desired dissociation rate (k_{off}^{mut}) and wild-type (k_{off}^{wt}) is at a maximum (Boder and Wittrup 1998):

$$t_{opt} = \frac{0.293 + 2.05\log\left(\frac{k_{off}^{wt}}{k_{off}^{mut}}\right) + \left[2.30 - 0.759\frac{1}{\left(\frac{k_{off}^{wt}}{k_{off}^{mut}}\right)}\right]\log\left(\frac{F_o}{F_\infty}\right)}{k_{off}^{wt}}.$$

Expected cell fluorescence ratios show distinct optimal competition reaction times, demonstrating the importance of selecting an appropriate finite competition time for isolating antibodies of specific target affinities, particularly when the desired improvement is small (threefold or less), as is often the case for antibodies with high wild-type affinities (Boder and Wittrup 1998).

A kinetic screen using competing unlabeled antigen has previously been used to affinity mature a yeast-displayed scFv for femtomolar binding affinity ($K_d = 48$ fM) to fluorescein (Boder, Midelfort, and Wittrup 2000). The astounding affinity of the engineered anti-fluorescein scFv was entirely attributable to a 1000-fold decrease in the dissociation rate compared to that of the original scFv. Conversely, the mutants isolated from this kinetic screen had antigen association rates that were either unchanged or slightly decreased (14-fold slower) compared to the original scFv (Boder, Midelfort, and Wittrup 2000).

For tumor-targeting applications, it is often most beneficial to increase the antibody retention time in the tumor relative to the normal tissue (Graff et al. 2004). Increased tumor retention may be accomplished by increased affinity, as indicated by experimental observations and mathematical modeling (Graff et al. 2004; Graff and Wittrup 2003). More precisely, the dissociation rate constant is a critical determinant of tumor retention time, although the extent to which retention can be improved is also determined by such variables as cellular internalization, recycling, and degradation rates. The therapeutic potential of an anti-carcinoembryonic antigen (CEA) scFv was improved by sorting successive yeast-displayed libraries using equilibrium binding screens at physiological temperature, followed by kinetic screens with competing antigen (Graff et al. 2004). Anti-CEA clones were isolated with approximate binding affinities of 80 pM, a 100-fold improvement over the original affinity. Notably, the best clone isolated by this method had the slowest reported dissociation rate for an engineered antibody against a protein antigen, corresponding to a dissociation half-life of several days at physiological temperature (Graff et al. 2004).

9.3.1.3 Kinetic Binding Screen: Association Rate

Although the majority of improvements in the affinity of engineered antibodies result from mutations corresponding to decreased dissociation rates, the association rate may also be of great functional importance for therapeutic or biotechnology applications. For example, a fast association rate will provide rapid and sensitive detection by engineered antibodies used as biosensors and diagnostic devices. The association rate constant (k_{on}) can be increased through a directed evolution strategy that uses a modified equilibrium binding screen, wherein experimental parameters are designed such that the time to reach equilibrium is dominated by the rate of association. With this screen, mutants with improved association rates can be selected by varying the antigen incubation time (Razai et al. 2005). Clones with faster association rates will achieve greater levels of antigen binding over the allotted incubation time, allowing for the identification and isolation of higher-affinity mutants.

For the reversible binding of a monovalent antigen to an antibody displayed on the surface of yeast, the fraction of antibody molecules bound to antigen (y) has been mathematically described by VanAntwerp and Wittrup (2002):

$$y = \frac{n \dfrac{k_{on}}{k_{off}}[R]}{1 + \dfrac{k_{on}}{k_{off}}[R]}$$

where n is fluorescence intensity when binding is fully saturated and $[R]$ is the concentration of displayed antibody. The reaction time (t_θ) needed to ensure equilibrium binding has been described by Razai et al. (2005):

$$t_\theta = \frac{-\ln(1-\theta)}{k_{on}[L] + k_{off}}$$

where $[L]$ is the labeled antigen concentration and θ is the percent equilibrium. When $k_{on}[L] >> k_{off}$, the time to equilibrium is dominated by the rate of association (Razai et al. 2005). Thus, by optimizing the incubation time and ligand concentration, a pre-equilibrium binding screen can be used to isolate mutants with enhanced association rates from a wild-type background (Razai et al. 2005).

A combined strategy employing a pre-equilibrium binding screen to increase k_{on} followed by a kinetic screen with competitive antigen to decrease k_{off} has been used to engineer scFvs for sensitive detection of botulinum neurotoxin type A. The engineered scFvs showed a 37- to 45-fold increase in affinity due both to increased association rate and decreased dissociation rate constants (Razai et al. 2005). The best evolved scFv allowed botulinum neurotoxin to be detected at concentrations as low as 0.1 pM, with a sensitivity equal to that commonly obtained in an enzyme-linked immunoadsorbant assay (ELISA).

9.3.2 Engineering Specificity

The yeast surface display platform can also be used to engineer antibody-antigen binding specificity via directed evolution. Antigen recognition can be engineered by designing yeast library screening strategies that select for antibody binding to particular domains, conformations, or subtypes of target antigens, while excluding others.

One strategy for isolating antibodies that target a specific or novel epitope involves a technique known as complementation screening (Siegel et al. 2004). A library of yeast-displayed antibodies previously enriched for antigen binding is incubated with fluorescently labeled target antigen for a sufficient amount of time to achieve binding equilibrium. A well-characterized monoclonal antibody (mAb) is then added to the bound antibody-antigen complexes, and multicolor flow cytometry is used to select yeast-displayed antibody clones that bind antigen while also allowing for mAb binding. Using this approach, antibodies that bind distinct or novel epitopes on the antigen can be identified and isolated. This screening strategy has been used to generate scFvs that bind to mutually exclusive epitopes on epidermal growth factor (Siegel et al. 2004). In addition, complementation screening allows for the rapid characterization of multiple antibody clones that often result from selection using a target antigen.

Conversely, antibody-binding specificity can also be broadened using yeast surface display. Antibody fragments with cross-reactivity for two subtypes of botulinum neurotoxin type A (BoNT/A) have been generated by screening yeast libraries for mutants that bind to both of these distinct antigenic targets (Garcia-Rodriquez et al. 2007). An scFv library, based on an antibody with high affinity for BoNT/A1 and low affinity for BoNT/A2, was constructed by introducing mutations into solvent-accessible positions along the antigen-binding loops. This yeast-displayed library was co-selected through equilibrium binding of soluble BoNT/A1 and BoNT/A2 antigens, to identify an scFv with a 1250-fold increased affinity for BoNT/A2 (87 pM) while retaining high affinity for BoNT/A1 (115 pM). Interestingly, the increase in cross-reactivity resulted entirely from mutations located in framework regions and a single antigen-binding loop. Furthermore, the engineered scFv had sufficiently high affinity for therapeutic use, and has been shown to boost BoNT/A2 neutralization 100-fold when combined with other human BoNT/A antibodies (Garcia-Rodriquez et al. 2007).

Yeast surface display can also be used to evolve antibodies for conformation-specific recognition of antigens. A yeast-displayed antibody library can first be enriched for antigen binding without initial bias towards a specific conformation. After an enriched antigen-binding population has been obtained, the selection strategy is modified such that equilibrium binding screens are performed against antigen in a specific conformation. This strategy was developed to isolate antibody fragments against the calcium-dependent signaling molecule calmodulin (CaM), either when bound to calcium (Ca^{2+}-CaM) or in the absence of calcium (apo-CaM) (Weaver-Feldhaus et al. 2005). A naïve yeast-displayed scFv library was first subjected to equilibrium sorting against untreated CaM that presumably existed in both conformations. Then the enriched pool of scFvs was sorted against CaM pretreated with calcium or a chelating agent to isolate antibody fragments that bound Ca^{2+}-CaM or apo-CaM, respectively (Weaver-Feldhaus et al. 2005). In this way, conformation-specific CaM-binding clones were identified. Importantly, a Ca^{2+}-CaM-specific scFv of high affinity ($K_d = 800$ pM) and specificity (>1000-fold compared to apo-CaM) and an antibody fragment recognizing apo-CaM, albeit with lower affinity and specificity, were isolated. Further rounds of directed evolution yielded an antibody fragment specific for apo-CaM with improved affinity ($K_d = 1$ nM) and specificity (>300-fold compared to Ca^{2+}-CaM) than the original antibody fragment (Weaver-Feldhaus et al. 2005).

9.3.3 Engineering Stability

Thermal and proteolytic stability is an important functional requirement for the use of antibodies and antibody fragments in medicine and biotechnology applications, or for structural studies (Orr et al. 2003). Therapeutic or diagnostic antibodies must retain their biological activity under physiological conditions (37°C in the presence of serum, which contains proteases) for several hours or days without precipitation or degradation. Stability requirements can be even greater for many non-pharmaceutical applications, in which antibodies must preserve their function in adverse environments such as nonpolar solvents, surfactants, proteases, and high temperatures. Because many antibodies or antibody fragments are not stable under such conditions, they have limited practical applications, despite having favorable antigen-binding properties (Worn and Pluckthun 2001). This instability can often be attributed to particular amino acid sequences rather than to an intrinsic limitation of the antibody format, indicating that antibodies with enhanced stability can be generated through protein engineering.

Structural stability is well correlated with *in vivo* activity for both extracellular and intracellular applications of antibodies (Worn and Pluckthun 2001; Worn et al. 2000). However, it is difficult to predict the stability of an antibody *a priori*, as the complex molecular and thermodynamic interactions governing protein stability are not well understood. Furthermore, the overall stability of multidomain antibodies depends not only on the stability of the individual domains, but also on that of the domain interfaces. Despite this complexity, yeast surface display has been used to engineer antibody fragments and other proteins for enhanced stability. The collective efficiency of processes involved in yeast protein production and surface display, including membrane translocation, protein folding, quality control, and vesicular transport, is well predicted by thermal stability (Shusta et al. 2000). Therefore, engineering an antibody for increased yeast surface display levels can be a way to improve both its stability and its soluble expression yield.

A common strategy for enhancing antibody stability through directed evolution is the replacement of structurally important cysteine residues with alanine or valine to create a library of clones devoid of intramolecular disulfide bonds (Proba et al. 1998). Such a library can be subjected to an equilibrium binding screen to select for mutants that retain native structure despite the lack of a disulfide bond. Mutations identified by this process are likely to enhance the stability of the protein upon the reintroduction of the native cysteine residues. This strategy was used to improve the stability of scFvs previously engineered for high-affinity antigen interactions (Graff et al. 2004). By screening a yeast library for clones with the highest surface display levels, a single stabilizing mutation was identified. Incorporation of this stabilizing mutation into previously engineered high-affinity constructs led to increased levels of protein expression without affecting antigen-binding affinity (Graff et al. 2004). The stabilized scFvs showed increased thermal stability, as demonstrated by retention of antigen-binding affinity after incubation at 37°C for nine days (Graff et al. 2004).

Yeast surface display has also been used to engineer single-chain T-cell receptors (scTCRs) for enhanced stability. ScTCRs are structurally homologous to scFvs, as they contain an Ig-like fold and variable CDRs that specifically bind to ligands. The use of scTCRs for practical applications in immunology has historically been limited, since they possess low ligand-binding affinities and are relatively unstable, with poor solubility and a high propensity to aggregate. To address these limitations, an scTCR with high thermal stability, solubility, and protein expression yields was engineered by directed evolution using yeast surface display (Shusta et al. 2000). A temperature-based library screening approach, in which selections were performed at elevated temperatures (37°C and 42°C), was used to identify a small number of mutations that enhanced thermal stability. Remarkably, combining these mutations resulted in an scTCR that was stable at 65°C for over an hour, with solubility and expression levels comparable to those for stable scFvs. In addition, enhanced yeast surface display levels of selected clones at 20°C corresponded with increased thermal stability, suggesting that protein folding and expression may be dictated by the *in vitro* stability of the protein (Shusta et al. 1999, 2000). The scTCR library screening strategies described above can be applied directly to engineer antibodies for enhanced stability by yeast surface display.

9.3.4 Engineering Expression

Yeast surface display is a useful platform to engineer proteins with enhanced soluble expression yields. Expression in yeast is an attractive option for recombinant antibody production, since its eukaryotic secretion machinery is well equipped to express disulfide-bonded proteins and its use in large-scale fermentation is well developed. Along the yeast secretory pathway, nascent polypeptide chains are folded and subjected to quality control in the endoplasmic reticulum (ER), wherein soluble and membrane-bound chaperones and foldases ensure the fidelity of the secreted protein species (Kowalski et al. 1998; Ellgaard and Helenius 2003). In this way, only properly folded proteins should be displayed as Aga2p fusions on the yeast cell surface (Shusta et al. 1999; Rakestraw and Wittrup 2006; Hammond and Helenius 1995). However, it has recently been found that the yeast ER quality control mechanism is unable to recognize some misfolded heterologous proteins, allowing for their efficient secretion despite structural inaccuracies (Kim et al. 2006; Park et al. 2006).

The secretion efficiency of eukaryotic proteins can vary widely due to intrinsic differences that affect folding kinetics, thermodynamic stability, susceptibility to proteolysis, and association with the quality control apparatus of the ER. Previous studies with scFvs, scTCRs, and other proteins have shown that thermal stability predominates in the modulation of yeast secretion levels (Shusta et al. 1998, 2000; Kowalski et al. 1998; Rakestraw and Wittrup 2006; Kondo and Ueda 2004; Kowalski, Parekh, and Wittrup 1998). A protein's thermodynamic stability determines secretion efficiency by defining a rapidly equilibrated partition between un- or misfolded proteins that are degraded, and properly folded forms that are exported (Kowalski et al. 1998). Hence, protein folding is often the rate-limiting step since proteins are exported only after proper folding, subunit assembly, and passing quality control in the ER (Kowalski et al. 1998; Kowalski, Parekh, and Wittrup 1998). In addition, because thermal denaturation kinetics are a good predictor of secretion efficiency (Shusta et al. 1998, 1999), proteins displayed on the surface of yeast can be engineered for increased secretion levels by screening for clones with increased thermodynamic stability. As discussed in the previous section, mutants with increased stability can be obtained by screening libraries for clones with increased yeast surface display levels. Indeed, previous studies have shown that increased protein secretion levels directly correlate with increased thermal stability and yeast display expression levels (Shusta et al. 1998, 1999).

ScFvs engineered for increased stability using temperature-based screens and yeast surface display levels had, in addition to increased thermal stability, enhanced protein secretion levels on the order of 10 to 20 mg/mL, a 100-fold improvement over the original high-affinity scFvs (Graff et al. 2004). Similar trends were found with scTCRs engineered for thermal stability by yeast surface display; the most stable scTCR had a solubility of over 4 mg/mL and protein secretion levels of 7.5 mg/L (Shusta et al. 2000). A myriad of other proteins have been engineered by yeast surface display for increased secretion levels, including cancer-testis antigen NY-ESO1 (Piatesi et al. 2006), green fluorescent protein (Huang and Shusta 2005), tumor necrosis factor receptor (Schweickhardt et al. 2003), and epidermal growth factor receptor (Kim et al. 2006).

9.4 GENERATING YEAST-DISPLAYED ANTIBODY LIBRARIES

9.4.1 Library Size

The number of distinct clones present in a combinatorial library is an important determinant for antibody engineering. Initially, *E. coli* mutator strains or plasmid ligation and chemical transformation methods were used to create yeast libraries of 10^5 to 10^6 clones (Boder and Wittrup 1997; Keike et al. 1997). However, yeast libraries on the order of 10^7 to 10^8 transformants are now routinely generated using homologous recombination (Kim et al. 2006; Piatesi et al. 2006; Lipovsek et al. 2007a), and libraries of 10^9 have been created with collaborative efforts (Feldhaus et al. 2003). In addition, yeast mating methods have been used to produce combinatorial libraries on the order of 10^9 clones (Blaise et al. 2004; Weaver-Feldhaus et al. 2004).

Significantly larger library sizes can be generated with other *in vitro* display platforms that are not limited by yeast transformation efficiency (phage display: 10^6 to 10^{11} clones, mRNA and ribosome display: 10^{11} to 10^{13} clones; Weaver-Feldhaus et al. 2004). Yet, all *in vitro* display methods grossly undersample the theoretical sequence space of the antibody to be engineered (CDR diversity $\approx 10^{80}$; Chao et al. 2006). Moreover, it is difficult to assess the true functional diversity of *in vitro* display libraries. However, the quality control mechanisms of the yeast ER could provide a protein folding or expression advantage that compensates for smaller library size. The effects of improved library quality were recently explored by using the same genetic material to create both phage-displayed and yeast-displayed immune scFv libraries that were screened for binders against the HIV-1 antigen gp120 (Bowley et al. 2007). Yeast surface display was found to more fully sample the antibody repertoire, resulting in the identification of all 6 scFvs selected by phage display and 12 additional scFvs (Bowley et al. 2007). The most influential factor accounting for the difference in the number of isolated clones appeared to be the lower level of correctly folded scFvs on the surface of phage compared to those displayed on yeast. Therefore, this direct comparison suggests that library quality, not size, more strongly influences repertoire sampling efficiency (Bowley et al. 2007).

9.4.2 Genetic Diversity

Antibody library diversity can be generated by cloning naturally occurring human antibody repertoires (Feldhaus et al. 2003; Lee et al. 2006; Bowley et al. 2007). Alternatively, when engineering a known antibody for desired properties, many methods have been used to create genetic diversity, including bacterial mutator strains (Boder, Midelfort, and Wittrup 2000; Boder and Wittrup 1997), combinatorial enumeration (Lipovsek et al. 2007a, 2007b), site-specific recombination (Sblattero and Bradbury 2000; Waterhouse et al. 1993), error-prone PCR (Griffin et al. 2001; Colby et al. 2004a; Holler et al. 2000), DNA shuffling (Graff et al. 2004), and shotgun ortholog scanning mutagenesis (Cochran et al. 2006). These general mutagenesis methods can be used to create antibody libraries for yeast, phage, *E. coli*, mRNA, and ribosome display with diversities ranging from clones that differ in sequence by one or more point mutations to clones with radically reconstructed regions. In addition, homologous recombination (Swers, Kellogg, and Wittrup 2004) and cell mating (Blaise et al. 2004; Weaver-Feldhaus et al. 2004) approaches have been developed specifically for the yeast display platform, and will be discussed in detail below.

9.4.2.1 *In Vivo Homologous Recombination* Homologous recombination, an endogenous DNA repair mechanism, can be harnessed for the efficient construction of plasmids in yeast following transformation (Ma et al. 1987). This method of plasmid construction utilizes the yeast gap-repair machinery, which is highly sensitive to double-strand breaks in DNA, to repair a linearized plasmid by recombination with an insert containing 5′ and 3′ homology. Because the gap-repair mechanism in yeast is highly efficient, plasmid and insert digestion with restriction enzymes can be used to create homology and ensure the correct directional recombination of the insert. In addition, *in vivo* homologous recombination has been used to introduce diversity into yeast-displayed antibody libraries (Fig. 9.4; Swers, Kellogg, and Wittrup 2004). With this method, transformation of multiple PCR products that share significant internal homology can result in recombination events that yield chimeric gene products, effectively shuffling the starting library *in vivo*. Two scFv antibodies with $\sim$89 percent sequence homology were shuffled *in vivo* by this homologous recombination method (Swers, Kellogg, and Wittrup 2004). The majority of the resulting scFv chimeras were the products of single crossovers. However, double and triple crossovers were also found, and the number of clones that were isolated with a given number of crossovers was well represented by a Poisson distribution.

Yeast surface display is uniquely suited to library generation by *in vivo* homologous recombination compared to other display platforms. For example, homologous recombination in *E. coli* (whose machinery is utilized in phage display for propagation and production) requires cooperation between two proteins, one of which is a vigorous exonuclease, thereby impeding the use of linear DNA.

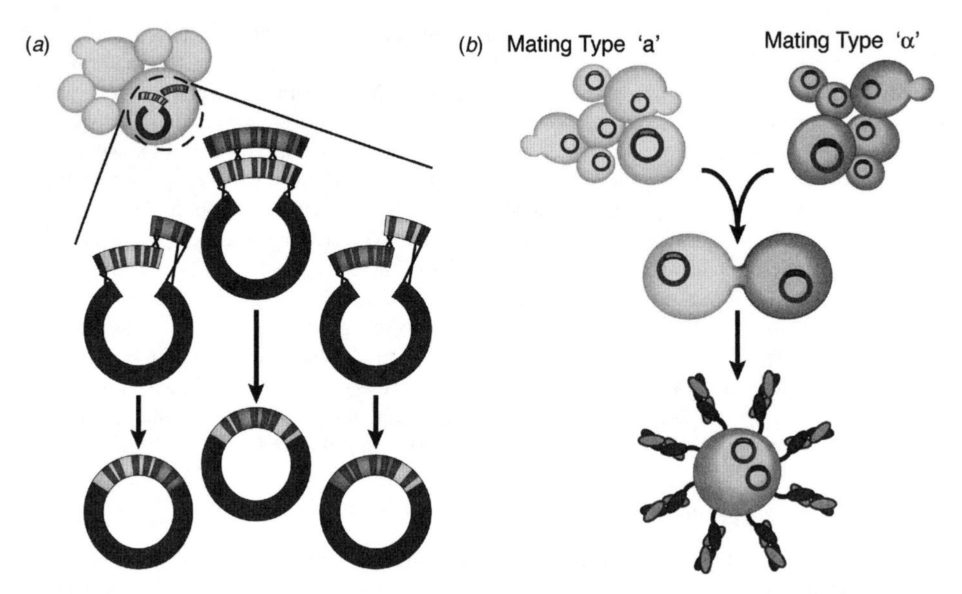

Figure 9.4 Methods of library generation that are uniquely suited for yeast surface display. (a) *In vivo* homologous recombination (Swers, Kellogg, and Wittrup 2004). PCR products that encode for mutant proteins are transformed into yeast, along with linearized vector. These inserts have regions of 3′ and 5′ homology to the yeast display plasmid backbone and shared internal regions of homology. The yeast gap-repair machinery uses these regions of homology to reconstruct whole plasmids, which in the process generates increased diversity from crossovers (denoted by an **x**) at regions of homology between inserts. (b) Combinatorial yeast mating (Blaise et al. 2004; Weaver-Feldhaus et al. 2004). Haploid yeast libraries of opposite mating type encoding secreted V_L domains and Aga2p-V_H fusion proteins are mated, resulting in a yeast-displayed combinatorial Fab library.

Importantly, because yeast is highly proficient at homologous recombination, as little as two nucleotides of homology are enough to produce crossovers, allowing diverse antibody libraries to be generated. Furthermore, this method provides a simple alternative for generating yeast-displayed antibody libraries that does not require the extensive PCR steps of DNA shuffling, pre-engineering of recombination sites, or ligation.

9.4.2.2 Yeast Mating To further address limitations of library size and to create antibody repertoires that better mimic the vast natural diversity of the human immune system, large combinatorial Fab libraries have been created by using yeast mating to bring together individual heavy and light chain libraries (Fig. 9.4; Blaise et al. 2004; Weaver-Feldhaus et al. 2004). The Fab antibody scaffold is uniquely suited for recombination by mating since the Fab heterodimer is composed of these two distinct polypeptide chains (Fig. 9.2). Libraries of each Fab domain are encoded within a different vector with distinct selectable markers in yeast strains of opposite mating types (a and α). The mating of a haploid library encoding secreted Fab light chains with a haploid library encoding Fab heavy chains fused to the Aga2p anchoring protein can generate a large combinatorial diploid Fab library with theoretical diversity on the order of 10^{14} and actual diversity of 10^9 (Blaise et al. 2004; Weaver-Feldhaus et al. 2004). Besides increasing library size, yeast mating can be used to effectively shuffle Fab heavy and light chains, creating new combinations for further affinity maturation and allowing for the selection of optimal V_H/V_L pairings. Furthermore, one or both of the polypeptide libraries can be diversified by common mutagenesis techniques to fine-tune the Fab antigen-binding affinity or for antibody humanization. In addition to affinity maturation, this yeast mating approach can be used to generate naïve libraries for isolating initial binders against an antigen target.

9.5 SCREENING YEAST-DISPLAYED ANTIBODY LIBRARIES

The choice of screening method (Fig. 9.5) is critically important for isolating improved antibodies from libraries of yeast-displayed mutants. As described below, there are several common techniques for isolating improved clones from antibody libraries using flow cytometry, magnetic particles, or mammalian cell-based selections.

9.5.1 FACS

Fluorescence-activated cell sorting (FACS) using labeled antigen enables quantitative screening of yeast-displayed antibody libraries by allowing the antigen-binding signal to be normalized for antibody display levels (Section 9.3). The elimination of binding artifacts due to differential expression between library clones allows for fine-affinity discrimination between mutants (Fig. 9.3; Chao et al. 2006; Kieke et al. 1997; VanAntwerp and Wittrup 2000). Moreover, FACS is an efficient method of screening yeast-displayed libraries; previous studies have shown that 100-fold enrichment ratios can be obtained after a single round of flow cytometry, even for mutants differing only twofold in affinity (VanAntwerp and Wittrup 2000).

The ability to measure yeast surface display levels by indirect immunofluorescent labeling of either the N- or C-terminal epitope tags is important for engineering antibodies for improved stability, since yeast cell surface expression is well correlated with thermal stability and soluble expression levels (Chao et al. 2006; Shusta et al. 1999). Furthermore, by using a two-color labeling scheme and two-channel flow cytometry in which one fluorophore tracks expression and the other antigen binding, directed evolution of enhanced stability and affinity can be conducted simultaneously. Finally, once library screening is complete, the antigen-binding affinity and thermal stability of the protein can be determined by titration experiments while the antibodies are still tethered to the yeast surface, eliminating the need for the expression and purification of individual clones (Chao et al. 2006) (Section 9.6.3).

However, the use of FACS for initial sorting rounds of large yeast-displayed libraries (theoretical diversity $> 10^7$ to 10^8) is limited due to technical considerations of the flow cytometer. A typical flow cytometer can sort 10^7 to 10^8 cells per hour. It is important to sort 10 times the library diversity,

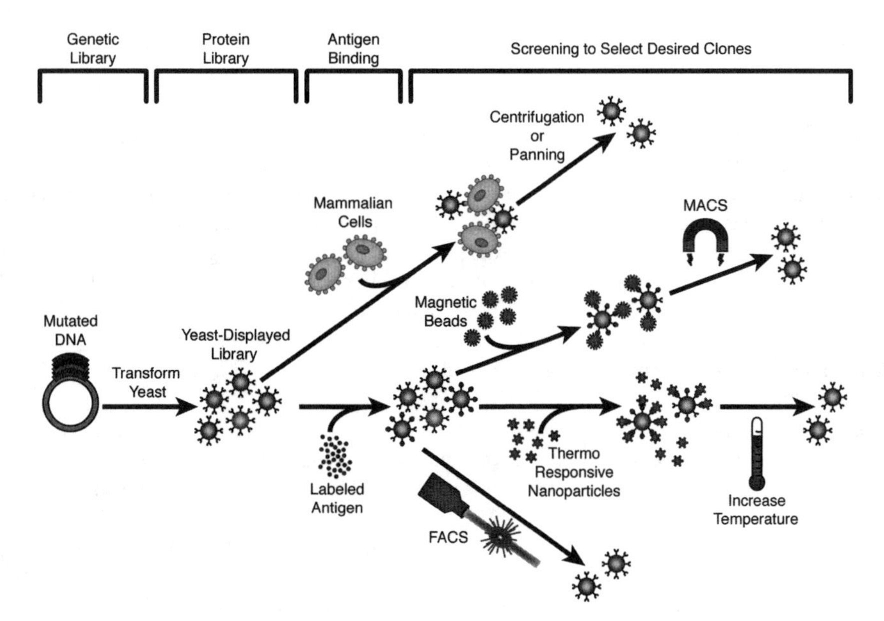

Figure 9.5 Schematic of methods used to screen yeast-displayed antibody libraries.

which statistically ensures that 90 percent of the clones will be present at least once in the sorting aliquot. Thus, it is often necessary to reduce the library size using other sorting methodologies before screening with FACS.

9.5.2 MACS

An alternate sorting method to FACS for large library sizes is magnetic-activated cell sorting (MACS). MACS provides a convenient and simple method to screen yeast-displayed libraries before using more quantitative sorting approaches. MACS can efficiently reduce cell number and library diversity while enriching the library several hundred fold in a few rounds of sorting (Siegel et al. 2004). Traditionally, streptavidin-coated magnetic beads are used to capture yeast cells displaying clones that bind to biotinylated target ligand. When used in combination with an equilibrium binding screen, single-pass enrichment of 10^4-fold was obtained; however, only 15 percent of binding clones were recovered (Yeung and Wittrup 2002). Upon subjecting the library to a kinetic screen, a single-pass enrichment of 600-fold was achieved, with a 25 percent probability of recovering binding clones (Yeung and Wittrup 2002).

A significant limitation of MACS is the high probability of losing binding clones, which must be compensated for by oversampling of library diversity. Compensation for typical recovery rates of binding clones on the order of 5 to 25 percent is of particular importance in early rounds of sorting where an individual clone is only represented between 1 and 100 times in the initial library aliquot (Feldhaus et al. 2003; Siegel et al. 2004; Yeung and Wittrup 2002). In addition to oversampling of library diversity, the impact of clonal loss can be minimized through optimization of practical parameters such as density of cell suspension, incubation time, and absence of a magnetic field during wash steps (Yeung and Wittrup 2002).

9.5.3 Mammalian Cell Surface Selection

While purified soluble proteins are most commonly used for screening processes in directed evolution, some antigens can be difficult to obtain, including transmembrane proteins, cell surface receptors, or targets existing in specific tissue types or diseased states (whose identity may not be known). To address this practical limitation, yeast displaying antibodies that bind to mammalian cell surface proteins can be rapidly selected as yeast-mammalian cell conjugates by both density centrifugation (Richman et al. 2006) and cell panning techniques (Wang and Shusta 2005).

9.5.3.1 Density Centrifugation For density centrifugation screens, yeast-displayed antibody libraries are incubated with mammalian cells expressing the target antigen on their surface. Yeast-displayed clones that bind to the target cell-surface antigen can be separated from nonbinding yeast on the basis of the density differential between yeast and mammalian cells. Yeast-mammalian cell conjugates are isolated by centrifugation through a discontinuous density gradient; these cell conjugates are held afloat by the larger buoyancy of mammalian cells, while unbound yeast form a pellet at the bottom of the tube (Richman et al. 2006). Remarkably, single-pass enrichments of 1000-fold can be achieved, making density centrifugation a rapid method for discriminating mutants with enhanced target binding affinity (Richman et al. 2006).

9.5.3.2 Biopanning An alternate cell-based selection strategy is to pan yeast-displayed antibody libraries against a monolayer of adherent mammalian cells (Wang and Shusta 2005; Wang, Cho, and Shusta 2007). The successful application of this strategy led to a three-round enrichment ratio of 10^6 using adherent cells expressing as few as 1700 ligands per cell (Wang and Shusta 2005). Yeast cells are negatively charged due to high concentrations of mannosylphosphate residues on the cell wall, and the strain used for yeast surface display is flocculin-deficient. These attributes lead to inherently low background binding in cell panning assays, giving good signal-to-noise ratios and allowing for rapid clonal enrichment (Wang and Shusta 2005).

Mammalian cell surface selection approaches, while offering less quantitative control than alternative methods and being limited to use only for target antigens expressed on the cell surface, have the advantage of presenting the target in a biologically relevant context. Additionally, sorting yeast-displayed libraries using mammalian cell surface selection methods allows selection of antibodies with high affinity for tissue-specific or tumor cell surface antigens (Richman et al. 2006; Wang and Shusta 2005; Wang, Cho, and Shusta 2007). However, care must be taken to preserve binding specificity, as the yeast-displayed library is exposed to all proteins present on the surface of the mammalian cells, and not just the target antigen.

9.5.4 Thermoresponsive Magnetic Particles

Magnetic particles greater than 1 μm cannot be used for automated sorting applications, since they do not readily disperse in solution. To address this, thermoresponsive magnetic nanoparticles that undergo a reversible transition between flocculation and dispersion around 30°C have been developed and used to screen a model yeast-displayed library (Furukawa et al. 2003). A yeast-displayed library of IgG binding proteins was sequentially incubated with biotinylated antibody, followed by soluble avidin and biotinylated magnetic nanoparticles. Yeast cells displaying the target protein were bound by the magnetic nanoparticles via a biotin-avidin-biotin bridge. Above the transition temperature, the nanoparticles are insoluble and flocculate, forming yeast-nanoparticle aggregates of sufficient size to allow for magnetic separation. Using this system, yeast cells displaying target protein with increased affinity were enriched from 0.001 to 100 percent after four rounds of sorting (Furukawa et al. 2003).

9.6 APPLICATIONS OF YEAST SURFACE DISPLAY

Yeast surface display has been used to engineer antibodies for a variety of applications. These engineered antibodies are being developed as therapeutics and analytical reagents, and are being used in bioseparation and epitope mapping techniques. Furthermore, the yeast strain *S. cerevisiae* has a "generally regarded as safe" status, meaning it can be used to solubly produce engineered antibodies for pharmaceutical and biotechnology applications (Kondo and Ueda 2004).

9.6.1 Therapeutics

Therapeutic antibodies offer many advantages over traditional small molecule drugs and are one of the fastest growing areas of the pharmaceutical industry. Currently, mAbs are used as clinical diagnostics and reagents, and play an important role in the treatment of cancer, autoimmune, cardiovascular, and infectious diseases (Nissim and Chernajovsky 2008). Potential clinical uses of engineered antibodies in therapeutics and diagnostics are vast and include immunotherapy, targeted drug or cytotoxin delivery, effector cell retargeting, and fluorescent and radioimaging (Rapley 1995). Given the long length of time from inception to translation for therapeutics and diagnostics, there are currently no scFvs approved by the U.S. Food and Drug Administration that were developed by yeast display. Antibodies engineered by phage display, an earlier established directed evolution platform, are just now making their way into clinical use. However, display technologies will be essential for the future development of antibody therapeutics and diagnostics with improved biophysical characteristics and clinical performance. Moreover, these methods can be used to engineer therapeutic antibodies to address current limitations on *in vivo* efficacy, such as induced allotypic immune response, limited tissue penetrance, and high clearance rates.

As illustrated above, yeast surface display is an efficient method for engineering clinically relevant single-chain antibody fragments, as well as hetero-oligomeric antibodies and antibody-like proteins, for high antigen-binding affinities and specificities and increased stability and soluble expression levels. Several additional examples applying yeast surface display to engineer antibodies as therapeutic agents are described below.

9.6.1.1 Engineered Antibodies for Specific Tumor Cell Targeting The epidermal growth factor receptor (EGFR) is a molecular target commonly overexpressed on tumor cells. Yeast surface display was used to increase the affinity of an scFv that binds to the EGFR and is subsequently internalized (Zhou et al. 2007). High-affinity engineered scFvs resulting from combined single mutations had affinities ranging from 1 to 10 nM, sufficiently high for therapeutic use. The engineered scFvs were then covalently attached to the surface of immunoliposome nanoparticles (ILs) containing fluorescent dye. Tumor cells overexpressing EGFR could specifically uptake these scFv-conjugated ILs, as measured by fluorescence microscopy (Fig. 9.6). In addition, it was found that scFv antibodies with affinities in the high nanomolar range were capable of effectively mediating the uptake of armed nanoparticles, due to avidity effects (Zhou et al. 2007). This technology could potentially be expanded to deliver therapeutic agents into the cytosol of tumor cells. Nanocarriers such as ILs could be armed with chemotherapeutic drugs or radioisotopes and targeted to specific cell types by incorporating engineered antibodies that are endocytosed after binding to surface receptors (Zhou et al. 2007).

Similarly, mAb AF-20, an antibody that binds a glycoprotein expressed on the surface of human hepatocellular carcinoma (HCC) cells and is internalized, was converted from a full-scale IgG format to an scFv format using yeast surface display (Mohr et al. 2004). The engineered AF-20 scFv retained the same binding specificity and internalization capabilities as mAb AF-20. Furthermore, the soluble AF-20 scFv produced in yeast was shown to bind to an HCC cell line and undergo internalization.

9.6.1.2 Engineered Antibodies for Intracellular Applications Another promising therapeutic application is the use of single domain intracellular antibodies (intrabodies) to target antigens normally found within the cell. Intrabodies can serve several potential therapeutic functions, such as the binding and inactivation of viral proteins and oncogenes. Additionally, intrabodies can be expressed within mammalian cells and directed to specific organelles, where they could potentially alter the folding, interaction, modification, or subcellular localization of their target antigens (Miller and Messer 2005). However, the efficacy of intrabodies is inherently limited due to their inability to form stabilizing disulfide bonds in the reducing environment of the cytoplasm. The lack of disulfide bonds, which typically contribute 4 to 5 kcal/mol to the lowest-energy protein conformation, results in reduced functional expression levels and a high propensity to aggregate (Worn and Pluckthun 2001). Moreover, the affinity and solubility of single domain antibodies are often reduced compared to other antibody fragments (Harmsen and De Haard 2007). To address these limitations, yeast surface display was used to engineer single domain antibody fragments (Fig. 9.2) devoid of disulfide bonds for use as therapeutic intrabodies. An engineered stable, high-affinity single domain intrabody against the huntingtin protein was capable of inhibiting huntingtin aggregation both *in vitro* and in a mammalian cell culture model of Huntington's disease (Fig. 9.6; Colby et al. 2004a, 2004b).

9.6.2 Epitope Mapping and Characterization

Yeast surface display has been used for antibody epitope mapping and characterization. These techniques, based on yeast cell surface expression and immunofluorescent staining, are considerably less time intensive than traditional approaches, which involve lengthy processes of soluble protein production and purification. Epitope mapping and characterization have important clinical relevance, as the efficacy of antibodies specific for therapeutic molecular targets is correlated with the location of interaction and ability to block ligand binding (Chowdhury and Wu 2005; Cochran et al. 2004). Additionally, identifying neutralizing epitopes on therapeutically relevant molecular targets is an important step in vaccine development (Levy et al. 2007).

9.6.2.1 Domain-Level Epitope Characterization Yeast surface display has been used to identify stable, functional protein domains and to characterize antibody-binding domains. This was first shown using EGFR as a model system (Cochran et al. 2004). In this study, EGFR ectodomain fragments were displayed on the surface of yeast, and were analyzed by flow cytometry using

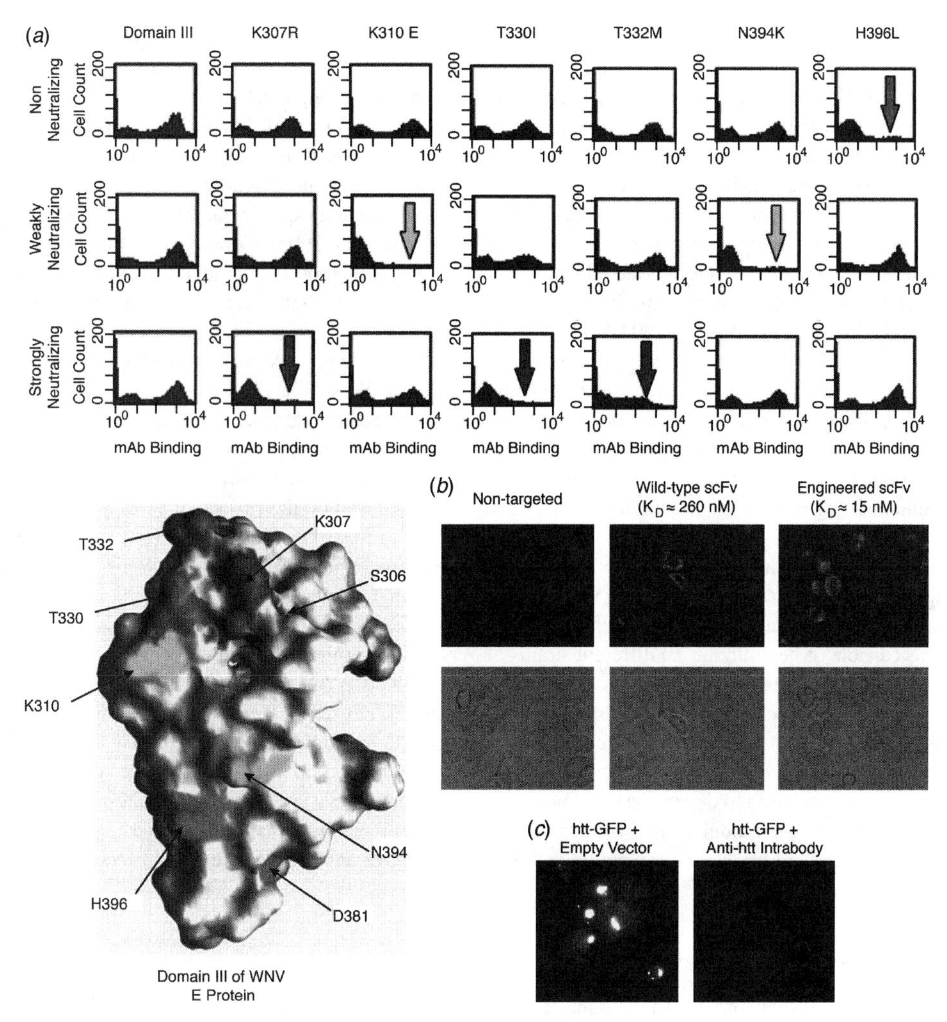

Figure 9.6 Antibody engineering applications of yeast surface display. (a) Fine epitope mapping of neutralizing and non-neutralizing monoclonal antibodies (mAb) against domain III of the West Nile virus E protein (WNV DIII). (Top) Flow cytometry histograms depict mAb immunoreactivity, with arrows highlighting point mutations that abolish binding of mAbs to yeast-displayed WNV DIII. (Bottom) Mapping of antibody binding epitopes on the surface of WNV DIII. From Oliphant et al. (2005). Reprinted by permission from Macmillan Publishers, Ltd. © 2005. (b) Tumor cell targeting and internalization is mediated by scFv-coated immunoliposomes (ILs). Wild-type and mutant anti-epidermal growth factor receptor scFvs attached to the surface of ILs loaded with fluorescent dye mediate internalization into tumor cells overexpressing the receptor. From Zhou et al. (2007). Reprinted with permission from Elsevier. © 2007. (c) An engineered single domain intrabody blocks intracellular huntingtin protein (htt) aggregation. Fluorescence microscopy images showing cells without (left) or treated with (right) the engineered anti-htt intrabody (Colby et al. 2004a). © 2004 National Academy of Sciences, USA. (See color insert.)

conformationally specific antibodies to discern properly folded EGFR domains from receptor truncations that resulted in aberrant protein folding. Domain-level antibody mapping was then performed, using the functionally displayed EGFR fragments to determine the receptor-binding location of several commercially available and clinically relevant antibodies. Next, heat denaturation studies were used to characterize antibody-binding epitopes as linear or conformational. Finally, antibodies were assayed for their ability to compete with ligand binding, and overlapping epitopes between antibodies were noted (Cochran et al. 2004).

9.6.2.2 Fine-Level Epitope Characterization Following domain-level epitope mapping of EGFR-specific antibodies, fine epitope mapping was performed using a yeast-displayed library based on a stable, functional EGFR ectodomain fragment (Chao, Cochran, and Wittrup 2004). Several therapeutically relevant antibodies were screened against this library to identify EGFR mutant fragments with decreased binding compared to the wild-type EGFR fragment. Sequence analysis and molecular modeling were used to propose the importance and potential roles of the mutated EGFR residues in antibody binding. Notably, these epitope mapping experiments have been confirmed in structural studies by other researchers (Johns et al. 2004; Li et al. 2005). Collectively, combinatorial library screening using yeast surface display provides a novel method for the identification of linear, discontinuous, and heat-denaturable antibody-binding epitopes with residue-specific resolution (Chao, Cochran, and Wittrup 2004).

Consistent with the methods applied to the model EGFR system, yeast surface display was used to identify contact residues of novel mAbs generated against the West Nile virus (WNV) (Oliphant et al. 2005). Briefly, domain III of the WNV E protein or the entire ectodomain was displayed on the surface of yeast. These WNV E proteins were used to identify mAbs that bound domain III, as convalescent antibodies from individuals who recovered from WNV infection were previously mapped to this domain. Next, a WNV E protein domain III library was displayed on the surface of yeast and labeled with select mAbs. Using flow cytometry, it was possible to identify point mutations that abolished binding of individual mAbs, thus allowing for the fine-level epitope mapping of WNV E protein domain III (Fig. 9.6). Interestingly, it was found that all strongly neutralizing mAbs, as well as the convalescent antibodies, mapped to an epitope on the lateral face of domain III. The best identified mAb was capable of neutralizing 10 different strains of WNV with significant therapeutic efficacy in mice, even when administered as a single dose five days after infection (Oliphant et al. 2005).

Similar approaches for fine and domain-level mapping were used to identify neutralizing epitopes on botulinum neurotoxin (BoNT), the most poisonous substance known (Levy et al. 2007). Individual domains of BoNT serotype A, and combinatorial libraries of these domains, were displayed on the surface of yeast and used to determine binding epitopes of human BoNT-specific mAbs. Epitope mapping results were used to model the interactions of three neutralizing antibodies and to propose a mechanism for the synergy observed on toxin neutralization when the three antibodies were administered together *in vivo* (Levy et al. 2007).

9.6.3 Biophysical Characterization

Yeast surface display provides a rapid means to quantitatively examine the antigen-binding affinity and thermal stability of individual antibody fragments, and engineered mutants. The approach is particularly advantageous, as alternative methods require soluble production, purification, and characterization of each protein to be analyzed, and thus are extremely limited in throughput.

Equilibrium binding and dissociation rate constants can be measured by flow cytometric detection of fluorescently labeled antigens while the antibodies are still tethered to the yeast cell surface. Binding constants obtained for yeast-displayed antibody fragments and antibody-like proteins are in close agreement with those determined by surface plasmon resonance or ELISA using soluble forms of the proteins (Gai and Wittrup 2007). In addition, the irreversible denaturation of yeast-displayed antibodies can be measured by flow cytometry (Orr et al. 2003). Thermal denaturation curves can be obtained by using fluorescently labeled antigens to detect the amount of functional antibody remaining after incubation over a range of temperatures. Using these methods, the relative antigen-binding affinities and thermal stabilities of yeast-displayed antibodies can be correlated with primary sequence information.

In some cases, antibody fragments can be tested directly for biological effects while still tethered to the yeast surface. This was demonstrated with an scFv engineered by yeast surface display to bind CD152, a T-cell protein, with high affinity (Griffin et al. 2001). The engineered high-affinity scFv was used in a novel immunotherapeutic strategy targeting T-cell activation in which surface-displayed scFvs were engineered to initiate signaling through individual T-cell proteins during cell–cell interactions.

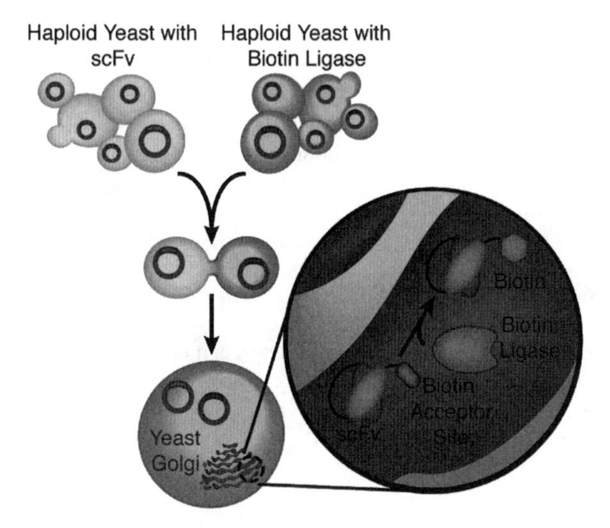

Figure 9.7 *In vivo* biotinylation of antibodies by yeast mating (Scholler et al. 2006). Haploid yeast populations of opposite mating type, one carrying a plasmid encoding a soluble scFv linked to a biotin acceptor peptide and the other with a plasmid encoding a Golgi-targeted biotin ligase enzyme, are fused. The resulting diploid yeast population is capable of secreting scFvs specifically biotinylated at their C-terminus.

9.6.4 Biotechnology

A novel technology has recently emerged that allows *in vivo* biotinylation of antibodies through yeast mating (Fig. 9.7; Scholler et al. 2006). Biotin has an extraordinarily high binding affinity and slow dissociation rate for streptavidin. Therefore, biotinylation is often used to label antibodies for diagnostic and biophysical applications. However, chemical methods for biotinylation nonspecifically modify surface-exposed lysine side chains, and can potentially disrupt residues important for antibody structure or function. To avoid this complication, yeast have been manipulated to perform *in vivo* biotinylation as a highly specific post-translational modification. ScFvs of interest are cloned into a yeast expression vector for soluble protein production. In this construct, the scFv is encoded in a cassette driven by a galactose-inducible promoter and is targeted for secretion by the inclusion of an alpha pre-pro leader sequence. The secreted scFv contains a hexahistidine tag at its carboxy terminus for protein purification, followed by an IgA1 hinge that connects it to a peptide substrate for biotin attachment through biotin ligase (BirA). Haploid yeast expressing the scFv-peptide fusion are then mated with haploid yeast of the opposite mating type engineered to contain Golgi-localized BirA enzyme, targeted to the yeast secretion pathway by a Kex2 leader sequence. This process results in diploid yeast capable of secreting biotinylated scFvs (termed *biobodies*). Biobodies obtained by this process retained nanomolar binding affinity for the target antigen. Moreover, the biotin tag provided practical functionality, allowing for immobilization, multimerization, and fluorescent labeling with streptavidin reagents (Scholler et al. 2006). Thus, this methodology allows efficient generation of biotin-functionalized antibodies for use in many biotechnology applications, such as flow cytometry, microscopy, and ELISA.

9.7 SUMMARY

Yeast display has accelerated the identification and engineering of antigen-specific antibodies and antibody-like proteins with high affinity, stability, and soluble expression yields. First, the use of a eukaryotic host offers the important advantage of post-translational processing and modification of mammalian proteins. Second, yeast-displayed libraries can be quantitatively screened by fluorescence activated cell sorting, providing unparalleled quantitative discrimination between clones and eliminating artifacts due to host expression biases. Third, antigen-binding properties, and epitope mapping and characterization of individual clones, can rapidly be performed by flow cytometry while the antibodies are still tethered to the yeast surface, eliminating the need for laborious soluble production and purification steps. These characteristics make yeast surface display an ideal platform for the directed evolution of antibodies and other proteins. Antibodies engineered through yeast surface display are being further developed for a myriad of clinical and biotechnology applications.

ACKNOWLEDGMENTS

We would like to thank Adam Silverman, Douglas Jones, and Ginger Chao for helpful feedback and comments on this manuscript. JLL has received support from a National Institutes of Health Interdisciplinary Regenerative Medicine training grant (T90 DK070103) and a California Breast Cancer Research Program fellowship (13GB-0161).

REFERENCES

Blaise, L., A. Wehnert, M.P. Steukers, T. van den Beucken, H.R. Hoogenboom, and S.E. Hufton. 2004. Construction and diversification of yeast cell surface displayed libraries by yeast mating: Application to the affinity maturation of Fab antibody fragments. *Gene* 342(2):211–218.

Boder, E.T., K.S. Midelfort, and K.D. Wittrup. 2000. Directed evolution of antibody fragments with monovalent femtomolar antigen-binding affinity. *Proc. Natl. Acad. Sci. USA* 97(20):10701–10705.

Boder, E.T., and K.D. Wittrup. 1997. Yeast surface display for screening combinatorial polypeptide libraries. *Nature Biotechnol.* 15(6):553–557.

Boder, E.T., and K.D. Wittrup. 1998. Optimal screening of surface-displayed polypeptide libraries. *Biotechnol. Prog.* 14(1):55–62.

Bowley, D.R., A.F. Labrijn, M.B. Zwick, and D.R. Burton. 2007. Antigen selection from an HIV-1 immune antibody library displayed on yeast yields many novel antibodies compared to selection from the same library displayed on phage. *Protein Eng. Des. Sel.* 20(2):81–90.

Cappellaro, C., C. Baldermann, R. Rachel, and W. Tanner. 1994. Mating type-specific cell-cell recognition of *Saccharomyces cerevisiae*: Cell wall attachment and active sites of a- and alpha-agglutinin. *EMBO J.* 13(20):4737–4744.

Chao, G., J.R. Cochran, and K.D. Wittrup. 2004. Fine epitope mapping of anti-epidermal growth factor receptor antibodies through random mutagenesis and yeast surface display. *J. Mol. Biol.* 342(2):539–550.

Chao, G., W.L. Lau, B.J. Hackel, S.L. Sazinsky, S.M. Lippow, and K.D. Wittrup. 2006. Isolating and engineering human antibodies using yeast surface display. *Nature Protoc.* 1(2):755–768.

Chowdhury, P.S., and H. Wu. 2005. Tailor-made antibody therapeutics. *Methods* 36(1):11–24.

Cochran, J.R., Y.S. Kim, S.M. Lippow, B. Rao, and K.D. Wittrup. 2006. Improved mutants from directed evolution are biased to orthologous substitutions. *Protein Eng. Des. Sel.* 19(6):245–253.

Cochran, J.R., Y.S. Kim, M.J. Olsen, R. Bhandari, and K.D. Wittrup. 2004. Domain-level antibody epitope mapping through yeast surface display of epidermal growth factor receptor fragments. *J. Immunol. Methods* 287(1–2):147–158.

Colby, D.W., Y. Chu, J.P. Cassady, et al. 2004a. Potent inhibition of huntingtin aggregation and cytotoxicity by a disulfide bond-free single-domain intracellular antibody. *Proc. Natl. Acad. Sci. USA* 101(51):17616–17621.

Colby, D.W., P. Garg, T. Holden, et al. 2004b. Development of a human light chain variable domain (V(L)) intracellular antibody specific for the amino terminus of huntingtin via yeast surface display. *J. Mol. Biol.* 342(3):901–912.

Colby, D.W., B.A. Kellogg, C.P. Graff, Y.A. Yeung, J.S. Swers, and K.D. Wittrup. 2004c. Engineering antibody affinity by yeast surface display. *Methods Enzymol.* 388:348–358.

Ellgaard, L., and A. Helenius. 2003. Quality control in the endoplasmic reticulum. *Nature Rev. Mol. Cell. Biol.* 4(3):181–191.

Feldhaus, M., and R. Siegel. 2004a. Flow cytometric screening of yeast surface display libraries. *Methods Mol. Biol.* 263:311–332.

Feldhaus, M.J., and R.W. Siegel. 2004b. Yeast display of antibody fragments: A discovery and characterization platform. *J. Immunol. Methods* 290(1–2):69–80.

Feldhaus, M.J., R.W. Siegel, L.K. Opresko, et al. 2003. Flow-cytometric isolation of human antibodies from a nonimmune *Saccharomyces cerevisiae* surface display library. *Nature Biotechnol.* 21(2):163–170.

Furukawa, H., R. Shimojyo, N. Ohnishi, H. Fukuda, and A. Kondo. 2003. Affinity selection of target cells from cell surface displayed libraries: A novel procedure using thermo-responsive magnetic nanoparticles. *Appl. Microbiol. Biotechnol.* 62(5–6):478–483.

Gai, S.A., and K.D. Wittrup. 2007. Yeast surface display for protein engineering and characterization. *Curr. Opin. Struct. Biol.* 17(4):467–473.

Garcia-Rodriguez, C., R. Levy, J.W. Arndt, et al. 2007. Molecular evolution of antibody cross-reactivity for two subtypes of type A botulinum neurotoxin. *Nature Biotechnol.* 25(1):107–116.

Graff, C.P., K. Chester, R. Begent, and K.D. Wittrup. 2004. Directed evolution of an anti-carcinoembryonic antigen scFv with a 4-day monovalent dissociation half-time at 37 degrees C. *Protein Eng. Des. Sel.* 17(4):293–304.

Graff, C.P., and K.D. Wittrup. 2003. Theoretical analysis of antibody targeting of tumor spheroids: Importance of dosage for penetration, and affinity for retention. *Cancer Res.* 63(6):1288–1296.

Griffin, M.D., P.O. Holman, Q. Tang, et al. 2001. Development and applications of surface-linked single chain antibodies against T-cell antigens. *J. Immunol. Methods* 248(1–2):77–90.

Hammond, C., and A. Helenius. 1995. Quality control in the secretory pathway. *Curr. Opin. Cell Biol.* 7(4):523–529.

Harmsen, M.M., and H.J. De Haard. 2007. Properties, production, and applications of camelid single-domain antibody fragments. *Appl. Microbiol. Biotechnol.* 77(1):13–22.

Holler, P.D., P.O. Holman, E.V. Shusta, S. O'Herrin, K.D. Wittrup, and D.M. Kranz. 2000. In vitro evolution of a T cell receptor with high affinity for peptide/MHC. *Proc. Natl. Acad. Sci. USA* 97(10):5387–5392.

Hoogenboom, H.R. 2005. Selecting and screening recombinant antibody libraries. *Nature Biotechnol.* 23(9):1105–1116.

Huang, D., and E.V. Shusta. 2005. Secretion and surface display of green fluorescent protein using the yeast *Saccharomyces cerevisiae. Biotechnol. Prog.* 21(2):349–357.

Huang, J., A. Koide, K.W. Nettle, G.L. Greene, and S. Koide. 2006. Conformation-specific affinity purification of proteins using engineered binding proteins: Application to the estrogen receptor. *Protein Expr. Purif.* 47(2):348–354.

Johns, T.G., T.E. Adams, J.R. Cochran, et al. 2004. Identification of the epitope for the epidermal growth factor receptor-specific monoclonal antibody 806 reveals that it preferentially recognizes an untethered form of the receptor. *J. Biol. Chem.* 279(29):30375–30384.

Karatan, E., M. Merguerian, Z. Han, M.D. Scholle, S. Koide, and B.K. Kay. 2004. Molecular recognition properties of FN3 monobodies that bind the Src SH3 domain. *Chem. Biol.* 11(6):835–844.

Kieke, M.C., B.K. Cho, E.T. Boder, D.M. Kranz, and K.D. Wittrup. 1997. Isolation of anti-T cell receptor scFv mutants by yeast surface display. *Protein Eng.* 10(11):1303–1310.

Kim, Y.S., R. Bhandari, J.R. Cochran, J. Kuriyan, and K.D. Wittrup. 2006. Directed evolution of the epidermal growth factor receptor extracellular domain for expression in yeast. *Proteins* 62(4):1026–1035.

Koide, A., S. Abbatiello, L. Rothgery, and S. Koide. 2002. Probing protein conformational changes in living cells by using designer binding proteins: Application to the estrogen receptor. *Proc. Natl. Acad. Sci. USA* 99(3):1253–1258.

Koide, A., C.W. Bailey, X. Huang, and S. Koide. 1998. The fibronectin type III domain as a scaffold for novel binding proteins. *J. Mol. Biol.* 284(4):1141–1151.

Koide, A., and S. Koide. 2007. Monobodies: Antibody mimics based on the scaffold of the fibronectin type III domain. *Methods Mol. Biol.* 352:95–109.

Kondo, A., and M. Ueda. 2004. Yeast cell-surface display: Applications of molecular display. *Appl. Microbiol. Biotechnol.* 64(1):28–40.

Kowalski, J.M., R.N. Parekh, J. Mao, and K.D. Wittrup. 1998. Protein folding stability can determine the efficiency of escape from endoplasmic reticulum quality control. *J. Biol. Chem.* 273(31):19453–19458.

Kowalski, J.M., R.N. Parekh, and K.D. Wittrup. 1998. Secretion efficiency in *Saccharomyces cerevisiae* of bovine pancreatic trypsin inhibitor mutants lacking disulfide bonds is correlated with thermodynamic stability. *Biochemistry* 37(5):1264–1273.

Lee, H.W., S.H. Lee, K.J. Park, J.S. Kim, M.H. Kwon, and Y.S. Kim. 2006. Construction and characterization of a pseudo-immune human antibody library using yeast surface display. *Biochem. Biophys. Res. Commun.* 346(3):896–903.

Levy, R., C.M. Forsyth, S.L. LaPorte, I.N. Geren, L.A. Smith, and J.D. Marks. 2007. Fine and domain-level epitope mapping of botulinum neurotoxin type A neutralizing antibodies by yeast surface display. *J. Mol. Biol.* 365(1):196–210.

Li, S., K.R. Schmitz, P.D. Jeffrey, J.J. Wiltzius, P. Kussie, and K.M. Ferguson. 2005. Structural basis for inhibition of the epidermal growth factor receptor by cetuximab. *Cancer Cell* 7(4):301–311.

Lin, Y., T. Tsumuraya, T. Wakabayashi, et al. 2003. Display of a functional hetero-oligomeric catalytic antibody on the yeast cell surface. *Appl. Microbiol. Biotechnol.* 62(2–3):226–232.

Lipovsek, D., E. Antipov, K.A. Armstrong, et al. 2007a. Selection of horseradish peroxidase variants with enhanced enantioselectivity by yeast surface display. *Chem. Biol.* 14(10):1176–1185.

Lipovsek, D., S.M. Lippow, B.J. Hackel, et al. 2007b. Evolution of an interloop disulfide bond in high-affinity antibody mimics based on fibronectin type III domain and selected by yeast surface display: Molecular convergence with single-domain camelid and shark antibodies. *J. Mol. Biol.* 368(4):1024–1041.

Ma, H., S. Kunes, P.J. Schatz, and D. Botstein. 1987. Plasmid construction by homologous recombination in yeast. *Gene* 58(2–3):201–216.

Midelfort, K.S., H.H. Hernandez, S.M. Lippow, B. Tidor, C.L. Drennan, and K.D. Wittrup. 2004. Substantial energetic improvement with minimal structural perturbation in a high affinity mutant antibody. *J. Mol. Biol.* 343(3):685–701.

Midelfort, K.S., and K.D. Wittrup. 2006. Context-dependent mutations predominate in an engineered high-affinity single chain antibody fragment. *Protein Sci.* 15(2):324–334.

Miller, T.W., and A. Messer. 2005. Intrabody applications in neurological disorders: Progress and future prospects. *Mol. Ther.* 12(3):394–401.

Mohr, L., A. Yeung, C. Aloman, D. Wittrup, and J.R. Wands. 2004. Antibody-directed therapy for human hepatocellular carcinoma. *Gastroenterology* 127(5, Suppl 1):S225–231.

Nissim, A., and Y. Chernajovsky. 2008. Historical development of monoclonal antibody therapeutics. *Handb. Exp. Pharmacol.* 181:3–18.

Oliphant, T., M. Engle, G.E. Nybakken, et al. 2005. Development of a humanized monoclonal antibody with therapeutic potential against West Nile virus. *Nature Med.* 11(5):522–530.

Orr, B.A., L.M. Carr, K.D. Wittrup, E.J. Roy, and D.M. Kranz. 2003. Rapid method for measuring ScFv thermal stability by yeast surface display. *Biotechnol. Prog.* 19(2):631–638.

Park, S., Y. Xu, X.F. Stowell, F. Gai, J.G. Saven, and E.T. Boder. 2006. Limitations of yeast surface display in engineering proteins of high thermostability. *Protein Eng. Des. Sel.* 19(5):211–217.

Parker, M.H., Y. Chen, F. Danehy, et al. 2005. Antibody mimics based on human fibronectin type three domain engineered for thermostability and high-affinity binding to vascular endothelial growth factor receptor two. *Protein Eng. Des. Sel.* 18(9):435–444.

Piatesi, A., S.W. Howland, J.A. Rakestraw, et al. 2006. Directed evolution for improved secretion of cancer-testis antigen NY-ESO-1 from yeast. *Protein Expr. Purif.* 48(2):232–242.

Proba, K., A. Worn, A. Honegger, and A. Pluckthun. 1998. Antibody scFv fragments without disulfide bonds made by molecular evolution. *J. Mol. Biol.* 275(2):245–253.

Rakestraw, A., and K.D. Wittrup. 2006. Contrasting secretory processing of simultaneously expressed heterologous proteins in *Saccharomyces cerevisiae*. *Biotechnol. Bioeng.* 93(5):896–905.

Rapley, R. 1995. The biotechnology and applications of antibody engineering. *Mol. Biotechnol.* 3(2):139–154.

Razai, A., C. Garcia-Rodriguez, J. Lou, et al. 2005. Molecular evolution of antibody affinity for sensitive detection of botulinum neurotoxin type A. *J. Mol. Biol.* 351(1):158–169.

Richman, S.A., S.J. Healan, K.S. Weber, et al. 2006. Development of a novel strategy for engineering high-affinity proteins by yeast display. *Protein Eng. Des. Sel.* 19(6):255–264.

Sblattero, D., and A. Bradbury. 2000. Exploiting recombination in single bacteria to make large phage antibody libraries. *Nature Biotechnol.* 18(1):75–80.

Scholler, N., B. Garvik, T. Quarles, S. Jiang, and N. Urban. 2006. Method for generation of in vivo biotinylated recombinant antibodies by yeast mating. *J. Immunol. Methods* 317(1–2):132–143.

Schweickhardt, R.L., X. Jiang, L.M. Garone, and W.H. Brondyk. 2003. Structure-expression relationship of tumor necrosis factor receptor mutants that increase expression. *J. Biol. Chem.* 278(31):28961–28967.

Shusta, E.V., P.D. Holler, M.C. Kieke, D.M. Kranz, and K.D. Wittrup. 2000. Directed evolution of a stable scaffold for T-cell receptor engineering. *Nature Biotechnol.* 18(7):754–759.

Shusta, E.V., M.C. Kieke, E. Parke, D.M. Kranz, and K.D. Wittrup. 1999. Yeast polypeptide fusion surface display levels predict thermal stability and soluble secretion efficiency. *J. Mol. Biol.* 292(5):949–956.

Shusta, E.V., R.T. Raines, A. Pluckthun, and K.D. Wittrup. 1998. Increasing the secretory capacity of *Saccharomyces cerevisiae* for production of single-chain antibody fragments. *Nature Biotechnol.* 16(8):773–777.

Siegel, R.W., J.R. Coleman, K.D. Miller, and M.J. Feldhaus. 2004. High efficiency recovery and epitope-specific sorting of an scFv yeast display library. *J. Immunol. Methods* 286(1–2):141–153.

Skerra, A. 2007. Alternative non-antibody scaffolds for molecular recognition. *Curr. Opin. Biotechnol.* 18(4):295–304.

Swers, J.S., B.A. Kellogg, and K.D. Wittrup. 2004. Shuffled antibody libraries created by in vivo homologous recombination and yeast surface display. *Nucleic Acids Res.* 32(3): e36.

VanAntwerp, J.J., and K.D. Wittrup. 2000. Fine affinity discrimination by yeast surface display and flow cytometry. *Biotechnol, Prog.* 16(1):31–37.

van den Beucken, T., H. Pieters, M. Steukers, et al. 2003. Affinity maturation of Fab antibody fragments by fluorescent-activated cell sorting of yeast-displayed libraries. *FEBS Lett.* 546(2–3):288–294.

van der Linden, R.H., B. de Geus, G.J. Frenken, H. Peters, and C.T. Verrips. 2000. Improved production and function of llama heavy chain antibody fragments by molecular evolution. *J. Biotechnol.* 80(3):261–270.

Wang, X.X., Y.K. Cho, and E.V. Shusta. 2007. Mining a yeast library for brain endothelial cell-binding antibodies. *Nature Methods* 4(2):143–145.

Wang, X.X., and E.V. Shusta. 2005. The use of scFv-displaying yeast in mammalian cell surface selections. *J. Immunol. Methods* 304(1–2):30–42.

Waterhouse, P., A.D. Griffiths, K.S. Johnson, and G. Winter. 1993. Combinatorial infection and in vivo recombination: A strategy for making large phage antibody repertoires. *Nucleic Acids Res.* 21(9):2265–2266.

Weaver-Feldhaus, J.M., J. Lou, J.R. Coleman, R.W. Siegel, J.D. Marks, and M.J. Feldhaus. 2004. Yeast mating for combinatorial Fab library generation and surface display. *FEBS Lett.* 564(1–2):24–34.

Weaver-Feldhaus, J.M., K.D. Miller, M.J. Feldhaus, and R.W. Siegel. 2005. Directed evolution for the development of conformation-specific affinity reagents using yeast display. *Protein Eng. Des. Sel.* 18(11):527–536.

Worn, A., A. Auf der Maur, D. Escher, A. Honegger, A. Barberis, and A. Pluckthun. 2000. Correlation between in vitro stability and in vivo performance of anti-GCN4 intrabodies as cytoplasmic inhibitors. *J. Biol. Chem.* 275(4):2795–2803.

Worn, A., and A. Pluckthun. 2001. Stability engineering of antibody single-chain Fv fragments. *J. Mol. Biol.* 305(5):989–1010.

Xu, L., P. Aha, K. Gu, et al. 2002. Directed evolution of high-affinity antibody mimics using mRNA display. *Chem. Biol.* 9(8):933–942.

Yeung, Y.A., and K.D. Wittrup. 2002. Quantitative screening of yeast surface-displayed polypeptide libraries by magnetic bead capture. *Biotechnol. Prog.* 18(2):212–220.

Zhou, Y., D.C. Drummond, H. Zou, et al. 2007. Impact of single-chain Fv antibody fragment affinity on nanoparticle targeting of epidermal growth factor receptor-expressing tumor cells. *J. Mol. Biol.* 371(4):934–947.

CHAPTER 10

Ribosomal Display

GEORGE THOM

ABSTRACT

Ribosome display is a powerful *in vitro* technology for the selection and directed evolution of proteins. This technology exploits cell-free translation to achieve coupling of phenotype and genotype by the production of stabilized ribosome complexes whereby translated proteins and their cognate mRNA remain attached to the ribosome. The *Escherichia coli* S30 extract for *in vitro* display of an mRNA library has proven to be very successful for the evolution of high affinity antibodies and the optimization of defined protein characteristics. However, this technology has so far been perceived as being technically challenging due to comparatively difficult protocols as well as the absence of tailored commercial reagents, particularly when using prokaryotic cell-free expression systems. This technical update, along with providing the S30 methods, also outlines new advances in prokaryotic and eukaryotic ribosome display technologies that make them more easily accessible for the end user.

Therapeutic Monoclonal Antibodies: From Bench to Clinic. Edited by Zhiqiang An

10.1 INTRODUCTION

Ribosome display is a powerful *in vitro* technology for the selection of proteins with specific function and as such has been demonstrated for the discovery of high affinity monoclonal antibodies and peptides (Hanes et al. 1998, 2000; Matthaekis, Bhatt, and Dower 1994; Lamla and Erdmann 2001; Takahashi, Ebihara, and Mie 2002; Thom et al. 2006) and the optimization of defined protein attributes by directed evolution (Hanes et al. 2000; Jermutus et al. 2001; Amstutz, Pelletier, and Guggisberg 2002; Matsuura and Plückthun 2004; Zahnd et al. 2007).

Ribosome display is a completely *in vitro* display system and offers advantages over other selection technologies, for example, phage or yeast display, that have limited library sizes (10^7 to 10^{10} per microgram DNA; Dower and Cwirla 1992) due to the *in vitro* transformation steps required for these technologies. The successful selection of antibodies from displayed libraries relies heavily on the size of the library being used. In general, the probability of finding a given antibody becomes higher as the size of the library increases (Vaughan et al. 1996). Ribosome display allows selection from much larger libraries (10^{14}) and has the added advantage of permitting Darwinian evolution of the displayed protein due to introduction of polymerase chain reaction (PCR)-based mutations in each selection cycle.

The ribosome display process contains a series of steps that may be performed in an iterative manor as required (Fig. 10.1). First, mRNA is transcribed *in vitro* from DNA encoding a ligand-binding molecule or a library of molecules. This is then transcribed *in vitro* under conditions to produce stable tertiary mRNA-ribosome-protein complexes (ribosomal complexes) (Fig. 10.2) to achieve coupling of phenotype and genotype via a noncovalent link. Recovery of the selected mRNA is achieved by dissociation of ribosomal complexes, and subsequent reverse transcription (RT) and PCR to generate DNA for subsequent rounds of selection or for expression and screening. Recovery of sequence in

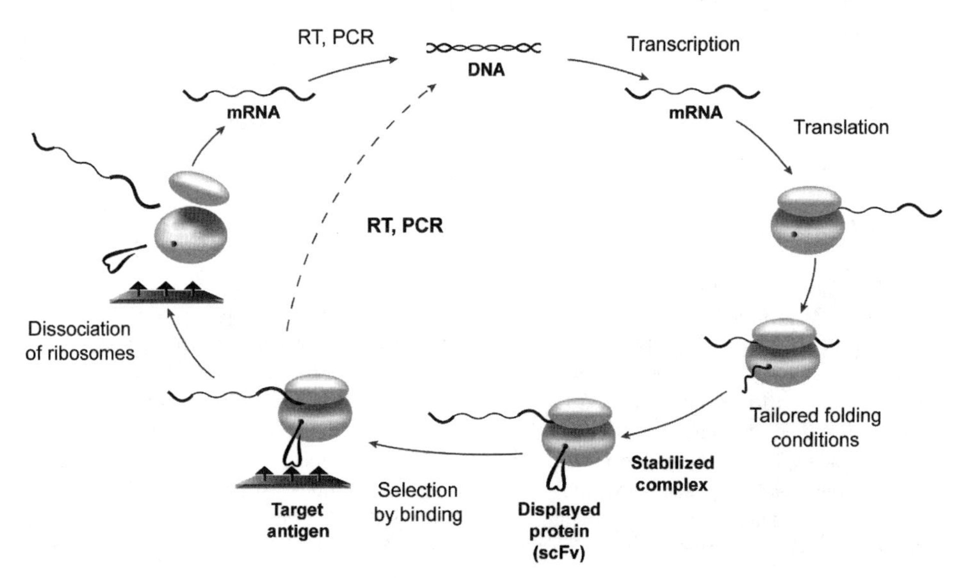

Figure 10.1 The ribosome display cycle. A library containing all features necessary for ribosome display is transcribed to mNRA. After its purification, mNRA is translated *in vitro*. Translation is stopped by cooling on water-ice, and the ribosome tertiary complexes are stabilized by increasing the magnesium concentration. Ribosomal complexes are affinity selected from the translation mixture, this can be via immobilized ligand or alternatively in solution with biotinylated ligand. Nonspecific ribosome complexes are removed by extensive washing, and mRNA is isolated from the bound ribosome complexes, reverse transcribed to cDNA, and then amplified by a separate PCR reaction. Processing of selected ribosome complexes directly into RT-PCR is also illustrated.

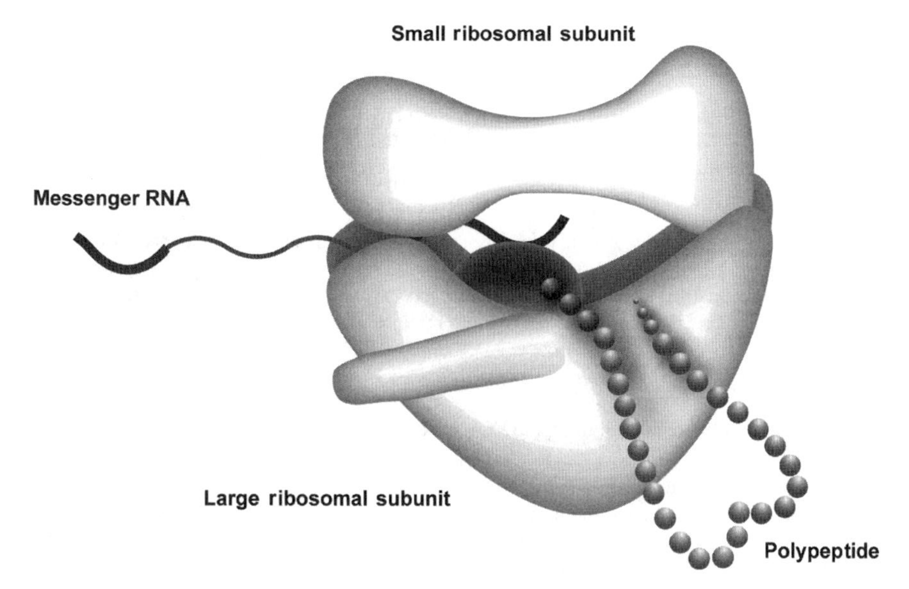

Figure 10.2 The stable tertiary mRNA-ribosome-protein complexes ("ribosomal complexes"). The two subunits of a translating ribosome have stalled at the C-terminus of the piece of mRNA with its cognate polypeptide remaining attached, achieving coupling of phenotype and genotype via a noncovalent link.

this way does not require the breakdown of the interaction between ligand and its binding partner; therefore, higher affinity interactions can be selected more efficiently, thus conferring an additional advantage of ribosome display over other *in vitro* display technologies, such as covalent mRNA display.

Current ribosome display systems that are well proven, by the evolution of high affinity antibodies and the optimization of defined protein characteristics, use an *E. coli* S30 cell extract (indicating the soluble fraction after centrifugation at 30,000 *g*) for *in vitro* translation and display of an mRNA library. This is a relatively crude cell extract that contains all enzymes and factors required for translation (Zubay 1973). Cell extracts suffer from a number of problems that limit their activity and overall protein yield compared to the rate of translation *in vivo*, the most significant of these being rapid energy depletion and mRNA degradation. Current ribosome display protocols overcome the problem of high endogenous RNase content in prokaryotic extracts, for example by the use of partially RNase-deficient *E. coli* strains, inclusion of RNase inhibitors, processing at low temperature, and by the addition of stem-loop sequences in mRNA templates (Jermutus, Ryabova, and Plückthun 1998).

A reconstituted cell-free translation system has been produced by combining recombinant *E. coli* protein factors with purified 70S ribosomes (Shimizu et al. 2001; Shimizu, Kanamori, and Ueda 2005). Soluble protein factors necessary for prokaryotic translation are prepared by overexpression in *E. coli* and subsequent purification via 6x histidine (HIS) tag. When combined with a highly purified 70S ribosome preparation and other requirements for translation, efficient cell-free synthesis of various proteins was demonstrated (Shimizu et al. 2001). This system has also been successful for cell-free expression and folding of a single chain antibody fragment (scFv), the solubility and function of which was improved when chaperone proteins were included (Ying et al. 2004). Such a reconstituted translation system is very attractive for cell-free translation-based display technologies, such as ribosome display, since nuclease and protease enzymes that are inherent in traditional cell extracts are present at very low levels (Ohashi et al. 2007). Using ribosome display in a reconstituted translation system may achieve higher display and recovery levels than present methods due to reduced degradation of mRNA, displayed protein, and target protein. Therefore, these technologies were combined (Villemagne, Jackson, and Douthwaite 2006) to improve the outcome of a ribosome display selection

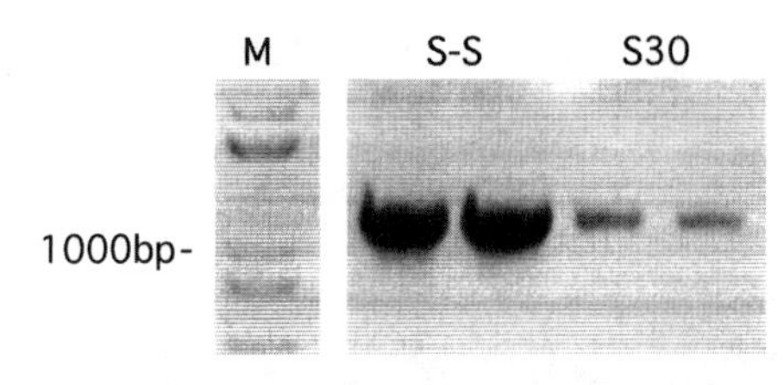

Figure 10.3 PUREsystem naïve selection. Ribosome display selection of a naïve scFv library performed using a reconstituted translation system (PUREsystem S-S) or an *E. coli* S30 extract for the generation of stabilized ribosome complexes; 100 nM biotinylated insulin was used as a model target antigen for selection. Agarose gel electrophoresis of reamplified duplicate first round selection outputs prior to a second round of ribosome display is shown. M is 1 kB^+ DNA ladder (Villemagne, Jackson, and Douthwaite 2006).

and this research demonstrated that higher cDNA yields were recovered from ribosome display selections (Fig. 10.3).

The use of a eukaryotic translation system for generation of stable ribosome complexes for selection has also been described and these methods propose an alternative mechanism of sequence recovery, where cDNA is recovered from mRNA without disruption of the ribosome complex (He and Taussig 1997, 2003, 2005). The eukaryotic system is simpler to set up and run due to the commercial availability of suitable *in vitro* translation reagents. In this report, various aspects of the eukaryotic ribosome display protocol have been altered, focusing attention on the proposed mechanism of mRNA recovery. Real-time PCR analysis has been used to determine relative efficiencies of ribosome complex disruption at the cDNA recovery stage. An alternative method was developed (Douthwaite et al. 2006) and is described in Section 10.3.6.2.

10.2 MATERIALS

Plasmid midiprep kit (Qiagen)
Qiaquick gel extraction kit (Qiagen)
2x PCR master mix (ABgene)
Nuclease-free water (Promega)
SYBRsafe stain (Invitrogen)
Ribomax large scale mRNA production kit (Promega)
RNase-free microcentrifuge tubes 1.5/0.5 mL (Ambion)
Superscript II reverse transcription kit (Invitrogen)
Probequant G50 microcolumn (Amersham Biosciences)
Complete amino acid (Sigma)
Snakeskin pleated dialysis tubing 7,000 MWCO (Pierce)
PUREsystem (Post Genome Institute)
PD-10 columns (GE Healthcare)
M280 Steptavidin beads (Dynal)
Protein disulfide isomerase (Fluka)
Magnetic device Dynal MPC-S (Dynal)
S. cerevisiae RNA (Roche)
High pure mRNA isolation kit (Roche)
RNase-free DNase I (Boehringer Mannheim)

Immunosorb 96-well plates (Nunc)
Dulbecco's PBS (Gibco)
RNasin (Promega)
Ultrapure agarose (Gibco)
ABI 7700 (Applied Biosystems)
2x ABI Universal master mix (Applied Biosystems)
Flexi rabbit reticulocyte lysate (Promega)
1 kB$^+$ DNA ladder (Invitrogen)

All other chemicals used are molecular biology grade and supplied by Sigma.

10.3 METHODS

10.3.1 Ribosome Display Constructs

The DNA construct used for ribosome display and its assembly by PCR have been described in detail previously (Hanes, Jermutus, and Plückthun 2000). On the DNA level, the construct requires a T7 RNA polymerase promoter for strong *in vitro* transcription to generate mRNA. On the mRNA level, the construct possesses a ribosome-binding site (Shine-Dalgarno sequence or Kozak sequence) for prokaryotic or eukaryotic translational initiation, respectively. Immediately following is an open reading frame encoding the protein library without a stop codon, to avoid dissociation of synthesized peptide and its cognate mRNA from the ribosome. The protein of interest is followed by a spacer/tether sequence fused in-frame to the protein. The mRNA contains 5′ and 3′ stem loop structures for increased mRNA stability against RNases. The presence of stem loops is important, especially in the *E. coli* ribosome display system, because at least 5 of the 20 or so *E. coli* RNases have been shown to contribute to mRNA degradation (Hajnsdorf et al. 1996), and they are most likely all present in the S30 extract used for *in vitro* translation. In one of the most frequently used ribosome display constructs the 5′-untranslated region of the mRNA, including the 5′ stem loop and the ribosome-binding site, is derived from gene 10 of phage T7 (Studier et al. 1990) and the 3′ stem loop is derived from the early terminator of phage T3 (Reynolds, Bermudez-Cruz, and Chamberlin 1992). One successful spacer/tether sequence is derived from the geneIII sequence of filamentous phage M13, spanning amino acids 130–204 (SwissProt P03622). Alternatively, spacers derived from other *E. coli* genes can be used. The spacer is essentially used to provide an unstructured portion at the C-terminus of the protein to be folded, such that the ribosomal tunnel can cover at least 20 to 30 amino acid residues of the emerging polypeptide without interfering with the folding of the protein of interest.

10.3.2 General Principles before Starting Experimental Work

(1) Ensure that all consumables are RNase free.
(2) Avoid sharing of reagents and equipment with scientists working on other ribosome display constructs.
(3) Use filter pipette tips at all times.
(4) Clean working area with detergent and 70 percent ethanol and irradiate pipettes with UV before commencing experiments each day.
(5) Prepare all PCR reactions in a clean environment to prevent DNA contamination.
(6) Always grow clones from different samples in separate plates.
(7) Perform inoculation and incubations of untransformed cells in defined areas, preferably in a laminar flow hood.

(8) Leave one lane free between samples when running PCR products on agarose gels.
(9) Use separate blades for each band being excised from an agarose gel.
(10) For gel extraction, always spin down the dissolved band before opening the eppendorf lid.
(11) Use Tris-based buffers to elute DNA from purification columns for greater long-term storage stability (e.g., 10 mM Tris, pH 8.0).

10.3.3 Generation of Ribosome Display Construct

Ribosome display constructs are easily prepared, completely *in vitro*, by assembly PCR. ScFv coding regions contained within the N*co*I/N*ot*I sites of the plasmid midiprep (Qiagen) preparation of pCANTAB6 (McCafferty et al. 1994) are amplified using the forward primer SD (AGACCACAACGGTTTCCCTCTAGAAATAATTTTGTTTAACTTTAAGAAGGAGATATATCC **ATG**NNNNNNNNN, where N is scFv-specific sequence, the underlined region is the Shine-Dalgarno sequence and the (**ATG**) in bold are the start codon at the beginning of the scFv coding region) or T7KOZ (GCAGCTAATACGACTCACTATAGGAACAGACCACC**ATG**NNNNNNNNN where N is scFv-specific sequence and the underlined regions are T7 RNA polymerase promoter and Kozak sequence), for prokaryotic and eukaryotic ribosome display, respectively, and the reverse primer MycR (ATTCAGATCCTCTTCTGAGATGAG). A separate PCR reaction using the forward primer MycG3 (ATCTCAGAAGAGGATCTGAATGGTGGCGGCTCCGGTTCCGGTGAT) and reverse primer G3rev (CCGTCACCGACTTGAGCC) is performed using pCANTAB6 as template to amplify a region of the geneIII coding sequence for use as the ribosome display spacer/tether.

Set up the following PCR reaction to either amplify the scFv region or geneIII from pCANTAB6. Also include a no template control. Each reaction should be performed in a 100 μL volume, containing the following reagents: 50 μL 2x PCR master mix; 2 μL SD, T7KOZ or MycG3 (10 μM); 2 μL MycR or G3rev (10 μM); 1 μL pCANTAB6 (~10 μg/mL); 45 μL nuclease-free water. Sequences are amplified using the following PCR conditions: 94°C 3 min; (94°C 30 sec, 55°C 30 sec, 72°C 105 sec) 25 cycles; 72°C 5 min and 10°C hold. PCR products are separated by agarose gel electrophoresis on a 1 percent agarose gel containing SYBRSafe stain (from ×10,000 stock solution). This stain is light sensitive and must be visualized under blue light. Subsequently, the correct DNA band is excised from the gel and DNA purified from the matrix via QIAquick Gel Extraction Kit. The DNA is eluted into elution buffer (containing 10 mM Tris/HCl pH 8.0) or nuclease-free water. The two purified PCR products are assembled by overlap PCR and reamplified with the forward primer T7B (ATACGA TAATACGACTCACTATAGGGAGACCACAACGG, underlined region is T7 RNA polymerase promoter) or T7KOZ, for prokaryotic and eukaryotic ribosome display respectively, and the reverse primer RT-1 (CCGCACACCAGTAAGGTGTGCGGTATCACCAGTAGCACCATTACCA TTAGCAAG).

Set up the following PCR reaction to assemble the scFv coding sequence with the geneIII tether and amplify the full-length construct. Also include a no template control. Each reaction should contain the following: 50 μL 2x PCR master mix; X μL gel-purified SD or T7KOZ/MycR product; Y μL gel-purified geneIII tether; Z μL nuclease-free water. X and Y volumes are added so that the DNA is in a 1 : 1 molar ratio. Z represents the volume of water required to make a final volume of 100 μL.

Sequences are amplified using the following PCR conditions: 94°C 3 min (94°C 30 sec, 50°C 30 sec, 72°C 105 sec) 5 cycles. During the fifth annealing step, that is, at 50°C: Pause the PCR block, open the tubes, and add 2 μL of mixed primers T7B or T7KOZ and RT-1 (at 10 μM each). Close the PCR tubes again and resume the program as follows: (94°C 30 sec, 35°C 30 sec, 72°C 105 sec) 3 cycles; (94°C 30 sec, 50°C 30 sec, 72°C 105 sec) 15 cycles; 72°C 5 min; 10°C hold.

PCR products are used directly for *in vitro* transcription of mRNA (Note: do not purify PCR products before use in the transcription reaction).

Libraries can also be built from this initial template DNA. These libraries can either be created for directed evolution, for example, NNS libraries directed to the complementarity determining regions (CDRs) of the variable region of an IgG or a binding region of any protein wishing to be displayed. Alternatively, libraries can be generated in a nondirected fashion, for example, using error prone

PCR and/or non-proof reading Taq polymerase during PCR steps of the process. These libraries are then selected to identify clones that have favorable specific binding characteristics for a particular antigen. These selections are usually, although not exclusively, affinity selections and are used to isolate clones with improved kinetics towards the antigen.

10.3.4 *In Vitro* Transcription of mRNA

DNA templates are prepared as PCR products for mRNA production. It is essential, to obtain the best ribosome display performance, that the quality of the PCR product used for transcription be extremely high. The DNA sample should contain a single PCR product and the DNA should have a concentration of at least 50 ng/μL. mRNA is prepared by T7 RNA polymerase directed *in vitro* transcription using the RiboMax Large Scale mRNA Production kit.

Assemble the transcription reaction as follows in a nonstick, RNAse-free microcentrifuge tube at room temperature in the order listed. If performing multiple reactions, a master mix can be prepared. Each sample should have a final reaction volume of 50 μL: 10 μL 5x Transcription Buffer, 15 μL rNTP mix (ATP, CTP, GTP, and UTP), 25 mM each; 20 μL linear DNA template (PCR reaction) and 5 μL T7 enzyme mix.

Mix well by pipetting or gentle vortexing and pulse spin to ensure all contents are at the bottom of the tube. Incubate at 37°C for at least 2 h and no more than 2.5 h. Pulse spin to collect condensation from the tube lid. Purify mRNA using a ProbeQuant G50 microcolumn according to manufacturer's instructions: Vortex column to resuspend the matrix. Loosen cap, add to a 2 ml RNase free tube, spin 1 min, 3000 rpm to remove storage buffer. Discard flow-through. Add 50 μL transcription reaction per column. Loosen cap, spin 2 min at 3000 rpm into fresh 1.5 mL RNAse-free microcentrifuge tube. Place mRNA on ice and quantify by measuring OD A_{260} in a spectrophotometer of 1/150 dilution in duplicate. If the OD A_{260} reading does not lie within the range 0.1 to 1.0 then repeat the analysis with a more suitable dilution. If using immediately, then dilute to 1 μg/μL ready for translation (alternatively aliquot mRNA, snap-freeze and store at −80°C *undiluted* and prepare dilution when required). For successful ribosome display, the mRNA quality is important. This can be checked by agarose gel electrophoresis and the sample should contain at least 90 percent full-length mRNA.

10.3.5 S30 *E. coli* Extract Preparation

Prokaryotic *in vitro* translation is performed by using an S-30 *E. coli* extract, prepared from the strain MRE600 by a modified protocol of Lesley (1995); based on the procedure of Chen and Zubay (1983). This protocol for the preparation of S-30 *E. coli* extract differs from the published protocols by the fact that DTT, 2-mercaptoenthanol, and *p*-toluene sulfonyl fluoride are omitted from all of the solutions. This is done to allow for the screening of libraries of proteins that contain disulfide bonds. However, if disulfide bonds are not present in the protein to be displayed, then the extract can be prepared in the presence of DTT and 2-mercaptoethanol. The quality of the S-30 extract being produced is of utmost importance for the efficiency of ribosome display and to that end the cells must be harvested during the exponential growth phase, where active ribosomes are at their most abundant. The cells should not be harvested at OD A_{600} higher than 0.6 when extracts are being produced from flask cultures (OD absorbance reading can be higher for fermenter cultures, for example OD A_{600} of 1.0 to 3.0).

A 100 mL *E. coli* MRE600 starter culture is grown overnight at 37°C in growth media (5.6 g of KH_2PO_4 [anhydrous], 28.9 g of K_2HPO_4 [anhydrous], 10 g yeast extract, 15 mg of thiamine, and 25 mL of 40 percent (w/v) glucose [added after separate sterilization]: per liter of distilled water) with shaking at 200 rpm. The following day, inoculate 500 mL of growth medium in a 2.5 L shaker flask with the 10 mL overnight culture and grow at 37°C with shaking at 250 rpm. Harvest the cells at OD A_{600} of 0.6 to 0.8 (early exponential growth phase) by centrifugation at 3500 *g* for 15 min at 4°C. Discard the supernatant and wash the pellet three times with 50 mL ice-cold S30 buffer (10 mM Tris-acetate [pH 7.5 at 4°C], 14 mM magnesium acetate, 60 mM potassium acetate) per liter of culture. The cell pellet can be frozen at −80°C at this point for up to two days. The pellet is thawed on ice before weighing and subsequent resuspension in ice-cold S30 buffer at a ratio of 4 mL of buffer per gram of wet cells. The cells are lysed by a single passage through a chilled French pressure cell at 12000 psi (more than one passage through the French press results in decreased translational activity of the S30 extract). The lysed cells are immediately centrifuged at 30,000 *g* for 30 min at 4°C. The supernatant is transferred to a clean centrifuge tube and centrifuged again at 30,000 *g* at 4°C for 30 min. The supernatant is once again transferred to a second clean centrifuge pot, and for each 6.5 mL of S-30 extract, 1 mL of preincubation mix is added (3.75 mL 2 M Tris-acetate [pH 7.5 at 4°C], 71 μL 3 M magnesium acetate, 75 μL amino acid mix [10 mM of each of the 20 amino acids; Sigma], 0.3 mL 0.2 M ATP, 0.2 g phosphenolpyruvate, 50 units of pyruvate kinase). The preincubation mix must be prepared fresh, immediately before use. Mix the solution slowly for 1 hour at room temperature. During this stage all endogenous mRNA will finish translating and the endogenous mRNA and DNA will be degraded by nucleases present in the cell extract. Afterward, the S30 extract is transferred to dialysis tubing and dialyzed in the cold room three times against chilled S30 buffer (500 mL buffer per aliquot of extract prepared from 1 L of culture). Each dialysis solution is replaced after 4 hours. The extract is snap frozen in 100 to 500 μL aliquots using liquid nitrogen. Store at −80°C. The extract can be stored for many months without losing activity. It is possible to refreeze the extract and reuse after thawing; however, the extract cannot be thawed more than twice without severely affecting its activity.

10.3.6 *In Vitro* Translation for Generation of Stable Ternary Ribosome Complexes

mRNA is translated *in vitro* using either an *E. coli* S30 extract or the PUREsystem reconstituted translation reagents (Post Genome Institute, Tokyo, Japan) to generate stabilized ribosome complexes for selection by prokaryotic ribosome display. For eukaryotic studies follow the method in Section 10.3.10 (Douthwaite et al. 2006).

10.3.6.1 E. coli S30 Extract Translation Reactions The *in vitro* translation step during ribosome display is predominantly performed at 37°C for anywhere from 6 to 10 minutes. The translating time needs to be optimized for each batch of S30 extract. This is done to minimize decay of mRNA-ribosome-protein complexes by proteases and mRNA by RNases, while at the same time long enough to generate ample full-length translated protein (Hanes, Jermutus, and Plückthun 2000). The synthesis of large proteins with a molecular weight greater than 70,000 Da is not efficient due to premature termination of translating ribosomes (Ramachandiran, Kramer, and Hardesty 2000).

When translating proteins that form disulfide bridges, it is imperative to have an oxidizing environment. There should be minimal or no DTT (or other reducing agents) present in the translation reaction. The *in vitro* transcription and translation steps are performed separately as T7 RNA polymerase requires the presence of reducing agents for stability. Protein disulfide isomerase (PDI), a eukaryotic chaperone that catalyzes disulfide bond formation (Freedman, Hawkins, and McLaughlin 1995), greatly improves the display of some proteins, for example, the display of antibody fragments by threefold (Hanes and Plückthun 1997).

One of the main contributing factors to successful ribosome display selections is to perform the entire procedure from *in vitro* translation to RNA isolation, on ice or water-ice and to keep all reagents and materials necessary for this part of the experiment (e.g., pipette tips, tubes, etc.) ice-cold. The best way to achieve this is to perform selections in a cold room where all materials are already chilled.

There are several preparatory steps to perform for optimal efficiency of the *in vitro* translation process.

(1) *Prechill Centrifuge*: Precool a refrigerated bench-top centrifuge to 4°C.
(2) *Prepare Selection Buffer*: Combine 5 mL 10x *E. coli* wash (0.5 M Tris-acetate pH 7.5, 1.5 M NaCl, 0.5 M magnesium acetate, 1 percent Tween-20), 45 mL filter sterilized high quality deionized water and 625 μL heparin (200 mg/mL) in a 50 mL Falcon tube. Keep on ice.

(3) *Prechill Microfuge Tubes*: For each clone/library to be selected, prechill 4x 1.5 mL microfuge tubes on ice (two for translation and two for selection). Prepare an additional 1.5 mL microfuge tube on ice containing 880 μL selection buffer.

(4) *Prepare Streptavidin-coated Beads*: Prepare 1 μL of beads per nanomolar of biotinyated antigen, down to a minimum of 20 μL (use 20 μL for 10 nM or less). Wash beads three times in selection buffer and resuspend in selection buffer to the original volume.

(5) *Prepare Biotinylated Antigen*: The antigen is body labeled with biotin with an average of two- to fourfold excess of biotin per antigen molecule for 20 min. Biotinylation success is observed via mass spectrometry, achieving an average of one to two biotinylation events per antigen molecule. Buffer exchange biotinylated antigen with PBS using PD-10 columns and test for activity, before selections take place.

(6) *Prepare De-biotinylated, Sterile Milk for Selection*: Add 100 μL streptavidin-coated beads to 1 mL of 10 percent autoclaved nonfat dried milk (Marvel) in water. Incubate with end over end mixing for at least 5 min. Collect beads using a magnetic particle concentrator and transfer milk to a fresh tube. Store on ice until required. Milk must be sterilized to remove intrinsic RNase activity.

(7) *Prepare Reagents*: Thaw Premix buffer (250 mM Tris-acetate pH 7.5, 1.75 mM of each amino acid, 10 mM ATP, 2.5 mM GTP, 5 mM cAMP, 150 mM acetylphosphate, 2.5 mg/mL *E. coli* tRNA, 0.1 mg/mL folinic acid, 7.5 percent PEG 8000) and PDI on ice. Remove S30 extract and mRNA for clones/libraries from the −80°C freezer. *Keep frozen on dry (card) ice until required.*

Selections are ideally performed in nonstick, RNase free microfuge tubes (1.5 mL). RNase free microtubes may be used without pretreatment to get rid of nucleases. Prepare a translation "master mix" for the total number of reactions required by combining the reagents in the order listed. 17.4 μL nuclease-free water; 11 μL 2 M potassium glutamate; 7.6 μL 0.1 M magnesium acetate; 2.0 μL 5 mg/mL PDI; 22 μL premix buffer; 40 μL *E. coli* S30 extract and 10 μL mRNA. Please note that mixing should be gentle by pipetting after addition of Premix buffer. Thaw the S30 extract at the latest possible moment (do not vortex). After addition of the S30 extract, mix the complete reaction again by pipetting gently. Subsequently, an aliquot of the mRNA clone/library should be thawed quickly and placed immediately on ice when thawed. Ice-cold library mRNA (10 μg or approximately 2×10^{13} molecules of a scFv library) in 10 μL of RNase free water is added to the mixture, gently mix and immediately place at 37°C. Unused mRNA should be immediately snap frozen.

Translation is performed at 37°C for 6 to 10 min, after which ribosome complexes are stabilized by dilution in ice-cold selection buffer [50 mM Tris acetate (pH 7.5), 150 mM NaCl, 50 mM magnesium acetate, 0.1 percent Tween 20, 2.5 mg/mL heparin) to a final volume of 550 μL per sample and mix. Centrifuge samples at 13,000 rpm at 4°C for 5 min in a prechilled rotor (to remove insoluble material) and replace on ice.

10.3.6.2 PUREsystem E. coli Translation Reactions Translation reactions can be performed using PUREsystem S-S (PGI) containing 10 μg mRNA, 90 μg/mL PDI, solutions A and B as recommended by the manufacturer in a total volume of 50 μL. Translation is performed at 37°C for 30 min, after which ribosome complexes are stabilized by dilution in ice-cold selection buffer to a final volume of 550 μL. The ribosomal complexes from the PUREsystem proceed through either soluble or surface ribosome display selections exactly as for complexes produced from S30 *E. coli* extract.

10.3.7 Soluble Selection and Capture of Specifically Bound Ribosomal Tertiary Complexes

It may be necessary to introduce a preselection step before the affinity selection stage, to reduce non-specific binding of the ribosomal tertiary complexes with debiotinylated milk or protein carriers that might be present in the antigen preparation (preferable to use carrier free antigens, where possible). The translation mixture is first incubated in debiotinylated milk-coated (or carrier-coated) microfuge tubes, where sticky ribosomal tertiary complexes can bind and be sequestered away from the main affinity selection process. The supernatant from this preselection step is then used for the subsequent affinity selection phase.

Transfer 500 μL stabilized tertiary complexes to each of two prechilled RNase free selection tubes (+ and −) per clone/library. Add 50 μL debiotinylated milk to both tubes and add biotinylated antigen at the required concentration to the positive antigen selections only. Incubate the selections at 4°C for the required time (2 h to overnight depending on the anticipated off-rate) with gentle end-over-end rotation. Capture selected complexes by addition of selection buffer-washed streptavidin-coated M280 paramagnetic beads. Incubate at 4°C for 5 min. Wash selections to remove nonspecifically bound complexes (this step can be performed manually or automated). To perform manual washing, collect the complexes by placing tubes on a magnet and allow beads to collect for 1 min. Subsequently, remove supernatant and add 1 mL selection buffer. Repeat three times. After the third wash, remove tubes from the magnet and add 220 μL elution buffer (50 mM Tris-acetate pH 7.5, 150 mM NaCl, 20 mM EDTA) containing *S. cerevisiae* mRNA (10 ng/mL final). The EDTA will chelate the Mg^{2+} ions, resulting in the dissociation of ribosomal tertiary complexes and elution of the mRNA. An excess of *S. cerevisiae* mRNA is used to sequester away any RNases that might be present, from the eluted selected mRNA, and hence slowing degradation. Incubate for 10 min at 4°C before placing the tubes on the magnet to collect the streptavidin beads. The mRNA in the supernatant was purified using a commercial mRNA purification kit (High Pure RNA Isolation Kit). For prokaryotic selections, the DNase I digestion option of the kit was performed. The purified mRNA is eluted with 40 μL nuclease-free water in a chilled microfuge. The mRNA is placed on ice and used immediately in a reverse transcription reaction followed by PCR. The resulting DNA is used for another round of ribosome display or for cloning and subsequent analysis of single clones by ELISA or other biochemical assay. Alternatively, the mRNA is immediately snap frozen on dry (card) ice for future analysis.

10.3.8 Surface Selection of Ribosomal Tertiary Complexes

The protocol for selection on a surface is very similar to that previously described for soluble selection (Section 10.3.7). Therefore, this section only concentrates on the differences between the two processes. All selection and elution buffers remain the same. All reagents and materials should be at 4°C, unless stated otherwise. Selections are performed on Nunc immunosorb 96-well plates. The wells are coated overnight at 4°C with 100 μL antigen in PBS solution. The following day, the coated wells are washed with PBS with the same number of uncoated wells, blocked with 4 percent (v/v) skimmed milk in PBS for 1 h at room temperature with shaking. (Each translation sample is split across four wells of the plate, i.e., two in presence of antigen and two in absence of antigen.) After blocking, the wells are washed with PBS and then with selection buffer (250 μL per well for each wash step). The nunc plate is placed on ice in the cold room before the addition of diluted translation mixture. The diluted translation mixture is mixed with 1/10th volume, 20 percent (v/v) skimmed milk in selection buffer, pipetted into milk-blocked wells (no antigen) for prebinding. The wells are incubated for 30 min on ice at 4°C. The content from the prebinding wells are then transferred to the ligand-coated and milk-blocked wells (see above), shaken gently and selected for at least 1 h. After five washes with ice-cold selection buffer, the bound ribosomal tertiary complexes are dissociated with 220 μL elution buffer. The remainder of the elution proceeds as described previously (Section 10.3.7).

10.3.9 Reverse Transcription and Polymerase Chain Reaction of Selection Outputs

Prepare a "master mix" for the required number of RT reactions to be performed. For prokaryotic and eukaryotic samples, prepare 20 μL reactions containing 4 μL 5x first strand buffer, 2 μL 0.1 M DTT, 0.25 μL 100 μM RT primer, 0.5 μL 25 mM PCR nucleotide mix, 0.5 μL RNasin, 0.5 μL

Superscript II and 12.25 μL mRNA. Incubate reactions at 50°C for 30 min. RT is performed with either the primer RT-1 (5′CCGCACACCAGTAAGGTGTGCGGTATCACCAGTAGCACCATTACCATT AGCAAG′3) which anneals at the 3′ end is hypothesized to recover mRNA only from disrupted ribosome complexes, or primer RT-2 (5′CCTTATTAGCGTTTGCCATTTTTTCATAATCAAAATCAC CGGA′3) which anneals 118 nucleotides upstream and is hypothesized to recover mRNA from all ribosome complexes. Aliquot 7.75 μL RT master mix into the bottom of an RNase free 0.5 mL microfuge tube or a thin-walled 0.2 mL PCR tube. Mix the eluted mRNA by pipetting and add 12.25 μL mRNA into each prepared RT tube. Mix well. Incubate RT reactions at temperature and length of time prescribed earlier in this section and transfer to ice when complete. During the incubation step prepare the PCR master mix.

10.3.9.1 End-Point PCR End-point PCR is performed to visualize amplification of the selected mRNA. During the initial round of selection it can be difficult to predict the number of amplification cycles necessary for PCR, as it depends on the amount of isolated mRNA. A 5 μL sample of each reverse transcription reaction is amplified with a PCR master mix containing 5 percent DMSO and 0.25 μM of forward primer (T7B or T7KOZ for prokaryotic and eukaryotic experiments, respectively), 0.25 μM RT primer (same as one used in reverse transcription reaction). Thermal cycling 94°C for 3 min, then 94°C for 30 sec, 50°C for 30 sec, and 72°C for 1.5 min, for 25 to 30 cycles, with a final step at 72°C for 5 min. A small sample (5 to 10 μL of the 100 μL reaction) of the PCR product are visualized by electrophoresis on a SYBRsafe stained 1 percent agarose gel under blue light. Usually, the more ribosome display cycles that have been carried out with a library the more mRNA encoding a cognate product to the antigen is isolated and therefore fewer PCR cycles are necessary during amplification. It is usually better to perform a test PCR first with fewer cycles, and if necessary to add further cycles, after analysis of PCR products by agarose gel electrophoresis. No additional PCR components need to be added, just incubate the samples for a further 5 to 10 cycles. It is not good to over amplify DNA. The PCR products become too smeary, background amplification of nonspecific DNA will become a problem, and the resulting RNA quality for subsequent rounds of selection will not be sufficient for a successful outcome. PCR products are gel purified using QiaQuick gel extraction kit according to manufacturer's instructions and reamplified using the PCR conditions described above. PCR products are then used directly for *in vitro* transcription as described in Section 10.3.4.

10.3.9.2 Real-Time PCR Real-time PCR can be performed to provide relative quantification of selection outputs. The technique allows accurate quantification of the amount of a specific product (cDNA) present in a sample. By comparing quantities of cDNA present following selection on the antigen of choice with selection in the absence of antigen it is possible to determine whether the selection has led to enrichment, and therefore been successful. Real-time PCR also allows estimation of the number of PCR cycles required to achieve amplification of specific outputs without amplification of background. This technique can be used to test clone function in ribosome display format, and is also useful for troubleshooting in the event that no end-point PCR product is seen.

Relative quantification of cDNA is performed by real-time PCR using the 5′ nuclease assay (Taqman). Real-time allows quantification to be performed at a point during the PCR reaction where amplification is proceeding exponentially (i.e., before amplification is limited by exhaustion of reactions or inhibited by accumulation of product). This technique allows more accurate quantification of cDNA compared to end-point PCR and gel analysis, is more sensitive, and is linear over a wide range of input cDNA levels. The Taqman assay employs two PCR primers and a dual labeled (reporter and quencher) primer probe that is specific for a common portion of the ribosome display construct. This assay is targeted to the geneIII tether sequence in a region 5′ to the annealing region of both RT primers. Real-time PCR reactions are performed in an ABI 7700 in 25 μL reactions containing 12.5 μL 2x ABI Universal Master Mix, 300 nM each of forward (5′GCCGCAGAACAAAAAC TCATCT′3) and reverse (5′AATCAAAATCACCGGAACCG′3) primer, 200 nM dual labeled (5′FAM, 3′TAMRA) probe (AGCCGCCACCATTCAGATCCTCTTCT) and 5 μL reverse transcription reaction.

Analyze the data using CT (cycle to threshold) values to calculate relative cDNA quantities based on delta CT value as follows:

$$\text{Relative quantitative cDNA level} = 2^{-\text{delta CT}}$$

where delta CT is the difference in CT between each sample and a control sample.

10.3.10 Eukaryotic Ribosome Display

Eukaryotic ribosome display has potentially been a more accessible alternative because of the availability of suitable commercial reagents. However, to date this method has been less widely used. The eukaryotic display process has been examined extensively (Douthwaite et al. 2006) and protocol modifications were identified to make it mechanistically comparable to the prokaryotic method. These changes resulted in a more efficient process. This should increase the ease of operating ribosome display technology, and make it more accessible to the scientific community.

At all stages of the display process, where stabilized eukaryotic ribosome complexes must be maintained, the presence of 5 mM magnesium is recommended and that samples are maintained at 4°C or on ice. Methods that include postselection steps lacking magnesium and incorporate DNase I digestion at 37°C on intact ribosome complexes are likely to result in the loss of selected complexes by spontaneous disruption. DNase I digestion is performed on purified mRNA where it does not impact on the selection efficiency. In addition, such relocation of the DNase I digestion step allows for more simple postselection washing. It is recommended that mRNA recovery is by ribosome complex disruption. The first overcomes the problem that the direct RT-PCR approach is not possible for certain selection scenarios, for example, selection on cell monolayers or on target protein that is immobilized on a solid surface, and the second allows mRNA to be purified prior to RT. Performing RT-PCR directly on selected ribosome complexes may be limited in terms of the capacity or efficiency of the reverse transcription step, since here RT is performed in the presence of magnetic beads, ribosomal and other proteins and "background" RNA and DNA. It would be preferable to produce a highly pure and concentrated sample of mRNA to allow maximum cDNA synthesis in an optimal reverse transcription reaction.

In vitro transcription proceeds exactly as described previously in Section 10.3.4 as for prokaryotic ribosome display.

10.3.10.1 In Vitro Translation Using the Flexi Rabbit Reticulocyte Lysate (RRL) system, prepare 100 μL translation reaction containing: 16.7 μL nuclease-free water, 2 μL complete amino acid mix (prepared by pooling all three amino acid mixtures), 1.6 μL 2.5 M KCl, 2 μL 5 mg/mL protein disulfide isomerase, 66 μL rabbit reticulocyte lysate, and 10 μL mRNA at a concentration of 1 μg/mL. Mix gently by pipetting and incubate at 30°C for 20 min.

10.3.10.2 Selection Transfer the translation reaction to a precooled microfuge tube on ice containing 100 μL PBS and mix gently by pipetting. Add 2 μL 10 percent debiotinylated autoclaved skimmed milk (prepared in water) to block nonspecific binding if required. Add the target antigen, suitably labeled for capture, for example coupled to biotin. Incubate at 4°C for 2 h to overnight (as appropriate) with end-over-end mixing.

10.3.10.3 Capture of Selected Complexes For biotinylated antigens, add 10 to 100 μL streptavidin M280 beads, prewashed three times in wash buffer (PBS, 5 mM magnesium acetate, 1 percent Tween 20). Incubate at 4°C for a further 2 min with gentle end-over-end mixing.

10.3.10.4 Postselection Washes Collect the beads on a magnetic particle concentrator, remove the supernatant, and gently resuspend in 500 μL wash buffer. Repeat a further two times (or more if required to improve signal-to-noise ratio). Ensure complete removal of buffer after last wash.

10.3.10.5 Ribosome Complex Disruption for mRNA Recovery Add 200 μL PBS containing 10 μg/mL *S. cerevisiae* RNA and incubate at 50°C for 5 min, mix by vortexing occasionally. Collect the beads and process the supernatant for RNA purification. Alternatively, the cell lysis step of the RNA purification kit may be used to recover mRNA as follows: resuspend beads in 200 μL PBS containing 10 μg/mL *S. cerevisiae* RNA and transfer the bead suspension to 400 μL cell lysis buffer for RNA purification (provided in the RNA isolation kit). Vortex mix well. Collect beads on a magnetic particle concentrator before processing the lysate for mRNA purification.

10.3.10.6 mRNA Purification Use a commercial kit, such as the High Pure RNA Isolation Kit. Include the DNase I digestion step and elute mRNA in 40 μL nuclease-free water. Keep mRNA on ice and proceed to the next step as quickly as possible.

Reverse transcription and PCR amplification steps are as described in Section 10.3.9. This protocol has been performed in parallel to the previously published methods (He and Taussig 1997, 2003) and cDNA yields measured as an indication of the selection efficiency. End-point PCR products generated following selections (Fig. 10.4) clearly indicate that selection efficiency is higher with the method described above and the real-time analysis of the selection outputs shows the improvement to be a

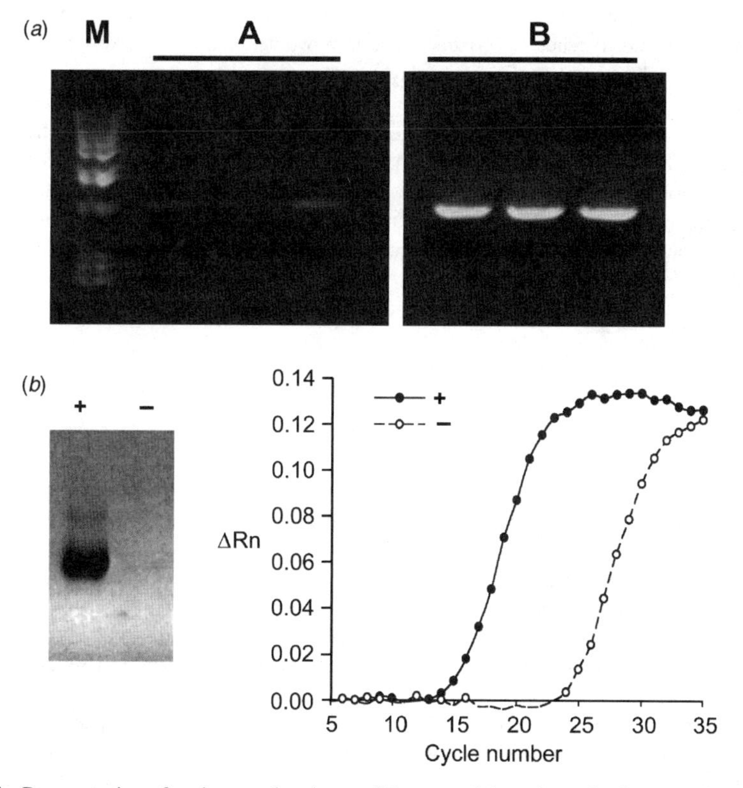

Figure 10.4 Demonstration of an improved and accessible protocol for eukaryotic ribosome display based on commercial reagents. A model scFv selection for fluorescein was used for selection on 50 nM biotin-fluorescein. (a) Comparison of eukaryotic ribosome display methods. A is the previously published methodologies and B is the new recommended (performed as described in methods). End-point PCR analysis (25 cycles) of selections performed in triplicate is shown. M is 1 Kb^{+} DNA ladder. (b) Demonstration of specificity of selection using this improved method. (+) is selection on 50 nM biotin-fluorescein; (−) is selection in the absence of biotin-fluorescein. Left panel is end-point PCR analysis (30 cycles); right panel is real-time PCR amplification. ΔRn is the normalized reporter signal (Douthwaite et al. 2006).

20.8-fold increase over the previous method. Figure 10.4 illustrates that cDNA recovery is specific to selection on antigen.

10.4 SUMMARY

Ribosome display has enormous potential in the areas of isolation and optimization of specific, high potency antibodies, as well as the improvement of defined protein characteristics. This potential is rapidly being realized and applicability will continue to rise with the evolution of the ribosome display protocols and the advent of tailored commercial reagents.

The *E. coli* S30 extract for *in vitro* display of an mRNA library have proven to be very successful, by the evolution of high-affinity antibodies and the optimization of defined protein characteristics. However, this technology has so far been perceived as being technically challenging due to comparatively difficult protocols as well as the absence of tailored commercial reagents, particularly when using prokaryotic cell-free expression systems.

Selection conditions developed with the S30 prokaryotic ribosome display system has proven to be directly transferable to the new reconstituted translation system "PUREsystem" to generate stabilized ribosome complexes. In a number of different ribosome display selection scenarios, it has been found that the reconstituted translation reagents are highly efficient and offer significant benefits in terms of DNA yields, compared to crude S30 extract. The improvements are most likely a reflection of the improved mRNA stability and quality of displayed proteins.

Eukaryotic ribosome display protocols were also investigated. An improved eukaryotic ribosome display protocol has been developed that is mechanistically similar to the prokaryotic method, but has been tailored for the specific requirements of eukaryotic ribosomes. The optimized protocol has resulted in a 20-fold improvement in selection outputs over previously published methods.

With the improvements described, the prokaryotic and eukaryotic ribosome display systems are now *in vitro* evolution technologies that are easily accessible owing to the availability of suitable commercial translation reagents.

REFERENCES

Amstutz, P., J.N. Pelletier, and A. Guggisberg. 2002. *In vitro* selection for catalytic activity with ribosome display. *J. Am. Chem. Soc.* 124:9396–9403.

Chen, H.Z., and G. Zubay. 1983. Prokaryotic coupled transcription-translation. *Methods Enzymol.* 101:674–690.

Douthwaite, J.A., M.A. Groves, P. Dufner, and L. Jermutus. 2006. An improved method for an efficient and easily accessible eukaryotic ribosome display technology. *Protein Eng. Design Selection* 19:85–90.

Dower, W.J., and S.E. Cwirla. 1992. Creating vast peptide expression libraries: Electroporation as a tool to construct plasmid libraries of greater than 10^9 recombinants. In *Guide to electroporation and electrofusion*, ed. D.C. Chang et al., 291–301. San Diego: Academic Press.

Freedman, R.B., H.C. Hawkins, and S.H. McLaughlin. 1995. Protein disulfide-isomerase. *Methods Enzymol.* 251:397–406.

Hajnsdorf, E., F. Braun, J. Haugel-Nielsen, J. Le Derout, and P. Régnier. 1996. Multiple degradation pathways of the rpsO mRNA of *Escherichia coli*. RNase E interacts with the 5′ and 3′ extremities of the primary transcript. *Biochimie* 78:416–424.

Hanes, J., L. Jermutus, and A. Plückthun. 2000. Selection and evolving functional proteins *in vitro* by ribosome display. In: Abelson J.N. and M.I. Simon (eds), Methods Enzymol., Academic Press, San Diego, p. 404.

Hanes, J., L. Jermutus, S. Weber-Bornhauser, H.R. Bosshard, and A. Plückthun. 1998. Ribosome display efficiently selects and evolves high-affinity antibodies *in vitro* from immune libraries. *Proc. Natl. Acad. Sci. USA* 95:14130–14135.

Hanes, J., and A. Plückthun. 1997. *In vitro* selection and evolution of functional proteins by using ribosome display. *Proc. Natl. Acad. Sci. USA* 94:4937–4942.

Hanes, J., C. Schaffitzel, A. Knappik, and A. Plückthun. 2000. Picomolar affinity antibodies from a fully synthetic naïve library selected and evolved by ribosome display. *Nature Biotechnol.* 18:1287–1292.

He, M., and M.J. Taussig. 1997. Antibody-ribosome-mRNA (ARM) complexes as efficient selection particles for *in vitro* display and evolution of antibody combining sites. *Nucleic Acids Res.* 25:5132–5134.

He, M., and M.J. Taussig. 2003. Ribosome complexes as selection particles for *in vitro* display and evolution of proteins. European Patent EP985032B1.

He, M., and M.J. Taussig. 2005. Ribosome display of antibodies: Expression, specificity and recovery in a eukaryotic system. *J. Immunol. Methods* 297:73–82.

Jermutus, L., A. Honeggar, F. Schwesinger, J. Hanes, and A. Plückthun. 2001. Tailoring *in vitro* evolution for protein affinity or stability. *Proc. Natl. Acad. Sci. USA* 98:75–80.

Jermutus, L., L.A. Ryabova, and A. Plückthun. 1998. Recent advances in producing and selecting functional proteins by using cell-free translation. *Curr. Opin. Biotechnol.* 9:534–548.

Lamla, T., and V.A. Erdmann. 2001. *In vitro* selection of other proteins than antibodies by means of ribosome display. *FEBS Lett.* 502:35–40.

Lesley, S.A. 1995. Preparation and use of *E. coli* S-30 extracts. *Methods Mol. Biol.* 37:265–278.

Matsuura, T., and A. Plückthun. 2004. Strategies for selection from protein libraries composed of *de novo* designed secondary structure modules. *Orig. Life Evol. Biosph.* 34:151–157.

Mattheakis, L.C., R.R. Bhatt, and W.J. Dower. 1994. An *in vitro* polysome display system for identifying ligands from very large peptide libraries. *Proc. Natl. Acad. Sci. USA* 91:9022–9026.

McCafferty, J., K.J. Fitzgerald, J. Earnshaw, D.J. Chiswell, J. Link, R. Smith, and J. Kenten. 1994. Selection and rapid purification of murine antibody fragments that bind a transition-state analog by phage display. *Appl. Biochem. Biotechnol.* 47:157–171.

Ohashi, H., Y. Shimizu, B.W. Ying, and T. Ueda. 2007. Efficient protein selection based on ribosome display system with purified components. *Biochem. Biophys. Res. Commun.* 352:270–276.

Ramachandiran, V., G. Kramer, and B. Hardesty. 2000. Expression of different coding sequences in cell-free bacterial and eukaryotic systems indicates translational pausing on *Escherichia coli* ribosomes. *FEBS Lett.* 482:185–188.

Reynolds, R., R.M. Bermudez-Cruz, and M.J. Chamberlin. 1992. Parameters affecting transcription termination by *Escherichia coli* RNA polymerase. I. Analysis of 13 rho-independent terminators. *J. Mol. Biol.* 224:31–51.

Shimizu, Y., A. Inoue, Y. Tomari, T. Suzuki, T. Yokogawa, K. Nishikawa, and T. Ueda. 2001. Cell-free translation reconstituted with purified components. *Nature Biotechnol.* 19:751–755.

Shimizu, Y., T. Kanamori, and T. Ueda. 2005. Protein synthesis by pure translation systems. *Methods* 36:299–304.

Studier, F.W., A.H. Rosenberg, J.J. Dunn, and J.W. Dubendorff. 1990. Use of T7 RNA polymerase to direct expression of cloned genes. *Methods Enzymol.* 185:60–89.

Takahashi, F., T. Ebihara, and M. Mie. 2002. Ribosome display for selection of active dihydrofolate reductase mutants using immobilized methotrexate on agarose beads. *FEBS Lett.* 514:106–110.

Thom, G., A.C. Cockroft, A.G. Buchanan, C.J. Candotti, E.S. Cohen, D. Lowne, P. Monk, C.P. Shorrock-Hart, L. Jermutus, and R.R. Minter. 2006. Probing a protein-protein interaction by *in vitro* evolution. *Proc. Natl. Acad. Sci. USA* 103:7619–7624.

Vaughan, T.J., A.J. Williams, K. Pritchard, J.K. Osbourn, A.R. Pope, J.C. Earnshaw, J. McCafferty, R.A. Hodits, J. Wilton, and K.S. Johnson. 1996. Human antibodies with sub-nanomolar affinities isolated from a large non-immunized phage display library. *Nature Biotechnol.* 14:309–314.

Villemagne, D., R. Jackson, and J.A. Douthwaite. 2006. Highly efficient ribosome display selection by use of purified components for *in vitro* translation. *J. Immunol. Methods* 30:140–148.

Ying, B.-W., H. Taguchi, H. Ueda, and T. Ueda. 2004. Chaperone-assisted folding of a single-chain antibody in a reconstituted system. *Biochem. Biophys. Res. Commun.* 320:1359–1364.

Zahnd, C., E. Wyler, J.M. Schwenk, D. Steiner, M.C. Lawrence, N.M. McKern, F. Pecorari, C.W. Ward, T.O. Joos, and A. Plückthun. 2007. A designed ankyrin repeat protein evolved to picomolar affinity to Her2. *J. Mol. Biol.* 369:1015–1028.

Zubay, G. 1973. *In vitro* synthesis of protein in microbial systems. *Annu. Rev. Genet.* 7:267–287.

CHAPTER 11

Bacterial Display of Antibodies

THOMAS J. VAN BLARCOM and BARRETT R. HARVEY

11.1 INTRODUCTION

The ability to express a functional binding domain of an antibody in bacteria was first reported in the late 1980s following the development of sufficient molecular biology techniques for the facile cloning and expression of recombinant antibody fragments (Skerra and Plückthun 1988). This has since led to the use of a number of display technologies, designed to isolate and engineer antibodies *in vitro* by striving to mimic the *in vivo* generation and selection of antibody diversity accomplished by the immune system (Foote and Eisen 1995, 2000; Holliger and Hudson 2005). In general, these technologies apply three common methods: a link between genotype and phenotype, application of a selective pressure, and amplification of the selected population (Hoogenboom 2005). They are often grouped by the method in which they achieve the genotype to phenotype linkage: how the antibody is displayed on a platform that retains its nucleic acid sequence while allowing association with antigen. Display has been achieved via a number of platforms, including phage display, ribosome display, mRNA display, microbead display by compartmentalization, bacteria display, yeast display, and mammalian cell display. Examples of antibody display in the majority of these systems have utilized the binding

Therapeutic Monoclonal Antibodies: From Bench to Clinic. Edited by Zhiqiang An

domain of the antibody in a single chain variable fragment (scFv) or antigen binding fragment (Fab) format, although systems have reported full length immunoglobulin G (IgG) display (Akamatsu et al. 2007; Mazor et al. 2007).

The first platform to demonstrate display of an active antibody-binding domain was bacteriophage display (McCafferty et al. 1990). Still today, it is the most universally used screening platform due to its robustness and ease of use. Of the competing technologies, all have their respective advantages and disadvantages. One of the distinct advantages of cells, namely mammalian cells, yeast, and bacteria over bacteriophage as antibody display platforms, is their suitability for screening by fluorescence activated cell sorting, or FACS (Daugherty et al. 2000; Hayhurst and Georgiou 2001). FACS is a high throughput screening methodology based on flow cytometry that incorporates real-time quantitative multiparameter analysis of each member of a population. This allows for precise control of what members of the cellular population are collected based on a multitude of their individual properties. In the 1990s, the technology behind FACS improved to the point where it was suitable for the screening of large libraries. As a result, cell display technologies were developed during this time. Flow cytometry coupled with the ease of genetic manipulation, fast growth rates, high DNA transformation efficiency, and low costs of bacteria culture, have made them an attractive choice for protein display (W. Chen and Georgiou 2002; Stahl and Uhlen 1997). This chapter will highlight the technologies available for bacterial display of antibodies, their applications, and their limitations.

The composition of the cell wall is used to define a bacterium as either Gram-negative or Gram-positive. In general, Gram-negative bacteria have an inner membrane, a thin cell wall composed of a few peptidoglycan layers, and an outer membrane. Gram-positive bacteria have an inner membrane, a much thicker cell wall composed of many peptidoglycan layers, and do not contain an outer membrane. The composition of the membranes and cell wall are a determining factor in how antibodies are displayed in bacteria. As a result of these differences, the way in which proteins are displayed on the surface of Gram-positive and Gram-negative bacteria vary tremendously and will be described independently of each other.

11.2 SYSTEMS FOR GRAM-NEGATIVE BACTERIA

Proteins to be displayed on the surface of Gram-negative bacteria must be transported across the bacterial envelope comprised of the inner membrane, the periplasm, and the outer membrane. The inner membrane is a phospholipid bilayer, the periplasm contains the peptidoglycan matrix, and the outer membrane consists of a phospholipid inner leaflet and a lipopolysaccharide outer leaflet (Earhart 2000). To target a protein across the envelope and anchor it such that it is freely displayed generally requires both a signal sequence to direct the protein out of the cytoplasm and a membrane anchoring domain. Numerous approaches have been used to display a variety of proteins in this heterologous manner and have been extensively reviewed elsewhere (Benhar 2001; W. Chen and Georgiou 2002; Samuelson et al. 2002). However, few have been shown to be compatible with the display of antibodies. These are defined in Table 11.1.

The majority of systems that have been engineered for the display of whole antibodies or antibody fragments in Gram-negative bacteria have utilized *E. coli*, likely due to the extensive understanding of the organism and the molecular biology tools available for its manipulation. Display of antibody fragments has been accomplished through heterologous fusions with the anchoring motifs of lipoproteins of both the inner and outer membrane, through fusions to endogenous and exogenous outer membrane proteins, through insertion into protein components of the flagella, and through soluble expression to the periplasm, each designed to maintain association of antibody with the cell (Bassi et al. 2000; G. Chen et al. 2001; Ezaki S, Takagi, and Imanaka 1998; Francisco et al. 1993; Fuchs et al. 1991; Harvey et al. 2004; K.J. Jeong et al. 2007).

All of the systems described in this chapter have demonstrated the ability to display some form of an antibody, typically a scFv. However, many of the systems require further evidence to demonstrate that they are useful as antibody engineering platforms. Considerations in evaluating each platform

TABLE 11.1 Systems Used for the Bacterial Display of Antibodies

Carrier Protein (Organism)	Protein Displayed	Year	Host	References
	Gram-Negative Bacteria			
PAL (*E. coli*)	Anti-chicken lysozyme scFv*	1991	*E. coli*	(Fuchs et al. 1991)
	Anti-phOx scFv	1996	*E. coli*	(Fuchs et al. 1996)
	Anti-c-myc peptide scFv	1996	*E. coli*	(Fuchs et al. 1996)
	Anti-atrazine scFv	1999	*E. coli*	(Dhillon, Drew, and Porter 1999)
Lpp-OmpA (*E. coli*)	β-lactamase*	1992	*E. coli*	(Francisco, Earhart, and Georgiou 1992)
	Anti-digoxin scFv	1993	*E. coli*	(Francisco et al. 1993)
	Anti-CEA	2007	*S. typhimurium*	(Bereta et al. 2007)
INP (*P. syringae*)	LevU*	1998	*E. coli*	(H.C. Jung, Lebeault, and Pan 1998)
	CMCase*	1998	*E. coli*	(H.C. Jung et al. 1998)
	Anti-c-myc scFv	2000	*E. coli*	(Bassi et al. 2000)
FliC (*E. coli*)	HEL-derived peptide*	1988	*E. coli*	(Kuwajima G et al. 1988)
	Anti-porphyrin scFv	1998	*E. coli*	(Ezaki S, Takagi, and Imanaka 1998)
IgA Protease (*N. gonorrhoeae*)	CTB*	1990	*S. typhimurium*	(Klauser, Pohlner, and Meyer 1990)
	Anti-polyhistidine	1999	*E. coli*	(Veiga, de Lorenzo, and Fernandez 1999)
	Anti-levan scFv	2004	*E. coli*	(Veiga, de Lorenzo, and Fernandez 2004)
	Anti-α-amylase dAb and polydAb	2004	*E. coli*	(Veiga, de Lorenzo, and Fernandez 2004)
NlpA (*E. coli*) or pIII (M13 bacteriophage)	Anti-PA scFv*	2004	*E. coli*	(Harvey et al. 2004)
	Anti-digoxin scFv*	2004	*E. coli*	(Harvey et al. 2004)
	Anti-methamphetamine	2006	*E. coli*	(Harvey et al. 2006)
	Anti-PA Fab (NlpA-V_L-C_κ, V_H-C_{H1} captured)*	2007	*E. coli*	(K.J. Jeong et al. 2007)
	Anti-PA IgG (NlpA-ZZ, IgG captured)*	2007	*E. coli*	(Mazor et al. 2007)
	Anti-digoxin IgG (NlpA-ZZ, IgG captured)*	2007	*E. coli*	(Mazor et al. 2007)
"Display-Less"	Anti-digoxin scFv*	2001	*E. coli*	(G. Chen et al. 2001)
	Gram-Positive Bacteria			
VEGETATIVE				
Protein A (*S. aureus*)	ABP*	1992	*S. xylosus*	(Hansson et al. 1992)
	Anti-human IgE scFv	1996	*S. xylosus*	(Gunneriusson et al. 1996)
	Anti-human IgE scFv	1996	*S. carnosus*	(Gunneriusson et al. 1996)
	Anti human IgA affibodies	1999	*S. carnosus*	(Gunneriusson et al. 1999)
	Anti human IgE affibodies	1999	*S. carnosus*	(Gunneriusson et al. 1999)
M6 (*S. pyogenes*)	E7 protein*	1992	*S. gordonii*	(Pozzi et al. 1992)
	Anti-idiotype scFv	2000	*S. gordonii*	(Beninati et al. 2000)
	Anti-SA I/II scFv	2000	*S. gordonii*	(Beninati et al. 2000)
Proteinase P (*L. paracasei*)	TTFC*	1999	*L. paracasei*	(Maassen et al. 1999)
	Anti-SA I/II scFv	2002	*L. paracasei*	(Kruger et al. 2002)
	Anti-RgpA protease scFv	2006	*L. paracasei*	(Marcotte et al. 2006)
	Anti-ROCP dAb	2006	*L. paracasei*	(Pant et al. 2006)
	Anti-MCP dAb	2007	*L. paracasei*	(Hultberg et al. 2007)

(*Continued*)

TABLE 11.1 *Continued*

Carrier Protein (Organism)	Protein Displayed	Year	Host	References
ENDOSPORES				
CotB (*B. subtilis*)	TTFC*	2001	*B. subtilis*	(Isticato et al. 2001)
CotC (*B. subtilis*)	TTFC*	2004	*B. subtilis*	(Mauriello et al. 2004)
	LT-B*	2004	*B. subtilis*	(Mauriello et al. 2004)
Protoxin (*B. thuringiensis*)	Anti-phOx scFv	2005	*B. thuringiensis*	(Du et al. 2005)

*Indicates first protein displayed using this carrier protein. PAL, peptidoglycan associated lipoprotein; phOx, 2-phenyloxazol-5-one; Lpp-OmpA, major lipoprotein-outer membrane protein A; CEA, carcinoembryonic antigen; INP, ice nucleation protein; LevU, levansucrase; CMCase, carboxymethylcellulase; CTB, cholera toxin B subunit; NlpA, new lipoprotein A; pIII, M13 bacteriophage gene III coat protein; PA, protective antigen from *Bacillus anthracis*; V_L, variable light chain; C_κ, light chain kappa constant; V_H, variable heavy chain; C_{H1}, IgG heavy chain constant domain 1; ZZ, synthetic analog of the IgG-binding domain of *Staphylococcus aureus* protein A; FliC, flagellin; HEL, hen egg-white lysozyme; ABP, albumin-binding protein from Streptococcal protein G; SA I/II, Streptococcal antigen I/II from *S. mutans*; TTFC, tetanus toxin fragment C; ROCP, rotavirus outer capsid proteins; MCP, major capsid protein from Lactococcal bacteriophage P2; LT-B, B subunit of the heat-labile enterotoxin of *E. coli*.

discussed will include whether they have demonstrated antigen-binding activity, the size of the antigen, compatibility with high throughput fluorescence activated cell sorting (FACS), unique features which contribute to the systems range of applications, and whether they have been utilized for antibody discovery or property enhancement.

11.2.1 PAL

The first antibody displayed on a bacterium was performed by Little and coworkers in 1991 (Fuchs et al. 1991; Fig. 11.1). They exploited the peptidoglycan associated lipoprotein (PAL) to display an scFv specific for chicken lysozyme on the surface of *E. coli.* PAL is normally secreted via its own signal sequence to the periplasm where it associates with the peptidoglycan layer via a region located on its C-terminus and to the outer membrane via an N-terminal cysteine following modification by a lipid moiety. By removing the N-terminal cysteine from PAL and introducing a leader peptide from the enzyme pectate lyase (pelB) followed by a gene encoding a scFv, the fusion protein remains bound to the cell via the peptidoglycan associated component. Fluorescence microscopy revealed that cells displaying the scFv-PAL fusion were specifically labeled with an antibody that recognizes an epitope tag in the scFv linker region. Specific binding to lysozyme was demonstrated with biotinylated lysozyme using a fluorescent streptavidin conjugate for detection.

Little and coworkers revisited the PAL system in 1996, further characterizing its utility for antibody engineering and demonstrating that scFvs displayed using PAL are indeed amenable to FACS (Fuchs et al. 1991, 1996). An scFv that recognizes the oncoprotein c-myc was displayed and specifically labeled with fluorescently conjugated c-myc peptide as determined by flow cytometry. Another scFv that recognizes the hapten 4-ethoxymethylene 2-phenyl 2-oxazolin 5-one (phOx) was also displayed, but only specifically bound a fluorescent conjugate of phOx following pretreatment with EDTA. This pretreatment permeabilizes the outer membrane of bacteria, but decreases the cell viability. However, increasing the permeability of the outer membrane to allow the interaction of the displayed antibody with the antigen is a recurring theme in Gram-negative bacterial display systems (Bassi et al. 2000; Fuchs et al. 1996; Harvey et al. 2004; Leive 1968). To further investigate the effect of antigen size on displayed antibody binding, the hapten phOx was conjugated to a variety of different proteins ranging up to 225 kDa. The only protein conjugated to the hapten which allowed for specific binding to the displayed scFv was the 45 kDa protein ovalbumin. Larger antigens were unable to specifically bind and smaller antigens were found to nonspecifically bind, perhaps due to surface charge or hydrophobicity.

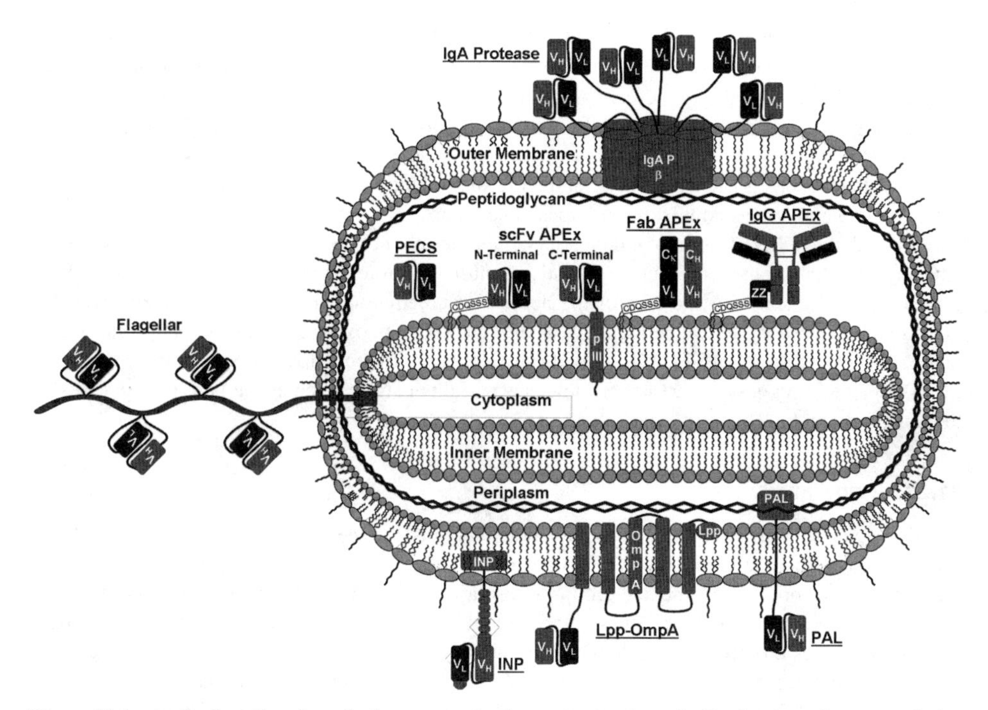

Figure 11.1 Antibody cell-surface display systems in Gram-negative bacteria. Depicted are the mature fusion proteins following processing and anchoring. In the peptidoglycan-associated lipoprotein (PAL) display system, the C-terminal domain of PAL from *E. coli* is fused to the C-terminus of an scFv (V_H-V_L or V_L-V_H), which is then displayed on the surface of the cell (Fuchs et al. 1991). The Lpp-OmpA system utilizes components from the major lipoprotein (Lpp) and outer membrane protein A (OmpA) of *E. coli* to display an scFv on the surface of a cell (Francisco, Earhart, and Georgiou 1992; Francisco et al. 1993). The first nine amino acids of Lpp properly target the fusion while the transmembrane helices of OmpA properly display the C-terminal scFv (Earhart 2000). The ice nucleation protein (INP) system utilizes the INP from *Pseudomonas syringae* as a fusion partner (H.C. Jung et al. 1998). The scFv is inserted either in or prior to the hydophilic C-terminal domain of the INP to display an scFv on the surface of *E. coli* (Bassi et al. 2000). Flagella display utilizes a major structural component of the flagellum, encoded by FliC, as a fusion partner for heterologous display (Kuwajima G et al. 1988). The scFv is inserted in an interior domain of fliC that is not necessary for the assembly of flagella and expressed on the surface of *E. coli* carrying deletions of the native *fliC* gene (Ezaki S, Takagi, and Imanaka 1998). Immunoglobulin A protease display utilizes the cell membrane domain of the autotransporter IgA protease from *Neisseria gonorrhoeae* as a fusion partner (Klauser, Pohlner, and Meyer 1990). The C-terminal β-domain of IgA Protease (IgA P β) is fused to the C-terminus of a scFv and used to anchor and transport the scFv to the surface of *E. coli* (Veiga, de Lorenzo, and Fernandez 1999). Anchored periplasmic expression (APEx) can be used to display scFvs, Fabs, and IgGs on the periplasmic face of the inner membrane in *E. coli* (Harvey et al. 2004; K.J. Jeong et al. 2007; Mazor et al. 2007). ScFvs are displayed as either N-terminal fusions with the first six amino acids (CDQSSS) of *E. coli* new lipoprotein A (NlpA) or as C-terminal fusions with the M13 bacteriophage gene 3 minor coat protein (p^{III}) (Harvey et al. 2004). Fabs are displayed by anchoring the entire light chain to the inner membrane using the NlpA format (V_L-C_κ), expressing the Fab heavy chain in the periplasm and allowing the heavy chain (V_H-C_H) to associate with the anchored light chain (K.J. Jeong et al. 2007). IgGs are displayed by expressing the entire full-length heavy and light chains in the periplasm where they assemble in aglycosylated full-length IgG and then capturing them with a synthetic Fc binding protein (ZZ) that is anchored using the NlpA format (Mazor et al. 2007). Periplasmic expression with cytometric screening (PECS) is a "display-less" system based on expressing an scFv in the periplasm of *E. coli* under appropriate conditions and temporarily shocking the cell to allow for free diffusion of antigen across the outer membrane (G. Chen et al. 2001).

As a first step in demonstrating the utility of PAL for library screening via FACS, cells displaying an scFv specific for either phOx or c-myc were enriched by FACS from a 1 : 1 mixture of each other following labeling with the fluorescently conjugated c-myc peptide. Enrichment was demonstrated through amplification of the scFv DNA by whole cell polymerase chain reaction (PCR) followed by restriction mapping. No future attempts using more stringent enrichment conditions were ever reported to demonstrate that the system has the potential to isolate clones with desired specificity from a library containing a large excess population of undesirable clones.

Although the utility of PAL for engineering antibodies has not been demonstrated, it was shown to have the potential for use as a bacterial bioadsorbant for environmental applications (Dhillon, Drew, and Porter 1999). An scFv specific for the herbicide and environmental pollutant atrazine was displayed and detected by whole cell ELISA using an alkaline phosphatase-atrazine conjugate. It was noted, however, that an anti-FLAG antibody did not strongly recognize an N-terminally displayed FLAG tag, again suggesting that larger molecular weight analytes, such as the IgG, may not be able to gain access to the PAL displayed scFv.

11.2.2 Lpp-OmpA

The display of proteins through outer membrane protein fusions seems to be the most obvious choice for a surface display system using Gram-negative bacteria. In 1986, Hofnung and coworkers and Henning and coworkers were the first groups to display peptides inserted in external loops of bacterial outer membrane proteins (Charbit et al. 1986; Freudl et al. 1986). However, inserting large proteins such as antibodies into the external loops often interferes with localization of the fusion protein and therefore unsuccessful display (Georgiou et al. 1997). This problem was overcome by Georgiou and coworkers in 1992 with the introduction of the Lpp-OmpA system (Francisco, Earhart, and Georgiou 1992).

The Lpp-OmpA tribrid system was designed based on the specific properties of the *E. coli* major lipoprotein (Lpp) and the outer membrane protein A (OmpA) (Earhart 2000). Lpp is an outer membrane protein localized to the periplasmic face of the outer membrane (Earhart 2000). OmpA is also an outer membrane protein, but it consists of eight membrane-spanning antiparallel β-strands forming a β-barrel structure (Earhart 2000). Independently, neither system appropriately displays fusion proteins on the bacterial surface (Earhart 2000; Ghrayeb and Inouye 1984). However, Georgiou and coworkers combined regions of each protein to successfully display an enzyme on the surface of *E. coli* as an Lpp-OmpA fusion (Francisco, Earhart, and Georgiou 1992).

The Lpp domain is responsible for targeting and anchoring of the heterologous protein to the outer membrane while the OmpA domain is responsible for localizing the protein to be displayed on the surface of the cell (Earhart 2000). The system is comprised of the signal sequence and the first nine amino acids of Lpp, a truncated form of OmpA containing five of the eight transmembrane helices and the protein of interest to be displayed. The Lpp sequence was found to be critical for efficient localization of the heterologous protein to the outer membrane. Additional studies revealed that further truncations of OmpA are also suitable for surface display of proteins (Georgiou et al. 1996). Since OmpA is highly conserved among Enterobacteria (Cole et al. 1982), the Lpp-OmpA system is amenable for use with other bacteria.

The initial work using the Lpp-OmpA system displayed the enzyme β-lactamase in a functional form. Following this, Georgiou and coworkers functionally displayed the enzyme cellulose (Francisco et al. 1993). An interesting aspect of cellulose is that it contains a catalytic domain as well as a cellulose-binding domain. Remarkably, display of the cellulose-binding domain alone was able to bind cellulose and this paved the way for the surface display of an antibody using this system.

In 1993, Georgiou and coworkers were the first to demonstrate the potential of bacterial surface display as a means to isolate antibodies of desired specificity (Francisco et al. 1993). They incorporated the use of flow cytometry with the Lpp-OmpA system to show the functional display of an scFv specific for digoxin on the surface of *E. coli.* As compared to cells displaying an enzyme, cells displaying the scFv were specifically labeled with a fluorescent conjugate form of the hapten digoxin. The scFv displaying cells were then mixed with a 100,000-fold excess of control bacteria lacking the fusion protein. The mixed population was fluorescently labeled with the digoxin conjugate and subjected to FACS followed by growth amplification of the sorted cells. Following two rounds of FACS, cells expressing the scFv were enriched to near homogeneity from this 100,000-fold excess of nonexpressing cells. For the first time, flow cytometry was used in conjunction with surface display of antibodies for the selection of antibodies with desired properties from a pool of undesired cells.

The signals obtained on the flow cytometer using the Lpp-OmpA system were encouraging for future use of the system in engineering applications. Scatchard analysis determined the presence of 60,000 scFvs displayed on the surface of each cell (G. Chen et al. 1996). This level of surface display is higher than other reported systems and partially explains the excellent signals previously observed on the flow cytometer (Francisco et al. 1993). The analysis also revealed that all the scFvs are functional and the system was used to measure the affinity of the scFv within twofold of biosensor measurements (G. Chen et al. 1996). The scFv expressing cells could also be used as a simple immunoprecipitation reagent by incubating the cells in a sample containing the antigen of interest followed by centrifugation (G. Chen et al. 1996).

Georgiou and coworkers' follow-up work in 1998 demonstrated the utility of the system in an actual antibody engineering application (Daugherty et al. 1998). First, they showed the system could be used quantitatively by discriminating between a few point mutants of the previously used anti-digoxin scFv that varied in affinity. Based on these results, they attempted to affinity mature the anti-digoxin scFv using the system. An scFv library was created by randomizing certain light chain residues. The library of scFvs displayed as Lpp-OmpA fusions was subjected to just one round of FACS under kinetic constrained conditions to isolate higher affinity scFvs. The highest affinity scFv isolated showed a threefold increase in affinity resulting from two point mutations located in the light chain complementarity determining region 3 (CDR 3).

Further experiments investigated the effect of mutation frequency in the affinity maturation of an antibody (Daugherty et al. 2000). Libraries were constructed by error-prone PCR of the scFv gene using varying mutational frequencies and subjected to FACS. Libraries with moderate to high error rates, in the range of 3.8 to 22.5 mutations per gene, resulted in higher affinity clones than lower error rates of 1.7 mutations per gene. The results indicated that hypermutated libraries contain unexpectedly high numbers of functional clones, higher affinity clones are well represented in these libraries, and the majority of mutations resulting in increased scFv affinity are located at positions distant from the binding site.

Despite the thorough evaluation of the Lpp-OmpA system for display and subsequent enrichment and maturation of anti-digoxin binding antibodies, no literature to date evaluates the display of antibodies using Lpp-OmpA on *E. coli* to target larger antigens, a requirement for engineering antibodies targeting the majority of protein antigens of interest. However, an antibody specific for the protein antigen carcinoma marker, carcinoembryonic antigen (CEA), was recently displayed on the surface of the Gram-negative bacteria *Salmonella typhimurium*, via the Lpp-OmpA system and used for tumor targeting *in vivo* (Bereta et al. 2007). Salmonella displaying the anti-CEA scFv were specifically labeled by incubation with a soluble CEA fluorophore conjugate. However, whether or not the anti-CEA scFv displayed on *E. coli* would recognize the protein antigen was not within the scope of the study and was therefore not determined.

Even though only a select few antibodies have been displayed using the Lpp-OmpA system, a variety of other proteins have been displayed, including the ones previously mentioned (Francisco, Earhart, and Georgiou 1992; Francisco et al. 1993). These include enzymes for use in biocatalysis (Curnow et al. 2005; Richins et al. 1997; Wan, Chang, and Lin 2002), peptides for vaccine delivery (Burnett et al. 2000), and bioaccumulation of heavy metals (Bae et al. 2000, 2001), green fluorescent protein (Shi and Wen Su 2001), and various binding domains (Francisco et al. 1993; Stephens, Choe, and Earhart 1995; A.A. Wang, Mulchandani, and Chen 2001; J.Y. Wang and Chao 2006). The utility of the system for various applications has been demonstrated, but future work is needed to determine the robustness of the system for broader antibody engineering applications.

11.2.3 Ice Nucleation Protein (INP)

The ice nucleation protein (INP) is a glycosyl phosphatidylinositol anchored outer membrane protein found in certain Gram-negative bacteria that catalyzes the formation of ice crystals from supercooled water (H.C. Jung, Lebeault, and Pan 1998; H.C. Jung et al. 1998). It contains a GPI-like anchor which is typically used for the attachment of surface proteins in eukaryotic cells (H.C. Jung et al. 1998; Nosjean 1998). An example of this is the yeast surface protein α-agglutinin, which is a protein previously used for anchoring in the yeast surface display system (Boder and Wittrup 1997). The INP from *Pseudomonas syringae* was first demonstrated as a surface display anchor by Pan and coworkers in 1998 (H.C. Jung et al. 1998).

There are several INP genes associated with the different species of *Pseudomonas*, but the one most frequently used for surface display is from *P. syringae* (H.C. Jung et al. 1998; Shimazu, Mulchandani, and Chen 2001a, 2001b). The INP system is simple in that it is comprised of only the INP as an N-terminal fusion to the protein of interest to be displayed. This is possible since the N-terminus of the INP contains an internal signal for translocation and localization and the C-terminus is exposed on the cell surface. The INP also contains an internal repeat domain that is required for nucleation activity, but not required for surface display of the fusion protein (H.C. Jung et al. 1998).

Pan and coworkers first displayed two different enzymes on the surface of *E. coli* and the activity of both the INP and the enzymes were maintained (H.C. Jung, Lebeault, and Pan1998; H.C. Jung et al. 1998). These enzymes were levansucrase (LevU) and carboxymehtylcellulase (CMCase) from *Zymomonas mobilis* and *Bacillus subtilis*, respectively. Surface display of the fusion proteins was verified by measuring the ice nucleation activity of the INP, the native activity of the displayed enzymes, as well as with electron microscopy and flow cytometry (H.C. Jung, Lebeault, and Pan 1998; H.C. Jung et al. 1998). The flow cytometric studies were based on detection of surface level expression of the enzymes using an antibody specific for the displayed enzyme (H.C. Jung et al. 1998). Fluorescent signals obtained were distinguishable from background and showed little overlap with the negative controls. These results were encouraging for potential use of the systems with antibodies for two reasons. First, detection antibodies were able to specifically interact with the displayed enzyme. If a 150 kDa IgG can recognize the displayed enzyme, displayed antibodies may be able to interact with and bind antigens of comparable size. Second, the level of display is high enough to allow for detection by flow cytometry.

The INP system was first shown to be a compatible platform for the bacterial surface display of antibodies by Margaritis and coworkers in 2000 (Bassi et al. 2000). They successfully expressed a functional scFv specific for the oncoprotein c-myc on the surface of *E. coli* using two versions of the INP system. In these systems, the gene encoding the scFv was inserted either prior to or in the hydrophilic C-terminal domain of the INP gene. Specific binding of the scFv was demonstrated using flow cytometry with a fluorescent conjugate of the c-myc peptide. However, binding was dependent on pretreatment of the cells with EDTA as is seen with other Gram-negative bacteria surface display systems (Bassi et al. 2000; Fuchs et al. 1996; Harvey et al. 2004; Leive 1968). Although the cells expressing the fusion protein seemed to be specifically labeled, there was also some nonspecific labeling of cells expressing just the INP. Further, the flow cytometry signals were very broad for the C-terminal insertion system and bimodal for the other system. This resulted in signal from cells expressing the fusion protein overlapping considerably with cells expressing the INP alone, which could complicate enrichment by FACS.

Although only one antibody has been successfully displayed on the surface of *E. coli* using the INP system, a variety of non-antibody proteins have been displayed in addition to the previously mentioned ones (H.C. Jung, Lebeault, and Pan 1998; H.C. Jung et al. 1998). These include enzymes for biocatalysis (P.H. Wu, Tsai, and Chen 2006; Yim et al. 2006), bioremediation (Li, Kang, and Cha 2004; Shimazu, Mulchandani, and Chen 2001a; Zhang et al. 2004), antifungal activity (M.L. Wu, Tsai, and Chen 2006), and other purposes (H. Jeong, Yoo, and Kim 2001). Further, proteins have been displayed using the INP system for vaccine delivery (E.J. Kim and Yoo 1999; Kwak, Yoo, and Kim 1999) and as diagnostic tools (Kang et al. 2003). The system has also been used for the directed evolution of two different enzymes (H.C. Jung et al. 2003; Y.S. Kim, Jung, and Pan 2000). The INP can also be expressed on the surface of many different Gram-negative bacteria and has already been used to display various antigens and enzymes to the cell surface of *Salmonella*, *Moraxella*, and other strains of *Pseudomonas* for a variety of applications (Drainas, Vartholomatos, and Panopoulos 1995; H.C. Jung, Kwon, and Pan 2006; J.S. Lee et al. 2000; Lei et al. 2005; Shimazu, Mulchandani, and Chen 2001b; Shimazu et al. 2003; Yang et al. 2008).

11.2.4 Flagella

Flagella have been used for the display of proteins on the surface of *E. coli* by several groups since 1988 (Ezaki S, Takagi, and Imanaka 1998; Kuwajima G et al. 1988; Lu et al. 1995; Westerlund-Wikstrom et al. 1997). Most of these engineered systems utilize the major structural component of the flagellum, flagellin (which is also known as FliC and formerly known as *hag*), as a fusion partner for heterologous display. Although a variety of proteins have been displayed using these systems, there is only one example of using flagella to display a scFv. In 1998, Imanaka and coworkers successfully engineered a version of the FliC system to display several different fusion proteins on the surface of *E. coli*, ranging in size from 1.2 kDa to 49.4 kDa, including an anti-porphyrin scFv (Ezaki S, Takagi, and Imanaka 1998).

The fusions were created by inserting the protein to be displayed into the dispensable interior D3 domain of FliC, which is not necessary for the assembly of flagella, and subsequent expression in a flagellin-deficient host strain. Expression and display were determined by Western blot analysis of the flagellar purified fraction (Ezaki S, Takagi, and Imanaka 1998). The smaller fusions were present on the flagellar surface in levels similar to the native FliC, but the scFv and other large fusions were present at much lower levels. It is unknown if the displayed scFv was functional since this was never tested. However, the 302 amino acid collagen binding domain of *Yersinia enterocolitica* has been functionally displayed as a FliC fusion, demonstrating that larger size fusions can be displayed without loss of function (Westerlund-Wikstrom et al. 1997).

One potential advantage of the flagella surface display system is the flagellar filament is formed by the assembly of approximately 20,000 units of FliC. If a large percentage of these can incorporate antibody fusions, the level of display might be suitable for use with flow cytometry. However, it has yet to be shown if the FliC displayed scFv is capable of specifically binding antigen. Further, there is no flow cytometry evidence that the scFv can be detected on the surface for expression or binding.

Much more work needs to be done before flagella display can be considered a practical antibody display platform. So far, only one of the flagella display systems has been used with an antibody and functionality was not demonstrated. However, a variety of non-antibody proteins have been functionally displayed using variations of the system in *E. coli* (Westerlund-Wikstrom 2000). These include peptides for vaccine delivery (Kuwajima G et al. 1988), peptides for binding studies (Lu et al. 1995), and other binding domains (Westerlund-Wikstrom et al. 1997). Various species of bacteria contain flagella and therefore have the potential to be exploited for the display of heterologous proteins, including antibodies. For example, the flagella from *Salmonella* have been repeatedly engineered to display peptides from various toxins and utilized as vaccine delivery systems (McEwen et al. 1992; Newton, Jacob, and Stocker 1989; Newton et al. 1991; J.Y. Wu et al. 1989).

11.2.5 IgA Protease

Certain species of Gram-negative bacteria contain specialized outer membrane proteins known as autotransporters (Dautin and Bernstein 2007). Autotransporters are composed of an outer membrane localized C-terminal β domain that facilitates the transport of an N-terminal passenger domain into the extracellular milieu (Dautin and Bernstein 2007). In 1990, Meyer and coworkers were the first to utilize an autotransporter for the display of a heterologous protein (Klauser, Pohlner, and Meyer 1990). They fused the β subunit of the cholera toxin between the N-terminal signal sequence and C-terminal β domain of the immunoglobulin A protease from *Neisseria gonorrhoeae* and expressed it on the surface of *Salmonella typhimurium*. Since then, various proteins have been expressed as fusions with the IgA protease β domain in Gram-negative bacteria, including *E. coli*.

These initial studies suggested the passenger protein needed to be maintained as a linear polypeptide devoid of disulfide bonds prior to transport across the outer membrane (Klauser, Pohlner, and Meyer 1990). Transport was thought to occur through a narrow pore created from a monomeric β domain (Maurer, Jose, and Meyer 1997). Therefore, autotransporter display of antibody fragments did not seem feasible since the formation of the intramolecular disulfide bonds in the oxidizing periplasm prior to transport is a requirement for antibody stability and function (Veiga, deLorenzo, and Fernandez 1999). However, de Lorenzo and coworkers disproved this theory and were able to display a functional scFv as the passenger protein on the surface of *E. coli* (Veiga, deLorenzo, and Fernandez 1999).

Although this effort was directed at better understanding the relationship of protein secretion and disulfide bond formation, rather than a platform for antibody engineering, it was successful at displaying correctly folded scFv, albeit at a low level. They fused an scFv specific for a polyhistidine tag between the pelB leader peptide and the β domain of IgA protease from *N. gonorrhoeae*. *E. coli* expressing the scFv fusion were shown to specifically bind immobilized dihydrofolate reductase containing a polyhistidine tag via an enzyme linked immunosorbent assay (ELISA). This suggests that the scFv is displayed in such a way that allows for ELISA surface interaction. The main apparent limitation of this system was that the presence of correctly folded scFv was low due to the inability of the system to efficiently export correctly folded disulfide containing scFv across the outer membrane. The level of surface display was increased approximately threefold by preventing the disulfide bonds from forming in the periplasm, but the displayed antibody was not functional. Subsequent work revealed that this antibody has a low thermodynamic stability and folding yield, which may have contributed to the display issues encountered (Veiga, deLorenzo, and Fernandez 2004).

Further work by de Lorenzo and coworkers revealed that a translocation pore was formed by an oligomeric ring-shaped structure containing at least six monomers (Veiga et al. 2002). This disproved previous theories that each monomeric autotransporter was responsible for transport of its N-terminal passenger and explains how scFvs can be transported following folding and disulfide bond formation in the periplasm. The pore size was shown to be large enough for transport of folded immunoglobulin domains and suggested transport should be more efficient than they originally demonstrated.

de Lorenzo and coworkers demonstrated the system can indeed efficiently transport not only scFvs, but single domain antibodies (dAb) as well as chains of single domains (Veiga, deLorenzo, and Fernandez 2004). By utilizing an scFv engineered for enhanced stability, it was shown that antibodies folded in the periplasm prior to export retained a higher degree of their innate binding function than those expressed in cells lacking chaperones to prevent aggregation. This effect was more dramatic when displaying dAbs that were not engineered for increased stability.

To date, the utility of autotransporters for combinatorial protein library screening applications has not been demonstrated. However, the system has been utilized to display a variety of different proteins for other applications. These include an scFv specific for transmissible gastroenteritis coronavirus (TGEV) that demonstrated viral neutralization properties in a cell-based assay (Veiga, deLorenzo, and Fernandez 2003b), a heavy metal binding protein for use in bioremediation (Valls et al. 2000), and binding proteins for cellular adhesion (Veiga, deLorenzo, and Fernandez 2003a).

11.2.6 Inner Membrane Display

Outer membrane proteins are the most obvious choice for displaying antibodies on the surface of Gram-negative cells for exogenous labeling. However, Georgiou and coworkers described a system in 2004 that is based on displaying antibodies on the inner membrane (Harvey et al. 2004). Anchored Periplasmic Expression (APEx) is an *E. coli* protein display platform based on anchoring proteins to the periplasmic face of the inner membrane. Following expression of the displayed protein, the outer membrane and peptidoglycan layer are disrupted by chemical and enzymatic means; the resulting spheroplasts are specifically labeled by exogenous fluorescently labeled molecules and subsequently analyzed by flow cytometry.

Proteins are displayed on the inner membrane by creating either N-terminal or C-terminal fusions with the first six amino acids of the *E. coli* mature lipoprotein, new lipoprotein A (NlpA), or the M13 bacteriophage gene 3 minor coat protein (g3p), respectively. Although better fluorescent discrimination is seen using the NlpA anchoring approach, pIII anchoring is sufficient to differentiate between positive and negative events and would allow existing phagemid libraries to be utilized in the APEx system (Harvey et al. 2004). It was later demonstrated that truncations of inner membrane proteins, such as TatC or MalF, could also be used for antibody fragment anchoring, although the fluorescent distribution of signal was not as defined as with NlpA or g3p anchoring (S.T. Jung et al. 2007; Ki et al. 2004).

APEx has been used to display antibodies specific for a protein as well as for two different haptens. Functionality was determined by labeling the cells with fluorescently conjugated forms of their respective antigens and subsequent analysis by flow cytometry. The specifically labeled cells had tight signals with clean discrimination from cells displaying a nonspecific antibody control. Further, a digoxin–phycoerythrin conjugate of 240 kDa was specifically recognized by the previously used anti-digoxin scFv which demonstrates the compatibility of the system with large antigens (G. Chen et al. 2001; Daugherty et al. 1998, 1999, 2000; Francisco et al. 1993; Harvey et al. 2004; Ribnicky, Van Blarcom, and Georgiou 2007). Under the proper conditions, spheroplasts displaying antibodies are even suitable for specific binding to immobilized surface antigen (S.T. Jung et al. 2007).

The APEx system has been utilized for the affinity maturation of two different scFvs to date (Harvey et al. 2004, 2006). One scFv known as 14B7 is specific for the 83 kDa protective antigen (PA) component of the *Bacillus anthracis* endotoxin and the other is specific for the hapten methamphetamine. Both scFv libraries used were constructed by error-prone PCR and subjected to two rounds of FACS. For the 14B7 library, the scFv affinity increased by over 100-fold from 4 nM to less than 40 pM, demonstrating affinity maturation of an antibody to a protein antigen utilizing a bacterial display platform. For the anti-methamphetamine library, the scFv affinity increase was a more moderate 3 fold.

A potential advantage of the APEx system is that the displayed antibody is located in the periplasm. This enables the antibody to interact with other proteins expressed in the periplasm. This concept resulted in the birth of the APEx two-hybrid system. The first demonstration of this system relied on expressing both an scFv in the APEx format along with the soluble expression of a fusion between a peptide antigen and a green fluorescent protein in the periplasm (Harvey et al. 2004). Following expression of both the scFv and the fusion, the cells were converted to spheroplasts to allow the unbound fusion to be released. Cells expressing an scFv specific for the peptide exhibited a higher fluorescent signal than cells expressing an scFv specific for an unrelated antigen. Although the signal was not as clean as compared to cells labeled with exogenous antigen, this may be the result of the particular antibody antigen pair being used instead of the system.

A variation of this concept was demonstrated in which the scFv is anchored in the APEx format while the antigen it is specific for is coexpressed as a soluble protein in the periplasm with an epitope tag on its C-terminus. Following conversion to spheroplasts, unassociated antigen is released and associated antigen is specifically labeled via an anti-epitope tag antibody conjugated to a fluorophore. This system was used to increase the affinity of the 14B7 antibody by more than 10-fold. Although the flow cytometry signals were rather broad as was seen previously with the peptide-GFP reporter, this was less of an issue with the high affinity M18 variant of 14B7 previously isolated using the original APEx system (Harvey et al. 2004; K.J. Jeong et al. 2007).

Another variation of the APEx platform demonstrated its compatibility with antibody-binding domains in the Fab format. Here, the light chain of the Fab (V_L-C_κ) is anchored in the APEx format while the heavy chain of the Fab (V_H-C_{H1}) is expressed as a soluble protein in the periplasm and allowed to dimerize with the light chain. Following conversion to spheroplasts, the Fab was labeled with both a fluorescently labeled antigen to detect antigen binding as well as a fluorescently labeled secondary antibody specific for an epitope tag located on the C-terminus of the heavy chain to detect assembly. This system was used to increase the expression of an anti-PA Fab at 37°C by more than fivefold. The signals obtained were tighter than those found when both the antibody and antigen were endogenously expressed, but they were still broader than desired (K.J. Jeong et al. 2007).

Recent engineering modifications to the APEx technology have led to the first full-length antibody display system in *E. coli* that is applicable to the screening of antibody libraries (Mazor et al. 2007). Designated the *E*-clonal system, full-length heavy and light chains are secreted into the periplasm of *E. coli* where they assemble into aglycosylated full-length IgG. The assembled IgG are then captured by an Fc-binding protein known as the ZZ domain that is anchored to the inner membrane using the APEx format. The cells are converted to spheroplasts, specifically labeled with fluorescently conjugated antigen and analyzed by flow cytometry. The system was used to isolate an array of antibodies from a mouse immune library directed against the PA toxin. Five antibodies were isolated ranging from 20 nM to 500 nM in affinity, with the majority less than 60 nM. The fluorescent signals obtained through specific interaction with the antigen fluorophore were strong and similar to those obtained using the original APEx system (Harvey et al. 2004, 2006; Mazor et al. 2007).

Potential advantages of the *E*-clonal system over previous display systems have been discussed (Mazor et al. 2007). First, selecting a protein as a fusion partners can alter its innate stability (Hayhurst 2000). This topic is further discussed in Section 11.2.7 below (G. Chen et al. 2001). This is not a concern in the *E*-clonal system since the IgG can be preassembled in the periplasm prior to being captured by the Fc-binding protein. Second, the avidity associated with the bivalent display of IgGs might aid in the identification of antibodies of lower affinity that may not be detectable using monovalent scFv or Fab display technologies. Third, although bacterial expressed IgG is aglycosylated, the serum half-life is not affected by the aglycosylated form of an IgG and other IgG effector functions are not a requirement for some antibody therapeutic applications (Mazor et al. 2007; Reichert et al. 2005; Tao and Morrison 1989). Therefore, selecting for full-length IgG circumvents the need to convert antibody fragments to IgGs and *E. coli*-produced IgG could potentially be used directly to test for functionality in some applications. This is most apparent for therapeutic purposes where the vast majority of the FDA-approved antibodies are full length (Benhar 2007; Carter 2006; Holliger and Hudson 2005).

Regardless of the APEx system used, it is important to note that *E. coli* spheroplasts are not viable and cannot be grown following a round of FACS. Similar viability issues also arose with the PAL display platform (Fuchs et al. 1996). Therefore, the DNA encoding the antibody, whether it is an scFv, Fab, or IgG, must be recovered by PCR following each round of FACS and recloned into the appropriate vector for subsequent rounds of FACS or soluble monoclonal screening. Although this process does increase the duration of each round of sorting, there are potential advantages (Harvey et al. 2004). Antibodies produced that have a toxic effect on the host organism, which may otherwise be lost as a result of growth disadvantages, have a better chance of being recovered. In addition, each PCR recovery could be performed under mutagenic conditions to promote evolution in each round by coupling mutagenesis and selection.

11.2.7 "Display-Less" System

In order to display an antibody on the surface of a biological particle, it is typically required that a fusion between the antibody and an anchoring motif be created. As has already been discussed, this is accomplished by creating either an N- or C-terminal fusion. Antibody fusions can demonstrate increased stability, which is not necessarily beneficial since the antibody will typically need to function as a soluble non-fusion for future applications (Hayhurst et al. 2003). For example, it is frequently observed that scFvs isolated as fusions using phage display do not function when expressed on their own, although there is not much literature available on this subject (Jensen et al. 2002). As a result, it can be beneficial to have a display system that does not rely on the creation of a fusion protein and at the same time retains the link between genotype and phenotype.

An innovative "display-less" system referred to as periplasmic expression with cytometric screening (PECS) was created by Georgiou and coworkers in 2001 (G. Chen et al. 2001). Following expression of an scFv in the periplasm of *E. coli*, the cells are specifically labeled with fluorescently conjugated antigen up to 10 kDa in size and are detectable by flow cytometry. By using the appropriate combination of bacterial strains and growth conditions followed by incubation with antigen under high osmotic conditions, they were able to increase the diffusion limitation of the outer membrane by over an order of magnitude while minimally compromising cell viability (G. Chen et al. 2001; Ribnicky, Van Blarcom, and Georgiou 2007). Depending on the antibody-antigen pair used, the signal from specifically labeling cells expressing scFv with antigen-conjugated fluorophore varied from relatively broad with significant background overlap to impressively tight with minimal background overlap.

Using this system, an affinity matured variant of the anti-digoxin scFv used previously with the Lpp-OmpA surface display system was isolated (G. Chen et al. 1996; Daugherty et al. 1998, 1999, 2000; Francisco et al. 1993). Based on previous results (Daugherty et al. 1998), only three residues in the light chain CDR 3 were randomized during library construction, which resulted in a modest twofold improvement after two rounds of sorting.

A convenient aspect of PECS is its amenability for use with existing phage display libraries. Many phage display vectors contain an amber codon located between the 3′ end of the scFv gene and the 5′ end of the phage coat protein pIII gene used for anchoring the scFv to the phage particle. Incomplete suppression of the amber codon results in early translation termination following scFv synthesis leading to the presence of soluble scFv located in the periplasm suitable for screening by PECS. This was demonstrated by isolating anti-hapten antibodies from the semisynthetic Griffin library (G. Chen et al. 2001). Standard phage panning was used in combination with PECS to isolate anti-digoxin specific scFvs. Of particular interest was the isolation of a high expressing scFv by PECS that behaved poorly when expressed as a pIII fusion on phage. This result demonstrates how the behavior of an scFv can be adversely affected when expressed as a fusion partner in a display system and how "display-less" approaches may improve antibody discovery.

The PECS system has recently been used in an innovative genetic selection to isolate a faster folder scFv variant (Ribnicky, Van Blarcom, and Georgiou 2007). This was accomplished by exploiting the proofreading properties of the twin arginine translocation pathway (Tat). Unlike the Sec pathway for protein export, the Tat pathway requires proteins to be folded prior to export from the cytoplasm (P.A. Lee, Tullman-Ercek, and Georgiou 2006). By fusing a Tat competent leader peptide to the N-terminus of the anti-digoxin scFv and expressing this fusion in an *E. coli* cell line with an oxidizing cytoplasm, the scFv was found to be exported to the periplasm. These cells could also be specifically labeled using the PECS conditions and detected by flow cytometry, but with approximately 10 percent of the scFv export obtained using a Sec leader. In an attempt to isolate a more efficiently exported mutant of this scFv, an error-prone library was made and subjected to three rounds of FACS. This resulted in the isolation of a mutant scFv that exhibited faster folding which in turn led to higher export efficiency, providing an example of how existing systems can be used to isolate antibodies with unique properties.

11.3 SYSTEMS FOR GRAM-POSITIVE BACTERIA

As is the case with Gram-negative bacteria, the display of proteins on the surface of Gram-positive bacteria has numerous biotechnology applications (Georgiou et al. 1993; Samuelson et al. 2002). Differences in the cell structure and composition of Gram-positive bacteria make them an excellent alternative to Gram-negative bacteria for surface display of proteins (Schneewind, Mihaylova-Petkov, and Model 1993; Wernerus and Stahl 2002). Without an outer membrane, translocation from the cytoplasm to the surface of the cell is in theory less invasive (Little et al. 1994). Many of the heterologous proteins displayed to date have been covalently linked to the peptidoglycan layer

(Schneewind, Model, and Fischetti 1992; Schneewind, Mihaylova-Petkov, and Model 1993). This linkage is catalyzed by a sortase enzyme which recognizes a common peptidoglycan layer anchoring motif, a C-terminal LPXTG domain (Marraffini, Dedent, and Schneewind 2006; Paterson and Mitchell 2004). Alternatively, some have relied on using the coat proteins of endospores (Beninati et al. 2000; Du et al. 2005; Gunneriusson et al. 1996, 1999; Hansson et al. 1992; Kruger et al. 2002; Liljeqvist et al. 1997, 1999; Lofblom, Wernerus, and Stahl 2005; Lofblom et al. 2007; Nord et al. 1997; Pant et al. 2006; Pozzi et al. 1992; Wernerus et al. 2001).

Several species of Gram-positive bacteria, including Staphylococci, Streptococci, Lactococci, and Lactobacillus, have been engineered to display various proteins using different targeting and anchoring strategies (Benhar 2001; Leenhouts, Buist, and Kok 1999; Wernerus and Stahl 2002). The major use of these systems has been for mucosal vaccine delivery (Fischetti, Medaglini, and Pozzi 1996). However, a subset of these systems have been shown to be compatible with the display of an antibody or other binding proteins (Beninati et al. 2000; Du et al. 2005; Gunneriusson et al. 1996, 1999; Hansson et al. 1992; Kruger et al. 2002; Liljeqvist et al. 1997, 1999; Lofblom, Wernerus, and Stahl 2005; Lofblom et al. 2007; Nord et al. 1997; Pant et al. 2006; Pozzi et al. 1992; Wernerus et al. 2001). In addition to scFvs and dAbs we have also included in our discussions affibodies, an alternative scaffold consisting of a 58 amino acid residue domain from staphylococcal protein A, engineered for specific antigen binding. Recent success with an affibody selection strategy represents the first use of Gram-positive display for the isolation of affinity proteins (Kronqvist et al. 2008). Gram-positive display for antibody isolation has yet to be demonstrated, although the potential for antibody engineering applications has been shown (Lofblom, Wernerus, and Stahl 2005; Wernerus, Samuelson, and Stahl 2003). Display platforms discussed will be evaluated for their ability to display functional antibodies, their compatibility with high throughput FACS, and how each system's unique features contribute to its range of applications.

11.3.1 Protein A Display

Two different systems were reported in parallel in 1992 that were the first to demonstrate the use of Gram-positive bacteria for the surface display of proteins (Hansson et al. 1992). One system reported by Uhlen and coworkers utilized the surface attachment regions from *Staphylococcus aureus* protein A (Goding 1978). Protein A plays a role in the pathogenicity of *S. aureus* by helping it avoid host organism immunological defenses. It has long been exploited in various ways due to its ability to bind the Fc domains of an immunoglobulin (Robert et al. 1996). Protein A contains a signal sequence that is processed during secretion, five surface exposed immunoglobulin domains, a cell wall spanning region, and a cell-wall anchoring region (Robert et al. 1996; Schneewind, Model, and Fischetti 1992). The anchoring region is a common tripartite domain containing the LPXTG motif, a hydrophobic region, and a charged tail (Robert et al. 1996; Schneewind, Model, and Fischetti 1992). Following proteolytic cleavage of the LPXTG motif, protein A is covalently linked to the cell wall by a sortase enzyme (Hansson et al. 1992). By utilizing the signal sequence and the C-terminal region of protein A, heterologous proteins were displayed on the surface of *Staphylococcus xylosus*, which does not normally express protein A (Gotz 1990; Liebl and Gotz 1986; Samuelson et al. 1995; Fig. 11.2).

The first fusion protein displayed was an albumin-binding protein (ABP) derived from streptococcal protein G and it was inserted between the signal peptide and the cell wall anchoring region of protein A (Hansson et al. 1992). Approximately half the cells containing the fusion protein could be specifically labeled based on fluorescent and electron microscopy performed with anti-sera. This was encouraging since it indicated that large proteins can interact with proteins displayed on the cell surface using this system. Additionally, the ABP and a peptide were inserted in series between the signal peptide and the cell wall anchoring region. While approximately half the cells were again specifically labeled by the terminal ABP, no more than 15 percent were specifically labeled with antibodies specific for the internal peptide. Although it makes sense for the more external protein to be more accessible and hence easier to bind, it does not explain why half the cells do now show any specific labeling.

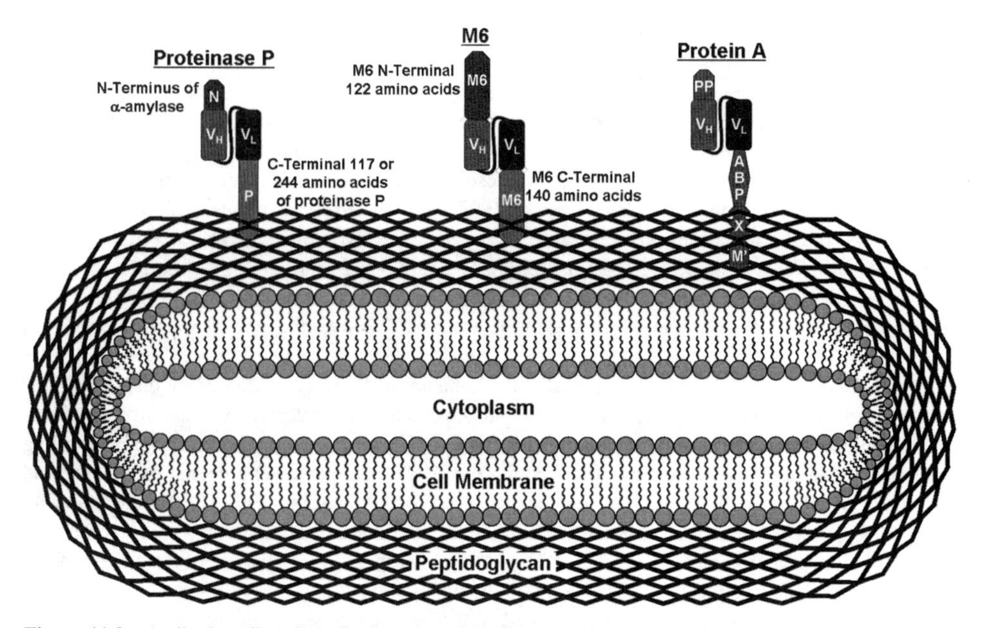

Figure 11.2 Antibody cell-surface display systems in Gram-positive bacteria. Depicted are the mature fusion proteins following processing and anchoring. Protein A display is based on using *Staphylococcus aureus* protein A (Gunneriusson et al. 1996; Samuelson et al. 1995; Wernerus and Stahl 2002). The mature fusion protein is comprised of the propeptide from *S. hyicus* (PP), the scFv, 198 amino acids from *streptococcal* protein G serum albumin binding region (ABP), the cell wall spanning region of protein A (X), and the processed form of the M sequence of protein A that is covalently anchored to the cell wall (M′). Note that the propeptide is processed in *Staphylococcus xylosus* but not *Staphylococcus carnosus*. M6 display utilizes the M6 protein from *Streptococcus pyogenes* as a fusion partner (Giomarelli et al. 2004; Oggioni et al. 1999). The displayed protein is composed of the N-terminal 122 amino acids of M6, the scFv, and the C-terminal 140 amino acids of M6 which contains the cell wall anchoring domain. Proteinase P display incorporates the cell wall anchoring domain of proteinase P from *Lactobacillus zeae* or *casei* (Kruger et al. 2002; Maassen et al. 1999). The displayed scFv depicted here utilized the N-terminal 26 amino acids of α-amylase from *Lactobacillus casei* (N) and either the C-terminal 117 or 244 amino acids of proteinase P from *L. zeae* (P).

In Stahl and coworkers' 1995 follow-up work, they engineered an alternative display system by modifying the original protein A system. They switched from using *S. xylosus* to *Staphylococcus carnosus* and replaced the promoter and signal sequence from protein A with a promoter, signal sequence, and propeptide region from a *Staphylococcus hyicus* ligase, previously optimized for expression in *S. carnosus* (Samuelson et al. 1995). The propeptide was included since it was previously demonstrated to be essential for secretion in *S. carnosus*. They continued to use ABP, but used it as a spacer to extend the displayed protein, in this case the same peptide used previously, away from the cell wall (Gunneriusson et al. 1996). Based on fluorescent and electron microscopy, nearly all the cells were specifically labeled by both the external peptide and the internal ABP spacer using antibodies specific for each. Flow cytometric analysis with ABP-specific antibodies confirmed the presence of the ABP spacer with and without the additional peptide. Although there was some nonspecific binding associated with cells not expressing either construct, both display constructs gave very tight signals highly distinguishable from background. Unfortunately, they did not demonstrate specific labeling with albumin or perform any FACS enrichment experiments.

The protein A system in both *S. xylosus* and *S. carnosus* were finally demonstrated to be compatible with the display of antibodies by Stahl and coworkers in 1996 (Gunneriusson et al. 1996; Samuelson et al. 1995). Based on previous work with the *S. carnosus* system, they modified the *S. xylosus* display system by utilizing the ABP as a spacer to separate the scFv from the surface of the cell (Gunneriusson et al. 1996). The *S. carnosus* system remained unchanged, with the exception of exchanging the displayed peptide for an scFv (Andreoni et al. 1997; Robert et al. 1996). Both systems were demonstrated to functionally display an scFv specific for human IgE based on ELISA. Compared to control cells, the *S. carnosus* cells displaying the scFv specifically bound human IgE significantly better than the *S. xylosus* displaying cells. This result is likely the consequence of increased levels of surface displayed protein (Andreoni et al. 1997). Follow-up experiments with non-antibody proteins confirmed a higher level of displayed proteins on the surface of *S. carnosus* (Gunneriusson et al. 1999; Liljeqvist et al. 1997, 1999; Lofblom, Wernerus, and Stahl 2005; Lofblom et al. 2007; Nord et al. 1997; Wernerus et al. 2001). However, following removal of the scFv, both systems demonstrated a high degree of nonspecific binding to the antigen. This was unexpected and no explanation was offered.

Although no additional publications exist in the literature with regard to the surface display of antibodies on either *S. xylosus* or *S. carnosus*, there are several articles that pertain to the display of affibodies and other protein-binding domains (Skerra 2007). Affibodies are one of several protein-binding scaffolds that are alternatives to antibodies (Nord et al. 1997). They were engineered by Nygren and coworkers in 1997 and first displayed on the surface of *S. carnosus* in 1999 (Gunneriusson et al. 1999). They are based on the Z domain, which is a 58 amino acid synthetic analog of one domain from staphylococcal protein A (Nord et al. 1997). Affibodies specific for a variety of different proteins have been isolated using phage display by randomizing 13 solvent-exposed residues.

The expression levels of affibodies on the cell surface is homogeneous and allows for precise affinity ranking among affibodies of similar affinities (Gunneriusson et al. 1999; Lofblom, Wernerus, and Stahl 2005; Wernerus, Samuelson, and Stahl 2003). Enrichment experiments using FACS have been performed that demonstrated the potential of this system for the high throughput isolation or affinity maturation of affibodies (Lofblom, Wernerus, and Stahl 2005). In an affinity maturation enrichment experiment, a Z domain specific for human IgG was enriched 140-fold from a 1000-fold excess of a mutant Z domain with an 8-fold lower affinity following one round of FACS (Lofblom, Wernerus, and Stahl 2005). The system was also utilized to enrich a population of a specific binder to one in four from a 100,000-fold excess of nonspecific binders following just two rounds of FACS and subsequent amplification by growth in liquid media (Wernerus, Samuelson, and Stahl 2003). Utility of this system was recently demonstrated in the selection of affibodies against human tumor necrosis factor-alpha (TNF-α). Following one round of phage display selection, a library of 1×10^6 variants was enriched using three rounds of FACS sorting and generated three high affinity affibody binders.

The success of affibody Staphylococcal display for enrichment and selection (Kronqvist et al. 2008; Lofblom, Wernerus, and Stahl 2005; Wernerus, Samuelson, and Stahl 2003), brings into question why similar success has not been seen with antibody fragment display. Since affibodies are staphylococcal derived, Staphylococci are an obvious choice for display. Size may contribute to affibody success, as they are only a fifth the size of an scFv. Antibody fragments generally require disulfide bridges for correct folding, a feature not required by the stable alpha helical domain of the affibody. Recent demonstration that disulfide bond protein A (DsbA) of Gram-positive *Staphylococcus aureus* has a different mechanism for forming disulfide bridges in secreted proteins than Gram-negative *E. coli* DsbA (Heras et al. 2008), may aid in our understanding of limitations to Gram-positive antibody display.

11.3.2 M6 Display

In parallel to the protein A system developed by Uhlen and coworkers in 1992, Fischetti and coworkers engineered nonpathogenic *Streptococcus gordonii* for the surface display of proteins using the M6 surface protein of *Streptococcus pyogenes* (Oggioni et al. 1999; Pozzi et al. 1992). The M proteins are fibrillar molecules located on the surface of streptococci and are a primary virulence factor that exhibit vast antigenic variation while maintaining the ability to avoid clearance by phagocytes. The M6 protein from *S. pyogenes* was chosen for anchoring because earlier studies revealed that it contains surface exposed regions and an anchoring domain (Pozzi et al. 1992; Schneewind, Mihaylova-Petkov, and

Model 1993). As with protein A and many surface proteins from Gram-positive bacteria, M6 contains a signal sequence that is processed during secretion, a surface exposed region, and a cell wall anchoring region containing the common LPXTG motif, a hydrophobic region, and a charged tail (Robert et al. 1996). Following proteolytic cleavage of the LPXTG motif, the M6 is covalently linked to the cell wall (Pozzi et al. 1992).

Fischetti and coworkers first displayed the E7 protein of human papillomavirus type 16 using the M6 system (Pozzi et al. 1992). *S. gordonii* was chosen as the host organism for display since it is highly competent for genetic transformation and is capable of efficient, stable chromosomal integration (Hollingshead, Fischetti, and Scott 1986). This was an important feature since the application of this initial work was for vaccine delivery. Surface display was determined by Western blot analysis and fluorescence microscopy. It was further demonstrated to be immunogenic and suitable for live recombinant vaccine delivery.

The system developed by Fischetti and coworkers is comprised of the protein to be displayed inserted between the M6 N-terminal signal sequence and the C-terminus containing the cell wall anchoring region (Oggioni et al. 1999). Using this configuration, various modifications of the system have been created and are suitable for display (Hansson et al. 1992; Oggioni et al. 1999; Xu and Li 2007). The system has been utilized to display proteins ranging from 15 to almost 500 amino acids on the surface of *S. gordonii*, which does not normally express M6 (Giomarelli et al. 2004). Two different scFvs have been displayed on the surface of *S. gordonii* using this system and the results are encouraging for future use of the system in antibody engineering (Beninati et al. 2000; Giomarelli et al. 2004).

In 2000, Polonelli and coworkers were the first to use the M6 system to display an scFv anti-idiotype antibody with microbicidal properties on the surface of *S. gordonii* (Beninati et al. 2000). Display of the scFv was confirmed by Western blot and functionality was demonstrated using both an *in vitro* candidacidal assay and an *in vivo* vaginal candidiasis model. Although the displayed scFv was used successfully as a mucosal pathogen therapy, the secreted scFv was twice as potent, potentially due to increased availability of the soluble product to *Candida albicans*.

A second scFv, specific for the surface antigen SA I/II of *Streptococcus mutans*, was displayed using the M6 system by Pozzi and coworkers in 2004 (Giomarelli et al. 2004). This scFv, known as Guy's 13, was previously displayed using the proteinase P system (described below) on the surface of *Lactobacillus casei* (Kruger et al. 2002). Using the M6 system, display of the scFv was confirmed by both Western blot and flow cytometry using antibodies specific for an internal epitope tag (Pouwels, Leer, and Boersma 1996). The flow cytometry signal of the cells displaying the scFv was distinguishable from cells expressing neither the fusion protein nor M6, but there was a noticeable amount of overlap. Unfortunately, the flow cytometry studies were not performed with a fluorescent conjugated form of the antigen, which is standard for library screenings by FACS. The combination of this and the overlap of the signals obtained may have contributed to the lack of FACS enrichment experiments being included in the study. However, an innovative approach was used to confirm functionally displayed scFvs. Whole cells displaying scFvs demonstrated specific binding when analyzed by surface plasmon resonance, evidence that the scFvs are surface exposed and available to antigen. The cell's ability to bind to the immobilized antigen was also inhibited by the presence of soluble antigen in a concentration-dependent manner, which further suggests specific binding.

11.3.3 Proteinase P Display

A surface display system for *Lactobacillus casei* was first described by Boersma and coworkers in 1996 (Pouwels, Leer, and Boersma 1996). However, it wasn't until their follow-up work in 1999 that they successfully demonstrated the surface display of a fusion protein (Maassen et al. 1999). They displayed the tetanus toxin fragment C by utilizing two different versions of a system dependent on the C-terminal cell-wall anchoring domain of proteinase P from *L. paracasei* (previously referred to as *L. casei* or *L. zeae*; Siezen 1999). The difference between the two systems was the use of either a regulatable promoter, signal sequence, and N-terminus of α-amylase from *L. amylovorus* or the constitutive promoter of lactate dehydrogenase from *L. paracasei* followed by the signal sequence and N-terminus of proteinase P. As with other Gram-positive cell wall anchoring systems, the cell wall anchoring region of proteinase P contains a cell wall spanning region, the common LPXTG motif, a hydrophobic region, and a charged tail (Siezen 1999).

Surface expression of the toxin was determined by flow cytometry using an antibody specific for the toxin (Maassen et al. 1999). Although the signals obtained from each version of the system were tight and distinguishable from background, the system using the constitutive promoter gave a higher signal. It was estimated that 1400 and 3900 molecules of toxin were displayed on the surface of cells using the regulatable and constitutive promoters, respectively.

Additional variations of the systems have been used to display several antibodies to date, namely two scFvs as well as two different llama single variable domain antibodies (dAbs) (Kruger et al. 2002). In 2002, Hammarstrom and coworkers successfully displayed the first scFv using a version of this system (Kruger et al. 2002; Maassen et al. 1999). They utilized the α-amylase version of the system to display the Guy's 13 scFv specific for the surface antigen SAI/II of *Streptococcus mutans* on the surface of *L. paracasei*. Display was confirmed by both Western blot and flow cytometry using antibodies specific for an internal epitope tag. The flow cytometry signal of the cells displaying the scFv using a longer portion of the proteinase P C-terminus was tight and approximately 100-fold higher than cells secreting the scFv. The flow cytometry studies were not performed with a fluorescent-conjugated form of the antigen for functional evaluation. The functionality of the scFv was instead demonstrated by using the purified fusion protein and a cell-based assay using surface-displayed scFv. Further *in vivo* studies in rats indicated the surface-displayed scFv specific for the surface antigen SAI/II of *Streptococcus mutans* significantly lowered oral cavity colonization levels.

In their follow-up work in 2006, an scFv specific for the RgpA protease of *Porphyromonas gingivalis* and the previously used anti-SAI/II scFv were displayed using a modified version of the proteinase P system (Marcotte et al. 2006). The controllable α-amylase promoter, signal sequence, and N-terminus were replaced by the equivalent parts from the lactate dehydrogenase of *L. paracasei*. This was done since the lactate dehydrogenase promoter is constitutive and better suited for *in vivo* applications (Pant et al. 2006). Although the flow cytometry data for this system was not as impressive as the original work, the signal for both scFvs was highly distinguishable from background using the same anti-epitope antibody used previously. The Guy's 13 scFv resulted in higher flow cytometry signals than the anti-RgpA scFv, but it was lower and more dispersed than that seen using the original system. These results indicate that the approximately 850 anti-RgpA scFvs displayed is less than that of the anti-SAI/II scFv. The functionality of the scFv was demonstrated using the cell extracts of the fusion protein and an *in vitro* agglutination assay.

In 2006, the system was further modified by Hammarstrom and co-workers and used to display a llama dAb specific for a rotavirus outer capsid protein (Pant et al. 2006). The lactate dehydrogenase promoter was again used, but this time with the signal sequence and N-terminus of proteinase P (Kruger et al. 2002; Marcotte et al. 2006; Pant et al. 2006). Surface expression was similar to previous results and was verified by flow cytometry (Hultberg et al. 2007). Functionality was verified using the fusion protein isolated from cell lysate and in a cell-based assay using the surface-displayed dAb and rhesus rotavirus. Both an *in vitro* neutralization assay and an *in vivo* prophylactic assay further confirmed the functionality of the displayed dAb.

Most recently, the original proteinase P system was used to display a dAb specific for the major capsid protein of Lactococcal bacteriophage P2 (Hultberg et al. 2007; Kruger et al. 2002; Marcotte et al. 2006; Pant et al. 2006). Surface expression was similar to previous results and was verified by flow cytometry (Xu and Li 2007). Functionality was verified using the fusion protein isolated from cell lysate and in a cell-based assay using the surface-displayed dAb and bacteriophage P2. An *in vivo* neutralization assay furthered verified functionality of the display dAb, but flow cytometry studies were not included.

The proteinase P *Lactobacillus* display system has been used to successfully display four different antibody fragments as well as additional proteins (Hultberg et al. 2007; Kruger et al. 2002, 2005; Marcotte et al. 2006; Pant et al. 2006). Subtle variations to the system have been used and all of them have successfully displayed the antibody fragments to similar levels. They have been used successfully to display antibodies that neutralize various pathogens *in vitro* and *in vivo* (Henriques and Moran 2007). Although the system has yet to be demonstrated for use in antibody engineering, the potential may be there. Antibodies displayed on the surface need to be shown to bind fluorescently conjugated forms of their respective antigens and subjected to FACS-based enrichment experiments to determine if this is a realistic application of the system.

11.3.4 Spore Display

Certain types of bacteria, most of which are Gram-positive, are known to form endospores to ensure their survival during periods of environmental stress. During this process, the vegetative cells are converted to dormant endospores that remain at rest until conditions that again favor growth are restored (Henriques and Moran 2007). The general features of all endospores are the same. They consist of a spore core, a spore cortex, and a spore coat. The cortex is a thick peptidoglycan layer that surrounds the core and it is surrounded by the coat, which is composed primarily of proteins (Henriques and Moran 2007). The coat of *Bacillus subtilis* spores is organized into two layers and is composed of more than 70 different proteins (Isticato et al. 2001). One of these proteins, CotB, was first reported in 2001 by Ricca and coworkers as a fusion partner for the surface display of a C-terminal fragment of the tetanus toxin fragment C (TTFC) on *B. subtilis* spores (Isticato et al. 2001). This was the first example of using a spore as a protein display system.

Although the CotB system was not utilized with an antibody, some knowledge relevant to the display of antibodies was uncovered and is worth discussing. The display of a fusion to CotB was dependent on the presence of endogenous CotB (Isticato et al. 2001). Even though this reduced the total number of displayed fusions to 1500, it was detectable by flow cytometry (Isticato et al. 2001). The flow cytometry signals were poor, but the displayed protein was 51.8 kDa and was detected using a primary and secondary antibody (Mauriello et al. 2004).

In some follow-up work, the coat protein CotC of *B. subtilis* spores was utilized to display either the same C-terminal fragment of TTFC or the B subunit of the heat-labile enterotoxin of *E. coli* (LT-B) on the surface of spores (Mauriello et al. 2004). While expression of the CotC fusions was not dependent on endogenous CotC expression as was the case with the CotB system, they did increase with coexpression (Isticato et al. 2007; Mauriello et al. 2004). Alternative fusion strategies to CotC were also shown to increase display efficiency (le Duc 2007; Isticato et al. 2001; Mauriello et al. 2004, 2007). Unfortunately the authors did not include any data utilizing flow cytometry. Both the CotB and CotC systems have been shown as very efficient vaccine delivery systems and additional work has already followed (Du et al. 2005). The current body of work indicates that both systems have the potential for the surface display of antibodies, but this has yet to be shown.

Several years later, *Bacillus thuringiensis* spores were used by Nickerson and coworkers to surface display an scFv (Du et al. 2005). An insecticidal protoxin that is a major component of the spore coat was utilized as a fusion partner to display an scFv specific for the hapten phOx, the same hapten used with the PAL system (Fuchs et al. 1996). Successful display was determined using fluorescence microscopy with BSA conjugated with phOx and a fluorophore. The overall molecular weight of the conjugated BSA was over 66 kDa, which indicates steric hindrance from other surface-exposed molecules should not interfere with the binding of large antigens. Unfortunately, there was no indication of the system's compatibility with flow cytometry.

11.4 SUMMARY

Perhaps the biggest driving force for the bacterial surface display and successful engineering of antibodies is their suitability for screening by flow cytometry. Flow cytometry enables quantitative multiparameter analysis to be performed simultaneously. This in turn allows for multiple antibody properties to be detected simultaneously, including relative affinity to different antigens, expression, and assembly. In order for a signal to be detectable by flow cytometry, there needs to be a minimal number of fluorescent molecules associated with the cell. This number is obviously dependent on the properties of the fluorescent molecule used. Be that as it may and given there are ways to amplify the fluorescent signal associated with each antibody antigen binding event, there is still a minimal threshold of functionally expressed antibodies required to be displayed on the surface in an accessible manner required for successful flow cytometric screening.

The ideal bacterial display system should possess the following features: (1) antibody expression does not affect cell growth rate or viability; (2) cells express a large number of functional and accessible antibodies that remain associated with the cell; (3) flow cytometric signals are distinguishable from background, narrow, and highly dependent on affinity; (4) cells are viable following FACS; (5) the system is amenable to multiple antibody formats. Although the systems discussed in this chapter possess some of these requirements, none of them possesses all of them. Therefore, there is still considerable work to be done in this field to truly take advantage of all the potential that bacterial display of antibodies has to offer.

ACKNOWLEDGMENTS

We thank Dr. George Georgiou for critical reading of this chapter. T.V.B. was supported by The Foundation for Research.

REFERENCES

Akamatsu, Y., K. Pakabunto, Z. Xu, Y. Zhang, and N. Tsurushita. 2007. Whole IgG surface display on mammalian cells: Application to isolation of neutralizing chicken monoclonal anti-IL-12 antibodies. *J. Immunol. Methods* 327:40–52.

Andreoni, C., L. Goetsch, C. Libon, P. Samuelson, T.N. Nguyen, A. Robert, M. Uhlen, H. Binz, and S. Stahl. 1997. Flow cytometric quantification of surface-displayed recombinant receptors on staphylococci. *Biotechniques* 23:696–702, 704.

Bae, W., W. Chen, A. Mulchandani, and R.K. Mehra. 2000. Enhanced bioaccumulation of heavy metals by bacterial cells displaying synthetic phytochelatins. *Biotechnol. Bioeng.* 70:518–524.

Bae, W., R.K. Mehra, A. Mulchandani, and W. Chen. 2001. Genetic engineering of *Escherichia coli* for enhanced uptake and bioaccumulation of mercury. *Appl. Environ. Microbiol.* 67:5335–5338.

Bassi, A.S., D.N. Ding, G.B. Gloor, and A. Margaritis. 2000. Expression of single chain antibodies (ScFvs) for c-myc oncoprotein in recombinant *Escherichia coli* membranes by using the ice-nucleation protein of *Pseudomonas syringae*. *Biotechnol. Prog.* 16:557–563.

Benhar, I. 2001. Biotechnological applications of phage and cell display. *Biotechnol. Adv.* 19:1–33.

Benhar, I. 2007. Design of synthetic antibody libraries. *Expert Opin. Biol. Ther.* 7:763–779.

Beninati, C., M.R. Oggioni, M. Boccanera, M.R. Spinosa, T. Maggi, S. Conti, W. Magliani, F. De Bernardis, G. Teti, A. Cassone, G. Pozzi, and L. Polonelli. 2000. Therapy of mucosal candidiasis by expression of an anti-idiotype in human commensal bacteria. *Nature Biotechnol.* 18:1060–1064.

Bereta, M., A. Hayhurst, M. Gajda, P. Chorobik, M. Targosz, J. Marcinkiewicz, and H.L. Kaufman. 2007. Improving tumor targeting and therapeutic potential of *Salmonella* VNP20009 by displaying cell surface CEA-specific antibodies. *Vaccine* 25:4183–4192.

Boder, E.T., and K.D. Wittrup. 1997. Yeast surface display for screening combinatorial polypeptide libraries. *Nature Biotechnol.* 15:553–557.

Burnett, M.S., N. Wang, M. Hofmann, and G. Barrie Kitto. 2000. Potential live vaccines for HIV. *Vaccine* 19:735–742.

Carter, P.J. 2006. Potent antibody therapeutics by design. *Nature Rev. Immunol.* 6:343–357.

Charbit, A., J.C. Boulain, A. Ryter, and M. Hofnung. 1986. Probing the topology of a bacterial membrane protein by genetic insertion of a foreign epitope: Expression at the cell surface. *EMBO J* 5:3029–3037.

Chen, G., J. Cloud, G. Georgiou, and B.L. Iverson. 1996. A quantitative immunoassay utilizing *Escherichia coli* cells possessing surface-expressed single chain Fv molecules. *Biotechnol. Prog.* 12:572–574.

Chen, G., A. Hayhurst, J.G. Thomas, B.R. Harvey, B.L. Iverson, and G. Georgiou, 2001. Isolation of high-affinity ligand-binding proteins by periplasmic expression with cytometric screening (PECS). *Nature Biotechnol.* 19:537–542.

Chen, W., and G. Georgiou. 2002. Cell-surface display of heterologous proteins: From high-throughput screening to environmental applications. *Biotechnol. Bioeng.* 79:496–503.

Cole, S.T., I. Sonntag, and U. Henning. 1982. Cloning and expression in *Escherichia coli* K-12 of the genes for major outer membrane protein OmpA from *Shigella dysenteriae*, *Enterobacter aerogenes*, and *Serratia marcescens*. *J. Bacteriol.* 149:145–150.

Curnow, P., P.H. Bessette, D. Kisailus, M.M. Murr, P.S. Daugherty, and D.E. Morse. 2005. Enzymatic synthesis of layered titanium phosphates at low temperature and neutral pH by cell-surface display of silicatein-alpha. *J. Am. Chem. Soc.* 127:15749–15755.

Daugherty, P.S., G. Chen, B.L. Iverson, and G. Georgiou. 2000. Quantitative analysis of the effect of the mutation frequency on the affinity maturation of single chain Fv antibodies. *Proc. Natl. Acad. Sci. USA* 97:2029–2034.

Daugherty, P.S., G. Chen, M.J. Olsen, B.L. Iverson, and G. Georgiou. 1998. Antibody affinity maturation using bacterial surface display. *Protein Eng.* 11:825–832.

Daugherty, P.S., B.L. Iverson, and G. Georgiou. 2000. Flow cytometric screening of cell-based libraries. *J. Immunol. Methods* 243:211–227.

Daugherty, P.S., M.J. Olsen, B.L. Iverson, and G. Georgiou. 1999. Development of an optimized expression system for the screening of antibody libraries displayed on the *Escherichia coli* surface. *Protein Eng.* 12:613–621.

Dautin, N., and H.D. Bernstein. 2007. Protein secretion in Gram-negative bacteria via the autotransporter pathway. *Annu. Rev. Microbiol.* 61:89–112.

Dhillon, J.K., P.D. Drew, and A.J. Porter. 1999. Bacterial surface display of an anti-pollutant antibody fragment. *Lett. Appl. Microbiol.* 28:350–354.

Drainas, C., G. Vartholomatos, and N.J. Panopoulos. 1995. The ice nucleation gene from *Pseudomonas syringae* as a sensitive gene reporter for promoter analysis in *Zymomonas mobilis*. *Appl. Environ. Microbiol.* 61:273–277.

Du, C., W.C. Chan, T.W. McKeithan, and K.W. Nickerson. 2005. Surface display of recombinant proteins on *Bacillus thuringiensis* spores. *Appl. Environ. Microbiol.* 71:3337–3341.

Earhart, C.F. 2000. Use of an Lpp-OmpA fusion vehicle for bacterial surface display. *Methods Enzymol.* 326:506–516.

Ezaki S, T.M., M. Takagi, and T. Imanaka. 1998. Display of heterologous gene products on the *Escherichia coli* cell surface as fusion proteins with flagellin. *J. Ferment. Bioeng.* 86:500–503.

Fischetti, V.A., D. Medaglini, and G. Pozzi. 1996. Gram-positive commensal bacteria for mucosal vaccine delivery. *Curr. Opin. Biotechnol.* 7:659–666.

Foote, J., and H.N. Eisen. 1995. Kinetic and affinity limits on antibodies produced during immune responses. *Proc. Natl. Acad. Sci. USA* 92:1254–1256.

Foote, J., and H.N. Eisen. 2000. Breaking the affinity ceiling for antibodies and T cell receptors. *Proc. Natl. Acad. Sci. USA* 97:10679–10681.

Francisco, J.A., R. Campbell, B.L. Iverson, and G. Georgiou. 1993. Production and fluorescence-activated cell sorting of *Escherichia coli* expressing a functional antibody fragment on the external surface. *Proc. Natl. Acad. Sci. USA* 90:10444–10448.

Francisco, J.A., C.F. Earhart, and G. Georgiou. 1992. Transport and anchoring of beta-lactamase to the external surface of *Escherichia coli*. *Proc. Natl. Acad. Sci. USA* 89:2713–2717.

Francisco, J.A., C. Stathopoulos, R.A. Warren, D.G. Kilburn, and G. Georgiou. 1993. Specific adhesion and hydrolysis of cellulose by intact *Escherichia coli* expressing surface anchored cellulase or cellulose binding domains. *Biotechnology (NY)* 11:491–495.

Freudl, R., S. MacIntyre, M. Degen, and U. Henning. 1986. Cell surface exposure of the outer membrane protein OmpA of *Escherichia coli* K-12. *J. Mol. Biol.* 188:491–494.

Fuchs, P., F. Breitling, S. Dubel, T. Seehaus, and M. Little. 1991. Targeting recombinant antibodies to the surface of *Escherichia coli*: Fusion to a peptidoglycan associated lipoprotein. *Biotechnology (NY)* 9:1369–1372.

Fuchs, P., W. Weichel, S. Dubel, F. Breitling, and M. Little. 1996. Separation of *E. coli* expressing functional cell-wall bound antibody fragments by FACS. *Immunotechnology* 2:97–102.

Georgiou, G., H.L. Poetschke, C. Stathopoulos, and J.A. Francisco. 1993. Practical applications of engineering Gram-negative bacterial cell surfaces. *Trends Biotechnol.* 11:6–10.

Georgiou, G., C. Stathopoulos, P.S. Daugherty, A.R. Nayak, B.L. Iverson, and R. Curtiss III. 1997. Display of heterologous proteins on the surface of microorganisms: From the screening of combinatorial libraries to live recombinant vaccines. *Nature Biotechnol.* 15:29–34.

Georgiou, G., D.L. Stephens, C. Stathopoulos, H.L. Poetschke, J. Mendenhall, and C.F. Earhart. 1996. Display of beta-lactamase on the *Escherichia coli* surface: Outer membrane phenotypes conferred by Lpp′-OmpA′-beta-lactamase fusions. *Protein Eng.* 9:239–247.

Ghrayeb, J., and M. Inouye. 1984. Nine amino acid residues at the NH2-terminal of lipoprotein are sufficient for its modification, processing, and localization in the outer membrane of *Escherichia coli*. *J. Biol. Chem.* 259:463–467.

Giomarelli, B., T. Maggi, J. Younson, C. Kelly, and G. Pozzi. 2004. Expression of a functional single-chain Fv antibody on the surface of *Streptococcus gordonii*. *Mol. Biotechnol.* 28:105–112.

Goding, J.W. 1978. Use of staphylococcal protein A as an immunological reagent. *J. Immunol. Methods* 20:241–253.

Gotz, F. 1990. *Staphylococcus carnosus*: A new host organism for gene cloning and protein production. *Soc. Appl. Bacteriol. Symp. Ser.* 19:49S–53S.

Gunneriusson, E., P. Samuelson, J. Ringdahl, H. Gronlund, P.A. Nygren, and S. Stahl. 1999. Staphylococcal surface display of immunoglobulin A (IgA)- and IgE-specific in vitro-selected binding proteins (affibodies) based on *Staphylococcus aureus* protein A. *Appl. Environ. Microbiol.* 65:4134–4140.

Gunneriusson, E., P. Samuelson, M. Uhlen, P.A. Nygren, and S. Stahl. 1996. Surface display of a functional single-chain Fv antibody on staphylococci. *J. Bacteriol.* 178:1341–1346.

Hansson, M., S. Stahl, T.N. Nguyen, T. Bachi, A. Robert, H. Binz, A. Sjolander, and M. Uhlen. 1992. Expression of recombinant proteins on the surface of the coagulase-negative bacterium *Staphylococcus xylosus*. *J. Bacteriol.* 174:4239–4245.

Harvey, B.R., G. Georgiou, A. Hayhurst, K.J. Jeong, B.L. Iverson, and G.K. Rogers. 2004. Anchored periplasmic expression, a versatile technology for the isolation of high-affinity antibodies from *Escherichia coli*-expressed libraries. *Proc. Natl. Acad. Sci. USA* 101:9193–9198.

Harvey, B.R., A.B. Shanafelt, I. Baburina, R. Hui, S. Vitone, B.L. Iverson, and G. Georgiou. 2006. Engineering of recombinant antibody fragments to methamphetamine by anchored periplasmic expression. *J. Immunol. Methods* 308:43–52.

Hayhurst, A. 2000. Improved expression characteristics of single-chain Fv fragments when fused downstream of the *Escherichia coli* maltose-binding protein or upstream of a single immunoglobulin-constant domain. *Protein Expr. Purif.* 18:1–10.

Hayhurst, A., and G. Georgiou. 2001. High-throughput antibody isolation. *Curr. Opin. Chem. Biol.* 5:683–689.

Hayhurst, A., S. Happe, R. Mabry, Z. Koch, B.L. Iverson, and G. Georgiou. 2003. Isolation and expression of recombinant antibody fragments to the biological warfare pathogen *Brucella melitensis*. *J. Immunol. Methods* 276:185–196.

Henriques, A.O., and C.P. Moran, Jr. 2007. Structure, assembly, and function of the spore surface layers. *Annu. Rev. Microbiol.* 61:555–588.

Heras, B., M. Kurz, R. Jarrott, S.R. Shouldice, P. Frei, G. Robin, M. Cemazar, L. Thony-Meyer, R. Glockshuber, and J.L. Martin. 2008. *Staphylococcus aureus* DsbA does not have a destabilizing disulfide. A new paradigm for bacterial oxidative folding. *J. Biol. Chem.* 283:4261–4271.

Holliger, P., and P.J. Hudson. 2005. Engineered antibody fragments and the rise of single domains. *Nature Biotechnol.* 23:1126–1136.

Hollingshead, S.K., V.A. Fischetti, and J.R. Scott. 1986. Complete nucleotide sequence of type 6 M protein of the group A Streptococcus. Repetitive structure and membrane anchor. *J. Biol. Chem.* 261:1677–1686.

Hoogenboom, H.R. 2005. Selecting and screening recombinant antibody libraries. *Nature Biotechnol.* 23: 1105–1116.

Hultberg, A., D.M. Tremblay, H. de Haard, T. Verrips, S. Moineau, L. Hammarstrom, and H. Marcotte. 2007. Lactobacilli expressing llama VHH fragments neutralise Lactococcus phages. *BMC Biotechnol.* 7:58.

Isticato, R., G. Cangiano, H.T. Tran, A. Ciabattini, D. Medaglini, M.R. Oggioni, M. De Felice, G. Pozzi, and E. Ricca. 2001. Surface display of recombinant proteins on *Bacillus subtilis* spores. *J. Bacteriol.* 183:6294–6301.

Isticato, R., D.S. Di Mase, E.M. Mauriello, M. De Felice, and E. Ricca. 2007. Amino terminal fusion of heterologous proteins to CotC increases display efficiencies in the *Bacillus subtilis* spore system. *Biotechniques* 42:151–152, 154, 156.

Jensen, K.B., M. Larsen, J.S. Pedersen, P.A. Christensen, L. Alvarez-Vallina, S. Goletz, B.F. Clark, and P. Kristensen. 2002. Functional improvement of antibody fragments using a novel phage coat protein III fusion system. *Biochem. Biophys. Res. Commun.* 298:566–573.

Jeong, H., S. Yoo, and E. Kim. 2001. Cell surface display of salmobin, a thrombin-like enzyme from *Agkistrodon halys* venom on *Escherichia coli* using ice nucleation protein. *Enzyme Microb. Technol.* 28:155–160.

Jeong, K.J., M.J. Seo, B.L. Iverson, and G. Georgiou. 2007. APEx 2-hybrid, a quantitative protein–protein interaction assay for antibody discovery and engineering. *Proc. Natl. Acad. Sci. USA* 104:8247–8252.

Jung, H.C., S. Ko, S.J. Ju, E.J. Kim, M.K. Kim, and J.G. Pan. 2003. Bacterial cell surface display of lipase and its randomly mutated library facilitates high-throughput screening of mutants showing higher specific activities. *J. Mol. Catalysis B: Enzymatic* 26:177–184.

Jung, H.C., S.J. Kwon, and J.G. Pan. 2006. Display of a thermostable lipase on the surface of a solvent-resistant bacterium, *Pseudomonas putida* GM730, and its applications in whole-cell biocatalysis. *BMC Biotechnol.* 6:23.

Jung, H.C., J.M. Lebeault, and J.G. Pan. 1998. Surface display of *Zymomonas mobilis* levansucrase by using the ice-nucleation protein of *Pseudomonas syringae*. *Nature Biotechnol.* 16:576–580.

Jung, H.C., J.H. Park, S.H. Park, J.M. Lebeault, and J.G. Pan. 1998. Expression of carboxymethylcellulase on the surface of *Escherichia coli* using *Pseudomonas syringae* ice nucleation protein. *Enzyme Microb. Technol.* 22:348–354.

Jung, S.T., K.J. Jeong, B.L. Iverson, and G. Georgiou. 2007. Binding and enrichment of *Escherichia coli* spheroplasts expressing inner membrane tethered scFv antibodies on surface immobilized antigens. *Biotechnol. Bioeng.* 98:39–47.

Kang, S.M., J.K. Rhee, E.J. Kim, K.H. Han, and J.W. Oh. 2003. Bacterial cell surface display for epitope mapping of hepatitis C virus core antigen. *FEMS Microbiol. Lett.* 226:347–353.

Ki, J.J., Y. Kawarasaki, J. Gam, B.R. Harvey, B.L. Iverson, and G. Georgiou. 2004. A periplasmic fluorescent reporter protein and its application in high-throughput membrane protein topology analysis. *J. Mol. Biol.* 341:901–909.

Kim, E.J., and S.K. Yoo. 1999. Cell surface display of hepatitis B virus surface antigen by using *Pseudomonas syringae* ice nucleation protein. *Lett. Appl. Microbiol.* 29:292–297.

Kim, Y.S., H.C. Jung, and J.G. Pan. 2000. Bacterial cell surface display of an enzyme library for selective screening of improved cellulase variants. *Appl. Environ. Microbiol.* 66:788–793.

Klauser, T., J. Pohlner, and T.F. Meyer. 1990. Extracellular transport of cholera toxin B subunit using *Neisseria* IgA protease beta-domain: Conformation-dependent outer membrane translocation. *EMBO J.* 9:1991–1999.

Kronqvist, N., J. Lofblom, A. Jonsson, H. Wernerus, and S. Stahl. 2008. A novel affinity protein selection system based on staphylococcal cell surface display and flow cytometry. *Protein Eng. Des. Sel.* 21:247–255.

Kruger, C., Y. Hu, Q. Pan, H. Marcotte, A. Hultberg, D. Delwar, P.J. van Dalen, P.H. Pouwels, R.J. Leer, C.G. Kelly, C. van Dollenweerd, J.K. Ma, and L. Hammarstrom. 2002. In situ delivery of passive immunity by lactobacilli producing single-chain antibodies. *Nature Biotechnol.* 20:702–706.

Kruger, C., A. Hultberg, C. van Dollenweerd, H. Marcotte, and L. Hammarstrom. 2005. Passive immunization by lactobacilli expressing single-chain antibodies against *Streptococcus mutans*. *Mol. Biotechnol.* 31:221–231.

Kuwajima G, A.-I., T. Fujiwara, T. Furiwara, K. Nakano, and E. Kondoh. 1988. Presentation of an antigenic determinant from hen egg-white lysoszyme on the flagellar filament of *Escherichia coli. Nature Biotechnol.* 6:1080–1083.

Kwak, Y.D., S.K. Yoo, and E.J. Kim. 1999. Cell surface display of human immunodeficiency virus type 1 gp120 on *Escherichia coli* by using ice nucleation protein. *Clin. Diagn. Lab. Immunol.* 6:499–503.

Le Duc, H., H.A. Hong, H.S. Atkins, H.C. Flick-Smith, Z. Durrani, S. Rijpkema, R.W. Titball, and S.M. Cutting. 2007. Immunization against anthrax using *Bacillus subtilis* spores expressing the anthrax protective antigen. *Vaccine* 25:346–355.

Lee, J.S., K.S. Shin, J.G. Pan, and C.J. Kim. 2000. Surface-displayed viral antigens on *Salmonella* carrier vaccine. *Nature Biotechnol.* 18:645–648.

Lee, P.A., D. Tullman-Ercek, and G. Georgiou. 2006. The bacterial twin-arginine translocation pathway. *Annu. Rev. Microbiol.* 60:373–395.

Leenhouts, K., G. Buist, and J. Kok. 1999. Anchoring of proteins to lactic acid bacteria. *Antonie Van Leeuwenhoek* 76:367–376.

Lei, Y., A. Mulchandani, and W. Chen. 2005. Improved degradation of organophosphorus nerve agents and p-nitrophenol by *Pseudomonas putida* JS444 with surface-expressed organophosphorus hydrolase. *Biotechnol. Prog.* 21:678–681.

Leive, L. 1968. Studies on the permeability change produced in coliform bacteria by ethylenediaminetetraacetate. *J. Biol. Chem.* 243:2373–2380.

Li, L., D.G. Kang, and H.J. Cha. 2004. Functional display of foreign protein on surface of *Escherichia coli* using N-terminal domain of ice nucleation protein. *Biotechnol. Bioeng.* 85:214–221.

Liebl, W., and F. Gotz. 1986. Studies on lipase directed export of *Escherichia coli* beta-lactamase in *Staphylococcus carnosus*. *Mol. Gen. Genet.* 204:166–173.

Liljeqvist, S., F. Cano, T.N. Nguyen, M. Uhlen, A. Robert, and S. Stahl. 1999. Surface display of functional fibronectin-binding domains on *Staphylococcus carnosus*. *FEBS Lett.* 446:299–304.

Liljeqvist, S., P. Samuelson, M. Hansson, T.N. Nguyen, H. Binz, and S. Stahl. 1997. Surface display of the cholera toxin B subunit on *Staphylococcus xylosus* and *Staphylococcus carnosus*. *Appl. Environ. Microbiol.* 63:2481–2488.

Little, M., F. Breitling, B. Micheel, and S. Dubel. 1994. Surface display of antibodies. *Biotechnol. Adv.* 12:539–555.

Lofblom, J., J. Sandberg, H. Wernerus, and S. Stahl. 2007. Evaluation of staphylococcal cell surface display and flow cytometry for postselectional characterization of affinity proteins in combinatorial protein engineering applications. *Appl. Environ. Microbiol.* 73:6714–6721.

Lofblom, J., H. Wernerus, and S. Stahl. 2005. Fine affinity discrimination by normalized fluorescence activated cell sorting in staphylococcal surface display. *FEMS Microbiol. Lett.* 248:189–198.

Lu, Z., K.S. Murray, V. Van Cleave, E.R. LaVallie, M.L. Stahl, and J.M. McCoy. 1995. Expression of thioredoxin random peptide libraries on the *Escherichia coli* cell surface as functional fusions to flagellin: A system designed for exploring protein–protein interactions. *Biotechnology (NY)* 13:366–372.

Maassen, C.B., J.D. Laman, M.J. den Bak-Glashouwer, F.J. Tielen, J.C. van Holten-Neelen, L. Hoogteijling, C. Antonissen, R.J. Leer, P.H. Pouwels, W.J. Boersma, and D.M. Shaw. 1999. Instruments for oral disease-intervention strategies: Recombinant *Lactobacillus casei* expressing tetanus toxin fragment C for vaccination or myelin proteins for oral tolerance induction in multiple sclerosis. *Vaccine* 17:2117–2128.

Marcotte, H., P. Koll-Klais, A. Hultberg, Y. Zhao, R. Gmur, R. Mandar, M. Mikelsaar, and L. Hammarstrom. 2006. Expression of single-chain antibody against RgpA protease of *Porphyromonas gingivalis* in *Lactobacillus*. *J. Appl. Microbiol.* 100:256–263.

Marraffini, L.A., A.C. Dedent, and O. Schneewind. 2006. Sortases and the art of anchoring proteins to the envelopes of gram-positive bacteria. *Microbiol. Mol. Biol. Rev.* 70:192–221.

Maurer, J., J. Jose, and T.F. Meyer. 1997. Autodisplay: One-component system for efficient surface display and release of soluble recombinant proteins from *Escherichia coli*. *J. Bacteriol.* 179:794–804.

Mauriello, E.M., G. Cangiano, F. Maurano, V. Saggese, M. De Felice, M. Rossi, and E. Ricca. 2007. Germination-independent induction of cellular immune response by *Bacillus subtilis* spores displaying the C fragment of the tetanus toxin. *Vaccine* 25:788–793.

Mauriello, E.M., H. le Duc, R. Isticato, G. Cangiano, H.A. Hong, M. De Felice, E. Ricca, and S.M. Cutting. 2004. Display of heterologous antigens on the *Bacillus subtilis* spore coat using CotC as a fusion partner. *Vaccine* 22:1177–1187.

Mazor, Y., T. Van Blarcom, R. Mabry, B.L. Iverson, and G. Georgiou. 2007. Isolation of engineered, full-length antibodies from libraries expressed in *Escherichia coli*. *Nature Biotechnol.* 25:563–565.

McCafferty, J., A.D. Griffiths, G. Winter, and D.J. Chiswell. 1990. Phage antibodies: Filamentous phage displaying antibody variable domains. *Nature* 348:552–554.

McEwen, J., R. Levi, R.J. Horwitz, and R. Arnon. 1992. Synthetic recombinant vaccine expressing influenza haemagglutinin epitope in *Salmonella* flagellin leads to partial protection in mice. *Vaccine* 10:405–411.

Newton, S.M., C.O. Jacob, and B.A. Stocker. 1989. Immune response to cholera toxin epitope inserted in *Salmonella* flagellin. *Science* 244:70–72.

Newton, S.M., M. Kotb, T.P. Poirier, B.A. Stocker, and E.H. Beachey. 1991. Expression and immunogenicity of a streptococcal M protein epitope inserted in *Salmonella* flagellin. *Infect. Immun.* 59:2158–2165.

Nord, K., E. Gunneriusson, J. Ringdahl, S. Stahl, M. Uhlen, and P.A. Nygren. 1997. Binding proteins selected from combinatorial libraries of an alpha-helical bacterial receptor domain. *Nature Biotechnol.* 15:772–777.

Nosjean, O. 1998. No prokaryotic GPI anchoring. *Nature Biotechnol.* 16:799.

Oggioni, M.R., D. Medaglini, T. Maggi, and G. Pozzi. 1999. Engineering the gram-positive cell surface for construction of bacterial vaccine vectors. *Methods* 19:163–173.

Pant, N., A. Hultberg, Y. Zhao, L. Svensson, Q. Pan-Hammarstrom, K. Johansen, P.H. Pouwels, F.M. Ruggeri, P. Hermans, L. Frenken, T. Boren, H. Marcotte, and L. Hammarstrom. 2006. Lactobacilli expressing variable domain of llama heavy-chain antibody fragments (lactobodies) confer protection against rotavirus-induced diarrhea. *J. Infect. Dis.* 194:1580–1588.

Paterson, G.K., and T.J. Mitchell. 2004. The biology of Gram-positive sortase enzymes. *Trends Microbiol.* 12:89–95.

Pouwels, P.H., R.J. Leer, and W.J. Boersma. 1996. The potential of *Lactobacillus* as a carrier for oral immunization: development and preliminary characterization of vector systems for targeted delivery of antigens. *J. Biotechnol.* 44:183–192.

Pozzi, G., M. Contorni, M.R. Oggioni, R. Manganelli, M. Tommasino, F. Cavalieri, and V.A. Fischetti. 1992. Delivery and expression of a heterologous antigen on the surface of streptococci. *Infect. Immun.* 60:1902–1907.

Reichert, J.M., C.J. Rosensweig, L.B. Faden, and M.C. Dewitz. 2005. Monoclonal antibody successes in the clinic. *Nature Biotechnol.* 23:1073–1078.

Ribnicky, B., T. Van Blarcom, and G. Georgiou. 2007. A scFv antibody mutant isolated in a genetic screen for improved export via the twin arginine transporter pathway exhibits faster folding. *J. Mol. Biol.* 369:631–639.

Richins, R.D., I. Kaneva, A. Mulchandani, and W. Chen. 1997. Biodegradation of organophosphorus pesticides by surface-expressed organophosphorus hydrolase. *Nature Biotechnol.* 15:984–987.

Robert, A., P. Samuelson, C. Andreoni, T. Bachi, M. Uhlen, H. Binz, T.N. Nguyen, and S. Stahl. 1996. Surface display on staphylococci: A comparative study. *FEBS Lett.* 390:327–333.

Samuelson, P., E. Gunneriusson, P.A. Nygren, and S. Stahl. 2002. Display of proteins on bacteria. *J. Biotechnol.* 96:129–154.

Samuelson, P., M. Hansson, N. Ahlborg, C. Andreoni, F. Gotz, T. Bachi, T.N. Nguyen, H. Binz, M. Uhlen, and S. Stahl. 1995. Cell surface display of recombinant proteins on *Staphylococcus carnosus*. *J. Bacteriol.* 177:1470–1476.

Schneewind, O., D. Mihaylova-Petkov, and P. Model. 1993. Cell wall sorting signals in surface proteins of gram-positive bacteria. *EMBO J.* 12:4803–4811.

Schneewind, O., P. Model, and V.A. Fischetti. 1992. Sorting of protein A to the staphylococcal cell wall. *Cell* 70:267–281.

Shi, H., and W. Wen Su. 2001. Display of green fluorescent protein on *Escherichia coli* cell surface. *Enzyme Microb. Technol.* 28:25–34.

Shimazu, M., A. Mulchandani, and W. Chen. 2001a. Cell surface display of organophosphorus hydrolase using ice nucleation protein. *Biotechnol. Prog.* 17:76–80.

Shimazu, M., A. Mulchandani, and W. Chen. 2001b. Simultaneous degradation of organophosphorus pesticides and p-nitrophenol by a genetically engineered *Moraxella* sp. with surface-expressed organophosphorus hydrolase. *Biotechnol. Bioeng.* 76:318–324.

Shimazu, M., A. Nguyen, A. Mulchandani, and W. Chen. 2003. Cell surface display of organophosphorus hydrolase in *Pseudomonas putida* using an ice-nucleation protein anchor. *Biotechnol. Prog.* 19:1612–1614.

Siezen, R.J. 1999. Multi-domain, cell-envelope proteinases of lactic acid bacteria. *Antonie Van Leeuwenhoek* 76:139–155.

Skerra, A. 2007. Alternative non-antibody scaffolds for molecular recognition. *Curr. Opin. Biotechnol.* 18:295–304.

Skerra, A., and A. Plückthun. 1988. Assembly of a functional immunoglobulin Fv fragment in *Escherichia coli. Science* 240:1038–1041.

Stahl, S., and M. Uhlen. 1997. Bacterial surface display: Trends and progress. *Trends Biotechnol.* 15:185–192.

Stephens, D.L., M.D. Choe, and C.F. Earhart. 1995. *Escherichia coli* periplasmic protein FepB binds ferrienterobactin. *Microbiology* 141(Pt 7):1647–1654.

Tao, M.H., and S.L. Morrison. 1989. Studies of aglycosylated chimeric mouse-human IgG. Role of carbohydrate in the structure and effector functions mediated by the human IgG constant region. *J. Immunol.* 143:2595–2601.

Valls, M., S. Atrian, V. de Lorenzo, and L.A. Fernandez. 2000. Engineering a mouse metallothionein on the cell surface of *Ralstonia eutropha* CH34 for immobilization of heavy metals in soil. *Nature Biotechnol.* 18:661–665.

Veiga, E., V. de Lorenzo, and L.A. Fernandez. 1999. Probing secretion and translocation of a beta-autotransporter using a reporter single-chain Fv as a cognate passenger domain. *Mol. Microbiol.* 33:1232–1243.

Veiga, E., V. de Lorenzo, and L.A. Fernandez. 2003a. Autotransporters as scaffolds for novel bacterial adhesins: Surface properties of *Escherichia coli* cells displaying Jun/Fos dimerization domains. *J. Bacteriol.* 185:5585–5590.

Veiga, E., V. De Lorenzo, and L.A. Fernandez. 2003b. Neutralization of enteric coronaviruses with *Escherichia coli* cells expressing single-chain Fv-autotransporter fusions. *J. Virol.* 77:13396–13398.

Veiga, E., V. de Lorenzo, and L.A. Fernandez. 2004. Structural tolerance of bacterial autotransporters for folded passenger protein domains. *Mol. Microbiol.* 52:1069–1080.

Veiga, E., E. Sugawara, H. Nikaido, V. de Lorenzo, and L.A. Fernandez. 2002. Export of autotransported proteins proceeds through an oligomeric ring shaped by C-terminal domains. *EMBO J.* 21:2122–2131.

Wan, H.M., B.Y. Chang, and S.C. Lin. 2002. Anchorage of cyclodextrin glucanotransferase on the outer membrane of *Escherichia coli. Biotechnol. Bioeng.* 79:457–464.

Wang, A.A., A. Mulchandani, and W. Chen. 2001. Whole-cell immobilization using cell surface-exposed cellulose-binding domain. *Biotechnol. Prog.* 17:407–411.

Wang, J.Y., and Y.P. Chao. 2006. Immobilization of cells with surface-displayed chitin-binding domain. *Appl. Environ. Microbiol.* 72:927–931.

Wernerus, H., J. Lehtio, T. Teeri, P.A. Nygren, and S. Stahl. 2001. Generation of metal-binding staphylococci through surface display of combinatorially engineered cellulose-binding domains. *Appl. Environ. Microbiol.* 67:4678–4684.

Wernerus, H., P. Samuelson, and S. Stahl. 2003. Fluorescence-activated cell sorting of specific affibody-displaying staphylococci. *Appl. Environ. Microbiol.* 69:5328–5335.

Wernerus, H., and S. Stahl. 2002. Biotechnological applications for surface-engineered bacteria. *Biotechnol. Appl. Biochem.* 40:209–228.

Westerlund-Wikstrom, B. 2000. Peptide display on bacterial flagella: Principles and applications. *Int. J. Med. Microbiol.* 290:223–230.

Westerlund-Wikstrom, B., J. Tanskanen, R. Virkola, J. Hacker, M. Lindberg, M. Skurnik, and T.K. Korhonen. 1997. Functional expression of adhesive peptides as fusions to *Escherichia coli* flagellin. *Protein Eng.* 10:1319–1326.

Wu, J.Y., S. Newton, A. Judd, B. Stocker, and W.S. Robinson. 1989. Expression of immunogenic epitopes of hepatitis B surface antigen with hybrid flagellin proteins by a vaccine strain of *Salmonella. Proc. Natl. Acad. Sci. USA* 86:4726–4730.

Wu, M.L., C.Y. Tsai, and T.H. Chen. 2006. Cell surface display of Chi92 on *Escherichia coli* using ice nucleation protein for improved catalytic and antifungal activity. *FEMS Microbiol. Lett.* 256:119–125.

Wu, P.H., R. Giridhar, and W.T. Wu. 2006. Surface display of transglucosidase on *Escherichia coli* by using the ice nucleation protein of *Xanthomonas campestris* and its application in glucosylation of hydroquinone. *Biotechnol. Bioeng.* 95:1138–1147.

Xu, Y., and Y. Li. 2007. Induction of immune responses in mice after intragastric administration of *Lactobacillus casei* producing porcine parvovirus VP2 protein. *Appl. Environ. Microbiol.* 73:7041–7047.

Yang, C., N. Cai, M. Dong, H. Jiang, J. Li, C. Qiao, A. Mulchandani, and W. Chen. 2008. Surface display of MPH on *Pseudomonas putida* JS444 using ice nucleation protein and its application in detoxification of organophosphates. *Biotechnol. Bioeng.* 99:30–37.

Yim, S.K., H.C. Jung, J.G. Pan, H.S. Kang, T. Ahn, and C.H. Yun. 2006. Functional expression of mammalian NADPH-cytochrome P450 oxidoreductase on the cell surface of *Escherichia coli. Protein Expr. Purif.* 49:292–298.

Zhang, J., W. Lan, C. Qiao, H. Jiang, A. Mulchandani, and W. Chen. 2004. Bioremediation of organophosphorus pesticides by surface-expressed carboxylesterase from mosquito on *Escherichia coli. Biotechnol. Prog.* 20:1567–1571.

CHAPTER 12

Antibody Selection from Immunoglobulin Libraries Expressed in Mammalian Cells

ERNEST S. SMITH and MAURICE ZAUDERER

ABSTRACT

Vaccinex has developed an antibody discovery technology that enables efficient selection of fully functional IgG antibodies from highly diverse immunoglobulin gene libraries expressed in mammalian cells. Because antibodies are expressed and selected in mammalian cells, the Vaccinex technology avoids problems associated with alterations in the folding and activity of antibodies expressed in

Therapeutic Monoclonal Antibodies: From Bench to Clinic. Edited by Zhiqiang An

bacteria and on the surface of phage particles. Unlike mouse-based humanization and transgenic selection systems, Vaccinex's *in vitro* antibody selection technology does not suffer the limitations of immunological tolerance to highly conserved proteins. In addition, the Vaccinex technology can be used to efficiently convert nonhuman monoclonal antibodies into fully human monoclonal antibodies.

12.1 INTRODUCTION

A quarter of a century after their debut, monoclonal antibodies have become the most rapidly expanding class of pharmaceuticals for treating a wide variety of human diseases, including cancer (Adams and Weiner 2005; Reichert et al. 2005). Antibody-based therapeutics bind to and either directly or indirectly influence the activity of cells or molecules involved in a given disease. The clinical success of a number of recent antibody-based therapies such as Rituxan®, Remicade®, and Herceptin®, as well as the regulatory acceptance of antibody-based therapeutics by regulatory agencies, has caused many leading pharmaceutical and biotechnology companies to increase their commercialization focus on the development of monoclonal antibody-based therapeutics. It is estimated that 25 to 30 percent of all products currently in clinical development are antibodies.

Previously, four general strategies have been employed to produce immunoglobulin molecules for drug development. In one approach, the variable domains from rodent antibodies are isolated and cloned in frame with human immunoglobulin constant domains. This process creates chimeric antibodies that preserve the antigen-binding specificity and affinity of the original rodent antibody, but now substitute human constant domains for both enhanced effector function and reduced immunogenicity in humans (Morrison et al. 1984). Although straightforward in application, chimeric antibodies still contain on average 30 percent nonhuman sequence. The rodent variable domains often result in immunogenicity in humans.

In a related approach, rodent antibody sequences have been converted into human antibody sequences, by grafting the specialized complementarity-determining regions (CDR) that comprise the antigen-binding site of a selected rodent monoclonal antibody onto the framework regions of a human antibody (Tsurushita and Vasquez 2003). In this approach, which has been termed antibody humanization, the three CDR loops of each rodent immunoglobulin heavy and light chain are grafted into homologous positions of the four framework regions of a corresponding human immunoglobulin chain (Jones et al. 1986). Because some of the framework residues also contribute to antibody affinity, the structure must, in general, be further refined by additional framework substitutions to enhance affinity (Queen et al. 1989). Although widely employed, this technology has two limitations. First, humanization is only possible when a specific, functional, high affinity nonhuman monoclonal antibody is available as starting material. In many cases mice are tolerant to the antigen or epitope of interest. Self-tolerance makes it difficult to generate the essential murine starting material. Without this initial antibody it is not possible to undertake humanization. Second, humanization results in antibodies that are on average only 95 percent human. The remaining mouse sequences can prove to be immunogenic when injected into humans. Clearly, this immunogenicity can be disastrous for clinical development of these antibodies, and it is not possible to predict which humanized antibodies will be immunogenic short of a clinical trial.

More recently, transgenic mice have been generated that express human immunoglobulin sequences (Green 1999; Lonberg 2005). While this strategy has been successfully applied to select human antibodies to multiple target antigens, it shares with the antibody humanization approach the limitation that antibodies are selected from the available mouse repertoire which has been shaped by proteins encoded in the mouse genome. This biases the epitope specificity of antibodies selected in response to a specific antigen. For example, immunization of mice with a human protein for which a mouse homolog exists might be expected to result predominantly in antibodies specific for those epitopes that are different in humans and mice. These may, however, not be the optimal target epitopes.

An alternative approach, which does not suffer this same limitation of self-tolerance, is to screen recombinant human antibody fragments displayed on bacteriophage, yeast, or bacterial cells (de Haard et al. 1999; Feldhaus et al. 2003; Hoogenboom 2005; Knappik et al. 2000; Marks et al. 1991; Mazor et al. 2007; Osbourn, Groves, and Vaughan 2005; Sheets et al. 1998). The most widely employed of these approaches is phage display. In phage display methods, functional immunoglobulin domains are displayed on the surface of a phage particle which carries polynucleotide sequences encoding them. In typical phage display methods, immunoglobulin fragments, for example, Fab or scFv, are displayed as fusion proteins, that is, fused to a phage surface protein. Antibody fragments with the desired binding specificity are then isolated by panning on antigen-coated tubes or by fluorescence activated cell sorting (FACS). Although this strategy does not suffer from an intrinsic repertoire limitation, it requires that variable domains of the expressed immunoglobulin fragment be synthesized and fold properly in bacterial cells. Many antigen-binding regions, however, are difficult to assemble correctly as a fusion protein in bacterial cells. In addition, the protein will not undergo normal eukaryotic posttranslational modifications. As a result, this method imposes a different selective filter on the antibody specificities that can be obtained.

As described above, each of the predominant platform technologies being used today has inherent limitations that can interfere with successfully selecting antibodies with the desired specificities or affinities. There was a need, therefore, for an alternative method to identify immunoglobulin molecules from an unbiased immunoglobulin repertoire that can be synthesized, properly glycosylated, and correctly assembled in mammalian cells.

12.2 IMMUNOGLOBULIN EXPRESSION LIBRARIES CONSTRUCTED IN A POXVIRUS VECTOR FOR EXPRESSION IN MAMMALIAN CELLS

In the following sections, we describe the Vaccinex ActivMAb® technology for selection of fully human antibodies from immunoglobulin gene libraries constructed in a poxvirus vector and expressed in mammalian cells. Different embodiments of this technology are based on construction and expression of either secreted or membrane-associated antibody libraries and support either *de novo* selection of a novel antibody, affinity improvement of an existing human antibody, or conversion of a mouse antibody to a fully human antibody specific for either soluble or membrane-associated target antigens. The principles of library construction will be presented first followed by applications of the secreted antibody platform to *de novo* selection, affinity improvement, and conversion of mouse antibodies. The advantages and limitations of the membrane antibody platform for these same applications will then be discussed.

12.2.1 Poxvirus Vectors

The ease of cloning and propagation in a variety of mammalian host cells has led to the widespread use of poxvirus vectors for expression of foreign proteins and as delivery vehicles for vaccine antigens (Moss 1991). A poxvirus-based library vector would have several advantages relative to more common plasmid or retrovirus-based vectors. Poxvirus replicates and is packaged into fully infectious particles in the cell cytoplasm and, as a result of its high infectivity, specific recombinants can be readily recovered even from very small numbers of selected cells, perhaps as few as a single cell. In addition, unlike plasmid- or retrovirus-based vectors, recombinant genes in a poxvirus vector can be recovered efficiently even from cells that have ceased to divide or that have died as a result of expression of the selected recombinant gene. Generally, the target protein coding sequence is cloned under the control of a vaccinia promoter in a plasmid transfer vector. The promoter and insert are flanked by sequences homologous to a nonessential region in the poxvirus, often the thymidine kinase (*tk*) gene, so that the plasmid intermediate can be introduced into the viral genome by homologous recombination at that locus. Recombinant virus can then be recovered based on a tk negative phenotype or other selectable marker. The frequency of recombinants derived in this fashion is of the order of 0.1 percent. This is sufficient to recover recombinants of a specific DNA clone but far too low to permit construction of a large, representative cDNA library. Initially we attempted to generate diverse cDNA libraries in poxvirus employing a direct ligation method (Merchlinksi et al. 1997). Although this method did select for a higher frequency of recombinants, relatively low viral titers were generated. We have developed a more efficient recombination method, termed Trimolecular Recombination, which generates recombinant vaccinia virus at high frequency.

12.2.2 Trimolecular Recombination

The rationale for this strategy is that a high frequency of recombinants would be obtained if cells were transfected with defective vaccinia DNA that could be packaged into infectious particles only if it had undergone recombination. One way to accomplish this is to cut the vaccinia DNA in the middle of the *tk* gene. Since there is no homology between the two *tk* gene fragments, the two vaccinia arms cannot be linked by homologous recombination except by bridging through the homologous tk sequences that flank the insert in a recombinant transfer plasmid. Because naked vaccinia DNA is not itself infectious, production of infectious particles requires that transfection be carried out in cells that are infected with a helper virus. As previously described, fowlpox virus (FPV) does not productively infect mammalian cells but provides the necessary helper functions required for replication and packaging of mature wild-type or recombinant vaccinia virus particles (Scheiflinger, Dorner, and Falkner 1992; Somogyi, Frazier, and Skinner 1993).

A vaccinia vector, v7.5/tk, was constructed that incorporates the early/late 7.5 k vaccinia promoter as well as unique NotI and ApaI restriction sites downstream of the promoter (Merchlinksi et al. 1997). Digestion with NotI and ApaI restriction endonucleases gives rise to two large fragments approximately 80 kilobases and 100 kilobases in size. Each of these arms includes a nonoverlapping fragment of the *tk* gene for bridging by a transfer plasmid. FPV-infected cells triply transfected with the two vaccinia arms and with DNA from a cDNA library constructed in a transfer plasmid gives rise to infectious vaccinia virus that is almost 100 percent recombinant (Smith et al. 2001). This method works very reliably and has been used to construct vaccinia cDNA libraries from many different cell lines and normal tissues (Smith, Shi, and Zauderer 2004).

12.2.3 Antibody Libraries Expressed in Mammalian Cells

The Vaccinex human monoclonal antibody discovery platform is based on the monoclonal expression of recombinant antibodies in mammalian cells. Separate libraries of human heavy and light chain immunoglobulin variable genes have been constructed in a vaccinia virus-based vector by Trimolecular Recombination.

Plasmid vectors incorporating constant regions of human heavy chain secreted gamma 1, human heavy chain membrane bound gamma 1, or human light chain were constructed to accommodate cloning human Ig variable region gene segments (VH, Vκ, and Vλ) in frame. These vectors are based on the plasmid pH5/tk, in which a vaccinia early/late promoter and multiple cloning sites were inserted into the viral *tk* gene. The multiple cloning sites were modified appropriately in order to clone in a modified Ig secretory signal peptide and the constant regions of γ1-secreted or γ1-membrane immunoglobulin heavy chains, and κ and λ immunoglobulin light chains. The resulting vectors retain unique cloning sites for inserting the VH, Vκ, and Vλ variable region genes. Throughout this manuscript Vκ and Vλ are collectively referred to as VL.

In order to take advantage of the increased diversity generated by random pairing of different heavy and light chains, we elected to construct independent libraries of heavy and light chains. The overall goal was to generate libraries of sufficient complexity that appropriate VH/VL combinations will occur at a frequency that permits efficient isolation of antibodies with a desired specificity and affinity. The independent assortment of germline V (D) and J segments, as well as the random combinatorial association of VL and VH, provides substantial diversity. Further diversification occurs during the response to antigen by the process of somatic mutation. To take advantage of all diversification processes, we have produced our libraries from four different human B-cell sources: (1) commercially obtained bone marrow-derived mRNA from large donor pools, (2) commercially available peripheral blood B-cells isolated from cancer patients, (3) commercially available peripheral blood B-cells and bone marrow from autoimmune patients (ex. Lupus), and (4) tonsil-derived germinal center B-cells. Because heavy and light chains are randomly reassorted in our system, it is possible to generate novel specificities that are more diverse than those of the antigen-driven B-cells from which these V genes derive. Somatic hypermutation in the germinal centers and selection resulting from the disease states of the B-cell donors contributes greatly to V gene diversity.

Immunoglobulin gene libraries were first constructed in the Ig-H and Ig-L plasmid vectors described above. The plasmid libraries were then used to construct immunoglobulin gene libraries in vaccinia virus by Trimolecular Recombination. It is reasonable to assume that the more VH/VL combinations that can be screened, the more likely it is that a monoclonal antibody with the desired epitope specificity and affinity will be isolated from an antibody library. Our current libraries contain 10 million Ig-H in both secreted and membrane format, and 10 million Ig-L. This calculates to a theoretical complexity of 1×10^{14} unique combinations. This complexity exceeds the size of most phage display libraries by several orders of magnitude. This theoretical complexity, however, vastly exceeds the number of recombinant antibodies that can be screened in mammalian cells. As described below, the availability of separate heavy and light chain libraries allows for sequential selection strategies that make it possible to tap into this enormous antibody diversity without needing to sample all of the 10^{14} possible VH/VL combinations directly.

12.2.4 Expression of Recombinant Human MAb

A number of mammalian cell lines were tested to identify an optimal cell line for MAb expression. Although vaccinia virus has a wide host cell range, vaccinia does not infect CHO cells or lymphocytes with high efficiency (Baixeras et al. 1998). As a result, we focused our testing on a variety of human and nonhuman epithelial cell lines. We found that HeLa cells, which are readily infected by vaccinia virus, express relatively high levels of vaccinia encoded immunoglobulins. It has been demonstrated that vaccinia virus can productively infect CHO cells if the cowpox CHO HR gene CP77 is provided by co-infection with CP77 recombinant vaccinia virus (Ramsey-Ewing and Moss 1996). We compared MAb expression levels in Hela cells versus in CP77 complemented CHO cells and determined that the expression levels were similar in the two cell lines. Because HeLa cells do not require complementation, HeLa was selected as the production line for the majority of our antibody selection projects. It is, however, useful in some instances in which a target antigen is expressed in HeLa cells to employ alternative host cell lines for antibody selection.

12.2.5 Selection of Specific Recombinant MAb

Mammalian cells infected with the vaccinia immunoglobulin gene recombinant vectors produce fully functional, bivalent antibodies. As outlined above, we have generated Ig-H libraries in a vaccinia expression vector that encodes the secretory form of the human gamma 1 heavy chain constant region. Co-infection of cells with these immunoglobulin heavy chain gene libraries and light chain libraries results in expression, assembly, and secretion of bivalent IgG1/L antibodies, permitting screening by ELISA or other functional assay. By infecting host cells with multiplicity of infection (moi) = 1 for both Ig-H and Ig-L vaccinia recombinants, each cell is on average infected with one Ig-H and one Ig-L recombinant vaccinia virus and thus expresses a single monoclonal antibody. We have also generated Ig-H libraries in a vaccinia expression vector that encodes the membrane-bound form of the gamma 1 heavy chain constant region. Co-infection of cells with these immunoglobulin heavy chain gene libraries and light chain libraries results in expression of bivalent antibody on the cell surface. The light chain libraries are the same as those used for the secreted antibody approach. By controlling moi = 1, each cell will express an average of one antibody specificity per cell. These cells can be "stained" with fluorescent labeled antigen and specific antibody producing cells selected using a combination of high throughput magnetic bead technology and cell sorting.

12.3 SELECTION OF ANTIBODIES IN SECRETED IgG FORMAT

Cells co-infected with secretory Ig-H and Ig-L recombinant vaccinia virus secrete bivalent IgG1/L, permitting screening by ELISA or functional assay. Secreted antibody has the advantage of being a very clean and specific reagent, which is particularly important for selection of antibodies to membrane-associated target antigens (see below). However, since secreted antibody is not associated with the producing cell, the selection strategy requires that small subsets of producing cells be isolated in separate wells for subsequent cloning. An efficient strategy with quantitative parameters is outlined below.

12.3.1 *De Novo* Selection of Fully Functional Antibodies in Mammalian Cells

We employ a standardized, highly sensitive ELISA for screening antibody libraries. Although each new target possesses unique characteristics, all other reagents are well characterized. The standard screening ELISA is a direct binding assay optimized for plates coated with antigen in the range of 0.1 to 1 μg/ml. Alternatively antigen can be coated onto an ELISA plate using a capture antibody. In order to achieve reliable discrimination of specific binding, it is desirable that there be at least 100 cells producing each specific antibody in an assay well. We routinely use 100,000 Hela cells for antibody production in each well, allowing 1000 different antibody combinations to be screened per well.

For screening, we generate multiple arrays in microtiter plates of mini Ig-L and Ig-H libraries, each containing a pool of 10 or 100 individual Ig gene recombinants, by amplifying 10 or 100 plaque forming units (pfu) from the parent Ig-L and Ig-H libraries, respectively, on feeder cells (monkey BSC1 cells) in individual wells of 96-well plates. Each of the resulting minilibraries contains 10 or 100 different Ig genes at titers of approximately 2×10^6 pfu/ml.

A standard *de novo* library screen is diagrammed in Figure 12.1. The Ig-H_{100} minilibraries are screened in combination with the Ig-L_{10} minilibraries. Thus, in a single well there are nominally 1000 distinct VH/VL combinations producing 1000 antibody clonotypes. Each assay well contains approximately 100,000 cells infected at a multiplicity of infection (moi) = 1 for both Ig-H and Ig-L recombinant virus. By calculation, there are nominally 100 cells producing each clonotype per well. Each 96-well plate, therefore, corresponds to approximately 10^5 distinct clonotypes. Because the independent assortment of heavy and light chains follows a Poisson distribution, the actual yield is somewhat less than this.

Following an incubation of 72 to 96 hours to allow for MAb secretion, culture supernatants are sampled and tested by ELISA for capacity to bind antigen. Specific antigens are coated onto duplicate plates, and each supernatant is also tested for binding to a control antigen. When an antigen-binding well is identified, the specific Ig-H and Ig-L are isolated from the parent minilibrary well by limiting dilution cloning and rescreening. In this process, mini Ig-H and Ig-L libraries corresponding to positive wells are sampled from the master plate and replated in 96-well plate format at limiting dilution (e.g., 10 or 1 pfu/well) and amplified on BSC1 feeder cells, creating Ig-H_{LD} and Ig-L_{LD} plates. In parallel, the original positive wells (Ig-H_{100} and Ig-L_{10}) are amplified to create high titer stocks that contain the original diversity of Ig-H and Ig-L recombinants. After amplification the Ig-H_{LD} plate is co-infected with the amplified original Ig-L_{10} stock into Hela cells and the Ig-L_{LD} plate is co-infected into Hela cells with the amplified original Ig-H_{100} stock. Following an incubation of 72 to 96 hours to allow for MAb secretion, culture supernatants are sampled and again tested by ELISA for capacity to bind antigen. In general, this screening identifies a number of positive wells in both the Ig-H_{LD} and Ig-L_{LD} plates. Following this second round of screening, individual plaques from a positive Ig-H_{LD} and an Ig-L_{LD} well are picked and amplified using standard vaccinia virus procedures. This step generates monoclonal Ig-H and Ig-L. We generally pick 12 Ig-H plaques and 6 Ig-L plaques. HeLa cell monolayers are then co-infected with each monoclonal Ig-H and Ig-L pair (12 Ig-H $\times$ 6 Ig-L = 72 combinations). The resulting supernatants are tested for antigen binding to verify that correct pairs of Ig-H and Ig-L that encode for specific antibody were isolated. The VH and VL genes are PCR

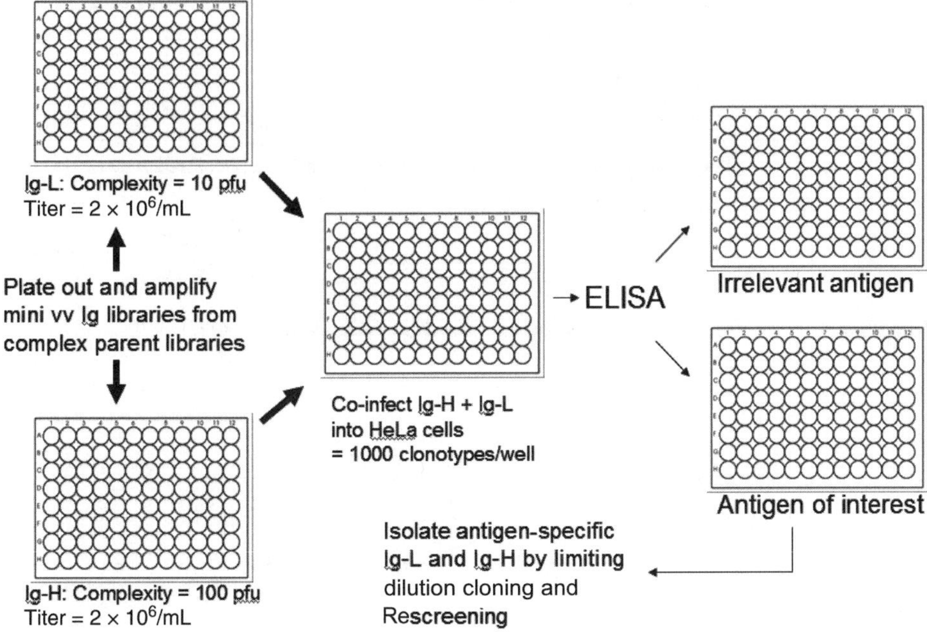

Figure 12.1 Selection of fully human antibodies using the secreted antibody platform. Separate master Ig libraries containing diverse Ig heavy and light chain genes are constructed in a vaccinia-based vector. Arrays of minilibraries of 100 Ig-H chain and 10 Ig-L chain clones are assembled and amplified from the master libraries. These minilibraries are co-infected into human cells which secrete fully assembled antibodies into the culture medium. Specific antibodies are detected by ELISA. Virus encoding the specific heavy and light chain genes are recovered from the corresponding minilibrary by limiting dilution cloning and rescreening.

amplified and subcloned into mammalian expression vectors containing human gamma 1 constant and human light chain constant domains. These expression plasmids are then transfected into CHO cells for high level expression and purification of full length IgG1/L and further analysis. Preliminary testing includes specificity testing by ELISA and affinity and functional testing.

12.3.1.1 Example of De Novo Antibody Selection Using Secreted IgG We have applied this approach to discovering MAbs to a range of different types of antigens. One such antigen, C35, is a protein of unknown function that we have determined is over-expressed in approximately 65 percent of breast tumors (Evans et al. 2006). The C35 protein can be readily expressed and purified from *Escherichia coli* as a 6X-his tagged protein. Results of a *de novo* screen for antibodies specific for C35 are shown in Figure 12.2a. The data shows ELISA results from one infection plate assayed in duplicate, as well as the same plate assayed on an irrelevant antigen. A strongly positive well was identified in this screen. The ELISA optical densities (ODs) are low but reliable. ODs increase significantly as the library complexity is reduced during second round and plaque screening. Additional screens in 200 plates identified 10 additional positive wells. The minilibraries corresponding to these positive wells were plated out at limiting dilution, rescreened as described above, and following plaque screening, pure monoclonal antibodies were obtained (Figs. 12.2b and 12.2c). The VH and VL genes from each clone were PCR amplified and subcloned into mammalian expression vectors containing human gamma 1 constant and human light chain constant domains. These expression plasmids

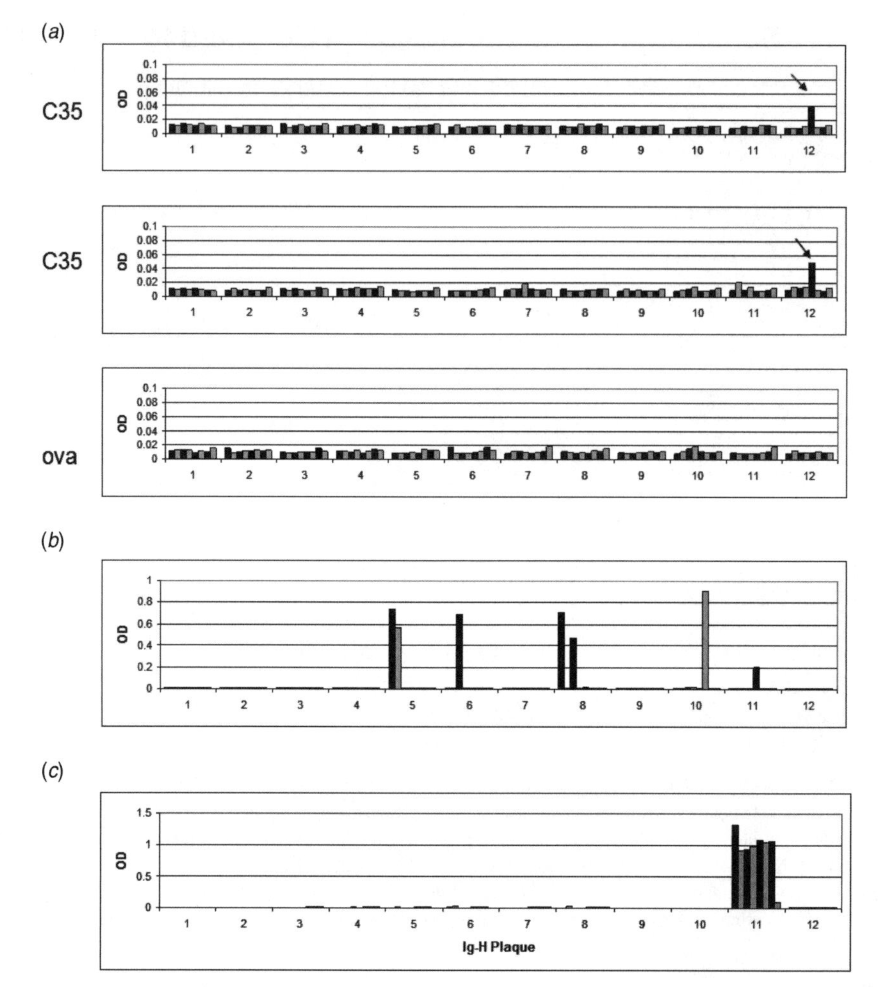

Figure 12.2 *De novo* selection of C35-specific monoclonal antibodies. Human antibodies specific for the "C35" breast cancer antigen were identified using the secreted antibody platform as diagrammed in Figure 12.1. (a) ELISA results from one 96-well infection plate. Supernatant from each well of one infection plate was assayed by ELISA for binding to C35-coated plates in duplicate, and a control antigen (ovalbumin). The well marked with an arrow indicates a well that contains a C35-specific antibody. (b) The positive well in **A** was subcloned by limiting dilution and rescreened as described in the text. A number of positive wells were identified. (c) Twelve individual Ig-H plaques were each paired with 8 Ig-L plaques. Each of the 96 plaque $\times$ plaque combinations were co-infected into Hela cells and the antibody produced was tested for binding to C35 by ELISA. The Ig-H clone is shown as a number, and each bar represents the pairing of an Ig-H clone with a different Ig-L clone. In this example, Ig-H clone 11 paired with Ig-L clones 1 to 5 created C35-specific antibodies. Sequencing of individual Ig-L clones confirmed that all five clones contained an identical Ig-L sequence.

were then transfected into CHO cells for high level expression and purification of full length IgG1/L and further analysis. Preliminary testing includes specificity testing by ELISA and affinity and functional testing.

At our current screening standard of 1000 clonotypes per well, it is possible using robotic pipeting stations (Biotek) and plate stackers for one individual to screen five million clonotypes per week. Although this number may seem low when compared to phage display throughput, we have found it to be more than adequate. In almost all cases, a workable initial panel of specific antibodies of moderate to high affinity can be isolated after screening between 2×10^7 and 10^8 clonotypes. The affinities of these selected antibodies can vary widely. Biacore analysis indicates that the affinities of these primary antibodies are often in the 10 nM range. As discussed below, the affinity of the initially selected antibodies can easily be improved by a process of V gene replacement.

12.3.2 Affinity Improvement by V Gene Replacement Selection

Whenever necessary, affinity maturation is accomplished by a method that we term V gene replacement. In this method the antigen-specific immunoglobulin light chain and/or heavy chains are fixed and used to screen larger libraries of complementary chains (e.g., use antigen-specific Ig-L to screen larger libraries of Ig-H clones). The format for this selection is very similar to the strategy employed for *de novo* screening, except in this method one chain is fixed (Fig. 12.3a). Having separate Ig-L and Ig-H recombinant vaccinia libraries and clones greatly facilitates this approach because the antigen-specific clones are ready for V gene replacement without any further manipulation. The rationale for this approach is that during the initial selection round, the antigen-specific Ig-L was only given the opportunity to pair with 100 Ig-H clones, and the antigen-specific Ig-H clone was only given the opportunity to pair with 10 Ig-L clones. It stands to reason that if these chains are given the opportunity to pair with hundreds of thousands or millions of complementing chains, then these new pairings will create higher affinity MAbs. New higher affinity combinations are readily detected by coating ELISA plates with reduced antigen concentration. When an antigen binding well is identified, the specific Ig-H or Ig-L is isolated from the parent minilibrary well by limiting dilution cloning and rescreening as described above.

12.3.2.1 Example of V Gene Replacement A C35 specific MAb was selected with affinity of 40 nM. Affinity of this MAb was improved by V gene replacement. The Ig-L from this clone was fixed and used to screen Ig-H libraries at both the original (1 μg/ml) and reduced (0.25 μg/ml) antigen concentrations. Figure 12.3b shows the ELISA results for four new clones and the original MAb tested at these two antigen-coating concentrations. The original selected antibody did not bind strongly to wells coated with the lower concentration of antigen, while the new clones still bound as well at the lower concentration. The affinities of the new MAbs were determined to be in the low nM range. This indicated that after one round of affinity improvement clones were identified that had 10- to 30-fold higher affinity than the original MAb by Biacore analysis. Multiple rounds of affinity improvement are possible, alternating H and L chains and incorporating mutagenesis in construction of sublibraries.

12.3.3 Fixed Ig-L Screening

A number of studies have demonstrated that the antibody heavy chain often makes a greater contribution to binding than the light chain. In order to increase the throughput of heavy chain screening, we have developed a screening strategy where a small number of defined light chains are fixed, and then used to screen for heavy chains. Similar to the results of others, we have determined that a limited diversity of light chains can be used to select antibodies specific for different antigens (Merchant et al. 1998; Nissim et al. 1994). The format for this screening is very similar to that used for V gene replacement (Fig. 12.3a). Once the initial antibody is selected, the heavy chain is then fixed and used to select new optimal light chains. This approach simplifies the manipulations involved, since only one new chain is isolated, and permits more rapid screening of the Ig-H libraries.

12.3.4 Conversion of Nonhuman Monoclonal Antibodies into Human Monoclonal Antibodies

Hybridoma technology has been used to identify a number of rodent antibodies with specificity, affinity, and functional activity towards important drug targets. For drug development these antibodies are often chimerized or humanized, with the attendant risk of immunogenicity and potential loss of

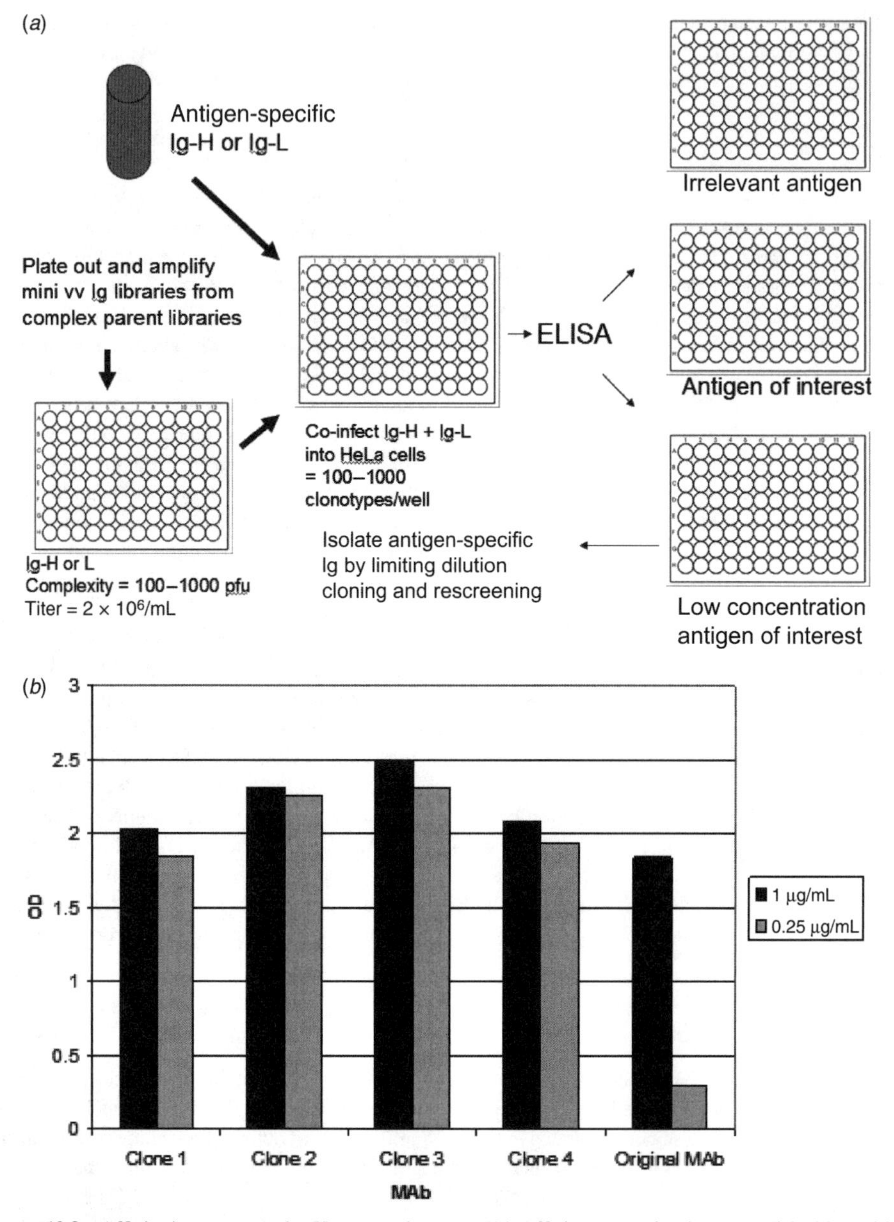

Figure 12.3 Affinity improvement by V gene replacement. (a) Affinity maturation is accomplished by pairing either the initial immunoglobulin light chain or heavy chain with a much larger selection of complementary chains. New higher affinity combinations are readily detected by coating ELISA plates with reduced antigen concentration. (b) Example of affinity improved antibodies selected by coating with a lower concentration of antigen.

affinity. We have applied the V gene replacement strategy of using vaccinia virus expressed antibody libraries for conversion of rodent antibodies into fully human antibodies.

The concept of V gene replacement in this application is to use a nonhuman antibody as a template and through a two-step process to identify human V genes that can replace the nonhuman V genes, while still retaining affinity and epitope specificity. The V gene replacement method is thus an alternative to traditional CDR grafted humanization. This method has several advantages compared to the more traditional humanization methods:

(1) V gene replacement results in the selection of fully human antibodies, while retaining the epitope specificity of the nonhuman MAb. In principle, these antibodies should have a lower risk of immunogenicity compared to CDR grafted and framework modified antibodies that retain significant amounts of murine sequences.
(2) V gene replacement results in the selection of multiple antibodies. This allows for the selection of lead antibodies derived from distinct VH and VL germline genes with different biochemical properties, including CDR sequences, expression levels, pI, etc.
(3) V gene replacement can result in the selection of antibodies with better affinity and functional activity than the original nonhuman antibody.

12.3.4.1 Summary of the Method In the first step in the method, the V genes from the nonhuman antibody are isolated and engineered to create chimeric heavy and light chains. The nonhuman Ig-H is paired with a library of human Ig-L and screened for specific binding to antigen. This initial selection yields a panel of hybrid antibodies comprising chimeric Ig-H and human Ig-L. The selected human Ig-Ls are then paired with a library of human Ig-H and selected for binding to antigen. Parallel selections can also be carried out starting with the nonhuman Ig-L to select human Ig-H, and then using the selected Ig-H to select human Ig-L. The human Ig-L selected with the chimeric Ig-H, and the human Ig-H selected with the chimeric Ig-L can also be cross-paired. The end result of these selection strategies is that panels of human antibodies that bind to the same antigen as the original nonhuman antibody are isolated. In most cases, the selected human antibodies recognize the same epitope as the original nonhuman antibody. If necessary, the first generation human antibodies can be affinity improved through either additional rounds of V gene replacement, or through mutagenesis.

12.3.4.2 Example of V Gene Replacement to Convert IL-6 Specific Mouse MAb into Human MAbs Interleukin-6 (IL-6) is a 23 kDa protein (212 amino acids) lymphokine that stimulates both B- and T-cell functions. IL-6 is believed to play an important role in the development and progression of rheumatoid arthritis (RA) (Gabay 2006; Lipsky 2006). IL-6, in conjunction with the soluble IL-6 receptor (sIL-6Rα), has been shown to activate endothelial cells to produce inflammatory chemokines and also to upregulate expression of adhesion molecules, contributing directly to recruitment of leukocytes at inflammatory sites. In addition, IL-6 can stimulate synoviocyte proliferation and osteoclast maturation and activation, suggesting a role in synovial pannus formation and in bone resorption in the inflamed joints. Neutralization of IL-6 function by blocking its receptor (IL-6R) with an IL-6R-specific monoclonal antibody has shown considerable promise in clinical trials of patients with RA (Smolen and Maini 2006).

Preliminary clinical trials have shown that treatment of late stage cancer patients with the murine anti-IL-6 MAb BE8 can block the *in vivo* proliferation of tumor cells and reduce IL-6-related toxicities (e.g., fever, cachexia; Haddad et al. 2001; Rossi et al. 2005; Wijdenes et al. 1991). Clinical use of a murine antibody is not ideal because the murine antibodies have short half-lives in humans, requiring frequent dosing, and murine antibodies are usually immunogenic, preventing long term treatment.

We have used the vaccinia library-based antibody discovery technology to convert the murine BE8 MAb into a fully human MAb. The goal of this work was to create a human MAb that was functionally similar to BE8, but that by virtue of being human, would have a superior half-life and reduced immunogenicity compared to BE8.

A detailed description of the discovery and characterization of human anti-IL-6 antibodies will be reported elsewhere. The following is a brief description of the antibody discovery process that we followed as an example of how the vaccinia library technology is used to convert mouse MAbs into human MAbs.

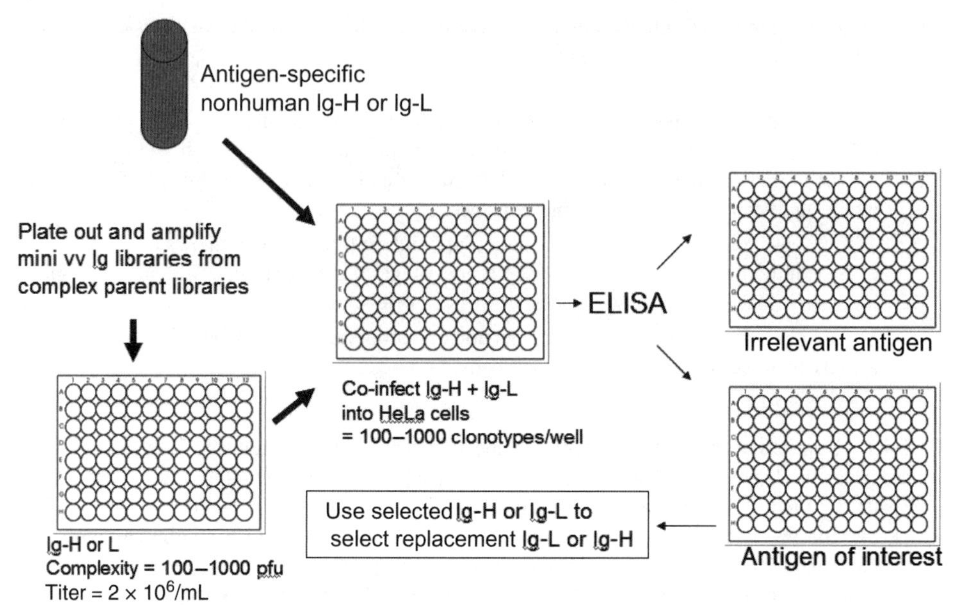

Figure 12.4 Conversion of mouse monoclonal antibody into human monoclonal antibodies. The antigen-specific mouse heavy chain is engineered so that it is expressed as a chimeric secreted heavy chain in vaccinia virus. A master Ig library containing diverse Ig light chain genes is constructed in a vaccinia-based vector. Arrays of minilibraries of 100 to 1000 Ig-L clones are assembled and amplified from the master library. The antigen-specific Ig-H and the Ig-L minilibraries are co-infected into human cells which secrete fully assembled antibodies into the culture medium. Specific antibodies are detected by ELISA. Virus encoding the specific light chain gene is recovered from the corresponding minilibrary by limiting dilution cloning and rescreening. To select human Ig-H that can pair with the selected human Ig-Ls, the process is reversed. The selected human Ig-Ls are used to screen minilibraries of human Ig-H gene libraries, each containing a pool of 100 to 1000 individual Ig-H gene recombinants. Specific antibodies are detected by ELISA. Virus encoding the specific heavy chain gene is recovered from the corresponding minilibrary by limiting dilution cloning and rescreening. Immunoglobulin genes encoding the specific antibody of interest are isolated and characterized.

The VH and VK genes from murine BE8 were engineered into full-length human Ig-H and human Ig-L in vaccinia virus so that they could be expressed as chimeric human IgG1/kappa. The chimeric BE8 Ig-H was screened in combination with human Ig-L minilibraries. A standard screen is diagrammed in Figure 12.4. For screening, we generate sets of mini Ig-L gene libraries, each containing a pool of 100 to 1000 individual Ig-L gene recombinants by amplification of 100 to 1000 pfu from the parent Ig-L libraries in individual wells of 96-well plates. Each of the resulting minilibraries carried 100 to 1000 different Ig-L genes at titers of approximately 2×10^6 pfu/mL. HeLa cells were co-infected with each of these Ig-L minilibraries along with chimeric BE8 Ig-H at moi = 1. Thus, in a single well there are nominally 100 to 1000 different Ig-L that could pair with chimeric BE8 Ig-H. Following incubation for 72 to 96 hours, culture supernatants are sampled and tested by ELISA for capacity to bind to IL-6 that had been captured onto an ELISA plate using the IL-6-specific BE4 MAb (specific for a BE8-independent epitope). When an antigen-binding well was identified, the specific Ig-L was isolated from the original minilibrary well by limiting dilution cloning and rescreening as described above. Monoclonal VL genes were then PCR amplified from the vaccinia clones and subcloned into mammalian expression vectors for high level expression and further analysis.

To select human Ig-H that can pair with the human Ig-Ls the process was reversed. The selected human Ig-Ls were used to screen minilibraries of human Ig-H gene libraries, each containing a pool of 100 to 1000 individual Ig-H recombinants as described above. We also performed parallel selections in which we started with the chimeric Ig-L to select human Ig-H, and then used the selected human Ig-H to select human Ig-Ls.

The selected VH and VL genes were PCR amplified and subcloned into mammalian expression vectors containing human gamma 1 constant and human kappa constant domains. These expression plasmids were then transfected into CHO cells for high level expression and purification of full length IgG1/L and further analysis. Preliminary testing includes specificity testing by ELISA and affinity and functional testing. By cross-pairing all of the selected Ig-H and Ig-L, we identified over 30 human MAbs that had low nM or sub-nM affinity for IL-6.

12.3.5 Screening Antibody Libraries on Whole Cells

A number of multipass membrane proteins, such as G-protein coupled receptors (GPCRs), are key mediators of signal transduction (Flower 1999; Nambi and Aiyar 2003). Members of the GPCR family contain a conserved heptahelical structure that results in the proteins making seven passes through the plasma membrane. Members of the GPCR family play critical roles in numerous physiological functions and often have tissue-specific expression patterns. The critical physiological role coupled with their tissue-specific expression profile has made GPCRs attractive targets for drug development. More than 50 percent of all marketed drugs target a member of the GPCR family. Despite this high rate of development as small molecule drug targets, however, there are very few monoclonal antibodies that have been developed to target GPCRs.

The paucity of monoclonal antibodies developed to target GPCRs is mainly due to the fact that it has proven to be extremely difficult to select antibodies that are specific for GPCRs. This difficulty is due to two main factors: (1) GPCRs are hydrophobic proteins that are extremely difficult to purify in biologically active form. This requires antibody screening on whole cells. (2) Many GPCRs are highly conserved between humans and rodents and, therefore, subject to immune self-tolerance. This is an obstacle to induction of high affinity antibody responses in immunized rodents. Several approaches employing various negative and positive selection strategies have been reported for using phage display and whole cell panning for the isolation of membrane receptor-specific antibodies. Despite these isolated successes, panning of phage expressed antibody libraries on whole cells has proved to be extremely difficult. One of the main challenges results from the tendency of phage to bind nonspecifically to cells. This background binding requires precise optimization of binding and washing steps. Even when specific antibodies have been isolated by phage display, in many cases these antibodies are of very poor affinity and require extensive optimization in order to generate clinical grade antibodies.

We have developed the secreted antibody platform so as to enable screening on whole cells for the selection of fully human antibodies specific for multipass membrane proteins. This approach screens full length antibodies for binding to whole cells in the absence of interference from viral coat proteins. This approach is very similar to the approach used to screen by ELISA, except that rather than screen by ELISA, we screen using the ABI 8200 (FMAT), which was developed to allow for the screening of hybridomas that make antibodies specific for membrane antigens in mix and read format: Using this instrument a mix of soluble antibody, cells expressing the antigen of interest, and fluorescently labeled secondary reagents are added in 96-well or 384-well format, mixed, incubated, and scanned in the ABI 8200. No washing steps are required, improving throughput. With this device it is possible to analyze 60 microtiter plates in 96-well or 384-well format per day. The major limitation of this instrument is that it was designed to sample hybridomas that are present at a diversity of one antibody per well.

12.3.5.1 Optimization of the Platform We elected CD20 as a model system to develop this antibody selection method. CD20 is a 35,000 kDa nonglycosylated tetraspanning cell membrane phosphoprotein. The antigen is located on normal pre-B and mature B lymphocytes and is found on most B-cell non-Hodgkin's lymphomas, but is not found on stem cells, pro-B cells, normal plasma cells, or other normal tissues. This high expression on normal and malignant B-cells has made CD20 an attractive target for immunotherapeutic depletion of B-cells. Rituxan®/MAbThera®

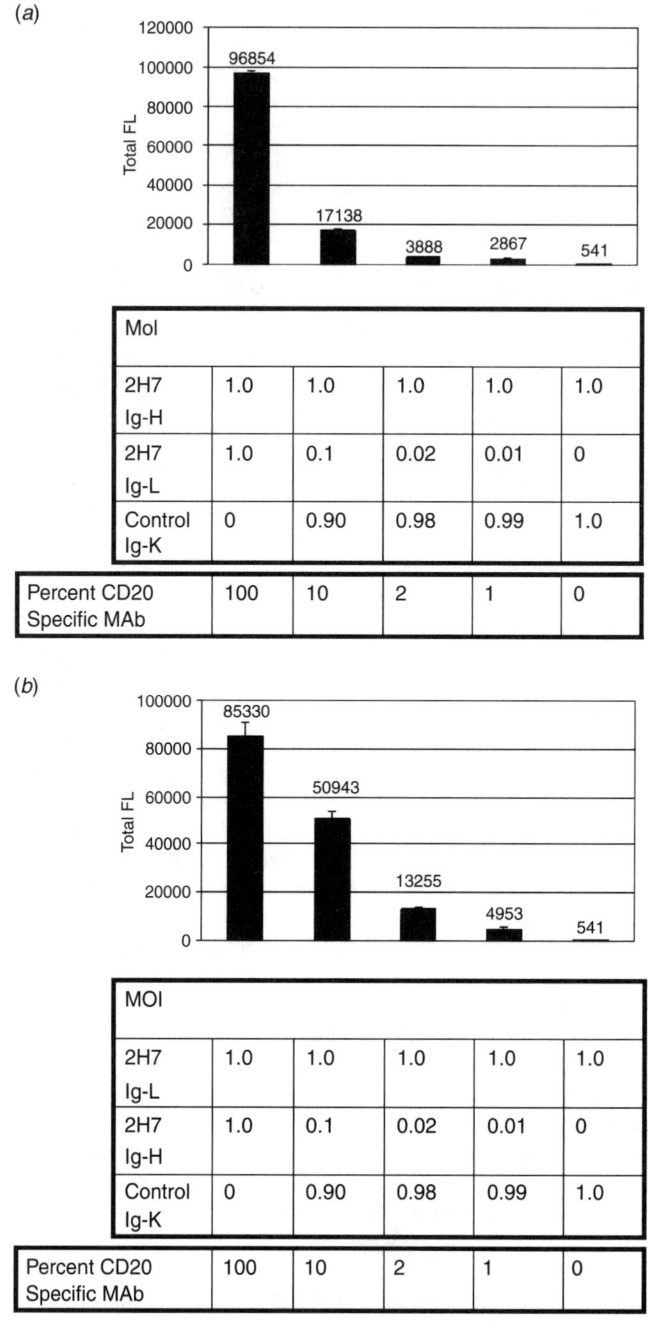

(a)

MoI					
2H7 Ig-H	1.0	1.0	1.0	1.0	1.0
2H7 Ig-L	1.0	0.1	0.02	0.01	0
Control Ig-K	0	0.90	0.98	0.99	1.0
Percent CD20 Specific MAb	100	10	2	1	0

(b)

MOI					
2H7 Ig-L	1.0	1.0	1.0	1.0	1.0
2H7 Ig-H	1.0	0.1	0.02	0.01	0
Control Ig-K	0	0.90	0.98	0.99	1.0
Percent CD20 Specific MAb	100	10	2	1	0

Figure 12.5 Detecting secreted antibody by FMAT. The evaluation of FMAT sensitivity using vaccinia virus spiking experiments. Vaccinia virus recombinant for the 2H7 Ig-H was fixed at moi = 1, and paired with dilutions of 2H7 Ig-L recombinant vaccinia virus (a) or the 2H7 Ig-L was fixed at moi = 1 and paired with dilutions of the 2H7 Ig-H (b). In both cases the specific 2H7 virus was diluted with vaccinia virus recombinant for unselected Ig V genes. CD20-specific binding was determined by adding antibody containing culture supernatant and anti-human IgG-APC secondary reagent to wells containing CHO.CD20 or CHO.vector cells. In both cases CD20-specific antibody was detected when it was present at as little as 1 percent of the total Ig. (c) Visual examination of the wells confirms the presence of specific binding.

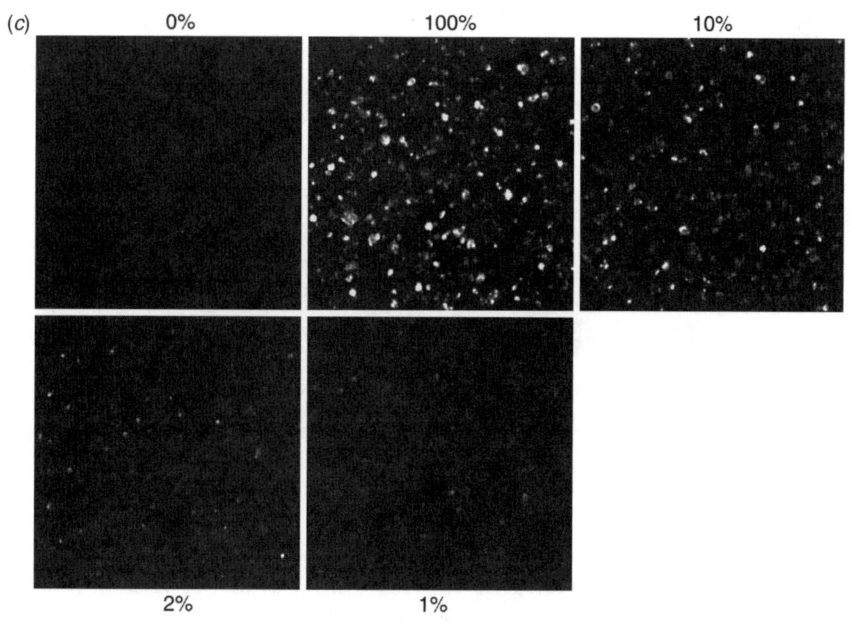

Figure 12.5 (*Continued*).

(rituximab) is licensed and commercialized in Europe and the United States for the treatment of relapsed or refractory, low-grade or follicular, CD20 antigen-positive, B-cell non-Hodgkin's lymphoma (NHL).

The mouse 2H7 MAb is specific for CD20 (Liu et al. 1987). A humanized version of this antibody is currently in late stage clinical testing (Vugmeyster et al. 2005). We cloned the CD20-specific 2H7 V genes from the 2H7 hybridoma (ATCC HB-9303) into mammalian expression vectors and into vaccinia transfer vectors for production of recombinant vaccinia virus. We also transfected CHO cells with the CD20 cDNA and selected a stable expressing clone. We used these tools: Chimeric 2H7 MAb, vaccinia virus recombinant for 2H7 Ig-H and Ig-L, and CHO.CD20 transfectants to develop FMAT conditions to detect CD20-specific Abs. Sensitivity by FMAT is a balance between signal and noise. The instrument quantitates fluorescence on each cell in a well at a specific depth of focus while subtracting out non-cell-bound background fluorescence. This enables the mix and read format because background is automatically subtracted in each well. We varied a number of conditions, including binding buffer, number of cells per well, choice of secondary reagents, binding time and temperature in order to increase the specific signal. Spiking experiments were performed to determine the limit of detection under the various conditions. In these spiking experiments, either the 2H7 Ig-H was fixed and paired with dilutions of 2H7 Ig-L, or the 2H7 Ig-L was fixed and paired with dilutions of the 2H7 Ig-H. In both cases, the specific 2H7 virus was diluted with vaccinia virus recombinant for unselected Ig-H or Ig-L genes. The data read out can be cell counts, fluorescence, or total fluorescence (counts $\times$ fluorescence) and a picture of each assay well is taken to allow for visual confirmation of positive results. As shown in Figure 12.5, in both cases we were able to detect CD20-specific MAb when it was present at as little as 1 percent of the total Ig. This translates into the ability to screen approximately 100 MAbs/well in 96 well plate format. Similar sensitivities have been observed in other antibody–antigen combinations.

Although this sensitivity limit of 100 clonotypes per well is somewhat low relative to 1000 clonotypes per well by ELISA, it is sufficient for MAb selection. We have successfully used with this approach for the conversion of mouse MAbs into human MAbs. With this approach secreted, fully human antibody can be screened on whole cells, removing the requirement for purification of

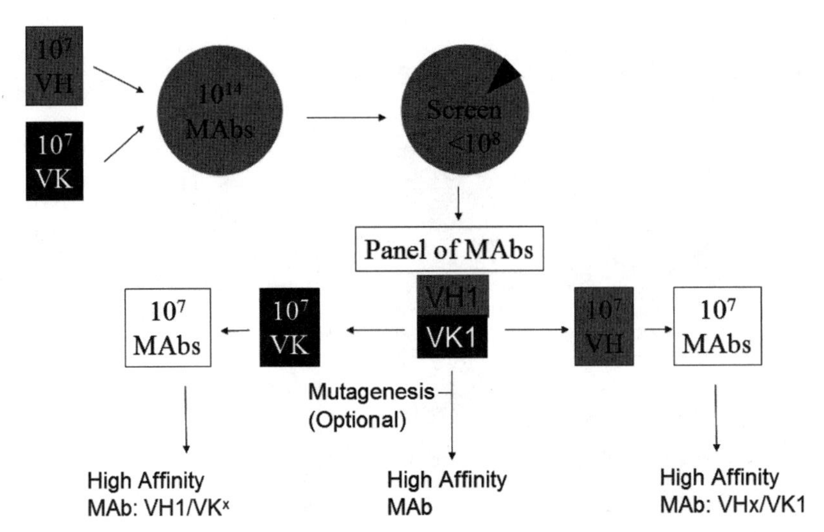

Figure 12.6 Summary of the secreted antibody platform. Employing the secreted antibody platform, 20 to 100 million recombinant antibody combinations can be screened using existing methods. By increasing automation, a larger number of antibodies could be screened. This level of diverse antibody combinations is sufficient to select a panel of fully human MAbs that have at least moderate affinity for the antigen of interest. Once an antibody of functional interest is identified, if necessary, V gene replacement screening can be employed to improve MAb affinity by rapidly sampling 10^6 to 10^7different complementary chains for both the heavy and light chains.

hydrophobic membrane proteins, and allows for the screening of antibodies against a target protein that is expressed in its natural milieu.

12.3.6 Summary of the Secreted Antibody Platform

A summary of the secreted antibody platform is shown in Figure 12.6. Employing the secreted antibody platform, it is possible to screen from 20 to 100 million recombinant antibody combinations using standard ELISA methods. By increasing automation, a larger number of antibodies can be screened. However, in our experience, this level of diverse antibody combinations is sufficient to select a panel of fully human MAbs that have at least moderate affinity for the antigen of interest and whose functional properties in the form of secreted IgG can be characterized quickly. Once an antibody of functional interest is identified, we use V gene replacement screening to improve MAb affinity by rapidly sampling 10^6 to 10^7 different complementary chains for either or both the heavy and light chains. This strategy of V gene replacement mimics the strategy of the natural immune system, which is to first select a moderate affinity antibody and then improve it. Selected MAbs can also be affinity improved by mutagenesis. This platform has been used to generate high affinity MAbs specific for a number of different types of target antigens. This technology also affords a robust method to convert mouse MAbs into human MAbs, and can be used to screen soluble antibodies for binding to membrane antigens expressed by whole cells.

12.4 MEMBRANE ANTIBODY PLATFORM

We have, in parallel, developed a platform to allow for the expression of a library of full length immunoglobulin on the surface of mammalian cells. In this embodiment the heavy chain immunoglobulin library is created in a vector that contains the human gamma 1 heavy chain containing a transmembrane domain. When cells are co-infected with heavy chain and light chain recombinant vaccinia, the antibody is expressed, assembled, and trafficked to the cell membrane. Cells expressing antigen-binding antibodies can be isolated by staining with labeled antigen and then isolating the antigen-binding cells using a combination of magnetic bead isolation and FACS. The affinities of selected antibodies can be driven by varying antigen concentrations employed for staining and selection. This method provides a rapid quantitative method to isolate specific high affinity monoclonal antibodies.

Despite these advantages, this method of antibody selection has a number of challenges associated with it. In order to take advantage of the increased diversity generated by random pairing of different heavy and light chains, we elected to construct independent libraries of heavy and light chains. While the independent assortment of individual Ig-H and Ig-L is a great asset during V gene replacement projects, it creates additional challenges when two independent libraries must be manipulated and the selected Ig-H and Ig-L must be maintained together in order to confirm the binding properties of the resulting antibody. As mentioned above, we have found by *de novo* screening that for many antigens antibodies of at least moderate affinity occur at a frequency of between 1 in a million to 1 in 5 million clonotypes. Isolating specific cells at this frequency by MACS and FACS is extremely challenging. Because of the challenges of co-isolating independent chains and the low hit rate, we have focused our efforts with this platform on the affinity improvement of selected MAbs and the conversion of mouse MAbs into human MAbs by V gene replacement. Both of these approaches are well suited to the cell surface display strategy because when one immunoglobulin chain is fixed, the hit rate for a complementary chain increases to on the order of 1 in 100,000.

12.4.1 Conversion of Nonhuman MAbs into Human MAbs Using the Membrane Antibody Platform

The method to use the membrane antibody platform for conversion of mouse MAbs into human MAbs is shown in Figure 12.7. This follows a similar two-step process as described above for the secreted

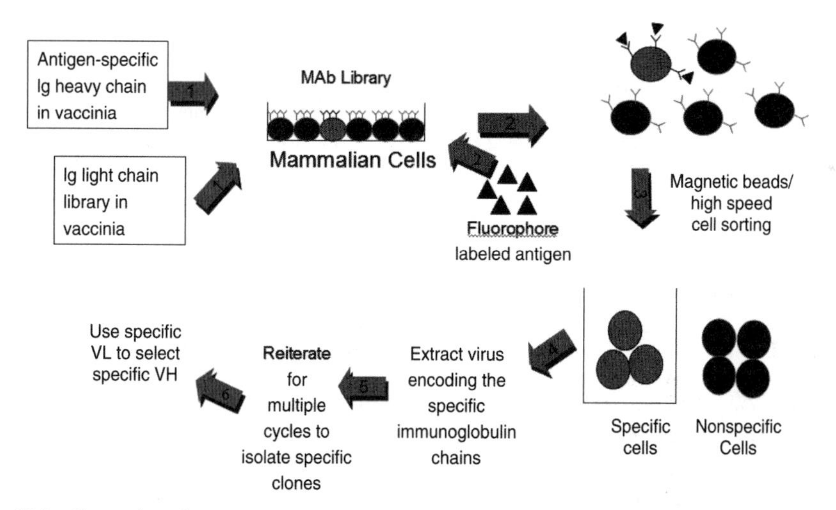

Figure 12.7 Conversion of a mouse monoclonal antibody into human monoclonal antibodies using the membrane antibody platform. The antigen-specific mouse heavy chain is engineered so that it is expressed in vaccinia virus as a chimeric membrane-bound heavy chain. Mammalian cells are co-infected with this specific recombinant heavy chain and with a recombinant library of human light chains. After overnight infection, cells are harvested and stained with fluorescent labeled antigen. Unbound antigen is washed away, and the minority of cells that express a specific antibody are isolated using magnetic bead separation and high speed cell sorting. Vaccinia virus is readily recovered from a small number of selected cells by freeze/thaw and can be rapidly expanded. If necessary initial rounds of selection can be performed until specific replacement Ig-L are identified. Once an antigen-specific fully human light chain is isolated, the chimeric heavy chain is set aside and the cycle is repeated to select complementary fully human heavy chains. This results in selection of fully human antibodies that bind the same antigen as the original mouse antibody.

antibody platform. The antigen-specific mouse heavy chain is engineered so that it is expressed as a chimeric membrane-bound heavy chain. Mammalian cells are co-infected with this specific recombinant heavy chain and with a recombinant library of human light chains. In order to simplify selection of the unknown replacement Ig-L clones, infections are performed using "inactivated" Ig-H. Inactivation is achieved by psoralen/UV treatment which blocks virus replication but permits viral entry and expression of viral and recombinant genes (Tsung et al. 1996). After overnight infection, cells are harvested and stained with fluorescent labeled antigen. Unbound antigen is washed away, and the minority of cells that express a specific antibody are isolated using magnetic bead separation and/or high speed cell sorting. This method has very high throughput. Specific antibodies can be selected from 10^8 to 10^9 recombinant antibody-expressing cells in one day. We screen our libraries at a redundancy of 100 infected cells per antibody combination, so the throughput of 10^8 to 10^9 cells translates into the ability to screen 10^6 to 10^7 Ig-L or Ig-H clones in each screening. Vaccinia virus is readily recovered from a small number of selected cells by simple freeze/thaw cycles and can be rapidly expanded by amplification on BSC1 cells. The psoralen/UV inactivated Ig-H virus will not replicate and so only the selected Ig-L recombinant vaccinia is recovered after amplification. After amplification, an aliquot of the selected pool of Ig-L can be paired with fresh chimeric Ig-H (cIg-H), stained with the antigen of interest, and analyzed by flow cytometry in order to estimate the percent of Ig-L that are able to pair with the cIg-H and bind antigen. If the percent of antigen-specific Ig-L is low, for example <10 percent, then the pooled Ig-L from the first sort can be subjected to additional rounds of sorting and analysis until a sufficient level of enrichment has been achieved. Once the percentage of antigen-specific Ig-L is high enough, individual vaccinia plaques are picked and amplified following standard vaccinia methods. The Ig-L plaques are then paired individually with the cIg-H and analyzed for antigen binding by flow cytometry. This step identifies clonal human Ig-L that can replace the cIg-L for pairing with the cIg-H and binding to antigen. Once an antigen-specific human light chain is isolated, the chimeric heavy chain is set aside and the process is repeated using the human Ig-L to select complementary human heavy chains. This results in selection of human antibodies that bind the same antigen as the original mouse antibody. Parallel selection strategies can be carried out starting with the chimeric Ig-L to select replacement human Ig-H, and then using the human Ig-H to select human Ig-L. The selected VH and VL genes are then PCR amplified from vaccinia and subcloned into mammalian expression vectors containing human gamma 1 constant and human kappa/lambda constant domains. These expression plasmids are then transfected into CHO cells for high level expression and purification of full length IgG1/L and further analysis. Preliminary testing includes specificity testing by ELISA, affinity measurements, and functional assays. If necessary, these antibodies can be affinity improved either through V gene replacement or through mutagenesis.

12.4.1.1 Example of Conversion of Mouse MAb into Human MAb Using the Membrane Platform The 3C9 hybridoma produces a mouse MAb that binds with high affinity to the proprietary VX5 antigen. We used the membrane antibody platform to convert 3C9 into a human MAb. In the first step the V genes from the 3C9 hybridoma were isolated and used to generate vaccinia virus recombinant for membrane bound chimeric Ig-H and chimeric Ig-L Next, we co-infected Hela cells at moi = 1 with the c3E10 Ig-L and a library of human Ig-H. After overnight infection, the cells were harvested and incubated with 50 ng/ml VX5 for 30 minutes on ice. Following this incubation unbound VX5 was washed away and the cells were incubated with biotinylated polyclonal anti-VX5 antibody (pAb) for 30 minutes on ice, washed, and incubated with streptavidin-APC for 20 minutes on ice, washed, and then cells binding to VX5 were isolated by cell sorting. Virus was released from selected cells by three cycles of freeze/thaw and selected virus was amplified on BSC1 feeder cells in one well of a six-well plate. After amplification and titration, the selected virus was used as input for a second round of selection as described above. Following the second round of selection, individual Ig-H plaques were picked and analyzed for the ability to pair with c3C9 Ig-L and bind to VX5. As shown in Figure 12.8a, a panel of Ig-H clones was isolated. Preliminary characterization of hybrid antibody (cIg-L/Ig-H) allowed the selection of a panel of

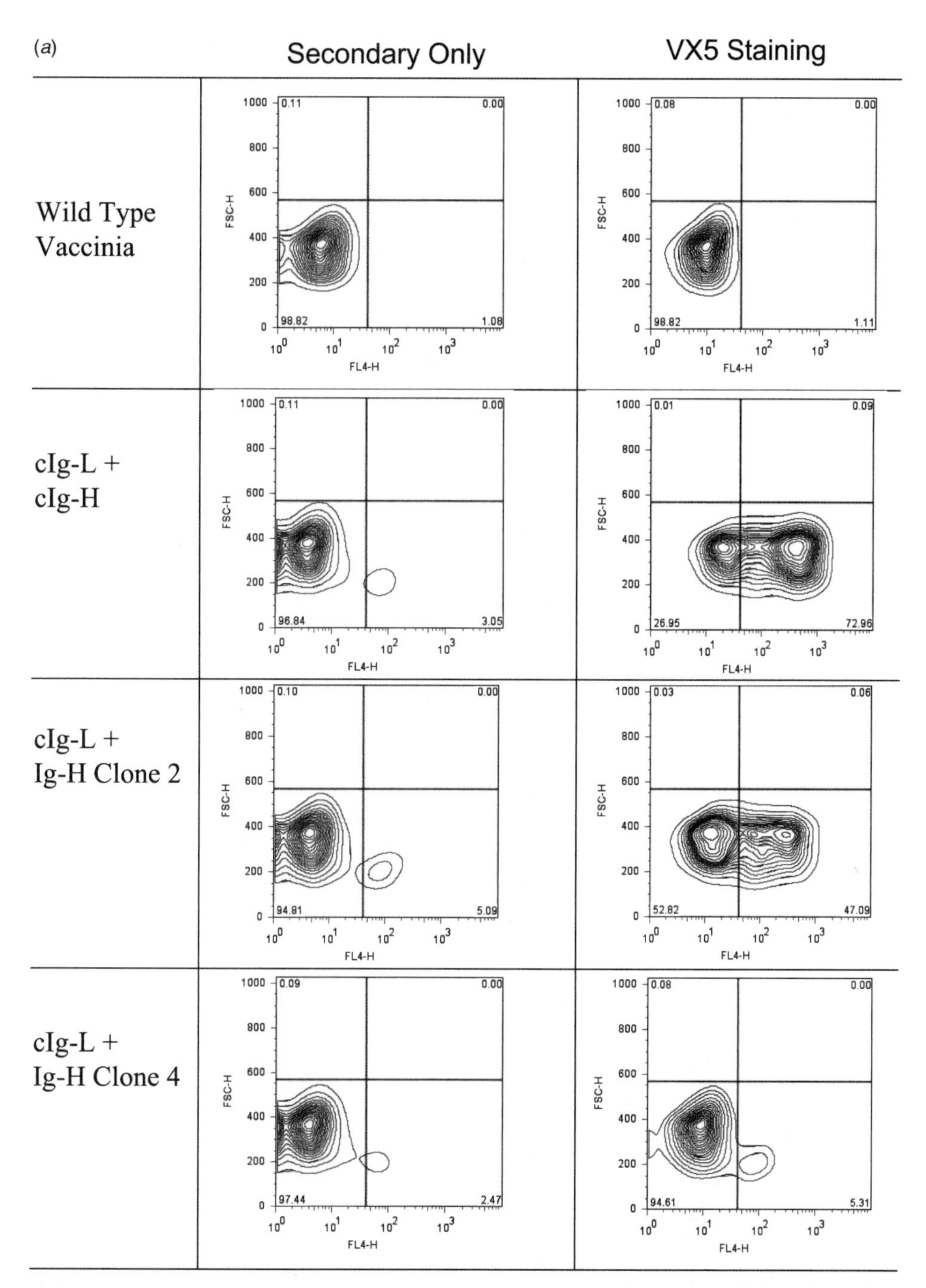

Figure 12.8 Flow cytometric analysis of antibodies selected with the membrane antibody platform. (a) Chimeric VX5 specific Ig-L was used to select replacement human Ig-H as described in Figure 12.7. Following two cycles of selection individual human Ig-H clones were tested for the ability to pair with the Chimeric Ig-L and bind to antigen. Ig-H clone 2 is a representative antigen specific clone, while clone 4 is a representation clone that does not bind to VX5.

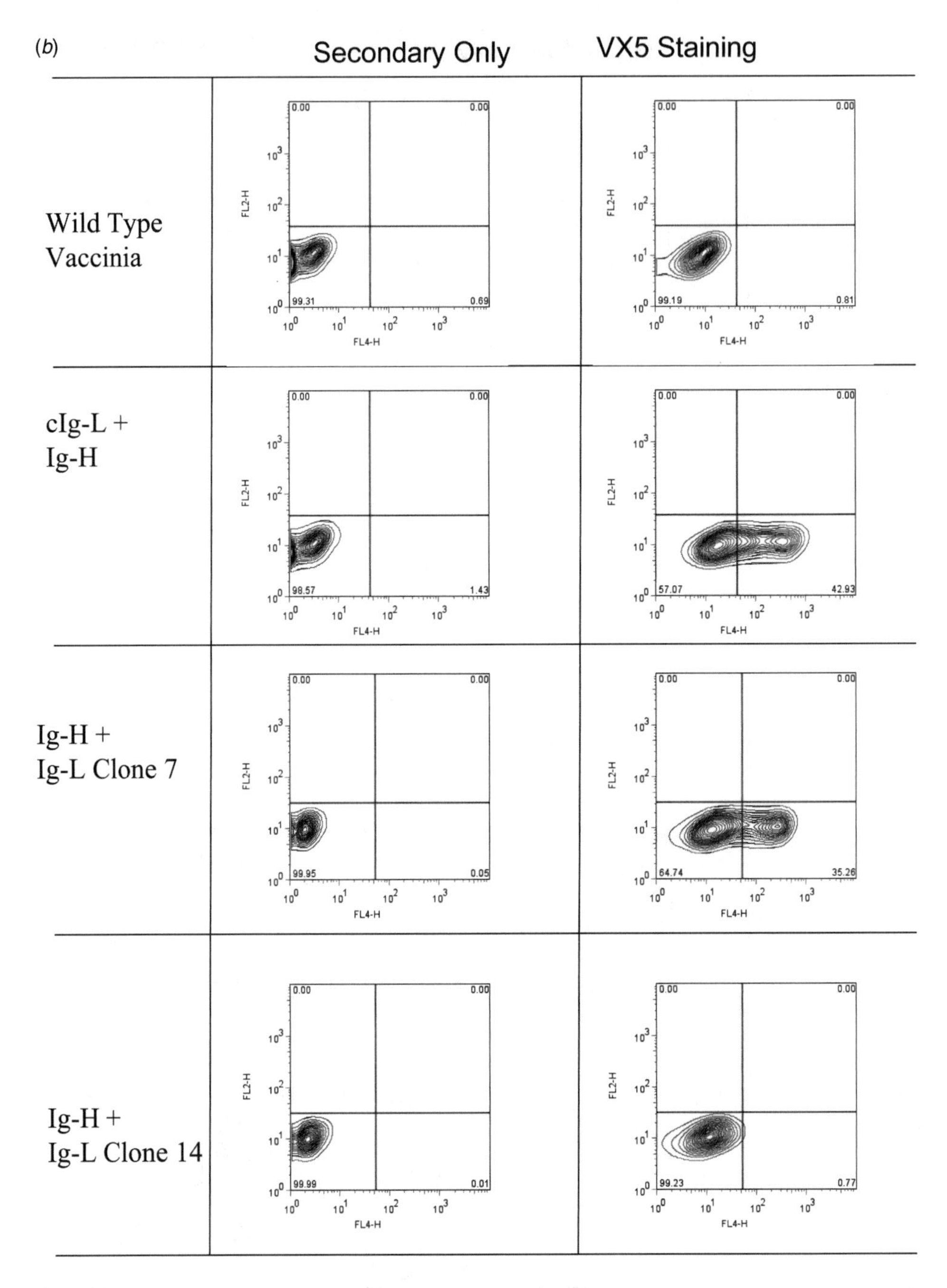

Figure 12.8 (*Continued*) (b) The human Ig-H was used to select human Ig-L. Clones were analyzed for the ability to pair with the Ig-H and bind to VX5. Ig-L clone 7 is a representative antigen specific clone, while clone 14 is a representation clone that does not bind to VX5.

promising Ig-Hs. Next, each human Ig-H was fixed and used to screen a library of human Ig-L. After two rounds of selection a panel of human Ig-L was isolated (Fig. 12.8b). The selected VH and VL genes were then PCR amplified and subcloned into mammalian expression vectors containing human gamma 1 constant and human kappa/lambda constant domains. These expression plasmids were then transfected into CHO cells for high level expression and purification of full length IgG1/Ls. Preliminary testing includes specificity testing by ELISA and affinity and functional testing. Selected antibodies had affinity (1 to 10 nM) and functional activity that was similar to that of chimeric 3C9.

12.4.2 V Gene Replacement with the Membrane Antibody Platform

The membrane antibody platform can be used to affinity improve antibodies by V gene replacement. These V genes can come from the conversion of mouse MAbs into human MAbs, from the secreted antibody platform, or from any other source such as phage display. This approach follows a similar strategy as described above for the secreted antibody platform and for the conversion of mouse MAbs into human MAbs using the membrane platform. An antigen-specific chain is fixed and used to interrogate a large number of complementary chains. The advantages of using the membrane format for affinity selection are twofold. First, this method has a very high throughput, with more than one million chains being screened at once. Second, since antibody-antigen binding is occurring in solution, affinity selection strategies have been developed where the concentration of selecting antigen is reduced from initial selection rounds to select for only higher affinity binders (Van den Buecken et al. 2003). By iterative bootstrapping with this strategy we can mimic the effect, although not the mechanism, of *in vivo* affinity maturation to isolate high affinity MAb.

12.4.2.1 Example of V Gene Replacement with the Membrane Antibody Platform Asdescribed above, we used the secreted antibody platform to convert the IL-6 specific BE8 MAb into a panel of human MAbs. One of these antibodies, MAb 190, had good functional activity in the *in vitro* assays, but had an affinity of only 6 nM. We elected to affinity improve MAb 190 using the membrane antibody platform. For affinity improvement, we fixed the Ig-L (L172) from MAb 190 and used it to screen a library of Ig-H. Because Ig-L has the same format for secreted and membrane-bound antibodies, the vaccinia clone selected with the secreted platform could be directly used for selection of heavy chain in the membrane platform. HeLa cells were infected with psoralen/UV inactivated Ig-L (L172) and a library of Ig-H. Following an overnight infection the cells were harvested and stained with 1 nM IL-6 for 30 minutes on ice. After washing, bound IL-6 was detected using the IL-6-specific B-F6 antibody which recognizes a different epitope than BE8/MAb 190. The B-F6 MAb is a mouse IgG1 MAb, and bound B-F6 was detected by staining the cells with goat-anti-mIgG1-APC polyclonal antibody. The brightest fluorescent cells were sorted, vaccinia virus was extracted from these cells by three cycles of freeze/thaw, and the selected virus was amplified on BSC1 feeder cells and titered. HeLa cells were co-infected with an aliquot of the amplified bulk sorted virus, Ig-H_1, and L172. The cells were stained for IL-6 binding as described above and analyzed by flow cytometry. As shown in Figure 12.9, a significant percentage of cells stained positive for IL-6 binding, indicating that a significant percentage of the Ig-H clones contain an IL-6-specific VH. Twenty vaccinia clones were picked from Ig-H_1, amplified, paired with L172, and analyzed for IL-6-specific binding. Eleven of the twenty clones produced an Ig-H that paired with L172 and bound to IL-6. Figure 12.9 shows representative data from several clones. The selected VH genes were then PCR amplified and subcloned into a mammalian expression vector containing human gamma 1 constant domain. The Ig-Hs were then cotransfected along with the L172 expression plasmid into CHO cells. MAb was purified and characterized for specificity, affinity, and functional activity. Affinity data demonstrated that five of these new MAbs had an affinity that was improved compared to MAb 190, with the new affinities ranging from 0.6 nM to 1.1 nM. This data demonstrates that with a single round of affinity improvement we had generated a number of new MAbs with 5- to 10-fold improvement in affinity.

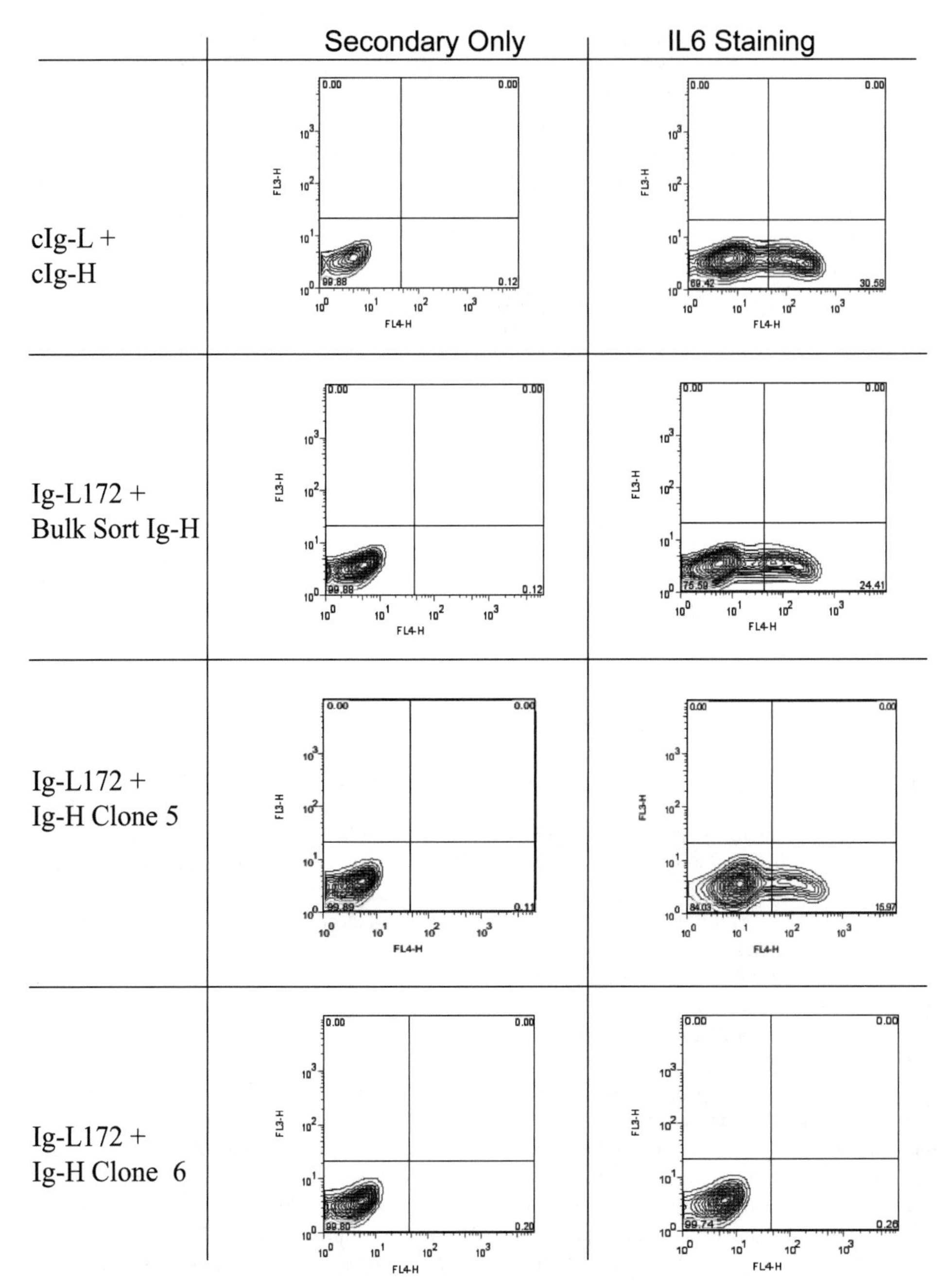

Figure 12.9 Affinity improvement by V gene replacement. An IL6 specific MAb was affinity improved by fixing the antigen specific Ig-L, and using this Ig-L to screen a library of human Ig-H. This approach is very similar to the approach outlined in Figure 12.7, except a human Ig-L is fixed instead of a non-human Ig. After one round of sorting, individual Ig-H clones were tested for the ability to pair with the Ig-L and bind to IL6. As shown in the figure, clone 5 is antigen specific, while clone 6 is not.

TABLE 12.1 Advantages of Expressing Immunoglobulin Libraries in Mammalian Cells

Existing Limitation	ActivMab Technology Advantage
Reengineering of scFv selected from phage and yeast display libraries	Engineering takes time and can create unplanned difficulties in production. Vaccinex directly expresses complete MAbs
Manufacturing	Intrinsic selection for high expression in mammalian cell lines, easily adaptable to manufacturing
Conversion of nonhuman antibodies	Many nonhuman MAbs exist, Vaccinex technology allows such antibodies to be converted into 100% human antibodies
Tolerance	Broader target range than mouse-based platforms since no homology issues
Difficult membrane targets	No interference in binding of soluble, secreted IgG to membrane targets

12.5 ADVANTAGES OF ANTIBODY SELECTION FROM IMMUNOGLOBULIN LIBRARIES EXPRESSED IN MAMMALIAN CELLS

The ability to select fully human monoclonal antibodies from immunoglobulin gene libraries expressed in mammalian cells has a number of advantages. First, using any of the variations of this technology, antibodies are selected as full length IgG antibodies. Other library technologies express and select antibody fragments, usually scFv or Fabs, which must be reengineered into full length IgG antibodies. Unanticipated difficulties can sometimes arise when these antibody fragments are removed from their selected environment and expressed as full length IgG antibodies. Second, employing this technology, there is a built in selection for antibodies that are efficiently expressed in mammalian cells. Any antibodies that do not express well will likely not be produced at sufficient concentration in the cell supernatant or on the cell surface to enable discovery with this technology. To date every antibody selected with this ActivMAb technology was efficiently expressed when transfected in CHO cells. Third, this technology has proven to be a robust method to convert nonhuman antibodies into human antibodies. Multiple potential lead antibodies are selected, and these human antibodies reproduce the affinity and functional activity of the parental murine antibody. Fourth, antibody selection with this technology is not restricted by self-tolerance. Similar to other library technologies this technology enables the selection of antibodies to virtually any target antigen. Fifth, secreted antibodies can be generated from our libraries and screened for binding to whole cells. This enables discovery of antibodies to proteins that cannot be readily purified. A summary of the advantages of this technology is shown in Table 12.1.

ACKNOWLEDGMENTS

This work was supported in part by a grant from the Advanced Technology Program (ATP) of the National Institute of Standards and Technology (NIST).

The authors would like to acknowledge and thank the large number of people who made important contributions in getting this technology established. In particular, we would like to thank Sumedha Bhagat, Holm Bussler, Leslie Croy, Elizabeth Evans, Terrence Fisher, Katya Klimatcheva, Richard Hall, He Huang, Loretta Mueller, Frank Murante, Tracy Pandina, Mark Paris, Christine Reilly, Maria Scrivens, Shuying Shi, and Wei Wang for providing data and scientific contributions to this work.

REFERENCES

Adams, G.P., and L.M. Weiner. 2005. Monoclonal antibody therapy of cancer. *Nature Biotechnol.* 23(9):1147–1157.

Baixeras, E., A. Cebrian, J.P. Albar, et al. 1998. Vaccinia virus-induced apoptosis in immature B lymphocytes: Role of cellular Bcl-2. *Virus Res.* 58(1–2):107–113.

Bataille, R., B. Barlogie, Z.Y. Lu, et al. 1995. Biologic effects of anti-interleukin-6 murine monoclonal antibody in advanced multiple myeloma. *Blood* 86(2):685–691.

de Haard, H.J., N. van Neer, A. Reurs, et al. 1999. A large non-immunized human Fab fragment phage library that permits rapid isolation and kinetic analysis of high affinity antibodies. *J. Biol. Chem.* 274(26):18218–18230.

Evans, E.E., A.D. Henn, A. Jonason, et al. 2006. 35 (C17orf37) is a novel tumor biomarker abundantly expressed in breast cancer. *Mol. Cancer Ther.* 5(11):2919–2930.

Feldhaus, M.J., R.W. Siegel, L.K. Opresko, et al. 2003. Flow-cytometric isolation of human antibodies from a nonimmune *Saccharomyces cerevisiae* surface display library. *Nature Biotechnol.* 21(2):163–170.

Flower, D.R. 1999. Modelling G-protein-coupled receptors for drug design. *Biochim. Biophys. Acta* 1422(3): 207–234.

Gabay, C. 2006. Interleukin-6 and chronic inflammation. *Arthritis Res. Ther.* 8(Suppl 2):S3.

Green, L.L. 1999. Antibody engineering via genetic engineering of the mouse: XenoMouse strains are a vehicle for the facile generation of therapeutic human monoclonal antibodies. *J. Immunol. Methods* 231(1–2):11–23.

Haddad, E., S. Paczesny, V. Leblond, et al. 2001. Treatment of B-lymphoproliferative disorder with a monoclonal anti-interleukin-6 antibody in 12 patients: A multicenter phase 1-2 clinical trial. *Blood* 97(6):1590–1597.

Hoogenboom, H.R. 2005. Selecting and screening recombinant antibody libraries. *Nature Biotechnol.* 23(9):1105–1116.

Jones, P.T., P.H. Dear, J. Foote, M.S. Neuberger, and G. Winter. 1986. Replacing the complementarity-determining regions in a human antibody with those from a mouse. *Nature* 321(6069):522–525.

Klein, B., J. Wijdenes, X.G. Zhang, et al. 1991. Murine anti-interleukin-6 monoclonal antibody therapy for a patient with plasma cell leukemia. *Blood* 78(5):1198–1204.

Knappik, A., L. Ge, A. Honegger, et al. 2000. Fully synthetic human combinatorial antibody libraries (HuCAL) based on modular consensus frameworks and CDRs randomized with trinucleotides. *J. Mol. Biol.* 296(1):57–86.

Lipsky, P.E. 2006. Interleukin-6 and rheumatic diseases. *Arthritis Res. Ther.* 8(Suppl 2):S4.

Liu, A.Y., R.R. Robinson, E.D. Murray, Jr., J.A. Ledbetter, I. Hellstrom, and K.E. Hellstrom. 1987. Production of a mouse-human chimeric monoclonal antibody to CD20 with potent Fc-dependent biologic activity. *J. Immunol.* 139(10):3521–3526.

Lonberg, N. 2005. Human antibodies from transgenic animals. *Nature Biotechnol.* 23(9):1117–1125.

Marks, J.D., H.R. Hoogenboom, T.P. Bonnert, J. McCafferty, A.D. Griffiths, and G. Winter. 1991. By-passing immunization. Human antibodies from V-gene libraries displayed on phage. *J. Mol. Biol.* 222(3):581–597.

Mazor, Y., T. Van Blarcom, R. Mabry, B.L. Iverson, and G. Georgiou. 2007. Isolation of engineered, full-length antibodies from libraries expressed in *Escherichia coli. Nature Biotechnol.* 25(5):563–565.

Merchant, A.M., Z. Zhu, J.Q. Yuan, et al. 1998. An efficient route to human bispecific IgG. *Nature Biotechnol.* 16(7):677–681.

Merchlinsky, M., D. Eckert, E. Smith, and M. Zauderer. 1997. Construction and characterization of vaccinia direct ligation vectors. *Virology* 238(2):444–451.

Morrison, S.L., M.J. Johnson, L.A. Herzenberg, and V.T. Oi. 1984. Chimeric human antibody molecules: Mouse antigen-binding domains with human constant region domains. *Proc. Natl. Acad. Sci. USA* 81(21):6851–6855.

Moss, B. 1991. Vaccinia virus: A tool for research and vaccine development. *Science* 252(5013):1662–1667.

Nambi, P., and N. Aiyar. 2003. G protein-coupled receptors in drug discovery. *Assay Drug Dev. Technol.* 1(2):305–310.

Nissim, A., H.R. Hoogenboom, I.M. Tomlinson, et al. 1994. Antibody fragments from a "single pot" phage display library as immunochemical reagents. *EMBO J.* 13(3):692–698.

Osbourn, J., M. Groves, and T. Vaughan. 2005. From rodent reagents to human therapeutics using antibody guided selection. *Methods* 36(1):61–68.

Queen, C., W.P. Schneider, H.E. Selick, et al. 1989. A humanized antibody that binds to the interleukin 2 receptor. *Proc. Natl. Acad. Sci. USA* 86(24):10029–10033.

Ramsey-Ewing, A., and B. Moss. 1996. Recombinant protein synthesis in Chinese hamster ovary cells using a vaccinia virus/bacteriophage T7 hybrid expression system. *J. Biol. Chem.* 271(28):16962–16966.

Reichert, J.M., C.J. Rosensweig, L.B. Faden, and M.C. Dewitz. 2005. Monoclonal antibody successes in the clinic. *Nature Biotechnol.* 23(9):1073–1078.

Rossi, J.F., N. Fegueux, Z.Y. Lu, et al. 2005. Optimizing the use of anti-interleukin-6 monoclonal antibody with dexamethasone and 140 mg/m^2 of melphalan in multiple myeloma: Results of a pilot study including biological aspects. *Bone Marrow Transplant.* 36(9):771–779.

Scheiflinger, F., F. Dorner, and F.G. Falkner. 1992. Construction of chimeric vaccinia viruses by molecular cloning and packaging. *Proc. Natl. Acad. Sci. USA* 89(21):9977–9981.

Sheets, M.D., P. Amersdorfer, R. Finnern, et al. 1998. Efficient construction of a large nonimmune phage antibody library: The production of high-affinity human single-chain antibodies to protein antigens. *Proc. Natl. Acad. Sci. USA* 95(11):6157–6162.

Smith, E.S., A. Mandokhot, E.E. Evans, et al. 2001. Lethality-based selection of recombinant genes in mammalian cells: Application to identifying tumor antigens. *Nature Med.* 7(8):967–972.

Smith, E.S., S. Shi, and M. Zauderer. 2004. Construction of cDNA libraries in vaccinia virus. *Methods Mol. Biol.* 269:65–76.

Smolen, J.S., and R.N. Maini. 2006. Interleukin-6: A new therapeutic target. *Arthritis Res. Ther.* 8(Suppl 2):S5.

Somogyi, P., J. Frazier, and M.A. Skinner. 1993. Fowlpox virus host range restriction: gene expression, DNA replication, and morphogenesis in nonpermissive mammalian cells. *Virology* 197(1):439–444.

Tsung, K., J.H. Yim, W. Marti, R.M. Buller, and J.A. Norton. 1996. Gene expression and cytopathic effect of vaccinia virus inactivated by psoralen and long-wave UV light. *J. Virol.* 70(1):165–171.

Tsurushita, N., and M. Vasquez. 2003. Humanization of monoclonal antibodies. In: T. Honjo, F.W. Alt, and M. Neuberger (eds), *Molecular Biology of B Cells*, Academic press. 533–546.

Van den Buecken, T., H. Pieters, M. Steukers, et al. 2003. Affinity maturation of Fab antibody fragments by fluorescent-activated cell sorting of yeast-displayed libraries. *FEBS Lett.* 546(2–3):288–294.

Vugmeyster, Y., J. Beyer, K. Howell, et al. 2005. Depletion of B cells by a humanized anti-CD20 antibody PRO70769 in *Macaca fascicularis*. *J. Immunother.* 28(3):212–219.

Wijdenes, J., C. Clement, B. Klein, et al. 1991. Human recombinant dimeric IL-6 binds to its receptor as detected by anti-IL-6 monoclonal antibodies. *Mol. Immunol.* 28(11):1183–1192.

PART IV

ANTIBODY ENGINEERING

CHAPTER 13

Antibody Engineering: Humanization, Affinity Maturation, and Selection Techniques

JUAN C. ALMAGRO and WILLIAM R. STROHL

ABSTRACT

Antibody optimization is a key process used to modify antibodies obtained from a primary antibody source, such as a murine hybridoma or human phage displayed antibody library, into molecules that possess properties desired in therapeutic antibodies. Strategies for humanization of nonhuman sequences to generate human-like antibodies are described in detail, as are strategies and basic methods for affinity optimization of antibodies.

13.1 INTRODUCTION

More than 20 antibodies have been approved by the U.S. Food and Drug Administration (FDA) for therapeutic applications in humans (see Table 1.1, Chapter 1 of this book), and a similar number is

Therapeutic Monoclonal Antibodies: From Bench to Clinic. Edited by Zhiqiang An

in the late stage of clinical trials to be approved in the near future (Riley 2006; see also Chapter 1 of this book). This remarkable progress by the biopharmaceutical industry has been made possible by several key antibody optimization technologies: (1) humanization of rodent antibodies that enhanced safety via minimization of undesired anti-rodent human immune responses, (2) affinity maturation strategies that improved their potency and efficacy, and (3) selection methodologies that facilitated humanization methods, revolutionized optimization of antibodies with promising therapeutic value, and enabled isolation of fully human antibodies.

The first key enablers of these technologies were the construction of human-murine antibody chimeras, which was then followed by methods for humanization of those chimeras. The first therapeutic antibody, Muronomab OKT3®, brought to the market in 1984 by Ortho Biotech, was a fully murine antibody that elicited substantial immune responses in people ranging from 38 to 83 percent, depending on the patient population (Kimball et al. 1995; Table 13.1). Subsequent chimeric antibodies such as ReoPro®, Rituxan®, Remicade®, and Erbitux®, which contain roughly 30 percent murine sequences, elicit anti-drug immune responses in approximately 5 to 15 percent of the patients treated with them (Table 13.1). Humanized antibodies such as Synagis®, Herceptin®, Campath®, Raptiva®, Xolair®, Tysabri®, and Soliris®, which retain significantly fewer murine amino acid sequences (usually less than 5 percent), elicited anti-drug responses in the range of 0.1 to 9 percent (Table 13.1). While antibody sequence is only one of many factors that can result in immunogenicity (other factors include, for example, route of administration, drug solubility and aggregation characteristics, patient population, soluble versus receptor target, contaminants, degradation, dosing concentration, and frequency), it is clear that humanization of antibodies has resulted in therapeutic antibodies that are tolerated significantly better than nonhuman antibodies would be (Koren, Zuckerman, and Mire-Sluis 2002; Pendley, Schantz, and Wagner 2003; Hwang and Foote 2005; Mukovosov et al. 2008). Thus, it is likely that for humanized and fully human antibodies, factors other than sequence will probably contribute most significantly to immunogenicity. The key is to make potent, well-behaved antibodies that are as human-like as possible.

In the first section of this chapter, we describe the two main modalities of humanization methods, namely, rational and empirical methods (Almagro and Fransson 2008). Rational methods have in common the design of few humanized antibody variants to be tested for binding or any other property of interest. If the designed variants prove to be unsatisfactory, a new design cycle and binding assessment is initiated. In contrast, empirical methods make few or no hypotheses on the impact of changes on the antibody structure and/or its binding profile. Instead, these methods rely on the power of selection techniques to isolate molecules with the desired functional and biophysical profile.

As an example of rational methods to humanize antibodies we describe the process for grafting of complementarity-determining regions (CDRs), the central paradigm for most humanization methods, followed by two variations of the original method: specificity-determining residues (SDR) grafting (Kashmiri et al. 2005) and superhumanization (Tan et al. 2002). Within empirical methods, we discuss guided selection (Osbourn, Groves, and Vaughan 2005), which enabled the isolation of Humira® (adalimumab), the first human antibody approved by the FDA (Table 13.1).

The second and third sections of this chapter describe affinity maturation strategies. *In vitro* affinity maturation consists of two well-defined components: (1) a strategy to generate libraries of antibody variants, and (2) a selection or screening strategy to enable isolation of variants of interest. The strategies to generate libraries of antibody variants include random and rational mutagenesis. Rational strategies generate diversity in predefined segments of the antibody V region, whereas random strategies introduce point mutations, insertion/deletions (indels) or homologous fragments containing more than one amino acid replacement all along the V region.

As an example of random strategies, we discuss in the second section of this chapter error-prone polymerase chain reaction (PCR) (Neylon 2004) and DNA shuffling (Stemmer 1994). As examples of rational methods, CDR walking mutagenesis (Yang et al. 1995) and rational design (Lippow, Wittrup, and Tidor 2007) are described. A review on *in vivo* strategies to optimize the affinity of antibodies has recently been published by Wark and Hudson (2006).

TABLE 13.1 Some of the Therapeutic Antibodies Approved by the FDA, With the Rate of Immunogenicity as Reported in the Prescribing Information Included With the Packaging, Unless Otherwise Noted

Antibody	Brand Name	Target	Type	Immunogenicity	Approval
Muromonab-CD3	Orthoclone OKT3	T-cell CD3 receptor	Murine	17%–63%, depending on patient population; Kimball et al. (1995)	1986
Abciximab	ReoPro	Glycoprotein IIb/IIIa	Chimeric	ca. 5%–6%; 24% upon readministration	1994
Daclizumab	Zenapax	IL-2 receptor A	Humanized	14%	1997
Rituximab	Rituxan, Mabthera	CD20	Chimeric	11% (118/1053 patients); 12% upon readministration	1997
Basiliximab	Simulect	IL-2 receptor A	Chimeric	1% (4/339)	1998
Infliximab	Remicade	TNF-α	Chimeric	ca. 10% over 1–2 years; 15% in psoriatic arthritis population	1998
Palivizumab	Synagis	RSV	Humanized	0.7% (in second season, 1/56; low titer only)	1998
Trastuzumab	Herceptin	ErbB2	Humanized	0.1% (1/903)	1998
Gemtuzumab ozogamicin	Mylotarg	CD33	Humanized	No data reported	2000
Alemtuzumab	Campath	CD52	Humanized	8.3% (11/133)	2001
Adalimumab	Humira	TNF-α	Human	5% (58/1062) overall; as monotherapy, 12%; with methotrexate, ~1%	2002
Efalizumab	Raptiva	CD11a	Humanized	6.3% (67/1063)	2002
Ibritumomab tiuxetan	Zevalin	CD20	Murine	1.3%–2.5% in multiple studies; mostly transient	2002
Tositumomab	Bexxar	CD20	Murine	7% at 6 months; 12%–13% at 12–18 months	2003
Bevacizumab	Avastin	VEGF	Humanized	No data reported	2004
Cetuximab	Erbitux	EGFR	Chimeric	5% (49/1001); non-neutralizing	2004
Omalizumab	Xolair	IgE	Humanized	<0.1% (1/1723)	2004
Natalizumab	Tysabri	VLA4	Humanized	~9%; significant number of neutralizing Abs	2006
Panitumumab	Vectibix	EGFR	Human	<1% (2/612) by ELISA; 4% (25/610) by BIAcore; 8/604 had neutralizing Abs	2006
Ranibizumab	Lucentis	VEGF	Humanized	1%–6% in patients	2006
Eculizumab	Soliris	Complement protein C5	Humanized	2% (3/196)	2007

Methods for selection of optimized clones are briefly discussed in the third section. One of the most robust and widely used selection techniques is phage display, a method first described by George Smith in the mid-1980s (Smith 1985) and applied to antibody engineering projects since the beginning of the 1990s. Additionally, *in vitro* display technologies (e.g., ribosome display), yeast display, and bacterial display all have been used to select improved antibodies. Reviews on *de novo* discovery of antibodies using phage display have been published by Hoogenboom (2005) and Sidhu and Fellouse (2006) and will not be covered here.

13.2 HUMANIZATION METHODS

The IgG isotype is the most abundant form of circulating antibodies and the molecular format of choice for most therapeutic antibodies thus far. IgGs are composed of two identical polypeptide, H and L chains. Each H chain has one V_H domain and three C domains, C_H1 to C_H3, counted from the amino terminus of the polypeptide chain. The L chain has one V_L domain at the amino terminus and only one C domain, C_L.

V domains define the antibody specificity and are composed of two β-sheets and loops connecting the β-strands (Amzel and Poljak 1979). Three of the loops are highly diverse in length and/or amino acid composition and are referred to as hypervariable (HV) loops (Chothia and Lesk 1987). The remaining non-HV loops and two β-sheets are referred to as framework regions (FRs). HV loops, denoted H1, H2, and H3 for V_H, and L1, L2, and L3 for V_L, are brought together by noncovalent association of V_H and V_L to accommodate the antigen-binding site at the N-terminal region of the Fv fragment (Fig. 13.1).

The antigen-binding site, or paratope, is defined by the set of residues in contact with the antigen, which are determined in the experimental structure of the antibody in complex with its corresponding antigen. By transferring the paratope from a nonhuman antibody into a human scaffold, the specificity of the nonhuman antibody should, in principle, be transferred into a human context and thus a humanized version of the nonhuman antibody is obtained.

Rational methods to humanize antibodies rely on the so-called "design cycle." It consists of designing one or few humanized variants, synthesizing the genes, expressing their products as Fv, scFv, Fabs, or IgGs, and assessing their binding profile. If the binding profile of the designed humanized variants is not satisfactory, a new design cycle is initiated. The key factor to succeed in the design cycle is the structural and physicochemical compatibility between residues targeted for transferring the specificity from a given nonhuman antibody and the human context.

13.2.1 CDR Grafting

In the mid-1980s, Greg Winter and his colleagues at the Medical Research Council (MRC) envisioned complementarity-determining region (CDR) grafting as a method to humanize antibodies (Jones et al. 1986). CDR grafting has three fundamental components: (1) delimitation of CDRs as regions determining the specificity of the nonhuman antibody and hence target for grafting, (2) selection of human FRs to be combined with the nonhuman CDRs, and (3) assessment of conflicts between nonhuman CDRs and human FRs and design of back mutations to prevent a loss of affinity in the final product.

Wu and Kabat (1970) defined the CDRs prior to resolution of the first antibody structures as the regions containing the antigen-binding site or paratope. CDRs were identified as the regions with highest variability values in a multiple alignment of antibody sequence and were predicted to contain the paratope. Subsequent resolution of antigen–antibody complexes (Amzel and Poljak 1979) validated Wu and Kabat's predictions. Figure 13.2 indicates the location of CDRs in the V regions. A set of simple rules to delimit the CDRs by looking at the V sequences can be found at Andrew Martin's webpage (URL: http://bioinf.org.uk/abs/).

Once the CDRs of the nonhuman antibody have been identified, the next step in CDR grafting is to select human sequences that will serve as FR receptors of nonhuman CDRs. Three sources of human sequences have been utilized in various CDR grafting protocols: (1) consensus sequences, (2) mature sequences, and (3) germline genes. Consensus sequences are created by artificial combination of different genes and thus being non-natural products could be immunogenic. Mature sequences, on the other hand, are products of immune responses and thus more often than not carry somatic mutations. Since these mutations are not under the species selection, they also may potentially be immunogenic. To minimize potential immunogenicity, human germline genes have increasingly been utilized as source of human sequences (Tan et al. 2002; Gonzales et al. 2004).

The physical maps of the human H and L chain loci were elucidated in the late 1990s and the functional germline gene repertoires they encode have been thoroughly characterized since then.

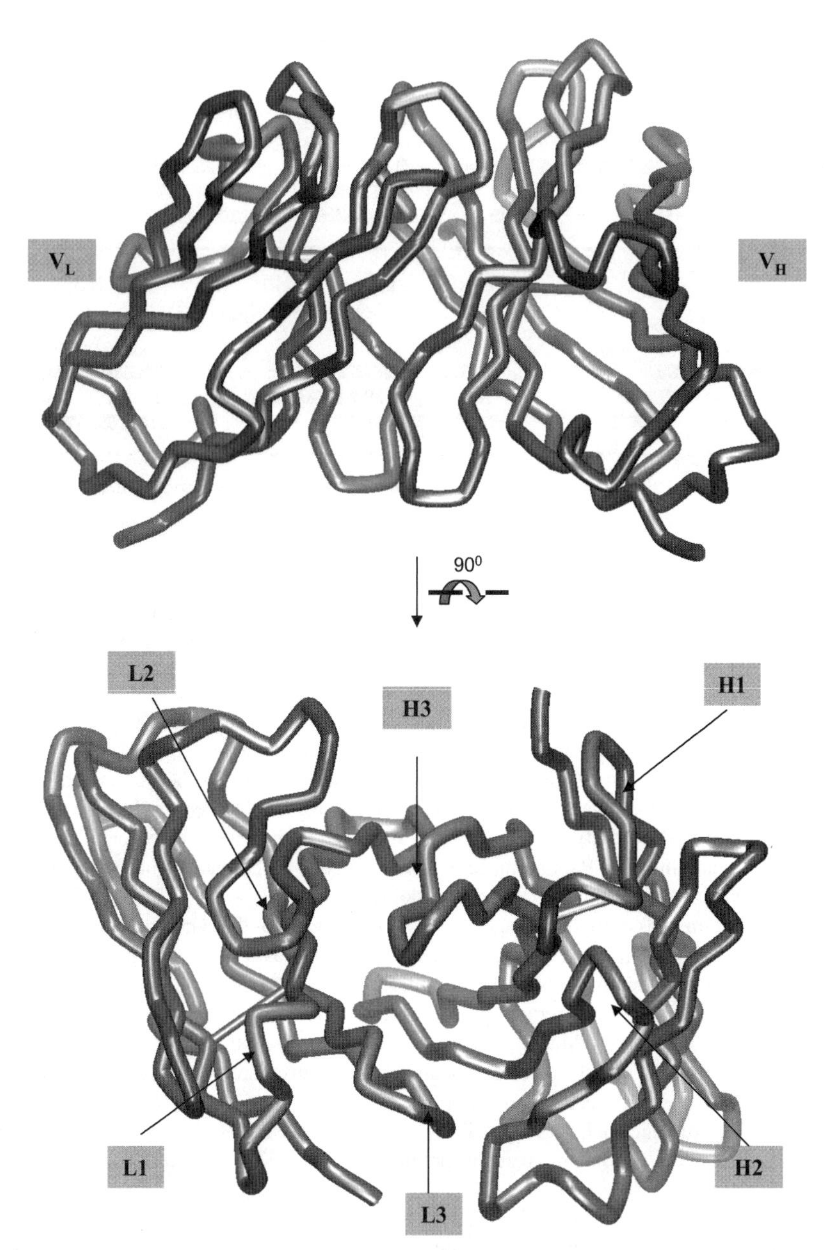

Figure 13.1 Fv fragment showing hypervariable loops and framework regions.

This information has been compiled in the Immunogenetics Database (IMGT; http://imgt.cines.fr/). The repertoire of human functional IGHV germline genes is composed of 49 genes distributed in seven IGHV families (Table 13.2). The human IGKV germline genes repertoire is composed of 34 functional genes distributed in six IGKV families (Table 13.3).

To select human germline genes to be used as FR donors, the amino acid sequence of the nonhuman sequence is compared to the sequences listed in Tables 13.2 and 13.3 and short lists with the highest homologous genes are built. For the IGHJ and IGLJ genes or FR-4 (Table 13.4) a similar procedure is carried out. This method of selection of human FRs based on sequence homology with the nonhuman

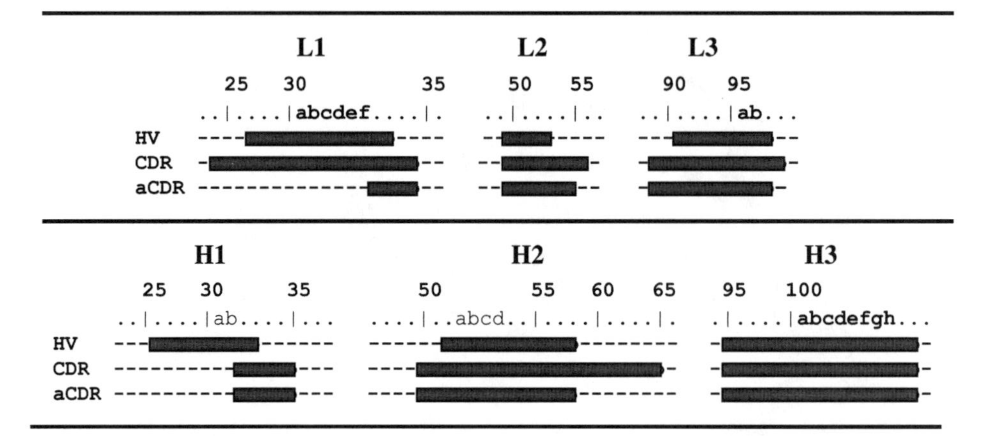

Figure 13.2 Location of the CDRs in the V regions and comparison to hypervariable loops and aCDRs, as defined in the text.

V regions is known in the field as best fit. Another modality of best fit in which the germline genes are selected by comparison to human germline gene CDRs alone, regardless of the FR homology, is described in Section 13.2.3.

The next step in the CDR grafting protocol is to combine the FRs of the chosen human genes compiled in the short lists noted above with the nonhuman CDRs. Of course, the higher the homology between the nonhuman and human genes, in particular at key residues to maintain the structure of the CDRs, the fewer the number of potential incompatibilities between the nonhuman CDRs and human FRs; hence, the fewer the number of back mutations that need to be introduced into the humanized product to retain the affinity of the nonhuman antibody. In the case of incompatibilities between nonhuman CDRs and human FRs, the impact of potential back mutations can be assessed in a three-dimensional model of the resulting CDR grafted molecules.

Antibody modeling has been highly facilitated by the discovery of the canonical structures (Chothia and Lesk 1987; Al-Lazikani, Lesk, and Chothia 1997), as well as by the large number of antibody crystal structures available in the Protein Data Bank (http://www.rcsb.org/). As of June 2008, there was 937 pdb entries (IMGT; 3D repertoire of antibodies). A website for automated antibody modeling called WAM (Web of Antibody Modeling) can be found at http://antibody.bath.ac.uk. WAM builds three-dimensional models of antibodies based on canonical structure patterns. The H3 structure is modeled using CONGEN (Bruccoleri and Karplus 1987). Current modeling software such as WAM and free modeling software such as Swiss PDB Viewer generate models with a reasonable precision at V_L and most of the V_H domain. Accurate prediction of the H3 loop conformation still remains troublesome.

Where should one look in the model? Any mutation introduced in the V region might, in principle, affect the antibody conformation and thus the antibody-binding profile. For example, Wedemayer and coworkers (1997) have shown that the affinity of an anti-hapten antibody could be increased by more than four orders of magnitude through somatic mutations located 10 Å away from the hapten-biding site. Moreover, mutations in residues in direct contact with the CDRs, collectively called the "Vernier zone" (Foote and Winter 1992), or individually as "Vernier residues," can have a substantial impact on the conformation and positioning of CDRs and thus the antibody affinity. Also, residues involved in the $V_H:V_L$ interface and H3 packing have been found to play an important role in affinity modulation (Fig. 13.3). Jose Saldanha has compiled a list of examples of humanized antibodies and residues modified during humanization protocols, which can be found at the website http://people.cryst.bbk.ac.uk/~ubcg07s/. This compilation can serve as a guide when designing back mutations in CDR grafting protocols.

Conflicts detected during examination of the humanized antibody three-dimensional models and potential back mutations to solve them should ideally be tested experimentally. Nonetheless, time and resources could be a limiting factor in assessing all variants, in particular if more than one human germline gene is considered as an FR donor. An alternative is to generate combinatorial libraries of variants at residues in conflict in the humanized variants and select the best variants by using a selection methodology (Section 13.4). This strategy was implemented by Baca et al. (1997) when they humanized the anti-vascular endothelial growth factor (anti-VEGF) monoclonal antibody A4.6.1. These authors grafted A4.6.1 CDRs onto a human V_L1 (κ) and V_H3 FRs and built a combinatorial library in 11 Vernier residues. Following phage display panning (Section 13.4), an optimized FR was selected that increased by 125-fold the affinity of the antibody used as starting point after CDR grafting.

13.2.2 SDR Grafting

SDR grafting was proposed by Tamura et al. (2000) as a modification of CDR grafting to minimize the content of nonhuman residues to be transferred into the human context. The rationale for SDR grafting was provided by analyses of antigen–antibody complexes conducted by Padlan, Abergel, and Tipper (1995), which indicated that only one-third of the CDR sequences actually contact the antigen. Comparison of the residues in contact with antigens and sequence variability values pinpointed residues in contact with the antigen as the most variable ones. These residues were defined as SDRs and the regions containing the SDRs were termed abbreviated-CDRs (aCDRs). Figure 13.2 shows a comparison of the CDRs, aCDRs, and HV loops.

Once the SDRs are identified in the nonhuman antibody, the remaining two steps, for example, human FR selection and design of mutants or antibody variants to prevent a loss in affinity, are implemented, the same as in CDR grafting (Section 13.2.1). De Pascalis et al. (2002) compared CDR-grafted and aCDR-grafted variants of the murine antibody COL-1 specific for carcinoembryonic antigen (CEA). No significant differences were found in binding to CEA between the CDR-grafted and SDR-grafted variants. Compared with CDR-grafted, the SDR-grafted COL-1 variant showed, however, lower reactivity to sera of patients carrying anti-V region antibodies to COL-1. A detailed description of the SDR grafting method has been published (Kashmiri et al. 2005).

13.2.3 Superhumanization

Superhumanization is a modification of best fit for selection of FR in the CDR grafting protocols. As mentioned in Section 13.2.1, the best fit approach to select FR is based on comparison of the nonhuman and human sequences along the entire V gene. Superhumanization, proposed by Jefferson Foote and coworkers (Tan et al. 2002), is based on comparison of CDRs; in this approach, FR homology is irrelevant. A detailed description of the method has recently been published by Hwang and Foote (2005).

Superhumanization consists of two *in silico* filters to select human FRs. First, the canonical structure class of the nonhuman antibody are determined and compared against the canonical structure classes encoded in the human germline genes (first columns of Tables 13.2 and 13.3). Those human genes encoding the same or structurally related canonical structures are selected. The second filter consists of choosing, within the genes sharing the canonical structures with the nonhuman antibody, those with highest homology in the CDRs. Once the human FRs are selected, the same steps as in CDR grafting to evaluate incompatibilities between the nonhuman CDRs and human FRs are followed. It should be noted that superhumanization could also be combined with SDR grafting to minimize the number of nonhuman residues transferred into the human FR.

Application of superhumanization to murine antibody 9.3, as the first case study (Tan et al. 2002), yielded a humanized antibody 9.3 V_L with insignificant loss in affinity. The loss in affinity of humanized V_H, on the other hand, was 20-fold. The fully superhumanized antibody lost 30-fold affinity but retained biological activity. Hwang et al. (2005) published the superhumanization of antibody D1.3 as a second case study. The affinity loss of superhumanized D1.3 was sixfold, as compared to

TABLE 13.2 Immunoglobulin Heavy Chain Germline Sequences

<table>
<tr><th></th><th></th><th>1   10   20   30   40   50   60   70   80   90</th></tr>
<tr><th>H1-H2</th><th>Name-allele</th><th>|...|....|....|....|....|....|.ab...|....|....|....|..abc..|....|....|....|....|....|..abc..|....|....</th></tr>
<tr><td>1-3</td><td>IGHV1-2*01</td><td>QVQLVQSGAEVKKPGASVKVSCKASGYTFTG--YYMHWVRQAPGQGLEWMGRINP--NSGGTNYAQKFQGRVTSTRDTSISTAYMELSRLRSDDTVVYYCAR</td></tr>
<tr><td>1-3</td><td>IGHV1-3*01</td><td>QVQLVQSGAEVKKPGASVKVSCKASGYTFTS--YAMHWVRQAPGQRLEWMGWINA--GNGNTKYSQKFQGRVTITRDTSASTAYMELSSLRSEDTAVYYCAR</td></tr>
<tr><td>1-3</td><td>IGHV1-8*01</td><td>QVQLVQSGAEVKKPGASVKVSCKASGYTFTS--YDINWVRQATGQGLEWMGWMNP--NSGNTGYAQKFQGRVTMTRNTSISTAYMELSSLRSEDTAVYYCAR</td></tr>
<tr><td>1-2</td><td>IGHV1-18*01</td><td>QVQLVQSGAEVKKPGASVKVSCKASGYTFTS--YGISWVRQAPGQGLEWMGWISA--YNGNTNYAQKLQGRVTMTTDTSTSTAYMELRSLRSDDTAVYYCAR</td></tr>
<tr><td>1-2</td><td>IGHV1-24*01</td><td>QVQLVQSGAEVKKPGASVKVSCKVSGYTLTE--LSMHWVRQAPGKGLEWMGGFDP--EDGETIYAQKFQGRVTMTEDTSTDTAYMELSSLRSEDTAVYYCAT</td></tr>
<tr><td>1-3</td><td>IGHV1-45*01</td><td>QMQLVQSGAEVKKTGSSVKVSCKASGYTFTY--RYLHWVRQAPGQALEWMGWITP--FNGNTNYAQKFQDRVTITRDRSMSTAYMELSSLRSEDTAMYYCAR</td></tr>
<tr><td>1-3</td><td>IGHV1-46*01</td><td>QVQLVQSGAEVKKPGASVKVSCKASGYTFTS--YYMHWVRQAPGQGLEWMGIINP--SGGSTSYAQKFQGRVTMTRDTSTSTVYMELSSLRSEDTAVYYCAR</td></tr>
<tr><td>1-3</td><td>IGHV1-58*01</td><td>QMQLVQSGPEVKKPGTSVKVSCKASGFTFTS--SAVQWVRQARGQRLEWIGWIVV--GSGNTNYAQKFQERVTITRDMSTSTAYMELSSLRSEDTAVYYCAA</td></tr>
<tr><td>1-2</td><td>IGHV1-69*01</td><td>QVQLVQSGAEVKKPGSSVKVSCKASGGTFSS--YAISWVRQAPGQGLEWMGGIIP--IFGTANYAQKFQGRVTITADESTSTAYMELSSLRSEDTAVYYCAR</td></tr>
<tr><td>1-2</td><td>IGHV1-f*01</td><td>EVQLVQSGAEVKKPGATVKISCKVSGYTFTD--YYMHWVQQAPGKGLEWMGLVDP--EDGETIYAEKFQGRVTITADTSTDTAYMELSSLRSEDTAVYYCAT</td></tr>
<tr><td>3-1</td><td>IGHV2-5*01</td><td>QITLKESGPTLVKPTQTLTLTCTFSGFSLSTSGVGVGWIRQPPGKALEWLALIY---WNDDKRYSPSLKSRLTITKDTSKNQVVLTMTNMDPVDTATYYCAH</td></tr>
<tr><td>3-1</td><td>IGHV2-26*01</td><td>QVTLKESGPVLVKPTETLTLTCTVSGFSLSNARMGVSWIRQPPGKALEWLAHIF---SNDEKSYSTSLKSRLTISKDTSKSQVVLTMTNMDPVDTATYYCAR</td></tr>
<tr><td>3-1</td><td>IGHV2-70*01</td><td>QVTLRESGPALVKPTQTLTLTCTFSGFSLSTSGMCVSWIRQPPGKALEWLALID---WDDDKYYSTSLKTRLTISKDTSKNQVVLTMTNMDPVDTATYYCAR</td></tr>
<tr><td>1-3</td><td>IGHV3-7*01</td><td>EVQLVESGGGLVQPGGSLRLSCAASGFTFSS--YWMSWVRQAPGKGLEWVANIKQ--DGSEKYYVDSVKGRFTISRDNAKNSLYLQMNSLRAEDTAVYYCAR</td></tr>
<tr><td>1-3</td><td>IGHV3-9*01</td><td>EVQLVESGGGLVQPGRSLRLSCAASGFTFDD--YAMHWVRQAPGKGLEWVSGISW--NSGSIGYADSVKGRFTISRDNAKNSLYLQMNSLRAEDTALYYCAK</td></tr>
<tr><td>1-3</td><td>IGHV3-11*01</td><td>QVQLVESGGGLVKPGGSLRLSCAASGFTFSD--YYMSWIRQAPGKGLEWVSYISS--SGSTIYYADSVKGRFTISRDNAKNSLYLQMNSLRAEDTAVYYCAR</td></tr>
<tr><td>1-1</td><td>IGHV3-13*01</td><td>EVQLVESGGGLVQPGGSLRLSCAASGFTFSS--YDMHWVRQATGKGLEWVSAIG---TAGDTYYPGSVKGRFTISRENAKNSLYLQMNSLRAGDTAVYYCAR</td></tr>
<tr><td>1-4</td><td>IGHV3-15*01</td><td>EVQLVESGGGLVKPGGSLRLSCAASGFTFSN--AWMSWVRQAPGKGLEWVGRIKSKTDGGTTDYAAPVKGRFTISRDDSKNTLYLQMNSLKTEDTAVYYCTT</td></tr>
<tr><td>1-3</td><td>IGHV3-20*01</td><td>EVQLVESGGGVVRPGGSLRLSCAASGFTFDD--YGMSWVRQAPGKGLEWVSGINW--NGGSTGYADSVKGRFTISRDNAKNSLYLQMNSLRAEDTALYHCAR</td></tr>
<tr><td>1-3</td><td>IGHV3-21*01</td><td>EVQLVESGGGLVKPGGSLRLSCAASGFTFSS--YSMNWVRQAPGKGLEWVSSISS--SSSYIYYADSVKGRFTISRDNAKNSLYLQMNSLRAEDTAVYYCAR</td></tr>
<tr><td>1-3</td><td>IGHV3-23*01</td><td>EVQLLESGGGLVQPGGSLRLSCAASGFTFSS--YAMSWVRQAPGKGLEWVSAISG--SGGSTYYADSVKGRFTISRDNSKNTLYLQMNSLRAEDTAVYYCAK</td></tr>
<tr><td>1-3</td><td>IGHV3-30*01</td><td>QVQLVESGGGVVQPGRSLRLSCAASGFTFSS--YAMHWVRQAPGKGLEWVAVISY--DGSNKYYADSVKGRFTISRDNSKNTLYLQMNSLRAEDTAVYYCAR</td></tr>
<tr><td>1-3</td><td>IGHV3-30-3*01</td><td>QVQLVESGGGVVQPGRSLRLSCAASGFTFSS--YAMHWVRQAPGKGLEWVAVISY--DGSNKYYADSVKGRFTISRDNSKNTLYLQMNSLRAEDTAVYYCAR</td></tr>
<tr><td>1-3</td><td>IGHV3-33*01</td><td>QVQLVESGGGVVQPGRSLRLSCAASGFTFSS--YGMHWVRQAPGKGLEWVAVIWY--DGSNKYYADSVKGRFTISRDNSKNTLYLQMNSLRAEDTAVYYCAR</td></tr>
<tr><td>1-3</td><td>IGHV3-43*01</td><td>EVQLVESGGVVVQPGGSLRLSCAASGFTFDD--YTMHWVRQAPGKGLEWVSLISW--DGGSTYYADSVKGRFTISRDNSKNSLYLQMNSLRTEDTALYYCAK</td></tr>
<tr><td>1-3</td><td>IGHV3-48*01</td><td>EVQLVESGGGLVQPGGSLRLSCAASGFTFSS--YSMNWVRQAPGKGLEWVSYISS--SSSTIYYADSVKGRFTISRDNAKNSLYLQMNSLRAEDTAVYYCAR</td></tr>
</table>

1-4	IGHV3-49*01	EVQLVESGGGLVQPGRSLRLSCTASGFTFGD--YAMSWFRQAPGKGLEWVGFIRSKAYGGTTEYTASVKGRFTISRDGSKSIAYLQMNSLKTEDTAVYYCTR
1-1	IGHV3-53*01	EVQLVESGGGLIQPGGSLRLSCAASGFTVSS--NYMSWVRQAPGKGLEWVSVIY---SGGSTYYADSVKGRFTISRDNSKNTLYLQMNSLRAEDTAVYYCAR
1-3	IGHV3-64*01	EVQLVESGGGLVQPGGSLRLSCAASGFTFSS--YAMHWVRQAPGKGLEYVSAISS--NGGSTYYANSVKGRFTISRDNSKNTLYLQMGSLRAEDMAVYYCAR
1-1	IGHV3-66*01	EVQLVESGGGLVQPGGSLRLSCAASGFTVSS--NYMSWVRQAPGKGLEWVSVIY---SGGSTYYADSVKGRFTISRDNSKNTLYLQMNSLRAEDTAVYYCAR
1-4	IGHV3-72*01	EVQLVESGGGLVQPGGSLRLSCAASGFTFSD--HYMDWVRQAPGKGLEWVGRTRNKANSYTTEYAASVKGRFTISRDDSKNSLYLQMNSLKTEDTAVYYCAR
1-4	IGHV3-73*01	EVQLVESGGGLVQPGGSLKLSCAASGFTFSG--SAMHWVRQASGKGLEWVGRIRSKANSYATAYAASVKGRFTISRDDSKNTAYLQMNSLKTEDTAVYYCTR
1-3	IGHV3-74*01	EVQLVESGGGLVQPGGSLRLSCAASGFTFSS--YWMHWVRQAPGKGLVWVSRINS--DGSSTSYADSVKGRFTISRDNAKNTLYLQMNSLRAEDTAVYYCAR
1-6	IGHV3-d*01	EVQLVESRGVLVQPGGSLRLSCAASGFTVSS--NEMSWVRQAPGKGLEWVSSI----SGGSTYYADSRKGRFTISRDNSKNTLHLQMNSLRAEDTAVYYCKK
2-1	IGHV4-4*01	QVQLQESGPGLVKPPGTLSLTCAVSGGSISSS-NWWSWVRQPPGKGLEWIGEIY---HSGSTNYNPSLKSRVTISVDKSKNQFSLKLSSVTAADTAVYCCAR
2-1	IGHV4-28*01	QVQLQESGPGLVKPSDTLSLTCAVSGYSISSS-NWWGWIRQPPGKGLEWIGYIY---YSGSTYYNPSLKSRVTMSVDTSKNQFSLKLSSVTAVDTAVYYCAR
3-1	IGHV4-30-2*01	QLQLQESGSGLVKPSQTLSLTCAVSGGSISSGGYSWSWIRQPPGKGLEWIGYIY---HSGSTYYNPSLKSRVTISVDRSKNQFSLKLSSVTAADTAVYYCAR
3-1	IGHV4-30-4*01	QVQLQESGPGLVKPSQTLSLTCTVSGGSISSGDYYWSWIRQPPGKGLEWIGYIY---YSGSTYYNPSLKSRVTISVDTSKNQFSLKLSSVTAADTAVYYCAR
3-1	IGHV4-31*01	QVQLQESGPGLVKPSQTLSLTCTVSGGSISSGGYYWSWIRQHPGKGLEWIGYIY---YSGSTYYNPSLKSLVTISVDTSKNQFSLKLSSVTAADTAVYYCAR
3-1	IGHV4-34*01	QVQLQQWGAGLLKPSETLSLTCAVYGGSFSG--YYWSWIRQPPGKGLEWIGEIN---HSGSTNYNPSLKSRVTISVDTSKNQFSLKLSSVTAADTAVYYCAR
1-1	IGHV4-39*01	QLQLQESGPGLVKPSETLSLTCTVSGGSISSSSYYWGWIRQPPGKGLEWIGSIY---YSGSTYYNPSLKSRVTISVDTSKNQFSLKLSSVTAADTAVYYCAR
1-1	IGHV4-59*01	QVQLQESGPGLVKPSETLSLTCTVSGGSISS--YYWSWIRQPPGKGLEWIGYIY---YSGSTNYNPSLKSRVTISVDTSKNQFSLKLSSVTAADTAVYYCAR
3-1	IGHV4-61*01	QVQLQESGPGLVKPSETLSLTCTVSGGSVSSGSYYWSWIRQPPGKGLEWIGYIY---YSGSTNYNPSLKSRVTISVDTSKNQFSLKLSSVTAADTAVYYCAR
2-1	IGHV4-b*01	QVQLQESGPGLVKPSETLSLTCAVSGYSISSG-YYWGWIRQPPGKGLEWIGSIY---HSGSTYYNPSLKSRVTISVDTSKNQFSLKLSSVTAADTAVYYCAR
1-2	IGHV5-51*01	EVQLVQSGAEVKKPGESLKISCKGSGYSFTS--YWIGWVRQMPGKGLEWMGIIYP--GDSDTRYSPSFQGQVTISADKSISTAYLQWSSLKASDTAMYYCAR
1-2	IGHV5-a*01	EVQLVQSGAEVKKPGESLRISCKGSGYSFTS--YWISWVRQMPGKGLEWMGRIDP--SDSYTNYSPSFQGHVTISADKSISTAYLQWSSLKASDTAMYYCAR
3-1	IGHV6-1*01	QVQLQQSGPGLVKPSQTLSLTCAISGDSVSSNSAAWNWIRQSPSRGLEWLGRTYYR-SKWYNDYAVSVKSRITINPDTSKNQFSLQLNSVTPEDTAVYYCAR
3-5	IGHV6-1*02	QVQLQQSGPGLVKPSQTLSLTCAISGDSVSSNSAAWNWIRQSPSRGLEWLGRTYYR-SKWYNDYAVSVKSRITINPDTSKNQFSLQLNSVTPEDTAVYYCAR
1-2	IGHV7-4-1*01	QVQLVQSGSELKKPGASVKVSCKASGYTFTS--YAMNWVRQAPGQGLEWMGWINT--NTGNPTYAQGFTGRFVFSLDTSVSTAYLQICSLKAEDTAVYYCAR

TABLE 13.3 Immunoglobulin Kappa Light Chain Germline Sequences

```
                1         10        20        30              40        50        60        70        80        90
L1-L2-L3 Name/allele     |...|....|....|....|....|....|abcdef....|....|....|....|....|....|....|....|....|....|....|....|....|
 2-1-1   IGKV1-12*01     DIQMTQSPSSVSASVGDRVTITCRASQGIS------SWLAWYQQKPGKAPKLLIYAASSLQSGVPSRFSGSGSGTDFTLTISSLQPEDFATYYCQQANSFP
 2-1-1   IGKV1-16*01     DIQMTQSPSSLSASVGDRVTITCRASQGIS------NYLAWFQQKPGKAPKSLIYAASSLQSGVPSRFSGSGSGTDFTLTISSLQPEDFATYYCQQYNSYP
 2-1-1   IGKV1-17*01     DIQMTQSPSSLSASVGDRVTITCRASQGIR------NDLGWYQQKPGKAPKRLIYAASSLQSGVPSRFSGSGSGTEFTLTISSLQPEDFATYYCLQHNSYP
 2-1-1   IGKV1-27*01     DIQMTQSPSSLSASVGDRVTITCRASQGIS------NYLAWYQQKPGKVPKLLIYAASTLQSGVPSRFSGSGSGTDFTLTISSLQPEDVATYYCQKYNSAP
 2-1-1   IGKV1-33*01     DIQMTQSPSSLSASVGDRVTITCQASQDIS------NYLNWYQQKPGKAPKLLIYDASNLETGVPSRFSGSGSGTDFTFTISSLQPEDIATYYCQQYDNLP
 2-1-1   IGKV1-39*01     DIQMTQSPSSLSASVGDRVTITCRASQSIS------SYLNWYQQKPGKAPKLLIYAASSLQSGVPSRFSGSGSGTDFTLTISSLQPEDFATYYCQQSYSTP
 2-1-1   IGKV1-5*01      DIQMTQSPSTLSASVGDRVTITCRASQSIS------SWLAWYQQKPGKAPKLLIYDASSLESGVPSRFSGSGSGTEFTLTISSLQPDDFATYYCQQYNSYS
 2-1-1   IGKV1-6*01      AIQMTQSPSSLSASVGDRVTITCRASQGIR------NDLGWYQQKPGKAPKLLIYAASSLQSGVPSRFSGSGSGTDFTLTISSLQPEDFATYYCLQDYNYP
 2-1-1   IGKV1-9*01      DIQLTQSPSFLSASVGDRVTITCRASQGIS------SYLAWYQQKPGKAPKLLIYAASTLQSGVPSRFSGSGSGTEFTLTISSLQPEDFATYYCQQLNSYP
 2-1-1   IGKV1D-12*01    DIQMTQSPSSVSASVGDRVTITCRASQGIS------SWLAWYQQKPGKAPKLLIYAASSLQSGVPSRFSGSGSGTDFTLTISSLQPEDFATYYCQQANSFP
 2-1-1   IGKV1D-16*01    DIQMTQSPSSLSASVGDRVTITCRASQGIS------SWLAWYQQKPEKAPKSLIYAASSLQSGVPSRFSGSGSGTDFTLTISSLQPEDFATYYCQQYNSYP
 2-1-1   IGKV1D-17*01    NIQMTQSPSAMSASVGDRVTITCRARQGIS------NYLAWFQQKPGKVPKHLIYAASSLQSGVPSRFSGSGSGTEFTLTISSLQPEDFATYYCLQHNSYP
 2-1-1   IGKV1D-33*01    DIQMTQSPSSLSASVGDRVTITCQASQDIS------NYLNWYQQKPGKAPKLLIYDASNLETGVPSRFSGSGSGTDFTFTISSLQPEDIATYYCQQYDNLP
 2-1-1   IGKV1D-39*01    DIQMTQSPSSLSASVGDRVTITCRASQSIS------SYLNWYQQKPGKAPKLLIYAASSLQSGVPSRFSGSGSGTDFTLTISSLQPEDFATYYCQQSYSTP
 2-1-1   IGKV1D-43*01    AIRMTQSPFSLSASVGDRVTITCWASQGIS------SYLAWYQQKPAKAPKLFIYYASSLQSGVPSRFSGSGSGTDYTLTISSLQPEDFATYYCQQYYSTP
 2-1-1   IGKV1D-8*01     VIWMTQSPSLLSASTGDRVTISCRMSQGIS------SYLAWYQQKPGKAPELLIYAASTLQSGVPSRFSGSGSGTDFTLTISCLQSEDFATYYCQQYYSFP

 3-1-1   IGKV2-40*01     DIVMTQTPLSLPVTPGEPASISCRSSQSLLDSDDGNTYLDWYLQKPGQSPQLLIYTLSYRASGVPDRFSGSGSGTDFTLKISRVEAEDVGVYYCMQRIEFP
 3-1-1   IGKV2D-40*01    DIVMTQTPLSLPVTPGEPASISCRSSQSLLDSDDGNTYLDWYLQKPGQSPQLLIYTLSYRASGVPDRFSGSGSGTDFTLKISRVEAEDVGVYYCMQRIEFP
 4-1-1   IGKV2-24*01     DIVMTQTPLSSPVTLGQPASISCRSSQSLVH-SDGNTYLSWLQQRPGQPPRLLIYKISNRFSGVPDRFSGSGAGTDFTLKISRVEAEDVGVYYCMQATQFP
 4-1-1   IGKV2-28*01     DIVMTQSPLSLPVTPGEPASISCRSSQSLLH-SNGYNYLDWYLQKPGQSPQLLIYLGSNRASGVPDRFSGSGSGTDFTLKISRVEAEDVGVYYCMQALQTP
 4-1-1   IGKV2-30*01     DVVMTQSPLSLPVTLGQPASISCRSSQSLVY-SDGNTYLNWFQQRPGQSPRRLIYKVSNRDSGVPDRFSGSGSGTDFTLKISRVEAEDVGVYYCMQGTHWP
 4-1-1   IGKV2D-26*01    EIVMTQTPLSLSITPGEQASISCRSSQSLLH-SDGYTYLYWFLQKARPVSTLLIYEVSNRFSGVPDRFSGSGSGTDFTLKISRVEAEDFGVYYCMQDAQDP
 4-1-1   IGKV2D-28*01    DIVMTQSPLSLPVTPGEPASISCRSSQSLLH-SNGYNYLDWYLQKPGQSPQLLIYLGSNRASGVPDRFSGSGSGTDFTLKISRVEAEDVGVYYCMQALQTP
 4-1-1   IGKV2D-29*01    DIVMTQTPLSLSVTPGQPASISCKSSQSLLH-SDGKTYLYWYLQKPGQPPQLLIYEVSNRFSGVPDRFSGSGSGTDFTLKISRVEAEDVGVYYCMQSIQLP
 4-1-1   IGKV2D-30*01    DVVMTQSPLSLPVTLGQPASISCRSSQSLVY-SDGNTYLNWFQQRPGQSPRRLIYKVSNWDSGVPDRFSGSGSGTDFTLKISRVEAEDVGVYYCMQGTHWP

 6-1-1   IGKV3-20*01     EIVLTQSPGTLSLSPGERATLSCRASQSVSS-----SYLAWYQQKPGQAPRLLIYGASSRATGIPDRFSGSGSGTDFTLTISRLEPEDFAVYYCQQYGSSP
 6-1-1   IGKV3D-20*01    EIVLTQSPATLSLSPGERATLSCGASQSVSS-----SYLAWYQQKPGLAPRLLIYDASSRATGIPDRFSGSGSGTDFTLTISRLEPEDFAVYYCQQYGSSP
 6-1-1   IGKV3D-7*01     EIVMTQSPATLSLSPGERATLSCRASQSVSS-----SYLSWYQQKPGQAPRLLIYGASTRATGIPARFSGSGSGTDFTLTISSLQPEDFAVYYCQQDYNLP
 2-1-1   IGKV3-11*01     EIVLTQSPATLSLSPGERATLSCRASQSVS------SYLAWYQQKPGQAPRLLIYDASNRATGIPARFSGSGSGTDFTLTISSLEPEDFAVYYCQQRSNWP
 2-1-1   IGKV3-15*01     EIVMTQSPATLSVSPGERATLSCRASQSVS------SNLAWYQQKPGQAPRLLIYGASTRATGIPARFSGSGSGTEFTLTISSLQSEDFAVYYCQQYNNWP
 2-1-?   IGKV3D-11*01    EIVLTQSPATLSLSPGERATLSCRASQGVS------SYLAWYQQKPGQAPRLLIYDASNRATGIPARFSGSGPGTDFTLTISSLEPEDFAVYYCQQRSNWH
 2-1-1   IGKV3D-15*01    EIVMTQSPATLSVSPGERATLSCRASQSVS------SNLAWYQQKPGQAPRLLIYGASTRATGIPARFSGSGSGTEFTLTISSLQSEDFAVYYCQQYNNWP

 3-1-1   IGKV4-1*01      DIVMTQSPDSLAVSLGERATINCKSSQSVLYSSNNKNYLAWYQQKPGQPPKLLIYWASTRESGVPDRFSGSGSGTDFTLTISSLQAEDVAVYYCQQYYSTP

 2-1-1   IGKV5-2*01      ETTLTQSPAFMSATPGDKVNISCKASQDIDD------DMNWYQQKPGEAAIFIIQEATTLVPGIPPRFSGSGYGTDFTLTINNIESEDAAYYFCLQHDNFP
```

TABLE 13.4 Immunoglobulin Heavy and Light J Chains

	L3		H3
	-		------
	CDR3		CDR3
	--		--------
	10		10 11
	678901234567		4567890123456789 0123
JK1	WTFGQGTKVEIK	JH1	---AEYFQHWGQGTLVTVSS
JK2	YTFGQGTKLEIK	JH2	---YWYFDLWGRGTLVTVSS
JK3	FTFGPGTKVDIK	JH3	-----AFDIWGQGTMVTVSS
JK4	LTFGGGTKVEIK	JH4	-----YFDYWGQGTLVTVSS
JK5	ITFGQGTRLEIK	JH5	----NWFDPWGQGTLVTVSS
		JH6	YYYYYGMDVWGQGTTVTVSS

70-fold loss in affinity of a CDR-grafted version of D1.3 (Hwang et al. 2005). Another example published by Hu et al. (2007) in which they superhumanized the murine antibody 1A4A1, a neutralizing antibody against the Venezuelan equine encephalitis virus (VEEV), demonstrating retention of antigen-binding specificity and neutralizing activity.

13.2.4 Guided Selection

Guided selection is an empirical approach to replace a murine antibody sequence with a fully human antibody sequence (and hence, is not truly a method for "humanization," but rather, it is a chain substitution method, as explained below). No assumptions at all are made on the impact of mutations on the antibody structure. Guided selection consists of three components: (1) the V regions of the nonhuman antibody to be humanized, (2) libraries of human V_H and V_L chains, and (3) a selection method. In a first step, the V_H or V_L chain of the nonhuman antibody is paired with a library human V_H or V_L chain to generate a nonhuman : human V chimerical library. The library is expressed and selected against the antigen recognized by the nonhuman antibody. The nonhuman V_H or V_L thus serves as a guide to select human V regions specific for the epitope recognized by the nonhuman antibody. In a second step, the specific human V chain identified in step 1 is paired with a library of human V_H or V_L chains to generate a library of fully human V regions and a fully human Fv product is then isolated through additional rounds of selection against the antigen of interest. A variation of this basic procedure is to select nonhuman : human V_H and V_L chimerical libraries in parallel and combine the outcome of independent selections to yield a human Fv. This was the approach used to isolate the human sequence for Humira®, the first fully human antibody to be marketed from the Cambridge Antibody Therapeutics phage displayed human antibody library. A detailed description of the method has been published by Osbourn, Groves, and Vaughan (2005).

It is evident that the higher the diversity of human V_H and V_L chain libraries, the higher the chance of isolating diverse human V_H and/or V_L chains against the epitope of interest, as well as the higher the chance of selecting higher affinity antibodies. The human V chains could be isolated from natural sources (Marks et al. 1991) or synthetically generated (Almagro et al. 2006; Sidhu and Fellouse 2006). The former has the advantage over the latter than V regions from circulating antibodies have undergone selection and thus the quality of the library could be superior to synthetically generated V regions. Nonetheless, like in the selection of human FRs for CDR grafting (Section 13.2.1), antibodies that are the product of immune responses could carry somatic mutations and thus potentially immunogenic residues. Synthetic libraries designed on well-characterized human germline genes could potentially mitigate such risk.

A potential drawback of guided selection is that shuffling of one of the V chains while keeping the other constant could result in epitope drift. Indeed, several studies have reported the isolation of binders with a change in fine-specificity (Watzka, Pfizenmaier, and Moosmayer 1998; Kuepper et al. 2005). In order to maintain the epitope recognized by the nonhuman antibody, CDR retention

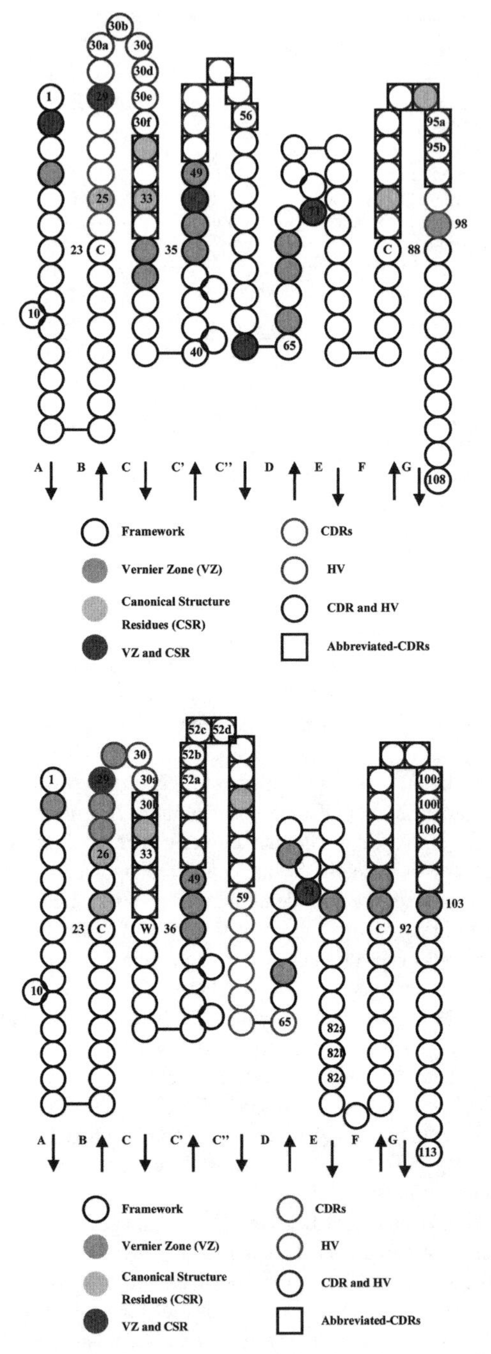

Figure 13.3 Important residues in modulation of antibody affinity.

can be included in the library design (Wang et al. 2000). CDR retention consists of combining the non-human CDR with the library of human genes. CDR-H3 is the most commonly retained CDR, as this region of the antigen-binding site plays the most important role in defining antibody specificity. Retention of CDR-H3 and CDR-L3, as well as CDR-H3, CDR-L3, and CDR-L2, has also been reported (Steinberger et al. 2000).

13.2.5 Other Humanization Methods

One of the early methods described for humanization was that of resurfacing, or veneering, of murine residues that were surface exposed (Daugherty et al. 2004; Roguska et al. 1994, 1994; Delagrave et al. 1999; Staelens et al. 2006). For this approach, no CDRs were grafted. Instead, the murine sequence itself was modified based on a combination of the following procedures: (1) comparing the source murine sequence with the closest human framework, (2) modeling the murine sequence based on the closest structural counterpart in the PDB database to identify surface potentially exposed residues, (3) converting those exposed residues to the human residues based on the most similar human framework sequence, and (4) restoration of any affinity lost via maturation, the same as with CDR grafting. Roguska et al. (1994) compared the results of humanization by standard CDR grafting with their resurfacing humanization strategy, and concluded that, while the resultant sequences were somewhat different, the overall humanness of the antibodies was similar and similar affinities to the antigen were retained. One of the antibodies resurfaced in that paper, the anti-CD56 Mab N901, is now in Phase II clinical trials as a mytansanoid-Mab conjugate, HuN-901-DM1.

Several other variations on the themes of CDR grafting, superhumanization, and resurfacing have been used in efforts to convert murine VH and VL sequences to those more human-like. While these are beyond the scope of this chapter, we will highlight one recent method described by Lazar et al. (2007), which is based on the principle of antibody humanness they termed "human string content" (HSC). This method relies on the use of multiple frameworks, and local sequence variation, to guide substitution of human amino acid residues for murine residues. The principle is that local (or short) sequences, which they call "human strings," possess the key information for binding to MHC-II and recognition by T-cells (Lazar et al. 2007). Because local sequences of multiple different germlines were used in their protocols, Lazar et al. (2007) also utilized modeling as a filter in their process to help ensure that three-dimensional structure was retained. The four antibodies humanized by Lazar et al. (2007) retained the biological activity of their murine parents, and in some cases, modest affinity optimization was achieved.

In summary, no matter which humanization, or human-adaptation, method is utilized, the keys to success remain the same: (1) generate antibodies with sequences that are the most human-like possible; (2) retain, recover, or in the best case, improve, affinity and functionality of the parental murine antibody; and (3) generate antibodies that possess superior folding, production, and solubility characteristics, while minimizing aggregation and other properties undesirable for antibodies that are intended to be developed as therapeutics.

13.3 AFFINITY MATURATION

Affinity, usually reported as equilibrium dissociation constant (K_D), is defined as the strength with which a given antibody binds to its correspondent antigen. K_D is governed by two kinetic constants, namely, association (k_{on} or ka) and dissociation (k_{off} or kd) constants. The former dictates how fast the antibody will bind the antigen. The latter defines for how long the antigen will last in complex with the antibody. Several methods to measure the kinetic constants and thus estimate K_D or measure the antigen–antibody affinity directly have been developed, the most commonly utilized being surface plasmon resonance (SPR, the most common form of which is BIAcore) and solution based kinetics exclusion assay (KinExA). The accuracy of these techniques depends on the K_D range to be measured. Values above 100 pM can be measured by BIAcore, while KinExA is the instrument of choice for measuring K_D values below 100 pM (Rathanaswami et al. 2005; Luginbühl et al. 2006).

It has been hypothesized that antibodies generated *in vivo* cannot reach affinities below 100 pM (Foote and Eisen 1995), whereas affinity of antibodies isolated *in vitro* using recombinant or synthetic antibody libraries ranges between 100 nM and 1 nM (Knappick et al. 2000; Söderlind et al. 2000; Hoet et al. 2005), with some exceptional binders reaching the high picomolar range. These hypotheses have proven to be over-generalizations, since antibodies with exceptionally high affinity have been isolated directly from murine hybridomas. Nevertheless, many antibodies humanized from hybridomas,

TABLE 13.5 Examples of Affinity Matured Antibodies Using *in Vitro* Technologies

Antibody	Target	Method	Selection	Affinity Increase	Final Affinity (pM)	References
b4/12	gp120	CDR walking	Phage display	420×	15	Yang et al. (1995)
C6.5	c-erbB-2	CDR-H3 and L3 mutagenesis	Phage display	1,230×	13	Schier et al. (1996)
4-4-20	FL-bio	Error-prone PCR DNA shuffling	Yeast display	10,000×	0.048	Boder, Midelfort, and Wittrup (2000)
H6	GCN4	Error-prone PCR DNA shuffling	Ribosome display	500×	5	Zahnd et al. (2004)
smE3	CEA	Error-prone PCR DNA shuffling	Yeast display	285×	30	Graff et al. (2004)
14B7	Anthrax toxin	Error-prone PCR	Bacteria display	200×	21	Harvey et al. (2004)
Synagis	RSV	Focused CDRs	Screening	125×	27	Wu et al. (2005)
PrP-SC	P Fab	Error-prone PCR DNA shuffling	Ribosome display Screening	300×	1	Luginbühl et al. (2006)
BAK1	IL-13	Error prone PCR CDR-3 mutagenesis CDR walking	Ribosome display Phage display	167×	81	Thom et al. (2006)
Erbitux	EGFR	Rational design	Screening	9×	52	Lippow, Wittrup, and Tidor (2007)
D44.1	Lysozyme	Rational design	Screening	140×	30	Lippow, Wittrup, and Tidor (2007)

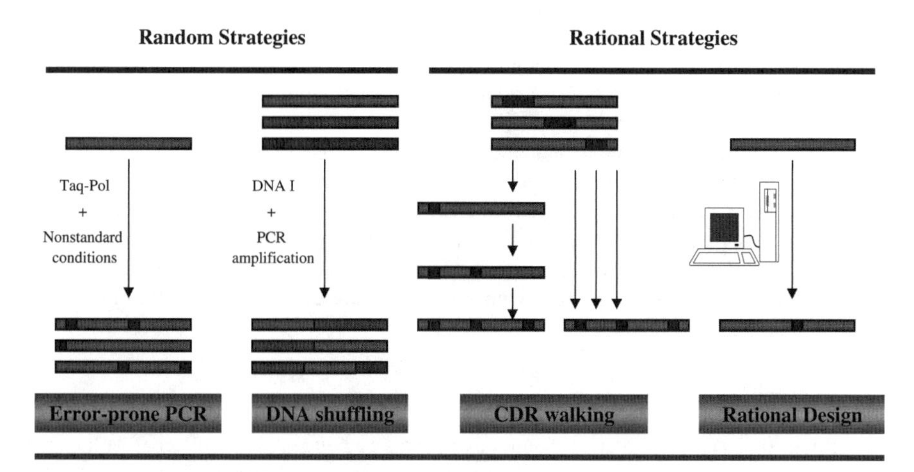

Figure 13.4 Schematic comparison of random strategies versus rational strategies for design of antibody optimization libraries, as described in the text.

transgenic humanized mice, and human antibody libraries alike still require affinity optimization. Table 13.5 compiles examples of antibodies that have undergone *in vitro* affinity maturation, reaching affinities beyond the ceiling imposed *in vivo* and two to three orders of magnitude below the affinity of antibodies isolated from recombinant or synthetic libraries. Of note is the report by Boder et al. (2000), who combined several mutagenesis strategies to enhance the affinity of an anti-hapten antibody (4-4-20) by more than four orders of magnitude. These authors produced a 48 fM binder; up to our knowledge, the highest affinity antibody reported. Affinity maturation of several of these antibodies reported in Table 13.5 has been translated into more potent antibodies (see references in Table 13.5), with the potential impact on safety, dosage, and hence production costs. Thus, demand for affinity maturation methods has been increasing in the last years, resulting in an increase and diversification of the strategies used for design of affinity maturation libraries (Fig. 13.4).

It should be noted here that while high affinity antibodies are desired or required for a great number of potential therapies, other factors also must be taken into account when designing the criteria for, and selecting, the therapeutic antibodies for any given target. These other criteria include epitope, target-based functionality, off-target activity, Fc functionality, and requirement for tissue penetration, among other factors.

13.3.1 Error-Prone PCR

Due to its simplicity, error-prone PCR is perhaps the most commonly used strategy (Table 13.5) to generate libraries of antibodies. This method relies on a low-fidelity DNA polymerase, such as *Thermus aquaticus* polymerase (Taq-Pol), to introduce errors all along the V gene during its amplification process. The error rate of Taq-Pol is enhanced using nonstandard PCR conditions such as Mn^{2+} instead of Mg^{+}, propanol, unbalanced dNTP, or unnatural mutagenic bases (Neylon 2004).

Limitations in the use of Taq-Pol are often associated with poor yields, low levels of mutation and, critically, a biased mutation spectrum. An alternative to the use of Tag-Pol to generate random mutations was reported by Biles and Connolly (2004). By mutating the region responsible for binding the incoming dNTPs, these authors converted the *Pyrococcus furiosus* polymerase (Pfu-Pol) from an extremely accurate wild-type polymerase into a variant with low fidelity. The Pfu-Pol mutant can be applied in error-prone PCR under the same conditions used for standard high-fidelity PCR, and large quantities of amplified product, with a high frequency of nearly indiscriminate mutations, are produced.

The effects of mutation frequency on affinity maturation generated by error-prone PCR have been assessed by Daugherty et al. (2000). Libraries containing different mutation rates were displayed on *Escherichia coli* and mutant populations were analyzed by flow cytometry. Specifically, three libraries were tested, one with approximately 2 mutations per gene, a second with approximately 4 mutations per gene, and a third with a high error rate of about 22 mutations per gene. After several rounds of enrichment, each of the three libraries yielded clones with improved affinity, with the libraries exhibiting moderate and high error rate yielding clones with greatest affinity improvement. The authors concluded that functional clones are well represented in such libraries despite the high random mutation rate. Interestingly, the majority of the mutations leading to higher affinity correspond to residues distant from the antigen-binding site, consistent with previous analyses of antibodies optimized *in vivo* (Wedemayer et al. 1997).

13.3.2 DNA Shuffling

DNA shuffling could generate libraries of antibody variants carrying segments of genes with more than one mutation (Fig. 13.4). This method (Stemmer 1994), is based on digestion of a gene with DNase I and pooling random DNA fragments. The fragments are reassembled into a full-length gene by repeated cycles of annealing in the presence of DNA polymerase. The fragments prime each other based on homology, and recombination occurs when fragments from one copy of a gene prime on another copy, causing a template switch.

Chodorge and coworkers (2008) assessed the impact of DNA recombination on affinity maturation. Parallel mutagenesis strategies combining error-prone PCR with and without a recombination step were carried out. As expected, both strategies improved affinity but the recombination step resulted in an increased population of affinity-improved variants. Moreover, the most improved variant, with a 22-fold affinity enhancement, was isolated only from the recombination-based approach. An analysis of mutations preferentially selected in the recombined population demonstrated strong cooperative effects. These results underscore the ability of combinatorial library approaches to explore larger regions of sequence space than error-prone PCR.

13.3.3 CDR Walking Mutagenesis

Overall, rational strategies to generate libraries of antibodies have the advantage over random mutagenesis, in that one has more control over the consequences of changes introduced into the V gene. In addition, by focusing the variation in a given region of the gene, a set of positions can be explored using saturation mutagenesis, e.g., the 20 natural amino acids. Nonetheless, the universe of variants that can be generated using saturation mutagenesis is reduced to a few positions, as the number of variants that can be sampled using the current selection methodologies (Section 13.4) is very limited. For instance, a library built on the common NNK diversification scheme introduces 32 codons in every position and thus grows by 32^n for every n number of residues. Phage display, as one of the most robust and widely used methods to search the sequence space, is limited to sampling libraries of 10^9 to 10^{10} variants, implying that only six to seven residues can be targeted for variegation if full sequence coverage is to be achieved in the library. Thus, the challenge of rational methods is to identify key positions to be variegated and the regime of diversification to be applied to these positions.

Comparisons of germline genes and sequences that have naturally been affinity matured *in vivo* indicate that although somatic hypermutation occurs all along the V gene, amino acid replacements are concentrated at CDRs (Tomlinson et al. 1996; Ramirez-Benitez and Almagro 2001). Mutational enrichment at CDRs has been explained in terms of presence of mutational hot and cold spots at CDRs and FRs, respectively, as well as antigenic selection. These observations, together with the fact that mutagenesis experiments in which CDR positions have been mutated to all amino acids have led to increases in affinity, fueled the concept of creating focused CDRs libraries.

CDRs are, on average, nine residues long (Fig. 13.2) and thus the sequence space generated by saturation mutagenesis in the six CDRs altogether cannot be exhaustively explored. Yang et al. (1995)

implemented a strategy called CDR walking (Fig. 13.4) to circumvent this limitation. In this strategy, NNK-randomized CDRs are selected in parallel, and the best variants combined and screened for binding improvement. Alternatively, a sequential search of the sequence space is conducted by optimizing one CDR at a time and taking the best variant to optimize the next CDR. Sequential CDR walking strategy has consistently yielded variants of improved affinity. Affinity improvements based on additivity effects by combining independently optimized CDRs have been unpredictable but have led to a modest improvement in affinity. The highest affinity binder reported to be obtained through this strategy was 15 pM, resulting in a 420-fold improvement with respect to the affinity of the starting antibody (Table 13.1).

A more focused strategy was implemented by Schier et al. (1996), in which only the sequence space of the CDR-3 regions was explored. The rationale was that CDR-3 regions are located at the center of the antigen-binding site (Fig. 13.1) and thus play a major role in determining the specificity and affinity of antibodies. Specifically, the CDR-3 of VH is by far the most diverse CDR of the antigen-binding site in terms of length, amino acid composition and conformation, as this region is naturally generated by multiple diversification mechanisms (Tonegawa 1983). First, the CDR-H3 is generated by recombination of the 5′ region of the VH region, the 3′ region of JH genes, and a repertoire of Diversity (D) genes estimated in humans to be composed of 27 gene segments (Corbett et al. 1997). Second, additional diversity is furnished by deletion of nucleotides at the VH, D, and JH genes. Also, additional noncoded nucleotides are added into the VH-D and D-JH junctions, thus contributing to more diversification. Finally, somatic mutation is instrumental in generating CDR-H3 variants. These mechanisms produce an enormous variation in the CDR-H3 that plays a major role in defining the binding profile of antibodies. Focusing on CDR-3s, Schier and coworkers (Table 13.1) were able to enhance the affinity of an anti-erbB-2 antibody by more than three orders of magnitude, down to 13 pM. These authors optimized the CDR-L3 first, followed by CDR-H3 and combined mutants with improved affinity from parallel selections.

13.3.4 Rational Design

Rational design enables a greater ability to explore sequence space than any experimental method. More exhaustive explorations of the sequence space hold the promise of reaching solutions beyond currently known limits for affinity. Moreover, rational design should simplify the library making by suggesting fewer changes than random mutagenesis strategies or CDR walking, hence minimizing time and costs. Thus, rational design has been a theme of constant research, although to date it has attained limited success, in particular when redesigning nanomolar antibodies to reach picomolar affinities, where a fine free-energy discrimination is required (Lippow and Tidor 2007).

Based on rational design, Lippow et al. (2007) recently enhanced the affinity of two antibodies down to the picomolar range (Table 13.5), thus infusing new expectations into the field. These authors explored single mutations at each of 60 CDR positions to the 20 common side chains, excluding proline and cysteine (10^{75} variants!), using a physics-based energy function. In a first step, they fixed the backbone conformation and conducted a discrete search of side chain conformations based on rotamers. It was followed by a second search with a more demanding computation protocol, including energy minimization. By combining six designed mutations in D44.1 and three in the Erbitux® (Table 13.1) medium picomolar affinities were achieved (Table 13.5). This exercise also provided valuable suggestions for future improvement of protocols aiming to rationally design variants for affinity maturation, including focus the energy function on the electrostatic term as a better predictor of improved binding and avoid destabilizing mutations based on calculating folding stability.

13.4 SELECTION TECHNIQUES

In this section, we will briefly discuss different selection techniques that have been used in affinity maturation, including display procedures using phage, yeast (Gai and Wittrup 2007), bacteria

(Harvey et al. 2004), mammalian cells (Ho, Nagata, and Pastan 2006; Akamatsu et al. 2007), or ribosomes (Lipovsek and Plückthun 2004) to couple phenotype (the antibody displayed) with genotype (the DNA encoding that antibody) (Hoogenboom 2005). Display of full IgGs also has been described (Akamatsu et al. 2007), which may in the future lead to methods for the generation of better "fit" antibodies (i.e., those with better folding, stability, and solubility features) using full IgG display procedures. Below, the three most widely used selection techniques, phage display, ribosome display, and yeast display, will be highlighted briefly.

Phage display using M13 was invented by George Smith in 1985 (Smith 1985) and has been applied to selection of antibody libraries since 1990, when McCafferty et al. (1990) first described antibody selections using phage display. The technology (Fig. 13.5) is based on the display of an antibody fragment, fused to one of the coat proteins of derivatives of the filamentous phage M13 or Fd. Most of the literature to date describes the use of protein III (pIII) as the fusion partner, although other coat proteins such as pVIII and pIX have been used successfully to display antibody fragments (Gao et al. 1999). In its genome, the phage carries the genes coding for the antibody fragment to be displayed, thus linking the phenotype to the genotype (Fig. 13.5). The strengths of phage display include ease of use, robustness, and stability of the phage particles. Because of these properties, phage display lends itself not only to panning on proteins, but also to various cell-based panning strategies which are now allowing for isolation of antibodies against active conformers of cell surface proteins (Schwartz et al. 2004). One potential negative attribute of phage display is that selection of antibody fragments on the surface of phages may not provide the best pressures for selecting the best-behaving (i.e., folding, solubility) antibody fragments.

Due to technical limitations of library construction based on transformation efficiency, a typical antibody library displayed on phage consists of 10^9 to 10^{10} variants (Dufner, Jermatus, and Minter 2006). Some of the strategies and methods for design of the libraries were described in Section 13.3 of this chapter, so will not be readdressed here. The variants are displayed on the phage surface and

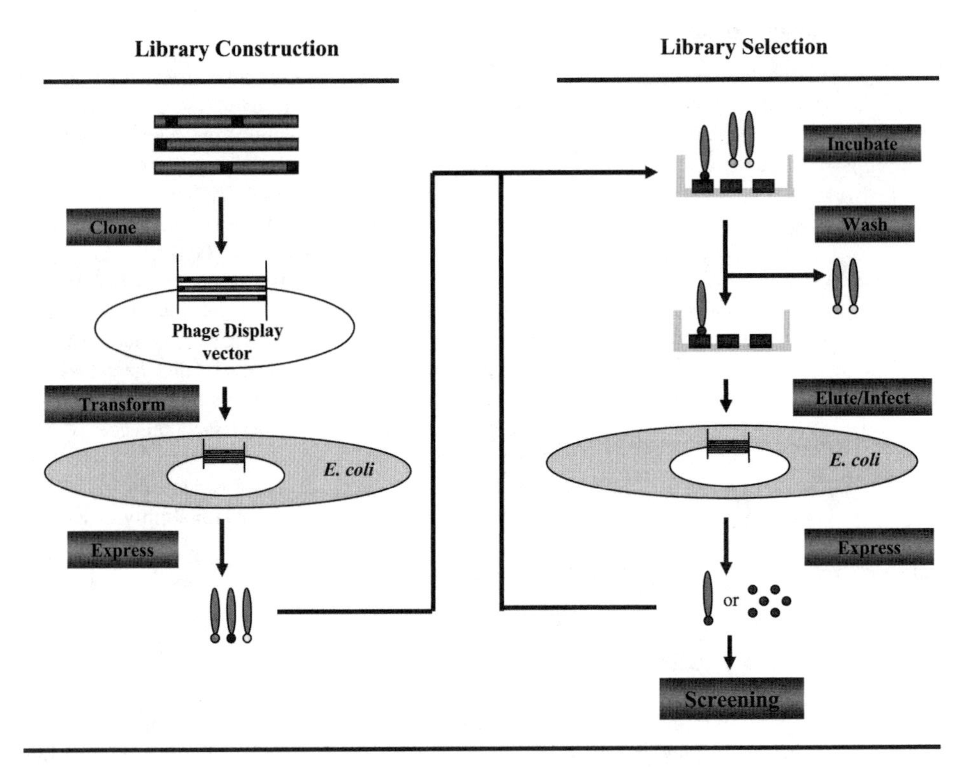

Figure 13.5 A schematic of phage display methodology, as described in the text.

then selected against a target antigen. The selection consists of incubation with the antigen of interest in such a way that phages bearing specific antibodies bind the antigen. Unbound phages are then removed through washing steps, followed by infection of *E. coli* cells and amplification of the bound phages. The procedure is usually repeated two to four times until a significant population of positive clones is obtained. Designing the panning strategy for affinity maturation is critical for an optimization project, as it can dramatically change the outcome of the selections (Hoogenboom 2005; Dufner, Jermatus, and Minter 2006).

Ribosome display, or other similar *in vitro* display techniques (e.g., mRNA display), have been hailed as important alternatives to phage display (Lipovsek and Plückthun 2004; Hoogenboom 2005; He and Khan 2005; Dufner, Jermatus, and Minter 2006; Groves et al. 2006; Fukuda et al. 2006; Yan and Xu 2006; Zahnd, Amstutz, and Plückthun 2007). Some of the advantages of ribosome or other *in vitro* display methods include speed, flexibility, library size (which, not limited by transformation frequency, can theoretically reach 10^{15}), protein folding in solution, absence of negative selection features, and the use of PCR in each round of selection, which enhances Darwinian selection (Dufner, Jermatus, and Minter 2006). Limitations of ribosome display include the inability to use it with cell-based protocols or impure antigens, and the requirement that the display molecule be in single chain format. Figure 13.6 shows a schematic for a generic ribosome display process (1) designing and constructing a library, (2) tethering of the library to the ribosome through initiation of the translation process, (3) incubation of the library-ribosome complex with the antigen of choice to enable binding, (4) washing of the antigen-mRNA-ribosome complex to remove phages that do not bind tightly, (5) dissociation of the complex to free up the mRNA, and (6) error-prone-based PCR amplification to regenerate a new library. The amplicons at the end of each process also can be screened using phage display or other methods for binding and functional characteristics (Fig. 13.6).

Yeast display is a technology that has become significantly more popular in the past few years, much of this owed to the success showed by Dane Wittrup and his colleagues (reviewed in Boder and Wittrup

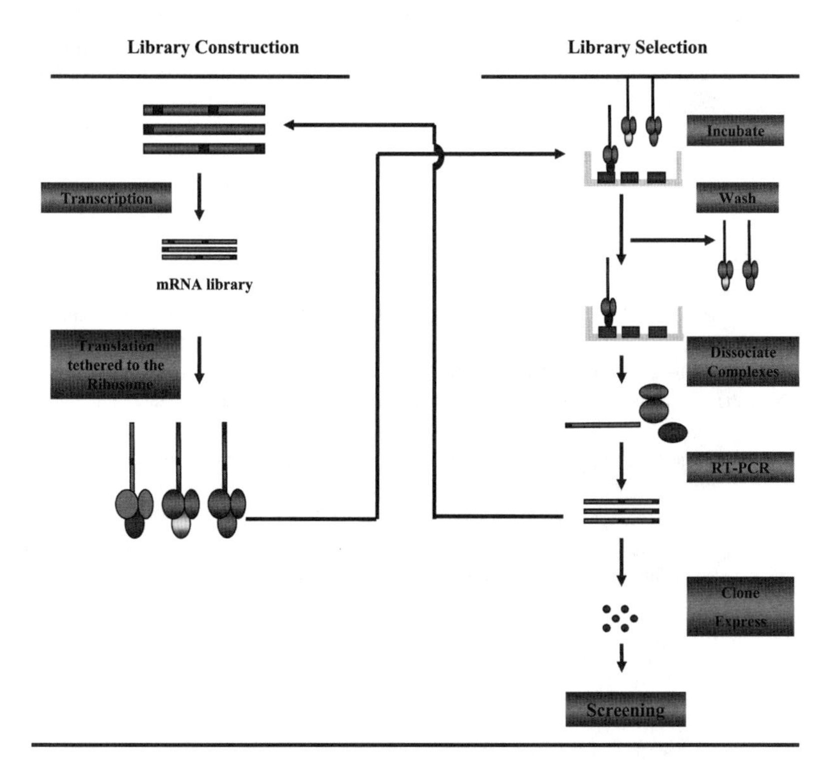

Figure 13.6 A schematic of typical ribosome display methodology as described in the text.

2000; Gai and Wittrup 2007), who have demonstrated the incredible power of yeast display in isolation of optimized antibodies with *Kd* values in the subpicomolar range (Boder, Midelfort, and Wittrup 2000; Table 13.2). Yeast display is inherently no different from any other display mechanism, with the exception that the antibody is displayed on the surface of the yeast, bound to a surface protein. The inherent limitation to yeast display is that the transformation frequency of yeast limits library sizes to the 10^7 range. The keys to the success of yeast display, on the other hand, are the ability to mate two different yeast strains, each carrying one of the chains of a Fab, hence multiplying the library size and overcoming the transformation frequency issue (Weaver-Feldhaus et al. 2004; Blaise et al. 2004), and the ability to use multicolor FACS selections and sorting to identify and select for the best binders (as well as best expressing Fabs; van den Beucken et al. 2003; Gai and Wittrup 2007). As mentioned previously, affinity-optimized antibodies in the subpicomolar range have been isolated using yeast display, demonstrating the power of the selection processes offered by the FACS selection and sorting methodology.

13.5 FUTURE DIRECTIONS

Despite the tremendous technological advances in the antibody engineering field during the last two decades, there is still room for improvement. The main drivers for future improvements in antibody optimization are to: (1) expedite and, where possible, combine the discovery and optimization process, (2) optimize for more potent antibodies with enhanced biophysical profiles, (3) generate humanized antibodies that possess the highest possible degree of humanness (Abhinandan and Martin 2007), and (4) continue to develop better and higher throughput means for co-selecting for both affinity and other desirable "developability" properties such as solubility, protein stability, proper folding, and lack of aggregation, a process that has been pioneered recently using domain antibodies by Greg Winter and his colleagues at the MRC (Jespers et al. 2004; Christ, Famm, and Winter 2007; Famm et al. 2008).

One advantage for improving the processes for engineering monoclonal antibodies these days is to use structural, functional, immunogenicity, and where known, solubility, information from the myriad of antibodies that have now entered the clinic (see Chapter 1); this includes 26 marketed monoclonal antibodies and Fc fusion proteins, 35 Phase III candidates, and another >140 known Phase I and Phase II candidates. This information may be able to be fed back into the various algorithms that drive humanization and affinity optimization paradigms.

REFERENCES

Abhinandan, K.R., and A.C. Martin. 2007. Analyzing the "degree of humanness" of antibody sequences. *J. Mol. Biol.* 369:852–862.

Akamatsu, Y., K. Pakabunto, Z. Xu, Y. Zhang, and N. Tsurushita. 2007. Whole IgG surface display on mammalian cells: Application to isolation of neutralizing chicken monoclonal anti-IL-12 antibodies. *J. Immunol. Methods* 327:40–52.

Al-Lazikani, B., A.M. Lesk, and C. Chothia. 1997. Standard conformations for the canonical structures of immunoglobulins. *J. Mol. Biol.* 273:927–948.

Almagro, J.C., and J. Fransson. 2008. Humanization of antibodies. *Front. Biosci.* 13:1619–1633.

Almagro, J.C., V. Quintero-Hernandez, M. Ortiz-Leon, A. Velandia, S.L. Smith, and B. Becerril. 2006. Design and validation of a synthetic VH repertoire with tailored diversity for protein recognition. *J. Mol. Recognit.* 19:413–422.

Amzel, L.M., and R.J. Poljak. 1979. Three-dimensional structure of immunoglobulins. *Annu. Rev. Biochem.* 48:961–997.

Baca, M., L.G. Presta, S.J. O'Connor, and J.A. Wells. 1997. Antibody humanization using monovalent phage display. *J. Biol. Chem.* 272:10678–10684.

Biles, B.D., and B.A. Connolly. 2004. Low-fidelity *Pyrococcus furiosus* DNA polymerase mutants useful in error-prone PCR. *Nucleic Acids Res.* 32:e176.

Blaise, L., A. Wehnert, M.P.G. Steukers, T. van den Beucken, H.R. Hoogenboom, and S.E. Hufton. 2004. Construction and diversification of yeast cell surface displayed libraries by yeast mating: Application to the affinity maturation of Fab antibody fragments. *Gene* 342:211–218.

Boder, E.T., and K.D. Wittrup. 2000. Yeast surface display for directed evolution of protein expression, affinity, and stability. *Methods Enzymol.* 328:430–444.

Boder, E.T., K.S. Midelfort, and K.D. Wittrup. 2000. Directed evolution of antibody fragments with monovalent femtomolar antigen-binding affinity. *Proc. Natl. Acad. Sci. USA* 97:10701–10705.

Bruccoleri, R.E., and M. Karplus. 1987. Prediction of the folding of short polypeptide segments by uniform conformational sampling. *Biopolymers* 26:137–168.

Chodorge, M., L. Fourage, G. Ravot, L. Jermutus, and R. Minter. 2008. In vitro DNA recombination by L-shuffling during ribosome display affinity maturation of an anti-Fas antibody increases the population of improved variants. *Protein Eng. Des. Sel.* 21:343–351.

Chothia, C., and A.M. Lesk. 1987. Canonical structures for the hypervariable regions of immunoglobulins. *J. Mol. Biol.* 196:901–917.

Christ, D., K. Famm, and G. Winter. 2007. Repertoires of aggregation-resistant human antibody domains. *Protein Eng. Des. Sel.* 20:413–416.

Corbett, S.J., I.M. Tomlinson, E.L. Sonnhammer, D. Buck, and G. Winter. 1997. Sequence of the human immunoglobulin diversity (D) segment locus: A systematic analysis provides no evidence for the use of DIR segments, inverted D segments, "minor" D segments or D-D recombination. *J. Mol. Biol.* 270:587–597.

Daugherty, P.S., G. Chen, B.L. Iverson, and G. Georgiou. 2000. Quantitative analysis of the effect of the mutation frequency on the affinity maturation of single chain Fv antibodies. *Proc. Natl. Acad. Sci. USA* 97:2029–2034.

Daugherty, B.L., G.E. Mark III, and E.A. Padlan. 2004. Method for reducing the immunogenicity of antibody variable domains. 45 pp. US 6,797,492 B2.

Delagrave, S., J. Catalan, C. Sweet, G. Drabik, A. Henry, A. Rees, T.P. Monath, and F. Guirakhoo. 1999. Effects of humanization by variable domain resurfacing on the antiviral activity of a single chain antibody against respiratory syncytial virus. *Protein Eng.* 12:357–362.

De Pascalis, R., M. Iwahashi, M. Tamura, E.A. Padlan, N.R. Gonzales, A.D. Santos, M. Giuliano, P. Schuck, J. Schlom, and S.V. Kashmiri. 2002. Grafting of "abbreviated" complementarity-determining regions containing specificity-determining residues essential for ligand contact to engineer a less immunogenic humanized monoclonal antibody. *J. Immunol.* 169:3076–3084.

Dufner, P., L. Jermatus, and R.R. Minter. 2006. Harnessing phage and ribosome display for antibody optimization. *Trends Biotechnol.* 24:523–529.

Famm, K., L. Hansen, D. Christ, and G. Winters. 2008. Thermodynamically stable aggregation-resistant antibody domains through directed evolution. *J. Mol. Biol.* 376:926–931.

Foote, J., and H.N. Eisen. 1995. Kinetic and affinity limits on antibodies produced during immune responses. *Proc. Natl. Acad. Sci. USA* 92:1254–1256.

Foote, J., and G. Winter. 1992. Antibody framework residues affecting the conformation of the hypervariable loops. *J. Mol. Biol.* 224:487–499.

Fukuda, I., K. Kojoh, N. Tabata, N. Doi, H. Takashima, E. Miyamoto-Sato, and H. Yanagawa. 2006. In vitro evolution of single-chain antibodies using mRNA display. *Nucleic Acids Res.* 34:e127, 1–8.

Gai, S.A., and K.D. Wittrup. 2007. Yeast surface display for protein engineering and characterization. *Curr. Opin. Struct. Biol.* 17:467–473.

Gao, C., S. Mao, C.H. Lo, P. Wirsching, R.A. Lerner, and K.D. Janda. 1999. Making artificial antibodies: A format for phage display of combinatorial heterodimeric arrays. *Proc. Natl. Acad. Sci. USA* 96:6025–6030.

Gonzales, N.R., E.A. Padlan, R. De Pascalis, P. Schuck, J. Schlom, and S.V. Kashmiri. 2004. SDR grafting of a murine antibody using multiple human germline templates to minimize its immunogenicity. *Mol. Immunol.* 41:863–872.

Graff, C.P., K. Chester, R. Begent, and K.D. Wittrup. 2004. Directed evolution of an anti-carcinoembryonic antigen scFv with a 4-day monovalent dissociation time at 37 degrees C. *Protein Eng. Des. Sci.* 17:293–304.

Groves, M., S. Lane, J. Douthwaite, D. Lowne, D.G. Rees, B. Edwards, and R.H. Jackson. 2006. Affinity maturation of phage display antibody populations using ribosome display. *J. Immunol. Methods* 313:129–139.

Harvey, B.R., G. Georgiou, A. Hayhurst, K.J. Jeong, B.L. Iverson, and G.K. Rogers. 2004. Anchored periplasmic expression, a versatile technology for the isolation of high-affinity antibodies from *Escherichia coli*-expressed libraries. *Proc. Natl. Acad. Sci USA* 101:9193–9198.

He, M., and F. Khan. 2005. Ribosome display: Next generation display technologies for production of antibodies in vitro. *Expert Rev. Proteomics* 2:421–430.

Ho, M., S. Nagata, and I. Pastan. 2006. Isolation of anti-CD22 Fv with high affinity by Fv display on human cells. *Proc. Natl. Acad. Sci. USA* 103:9637–9642.

Hoet, R.M., E.H. Cohen, R.B. Kent, K. Rookey, S. Schoonbroodt, S. Hogan, L. Rem, N. Frans, M. Daukandt, H. Pieters, R. van Hegelsom, N.C. Neer, H.G. Nastri, I.J. Rondon, J.A. Leeds, S.E. Hufton, L. Huang, I. Kashin, M. Devlin, G. Kuang, M. Steukers, M. Viswanathan, A.E. Nixon, D.J. Sexton, H.R. Hoogenboom, and R.C. Ladner. 2005. Generation of high-affinity human antibodies by combining donor-derived and synthetic complementarity-determining-region diversity. *Nature Biotechnol.* 23:344–348.

Hoogenboom, H.R. 2005. Selecting and screening recombinant antibody libraries. *Nature Biotechnol.* 23:1105–1116.

Hu, W.G., D. Chau, J. Wu, S. Jager, and L.P. Nagata. 2007. Humanization and mammalian expression of a murine monoclonal antibody against Venezuelan equine encephalitis virus. *Vaccine* 25:3210–3214.

Hwang, W.Y., and J. Foote. 2005. Immunogenicity of engineered antibodies. *Methods* 36:3–10.

Hwang, W.Y., J.C. Almagro, T.N. Buss, P. Tan, and J. Foote. 2005. Use of human germline genes in a CDR homology-based approach to CDR grafting. *Methods* 36:35–42.

Jespers, L.S., A. Roberts, S.M. Mahler, G. Winter, and H.R. Hoogenboom. 1994. Guiding the selection of human antibodies from phage display repertoires to a single epitope of an antigen. *Biotechnology (NY).* 12:899–903.

Jespers, L., O. Schon, K. Famm, and G. Winter. 2004. Aggregation-resistant domain antibodies selected on phage by heat denaturation. *Nature Biotechnol.* 22:1161–1165.

Jones, P.T., P.H. Dear, J. Foote, M.S. Neuberger, and G. Winter. 1986. Replacing the complementarity-determining regions in a human antibody with those from a mouse. *Nature* 321:522–525.

Kashmiri, S.V., R. De Pascalis, N.R. Gonzales, and J. Schlom. 2005. SDR grafting: A new approach to antibody humanization. *Methods* 36:25–34.

Kimball, J.A., D.J. Norman, C.F. Shield, T.J. Schroeder, P. Lisi, M. Garovoy, J.B. O'Connell, F. Stuart, S.V. McDiarmid, and W. Wall. 1995. The OKT3 antibody response study: A multicentre study of human anti-mouse antibody (HAMA) production following OKT3 use in solid organ transplantation. *Transpl. Immunol.* 3(3):212–221.

Knappik A, L. Ge, A. Honegger, P. Pack, M. Fischer, G. Wellnhofer, A. Hoess, J. Wölle, A. Plückthun, and B. Virnekäs. 2000. Fully synthetic human combinatorial antibody libraries (HuCAL) based on modular consensus frameworks and CDRs randomized with trinucleotides. *J. Mol. Biol.* 296:57–86.

Koren, E., L.A. Zuckerman, and A.R. Mire-Sluis. 2002. Immune responses to therapeutic proteins in humans: Clinical significance, assessment and prodiction. *Curr. Pharmaceut. Biotechnol.* 3:349–368.

Kuepper, M.B., M. Huhn, H. Spiegel, J.K. Ma, S. Barth, R. Fischer, and R. Finnern. 2005. Generation of human antibody fragments against *Streptococcus mutans* using a phage display chain shuffling approach. *BMC Biotechnol.* 5(4):1–12.

Lazar, G.A., J.R. Desjarlais, J. Jacinto, S. Karki, and P.W. Hammond. 2007. A molecular immunology approach to antibody humanization and functional optimization. *Mol. Immunol.* 44:1986–1998.

Lee, C.V., W.C. Liang, M.S. Dennis, C. Eigenbrot, S.S. Sidhu, and G. Fuh. 2004. High-affinity human antibodies from phage-displayed synthetic Fab libraries with a single framework scaffold. *J. Mol. Biol.* 340:1073–1093.

Lipovsek, D., and A. Plückthun. 2004. In-vitro protein evolution by ribosome display and mRNA display. *J. Immunol. Methods* 290:51–67.

Lippow, S.M., and B. Tidor. 2007. Progress in computational protein design. *Curr. Opin. Biotechnol.* 18:305–311.

Lippow, S.M., K.D. Wittrup, and B. Tidor. 2007. Computational design of antibody-affinity improvement beyond in vivo maturation. *Nature Biotechnol.* 25:1171–1176.

Luginbühl, B., Z. Kanyo, R.M. Jones, R.J. Fletterick, S.B. Prusiner, F.E. Cohen, R.A. Williamson, D.R. Burton, and A. Plückthun. 2006. Directed evolution of an anti-prion protein scFv fragment to an affinity of 1 pM and its structural interpretation. *J. Mol. Biol.* 363:75–97.

Marks, J.D., H.R. Hoogenboom, T.P. Bonnert, J. McCafferty, A.D. Griffiths, and G. Winter. 1991. By-passing immunization. Human antibodies from V-gene libraries displayed on phage. *J. Mol. Biol.* 222:581–597.

McCafferty, J., A.D. Griffiths, G. Winter, and D.J. Chiswell. 1990. Phage antibodies: Filamentous phage displaying antibody variable domains. *Nature* 348:552–554.

Mukovosov, I., T. Sabljic, G. Hortelano, and F. Ofosu. 2008. Factors that contribute to the immunogenicity of therapeutic recombinant human proteins. *Thromb. Haemost.* 99:874–882.

Neylon, C. 2004. Chemical and biochemical strategies for the randomization of protein encoding DNA sequences: library construction methods for directed evolution. *Nucl. Acids. Res.* 32:1448–1459.

Osbourn, J., M. Groves, and T. Vaughan. 2005. From rodent reagents to human therapeutics using antibody guided selection. *Methods* 36:61–68.

Padlan, E.A., C. Abergel, and J.P. Tipper. 1995. Identification of specificity-determining residues in antibodies. *FASEB J.* 9:133–139.

Pendley, C., A. Schantz, and C. Wagner. 2003. Immunogenicity of therapeutic monoclonal antibodies. *Curr. Opin. Mol. Therap.* 5:172–179.

Queen, C., W.P. Schneider, H.E. Selick, P.W. Payne, N.F. Landolfi, J.F. Duncan, N.M. Avdalovic, M. Levitt, R.P. Junghans, and T.A. Waldmann. 1989. A humanized antibody that binds to the interleukin 2 receptor. *Proc. Natl. Acad. Sci. USA* 86, 10029–10033.

Ramirez-Benitez, M.C., and J.C. Almagro. 2001. Analysis of antibodies of known structure suggests a lack of correspondence between the residues in contact with the antigen and those modified by somatic hypermutation. *Proteins* 45:199–206.

Rathanaswami, P., S. Roalstad, L. Roskos, Q.J. Su, S. Lackie, and J. Babcook. 2005. Demonstration of an in vivo generated sub-picomolar affinity fully human monoclonal antibody to interleukin-8. *Biochem. Biophys. Res. Commun.* 334:1004–1013.

Riley, S. 2006. In *The Future of Monoclonal Antibodies*. Business Insight Ltd., London.

Roguska, M.A., J.T. Pedersen, A.H. Henry, S.M.J. Searle, C. M. Roja, B. Avery, M. Hoffee, S. Cook, J.M. Lambert, W.A. Blaettler, A.R. Rees, and B.C. Guild. 1994. A comparison of two murine monoclonal antibodies humanized by CDR-grafting and variable domain resurfacing. *Protein Eng.* 9:895–904.

Roguska, M.A., J.T. Pedersen, C.A. Keddy, A.H. Henry, S.J. Searle, J.M. Lambert, V.S. Goldmacher, W.A. Blaettler, A.R. Rees, and B.C. Guild. 1994. Humanization of murine monoclonal antibodies through variable domain resurfacing. *Proc. Natl. Acad. Sci. USA* 91:969–973.

Schier, R., A. McCall, G.P. Adams, K.W. Marshall, H. Merritt, M. Yim, R.S. Crawford, L.M. Weiner, C. Marks, and J.D. Marks. 1996. Isolation of picomolar affinity anti-c-erbB-2 single-chain Fv by molecular evolution of the complementarity determining regions in the center of the antibody binding site. *J. Mol. Biol.* 263:551–567.

Schwartz, M., P. Boettgen, Y. Takada, F. Le Gall, S. Knackmuss, N. Bassler, C. Buettner, M. Little, C. Bode, and K. Peter. 2004. Single-chain antibodies for the conformation-specific blockade of activated platelet integrin alpha-II beta-3 designed by subtractive selection from naïve human phage libraries. *FASEB J.* 18:1704–1706.

Sidhu, S.S., and F.A. Fellouse. 2006. Synthetic therapeutic antibodies. *Nature Chem. Biol.* 2:682–688.

Smith, G.P. 1985. Filamentous fusion phage: Novel expression vectors that display cloned antigens on the virion surface. *Science* 228:1315–1317.

Söderlind, E., L. Strandberg, P. Jirholt, N. Kobayashi, V. Alexeiva, A.M. Aberg, A. Nilsson, B. Jansson, M. Ohlin, C. Wingren, L. Danielsson, R. Carlsson, and C.A. Borrebaeck. 2000. Recombining germline-derived CDR sequences for creating diverse single-framework antibody libraries. *Nature Biotechnol.* 18:852–856.

Staelens, S., J. Desmet, T.H. Ngo, S. Vauterin, I. Pareyn, P. Barbeaux, I. Van Rompaey, J.-M. Stassen, H. Deckmyn, and K. Vanhoorelbeke. 2006. Humanization by variable domain resurfacing and grafting on a human IgG4, using a new approach for determination of non-human like surface accessible framework residues based on homology modeling of variable domains. *Mol. Immunol.* 43:1243–1257.

Steinberger, P., J.K. Sutton, C. Rader, M. Elia, and C.F. Barbas. 2000. Generation and characterization of a recombinant human CCR5-specific antibody. A phage display approach rabbit antibody humanization. *J. Biol. Chem.* 275:36073–36078.

Stemmer, W.P. 1994. DNA shuffling by random fragmentation and reassembly: In vitro recombination for molecular evolution. *Proc. Natl. Acad. Sci. USA* 91:10747–10751.

Tamura, M., D.E. Milenic, M. Iwahashi, E. Padlan, J. Schlom, and S.V. Kashmiri. 2000. Structural correlates of an anticarcinoma antibody: Identification of specificity-determining residues (SDRs) and development of a minimally immunogenic antibody variant by retention of SDRs only. *J. Immunol.* 164:1432–1441.

Tan, P., D.A. Mitchell, T.N. Buss, M.A. Holmes, C. Anasetti, and J. Foote. 2002. "Superhumanized" antibodies: Reduction of immunogenic potential by complementarity-determining region grafting with human germline sequences—application to an anti-CD28. *J. Immunol.* 169:1119–11125.

Thom, G., A.C. Cockroft, A.G. Buchanan, C.J. Candotti, E.S. Cohen, D. Lowne, P. Monk, C.P. Shorrock-Hart, L. Jermatus, and R.R. Minter. 2006. Probing a protein–protein interaction by *in vitro* evolution. *Proc. Natl. Acad. Sci. USA* 103:7619–7624.

Tonegawa, S. 1983. Somatic generation of antibody diversity. *Nature* 302:575–581.

Tomlinson, I.M., G. Walter, P.T. Jones, P.H. Dear, E.L. Sonnhammer, and G. Winter. 1996. The imprint of somatic hypermutation on the repertoire of human germline V genes. *J. Mol. Biol.* 256:813–817.

Van den Beucken, T., H. Pieters, M. Steukers, M. van der Vaart, R.C. Ladner, H.R. Hoogenboom, and S.E. Hufton. 2003. Affinity maturation of Fab antibody fragments by fluorescent-activated cell sorting of yeast-displayed libraries. *FEBS Lett.* 546:288–294.

Wang, Z., Y. Wang, Z. Li, J. Li, and Z. Dong. 2000. Humanization of a mouse monoclonal antibody neutralizing TNF-alpha by guided selection. *J. Immunol. Methods* 241:171–184.

Wark, K.L., and P.J. Hudson. 1996. Latest technologies for the enhancement of antibody affinity. *Adv. Drug Deliv. Rev.* 58:657–670.

Watzka, H., K. Pfizenmaier, and D. Moosmayer. 1998. Guided selection of antibody fragments specific for human interferon γ receptor 1 from a human VH- and VL-gene repertoire. *Immunotechnology* 3:279–291.

Weaver-Feldhaus, J.M., J. Lou, J.R. Coleman, R.W. Siegel, J.D. Marks, and M.J. Feldhaus. 2004. Yeast mating for combinatorial Fab library generation and surface display. *FEBS Lett.* 564:24–34.

Wedemayer, G.J., P.A. Patten, L.H. Wang, P.G. Schultz, and R.C. Stevens. 1997. Structural insights into the evolution of an antibody combining site. *Science* 276:1665–1669.

Wu., H., D.S. Pfarr, Y. Tang, L.L. An, N.K. Patel, J.D. Watkins, W.D. Huse, P.A. Kiener, and J.F. Young. 2005. Ultra-potent antibodies against respiratory syncytial virus: effects of binding kinetics and binding valence on viral neutralization. *J. Mol. Biol.* 350:126–144.

Wu, T.T., and E.A. Kabat. 1970. An analysis of the sequences of the variable regions of Bence Jones proteins and myeloma light chains and their implications for antibody complementarity. *J. Exp. Med.* 132:211–250.

Yan, X., and Z. Xu. 2006. Ribosome-display technology: Applications for directed evolution of functional proteins. *Drug Discovery Today* 11:911–916.

Yang, W.P., K. Green, S. Pinz-Sweeney, A.T. Briones, D.R. Burton, and C.F. Barbas III. 1995. CDR walking mutagenesis for the affinity maturation of a potent human anti-HIV-1 antibody into the picomolar range. *J. Mol. Biol.* 254:392–403.

Zahnd, C., P. Amstutz, and A. Plückthun. 2007. Ribosome display: Selecting and evolving proteins in vitro that specifically bind to a target. *Nature Methods* 4:269–279.

Zahnd, C., S. Spinelli, B. Luginbühl, P. Amstutz, C. Cambillau, and A. Plückthun. 2004. Directed in vitro evolution and crystallographic analysis of a peptide-binding single chain antibody fragment (scFv) with low picomolar affinity. *J. Biol. Chem.* 279:18870–18877.

CHAPTER 14

Modulation of Serum Protein Homeostasis and Transcytosis by the Neonatal Fc Receptor

WILLIAM F. DALL'ACQUA and HERREN WU

ABSTRACT

The neonatal Fc receptor (FcRn) plays an essential role in controlling the homeostasis and transcytosis of immunoglobulin G (IgG) and albumin in mammals. Engineering the Fc portion of IgGs for increased or decreased binding to FcRn constitutes a sound strategy to modulate their serum pharmacokinetics properties and ability to be transported across select tissues. In particular, several recent and independent studies have shown that human IgGs engineered for increased binding to human FcRn exhibit significantly longer serum half-life in nonhuman primates when compared with their unmutated counterparts. The ability to tailor the properties of a given IgG in terms of its serum persistence, maternofetal transfer, tissue biodistribution, and/or effect on endogenous IgG levels promises to deliver a novel class of antibody-related medicines. Such advances will provide valuable additions to the fields of antibody therapy and diagnosis. It may also be possible to improve the pharmacokinetics properties of albumin by engineering its interaction with FcRn. Although not as advanced as when applied to IgGs, this engineering strategy may ultimately prove useful in developing novel albumin-based therapies or drug carriers.

Therapeutic Monoclonal Antibodies: From Bench to Clinic. Edited by Zhiqiang An

14.1 MODULATING THE SERUM HALF-LIFE AND TRANSCYTOSIS OF ANTIBODIES

14.1.1 The Neonatal Fc Receptor Controls IgG Recycling and Transport

Immunoglobulin G molecules exhibit two remarkable features that differentiate them from other immunoglobulin classes in humans. First, they are actively transferred across the placenta to confer passive immunity from mother to the fetus or neonate (Brambell 1966; Firan et al. 2001). Second, their serum half-life is significantly prolonged when compared with other serum proteins, a direct consequence of their Fc portion binding to a γ-globulin receptor (Brambell, Hemmings, and Morris 1964). The neonatal Fc receptor (FcRn) lies at the heart of these two essential, tightly controlled processes (Junghans and Anderson 1996; Leach et al. 1996; Kristoffersen 1996). The structure, function, and properties of this molecule have previously been described and reviewed in detail elsewhere (Ghetie and Ward 2000; Roopenian and Akilesh 2007). In short, FcRn is homologous to the class I major histocompatibility complex both in terms of sequence (Simister and Mostov 1989; Ahouse et al. 1993) and three-dimensional structure (Burmeister et al. 1994; Burmeister, Huber, and Bjorkman 1994). Expressed diffusely in the endothelial cells located throughout the adult body (Ghetie et al. 1996; Borvak et al. 1998), it is believed that FcRn acts as a salvage receptor for IgGs in mammals. Once pinocytosed and bound to FcRn, IgG molecules are transported across or recycled within endothelial cells and rescued from a degradative pathway. This process is illustrated in Figure 14.1. More precisely, IgG transcytosis refers to an FcRn-driven transport process occurring across the epithelium of various tissues such as placenta (Firan et al. 2001; Leach et al. 1996; Story, Mikulska, and Simister 1994), neonatal or adult intestine (Wallace and Rees 1980; Medesan et al. 1996; Dickinson et al. 1999), kidney (Haymann et al. 2000), lung (Spiekermann et al. 2002; Bitonti et al. 2004), yolk sac (Medesan et al. 1996), and mammary gland (Cianga et al. 1999). The control of IgG homeostasis refers to an FcRn-driven recycling mechanism, thought to occur mostly within the endothelial cells

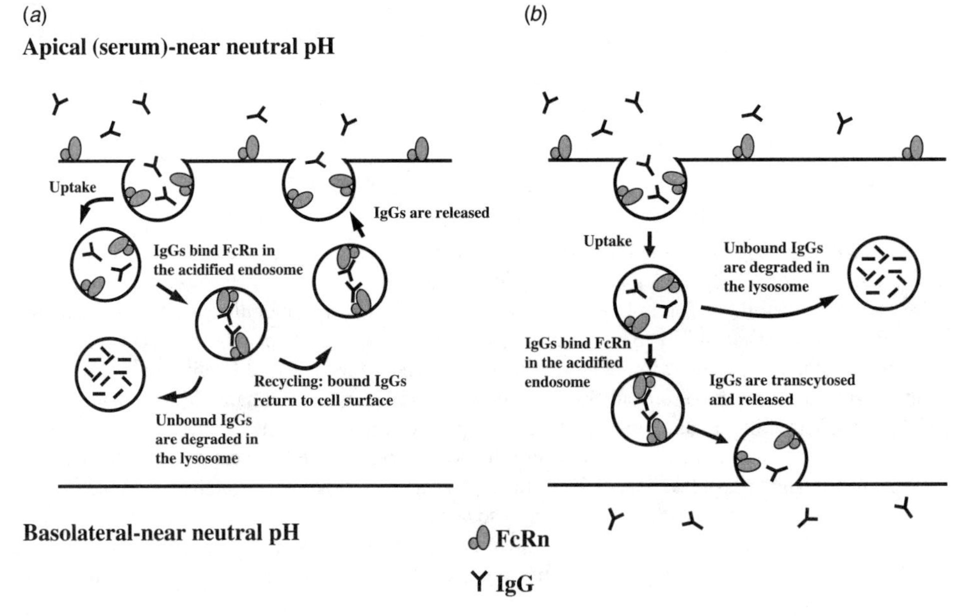

Figure 14.1 Model of IgG recycling and transcytosis viewed from a cross section of an FcRn-expressing endothelial cell. Upon pinocytosis, IgG molecules enter acidic endosomes and interact with FcRn. Once bound, IgG molecules are diverted from the lysosomal degradation pathway. Vesicles then return or migrate to the cell surface where the IgG molecules are released at physiological pH, leading to their (a) recycling, or (b) transcytosis. The IgG transcytosis process may possibly occur in a bidirectional manner (Ghetie and Ward 2000).

of small vessels and capillaries (Borvak et al. 1998). The latter phenomenon accounts for the comparatively long serum half-lives of IgGs, which range from approximately seven days to over three weeks (Morell, Terry, and Waldmann 1970).

One early pharmacokinetics study which provided direct evidence of a link between control of IgG homeostasis and presence of FcRn was carried out in the context of β2-microglobulin-deficient mice (Ghetie et al. 1996). Because the heterodimeric FcRn comprises β2-microglobulin in addition to a heavy (α) chain, these animals do not express a functional receptor. When dosed in such hosts, murine IgG1s as well as their Fc-hinge fragments exhibit an about sixfold shorter serum half-life relative to wild-type mice. The same mice also exhibit abnormally low serum levels of endogenous IgGs. Taken together, these data strongly suggest that FcRn is in direct control of IgG homeostasis.

14.1.2 Major Characteristics of the IgG-FcRn Complex

As a preamble and for comparison purposes, all antibody amino acid positions mentioned for the remainder of the chapter will be in accordance with the EU numbering system (Kabat et al. 1991).

One hallmark of the IgG/FcRn interaction is its acute pH sensitivity. Indeed, across various species from rodent to man, the corresponding binding affinities are at their strongest at slightly acidic pH and virtually abolished at neutral pH (Rodewald 1976; Raghavan et al. 1995; Ober et al. 2001). This observation is an agreement with co-localization studies (Kristoffersen and Matre 1996) and strongly suggests that the IgG/FcRn complexes form within acidified endosomes. The molecular mechanism responsible for this pH dependence is generally attributed to the titration of two histidine residues on the Fc (H310/H433; Raghavan et al. 1995; Vaughan and Bjorkman 1998) and FcRn (H250/H251; Raghavan et al. 1995) moieties. As will be discussed later in Section 14.1.4, this critical property must be conserved in order to engineer IgGs exhibiting extended serum half-life.

A plethora of mutagenesis studies have identified several human or mouse Fc residues important for the interaction of IgGs with their cognate FcRn. These elegant approaches have shown that essential components of FcRn functional epitope span both Fc C_H2 and C_H3 domains. Most notably among these, positions I253, H310, H435, and H436 were found to be particularly important for mouse Fc binding to mouse FcRn (Medesan et al. 1997). Likewise, the analysis of the human IgG/human FcRn interaction revealed the important role played by Fc positions I253, S254, H435, and Y436 (Firan et al. 2001; Shields et al. 2001). This is in very good agreement with the crystal structure of the complex between rat FcRn with rat Fc, which located FcRn structural epitope at the interface between Fc C_H2 and C_H3 domains (Burmeister, Huber, and Bjorkman 1994). However, we note that the nature of the human light chain (namely κ versus λ) in otherwise identical chimeric IgGs has been shown to influence their kinetics of binding to murine FcRn, suggesting that non-Fc areas could also play an indirect role in the interaction between the two partners (Gurbaxani et al. 2006).

Despite significant similarities in terms of amino acid sequences and three-dimensional structures, differences in the interaction between IgG and FcRn can be seen across various species. For instance, the strength of the IgG/FcRn interaction at pH 6.0 ranges from micromolar in human (Firan et al. 2001; Dall'Acqua et al. 2002; Dall'Acqua, Kiener, and Wu 2006) down to low micromolar affinity in mouse (Ghetie et al. 1997; Datta-Mannan et al. 2007b). It is also worth noting that human and mouse FcRn molecules exhibit different levels of promiscuity when binding to IgGs from different species (Ober et al. 2001). In fact, human FcRn does not significantly bind to rat or mouse IgG, whereas mouse FcRn binds to IgG from many different species, including human. These observations highlight the inherent engineering challenges in terms of selecting relevant animal models for pharmacokinetic and pharmacodynamic analyses. In what follows, we will attempt to summarize the major Fc engineering efforts that have aimed at down- or up-modulating the IgG/FcRn interaction, and discuss their respective impacts on IgG homeostasis, transport, or tissue distribution.

14.1.3 Down-Modulation of the IgG-FcRn Interaction

A strong correlation exists between decrease in the affinity of Fc for FcRn at acidic pH and worsening of the serum half-life of the corresponding antibodies (or antibody fragments). A first natural example relates to the γ3 isotype of human IgGs. More precisely, human IgG3s display a shorter serum half-life when compared to that of human IgG1 in mice (~107 versus ~178 hours; Kim et al. 1999). This unusual property is to be related to the approximately threefold lower binding affinity of human IgG3 for murine FcRn when compared with human IgG1 (Kim et al. 1999). Naturally occurring amino acid sequence variations provide a possible explanation for this difference. Indeed, human IgG3 molecules differ from other isotypes at position 435 where they encode an arginine compared to a histidine in human IgG1,2,4 (with the exception of the uncommon G3 $m^{s,t}$ allotype, which encodes a histidine; Matsumoto et al. 1983). The relevance of this observation is illustrated by the approximately twofold reduction in mouse FcRn binding to a human Fc(γ1)-hinge fragment containing the H435R mutation (Kim et al. 1999). It is worth noting that human IgG3s also exhibit a shorter serum half-life in human when compared with all other IgG isotypes (~7 versus ~21 days; Morell, Terry, and Waldmann 1970), a property likely related to the different ability of the various human IgG3 allotypes to interact with human FcRn (West and Bjorkman 2000 and our unpublished observations).

Extensive mutagenesis studies have provided additional support for such a correlation. In particular, Fc-hinge fragments derived from a mouse IgG1 and containing either one of the I253A, H310A, H435A, or H436A substitutions see their binding affinity to murine FcRn at pH 6.0 significantly reduced (approximately fourfold). When tested in mice, such mutated fragments exhibit an about 2.5- to 7-fold reduction in serum half-life when compared with their unmutated counterpart (Kim et al. 1994; Medesan et al. 1997). Similar results were observed when the H435A substitution was introduced into the Fc portion of a humanized IgG1. In this situation, binding of the modified antibody to murine FcRn was ablated at pH 6.0, and its serum half-life in mice reduced by a factor of nearly 10-fold (Firan et al. 2001). In a related series of experiments, when the I253A, H310A, and H435A mutations were introduced separately into the Fc-hinge fragment derived from a human IgG1, the resulting affinities to mouse FcRn at acidic pH were reduced by about 5-, 14-, and 13-fold, respectively (Kim et al. 1999). This loss translated to an about 2.5-, 3-, and 3-fold reduction, respectively, in the fragments serum half-life in mice (Kim et al. 1999). Finally, a humanized IgG1 containing the I253A mutation was described, whose binding to human FcRn was substantially reduced although not quantified (Petkova et al. 2006). In agreement with all previous data, this molecule exhibited a significantly decreased serum half-life in transgenic mice expressing human but not mouse FcRn (up to sixfold when compared with the same unmutated IgG1; Petkova et al. 2006).

The ability to generate and develop such short-lived molecules could prove important in the diagnosis field, where reduction of the patient's overall exposure to the radioactive moiety of imaging compounds is seen as beneficial. In fact, various ^{124}I-labeled forms of an anti-carcinoembryonic antigen (CEA) single chain Fv (scFv) fragment fused to a human Fc moiety engineered for decreased binding to FcRn have been described (Kenanova et al. 2005). In this situation, faster serum clearance rates were shown to correlate with quick tumor localization of the scFv-Fc and clear images in xenografted mice. Additionally, antibodies or antibody fragments specifically engineered for fast serum clearance could be desirable to reduce the overall body exposure and potential toxicity of therapeutics such as immunotoxins and radioconjugates (Sharkey and Goldenberg 2006).

14.1.4 Up-Modulation of the IgG-FcRn Interaction

Contrary to when the affinity of IgG for FcRn is decreased (see Section 14.1.3), strengthening of this interaction correlates with improvements in IgG serum half-life. Independent engineering strategies have identified various Fc substitutions resulting in increased affinity of IgG or Fc to FcRn, as shown in Figure 14.2. Although molecular modeling procedures have been described for this purpose (Kamei et al. 2005), most of the currently available body of data was generated from experimental, mutagenesis-based approaches. For instance, a full-scale alanine scanning campaign targeting solvent-exposed amino acids on a human IgG1 Fc led to the identification of over a dozen beneficial single variants, the most improved of which (N434A) exhibited an about 3.5-fold increase in its binding to human FcRn at pH 6.0 (Shields et al. 2001). When some of these beneficial mutations were

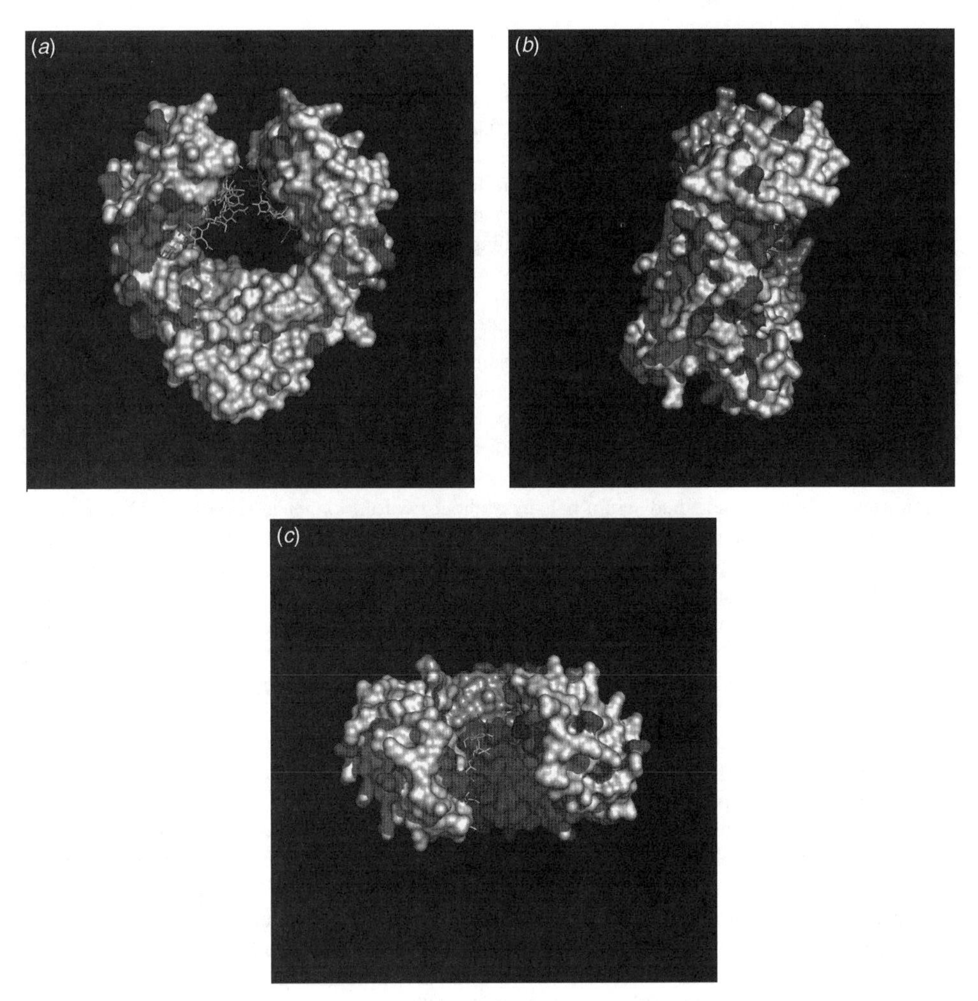

Figure 14.2 Space-filling model of an Fc fragment derived from a human IgG1, based on the x-ray structure corresponding to protein database ID number 1FC1 (Deisenhofer 1981). The Fc fragment is oriented so that its front (a), side (b), or top (c) points toward the reader. Residues in red (see color insert) correspond to the various positions where single or multiple substitutions were shown to improve the affinity of the human IgG/FcRn interaction at acidic pH. Carbohydrates are shown as sticks. (See color insert.)

combined, further increases in human IgG1 binding to human FcRn were attained; most notably, an about 12-fold affinity improvement at pH 6.0 was seen for a human IgG1 containing the T307A/E380A/N434A triple substitution (Shields et al. 2001). All variants retained good pH sensitivity, namely no or very poor binding to FcRn at or near neutral pH. Importantly, when dosed in transgenic mice expressing human FcRn, a human IgG1 containing the N434A or T307A/E380A/N434A mutations exhibited an about two- to threefold increase in its serum half-life when compared with the same unmutated antibody (Petkova et al. 2006). Similar results had previously been observed with a murine IgG1 Fc-hinge fragment randomly mutated at select positions and selected for increased binding to murine FcRn (Ghetie et al. 1997). In this set of experiments, a triple mutant (T252L/T254S/T256F) exhibited significant increases in both binding to murine FcRn at pH 6.0 (by 3.5-fold) and serum half-life in mice (up to 1.6-fold) when compared with the wild-type fragment. This variant still retained a strong pH dependency of binding to murine FcRn.

Other recent studies have aimed at discovering and characterizing beneficial Fc mutations in the context of nonhuman primates. In particular, a combination of rational design and phage display-based library screening succeeded in the identification of several humanized IgG1 Fc mutants whose binding to human and murine FcRn at pH 6.0 was improved by up to 60- and 30-fold, respectively (Dall'Acqua et al. 2002). One particular set of mutations (M252Y/S254T/T256E) was further characterized (Dall'Acqua, Kiener, and Wu 2006). In short, the binding of a M252Y/S254T/T256E-modified humanized IgG1 to both human and cynomolgus monkey FcRn was improved by ~10-fold at pH 6.0 while a strong pH dependency of binding towards FcRn of both species was maintained. Pharmacokinetic studies in cynomolgus monkeys indicated that the serum half-life of this mutated humanized IgG1 was increased by nearly fourfold when compared with its unmutated counterpart. Importantly, the authors also showed that the interaction of cynomolgus monkey FcRn with endogenous (cynomolgus) or human IgG was similar. This further established the relevance of this model when compared with human. Likewise, human IgG2 mutants with increased affinity to human FcRn have been described (Hinton et al. 2004). Among those, M428L and T250Q/M428L showed an increase in binding to human FcRn at pH 6.0 of 7- and 28-fold, respectively, and no binding to the receptor at near neutral pH. Here again, significantly improved pharmacokinetics parameters were observed in primates, as both mutants had serum half-lives about twofold longer in rhesus monkeys than the wild-type form of the same human IgG2. The T250Q/M428L double mutation was also introduced into a human IgG1 background and resulted in an about 30- and 40-fold pH-dependent increase in IgG binding to human and rhesus FcRn, respectively (Hinton et al. 2005). This correlated with a 2.5-fold extension in serum half-life in rhesus monkeys when compared with the unmutated parental molecule.

Although the stoichiometry of the IgG/FcRn interaction or the molecule's relative arrangements in the complex are still unclear (as reviewed in Ghetie and Ward 2000), the examples mentioned above seem to demonstrate a clear correlation between increasing the affinity of IgGs for FcRn and extending their serum persistence. However, additional aspects must be taken into consideration, which complicate this deceptively simple picture.

In the first place, the pH dependence that characterizes the IgG/FcRn interaction should be maintained. Indeed, it has been shown that humanized IgG1 Fc mutants that exhibit an affinity increase towards murine FcRn at pH 6.0 also generally show significantly increased binding to the same receptor at pH 7.4 (Dall'Acqua et al. 2002). In this situation, humanized IgG1 mutants whose affinities to mouse FcRn were improved by about 1.5- to 20-fold at pH 6.0 showed significantly lower serum concentrations than the wild-type antibody in mice. Similarly, a human IgG1 triple mutant (T307A/E380A/N434A) whose binding affinity towards murine FcRn was increased at both pH 6.0 and neutral pH has been described (Guraxani et al. 2006). Its serum half-life in mice was indistinguishable from the unmutated molecule. As suggested before (Dall'Acqua et al. 2002), it is possible that retention of IgG binding to FcRn around neutral pH leads to the intracellular sequestration and degradation of the corresponding IgG molecules in FcRn-containing tissues. Regardless of their own fate, the increased stickiness of certain engineered antibodies to FcRn at near neutral pH can have an indirect effect, namely, the ability to outcompete the endogenous IgGs for binding to the receptor. This interesting phenomenon can result in an increased clearance of the host's own antibodies from the serum, as shown in mice (Vaccaro et al. 2005).

Second, a certain threshold in affinity improvement at acidic pH may need to be overcome. In agreement with this notion, a triple mutant of a mouse IgG1 Fc-hinge fragment (T252A/T254S/T256A) with an about twofold increase in binding to murine FcRn at pH 6.0 and no binding at pH 7.4 failed to show improved serum half-life in mice (Ghetie et al. 1997).

Third, parameters other than the affinity of Fc binding to FcRn at pH 6.0 or neutral pH may be important. This is illustrated by the fact that the >50-fold pH-dependent increase in the binding affinity of various humanized IgG1 variants (P257I/N434H, D376V/N434H, and P257I/Q311I) to cynomolgus monkey FcRn did not significantly impact their serum half-life in the same host (Datta-Mannan et al. 2007a). That the enhanced affinities of these mutants to cynomolgus monkey FcRn were predominantly driven by an increase in the corresponding association rates (k_{on}) seems to imply that the rates of dissociation (k_{off}) may play a major role in determining the fate of the recycled antibodies. Essentially similar conclusions were achieved when a human IgG1 containing the T250Q/M428L mutations was dosed in cynomolgus monkeys (Datta-Mannan et al. 2007b). Here again, no significant pharmacokinetic changes were seen for the T250Q/M428L mutant, despite its 40-fold pH-dependent, k_{on}-driven affinity increase to cynomolgus FcRn. Conversely, the same mutated human IgG1 exhibited a serum clearance about 2-fold slower than the unmodified antibody in mice, in accordance with its 500-fold pH-dependent, k_{off}-driven affinity increase to mouse FcRn.

In conclusion, for parts 14.1.3 and 14.1.4, the authors would like to note that our review of the literature indicates an (understandable) lack of normalization in terms of characterizing the previously described engineered IgGs. This obviously complicates the interpretation of the corresponding data and precludes accurate comparison between studies. Additional information would likely be gained if similar IgG/FcRn binding assays were used, the kinetic rates of the various IgG/FcRn interactions determined, the binding of engineered IgGs to components of the host(s) serum analyzed (to rule out any potential carrier effect), and the competition for FcRn binding between the engineered and host's endogenous IgGs assessed.

14.1.5 Future Prospects: Beyond the Control of IgG Homeostasis

How an increase or decrease in IgG binding affinity to FcRn would affect (if at all) the FcRn-related transport of IgGs or their tissue distribution *in vivo* remains largely undetermined. However, three studies have started to shed light on this process. In the first, a human IgG1 engineered for an about 15-fold pH-dependent increase in binding to human FcRn at pH 6.0 was transported significantly more efficiently than its wild-type counterpart across human placenta *ex vivo* (by about 2-fold; Vaccaro et al. 2006). In a second study, a humanized IgG1 engineered for an about 10-fold pH-dependent increase in binding to cynomolgus monkey FcRn was shown to exhibit an $\sim$4-fold increase in lung bioavailability in cynomolgus monkeys relative to its unmutated counterpart (Dall'Acqua, Kiener, and Wu 2006). In the third study, the *ex vivo* transport across human placenta of a human IgG1 engineered for lack of binding to human FcRn was reduced to background levels (Firan et al. 2001). In addition, FcRn also seems to play a major role in IgG transport from the lumen side of the lungs to the systemic circulation in both mouse and cynomolgus monkeys (Spiekermann et al. 2002; Bitonti et al. 2004), as well as in the distribution of IgGs across the epithelial barrier of human intestine (Dickinson et al. 1999). Thus, conceivably, up- or down-modulating the interaction of IgG with FcRn may provide an opportunity to optimize the distribution or transcytosis of therapeutic antibodies to or across specific tissues such as placenta, lungs, and intestine. In particular, the modulation of IgG placental transport could have implications for neonatal medicine when therapeutic antibodies need to be efficiently transferred from the mother to the fetus. It is also possible that increased IgG transcytosis from the lungs or intestine to the serum can be achieved, a direct consequence of which would be the development of novel and more convenient IgG routes of administration (such as oral or inhalation).

14.1.6 Summary

The rationale behind engineering the interaction of IgG with the neonatal Fc receptor has now been well established. Significant increases or decreases in antibody serum half-life have been achieved in rodents and primates through the identification of select Fc mutations. Studies are ongoing to better understand the detailed relationship between the molecular mechanisms of the IgG/FcRn interaction and the corresponding IgG pharmacokinetics/pharmacodynamics parameters. Despite this fact, the successful demonstration that human IgGs can be rationally engineered for improved pharmacokinetics properties in both cynomolgus (Dall'Acqua, Kiener, and Wu 2006) and rhesus (Hinton et al. 2004, 2005) monkeys bodes well for their future therapeutic use in human. Among the many benefits that could be derived from immunoglobulins exhibiting very long serum persistence in the human population, the possibility to decrease their administration frequency or dosing requirements while maintaining (or improving) their efficacy comes first. Economic benefits, in the form of a decrease in the costs of the drugs, could follow. However, because of their prolonged retention in serum and increase in overall exposure, the potential toxicity of such enhanced molecules will need to be closely monitored on a case by case basis. Conversely, the ability to reduce the serum persistence of antibodies or antibody fragments will likely prove useful in situations where an acute effect is desired (such as for imaging purposes). The possibility to control IgG transport or tissue distribution may enhance the efficacy and improve the dosing protocols of therapeutic antibodies. Finally, novel opportunities to reduce IgG levels in antibody-mediated diseases may be created in light of the observed outcompetition of endogenous IgGs by antibodies engineered for high-affinity, pH-*independent* binding to FcRn (Vaccaro et al. 2005). Various therapeutic areas are likely to benefit from this new field of protein engineering, including oncology, autoimmunity, inflammation, respiratory diseases, and neonatal medicine.

14.2 HOMEOSTATIC REGULATION OF ALBUMIN

14.2.1 FcRn-Mediated Recycling of Albumin

Albumin ranks among the most abundant plasma proteins. With a molecular weight of about 67 kDa, it exhibits circulation levels of ~50 mg/mL and has a serum half-life of about three weeks in human (Peters 1985, 1996). Albumin contributes to the maintenance of osmotic pressure and plasma pH, and is an important natural circulating carrier for various substances such as hormones, vitamins, fatty acids, and ions. The process by which it is synthesized by the liver has been well documented over the past decades; however, its degradation mechanism has remained unknown for most of that time. About 40 years ago, Schultze and Heremans (1966) suggested that Brambell's hypothesis pertaining to the protection of IgG from degradation via binding to an Fc receptor (Brambell, Hemmings, and Morris 1964) could be expanded to the homeostatic regulation of albumin. However, no further work to confirm this hypothesis was carried out. Recently, however, Brambell's hypothesis for IgG recycling has been extensively studied and validated (as reviewed in Section 14.1). This may have triggered interest in deciphering the albumin recycling molecular mechanism, the nature of which will be reviewed in what follows.

Several similarities can be found when comparing albumin with IgG. For instance, they both exhibit about a three week serum half-life. In addition, both are the two most abundant plasma proteins. It has also been shown that a direct relationship exists between the fractional catabolic rate of albumin and its serum concentration (Freeman and Gordon 1965). The latter property is characteristic of IgG, but not of other serum proteins such as IgA, IgM, transferrin, or fibrinogen (Schultze and Heremans 1966; Waldmann and Strober 1969). Thus, since the mechanism by which IgGs are recycled has been shown to be directly mediated by FcRn, one might speculate that the same receptor protects albumin from degradation [although Schultze and Heremans (1966) had initially proposed that receptor(s) different from Brambell's Fc receptor may be involved]. The first direct evidence linking albumin to FcRn was a result of serendipity (Chaudhury et al. 2003; Anderson et al. 2006). While the authors were attempting to purify recombinant human FcRn from culture medium using an IgG affinity column, they noted that bovine albumin co-eluted with FcRn. This led to further investigation, which eventually demonstrated that albumin, similarly to IgG, binds to FcRn in a pH-dependent manner (Chaudhury et al. 2003). The same authors also showed that albumin serum half-life was shortened in FcRn-deficient mice. Furthermore, the same hosts exhibited substantially lower plasma albumin levels when compared with wild-type mice. Taken together, these results clearly confirmed the receptor-mediated hypothesis for protecting albumin from degradation. They also united the mechanisms controlling IgG and albumin catabolism via their binding to FcRn in a pH-dependent fashion.

14.2.2 Characterization of the Albumin-FcRn Interaction

Recent data have shown that FcRn is capable of binding IgG and albumin simultaneously to form a trimolecular complex (Chaudhury et al. 2003). In addition, both ligands bind noncooperatively to distinct sites on FcRn (Chaudhury et al. 2006). The binding of albumin to FcRn relies on a "fast on/fast off" kinetic interaction (Chaudhury et al. 2006), similar to that seen for the IgG-FcRn complex (Datta-Mannan et al. 2007b). More precisely, under acidic conditions, the association and dissociation rates of both ligands to FcRn were estimated at $\sim 10^4$ M^{-1} s^{-1} and $\sim 10^{-2}$ to 10^{-3} M s^{-1}, respectively. At pH 6, the dissociation constant of human FcRn to immobilized human serum albumin (HSA) is 0.9 μM, a value comparable to that of immobilized human IgG1 (1.3 to 2.5 μM; Dall'Acqua et al. 2002; Dall'Acqua, Kiener, and Wu 2006). Despite such similarities, differences between the IgG-FcRn and albumin-FcRn interactions can be found. For instance, the albumin-FcRn interaction is characterized by hydrophobic interactions, which result in a large positive entropy change counteracted by an unfavorable enthalpy change (Chaudhury et al. 2006). In contrast, the IgG-FcRn interaction is driven by a favorable enthalpy change (Huber et al. 1993). Structurally, both ligands are also quite different since albumin contains about 67 percent α-helix and no β-sheet (He and Carter 1992), whereas IgG Fc fragments contain four β-barrel domains (Harris et al. 1992).

Efforts to characterize the albumin region binding to FcRn have been carried out. Initial results have identified HSA domain III as a major contributor (Chaudhury et al. 2006). The characteristic pH dependence of the albumin-FcRn interaction has been attributed to a conserved histidine at position 166 (H166) in FcRn α-chain using sequence comparison among 11 different species as well as mutagenesis studies (Andersen, Dee Qian, and Sandlie 2006). More precisely, H166 is located near a pocket of the peptide cleft directly opposite to the main IgG binding site. Finally, an attempt to modify HSA to alter its pharmacokinetic properties showed that variants with a +2 change in net charge exhibited a slightly prolonged half-life in mice, whereas those engineered for increased hydrophobicity had an opposite behavior (Iwao et al. 2007). However, in general, no strong correlation could be found between the nature of mutations and their pharmacokinetic consequences. This might be attributed to various factors, such as differential alterations of (1) the stability and resistance of the HSA variants to plasma enzymes, (2) the FcRn-mediated albumin recycling mechanism, and (3) the liver/kidney HSA uptake and clearance efficiency.

14.2.3 Future Prospects

Albumin has a remarkably long serum half-life. Along with its extensive *in vivo* biodistribution and lack of immunological or enzymatic activities, it serves as a very attractive carrier for various therapeutic drugs (Stehle et al. 1999; Chuang, Kragh-Hansen, and Otagiri 2002), and has been used as a fusion partner with interferon-α (Subramanian et al. 2007), hirudin (Syed et al. 1997), and CD4 (Yeh et al. 1992). HSA is also used clinically to treat severe hypoalbuminemia or traumatic shock (Peters 1996). It would therefore be highly desirable to improve its serum half-life. However, very little effort has been invested in this area. In contrast, there has been great progress in the area of FcRn-mediated IgG serum half-life extension (see Section 14.1). By engineering the Fc portion of IgGs for increased pH-dependent binding to FcRn, one could extend the half-lives of antibodies or Fc fusion proteins in human. This was shown in multiple animal models, which included nonhuman primates. One of the most advanced set of mutations (M252Y/S254T/T256E) is currently being investigated in a human clinical trial for its potential in increasing the circulating half-life of motavizumab (Dall'Acqua, Kiener, and Wu 2006; Wu et al. 2008). Perhaps, if promising clinical results are obtained, more interest in trying to improve the pharmacokinetic properties of albumin will emerge. Conceivably, the experience gained from engineering IgG Fc may help guide the engineering of albumin. Such studies will hopefully lead to a detailed understanding of both molecule's recycling and transcytotic pathways.

ACKNOWLEDGMENTS

We thank Vaheh Oganesyan for his help in preparing the illustrations in this chapter.

REFERENCES

Ahouse, J.J., C.L. Hagerman, P. Mittal, D.J. Gilbert, N.G. Copeland, N.A. Jenkins, and N.E. Simister. 1993. Mouse MHC class I-like Fc receptor encoded outside the MHC. *J. Immunol.* 151(11):6076–6088.

Andersen, J.T., J. Dee Qian, and I. Sandlie. 2006. The conserved histidine 166 residue of the human neonatal Fc receptor heavy chain is critical for the pH-dependent binding to albumin. *Eur. J. Immunol.* 36(11):3044–3051.

Anderson, C.L., C. Chaudhury, J. Kim, C.L. Bronson, M.A. Wani, and S. Mohanty. 2006. Perspective: FcRn transports albumin—relevance to immunology and medicine. *Trends Immunol.* 27(7):343–348.

Bitonti, A.J., J.A. Dumont, S.C. Low, R.T. Peters, K.E. Kropp, V.J. Palombella, J.M. Stattel, Y. Lu, C.A. Tan, J.J. Song, A.M. Garcia, N.E. Simister, G.M. Spiekermann, W.I. Lencer, and R.S. Blumberg. 2004. Pulmonary delivery of an erythropoietin Fc fusion protein in non-human primates through an immunoglobulin transport pathway. *Proc. Natl. Acad. Sci. USA* 101(26):9763–9768.

Borvak, J., J. Richardson, C. Medesan, F. Antohe, C. Radu, M. Simionescu, V. Ghetie, and E.S. Ward. 1998. Functional expression of the MHC class I-related receptor, FcRn, in endothelial cells of mice. *Int. Immunol.* 10(9):1289–1298.

Brambell, F.W. 1966. The transmission of immunity from mother to young and the catabolism of immunoglobulins. *Lancet* 2(7473):1087–1093.

Brambell, F.W., W.A. Hemmings, and I.G. Morris. 1964. A theoretical model of gamma-globulin catabolism. *Nature* 203:1352–1354.

Burmeister, W.P., L.N. Gastinel, N.E. Simister, M.L. Blum, and P.J. Bjorkman. 1994. Crystal structure at 2.2 Å resolution of the MHC-related neonatal Fc receptor. *Nature* 372(6504):336–343.

Burmeister, W.P., A.H. Huber, and P.J. Bjorkman. 1994. Crystal structure of the complex of rat neonatal Fc receptor with Fc. *Nature* 372(6504):379–383.

Chaudhury, C., C.L. Brooks, D.C. Carter, J.M. Robinson, and C.L. Anderson. 2006. Albumin binding to FcRn: Distinct from the FcRn-IgG interaction. *Biochemistry* 45(15):4983–4990.

Chaudhury, C., S. Mehnaz, J.M. Robinson, W.L. Hayton, D.K. Pearl, D.C. Roopenian, and C.L. Anderson. 2003. The major histocompatibility complex-related Fc receptor for IgG (FcRn) binds albumin and prolongs its lifespan. *J. Exp. Med.* 197(3):315–322.

Chuang, V.T., U. Kragh-Hansen, and M. Otagiri. 2002. Pharmaceutical strategies utilizing recombinant human serum albumin. *Pharm. Res.* 19(5):569–577.

Cianga, P., C. Medesan, J.A. Richardson, V. Ghetie, and E.S. Ward. 1999. Identification and function of neonatal Fc receptor in mammary gland of lactating mice. *Eur. J. Immunol.* 29(8):2515–2523.

Dall'Acqua, W.F., P.A. Kiener, and H. Wu. 2006. Properties of human IgG1s engineered for enhanced binding to the neonatal Fc receptor (FcRn). *J. Biol. Chem.* 281(33):23514–23524.

Dall'Acqua, W.F., R.M. Woods, E.S. Ward, S.R. Palaszynski, N.K. Patel, Y.A. Brewah, H. Wu, P.A. Kiener, and S. Langermann. 2002. Increasing the affinity of a human IgG1 for the neonatal Fc receptor: Biological consequences. *J. Immunol.* 169(9):5171–5180.

Datta-Mannan, A., D.R. Witcher, Y. Tang, J. Watkins, W. Jiang, and V.J. Wroblewski. 2007a. Humanized IgG1 variants with differential binding properties to the neonatal Fc receptor: Relationship to pharmacokinetics in mice and primates. *Drug Metab. Dispos.* 35(1):86–94.

Datta-Mannan, A., D.R. Witcher, Y. Tang, J. Watkins, and V.J. Wroblewski. 2007b. Monoclonal antibody clearance: Impact of modulating the interaction of IgG with the neonatal Fc receptor. *J. Biol. Chem.* 282(3):1709–1717.

Deisenhofer, J. 1981. Crystallographic refinement and atomic models of a human Fc fragment and its complex with fragment B of protein A from *Staphylococcus aureus* at 2.9- and 2.8-Å resolution. *Biochemistry* 20(9):2361–2370.

Dickinson, B.L., K. Badizadegan, Z. Wu, J.C. Ahouse, X. Zhu, N.E. Simister, R.S. Blumberg, and W.I. Lencer. 1999. Bidirectional FcRn-dependent IgG transport in a polarized human intestinal epithelial cell line. *J. Clin. Invest.* 104(7):903–911.

Firan, M., R. Bawdon, C. Radu, R.J. Ober, D. Eaken, F. Antohe, V. Ghetie, and E.S. Ward. 2001. The MHC class I-related receptor, FcRn, plays an essential role in the maternofetal transfer of gamma-globulin in humans. *Int. Immunol.* 13(8):993–1002.

Freeman, T., and A.H. Gordon. 1965. Albumin catabolism in hypoproteinaemic states studies with ^{131}I-albumin. *Bibl. Haematol.* 23:1108–1115.

Ghetie, V., J.G. Hubbard, J.K. Kim, M.F. Tsen, Y. Lee, and E.S. Ward. 1996. Abnormally short serum half lives of IgG in beta2-microglobulin-deficient mice. *Eur. J. Immunol.* 26(3):690–696.

Ghetie, V., S. Popov, J. Borvak, C. Radu, D. Matesoi, C. Medesan, R.J. Ober, and E.S. Ward. 1997. Increasing the serum persistence of an IgG fragment by random mutagenesis. *Nature Biotechnol.* 15(7):637–640.

Ghetie, V., and S. Ward. 2000. Multiple roles for the major histocompatibility complex class I-related receptor FcRn. *Annu. Rev. Immunol.* 18:739–766.

Gurbaxani, B., L.L. Dela Cruz, K. Chintalacharuvu, and S.L. Morrison. 2006. Analysis of a family of antibodies with different half-lives in mice fails to find a correlation between affinity for FcRn and serum half-life. *Mol. Immunol.* 43(9):1462–1473.

Harris, L.J., S.B. Larson, K.W. Hasel, J. Day, A. Greenwood, and A. McPherson. 1992. The three-dimensional structure of an intact monoclonal antibody for canine lymphoma. *Nature* 360(6402):369–372.

Haymann, J.P., J.P. Levraud, S. Bouet, V. Kappes, J. Hagège, G. Nguyen, Y. Xu, E. Rondeau, and J.D. Sraer. 2000. Characterization and localization of the neonatal Fc receptor in adult human kidney. *J. Am. Soc. Nephrol.* 11(4):632–639.

He, X.M., and D.C. Carter. 1992. Atomic structure and chemistry of human serum albumin. *Nature.* 358(6383):209–215.

Hinton, P.R., M.G. Johlfs, J.M. Xiong, K. Hanestad, K.C. Ong, C. Bullock, S. Keller, M.T. Tang, J.Y. Tso, M. Vásquez, and N. Tsurushita. 2004. Engineered human IgG antibodies with longer serum half-lives in primates. *J. Biol. Chem.* 279(8):6213–6216.

Hinton, P.R., J.M. Xiong, M.G. Johlfs, M.T. Tang, S. Keller, and N. Tsurushita. 2005. An engineered human IgG1 antibody with longer serum half-life. *J. Immunol.* 176(1):346–356.

Huber, A.H., R.F. Kelley, L.N. Gastinel, and P.J. Bjorkman. 1993. Crystallization and stoichiometry of binding of a complex between a rat intestinal Fc receptor and Fc. *J. Mol. Biol.* 230(3):1077–1083.

Iwao, Y., M. Hiraike, U. Kragh-Hansen, K. Mera, T. Noguchi, M. Anraku, K. Kawai, T. Maruyama, and M. Otagiri. 2007. Changes of net charge and alpha-helical content affect the pharmacokinetic properties of human serum albumin. *Biochim. Biophys. Acta.* 1774(12):1582–1590.

Junghans, R.P., and C.L. Anderson. 1996. The protection receptor for IgG catabolism is the beta2-microglobulin-containing neonatal intestinal transport receptor. *Proc. Natl. Acad. Sci. USA* 93(11):5512–5516.

Kabat, E.A., T.T. Wu, H.M. Perry, K.S. Gottesman, and C. Foeller. 1991. *Sequences of Proteins of Immunological Interest.* Washington, D.C.: U.S. Public Health Service, National Institutes of Health.

Kamei, D.T., B.J. Lao, M.S. Ricci, R. Deshpande, H. Xu, B. Tidor, and D.A. Lauffenburger. 2005. Quantitative methods for developing Fc mutants with extended half-lives. *Biotechnol. Bioeng.* 92(6):748–760.

Kenanova, V., T. Olafsen, D.M. Crow, G. Sundaresan, M. Subbarayan, N.H. Carter, D.N. Ikle, P.J. Yazaki, A.F. Chatziioannou, S.S. Gambhir, L.E. Williams, J.E. Shively, D. Colcher, A.A. Raubitschek, and A.M. Wu. 2005. Tailoring the pharmacokinetics and positron emission tomography imaging properties of anti-carcinoembryonic antigen single-chain Fv-Fc antibody fragments. *Cancer Res.* 65(2):622–631.

Kim, J.K., M. Firan, C.G. Radu, C.H. Kim, V. Ghetie, and E.S. Ward. 1999. Mapping the site on human IgG for binding of the MHC class I-related receptor, FcRn. *Eur. J. Immunol.* 29(9):2819–2825.

Kim, J.K., M.F. Tsen, V. Ghetie, and E.S. Ward. 1994. Identifying amino acid residues that influence plasma clearance of murine IgG1 fragments by site-directed mutagenesis. *Eur. J. Immunol.* 24(3):542–548.

Kristoffersen, E.K. 1996. Human placental Fc gamma-binding proteins in the maternofetal transfer of IgG. *APMIS Suppl.* 64:5–36.

Kristoffersen, E.K., and R. Matre. 1996. Co-localization of the neonatal Fc gamma receptor and IgG in human placental term syncytiotrophoblasts. *Eur. J. Immunol.* 26(7):1668–1671.

Leach, J.L., D.D. Sedmak, J.M. Osborne, B. Rahill, M.D. Lairmore, and C.L. Anderson. 1996. Isolation from human placenta of the IgG transporter, FcRn, and localization to the syncytiotrophoblast: Implications for maternal-fetal antibody transport. *J. Immunol.* 157(8):3317–3322.

Matsumoto, H., S. Ito, T. Miyazaki, and T. Ohta. 1983. Structural studies of a human gamma 3 myeloma protein (Jir) bearing the allotypic marker Gm(st). *J. Immunol.* 131(4):1865–1870.

Medesan, C., D. Matesoi, C. Radu, V. Ghetie, and E.S. Ward. 1997. Delineation of the amino acid residues involved in transcytosis and catabolism of mouse IgG1. *J. Immunol.* 158(5):2211–2217.

Medesan, C., C. Radu, J.K. Kim, V. Ghetie, and E.S. Ward. 1996. Localization of the site of the IgG that regulates maternofetal transmission in mice. *Eur. J. Immunol.* 26(10):2533–2536.

Morell, A., W.D. Terry, and T.A. Waldmann. 1970. Metabolic properties of IgG subclasses in man. *J. Clin. Invest.* 49(4):673–680.

Ober, R.J., C.G. Radu, V. Ghetie, and E.S. Ward. 2001. Differences in promiscuity for antibody-FcRn interactions across species: Implications for therapeutic antibodies. *Int. Immunol.* 13(12):1551–1559.

Peters, T., Jr. 1985. Serum albumin. *Adv. Protein Chem.* 37:161–245.

Peters, T., Jr. 1996. *All About Albumin: Biochemistry, Genetics, and Medical Applications*. New York: Academic Press.

Petkova, S.B., S. Akilesh, T.J. Sproule, G.J. Christianson, H. Al Khabbaz, A.C. Brown, L.G. Presta, Y.G. Meng, and D.C. Roopenian. 2006. Enhanced half-life of genetically engineered human IgG1 antibodies in a humanized FcRn mouse model: Potential application in humorally mediated autoimmune disease. *Int. Immunol.* 18(12):1759–1769.

Raghavan, M., V.R. Bonagura, S.L. Morrison, and P.J. Bjorkman. 1995. Analysis of the pH dependence of the neonatal Fc receptor/immunoglobulin G interaction using antibody and receptor variants. *Biochemistry.* 34(45):14649–14657.

Rodewald, R. 1976. pH-dependent binding of immunoglobulins to intestinal cells of the neonatal rat. *J. Cell. Biol.* 71(2):666–669.

Roopenian, D.C., and S. Akilesh. 2007. FcRn: The neonatal Fc receptor comes of age. *Nature Rev. Immunol.* 7(9):715–725.

Schultze, H.E., and J.F. Heremans. 1966. Molecular biology of human proteins: With special reference to plasma proteins. Vol. 1. In Nature and Metabolism of Extracellular Proteins. New York: Elsevier.

Sharkey, R.M., and D.M. Goldenberg. 2006. Targeted therapy of cancer: New prospects for antibodies and immunoconjugates. *CA Cancer J. Clin.* 56(4):226–243.

Shields, R.L., A.K. Namenuk, K. Hong, Y.G. Meng, J. Rae, J. Briggs, D. Xie, J. Lai, A. Stadlen, B. Li, J.A. Fox, and L.G. Presta. 2001. High resolution mapping of the binding site on human IgG1 for Fc gamma RI, Fc gamma RII, Fc gamma RIII, and FcRn and design of IgG1 variants with improved binding to the Fc gamma R. *J. Biol. Chem.* 276(9):6591–6604.

Simister, N.E., and K.E. Mostov. 1989. An Fc receptor structurally related to MHC class I antigens. *Nature* 337(6203):184–187.

Spiekermann, G.M., P.W. Finn, E.S. Ward, J. Dumont, B.L. Dickinson, R.S. Blumberg, and W.I. Lencer. 2002. Receptor-mediated immunoglobulin G transport across mucosal barriers in adult life: Functional expression of FcRn in the mammalian lung. *J. Exp. Med.* 196(3):303–310.

Stehle, G., A. Wunder, H.H. Schrenk, G. Hartung, D.L. Heene, and H. Sinn. 1999. Albumin-based drug carriers: Comparison between serum albumins of different species on pharmacokinetics and tumor uptake of the conjugate. *Anticancer Drugs* 10(8):785–790.

Story, C.M., J.E. Mikulska, and N.E. Simister. 1994. A major histocompatibility complex class 1-like Fc receptor cloned from human placenta: Possible role in transfer of immunoglobulin G from mother to fetus. *J. Exp. Med.* 180(6):2377–2381.

Subramanian, G.M., M. Fiscella, A. Lamousé-Smith, S. Zeuzem, and J.G. McHutchison. 2007. Albinterferon alpha-2b: A genetic fusion protein for the treatment of chronic hepatitis C. *Nature Biotechnol.* 25(12):1411–1419.

Syed, S., P.D. Schuyler, M. Kulczycky, and W.P. Sheffield. 1997. Potent antithrombin activity and delayed clearance from the circulation characterize recombinant hirudin genetically fused to albumin. *Blood* 89(9):3243–3252.

Vaccaro, C., R. Bawdon, S. Wanjie, R.J. Ober, and E.S. Ward. 2006. Divergent activities of an engineered antibody in murine and human systems have implications for therapeutic antibodies. *Proc. Natl. Acad. Sci. USA* 103(49):18709–18714.

Vaccaro, C., J. Zhou, R.J. Ober, and E.S. Ward. 2005. Engineering the Fc region of immunoglobulin G to modulate in vivo antibody levels. *Nature Biotechnol.* 23(10):1283–1288.

Vaughn, D.E., and P.J. Bjorkman. 1998. Structural basis of pH-dependent antibody binding by the neonatal Fc receptor. *Structure* 6(1):63–73.

Waldmann, T.A., and W. Strober. 1969. Metabolism of immunoglobulins. *Prog. Allergy* 13:1–110.

Wallace, K.H., and A.R. Rees. 1980. Studies on the immunoglobulin-G Fc-fragment receptor from neonatal rat small intestine. *Biochem. J.* 188(1):9–16.

West, A.P., Jr., and P.J. Bjorkman. 2000. Crystal structure and immunoglobulin G binding properties of the human major histocompatibility complex-related Fc receptor. 39(32):9698–9708.

Wu, H., D.S. Pfarr, G.A. Losonsky, and P.A. Kiener. 2008. Immunoprophylaxis of RSV infection: Advancing from RSV-IGIV to palivizumab and motavizumab. *Curr. Top. Microbiol. Immunol.* 317:103–123.

Yeh, P., D. Landais, M. Lemaître, I. Maury, J.Y. Crenne, J. Becquart, A. Murry-Brelier, F. Boucher, G. Montay, R. Fleer, P.H. Hirel, J.F. Mayaux, and D. Klatzmann. 1992. Design of yeast-secreted albumin derivatives for human therapy: Biological and antiviral properties of a serum albumin-CD4 genetic conjugate. *Proc. Natl. Acad. Sci. USA* 89(5):1904–1908.

CHAPTER 15

Engineering the Antibody Fc Region for Optimal Effector Function

GREG A. LAZAR and JOHN R. DESJARLAIS

ABSTRACT

The Fc region mediates a spectrum of powerful effector functions that monoclonal antibodies can use against tumors and infectious pathogens. A growing set of modifications provide antibody drug developers with the capacity to control Fc interactions with Fc gamma receptors (FcγRs) and complement in order to enhance effector functions and remove unwanted Fc-mediated effects. We discuss the motivations behind these developments, the current state of Fc engineering capabilities, and their *in vivo* validation in animal models and clinical trials.

Therapeutic Monoclonal Antibodies: From Bench to Clinic. Edited by Zhiqiang An

15.1 INTRODUCTION

As the bridge between targeted antigen and the body's immune system, the Fc region is central to the versatile array of weapons that monoclonal antibodies (mAbs) can use against tumors and infectious pathogens. Whereas the approach to the Fc region throughout most of the history of antibody drug development has been to keep it fixed, recently there has been both a change in mindset about its modification and a revolution in capabilities to do so. This shift has been brought about by a more mature understanding of the detailed role of immune receptors in antibody therapy, and the development of modifications for controlling their interactions with the antibody. In this chapter we discuss these two aspects of Fc engineering in the context of enhancing Fc gamma receptor (FcγR)-mediated effector functions, improving complement-mediated effector functions, and removal of unwanted Fc-mediated effects.

15.2 ANATOMY OF THE Fc REGION

Fc is a historical name derived from the crystallizable fragment obtained by cleavage with the broad specificity protease papain, and as a result there is no strict sequence definition. The region is generally considered to include the C-terminal portion of the hinge, often referred to as the lower hinge, and the CH2 and CH3 domains, extending approximately from one of the hinge cysteines at positions 226 or 230 to the C-terminus. Positions are most commonly numbered according to a scheme called the EU index or EU numbering system, which is based on the sequential numbering of the first human IgG1 sequenced (the EU antibody; Edelman et al. 1969). Because the most common reference for this convention is the Kabat sequence manual (Kabat et al. 1991), the EU index is sometimes erroneously used synonymously with the Kabat index. The EU index does not provide insertions and deletions, and thus in some cases comparisons of IgG positions across IgG subclass and species can be unclear, particularly in the hinge region. Nonetheless, the convention has sufficed at enabling straightforward communication among the numerous Fc structure/function studies.

The Fc region mediates binding of the antibody to all endogenous receptors other than target antigen. The natural human receptors with which Fc interacts can be divided into three groups: FcγRs, complement protein C1q, and the neonatal Fc receptor FcRn (Fig. 15.1). The family of FcγRs all bind to essentially the same site on Fc, specifically the lower hinge and proximal CH2 region (Sondermann, Kaiser, and Jacob 2001). C1q binds at a region that is separate from but overlapping with the FcγR-binding site. FcRn binds Fc in the region between the CH2 and CH3 domains, a site that also mediates binding to microbial proteins A and G that are important purification reagents for large-scale manufacturing. The interaction of Fc with FcRn is the subject of the accompanying chapter by Dall'Acqua and Wu (Chapter 14) and is therefore not covered here. The collection of endogenous Fc-binding partners is referred to most commonly as Fc receptors or Fc ligands, although often the term Fc receptor is used to describe only the FcγR family.

The binding stoichiometries of Fc to each Fc ligand are well established. Whereas C1q (Michaelsen et al. 2006) and FcRn (West and Bjorkman 2000) have a distinct binding site on each Fc monomer and thus are able to bind as dimers, the FcγRs bind 1 : 1 asymmetrically to the Fc homodimer (Sondermann et al. 2000; Radaev et al. 2001). Although the majority of data indicate that FcγRs and complement bind to the hinge and proximal CH2 region, some mutagenesis studies have shown that distal residues in CH2, CH3, and even the Fab region can impact binding. These results suggest quaternary interactions across the full-length antibody that are not well understood, and highlight the complexity of the molecule. Indeed the only crystal structure of a full-length antibody with a fully intact hinge indicates long-range interactions between the various regions of the molecule (Saphire et al. 2002). Whether these particular interactions or other long-range interactions exist in solution, and contribute to antibody function, remains to be determined.

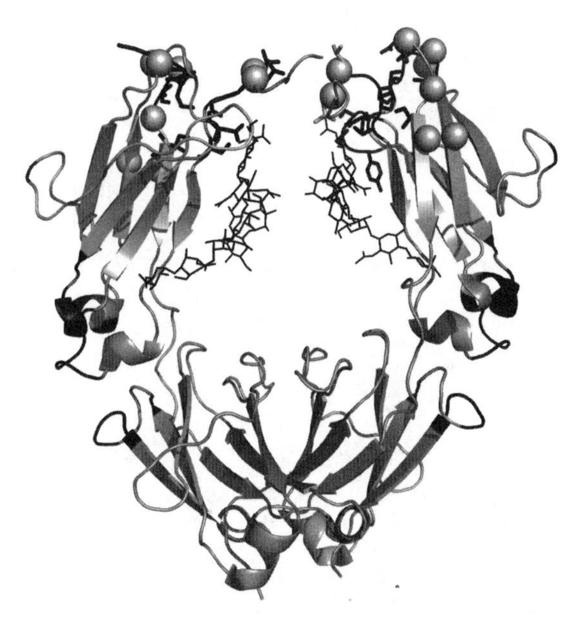

Figure 15.1 Anatomy of the Fc region. The structure is taken from pdb accession 1E4K (Sondermann et al. 2000) minus the bound FcγR, which provides the most structurally defined Fc hinge region. Thin black sticks represent the carbohydrate attached at N297. Thick black sticks represent the binding site for FcγRs based on visual inspection of the 1E4K and 1IIS pdb structures (Sondermann et al. 2000; Radaev et al. 2001), and include positions 234, 235, 236, 237, 239, 265, 326, 327, 328, 329, 330, and 332 on the left (A) chain, and positions 234, 235, 236, 237, 238, 239, 265, 267, 268, 269, 270, 296, 297, 298, 299, 325, 327, and 332 on the right (B) chain. Gray spheres represent the Cα carbons of the residues that have been shown to affect complement binding and/or CDC activity when mutated, including 234, 235, 270, 322, 326, 329, 331, and 333. Black ribbon indicates positions that contact FcRn, determined by visual inspection of a model of the human Fc/FcRn complex generated from the available rat complex structure (Martin et al. 2001). FcRn contact residues include 252, 253, 254, 255, 256, 286, 307, 309, 310, 311, 312, 314, 428, 433, 434, 435, and 436.

15.3 ENGINEERING Fc FOR IMPROVED AFFINITY TO FcγRs

15.3.1 Background on FcγR Biology

The family of human FcγRs consists of six known members in three subgroups, including FcγRI (CD64), FcγRIIa,b,c (CD32a,b,c), and FcγRIIIa,b (CD16a,b). The differences between the receptors in expression, signaling, and affinities for the IgG isotypes (Fig. 15.2) make this a versatile and highly regulated biological system. Four of the receptors are activating due to their possession of a cytoplasmic immunoreceptor tyrosine-based activation motif (ITAM), which is either genetically encoded (FcγRIIa,c) or gained by association with a common ITAM γ-chain (FcγRI and FcγRIIIa). In contrast, FcγRIIb possesses an inhibitory motif (ITIM) in its cytoplasmic domain, eliciting negative intracellular signals that downregulate effector functions. FcγRIIIb does not signal because it is linked to the membrane with a glycosyl phosphatidyl inositol (GPI) anchor. The generally accepted mechanism for ITAM/ITIM signaling is based on receptor clustering induced by binding to immune complexes, a trigger that does not occur for monomeric IgG due to the low affinity (10^{-5} to 10^{-7}) of the FcγRII and FcγRIII receptors. FcγRI binds with high affinity (10^{-10} M) to monomeric IgG, and as a result it is poor at distinguishing between unbound IgG and immune-complexed antigen.

A variety of allelic forms of these receptors exist that impact their interaction with IgG or signaling properties (Fig. 15.2; van Sorge, van der Pol, and van de Winkel 2003). Because these polymorphisms have functional effects on immune response, a large number of studies have characterized their

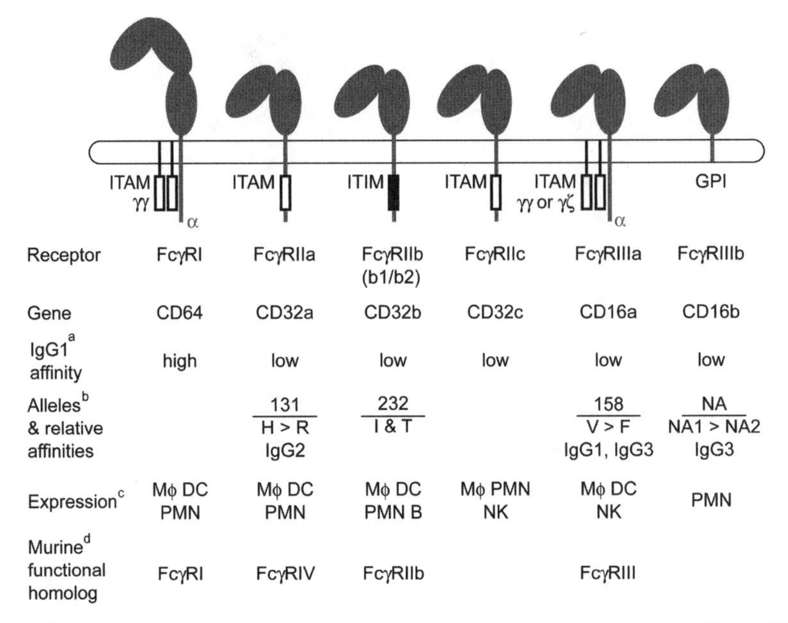

Figure 15.2 The human FcγR family. Whereas the FcγRIIs encode activating ITAMs or inhibitory ITIM genetically, FcγRs I and IIIa associate intracellularly with homodimeric γγ (FcγRI and FcγRIIIa) or heterodimeric γζ (FcγRIIIa). FcγRIIIb is GPI-linked and does not signal. FcγRIIb is expressed as one of two spliceforms, referred to as FcγRIIb1 and FcγRIIb2.

[a]FcγRI binds IgG1 with ~0.1 nM affinity. The FcγRII and FcγRIII receptors bind in the 10 to 0.1 μM affinity range.

[b]H131 FcγRIIa binds IgG2 more tightly than the R131 form; V158 FcγRIIIa binds IgG1 and IgG3 more tightly than the F158 form; the NA1 allele of FcγRIIIb binds IgG3 more tightly, and mediates neutrophil phagocytosis by IgG1 and IgG3 better than the NA2 form. Position 232 of FcγRIIb is in the transmembrane domain, and as a result the I/T polymorphism affects signaling but not IgG affinity.

[c]Mϕ = macrophage, DC = dendritic cell, PMN = neutrophil, and NK = NK cell.

[d]Murine functional homologs are based on similarity of expression patterns, not on sequence homology.

relationship with susceptibility to autoimmune and infectious diseases (van Sorge, van der Pol, and van de Winkel 2003; Lehmbecher et al. 1999). A polymorphism at position 131 in FcγRIIa results in either a histidine or arginine. Only the H131 form is capable of binding IgG2 and carrying out IgG2-mediated phagocytosis by neutrophils and monocytes (Parren et al. 1992; Salmon et al. 1992; Sanders et al. 1995). FcγRIIb is polymorphic at position 232 in the transmembrane region. Conflicting results indicate that the T232 allele shows greater or reduced inhibition of BCR signaling relative to the I232 form (X. Li et al. 2003; Kono et al. 2005). A polymorphism at position 158 of FcγRIIIa results in either a low affinity phenylalanine form, or a higher affinity valine form that binds approximately fivefold more tightly to IgG1 and IgG3 (Koene et al. 1997). Finally, there are two neutrophil antigen (NA) polymorphisms of FcγRIIIb, referred to as NA1 and NA2, that differ by four amino acids. This variation results in differences in glycosylation, which translates into a tighter binding affinity of NA1 to IgG3 and greater IgG1- and IgG3-mediated neutrophil phagocytosis relative to the NA2 form (Bredius et al. 1994; Nagarajan et al. 1995; Salmon et al. 1995). It should be kept in mind that the numbering conventions of the receptors and these polymorphic positions vary in the literature, often due to inconsistencies of whether numbering is sequential or based on an alignment, and whether leader sequences are counted.

Fundamental to their diverse ability to control cellular immune response are the differences in receptor expression on the various immune cells (Fig. 15.2). The most relevant for possible therapeutic effects include natural killer (NK) cells, monocytes/macrophages, dendritic cells (DCs), and neutrophils. A variety of other FcγR-expressing cells, including basophils, eosinophils, mast cells, B-cells, and γδ T-cells, are presumed less relevant with respect to the *in vivo* cytotoxic activity of antibody drugs, and therefore will not be discussed. Engagement of immune-complexed antibody with FcγRs can mediate an array of antibody-dependent effector functions, including cytolysis, phagocytosis, release of cytokines and chemokines, major histocompatibility complex (MHC) presentation, and other more complex activities (Fig. 15.3). For all of these immune cell types, expression of FcγRs can be modulated by cytokines, and likewise FcγR activation can induce cytokine production.

NK cells typically express only the activating receptor FcγRIIIa, and in this way are unique among effector cells. Due to allelic variation some individuals express NK cell FcγRIIc, an activating receptor highly homologous to FcγRIIb (Ernst et al. 2002). The major FcγR-induced effector activities of NK cells are cytolysis of target cells via lytic granule release, induction of target cell apoptosis through secretion of TNF family ligands (e.g., TNF, FasL; Kashii et al. 1999), and production of cytokines such as IFNγ. Cells of the myeloid lineage, including monocytes, macrophages, DCs, and neutrophils, have more complex FcγR expression profiles (Boruchov et al. 2005; Michon et al. 1998; Pricop et al. 2001; Schakel et al. 1998). All express FcγRIIa and at least one splice form (b1 or b2) of the inhibitory receptor FcγRIIb. Macrophages and DCs also express FcγRI and FcγRIIIa depending on their source and activation state. Neutrophils express FcγRIIIb rather than FcγRIIIa, and FcγRI when activated by G-CSF. FcγRs I, IIa, and IIIa are capable of activating cellular effector functions via their ITAM

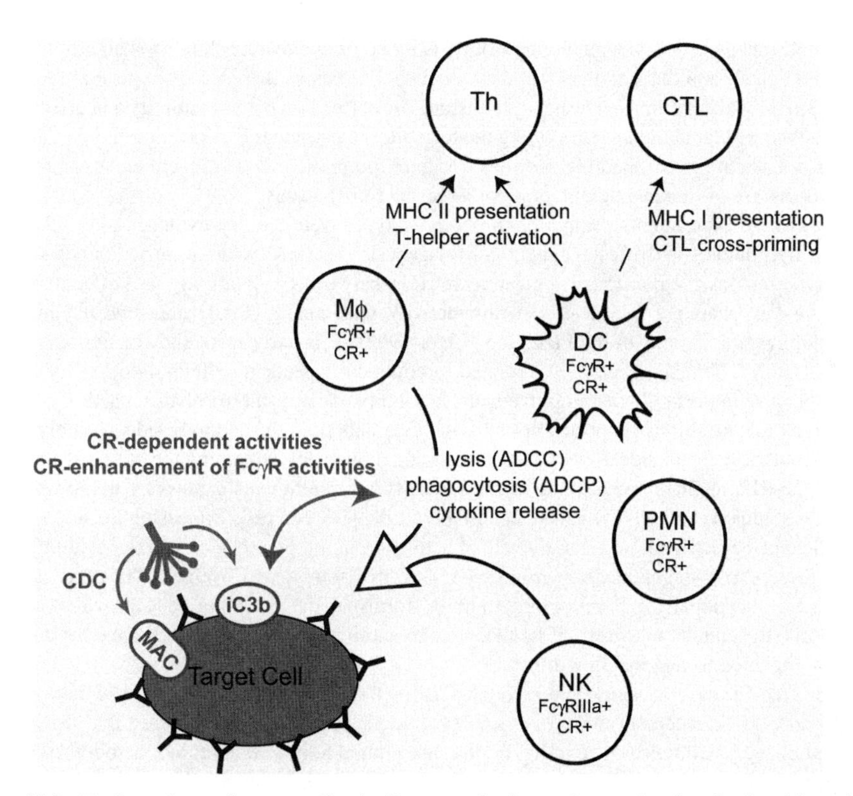

Figure 15.3 FcγR- and complement-mediated effector mechanisms of monoclonal antibodies. Macrophages (Mϕ), dendritic cells (DC), neutrophils (PMN), and NK cells (NK) are all capable of carrying out FcγR-mediated effects (ADCC, ADCP, and cytokine/chemokine release) against target tumor or microbial cells. Macrophages and DCs can present engulfed antigen on class II MHC to generate $CD4^+$ helper T-cells (Th), and DCs also have the capacity to cross-prime $CD8^+$ cytotoxic T-lymphocytes (CTL). Complement-mediated mechanisms include non-cellular MAC-mediated lysis of target cells (CDC), and cellular pathways mediated by interaction between complement receptors (CR1, CR3, CR4) and opsonic iC3b (as well as C3b and C4b, not shown). These cellular pathways include CR-dependent phagocytic and cytotoxic activities, as well as CR enhancement of FcγR-mediated effector functions.

signaling components. Although FcγRIIIb has no intrinsic signaling capacity, some studies have shown that it can mediate neutrophil effector functions through direct cooperation with FcγRIIa or complement receptor CR3 (discussed below), or via indirect effects on signaling molecules by localizing complexes into lipid rafts (Edberg et al. 1998; Fernandes et al. 2006; Stockinger 1997). Macrophages and neutrophils can phagocytose opsonized target cells upon FcγR engagement, lyse them through release of cytolytic granules, or induce apoptosis via release of reactive nitrogen (RNI) and oxygen (ROI) intermediates. Macrophages and DCs are professional antigen-presenting cells (APCs), and FcγR-mediated phagocytosis can not only result in target cell destruction, but also lead to presentation of target cell-derived antigens to T-cells. Both macrophages and DCs express surface class II MHC, and thus can present to $CD4^+$ T-helper cells (Fig. 15.3). Additionally, when tumor uptake is mediated by FcγRs, DCs can present tumor-derived antigens on class I MHC, thus acquiring the ability to activate $CD8^+$ cytotoxic T-lymphocytes (CTLs), a process called cross-priming (Dhodapkar et al. 2002). This mechanism can potentially lead to long-lasting adaptive antitumor immunity, and these effects are sometimes invoked to explain the long-term responses observed in lymphoma patients after therapy with the anti-CD20 antibody rituximab (Rituxan®; Selenko et al. 2002).

15.3.2 Support for Engineering Fc for Improved Affinity to FcγRs

The central strategy of all Fc engineering efforts is based on the relationship between affinity for the various Fc ligands and the activities that they mediate. For FcγRs there are three general sets of data that support this strategy: *in vitro* results establishing the correlation between affinity and effector function, results from animal models supporting the relevance of particular FcγRs and cell types, and clinical data documenting correlations between receptor polymorphisms and either susceptibility to infectious disease or therapeutic outcome for anticancer mAb drugs.

Early *in vitro* experiments comparing antibody isotypes were the first evidence for a relationship between FcγR binding and effector function. Mutagenesis experiments, often carried out by swapping IgG isotype regions, demonstrated a clear correlation between FcγR affinity and effector function activity, and generated a wealth of structure-activity relationship (SAR) data that elucidated the FcγR binding determinants of the Fc region (Clark 1997). This early work showed not only that the affinity of an IgG for Fc receptors was related to antibody-dependent cellular cytotoxicity (ADCC) *in vitro*, but also importantly that improvements in affinity could result in enhanced activity. The affinity/activity relationship is supported not only with variants from the antibody side, but polymorphic variants from the receptor side as well. Neutrophils and monocytes expressing the higher affinity H131 allele of FcγRIIa mediate higher phagocytosis activity relative to cells expressing the lower affinity R131 form (Salmon et al. 1992; Sanders et al. 1995). Likewise, NK cells expressing the higher affinity V158 allele of FcγRIIIa show greater levels of activation in the presence of IgG1-coated target cells relative those expressing the lower affinity F158 form (Bowles and Weiner 2005). As described below, these functionally difference polymorphic forms of the receptors have provided a unique window into the clinical relevance of Fc receptors to antitumor activity, and thus have become a significant motivation for engineering efforts.

Animal models have been invaluable for elucidating the details of FcγR biology and demonstrating the relevance of Fc receptors to *in vivo* activity. Genetic knockout mice showed that the common γ-chain, utilized by all murine activating FcγRs, is essential to *in vivo* antitumor activity for multiple antigen/antibody systems, including the anti-Her2 antibody trastuzumab (Herceptin®) and the anti-CD20 antibody rituximab (Rituxan®) (Clynes et al. 1998, 2000). Consistent with these observations, later studies showed that depletion of normal B-cells by a panel of murine anti-CD20 antibodies is dependent on the antibody isotype and the common γ-chain (Hamaguchi et al. 2006; Uchida et al. 2004). It is important to caution that there are substantial differences between mouse and human Fc receptor biology and IgG isotypes that preclude direct extrapolation of results from murine models. These include not only differences in sequence homology, but also in isotype affinities, expression patterns, and signaling (Clynes 2006; Desjarlais et al. 2007). Nonetheless, improvements in activity have been demonstrated in standard SCID xenograft models (Horton et al. 2008). More sophisticated models that more accurately represent human FcγR biology include mice engrafted transiently with human immune cells, mice engineered transgenically with human receptors, and cynomolgus monkeys. Notably, evidence that the affinity/activity relationship observed *in vitro* does translate to greater antibody activity *in vivo* has been obtained from experiments studying Fc-engineered antibodies (discussed below) in mice engrafted with human peripheral blood mononuclear cells (PBMCs) (Niwa et al. 2004), mice transgenic for human FcγRIIIa (Stavenhagen et al. 2007), and a monkey B-cell depletion model (Lazar et al. 2006; Zalevsky et al. 2009).

Perhaps the most compelling support for the engineering of improved interaction with FcγRs has come from studies correlating human clinical outcome with functionally relevant receptor polymorphisms. These studies are based on the concept that possession of the higher affinity allele for an activating FcγR should provide greater clinical benefit. FcγR polymorphism associations have long been studied as predictors of autoimmune disease incidence and susceptibility to microbial infections (van Sorge, van der Pol, and van de Winkel 2003; Lehmbecher et al. 1999; Karassa, Trikalinos, and Ioannidis 2004). R131 FcγRIIa is associated with greater susceptibility to infectious disease, a relationship that is hypothesized to be due to the critical role of IgG2 in fighting pathogens and the capacity of this isotype to mediate monocyte and neutrophil effector function only with the H131 form (van Sorge, van der Pol, and van de Winkel 2003; Salmon et al. 1992; Sanders et al. 1995). More recently, polymorphism analysis has been extended to studying clinical response to administered anticancer antibody drugs (Desjarlais et al. 2007). Significant response differences between the V158 and F158 alleles of FcγRIIIa have been observed with anti-CD20 rituximab for treatment of follicular non-Hodgkin's lymphoma (NHL), diffuse large B-cell lymphoma (DLBCL), and Waldenström's macroglobulinemia (WM) (Cartron et al. 2002; Kim et al. 2006; Weng and Levy 2003, 2005; Treon et al. 2005), as well as with anti-Her2 trastuzumab for treatment of metastatic breast cancer (Musolino et al. 2007). It should be noted that not all such studies have documented significant correlations. For anti-CD20 therapy, no correlation was observed between receptor polymorphism and outcome for chronic lymphocytic leukemia (CLL; Farag et al. 2004), and there are conflicting results for DLBCL (Kim et al. 2006; Mitrovic et al. 2007).

A number of these same studies have also documented significant correlations with the high affinity H131 allele of FcγRIIa (Weng and Levy 2003, 2005; Musolino et al. 2007; Zhang et al. 2007). These results are perplexing given that IgG1 (in contrast to IgG2) shows no affinity difference between these polymorphic forms (Richards et al. 2008). A recent study suggests that these correlations are due to linkage disequilibrium between FcγRIIa and FcγRIIIa (Hatjiharissi et al. 2007) rather than mechanism. Inconsistent with these data, and a particularly confusing result, is the correlation observed between the F158 form of FcγRIIIa and the H131 form of FcγRIIa and progression-free survival in metastatic colorectal cancer in response to therapy with the anti-EGFR cetuximab (Erbitux®; Zhang et al. 2007). The strongest direct support for FcγRIIa-mediated antitumor effects comes from a study with a murine IgG3 antibody, which binds with greater affinity to the R131 allele relative to the H131 form (this is opposite to the preference of human IgG2). In this study, anti-GD2 therapy for neuroblastoma provided greater progression-free survival in patients having the higher affinity R/R131 genotype (Cheung et al. 2006). Given that FcγRIIa is expressed on phagocytes such as neutrophils, macrophages, and DCs, the results implicate at least one of these cell types for antibody efficacy against solid tumors. This point raises another key set of clinical data that supports the strategy of enhancing FcγR affinity to improve therapeutic outcome, namely the infiltration of various effector populations into tumor tissues. The presence of neutrophils (van Egmond 2008; Di Carlo et al. 2001), macrophages (Mantovani et al. 2002), DCs (Guiducci et al. 2005), and in some cases even NK cells (Arnould et al. 2006) in the tumor microenvironment suggests that FcγR-mediated effector functions are indeed accessible mechanisms of action by administered antibody drugs.

15.3.3 Target FcγR Profiles for Improving Therapeutic mAb Activity

A variety of immune cells can carry out FcγR-mediated effector functions (Fig. 15.3), and thus are capable of contributing to the *in vivo* activity of monoclonal antibodies. The specific roles of each cell type in the destroying tumor or pathogenic target cells are not well understood, and indeed may vary with indication, antigen, and antibody. Moreover, the specific FcγR dependencies of each of these cell types are not well defined. Whether a single optimal FcγR selectivity profile exists for cytotoxic mAbs, or whether practically the best set of receptor affinities is a compromise, is a major knowledge gap for Fc engineering efforts. The parameters that determine the answer to this question are the capacity of each effector cell type to affect a target cell, the accessibility of each effector cell to the target cells, and finally what specific FcγR profiles are optimal for each effector cell type.

NK cells have developed a prominent reputation as target cell killers. When PBMCs are used as effectors, lysis is mediated almost completely by the NK cell population (Abdullah et al. 1999). Their efficient ADCC of antibody-opsonized tumor cell lines *in vitro* has contributed to the perception that their activity is necessarily relevant *in vivo*. Because NK cells typically express only the activating receptor FcγRIIIa, the optimal FcγR profile is simply high affinity for this receptor. Accordingly, the correlations observed between clinical response to rituximab and FcγRIIIa polymorphism have foremost been attributed to the involvement of NK cells, a plausible hypothesis given the access of NK cells to blood. The role of NK cells in antibody-mediated destruction of solid tumors is less established, as NK cells are generally considered poor solid tumor infiltrators (Albertsson et al. 2003). Notably, however, a recent study demonstrated that tumor infiltration of NK cells increases after treatment with trastuzumab (Arnould et al. 2006).

FcγR-positive effector populations with stronger reputations for infiltrating tumors are the monocytic phagocytes, including macrophages (Mantovani et al. 2002) and DCs (Guiducci et al. 2005), as well as neutrophils (van Egmond 2008; Di Carlo et al. 2001). Although there is evidence that in some cases tumor-associated macrophages (TAMs) can either promote tumor progression or enhance antitumor immunity depending on local environment and their stage of development (Pollard 2004; Van Ginderachter et al. 2006), their presence in the tumor microenvironment naturally implicates them as key mediators of antibody efficacy. This is consistent with observations in mouse models that monocyte populations are critical for antibody-mediated target cell depletion (Hamaguchi et al. 2006; Uchida et al. 2004). Macrophages and DC have more complex FcγR expression than NK cells, expressing various levels of all the activating and inhibitory FcγRs, subject to regulation by cytokines. Macrophages and DCs express FcγRIIIa, and the correlations observed between polymorphism of this receptor and clinical outcome apply equally to these cell types as they do to NK cells. Recent data have indicated that FcγRIIa in particular is a key receptor for macrophages, with more minor contributions from FcγRI and FcγRIIIa and no impact from FcγRIIb (Richards et al. 2008). Dendritic cells have also been shown to be dependent on FcγRIIa for immune complex-stimulated maturation (Boruchov et al. 2005), an important step in cross-presentation of tumor-derived antigens to antitumor CTLs (Dhodapkar et al. 2005). Thus, FcγRIIa appears to be a critical link between the humoral and adaptive arms of the immune system. Neutrophil effector function is also known to be strongly dependent on FcγRIIa, with a cooperative role for FcγRIIIb (Edberg et al. 1998; Fernandes et al. 2006; Nagarajan et al. 2000). Given the strong involvement of FcγRIIa in their effector functions and their capacity to infiltrate tumors, the observed correlations between FcγRIIa polymorphism and clinical outcome (Cheung et al. 2006) would seem to support a role for macrophages, DCs, and/or neutrophils in antibody efficacy.

Among the receptors expressed on these cell types, the current data support the greatest roles for FcγRIIa, FcγRIIb, and FcγRIIIa. Although FcγRI is an activating receptor, there is no evidence to suggest that increasing mAb affinity to it will improve effector function; its high affinity for monomeric IgG makes it poor at distinguishing between unbound antibody and immune complexes, and no impact on antibody activity was observed in FcγRI$^{-/-}$ mice (Nimmerjahn and Ravetch 2005). The absence of cytoplasmic signaling by FcγRIIIb, and the variable expression and poor understanding of FcγRIIc have placed these receptors in the back seat as well. Of course, the other strike against these two receptors is the lack of functional orthologs in mice. Likewise, there is no polymorphism data supporting the relevance of FcγRs I, IIc, or IIIb for antibody therapy. It should be cautioned that absence of such data does not conclude the irrelevance of these receptors. Nonetheless, the approach thus far with these receptors has been to leave them be, and the emphasis of Fc engineering has been on improving interaction profiles with FcγRIIa, FcγRIIb, and FcγRIIIa.

	1IIS	21	22	88	89	90	113	114	115	116	117	119	120	122	128	129	130	131	132	133	134	135	137	155	157	158	159	160	161	162	163	164
	1E4K	18	19	85	86	87	110	111	112	113	114	116	117	119	125	126	127	128	129	130	131	132	134	152	154	155	156	157	158	159	160	161
Human	I	Q	E	R	G	W	W	K	D	K	L	Y	N	L	K	A	F	K	F	F	H	W	S	S		M	G	K	H	R	Y	T
Human	IIa	Q	E	S	E	W	W	K	D	K	P	V	K	T	K	S	Q	K	F	S	H/R	L	P	T	N	I	G	Y	T	L	F	S
Human	IIb	Q	E	S	E	W	W	K	D	K	P	V	K	T	K	S	K	K	F	S	R	S	P	T	N	I	G	Y	T	L	Y	S
Human	IIc	Q	E	S	E	W	W	K	D	K	P	V	K	T	K	S	K	K	F	S	R	S	P	T	N	I	G	Y	T	L	Y	S
Human	IIIa	E	K	I	G	W	W	K	N	T	A	H	K	T	K	G	R	K	Y	F	H	H	S	R	L	V/F	G	S	K	(N)	V	S
Human	IIIb	E	K	I	G	W	W	K	N	T	A	H	K	T	K	D	R	K	Y	F	H	H	S	R	L	V	G	S	K	(N)	V	S
Mouse	I	Q	K	N/E	D	W	W	K	N	K	L	Y	N	V	K	S	F	Q/K	F		S	S	S	S	T/M	G	R	H	R	Y	T	S
Mouse	II	K	E	S	D	W	W	R	N	K	L/P	N	R	S	K	S	V	R	Y	H	H	Y	S	K	S	L	G	R	T	L/Q	H	Q
Mouse	III	K	E	S	D	W	W	R	N	K	L	N	R	S	K	S	V	R	Y	H	H	Y	S	K	S	L	G	S	T	Q	H	Q
Mouse	IV	E	E	M	G	W	W	Q	N	R	P	R	K	T	K	G	K	K	Y	F	H	E	S	R	L	I	G	H	N	(N)	K	S

Mutational Differences		Human I	Human IIa	Human IIb	Human IIc	Human IIIa	Human IIIb	Mouse I	Mouse II	Mouse III	Mouse IV
Human	I	0	19	19	19	21	21	18	22	22	21
Human	IIa		0	4	4	21	21	22	20	21	19
Human	IIb			0	0	22	22	21	21	22	19
Human	IIc				0	22	22	21	21	22	19
Human	IIIa					0	1	23	23	22	12
Human	IIIb						0	23	23	22	13
Mouse	I							0	22	22	24
Mouse	II								0	2	22
Mouse	III									0	22
Mouse	IV										0

Figure 15.4 FcγR sequence similarity at positions that bind the antibody Fc region. Mouse sequences are highlighted in gray. Polymorphic amino acids are separated by a slash. Positions are numbered according to that provided in the two available Fc/FcγRIIIb complex structures 1E4K (Sondermann et al. 2000) and 1IIS (Radaev et al. 2001). These structures differ in the numbering convention used, which results in differences in positional references for the FcγRIIa 131 and FcγRIIIa 158 polymorphisms. Circles indicate Asn162 in the human FcγRIIIa/b and mouse FcγRIV receptors. N-linked glycosylation at this site is the most well-supported mechanism for the affinity enhancements provided by afucosylation. The mutational differences in the right panel indicate the number of differences between the receptors at the Fc interface out of the 31 positions listed.

At first glance, the ideal FcγR selectivity profile of an engineered antibody would be high affinity for FcγRIIa and FcγRIIIa but low affinity for the inhibitory FcγRIIb. A significant obstacle to this profile is the high homology between FcγRIIa and FcγRIIb. There are only four differences between these receptors at the Fc interface (Fig. 15.4), three when the polymorphic position 131 is excluded (FcγRIIb has an arginine at 131). Not only does this high similarity make engineering selectivity between these receptors a challenge, it suggests that a possible compromise between the three receptors may need to be made. A corollary of the high homology between FcγRIIa and FcγRIIb is that a modification that provides a reduction in FcγRIIb affinity, for example in order to improve IIIa : IIb ratio, will likely also reduce interaction with FcγRIIa. This nuance between the three receptors puts pressure on the hypothesis that activating to inhibitory ratios is a critical determinant of effector function. Despite support for the relevance of the inhibitory receptor for antibody activity in mice (Clynes et al. 1998; Hamaguchi et al. 2006) and the opposed biochemistry of FcγRIIa and FcγRIIb for DCs (Dhodapkar et al. 2005), as of yet there is no evidence that reduced binding to FcγRIIb is preferable to, and therefore should come at the expense of, maximal binding to FcγRIIa or FcγRIIIa.

15.3.4 Amino Acid Engineering to Improve FcγR-Mediated Effector Functions

Amino acid modification has provided a versatile approach for controlling FcγR binding. Significant progress has been made in improving overall affinity, as well as in tuning the Fc for particular FcγR profiles. Principal challenges have included the virtually identical binding site of the different Fc receptors, and the asymmetric binding of the Fc homodimer to the monomeric receptor. A consequence of this latter issue is that a designed Fc substitution is actually two mutations, each in a distinct structural environment on opposite sides of the interface. Efforts have been aided by available crystal structures of the Fc/FcγRIIIb complex (Sondermann et al. 2000; Radaev et al. 2001). Efficient screens have also been critical, and a variety of approaches have been used to select favorable mutations, including alanine scanning and site-directed mutagenesis (Shields et al. 2001), computational structure-based design (Lazar et al. 2006; Richards et al. 2008), and selection-based methods (Stavenhagen et al. 2007). Primary screens/selections have for the most part been based on receptor binding, with subsequent evaluation of mutational hits in cell-based effector function assays, typically ADCC. Favorable substitutions have generally been additive in subsequent engineering rounds, with some exceptions. The greatest improvements in the published literature have been provided by substitutions at positions 298, 333, and 334 (Shields et al. 2001), 239, 332, 330, and 236 (Lazar et al. 2006; Richards et al. 2008), and 243, 292, 300, 305, and 396 (Stavenhagen et al. 2007). However, a much larger number of variants, on the order of thousands, have been disclosed in patents, and the Fc region at the FcγR-binding interface has been virtually saturated with all possible mutations. The majority of improvements have been afforded by substitutions at the interface (Fig. 15.1), although mutations at sites distal to it have also impacted binding (Stavenhagen et al. 2007). It is not clear whether these results are artifacts of the binding assays, or reflect a more complex dependence of FcγR binding on Fc tertiary structure.

The early goal of engineering efforts was improved FcγRIIIa affinity, motivated by the clearer relevance of this receptor from polymorphism data and the dominance of NK cells among other PBMC components in ADCC assays. Between one and two orders of magnitude greater affinity for this receptor has been obtained with a small number of mutations, providing substantial enhancements in both ADCC potency and efficacy (Stavenhagen et al. 2007; Lazar et al. 2006; Shields et al. 2001). Experiments exploring the impact of some of these Fc variants in mice transgenic for human FcγRIIIa and in a monkey B-cell depletion model have confirmed their improved activity *in vivo* (Stavenhagen et al. 2007; Lazar et al. 2006; Zalevsky et al. 2009). The first generation of Fc-engineered therapeutic mAbs have been geared primarily towards consideration of FcγRIIIa, and these candidates are just beginning to make their entry into clinical trials (Desjarlais et al. 2007).

Variants with up to 70-fold improved affinity to FcγRIIa have also been engineered, providing substantial enhancements to macrophage phagocytosis (Richards et al. 2008). More challenging has been selective affinity enhancement to activating receptors relative to the inhibitory receptor FcγRIIb. Variants that improve Fc affinity for FcγRIIIa relative to FcγRIIb have been generated (Stavenhagen et al. 2007; Lazar et al. 2006; Shields et al. 2001), although as of yet no clear effector function benefit has been conclusively attributed to this selectivity. Variants have also been engineered that provide selective enhancement to FcγRIIa relative to FcγRIIb (Richards et al. 2008), despite their near identical extracellular domains (Fig. 15.4). Interestingly, the greater macrophage phagocytosis of these variants depended primarily on their absolute affinity for FcγRIIa, with no impact from FcγRIIb binding, a result that is counter to the perceived role of FcγRIIb for this cell type. These results highlight the value of Fc variants with a diverse set of FcγR selectivities as tools for helping define target profiles for relevant immune cell types.

15.3.5 Glycoengineering to Improve FcγR-Mediated Effector Functions

The observation that ADCC activity was affected by different Fc glycosylation patterns, produced by altering cell line and culture conditions, (Lifely et al. 1995) led to the development of glycoform engineering as a strategy for improving FcγR affinity and effector function. Originally there was some confusion as to the relevance of particular structural aspects of the glycoform at Asn297, specifically the roles of afucosylation and bisecting *N*-acetylglucosamine (GlcNAc), both of which are achieved by overexpression of β(1,4)-*N*-acetylglucosaminyltransferase III (GnTIII) (Umana et al. 1999; Davies et al. 2001). Later work, however, showed that the absence of fucose is the sole determinant of enhanced FcγR affinity and effector function (Shinkawa et al. 2003; Shields et al. 2002), indicating that the impact of GnTIII on FcγR binding was due not to the generation of bisecting GlcNAc, but rather the preclusion of subsequent modification by α1,6-fucosyltransferase.

The Lec13 CHO and the YB2/0 rat hybridoma cell lines naturally generate fucose-deficient antibodies, and much of the early work studying the impact of glycoforms on antibody activity has been carried out with these systems. However, these cell lines are impractical for large-scale mAb manufacturing because they grow poorly and produce mixtures of fucosylated and nonfucosylated antibodies. CHO cell lines that more robustly produce afucosylated antibodies and are more commercially viable have been glycoengineered using inducible expression of the enzyme GnTIII (Umana et al. 1999), and genetic knockout of the enzyme α-1,6-fucosyltransferase (FUT8; Yamane-Ohnuki et al. 2004). Spurred by the success of these efforts at improving FcγRIIIa affinity and ADCC, subsequent work has been carried out in nonmammalian expression systems. Deletion of fucosylation pathways has been accomplished in yeast, plants, and moss (Cox et al. 2006; H. Li et al. 2006; Nechansky et al. 2007), often accompanying efforts to remove glycosylation pathways that produce nonhuman and thus potentially immunogenic glycoforms. Afucosylated antibodies, in general, provide one to two orders of magnitude greater affinity for both iosoforms of FcγRIIIa (Ferrara et al. 2006; Okazaki et al. 2004). Improved antitumor activity has been demonstrated for an afucosylated antibody using mice engrafted with human PBMCs (Niwa et al. 2004). Clinical candidates based on some of these technologies are currently being developed, the most advanced of which is an anti-GD3 antibody being tested for metastatic melanoma (Forero et al. 2006).

The mechanism by which lack of fucose increases FcγR affinity and ADCC has been the subject of debate. Structural work has suggested that improved FcγR binding is due to subtle conformational alterations in the Fc caused by the absence of fucose (Matsumiya et al. 2007). In contrast, other work has indicated that enhancement occurs via a steric interaction between Fc and a receptor carbohydrate at Asn162 that is relieved upon removal of Fc fucose (Ferrara et al. 2006). The latter model is more consistent with two key sets of data. First, no improvement by afucosylation is observed when receptor carbohydrate is removed enzymatically (Ferrara et al. 2006). Second, the requisite for receptor glycosylation at position 162 correctly predicts that affinity should only be improved for FcγRIIIa/b and mouse FcγRIV (Ferrara et al. 2006; Masuda et al. 2007), which are the only receptors in these organisms that possess an asparagine at that position (as well as the requisite serine/threonine at the $i + 2$ position) (Fig. 15.4). It is noted, however, that some data show subtle binding increases of afucosylated mAbs to human FcγRIIb, mouse FcγRIIb, and mouse FcγRIII (Nimmerjahn and Ravetch 2005; H. Li et al. 2006; Siberil et al. 2006); whether these differences are significant, and if so whether they reflect another mechanism, requires further investigation. For the moment, the receptor carbohydrate mechanism has several ramifications for the glycoengineering approach. First, it reaffirms that lack of fucose is the most important if not the sole factor for enhancement. Second, it indicates that variability in FcγRIIIa glycosylation, known to occur for some effector cell types (Drescher, Witte, and Schmidt 2003; Edberg and Kimberly 1997) could impact *in vivo* benefit. Finally, it suggests that afucosylation is afucosylation irrespective of the method used to generate it. Thus the only difference between the various CHO and nonmammalian expression systems with respect to ADCC enhancement should be percentage of afucosylated antibody and manufacturing efficiency. However, it must be cautioned that whereas lack of fucose appears to be the only modification that improves FcγRIIIa affinity, other sugar structures can adversely affect antibody Fc properties (Kanda et al. 2007; Scallon et al. 2007).

15.3.6 Combination of Amino Acid and Glycoform Engineering

The affinity improvement by glycoengineering to only FcγRIIIa/b among the human receptors highlights a key distinction from amino acid modifications, which create a more diverse range of effects. An experiment combining the two approaches resulted in an additive effect on FcγRIIIa affinity, but no consequent benefit in ADCC (Masuda et al. 2007). This result suggests a maximal threshold of ADCC exists that each method alone is sufficient to induce. Whether such a ceiling exists *in vivo* remains to be seen. More importantly and as previously discussed, NK cell FcγRIIIa-mediated ADCC is not the only, nor necessarily the most relevant, effector function for improving therapeutic antibody potency. Indeed combination of afucosylation with amino acid modifications selectively enhanced for FcγRIIa affinity (Richards et al. 2008) results in additive FcγRIIa and FcγRIIIa profiles that do provide additive macrophage phagocytosis (Xencor, unpublished results). Currently, the full utility of combining the two technologies remains relatively unexplored.

15.4 COMPLEMENT

15.4.1 Background on Complement

There are multiple cytotoxic mechanisms accessible by the classical (antibody-dependent) complement pathway (the alternative and lectin pathways are not activated by opsonic antibodies and are therefore not discussed). The classical pathway includes both noncellular and cellular mechanisms, as well as significant crosstalk between complement and FcγR pathways (Fig. 15.3). Binding of the C1q component of the C1 complex to the antibody Fc region activates a cascade of events involving numerous complement proteins, ultimately resulting in recruitment of the membrane attack complex (MAC, also called C5b-9), and deposition of opsonic C3 and C4 components on the target cell surface. MAC is responsible for mediating complement-dependent cytotoxicity (CDC), a noncellular activity and the most widely recognized mechanism of target cell destruction. Opsonic C3b, iC3b, and C4b proteins on target cells result in two cellular complement activities, both of which are mediated by interaction with complement receptors CR1 (CD35), CR3 (CD11b-CD18), and CR4 (CD11c-CD18) expressed on phagocytes and NK cells. CR-dependent phagocytosis and cytotoxicity, also referred to as complement-dependent cellular cytotoxicity (CDCC), are analogs of the FcγR-dependent mechanisms of similar name, except that effector function is mediated directly by binding of CR to opsonin. These processes are activated by cell wall β-glucan, and therefore presumed to be relevant for pathogenic target cells but not tumors. The other mechanism involves enhancement of FcγR-mediated effector functions by the CR/opsonin interaction. This synergy does not require microorganism danger signals, making it potentially applicable for both antitumor and anti-infectious disease antibodies. CR enhancement of cellular effector functions is activated by opsonic complement protein C5a, which is not only chemotactic for effector cells, but also selectively increases macrophage expression of activating FcγRs relative to FcγRIIb (Godau et al. 2004; Konrad et al. 2006), providing additional crosstalk between FcγR and complement effector arms. Complement activation is tightly regulated by a number of complement regulatory proteins (CRPs), including both soluble proteins, and membrane-bound proteins that are expressed on most cell types.

15.4.2 Support for Engineering Fc for Improved Complement

Given the activation of complement pathways by microbial surfaces, the most intuitive applications for complement-enhanced mAbs are infectious disease indications. The capacity of mAbs to destroy microbes using complement mechanisms is supported by both *in vitro* (Kelly-Qunitos et al. 2006; Preston et al. 1998) and *in vivo* (Wells et al. 2006; Han et al. 2001) data. Unfortunately little clinical data is available concerning the mechanisms of action of antipathogenic mAbs, due principally to the low number of such drugs that have progressed through clinical trials (Baker 2006). It remains to be seen whether complement plays a role in the therapeutic activity of antipathogen antibodies, and accordingly whether enhanced complement-mediated effector functions are a valid strategy for improving them.

The role of complement in anticancer mAb activity is less clear, and has generally been in doubt due to the overexpression of complement inhibitory proteins on some tumors that protect them from complement-mediated injury (Gelderman et al. 2004; L. Li et al. 2001). The strongest support for a relationship between complement activity and mAb anticancer efficacy exists for rituximab. However, although some experiments in mice have demonstrated a dependence of anti-CD20 activity on complement (Cragg and Glennie 2004; Di Gaetano et al. 2003; Golay et al. 2006), others have not (Uchida et al. 2004; Hamaguchi et al. 2005). *In vivo* experiments with an anti-GD2 antibody have demonstrated that complement-dependent mechanisms are effective against tumor cells *in vivo*, even when the tumor cell expresses high levels of complement inhibitors (Imai et al. 2005). The best clinical support comes from observed correlations between expression of CRPs and lower patient response to rituximab (Bannerji et al. 2003; Treon et al. 2001), and the consumption of complement by rituximab in chronic lymphocytic leukemia (CLL) patients (Kennedy et al. 2004). However, no differences in complement-mediated cytotoxicity *in vitro* were observed using tumor cells from the different response groups (Weng and Levy 2001). Recent work has investigated relationships between complement polymorphisms and clinical response, similar to the FcγR studies, and found an association between C1q polymorphism and breast cancer metastasis (Racilla et al. 2006).

A contributing factor to the lower emphasis on complement relative to FcγRs for mAb cancer therapy, and a practical issue for engineering, is that whereas most mAbs mediate ADCC *in vitro*, few are capable of mediating complement activity (Xia, Hale, and Waldmann 1993; Cragg et al. 2003). The suggested reason for this difference is that complement requires high antigen/antibody surface density, which is consistent with the fact that pentameric IgM is the most active isotype for complement. The importance of surface density may be related to the epitope dependence of complement activity for antibodies against the same target (Cragg and Glennie 2004). The importance of both of these parameters is supported by the greater complement activity observed using antibodies with higher valency and heterogenous epitopes (Meng et al. 2004; Macor et al. 2006; Spiridon et al. 2002). Despite the sensitivities of complement activity *in vitro*, a correlation between C1q affinity and complement activity has been established in a number of antigen/antibody systems (Dall'Acqua et al. 2006; Idusogie et al. 2001; Natsume et al. 2008). Importantly, this relationship serves as a foundation for an Fc engineering approach to improve complement-mediated effector functions.

15.4.3 Amino Acid Engineering to Improve Complement-Mediated Effector Functions

Although a structure of the Fc/C1q complex is unavailable, mapping of the interaction by mutagenesis (Hezareh et al. 2001; Idusogie et al. 2000; Thommesen et al. 2000; Redpath et al. 1998; Sensel, Kane, and Morrison 1997; Tao, Canfield, and Morrison 1991; Tao, Smith, and Morrison 1993) has determined the importance to binding of the hinge and Fc positions 234, 235, 270, 322, 326, 329, 331, and 333 (Fig. 15.1). This site is separate from but overlapping with the binding site for FcγRs (Fig. 15.1), and this overlap is an important consideration for Fc engineering work in general. Capitalizing on this SAR, as well as the known differences in complement activity between IgG isotypes, several groups have successfully engineered mutations that improve both mAb/complement interactions and *in vitro* CDC. These include substitutions at 326 and 333 (Idusogie et al. 2001), hinge modifications (Dall'Acqua et al. 2006), and IgG1/IgG3 isotype switch variants (Natsume et al. 2008). In addition, up to 200-fold greater CDC has been achieved by engineering a C-terminal disulfide to generate covalent dimers (Shopes 1992), consistent with the influence of valency on complement activity. Importantly, data from a monkey B-cell depletion study using a variant anti-CD20 demonstrated that improved C1q affinity and CDC *in vitro* can translate into improved mAb activity *in vivo* (Natsume et al. 2008).

In the future it will be necessary to evaluate the impact of these modifications not only on MAC-mediated CDC activity, but also on CR-dependent enhancement of cellular effector functions. Of particular importance is how improved complement activity may synergize with FcγR-mediated pathways. In this regard, variants with greater complement activity, as with those enhancing FcγR binding, will make useful reagents to study the role and mechanisms of complement in mAb therapeutic efficacy. As of yet there are no mAb candidates in clinical development that contain Fc modifications that improve complement activity. Obviously the first such antibody drug will be an important step in the maturity of this particular area of Fc engineering.

15.5 EFFECTOR FUNCTION SILENT Fc REGIONS

15.5.1 Rationale and Support for Using Effector Function Silent Fcs

A final area of Fc engineering, and one that is often overlooked as a means for improving the clinical properties of antibodies, is the removal of FcγR- and/or complement-mediated effector functions. Knockout or silent Fc regions are valuable when FcγR- and/or complement-mediated effects are definitively not part of mechanism of action, plus one of several scenarios: (1) when effector function results in off-mechanism toxicity, (2) when the goal is to block a surface antigen but not deplete the target cell; (3) when the primary application for an mAb that is not directly cytotoxic is cotherapy with one that is, for example, an anti-angiogenic or immunomodulatory mAb with a tumor-targeting mAb. Obviously a critical component of these criteria is a confident understanding of the mechanism of a particular antibody, specifically the knowledge that effector functions and Fc ligand interactions are not relevant or are harmful. Examples of antibodies that have met these criteria are few. The most well-characterized cases in which Fc interactions contribute only side effects involve mAbs targeting CD3 (Chatenoud 2004) and CD4 (Newman et al. 2001). Ironically, a complete understanding of the mechanism of action of many approved antibody drugs is lacking. In practice, acceleration of candidates into development typically occurs based on *in vitro* assays that do not capture Fc interactions, or animal models that do not accurately mimic the human receptor biology. In this context it is important to keep in mind that these differences not only can misrepresent cellular and complement effector functions but, because of the avidity changes brought about by FcγR and complement cross-linking, can also potentially impact Fv-mediated effects. In addition, there are data suggesting that interaction with effector molecules can alter half-life and biodistribution, and in some cases reduction of these interactions may provide improved pharmacokinetic properties (Gillies et al. 1999; Hutchins et al. 1995). Overall these issues highlight the need for greater characterization of antibody mechanism at the research stage. With respect to Fc engineering, the inability to definitively meet the above criteria, coupled with the additional risk of changing a drug or development process, has meant that use of silent Fcs for clinical candidates has thus far not been a common practice.

15.5.2 IgG Isotypes versus Fc Engineering

One apparent approach to removing effector functions is to use weaker IgG isotypes, specifically IgG2 and IgG4. However, the view that there are two high (IgG1 and IgG3) and two low (IgG2 and IgG4) effector function IgGs is overly simplistic and misleading. IgG2 binds with significant affinity to FcγRIIa, particularly the H131 allele (Fig. 15.2), providing it with phagocytic capacity by neutrophils and monocytes (Salmon et al. 1992; Sanders et al. 1995). IgG4 binds with high affinity to FcγRI and weak but significant affinity to FcγRIIa/b, and is capable of Fc-mediated effects (Isaacs et al. 1996). The engagement of Fc receptors by IgG4 has been suggested as a possible explanation for the disastrous clinical outcome of TGN1412 (Wise, Gallimore, and Godkin 2006), an anti-CD28 antibody that caused a dangerous cytokine storm in six phase I patients. The cross-linking mediated by Fc receptor binding and the resulting increased avidity may have resulted in activities that were not observed *in vitro* in the absence of effector cells. Nor were they observed at the equivalent dose in nonhuman primates, which may differ in their receptor affinities for human IgG4. Although this hypothesis is one of several that are currently being evaluated, it nonetheless highlights the importance of characterizing the impact of Fc engagement in a manner that enables cross-species comparisons. Altogether, the perception that IgG2 and IgG4 are inactive is not valid, highlighted by the superior bactericidal activity of these allotypes relative to the IgG1 and IgG3 in a *Cryptococcus neoformans* infection model (Beenhouwer et al. 2007). Another aspect of the IgG2 and IgG4 isotypes that requires consideration is their sometimes problematic solution properties relative to IgG1. It is well established that IgG4 heavy chains readily exchange with one another. Because this process occurs irrespective of antigen specificity, exchange results in heterogeneous mixtures of homodimeric (monospecific) and heterodimeric (bispecific) antibodies (van der Neut Kolfschoten et al. 2007). There is also some indication that IgG2 can form covalent dimers, a property apparently related to its disulfide pairing (Yoo et al. 2003). Overall the differences between the IgG isotypes are not simplistic, and selection of which isotype best suits a clinical candidate requires careful consideration (Salfeld 2007).

The lack of complete reduction of Fc-mediated effects by IgG2 and IgG4, and their suboptimal solution properties, have motivated engineering efforts to generate silent Fcs using amino acid modifications. Studies aimed at generating generic variants for nonablative applications while minimizing the creation of new epitopes have explored intersequence variants of the IgG1, IgG2, and IgG4 isotypes (Armour et al. 1999; Strohl 2006). Other work has been directed at engineering out the unfavorable exchange behavior of IgG4 (Angal et al. 1993), although it should be noted that this property is a result not only of the hinge cysteine arrangement but also the CH3 domain (van der Neut Kolfschoten et al. 2007). The most well-characterized examples of engineering Fc silent modifications have been aimed at fixing unwanted clinical properties of specific mAbs. The most notable of these is muromonab-CD3 (Orthoclone OKT3), a murine anti-CD3 antibody for transplant rejection and the first clinically approved mAb. Muromonab-CD3 is the clearest case of an antibody whose beneficial Fv-mediated activity is corrupted by off-mechanism Fc-mediated effects, specifically the rapid induction of a cytokine storm upon administration (Chatenoud 2004; Raasveld et al. 1993; Vallhonrat et al. 1999). A variety of engineering approaches have been used to address this problem, including the use of an aglycosyl (N297A) IgG1 (Bolt et al. 1993), an L234A/L235A variant of IgG1 and IgG4 isotypes (Xu et al. 2000), and a V234A/G237A variant of IgG2 (Cole, Anasetti, and Tso 1997). Phase I clinical trials with these variant mAbs have demonstrated that the engineered modifications successfully reduced the acute clinical syndrome while maintaining Fv-mediated immusuppressive activity (Friend et al. 1999; Norman et al. 2000; Woodle et al. 1999). Another example of engineering to remove unwanted Fc-mediated effects is anti-CD4. In this case an IgG4 isotype was mutated with L235E to reduce residual FcγRI binding, and S228P to stabilize the hinge disulfides and reduce IgG exchange (Newman et al. 2001; Reddy et al. 2000). This antibody, referred to as IgG4-PE, retained the Fv-mediated capacity to inhibit CD4 interaction with MHC II, but did not deplete CD4+ T-cells in a chimpanzee model (Newman et al. 2001; Reddy et al. 2000). More precise tuning has recently been explored in an anti-HLA-DR mAb to selectively reduce complement activity, which was blamed for the severe infusion reactions caused by this antibody upon administration. P331S and K322A modifications in IgG1, which ablated complement- but not FcγR-mediated effector functions, successfully reduced toxicity in rats and monkeys while retaining antitumor activity (Tawara et al. 2008). Together, the preclinical and clinical results with these modified anti -CD3, -CD4, and -HLA-DR antibodies illustrate how simple but well-designed amino acid modifications can dramatically improve the *in vivo* properties of therapeutic mAbs.

15.6 SUMMARY AND FUTURE DIRECTIONS

The exciting developments in antibody Fc optimization in recent years are the product of decades of progress in immunobiology and protein engineering. The convergence of this work has generated Fc technologies that have the potential to positively impact therapeutic mAbs to the extent that hybridoma methodology and chimerization/humanization once did. The next few years will see clinical data emerge from the first wave of Fc engineered mAbs, providing critical validation and feedback. As the field develops, a number of gaps in knowledge and capability remain to be addressed. These include, foremost, understanding at finer resolution the roles of different cell populations and complement activities in the therapeutic effects of anticancer and antipathogen mAbs. In addition, there is an urgent need for greater definition of the optimal FcγR profiles for each immune cell type, as well as for modifications that more precisely tune the Fc region to these target specificities. Finally, the capacity to improve antibodies by altering their Fc regions will put more pressure on antibody development in early research stages to understand with confidence the mechanism(s) of action of a given mAb.

ACKNOWLEDGMENTS

We thank Bassil Dahiyat, Bill Strohl, and Zhiqiang An for encouragement.

REFERENCES

Abdullah, N., J. Greenman, A. Pimenidou, K.P. Topping, and J.R. Monson. 1999. The role of monocytes and natural killer cells in mediating antibody-dependent lysis of colorectal tumour cells. *Cancer Immunol. Immunother.* 48(9):517–524.

Albertsson, P.A., P.H. Basse, M. Hokland, et al. 2003. NK cells and the tumour microenvironment: Implications for NK-cell function and anti-tumour activity. *Trends Immunol.* 24(11):603–609.

Angal, S., D.J. King, M.W. Bodmer, et al. 1993. A single amino acid substitution abolishes the heterogeneity of chimeric mouse/human (IgG4) antibody. *Mol. Immunol.* 30(1):105–108.

Armour, K.L., M.R. Clark, A.G. Hadley, and L.M. Williamson. 1999. Recombinant human IgG molecules lacking Fcgamma receptor I binding and monocyte triggering activities. *Eur. J. Immunol.* 29(8):2613–2624.

Arnould, L., M. Gelly, F. Penault-Llorca, et al. 2006. Trastuzumab-based treatment of HER2-positive breast cancer: An antibody-dependent cellular cytotoxicity mechanism? *Br. J. Cancer* 94(2):259–267.

Baker, M. 2006. Anti-infective antibodies: Finding the path forward. *Nature Biotechnol.* 24(12):1491–1493.

Bannerji, R., S. Kitada, I.W. Flinn, et al. 2003. Apoptotic-regulatory and complement-protecting protein expression in chronic lymphocytic leukemia: rRelationship to in vivo rituximab resistance. *J. Clin. Oncol.* 21(8):1466–1471.

Beenhouwer, D.O., E.M. Yoo, C.W. Lai, M.A. Rocha, and S.L. Morrison. 2007. Human immunoglobulin G2 (IgG2) and IgG4, but not IgG1 or IgG3, protect mice against *Cryptococcus neoformans* infection. *Infect. Immun.* 75(3):1424–1435.

Bolt, S., E. Routledge, I. Lloyd, et al. 1993. The generation of a humanized, non-mitogenic CD3 monoclonal antibody which retains in vitro immunosuppressive properties. *Eur. J. Immunol.* 23(2):403–411.

Boruchov, A.M., G. Heller, M.C. Veri, E. Bonvini, J.V. Ravetch, and J.W. Young. 2005. Activating and inhibitory IgG Fc receptors on human DCs mediate opposing functions. *J. Clin. Invest.* 115(10):2914–2923.

Bowles, J.A., and G.J. Weiner. 2005. CD16 polymorphisms and NK activation induced by monoclonal antibody-coated target cells. *J. Immunol. Methods* 304(1–2):88–99.

Bredius, R.G., C.A. Fijen, M. De Haas, et al. 1994. Role of neutrophil Fc gamma RIIa (CD32) and Fc gamma RIIIb (CD16) polymorphic forms in phagocytosis of human IgG1- and IgG3-opsonized bacteria and erythrocytes. *Immunology* 83(4):624–630.

Cartron, G., L. Dacheux, G. Salles, et al. 2002. Therapeutic activity of humanized anti-CD20 monoclonal antibody and polymorphism in IgG Fc receptor FcgammaRIIIa gene. *Blood* 99(3):754–758.

Chatenoud, L. 2004. Anti-CD3 antibodies: Towards clinical antigen-specific immunomodulation. *Curr. Opin. Pharmacol.* 4(4):403–407.

Cheung, N.K., R. Sowers, A.J. Vickers, I.Y. Cheung, B.H. Kushner, and R. Gorlick. 2006. FCGR2A polymorphism is correlated with clinical outcome after immunotherapy of neuroblastoma with anti-GD2 antibody and granulocyte macrophage colony-stimulating factor. *J. Clin. Oncol.* 24(18):2885–2890.

Clark, M.R. 1997. IgG effector mechanisms. *Chem. Immunol.* 65:88–110.

Clynes, R. 2006. Antitumor antibodies in the treatment of cancer: Fc receptors link opsonic antibody with cellular immunity. *Hematol. Oncol. Clin. North Am.* 20(3):585–612.

Clynes, R., Y. Takechi, Y. Moroi, A. Houghton, and J.V. Ravetch. 1998. Fc receptors are required in passive and active immunity to melanoma. *Proc. Natl. Acad. Sci. USA* 95(2):652–656.

Clynes, R.A., T.L. Towers, L.G. Presta, and J.V. Ravetch. 2000. Inhibitory Fc receptors modulate in vivo cytotoxicity against tumor targets. *Nature Med.* 6(4):443–446.

Cole, M.S., C. Anasetti, and J.Y. Tso. 1997. Human IgG2 variants of chimeric anti-CD3 are nonmitogenic to T cells. *J. Immunol.* 159(7):3613–3621.

Cox, K.M., J.D. Sterling, J.T. Regan, et al. 2006. Glycan optimization of a human monoclonal antibody in the aquatic plant *Lemna minor. Nature Biotechnol.* 24(12):1591–1597.

Cragg, M.S., and M.J. Glennie. 2004. Antibody specificity controls in vivo effector mechanisms of anti-CD20 reagents. *Blood* 103(7):2738–2743.

Cragg, M.S., S.M. Morgan, H.T. Chan, et al. 2003. Complement-mediated lysis by anti-CD20 mAb correlates with segregation into lipid rafts. *Blood* 101(3):1045–1052.

Dall'Acqua, W.F., K.E. Cook, M.M. Damschroder, R.M. Woods, and H. Wu. 2006. Modulation of the effector functions of a human IgG1 through engineering of its hinge region. *J. Immunol.* 177(2):1129–1138.

Davies, J., L. Jiang, L.Z. Pan, M.J. LaBarre, D. Anderson, and M. Reff. 2001. Expression of GnTIII in a recombinant anti-CD20 CHO production cell line: Expression of antibodies with altered glycoforms leads to an increase in ADCC through higher affinity for FC gamma RIII. *Biotechnol. Bioeng.* 74(4):288–294.

Desjarlais, J.R., G.A. Lazar, E.A. Zhukovsky, and S.Y. Chu. 2007. Optimizing engagement of the immune system by anti-tumor antibodies: An engineer's perspective. *Drug Discov. Today* 12(21–22):898–910.

Dhodapkar, K.M., J.L. Kaufman, M. Ehlers, et al. 2005. Selective blockade of inhibitory Fcgamma receptor enables human dendritic cell maturation with IL-12p70 production and immunity to antibody-coated tumor cells. *Proc. Natl. Acad. Sci. USA*. 102(8):2910–2915.

Dhodapkar, K.M., J. Krasovsky, B. Williamson, and M.V. Dhodapkar. 2002. Antitumor monoclonal antibodies enhance cross-presentation of cellular antigens and the generation of myeloma-specific killer T cells by dendritic cells. *J. Exp. Med.* 195(1):125–133.

Di Carlo, E., G. Forni, P. Lollini, M.P. Colombo, A. Modesti, and P. Musiani. 2001. The intriguing role of polymorphonuclear neutrophils in antitumor reactions. *Blood* 97(2):339–345.

Di Gaetano, N., E. Cittera, R. Nota, et al. 2003. Complement activation determines the therapeutic activity of rituximab in vivo. *J. Immunol.* 171(3):1581–1587.

Drescher, B., T. Witte, and R.E. Schmidt. 2003. Glycosylation of FcgammaRIII in N163 as mechanism of regulating receptor affinity. *Immunology* 110(3):335–340.

Edberg, J.C., and R.P. Kimberly. 1997. Cell type-specific glycoforms of Fc gamma RIIIa (CD16): Differential ligand binding. *J. Immunol.* 159(8):3849–3857.

Edberg, J.C., J.J. Moon, D.J. Chang, and R.P. Kimberly. 1998. Differential regulation of human neutrophil FcgammaRIIa (CD32) and FcgammaRIIIb (CD16)-induced Ca2+ transients. *J. Biol. Chem.* 273(14): 8071–8079.

Edelman, G.M., B.A. Cunningham, W.E. Gall, P.D. Gottlieb, U. Rutishauser, and M.J. Waxdal. 1969. The covalent structure of an entire gammaG immunoglobulin molecule. *Proc. Natl. Acad. Sci. USA* 63(1):78–85.

Ernst, L.K., D. Metes, R.B. Herberman, and P.A. Morel. 2002. Allelic polymorphisms in the FcgammaRIIC gene can influence its function on normal human natural killer cells. *J. Mol. Med.* 80(4):248–257.

Farag, S.S., I.W. Flinn, R. Modali, T.A. Lehman, D. Young, and J.C. Byrd. 2004. Fc gamma RIIIa and Fc gamma RIIa polymorphisms do not predict response to rituximab in B-cell chronic lymphocytic leukemia. *Blood* 103(4):1472–1474.

Fernandes, M.J., E. Rollet-Labelle, G. Pare, et al. 2006. CD16b associates with high-density, detergent-resistant membranes in human neutrophils. *Biochem. J.* 393(Pt 1):351–359.

Ferrara, C., F. Stuart, P. Sondermann, P. Brunker, and P. Umana. 2006. The carbohydrate at FcgammaRIIIa Asn-162: An element required for high affinity binding to non-fucosylated IgG glycoforms. *J. Biol. Chem.* 281(8):5032–5036.

Forero, A., J. Shah, R. Carlisle, et al. 2006. A phase I study of an anti-GD3 monoclonal antibody, KW-2871, in patients with metastatic melanoma. *Cancer Biother. Radiopharm.* 21(6):561–568.

Friend, P.J., G. Hale, L. Chatenoud, et al. 1999. Phase I study of an engineered aglycosylated humanized CD3 antibody in renal transplant rejection. *Transplantation* 68(11):1632–1637.

Gelderman, K.A., S. Tomlinson, G.D. Ross, and A. Gorter. 2004. Complement function in mAb-mediated cancer immunotherapy. *Trends Immunol.* 25(3):158–164.

Gillies, S.D., Y. Lan, K.M. Lo, M. Super, J. Wesolowski. 1999. Improving the efficacy of antibody-interleukin 2 fusion proteins by reducing their interaction with Fc receptors. *Cancer Res.* 59(9):2159–2166.

Godau, J., T. Heller, H. Hawlisch, et al. 2004. C5a initiates the inflammatory cascade in immune complex peritonitis. *J. Immunol.* 173(5):3437–3445.

Golay, J., E. Cittera, N. Di Gaetano, et al. 2006. The role of complement in the therapeutic activity of rituximab in a murine B lymphoma model homing in lymph nodes. *Haematologica* 91(2):176–183.

Guiducci, C., A.P. Vicari, S. Sangaletti, G. Trinchieri, and M.P. Colombo. 2005. Redirecting in vivo elicited tumor infiltrating macrophages and dendritic cells towards tumor rejection. *Cancer Res.* 65(8):3437–3446.

Hamaguchi, Y., J. Uchida, D.W. Cain, et al. 2005. The peritoneal cavity provides a protective niche for B1 and conventional B lymphocytes during anti-CD20 immunotherapy in mice. *J. Immunol.* 174(7):4389–4399.

Hamaguchi, Y., Y. Xiu, K. Komura, F. Nimmerjahn, and T.F. Tedder. 2006. Antibody isotype-specific engagement of Fcgamma receptors regulates B lymphocyte depletion during CD20 immunotherapy. *J. Exp. Med.* 203(3):743–753.

Han, Y., T.R. Kozel, M.X. Zhang, R.S. MacGill, M.C. Carroll, and J.E. Cutler. 2001. Complement is essential for protection by an IgM and an IgG3 monoclonal antibody against experimental, hematogenously disseminated candidiasis. *J. Immunol.* 167(3):1550–1557.

Hatjiharissi, E., M. Hansen, D.D. Santos, et al. 2007. Genetic linkage of Fc gamma RIIa and Fc gamma RIIIa and implications for their use in predicting clinical responses to CD20-directed monoclonal antibody therapy. *Clin. Lymphoma Myeloma* 7(4):286–290.

Hezareh, M., A.J. Hessell, R.C. Jensen, J.G. van de Winkel, and P.W. Parren. 2001. Effector function activities of a panel of mutants of a broadly neutralizing antibody against human immunodeficiency virus type 1. *J. Virol.* 75(24):12161–12168.

Horton, H.H., M.J. Bernett, E.K. Pong, et al. 2008. Potent in vitro and in vivo activity of an Fc-engineered anti-CD19 monoclonal antibody against lymphoma and leukemia. *Cancer Res.* 68(19):8049–8057.

Hutchins, J.T., F.C. Kull, Jr., J. Bynum, V.C. Knick, L.M. Thurmond, and P. Ray. 1995. Improved biodistribution, tumor targeting, and reduced immunogenicity in mice with a gamma 4 variant of Campath-1H. *Proc. Natl. Acad. Sci. USA* 92(26):11980–11984.

Idusogie, E.E., L.G. Presta, H. Gazzano-Santoro, et al. 2000. Mapping of the C1q binding site on rituxan, a chimeric antibody with a human IgG1 Fc. *J. Immunol.* 164(8):4178–4184.

Idusogie, E.E., P.Y. Wong, L.G. Presta, et al. 2001. Engineered antibodies with increased activity to recruit complement. *J. Immunol.* 166(4):2571–2575.

Imai, M., C. Landen, R. Ohta, N.K. Cheung, and S. Tomlinson. 2005. Complement-mediated mechanisms in anti-GD2 monoclonal antibody therapy of murine metastatic cancer. *Cancer Res.* 65(22):10562–10568.

Isaacs, J.D., M.G. Wing, J.D. Greenwood, B.L. Hazleman, G. Hale, and H. Waldmann. 1996. A therapeutic human IgG4 monoclonal antibody that depletes target cells in humans. *Clin. Exp. Immunol.* 106(3):427–433.

Kabat, E.A., T.T. Wu, H.M. Perry, K.S. Gottesman, and C. Foeller. 1991. *Sequences of Proteins of Immunological Interest.* Bethesda, MD: U.S. Department of Health and Human Services.

Kanda, Y., T. Yamada, K. Mori, et al. 2007. Comparison of biological activity among nonfucosylated therapeutic IgG1 antibodies with three different N-linked Fc oligosaccharides: The high-mannose, hybrid, and complex types. *Glycobiology* 17(1):104–118.

Karassa, F.B., T.A. Trikalinos, and J.P. Ioannidis. 2004. The role of FcgammaRIIA and IIIA polymorphisms in autoimmune diseases. *Biomed. Pharmacother.* 58(5):286–291.

Kashii, Y., R. Giorda, R.B. Herberman, T.L. Whiteside, and N.L. Vujanovic. 1999. Constitutive expression and role of the TNF family ligands in apoptotic killing of tumor cells by human NK cells. *J. Immunol.* 163(10): 5358–5366.

Kelly-Quintos, C., L.A. Cavacini, M.R. Posner, D. Goldmann, and G.B. Pier. 2006. Characterization of the opsonic and protective activity against *Staphylococcus aureus* of fully human monoclonal antibodies specific for the bacterial surface polysaccharide poly-N-acetylglucosamine. *Infect. Immun.* 74(5):2742–2750.

Kennedy, A.D., P.V. Beum, M.D. Solga, et al. 2004. Rituximab infusion promotes rapid complement depletion and acute CD20 loss in chronic lymphocytic leukemia. *J. Immunol.* 172(5):3280–3288.

Kim, D.H., H.D. Jung, J.G. Kim, et al. 2006. FCGR3A gene polymorphisms may correlate with response to front-line R-CHOP therapy for diffuse large B-cell lymphoma. *Blood* 108(8):2720–2725.

Koene, H.R., M. Kleijer, J. Algra, D. Roos, A.E. von dem Borne, and M. de Haas. 1997. Fc gammaRIIIa-158V/F polymorphism influences the binding of IgG by natural killer cell Fc gammaRIIIa, independently of the Fc gammaRIIIa-48L/R/H phenotype. *Blood* 90(3):1109–1114.

Kono, H., C. Kyogoku, T. Suzuki, et al. 2005. FcgammaRIIB Ile232Thr transmembrane polymorphism associated with human systemic lupus erythematosus decreases affinity to lipid rafts and attenuates inhibitory effects on B cell receptor signaling. *Hum. Mol. Genet.* 14(19):2881–2892.

Konrad, S., U. Baumann, R.E. Schmidt, and J.E. Gessner. 2006. Intravenous immunoglobulin (IVIG)-mediated neutralisation of C5a: A direct mechanism of IVIG in the maintenance of a high Fc gammaRIIB to Fc gammaRIII expression ratio on macrophages. *Br. J. Haematol.* 134(3):345–347.

Lazar, G.A., W. Dang, S. Karki, et al. 2006. Engineered antibody Fc variants with enhanced effector function. *Proc. Natl. Acad. Sci. USA* 103(11):4005–4010.

Lehrnbecher, T., C.B. Foster, S. Zhu, et al. 1999. Variant genotypes of the low-affinity Fcgamma receptors in two control populations and a review of low-affinity Fcgamma receptor polymorphisms in control and disease populations. *Blood* 94(12):4220–4232.

Li, H., N. Sethuraman, T.A. Stadheim, et al. 2006. Optimization of humanized IgGs in glycoengineered *Pichia pastoris. Nature Biotechnol.* 24(2):210–215.

Li, L., I. Spendlove, J. Morgan, and L.G. Durrant. 2001. CD55 is over-expressed in the tumour environment. *Br. J. Cancer* 84(1):80–86.

Li, X., J. Wu, R.H. Carter, et al. 2003. A novel polymorphism in the Fcgamma receptor IIB (CD32B) transmembrane region alters receptor signaling. *Arthritis Rheum.* 48(11):3242–3252.

Lifely, M.R., C. Hale, S. Boyce, M.J. Keen, and J. Phillips. 1995. Glycosylation and biological activity of CAMPATH-1H expressed in different cell lines and grown under different culture conditions. *Glycobiology* 5(8):813–822.

Macor, P., D. Mezzanzanica, C. Cossetti, et al. 2006. Complement activated by chimeric anti-folate receptor antibodies is an efficient effector system to control ovarian carcinoma. *Cancer Res.* 66(7):3876–3883.

Mantovani, A., S. Sozzani, M. Locati, P. Allavena, and A. Sica. 2002. Macrophage polarization: tumor-associated macrophages as a paradigm for polarized M2 mononuclear phagocytes. *Trends Immunol.* 23(11):549–555.

Martin, W.L., A.P. West, Jr., L. Gan, and P.J. Bjorkman. 2001. Crystal structure at 2.8 Å of an FcRn/heterodimeric Fc complex: Mechanism of pH-dependent binding. *Mol. Cell* 7(4):867–877.

Masuda, K., T. Kubota, E. Kaneko, et al. 2007. Enhanced binding affinity for FcgammaRIIIa of fucose-negative antibody is sufficient to induce maximal antibody-dependent cellular cytotoxicity. *Mol. Immunol.* 44(12):3122–3131.

Matsumiya, S., Y. Yamaguchi, J. Saito, et al. 2007. Structural comparison of fucosylated and nonfucosylated Fc fragments of human immunoglobulin G1. *J. Mol. Biol.* 368(3):767–779.

Meng, R., J.E. Smallshaw, L.M. Pop, et al. 2004. The evaluation of recombinant, chimeric, tetravalent antihuman CD22 antibodies. *Clin. Cancer Res.* 10(4):1274–1281.

Michaelsen, T.E., J.E. Thommesen, O. Ihle, et al. 2006. A mutant human IgG molecule with only one C1q binding site can activate complement and induce lysis of target cells. *Eur. J. Immunol.* 36(1):129–138.

Michon, J.M., A. Gey, S. Moutel, et al. 1998. In vivo induction of functional Fc gammaRI (CD64) on neutrophils and modulation of blood cytokine mRNA levels in cancer patients treated with G-CSF (rMetHuG-CSF). *Br. J. Haematol.* 100(3):550–556.

Mitrovic, Z., I. Aurer, I. Radman, R. Ajdukovic, J. Sertic, and B. Labar. 2007. FCgammaRIIIA and FCgammaRIIA polymorphisms are not associated with response to rituximab and CHOP in patients with diffuse large B-cell lymphoma. *Haematologica* 92(7):998–999.

Musolino, A., N. Naldi, B. Bortesi, et al. 2007. Immunoglobulin G fragment C receptor polymorphisms and response to trastuzumab-based treatment in patients with HER-2/neu-positive metastatic breast cancer. Presented at the AACR Annual Meeting Los Angeles.

Nagarajan, S., S. Chesla, L. Cobern, P. Anderson, C. Zhu, and P. Selvaraj. 1995. Ligand binding and phagocytosis by CD16 (Fc gamma receptor III) isoforms. Phagocytic signaling by associated zeta and gamma subunits in Chinese hamster ovary cells. *J. Biol. Chem.* 270(43):25762–25770.

Nagarajan, S., K. Venkiteswaran, M. Anderson, U. Sayed, C. Zhu, and P. Selvaraj. 2000. Cell-specific, activation-dependent regulation of neutrophil CD32A ligand-binding function. *Blood* 95(3):1069–1077.

Natsume, A., M. In, H. Takamura, et al. 2008. Engineered antibodies of IgG1/IgG3 mixed isotype with enhanced cytotoxic activities. *Cancer Res.* 68(10):3863–3872.

Nechansky, A., M. Schuster, W. Jost, et al. 2007. Compensation of endogenous IgG mediated inhibition of antibody-dependent cellular cytotoxicity by glyco-engineering of therapeutic antibodies. *Mol. Immunol.* 44(7):1815–1817.

Newman, R., K. Hariharan, M. Reff, et al. 2001. Modification of the Fc region of a primatized IgG antibody to human CD4 retains its ability to modulate CD4 receptors but does not deplete CD4(+) T cells in chimpanzees. *Clin. Immunol.* 98(2):164–174.

Nimmerjahn, F., and J.V. Ravetch. 2005. Divergent immunoglobulin G subclass activity through selective Fc receptor binding. *Science* 310(5753):1510–1512.

Niwa, R., E. Shoji-Hosaka, M. Sakurada, et al. 2004. Defucosylated chimeric anti-CC chemokine receptor 4 IgG1 with enhanced antibody-dependent cellular cytotoxicity shows potent therapeutic activity to T-cell leukemia and lymphoma. *Cancer Res.* 64(6):2127–2133.

Norman, D.J., F. Vincenti, A.M. de Mattos, et al. 2000. Phase I trial of HuM291, a humanized anti-CD3 antibody, in patients receiving renal allografts from living donors. *Transplantation* 70(12):1707–1712.

Okazaki, A., E. Shoji-Hosaka, K. Nakamura, et al. 2004. Fucose depletion from human IgG1 oligosaccharide enhances binding enthalpy and association rate between IgG1 and FcgammaRIIIa. *J. Mol. Biol.* 5 336(5):1239–1249.

Parren, P.W., P.A. Warmerdam, L.C. Boeije, et al. 1992. On the interaction of IgG subclasses with the low affinity Fc gamma RIIa (CD32) on human monocytes, neutrophils, and platelets. Analysis of a functional polymorphism to human IgG2. *J. Clin. Invest.* 90(4):1537–1546.

Pollard, J.W. 2004. Tumour-educated macrophages promote tumour progression and metastasis. *Nature Rev. Cancer.* 4(1):71–78.

Preston, M.J., A.A. Gerceker, M.E. Reff, and G.B. Pier. 1998. Production and characterization of a set of mouse-human chimeric immunoglobulin G (IgG) subclass and IgA monoclonal antibodies with identical variable regions specific for *Pseudomonas aeruginosa* serogroup O6 lipopolysaccharide. *Infect. Immun.* 66(9): 4137–4142.

Pricop, L., P. Redecha, J.L. Teillaud, et al. 2001. Differential modulation of stimulatory and inhibitory Fc gamma receptors on human monocytes by Th1 and Th2 cytokines. *J. Immunol.* 166(1):531–537.

Raasveld, M.H., F.J. Bemelman, P.T. Schellekens, et al. 1993. Complement activation during OKT3 treatment: A possible explanation for respiratory side effects. *Kidney Int.* 43(5):1140–1149.

Racila, E., D.M. Racila, J.M. Ritchie, C. Taylor, C. Dahle, and G.J. Weiner. 2006. The pattern of clinical breast cancer metastasis correlates with a single nucleotide polymorphism in the C1qA component of complement. *Immunogenetics* 58(1):1–8.

Radaev, S., S. Motyka, W.H. Fridman, C. Sautes-Fridman, and P.D. Sun. 2001. The structure of a human type III Fcg receptor in complex with Fc. *J. Biol. Chem.* 276(19):16469–16477.

Reddy, M.P., C.A. Kinney, M.A. Chaikin, et al. 2000. Elimination of Fc receptor-dependent effector functions of a modified IgG4 monoclonal antibody to human CD4. *J. Immunol.* 164(4):1925–1933.

Redpath, S., T. Michaelsen, I. Sandlie, and M.R. Clark. 1998. Activation of complement by human IgG1 and human IgG3 antibodies against the human leucocyte antigen CD52. *Immunology* 93(4):595–600.

Richards, J.O., S. Karki, G.A. Lazar, H. Chen, W. Dang, and J.R. Desjarlais. 2008. Optimization of antibody binding to FcgammaRIIa enhances macrophage phagocytosis of tumor cells. *Mol. Cancer Ther.* 7(8):2517–2527.

Salfeld, J.G. 2007. Isotype selection in antibody engineering. *Nature Biotechnol.* 25(12):1369–1372.

Salmon, J.E., J.C. Edberg, N.L. Brogle, and R.P. Kimberly. 1992. Allelic polymorphisms of human Fc gamma receptor IIA and Fc gamma receptor IIIB. Independent mechanisms for differences in human phagocyte function. *J. Clin. Invest.* 89(4):1274–1281.

Salmon, J.E., S.S. Millard, N.L. Brogle, and R.P. Kimberly. 1995. Fc gamma receptor IIIb enhances Fc gamma receptor IIa function in an oxidant-dependent and allele-sensitive manner. *J. Clin. Invest.* 95(6):2877–2885.

Sanders, L.A., R.G. Feldman, M.M. Voorhorst-Ogink, et al. 1995. Human immunoglobulin G (IgG) Fc receptor IIA (CD32) polymorphism and IgG2-mediated bacterial phagocytosis by neutrophils. *Infect. Immun.* 63(1):73–81.

Saphire, E.O., R.L. Stanfield, M.D. Crispin, et al. 2002. Contrasting IgG structures reveal extreme asymmetry and flexibility. *J. Mol. Biol.* 319(1):9–18.

Scallon, B.J., S.H. Tam, S.G. McCarthy, A.N. Cai, and T.S. Raju. 2007. Higher levels of sialylated Fc glycans in immunoglobulin G molecules can adversely impact functionality. *Mol. Immunol.* 44(7):1524–1534.

Schakel, K., E. Mayer, C. Federle, M. Schmitz, G. Riethmuller, and E.P. Rieber. 1998. A novel dendritic cell population in human blood: One-step immunomagnetic isolation by a specific mAb (M-DC8) and in vitro priming of cytotoxic T lymphocytes. *Eur. J. Immunol.* 28(12):4084–4093.

Selenko, N., O. Majdic, U. Jager, C. Sillaber, J. Stockl, and W. Knapp. 2002. Cross-priming of cytotoxic T cells promoted by apoptosis-inducing tumor cell reactive antibodies? *J. Clin. Immunol.* 22(3):124–130.

Sensel, M.G., L.M. Kane, and S.L. Morrison. 1997. Amino acid differences in the N-terminus of C(H)2 influence the relative abilities of IgG2 and IgG3 to activate complement. *Mol. Immunol.* 34(14):1019–1029.

Shields, R.L., J. Lai, R. Keck, et al. 2002. Lack of fucose on human IgG1 N-linked oligosaccharide improves binding to human Fcgamma RIII and antibody-dependent cellular toxicity. *J. Biol. Chem.* 277(30):26733–26740.

Shields, R.L., A.K. Namenuk, K. Hong, et al. 2001. High resolution mapping of the binding site on human IgG1 for Fc gamma RI, Fc gamma RII, Fc gamma RIII, and FcRn and design of IgG1 variants with improved binding to the Fc gamma R. *J. Biol. Chem.* 276(9):6591–6604.

Shinkawa, T., K. Nakamura, N. Yamane, et al. 2003. The absence of fucose but not the presence of galactose or bisecting N-acetylglucosamine of human IgG1 complex-type oligosaccharides shows the critical role of enhancing antibody-dependent cellular cytotoxicity. *J. Biol. Chem.* 278(5):3466–3473.

Shopes, B. 1992. A genetically engineered human IgG mutant with enhanced cytolytic activity. *J. Immunol.* 148(9):2918–2922.

Siberil, S., C. de Romeuf, N. Bihoreau, et al. 2006. Selection of a human anti-RhD monoclonal antibody for therapeutic use: Impact of IgG glycosylation on activating and inhibitory Fc gamma R functions. *Clin. Immunol.* 118(2–3):170–179.

Sondermann, P., R. Huber, V. Oosthuizen, and U. Jacob. 2000. The 3.2-Å crystal structure of the human IgG1 Fc fragment-FcgRIII complex. *Nature* 406(6793):267–273.

Sondermann, P., J. Kaiser, and U. Jacob. 2001. Molecular basis for immune complex recognition: A comparison of Fc-receptor structures. *J. Mol. Biol.* 309(3):737–749.

Spiridon, C.I., M.A. Ghetie, J. Uhr, et al. 2002. Targeting multiple Her-2 epitopes with monoclonal antibodies results in improved antigrowth activity of a human breast cancer cell line in vitro and in vivo. *Clin. Cancer Res.* 8(6):1720–1730.

Stavenhagen, J.B., S. Gorlatov, N. Tuaillon, et al. 2007. Fc optimization of therapeutic antibodies enhances their ability to kill tumor cells in vitro and controls tumor expansion in vivo via low-affinity activating Fc gamma receptors. *Cancer Res.* 67(18):8882–8890.

Stockinger, H. 1997. Interaction of GPI-anchored cell surface proteins and complement receptor type 3. *Exp. Clin. Immunogenet.* 14(1):5–10.

Strohl, W.R. 2006 (Oct. 17). Non-immunostimulatory antibody and compositions containing the same. USSN 11/581,931.

Tao, M.H., S.M. Canfield, and S.L. Morrison. 1991. The differential ability of human IgG1 and IgG4 to activate complement is determined by the COOH-terminal sequence of the CH2 domain. *J. Exp. Med.* 173(4):1025–1028.

Tao, M.H., R.I. Smith, and S.L. Morrison. 1993. Structural features of human immunoglobulin G that determine isotype-specific differences in complement activation. *J. Exp. Med.* 178(2):661–667.

Tawara, T., K. Hasegawa, Y. Sugiura, et al. 2008. Complement activation plays a key role in antibody-induced infusion toxicity in monkeys and rats. *J. Immunol.* 180(4):2294–2298.

Thommesen, J.E., T.E. Michaelsen, G.A. Loset, I. Sandlie, and O.H. Brekke. 2000. Lysine 322 in the human IgG3 C(H)2 domain is crucial for antibody dependent complement activation. *Mol. Immunol.* 37(16): 995–1004.

Treon, S.P., M. Hansen, A.R. Branagan, et al. 2005. Polymorphisms in FcgammaRIIIA (CD16) receptor expression are associated with clinical response to rituximab in Waldenstrom's macroglobulinemia. *J. Clin. Oncol.* 23(3):474–481.

Treon, S.P., C. Mitsiades, N. Mitsiades, et al. 2001. Tumor cell expression of CD59 is associated with resistance to CD20 serotherapy in patients with B-cell malignancies. *J. Immunother.* 24(3):263–271.

Uchida, J., Y. Hamaguchi, J.A. Oliver, et al. 2004. The innate mononuclear phagocyte network depletes B lymphocytes through Fc receptor-dependent mechanisms during anti-CD20 antibody immunotherapy. *J. Exp. Med.* 199(12):1659–1669.

Umana, P., J. Jean-Mairet, R. Moudry, H. Amstutz, and J.E. Bailey. 1999. Engineered glycoforms of an antineuroblastoma IgG1 with optimized antibody-dependent cellular cytotoxic activity. *Nature Biotechnol.* 17(2):176–180.

Vallhonrat, H., W.W. Williams, A.B. Cosimi, et al. 1999. In vivo generation of C4d, Bb, iC3b, and SC5b-9 after OKT3 administration in kidney and lung transplant recipients. *Transplantation* 67(2):253–258.

van der Neut Kolfschoten, M., J. Schuurman, M. Losen, et al. 2007. Anti-inflammatory activity of human IgG4 antibodies by dynamic Fab arm exchange. *Science* 317(5844):1554–1557.

Van Egmond, M. 2008. Neutrophils in antibody-based immunotherapy of cancer. *Expert Opin. Biol. Ther.* 8(1):83–94.

Van Ginderachter, J.A., K. Movahedi, G. Hassanzadeh Ghassabeh, et al. 2006. Classical and alternative activation of mononuclear phagocytes: Picking the best of both worlds for tumor promotion. *Immunobiology* 211(6–8):487–501.

von Sorge, N.M., W.L. von der Pol, and J.G. von de Winkel. 2003. FcgammaR polymorphisms: Implications for function, disease susceptibility and immunotherapy. *Tissue Antigens* 61(3):189–202.

Wells, J., C.G. Haidaris, T.W. Wright, and F. Gigliotti. 2006. Complement and Fc function are required for optimal antibody prophylaxis against *Pneumocystis carinii* pneumonia. *Infect. Immun.* 74(1):390–393.

Weng, W.K., and R. Levy. 2001. Expression of complement inhibitors CD46, CD55, and CD59 on tumor cells does not predict clinical outcome after rituximab treatment in follicular non-Hodgkin lymphoma. *Blood* 98(5):1352–1357.

Weng, W.K., and R. Levy. 2003. Two immunoglobulin G fragment C receptor polymorphisms independently predict response to rituximab in patients with follicular lymphoma. *J. Clin. Oncol.* 21(21):3940–3947.

Weng, W.K., and R. Levy. 2005. Genetic polymorphism of the inhibitory IgG Fc receptor Fc gamma RIIb is not associated with clinical outcome of rituximab treated follicular lymphoma patients. ASH Annual Meeting, Atlanta. Vol. 106, 2430.

West, A.P., Jr., and P.J. Bjorkman. 2000. Crystal structure and immunoglobulin G binding properties of the human major histocompatibility complex-related Fc receptor(,). *Biochemistry* 39(32):9698–9708.

Wise, M.P., A. Gallimore, A. Godkin. 2006. T-cell costimulation. *N. Engl. J. Med.* 355(24):2594–2595; author reply 2595.

Woodle, E.S., D. Xu, R.A. Zivin, et al. 1999. Phase I trial of a humanized, Fc receptor nonbinding OKT3 antibody, huOKT3gamma1(Ala-Ala) in the treatment of acute renal allograft rejection. *Transplantation* 68(5):608–616.

Xia, M.Q., G. Hale, and H. Waldmann. 1993. Efficient complement-mediated lysis of cells containing the CAMPATH-1 (CDw52) antigen. *Mol. Immunol.* 30(12):1089–1096.

Xu, D., M.L. Alegre, S.S. Varga, et al. 2000. In vitro characterization of five humanized OKT3 effector function variant antibodies. *Cell Immunol.* 200(1):16–26.

Yamane-Ohnuki, N., S. Kinoshita, M. Inoue-Urakubo, et al. 2004. Establishment of FUT8 knockout Chinese hamster ovary cells: An ideal host cell line for producing completely defucosylated antibodies with enhanced antibody-dependent cellular cytotoxicity. *Biotechnol. Bioeng.* 87(5):614–622.

Yoo, E.M., L.A. Wims, L.A. Chan, and S.L. Morrison. 2003. Human IgG2 can form covalent dimers. *J. Immunol.* 170(6):3134–3138.

Zalevsky, J., I.W. Leung, S. Karki, et al. 2009. The impact of Fc engineering on an anti-CD19 antibody: increased Fc gamma receptor affinity enhances B-cell clearing in nonhuman primates. *Blood* 113:3735–3743.

Zhang, W., M. Gordon, A.M. Schultheis, et al. 2007. FCGR2A and FCGR3A polymorphisms associated with clinical outcome of epidermal growth factor receptor expressing metastatic colorectal cancer patients treated with single-agent cetuximab. *J. Clin. Oncol.* 25(24):3712–3718.

PART V

PHYSIOLOGY AND *IN VIVO* BIOLOGY

CHAPTER 16

Antibody-Complement Interaction

KILEEN L. MERSHON and SHERIE L. MORRISON

ABSTRACT

The complement cascade is a group of serum proteins that plays a major role in both innate and adaptive immunity. Complement can be activated through any of three pathways: the classical, alternative, and lectin, but only the classical pathway is antigen and antibody dependent. Mouse IgG and human IgG and IgM are the only isotypes that activate complement through the classical pathway, with the level of activation varying among the isotypes. Recombinant IgA can activate through the alternative pathway. Site-directed mutagenesis has been used to identify key amino acids in antibody-complement interaction. In murine IgG, E318, G320, and K322 are essential for complement activation. Complement activation by human isotypes is more complicated, but there is a general dependence on the amino acids in the C_H2 and hinge region of IgG. Complement functions through a variety of mechanisms, including cell lysis, opsonization and phagocytosis, and generation of anaphylatoxins.

16.1 DISCOVERY OF COMPLEMENT

In the late 1890s, Jules Bordet of the Institut Pasteur in Paris demonstrated sheep antiserum to *Vibrio cholerae* was capable of lysing the bacteria (Bordet 1895). In subsequent experiments, when serum from a rabbit immunized with sheep red blood cells (SRBC) was mixed with SRBC, lysis was observed. However, if the serum was heated at 56°C for 30 minutes prior to addition to the SRBC,

Therapeutic Monoclonal Antibodies: From Bench to Clinic. Edited by Zhiqiang An

lysis was abrogated. Serum from a nonimmunized guinea pig did not lyse SRBC, but when added to the heat-treated immune rabbit serum, the mixture was capable of lysis. Bordet concluded bacteriolytic activity requires two factors: antibodies specific for the bacteria, which are not heat sensitive, and a nonspecific heat-sensitive component. It was Paul Ehrlich, of Berlin, who coined the term *complement* (C′), after independently carrying out similar experiments. Ehrlich defined complement as "the activity of blood serum that completes the action of antibody" (Ehrlich 1899/1900).

16.2 THE THREE PATHWAYS OF COMPLEMENT AND THEIR FUNCTIONS

Complement is a cascade of more than 30 serum proteins, which are made mainly in the liver or in macrophages. There are three pathways of complement activation, the classical, the alternative, and the lectin pathways, consisting of unique and shared complement components. All three pathways converge at a single point, the cleavage of C3 to C3a and C3b, which is the central event of complement activation (Fig. 16.1).

The classical pathway, which was discovered by Bordet, is part of the adaptive immune response. This pathway is antibody (Ab) dependent and therefore antigen (Ag) specific. The pathway is initiated by the binding of C1q to the Fc portion of an immobilized antibody, either as part of an antigen-antibody immune complex, or bound to a surface. C1q possesses six legs, each with a binding site for Fc. The affinity of a single binding site for Fc regions is low, such that multiple binding events are required for complement activation. Once C1q has bound multiple Fc regions, the associated C1r and C1s proteases are activated. Activation of C1 results in cleavage of C4 followed by C2, forming C4b2a, which is also a protease. This new protease cleaves C3 into C3a and C3b, forming the protease C4b2a3b, which now cleaves C5. C5b associates with C6, C7, C8, and C9 to form

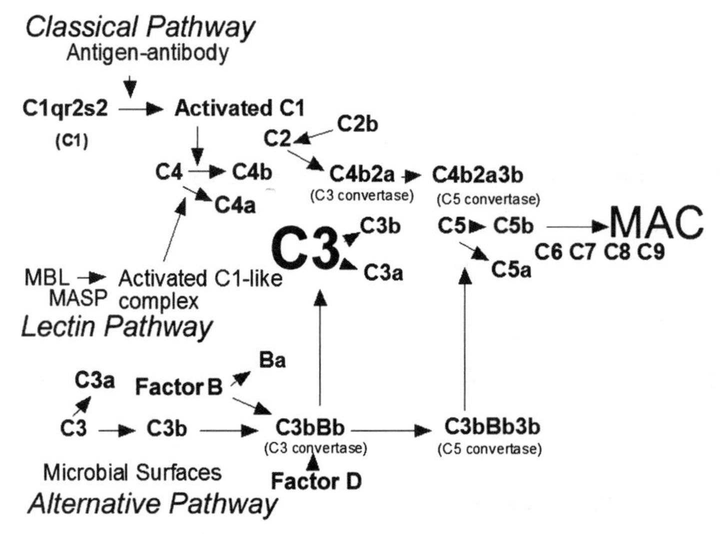

Figure 16.1 Schematic of the complement cascade. The three activation pathways of the complement cascade all converge at the cleavage of C3 into C3a and C3b. In the classical pathway $C1qr_2s_2$ binds an antigen–antibody complex, activating the C1q-associated proteases which cleave C4 and C2 that then form the C3 convertase, C4b2a. In the lectin pathway, mannose-binding lectin (MBL) or other lectins such as the ficolins, bind to microbial surfaces activating the MBL-associated proteases (MASP) which then cleave C4 and C2 to form the C3 convertase, C4b2a. The alternative pathway relies on spontaneous low-level hydrolysis of C3 into C3a and C3b, which then binds to microbial surfaces and associates with factor B. Subsequent cleavage of factor B by factor D results in formation of the C3 convertase, C3bBb. C3 cleavage products are used to form C5 convertases which cleave C5 into C5a and C5b. C5b initiates the assembly of the membrane attack complex (MAC).

the membrane attack complex (MAC). The MAC inserts itself into the lipid bilayer, disrupting the cell membrane and causing osmotic lysis and necrotic cell death.

Although traditionally the classical pathway initiated using C1 has been described as Ab dependent, it has recently been discovered that the classical pathway can also be triggered by the binding of C1 by SIGN-R1, a transmembrane C-type lectin that binds to microbial polysaccharides in the spleen (Kang et al. 2006). It is proposed that in the SIGN-R1 pathway the lectin binds the same C1 globular heads as does immunoglobulins, resulting in activation of the C1r and C1s proteases leading to the capture and proteolysis of C4 and C2 and formation of a C3 convertase.

The lectin pathway is part of the innate immune system. It is similar to the classical pathway, except initiation occurs via interaction of targets with lectins such as ficolins or mannose-binding lectin (MBL), both of which resemble C1q in structure. MBL, also called mannose-binding protein, is a C-type lectin that binds to mannose residues while the ficolins recognize different carbohydrates (reviewed in Holmskov et al. 2003). Following carbohydrate binding, the proteases called mannan-binding lectin associated serine protease (MASP), although they are also associated with the ficolins, activate C4 and C2. After the initial step, complement activation follows the same route as the classical pathway. Recent studies have shown that MBL can also activate complement utilizing components of the alternative pathway in the absence of C4 and/or C2, although this is presumably a minor bypass pathway (Selander et al. 2006).

The alternative pathway is also part of the innate immune response and takes advantage of the continuous low-level C3 cleavage or "tickover" that occurs independently of antigen and antibody. Normally the hydrolyzed C3 is broken down prior to interaction with other complement components; however, if hydrolyzed C3 is deposited on a microbial surface it can bind to factor B. A serum protease, factor D, cleaves the complex to form C3bBb, which is a C3 convertase, creating an amplification loop at the cell surface. As the supply of factor B diminishes, complexes of C3bBb form and act as a C5 convertase, cleaving C5 to C5a and C5b. C5b binds to the cell surface and initiates MAC formation.

16.3 ACTIVITIES OF COMPLEMENT

The cleavage products of complement play a critical role in its function. C3a, C4a, and C5a are anaphylatoxins, small peptides that cause smooth muscle contraction, mast cell and basophil degranulation, and increased vascular permeability. They also induce synthesis of pro-inflammatory cytokines. Anaphylatoxins contribute to multiple diseases and conditions, including hypersensitivity reactions, endotoxin shock, multiple organ failure, and respiratory distress syndrome. Receptors for anaphylatoxins are located on circulating leukocytes, mast cells, macrophages, hepatocytes, lung epithelial cells, endothelial cells, astrocytes, and brain microglial cells. C5a also functions as a chemoattractant and white blood cell activator. Receptors for C3b are present on macrophages and polymorphonuclear leukocytes and the coating of the cell with C3b opsonizes them, marking them for phagocytosis.

16.4 REGULATION OF COMPLEMENT

In general, complement components are highly labile and spontaneously inactivate if they do not interact with other components of the complement cascade. This limits the time complement can function and restricts reactive species to the activation site. There are also complement regulators and inhibitors that function by binding to complement components, preventing their subsequent binding to the next component in the activation cascade.

C1 is regulated by C1 inhibitor (C1Inh), which disrupts the $C1r_2s_2$ complex prior to C4 and C2 activation. The soluble C4b-binding protein (C4bBP) and the membrane-bound complement receptor type 1 (CR1) and membrane cofactor protein (MCP) all bind to C4b to prevent subsequent binding to C2a. After C4b has been bound by a regulatory protein, factor I cleaves C4b into C4d and C4c, which is soluble. C3b can be bound by CR1, MCP, or factor H, another regulatory protein. Bound C3b can no longer interact with factor B, and factor I cleaves C3b into C3f, a soluble component, and iC3b, which remains bound. Factor I can cleave iC3b further into C3c, which is soluble, and C3dg, which is membrane bound. The C3 convertase can also be a source of regulation. C4bBP, CR1, factor H, and decay-accelerating factor (DAF or CD55) can each dissociate the C3 convertase by releasing the enzymatically active component (C2a or Bb) from the surface-bound component (C4b or C3b), which is then cleaved by factor I.

Two proteins regulate the MAC, helping to prevent nonspecific complement-mediated lysis. Homologous restriction factor (HRF) and membrane inhibitor of reactive lysis (MIRL or CD59) bind C8 preventing the subsequent binding of poly-C9 and the assembly of the MAC in the plasma membrane. Inhibition is only seen if the complement components and the target cells are from the same species.

One of the most important actions of complement is to facilitate the uptake and destruction of pathogens by phagocytic cells. Opsonization of pathogens is a major function of C3b and its proteolytic products. C4b also acts as an opsonin, but generally has a minor role in part because less C4b is produced. There are five known complement receptors that bind C3 fragments. CR1 (CD35), as described above, binds to C3b and C4b and functions both to promote C3b and C4b decay and to stimulate phagocytosis. CR2 (CD21) binds to C3d, C3dg, or iC3b and is part of the B cell co-receptor. CR3 (Mac-1 or CD11b/18) and CR4 (CD11c/CD18) bind to iC3b and enhance phagocytosis. Recently a receptor on Kupffer cells that binds C3b and iC3b has been identified (Helmy et al. 2006). This receptor, CRIg, which is a member of the immunoglobulin superfamily, is required for efficient binding and phagocytosis of C3-opsonized particles. Whereas CRIg is localized on constitutively recycling endosomes, CR1, CR3, and CR4 are located on secretory vesicles that internalize ligand through a macropinocytotic process only after cross-linking.

16.5 ANTIBODY-DEPENDENT COMPLEMENT ACTIVATION

The classical pathway of complement activation is dependent on antibodies for pathogen specificity and initiation of the complement cascade. However, not all isotypes activate complement. As discussed above, each binding site on C1q interacts with Fc with low affinity. Therefore, binding of multiple heads of C1q at the same time is required to initiate the complement cascade. Multiple Fcs are present in IgM because of its polymeric structure and binding of one polymeric IgM molecule is sufficient for complement activation. IgM activates the classical pathway most effectively at higher epitope densities and at equivalence or Ab excess, and does not activate the alternative pathway (Lucisano Valim and Lachmann 1991). Activation of complement by IgG requires that two or more IgG molecules be bound to the Ag in close proximity so that C1q can bind to more than one Fc.

16.5.1 IgM and Complement Activation

IgM exists as a hexameric protein comprised of six H_2L_2 building blocks and as a pentameric protein with five H_2L_2 building blocks and J chain although pentamer lacking J chain can also be produced (Wiersma et al. 1998). Highly efficient complement activation is associated with hexameric IgM (Randall et al. 1990); hexameric IgM activates guinea pig complement 100-fold more efficiently than J chain-deficient pentamer, which in turn is more active than J chain-containing pentamer. Within murine IgM, interchain disulfides are present at positions 337, 414, and 575. Using site-directed mutagenesis, it was shown that the disulfide present at position 414 is required for effective complement activation (Davis et al. 1989). It is suggested that the formation of the Cys414-Cys414 disulfide bond rigidified the IgM and that this rigidity is necessary for activation of C1.

Cμ3 in mouse IgM is important for complement-dependent cytolytic activity. The mutations D432G, P434A, and P436S within Cμ3 disrupt cytolysis but not polymerization of IgM (Arya et al. 1994). However, using IgG2b/human IgM exchanged proteins it was demonstrated that Cμ3 alone was not responsible for complement activation by IgM, and that Cμ1, Cμ2, and Cμ4 were required for full complement activation activity (Poon et al. 1995). Using variable region identical antibodies, guinea pig and human C1 bound more efficiently to mouse IgM than to chimeric mouse/human IgM, whereas rat and rabbit C1 bound more efficiently to chimeric IgM (Boulianne et al. 1987). Therefore, differences between human and mouse μ constant regions define structures that are important in C1-IgM interactions and species-specific differences in C1 also influence this interaction.

One question is why does IgM not spontaneously activate complement since it is oligomeric with multiple C1q binding sites. The solution structure of murine and human IgM yields some insight (Feinstein, Richardson, and Taussig 1986; Perkins et al. 1991). It appears that in solution the $(Fab')_2$ and the Fc fragments form a planar structure, where access to the C1q binding site in $C\mu3$ is prevented. Upon antigen binding there is a conformational change so that the $(Fab')_2$ fragments are rotated out of the plane of Fc so that the C1q heads can now bind.

16.5.2 IgG and Complement Activation

Different isotypes of IgG show differing ability to activate complement. In mice, IgG3 is the best activator of complement, with IgG2a and IgG2b also capable of activation. C1q binds within the C_H2 domain and early studies identified the C1q binding site for mouse IgG2b as E318, G320, and K322 (Duncan and Winter 1988) on an exposed loop of C_H2. Murine IgG1, which has R322, does not activate complement.

The amino acid sequence of C_H2 contributes to the differing abilities of the different human IgG isotypes to activate complement. IgG1 and IgG3 are effective in complement activation while IgG2 activates only under selected conditions, and IgG4 does not activate complement through the classical pathway. IgG2 and IgG3 differ at four positions in the N-terminus of C_H2 and six in the C-terminus. Mutants of IgG2 containing the N-terminus of the C_H2 of IgG3 were as effective as wild-type IgG3 in C1q binding, complement activation, and MAC formation, but had reduced ability to effect complement-mediated lysis (Sensel et al. 1997). Changes in IgG1 at amino acids 233 to 236 to those of IgG2 abolished its ability to mediate cell lysis using human complement (Morgan et al. 1995). However, unlike what had been observed with murine IgG2b, a K320A mutation had no effect on IgG1-mediated complement lysis (Fig. 16.2). Studies by others also found that in the context of human IgG1, the K320A mutation as well as E318A did not abolish the ability to activate complement (Idusogie et al. 2000). However, D270A and P329A were deficient in complement activation and C1q binding. Human IgG3 also showed differences from murine IgG2b in that A318 and A320 did not affect antibody-dependent complement lysis; however, K322 was crucial for complement activation by IgG3 (Thommesen et al. 2000).

Amino acid differences in the carboxyl terminus of C_H2 are responsible for the differential ability of IgG1 and IgG4 to activate complement (Tao et al. 1991). S331 in IgG4 is critical for determining the inability of that isotype to bind C1q and activate complement and P331 is important for C1q binding and complement activation in human IgG1 (Tao et al. 1991; Brekke et al. 1994; Xu et al. 1994; Idusogie et al. 2000).

Site-directed mutagenesis at positions within C_H2 has also been used to produce IgG1 that is more effective in complement activation (Idusogie et al. 2001). K326A and E333A showed about a 50 percent increase in binding to C1q and increased complement-dependent cytotoxicity (CDC) activity. K326W gave a threefold increase in C1q binding and a twofold increase in CDC. E333S had a twofold increase in C1q binding and CDC. E326W/E333S showed a fivefold increase in C1q binding but only a 2.3-fold increase in CDC compared to the wild-type IgG1, Rituximab. In the context of IgG2, E333S increased its affinity for C1q while K326W had little effect. Although IgG2 with E333S bound C1q better than IgG1 and showed some CDC, the CDC was greatly reduced compared with IgG1 (see Table 16.1).

In addition to the C_H2 region of IgGs, the hinge also plays a role in complement activation. Within the IgG molecule, the hinge region serves both as a spacer, separating the Fab arms of the Ab from the Fc, as well as a connector between the two heavy chains, providing the interheavy chain disulfide bond. The segmental flexibility of the IgG molecule is correlated with upper hinge length. For human IgG, IgG3 is the most flexible, followed by IgG1, IgG4, and IgG2 (Dangl et al. 1988). When complement activation was examined using a complement consumption assay, the more flexible IgG3 was more

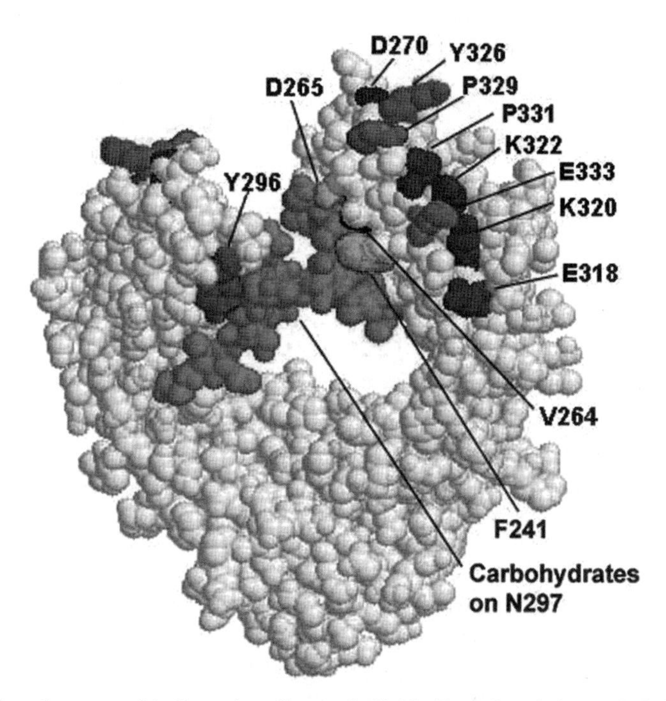

Figure 16.2 Crystal structure of the Fc portion of human IgG1. The Fc portion of a human IgG1 is shown in CPK space-fill form (PDB coordinates 1Fc1). Amino acids that are playing a role in complement binding and activation are colored as follows: F241 magenta; V264 black; D265 violet; D270 blue; Y296 orange; E318, K320, and K322 green-blue; Y326 cyan; P329 purple, P331 red; E333 yellow. Carbohydrates attached to asparagines 297 are colored green. Amino acids L234, L235, and G236 are not shown because they are not resolved in the crystal structure. (See color insert.)

potent than IgG1 (Dangl et al. 1988). When the hinge of IgG4 was used to replace that of IgG3, the molecule had the same ability to bind C1q, but showed decreased ability to activate complement in a complement consumption assay (Tan et al. 1990), although in a different assay system, chimeric IgG3 with the hinge of IgG4 showed improved ability to effect complement-mediated lysis of antigen-coated SRBCs (Norderhaug et al. 1991). Within the context of human IgG3, there was no correlation between proficiency in complement activation and segmental flexibility. Human IgG2, which is the most rigid Ab, was unable to activate human and guinea pig complement, but could consume rabbit complement (Dangl et al. 1988). Thus, as was seen for IgM, there are species-specific differences in complement activation.

Although segmental flexibility does not correlate with the ability to activate complement, an intact hinge is required for effective complement activation. Substitution of the two hinge cysteines of human IgG1 with serine resulted in the production of HL half molecules with reduced but significant ability to activate human complement (Gillies and Wesolowski 1990). However, IgG3 lacking a hinge region could not activate complement (Tan et al. 1990). When a cysteine was introduced between A231 and P232 in hinge deleted IgG3, some H_2L_2 moles were produced, but C1q binding and complement-mediated cytolysis were not observed (Coloma et al. 1997).

Complement activation is dependent on antigen and antibody concentrations, as well as epitope density. When complement fixation was investigated using a complement consumption assay, normal human serum, and anti-NIP (the hapten 4-hydroxy-3-iodo-5 nitrophenylactive acid) chimeric Abs, in general, complement activation was better in antibody excess and at equivalence than antigen excess, and better at high epitope density than at low epitope density (Lucisano Valim and Lachmann 1991). Human IgG1 and IgG3 are good activators of the classical pathway, but do not significantly

TABLE 16.1 Amino Acids on Human IgG that Affect Classical Pathway Complement Activation

Amino Acid Mutation	Isotype(s)	Effect on C′ Activation	References
233–236 to IgG2	IgG1	Decreased	Morgan et al. (1995)
L235E	IgG4	Decreased	Morgan et al. (1995)
V234L/A235L/−236G	IgG2	Greatly increased	Sensel et al. (1997)
F241A	IgG3	Decreased	Lund et al. (1996)
V264A	IgG3	Decreased	Lund et al. (1996)
D265A	IgG3	Decreased	Lund et al. (1996)
D270A	IgG1	Decreased	Idusogie et al. (2000)
D270E	IgG1	No change	Mershon (2008)
V296A	IgG3	Decreased	Lund et al. (1996)
N297A	IgG3	Decreased	Lund et al. (1996)
N297Q	IgG1	Decreased	Tao et al. (1993)
N297Q	IgG3	Decreased	Tao et al. (1993)
E318A	IgG1	No effect	Idusogie et al. (2000)
E318A	IgG3	No effect	Thommesen et al. (2000)
K320A	IgG1	No effect	Morgan et al. (1995); Idusogie et al. (2000)
K320A	IgG3	No effect	Thommesen et al. (2000)
K322A	IgG1	Decreased	Idusogie et al. (2000)
K322A	IgG3	Decreased	Idusogie et al. (2000)
K326A	IgG1	Increased	Idusogie et al. (2001)
K326W	IgG1	Increased	Idusogie et al. (2000)
P329A	IgG1	Decreased	Idusogie et al. (2000)
P331S	IgG1	Decreased	Idusogie et al. (2000)
P331S	IgG3	Decreased	Tao et al. (1993); Xu et al. (1994)
S331P	IgG4	Increased	Tao et al. (1993); Brekke et al. (1994); Xu et al. (1994)
E333A	IgG1	Increased	Idusogie et al. (2001)
E333S	IgG1	Increased	Idusogie et al. (2001)
E333S	IgG2	Increased	Mershon (2008)
K326W/E333S	IgG2	Increased	Idusogie et al. (2001)

activate the alternative pathway. IgG2 is a very poor activator of complement and activates the classical pathway only at high epitope density and at equivalence or antibody excess; however, IgG2 is the best IgG subclass for alternative pathway activation (Lucisano Valim and Lachmann 1991). IgG4 did not activate the classical pathway under any circumstances (Garred et al. 1989) and does not bind human C1q (Bindon et al. 1988).

Although patterns of complement activation change depending on the epitope density, IgG3 was found to bind more C1q than IgG1 or IgG2 regardless of the epitope density. While IgG3 bound C1q better than IgG1, IgG1 was more effective in the lysis of NIP-coated RBC using human serum (Bindon et al. 1988). The advantage of human IgG1 over IgG3 occurs at the stage of C4 activation with IgG1 causing considerably more C4b binding than IgG3 because C1s activated via IgG3 is very inefficient at activation of C4. IgG2 was able to cause RBC lysis using guinea pig, but not human complement (Bindon et al. 1988).

Fc glycosylation influences the ability of IgG to activate complement. In IgG the carbohydrate is not freely mobile, but is largely sequestered within the space enclosed between the C_H2 domains. Aglycosylated IgG1 and IgG3 were found to be inactive in a complement consumption assay (Tao and Morrison 1989). Many different glycoforms can be attached to the conserved glycosylation site at N297 and alterations in the structure of the conserved C_H2 carbohydrates can influence the ability of human IgG1 to bind C1q and activate complement (Wright and Morrison 1998). Alterations in the orientation of the N-glycans also can influence complement activation (White et al. 1997). In two murine IgG2b antibodies the glycans were of apparently the same composition, but exhibited different accessibility as assessed by lectin binding. The antibody with the more accessible glycan did not bind C1q as well (White et al. 1997). Within C_H2 there are multiple noncovalent interactions between the oligosaccharides and the amino acids of the peptide backbone and these noncovalent interactions are necessary for optimal recognition of IgG by C1q (Lund et al. 1996). In particular, replacement of F241, V264, or D265 with alanine in chimeric IgG3 results in reduced recognition by guinea pig complement and human C1q. In contrast to what had been observed earlier with a chimeric IgG3 using a complement consumption assay (Tao and Morrison 1989), in this study aglycosylated IgG3 (N297A) retained some ability to bind C1q and effect lysis of SRBCs derivatized with NIP. Rheumatoid arthritis is associated with a marked increase in IgG glycoforms that lack galactose in the Fc region and terminate in *N*-acetyl glucosamine (GlcNAc). These terminal GlcNAc residues become accessible for MBL binding and the bound MBL was shown to activate complement through the lectin pathway (Malhotra et al. 1995).

It is generally accepted that the constant region of an antibody should retain its ability to interact with the complement system irrespective of its binding specificity. However, when a mouse IgG2b anti-CD25 antibody that was able to lyse target cells in the presence of rabbit complement was converted to a chimeric IgG1 or IgG3, isotypes known to activate complement, it failed to activate CDC with rabbit complement and neither rabbit nor human C1q bound to the antibody (Junghans et al. 1990). Controls confirmed that the variable regions were intact and when transferred back to murine IgG2b once again in CDC to CD25 expressing cells was seen.

16.5.3 IgA and Complement Activation

IgA is a glycoprotein with N-linked carbohydrate present on its heavy chain. The J chain present in polymeric Igs also has one *N*-glycan. Secretory chain (SC), the cleavage product of the polymeric Ig receptor (pIgR), is associated with IgA at mucosal surfaces and is highly glycosylated, with seven *N*-glycan sites. The IgA subclasses differ in their carbohydrate content: IgA1 contains three to five O-linked glycans in its hinge and two N-linked carbohydrate moieties, whereas IgA2 lacks the O-linked glycans but contains either four [IgAm(1)] or five [IgA2 m(2) and IgA2(n)] N-linked carbohydrate moieties. About one-third of the Fab fragments isolated from IgA also contain an *N*-glycan (Mattu et al. 1998). In contrast to IgG, the *N*-glycans of both IgA1 and IgA2 are exposed on the surface (Mattu et al. 1998; Boehm et al. 1999; Herr et al. 2003a, 2003b; Furtado et al. 2004).

There is agreement that IgA does not bind C1q nor does it activate the classical pathway (Kerr 1990); in fact IgA can prevent bacterial lysis by IgGs (Hamadeh et al. 1995). However, there are conflicting reports as to whether IgA activates the lectin and/or alternative pathways. In one study, recombinant IgA did not activate the classical pathway in any circumstance but IgA was the best activator of the alternative pathway. IgA is relatively tolerant to epitope density and Ag : Ab ratio (Lucisano Valim and Lachmann 1991). Interpretation of the studies is frequently made difficult by the nature of the assays and the fact that for many studies the IgA is not in the biologically relevant conformation. An additional issue is the source of the complement used for the assay because, as discussed above, depending on the species from which the complement was obtained, different degrees of complement activation are observed.

A common feature of IgA that contributes to activation of the alternative pathway is the presence of carbohydrate. When immune complexes made using aglycosylated IgA2 produced by growing cells in the presence of tunicamycin were compared to those using glycosylated IgA2, the immune complexes made with the latter were at least five times as potent as those lacking carbohydrate in consumption of complement (Zhang and Lachmann 1994). However, the experiments did not distinguish between the lectin and alternative pathways.

The contribution of IgA *N*-glycans to complement activation was examined using recombinant IgA1 that lacked one or both of the N-glycosylation sites (Chuang and Morrison 1997). Wild-type and mutant IgAs were bound to antigen-coated plates and normal human serum was added to determine

C3 binding by ELISA. IgA1 lacking the C_H2 glycan showed C3 deposition similar to wild-type IgA1. However, IgA1 lacking the C_H3 glycan showed reduced C3 deposition and IgA1 lacking both *N*-glycans showed minimal C3 deposition. The assays were performed in the presence of 10 mM EGTA so activation of C3 should be solely through the alternative pathway.

Human IgA purified from pooled normal human serum has been reported to bind MBL and activate human complement, but for this study the IgA was directly coated on to microtiter plates, which can denature/distort the protein (Roos et al. 2001). Similarly, polymeric but not monomeric IgA from patients with IgA nephropathy coated on plastic bound MBL (Roos et al. 2006). In contrast, Terai et al. (2006) found that serum IgA and most myeloma IgA proteins did not bind MBL although one IgA2 m(2) myeloma did bind; enzymatic degalactosylation and denaturation led to binding by all. These results would appear to indicate that MBL cannot bind the glycans in the native protein but can either bind following the alteration of the structure of the glycans or denaturation of the protein. IgA1 made in baculovirus was reported to activate the alternative pathway, but the data presented did not exclude the MBL pathway and the structure of the attached glycans was not reported (Carayannopoulos et al. 1994).

In other species, 12 different rabbit IgAs produced recombinantly were found to activate the alternative, but not the classical pathway (Schneiderman et al. 1990), using rabbit complement. Activation occurred in the presence of EGTA, confirming that it was via the alternative, not the lectin, pathway.

16.6 SUMMARY

The three pathways of the complement cascade each play unique roles in the immune response. While the alternative and lectin pathways are both part of the innate immune system, the antibody-dependent classical pathway bridges innate and adaptive immunity. IgM and IgG but not IgA activate the classical pathway. Various amino acid mutations have made it possible to identify regions in the C_H2 domain and hinge of IgG that are essential for antibody-complement interactions. Complement activation is highly dependent on antibody isotype, as well as species from which serum is obtained. Complement activation provides protection through various mechanisms, including opsonization by complement fragments to promote phagocytosis, cell lysis by the MAC, and generation of anaphylatoxins.

REFERENCES

Arya, S., F. Chen, et al. 1994. Mapping of amino acid residues in the C mu 3 domain of mouse IgM important in macromolecular assembly and complement-dependent cytolysis. *J. Immunol.* 152(3):1206–1212.

Bindon, C.I., G. Hale, et al. 1988. Human monoclonal IgG isotypes differ in complement activating function at the level of C4 as well as C1q. *J. Exp. Med.* 168(1):127–142.

Boehm, M.K., J.M. Woof, et al. 1999. The Fab and Fc fragments of IgA1 exhibit a different arrangement from that in IgG: A study by X-ray and neutron solution scattering and homology modelling. *J. Mol. Biol.* 286(5):1421–1447.

Bordet, J.J.B.V. 1895. Les leucocytes et les proprietes actives du serum chez les vaccines. *Ann. Inst. Pasteur* 9:462.

Boulianne, G.L., D.E. Isenman, et al. 1987. Biological properties of chimeric antibodies: Interaction with complement. *Mol. Biol. Med.* 4(1):37–49.

Brekke, O., T. Michaelsen, et al. 1994. Human IgG isotype-specific amino acid residues affecting complement-mediated cell lysis and phagocytosis. *Eur. J. Immunol.* 24(10):2542–2547.

Carayannopoulos, L., E.E. Max, et al. 1994. Recombinant human IgA expressed in insect cells. *Proc. Natl. Acad. Sci. USA* 91(18):8348–8352.

Chuang, P.D., and S.L. Morrison. 1997. Elimination of N-linked glycosylation sites from the human IgA1 constant region: Effects on structure and function. *J. Immunol.* 158(2):724–732.

Coloma, M.J., K.R. Trinh, et al. 1997. The hinge as a spacer contributes to covalent assembly and is required for function of IgG. *J. Immunol.* 158(2):733–740.

Dangl, J.L., T.G. Wensel, et al. 1988. Segmental flexibility and complement fixation of genetically engineered chimeric human, rabbit and mouse antibodies. *EMBO J.* 7(7):1989–1994.

Davis, A.C., K.H. Roux, et al. 1989. Intermolecular disulfide bonding in IgM: Effects of replacing cysteine residues in the mu heavy chain. *EMBO J.* 8(9):2519–2526.

Duncan, A.R., and G. Winter. 1988. The binding site for C1q on IgG. *Nature* 332(6166):738–740.

Ehrlich, P. 1899/1900. Croonian Lecture: On immunity with special references to cell life. *Proc. R. Soc. (London)* 66:424–448.

Feinstein, A., N. Richardson, and M.J. Taussig. 1986. Immunogobulin flexibility in complement activation. *Immunol. Today* 7(6):169–174.

Furtado, P.B., P.W. Whitty, et al. 2004. Solution structure determination of monomeric human IgA2 by X-ray and neutron scattering, analytical ultracentrifugation and constrained modelling: A comparison with monomeric human IgA1. *J. Mol. Biol.* 338(5):921–941.

Garred, P., T.E. Michaelsen, et al. 1989. The IgG subclass pattern of complement activation depends on epitope density and antibody and complement concentration. *Scand. J. Immunol.* 30(3):379–382.

Gillies, S.D., and J.S. Wesolowski. 1990. Antigen binding and biological activities of engineered mutant chimeric antibodies with human tumor specificities. *Hum. Antibodies Hybridomas* 1(1):47–54.

Hamadeh, R.M., M.M. Estabrook, et al. 1995. Anti-Gal binds to pili of *Neisseria meningitidis*: The immunoglobulin A isotype blocks complement-mediated killing. *Infect. Immun.* 63(12):4900–4906.

Helmy, K.Y., K.J. Katschke, Jr., et al. 2006. CRIg: A macrophage complement receptor required for phagocytosis of circulating pathogens. *Cell* 124(5):915–927.

Herr, A.B., E.R. Ballister, et al. 2003. Insights into IgA-mediated immune responses from the crystal structures of human FcalphaRI and its complex with IgA1-Fc. *Nature* 423(6940):614–620.

Herr, A.B., C.L. White, et al. 2003. Bivalent binding of IgA1 to FcalphaRI suggests a mechanism for cytokine activation of IgA phagocytosis. *J. Mol. Biol.* 327(3):645–657.

Holmskov, U., S. Thiel, et al. 2003. Collections and ficolins: Humoral lectins of the innate immune defense. *Annu. Rev. Immunol.* 21:547–578.

Idusogie, E.E., L.G. Presta, et al. 2000. Mapping of the C1q binding site on Rituxan, a chimeric antibody with a human IgG1 Fc. *J. Immunol.* 164(8):4178–4184.

Idusogie, E.E., P.Y. Wong, et al. 2001. Engineered antibodies with increased activity to recruit complement. *J. Immunol.* 166(4):2571–2575.

Junghans, R.P., T.A. Waldmann, et al. 1990. Anti-Tac-H, a humanized antibody to the interleukin 2 receptor with new features for immunotherapy in malignant and immune disorders. *Cancer Res.* 50(5):1495–1502.

Kang, Y.S., Y. Do, et al. 2006. A dominant complement fixation pathway for pneumococcal polysaccharides initiated by SIGN-R1 interacting with C1q. *Cell* 125(1):47–58.

Kerr, M.A. 1990. The structure and function of human IgA. *Biochem. J.* 271(2):285–296.

Lucisano Valim, Y.M., and P.J. Lachmann. 1991. The effect of antibody isotype and antigenic epitope density on the complement-fixing activity of immune complexes: A systematic study using chimaeric anti-NIP antibodies with human Fc regions. *Clin. Exp. Immunol.* 84(1):1–8.

Lund, J., N. Takahashi, et al. 1996. Multiple interactions of IgG with its core oligosaccharide can modulate recognition by complement and human Fc gamma receptor I and influence the synthesis of its oligosaccharide chains. *J. Immunol.* 157(11):4963–4969.

Malhotra, R., M.R. Wormald, et al. 1995. Glycosylation changes of IgG associated with rheumatoid arthritis can activate complement via the mannose-binding protein. *Nature Med.* 1(3):237–243.

Mattu, T.S., R.J. Pleass, et al. 1998. The glycosylation and structure of human serum IgA1, Fab, and Fc regions and the role of N-glycosylation on Fc alpha receptor interactions. *J. Biol. Chem.* 273(4):2260–2272.

Mershon, K.L. 2008. Complement, passively administered antibodies, and disease dissemination in mice infected with *Cryptococcus neoformans* or *Cryptococcus gatti*. Dept. of Microbiology, Immunology & Molecular Genetics. Los Angeles, University of California. Doctoral Dissertation.

Morgan, A.J.N., A.M. Nesbitt, L. Chaplin, M.W. Bodmer, and J.S Emtage. 1995. The N-terminal end of the CH2 domain of chimeric human IgG1 anti-HLA-DR is necessary for C1q, Fc gamma RI and Fc gamma RIII binding. *Immunology* 86(2):319–324.

Norderhaug, L., O.H. Brekke, et al. 1991. Chimeric mouse human IgG3 antibodies with an IgG4-like hinge region induce complement-mediated lysis more efficiently than IgG3 with normal hinge. *Eur. J. Immunol.* 21(10):2379–2384.

Perkins, S.J., A.S. Nealis, et al. 1991. Solution structure of human and mouse immunoglobulin M by synchrotron X-ray scattering and molecular graphics modelling: A possible mechanism for complement activation. *J. Mol. Biol.* 221(4):1345–1366.

Poon, P.H., S.L. Morrison, et al. 1995. Structure and function of several anti-dansyl chimeric antibodies formed by domain interchanges between human IgM and mouse IgG2b. *J. Biol. Chem.* 270(15):8571–8577.

Randall, T.D., L.B. King, et al. 1990. The biological effects of IgM hexamer formation. *Eur. J. Immunol.* 20(9):1971–1979.

Roos, A., L.H. Bouwman, et al. 2001. Human IgA activates the complement system via the mannan-binding lectin pathway. *J. Immunol.* 167(5):2861–2868.

Roos, A., M.P. Rastaldi, et al. 2006. Glomerular activation of the lectin pathway of complement in IgA nephropathy is associated with more severe renal disease. *J. Am. Soc. Nephrol.* 17(6):1724–1734.

Schneiderman, R.D., T.F. Lint, et al. 1990. Activation of the alternative pathway of complement by twelve different rabbit-mouse chimeric transfectoma IgA isotypes. *J. Immunol.* 145(1):233–237.

Selander, B., U. Martensson, et al. 2006. Mannan-binding lectin activates C3 and the alternative complement pathway without involvement of C2. *J. Clin. Invest.* 116(5):1425–1434.

Sensel, M.G., L.M. Kane, et al. 1997. Amino acid differences in the N-terminus of C(H)2 influence the relative abilities of IgG2 and IgG3 to activate complement. *Mol. Immunol.* 34(14):1019–1029.

Tan, L.K., R.J. Shopes, et al. 1990. Influence of the hinge region on complement activation, C1q binding, and segmental flexibility in chimeric human immunoglobulins. *Proc. Natl. Acad. Sci. USA* 87(1):162–166.

Tao, M.H., S.M. Canfield, et al. 1991. The differential ability of human IgG1 and IgG4 to activate complement is determined by the COOH-terminal sequence of the CH2 domain. *J. Exp. Med.* 173(4):1025–1028.

Tao, M.H., and S.L. Morrison. 1989. Studies of aglycosylated chimeric mouse-human IgG: Role of carbohydrate in the structure and effector functions mediated by the human IgG constant region. *J. Immunol.* 143(8):2595–2601.

Tao, M.H., R.I. Smith, et al. 1993. Structural features of human immunoglobulin G that determine isotype-specific differences in complement activation. *J. Exp. Med.* 178(2):661–667.

Terai, I., K. Kobayashi, et al. 2006. Degalactosylated and/or denatured IgA, but not native IgA in any form, bind to mannose-binding lectin. *J. Immunol.* 177(3):1737–1745.

Thommesen, J.E., T.E. Michaelsen, et al. 2000. Lysine 322 in the human IgG3 C(H)2 domain is crucial for antibody dependent complement activation. *Mol. Immunol.* 37(16):995–1004.

White, K.D., R.D. Cummings, et al. 1997. Ig N-glycan orientation can influence interactions with the complement system. *J. Immunol.* 158(1):426–435.

Wiersma, E.J., C. Collins, et al. 1998. Structural and functional analysis of J chain-deficient IgM. *J. Immunol.* 160(12):5979–5989.

Wright, A., and S.L. Morrison. 1998. Effect of C2-associated carbohydrate structure on Ig effector function: Studies with chimeric mouse-human IgG1 antibodies in glycosylation mutants of Chinese hamster ovary cells. *J. Immunol.* 160(7):3393–3402.

Xu, Y., R. Oomen, et al. 1994. Residue at position 331 in the IgG1 and IgG4 CH2 domains contributes to their differential ability to bind and activate complement. *J. Biol. Chem.* 269(5):3469–3474.

Zhang, W., and P.J Lachmann. 1994. Glycosylation of IgA is required for optimal activation of the alternative complement pathway by immune complexes. *Immunology* 81(1):137–141.

CHAPTER 17

Bacteria Immunoglobulin-Binding Proteins: Biology and Practical Applications

LESLIE COPE and TESSIE McNEELY

ABSTRACT

Bacteria from several genera produce unique proteins which tightly bind mammalian immunoglobulin (immunoglobulin-binding proteins, IBP). These proteins are displayed on bacterial cell wall surfaces, in addition to being excreted. The best known of these are protein A from *Staphylococcus aureus* (SPA), protein G from groups G and C streptococcus (PG), and protein L from *Peptostreptococcus*

Therapeutic Monoclonal Antibodies: From Bench to Clinic. Edited by Zhiqiang An

magnus (PL). These proteins were originally assumed to act as virulence factors, protecting the bacteria during infection; however, this role is only recently beginning to be elucidated. Most of the research on these proteins centers on their ability to bind immunoglobulin and their uses in both the laboratory and industrial setting as immunological tools. A vast number of applications have been developed from these "virulence factors" for the advancement of science and medicine. The biology of SPA, PG, and PL will be discussed, as well as major developments in the uses of these proteins as immunological reagents or tools.

17.1 INTRODUCTION

The discovery of immunoglobulin-binding proteins (IBP) has been a boon to science. Without these molecules, a large degree of scientific progress and innovative technology would not have been possible. These molecules are still the mainstay of antibody purification and are also being used for newer and more creative protocols for both medicine and industry. The workhorses of this technology have come from proteins A, G, and L. This chapter will review the discovery and characterization of these proteins, followed by current uses for them in science. However, bacteria do not produce these proteins for the benefit of science. IBP play an important role in the survival and proliferation of the bacteria in the host. Their function of binding host immunoglobulin is a direct adaptation to host-parasite interactions.

There are a large number of bacterial species that produce IBP. One of the most interesting features of these molecules is the lack of primary sequence identity between them, yet they form similar secondary and tertiary structures, with similar functions. This is an example of convergent evolution in which a number of species have independently generated a protein with a common function that gives them a selective survival advantage (Frick et al. 1992). In addition to the three molecules that will be discussed, there are several additional IBP that are worth a brief mention. Proteins A, G, and L (Fig. 17.1) are produced by Gram positive organisms, yet Gram negative bacteria also employ IBP as virulence factors. *Haemophilus influenzae* and *Moraxella catarrhalis* both possess IgD-binding proteins. It has been postulated that these IgD-binding proteins can modify the immune response in

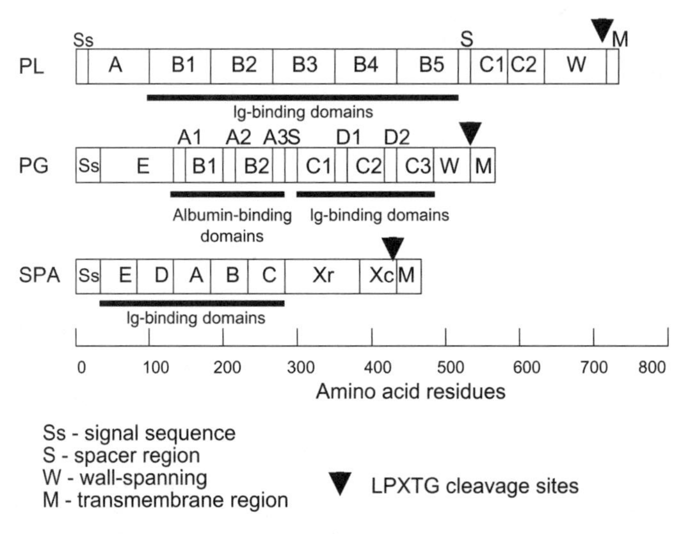

Figure 17.1 Comparison of amino acid sequences of proteins A, G, and L, and arrangements of subdomains. Domain designation is based on sequencing reports (J.P. Murphy et al. 1994; Kastern, Sjobring, and Bjorck 1992). These protein sequences have similar protein frameworks. They all have domains for signal sequences, wall spanning regions (termed Xr and Xc in SPA), LPXTG cleavage sites and transmembrane regions that are removed upon binding to peptidoglycan by a sortase A-like molecule. The immunoglobulin- and albumin-binding regions are further described in the text. The function of A, C1, and C2 domains of PL, the E domain of PG, and portions of the Xr and Xc domains in SPA still await further investigation.

pneumonia and otitis media by binding to B-cell receptors (BCR) in the local environment (Forsgren and Grubb 1979; Forsgren et al. 2001; M. Samuelsson et al. 2007). The association between bacterial IBP antibody specificity and bacterial location in the body was recognized by Goward et al. (1993). They observed that bacteria that are frequently found in the blood stream have IgG-binding activity. Bacteria associated with the mucosa, which has locally high IgA levels, have IgA-binding activity, for example, Group B streptococci (Russelljones, Gotschlich, and Blake 1984) and bacteria that bind IgD, for example, *Haemophilus* and *Moraxella*, are found in the lower respiratory tract and inner ear, which have higher local levels of IgD.

This chapter will describe the three most predominately used molecules for Ig purification: protein A from *Staphylococcus aureus* (SPA), protein G from groups G and C streptococcus (PG), and protein L from *Peptostreptococcus magnus* (PL). Each of these molecules has its advantages and disadvantages. SPA has received the greatest amount of scientific scrutiny, yet it cannot bind all subclasses of IgG. PG can bind all subclasses of IgG, but is not as stable as SPA. PL can bind all classes of Ig, but only those that have κ light chains. Even with these drawbacks, these proteins have met the majority of needs for antibody purification. Modification of the existing IBP appears to be the main avenue for increasing the utility of these molecules, rather than development of new IBP. With the knowledge gained from the study and utilization of these proteins, if a new and more advantageous IBP is found, scientists would be able to quickly adapt this molecule to add to our tool box for medical and scientific discovery.

17.2 IBP: BIOLOGICAL FUNCTION AND STRUCTURE

17.2.1 Protein A

SPA was first characterized in *Staphylococcus aureus* strain Cowen (Forsgren and Sjoquist 1966) and named protein A by Grov, Oeding, and Myklestad (1964), to distinguish it from polysaccharide A (Julianelle and Wieghard 1935). *S. aureus* is a Gram positive coccus that constitutes a part of the normal flora in the nose, axilla, and rectum (Hook and Foster 2000). This organism is normally a benign commensal, but may become an opportunistic pathogen if it gains entry into the body. It is the cause of numerous diseases, ranging from minor skin infections, such as carbuncles and furuncles, to more serious sequelae, such as septicemia, deep tissue infections, endocarditis, pneumonia, and septic arthritis. *S. aureus* produces multiple virulence factors. Among these are the polysaccharide capsule, which has been associated with resistance to phagocytosis, and numerous exotoxins, such as α-toxin, a cytolytic pore-forming toxin, and leukocidins (e.g., β-, δ-, γ-toxins, and Panton-Valentine). *S. aureus* also displays a large number of surface proteins involved in bacterial adherence, called MSCRAMMs (microbial surface components recognizing adhesive matrix molecules). These proteins are thought to enhance bacterial invasion by promoting binding to fibrinogen, collagen, and fibronectin.

The role of SPA in bacterial virulence has been investigated by numerous groups. Initially it was believed that SPA functions principally by binding IgG and thus coating and camouflaging the bacterium with host protein (Dossett et al. 1969). To investigate SPA function *in vivo*, several groups generated mutants by use of chemical mutagenesis (Forsgren et al. 1971; Jonsson et al. 1985); however, more definitive studies were performed with a well-defined SPA deletion mutant. Using this mutant, it was demonstrated that SPA has only a small effect on the ability of *S. aureus* to generate a subcutaneous skin infection or produce septic arthritis in a mouse model (Patel et al. 1987; Palmqvist et al. 2002); this is most likely due to compensation by other virulence factors.

On a cellular level, investigators have determined a number of potential virulence functions of SPA. SPA may interfere with phagocytosis of opsonized bacteria by competing with phagocytic cells for available IgG-Fc receptors. This serves to diminish the local IgG pool mediating opsonization (Peterson et al. 1977). SPA activates the classical complement pathway and induces nonspecific reduction of complement components by binding to IgG or the V_H3 region of IgM (Kozlowski et al. 1996; Stalenhe et al. 1973). SPA is also an adhesion molecule and was determined to be the von Willebrand factor (vWF) binding protein of *S. aureus* (Herrmann et al. 1997). Following endothelial damage, vWF mediates platelet adhesion to begin tissue clotting and repair. SPA recognition and binding to vWF presumably allows bacterial adhesion at sites of injury. Evidence for this function included use of an SPA deletion mutant, which was shown to have a reduced ability to bind vWF. *S. aureus* binding to vWF was competed by soluble vWF, IgG, and anti-SPA antibodies. Even though IgG binding to SPA blocks binding to vWF *in vitro*, *in vivo* bacteria may not be completely coated with IgG, and therefore some SPA should be available to bind vWF (Hartleib et al. 2000). SPA also binds platelet gC1q receptor (gC1qR). This protein is found in high levels on activated platelets and functions to bind the globular heads of the C1q complement component. SPA appears to bind to gC1qR outside of the C1q binding site. Studies indicate that the gC1qR binding site on SPA is the Fc binding domain. Binding between SPA and gC1qR is thought to help localize *S. aureus* to platelet surfaces (Nguyen, Ghebrehiwet, and Peerschke 2000). SPA was recently found to mimic the ability of TNF-α to activate tumor necrosis factor receptor (TNFR1) and induce shedding of TNFR1 from airway epithelium during pneumonia (Gomez et al. 2004). This release of TNFR1 neutralizes free TNF-α in the airway, which can presumably enhance the growth and spread of *S. aureus*. A second potential role of SPA in pneumonia pathogenesis is binding to epidermal growth factor receptor (EGFR), which can secondarily regulate the availability of TNFR1 on mucosal cells (Gomez, Seaghdha, and Prince 2007). SPA has been show to activate basophils and mast cells by cross-linking the IgE surface receptors; this in turn activates the release of histamine (Genovese et al. 2000). Recently, investigators have elucidated an important virulence property of SPA, which is to function as a B-cell superantigen (see Section 17.2.5).

SPA binding to Ig was first described by Forsgren and Sjoquist (1966). They found that the protein was able to bind to human IgG antibodies and the binding was both heavy chain and Fc specific. Further work characterized the heavy chain binding to the Cγ2-Cγ3 region which has been termed the "classical binding site" (Deisenhofer 1981). This region is the same domain that PG recognizes and the two molecules can compete with each other for IgG binding. However, SPA does not bind to the same subclasses as PG. SPA binds well to IgG1, IgG2, and IgG4, and poorly to IgG3 (Sauereriksson et al. 1995). Presumably the binding region is close but not exactly the same as PG, since the subclass binding is different between the two molecules. In addition to Fc binding, SPA was found to also bind to Fab by binding to the conserved portion of the V_H3 germline sequence (Inganas, Johansson, and Bennich 1980; Roben, Salem, and Silverman 1995). This V_H3 region has been found to be expressed in a large number of antibodies involved in the innate immune system (see B-cell superantigens, Section 17.2.5). The *spa* gene is composed of a number of distinct segments and produces a 42 kDa surface protein. After the signal sequence, there are five 56- to 61-amino acid (aa) Ig-binding domains (E, D, A, B, and C) (Fig. 17.1). Following this region, there are two cell wall spanning segments, X_r and X_c, the LPXTG peptidoglycan attachment site, and an 18- to 20-aa hydrophobic tail (Lofdahl et al. 1983; Uhlen et al. 1984). Attachment of this protein to the cell wall is mediated by the enzyme sortase A. This transpeptidase cleaves the sorting signal between the threonine and glycine of the LPXTG motif. The C-terminal threonine is then linked to peptidoglycan by an amide bond to cell wall cross bridges (Marraffini, Dedent, and Schneewind 2006). Recently the location of newly synthesized SPA was described by DeDent, McAdow, and Schneewind (2007). They showed that SPA appears on the bacterial surface in two to four distinct foci and is then spread across the surface from these sites. This molecule is characterized as an elongated, fibrous three-dimensional structure, similar to both PG and PL (Wikstrom et al. 1994). The secondary structure of the Ig-binding domain is different from PG and PL and consists of three α-helices (Deisenhofer 1981). Helices 1 and 2 bind to Fc primarily via hydrophobic interactions and are stabilized by four hydrogen bonds between the two molecules. Fab binding results from interactions with helices 2 and 3, which bind the V_H3 region of the Ig molecule (Sasso, Silverman, and Mannik 1991; Graille et al. 2000).

17.2.2 Protein G

Protein G (PG) was initially discovered in group G streptococcus (GGS), group C streptococcus (GCS), and group A streptococcus (GAS) (Kronvall 1973). PG from groups G and C streptococcus was described in 1984 (Bjorck and Kronvall 1984; Reis, Ayoub, and Boyle 1984), and cloned in 1986 (Guss et al. 1986; Fahnestock et al. 1986; Olsson et al. 1987). The distinction between the various groups of streptococci is based on their Lancefield grouping. Human isolates of GGS and GCS have been described as Gram positive β-hemolytic streptococci that produce large colonies on horse blood agar. Many taxonomic studies have resulted in a division of the GGS and GCS into human and animal origin. The human origin strains of GGS and GCS are classified as *S. dysgalactiae* subsp. *equisimilis*, and some GCS strains are also classified as *S. equi* subsp. *zooepidemicus* (Malke 2000).

GGS and GCS are found as opportunistic commensals on skin, nose, throat, vagina, and gastrointestinal tract (Gaunt and Seal 1987). Even though these strains have a number of virulence factors in common with GAS, they are not as pathogenic, probably due to the lack of pyrogenic exotoxins. These strains have been found, albeit at a low frequency, as the cause of pharyngitis and septicemia, and at an even lower frequency in bone and joint infections, respiratory tract infections, and soft tissue infections. There are a number of GGS/GCS surface proteins shown to be virulence factors. These strains have M and M-like proteins used for attachment and inhibition of phagocytosis. They have fibronectin-binding proteins, plasminogen-binding proteins, and C5a peptidase. The C5a peptidase cleaves complement-activated C5a anaphylatoxin, reducing this chemotactic signal. In addition to C5a peptidase, GGS have a fibrinogen-binding M-like protein, FOG, which has an antiphagocytic activity through formation of fibrinogen aggregates on the bacterial surface (Johansson, Morgelin, and Frick 2004). These organisms also have a number of extracellular proteins associated with virulence. Streptokinase (plasminogen activator), streptolysin L and S (cytolytic toxins), and streptodornase (DNase) are secreted and help the organism to survive *in vivo* (Malke 2000).

PG from GGS and GCS has an elongated, fibrous three-dimensional structure, similar to both SPA and PL (Wikstrom et al. 1994). Unlike SPA and PL, PG is not secreted into the culture supernatant. This appears to correlate to the lack of B-cell superantigen activity (see Section 17.2.5). This protein has been designated as a virulence factor due to its presence in human infectious isolates and not in animal isolates but there is no direct evidence that it enhances the pathogenesis of these strains. However, it does appear to help prevent initiation of the innate immune response by decreasing C1q binding to the bacteria (Nitsche-Schmitz et al. 2007). PG also has been found to bind α_2-macroglobulin and this ability is thought to enhance the virulence of these bacterial strains (see Section 17.2.4). PG is a bifunctional molecule. In addition to Ig binding, another domain of the protein has been shown to bind albumin (Bjorck et al. 1987). The albumin-binding portion of the molecule is thought to coat the bacterial cell with host protein and disguise it from host defenses. Another function for albumin binding by PG, proposed by Sjobring, is to alter the bacterial cell adherence properties and thus influence the establishment of infection (Sjobring 1992).

Studies have shown multiple forms of PG in both GGS and GCS strains (Sjobring, Bjorck, and Kastern 1991; Otten and Boyle 1991). The first strain sequenced, G148, had a 65 kDa sized PG protein that bound to IgG and human albumin (Olsson et al. 1987). The gene was composed of a number of repetitive regions. Following the signal sequence, an alanine-rich region, E, is then followed by the albumin-binding region. This region is composed of three A subunits divided by two B subunits such that the sequence is A1, B1, A2, B2, and A3. This albumin-binding region is separated by a small spacer region, S, from the IgG-binding portion. The IgG-binding region is also composed of two types of repetitive subunits. First are three C subunits divided by two D subunits in a similar orientation as the albumin-binding region, C1, D1, C2, D2, and C3. This functional region of the protein is followed by a wall spanning region, W, and finally the membrane anchor, M (Fig. 17.1). The PG sequence has an LPXTG domain that binds the protein to cell wall peptidoglycan. Two other PG proteins are found in the GGS and GCS population (Sjobring, Bjorck, and Kastern 1991). A 50.4 kDa protein, as exemplified by GCS strain C40 and a 31.5 kDa protein in GGS strain G53. Both of these proteins have reduced binding of radiolabeled human IgG. This is accounted for by the loss of a 210-aa fragment from the IgG-binding region of the protein, C2 and D2. Possessing only a single full length D and C region, D1 and C3, probably accounts for the lower binding of IgG. In strain G53, the further reduction in size is due to a loss of nearly all of the albumin-binding regions, A1,

B1, A2, and B2. This strain only retains the A3 domain and this portion is not able to mediate albumin binding. Otten and Boyle further characterized the PG molecules from GGS and GCS from a large number of human isolates by relating the binding of probes to the Ig-binding and albumin-binding sequences in a Southern blot with the molecular weight for the protein. They found that the majority of GGS strains (Stone et al. 1989) had three Ig-binding domains and three albumin-binding domains, and GCS strains had mainly two Ig-binding domains and two albumin-binding domains (Olsson et al. 1987; Sjobring 1992). They also found two GGS isolates that did not have albumin-binding but had either three or two Ig-binding domains, and one GGS isolate that had three Ig-binding domains and two albumin-binding domains (Otten and Boyle 1991). Overall, there appears to be a significant amount of variation in the composition of PG molecules, but none have been found that lack IgG binding. There appears to be selective pressure to retain a minimal level of Ig binding in these strains.

PG binds to human IgG heavy chains, but not the light chains. Unlike SPA, PG can bind to all isoforms of human and mouse IgG, which increases its usefulness for IgG purification. This molecule does not bind to IgM, IgA, IgD, or IgE (Table 17.1; Kronvall 1973). The minimal IgG-binding

TABLE 17.1 IBP Affinity for Ig from Various Animal Species

Species	Ig Class or Subclass	Protein A	Protein G	Protein L*
Human	IgG1	S	S	S
	IgG2	S	S	S
	IgG3	W	S	S
	IgG4	S	S	S
	IgM	W	NB	S
	IgD	NB	NB	S
	IgE	M	NB	S
	IgA1	W	NB	S
	IgA2	W	NB	S
	Fab	W	W	S
	scFv	W	NB	S
Mouse	IgG1	W	M	S
	IgG2a	S	S	S
	IgG2b	S	S	S
	IgG3	S	S	S
	IgM	NB	NB	S
Rat	IgG1	W	M	S
	IgG2a	NB	S	S
	IgG2b	NB	W	S
	IgG2c	S	S	S
Cow	Total IgG	W	S	NB
Goat	Total IgG	W	S	NB
Sheep	Total IgG	W	S	NB
Horse	Total IgG	W	S	–
Rabbit	Total IgG	S	S	W
Guinea pig	Total IgG	S	W	W
Pig	Total IgG	S	W	S
Dog	Total IgG	S	W	–
Hamster	Total IgG	M	M	S
Donkey	Total IgG	M	S	–
Cat	Total IgG	S	W	–
Monkey (rhesus)	Total IgG	S	S	–
Chicken	Total IgY	NB	NB	NB

*PL will only bind Ig with κ light chains, no binding to Ig with λ light chains.

Binding: W = weak binding; M = medium binding; S = strong binding; NB = no binding; – = no information.

Source: Adapted from *Pierce (Perbio) Applications Handbook and Catalog*, 2006, p. 459. Data from Bjorck 1984; Akerstrom 1985; Lindmark 1983; Langone 1982; Dechateau 1993.

region is composed of the C domain. The binding affinity of a single unit is $\sim 3 \times 10^7\ M^{-1}$ and two subunits is $\sim 5 \times 10^9\ M^{-1}$. The structure of this unit is similar to protein L. This domain consists of a central α-helix on top of a mixed four-stranded β-sheet. The loop between the α-helix and the third β strand is longer in PG than PL (Wikstrom et al. 1994). The portion of the IgG molecule that is recognized and bound by PG is the Cγ2/Cγ3 region of the heavy chain (Stone et al. 1989). This is close to the hinge region, and binding to this site is presumably why C1q cannot bind to the bacterial cell surface concurrently with IgG; the PG covers or sterically hinders the C1q binding site on the IgG. PG also has an affinity for the Cγ1 domain for Fab binding. There are physically separate regions of PG that bind to the Fc and the Fab portion of the IgG molecule. Fc binding involves PG residues from the α-helix, the extended loop, and the third β strand (Gronenborn and Clore 1993). Fab binding involves PG residues from the second β strand, C-terminal residues from the α-helix, and the C-terminus of the IgG molecule (Derrick and Wigley 1992; Lian et al. 1994).

17.2.3 Protein L

Peptostreptococcus magnus is a Gram positive anaerobic coccus that is the source for protein L (PL). This organism is the most commonly isolated and possibly the most pathogenic of the Gram positive anaerobic cocci (reviewed in Murdoch 1998). It is found in humans and other mammals. *P. magnus* was originally placed in the genus *Peptococcus* and later reclassified as a *Peptostreptococcus* based on the G + C content, and on DNA homology studies. *Peptostreptococcus* spp. are routinely found in humans as normal flora and are recovered from the skin, oropharynx, upper respiratory tract, gut, and urogenital tract. *P. magnus* along with *P. prevotti*, *P. tetradius*, and *P. asaccharolyticus* have been recovered from the vagina of 90 percent of pregnant women and *P. magnus* has been found to be a possible cause of bacterial vaginosis. *P. magnus* has been isolated from a number of infection sites, from skin abscesses to endocarditis. This species is also commonly isolated from soft tissue infections and bone/joint infections (Murdoch 1998). *P. magnus* appears to be aerotolerant and is found in mixed infections with aerobes. Animal studies demonstrated that *P. magnus* was able to synergize with both facultative and anaerobic bacteria to increase abscess formation in animal tissue and skin models (Brook 1987, 1988).

Several virulence factors have been identified in *P. magnus*; a human serum albumin-binding protein, PAB, was found in 42 percent of human isolates (Myhre 1984; Dechateau, Holst, and Bjorck 1996), a capsule was identified and determined to be an important virulence factor by Brook and Walker (1985), and PL was characterized as a human immunoglobulin-binding protein (Bjorck 1988). The importance of PL as a virulence factor in *P. magnus* infections has not been extensively studied. Kastern et al. (1990) showed that PL was only present in 4 out of 30 clinical strains. However, of the seven strains isolated from bacterial vaginosis, all four PL expressing strains were from this group. In addition, the strain used in the original description, strain 312, was also isolated from the urethra and vagina of a patient with bacterial vaginosis. The authors believed that the significant correlation between PL bearing strains and bacterial vaginosis implied that PL was a virulence factor. They postulated that host immunoglobulin binding to the bacterial surface would have an effect on the interaction between host and parasite. Due to the inability to establish an animal model specifically for *P. magnus*, the gene function was examined in a *Streptococcus gordonii* murine vaginal model (Ricci et al. 2001). Four of the five Ig-binding domains of PL were cloned into and replaced the middle of the M6 protein from *S. pyogenes;* the fusion protein was heterologously expressed in *S. gordonii* and shown to bind radiolabeled kappa light chains. This construct was further investigated in a vaginal colonization model in mice. The strain expressing the PL Ig-binding domain was able to colonize the vagina longer than the control strain in both Balb/c and C57BL/6 mice. The increase in colonization was postulated to be due to binding of surface-associated IgA, thus avoiding bacterial clearance. This would give this organism an advantage over the normal flora and thus the potential to cause vaginosis.

Another role for PL in pathogenesis was first described by Patella et al. (1990) working with basophils and human skin mast cells. Their report was further investigated by Genovese et al. (2000). These authors showed that PL was a potent inducer of histamine release by human heart mast cells. This release occurs when *P. magnus* or PL binds the IgE kappa light chain on the surface of basophils and mast cells. Binding triggers the release of histamine and other pro-inflammatory mediators. The amount of histamine released is proportional to the amount of PL present, and this activity can be competed by a kappa light chain-bearing IgE myeloma. This binding activity has been categorized as an Ig superantigen function (Genovese et al. 2000). The authors postulate that *P. magnus* cardiac abscesses and endocarditis infections may be enhanced by PL, yet PL producing strains have only been found in vaginosis infections. Human mast cells are spread throughout the body and found associated with all vascularized tissue (Galli 1993). The presence of mast cells in the vagina may contribute to the virulence and persistence of *P. magnus* in this location. In addition to Ig superantigen activity, PL can also function as a B-cell superantigen (see Section 17.2.5).

The purification and description of protein L (PL) from *Peptosteptococcus magnus* was published in 1985 (Myhre and Erntell 1985). They published the description of a heat stable surface protein that was able to bind to Ig kappa light chains. They found this activity on only one strain, 312, of the five evaluated. They showed that whole bacteria were able to bind radiolabeled human polyclonal IgG and 24 out of 40 purified human monoclonal antibodies. These antibodies consisted of all of the IgG subclasses, IgA, and IgM isolated from human myeloma sera. A predominance of kappa light chain was found in the monoclonals that bound the bacteria. They went on to show that light chains purified from polyclonal IgG were able to inhibit binding of radiolabeled IgG to bacteria, while purified heavy chains did not bind. This study was followed by a report by Bjorck, who named the 95 kDa surface protein from *P. magnus* protein L, to indicate the specificity of the binding substrate (Bjorck 1988). He further characterized the binding site on the light chain to the variable region.

PL is present on the surface of the bacteria and can also be found as a $\sim$76 kDa protein released into the culture medium. Since more PL can be found released from the bacteria than still bound to the cell surface, this molecule is apparently naturally released from the cell surface as part of its life cycle (Akerstrom and Bjorck 1989). The function of the secreted PL *in vivo* has not been addressed directly, but is thought to function as a B-cell superantigen.

The binding characteristics of PL are quite distinct from those of SPA or PG, for unlike those proteins, PL binds to the Ig light chain (Bjorck 1988; Akerstrom and Bjorck 1989). PL is has an elongated, fibrous three-dimensional structure, similar in structure to both SPA and PG, even though there is very little sequence identity (Akerstrom and Bjorck 1989; Wikstrom et al. 1994). PL generally consists of an N-terminal region A, that is followed by repetitive Ig-binding domains, B, a repetitive C domain, and ending with a wall spanning domain W, and a C-terminal transmembrane domain, M (Fig. 17.1). PL has five Ig-binding domains, each with 72 to 76 amino acids (Svensson, Hoogenboom, and Sjobring 1998). Each Ig-binding domain consists of a four-stranded β-sheet held by a central α-helix. The residues involved in the interaction between the κ chain and Ig-binding domain are concentrated along the second β strand, the C-terminal end of the α-helix and the loop between the α-helix and the third β strand. Ig binding is through the framework region of the κ light chain variable domain, primarily to the V_{κ} I, III, and IV subtypes (Nilson et al. 1992), which make up about 55 to 60 percent of the Igs in human serum (Solomon 1976). PL does not appear to bind to λ light chains or V_{κ} II light chains due to sequences that result in a secondary structure that prevents PL binding to the "classical binding site" (Nilson et al. 1992; Housden et al. 2003). There is an LPXTG (LPKAG) sequence motif just after the W domain that is used to bind this molecule to the cell wall peptidoglycan (Kastern, Sjobring, and Bjorck 1992). PL can interact with all of the Ig classes, since it binds to light chain variable regions (Wikstrom et al. 1995). For this reason, it also binds effectively to scFv (Wikstrom et al. 1995).

Two strains of *P. magnus* have been sequenced and show slightly different protein sequences. Strain 312 has five Ig-binding domains (Kastern, Sjobring, and Bjorck 1992), while strain 3316 has four Ig-binding domains (J.P. Murphy et al. 1994). Studies with strain 3316 have shown that each binding domain has the ability to bind to two κ light chains, with one site having a 25- to 55-fold higher affinity than the other site (Housden et al. 2003).

17.2.4 α_2 Macroglobulin Binding

Binding to α_2 macroglobulin (α_2M) appears to be a function common to all of the bacterial IBP (Sjobring et al. 1989). α_2M is a large glycoprotein (718 kDa), with several biological functions, as reviewed by Armstrong (2006). It is a broad spectrum proteinase inhibitor and can inhibit serine, thiol, aspartic, and metallo-type proteinases. The native protein consists of four identical subunits. Two forms of α_2M may be identified on nondenaturing PAGE gels. When α_2M is bound to a proteinase, it undergoes a conformational change, which results in a molecule that has faster migration on gels. This is called the fast form (f-α_2M), whereas the native, noncomplexed form is slower migrating, and is referred to as s-α_2M. Radiolabeled SPA, PG, and PL bind α_2M and kininogen (a cysteine proteinase inhibitor), which may be detected with Western blots. Disagreement exists as to which domain of PG recognizes α_2M. In one report, the C domain (Ig-binding domain) of PG was found to interact with α_2M (Sjobring et al. 1989), while another report indicates that the E domain (the alanine rich domain) binds α_2M. Different strains of GGS, which may have different PG phenotypes exhibiting variable numbers of binding domains, were used in these two studies. These differences may explain discrepancies in apparent binding sites.

Muller and Rantamaki (1995) found that PG binds the native (s-α_2M) form of α_2M. S-α_2M is not complexed with proteinases and binds PG in such a manner that any available proteinases can bind concurrently with PG to the s-α_2M. There is conflicting data on the amount of IgG binding to IBP in the presence of α_2M. Some researchers have shown a high degree of inhibition of IgG binding to IBP by α_2M, while others have shown none (Sjobring et al. 1989; Muller and Rantamaki 1995; Chhatwal, Muller, and Blobel 1983). In general, it appears that interference of IgG binding in the presence of α_2M is IgG concentration dependent, and possibly dependent on the number of Ig-binding domains on the IBP. Binding of s-α_2M to the bacterial cell surface may give bacteria a greater advantage *in vivo*, as the native form of α_2M is still capable of binding proteinases. The bacteria may use the α_2M to bind and inactivate host proteinases to protect themselves, or to lower the local concentration of proteinase inhibitor and thus increase tissue destruction by host proteins. This may aid in bacterial spreading.

17.2.5 Superantigen Activity

Several IBP have been classified as B-cell superantigens (SAg). The most extensively studied are SPA and PL (reviewed in Silverman and Goodyear 2006 and Zouali 2007). There is no evidence yet that PG is a B-cell SAg, but it does have the ability to bind to the Fab portion of antibodies, and therefore also to the B-cell receptor (BCR). PG is not a secreted protein, which correlates with its lack of reported SAg activity. Both SPA and PL appear to be secreted in culture, and presumably *in vivo*. Secretion is important, because B-cell SAgs have the greatest effect when they can circulate and interact with large numbers of host cells. As previously described, SPA has the ability not only to bind Ig via the Fc region; it can also bind to the Fab region through the γ heavy chain variable region. PL can bind to the Fab via the light chain variable regions. Due to binding specificity, SPA can only affect B-cells that have BCRs possessing the V_H3 domain. The frequency of BCRs with germline V_H3 is relatively high; one estimate puts it as high as 50 percent of circulating B-cells. PL binds to any κ-bearing light chain BCR and can affect up to 66 percent of B-cells, regardless of their heavy chain type.

B-cell SAgs have been shown to selectively delete marginal zone (MZ) cells, and B1a lineage B-cells in mice reconstituted with human Ig expression genes (Viau et al. 2004, 2005). Both of these B-cell populations are involved in the innate immune system, while B2 cells are more involved in the adaptive immune system. MZ B-cells are found in the extrafollicular regions outside the marginal sinuses of the spleen and have a lower threshold for BCR-based response to antigen. Thus, they react quickly to antigenic stimulation. B2, or follicular B-cells, are mature recirculating B-cells that react slower, and produce antibodies of narrow specificity that depend on MHC-class II mediated T-cell help. B1 cells can be placed into two groups, B1a and B1b, based on CD5 expression. B1a cells are a major source of constitutively expressed IgM. The B1 cells respond quickly to antigens to produce broad specificity Ig, and do not need MHC class II-mediated T-cell help. Both MZ and B1 cells are said to produce the majority of natural antibodies and are considered part of the innate immune defense. Interestingly both SPA and PL have been found to specifically delete the B1a set of B1 cells; the reason for this is still unclear (Zouali 2007).

The mechanism of action for B-cell SAg is to cross-link the BCR via the Fab region (Silverman and Goodyear 2006). This occurs with both SPA and PL because they have multiple binding sites per molecule. This cross-linking leads to patching of the BCR and can lead to four different outcomes; proliferation, activation, migration, and/or deletion. Proliferation can occur when either V_H3-bearing B-cells are cross-linked by soluble SPA, or when κ light chain-bearing B-cells are cross-linked by PL. The B-cells must also receive external stimulatory signals from CD3-specific-antibody-activated T-cells, or IL-2. B-cells can be activated when BCR and SPA or PL complexes recruit co-receptors CD19 and CD21. This leads to upregulation of the early activation markers, CD69 and CD86, followed later by increasing expression of additional markers CD40, CD54, CD80, and CD95. IBP SAgs can also induce the migration of mature B-cells from lymph nodes to the spleen, and the migration of splenic MZ B-cells to the follicular spaces. This leads to a transient increase of follicular B-cells in the spleen. Clonal deletion occurs after prolonged exposure. MZ B-cells are removed early, within 16 to 24 hours, followed by follicular B-cells, within 72 to 96 hours. The mechanism of cell death is through caspase-induced apoptosis.

Which path a susceptible B-cell follows is very dependent on environmental factors. If a small amount of SAg is present, and it binds a relatively small number of B-cells, these B-cells may also receive secondary signals from cytokine or cognate help (IL-4 or CD40L) which are required for the cells to survive. In this case, the B-cells may proliferate or become activated. If a larger concentration of SAg is present, then a larger number of B-cells will have SAg bound, and will be competing for the same amount of co-stimulatory help, resulting in a greater amount of B-cell death.

17.3 EARLY SCIENTIFIC USES

Although SPA was first described in 1940 as a major antigen of *S. aureus* (Verwey 1940), its IgG Fc-binding property was not described for another 26 years (Forsgren and Sjoquist 1966). The high affinity binding of SPA to IgG has led to the development of a multitude of applications for both laboratory- and industrial-scale use. Although other IBP are now in common use, SPA was the first described and was the sole reagent for laboratory immuno-techniques for many years. A wide range of laboratory applications for SPA were described in the 1970 to 1980s. These applications have continued to evolve during the last several decades, as technological advances have been made (covered in Sections 17.4 through 17.11).

After discovery of the Ig-binding property of SPA in the 1960s, multiple applications for this useful protein quickly followed (Goding 1978; Langone 1982a, 1982b, 1982c). Reflecting its normal biological expression as both a cell wall anchored protein and a secreted protein, SPA has been utilized both as an insoluble particulate (in the form of whole bacteria, and later covalently bound to Sepharose®), and as a soluble reagent. For many years, *S. aureus* strain Cowan I (SACI) was the particulate SPA reagent of choice, and it may still be obtained today under the brand name Pansorbin [note that PG-expressing nonviable bacteria are also marketed today under the name of ProACTR (ReceptorPro, Ann Arbor, Michigan)]. Early investigators easily cultured SACI in the lab for an inexpensive source of particulate SPA. Cowan I has a high level of SPA expression, and may be used intact after formalin fixation (1.5 percent, 1.5 hours, room temperature) and heating (80°C, 5 minutes, followed by rapid cooling) to ensure all bacteria are killed (Goding 1978). This reagent may then be used for precipitation of Ig, Ig-antigen complexes (Ig-Ag), and circulating immune complexes (CIC).

For investigating solubilized antigens, it was found that precipitation of Ig-Ag complexes was much more efficient when using particulate SPA (SACI) than the older method of using a second anti-Ig antibody. The SACI method gave fast, quantitative results and was widely adopted in the 1970s.

The investigation of surface expressed antigens was facilitated by using SACI to precipitate antibody-bound solubilized antigens. As described by Langone (1982a), there were four steps to this technique. Proteins on the mammalian cell surface were radiolabeled (metabolically or chemically), cells were solubilized, antigen-specific Ab followed by SACI were added to the mixture, and lastly the precipitated complexes were analyzed using SDS PAGE or other methods. This procedure was used to analyze multiple surface-expressed antigens on mouse, rabbit, human, porcine, and guinea pig cells. Many immunoassays reported in the 1970s to 1980s incorporated SACI precipitation to remove unbound Ag from Ig-Ag complexes. This procedure was included in radioactive ligand immunoassays (RIAs) and other immunoassays for multiple serum proteins, hormones, bacterial toxins, and autoantibodies (reviewed in Langone 1982a). Immobilized SPA was widely used for the affinity isolation of CIC, and in the investigation of Ig contained in CIC found in many diseases, such as systemic lupus erythematosus (SLE), Sjögren syndrome, and rheumatoid arthritis (see also Section 17.10).

SACI was widely used for cell separation techniques through the 1970s. Ig bearing cells formed rosettes with SACI which could be separated by density gradient centrifugation. In this way, lymphocytes were isolated from other peripheral blood cells. Alternatively, SACI was used to coat the bottom of wells in tissue culture plates, and Ig bearing cells specifically bound to the SACI through SPA affinity. Essentially any cells could be isolated by using SPA reactive, antigen-specific Ig bound to their cell surface target and then reacted with SACI-coated plates.

With its high affinity for IgG, one early application for SPA was to selectively adsorb IgG from human serum, leaving IgM and IgA relatively untouched. This procedure helped investigators attribute certain immunological responses in immune sera to IgG or IgM (Langone 1982b). The practical application of this principle was in diagnostic serology where it was used for uncovering relatively low levels of antigen-specific IgM. The low level of IgM could be masked *in vitro* by the large non-immune pool of IgG. This principle was used in the detection of antibodies to diagnose diseases such as rubella, hepatitis A, hepatitis B, herpes simplex virus (HSV)-1, and Coxsackie B.

Among the early applications for SPA was its use as an antibody detection reagent. For this purpose, SPA was used in place of secondary anti-Ig antibodies, which were often unavailable, or of inadequate specificity, or would bind nonspecifically to eukaryote cells (e.g., through Fc receptors; reviewed in Goding 1978). Purified SPA was labeled by one of a variety of techniques, such as fluorescein, ^{125}I, or other radiolabel. The labeled SPA could then be used as a tag for the presence of primary Ig bound to specific antigens on mammalian cells. Alternatively, the labeled SPA could be used to directly detect Ig-producing cells (i.e., lymphocytes).

17.4 IMMUNOGLOBULIN PURIFICATION

17.4.1 Immunoglobulin Isolation from Multiple Species

Early on, purified SPA covalently linked to Sepharose® was demonstrated to be useful as an affinity resin for isolation of immunoglobulin, and has been available commercially since at least 1978 (Hjelm, Hjelm, and Sjoquist 1972; Goding 1978). SPA is a highly stable protein and is very resistant to a wide range of pH, temperatures, and denaturing agents (Sjoholm 1975). Due to these properties, it makes a very stable resin, which may be regenerated up to 200 times (Lindmark, Thorentolling, and Sjoquist 1983). Much work has been devoted to characterizing the binding of Ig from many animal species to SPA (reviewed in Lindmark, Thorentolling, and Sjoquist 1983 and Langone 1982a; Table 17.1). As discussed in Section 17.2.1, the five extracellular homologous domains of SPA each mediate binding to Fc (Uhlen et al. 1984; Moks et al. 1986; Svensson, Hoogenboom, and Sjobring 1998; Fig. 17.1). The highest affinity binding of SPA is to the Fc region of the IgG heavy chain (Deisenhofer 1981). The SPA B domain binds human IgG Fc tightly with a dissociation constant of 2×10^{-8} M (Lindmark, Thorentolling, and Sjoquist 1983). Additionally there is weaker binding to the Fab portion of Ig (Lindmark, Thorentolling, and Sjoquist 1983); specifically, the V_H domain (Vidal and Conde 1985). A small proportion of human, rabbit, mouse, porcine, and guinea pig Fab is recognized by SPA (Proudfoot et al. 1992; Endresen 1979; Zikan 1980; Young et al. 1984), and some human $F(ab')_2$ (Inganas 1981; Biguzzi 1982). SPA binds with high affinity to essentially 100 percent of human IgG isotypes IgG_1, IgG_2, and IgG_4; all mouse IgG isotypes, rabbit IgG, and rhesus macaque IgG (Lindmark, Thorentolling, and Sjoquist 1983). Human IgG_3 is not bound by SPA, and only about 15, 13, and 40 percent of IgG Fab, IgA, and IgM, respectively, are bound (Svensson, Hoogenboom, and Sjobring 1998). Immunoglobulins from more than 140 different species of animals have been tested for binding to SPA (Lindmark, Thorentolling, and Sjoquist 1983). Ig, mainly IgG, from a wide range of mammals binds to SPA, whereas avian IgY, in general does not (Table 17.1).

Binding to SPA is dependent on buffer pH and other column conditions, such as SPA density and column flow rate (reviewed in Boyle 1990). Due to the differential affinity of IgG isotypes for SPA, chromatography over SPA-Sepharose with pH gradient elution may be used to separate and recover individual IgG isotypes, for example, murine IgG1 from IgG2a, IgG2b, and IgG3 (reviewed in Langone 1982a and Goding 1978), or human IgG1 and IgG2 completely free of other Ig, or containing only low levels of other subclasses (Duhamel et al. 1979). This principle also applies to subfractions of IgG, such as F(ab), F(ab′), and $F(ab')_2$ (Goding 1978).

Protein G (PG) has Ig-binding characteristics similar to but distinct from those of SPA (see Section 17.2.2). As a consequence, this IBP has been exploited for purification purposes not amenable with SPA. PG binds the Fc portion of Ig similarly to SPA, at the $C\gamma2/C\gamma3$ interface or the $C\gamma1$ domain (Akerstrom et al. 1994), but with higher affinity to some Ig types (Table 17.1). Of note, PG binds all murine IgG isotypes with higher affinity than SPA (Harlow and Lane 1999). PG binds human IgG subclasses IgG_1, IgG_2, and IgG_4, and, as opposed to SPA, binds IgG_3. It also binds weakly $F(ab')_2$ fragments, Fab′, and the IgG Fab region (Bjorck and Kronvall 1984; Erntell et al. 1988). Optimal binding of IgG to PG is reported to occur at pH 5 to 6, although it has been reported that pH 8 may be optimal for binding to PG-Sepharose (Pilcher et al. 1991; Boyle 1990).

As noted in Section 17.2.3, the binding characteristics of PL are quite distinct from those of SPA or PG, as PL binds to the Ig light chain (Bjorck 1988; Akerstrom and Bjorck 1989). PL binding to Ig light chains does not interfere with antigen binding (Akerstrom and Bjorck 1989) and is dependent on the three-dimensional structure of the Ig variable domain (Nilson et al. 1992). Ig binding is through the framework region of the κ light chain variable domain, primarily to the V_κ I, III, and IV subtypes (Nilson et al. 1992), which make up about 55 to 60 percent of the Igs in human serum (Solomon 1976). This molecule has a binding affinity of $\sim10^9$ to 10^{10} M^{-1} for kappa light chains (Nilson et al. 1992; Akerstrom and Bjorck 1989). PL binds to human, chimpanzee, mouse, rat, rabbit, and guinea pig Ig (Nilson et al. 1993; Dechateau et al. 1993; Table 17.1). Since binding of Ig by PL is not dependent on the heavy chain, it can be used to purify all Ig subgroups. In 1993, Nilson et al. (1993) demonstrated the utility of PL for Ig purification. They reported the isolation of human and mouse IgG, IgM, and IgA from serum in a single step, and the purification of human and mouse monoclonals, human IgG Fab, and Fv fragments, as well as recombinant human/mouse chimeras, using PL. Additionally these authors engineered a humanized mouse Ig in which mouse hypervariable antigen-binding regions were introduced into a human IgG with a κ light chain variable subtype III, to demonstrate introduction of a PL-binding region for purposes of monoclonal purification. This κ light chain framework allows isolation of monoclonals while avoiding the issues of introducing a potentially immunogenic affinity structure, such as C-terminal histidine residues. In 1997, Kouki et al. (1997) published methods to separate human IgG fragments such as free κ light chain, Fab or $F(ab')_2$, and the Fd fragment, using PL affinity chromatography.

17.4.2 Purification of Humanized and Chimeric IgG and Antibody Fragments

In the early 1990s, development of monoclonal and recombinant antibodies for treatment of cancer and other illnesses became an active and important area of research. Investigators needed efficient means to purify these new Ig constructs, preferably without the use of additional protein tags. The use of SPA, PG, and PL for the purification of these novel antibodies was pursued.

Early (murine) monoclonal antibodies developed for clinical use induced a human anti-animal Ig antibody (HAIA) response after multiple injections. To overcome HAIA response to murine monoclonals, chimeric monoclonals (mouse variable regions fused to human constant regions through use of recombinant DNA) were developed. Purification of recombinant mouse-human chimeric Fab′ and $F(ab')_2$ from Chinese hamster ovary (CHO) cell supernatant was described in 1992 (Proudfoot et al. 1992). Differential binding of these constructs to PG allowed the separation of functional Fab′ from the $F(ab')_2$.

With the recent development of single chain variable fragments (scFv) for research and clinical purposes, reliable methods of purification for these small molecules was explored (Akerstrom et al. 1994; Das, Allen, and Suresh 2005; S.L. Li et al. 2000; U.B. Nielsen et al. 2002). In an effort to define the binding characteristics of SPA, PG, protein H (PH), and PL for recombinant Ig fractions, Akerstrom and coworkers (1994) measured binding to a set of 34 human scFv antibodies, expressed in *Escherichia coli* and with known specificity and variable domain types. They found that PG and PH did not bind to the scFv, whereas SPA and PL bound about one-third of the constructs. SPA bound scFv with V_H1, 3, 4, and 5 domains. PL bound all scFv with light chain domains of $\kappa 1$, one with $\lambda 1$, and one with a $\lambda 3$ domain. SPA and PL immobilized on Sepharose® could be used to purify scFv from complex bacterial culture media. The authors noted that their characterization of the IBP agreed with the previously published findings that SPA binds to V_H domains and PL to V_L domains, whereas proteins G and H bind outside this region [note that PH binds through the IgG $C\gamma 2/C\gamma 3$ interface (Frick et al. 1992)]. Das and coworkers (2005) found PL to be an efficient method of purification of scFv, which could be used even in the absence of a His_6 tag. Their one-step purification of anti-CD19-c-myc-His_6-Cys scFv gave a higher yield, and purer product, than purification of this scFv using a Ni-NTA resin.

17.4.3 Molecular Modification of IBP for Enhanced Purification Purposes

The successful and increasing use of monoclonal Ab treatments for cancer, asthma, arthritis, and other diseases necessitates large-scale purification procedures for Ig. Affinity resins made from bacterial IBP are in demand for this purpose. PG, sold as G-Sepharose Fast Flow (Pharmacia, Uppsala, Sweden), contains recombinant PG from which the albumin-binding region of the native molecule was deleted. This resin is used for large-scale purification of human monoclonal mAb for clinical applications (Bill et al. 1995). An advantage of PG is that it binds human IgG_3; however, PG is much less stable than SPA. For this reason, the workhorse of mAb isolation at large scale is SPA-Sepharose® (McCormick 2005). Many manufacturers currently market SPA-based resins for scale-up purification, for example, GE Healthcare, Fermentech, Bio-Rad, Pierce, Tosoh Biosep, Millipore, Amersham Biosciences, and others. Head to head comparisons of these resins for industrial use have been conducted (reviewed in Hober, Nord, and Linhult 2007; Swinnen et al. 2007; Hahn et al. 2005). However, due to many potential evaluation parameters, some of which may be more important in one application than another, no clear recommendations have been made for one resin versus another. The use of SPA and PG resins has been well validated by the pharmaceutical industry over the last decade; however, they remain very expensive to purchase. At around $8,000 to $10,000 per liter, SPA resins are a costly part of the manufacturing of clinical mAbs. For a 300 L column of resin, the cost is as high as $3 million. Due to stringent cleaning-in-place procedures required for validated column regeneration, a resin may be used only 15 to 20 times (McCormick 2005). Improvements to SPA or the other IBP resins have fallen into three broad categories: fusion of individual Ig-binding proteins together, molecular substitutions to the aa sequences, and addition of coupling tags.

To create IBP with wider or improved binding spectra, several groups have made fusion proteins derived from SPA, PG, and PL (Kihlberg et al. 1992; Eliasson et al. 1988; Svensson, Hoogenboom, and Sjobring 1998). Fusion proteins with combinations of SPA and PG, PL and PG, and SPA and PL have been reported. The combination of SPA and PG, into fusion protein AG, optimized the binding properties of the parent proteins, and enlarged the number of IgG recognized and bound

(Eliasson et al. 1988; K. Nielsen et al. 2004). Fusion protein LG, reported by Kihlberg et al. (1992), was made by a combination of PL and PG. The Ig-binding regions of PL and PG were fused to form hybrid protein LG. This protein bound most intact human Igs, as well as Fc and Fab fragments, and Ig light chains. Binding of LG to Ig was specific, with high affinity, in the nanomolar range (Kihlberg et al. 1992). The fusion protein LA was made by fusing four of the Igκ light chain-binding domains of PL with four of the IgG Fc- and Fab-binding regions of SPA. Fusion protein LA maintained binding to κ light chains and IgG Fc, and it bound to ligands with enhanced affinity compared to the parent SPA or PL. Immobilized LA could be used to purify Ig ligands from human serum and saliva. Protein LA bound human Ig of different classes and IgG from a wide range of mammalian species, IgM and IgA, and also scFv.

During the late 1980s and 1990s researchers began to look for ways to modify the Ig-binding proteins to increase protein stability, and thus enhance purification of recombinant antibodies and antibody fragments. SPA and PG stability under harsh column elution and regenerating conditions were of major concern. It is common to use sodium hydroxide in concentrations ranging from 0.1 to 1 M for cleaning Ig affinity resins. Genetic modifications of single amino acids in the B domain of SPA were investigated for enhanced purification efficacy. A single mutation in the B domain at residue 29, from Gly to Ala, enhanced protein stability. The resultant B domain was denoted Z (Nord et al. 1995; Tolmachev et al. 2007). In another example, a single tyrosine residue of domain B was mutated to enhance pH responsiveness (Gore et al. 1992). Increasing or mutating aa in the loop region between helix 1 and 2 in domain B was observed to affect Ig binding, and resulted in variants with milder elution requirements. Using this approach, a variant with elution pH of 4.5 instead of pH 3.3 was reported (Gulich, Uhlen, and Hober 2000). Further improvements on protein stability were obtained by examination of individual asparagine residues, since asparagine structure is susceptible to low pH conditions. Individual asparagines were mutated to aa less susceptible to alkaline conditions, in the background of an SPA Z domain construct. Interestingly, some of the variants showed enhanced alkaline susceptibility, while some were more resistant. Exchange of asparagine 23 for a threonine greatly increased alkaline resistance, up to 0.5 M NaOH. A similar asparagine replacement study was performed for the C2 domain of PG (Gulich et al. 2002; Linhult et al. 2004). In that investigation, asparagine as well as glutamine and aspartate were exchanged for aa with increased alkaline stability. The double mutant $C2_{N7,36A}$ was the most stable to alkaline conditions, and importantly retained its secondary structure and high affinity binding to Ig Fc. The enhanced alkaline resistance of the Z domain and the C2 domain will greatly improve affinity resin stability under conditions of column cleaning and regeneration.

Coupling of Ig-binding proteins to the matrix of support resins has been improved. To increase per gram capacity, Ig-binding proteins should be displayed with their binding region free in solution, while the nonbinding region is used to covalently anchor the protein. By introduction of residues that may be targeted for covalent modification, Ig-binding proteins may be directed in the appropriate orientation on the matrix support (Ljungquist et al. 1989). Johnson and coauthors (2003) engineered a His_6 tag on the C-terminal of the SPA B domain, thus enabling oriented binding versus random binding of this domain to a Ni-NTA surface. In this fashion, the B domain exhibited a single homogeneous, high affinity population, at densities where steric crowding between bound IgG did not affect the apparent receptor affinity. The authors offer the utility of this construct for use in sensor design and purification of Ig, as well as Fc tagged proteins.

17.5 CELL ISOLATION TECHNIQUES

Use of SACI for mammalian cell separation/isolation has generally been supplanted by the advent of magnetic beads, a much cleaner system. The use of 1 μm iron oxide beads coated with SPA was reported in 1979 for the isolation of Ig-bearing cells (Widder et al. 1979). Isolation of the bead-bound cells was achieved using a bar magnet. Specific cells could be isolated from a heterogeneous mixture, such as peripheral blood.

Magnetic beads coated with Ig were developed for cell separation purposes during the 1980s. Use of magnetic beads for cell separation or immunoprecipitation has the benefits of being rapid, highly specific, and simple to perform. The drawback of using beads coated with antigen-specific Ig was that the coupling of Ig to beads could result in low binding efficiency for their target antigen or monoclonal antibody. This was due to nonselective or random coupling of Ig to the beads, through the Fab as well as the Fc region. Widjojoatmodjo and coauthors (1993) compared immunomagnetic beads coated with SPA, PG, or goat anti-mouse immunoglobulin (GαM-beads). These authors found that their SPA and PG magnetic beads had a higher binding capacity for Ig, than the GαM Ig-coated beads. For purposes of immunoseparation of bacteria, the SPA-coated beads were superior to both the PG and the GαM-beads. The authors suggest that this is due to the presence of five Ig-binding domains on the SPA versus only two on PG or GαM.

Magnetic separation of mammalian peripheral cells has recently been demonstrated using SPA expressing bacterial magnetic particles (BacMPs) isolated from *Magnetospirillum magneticum* AMB-1 (Kuhara et al. 2004; Matsunaga et al. 2006). BacMPs are nanometer sized particles containing a single magnetic domain of magnetite covered with a phospholipid membrane. MagA, an iron transporting protein, is embedded in the BacMP membrane, and protein Mms16 binds tightly to the BacMPs surface. SPA was engineered as a fusion protein to either MagA or Mms16. The fusion proteins were expressed on the surface of BacMPs, and maintained Fc-binding activity. The resultant BacMPs possessed good monodispersity, relatively high SPA-Ab-binding capacity and were easy to isolate from *M. magneticum*. The BacMPs were used to isolate with high purity four cell types, CD19+, CD20+, CD8+, and CD14+. BacMPs with the SPA-Mms16 fusion appeared to give better cell separation results, with respect to CD14+ cells. The magnetic particles did not adversely affect the separated cells.

17.6 IBP EXPRESSION BY HETEROLOGOUS MICROBES

Motivated by economic concerns, many groups have begun to search for low cost alternatives to Ig-binding resins. In a technological update of SACI and SPA-Sepharose®, several labs have described recombinant expression of SPA, PG, or the Ig-binding domains of these proteins on the surface of alternative microbe carriers, such as bacteriophage capsid proteins (Kushwaha et al. 1994; Y. Li, Cockburn, and Whitelam 1998), or bacterial S-layer proteins. One such construct was described by Nomellini and coauthors (2007). These authors utilized the surface expression system of *Caulobacter cresentus*, which synthesizes a 1026-amino acid protein RsaA as the sole component of its paracrystalline protein surface layer (S-layer). Approximately 40,000 subunits of RsaA are assembled into a lattice array with hexagonal symmetry on the cell surface. The monomers are secreted from the bacterium via a type I ABC transporter system (Nomellini et al. 2007). Nomellini and coworkers engineered RsaA as a fusion protein with up to three PG B1 binding domains surrounded by Muc-1 (a 20 aa sequence derived from human mucin) spacer sequences. The resultant *C. cresentus* expressed RsaA fusion protein in quantities equivalent to native RsaA in the wild-type bacterium. The ability to bind IgG increased as PG B1 units increased. Those with three PG B1 units bound twice as much rabbit IgG per cell as SACI (Pansorbin). The authors concluded that this recombinant *C. cresentus* is a low cost equivalent to PG-Sepharose® beads for immunoreactive functions, such as immunoprecipitation. Since PG binds a wider range of IgG isotypes and species than SPA, it would be a more versatile reagent than SACI.

Vollenkle et al. (2004) constructed a recombinant protein containing two SPA Z domains in tandem, fused with the S-layer protein of *Bacillus sphaericus* SbpA. The hybrid protein was found to self-assemble in solution (as an S-layer), to bind and recrystallize to its native bacterial support (secondary cell wall polymer), and to present the ZZ domain to the outermost surface of the protein lattice. The fusion protein was assembled on cellulose-based microbeads, a construct developed to purify autoantibodies from blood. Binding capacity of these beads was about 20 times higher than that reported for other commercial immunoadsorbent microbeads (Vollenkle et al. 2004).

In yet another example of technological improvements to SACI, Brockelbank, Peters, and Rehm (2006) engineered a fusion between the ZZ-binding domain of SPA and polyhydroxyalkanoate (PHA) synthase, the key enzyme of PHA biosynthesis. PHA biosynthesis leads to PHA granule formation inside bacteria. The PHA granule core is composed of PHA, and enclosed with a phospholipid membrane containing embedded protein. PHA synthase is covalently bound to the PHA granule core such that the protein is displayed on the surface of the granules (Brockelbank, Peters, and Rehm 2006). These authors made a genetic construct containing SPA ZZ peptide fused to *Cupriavidus necator* PHA synthase, which they heterologously expressed in *E. coli*. The resultant fusion protein was enzymatically active and mediated formation of granules exhibiting the Ig-binding domain ZZ on their surface. PHA/ZZ granules were used to isolate Ig from serum with the same purity as obtained using SPA-Sepharose®. The synthesis of these granules in *E. coli* allows for a low cost alternative to SPA-Sepharose®, for use in monoclonal purification or other immunoreactive functions.

In perhaps the most cost effective system, the binding domain of SPA has been engineered for expression on a tobamovirus, resulting in immunoabsorbent nanoparticles (Werner et al. 2006). In this construct, the coat protein of turnip vein clearing virus (TVCV) was fused to a 133-aa fragment of SPA (containing two of the five Ig-binding domains) with a 15-aa linker between. The fusion protein did not interfere with virus formation, thus viral particles were formed with a functional IgG-binding domain expressed on their coat protein. The particles were used for Ig purification, and a 50 percent yield of human IgG with >90 percent purity was obtained. This compared favorably to yields/purity obtained using SPA magnetic beads. Due to the high number of fusion protein copies per particle (>2100 copies per viral particle), the immunosorbent had a capacity of 2 g mAb per gram of particles. A yield of >3 g of TVCV per kilogram of leaf mass is possible, making this a very attractive economical alternative to SPA-Sepharose®.

17.7 FUSION PROTEINS

IBPs, or their smaller domains, have been used as fusion partners for many applications in addition to those discussed above (also reviewed in Stahl et al. 1993). Multiple expression and secretion systems for SPA fusion proteins have been developed for several bacteria (Goeddel 1990), including *E. coli* and *S. aureus*. A partial listing of fusion strategies includes fusion of an IBP or domain to a protein of interest to serve as an affinity tag for purification with Ig sepharose; fusion of the albumin-binding BB domain of PG to a protein of interest for purification with albumin-Sepharose; fusion of the SPA ZZ domain to a protein of interest for enhanced expression, solubility, and stability in heterologous bacteria (Hua, Jie, and Zhu 1994; Lobeck et al. 1998; E. Samuelsson et al. 1991); fusion of the SPA ZZ domain to fluorescent proteins to serve as detection reagents in immunoassays, immunoblotting, and immunofluorescence (Huang et al. 2006; Arai, Ueda, and Nagamune 1998); fusion of SPA to streptavidin for use in immuno-PCR assays (Sano, Smith, and Cantor 1992); fusion of PG to luciferase for immunoassays (Maeda et al. 1997); and fusion of PG BB to proteins for binding to albumin on a solid phase surface, such as a gold sensor chip (Baumann et al. 1998). In biosensor development, cysteine or other reactive tags have been engineered to SPA or PG for purposes of orienting them on a solid surface, which will in turn anchor/orient Ig (Lee et al. 2007; Briand et al. 2006). SPA and its ZZ domain are quite immunogenic, therefore fusion of SPA or ZZ to a protein of interest can be used to increase protein immunogenicity for purposes of making polyclonal Ab (Goeddel 1990; Leonetti et al. 1998). Finally, the D domain of SPA has been fused to cholera toxin for use as an adjuvant (Lycke 2001; Mowat et al. 2001). Undoubtedly the applications for these IBP-fusion proteins will continue to increase.

17.8 SPA-DERIVED AFFIBODY

Through investigation of the binding interaction between SPA and IgG Fc, the binding of SPA on a molecular level was solved (Jendeberg et al. 1996). The structure of the second, B, domain in a complex with the Fc fragment of IgG was determined, and the essential residues in the Fc-binding sites were elucidated (Takahashi et al. 2000; Deisenhofer 1981; Gouda et al. 1992; Lyons et al. 1993; Nilsson et al. 1987; Zheng, Aramini, and Montelione 2004). The B domain and its synthetic counterpart, the Z domain, have proved to be highly versatile, as this domain retains secondary structure (a three α-helix bundle), is easily manipulated genetically, and is highly structurally stable. The Z domain can be boiled at neutral pH without significant loss of activity after cooling (Nord et al. 1995). The Z domain is stable under a wide range of pH (4.5 to 9.3), and its structure is maintained without disulfide bonds; therefore, the domain can be exposed to harsh reducing environments (300-fold molar excess of dithiothreitol) without losing binding ability to target (Tolmachev et al. 2007). The Z domain is also resistant to degradation by proteases and peptidases.

Nord and coworkers have developed the SPA Z domain into binding peptides called affibodies (Nord et al. 1995, 1997, 2001). The region of the Z domain contacting Ig Fc involves two of three α-helices and covers a surface area of approximately 800 $Å^2$, this is analogous in size to the surfaces involved in many antigen–antibody binding sites (Nord et al. 1995), thus resulting in high affinity binding between affibodies and their binding target (10 μM to 22 pM K_d). Nord and coauthors speculated that using the stable scaffold of the Z domain would allow construction of binding peptide variants that would maintain high binding affinity. Use of peptides with highly stable constrained structures eliminates the major entropy loss associated with the ordering of flexible molecules such as linear peptides, thus increasing the potential binding affinity of these molecules. The 13 surface exposed amino acids of the Z domain were randomly simultaneously mutated (substitutions by all 20 possible amino acids) for the construction of an affibody library. Building on previous work by Djojonegoro and coworkers, Nord and coauthors constructed a phage display library with affibodies. (In 1994, Djojonegoro, Benedik, and Willson described the first construction of a bacteriophage with domain B from SPA displayed on the surface. It was the investigators' intent to use this method to eventually create a variant display library of the B domain of SPA in order to isolate mutated SPA with better attributes for Ig purification, for example, higher affinity for murine IgG_1, ease of elution of humanized IgG_1, and enhanced binding of Fab of certain Ig types.)

Using phage display selection technology, affibody variants with high affinity for particular targets have been identified using the panning technique originally developed for scFv (reviewed in Tolmachev et al. 2007 and Nord et al. 1995). In one application, affibodies were used in place of monoclonal or polyclonal Ig in a sandwich-type ELISA (Andersson et al. 2003). Through panning of an affibody library, an affibody that specifically bound IgA and one that specifically bound apolipoprotein A-1 were isolated. Each of these was used as the capture reagent in a sandwich ELISA, with the detecting Ab being either a goat polyclonal anti-human IgA, or a murine monoclonal anti-human lipoprotein A-1. These capture reagents worked comparably to using goat Ab or murine mAb as both the capture and detection Abs, even though the affinity of the affibodies for their ligands was not as high as the Abs for their ligands. The authors speculate that due to the small size (approximately 6 kDa) of the affibodies, they can bind the solid phase (ELISA plate) at much higher densities than full size Ab, and thus compensate for the reduced affinity for their ligands. Additionally the authors report that use of affibodies for this purpose overcame the problem of background, or false positive, titers in human sera due to the presence of human heterophilic anti-animal immunoglobulin antibodies (HAIA).

In another application, Tolmachev and coauthors (2007) describe the use of affibodies as potential affinity reagents for *in vivo* imaging or targeting of tumors, for the purpose of cancer therapy. Using an affibody library of 3×10^9 variants, affinity proteins that have a high specificity for individual cancer-associated antigens may be selected (Tolmachev et al. 2007; Wikman et al. 2004). The small size of the affibody molecules allows for easy tissue penetration with rapid tumor localization. Additionally, the affibody molecules are rapidly cleared from the blood, thus reducing background and off-target binding. Initial efforts in this area were directed to the HER2 protein expressed on the surface of multiple cancer cells, such as carcinomas of breast, ovary, and bladder. Upon panning the expression library, an affibody molecule was selected to carry forward for development. *In vitro* experiments demonstrated specific binding of this molecule to HER2-expressing SKBR-3 breast cancer cells. To enhance binding affinity, two approaches were utilized. In the first approach, the selected affibody was dimerized, which reduced the K_d from 50 nM to 3 nM (Steffen et al. 2005). In a second approach, the affibody was affinity matured using a new library of 3×10^8 new affibody variants, to a K_d of 22 pM. The affinity matured affibody demonstrated better tumor targeting *in vivo* than either the parent or the dimerized parent molecule. To enhance serum residence, the dimerized HER2 affibody was fused with an albumin-binding peptide derived from the albumin-binding domain of PG (ABD). This construct exhibited better biodistribution in tumor-bearing mice and specific uptake of the construct in HER2-expressing xenografts.

17.9 PROTEOMICS

One of the original applications for SPA was for the investigation of surface expressed proteins on mammalian cells (as discussed in Section 17.3). In those experiments, surface proteins were labeled (by isotope or other label), bound by specific Ig, solubilized, and immunoprecipitated using SACI. This method has now been updated through the use of modern genetic and analytical techniques. More recently, SPA is genetically fused to a protein of interest and used as an affinity tag for protein purification with immobilized IgG. It is known that cellular processes, for example, protein synthesis, protein degradation, intracellular signaling, etc., require the interaction of multiple proteins or protein complexes. Some of these complexes are stable and some are transient. The study of such protein-protein interactions has been facilitated by the elucidation of the complete genome of multiple animals/organisms, the use of highly sophisticated mass spectrometers, and the use of affinity tags such as SPA. These tools have allowed the dissection of many of the large interacting protein complexes (reviewed in Chang 2006). By use of affinity purification, protein complexes can be isolated. The isolated protein complexes can then be subjected to protease digestion followed by mass spectrometer analysis for protein identification. SPA has been utilized for such proteomics studies of human cancer cells, *Arabidopsis*, mouse cells, human cell transcription studies, yeast chromatin remodeling, and human TGF receptor signaling (Chang 2006). A method for the rapid elution of SPA-tagged proteins/complexes from Ig columns was reported (Strambio-de-Castillia et al. 2005). This method utilizes a 13-aa peptide mimicking the protein-protein binding interface of SPA for the hinge region on the Fc domain of human IgG. The peptide was N-terminally biotinylated to increase solubility. A one step purification of SPA fusion proteins which does not denature the protein complexes may be accomplished using this peptide.

Another approach utilizing IBP for proteomics incorporates the PG-expressing nonviable bacterial preparation, ProACTR. In this application, ProACTR is used as a platform to capture specific antigens and display them directly on a chip for analysis by SELDI-TOF mass spectrometry. This procedure consists of three steps: binding of antigen-specific Ab to ProACTR (through binding to PG on the ProACTR bacteria), interaction of the Ab-loaded ProACTR with specific antigen in some proteinaceous milieu, and removal of the antigen–antibody-loaded ProACTR for direct placement on a MS chip. This method was successful in identifying a secreted toxin (SpeB, from *S. pyogenes*) and several derivatives of the toxin in a complex bacterial broth. After placement on the chip, the SELDI-TOF laser intensity was adjusted to remove the displayed antigen for mass analysis, but not the Ab or antigens from the ProACTR. The technique did not work with PG- or SPA-Sepharose®, as these reagents did not bind directly and uniformly to the MS chip (Saouda, Romer, and Boyle 2002).

17.10 CLINICAL APPLICATIONS

The use of SPA as a medical reagent began in the 1970s. Early investigators explored its utility as a reagent to remove IgG or circulating immune complexes (CIC) from the circulation of patients with certain medical conditions, such as cancer (reviewed in R.M. Murphy 1989), or autoimmune diseases, such as idiopathic thrombocytopenic purpura (ITP).

The first described studies on the extracorporeal immunoperfusion of plasma from cancer patients and animals were particularly exciting (Bansal et al. 1978; Terman 1985). It appeared that plasma passage over SPA columns induced a reduction in cancer symptoms and even surprisingly a reduction of tumor size (Terman 1985; Korec et al. 1984). Research at the time indicated the presence of blocking factors in cancer patients' plasma, which were believed to be composed of CIC, consisting of tumor antigens bound to anti-tumor antibodies (reviewed in Ray 1985 and R.M. Murphy 1989). It was thought that the CIC blocked normal tumor clearance by the patient's immune system, and therefore removal of the complexes would allow tumor regression (Ciavarella 1992; Nand 1990). It was also known that SPA binds preferentially to the IgG of CIC versus free IgG (McDougal et al. 1979). Immunoperfusion led to the removal of CIC from patients' plasma. This procedure also induced acute symptoms, including chills, fever, nausea, rash, and respiratory complaints. The plasma adsorption procedure caused tumor regression in several patients with colon carcinoma and other human malignancies (Nand 1990; Ray 1985). Later work implied that the removal of CIC may not have been the mechanism inducing tumor regression, but rather it was due to the introduction of enterotoxins from impure SPA used to prepare the immunoperfusion columns (Terman 1985). *S. aureus* enterotoxins are potent T-cell activators and pyrogens. Alternatively, SPA, a B-cell superantigen, may have leached from the column and activated B-cells (Section 17.2.5). Immunoperfusion over SPA columns was variously reported to activate the complement system, increase the T-helper/suppressor ratio, increase blastogenic responses, and augment the natural killer (NK) cell activity (Nand 1990). After initial reports of the clinical benefit of SPA treatment in cancer patients, many uncontrolled clinical trials were initiated. However, trials varied in treatment protocols as well as cancer disease types and clinical stages of the patients (R.M. Murphy 1989). Results ranged from partial regression of tumors to no beneficial effect.

With the availability of a highly purified and standardized SPA affinity resin, controlled clinical trials could be conducted. In 1988, a large randomized clinical trial was conducted in which patients with a variety of neoplasms were enrolled (Messerschmidt et al. 1988, 1989). The IMRA PROSORBA® column made by Imre' (Seattle, Washington; approved by the FDA for marketing in 1987), was used for the immunoperfusion procedures in this trial. Results were somewhat disappointing in that only 22 of 101 patients had tumor size regression of >25 percent; however, some interesting evidence of treatment-induced immunomodulation was observed in a few patients (Snyder, Balint, and Jones 1989). At that time, the mechanism of action of SPA immunoperfusion was unclear (R.M. Murphy 1989). With the inability to reproduce and expand on the original observations of SPA benefits in cancer patients, this approach for treatment of cancer tumors was eventually abandoned.

In addition to tumor treatment, treatment of cancer-associated syndromes like hemolytic uremic syndrome with SPA immunoperfusion was explored. Immunoperfusion was successful in reducing symptoms associated with CIC deposition in the kidneys. This relief of symptoms was only temporary in some patients, while in others it lasted over 9 months post-treatment (Korec 1986). At the time of its early use, SPA immunoperfusion was among the few treatment options for C-HUS (cancer-associated hemolytic-uremic syndrome; Lesesne et al. 1989), a severe, often fatal, syndrome associated with some solid tumor cancers (e.g., adenocarcinoma or gastric carcinoma).

Although extracorporeal immunosorption for treatment of malignant tumors was not pursued, the therapy was explored and adopted for the treatment of various autoimmune disorders. By 2001, two SPA resins were approved for use by the FDA (reviewed in Bertram, Jones, and Balint 1996 and Matic, Bosch, and Ramlow 2001). Immunoadsorption may be performed with either of the two SPA-linked resins (either silica-based Prosorba®, Cypress Bioscience, San Diego, California, Fresenius Hemocare, Redmond, Washington, or beaded agarose, Immunosorba®, Fresenius Hemocare, St. Wendel, Germany). The Prosorba column is designed to be used one time and then discarded, while the Immunosorba column may be recycled. The Immunosorba column is used mainly in Europe, while the Prosorba column is used primarily in the United States (Bertram, Jones, and Balint 1996).

Licensed uses for the two SPA resins include treatment of idiopathic thrombocytopenic purpura (ITP), rheumatoid arthritis, and hemophilia with inhibitors (Matic, Bosch, and Ramlow 2001). Chronic ITP is an immune disease marked by circulating antiplatelet antibodies leading to the destruction of platelets, primarily in the spleen, and occurs in both adults and children. An early trial using

Prosorba® columns for treatment of this disease was published in 1992 (Snyder et al. 1992). Twenty-five percent of 72 patients had good responses, 21 percent fair responses, and 54 percent had poor responses. SPA immunoadsorption, although still available, is a rare treatment option for this disease. Rheumatoid arthritis is one of the most common autoimmune diseases. The Prosorba® system is approved for treatment of this debilitating disease (*Joint, Bone, Spine* 2005). Licensure for this indication, in 1999, was supported by a prospective, controlled, and double-blinded clinical study (Matic, Bosch, and Ramlow 2001). The amount of plasma treated with a Prosorba column for RA is limited, about 1250 mL, and only about 1.5 percent of serum IgG is removed. Therefore, removal of a significant amount of CIC cannot explain its beneficial effects. In the clinical trial for licensure, there was no effect on CD4/8 or TH1/2 lymphocyte profiles (Bosch 2003). There was a small amount of SPA released from the column (100 to 200 μg; Bosch 2003); however, it was not known if this had an affect on patients' health. Mechanisms of action of Prosorba® may include the induction of nonspecific immunomodulating effects, or complement activation without changes in C3 or C4 blood levels, or release of SPA into the bloodstream (*Joint, Bone, Spine* 2005). Other autoimmune diseases that have been explored for treatment with SPA immunoadsorption are thrombotic thrombocytopenic purpura, hemolytic uremic syndrome, paraneoplastic syndromes, Guillain-Barré syndrome, amyotrophic lateral sclerosis, and human immunodeficiency virus infection (Arbiser et al. 1995). Positive results were observed for treatment of myasthenia gravis using the Excorim column, and it was apparently safe and effective for this use (Batocchi et al. 2000). SPA immunosorption has also been used for treatment of HLA hyperimmunized transplantation candidates, and patients with rapidly progressive glomerulonephritis (G. Samuelsson 1999). Recently, successful treatment of pemphigus foliaceus and pemphigus vulgaris by immunoadsorption was reported (Schmidt et al. 2003) in an uncontrolled clinical trial. The treatment was found to be safe and highly effective at reducing clinical signs of disease.

17.11 *SPA* TYPING

In the last decade, SPA has been utilized for a clinical purpose unrelated to its Ig-binding property. *S. aureus* can cause very serious infections in both the hospital (it is an important cause of nosocomial infections) and in the community setting. Acquisition of a mobile cassette chromosome containing the *mec* gene has produced strains resistant to methicillin and other beta-lactams (methicillin-resistant *Staphylococcus aureus*, or MRSA). As a result of acquisition of *mec* and other drug-resistance genes, treatment options for *S. aureus* infections are becoming fewer in number. In order to track epidemic strains and monitor drug resistance, genotyping methods have been developed. SPA is expressed on the vast majority of *S. aureus* clinical isolates, and therefore a method of typing based on the X domain of the *spa* gene has been developed. The gold standard of typing for *S. aureus* is the pulsed field gel electrophoresis, PFGE, method (Murchan et al. 2003; Strommenger et al. 2006) in which the macro-restriction fragments of the *S. aureus* genome are evaluated. However, this method is difficult to standardize between laboratories, is technically challenging to perform, and is labor intensive. Multi-gene locus species typing (MLST), which evaluates DNA partial sequences of seven housekeeping genes, has been adopted as an alternative to PFGE. This method of typing gives reliable, unambiguous results, standardizable between labs (Enright et al. 2000, 2002; Enright and Spratt 1999). However, use of MLST is not amenable for quick routine typing due to high costs and labor associated with this method. It has been shown that a single gene sequence, the *spa* gene, can give sufficient discriminatory information similar to using the seven housekeeping genes (Hallin et al. 2007). Use of *spa* for this purpose (*spa* typing) was pioneered by Frenay and coworkers (1994, 1996). The polymorphic X region of *spa* contains multiple repeating regions (short sequence repeats, SSR). Genetic analysis of the X region of *spa* can monitor micro and macro genetic changes, allowing for both local and global epidemiological studies (Koreen et al. 2004). It was shown by calculation of Simpson's index of diversity that *spa* typing is nearly as discriminatory as PFGE (Aires-de-Sousa et al. 2006). While the function of the octapeptide of the SSR is not known, variations are clearly related to bacterial pathogenesis and virulence

(Ruppitsch et al. 2006). Software for assignment of *spa* types is now available (Ridom StaphType, Ridom GmbH, Würzburg, Germany; Harmsen et al. 2003), allowing worldwide assignment of *spa* types. More than 1400 spa types have been described (accessible at http://spaserver.ridom.de; Ruppitsch et al. 2006). In a study of 217 strains by the Belgian Reference Laboratory for Staphylococci, *spa* typing was highly concordant with PFGE and also showed good agreement with MLST methods (Hallin et al. 2007). In a large study by 10 laboratories from around the world, *spa* typing (range of 110 to 422 bp sequenced) was 100 percent reproducible between and within labs (Aires-de-Sousa et al. 2006). *Spa* typing has been used to type sporadically occurring clones, track hospital transmission of *S. aureus* isolates from health care workers to neonates (Matussek et al. 2007), and type and distinguish MRSA strains in a university hospital setting (Harmsen et al. 2003) or geographic region (Ruppitsch et al. 2006). For the best discriminatory power, it has been suggested that *spa* typing should be used in conjunction with another typing method, such as *clfB*, *mec*, or other virulence gene sequence (Kuhn, Francioli, and Blanc 2007; Hallin et al. 2007), and/or the clustering software algorithm Based Upon Repeat Pattern (BURP) (Strommenger et al. 2006).

17.12 SUMMARY

IBP have been the object of research, and a tool for researchers, for many decades. Their biological functions as bacterial virulence factors are slowly being elucidated. However, their utility as scientific and clinical tools has been expanding at a rapid rate. *In vivo* biological functions include surface binding of Ig, α2M, and other plasma proteins, and B-cell superantigen activity. As tools for clinical and research purposes, these proteins or their derivatives are utilized for Ig purification, a wide variety of immunoassay procedures, cell staining, proteomics, immunoprecipitation, cell separation, Ig adsorption and orientation on solid surfaces, and more.

REFERENCES

Aires-de-Sousa, M., K. Boye, H. de Lencastre, A. Deplano, M.C. Enright, J. Etienne, A. Friedrich, D. Harmsen, A. Holmes, X.W. Huijsdens, A.M. Kearns, A. Mellmann, H. Meugnier, J.K. Rasheed, E. Spalburg, B. Strommenger, M.J. Struelens, F.C. Tenover, J. Thomas, U. Vogel, H. Westh, J. Xu, and W. Witte. 2006. High interlaboratory reproducibility of DNA sequence-based typing of bacteria in a multicenter study. *Journal of Clinical Microbiology* 44:619–621.

Akerstrom, B., and L. Bjorck. 1989. Protein-l: An immunoglobulin light chain-binding bacterial protein—characterization of binding and physicochemical properties. *Journal of Biological Chemistry* 264:19740–19746.

Akerström, B., T. Brodin, K. Reis, and L. Björck. 1985. Protein G: a powerful tool for binding and detection of monoclonal and polyclonal antibodies. *Journal of Immunology* 135:2589–2592.

Akerstrom, B., B.H.K. Nilson, H.R. Hoogenboom, and L. Bjorck. 1994. On the interaction between single chain Fv antibodies and bacterial immunoglobulin-binding proteins. *Journal of Immunological Methods* 177:151–163.

Andersson, M., J. Ronnmark, I. Arestrom, P.A. Nygren, and N. Ahlborg. 2003. Inclusion of a non-immunoglobulin binding protein in two-site ELISA for quantification of human serum proteins without interference by heterophilic serum antibodies. *Journal of Immunological Methods* 283:225–234.

Arai, R., H. Ueda, and T. Nagamune. 1998. Construction of chimeric proteins between protein G and fluorescence-enhanced green fluorescent protein, and their application to immunoassays. *Journal of Fermentation and Bioengineering* 86:440–445.

Arbiser, J.L., J.S. Dzieczkowski, J.V. Harmon, and L.M. Duncan. 1995. Leukocytoclastic vasculitis following staphylococcal protein-a column immunoadsorption therapy: Two cases and a review of the literature. *Archives of Dermatology* 131:707–709.

Armstrong, P.B. 2006. Proteases and protease inhibitors: A balance of activities in host-pathogen interaction. *Immunobiology* 211:263–281.

Bansal, S.C., B.R. Bansal, H.L. Thomas, P.D. Siegel, J.E. Rhoads, D.R. Cooper, D.S. Terman, and R. Mark. 1978. Ex vivo removal of serum IgG in a patient with colon carcinoma: Some biochemical, immunological and histological observations. *Cancer* 42:1–18.

Batocchi, A.P., A. Evoli, C. Di Schino, and P. Tonali. 2000. Therapeutic apheresis in myasthenia gravis. *Therapeutic Apheresis* 4:275–279.

Baumann, S., P. Grob, F. Stuart, D. Pertlik, M. Ackermann, and M. Suter. 1998. Indirect immobilization of recombinant proteins to a solid phase using the albumin binding domain of streptococcal protein G and immobilized albumin. *Journal of Immunological Methods* 221:95–106.

Bertram, J.H., F.R. Jones, and J.P. Balint, Jr. 1996. Protein A immunoadsorption: Clinical potential. *Clinical Immunotherapeutics* 6:211–227.

Biguzzi, S. 1982. Fc-gamma-like determinants on immunoglobulin variable regions: Identification by staphylococcal protein-A. *Scandinavian Journal of Immunology* 15:605–618.

Bill, E., U. Lutz, B.M. Karlsson, M. Sparrman, and H. Allgaier. 1995. Optimization of protein G chromatography for biopharmaceutical monoclonal antibodies. *Journal of Molecular Recognition* 8:90–94.

Bjorck, L. 1988. Protein-l: A novel bacterial-cell wall protein with affinity for Ig L-chains. *Journal of Immunology* 140:1194–1197.

Bjorck, L., W. Kastern, G. Lindahl, and K. Wideback. 1987. Streptococcal protein-G, expressed by streptococci or by *Escherichia coli*, has separate binding-sites for human-albumin and IgG. *Molecular Immunology* 24:1113–1122.

Bjorck, L., and G. Kronvall. 1984. Purification and some properties of streptococcal protein-G, protein-A novel IgG-binding reagent. *Journal of Immunology* 133:969–974.

Bosch, T. 2003. Recent advances in therapeutic apheresis. *Journal of Artificial Organs* 6:1–8.

Boyle, M.D.P. 1990. The type I bacterial immunoglobulin-binding protein: staphylococcal protein A. In: Boyle, M.D.P. (Ed.). *Bacterial Immunoglobulin-Binding Proteins: Microbiology, Chemistry and Biology*. Academic Press, New York, pp. 17–28.

Briand, E., M. Salmain, C. Compere, and C.M. Pradier. 2006. Immobilization of protein A on SAMs for the elaboration of immunosensors. *Colloids and Surfaces B: Biointerfaces* 53:215–224.

Brockelbank, J.A., V. Peters, and B.H.A. Rehm. 2006. Recombinant *Escherichia coli* strain produces a ZZ domain displaying biopolyester granules suitable for immunoglobulin G purification. *Applied and Environmental Microbiology* 72:7394–7397.

Brook, I. 1987. Bacteremia and seeding of capsulate *Bacteroides* spp. and anaerobic cocci. *Journal of Medical Microbiology* 23:61–67.

Brook, I. 1988. Enhancement of growth of aerobic, anaerobic, and facultative bacteria in mixed infections with anaerobic and facultative Gram-positive cocci. *Journal of Surgical Research* 45:222–227.

Brook, I., and R.I. Walker. 1985. The role of encapsulation in the pathogenesis of anaerobic Gram-positive cocci. *Canadian Journal of Microbiology* 31:176–180.

Chang, I.F. 2006. Mass spectrometry-based proteomic analysis of the epitope-tag affinity purified protein complexes in eukaryotes. *Proteomics* 6:6158–6166.

Chhatwal, G.S., H.P. Muller, and H. Blobel. 1983. Characterization of binding of human alpha-2-macroglobulin to group-G streptococci. *Infection and Immunity* 41:959–964.

Ciavarella, D. 1992. The use of protein A columns in the treatment of cancer and allied diseases. *International Journal of Clinical and Laboratory Research* 21:210–213.

Das, D., T.M. Allen, and M.R. Suresh. 2005. Comparative evaluation of two purification methods of anti-CD19-c-myc-His6-Cys scFv. *Protein Expression and Purification* 39:199–208.

Dechateau, M., E. Holst, and L. Bjorck. 1996. Protein PAB, an albumin-binding bacterial surface protein promoting growth and virulence. *Journal of Biological Chemistry* 271:26609–26615.

Dechateau, M., B.H.K. Nilson, M. Erntell, E. Myhre, C.G.M. Magnusson, B. Akerstrom, and L. Bjorck. 1993. On the interaction between protein-l and immunoglobulins of various mammalian species. *Scandinavian Journal of Immunology* 37:399–405.

Dedent, A.C., M. McAdow, and O. Schneewind. 2007. Distribution of protein A on the surface of *Staphylococcus aureus*. *Journal of Bacteriology* 189:4473–4484.

Deisenhofer, J. 1981. Crystallographic refinement and atomic models of a human Fc fragment and its complex with fragment-B of protein-A from *Staphylococcus aureus* at 2.9-A and 2.8-A resolution. *Biochemistry* 20:2361–2370.

Derrick, J.P., and D.B. Wigley. 1992. Crystal structure of a streptococcal protein-G domain bound to an Fab fragment. *Nature* 359:752–754.

Djojonegoro, B.M., M.J. Benedik, and R.C. Willson. 1994. Bacteriophage surface display of an immunoglobulin-binding domain of *Staphylococcus aureus* protein-A. *Bio-Technology* 12:169–172.

Dossett, J.H., G. Kronvall, R.C. Williams Jr, and P.G. Quie. 1969. Antiphagocytic effects of staphylococfcal protein A. *Journal of Immunology* 103:1405–1410.

Duhamel, R.C., P.H. Schur, K. Brendel, and E. Meezan. 1979. pH gradient elution of human IgG1, IgG2 and IgG4 from protein A-Sepharose. *Journal of Immunological Methods* 31:211–217.

Joint Bone Spine. 2005. Editorial. Protein A-immunoadsorption (Prosorba column) in the treatment of rheumatoid arthritis. 72:101–103.

Eliasson, M., A. Olsson, E. Palmcrantz, K. Wiberg, M. Inganas, B. Guss, M. Lindberg, and M. Uhlen. 1988. Chimeric IgG-binding receptors engineered from staphylococcal protein-A and streptococcal protein-G. *Journal of Biological Chemistry* 263:4323–4327.

Endresen, C. 1979. Binding to protein A of immunoglobulin-G and of Fab and Fc fragments. *Acta Pathologica et Microbiologica Scandinavica Section C-Immunology* 87:185–189.

Enright, M.C., N.P.J. Day, C.E. Davies, S.J. Peacock, and B.G. Spratt. 2000. Multilocus sequence typing for characterization of methicillin-resistant and methicillin-susceptible clones of *Staphylococcus aureus. Journal of Clinical Microbiology* 38:1008–1015.

Enright, M.C., D.A. Robinson, G. Randle, E.J. Feil, H. Grundmann, and B.G. Spratt. 2002. The evolutionary history of methicillin-resistant *Staphylococcus aureus* (MRSA). *Proceedings of the National Academy of Sciences of the United States of America* 99:7687–7692.

Enright, M.C., and B.G. Spratt. 1999. Multilocus sequence typing. *Trends in Microbiology* 7:482–487.

Erntell, M., E.B. Myhre, U. Sjobring, and L. Bjorck. 1988. Streptococcal protein-G has affinity for both Fab-fragments and Fc-fragments of human-IgG. *Molecular Immunology* 25:121–126.

Fahnestock, S.R., P. Alexander, J. Nagle, and D. Filpula. 1986. Gene for an immunoglobulin-binding protein from a group-G streptococcus. *Journal of Bacteriology* 167:870–880.

Forsgren, A., M. Brant, A. Mollenkvist, A. Muyombwe, H. Janson, N. Woin, and K. Riesbeck. 2001. Isolation and characterization of a novel IgD-binding protein from *Moraxella catarrhalis. Journal of Immunology* 167:2112–2120.

Forsgren, A., and A.O. Grubb. 1979. Many bacterial species bind human IgD. *Journal of Immunology* 122:1468–1472.

Forsgren, A., K. Nordstro, L. Philipso, and J. Sjoquist. 1971. Protein-A mutants of *Staphylococcus aureus. Journal of Bacteriology* 107:245.

Forsgren, A., and J. Sjoquist. 1966. Protein A from *S. aureus.* I. Pseudo-immune reaction with human gamma-globulin. *Journal of Immunology* 97:822.

Frenay, H.M.E., A.E. Bunschoten, L.M. Schouls, W.J. van Leeuwen, C.M.J.E. VandenbrouckeGrauls, J. Verhoef, and F.R. Mooi. 1996. Molecular typing of methicillin-resistant *Staphylococcus aureus* on the basis of protein A gene polymorphism. *European Journal of Clinical Microbiology & Infectious Diseases* 15:60–64.

Frenay, H.M.E., J.P.G. Theelen, L.M. Schouls, C.M.J.E. VandenbrouckeGrauls, J. Verhoef, W.J. van Leeuwen, and F.R. Mooi. 1994. Discrimination of epidemic and nonepidemic methicillin-resistant *Staphylococcus aureus* strains on the basis of protein A gene polymorphism. *Journal of Clinical Microbiology* 32:846–847.

Frick, I., M. Wikstrom, S. Forsen, T. Drakenberg, H. Gomi, U. Sjobring, and L. Bjorck. 1992. Convergent evolution among immunoglobulin G-binding bacterial proteins. *Proceedings of the National Academy of Sciences USA* 89:8532–8536.

Galli, S.J. 1993. Seminars in Medicine of the Beth-Israel-Hospital, Boston: New concepts about the mast-cell. *New England Journal of Medicine* 328:257–265.

Gaunt, P.N., and D.V. Seal. 1987. Group-G streptococcal infections. *Journal of Infection* 15:5–20.

Genovese, A., J.P. Bouvet, G. Florio, B. Lamparter-Schummert, L. Bjorck, and G. Marone. 2000. Bacterial immunoglobulin superantigen proteins A and L activate human heart mast cells by interacting with immunoglobulin E. *Infection and Immunity* 68:5517–5524.

Goding, J.W. 1978. Use of staphylococcal protein A as an immunological reagent. *Journal of Immunological Methods* 20:241–253.

Goeddel, D.V. 1990. Systems for heterologous gene expression. *Methods in Enzymology* 185:3–7.

Gomez, M.I., A. Lee, B. Reddy, A. Muir, G. Soong, A. Pitt, A. Cheung, and A. Prince. 2004. *Staphylococcus aureus* protein A induces airway epithelial inflammatory responses by activating TNFR1. *Nature Medicine* 10:842–848.

Gomez, M.I., O. Seaghdha, and A.S. Prince. 2007. *Staphylococcus aureus* protein A activates TACE through EGFR-dependent signaling. *EMBO Journal* 26:701–709.

Gore, M.G., W.F. Ferris, A.G. Popplewell, M. Scawen, and T. Atkinson. 1992. Ph-sensitive interactions between IgG and a mutated IgG-binding protein based upon 2 B-domains of protein-A from *Staphylococcus aureus*. *Protein Engineering* 5:577–582.

Gouda, H., H. Torigoe, A. Saito, M. Sato, Y. Arata, and I. Shimada. 1992. Three-dimensional solution structure of the B-domain of staphylococcal protein-A: Comparisons of the solution and crystal structures. *Biochemistry* 31:9665–9672.

Goward, C.R., M.D. Scawen, J.P. Murphy, and T. Atkinson. 1993. Molecular evolution of bacterial cell-surface proteins. *Trends in Biochemical Sciences* 18:136–140.

Graille, M., E.A. Stura, A.L. Corper, B.J. Sutton, M.J. Taussig, J.B. Charbonnier, and G.J. Silverman. 2000. Crystal structure of a *Staphylococcus aureus* protein A domain complexed with the Fab fragment of a human IgM antibody: Structural basis for recognition of B-cell receptors and superantigen activity. *Proceedings of the National Academy of Sciences of the United States of America* 97:5399–5404.

Gronenborn, A.M., and G.M. Clore. 1993. Identification of the contact surface of a streptococcal protein G-domain complexed with a human Fc fragment. *Journal of Molecular Biology* 233:331–335.

Grov, A., P. Oeding, and B. Myklestad. 1964. Immunochemical studies on antigen preparations from *Staphylococcus aureus*. I. Isolation and chemical characterization of antigen A. *Acta Pathologica et Microbiologica Scandinavica* 61:588.

Gulich, S., M. Linhult, S. Stahl, and S. Hober 2002. Engineering streptococcal protein G for increased alkaline stability. *Protein Engineering* 15:835–842.

Gulich, S., M. Uhlen, and S. Hober. 2000. Protein engineering of an IgG-binding domain allows milder elution conditions during affinity chromatography. *Journal of Biotechnology* 76:233–244.

Guss, B., M. Eliasson, A. Olsson, M. Uhlen, A.K. Frej, H. Jornvall, J.I. Flock, and M. Lindberg. 1986. Structure of the IgG-binding regions of streptococcal protein-G. *EMBO Journal* 5:1567–1575.

Hahn, R., P. Bauerhansl, K. Shimahara, C. Wizniewski, A. Tscheliessnig, and A. Jungbauer. 2005. Comparison of protein A affinity sorbents. II. Mass transfer properties. *Journal of Chromatography A* 1093:98–110.

Hallin, M., A. Deplano, O. Denis, R. De Mendonca, R. De Ryck, and M.J. Struelens. 2007. Validation of pulsed-field gel electrophoresis and spa typing for long-term, nationwide epidemiological surveillance studies of *Staphylococcus aureus* infections. *Journal of Clinical Microbiology* 45:127–133.

Harlow, E., and D. Lane. 1999. *Using Antibodies: A Laboratory Manual.* Cold Spring Harbor Laboratory Press, Cold Spring Harbor, NY.

Harmsen, D., H. Claus, W. Witte, J. Rothganger, H. Claus, D. Turnwald, and U. Vogel. 2003. Typing of methicillin-resistant *Staphylococcus aureus* in a university hospital setting by using novel software for spa repeat determination and database management. *Journal of Clinical Microbiology* 41:5442–5448.

Hartleib, J., N. Kohler, R.B. Dickinson, G.S. Chhatwal, J.J. Sixma, O.M. Hartford, T.J. Foster, G. Peters, B.E. Kehrel, and M. Herrmann. 2000. Protein A is the von Willebrand factor binding protein on *Staphylococcus aureus*. *Blood* 96:2149–2156.

Herrmann, M., J. Hartleib, B. Kehrel, R.R. Montgomery, J.J. Sixma, and G. Peters. 1997. Interaction of von Willebrand factor with *Staphylococcus aureus*. *Journal of Infectious Diseases* 176:984–991.

Hjelm, H., K. Hjelm, and J. Sjoquist. 1972. Protein A from *Staphylococcus aureus*: Its isolation by affinity chromatography and its use as an immunosorbent for isolation of immunoglobulins. *FEBS Letters* 28:73–76.

Hober, S., K. Nord, and M. Linhult. 2007. Protein A chromatography for antibody purification. *Journal of Chromatography B Analytical Technologies in the Biomedical and Life Sciences* 848:40–47.

Hook, M., and T. Foster. 2000. Staphylococcal surface proteins. In *Gram-Positive Pathogens*, ed. V.A. Fishetti, R.P. Novick, J.J. Ferretti, D.A. Portnoy, and J.I. Rood, 386–391. Washington, D.C.: ASM Press.

Housden, N.G., S. Harrison, S.E. Roberts, J.A. Beckingham, M. Graille, E. Stura, and M.G. Gore. 2003. Immunoglobulin-binding domains: Protein L from *Peptostreptococcus magnus*. *Biochemical Society Transactions* 31:716–718.

Hua, Z.C., L. Jie, and D.X. Zhu. 1994. Expression of a biologically-active human granulocyte-macrophage colony-stimulating factor fusion protein in *Escherichia coli*. *Biochemistry and Molecular Biology International* 34:621–626.

Huang, Q.L., C. Chen, Y.Z. Chen, C.G. Gong, J. Wang, and Z.C. Hua. 2006. Fusion protein between protein ZZ and red fluorescent protein DsRed and its application to immunoassays. *Biotechnology and Applied Biochemistry* 43:121–127.

Inganas, M. 1981. Comparison of mechanisms of interaction between protein-A from *Staphylococcus aureus* and human monoclonal IgG, IgA and IgM in relation to the classical Fc-gamma and the alternative F(Ab′)2-epsilon protein-A interactions. *Scandinavian Journal of Immunology* 13:343–352.

Inganas, M., S.G.O. Johansson, and H.H. Bennich. 1980. Interaction of human polyclonal IgE and IgG from different species with protein-A from *Staphylococcus aureus*: Demonstration of protein-A-reactive sites located in the Fab′2 fragment of human-IgG. *Scandinavian Journal of Immunology* 12:23–31.

Jendeberg, L., M. Tashiro, R. Tejero, B.A. Lyons, M. Uhlen, G.T. Montelione, and B. Nilsson. 1996. The mechanism of binding staphylococcal protein A to immunoglobin G does not involve helix unwinding. *Biochemistry* 35:22–31.

Johansson, H.M., M. Morgelin, and I.M. Frick. 2004. Protein FOG: A streptococcal inhibitor of neutrophil function. *Microbiology-Sgm* 150:4211–4221.

Johnson, C.P., I.E. Jensen, A. Prakasam, R. Vijayendran, and D. Leckband. 2003. Engineered protein A for the orientational control of immobilized proteins. *Bioconjugate Chemistry* 14:974–978.

Jonsson, P., M. Lindberg, I. Haraldsson, and T. Wadstrom. 1985. Virulence of *Staphylococcus aureus* in a mouse mastitis model: Studies of alpha-hemolysin, coagulase, and protein-A as possible virulence determinants with protoplast fusion and gene cloning. *Infection and Immunity* 49:765–769.

Julianelle, L.A., and C.W. Wieghard. 1935. The immunological specificity of staphylococci. I. The occurrence of serological types. *Journal of Experimental Medicine* 62:11–21.

Kastern, W., E. Holst, E. Nielsen, U. Sjobring, and L. Bjorck. 1990. Protein-L, a bacterial immunoglobulin-binding protein and possible virulence determinant. *Infection and Immunity* 58:1217–1222.

Kastern, W., U. Sjobring, and L. Bjorck. 1992. Structure of peptostreptococcal protein-l and identification of a repeated immunoglobulin light chain-binding domain. *Journal of Biological Chemistry* 267:12820–12825.

Kihlberg, B.M., U. Sjobring, W. Kastern, and L. Bjorck. 1992. Protein LG: A hybrid molecule with unique immunoglobulin binding properties. *Journal of Biological Chemistry* 267:25583–25588.

Korec, S., P.S. Schein, F.P. Smith, J.R. Neefe, P.V. Woolley, R.M. Goldberg, and T.M. Phillips. 1986. Treatment of cancer-associated hemolytic uremic syndrome with staphylococcal protein A immunoperfusion. *Journal of Clinical Oncology* 4:210–215.

Korec, S., F.P. Smith, P.S. Schein, and T.M. Phillips. 1984. Clinical experiences with extracorporeal immunoperfusion of plasma from cancer patients. *J. Biol. Response Mod.* 3:330–335.

Koreen, L., S.V. Ramaswamy, E.A. Graviss, S. Naidich, J.A. Musser, and B.N. Kreiswirth. 2004. spa typing method for discriminating among *Staphylococcus aureus* isolates: Implications for use of a single marker to detect genetic micro- and macrovariation. *Journal of Clinical Microbiology* 42:792–799.

Kouki, T., T. Inui, H. Okabe, Y. Ochi, and Y. Kajita. 1997. Separation method of IgG fragments using protein L. *Immunological Investigations* 26:399–408.

Kozlowski, L.M., A.M. Soulika, G.J. Silverman, J.D. Lambris, and A.I. Levinson. 1996. Complement activation by a B cell superantigen. *Journal of Immunology* 157:1200–1206.

Kronvall, G. 1973. Surface component in group A, C, and G streptococci with non-immune reactivity for immunoglobulin-G. *Journal of Immunology* 111:1401–1406.

Kuhara, M., H. Takeyama, T. Tanaka, and T. Matsunaga. 2004. Magnetic cell separation using antibody binding with protein A expressed on bacterial magnetic particles. *Analytical Chemistry* 76:6207–6213.

Kuhn, G., P. Francioli, and D.S. Blanc. 2007. Double-locus sequence typing using clfB and spa, a fast and simple method for epidemiological typing of methicillin-resistant *Staphylococcus aureus*. *Journal of Clinical Microbiology* 45:54–62.

Kushwaha, A., P.S. Chowdhury, K. Arora, S. Abrol, and V.K. Chaudhary. 1994. Construction and characterization of M13 bacteriophages displaying functional IgG-binding domains of staphylococcal protein-A. *Gene* 151:45–51.

Langone, J.J. 1982a. Applications of immobilized protein-A in immunochemical techniques. *Journal of Immunological Methods* 55:277–296.

Langone, J.J. 1982b. Protein-A of staphylococcus aureus and related immunoglobulin receptors produced by streptococci and pneumococci. *Advances in Immunology* 32:157–252.

Langone, J.J. 1982c. Use of labeled protein A in quantitative immunochemical analysis of antigens and antibodies. *Journal of Immunological Methods* 51:3–22.

Lee, J.M., H.K. Park, Y. Jung, J.K. Kim, S.O. Jung, and B.H. Chung. 2007. Direct immobilization of protein G variants with various numbers of cysteine residues on a gold surface. *Analytical Chemistry* 79:2680–2687.

Leonetti, M., R. Thai, J. Cotton, S. Leroy, P. Drevet, F. Ducancel, J.C. Boulain, and A. Menez. 1998. Increasing immunogenicity of antigens fused to Ig-binding proteins by cell surface targeting. *Journal of Immunology* 160:3820–3827.

Lesesne, J.B., N. Rothschild, B. Erickson, S. Korec, R. Sisk, J. Keller, M. Arbus, P.V. Woolley, L. Chiazze, and P.S. Schein. 1989. Cancer-associated hemolytic-uremic syndrome: Analysis of 85 cases from a national registry. *Journal of Clinical Oncology* 7:781–789.

Li, S.L., S.J. Liang, N. Guo, A.M. Wu, and Y. Fujita-Yamaguchi. 2000. Single-chain antibodies against human insulin-like growth factor I receptor: Expression, purification, and effect on tumor growth. *Cancer Immunology Immunotherapy* 49:243–252.

Li, Y., W. Cockburn, and G.C. Whitelam. 1998. Filamentous bacteriophage display of a bifunctional protein A:scFv fusion. *Molecular Biotechnology* 9:187–193.

Lian, L.Y., I.L. Barsukov, J.P. Derrick, and G.C.K. Roberts. 1994. Mapping the interactions between streptococcal protein-G and the Fab fragment of IgG in solution. *Nature Structural Biology* 1:355–357.

Lindmark, R., K. Thorentolling, and J. Sjoquist. 1983. Binding of immunoglobulins to protein-A and immunoglobulin levels in mammalian sera. *Journal of Immunological Methods* 62:1–13.

Linhult, M., S. Gulich, T. Graslund, A. Simon, M. Karlsson, A. Sjoberg, K. Nord, and S. Hober. 2004. Improving the tolerance of a protein A analogue to repeated alkaline exposures using a bypass mutagenesis approach. *Proteins Structure Function and Bioinformatics* 55:407–416.

Ljungquist, C., B. Jansson, T. Moks, and M. Uhlen. 1989. Thiol-directed immobilization of recombinant IgG-binding receptors. *European Journal of Biochemistry* 186:557–561.

Lobeck, K., P. Drevet, M. Leonetti, C. Fromen-Romano, F. Ducancel, E. Lajeunesse, C. Lemaire, and A. Menez. 1998. Towards a recombinant vaccine against diphtheria toxin. *Infection and Immunity* 66:418–423.

Lofdahl, S., B. Guss, M. Uhlen, L. Philipson, and M. Lindberg. 1983. Gene for staphylococcal protein-A. *Proceedings of the National Academy of Sciences of the United States of America Biological Sciences* 80:697–701.

Lycke, N. 2001. The B-cell targeted CTA1-DD vaccine adjuvant is highly effective at enhancing antibody as well as CTL responses. *Current Opinion in Molecular Therapeutics* 3:37–44.

Lyons, B.A., M. Tashiro, L. Cedergren, B. Nilsson, and G.T. Montelione. 1993. An improved strategy for determining resonance assignments for isotopically enriched proteins and its application to an engineered domain of staphylococcal protein-A. *Biochemistry* 32:7839–7845.

Maeda, Y., H. Ueda, J. Kazami, G. Kawano, E. Suzuki, and T. Nagamune. 1997. Engineering of functional chimeric protein G-vargula luciferase. *Analytical Biochemistry* 249:147–152.

Malke, H. 2000. Genetics and pathogenicity factors of group C and G streptococci. In: Fischetti, V.A., R.P. Novick, J.J. Ferretti, D.A. Portnoy, J.I. Rood. (eds), *Gram-positive pathogens*. Washington, D.C.: ASM Press, pp. 163–176.

Marraffini, L.A., A.C. Dedent, and O. Schneewind. 2006. Sortases and the art of anchoring proteins to the envelopes of gram-positive bacteria. *Microbiology and Molecular Biology Reviews* 70:192.

Matic, G., T. Bosch, and W. Ramlow. 2001. Background and indications for protein A-based extracorporeal immunoadsorption. *Therapeutic Apheresis* 5:394–403.

Matsunaga, T., M. Takahashi, T. Yoshino, M. Kuhara, and H. Takeyama. 2006. Magnetic separation of CD14+ cells using antibody binding with protein A expressed on bacterial magnetic particles for generating dendritic cells. *Biochemical and Biophysical Research Communications* 350:1019–1025.

Matussek, A., J. Taipalensuu, I.M. Einemo, M. Tiefenthal, and S. Lofgren. 2007. Transmission of *Staphylococcus aureus* from maternity unit staff members to newborns disclosed through spa typing. *American Journal of Infection Control* 35:122–125.

McCormick, D. 2005. Artificial distinctions: Protein A mimetic ligands for bioprocess separations. *Pharmaceutical Technology* 29:58–62.

McDougal, J.S., P.B. Redecha, R.D. Inman, and C.L. Christian. 1979. Binding of immunoglobulin G aggregates and immune complexes in human sera to staphylococcal containing protein A. *Journal of Clinical Investigation* 63:627–635.

Messerschmidt, G.L., D.H. Henry, H.W. Snyder, Jr., J.H. Bertram, A. Mittelman, S. Ainsworth, J. Fiore, M.V. Viola, J. Louie, E. Ambinder, F.R. MacKintosh, D.J. Higby, K.D. O'Brien, M. Hamburger, J.P. Balint, L.D. Fisher, W. Perkins, C.M. Pinsky, and F.R. Jones. 1988. Protein A immunoadsorption in the treatment of malignant disease. *Journal of Clinical Oncology* 6:203.

Messerschmidt, G.L., D.H. Henry, H.W. Snyder, Jr., J.H. Bertram, A. Mittelman, S. Ainsworth, J. Fiore, M.V. Viola, J. Louie, E. Ambinder, F.R. MacKintosh, D.J. Higby, P. O'Brien, D. Kiprov, M. Hamburger, J.P. Balint, L.D. Fisher, W. Perkins, C.M. Pinsky, and F.R. Jones. 1989. Protein A immunotherapy in the treatment of cancer: An update. *Seminars in Hematology* 26(Suppl 1):19.

Moks, T., L. Abrahmsen, B. Nilsson, U. Hellman, J. Sjoquist, and M. Uhlen. 1986. Staphylococcal protein-A consists of Five IgG-binding domains. *European Journal of Biochemistry* 156:637–643.

Mowat, A.M.I., A.M. Donachie, S. Jagewall, K. Schon, B. Lowenadler, K. Dalsgaard, P. Kaastrup, and N. Lycke. 2001. CTA1-DD-immune stimulating complexes: A novel, rationally designed combined mucosal vaccine adjuvant effective with nanogram doses of antigen. *Journal of Immunology* 167:3398–3405.

Muller, H.P., and L.K. Rantamaki. 1995. Binding of native alpha(2)-macroglobulin to human group-G streptococci. *Infection and Immunity* 63:2833–2839.

Murchan, S., M.E. Kaufmann, A. Deplano, R. De Ryck, M. Struelens, C.E. Zinn, V. Fussing, S. Salmenlinna, J. Vuopio-Varkila, N. De Solh, C. Cuny, W. Witte, P.T. Tassios, N. Legakis, W. van Leeuwen, A. van Belkum, A. Vindel, I. Laconcha, J. Garaizar, S. Haeggman, B. Olsson-Liljequist, U. Ransjo, G. Coombes, and B. Cookson. 2003. Harmonization of pulsed-field gel electrophoresis protocols for epidemiological typing of strains of methicillin-resistant *Staphylococcus aureus*: A single approach developed by consensus in ten European laboratories and its application for tracing the spread of related strains. *Journal of Clinical Microbiology* 41:1574–1585.

Murdoch, D.A. 1998. Gram-positive anaerobic cocci. *Clinical Microbiology Reviews* 11:81–120.

Murphy, R.M., C.K. Colton, and M.L. Yarmush. 1989. Staphylococcal protein A adsorption in neoplastic disease: Analysis of physicochemical aspects. *Molecular Biotherapy* 1:186–207.

Murphy, J.P., C.J. Duggleby, M.A. Atkinson, A.R. Trowern, T. Atkinson, and C.R. Goward. 1994. The functional units of a peptostreptococcal protein-l. *Molecular Microbiology* 12:911–920.

Myhre, E.B. 1984. Surface receptors for human-serum albumin in *Peptococcus magnus* strains. *Journal of Medical Microbiology* 18:189–195.

Myhre, E.B., and M. Erntell. 1985. A non-immune interaction between the light chain of human immunoglobulin and a surface component of a *Peptococcus magnus* strain. *Molecular Immunology* 22:879–885.

Nand, S.M.R. 1990. Therapeutic plasmapheresis and protein A immunoadsorption in malignancy: A brief review. *Journal of Clinical Apheresis* 5:206–212.

Nguyen, T., B. Ghebrehiwet, and E.I.B. Peerschke. 2000. *Staphylococcus aureus* protein A recognizes platelet gC1qR/p33: A novel mechanism for staphylococcal interactions with platelets. *Infection and Immunity* 68:2061–2068.

Nielsen, K., P. Smith, W. Yu, P. Nicoletti, P. Elzer, A. Vigliocco, P. Silva, R. Bermudez, T. Renteria, F. Moreno, A. Ruiz, C. Massengill, Q. Muenks, K. Kenny, T. Tollersrud, L. Samartino, S. Conde, G. Draghi de Benitez,

D. Gall, B. Perez, and X. Rojas. 2004. Enzyme immunoassay for the diagnosis of brucellosis: Chimeric protein A-protein G as a common enzyme labeled detection reagent for sera for different animal species. *Veterinary Microbiology* 101:123–129.

Nielsen, U.B., D.B. Kirpotin, E.M. Pickering, K.L. Hong, J.W. Park, M.R. Shalaby, Y. Shao, C.C. Benz, and J.D. Marks. 2002. Therapeutic efficacy of anti-ErbB2 immunoliposomes targeted by a phage antibody selected for cellular endocytosis. *Biochimica et Biophysica Acta Molecular Cell Research* 1591:109–118.

Nilson, B.H.K., L. Logdberg, W. Kastern, L. Bjorck, and B. Akerstrom. 1993. Purification of antibodies using protein L-binding framework structures in the light chain variable domain. *Journal of Immunological Methods* 164:33–40.

Nilsson, B., T. Moks, B. Jansson, L. Abrahmsen, A. Elmblad, E. Holmgren, C. Henrichson, T.A. Jones, and M. Uhlen. 1987. A synthetic IgG-binding domain based on staphylococcal protein-A. *Protein Engineering* 1:107–113.

Nilson, B.H.K., A. Solomon, L. Bjorck, and B. Akerstrom. 1992. Protein-l from *Peptostreptococcus magnus* binds to the kappa-light chain variable domain. *Journal of Biological Chemistry* 267:2234–2239.

Nitsche-Schmitz, D.P., H.M. Johansson, I. Sastalla, S. Reissmann, I.M. Frick, and G.S. Chhatwal. 2007. Group G streptococcal IgG binding molecules FOG and protein G have different impacts on opsonization by C1q. *Journal of Biological Chemistry* 282:17530–17536.

Nomellini, J.F., G. Duncan, I.R. Dorocicz, and J. Smit. 2007. S-layer-mediated display of the immunoglobulin g-binding domain of streptococcal protein G on the surface of *Caulobacter crescentus*: Development of an immunoactive reagent. *Applied and Environmental Microbiology* 73:3245–3253.

Nord, K., E. Gunneriusson, J. Ringdahl, S. Stahl, M. Uhlen, and P.A. Nygren. 1997. Binding proteins selected from combinatorial libraries of an alpha-helical bacterial receptor domain. *Nature Biotechnology* 15:772–777.

Nord, K., J. Nilsson, B. Nilsson, M. Uhlen, and P.A. Nygren. 1995. A combinatorial library of an alpha-helical bacterial receptor domain. *Protein Engineering* 8:601–608.

Nord, K., O. Nord, M. Uhlen, B. Kelley, C. Ljungqvist, and P.A. Nygren. 2001. Recombinant human factor VIII-specific affinity ligands selected from phage-displayed combinatorial libraries of protein A. *European Journal of Biochemistry* 268:4269–4277.

Olsson, A., M. Eliasson, B. Guss, B. Nilsson, U. Hellman, M. Lindberg, and M. Uhlen. 1987. Structure and evolution of the repetitive gene encoding streptococcal protein-G. *European Journal of Biochemistry* 168:319–324.

Otten, R.A., and M.D.P. Boyle. 1991. Characterization of protein-G expressed by human group-C and group-G streptococci. *Journal of Microbiological Methods* 13:185–200.

Palmqvist, N., T. Foster, A. Tarkowski, and E. Josefsson. 2002. Protein A is a virulence factor in *Staphylococcus aureus* arthritis and septic death. *Microbial Pathogenesis* 33:239–249.

Patel, A.H., P. Nowlan, E.D. Weavers, and T. Foster. 1987. Virulence of protein-A-deficient and alpha-toxin-deficient mutants of *Staphylococcus aureus* isolated by allele replacement. *Infection and Immunity* 55:3103–3110.

Patella, V., V. Casolaro, L. Bjorck, and G. Marone. 1990. Protein-l: A bacterial Ig-binding protein that activates human basophils and mast cells. *Journal of Immunology* 145:3054–3061.

Peterson, P.K., J. Verhoef, L.D. Sabath, and P.G. Quie. 1977. Effect of protein-A on staphylococcal opsonization. *Infection and Immunity* 15:760–764.

Pilcher, J.B., V.C.W. Tsang, W. Zhou, C.M. Black, and C. Sidman. 1991. Optimization of binding-capacity and specificity of protein-G on various solid matrices for immunoglobulins. *Journal of Immunological Methods* 136:279–286.

Proudfoot, K.A., C. Torrance, A.D.G. Lawson, and D.J. King. 1992. Purification of recombinant chimeric B72.3 Fab′ and F(ab′)2 using streptococcal protein G. *Protein Expression and Purification* 3:368–373.

Ray, P.K. 1985. Immunosuppressor control as a modality of cancer treatment: Effect of plasma adsorption with *Staphylococcus aureus* protein A. In *Immune Complexes and Human Cancer*, ed. F.A. Salinas and M.G. Hanna, Jr., 147–211. New York: Plenum Press.

Reis, K.J., E.M. Ayoub, and M.D. Boyle. 1984. Streptococcal Fc receptors. I. Isolation and partial characterization of the receptor from a group C streptococcus. *Journal of Immunology* 132:3091–3097.

Ricci, S., D. Medaglini, H. Marcotte, A. Olsen, G. Pozzi, and L. Bjorck. 2001. Immunoglobulin-binding domains of peptostreptococcal protein L enhance vaginal colonization of mice by *Streptococcus gordonii. Microbial Pathogenesis* 30:229–235.

Roben, P.W., A.N. Salem, and G.J. Silverman. 1995. V(H)3 family antibodies bind domain-D of staphylococcal protein-A. *Journal of Immunology* 154:6437–6445.

Ruppitsch, W., A. Indra, A. Stoger, B. Mayer, S. Stadlbauer, G. Wewalka, and F. Allerberger. 2006. Classifying spa types in complexes improves interpretation of typing results for methicillin-resistant *Staphylococcus aureus*. *Journal of Clinical Microbiology* 44:2442–2448.

Russelljones, G.J., E.C. Gotschlich, and M.S. Blake. 1984. A surface-receptor specific for human IgA on group-B streptococci possessing the Ibc protein antigen. *Journal of Experimental Medicine* 160:1467–1475.

Samuelsson, E., H. Wadensten, M. Hartmanis, T. Moks, and M. Uhlen. 1991. Facilitated in vitro refolding of human recombinant insulin-like growth factor-I using a solubilizing fusion partner. *Bio-Technology* 9:363–366.

Samuelsson, G. 1999. What's happening? Protein A columns: Current concepts and recent advances. *Transfusion Science* 21:215–217.

Samuelsson, M., T. Hallstrom, A. Forsgren, and K. Riesbeck. 2007. Characterization of the IgD binding site of encapsulated *Haemophilus influenzae* serotype b. *Journal of Immunology* 178:6316–6319.

Sano, T., C.L. Smith, and C.R Cantor. 1992. Immuno-PCR: Very sensitive antigen detection by means of specific antibody-DNA conjugates. *Science* 258:120–122.

Saouda, M., T. Romer, and M.D.P. Boyle. 2002. Application of immuno-mass spectrometry to analysis of a bacterial virulence factor. *Biotechniques* 32:916.

Sasso, E.H., G.J. Silverman, and M. Mannik. 1991. Human-IgA and IgG F(Ab′)2 that bind to staphylococcal protein-A belong to the Vhiii subgroup. *Journal of Immunology* 147:1877–1883.

Sauereriksson, A.E., G.J. Kleywegt, M. Uhl, and T.A. Jones. 1995. Crystal structure of the C2 fragment of streptococcal protein-G in complex with the Fc domain of human IgG. *Structure* 3:265–278.

Schmidt, E., E. Klinker, A. Opitz, S. Herzog, C. Sitaru, M. Goebeler, B. Mansouri Taleghoni, E.B. Brocker, and D. Zillikens. 2003. Protein A immunoadsorption: A novel and effective adjuvant treatment of severe pemphigus. *British Journal of Dermatology* 148:1222–1229.

Silverman, G.J., and C.S. Goodyear. 2006. Confounding B-cell defences: Lessons from a staphylococcal superantigen. *Nature Reviews Immunology* 6:465–475.

Sjobring, U. 1992. Isolation and molecular characterization of a novel albumin-binding protein from group-G streptococci. *Infection and Immunity* 60:3601–3608.

Sjobring, U., L. Bjorck, and W. Kastern. 1991. Streptococcal protein-G: Gene structure and protein-binding properties. *Journal of Biological Chemistry* 266:399–405.

Sjobring, U., J. Trojnar, A. Grubb, B. Akerstrom, and L. Bjorck. 1989. Ig-binding bacterial proteins also bind proteinase-inhibitors. *Journal of Immunology* 143:2948–2954.

Sjoholm, I. 1975. Protein A from *Staphylococcus aureus*: Spectropolarimetric and spectrophotometric studies. *European Journal of Biochemistry* 51:55–61.

Snyder HW Jr, Balint JP, and Jones FR. 1989. Modulation of immunity in patients with autoimmune disease and cancer treated by extracorporeal immunoadsorption with Prosorba columns. *Seminars in Hematology*, 26[Suppl 1], 31–38.

Snyder, H.W., Jr., S.K. Cochran, J.P. Balint, Jr., J.H. Bertram, A. Mittelman, T.H. Guthrie, Jr., and F.R. Jones. 1992. Experience with protein A-immunoadsorption in treatment-resistant adult immune thrombocytopenic purpura. *Blood* 79:2237–2245.

Solomon, A. 1976. Bence-Jones proteins and light-chains of immunoglobulins. II. *New England Journal of Medicine* 294:91–98.

Stahl, S., P.A. Nygren, A. Sjolander, and M. Uhlen. 1993. Engineered bacterial receptors in immunology. *Current Opinion in Immunology* 5:272–277.

Stalenhe, G., O. Gotze, N.R. Cooper, J. Sjoquist, and H.J. Mullereb. 1973. Consumption of human complement components by complexes of IgG with protein A of *Staphylococcus aureus*. *Immunochemistry* 10:501–507.

Steffen, A.C., M. Wikman, V. Tolmachev, G.P. Adams, F.Y. Nilsson, S. Stahl, and J. Carlsson. 2005. In vitro characterization of a bivalent anti-HER-2 affibody with potential for radionuclide-based diagnostics. *Cancer Biotherapy and Radiopharmaceuticals* 20:239–248.

Stone, G.C., U. Sjobring, L. Bjorck, J. Sjoquist, C.V. Barber, and F.A. Nardella. 1989. The Fc binding-site for streptococcal protein-G is in the C-gamma-2-C-gamma-3 interface region of IgG and is related to the sites that bind staphylococcal protein-A and human rheumatoid factors. *Journal of Immunology* 143:565–570.

Strambio-de-Castillia, C., J. Tetenbaum-Novatt, B.S. Imai, B.T. Chait, and M.P. Rout. 2005. A method for the rapid and efficient elution of native affinity-purified protein A tagged complexes. *Journal of Proteome Research* 4:2250–2256.

Strommenger, B., C. Kettlitz, T. Weniger, D. Harmsen, A.W. Friedrich, and W. Witte. 2006. Assignment of staphylococcus isolates to groups by spa typing, SmaI macrorestriction analysis, and multilocus sequence typing. *Journal of Clinical Microbiology* 44:2533–2540.

Svensson, H.G., H.R. Hoogenboom, and U. Sjobring. 1998. Protein LA, a novel hybrid protein with unique single-chain Fv antibody- and Fab-binding properties. *European Journal of Biochemistry* 258:890–896.

Swinnen, K., A. Krul, I. Van Goidsenhoven, N. Van Tichelt, A. Roosen, and K. Van Houdt. 2007. Performance comparison of protein A affinity resins for the purification of monoclonal antibodies. *Journal of Chromatography B Analytical Technologies in the Biomedical and Life Sciences* 848:97–107.

Takahashi, H., T. Nakanishi, K. Kami, Y. Arata, and I. Shimada. 2000. A novel NMR method for determining the interfaces of large protein-protein complexes. *Nature Structural Biology* 7:220–223.

Terman, D.S. 1985. Protein A and staphylococcal products in neoplastic disease. *Critical Reviews in Oncology Hematology* 4:103–124.

Tolmachev, V., A. Orlova, F.Y. Nilsson, J. Feldwisch, A. Wennborg, and L. Abrahmsen. 2007. Affibody molecules: Potential for in vivo imaging of molecular targets for cancer therapy. *Expert Opinion on Biological Therapy* 7:555–568.

Uhlen, M., B. Guss, B. Nilsson, S. Gatenbeck, L. Philipson, and M. Lindberg. 1984. Complete sequence of the staphylococcal gene encoding protein-A: A gene evolved through multiple duplications. *Journal of Biological Chemistry* 259:1695–1702.

Verwey, W.F. 1940. A type-specific antigenic protein derived from the staphylococcus. *Journal of Experimental Medicine* 71:635–644.

Viau, M., N.S. Longo, P.E. Lipsky, L. Bjorck, and M. Zouali. 2004. Specific in vivo deletion of B-cell subpopulations expressing human immunoglobulins by the B-cell superantigen protein L. *Infection and Immunity* 72:3515–3523.

Viau, M., N.S. Longo, P.E. Lipsky, and M. Zouali. 2005. Staphylococcal protein A deletes B-1a and marginal zone B lymphocytes expressing human immunoglobulins: An immune evasion mechanism. *Journal of Immunology* 175:7719–7727.

Vidal, M.A., and F.P. Conde. 1985. Alternative mechanism of protein A-immunoglobulin interaction: The Vh-associated reactivity of a monoclonal human-IgM. *Journal of Immunology* 135:1232–1238.

Vollenkle, C., S. Weigert, N. Ilk, E. Egelseer, V. Weber, F. Loth, D. Falkenhagen, U.B. Sleytr, and M. Sara. 2004. Construction of a functional S-layer fusion protein comprising an immunoglobulin G-binding domain for development of specific adsorbents for extracorporeal blood purification. *Applied and Environmental Microbiology* 70:1514–1521.

Werner, S., S. Marillonnet, G. Hause, V. Klimyuk, and Y. Gleba. 2006. Immunoabsorbent nanoparticles based on a tobamovirus displaying protein A. *Proceedings of the National Academy of Sciences USA* 103:17678–17683.

Widder, K.J., A.E. Senyei, H. Ovadia, and P.Y. Paterson. 1979. Magnetic protein A microspheres: A rapid method for cell separation. *Clinical Immunology and Immunopathology* 14:395–400.

Widjojoatmodjo, M.N., A.C. Fluit, R. Torensma, and J. Verhoef. 1993. Comparison of immunomagnetic beads coated with protein A, protein G, or goat anti-mouse immunoglobulins Applications in enzyme immunoassays and immunomagnetic separations. *Journal of Immunological Methods* 165:11–19.

Wikman, M., A.C. Steffen, E. Gunneriusson, V. Tolmachev, G.P. Adams, J. Carlsson, and S. Stahl. 2004. Selection and characterization of HER2/neu-binding affibody ligands. *Protein Engineering Design & Selection* 17:455–462.

Wikstrom, M., T. Drakenberg, S. Forsen, U. Sjobring, and L. Bjorck. 1994. Three-dimensional solution structure of an immunoglobulin light chain-binding domain of protein-l: Comparison with the IgG-binding domains of protein-G. *Biochemistry* 33:14011–14017.

Wikstrom, M., U. Sjobring, T. Drakenberg, S. Forsen, and L. Bjorck. 1995. Mapping of the immunoglobulin light chain-binding site of protein-l. *Journal of Molecular Biology* 250:128–133.

Young, W.W., Y. Tamura, D.M. Wolock, and J.W. Fox. 1984. Staphylococcal protein-A binding to the Fab fragments of mouse monoclonal antibodies. *Journal of Immunology* 133:3163–3166.

Zheng, D., J.M. Aramini, and G.T. Montelione. 2004. Validation of helical tilt angles in the solution NMR structure of the Z domain of staphylococcal protein A by combined analysis of residual dipolar coupling and NOE data. *Protein Science* 13:549–554.

Zikan, J. 1980. Interactions of pig Fab-gamma fragments with protein-A from *Staphylococcus aureus*. *Folia Microbiologica* 25:246–253.

Zouali, M. 2007. B cell superantigens subvert innate functions of B cells. *Chem. Immunol. Allergy* 93:92–105.

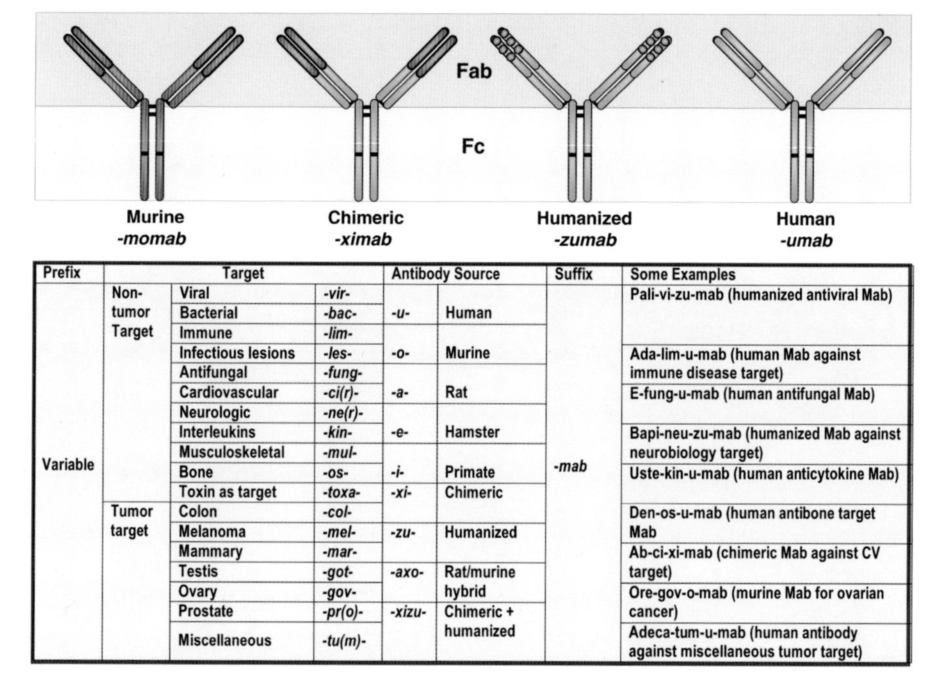

Prefix	Target			Antibody Source		Suffix	Some Examples
Variable	Non-tumor Target	Viral	*-vir-*	*-u-*	Human	*-mab*	Pali-vi-zu-mab (humanized antiviral Mab)
		Bacterial	*-bac-*				
		Immune	*-lim-*				
		Infectious lesions	*-les-*	*-o-*	Murine		Ada-lim-u-mab (human Mab against immune disease target)
		Antifungal	*-fung-*				
		Cardiovascular	*-ci(r)-*	*-a-*	Rat		E-fung-u-mab (human antifungal Mab)
		Neurologic	*-ne(r)-*				
		Interleukins	*-kin-*	*-e-*	Hamster		Bapi-neu-zu-mab (humanized Mab against neurobiology target)
		Musculoskeletal	*-mul-*				
		Bone	*-os-*	*-i-*	Primate		Uste-kin-u-mab (human anticytokine Mab)
		Toxin as target	*-toxa-*	*-xi-*	Chimeric		
	Tumor target	Colon	*-col-*				Den-os-u-mab (human antibone target Mab)
		Melanoma	*-mel-*	*-zu-*	Humanized		
		Mammary	*-mar-*				Ab-ci-xi-mab (chimeric Mab against CV target)
		Testis	*-got-*	*-axo-*	Rat/murine hybrid		
		Ovary	*-gov-*				Ore-gov-o-mab (murine Mab for ovarian cancer)
		Prostate	*-pr(o)-*	*-xizu-*	Chimeric + humanized		
		Miscellaneous	*-tu(m)-*				Adeca-tum-u-mab (human antibody against miscellaneous tumor target)

Figure 1.1 The different forms of therapeutic monoclonal antibodies that have been approved for marketing, including murine, chimeric, humanized, and fully human antibodies, as well as how the generic names are applied to each of them based on structure, source, and target.

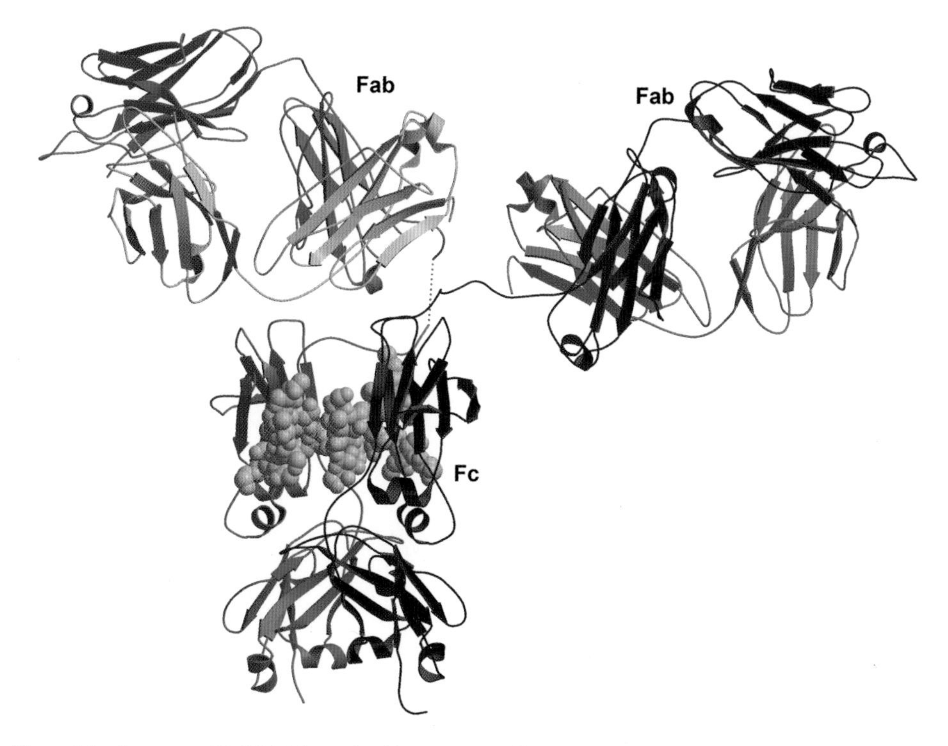

Figure 2.3 Structure of IgG b12 determined by x-ray crystallography. The heavy chains are shown in green and dark blue, and the light chains in yellow and cyan. Residues in the linker region of the green heavy chain that have no visible electron density are shown as a dotted line. Carbohydrate within the Fc domain is shown in a gray CPK representation. Rather than a rigid Y shape as depicted in Figure 2.2, in a real IgG, the Fab arms are very flexible with respect to the Fc domain, allowing for greater flexibility in binding antigen.

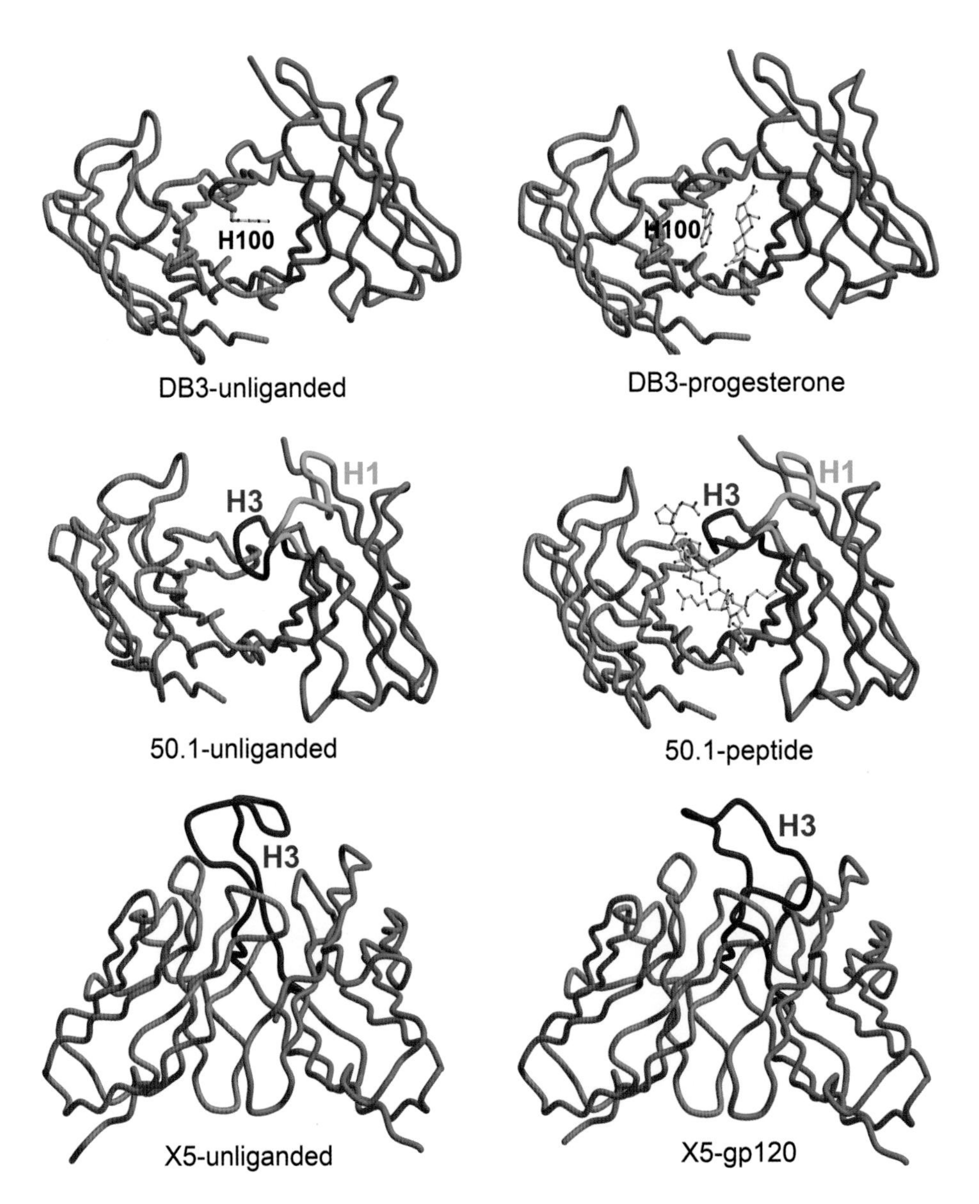

Figure 2.7 Examples of conformational changes in Fabs after binding antigen. Top left: The anti-progesterone Fab DB3 in its unliganded form, with Trp^{H100} filling its antigen-binding pocket. Top right: DB3 with progesterone (ball-and-stick) bound, and Trp^{H100} moved away from its position in the unliganded Fab. (Middle left) The anti-HIV-1 Fab 50.1 in its unliganded form. Middle right: 50.1 with bound peptide (ball-and-stick). The H3 and H1 CDRs are labeled. H3 undergoes a structural rearrangement, while H1 moves away from the binding site while maintaining its overall shape. Bottom left: The unliganded anti-HIV-1 Fab X5, with its long CDR H3 labeled. Bottom right: Fab X5 in its gp120-bound conformation, with H3 labeled. A large conformational change occurs in the CDR H3 between the unliganded and liganded forms.

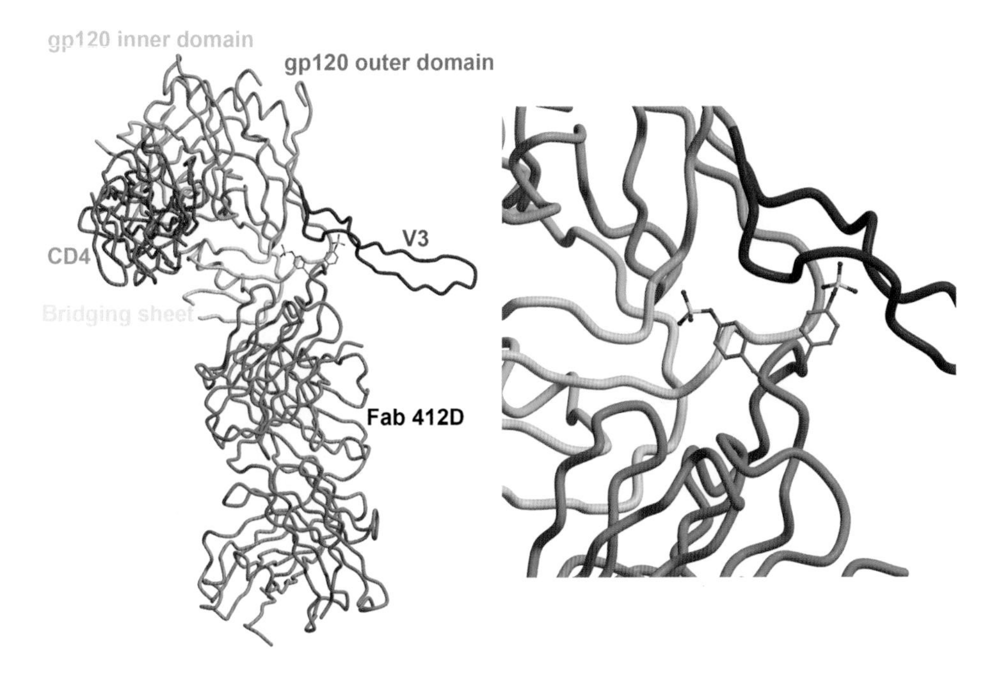

Figure 2.9 The anti-HIV-1 Fab 412D has sulfated tyrosine residues in CDR H3. Left: The complex of Fab 412d with CD4 and gp120. Right: enlargement of the sulfated tyrosine residues and their interaction with gp120. It is thought that the gp120 co-receptor (CCR5) binds to the bridging sheet and V3 regions of the gp120 molecule. CCR5 has four sulfated tyrosine residues in its N-terminal region, and at least two of these are thought to take part in the interaction with gp120. Several antibodies that recognize the same region of gp120 have also evolved to have sulfated tyrosine residues.

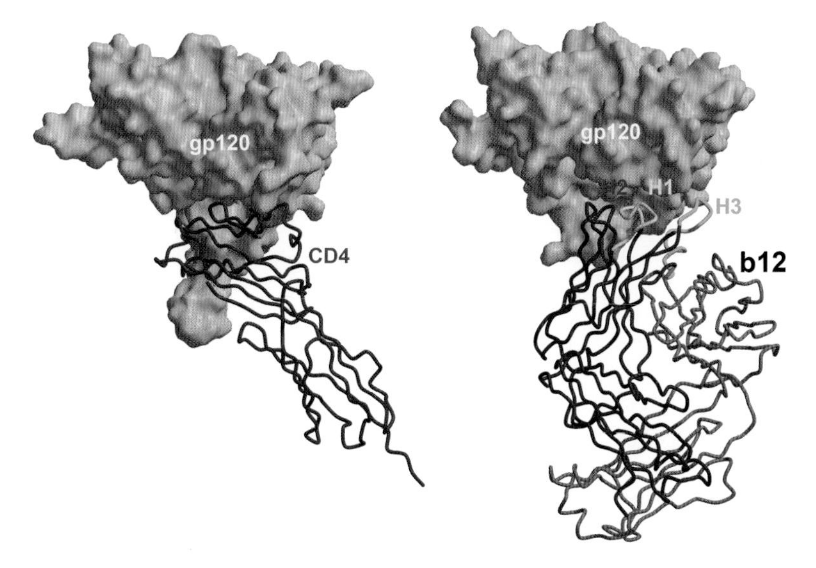

Figure 2.10 Broadly neutralizing anti-HIV-1 antibody b12 interactions with gp120. Left: The CD4 (tubes) binding site on gp120 (solid surface) is shown. Right: b12 binds to the CD4-binding site, accessing a deep cleft with its long CDR loop (H2) and clasping the CD4-binding loop between CDRs H2 and H1 on one side and H3 on the other side.

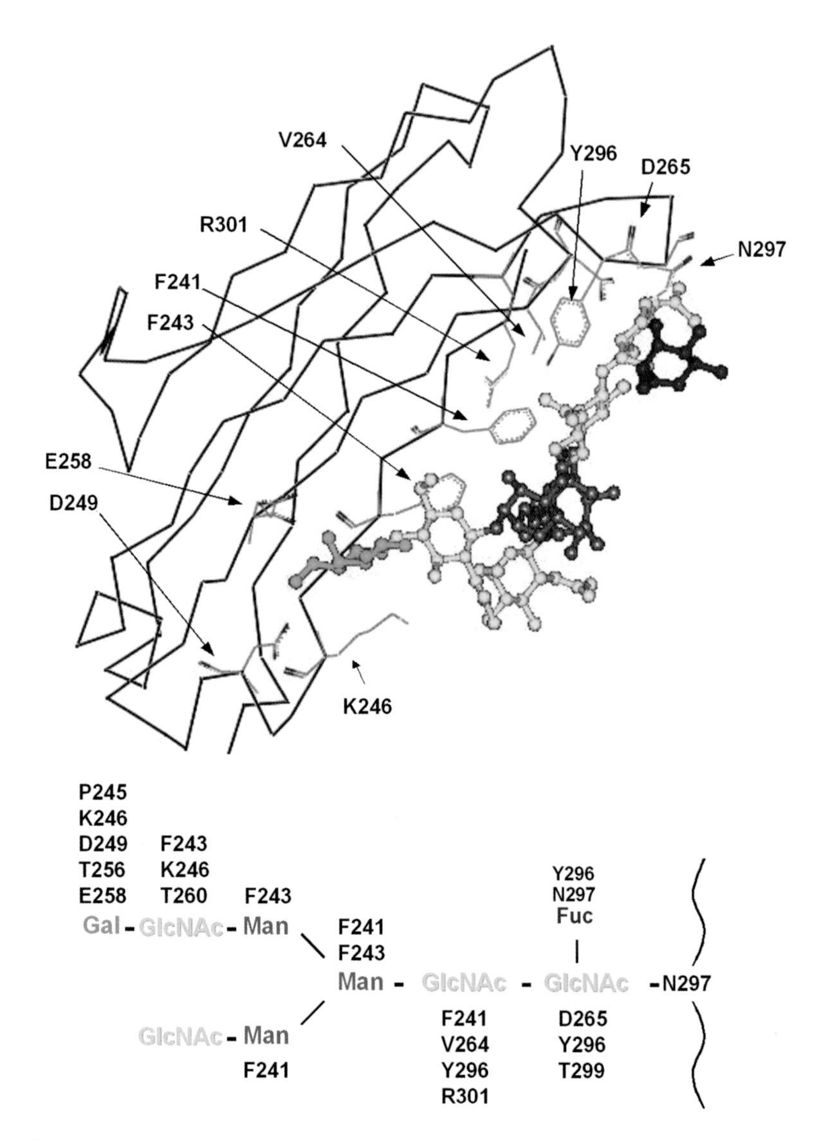

Figure 3.2 Contacts between the C_H2 domain and the oligosaccharide. The C^{α}-atoms of the polypeptide chain of the C_H2 domain is drawn with a line while the oligosaccharide chain is shown in colored ball-and-stick representation (PDB code 1Fc). Residues that make contacts with the oligosaccharide are labeled with the Eu numbering system.

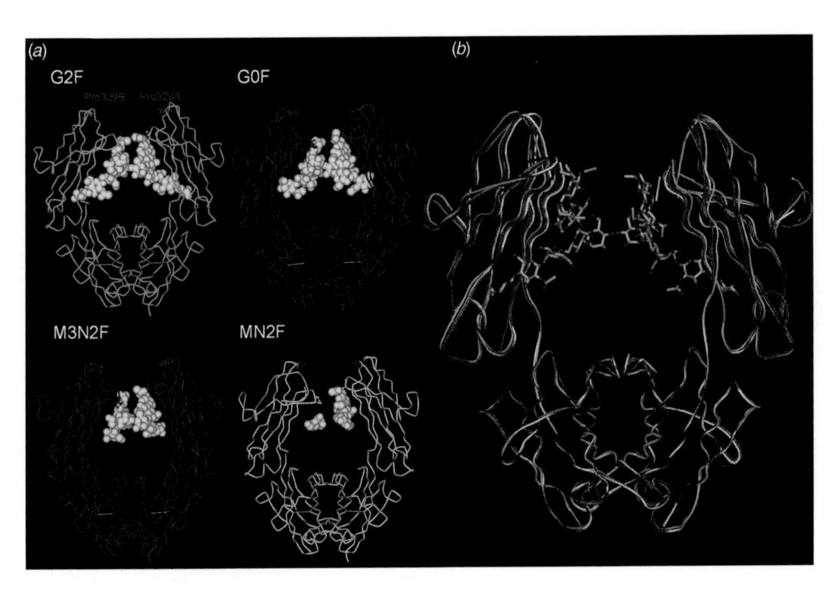

Figure 3.6 Crystal structures of homogeneous glycoforms of IgG-Fc Cri (a) and superimposed image (b). Structures of homogeneous Fc glycoforms G2F, G0F, M3N2F, and MN2F in space group $P2_12_12_1$ are shown (PDB codes 1H3V, 1H3X, 1H3U, and 1H3T, respectively). Pro329 residues are shown on the tips of the C_H2 domains. Note that the tetrasaccharide on the chain-B (left) of the MN2F glycoform is not completely defined but present, due to less/weaker crystal contacts. The C_H2 domains of the MN2F glycoform exhibit a higher B-factor and "softer" structure than in the other glycoforms. The superimposed image was created using PYMOL (www.pymol.org). (Courtesy of Dr. Peter Sondermann). Only the G2F glycan is shown in the superimposed image. Note that the conformational difference between the glycoforms is prominent in the hinge proximal region of the C_H2 domain in contrast to the C_H3 domain and C_H2/C_H3 domain interface.

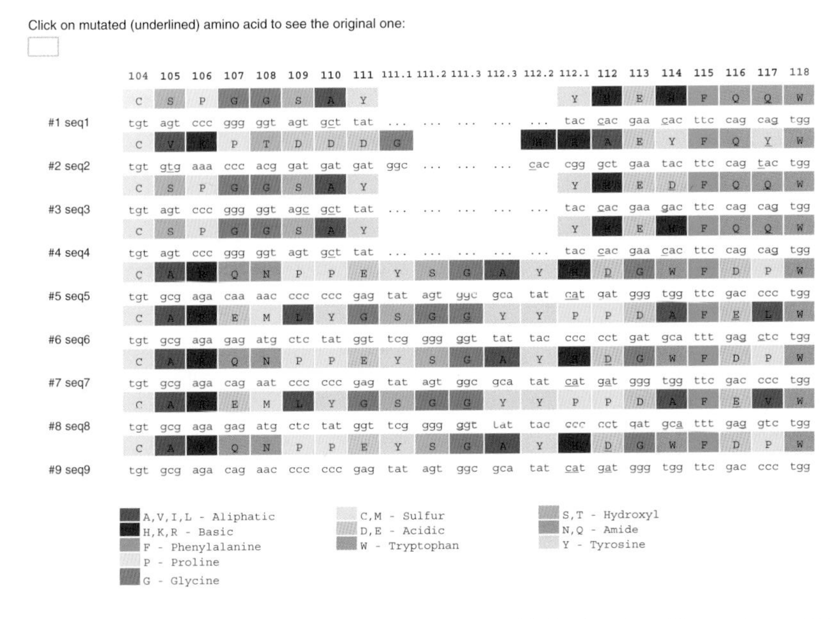

Translation of the JUNCTIONs

Click on mutated (underlined) amino acid to see the original one:

	104	105	106	107	108	109	110	111	111.1	111.2	111.3	112.3	112.2	112.1	112	113	114	115	116	117	118
	C	S	P	G	G	S	A	Y						Y	H	E	H	F	Q	Q	W
#1 seq1	tgt	agt	ccc	ggg	ggt	agt	gct	tat	...	...	...	...	...	tac	cac	gaa	cac	ttc	cag	cag	tgg
	C	V	K	P	T	D	D	D	G				H	R	A	E	Y	F	Q	Y	W
#2 seq2	tgt	gtg	aaa	ccc	acg	gat	gat	gat	ggc	...	...	...	cac	cgg	gct	gaa	tac	ttc	cag	tac	tgg
	C	S	P	G	G	S	A	Y						Y	H	E	D	F	Q	Q	W
#3 seq3	tgt	agt	ccc	ggg	ggt	agc	gct	tat	...	...	...	...	...	tac	cac	gaa	gac	ttc	cag	cag	tgg
	C	S	P	G	G	S	A	Y						Y	H	E	H	F	Q	Q	W
#4 seq4	tgt	agt	ccc	ggg	ggt	agt	gct	tat	...	...	...	...	...	tac	cac	gaa	cac	ttc	cag	cag	tgg
	C	A	R	Q	N	P	P	E	Y	S	G	A	Y	H	D	G	W	F	D	P	W
#5 seq5	tgt	gcg	aga	caa	aac	ccc	ccc	gag	tat	agt	ggc	gca	tat	cat	gat	ggg	tgg	ttc	gac	ccc	tgg
	C	A	R	E	M	L	Y	G	S	G	G	Y	Y	P	P	D	A	F	E	L	W
#6 seq6	tgt	gcg	aga	gag	atg	ctc	tat	ggt	tcg	ggg	ggt	tat	tac	ccc	cct	gat	gca	ttt	gag	ctc	tgg
	C	A	R	Q	N	P	P	E	Y	S	G	A	Y	H	D	G	W	F	D	P	W
#7 seq7	tgt	gcg	aga	cag	aat	ccc	ccc	gag	tat	agt	ggc	gca	tat	cat	gat	ggg	tgg	ttc	gac	ccc	tgg
	C	A	R	E	M	L	Y	G	S	G	G	Y	Y	P	P	D	A	F	E	V	W
#8 seq8	tgt	gcg	aga	gag	atg	ctc	tat	ggt	tcg	ggg	ggt	tat	tac	ccc	cct	gat	gca	ttt	gag	gtc	tgg
	C	A	R	Q	N	P	P	E	Y	S	G	A	Y	H	D	G	W	F	D	P	W
#9 seq9	tgt	gcg	aga	cag	aac	ccc	ccc	gag	tat	agt	ggc	gca	tat	cat	gat	ggg	tgg	ttc	gac	ccc	tgg

A,V,I,L - Aliphatic
H,K,R - Basic
F - Phenylalanine
P - Proline
G - Glycine
C,M - Sulfur
D,E - Acidic
W - Tryptophan
S,T - Hydroxyl
N,Q - Amide
Y - Tyrosine

Figure 4.5 IMGT/JunctionAnalysis for the analysis of the V-J and V-D-J junctions. The translation of the junctions is shown with amino acids colored according to the IMGT physicochemical characteristics classes (Pommié et al. 2004).

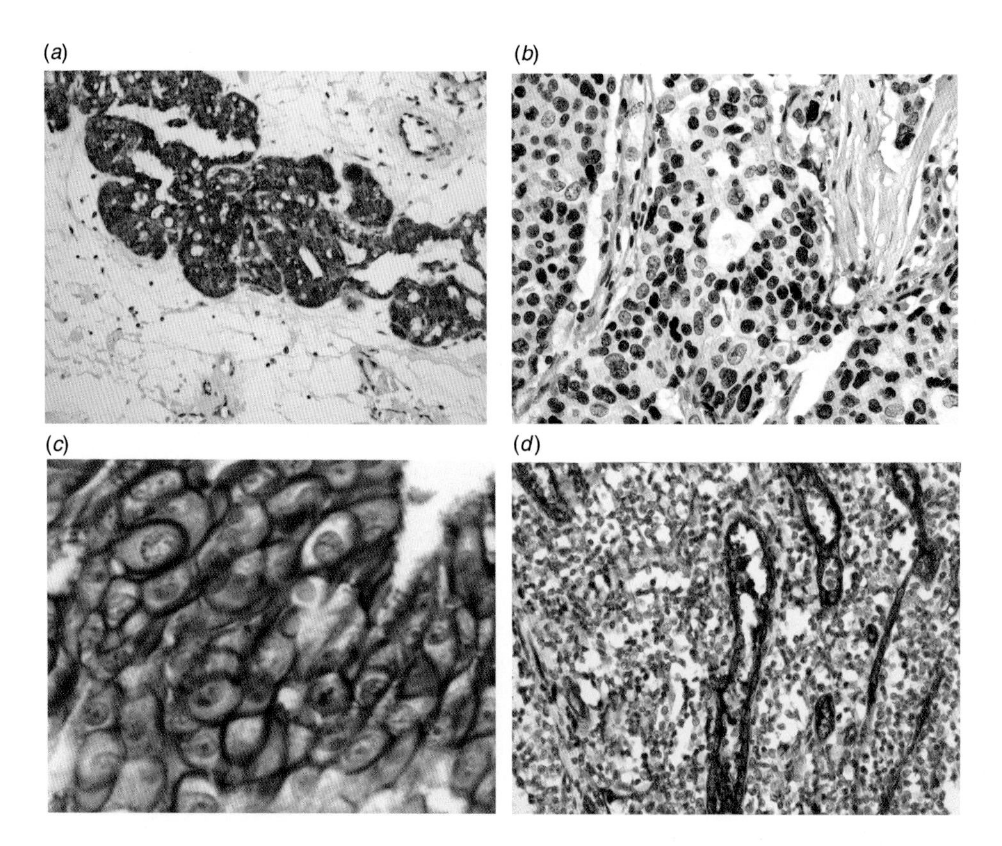

Figure 6.3 Immunohistochemistry staining of paraffin embedded tissues by RabMAbs. Human paraffin embedded tissues were deparaffinized by heating followed by a series of washes, with two xylene washes and two 100 percent ethanol rinses, followed by 95 percent ethanol, 70 percent ethanol, 50 percent ethanol, 30 percent ethanol, followed by H_2O and a TBST wash for 5 min on a shaker. Antigen retrieval was done using a rice cooker to cook in antigen retrieval solution for about 20 to 30 min. After slides were washed with TBST for 5 min on a shaker, endogenous peroxidase was inactivated by covering tissue with 3 percent hydrogen peroxide for 10 min. Slides were washed three times with TBST and blocked with blocking solution for one hour. Diluted primary antibody in the blocking buffer was applied to each section and incubated overnight in a humidified chamber (4°C). After washing slides three times with TBST, secondary HRP-conjugated anti-rabbit antibody diluted in the blocking solution was applied to incubate for one hour at room temperature. Slides were washed three times with TBST and freshly prepared DAB substrate was added to the tissue sections. The tissue sections with the substrate were incubated at room temperature until suitable staining developed, followed by rinsing with water and counterstaining with Hematoxylin. Slides were rinsed with water again and dehydrated using two rinses with 100 percent ethanol followed by two rinses with xylene. Slides were mounted with coverslips using permount medium and visualized under a microscope. (a) Ovarian adenocarcinoma tissue stained with anti-phospho-AKT RabMAb. (b) Breast ductal infiltrating carcinoma tissue stained with anti-PR RabMAb. (c) Breast ductal infiltrating carcinoma tissue stained with anti-Her2 RabMAb. (d) Vascular endothelial cells stained with anti-CD34 RabMAb.

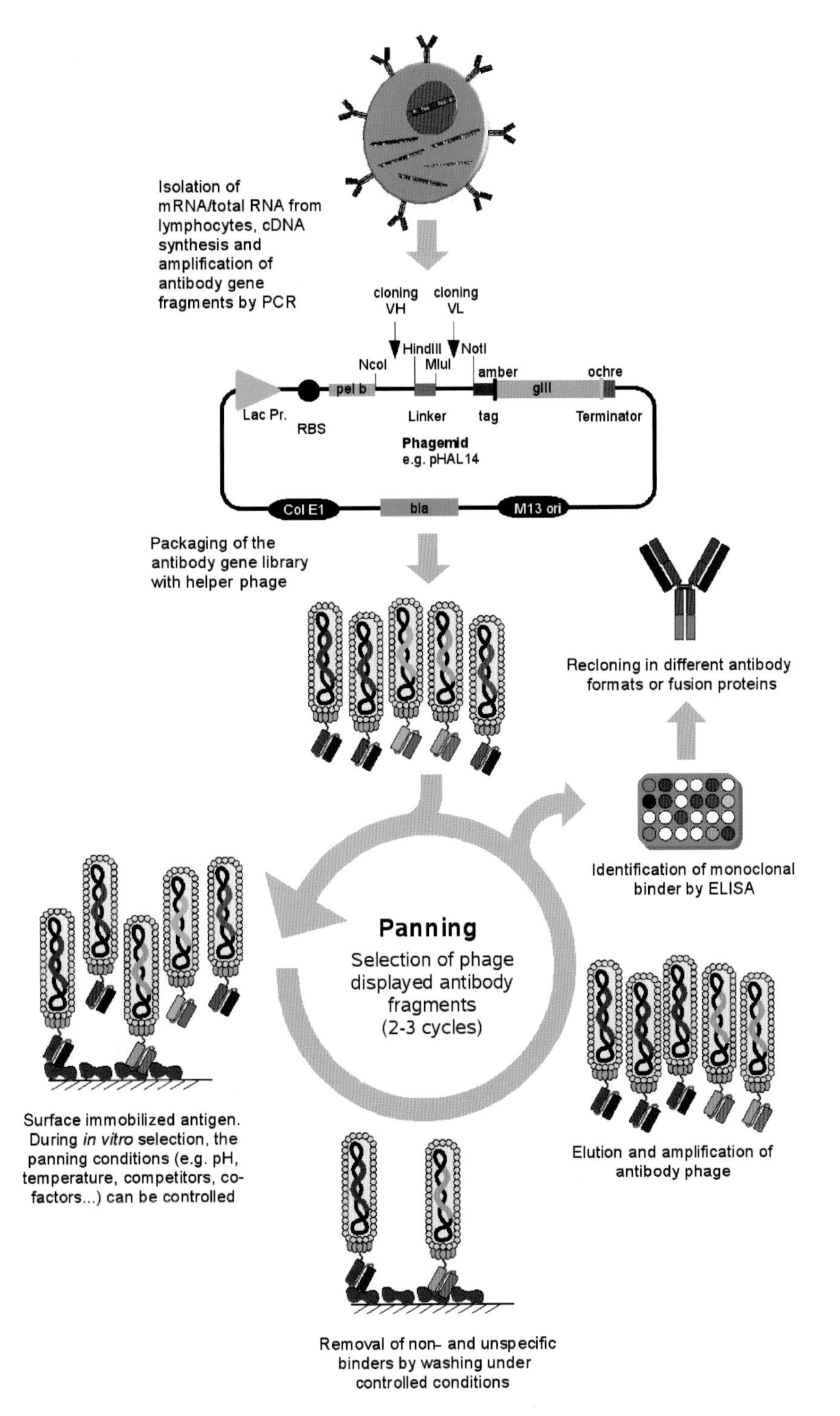

Figure 8.1 Construction of antibody gene libraries in the scFv format and selection of antibodies from antibody libraries ("panning") by phage display.

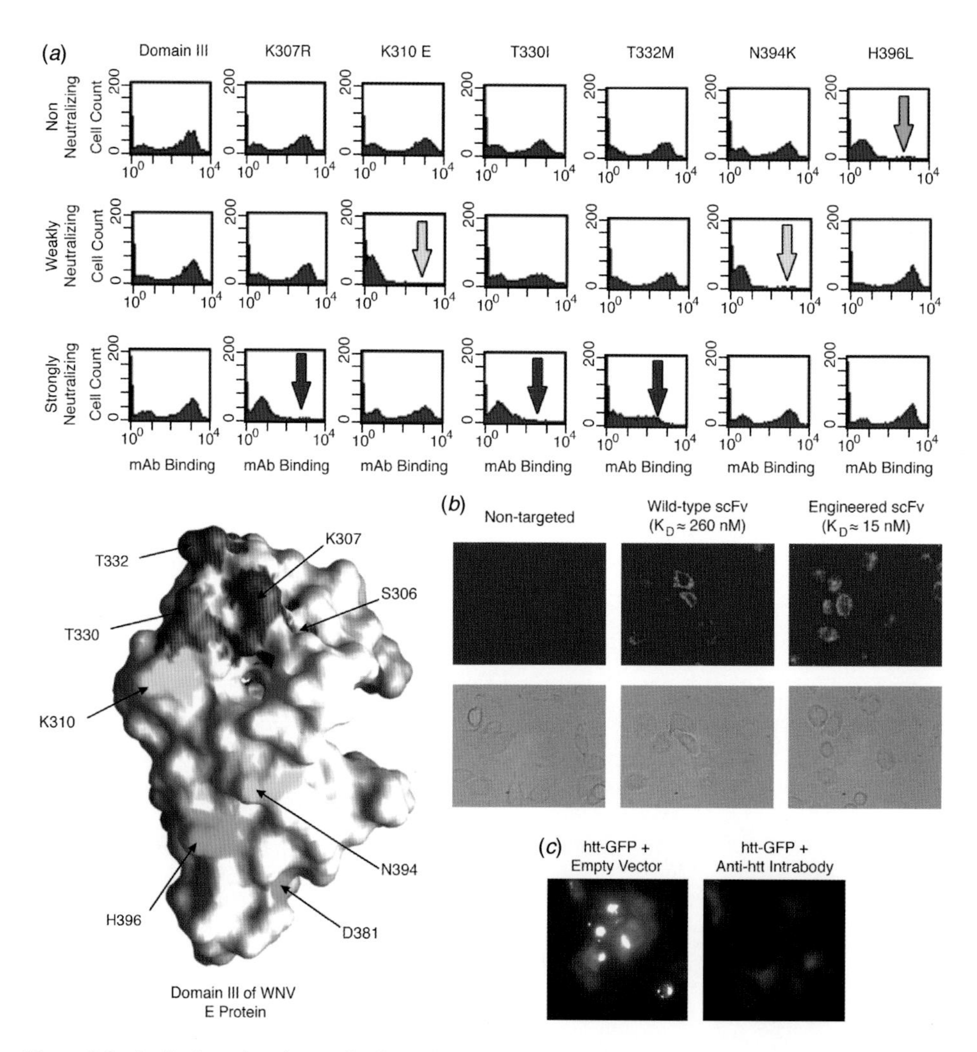

Figure 9.6 Antibody engineering applications of yeast surface display. (a) Fine epitope mapping of neutralizing and non-neutralizing monoclonal antibodies (mAb) against domain III of theWest Nile virus E protein (WNV DIII). (Top) Flow cytometry histograms depict mAb immunoreactivity, with arrows highlighting point mutations that abolish binding of mAbs to yeast-displayed WNV DIII. (Bottom) Mapping of antibody binding epitopes on the surface of WNV DIII. From Oliphant et al. (2005). Reprinted by permission from Macmillan Publishers, Ltd. © 2005. (b) Tumor cell targeting and internalization is mediated by scFv-coated immunoliposomes (ILs). Wild-type and mutant anti-epidermal growth factor receptor scFvs attached to the surface of ILs loaded with fluorescent dye mediate internalization into tumor cells overexpressing the receptor. From Zhou et al. (2007). Reprinted with permission from Elsevier. © 2007. (c) An engineered single domain intrabody blocks intracellular huntingtin protein (htt) aggregation. Fluorescence microscopy images showing cells without (left) or treated with (right) the engineered anti-htt intrabody (Colby et al. 2004a). © 2004 National Academy of Sciences, USA.

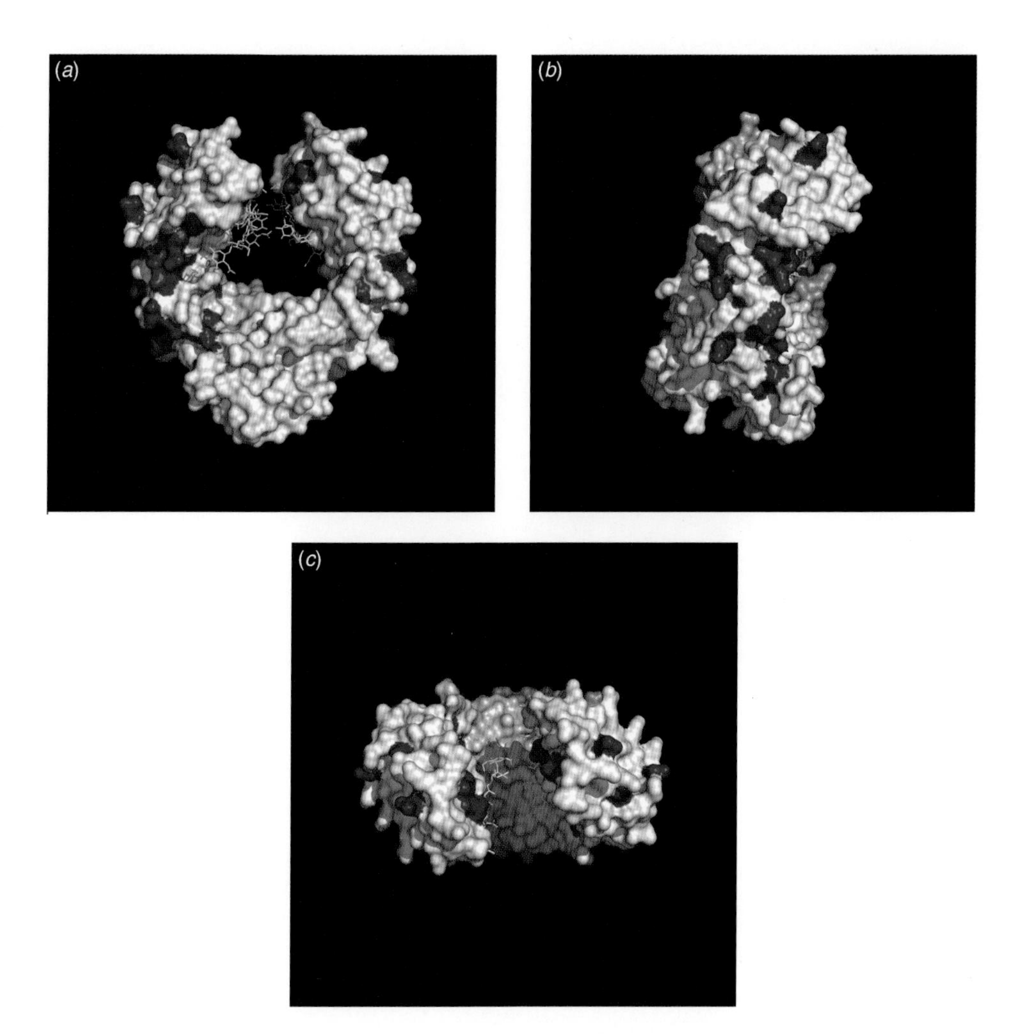

Figure 14.2 Space-filling model of an Fc fragment derived from a human IgG1, based on the x-ray structure corresponding to protein database ID number 1FC1 (Deisenhofer 1981). The Fc fragment is oriented so that its front (a), side (b), or top (c) points toward the reader. Residues in red correspond to the various positions where single or multiple substitutions were shown to improve the affinity of the human IgG/FcRn interaction at acidic pH. Carbohydrates are shown as sticks.

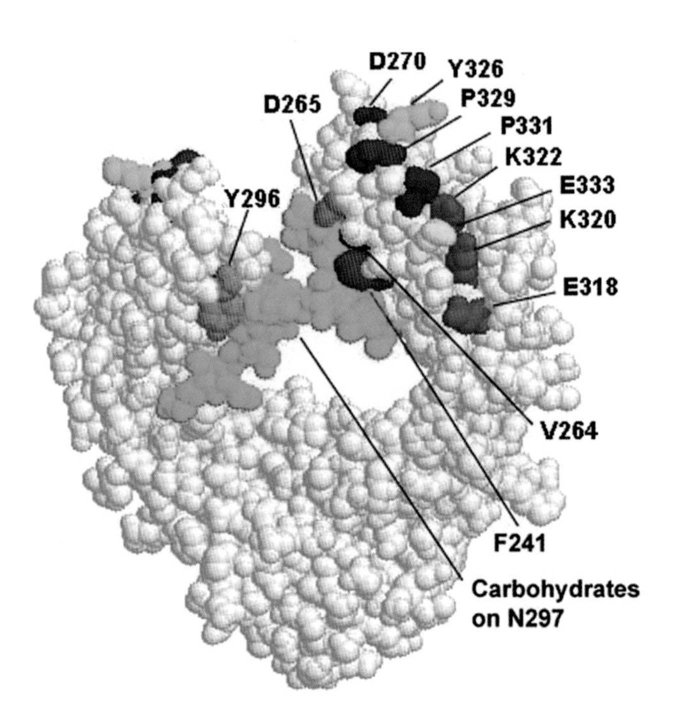

Figure 16.2 Crystal structure of the Fc portion of human IgG1. The Fc portion of a human IgG1 is shown in CPK space-fill form (PDB coordinates 1Fc1). Amino acids that are playing a role in complement binding and activation are colored as follows: F241 magenta; V264 black; D265 violet; D270 blue; Y296 orange; E318, K320, and K322 green-blue; Y326 cyan; P329 purple, P331 red; E333 yellow. Carbohydrates attached to asparagines 297 are colored green. Amino acids L234, L235, and G236 are not shown because they are not resolved in the crystal structure.

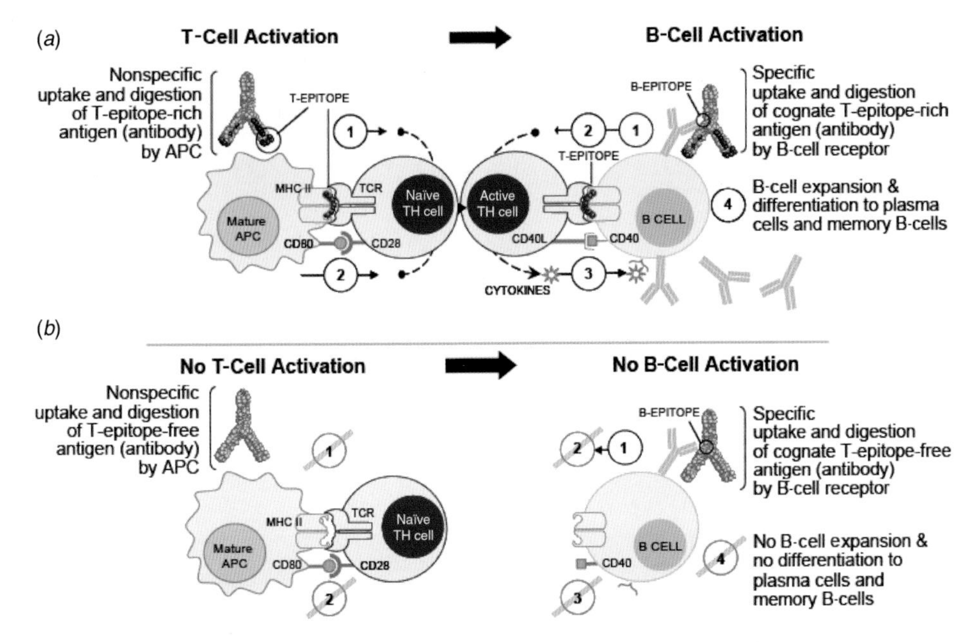

Figure 18.1 Activation of CD4 T-cells and the T-dependent antibody response. In order to induce naïve B-cells to produce antibodies against a protein antigen, several events must be coordinated, usually within specialized regions of secondary lymphoid organs (e.g., lymph nodes, spleen, etc.). The first step in this process is internalization of antigen by professional antigen-presenting cells (APC), such as dendritic cells (left side of figure). The mature APC processes antigen into peptides that are then presented to naïve T-cells in major histocompatibility complex (MHC) class II molecules on the APC surface. The interaction of a T-cell receptor (TCR) with this MHCII:peptide complex is identified as a T-cell signal 1 (T1). In order to fully activate the T-cell, T1 must be accompanied by additional signals from what are termed *co-stimulatory molecules*, such as CD80, CD86, etc. provided by the APC (T-cell signal 2, T2). In the absence of T2, T-cells may become anergic. Once fully activated, these CD4 T-cells divide and produce an array of cytokines with manifold activities. Activation of the naïve B-cell is initiated by an interaction between B-cell IgM and IgD receptors and cognate antigen. This is termed B-cell signal 1 (B1). Upon encountering a B-cell that has recognized antigen via a specific T-cell epitope-MHC:TCR interaction, the T-cell delivers cytokines that drive the B-cell to proliferate and mature towards an antibody-secreting plasma cell. A CD4 T-cell thus provides help to B-cells [B-cell signal 2 (B2)]. This interaction results in the engagement of CD40 and CD40 ligand communicating a further signal to the presenting B-cell [B-cell signal 3 (B3)] leading to antibody production via B-cell expansion and differentiation to plasma cells and memory B cells (B4). Adapted with permission from A.S. De Groot and L. Moise, *Current Opinion in Drug Discovery & Development* 10, no. 3 (2007):332–340. Copyright 2007, the Thomson Corporation.

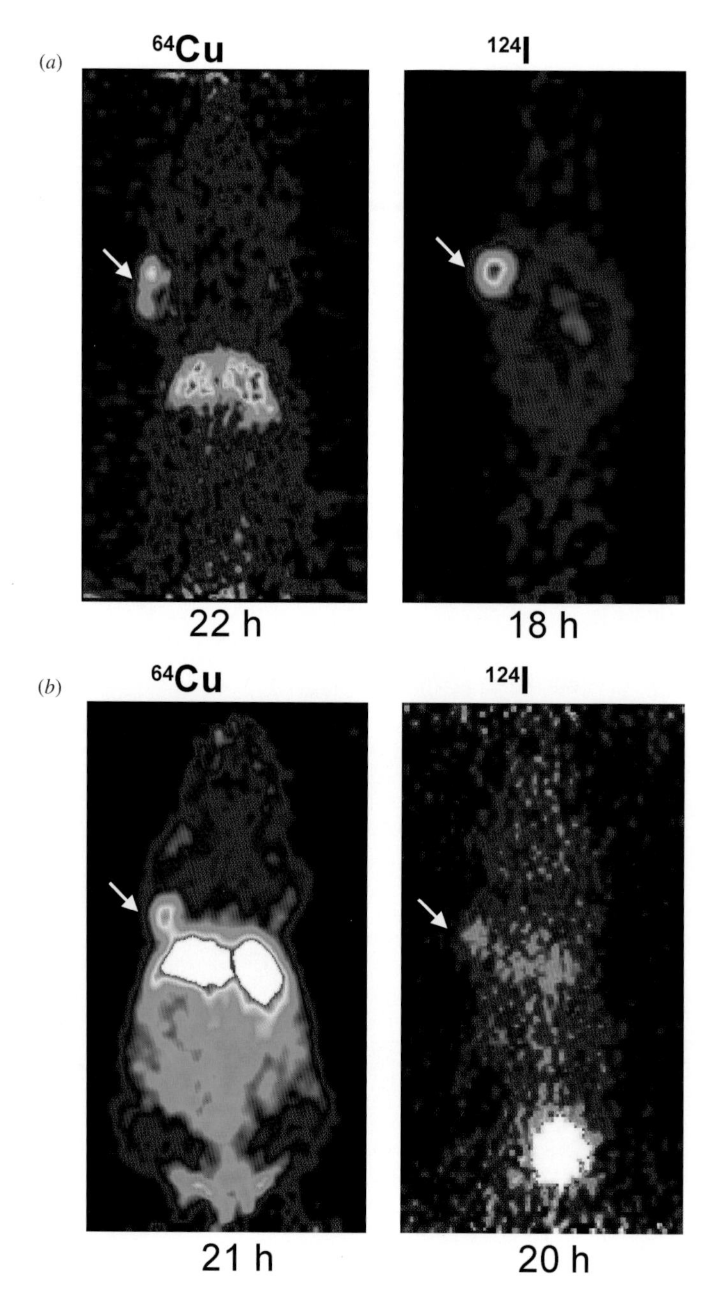

Figure 20.2 PET images of noninternalizing (a) versus internalizing (b) tumor cell surface targets with ^{64}Cu- and ^{124}I-labeled scFv-Fc fragments. (a) Coronal images of mice injected with anti-CEA scFv-Fc H310A/H435Q bearing LS174T xenografts (arrow). (b) Coronal images of mice injected with anti-HER2 scFv-Fc H310A/H435Q bearing MCF7/HER2 xenografts (arrow). The radiolabel is shown above each image and the time in hours after administration of the tracer is shown below. High liver activity is seen in mice injected with ^{64}Cu-labeled tracer in both tumor models, which is more or less absent in the mice injected with ^{124}I-labeled tracer. A higher contrast image is seen in the noninternalizing CEA system, which is further enhanced by blocking for thyroid and stomach uptakes.

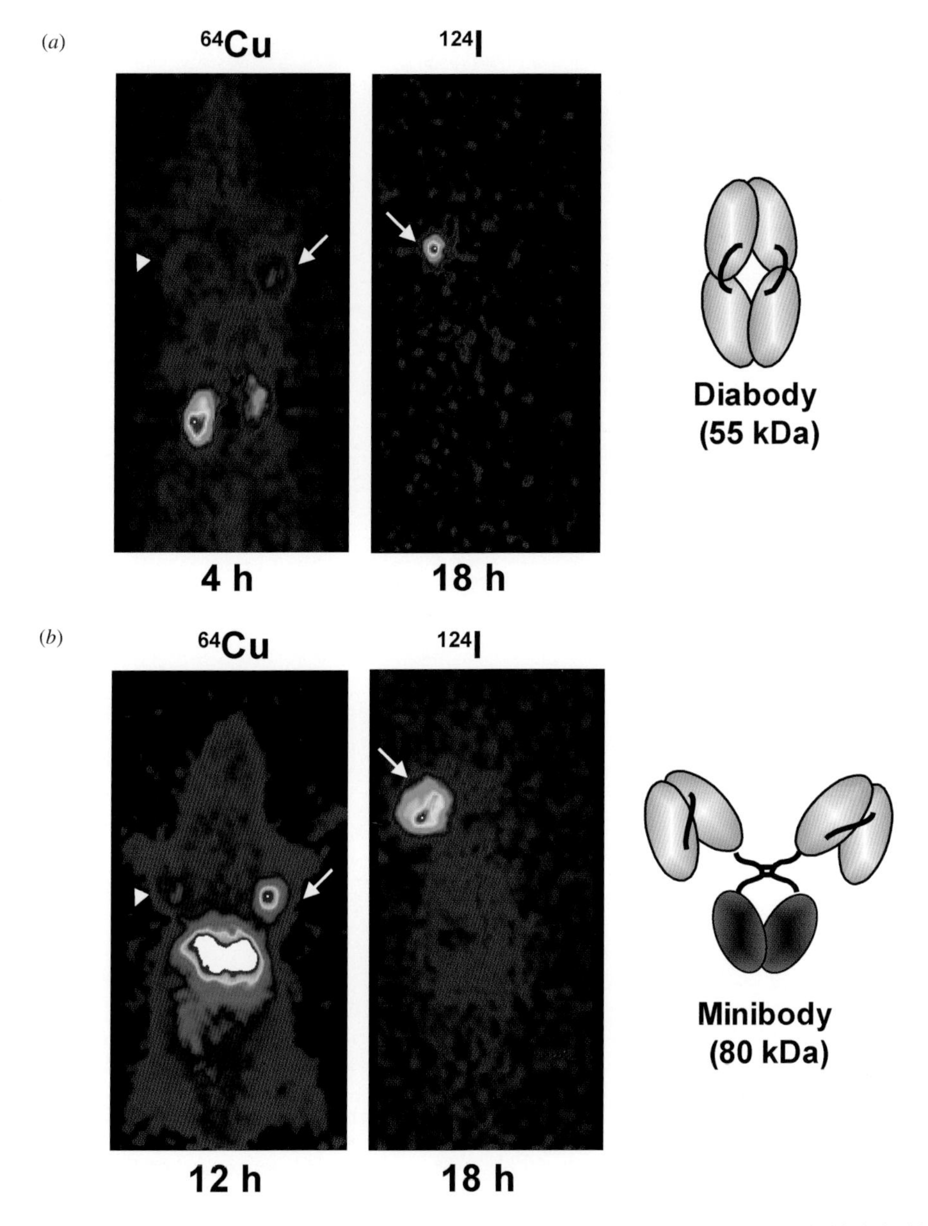

Figure 20.3 PET images of tumor-bearing mice injected with diabody (a) and minibody (b) radiolabeled with ^{64}Cu and ^{124}I. Positive tumor is indicated by an arrow, whereas the negative tumor is indicated by an arrowhead. The radiolabel is shown above each image and the time in hours after administration of the tracer is shown below. The primary excretion routes for the diabody and minibody are kidney and liver, respectively, which can be seen in the mice injected with ^{64}Cu-labeled fragments. In the mice injected with ^{124}I-labeled fragment, only tumor is visible at 18 hour with the diabody due to its rapid blood clearance, whereas the more background (blood pool) activity is seen in the mice injected with the minibody. However, both fragments produce excellent, high contrast images in mice (stomach and thyroid uptakes were blocked).

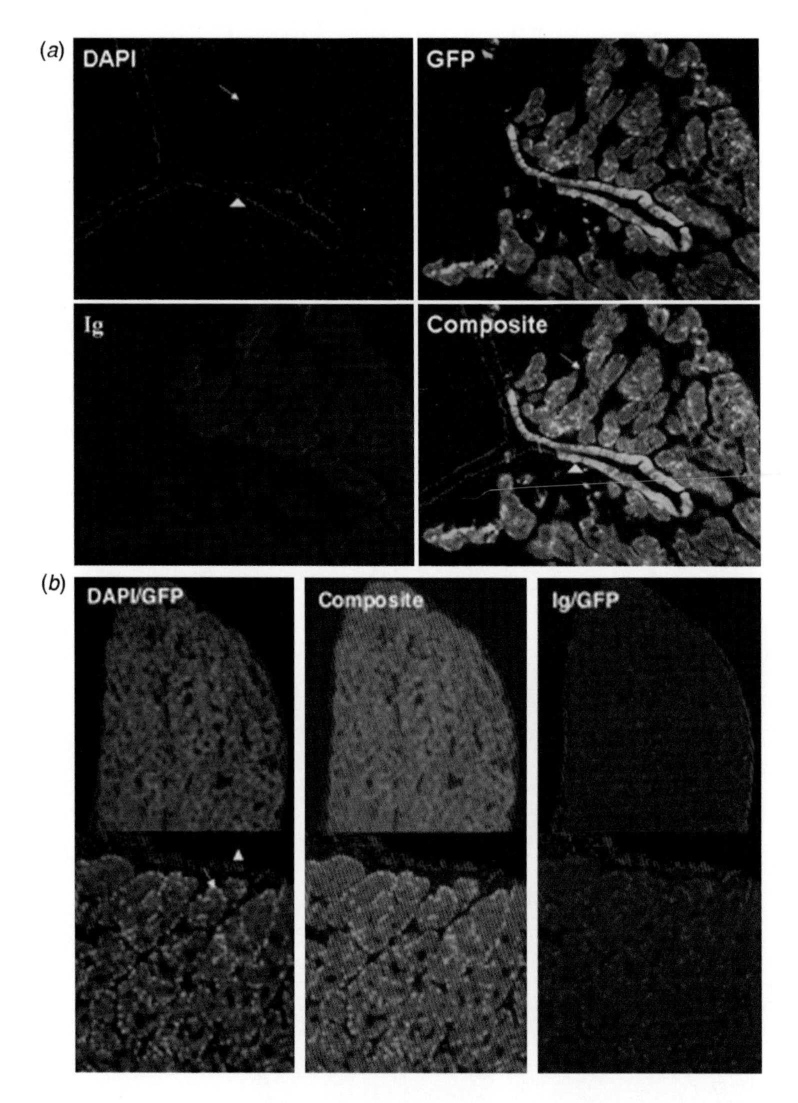

Figure 28.7 Expression of mAb was localized to the tubular gland cells in chimeras. (a) Frozen sections of the magnum portion of the oviduct from an estrogen-induced chimeric female chick produced from cES cells transfected with Ov7.5mAbdns. Images of a section under DAPI staining showing the location of the cells, EGFP fluorescence showing the contribution of cES cells in chimera, human Ig Ab staining showing the transgene expression and composite of all three images. (Adapted from Zhu, L. et al. 2005. *Nature Biotechnol.* 23:1159–1169.) (b) Expression of mAbF1 in the magnum portion of the oviduct from an estrogen-induced chimeric female chick produced from OvBACmAbF1transfected cES cell line OV61/42. Note the restricted transgene expression in donor cES cell-derived tubular glands (arrow) but not in epithelial cells (arrowhead). Top panels under low magnification showed extensive contribution of the cES cells to the tubular glands. Lower panels under higher magnification showed restricted expression of mAb in the tubular gland cells. (From Zhu, L. et al. unpublished data.)

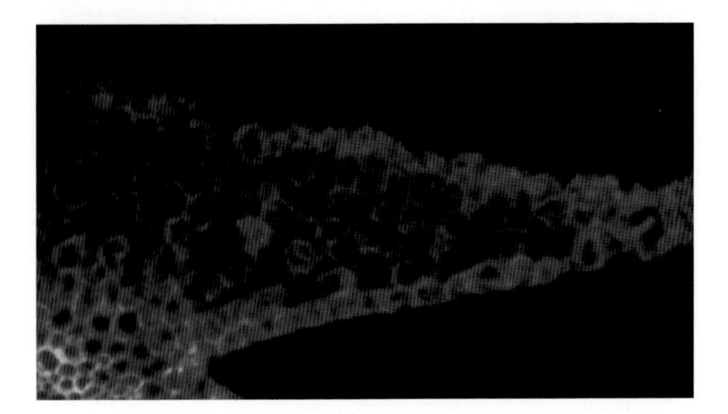

Figure 29.1 Immunolocalization of IgG1 mAb expressed in *Lemna* (cross section). IgG1 mAb expressed in a stable transgenic *Lemna* line; immunolocalized using a goat anti-human IgG1 followed by a donkey anti-goat flourescently labeled antibody. The recombinant antibody visualized in blue is localized to the apoplast of the *Lemna* frond and is expressed evenly throughout.

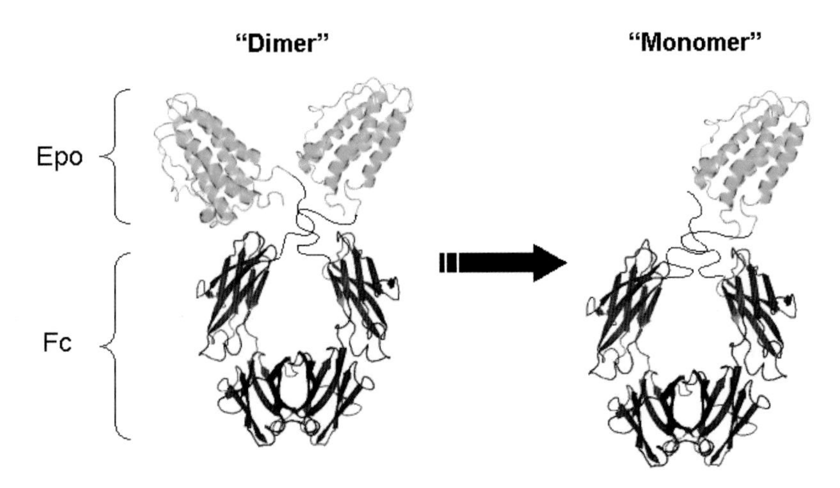

Figure 33.1 Schematic diagram for erythropoietin Fc (EpoFc) dimer and monomer. The dimer is comprised of two molecules of Epo joined to dimeric Fc and the monomer has a single molecule of Epo joined to dimeric Fc.

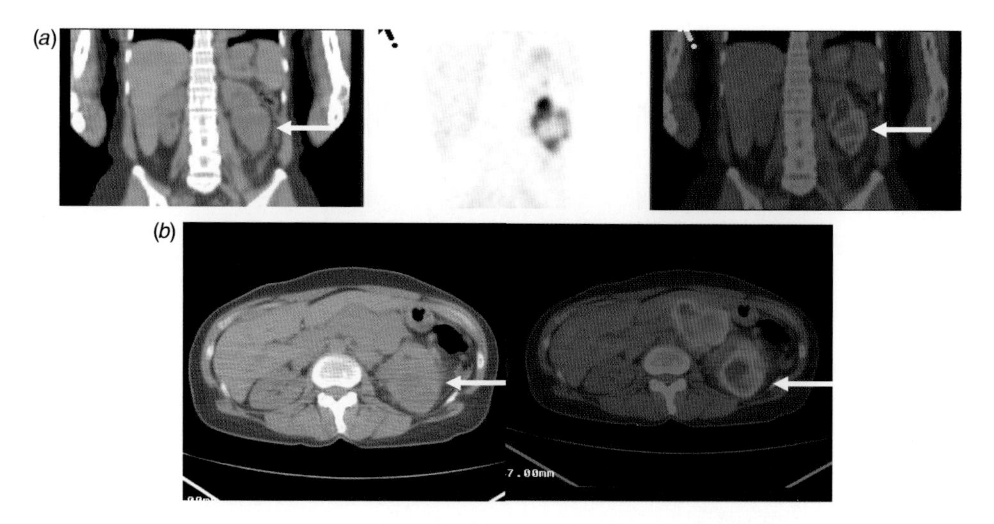

Figure 34.3 Coronal (a) and transaxial (b) images of a patient with clear-cell renal cell carcinoma following an injection with ^{124}I cG250 antibody. There is excellent targeting of antibody to tumor (white arrow).

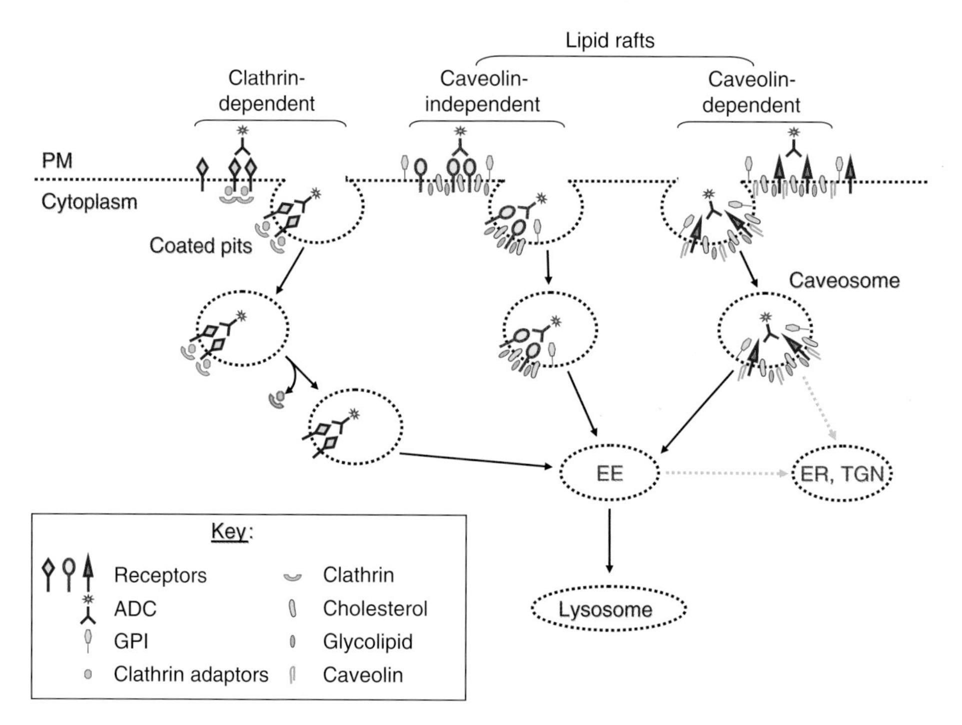

Figure 35.2 Major receptor-mediated endocytosis pathways available for internalization of antibody drug conjugates. Three pathways: clathrin-dependent endocytosis, clathrin-independent noncaveolar endocytosis, and clathrin-independent caveolar endocytosis have been described to mediate internalization of antibody drug conjugates. All three pathways can converge at the early endosomes and eventually end at the lysosomes. The green arrows indicate retrograde trafficking from the early endosomes or caveosome to the endoplasmic reticulum or trans-Golgi network bypassing the lysosomal compartment. Abbreviations: PM, plasma membrane; EE, early endosome; ER, endoplasmic reticulum; TGN, trans-Golgi network; ADC, antibody drug conjugate; GPI, glycophosphatidylinostol-linked proteins.

CHAPTER 18

Immunogenicity Screening Using *in Silico* Methods: Correlation between T-Cell Epitope Content and Clinical Immunogenicity of Monoclonal Antibodies

SI-HAN HAI, JULIE A. McMURRY, PAUL M. KNOPF, WILLIAM MARTIN, and ANNE S. DE GROOT

Therapeutic Monoclonal Antibodies: From Bench to Clinic. Edited by Zhiqiang An

ABSTRACT

One of the great surprises of the recombinant protein revolution has been the discovery that humanized monoclonal antibodies, and even those monoclonals of human origin, can cause immune responses when administered to immune-competent patients. Evaluation of the immunogenicity of monoclonals has been focused primarily on humoral immune response, and as a result the critical contribution of T-cells to the development of anti-monoclonal antibodies (also known as anti-drug antibodies or ADA) has been overlooked. This chapter addresses the relationship between T-cell epitopes contained in antibody sequences, T-cell response, and the clinical immunogenicity of monoclonals. An analysis of common monoclonals for T-cell-dependent immunogenicity is provided and contrasted with immunogenicity observed in the clinic. Tools for mapping T-cell epitopes and for deimmunizing monoclonals are also discussed.

18.1 INTRODUCTION

Monoclonal antibodies (mAbs) and antibody-like protein scaffolds hold significant promise for improving human health. While human immune responses to foreign proteins can be expected and are well understood, the basis for the development of responses to mAbs, especially those that are more human-like, is the subject of some debate. These anti-antibody responses result in the formation of immune complexes that neutralize the therapeutic effects of the mAb and alter its pharmacokinetics. To reduce immunogenicity, monoclonal antibody developers have typically concentrated on purity, glycosylation status, removal of aggregates, proper dosing, and the route of delivery. A number of mAb developers have explored antibody humanization, substituting human germline regions for framework segments in the variable regions that are not necessary for antigen binding. These approaches have led to significant reductions in the immunogenicity of mAbs, with some notable exceptions, which will be discussed in this chapter. One newer strategy to reduce B-cell immunogenicity is to focus on the underlying T-cell responses to T-cell epitopes present in the mAb. We review here some of the theories put forward to explain the T-cell-dependent immunogenicity of monoclonal antibody therapeutics and also describe emerging protein engineering approaches that may prevent the development of these "anti-drug" antibodies.

18.1.1 Factors that Influence Immunogenicity

Human immune responses can be described as the net effect of the various components of the immune system, which can be categorized as innate or adaptive, humoral or cellular. These arms of the immune system each play a role in the development of antibodies to protein therapeutics such as mAb drugs. Even though the immunogenicity of protein therapeutics is usually measured in terms of antibody titers (both neutralizing and non-neutralizing), helper (CD4+) T-cells can play an important role to accelerate antibody responses following exposure to the mAb drug, just as they do for any foreign proteins. T-cells recognize foreign proteins by binding to the MHC:T-cell epitope complex on the surface of antigen-presenting cells (APCs). This T-cell epitope is derived from the original protein by a series of processing and presentation steps performed by the APCs.

The critically important effect of T-cell recognition on antibody titers can be seen following vaccination with an antibody-directed vaccine such as influenza vaccine. The effect of CD4+ T-cells and their contribution to priming the immune response is clear, since titers of antibody usually increase following revaccination, due to T-cell recognition of the epitopes contained in the vaccine and the resultant cognate T-cell help. In fact, nonresponsiveness to conventional influenza vaccines is associated with certain major histocompatibility (MHC) haplotypes, suggesting that T-cell responses provide critically important support for the humoral immune response (Schneider and Van Regenmortel 1992; Gelder et al. 2002). Similarly, T-cell help is an important, if not critical, factor in the human immune response to protein therapeutics, and one sign of the role of T-cell help is that titers of antibodies usually increase following readministration of the therapeutic.

Although not addressed in this chapter, therapeutic protein dose, route of administration, purity, and aggregation are also important immunogenicity factors to consider. Larger doses, prolonged or recurrent administration, contamination of the product with bacterial or inorganic adjuvants, aggregated drug complexes, and the subcutaneous dosing route can generate more pronounced anti-drug antibody (ADA) responses, reducing the effectiveness of the therapeutic.

18.1.1.1 T-Independent and T-Dependent Immune Response The sequence of events that lead to the activation of B-cells, and the resultant production of antibodies, can be divided into T-cell independent (Ti) and T-cell dependent (Td) scenarios. Ti activation of B-cells occurs when structural features of certain molecules, such as polymeric repeats of an epitope, induce the "signals" required to stimulate activation of a B-cell subset. In addition, the complement component C3d may also play a role when a T-cell-dependent immune response cannot be activated, leading to the development of immune memory (Test et al. 2001). C3d enhances the immune response and lowers the affinity threshold for B-cell activation. Although Ti activation is often cited as a source of antibodies to protein therapeutics, Ti activation of B-cells generally does not lead to affinity maturation nor does it result in the generation of memory B-cells. More commonly, Td activation of B-cells results in a more robust antibody response, affinity maturation, isotype switching, and the development of B-cell memory. The induction of IgG class ADA that are measurable in standard immunogenicity assays normally implies that the therapeutic protein is a Td antigen that has led to isotype switching (Zubler 2001).

Since Td responses require T-cell recognition of epitopes contained in the protein drug, it can be assumed that (1) its constituent peptide epitopes (derived from internal processing by antigen-presenting cells) bind to MHCII (major histocompatibility complex class II) molecules and (2) that the epitope-HLA complex activates helper T-cells (Fig. 18.1a). In the absence of signals provided by the cytokines released by T-cell interactions with APCs, the naïve B-cell does not mature. Indeed, without T-cell help, activated antigen-specific B-cells may be rendered anergic or undergo apoptosis (Fig. 18.1b). As a result, T-cell recognition of the peptide epitopes derived from the antigen is a key determinant of Td antibody formation.

18.1.1.2 The Role of T-Cell Epitopes Presentation of T-cell epitopes in the context of MHC class II molecules is essential for robust B-cell activation with isotype switching, affinity maturation, and immunological memory. If T-cell epitopes can be derived from human proteins, as they appear to be when human proteins are used as therapeutic drugs, then it is worth considering what features of the protein therapeutic can be modified in order to minimize the pathology that would presumably follow.

Currently, there are several methods available for screening mAb sequences for T-cell epitopes. Over the last 10 years, a number of computer algorithms have been developed and put into use for detecting MHC class I- and class II-restricted T-cell epitopes within protein molecules of various origins (see review and a current list of immunoinformatics tools in De Groot and Berzofsky 2004). Such *in silico* predictions of T-helper epitopes have already been successfully applied to the design of vaccines (Ahlers et al. 2001; De Groot et al. 2002) and to the selection of epitopes in studies of autoimmunity (Inaba et al. 2006). The authors of this report currently use the EpiMatrix system, a suite of epitope mapping tools that has been validated after more than a decade of use for *in vitro* and *in vivo* studies (for example, De Groot et al. 1997; Bond et al. 2001; McMurry et al. 2005; Dong et al. 2004; Koita et al. 2006). The EpiMatrix algorithm is comprised of class I and class II HLA matrices that apply individual frequencies of all 20 amino acids (aa) in each HLA pocket position to the prediction of overlapping 9- and 10-mer peptides.

More recently, the EpiMatrix system has been utilized to measure the potential immunogenicity of whole proteins such as mAbs and their subregions (see Fig. 18.2). In this context, EpiMatrix assesses the epitope density of a given protein relative to that of randomly generated pseudo-protein sequences of similar size. Setting aside for a moment the well-established concept of tolerance (see below), EpiMatrix has been used to evaluate selected self-proteins (specifically those that are abundant in

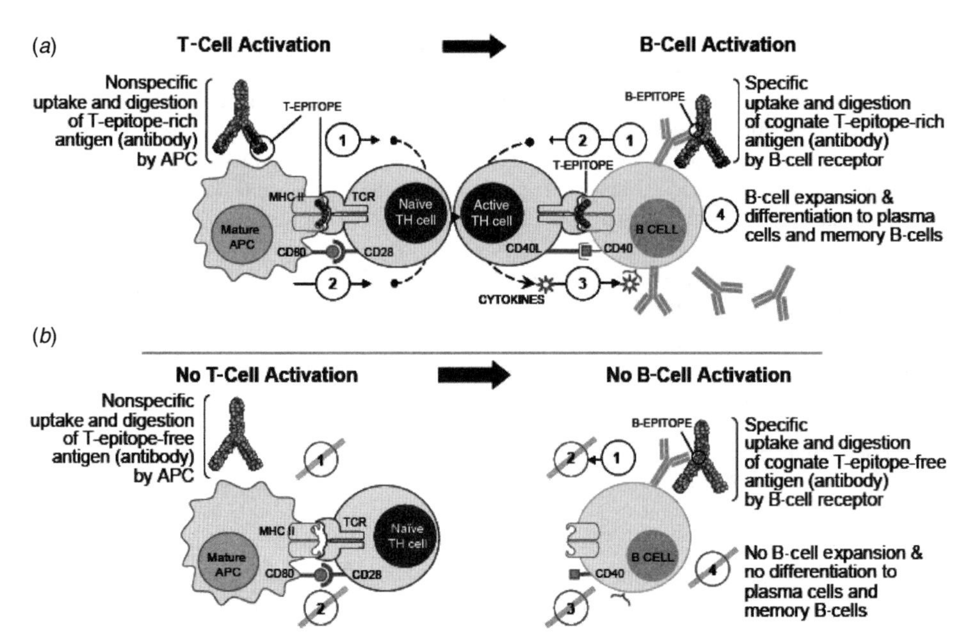

Figure 18.1 Activation of CD4 T-cells and the T-dependent antibody response. In order to induce naïve B-cells to produce antibodies against a protein antigen, several events must be coordinated, usually within specialized regions of secondary lymphoid organs (e.g., lymph nodes, spleen, etc.). The first step in this process is internalization of antigen by professional antigen-presenting cells (APC), such as dendritic cells (left side of figure). The mature APC processes antigen into peptides that are then presented to naïve T-cells in major histocompatibility complex (MHC) class II molecules on the APC surface. The interaction of a T-cell receptor (TCR) with this MHCII:peptide complex is identified as a T-cell signal 1 (T1). In order to fully activate the T-cell, T1 must be accompanied by additional signals from what are termed *co-stimulatory molecules*, such as CD80, CD86, etc. provided by the APC (T-cell signal 2, T2). In the absence of T2, T-cells may become anergic. Once fully activated, these CD4 T-cells divide and produce an array of cytokines with manifold activities. Activation of the naïve B-cell is initiated by an interaction between B-cell IgM and IgD receptors and cognate antigen. This is termed B-cell signal 1 (B1). Upon encountering a B-cell that has recognized antigen via a specific T-cell epitope-MHC:TCR interaction, the T-cell delivers cytokines that drive the B-cell to proliferate and mature towards an antibody-secreting plasma cell. A CD4 T-cell thus provides help to B-cells [B-cell signal 2 (B2)]. This interaction results in the engagement of CD40 and CD40 ligand communicating a further signal to the presenting B-cell [B-cell signal 3 (B3)] leading to antibody production via B-cell expansion and differentiation to plasma cells and memory B cells (B4). Adapted with permission from A.S. De Groot and L. Moise, *Current Opinion in Drug Discovery & Development* 10, no. 3 (2007):332–340. Copyright 2007, the Thomson Corporation. (See color insert.)

serum) and found that they have lower inherent class II epitope-restricted epitope densities, as compared to random proteins and to known antigens (De Groot 2006). This approach is gaining acceptance as a means of comparing one monoclonal to another, so as to select the best candidate to carry forward from preclinical to clinical studies, reducing the risk of failure due to immunogenicity. Several examples of this type of application are described herein.

18.1.2 Tolerance and the Regulation of Anti-Self Response

Immune response against self (i.e., autologous proteins) is relatively rare; this is thought to be because of the development of tolerance to self-antigens in the primary lymphoid organs (thymus and bone marrow), as well as in the periphery. For T-cells, initial self/nonself discrimination occurs in the thymus during neonatal development (continuing as the host matures) when medullary epithelial cells express tissue-specific self-proteins and present these epitopes to immature T-cells in the context

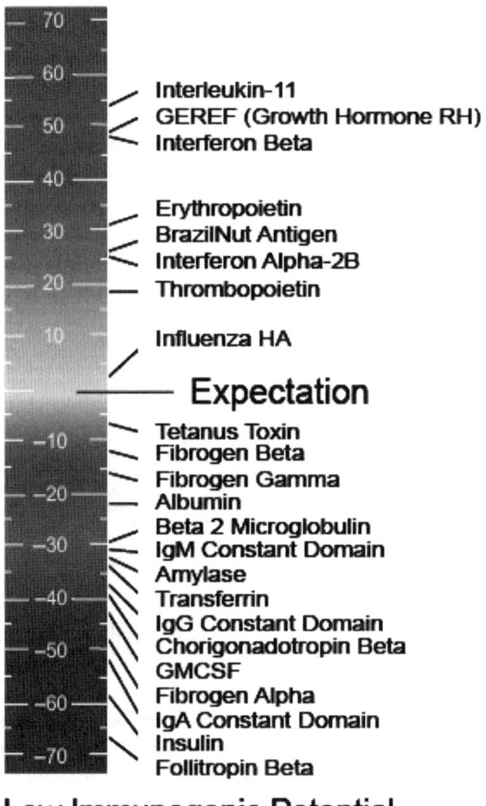

Figure 18.2 Immunogenicity scale. Well-known proteins shown as they rank and as compared to each other on the EpiMatrix immunogenicity scale.

of MHC. According to the most widely accepted theory regarding T-cell tolerance, T-cells with high affinity for self-antigens are deleted, whereas those with moderate affinity for self-antigens may escape deletion and may be reprogrammed to function as natural regulatory T-cells (Treg) (Bluestone and Abbas 2003). In fact, the potential to make antibodies to self-proteins is supported by the presence of B-cells with receptors for self-antigens. Anti-self immune responses may be ongoing and controlled by circulating Tregs. When tolerance to autologous proteins is broken, autoimmunity results, manifested as diseases such as rheumatoid arthritis, Grave's disease, and multiple sclerosis. The presence of both T-cells and B-cells specific for self-antigens and the absence of T-regulatory immune responses are the hallmarks of these autoimmune diseases.

It has been shown that a given peptide epitope can be recognized by regulatory T-cells and effector T-cells alike. For example, regulatory T-cell epitopes have been identified in insulin and GAD-65 (Congia et al. 1998; Oling et al. 2006). In the context of autoimmune (type 1) diabetes, these T-cell epitopes are recognized by effector T-cells instead of regulatory T-cells, reflecting a shift in the type of immune response to the identical epitope.

Peripheral tolerance can be overcome in a number of different contexts that most likely relate to the milieu in which the immune response is taking place (Jiang and Moudgil 2006). For example, inflammation caused by minute amounts of contaminants such as LPS, or bacterial DNA, both of which can act as toll-like receptor (TLR) agonists, may stimulate APCs to release sufficient amounts of cytokines and chemokines to overcome Treg immune responses in the periphery (Hacker 2000). In addition, autoimmune responses are believed to occur after a priming event such as a viral infection. Such infections may result in inflammatory reactions and activation of the innate immune system, leading to increased immune responsiveness by maturing dendritic cells (DC), modified cytokine expression profiles, and increased MHC expression. Viral infections may also create responses by stimulating cross-reactive T-cell help for B-cells and/or by "antigen mimicry" of self-proteins (Page, Scott, and Manabe 2006). Any or all of these conditions may also be present when monoclonal antibodies are administered, resulting in greater immune response than would be expected for a protein that is human-like in origin.

The heavy chain of antibodies has both a constant and a variable region. As the name implies, the constant region is conserved in all antibodies of the same isotype, whereas the variable region differs in each B-cell. The light chain also has one constant and one variable domain. The variable domains of each chain are most crucial since they bind to antigen and determine the specificity of the antibody. This variable domain [one genetic region per haploid DNA for heavy (H) and another for light (L)] is generated in primary lymphoid organs (bone marrow and thymus) prior to exogenous antigen exposure by a genetic cross-over event between two families of DNA segments (v and j for L and v, d, and j for H) recombining but also generating nucleotide sequence heterogeneity at the cross-over site, creating *junctional diversity* ($v \times j = v'/j' = VL$ and $v \times d \times j = v'/d'/j' = VH$) that further alters both affinity and specificity for each antigen–antibody interaction. The resulting novel heterogeneity within the V segments becomes the common theme for a small subset of B-cells that out-compete other clonally related B-cells by virtue of their superior antigen-binding capacity. There have been many explanations for the apparent lack of induction of T-cell responses to antibody variable regions, including T-regulatory cells and apparent T-cell tolerance to B-cell receptor variable regions (Eyerman and Wysocki 1994).

Recent evidence points to a role for regulatory T-cells (Fig. 18.3) in the suppression of anti-immunoglobulin immune responses. In recent work, the authors of this chapter identified regulatory T-cell epitopes within the heavy and light chains of Ig. Termed *Tregitopes*, these regulatory epitopes synthesized as peptides were able to suppress effector immune responses to coadministered antigen, both *ex vivo* and *in vivo*. In mouse models, the level of suppression was greater than that achieved by Fc, perhaps explaining the mechanism of IVIG suppression. The Tregitopes were found to upregulate CD25 and Foxp3 expression leading to the secretion of IL-10 and TGF-β (De Groot et al. 2008). These findings have important implications for the design of mAbs and indeed for therapeutic proteins in general.

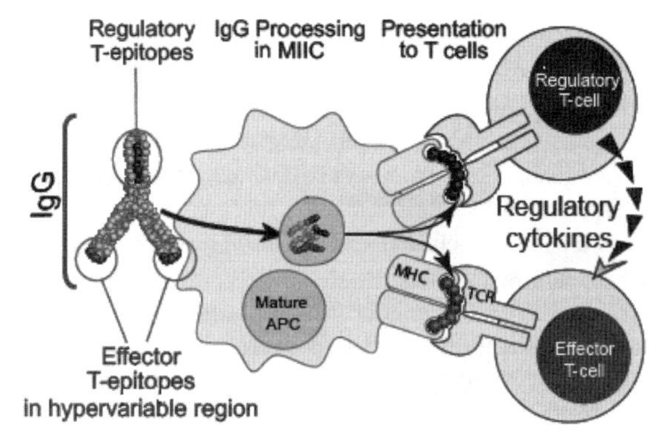

Figure 18.3 Hypothesized tolerizing mechanism of IgG. We have discovered conserved T-cell epitopes in IgG that engage natural regulatory T-cells. The data suggest that the antibody-derived Treg epitopes (dark epitope) activate regulatory T cells that lead to suppression of effector T-cells that recognize effector epitopes (grey epitope), like those of IgG hypervariable regions to which central tolerance does not exist.

The development of antibodies to the human components of recombinant autologous proteins—antibody or other—can be regarded as a breach of tolerance. Hypothetically, removal or alteration of regulatory T-cell epitopes by mAb developers in the humanization process may adversely affect the immune response to the mAbs. The link between T-cell (and HLA-restricted) immune response and the development of auto-antibodies is still being defined; early evidence points to the reduction of Treg immune responses and to induction of T-effector responses as significant contributors to human anti-human antibodies (HAHA) (Reveille 2006).

18.2 COMPARATIVE TD IMMUNOGENICITY OF DIFFERENT MONOCLONAL ANTIBODIES

Much effort has been applied to reengineering murine and mouse-human chimeric antibodies to reduce nonhuman structure as much as possible so as to minimize antitherapeutic effects. In some cases, modification of a murine monoclonal (foreign protein) to a less foreign protein sequence via replacement of mouse Ig constant domain (chimeric monoclonal) has resulted in the largest immunogenicity reduction. Humanization of variable domains further decreases the immune response against the therapeutic. There is clear evidence that chimeric antibodies are less immunogenic than murine monoclonal antibodies; however, humanization has not yet solved the problem of monoclonal antibody immunogenicity. One such example is the Campath antibody, which continued to induce an immune response despite humanization (Clark 2001; Hwang and Foote 2005).

18.2.1 Murine Monoclonal Antibodies

Monoclonal antibodies were developed initially by Köhler and Milstein (1976), who introduced hybridoma technology. By fusing an antibody-negative myeloma and an antibody-producing B-cell derived from mice, these researchers were able to produce the first murine monoclonal antibodies. Although murine monoclonal antibodies seemed promising, many therapeutics induced human anti-mouse antibodies (HAMA) that resulted in allergic reactions culminating in anaphylactic shock. OKT3, an immunosuppressive drug for use in transplantation, was the first monoclonal antibody to be approved for human use. OKT3 is a murine monoclonal IgG2a antibody that targets the ε chain of the T-cell antigen-recognition mechanism by reacting with the T-cell receptor:CD3 complex. Thus, mature circulating T-cell function and its associated cell lysis are blocked, reducing tissue-graft rejection (Cosimi 1981). However, chills and fever are associated with use of OKT3, and 86 percent of patients develop anti-drug antibodies due to the murine components of the monoclonal.

In contrast, Zevalin (ibritumomab tixuxetan), a form of murine monoclonal antibody radiotherapy for some forms of B-cell non-Hodgkin's lymphoma, shows markedly low immunogenicity. Licensed by the FDA in 2002, Zevalin uses an IgG1 anti-CD20 murine mAb conjugated to an yttrium-90 or indium-111-labeled tixuxetan chelator. This antibody binds to mature normal and malignant B-cells but spares precursor B-cells; it kills either through antibody-dependent cell cytotoxicity (ADCC) or by β-radiation from the attached isotope. In clinical trials, these patients, who may be immunosuppressed due to other marrow toxic agents, showed a <2 percent rate of HAMA responses, making Zevalin one of the least immunogenic murine monoclonal antibodies available today (Hwang and Foote). This low rate of response may have been due to the immunosuppression of patient populations receiving the drug.

18.2.2 Chimeric Antibodies

Chimeric antibodies have experienced limited success in chronic use due to the HAMA response to the murine variable regions. Examples of chimeric antibodies that induce antibody responses include Simulect (basiliximab), Erbitux (cetuximab), Remicade (infliximab), and Rituxan (rituximab). Although the constant domains of the light and heavy chain are replaced by human C-regions, the antibody is still recognized as foreign by the human immune system (Riechmann et al. 1988). Immune responses to these antibodies vary depending on their target and their indication. When used in immunocompromised subjects, the titer of antibody can be low to undetectable, due to the absence of effective T-cell or B-cell response. In addition, if the surface protein that is the target of the antibody is implicated in effective immune response, the antibody may have a direct effect, reducing antibody titers due to its inherent mechanism of action. For example, rituximab (Rituxan) targets CD20 for the treatment of B-cell non-Hodgkin's lymphoma and basiliximab (Simulect) targets CD25 and is used in the context of other immunosuppressive treatments in the context of graft-versus-host disease (GVHD) in stem cell transplant. These chimeric antibodies are generally effective but there have been cases of an anti-antibody response (ADA) during the course of treatment (Table 18.1; Hwang and Foote 2005). On a relative scale, however, chimeric antibodies are much less immunogenic than are murine monoclonal antibodies.

Rituximab (Rituxin) has been very successful in the treatment of non-Hodgkin's lymphoma. CD20 is a particularly attractive target for treatment of B-cell lymphomas due to the fact that it does not circulate freely in the plasma, is shed from cells after antigen binding, and is only present on mature B-cells (Einfeld et al. 1988). Perhaps due to rituximab's effect on mature B-cells, patients given weekly doses showed no human anti-murine antibody (HAMA) responses and approximately 50 percent experienced a therapeutic benefit (Piro et al. 1999). In contrast, subjects who receive this same drug for other off-label indications (dermatologic, for example) may develop anti-drug antibodies. At least for the approved indication, the safety and efficacy of rituximab show that chimeric monoclonal antibody therapies are much less immunogenic than their murine counterparts.

Remicade (Infliximab) and Embrel (etanercept) are examples of antibodies that have the same target but very different structures. Remicade is a chimeric antibody while Embrel is a fusion of human IgFc and two soluble TNF receptors. As can be expected, in an analysis of 233 cases, the fully human (but fused) Embrel (etanercept) is less immunogenic than is its chimeric anti-TNF counterpart (Ramos-Casals et al. 2007).

18.2.3 Humanized Antibodies

Developers of humanized antibodies have managed to substitute all but the antigen-binding site loci (the hypervariable region) of the original mouse antibody with human domains, resulting in a mostly human monoclonal antibody. There are, however, some structural limitations to the substitution of human for mouse. As humanized antibodies developed, scientists found that using a human framework that had close sequence homology to its murine counterpart generated the most effective humanized variant (Queen et al. 1989). These antibodies offer the advantage of high antigen affinity and limited HAMA response. As a result, many humanized antibodies have been approved by the FDA for clinical use in the United States.

Despite the success of humanization, some of these antibodies remain immunogenic for nonimmunocompromised subjects. For example, alemtuzumab (Campath-1H), which is a humanized monoclonal antibody, targets the human CD52 antigen expressed by many lymphoid neoplasms, and has been shown to be strongly immunogenic (up to 70 percent) in some studies. Alemtuzumab was originally approved by the FDA for the treatment of fludarabine-refractory chronic lymphocytic leukemia (CLL; Alinari et al. 2007) but is being considered for use in multiple sclerosis and rheumatoid arthritis (Dumont 2001). Although there are some adverse reactions when the first dose is administered intravenously, researchers have found that frequent administration subcutaneously reduces ADA responses without reducing the therapeutic value of the monoclonal. Since alemtuzumab and other humanized monoclonal antibodies have generally proven to be both safe and effective, they are among the most widely used monoclonal antibodies licensed by the FDA.

Synagis (palivizumab) is another humanized monoclonal antibody targeted for the treatment of respiratory syncytial virus (RSV) in infants. RSV at a young age causes long-term pulmonary disease. The development of a protein therapy against this childhood disease was a breakthrough in pediatric care. Humanized monoclonal antibodies containing the murine complementary determining regions

TABLE 18.1 Published Prevalence of Anti-Drug Antibodies (ADAs) for Selected FDA-Approved mAbs

Indication	Target	Source	Brand	Generic	Obs	Pred	Comment
Breast cancer	HER2	GenBank	Herceptin	Trastuzumab	0.1	0.0	
Prevention of respiratory syncytial virus	RSV	US 6818216	Synagis	Palivizumab	1.0	0.0	
Ulcerative colitis & Chrohn's	CD3	US 5934597	HUM291	Visilizumab	0.0	0.4	Concomitant immunosuppressive used
Severe plaque psoriasis	CD11-a	US 6703018	Raptiva	Efalizumab	4.7	1.3	Three trials: 2.3, 4, 6% averaged to 4.7%
Asthma	IgE	US 6914129	Xolair	Omalizumab	0.0	3.7	
Metastatic carcinoma of the colon or rectum	VEGF	US 6884879	Avastin	Bevacizumab	0.0	3.8	Concomitant immunosuppressive used
Transplant prophylaxis	IL-2 receptor	US 6180370	Zenapax	Daclizumab	14.0	4.2	8% when used w/concomitant immuosuppressive. 34% in children per package insert
Prostate cancer	CD3	US 5934597	HUJ591	HUJ591	0.0	4.6	
Ischemic stroke	CD11/CD18	US 5S54070	Leukarrest	Rovelizumab	0.0	4.8	Single dose
Crohn's	TNF	US 7012135	Cimzia	Certolizumab pegol	11.0	5.4	See Sandbom (10%) and also Schreiber (12%–24%) in NEJM July 19, 2007; lacks Treg epitope
Macular degeneration	VEGF	US 7060269	Lucentis	Ranibizumab	6.3	6.1	
Carcinoma	CD44v6	US 6972324	BIWI1 conjugate	Bivatuzumab mertansine	6.7	8.4	Single dose; 0 and 10 averaged to 6.7%
Transplant prophylaxis	IL-2 receptor	US 6521230	Simulect*	Basiliximab	1.4	9.1	Single dose
Arthritis	TNF	US 6258562	Humira	Adalimumab	12.0	9.6	12% without methotrexate, 1% with methotrexate
Arthritis and nerve root injury	TNF-alpha	US 5994510	Humicade	Adalimumab	7.0	10.4	Concomitant immunosuppressive used
Adult leukemia	CD33	US 5877296	Mylotarg	Gemtuzumab Ozogamicin	2.9	13.1	Acute myeloid leukemia
Multiple sclerosis	Integrin	US 5840299	Tsabri	Natalizumab	7.0	13.7	Concomitant immunosuppressive used
Plaque psoriasis, arthritis, Crohn's, ulcerative colitis, and ankylosing spondylitis	TNF-alpha	US 5656272	Remicade*	Infliximab	26.0	14.8	Package insert reports antibody rates from 10 to 51% with an average of 26%
Leukemia	CD52	US 6569430	Campath	Alemtuzumab	33.0	16.9	Seven trials: 63, 10, 75, 29, 53, 0, and 1.9% averaged to 33%
Colorectal cancer	EGF	US 7060808	Erbitux*	Cetuximab	5.0	17.8	Packaged insert states that observed estimate of 5% is unreliable
Rheumatoid arthritis and non-Hodgkin's lymphoma	CD20	US 6682734	Rituxan*	Rituximab	27.0	21.4	Four trials: 0, 0, 65, 27%; Cancer studies and dose escalation study omitted from observed value

(CDRs) to the F protein in RSV (a highly conserved protein over numerous strains) were incorporated into the light and heavy chains of a human IgG antibody (Young 2002). Phase III clinical trials showed that Synagis (palivizumab) reduced rates of infant hospitalization for respiratory illnesses and all adverse side effects were determined to be unrelated to the drug since incidence was equal in both the placebo and experimental groups. FDA approval was given in 1998 for use in the United States, followed by European approval in 1999 and approval in Japan in 2002 (Pollack and Groothuis 2002).

18.2.4 Fully Human Monoclonal Antibodies

Human antibodies have recently become available with the introduction of the transgenic mouse engineered with a humanized immune system. By introducing almost the entire human immunoglobulin loci into the germline of mice with an inactivated murine antibody system, these mice can be utilized as expression vectors for human antibodies against autologous antigens. Abgenix, a biotechnology company in California, has recently introduced the XenoMouse system that generates high-affinity, fully human antibodies to target the disease of choice (Jakobovits et al. 2007). In this system, the crucial cis-acting regions of the mouse immunoglobulin heavy and light chains (Jh and Ck) are deleted and, through crossbreeding, homozygous double-inactivated mice are generated (Jakobovits et al. 1993). In September 2006 the FDA approved panitumumab, a fully human monoclonal antibody against epidermal growth factor receptor (EGFR), an antigen over-expressed in most solid tumors. It is the first human antibody to be approved by the FDA for the treatment of EGFR-expressing colorectal cancer with disease progression. Phases I, II, and III clinical studies showed no significant host immune response against the monoclonal therapy and many patients showed reduced rates of cancer progression (Carteni et al. 2007). With the FDA approval of panitumumab, the XenoMouse shows great potential in the development of fully human monoclonal antibodies that may be used to treat other chronic diseases.

18.2.5 Epitope-Modified Monoclonal Antibodies

Since T-cell epitopes play a critical role in the development of Td antibody responses, it stands to reason that protein sequence modifications resulting in removal of potential T-cell epitopes from autologous recombinant therapeutic proteins could reduce the potential for immune response induction to the protein; sequence modification of this nature is known as *deimmunization*. Current literature is replete with evidence for the attenuating effect of epitope-sequence modification on T-cell response, particularly in reference to viral immune escape from class I- and class II-restricted immune responses. Loss of T-cell help removes signal 2 for antigen-specific B-cells and, in theory, could lead to B-cell apoptosis (Kappler, Roehm, and Marrack 1987). T-cell epitope modification may be thought of as the directed version of the process that occurs naturally when tumor cells (Scanlan and Jager) and pathogens (Mullbacher 1992; Hill et al. 1997) evolve to escape immune pressure by accumulating mutations that reduce the binding of their constituent epitopes to host HLA (Vossen et al. 2002). The epitope-modification approach may also be relevant to the design of novel replacement proteins as illustrated by the development of a deimmunized recombinant erythropoietin (rEPO; Tangri et al. 2005), recent progress on a novel FVIII (Hay et al. 2006), and the development of deimmunized monoclonal antibodies (Damschroder et al. 2007; Hellendoorn et al. 2004; Staelens et al. 2006). Successful deimmunization of therapeutic proteins has been demonstrated by workers at Genencor (Yeung et al. 2004) and EpiVax (De Groot, Knopf, and Martin 2005) using slightly different approaches. However, such modification could introduce new epitopes that could be presented by other MHC molecules or could create novel B-cell epitopes, possibilities that can be assessed in (HLA transgenic) animal studies or by reiterative T-cell epitope analysis (*in silico*). Most importantly, these modifications must maintain the biological function of the product while also reducing or eliminating immunogenicity.

Using this technique of T-cell epitope analysis and subsequent modification, the researchers at Biovation developed a monoclonal antibody therapy for chronic hepatitis C (HCV) infection. Scientists identified the most immunogenic regions of the IFN-α2b and engineered a variant containing six mutations within those regions. By conjugating the deimmunized IFN-α2b to the Fc portion of an IgG1 antibody, the half-life of the therapeutic was extended and the immunogenicity of the monoclonal was reduced (Jones et al. 2004).

18.2.6 Scaffolds

Novel protein scaffolds are a cost-effective strategy for developing protein therapeutics. Although monoclonal antibodies are a proven therapeutic, their high manufacturing costs make them difficult to commercialize. Wyeth Pharmaceuticals has developed AdNectins that are made from the 10FN3 domain of human fibronectin. A library of monoclonal therapeutics can be made easily by introducing antigen-recognition sequences into the three loops of the AdNectin that mimic the CDR domains of immunoglobulins. AdNectins generated can be selected for antigen binding and further modified to tailor the pharmacokinetics of the drug (Gill and Damle 2006). AdNexus developed Angiocept, an AdNectin specific for human vascular endothelial growth factor receptor 2 (VEGFR2) that entered Phase I clinical trials in 2006. As these molecules are only now being tested in the clinic, their potential for immunogenicity remains to be determined. However, depending on their epitope content and their homology to human proteins, they may have little immunogenicity and could represent a viable alternative to monoclonals.

18.2.7 Nonhuman Antibodies (Llama, Rabbit)

Llama mAbs are currently used for the purification of biological substances via affinity chromatography but some "second generation" monoclonals derived from llama antibodies are being considered for use in the clinic for selected target diseases. Llama antibodies are of great interest for hard-to-reach cellular targets because some of the antibodies lack the light chain. These specialized antibodies only contain one binding domain and are easily cloned and produced in *Escherichia coli* and *Saccharomyces cerevisiae* (Frenken et al. 2000). In addition, llama antibodies have very small antigen-binding domains (Muylermans 2001) and are said to be much more stable than other antibodies (Verheesen et al. 2003). Kooster et al. (2007) describe the use of high-affinity llama antibodies against HSA and IgG to bulk purify human serum and plasma.

Rabmabs (rabbit monoclonal antibodies) are produced by EpiTomics. These antibodies are also used for diagnostic kits but are being considered for use in humans. In the future, using epitope-driven modification approaches, it may be possible to develop rabbit or llama antibodies for human use that do not engender ADA.

18.3 IMMUNOGENIC EFFECTS OF MONOCLONAL ANTIBODIES

18.3.1 Neutralizing and Non-Neutralizing Antibodies

Both neutralizing and non-neutralizing antibodies can develop following therapeutic monoclonal antibody administration. Neutralizing antibodies can lead to the loss of activity of the monoclonal. By definition, neutralizing antibodies that interact with a ligand receptor inhibit the efficacy of all ligands in the same class. For example, antibodies that neutralize rEPO by interfering with the binding of the protein to its receptor may interfere with all forms of erythropoietin, leading to anemia. Neutralizing antibodies are of greatest concern when they develop in conjunction with conditions for which there are no treatment alternatives (Koren, Zuckerman, and Mire-Sluis 2002). Cross-reactive antibodies may also cause particular problems. In general, when neutralizing antibodies cross-react with an endogenous factor that has an essential biological function, there is a greater likelihood of adverse events.

18.3.2 HAMA Responses

Human anti-mouse (HAMA) responses can be classified into three categories: (1) anti-isotypic, (2) anti-idiotypic, and (3) anti-anti-idiotypic. Anti-isotypic antibodies are the most common, usually directed against the Fc region of the mouse immunoglobulin. These antibodies are nonspecific and often found in maternal serum. Anti-idiotypic antibodies are produced against specific idiotypes of the mouse antibody. By binding in or around the antigen-binding site, these antibodies can interfere with the efficacy of the protein therapeutic.

HAMA responses are characterized by a range of reactions from mild allergic responses to anaphylactic shock; compared to other antibodies, humanized antibodies induce the least frequent and least severe of ADA responses. Methods for the precise measurement of HAHA and HAMA and their effects are critical to rational design of future therapeutics. These methods have evolved rapidly over the past few years and have become the subject of two industry white papers. Assays for measuring immune responses to therapeutic proteins are discussed by Mire-Sluis et al. (2004). Currently, the only method of discriminating between neutralizing and non-neutralizing ADA is to perform a biological assay using cell lines that express the target and mimic the therapeutic effect of the drug. Measurement of the biological effect of anti-protein therapeutic antibodies is also discussed by Gupta et al. (2007).

18.3.3 HAHA Responses

Humanized antibodies can also generate potent anti-antibody responses when administered in high doses. These responses can be characterized as either type I or type II responses (Szolar et al. 2006). Type II responses pose the greatest threat since antibody titers continue to rise with repeated doses of the monoclonal therapeutic. In type I responses, the patient initially develops a limited anti-antibody response that stabilizes within weeks of treatment (Scott et al. 2005). Eventually, type I responses result in tolerance to the monoclonal, allowing effective treatment to continue.

18.4 T-CELL EPITOPE PREDICTION

A number of computational T-cell epitope-mapping tools are now available for use in the identification of T-cell epitopes contained within protein sequences (De Groot and Berzofsky 2004; Van Walle et al. 2007). Highly immunogenic proteins contain many T-cell epitopes or concentrated clusters of T-cell epitopes, whereas nonimmunogenic proteins tend to contain fewer epitopes. The authors hypothesize that T-helper (Th) epitope content may partially explain why slightly different versions of the same recombinant human protein result in different observed antibody responses (De Groot, Knopf, and Martin 2005). The EpiMatrix protein immunogenicity scale (see Fig. 18.2) allows protein sequences that are slight sequence variants of one another to be compared for epitope content (and potential for immunogenicity) facilitating the process of selecting the best candidate for clinical trials. Immunoinformatics, when combined with *in vitro* and *in vivo* methods, provides an efficient alternative to conventional epitope mapping using overlapping peptides; reductions in time and effort up to 700-fold have been shown (De Groot et al. 1997, 2001a, 2001b; De Groot, Rayner, and Martin 2003; Kast et al. 1994; Moutaftsi et al. 2006; Southwood et al. 1998). However, since *in silico* tools may vary (see Pollack and Groothuis 2002 for a review of epitope prediction tools), protein developers are advised to request or to perform comparison of prediction accuracy before selecting one T-cell epitope prediction tool. Four such tools are discussed below.

18.4.1 EpiMatrix (EpiVax)

The EpiMatrix method for ranking prospective epitopes has been published (De Groot et al. 1997; De Groot, Rayner, and Martin 2003). Three steps are involved in epitope mapping with EpiMatrix. In the following paragraphs these steps are outlined as they were applied to the discovery of Treg epitopes in IgG.

Step 1: *Searching for putative T-cell epitopes.* Evaluation is initiated by parsing the amino acid sequence of a given protein into all its possible 9-mer frames. The binding potential of each frame is then evaluated with respect to each of a panel of eight common class II alleles

Influenza Hemagglutinin AA 306-318

Frame Start	AA Sequence	Frame Stop	DRB1*0101 Z score	DRB1*0301 Z score	DRB1*0401 Z score	DRB1*0701 Z score	DRB1*0801 Z score	DRB1*1101 Z score	DRB1*1301 Z score	DRB1*1501 Z score	HITS	Cluster Score
306	PRYVKQNTL	314	1.34	1.40		2.06				1.28	1	Deviation from Expectation: 17.62
307	RYVKQNTLK	315										
308	YVKQNTLKL	316	3.33	1.97	3.15	3.27	1.96	1.99	2.37	2.36	8	
309	VKQNTLKLA	317						1.59	1.67		1	
310	KQNTLKLAT	318										

Figure 18.4 EpiBar, typical EpiMatrix analysis. *Z* score (Top 10%** Top 5% Top 1%) indicates potential of a 9-mer frame to bind to a given HLA allele. All *Z* scores in the top 5 percent (>1.64) are considered hits. **Though not hits, scores in the top 10 percent are considered elevated; scores below 10 percent are masked for simplicity. Frames containing four or more alleles scoring above 1.64 are colloquially referred to as EpiBars and are highlighted (see frame 308:YVKQNTLKL). This band-like pattern is characteristic of promiscuous epitopes. The influenza peptide scores are extremely high for all eight alleles in EpiMatrix; the deviation compared to expectation is +17.62.

(DRB1*0101, *0301, *0401, *0701, *0801, *1101, *1301, and *1501). Taken together these alleles cover the genetic backgrounds of most humans worldwide (Southwood et al. 1998). These alleles are not only common but their peptide-binding pockets are also representative of those found in most of the other known alleles, thus maximizing the clinical relevance of the prediction. In order to compare potential epitopes across multiple HLA alleles, EpiMatrix raw scores are converted to a normalized *Z* scale. Peptides scoring above 1.64 on the EpiMatrix *Z* scale (typically the top 5 percent of any given sample) are likely to be MHC ligands (see Fig. 18.4).

Step 2: *Searching for epitope clusters.* Potential immunogenicity is not randomly distributed throughout protein sequences but instead tends to cluster in immunogenic regions (which are often also immunodominant). ClustiMer maps the EpiMatrix motif matches along the length of a protein and calculates the density of motifs for the EpiMatrix panel of eight HLA alleles (Kast et al. 1994). Epitope-dense regions are simply referred to as *clusters*. A known promiscuous T-cell cluster (Influenza HA 306-318) is shown as an example of a typical EpiMatrix cluster report (Fig. 18.4).

Step 3: *Ranking regional immunogenic potential.* A T-cell epitope cluster usually ranges from 9 to about 25 amino acids in length and can contain anywhere from 4 to 40 binding motifs. Scores above 10 and, in particular, scores above 15 indicate significant immunogenic potential (De Groot 2006; Koren et al. 2007). The influenza epitope shown above has an immunogenicity score (standard deviation above random) of 17.62. Note the horizontal bar of high *Z* scores at position 308. Having observed this EpiBar pattern to be characteristic of promiscuous epitopes, the authors have integrated the pattern into the prospective selection of clusters. Sequences that contain EpiBars include tetanus toxin 825-850 (cluster score +16), influenza HA 306-318 (cluster score +18) and GAD65 557-567 (cluster score +19). Multiple prospective studies that confirm the accuracy of the algorithm are published in the vaccine literature (McMurry et al. 2007a, 2007b) and have also been presented or published in the biological therapeutics field (Koren et al. 2007; M. Moxness, personal communication). In two cases (for the GDNF and FPX proteins), clinical immunogenicity was predicted by EpiVax, using EpiMatrix, ClustiMer, and the immunogenicity scale, and subsequently confirmed by researchers at Amgen.

18.4.2 SYFPEITHI

SYFPEITHI is a database for peptide motifs and MHC ligands, including both class I and class II molecules of multiple species. This algorithm allows users to perform a web-based search for MHC alleles, MHC motifs, natural ligands and T-cell epitopes. Beginning at the first amino acid, the protein sequence is parsed into all possible 8-mer, 9-mer, and 10-mer frames; the amino acids that commonly appear in anchor positions are given a score of 10, those occurring less frequently are given a score of 8, and those appearing rarely or at auxiliary anchor positions are given a score of 6. Preferred amino acids are then given a coefficient of 1 to 4 and those that are not favored for binding are given scores of −3 to −1 (Rammensee et al. 1999). Higher scores indicate better MHC ligands and can be used to determine monoclonal therapies that may be highly immunogenic. Scores are not comparable across alleles; therefore, finding clusters is difficult unless it is performed manually.

18.4.3 Epibase (Algonomics)

The Epibase algorithm developed by Algonomics offers predictions of T-cell epitopes for rare HLA types for which information may not be readily available. Both class I and class II allotypes that are present in at least 3 percent of given populations can be addressed and epitopes with atypical sequence motifs can also be identified. The selection of HLA binding motifs for the purpose of finding clusters has not been published; therefore, it is difficult to determine whether cluster finding is possible and to evaluate the accuracy of this tool. Epibase uses the interaction of the peptide and its receptors to determine the binding affinity to multiple HLA alleles. The developers of this algorithm state that this algorithm may allow researchers to reduce immunogenicity of protein therapeutics and also optimize HLA promiscuity for vaccine development. Prospective evaluations of the algorithm have yet to be published.

18.4.4 Tepitope

Tepitope, an algorithm developed by Hammer and Sturniolo (Bian and Hammer 2004), is used to predict HLA class II T-cell epitopes, both promiscuous and allele specific. After inserting a protein sequence into the computer algorithm, the user can set the stringency of the prediction process and can choose multiple HLA alleles. Nonamers are examined for binding to the selected HLA class II alleles and multiple scans along the protein sequence can identify promiscuous regions or sequences. Candidate peptide frames are selected based on quantitative analysis of multiple peptide side chains for binding affinity to HLA class II alleles. Using this algorithm, researchers are able to determine promiscuous epitopes that may be good candidates for monoclonal antibody therapy. Prospective evaluations of the algorithm have yet to be published; however, internal comparisons performed by EpiVax have demonstrated that the Tepitope algorithm is less accurate than EpiMatrix for HLA DR alleles DRB1*0201, DRB1*0301, and DRB1*0401.

18.5 CONFIRMATION OF PREDICTED T-CELL EPITOPES

18.5.1 Peptide Binding Assays

These assays can be used to assess the affinities of therapeutic-derived epitope sequences for multiple HLA alleles. *In vitro* evaluation of MHC binding can be performed by quantifying the ability of exogenously added peptides to compete with a fluorescently labeled known MHC ligand (Steere et al. 2006). Competition-based HLA binding assays can be adapted for high throughput (McMurry et al. 2007a). EpiVax routinely uses these high-throughput HLA binding assays to confirm epitope predictions *in vitro*. A correlation between HLA binding and immunogenicity is often observed (McMurry et al. 2005).

18.5.2 Transgenic Mouse Models

Even though their functions may be similar, animal and human proteins often differ by as much as 20 percent of their total sequence at the amino acid level. Therefore, most proteins intended for therapeutic use in humans are, for animals, relatively foreign and therefore immunogenic. The degree of foreignness may depend on the number of amino acids that are different among peptides that are processed and presented by the animal MHC molecule. Haplotype differences may partially explain why different strains of mice (Blab/C, C57Bl/6) have different immune responses to therapeutic monoclonals (Klitgaard et al. 2006). Although sometimes similar, animal and human MHC differ due to restrictions in the amino acid side chain. Nonhuman primate MHCs are also dissimilar to human MHC; however, there may be a higher degree of sequence homology between the human product and the primate's native molecule to which the animal is tolerant.

Since mice present different T-cell epitopes than do humans, the clinical immunogenicity of mAbs cannot be accurately predicted in mice. An alternative means of evaluating the impact of epitope modifications on *de novo* T-cell response is to measure the immunogenicity of the proteins in mice that are transgenic for human HLA (HLA Tg mice; Depil et al. 2006). Fortunately, several transgenic mouse strains expressing the most common HLA DR molecules are available (Kong et al. 1996; Pan et al. 1998). To compare immunogenicity of wild-type and modified epitopes, mice are immunized with peptide epitopes in adjuvant. T-cell responses in infected humans correlate directly with T-cell responses in immunized HLA transgenic mice (Shirai et al. 1995; Man et al. 1995); thus, HLA transgenic mice are now routinely used to assay and optimize (human) epitope-driven vaccines in preclinical studies (Charo et al. 2001; Ishioka et al. 1999; Livingston et al. 2001). Despite the limited number of HLA for which Tg mice have been developed, comparisons of immunogenicity can be done to a high degree of accuracy in the mouse model for selected HLA class II alleles (HLA DRB1*0101, *1501, *0201, *0301, *0401).

18.6 APPLICATION: PROSPECTIVE PREDICTION OF IMMUNOGENICITY

Because the existing data support the clinical relevance of T-cell-dependent immunogenicity we sought to determine whether the Protein Immunogenicity Score obtained by analysis of the number of T-cell epitopes restricted by eight common HLA alleles could retrospectively predict the immunogenicity observed in the clinic. We have therefore performed a limited analysis of 21 monoclonal antibodies that have been tested in Phase I, II, or III trials and/or that are currently available for use in clinical settings. Vitaxin and HUA33 were excluded because immunogenicity data was not of the same quality as that for the other antibodies and because the sequences for these two antibodies could not be precisely determined from patent databases.

Twenty human and chimeric antibodies that had been studied for immunogenicity were selected from the literature (see Table 18.1). Where multiple studies were available, the scores were averaged. Amino acid sequences for the variable regions of the heavy and light chains were obtained from GenBank and the U.S. Patent and Trademark Office.

All sequences were scored with EpiMatrix and rated on the EpiMatrix immunogenicity scale (De Groot, Knopf, and Martin 2005). Each sequence was then scanned for the presence of putative Treg epitopes defined as 9-mer sequences conserved in more than 10 percent of observed antibodies and scoring at least 5 on the EpiMatrix cluster scale. This included only the most common variants on the Epibars (see Fig. 18.5) found in the regulatory T-cell epitopes previously identified as potentially tolerogenic (De Groot et al. 2008). Observed immunogenicity was then regressed against the immunogenicity (epitope per AA) score of the heavy chains. Heavy chain scores were found to be a significant predictor of immunogenicity (correlation coefficient = 0.58). Observed immunogenicity was also regressed against the immunogenicity score of the light chains. Light chain scores were found to be a significant predictor of immunogenicity (correlation coefficient = 0.6).

As a final test we summed the scores of the heavy and light chains and repeated the regression analysis. This combined score is a better predictor then either the light or heavy score (correlation coefficient = 0.71), showing that the pattern was common to both chains.

That said, the data set used for this pilot analysis is very small. A number of caveats apply to the interpretation of the data. For example, in some cases the data was only provided for subjects who were clinically immunosuppressed prior to having received the antibody therapy. In addition, the total sample size was limited to antibodies for which data was available ($n = 20$). In most cases, the

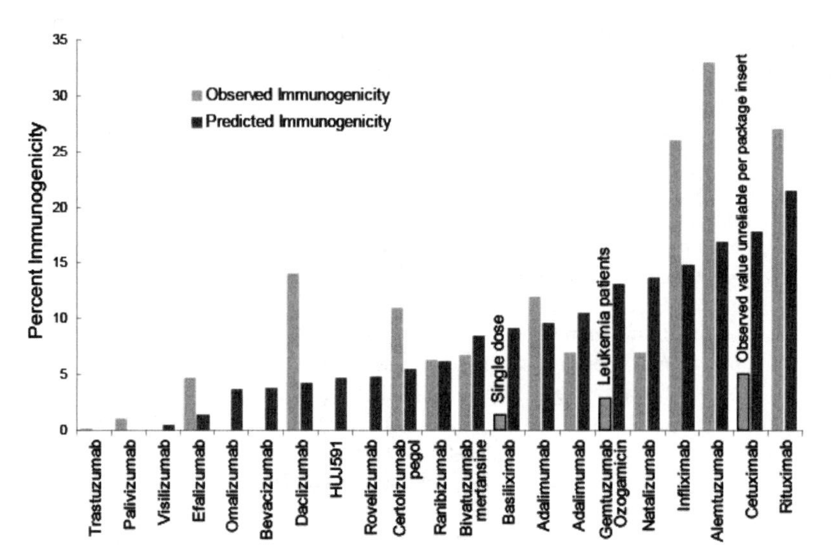

Figure 18.5 Predicted vs. observed Td immunogenicity. Twenty monoclonal antibody sequences were obtained from the literature or the U.S. Patent and Trademark Office (USPTO). Each sequence was then scanned both for epitopes (any epitope restricted by eight common HLA alleles) and also for the presence of putative Treg epitopes defined as 9-mers conserved in more than 50 percent of observed antibodies (>4000) and scoring at least 5 on the EpiMatrix cluster scale. Positive EpiMatrix scores for each of the antibodies were summed; scores for epitopes contained in putative T-regulatory epitope (as defined above) were set aside (not included in the final sum). The final regression is shown.

number of study subjects is small (20 to 30). Furthermore, some of the antibodies target receptors involved in human immune response. And finally, as stated previously, the presence or absence of MHC class II ligands is not the only factor necessary to the formation of robust humoral response.

The findings from this retrospective study can be summarized as follows:

- Predicted T-cell epitope content correlated with observed immunogenicity in this study.
- In particular, low predicted immunogenicity correlated very well with low observed immunogenicity.
- This data suggests that *immunogenicity in the clinic can be avoided if effector T-cell epitope content is reduced.*

While epitope content is not the only factor determining immune responses to mAb, T-cell epitope content appeared to be a solid predictor of observed immune response for this very limited data set.

18.7 INCORPORATING T-CELL IMMUNOGENICITY SCREENING INTO THE DEVELOPMENT PIPELINE

We have briefly described several studies in this chapter that confirm the utility of *in silico* methods for prescreening protein therapeutics and have described a retrospective study of monoclonals. In the context of vaccine studies, *in vitro* confirmation has validated the efficacy of immunogenicity screening using *in silico* methods. Because T-cell epitopes are necessary for a robust humoral response, predictions correlate to the actual response *in vivo*; however, because T-cell epitopes are not, in themselves, sufficient for a robust humoral response some of the predicted immune potential may not be realized *in vivo*. Taken together, these results suggest that modifying epitopes using epitope-mapping tools may prove to be the most efficient means of avoiding antidrug responses. Screening monoclonal antibodies in the preclinical stages, using epitope mapping, may allow researchers to avoid the development of HAHA and HAMA responses and also reduce the costs of monoclonal antibody drug development by eliminating possible candidates that are determined to be too immunogenic.

18.8 SUMMARY

Improved understanding of the relationship between chimeric and humanized monoclonal antibodies and T-cell-dependent immune response is emerging. T-cell responses to epitopes contained in the variable region of monoclonal antibodies contribute to the development of anti-monoclonal antibodies (also known as anti-drug antibodies or ADA). In this chapter, we described the relation between T-cell epitopes contained in antibody sequences, T-cell response, and the clinical immunogenicity of monoclonals. An epitope-based analysis of common monoclonal antibodies was provided and contrasted with immunogenicity observed in the clinic. The role of the T-cell epitope in regards to the development of ADA is now established. Screening and ranking new monoclonal antibodies by T-cell epitope content and T-cell epitope-directed deimmunization of monoclonals are methods that are likely to be integrated in the monoclonal development pathway.

REFERENCES

Ahlers, J.D., I.M. Belyakov, E.K. Thomas, and J.A. Berzofsky. 2001. High-affinity T helper epitope induces complementary helper and APC polarization, increased CTL, and protection against viral infection. *J. Clin. Invest.* 108:1677–1685.

Alinari, L., R. Lampalombella, L. Andritsos, R.A. Baiocchi, T.S. Lin, and J.C. Byrd. 2007. Alemtuzumab (Campath-1H) in the treatment of chronic lymphocytic leukemia. *Oncogene* 26:3644–3653.

Bian, H., and J. Hammer. 2004. Discovery of promiscuous HLA-II-restricted T cell epitopes with TEPITOPE. *Methods* 34(4):468–475.

Bluestone, J.A., and A.K. Abbas. 2003. Natural versus adaptive regulatory T cells. *Nature Rev. Immunol.* 3(3):253–257.

Bond, K.B., B. Sriwanthana, T.W. Hodge, A.S. De Groot, T.D. Mastro, N.L. Young, N. Promadej, J.D. Altman, K. Limpakarnjanarat, and J.M. McNicholl. 2001. An HLA-directed molecular and bioinformatics approach identifies new HLA-A11 HIV-1 subtype E cytotoxic T lymphocyte epitopes in HIV-1-infected Thais. *AIDS Res. Hum. Retroviruses* 20:703–717.

Carteni, G., R. Fiorentino, L. Vecchione, B. Chiurazzi, and C. Battista. 2007. Panitumumab a novel drug in cancer treatment. *Ann. Oncol.* Suppl. 6(vi):16–21.

Charo, J., M. Sundback, A. Geluk, T. Ottenhoff, and R. Kiessling. 2001. DNA immunization of HLA transgenic mice with a plasmid expressing mycobacterial heat shock protein 65 results in HLA class I- and II-restricted T cell responses that can be augmented by cytokines. *Hum. Gene Ther.* 12(14):1797–1804.

Clark, M. 2001. Antibody humanization: A case of the "Emperor's new clothes"? *Immunol. Today* 21:397–402.

Congia, M., S. Patel, A.P. Cope, S. De Virgiliis, and G. Sonderstrup. 1998. T cell epitopes of insulin defined in HLA-DR4 transgenic mice are derived from preproinsulin and proinsulin. *Proc. Natl. Acad. Sci. USA* 95(7):3833–3838.

Cosimi, A.B. 1981. Treatment of acute renal allograft rejection with OKT3 monoclonal antibody. *Transplantation* 32(6):535–539.

Damschroder, M.M., L. Widjaja, P.S. Gill, V. Krasnoperov, D. Jiang, W.F. Dall'Acqua, and H. Wu. 2007. Framework shuffling of antibodies to reduce immunogenicity and manipulate functional and biophysical properties. *Mol. Immunol.* 44(11):3049–3060.

De Groot, A.S. 2006. Immunomics: Discovering new targets for vaccine and therapeutics. *Drug Discovery Today* 11:203–209.

De Groot, A.S., and J.A. Berzofsky. 2004. From genome to vaccine: New immunoinformatics tools for vaccine design. *Methods* 34:425–428.

De Groot, A.S., A. Bosma, N. Chinai, J. Frost, B.M. Jesdale, M.A. Gonzalez, W. Martin, and C. Saint-Aubin. 2001a. From genome to vaccine: In silico predictions, ex vivo verification. *Vaccine* 19(31):4385–4395.

De Groot, A.S., D.S. Rivera, J.A. McMurry, S. Buus, and W. Martin. 2008. Identification of immunogenic HLA-B7 “achilles' heel” epitopes within highly conserved regions of HIV. *Vaccine* 26:3059–3071.

De Groot, A.S., B.M. Jesdale, E. Szu, J.R. Schafer, R.M. Chicz, and G. Deocampo. 1997. An interactive Web site providing major histocompatibility ligand predictions: Application to HIV research. *AIDS Res. Hum. Retroviruses* 13:529–5231.

De Groot, A.S., P.M. Knopf, and W. Martin. 2005. De-immunization of therapeutic proteins by T-cell epitope modification. *Dev. Biol. (Basel)* 122:137–160.

De Groot, A.S., J. Rayner, and W. Martin. 2003. Modeling the immunogenicity of therapeutic proteins using T cell epitope mapping. *Dev. Biol. (Basel)* 112:71–80.

De Groot, A.S., C. Saint-Aubin, A. Bosma, H. Sbai, J. Rayner, and W. Martin. 2001b. Rapid determination of HLA B*07 ligands from the West Nile virus NY99 genome. *Emerg. Infect. Dis.* 7(4):706–713.

De Groot, A.S., H. Sbai, C. Saint-Aubin, J.A. McMurry, and W. Martin. 2002. Immuno-informatics: Mining genomes for vaccine components. *Immunol. Cell. Biol.* 80:255–269.

Depil, S., G. Angyalosi, O. Morales, M. Delacre, N. Delhem, V. Francois, B. Georges, J. Hammer, B. Maillere, C. Auriault, and V. Pancre. 2006. Peptide-binding assays and HLA II transgenic Abeta degrees mice are consistent and complementary tools for identifying HLA II-restricted peptides. *Vaccine* 24:2225–2259.

Dong, Y., S. Demaria, X. Sun, F.R. Santori, B.M. Jesdale, A.S. De Groot, W.N. Rom, and Y. Bushkin. 2004. HLA-A2-restricted CD8+-cytotoxic-T cell responses to novel epitopes in *Mycobacterium tuberculosis* superoxide dismutase, alanine dehydrogenase, and glutamine synthetase. *Infect. Immun.* 72:2412–2415.

Dumont, F.J. 2001. Alemtuzuman (Millenium/ILEX). *Curr. Opin. Investig. Drugs* 2(1):139–160.

Einfeld, D.A., J.P. Brown, M.A. Valentine, E.A. Clark, and J.A. Ledbetter. 1988. Molecular cloning of the human B cell CD20 receptor predicts a hydrophobic protein with multiple transmembrane domains. *EMBO J.* 7(3):711–717.

Eyerman, M.C., and L. Wysocki. 1994. T cell recognition of somatically-generated Ab diversity. *J. Immunol.* 152:1569–1577.

Frenken, L.G., R.H. van der Linden, P.W. Hermans, J.W. Jos, R.C. Ruuls, B. de Geus, and C.T. Verrips. 2000. Isolation of antigen specific llama VHH antibody fragments and their high level secretion by *Saccharomyces cerevisiae*, *J. Biotechnol.* 78(1):11–21.

Gelder, C.M., R. Lambkin, K.W. Hart, D. Fleming, O.M. Williams, M. Bunce, K.I. Welsh, S.E. Marshall, and J. Oxford. 2002. Associations between human leukocyte antigens and nonresponsiveness to influenza vaccine. *J. Infect. Dis.* 185:114–117.

Gill, D.S., and N.K. Damle. 2006. Biopharmaceutical drug discovery using novel protein scaffolds. *Curr. Opin. Biotechnol.* 17(6):653–658.

Gupta, S., S.R. Indelicato, V. Jethwa, T. Kawabata, M. Kelley, A.R. Mire-Sluis, S.M. Richards, B. Rup, E. Shores, S.J. Swanon, and E. Wakshull. 2007. Recommendations for the design, optimization, and qualification of cell-based assays used for the detection of neutralizing antibody responses elicited to biological therapeutics. *J. Immunol. Methods* 32:11–18.

Hacker, H. 2000. Signal transduction pathways activated by CpG-DNA. *Curr. Top. Microbiol. Immunol.* 247:277–292.

Hay, C., M. Recht, M. Carcao, and B. Reipert. 2006. Current and future approaches to inhibitor management and aversion. *Semin. Thromb. Hemost.* 32(Suppl 2):15–21.

Hellendoorn, K., T. Jones, J. Watkins, M. Baker, A. Hamilton, and F. Carr. 2004. Limiting the risk of immunogenicity by identification and removal of T-cell epitopes (DeImmunisation™), Association for Immunotherapy of Cancer: Cancer Immunotherapy-2nd Annual Meeting Mainz, Germany. 6–7 May 2004. *Cancer Cell International* 4:S20.

Hill, A.V., A. Jepson, M. Plebanski, and S.C. Gilbert. 1997. Genetic analysis of host-parasite coevolution in human malaria. *Philos. Trans. R. Soc. London B Biol. Sci.* 352:1317–1325.

Hwang, W.Y., and J. Foote. 2005. Immunogenicity of engineered antibodies. *Methods* 36:3–10.

Inaba, H., W. Martin, A.S. De Groot, S. Qin, and L.J. De Groot. 2006. Thyrotropin receptor epitopes and their relation to histocompatibility leukocyte antigen-DR molecules in Grave's disease. *J. Clin. Endocrinol. Metab.* 91:2286–2294.

Ishioka, G.Y., J. Fikes, G. Hermanson, B. Livingston, C. Crimi, M. Qin, M.F. del Guercio, C. Oseroff, C. Dahlberg, J. Alexander, R.W. Chesnut, and A Sette. 1999. Utilization of MHC class I transgenic mice for development of minigene DNA vaccines encoding multiple HLA-restricted CTL epitopes. *J. Immunol.* 162(7):3915–3925.

Jakobovits, A., R.G. Amado, X. Yang, L. Roskos, and G. Schwab. 2007. From XenoMouse technology to panitumumab, the first fully human antibody product from transgenic mice. *Nature Biotechnol.* 25(10):1134–1143.

Jakobovits, A., G.J. Vergara, J.L. Kennedy, J.F. Hales, R.P. McGuinness, D.E. Casentini-Borocz, D.G. Brenner, and G.R. Otten. 1993. Analysis of homozygous mutant chimeric mouse: Deletion of the immunoglobulin heavy-chain joining region blocks B-cell development and antibody production. *Proc. Natl. Acad. Sci. USA* 90:2551–2555.

Jiang, X., and K.D. Moudgil. 2006. The unveiling of hidden T-cell determinants of a native antigen by defined mediators of inflammation: Implications for the pathogenesis of autoimmunity. *Scand. J. Immunol.* 63:338–346.

Jones, T.D., M. Hanlon, B.J. Smith, C.T. Heise, P.D. Nayee, D.A. Sanders, A. Hamilton, C. Sweet, E. Unitt, G. Alexander, K.M. Lo, S.D. Gillies, F.J. Carr, and M.P. Baker. 2004. The development of a modified human IFN-alpha2 linked to the Fc portion of human IgG1 as a novel potential therapeutic for the treatment of hepatitis C infection. *J. Interferon Cytokine Res.* 24(9):560–572.

Kappler, J.W., N. Roehm, and P. Marrack. 1987. T-cell tolerance by clonal elimination in the thymus. *Cell* 49:273–280.

Kast, W.M., R.M. Brandt, J. Sidney, J.W. Drijfhout, R.T. Kubo, H.M. Grey, C.J. Melief, and A. Sette. 1994. Role of HLA-A motifs in identification of potential CTL epitopes in human papillomavirus type 16 E6 and E7 proteins. *J. Immunol.* 152:3904–3912.

Klitgaard, J.L., V.W. Coljee, P.S. Andersen, L.K. Rasmussen, L.S. Nielsen, J.S. Haurum, and S. Bregenholt. 2006. Reduced susceptibility of recombinant polyclonal antibodies to inhibitory anti-variable domain antibody responses. *J. Immunol.* 177(6):3782–3790.

Kohler, G., and C. Milstein. 1976. Derivation of specific antibody-producing tissue culture and tumor lines by cell fusion. *Eur. J. Immunol.* 6(7):511–519.

Koita, O.A., D. Dabitao, I. Mahamadou, M. Tall, S. Dao, A. Tounkara, H. Guiteye, C. Noumsi, O. Thiero, M. Kone, D. Rivera, J.A. McMurry, W. Martin, and A.S. De Groot. 2006. Confirmation of immunogenic consensus sequence HIV-1 T cell epitopes in Bamako, Mali and Providence, Rhode Island. *Hum. Vaccin.* 2(3):119–128.

Kong, Y.C., L.C. Lomo, R.W. Motte, A.A. Giraldo, J. Baisch, G. Strauss, G.J. Hammerling, and C.S. David. 1996. HLA-DRB1 polymorphism determines susceptibility to autoimmune thyroiditis in transgenic mice: Definitive association with HLA-DRB1*0301 (DR3) gene. *J. Exp. Med.* 184:1167–1172.

Kooster, R., B.T. Maassen, J.C. Stam, P.W. Hermans, M.R. Ten Haaft, F.J. Detmers, H.J. de Haard, J., Post, and C. Theo Verrips. 2007. Improved anti-IgG and HAS affinity ligands: Clinical application of VHH antibody technology. *J. Immunol. Methods* 324(1-2):1–12.

Koren, E., A.S. De Groot, V. Jawa, K.D. Beck, D. Rivera, L. Li, D. Mytych, M. Kosec, D. Weeraratne, S. Swanson, and W. Martin. 2007. Clinical validation of the "in silico" prediction of immunogenicity of a human recombinant therapeutic protein. *Clin. Immunol.* 124(1):25–32.

Koren, E., L.A. Zuckerman, and A.R. Mire-Sluis. 2002. Immune responses to therapeutic proteins in humans: Clinical significance, assessment and prediction. *Curr. Pharm. Biotechnol.* 3:349–360.

Livingston, B.D., M. Newman, C. Crimi, D. McKinney, R.W. Chesnut, and A. Sette. 2001. Optimization of epitope processing enhances immunogenicity of multiepitope DNA vaccines. *Vaccine* 19(32):4652–4660.

Man, S., M.H. Newberg, V.L. Crotzer, C.J. Luckey, N.S. Williams, Y. Chen, E.L. Huczko, J.P. Ridge, and V.H. Engelhard. 1995. Definition of a human T-cell epitope from influenza A non-structural protein 1 using HLA-A2.1 transgenic mice. *Int. Immunol.* 7(4):597–605.

McMurry, J.A., S.H. Gregory, L. Moise, D. Rivera, S. Buus, and A.S. De Groot. 2007a. Diversity of *Francisella tularensis* Schu4 antigens recognized by T lymphocytes after natural infections in humans: Identification of candidate epitopes for inclusion in a rationally designed tularemia vaccine. *Vaccine* 25(16):3179–3191.

McMurry, J.A., S. Kimball, J.H. Lee, D. Rivera, W. Martin, D.B. Weiner, M. Kutzler, D.R. Sherman, H. Kornfeld, and A.S. De Groot. 2007b. Epitope-driven TB vaccine development: A streamlined approach using immunoinformatics, ELISpot assays and HLA transgenic mice. *Curr. Mol. Med.* 7(4):351–363.

McMurry, J., H. Sbai, M.L. Gennaro, E.J. Carter, W. Martin, and A.S. De Groot. 2005. Analyzing *Mycobacterium tuberculosis* proteomes for candidate vaccine epitopes. *Tuberculosis (Edinburgh)* 85:95–105.

Mire-Sluis, A.R., Y.C. Barrett, V. Devanarayan, E. Koren, H. Liu, M. Maia, T. Parish, G. Scott, G. Shankar, E. Shores, S.J. Swanson, G. Taniguchi, D. Wierda, and L.A. Zuckerman. 2004. Recommendations for the design and optimization of immunoassays used in the detection of host antibodies against biotechnology products. *J. Immunol. Methods* 289:1–16.

Moutaftsi, M., B. Peters, V. Pasquetto, D.C. Tscharke, J. Sidney, H.H. Bui, H. Grey, and A. Sette. 2006. A consensus epitope prediction approach identifies the breadth of murine T(CD8+)-cell responses to vaccinia virus. *Nature Biotechnol.* 24(7):817–819.

Mullbacher, A. 1992. Viral escape from immune recognition: Multiple strategies of adenoviruses. *Immunol. Cell. Biol.* 70:59–63.

Muyldermans, S. 2001. Single domain camel antibodies: Current status. *J. Biotechnol.* 74:277.

Oling, V., J. Marttila, J. Ilonen, W.W. Kwok, G. Nepom, M. Knip, O. Simell, and H. Reijonen. 2006. GAD65- and proinsulin-specific CD4+ T-cells detected by MHC class II tetramers in peripheral blood of type 1 diabetes patients and at-risk subjects. *J. Autoimmun.* 27(1):69.

Page, K.R., A.L. Scott, and Y.C. Manabe. 2006. The expanding realm of heterologous immunity: Friend or foe? *Cell Microbiol.* 8:185–196.

Pan, S., T. Trejo, J. Hansen, M. Smart, and C.S. David. 1998. HLA-DR4 (DRB1*0401) transgenic mice expressing an altered CD4- binding site: Specificity and magnitude of DR4-restricted T-cell response. *J. Immunol.* 161(6):2925–2929.

Piro, L.D., C.A. White, A.J. Grillo-López, N. Janakiraman, A. Saven, T.M. Beck, C. Varns, S. Shuey, M. Czuczman, J.W. Lynch, J.E. Kolitz, and V. Jain. 1999. Extended Rituximab (anti-CD20 monoclonal antibody) therapy for relapsed or refractory low-grade or follicular non-Hodgkin's lymphoma. *Ann. Oncol.* 10(6):655–661.

Pollack, P., and J.R. Groothuis. 2002. Development and use of palivizumab (Synagis): A passive immunoprophylactc agent for RSV. *J. Infect. Chemother.* 8:201–216.

Queen, C., W.P. Schneider, H.E. Selick, P.W. Payne, N.F. Landolfi, F.J. Duncan, N.M. Avdalovic, M. Levitt, R.P. Junghans, and T.A. Waldmann. 1989. A humanized antibody that binds to the interleukin 2 receptor. *Proc. Natl. Acad. Sci. USA* 86:10029–10033.

Rammensee, H., J. Bachmann, N.P. Emmerich, O.A. Bachor, and S. Stevanović. 1999. SYFPEITHI: Database for MHC ligands and peptide motifs. *Immunogenetics* 50(3-4):213–219.

Ramos-Casals, M., P. Brito-Zerón, S. Munoz, N. Soria, D. Galiana, L. Bertolaccini, M.J. Cuadrado, and M.A. Khamashta. 2007. Autoimmune diseases induced by TNF-targeted therapies: Analysis of 233 cases. Medicine (Baltimore) 86(4):242–251.

Reveille, J.D. 2006. The genetic basis of autoantibody production. *Autoimmun. Rev.* 5(6):389–398.

Riechmann, L., M. Clark, H. Waldmann, and G. Winter. 1988. Reshaping human antibodies for therapy. *Nature* 332:323–327.

Scanlan, M.J., and D. Jager. 2001. Challenges to the development of antigen-specific breast cancer vaccines. *Breast Cancer Res.* 3:95–98.

Schneider, C., and M.H. Van Regenmortel. 1992. Immunogenicity of free synthetic peptides corresponding to T helper epitopes of the influenza HA 1 subunit. Induction of virus cross reacting CD4+ T lymphocytes in mice. *Arch. Virol.* 125:103–119.

Scott, A.M., F.T. Lee, R. Jones, W. Hopkins, D. MacGregor, J.S. Cebon, A. Hannah, G.U.P. Chong, A. Papenfuss, A. Rigopoulos, S. Sturrock, R. Murphy, V. Wirth, C. Murone, F.E. Smyth, S. Knight, S. Welt, G. Ritter, E. Richards, E.C. Nice, A.W. Burgess, and L.J. Old. 2005. Phase I study of anticolon cancer humanized antibody A33: Biodistribution, pharmacokinetics and quantitative tumor uptake. *Clin. Cancer Res.* 11(13):4810–4817.

Shirai, M., T. Arichi, M. Nishioka, T. Nomura, K. Ikeda, K. Kawanishi, V.H. Engelhard, S.M. Feinstone, and J.A. Berzofsky. 1995. CTL responses of HLA-A2.1-transgenic mice specific for hepatitis C viral peptides predict epitopes for CTL of humans carrying HLA-A2.1. *J. Immunol.* 154(6):2733–2742.

Southwood, S., J. Sidney, A. Kondo, M.F. Del Guercio, E. Appella, S. Hoffman, R.T. Kubo, R.W. Chesnut, H.M. Grey, and A. Sette. 1998. Several common HLA-DR types share largely overlapping peptide binding repertoires. *J. Immunol.* 160(7):3363–3373.

Staelens, S., J. Desmet, T.H. Ngo, S. Vauterin, I. Pareyn, P. Barbeaux, I. Van Rompaey, J.M. Stassen, H. Deckmyn, and K. Vanhoorelbeke. 2006. Humanization by variable domain resurfacing and grafting on a human IgG4, using a new approach for determination of non-human like surface accessible framework residues based on homology modelling of variable domain. *Mol. Immunol.* 43:1243–1257.

Steere, A.C., W. Klitz, E.E. Drouin, B.A. Falk, W.W. Kwok, G.T. Nepom, and L.A. Baxter-Lowe. 2006. Antibiotic-refractory Lyme arthritis is associated with HLA-DR molecules that bind a *Borrelia burgdorferi* peptide. *J. Exp. Med.* 203(4):961–971.

Szolar, O.H., S. Stranner, I. Zinoecker, G.C. Mudde, G. Himmler, G. Waxenecker, and A. Nechansky. 2006. Qualification and application of a surface plasmon resonance-based assay for monitoring potential HAHA responses induced after passive administration of a humanized anti Lewis-Y antibody. *J. Pharm. Biomed. Anal.* 41(4):1347–1353.

Tangri, S., B.R. Mothé, J. Eisenbraun, J. Sidney, S. Southwood, K. Briggs, J. Zinckgraf, P. Bilsel, M. Newman, R. Chesnut, C. Licalsi, and A. Sette. 2005. Rationally engineered therapeutic proteins with reduced immunogenicity. *J. Immunol.* 174:3187–3196.

Test, S.T., J. Mitsuyoshi, C.C. Connolly, and A.H. Lucas. 2001. Increased immunogenicity and induction of class switching by conjugation of complement C3d to pneumococcal serotype 14 capsular polysaccharide. *Infect. Immun.* 69(5):3031–3040.

Van Walle, I., Y. Gansemans, P.W. Parren, P. Stas, and I. Lasters 2007. Immunogenicity screening in protein drug development. *Expert Opin. Biol. Ther.* 7:405–418.

Verheesen, P., M.R. Ten Haaft, N. Lindner, C.T. Verrips, and J.J. De Haard. 2003. Beneficial properties of single-domain antibody fragments for application in immunoaffinity purification and immuno-perfusion chromatography. *Biochem. Biophys. Acta* 1624:21.

Vossen, M.T., E.M. Westerhout, C. Soderberg-Naucler, and E.J. Wiertz. 2002. Viral immune evasion: A masterpiece of evolution. *Immunogenetics* 54:527–542.

Yeung, P., J. Chang, J. Miller, C. Barnett, M. Stickler, and F.A. Harding. 2004. Elimination of an immunodominant CD4 T-cell epitope in human IFN- does not result in an in vivo response directed at the subdominant epitope. *J. Immunol.* 172(11):6658–6665.

Young, J. 2002. Development of a potent respiratory syncytial virus-specific monoclonal antibody for the prevention of serious lower respiratory tract disease in infants. *Respir. Med. Suppl. B* S31–S35.

Zubler, R.H. 2001. Naive and memory B cells in T-cell-dependent and T-independent responses. *Springer Seminars Immunopathol.* 23:405–419.

CHAPTER 19

Monoclonal Antibody Pharmacokinetics and Pharmacodynamics

CHRISTOPHER R. GIBSON, PUNAM SANDHU, and WILLIAM D. HANLEY

ABSTRACT

This chapter describes the concepts of pharmacokinetics and pharmacodynamics relating to the absorption, distribution, elimination (target-mediated kinetics), and immunogenic properties of monoclonal antibodies. Examples are provided for of comparison of pharmacokinetics of monoclonal antibodies between preclinical species and humans. Pharmacokinetic and pharmacodynamic modeling approaches for antibodies and drug-drug interaction potential of antibodies are also discussed.

Therapeutic Monoclonal Antibodies: From Bench to Clinic. Edited by Zhiqiang An

19.1 INTRODUCTION: PHARMACOKINETICS AND PHARMACODYNAMICS IN DRUG DEVELOPMENT

Pharmacokinetics (PK) is the study of the time course of a drug in the body following its administration. The time course of a drug is determined by all processes involved in the absorption, distribution, metabolism, and excretion of the drug (i.e., the disposition of the drug in the body). In turn, the drug's desirable and undesirable effects on the body represent its pharmacodynamic (PD) properties. The integration of these fields constitutes the exposure-response relationship (or PK-PD), which allows a continuous description of the effect of the drug over time. The application of PK-PD concepts has become an important tool enhancing the decision-making process during drug development.

Despite the increasing number of therapeutic monoclonal antibodies (mAb) on the market and in development, certain aspects of the pharmacokinetic and pharmacodynamic properties of these molecules are not fully understood. Target-mediated disposition, in part, explains the characteristic non-linear PK and PD characteristics often observed for mAbs. The interindividual variability of mAb PK can be explained by several factors, including interindividual variability with respect to target antigen expression, and generation of immune responses against the administered mAb while, among other sources, genetics and clinical status may explain interindividual PD variability. The PK and PD and related variability may be explored through the use of PK-PD modeling approaches, which may be more complex than those for conventional drugs.

Thus far, 23 antibody products have been approved for use in various therapeutic areas, such as inflammation, oncology, cardiovascular, transplantation, antiviral, and macular degeneration. These include intact antibodies, antibody fragments, and antibody-fusion proteins (Mould and Sweeney 2007; Seitz and Zhou 2007). Of these 23 antibody products, 19 are monoclonal antibodies, and all are of the IgG class. Therapeutic monoclonal antibodies are designed to either target soluble antigens, such as circulating cytokines [e.g. IL-4, IL-5, IL-8, or vascular endothelial growth factor (VEGF)], or cellular membrane-associated antigens, such as epidermal growth factor receptor (EGFR) or human epidermal growth factor receptor 2 (HGR2) (Tabrizi, Tseng, and Roskos 2006). The pharmacokinetic properties of mAbs targeted against soluble antigens, in general, tend to display dose-proportional exposure with linear clearance. Monoclonal antibodies that target cellular antigens, on the other hand, display more complex non-linear pharmacokinetic properties, such that the half-life of these can be concentration- and time-dependent. This often leads to non-linear clearance due to clearance being mediated by formation of an antibody-antigen complex which undergoes subsequent internalization and degradation. This impacts mAb pharmacokinetics such that at lower antibody doses there is an increase in clearance. At higher antibody doses, the clearance decreases due to dose-dependent saturation of the antigen sink (Tabrizi, Tseng, and Roskos 2006).

19.2 PHARMACOKINETICS OF MONOCLONAL ANTIBODIES

19.2.1 General Overview of Pharmacokinetics

Monoclonal antibodies and biological compounds such as recombinant fusion proteins have in the recent past played an increasingly important role in the development of therapeutics. In humans, there are five major classes of immunoglobulins (Ig), namely, IgA, IgD, IgE, IgG, and IgM. IgG is the most abundant immunoglobulin in plasma and has several subclasses in humans (IgG1, 2, 3, and 4). The vast majority (14 out of 19) of therapeutic antibodies are of the IgG1 subtype (Mould and Sweeney 2007; Seitz and Zhou 2007). The half-life of IgG1, 2, and 4 in humans is approximately 21 days and that of IgG3 is approximately 7 days (Ternant and Paintaud 2005). The shorter half-life of IgG3 as compared to IgGs 1, 2, and 4 has been linked to the differences in binding to FcRn resulting from a single amino acid difference in the binding domain of FcRn (Fc-receptor of the neonate, also known as the Brambell receptor; Ghetie and Ward 2002). The other immunoglobulins (IgA, IgD, IgE, and IgM) have a shorter half-life, in the range of 2.5 to 6 days (Ternant and Paintaud 2005).

The half-life of mAbs generally increases with the degree of humanization: murine (1.5 days) < chimeric (10 days) < humanized (12 to 20 days) ≈ fully human (15 to 20 days) (Ternant and Paintaud 2005). The shorter half-life of murine mAbs has been attributed to lack of binding of murine IgG to human FcRn, which has been demonstrated to be responsible for the protection of IgG from systemic elimination (Tabrizi, Tseng, and Roskos 2006; Lobo, Hansen, and Balthasar 2004). The generation of human anti-mouse antibodies (HAMA) has also been, partly, attributed for the shorter half-life of murine mAbs as these tend to accelerate the clearance of the antibody by forming immunocomplexes. The generation of an anti-antibody response is governed by several factors, such as the dose of the antibody, the route of administration, and the number of times an antibody is administered. Other factors that can influence the generation of an anti-antibody response include the form and immunogenicity of the antibody as well as the immunocompetence of the patient (Kuus-Reichel et al. 1994).

19.2.2 Absorption, Distribution, Metabolism, and Excretion Properties of Monoclonal Antibodies

19.2.2.1 Absorption Various routes of administration have been employed for dosing antibodies. These have included intravenous (IV) as well as extravascular routes such as subcutaneous (SC) and intramuscular (IM) administration. The vast majority of the marketed antibodies are administered intravenously (Ternant and Paintaud 2005; Tang et al. 2004). The obvious advantages of IV administration are complete systemic availability, rapid achievement of high serum concentrations, as well as the ability to administer high doses if necessary. The IV route of administration, however, is not as convenient as a parenteral form of dosing as it often requires hospitalization of patients. Following SC or IM administration, systemic absorption of antibodies takes place through lymphatic vessels, thus absorption tends to be slow due to the low flow rate of lymph (Ternant and Paintaud 2005; Lobo, Hansen, and Balthasar 2004; Tang et al. 2004). The bioavailability of an antibody following extravascular dosing may be affected by the extent of presystemic degradation by proteolytic enzymes. This degradation can be saturable and can hence increase the bioavailability with increasing dose of antibody (Lobo, Hansen, and Balthasar 2004). In general, monoclonal antibodies administered via extravascular routes display high bioavailability (>50 percent; Tang et al. 2004; Scheinfeld 2003). A limitation of extravascular routes is that large volumes of antibody (greater than 5 mL) cannot be administered due to practical considerations such as discomfort associated at the site of administration and solubility limitations of the mAb.

The oral route of administration has been minimally explored as its utility has thus far been limited to the treatment of gastrointestinal diseases (Losonsky et al. 1985; Guarino et al. 1994). Antibodies are susceptible to proteolytic degradation in the gastrointestinal tract and can denature in the acidic environment of the stomach. The high polarity and large molecular size of antibodies can also hinder diffusion through the gastrointestinal epithelium (Lobo, Hansen, and Balthasar 2004). Due to these reasons, oral administration of antibodies is not a very attractive alternative as compared to the other extravascular routes.

Inhalation delivery of monoclonal antibodies has also been explored as it offers the advantage of ease of administration coupled with the presence of a large surface area for absorption and high vascularity (Tang et al. 2004). Administration of antibodies by inhalation, however, has not proven to demonstrate efficacy presumably due to the inability of the aerosol route to result in high enough concentrations in target tissues. It has also been speculated that the delivery of monoclonal antibodies by inhalation may cause a higher immunogenic response as compared to parenteral routes of administration (Fahy et al. 1999).

19.2.2.2 Distribution Distribution of antibodies depends on factors such as the rate and extent of antibody extravasation within tissue, as well as distribution and elimination from tissue (Lobo, Hansen, and Balthasar 2004). The apparent volume of distribution at steady state (V_{ss}) in which large molecules such as antibodies distribute following IV administration is generally in the range of 3 to 8 L. This is equal to two- to three-fold higher than the plasma volume. It should be noted that V_{ss} values reported in the literature for most antibody pharmacokinetics should be interpreted with caution as most of these calculations are based on noncompartmental approaches (described in more detail in Section 19.4) which assume that the site of antibody elimination is in rapid equilibrium with plasma and that all elimination is from the central compartment. Since antibodies undergo degradation in tissues, these assumptions are not necessarily always valid (Lobo, Hansen, and Balthasar 2004).

Preclinically, distribution of antibodies can be assessed via conduct of radiolabeled tissue distribution studies as has been demonstrated in rabbits for ^{125}I-rhumAb VEGF, a recombinant humanized monoclonal antibody against the vascular endothelial growth factor (Lin et al. 1999) and ^{125}I-rhumAb-E25 in monkeys (Fox et al. 1996). In both cases, there was no significant accumulation of antibody in tissues. These and other studies have demonstrated that, in general, for antibodies and antibody fragments, the tissue to blood ratio is in the range of 0.1 to 1.0 (Molthoff et al. 1992; Baxter et al. 1994). The tissue to blood concentration ratio though can be higher in cases where the antibody binds avidly to tissues. For example, the tissue to blood concentration ratio for several antibodies targeted against endothelial antigens was in the range of 4.8 to 23.5 in the lung, spleen, heart, kidney and, liver of rats (Danilov et al. 2001).

19.2.2.3 Metabolism/Elimination The elimination of monoclonal antibody drugs and endogenous immunoglobulins from the body is a complex multifactorial process. In general, the clearance process can be broken down into four distinct entities, namely, normal protein catabolism, interaction with the neonatal Fc-receptor (FcRn), target-mediated elimination (Section 19.2.2.3), and immunogenicity (Section 19.2.2.3). It is the overall contribution of each of the individual processes that determines the pharmacokinetic profile of a monoclonal antibody drug. Other factors that can contribute to the elimination of antibodies include proteolytic degradation (Gillies et al. 2002) and glycosylation (Meier et al. 1995) of the antibody.

Excretion of intact mAbs is a minor component of the overall elimination process from the body. Antibodies are primarily eliminated via catabolism and thus renal and hepatic excretion do not play as dominant a role as catabolic processes. Due to the large size and physicochemical properties, very little antibody is excreted in the urine or bile. The majority of the antibody that is transported into the urinary space of the glomerulus is reabsorbed in the proximal tubules of the nephron and either re-enters the systemic circulation or is catabolized in the tubular cells (Lobo, Hansen, and Balthasar 2004). Lower molecular weight antibody fragments, such as Fab (fragment antigen binding), are filtered by the glomerulus and subsequently reabsorbed and catabolized by proximal tubule cells. Biliary excretion has been shown to play a role in the elimination of IgA-type immunoglobulins, but the contribution to the overall clearance was minor at 3 percent (Delacroix et al. 1982). Catabolism to smaller peptides appears to be the major elimination mechanism of intact monoclonal antibody drugs. Catabolic enzymes are relatively ubiquitous in the human body. Carrier-mediated membrane transport as well as classical endocytosis/pinocytosis are responsible for movement of the antibodies into cells where they can be degraded into smaller peptides (Baumann 2006; Leslie 1982). The pathway for intracellular catabolism of a protein will differ depending on the cell type and physiological state (Ghetie and Ward 2002). Antibody catabolism is known to occur in the lysosome and is considered to be a slow and non-selective process. In contrast, antibodies that are released from endosomes into the cytosol may undergo rapid degradation via the ATP-dependent ubiquitin/proteasome system (Ghetie and Ward 2002).

Once inside the cell, it is the interaction with the FcRn that plays a very important role regarding the fate of the antibody. The elimination of IgG is known to be concentration dependent, with half-life decreasing as a function of increasing serum IgG concentrations. In 1964, Brambell and co-workers proposed that IgG may be protected from elimination by a transport protein that is saturable at high concentrations of IgG, thereby leading to an increase in the elimination rate of IgG (Brambell, Hemmings, and Morris 1964). These workers also hypothesized that the same mechanism may mediate the transport of maternal IgG to the fetus, and subsequent molecular biology efforts led to the identification and cloning of FcRn in rat (Lobo, Hansen, and Balthasar 2004; Brambell, Hemmings, and Morris 1964). The importance of FcRn on the pharmacokinetics of IgG was conclusively shown in FcRn knockout mice in which the IgG elimination rate was 10- to 15-fold faster than in the wild-type strain (Ghetie et al. 1996; Israel et al. 1996). Interestingly, the elimination of other immunoglobulins (IgA, D, E, and M) was not affected in the knockout mice. Thus, these data suggest that IgG-FcRn interactions are responsible for the characteristic lower clearance and longer half-life of IgGs, while FcRn may play a lesser role in the pharmacokinetics of other Ig subtypes.

The binding of IgG to FcRn was demonstrated to be pH dependent, with increased binding at acidic pH (6.0) and lower binding at physiologic neutral pH (Ternant and Paintaud 2005). As shown in Figure 19.1, IgGs are transported into cells either non-selectively by pinocytosis, or by specific membrane transporters, into endosomal pockets. As the pH in the endosome decreases (from 7.4 to 6), the affinity of FcRn for the Fc portion of IgG increases. Following a sorting event, the endosome is split into halves, one containing the FcRn-IgG bound complexes and the other containing free IgG and other proteins destined for catabolism. The endosome containing free IgG then fuses with a lysosome and its contents are released into the lysosome where they undergo catabolism (Ternant and Paintaud 2005). The endosome, containing IgG-FcRn complexes, fuses with the cell membrane and is exposed to physiologic pH (7.4), thereby releasing IgG back into the systemic circulation. It can be rationalized, based on this mechanism of FcRn elimination protection, how saturation can lead to an increased clearance of IgG. Once all available FcRn is bound to IgG, any additional antibody that is present will not be able to undergo the recycling mechanism afforded by the interaction with FcRn and will be subject to catabolic elimination.

FcRn is also expressed in the epithelial cells of the kidney glomerulus and the cells of the proximal convoluted tubule. This is noteworthy since intact IgG is too large to be filtered by the glomerulus. An interesting hypothesis to explain the presence of FcRn in the glomerulus is that it serves to keep the glomerular filter structures free from antibody deposition (Roopenian and Akilesh 2007). Once the antibody comes into contact with the podocyte cells of the glomerulus, it is transported into the urinary

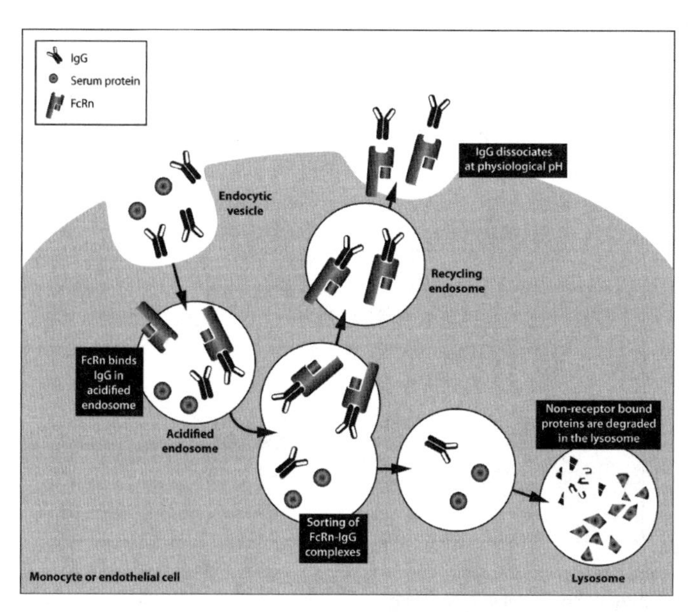

Figure 19.1 Following uptake into the cell, IgG binds to FcRn as the pH is decreased. After sorting, the IgG-FcRn complexes are returned to the cell surface where the IgG is released back into the interstitial space. Any IgG that does not bind FcRn upon entering the cell is destined for degradation. (Adapted from D.C. Roopenian and S. Akilesh. 2007. *Nature Rev. Immunol.* 7:715–725.)

space of the glomerulus instead of aggregating and reducing the filtering capacity of the kidney. The antibody could then be reabsorbed into the circulation by the FcRn expressed downstream in the proximal convoluted tubule cells.

Target-Mediated Elimination Kinetics One major difference between the pharmacokinetics of small molecules and monoclonal antibodies is the proportion of molecules that exhibit saturable elimination kinetics. This type of pharmacokinetic behavior is relatively common for monoclonal antibodies, especially for those with high affinity for cellular protein targets. In this case, the elimination process of the antibody can be viewed as one comprising high affinity to a target with low capacity (i.e., low relative abundance of target in the body) making it susceptible to saturation. In such instances, at low serum concentrations, the elimination rate (half-life) of the antibody is rapid and governed by binding of the antibody to its target antigen, whereas, at higher serum antibody concentrations, the available pool of target antigen is vastly reduced and excess antibody is eliminated more slowly by participating in FcRn-mediated processes.

A simple version of a typical saturable PK model for monoclonal antibodies is shown in Figure 19.2, which represents a two-compartment model where clearance of the antibody occurs only from the central compartment. The clearance term is represented as two separate processes, a saturable Michaelis–Menten term and a standard first-order elimination term. The Michaelis–Menten term represents saturable binding of antibody to its target. In this model, binding of the antibody to the target is considered an irreversible elimination process, such as binding of an antibody to a cellular receptor, and subsequent internalization and degradation of the antibody-receptor complex. The first-order elimination term represents the normal antibody catabolism rate and is the net effect of protein metabolism and protection from elimination by FcRn.

By considering the mathematics of this relatively simple model of target-mediated disposition, several important features of the system can be rationalized. First, apparent overall clearance is dependent on the serum concentration of the antibody. When the serum concentration of antibody is below the apparent K_m, target-mediated elimination is not saturated and the clearance and half-life of the antibody are not constant values and thus cannot be reported as such (Fig. 19.3). However, when the serum concentration of antibody is in excess of what is needed to saturate the available target (serum concentration approximately 10-fold higher than the apparent K_m), the saturable elimination term becomes negligible and first-order elimination becomes the primary determinant of the elimination rate. Accordingly, as the serum concentration of antibody continues to rise, the clearance and half-life approach a constant values. This is illustrated in Figure 19.4, which represents three simulated IV

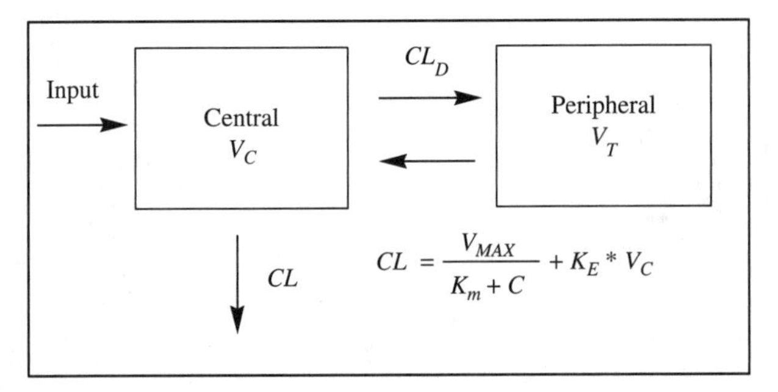

Figure 19.2 A classical two-compartment open model with parallel saturable (Michaelis–Menten) and first-order elimination occurring from the central compartment. CL = clearance, CL_D = distribution clearance resulting from drug movement between central and peripheral compartment, V_C = volume of central compartment, V_T = volume of peripheral compartment, K_E = first-order elimination rate constant, V_{MAX} = maximum rate of elimination from saturable pathway, K_m = serum concentration where saturable elimination rate is half maximal (Michaelis constant), C = serum concentration.

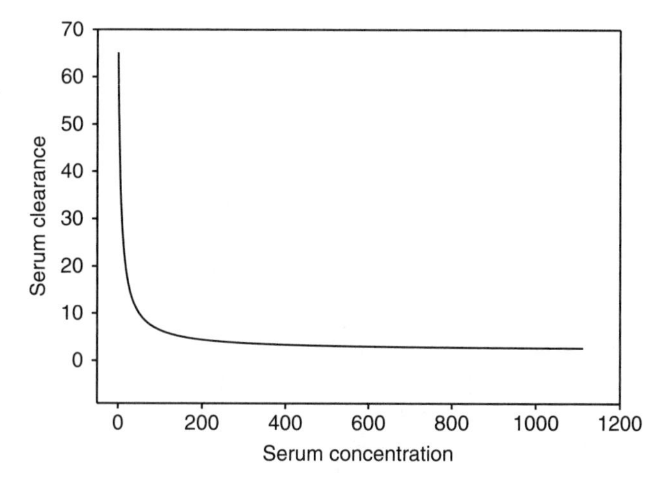

Figure 19.3 The relationship between clearance and serum concentration for a monoclonal antibody that has parallel saturable target-mediated and first-order elimination. At low serum concentrations, the saturable target-mediated clearance is the primary driver and results in large values for overall clearance of antibody. At higher serum concentrations, where most of the available target antigen has been saturated by antibody, the clearance decreases and approaches a value governed by the slower endogenous antibody catabolism rate and interaction with FcRn.

doses (high, medium, and low) of a hypothetical monoclonal antibody that behaves according to the pharmacokinetic model illustrated in Figure 19.2. When the serum concentration is above the apparent K_m value (set to 7 arbitrary concentration units for this example), the pharmacokinetic behavior displays a classical two-compartment model profile although the apparent terminal half-life increases with increasing dose. Therefore, at saturating concentrations for a monoclonal antibody that behaves according to this model, the half-life will depend on dose, by way of serum concentration, and will not have a discrete value for all doses. Once serum levels decline and approach the K_m value, there is a rapid increase in clearance as the saturable portion becomes more prominent in the overall elimination of the antibody.

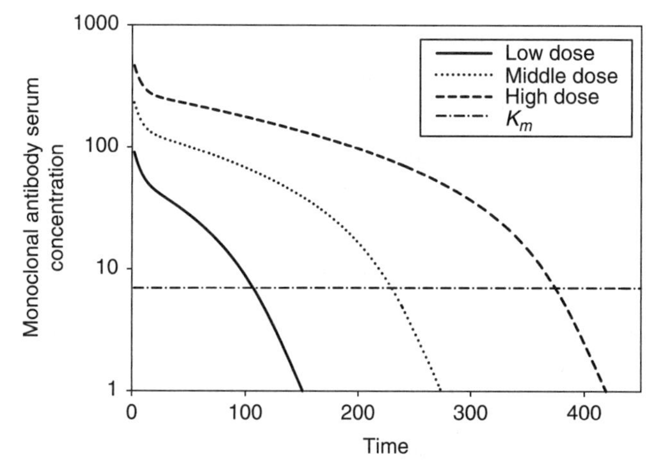

Figure 19.4 Simulated serum concentration-time profiles of a hypothetical monoclonal antibody, following three escalating doses, which behaves according to the pharmacokinetic model shown in Figure 19.2. For this example, the K_m was fixed to a value of 7 arbitrary concentration units.

For antibodies that have high affinity to targets that are present in low abundance, relative to the amount of antibody, it would be important to study the pharmacokinetics at several doses to determine whether saturable target-mediated elimination plays a role. It is of equal importance to rule out immunogenicity as a cause for a rapid increase in systemic clearance. The presence of an antidrug immune response could cause a rapid increase in clearance and would be independent of any target-mediated elimination. In cases where both target-mediated elimination and an immune response are occurring, an increasing half-life with increasing dose, prior to onset of the immune response, may be the only indicator of target-mediated elimination.

Immunogenicity and the Impact on Monoclonal Antibody Pharmacokinetics It is generally accepted that administration of any exogenous therapeutic antibody to humans can elicit an immune response and formation of human anti-human antibodies. The propensity for an mAb to promote an immune response has been reported to be dependent on several factors, including the dose, route of administration, amount of aggregation, and the similarity of the exogenous antibody to endogenous immunoglobulins (Lobo, Hansen, and Balthasar 2004). Early therapeutic mAbs were often entirely of murine origin; however, the intended pharmacological effect was often abrogated by the generation of human anti-mouse antibodies (HAMA), potentially leading to an increase in clearance of the mAb and occasionally adverse events. Consequently, much effort has been devoted to reducing the murine content of therapeutic mAbs in an attempt to reduce the immunogenicity. A thorough review of the literature reveals that the incidence of anti-antibody responses is roughly proportional to the residual amount of murine content in the mAb (Hwang and Foote 2005). Therapeutic mAbs that have been engineered to minimize the murine content may be categorized as follows: (1) chimeric, consisting of mAbs with mouse variable regions and human constant regions; (2) humanized, mAbs that retain only murine complementary-determining regions (CDRs); and (3) fully human, mAbs generated from phage libraries or using transgenic mice that are reported to be indistinguishable from native human antibodies. However, even fully human mAbs may illicit an immune response, potentially due to the unique antibody sequences responsible for antigen binding, or, perhaps because mAbs are mass-produced using non-human cell lines which can introduce non-native post-translational modifications (Pendley, Schantz, and Wagner 2003).

In addition to safety concerns (which can include injection site or infusion reactions, as well as allergic and anaphylactic reactions), the presence of an immune response may also substantially alter the pharmacokinetics of an administered mAb. The nature of the anti-antibody response, as well as the characteristics of the mAb itself (e.g. intact antibody vs. Fab fragment) can result in enhanced or diminished clearance of the therapeutic mAb. It has been suggested that the number of antigenic sites found on therapeutic proteins influences the PK when an immune response is triggered (Rehlaender and Cho 1998), but the location of the antigenic sites and the nature of the target (e.g. soluble vs. membrane-bound receptor) also likely impact whether the response will result in alterations in the clearance of the administered mAb.

It has been reported that one or two antigenic sites may lead to an increase in the half-life of a therapeutic protein following binding of an anti-antibody such that the half-life of the immune complex approaches that of endogenous IgG (e.g. approximately 21 days; Lobo, Hansen, and Balthasar 2004). This phenomenon was observed for the Fab fragment abciximab in primates, where the presence of an anti-Fab immune response led to an increase in the half-life of abciximab (Pendley, Schantz, and Wagner 2003). Although the mechanism for this increase is unknown, the Fc portion of the IgG from the anti-antibody may allow certain anti-antibody-Fab complexes to participate in FcRn-mediated protection from degradation (Busse et al. 2001). Although not related to an anti-antibody response, a similar effect was reported for omalizumab, an IgG1 monoclonal antibody targeted against IgE, a soluble target. Following administration of omalizumab, total IgE concentrations increased four- to five-fold (Casale et al. 1997), suggesting that the half-life increased from about 3 days for endogenous IgE, to approximately 15 days, which is closer to the reported half-life of IgG (Lobo, Hansen, and Balthasar 2004).

In contrast, where anti-antibodies are targeted against more than two sites, the clearance of the therapeutic protein often increases (Lobo, Hansen, and Balthasar 2004). However, the pharmacokinetic effect of an induced antibody response is very difficult to predict. Immune responses targeted against chimeric or humanized antibodies, such as basiliximab, palivizumab, and daclizumab, have resulted in minimal effects on the PK, while infliximab was reported to clear faster when anti-mAb antibodies were detected (Pendley, Schantz, and Wagner 2003). Although fewer antigenic sites would be expected for a fully human mAb, adalizumab also cleared faster in the presence of anti-antibodies [Humira® (adalizumab) Package Insert, Abbott Laboratories, 2003)].

For immune responses that result in enhanced clearance of the exogenous mAb, the rate and mechanism of clearance of the anti-antibody-mAb complex, as well as the extent of tissue distribution, may be mAb-specific and dependent on the size and nature of the immune complex (Rojas et al. 2005). The size of the complex likely depends on a number of factors, including the characteristics of the mAb/Fab, the concentration and stoichiometry of the immune complex components, and the number of immunogenic epitopes (Schifferli and Taylor 1989). In humans and non-human primates, soluble immune complexes may be cleared in the liver and spleen via a complement-mediated mechanism (Davies et al. 1995). Immune complexes that are sufficiently opsonized bind to complement receptor 1 (CR1, CD35) on erythrocytes via complement component C3b and are transported through the circulatory system to the liver. Here, immune complexes are phagocytized and stripped from erythrocytes by acceptor cells, which recognize immune complexes bound to CR1 via Fc receptors (most likely Fcγ receptor RI), and intact erythrocytes are returned to the circulation (Davies et al. 1990). In a complement-independent mechanism, immune complexes may also be cleared by macrophages via FcRn-mediated processes (Daeron 1997).

In one study, intentional administration of radiolabeled anti-infliximab antibodies following administration of infliximab resulted in an initial increase in the rate of clearance of infliximab in cynomolgus monkeys (test group), compared with administration of non-specific radiolabeled antibody (control) (Rojas et al. 2005). Gamma imaging indicated that a higher proportion of radioactivity was detected in the liver and kidneys of the test group. In addition, immunohistochemical analysis revealed that both infliximab and anti-infliximab antibodies were independently identified in these organs. However, erythrocytes and CR1 did not appear to play a prominent role in elimination of the induced immune complexes. Rather, these results are consistent with immune complex uptake by the mononuclear phagocytic system in the liver, followed by degradation and excretion of radiolabeled degradates into urine. It was also postulated that the immune complexes were potentially recognized by Fcγ receptors on Kupffer and liver endothelial cells, and subsequently internalized and enzymatically digested within lysosomes (Rojas et al. 2005).

In conclusion, antibodies generated from an immune response can impact the PK and PD of the administered mAb. Some product labels indicate that anti-mAb responses have led to neutralizing of the mAb affinity for its biological target, thus resulting in a loss of efficacy. Furthermore, immune responses can lead to alterations in the PK, which may also impact efficacy when anti-antibodies result in enhanced clearance of the administered mAb. The incidence of an immune response for a given exogenous mAb is difficult to predict, and it seems likely that all antibody therapeutics will possess some degree of immunogenicity. However, increasing the dose or dosing frequency may compensate for any detrimental impact of the PK or pharmacological efficacy. Advances in the identification of specific sequences or post-translational modifications associated with immune responses are likely to minimize the incidence of anti-antibody responses, and thus result in improved therapeutic agents.

19.2.2.4 Alteration of the Antibody-FcRn Interaction and the Effect on Pharmacokinetics Several groups have begun to experiment with IgG-FcRn interaction to influence the pharmacokinetic behavior of the mAb. The initial hypothesis was that increasing affinity of a mutant IgG for FcRn, relative to wild-type, would result in decreased systemic clearance and longer terminal half-life by shifting the equilibrium of mAb disposition towards the FcRn recycling pathway. Conversely, for therapeutic indications requiring rapid clearance of an mAb, an IgG could be engineered with less affinity for FcRn such that it is eliminated more rapidly. Kim et al. (1994) were the first to show that specific amino acid mutations in the Fc portion of mouse IgG, the region that binds to mouse FcRn, can influence the half-life and clearance of antibodies *in vivo*. Advances in computer modeling are being used to identify key residues in the IgG protein that interact with FcRn. Several groups are now performing site-directed mutagenesis studies of these amino acid residues with some success. Hinton et al. (2006) generated one such mutant IgG1 which had a 37-fold increased binding affinity for rhesus FcRn with an approximate 2.5-fold increase in serum AUC (area under the concentration curve) and half-life, relative to wild-type. The situation, however, may not be so straightforward, as studies from other groups have revealed that simply increasing affinity for FcRn may not be sufficient to reduce the clearance of an mAb since there is complex interplay between binding kinetics and pH influences which determine the protection from elimination afforded by the interaction with FcRn.

Dall'Acqua et al. (2002) conducted a series of *in vitro* and *in vivo* experiments with several human IgG1 mutants. The mutants had an increased affinity towards murine FcRn of more than five-fold at pH 6.0 and had increased binding at pH 7.4, relative to wild-type. The same binding properties were not evident when the mutants were studied with human FcRn since the variants which had an increased affinity (up to 10-fold) for human FcRn at pH 6.0 had poor binding at pH 7.4. This suggested a species-dependent binding interaction of the human IgG mutants to murine and human FcRn. Interestingly, when three of the IgG1 mutants with 14- to 19-fold relative increase in affinity for murine FcRn at pH 6.0 were administered as equal doses by intramuscular administration to BALB/c mice, they exhibited significantly lower serum concentration-time profiles than the wild-type IgG1. The lower AUC of the mutants, relative to the wild-type, could not be explained by proteolytic susceptibility in serum or whole blood, nor by excretion, nor by accumulation into different tissues. It was hypothesized that the advantage of having increased binding affinity for FcRn at pH 6.0 was negated by the analogous increase in interaction at physiologic pH by impeding the dissociation of the antibody from FcRn so that it could re-enter the circulation.

Based on these observations, it is clear that the binding affinity (K_D) of the antibody-FcRn interaction is insufficient to fully describe the effect of altered affinity on pharmacokinetics. As a result, other groups have studied, in more detail, the pH-dependent binding kinetics of the antibody-FcRn interaction. Datta-Mannan et al. (2007) studied a series of humanized anti-tumor necrosis factor IgG1 mutants with varying binding properties to human, murine, and cynomolgus monkey FcRn. The mutants showed substantial increase in affinity (15- to 197-fold) to human, cynomolgus monkey, and murine FcRn at pH 6.0, yet, had no direct binding at pH 7.4. Although the mutants showed no binding at pH 7.4, following formation of the antibody-murine FcRn complex at pH 6.0, there was a decrease in the proportion of the antibody released from the complex as the pH was increased. This phenomenon was not observed with either the human or cynomolgus monkey FcRn-antibody interaction and suggested that once the mutant antibody bound to murine FcRn at acidic pH *in vivo*, a certain population of protein-protein interactions was of sufficient affinity to withstand dissociation as pH increased. Following intravenous administration to cynomolgus monkeys, the mutant antibodies had similar pharmacokinetic profiles to the wild-type, suggesting that increased binding affinity for FcRn at pH 6.0 was insufficient to affect the systemic clearance. When administered intravenously to mice, however, the mutant antibodies showed drastically increased clearance, relative to wild-type. These experiments suggest that dissociation kinetics, and extent of dissociation upon raising pH, are important determinates for the *in vivo* pharmacokinetic profile of mutant IgGs.

Several mutant human-mouse chimeric IgGs and a mutant human IgG1 were characterized for their binding properties to mouse FcRn (Gurbaxani et al. 2006). BIAcore Surface Plasmon Resonance was used to study FcRn binding kinetics by immobilizing mouse FcRn to a biosensor chip. Based on the *in vitro* binding data, a new mathematical model of IgG-FcRn interaction called the dual bivalent analyte model (DBVA) was developed. This model assumes that there are two types of binding events to the immobilized mouse FcRn, one being high affinity and the other low affinity, and that both sides of the antibody constant region (Fc) may be involved in the binding. These mutant antibodies were administered to mice and their pharmacokinetic profiles were characterized. As noted in the previous studies mentioned above, the increased affinity for mouse FcRn at pH 6.0 was not associated with an increased serum half-life and there was no single kinetic parameter from the DBVA model that was predictive of the half-life *in vivo*. The mutant antibodies were shown to have low affinity for mouse FcRn at pH 7.4. Simulations using the DBVA model suggest that even low affinity of interaction of IgG for FcRn may result in a substantial population of IgG being bound to FcRn when normal physiologic blood levels of IgG (~60,000 nM) are considered. This low affinity interaction may be the primary determinant of sequestration inside cells.

Advances in transgenic animal models have allowed for interesting studies to be conducted in humanized FcRn mice. A recent study by Petkova et al. (2006) evaluated the pharmacokinetics of mutant human IgG1 antibodies in transgenic mice expressing human FcRn. The results indicated that the high affinity mutants had a modest but significant (up to 2.5-fold) increase in terminal half-life, relative to the wild-type antibody in transgenic mice. It was also observed that the level of expression of human FcRn was correlated to the degree of protection of IgG from elimination. In other words, terminal half-lives for the mutant antibodies were longer in a homozygote transgenic, relative to heterozygotes. These data provide promising results indicating that humanized mice may be a useful tool to discriminate the pharmacokinetics of human IgGs *in vivo* and should thus prove to be more useful than traditional rodent models for pharmacokinetic screening.

Changing the complex binding behavior of an engineered IgG to FcRn remains a realistic strategy to generate antibodies with desirable pharmacokinetic properties in humans. It is clear, however, that the interaction of IgG with FcRn is complex and that binding kinetics, and other factors such as glycosylation (Sinclair and Elliott 2005), need to be characterized in detail so that the mutant antibody will have the highest probability of possessing the desired pharmacokinetics *in vivo*. In this regard, computer aided modeling remains key to helping identify residues critical for the interaction. Furthermore, humanized FcRn animal models can provide an *in vivo* tool that will help select an IgG mutant with the best chance of having the desired pharmacokinetic profile in humans. For the purpose of pharmacokinetic screening, transgenic animal models appear to be more reliable than standard rodent models and may be ultimately less expensive than a colony of monkeys.

19.3 PRECLINICAL TO CLINICAL PHARMACOKINETIC COMPARISON

19.3.1 Preclinical to Clinical Comparisons Utilizing PK and PK-PD Approaches

The pharmacokinetics of keliximab and clenoliximab were evaluated in transgenic mice bearing human CD4 molecules (Sharma et al. 2000). Keliximab and clenoliximab both are monkey/human chimeric CD4 monoclonal antibodies of the IgG1 and IgG4 isotypes, respectively, and have been studied for treatment of autoimmune disorders. The pharmacokinetic model used for these studies can be described as a two-compartment model with saturable Michaelis–Menten-type elimination occurring from both the central and tissue compartments. The model assumes unidirectional flow of antibody from the central compartment to the tissue compartment. Incorporating saturable elimination events in both the central and tissue compartments is physiologically relevant as CD4 in both the circulation and tissues serves as a target for the mAbs. The concentration-time data for keliximab and clenoliximab in the transgenic mice were superimposable following single IV bolus injection of 5, 25, or 125 mg/kg indicating that the IgG isotype had no apparent effect on the pharmacokinetics. Accordingly, the concentration-time data for keliximab and clenoliximab were pooled and used to fit the pharmacokinetic model. The implication of these results is that target-mediated elimination (interaction with CD4) of the mAbs was the primary determinant of disposition, as opposed to interaction with FcRn which would presumably be different for each isotype. The results indicated that the estimated volume of distribution for the tissue compartment ($V_T = 25.6$ mL) was approximately 10-fold larger than the apparent volume of the central compartment ($V_C = 2.5$ mL), thus indicating that the mAbs had significant distribution to tissues containing CD4. Furthermore, the pharmacokinetics of clenoliximab were evaluated in rheumatoid arthritis patients following 0.05 to 15 mg/kg IV infusion using a model slightly different than what was used for the transgenic mice (Mould et al. 1999). The pharmacokinetic model used for the clinical study was a two-compartment model with saturable elimination from the central compartment only, and bidirectional flow of mAb from the central and tissue compartments. The K_m observed in patients (1290 ng/mL) was approximately four-fold lower than that observed in transgenic mice (5249 ng/mL). The difference in the fitted K_m values between mice and humans is curious since the mice expressed the human form of CD4. This difference may have been due to either the different modeling approaches used in the studies or due to a species difference in the location and/or extent of CD4 expression between the transgenic animals and humans.

Another example of a cross-species PK-PD comparison was published for the CD11a mAb efalizumab. Efalizumab is a humanized IgG1 antibody that has shown efficacy in treatment of psoriasis (Ng et al. 2005). Two separate pharmacokinetic models were evaluated to describe the plasma concentration-time profile in chimpanzees and humans (Bauer et al. 1999). The first approach (model A) was based on a classical two-compartment model with both first-order and Michaelis–Menten elimination from the central compartment. Again, this model was used to describe the apparent increase in clearance that was observed at low concentrations where elimination was due to interaction with the CD11a receptor and assumes that the total receptor level remains unchanged. The second model (model B) used a different approach in which the saturable clearance term was modified to consider the binding of the mAb to the CD11a receptor as a process that simultaneously eliminates both the mAb and the receptor. This model, termed the dynamic receptor-mediated clearance model (further described in Section 19.5.2.1), is theoretically more consistent with the actual dynamics of the *in vivo* disposition of both the mAb and its target.

The PK-PD of efalizumab were determined following a single intravenous administration to chimpanzees (0.5 to 10 mg/kg) and humans (0.03 to 10 mg/kg) (Bauer et al. 1999). Both model A and B provided reasonable fits to the experimental data, indicating that the available data could not be used to determine if the overall pool of CD11a receptors is decreased as a result of binding to efalizumab. The estimate for K_m from model A in both chimpanzees (0.116 μg/mL) and humans (0.0973 μg/mL) was similar to the affinity of efalizumab for CD11a as measured *in vitro*. There was some difference in the estimate for the maximum elimination rate (V_{max}) between chimpanzees (13.1 μg/kg/day) and humans (39.0 μg/kg/day), with humans having a larger value. At serum concentrations well above the K_m, the saturable component of clearance approaches zero and all the mAb elimination is attributed to the slow first-order component. As the serum concentration of efalizumab approaches the K_m, the clearance increases as the saturable elimination mechanism starts to play an increasingly important role. As with the other mAbs that show similar elimination behavior, the pharmacokinetics could not be attributed to the generation of an anti-efalizumab response in either chimpanzees or humans.

For the dynamic receptor-mediated clearance model, the saturable elimination term varies not only with mAb concentration but also with the target (CD11a) concentration. As such, the V_{max} term, designated V'_{max}, is normalized to the percent of circulating CD11a receptors at baseline. In general, the model parameters resulting from fitting the dynamic receptor-mediated clearance model did not differ from the parameter estimates obtained using a simple Michaelis–Menten-type model for saturable elimination. Overall, both pharmacokinetic models developed for efalizumab in chimpanzee were predictive of the pharmacokinetic behavior in humans. This is due, at least in part, to the similar binding potency of efalizumab in chimpanzees and humans and to a presumably similar pattern and extent of expression of the CD-11a in the two species.

19.3.2 Preclinical to Clinical Comparisons Utilizing Allometric Scaling

The pharmacokinetics of bevacizumab, a vascular endothelial growth factor mAb, has been studied in mouse, rat, and cynomolgus monkeys (Lin et al. 1999). Bevacizumab exhibited multiple-compartmental pharmacokinetics in preclinical species, consistent with other IgG antibodies and there was no evidence of saturable target-mediated clearance in monkeys or mice. There was, however, some evidence for dose-dependent pharmacokinetics in rats having a lower clearance and longer half-life following a 10 mg/kg intravenous dose relative to a 0.66 mg/kg dose. The resulting preclinical data was used to perform allometric scaling to predict human pharmacokinetics. Allometric scaling is a technique that uses animal *in vivo* data to predict the pharmacokinetics in humans (Mahmood 1999). It is based on a power function that relates a pharmacokinetic parameter to a physiologic parameter, usually body weight. Since there was dose-dependent pharmacokinetics in the rat, the values used for the allometric analysis were taken from the 10 mg/kg dose group since the saturable component of the clearance was negligible at this dose. Pharmacokinetic information from rats following a saturating dose of bevacizumab, along with the monkey and mouse data, was used to predict pharmacokinetics in humans. This approach seems reasonable since the clearance at a saturating dose approaches an asymptotic value that is not greatly affected by higher doses. In other words, at a saturating dose, the clearance is a constant value and approximates a first-order situation, making it possible to perform allometric analysis.

Using allometric scaling, the clearance of bevacizumab and terminal half-life in humans was predicted to be 4.3 mL/day/kg and 12 hours, respectively (Lin et al. 1999). Following intravenous infusion of bevacizumab to patients (0.3 to 10 mg/kg) the mean serum clearance was 3.8 ± 1.9 mL/day/kg and the mean terminal half-life was 21 hours (Gordon et al. 2001). The clearance of bevacizumab was well predicted from the allometric analysis of the preclinical data, whereas the terminal half-life was under-predicted by 1.7-fold. Interestingly, there were two patients who received a lower bevacizumab dose of 0.1 mg/kg who had a much higher clearance of approximately 16.5 mL/day/kg. This suggests the possibility of a saturable target-mediated elimination mechanism for bevacizumab at low doses in patients.

Allometric scaling has also been used to predict the clearance and volume of distribution of a CD4-IgG1 immunoadhesion molecule in man Mordenti et al. 1991). Similar to the situation with bevacizumab, rat, mouse, and cynomolgus monkey pharmacokinetics were used to fit the allometric model. The predicted clearance (2.6 mL/min) and volume of distribution (5.7 L) correlated well with the observed values of 2.6 mL/min and 6.3 L, respectively.

While the overall published data on the correlation of preclinical and clinical pharmacokinetics of monoclonal antibodies is lacking, there are some preliminary trends that are worthwhile noting. The available data suggests that allometric scaling may be a useful technique to predict the pharmacokinetics of a monoclonal antibody in humans from animal *in vivo* data in situations when there is no saturable elimination occurring. If saturable elimination is noted in the preclinical evaluation, allometric scaling should be used cautiously and evaluating the pharmacokinetics following saturating doses will need to be considered. In this case, the predicted human pharmacokinetics would probably reflect the disposition of the antibody following a saturating dose in humans and likely would not predict the pharmacokinetics at sub-saturating doses. More research needs to be published in this area, especially in the field of transgenic animal models, so that meaningful and robust conclusions can be made regarding the predictability of animal models for human pharmacokinetics.

19.4 NONCOMPARTMENTAL ANALYSIS

19.4.1 Noncompartmental Pharmacokinetic Analysis for Monoclonal Antibodies

Noncompartmental analysis has become a popular and useful tool for characterizing the pharmacokinetic properties of therapeutic agents. Using noncompartmental approaches, pharmacokinetic parameters (such as the clearance and apparent volume of distribution of a drug at steady state) are estimated by geometrically evaluating the area under the concentration vs. time curve (AUC) and area under the concentration $\times$ time vs. time, or moment, curve (AUMC). Although noncompartmental analysis has advantages for many drugs, it may lead to an inaccurate estimation of the volume of distribution. The volume of distribution (V) of a drug is the theoretical space into which drug distributes, as opposed to a real volume. It may be considered as a proportionality constant relating the amount of drug in the body at a given time (A) to the concentration in the plasma at the same time (C); thus $V = A/C$. For monoclonal antibodies that bind with high affinity to molecules within tissue sites, a substantial quantity of antibody may be bound to target antigens within tissues, and consequently a high volume of distribution would be expected. However, many reports of antibody pharmacokinetics cite that the V_{ss} is roughly equal to the volume of plasma. The source of the discrepancy lies in the use of noncompartmental analysis, which assumes that the elimination of drug is exclusively from the plasma or serum, and thus implicitly suggests that the site of drug elimination is in rapid equilibrium with the sampling compartment (Lobo, Hansen, and Balthasar 2004). Although this assumption may be valid for some antibodies, it is not valid for those with target-mediated disposition where antibody elimination occurs predominantly in tissue sites via receptor-mediated endocytosis, rather than from the systemic circulation (or central sampling compartment). In addition, noncompartmental analysis assumes first-order elimination, which may or may not be applicable to mAbs. For these reasons, the pharmacokinetic parameter values obtained from noncompartmental analysis may not be appropriate for further use in PK-PD modeling efforts. Therefore, although noncompartmental analysis is very useful for obtaining preliminary information on the PK of mAbs, it must be used with caution, and may not be appropriate if the ultimate goal is to define a predictive PK-PD relationship designed to represent the pharmacological effect of a drug.

19.5 PHARMACOKINETIC/PHARMACODYNAMIC RELATIONSHIPS

19.5.1 General Overview of Pharmacokinetic-Pharmacodynamic Relationships

The exposure-response relationship of a drug is characterized by the time course of the drug in the body (pharmacokinetics; PK) and the corresponding pharmacological effect of the drug (pharmacodynamics; PD). Understanding the relationship between a drug's concentration-time course and concentration-effect profile will help researchers design the most informative clinical studies and ultimately provide patients with dosing regimens that will have maximum efficacy. Furthermore, the integration and application of PK-PD concepts in drug discovery and development can help differentiate potential drug candidates, aid dose selection for proof-of-concept studies, and improve the design of clinical trials such that fewer or more efficient clinical studies are planned.

Monoclonal antibodies can exert pharmacological effects by blocking biological interactions via adhesion to key domains on ligands or their respective receptors, downregulating receptors in target tissues via receptor-mediated endocytosis, and/or eliciting effector functions. PK-PD concepts are equally applicable to monoclonal antibody therapeutics as they have been for traditional small molecule drugs. Monoclonal antibodies, however, pose unique challenges in the collection and interpretation of pharmacokinetic and pharmacodynamic data (e.g. target-mediated elimination and saturable clearance).

19.5.2 PK-PD Modeling Approaches for Monoclonal Antibodies

Many of the pharmacokinetic and pharmacodynamic principles that apply to small molecules also apply to monoclonal antibody therapeutics. However, there are a number of unique attributes of mAbs that require special considerations in order to appropriately apply these principles. As discussed in Section 19.2.2.3, the pharmacokinetic properties of mAbs are frequently governed by target-mediated disposition. The interaction of a therapeutic mAb with its target ligand serves as a pathway for elimination from the tissue interstitium. In situations where the number of target ligands is comparable to the amount of circulating mAbs, the pharmacokinetics and pharmacodynamics are interdependent processes. Consequently, simultaneous modeling of pharmacokinetic and pharmacodynamic data is frequently employed to derive exposure-response relationships. Additionally, the high target specificity of therapeutic mAbs often facilitates use of mechanistic PK-PD modeling approaches, although empirical relationships may be equally suitable under certain circumstances. While empirical models are mainly descriptive in nature, mechanism-based models have a structure that explicitly represents a phenomenon or process quantitatively through incorporation of first principles. The following section briefly describes the application of mechanism-based PK-PD modeling approaches for two mAbs: (1) efalizumab, a monoclonal antibody for the treatment of psoriasis targeted against a membrane-bound receptor on lymphocytes (CD11a), and (2) omalizumab, a monoclonal antibody for the treatment of asthma targeted against a soluble antigen (IgE).

19.5.2.1 Efalizumab Efalizumab is a recombinant humanized IgG1 κ monoclonal antibody that selectively binds to the alpha subunit of leukocyte function-associated antigen-1 (LFA-1, CD11a; Joshi et al. 2006; Gottlieb et al. 2002). LFA-1, which is expressed on activated T-lymphocytes, interacts with intracellular adhesion molecule 1 (ICAM-1) and facilitates the processes leading to the pathogenesis of psoriasis (Krueger 2002). Efalizumab exhibits dose-dependent, non-linear pharmacokinetics, which can be explained by its saturable binding to CD11a (Bauer et al. 1999; Joshi et al. 2006; Gottlieb et al. 2002). Following termination of dosing, the elimination phase of efalizumab was biphasic, with a slow phase followed by a rapid terminal elimination phase when drug levels fell below ~10 μg/mL. Previous studies demonstrated that administration of efalizumab resulted in a rapid reduction in CD11a expression on circulating lymphocytes, until efalizumab levels fell below 3 μg/mL. Subsequently, the drug was rapidly cleared from the circulation and CD11a expression returned to baseline within 7 to 10 days (Joshi et al. 2006).

A mechanism-based PK-PD model, which assumes that the clearance of efalizumab by CD11a receptors from the systemic circulation is a saturable process represented by a Michaelis–Menten function, has been used to describe the PK-PD relationship (Bauer et al. 1999; Wu et al. 2006). Absorption of efalizumab was assumed to be first-order following subcutaneous administration. A two-compartment model, with central and peripheral compartments, was used to characterize the pharmacokinetics. In this model, the clearance of mAb simultaneously results in dynamic changes in CD11a expression on T-lymphocytes, thus linking the PK and PD (Bauer et al. 1999; Wu et al. 2006). Figure 19.5 provides a schematic form of the PK-PD model.

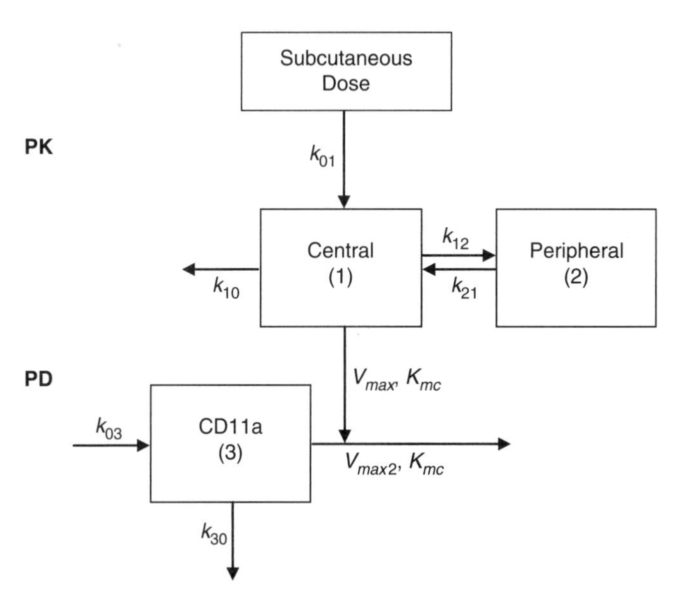

Figure 19.5 Schematic of a mechanism-based PK-PD model for efalizumab. A two-compartmental model characterizes the PK. Clearance of mAb from the central compartment (which is a saturable process represented in the model by a Michaelis–Menten expression) simultaneously results in dynamic changes in CD11a expression on T-lymphocytes. CD11a is also cleared via a saturable process initiated by its binding to efalizumab. (Adapted from R.J. Bauer et al. 1999. *J. Pharmacokinet. Biopharm.* 27(4):397–420.)

Mass balance principles can be used to derive the set of differential equations that describe the system:

$$\frac{dX_{SC}}{dt} = -k_{01} \cdot X_{SC} \tag{19.1}$$

$$\frac{dX_1}{dt} = Fk_{01} \cdot X_{SC} - k_{10} \cdot X_1 - k_{12} \cdot X_1 + k_{21} \cdot X_2 - \frac{V_{max1} \cdot X_1 \cdot \mathrm{CD11a}}{(K_{mc} + X_1)} \tag{19.2}$$

$$\frac{dX_2}{dt} = k_{12} \cdot X_1 - k_{21} \cdot X_2 \tag{19.3}$$

$$\frac{d\mathrm{CD11a}}{dt} = k_{03} - k_{30} \cdot CD11a - \frac{V_{max2} \cdot X_1 \cdot \mathrm{CD11a}}{(K_{mc} + X_1)} \tag{19.4}$$

where X_1, X_2, and X_{SC} are the amounts of efalizumab in plasma, and the peripheral and dosing compartments, respectively, F is the fraction that enters systemic circulation following subcutaneous administration (i.e., the bioavailability), k_{01} is the first-order absorption rate constant for subcutaneous dosing, k_{12} and k_{21} are the rates of distribution to and from the peripheral compartment, respectively, and k_{10} is the non-specific elimination rate constant of efalizumab from the plasma compartment. In addition, efalizumab is cleared by binding to CD11a receptor on T-lymphocytes. This saturable process is represented by the Michaelis–Menten expression

$$\frac{V_{max1} \cdot X_1 \cdot \mathrm{CD11a}}{(K_{mc} + X_1)}$$

where V_{max1} is the maximal rate of efalizumab elimination, and K_{mc} is the amount of efalizumab when this clearance is at half the maximal rate. Note that this expression varies with both efalizumab levels and CD11a expression.

Equation 19.4 describes the interaction between CD11a and efalizumab, where CD11a represents the total CD11a expression, k_{03} is the zero-order rate of CD11a synthesis, and k_{30} is the first-order elimination rate constant for non-specific degradation of CD11a. CD11a is also cleared via a saturable process initiated by its binding to efalizumab with its own V_{max} (or V_{max2}). The same K_m was used for the PK and PD processes, since defining a unique K_{m2} for CD11a elimination resulted in model over-specification and an inability for the model to converge in a prior adaptation (Bauer et al. 1999). Data were obtained using a competitive ELISA assay to quantify efalizumab levels in human serum, while a validated fluorescence activated cell sorter (FACS) assay was used to quantify CD11a expression on T-lymphocytes (Ng et al. 2005; Bauer et al. 1999).

The efalizumab PK-PD model was used to simulate CD11a expression vs. time, and suggested that IV doses of ~0.3 mg/kg would achieve maximal reduction in CD11a on T-lymphocytes (Joshi et al. 2006). This result correlated well with efficacy data, which demonstrated that 0.3 mg/kg/week resulted in improvements in patients with psoriasis. In particular, histological improvements in skin biopsies, including statistically significant decreases in epidermal thickness and dermal and epidermal T-lymphocytes at this dose were reported (Joshi et al. 2006).

19.5.2.2 Omalizumab Omalizumab is a recombinant humanized IgG1 κ monoclonal antibody that selectively binds to human immunoglobulin E (IgE). An allergic response is initiated when IgE bound to high-affinity FcεRI receptors expressed on basophils and mast cells is cross-linked by an allergen. This results in degranulation of the effector cells and release of inflammatory mediators such as histamine and leukotrienes (Hart 2001). Omalizumab interrupts the aforementioned allergic response by forming complexes with IgE, which prevents activation of the effector cells. A reduction in free circulating IgE also leads to downregulation of FcεRI (Owen 2002).

Omalizumab binds IgE in a second-order reversible reaction to form the omalizumab-IgE complex. All three components in the system (omalizumab, IgE, and the complex) are subject to first-order

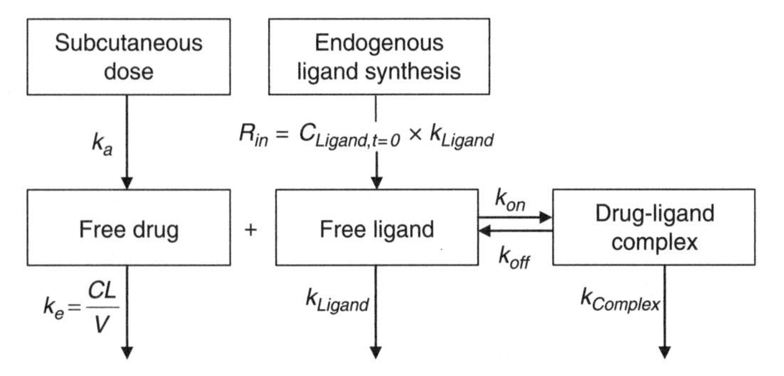

Figure 19.6 Schematic of a mechanism-based PK-PD model for omalizumab. Omalizumab binds to IgE in a second-order reversible reaction to form the omalizumab-IgE complex. All three components in the system (omalizumab, IgE, and the complex) are subject to first-order clearance by the reticuloendothelial system, either by the IgG (omalizumab) or the IgE pathways. (Adapted from G.M. Meno-Tetang and P.J. Lowe. 2005. *Basic Clin. Pharmacol. Toxicol.* 96(3):182–192.)

clearance by the reticuloendothelial system, either by the IgG (omalizumab) or the IgE pathways. The complex, in theory, can be cleared by both pathways, although the presence of omalizumab bound to IgE may cause some steric hindrance for the binding of IgE to its clearance receptor (Meno-Tetang and Lowe 2005). A schematic representing these processes is provided in Figure 19.6.

The model was implemented according to the following set of differential equations:

$$\frac{dA_{Dose}}{dt} = -k_a \cdot A_{Dose} \tag{19.5}$$

$$\frac{dC_{FreeDrug}}{dt} = \frac{k_a \cdot A_{Dose}}{V_{FreeDrug} \cdot M_W} - \frac{CL_{Drug} \cdot C_{FreeDrug}}{V_{FreeDrug}} - k_{on} \cdot C_{FreeDrug} \cdot C_{Ligand} + k_{off} \cdot C_{Ligand} \tag{19.6}$$

$$\frac{dC_{Ligand}}{dt} = k_{Ligand} \cdot C_{Ligand,t=0} - k_{Ligand} \cdot C_{Lignad} - k_{on} \cdot C_{FreeDrug} \cdot C_{Ligand} + k_{off} \cdot C_{Complex} \tag{19.7}$$

$$\frac{dC_{Complex}}{dt} = -k_{Complex} \cdot C_{Complex} + k_{on} \cdot C_{FreeDrug} \cdot C_{Ligand} - k_{off} \cdot C_{Complex} \tag{19.8}$$

where A_{Dose} is the amount of the subcutaneous dose remaining at the site of injection and k_a is the first-order absorption rate constant. $C_{FreeDrug}$, $V_{FreeDrug}$, and CL_{Drug} are the concentration of unbound omalizumab, its volume of distribution, and its clearance, respectively. k_{on} and k_{off} are the on and off rate constants for omalizumab binding to its target ligand, IgE, whose concentration is specified as C_{Ligand}. k_{Ligand} and $k_{Complex}$ are the first-order rate constants for elimination of the ligand (IgE) and the complex (IgG-IgE), respectively, while $C_{Complex}$ designates the concentration of the complex. The conversion from weight-based drug input in Equation 19.5 to molar concentrations for the remaining differential equations was made by division of the volume of distribution for the free drug ($V_{FreeDrug}$) and the molecular weight (M_W) in terms of Equation 19.6. Omalizumab and IgE were assayed by ELISA as previously described (Casale et al. 1997).

The model adequately described the observed PK data in patients and the PK and PD parameters were estimated precisely, with low standard errors. In addition to characterizing the PK of omalizumab, this model may be used to characterize the properties of the ligand, in this case endogenous IgE. The half-life of IgE, determined from the value for k_{Ligand}, was estimated to be 1.76 days, which is in good agreement with the value previously reported in the literature of 2.7 days (Waldmann et al. 1976). Furthermore, the estimated synthesis rate of IgE, 3.91 $\mu g/kg/day$, is also on the same order as that previously reported (3.77 $\mu g/kg/day$; Waldmann et al. 1976).

In conclusion, target-mediated phenomena represent an important consideration in characterizing the PK and PD for mAbs. The models presented in this section are of general applicability, and variations may be implemented to characterize the exposure-response relationships for other mAbs with target–mediated disposition. The application of mechanistic pharmacodynamic models to represent complex biological processes and the application of PK-PD modeling techniques have contributed to the overall understanding of the clinical activity of monoclonal antibodies. The application of PK-PD modeling is a powerful tool that will continue to facilitate the drug development process for biologics. Increased publication and awareness of PK-PD modeling strategies for mAbs will be needed to further develop the understanding and capabilities for development of mAbs as efficacious agents.

19.6 DRUG-DRUG INTERACTIONS

Unlike small molecules, the literature on drug-drug interactions (DDI) for therapeutic proteins and monoclonal antibodies is somewhat limited. This can be attributed partly to the fact that major metabolic enzymes, such as the cytochrome P450 and UDP-glucuronsyltransferase subfamilies, are not typically thought to be involved in the metabolism of monoclonal antibodies as metabolism for mAbs is primarily mediated by catabolic processes (described in Section 19.2.2.3). In addition, the underlying mechanisms for mAb-mediated DDIs remain unknown for most cases. Recently, a couple of reviews have attempted to compile available DDI information on therapeutic proteins and monoclonal antibodies (Seitz and Zhou 2007; Mahmood and Green 2007). A vast majority of the currently marketed mAbs target the disease areas of oncology and immunology. As a result, most of the pharmacokinetic DDI data on mAbs is in the presence of concomitantly administered chemotherapeutic drugs (e.g., paclitaxel, cisplatin, irinotecan, docetaxel, and carboplatin) or immunosuppressive agents (e.g., methotrexate, cyclosporine, azathioprine, and mycophenolate mofetil). Though few formal DDI studies have been conducted with monoclonal antibodies, useful information in this area can be gleaned from pharmacokinetic and pharmacodynamic data obtained from cross-study comparison trials. Examples of some mAbs where altered PK or PD was observed in the presence of concomitant administration of another drug are discussed below.

19.6.1 Drug-Drug Interactions Affecting Pharmacokinetics Following Administration of Monoclonal Antibodies with Concomitant Medications

Muromonab-CD3, a murine, IgG2 anti-CD3 mAb for treatment of acute rejection in renal transplant patients when coadministered with cyclosporine, an immunosuppressive agent, led to a significant increase in trough concentrations of cyclosporine on day 5 post-transplant (Vasquez and Pollak 1997). This observation led to lowering the cyclosporine dose for subsequent post-transplant days. Studies with another immunosuppressive mAb, basiliximab [a mouse-human chimeric IgG1 monoclonal antibody targeted against the interleukin-2 receptor (CD25) for prophylaxis against renal transplant rejection] also indicated the need for dose adjustment of cyclosporine when coadministered to pediatric patients (Seitz and Zhou 2007; Strehlau et al. 2000). A clinical study of basiliximab with another immunosuppressive agent, tacrolimus, indicated the need for close monitoring of tacrolimus levels post-transplantation. In this particular study, a 63 percent increase in tacrolimus trough levels was noted on day 3 post-transplantation in the basiliximab group as compared to the control group (Sifontis, Benedetti, and Vasquez 2002). These observations of altered trough levels of cyclosporine and tacrolimus were hypothesized to be mediated via cytokine-induced changes in the metabolism of these drugs at the level of cytochrome P450 expression. In a separate study, the serum clearance of basiliximab was lowered when coadministered with triple immunosuppressive therapies consisting of cyclosporine and corticosteroids in the presence of either mycophenolate mofetil or azathioprine (Seitz and Zhou 2007; Kovarik et al. 2001). The serum clearance of basiliximab was reduced by an average of 51 and 22 percent in the presence of mycophenolate mofetil and azathioprine, respectively, as compared to treatment of basiliximab with dual therapy in the absence of these drugs (Kovarik et al. 2001).

It has been observed that the PK of infliximab [an IgG1 mouse-human chimeric anti-tumor necrosis factor-alpha (TNF-α) mAb for the treatment of Crohn's disease, ulcerative colitis, rheumatoid and psoriatic arthritis, plaque psoriasis, and ankylosing spondylitis] can be affected by the formation of human anti-chimeric antibodies (HACA), thus limiting its efficacy. However, additive or synergistic efficacy was observed in the presence of methotrexate (a traditional disease-modifying antirheumatic drug) with a concomitant decrease in HACA formation (Maini et al. 1998). The results of this combination trial indicated that methotrexate may reduce serum clearance of infliximab. The full clinical significance of the role of HACA formation, however, remains unclear in terms of the efficacy of infliximab (Mori 2007). Studies with adalimumab, a fully human IgG1 anti-TNF-α mAb for the treatment of rheumatoid arthritis, have indicated that concomitant administration of methotrexate reduces the clearance of adalimumab by 29 to 44 percent (Seitz and Zhou 2007; Nestorov 2005).

19.6.2 Drug-Drug Interactions Affecting Pharmacodynamics Following Administration of Monoclonal Antibodies with Concomitant Medications

Kereiakes et al. (1996) studied the pharmacodynamic effects of xemilofiban (an oral nonpeptide GP IIb/IIIa antagonist) in the presence and absence of abciximab (a mouse-human chimeric IgG1 Fab and GP IIb/IIIa antagonist for percutaneous coronary intervention therapy). The results from this study indicated that in patients pretreated with abciximab, the magnitude and duration of inhibition of the *ex vivo* platelet aggregation response was enhanced to xemilofiban, thus suggesting that extended GP IIb/IIIa antagonism may confer longer-term clinical benefit through secondary prevention (Mahmood and Green 2007; Kereiakes et al. 1996). In another study, the efficacy of cetuximab (an IgG1 mouse-human chimeric epidermal growth factor receptor Mab) was compared in the presence and absence of irinotecan in patients with metastatic colorectal cancer that was refractory to treatment with irinotecan (Cunningham et al. 2004). The results indicated a beneficial effect of the combination therapy as compared to monotherapy with cetuximab alone. The median time for tumor progression was 4.1 vs. 1.5 months and the median survival time was 8.6 vs. 6.9 months for the combination therapy group vs. the monotherapy group, respectively. The results of this study thus suggest that EGFR inhibition by cetuximab may help overcome irinotecan resistance by abrogating the mechanisms involved in tumor resistance to irinotecan, such as suppression of apoptosis, enhanced DNA repair, and decreased intracellular levels due to active drug efflux (Mahmood and Green 2007; Cunningham et al. 2004; Xu and Villalona-Calero 2002).

In summary, even though DDI studies involving therapeutic proteins or mAbs have not been conducted as widely as for small molecules, useful information has been obtained when such studies were conducted with macromolecules. The conduct of drug interaction studies with macromolecules will become progressively more important due to the increasing desire to administer macromolecules with conventional small molecule drugs in order to improve efficacy for various disease conditions.

ACKNOWLEDGMENTS

The authors would like to thank Dr. Ronda Rippley (Clinical Pharmacokinetics and Pharmacodynamics, Merck Research Laboratories) and Dr. Jiunn Lin (Preclinical Drug Metabolism and Pharmacokinetics, Merck Research Laboratories) for the helpful suggestions during editing of this chapter. We also would like to thank Ms. Jill Williams (Visual Communications, Merck Research Laboratories) for her assistance in the preparation of this manuscript.

REFERENCES

Bauer, R.J., R.L. Dedrick, M.L. White, M.J. Murray, and M.R. Garovoy. 1999. Population pharmacokinetics and pharmacodynamics of the anti-CD11a antibody hu1124 in human subjects with psoriasis. *J. Pharmacokinet. Biopharm.* 27(4):397–420.

Baumann, A. 2006. Early development of therapeutic biologics: Pharmacokinetics. *Curr. Drug Metab.* 7(1): 15–21.

Baxter, L.T., H. Zhu, D.G. Mackensen, and R.K. Jain. 1994. Physiologically based pharmacokinetic model for specific and nonspecific monoclonal antibodies and fragments in normal tissues and human tumor xenografts in nude mice. *Cancer Res.* 54(6):1517–1528.

Brambell, F.W., W.A. Hemmings, and I.G. Morris. 1964. A theoretical model of gamma-globulin catabolism. *Nature* 203:1352–1354.

Busse, W., J. Corren, B.Q. Lanier, et al. 2001. Omalizumab, anti-IgE recombinant humanized monoclonal antibody, for the treatment of severe allergic asthma. *J. Allergy Clin. Immunol.* 108(2):184–190.

Casale, T.B., I.L. Bernstein, W.W. Busse, et al. 1997. Use of an anti-IgE humanized monoclonal antibody in ragweed-induced allergic rhinitis. *J. Allergy Clin. Immunol.* 100(1):110–121.

Cunningham, D., Y. Humblet, S. Siena, et al. 2004. Cetuximab monotherapy and cetuximab plus irinotecan in irinotecan-refractory metastatic colorectal cancer. *N. Engl. J. Med.* 351(4):337–345.

Daeron, M. 1997. Fc receptor biology. *Annu. Rev. Immunol.* 15:203–234.

Dall'Acqua, W.F., R.M. Woods, E.S. Ward, et al. 2002. Increasing the affinity of a human IgG1 for the neonatal Fc receptor: bBiological consequences. *J. Immunol.* 169(9):5171–5180.

Danilov, S.M., V.D. Gavrilyuk, F.E. Franke, et al. 2001. Lung uptake of antibodies to endothelial antigens: Key determinants of vascular immunotargeting. *Am. J. Physiol. Lung Cell. Mol. Physiol.* 280(6):L1335–L1347.

Datta-Mannan, A., D.R. Witcher, Y. Tang, J. Watkins, W. Jiang, and V.J. Wroblewski. 2007. Humanized IgG1 variants with differential binding properties to the neonatal Fc receptor: Relationship to pharmacokinetics in mice and primates. *Drug Metab. Dispos.* 35(1):86–94.

Davies, K.A., P.T. Chapman, P.J. Norsworthy, et al. 1995. Clearance pathways of soluble immune complexes in the pig. Insights into the adaptive nature of antigen clearance in humans. *J. Immunol.* 155(12):5760–5768.

Davies, K.A., V. Hird, S. Stewart, et al. 1990. A study of in vivo immune complex formation and clearance in man. *J. Immunol.* 144(12):4613–4620.

Delacroix, D.L., H.J. Hodgson, A. McPherson, C. Dive, and J.P. Vaerman. 1982. Selective transport of polymeric immunoglobulin A in bile. Quantitative relationships of monomeric and polymeric immunoglobulin A, immunoglobulin M, and other proteins in serum, bile, and saliva. *J. Clin. Invest.* 70(2):230–241.

Fahy, J.V., D.W. Cockcroft, L.P. Boulet, et al. 1999. Effect of aerosolized anti-IgE (E25) on airway responses to inhaled allergen in asthmatic subjects. *Am. J. Respir. Crit. Care Med.* 160(3):1023–1027.

Fox, J.A., T.E. Hotaling, C. Struble, J. Ruppel, D.J. Bates, and M.B. Schoenhoff. 1996. Tissue distribution and complex formation with IgE of an anti-IgE antibody after intravenous administration in cynomolgus monkeys. *J. Pharmacol. Exp. Ther.* 279(2):1000–1008.

Ghetie, V., J.G. Hubbard, J.K. Kim, M.F. Tsen, Y. Lee, and E.S. Ward. 1996. Abnormally short serum half-lives of IgG in beta 2-microglobulin-deficient mice. *Eur. J. Immunol.* 26(3):690–696.

Ghetie, V., and E.S. Ward. 2002. Transcytosis and catabolism of antibody. *Immunol. Res.* 25(2):97–113.

Gillies, S.D., K.M. Lo, C. Burger, Y. Lan, T. Dahl, and W.K. Wong. 2002. Improved circulating half-life and efficacy of an antibody-interleukin 2 immunocytokine based on reduced intracellular proteolysis. *Clin. Cancer Res.* 8(1):210–216.

Gordon, M.S., K. Margolin, M. Talpaz, et al. 2001. Phase I safety and pharmacokinetic study of recombinant human anti-vascular endothelial growth factor in patients with advanced cancer. *J. Clin. Oncol.* 19(3):843–850.

Gottlieb, A.B., J.G. Krueger, K. Wittkowski, R. Dedrick, P.A. Walicke, and M. Garovoy. 2002. Psoriasis as a model for T-cell-mediated disease: Immunobiologic and clinical effects of treatment with multiple doses of efalizumab, an anti-CD11a antibody. *Arch. Dermatol.* 138(5):591–600.

Guarino, A., R.B. Canani, S. Russo, et al. 1994. Oral immunoglobulins for treatment of acute rotaviral gastroenteritis. *Pediatrics* 93(1):12–16.

Gurbaxani, B., L.L. Dela Cruz, K. Chintalacharuvu, and S.L. Morrison. 2006. Analysis of a family of antibodies with different half-lives in mice fails to find a correlation between affinity for FcRn and serum half-life. *Mol. Immunol.* 43(9):1462–1473.

Hart, P.H. 2001. Regulation of the inflammatory response in asthma by mast cell products. *Immunol. Cell Biol.* 79(2):149–153.

Hinton, P.R., J.M. Xiong, M.G. Johlfs, M.T. Tang, S. Keller, and N. Tsurushita. 2006. An engineered human IgG1 antibody with longer serum half-life. *J. Immunol.* 176(1):346–356.

Hwang, W.Y., and J. Foote. 2005. Immunogenicity of engineered antibodies. *Methods* 36(1):3–10.

Israel, E.J., D.F. Wilsker, K.C. Hayes, D. Schoenfeld, and N.E. Simister. 1996. Increased clearance of IgG in mice that lack beta 2-microglobulin: Possible protective role of FcRn. *Immunology* 89(4):573–578.

Joshi, A., R. Bauer, P. Kuebler, et al. 2006. An overview of the pharmacokinetics and pharmacodynamics of efalizumab: A monoclonal antibody approved for use in psoriasis. *J. Clin. Pharmacol.* 46(1):10–20.

Kereiakes, D.J., J.P. Runyon, N.S. Kleiman, et al. 1996. Differential dose-response to oral xemilofiban after antecedent intravenous abciximab. Administration for complex coronary intervention. *Circulation* 94(5):906–910.

Kim, J.K., M.F. Tsen, V. Ghetie, and E.S. Ward. 1994. Identifying amino acid residues that influence plasma clearance of murine IgG1 fragments by site-directed mutagenesis. *Eur. J. Immunol.* 24(3):542–548.

Kovarik, J.M., M.D. Pescovitz, H.W. Sollinger, et al. 2001. Differential influence of azathioprine and mycophenolate mofetil on the disposition of basiliximab in renal transplant patients. *Clin. Transplant.* 15(2):123–130.

Krueger, J.G. 2002. The immunologic basis for the treatment of psoriasis with new biologic agents. *J. Am. Acad. Dermatol.* 46(1):1–23; quiz 23–26.

Kuus-Reichel, K., L.S. Grauer, L.M. Karavodin, C. Knott, M. Krusemeier, and N.E. Kay. 1994. Will immunogenicity limit the use, efficacy, and future development of therapeutic monoclonal antibodies? *Clin. Diagn. Lab. Immunol.* 1(4):365–372.

Leslie, R.G. 1982. Macrophage interactions with antibodies and soluble immune complexes. *Immunobiology* 161(3–4):322–333.

Lin, Y.S., C. Nguyen, J.L. Mendoza, et al. 1999. Preclinical pharmacokinetics, interspecies scaling, and tissue distribution of a humanized monoclonal antibody against vascular endothelial growth factor. *J. Pharmacol. Exp. Ther.* 288(1):371–378.

Lobo, E.D., R.J. Hansen, and J.P. Balthasar. 2004. Antibody pharmacokinetics and pharmacodynamics. *J. Pharm. Sci.* 93(11):2645–2668.

Losonsky, G.A., J.P. Johnson, J.A. Winkelstein, and R.H. Yolken. 1985. Oral administration of human serum immunoglobulin in immunodeficient patients with viral gastroenteritis. A pharmacokinetic and functional analysis. *J. Clin. Invest.* 76(6):2362–2367.

Mahmood, I. 1999. Allometric issues in drug development. *J. Pharm. Sci.* 88(11):1101–1106.

Mahmood, I., and M.D. Green. 2007. Drug interaction studies of therapeutic proteins or monoclonal antibodies. *J. Clin. Pharmacol.* 47(12):1540–1554.

Maini, R.N., F.C. Breedveld, J.R. Kalden, et al. 1998. Therapeutic efficacy of multiple intravenous infusions of anti-tumor necrosis factor alpha monoclonal antibody combined with low-dose weekly methotrexate in rheumatoid arthritis. *Arthritis Rheum.* 41(9):1552–1563.

Meier, W., A. Gill, M. Rogge, et al. 1995. Immunomodulation by LFA3TIP, an LFA-3/IgG1 fusion protein: Cell line dependent glycosylation effects on pharmacokinetics and pharmacodynamic markers. *Therap. Immunol.* 2(3):159–171.

Meno-Tetang, G.M., and P.J. Lowe. 2005. On the prediction of the human response: A recycled mechanistic pharmacokinetic/pharmacodynamic approach. *Basic Clin. Pharmacol. Toxicol.* 96(3):182–192.

Molthoff, C.F., H.M. Pinedo, H.M. Schluper, H.W. Nijman, and E. Boven. 1992. Comparison of the pharmacokinetics, biodistribution and dosimetry of monoclonal antibodies OC125, OV-TL 3, and 139H2 as IgG and F(ab')2 fragments in experimental ovarian cancer. *Br. J. Cancer* 65(5):677–683.

Mordenti, J., S.A. Chen, J.A. Moore, B.L. Ferraiolo, and J.D. Green. 1991. Interspecies scaling of clearance and volume of distribution data for five therapeutic proteins. *Pharm. Res.* 8(11):1351–1359.

Mori, S. 2007. A relationship between pharmacokinetics (PK) and the efficacy of infliximab for patients with rheumatoid arthritis: Characterization of infliximab-resistant cases and PK-based modified therapy. *Mod. Rheumatol.* 17(2):83–91.

Mould, D.R., C.B. Davis, E.A. Minthorn, et al. 1999. A population pharmacokinetic-pharmacodynamic analysis of single doses of clenoliximab in patients with rheumatoid arthritis. *Clin. Pharmacol. Ther.* 66(3):246–257.

Mould, D.R., and K.R. Sweeney. 2007. The pharmacokinetics and pharmacodynamics of monoclonal antibodies: Mechanistic modeling applied to drug development. *Curr. Opin. Drug. Discov. Develop.* 10(1):84–96.

Nestorov, I. 2005. Clinical pharmacokinetics of TNF antagonists: How do they differ? *Semin. Arthritis Rheum.* 34(5 Suppl 1):12–18.

Ng, C.M., A. Joshi, R.L. Dedrick, M.R. Garovoy, and R.J. Bauer. 2005. Pharmacokinetic-pharmacodynamic-efficacy analysis of efalizumab in patients with moderate to severe psoriasis. *Pharm. Res.* 22(7):1088–1100.

Owen, C.E. 2002. Anti-immunoglobulin E therapy for asthma. *Pulm. Pharmacol. Ther.* 15(5):417–424.

Pendley, C., A. Schantz, and C. Wagner. 2003. Immunogenicity of therapeutic monoclonal antibodies. *Curr. Opin. Mol. Ther.* 5(2):172–179.

Petkova, S.B., S. Akilesh, T.J. Sproule, et al. 2006. Enhanced half-life of genetically engineered human IgG1 antibodies in a humanized FcRn mouse model: Potential application in humorally mediated autoimmune disease. *Int. Immunol.* 18(12):1759–1769.

Rehlaender, B.N., and M.J. Cho. 1998. Antibodies as carrier proteins. *Pharm. Res.* 15(11):1652–1656.

Rojas, J.R., R.P. Taylor, M.R. Cunningham, et al. 2005. Formation, distribution, and elimination of infliximab and anti-infliximab immune complexes in cynomolgus monkeys. *J. Pharmacol. Exp. Ther.* 313(2):578–585.

Roopenian, D.C., and S. Akilesh. 2007. FcRn: The neonatal Fc receptor comes of age. *Nature Rev. Immunol.* 7(9):715–725.

Scheinfeld, N. 2003. Adalimumab (HUMIRA): A review. *J. Drugs Dermatol.* 2(4):375–377.

Schifferli, J.A., and R.P. Taylor. 1989. Physiological and pathological aspects of circulating immune complexes. *Kidney Int.* 35(4):993–1003.

Seitz, K., and H. Zhou. 2007. Pharmacokinetic drug-drug interaction potentials for therapeutic monoclonal antibodies: Reality check. *J. Clin. Pharmacol.* 47(9):1104–1118.

Sharma, A., C.B. Davis, L.A. Tobia, et al. 2000. Comparative pharmacodynamics of keliximab and clenoliximab in transgenic mice bearing human CD4. *J. Pharmacol. Exp. Ther.* 293(1):33–41.

Sifontis, N.M., E. Benedetti, and E.M. Vasquez. 2002. Clinically significant drug interaction between basiliximab and tacrolimus in renal transplant recipients. *Transplant Proc.* 34(5):1730–1732.

Sinclair, A.M., and S. Elliott. 2005. Glycoengineering: The effect of glycosylation on the properties of therapeutic proteins. *J. Pharm. Sci.* 94(8):1626–1635.

Strehlau, J., L. Pape, G. Offner, B. Nashan, and J.H. Ehrich. 2000. Interleukin-2 receptor antibody-induced alterations of ciclosporin dose requirements in paediatric transplant recipients. *Lancet* 356(9238):1327–1328.

Tabrizi, M.A., C.M. Tseng, and L.K. Roskos. 2006. Elimination mechanisms of therapeutic monoclonal antibodies. *Drug Discov. Today* 11(1–2):81–88.

Tang, L., A.M. Persky, G. Hochhaus, and B. Meibohm. 2004. Pharmacokinetic aspects of biotechnology products. *J. Pharm. Sci.* 93(9):2184–2204.

Ternant, D., and G. Paintaud. 2005. Pharmacokinetics and concentration-effect relationships of therapeutic monoclonal antibodies and fusion proteins. *Expert Opin. Biol. Ther.* 5 (Suppl 1):S37–S47.

Vasquez, E.M., and R. Pollak. 1997. OKT3 therapy increases cyclosporine blood levels. *Clin. Transplant.* 11(1):38–41.

Waldmann, T.A., A. Iio, M. Ogawa, O.R. McIntyre, and W. Strober. 1976. The metabolism of IgE. Studies in normal individuals and in a patient with IgE myeloma. *J. Immunol.* 117(4):1139–1144.

Wu, B., A. Joshi, S. Ren, and C. Ng. 2006. The application of mechanism-based PK/PD modeling in pharmacodynamic-based dose selection of muM17, a surrogate monoclonal antibody for efalizumab. *J. Pharm. Sci.* 95(6):1258–1268.

Xu, Y., and M.A. Villalona-Calero. 2002. Irinotecan: Mechanisms of tumor resistance and novel strategies for modulating its activity. *Ann. Oncol.* 13(12):1841–1851.

CHAPTER 20

Biodistribution and Imaging

TOVE OLAFSEN and ANNA M. WU

ABSTRACT

Molecular imaging using positron emission tomography (PET) is becoming a valuable tool for preclinical assessment of drugs before further development, leading to rapid evaluation of new tracers with a decreased workload. The majority of clinical PET imaging is currently performed with [^{18}F]-FDG, but due to its limitations there is ongoing investigation and development of additional radiopharmaceuticals. Antibodies are increasingly being recognized as a powerful class of molecular imaging probes that can be genetically modified for optimal *in vivo* imaging of surface markers (immunoPET).

In cancer diagnostics and therapy, immunoPET can potentially play an important role at many points along the cancer care spectrum, such as determining the extent of disease, and stratifying who will benefit from antibody therapy and respond to treatment. Overall, the broad implementation of PET imaging will bolster basic and preclinical research, enabling a translational tool for future drug discovery and clinical application.

Therapeutic Monoclonal Antibodies: From Bench to Clinic. Edited by Zhiqiang An

20.1 INTRODUCTION

The science of drug action (pharmacodynamics, PD) and the dynamics by which a drug is absorbed, distributed, metabolized, and eliminated (pharmacokinetics, PK) are essential to understand in drug discovery, since the objective of drug treatment is to produce therapeutic benefit while minimizing side effects. Thus, the precise relationship between PK and PD in the sense of how much of the drug will reach the target and how it acts is equally important in the development of any biotherapeutics, including antibodies. The use of monoclonal antibodies (mAbs) as targeted biotherapeutics is a rapidly expanding field as biotechnology and pharmaceutical companies are starting to see the advantages of the high affinity and specificity offered by mAbs. In addition, mAbs have endogenous effector functions such as antibody–dependent, cell-mediated cytotoxicity (ADCC) and complement-mediated cell lysis (CMC), and can be used as carriers of effector molecules such as radionuclides, enzymes, and toxins that will exert their therapeutic effect at the target site. In order to understand whether a particular antibody localizes to its target and can provide a therapeutic effect *in vivo*, it is useful to evaluate its biodistribution properties in an animal model that represents the particular disease under study. Two approaches can be used for evaluating new therapeutic compounds *in vivo*. These are conventional biodistribution studies and molecular imaging techniques using radiolabeled mAbs (immunoPET).

The purpose of biodistribution studies in small animal models is to track where compounds of interest travel in living subjects. In conventional biodistribution studies the compound is radiolabeled and injected intravenously into a group of animals (typically mice or rats). At certain time intervals, smaller groups (four to five) of the animals are sacrificed. The organs of interest, usually blood, liver, spleen, kidney, muscle, lung, gastrointestinal tract, bone, and tumor (if cancer is the disease under study) are harvested and the radioactive uptake in each organ is determined and expressed as a percentage of injected dose per gram (percent ID/g) tissue. A dynamic view (time-activity curve) of how the compound moves through the animal is obtained by plotting the radioactive uptake against time. The disadvantages of conventional biodistribution studies, however, are the workload and invasiveness, that is, number of animals that need to be sacrificed in order to generate reproducible data and good statistics.

Molecular imaging is defined as "non-invasive, quantitative and repetitive imaging of targeted macromolecules and biological processes in living organisms" (Herschman 2003, p. 606). There is a growing interest in using noninvasive molecular imaging for screening new drugs with regards to their distribution and effectiveness, in addition to studying cellular events and cell trafficking. The molecular imaging modalities that are utilized in the clinic include positron emission tomography (PET), single-photon emission computed tomography (SPECT), and to a certain extent magnetic resonance imaging (MRI); with anatomical information provided by MRI and computed tomography (CT). These imaging modalities are also available for small-animal imaging as well as optical imaging (fluorescence and bioluminescence) (reviewed in Massoud and Gambhir 2003). In preclinical stages, molecular imaging stands to become a high-content assay because data are generated repeatedly and noninvasively in the same animal over time. Thus, a time-activity curve of how a compound moves through the animal over time can be provided with a lower workload and fewer animals.

PET provides a highly sensitive as well as a quantitative molecular imaging technique that uses positron emitting radionuclides coupled to specific ligands. The metabolic tracer [^{18}F]-FDG (18-fluoro-2-deoxy-D-glucose) was approved by the U.S. Food and Drug Administration (FDA) in 1997, and is the most widely used tracer in oncology today. The principle of this tracer is based on the higher metabolic rate associated with neoplasia relative to normal cells. In cells, [^{18}F]-FDG becomes trapped when it is phosphorylated by hexokinase. Fast growing tumor cells that metabolize glucose faster will therefore accumulate more [^{18}F]-FDG and be differentiated from benign tissues. However, malignancies with low metabolic rates such as prostate cancers and low-grade lymphomas are poorly imaged with [^{18}F]-FDG. Also, the nonspecific nature of [^{18}F]-FDG makes it unable to distinguish tumor cells from normal immune cells present in infectious and inflammatory regions. In addition to its imaging abilities, PET enables quantitation of radioactive uptake in the tissues in three dimensions. The standardized uptake value (SUV) is a useful semiquantitative index for [^{18}F]-FDG accumulation in tissue (Sugawara et al. 1999), obtained by placing a region of interest (ROI) over the lesion and dividing the value (in μCi per cm^3) by the injected dose (in μCi) divided by the patient's body weight (in grams) (Rohren, Turkington, and Coleman 2004). Thus, the amount of radioactive uptake of [^{18}F]-FDG in a lesion gives a measurement that in itself can be used to predict the likelihood of malignancy.

20.1.1 Antibody-Based Imaging Agents

The discovery of novel cell surface targets has boosted the design and development of targeted pharmaceuticals. Of these, mAbs are the most rapidly expanding group of molecules being investigated for targeted therapy of several different diseases, with almost 30 approved by the FDA. Although the majority of these are approved for therapy, a few have also been developed for diagnostic imaging by SPECT. These are capromab pendetide (ProstaScint™, Cytogen, Princeton, NJ) for imaging prostate cancers, sulesomab (LeukoScan®, Immunomedics Europe, Darmstadt, Germany) for imaging inflammation and infection, and the no longer available agents satumomab pendetide (OncoScint™, Cytogen) for imaging metastatic disease associated with ovarian and colorectal cancers, arcitumomab (CEA-Scan™, Immunomedics, Morris Plains, NJ) for imaging colorectal cancers, and nofetumomab merpentan (Verluma™, Boehringer Ingelheim, Ingelheim, Germany) for imaging small cell lung cancer. In addition, a couple of agents have been withdrawn from the market due to serious side effects in some patients: imciromab pentetate (MyoScint™, Centocor, Leiden, The Netherlands) for imaging myocardial infarction and fanolesomab (NeutroSpec®, Palatin Technologies, Cranbury, NJ) for imaging of appendicitis. These agents represent the first generation of targeted diagnostic pharmaceuticals consisting of intact murine mAbs or murine Fab fragments radiolabeled with the γ-emitting radionuclides ^{111}In or ^{99m}Tc. As mentioned, only ProstaScint and LeukoScan remain in the clinic as the others have either been set aside by [^{18}F]-FDG PET or suspended from the market. Although the specificity of these mAbs/Fab fragments was excellent, the sensitivity and image resolution was poor because of low target to background ratios. Thus, the overall clinical impact of these early imaging compounds has not been impressive and there is still room for improvement.

Even though SPECT images are informative, the spatial resolution and the sensitivity are much lower than that of PET images. Since PET offers significant advantages over SPECT and many PET radionuclides are now commercially available, there has been a resurgence of interest in developing non-FDG-based ligands in order to overcome the limitations encountered with [^{18}F]-FDG. The use of mAbs as PET imaging probes is appealing due to their specificity and extensive knowledge gained since the introduction of hybridoma technology in 1975 (Kohler and Milstein 1975). Since then, means for redesigning mAbs and optimizing their characteristics without compromising their specificity for the target antigen have been provided by recombinant protein engineering technology. Today, reduction of immunogenicity by producing chimeric (mouse/human) and humanized mAbs is routine, and fully human mAbs derived from display libraries or from immunized transgenic mice carrying human immunoglobulin genes are commonly produced. Pharmacokinetics can be modified using engineered antibody fragments of smaller sizes that clear faster from the circulation and penetrate tumors better than intact mAbs (Fig. 20.1). Modifications of binding to the neonatal Fc receptor (FcRn) by domain deletion or site-specific mutations, generation of antibody-fusion proteins using human serum albumin (HSA), as well as PEGylation and changing the isoelectric point (pI) of an antibody have been investigated as approaches for controlling PK and clearance.

20.1.2 Advantages of immunoPET

The notion that mAbs could be used in PET (immunoPET) arose in the late 1980s and early 1990s. Intact mAbs, $F(ab')_2$ and Fab fragments radiolabeled with positron emitting nuclides such as ^{18}F, ^{64}Cu, ^{124}I, and ^{68}Ga were investigated by several groups (Anderson et al. 1992; Garg, Garg, and Zalutsky 1991; Larson et al. 1992; Otsuka et al. 1991). In addition to specific targeting,

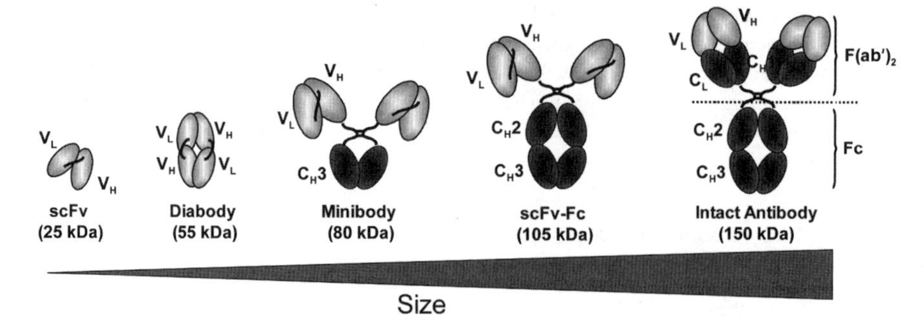

Figure 20.1 Schematic drawings of an intact antibody and engineered antibody fragments that can be derived from it. Single chain Fv (scFv) is shown with a linker between the variable domains. Molecular weights are indicated in parentheses. V_L = variable light; V_H = variable heavy, C_L = constant light; C_H = constant heavy.

immunoPET also enables quantification of radioactivity accumulation in the tumor. This can facilitate a personalized approach to identifying who may benefit most from FDA-approved mAb-based therapy. For instance, only 20 to 30 percent of patients with breast cancer overexpress HER2, the target for trastuzumab (Herceptin™, Genentech, San Fransisco, CA) and biopsies of tumors are required for confirmation of HER2 expression. However, the expression of HER2 in the primary tumor and metastatic lesions may differ over the course of the disease while the patient undergoes chemo- and/or hormonal therapy (Gancberg et al. 2002; Gong, Booser, and Sneige 2005; Rasbridge et al. 1994; Zidan et al. 2005). Using immunoPET to identify patients who will benefit from trastuzumab treatment would be less invasive than taking multiple or repetitive biopsies. The most serious complication in breast cancer patients treated with trastuzumab is congestive heart failure, especially when combined with anthracyclines (Behr, Behe, and Wormann 2001; Cobleigh et al. 1999; Eisenhauer 2001; Slamon et al. 2001). In a recent study, scintigraphy demonstrated myocardial overexpression of HER2 in 50 percent of the patients following anthracycline treatment (de Korte et al. 2007). Thus, imaging HER2 for evaluation of receptor expression and therapeutic efficacy of trastuzumab, as well as predicting cardiac toxicity, seems worthwhile for this patient group. Additional patient groups that may also benefit from immunoPET are those suffering from urological malignancies such as prostate and renal cancers (Larson and Schoder 2008), and hematological malignancies of low grade.

The potential quantification of molecular interactions in immunoPET is an attractive utility for assessing target expression of all lesions and normal tissues prior to therapy. Combined treatment modalities can then be monitored and adapted for optimal treatment efficacy. Until now, quantitative pretherapy imaging has mostly been applied for assessment of dosimetry prior to radioimmunotherapy (RIT) of intact mAbs where bone marrow toxicity is the dose-limiting factor (Verel, Visser, and van Dongen 2005). Thus, quantitative immunoPET imaging can become a valuable tool in the development and application of mAbs, especially against novel targets. In addition, fewer patients should be required to provide information for oncologists on optimal dosimetry, toxicity in normal organs, and variations between patients in PK and tumor targeting.

20.1.3 Radionuclides Used in immunoPET and Effect on Image Contrast

Several positron emitters for immunoPET are currently under investigation. These can be grouped according to their physical half-lives ($t_{1/2}$). Short-lived positron emitters are ^{68}Ga ($t_{1/2} = 1.13$ h) and ^{18}F ($t_{1/2} = 1.83$ h). Intermediate-lived positron emitters are ^{64}Cu ($t_{1/2} = 12.7$ h), ^{86}Y ($t_{1/2} = 14.7$ h), and ^{76}Br ($t_{1/2} = 16.2$ h). Long-lived positron emitters are ^{89}Zr ($t_{1/2} = 78.4$ h) and ^{124}I ($t_{1/2} = 100.3$ h). Positron-emitting radionuclides produce high-energy gamma rays through positron emission decay. A positron emitted from the nucleus will annihilate with a nearby electron to produce two 511 keV gamma rays that are emitted in opposite directions at an approximately 180 degree angle to each other. These gamma rays are detected by scintillation crystals arranged in a ring in the PET scanner. The technique depends on simultaneous or coincident detection (within a few nanoseconds) of the pair of gamma rays, making it possible to localize the gamma rays source along a straight line of response (LOR) in space. The unique ability of PET to measure radioactive concentrations in three-dimensional tissue volumes is a major advantage. We and others have validated the agreement between tumor/organ uptake by invasive biodistribution studies and the non-invasive PET scans (Cai et al. 2006, 2007; Olafsen et al. 2005; Wu et al. 2000; Zhang et al. 2006). PET can be combined with other imaging modalities such as MRI and CT for simultaneous registration of biological function and anatomy. Multimodality imaging in the same subject within the setting of a single examination is rapidly becoming commonplace with the development of SPECT-CT and PET-CT scanners.

Radiolabeling of mAbs can be direct or indirect. In direct labeling methods, oxidants such as iodogen, Chloramine-T (CAT), and hydrogen peroxide or peroxidase catalyzed reactions to make halogen ions positively charged are used to enable electrophilic substitution in the *ortho* position relative to the hydroxyl functional group of phenyl in tyrosine residues. The PET radionuclides that can be directly coupled to mAbs are ^{76}Br and ^{124}I. Problems associated with direct methods are loss of immunoreactivity following randomized incorporation of the radiolabel; unwanted oxidation of susceptible groups such as histidine, methionine, and tryptophan; and loss of label *in vivo* (dehalogenation and metabolism). The major metabolic product of conventionally radioiodinated proteins is iodinated tyrosines (mainly monoiodotyrosine) that are rapidly deiodinated and quickly released from the cells and excreted via the kidneys into the urine (Geissler et al. 1992; Xu et al. 1997). Thus, internalization of mAb after binding to the target cells will reduce the contrast in images. On the contrary, mAbs that are not internalized will produce high contrast images due to rapid reduction of activity in nonspecific organs such as kidneys and liver (Fig. 20.2). Moreover, the overall image can be enhanced further when thyroid and stomach uptakes of iodine are blocked (Figs. 20.2 and 20.3) as described (Sundaresan et al. 2003). This was indeed demonstrated with ^{124}I-labeled anti-CEA T84.66 diabody and a slightly larger fragment, the minibody (scFv-C_H3 dimer, 80 kDa) (Sundaresan et al. 2003). The harsh, oxidative conditions used in direct labeling procedures can be avoided by indirect labeling procedures using a precursor molecule such as Bolton-Hunter reagent. The labeled precursor is then coupled to the lysine residues of the mAb under slightly basic conditions.

Indirect labeling procedures are also used for labeling radiometals to mAbs. In these procedures, chelates such as *p*-isothiocyanatobenzyl-diethylenetriamine-pentaacetic acid (MX-DTPA), 1,4,7,10-tetraazacyclododecane-1,4,7,10-tetraacetic acid (DOTA), and 1,4,8,11-tetraazacyclododecane-1,4,8,11-tetraacetic acid (TETA) are initially coupled to the ε amino-group on lysine residues. The mAb carrying the chelate is then incubated with the radiometal under mild conditions. The advantage of this procedure is that the same mAb can be labeled with different radiometals. Moreover, the amino acids derivatives from radiometal-chelated mAbs are trapped in the lysosomes, leading to increased accumulation of radioactivity in the tissue. Thus, in addition to radiouptake in tumor, nonspecific accumulation in normal tissues such as the routes of excretion can be evaluated by PET (Figs. 20.2 and 20.3). The trapping of radioactivity in the tissue is particularly advantageous when evaluating tumor targeting of internalizing mAbs. For example, an anti-HER2 scFv-Fc fragment (105 kDa) modified to clear quickly from the circulation was generated from trastuzumab (Herceptin™; Olafsen et al. 2005). When the ^{64}Cu-DOTA scFv-Fc was evaluated by microPET in mice bearing MCF7/HER2 xenografts, high activity was seen in the tumor at 21 h. However, high accumulation of the radioactivity was also seen in the liver, which is the primary excretion organ for this fragment (Fig. 20.2). Thus, radiometal-labeled mAbs are not suitable for imaging lesions located in the liver, kidney, or gut areas. The mice were sacrificed and the percent injected dose per gram (percent ID/g) in the tumor was calculated to be 12.2 ($\pm$1.2; Olafsen et al. 2005). When the same fragment was labeled with ^{124}I, the activity in the positive tumor was only 1.5 ($\pm$0.2) percent ID/g at 20 h (unpublished) reflecting rapid dehalogenation and metabolism following internalization (Fig. 20.2).

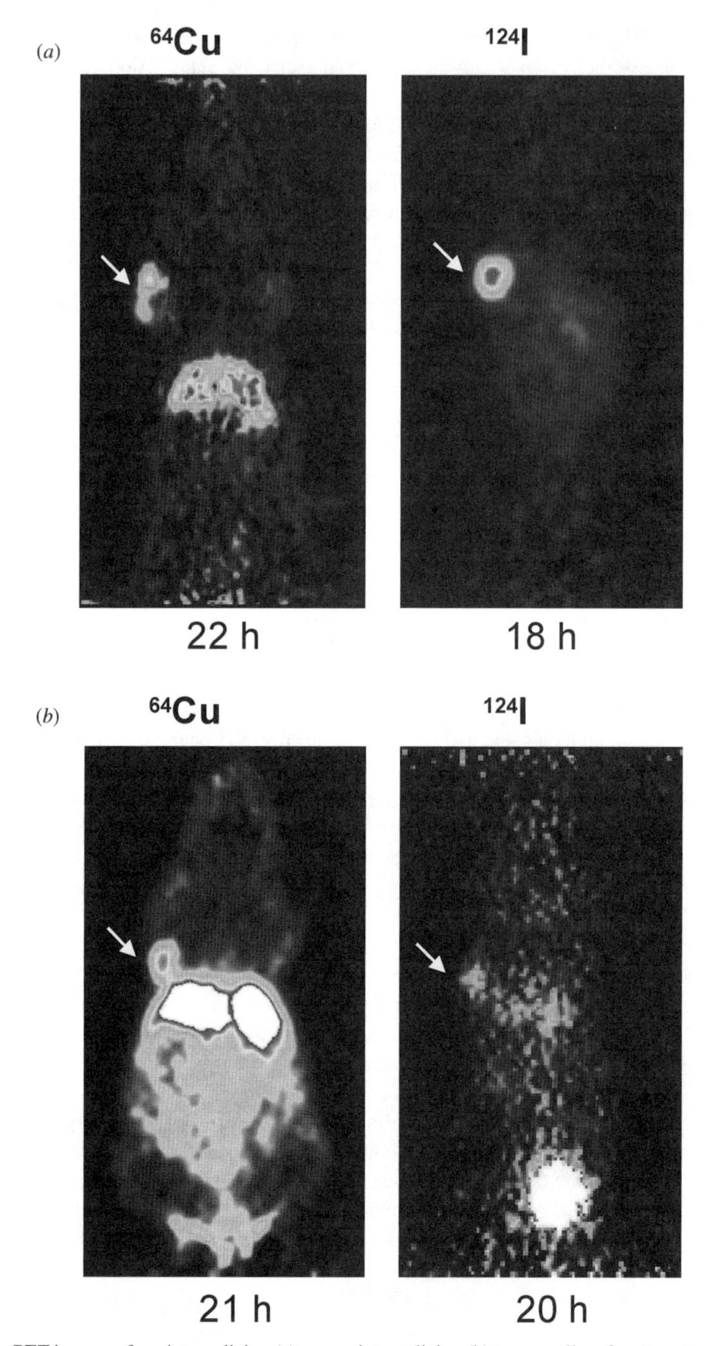

Figure 20.2 PET images of noninternalizing (a) versus internalizing (b) tumor cell surface targets with ^{64}Cu- and ^{124}I-labeled scFv-Fc fragments. (a) Coronal images of mice injected with anti-CEA scFv-Fc H310A/H435Q bearing LS174T xenografts (arrow). (b) Coronal images of mice injected with anti-HER2 scFv-Fc H310A/H435Q bearing MCF7/HER2 xenografts (arrow). The radiolabel is shown above each image and the time in hours after administration of the tracer is shown below. High liver activity is seen in mice injected with ^{64}Cu-labeled tracer in both tumor models, which is more or less absent in the mice injected with ^{124}I-labeled tracer. A higher contrast image is seen in the noninternalizing CEA system, which is further enhanced by blocking for thyroid and stomach uptakes. (See color insert.)

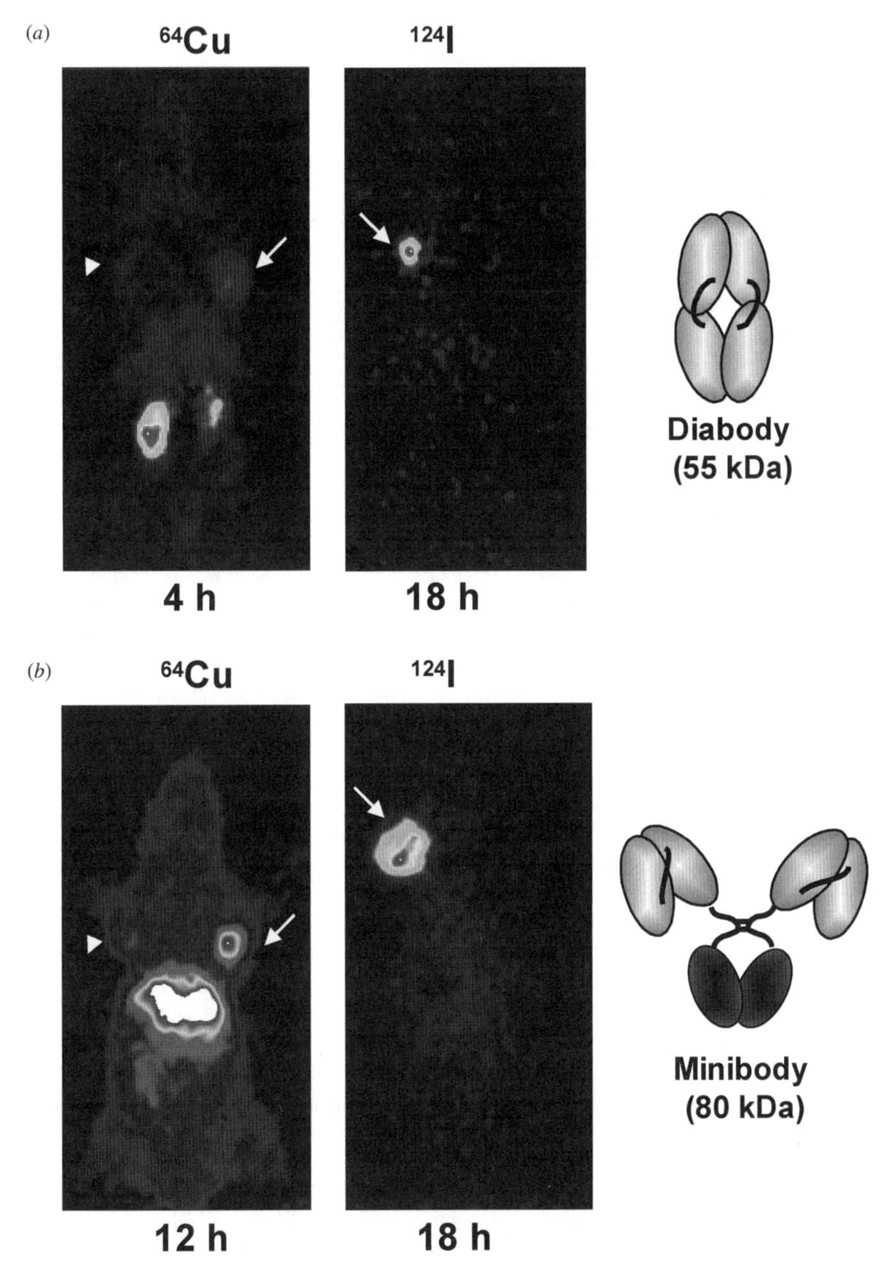

Figure 20.3 PET images of tumor-bearing mice injected with diabody (a) and minibody (b) radiolabeled with ^{64}Cu and ^{124}I. Positive tumor is indicated by an arrow, whereas the negative tumor is indicated by an arrowhead. The radiolabel is shown above each image and the time in hours after administration of the tracer is shown below. The primary excretion routes for the diabody and minibody are kidney and liver, respectively, which can be seen in the mice injected with ^{64}Cu-labeled fragments. In the mice injected with ^{124}I-labeled fragment, only tumor is visible at 18 hour with the diabody due to its rapid blood clearance, whereas the more background (blood pool) activity is seen in the mice injected with the minibody. However, both fragments produce excellent, high contrast images in mice (stomach and thyroid uptakes were blocked). (See color insert.)

20.2 IMPROVING ANTIBODY PHARMACOKINETICS (PK) FOR immunoPET

20.2.1 Clinical immunoPET Studies: Intact Antibodies

Of the current mAbs approved by FDA for therapy, almost all are intact immunoglobulins of 150 kDa in size. When contemplating using mAbs as imaging agents their blood clearance properties need to be considered. Ideally, an imaging agent should localize quickly to the target at high concentration and clear rapidly from the circulation. Intact antibodies exhibit long residence times (one to three weeks) in the circulation, which allows ample time for high accumulation in the tumor. However, days are required for the activity levels to drop sufficiently to allow good target to background ratios, making intact Abs unsuitable as radiolabeled imaging agents. Despite their prolonged circulation time, clinical immunoPET studies have all been performed with intact mAbs.

In an early Phase I/II study, 36 patients with suspected advanced primary or metastatic colorectal cancer were imaged with a ^{64}Cu-labeled anti-colorectal carcinoma 1A3 mAb, using bromoacetamido-benzyl-TETA as chelator (Philpott et al. 1995). Early imaging times, at 4 to 36 hours after injection, were chosen due to the short half-life of ^{64}Cu. In 29 patients, all 17 primary and recurrent sites were clearly visualized and 59 percent (23/39) of the metastatic sites were detected. In addition, 11 new occult tumor sites (<2 cm in diameter) were detected that were not detected by CT or MRI. Detection of metastases in lung and liver, the most important sites of this disease, was difficult because of the blood activity and accumulation of ^{64}Cu chelate complexes in the liver. There were no complications, but significant levels of human anti-mouse antibody (HAMA) titers were found in 28 percent of the 29 patients tested 1 to 12 months later. The high accumulation of Cu^{2+} in liver is facilitated by dissociation from the chelator followed by binding to copper-binding proteins such as superoxide dismutase (SOD), ceruloplasmin, and copper transporter 1 receptor (CTr 1). This has prompted the search for alternative ^{64}Cu chelates that produce less liver uptake (Li et al. 2008; Wadas et al. 2007), but these remain to be evaluated in the clinic.

The use of positron emitters with longer physical half-lives, such as ^{124}I and ^{89}Zr, is particularly suitable for immunoPET when used in combination with intact mAbs (Verel et al. 2003). In 1991, a study was done of nine patients with ductal breast carcinoma. They were imaged with ^{124}I-labeled mAbs for quantitative measurement of tumor uptakes (Wilson et al. 1991). Subsequently, one patient with neuroblastoma was scanned with ^{124}I-labeled 3F9 mAb for estimating tumor dosimetry during treatment planning for RIT (Larson et al. 1992). These early studies exemplified the potential of using ^{124}I-labeled mAbs in PET. With current commercial availability of ^{124}I, interest in ^{124}I-labeled mAbs has been renewed. In a Phase I trial, ^{124}I-HuMV833 (a humanized version of the mouse anti-VEGF MV833 mAb) was used to measure the antibody distribution and clearance from tissues in patients with a variety of progressive tumors (Jayson et al. 2002). Heterogeneous antibody distribution and clearance between and within patients and between and within individual tumors were observed, which may reflect the difference in available targets or variability in delivery of the mAb. ImmunoPET may also be beneficial for patients suffering from renal cancers since the specificity and sensitivity of [^{18}F]-FDG-PET seem less effective than CT (Aide et al. 2003; Kang et al. 2004). Thus, in a recent clinical pilot study, immunoPET was used for profiling renal cancer (Divgi et al. 2007). Here, the chimeric (mouse/human) G250 (cG250), which targets carbonic anhydrase-IX (CA-IX) that is over-expressed in clear-cell renal carcinoma, was labeled with ^{124}I and evaluated in 26 patients with renal masses. The ^{124}I-cG250 mAb was able to identify 15 of 16 clear-cell carcinomas accurately, and was negative for all non-clear, less aggressive cell renal masses. This study illustrates how molecular imaging can be used instead of biopsy for identification of aggressive tumors to aid in stratification and treatment decisions of patients with renal masses (personalized medicine). However, the usefulness of knowing this with regard to management of the patients has been questioned (Powles and Ell 2007).

Clinical translation of using ^{89}Zr in immunoPET has recently been achieved (Borjesson et al. 2006; Perk et al. 2006; Zalutsky 2006). In a pilot study, the clinical performance of ^{89}Zr-Zevalin was evaluated by whole-body PET imaging in one patient with indolent CD20^{+} non-Hodgkin's lymphoma (NHL) at 2 and 96 h after administration (Perk et al. 2006). At 2 h mainly blood pool activity was observed, whereas at 96 h clear uptake in all known tumor lesions previously identified by [^{18}F]-FDG-PET was seen. This study suggests that ^{89}Zr-Zevalin can be implemented to identify NHL patients who are suitable for high-dose ^{90}Y-Zevalin RIT. A more extensive PET imaging study was carried out with the chimeric (mouse/human) anti-CD44v6 U36 mAb (Borjesson et al. 2006). In this study, cU36 mAb was radiolabeled with ^{89}Zr and administered to 20 patients with head and neck squamous cell carcinoma (HNSCC) scheduled to undergo neck dissection with or without resection of the primary tumor. All patients were also examined by CT and/or MRI. ImmunoPET was performed up to six days after injection of ^{89}Zr-cU36 mAb. All 17 primary tumors as well as lymph node metastases in 18 of 25 positive neck regions were detected. It was concluded that the sensitivity and accuracy of immunoPET was at least as good as CT/MRI with optimal tumor uptakes at later imaging times.

20.2.2 Engineered Antibody Fragments

20.2.2.1 Preclinical Studies One strategy for improving and optimizing the targeting and clearance properties of anti-tumor antibodies is to produce smaller antibody fragments (Fig. 20.1). When the molecular size of the antibody is reduced from a bivalent F(ab$'$)$_2$ (110 kDa) to a monovalent Fab (55 kDa) fragment, there is faster clearance from the circulation. The early imaging studies with ^{18}F-, ^{64}Cu-^{124}I-labeled Fab and F(ab$'$)$_2$ demonstrated their feasibility as PET tracers (Anderson et al. 1992; Garg, Garg, and Zalutsky 1991; Page et al. 1994; Westera et al. 1991). Protein engineering techniques have enabled the formation of a variety of different sized antibody fragments (Fig. 20.1) (Wu and Senter 2005). For example, the small monovalent single chain (sc) Fv (25 kDa) fragment clears rapidly from the blood ($t_{1/2\beta} = 0.5$ to 2 h), but suffer from being unable to accumulate significant radioactivity in the tumor as a result. Therefore, scFv fragments are used as building blocks for fusion proteins and multivalent antibody fragments. Cross-paired scFv dimers (diabody; 55 kDa) are spontaneously formed when the linker between the variable domains is shortened. The slightly longer residence time in the blood by the diabodies ($t_{1/2\beta} = 3$ to 7 h) and their increased avidity due to their bivalency, account for the increase in tumor targeting and retention. A variety of protein domains has been used in order to increase mass and promote multivalency of scFv fragments. For example, the immunoglobulin C$_H$3 domain and Fc region (C$_H$2-C$_H$3 domains) have been used to make minibodies (scFv-C$_H$3 dimers, 80 kDa) and scFv-Fc fragments (105 kDa), respectively. The overall advantage of rapidly clearing antibody fragments is that higher tumors to blood ratios are obtained. However, at the same time less of the injected product reaches the tumor. Still, excellent tumor uptakes and high tumors to blood ratios have been achieved with intermediate-sized antibody fragments in animal models (Fig. 20.3).

The ability of ^{123}I-anti-CEA T84.66 diabody and minibody to rapidly localize to the tumor were initially evaluated by SPECT in athymic mice bearing LS174T colon cancer xenografts (Hu et al. 1996; Wu et al. 1999). The larger minibody fragment persists longer than diabodies in serum ($t_{1/2\beta} = 6$ to 11 h), allowing higher activity accumulation in tumors (Hu et al. 1996; Wu and Yazaki 2000; Yazaki and Wu 2003).

A preclinical immunoPET study with the anti-CEA minibody fragment was performed with ^{64}Cu, using DOTA as chelator (Wu et al. 2000). In this study, tumors could be detected readily at 2 to 24 hours post-injection. However, significant nonspecific uptake was seen in the kidney and liver region, hampering detection of lesion in this region. A similar uptake in the liver region was observed when a DOTA-conjugated minibody against anti-HER2 was radiolabeled with ^{64}Cu (Olafsen et al. 2005). In addition, much higher kidney activity was observed with this fragment. Immunohistochemical staining showed that localization to the kidneys could be attributed to the presence of a cross-reacting antigen.

The T84.66 diabody and minibody were also radiolabeled with ^{124}I and evaluated by PET in mice bearing LS174T xenografts (Sundaresan et al. 2003). Both diabody and minibody demonstrated excellent uptake in the tumors and minimal activity in the normal tissues, enhancing the overall images. The HER2 tyrosine kinase receptor has been targeted successfully by the C6.5 diabody (Adams et al. 1998). Excellent tumor uptake was achieved when the C6.5 diabody was radiolabeled with ^{124}I and evaluated by PET in SCID mice bearing HER2-positive human ovarian carcinoma (SK-OV-3) xenografts (Gonzalez Trotter et al. 2004). However, in this study the thyroid and stomach were not blocked, resulting in significant uptake of iodine at 24 and 48 h post-injection. High contrast images have also been obtained with ^{124}I-anti-CD20 minibodies (Olafsen et al. 2006).

Recently, an anti-angiogenesis minibody variant (scFv-$C_H4_{\varepsilon\text{-}S2}$) referred to as L19-SIP that targets the extra domain B (ED-B) of fibronectin (Borsi et al. 2002) was labeled with ^{76}Br and evaluated in tumor-bearing mice bearing solid F9 tumors (Rossin et al. 2007). The tumors were clearly visible from 5 to 46 h after injection. However, the long retention of the radioactivity in the blood and very slow renal excretion resulted in low target to nontarget ratios that were explained to be partially due to debromination. Thus, the prolonged clearance observed with the radiobromide from blood (Lovqvist et al. 1997) may render ^{76}Br less suitable for immunoPET with fast clearing engineered antibody fragments.

20.2.2.2 Clinical Studies Engineered antibody fragments have progressed into clinical evaluation by SPECT. Tumor targeting of the anti-CEA ^{123}I-MFE-23 scFv ($\sim$27 kDa) was evaluated in 10 patients, who were imaged at 1, 4, and 22 h after administration (Begent et al. 1996). Rapid blood clearance through the kidneys with a β phase half-life of 5.32 h was observed for this fragment. The highest median uptake in tumor was seen at 1 h (2.4 percent ID/kg) and the tumor to blood ratios were 1.5 at 4 h and 5.6 at 22 h. The kidneys retained activity at levels similar to those of the tumors. This study demonstrated the applicability of using an scFv as a tumor-targeting agent for imaging. The MFE-23-his scFv was later evaluated for radioimmunoguided surgery (RIGS) in a Phase I clinical study, involving 35 patients scheduled for surgery (Mayer et al. 2000). In this study, the ^{125}I-MFE-23-his scFv was reported to clear from the blood with a $t_{1/2\beta} = 10.95$ h. The scFv showed localization to 84 percent of the primary colorectal cancer and metastases when compared with histology and could be detected at 24 to 96 h after administration. In another study, an anti-TAG-72 ^{123}I-scFv fragment, also targeting colorectal carcinoma, was evaluated in five patients with metastatic lesions in liver who were scheduled for surgery (Larson et al. 1997). The scFv exhibited rapid clearance from the blood ($t_{1/2\beta} =$ 10.5 h; $n = 4$) and imaged both primary and metastatic lesions in all five patients at four to six hours. Although image quality was suboptimal, the study demonstrated that early, same-day imaging of both primary and metastatic tumors was feasible with this antibody fragment.

A dimeric L19 scFv fragment [L19$(scFv)_2$] of about 57 kDa in size (Borsi et al. 2002) was evaluated for tumor targeting of primary and metastatic lesions in 20 patients with lung, colorectal, or brain cancer (Santimaria et al. 2003). Rapid clearance ($t_{1/2\beta} = 5.2$ h) via the kidneys was observed, and selective localization to both primary tumors and metastases as early as 6 h after administration of ^{123}I-L19 $(scFv)_2$ in 16 patients was seen. Of the four negative scans, one was a patient with low grade astrocytoma that does not express ED-B fibronectin. Since ED-B fibronectin is a marker of angiogenesis it was indicated that the tumors in the other three patients were possibly in a quiescent phase. The study demonstrated that L19$(scFv)_2$ could potentially be suitable for diagnosis and delivery of therapeutic agents. The ^{123}I- L19$(scFv)_2$ was recently evaluated in five patients with HNSCC in a Phase I/II clinical study (Birchler et al. 2007). Successful imaging comparable to [^{18}F]-FDG PET was achieved in four of five patients. The fifth patient appeared not to have a viable tumor. The authors conclude that the ^{123}I-L19 $(scFv)_2$ is less suitable for diagnostic imaging of HNSCC due to free iodine in thyroid, salivary, parotid, and submandibular glands, but may be more suitable for selective delivery of therapeutic compounds.

In another clinical pilot study, the T84.66 anti-CEA ^{123}I-minibody (80 kDa) was evaluated in 10 patients with colorectal cancer scheduled for surgery (Wong et al. 2004). The mean clearance of the minibody from the blood was 29.8 hours, but a wide range was observed (10.9 to 65.4 h). In seven of the eight patients not receiving neoadjuvant therapy, the minibody imaged 8 of 10 lesions that were $\geq$1.0 cm in size, whereas CT imaged 5. The minibody could not detect any tumor in the two remaining patients who received therapy prior to imaging. At surgery, only a 2 mm residual focus was seen in one patient and no tumor was found in the second patient. The study demonstrates the sensitivity and specificity of the minibody in detecting infiltrative and diffuse lesions that could not be detected by CT.

In addition to blood clearance properties, the off-rate of an antibody needs to be considered when contemplating the use of mAbs as imaging agents. Ideally, the off-rate of the antibody following binding to target should be slower than the elimination from the blood. The physical half-life of the radionuclide must also be considered. A numerical criterion, called the imaging figure of merit (IFOM), for comparing radiolabeled mAbs and their fragments has been described (Williams et al. 1995, 2001). This criterion estimates from biodistribution data how long it takes to statistically distinguish tumor from blood pool background. High IFOM values imply shorter imaging times. Thus, for a given radionuclide, IFOM can provide prediction as to which fragment is most suitable and the optimal time interval for imaging. Analysis of an mAb and four related engineered fragments confirmed the notion that optimal results are obtained when the physical half-life of the radionuclide match the biological half-life of the mAb or fragments (Williams et al. 2001). For example, the physical half-life of ^{18}F is well matched to the biologic half-life of the diabody. Recently, the anti-CEA T84.66 diabody was labeled with ^{18}F and evaluated by PET in tumor-bearing mice (Cai et al. 2007; Shively 2007). In this study, IFOM calculations were supported by the experimental data that PET imaging, using ^{18}F-diabody, can begin as early as 1 hour after injection.

20.2.3 Pretargeting Approaches

In a two-step pretargeting approach, an unlabeled macromolecular targeting agent such as an antibody is administered. Sufficient time is allowed for target localization and blood clearance followed by injection of a small radiolabeled hapten. In order for this method to be successful, the macromolecule must clear sufficiently from the blood and normal tissue, and a three-step protocol is often employed to facilitate this. In three-step pretargeting, a synthetic clearing agent that is capable of removing residual macromolecules from the blood is administered before the small radiolabeled hapten. Several antibody-based macromolecules have been investigated for pretargeting; these are bispecific mAbs (Cao and Suresh 1998), streptavidin- or avidin-conjugated mAbs (Hnatowich, Virzi, and Rusckowski 1987; Kalofonos et al. 1990), biotinylated mAbs (Paganelli et al. 1991; Saga et al. 1994; Wilbur et al. 1996), and mAbs conjugated to DNA (Kuijpers et al. 1993).

Studies involving streptavidin- or avidin-based pretargeting approaches for RIT are attractive because of their high affinity for biotin (10^{-13} to $10^{-15}\,M^{-1}$). However, a major limitation of using streptavidin is that it is highly immunogenic, which limits repeated administration of therapeutic doses. In addition, renal radiotoxicity due to high and prolonged retention of radioactivity in the kidneys has been described with this approach (Alvarez-Diez, Polihronis, and Reilly 1996; Breitz et al. 2000; Yao et al. 2004). When the electrical charge (pI) of an scFv-streptavidin conjugate through chemical modification (succinylation) was changed, the renal dose was significantly reduced ($\sim$30 percent) without affecting the tumor activity (Forster et al. 2006). For PET and RIT applications, the biodistribution, clearance, and tumor targeting of a directly labeled mAb and mAb-streptavidin pretargeting were compared in mice bearing human colorectal carcinoma xenografts using ^{64}Cu as tracer (Lewis et al. 2003). The antibody pretargeting strategy with ^{64}Cu-DOTA-biotin displayed a more rapid tumor uptake, substantially faster clearance, and superior tumor to normal tissue ratios. Thus, this study suggests that the pretargeting approach will be superior in PET imaging and exhibit considerably greater efficacy in RIT.

Although pretargeting methods are more complicated than a single-agent targeting method, preclinical data using bispecific (bs) mAbs suggest that superior image quality to that of a directly labeled antibody can be achieved (McBride et al. 2006; Sharkey et al. 2005). Production of bs-mAbs was initially made by chemically linking an anti-tumor Fab′ or $F(ab')_2$ fragment to an antichelate Fab′ fragment (Goodwin et al. 1986; Reardan et al. 1985; Schuhmacher et al. 1995). The hapten was initially the chelate-radiometal complex by itself, which was used until the development of the affinity enhancement system (AES) (Le Doussal et al. 1989). The utility of pretargeting for molecular imaging using bs-mAbs has been evaluated by several groups. For example, the use of different antitumor mAbs for bs-mAb preparation and ^{68}Ga-labeled chelate enhanced the sensitivity of tumor detection at 1 h in mice with rat pancreatic carcinoma (Schuhmacher et al. 1995) and human colon carcinoma

(Klivenyi et al. 1998) xenografts. The first clinical PET imaging pretargeting study using ^{68}Ga-chelate and bs-mAbs was carried out in 10 patients with primary breast cancer targeting MUC1 (Schuhmacher et al. 2001). Fourteen of 17 known lesions were clearly visualized at 60 to 90 min after injection of ^{68}Ga-chelate in patients pretreated with bs-mAb and a clearing agent. Although there were no false-positive readings, three false-negative readings were obtained. Thus, detection of breast cancer with this particular system was not optimal due to the nonspecific distribution of shed antigen and the fast tumor kinetics of the bs-mAb.

Bs-mAbs based on Fab′ fragments that range from ~50 to 157 kDa have been made through molecular engineering using the so-called dock-and-lock method (Goldenberg et al. 2008). A flexible bs-mAb pretargeting system based on using an anti-hapten antibody specific for a synthetic compound, histamine-succinyl-glycine (HSG), which can be labeled with any radionuclide, has also been developed (McBride et al. 2006; Sharkey et al. 2003). McBride et al. demonstrated in mice bearing colon xenografts that tumor uptake in pretargeted mice was about threefold higher with divalent ^{124}I-HSG than in those injected with [^{18}F]-FDG at 1 h. In addition, the background activity in normal tissues was less in the pretargeted animals. Using the dock-and lock-module system, a bispecific complex consisting of three Fab′ fragments of which two bound to CEA and one bound to divalent-HSG peptide, was radiolabeled with ^{124}I and evaluated for PET imaging a disseminated human colorectal cancer model (Sharkey et al. 2007, 2008). In this study, tumors less than 0.3 mm in diameter were detected in pretargeted mice with divalent ^{124}I-HSG peptide that were not seen with [^{18}F]-FDG. Thus, pretargeting appears to be highly applicable for PET imaging. One should note, though, that rapid internalization of target upon binding of bs-mAb would not be a good candidate for pretargeting applications.

20.2.4 Altering PK through Neonatal Fc Receptor (FcRn) Binding

The MHC class 1-like protein, FcRn, plays a central role in prenatal IgG transfer and protection of IgG from catabolism in adults (reviewed in Roopenian and Akilesh 2007). Binding of Fc to FcRn is pH specific and takes place at pH 6 in the early endosomes where the pino- and endocytosed IgG is salvaged from degradation by FcRn and recycled to the cell surface and released into the extracellular space at pH 7.4. Unbound IgG will enter the lysosomal pathway and become degraded (Ober et al. 2004). The residues in the Fc region that are involved in the FcRn binding have been mapped and extensively characterized. Kim et al. (1999) showed that single residue mutation of 253-Ile, 310-His, and 435-His reduced the half-life of an Fc fragment. Larger engineered antibody fragments consisting of scFv fused to Fc region (scFv-Fc, 105 kDa) exhibit similar PK properties to those of intact mAbs (Slavin-Chiorini et al. 1995). However, mutating the residues binding to FcRn in the Fc region of an anti-CEA scFv-Fc fragment resulted in variants with distinct clearance profiles in mice (Kenanova et al. 2005). Each of these variants was radiolabeled with ^{124}I and evaluated by PET in mice bearing LS174T xenografts. Imaging showed that there was no difference between the fragments at 3 to 4 h, whereas a distinct difference between the fast and the slow clearing fragments was obvious at 16 to 18 h and even more distinct at 48 to 90 h. The highest tumor to soft tissue ratio was achieved with the fastest clearing fragments. The scFv-Fc variant that contained two mutations (H310A/H435Q) exhibited blood clearance profile similar to that of the C_H2 deleted antibody variant (i.e. minibody). This fragment was subsequently evaluated by PET in two other tumor systems targeting CD20 antigen (B-cell lymphoma) and HER2 (breast cancer). The ^{124}I-labeled anti-CD20 scFv-Fc H310A/H435Q antibody fragment successfully targeted and imaged CD20-positive xenografts in mice (Olafsen et al. 2006). The internalizing anti-HER2 scFv-Fc H310A/H435Q fragment was radiometal-labeled using ^{64}Cu and DOTA as chelator and evaluated in mice bearing human MCF7/HER2 breast cancer xenografts (Olafsen et al. 2005). Although high accumulation of radioactivity was seen in the tumor at 21 h, high accumulation of radioactivity was also seen in the liver as expected for this fragment and label. PET images using the CEA- and HER2-specific scFv-Fc H310A/H435Q fragments labeled with ^{64}Cu and ^{124}I are shown in Figure 20.3. Overall, this approach provides means to tailor the PK of mAbs and enable selection of versions that are optimized for imaging or therapy.

In addition to shortening the plasma half-life of mAbs there is also a trend to mutate the FcRn-binding site so that binding is increased at pH 6.0, but not at pH 7.4 (Dall'Acqua et al. 2002; Ghetie et al. 1996; Hinton et al. 2004, 2006). Blocking FcRn binding by high affinity mAb has been shown to affect the concentrations of endogenous IgG *in vivo* by rapid clearance (Vaccaro et al. 2005). A recent study demonstrated that rapid clearance of radiolabeled mAbs was achieved in both mice and humans by administration of high doses of polyclonal IgG (1 g/kg) without altering tumor uptake (Jaggi et al. 2007). PET imaging was performed using ^{124}I-labeled cG250 and A33 mAbs in SK-RC-52 and SW1222 xenograft-bearing athymic mice, respectively. Both untreated mice and IgG-treated mice showed comparable tumor uptakes by 40 h post-injection, but the IgG-treated mice showed marked reduction in the blood. The tumor to blood contrast with IgG-treated mice was even higher in the ^{124}I-A33 imaging studies, in which tumors were visible as early as 4 h after injection. Although significant blood clearance with ^{124}I-A33 was also seen in humans (at 160 h post-injection), it was not as pronounced as that seen in the mice. This was partially explained by the 8-fold lower affinity of human FcRn for human IgG and because humans may require administration of higher doses (1 g/kg twice a day) in order to achieve the effects seen in mice. The study describes a novel approach that results in enhanced tumor contrast in imaging studies where intact mAbs are used.

20.2.5 Other Methods to Extend Half-Life of Antibody Fragments

20.2.5.1 Human Serum Albumin (HSA) Albumin (67 kDa), the most abundant protein in the plasma (5 g/L), has a half-life of 19 days in humans (Peters 1985). This long serum half-life is due to a recycling process mediated by FcRn, similar to that observed for IgG molecules (Chaudhury et al. 2003, 2006). HSA has been used as carrier molecule for drugs for therapeutic or diagnostic purpose (Chuang, Kragh-Hansen, and Otagiri 2002). In addition, HSA has been fused to hormones and cytokines in order to modulate their PKs and reduce their immunogenicity (Duttaroy et al. 2005; Melder et al. 2005; Osborn et al. 2002a, 2002b; Sung et al. 2003). Smith and coworkers (2001) showed that Fab′ and F(ab′)$_2$ fragments associated with albumin by conjugation, fusion, or noncovalent binding resulted in extended persistence close to that of the albumin itself in plasma *in vivo*. Two imaging studies involving an antibody fragment and HSA have recently been published. In one study, an albumin-binding peptide sequence fused to an anti-HER2 Fab′ was evaluated for targeting by SPECT using ^{111}In as tracer and DOTA as chelator, in mice bearing HER2-positive xenografts (Dennis et al. 2007). It was shown that similar tumor deposition was achieved with the Fab′ albumin-binding peptide fragment and the intact trastuzumab (Herceptin^{TM}) mAb at 48 h. However, the albumin-binding Fab′ fragment targeted tumor more quickly and cleared faster from the blood, resulting in enhanced contrast. In the other study, an anti-CEA T84.66 scFv-HSA fusion protein referred to as immunobumin was evaluated by both SPECT and PET in tumor-bearing mice (Yazaki et al. 2008). SPECT images using ^{125}I and ^{111}In-DOTA-labeled immunobumin demonstrated localization to tumor and rapid clearance from the blood. Higher tumor and lower normal tissue uptakes were obtained with ^{111}In-DOTA immunobumin relatively to ^{111}In-DOTA minibody. When the DOTA-immunobumin was evaluated by PET in one mouse using ^{64}Cu, tumor localization was evident at 4 h and reached highest intensity at 24 h. Thus, according to the authors, the improved tumor to normal organ ratio achieved with immunobumin supports the development of a new imaging agent using this approach.

20.2.5.2 PEGylation The conjugation of polyethylene glycol (PEG) polymers to drugs has become one of the best validated drug delivery methods for extension of serum half-life (Bailon et al. 2001; Greenwald et al. 2003) and several PEGylated protein therapeutics are on the market (Harris and Chess 2003). In addition to increased retention of the drugs in circulation (especially α phase), PEGylation of antibody fragments reduces immunogenicity, increases solubility, protects against enzymatic digestion, and slows filtration by the kidneys (Chapman 2002). Several recent studies have investigated PEGylation of scFv, Fab, or other domain conjugates (Albrecht, Denardo, and Denardo 2006; Germershaus et al. 2006; Li et al. 2006). However, PEGylation frequently leads to loss or reduction of antigen binding due to steric interference with the antibody–antigen binding interaction. Several approaches of conjugating antibody fragment to PEG have been reported (Chapman et al. 1999; Lee et al. 1999; Pedley et al. 1994; Yang et al. 2003). A general strategy for creating tailored site-specific PEGylated single-chain antibodies using thiol chemistry was developed by Yang and coworkers (2003). They showed that conjugated scFv fragments with 5, 20, and 40 kDa maleimide-PEG polymers demonstrated prolongation of biological half-lives that correlated with polymer molecular mass *in vivo* with the 40 kDa PEG-scFv derivatives exhibiting half-lives similar to intact antibodies. Thus, the PK of scFv fragments can be tailored through polymer selection.

The PK of PEGylated antibody fragments has also been investigated in oncology. Several studies have shown that PEGylated antibody fragments with prolonged plasma half-life have increased accumulation in tumor (Delgado et al. 1996; Kubetzko et al. 2006; Lee et al. 1999; Pedley et al. 1994). In a recent study, the T84.66 anti-CEA diabody was conjugated to a bifunctional PEG-3400 derivative followed by reaction with cysteinyl-DOTA and radiolabeled with ^{111}In (Li et al. 2006). Due to the apparent increase in Stokes radius, a three- to fourfold increase in tumor uptake and more than a fourfold reduction in kidney uptake were achieved with the ^{111}In-cysteinyl-DOTA-PEG3400. In SPECT, excellent tumor images similar to that of the larger antibody fragment minibody (Hu et al. 1996) were achieved, but with improved tumor to liver ratios. The authors suggest that this agent can potentially be suitable for imaging liver metastases. To date, none of the PEGylated antibody fragments generated have been evaluated by PET imaging.

20.3 APPLICATIONS OF immunoPET IN ONCOLOGY

The high sensitivity and resolution of the PET scanner makes this a very useful tool for detection of malignancies. The primary application of immunoPET is therefore to assess the presence and extent of disease by targeting cell surface phenotype. In addition, target expression, tumor type, and tissue of origin can be determined by immunoPET depending on the surface phenotype. Once the diagnosis has been established the aim of evaluation turns to staging and treatment planning. Staging is a means to categorize the extent of the disease. The common elements considered in most staging systems are the location of the primary tumor, tumor size and number of tumors, lymph node involvement, cell type and tumor grade (how closely the cancer cells resemble normal tissue), and presence or absence of metastases. Thus, staging is used as a predictor of prognosis and treatment planning and is largely dependent on which stage the patient presents at the time of diagnosis. For example, patients who are candidates for Ab therapy could be stratified based on cell surface phenotype expression identified by immunoPET. A potential application might be breast cancer patients, of whom only 20 to 30 percent over-express HER2 and are eligible for trastuzumab (Herceptin^{TM}) therapy. ImmunoPET can be used to evaluate the biodistribution of the therapeutic mAb, that is, whether it targets tumor and/or normal organs such as heart in breast cancer patients pretreated with anthracyclines (Behr, Behe, and Wormann 2001; de Korte et al. 2007).

In individualized medicine, variations in drug PK need to be considered in order to calculate a drug dose that is adapted specifically to the patient. In order to establish patient-specific biokinetics, a small amount of radiolabeled pharmaceutical is administered and serial quantitative imaging is performed (Siegel et al. 1999). Assuming that isotopes have the same biokinetic behavior, PET can be used for estimating dosimetry prior to RIT by using a positron emitter to simulate another radionuclide with the same atomic number for which activity measurements and absorbed dose estimates are required. Examples of PET/RIT isotope pairs are ^{64}Cu/^{67}Cu, ^{124}I/^{131}I, ^{86}Y/^{90}Y, ^{89}Zr/^{177}Lu, and ^{89}Zr/^{90}Y. PET dosimetry calculations can also be useful following RIT to verify the predicted absorbed dose distribution and to analyze outcome (dose-response) of the therapy. By assessing the early effects of therapy, changes to the treatment protocol can be made much sooner, if necessary. In addition, identification of residual tumor after chemotherapy, radiation therapy, or surgery can be made by PET. One study compared immunoPET targeting HER2 and [^{18}F]-FDG for imaging tumor response following Hsp90 therapy (Smith-Jones et al. 2006). Within 24 h a significant decrease in HER2 was observed whereas [^{18}F]-FDG PET was unchanged, demonstrating that targeting a surface marker is more sensitive than measuring glycolysis. Similarly, pretreatments interfering with neoangiogenesis or signaling pathways and their possible effect on tumor surface target expression could also be monitored by immunoPET. This sensitivity makes immunoPET also potentially suitable for evaluating receptor occupancy, that is, whether a linear relationship between dose and target occupancy exists.

20.4 SUMMARY

Molecular imaging using PET is becoming a valuable tool for preclinical assessment of drugs before further development, leading to rapid evaluation of new tracers with a decreased work load. Advances have been made regarding imaging instrumentation and the widespread clinical implementation of SPECT/CT and PET/CT. The vast majority of clinical PET imaging is currently performed with [^{18}F]-FDG. However, with its limitations, [^{18}F]-FDG PET represents only a fraction of the potential of PET imaging and there is ongoing investigation and development of additional radiopharmaceuticals. One such group of pharmaceuticals is antibodies that are increasingly being recognized as a powerful class of molecular imaging probes for *in vivo* imaging of surface markers. Several approaches have been developed to optimize their *in vivo* properties as imaging agents. With the improved positron emitting radionuclides availability, radiometal chelates, and conjugation chemistry, radioimmunoconjugates with optimal properties have become available. Thus, immunoPET might become a powerful tool for imaging cell surface phenotypes *in vivo*—in patients—in cancer diagnostics and therapy. This can potentially play an important role at many points along the cancer care spectrum, such as determining the extent of disease, stratifying who will benefit from antibody therapy, and response to treatment. Overall, the broad implementation of cell-surface imaging will bolster basic and preclinical research, enabling a translational tool for drug discovery and clinical application in future molecular imaging.

REFERENCES

Adams, G.P., R. Schier, A.M. McCall, R.S. Crawford, E.J. Wolf, L.M. Weiner, and J.D. Marks. 1998. Prolonged in vivo tumour retention of a human diabody targeting the extracellular domain of human HER2/neu. *Br. J. Cancer* 77(9):1405–1412.

Aide, N., O. Cappele, P. Bottet, H. Bensadoun, A. Regeasse, F. Comoz, F. Sobrio, G. Bouvard, and D. Agostini. 2003. Efficiency of [(^{18})F]FDG PET in characterising renal cancer and detecting distant metastases: A comparison with CT. *Eur. J. Nuclear Med. Mol. Imaging* 30(9):1236–1245.

Albrecht, H., G.L. Denardo, and S.J. Denardo. 2006. Monospecific bivalent scFv-SH: Effects of linker length and location of an engineered cysteine on production, antigen binding activity and free SH accessibility. *J. Immunol. Methods* 310(1–2):100–116.

Alvarez-Diez, T.M., J. Polihronis, and R.M. Reilly. 1996. Pretargeted tumour imaging with streptavidin immunoconjugates of monoclonal antibody CC49 and ^{111}In-DTPA-biocytin. *Nuclear Med. Biol.* 23(4):459–466.

Anderson, C.J., J.M. Connett, S.W. Schwarz, P.A. Rocque, L.W. Guo, G.W. Philpott, K.R. Zinn, C.F. Meares, and M.J. Welch. 1992. Copper-64-labeled antibodies for PET imaging. *J. Nuclear Med.* 33(9):1685–1691.

Bailon, P., A. Palleroni, C.A. Schaffer, C.L. Spence, W.J. Fung, J.E. Porter, G.K. Ehrlich, W. Pan, Z.X. Xu, M.W. Modi, A. Farid, W. Berthold, and M. Graves. 2001. Rational design of a potent, long-lasting form of interferon: A 40 kDa branched polyethylene glycol-conjugated interferon alpha-2a for the treatment of hepatitis C. *Bioconjug. Chem.* 12(2):195–202.

Begent, R.H., M.J. Verhaar, K.A. Chester, J.L. Casey, A.J. Green, M.P. Napier, L.D. Hope-Stone, N. Cushen, P.A. Keep, C.J. Johnson, R.E. Hawkins, A.J. Hilson, and L. Robson. 1996. Clinical evidence of efficient tumor targeting based on single-chain Fv antibody selected from a combinatorial library. *Nature Med.* 2(9):979–984.

Behr, T.M., M. Behe, and B. Wormann. 2001. Trastuzumab and breast cancer. *N. Engl. J. Med.* 345(13):995–996.

Birchler, M.T., C. Thuerl, D. Schmid, D. Neri, R. Waibel, A. Schubiger, S.J. Stoeckli, S. Schmid, and G.W. Goerres. 2007. Immunoscintigraphy of patients with head and neck carcinomas, with an anti-angiogenetic antibody fragment. *Otolaryngol. Head Neck Surg.* 136(4):543–548.

Borjesson, P.K., Y.W. Jauw, R. Boellaard, R. de Bree, E.F. Comans, J.C. Roos, J.A. Castelijns, M.J. Vosjan, J.A. Kummer, C.R. Leemans, A.A. Lammertsma, and G.A. van Dongen. 2006. Performance of immuno-positron emission tomography with zirconium-89-labeled chimeric monoclonal antibody U36 in the detection of lymph node metastases in head and neck cancer patients. *Clin. Cancer Res.* 12(7 Pt 1):2133–2140.

Borsi, L., E. Balza, M. Bestagno, P. Castellani, B. Carnemolla, A. Biro, A. Leprini, J. Sepulveda, O. Burrone, D. Neri, and L. Zardi. 2002. Selective targeting of tumoral vasculature: Comparison of different formats of an antibody (L19) to the ED-B domain of fibronectin. *Int. J. Cancer* 102(1):75–85.

Breitz, H.B., P.L. Weiden, P.L. Beaumier, D.B. Axworthy, C. Seiler, F.M. Su, S. Graves, K. Bryan, and J.M. Reno. 2000. Clinical optimization of pretargeted radioimmunotherapy with antibody-streptavidin conjugate and ^{90}Y-DOTA-biotin. *J. Nucl. Med.* 41(1):131–140.

Cai, W., T. Olafsen, X. Zhang, Q. Cao, S.S. Gambhir, L.E. Williams, A.M. Wu, and X. Chen. 2007. PET imaging of colorectal cancer in xenograft-bearing mice by use of an ^{18}F-labeled T84.66 anti-carcinoembryonic antigen diabody. *J. Nuclear Med.* 48(2):304–310.

Cai, W., Y. Wu, K. Chen, Q. Cao, D.A. Tice, and X. Chen. 2006. In vitro and in vivo characterization of ^{64}Cu-labeled Abegrin, a humanized monoclonal antibody against integrin alpha v beta 3. *Cancer Res.* 66(19):9673–9681.

Cao, Y., and M.R. Suresh. 1998. Bispecific antibodies as novel bioconjugates. *Bioconjug. Chem.* 9(6):635–644.

Chapman, A.P. 2002. PEGylated antibodies and antibody fragments for improved therapy: A review. *Adv. Drug Deliv. Rev.* 54(4):531–545.

Chapman, A.P., P. Antoniw, M. Spitali, S. West, S. Stephens, and D.J. King. 1999. Therapeutic antibody fragments with prolonged in vivo half-lives. Nature Biotechnol. 17(8):780–783.

Chaudhury, C., C.L. Brooks, D.C. Carter, J.M. Robinson, and C.L. Anderson. 2006. Albumin binding to FcRn: Distinct from the FcRn-IgG interaction. *Biochemistry* 45(15):4983–4990.

Chaudhury, C., S. Mehnaz, J.M. Robinson, W.L. Hayton, D.K. Pearl, D.C. Roopenian, and C.L. Anderson. 2003. The major histocompatibility complex-related Fc receptor for IgG (FcRn) binds albumin and prolongs its life-span. *J. Exp. Med.* 197(3):315–322.

Chuang, V.T., U. Kragh-Hansen, and M. Otagiri. 2002. Pharmaceutical strategies utilizing recombinant human serum albumin. *Pharm. Res.* 19(5):569–577.

Cobleigh, M.A., C.L. Vogel, D. Tripathy, N.J. Robert, S. Scholl, L. Fehrenbacher, J.M. Wolter, V. Paton, S. Shak, G. Lieberman, and D.J. Slamon. 1999. Multinational study of the efficacy and safety of humanized anti-HER2 monoclonal antibody in women who have HER2-overexpressing metastatic breast cancer that has progressed after chemotherapy for metastatic disease. *J. Clin. Oncol.* 17(9):2639–2648.

Dall'Acqua, W.F., R.M. Woods, E.S. Ward, S.R. Palaszynski, N.K. Patel, Y.A. Brewah, H. Wu, P.A. Kiener, and S. Langermann. 2002. Increasing the affinity of a human IgG1 for the neonatal Fc receptor: Biological consequences. *J. Immunol.* 169(9):5171–5180.

de Korte, M.A., E.G. de Vries, M.N. Lub-de Hooge, P.L. Jager, J.A. Gietema, W.T. van der Graaf, W.J. Sluiter, D.J. van Veldhuisen, T.M. Suter, D.T. Sleijfer, and P.J. Perik. 2007. ^{111}Indium-trastuzumab visualises myocardial human epidermal growth factor receptor 2 expression shortly after anthracycline treatment but not during heart failure: A clue to uncover the mechanisms of trastuzumab-related cardiotoxicity. *Eur. J. Cancer* 43(14):2046–2051.

Delgado, C., R.B. Pedley, A. Herraez, R. Boden, J.A. Boden, P.A. Keep, K.A. Chester, D. Fisher, R.H. Begent, and G.E. Francis. 1996. Enhanced tumour specificity of an anti-carcinoembryonic antigen Fab' fragment by poly(ethylene glycol) (PEG) modification. *Br. J. Cancer* 73(2):175–182.

Dennis, M.S., H. Jin, D. Dugger, R. Yang, L. McFarland, A. Ogasawara, S. Williams, M.J. Cole, S. Ross, and R. Schwall. 2007. Imaging tumors with an albumin-binding Fab, a novel tumor-targeting agent. *Cancer Res.* 67(1):254–261.

Divgi, C.R., N. Pandit-Taskar, A.A. Jungbluth, V.E. Reuter, M. Gonen, S. Ruan, C. Pierre, A. Nagel, D.A. Pryma, J. Humm, S.M. Larson, L.J. Old, and P. Russo. 2007. Preoperative characterisation of clear-cell renal carcinoma using iodine-124-labelled antibody chimeric G250 (^{124}I-cG250) and PET in patients with renal masses: A phase I trial. *Lancet Oncol.* 8(4):304–310.

Duttaroy, A., P. Kanakaraj, B.L. Osborn, H. Schneider, O.K. Pickeral, C. Chen, G. Zhang, S. Kaithamana, M. Singh, R. Schulingkamp, D. Crossan, J. Bock, T.E. Kaufman, P. Reavey, M. Carey-Barber, S.R. Krishnan, A. Garcia, K. Murphy, J.K. Siskind, M.A. McLean, S. Cheng, S. Ruben, C.E. Birse, and

O. Blondel. 2005. Development of a long-acting insulin analog using albumin fusion technology. *Diabetes* 54(1):251–258.

Eisenhauer, E.A. 2001. From the molecule to the clinic: Inhibiting HER2 to treat breast cancer. *N. Engl. J. Med.* 344(11):841–842.

Forster, G.J., E.B. Santos, P.M. Smith-Jones, P. Zanzonico, and S.M. Larson. 2006. Pretargeted radioimmunotherapy with a single-chain antibody/streptavidin construct and radiolabeled DOTA-biotin: Strategies for reduction of the renal dose. *J. Nucl. Med.* 47(1):140–149.

Gancberg, D., A. Di Leo, F. Cardoso, G. Rouas, M. Pedrocchi, M. Paesmans, A. Verhest, C. Bernard-Marty, M.J. Piccart, and D. Larsimont. 2002. Comparison of HER-2 status between primary breast cancer and corresponding distant metastatic sites. *Ann. Oncol.* 13(7):1036–1043.

Garg, P.K., S. Garg, and M.R. Zalutsky. 1991. Fluorine-18 labeling of monoclonal antibodies and fragments with preservation of immunoreactivity. *Bioconjug. Chem.* 2(1):44–49.

Geissler, F., S.K. Anderson, P. Venkatesan, and O. Press. 1992. Intracellular catabolism of radiolabeled anti-mu antibodies by malignant B-cells. *Cancer Res.* 52(10):2907–2915.

Germershaus, O., T. Merdan, U. Bakowsky, M. Behe, and T. Kissel. 2006. Trastuzumab-polyethylenimine-polyethylene glycol conjugates for targeting Her2-expressing tumors. *Bioconjug. Chem.* 17(5):1190–1199.

Ghetie, V., J.G. Hubbard, J.K. Kim, M.F. Tsen, Y. Lee, and E.S. Ward. 1996. Abnormally short serum half-lives of IgG in beta 2-microglobulin-deficient mice. *Eur. J. Immunol.* 26(3):690–696.

Goldenberg, D.M., E.A. Rossi, R.M. Sharkey, W.J. McBride, and C.H. Chang. 2008. Multifunctional antibodies by the dock-and-lock method for improved cancer imaging and therapy by pretargeting. *J. Nucl. Med.* 49(1):158–163.

Gong, Y., D.J. Booser, and N. Sneige. 2005. Comparison of HER-2 status determined by fluorescence in situ hybridization in primary and metastatic breast carcinoma. *Cancer* 103(9):1763–1769.

Gonzalez Trotter, D.E., R.M. Manjeshwar, M. Doss, C. Shaller, M.K. Robinson, R. Tandon, G.P. Adams, and L.P. Adler. 2004. Quantitation of small-animal (124)I activity distributions using a clinical PET/CT scanner. *J. Nucl. Med.* 45(7):1237–1244.

Goodwin, D.A., C.F. Mears, M. McTigue, and G.S. David. 1986. Monoclonal antibody hapten radiopharmaceutical delivery. *Nuclear Med. Commun.* 7(8):569–580.

Greenwald, R.B., H. Zhao, J. Xia, and A. Martinez. 2003. Poly(ethylene glycol) transport forms of vancomycin: A long-lived continuous release delivery system. *J. Med. Chem.* 46(23):5021–5030.

Harris, J.M., and R.B. Chess. 2003. Effect of pegylation on pharmaceuticals. *Nature Rev. Drug Discov.* 2(3):214–221.

Herschman, H.R. 2003. Molecular imaging: Looking at problems, seeing solutions. *Science* 302(5645):605–608.

Hinton, P.R., M.G. Johlfs, J.M. Xiong, K. Hanestad, K.C. Ong, C. Bullock, S. Keller, M.T. Tang, J.Y. Tso, M. Vasquez, and N. Tsurushita. 2004. Engineered human IgG antibodies with longer serum half-lives in primates. *J. Biol. Chem.* 279(8):6213–6216.

Hinton, P.R., J.M. Xiong, M.G. Johlfs, M.T. Tang, S. Keller, and N. Tsurushita. 2006. An engineered human IgG1 antibody with longer serum half-life. *J. Immunol.* 176(1):346–356.

Hnatowich, D.J., F. Virzi, and M. Rusckowski. 1987. Investigations of avidin and biotin for imaging applications. *J. Nucl. Med.* 28(8):1294–1302.

Hu, S., L. Shively, A. Raubitschek, M. Sherman, L.E. Williams, J.Y. Wong, J.E. Shively, and A.M. Wu. 1996. Minibody: A novel engineered anti-carcinoembryonic antigen antibody fragment (single-chain Fv-C_H3) which exhibits rapid, high-level targeting of xenografts. *Cancer Res.* 56(13):3055–3061.

Jaggi, J.S., J.A. Carrasquillo, S.V. Seshan, P. Zanzonico, E. Henke, A. Nagel, J. Schwartz, B. Beattie, B.J. Kappel, D. Chattopadhyay, J. Xiao, G. Sgouros, S.M. Larson, and D.A. Scheinberg. 2007. Improved tumor imaging and therapy via i.v. IgG-mediated time-sequential modulation of neonatal Fc receptor. *J. Clin. Invest.* 117(9):2422–2430.

Jayson, G.C., J. Zweit, A. Jackson, C. Mulatero, P. Julyan, M. Ranson, L. Broughton, J. Wagstaff, L. Hakannson, G. Groenewegen, J. Bailey, N. Smith, D. Hastings, J. Lawrance, H. Haroon, T. Ward, A.T. McGown, M. Tang, D. Levitt, S. Marreaud, F.F. Lehmann, M. Herold, and H. Zwierzina. 2002. Molecular imaging and biological evaluation of HuMV833 anti-VEGF antibody: Implications for trial design of antiangiogenic antibodies. *J. Natl. Cancer Inst.* 94(19):1484–1493.

Kalofonos, H.P., M. Rusckowski, D.A. Siebecker, G.B. Sivolapenko, D. Snook, J.P. Lavender, A.A. Epenetos, and D.J. Hnatowich (1990). Imaging of tumor in patients with indium-111-labeled biotin and streptavidin-conjugated antibodies: Preliminary communication. *J. Nuclear Med.* 31(11):1791–1796.

Kang, D.E., R.L. White, Jr., J.H. Zuger, H.C. Sasser, and C.M. Teigland. 2004. Clinical use of fluorodeoxyglucose F 18 positron emission tomography for detection of renal cell carcinoma. *J. Urol.* 171(5):1806–1809.

Kenanova, V., T. Olafsen, D.M. Crow, G. Sundaresan, M. Subbarayan, N.H. Carter, D.N. Ikle, P.J. Yazaki, A.F. Chatziioannou, S.S. Gambhir, L.E. Williams, J.E. Shively, D. Colcher, A.A. Raubitschek, and A.M. Wu. 2005. Tailoring the pharmacokinetics and positron emission tomography imaging properties of anti-carcinoembryonic antigen single-chain Fv-Fc antibody fragments. *Cancer Res.* 65(2):622–631.

Kim, J.K., M. Firan, C.G. Radu, C.H. Kim, V. Ghetie, and E.S. Ward. 1999. Mapping the site on human IgG for binding of the MHC class I-related receptor, FcRn. *Eur. J. Immunol.* 29(9):2819–2825.

Klivenyi, G., J. Schuhmacher, E. Patzelt, H. Hauser, R. Matys, M. Moock, T. Regiert, and W. Maier-Borst. 1998. Gallium-68 chelate imaging of human colon carcinoma xenografts pretargeted with bispecific anti-CD44V6/anti-gallium chelate antibodies. *J. Nuclear Med.* 39(10):1769–1776.

Kohler, G., and C. Milstein. 1975. Continuous cultures of fused cells secreting antibody of predefined specificity. *Nature* 256(5517):495–497.

Kubetzko, S., E. Balic, R. Waibel, U. Zangemeister-Wittke, and A. Pluckthun. 2006. PEGylation and multimerization of the anti-p185HER-2 single chain Fv fragment 4D5: Effects on tumor targeting. *J. Biol. Chem.* 281(46):35186–35201.

Kuijpers, W.H., E.S. Bos, F.M. Kaspersen, G.H. Veeneman, and C.A. van Boeckel. 1993. Specific recognition of antibody-oligonucleotide conjugates by radiolabeled antisense nucleotides: A novel approach for two-step radioimmunotherapy of cancer. *Bioconjug. Chem.* 4(1):94–102.

Larson, S.M., A.M. El-Shirbiny, C.R. Divgi, G. Sgouros, R.D. Finn, J. Tschmelitsch, A. Picon, M. Whitlow, J. Schlom, J. Zhang, and A.M. Cohen. 1997. Single chain antigen binding protein (sFv CC49): First human studies in colorectal carcinoma metastatic to liver. *Cancer* 80(12 Suppl):2458–2468.

Larson, S.M., K.S. Pentlow, N.D. Volkow, A.P. Wolf, R.D. Finn, R.M. Lambrecht, M.C. Graham, G. Di Resta, B. Bendriem, F. Daghighian, et al. 1992. PET scanning of iodine-124-3F9 as an approach to tumor dosimetry during treatment planning for radioimmunotherapy in a child with neuroblastoma. *J. Nuclear Med.* 33(11):2020–2023.

Larson, S.M., and H. Schoder. 2008. Advances in positron emission tomography applications for urologic cancers. *Curr. Opin. Urol.* 18(1):65–70.

Le Doussal, J.M., M. Martin, E. Gautherot, M. Delaage, and J. Barbet. 1989. In vitro and in vivo targeting of radiolabeled monovalent and divalent haptens with dual specificity monoclonal antibody conjugates: Enhanced divalent hapten affinity for cell-bound antibody conjugate. *J. Nucl. Med.* 30(8):1358–1366.

Lee, L.S., C. Conover, C. Shi, M. Whitlow, and D. Filpula. 1999. Prolonged circulating lives of single-chain Fv proteins conjugated with polyethylene glycol: A comparison of conjugation chemistries and compounds. *Bioconjug. Chem.* 10(6):973–981.

Lewis, M.R., M. Wang, D.B. Axworthy, L.J. Theodore, R.W. Mallet, A.R. Fritzberg, M.J. Welch, and C.J. Anderson. 2003. In vivo evaluation of pretargeted ^{64}Cu for tumor imaging and therapy. *J. Nucl. Med.* 44(8):1284–1292.

Li, L., J. Bading, P.J. Yazaki, A.H. Ahuja, D. Crow, D. Colcher, L.E. Williams, J.Y. Wong, A. Raubitschek, and J.E. Shively. 2008. A versatile bifunctional chelate for radiolabeling humanized anti-CEA antibody with In-111 and Cu-64 at either thiol or amino groups: PET imaging of CEA-positive tumors with whole antibodies. *Bioconjug. Chem.* 19(1):89–96.

Li, L., P.J. Yazaki, A.L. Anderson, D. Crow, D. Colcher, A.M. Wu, L.E. Williams, J.Y. Wong, A. Raubitschek, and J.E. Shively. 2006. Improved biodistribution and radioimmunoimaging with poly(ethylene glycol)-DOTA-conjugated anti-CEA diabody. *Bioconjug. Chem.* 17(1):68–76.

Lovqvist, A., A. Sundin, H. Ahlstrom, J. Carlsson, and H. Lundqvist. 1997. Pharmacokinetics and experimental PET imaging of a bromine-76-labeled monoclonal anti-CEA antibody. *J. Nucl. Med.* 38(3):395–401.

Massoud, T.F., and S.S. Gambhir. 2003. Molecular imaging in living subjects: Seeing fundamental biological processes in a new light. *Genes Dev.* 17(5):545–580.

Mayer, A., E. Tsiompanou, D. O'Malley, G.M. Boxer, J. Bhatia, A.A. Flynn, K.A. Chester, B.R. Davidson, A.A. Lewis, M.C. Winslet, A.P. Dhillon, A.J. Hilson, and R.H. Begent. 2000. Radioimmunoguided surgery in colorectal cancer using a genetically engineered anti-CEA single-chain Fv antibody. *Clin. Cancer Res.* 6(5):1711–1719.

McBride, W.J., P. Zanzonico, R.M. Sharkey, C. Noren, H. Karacay, E.A. Rossi, M.J. Losman, P.Y. Brard, C.H. Chang, S.M. Larson, and D.M. Goldenberg. 2006. Bispecific antibody pretargeting PET (immunoPET) with a ^{124}I-labeled hapten-peptide. *J. Nucl. Med.* 47(10):1678–1688.

Melder, R.J., B.L. Osborn, T. Riccobene, P. Kanakaraj, P. Wei, G. Chen, D. Stolow, W.G. Halpern, T.S. Migone, Q. Wang, K.J. Grzegorzewski, and G. Gallant. 2005. Pharmacokinetics and in vitro and in vivo anti-tumor response of an interleukin-2-human serum albumin fusion protein in mice. *Cancer Immunol. Immunother.* 54(6):535–547.

Ober, R.J., C. Martinez, C. Vaccaro, J. Zhou, and E.S. Ward. 2004. Visualizing the site and dynamics of IgG salvage by the MHC class I-related receptor, FcRn. *J. Immunol.* 172(4):2021–2029.

Olafsen, T., D. Betting, V.E. Kenanova, A.A. Raubitschek, J.M. Timmerman, and A.M. Wu. 2006. MicroPET imaging of CD20 lymphoma Xenografts using engineered antibody fragments. *J. Nuclear Med.* 47(5, S1):48, Abstract 140.

Olafsen, T., V.E. Kenanova, G. Sundaresan, A.L. Anderson, D. Crow, P.J. Yazaki, L. Li, M.F. Press, S.S. Gambhir, L.E. Williams, J.Y. Wong, A.A. Raubitschek, J.E. Shively, and A.M. Wu. 2005. Optimizing radiolabeled engineered anti-p185HER2 antibody fragments for in vivo imaging. *Cancer Res.* 65(13):5907–5916.

Osborn, B.L., H.S. Olsen, B. Nardelli, J.H. Murray, J.X. Zhou, A. Garcia, G. Moody, L.S. Zaritskaya, and C. Sung. 2002a. Pharmacokinetic and pharmacodynamic studies of a human serum albumin-interferon-alpha fusion protein in cynomolgus monkeys. *J. Pharmacol. Exp. Ther.* 303(2):540–548.

Osborn, B.L., L. Sekut, M. Corcoran, C. Poortman, B. Sturm, G. Chen, D. Mather, H.L. Lin, and T.J. Parry. 2002b. Albutropin: A growth hormone-albumin fusion with improved pharmacokinetics and pharmacodynamics in rats and monkeys. *Eur. J. Pharmacol.* 456(1–3):149–158.

Otsuka, F.L., M.J. Welch, M.R. Kilbourn, C.S. Dence, W.G. Dilley, and S.A. Wells, Jr. 1991. Antibody fragments labeled with fluorine-18 and gallium-68: in vivo comparison with indium-111 and iodine-125-labeled fragments. *Int. J. Rad. Appl. Instrum. B* 18(7):813–816.

Paganelli, G., P. Magnani, F. Zito, E. Villa, F. Sudati, L. Lopalco, C. Rossetti, M. Malcovati, F. Chiolerio, E. Seccamani, et al. 1991. Three-step monoclonal antibody tumor targeting in carcinoembryonic antigen-positive patients. *Cancer Res.* 51(21):5960–5966.

Page, R.L., P.K. Garg, S. Garg, G.E. Archer, O.S. Bruland, and M.R. Zalutsky. 1994. PET imaging of osteosarcoma in dogs using a fluorine-18-labeled monoclonal antibody Fab fragment. *J. Nuclear Med.* 35(9):1506–1513.

Pedley, R.B., J.A. Boden, R. Boden, R.H. Begent, A. Turner, A.M. Haines, and D.J. King. 1994. The potential for enhanced tumour localisation by poly(ethylene glycol) modification of anti-CEA antibody. *Br. J. Cancer* 70(6):1126–1130.

Perk, L.R., O.J. Visser, M. Stigter-van Walsum, M.J. Vosjan, G.W. Visser, J.M. Zijlstra, P.C. Huijgens, and G.A. van Dongen. 2006. Preparation and evaluation of (89)Zr-Zevalin for monitoring of (90)Y-Zevalin biodistribution with positron emission tomography. *Eur. J. Nuclear Med. Mol. Imaging* 33(11):1337–1345.

Peters, T, Jr. 1985. Serum albumin. *Adv. Protein Chem.* 37:161–245.

Philpott, G.W., S.W. Schwarz, C.J. Anderson, F. Dehdashti, J.M. Connett, K.R. Zinn, C.F. Meares, P.D. Cutler, M.J. Welch, and B.A. Siegel. 1995. RadioimmunoPET: Detection of colorectal carcinoma with positron-emitting copper-64-labeled monoclonal antibody. *J. Nuclear Med.* 36(10):1818–1824.

Powles, T., and P.J. Ell. 2007. Does PET imaging have a role in renal cancers after all? *Lancet Oncol.* 8(4):279–281.

Rasbridge, S.A., C.E. Gillett, A.M. Seymour, K. Patel, M.A. Richards, R.D. Rubens, and R.R. Millis. 1994. The effects of chemotherapy on morphology, cellular proliferation, apoptosis and oncoprotein expression in primary breast carcinoma. *Br. J. Cancer* 70(2):335–341.

Reardan, D.T., C.F. Meares, D.A. Goodwin, M. McTigue, G.S. David, M.R. Stone, J.P. Leung, R.M. Bartholomew, and J.M. Frincke. 1985. Antibodies against metal chelates. *Nature* 316(6025):265–268.

Rohren, E.M., T.G. Turkington, and R.E. Coleman. 2004. Clinical applications of PET in oncology. *Radiology* 231(2):305–332.

Roopenian, D.C., and S. Akilesh. 2007. FcRn: The neonatal Fc receptor comes of age. *Nature Rev. Immunol.* 7(9):715–725.

Rossin, R., D. Berndorff, M. Friebe, L.M. Dinkelborg, and M.J. Welch. 2007. Small-animal PET of tumor angiogenesis using a (76)Br-labeled human recombinant antibody fragment to the ED-B domain of fibronectin. *J. Nuclear Med.* 48(7):1172–1179.

Saga, T., J.N. Weinstein, J.M. Jeong, T. Heya, J.T. Lee, N. Le, C.H. Paik, C. Sung, and R.D. Neumann. 1994. Two-step targeting of experimental lung metastases with biotinylated antibody and radiolabeled streptavidin. *Cancer Res.* 54(8):2160–2165.

Santimaria, M., G. Moscatelli, G.L. Viale, L. Giovannoni, G. Neri, F. Viti, A. Leprini, L. Borsi, P. Castellani, L. Zardi, D. Neri, and P. Riva. 2003. Immunoscintigraphic detection of the ED-B domain of fibronectin, a marker of angiogenesis, in patients with cancer. *Clin. Cancer Res.* 9(2):571–579.

Schuhmacher, J., S. Kaul, G. Klivenyi, H. Junkermann, A. Magener, M. Henze, J. Doll, U. Haberkorn, F. Amelung, and G. Bastert. 2001. Immunoscintigraphy with positron emission tomography: Gallium-68 chelate imaging of breast cancer pretargeted with bispecific anti–MUC1/anti-Ga chelate antibodies. *Cancer Res.* 61(9):3712–3717.

Schuhmacher, J., G. Klivenyi, R. Matys, M. Stadler, T. Regiert, H. Hauser, J. Doll, W. Maier-Borst, and M. Zoller. 1995. Multistep tumor targeting in nude mice using bispecific antibodies and a gallium chelate suitable for immunoscintigraphy with positron emission tomography. *Cancer Res.* 55(1):115–123.

Sharkey, R.M., T.M. Cardillo, E.A. Rossi, C.H. Chang, H. Karacay, W.J. McBride, H.J. Hansen, I.D. Horak, and D.M. Goldenberg. 2005. Signal amplification in molecular imaging by pretargeting a multivalent, bispecific antibody. *Nature Med.* 11(11):1250–1255.

Sharkey, R.M., H. Karacay, W.J. McBride, E.A. Rossi, C.H. Chang, and D.M. Goldenberg. 2007. Bispecific antibody pretargeting of radionuclides for immuno single-photon emission computed tomography and immuno positron emission tomography molecular imaging: An update. *Clin. Cancer Res.* 13(18 Pt 2):5577s–5585s.

Sharkey, R.M., H. Karacay, S. Vallabhajosula, W.J. McBride, E.A. Rossi, C.H. Chang, S.J. Goldsmith, and D.M. Goldenberg. 2008. Metastatic human colonic carcinoma: Molecular imaging with pretargeted SPECT and PET in a mouse model. *Radiology* 246(2):497–507.

Sharkey, R.M., W.J. McBride, H. Karacay, K. Chang, G.L. Griffiths, H.J. Hansen, and D.M. Goldenberg. 2003. A universal pretargeting system for cancer detection and therapy using bispecific antibody. *Cancer Res.* 63(2):354–363.

Shively, J.E. 2007. ^{18}F labeling for immuno-PET: Where speed and contrast meet. *J. Nuclear Med.* 48(2):170–172.

Siegel, J.A., S.R. Thomas, J.B. Stubbs, M.G. Stabin, M.T. Hays, K.F. Koral, J.S. Robertson, R.W. Howell, B.W. Wessels, D.R. Fisher, D.A. Weber, and A.B. Brill. 1999. MIRD pamphlet no. 16: Techniques for quantitative radiopharmaceutical biodistribution data acquisition and analysis for use in human radiation dose estimates. *J. Nucl. Med.* 40(2):37S–61S.

Slamon, D.J., B. Leyland-Jones, S. Shak, H. Fuchs, V. Paton, A. Bajamonde, T. Fleming, W. Eiermann, J. Wolter, M. Pegram, J. Baselga, L. Norton. 2001. Use of chemotherapy plus a monoclonal antibody against HER2 for metastatic breast cancer that overexpresses HER2. *N. Engl. J. Med.* 344(11):783–792.

Slavin-Chiorini, D.C., S.V. Kashmiri, J. Schlom, B. Calvo, L.M. Shu, M.E. Schott, D.E. Milenic, P. Snoy, J. Carrasquillo, K. Anderson, et al. 1995. Biological properties of chimeric domain-deleted anticarcinoma immunoglobulins. *Cancer Res.* 55(23 Suppl):5957s–5967s.

Smith, B.J., A. Popplewell, D. Athwal, A.P. Chapman, S. Heywood, S.M. West, B. Carrington, A. Nesbitt, A.D. Lawson, P. Antoniw, A. Eddelston, and A. Suitters. 2001. Prolonged in vivo residence times of antibody fragments associated with albumin. *Bioconjug. Chem.* 12(5):750–756.

Smith-Jones, P.M., D. Solit, F. Afroze, N. Rosen, and S.M. Larson. 2006. Early tumor response to Hsp90 therapy using HER2 PET: Comparison with ^{18}F-FDG PET. *J. Nuclear Med.* 47(5):793–796.

Sugawara, Y., K.R. Zasadny, A.W. Neuhoff, and R.L. Wahl. 1999. Reevaluation of the standardized uptake value for FDG: Variations with body weight and methods for correction. *Radiology* 213(2):521–525.

Sundaresan, G., P.J. Yazaki, J.E. Shively, R.D. Finn, S.M. Larson, A.A. Raubitschek, L.E. Williams, A.F. Chatziioannou, S.S. Gambhir, and A.M. Wu. 2003. ^{124}I-Labeled engineered anti-CEA minibodies and diabodies allow high-contrast, antigen-specific small-animal PET imaging of xenografts in athymic mice. *J. Nuclear Med.* 44(12):1962–1969.

Sung, C., B. Nardelli, D.W. LaFleur, E. Blatter, M. Corcoran, H.S. Olsen, C.E. Birse, O.K. Pickeral, J. Zhang, D. Shah, G. Moody, S. Gentz, L. Beebe, and P.A. Moore. 2003. An IFN-beta-albumin fusion protein that displays improved pharmacokinetic and pharmacodynamic properties in nonhuman primates. *J. Interferon Cytokine Res.* 23(1):25–36.

Vaccaro, C., J. Zhou, R.J. Ober, and E.S. Ward. 2005. Engineering the Fc region of immunoglobulin G to modulate in vivo antibody levels. *Nature Biotechnol.* 23(10):1283–1288.

Verel, I., G.W. Visser, O.C. Boerman, J.E. van Eerd, R. Finn, R. Boellaard, M.J. Vosjan, M. Stigter-van Walsum, G.B. Snow, and G.A. van Dongen. 2003. Long-lived positron emitters zirconium-89 and iodine-124 for scouting of therapeutic radioimmunoconjugates with PET. *Cancer Biother. Radiopharm.* 18(4):655–661.

Verel, I., G.W. Visser, and G.A. van Dongen. 2005. The promise of immuno-PET in radioimmunotherapy. *J. Nuclear Med.* 46 (Suppl 1):164S–171S.

Wadas, T.J., E.H. Wong, G.R. Weisman, and C.J. Anderson. 2007. Copper chelation chemistry and its role in copper radiopharmaceuticals. *Curr. Pharm. Design* 13(1):3–16.

Westera, G., H.W. Reist, F. Buchegger, C.H. Heusser, N. Hardman, A. Pfeiffer, H.L. Sharma, G.K. von Schulthess, and J.P. Mach. 1991. Radioimmuno positron emission tomography with monoclonal antibodies: A new approach to quantifying in vivo tumour concentration and biodistribution for radioimmunotherapy. *Nuclear Med. Commun.* 12(5):429–437.

Wilbur, D.S., D.K. Hamlin, R.L. Vessella, J.E. Stray, K.R. Buhler, P.S. Stayton, L.A. Klumb, P.M. Pathare, and S.A. Weerawarna. 1996. Antibody fragments in tumor pretargeting. Evaluation of biotinylated Fab' colocalization with recombinant streptavidin and avidin. *Bioconjug. Chem.* 7(6):689–702.

Williams, L.E., A. Liu, A.M. Wu, T. Odom-Maryon, A. Chai, A.A. Raubitschek, and J.Y. Wong. 1995. Figures of merit (FOMs) for imaging and therapy using monoclonal antibodies. *Med. Phys.* 22(12):2025–2027.

Williams, L.E., A.M. Wu, P.J. Yazaki, A. Liu, A.A. Raubitschek, J.E. Shively, and J.Y. Wong. 2001. Numerical selection of optimal tumor imaging agents with application to engineered antibodies. *Cancer Biother. Radiopharm.* 16(1):25–35.

Wilson, C.B., D.E. Snook, B. Dhokia, C.V. Taylor, I.A. Watson, A.A. Lammertsma, R. Lambrecht, J. Waxman, T. Jones, A.A. Epenetos. 1991. Quantitative measurement of monoclonal antibody distribution and blood flow using positron emission tomography and ^{124}iodine in patients with breast cancer. *Int. J. Cancer* 47(3):344–347.

Wong, J.Y., D.Z. Chu, L.E. Williams, D.M. Yamauchi, D.N. Ikle, C.S. Kwok, A. Liu, S. Wilczynski, D. Colcher, P.J. Yazaki, J.E. Shively, A.M. Wu, and A.A. Raubitschek. 2004. Pilot trial evaluating a ^{123}I-labeled 80-kilodalton engineered anticarcinoembryonic antigen antibody fragment (cT84.66 minibody) in patients with colorectal cancer. *Clin. Cancer Res.* 10(15):5014–5021.

Wu, A.M., and P.D. Senter. 2005. Arming antibodies: Prospects and challenges for immunoconjugates. *Nature Biotechnol.* 23(9):1137–1146.

Wu, A.M., L.E. Williams, L. Zieran, A. Padma, M. Sherman, G.G. Bebb, T. Odom-Maryon, J.Y.C. Wong, J.E. Shively, and A.A. Raubitschek. 1999. Anti-carcinoembryonic antigen (CEA) diabody for rapid tumor targeting and imaging. *Tumor Targeting* 4:47–58.

Wu, A.M., and P.J. Yazaki. 2000. Designer genes: Recombinant antibody fragments for biological imaging. *Q. J. Nuclear Med.* 44(3):268–283.

Wu, A.M., P.J. Yazaki, S. Tsai, K. Nguyen, A.L. Anderson, D.W. McCarthy, M.J. Welch, J.E. Shively, L.E. Williams, A.A. Raubitschek, J.Y. Wong, T. Toyokuni, M.E. Phelps, and S.S. Gambhir. 2000. High-resolution microPET imaging of carcinoembryonic antigen-positive xenografts by using a copper-64-labeled engineered antibody fragment. *Proc. Natl. Acad. Sci. USA* 97(15):8495–8500.

Xu, F.J., Y.H. Yu, D.S. Bae, X.G. Zhao, S.K. Slade, C.M. Boyer, R.C. Bast, Jr., and M.R. Zalutsky. 1997. Radioiodinated antibody targeting of the HER-2/neu oncoprotein. *Nuclear Med. Biol.* 24(5):451–459.

Yang, K., A. Basu, M. Wang, R. Chintala, M.C. Hsieh, S. Liu, J. Hua, Z. Zhang, J. Zhou, M. Li, H. Phyu, G. Petti, M. Mendez, H. Janjua, P. Peng, C. Longley, V. Borowski, M. Mehlig, and D. Filpula. 2003. Tailoring structure-function and pharmacokinetic properties of single-chain Fv proteins by site-specific PEGylation. *Protein Eng.* 16(10):761–770.

Yao, Z., W. Dai, J. Perry, M.W. Brechbiel, and C. Sung. 2004. Effect of albumin fusion on the biodistribution of interleukin-2. *Cancer Immunol. Immunother.* 53(5):404–410.

Yazaki, P.J., T. Kassa, C.W. Cheung, D.M. Crow, M.A. Sherman, J.R. Bading, A.L. Anderson, D. Colcher, and A. Raubitschek. 2008. Biodistribution and tumor imaging of an anti-CEA single-chain antibody-albumin fusion protein. *Nuclear Med. Biol.* 35(2):151–158.

Yazaki, P.J., and A.M. Wu. 2003. Construction and characterization of minibodies for imaging and therapy of colorectal carcinomas. *Methods Mol. Biol.* 207:351–364.

Zalutsky, M.R. 2006. Potential of immuno-positron emission tomography for tumor imaging and immunotherapy planning. *Clin. Cancer Res.* 12(7 Pt 1):1958–1960.

Zhang, X., Z. Xiong, Y. Wu, W. Cai, J.R. Tseng, S.S. Gambhir, and X. Chen. 2006. Quantitative PET imaging of tumor integrin alphavbeta3 expression with ^{18}F-FRGD2. *J. Nuclear Med.* 47(1):113–121.

Zidan, J., I. Dashkovsky, C. Stayerman, W. Basher, C. Cozacov, A. Hadary. 2005. Comparison of HER-2 over-expression in primary breast cancer and metastatic sites and its effect on biological targeting therapy of metastatic disease. *Br. J. Cancer* 93(5):552–556.

CHAPTER 21

Antibodies and the Blood-Brain Barrier

ANGELA R. JONES and ERIC V. SHUSTA

ABSTRACT

A formidable obstacle for drug delivery to the brain is the blood-brain barrier (BBB). The BBB is necessary for maintaining brain homeostasis and healthy function, but because of its barrier properties it also prevents the brain uptake of neuropharmaceuticals, including antibodies, from the bloodstream. Though various invasive procedures, such as convection-enhanced delivery, may provide some limited antibody access to brain tissue, noninvasive alternatives that circumvent the BBB would be preferred. Cationization and polyamination of antibodies can mediate nonspecific transport across the BBB; however, antibodies targeted to specific receptors at the BBB that undergo transcytosis likely represent key players in the future of brain drug delivery. These antibodies can act as transport vectors to deliver small molecule drugs as well as large biopharmaceuticals like neurotrophins or therapeutic antibodies. This chapter focuses on antibody permeability at the BBB, approaches for delivery of therapeutic antibodies to the brain, and the use of antibodies as specific transport vectors for brain drug delivery.

Therapeutic Monoclonal Antibodies: From Bench to Clinic. Edited by Zhiqiang An

21.1 INTRODUCTION

There are more than 600 disorders of the nervous system (National Institute of Neurological Disorders and Stroke 2005). Based on the final statistics for 2004 released in 2007 through the National Center for Health Statistics, 3 of the top 15 causes of death in the United States were neurological disorders: stroke (ranked third), Alzheimer's disease (ranked seventh), and Parkinson's disease (ranked fourteenth; Minino et al. 2007). Moreover, central nervous system (CNS)-based cancers comprised 2.3 percent of cancer deaths (Minino et al. 2007). Given the aging population, these numbers are sure to grow and, as an example, the incidence of Alzheimer's disease (AD) is expected to almost triple in the United States by 2050 (Hebert et al. 2003). Due to the prevalence of these neurological diseases, it is imperative to develop new treatments. However, for systemic delivery, one obstacle that any new treatment must overcome is the blood-brain barrier (BBB). The BBB provides the brain with nutrients, prevents the introduction of harmful blood-borne substances, and restricts the movement of ions and fluid to ensure an optimal environment for brain function (Abbott, Ronnback, and Hansson 2006). Consequently, the BBB also prevents the movement of drugs from the blood into the brain, severely limiting the human brain microvasculature as a distribution route for noninvasive brain drug delivery. As a general rule, only those drugs that are lipid soluble with a molecular weight of 400 to 600 Da can readily diffuse through the endothelial cells that comprise the BBB and enter the brain (Pardridge 2002; Fig. 21.1). Examples of this restrictive class of lipid-soluble neuropharmaceuticals are codeine (for headaches), diazepam (for anxiety disorders), and chlorpromazine (for schizophrenia) (Pardridge 1991). Biopharmaceuticals like antibodies, proteins, and genes, which are much larger in size, do not appreciably cross the BBB by diffusion.

Thus, new modes of delivery via the bloodstream are necessary and are currently being developed to take advantage of the ~400 miles of capillaries and large drug transport surface area of ~20 square meters found in the human brain (Pardridge 2002). In this chapter, we will briefly highlight critical features of the BBB, and we will discuss how these attributes result in limited antibody permeability. Next, strategies for circumventing the BBB to increase the potency of disease-targeting antibodies for treatment of CNS disorders will be covered. Finally, a promising application for antibodies includes

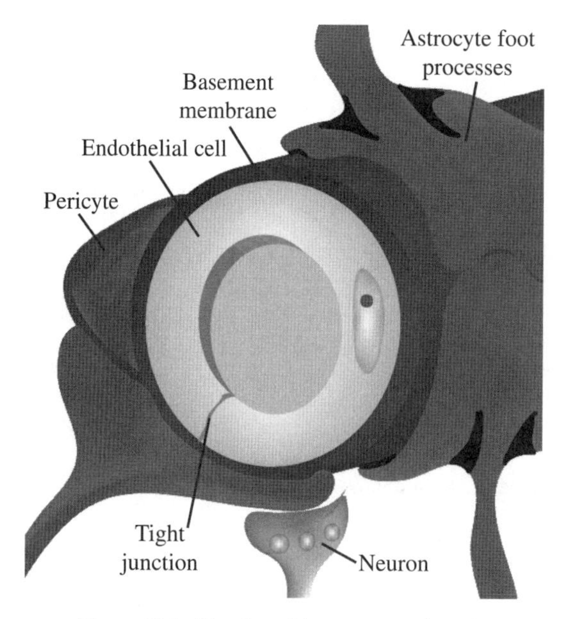

Figure 21.1 Drawing of the neurovascular unit.

their use as BBB transporting agents for shuttling of drug cargo into the brain, and these approaches will be extensively reviewed.

21.2 ARCHITECTURE OF THE BBB

The BBB is formed by the endothelium of the brain microvasculature (Fig. 21.1; Ge, Song, and Pachter 2005). The endothelial cells (EC) have specialized intercellular tight junctions which prevent free diffusion of substances, including drugs, between the circulating blood and the brain (Pardridge 1991). In contrast, the endothelia of peripheral tissue are not joined as tightly and therefore do not present as great an obstacle to drug delivery (Pardridge 1991). Adding to the intrinsic impermeability of the BBB, the endothelial cells also express efflux transporters such as P-glycoprotein (P-gp) that can remove small lipophilic drugs that have managed to penetrate the endothelial cells of the brain by pumping them back to the lumen of the blood vessel (Endicott and Ling 1989; Cordon-Cardo et al. 1989; Sugawara et al. 1990; Tsuji et al. 1992). Tight junctions among other unique attributes are elicited by close contact with other brain cell types that form an inductive microenvironment. For example, astrocyte foot processes, which cover nearly 99 percent of the capillary surface (Pardridge 2002), play a prominent role in regulating barrier tightness (Abbott 2002). Simple co-culture of brain EC with astrocytes, or EC culture in astrocyte-conditioned media, can improve barrier properties of an *in vitro* BBB model (Tao-Cheng, Nagy, and Brightman 1987; Raub, Kuentzel, and Sawada 1992; Arthur, Shivers, and Bowman 1987; Rubin et al. 1991). Pericytes, another microvascular cell type, control the growth and migration of EC (Dohgu et al. 2005), stabilize the brain capillary structure (Ramsauer, Krause, and Dermietzel 2002), and play a role in immune surveillance (Pardridge et al. 1989). Little is known about the role of neurons on the BBB *in vivo*, but when co-cultured with brain microvascular endothelial cells (BMEC) *in vitro*, the endothelial permeability has been shown to decrease (Cestelli et al. 2001). Taken together, this multicellular composite, also known as the neurovascular unit (Fig. 21.1), controls the bidirectional transport of material from the blood to the brain. There are other potential portals of drug entry into the brain at the blood-cerebrospinal fluid interface of the choroid plexus (blood-CSF barrier) and in locations where fenestrated capillaries perfuse circumventricular organs (CVOs). However, the total surface areas of the CVOs and choroid plexus are at least three orders of magnitude smaller than the BBB (Pardridge 1991), limiting their potential as drug transport interfaces. The BBB will therefore be the sole focus of further discussion.

21.3 INTRINSIC ANTIBODY TRANSPORT AT THE BBB

Antibodies and other biopharmaceuticals do not readily cross the BBB. In order to determine a drug's potential for brain delivery, the permeability, also known as the permeability surface area (PS product = permeability coefficient × surface area), of the drug is often calculated. The percent of the injected dose of the drug (% ID = PS × area under the concentration curve, AUC) is also an indicator of BBB permeability. Table 21.1 compares the PS products and/or % ID/brain of various antibodies and natural ligands. The PS product for antibody (IgG) uptake into the rat brain is 0.05 to 0.06×10^{-6} mL/g/s which is on the same order of a circulating serum protein, albumin (Table 21.1). In some experiments, IgG uptake is not even detected (Table 21.1). For comparison, the average PS product of transferrin and insulin, proteins known to cross the BBB by receptor mediation, are 2.432×10^{-6} mL/g/s and 18.50×10^{-6} mL/g/s, respectively (Poduslo, Curran, and Berg 1994), or 40 to 300 times higher than antibodies. Thus, while passive protein transport at the BBB either by diffusion or nonspecific fluid phase endocytosis is nonzero, antibody uptake levels are likely too low for most therapeutic applications without the use of excessively large doses. Therefore, strategies that can augment IgG uptake into the brain will be the main focus of the remainder of the chapter.

TABLE 21.1 Pharmacokinetic Data for Transport of Antibodies and Natural Ligands at the BBB

Vector	PS Product (mL/g-s × 10^6) (% ID/Brain)	Transporter (Mechanism)	Animal	Ref.
Human IgG	0.062 (ND)		Rat	Poduslo, Curran, and Berg (1994)
Albumin[a]	0.097 {0.062}[b] (ND)		Rat	Poduslo, Curran, and Berg (1994)
Tf[c]	2.432 {0.062}[d] (ND)	TfR (RMT)	Rat	Poduslo, Curran, and Berg (1994)
Insulin[a]	18.50 {0.062}[b] (ND)	IR (RMT)	Rat	Poduslo, Curran, and Berg (1994)
Cationized IgG	9.5 (ND)	NS (AMT)	Rat	Triguero, Buciak, and Pardridge (1991)
Cationized D146 MAb	ND (0.07) {0}	NS (AMT)	Mouse	Pardridge et al. (1995b)
Glycated IgG[a]	0.19 {0.05}[b] (ND)	NS (Unknown)	Rat	Poduslo and Curran (1994)
Polyaminated IgG[a]	9.69 {0.06}[b] (ND)	NS (likely AMT)	Rat	Poduslo and Curran (1996)
MAb OX26	27 (0.44) {0}[c]	Rat TfR (RMT)	Rat	Bickel, Yoshikawa, and Pardridge (1993); Friden et al. (1991)
MAb OX26	0.77 (0.03)[c]	Rat TfR (RMT)	Mouse	Lee et al. (2000)
MAb RI7-127	20 (0.8)[c]	Mouse TfR (RMT)	Mouse	Lee et al. (2000)
MAb 8D3	25 (1.5)[c]	Mouse TfR (RMT)	Mouse	Lee et al. (2000)
MAb 128.1	ND (0.3%) {0.06}	Human TfR (RMT)	Monkey	Friden et al. (1996)
MAb Z35.2	ND (0.2%) {0.03}	Human TfR (RMT)	Monkey	Friden et al. (1996)
MAb 83-14	88–90 (2.5–3.8) {0.06}	HIR (RMT)	Monkey	Pardridge et al. (1995a); Coloma et al. (2000)
Chimeric MAb 83-14	28 (2)	HIR (RMT)	Monkey	Coloma et al. (2000)
Humanized MAb 83-14	ND (1)	HIR (RMT)	Monkey	Boado et al. (2007b)
Phage-displayed FC44	ND (2.3)[c]	Unknown (RMT)	Mouse	Muruganandam et al. (2002)
Phage-displayed FC5	ND (1.5)[c]	Unknown (RMT)	Mouse	Muruganandam et al. (2002)

ND = Not determined or not available in reference; NS = nonspecific.

[a]Average PS product from various brain regions determined in referenced article.

[b]Numbers in { } are values for isotype control antibodies and are indicators of general antibody permeability.

[c]Values were calculated from reported % ID/g, assuming representative mass of animal brain (100 g for monkey, 1 g for rat, and 0.5 g for mouse).

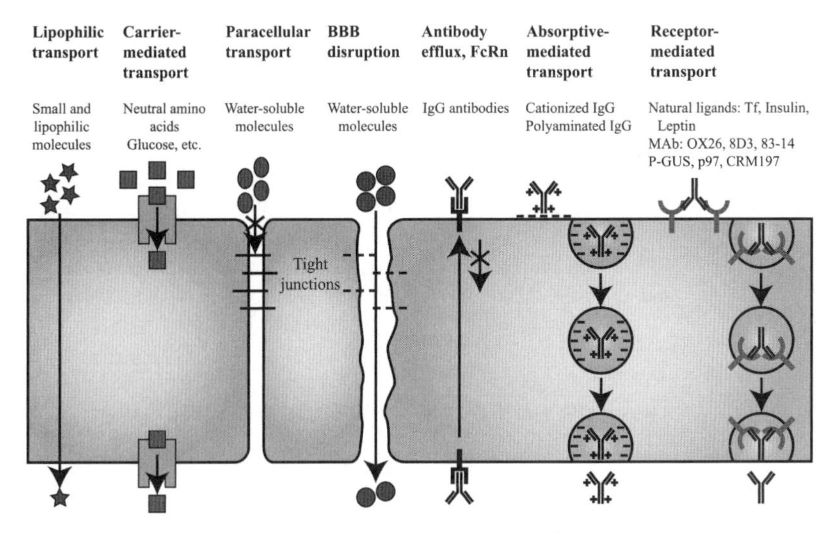

Figure 21.2 BBB transport mechanisms.

Complicating matters further, antibodies have an additional obstacle to delivery through the BBB due to the presence of the neonatal Fc receptor (FcRn) at the BBB endothelium. The FcRn recognizes the constant region (Fc) of an IgG and can transport IgG across cellular interfaces, such as from the mother to the fetus across the placenta (Roopenian and Akilesh 2007). In addition, FcRn has recently been shown to contribute to the efflux of antibodies at the BBB (Fig. 21.2). In contrast to the limited blood-to-brain transport of antibodies discussed above, it was shown that IgG molecules rapidly efflux from brain to blood following intracerebral injection (Zhang and Pardridge 2001a). Ninety minutes after intracerebral injection of radiolabeled 70 kDa dextran, anti-rat transferrin monoclonal antibody OX26, or mouse IgG2a, the level of molecule remaining within the brain was 93 percent, 31 percent, and 29 percent of the injected dose, respectively (Zhang and Pardridge 2001a). Efflux was competitively inhibited by Fc fragments and not $F(ab')_2$ fragments, indicating Fc-dependent efflux (Zhang and Pardridge 2001a). Using confocal microscopy, it was shown that rat FcRn was co-localized with the glucose transporter (GLUT1), known to be expressed on the capillary endothelium of the brain (Schlachetzki, Zhu, and Pardridge 2002). Thus, the potential of antibody efflux by FcRn-mediated transport in the brain to blood direction must also be taken into consideration when developing antibody-based therapeutic strategies for brain disease.

21.4 ANTIBODY MODIFICATIONS FOR IMPROVING THE TRANSPORT OF NON-BBB TARGETING ANTIBODIES

Various chemical modifications of antibodies have been tested for their ability to improve BBB permeability. Early attempts include the cationization of antibodies. The idea is that the positive charge of the cationized antibodies will bind to negative charges on a cell surface signaling for absorptive-mediated endocytosis, thereby increasing cellular uptake (Triguero et al. 1989; Triguero, Buciak, and Pardridge 1991; Fig. 21.2). Cationization of surface carboxyl groups on the antibody by covalent linkage to primary amino groups resulted in increased uptake of radiolabeled antibody into the brain parenchyma relative to noncationized antibody, with a PS product of 9.5×10^{-6} mL/g/s (Triguero, Buciak, and Pardridge 1991; Table 21.1). In another study, an antibody targeted to the Asp^{13} point mutation of the *ras* proto-oncogenic p21 protein (D146) was similarly cationized, and the % ID/brain increased from undetectable to 0.07 percent (Pardridge et al. 1995b). Another technique for cationization is the polyamination of the antibody, which involves covalently attaching polyamines like putrescine, spermidine, or spermine to carboxyl groups on the antibody (Poduslo and Curran 1996). Through the addition of putrescine to an antibody, the PS product was increased 111- to 313-fold throughout the brain relative to unaltered antibody (Poduslo and Curran 1996; Table 21.1). Another antibody modification used is glycation, which is the covalent linkage of sugar molecules to the primary amino groups in antibodies (Poduslo and Curran 1994). This modification resulted in an increased PS product roughly three to five times greater than the nonglycated counterpart although the mechanism is not clear (Poduslo and Curran 1994). While improved brain delivery of antibodies can be achieved through cationization and glycation, these approaches may not be the most desirable because cationized proteins will indiscriminately cross plasma membranes of cells throughout the body, and therefore lack ideal targeting (brain specificity) and uptake attributes (% ID/brain).

21.5 INVASIVE MODES FOR IMPROVING ANTIBODY DELIVERY

Intraventricular administration of drugs, a means of delivery via CSF, requires highly invasive techniques like intracerebroventricular (ICV) infusion, and drugs delivered to the CSF are cleared rather rapidly as CSF turns over four to five times daily in humans (Pardridge 1991; Davson, Welch, and Segal 1987). Thus, transport via this mechanism results in local delivery of drugs to the brain surface as was demonstrated with limited penetration (1 to 2 mm) of brain-derived neurotrophic factor (BDNF) or horseradish peroxidase from the ventricles into brain tissue (Yan et al. 1994; Brightman, Reese, and Feder 1970). If one is targeting diseases localized at any of the brain surfaces, for example, meningeal leukemia (Fleischhack, Jaehde, and Bode 2005) or bacterial and viral meningitis (Tunkel et al. 2004; Falagas, Bliziotis, and Tam 2007; Mellouli et al. 2003), this might be a useful technique, but it will not be particularly effective for drug delivery to the majority of brain tissue. Another highly invasive treatment relies on direct injection of drug into brain tissue or implantation of drug-loaded particles. As is the case for ICV infusion, drug administered in this way does not diffuse more than 1 mm away from the injection/implantation site (Mak et al. 1995; Pardridge 2001). Convection-enhanced delivery, the most effective of these invasive surgery-based treatments for delivery to the brain parenchyma, has been used to treat human brain tumors, and involves infusion of fluid at a rate of 10 μL/min in humans (Laske, Youle, and Oldfield 1997) and leads to a larger treatment volume than direct injection alone (Pardridge 2001; P.F. Morrison et al. 1994). In this way, therapeutic antibodies have been delivered to the brain for the treatment of malignant gliomas. Local delivery of a monoclonal antibody (MAb) to the tumor by injection directly into the tumor has provided some hope of treatment for malignant gliomas as it bypasses the BBB. For example, after a single intratumoral injection of an antibody against the common mutation of the epidermal growth factor receptor (EGFRvIII), mice implanted with tumors expressing this receptor had an increased median survival of 286 percent (Sampson et al. 2000; Gerber and Laterra 2007). In another study using a different anti-EGFRvIII MAb and a rat model with implanted EGFRvIII-expressing gliomas, 7.7 percent of injected dose per gram tissue (% ID/g) found its way into the tumor 24 hours after intravenous injection, while 59.8% ID/g made it 24 hours after convection-enhanced delivery (Gerber and Laterra 2007; Yang et al. 2005). In a safety and feasibility study, Cotara, a radiolabeled MAb against an intracellular antigen found in the necrotic core of malignant solid tumors, was delivered to patients with glioblastoma multiforme and anaplastic astrocytomas via convection-enhanced delivery and proved to have an acceptable safety profile (Gerber and Laterra 2007; Patel et al. 2005). We refer the reader to Gerber and Laterra (2007) for a more comprehensive discussion of antibody treatment of malignant gliomas.

Another delivery mechanism that allows nonselective delivery through the BBB involves opening the tight junctions. In this way, blood-resident drugs such as antibodies could gain entry into brain tissue as a result of removing the physical barrier to diffusion and increasing the paracellular movement of fluid (Fig. 21.2). BBB disruption can be therapeutically induced by intracarotid infusion of hyperosmolar solutions like mannitol or with vasoactive agents like bradykinin peptides. Hyperosmolar infusions "shrink" the endothelium, rupturing the tight junctions to allow drug entrance (Kemper et al. 2004; Neuwelt et al. 1982). Bradykinin is a vasoactive peptide that disrupts the BBB (Unterberg, Wahl, and Baethmann 1984; Black et al. 1997), and at low concentrations its permeability effect is somewhat specific for tumor vasculature because bradykinin binds to receptors that are preferentially expressed on the leaky tumor vasculature (Kemper et al. 2004; Kroll et al. 1998). Upon comparison of disruption due to osmotic solutions and bradykinin peptides, the delivery of small (methotrexate) and large (dextran 70, MW 70 kDa) molecules using osmotic solutions was increased 3- to 6-fold in the tumor and 4- to 10-fold in surrounding brain tissue relative to saline controls, but the greatest increase shown with bradykinin was less than 60 percent over controls (Kroll and Neuwelt 1998). Although BBB disruption is reversible, seizures are the most common neurological side effect (Marchi et al. 2007; McAllister et al. 2000), but disruption can also cause transient obtundation, stroke, and cerebral herniation (McAllister et al. 2000). Because of these attributes, BBB disruption is probably not the preferred mode of antibody delivery. In addition, direct injection or implantation strategies would be most useful for highly localized delivery, such as to a tumor mass. To achieve more widespread delivery, other methods will be required.

21.6 NONINVASIVE MODES FOR IMPROVING ANTIBODY DELIVERY

Due to the delivery shortcomings of invasive strategies, a variety of noninvasive brain drug delivery methods utilizing natural transport mechanisms (Fig. 21.2) that make use of the brain blood vessel network to gain widespread distribution have been investigated. Nutrients and other water-soluble compounds necessary for brain function are transported to the brain by three different classes of transport systems embedded in the plasma membrane of the BBB: carrier-mediated, absorptive-mediated, and receptor-mediated transport. Carrier-mediated transport utilizes highly stereospecific carriers that selectively transport small molecules like ions and amino acids in both the blood to brain and brain to blood directions (Fig. 21.2). For example, a hexose carrier transports D-glucose, and a neutral amino acid carrier transports 13 neutral amino acids (Pardridge 1991). Though therapeutics have been conjugated to natural substrates of carrier-mediated transport systems, this approach has not proven effective for brain delivery of large-molecule therapeutics, like antibodies (Pardridge 2002). More relevant transport systems for large molecule drug delivery are absorptive-mediated (AMT) and receptor-mediated (RMT) transcytosis, which transport material across the BBB using the vesicular trafficking machinery of the EC. AMT is initiated by polycationic molecules binding to negative charges on the plasma membrane (Pardridge 1991; Fig. 21.2). As described earlier, the cationization of antibodies can be used to improve delivery of antibodies through the BBB via AMT. Unlike AMT, RMT depends on the interaction of a ligand to a *specific* receptor on the blood side of the EC. Nutrients such as iron (Jefferies et al. 1984), insulin (Duffy and Pardridge 1987), and leptin (Golden, Maccagnan, and Pardridge 1997) are delivered to the brain via this mechanism. In RMT, the ligand binds to a particular endocytosing receptor that clusters in clathrin-coated pits or caveolae. The receptor-ligand complex then invaginates into an intracellular vesicle (Brown and Greene 1991). Vesicles fuse together to form endosomes in which the receptor-ligand complex can be sorted toward different intracellular pathways including transcytosis. As Figure 21.2 depicts for transcytosis, the receptor-ligand complex is sent to the opposite side of the EC, at which time the ligand is released (Brown and Greene 1991). Interestingly proteins, including antibodies, that bind to known BBB transcytosing receptors can also promote transcytosis and act as a vector for trans-BBB drug delivery of therapeutics.

21.6.1 Antibodies as Trans-BBB Delivery Vectors

If appropriately designed, antibodies themselves are well qualified to act as BBB targeting vectors, as their intrinsic properties allow them to bind tightly to a specific antigen. In this way, antibodies that target known transcytosing receptors can piggy-back across the BBB allowing noninvasive entry of linked drug cargo into the brain. There are several such antibodies used as delivery vectors in a research capacity today. The two most frequently used antibody classes recognize the transferrin and insulin receptors of EC, and more recent studies have employed recombinant antibody library technology to identify additional novel transcytosing antibody vectors.

21.6.1.1 Antibodies to the Transferrin Receptor The transferrin receptor (TfR) mediates the delivery of iron into the brain and is highly expressed by brain capillaries (Jefferies et al. 1984). The TfR is a disulfide-linked homodimer that binds to transferrin (Tf), an iron-carrying protein (Pardridge, Eisenberg, and Yang 1987; Moos and Morgan 2000; Qian et al. 2002). Since the TfR is nearly saturated with endogenous Tf in the blood, any Tf-targeted drug would have to compete with the natural ligand (Qian et al. 2002; Bickel, Yoshikawa, and Pardridge 2001). As an alternative, antibodies against a distinct epitope on the TfR have been utilized for transport of drugs into the brain.

The mouse MAb against the rat TfR, OX26, has been the most commonly used anti-TfR antibody (Table 21.1). The OX26 antibody and Tf have different binding sites on TfR, which allows for minimal inhibition of normal Tf transport (Lee et al. 2000). This antibody has been used successfully to deliver conjugated therapeutics through the BBB and produce desired pharmacological results. As one example for a potential treatment of stroke, brain-derived neurotrophic factor (BDNF) was conjugated to OX26 (Zhang and Pardridge 2001b). Unconjugated BDNF or the PEGylated BDNF conjugated to OX26 was injected into the femoral vein of a rat one hour after middle cerebral artery occlusion (MCAO). Unconjugated BDNF provided no neuroprotection at either 24 hours or 7 days after MCAO. However, the OX26-conjugated BDNF provided a 68 percent reduction in cortical stroke volume at 24 hours and a 70 percent reduction in cortical stroke volume at 7 days after administration (Zhang and Pardridge 2001b). Subsequent studies further corroborated the neuroprotective effects of the OX26-BDNF conjugate on stroke (Zhang and Pardridge 2006). This MAb has also been conjugated to many other drugs for noninvasive delivery. Like BDNF, human basic fibroblast growth factor (bFGF) was conjugated to OX26 and delivered intravenously to rats to treat permanent MCAO (Song et al. 2002). Similarly, a nerve growth factor (NGF)-OX26 conjugate was used to prevent degeneration of neurons in a rat model of Huntington's disease (HD; Kordower et al. 1994). Most recently, it has been shown that OX26 can be used to deliver short interfering RNA (siRNA) to the rat brain after intravenous injection (Xia et al. 2007). Rat C6 glioma cells or rat glial RG-2 cells permanently transfected with a luciferase expression plasmid were implanted in the brain of adult rats, and the tumor cells were allowed to grow. Anti-luciferase siRNA conjugated to OX26 was injected into the C6 and RG-2 implanted rats, and 69 percent and 81 percent inhibition, respectively, of luciferase enzyme activity was observed after 48 hours (Xia et al. 2007). The MAb OX26 has also been used to deliver small molecules like methotrexate (Friden et al. 1991) or daunomycin (Cerletti et al. 2000; Huwyler, Wu, and Pardridge 1996; Huwyler, Yang, and Pardridge 1997), plasmid DNA encoding exogenous genes (Shi, Boado, and Pardridge 2001), antisense oligonucleotides (ODN) and peptide nucleic acids (PNA) as therapeutics (Boado, Tsukamoto, and Pardridge 1998; Shi, Boado, and Pardridge 2000; Suzuki et al. 2004; Penichet et al. 1999).

The monoclonal antibody OX26 is species specific to the rat TfR, and another MAb, 8D3, has been used for targeting the mouse TfR (Lee et al. 2000; Table 21.1). Examples of drug delivery to the mouse brain using 8D3 are similar to those using OX26. Proof-of-concept experiments include using exogenous plasmid DNA encoding the enzymes β-galactosidase and luciferase encapsulated in PEGylated liposomes with 8D3 as a targeting vector (Shi et al. 2001). β-galactosidase activity was present in the brain as well as other TfR-rich organs like the spleen, liver, and lung, under the constitutive simian virus 40 (SV40) promoter. Further refinement can be gained through the use of tissue-specific promoters to attain tissue-localized gene expression. For example, brain specificity of β-galactosidase expression was attained when the brain-specific glial fibrillary acidic protein (GFAP) promoter was employed (Shi et al. 2001). The MAb RI7-127 also targets the mouse TfR and undergoes RMT through the mouse BBB. However, on comparison, 8D3 exhibited a higher plasma area under the curve (AUC), brain uptake, and % ID/brain than RI7-127 (Table 21.1). Interestingly, brain selectivity was better attained with RI7-127 than 8D3 as there was no measurable uptake in the mouse kidney or liver (Lee et al. 2000). It was hypothesized that the RI7-127 antibody may target an epitope on the mouse BBB TfR that is not accessible in the peripheral organs (Lee et al. 2000). Moreover, from this data, it is clear that though antibodies may target the same receptor, their pharmacokinetic behavior may be dramatically different.

Although antibodies targeting the TfR have PS products that are considerably higher than a non-transporting IgG (Table 21.1) and are roughly two- to threefold higher than the cationized and polyaminated antibodies, they do have their limitations. First, the TfR is expressed ubiquitously on peripheral organs, and as a result only 0.44 percent of the injected dose of OX26 reaches the rat brain after 24 hours (Friden et al. 1991). Similarly, the percent injected dose to the mouse brain of 8D3 and RI7-127 was approximately 1.5 percent (3.1% ID/g) and 0.8 percent (1.6% ID/g), respectively, after one hour (Lee et al. 2000). To promote drug action restricted to the CNS, secondary targeting strategies such as the brain-specific promoter utilized in the β-galactosidase study mentioned earlier could be used for some therapeutic approaches (Shi et al. 2001). As a second drawback, these antibodies are species specific and, consequently, do not recognize the human TfR. Antibodies against the human TfR have exhibited transport into the primate brain (Friden et al. 1996), but these antibodies are mouse derived and would likely require humanization for therapeutic application.

21.6.1.2 Antibodies to the Insulin Receptor The insulin receptor also undergoes RMT through the endothelial cells that make up the BBB (Duffy and Pardridge 1987). Like the TfR, the insulin receptor is found on the blood-side membrane of brain capillary endothelial cells and at the plasma membrane of other brain cells (Havrankova, Brownstein, and Roth 1981; Smith and Gumbleton 2006). The insulin receptor is made up of two α-subunits and two β-subunits with each α- and β-subunit forming a cylindrical structure joined by a disulfide bond (Gaillard, Visser, and de Boer 2005; Ullrich et al. 1985). Insulin binding triggers a conformational change that leads to receptor internalization as well as tyrosine kinase activity (Gaillard, Visser, and de Boer 2005; Ullrich et al. 1985). Like the TfR, antibodies against the insulin receptor have been used for targeted delivery to the brain. The most studied is the mouse MAb against the human insulin receptor (HIR), 83-14, which has been used in Old World primates to model behavior of this targeting vector in humans (Pardridge et al. 1995a; Table 21.1).

When compared to the transport of an anti-human TfR MAb (0.3 percent brain uptake at 24 hours; Friden et al. 1996), the transport of 83-14 across the primate BBB is approximately 10-fold greater, with uptake nearing 4 percent of the injected dose three hours after injection (Table 21.1; Pardridge et al. 1995a). Though 83-14 is effective at delivering to the primate brain, it is also a mouse-derived antibody, which could lead to immunogenic effects if used in humans, so a chimeric antibody (Coloma et al. 2000) and a fully humanized form of the antibody (Boado et al. 2007b) have been created to address this problem. To create the chimeric MAb, the mouse sequences encoding the variable heavy and light chains of the antibody were grafted onto a human immunoglobulin scaffold, producing an antibody that is composed of 85 percent human sequence and 15 percent mouse sequence (Coloma et al. 2000; S.L. Morrison et al. 1984). For the fully humanized antibody, only the CDR loops of the mouse 83-14 antibody were grafted onto the variable chain framework regions of a homologous human antibody (B43 IgG heavy chain and the human REI kappa light chain), and five human residues in the variable heavy region were replaced with original mouse residues to improve antibody secretion from myeloma cells (Boado et al. 2007b). The chimeric antibody retained similar activity to that of the parent antibody in terms of binding to human brain capillaries as well as uptake into the rhesus monkey brain (Coloma et al. 2000; Boado et al. 2007b). The fully humanized antibody, which could be less immunogenic than the chimeric antibody (Boado et al. 2007b; Hwang and Foote 2005), showed a 27 percent decrease in affinity relative to the parent mouse antibody (Boado et al. 2007b), and intravenous injection led to a twofold decrease in % ID/brain to the rhesus monkey relative to the chimeric antibody (Table 21.1; Pardridge et al. 1995a; Coloma et al. 2000; Boado et al. 2007b).

Experiments involving the delivery of therapeutic cargo using 83-14 are similar to those using the antibodies against the TfR, and a few representative examples follow. Proof-of-concept experiments once again included the use of a β-galactosidase expression plasmid. PEGylated liposomes targeted with 83-14 encapsulating the expression plasmid were intravenously injected into the rhesus monkey, and the immunoliposome crossed both the BBB and since HIR is also found on the plasma membrane of neurons, it mediated plasmid delivery to the neurons where the enzyme was expressed at high levels (Zhang, Schlachetzki, and Pardridge 2003). To attain tissue specificity, an opsin promoter was used to control β-galactosidase gene expression so that the enzyme was only expressed in the primate eye (Zhang et al. 2003). Therapeutic cargo examples include human BDNF linked to the chimeric 83-14 MAb. The level of BDNF after intravenous administration into rhesus monkeys was more than 10-fold greater than endogenous levels (Boado, Zhang, and Pardridge 2007). As a means of treating lysosomal storage disorders, a fusion protein of recombinant α-L-iduronidase (IDUA) and the chimeric HIR MAb was created. Intravenous injection into rhesus monkeys resulted in 1.05% ID/brain, which would be adequate to elicit therapeutic effects (Boado et al. 2008).

21.6.1.3 Antibodies to Novel BBB Transport Targets Although antibodies against the transferrin and insulin receptors are capable of delivering therapeutic cargo to the brain in an appreciable amount, they target receptors that are ubiquitously expressed, and generally only 1 to 4 percent of the injected dose actually reaches the brain (Friden et al. 1991; Pardridge et al. 1995a; Coloma et al. 2000; Kang, Bickel, and Pardridge 1994). While these approaches can clearly result in a therapeutic drug concentration within the brain tissue, the loss of drug to clearance and/or other tissues could hinder further development as a result of unwanted side effects. In light of the fact that at least 700 protein- and gene-based therapeutics are in various stages of clinical trials (Pissarra 2004), production costs must also be considered in the case of antibody-mediated brain delivery since both the antibody and drug cargo (protein, gene, another antibody, etc.) are expensive drugs to manufacture. Thus, one of the goals for extending antibody-based delivery modalities has been to identify BBB receptors with a high transport capacity and selective or specific expression at the BBB. To this end, several groups in the BBB field are working to identify new, and potentially better, transcytosing antibodies which could ultimately lead to the development of improved BBB transporting vectors.

As one promising approach, combinatorial antibody library technology has been applied to these goals. Combinatorial antibody libraries are large diverse pools of $\sim 10^8$ to 10^{12} antibodies that can be searched for antibodies capable of binding, internalizing, and transcytosing across the capillary endothelium of the brain. Such selections are performed with the assistance of various antibody display technologies. Typically, antibody fragments consisting of solely V_H domains (single-domain antibody, sdAb), V_H and V_L domains (single-chain variable fragment antibodies, scFv), or Fab fragments are displayed on the surface of a host particle, be it bacteriophage, yeast, or bacteria. These antibody-decorated particles can then be mined for desirable properties such as BBB binding and transport.

One study relied on the display of a library of naïve llama sdAb on the surface of phage particles to identify antibodies that trigger transcytosis across a human *in vitro* BBB model (Muruganandam et al. 2002). Since sdAb, like scFv, lack the Fc domain of a full antibody, the undesired uptake in organs and cells that highly express Fc receptors can be lowered (Muruganandam et al. 2002). Furthermore, it has been contended that due to high homology between the camelid heavy chain variable region of the sdAb and human V_H sequences, sdAb would have lowered immunogenicity as human therapeutics (Cortez-Retamozo et al. 2004; Saerens et al. 2005). Through subtractive panning of the sdAb phage library, two antibodies, FC5 and FC44, produced a desired response of tissue selectivity as well as transcytosis capacity through human cerebromicrovascular endothelial cells (Muruganandam et al. 2002). Both phage-displayed and soluble FC44 and FC5 accumulated in the mouse brain, kidney, and liver after intravenous injection with brain levels of 4.5% ID/g ($\sim$2.3% ID/brain) and 2.9% ID/g ($\sim$1.5% ID/brain), respectively for the phage-displayed antibodies (Table 21.1; Muruganandam et al. 2002). Rigorous analysis to determine the transport mechanism suggests that the FC5 target undergoes receptor-mediated transport, and the antibody binds to an α(2,3)-sialoglycoprotein on the luminal (blood) surface of the human brain endothelial cells (Abulrob et al. 2005).

For human therapeutic application, fully human antibodies would be preferable to antibodies raised in other species or to partially humanized antibodies as they could elicit undesirable immune responses (Holliger and Hudson 2005). Recently, a library of nonimmune human scFv was displayed on the surface of yeast particles and screened using a biopanning technique for antibodies that targeted receptors on the surface of cells in an *in vitro* model of the BBB (Wang, Cho, and Shusta 2007). From this yeast library, 34 unique BBB-binding antibody clones were identified, and some recognized receptors that trigger endocytosis (Wang, Cho, and Shusta 2007). Immunoprecipitation analysis suggested that these internalizing receptors were neither the insulin nor transferrin receptors and, therefore, may represent novel BBB-transport targets (Wang, Cho, and Shusta 2007).

Use of these antibody library techniques have resulted in new antibody vectors for drug delivery to the brain. Further analysis is needed to determine if they outperform the current antibody vectors against the transferrin and insulin receptors in brain tissue specificity and/or rate of delivery to the brain. As described in the discussion of the antibodies to the mouse TfR, 8D3 and RI7-127 antibodies both bind and undergo transcytosis but have different advantages regarding drug delivery in terms of biodistribution and brain uptake levels (Lee et al. 2000). Thus, at the very least, newly identified antibody vectors might provide additional options to fit varying applications in drug delivery.

21.6.2 Non-Antibody Brain Delivery Vectors

Given the description above, one could envision delivery of disease-specific therapeutic antibodies by conjugating to BBB-transporting antibodies. Alternatively, other non-antibody proteins and peptide transporting vectors could also be used for delivery of therapeutic antibodies to the brain. For example, the mannose 6-phosphate (M6P) receptor, also known as the insulin-like growth factor II (IGF-II) receptor, binds to the M6P component of lysosomal enzymes leading to internalization and subsequent sorting to the lysosome and has been indicated to undergo transport across the BBB of neonatal mice (Urayama et al. 2004). Phosphorylated β-glucuronidase (P-GUS) and, to a lesser extent, nonphosphorylated β-glucuronidase undergo RMT via the M6P/IGF-II receptor early in postnatal life but not in adults (Urayama et al. 2004). This receptor may provide a new way to target the neonatal brain parenchyma for therapeutic antibody delivery. Alternatively, low density lipoprotein receptor-related proteins 1 (LRP1) and 2 (LRP2) can also be used for BBB-targeted delivery as their receptors are multifunctional RMT systems with multiple ligands (Gaillard, Visser, and de Boer 2005). Proteins such as melanotransferrin, also known as human melanoma antigen p97 (Demeule et al. 2002), and receptor-associated protein (RAP; Pan et al. 2004) using the LRP1 and LRP2 transporters, respectively, could also be used as drug delivery vectors across the BBB. Interestingly, nanoparticles coated with the surfactant polysorbate 80 (Tween 80) have been used for brain delivery of drugs such as dalargin (Schroder and Sabel 1996), doxorubicin (Gulyaev et al. 1999; Ambruosi, Yamamoto, and Kreuter 2005; Ambruosi et al. 2006; Steiniger et al. 2004), loperamide (Alyautdin et al. 1997), and methotrexate (Gao and Jiang 2006). The mechanism of transport has yet to be definitively elucidated, but a pharmacologic effect is clearly elicited, and it is hypothesized that these coated nanoparticles might undergo receptor-mediated endocytosis using mechanisms similar to lipoproteins via the LDL receptor family (Kreuter et al. 2002). Additionally, CRM197, a nontoxic mutant of diphtheria toxin, may have the potential to serve as a targeting vector for drug delivery to the brain. It has been shown to endocytose after binding to the diphtheria toxin receptor, also known as the membrane-bound precursor of heparin-binding epidermal growth factor-like growth factor (HB-EGF; Gaillard, Brink, and de Boer 2005). For a complete review of BBB delivery vectors, see the review by Jones and Shusta (2007). Thus, there are multiple delivery vectors, both antibody and otherwise, that have the potential to deliver a drug cargo, whether it is a small molecule or antibody therapeutic.

21.7 ANTIBODIES FOR SECONDARY TARGETING TO SITE(S) OF ACTION WITHIN THE CENTRAL NERVOUS SYSTEM

In drug delivery to the brain, crossing the BBB is only part of the battle. Further targeting may be needed to deliver the payload to specific cells like neurons, glia, or cancerous cell populations. Such secondary targeting can also be performed by antibodies.

21.7.1 Antibodies to Transporter Present at Both the BBB and Target Cell Population

The best-case scenario for secondary targeting is if the vector that crosses the BBB can also selectively target the diseased cell population. As mentioned earlier, the TfR is ubiquitously expressed in other tissues in the body aside from the BBB. Conveniently, it is expressed on neurons (Giometto et al. 1990; Mash et al. 1991) and over-expressed in tumor tissue (Kratz and Beyer 1998). This fact has been exploited for the development of a therapeutic for Parkinson's disease (PD; Zhang et al. 2003, 2004a). General consensus is that the symptoms of PD are caused by a deficiency of dopamine in the striatum as a result of the degeneration of dopaminergic neurons in the substantia nigra pars compacta and possibly decreased biosynthesis of dopamine from the remaining neurons (Chinta and Andersen 2005; Haavik and Toska 1998). The rate-limiting step of the biosynthesis of dopamine is the hydroxylation of L-tyrosine to L-DOPA, which is catalyzed by tyrosine hydroxylase (TH), and a deficiency in this enzyme in the striatum can cause the symptoms associated with PD (Haavik and Toska 1998). By encapsulating a TH expression plasmid in a PEGylated liposome decorated with the OX26 antibody against rat TfR, the liposome can deliver the gene therapeutic through the BBB via RMT and across the neuronal plasma membrane via receptor-mediated endocytosis (Zhang et al. 2003, 2004a). After intravenous injection of this immunoliposome into the 6-hydroxydopamine rat model of PD, striatal TH activity was normalized and apomorphine-induced rotation behavior was significantly reduced (Zhang et al. 2003, 2004a). Furthermore, this methodology has also been used for imaging of diseased tissue by delivering radiolabeled antisense selective for affected tissue. Caveolin-1 is an integral membrane protein known to localize in caveolae (Rothberg et al. 1992) and has implications in tumor biology. In particular, the mRNA and protein levels of caveolin-1 in rat C6 glioma cells are 2-fold and 1.5-fold greater, respectively, than in type 1 astrocytes, making it a potential marker for tumor cells, provided one can bypass the BBB (Cameron et al. 2002). To capitalize on this, OX26-targeted and radiolabeled peptide nucleic acid (PNA) antisense to the rat caveolin-1α mRNA was used to image brain cancer in rats implanted with RG-2 rat glioma cells (Suzuki et al. 2004). After intravenous injection, the radiolabeled PNA-OX26 conjugate transcytosed the BBB and endocytosed into glioma cells, where it hybridized to the target mRNA, allowing imaging of the tumor (Suzuki et al. 2004).

21.7.2 Antibodies Used for Sequential Targeting

It is not always possible to target the subpopulation of interest within the CNS with a single delivery/targeting reagent. For this task, secondary targeting antibodies in conjunction with the BBB transport vector can be used to selectively deliver the therapeutic cargo to its specific destination within the CNS. Sequential targeting has been proven feasible in mice implanted with human U87 glial brain tumors. Every week these mice were injected intravenously with PEGylated liposomes decorated with two different antibodies, one for BBB transport and one for tumor targeting. Liposomes were loaded with either a nonviral expression plasmid encoding antisense mRNA against the human epidermal growth factor receptor (EGFR) gene (Zhang, Zhu, and Pardridge 2002) or an expression plasmid encoding a short interfering hairpin RNA (siRNA) directed against the human EGFR gene (Zhang et al. 2004b). The gene therapeutics within the liposomes first crossed the mouse BBB by RMT using anti-mouse TfR MAb 8D3 and subsequently traversed the plasma membrane of the human glioma cells via endocytosis of the anti-HIR 83-14 MAb. Continuing weekly injection of the antisense gene therapy increased the survival time by 100 percent (Zhang, Zhu, and Pardridge 2002), and the siRNA gene therapy increased survival time by 88 percent (Zhang et al. 2004b).

21.7.3 Selective Action within the CNS

In light of the ubiquitous expression of some of the receptor targets discussed above that lead to CNS delivery to all cell types (e.g. insulin and transferrin receptors), a form of secondary targeting can instead be attained through the design of therapeutics that elicit an action only on a select subset of cells that receive the therapeutic cargo. For instance, gene therapeutics can be controlled to act on a particular cell type by using gene promoter elements that limit gene expression to the target cell population, although the genes themselves would be delivered to many different cells and tissues. Using this mode of secondary targeting, enzyme activity from the delivery of the TH expression plasmid contained within the OX26-targeted immunoliposome in the Parkinson's study detailed above was restricted to the CNS through the use of the brain-restrictive GFAP promoter, thereby eliminating undesirable expression of TH in the liver (Zhang et al. 2004a). Additionally, only the target population of nigral-striatal neurons produces the necessary cofactor for TH activity, further restricting enzyme activity to the target neuron population even though the GFAP promoter drives expression in both neurons and astrocytes. As a result of the administration of this targeted immunoliposome into a rat model of PD, the enzyme activity was specifically restored to neurons (Zhang et al. 2004a). As another example, a potential treatment of Alzheimer's disease was recently demonstrated (Boado et al. 2007a). In this study, an anti-beta amyloid (anti-Aβ) scFv designed to bind and disaggregate amyloid plaques within the brain tissue was fused directly to the 83-14 anti-HIR MAb, to mediate brain uptake of the scFv. Because of the presence of FcRn at the BBB, antibodies can efflux in the brain to blood direction. Thus, in theory, any amyloid bound and disaggregated by the anti-Aβ scFv portion of the fusion protein would be actively carried out of the brain as the FcRn effluxes the MAb. Indeed, intracerebral injection of the fusion protein to a transgenic mouse model of Alzheimer's disease was shown to reduce the plaque burden of the mice (Boado et al. 2007a). This proof-of-concept study illustrates the potential use of bispecific antibody constructs that can serve both transport and therapeutic roles.

21.8 IMMUNIZATION FOR ALZHEIMER'S DISEASE

Though therapeutic antibodies do not appreciably cross the BBB on their own, it has been hypothesized that one can immunize against certain brain disorders. As one example, this approach has been tried with Alzheimer's disease (AD), progressing all the way to human clinical trials. Early experiments showed promise using active immunization by injecting PDAPP transgenic mice, which over-express mutant human amyloid precursor protein, with $A\beta^{1\text{-}42}$, the predominant form of beta amyloid (Aβ) peptide found in the amyloid plaques of AD (Schenk et al. 1999). Immunization of the young mice prior to plaque occurrence resulted in almost complete prevention of AD-like pathology, and immunization of the older mice having substantial plaque burden showed marked improvement (Schenk et al. 1999). The $A\beta^{1\text{-}42}$ immunization therapy (drug name: AN1792) progressed to Phase II clinical trials at which time treatment was halted because of meningoencephalitis in 6 percent of the patients administered the drug (Orgogozo et al. 2003), most likely due to T-cell activation (Nicoll et al. 2003).

As a means of reducing this immune response, passive immunization using anti-Aβ antibodies was subsequently attempted. In this form of therapy, anti-Aβ antibodies are administered rather than relying on the immune system to produce its own anti-Aβ antibodies. A proof-of-concept example of passive immunization for AD has already been discussed in the previous section with the administration of a bifunctional fusion protein consisting of an anti-Aβ scFv fused to the 83-14 MAb (Boado et al. 2007a). Elan and Wyeth Pharmaceuticals, the sponsors of AN1792, are currently attempting such passive immunization in Phase II clinical trails for AD using a humanized monoclonal antibody, AAB-001 (Elan Corporation n.d.), and other companies are also currently exploring this mode of treatment.

The major question that arises in light of the antibody permeability data presented in this chapter is how the antibodies from active or passive immunization cross the BBB at levels necessary to act on the amyloid plaques. The exception of course is the aforementioned bifunctional antibody that crosses the BBB via its 83-14 MAb module. There are at least three hypotheses as to the plaque clearance mechanism; two require the crossing of the BBB while the third does not [see Dodel, Hampel, and Du (2003) for an expanded discussion]. The first mechanism assumes the antibody enters the CNS, whereby it can bind to plaques and induce Aβ clearance via Fcγ-receptor-mediated phagocytosis and hydrolysis in microglia. This hypothesis is supported by Fc-mediated microglial phagocytosis of Aβ in *ex vivo* experiments with brain tissue sections of PDAPP mouse or human AD brains (Bard et al. 2000). The second mechanism once again requires the antibody to cross the BBB and bind to Aβ peptides, leading to disaggregation of the amyloid plaques. This is supported by *in vitro* experiments where antibodies raised against the N-terminal region of the Aβ peptide bind to the Aβ fibrils, leading to disaggregation of the fibrils (Solomon et al. 1997). A study involving the age-dependent clearance of Aβ found that the FcRn participated in the clearance of antibody-Aβ immune complexes in older mouse brains upon direct brain injection of anti-Aβ antibodies (Deane et al. 2005). Neither of these clearance hypotheses explain how the antibodies get into the brain, but studies suggest that Aβ peptides cross the BBB via receptor mediation (Poduslo et al. 1999; Zlokovic et al. 1996), and therefore indicate that it might be possible for the antibody to cross the BBB as an Aβ-antibody immune complex (Poduslo and Curran 2001). The third mechanism of clearance, the peripheral sink hypothesis, does not require the antibodies to cross the BBB. An intravenously administered MAb directed against a central domain of Aβ, m266, did not bind to plaques but rather disrupted the equilibrium of Aβ between the blood and CNS, inducing a rapid efflux of Aβ from brain to blood and CSF and a marked reduction of Aβ deposition in the brain (DeMattos et al. 2001). Although the therapeutic mechanism of passive immunization with Aβ antibodies remains to be fully elucidated, it is likely that the ability of Aβ peptides to transport across the BBB is a key component to the success of the strategy. As such, the treatment of neurological disease with peripherally administered antibodies is likely to be an exception, rather than the rule.

21.9 SUMMARY

Given the prevalence of neurological disease and the potential therapeutic benefit of antibodies, both from a delivery and treatment perspective, it is anticipated that antibodies will play a crucial role in the treatment of brain disease. Humanization of antibodies like 83-14 and the discovery of novel BBB transport system-antibody conjugates could offer drug developers the ability to choose among delivery systems to elicit the controlled and desired response in treating brain disease in humans. For now, antibody usage in the CNS as been limited predominately to the research realm, but clinical trials using Aβ antibodies in passive immunization are paving the way for further attempts at treating the brain with large biopharmaceuticals.

ACKNOWLEDGMENTS

This chapter was supported in part by the National Institutes of Health (NS052649, EY018506). A.R.J. is the recipient of a National Science Foundation Graduate Research Fellowship. We would also like to acknowledge Robert Gonzales for his help in the artistic rendering of the neurovascular unit (Fig. 21.1).

REFERENCES

Abbott, N.J. 2002. Astrocyte-endothelial interactions and blood-brain barrier permeability. *J. Anat.* 200(6):629–638.

Abbott, N.J., L. Ronnback, and E. Hansson. 2006. Astrocyte-endothelial interactions at the blood-brain barrier. *Nature Rev. Neurosci.* 7(1):41–53.

Abulrob, A., H. Sprong, P. Henegouwen, and D. Stanimirovic. 2005. The blood-brain barrier transmigrating single domain antibody: Mechanisms of transport and antigenic epitopes in human brain endothelial cells. *J. Neurochem.* 95(4):1201–1214.

Alyautdin, R.N., V.E. Petrov, K. Langer, A. Berthold, D.A. Kharkevich, and J. Kreuter. Delivery of loperamide across the blood-brain barrier with polysorbate 80-coated polybutylcyanoacrylate nanoparticles. *Pharm. Res.* 1997; 14(3):325–328.

Ambruosi, A., A.S. Khalansky, H. Yamamoto, S.E. Gelperina, D.J. Begley, and J. Kreuter. 2006. Biodistribution of polysorbate 80-coated doxorubicin-loaded [C-14]-poly(butyl cyanoacrylate) nanoparticles after intravenous administration to glioblastoma-bearing rats. *J. Drug Target.* 14(2):97–105.

Ambruosi, A., H. Yamamoto, and J. Kreuter. 2005. Body distribution of polysorbate-80 and doxorubicin-loaded [C-14]poly(butyl cyanoacrylate) nanoparticles after i.v. administration in rats. *J. Drug Target.* 13(10):535–542.

Arthur, F.E., R.R. Shivers, and P.D. Bowman. 1987. Astrocyte-mediated induction of tight junctions in brain capillary endothelium: An efficient in vitro model. *Dev. Brain Res.* 36(1):155–159.

Bard, F., C. Cannon, R. Barbour, et al. 2000. Peripherally administered antibodies against amyloid beta-peptide enter the central nervous system and reduce pathology in a mouse model of Alzheimer disease. *Nature Med.* 6(8):916–919.

Bickel, U., T. Yoshikawa, and W.M. Pardridge. 1993. Delivery of peptides and proteins through the blood-brain barrier. *Adv. Drug Deliv. Rev.* 10(2–3):205–245.

Bickel, U., T. Yoshikawa, and W.M. Pardridge. 2001. Delivery of peptides and proteins through the blood-brain barrier. *Adv. Drug Deliv. Rev.* 46(1–3):247–279.

Black, K.L., T. Cloughesy, S.C. Huang, et al. 1997. Intracarotid infusion of RMP-7, a bradykinin analog, and transport of gallium-68 ethylenediamine tetraacetic acid into human gliomas. *J. Neurosurg.* 86(4):603–609.

Boado, R.J., H. Tsukamoto, and W.M. Pardridge. 1998. Drug delivery of antisense molecules to the brain for treatment of Alzheimer's disease and cerebral AIDS. *J. Pharm. Sci.* 87(11):1308–1315.

Boado, R.J., Y. Zhang, and W.M. Pardridge. 2007. Genetic engineering, expression, and activity of a fusion protein of a human neurotrophin and a molecular Trojan horse for delivery across the human blood-brain barrier. *Biotechnol. Bioeng.* 97(6):1376–1386.

Boado, R.J., Y. Zhang, C.F. Xia, and W.M. Pardridge. 2007a. Fusion antibody for Alzheimer's disease with bidirectional transport across the blood-brain barrier and A-beta fibril disaggregation. *Bioconjug. Chem.* 18(2):447–455.

Boado, R.J., Y.F. Zhang, Y. Zhang, and W.M. Pardridge. 2007b. Humanization of anti-human insulin receptor antibody for drug targeting across the human blood-brain barrier. *Biotechnol. Bioeng.* 96(2):381–391.

Boado, R.J., Y. Zhang, Y.F. Zhang, C.F. Xia, Y.T. Wang, W.M. Pardridge. 2008. Genetic engineering of a lysosomal enzyme fusion protein for targeted delivery across the human blood-brain barrier. *Biotechnol. Bioeng.* 99(2):475–484.

Brightman, M.W., T.S. Reese, and N. Feder. 1970. Assessment with the electron microscope of the permeability to peroxidase of cerebral endothelium and epithelium in mice and sharks. In *Capillary Permeability*, ed. C. Crone and N.A. Lassen. Copenhagen: Munksgaard.

Brown, V.I., and M.I. Greene. 1991. Molecular and cellular mechanisms of receptor-mediated endocytosis. *DNA Cell Biol.* 10(6):399–409.

Cameron, P.L., C.D. Liu, D.K. Smart, S.T. Hantus, J.R. Fick, and R.S. Cameron. 2002. Caveolin-1 expression is maintained in rat and human astroglioma cell lines. *Glia* 37(3):275–290.

Cerletti, A., J. Drewe, G. Fricker, A.N. Eberle, and J. Huwyler. 2000. Endocytosis and transcytosis of an immunoliposome-based brain drug delivery system. *J. Drug Target.* 8(6):435–436.

Cestelli, A., C. Catania, S. D'Agostino, et al. 2001. Functional feature of a novel model of blood brain barrier: Studies on permeation of test compounds. *J. Control. Release* 76(1–2):139–147.

Chinta, S.J., and J.K. Andersen. 2005. Dopaminergic neurons. *Int. J. Biochem. Cell Biol.* 37(5):942–946.

Coloma, M.J., H.J. Lee, A. Kurihara, et al. 2000. Transport across the primate blood-brain barrier of a genetically engineered chimeric monoclonal antibody to the human insulin receptor. *Pharm. Res.* 17(3):266–274.

Cordon-Cardo, C., J.P. Obrien, D. Casals, et al. 1989. Multidrug-resistance gene (P-glycoprotein) is expressed by endothelial cells at blood-brain barrier sites. *Proc. Natl. Acad. Sci. USA* 86(2):695–698.

Cortez-Retamozo, V., N. Backmann, P.D. Senter, et al. 2004. Efficient cancer therapy with a nanobody-based conjugate. *Cancer Res.* 64(8):2853–2857.

Davson, H., K. Welch, and M.B. Segal. 1987. Secretion of the cerebrospinal fluid. In *The Physiology and Pathophysiology of the Cerebrospinal Fluid*, ed. E.A. Neuwelt. London: Churchill Livingstone.

Deane, R., A. Sagare, K. Hamm, et al. 2005. IgG-assisted age-dependent clearance of Alzheimer's amyloid beta peptide by the blood-brain barrier neonatal Fc receptor. *J. Neurosci.* 25(50):11495–11503.

DeMattos, R.B., K.R. Bales, D.J. Cummins, J.C. Dodart, S.M. Paul, and D.M. Holtzman. 2001. Peripheral anti-A beta antibody alters CNS and plasma A beta clearance and decreases brain A beta burden in a mouse model of Alzheimer's disease. *Proc. Natl. Acad. Sci. U.S.A.* 98(15):8850–8855.

Demeule, M., J. Poirier, J. Jodoin, et al. 2002. High transcytosis of melanotransferrin (P97) across the blood brain barrier. *J. Neurochem.* 83(4):924–933.

Dodel, R.C., H. Hampel, and Y.S. Du. 2003. Immunotherapy for Alzheimer's disease. *Lancet Neurol.* 2(4):215–220.

Dohgu, S., F. Takata, A. Yamauchi, et al. 2005. Brain pericytes contribute to the induction and up-regulation of blood-brain barrier functions through transforming growth factor-beta production. *Brain Res.* 1038(2):208–215.

Duffy, K.R., and W.M. Pardridge. 1987. Blood-brain barrier transcytosis of insulin in developing rabbits. *Brain Res.* 420(1):32–38.

Elan Corporation. n.d. *Elan: Alzheimer's Research* (http://www.elan.com/research_development/Alzheimers/Default.asp. Accessed April 21, 2008.)

Endicott, J.A., and V. Ling. 1989. The biochemistry of P-glycoprotein-mediated multidrug resistance. *Annu. Rev. Biochem.* 58:137–171.

Falagas, M.E., I.A. Bliziotis, and V.H. Tam. 2007. Intraventricular or intrathecal use of polymyxins in patients with Gram-negative meningitis: A systematic review of the available evidence. *Int. J. Antimicrob. Agents* 29(1):9–25.

Fleischhack, G., U. Jaehde, and U. Bode. 2005. Pharmacokinetics following intraventricular administration of chemotherapy in patients with neoplastic meningitis. *Clin. Pharmacokinet.* 44(1):1–31.

Friden, P.M., T.S. Olson, R. Obar, L.R. Walus, and S.D. Putney. 1996. Characterization, receptor mapping and blood-brain barrier transcytosis of antibodies to the human transferrin receptor. *J. Pharmacol. Exp. Ther.* 278(3):1491–1498.

Friden, P.M., L.R. Walus, G.F. Musso, M.A. Taylor, B. Malfroy, and R.M. Starzyk. 1991. Antitransferrin receptor antibody and antibody-drug conjugates cross the blood-brain barrier. *Proc. Natl. Acad. Sci. U.S.A.* 88(11):4771–4775.

Gaillard, P., A. Brink, and A.G. de Boer. 2005. Diphtheria toxin receptor-targeted brain drug delivery. *Int. Congres. Series* 1277:185–198.

Gaillard, P.J., C.C. Visser, and A.G. de Boer. 2005. Targeted delivery across the blood-brain barrier. *Expert Opin. Drug Deliv.* 2(2):299–309.

Gao, K., and X. Jiang. 2006. Influence of particle size on transport of methotrexate across blood brain barrier by polysorbate 80-coated polybutylcyanoacrylate nanoparticles. *Int. J. Pharm.* 310(1–2):213–219.

Ge, S.J., L. Song, and J.S. Pachter. 2005. Where is the blood-brain barrier ... really? *J. Neurosci. Res.* 79(4):421–427.

Gerber, D.E., and J. Laterra. 2007. Emerging monoclonal antibody therapies for malignant gliomas. *Expert Opin. Investig. Drugs* 16(4):477–494.

Giometto, B., F. Bozza, V. Argentiero, et al. 1990. Transferrin receptors in rat central nervous system. An immunocytochemical study. *J. Neurol. Sci.* 98(1):81–90.

Golden, P.L., T.J. Maccagnan, and W.M. Pardridge. 1997. Human blood-brain barrier leptin receptor. Binding and endocytosis in isolated human brain microvessels. *J. Clin. Invest.* 99(1):14–18.

Gulyaev, A.E., S.E. Gelperina, I.N. Skidan, A.S. Antropov, G.Y. Kivman, and J. Kreuter. 1999. Significant transport of doxorubicin into the brain with polysorbate 80-coated nanoparticles. *Pharm. Res.* 16(10):1564–1569.

Haavik, J., and K. Toska. 1998. Tyrosine hydroxylase and Parkinson's disease. *Mol. Neurobiol.* 16(3):285–309.

Havrankova, J., M. Brownstein, and J. Roth. 1981. Insulin and insulin receptors in rodent brain. *Diabetologia* 20 (Suppl 1):268–273.

Hebert, L.E., P.A. Scherr, J.L. Bienias, D.A. Bennett, and D.A. Evans. 2003. Alzheimer disease in the U.S. population: Prevalence estimates using the 2000 census. *Arch. Neurol.* 60(8):1119–1122.

Holliger, P., and P.J. Hudson. 2005. Engineered antibody fragments and the rise of single domains. *Nature Biotechnol.* 23(9):1126–1136.

Huwyler, J., D.F. Wu, and W.M. Pardridge. 1996. Brain drug delivery of small molecules using immunoliposomes. *Proc. Natl. Acad. Sci. U.S.A.* 93(24):14164–14169.

Huwyler, J., J. Yang, and W.M. Pardridge. 1997. Receptor mediated delivery of daunomycin using immunoliposomes: Pharmacokinetics and tissue distribution in the rat. *J. Pharmacol. Exp. Ther.* 282(3):1541–1546.

Hwang, W.Y., and J. Foote. 2005. Immunogenicity of engineered antibodies. *Methods* 36(1):3–10.

Jefferies, W.A., M.R. Brandon, S.V. Hunt, A.F. Williams, K.C. Gatter, and D.Y. Mason. 1984. Transferrin receptor on endothelium of brain capillaries. *Nature* 312(5990):162–163.

Jones, A.R., and E.V. Shusta. 2007. Blood-brain barrier transport of therapeutics via receptor-mediation. *Pharm. Res.* 24(9):1759–1771.

Kang, Y.S., U. Bickel, and W.M. Pardridge. 1994. Pharmacokinetics and saturable blood-brain barrier transport of biotin bound to a conjugate of avidin and a monoclonal antibody to the transferrin receptor. *Drug Metab. Dispos.* 22(1):99–105.

Kemper, E.M., W. Boogerd, I. Thuis, J.H. Beijnen, and O. van Tellingen. 2004. Modulation of the blood-brain barrier in oncology: Therapeutic opportunities for the treatment of brain tumours? *Cancer Treat. Rev.* 30(5):415–423.

Kordower, J.H., V. Charles, R. Bayer, et al. 1994. Intravenous administration of a transferrin receptor antibody nerve growth-factor conjugate prevents the degeneration of cholinergic striatal neurons in a model of Huntington disease. *Proc. Natl. Acad. Sci. U.S.A.* 91(19):9077–9080.

Kratz, F., and U. Beyer. 1998. Serum proteins as drug carriers of anticancer agents: A review. *Drug Deliv.* 5(4):281–299.

Kreuter, J., D. Shamenkov, V. Petrov, et al. 2002. Apolipoprotein-mediated transport of nanoparticle-bound drugs across the blood-brain barrier. *J. Drug Target.* 10(4):317–325.

Kroll, R.A., and E.A. Neuwelt. 1998. Outwitting the blood-brain barrier for therapeutic purposes: Osmotic opening and other means. *Neurosurgery.* 42(5):1083–1099.

Kroll, R.A., M.A. Pagel, L.L. Muldoon, S. Roman-Goldstein, S.A. Fiamengo, and E.A. Neuwelt. 1998. Improving drug delivery to intracerebral tumor and surrounding brain in a rodent model: A comparison of osmotic versus bradykinin modification of the blood-brain and/or blood-tumor barriers. *Neurosurgery* 43(4):879–886.

Laske, D.W., R.J. Youle, and E.H. Oldfield. 1997. Tumor regression with regional distribution of the targeted toxin TF-CRM107 in patients with malignant brain tumors. *Nature Med.* 3(12):1362–1368.

Lee, H.J., B. Engelhardt, J. Lesley, U. Bickel, and W.M. Pardridge. 2000. Targeting rat anti-mouse transferrin receptor monoclonal antibodies through blood-brain barrier in mouse. *J. Pharmacol. Exp. Ther.* 292(3):1048–1052.

Mak, M., L. Fung, J.F. Strasser, and W.M. Saltzman. 1995. Distribution of drugs following controlled delivery to the brain interstitium. *J. Neurooncol.* 26(2):91–102.

Marchi, N., L. Angelov, T. Masaryk, et al. 2007. Seizure-promoting effect of blood-brain barrier disruption. *Epilepsia* 48(4):732–742.

Mash, D.C., J. Pablo, B.E. Buck, J. Sanchezramos, and W.J. Weiner. 1991. Distribution and number of transferrin receptors in Parkinson's disease and in MPTP-treated mice. *Exp. Neurol.* 114(1):73–81.

McAllister, L.D., N.D. Doolittle, P.E. Guastadisegni, et al. 2000. Cognitive outcomes and long-term follow-up results after enhanced chemotherapy delivery for primary central nervous system lymphoma. *Neurosurgery* 46(1):51–60.

Mellouli, E., Z. Arrouji, M. Debre, and M. Bejaoui. 2003. Successful treatment of Echovirus 27 meningoencephalitis in agammaglobulinaemia with intraventricular injection of gammaglobulin. A case report. *Arch. Pediatr.* 10(2):130–133.

Minino, A.M., M.P. Heron, S.L. Murphy, and K.D. Kochanek. 2007. Deaths: Final Data for 2004. *National Vital Statistics Reports*, Vol 55. Hyattsville, MD: National Center for Health Statistics.

Moos, T., and E.H. Morgan. 2000. Transferrin and transferrin receptor function in brain barrier systems. *Cell Mol. Neurobiol.* 20(1):77–95.

Morrison, P.F., D.W. Laske, H. Bobo, E.H. Oldfield, and R.L. Dedrick. 1994. High-flow microinfusion: Tissue penetration and pharmacodynamics. *Am. J. Physiol.* 266(1):R292–R305.

Morrison, S.L., M.J. Johnson, L.A. Herzenberg, and V.T. Oi. 1984. Chimeric human antibody molecules: Mouse antigen-binding domains with human constant region domains. *Proc. Natl. Acad. Sci. U.S.A.* 81(21):6851–6855.

Muruganandam, A., J. Tanha, S. Narang, and D. Stanimirovic. 2002. Selection of phage-displayed llama single-domain antibodies that transmigrate across human blood-brain barrier endothelium. *FASEB J.* 16(2):240–242.

National Institute of Neurological Disorders and Stroke. 2005. *NINDS Overview* (April 15, 2005; http://www.ninds.nih.gov/about_ninds/ninds_overview.htm. Accessed April 10, 2008).

Neuwelt, E.A., P.A. Barnett, D.D. Bigner, and E.P. Frenkel. 1982. Effects of adrenal-cortical steroids and osmotic blood-brain barrier opening on methotrexate delivery to gliomas in the rodent: The factor of the blood-brain barrier. *Proc. Natl. Acad. Sci. U.S.A.* 79(14):4420–4423.

Nicoll, J.A.R., D. Wilkinson, C. Holmes, P. Steart, H. Markham, and R.O. Weller. 2003. Neuropathology of human Alzheimer disease after immunization with amyloid-beta peptide: A case report. *Nature Med.* 9(4):448–452.

Orgogozo, J.M., S. Gilman, J.F. Dartigues, et al. 2003. Subacute meningoencephalitis in a subset of patients with AD after A beta 42 immunization. *Neurology* 61(1):46–54.

Pan, W.H., A.J. Kastin, T.C. Zankel, P. van Kerkhof, T. Terasaki, and G.J. Bu. 2004. Efficient transfer of receptor-associated protein (RAP) across the blood-brain barrier. *J. Cell. Sci.* 117(21):5071–5078.

Pardridge, W.M. 1991. *Peptide Drug Delivery to the Brain*. New York: Raven Press.

Pardridge, W.M. 2001. *Brain Drug Targeting: The Future of Brain Drug Development*. Cambridge, UK: Cambridge University Press.

Pardridge, W.M. 2002. Drug and gene targeting to the brain with molecular Trojan horses. *Nature Rev. Drug Discov.* 1(2):131–139.

Pardridge, W.M., J. Eisenberg, and J. Yang. 1987. Human blood-brain barrier transferrin receptor. *Metabolism* 36(9):892–895.

Pardridge, W.M., Y.S. Kang, J.L. Buciak, and J. Yang. 1995a. Human insulin receptor monoclonal antibody undergoes high affinity binding to human brain capillaries in vitro and rapid transcytosis through the blood-brain barrier. *Pharm. Res.* 12(6):807–816.

Pardridge, W.M., Y.S. Kang, J. Yang, and J.L. Buciak. 1995b. Enhanced cellular uptake and in vivo biodistribution of a monoclonal antibody following cationization. *J. Pharm. Sci.* 84(8):943–948.

Pardridge, W.M., J. Yang, J. Buciak, and W.W. Tourtellotte. 1989. Human brain microvascular DR-antigen. *J. Neurosci. Res.* 23(3):337–341.

Patel, S.J., W.R. Shapiro, D.W. Laske, et al. 2005. Safety and feasibility of convection-enhanced delivery of cotara for the treatment of malignant glioma: Initial experience in 51 patients. *Neurosurgery* 56(6):1243–1252.

Penichet, M.L., Y.S. Kang, W.M. Pardridge, S.L. Morrison, and S.U. Shin. 1999. An antibody-avidin fusion protein specific for the transferrin receptor serves as a delivery vehicle for effective brain targeting: Initial applications in anti-HIV antisense drug delivery to the brain. *J. Immunol.* 163(8):4421–4426.

Pissarra, N. 2004. Changes in the business of culture. *Nature Biotechnol.* 22:1355–1356.

Poduslo, J.F., and G.L. Curran. 1994. Glycation increases the permeability of proteins across the blood-nerve and blood-brain barriers. *Mol. Brain Res.* 23(1–2):157–162.

Poduslo, J.F., and G.L. Curran. 1996. Polyamine modification increases the permeability of proteins at the blood-nerve and blood-brain barriers. *J. Neurochem.* 66(4):1599–1609.

Poduslo, J.F., and G.L. Curran. 2001. Amyloid beta peptide as a vaccine for Alzheimer's disease involves receptor-mediated transport at the blood-brain barrier. *Neuroreport* 12(15):3197–3200.

Poduslo, J.F., G.L. Curran, and C.T. Berg. 1994. Macromolecular permeability across the blood-nerve and blood-brain barriers. *Proc. Natl. Acad. Sci. U.S.A.* 91(12):5705–5709.

Poduslo, J.F., G.L. Curran, B. Sanyal, and D.J. Selkoe. 1999. Receptor-mediated transport of human amyloid beta-protein 1-40 and 1-42 at the blood-brain barrier. *Neurobiol. Dis.* 6(3):190–199.

Qian, Z.M., H.Y. Li, H.Z. Sun, and K. Ho. 2002. Targeted drug delivery via the transferrin receptor-mediated endocytosis pathway. *Pharmacol. Rev.* 54(4):561–587.

Ramsauer, M., D. Krause, and R. Dermietzel. 2002. Angiogenesis of the blood-brain barrier in vitro and the function of cerebral pericytes. *FASEB J.* 16(8):1274–1276.

Raub, T.J., S.L. Kuentzel, and G.A. Sawada. 1992. Permeability of bovine brain microvessel endothelial cells in vitro: Barrier tightening by a factor released from astroglioma cells. *Exp. Cell. Res.* 199(2):330–340.

Roopenian, D.C., and S. Akilesh. 2007. FcRn: The neonatal Fc receptor comes of age. *Nature Rev. Immunol.* 7(9):715–725.

Rothberg, K.G., J.E. Heuser, W.C. Donzell, Y.S. Ying, J.R. Glenney, and R.G.W. Anderson. 1992. Caveolin, a protein-component of caveolae membrane coats. *Cell* 68(4):673–682.

Rubin, L.L., D.E. Hall, S. Porter, et al. 1991. A cell culture model of the blood-brain barrier. *J. Cell Biol.* 115(6):1725–1735.

Saerens, D., M. Pellis, R. Loris, et al. 2005. Identification of a universal VHH framework to graft non-canonical antigen-binding loops of camel single-domain antibodies. *J. Mol. Biol.* 352(3):597–607.

Sampson, J.H., L.E. Crotty, S. Lee, et al. 2000. Unarmed, tumor-specific monoclonal antibody effectively treats brain tumors. *Proc. Natl. Acad. Sci. U.S.A.* 97(13):7503–7508.

Schenk, D., R. Barbour, W. Dunn, et al. 1999. Immunization with amyloid-beta attenuates Alzheimer disease-like pathology in the PDAPP mouse. *Nature* 400(6740):173–177.

Schlachetzki, F., C.N. Zhu, and W.M. Pardridge. 2002. Expression of the neonatal Fc receptor (FcRn) at the blood-brain barrier. *J. Neurochem.* 81(1):203–206.

Schroder, U., and B.A. Sabel. 1996. Nanoparticles, a drug carrier system to pass the blood-brain barrier, permit central analgesic effects of i.v. dalargin injections. *Brain Res.* 710(1–2):121–124.

Shi, N., R.J. Boado, and W.M. Pardridge. 2000. Antisense imaging of gene expression in the brain in vivo. *Proc. Natl. Acad. Sci. U.S.A.* 97(26):14709–14714.

Shi, N.Y., R.J. Boado, and W.M. Pardridge. 2001. Receptor-mediated gene targeting to tissues in vivo following intravenous administration of pegylated immunoliposomes. *Pharm. Res.* 18(8):1091–1095.

Shi, N.Y., Y. Zhang, C.N. Zhu, R.J. Boado, and W.M. Pardridge. 2001. Brain-specific expression of an exogenous gene after i.v. administration. *Proc. Natl. Acad. Sci. U.S.A.* 98(22):12754–12759.

Smith, M.W., and M. Gumbleton. 2006. Endocytosis at the blood-brain barrier: From basic understanding to drug delivery strategies. *J. Drug Target.* 14(4):191–214.

Solomon, B., R. Koppel, D. Frankel, and E. HananAharon. 1997. Disaggregation of Alzheimer beta-amyloid by site-directed mAb. *Proc. Natl. Acad. Sci. U.S.A.* 94(8):4109–4112.

Song, B.W., H.V. Vinters, D.F. Wu, and W.M. Pardridge. 2002. Enhanced neuroprotective effects of basic fibroblast growth factor in regional brain ischemia after conjugation to a blood-brain barrier delivery vector. *J. Pharmacol. Exp. Ther.* 301(2):605–610.

Steiniger, S.C.J., J. Kreuter, A.S. Khalansky, et al. 2004. Chemotherapy of glioblastoma in rats using doxorubicin-loaded nanoparticles. *Int. J. Cancer* 109(5):759–767.

Sugawara, I., H. Hamada, T. Tsuruo, and S. Mori. 1990. Specialized localization of P-glycoprotein recognized by Mrk 16 monoclonal antibody in endothelial cells of the brain and the spinal cord. *Jpn. J. Cancer Res.* 81(8):727–730.

Suzuki, T., D.F. Wu, F. Schlachetzki, J.Y. Li, R.J. Boado, and W.M. Pardridge. 2004. Imaging endogenous gene expression in brain cancer in vivo with In-111-peptide nucleic acid antisense radiopharmaceuticals and brain drug-targeting technology. *J. Nuclear Med.* 45(10):1766–1775.

Tao-Cheng, J.H., Z. Nagy, and M.W. Brightman. 1987. Tight junctions of brain endothelium in vitro are enhanced by astroglia. *J. Neurosci.* 7(10):3293–3299.

Triguero, D., J.L. Buciak, and W.M. Pardridge. 1991. Cationization of immunoglobulin G results in enhanced organ uptake of the protein after intravenous administration in rats and primate. *J. Pharmacol. Exp. Ther.* 258(1):186–192.

Triguero, D., J.B. Buciak, J. Yang, and W.M. Pardridge. 1989. Blood-brain barrier transport of cationized immunoglobulin G enhanced delivery compared to native protein. *Proc. Natl. Acad. Sci. U.S.A.* 86(12): 4761–4765.

Tsuji, A., T. Terasaki, Y. Takabatake, et al. 1992. P-glycoprotein as the drug efflux pump in primary cultured bovine brain capillary endothelial-cells. *Life Sci.* 51(18):1427–1437.

Tunkel, A.R., B.J. Hartman, S.L. Kaplan, et al. 2004. Practice guidelines for the management of bacterial meningitis. *Clin. Infect. Dis.* 39(9):1267–1284.

Ullrich, A., J.R. Bell, E.Y. Chen, et al. 1985. Human insulin receptor and its relationship to the tyrosine kinase family of oncogenes. *Nature* 313(6005):756–761.

Unterberg, A., Wahl, M., and A. Baethmann. 1984. Effects of bradykinin on permeability and diameter of pial vessels in vivo. *J. Cereb. Blood Flow Metab.* 4(4):574–585.

Urayama, A., J.H. Grubb, W.S. Sly, and W.A. Banks. 2004. Developmentally regulated mannose 6-phosphate receptor-mediated transport of a lysosomal enzyme across the blood-brain barrier. *Proc. Natl. Acad. Sci. U.S.A.* 101(34):12658–12663.

Wang, X.X., Y.K. Cho, and E.V. Shusta. 2007. Mining a yeast library for brain endothelial cell-binding antibodies. *Nature Methods* 4(2):143–145.

Xia, C.F., Y. Zhang, Y. Zhang, R.J. Boado, and W.M. Pardridge. 2007. Intravenous siRNA of brain cancer with receptor targeting and avidin-biotin technology. *Pharm. Res.* 24(12):2309–2316.

Yan, Q., C. Matheson, J. Sun, M.J. Radeke, S.C. Feinstein, and J.A. Miller. 1994. Distribution of intracerebral ventricularly administered neurotrophins in rat brain and its correlation with trk receptor expression. *Exp. Neurol.* 127(1):23–36.

Yang, W.L., R.F. Barth, G. Wu, et al. 2005. Development of a syngeneic rat brain tumor model expressing EGFRvIII and its use for molecular targeting studies with monoclonal antibody L8A4. *Clin. Cancer Res.* 11(1):341–350.

Zhang, Y., F. Calon, C.N. Zhu, R.J. Boado, and W.M. Pardridge. 2003. Intravenous nonviral gene therapy causes normalization of striatal tyrosine hydroxylase and reversal of motor impairment in experimental parkinsonism. *Hum. Gene Ther.* 14(1):1–12.

Zhang, Y., and W.M. Pardridge. 2001a. Mediated efflux of IgG molecules from brain to blood across the blood-brain barrier. *J. Neuroimmunol.* 114(1–2):168–172.

Zhang, Y., and W.M. Pardridge. 2001b. Neuroprotection in transient focal brain ischemia after delayed intravenous administration of brain-derived neurotrophic factor conjugated to a blood-brain barrier drug targeting system. *Stroke* 32(6):1378–1383.

Zhang, Y., and W.M. Pardridge. 2006. Blood-brain barrier targeting of BDNF improves motor function in rats with middle cerebral artery occlusion. *Brain Res.* 1111(1):227–229.

Zhang, Y., F. Schlachetzki, J.Y. Li, R.J. Boado, and W.M. Pardridge. 2003. Organ-specific gene expression in the rhesus monkey eye following intravenous non-viral gene transfer. *Mol. Vis.* 9:465–472.

Zhang, Y., F. Schlachetzki, and W.M. Pardridge. 2003. Global non-viral gene transfer to the primate brain following intravenous administration. *Mol. Ther.* 7(1):11–18.

Zhang, Y., F. Schlachetzki, Y.F. Zhang, R.J. Boado, and W.M. Pardridge. 2004a. Normalization of striatal tyrosine hydroxylase and reversal of motor impairment in experimental parkinsonism with intravenous nonviral gene therapy and a brain-specific promoter. *Hum. Gene Ther.* 15(4):339–350.

Zhang, Y., Y.F. Zhang, J. Bryant, A. Charles, R.J. Boado, and W.M. Pardridge. 2004b. Intravenous RNA interference gene therapy targeting the human epidermal growth factor receptor prolongs survival in intracranial brain cancer. *Clin. Cancer Res.* 10(11):3667–3677.

Zhang, Y., C.N. Zhu, and W.M. Pardridge. 2002. Antisense gene therapy of brain cancer with an artificial virus gene delivery system. *Mol. Ther.* 6(1):67–72.

Zlokovic, B.V., C.L. Martel, E. Matsubara, et al. 1996. Glycoprotein 330/megalin: Probable role in receptor-mediated transport of apolipoprotein J alone and in a complex with Alzheimer disease amyloid beta at the blood-brain and blood-cerebrospinal fluid barriers. *Proc. Natl. Acad. Sci. U.S.A.* 93(9):4229–4234.

PART VI

ANTIBODY CHARACTERIZATION

CHAPTER 22

Determination of Equilibrium Dissociation Constants

ROBIN E. ERNST, KATRINA N. HIGH, TOM R. GLASS, and QINJIAN ZHAO

22.1 INTRODUCTION

Binding specificities of monoclonal antibodies (mAb) against virtually any antigen can be generated whether with conventional hybridoma technology or with antibody libraries displayed on filamentous phage or other display systems. The determination of antibody–antigen affinity constants or equilibrium dissociation constants (K_D values) is a key parameter for understanding antibody–antigen

Therapeutic Monoclonal Antibodies: From Bench to Clinic. Edited by Zhiqiang An

interactions. It can be predictive of antibody performance in biotechnological applications, such as in immunochemical and bioanalytical assays, in diagnostic assays, and as therapeutic drugs.

During the early discovery stage, affinity maturation using protein engineering techniques of a prototype mAb is often necessary to achieve the level of binding that would result in a biological effect. The quantitative determination of K_D values of the newly obtained mAbs through affinity maturation is a critical benchmark for tracking progress against a putative affinity goal to incur a certain biological effect. If the total concentrations of antibody (Ab) and antigen (Ag) are well below the K_D value, only a small fraction of them will exist in the bound form as Ab–Ag complex. Under these conditions, it is unlikely any biological effects will be triggered. Conversely, at total concentrations of antibody (Ab) and antigen (Ag) well above the K_D, a large proportion of one or both of the binding partners will be in the associated form. Therefore, K_D can be seen as a useful parameter to predict whether the Ab or Ag is largely in free or associated form under certain conditions. It is useful to determine the K_D value of an mAb obtained via a library panning or after some rounds of affinity maturation, because this information, along with the circulating Ag level or the surface receptor level (if the Ag is a membrane-associated receptor), will give us a rough idea as to whether this mAb will have a chance to engage the intended target *in vivo* and cause desired biological effects.

Once a given mAb has passed the preclinical development stages with established safety and efficacy, the determination of K_D values and active concentrations, for different lots of that mAb from different manufacturing scales and processes, still remains a critical task. The K_D values can be used for demonstrating the integrity and functional activity of the mAb during toxicological studies and various phases of clinical trials and post-marketing. There are a few different ways one can measure the K_D value of a given Ab–Ag pair and which technique one uses depends on the kinetic rate characteristics between the Ab and Ag. The methods fall into two categories: (1) equilibrium saturation analysis when the free and bound concentrations of one of the binding partner need to be measured; and (2) kinetics-based determination of association (k_a or k_{on}) and dissociation rate constants (k_d or k_{off}) from which K_D values can be derived from the ratio of the two. For determination of K_D at equilibrium, the measurement and/or the separation (via dialysis, centrifugation and chromatography, etc) of free and bound forms need to be achieved, with the assumption that the dissociation of the Ab–Ag complex is not significant during the time of separation, before the measurements are made. For kinetics-based measurements, label-free and real-time biosensor-based technology is employed, such as surface plasmon resonance (SPR)-based Biacore or other SPR-based technologies. The characterization of high affinity binding pairs using kinetics-based and equilibrium-based methods has been reviewed (Drake, Myszka, and Klakamp 2004).

In this chapter, we introduce the Biacore technology and its application for determination of K_D based on the measurements of association (k_a or k_{on}) and dissociation rate constants (k_d or k_{off}) with some examples using several human mAbs. Subsequently, two equilibrium-based K_D determination methods, fluorescence ELISA and KinExA, are presented, with some examples. Finally, the application of the KinExA technique in the determination of K_D involving whole cells is discussed.

22.2 KINETIC-BASED DISSOCIATION CONSTANT DETERMINATION BY SPR-BASED TECHNOLOGY

22.2.1 Surface Plasmon Resonance-Based Technology

BIAcore™ is the most commonly used SPR-based technology for characterization of real-time biomolecular interactions. It has been used for kinetic characterization of engineered antibodies during early and advanced stages of the development process. An SPR-based optical biosensor is designed to measure the change in the refractive index of a solution near a thin metal surface that occurs as a result of molecular complex formation or dissociation (Huber and Mueller 2006). The response detected has been shown to be proportional to the mass bound or deposited at the surface (Stenberg et al. 1991). SPR detection allows one to directly measure the binding of a molecule in solution to a

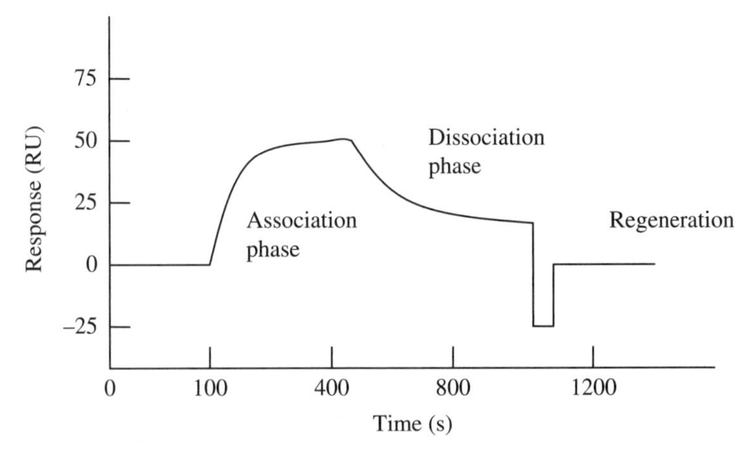

Figure 22.1 An example of a sensorgram that shows real-time information about the antibody–antigen interaction. The association and dissociation phases are measured in resonance units (RU) as a function of time.

surface immobilized binding partner and determine both the kinetics of that interaction, on-rate and off-rate, and the dynamics of the interaction or affinity (equilibrium dissociation constant, K_D). The advantages of these technologies include no labeling of the molecules, small sample volumes, and analysis of crude sample preparations. The majority of the work published in the antibody field originates from real-time binding analysis based on biosensor instruments produced by BIAcore (GE Biosciences) although several other instrumentation manufacturers have entered the field (Rich and Myszka 2007a).

Biacore has gained its popularity partly due to its automated liquid handling capacity with the integrated fluid control unit. A basic diagram of biosensor output or sensorgram is illustrated in Figure 22.1. The response unit (RU) or RU changes represents changes in the refractive index at the interface between the sensor surface and solution. RU changes are proportional to the deposited mass changes or binding events at the Ag or Ab modified gold surface measured in real time.

22.2.2 Kinetic Analysis by SPR via Real Time Monitoring

The changes in the detected SPR signal are expressed as resonance/response units or RU and are followed as a function of time. The sensorgram in Figure 22.1 outlines the five basic phases during the binding of an Ag to an immobilized Ab or vice versa. The first phase is the baseline signal or the pre-injection phase of buffer only, which is followed by the injection of the Ag or association phase. Once the association phase is finished, the Ag solution is replaced with buffer and the dissociation phase begins, which monitors the dissociation of the Ag from the Ab or dissociation of the complex over time. The final phases are for the regeneration of the sensor surface, to remove all bound Ag using predetermined regeneration reagents and conditions, followed by a stabilization phase where only buffer is flowing and maintaining the surface in preparation for the next round of Ag injection. In a typical experiment a set of varying Ag concentrations are injected sequentially onto the specifically derivatized surface for K_D determination. The kinetic-based equilibrium dissociation constant or K_D is the ratio of off-rate over on-rate for a given bimolecular interaction. K_D is expressed in molar units, M, and describes strength of the binding.

Once the surface is prepared and stabilized, a series of serially diluted Ag or Ab samples can be injected onto the surface with regeneration between cycles of injections. The kinetic data analysis and curve fitting calculations can be done by evaluation software supplied by the manufacturer of the biosensor instrument, such as BIA evaluation (from Biacore). These software programs also have multiple binding models that consider simple 1 : 1 interactions, binding due to mass transport limitation, heterogeneous ligand model where bound antibody may be present in multiple forms, bivalent and heterogeneous analyte models, and more complicated two state and competing reactions models (Karlsson, Roos, and Fägerstam 1994; Karlsson and Fält 1997; Alfthan 1998). The use of these alternative models requires previous knowledge of such interactions that deviate from a 1 : 1 interaction.

To generate reliable and accurate kinetic measurements for an Ag–Ab interaction, one must have well-characterized starting material in which the concentration, purity, and activity have been determined by multiple methods if possible. Most of these interactions are at first considered to follow a simple 1 : 1 kinetic interaction model and one is encouraged to set up the experiment to support that model, so it is important that the molecule in solution or Ag is monovalent to prevent avidity effects that cause a deviation from the true rate constants. In most cases, this means that the Ag will be the analyte in solution and the bivalent Ab (IgG) will be immobilized on the chip surface either directly or indirectly via a capture molecule. The manner in which the Ab is immobilized depends on its activity when directly bound to the chip surface or if stability is an issue or one wishes to re-use the chip for several different IgGs, a capture molecule such as protein A or an anti-IgG antibody can be used (Darling et al. 2002; Canziani, Klakamp, and Myszka 2004). While there are several kinds of chip surfaces available for the covalent coupling of a protein, the most common chips used in these studies are coated with a carboxymethylated dextran matrix like the Biacore CM5 chip. Immobilization via amine groups is the most popular method used for surface immobilization. The coupling normally occurs between the primary amine group of lysine residues on the Ab (or Ag) and free carboxylic acid groups on the surface which are generated by treatment with 1-ethyl-3-(3-dimethylaminopropyl) carbodiimide (EDC) and *N*-hydroxysuccinimide (NHS). The preparation of a surface that can be regenerated for multiple cycles of analysis is crucial for kinetic analysis using a series of Ab or Ag concentrations in solution for binding kinetic studies.

22.2.3 Kinetic Analysis of a Human mAb (IgG) with Its Antigen (a Receptor)

To study the binding kinetics and affinity for a given binding pair, different ways of surface immobilization need to be considered. The surface can be chemically derivatized by immobilizing a specific Ab (as a specific capture agent, in the case of Fig. 22.2, or directly as the test article) or an Ag (e.g., the extracellular domain of IL-13R in Fig. 22.3). The regeneration of the immobilized molecule—Ab, Ag, or a specific captured Ab—is critical for subsequent regeneration of the surface after each cycle. In addition, one important consideration for kinetic analysis is the amount of ligand to immobilize on a sensor flow cell surface. For kinetic analysis it is recommended that the immobilized ligand density be kept low to prevent steric hindrance, aggregation, rebinding of the analyte during dissociation and mass transport limited binding. The magnitude of the SPR response is proportional to the mass concentration at the sensor surface and the binding capacity of the surface is related to the amount of ligand immobilized. It is recommended that a ligand density be used that produces a RU_{max} between 20 and 150 RU depending on the instrument model used. Once the chip surface has been prepared, the kinetics experiment may be set up and run. It is important to consider the estimated K_D of the Ab–Ag interaction when determining the concentration titration of Ag to use. The Ag or analyte concentration dilution series should range from approximately 10-fold above to 10-fold below the estimated K_D value. The example in Figure 22.2 illustrates data output and curve fitting for binding of an Ag in solution to a surface-captured mAb.

In this experiment, a goat anti-human Fcγ IgG specific polyclonal antibody was immobilized to both an experimental and control reference flow cell surface via amine coupling. During each cycle of a run, the mAb was captured on the experimental flow cell surface via its Fcγ at a density of approximately 200 RU at a low flow rate. After capture of the mAb, the Ag is injected over both the experimental and reference flow cell surfaces at an increased flow rate of 60 μL/min. The reference-subtracted signal represents binding of the Ag to the mAb and discounts binding to the anti-human Fc IgG only flow cell. The recombinant human IgG tested was originally cloned from a mouse hybridoma. The mAb that was produced by the hybridoma was shown to have a K_D of 20 to 30 nM. Based on this estimated K_D, the antigen series titration ranged from 200 nM to

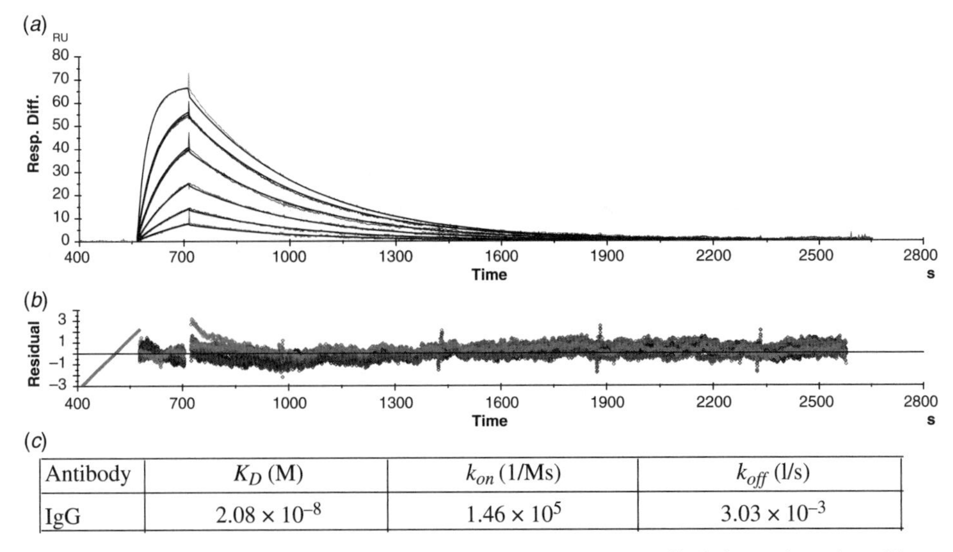

Antibody	K_D (M)	k_{on} (1/Ms)	k_{off} (l/s)
IgG	2.08×10^{-8}	1.46×10^{5}	3.03×10^{-3}

Figure 22.2 Example of kinetic data output and analysis for an antigen–antibody interaction using a Biacore 3000 instrument. In this experiment, the IgG to be tested was captured on a goat anti-human Fc IgG surface using a CM5 Biacore chip. The binding data was generated by injecting (six) 2-fold dilutions of the antigen in duplicate. The model used to fit the experimental data was 1 : 1 Langmuir using global analysis. (a) The overlaid fitted curve for each duplicated antigen concentration titration curve. (b) The curve fitting residuals, comparing the calculated and experimental point values for each of the experimental curves. The precision of the overall fit is reported as the statistical value χ^2 and for each parameter (R_{max}, k_{on}, and k_{off}) the standard error is reported estimates of how sensitive the fitting is to changes in the parameter. For this data set, the χ^2 was <1 and the T-value for each parameter was >700. (c) The calculated rate constants and the dissociation constant or affinity of the IgG for the antigen.

6.25 nM. The model used to analyze the kinetic data was a 1 : 1 Langmuir global analysis done by the manufacturer's installed software. Figure 22.2 shows the fitted sensorgram curves for each Ag concentration, as well as the calculated residuals along those curves. Under these experimental conditions, the apparent affinity of the recombinant IgG was 20.8 nM, very similar to the K_D value obtained with the original hybridoma IgG material. This is an ideal case for kinetic studies by Biacore as the relatively fast off-rate allows the monitoring of the complete dissociation phase down to baseline in about 30 min. It is not always the case for high affinity binding pairs and for those with multivalent interactions.

22.2.4 Kinetic Analysis of High Affinity Interactions

While the previous section presented an ideal case for SPR-based kinetic studies, analyzing high affinity binding pairs presents many challenges. When the affinity is high, such as subnanomolar down to picomolar, the dissociation rate is often much slower, thus causing difficulties in observing the dissociation phase. High affinity mAbs are very common, and as a matter of fact, these are highly desired for achieving biological effects. One of the main goals of recombinant antibody discovery and optimization is to develop therapeutics with enhanced efficacy. This can be achieved with higher affinity for the target molecule while preserving the specificity. Molecular evolution-based technologies such as phage display can generate large libraries of variant antibody fragments with optimized properties such as overall affinity for the target and improved functional characteristics. SPR-based screening has been used as a tool to identify and characterize these improved antibodies that have been isolated by selection against the target molecule. During affinity maturation, the kinetic rate constants of the variant antibodies can be compared to the original antibody. Figure 22.3 is an example of variant clones isolated from combined light chain CDR3 libraries covering the complete sequence diversity

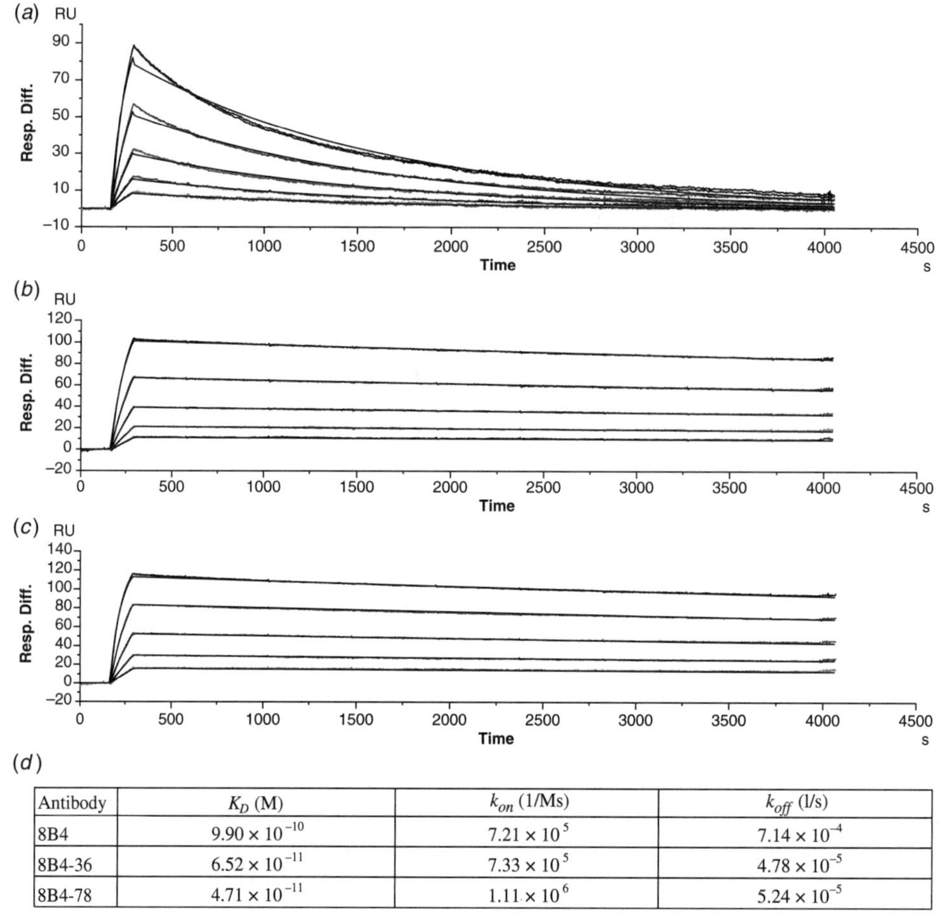

Antibody	K_D (M)	k_{on} (1/Ms)	k_{off} (l/s)
8B4	9.90×10^{-10}	7.21×10^{5}	7.14×10^{-4}
8B4-36	6.52×10^{-11}	7.33×10^{5}	4.78×10^{-5}
8B4-78	4.71×10^{-11}	1.11×10^{6}	5.24×10^{-5}

Figure 22.3 Affinity characterization of recombinant Fab antibody fragments recovered from phage display libraries generated from the variable regions of a recombinant antibody specific for IL13Rα1 receptor using a Biacore 3000 instrument. Panels A to C show the fitted curves for each Fab fragment (A = parental antibody 8B4, B and C are the affinity matured antibody fragments, 8B4-36 and 8B4-78, respectively). The IL13Rα1 extracellular domain was immobilized on a CM5 chip surface at a low density using amine coupling. The kinetic constants were calculated by injecting (five) 2-fold dilutions of the Fab fragments for 1 minute at a flow rate of 60 μL/min and the dissociation was measured for 60 minutes. The curves were fitted using 1 : 1 Langmuir model. Panel D summarizes the kinetic data and shows an overall affinity improvement of 15- to 20-fold for the affinity matured Fab antibody fragments.

by Fab phage display that have an affinity improvement for the target protein human IL13RαI extracellular domain.

The kinetic analysis shows an affinity improvement of 15- to 20-fold compared to the parental antibody. The binding improvement appears to be mainly due to a decrease in the dissociation rate of the Ag–Ab complex with limited effects on the association rate. The experimental design of SPR-based kinetic analyses is an evolving process and careful review of the literature is necessary to apply current practical and theoretical considerations to a specific set of experiments (Rich and Myszka 2007b). One of the issues with using Biacore in kinetic analysis is the slow dissociation of mAbs with high affinity (Drake, Myszka, and Klakamp 2004). It is evident from Figure 22.3 that for the two affinity matured Fab (8B4-36 and 8B4-78) only a very small fraction was observed to dissociate in approximately one hour or 4000 s. Based on the calculated k_{off} value of $5 \times 10^{-5}\,s^{-1}$, the half-life of the complex is 1.2×10^4 s, which is ~3.3 hours. This means, if one desires to monitor 50 percent of the immunocomplex to dissociate to gain enough confidence for k_{off} analysis, one has to monitor for 3.3 hours, which makes it impractical. In the following sections, we will look at the techniques that allow prolonged incubation of the Ab and Ag mixtures to completely reach equilibrium, and then determine the equilibrium dissociation constants by quantitating the fraction of free versus bound concentrations. While there is no information on kinetic parameters being obtained, much higher confidence is achieve for K_D determination as the measurements take place on the fully equilibrated mixtures.

22.3 EQUILIBRIUM-BASED DISSOCIATION CONSTANT DETERMINATION BY FLUORESCENCE ELISA (FL-ELISA)

22.3.1 Relationship between Kinetic- and Equilibrium-Based K_D Determination

One of the most fundamental ways to quantitatively characterize the interaction between two molecules as defined in Equation 22.1 is to determine the dissociation constant (K_D), commonly referred to as *affinity constant* in most literature. The affinity constant is defined as the ratio of the rate constants (Kinetic) or the ratio of concentrations at equilibrium, when k_{on} is equal to k_{off}, for a two-phase reversible interaction as defined in Equation 22.2. Affinity constants give quantitative meaning to phrases such as tight binding and weak interaction, and refers to the stability of the bimolecular complex (Goodrich and Kugel 2007). Most often bimolecular interactions are dynamic processes that occur in solution and include multiple association and dissociation phases. The dissociation constant (K_D) is the intrinsic property or measure of the binding energetics of a given binding pair that can be measured experimentally.

$$\mathrm{mAb} + \mathrm{L} \underset{k_{off}(\mathrm{s}^{-1})}{\overset{k_{on}(\mathrm{M}^{-1}\cdot\mathrm{s}^{-1})}{\rightleftharpoons}} \mathrm{mAb}\cdot\mathrm{L} \tag{22.1}$$

$$K_D = \frac{[\mathrm{mAb}]\cdot[\mathrm{L}]}{[\mathrm{mAb}\cdot\mathrm{L}]}(\mathrm{M}) \quad \text{or} \quad K_D = \frac{k_{off}}{k_{on}}(\mathrm{M}) \tag{22.2}$$

In the previous section, we discussed K_D determination using methodology that determined the kinetic rate constants for the interaction. While solution affinity at equilibrium can be determined using the Biacore platform, K_D values less than 100 pM are difficult to measure. ELISA with fluorescence detection (FL)-ELISA and KinExA technologies are able to measure subpicomolar affinity values.

22.3.2 ELISA Sensitivity: Determining Factor for Low K_D Analysis (High Affinity Interactions)

There are various methods available and in practice for measuring K_D values, including equilibrium dialysis, protein affinity chromatography, sedimentation through gradient, gel filtration, surface plasmon resonance, KinExA, and ELISA (Goldberg and Ohaniance 1993). FL-ELISA is one of the most convenient and sensitive techniques for low level analyte quantitation without having to resort to radiolabeling of Ab (or Ag) for increased sensitivity. The lower limit of K_D determination for a sensitive binding system correlates to the ELISA sensitivity.

The determining factors for the FL-ELISA sensitivity are the choices of enzyme conjugates and substrates used. The FL-ELISA is more sensitive than the colorimetric-based assays due to use of fluorescence substrate in place of the colorimetric substrate (Meng et al. 2005; Towne et al. 2004). With fluorescence detection in ELISA a 5- to 10-fold enhancement in sensitivity may be achieved over commonly used colorimetric substrates. K_D values as low as 1 nM have been reported using ELISA with colorimetric detection (Friguet et al. 1985). The FL-ELISA method was shown to determine solution K_D values as low as 10 pM (Friguet et al. 1985; High et al. 2005). The key to the FL-ELISA method is that it is solution based so there are no modifications or surface adsorption of either the Ab or Ag and the method is sensitive due to fluorescence detection.

22.3.3 Determination of Free and Bound Components in Ab–Ag Mixtures

In order to determine the K_D of a bimolecular interaction using the equilibrium-based FL-ELISA method, the free and bound fractions of one of the binding components must be determined in the reaction mixture at equilibrium. In fact, the first step in K_D determination is to quantitate the free and bound fractions of the Ab and Ag in the reaction mixtures at equilibrium. For example, an Ag-coated plate can be used to capture the free Ab in solution at equilibrium. This is referred to as indirect ELISA, where the Ag–Ab mixture is transferred to an Ag-coated plate to quantitate the free Ab concentrations in the mixtures. This in-direct method works only when the dissociation of the Ab–Ag complex is insignificant over the time allowed for the capture step. Thus, it is advantageous for the complex to have a slow off-rate. After the Ab–Ag equilibrium mixtures are transferred to Ag-coated plate for a short incubation period; the plates should be washed to remove any unbound material and a secondary antibody enzyme conjugate should be applied to detect the free Ab captured on the Ag-coated plate. As mentioned earlier, the free Ab levels must be detectable and quantifiable.

The indirect FL-ELISA method experimental set up demands the Ab concentration to be held constant at 2- to 5-fold below K_D while the Ag concentration range is from 5- to 10-fold below and 5- to 10-fold above the K_D value. With the FL-ELISA method, being particularly useful for determining K_D values of high affinity interactions, the expectation is that very little free Ab will be available so the detection system must be very sensitive. After the enzyme conjugate incubation for detection of the Ag–Ab complex, the plates should be thoroughly washed and a horseradish peroxidase sensitive substrate like ADHP should be used as the detection system. ADHP, 10-acetyl-3,7-dihydroxyphenoxazine, is a resorufin-based substrate that acts on the HRP enzyme producing an oxidation product (High et al. 2005). The horseradish peroxidase with ADHP detection system was shown to offer lower range detection for sensitive systems (Meng et al. 2005).

Once the data are obtained with FL-ELISA for the Ab–Ag mixtures, using a set of Ab-only dilution series as a calibration curve, the raw FL-ELISA readouts were transformed into the free Ab concentrations in the mixtures (Fig. 22.4). This is one of two curve-fitting steps needed for the K_D determination. This very important step uses the calibrations process that allows elimination of potential errors in the curve fittings that rely on raw signal outputs. And thus, the final K_D result does not rely on raw signal intensities, as found with many other methods where nonlinearity (assay signal vs. analyte) could potentially skew the K_D values (Azimzadeh, Pellequer, and Van Regenmortel 1992; Goldberg and Ohaniance 1993). The nonlinearity in the raw signal outputs versus fraction of bound species can be removed in this step (Fig. 22.4).

22.3.4 K_D Determination of Affinity Macromolecules with FL-ELISA

True solution-based K_D values for Ab–Ag binding pairs can be measured using an indirect FL-ELISA method with both binding partners in the solution phase. The indirect FL-ELISA method, in addition to being a more practical application for measuring true solution-based K_D values, more importantly allows the analysis of high affinity macromolecules. While some of the methods used for measuring K_D values mentioned previously are reliable methods to determine the K_D values for a small molecule binding to a large molecule, measuring the K_D value for two interacting macromolecules in solution is not trivial (Friguet et al. 1985). It is even more challenging to measure subnanomolar K_D values for extremely tight-binding pairs due to the level of detection sensitivity required. In cases like this, the FL-ELISA method is the most suitable technique for determining K_D values.

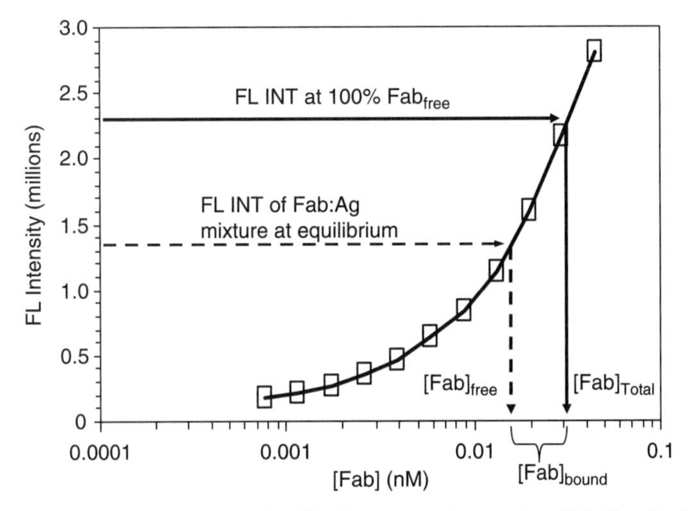

Figure 22.4 A representative Fab only ELISA calibration curve and calculation of $[\mathrm{Fab}]_{\mathrm{free}}$ (in nM) from the fluorescence intensity (FL INT) in a given Ab–Ag mixture. (Note: although intact human IgG is used, the binding-site concentration of Ab is expressed in the putative Fab concentration, that is, twice that of the IgG concentration as there are two Fabs per IgG molecule.) The fluorescence readings are used to generate the Ab only calibration curve, from which $[\mathrm{Fab}]_{\mathrm{free}}$ at equilibrium was determined using a four-parameter logistic curve fit. This information is used to calculate $[\mathrm{Fab}]_{\mathrm{bound}}$ and ultimately K_D values (see Fig. 22.5).

Some of the traditional methods for measuring K_D values for tight-binding pairs were found to be unsatisfactory due to chemical modification of one binding partner or due to conformational heterogeneity induced by immobilization onto a sensor chip or a solid surface (Friguet et al. 1985; Azimzadeh, Pellequer, and Van Regenmortel 1992). In addition, adsorption to a solid surface could potentially lead to undesired conformational changes or prevent important induced fit-based conformational changes, thus leading to deceptive or biased K_D values for some bimolecular interactions (Goldberg and Ohaniance 1993). For the methods involving chemically modifying either the Ab or Ag for detection purposes, potential modification of residues in or near the binding site could alter the interactions between Ab and Ag (Goldberg and Ohaniance 1993). It has also been shown that binding to sites outside the variable regions of the Ab, such as the Fc portion, by protein A or anti-Fc antibodies can change the K_D (Stevens 1987). Therefore, probing Ab–Ag interactions in solution with an indirect FL-ELISA is very attractive since it allows free interaction between the Ab and Ag, reducing the chances for any undesired conformational changes or other molecular heterogeneities such as oligomerization or aggregation and the solution-based interaction poses no size limitation (Friguet et al. 1985).

The details of a method for measuring K_D values down to picomolar revels using the indirect FL-ELISA were reported previously (High et al. 2005). Examples of solution based equilibrium K_D values for human IgGs binding their target antigen determined using the indirect FL-ELISA are described below. For a typical experiment, there are two interacting species, namely the Ab and Ag, one of which will be used for detection and thus will require a means of detection. ELISA assays are typically performed on 96-well microtiter plates and there are two microtiter plates needed in the assay. One plate should be a high binding plate where molecules can adsorb to the plate passively over time. The high binding plate should be coated with the Ag (note the Ab could also be used for coating depending on assay set up). The second plate should be a low binding microtiter plate. The low binding plate is used to prepare the equilibrium binding state of the Ab and Ag. Low binding 96-well plate set up should consist of the calibration standard and the Ab–Ag mixture referenced in Figure 22.4.

The calibration curve should represent the molecule in the system that will be detected in the indirect FL-ELISA. The calibration curve in this example should be Ab only, tested in duplicate and titrated across the columns of the microtiter plate leaving a couple of wells to serve as the plate blank (no Ag or Ab). The remaining rows of the microtiter plate can be used to prepare the Ab–Ag mixture. Prepare replicate titrations of each antigen sample covering an appropriate concentration range above and below the anticipated K_D discussed previously. Add a constant concentration of Ab 2- to 5-fold below the K_D value if feasible to the Ag serial dilution series. Having the concentration of Ab in solution well below the K_D allows the assumption that only a small portion of the Ag will be in the bound form, which simplifies future calculations for deriving the K_D (refer back to Equation 22.1). Both plates, the one containing the coated Ag and the one containing the calibration curves and the Ab–Ag mixture should be incubated for 16 to 20 hours, which is assumed to be sufficient time to allow for solution-based Ab–Ag binding equilibria. At equilibrium, net changes in the concentrations of Ab, Ag, or Ab–Ag complex should be negligible. The coated plates should probably be kept at 4°C during the incubation so that the Ag is not at ambient temperature for an extended time while adsorbing onto the plate. The remaining steps of the FL-ELISA were described in Section 22.3.3 for determining the free and bound components in Ab–Ag mixtures.

It is important to emphasize that the key to determining K_D values using the FL-ELISA method described in this section relies on a series of reactions in which the concentration of Ab is kept constant, the concentration of Ag is varied, and the concentration of Ab_{free} is determined and used to derive the K_D. Therefore, in the experimental design it is important that the Ag dilution series span a concentration range ideally from 10-fold below to 10-fold above the K_D value and if possible the [Ab] should be kept well below the K_D, ideally 100-fold, which simplifies later K_D calculations. With the $[Ab] \ll K_D$, the concentration of Ag in the Ab–Ag complex will be small and Ag_{free} will always be much greater, thus approximating Ag_{total} which should be a known value. Also, 1 : 1 binding is assumed between the Ab and the Ag in the equations used to derive the K_D. Higher-order binding involving multivalent species require complex fitting models, which will not be discussed here (Blake et al. 2005).

For data analysis, two independent curve-fitting steps are required. The first curve-fitting step, discussed earlier, is used to determine free Ab concentrations in the mixtures (Fig. 22.4) from the Ab only calibration curve. For calculation purposes, although intact IgG is used, the binding-site concentration of Ab is expressed in the putative Fab concentration, that is, twice that of the IgG concentration as there are two Fabs per IgG molecule available to bind antigen. After calculating the free Fab concentrations ($[Fab]_{free}$) in various Ab–Ag mixtures, assuming two binding sites per Ab and 1 : 1 binding between

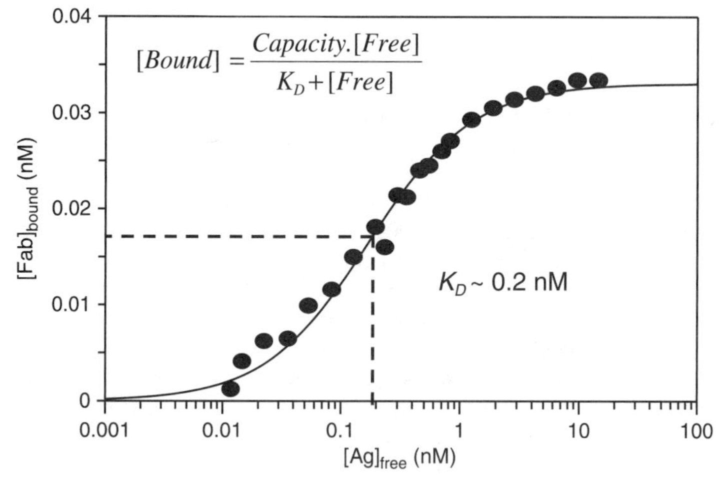

Figure 22.5 Bound vs. free curve fitting for deriving K_D values for an Ab–Ag binding pair. The equation reflects a single binding-site model (1 : 1) from the commercial software Grafit. The equation shown in the figure is a transformed version of Equation 22.2 with varying free Ag concentrations.

TABLE 22.1 Experimentally Determined K_D Measurements for Human IgG: Receptor I Binding

K_D Determinations for Human IgG Batches	Average K_D, nM ($\pm$ SD)
Batch A	0.42 ± 0.02
Batch B	0.40 ± 0.04
Batch C	0.22 ± 0.07
Batch D	0.24 ± 0.03
Batch E	0.27 ± 0.04
Batch F	0.21 ± 0.03

SD = standard deviation; $n = 3$ unless otherwise stated.

Fab and Ag, the $[\mathrm{Ag}]_{\mathrm{bound}}$, $[\mathrm{Fab}]_{\mathrm{bound}}$, and $[\mathrm{Fab}]_{\mathrm{free}}$ in the equilibrated mixtures can be derived. However, in order to obtain K_D values, the concentrations of total Ag ($[\mathrm{Ag}]_{\mathrm{total}}$) and total Fab ($[\mathrm{Fab}]_{\mathrm{total}}$) need to be known in the mixtures. The second curve fitting is used to derive the K_D values from the calculated $[\mathrm{Ag}]_{\mathrm{free}}$ versus $[\mathrm{Fab}]_{\mathrm{bound}}$ plot (Fig. 22.5). Table 22.1 lists K_D values obtained by the FL-ELISA for various human IgG batches prepared at different times, on different scales, and by different manufacturing processes. This data shows that over the different changes being made to the scale and manufacturing process the K_D values for the final human IgG sample did not impact binding to the antigen receptor and the batches were comparable. This is valuable information needed to see how changing manufacturing components impact final products.

22.4 EQUILIBRIUM-BASED DISSOCIATION CONSTANT DETERMINATION BY KinExA

22.4.1 KinExA Method for K_D Determination of High Affinity Binding Partners

KinExA, which is a contraction of Kinetic Exclusion Assay, refers to the technique for measuring the concentration of one of the reactants in a two phase reversible reaction mixture without perturbing the equilibrium of the solution-bound components. The assumption underlying Kinetic Exclusion is that the time of contact between the mixture and the solid phase is short enough that there is not time for significant dissociation of the solution-bound component to occur, and the captured portion of free component thus provides a direct measure of the amount free at equilibrium (Darling and Brault 2004; Blake, Pavlov, and Blake 1999).

22.4.2 Determination of Free and Bound Components in Ab–Ag Mixtures

The heart of the KinExA instrument is a small column with a fixed screen or frit which is called a flow cell. Solid phase particles, for example, polymethyl methacrylate (PMMA), polystyrene, sepharose, or azlactone, are precoated with a molecule used to capture the free molecule and put in a reservoir attached to the instrument. For each measurement a new bead pack is generated by suspending the beads with a mechanical stirrer. A precise volume of bead suspension controlled by a syringe pump is aspirated into the flow cell where it is trapped by the screen. There are 13 input lines where samples are then drawn through the flow cell at rates ranging from 0.25 to 1.5 mL/min which is specified by the user through the dedicated software. After the samples are drawn through, the molecule captured on the solid phase is labeled by flowing a fluorescently labeled antibody through the bead pack. The fluorescent signal is read in place from the beads and the fluorescent signal bound to the beads after washing away unbound labeled antibodies is interpreted as directly proportional to the free receptor present in the sample solution. Figure 22.6 shows a schematic view of the KinExA instrumentation.

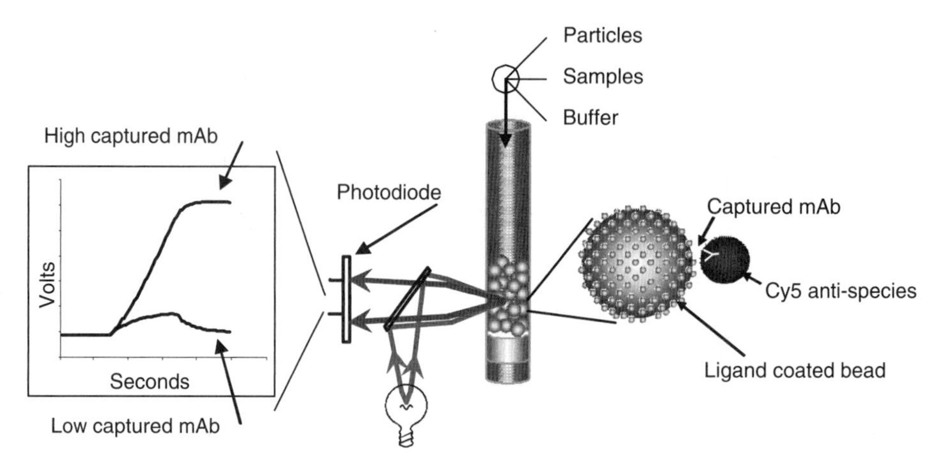

Figure 22.6 Schematic view of basic KinExA instrumentation and signals.

22.4.3 Curve Fitting and Standard K_D Analysis of a High Affinity IgG

As described previously (Darling and Brault 2004; Blake, Pavlov, and Blake 1999), the signal in KinExA is fitted with the following equation:

$$\text{Signal} = \frac{Sig_{100\%} - Sig_{0\%}}{2 \cdot A_T} \times \left[(A_T - K_D - S_T) + \sqrt{A_T{}^2 + 2 \cdot A_T \cdot S_T + K_D{}^2 + 2 \cdot S_T \cdot K_D + (S_T)^2} \right] + Sig_{0\%} \quad (22.3)$$

where A_T = total receptor (often but not always an antibody); S_T = total solution ligand or antigen; $Sig_{100\%}$ = signal generated by A_T in the absence of S; and $Sig_{0\%}$ = residual signal (NSB) generated by A_T in the presence of saturating S.

In a typical experiment, a series of antigen concentrations are prepared, each with identical antibody concentration. The antigen concentrations are chosen to span the range from zero to fully saturating all antibody binding sites. Equation 22.3 is then fitted to the data by varying $Sig_{100\%}$, $Sig_{0\%}$, A_T, and K_D using an iterative least squares algorithm. Including the total receptor (A_T) as a fit parameter means that all the analysis is tied to one concentration (the ligand concentration) supplied by the user. This is advantageous because usually the active concentration of biomolecules is not equal to the total protein concentration as determined by A_{280nM} and specifying both concentrations results in poorer fits to the data and less accuracy in the determination of K_D. The KinExA software allows the user to specify either A_T or S_T, whichever they have more confidence in, and then fits the data to determine the other. Equation 22.3 is fit directly to the signal data. This is possible because the KinExA response is linear with respect to free antibody concentration up to the maximum concentration of antibody used (A_T). This fact can be easily confirmed by using a standard curve but the standard curve thus constructed is not used in the fitting data (Darling and Brault 2004; Blake, Pavlov, and Blake 1999). Including the NSB or $Sig_{0\%}$ in the fit to the signal data instead of fitting it in a separate standard curve improves accuracy because it allows for the possibility of nonspecific binding of the primary antibody and/or solution antigen in the experiment. Fitting in a single step also avoids the potential for compounding of errors through propagation.

Once the parameters resulting in optimum fit have been identified, error curves are computed by moving (first) the K_D away from its optimum value and then re-optimizing the other three parameters. This process is repeated for a series of different K_D values bracketing the optimum value, resulting in a graph of minimum residual error versus K_D. The value of such a graph is that it is immediately obvious whether there is really a single well-defined K_D value resulting in an optimum fit or conversely whether there are actually many K_D values all resulting in similar or equivalent fits to the data. Which of these situations occurs in any given experiment depends primarily on the ratio of the receptor concentration to the K_D. If this ratio is much greater than one (i.e., the experiment is performed with a receptor concentration many times greater than the K_D) then for ligand concentrations less than the receptor concentration nearly all ligand is bound. Put another way, as ligand is added, every ligand binds in solution until the receptor is depleted. In this situation the response curve is steep and it can be fit by a wide variety of different K_Ds. Conversely, experiments conducted with ratios much less than one give tight bounds for the K_D but are usually one sided for the A_T.

A typical KinExA standard K_D analysis is shown in Figure 22.7 for an affinity matured mAb, in two separate experiments using different fixed concentrations of Ab. Panels A and D show the signal data, plotted as percent free Ab versus antigen concentration and the curves represent the best fit of Equation 22.3. As described above, best fit error curves are generated for K_D and A_T values and these are summarized in panel G for both curves. Taking a closer look at a single example, the overall best fit K_D value for curve 1 is 16.24 pM but any value between 10.3 pM and 23.53 pM gives an excellent fit. The calculated active binding concentration, A_T, for this relatively low ratio ($A_T/K_D = 10$) binding curve, is 169.52 pM, which is 1.8-fold less than the nominal Ab concentration that was determined based on A_{280nM} measurements.

In the section above on standard analysis, the ligand concentrations were treated as known values in the fit algorithm, and the values of K_D and A_T determined are both directly dependent on the accuracy of the supplied ligand concentration. This is reasonable in cases where investigators have more confidence in their knowledge of the ligand concentration than in the receptor concentration. In cases where investigators have more confidence in the receptor concentration than in the ligand concentration, the KinExA software supports an alternate analysis (unknown ligand) in which the A_T parameter is replaced with an LCM (ligand concentration multiplier) parameter which, as the name suggests, provides a concentration correction for the ligand based on the known receptor concentration. In actual fact there is usually some uncertainty in both of these concentrations, due to factors such as misfolded or inactive proteins. Investigators are encouraged to think carefully about which concentration they trust more and base their results on that.

There are two cautions associated with the unknown ligand analysis. First, if the goal of the study is to compare the K_Ds of two or more receptors against the same ligand, then it is critical that both receptor K_D values be referenced to the same concentration. If both are fitted using the unknown ligand method then each is referenced to its own concentration and concentration errors may introduce additional uncertainty in the comparison. The ideal situation is to use the standard analysis with a well-known ligand concentration. If the ligand concentration is unknown then it may be determined using the unknown ligand method and *one* of the receptors (the one whose concentration is most certain), but subsequent receptors should then be analyzed using the standard analysis and the corrected (nominal value times the LCM) ligand concentration.

The second caution is that for very low ratios, the unknown ligand method is indeterminate for both the K_D and LCM. If this occurs, both error graphs will have only a lower bound. The fix for this is to do a second experiment at a higher receptor concentration and analyze the two curves in the n-curve analysis described below.

22.4.4 K_D Analysis of High Affinity Ab–Ag Interactions

The sensitivity of the KinExA platform allows one to determine the dissociation constants of high affinity recombinant antibodies now routinely generated via directed evolution technologies and other methods (Drake, Myszka, and Klakamp 2004; Luginbühl et al. 2006; Rathanaswami et al. 2005). When setting up an experiment one needs to determine which component will be measured in solution and how it will be detected. Normally, the free antibody concentration is measured by capture on the solid phase bead column which has been coated with ligand and detected by a fluorescently labeled anti-IgG antibody. After the conditions for coupling the ligand to the beads have been determined

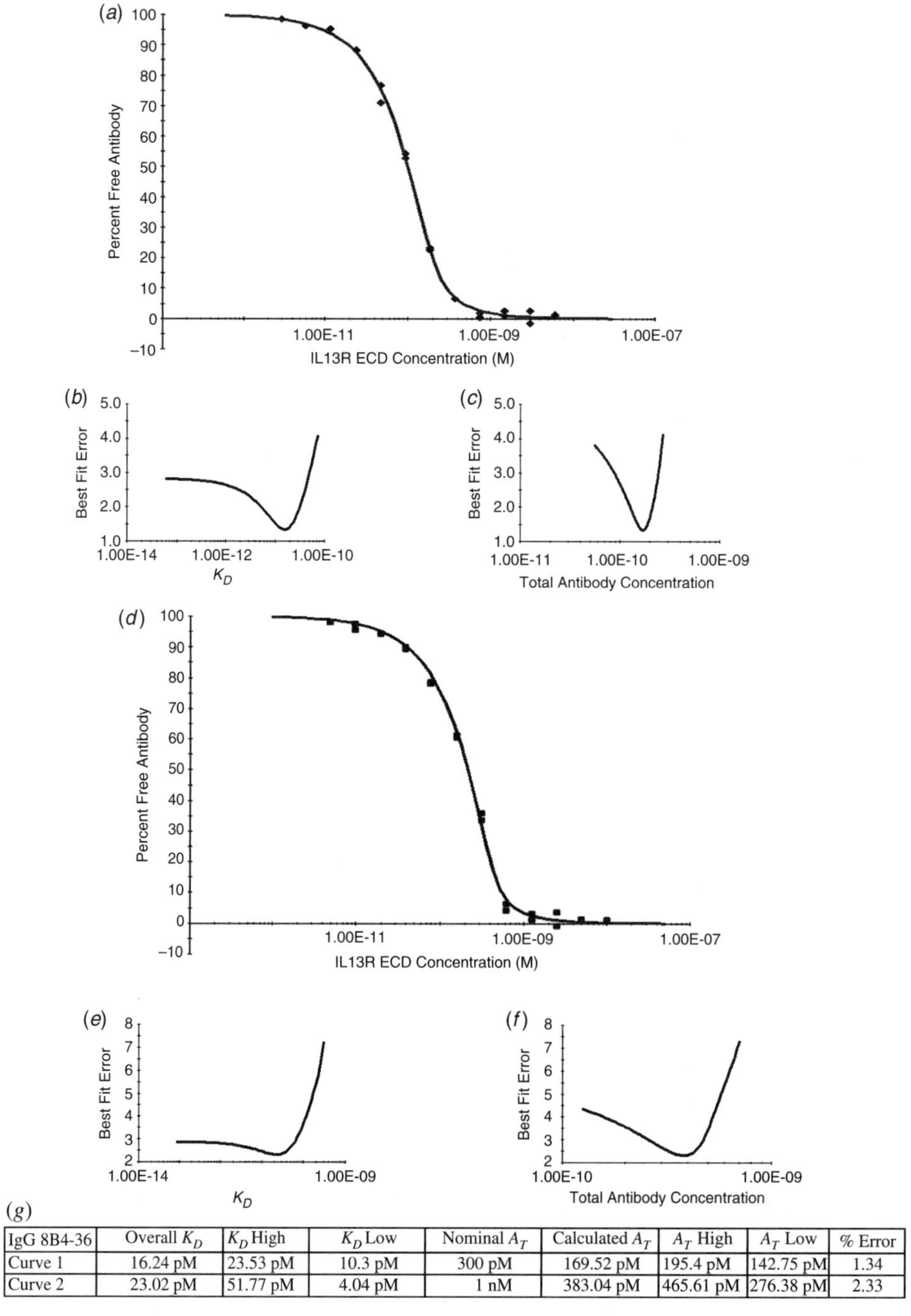

(g)

IgG 8B4-36	Overall K_D	K_D High	K_D Low	Nominal A_T	Calculated A_T	A_T High	A_T Low	% Error
Curve 1	16.24 pM	23.53 pM	10.3 pM	300 pM	169.52 pM	195.4 pM	142.75 pM	1.34
Curve 2	23.02 pM	51.77 pM	4.04 pM	1 nM	383.04 pM	465.61 pM	276.38 pM	2.33

Figure 22.7 Measurement of the K_D of an anti-human IL13Rα1 extracellular domain (ECD) recombinant human mAb. Two standard K_D binding curves were run using different antibody concentrations for an affinity matured mAb 8B4-36. Human IL13Rα1 ECD was serially diluted in PBS pH 7.4 containing 1 mg/mL bovine serum albumin with a fixed amount of mAb and incubated at room temperature for 12 hours. The free mAb was captured over a human IL13Rα1 ECD azlactone bead pack and detected using a Cy5-labeled goat anti-human Fcγ IgG. Panels A (300 pM mAb curve) and D (1 nM mAb curve) show the signal data plotted as percent free antibody versus IL13Rα1 ECD concentration (molar units). The error graphs for best fit KD and ABC or A_T values are shown in panels B and C for the 300 pM antibody concentration curve and panels E and F for the 1 nM antibody concentration curve. Panel G summarizes the best fit values for the K_D and ABC or A_T for each curve, as well as the 95% confidence interval values (high/low) for each.

and shown to remain active and recognized by the mAb, a series of signal tests are performed to determine the instrument settings such as flow rate and binding signal to demonstrate the linearity and reproducibility of the response as well as the signal to noise ratio.

A typical experimental set up includes determining the range of antigen concentrations and fixed antibody concentration to be used to generate a K_D curve. The antigen dilutions are prepared in a buffer containing a fixed amount of antibody and the volume needed per sample is based on the signal testing results performed as described above. The instrument allows one to set up a 13 point curve and, as discussed in Section 22.3.3, one wants conditions where the signal ranges from 100 percent bound to 100 percent free antibody. It is important to have a general idea of the expected K_D for a given Ag–Ab pair, and this can be done by setting up a K_D range find experiment, over a broad range of antigen concentrations. Based on these results, a standard K_D curve can be generated using a tighter range of Ag concentrations and optimally fixed Ab concentrations that are both above and below the expected K_D value.

As discussed in Section 22.4.3 above, a single experimental curve with a high ratio is generally inconclusive with regard to K_D determination and a single curve with a very low ratio has the converse problem. KinExA software also supports analyzing both curves together in an n-curve analysis. This analysis method is used in cases where multiple experiments with the same antibody–antigen pair have been performed at different antibody concentrations. The strength of this analysis is that both the K_D and the A_T can usually be determined with high confidence. Figure 22.8 shows the n-curve analysis for mAb 8B4-36 as determined from the two K_D curves generated using different Ab concentrations (from Fig. 22.7). In this case the A_T is given for one of the curves, curve 1, and the other curve is computed

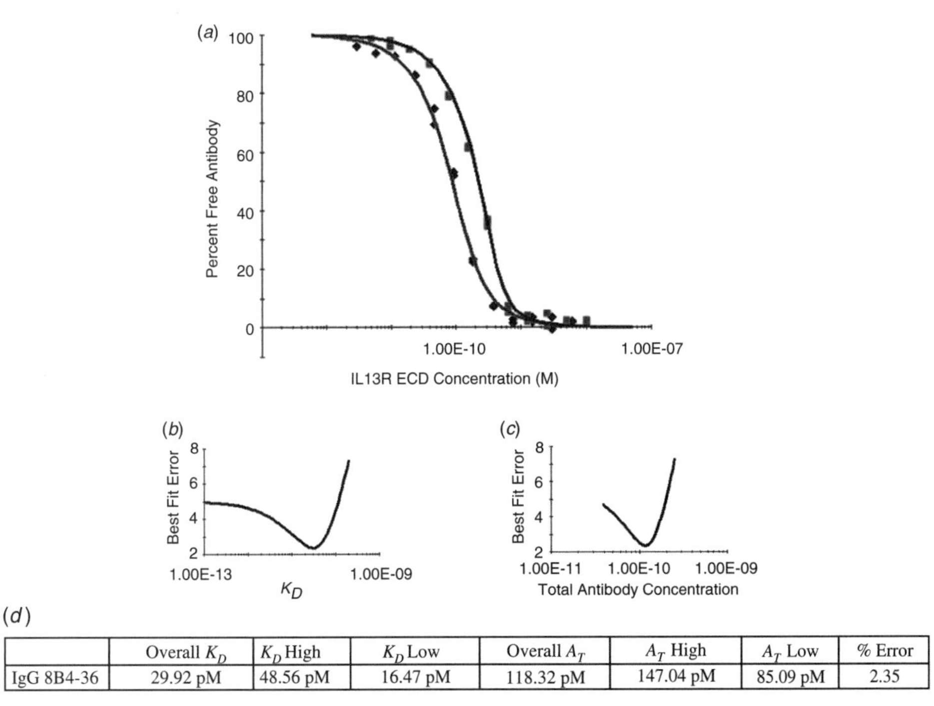

	Overall K_D	K_D High	K_D Low	Overall A_T	A_T High	A_T Low	% Error
IgG 8B4-36	29.92 pM	48.56 pM	16.47 pM	118.32 pM	147.04 pM	85.09 pM	2.35

Figure 22.8 Determination of overall K_D using n-curve analysis for an anti-IL-13α1 mAb. The standard K_D curves shown in Figure 22.7 were analyzed simultaneously to determine the optimal K_D and ABC or A_T values for mAb 8B4-36. Panel A shows the equilibrium titration curves for 300 pM (diamonds) and 1 nM (squares) fixed IgG concentrations as the percent free antibody plotted against IL13R ECD concentration (molar units). The best fit error plots for K_D and ABC or A_T are shown in panels B and C. The n-curve analysis results are shown in panel D. The n-curve analysis calculated a best fit K_D value of 29.92 pM.

TABLE 22.2 Equilibrium Dissociation Constants for anti-IL13Rα1 IgG and Fab Using KinExA n-Curve Analysis

	IgG				Fab			
Antibody	K_D (pM)	K_D High[a]	K_D Low[a]	% Error	K_D (pM)	K_D High[a]	K_D Low[a]	% Error
8B4	461.7	550.7	384.22	1.16	740.29	1080	499.21	1.24
8B4-36	29.92	48.56	16.47	2.35	20.06	33.99	9.33	1.54
8B4-78	8.02	15.99	1.93	2.36	23.2	35.18	13.8	1.46

[a]The K_D high and K_D low values are the 95% confidence intervals from the multiple curve fitting analysis done by the instrument's evaluation software. For each antibody sample, two standard K_D curves were generated using different antibody concentrations.

from their nominal concentrations using the same factor. The best fit K_D value for this analysis is 29.92 pM with an acceptable fit range of 16.47 to 48.56 pM. The K_D/A_T ratios calculated were 4 and 13 times the overall K_D of the system. Optimally one would want ratios $\approx < K_D <$ 10-fold for a given series of experimental curves but it may not be feasible when characterizing a series of mAbs for a given program.

Once the experimental conditions are established for an Ag–Ab pair, they can be used to identify optimized engineered antibodies. Table 22.2 is a summary of experiments performed to characterize the binding strength of a series of anti-human IL13Rα1 extracellular domain (ECD) mAbs engineered and selected to have higher affinities than the original mAb, which had an apparent K_D value of 1 nM as determined by preliminary kinetic-based and equilibrium-based analysis.

These antibodies were originally identified during a Fab phage display affinity maturation selection process. The overall K_D values were determined for both the Fab and IgG for each clone using dual curve analysis as outlined in the paragraph above for one of the clones, 8B4-36. All data was generated using the same antigen lot and normalized to a single antigen concentration based on curves generated with a single antibody. Overall the K_D values are similar for the Fab and IgG for each clone. The affinity matured clones 36 and 78 show a marked affinity improvement compared to the original 8B4 antibody. These studies can be used to benchmark the progress of an affinity maturation program, as well as determine activity levels of future preparations of IgG and antigen material.

22.5 EQUILIBRIUM DISSOCIATION CONSTANT DETERMINATION FOR IgG TO WHOLE CELLS

In many cases of therapeutic interest, the binding target of a monoclonal antibody is a cellular protein or receptor. In some cases, the binding target can be isolated from the cell membrane, purified, and the binding constant for antibody-target interaction can be measured for the soluble target. In other cases, the target cannot be purified, cannot be solubilized, or is so denatured when removed from its natural environment that solution measurements are not possible. Even in cases where the solution measurement can be made, the question generally remains open whether the binding constant measured for the soluble molecule accurately reflects the K_D for the membrane bound molecule. In addition, many therapeutic antibody candidates are bivalent IgG molecules and there is a possibility that they may interact with two membrane-bound target molecules simultaneously. In these cases, the affinity (which characterizes the single site binding) of the binding interaction may be less relevant than the avidity of the binding. For all of these reasons, a methodology for measuring the effective K_D of the binding of the antibody to the cell itself is highly desirable. The discussion that follows describes measurements made for binding of IgG to cell surface ligands.

Direct measurement of an equilibrium K_D always involves concentration measurements, either of the bound or of the free concentration at equilibrium. The trick is to make this measurement without perturbing the equilibrium. In the discussion of KinExA above, the measurement is made by briefly exposing the equilibrated solution to the solid phase, keeping the time of contact too short to allow significant competition between the solid phase and solution phase ligands to occur. It is not generally feasible to treat the whole cells like a solution antigen and flow them through the solid phase in the KinExA column. Instead, an alternative methodology has been developed in which a fixed concentration of the antibody is incubated with serially diluted cell suspensions. The highest cell concentration is chosen to bind all of the antibody and the titration extends downward far enough to bind little or no antibody. After equilibrium is reached, the cells are briefly centrifuged and the concentration of antibody in the cell supernatant is measured. This separation step is analogous to that done using titrated radiolabeled ligands and a fixed concentration of cells. It is generally believed that the separation step does not significantly shift the equilibrium of the bound ligand in these radiolabeled experiments (Limbird 1996). The KinExA experiment differs from the well-known radiolabel technique in that the cells are titrated rather than the ligand, and the unbound ligand in the supernatant is measured. There is at least the possibility that when the cells are pelleted the resulting higher effective receptor concentration could lead to increased binding on the cell surface but in order for this to occur there would also have to be enough time for the antibody to diffuse from the supernatant into the pellet. Detailed studies are not yet available but this is unlikely to be a significant effect at the relatively low solution antibody concentrations typical in these experiments.

The measurement of the free antibody in the supernatant can be accomplished using the KinExA in at least two different ways. In the first, the solid phase particles are coated with a solubilized version of the receptor which is then used to capture specific antibodies from the supernatant. This approach has the advantage that antibody need not necessarily be completely pure because the capture step is also specific. Results using this approach have been published measuring the effective K_D for both whole IgG and Fab fragments binding to human insulin-like growth factor receptor (hIGFR; Xie et al. 2005). The second approach uses particles coated with an antispecies antibody to capture the antibody present in the supernatant (Rathanaswami, Babcock, and Gallo 2008). This approach has the advantage of not requiring a solubilized version of the cell surface ligand.

When soluble receptor is available at a known concentration, it can be used to establish the active concentration of the antibody. This can be done on the KinExA instrument using a high ratio experiment and the A_T (active binding concentration) parameter as described above. This capability is valuable because the active concentration of antibody in a solution is rarely equal to the nominal value. The discrepancy is thought to arise because of incomplete purification or denaturation of a portion of the antibody during the purification process.

As described above, in KinExA (and in every other affinity measurement) the K_D measured is ultimately referenced to a concentration and will generally be in error to the same extent that the concentration is in error. In the most common solution case, KinExA measurements are referenced to the ligand concentration but the software also supports referencing to the antibody concentration instead in cases where that is believed to be more accurately known. For soluble receptor it is often easy to get at least a nominal concentration. For cellular bound receptors, it may be possible to make some estimate of the expression level (ligands per cell) from fluorescence activated cell sorting (FACS) data or there may be no estimate available. In the work cited, Xie et al. (2005) used the nominal concentration of the commercially supplied soluble hIGFR to estimate the active concentration of their antibodies. They then used the active concentration of the antibody to estimate the effective solution concentration of the cells and then used this in conjunction with the cells per milliliter to estimate the expression level of hIGFR on the cells.

Xie et al. (2005) measured K_D values for Fab and whole IgG to both soluble and cell bound hIGFR. The K_Ds measured ranged from 4 pM (for both whole IgG and Fab binding to soluble receptor) to 10 pM (Fab binding to cellular receptor). Because of overlap in the 95% confidence intervals the authors concluded that the measured K_Ds were not significantly different from one another.

Rathanaswami and coworkers (2008) measured the K_D of a monoclonal anti-TNFα (both whole IgG and Fab) against membrane-expressed TNFα. In their work they reproduced the method of Xie et al. (immobilizing soluble TNFα on the beads) and then compared this to results obtained using an immobilized anti-species antibody on the KinExA solid phase. Their cell experiments were identical in concept to Xie's in that a fixed concentration of antibody was incubated to equilibrium with a serial dilution of suspended cells. In their case, in contrast to the results of Xie et al., they found that the K_D of the Fab was three (solution) to five (membrane expressed) times higher (weaker binding) than the K_D of the whole IgG. The factor was similar enough that they conclude that the differences in binding are due to changes in structure associated with cleaving the IgG rather than to valency effects of the IgG binding to membrane-expressed TNFα.

22.6 PRECAUTIONS AND ARTIFACTS IN K_D DETERMINATION

22.6.1 Active Concentrations

All methods for K_D determination require a precise knowledge of the concentration of at least one of the binding partners. Therefore, the accuracy of the K_D determination is highly dependent on the errors in the estimation or measuring of the active concentration of the mAb preparations. In general, the concentration of a purified mAB preparation can be determined by absorption at 280 nm (see also Chapter 23). In addition to the overall mass and molar concentration, the knowledge of concentration of basic binding sites (or number of binding sites per IgG or per Ag molecules) is also essential. The active concentration, however, may or may not be the same as the total mAb concentration as determined by 280 nm absorbance. In other words, a fraction of the mAb can exist in a denatured inactive form, there may be non-mAb proteins in the solution, or the mAb may form aggregates with complementary determinant regions (also referred to as paratope) partially or completely masked, rendering it inactive in binding analysis. When estimating the K_D values using whole cells, it is very challenging to get precise concentrations of both binding partners, the mAb and the surface bound antigens/receptors. Binding of an mAb with a cell surface bound Ag may induce certain conformational changes resulting in a significant change in their binding interface, thus a modulation in the dissociation rate. To make things more complicated, internalization of the receptor (if Ag is a receptor) and/or the receptor-mAb complex takes place, leading to irreversible changes in the active concentrations of both the receptor and the mAb.

22.6.2 Multivalency, Avidity, and Rebinding Effects

More than one active binding site on the mAb and/or on the Ag could complicate the data analysis and interpretation. Immobilizing one binding partner on the surface through chemical coupling or specific capture would further complicate the situation. Multiple on- and off-rates would be exhibited based on the binding order and/or the closeness to the surface and the freedom of solution-like mobility of the binding interactions. Dissociated molecules can rebind to a nearby site on the surface, resulting in a much higher apparent affinity or slower off-rate. In general, a thermodynamically meaningful affinity constant describes the true strength of the interaction between two binding partners in solution. Whenever one of the binding partners is present in multiple copies and in a certain restricted manner on a support such as a cellular membrane, a sensor chip, or a microtiter plate, the observed affinity constant represents more of the functional affinity or avidity.

22.6.3 Solution versus Surface-Bound Interaction

When two molecules interact freely in solution, their intrinsic dynamics and flexibility during the docking and undocking process of two biomolecules play major roles in defining the true dissociation constants with freely interchanging on and off processes at equilibrium. This equilibrium dissociation (or affinity) constant with monovalent interaction in solution is truly thermodynamically meaningful. If one of the binding partners is restricted on a surface, we cannot apply or derive a concentration term for that partner. The density of the functional partner on the surface would significantly impact the off-rate, the avidity, and the observed affinity constants. Due to mass transport limitation, surface-induced conformational changes, and other factors affecting the readout signals, the signal intensities on the surface may not faithfully reflect the changes in the fractions of bound versus free of a given binding partner in solution. Many methods relying on the raw signal intensities for curve fitting may introduce some biases due to the nonlinearity between signals versus the free (or bound) concentrations being used for the curve fitting to derive K_D. One of the two equilibrium-based methods introduced here, FL-ELISA in Section 22.3, uses a standard curve to convert the observed signal intensities into free Ag (or Ab) concentrations in solution. The nonlinearity, the equipment variations, signal amplification, etc., are all eliminated after converting the signals into concentrations with standard curves obtained in the same run or in an adjacent run. The second solution method discussed, KinExA in Section 22.4, avoids the issue by using a densely coated solid phase to assure linearity of response in the free receptor range used. However, it is hard for any other techniques to adopt a similar approach, such as fluorescence quenching or Biacore, because it is very difficult to simulate what happens in a complex without forming one or simulating the surface mass deposits with solution concentrations.

22.6.4 Buffer versus Blood

The binding strength between an Ab–Ag pair may be dramatically altered when going from a simple buffer such as phosphate buffered saline to complex human or animal blood with high concentrations of albumin, immunoglobulins, fibrinogen, and other blood components. Thus, it is essential to take this into consideration during the early discovery phase when modeling the affinity needed for a certain target engagement and bench marking the progress of affinity maturation.

22.7 CONCLUSION

In this chapter, several methods utilizing different technologies for determining the dissociation constants are introduced and illustrated with examples. Their advantages and disadvantages are discussed. The sensor chip-based Biacore and other technologies offer potential higher throughput automation for kinetic rate constant-based screening and less sample volume input. Yet, the nature of the surface interaction for the analysis complicates the data interpretation and makes it susceptible to some surface-induced changes in conformation or multivalency issues that could be avoided in a solution-based technique. For high affinity mAbs, the slow dissociation often limits the use of kinetic-based K_D determination as only a very small fraction of Ab–Ag complex would dissociate in the reasonable 30 to 90 min monitoring period. Introduction of heterogeneities in the molecular association and (more so) the dissociation on the surface-bound molecules/complexes is also worth noting because most fitting algorithms assume homogeneous dissociate in a pseudo first order kinetics. Solution-based ELISA and KinExA introduce fewer artifacts as compared to methods monitoring the interaction on a surface.

REFERENCES

Alfthan, K. 1998. Surface plasmon resonance biosensors as a tool in antibody engineering. *Biosensor Bioelectronics* 13:653–663.

Azimzadeh, A., J.L. Pellequer, and H.V. Van Regenmortel. 1992. Operational aspects of antibody affinity constants measured by liquid-phase and solid phase assays. *J. Mol. Recogn.* 5:9–18.

Blake, R.C., N. Ohmura, S.J. Lackie, et al. 2005. In *Trends in Monoclonal Antibody Research*, ed. M.A. Simmons, 1–36. New York: Nova Biomedical Books.

Blake, R.C. II, A.R. Pavlov, and D.A. Blake. 1999. Automated kinetic exclusion assays to quantify protein binding interaction in homogeneous solution. *Anal. Biochem.* 272:123–134.

Canziani, G.A., S. Klakamp, and D.G. Myszka. 2004. Kinetic screening of antibodies from crude hybridoma samples using Biacore. *Anal. Biochem.* 325:301–307.

Darling, R.J., and P.-A. Brault. 2004. Kinetic exclusion assay technology: Characterization of molecular interactions. *Assay Drug Development Tech.* 2:647–657.

Darling, R.J., U. Kuchibhotla, W. Glaesner, et al. 2002. Glycosylation of erythropoietin affects binding kinetics: Role of electrostatic interactions. *Biochemistry* 41:14524–14531.

Drake, A.W., D.G. Myszka, and S.L. Klakamp. 2004. Characterizing high-affinity antigen/antibody complexes by kinetic- and equilibrium-based methods. *Anal. Biochem.* 328:35–43.

Friguet, B., A.F. Chaffotte, L.D. Ohaniance, and M.E. Goldberg. 1985. Measurements of the true affinity constant in solution of antigen–antibody complexes by enzyme-linked immunosorbent assay. *J. Immunol. Methods* 77:305–319.

Goldberg, M.E., and L.D. Ohaniance. 1993. Methods for measurement of antibody/antigen affinity based on ELISA and RIA. *Curr. Opin. Immunol.* 5:278–281.

Goodrich, J.A., and J.F. Kugel. 2007. *Binding and Kinetics for Molecular Biologists: Affinity Constants.* New York: Cold Spring Harbor Laboratory Press.

High, K., Y. Meng, M. Washabaugh, and Q. Zhao. 2005. Determination of picomolar equilibrium dissociation constants in solution by enzyme-linked immunosorbent assay with fluorescence detection. *Anal. Biochem.* 347:159–161.

Huber, W., and F. Mueller. 2006. Biomolecular interaction analysis in drug discovery using surface plasmon resonance technology. *Curr. Pharm. Design* 12:3999–4021.

Karlsson, R., and A. Fält. 1997. Experimental design for kinetic analysis of protein-protein interactions with surface plasmon resonance biosensors. *J. Immunol. Methods* 200:121–131.

Karlsson, R., H. Roos, and L. Fägerstam. 1994. Kinetic and concentration analysis using BIA technology. *Methods Companion Methods Enzymol.* 6:99–110.

Limbird, L.E. 1996. *Cell Surface Receptors: A Short Course on Theory and Methods*, 2nd edition. New York: Springer.

Luginbühl, B., Z. Kanyo, R.M. Jones, et al. 2006. Directed evolution of an anti-prion protein scFv fragment to an affinity of 1 pM and its structural interpretation. *J. Mol. Biol.* 363:75–97.

Meng, Y., K. High, J. Antonello, M. Washabaugh, and Q. Zhao. 2005. Enhanced sensitivity and precision in an enzyme-linked immunosorbent assay with fluorogenic substrates compared with commonly used chromogenic substrates. *Anal. Biochem.* 345:227–236.

Rathanaswami, P., J. Babcock, and M. Gallo. 2008. High-affinity binding measurements of antibodies to cell-surface expressed antigens. *Anal. Biochem.* 373:52–60.

Rathanaswami, P., S. Roalstad, L. Roskos, et al. 2005. Demonstration of an in vivo generated sub-picomolar affinity fully human monoclonal antibody to interleukin-8. *Biochem. Biophys. Res. Commun.* 334:1004–1013.

Rich, R.L., and D.G. Myszka. 2007a. Higher-throughput, label-free, real-time molecular interaction anlaysis. *Anal. Biochem.* 361:1–6.

Rich, R.L., and D.G. Myszka. 2007b. Survey of the year 2006 commercial optical biosensor literature. *J. Mol. Recognit.* 20:300–366.

Stenberg, E., B. Persson, H. Roos, et al. 1991. Quantitative determination of surface concentration of protein with surface plasmon resonance using radiolabelled proteins. *J. Colloid Interface Sci.* 143:513–526.

Stevens, F. 1987. Modification of an ELISA-based procedure for affinity determination: Correction necessary for use with bivalent antibody. *Mol. Immunol.* 24:1055–1060.

Towne, V., M. Will, B. Oswald, and Q. Zhao. 2004. Complexities in horseradish peroxidase-catalyzed oxidation of dihydroxyphenoxazine derivatives: Appropriate ranges for pH values and hydrogen peroxide concentrations in quantitative analysis. *Anal. Biochem.* 334:290–296.

Xie, L., M. Jones, T.R. Glass, et al. 2005. Measurement of the functional affinity constant of a monoclonal antibody for cell surface receptors using kinetic exclusion fluorescence immunoassay. *J. Immunol. Methods* 304:1–14.

CHAPTER 23

Molecular and Functional Characterization of Monoclonal Antibodies

QINJIAN ZHAO, TERRANCE A. STADHEIM, LORENZO CHEN, and
MICHAEL W. WASHABAUGH

23.1 INTRODUCTION

There are over a dozen therapeutic monoclonal antibodies (mAbs) on the market, with many more being tested in clinical trials. Most of them are chimeric, humanized, or fully human IgG1 molecules. Analytical characterization plays a key role in the development of an IgG therapeutic with high specificity and affinity to its target. Moreover, analytical characterization of an mAb molecule is critical during all stages of development, including discovery, research and development, and the whole life cycle of a therapeutic mAb as a drug. Throughout the product development lifecycle (discovery through post-marketing phases), analytical characterization will focus on different aspects of the properties of the mAb.

Therapeutic Monoclonal Antibodies: From Bench to Clinic. Edited by Zhiqiang An

The initial goals during discovery and the preclinical stages are to ensure, with the best efforts and knowledge, that the mAb's affinity and specificity to its target(s) is sufficient, and to ensure reliability of animal models for predicting safety and efficacy in humans. Assays should be performed to show the specific binding *in vitro* for the specific pairs of interacting partners, on relevant tissue level and ideally on the animal as a whole off-target binding is always a concern because at the proposed dose for clinical trials, weak nonspecific binding might result in substantial undesired off-target binding *in vivo*, thus causing undesired biological events or rendering the drug nonefficacious.

Once an mAb candidate has passed the preclinical stage (i.e., entering Phase I and beyond), comparability of the pivotal lots made differently, either in scale and/or processes, for different phases needs to be demonstrated. The goal of the comparability exercise is to ensure consistency in the quality, safety, and efficacy of mAb products produced by a manufacturing process at different scales or with certain process changes being implemented. Comparability can be determined using a combination of *in vitro* analytical testing, including certain biological (or cell-based) assays, and, in some cases, some preclinical *in vivo* pharmacokinetics and pharmacodynamics data, as well as some limited clinical bridging data if necessary.

In general, the primary sequence should be deduced by DNA sequencing of the plasmid encoding the heavy chain and light chain of the mAb and confirmed experimentally by peptide mapping and amino acid sequencing. This should be done during the clone-selection stage as well as in the purified final bulk/container stage. Approximately 95 percent of the amino acid sequence of human IgG1 molecules is conserved (i.e., the constant regions of an IgG1). Although with defined amino acid sequence of the heavy chain and light chain, certain degrees of molecular heterogeneity are common for all IgG1 molecules when over-expressed in commonly used mammalian cells. These include heterogeneity in N-linked glycans at Asn297 of the heavy chain, different degrees of carboxypeptidase processing of C-terminal Lys residues, or conversion of heavy chain N-terminal glutamine residue into a cyclic pyroglutamate. The variable regions of the primary structure, $\sim$5 percent of the whole IgG1, confer antigen (or target) specificity. This is the portion of an IgG molecule that can introduce variations in molecular properties as compared to other well-studied IgG molecules. In addition to evaluating antigen binding and other biological effects for the new mAb candidate using different soluble ligands or cells, other analytical methods may need to be reoptimized to accommodate these amino acid changes in the complementary determining regions (CDRs) for achieving comparable assay performance.

The following sections describe key analytical methods for assessing the molecular properties and functional characteristics of an mAb and for defining its structure and activity. Also discussed are applications of these physicochemical, biochemical, and functional assays during different stages of product development in the context of bridging different mAb preparations and assessing comparability.

23.2 MOLECULAR STRUCTURAL ANALYSIS OF mAb BY PHYSICOCHEMICAL METHODS

Table 23.1 shows a list of essential assays for characterizing biological molecules. With these assays, one can define a well-characterized biological—bringing it closer to a small molecule pharmaceutical from a regulatory perspective than the classical process-defined biologic.

In the following sections, we will review some of these analytical methods and the molecular characteristics revealed by those techniques.

23.2.1 Purity and Concentration Determination

It is common practice, after the mAb is >95 percent pure, to use the extinction coefficient and UV absorbance at 280 nm to determine mAb concentration. The extinction coefficient for the antibody in question can be calculated as described by Pace et al. (1995) and references therein, where the molar extinction coefficients of Trp, Tyr, Phe residues and disulfide bonds can be summed for a given IgG from its sequence information. These calculated values should correlate with the results

TABLE 23.1 Physicochemical Characterization Tests of Biological Molecules

Parameter	Analytical Tests
Primary structure	Amino acid composition analysis, MS, N/C-terminal sequencing, TPM and sequencing/MS
Higher-order structure	CD, NMR, immunoreactivity with conformational-dependent antibodies, TPM/MS, biological assays
Size	AUC, field flow fractionation, MALDI-TOF MS, LC-ESI, SDS-PAGE, SEC-HPLC
Charge	CE, IEC, IEF
Hydrophobicity	HIC-HPLC, RP-HPLC
Immunoreactivity	Immunoprecipitation, Western blot analysis
For glycosylated products	
Glycosylation pattern/sequence	CE, HPAEC-PAD, LC-ESI, MALDI-TOF MS, RP-HPLC
Identification of glycosylation sites	TPM/MS

AUC, analytical ultracentrifugation; CD, circular dichroism; CE, capillary electrophoresis; ESI, electrospray ionization mass spectrometry; HPAEC-PAD, high pH anion exchange chromatography with pulsed amperometric detection; HIC, hydrophobic interaction column; IEC, ion-exchange chromatography; IEF, isoelectric focusing gel electrophoresis; LC, liquid chromatography; MS, mass spectrometry; MALDI-TOF MS, matrix-assisted laser desorption/ionization-time of flight; NMR, nuclear magnetic resonance; RP-HPLC, reversed-phase high performance liquid chromatography; SEC, size exclusion chromatography; TPM, trypsin peptide mapping.

From Chirino, A.J. and Mire-Sluis, A. 2004. Characterizing biological products and assessing comparability following manufacturing changes. *Nature Biotechnology* 22:1383–1391.

from amino acid analyses, one of the most commonly used techniques for quantitation of proteins and peptides that does not depend on specific structural features of the polypeptide (Anders et al. 2003; Macchi et al. 2001). The accuracy achieved with this approach is sufficient for product development and consistency.

23.2.2 Primary Structure

Characterization should start with the determination of class, subclass, light and heavy chain composition, and in certain cases, such as when the monoclonal antibody belongs to the IgG4 subclass, the number of half-antibody molecules.

Obviously, at the initial stages of development, it is important to establish that the protein in production is the desired one. DNA sequencing of the stable, overproducing clone sets the stage, followed by determination of the amino acid sequence of the protein for confirmation of the sequence. Confirmation of the expected amino acid sequence can take several forms. Currently, the application of mass spectrometry for the determination of the intact mass of either the whole mAb or the light and heavy chains after reduction of the intermolecular disulfide bonds is a good initial step. Commercially available mass spectrometers can provide data with a precision within 1 to 2 Da of the expected masses of these components. This is illustrated in Figure 23.1, where the light and heavy chain masses for a typical IgG1 was determined. The mass of the light chain conforms with the expected mass, but in Figure 23.1b, the expected heavy chain mass is not observed because of glycosylation in the Fc region. Instead, one observes several different masses that conform to a mixed population of glycans that are attached to the heavy chain. To ascertain the aglycosylated heavy chain mass, methods employing enzymatic release of N-linked glycans can be used, such as PNGase F treatment, followed by mass spectrometry with the consideration of the resulting 1 Da mass change due to the conversion of asparagine to aspartate during the deglycosylation reaction.

Although there exists the possibility for errors in protein translation to result in isometric mutations, the data obtained by this method, in conjunction with ligand binding and cell-based functional assays, can be an additional screen to ensure that the DNA sequence in the plasmid was accurately translated to generate the protein/mAb of choice. Intact mass determination of the light and heavy chains can also be implemented during clonal selection and fermentation process optimization, since it not

(a) 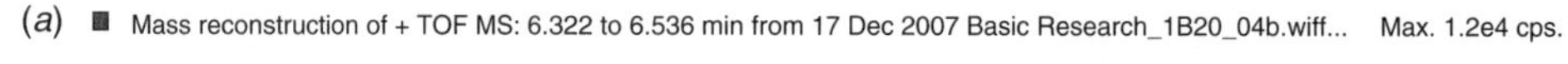

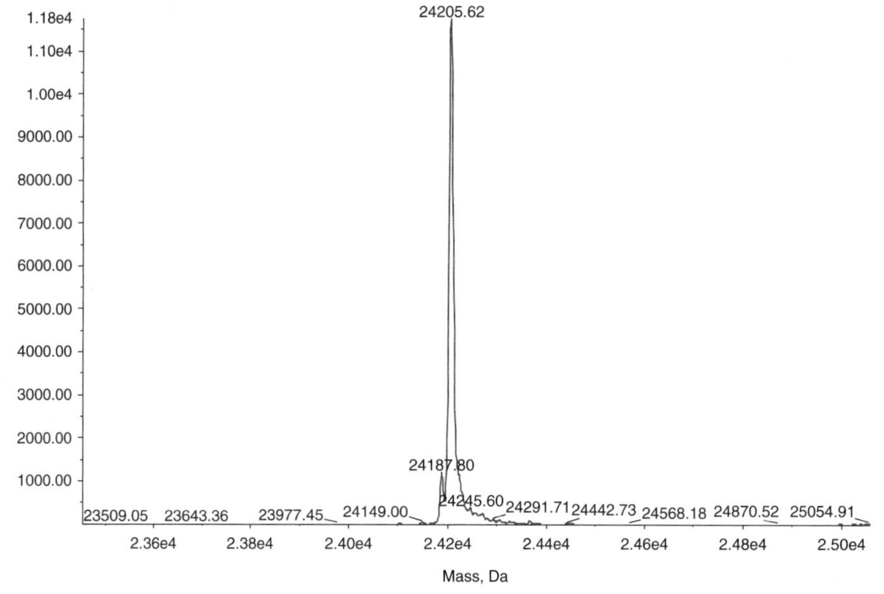

(b) Mass reconstruction of + TOF MS: 7.872 to 8.121 min from 17 Dec 2007 Basic Research_1B20_04b.wiff... Max. 3872.9 cps.

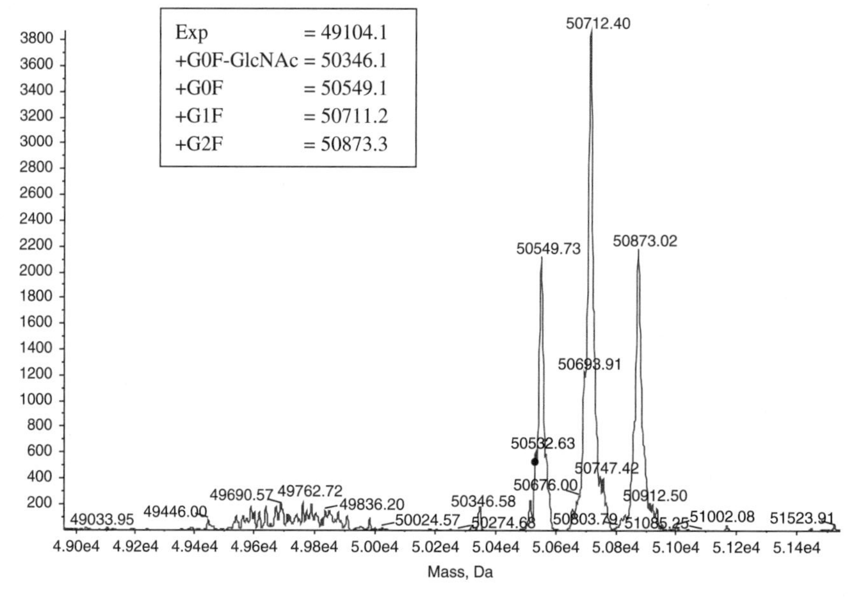

Figure 23.1 Determination of the (a) light and (b) heavy chains of a typical IgG1 molecule by mass spectrometry. The sample was reduced with DTT at 75°C for 15 minutes and the reduction reaction quenched by the immediate addition of 0.1 percent formic acid. The samples were analyzed by LC-MS with an Agilent binary capillary pump coupled to an Agilent MSD-TOF, an electrospray ionization time-of-flight (ESI-TOF) mass spectrometer. A PSDVB column (0.5 × 50 mm) maintained at 80°C and flow rate of 20 μL/min was used in the LC separation. Chromatographic separation was achieved using mobile phases of 0.1 percent formic acid in H_2O and 0.1 percent formic acid in CH_3CN. The masses of the analytes were determined from the deconvolution of the acquired spectra using the TOF Protein Confirmation software (Agilent). The light chain molecular weight conforms to the expected molecular weight calculated based on the amino acid sequence.

TABLE 23.2 Application of Intact Mass Determination of Antibody Light Chain by ESI- TOF during Clonal Selection

Theoretical	23946.1	Δ, Da
Clone 1	23946.2	+0.1
Clone 2	23945.9	−0.2
Clone 3	23946.2	+0.1
Clone 4	23946.1	0
Clone 5	23945.8	−0.3
Clone 6	23945.8	−0.3
Clone 7	23957.0	+10.9
Clone 8	23957.5	+11.4

ESI-TOF = Electrospray ionization-time-of-flight mass spectrometry.

only provides mass confirmation but also a glimpse of some aspects of posttranslational modifications without much sample manipulation. Compared to the aforementioned binding and functional assays, mass spectrometry provides a quick screen to determine the suitability of a clone. An example is shown in Table 23.2, where the light chain masses of several clones were determined. With the expected molecular weight of 23946.1 Da, the intact light chain masses from six clones (clones 1 to 6) were within ±0.3 Da of the expected mass. The observed light chain molecular weight of the remaining two clones (clones 7 and 8) deviated significantly from the expected value, leading to a reevaluation of the DNA sequence of the latter two clones. Subsequent experiments revealed that there was a mistranslation in the open reading frame, resulting in a single mutation due to frameshift.

Further confirmation of the amino acid sequences of the light and heavy chains can be carried out by peptide mapping at the protein level. The mAb is digested with various endoproteases with known specificity, and these peptide fragments are subsequently analyzed by a combination of reversed phase chromatography and mass spectrometry. In the latter methodology, particular emphasis is placed on the confirmation of the various complementarity determining regions (CDR) because these are key to the specificity of the mAb. The combination of the intact mass analyses and peptide mapping provides confirmation of the expected amino acid sequences.

The generation of these molecules in living organisms, as well as the physical and chemical constraints imparted on them during the course of fermentation and purification development, can result in various posttranslational modifications to the molecules. Some of these changes result in an increase in molecular mass large enough to be determined by the above method (intact mass analyses). These modifications include molecular eliminations such as pyroglutamate formation at the N-termini of light and heavy chains containing glutamic acid residues. Some of these changes appear to be spontaneous, in that the observed mass of the intact antibody is consistent with the presence of pyroglutamate instead of glutamate. In other cases, the cyclization appears to be a function of the structure, storage, or perhaps formulation conditions of the mAb (Chelius et al. 2006). Examples of additional modifications such as methionine oxidation abound in the current literature (such as described in Liu et al. 2006; Chumsae et al. 2007). Several recent reviews have been published on protein modifications like these, as well as on more subtle changes like deamidation and isoaspartate formation, which are difficult to observe by intact mass analyses of antibodies, yet they contribute to the observed heterogeneity (Liu H. et al. 2008; and see Chapter 24).

Reverse phase-HPLC of peptide enzyme-digested protein (peptide mapping) with or without MS detection is useful for obtaining peptide fractions/fragments to confirm amino acid sequences and potential modification sites, especially those that are prone to deamidation and/or isoaspartate formation. Two different types of genetic heterogeneity (mutation and recombination) have been identified by peptide mapping (Harris et al. 1993; Wan et al. 1999), thus justifying the use of this technique for demonstrating cell line comparability (e.g., transition from a hybridoma to mammalian cell expression system) and cell line stability across subsequent passages and throughout the cell banking process. Peptide mapping is also a valuable tool in showing the comparability of the purified protein product for lot-to-lot consistency and particularly for comparability exercises where certain changes in process or cell bank are implemented during different phases of clinical developmental phases or during manufacturing. Once the peaks have been identified by MS, HPLC with a UV detector at ~210 nm for amide bonds can be a very useful tool for demonstrating product consistency.

Using RP-HPLC without enzyme digestion and under reducing conditions, the light chain can produce a single sharp peak, but the heavy chain is often a broad peak without well-resolved forms. Increasing the temperature improves RP-HPLC resolution (Battersby et al. 2001); however, this may facilitate the cleavage of an acid-labile Asp/Pro bond in the heavy chain. Therefore, without enzyme digestion, RP-HPLC has very limited use except for checking the overall purity and integrity of heavy chain and light chain under denaturing conditions. In addition, hydrophobic interaction chromatography (HIC) of intact, pepsin-digested, or papain-digested antibodies can be a useful alternative to RP-HPLC. For example, Fab forms of a recombinant IgG1 antibody with Asp, isoAsp, or the succinimide at a site in light chain CDR1 were resolved by HIC after papain cleavage (Cacia et al. 1996).

23.2.3 Secondary Structures

Circular dichroism (CD) and Fourier transformed infrared (FT-IR) are two commonly used tools for determining the secondary structure of a protein, because they can derive the approximate percentage of the α-helix, β-sheet, and random coils. The CD technique and its application to protein analysis were reviewed in Kelly and Price (2000). CD is sensitive to chirality and asymmetry of the molecules being tested. Far UV CD examines the peptide backbone and can be used to estimate the fraction of secondary structure, whereas near UV CD spectra can be used to characterize the environment around aromatic residues (dominated by Trp residues) and disulfide pairings, and it is particularly sensitive at detecting subtle changes in protein structure.

Fluorescence spectroscopy is used to show the proper folding by measuring the local environments of Trp residues, which are linked to the proper folding of secondary and tertiary structures.

23.2.4 Identification of Posttranslational Modifications by Mass Spectrometry

With rapid and significant advances in both instrumentation and software, mass spectrometry has become almost the application of choice in the investigation of the molecular structure of antibodies and large proteins. A recent review by Srebalus Barnes and Lim (2007) cites many examples in which this tool shed light on posttranslational modifications, as well as modifications that could have arisen during processing or storage of the mAb. Additional examples include the elucidation of inter- and intramolecular disulfide linkages, deamidation, and other chemical changes to the mAb which all result in the heterogeneity so frequently observed and reported.

Many changes to the primary structure are benign with respect to effects on potency and stability, but others can lead to some degree of protein misfolding, which has been implicated in changes in the immunogenicity of mAbs (Maas et al. 2007). Variations in glycosylation patterns and site occupancy, primarily of N-linked sites, are also readily detected by mass spectrometric analyses (Lim et al. 2008). Mass spectrometry can be applied either as a primary or orthogonal method in the development of pharmacologically relevant mAbs, from clonal selection to formulation development to assessments of the mechanisms of mAb degradation (Cordoba et al. 2005; Cohen, Price, and Vlasak 2007).

Several molecular variations that introduce charge heterogeneity have been reported. Due to variations in cell culture conditions, over-expression of the mAb in the fermentation process, and physical and chemical stresses of the downstream purification processes, antibodies are also subject to incomplete disulfide formation (Zhang and Czupryn 2002; Zhang et al. 2002). Again, application of mass spectrometry with a good knowledge of protein biochemistry can identify disulfide linkages. Unpaired cysteines are detected using Ellman's reagent (Riddles, Blakely, and Zerner 1979) or other thiol-reactive compounds, and assignment of unpaired cysteine sites may be possible using thiol-specific reagents.

Gorman, Wallis, and Pitt (2002) discuss many aspects of the evaluation and determination of disulfide linkages in proteins and mAbs in a review of methods and considerations for disulfide bond determination in general. The major considerations in these experiments are artifacts that could be introduced from disulfide shuffling and how to mitigate these effects. Disulfide interchange can decrease significantly at lower pH, but not many proteases are efficient at low pH. Pepsin has favorable properties at low pH, but its lack of specificity adds a degree of complexity to data interpretation. Another approach is to include low concentrations of an alkylating agent during the digestion process to scavenge any free thiols. Another approach entails sequential labeling using different alkylating agents, such as *N*-ethylmaleimide and iodoacetamide (Wenger et al. 2006). Oxidation of methionine residues to the sulfoxide can occur in solution, particularly if peroxide-contaminated polysorbates are used as excipients (Herman, Boone, and Lu 1996). Oxidation of IgG may lead to a dramatic drop in potency (Rao and Kroon 1993) or even enhanced immunogenicity. In IgG1 constant regions, the methionines that are the most susceptible to oxidation are found in the Fc region at Met252 and Met428; oxidation of these residues can be monitored by hydrophobic interaction chromatography after papain digestion (Shen et al. 1996). Minimizing air/oxygen contact, such as filling the vials with inert gas in the headspace and minimizing light exposure using amber vials, helps to stabilize oxidation-prone IgG. This is particularly important if there is a susceptible site in the CDR regions (Rao and Kroon 1993).

Covalent modifications and fragmentation can also be observed after long-term storage and can become apparent in accelerated stress studies. The hinge between the Fab domain and the Fc domain is a susceptible region for proteolytic and nonenzymatic cleavage. There is already a high degree of heterogeneity of IgG due to the variations in glycans and inconsistent clipping of C-terminal Lys residues of the heavy chain (see Section 23.2.3), and the undesired fragmentations at the hinge region further exacerbate the molecular heterogeneity. For example, IgG1 monoclonal antibodies that are stored long term in solution are known to undergo gradual nonenzymatic cleavage in the hinge region (Cordoba et al. 2005) and form thioether-linked antibodies (Tous et al. 2005). Recently, a detailed analytical characterization (Cohen, Price, and Vlasak 2007) revealed that two distinct mechanisms are responsible for the hinge cleavage, one of which involves beta-elimination of the light chain-heavy chain disulfide bridge. The Cohen et al. study indicated that the β-elimination mechanism also led to an uncommon covalent modification (specifically a pyruvoyl group capping the N-terminal of the Fc fragment) and it provided a plausible explanation of the formation of the frequently observed thioether-linked antibody.

23.3 GLYCOSYLATION AND GLYCAN ANALYSIS

Typically, mAbs have only one N-glycosylation site on each heavy chain in the Fc region and no sites on the light chain. N-linked glycosylation in the Fc portion of therapeutic mAbs has emerged as an important contributor to the overall effector function of the molecule. Those derived from mammalian cells contain N-glycosylation profiles that are predominantly composed of complex, fucosylated glycans of the G0f, G1f, and G2f types. However, a variety of glycan structures can be identified from an mAb, including G0, G1, and G2 structures without fucose, hybrid-type glycans, and human high mannose (Man5-Man9). It is well understood that glycosylation at the conserved consensus site in the Fc region is critical for binding to Fc receptors, including FcγRI, FcγRII, FcγRIII, and C1q (Wright and Morrison 1997). Not only is the presence of glycosylation in the Fc region important, the composition of the particular oligosaccharide can also act to modulate relative binding to various Fc receptors. One of the best studied examples involves the FcγRIIIa receptor, where it has been shown by several groups that the presence of core α-1,6 fucose on the Fc oligosaccharide affects receptor binding, with a lack of fucose conferring better binding relative to the presence of fucose (Okazaki et al. 2004; Shields et al. 2002; Li et al. 2006; Rothman et al. 1989). As such, the evaluation of N-linked glycosylation in terms of both lot-to-lot consistency and composition are important metrics towards the establishment and implementation of a robust mAb production process (Harris, Shire, and Winter 2004).

The characterization of the N-glycosylation profile for an mAb is an important determinant for clonal selection and for evaluating process consistency. Several accepted approaches are commonly used to evaluate the glycosylation profile, including total glycoprotein quality and glycan analysis following chemical or enzymatic release. The latter approach, or the analysis of released glycans, will be discussed in the subsequent sections.

23.3.1 Enzymatic Release of N-Linked Glycans

Releasing N-linked glycans is a common method for evaluating the oligosaccharide profile of an mAb sample. Although the release of glycans by chemical means can be achieved, such as through the use of hydrazine, many investigators prefer the use of specific enzymes like PNGase F (Tarentino, Gomez, and Plummer 1985) to separate N-linked oligosaccharides after cleavage from the protein. A convenient method established at Genentech involves the binding of a sample protein to a polyvinylidene fluoride (PVDF) membrane in a 96-well plate, followed by reducing the protein with dithiothreitol reduction and then denaturation with a chaotropic buffer (Papac et al. 1998). After extensive washing, the denatured protein is treated with PNGase F. Released glycans can then be retrieved by removing the buffer contents in the well or by centrifuging and collecting the buffer contents into a 96-well sample collection plate. While either method results in the collection of free glycans, the latter provides for a cleaner sample because the PNGase F enzyme is retained in the sample well. The collected glycans can then be analyzed in a qualitative/semiquantitative fashion using matrix-assisted, laser-desorption, ionization time of flight, or MALDI-TOF. Relative quantitation of N-linked oligosaccharides is obtained by fluorescently tagging the reducing terminus, followed by normal- or reversed-phase high performance liquid chromatography (HPLC) or capillary electrophoresis (CE) methods.

23.3.2 Oligosaccharide Profiling

One of the most convenient and fastest ways to assess N-glycosylation involves the use of MALDI-TOF. This method provides mass analysis of free glycans using a matrix, most commonly dihydroxybenzoic acid (DHB). Sample preparation procedures in 96- and 384-well formats make this an attractive assay for screening N-linked glycan profiles across a large number of protein samples. MALDI-TOF analysis produces a spectrum of discrete peaks that can be cross-referenced to a predicted N-linked oligosaccharide mass. Glycan assignments to an observed mass can be confirmed by further analysis, including NMR. Alternatively, the sample can be treated with specific glycosidase enzymes or specific glycosyltransferases with the appropriate charged monosaccharide followed by MALDI-TOF analysis. With this latter technique, the resultant masses are compared with the expected theoretical masses to confirm the correct structural assignment. This qualitative/semiquantitative method is very useful when rapid turnaround times for a large number of samples are required, such as with initial clone selection, media optimization, and other downstream process development activities.

23.3.3 N-Linked Oligosaccharide Quantitation

While mass spectrometry methods offer highly specific resolution and identification assignments for oligosaccharides, other methods can be used to quantify the relative abundance of specific oligosaccharides. One of the most common methods for quantifying N-linked glycosylation without using labeling methods is high performance anion exchange chromatography with pulsed amperometric detection, or HPAEC/PAD (Weitzhandler et al. 1994). This method can also be used to quantify monosaccarides in a protein sample. For example, N-glycans can be enzymatically released, followed by an acid hydrolysis step. The content of the individual monosaccharide species can be estimated by HPAEC/PAD.

By labeling the reducing end of an oligosaccharide with a fluorescent molecule, such as 2-aminobenzamide (2-AB) or 2-anthranilic acid (2-AA), the relative content of individual oligosaccharides can be quantified using either normal- or reverse-phase high performance liquid

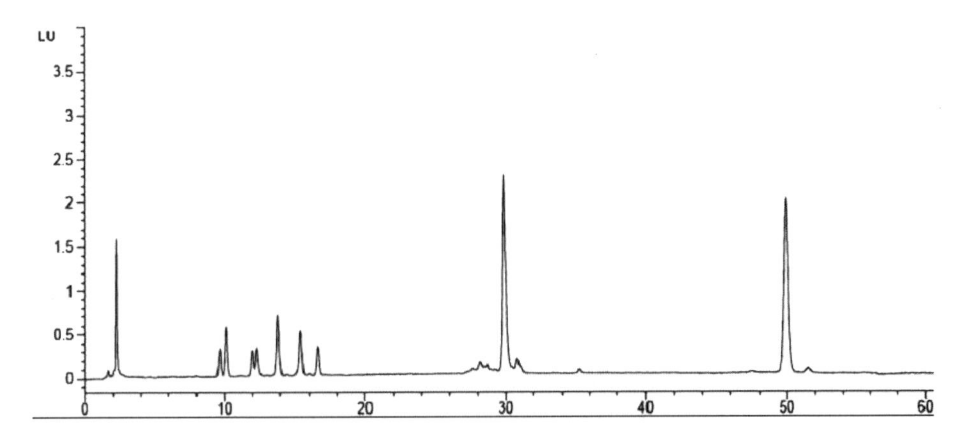

N-glycan	Retention time (min)
GlcNAc2Man3GlcNAc2~2AB	9.645
Man5GlcNAc2~2AB	10.099
Man6GlcNAc2~2AB	12.011
Gal2GlcNAc2Man3GlcNAc2~2AB	12.340
Man7GlcNAc2~2AB	13.828
Man8GlcNAc2~2AB	15.427
Man9GlcNAc2~2AB	16.643
NeuAcGal2GlcNAc2Man3GlcNAc2~2AB	29.854
NeuAc2Gal2GlcNAc2Man3GlcNAc2~2AB	49.912

Figure 23.2 Separation of N-linked oligosaccharides by normal-phase HPLC. Authentic N-linked glycan standards were labeled with 2-aminobenzamide (2-AB) and separated by normal-phase HPLC using an Alltech Prevail Carbohydrate ES column and an Agilent 1200 series HPLC instrument.

chromatography equipped with a fluorescence detector (Goldman et al. 1998; Bigge et al. 1995). This method allows for the profiling of glycans released from a glycoprotein using mole percent values, because each oligosaccharide contains only one fluorescent label. Of particular utility is the ability of the method to evaluate both neutral and acidic glycoforms in the same HPLC run (Fig. 23.2). Coupled with a highly qualitative and rapid turnaround method such as MALDI-TOF mass spectrometry, HPLC estimation of oligosaccharides species presents a powerful strategy for the support of any stage of glycoprotein drug development.

23.3.4 N-Linked Oligosaccharide Occupancy at Asn297

The extent to which N-linked oligosaccharides are found at Asn297 in an mAb is known as macroheterogeneity or occupancy. This glycoprotein characteristic can be an important metric for process development and can be included as part of a physicochemical characterization package. Since an mAb is composed of two identical heavy chains covalently linked by disulfide bonds, there exist three theoretical glycosylation states: (1) fully occupied, with Asn297 glycosylation on both heavy chains, (2) partially occupied, with Asn297 glycosylation on one of the two heavy chains, and (3) unoccupied, with no glycosylation present on either of the two heavy chains. A number of methods can be used to monitor Asn297 glycosylation, including mass spectrometry, SDS-PAGE, CE, and HPLC.

A particularly convenient method for assessing the overall degree of heavy chain N-glycosylation is CE. This method can be used under both nonreducing and reducing conditions. Under reducing conditions, the electropherogram shows two prominent peaks, light chain and heavy chain. In some samples, the heavy chain exists as two peaks, one glycosylated and one aglycosylated. These peaks can be identified by the use of PNGase F to remove the N-linked oligosaccharide from the heavy chain. By quantifying these two peaks, a relative occupancy can be estimated. Minimal sample preparation and the quantitative nature of this assay make it well suited for guiding clone selection during early development and assessing comparability throughout development.

23.4 MOLECULAR HETEROGENEITY

mAb molecules have significant intrinsic heterogeneities at different levels. In addition to the above-mentioned heterogeneities in the glycans, which is made worse when two heavy chains are paired together, mAbs are also subject to posttranslational modifications or degradation at several independent sites. Such modifications may result in the presence of many different species in crude mAB preparations and in final, purified products. They therefore display considerable heterogeneity that can be characterized by several orthogonal methods, such as isoelectric focusing (IEF), ion exchange chromatography (IEC), and capillary electrophoresis (CE). The mAb should be characterized as much as possible using relevant methods, and batch-to-batch consistency with respect to heterogeneity should be shown. Because the immunoglobulin molecule is relatively robust, many of these species will have full bioactivity *in vitro*. When using certain methods, such as state-of-the art liquid chromatography (LC) or CE, these species can be seen as many poorly resolved peaks.

One should characterize the possible discrete modifications and the major peaks seen in chromatograms; however, a full identification of all the different minor species will often not be feasible. A form of heterogeneity very specific for mAbs is C-terminal charge heterogeneity. Lysine residues from the C-termini are often partially or completely removed by a carboxypeptidase B-like activity. The extent of Lys removal should be addressed. Covalent or noncovalent association between IgG molecules during purification and/or storage also leads to reversible or irreversible aggregations, further increasing the degree of heterogeneity of mAb. More indepth discussion of mAb heterogeneity is presented in Chapter 24.

23.5 FUNCTIONAL ANALYSES OF mAb CANDIDATES

The mode of action and immunological properties of the antibody should be defined and demonstrated. During preclinical development, the desired biological effects of the mAb should be demonstrated in convincing animal models, including nonhuman primates. On the organ or tissue level, imaging or classical histological techniques can be useful. If feasible, mechanism-based localization of the administered mAb should be demonstrated. Based on the targeted and immobilized ligand(s), the mAb should bind to specific tissues (e.g., cancer cells for oncology mAbs) or organs by local imaging or whole animal imaging. Histological study is also a way to demonstrate the desired specificity, particularly for anti-cancer mAbs targeting certain cell surface molecules on different types of cancer cells or tissues.

23.5.1 Cell-Based Functional Analysis

Most mAbs are designed to bind a soluble cytokine or a receptor on the surface of a cell, subsequently triggering a series of downstream signaling events, such as (auto)phosphorylation of a receptor or cytokine(s). While the binding specificity of the IgG is determined by the amino acid sequence in several segments in CDRs of the Fab fragment, the Fc portion of IgG molecule is critical in certain cases for IgG function at the cell level and for its metabolic fate. The ability for complement binding and activation and for other effector functions of an IgG needs to be studied. In addition, cytotoxic effector properties like complement-dependent cytotoxicity (CDC) and antibody-dependent cellular cytotoxicity (ADCC) should be characterized as to whether these activities are linked to the mAbs function or to undesired adverse effects. It has been shown that differential sialylation, galactosylation, and fucosylation can potentially affect antigen binding, CDC, ADCC, and pharmacokinetics (Wright and Morrison 1997; Jefferis 2001). Cell-based potency analysis is discussed in Chapter 20 in more detail.

23.5.2 Functional Characterization on the Molecular Level

To elucidate the molecular properties of the IgG and to link them to its mode of action, the following aspects need to be included for characterizing the mAb: (1) antigenic specificity/ligand binding, including the characterization of the epitope that the antibody recognizes; (2) the paratope (the part of the mAb that recognizes and binds to the epitope) should be identified; (3) the equilibrium dissociation constant (affinity), K_D; and (4) the immunoreactivity of the antibody (ELISA, Western blot, or immuoprecipitation).

The K_D determination of a given antibody–antigen pair with different methods, including Biacore, Kinexa, and ELISA, are discussed in detail in Chapter 22.

The specific activity (i.e., binding to the ligand or receptor) of the purified mAb should be determined (units of activity/mass of product). This activity can also be measured on a relative scale once a reference lot is identified. In the meantime, whether the mAb can fully or partially block the binding of intrinsic ligand to the target molecule needs to be characterized to further aid the understanding of its mode of action.

In order to assess the *in vitro* binding activity of the mAb drug candidate, a soluble, stable antigen is necessary. This antigen will act as a surrogate for the *in vivo* target of the mAb. This antigen should be available during the early development phase to enable development of assays that are important for product development support. Two of the more popular platforms for binding analysis methods are ELISA and surface plasmon resonance-based assays. Here we will focus on the ELISA method, where certain labels (biotin, SPR, or AP) that covalently link to the primary or secondary detection reagents are needed.

When a well-characterized antigen is available, there are a few options for analyzing *in vitro* binding activity. The true solution affinity constant (or equilibrium dissociation constant, K_D, generally based on molarity) is a key attribute of IgG in order for it to function in the biological system as well as to demonstrate lot-to-lot consistency during manufacturing. This is discussed in more detail in Chapter 22. Here, we will discuss four different formats, as illustrated in Figure 23.3, where the relative binding activity to the receptor (as an example) is needed to analyze different preparations of the mAb during different developmental stages. These include the discovery stage, preclinical stage, preclinical stage for animal studies, and later for preparations of clinical supplies or during routine manufacturing after licensure.

During early development, Format A (in Fig. 23.3)—the relative EC_{50}, the effective concentration achieving 50 percent of maximal binding—is preferred because only a limited amount of the purified IgG is usually available at this stage. In addition, Format A can be used as a titer assay for culture supernatant and for process intermediate. At a later stage of drug development, Format C, IC_{50} (inhibition concentration achieving 50 percent binding) for solution activity analysis of the test article is preferred as the activity is probed in solution without going through poorly controlled wash steps, unlike in the other three formats (see Fig. 23.3). These four different ligand binding assays are compared in Table 23.3 and their advantages and disadvantages are listed.

One of these assays should be a release and stability test for the IgG during clinical development of the drug or during routine manufacturing. A functional ELISA as described here is the most mechanism-driven assay for IgG product release and characterization. In addition to this ligand binding assay, a cell-based potency assay, normally analyzing a downstream biochemical marker, is also needed for demonstrating the product potency. Together, these stability indicating potency assays should be eventually linked to clinical efficacy.

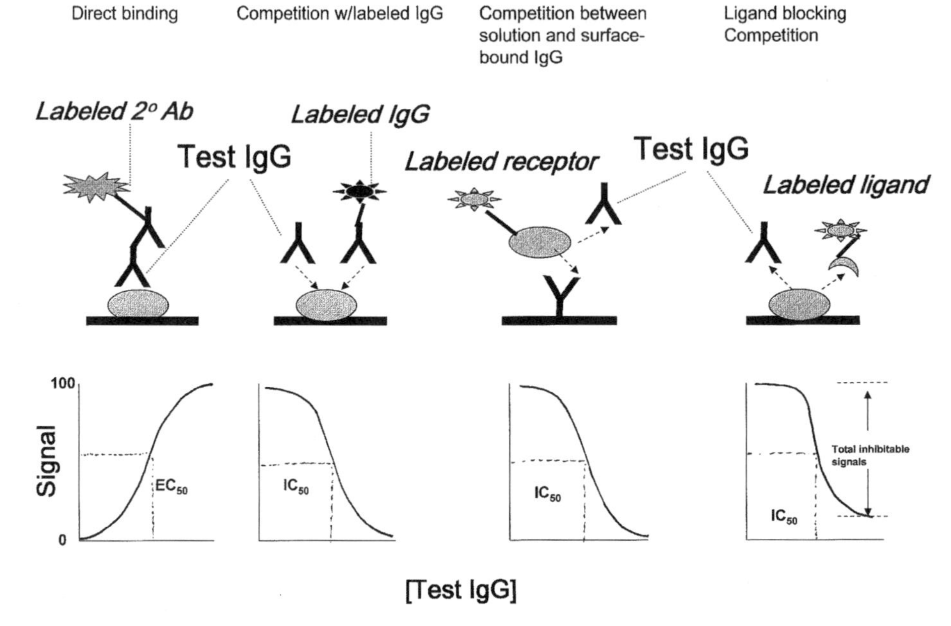

Figure 23.3 Four different formats of ELISA assays (Biacore assays can be designed in a similar way) for ligand binding analyses. The examples are illustrated with a receptor (with an oval shape) as the antigen for a therapeutic IgG. Assuming the soluble receptor (in most cases, the soluble extracellular domain) could be over-expressed and stabilized in solution, one would utilize that as a key reagent in the *in vitro* binding assays for the IgG being developed as a drug. Format A is the most straightforward way to derive the binding activity to the intended receptor in a simple ELISA (or Biacore) format. Formats B and C are competitive ELISA formats; Format C is highly preferred for purified IgG as the test IgG (also referred to as the test article) is never exposed to the subsequent wash cycles, unlike the other three formats. Format D is a ligand-blocking assay in a competitive ELISA format.

23.5.3 Specificity, Ligand Blocking, and Cross-Reactivity

The epitope (amino acid sequence or equivalent structural moiety) of the targeted antigen that is recognized by the mAb should be determined. The analysis should further include the determination of unintentional reactivity with or cytotoxicity for human tissues distinct from the intended target, and cross-reactivity with a range of human tissues by immunohistochemical procedures. After the binding specificity to the intended antigen is established, binding to some other analogous antigens should be tested for potential cross-reactivity. If there is an intrinsic ligand to the target antigen (e.g., a receptor), a blocking assay (ELISA or Biacore) should be used to demonstrate the effectiveness and completeness of ligand blocking by the mAb (see Format D in Fig. 23.3). Depending on the epitope, partial or complete ligand blocking can be achieved.

It is common and has come to be expected in the current regulatory environments, that a ligand binding assay and another cell-based potency assay be used for product release and stability testing. Though both assays are relevant, the binding assay may provide the best precision for process and formulation development, whereas the cell-based assay is good for indicating if changes are either enhanced or tolerated in a biological system. By the end of Phase III, and provided there is a good correlation between the two different assays, it is possible that one can be dropped from routine testing. In both the ligand binding assay and the cell-based potency assay, the ultimate readout is often some sort of immunoassay. Although immunoassay is a very powerful analytical tool and is widely used, assay variation can differ dramatically based on the assay format, choice of substrates and hardware for liquid handling and plate reading, standard curve design, data processing and subsequent analysis,

TABLE 23.3 Comparison of the Four Different Formats for Analyzing the *in Vitro* Binding Activity of the Human IgG to its Intended Receptor

	On the Surface	Requirements	Advantages	Disadvantages
Format A[a]	Receptor		Works in crude preps	Test IgG gores through multiple wash cycles
Format B	Receptor	Labeled IgG is available		% of active IgG between labeled and test IgG needs to be comparable
Format C[b]	IgG	Labeled receptor is available	Test IgG does not go through multiple was steps	Need to label the ligand
Format D[c]	Receptor	Labeled ligand is available	Relevant to mode of action information of affinity $(IC_{50})^{d}$ as well as footprint (% inhibitable)	Need to have soluble and labeled ligand Trimolecular interactions

[a]Format A is used most commonly due to its simplicity. In addition, the advantage for this format is that it can be used in the crude preparation as it functions in a direct binding mode and anything with no specific binding to the receptor is being washed off the surface.
[b]Format C is the most preferred format as it probes the interaction of the test IgG in solution. This is critical as the binding activity of IgG is probed in solution as a snapshot without subjecting the test article to harsh steps where the degree of immunocomplex loss is poorly understood.
[c]Format D is important to the understanding of the mode of action of the IgG as it shows the effectiveness of the IgG to inhibit the binding of the intrinsic ligand of the receptor to the receptor (the IC_{50}) and the degree of completeness of ligand blocking by the administered recombinant human IgG—as indicated by the percentage of inhibition of the ligand-receptor binding by saturating concentration of IgG.
[d]IC_{50} = median inhibitory concentration at 50% of maximal binding.

etc. (Meng et al. 2005; High et al. 2005). Therefore, it is not hard to imagine that the assay variations can be quite different when similar assays are designed and performed in different labs.

23.6 SUMMARY AND CONCLUDING REMARKS

Recombinant monoclonal antibodies and antibody-related biologics need to go through extensive biochemical, biophysical, biological, immunochemical, and immunological characterizations. Detailed molecular characterization of mAbs was described for primary, secondary, tertiary, and quaternary structures, micro- and macro-heterogeneity, affinity, epitope, ligand blocking, and glycan structures, and their potential implications for mAb functions. Most of these analyses should be performed during discovery and preclinical development. Starting from early clinical studies, a state-of-the-art characterization test panel should be performed that is aligned with the ICH Q6B guideline, which addresses the primary and higher-order structures as well as the physicochemical and functional properties of the product. A comprehensive characterization package is essential for demonstrating process and product consistency and for supporting future scale, strain, and process changes.

For product development, it is desirable to adopt an early development manufacturing process and analytical platform strategy to support a high throughput of development candidates. Later, a comparability strategy must be developed to seamlessly bridge early- and late-development processes (process comparability) and product characteristics (product comparability) relevant to the quality, safety, and efficacy of the product. The key element of this strategy is to perform extensive product characterization as early as possible for an accurate product profile. This helps to ensure equivalence between pivotal toxicology supplies and clinical supplies, improve understanding of analytical capability and process capability throughout the full development cycle, and to maintain good links between meeting regulatory requirements and performing the right experiments. Advances in the analytical technologies described in this chapter are crucial for enabling the concept of well-characterized biologicals.

The subtle differences detected using these modern analytical techniques are meaningful and should be weighted based on their impacts on pharmacokinetics, pharmacodynamics, and immunogenicity assays using relevant animal models. Understanding the mode of action helps to link the analytical results to product safety and efficacy, and facilitates the choice of meaningful assays in defining the product profiles early on in the developmental program.

ACKNOWLEDGMENTS

The authors thank Drs. Steven L. Cohen and Yang Wang for critical reading of the manuscript, and Mrs. Tracy Janus for her editorial assistance.

REFERENCES

Anders, J.C., B.F. Parten, G.E. Petrie, R.L. Marlowe, and J.E. McEntire. 2003. Using amino acid analysis to determine absorptivity constants. *BioPharm. Int.* 16:30–37.

Battersby, J.E., B. Snedecor, C. Chen, K.M. Champion, L. Riddle, and M. Vanderlaan. 2001. Affinity-reversed phase liquid chromatography assay to quantitate recombinant antibodies and antibody fragments in fermentation broth. *J. Chromatogr. A* 927:61–76.

Bigge, J.C., T.P. Patel, J.A. Bruce, P.N. Goulding, S.M. Charles, and R.B. Parekh. 1995. Nonselective and efficient fluorescent labeling of glycans using 2-amino benzamide and anthranilic acid. *Anal. Biochem.* 230(2):229–238.

Cacia, J., R. Keck, L.G. Presta, and J. Frenz. 1996. Isomerization of an aspartic acid residue in the complementarity-determining regions of a recombinant antibody to human IgE: Identification and effect on binding affinity. *Biochemistry* 35:1897–1903.

Chelius, D., et al. 2006. Formation of pyroglutamic acid from N-terminal glutamic acid in immunoglobulin gamma antibodies. *Anal. Chem.* 78(7):2370–2376.

Chirino, A.J., and Mire-Sluis, A. 2004. Characterizing biological products and assessing comparability following manufacturing changes. *Nature Biotechnol.* 22:1383–1391.

Chumsae, C., et al. 2007. Comparison of methionine oxidation in thermal stability and chemically stressed samples of a fully human monoclonal antibody. *J. Chromatogr. B* 850(1–2):285–294.

Cohen, S.L., C. Price, and J. Vlasak. 2007. Beta-elimination and peptide bond hydrolysis: Two distinct mechanisms of human IgG1 hinge fragmentation upon storage. *J. Am. Chem. Soc.* 129:6976–6977.

Cordoba, A.J., B.J. Shyong, D. Breen, and R.J.J. Harris. 2005. Non-enzymatic hinge region fragmentation of antibodies in solution. *Chromatogr. B Analyt. Technol. Biomed. Life Sci.* 818:115–121.

Goldman, M.H., D.C. James, M. Rendall, A.P. Ison, M. Hoare, and A.T. Bull. 1998. Monitoring recombinant human interferon-gamma N-glycosylation during perfused fluidized-bed and stirred-tank batch culture of CHO cells. *Biotechnol. Bioeng.* 60(5):596–607.

Gorman, J.J., T.P. Wallis, and J.J. Pitt. 2002. Protein disulfide bond determination by mass spectrometry. *Mass Spectrometry Rev.* 21(3):183–216.

Harris, R.J., A.A. Murnane, S.L. Utter, K.L. Wagner, E.T. Cox, G. Polastri, J.C. Helder, and M.B. Sliwkowski. 1993. Assessing genetic heterogeneity in production cell lines: Detection by peptide mapping of a low level Tyr to Gln sequence variant in a recombinant antibody. *Biotechnology* 11:1293–1297.

Harris, R.J., S.J. Shire, and C. Winter. 2004. Commercial manufacturing scale formulation and analytical characterization of therapeutic recombinant antibodies. *Drug Dev. Res.* 61:17.

Herman, A.C., T.C. Boone, and H.S. Lu. 1996. Characterization, formulation, and stability of Neupogen (filgrastim), a recombinant human granulocyte colony-stimulating factor. In *Formulation, Characterization, and Stability of Protein Drugs*, ed. R. Pearlman and Y.J. Wang, 303–328. New York: Plenum Press.

High, K., Y. Meng, M. Washabaugh, and Q. Zhao. 2005. Determination of picomolar equilibrium dissociation constants in solution by enzyme-linked immunosorbent assay with fluorescence detection. *Anal. Biochem* 347:159–161.

Jefferis, R. 2001. Glycosylation of human IgG antibodies: Relevance to theraputic applications. *BioPharm (International)* 14:19–26.

Kelly, S.M., and N.C. Price. 2000. The use of circular dichroism in the investigation of protein structure and function. *Curr. Protein Pept. Sci.* 1:349–384.

Li, H., N. Sethuraman, T.A. Stadheim, et al. 2006. Optimization of humanized IgGs in glycoengineered *Pichia pastoris. Nature Biotechnol.* 24(2):210–215.

Lim, A., A. Reed-Bogan, and B.J. Harmon. 2008. Glycosylation profiling of a therapeutic recombinant monoclonal antibody with two N-linked glycosylation sites using liquid chromatography coupled to a hybrid quadrupole time-of-flight mass spectrometer. *Anal. Biochem.* 375(2):163–172.

Liu, H., et al. 2006. Effect of posttranslational modifications on the thermal stability of a recombinant monoclonal antibody. *Immunol. Lett.* 106(2):144–153.

Liu, H., D. Faldu, C. Chumsae, and J. Sun. 2008. Heterogeneity of monoclonal antibodies. *J. Pharm. Sci.* 97(7):2426–2447.

Maas, C., et al. 2007. A role for protein misfolding in immunogenicity of biopharmaceuticals. *J. Biol. Chem* 282(4):2229–2236.

Macchi, F.D., F.J. Shen, R.G. Keck, and R.J. Harris. 2001. The continuing role of amino acid analysis in a biotechnology laboratory. In *Amino Acid Analysis Protocols*, ed. C. Cooper, N. Packer, and K. Williams, 9–30. *Methods in Molecular Biology* no. 159. Totowa, NJ: Humana Press.

Meng, Y., K. High, J. Antonello, M. Washabaugh, and Q. Zhao. 2005. Enhanced sensitivity and precision in an enzyme-linked immunosorbent assay with fluorogenic substrates compared with commonly used chromogenic substrates. *Anal. Biochem.* 345:227–236.

Okazaki, A., E. Shoji-Hosaka, K. Nakamura, et al. 2004. Fucose depletion from human IgG1 oligosaccharide enhances binding enthalpy and association rate between IgG1 and FcgammaRIIIa. *J. Mol. Biol.* 336(5): 1239–1249.

Pace, C.N., et al. 1995. How to measure and predict the molar absorption coefficient of a protein. *Protein Sci.* 4(11):2411–2423.

Papac, D.I., J.B. Briggs, E.T. Chin, and A.J. Jones. 1998. A high-throughput microscale method to release N-linked oligosaccharides from glycoproteins for matrix-assisted laser desorption/ionization time-of-flight mass spectrometric analysis. *Glycobiology* 8(5):445–454.

Rao, P.E., and D.J. Kroon. 1993. Orthoclone OKT3. Chemical mechanisms and functional effects of degradation of a therapeutic antibody. In *Stability and Characterization of Protein and Peptide Drugs: Case Histories*, ed. Y.J. Wang and R. Pearlman, 135–158. New York: Plenum Press.

Riddles, P.W., R.L. Blakely, and B. Zerner. 1979. Ellman's reagent: 5,5′-dithiobis(2-nitrobenzoic acid): A reexamination. *Anal. Biochem.* 94:75–81.

Rothman, R.J., B. Perussia, D. Herlyn, and L. Warren. 1989. Antibody-dependent cytotoxicity mediated by natural killer cells is enhanced by castanospermine-induced alterations of IgG glycosylation. *Mol. Immunol.* 26(12):1113–1123.

Shen, F.J., M.Y. Kwong, R.G. Keck, and R.J. Harris. 1996. The application of tert-butylhydroperoxide oxidation to study sites of potential methionine oxidation in a recombinant antibody. In *Techniques in Protein Chemistry*, vol. 7, ed. D. Marshak, 275–284. San Diego: Academic Press.

Shields, R.L., J. Lai, R. Keck, et al. 2002. Lack of fucose on human IgG1 N-linked oligosaccharide improves binding to human Fcgamma RIII and antibody-dependent cellular toxicity. *J. Biol. Chem.* 277(30):26733–26740.

Srebalus Barnes, C.A., and A. Lim. 2007. Applications of mass spectrometry for the structural characterization of recombinant protein pharmaceuticals. *Mass Spectrometry Rev.* 26(3):370–388.

Tarentino, A.L., C.M. Gomez, and T.H. Plummer, Jr. 1985. Deglycosylation of asparagine-linked glycans by peptide: N-glycosidase F. *Biochemistry* 24(17):4665–4671.

Tous, G.I., Z. Wei, J. Feng, S. Bilbulian, S. Bowen, J. Smith, R. Strouse, P. McGeehan, J. Casas-Finet, and M.A. Schenerman. 2005. Characterization of a novel modification to monoclonal antibodies: Thioether cross-link of heavy and light chains. *Anal. Chem.* 77:2675–2682.

Wan, M., F.Y. Shiau, W. Gordon, and G. Wang. 1999. Variant antibody identification by peptide mapping. *Biotechnol. Bioeng.* 62:485–488.

Weitzhandler, M., M. Hardy, M.S. Co, and N. Avdalovic. 1994. Analysis of carbohydrates on IgG preparations. *J. Pharm. Sci.* 83(12):1670–1675.

Wenger, M., P. DePhillips, and L. Chen. 2006. Evaluating the Disulfide Map of a Recombinant Membrane Protein by LC/MS/MS. (54th ASMS Conference on Mass Spectrometry and Allied Topics, Seattle, WA.)

Wright, A., and S.L. Morrison. 1997. Effect of glycosylation on antibody function: Implications for genetic engineering. *Trends Biotechnol.* 15(1):26–32.

Zhang, W., and M.J. Czupryn. 2002. Free sulfhydryl in recombinant monoclonal antibodies. *Biotechnol. Prog.* 18:509–513.

Zhang, W., L.A. Marzilli, J.C. Rouse, and M.J. Czupryn. 2002. Complete disulfide bond assignment of a recombinant immunoglobulin G4 monoclonal antibody. *Anal. Biochem.* 311(1):1–9.

CHAPTER 24

Characterization of Heterogeneity in Monoclonal Antibody Products

YANG WANG, MICHAEL W. WASHABAUGH, and QINJIAN ZHAO

24.1 INTRODUCTION

Unlike small molecule drugs, most therapeutic proteins, including monoclonal antibody (mAb)-based biologics are produced in live cells, followed by bioprocessing and formulation. In addition to naturally occurring posttranslational modifications, other chemical modification or degradation may take place during protein expression/production, purification, formulation, and storage stages. Though the primary structure of an mAb can be maintained from the perspective of cell expression and confirmed to be homogeneous by mass spectrometry, all antibody products studied or marketed are heterogeneous in nature from other structural perspectives. Furthermore, the presence of heterogeneity at the molecular and structural levels and their contributions to biological functionality are being increasingly appreciated and understood with more preclinical and clinical data. This is also facilitated by advances in the introduction and application of new *in vitro* analytical techniques to these already well-understood biomolecules and thus facilitating the correlation between variations in structures (such

Therapeutic Monoclonal Antibodies: From Bench to Clinic. Edited by Zhiqiang An

as a change in *N*-glycan pattern) in IgG and its desired and undesired biological effects. Different factors in bioprocessing, formulation, and during storage may result in the presence of many different species (discussed in more detail below) in the final product, which need to be characterized to the best of one's ability for use in the clinic, since such differences can influence product consistency or equivalency. Better understanding of the heterogeneity issue and its correlation to the intended function of the product is important to obtaining product licensure for both original innovators and follow-on manufacturers.

24.2 HETEROGENEITY OF mAb: HOW IT IS FORMED AND WHAT MAKES IT WORSE

Heterogeneity in monoclonal antibody product can be introduced in every step of the protein expression, recovery/purification, formulation preparation, storage, and even during analysis (Fig. 24.1). Intracellularly, antibodies are subjected to both multiple potential protein folding pathways or folding mistakes, and various posttranslational modifications. Extracellularly, exposure to culture media and culture processing conditions (i.e., pH, temperature, metal ions, oxidatives, and other stresses) can introduce additional modifications or degradations. Conformational search and disulfide bond pairings may introduce a certain degree of heterogeneity as not all molecules can end up in the ideal structures. The purification process can not only apply certain stress on the product, but also is dependent of some structural features, thus enriching certain populations of minor species along with the main product. Certain formulation process, with either intended manipulation or storage condition or deliberated stress as part of development, can further augment the heterogeneity for better understanding purpose. Finally, it is critically important to realize that certain analytical procedures or techniques could provide additional perturbation to the molecular heterogeneity. The nature and characteristics of the heterogeneity introduced in different steps and the techniques used for analyzing them are outlined as examples in Table 24.1. These techniques are useful in characterizing the three different kinds of mAb heterogeneities: (1) heterogeneities that are intrinsic to human antibodies in blood, (2) naturally occurring heterogeneities during cell culture and downstream processing; and (3) additional heterogeneities formed during storage and or the purposely stressed samples.

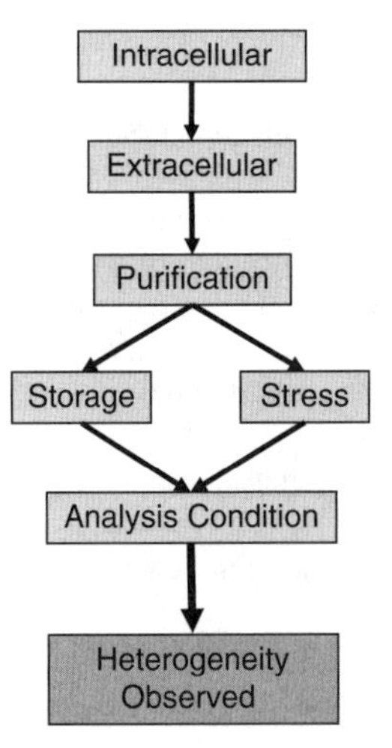

Figure 24.1 Various molecular heterogeneities can be introduced throughout the processes, including cellular functions during cell culture, downstream processing, and during analytical procedures.

TABLE 24.1 Examples of Common Analytical Techniques for Characterization of Molecular Heterogeneity or Product-Related Impurities in Monoclonal Antibody Products

Structural Characteristics	Analytical Technique
Primary structure	Amino acid analysis
	Mass spectrometry/peptide mapping (MS)
	N- and C-terminal sequencing
	SDS-PAGE (reduced for truncation and nonreduced for S-S pairing)
	Western blotting
Secondary structure	Circular dichroism (CD)
	Infrared spectroscopy (IR)
	Direct binding and biological assays
Tertiary and higher-order structure	Size exclusion chromatography (SEC)
	Analytical ultracentrifugation (AUC)
	Field-flow fractionation (FFF)
Charge variants	Ion exchange chromatography (IEC)
	Isoelectric focusing (IEF)
	Hydrophobic interaction chromatography (HIC)
Glycosylation	Mass spectrometry
	NP/RP-HPLC
	Capillary electrophoresis post protein removal
Chemical modification (oxidation, photon cross-linking, glycation, etc.)	Mass spectrometry/peptide mapping
	SDS-PAGE (reduced and nonreduced)
	Size exclusion chromatography
	Ion exchange chromatography

24.3 NATURE OF HETEROGENEITY IN mAb PRODUCTS

Variations in the form and quantity of *N*-glycans are the most common structural heterogeneity at the molecular level. Monoclonal antibodies made of four polypeptide chains (two heavy chains and two light chains) can display considerable heterogeneity in individual chains or as a result of different combination of the four chains in a full IgG molecule. There are several common modifications leading to antibody charge variants on the peptide chains, some of which occur enzymatically, such as N-terminal pyroglutamation (though a nonenzymatic pathway exists), C-terminal lysine truncation, and proteolytic fragmentation, while others happen nonenzymatically, such as deamidation, partial structural unfolding, and molecular aggregation or complex formation with other host cell or process impurities.

Different combinations of the various forms of the heterogeneity in glycans and in peptide chains would make the overall heterogeneity significantly worse in a whole IgG molecule. The different species, considering two heavy chains and two light chains in an IgG, can theoretically add up to thousands even considering only the common ones like pyroglutamation (Walsh and Jefferis 2006), deamidation (Hunt, Hotaling, and Chen 1998), methionine oxidation (Harris et al. 2001), C-terminal lysine truncation (Perkins et al. 2000), disulfide mispairing (Harris 1995), and up to six glycan variants (Dick et al. 2007). Characterization of heterogeneity due to most of the chemical modifications by mass spectrometry can be accomplished readily as discussed in the previous section. Heterogeneity due to differences in glycosylation represents one of the most common and complex posttranslational modifications taking place in the cell. A certain degree of variation in *N*-glycans is characteristic for all cell lines used for producing mAbs for human use. The molecular heterogeneity (also referred to as product-related impurities) can also be categorized based on protein structural characteristics. The two main classes of heterogeneities are those at the amino acid or peptide level, typically chemical modifications, and those at the secondary and tertiary structural levels, typically physical alterations, although it is often common that changes at one level will result in changes at the other level. For example, a charge variant caused by deamidation (introducing a negative charge) or oxidation, could lead to local protein conformational changes resulting in elevated tendency of the modified molecules to aggregate.

Analyzing and fully understanding mAb heterogeneity can be quite challenging in some cases as one is dealing with the trace rather than bulk properties of the mAb. Identification and, if feasible, quantitation of the minor species present in the mAb help to shed light on the degradation pathways. The more insightful understanding of these variations in turn is critical in developing formulation prolonging the shelf life of the mAb product and enhancing the process consistency during production. In Table 24.1, we provide an overview of the common heterogeneities and how they can be characterized with typical analytical methods. The two common heterogeneities related to charge variants and aggregations are further discussed in more details in the following sections with examples of using several orthogonal analytical methods. Specific examples are provided to illustrate how each analytical method can provide quantitative information on antibody heterogeneity.

24.4 ANALYSIS OF CHARGE-RELATED HETEROGENEITY

24.4.1 Solution-Based Isoelectric Focusing

In order to avoid assay-induced heterogeneity, solution-based techniques applicable to protein in native state are thought to cause minimal perturbation to the mAb structure. Isoelectric focusing (IEF), running in many different formats such as imaging based capillary-base IEF from Convergence Bioscience or other forms, is one of the commonly used solution-based methods for analysis of protein charge variants. IEF separates different charge variants based on the pH where the total net charge of a molecule in solution is zero, that is, their isoelectric points. With no surface adsorption and no resin interaction required, IEF is viewed to be less intrusive than other methods that involve resin interactions or surface adsorption where stress could occur. In addition, IEF can be run under native conditions with all structures of mAbs retained during IEF measurement, as well as under fully denaturing conditions with a neutral denaturant (such as urea) where the higher-order structural features are completely eliminated. The comparison of the native and the denatured protein forms can provide structural information in the presence or absence of a denaturant, thus elucidating some interactions or shielding effects due to exposure to ionic species in buffer and charge-charge interactions, with some of them being made possible only by three-dimensional structure under native conditions. Another advantage of IEF is that the peak intensities can be monitored directly with 280 nm absorbance, where signals are proportional to protein concentrations/amounts in these peaks.

24.4.2 Chromatographic Separation of Charge Variants

Ion exchange chromatography (IEC) operating in either anion (AIEC) or cation (CIEC) exchange mode for separation is another commonly used column-based technique for assessing charge-related heterogeneity. Mechanistically, IEC probes differences in molecular charge distribution on the surface of an antibody that is accessible to opposite charges on the resin employed in the chromatography. Therefore, IEC results can sometimes contain certain structural information, such as the accessibility of N-terminal and C-terminal peptide segments to the bulk media or chromatographic resin. Due to the structurally sensitive nature of the IEC methods, it is less relevant to apply the IEC technique to fully denatured antibody where both heavy chain and light chain are linearized into random coils. On the other hand, IEF can be readily applied to both native and denatured proteins for assessing the structural impacts on the isoelectric points. Due to the chromatographic nature of IEC, it is less sensitive to the buffer matrix as compared to IEF since there are adsorption and desorption processes to the resin where matrix effects can be minimized. Utilization of different elution buffers with different gradients may significantly improve the resolution. On the contrary, IEF cannot tolerate much salt as the matrix needs to be compatible with the electrolytes for the IEF analysis. The applications of and correlation between results obtained with IEC and IEF methods will be further discussed in the following sections.

24.4.3 Applications and Correlation of Different Methods for Analyzing Charge Variants

Figures 24.2a and b show, respectively, the typical data profiles for a human IgG1 obtained with cation exchange chromatography (CIEC) and isoelectric focusing (IEF). At least four components are present in each example using both bioanalytical techniques. The main component in both the CIEC and IEF profiles is from the monomeric antibody with unmodified amino acid sequence and uncharged glycans except that the C-terminal lysine on each of the heavy chains is removed. The two basic peaks are from either the antibody with the C-terminal lysine removed from one heavy chain or not at all. Confirmation of the assignment can be readily achieved with the treatment of carboxypeptidase B (Wright 1992). Additional basic peaks present after the carboxypeptidase B treatment may be related to N-terminal pyroglutamation, which adds a positive charge to a polypeptide (by eliminating a negative charge of the side chain of glutamic acid; Lebowitz, Lewis, and Schuck 2002). The presence or absence of

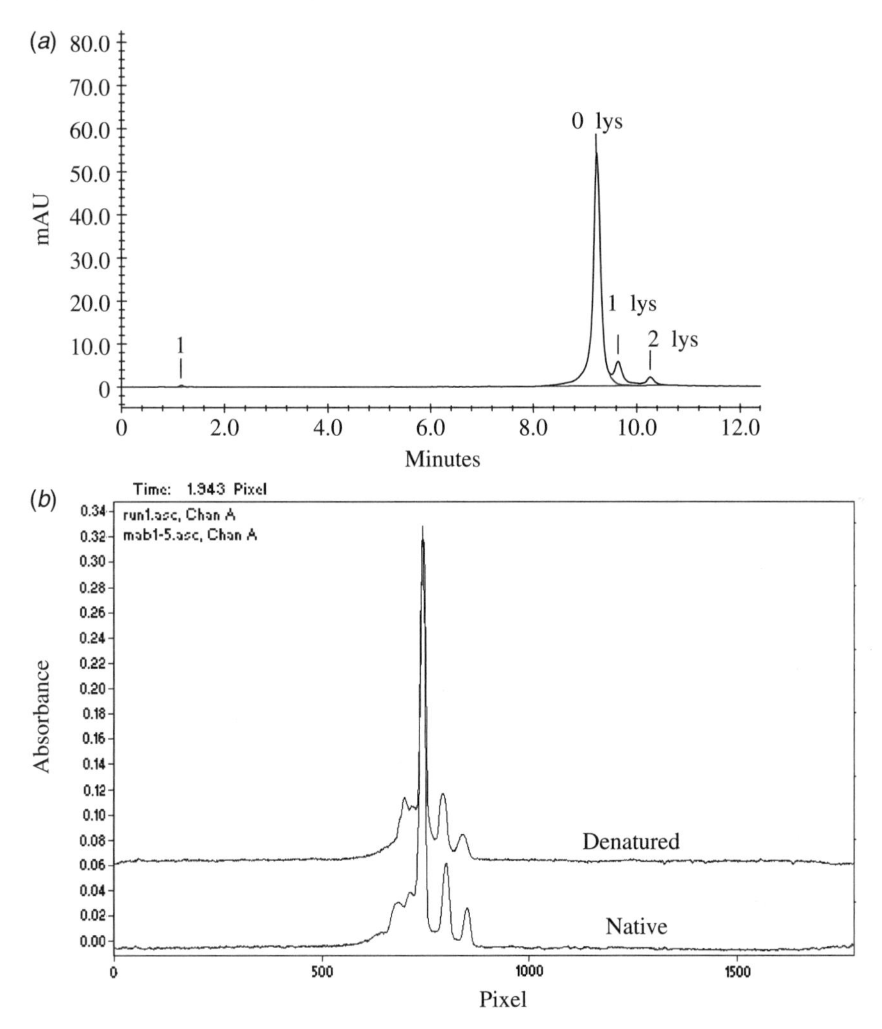

Figure 24.2 Charge heterogeneity assessed by (a) cation exchange chromatography (CIEC) and (b) isoelectric focusing (IEF). The two minor basic components shown on the left side of the main peak in both CIEC and IEF profiles are due to lysine variants. There is little difference in the IEF profiles for an IgG1 under native (b: bottom trace) and denatured (b: top trace) conditions. The denaturation was achieved with urea.

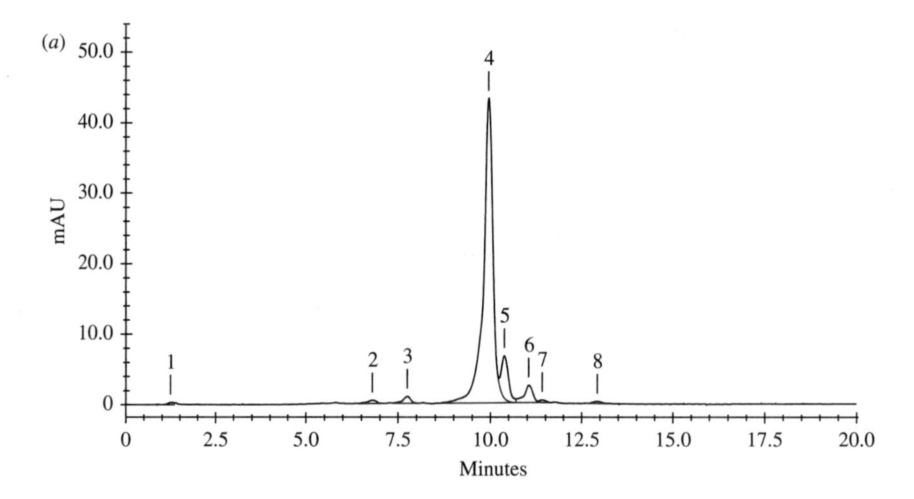

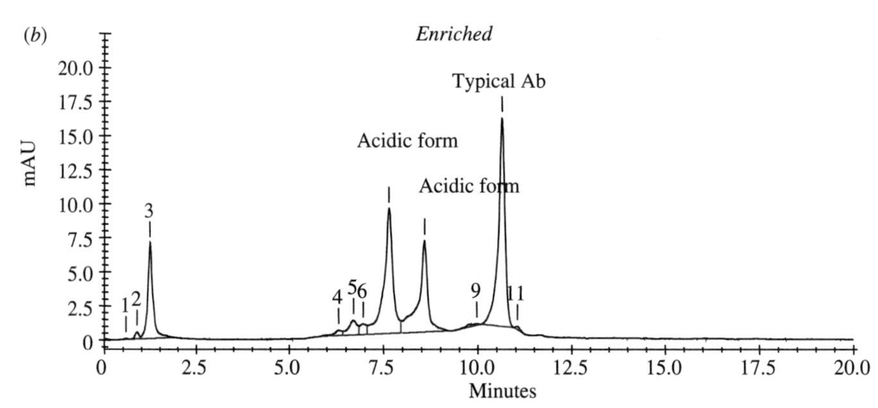

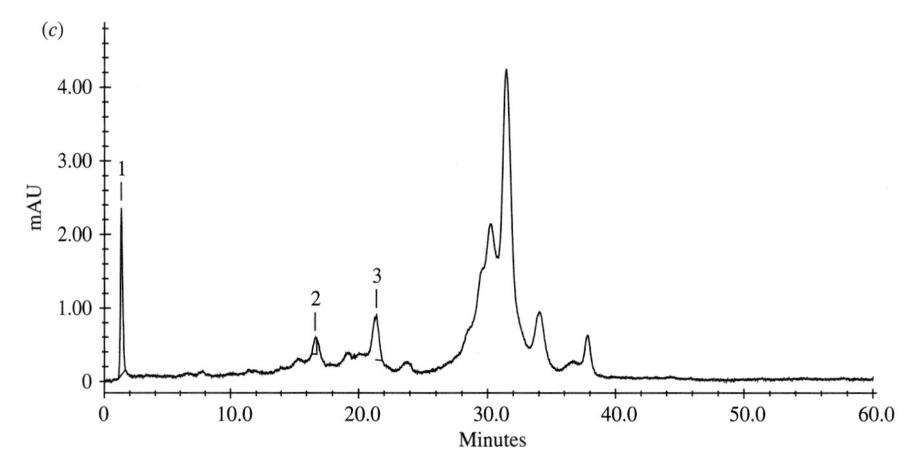

Figure 24.3 Cation exchange chromatographic profiles of bioprocess intermediates of an IgG1:intermediate from baseline purification process with normal sample collection (a), process intermediate from an aggressive elution cut (b), and stressed process intermediate at 37°C for a week (c).

C-terminal lysine residue(s) and N-terminal pyroglutamation is not usually a regulatory issue, as such processing occurs in nature with endogenous antibodies in human plasma.

Various modifications contributing to the observed mAb heterogeneity presented as the acidic components can be complicated. The two major contributing modifications are variations in the attachment of charged glycan motifs and deamidation of asparagine residues (Liu, Andya, and Shire 2006). In theory, structural changes can introduce additional acidic or basic variants in either IEC or IEF. A comparison of the IEF profiles in native and denatured states can further help in assessing the extent of structural contributions to the isoelectric points of different species. In the example shown in Figure 24.2b, very few changes were observed between native and denatured mAb beyond assay capability. This result suggests that the secondary and tertiary structures play a limited role in affecting the isoelectric point of the IgG molecule.

During antibody drug development, in addition to thorough assessment of charge-variant-related heterogeneity in the final product, both purification and formulation process developments need to be studied in great detail with respect of heterogeneity of mAb, thus searching for the best conditions to minimize it and increase the homogeneity and stability of the formulated product. An example is given in Figure 24.3, where different purification processes can lead to different amount of acidic variants in the final product (Figs. 24.3a and 3b), while stress under elevated temperature can lead to formation of acidic variants (Fig. 24.3c). Acidic variants formed from heat stress are different from those derived from the purification process. Therefore, minimizing those acidic variants is a good benchmark for developing ideal formulation conditions with respect to the long-term mAb stability.

A major advantage of the IEC method is its ability to isolate or enrich minor charge variants for further characterization after collecting the fractions of interests. However, one of its limitations may

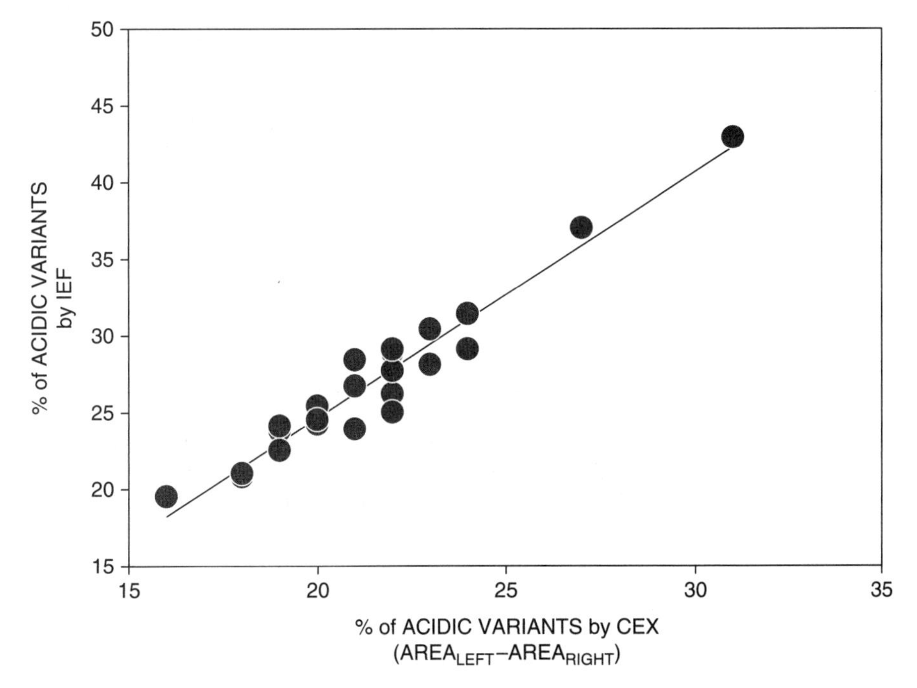

Figure 24.4 Correlation between results obtained with cation exchange chromatography (CIEC) and isoelectric focusing (IEF) for a set of process development samples. The IEF data were obtained by a drop-line integration approach for all samples, and the CIEC data were obtained by analyzing the difference between the total areas on the left and right sides of a fixed line at the maximum of the main peak (for example, the highest peak in Figs. 24.3a to c).

be the relatively poor resolution because acidic components showed as shoulders on the main peak or just poorly resolved from the main peak. Even with limited resolution, IEC has been demonstrated to be a high throughput analytical method for supporting process and formulation development. Using the relative purity approach, the molecular heterogeneity as resolved in IEC profiles can be analyzed and compared quantitatively and reproducibly by dividing individual chromatograms at the maximum of the main peak. Analyzing the difference between the total contributions from the two components as a function of process or formulation or stress parameters can provide insights of the root cause for these heterogeneities, thus aiding the search for a solution to the problem of minimizing them for mAb homogeneity and stability.

Furthermore, in most cases, the correlation between the results from IEC and IEF methods can be established. If so, one method can be used predominantly and in high throughput format for screening process and formulation conditions, while the other method can be used only for the confirmation purpose. An example is illustrated in Figure 24.4; acidic variants of a set of process development samples were assessed by both IEF using acidic peak integration and IEC by difference in peak areas on the left and right of the main peak. Both methods offer similar quantitative information and good correlation was observed from the results obtained with these two different methods.

24.5 ANALYSIS OF MOLECULAR HETEROGENEITY RELATED TO SIZE

Size distribution or mono-dispersity of an mAb product is important in terms of both safety and efficacy. There are several common causes for molecular heterogeneity in size. Components smaller than the molecular weight of an intact antibody often result from enzymatic or nonenzymatic (chemical) cleavage, in addition to incomplete formation of or mispaired disulfide bridges. In the latter case, the entire light or heavy chain (two light chains and two heavy chains in an IgG molecule) may not be incorporated into the IgG molecule in either native or denatured state.

The size-related heterogeneity due to components larger than the molecular weight of an individual antibody is often a result of molecular association, oligomer formation, aggregation, or even precipitation. The root cause of the aggregation can be different, ranging from intracellular misfolding of the protein to extracellularly stress-induced. As a result, a full spectrum of species, from molecular dimer, oligomer, higher-order aggregates, to precipitation may be present in the purified/recovered mAb preparations. Furthermore, they can form by either noncovalent association or covalent linkage with or without disulfide bonds. Additional complexity is introduced with respect to whether the noncovalent association is partially or completely reversible and to the time scales and concentration dependence of those reversible associations.

24.5.1 Commonly Used HPSEC, SDS-PAGE, and SDS-CE

High performance size exclusion chromatography (HPSEC) and sodium dodecyl sulfate polyacrylamide gel electrophoresis (SDS-PAGE) are the most commonly used bioanalytical techniques for assessing size-related heterogeneity (Table 24.1). The HPSEC can be developed for both native and denatured antibodies to assess whether association is covalent or noncovalent. It is also useful in elucidating whether cleaved components are incorporated into monomeric molecules under native condition. The SDS-PAGE method is used mostly for assessing molecular integrity of the light chain and heavy chain under denaturing conditions. SDS-PAGE results with or without reducing agents can shed light on the correct or incorrect covalent linkages due to disulfide bond between peptide chains.

24.5.2 Other Biophysical Methods

In recent years, heterogeneity of mAb products is being better appreciated. As a result, more and more new and improved methods, including the new algorithms for data analysis, have been developed to address this issue, particularly for products in clinical trials or in post-market stage. In addition to

HPSEC and SDS-PAGE, other biophysical techniques used to assess the size of molecular species include analytical ultracentrifugation (AUC; Bermudez and Forciniti 2004), field-flow fractionation (FFF) (Liu et al. 2006), and dynamic light scattering (DLS) methods (Murphy 1997). These techniques are being used for analyzing mAb products more and more frequently. Their advantages, as well as their limitations and potential artifacts introduced by certain techniques, should be fully appreciated when an mAb is being studied by a set of orthogonal methods.

Though it is common to have antibody aggregations proceed from more abundant smaller species, such as IgG dimer to larger aggregation (example shown in Fig. 24.5a), different distribution of larger antibody aggregations can occur under special conditions (example shown in Fig. 24.5b). Analytically, it is important to track the performance of the HPSEC column in terms of both mass balance (i.e., recovery, typically by peak area expected for the testing articles) and resolution, which can readily be accomplished with running molecular weight standards. Both relative peak areas and peak widths can be evaluated quantitatively to ensure consistent instrumentation performance and buffer properties.

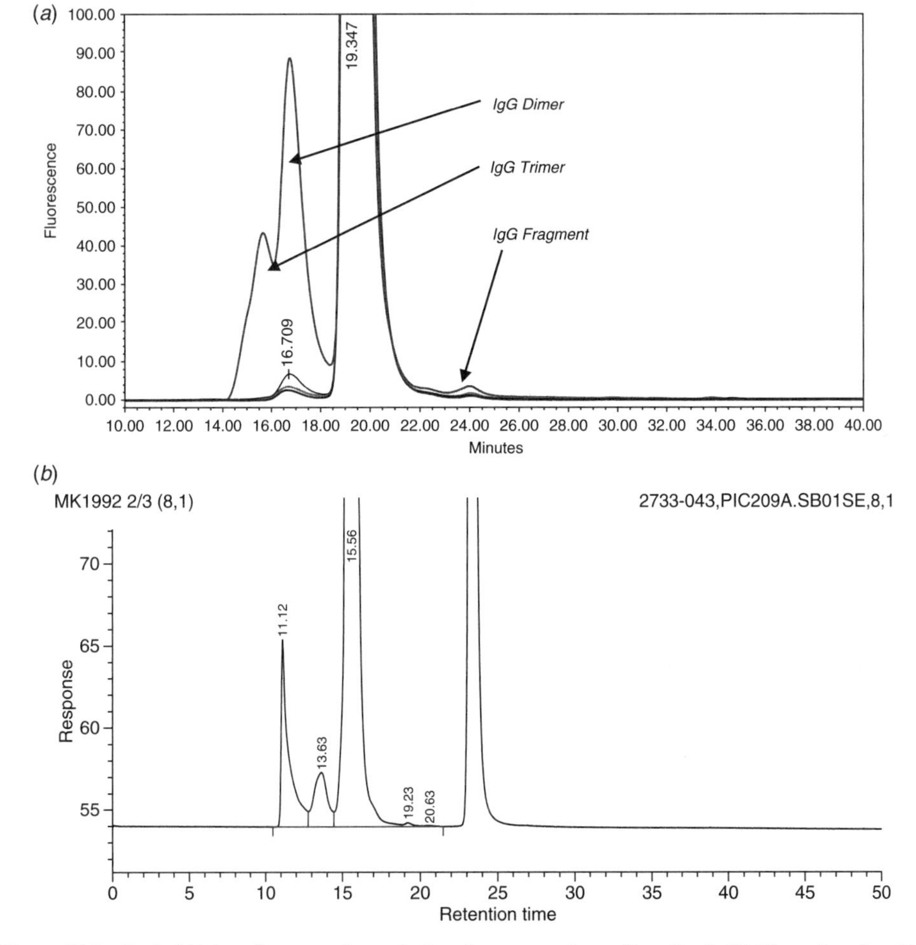

Figure 24.5 Typical high performance size exclusion chromatography profiles of an IgG1. The series of profiles in (a) represent storage stress of the same IgG1 sample. The example trace in (b) represents unique, process-induced high-order aggregate of IgG molecules. The amount of the high-order aggregate is a function of specific process parameter, while the amount of the IgG dimer stays relatively constant for the same range of the process parameter studied.

24.5.3 Concentration Dependence and Reversibility of mAb Aggregation

One aspect of soluble antibody size heterogeneity that needs to be addressed is its reversibility. From the analysis perspective, HPSEC has the capability to assess aggregation that is either not reversible or reversible with kinetics longer than a couple of hours. Typically, the initial investigation could involve

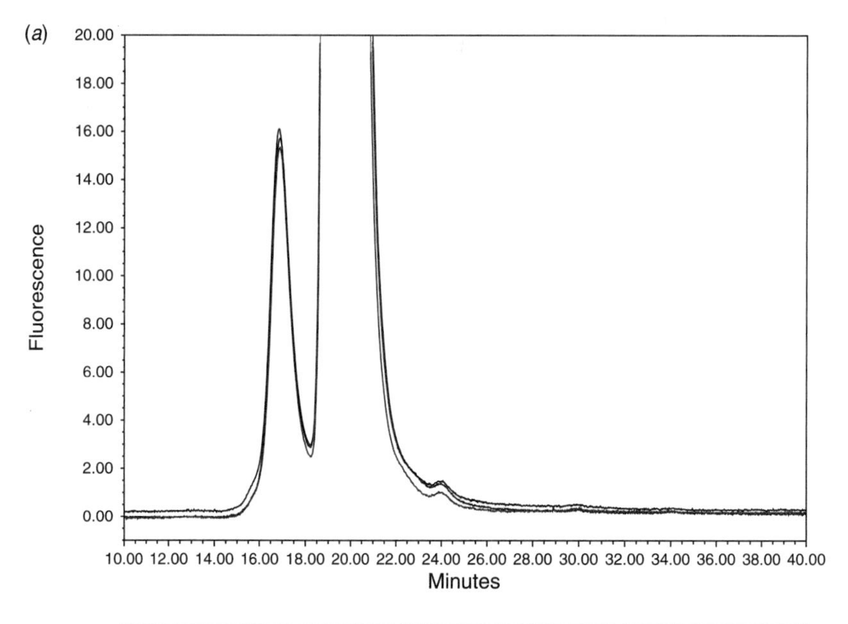

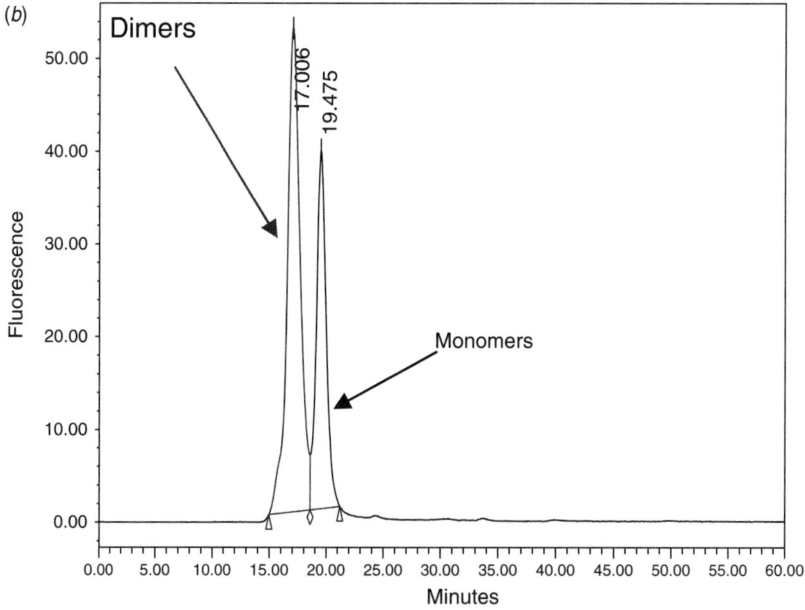

Figure 24.6 Examples for using the HPSEC method to study reversibility of an antibody size heterogeneity. The IgG1 dimer content represents less than 1 percent of the total antibody (a), which can be enriched through a fractional collection from an HPSEC run (b). The relative content of IgG1 monomer and dimer can be assessed as a function of time at room temperature (c). In this case, the process-derived IgG1 dimer is tightly associated, although not covalently linked (SDS-PAGE data not shown).

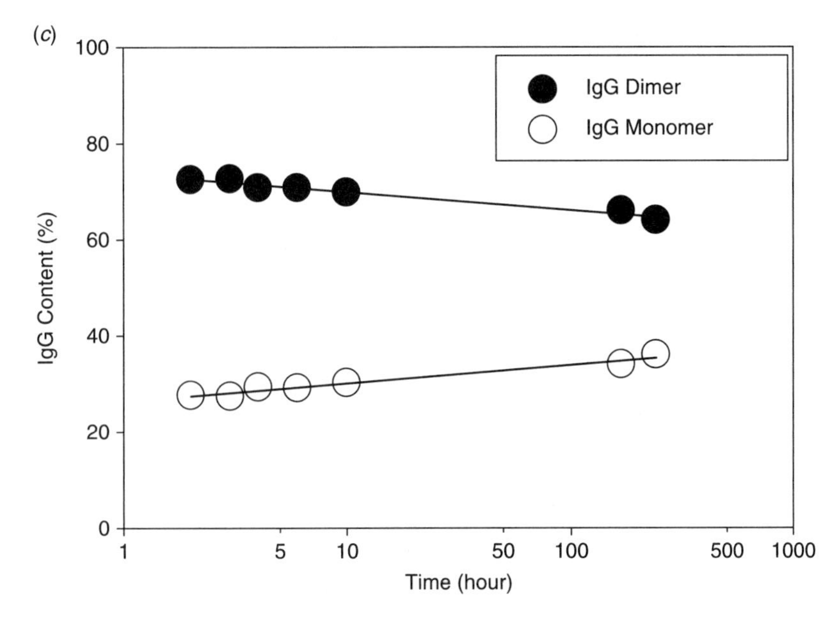

Figure 24.6 *(Continued)*.

a concentration-dependence study with HPSEC as well as other techniques such as AUC, FFF, and DLS, which are more compatible with higher antibody concentration samples. It is important to assess the status of molecular associations at elevated concentration even if HPSEC did not reveal their presence, since such species could in theory lead to more heterogeneity upon prolonged storage, impacting the product stability as well as immunogenicity depending on their dissociation kinetics and sensitivity to other environmental factors in a living system. An example is provided in Figure 24.6 where the HPSEC method is used to assess the properties of molecular association of an IgG1. The technique offers greater flexibility in isolating or enriching the aggregated component (Fig. 24.6b) even though it may be in trace amount (~1 percent) in the final product (Fig. 24.6a). Once enriched, one can further assess the stability of the molecular aggregate (such as the IgG1 dimer, in the example provided, Fig. 24.6c).

Molecular heterogeneity in size can also be readily investigated with analytic ultracentrifugation (AUC). As shown in Figure 24.7, the technique is complementary to HPSEC, and in these examples, the results correlate quantitatively with those obtained from HPSEC in terms of amount of IgG1 dimer detected (0.3 percent in Fig. 24.7a and 1.0 percent in Fig. 24.7b; HPSEC data not shown). However, the AUC-based methods are technically more complex, computationally intensive, and intrinsically lower throughput. Therefore, it is often used in confirmatory studies to support the validity of antibody heterogeneity observed with the HPSEC method which is considered the work horse for characterizing protein aggregation in the industry. Other methods, such as FFF and DLS, have advantages and disadvantages as well, and are commonly used in supportive mode. For example, the DLS method can provide sizing information in a highly concentrated sample (>100 mg/mL), while a single technique like FFF can be used to quantitatively assess molecular size ranging from ~10 nm to 10 μm with tuning of the cross flow rate (Levin 1991).

There exists additional size-related heterogeneity that can be masked by the molecular structure in native state. They include clipping in the polypeptide chains, and undesired covalent bonds either by wrong disulfide pairing or other chemical cross-link. Though it is possible to extract the information with the HPSEC or AUC methods modified for denatured and reduced conditions, the most effective way is to use SDS-PAGE, either the slap gel or capillary electrophoresis (CE)-based analytical technique under both nonreduced and reduced conditions. Under nonreduced conditions, one can evaluate

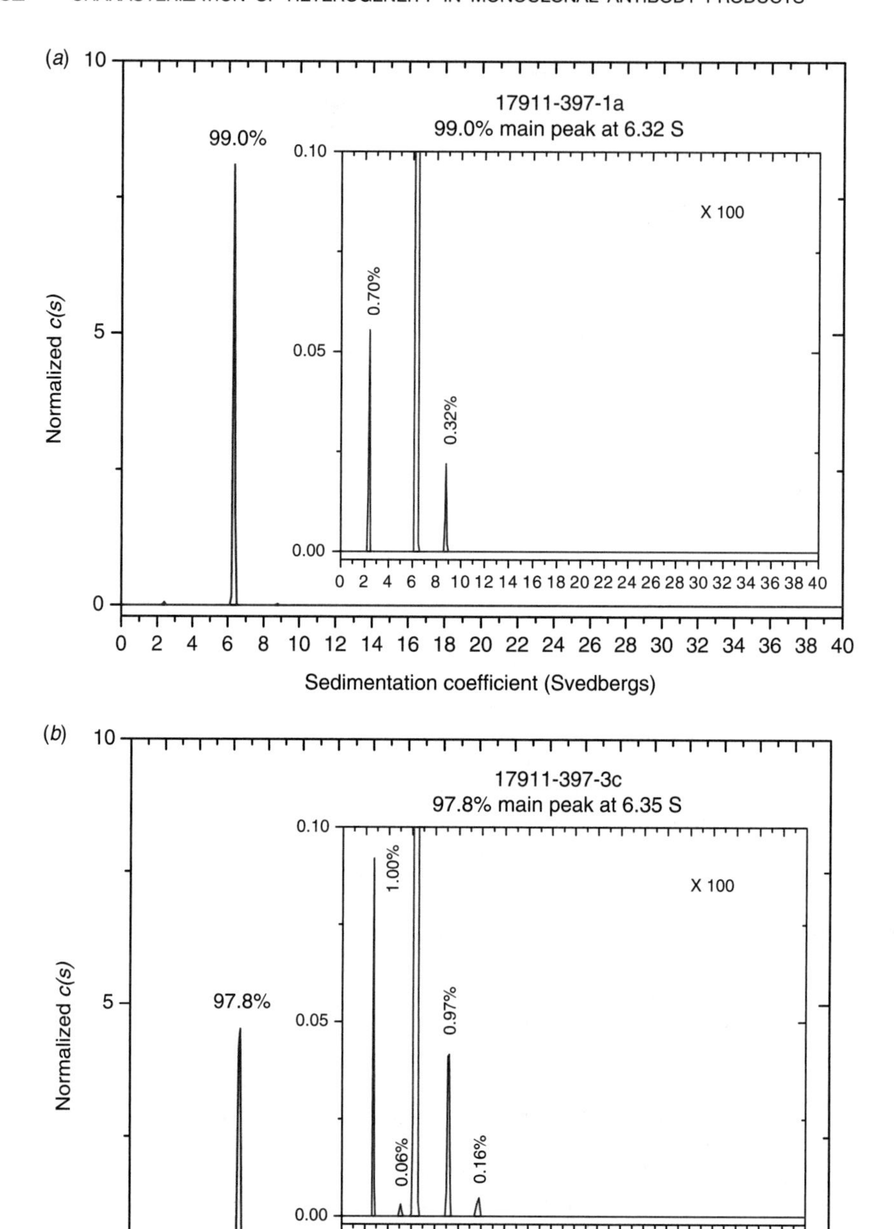

Figure 24.7 Examples of calculated molecular size distribution profiles from the sedimentation velocity measured by analytical ultracentrifugation technique. Sample shown in (b) has slightly elevated IgG1 dimer content compared to the samples shown in (a), which is consistent with the data obtained with HPSEC (data not shown). The peaks near 2.5 Svedbergs in both traces are due to the polysorbate 80 introduced as part of formulation.

if there are clippings and either absence of correct or presence of mistakes in disulfide bonds as well as whether molecular aggregates are covalently associated or not. Under reduced conditions, one can further assess if additional clipping is present that is not detectable due to disulfide link, and whether chemical cross-linking not involving disulfide bond exists.

In the example provided in Figure 24.8a, the heterogeneity observed under nonreduced conditions include two heavy chains, one heavy and one light chain (half antibody), and individual heavy and light chains as well as other fragments. Many of these size heterogeneous components are not detectable in standard native HPSEC assay. Upon denaturation in the SDS-PAGE assay, the identification or assignment of the bands (for full length or fragmented peptide chains) can be readily achieved with the technique of Western blotting using, for example, anti-human light chain and anti-human heavy chain or anti-human Fc antibodies. The example of molecular aggregation involving mispairing disulfide bonds

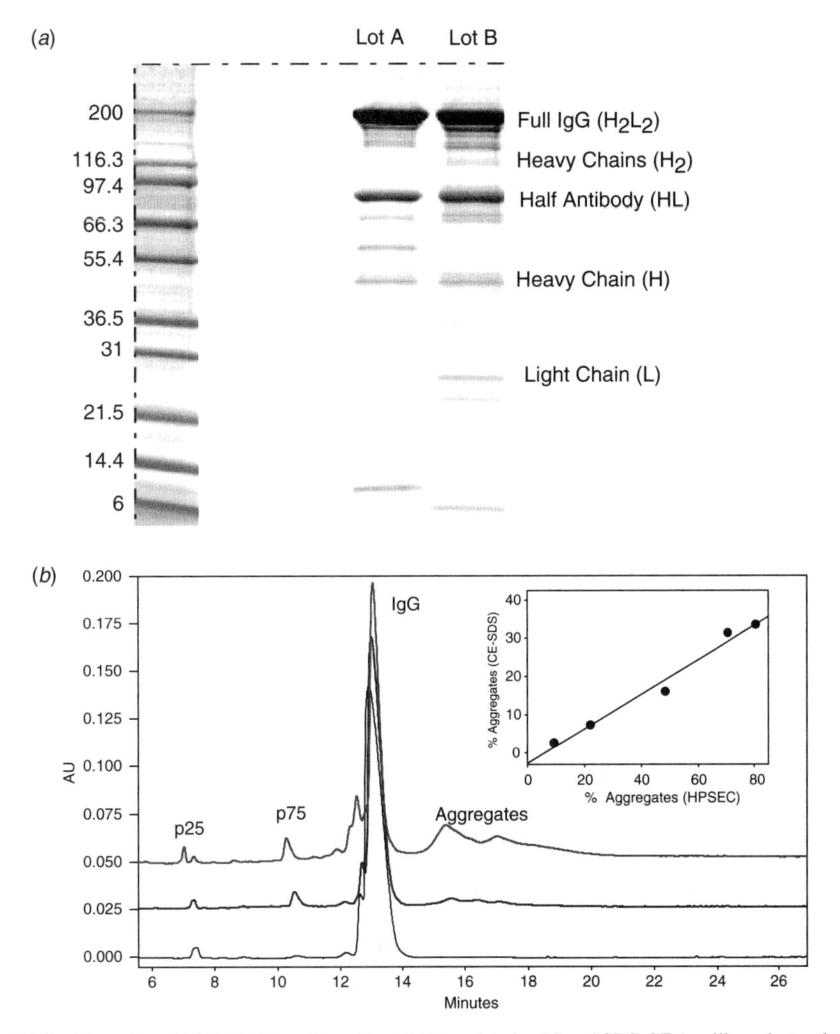

Figure 24.8 Nonreduced SDS-PAGE profiles of two IgG4 molecules (a) and SDS-CE (capillary electrophoresis) profiles of process fractions of a modified IgG2 molecule (b). The identity assignment shown in (a) was made based on Western blot analysis (data not shown). The inset in (b) shows the correlation between aggregations detected by HPSEC and by SDS-CE. HPSEC detects more aggregation, suggesting that not all aggregations are covalently linked.

is shown with a different technique (SDS capillary electrophoresis) in Figure 24.8b. As discussed previously, attention needs to be paid to careful experimental assessment of artifacts potentially introduced during the sample treatment for analysis to avoid two major and common artifacts for increased heterogeneity, the additional clipping and disulfide shuffling. Screening of the experimental conditions in terms of incubation temperature and duration as well as proper use of free-thiol capping reagents (such as iodoacetamide or iodoacetic acid) is often necessary.

24.6 CONCLUSION

In summary, heterogeneity is intrinsic to protein based biopharmaceuticals, including mAb-based drugs. The issue needs to be addressed early in a developmental program as relative heterogeneity among different mAbs can be one of the key criteria for picking mAb candidates for advancing to preclinical and clinical development. The heterogeneity is both naturally occurring and externally induced, and can be pronounced in certain special cases with proteins having "hot" spots. Analytical methods exist for a thorough characterization of most, if not all, of these heterogeneous components when the presence is above a meaningful level. Charge- and size-based heterogeneity can be fully characterized from a quantitative perspective by ion exchange chromatography and isoelectric focusing, and high performance size exclusion chromatography and SDS-PAGE methods, respectively, with additional information obtained with some newly introduced and more complex analytical methods (such as FFF and AUC). While each method has merits of its own, precautions should be taken in the data interpretation since assay-induced changes are possible when certain stress or surface adsorption/interactions are necessary for the technique. Applications of these methods and the accumulative knowledge about the product-related heterogeneity have played and will continue to play important roles in the development of mAb-based biopharmaceuticals.

REFERENCES

Bermudez, O., and D. Forciniti. 2004. Aggregation and denaturation of antibodies: A capillary electrophoresis, dynamic light scattering, and aqueous two-phase partitioning study. *J. Chromatogr. B* 807:17–24.

Dick, L.W., C. Kim, D. Qiu, and K.C. Cheng. 2007. Determination of the origin of the N-terminal pyro-glutamate variation in monoclonal antibodies using model peptides. *Biotechnol. Bioeng.* 97:544–553.

Harris, R. 1995. Processing of C-terminal lysine and arginine residues of proteins isolated from mammalian cell culture. *J. Chromatogr. A* 705:129–134.

Harris, R., et al. 2001. Identification of multiple sources of charge heterogeneity in a recombinant antibody. *J. Chromatogr. B* 752:233–245.

Hunt, G., T. Hotaling, and A.B. Chen. 1998. Validation of a capillary isoelectric focusing method for the recombinant monoclonal antibody C2B8. *J. Chromatogr. A* 800:355–367.

Lebowitz, J., M. Lewis, and P. Schuck. 2002. Moden analytical ultracentrifugation in protein sciences: A tutorial review. *Protein Sci.* 11:2067–2079.

Levin, S. 1991. Field-flow fractionation in biomedical analysis. *Biomedical chromatography* 5:133–137.

Liu, J., J. Andya, and S. Shire. 2006. Critical review of analytical ultracentrifugation and field-flow fractionation methods for measuring protein aggregation. *AAPS Journal* 8:E580–E589.

Murphy, R.M. 1997. Static and dynamic light scattering of biological macromolecules: What can we learn? *Current Opinion in Biotechnology* 8:25–30.

Perkins, M., T. Theiler, S. Lunte, and M. Jeschke. 2000. Determination of the origin of charge heterogeneity in a murine monoclonal antibody. *Pharm. Res.* 17:1110–1117.

Walsh, G., and R. Jefferis. 2006. Post-translational modifications in the contest of therapeutic proteins. *Nature Biotechnology* 24:1241–1252.

Wright, H.T. 1992. Nonenxymatic deamidation of asparaginyl and glutaminyl residues in proteins. *Crit. Rev. Biochem. Mol. Biol.* 26:1–52.

PART VII

ANTIBODY EXPRESSION

CHAPTER 25

Antibody Expression in Mammalian Cells

FUBAO WANG, LORENZO CHEN, NEAL CONNORS, and HENRYK MACH

ABSTRACT

The expression of recombinant antibodies in mammalian cells provides speed and flexibility for early phase antibody drug discovery via transient expression and sustained high yield for GMP (Good Manufacturing Practice) production via stable cell lines. In this chapter, generic experimental procedure and conditions are described which include the general features of vectors used for antibody expression, cell substrates, cell culture conditions, transfection agents, cell culture transfection and antibody production, harvest, and purification. With purified antibodies, several biophysical and analytical characterization methodologies are described for dose, purity, identity, aggregation, thermal stability, and antibody modifications such as glycosylation. The chapter provides a general description of transient production of antibodies and a toolbox of assays with which the purified antibodies are analyzed.

Therapeutic Monoclonal Antibodies: From Bench to Clinic. Edited by Zhiqiang An

25.1 INTRODUCTION

Antibody discovery has become a prominent phenomenon in the pharmaceutical and biotech industries. Recent market research suggests antibody-based therapeutics will be one of the major drivers for the continued growth of the industry during the next five years and beyond (Riley 2006). So far, the predominant molecules on the market and in the published pipelines are full-length immunoglobulins subclass G (IgG). How to produce IgG-based molecules in desired quantities has become a critical challenge. While stably transformed, clonal mammalian cell lines such as Chinese hamster ovary (CHO) or NS0 are used for the commercial production of biopharmaceuticals, the time and resources required to establish these cell lines are substantial. In the case of discovery and development of monoclonal antibody (mAb)-based therapeutics, there is a real need to rapidly and efficiently produce tens or hundreds of milligrams for *in vitro* and *in vivo* evaluation. Because of the low probability of success associated with an early phase project, the investment in the production of a stable cell line is not warranted. During the time window from the inception of a program to the selection of a candidate for clinical development, a more flexible and speedy way is needed to meet the challenge. Transient expression of antibody provides such a speed and flexibility and can provide gram quantities of antibodies for early lead study before the materials generated from the stable cell line are available.

Mammalian cell lines are typically used for transient expression because of their transfectability and the ability to carry out posttranslational modifications critical to the function of the protein of interest (e.g., glycosylation and sialylation). Transient expression relies on the principle that certain transfection reagents can mediate the introduction of DNAs into cells. Once the DNA is inside the cells, the genes driven by the corresponding promoter and polyadenylation signal can be transcribed and mRNA is in turn translated into protein. Based on the design, the expressed protein can be guided through the secretion pathway through the endoplasmic reticulum (ER) and Golgi apparatus and eventually released to the culture medium. The secreted antibody is harvested and subject to purification, usually affinity chromatography, such as with protein A or G. The yield is dependent on several factors: the nature of the DNA sequence (codon usage), the nature of the protein (good or poor expressor), the longevity of the transfected DNA inside cells (oriP/EBNA; Lindner and Sugden 2007), the ratio between cells and plasmid DNA, transfection agents, the amount of DNA transfected, expression time length, quality of DNA, transfection protocol, etc.

The demand for the antibody yield is another significant factor to be considered. In the early stages, 0.1 to 1 mg per antibody is sufficient for most *in vitro* biochemical, biophysical characterization, and functional assays. At later stages, 0.1 to 1 g per antibody could be required for various proposed animal models. One typical scenario is a complementary triangular relationship between the number of antibodies to be produced and the quantity per antibody. During the early stages, a large number of antibodies is produced in a small quantity and at later stages, a small number of antibodies is needed in large quantity. Transient expression takes about two to three weeks including purification. This is a great time saving compared to the stable cell line. Transient expression can be expensive if repeated using the same molecule. In this case, one should consider alternative methods, such as the stable cell line or other expression system. During preclinical and clinical development of antibody therapeutics, stable cell lines are generated for the production of gram to kilogram quantities of antibodies. CHO and NS0 are the most commonly used cell lines for stable cell line generation. Standard protocols are well established for industrial stable antibody producing cell lines. This chapter focuses on the transient expression of research quantities of drugs during the discovery stage of antibody therapeutic programs.

25.2 VECTORS FOR IMMUNOGLOBULIN EXPRESSION

The immunoglobulin heavy and light chains can usually be expressed using various vectors such as plasmid DNA and viral vectors. Although the viral vector-based antibody expression systems may offer benefits under certain circumstances, it takes considerable effort to construct viral vectors and

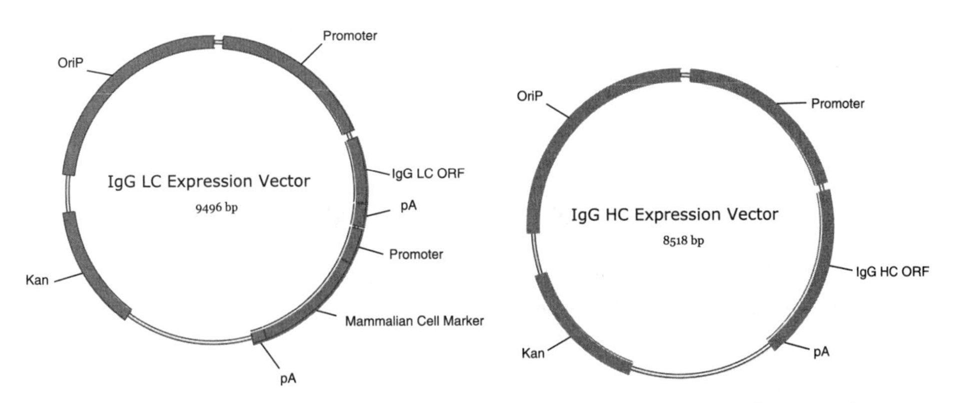

Figure 25.1 Schematic representation of plasmid vectors used for antibody transient expression.

this may not be the most obvious choice during the stage where transient expression offers maximal benefits. In this chapter, we will focus on the most widely used transient expression vector, the plasmid DNA-based antibody transient expression. For a transient expression system with two plasmids, typical features of the two plasmids are shown in Figure 25.1. For transient expression, the light and heavy chain genes can be encoded by plasmids, separate or combined. The convenience of a two plasmid system sometimes makes it possible for a change in the ratio between the two plasmids to compensate for the expression deficiency of either chain. Both genomic and cDNA for antibody open reading frames are used for expression. In the case of genomic DNA, the vector will be a bit larger due to the presence of the intron sequences. On the other hand, the presence of intron sequences allows easier manipulation of the vectors to facilitate the insertion of cloning sequences without the potential change of antibody coding regions. In contrast, the cDNA containing expression vector is, relatively speaking, more compact.

Both cDNA and genomic DNA-based vectors give rise to good transient expression yields. The structure of the coding region in the vector usually is designed to contain the leader sequence for the secretion of the expressed antibody molecule into the culture medium and the constant region of defined isotype such as IgG1 and IgG2. Between the two coding regions, intervening DNA sequences contain the cloning sites for the insertion of the variable regions. Once the variable region is inserted, the leader sequence, variable region, and constant region form one open reading frame for the expression. The open reading frame is driven by a promoter and followed by a polyadenylation signal. The most commonly used ones are human cytomegalovirus (Ghazal, Lubon, and Hennighausen 1988a, b) and elongation factor 1 alpha promoter, SV40 and human growth hormone polyadenylation signal sequences. All the vectors carry the antibiotic resistance marker and replication origin for the manipulation and amplification of the plasmids in the bacterium *Escherichia coli*. One of the vectors may also contain the expression cassette for a selection marker for mammalian cells, such as hygromycin. In addition, many expression vectors include the Epstein-Barr virus (EBV) oriP sequence (Lindner and Sugden 2007) for extended expression in cells that provide the EBV EBNA protein. With all the elements included, the average size of the expression vector is about 8 to 10 kb. The purity of the plasmid DNA plays a role in the expression level. Ideally, DNA for transfection should contain low or no endotoxin. In all plasmid designs, vectors used for transient expression have all the genetic elements necessary for a high level of expression in mammalian cell lines and high copy number replication in *E. coli*. For example, the pCEP4 plasmid (Invitrogen Corp., Carlsbad, California; www.invitrogen.com) possesses the strong eukaryotic promoter CMV, a multiple cloning site for inserting the gene of interest, the SV40 polyadenylation signal, and the EBV origin of replication (OriP) to work in concert with the EBNA sequence in the HEK-293 host cell line. While not utilized for transient gene expression this vector also has the hygromycin resistance gene for the selection of stable cell lines utilizing the TK promoter and polyadenylation signal. The amounts of plasmid required for transient gene expression are typically generated by growth of *E. coli* and thus plasmids have an antibiotic resistance determinant and a pUC origin of replication. For large scale transient expression batches (10 to 100 L), multi-milligram quantities of each plasmid are required. This typically requires 10 to 20 L fermentations of *E. coli* to generate the amounts of plasmid required. Purification of plasmid DNA can be carried out using commercially available kits but for larger amounts a larger scale process is required.

25.3 ANTIBODY PRODUCTION BY TRANSIENT EXPRESSION

The antibody transient production starts with the available plasmid DNAs ready to be transfected and cells seeded for transfection. The transiently expressed antibodies are secreted into medium and the medium is harvested and subject to purification through typically affinity chromatography using protein A or G affinity column. The purified antibodies are characterized for biophysical and analytical characters before use. The whole process is outlined in Figure 25.2.

25.3.1 Cell Lines Used for Transient Expression

HEK-293 (human embryonic kidney; Thomas and Smart 2005) and CHO (Chinese hamster ovary; Derouazi et al. 2004) are the predominant cell lines used for transient expression of mAbs. COS (African green monkey kidney; Blasey et al. 1996) and BHK (baby hamster kidney; Wurm and Bernard 1999) are used to a much lesser extent and will not be covered here. However, factors affecting transient gene expression are common for all cells lines and protein types.

25.3.1.1 HEK-293 HEK-293 cell lines are the most widely used for transient expression and, in addition to the production of mAbs, have been used successfully for the expression of receptors and a variety of ion channels (Thomas and Smart 2005). The original HEK-293 cell line was generated in 1977 by transforming human embryonic kidney (HEK) cells with calcium phosphate precipitated

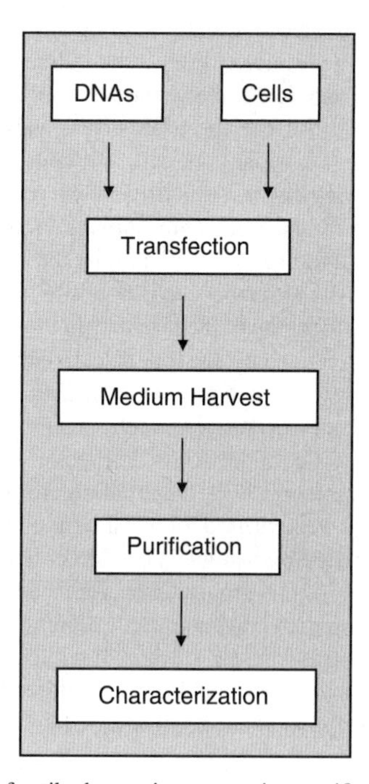

Figure 25.2 Flowchart of antibody transient expression, purification, and characterization.

sheared fragments of human adenovirus type 5 (Ad5), resulting in the incorporation of Ad5 into the HEK genome (Graham et al. 1977). Additional cell lines have been derived from the original HEK-293 line: HEK-293 EBNA expressing the Epstein-Barr virus nuclear antigen I (Cachianes et al. 1993), and the HEK-293T expressing the SV40 large T antigen (Kim, Oh, and Lee 1997). Both modifications allow for episomal replication of plasmids containing the Epstein-Barr virus or SV40 origin of replication. Other virtues of the HEK-293 cell lines include the ease of cultivation as anchorage-dependent or suspension-adapted, growth in the presence or absence of fetal bovine serum (FBS), and growth at scale in bioreactors. Moreover, high transfectability of the HEK-293 EBNA cell line was demonstrated by Durocher, Perret, and Kamen (2002). In the presence of 1 to 2 percent FBS, 60 to 70 percent of cells were green fluorescent protein (GFP)—positive when transected with a GFP gene-containing plasmid (Durocher, Perret, and Kamen 2002).

25.3.1.2 CHO The advantage of using CHO cell lines (e.g., DG44 or K1) for transient gene expression is that this host cell line is typically used to generate high-producing, stably transformed cell lines for commercial manufacturing. Therefore, proteins produced by transient expression in CHO cells should be the same as those produced by a corresponding stable CHO cell line. By contrast, HEK-293 cells can process *N*-glycans differently compared to CHO cells (Van den Nieuwenhof et al. 2000), which could lead to differences in bioactivity (Haack et al. 1999). The benefits of using CHO cells for transient expression are offset by the generally held belief that volumetric productivity is lower compared to HEK-293 cells (Muller et al. 2007). However, these comparisons do not take into account differences in transfection reagents, culture media, bioreactor scale, etc. Moreover, it is well accepted that for a given set of conditions for transient expression using a single cell line, volumetric productivity can vary widely as a function of the proteins being expressed. Nevertheless, a great deal of progress has been made in developing procedures for transient expression in CHO cells.

25.3.2 Transfection Reagents

Under normal circumstances, mammalian cells do not take up and express exogenous DNA readily. In order to introduce DNA into the cell for transient expression, a vehicle is required to get the DNA through the plasma membrane. Chemical reagents such as inorganic compounds, cationic polymers, or cationic lipids are efficient vehicles for introducing plasmid DNA into mammalian cells. These reagents are mixed with the plasmid to form a precipitate or complex and then added to the cell culture which has been grown to a target cell density. Each reagent has attributes and limitations relative to scale, transfection efficiency, and cost. Most of the early development work done with these different reagents has been carried out with single open reading frame (i.e., non-mAb) reporter proteins, such as green fluorescent protein, alkaline phosphatase, or ß-galactosidase. However, all the parameters identified for efficient transient expression of these reporter proteins are also important for the production of mAbs.

25.3.2.1 Inorganic Compounds Calcium phosphate is widely used as a transfection reagent for transient expression and for the generation of stable cell lines. This reagent is low cost, which makes it appealing for large scale transient expression. A calcium/DNA solution and DNA/phosphate solution are mixed and the resulting calcium phosphate precipitate traps the DNA. When added to cell culture, the DNA and calcium phosphate particles are then taken up by the cells. This process can be difficult to control and is highly dependent on DNA concentration, time, and temperature. Jordan and coworkers (1996) carried out a systematic evaluation of the parameters affecting transfection of HEK-293 and CHO cell lines. The coprecipitation reaction occurs within 1 minute, with no DNA being present in the supernatant based on a DNA concentration of 25 μg/mL. This short reaction time resulted in a very large number of small DNA-calcium phosphate particles, which are readily taken up by the cells. Using a ß-galactosidase-containing plasmid and X-gal staining, up to 60 percent of CHO cells and 40 percent of HEK-293 cells were transfected. When the precipitation reaction was carried out for as long as 20 minutes, larger particles were formed and the transfection efficiencies for CHO and HEK-293 cells dropped approximately 10-fold.

25.3.2.2 Cationic Polymers The use of polyethylenimines (PEI, 800 kDa and 50 kDa) was first described by Boussif et al. (1995) where it was used to deliver a luciferase reporter gene to different cell lines and embryonic neurons. PEI forms a complex with DNA as a result of its high positive charge density. The complexes bind to negatively charged residues on the plasma membrane surface and are then taken up by the cell. Moreover, PEIs have a large buffering capacity over a wide pH range. This property led this group of investigators to speculate that the transfection efficiency of PEI lies in the ability to provide buffering in the lysosome, thus protecting the DNA from degradation. A great deal of what is known about this class of transfection reagents has come from the gene delivery field (Godbey, Wu, and Mikos 1999). PEIs are available in a wide range of molecular weights and polydispersities which is critical to transfection efficiency; the most widely used PEI for the transient expression to produce antibodies is the 25 kDa branched or unbranched isoforms. The ratio of PEI to plasmid (w/w) has proven to be a critical parameter for transfection efficiency using this reagent and must be determined experimentally for each protein of interest to maximize productivity. Lipids or dextrans that are part of the medium or polymers such as glycosaminoglycans (Belting and Peterson 1999) produced by the cells are known to reduce the transfection efficiency of cationic polymer transfection reagents.

25.3.2.3 Cationic Lipids Because of cost, the cationic lipids are typically not used for large scale transient transfection. They are, however, useful for small scale (e.g., tissue culture flask) transfection. The cationic lipids are very versatile with regards to types of cells transfected. These include cell lines along with primary cells and stem cells. Similar to the two reagents described above, cationic lipids complex with plasmid DNA. For this class of reagent, a liposome vesicle is formed when a cationic lipid and a neutral lipid are mixed. The net positive charge that these vesicles have adsorbs the plasmid and the complex is taken up through the plasma membrane. A number of next-generation reagents have been developed called cationic lipid transfection reagents, which consist of lipids and proprietary components, including ones that have been developed specifically for certain cell types (Invitrogen, www.invitrogen.com; Bio-Rad, www.bio-rad.com; and Roche, www.roche.com).

25.3.3 Culture Media, Conditions, and Additives Used for Transient Expression

Essentially all commercially available media used for cultivating HEK-293 or CHO cells can be used for transient expression although some do work better than others. Special formulations such as FreeStyle (Invitrogen) or Excell-293 (JRH Biosciences, Lenexa, Kansas) have been developed specifically for transient expression in HEK-293 cells although cost limits their utility at large scale. Many biopharmacuetical companies have developed their own proprietary media. While cell growth to a target density is a trivial matter for these robust cell lines, the makeup of the medium at the time of transfection is critical to gene delivery and expression. For example, Baldi et al. (2005) carried out a complete Excell-293 medium change (by centrifugation) 4 hours post-calcium-mediated HEK-293 transfection (20). While this is not feasible for cell culture batches greater than a few liters, it is practical to seed fresh medium at a viable cell density of up to 1×10^6 (Galbraith et al. 2006). Both HEK-293 and CHO cells grow well in suspension in the absence of FBS. However, reports are mixed regarding the impact of FBS on transfection efficiency. Titers of over 20 mg/L at the multiliter scale have been achieved using CHO cells, serum-free ProCHO5 CDM medium, and PEI and the transfection reagent (Muller et al. 2007). In a report by Baldi et al. (2005) both serum-containing (1 percent v/v) and serum-free medium were used for HEK-293 transient expression. Durocher and coworkers (2002) demonstrated a dose-dependent increase in gene expression when FBS concentrations were increased from 0 to 5 percent. When serum was added to a final concentration of 1 percent 3 hours post-transfection, expression levels were twofold higher than in serum-free medium but only 50 percent of the expression level when 1 percent serum was present at the time of transfection. This suggests an important role for serum in the transfection process. In an effort to eliminate serum or other animal-derived proteins from culture medium, plant peptones and hydrolysates have been evaluated for their impact on transient expression. Pham et al. (2003) evaluated a number of different peptones at 1 percent w/v and demonstrated an increase in expression by HEK-293 cells of secreted alkaline phosphatase compared to serum-free medium. Control cultures with 1 percent v/v FBS expressed more protein than the peptone-containing cultures (Pham et al. 2003). Another interesting approach to improving expression levels was carried out by Tait and coworkers (2004), who utilized the microtubule depolymerization agent nocodazole to synchronize CHO cells in the G2/M phase of the cell cycle. This treatment resulted in increases in transfection efficiency, a fivefold increase in the rate of secreted alkaline phosphatase production and a twofold overall increase in protein production. Using the nocodazole treatment, a twofold increase in the expression of an IgG4 mAb with no change in glycosylation was observed.

25.3.4 Production of mAbs by Transient Expression

Unlike many proteins expressed by transient expression, mAbs are heterotetramers composed of two heavy and two light chain polypeptides associated by disulfide bonds. Production of full length mAbs by transient expression requires either the transfection of a single plasmid containing both the heavy and light chain coding sequences or co-transfection of separate vectors each containing either the heavy chain or light chain coding sequence. For single plasmids, the total DNA concentration introduced to the culture and the ratio of DNA to transfection reagent (w/w) are key parameters. For PEI-mediated transfections, a typical reagent to DNA ratio is 3 : 1 and final DNA concentration of 1 to 3 mg/L of culture volume. Alternatively the reagent to DNA ratio is expressed as PEI nitrogen:DNA phosphate ranging from 10 : 1 to 20 : 1. These parameters are also impacted by cell type, cell density, and culture age. While transient expression from a single plasmid is easier, the fine control of heavy and light chain expression is lost unless promoters are tuned to allow different stoichiometries. In the early phases of an mAb discovery and development program, it is typical that the heavy and light chain coding sequences are expressed from different plasmids that have all the genetic elements previously described. In addition to determining the optimal total plasmid DNA concentration (heavy chain + light chain plasmids) added to the culture and the plasmid DNA to transfection reagent ratio, the heavy chain to light chain (hc:lc) plasmid ratio is critical to a high level of expression of full length mAbs. As a first approximation, most investigators use a hc:lc ratio of 1 : 1 consistent with the stoichiometry of the heterotetramer. Schlatter and colleagues (2005) demonstrated maximum production of a recombinant IgG_4 mAb at an hc:lc ratio of 3 : 2 for PEI-mediated transfection of CHO cells. While empirical and mathematical modeling techniques were used in this study, a more practical approach would be to utilize statistical experimental design techniques to optimize all critical parameters concurrently. It is very likely that different subclasses of mAbs (IgG1, IgG2m4; kappa or lambda light chains) produced in either HEK-293 or CHO cell lines will have different optimal transfection parameters. Commercially available software packages such as JMP (SAS Institute, Cary, North Carolina; www.jmp.com), Minitab (Minitab, Inc., State College, Pennsylvania; www.minitab.com), or Design-Expert (Stat-Ease, Inc., Minneapolis, Minnesota; www.statease.com) make the optimization of final DNA concentration, DNA:transfection reagent, hc:lc, cell density, etc. a straightforward undertaking.

25.3.4.1 *Well Plates and Tissue Culture Flasks* Transient expression can be carried out at all scales, as static cultures in tissue culture flask and well plates or in mixed bioreactors, depending on the amount of material required and the throughput. The nature of the application may require that the cells be grown in an anchorage-dependent fashion, for example, in polylysine-coated plates. Well plates and tissue culture flasks are the simplest and most straightforward approach to transient expression. Six- and 24-well plates are very useful when small amounts of a large number of antibodies are required for evaluation in an *in vitro* or cell-based assay. Tissue culture flasks are directly scalable and a few liters of conditioned medium can be generated if enough traditional tissue culture flasks are used. Well-plate and tissue culture flasks are readily automated using a number of off-the-shelf systems such as those available from Tecan (www.tecan.com). The drawback of this approach is that cell growth is limited by the surface area of the vessel and gas transfer into the static liquid from the headspace. Newly designed flasks such as the HYPER*Flask*™ made by Corning (Corning, New York; www.corning.com) have a footprint similar to a standard tissue culture flask but have multiple gas-permeable layers. The increase in surface area and gas exchange for cell growth allows for up to 500 mL of conditioned medium to be generated per flask. The process can be automated, and several automated cell culture systems are on the market. These flasks have been used by the Automation Partnership (Royston, Herdfordshire, UK; www.automationpartnership.com) to automate cell culture manipulations. In particular the SelectT system has been billed as capable of automating transient expression.

25.3.4.2 Shaking Flasks and Tubes These vessels are also very straightforward to use and require only an investment in a CO_2 incubator with orbital shaking capability. Several liters of conditioned medium can be generated in 2 L flasks with a final culture volume of 500 mL. There is little opportunity to automate transient expression in shake flasks but the amount generated from several flasks can be on the order of 10 to 100 mg of mAb, typically enough for a small rodent-based study. Recently, 50 mL ventilated centrifuge tubes have become available from TPP (Trasadingen, Switzerland; www.tpp.ch) and have proven useful for mAb transient expression studies (Muller et al. 2007). They are easy to manipulate, have the potential for automated manipulation, and provide a scale-down model for suspension-adapted cell lines in batch or fed-batch mode. These vessels are the perfect scale to carry out statistical experimental designs for optimizing transfection parameters, screening media, or media additives.

25.3.4.3 Bioreactors Different types of bioreactors are fully effective for transient expression at the 10 to 50 L scale. Traditional stirred tank bioreactors with marine-style impellors and gas blending for introducing CO_2 to the culture have been a mainstay in the cell culture field for quite some time. Many companies prefer these fixed bioreactors because they can be retrofitted back and forth between cell culture and microbial fermentation. Disposable bioreactors are the perfect solution for transient expression mAb or other proteins. These reactors eliminate the downside of fixed bioreactors, which require time and resources to prepare, clean, and turn-around this equipment. The most common type of disposable bioreactor is the rocking-style, such as the Wave™ (Wave Biotech, www.wavebiotech.com) or the CultiBag (Satorius, www.bbiserver.de). The culture is grown in a specially designed sterile plastic bag. Medium and inoculum are added to the culture bag by welding the tubing on the vessel containing the cells or medium to the tubing on the culture bag. The transfection mixture or any feeds are added the same way. The bag rocks back and forth on a mechanical platform and the waves that are produced provide mixing and gas exchange. Temperature control is achieved through a heating element on the platform, which is positioned under the bag. Aeration is provided by introducing the air/CO_2 mixture into the head space of the bag. Rocking speed, rocking angle, and gas flow are key parameters for maintaining an adequate dissolved oxygen concentration. Dissolved oxygen and pH can be monitored on- or off-line using a conventional blood gas analyzer. Cell densities of 1×10^6 to 1×10^7 cells/mL can be achieved in these bioreactors. Haldankar et al. (2006) describe a PEI-mediated transient transfection of CHO cells to produce 9.4 mg/L of a human/mouse chimeric IgG1.

25.4 PURIFICATION OF TRANSIENTLY EXPRESSED ANTIBODIES

The next step of mAb production by transient expression is purification of the product. This involves removing the cells, affinity chromatography, removal of endotoxins if the level is high, and exchange into a buffer suitable for the indented purpose.

25.4.1 Primary Recovery (Cell Removal and Concentration)

Depending on the batch size, cells can be removed by batch centrifugation, continuous centrifugation, or microfiltration resulting in a clarified conditioned medium. To facilitate the subsequent chromatography steps, the mAb in the conditioned medium must be concentrated. If the volume of conditioned medium is low, this step is unnecessary. The concentration step is typically accomplished via ultrafiltration using a 30 kDa molecular weight cut-off membrane. As the conditioned medium is passed over the membrane, the mAb (molecular weight 150 kDa) is retained while the bulk medium passes through the membrane. Depending on the concentration of protein in the medium (e.g., if serum is used), the protein can be over-concentrated and the proteins can precipitate out of solution. Typically a 10- to 20-fold concentration is targeted.

25.4.2 Affinity Chromatography

The primary method for purification of mAbs is protein A chromatography. Protein A binds specifically to the Fc portion of an antibody at physiological pH. Binding capacities for protein A range from 20 to 40 mg mAb/mL resin. GE Healthcare Life Sciences (Uppsala, Sweden; www.gehealthcare.com) and Applied Biosystems (Foster City, California; www.appliedbiosystems.com) are two examples of the major suppliers of bulk protein A resin. Protein G affinity chromatography is an alternative method for purifying mAbs that bind weakly to protein A. Elution from protein G requires lower pH elution buffers (e.g., 2.5) compared to protein A, which may compromise mAb stability. The concentrated conditioned medium, 1 to 2 L from a multiliter cultivation, is applied to the column and with a flow rate adjusted to achieve a 4 to 6 minute residence time [resin volume (mL)/residence time (min) = target flow rate (mL/min)]. The residence time, which is a key variable for optimizing capture chromatography, allows for an adequate interaction of the antibody with the resin-bound protein A. After loading, the column is washed extensively with phosphate buffer saline at neutral pH to eliminate any nonspecifically bound proteins. The antibody is eluted using 1 to 2 column volumes of a pH 3.0 to 3.5 buffer, typically citrate. Although mAbs tolerate low pH, the pH of the mAb solution should be adjusted promptly to neutral using a strong buffering solution (e.g., 1 M Tris, pH 8.0) at roughly 15 percent (v/v) relative to the mAb eluate volume. Purity after the protein A chromatography step is typically 90 to 95 percent, with yields of 80 to 90 percent. This is highly dependent on the antibody and the nature of the conditioned medium, for example, the presence of serum.

For larger numbers of small preps (e.g., 30 to 300 mL of conditioned medium)—typical for a screening experiment—the mAb can be bound in a batch mode without concentration by adding pre-equilibrated resin directly to the condition medium, incubating for a period of time to allow binding of the antibody to the resin, decanting the supernatant after settling of the resin, and then transferring the resin to empty columns for elution by gravity. This method is simple and allows for a number of samples to be processed in parallel. The downside is that yields are not always as high compared to standard column chromatography methods but the amount of antibody generated is likely to be adequate for small scale studies. Other approaches to small scale purification include columns that are loaded and eluted by centrifugation or vacuum (see www.nuncbrand.com).

25.4.3 Endotoxin Removal

Knowing the endotoxin levels and extent of mAb aggregation (covered in a following section) are important in particular where the mAb is used *in vivo* or in a cell-based assay. Endotoxin levels significantly greater than 5 EU/mL can dictate the outcome of a rodent study rather than the effectiveness of the mAb. Endotoxins and their removal from biological preparations are reviewed by Magalhães and coworkers (2007). Endotoxin removal can be accomplished using a couple of chromatography techniques. Endotoxins are highly negatively charged and bind strongly to anion exchange resins in low ionic strength buffers, while the mAb is not bound and is present in the flow through. This is typically a very effective and high yielding step. This step can be performed using columns or, if the volume of the sample is small, membrane chromatography can be effective (see www.sartorius.co.uk). This technique is dependent on the isoelectric point of the mAb. Where the isoelectric point of the protein is not known, Polymixin-B chromatography is a suitable alternative (Rogers and Cohen 1986).

25.4.4 Buffer Exchange

Prior to experimental use, the buffer that the mAb is in, protein A or ion exchange buffers, must be exchanged for a buffer that is amenable to use *in vivo* or *in vitro*. For analytical uses, phosphate buffered saline is typical. For *in vivo* or cell-based assays, a histidine-based buffer can be used. Moreover, the choice of buffer and any additional components in the formulation contribute to the stability of the mAb, reduce aggregate formation, etc. Ultrafiltration using a 30 kDa molecular weight cutoff membrane is the appropriate option for large scale samples >500 mL. Approximately 3 to 4 buffer volumes are required and the operation takes approximately 3 to 4 hours depending on the concentration of the protein. One drawback to ultrafiltration is the length of time required for set up and clean out of the unit in addition to the time required to do the actual diafiltration. An alternative for 100 to 500 mL samples is the use of centrifugal devices (e.g., see www.millipore.com) which come with several molecular weight cut-off membranes and have capacities up to 80 mL. These units require a standard swinging-bucket centrifuge to force the sample through the membrane, retaining the antibody. The buffer exchange is carried out by sequential centrifugations and volume adjustments back to the original sample volume with the buffer of interest. Typically three to four cycles of 15 to 20 minutes each is adequate for ≈95 percent buffer exchange. Several units can be used to adequately buffer exchange the entire sample. The major benefit to these devices is that they are essentially ready to use right out of the package. Moreover, they are disposable and do not require time for clean out and set up.

25.5 ANALYTICAL AND BIOPHYSICAL CHARACTERIZATION OF PURIFIED ANTIBODIES

With the antibodies purified, one would often like to know the profile of the antibodies in several aspects: dose/concentration, purity, mass, aggregation, stability, and *in vitro* and sometimes *in vivo* potency (Table 25.1). Critical parameters need to be evaluated at every stage in the generation of the antibody of choice both for an understanding of the product, as well as to establish a baseline for future comparability studies as the product matures toward eventual licensure. In addition, product characterization such as the determination of isoforms and posttranslational modifications should be accomplished.

25.5.1 Dose/Concentration

The accurate concentration measurements of monoclonal antibody preparations are essential in the antibody development cycle, starting from the initial small scale experimental preparations that are used for *in vitro* and *in vivo* potency tests, and ending with the large scale production batches. Initially, when the purified protein is scarce, the concentration can be evaluated by nondestructive methods, such as near-UV spectroscopy. The protein concentration can be accurately measured by near-UV absorption spectroscopy, provided that the value of the extinction coefficient is known. In the initial stages of the development process, use of traditional methods of extinction coefficient determination, such as dry weight determination (Wetlaufer 1954), quantitative amino acid analysis, or colorimetric methods (Bradford 1976; Lowry et al. 1953) is not feasible. Since the sequence and thus the amino acid composition are known, the molar extinction coefficient can be calculated. However, initial efforts produced error often exceeding 10 percent due to the use of extinction coefficient values of free tryptophan and tyrosine amino acids (Gill and von Hippel 1989). Examination of large experimental

TABLE 25.1 Biophysical and Analytical Assays for Antibody Characterization

Character	Assay(s)	Example(s)
Dose/concentration	UV spectroscopy Colorimetric assay(s) Size exclusion chromatography (HP-SEC) Biosensor-based quantitation assay	Multicomponent analysis of second derivative UV spectra, BCA, surface plasmon resonance (SPR)
Purity	SDS-PAGE Western blot	Reducing and nonreducing
Identity/mass	LC-MS	MWs and modifications, e.g., pyroglutamate formation
	Cation exchange chromatography (CEX)	Glycoform analysis
	Capillary isoelectric focusing (cIEF)	Deamidation
	Reverse-phase HPLC	Oxidation
Aggregation/stability	HP-SEC Dynamic light scattering Capillary differential scanning calorimetry (cDSC)	HP-SEC with light scattering and photodiode-array detectors
Potency	Biological assays Surrogate assays	ELISA or SPR

data sets revealed that the hydrophobic environments in most proteins result in increased absorptivity of tyrosine residues (Mach, Middaugh, and Lewis 1992; Pace et al. 1995). Consequently, the following equation was shown to produce accurate results, typically within 2 percent:

$$E(280\ \text{nm}) = \#\text{Trp}^{*}5500 + \#\text{Tyr}^{*}1490 + \#\text{cystine}^{*}125$$

Since the monoclonal antibody has a molecular weight in the range of 150 kDa, a small optical density resulting from light scattering is present in the UV spectra. If the UV instrument used does not have a light scattering correction algorithm implemented, a linear extrapolation of 320 nm and 350 nm values, or even subtraction of 320 nm optical density values, for nonaggregated antibody samples is sufficient. The extent of the potential error can be estimated by comparing UV absorbance values at 280 nm, 320 nm, and 350 nm. Once the antibody spectrum is measured and the concentration of the sample is determined, the highest precision and convenience of subsequent measurements is achieved by employing a method of multicomponent analysis of the second derivative spectra that uses the prerecorded spectrum of known concentration as a standard (Mach, Thomson, and Middaugh 1989; Mach et al. 1997). The multicomponent analysis approach allows analysis of other UV-absorbing components, provided that their standard spectra are previously recorded. In the early stages of development in the basic research environment, however, where properties of newly expressed antibodies are evaluated, incomplete antibodies or antibody fragments result in overestimation of the amount of the active form. An appropriate correction to the concentration value can be derived from other characterization tools, such as size-exclusion chromatography (Hjerten and Mosbach 1962; Hjerten and Eriksson 1984; Wehr and Rodriguez-Diaz 2005) or SDS-PAGE (Laemmli 1970). In practice, selectivity of the biosensor instruments that rely on the binding of the mAb Fc domain to immobilized protein A also allows accurate estimation of mAb concentration (www.fortebio.com).

25.5.2 Purity

The purity of proteins, whether they are mAbs or other recombinant species, is routinely assessed through SDS-PAGE, with either colorimetric (e.g., silver, Coomassie-blue) or fluorometric (e.g., SYPRO-Ruby) staining reagents. While this is the standard method of evaluation, SDS-capillary electrophoresis is another option. Advantages to this latter technique include higher sensitivity, throughput, faster run times, and perhaps more uniformity from run to run. A major disadvantage is economic, although this dynamic is changing. In the early stages of development, and especially in products arising from transiently transfected cells, additional protein species are sometimes observed by SDS-PAGE. In addition, there may be additional unexpected higher molecular weight species which may be dimers or contaminants. The application of Western blotting with product-directed antibodies can distinguish whether these species are product- or host cell-derived that have simply been co-purified with the antibody. This is an important aspect, since these observations may provide insights into the stability of either the construct itself or the product in particular. As another example, the hinge regions of the antibodies are exposed to solvent and susceptible to strain and proteolytic clipping. These observations speak to the purity of the transiently expressed materials; they are readily evaluated by SDS-PAGE, and a Western blot would indicate whether these fragments or other species observed on the gels are product derived.

25.5.3 Antibody Mass and Modifications

25.5.3.1 Mass The application of SDS-PAGE to evaluate the molecular weight of the intact antibody, as well as the constituent light and heavy chains, is now used more to evaluate purity than as an estimate of the mass. Liquid chromatography mass spectrometry (LC-MS) is becoming an important tool in these early stages of development, providing significantly more accuracy and precision than SDS-PAGE with which to assess the integrity of the expressed protein. The technology has advanced significantly and manufacturers are providing more user friendly packages with which to carry out these analyses. What emerges from this is that an investigator can readily evaluate whether the antibody with the correct amino acid sequences in the light and heavy chains is being produced. Furthermore, the resolving power of current instrumentation provides insights into several of the expected posttranslational modifications occurring in the transiently expressed product. For example, current LC-MS instrumentation with electrospray ionization (ESI) and time-of-flight (TOF) detector can determine if the light chain molecular weight is within 1 or 2 Da of the expected mass based on the amino acid sequence. While the possibility of a mutation that results in either an isomeric light and/or heavy chain does exist, such a mutation would also presumably be discovered during the DNA sequencing of the plasmid.

25.5.3.2 Pyroglutamate Formation One modification that is sometimes observed in the light or heavy chain is pyroglutamate formation at the N-terminus if this residue is a glutamine. This would result in a mass shift of -17 Da ($\Delta = -17$ Da), and can be observed with current LC-MS technology. Pyroglutamate formation from N-terminal glutamic acid residues is less frequent, but also observed. This type of resolving power is not possible in SDS-PAGE analysis. The determination of the heavy chain mass is similar to that of the light chain, but complicated by the likely presence of several glycans on the Fc region of the heavy chain; the partial excision of the C-terminal lysine residue ($\Delta = -128$ Da) also occurs. Incomplete intra- or interchain disulfide bond formation also results in some of the observed heterogeneity. Populations of molecules lacking interchain disulfides can be observed both by SDS-PAGE and by mass spectrometry. These modifications combine to introduce significant heterogeneity in the heavy chain but are well resolved by mass spectrometry. The mass of the intact mAb ($\sim$150,000 Da) can also be determined with the current LC-MS technologies. The information that is provided gives a glimpse of the heterogeneity that exists in the product (due to glycan variability, discussed below) and a rough estimate as to the relative ratios of the various glycoforms produced.

25.5.3.3 Glycoform Analysis The antibodies that are produced by transient transfections display a wide array of glycan occupancy in the Fc, ranging from zero occupancy (no glycans attached to either of the mAb Fc) to various stages of glycosylation. In CHO cells, the expected glycoforms range from G0F (Liu et al. 2008) to higher ordered populations. Several techniques exist for the analysis of these glycans. As stated above, LC-MS is now commonly utilized to evaluate these N-linked glycans. Quantitation of the ratios of the various glycans can be achieved by application of a glycanase such as PNGase to the antibody, recovery of the excised glycans, and evaluation by capillary electrophoresis. Application of a commercial standard kit provides identity and quantitation of these forms.

25.5.4 Aggregation State

The information about aggregation state is critical in the early development of monoclonal antibodies. It provides a first glimpse into the intrinsic structural stability, allowing selection of the most stable candidate. For the initial animal potency studies, aggregation can result in the underestimation of the concentration determination, decreased potency due to lack of activity of the aggregated fraction, and/or undesirable immunogenic effects. Aggregation-detecting techniques that are rapid, nondestructive, and require small amounts of sample are critical in the early development stages, since they allow minimizing of protein loss through pH and buffer component optimization during expression and purification, as well as allowing accelerated micro-scale chemical and structural stability studies to be performed that can be used to predict future dosing and shelf-life limitations and thus aid in the selection process. The first obvious estimation of the aggregation state comes from ultraviolet spectra that are measured for the purpose of concentration determination. Relating the amount of turbidity originating from light scattering, most frequently defined as the difference of absorbance at 350 nm and 550 nm (European Pharmacopeia 2002; U.S. Pharmacopeia 2003) to the concentration as determined by the corrected absorbance at 280 nm yields a useful and universal parameter that can be used to evaluate aggregate content. Light scattering models specific for the instruments used are often constructed to streamline sample analysis and result reporting. Size exclusion chromatography (SEC) provides an insight into the size distribution of antibody samples, revealing the presence and amount of aggregated species (for review, see Wehr and Rodriguez-Diaz 2005). The method is frequently used in early development stages due to its low sample requirement and high throughput capability. Since the largest aggregates are typically retained by the column, filtration through a 0.22 μm filter is advised, followed by a second UV measurement, to ascertain the relative amount of the aggregates that are too large to be eluted from the SEC column. Increase of ionic strength and addition of running phase excipients such as arginine frequently improve sample recovery (Arakawa et al. 2006). The small amount of largest aggregates has the most profound effect on the results obtained from dynamic (quasielastic) light scattering, a rapid and nondestructive method that requires a small amount of sample (Phillies 1990; Koppel 1972). Since the intensity of the scattered light at right angle is approximately proportional to the molecular weight for submicron sized particles (Van Holde 1985), the relative weight-based contributions of particles of various sizes is a function of the cube of the diameter, making it the most sensitive method of the detection of large aggregates. If the aggregates are present, the results of a regularization fitting routine, presented as weight-based distribution of several peaks, that is, corrected for variable intrinsic scattering intensity of various sizes (Malvern Instruments, 2005; Wyatt Technology, 2007), are used to estimate the amount of nonaggregated material.

25.5.5 Structural Stability

Structural instability, while having a direct effect on potency, affects the quality of the product mostly through opening a pathway to irreversible aggregation (Chen et al. 1994). At early stages of the development, the intrinsic stability is evaluated by observing the resistance to forced destabilization. Thermal unfolding is typically a method of choice due to low sample requirement and availability of automated instruments. The onset of unfolding can be detected with high sensitivity using fluorescent dyes sensitive to the presence of hydrophobic surfaces, such as ANS (Cummings, Farnum, and Nelen 2006) and Sypro-Orange (Epps, Sarver, and Rogers 2001). Since unfolding of the less aggregation-prone Fc region frequently occurs at temperatures lower than unfolding of the Fab region, onset of aggregation-related turbidity is better detected by observing the light scattering at a wavelength overlapping with the excitation peak of the fluorometer. More detailed information about unfolding temperatures of the antibody constituent domains is obtained from capillary differential scanning calorimetry (Microcal, Inc.), which allows accurate measurements at protein concentrations of less than 1 mg/mL (Harn et al. 2007). The selection of structurally stable variants can also be achieved through *in vitro* proteolysis during the phage propagation cycle (Sieber, Pluckthun, and Schmid 1998). The evaluation of long-term structural stability in selected formulation can be performed at the early development stage through accelerated stability studies, subjecting samples to incubation at elevated temperatures (e.g., 40 to 50°C) for relatively short periods of time (weeks), followed by analysis with high throughput, sensitive methods such as SEC that reveal both the aggregation and fragmentation of the antibody.

25.5.6 Potency

Finally, a potency assay is essential to determine the *in vitro* functionality of the early leads, which can be used to rank order them for lead selection. A potency assay is typically a biological assay in which the change caused by the addition of antibodies can be registered and reported quantitatively. The assay can also be used to establish comparability between the materials generated by transient transfection and a later, stable, over-expressing clone. This assay is but a surrogate for the proposed interaction between the mAb and its intended target, since what is ultimately desired is a clinical marker to determine its efficacy. Sometimes, a binding assay, either in an ELISA format or in another form such as an SPR response, is also necessary to corroborate the antibody potency and bridge from antibodies produced by transient transfection to those from a stable clone.

REFERENCES

Arakawa, T., J.S. Philo, D. Ejima, K. Tsumoto, and F. Arisaka. 2006. Aggregation analysis of therapeutic proteins. Part 1. *BioProcess Int.* 4:42–43.

Baldi, L., N. Muller, S. Picasso, R. Jacquet, P. Girard, H.P. Thanh, E. Derow, and F.M. Wurm. 2005. Transient gene expression in suspension HEK-293 cells: Application to large-scale protein production. *Biotechnol. Progress* 21:148–153.

Belting, M., and P. Peterson. 1999. Intracellular accumulation of secreted proteoglycans inhibits cationic lipid-mediated gene transfer: Co-transfer of glycosaminoglycans to the nucleus. *J. Biol. Chem.* 274:19375–19382.

Blasey, H.D., J.P. Aubry, G.J. Mazzei, and A.R. Bernard. 1996. Large-scale transient expression with COS cells. *Cytotechnology* 18:138–142.

Boussif, O., F. Lezoualch, M.A. Zanta, M.D. Mergny, D. Scherman, B. Demenex, and J.P. Behr. 1995. A versatile vector for gene and oligonucleotide transfer into cells in culture and in vivo: Polyethylenimine. *Proc. Nat. Acad. Sci. U.S.A.* 92:7297–7301.

Bradford, M.M. 1976. A rapid and sensitive method for the quantitation of microgram quantities of protein utilizing the principle of protein-dye binding. *Anal. Biochem.* 72:248–254.

Cachianes, G., C. Ho, R.F. Weber, S.R. Williams, D.V. Goeddel, and D.W. Leung. 1993. Epstein-Barr virus-derived vectors for transient and stable expression of recombinant proteins. *Bio/Techniques* 15:255–259.

Chen, B. et al. 1994. Strategies to suppress aggregation of recombinant keratinocyte growth factor during liquid formulation development. *J. Pharm. Sci.* 83:1657–1661.

Cummings, M.D., M.A. Farnum, and M.I. Nelen. 2006. Universal screening methods and applications of ThermoFluor®. *J. Biomol. Screening* 11:854–863.

Derouazi, M., P. Girard, F.V. Tilborgh, et al. 2004. Serum-free large-scale transient transfection of CHO cells. *Biotechnol. Bioeng.* 87:537–545.

Durocher, Y., S. Perret, and A. Kamen. 2002. High-level and high throughput recombinant protein production by transient transfection of suspension-growing human 293-EBNA1 cells. *Nucleic Acids Res.* 30:E9.

Epps, D.E., R.W. Sarver, and J.M. Rogers. 2001. The ligand affinity of proteins measured by isothermal denaturation kinetics. *Anal. Biochem.* 292:40–50.

European Pharmacopeia. 2002. Particulate Contamination: Subvisible Particles, 2915–2918.

Galbraith, D.J., A.S. Tait, A.J. Racher, J.R. Birch, and D.C. James. 2006. Control of culture environment for improved polyethylenimine-mediated transient production of recombinant monoclonal antibodies by CHO cells. *Biotechnol. Progress* 22:753–762.

Ghazal, P., H. Lubon, and L. Hennighausen. 1988a. Specific interactions between transcription factors and the promoter-regulatory region of the human cytomegalovirus major immediate-early gene. *J. Virol.* 62:1076–1079.

Ghazal, P., H. Lubon, and L. Hennighausen. 1988b. Multiple sequence-specific transcription factors modulate cytomegalovirus enhancer activity *in vitro*. *Mol. Cell. Biol.* 8:1809–1811.

Gill, S.C., and P.H. von Hippel. 1989. Calculation of protein extinction coefficients from amino acid sequence data. *Anal. Biochem.* 182:319–326.

Godbey, W.T., K.K. Wu, and A.G. Mikos. 1999. Poly(ethylenimine) and its role in gene delivery. *J. Controlled Release* 60:149–160.

Graham, F.L., J. Smiley, W.C. Russell, and R. Nairn. 1977. Characteristics of a human cell line transformed by DNA from a human adenovirus type 5. *J. Gen. Virol.* 36:59–74.

Haack, A., C. Schmitt, W. Poller, J. Oldenburg, P. Hanfland, H.H. Brackmann, and R. Schwaab. 1999. Analysis of expression kinetics and activity of a new B-domain truncated and full-length FVIII protein in three different cell lines. *Ann. Hematol.* 78:111–116.

Haldankar, R., D. Li, Z. Saremi, C. Baikalov, and R. Deshpande. 2006. Serum-free suspension large-scale transient transfection of CHO cells in WAVE bioreactors. *Mol. Biotechnol.* 34:191–199.

Harn, N., C. Allan, C. Oliver, and C.R. Middaugh. 2007. Highly concentrated monoclonal antibody solutions: Direct analysis of physical structure and thermal stability. *J. Pharm. Sci.* 96:532–546.

Hjerten, S., and K.O. Eriksson. 1984. High-performance molecular sieve chromatography of protein on agarose columns: The relations between concentration and porosity of the gel. *Anal. Biochem.* 137:313–317.

Hjerten, S., and R. Mosbach. 1962. "Molecular-sieve" chromatography of proteins on columns of cross-linked polyacrylamide. *Anal. Biochem.* 3:109–118.

Jordan, M., A. Schallhorn, and F.M. Wurm. 1996. Transfecting mammalian cells: Optimization of critical parameters affecting calcium-phosphate precipitate formation. *Nucleic Acids Res.* 24:596–601.

Kim, C.H., Y. Oh, and T.H. Lee. 1997. Codon optimization for high-level expression of human erythropoietin (EPO) in mammalian cells: Human or yeast codon usage effect on over-expression in 293T cell culture. *Gene* 199:293–301.

Koppel, D.E. 1972. Analysis of macromolecular polydispersity in intensity correlation spectroscopy: The method of cumulants. *J. Chem. Phys.* 37:4814–4820.

Laemmli, U.K. 1970. Cleavage of structural proteins during the assembly of the head of bacteriophage T4. *Nature* 227:680–685.

Lindner, S.E., and B. Sugden. 2007. The plasmid replicon of Epstein-Barr virus: Mechanistic insights into efficient, licensed, extrachromosomal replication in human cells. *Plasmid* 58:1–12.

Liu, H., G. Gaza-Bulseco, D. Faldu, C. Chumsae, and J. Sun. 2008. Heterogeneity of monoclonal antibodies. *J. Pharm. Sci.* 97:2426–2447.

Lowry, O.H., N.J. Rosenbrough, A.L. Farr, and R.J. Randall. 1953. Protein measurement with the folin phenol reagent. *J. Biol. Chem.* 193:265–275.

Mach, H., C.R. Middaugh, and R.V. Lewis. 1992. Statistical determination of the average values of the extinction coefficients of tryptophan and tyrosine in native proteins. *Anal. Biochem.* 200:74–80.

Mach, H., G. Sanyal, D.B. Volkin, and C.R. Middaugh. 1997. Applications of ultraviolet absorption spectroscopy to the analysis of biopharmaceuticals. In *Therapeutic Protein and Peptide Formulation and Delivery*, ed. Z. Shahrokh et al. ACS Symposium Series. Washington, D.C.: American Chemical Society.

Mach, H., J.A. Thomson, and C.R. Middaugh. 1989. Quantitative analysis of protein mixtures by second derivative absorption spectroscopy. *Anal. Biochem.* 181:79–85.

Magalhães, P.O., A.M. Lopes, P.G. Mazzola, C. Rangel-Yagui, T.C.V. Penna, and A. Pessoa. 2007. Methods of endotoxin removal from biological preparations: A review. *J. Pharmacy Pharm. Sci.* 14:388–404.

Muller, N., M. Derouazi, F. Van Tilborgh, S. Wulhfard, D.L. Hacker, M. Jordan, and F.M. Wurm. 2007. Scaleable transient gene expression in Chinese hamster ovary cells in instrumented and non-instrumented cultivation systems. *Biotechnol. Lett.* 29:703–711.

Pace, C.N., F. Vajdos, L. Fee, G. Grimsley, and T. Gray. 1995. How to measure and predict the molar absorption coefficient of a protein. *Protein Sci.* 4:2411–2423.

Pham, P.L., S. Perret, H.C. Doan, B. Cass, G. St.-Laurent, A. Kamen, and Y. Durocher. 2003. Large-scale transient transfection of serum-free suspension-growing HEK-293 EBNA1 cells: Peptone additives improve cell growth and transfection efficiency. *Biotechnol. Bioeng.* 84:332–342.

Phillies, G.D.J. 1990. Quasielastic light scattering. *Anal. Chem.* 62:1049–1057.

Riley, S. 2006. The future of the monoclonal antibody therapeutics: Innovation in antibody engineering, key growth strategies and forecasts to 2011. *Business Insights.* http://www.globalbusinessinsights.com.

Rogers, M.J., and J. Cohen. 1986. Comparison of the binding of gram-negative bacterial endotoxin by polymyxin B sulphate, colistin sulphate and colistin sulphomethate sodium. *Infection* 14:79–81.

Schlatter, S., S.H. Stansfield, D.M. Dinnis, A.J. Racher, J.R. Birch, and D.C. James. 2005. On the optimal ratio of heavy to light chain genes for efficient recombinant antibody production by CHO cells. *Biotechnol. Progress* 21:122–138.

Sieber, V., A. Pluckthun, and F.X. Schmid. 1998. Selecting proteins with improved stability by a phage-based method. *Nature Biotechnol.* 16:955–960.

Tait, A.S., C.J. Brown, D.J. Galbraith, M.J. Hines, M. Hoare, J.R. Birch, and D.C. James. 2004. Transient production of recombinant proteins by Chinese hamster ovary cells using polyethyleneimine/DNA complexes in combination with microtubule disrupting anti-mitotic agents. *Biotechnol. Bioeng.* 88:707–721.

Thomas, P., and T.G. Smart. 2005. HEK293 cell line: A vehicle for the expression of recombinant proteins. *J. Pharmacol. Toxicol. Methods* 51:187–200.

U.S. Pharmacopeia. 2003. Particulate Matter in Injections, 2189–2196.

Van den Nieuwenhof, I.M., H. Koistinen, R.L. Easton, R. Koistinen, M. Kamarainen, H.R. Morris, I. van Die, M. Seppala, A. Dell, and D.H. Van den Eijnden. 2000. Recombinant glycodelin carrying the same type of glycan structures as contraceptive glycodelin-A can be produced in human kidney 293 cells but not in Chinese hamster ovary cells. *Eur. J. Biochem.* 267:4753–4762.

Van Holde, K.E. 1985. In *Physical Biochemistry*, 209–227. Englewood Cliffs, NJ: Prentice-Hall.

Wehr, T., and R. Rodriguez-Diaz. 2005. Use of size exclusion chromatography in biopharmaceutical development. In *Analytical Techniques for Biopharmaceutical Development*, 1st edition, ed. R. Rodriguez-Diaz, T. Wehr, and S. Tuck, 95–112. New York: Marcel Dekker.

Wetlaufer, D.B. 1954. Ultraviolet spectra of proteins and amino acids. *Adv. Protein Chem.* 17:303–390.

Wurm, F.M., and A. Bernard. 1999. Large-scale transient expression in mammalian cells for recombinant protein production. *Curr. Opin. Biotechnol.* 313:156–159.

CHAPTER 26

Production of Antibodies in *Pichia pastoris*

JUERGEN H. NETT

ABSTRACT

In recent years, the methylotropic yeast *Pichia pastoris* has become one of the most commonly used organisms for production of both secreted and intracellular proteins. This chapter provides an overview of the general principles for expression vector and strain construction and screening and selection of production strains. It also compiles and reviews published examples of antibody expression and provides a discussion on parameters that can be used for optimizing expression in this host.

26.1 INTRODUCTION

26.1.1 Background

Pichia pastoris is a methylotropic yeast that now has been used for more than two decades to successfully produce a large number of recombinant proteins (Lin Cereghino and Cregg 2000; Macauley-Patrick et al. 2005). See Figure 26.1 for an electron microscope picture of a *Pichia pastoris* cell.

Therapeutic Monoclonal Antibodies: From Bench to Clinic. Edited by Zhiqiang An

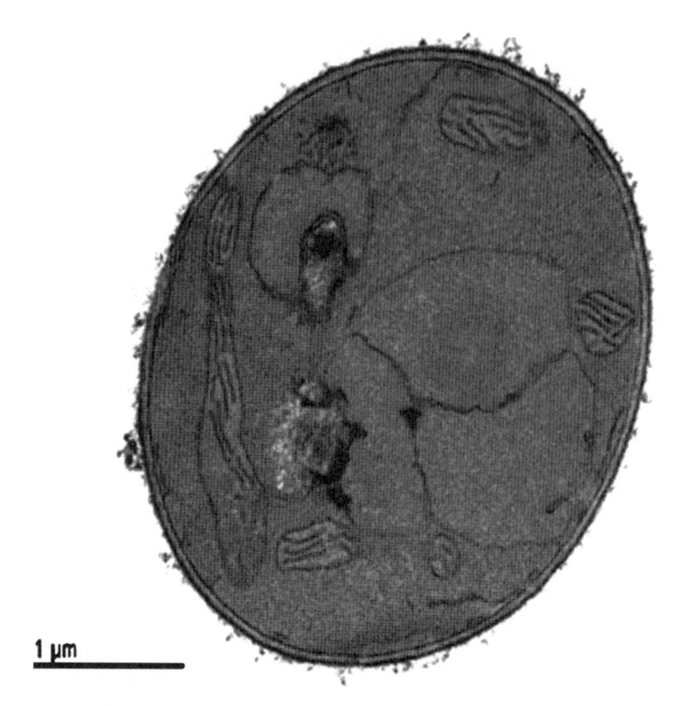

Figure 26.1 Electron micrograph of a *Pichia pastoris* cell.

Initially it was developed by Phillips Petroleum Company of Bartlesville, Oklahoma, during the early 1970s as a single cell protein (SCP) for use as animal feed (Wegner 1990). However during the oil crisis of the 1970s the cost of petroleum-based products, including methanol, increased drastically and therefore the economics for production of SCP from methanol were never economically favorable.

In the early 1980s Phillips Petroleum Company engaged the Salk Institute Biotechnology/ Industrial Associates, Inc. (SIBIA) in a collaboration to develop *P. pastoris* as an organism for foreign gene expression.

Some of the key factors contributing to the popularity of *P. pastoris* are (1) the fact that Phillips Petroleum licensed Invitrogen Corporation of Carlsbad, California, to sell a commercial kit containing components of the system; (2) the simplicity of genetic manipulation and large scale fermentation; (3) an unusually efficient and tightly regulated promoter from the alcohol oxidase I gene (*AOX1*), enabling *P. pastoris* to produce heterologous proteins either intracellularly or extracellularly at high levels; (4) its capability to perform eukaryotic posttranslational modifications like folding, disulfide bond formation, proteolytic processing, and N- and O-linked glycosylation; and (5) its preference for a respiratory rather than a fermentative growth, enabling it to reach ultra-high cell densities of $\geq$400 g/L wet cell weight during growth in fermenters.

26.1.2 Host Strains

All currently available *P. pastoris* strains are derived from the wild-type strain NRRL-Y 11430 (Northern Regional Research Laboratories, Peoria, Illinois). Several research groups have described the isolation or construction of whole sets of auxotrophic *P. pastoris* strains that can be used employing expression plasmids with the appropriate selectable marker genes (Gould et al. 1992; Lin Cereghino et al. 2001; Nett and Gerngross 2003; Thor et al. 2005; Nett et al. 2005). In addition, the use of several dominant drug (Clare et al. 1991; Scorer et al. 1994; Kimura et al. 1994; Higgins and Cregg 1998; Wan et al. 2004) and nutritional markers (Sreekrishna, Tschopp, and Fuke 1987; Sunga and Cregg 2004) has been reported. The multitude of these very simple and easy to use selection systems coupled with the propensity of *P. pastoris* for homologous recombination and the possibility to use standard yeast genetic methods similar to those in use with *Saccharomyces cerevisiae* has opened up almost limitless possibilities for genetic modification of this organism, as evidenced by reports of extensive genetic engineering projects in the recent past (Hamilton et al. 2006).

Some of the earliest genetic modifications were the generation of protease-deficient strains (Gleeson et al. 1998). Since many secreted or intracellularly expressed heterologous proteins are susceptible to degradation by proteases, the use of these strains deficient in proteinase A (encoded by the *PEP4* gene) or proteinase B (encoded by the *PRB1* gene) can often significantly improve protein yield (White et al. 1995; Weiss et al. 1995).

26.1.3 Expression Vectors

All *Pichia pastoris* expression vectors to date have been constructed as *Escherichia coli* shuttle vectors, using an *E. coli* origin of replication and selectable markers functional in one or both organisms. Because *P. pastoris* has no stable episomal vectors, all expression vectors are designed to integrate into the genome. Based on their mode of integration, the vectors can be divided into single crossover vectors and double crossover vectors.

The single crossover vectors are commonly digested at a unique restriction site within the integration locus contained on the vector and then recombine homologously at the cut locus via a single crossover event once transformed into an appropriate *P. pastoris* strain (Lin Cereghino et al. 2001). This so-called roll-in allows for the integration of multiple copies of the vector into a predetermined site of the genome, but also leads to the duplication of the integration locus and therefore to the potential excision of the plasmid sequences after homologous recombination of these direct repeats (Nett and Gerngross 2003).

The double crossover vectors integrate via gene replacement events, leading to the knockout of the integration locus (Nett and Gerngross 2003; Nett et al. 2005). While this way of integration into the genome is much more stable, it generally only allows for a single copy of the vector to be integrated and also knocks out the integration locus at the same time (Nett and Gerngross 2003). In the case of gene replacement at the *AOX1* locus this leads to strains that metabolize methanol at a greatly reduced rate. This is not necessarily a bad thing, since in some cases these so-called Mut^s strains express higher levels of heterologous proteins than do wild-type (also called Mut^+) strains (Tschopp et al. 1987; Cregg et al. 1987). Both types of vectors also contain one or more expression cassettes composed of a *P. pastoris* promoter, a multiple cloning site, and a transcriptional terminator.

The two most commonly used promoters are the strongly inducible *AOX1* promoter (Ellis et al. 1985) and the constitutive *GAP* promoter (Waterham et al. 1997). As transcriptional terminators, the use of the *AOX1* (Lin Cereghino et al. 2001) and the heterologous *ScCYC1* (Nett and Gerngross 2003) terminator have been described.

Additionally some expression vectors may also contain a secretion signal, which, when fused in frame to the protein to be expressed, targets the protein through the cell's secretion pathway into the supernatant. The most commonly used secretion signal is derived from the *S. cerevisiae* α-mating factor (Kurjan and Herskowitz 1982), but the use of signals from the *P. pastoris* acid phosphatase (*PHO1*; Laroche et al. 1994), the *P. pastoris KAR2* gene product (Li et al. 2006), chicken lysozyme (Li et al. 2006), and the *S. cerevisiae* invertase (*SUC2*; Freyre et al. 2000) have also been published.

The construction of several more or less specialized antibody expression plasmids has been reported by different groups (Liu et al. 2003; Gurkan, Symeonides, and Ellar 2004; Braren et al. 2007). Figure 26.2 depicts the plasmid map of a more generalized antibody expression vector. It consists of two expression cassettes for the individual immunoglobulin chains, a selectable marker, and an integration locus for targeting into the *P. pastoris* genome.

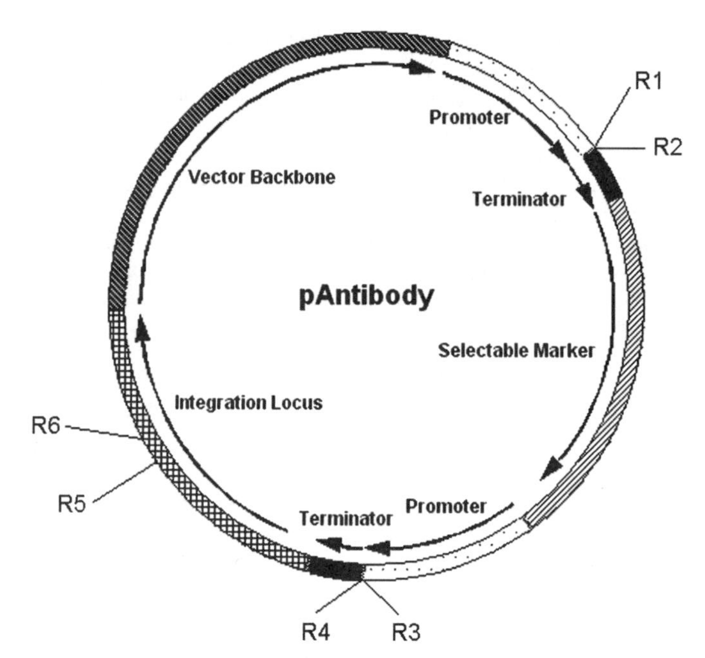

Figure 26.2 Plasmid map of a generalized *Pichia pastoris* antibody expression vector. R1 to R6 designate unique restriction sites.

26.1.4 Construction of Expression Strains

The third step after selection of the appropriate host strain and construction of a suitable expression vector is the introduction of the expression vector into the *P. pastoris* genome and subsequent selection and identification of strains expressing the gene of interest. In some cases it is advantageous to employ a selection scheme that allows for construction or selection of strains integrating multiple copies of the gene of interest into the genome (for a review see Lin Cereghino and Cregg 2000), but for some less efficiently assembled or secreted proteins a higher copy number might actually lead to reduced secretion and intracellular accumulation of protein (Romanos 1995). In order to consistently find a strain that is expressing the protein of interest at optimal levels, it is therefore recommended to follow a stepwise strain identification and verification protocol (see Fig. 26.3).

To start out, a relatively large number of transformants should be grown up and induced in 96- or 24-well plates and analyzed for expression of the desired protein using immuno dot blots or ELISA. In the second step, 5 to 10 promising strains are then grown in small shake flask cultures to determine which strain produces the foreign protein at the highest level. If the highest possible level of expression is desired, it is generally also recommended to transfer the highest expressing strain in a fermenter and there further increase expression by optimizing the multitude of parameters that can be controlled during this process (for a review see Lin Cereghino et al. 2002).

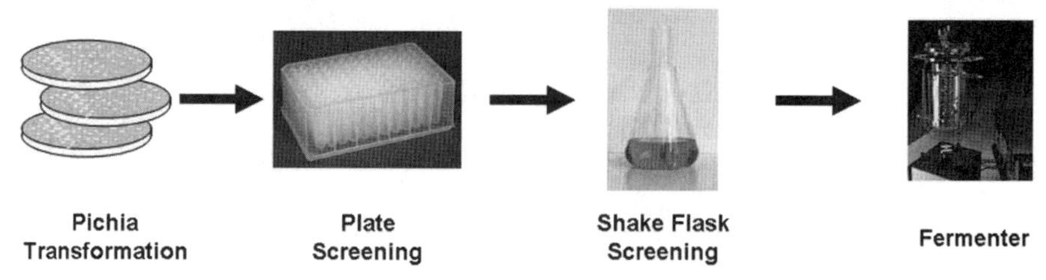

Figure 26.3 Scheme for selection of *Pichia pastoris* production strains.

26.2 EXAMPLES OF ANTIBODY EXPRESSION IN *Pichia pastoris*

Since the emergence of monoclonal antibodies as potential diagnostic and therapeutic agents in the late 1970s, the quick and cheap expression of whole antibodies or antibody fragments in recombinant form has been one of the main challenges of antibody technology. While the expression in *E. coli* can yield titers of up to grams per liter in fermenter cultures (Horn et al. 1996, and reviewed in Harrison and Keshavarz-Moore 1996) there are also several drawbacks associated with expression in this prokaryotic host. First, the secretion of immunoglobulins into the culture medium or the periplasmic space, where folding and assembly of the light and heavy chains is facilitated, can often be accompanied by cell lysis and subsequent loss of product. Second, expression of the desired product in inclusion bodies, often resulting in insoluble aggregates, necessitates time-consuming and often inefficient *in vitro* refolding followed by purification, and frequently only a small fraction of product can be recovered in functional form. Third, the fact that bacteria are unable to carry out eukaryotic posttranslational modifications makes them unsuitable in cases where glycosylation of antibodies or fusion proteins is required for functional activity.

Because yeasts are able to efficiently secrete heterologous proteins into the culture medium and also possess a functional glycosylation machinery, researchers from early on explored these model organisms as alternatives to *E. coli* for expression of antibodies. The first successful production of functional antibody in *Saccharomyces cerevisiae* was reported by Wood and colleagues in 1985 and Horwitz and colleagues in 1988. It took several more years until Ridder and collaborators (1995) published the first report of expression of antibody fragments in *Pichia pastoris*. Using the α-mating factor signal sequence as secretion signal they managed to produce more than 100 mg/L of pure and functional rabbit single-chain Fv against recombinant human leukemia inhibitory factor in shake flask cultures. Since then almost 50 reports have been published reporting the successful expression of antibodies in *P. pastoris*, mostly scFv fragments and fusions thereof (for a list see Table 26.1).

26.2.1 Single Domain Antibody Fragments

To date there have only been two published reports of production of so-called single domain antibodies (VHH).

Rahbarizadeh and coworkers (2006) expressed two VHH constructs against cancer associated mucin, MUC1, in a Mut^S strain of *P. pastoris*. They found that supplementation of the induction medium with casamino acids or EDTA increased production up to twofold, and that preinduction osmotic stress using media containing 1 M sorbitol also increased product yield. Under optimal conditions they were able to achieve yields in the range of 10 to 15 mg/L. They also showed that the specific activity in an ELISA of the purified yeast VHH was higher than that of a bacterial periplasmic counterpart.

In a rare example of intracellular expression, Omidfar et al. (2007) reported the expression and purification from the cell lysate of *P. pastoris* of a VHH antibody recognizing the external domain of the mutant epidermal growth factor receptor EGFR VIII at levels of 8 to 10 mg/L after purification.

26.2.2 Single Chain Fv Antibody Fragments

Most of the early publications describing the expression of scFv fragments reported yields ranging from a few milligrams to a few hundred milligrams, but in 2000 Freyre and colleagues achieved levels of 1.2 g/L of a biologically active anti-carcinoembryonic antigen scFv secreted into the culture supernatant of a *P. pastoris* fermenter culture. Contrary to the work of most other groups, they used a *P. pastoris* strain that had been developed in house, employed a *HIS3* auxotrophic marker for selection and an *S. cerevisiae* sucrose invertase secretion signal for expression.

Using minimal medium and feedback control of the methanol concentration in their fermentation strategy, Pla and coworkers (2006) showed several years later that the A33 scFv antibody fragment can be secreted extracellularly at levels of over 4 g/L. They also detailed that in addition to optimized

TABLE 26.1 Antibody Fragments and Antibodies Expressed in *Pichia pastoris*

Antibody Format	Antigen	Secretion Signal	Shake Flask/ Fermenter	Titer [mg/L]	Reference
VHH	MUC1	Sc α-mating factor	Shake flask	10–15	Rahbarizadeh et al. (2006)
VHH	External domain of EGFR VIII	None	Shake flask	8–10	Omidfar et al. (2007)
scFv	Leukemia inhibitory factor	Sc α-mating factor	Shake flask	≥100	Ridder et al. (1995)
scFv	Squamous carcinoma	Sc α-mating factor	Shake flask	10–50	Luo et al. (1995, 1996)
scFv	Desipramine	Sc α-mating factor/PpPHO1	Shake flask	60	Eldin et al. (1997)
scFv	CD7	Sc α-mating factor/PpPHO1	Shake flask	100–250	Eldin et al. (1997)
scFv	Thomsen–Friedenreich (pan-adenocarcinoma)	Sc α-mating factor/PpPHO1	Shake flask	0–200	Luo et al. (1997b)
scFv	Human type V acid phosphatase	PpPHO1	Shake flask	0.31–0.44	Cupit et al. (1999)
scFv	*Pseudomonas aeruginosa* lipoprotein I	PpPHO1	Shake flask	1.7	Cupit et al. (1999)
scFv	Carcinoembryonic antigen	ScSUC2	Fermenter	1200	Freyre et al. (2000)
scFv	Rat transferrin receptor (OX26)	Sc α-mating factor/PpPHO1	Shake flask	0.62	Boado, Ji, and Pardridge (2000)
scFv	Same as mouse MAb4813	Sc α-mating factor	Fermenter	40–45	Hellwig et al. (2001)
scFv	ED-B domain of B-fibronectin isoform	Sc α-mating factor	Fermenter	5–20	Marty et al. (2001)
scFv	Prostate-specific antigen	Sc α-mating factor	Shake flask	≥15	Y. Wang et al. (2001)
scFv	Botulinum A neurotoxin	Sc α-mating factor	Shake flask	Retained intracellularly	Mah et al. (2003)
scFv	*Mamestra configurata* serpins	Sc α-mating factor	Shake flask	25	Shi et al. (2003)
scFv	ED-B domain of B-fibronectin isoform	Sc α-mating factor	Fermenter	19	Cunha et al. (2004)
scFv	Fimbriae of enterotoxigenic *E. coli* F4	Sc α-mating factor	Fermenter	106.5	Trentmann, Khatri, and Hoffmann (2004)
scFv	$p185^{HER\text{-}2}$	Sc α-mating factor	Shake flask	70	Gurkan, Symeonides, and Ellar (2004)
scFv	CD33	Sc α-mating factor	Shake flask	11–48	Emberson et al. (2005)
scFv	Human epidermal growth factor	Sc α-mating factor	Shake flask	7.3	Miller et al. (2005)
scFv	*Shewanella oneidensis* protein phosphotyrosine phosphatase	Sc α-mating factor	Shake flask	0.42	Miller et al. (2005)

scFv	*Shewanella oneidensis* Gfo/Idh/MocA family oxidoreductase	Sc α-mating factor	Shake flask	5.6	Miller et al. (2005)
scFv	Fimbriae of enterotoxigenic *E. coli* F4	Sc α-mating factor	Fermenter	300–400	Khatri and Hoffmann (2005)
scFv	CD11bA domain of CR3	Sc α-mating factor	Shake flask	40	Ben Tanfous et al. (2006)
scFv	ErbB2	Sc α-mating factor	Shake flask	15	Hu et al. (2006)
scFv	HIV envelope glycoproteins	Sc α-mating factor/native	Shake flask	Not reported	Gasser et al. (2006)
scFv	A33 antigen	Sc α-mating factor	Fermenter	4000–6000	Pla et al. (2006)
scFv	A33 antigen	Sc α-mating factor	Fermenter	≥8000	Damasceno et al. (2007)
Glycosylated scFv (Mannabodies)	TAG-72	PpPHO1	Shake flask	50–100	M. Wang et al. (1998)
Bivalent scFv	CA 125 ovarian-carcinoma	Sc α-mating factor	Shake flask	100	Luo et al. (1997a)
Bivalent scFv	TAG-72	Sc α-mating factor	Shake flask	15–20	Goel et al. (2000a)
Tetravalent scFv	TAG-72	Sc α-mating factor	Shake flask	2.0–3.5	Goel et al. (2000b)
Cross-linked bivalent diabody	Carcinoembryonic antigen (CEA)	Sc α-mating factor	Shake flask	1	FitzGerald, Hollinger, and Winter (1997)
Cross-linked bispecific diabody	CEA/CD3	Sc α-mating factor	Shake flask	1	FitzGerald, Hollinger, and Winter (1997)
scFv fusion with B7-2	ErbB2/CD28	Sc α-mating factor	Shake flask	0.5	Gerstmayer et al. (1997)
scFv-fusion with biotin mimetic sequence	CA125 (ovarian carcinoma)	Sc α-mating factor	Shake flask	Not reported	Luo et al. (1998)
scFv fusion with protein A	Z-form of DNA	Sc α-mating factor	Shake flask	5	De Andrade et al. (2000)
scFv fusion with carboxypeptidase G2	Carcinoembryonic antigen (CEA)	Sc α-mating factor	Fermenter	65–110	Medzihradszky et al. (2004); Sharma et al. (2005)
scFv fusion with Fc fragment of human IgG1	Z-form of DNA	Sc α-mating factor	Shake flask	1.5	De Andrade et al. (2000)
scFv fusion with Fc fragment of human IgG1	Botulinum A neurotoxin	Sc α-mating factor	Shake flask	2	Powers et al. (2001)
scFv fusion with Fc fragment of human IgG1	Her2/*neu* oncoprotein	Sc α-mating factor	Shake flask	0.3	Powers et al. (2001)
scFv fusion with Fc fragment of mouse IgG1	*Schistosoma japonicum* glutathione-*S*-transferase (GST)	Sc α-mating factor	Fermenter	10–30	Liu et al. (2003)

(*Continued*)

TABLE 26.1 *Continued*

Antibody Format	Antigen	Secretion Signal	Shake Flask/ Fermenter	Titer [mg/L]	Reference
scFv fusion with Fc fragment of human IgG1	Hen egg lysozyme	Sc α-mating factor	Shake flask	1–10	Braren et al. (2007)
scFv fusion with C_H1-3 of human IgG1	Hen egg lysozyme	Sc α-mating factor	Shake flask	1–10	Braren et al. (2007)
Fab	FcεRI	PpPHO1	Shake flask	10–40	Takahashi et al. (2000)
Fab	FcεRI	Native	Shake flask	0.5–1.5	Takahashi et al. (2000)
Fab	Atrazine	Sc α-mating factor	Fermenter	40	Lange, Schmitt, and Schmid (2001)
Fab	Hepatitis B surface antigen	Sc α-mating factor	Shake flask	50	Ning et al. (2003)
Fab	Hepatitis B surface antigen	Sc α-mating factor	Fermenter	420–458	Ning et al. (2005)
Fab	HIV envelope glycoproteins	Sc α-mating factor/native	Shake flask/ fermenter	21.8/41.2	Gasser et al. (2006)
Fab	Human monoclonal antibody 2F5	Sc α-mating factor	Fermenter	260	Gach et al. (2007)
Fab	Tumor-associated glycoprotein-72 (TAG-72)	Sc α-mating factor	Shake flask	1–2	Tang et al. (2007)
scFab	Hen egg lysozyme	Sc α-mating factor	Shake flask	Not reported	Hust et al. (2007)
Murine MAb	2,3,7,8-tetrachloro-dibenzo-*p*-dioxin	Sc α-mating factor	Shake flask	36	Ogunjimi et al. (1999)
Chimeric MAb	CD20	PpKAR2 Chicken lysozyme	Fermenter	Not reported	Li et al. (2006)

fermentation conditions the choice of Mut^+ and Mut^S phenotype strains can have a significant impact on product yields. Improving on this work even further, Damasceno et al. (2007) subsequently showed that overexpression of the *P. pastoris* immunoglobulin binding protein (BiP) was able to further increase secretion of A33scFv by approximately threefold in small scale cultures and twofold in benchtop fermenter cultures yielding over 8 g/L of secreted product. However, when the same group overexpressed protein disulfide isomerase (PDI) either alone or in addition to BiP, this had no apparent effect on secretion of A33scFv. Somewhat contrary to these findings, Gasser and coworkers (2006) reported that the overexpression of the unfolded protein response transcription factor *S. cerevisiae* HAC1 led to an increase in Fab secretion of a monoclonal anti-HIV-1 antibody of 1.3-fold and the overexpression of *S. cerevisiae* PDI enabled an increase of the Fab level of 1.9-fold. These results reflect probably one of the biggest challenges and opportunities to further improve the *P. pastoris* system, that is, by optimization of the chaperone and folding pathway in the endoplasmatic reticulum of this yeast.

Shi and colleagues (2003) took a different approach to maximizing product titer by optimization of expression conditions such as protease activity, temperature, cell density, osmotic stress, medium composition, pH, and reiterative induction. During expression of an scFv targeted against *Mamestra configurata* (bertha armyworm) serpins, they found that medium composition, preinduction osmotic stress, and temperature had the greatest effects on protein production. Supplementation of the induction medium with arginine, casamino acids, or EDTA increased product yield by several fold, as did cultivation under osmotic stress conditions during preinduction biomass accumulation. Incubation at 15°C versus 30°C extended the period whereby cells were capable of producing scFv from one to seven days, with the possibility of performing up to five reiterative inductions.

Yet another approach was taken by Hu et al. (2006). They designed an anti-ErbB2 scFv, choosing the *P. pastoris* codon bias while keeping the G + C content at a relatively low level. This codon optimization increased the scFv level three- to fivefold, and Northern blotting confirmed that this increase of scFv expression was mainly due to enhanced translation efficiency. However, there are also instances where codon optimization gave no improvement of specific productivity. Gach and colleagues (2007), for example, reported that codon optimization of a Fab directed against the HIV-1 broadly neutralizing antibody 2F5 did not result in an increase in expression of the desired product from the 260 mg/L level they had achieved in fermenter cultures.

26.2.3 Modified scFv Fragments and scFv Fusions

Numerous approaches have been reported to further increase the usefulness of scFv fragments by either fusing them to other protein fragments or simple multimerization. In early 1997 Luo and coworkers (Luo et al. 1997a) described the expression in *P. pastoris* of scFv fragments targeted against a unique epitope of CA125 expressed on ovarian carcinomas that were able to dimerize due to additional C-terminal extensions containing one or two cysteines. They were able to secret these bivalent scFv fragments at a level of 100 mg/L in a shake flask and show that they exhibited specific antigen binding activity which was almost the same as the parental MAbs.

In the same year, FitzGerald, Hollinger, and Winter (1997) reported the expression of disulfide-linked bivalent and bispecific diabodies in *E. coli* and *P. pastoris*. Although they were only able to secret the products from *P. pastoris* in a shake flask at a level of 1 mg/L using a GAPDH promoter-driven process, they concluded that disulfide-linked diabodies could greatly benefit from expression by *P. pastoris* due to the higher thermal stability and improved cross-linking and tumor-localizing properties when compared to expression in the *E. coli* system.

To improve further the *in vitro* and ultimately the *in vivo* performance of scFvs, Goel and coworkers (2000a, 2000b) expressed bivalent and tetravalent scFvs recognizing the tumor-associated glycoprotein TAG-72 in *P. pastoris*. They showed that the expression of divalent noncovalent and covalent $sc(Fv)_2$ in *P. pastoris* was 30- to 40-fold higher than in *E. coli*. They were also able to show that the tetravalent scFv formed by a noncovalent association of the covalent dimer showed improved *in vitro* binding properties as compared with $sc(Fv)_2$ and the respective IgG called CC49. The tetravalent scFv exhibited an increased tumor uptake rate, and since the larger molecular size exceeds the renal threshold for the first pass elimination, this also translated into an improved biological half-life.

In order to generate a conjugation site for drugs and other molecules, M. Wang and colleagues (1998) engineered an scFv with a glycan attached to the C-terminus. After expression in *P. pastoris* shake flask cultures at levels of up to 100 mg/L they were able to show that the glycosylated scFv mannose chains of these mannabodies could be effectively conjugated to polyethylene glycol and that the resulting conjugate displayed a 10-fold increased circulating life in mice.

In a more direct approach of fusing a costimulary protein to scFv in order to generate a potential tumor vaccine for cancer immunotherapy, Gerstmayer et al. (1997) constructed a chimeric fusion protein consisting of the extracellular domain of human B7-2 (CD86) fused to an scFv specific for the ErbB2 protein, a type I growth factor overexpressed in a high percentage of human adenocarcinomas. After expression in *P. pastoris* shake flask cultures and subsequent purification, albeit only at yields of 0.5 mg/L, this biologically active fusion protein localized specifically to the surface of ErbB2-expressing target cells, thereby providing a costimulatory signal that resulted in enhanced proliferation of syngeneic T-cells.

In an attempt to produce an agent suitable for use in antibody-directed enzyme prodrug therapy (ADEPT), Medzihradszky and coworkers (2004), and Sharma and coworkers (2005) expressed the recombinant antibody-enzyme fusion protein MFE-CP (a fusion protein of carboxypeptidase G2 with an anti-carcinoembryonic antigen scFv) in *P. pastoris* fermenter cultures at levels of 65 to 110 mg/L. They proposed that its therapeutic potential could be improved by N-glycosylation obtained by expression in *P. pastoris*. They also postulated that this glycosylation could enhance clearance from healthy tissue and might result in better tumor-normal tissue ratios. After purification of MFE-CP from the fermenter cultures they went on to show that the fusion protein was enzymatically active, showed enhanced clearance from normal tissues *in vivo*, and also exhibited effective tumor localization. This led to reproducible tumor growth delay in two models of human colon carcinoma xenografts after a single ADEPT cycle with minimal toxicity. The authors concluded that the carbohydrate structures resulting from expression in *P. pastoris* were responsible for this improvement and also developed a method suitable for characterization of these carbohydrates.

Several groups (see Table 26.1) have reported the expression in *P. pastoris* of fusions of scFv with Fc fragment of human IgG1 either to simplify detection and purification or to prolong the protein's serum half-life *in vivo* and mediate antibody-dependent cellular cytotoxicity against tumor target cells. All of the reports were in agreement that the binding activity of the fusion proteins was comparable to that of the scFV molecules. However, while the expression levels in shake flask cultures are generally reported to be in the single-digit milligram per liter range, Liu et al. (2003) were able to show that the expression yield in fermenters can reach 10 to 30 mg/L of culture medium.

26.2.4 Fab Antibody Fragments

In addition to the examples already mentioned above of Fab expression in *P. pastoris*, several additional groups have reported the expression of Fabs or scFabs in this system.

Takahashi and collaborators for the first time demonstrated the expression of a Fab fragment in *P. pastoris* in 2000 (Takahashi et al. 2000). They showed that the replacement of the native signal sequence with that from the *P. pastoris PHO1* gene increased the production of the Fab fragment by a Mut^S *P. pastoris* strain by more than 10-fold to a titer of up to 40 mg/L in a shake flask culture. They also demonstrated that when only the light or heavy chain was introduced into *P. pastoris* individually neither chain was secreted into the medium and the product was hardly detectable in the cells.

In 2001 Lange, Schmitt, and Schmid published the expression of pure, atrazine-specific Fab fragments using a Mut^+ *P. pastoris* strain. They were able to significantly increase the product yield to up to 40 mg/L by switching from a shake flask process to a 5 L fed-batch process. Although the IC_{50} value determined by means of a competitive ELISA of the Fab produced in *P. pastoris* was one magnitude higher than the value obtained with the parental MAb it still equaled that obtained with Fab expressed in *E. coli* cells.

Another comparison of cultures grown in shake flasks versus fermenter cultures was made by Ning and colleagues (2003, 2005). They report the expression of an anti-hepatitis B Fab fragment in shake flasks to reach levels of 40 to 50 mg/L employing a Mut^S *P. pastoris* strain as expression host, and were able to later increase this titer to up to 450 mg/L using a 5 L fed-batch process. After purification, they also determined that the recombinant Fab was able to neutralize the hepatitis B surface antigen with high affinity.

In a more recent study, Tang et al. (2007) generated a recombinant Fab of the TAG-72 monoclonal antibody CC49, expressed it in *P. pastoris* shake flask cultures at a level of 1 to 2 mg/L and, after purification, directly compared its tumor and normal tissue localization properties with those of enzymatically generated CC49 Fab in athymic mice implanted with LS174T human colon cancer xenografts. The authors concluded that the specific accumulation of ^{123}I-labeled CC49 rFab in these xenografts and its rapid clearance from the blood and most normal tissues suggest that it would be useful for imaging colorectal cancer and other tumors expressing TAG-72.

26.2.5 Full Length MAbs

To date there have only been two published reports of expression of full length MAbs in *P. pastoris*. In 1999 Ogunjimi and coworkers reported the production of a functional murine antibody to dioxin at a level of 36 mg/mL in shake flask cultures. The two individual antibody chains were expressed from two separate expression cassettes on a single plasmid that had been transformed into a *pep4* protease knockout strain of *P. pastoris*. After preliminary screening using colonies replicated to nitrocellulose membranes, the authors selected one Mut^+ strain and one Mut^S strain and analyzed the ability of these strains to express antibody. Using results from Western blot and slot-immunoblot analysis they concluded that *P. pastoris* was able to secrete a major portion (about 90 percent) of the produced antibody into the supernatant and that the antibody was detectable between 12 h and 120 h of induction, with the highest levels detected between 72 and 108 h.

Li et al. (2006) used several glycoengineered lines of *P. pastoris* to produce a library of seven different glycoforms of the anti-CD20 antibody rituximab and compared the receptor-binding properties of these proteins to the mammalian cell-derived commercial counterpart Rituxan. Fcγ receptor-binding assays showed that changing the glycan structures of the antibody improved binding affinities for the V158 variant of the FcγRIIIa receptor at least 10-fold, and the binding to the low affinity variant FcγRIIIa-F158 over 100-fold compared to commercial rituximab. It was also reported that the glycoengineered variants showed improved antibody-mediated killing in B-cell depletion assays when compared to the mammalian cell culture-derived product. It is noteworthy that the expression constructs described did not use any of the standard signal sequences to drive secretion, but a Kar2 signal peptide and a chicken lysozyme-derived signal sequence.

These results underscore the ease with which *P. pastoris* can be genetically modified to express antibodies with improved functionalities with reduced cycle times and favorable cost of goods.

26.3 SUMMARY

During the last decade a multitude of immunoglobulins have been successfully expressed in *P. pastoris*. While there seems to be a general consensus that the antigen-binding activity of these proteins is as good as or better than that of their counterparts expressed in *E. coli* and sometimes even exceeding that of the proteins obtained from mammalian cell culture, various approaches have been described attempting the optimization of expression in *P. pastoris*. Despite the general simplicity with which proteins can be expressed in this yeast, there are a multitude of parameters that can be varied in order to achieve optimal expression of a protein of interest.

- *Choice of Site and Mode of Integration*. This can determine the methanol utilization phenotype (Mut^+ or a Mut^S) and, through choice of the appropriate selectable marker, also the number of copies of expression cassette that are integrated into the genome. Whereas there doesn't seem to exist a strong preference for either Mut^{+} or Mut^{S} strains, expression of especially the smaller antibody fragments seems to be increased when multicopy integrants are selected for.

- *Choice of Host Strain.* The availability of protease knockout strains facilitates the determination of whether degradation through proteases has a significant effect on product titer. It seems that in isolated cases the use of protease knockouts can lead to titer increases of several fold; however, it should also be considered that some of these host strains grow less vigorously than their wild-type counterparts and therefore their use should be evaluated on an individual basis.
- *Codon Optimization.* There are only very few published examples where the choice of codon bias has had a significant influence on the expression level of antibodies. However, considering the ease with which codon optimized genes can nowadays be obtained (Codon Devices, DNA2.0, GENEART), it should be standard practice to start any kind of expression study with a for *P. pastoris* codon optimized version of the gene.
- *Choice of Promoter.* The availability of the inducible AOX1 and the constitutive GAPDH promoter affords the investigator the choice of two different ways of protein induction. While there have been isolated reports of optimal protein expression using the GAPDH system, the vast majority of cases find the AOX1 promoter to be more effective. However, the fact that methanol, a highly flammable and toxic liquid, has to be used to induce the AOX1 promoter can become a consideration once larger fermenter volumes are used for protein expression.
- *Intracellular versus Secreted Expression.* Given the fact that secretion into the medium can already be considered a first purification step, due to the fact that *P. pastoris* secretes only a very small number of endogenous proteins, secreted expression is commonly the method of choice. However, in instances where this attempt is not successful, there are also ways of efficiently expressing heterologous proteins intracellularly in this yeast.
- *Choice of Secretion Signal.* This is one of the most important variables when it comes to optimizing the expression of a given protein. For most proteins the native signal sequence is usually a suboptimal choice for expression in *P. pastoris*. Whereas the majority of published reports use the *S. cerevisiae* alpha mating factor signal sequence that is contained in most commercially available *P. pastoris* expression vectors, there are instances where a switch to other secretion signals results in a significant increase in product titer.
- *Fermentation Parameters.* It is generally accepted that titers in a fermentation process can reach levels several fold higher than those achieved in shake flask cultures. This is due to the higher cell densities that can be reached in fermenters, but also due to the fact that a multitude of factors can be tightly controlled and optimized in these vessels. Some of these factors are, for example, culture pH, oxygen uptake rate, and methanol feeding strategy. Whereas the high cell densities in the fermenters can sometimes lead to cell lysis and release of vacuolar proteases, this can often be overcome by using protease knockout strains and supplementation of the medium with protease inhibitors or enhancers of product stability such as casamino acids or arginine.
- *Genetic Strain Modifications.* Multiple reports of increased antibody titers generated in genetically modified strains demonstrate that this method of overcoming bottlenecks for antibody assembly and secretion is a viable strategy for the generation of strains with improved antibody production capabilities. However, the complexity of the folding and secretion pathway in *P. pastoris* and other yeasts precludes these kinds of optimizations to be done on a routine basis. Until there is further evidence which genetic modifications lead to reliable and significant increases in antibody titer, attempts to optimize product yields this way should be evaluated on a case by case basis.

In conclusion, the ease of manipulation and the availability of a multitude of optimization parameters make *P. pastoris* an attractive alternative to bacterial and mammalian systems for the expression of antibodies.

REFERENCES

Ben Tanfous, N.G., H. Kallel, M.A. Jarboui, and D.M. Fathallah. 2006. Expression in *Pichia pastoris* of a recombinant scFv form of MAb 107, an anti human CD11b integrin antibody. *Enzyme Microb. Tech.* 38:636–642.

Boado, R.J., A. Ji, and W.M. Pardridge. 2000. Cloning and expression in *Pichia pastoris* of a genetically engineered single chain antibody against the rat transferring receptor. *J. Drug Target.* 8:403–412.

Braren, I., K. Greunke, O. Umland, S. Deckers, R. Bredehorst, and E. Spillner. 2007. Comparative expression of different antibody formats in mammalian cells and *Pichia pastoris*. *Biotechnol. Appl. Biochem.* 47:205–214.

Clare, J.J., F.B. Rayment, S.P. Ballantine, K. Sreekrishna, and M.A. Romanos. 1991. High-level expression of tetanus toxin fragment c in *Pichia pastoris* strains containing multiple tandem integrations of the gene. *Nature Biotechnol.* 9:455–460.

Cregg, J.M., J.F. Tschopp, C. Stillman, R. Siegel, M. Akong, W.S. Craig, R.G. Buckholz, K.R. Madden, P.A. Kellaris, G.R. Davis, B.L. Smiley, J. Cruze, R. Torregrossa, G. Veliçelebi, and G.P. Thill. 1987. High-level expression and efficient assembly of hepatitis B surface antigen in the methylotropic yeast *Pichia Pastoris*. *Bio/Technol.* 5:479–485.

Cunha, A.E., J.J. Clemente, R. Gomes, F. Pinto, M. Thomaz, S. Miranda, R. Pinto, D. Moosmayer, P. Donner, M.J.T. Carrondo. 2004. Methanol induction optimization for scFv antibody fragment production in *Pichia pastoris*. *Biotechnol. Bioeng.* 86:458–467.

Cupit, P.M., J.A. Whyte, A.J. Porter, M.J. Browne, S.D. Holmes, W.J. Harris, and C. Cunningham. 1999. Cloning and expression of single chain antibody fragments in *Escherichia coli* and *Pichia pastoris*. *Lett. Appl. Microbiol.* 29:273–277.

Damasceno, L.M., K.A. Anderson, G. Ritter, J.M. Cregg, L.J. Old, and C.A. Batt. 2007. Cooverexpression of chaperones for enhanced secretion of a single-chain antibody fragment in *Pichia pastoris*. *Appl. Microbiol. Biotechnol.* 74:381–389.

De Andrade, E.V., F.C. de Albuquerque, L.M.P. de Moraes, M. de Macedo Brígido, and M.A. Santos-Silva. 2000. Single-chain Fv fragment of the human IgG1 Tag: Construction, *Pichia pastoris* expression and antigen binding characterization. *J. Biochem.* 128:891–895.

Eldin, P., M.E. Pauza, Y. Hieda, G. Lin, M.P. Murtaugh, P.R. Pentel, and C.A. Pennell. 1997. High-level secretion of two antibody single chain Fv fragments by *Pichia pastoris*. *J. Immunol. Methods* 201:67–75.

Ellis, S.B., P.F. Brust, P.J. Koutz, A.F. Waters, M.M. Harpold, and T.R. Gingeras. 1985. Isolation of alcohol oxidase and two other methanol regulatable genes from the yeast *Pichia pastoris*. *Mol. Cell. Biol.* 5:1111–1121.

Emberson, L.M., A.J. Trivett, P.J. Blower, and P.J. Nicholls. 2005. Expression of an anti-CD33 single-chain antibody by *Pichia pastoris*. *J. Immunol. Methods* 305:135–151.

Fischer, R., J. Drossard, N. Emans, U. Commandeur, and S. Hellwig. 1999. Towards molecular farming in the future: *Pichia pastoris*-based production of single-chain antibody fragments. *Biotechnol. Appl. Biochem.* 30:117–120.

FitzGerald, K., P. Hollinger, and G. Winter. 1997. Improved tumour targeting by disulphide stabilized diabodies expressed in *Pichia pastoris*. *Protein Eng.* 10:1221–1225.

Freyre, F.M., J.E. Vazquez, M. Ayala, L. Canaan-Haden, H. Bell, I. Rodriguez, A. Gonzalez, A. Cintado, and J.V. Gavilondo. 2000. Very high expression of an anti-carcinoembryonic antigen single chain Fv antibody fragment in the yeast *Pichia pastoris*. *J. Biotechnol.* 76:157–163.

Gach, J.S., M. Maurer, R. Hahn, B. Gasser, D. Mattanovich, H. Katinger, and R. Kunert. 2007. High level expression of a promising anti-idiotypic antibody fragment vaccine against HIV-1 in *Pichia pastoris*. *J. Biotechnol.* 128:735–746.

Gasser, B., M. Maurer, J. Gach, R. Kunert, and D. Mattanovich. 2006. Engineering of *Pichia pastoris* for improved production of antibody fragments. *Biotechnol. Bioeng.* 94:353–361.

Gerstmayer, B., U. Altenschmidt, M. Hoffmann, and W. Wels. 1997. Costimulation of T cell proliferation by a chimeric B7-2 antibody fusion protein specifically targeted to cells expressing the *erb*B2 proto-oncogene. *J. Immunol.* 158:4584–4590.

Gleeson, M.A.G., C.E. White, D.P. Meininger, and E.A. Komives. 1998. Generation of protease-deficient strains and their use in heterologous protein expression. In *Methods in Molecular Biology: Pichia Protocols*, ed. D.R. Higgins and J.M. Cregg, 81–94. Totowa, NJ: Humana Press.

Goel, A., G.W. Beresford, D. Colcher, G. Pavlinkova, B.J.M. Booth, J. Baranowska-Kortylewicz, and S.K. Batra. 2000. Divalent forms of CC49 single-chain antibody constructs in *Pichia pastoris*: Expression, purification and characterization. *J. Biochem.* 127:829–836.

Goel, A., D. Colcher, J. Baranowska-Kortylewicz, S. Augustine, B.J.M. Booth, G. Pavlinkova, and S.K. Batra. 2000. Genetically engineered tetravalent single-chain Fv of the pancarcinoma monoclonal antibody CC49: Improved biodistribution and potential for therapeutic application. *Cancer Res.* 60:6964–6971.

Gould, S.J., D. McCollum, A.P. Spong, J.A. Heyman, and S. Subramani. 1992. Development of the yeast *Pichia pastoris* as a model organism for a genetic and molecular analysis of peroxisome assembly. *Yeast* 8:613–628.

Gurkan, C., S.N. Symeonides, and D.J. Ellar. 2004. High-level production in *Pichia pastoris* of an anti-pI85^{HER-2} single-chain antibody fragment using an alternative secretion expression vector. *Biotechnol. Appl. Biochem.* 39:115–122.

Hamilton, S.R., R.C. Davidson, N. Sethuraman, J.H. Nett, Y. Jiang, S. Rios, P. Bobrowicz, T.A. Stadheim, H. Li, B.-K. Choi, D. Hopkins, H. Wischnewski, J. Roser, T. Mitchell, R.R. Strawbridge, J. Hoopes, S. Wildt, and T.U. Gerngross. 2006. Humanization of yeast to produce complex terminally sialated glycoproteins. *Science* 313:1441–1443.

Harrison, J.S., and E. Keshavarz-Moore. 1996. Production of antibody fragments in *Escherichia coli*. *Ann. NY Acad. Sci.* 782:143–158.

Hellwig, S., F. Emde, N.P.G. Raven, M. Henke, P. van der Logt, and R. Fischer. 2001. Analysis of single-chain antibody production in *Pichia pastoris* using on-line methanol control and mixed-feed fermentations. *Biotechnol. Bioeng.* 74:344–352.

Higgins, D.R., and J.M. Cregg. 1998. Introduction to *Pichia pastoris*. In *Methods in Molecular Biology: Pichia Protocols*, ed. D.R. Higgins and J.M. Cregg, 1–15. Totowa, NJ: Humana Press.

Horn, U., W. Strittmatter, A. Krebber, U. Knupfer, M. Kujau, R. Wenderoth, K. Muller, S. Matzku, A. Pluckthun, and D. Riesenberg. 1996. High volumetric yields of functional dimeric miniantibodies in *Escherichia coli*, using an optimized expression vector and high-cell-density fermentation under nonlimited growth conditions. *Appl. Microbiol. Biotechnol.* 46:524–532.

Horwitz, A.H., C.P. Chang, M. Better, K.E. Hellstrom, and R.R. Robinson. 1988. Secretion of functional antibody and Fab fragment from yeast cells. *Proc. Natl. Acad. Sci. U.S.A.* 85:8678–8682.

Hu, S., L. Li, J. Qiao, Y. Guo, L. Cheng, and J. Liu. 2006. Codon optimization, expression, and characterization of an internalizing anti-ErbB2 single-chain antibody in *Pichia pastoris*. *Protein Expres. Purif.* 47:249–257.

Hust, M., T. Jostock, C. Menzel, B. Voedisch, A. Mohr, M. Brenneis, M.I. Kirsch, D. Meier, and S. Dübel. 2007. Single chain Fab (scFab) fragment. *BMC Biotechnol.* 7:14.

Khatri, N.K., and F. Hoffmann. 2005. Impact of methanol concentration on secreted protein production in oxygen-limited cultures of recombinant *Pichia pastoris*. *Biotechnol. Bioeng.* 93:871–879.

Kimura, M, T. Kamakura, Q.Z. Tao, I. Kaneko, and I. Yamaguchi. 1994. Cloning of the blasticidin S deaminase gene (*BSD*) from *Aspergillus terreus* and its use as a selectable marker for *Schizosaccharomyces pombe* and *Pyricularia oryzae*. *Mol. Gen. Genet.* 242:121–129.

Kurjan, J., and I. Herskowitz. 1982. Structure of a yeast pheromone gene (MFα): A putative α-factor precursor contains four tandem copies of mature α-factor. *Cell* 30:933–943.

Lange, S., J. Schmitt, and R.D. Schmid. 2001. High-yield expression of the recombinant, atrazine-specific Fab fragment K411B by the methylotropic yeast *Pichia pastoris*. *J. Immunol. Methods* 255:103–114.

Laroche, Y., V. Storme, J. De Meutter, J. Messens, and M. Lauwereys. 1994. High-level secretion and very efficient isotopic labeling of tick anticoagulant peptide (TAP) expressed in the methylotropic yeast *Pichia pastoris*. *Nature Biotechnol.* 12:1119–1124.

Li, H., N. Sethuraman, T.A. Stadheim, D. Zha, B. Prinz, N. Ballew, P. Bobrowicz, B.-K. Choi, W.J. Cook, M. Cukan, N.R. Houston-Cummings, R. Davidson, B. Gong, S.R. Hamilton, J.P. Hoopes, Y. Jiang, N. Kim, R. Mansfield, J.H. Nett, S. Rios, R. Strawbridge, S. Wildt, and T.U. Gerngross. 2006. Optimization of humanized IgGs in glycoengineered *Pichia pastoris*. *Nature Biotechnol.* 24:210–215.

Lin Cereghino, J., and J.M. Cregg. 2000. Heterologous protein expression in the methylotropic yeast *Pichia pastoris*. *FEMS Microbiol. Rev.* 24:45–66.

Lin Cereghino, G.P., J. Lin Cereghino, C. Ilgen, and J.M. Cregg. 2002. Production of recombinant proteins in fermenter cultures of the yeast *Pichia pastoris*. *Curr. Opin. Biotechnol.* 13:329–332.

Lin Cereghino, G.P., J. Lin Cereghino, A.J. Sunga, M.A. Johnson, M. Lim, M.A. Gleeson, and J.M. Cregg. 2001. New selectable marker/auxotrophic host strain combinations for molecular genetic manipulation of *Pichia pastoris*. *Gene*. 263:159–169.

Liu, J., D. Wei, F. Quian, Y. Zhou, J. Wang, Y. Ma, and Z. Han. 2003. pPIC9-Fc: A vector system for the production of single-chain Fv-Fc fusions in *Pichia pastoris* as detection reagents *in vitro*. *J. Biochem.* 134:911–917.

Luo, D., M. Geng, A.A. Noujaim, and R. Madiyalakan. 1997a. An engineered bivalent single-chain antibody fragment that increases antigen binding activity. *J. Biochem.* 121:831–834.

Luo, D., M. Geng, B. Schultes, J. Ma, D.Z. Xu, N. Hamza, W. Qi, A.A. Noujaim, and R. Madiyalakan. 1998. Expression of a fusion protein of scFv-biotin mimetic peptide for immunoassay. *J. Biotechnol.* 65:225–228.

Luo, D., N. Mah, M. Krantz, K. Wilde, D. Wishart, Y. Zhang, F. Jacobs, and L. Martin. 1995. Vl-linker-Vh orientation-dependent expression of single chain Fv containing an engineered disulfide-stabilized bond in the framework regions. *J. Biochem.* 118:825–831.

Luo, D., N. Mah, M. Krantz, D. Wishart, F. Jacobs, and L. Martin. 1997b. High level secretion of single-chain antibody in *Pichia pastoris* expression system. *Biotechnol. Tech.* 11:759–761.

Luo, D., N. Mah, D. Wishart, Y. Zhang, F. Jacobs, and L. Martin. 1996. Construction and expression of bi-functional proteins of single-chain Fv with effector domains. *J. Biochem.* 120:229–232.

Macauley-Patrick, S., M.L. Fazenda, B. McNeil, and L.M. Harvey. 2005. Heterologous protein production using the *Pichia pastoris* expression system. *Yeast* 22:249–270.

Mah, D.C.W., W.G. Hu, J.K. Pon, S.A. Masri, R.E. Fulton, P.L. Monette, and L.P. Nagata. 2003. Recombinant anti-botulinum neurotoxin A single-chain variable fragment antibody generated using a phage display system. *Hybrid Hybridomics* 22(5):277–283.

Marty, C., P. Scheidegger, K. Ballmer-Hofer, R. Klemenz, and R.A. Schwendener. 2001. Production of functionalized single-chain Fv antibody fragments binding to the ED-B domain of the B-isoform of fibronectin in *Pichia pastoris*. *Protein Expres. Purif.* 21:156–164.

Medzihradszky, K.F., D.I.R. Spencer, S.K. Sharma, J. Bhatia, R.B. Pedley, D.A. Read, R.H.J. Begent, and K.A. Chester. 2004. Glycoforms obtained by expression in *Pichia pastoris* improve cancer targeting potential of a recombinant antibody-enzyme fusion protein. *Glycobiology* 14:27–37.

Miller, K.D., J. Weaver-Feldhaus, S.A. Gray, R.W. Siegel, and M.J. Feldhaus. 2005. Production, purification, and characterization of human scFv antibodies expressed in *Saccharomyces cerevisiae, Pichia pastoris*, and *Escherichia coli*. *Protein Expres. Purif.* 42:255–267.

Nett, J.H., and T.U. Gerngross. 2003. Cloning and disruption of the *PpURA5* gene and construction of a set of integration vectors for the stable genetic modification of *Pichia pastoris*. *Yeast* 20:1279–1290.

Nett, J.H., N. Hodel, S. Rausch, and S. Wildt. 2005. Cloning and disruption of the *Pichia pastoris ARG1, ARG2, ARG3, HIS1, HIS2, HIS5, HIS6* genes and their use as auxotrophic markers. *Yeast* 22:295–304.

Ning, D., X. Junjian, W. Xunzhang, C. Wenyin, Z. Quing, S. Kuanyuan, R. Guirong, R. Xiangrong, L. Quingxin, and Y. Zhouyao. 2003. Expression, purification, and characterization of humanized anti-HBs Fab fragment. *J. Biochem.* 134:813–817.

Ning, D., X. Junjian, Q. Zhang, X. Sheng, C. Wenyin, R. Guirong, and W. Xunzhang. 2005. Production of recombinant humanized anti-HBsAg Fab fragment from *Pichia pastoris* by fermentation. *J. Biochem. Mol. Biol.* 38:294–299.

Ogunjimi, A.A., J.M. Chandler, C.M. Gooding, A. Recinos III, and P.V. Choudary. 1999. High-level secretory expression of immunologically active intact antibody from the yeast *Pichia pastoris*. *Biotechnol. Lett.* 21:561–567.

Omidfar, K., M.J. Rasaee, S. Kashanian, M. Paknejad, and Z. Bathaie. 2007. Studies of thermostability in *Camelus bactrianus* (Bactrian camel) single-domain antibody specific for the mutant epidermal-growth-factor receptor expressed by *Pichia*. *Biotechnol. Appl. Biochem.* 46:41–49.

Pennel, C.A., and P. Eldin. 1998. *In vitro* production of recombinant antibody fragments in *Pichia pastoris*. *Res. Immunol.* 149:599–603.

Pla, I.A., L.M. Damasceno, T. Vannelli, G. Ritter, C.A. Batt, and M.L. Shuler. 2006. Evaluation of Mut^{+} and Mut^{s} *Pichia pastoris* phenotypes for high level extracellular scFv expression under feedback control of the methanol concentration. *Biotechnol. Prog.* 22:881–888.

Powers, D.B., P. Amersdorfer, M.A. Poul, U.B. Nielsen, M.R. Shalaby, G.P. Adams, L.M. Weiner, and J.D. Marks. 2001. Expression of single-chain Fv-Fc fusions in *Pichia pastoris*. *J. Immunol. Methods* 251:123–135.

Rahbarizadeh, F., M.J. Rasaee, M. Forouzandeh, and A.A. Allameh. 2006. Over expression of anti-MUC1 single-domain antibody fragments in the yeast *Pichia pastoris*. *Mol. Immunol.* 43:426–435.

Ridder, R., R. Schmitz, F. Legay, and H. Gram. 1995. Generation of rabbit monoclonal antibody fragments from a combinatorial phage library and their production in the yeast *Pichia pastoris*. *Nature Biotechnol.* 13:255–260.

Romanos, M. 1995. Advances in the use of *Pichia pastoris* for high-level gene expression. *Curr. Opin. Biotechnol.* 6:527–533.

Scorer, C.A., J.J. Clare, W.R. McCombie, M.A. Romanos, and K. Sreekrishna. 1994. Rapid selection using G418 of high copy number transformants of *Pichia pastoris* for high-level foreign gene expression. *Nature Biotechnol.* 12:181–184.

Sharma, S.K., R.B. Pedley, J. Bhatia, G.M. Boxer, E. El-Emir, U. Qureshi, B. Tolner, H. Lowe, N.P. Michael, N. Minton, R.H.J. Begent, and K.A. Chester. 2005. Sustained tumor regression of human colorectal cancer xenografts using a multifunctional mannosylated fusion protein antibody-directed enzyme prodrug therapy. *Clin. Cancer Res.* 11:814–825.

Shi, X., T. Karkut, M. Chamankhah, M. Alting-Mees, S.M. Hemmingsen, and D. Hegedus. 2003. Optimal conditions for the expression of a single-chain antibody (scFv) gene in *Pichia pastoris*. *Protein Expres. Purif.* 28:321–330.

Sreekrishna, K., J.F. Tschopp, and M. Fuke. 1987. Invertase gene (*SUC2*) of *Saccharomyces cerevisiae* as a dominant marker for transformation of *Pichia pastoris*. *Gene.* 59:115–125.

Sunga, A.J., and J.M. Cregg. 2004. The *Pichia pastoris* formaldehyde dehydrogenase gene (*FLD1*) as a marker for selection of multicopy expression strains of *P. pastoris*. *Gene* 330:39–47.

Takahashi, K., T. Yuuki, T. Takai, C. Ra, K. Okumura, T. Yokota, and Y. Okumura. 2000. Production of humanized Fab fragment against human high affinity IgE receptor in *Pichia pastoris*. *Biosci. Biotechnol. Biochem.* 64:2138–2144.

Tang, Y., S. Yang, J. Gariépy, D.A. Scollard, and R.M. Reilly. 2007. Construction and evaluation of the tumor imaging properties of ^{123}I-labeled recombinant and enzymatically generated Fab fragments of the TAG-72 monoclonal antibody CC49. *Bioconjugate Chem.* 18:677–684.

Thor, D., S. Xiong, C.C. Orazem, A.C. Kwan, J.M. Cregg, J. Lin Cereghino, and G.P. Lin Cereghino. 2005. Cloning and characterization of the *Pichia pastoris MET2* gene as a selectable marker. *FEMS Yeast Res.* 5:935–942.

Trentmann, O., N.K. Khatri, and F. Hoffmann. 2004. Reduced oxygen supply increases process stability and product yield with recombinant *Pichia pastoris*. *Biotechnol. Prog.* 20:1766–1775.

Tschopp, J.F., G. Sverlow, R. Kosson, W. Craig, and L. Grinna. 1987. High-level secretion of glycosylated invertase in the methylotropic yeast *Pichia pastoris*. *Nature Biotechnol.* 5:1305–1308.

Wan, H., M. Sjoelinder, H.U. Schairer, and A. Leclerque. 2004. A new dominant selection marker for transformation of *Pichia pastoris* to soraphen A resistance. *J. Microbiol. Meth.* 57:33–39.

Wang, M., L. Stanford Lee, A. Nepomich, J.D. Yang, C. Conover, M. Whitlow, and D. Filpula. 1998. Single chain Fv with manifold *N*-glycans as bifunctional scaffolds for immunomolecules. *Protein Eng.* 11:1277–1283.

Wang, Y., K. Wang, D.C. Jette, and D.S. Wishart. 2001. Production of an anti-prostate-specific antigen single-chain antibody fragment from *Pichia pastoris*. *Protein Expres. Purif.* 23:419–425.

Waterham, H.R., M.E. Digan, P.J. Koutz, S.V. Lair, and J.M. Cregg. 1997. Isolation of the *Pichia pastoris* glyceraldehyde-3-phosphate dehydrogenase gene and regulation and use of its promoter. *Gene.* 186:37–44.

Wegner, G. 1990. Emerging applications of the methylotropic yeasts. *FEMS Microbiol. Rev.* 87:279–284.

Weiss, H.M., W. Haase, H. Michel, and H. Reilaender. 1995. Expression of functional mouse 5-HT_{5A} serotonin receptor in the methylotropic yeast *Pichia pastoris*: pharmacological characterization and localization. *FEBS Lett.* 377:451–456.

White, C.E., M.J. Hunter, D.P. Meininger, L.R. White, and E.A. Komives. 1995. Large-scale expression purification and characterization of small fragments of thrombomodulin: The roles of the sixth domain and of methionine 388. *Protein Eng.* 8:1177–1187.

Wood, C.R., M.A. Boss, J.H. Kenten, J.E. Calvert, N.A. Roberts, and J.S. Emtage. 1985. The synthesis and *in vivo* assembly of functional antibodies in yeast. *Nature* 314:446–449.

CHAPTER 27

Production of Antibody Fab′ Fragments in *E. coli*

DAVID P. HUMPHREYS and LEIGH BOWERING

Therapeutic Monoclonal Antibodies: From Bench to Clinic. Edited by Zhiqiang An

ABSTRACT

Proteolytic and engineered Fab′ fragments are useful as research and diagnostic tools in the laboratory and beneficial as therapeutic and imaging agents in the clinic. Successful clinical use has been demonstrated with the proteolytically derived CroFab®, DigiFab®, Digibind®, and ReoPro® and the microbially produced Lucentis® and Cimzia®. This chapter deals with all aspects of the production and purification of recombinant Fab′ fragments from *Escherichia coli*, including protein engineering, host strain selection, expression vector design, small and large scale expression methods, product harvest, and purification. Some of the most useful end uses and secondary modifications such as PEGylation, multimerization, and conjugation are also highlighted.

27.1 INTRODUCTION

Antibody fragments and *E. coli* production are natural partners and have been in a stable and mutually beneficial and instructive relationship for many years. Fab, Fab′ and $F(ab')_2$ fragments have been produced by proteolytic cleavage of purified polyclonal or monoclonal antibodies since the 1930s (Pope 1939; Porter 1959; Parham et al. 1982). Papain can be used to cleave above the hinge to produce Fab. Pepsin or bromelain can be used to cleave below the hinge to produce $F(ab')_2$ fragments. When these are reduced, Fab′ fragments with two or more free cysteines can also be produced. These methods are still used in research laboratories around the world as a convenient and time efficient means of generating fragments from preexisting purified Ig. The method is also employed on a commercial scale. For example, monoclonal (e.g., ReoPro®) and polyclonal (e.g., DigiFab®) therapeutic and imaging products (e.g., CEA-Scan®) are made using these methods. These processes are multistep, slow, and somewhat wasteful but clearly scalable. The implementation of molecular biology techniques and use of host organisms amenable to fragment expression make it seem unlikely however, that these methods will be used again for the commercial production of monoclonal fragments.

The major strengths of *E. coli* as a production host are the simplicity, variety, and practicality of its expression vectors, its rapid growth (doubling time ~20 to 30 minutes), and ability to be grown on a very large scale (e.g., 20 to 100,000l) in a wide variety of relatively inexpensive defined media. *E. coli* was originally poorly suited to the expression of relatively large, disulfide-rich heterodimeric proteins but many of these problems have now largely been minimized or overcome.

Successful expression of Fab′ fragments in *E. coli* requires a good fit between some basic considerations. Use of a single host cell for both polypeptides, use of a single expression plasmid rather than two, and aiming for soluble periplasmic expression rather than insoluble cytoplasmic expression by fusion to bacterial signal peptides were critical first steps forward toward meaningful expression methods (Boss et al. 1984; Cabilly et al. 1984; Wood et al. 1984). Striking a balance between plasmid copy number, promoter strength, and relative expression of light chain and heavy chain were all simplified after the inception of dicistronic systems (Better et al. 1988; Cabilly 1989; Shibui and Nagahari 1992). Finally, the realization that not all antibodies are equal with regard to expression, as shown by the increase in yield with the move from IgM to IgG Fab, murine to chimeric and then to human/humanized sequences. In contrast to the proteolytic methods, employment of molecular biology techniques also enabled scientists to define precisely the molecular structure of the hinge region of the fragment (Better et al. 1993; Carter et al. 1992; Humphreys et al. 1998; Shibui et al. 1993), thereby introducing the concept of DNA-encoded product homogeneity.

The following years have involved fine tuning and optimization with the general aim of increasing product functionality, yield, and quality or process reliability in order to make *E. coli* expression more robust, user friendly, and cost effective. The bulk of this chapter covers many of these key practical aspects with the dual aims of informing and instructing the reader who might be interested in *E. coli* expression of antibody fragments.

27.2 DESCRIPTION OF ANTIBODY Fab′ FRAGMENTS

Antibody Fab fragments are defined as the antigen binding fragment of an antibody, while the "prime" of Fab′ was originally used to describe Fab containing hinge cysteines. However, with molecular biology and heterologous expression overtaking proteolysis, Fab′ is most commonly used to denote Fab with any form of modified hinge. Dimeric structures are interchangeably described as $F(ab')_2$, di-Fab, or DFM (di-Fab maleimide), depending on their source and method of generation. Fab′ are ~50 kDa heterodimeric proteins composed of one light chain (LC) and one heavy chain (HC). They have a variable N-terminal region (fragment variable, Fv) composed of two immunoglobulin folds, V_L and V_H, respectively. The variable regions each have three hypervariable loops called CDRs (complementarity determining regions) at one end of the protein that together form a common exposed antigen-binding surface. The Fv region is followed by the constant regions, which provide some physical support for the Fv to ensure that the antigen-binding Fv region is kept in the correct tertiary structure—a feature and characteristic that distinguishes Fab′ from Fv and scFv. Light chains can be either of lambda or kappa sequence origin. Both light chain constant regions can be expressed in *E. coli* although the kappa light chain appears to be the most prevalent, possibly reflecting the output from antibody v-region selection strategies. The CH_1 region is antibody class and isotype specific and so can be drawn from the μ, α, γ, δ, or ε classes and their isotypes, $\gamma 1$, $\gamma 2$, $\gamma 3$, and $\gamma 4$. In practice, however, engineered Fab′ fragments are $\gamma 1$. The lack of Fc and hinge encoded activities make all Fab inactive and the linear arrangement of cysteines in $\gamma 1$ Fab′ is found to confer better expression (Humphreys et al. 1996).

27.3 EFFECT OF ANTIBODY SEQUENCE AND STABILITY ON EXPRESSION

It is clear that not all antibodies are equal in terms of their expression and purification. Detailed study has improved our understanding of this area considerably, but this information may be of limited practical use to some researchers. Many will find themselves in the situation of having very few or even only a single antibody that has the desired functional properties and hence this sequence must be produced regardless. Others will find themselves with the relative luxury of being able to select the most appropriate antibody sequence from a panel or even modify a particular sequence armed with this information. Furthermore, since sequence composition may exert more influence over yield in *E. coli* (than hosts such as mammalian cells, for example), the basic factors underlying this are worth highlighting.

At the most basic level, all reports to date suggest that human (or humanized) Fab′s are expressed at higher levels in *E. coli* than murine Fab′s or chimeric murine:human Fab′s. It is not yet entirely clear how much of this apparent difference is due to the fact that a disproportionate amount of effort has been expended in the study, optimization, and fermentation expression of human rather than rodent antibody sequences. As an approximation, human-based Fab′ can be expressed at the 500 to 1000 mg/L level fairly routinely in *E. coli* fermentations (Better et al. 1993; Carter et al. 1992; Humphreys et al. 1998) and as high as 2.4 g/L (Chen et al. 2004), but has also been reported in ~15 mg/L quantities in shake flasks (Corisdeo and Wang 2004; Quintero-Hernandez et al. 2007). In contrast, reports for expression of murine and chimeric Fab tend to be in the 5 to 15 mg/L range from shake flasks and fermentations (Better et al. 1988; Cabilly 1989; Nadkarni, Kelley, and Momany 2007).

Antibody v-regions can be divided into various subgroups based on their sequence relatedness. There are clear differences in the relative stability and expression of these subgroups and this information is of particular importance to researchers who humanize and hence have the opportunity to select a framework for CDR grafting. V_H3 and $V\kappa3$ have the highest stabilities as individual domains but as combinations (Fv) there are several pairs that express well, including combinations of V_H1, V_H3, $V\kappa3$, and $V_\lambda1$-3. Other subgroups such as V_H2, 4, 6 and $V\kappa2$, 4 may in general be less useful for expression in *E. coli*. The CDRs are also known to contribute to the expression/stability profile of Fab′s in *E. coli* but this may be more difficult to alter (Ewert et al. 2003). The interaction of the v-regions with the constant regions and the constant regions with each other has also been studied and has been shown to contribute significantly to Fab′ stability and so potentially to also influence Fab′ yield (Röthlishberger, Honegger, and Plückthun 2005).

A number of studies have demonstrated that the stability of v-regions and consequently Fab′s can be improved. A wide variety of methods, including those based on structural knowledge, random diversity generation followed by selection for improved properties, selection after removal of intradomain disulfide bonds in the reducing *E. coli* or yeast cytoplasm, or combined use of sequence comparisons with targeted mutagenesis have all proved to be successful (Ewert, Honegger, and Plückthun 2003; Jung, Honegger, and Plückthun 1999; Jermutus et al. 2001; Martineau and Betton 1999; Auf Der Maur, Tissot, and Barberis 2004; Worn and Plückthun 2001; Demarest 2006). A positive correlation between stability and yield for non-antibody proteins expressed in *Pichia pastoris* supports this link (Kumita et al. 2006). Researchers with large panels of antibodies might benefit from identification of their v-region subgroups and measurement of their *in vitro* stabilities to enable selection of the most user-friendly molecule. If a limited choice of v-regions is available and they are found to express poorly then modification of the sequence/structure of the Fab′ should be considered seriously.

27.4 EXPRESSION ALTERNATIVES

There are a number of theoretical possibilities for the expression of Fab′ fragments in *E. coli*, one based on a soluble versus insoluble partition and another based on subcellular location. Insoluble expression can be achieved in both the cytoplasm and periplasm, and there are three possible locations for soluble expression: cytoplasm, periplasm, and secretion into the extracellular medium.

Insoluble expression benefits from very high expression yield per cell and per unit time and the fact that inclusion bodies can be purified virtually free from contaminating host proteins. However, this approach necessarily involves *in vitro* refolding, which can be costly in terms of time, reagent, and percent product recovery. Furthermore, the cytoplasm is the site for the highest yielding inclusion body formation and so even after successful refolding the protein may not have the desired N-terminus.

Soluble expression is most often preferred for these reasons. However, one is still faced with choosing the subcellular site for expression. All three locations have strengths and weaknesses but so far periplasmic expression has been shown to have the best overall characteristics for the reasons described below.

27.4.1 Fab′ Fragments Contain Disulfide Bonds

Fab′ fragments contain four structural intramolecular and one dispensable intermolecular disulfide bond. Since the periplasm is the naturally oxidizing subcellular location, peptides must be directed to or through the periplasm if they are to acquire these bonds. *E. coli* strains that are mutant for the cytoplasmic proteins thioredoxin (*trxB*) and glutaredoxin (*gor*) are capable of disulfide formation in the cytoplasm (Prinz et al. 1997). These strains suffer a serious reduction in growth rate unless the double *trxB*/*gor* mutation is accompanied by a compensating mutation in the *ahpC* gene that encodes for a peroxiredoxin (Bessette et al. 1999). Even so these strains are found to be incompatible with use of tetracycline as a selection agent and have generally been found to be low yielding even for expression of scFv (Jurado et al. 2002; Heo et al. 2006). Coexpression of various chaperones or foldases has variously been found to be essential for even low Fab′ yield, ~800 $\mu g/L/OD_{600}$ Fab′ in shake flasks (Levy et al. 2001). The highest yield for Fab′ expression was from the FÅ113 strain, estimated to be ~10 to 30 mg/L in shake flask experiments (Venturi, Seifert, and Hunte 2002).

27.4.2 Retention of a Correct N-Terminus

Expression in the cytoplasm results in the variable retention of the N-terminal initiator methionine (Hexham et al. 2001). This may be a problem for antibody fragments destined for commercial or therapeutic uses or where it affects antigen binding or immunogenicity. Variability can be reduced with additional culturing and enzymatic procedures but both add to process cost and analytical complexity (Sandman, Grayling, and Reeve 1995). Peptides directed to the periplasm by an accurately cleaved N-terminal extension signal peptide avoid this problem. Further information on signal peptide function and choice is given later in this chapter.

27.4.3 Exposure of the Protein to Extracellular Chemical Agents

The periplasm is accessible to media components ≤500 Da in size; hence, the protein can be exposed to chemical agents such as salts, pH, redox molecules, and mild chaotropes. These can be useful for aiding the folding, modification, or experimental investigation of the protein of interest. Additional information on the use of media additions is given later in this chapter.

27.4.4 Simplification of Protein Purification

Poorly expressed proteins become diluted when secreted into the culture medium. This results in the need for high volume capture or filtration/concentration steps. Also, some proteins can become physically damaged by the shear forces generated by some microbial fermentation processes (Harrison, Gill, and Hoare 1998; Mukherjee et al. 2004; Chou et al. 2005). Hence, the periplasm can conveniently act as a means of product encapsulation. Simple harvest of the cells by centrifugation or filtration can result in a significant volumetric concentration. Following harvest, the periplasmic proteins can be released with varying degrees of selectivity by chemical or mechanical extraction methods. The periplasm is the site of expression for approximately one tenth of the soluble proteins of *E. coli* and does not contain any DNA or RNA, which can also interfere with purification. Therefore, combination of periplasmic expression with a carefully considered harvest and extraction regime can result in a very useful Fab′ concentration and partial purification/enrichment.

Secretion into the medium has strong theoretical attractions that have yet to be demonstrated in *E. coli* in a reproducible and practical manner for an ~50 kDa Fab. Secretion of small ~3.5 kDa polypeptides has been demonstrated (Ray et al. 2002) but *E. coli* is not considered to be a professional secretor of large proteins. Proteins such as scFv have been secreted into the medium at high levels by other microbes such as *Saccharomyces cerevisiae* and *Pichia pastoris*, in which case the product may constitute >80 percent of total medium protein. This can greatly simplify the primary capture step. Various attempts have been made to overcome this hurdle in *E. coli*, for example by utilization of the Tat secretion pathway for folded proteins (De Lisa, Tullman, and Geogiou 2003), the α-hemolysin secretion (Fernandez et al. 2000; Sugamata and Shiba 2004) and flagellar secretion systems (Majander et al. 2005). To date these have resulted only in low level secretion of small proteins (~0.1 to 15 mg/L), often while attached to a secretion polypeptide.

27.5 EXPRESSION VECTORS

E. coli fermentations are relatively short in duration (~3 to 4 days), and many plasmids are stably inherited even in the absence of selection. Hence, extrachromosomal plasmids are the normal choice as the vehicle for Fab′ expression. The choice of plasmid components is affected by the characteristics of the Fab′ and the type of expression regime and are detailed below.

27.5.1 Copy Number/Origin of Replication

Small scale (5 to 1000 mL) and rapid induction expression methods (a few hours to overnight) may be best suited to high copy number plasmids (100 to 500 copies per cell, e.g., pUC-based plasmids). In contrast, expression regimes which are longer in duration (such as for fermentations) or where the intention is to express a Fab′ that may be somewhat toxic to the cell, lower copy number plasmids may be favored (20 to 50 copies per cell, e.g., pACYC; Chang and Cohen 1978). In extreme conditions a copy number of 1 can be achieved by integration of the expression cassette into the F′ plasmid. At the opposite extreme, runaway replication plasmids can be used to give copy numbers of ~1000. These may be useful where the Fab′ is to be made as an inclusion body (Nordström and Uhlin 1992). Very high copy number plasmids can be especially useful when combined with leaky promoters such as *trp* (Hallewell and Emtage 1980) that do not require the addition of exogenous chemical inducer.

27.5.2 Promoter Choice

This is perhaps the most obvious plasmid variable affecting expression. Since an excellent and detailed review of promoters can be found elsewhere (Makrides 1996) we will limit ourselves to discussion of the general concepts surrounding promoter choice.

Promoters can be divided into those that are constitutively expressed and those that are inducible. Inducible promoters are generally preferred since biomass can be accumulated prior to protein expression, thereby increasing volumetric yield. Inducible promoters can be further divided into those induced by nonmetabolizable agents (e.g., IPTG, IAA, benzoic acid, Hg^{2+}), those induced by metabolizable agents (e.g., lactose, arabinose) and those induced by agents whose concentration may fluctuate during the fermentation due to volatility or oxidation (e.g., tetracycline, propionate, dicyclopropylketone). It is also possible to use culture conditions such as change in temperature, pH, osmotic stress, depletion of media component such as tryptophan or phosphate to induce protein production.

In practice, one finds that each promoter has its own strengths and weaknesses. For example, metabolizable components generally have the benefits of being natural, safe, and cheap compounds but since they are consumed during the induction process their levels may need to be carefully monitored or controlled. This can add to induction (yield) variability at small scale or process complexity (cost) at production scale. Nonmetabolizable agents may not suffer in this regard but the molecules themselves can be mildly toxic to bacteria. Induction by change of culture conditions can also be problematic. For example where depletion of a medium component is used for induction, timing or coordination with linked process changes can be an issue. The control of physical parameters such as temperature, pH, or salt concentration enable greater temporal control but such approaches may be difficult to implement satisfactorily at large scale and may induce undesirable or variable stress responses. For example, the rapid heating or cooling of large volume fermenters can be difficult from an engineering perspective. Rapid or pronounced changes in critical process parameters such as pH or ionic strength could have profound effects on cell health and product characteristics.

On balance the combined use of the *lac* or *tac* promoters with either lactose or IPTG as inducers is probably the most widely used for Fab′ production. The *lac* promoter uses the wild-type −35 and −10 sequence of the lactose operon, whereas the stronger *tac* promoter uses the −35 sequence of *trp* and the −10 sequence of *lac*. There are two different *tac* sequences, which differ slightly in sequence and strength of induction: *tac*I and *tac*II being 11 times and 7 times stronger, respectively, than *lac* and about 3 times stronger than *trp* but either can suffice for Fab′ expression (De Boer, Comstock, and Vasser 1983). The *phoA* promoter along with depletion of phosphate has also been used to good effect for the production of antibody fragments (Carter et al. 1992; Simmons et al. 2002). However, depletion or reduction of free PO_4 to 0.1–600 μM induces the expression of the proteins in the *pho* regulon (Torriani 1990), which may indirectly increase the metabolic burden on the cells.

PO_4 is also required for many cellular processes and so its depletion may directly result in reduced cell fitness. Amino acid substitutions in the periplasmic scavenger of free PO_4 (PhoS/PstS) are able to desensitize this system such that induction occurs at higher concentrations of free PO_4, thereby potentially alleviating the detrimental effects of PO_4 starvation (Bass and Swartz 1994).

If the Fab′ is poorly tolerated by the cell then the primary need might be for tight preinduction promoter control. Suitable promoters include *tetA* (tetracycline; Skerra 1994), *araBAD* (arabinose; Greenfield, Boone, and Wilcox 1978; Guzman et al. 1995; Morgan-Kiss, Wadler, and Cronan 2002) and the T7/pLys/BL21 (DE3) system (IPTG; Studier 1991), and propionate (Lee and Keasling 2005).

Very strong and rapid induction using the powerful T7 promoter might be preferred for the deliberate production of Fab′ in inclusion bodies. The gene to be expressed is usually under the direct control of a plasmid-borne T7 promoter, whereas the gene encoding for the T7 RNA polymerase is under the control of a lacUV5 promoter placed on the chromosome of a recipient strain. IPTG is used to induce the production of the T7 RNA polymerase and hence indirectly the gene of interest in an amplified manner. In further adaptations the T7 RNA polymerase is under the control of the heat-inducible λP_L and λP_R tandem promoter or the *ara*BAD promoter (Chao et al. 2002; Chao, Chiang, and Hung 2002). Induction in response to a decrease in temperature from $\sim 37°C$ to $28°C$ may be especially useful for automated high throughput Fab expression systems (Trepod and Mott 2002).

27.5.3 Resistance/Selection Marker

Fermentations for the production of therapeutic Fab are generally performed in the absence of antibiotic. However, plasmid selection may be preferred while preparing cell banks and for production runs for nontherapeutic Fab′. The use of chloramphenicol or ampicillin/carbenicillin resistance genes is not favored where large scale production is contemplated. β-lactamase and chloramphenicol acetyl-transferase are produced at very high levels and both are found to accumulate in the medium (probably due to outer membrane leakiness and/or cell lysis) to levels that can deplete medium antibiotics. Bacteriostatic antibiotics such as ampicillin are further penalized by the inability to kill cells that have lost the plasmid or resistance marker. Hence, when antibiotic is depleted from the medium plasmid-free cells can outgrow plasmid-bearing cells, causing a reduction in product yield. The bactericidal antibiotics kanamycin and tetracycline may be preferred due to a combination of these reasons.

27.5.4 Choice of Signal Peptide/Signal Peptide Coding Region

Signal peptides are the mechanism for directing the LC and HC into the periplasm. The important features of signal peptides of the *sec* type are well understood and include:

- Overall length of ~19 to 21 amino acids.
- A short ~5 to 6 amino acid, positively charged N-terminal region due to the amino terminus and the presence of lysine or arginine residues.
- A longer central hydrophobic region of ~8 to 12 amino acids. The C-terminal end of the hydrophobic stretch is often punctuated by a turn-promoting residue such as glycine or proline.
- A C-terminal 5 to 6 amino acid cleavage site with a reasonable degree of sequence conservation. Cleavage occurs after the most C-terminal residue of the tripeptide ^{N}small-X-small^{C} such as ^{N}A-X-A^{C} or ^{N}A-X-G^{C} (Nielsen et al. 1997).

The first mature amino acid after the signal peptide can detrimentally affect the efficiency of cleavage of the signal peptide and hence protein yield (Von Heijne 1990; Laforet and Kendall 1991). Small or acidic side chain amino acids are preferred at this position, whereas large hydrophobic and basic side chains are problematic (Nielsen et al. 1997). Since antibodies are secreted proteins they tend to have reasonably effective mature N-terminal residues. The presence of certain amino acids, particularly arginine and lysine, in the first 20 to 30 amino acids of the mature domain may reduce the efficiency

TABLE 27.1 Protein Sequences of Useful Signal Peptides

Protein	Subcellular Location	Signal Peptide Sequence
β-lactamase, bla	Periplasmic	MSIQHFRVALIPFFAAFCLPVFA
Alkaline phosphatase, AP/PhoA	Periplasmic	MKQSTIALALLPLLFTPVTKA
OmpE, PhoE	Periplasmic	MKKSTLALVVMGIVASASVQA
Phosphate binding protein, PhoS	Periplasmic	MKVMRTTVATVVAATLSMSAFSVFA
M13 major coat protein	Inner membrane	MKKSLVLKASVAVATLVPMLSFA
λ-receptor, LamB	Outer membrane	MMITLRKLPLAVAVAAGVMSAQAMA
Maltose binding protein, MalE	Periplasmic	MKIKTGARILALSALTTMMFSASALA
Outer membrane protein A, OmpA	Outer membrane	MKKTAIAIAVALAGFATVAQA
Outer membrane protein F, OmpF	Outer membrane	MMKRNILAVIVPALLVAGTANA
Pectate lyase B, PelB	Periplasmic/secreted	MKYLLPTAAAGLLLLAAQPAMA
Heat stable toxin II, stII	Excreted	MKKNIAFLLASMFVFSIATNAYA

of secretion to the periplasm and should be examined as a potential cause for low Fab′ yield (Li, Beckwith, and Inouye 1988; Andersson and Von Heijne 1991; Kajava et al. 2000). Similarly, the presence of a double arginine motif that may resemble the Tat secretion motif (S/T-RRXFLK) might also cause reductions of secretion efficiency (Cristobal et al. 1999). If the N-terminus of the mature Fab′ sequence particularly disfavors signal peptide cleavage then reengineering of the mature N-terminus may be tested for retention of antigen affinity. Alternatively, coexpression of signal peptidase I (Lep) may be helpful (Van Dijl et al. 1991).

There are many sequences that can be used as signal peptides. Examples of commonly used signal peptides are shown in Table 27.1. They are generally highly expressed proteins of the periplasm or outer membrane of *E. coli* or the closely related organisms *Erwinia* and *Salmonella* (see Sjöström et al. 1987 for a review). Most commonly, two copies of the same signal peptide are used within one plasmid but examples of using a different signal peptides for LC and HC have been demonstrated (Corisdeo and Wang 2004; Nadkarni, Kelley, and Momany 2007). Due to the high degree of sequence and functional conservation signal peptides can sometimes be interchanged between species (Humphreys et al. 2000b; Tan, Ho, and Ding 2002). Of equal importance to the choice of signal peptide amino acid sequence is the nucleotide composition of the signal peptide. Various factors conspire together to make certain aspects of the nucleotide composition critical to achieving the very highest levels of Fab′ expression in *E. coli* and these are described in more detail below.

27.5.5 Optimization of Translation

A good translation unit is essential if one is to achieve high level Fab′ expression. The consensus sequence for ribosome binding (Shine–Dalgarno site, TAAGGAGGTGATC) and its spacing (5–13 nt) to the initiator codon (ATG $>$ GTG $\gg$ TTG) are well established (Shine and Dalgarno 1974; Ringquist et al. 1992; Gold 1988). Furthermore, codon usage differs between species (Wada et al. 1991) and since genes encoding Fab′ fragments are of non *E. coli* origin (e.g., mouse, rat, rabbit, or synthetic) one might consider codon optimization.

Of primary importance is the codon choice at the 5′ end of the coding region, which in the case of periplasmic Fab′ is the signal peptide coding region. The effect of codon usage here is twofold: ribosomal stalling events caused by rare codons toward the 5′ have a strong impact on the translation rate of the whole mRNA (Goldman et al. 1995) and occlusion of the Shine–Dalgarno site by hairpin loops encoded within the 5′-untranslated region (UTR)-signal peptide region (Wood et al. 1984). It is for the latter reason that 5′ UTRs and 5′ signal peptide codons tend to be rich in A and T nucleotides. The importance and difficulty in the prediction of good signal peptide coding regions have been demonstrated several times by random mutagenesis/selection regimes (Humphreys et al. 2000b; Stemmer et al. 1993; Le Calvez, Green, and Baty 1996; Simmons and Yansura 1996). One should also try to avoid problems caused by multiple repeats of rare codons elsewhere in the coding sequence (see Kane 1995 for a review).

Finally it is worth highlighting that there are also differences in the prevalence and efficiency of the three termination codons, with the order TAA > TAG > TGA generally being observed. These observations can be rationalized by the fact that there are two release factors in *E. coli*, each of which recognizes two stop codons: RF-1 recognizes TAA and TAG, while RF-2 recognizes TAA and TGA. TGA is observed to be the least efficient stop codon and may lead to small amounts of read through and frame shift errors (Macbeath and Kast 1998; Wenthzel, Stancek, and Isaksson 1998). Experiments have also shown that the nucleotides following the stop codon are nonrandom and can exert an additional effect on termination efficiency. TAA(T) appears to be the overall best combination for *E. coli* (Poole, Brown, and Tate 1995; Björnsson, Mottagui-Tabar, and Isaksson 1996).

27.6 EXPRESSION OF MULTIMERIC Fab′ PROTEINS

Multimeric proteins such as Fab′ add an extra level of complexity to the expression problem. The peptides involved in the final 4° structure need to be presented for association in the periplasm in approximately equal functional quantities. The control of the relative expression of two polypeptides can be achieved in three ways.

Dual plasmid expression can be achieved where compatible origins of replication and resistance markers are used and one protein of interest is encoded on each plasmid. For example, the colE1 and pACYC origins can be used with the ampicillin and chloramphenicol resistance markers (Boss et al. 1984; Cabilly et al. 1984; Humphreys et al. 1996). The individual copy number of compatible plasmids present simultaneously within a cell can vary during a culture and antibiotic selection must be maintained at all times—neither of which are attractive features for Fab′ expression.

Dual promoter plasmids encode both LC and HC polypeptides on the same plasmid but under the control of two separate promoters. In one example both polypeptides are under the control of the *tac* promoter (Humphreys et al. 2002) while in another polypeptides are under the control of the *phoA* promoter (Simmons et al. 2002).

Dicistronic plasmids are where the translation units for both LC and HC polypeptides are encoded on the same mRNA under the control of a single promoter. The dicistronic approach has been widely used for Fab′ expression (Better et al. 1993; Carter 1992). See Corisdeo and Wang (2004) for a comparison with dual promoters.

Dicistronic approaches do not allow independent temporal or strength control over the induction of transcription of each polypeptide and this is in contrast to dual promoter approaches. Both polypeptides are expressed simultaneously but some control over the relative expression of each polypeptide can be achieved by engineering of the signal peptide coding regions, intergenic sequence, or other coding elements.

27.7 INFLUENCE OF THE HOST ENVIRONMENT: STRAIN SELECTION AND IMPROVEMENT

Judicious choice and engineering of v-region pairs and use of a sensible expression plasmid should result in useful levels of Fab′ expression in any *E. coli* strain. Experience shows that every Fab′ has its own subtleties which may be managed by careful choice or engineering of the host strain. Many of the gene products involved in the catalysis, assistance, or interference of protein translocation and folding in *E. coli* have been identified. These include proteins involved directly or indirectly in translocation, such as trigger factor, SecA, SecB, SecYEG, SecD, F and YajC (see Mergulhao, Summers, and Monteiro 2005 for a review). Host proteins are also intimately involved in the folding and prevention of aggregation of periplasmic proteins. These include at least seven members of the Dsb protein family which are involved in the formation and isomerization of disulfide bonds, PPIases that catalyze the slow *trans* → *cis* isomerization of peptidyl-prolyl bonds, general chaperones such as FkpA (Richarme and Caldas 1997) and specific chaperones such as Skp and PapD (Jones et al. 1997). A practical review of plasmid resources available for coexpression studies has been compiled (Baneyx and Palumbo 2003).

27.7.1 Influence of Disulfides

Examples which illustrate the importance of efficient disulfide formation and the potential utility of disulfide catalyst expression include the periplasmic expression of Fab′ of the $\gamma 4$ isotype and full length IgG1. Expression yield of $\gamma 4$ Fab′ is lower than that of a comparable $\gamma 1$ Fab′. These two isotypes have different relative positions of their inter-domain and hinge cysteines in the 1° sequence. The lower yield can be rescued by co-expression of human Protein Disulfide Isomerase (PDI), suggesting that disulfide bond isomerization may at times be an issue for *E. coli* (Humphreys et al. 1996). Yields of IgG1 have also been improved by coexpression of DsbA + C (Simmons et al. 2002; Reilly and Yansura 2004).

27.7.2 Periplasmic Chaperones

Skp (OmpH) is a small trimeric protein whose *in vivo* function appears to be the chaperoning and transport of outer membrane proteins (OMPs) across the periplasmic space to the outer membrane (R. Chen and Henning 1996; Schafer, Beck, and Muller 1999). It has been widely used in coexpression studies where the effect on the yield of the protein of interest has been mixed (Bothman and Plückthun 1998; Hayhurt and Harris 1999; Mavrangelos et al. 2001). Coexpression of Skp does, however, commonly result in an increase in the general health of the cells as evidenced by the attainment of higher OD_{600}. This feature might itself be of some practical use when trying to express Fab′ fragments in flask cultures.

27.7.3 Periplasmic PPlases

The presence of *cis* prolines within a Fab′ may reduce its expression since Xaa-Pro peptide bonds are uniformly found in the *trans* form after exiting the ribosome. Seven percent of all prolyl-peptide bonds in folded proteins are *cis* (see Schmid 2002 for a review) and antibodies are known to contain *cis* bonds. Rotation around this bond is sterically restricted and so it can form the rate limiting step during Fab′ folding (Skerra and Plückthun 1991; Jager and Plückthun 1997; Thies et al. 1999). The enzymes that catalyze this rearrangement are called PPIases (peptidyl-prolyl *cis-trans* isomerases) and the periplasm has four such proteins: three soluble proteins (FkpA, SurA, and PpiA) and one inner membrane protein (PpiD).

FkpA and SurA are both known to have dual PPIase and chaperone functions (Missiakas, Betton, and Raina 1996; Ramm and Plückthun 2000; Behrens et al. 2001). Coexpression of FkpA with antibody fragments has been found to be beneficial in some instances (Simmons et al. 2002; Z. Zhang et al. 2003) even when the protein expressed lacks *cis* peptidyl-prolyl bonds (Bothmann and Plückthun 2000). This suggests that the major benefit from FkpA with regard to antibody fragment expression arises from its general chaperone function rather than its PPIase activity. Little is known about the *in vivo* function of PpiA (RotA) although its gene appears to be dispensable without causing perturbation of the periplasmic environment (Kleerebezem, Heutink, and Tommassen 1995). Coexpression of PpiA with scFv was found to have no effect on yield (Knappik, Krebber, and Plückthun 1993). No information about the utility of SurA or PpiD in the context of Fab′ expression is available.

27.7.4 Proteases

Proteins that fold incorrectly in the periplasm maybe subject to proteolysis. In these situations, strains deficient in certain proteases may help to increase the yield of both soluble and insoluble protein. Key proteases have been disrupted specifically to help increase the expression of antibody fragments: the outer membrane OmpT/Protease VII (*ompT*), the periplasmic proteases; Protease III (*ptr3*), DegP (*htrA*), and Tail specific protease/Prc (*tsp*) and even the cytoplasmic Lon protease (*lon*). DegP has dual protease and chaperone functions that can be separated by temperature or a single active site S210A mutation (Spiess, Beil, and Ehrmann 1999). In practice, the deletion of proteases from the genome causes a loss of fitness to the cells that can limit their practical use. This underlines the positive function that proteases perform within the periplasm (Meerman and Georgiou 1994; Park, Georgiou, and Lee 1999). In one Fab′ related example, a spontaneous suppressor mutation (*spr*) was required to overcome such a problem after deletion of the Tsp protease (C. Chen et al. 2004). There are a number of widely available protease-deficient strains that may be of practical use for those experiencing severe expression problems. For example, wild-type B-strains are deficient for the Lon protease and BL-21, which is derived from a B-strain, is additionally deficient in OmpT (Meerman and Georgiou 1994; Studier et al. 1990; Casali 2003).

27.8 ENGINEERED FUNCTIONALITY OF Fab′ FRAGMENTS

While engineering an antibody fragment for expression in *E. coli*, it is also worth considering methods for their detection, purification, multimerization, and for the addition of other functionalities.

27.8.1 Hinge Cysteines, Tags, and Tag Cleavage

The ability to control the number and position of hinge cysteines through protein engineering is a key feature of Fab′. They have proved to be invaluable for the attachment of other functionalities, such as PEG, macrocycles (for the chelation of radionuclides), and other Fab′ [to make homo- or hetero-bifunctional $F(ab')_2$; Humphreys et al. 1998; Glennie et al. 1987], or other proteins, such as enzymes. Some reports suggest that the presence of free cysteines (in scFv) can result in reduced periplasmic yields (Kipriyanov, Moldenhauer, and Little 1997; Schmiedl et al. 2000), while other reports show that free cysteines in the hinge of Fab′ do not affect yield (Carter et al. 1992; Humphreys et al. 1997, 1998, 2007).

If the end use of the Fab′ is nontherapeutic it may be sensible to engineer in N- or C-terminal tags or fusion partners. These can aid with the expression, solubilization, immunodetection, and purification of Fab′. This can be especially useful when the Fab′ is being expressed as part of a high throughput strategy where a uniform and dependable purification and detection is required. There are a wide variety of tags available that range in size and conferred functionality; a list is shown in Table 27.2 (reviewed by Lichty et al. 2005; Esposito and Chatterjee 2006; Leder et al. 2007). If tags are used with Fab′ to be used *in vivo* or for crystallization then it may be necessary to remove the translational fusion following purification but this adds additional steps and cost. Examples of potential cleavage methods are shown in Table 27.3. Residual cleavage site amino acids remain an issue with some of these approaches. An enzymatic procedure using a dipeptidyl peptidase I (DAPase) may be especially well suited to the removal of N-terminal His tags (Arnau et al. 2006). Site-specific chemical cleavage mechanisms can be used, such as CNBr (Haught et al. 1998), *o*-iodobenzoic acid (Hara and Yamakawa 1996), and Cu^{2+} ions (Humphreys et al. 2000a) but they have unresolved technical issues. Self-cleavable inteins use less harsh conditions but the requirement for a cysteine in the cleavage site and exposure to ~50 mM DTT for 16 to 40 h may not be entirely compatible with Fab expression and purification (Chong et al. 1998).

27.9 MEDIA EFFECTS

The outer membrane is porous to solutes ≤500 Da in size. This means that media components can have a direct physical effect on the folding of the polypeptide. A wide range of solutes have been investigated for their ability to improve the expression of heterologous proteins in the periplasm,

TABLE 27.2 Examples of Translational Fusion Partners for Periplasmic Proteins

Name	Uses[a]	Comments	References
Albumin binding motifs	ID, SSP	22–46 amino acid (aa) C-terminal ABD motifs can be used to bind to albumin immobilized on ELISA plates or columns.	Konig and Skerra (1998); Dennis et al. (2002)
Alkaline phosphatase	ID, MM	47 kDa periplasmic protein dimerizes into its active form and so can be used for direct substrate detection.	Wels et al. (1992); Dennis et al. (2002)
c- myc tag	ID, AP	Ab 9E10 binds to EQKLISEEDL	Munro and Pelham (1986); Evan et al. (1985)
Calmodulin binding peptide	ID, SSP	Binds to immobilized calmodulin in the presence of Ca^{2+}, eluted with EGTA; 27 aa minimal binding sequence.	Stofko-Hahn, Carr, and Scott (1992)
Cellulose binding domain	SSP	CBP is a ~45 kDa bacterial exoenzyme that binds to cellulose, eluted with H_2O.	Ong et al. (1989)
Chitin binding domain	SSP	5 kDa binding domain from *Bacillus circulans* binds to immobilized chitin beads.	Chong et al. (1997); Blank et al. (2002)
DsbA	SM	21 kDa periplasmic protein used as a solubilization motif.	Collins-Racie et al. (1995); Y. Zhang et al. (1998)
DsbC	SM	25.6 kDa periplasmic protein used as a solubilization motif.	Z. Zhang et al. (2002)
Ecotin	SM, AP	17 kDa periplasmic protein that can be purified on trypsinogen columns.	Malik et al. (2007)
FLAG tag	ID, AP	DYLDDDDK for "FLAG" or DYKDHDGDYKDHDIDYKDDDDK for "3 × FLAG"; encodes enterokinase cleavage site. Ab M1 sees N-terminal FLAG in a Ca^{2+}-dependent manner, Ab M2 sees N- and C-terminal FLAG and is Ca^{2+} independent.	Brizzard, Chubet, and Vizard (1994); Knappik and Plückthun (1994)
g3 peptide tag	ID, IP, AP	ATDYGAAIDGF bound by Mab 10C31 at 5×10^{-7} M.	Beckmann et al. (1998)
GFP	ID	~30 kDa GFP does not seem to be well expressed or attain its fully fluorescent form in the periplasm but is useful for real time ID.	Yi et al. (2004); Griep et al. (1999)
HA tag	ID, IP, AP	YPYDVPDYA is the target for ID antibodies.	Nice, McInerney, and Jackson (1996)
His tag	ID, AP, SSP, LSP	Inexpensive and rechargeable Ni^{2+} or Cu^{2+} columns are ideal for purifications and antibodies are available for immunodetection.	Xiang et al. (2002); Hochuli et al. (1988)
Lectin	AP, MM	11.8 kDa tetrameric that can be purified on D-mannose agarose.	Tielker et al. (2006)
Leucine zippers and coiled coils	MM	Helical motifs for di- and tetramerization from Fos/Jun and GCN4.	Kostelny, Cole, and Tso (1992); Pack and Plückthun (1992); Pack et al. (1995); Arndt, Muller, and Plückthun (2001)
Maltose binding protein	ID, IP, SSP, SM	MBP is a ~40 kDa periplasmic protein. Binds to immobilized starch, eluted with maltose.	Di Guan et al. (1988); Hayhurst (2000)

(*Continued*)

TABLE 27.2 *Continued*

Name	Uses[a]	Comments	References
PDI	SM	58 kDa endoplasmic reticulum protein	Liu et al. (2005)
Peptide 38 tag	ID, IP, AP	LPSDR motif recognized by antibody 2E11 with 4.9×10^{-6} affinity.	Boldicke et al. (2000)
Polycysteine	SSP, LSP	Tetra-Cys. Potential for disulfide interference or capping.	Persson et al. (1988)
Polyionic peptides	MM	Poly-Lys and Poly-Glu for dimerization.	Richter et al. (2001)
Polyphenylalanine	SSP, LSP	$11\times$ F. Hydrophobicity expression issue.	Persson et al. (1988)
Protein A binding domain	SSP	Z domain of staphylococcal protein A is encoded by ~58 aa in two antiparallel helices. Binds between CH2 and CH3 of Fc with 2×10^{-8} M affinity.	Nilsson et al. (1987)
Protein D	SM	11.6 kDa phage head protein for cytoplasmic expression.	Forrer and Jaussi (1998)
Skp	SM	Soluble 17 kDa periplasmic chaperone	Chatterjee and Esposito (2006)
Strep tag	ID, AP	9 aa StrepI tag AYRHPQFGG and 10 aa StrepII tag SNWSHPQFEK. Affinity for streptavidin 2.7×10^{-4}.	Schmidt and Skerra (1993); Voss and Skerra (1997)
Thermally responsive polypeptides	SSP, LSP	(VPGXG)n motif has a sharp precipitation point above ~32°C which allows high volume enrichment/purification.	Meyer and Chilkoti (1999); Fong et al. (2002)
V5 tag	ID, IP, AP	GKPIPNPLLGLDST is the target for ID antibodies.	Dunn, O'Dowd, and Randall (1999)

[a]ID = immunodetection, IP = immunoprecipitation, AP = affinity purification, SSP = small scale purification, LSP = larger scale purification, SM = solubilization motif, MM = multimerization motif.

including sucrose, raffinose, glycine, betaine, sorbitol, L-arginine, ethylurea, and acetamide. Redox reagents such as glutathione, glutaredoxin, cysteine, cystine, β-mercaptoethanol, and dithiothreitol can be used to alter the redox potential of the periplasm and hence alter the spectrum of disulfide formation and oxidation (Humphreys et al. 1995; Veiga, De Lorenzo, and Fernandez 1999). Hence, special media additions might help to alleviate problems with small scale periplasmic expression of Fab fragments.

27.10 EXTRACTION OF PROTEINS FROM THE PERIPLASM

When desirable levels of expression have been achieved consideration must be given to how best to recover the protein of interest from the periplasm. Advantages of periplasmic expression over extracellular expression are the ability to concentrate the product simply by harvest of the cells by centrifugation or filtration and avoiding release of large quantities of DNA, RNA, and cytoplasmic proteins that may complicate the purification process. The aim of the extraction process is therefore to release the maximum amount of heterologous protein while minimizing the release of other host proteins and biomolecules.

In practice two groups of methods have been employed to try and disrupt or destabilize the outer membrane and cell wall while leaving the inner membrane largely intact: mechanical disruption, such as by sonication, passage through a pressure system such as a French press or Dounce device, freeze-thaw fracture or osmotic shock; or chemical disruption with agents that specifically target the outer membrane and cell wall. Key classes of such molecules include detergents, buffers such as Tris, and peptides that interact with and disrupt the lipid leaflet, metal ion chelating agents that destabilize the outer membrane and cell wall, enzymes such as lysozyme that destroy key structural bonds

TABLE 27.3 Examples of Fusion Partners and Cleavage Regimes for Periplasmic Expression of Antibody Fragments

Name	Cleavage Site	Comments	References
2A protease	KGDIKSY/G	Requires DTT, leaves 1 C-terminal post-cleavage aa.	Walker et al. (1994)
3C protease	E(T or V)LFQ/GP	Requires DTT, leaves 2 C-terminal post-cleavage aa.	Walker et al. (1994)
Cu^{2+} ions	Cleaves at DK/TH or DK/SH	Slow cleavage improved by warm alkaline conditions. DKTH is a natural upper hinge antibody sequence.	Humphreys et al. (2000a)
Cyanogen bromide	Cleaves at Met	Safety issues, derivatization of His, poor specificity.	Haught et al. (1998)
DAPase	N-terminal -XX-	Processive N-terminal exopeptidase for His removal.	Arnau et al. (2006); Arnau, Lauritzen, and Pedersen (2006)
Enterokinase	DDDDK/	Encoded within the FLAG sequence, does not leave C-terminal post-cleavage aa.	Collins-Racie et al. (1995)
Factor Xa protease	I(EorD)GR/X	Commonly used enzyme leaves 1 C-terminal post-cleavage aa.	Nagai and Thogersen (1984)
Inteins	X/C(X)nQ/NA	Self-cleavable under mild reaction conditions, activated by thiol agents.	Chong et al. (1998); Sydor et al. (2002)
o-Iodobenzoic acid	Cleaves at Trp	Very poor specificity.	Hara and Yamakawa (1996)
Subtilisin (modified)	PGAAHY/X	Mutation H64A of subtilisin narrowed the site specifity.	Carter and Wells (1987)
Tev protease	EXXYXQS/	Tobacco etch virus protein does not leave C-terminal post-cleavage aa.	Dougherty et al. (1988); Mondigler and Ehrmann (1996)
Thrombin	LVPR/GS	Leaves 2 C-terminal post-cleavage aa.	McKenzie et al. (1991)

with the cell wall, pH and monovalent cations (for reviews see Harrison, Dennis, and Cahse 1991; Vaara 1992). The most commonly used agents are Tris, EDTA, sucrose, and lysozyme.

A brief overview of some useful methods is shown in Table 27.4. Periplasmic extracts resulting from some of these procedures may not be suitable for use directly in the first stage of the purification or capture step. For example, the presence of excess EDTA may strip Ni^{2+} or Cu^{2+} off columns used for purification of His tagged proteins, although this can be turned to advantage by inclusion of correct amounts of Mg^{2+} (Xiang et al. 2002). The pH or ionic content of the extract may disable binding of the protein to ion exchange matrices and the presence of trace amounts of cell wall debris and nucleic acid can block columns or filter disks. Therefore, certain pretreatments such as dilution, dialysis, buffer exchange, pH adjustment, centrifugation, filtration, or treatment with DNaseI might be necessary before continuing with a purification step. In general these are easy to perform on a laboratory scale, but less so on an industrial scale. Fab′ demonstrates high thermal stability; incubation of the extraction at elevated temperatures as high as ~65°C can preclear a substantial percentage of host proteins, thereby achieving a useful enrichment of heterologous protein (Weir and Bailey 1997).

27.10.1 Tris/EDTA

This method is probably the best to use with large volumes and to achieve a partial enrichment of heterologous over host proteins. Resuspend cells in buffer (100 mM Tris pH 7.4, 10 mM EDTA) and incubate with rapid agitation overnight or for at least 4 hours. Agitation can be performed at

TABLE 27.4 Comparison of Periplasmic Extraction Methods

Method	Volume	Speed	Handling Time	Physical Insult	Notes	References
Tris/EDTA	~0.5 mL–000s L	Slow (hours)	Few minutes	Mild (Overnight agitation)	Can be combined with high temp.	Weir and Bailey (1997)
Sucrose/EDTA	~1–100 mL	Medium (30–60 min)	Few minutes	Gentle (Stand/agitate on ice)	Requires ≥13,000 g centrifugation	Oliver and Beckwith (1982)
Cold osmotic shock	0.5–50 mL	Fast (sample number dependent)	5–15 min	Bursts outer membrane, speed of resuspension limits volume	Speed of pellet resuspension limits volume	Manoil and Beckwith (1986)
Homogenization	~100 mL–000s L	Slow (volume related)	Volume related	Pressure dependent	Requires capital equipment	Harrison, Dennis, and Cahse (1991)

temperatures from 25°C to 65°C to alter the profile of proteins recovered in the periplasmic fraction. Collect the periplasmic supernatant fraction after centrifugation at 13,000 rpm for 10 minutes in microfuge tubes or up to an hour for larger volumes.

27.10.2 Sucrose/EDTA

This method combines mild conditions and minimal manipulation times and so is amenable for medium and large volumes. Wash the pellet from 1 volume of culture with 1 vol. ice cold 10 mM Tris.Cl pH 8.0 and then resuspend in 0.1 to 1 vol. ice cold 30 mM Tris pH 8.0, 20 percent sucrose, 10 mM EDTA. Add 25 μL of lysozyme at 2 mg/mL. Incubate on ice for 30 to 60 minutes with gentle agitation. Centrifuge at 13,000 rpm/4°C in a microfuge for 3 minutes and take the supernatant off as the periplasmic fraction. A cytoplasmic fraction can then be prepared by resuspending the pellet in 500 μL 10 mM Tris.Cl pH 8.0, 100 mM NaCl, 1 mM $MgCl_2$ with fresh DnaseI added to 1 μg/mL. The resuspension is freeze/thawed three times, before centrifugation at 13,000 rpm/4°C in a microfuge for 3 minutes and take supernatant off as cytoplasmic fraction.

27.10.3 Cold Osmotic Shock

This is the best method for getting very clean separation of periplasmic, cytoplasmic, and insoluble (membrane) fractions but it is relatively labor intense and can only reliably be used on small volumes. Attempts have been made to adapt this method to large scale with ~65 percent product recovery (Rathore, Bilbrey, and Steinmeyer 2003). Pellet 1 ml of culture in a microfuge tube for 1 minute and then resuspend in 150 μL of ice cold spheroplast buffer (100 mM Tris.Cl pH 8.0, 500 mM sucrose, 0.5 mM EDTA). Store on ice for 5 minutes before taking a 50 μL whole cell sample. Pellet with 1 minute centrifugation, discard the supernatant, removing the last few drops of liquid with a pipette before resuspending with vigorous pipetting or vortex mixing in 100 μL of ice cold deionized water for 30 seconds. Add 5 μL of ice cold 20 mM $MgCl_2$, mix, then pellet by centrifugation for 2 minutes. Take off the supernatant as periplasmic fraction before resuspending the pellet in 150 μL of ice cold spheroplast buffer and adding 15 μL of fresh lysozyme from a 2 mg/mL stock. Incubate on ice for 5 minutes and then centrifuge for 6 minutes. Carefully remove and discard the supernatant before resuspending the pellet in 600 μL 10 mM Tris.Cl pH 8.0. Freeze thaw the resuspension three times, add

20 μL 1 M $MgCl_2$ and 6 μL DNAseI (fresh stock at 1 mg/mL). Mix to degrade DNA then centrifuge for 25 minutes. The supernatant is removed as the cytoplasmic fraction and the pellet is resuspended in 150 μL of 10 mM Tris.Cl pH 8.0 as the membrane/insoluble fraction.

27.11 EXPRESSION METHODS

Shake flasks provide a convenient means for assessing the yield of recombinant protein achievable from different host strains and expression vectors, and for generating small quantities of material for research use. Growth in shake flasks is generally performed using complex medium containing yeast extract and a peptone such as tryptone or phytone. The medium contains all the nutrients required for *E. coli* growth, and culture in shake flasks can generate cell densities in the range of 2 to 3 OD_{600}. Growth to higher biomass concentrations is limited by oxygen transfer, and the culture will eventually enter stationary phase as growth inhibitory by-products accumulate or as specific nutrients are metabolized and deplete from the medium.

The generation of larger quantities of material requires the use of fermenters, which increase biomass concentration and specific productivity through control of nutrient concentrations and key process parameters such as pH, temperature, and dissolved oxygen concentration (DO; Lin and Swartz 1992). Fermenters can be operated in batch mode, where all the nutrients are included in the medium at the start of the fermentation, or fed-batch mode, where an initial batch phase of growth is followed by the feeding of a concentrated nutrient solution at a controlled rate.

Batch fermentations typically generate cell concentrations of 5 to 10 grams dry cell weight per liter (g DCW/L; Yee and Blanch 1992); however, maximum cell densities are still limited by availability of nutrients provided within the basic culture medium. Complex media are commonly used in batch fermentation and generally support higher specific growth rates than chemically defined media. The nutrient composition and quality of complex media can vary, however, and this can provide a source of process variability.

To generate the high cell densities required to increase productivity to levels needed for commercial manufacture, fed-batch techniques are frequently employed (Yee and Blanch 1992; Lee 1996; Riesenberg and Guthke 1999). In such processes a chemically defined medium (minimal medium) is generally used since the nutrient concentrations are user defined and can be controlled at concentrations optimal for growth or product formation throughout the fermentation. The medium must contain all the necessary nutrients, including sources of carbon, nitrogen, and trace elements essential for *E. coli* metabolism. Glucose or glycerol is generally used as the carbon source and nitrogen may be provided in the form of ammonium hydroxide used for pH control. Examples of defined media that have been used for the production of antibody fragments in *E. coli* are presented in Table 27.5; the final design of the medium will be influenced by the host strain, the target cell density and product yield, the induction mechanism, and the potential impact of specific medium components on product quality. The effect the medium composition can have on the product itself has been shown with Il-2 produced in minimal medium where the encoded methionine residues had been replaced by norleucine (Tsai et al. 1988). A decrease in norleucine misincorporation was observed upon supplementation of the minimal medium with leucine or methionine.

In the initial batch phase of a fed-batch culture, the culture grows at the maximum specific growth rate for the specific strain under the growth conditions employed. In the fed-batch phase growth is often limited by the rate of addition of a key nutrient (normally the carbon source). Such fed-batch strategies can be used to alleviate a number of challenges associated with growth of *E. coli* cultures to high cell density, including substrate inhibition (where nutrients become inhibitory above certain concentrations), the formation of growth inhibitory by-products (such as acetate in *E. coli* cultures growing in the presence of excess glucose or under anaerobic conditions), and limited oxygen transfer or cooling capacity in high cell density cultivations.

A range of feeding strategies have been developed to optimize biomass accumulation and productivity in fed-batch fermentation (reviewed in Yee and Blanch 1992; Lee 1996), and using

TABLE 27.5 Examples of Defined (Minimal) Media Used for Cultivation of *E. coli*

M9 Minimal Media		SM6 Minimal Media		Minimal Media Supplemented with Casein	
Component	Concentration	Component	Concentration	Component	Concentration
Na_2HPO_4	12.8 g L^{-1}	$(NH_4)_2SO_4$	5.2 g L^{-1}	Digested casein	12 g L^{-1}
KH_2PO_4	3.0 g L^{-1}	$NaH_2PO_4 \cdot H_2O$	4.14 g L^{-1}	Glucose	17 mM
NaCl	0.5 g L^{-1}	KCl	4.025 g L^{-1}	Isoleucine hydrochloride	2.4 mM
NH_4Cl	1.0 g L^{-1}	$MgSO_4$	1.04 g L^{-1}	$(NH_4)_2SO_4$	47 mM
$MgSO_4 \cdot 7H_2O$	0.494 g L^{-1}	Citric acid	5.20 g L^{-1}	NaH_2PO_4	10 mM
$CaCl_2 \cdot 2H_2O$	0.0152 g L^{-1}	$CaCl_2 \cdot 2H_2O$	0.0522 g L^{-1}	K_2HPO_4	18 mM
Thiamine	0.01 g L^{-1}	$ZnSO_4 \cdot 7H_2O$	0.0206 g L^{-1}	Trisodium citrate	4.1 mM
$FeSO_4 \cdot 7H_2O$	0.01 g L^{-1}	$MnSO_4 \cdot 4H_2O$	0.0272 g L^{-1}	$MgSO_4$	12 mM
Glucose	2.0 g L^{-1}	$CuSO_4 \cdot 5H_2O$	0.0081 g L^{-1}	$FeCl_3$	125 μM
Common laboratory medium for growth of *E. coli* cultures at small scale to low cell densities		$CoSO_4 \cdot 7H_2O$	0.0042 g L^{-1}	$ZnSO_4$	20 μM
		$FeCl_3 \cdot 6H_2O$	0.1006 g L^{-1}	$MnSO_4$	20 μM
		H_3BO_3	0.0003 g L^{-1}	$CuSO_4$	20 μM
		$Na_2MoO_4 \cdot 2H_2O$	0.0002 g L^{-1}	$CoCl_2$	20 μM
		Glycerol	31.1 g L^{-1}	H_3BO_3	20 μM
				$NaMoO_4$	20 μM
		Fermentation medium. Culture batch fed with glycerol and lactose; final cell density ~90–100 OD_{600}		Fermentation medium. Culture fed with glucose; final cell density ~120–150 OD_{600}	
Reference: Sambrook Fritsch, and Maniatis (1987)		Reference: Humphreys et al. (1998)		Reference: Carter et al. (1992)	

fed-batch techniques cell concentrations as high as 145 g DCW/L have been reported (Horn et al. 1996). A carbon limiting feed may be applied at a constant or increasing (linear, stepwise, or exponential) rate. A constant feed, although simple to operate, will result in a decreasing growth rate, as biomass concentration and culture volume increase. An exponential feed has the advantage that it can be used to control the culture at a constant, optimal specific growth rate. More complex feeding strategies that couple feeding with measurement of physical parameters such as dissolved oxygen (DO) and pH have also been developed. The DO-stat method is based on the observation that DO increases when substrate is depleted. The substrate concentration is maintained within a desired range by the automatic addition of a predetermined amount of nutrient when the DO rises above the preset value (Seo et al. 1992). Similarly, with the pH-stat method, nutrient is fed when pH rises as a result of depletion of the principal carbon source (Kim et al. 2004).

Expression of recombinant proteins imparts a metabolic burden on the host cell, which can result in decreased growth rate and decreased target protein production (Neubauer, Lin, and Mathiszik 2003). The feeding of complex medium components, amino acids, tricarboxylic acid (TCA) cycle intermediates, or vitamins (Ramirez and Bentley 1993; Ramchuran, Holst, and Karlsson 2005) may facilitate recombinant protein expression and improve yields. The most beneficial amino acids to feed may be estimated by comparing the amino acid composition of the *E. coli* host cell with that of the highly expressed product.

As described previously, a number of options are available for the expression of Fab′ fragments in *E. coli*, including expression as insoluble inclusion bodies in the cytoplasm and secretion as soluble protein to the periplasm or supernatant. The choice of expression route will influence the design of both the production strain and the fermentation process. For Fab′ fragments, the preferred expression route is most often expression to the periplasm, where the oxidizing environment promotes correct protein folding and formation of disulfide bonds, and cleavage of the signal peptide gives a consistent and authentic N-terminus. Furthermore the periplasm provides a level of protection from the harsh extracellular environment where shear forces in the fermenter and protease activity may lead to product degradation.

Fermentation processes for the production of antibody fragments in *E. coli* typically involve growing cultures to high cell densities before inducing recombinant protein expression. This allows the culture to direct all activities towards biomass production during the growth phase, and towards product expression during the induction phase. To maximize expression of soluble, correctly folded periplasmic Fab′, it is necessary to tightly control both culture growth and the expression of recombinant protein during the induction phase. Control of these parameters can be achieved through application of a carbon- or nitrogen-limiting feed or through depletion of a nutrient such as phosphate from the medium. This leads to a gradual accumulation of soluble product throughout an extended induction period. Protein folding and soluble product formation can be improved by reducing the fermentation temperature during induction (Cabilly 1989; Shibui and Nagahari 1992). Control of growth and recombinant protein expression has also been shown to increase segregational plasmid stability (Neubauer, Lin, and Mathiszik 2003), which is important for industrial fermentations where antibiotic selection is often avoided in the final production vessel.

Examples of two fermentation processes for the production of soluble Fab′ in the periplasm of *E. coli* are shown in Figure 27.1. In the first example (Humphreys et al. 1998), the culture is initially grown to high cell density in batch mode using glycerol as the carbon source; glycerol is provided both in the initial culture medium and through further batch additions made during the growth phase of the cultivation. At an OD_{600} of ~80, lactose is added to the fermentation medium to induce expression of Fab′ from a *tac* promoter. The process is designed such that carbon utilization switches from glycerol to lactose and growth is controlled by phosphate depletion during the induction phase. The lactose functions as both the carbon source and inducer during the induction phase of the fermentation, and hence

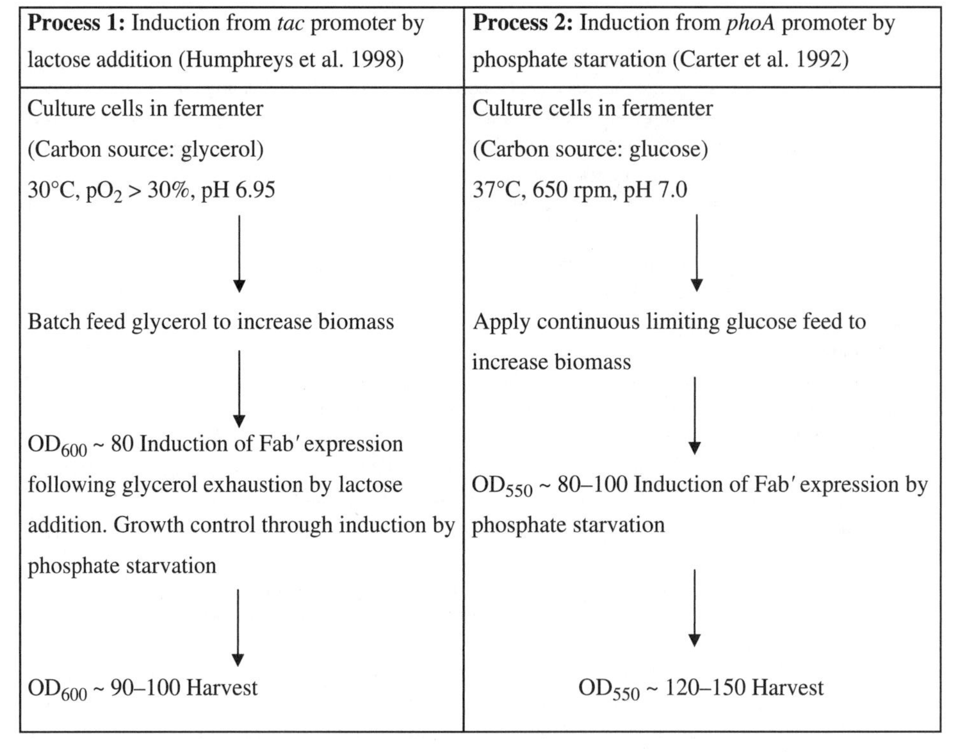

Figure 27.1 Examples of fermentation processes for the production of soluble Fab′ in *E. coli.*

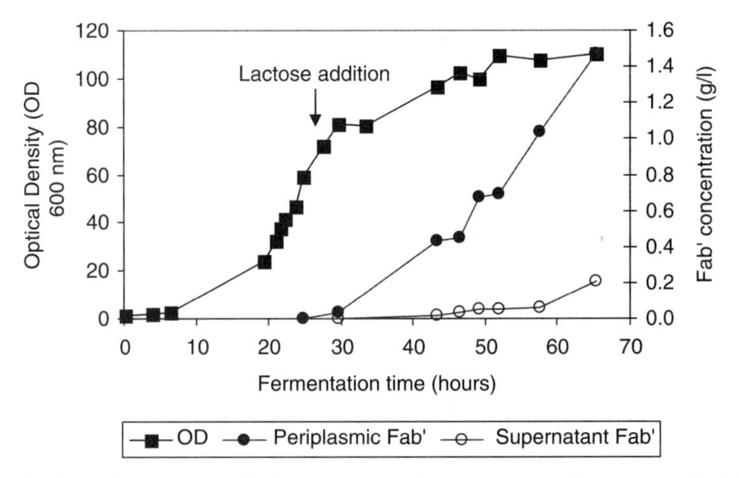

Figure 27.2 Typical culture growth and Fab′ expression profile generated using Process 1 (see Fig. 27.1). Culture initially grown to high cell density in defined medium using glycerol as the carbon source. Induction of Fab′ expression from *tac* promoter by lactose addition at ~80 OD_{600}. Lactose also utilized as sole carbon source during the induction phase of the fermentation.

further additions of lactose are required throughout the induction period. Phosphate starvation during induction results in reduced rates of growth and expression, which facilitate the formation of soluble product. The Fab′ accumulates slowly over an extended induction phase (24 to 36 hours). A typical fermentation profile for this process, showing oxygen uptake rate (OUR), culture growth, and product expression is presented in Figure 27.2.

In the second example (Carter et al. 1992), cells are grown to high biomass using a glucose fed-batch strategy. At an OD_{550} of approximately 80 to 100 expression of Fab′ is induced from a *phoA* promoter by phosphate starvation. Again Fab′ accumulates during induction to give periplasmic yields of 1 to 2 g/L from cell densities of 120 to 150 OD_{550}.

The design of fermentation processes for the production of therapeutic proteins must take into consideration any constraints associated with the intended final scale of operation. High cell density cultures have high oxygen and cooling demands that may be difficult to satisfy at large scale, and reduced mixing efficiency in large vessels may result in culture heterogeneity (Enfors et al. 2001) and CO_2 toxicity. Accurate control of medium composition is more difficult to achieve when large volumes are being prepared, and this may lead to process variability where, for example, depletion of a medium component is used for induction and hence the point of induction is dependent on the initial medium concentration of that component. The requirement to add metabolizable inducers such as lactose may also be problematic at scale as lactose solutions must be kept hot (>60°C) to avoid crystallization in tanks and pipework. Cost of goods (COGs) is another important consideration for commercial processes; this will be impacted by final product titer, time in the production vessel, compatibility with both large scale operation and low cost purification techniques, and effective integration with downstream processing to maximize process yield while minimizing overall process cycle time.

27.12 RECOVERY AND PURIFICATION

27.12.1 Primary Recovery

The methods and sequence of operations used in primary recovery will depend on the subcellular location of the product. For antibody fragments secreted to the extracellular medium, the initial stage of primary recovery will be removal of cells from the process stream. For fragments expressed in the periplasm or as intracellular inclusion bodies, primary recovery will involve cell harvest followed by cell disruption or selective release of periplasmic proteins and subsequently the removal of cells and cell debris.

Separation of cells from the fermentation medium can be achieved using centrifugation and/or filtration. Industrial scale centrifuges can be operated in continuous mode; however, they are less effective at solid-liquid separation than laboratory batch centrifuges. Hence, separation using a continuous disk stack centrifuge will generate a concentrated cell suspension rather than a dry pellet, resulting in carry over of culture supernatant and associated process contaminants during cell harvest, or product losses during cell removal. Furthermore, where centrifugation is being used to remove cells from the process stream prior to application to the initial capture column, continuous centrifugation is unlikely to provide sufficient particulate removal and a further filtration step will be required to protect the column from blocking. The high shear forces generated within the feed zone of continuous centrifuges and upon solids discharge may also damage the product or cause cell breakage and release further unwanted intracellular contaminants into the process stream. For Fab′s expressed as insoluble inclusion bodies, differential centrifugation can be used to separate the inclusion bodies from cell homogenates (Van Hee et al. 2004).

An alternative method to centrifugation for cell removal is tangential flow (or cross-flow) microfiltration. In this mode of operation the feed stream is pumped parallel to the membrane surface and resulting shear forces reduce fouling of the membrane with suspended solids. A number of different formats are available including hollow fiber membranes and flat bed sheets (cassettes). Pore size and membrane composition are important in achieving good separation and membrane pore sizes of 0.1 to 0.2 μm are typically used for the separation of cells or cell debris from fermentation broths and homogenates.

Microfiltration operations typically employ two phases; an initial concentration phase where supernatant is removed and cell concentration increases, followed by a diafiltration stage in which the concentrated cells are washed with buffer or water to improve product recovery. The degree of concentration that can be achieved is limited by the viscosity of the concentrated cell solution, and in the case of high cell density feed streams only a twofold concentration may be achievable (Bowering, unpublished results). Product recovery will improve with increasing diafiltration volumes, although this will result in dilution of the process stream. Optimization of operating parameters such as concentration factor, diafiltration volumes, cross-flow velocity, and transmembrane pressure, and solution conditions such as pH and ionic strength, will be the key to ensuring effective process operation. The impact of membrane costs (and potential for membrane reuse), buffer consumption and process time on overall process time and costs will also need to be taken into consideration when evaluating microfiltration as a clarification step.

As with centrifugation, microfiltration may provide insufficient particulate removal for application of the process stream to a chromatography column. A wide range of filters are available for final clarification, differing in material type, structure, surface charge, and pore size. The final clarification is often best achieved using a combination of both depth and membrane filters; however, the large membrane areas required for certain processes, in combination with difficulties in operator handling of some filter systems, may prove prohibitive for large scale operations (10,000 L and above). Depth and membrane filters are normally single use and hence the requirement for large filter areas can add significant cost to the process. However, use of disposable filter/filter housing units can increase process flexibility and alleviate or remove some of the operator handling difficulties and cleaning issues associated with the use of stainless steel filter housings.

Complete cell disruption or the selective release of periplasmic proteins can be achieved using a variety of methods at small scale, as reviewed in Section 27.10. The simplest methods compatible with large scale operation include homogenization and Tris/EDTA extraction. High pressure homogenization is used for release of intracellular products; however, complete cellular disruption generates a complex mixture of proteins and nucleic acids for subsequent purification. Tris/EDTA extraction disrupts the outer membrane and releases periplasmic proteins without overly compromising inner membrane integrity. This results in a process stream with significantly fewer contaminating host cell proteins and nucleic acids, and simplifies the subsequent purification. Inclusion of a heat treatment step can further enhance purification; correctly assembled Fab′s are stable at high temperatures up to ~80°C in pure form, whereas many contaminating *E. coli* proteins are less stable and form precipitates and/or aggregates that can be removed during centrifugation and/or filtration operations (Weir and Bailey 1997). The efficiency of methods for specific periplasmic release can be improved by combination with a low pressure homogenization step, which assists in destabilizing the outer membrane without completely disrupting the cells, and may also separate cell clumps, allowing the buffer greater access to the cell surface area.

Further conditioning of the process stream for the first purification step may also be required; for example, adjustment of pH and/or conductivity to provide conditions that favor binding of product to the capture adsorbent. This may present challenges at scale, particularly where large volumes need to be conditioned, possibly requiring large tanks for dilutions or adjustments, inline dilution, or diafiltration steps. For fragments expressed as insoluble intracellular inclusion bodies, extensive solubilization and renaturation stages will be required before further purification.

27.12.2 Purification

The objective of the purification process is to recover, concentrate, and purify the desired product to an extent suitable for its intended use. The product must conform to predefined specifications of identity and purity with regard to both process-related and product-related impurities. Typical process-related impurities that must be removed from *E. coli* feed streams include host cell protein (HCP), DNA, and endotoxin. Product-related impurities are molecular variants of the desired product that do not show comparable activity, efficacy, or safety to the desired product. Such impurities may be generated during biosynthesis (through, for example, mistranslation or misincorporation) or may result from degradation (such as proteolytic activity, deamidation, oxidation, or aggregation). These impurities may have very similar physiochemical characteristics to the desired product, and hence their removal can present a challenge.

The design of the purification process will be influenced by the physiochemical properties of the product (including size, charge, and hydrophobicity) and the nature of the process- and product-related impurities. Consideration must also be given to practical and economic considerations, such as affinity and capacity of the resin for the target product, resin cost, resin stability, and the intended final scale of operation. Downstream processing costs usually represent a significant proportion of the total production costs for therapeutic antibodies. Consequently it is desirable to minimize purification costs through use of high capacity, low cost resins that are stable under stringent operation and cleaning conditions and hence compatible with multiple cycles of reuse.

Fab′ fragments can be effectively purified from *E. coli* feed streams using a combination of ion exchange resins that separate molecules based on surface charge differences. Many Fab′ fragments are basic, making cation exchange a suitable choice for the first capture step. Binding and elution conditions will depend on the isoelectric point of the Fab′ and will need to be determined experimentally for each antibody fragment. Conditioning (adjustment of pH and conductivity) of the feed stream may be required to ensure optimal binding to the charged resin. Use of a carefully optimized salt or pH gradient elution can give effective separation of HCP and product variants.

Initial capture is generally performed in packed bed mode using a clarified feed stream. A potentially more efficient alternative is the use of expanded bed adsorption (EBA) to capture product directly from unclarified fermentation broth or cell extract. EBA has the potential to replace multiple process steps, including centrifugation, filtration, concentration, and primary capture, with a single process step, resulting in process simplification and a reduction in cycle time. EBA is currently being used in the commercial manufacture of Cimzia® to recover Fab′ from unclarified cell extracts. The technology is limited, however, by column size (restricted to 1 meter diameter) and the considerable buffer volume requirements at large scale.

Anion exchange chromatography is commonly used as the second step in purification after cation exchange, and is frequently performed in negative or flow through mode. DNA and endotoxin bind strongly to the anion exchange resin while the basic Fab′ is collected in the flow through. Anion exchange membranes have recently been introduced and may be used as an alternative to packed bed chromatography for DNA and endotoxin removal. The membranes provide more rapid contaminant removal due to reduced mass transfer limitations and hence more efficient binding of large molecules.

It is possible to generate pure Fab′ from *E. coli* systems in just two column steps. In cases where further purification is required, hydrophobic interaction chromatography (HIC) can be employed. HIC separates proteins based on differences in hydrophobicity, and a wide variety of resins with different base matrices, ligands, and ligand densities are available. Appropriate choice of matrix and optimization of buffer composition and elution conditions can provide excellent HCP and product variant separation. Salts (such as ammonium sulfate) are usually added to the process stream to achieve binding to hydrophobic resins. However, some Fab′s have a tendency to precipitate or lose binding affinity at high salt concentrations and hence HIC may not be suitable for all antibody fragments.

Size exclusion chromatography (SEC) can be performed as a final polishing step to remove protein aggregates and for buffer exchange into the final drug substance buffer, although this technique can be time consuming and is not readily scalable. Therefore it is preferential to optimize earlier column steps (CEX and HIC) for removal of multimeric protein forms and aggregates, and perform final concentration and buffer exchange using ultrafiltration/diafiltration (UF/DF) operations. UF and DF steps are also frequently positioned between chromatographic separations for protein concentration, buffer exchange, or salt removal.

Some Fab′s can be purified to homogeneity in one step using protein A affinity chromatography, although this is sequence and V-region subgroup dependent (V_H3). Almost all Fab′s can be purified with protein G or protein L ligands. These ligands are ideal for rapid purification in 10 to 500 mg quantities but are not so well suited to large scale commercial production due to their relatively high cost.

Use of synthetic peptide ligands that reversibly bind Fab′ and/or mimic existing antibody-protein interactions (mimotopes) could provide affinity purification matrices that may be cheaper and more robust than protein A/G while also offering a reduction in processing time over multistep nonaffinity methods. Such peptides can be identified by screening phage-displayed, combinatorial peptide libraries with immobilized target protein (Zwick, Shen, and Scott 1998; McConnell et al. 1998). Synthetic versions of the selected peptide can be immobilized on a suitable matrix (Murray et al. 1997). Alternatively combinatorial chemistry libraries can be screened for IgG biomimetic ligands that can also be immobilized on conventional matrices (Teng et al. 2000; Sproule et al. 2000). Affinity columns based on chemical ligands are likely to be more cost effective and robust than those based on synthetic peptides. Development of improved chromatography matrices is, however, largely driven by commercial interests. A reflection of this is that although there have been multiple media improvements in the competitive and lucrative protein A/IgG area, comparatively little progress has been made in media developments aimed specifically at Fab′ purification.

Thiophilic adsorption chromatography has been shown to exert some considerable degree of selectivity for antibodies (Porath, Maisano, and Belew 1985; Belew et al. 1987), and this method has also been adapted for the one step purification of Fab′ (Fiedler and Skerra). Immobilized metal affinity chromatography (IMAC) of Fab containing His tags has traditionally only been performed at small scale. This is in part because large scale periplasmic extraction procedures tend to contain EDTA which may interfere with the chromatography. However, it is possible to use this EDTA to increase the specificity of the binding by including Mg^{2+} in the binding step (Xiang et al. 2002).

27.13 QUALITY ISSUES

27.13.1 Posttranslational Modifications

All proteins, whether they be heterologously or natively expressed, can be subject to a wide variety of posttranslational modifications (PTMs). *E. coli* lacks major enzymatic (and therefore highly efficient) PTMs, such as N-linked and O-linked glycosylation (see Chapter 3). Therefore, the events witnessed

TABLE 27.6 Examples of Potential Posttranslational Modifications of *E. coli* Produced Proteins

PTM	Target Sequence	Notes	±MW	References
N-terminal Met heterogeneity	Met	Met of cytoplasmically expressed Fab may or may not be removed.	−131 Da	Hexham et al. (2001); Sandman, Grayling, and Reeve (1995)
Disulfide isomerization	Cys	Fab′ containing hinge thiols can form an inappropriate LC-hinge disulfide bond or intermolecular disulfides.	0	Humphreys et al. (2007); Kamat, Tolman, and Brown (1996)
Deamidation	Asn > Gln (NG, NS)	Change of Asn → Asp/isoAsp and Gln → Glu results in a change in pI but only a very small one in MW. Rate influenced by sequence, location, pH, and temperature. A major cause of acidic species.	+1	Robinson (2002); Robinson and Robinson (2001); Perkins et al. (2000)
Asp isomerization	Asp	Asp → isoAsp no charge or mass change; can form directly or via succinimide intermediate.	0	Cacia et al. (1996); Harris et al. (2001)
Oxidation	Met > Cys, Trp, Tyr, His	e.g., Methionine R-S-CH_3 → R-S=(O)-CH_3 methyl sulfoxide. Accelerated by light, surfactants (Tween 20), and metal ions, reduced by antioxidants.	+16 Da	Usami et al. (1996); Lam, Yang, and Cleland (1997)
Cysteinylation	Cys	Covalent attachment of reducible cysteine.	+119 Da	Gadgil et al. (2006)
Thioether	Cys-Cys disulfides	Heat-accelerated formation of a nonreducible bond between two cysteines.	−32 Da	Tous et al. (2005)
Acetylation	Surface Lys	Nonenzymatic and sequence-dependent addition of acetyl group.	+42 Da	D'Alayer, Expert-Bezancon, and Beguin (2007)
Glycation	Surface Lys	Nonenzymatic and sequence-dependent addition of glucose and lactose. Loss of Lys +ve charge causes acidic species.	+162 Da	Brady, Martinez, and Balland (2007); Andya et al. (1999)
N-terminal clipping		Incomplete signal peptide removal/aminopeptidase or nonenzymatic cyclization of XXG- to DKP and cleavage at mature N-terminus.	—	Battersby et al. (1994)
N-terminal acylation	His tag	Spontaneous cytoplasmic addition of 6-phosphogluconoyl to GSSHHHHHH tag.	+178 Da +258 Da	Geoghegan et al. (1999)
N-terminal pyroGlu	Gln	Gln → pyroGlu, considered a natural by-product and a cause of basic species.	−18 Da	Powell et al. (1996)
C-terminal clipping	Ala	Removal of one or more amino acids by peptidases or cleavage of hinge region (DKTH) by metal ions.	—	Humphreys et al. (2000a); Smith et al. (1996); Cordoba et al. (2005)
Phosphorylation	Ser, Thr > Tyr, His, Asp	No reports in IgG or Fab′. Possible cause of acidic species.	+80	

with Fab′ tend to be very low abundance, typically <1 percent. As such, they are largely irrelevant to researchers interested in expression of Fab′ for their own *in vitro* research needs. PTMs can be of interest when making very large batches of protein where knowledge about the causes of batch to batch variation can be of commercial interest and will certainly be of interest to those engaged in the production of Fab destined for human *in vivo* uses.

The cause and analysis of PTMs is a very detailed and complex area of its own, some of which has been described in Chapter 23 (see Powell et al. 1996; Wang et al. 2007 for reviews). Free Web-based PTM predictions can also be useful, such as that found at http://ams2.bioinfo.pl/.

However, it is worth highlighting the principal PTMs that have the potential for being observed during the expression, purification, and storage of Fab′ from *E. coli*, shown in Table 27.6.

27.13.2 GMP and Regulatory Considerations

An understanding of the quality control systems and regulatory requirements surrounding production of proteins for human therapeutic use is a necessity in the biopharmaceutical industry. Manufacture of human therapeutics must be performed with strict adherence to Good Manufacturing Practice (GMP) guidelines, and it must be demonstrated that the process is sufficiently controlled to consistently deliver product that meets predefined specifications of identity and purity. Quality control considerations must be given to raw materials and reagents used during the process as these can impact on process performance and product quality. The use of animal-derived raw materials must be avoided wherever possible. Cell banks for use in manufacture must be prepared in accordance with the principles of GMP, and must be characterized for identity, purity, and genetic stability. Process-related impurities and product-related variants must be identified and removal (clearance) through the process must be demonstrated. Furthermore a detailed understanding of the impact of deviations from predefined process operating ranges on both process performance and final product quality must be demonstrated, and key process parameters affecting product quality identified. This will require the development of appropriate analytical tools, and detailed characterization of both the desired product and product-related variants.

27.14 SUMMARY AND PERSPECTIVES

E. coli has been used as the global model for biochemical study and the vehicle for cloning and sequencing since the founding of molecular microbiology. As such it has always been used as a simple model for the study of basic cell biology processes. This resulted in a deep understanding of its biology and ultimately how it could best be put to use as an expression host. From the outset of the biotech era *E. coli* was therefore used for the safe, dependable, and productive expression of research and therapeutic proteins. *E. coli* is currently used for the production of at least 24 approved drugs covering a diverse range of protein classes, including growth hormones, insulins, enzymes, interferons, cytokines, and antibody Fab′ fragments. The expression methods employed cover all of those described in this chapter; insoluble/refolding, soluble cytoplasmic, soluble periplasmic and media/secretion. The most recent additions to this list, Lucentis®, a naked Fab′, and Cimzia®, a PEGylated Fab′, are both made as soluble proteins in the periplasm followed by multistep ion exchange-based column chromatography. They represent the upper end both in terms of size and molecular complexity that has been achieved so far for therapeutic products from *E. coli* and fully qualify the potential that the organism has for delivery of high quality, cost-effective protein.

REFERENCES

Andersson, H., and G. Von Heijne. 1991. A 30-residue-long “export initiation domain” adjacent to the signal sequence is critical for protein translocation across the inner membrane of *Escherichia coli*. *Proc. Natl. Acad. Sci. U.S.A.* 88:9751–9754.

Andya, J.D., Y.F. Maa, H.R. Costantino, et al. 1999. The effect of formulation exipients on protein stability and aeosol performance of spray-dried powders of a recombinant humanized anti-IgE monoclonal antibody. *Pharmaceut. Res.* 16:350–358.

Arnau, J., C. Lauritzen, and J. Pedersen. 2006. Cloning strategy, production and purification of proteins with exopeptidase-cleavable His-tags. *Nature Protocols* 1:2326–2332.

Arnau, J., C. Lauritzen, G.E. Petersen, and J. Pedersen. 2006. Current strategies for the use of affinity tags and tag removal for the purification of recombinant proteins. *Protein Exp. Purif.* 48:1–13.

Arndt, K.M., K. Muller, and A. Plückthun. 2001. Helix-stabilized Fv (hsFv) antibody fragments: Substituting the constant domains of a Fab fragment for a heterodimeric coiled-coli domain. *J. Mol. Biol.* 312:221–228.

Auf Der Maur, A., K. Tissot, and A. Barberis. 2004. Antigen-independent selection of intracellular stable antibody frameworks. *Methods* 34:215–224.

Baneyx, F., and J.L. Palumbo. 2003. Improving heterologous protein folding via molecular chaperone and foldase co-expression. *Meth. Mol. Biol.* 205:171–197.

Bass, S., and J.R. Swartz. 1994. Method of controlling polypeptide production in bacterial cells. U.S. Patent 5304472.

Battersby, J.E., W.S. Hancock, E. Canova-Davis, et al. 1994. Diketopiperazine formation and N-terminal degradation in recombinant human growth hormone. *Intl. J. Pep. Prot. Res.* 44:215–222.

Beckmann, C., B. Haase, K.N. Timmis, and M. Tesar. 1998. Multifunctional g3p-peptide tag for current phage display systems. *J. Immunol. Meth.* 212:131–138.

Behrens, S., R. Maier, H. De Cock, et al. 2001. The SurA periplasmic PPIase lacking its parvulin domains functions *in vivo* and has chaperone activity. *EMBO J.* 20:285–294.

Belew, M., N. Juntti, A. Larsson, and J. Porath. 1987. A one step purification method for monoclonal antibodies based on salt-promoted adsorption chromatography on a "thiophilic" adsorbent. *J. Immunol. Meth.* 102:173–182.

Bessette, P.H., J.J. Cotto, H.F. Gilbert, and G. Georgiou. 1999. *In vivo* and *in vitro* function of the *Escherichia coli* periplasmic cysteine oxidoreductase DsbG. *J. Biol. Chem.* 274:7784–7792.

Better, M., S.L. Bernhard, S.P. Lei, et al. 1993. Potent anti-CD5 ricin A chain immunoconjugates from bacterially produced Fab′ and $F(ab')_2$. *Proc. Natl. Acad. Sci. U.S.A.* 90:457–461.

Better, M., C.P. Chang, R.R. Robinson, and H. Horwitz. 1988. *Escherichia coli* secretion of an active chimeric antibody fragment. *Science* 240:1041–1043.

Björnsson, A., S. Mottagui-Tabar, and L.A. Isaksson. 1996. Structure of the C-terminal end of the nascent peptide influences translation termination. *EMBO J.* 15:1696–1704.

Blank, K., P. Lindner, B. Diefenbach, and A. Plückthun. 2002. Self-immobilizing recombinant antibody fragments for immunoaffinity chromatography: Generic, parallel, and scalable protein purification. *Protein Exp. Purif.* 24:313–322.

Boldicke, T., F. Struck, F. Schaper, et al. 2000. A new peptide-affinity tag for the detection and affinity purification of recombinant proteins with a monoclonal antibody. *J. Immunol. Meth.* 240:165–183.

Boss, M.A., J.H. Kenten, C.R. Wood, and J.S. Emtage. 1984. Assembly of functional antibodies from immunoglobulin heavy and light chains synthesised in *E. coli*. *Nucleic Acids Res.* 12:3791–3806.

Bothman, H., and A. Plückthun. 1998. Selection for a periplasmic factor improving phage display and functional periplasmic expression. *Nature Biotechnol.* 16:376–380.

Bothmann, H., and A. Plückthun. 2000. The periplasmic *Escherichia coli* peptidylprolyl *cis,trans*-isomerase FkpA. *J. Biol. Chem.* 275:17100–17105.

Brady, L.J., T. Martinez, and A. Balland. 2007. Characterization of nonenzymatic glycation on a monoclonal antibody. *Anal. Chem.* 79:9403–9413.

Brizzard, B.L., R.G. Chubet, and D.L. Vizard. 1994. Immunoaffinity purification of FLAG epitope-tagged bacterial alkaline phosphatase using a novel monoclonal antibody and peptide elution. *BioTechniques* 16:730–734.

Cabilly, S. 1989. Growth at sub-optimal temperatures allows the production of functional, antigen-binding Fab fragments in *Escherichia coli*. *Gene* 85:553–557.

Cabilly, S., A.D. Riggs, H. Pande, et al. 1984. Generation of antibody activity from immunoglobulin polypeptide chains produced in *Escherichia coli*. *Proc. Natl. Acad. Sci. U.S.A.* 81:3273–3277.

Cacia, J., R. Keck, L.G. Presta, and J. Frenz. 1996. Isomerization of an aspartic acid residue in the complementarity-determining regions of a recombinant antibody to human IgE: Identification and effect on binding affinity. *Biochemistry* 35:1897–1903.

Carter, P., R.F. Kelley, M.L. Rodrigues, et al. 1992. High level *Escherichia coli* expression and production of a bivalent humanized antibody fragment. *Bio/Technol.* 10:163–167.

Carter, P., and J.A. Wells. 1987. Engineering enzyme specificity by "substrate-assisted catalysis." *Science* 237:394–399.

Casali, N. 2003. *Escherichia coli* host strains. *Meth. Mol. Biol.* 235:27–48.

Casey, J.L., A.M. Coley, et al. 2000. Green fluorescent antibodies: Novel *in vitro* tools. *Protein Eng.* 13:445–452.

Chang, A.C.Y., and S.N. Cohen. 1978. Construction and characterization of amplifiable multicopy DNA cloning vehicles derived from the p15A cryptic miniplasmid. *J. Bacteriol.* 134:1141–1156.

Chao, Y.P., C.J. Chiang, and W.B. Hung. 2002. Stringent regulation and high level expression of heterologous genes in *Escherichia coli* using T7 system controllable by the araBAD promoter. *Biotechnol. Prog.* 18:394–400.

Chao, Y.P., W.S. Law, P.T. Chen, and W.B. Hung. 2002. High production of heterologous proteins in *Escherichia coli* using the thermo-regulated T7 expression system. *Appl. Microbial Biotechnol.* 58:446–453.

Chatterjee, D.K., and D. Esposito. 2006. Enhanced soluble protein expression using two new fusion tags. *Protein Exp. Purif.* 46:122–129.

Chen, C., B. Snedecor, J.C. Nishihara, et al. 2004. High-level accumulation of a recombinant antibody fragment in the periplasm of *Escherichia coli* requires a triple-mutant (*degP prc spr*) host strain. *Biotechnol. Bioeng.* 85:463–474.

Chen, R., and U. Henning. 1996. A periplasmic protein (Skp) of *Escherichia coli* selectively binds a class of outer membrane proteins. *Mol. Microbiol.* 19:1287–1294.

Chong, S., F.B. Mersha, D.G. Comb, et al. 1997. Single-column purification of free recombinant proteins using a self-cleavable affinity tag derived from a protein splicing element. *Gene* 192:271–281.

Chong, S., G.E. Montello, A. Zhang, et al. 1998. Utilizing the C-terminal cleavage activity of a protein splicing element to purify recombinant proteins in a single chromatographic step. *Nucleic Acids Res.* 26:5109–5115.

Chou, D.K., R. Krishnamurthy, T.W. Randolph, J.F. Carpenter, and M.C. Manning. 2005. Effect of Tween20 and Tween80 on the stability of albutropin during agitation. *J. Pharmaceut. Sci.* 94:1368–1381.

Collins-Racie, L.A., J.M. McColgan, K.L. Grant, et al. 1995. Production of recombinant bovine enterokinase catalytic subunit in *Escherichia coli* using the novel secretory fusion partner DsbA. *Bio/Technol.* 13:982–987.

Cordoba, A.J., B.J. Shyong, D. Breen, and R.J. Harris. 2005. Non-enzymatic hinge region fragmentation of antibodies in solution. *J. Chromatogr.* 818:115–121.

Corisdeo, S., and B. Wang. 2004. Functional expression and display of an antibody Fab fragment in *Escherichia coli*: Study of vector designs and culture conditions. *Protein Exp. Purif.* 34:270–279.

Cristobal, S., J.W. De Gier, H. Nielsen, and G. Von Heijne. 1999. Competition between Sec and TAT dependent protein translocation in *Escherichia coli*. *EMBO J.* 18:2982–2990.

D'Alayer, J., N. Expert-Bezancon, and P. Beguin. 2007. Time- and temperature-dependent acetylation of the chemokine RANTES produced in recombinant *Escherichia coli*. 55:9–16.

De Boer, H.A., L.J. Comstock, and M. Vasser. 1983. The *tac* promoter: A functional hybrid derived from the *trp* and *lac* promoters. *Proc. Natl. Acad. Sci. U.S.A.* 80:21–25.

De Lisa, M.P., D. Tullman, and G. Geogiou. 2003. Folding quality control in the export of proteins by the bacterial twin-arginine translocation pathway. *Proc. Natl. Acad. Sci. U.S.A.* 100:6115–6120.

Demarest, S.J., G. Chen, B.E. Kimmel, D. Gustafson, J. Wu, J. Salbato, J. Poland, M. Elia, X. Tan, K. Wong, J. Short, and G. Hansen. 2006. Engineering stability into *Escherichia coli* secreted Fabs leads to increased functional expression. *Protein Eng. Design Select.* 19:325–336.

Dennis, M.S., M. Zhang, Y.G. Meng, et al. 2002. Albumin binding as a general strategy for improving the pharmacokinetics of proteins. *J. Biol. Chem.* 277:35035–35043.

Di Guan, C., P. Li, P.D. Riggs, and H. Inouye. 1988. Vectors that facilitate the expression and purification of foreign peptides in *Escherichia coli* by fusion to maltose-binding protein. *Gene* 67:21–30.

Dougherty, W.G., J.C. Carrington, et al. 1988. Biochemical and mutational analysis of a plant virus polyprotein cleavage site. *EMBO J.* 7:1281–1287.

Ducancel, F., D. Gillet, A. Carrier, et al. 1993. Recombinant colorimetric antibodies: Construction and characterization of a bifunctional F(ab)2/alkaline phosphate conjugate produced in *Escherichia coli. Bio/Technol.* 11:601–605.

Dunn, C., A. O'Dowd, and R.E. Randall. 1999. Fine mapping of the binding sites of the monoclonal antibodies raised against the Pk tag. *J. Immunol. Meth.* 224:141–150.

Enfors, S.O., M. Jahic, A. Rozkov, et al. 2001. Physiological responses to mixing in large scale bioreactors. *J. Biotechnol.* 85:175–185.

Esposito, D., and D.K. Chatterjee. 2006. Enhancement of soluble protein expression through the use of fusion tags. *Curr. Opin. Biotechnol.* 17:353–358.

Evan, G.I., G.K. Lewis, G. Ramsay, and J.M. Bishop. 1985. Isolations of monoclonal antibodies specific for human-myc proto-oncogene product. *Mol. Cell. Biol.* 5:3610–3616.

Ewert, S., A. Honegger, and A. Plückthun. 2003. Structure-based improvement of the biophysical properties of immunoglobulin VH domains with a generalizable approach. *Biochemistry* 42:1517–1528.

Ewert, S., T. Huber, A. Honegger, and A. Plückthun. 2003. Biophysical properties of human antibody variable regions. *J. Mol. Biol.* 325:531–553.

Fernandez, L.A., I. Sola, L. Enjuanes, and V. De Lorenzo. 2000. Specific secretion of active single-chain Fv antibodies into the supernatants of *Escherichia coli* cultures by use of the hemolysin system. *Appl. Environment. Microbiol.* 66:5024–5029.

Fiedler, M., and A. Skerra. 1999. Use of thiophilic adsorption chromatography for the one-step purification of a bacterially produced antibody Fab fragment without the need for an affinity tag. *Protein Exp. Purif.* 17:421–427.

Fong, R.B., Z. Ding, A.S. Hoffman, and P.S. Stayton. 2002. Affinity separation using an Fv antibody fragment-"smart" polymer conjugate. *Biotech. Bioeng.* 79:271–276.

Forrer, P., and R. Jaussi. 1998. High-level expression of soluble heterologous proteins in the cytoplasm of *Escherichia coli* by fusion to the bacteriophage Lambda head protein D. *Gene* 224:45–52.

Gadgil, H.S., P.V. Bondarenko, G.D. Pipes, et al. 2006. Identification of cysteinylation of a free cysteine in the fab region of a recombinant monoclonal IgG1 antibody using Lys-C limited proteolysis coupled with LC/MS analysis. *Anal. Biochem.* 355:165–174.

Geoghegan, K.F., H.B.F. Dixon, P.J. Rosner, et al. 1999. Spontaneous a-N-6-phosphogluconoylation of a "His-tag" in *Escherichia coli*: The cause of extra mass of 258 or 178Da in fusion proteins. *Anal. Biochem.* 267:169–184.

Glennie, M.J., H.M. McBride, A.T. Worth, and G.T. Stevenson. 1987. Preparation and performance of bispecific $F(ab'\gamma)_2$ antibody containing thioether-linked $Fab'\gamma$ fragments. *J. Immunol.* 139:2367–2375.

Gold, L. 1988. Posttranslational regulatory mechanisms in *Escherichia coli. Annu. Rev. Biochem.* 57:199–233.

Goldman, E., A.H. Rosenberg, G. Zubay, and F.W. Studier. 1995. Consecutive low-usage leucine codons block translation only when near the 5′ end of a message in *Escherichia coli. J. Mol. Biol.* 245:467–473.

Greenfield, L., T. Boone, and G. Wilcox. 1978. DNA sequence of the *araBAD* promoter in *Escherichia coli* B/r. *Proc. Natl. Acad. Sci. U.S.A.* 75:4724–4728.

Griep, R.A., C. Van Twisk, et al. 1999. Fluobodies: Green fluorescent single-chain Fv fusion proteins. *J. Immunol. Meth.* 230:121–130.

Guzman, L.M., D. Belin, M.J. Carson, and J. Beckwith 1995. Tight regulation, modulation, and high-level expression by vectors containing the arabinose P_{bad} promoter. *J. Bacteriol.* 177:4121–4130.

Hallewell, R.A., and S. Emtage. 1980. Plasmid vectors containing the tryptophan operon promoter suitable for efficient regulated expression of foreign genes. *Gene* 9:27–47.

Hara, S., and M. Yamakawa. 1996. Production in *Escherichia coli* of Moricin, a novel type antibacterial peptide from the silkworm, *Bombyx mori. Biochem. Biophys. Res. Commun.* 220:664–669.

Harris, R.J., B. Kabakoff, F.D. Macchi, et al. 2001. Identification of multiple sources of charge heterogeneity in a recombinant antibody. *J. Chromatogr. B* 752:233–245.

Harrison, J.S., A. Gill, and M. Hoare. 1998. Stability of a single-chain Fv antibody fragment when exposed to a high shear environment combined with air-liquid interfaces. *Biotech. Bioeng.* 59:517–519.

Harrison, S.T.L., J.S. Dennis, and H.A. Cahse. 1991. Combined chemical and mechanical processes for the disruption of bacteria. *Bioseparation* 2:95–105.

Haught, C., G.D. Davis, R. Subramanian, et al. 1998. Recombinant production and purification of novel antisense antimicrobial peptide in *Escherichia coli*. *Biotech. Bioeng.* 57:55–61.

Hayhurst, A. 2000. Improved expression characteristics of single-chain Fv fragments when fused downstream of the *Escherichia coli* maltose-binding protein or upstream of a single immunoglobulin-constant domain. *Protein Exp. Purif.* 18:1–10.

Hayhurst, A., and W.J. Harris. 1999. *Escherichia coli* Skp chaperone coexpression improves solubility and phage display of single-chain antibody fragments. *Protein Exp. Purif.* 15:336–343.

Heo, M.A., S.H. Kim, S.Y. Kim, et al. 2006. Functional expression of single-chain variable fragment antibody against c-Met in the cytoplasm of *Escherichia coli*. *Protein Exp. Purif.* 47:203–209.

Hexham, J.M., V. King, D. Dudas, et al. 2001. Optimization of the anti-(human CD3) immunotoxin DT389-scFv(UCHT1) N-terminal sequence to yield a homogeneous protein. *Biotechnol. Appl. Biochem.* 34:183–187.

Hochuli, E., W. Bannwarth, H. Dobeli, et al. 1988. Genetic approach to facilitate purification of recombinant proteins with a novel metal chelate adsorbent. *Bio/Technol.* 6:1321–1325.

Horn, U., W. Strittmatter, A. Krebber, et al. 1996. High volumetric yields of functional dimeric miniantibodies in *Escherichia coli*, using an optimized expression vector and high-cell-density fermentation under non-limited growth conditions. *Appl. Microbiol. Biotechnol.* 46:524–532.

Humphreys, D.P., B. Carrington, L.C. Bowering, et al. 2002a. A plasmid system for optimization of Fab′ production in *Escherichia coli*: Importance of balance of heavy chain and light chain synthesis. *Protein Exp. Purif.* 26:309–320.

Humphreys, D.P., A.P. Chapman, D.G. Reeks, et al. 1997. Formation of dimeric Fabs in *Escherichia coli*: Effect of hinge size and isotype, presence of interchain disulphide bond, Fab′ expression levels, tail piece sequences and growth conditions. *J. Immunol. Meth.* 209:193–202.

Humphreys, D.P., S.P. Heywood, A. Henry, et al. 2007. Alternative antibody Fab′ fragment PEGylation strategies: Combination of strong reducing agents, disruption of the interchain disulphide bond and disulphide engineering. *Protein Eng. Design Select.* 20:227–234.

Humphreys, D.P., L.M. King, S.M. West, et al. 2000. Improved efficiency of site-specific copper(II) ion-catalysed protein cleavage effected by mutagenesis of cleavage site. *Protein Eng.* 3:201–206.

Humphreys, D.P., M. Sehdev, A.P. Chapman, et al. 2000b. High-level periplasmic expression in *Escherichia coli* using a eukaryotic signal peptide: Importance of codon usage at the 5′ end of the coding sequence. *Protein Exp. Purif.* 20:252–264.

Humphreys, D.P., O.M. Vetterlein, A.P. Chapman, et al. 1998. $F(ab')_2$ molecules made from *E. coli* produced Fab′ with hinge sequences conferring increased serum permanence times in an animal model. *J. Immunol. Meth.* 217:1–10.

Humphreys, D.P., N. Weir, A. Lawson, A. Mountain, and P.A. Lund. 1996. Co-expression of human protein disulphide isomerase (PDI) can increase the yield of an antibody Fab′ fragment expressed in *Escherichia coli*. *FEBS Letts.* 380:194–197.

Humphreys, D.P., N. Weir, A. Mountain, and P.A. Lund. 1995. Human protein disulfide isomerase functionally complements a *dsbA* mutation and enhances the yield of pectate lyase C in *Escherichia coli*. *J. Biol. Chem.* 270:28210–28215.

Jager, M., and A. Plückthun. 1997. The rate-limiting steps for the folding of an antibody scFv fragment. *FEBS Letts.* 418:106–110.

Jermutus, L., A. Honegger, F. Schwesinger, et al. 2001. Tailoring *in vitro* evolution for protein affinity or stability. *Proc. Natl. Acad. Sci. U.S.A.* 98:75–80.

Jones, C.H., P.N. Danese, J.S. Pinkner, et al. 1997. The chaperone-assisted membrane release and folding pathway is sensed by two signal transduction systems. *EMBO J.* 16:6394–6406.

Jung, S., A. Honegger, and A. Plückthun. 1999. Selection for improved protein stability by phage display. *J. Mol. Biol.* 294:163–180.

Jurado, P., D. Ritz, J. Beckwith, V. De Lorenzo, and L.A. Fernandez. 2002. Production of functional single-chain Fv antibodies in the cytoplasm of *Escherichia coli*. *J. Mol. Biol.* 320:1–10.

Kajava, A.V., S.N. Zolov, A.E. Kalinin, and M.A. Nesmeyanova. 2000. The net charge of the first 18 residues of the mature sequence affects protein translocation across the cytoplasmic membrane of Gram-negative bacteria. *J. Bacteriol.* 182:2163–2169.

Kamat, M.S., G.L. Tolman, and J.M. Brown. 1996. Formulation development of an antifibrin monoclonal antibody radiopharmaceutical. *Pharmaceut. Biotechnol.* 9:343–364.

Kane, J.F. 1995. Effects of rare codon clusters on high-level expression of heterologous proteins in *Escherichia coli. Curr. Biol.* 6:494–500.

Kim, B.S., S. Lee, S.Y. Lee, et al. 2004. High cell density fed-batch cultivation of *Escherichia coli* using exponential feeding combined with pH-stat. *Bioprocess Biosys. Eng.* 26:147–150.

Kipriyanov, S.M., G. Moldenhauer, and M. Little. 1997. High level production of soluble single chain antibodies in small-scale *Escherichia coli* cultures. *J. Immunol. Meth.* 200:69–77.

Kleerebezem, M., M. Heutink, and J. Tommassen. 1995. Characterization of an *Escherichia coli* rotA mutant, affected in periplasmic peptidyl-prolyl *cis/trans* isomerase. *Mol. Microbiol.* 18:313–320.

Kostelny, S.A., M.S. Cole, and J.Y. Tso. 1992. Formation of a bispecific antibody by the use of leucine zippers. *J. Immunol.* 148:1547–1553.

Knappik, A., C. Krebber, and A. Plückthun. 1993. The effect of folding catalysts on the *in vivo* folding process of different antibody fragments expressed in *Escherichia coli. Bio/Technol.* 11:77–83.

Knappik, A., and A. Plückthun. 1994. An improved affinity tag based on the FLAG peptide for the detection and purification of recombinant antobody fragments. *BioTechniques* 17:754–761.

Konig, T., and A. Skerra. 1998. Use of an albumin-binding domain for the selective immobilisation of recombinant capture antibody fragments on ELISA plates. *J. Immunol. Meth.* 218:73–83.

Kumita, J.R., R.J.K. Johnson, M.J.C. Alcocer, et al. 2006. Impact of the native-state stability of human lysozyme variants on protein secretion by *Pichia pastoris. FEBS J.* 273:711–720.

Laforet, G.A., and D.A. Kendall. 1991. Functional limits of conformation, hydrophobicity, and steric constraints in prokaryotic signal peptide cleavage regions. *J. Biol. Chem.* 266:1326–1334.

Lam, X.M., J.Y. Yang, and J.L. Cleland. 1997. Antioxidants for prevention of methionine oxidation in recombinant monoclonal antibody HER2. *J. Pharmaceut. Sci.* 86:1250–1255.

Le Calvez, H., J. Green, and D. Baty. 1996. Increased efficiency of alkaline phosphatase production levels in *Escherichia coli* using a degenerate PelB signal sequence. *Gene* 170:51–55.

Leder, L., F. Freuler, M. Forstner, and L.M. Mayr. 2007. New methods for efficient protein production in drug discovery. *Curr. Opin. Drug Disc. Devel.* 10:193–202.

Lee, S.Y. 1996. High cell density culture of *Escherichia coli. Tibtech.* 14:98–105.

Lee, S.K., and J.D. Keasling. 2005. Propionate-regulated high-yield protein production in *Escherichia coli. Biotech. Bioeng.* 93:912–918.

Levy, R., R. Weiss, G. Chen, et al. 2001. Production of correctly folded Fab antibody fragment in the cytoplasm of *Escherichia coli* trxB gor mutants via the coexpression of molecular chaperones. *Protein Exp. Purif.* 23:338–347.

Li, P., J. Beckwith, and H. Inouye. 1988. Alteration of the amino terminus of the mature sequence of a periplasmic protein can severely affect protein export in *Escherichia coli. Proc. Natl. Acad. Sci. U.S.A.* 85:7685–7689.

Lichty, J.J., J.L. Malecki, H.D. Agnew, et al. 2005. Comparison of affinity tags for protein purification. *Protein Exp. Purif.* 41:98–105.

Lin, N.S., and J.R. Swartz. 1992. Production of heterologous proteins from recombinant DNA *Escherichia* in bench fermentors. *Methods Compan. Meth. Enzymol.* 4:159–168.

Liu, Y., T.J. Zhao, et al. 2005. Increase of soluble expression in *Escherichia coli* cytoplasm by a protein disulfide isomerase gene fusion system. *Protein Exp. Purif.* 44:155–161.

Macbeath, G., and P. Kast. 1998. UGA read through artifacts: When popular gene expression systems need a pATCH. *BioTechniques* 24:789–794.

Majander, K., L. Anton, J. Antikainen, et al. 2005. Extracellular secretion of polypeptides using a modified *Escherichia coli* flagellar secretion apparatus. *Nature Biotechnol.* 23:475–481.

Makrides, S.C. 1996. Strategies for achieving high-level expression of genes in *Escherichia coli. Microbiol. Rev.* 60:512–538.

Malik, A., M. Jenzsch, A. Lubbert, et al. 2007. Periplasmic production of native human proinsulin as a fusion to *E. coli* ecotin. *Protein Exp. Purif.* 55:100–111.

Manoil, C., and J. Beckwith. 1986. A genetic approach to analysing membrane protein topology. *Science* 233:1403–1408.

Martineau, P., and J.M. Betton. 1999. *In vitro* folding and thermodynamic stability of an antibody fragment selected in vivo for high expression levels in *Escherichia coli* cytoplasm. *J. Mol. Biol.* 292:921–929.

Mavrangelos, C., M. Thiel, P.J. Adamson, et al. 2001. Increased yield and activity of soluble single-chain antibody fragments by combining high-level expression and the Skp chaperonin. *Protein Expression Purific.* 23:289–295.

McConnell, S.J., T. Dinh, M.H. Le, et al. 1998. Isolation of erythropoietin receptor agonist peptides using evolved phage libraries. *Biol. Chem.* 379:1279–1286.

McKenzie, K.R., E. Adams, W.J. Britton, et al. 1991. Sequence and immunogenicity of the 70-kDa heat shock protein of *Mycobacterium leprae*. *J. Immunol.* 147:312–319.

Meerman, H.J., and G. Georgiou. 1994. Construction and characterization of a set of *E. coli* strain deficient in all known loci affecting the proteolytic stability of secreted recombinant proteins. *Bio/Technol.* 12:1107–1110.

Mergulhao, F.J.M., D.K. Summers, and G.A. Monteiro. 2005. Recombinant protein secretion in *Escherichia coli*. *Biotechnol. Adv.* 23:177–202.

Meyer, D.E., and A. Chilkoti. 1999. Purification of recombinant proteins by fusion with thermally-responsive polypeptides. *Nature Biotechnol.* 17:1112–1115.

Missiakas, D., J.M. Betton, and S. Raina. 1996. New components of protein folding in extracytoplasmic compartments of *Escherichia coli* SurA, FkpA and Skp/OmpH. *Mol. Microbiol.* 21:871–884.

Mondigler, M., and M. Ehrmann. 1996. Site-specific proteolysis of the *Escherichia coli* SecA protein in vivo. *J. Bacteriol.* 178:2986–2988.

Morgan-Kiss, R.M., C. Wadler, and J.E. Cronan, Jr. 2002. Long-term and homogeneous regulation of the *Escherichia coli araBAD* promoter by use of a lactose transporter of relaxed specificity. *Proc. Natl. Acad. Sci. U.S.A.* 99:7373–7377.

Mukherjee, K.J., D.C.D. Rowe, N.A. Watkins, and D.K. Summers. 2004. Studies of single-chain antibody expression in quiescent *Escherichia coli*. *Appl. Environment. Microbiol.* 70:3005–3012.

Munro, S., and H.R. Pelham. 1986. An hsp70-like protein in the ER: Identity with the 78 kDa glucose-regulated protein and immunoglobulin heavy chain binding protein. *Cell* 46:291–300.

Murray, A., M. Sekowski, D.I.R. Spencer, et al. 1997. Purification of monoclonal antibodies by epitope and mimotope affinity chromatography. *J. Chromatogr. A* 782:49–54.

Nadkarni, A., L.L.C. Kelley, and C. Momany. 2007. Optimization of a mouse recombinant antibody fragment for efficient production from *Escherichia coli*. *Protein Expression Purific.* 52:219–229.

Nagai, K., and H.C. Thogersen. 1984. Generation of b-globin by sequence-specific proteolysis of a hybrid protein produced in *Escherichia coli*. *Nature* 309:810–812.

Neubauer, P., H.Y. Lin, and B. Mathiszik. 2003. Metabolic load of recombinant protein production: Inhibition of cellular capacities for glucose uptake and respiration after induction of a heterologous gene in *Escherichia coli*. *Biotech. Bioeng.* 83:53–64.

Nice, E., T.L. McInerney, and D.C. Jackson. 1996. Analysis of the interaction between a synthetic peptide of influenza virus hemagglutinin and monoclonal antibodies using an optical biosensor. *Mol. Immunol.* 33:659–670.

Nielsen, H., J. Engelbrecht, S. Brunak, and G. Von Heijne. 1997. Identification of prokaryotic and eukaryotic signal peptides and prediction of their cleavage sites. *Protein Eng.* 10:1–6.

Nilsson, B., T. Moks, B. Jansson, et al. 1987. A synthetic IgG-binding domain based on staphylococcal protein A. *Protein Eng.* 1:107–113.

Nordström, K., and B.E. Uhlin. 1992. Runaway-replication plasmids as tools to produce large quantities of proteins from cloned genes in bacteria. *Bio/Technol.* 10:661–666.

Oliver, D.B., and J. Beckwith. 1982. Regulation of a membrane component required for protein secretion in *Escherichia coli*. *Cell* 30:311–319.

Ong, E., N.R. Gilkes, R.A.J. Warren, et al. 1989. Enzyme immobilization using the cellulose-binding domain of a Cellulomonas fimi exoglucanase. *Bio/Technol.* 7:604–607.

Pack, P., K. Muller, R. Zahn, and A. Plückthun. 1995. Tetravalent miniantibodies with high avidity assembling in *Escherichia coli*. *J. Mol. Biol.* 246:28–34.

Pack, P., and A. Plückthun. 1992. Miniantibodies: Use of amphipathic helices to produce functional, flexibly linked dimeric Fv fragments with high avidity in *Escherichia coli*. *Biochemistry* 31:1579–1584.

Parham, P., M.J. Androlewicz, F.M. Brodsky, et al. 1982. Monoclonal antibodies: Purification, fragmentation and application to structural and functional studies of class I MHC antigens. *J. Immunol. Meth.* 53:133–177.

Park, S.J., G. Georgiou, and S.Y. Lee. 1999. Secretory production of recombinant protein by a high cell density culture of a protease negative mutant *Escherichia coli* strain. *Biotechnol. Prog.* 15:164–167.

Perkins, M., R. Theiler, S. Lunte, and M. Jeschke. 2000. Determination of the origin of charge heterogeneity in a murine monoclonal antibody. *Pharmaceut. Res.* 17:1110–1117.

Persson, M., M.G. Bergstrand, L. Bulow, and K. Mosbach. 1988. Enzyme purification by genetically attached polycysteine and polyphenylalanine affinity tails. *Anal. Chem.* 172:330–337.

Poole, E.S., C.M. Brown, and W.P. Tate. 1995. The identity of the base following the stop codon determines the efficiency of an *in vivo* translational termination in *Escherichia coli. EMBO J.* 14:151–158.

Pope, C.G. 1939. The action of proteolytic enzymes on the antitoxins and proteins in immune sera. I. True digestion of the proteins. *Br. J. Exper. Pathol.* 20:132–149.

Porath, J., F. Maisano, and M. Belew. 1985. Thiophilic adorption: A new method for protein fractionation. *FEBS Letts.* 185:306–310.

Porter, R.R. 1959. The hydrolysis of rabbit γ-globulin and antibodies by crystalline papain. *Biochem. J.* 73:119–127.

Powell, M.F., et al. 1996. A compendium and hydropathy/flexibility analysis of common reactive sites in proteins: Reactivity at Asn, Asp, Gln, and Met motifs in neutral pH solution. *Pharmaceut. Biotechnol.* 9:1–140.

Prinz, W.A., F. Aslund, A. Holmgren, and J. Beckwith. 1997. The role of the thioredoxin and glutaredoxin pathways in reducing protein disulfide bonds in the *Escherichia coli* cytoplasm. *J. Biol. Chem.* 272:15661–15667.

Quintero-Hernandez, V., V.R. Juarez-Gonzalez, et al. 2007. The change of the scFv into the Fab format improves the stability and *in vivo* toxin neutralization capacity of recombinant antibodies. *Mol. Immunol.* 44:1307–1315.

Ramchuran, S.O., O. Holst, and E.N. Karlsson. 2005. Effect of postinduction nutrient feed composition and use of lactose as inducer during production of thermostable xylanase in *Escherichia coli* glucose limited fed-batch cultivations. *J. Biosci. Bioeng.* 99:477–484.

Ramirez, D.M., and W.E. Bentley. 1993. Enhancement of recombinant protein synthesis and stability via coordinated amino acid addition. *Biotech. Bioeng.* 41:557–565.

Ramm, K., and A. Plückthun. 2000. The periplasmic *Escherichia coli* peptidylprolyl *cis,trans*-isomerase FkpA. *J. Biol. Chem.* 275:17106–17113.

Rathore, A.S., R.E. Bilbrey, and D.E. Steinmeyer. 2003. Optimization of an osmotic shock procedure for isolation of a protein product expressed in *E. coli. Biotechnol. Prog.* 19:1541–1546.

Ray, M.V.L., C.P. Meenan, A.P. Consalvo, et al. 2002. Production of salmon calcitonin by direct expression of a glycine-extended precurser in *Escherichia coli. Protein Expression Purific.* 26:249–259.

Reilly, D., and D.G. Yansura. 2004. Methods and composition for increasing antibody production. Patent WO2004042017.

Richarme, G., and T.D. Caldas. 1997. Chaperone properties of the bacterial periplasmic substrate-binding proteins. *J. Biol. Chem.* 272:15607–15612.

Richter, S.A., K. Stubenrauch, H. Lilie, and R. Rudolph. 2001. Polyionic fusion peptides function as specific dimerization motifs. *Protein Eng.* 14:775–783.

Riesenberg, D., and R. Guthke. 1999. High cell density cultivation of microorganisms. *Appl. Environment. Microbiol.* 51:422–430.

Ringquist, S., S. Shinedling, D. Barrick, et al. 1992. Translation initiation in *Escherichia coli*: Sequences within the ribosome-binding site. *Mol. Microbiol.* 6:1219–1229.

Robinson, N.E. 2002. Protein deamidation. *Proc. Natl. Acad. Sci. U.S.A.* 99:5283–5288.

Robinson, N.E., and A.B. Robinson. 2001. Prediction of protein deamidation rates from primary and three-dimensional structure. *Proc. Natl. Acad. Sci. U.S.A.* 98:4367–4372.

Röthlishberger, D., A. Honegger, and A. Plückthun. 2005. Domain interactions in the Fab fragment: A comparative evaluation of the single-chain Fv and Fab format engineered with variable domains of different stability. *J. Mol. Biol.* 347:773–789.

Sambrook, J., E.F. Fritsch, and T. Maniatis. 1987. *Molecular Cloning: A Laboratory Manual.* Cold Spring Harbour, NY: Cold Spring Harbour Laboratory Press.

Sandman, K., R.A. Grayling, and J.N. Reeve. 1995. Improved N-terminal processing of recombinant proteins synthesized in *Escherichia coli*. *Bio/Technol.* 13:504–506.

Schafer, U., K. Beck, and M. Muller. 1999. Skp, a molecular chaperone of Gram-negative bacteria, is required for the formation of soluble periplasmic intermediates of outer membrane proteins. *J. Biol. Chem.* 274:24567–24575.

Schmid, F.X. 2002. Prolyl isomerases. *Adv. Protein Chem.* 59:243–282.

Schmidt, T.G.M., and A. Skerra. 1993. The random peptide library-assisted engineering of a C-terminal affinity peptide, useful for the detection and purification of a functional Ig Fv fragment. *Protein Eng.* 6:109–122.

Schmiedl, A., F. Breitling, C.H. Winter, et al. 2000. Effects of unpaired cysteines on yield, solubility and activity of different recombinant antibody constructs expressed in *E. coli*. *J. Immunol. Meth.* 242:101–114.

Seo, D.J., B.H. Chung, Y.B. Hwang, and Y.H. Park. 1992. Glucose limited fed batch culture of *Escherichia coli* for production of recombinant human interleukin-2 with the DO-stat method. *J. Ferment. Bioeng.* 74:196–198.

Shibui, T., K. Munakata, R. Matsumoto, et al. 1993. High-level production and secretion of a mouse-human chimeric Fab fragment with specificity to human carcinoembryonic antigen in *Escherichia coli*. *Appl. Microbiol. Biotechnol.* 38:770–775.

Shibui, T., and K. Nagahari. 1992. Secretion of a functional Fab fragment in *Eschericheria coli* and the influence of culture conditions. *Appl. Microbiol. Biotechnol.* 37:352–357.

Shine, J., and L. Dalgarno. 1974. The 3′ terminal sequence of *Escherichia coli* 16S ribosomal RNA: Complementarity to nonesense triplets and ribosome binding sites. *Proc. Natl. Acad. Sci. U.S.A.* 71:1342–1346.

Simmons, L., L. Klimowski, D.E. Reilly, and D.G. Yansura. 2002. Prokaryotically produced antibodies and uses thereof. WO 02061090.

Simmons, L.C., D. Reilly, L. Klimowski, et al. 2002. Expression of full-length immunoglobulins in *Escherichia coli*: Rapid and efficient production of aglycosylated antibodies. *J. Immunol. Meth.* 263:133–147.

Simmons, L.C., and D.G. Yansura. 1996. Translational level is a critical factor for the secretion of heterologous proteins in *Escherichia coli*. *Nature Biotechnol.* 14:629–634.

Sjöström, M., S. Wold, A. Wieslander, and L. Rilfors. 1987. Signal peptide amino acid sequences in *Escherichia coli* contain information related to final protein localization. A multivariate data analysis. *EMBO J.* 6:823–831.

Skerra, A. 1994. Use of the tetracycline promoter for the tightly regulated production of a murine antibody fragment in *Escherichia coli*. *Gene* 151:131–135.

Skerra, A., and A. Plückthun. 1991. Secretion and *in vivo* folding of the Fab fragment of the antibody McPC603 in *Escherichia coli*: Influence of disulphides and cis-prolines. *Protein Eng.* 4:971–979.

Smith, M.A., M. Easton, P. Everett, et al. 1996. Specific cleavage of immunoglobulin G by copper ions. *Int. J. Pep. Prot. Res.* 48:48–55.

Spiess, C., A. Beil, and M. Ehrmann. 1999. A temperature-dependent switch from chaperone to protease in a widely conserved heat shock protein. *Cell* 97:339–347.

Sproule, K., P. Morrill, J.C. Pearson, et al. 2000. New strategy for the design of ligands for the purification of pharmaceutical proteins by affinity chromatography. *J. Chromatogr. B* 740:17–33.

Stemmer, W.P.C., S.K. Morris, C.R. Kautzer, and B.S. Wilson. 1993. Increased antibody expression from *Escherichia coli* through wobble-base library mutagenesis by enzymatic inverse PCR. *Gene* 123:1–7.

Stofko-Hahn, R.E., D.W. Carr, and J.D. Scott. 1992. A single step purification for recombinant proteins. *FEBS Letts.* 302:274–278.

Studier, F.W. 1991. Use of bacteriophage T7 lysozyme to improve an inducible T7 expression system. *J. Mol. Biol.* 219:37–44.

Studier, F.W., A.H. Rosenberg, J.J. Dunn, and J.W. Dubendorff. 1990. Use of T7 RNA polymerase to direct expression of cloned genes. *Meth. Enzymol.* 185:60–89.

Sugamata, Y., and T. Shiba. 2004. Improved secretory production of recombinant proteins by random mutagenesis of hlyB, and alph-hemolysin transporter from *Escherichia coli*. *Appl. Environment. Microbiol.* 2005:656–662.

Sydor, J.R., M. Mariano, S. Sideris, and S. Nock. 2002. Establishment of intein-mediated protein ligation under denaturing conditions: C-terminal labelling of a single-chain antibody for biochip screening. *Bioconj. Chem.* 13:707–712.

Tan, N.S., B. Ho, and J.L. Ding. 2002. Engineering of a novel secretion signal for cross-host recombinant protein expression. *Protein Eng.* 15:337–345.

Teng, S.F., K. Sproule, A. Husain, and C.R. Lowe. 2000. Affinity chromatography on immobilized "biomimetic" ligands. Synthesis, immobilization and chromatographic assessment of an immunoglobulin G-binding ligand. *J. Chromatogr. B* 740:1–15.

Thies, M.J.W., J. Mayer, J.G. Augustine, C.A. Frederick, H. Lilie, and J. Buchner. 1999. Folding and association of the antibody domain CH3: Prolyl isomerization precedes dimerization. *J. Mol. Biol.* 293:67–79.

Tielker, D., F. Rosenau, K.M. Artels, et al. 2006. Lectin-based affinity tag for one-step protein purification. *BioTechniques* 41:327–332.

Torriani, A. 1990. From cell membrane to nucleotides: The phosphate regulon in *Escherichia coli. BioEssays* 12:371–376.

Tous, G.I., Z. Wei, J. Feng, et al. 2005. Characterization of a novel modification to monoclonal antibodies: Thioether cross-link of heavy and light chains. *Anal. Chem.* 77:2675–2682.

Trepod, C.M., and J.E. Mott. 2002. A spontaneous runaway vector for production-scale expression of bovine somatotropin from *Escherichia coli. Appl. Microbiol. Biotechnol.* 58:84–88.

Tsai, L.B., H.S. Lu, W.C. Kenney, et al. 1988. Control of misincorporation of *de novo* synthesized norleucine into recombinant interleukin-2 in *E. coli. Biochem. Biophys. Res. Commun.* 156:733–739.

Usami, A., A. Ohtsu, S. Takahama, and T. Fujii. 1996. The effect of pH, hydrogen peroxide and temperature on the stability of human monoclonal antibody. *J. Pharmaceut. Biomed. Anal.* 14:1133–1140.

Vaara, M. 1992. Agents that increase the permeability of the outer membrane. *Microbiol. Rev.* 56:395–411.

Van Dijl, J.M., A. De Jong, H. Smith, et al. 1991. Signal peptidase I overproduction results in increased efficiencies of export and maturation of hybrid secretory proteins in *Escherichia coli. Mol. Gen. Genetics* 227:40–48.

Van Hee, P., A.P.J. Middelberg, et al. 2004. Relation between cell disruption conditions, cell debris particle size, and inclusion body release. *Biotech. Bioeng.* 88:100–110.

Veiga, E., V. De Lorenzo, and L.A. Fernandez. 1999. Probing secretion and translocation of a β-autotransporter using a reporter single-chain Fv as a cognate passenger domain. *Mol. Microbiol.* 33:1232–1243.

Venturi, M., C. Seifert, and C. Hunte. 2002. High level production of functional antibody Fab fragments in an oxidising bacterial cytoplasm. *J. Mol. Biol.* 315:1–8.

Von Heijne, G. 1990. The signal peptide. *J. Memb. Biol.* 115:195–201.

Voss, S., and A. Skerra. 1997. Mutagenesis of a flexible loop in streptavidin leads to higher affinity for the Strep-tag II peptide and improved performance in recombinant protein purification. *Protein Eng.* 10:975–982.

Wada, K.N., Y. Wada, H. Doi, et al. 1991. Codon usage tabulated from the GenBank genetic sequence data. *Nucleic Acids Res.* 19:1981–1986.

Walker, P.A., L.E.C. Leong, P.W.P. Ng, et al. 1994. Efficient and rapid affinity purification of proteins using recombinant fusion proteases. *Bio/Technol.* 12:601–605.

Wang, W., S. Singh, D.L. Zeng, K. King, and S. Nema. 2007. Antibody structure, instability, and formulation. *J. Pharmaceut. Sci.* 96:1–26.

Weir, A.N.C., and N.A. Bailey. 1997. Process for obtaining antibodies utilizing heat treatment, Celltech Therapeutics Ltd. U.S. Patent 5665866.

Wels, W., I.M. Harwerth, M. Zwickl, et al. 1992. Construction, bacterial expression and characterization of a bifunctional single-chain antibody-phosphatase fusion protein targeted to the human erbB-2 receptor. *Bio/Technol.* 10:1128–1132.

Wenthzel, A.M.K., M. Stancek, and L.A. Isaksson. 1998. Growth phase dependent stop codon read through and shift of translation reading frame in *Escherichia coli. FEBS Letts.* 421:237–242.

Wood, C.R., M.A. Boss, T.P. Patel, and J.S. Emtage. 1984. The influence of messenger RNA secondary structure on expression of an immunoglobulin heavy chain in *Escherichia coli. Nucleic Acids Res.* 12:3937–3950.

Worn, A., and A. Plückthun. 2001. Stability engineering of an antibody single-chain Fv fragments. *J. Mol. Biol.* 305:989–1010.

Xiang, H., R. Wynn, L.H.T. Nguyen, et al. 2002. Immobilized metal-ion affinity chromatography of recombinant Fab protein OPG C11 in the presence of EDTA-Mg(II). *J. Chromatogr. A* 978:153–164.

Yee, L., and H.W. Blanch. 1992. Recombinant protein expression in high cell density fed-batch cultures of *Escherichia coli. Bio/Technol.* 10:1550–1556.

Yi, K.S., J. Chung, K.H. Park, et al. 2004. Expression system for enhanced green fluorescence protein conjugated recombinant antibody fragment. *Hybridoma and Hybridomics* 23:279–286.

Zhang, Y., D.R. Olsen, K.B. Nguyen, et al. 1998. Expression of eukaryotic proteins in soluble form in *Escherichia coli. Protein Exp. Purif.* 12:159–165.

Zhang, Z., Z.H. Li, F. Wang, et al. 2002. Overexpression of DsbC and DsbG markedly improves soluble and functional expression of single-chain Fv antibodies in *Escherichia coli. Protein Exp. Purif.* 26:218–228.

Zhang, Z., L.P. Song, M. Fang, et al. 2003. Production of soluble and functional engineered antibodies in *Escherichia coli* improved by FkpA. *BioTechniques* 35:1032–1042.

Zwick, M.B., J. Shen, and J.K. Scott. 1998. Phage-displayed peptide libraries. *Curr. Opin. Biotechnol.* 9:427–436.

CHAPTER 28

Production of Human Therapeutic Monoclonal Antibodies in Chicken Eggs

LEI ZHU and ROBERT J. ETCHES

ABSTRACT

We have produced a human therapeutic monoclonal antibody in chicken eggs using the regulatory sequences of the ovalbumin locus and coding regions derived from the human IgG1 locus. The antibody is produced in the tubular gland cells of the chicken reproductive tract and secreted into egg white as the egg is formed. Glycosylation of the antibody lacks fucose and this attribute confers enhanced antibody-dependent cellular cytotoxicity (ADCC). All other sugar residues in the antibody are shared with native human antibodies, indicating that the material is unlikely to be antigenic in patients. Up to 6 mg of antibody is deposited per egg, which is purified using precipitation and a protein A

Therapeutic Monoclonal Antibodies: From Bench to Clinic. Edited by Zhiqiang An

affinity column. The antibody transgene was inserted into the genome of chicken embryonic stem cells (see van de Lavoir et al. 2006b) and high-grade somatic chimeras were made to produce the antibody. Greater yields of antibody can now be made by inserting the antibody transgene into chicken primordial germ cells to make transgenic chickens in which the entire population of tubular gland cells secretes the therapeutic antibody product (see van de Lavoir et al. 2006a).

28.1 INTRODUCTION

The emergence of new technologies to rapidly generate human antibodies and the success of antibodies as therapeutics have led to the realization of the tremendous potential for monoclonal antibodies (mAbs) as a class of human therapeutics. Currently, 21 therapeutic mAbs or fragments thereof are approved by the U.S. Food and Drug Administration and more than 150 therapeutic mAbs are in the clinical study pipeline. As more antibodies are approved, the volume of annual sales is expected to surpass the current value of $20 billion (Sheridan 2007). The demand for mAbs is likely to increase at a rate that exceeds the capacity of traditional production systems such as Chinese hamster ovary (CHO) cells. Furthermore, the financial and logistical challenges of constructing large cell culture facilities create a very high barrier to the entry of potential therapeutic mAbs into the advanced stages of the approval process. Many technologies have been proposed for production of therapeutic mAbs (see Chapters 26, 27, and 29; Pollock et al. 1999). Here, we will discuss the advantages of using the chicken tubular gland cells to produce mAbs as a component of egg white.

28.2 EGG WHITE AND THE OVALBUMIN PROMOTER

The egg white component of the average chicken egg contains approximately 3 g of protein. Slightly more that half of this protein is ovalbumin (Ov) which is a member of the serpin family of proteins (Benarafa and Remold-O'Donnell 2005). Ov is produced by the tubular gland cells of the chicken oviduct (see Fig. 28.1),where it is stored in secretory granules. As the yolk traverses the magnum region of the oviduct, the tubular gland cells release their contents and 3 g of thick albumen is deposited around the yolk. Chickens lay 300 to 320 eggs annually and therefore, the tubular gland cells produce approximately 1 kg of egg white protein per year, of which approximately 500 g is Ov. This prodigious production of protein provides a unique opportunity to harness the capacity to direct large amounts of tissue-specific protein synthesis into the egg to produce novel products. In this chapter we describe the molecular and cellular biology that we employed to generate chickens producing mAbs in their tubular gland cells and the properties of the antibodies that affect their utility as therapeutics.

The Ov locus is located in a duplicated region of chromosome 2 that includes the coding and regulatory regions of several serpins (Fig. 28.2). Within this family, the Ov locus appears to have been duplicated to form the Y and X loci that lie upstream from the Ov locus. Precise regulation of the Ov locus is unknown although several studies have identified components that contribute to tissue specific and hormonally induced expression. For example, high level, tissue-specific expression of the Ov gene in laying hens is stimulated by the administration of both steroid hormones (i.e., estrogens, progestins, androgens, and glucocorticoids) and nonsteroid hormones (i.e., insulin) (Evans and McKnight 1984; McKnight 1978; Sanders and McKnight 1985). Regulation of Ov gene expression was studied in detail by DNase I hypersensitivity analysis and promoter assays in primary tubular gland cell cultures. Four DNase I hypersensitive (DH-I, -II, -III, and -IV) sites were identified at -0.15, -0.8, -3.2, and -6.0 kb from the transcription start site of the Ov gene (Kaye et al. 1984, 1986). The DH-I site contains a negative-response element (NRE) that represses transcription in the absence of steroids and activates transcription in the presence of steroids by cooperating with a steroid-dependent response element (SDRE) present in the DH-II site (Kaye et al. 1986). An estrogen responsive element (ERE) that consists of four half-palindromic motifs was identified in the DH-III site located approximately 3.3 kb upstream from the transcription start site (Kato et al. 1992) and

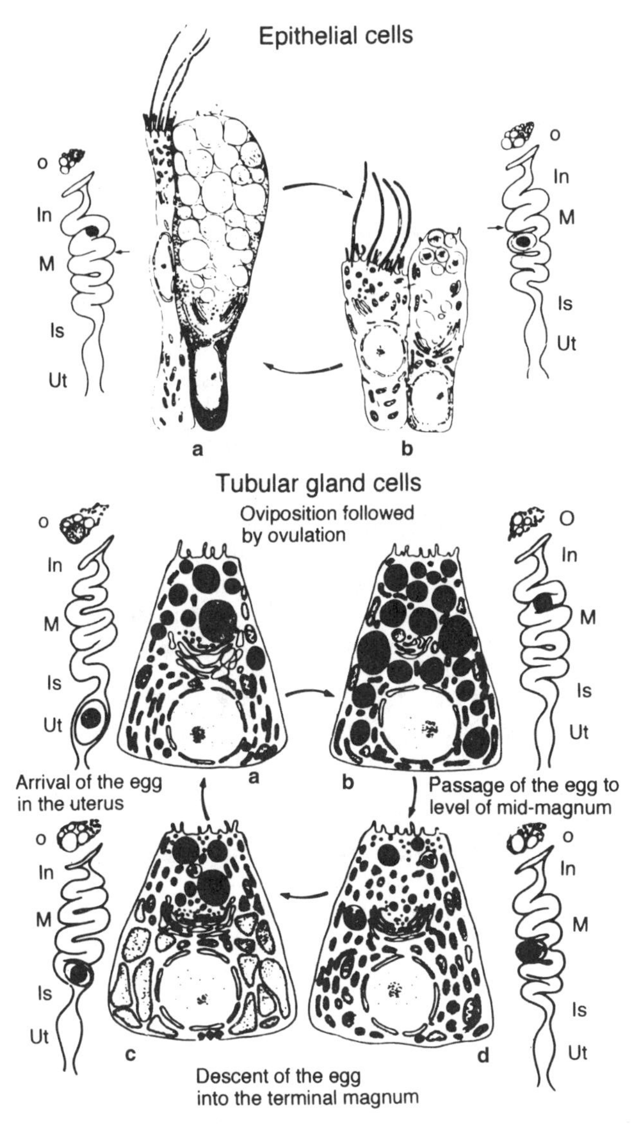

Figure 28.1 A diagrammatic representation of the changes in the secretory cells of the epithelium (upper panel) and the tubular gland cells (lower panel) of the magnum throughout the ovulatory cycle. As the egg enters the magnum, the epithelial cells are 20 to 30 μm in height (upper panel, **a**) and the nonsecretory ciliated cells are compressed between the nonciliated cells, which are packed with secretory granules at their apical surface. When an egg has traversed the magnum (upper panel, **b**), the secretory granules have discharged most of their contents and the epithelium is 13 to 18 μm in height. The histological characteristics of the cells in the tubular glands indicate that they produce egg white proteins continuously throughout the ovulatory cycle and store the proteins in secretory granules (shown in black in the lower panel, **a** and **b**). Most of the preformed protein is released as the egg passes through the magnum (lower panel, **d**) and the rough endoplasmic reticulum (ER) becomes more conspicuous within an hour after the egg leaves this region (lower panel, **c**). The proteins produced in the ER are packaged by the Golgi apparatus into secretory granules, which are most abundant as the egg enters the magnum. (Taken from Etches, R.J. 1996. *Reproduction in Poultry*. Wallingford, UK: CAB International. Used with permission.)

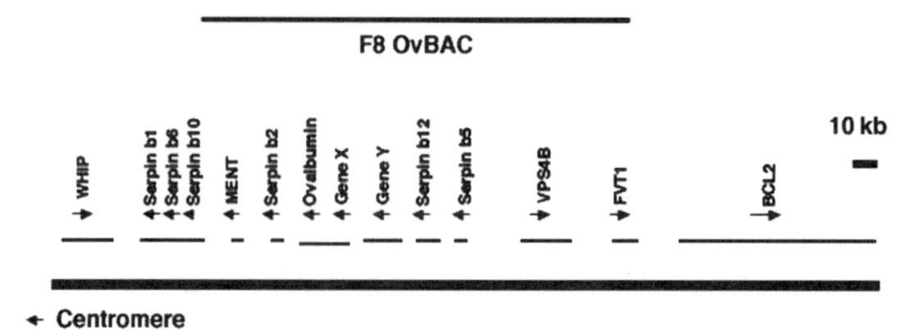

Figure 28.2 The chicken ov-serpin locus and the relative location of F8 OvBAC in the genome. The 10 Ov-serpins and flanking genes with their respective orientation on chicken chromosome 2q are shown. The serpin cluster is flanked downstream by the Werner helicase-interacting protein (*WHIP*) and upstream by the vacuolar sorting protein 4b (*VPS4B*), follicular lymphoma variant translocation 1 (*FVT1*), and B-cell leukemia/lymphoma 2 (*BCL2*). (Adapted from Benefara, C., and E. Remold-O'Donnell. 2005. *Proc. Natl. Acad. Sci. U.S.A.* 102: 11367–11372.)

this ERE was included in one of the vectors used by Lillico et al. (2007; see Section 28.4). An additional tissue-specific silencer-like element residing between −3.2 and −2.8 kb of the promoter region was identified by gene gun-mediated *in vivo* analysis of the Ov promoter in the liver and oviduct of laying hens (Muramatsu et al. 1998). In summary, these regions contribute to oviduct specificity and steroid induction of Ov gene expression in laying hens. Additional transcriptional regulatory information could also reside in the 3′ flanking region and in introns but a precise analysis of these elements has not been undertaken. Both 5′- and 3′-untranslated regions (UTR) often contain translational regulatory elements such as the Kozak consensus sequence in the 5′ UTR for efficient translation (Kozak 1990). While regulatory sequences upstream from the DH-IV site situated 6.0 kb upstream from the transcription start site are likely to exist, they have not yet been identified.

28.3 EVOLUTION OF TRANSGENIC TECHNOLOGY IN CHICKENS

In the absence of a precise understanding of the Ov regulatory regions and the inability to insert large amounts of genomic DNA into transgenic chickens, the majority of studies during the past three decades have focused on small transgenes that could be packaged into viral vectors. Expression vectors derived from reticuloendothelial virus, spleen necrosis virus, and avian leukosis virus are restricted to applications using relatively small, ubiquitous promoters for transgene expression since the maximum size of the transgene that can be accommodated in these vectors is approximately 8 kb (Bosselman et al. 1989; Cook et al. 1993; A.J. Harvey et al. 2002a, 2002b; Hippenmeyer, Krivi, and Highkin 1988; Mozdziak et al. 2003; Petropoulos et al. 1992; Rapp et al. 2003; Temin 1989; Vile, Tuszynski, and Castleden 1996). Typically, these retroviral vectors produced transgenic chickens at low frequency with germline transmission rates around 1 percent. Transgenes in these studies were either not expressed or expressed at relatively low levels in a ubiquitous manner. Another approach that generated transgenic chickens is direct injection of DNA into the germinal disc of newly fertilized eggs (Love et al. 1994; Sang and Perry 1989; Sherman et al. 1998). Although pronuclear injection is used routinely to generate transgenic mice and several other animal species, injection into the pronuclei of the chicken ovum is impossible due to the fact that the pronuclei are masked by yolk in the cytoplasm and cannot be easily distinguished from the supernumerary sperm nuclei arising from polyspermic fertilization (Sang et al. 1993). Although injection into the germinal disc has successfully produced transgenic chickens, the frequency is very low and to date, only 0.4 percent of injected zygotes give rise to germline transgenics, and none of the transgenes are transcribed or translated.

Progress has been made in recent years in generating transgenic chickens with transgene expression at levels that are competitive with other systems for the production of mAb. Stable integration and expression of GFP transgenes in chickens were reported using either an HIV-1-derived lentiviral vector approach (Chapman et al. 2005) or a Moloney murine leukemia virus (MoMLV)-based retroviral vector system (Koo et al. 2006) although germline transmission rates remained relatively low, at less than 1 percent, in these studies. More recently, improvements to the viral vectors and better characterization of the immediate 5′ regulatory region of the Ov promoter have yielded transgenic chickens at higher frequencies that deposit larger amounts of antibody in their eggs (see Section 28.4). Production of proteins in the eggs of chickens at levels that are commercially relevant has also been made possible by the ability to insert large amounts of DNA into chicken embryonic stem (cES) cells (van de Lavoir et al. 2006b; Zhu et al. 2005) and chicken primordial germ cells (PGCs; van de Lavoir et al. 2006a).

28.4 PRODUCTION OF mAb FRAGMENTS IN TRANSGENIC CHICKENS MADE BY RETROVIRAL VECTORS AND ITS LIMITATIONS

In their initial report, Mizuarai et al. (2001) observed high rates of germline transmission (in excess of 80 percent) when a MoMLV-based viral vector was injected into chicken embryos at the blastodermal stage, although expression of the transgene was very low. Subsequently Koo et al. (2006) overcame transgene silencing during early embryo development by introducing a second promoter in the viral vector sequence. In their report, Kamihira et al. (2005) tested the efficiency of transgenesis and level of transgene expression by injecting either a mouse stem cell virus (MSCV)-based or MoMLV-based retroviral vector into embryos at various developmental stages. These experiments yielded maximal transgene expression when the viral vectors were injected into the heart of developing embryos at Hamburger and Hamilton stage 14 to 16 after 55 hours of incubation. Using this method, they produced transgenic chickens that carried antiprion single-chain Fv fused with the Fc region of human immunoglobulin G1 (scFv-Fc). Since the transgene was driven by the ubiquitous chicken β-actin promoter, scFv-Fc was produced in many tissues of the G_0 mosaic chickens and was secreted into serum, egg white, and egg yolk at high levels, with up to 5.6 mg/mL in egg white. Transgenic G_1 chickens were produced with a frequency of 3.3 percent. However, the transgene expression was suppressed to 0.2–1.5 mg/mL of egg white in the G_1 offspring and subsequently suppressed again in the G_2. The expression level of scFv-Fc in egg white varied among G_1 hens at a concentration between 0.2 and 1.5 mg/mL; this variability was attributed to the positional effect of chromosomal insertion sites of the transgene. The author speculated that the drop in expression level from G_0 to G_1 could be due to the toxicity of the ubiquitous high level expression of the transgene and the mosaic insertion of the transgene in G_0 birds. The data indicated that some kind of transgene suppression was still present although it was not as severe as the gene silencing observed when a retroviral vector was injected into early-stage embryos. The antigen recognition ability of the scFv-Fc produced in egg white was confirmed by Western blot analysis with natural prion proteins from a mouse brain homogenate and glutathione *S*-transferase tagged epitope peptide of human prion protein as antigens. However, the antigen-binding affinity and other functional activities of scFv-Fc produced by the chickens were not discussed in comparison to that of the original mAb fragment.

McGrew et al. (2004) and Lillico et al. (2007) described the use of lentiviral vectors to deliver transgene constructs. Initially they reported generating germline transgenic chickens using equine infectious anemia virus (EIAV)-derived lentiviral vectors and showed that the efficiency of production of transgenic chickens was at least two orders of magnitude greater than that of other retroviral vectors (McGrew et al. 2004). The germline transmission rate in 10 founder males ranged from 4 to 45 percent. They attributed the high efficiency to the ability of lentiviral vectors to transduce nondividing cells, to the use of the VSV-G pseudotype and to the high viral titers. The transgene (EGFP or LacZ) was under the control of human cytomegalovirus (CMV) immediate-early enhancer/promoter. Therefore the transgene expression was largely ubiquitous. In their most recent publication, Lillico et al. (2007) extended their studies by expressing two therapeutic proteins [a humanized ScFv-Fc miniantibody (miR24) derived from a mouse mAb that has potential in treating malignant melanoma, and human

IFN-β-1a (hIFNβ1a)] under truncated regulatory sequences from the Ov gene to direct synthesis of proteins to the tubular gland cells. Because of the limitation of the size of DNA a viral vector can carry, the Ov sequence used here contained only 2.8 kb of sequence immediately 5′ to the translation start site of the Ov gene (which means the presence of around 1.1 kb of promoter sequence). A 675 bp fragment containing an estrogen-responsive enhancer element (ERE) that normally is located further upstream was artificially added in the 5′ end of the 2.8 kb Ov sequence in some transgenic lines. Tissue samples were collected from the magnum portion of the oviduct, pancreas, brain, intestine, liver, heart, and breast muscles from adult transgenic hens in lay. Northern blot analysis of the tissue samples indicated transgene expression from the truncated Ov promoter was restricted to the magnum region of the oviduct regardless of the presence of the ERE. Coupled with localized protein expression to tubular gland cells identified by immunohistochemistry on magnum sections, the data suggested that 2.8 kb of sequence immediately 5′ to the translation start site of the Ov gene was sufficient to confer tissue specificity in these transgenic chickens. Deposition of recombinant miR24 or hIFNβ1a in egg white from transgenic hens was in the range of 15 to 50 μg/mL and 3.5 to 100 μg/mL, respectively. Again, the presence of ERE in the truncated Ov regulatory sequence made no difference in the level of transgene expression. The concentration of the recombinant protein in eggs was also variable among different hens presumably due to differences in the integration site of the transgene. The recombinant hIFNβ1a was found to be biologically active in a standard antiviral/cytopathic effect assay. However, neither the structure nor the functionality of the ScFv-Fc miniantibody (miR24) was described in the report.

Taken together, these groups have made transgenic chickens producing ScFv-Fc antibody fragments in egg white at a reasonable level. However, transgene expression is subject to positional effects and further studies of the characteristics and biological activities of the proteins are necessary to validate the commercial value of the retroviral-based transgenic production systems.

28.5 PRODUCTION OF COMPLETE mAb MOLECULES IN CHIMERIC CHICKENS MADE BY cES CELL-MEDIATED TECHNOLOGY

The ES cell-mediated transgenic approach has been used extensively in the mouse and was widely advocated (although rarely implemented) for other species. The initial studies with chickens showed that cells derived from the area pellucida of newly laid, fertile, unincubated eggs and blastodermal cells that were cultured for short periods contributed to somatic tissues and the germline of chimeras (Pain et al. 1996; Petitte et al. 1990). The ability of cultured cells to contribute to somatic tissues and the germline, however, diminished after one week and was nonexistent after more than 21 days in culture (Petitte, Liu, and Yang 2004). The concept of using embryonic stem cells to make transgenic somatic chimeras was demonstrated for the first time when Zhu et al. (2005) produced mAbs with therapeutic potential in eggs. Subsequently, van de Lavoir et al. (2006a) reported the development of cES cell lines which could be cultured indefinitely and genetically modified with both small and large transgenes (van de Lavoir et al. 2006b). After more than nine months in culture, the genetically modified cES cells retain their ability to contribute extensively to the somatic tissues of chimeras. Although these cells contribute extensively to the somatic tissues, germline transmission of a transgene has not yet been observed.

The ES cell approach to transgenesis has the potential to overcome the aforementioned size limitation inherent in retroviral-based strategies. Relatively large regulatory sequences can be included to ensure tissue-specific and high level expression of the transgene with minimized positional effects. Since several reports have shown that genomic DNA increases the level of transgene expression (Choi et al. 1991; Le Hir, Nott, and Moore 2003; Webster et al. 1997; Whitelaw et al. 1991), lengthy transgenes that include genomic DNA sequences encoding recombinant proteins can be used instead of short cDNA sequences, along with large known and unidentified regulatory sequences for directing tissue-specific expression in transgenes destined for incorporation into the genome of ES cells or PGCs . In addition, for proteins assembled from two subunits (e.g., the immunoglobulin heavy and light chains), two subunits can be expressed from the same vector by utilizing an internal ribosome entry site (IRES) bicistronic expression system (Jang et al. 1988, 1989, 1990; Jang and Wimmer 1990).

28.5.1 Construction of Ov Expression Vectors

Based on published information about the Ov promoter (see Section 28.2), it is preferable to clone at least 7.4 kb of the 5′ flanking region, including the transcription start site to include all the regulatory elements known so far. Furthermore, it may be beneficial to include additional 5′ regions to include the matrix attachment regions (MARs) that reside 8 kb 5′ of the Ov gene (Ciejek, Tsai, and O'Malley 1983) since these elements have the ability to insulate transgene sequences from the positional effects of flanking chromatin (McKnight et al. 1992).

Unlike the size limitation imposed by the retroviral approach, cES cell-based technology allows exploration of the use of large regulatory sequences of the Ov gene to direct transgene expression. In order to include all of the necessary regulatory elements of the Ov gene for proper gene expression, a BAC clone (F8) was identified by screening a chicken BAC library (Texas A & M University, BAC center; Crooijmans et al. 2000). This clone contained approximately 125 kb of 5′ flanking DNA and 25 kb of 3′ flanking DNA in addition to the Ov coding sequence. Therefore, Ov expression vectors varying in size were constructed from the F8BAC (Fig. 28.3a). Ov7.5 and Ov15 contained 7.5 and 15 kb of promoter sequence, respectively, in addition to the 5′ UTR, the first intron, and 15.5 kb of the 3′ flanking sequence, including the 3′ UTR (Zhu et al. 2005). Both vectors contained a unique restriction site for cloning any transgene to be expressed. For expression of mAb, the vectors were designed in such a way that unique V genes from an mAb can be easily inserted into the vectors. A chimeric human-mouse mAb specific for the dansyl hapten (mAbdns) or a fully human mAb specific for prostate specific membrane antigen (mAbF1) was used to demonstrate the production of mAb in eggs from chimeras. To assemble these transgenes, genomic DNA encoding mAb was used instead of cDNA. A third expression vector OvBAC was developed to increase the amount of both 5′ and 3′ regulatory sequence by using all the sequences present in F8 BAC clone to increase the expression level of mAb in eggs (Zhu et al. unpublished data). The OvBAC vector was assembled using a novel bacterial recombineering technology developed in Dr. Neal Copeland's laboratory at the National Institutes of Health (Copeland, Jenkins, and Court 2001; Fig. 28.3b). First, the F8 BAC was retrofitted with Cxneo from a modified pRetro vector carrying a Cxneo cassette by loxP-mediated recombination (Wang et al. 2001). Second, the retrofitted F8 BAC was targeted using a targeting vector in an *Escherichia coli* strain with a defective λ prophage that efficiently facilitates homologous recombination. The targeting vector carried DNA fragments corresponding to 125 bp of DNA immediately upstream of the Ov translation start site and 125 bp of DNA sequence immediately downstream of the Ov translation stop site as homologous arms flanking the mAbF1 cassette. Homologous arms can be as short as 40 to 50 bp, but efficiency increases with longer homologies (Copeland, Jenkins, and Court 2001). The resulting OvBAC vector contained around 125 kb of the 5′ flanking sequence and 25 kb of the 3′ flanking sequence with the mAbF1 cassette replacing the coding sequence of the Ov gene.

28.5.2 Establishment of cES Cell Lines and Production of Chimeras Carrying Ov Expression Vectors

We have derived a number of male and female cES cell lines that contribute extensively to somatic tissues (van de Lavoir et al. 2006b). Briefly, these cell lines are derived from embryos in newly laid, fertile eggs and can be maintained in culture for extended periods. Chicken ES cells are diploid and can form chimeras with extensive contributions from the cES cells and minimal contributions from the recipient embryo. The male cES clone OVA and female cES cell clone OVH were established after transfection of Ov7.5mAbdns and Ov15mAbF1, respectively. Stable integration was confirmed by Southern blot analysis (Fig. 28.4 for OVH). Chimeras were made with OVA or OVH cells by injection into recipient embryos as described by van de Lavoir and Mather-Love (2006). The cES cells that were used to make OvBAC chimeric hens also carried an Ov7.5EGFP transgene that was previously

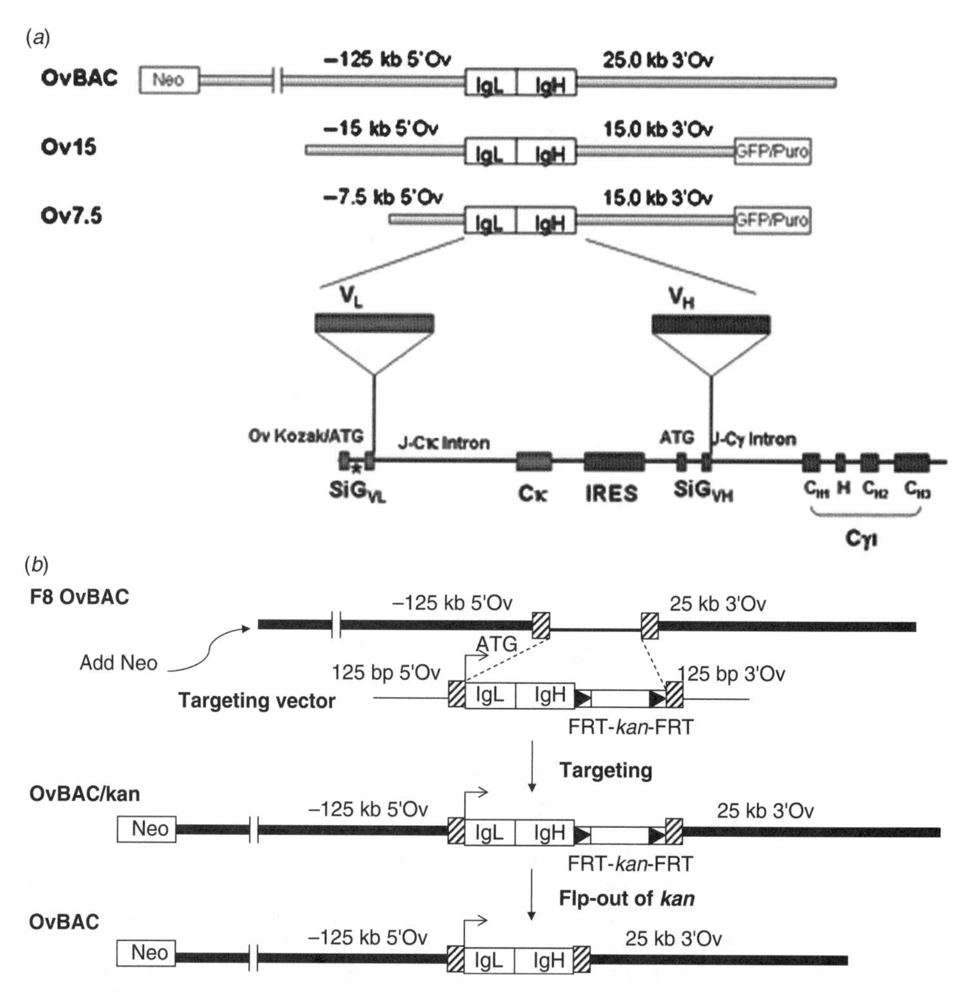

Figure 28.3 (a) Diagram of the Ov7.5mAbdns, Ov15mAbF1, and OvBACmAbF1 expression vectors. Positions of the 5′ and 3′ regulatory sequences of Ov are indicated. Included in the 3′ end of the Ov7.5 and Ov15 mAb expression vectors is a cassette with both EGFP and puromycin (Puro) driven by a promoter (Cx) that functions in all cell types to facilitate the isolation of stably transfected lines of cES cells and the identification of the cES cell contribution in the chimeras. The strategy for construction the OvBAC mAb expression vector is illustrated in the upper part of the diagram. A neo selection cassette is included in the 5′ end of the vector to allow the isolation of stably transfected cES cell clones. Details of the mAb cassette are illustrated in the lower part of the diagram. The thin black line represents intron sequences or nontranslated sequences derived from the human L and H chains. SiG_{VL}: sequence of the signal peptide of the L chain, VL: sequence of the V gene of the L chain, Cκ: sequence of the constant region of the Kappa L chain, IRES: internal ribosomal entry sequence, SiG_{VH}: sequence of the signal peptide of the H chain, VH: sequence of the V gene of the H chain, C_{H1}, H, C_{H2} and C_{H3}: coding sequences of the C_H1, Hinge, C_H2 and C_H3 domains of the gamma1 H chain. *intron sequence present in mAbdns but not in mAbF1. (Adapted from Zhu, L. et al. 2005. *Nature Biotechnol.* 23:1159–1169.) (b) Schematic description of the construction of OvBACmAbF1 expression vector. A neo selection cassette driven by Cx promoter is introduced at the 5′ end of F8 OvBAC. The black line in F8 OvBAC represents the coding region of the Ov gene. The 5′ and 3′ 125 bp homologous sequences in the targeting vector are indicated by the striped rectangles. The black triangles indicate the location of FRT sequences. After flipping out the *kan* cassette, one FRT is still present in the vector, which is not illustrated in the diagram.

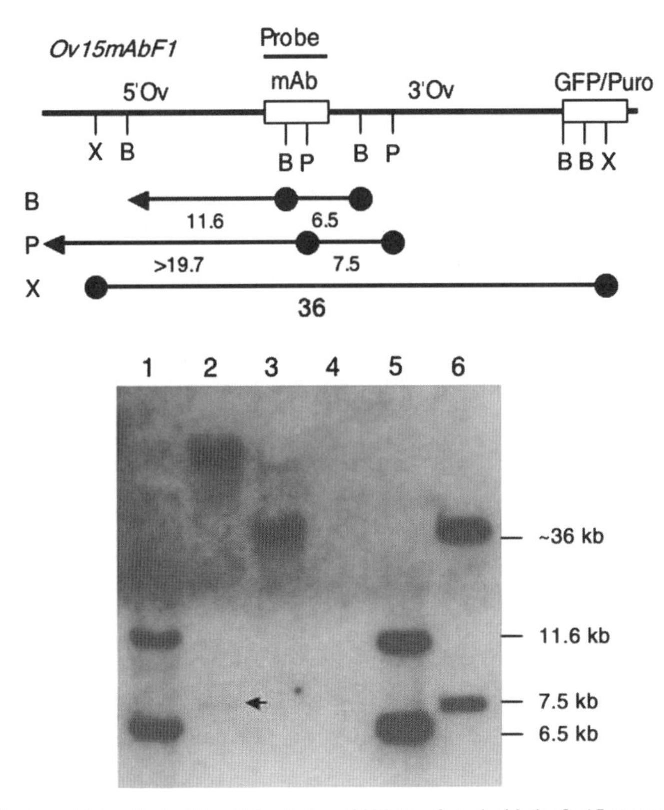

Figure 28.4 Southern blot analysis of the cES cell clone OVH transfected with the Ov15 expression vector; 10 μg of digested genomic DNA from OVH was run on a 1 percent agarose gel. The probe was a NotI fragment of the mAbF1 cassette. The expected sizes of DNA fragments are indicated in the diagram. The presence of a 36 kb band in the P digestion suggests stable integration of the transgene. The presence of a 36 kb band in the X digestion suggests the presence of at least a 10 kb 5′Ov promoter sequence and all of the 15.5 kb 3′Ov flanking sequences. 1: OVH cut with B; 2: OVH cut with P; 3: OVH cut with X; 4: empty; 5: Ov15MAbF1 plasmid DNA cut with B; 6: Ov15MAbF1 plasmid DNA cut with P. In lanes 5 and 6, genomic DNA from a nonchimera was used as the carrier. The arrow points to an internal fragment detected in lane 2. B: BamHI, P: PmeI, X: XhoI. (Adapted from Zhu, L. et al. 2005. *Nature Biotechnol.* 23:1159–1169.)

inserted into the cES cell genome in addition to the OvBACmAb. Four different parental cES cell lines were transfected first with an Ov7.5EGFP construct so that GFP was exclusively expressed in every cES cell-derived tubular gland cell in the magnum. This was done to avoid GFP expression in other irrelevant tissues but still allow the identification of cES-derived tubular gland cells in a chimera. Ov7.5EGFP transfected cES cell lines in different parental backgrounds were then selected for transfection of OvBACmAbF1. In total, 22 independent cell lines were established and 11 of them probably contained an intact copy of the transgene. The transgene was truncated at the 3′ end in one of the cell lines, seven cell lines had various truncations in the 5′ end, and three cell lines carried large truncations as indicated by Southern blot analysis (Fig. 28.5). Fifteen OvBACmAbF1 cES cell lines (10 with an intact copy and 5 with a truncated copy) were used to generate chimeras.

28.5.3 Tissue-Restricted Expression in Chimeras Carrying Ov Expression Vectors

We have emphasized the need for tissue-specific expression to optimize animal health and product quality (Etches 2006) and designed transgenes accordingly. In terms of animal health, high levels of

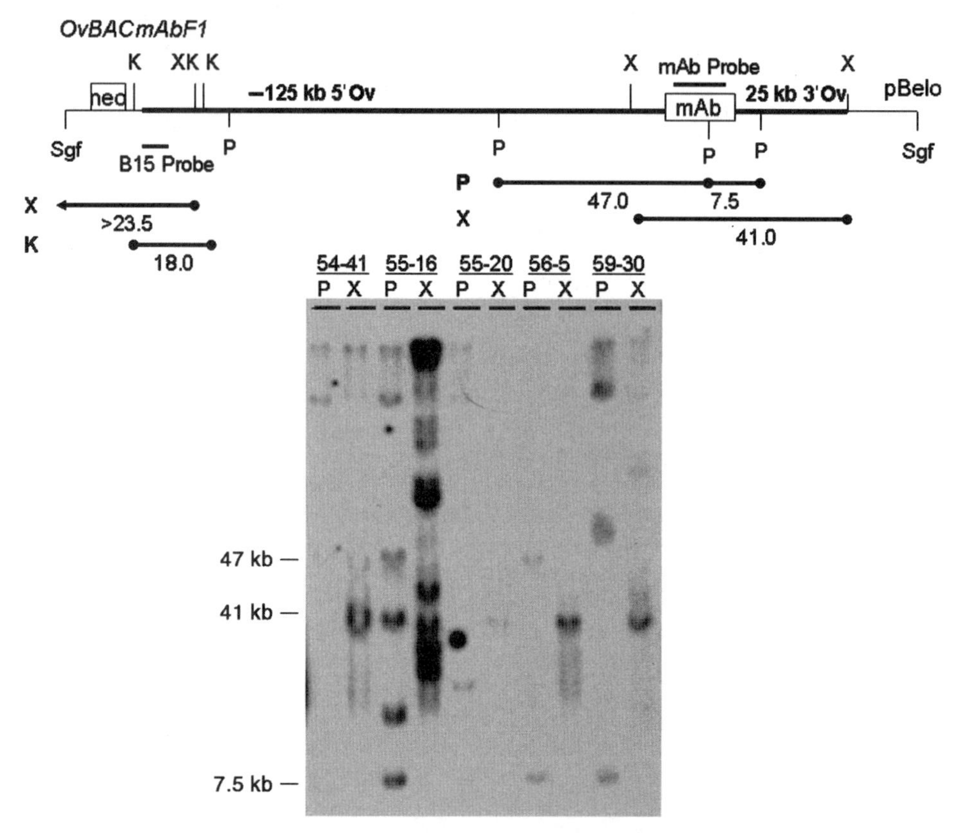

Figure 28.5 Southern blot analysis of the cES cell clones transfected with the OvBAC expression vector; 10 μg each of digested genomic DNA from OVBAC cES cell clones 54-41, 55-16, 55-20, 56-5, and 59-30 were run on a 1 percent agarose gel. The probe was a NotI fragment of the mAbF1 cassette. The expected sizes of DNA fragments are indicated in the diagram. The presence of a 47 kb band in the P digestion suggests the inclusion of a 44 kb 5′Ov promoter sequence. The presence of a 41 kb band in the X digestion suggests stable integration of at least 10 kb of the proximal 5′Ov promoter sequence and all of the 25 kb proximal 3′Ov flanking sequences. In addition to the mAb probe, a second probe (B15) in the most 5′ end of the OvBAC was used to evaluate stable integration of the transgene (data not shown). The presence of an 18 kb band in the K digestion indicates the presence of the most 5′ end of the OvBAC sequence. The presence of a band bigger than 23.5 kb in the X digestion supports the conclusion that the transgene is stably integrated into the genome. K: KpnI; P: PmeI; X: Xho I.

ubiquitous expression of a foreign protein are likely to have unintended consequences that compromise the well-being of the animal. Hence, we have sought regulatory sequences that restrict production of ectopic proteins to the tubular gland cells of the oviduct. We have also recognized that glycosylation of proteins is tissue specific and therefore, proteins derived from different tissues will be a mixture of glycoforms with different biological properties. For example, mAbs derived from the tubular gland cells are nonfucosylated (Zhu et al. 2005) whereas those derived from B-cells contain a fucose residue at the branching GlcNAc (Raju et al. 2000). The nonfucosylated mAb has enhanced antibody-dependent cellular cytotoxicity (ADCC) (see below). Products destined for therapeutic application, therefore, should be derived from the tissue that produces proteins with the desired therapeutic profile.

Tissue-specific and hormone-inducible expression of the transgene was assessed in two-week-old female chimeric chicks injected with estrogen to induce transcription and protein production in the tubular gland cells. In both Ov7.5mAbdns- and Ov15mAbF1-derived chimeras, RT-PCR revealed that both the heavy (H) and light (L) chain transcripts were detected in samples from the magnum

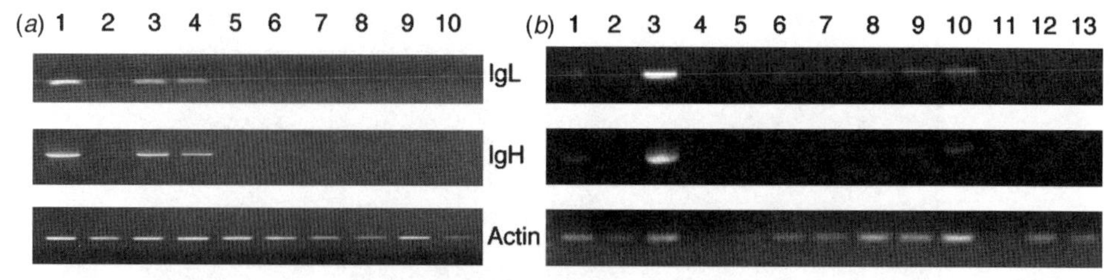

Figure 28.6 Tissue-restricted expression of mAb from Ov7.5mAbdns and Ov15mAbF1 vectors. (a) RT-PCR of GFP-positive tissue samples from estrogen-induced Ov7.5mAbdns chimeric chicks for IgH, IgL, and β-actin. 1: #OV10-17 magnum; 2: #OV10-55 magnum; 3: #OV10-66 magnum; 4: #OV10-79 magnum; 5: #OV10-102 magnum; 6: #OV10-55 gut; 7: #OV10-79 pancreas; 8: #OV10-79 brain; 9: #OV10-102 brain; 10: #OV10-102 leg muscle. (b) RT-PCR of GFP-positive tissue samples from estrogen-induced Ov15mAbF1 chimeric chicks for IgH, IgL, and β-actin. 1: #OV15-34 magnum; 2: #OV15-76 magnum; 3: #OV15-83 magnum; 4: #OV15-121 magnum; 5: #OV15-76 stomach; 6: #OV15-76 duodenum; 7: #OV15-76 jejunum; 8: #OV15-76 ileum; 9: #OV15-76 cecum; 10: #OV15-76 colon; 11: #OV15-111 brain; 12: #OV15-111 liver; 13: #OV15-34 muscle. Note the detection of both full-length L and H transcripts. (Adapted from Zhu, L. et al. 2005. *Nature Biotechnol.* 23:1159–1169.)

portions of the oviducts (where the tubular gland cells are located; Fig. 28.6). RT-PCR was also performed on magnum samples collected from chimeras made from eight independent OvBACmAbF1 cES cell lines. We detected H and L chain transcripts in six lines (data not shown). Of the two nonexpressing cell lines, one was known to have a 5′ truncation and the other did not have any contribution to the magnum in the chimeras examined. Neither transcript for H or L chain was detected in cES cell-derived, GFP-expressing brain, pancreas, gut, and muscle samples taken from chimeras carrying Ov7.5mAbdns transgene (Fig. 28.6a). While L and H chain transcripts were not present in muscle, brain, and liver samples, low levels of expression were detected in samples from the ileum, cecum, and colon of chimeras carrying the Ov15mAbF1 transgene (Fig. 28.6b). Whether mAb transcripts were detectable in tissues other than tubular glands in chimeras carrying the OvBACmAbF1 transgene was not investigated since the GFP transgene used to identify cES contribution was only expressed in the tubular gland cells of these chimeras.

As additional evidence for tissue specificity, immunohistochemistry of sections from estrogen-induced magnums confirmed the expression of mAb in cES cell-derived (i.e., GFP-positive) tubular gland cells but not in the adjacent cES cell-derived epithelial cells in chimeras carrying Ov7.5mAbdns (Fig. 28.7a) or Ov15mAbF1 vectors (data not shown). Similarly, mAb was localized in the GFP-expressing tubular gland cells in chimeras derived from five independent OvBACmAbF1 cES cell lines (Fig. 28.7b). Here both GFP (controlled by Ov7.5 promoter) and mAb (controlled by OvBAC promoter) were only expressed in the tubular gland cells but not in other cells and tissues in this region of the oviduct.

The 5′ regulatory regions that confer tissue specificity can also be inferred from 15 Ov7.5EGFP cES cell lines that were used to identify cES cell-derived tubular gland cells (Zhu et al. unpublished data). Southern analysis revealed that seven lines contained at least 5 kb of promoter sequence while eight others had less than 5 kb of promoter sequence (data not shown). GFP expression in all seven cES cell lines with more than 5 kb of promoter sequence was restricted to the tubular gland cells in two-week-old estrogen-induced female chimeras, whereas GFP expression was restricted to the tubular gland cells in only three out of eight cES cell lines carrying less than 5 kb of the Ov promoter sequence. No expression was observed in the remaining five lines. Ectopic GFP expression was not observed in chimeras derived from any of the Ov7.5EGFP transfected cES cell lines. These data suggest that the first 5 kb of the Ov genomic regulatory sequence contains elements that confer both tissue specificity and binary expression.

Taken together, these data demonstrated that the regulatory sequences present in Ov7.5, Ov15, and OvBAC vectors can direct mAb expression in the tubular gland cells of the oviduct. While data

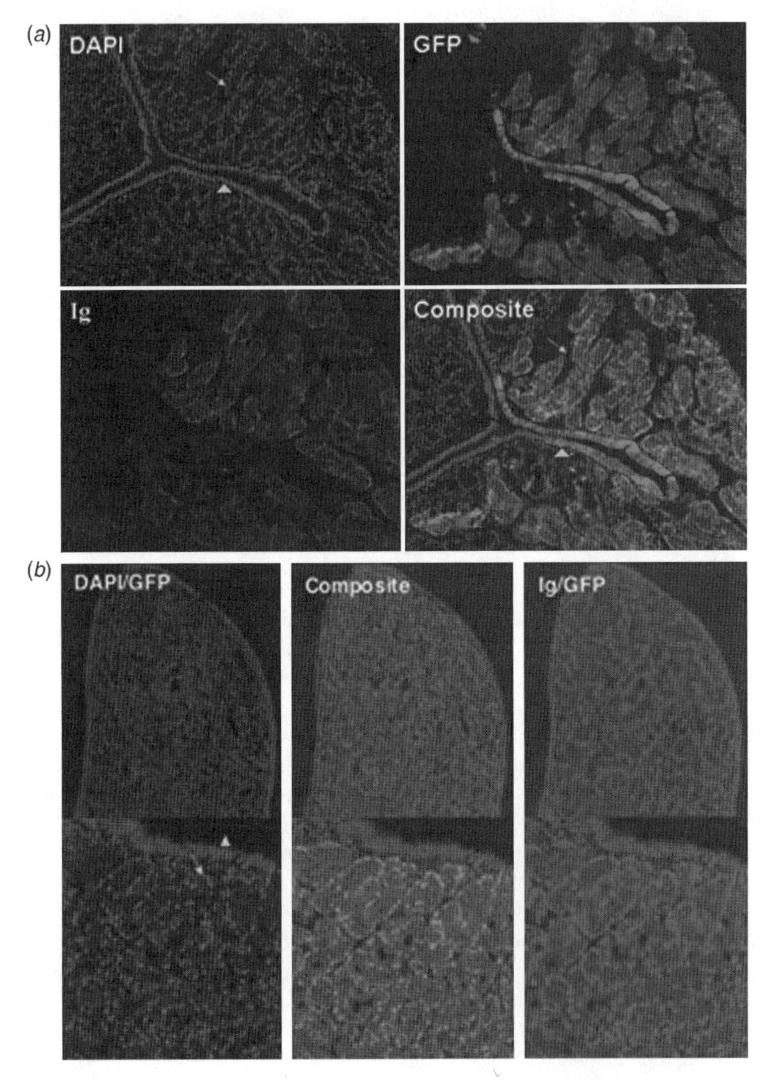

Figure 28.7 Expression of mAb was localized to the tubular gland cells in chimeras. (a) Frozen sections of the magnum portion of the oviduct from an estrogen-induced chimeric female chick produced from cES cells transfected with Ov7.5mAbdns. Images of a section under DAPI staining showing the location of the cells, EGFP fluorescence showing the contribution of cES cells in chimera, human Ig Ab staining showing the transgene expression and composite of all three images. (Adapted from Zhu, L. et al. 2005. *Nature Biotechnol.* 23:1159–1169.) (b) Expression of mAbF1 in the magnum portion of the oviduct from an estrogen-induced chimeric female chick produced from OvBACmAbF1transfected cES cell line OV61/42. Note the restricted transgene expression in donor cES cell-derived tubular glands (arrow) but not in epithelial cells (arrowhead). Top panels under low magnification showed extensive contribution of the cES cells to the tubular glands. Lower panels under higher magnification showed restricted expression of mAb in the tubular gland cells. (From Zhu, L. et al. unpublished data.) (See color insert.)

obtained from Ov7.5mAbdns and Ov7.5EGFP (see above) suggested that a 7.5 kb Ov promoter was sufficient to direct both tissue-restricted and hormonally induced transgene expression, the low level of ectopic expression detected in the gut of chimeras carrying Ov15mAbF1 transgene suggested that 10 to 15 kb of the 5′ regulatory sequences of the Ov gene may not be sufficient to confer fully tissue specific expression of the transgene in a positional independent manner. This level of ectopic expression may also be present in wild-type birds because the plasma concentration of Ov protein is approximately 50 μg/mL (Zhu et al. unpublished data). Since the concentration gradient between egg white and blood is approximately 1000-fold, Ov protein may be transferred to blood by diffusion although the absence of a correlation between the concentration of mAb in egg white and blood (Zhu et al. unpublished data) argues against this possibility. In light of the presence of Ov protein in plasma of wild-type birds, the pronouncement of absolute tissue specificity using 7.5 kb of Ov regulatory sequence (Zhu et al. 2005) or 1.1 kb of Ov regulatory sequence (Lillico et al. 2007) should be tempered by the relatively low expression levels that are induced by these transgenes, the exquisite sensitivity of PCR, and the lack of sensitivity of Northern analysis.

28.5.4 Deposition of Human mAb in Eggs from Chimeras Carrying Ov Expression Vectors

Nine out of the twelve chimeric females carrying the Ov7.5mAbdns transgene commenced laying when they were reared to sexual maturity. Low levels of mAb in the range of 1 to 11 ng/mL as determined by ELISA for human IgG were detected in eggs from three chimeric hens, indicating that the transgene product was exported from the tubular gland cells and deposited in egg white. This low level of expression was attributed to the very low donor contributions to the tubular gland cells, masculinization of the hens due to the male cES cell line used for transfection, and an aberrant splicing in the L chain mRNA. All of these deficiencies were corrected in the subsequent two Ov expression vector systems.

Thirty chimeric hens generated from OVH cES cells carrying the Ov15mAbF1 were raised to sexual maturity and were evaluated for the deposition of the fully human mAb into eggs by ELISA. The concentration and amount of mAb in eggs varied among the chimeric hens, presumably due to the varying degree of chimerism in the magnum. Eight chimeric hens laid eggs that contained more than 1 mg of mAb per egg and the maximal level among the 30 chimeras was 3.4 mg of mAb per egg. Four chimeric hens laid eggs whose mean level of mAb was between 1.2 and 1.6 mg per egg. Six chimeric hens laid eggs whose mean level of mAb was between 0.5 and 1.0 mg per egg.

One hundred thirty-seven chimeric hens were produced and reared to sexual maturity from 15 independent cES cell lines carrying OvBACmAbF1 transgene. Monoclonal Ab production in eggs was analyzed from all the hens by ELISA. Out of the 15 cell lines, 9 cell lines (7 with intact transgene, 2 with truncations) had at least 1 mg per egg. The concentration of mAb in the two highest expressing cell lines was in the range of 4 to 6 mg per egg in chimeras. The concentration of human mAb in eggs of chimeras derived from one cell line (OV61-42) is shown in Table 28.1. In chimeric hens made from this cell line, seven had more than 3 mg of mAb per egg, two had between 2 and 3 mg mAb per egg, and another seven had between 1 and 2 mg mAb per egg.

MAbF1 produced in egg white was further analyzed by Western blot with a rabbit anti-human IgG (H + L) that recognizes both the H and L chains. As shown in Figure 28.8, full-length L and H chains were detected under reducing conditions in egg white samples from five Ov15mAbF1 chimeric hens.

TABLE 28.1 Concentration of Human IgG mAb in Eggs from OvBACmAbF1 Chimeras Made from Clone OV61-42

Mean of mAb/Egg (mg)	No. of Hens	Mean Extent of Chimerism (%)	Range of [mAb] in Egg (μg/mL)	Range of mAb/ Egg (mg)
>3.0	7	89	81.2–287.6	2.0–5.8
>2.0–3.0	2	57	84.4–116.8	2.1–3.2
>1.0–2.0	7	82	29.7–97.4	0.8–2.8
>0.5–1.0	3	75	21.2–37.3	0.6–0.9
>0.1–0.5	2	60	8.6–19.5	0.2–0.5
0.1 or less	5	53	Up to 4.2	Up to 0.2

The mean value of MAb/egg from each hen was an average of values from 2 to 9 eggs. The extent of chimerism was scored by comparing the extent of black down relative to that of the Barrred Plymouth Rock from which chicken ES cells were derived.

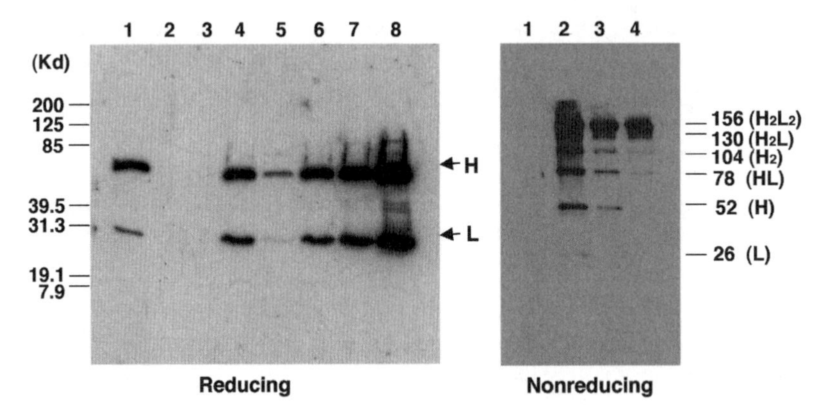

Figure 28.8 Western blot analysis of egg white samples from Ov15mAbF1 chimeric hens under reducing (left) and nonreducing conditions (right) showing deposition of mAbF1 into eggs. (left) 1: 4 ng of purified human IgG1, κ (Sigma) as a positive control; 2: M.W marker; 3: egg white from a wild-type White Leghorn hen as negative control; 4: egg white from OV15-17; 5: egg white from OV15-29; 6: egg white from OV15-71; 7: egg white from OV16-34; 8: egg white from OV16-42. (right) 1: egg white sample from a wild-type White Leghorn hen; 2: egg white sample from OV16-42; 3: egg white sample from OV16-34; 4: human IgG1, κ control (8 ng, Sigma). Note the presence of L and H chains of the correct size in all samples. HRP-labeled goat anti-human IgG (Southern Biotechnology) was used for detection. (Adapted from Zhu, L. et al. 2005. *Nature Biotechnol.* 23:1159–1169.)

Fully assembled H_2L_2 was seen under nonreducing conditions as well as small amounts of assembly intermediates (Fig. 28.8).

Variation of the mAb concentration deposited in egg white was monitored in eggs from five different Ov15mAbF1 chimeric hens over a period of three to four months. In three hens, the amount of mAb peaked during the first few weeks, then dropped approximately twofold and remained stable over the next few months (Fig. 28.9). In two hens, the expression level remained relatively stable over

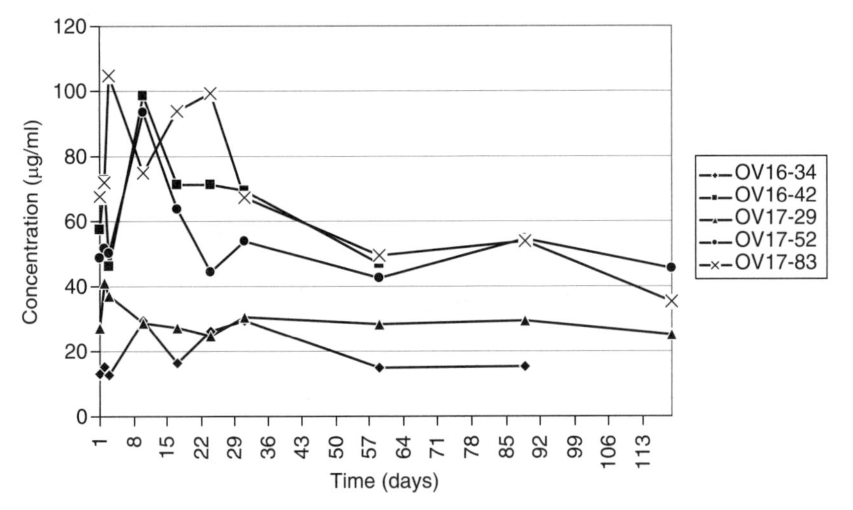

Figure 28.9 Concentration of mAb in eggs from representative chimeric hens during the first 120 days after the first oviposition. At each time point, the mean values were calculated from at least two independent preparations of egg white that were assayed in two independent ELISA assays. (Adapted from Zhu, L. et al. 2005. *Nature Biotechnol.* 23:1159–1169.)

the time course (Fig. 28.9). Similar variation was also observed in eggs from chimeras carrying the OvBACmAbF1 transgene (data not shown).

28.5.5 Purification and Physical Characteristics of mAb Produced in Tubular Gland Cells

MAb produced in tubular gland cells and deposited in egg white was purified using a protein A Sepharose column. Egg white was first separated from yolk. The egg white was then diluted with three volumes of water and approximately 3 percent of the egg white protein, containing mostly ovomucin, was removed after precipitation by adjusting pH to 6.0 with phosphoric acid and centrifugation. The supernatant was loaded onto a protein A-Sepharose FF column and the final elution contained purified mAb. The majority of the purified material was a fully assembled H_2L_2 (Fig. 28.10). As determined by ELISA and A280, the purity of the material was greater than 90 percent. The yield of purified mAb was estimated in excess of 85 percent and sufficient mAbF1 was purified from eggs of transgenic chimeras for analysis of functionality and glycosylation (see below).

Characterization of mAbF1 produced in chicken was conducted in several assays in parallel with mAbF1 produced in conventional CHO cell culture. Monoclonal Ab produced in the chicken tubular gland cells and CHO cells shared a number of similarities in their protein characteristics (Zhu et al. 2005). They ran similarly on SDS-PAGE (data not shown), their approximate molecular weights were similar (data not shown), and both were more than 90 percent IgG monomer as determined by

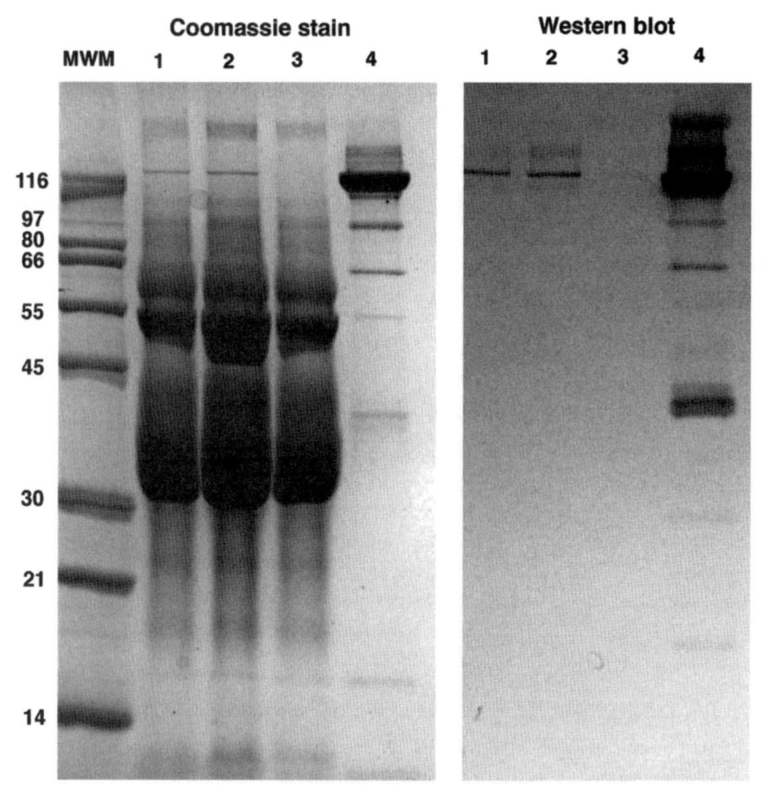

Figure 28.10 Purification of mAbF1 from egg white. 1: Starting material; 2: protein A load; 3: protein A flow through; 4: protein A eluate. Both Coomassie stain and Western blot are under nonreducing conditions. (Adapted from Zhu, L. et al. 2005. *Nature Biotechnol.* 23:1159–1169.)

TABLE 28.2 SEC-HPLC Analysis of mAbF1 Produced by Chicken Tubular Gland Cells and CHO Cells

Production Systems	IgG Monomer (% Peak Area)	IgG Aggregate (% Peak Area)	IgM (% Peak Area)
F1 CHO	97.2	2.8	0.0
F1 Chicken	100.0	0.0	0.0

size-exclusion high-performance liquid chromatography (SEC-HPLC) analysis (Table 28.2). The lack of aggregates in the chicken produced mAbF1, presumably due to the inhibition of proteases in egg white, makes the chicken a unique system for producing proteins. No differences in the primary amino acid sequence were identified between the two mAb preparations by analysis of corresponding tryptic peptides by nano liquid chromatograph-tandem mass spectrometry (LC-MS/MS). Furthermore, L chains from both mAb preparations had identical mass (± 3 Da) in liquid chromatograph-mass spectrometric (LC-MS) analysis. Some minor differences were also identified between the mAbs produced in the two systems. For example, while both mAb preparations had similar isoelectic points (pI) as identified in capillary isoelectric focusing analysis, CHO-derived mAbF1 had greater charge heterogeneity than chicken-derived mAbF1 (data not shown). Charge micro-heterogeneity for mAb is believed to arise from C-terminal lysine variants, N-terminal pyroglutamate formation, deamidation of asparagine residues, and carbohydrate composition. Here the differences in carbohydrate composition could be the main reason for the observed micro-heterogeneity (see below).

Another difference was observed in the thermal stabilities of mAbF1 produced in the chicken and CHO expression systems. Melting temperatures (T_m) were obtained from both Fc glycosylated and deglycosylated forms of mAbF1 produced in the two systems using differential scanning calorimetry (DSC; Ghirlando et al. 1999). It was found that the unfolding profiles of the glycosylated mAbs derived from CHO and chicken systems were quite different, whereas the unfolding profiles of the Fc-deglycosylated mAbs were almost identical (Fig. 28.11). Therefore, it was evident that the differences in thermal stabilities of the CHO- and chicken-derived mAbs were primarily due to the different glycosylation patterns present in their respective Fc domains. Of the several melting transition points, the first one is most critical as it determines the overall stability of the antibodies at lower temperatures. The higher T_{m1} of chicken-derived mAb (63.8°C) compared to that of CHO-derived mAb (62.7°C), predicted that the chicken-derived antibody should be more stable under typical storage conditions. This attribute may be beneficial as mAbs are formulated for therapeutic applications.

28.5.6 Glycosylation of mAb Produced in Chicken

As indicated above, glycosylation of mAbF1 produced in our chicken expression system differed from that of mAbF1 produced in CHO cells. As shown in Figure 28.12, both CHO cells and chicken tubular gland cells glycosylated mAbF1 in both the Fab and Fc regions. Although most mAb are glycosylated in the Fc region only, mAbF1 is glycosylated in both the Fc and the Fab regions. However, capillary zone electrophoresis with laser induced fluorescence (CE-LIF) revealed that the oligosaccharides released from samples of mAbF1 produced in the chicken are quite different from those found on mAbF1 produced in CHO cells (Fig. 28.13). Differences in carbohydrate composition were identified by monosaccharide analysis of chicken- and CHO-produced mAbF1 with high-performance anion exchange chromatography with pulsed amperometric detection (HPAE-PAD) (Table 28.3). The most striking features were the absence of fucose residues, low content of galactose residues, and the high content of mannose residues in mAbF1 produced in chicken. Exoglycosidase analysis of glycans by capillary electrophoresis (CE) and mass spectrometry to identify the terminal sugar residues confirmed the above observation. Structures of major glycans in chicken-derived mAbF1 were analyzed by MALDI time-of-flight (TOF) mass spectrometry. The deduced carbohydrate composition and possible glycosidic linkages are summarized in Table 28.4. The oligosaccharide structures

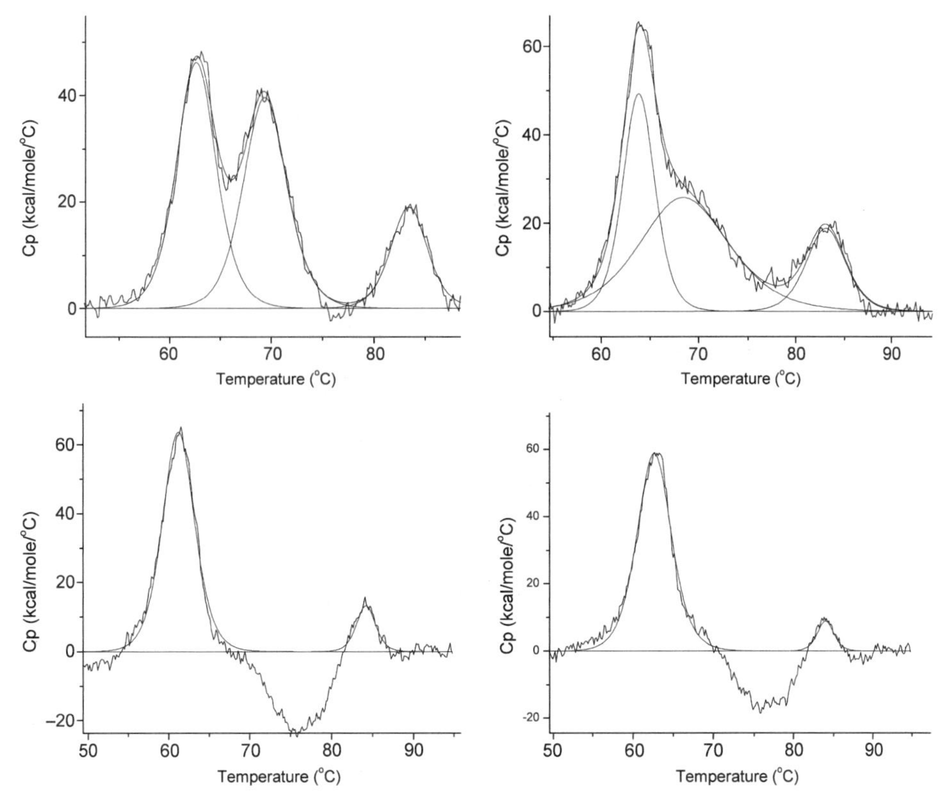

Figure 28.11 Thermal stabilities of chicken- and CHO cell-derived MAbF1. DSC data of MAbF1 derived from CHO cells (top left) and chicken tubular gland cells (top right). Data is deconvoluted to fit a two-state transition model with three peaks in both cases, yielding T_m values of 62.7°C, 69.4°C, and 83.4°C for antibody expressed in CHO cells and T_m values of 63.8°C, 68.5°C, and 83.1°C for the chicken-derived antibody. The DSC data of the respective Fc-deglycosylated antibodies (bottom left and bottom right, respectively), were also fitted to a two-state transition model, but with two peaks in each case, which yields T_m values of 61.2°C and 84.0°C for the CHO-expressed antibody and 62.6°C and 84.0°C for the chicken-derived antibody. The different profiles of the intact antibodies and the similar profiles of the Fc-deglycosylated antibodies indicate that the different thermal stabilities are the consequence of the difference in glycosylation of MAb produced in tubular gland cells and CHO cells. (Adapted from Zhu, L. et al. 2005. *Nature Biotechnol.* 23:1159–1169.)

contained high-mannose type, complex type, and hybrid type *N*-glycans. These data provided confirmation of the differences in glycosylation between chicken tubular gland-produced and CHO-produced mAbF1 that were revealed by global monosaccharide analysis. All of these residues are shared with those on human proteins and therefore are unlikely to be antigenic in patients receiving therapeutic antibodies produced in egg white.

28.5.7 mAb Produced in Chicken Tubular Gland Cells Is Biologically Active with Enhanced ADCC Activity

The antigen-binding property of mAbF1 produced in chicken tubular gland cells was compared with that of mAbF1 produced in CHO cells. Both antibody preparations produced nearly identical binding curves when PSMA expressed on LNCaP cells (ATCC) was used as antigen to assay for binding (Fig. 28.14). The EC_{50} value of the two preparations of chicken-produced mAbF1 was 1.03 nM and 0.81 nM, and was 1.04 nM for the CHO cell-produced mAbF1. The data showed that while the

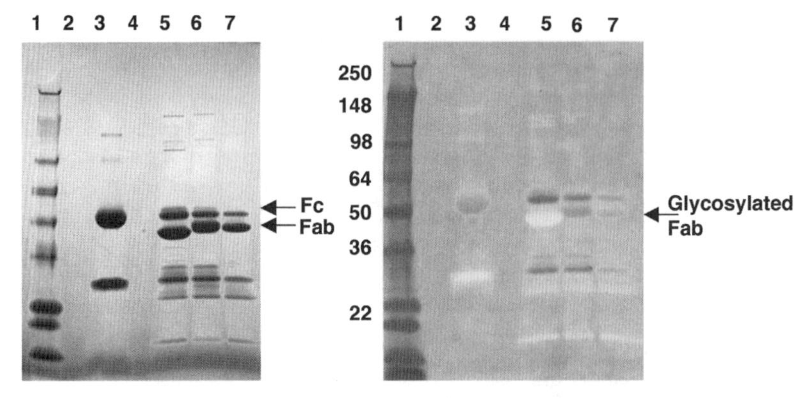

Figure 28.12 Analysis of Fab glycosylation of mAbF1 protein produced in chicken tubular gland cells. (left) SDS-PAGE with protein-specific staining showing both Fc and Fab fragments. (right) Western blot with a carbohydrate-specific stain showing that both Fc and Fab are glycosylated. 1: MW standard; 2: empty; 3: anti-CTLA4 mAb control, reduced; 4: empty; 5: anti-CTLA4 mAb control papain digest, nonreduced; 6: mAbF1 CHO papain digest, nonreduced; 7: mAbF1 chicken papain digest, nonreduced. The anti-CTLA4 mAb is glycosylated only in the Fc region. (Adapted from Zhu, L. et al. 2005. *Nature Biotechnol.* 23:1159–1169.)

chicken-derived and CHO-derived antibodies are glycosylated differently, they recognize and bind antigen equally efficiently.

A potential strategy for therapeutic use of mAbF1 relies on internalization of the antibody following binding to the PSMA antigen (Liu et al. 1998). In this application, the mAbF1 could be conjugated with cytotoxins in order to target and destroy PSMA-expressing tumor cells. The internalization of mAbF1

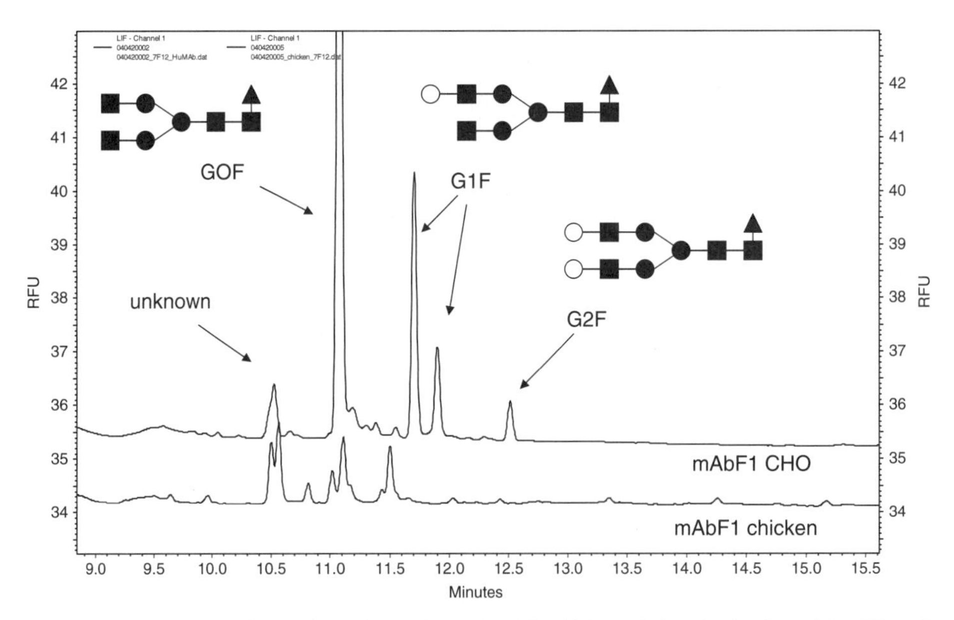

Figure 28.13 Oligosaccharide profiles of mAbF1 produced in chicken tubular gland cells and in CHO cells. APTS-labeled oligosaccharides were separated by capillary electrophoresis with LIF detection. Known glycan structures of the CHO-produced mAb are indicated. Glycan structures were identified based on the retention time of oligosaccharide standards; glycan structures were also confirmed by MS analysis. Mannose is shown as a black circle; galactose is shown as an open circle; *N*-acetyl glucosamine is shown as a black square; fucose is shown as a black triangle. (Adapted from Zhu, L. et al. 2005. *Nature Biotechnol.* 23:1159–1169.)

TABLE 28.3 Monosaccharide Analysis of Chicken-Produced and CHO-Produced MAb7F12

Monosaccharide	CHO HuMAb pmol (% Total)	Chicken HuMAb pmol (% Total)
Fucose	692 (18)	0
Glucosamine	1536 (40)	1571 (52)
Galactose	671 (17)	43 (1)
Mannose	940 (25)	1513 (47)
Total	3839 (100)	3127 (100)

TABLE 28.4 Carbohydrate Composition of mAbFl Produced in Chicken Eggs

Observed Mass + Na+	Theoretical Mass + Na+	Composition	Possible Structure
1136.4	1136.5	$(Man)_3(GlcNAc)_3$	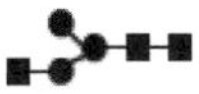
1257.4	1257.6	$(Hex)_2(Man)_3(GlcNAc)_2$	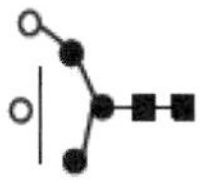
1298.4	1298.44	$(Hex)(Man)_3(GlcNAc)_3$	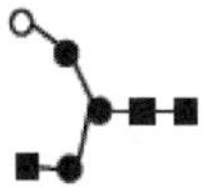
1339.46	1339.5	$(Man)_3(GlcNAc)_4$	
1460.49	1460.5	$(Hex)_2(Man)_3(GlcNAc)_3$	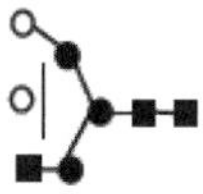
1501.53	1501.5	$(Hex)(Man)_3(GlcNAc)_4$	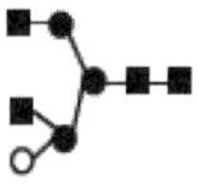
1542.56	1542.6	$(Man)_3(GlcNAc)_5$	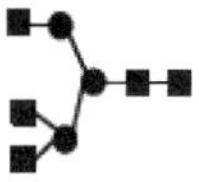
1663.6	1663.58	$(Hex)_2(Man)_3(GlcNAc)_4$	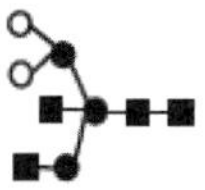 or 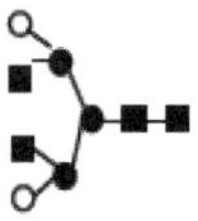
1704.6	1704.6	$(Hex)(Man)_3(GlcNAc)_0$	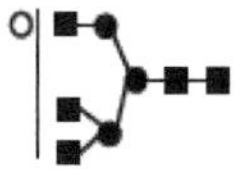
1745.66	1745.6	$(Man)_3(GlcNAc)_5$	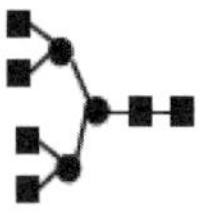
1866.7	1866.7	$(Hex)_2(Man)_3(GlcNAc)_5$	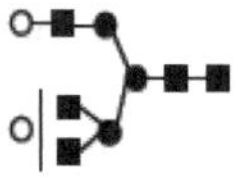
1948.7	1948.68	$(Man)_3,(GlcNAc)_7$	Not determined

Man, mannose, shown as black circle; Hex, hexose (mannose or galactose), shown as open circle; GlcNAc, *N*-acetyl glucosamine, shown as black square. The vertical line indicates that the last hexose could be connected to any one of Man or GlcNAc residues along the line.

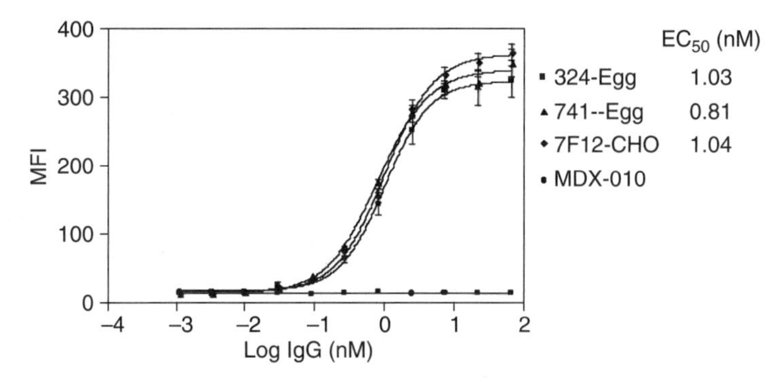

Figure 28.14 Binding of chicken tubular gland cell or CHO cell produced mAbF1 to PSMA expressed on LNCaP cells. 324-Egg and 741-Egg are two different preparations of mAbF1 isolated from eggs. F1-CHO is mAbF1 produced by CHO cells. Anti-CTLA4 is a human IgG1 control mAb not recognizing antigen on LNCaP cells. (Adapted from Zhu, L. et al. 2005. *Nature Biotechnol.* 23:1159–1169.)

was evaluated with the HumZap assay in which cells are killed when the mAbF1/Hum-Zap complex bound to PSMA on the cell surface is internalized. In the presence of antibody alone or Hum-Zap alone, LNCaP cells survive. As shown in Figure 28.15, internalization of mAbF1 preparations from both chicken tubular gland cells and CHO cells was similar. When titrated over a range of antibody concentrations, the EC_{50} values for internalization of both the chicken-derived and CHO-derived mAbF1 were 0.49 nM.

Another important benchmark for evaluating mAb is the rate of serum clearance. In general, longer clearance rates are preferred to reduce the dosage of the drug and to reduce the frequency of administration. The *in vivo* half-life of the chicken-produced mAbF1 was analyzed in parallel with the CHO-produced antibody in BALB/c mice by intravenous injection of radiolabeled antibodies. Residual radioactivity was determined by whole body counting over the time course. While mAbF1 produced by CHO cells cleared with a half-life ($t_{1/2}$) of 207.5 $\pm$ 18.3 hours, mAbF1 produced by chicken tubular gland cells cleared more quickly, with a half-life of 102.4 $\pm$ 0.9 hours (Fig. 28.16). Although the half-life of the chicken-produced material is reduced by approximately twofold relative to the half-life of the CHO-produced material, it is within the range for therapeutic application. The shorter serum half-life in mice of mAbF1 produced in the chicken oviduct is presumably due to the higher content of high-mannose type *N*-glycans. However, serum half-life of the chicken-produced mAbF1 remains to be evaluated in primates and ultimately in humans. It is expected that

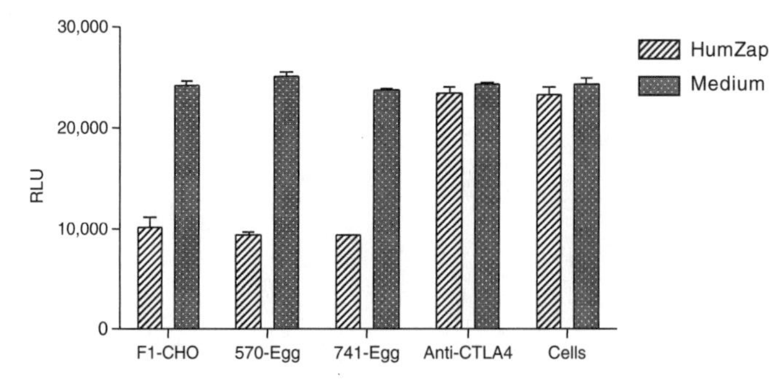

Figure 28.15 Internalization assay of mAbF1 produced in chicken tubular gland cells and CHO cells. 570-Egg and 741-Egg were two preparations of mAbF1 isolated from eggs; 300 ng per well each of mAb was added either with Hum-Zap (anti-Human IgG, Saporin conjugate) or cell culture medium to LNCaP Cells. Cell viability was determined 48 hours after addition of mAb and HumZap. The Y axis is relative luminescence units (RLU). (Adapted from Zhu, L. et al. 2005. *Nature Biotechnol.* 23:1159–1169.)

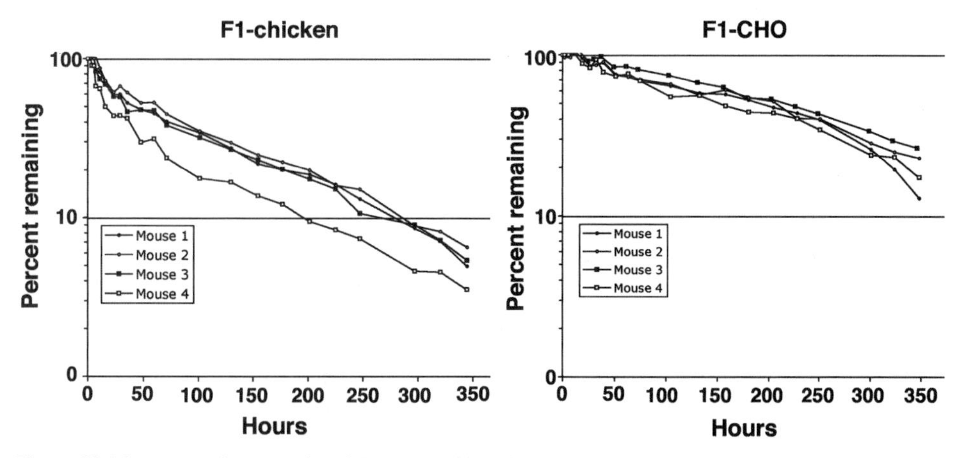

Figure 28.16 *In vivo* clearance of mAbF1 produced in chicken tubular gland cells (F1-chicken) and CHO cells (F1-CHO). Four BALB/c mice were injected with each radiolabeled antibody preparation. The residual radioactivity was measured over a period of 350 hours using a whole-body counter. Data are expressed as the percentage of the injected dose remaining. (Adapted from Zhu, L. et al. 2005. *Nature Biotechnol.* 23:1159–1169.)

longer serum half-life would be observed when the fully human mAbF1 is evaluated in a human volunteer or patient.

Antibody-dependent cellular cytotoxicity (ADCC) is implicated in many applications of therapeutic mAbs and is mediated through binding to the CD16 (FcγRIIIA) receptor. CHO-derived mAbF1 induced a dose-dependent cell lysis, which reached a plateau at 38% lysis, with an EC_{50} of 0.11 μg/ml in a modified ^{51}Cr ADCC assay (Fig. 28.17). In contrast, the maximum ability of mAbF1 derived from chicken eggs to lyse target cells was increased to 60% and the EC_{50} was 0.018 μg/mL (Fig. 28.17). This indicated that chicken-derived mAbF1 had higher potency and efficacy in ADCC when compared to the CHO-produced mAbF1.

Blockade of antibody binding to CD16 with anti-CD16 antibody eliminated ADCC activity of mAbF1 derived from both chicken and CHO cells (Fig. 28.18). The binding of the two antibody preparations to CD16-Phe and CD16-Val (two isotypes of the FcγRIIIA) showed the dissociation constant (KD) of chicken-derived mAbF1 was about 10-fold lower for both types of the CD16 as compared to

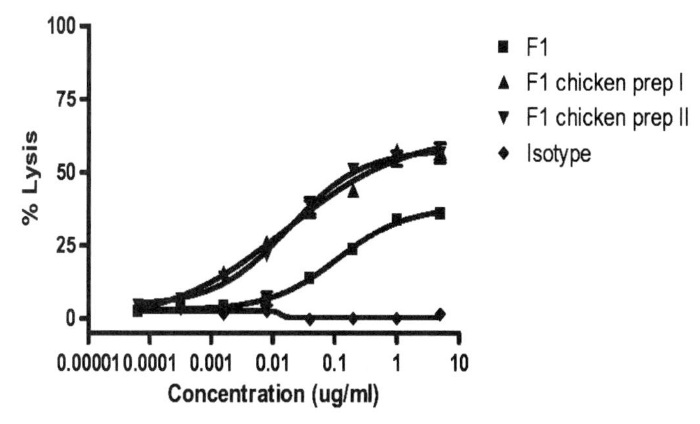

Figure 28.17 Antibody-dependent cellular cytotoxicity (ADCC) assay of mAbF1 produced in chicken tubular gland cells (F1 chicken prep I and II) and CHO cells (F1). A human IgG1 not recognizing antigen on LNCaP cells was used as isotype control. (Adapted from Zhu, L. et al. 2005. *Nature Biotechnol.* 23:1159–1169.)

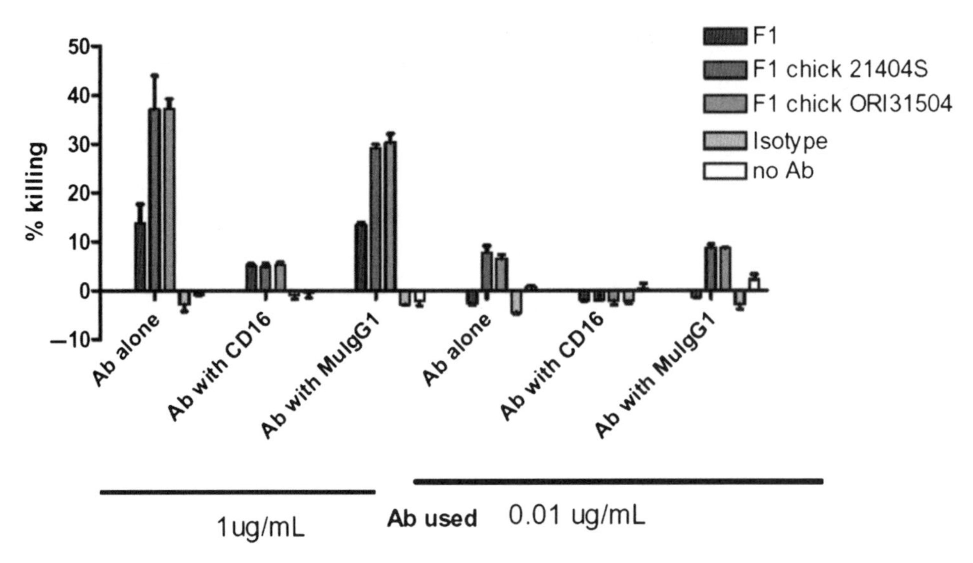

Figure 28.18 Blockade of ADCC with anti-CD16 antibodies. Percent of cell lysis by different mAbF1 preparations at a saturating dose (1 μg/mL) or a suboptimal dose (0.01 μg/mL) was measured in the absence or presence of 5 μg/mL of anti-CD16 antibody 3G8, or isotype control antibody (MuIgG1). The percent lysis was greatly reduced in the presence of anti-CD16 antibody while isotype control antibody had no effect. F1: mAbF1 from CHO; F1 chick 21404S and F1 chick ORI31504: two different batches of mAbF1 produced in chicken eggs; isotype: a nonspecific IgG1 control; no Ab: no antibody added. (From B.T. Preston and P.M. Cardarelli unpublished data.)

the corresponding CHO-derived mAb (Fig. 28.19). The enhanced affinity of chicken-derived mAbF1 to both isotypes of CD16 could be useful if a mAb is only effective in a group of patients with one particular CD16 isotype. For example, Rituxan was found to be more effective for non-Hodgkin lymphoma patients with CD16-Val isotype compared with patients with CD16-Phe isotype (Cartron et al. 2002). The use of a defucosylated form of Rituxan might potentially improve the therapeutic effects in all patients independent of their CD16 isotypes.

The absence of the α1-6 fucose residue on the Asn297 linked *N*-acetyl glucosamine residue of human immunoglobulins is shown to substantially increase ADCC by 10- to 100-fold (Niwa et al. 2004b; Okazaki et al. 2004; Shields et al. 2002; Shinkawa et al. 2003). In some therapeutic mAbs when ADCC is of therapeutic importance (e.g., Rituxan and Herceptin; Clynes et al. 2000; Niwa

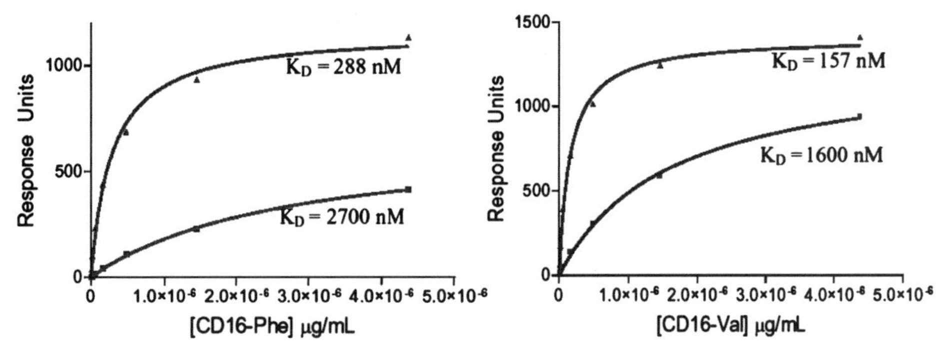

Figure 28.19 Equilibrium binding of chicken- (top, filled triangles) and CHO- (bottom, filled squares) derived mAbF1 with CD16-Phe (left panel) and CD16-Val (right panel). Estimated values of dissociation constants (K_D) are shown near corresponding fits. (From M. Srinivasan unpublished data.)

et al. 2004a, 2004b), this attribute is very attractive. Although the immunoglobulin molecules produced in B-cells of birds possess α1-6 fucose residues (Raju et al. 2000; Suzuki et al. 2003), the α1-6 fucose linkage is absent from the major egg white proteins (D.J. Harvey et al. 2000; Hase et al. 1986; Iwase, Kato, and Hotta 1984; Kuster, Naven, and Harvey 1996; Saba et al. 2001; Takahashi et al. 1993, 2001). MAbF1 was not fucosylated when produced in the chicken tubular gland cells and this property was associated with a significant increase in ADCC. Currently, there are several *in vitro* systems to produce mAb with little or no fucose. A CHO cell line with a FUT8 (alpha-1,6-fucosyltransferase) knockout is used to produce completely defucosylated mAb (Yamane-Ohnuki et al. 2004) and mAb with a low fucose content is produced by a mutant CHO cell line (Lec13) that is deficient in GDP-mannose 4,6-dehydratase which is required for converting mannose to fucose (Ripka, Adamany, and Stanley 1986; Shields et al. 2002). A rat hybridoma cell line with low level of FUT8 fucosyltransferase produces mAb that is 30 to 90 percent defucosylated (Niwa et al. 2004b). In each case, the ADCC of the fucose-free or fucose-reduced antibody is enhanced. Expression of mAb in chicken tubular gland cells thus offers an alternative system to obtain large amounts of defucosylated mAb. It would be interesting to compare the ADCC activity of the same mAb produced in these different systems.

28.6 OTHER TRANSGENIC SYSTEMS, ADVANTAGES OF THE TRANSGENIC CHICKEN SYSTEM, AND ITS FUTURE OUTLOOK FOR THE PRODUCTION OF mAb IN EGGS

Transgenic technology has become increasingly attractive in large-scale production of biologically active recombinant proteins, including mAbs as the technology has matured. However, the slow passage of products from transgenic animals through the regulatory process has been a major concern to investors. The recent approval of recombinant human antithrombin produced in transgenic dairy goats by the European Commission has demonstrated that protein therapeutics from transgenic animals can be approved by the regulatory agencies and should allay some of this concern (Echelard, Ziomek, and Meade 2006).

Transgenic plants and transgenic dairy animals, including cows and goats, have been used to produce antibodies. Production of mAbs in plants such as tobacco has the advantages of low cost, relatively short generation time, and easy scale-up (Ma, Drake, and Christou 2003). However, the presence of immunogenic plant-specific glycans and biosafety issues regarding containment need to be addressed (Ma, Drake, and Christou 2003). Transgenic goats can be used to produce recombinant proteins in milk at concentrations between 1 and 5 g/L and the proteins are glycosylated with patterns typical of mammalian proteins although increased presence of oligomannose structures were observed on protein produced in transgenic goat milk (Echelard 1996; Edmunds et al. 1998; Pollock et al. 1999). Furthermore, heterogeneity in glycosylation was observed in human recombinant antithrombin produced in the mammary gland of transgenic goats among proteins derived from different founder goats (Zhou et al. 2005). This system is hindered by the relatively long time frame required for developing new products (18 months for goat and 3 years for cow from microinjection to full lactation).

In comparison to plant and mammalian transgenic systems, transgenic chickens that produce recombinant mAbs in their eggs offer several advantages. Products can be obtained from cES cell-derived chimeras in 8 months and from fully transgenic chickens (see below) in 14 months. Another advantage of the transgenic chicken expression system is that production can be easily and rapidly scaled up by increasing flock size to meet market demand. The capacity of the poultry industry to produce large quantities of eggs makes protein production in eggs a promising alternative to transgenic dairy animal systems in many applications. In addition, chickens can be contained to prevent escape of the genotype and the egg is a sterile and compact compartment providing a good starting material for isolation and purification. The glycosylation profile of proteins produced in chicken eggs is shared by human proteins and, therefore, chicken-derived products are unlikely to be antigenic.

The absence of aggregates in chicken-produced mAbs further reduces the likelihood of immunological reactions in patients. Protein production from transgenic chickens made with a defined cES clone coupled with the easy scalability from one founder can also overcome the glycosylation heterogeneity observed in proteins derived from different founder transgenic goats (Zhou et al. 2005). Good manufacturing practices (GMP) conditions are well known in poultry production for the rearing of elite breeding stock and for the production of eggs that are used for vaccine production.

We have obtained up to 6 mg of MAb per egg from cES cell-derived chimeras and we predict that the level of mAb will be higher in eggs from germline transgenic hens when every cell in the tubular glands expresses the mAb transgene. The recent development of long-term culture of chicken primordial germ cells (PGCs) which retain the ability to colonize the germ line after genetic modification (van de Lavoir et al. 2006a), creates the opportunity of generating fully transgenic hens for the production of human mAb with beneficial features in chicken eggs. This improvement will yield higher amounts of mAb per egg and it will simplify scalability. Typically, a cohort of 10 PGC chimeras contains 3 males that transmit the transgene to the next generation at rates exceeding 25 percent (van de Lavoir et al. 2006a). The remarkable fecundity of chickens can then be used to produce large numbers of female offspring carrying the transgene and depositing mAb in their eggs. For example, each male PGC chimera could be bred to 140 hens per week (each male is ejaculated daily and each ejaculate can be used to inseminate 20 hens), and each hen can then lay eggs daily for a week after being inseminated at seven day intervals. The rate of lay of the flock would typically be 90 percent and approximately 80 percent of the eggs will produce a chick. Under these conditions three PGC chimeric males transmitting at 25 percent or higher would produce a flock of approximately 3000 G_1 hens in three months. If each hen produces 50 mg of mAb per day, the G_1 flock yields 45 kg of mAb annually.

Adoption of an alternate system for the manufacture of therapeutic mAbs is dependent in part on reducing the cost of bringing a new product to market. A significant part of this cost using current CHO cell-based technology is the construction of the cell culture facility. We have calculated that the cost of establishing a chicken-based production system is approximately 10 percent of that of constructing a CHO culture facility. In addition, the ongoing personnel costs are reduced to about 10 percent in a chicken-based facility. Assuming equal purification costs, the financial risk to bringing a new product to the marketplace is dramatically reduced by isolating mAb from chicken eggs.

28.7 SUMMARY

Taken together, this work demonstrates the potential of producing therapeutic proteins in chicken eggs to be a viable alternative to mammalian cell culture system. Introduction of genetic modifications with transgenes up to 150 kb via cES cells overcomes the limitation of viral expression systems on the size of the foreign DNA that can be introduced. Therefore, large portions of the regulatory sequence can be included to yield high expression levels while maintaining the expression pattern in a tissue-restricted, position-independent manner. Glycosylation of the mAb is consistent with that of humans indicating the absence of antigenicity. The absence of fucose increases ADCC which is advantageous in some applications. Purification of mAb from egg white is straightforward and yields product that is free of aggregates. The infrastructure costs to establish a production facility and the timeline for scaling up are modest in comparison to a CHO cell culture facility making it possible to increase the number of therapeutic candidates in late stage testing.

ACKNOWLEDGMENTS

This work was supported by SBIR grants from the National Institutes of Health.

REFERENCES

Benarafa, C., and E. Remold-O'Donnell. 2005. The ovalbumin serpins revisited: Perspective from the chicken genome of clade B serpin evolution in vertebrates. *Proc. Natl. Acad. Sci. U.S.A.* 102:11367–11372.

Bosselman, R.A., R.-Y. Hsu, T. Boggs, S. Hu, J. Bruszewski, S. Ou, L. Kozar, F. Martin, C. Green, F. Jacobsen, et al. 1989. Germline transmission of exogenous genes in the chicken. *Science* 243:533–535.

Cartron, G., L. Dacheux, G. Salles, P. Solal-Celigny, P. Bardos, P. Colombat, and H. Watier. 2002. Therapeutic activity of humanized anti-CD20 monoclonal antibody and polymorphism in IgG Fc receptor FcgammaRIIIa gene. *Blood* 99:754–758.

Chapman, S.C., A. Lawson, W.C. Macarthur, R.J. Wiese, R.H. Loechel, M. Burgos-Trinidad, J.K. Wakefield, R. Ramabhadran, T.J. Mauch, and G.C. Schoenwolf. 2005. Ubiquitous GFP expression in transgenic chickens using a lentiviral vector. *Development* 132:935–940.

Choi, T., M. Huang, C. Gorman, and R. Jaenisch. 1991. A generic intron increases gene expression in transgenic mice. *Mol. Cell Biol.* 11:3070–3074.

Ciejek, E.M., M.J. Tsai, and B.W. O'Malley. 1983. Actively transcribed genes are associated with the nuclear matrix. *Nature* 306:607–609.

Clynes, R.A., T.L. Towers, L.G. Presta, and J.V. Ravetch. 2000. Inhibitory Fc receptors modulate in vivo cytoxicity against tumor targets. *Nature Med.* 6:443–446.

Cook, R.F., S.J. Cook, S. Savon, M. McGrane, M. Hartitz, R.W. Hanson, and C.P. Hodgson. 1993. Liver-specific expression of a phosphoenolpyruvate carboxykinase-neo gene in genetically modified chickens. *Poult. Sci.* 72:554–567.

Copeland, N.G., N.A. Jenkins, and D.L. Court. 2001. Recombineering: A powerful new tool for mouse functional genomics. *Nature Rev. Genet.* 2:769–779.

Crooijmans, R.P., J. Vrebalov, R.J. Dijkhof, J.J. van der Poel, and M.A. Groenen. 2000. Two-dimensional screening of the Wageningen chicken BAC library. *Mamm. Genome* 11:360–363.

Echelard, Y. 1996. Recombinant protein production in transgenic animals. *Curr. Opin. Biotechnol.* 7:536–540.

Echelard, Y., C. Ziomek, and H. Meade. 2006. Production of recombinant therapeutic proteins in the milk of transgenic animals. *BioPharm Int.* August, 36–46.

Edmunds, T., S.M. Van Patten, J. Pollock, E. Hanson, R. Bernasconi, E. Higgins, P. Manavalan, C. Ziomek, H. Meade, J.M. McPherson, and E.S. Cole. 1998. Transgenically produced human antithrombin: Structural and functional comparison to human plasma-derived antithrombin. *Blood* 91:4561–4571.

Etches, R.J. 1996. *Reproduction in Poultry.* Wallingford, UK: CAB International.

Etches, R.J. 2006. The hard cell(s) of avian transgenesis. *Transgenic Res.* 15:521–526.

Evans, M.I., and G.S. McKnight. 1984. Regulation of the ovalbumin gene: Effects of insulin, adenosine 3′,5′-monophosphate, and estrogen. *Endocrinology* 115:368–377.

Ghirlando, R., J. Lund, M. Goodall, and R. Jefferis. 1999. Glycosylation of human IgG-Fc: Influences on structure revealed by differential scanning micro-calorimetry. *Immunol. Lett.* 68:47–52.

Harvey, A.J., G. Speksnijder, L.R. Baugh, J.A. Morris, and R. Ivarie. 2002a. Expression of exogenous protein in the egg white of transgenic chickens. *Nature Biotechnol.* 20:396–399.

Harvey, A.J., G. Speksnijder, L.R. Baugh, J.A. Morris, and R. Ivariet. 2002b. Consistent production of transgenic chickens using replication-deficient retroviral vectors and high-throughput screening procedures. *Poult. Sci.* 81:202–212.

Harvey, D.J., D.R. Wing, B. Kuster, and I.B. Wilson. 2000. Composition of N-linked carbohydrates from ovalbumin and co-purified glycoproteins. *J. Am. Soc. Mass Spectrom.* 11:564–571.

Hase, S., T. Sugimoto, H. Takemoto, T. Ikenaka, and K. Schmid. 1986. The structure of sugar chains of Japanese quail ovomucoid. The occurrence of oligosaccharides not expected from the classical biosynthetic pathway for N-glycans; a method for the assessment of the structure of glycans present in picomolar amounts. *J. Biochem. (Tokyo)* 99:1725–1733.

Hippenmeyer, P.J., G.G. Krivi, and M.K. Highkin. 1988. Transfer and expression of the bacterial NPT-II gene in chick embryos using a Schmidt-Ruppin retrovirus vector. *Nucleic Acids Res.* 16:7619–7632.

Iwase, H., Y. Kato, and K. Hotta. 1984. Comparative study of the carbohydrate chain of ovalbumin from various avian species by high pressure liquid chromatography. *Comp. Biochem. Physiol. B* 77B:743–747.

Jang, S.K., M.V. Davies, R.J. Kaufman, and E. Wimmer. 1989. Initiation of protein synthesis by internal entry of ribosomes into the 5′ nontranslated region of encephalomyocarditis virus RNA in vivo. *J. Virol.* 63:1651–1660.

Jang, S.K., H.G. Krausslich, M.J. Nicklin, G.M. Duke, A.C. Palmenberg, and E. Wimmer. 1988. A segment of the 5′ nontranslated region of encephalomyocarditis virus RNA directs internal entry of ribosomes during in vitro translation. *J. Virol.* 62:2636–2643.

Jang, S.K., T.V. Pestova, C.U. Hellen, G.W. Witherell, and E. Wimmer. 1990. Cap-independent translation of picornavirus RNAs: Structure and function of the internal ribosomal entry site. *Enzyme* 44:292–309.

Jang, S.K., and E. Wimmer. 1990. Cap-independent translation of encephalomyocarditis virus RNA: Structural elements of the internal ribosomal entry site and involvement of a cellular 57-kD RNA-binding protein. *Genes Dev.* 4:1560–1572.

Kamihira, M., K. Ono, K. Esaka, K. Nishijima, R. Kigaku, H. Komatsu, T. Yamashita, K. Kyogoku, and S. Iijima. 2005. High-level expression of single-chain Fv-Fc fusion protein in serum and egg white of genetically manipulated chickens by using a retroviral vector. *J. Virol.* 79:10864–10874.

Kato, S., L. Tora, J. Yamauchi, S. Masushige, M. Bellard, and P. Chambon. 1992. A far upstream estrogen response element of the ovalbumin gene contains several half-palindromic 5′-TGACC-3′ motifs acting synergistically. *Cell* 68:731–742.

Kaye, J.S., M. Bellard, G. Dretzen, F. Bellard, and P. Chambon. 1984. A close association between sites of DNase I hypersensitivity and sites of enhanced cleavage by micrococcal nuclease in the 5′-flanking region of the actively transcribed ovalbumin gene. *EMBO J.* 3:1137–1144.

Kaye, J.S., S. Pratt-Kaye, M. Bellard, G. Dretzen, F. Bellard, and P. Chambon. 1986. Steroid hormone dependence of four DNase I-hypersensitive regions located within the 7000-bp 5′-flanking segment of the ovalbumin gene. *EMBO J.* 5:277–285.

Koo, B.C., M.S. Kwon, B.R. Choi, J.H. Kim, S.K. Cho, S.H. Sohn, E.J. Cho, H.T. Lee, W. Chang, I. Jeon, et al. 2006. Production of germline transgenic chickens expressing enhanced green fluorescent protein using a MoMLV-based retrovirus vector. *FASEB J.* 20:2251–2260.

Kozak, M. 1990. Downstream secondary structure facilitates recognition of initiator codons by eukaryotic ribosomes. *Proc. Natl. Acad. Sci. U.S.A.* 87:8301–8305.

Kuster, B., T.J. Naven, and D.J. Harvey. 1996. Rapid approach for sequencing neutral oligosaccharides by exoglycosidase digestion and matrix-assisted laser desorption/ionization time-of-flight mass spectrometry. *J. Mass Spectrom.* 31:1131–1140.

Le Hir, H., A. Nott, and M.J. Moore. 2003. How introns influence and enhance eukaryotic gene expression. *Trends Biochem. Sci.* 28:215–220.

Lillico, S.G., A. Sherman, M.J. McGrew, C.D. Robertson, J. Smith, C. Haslam, P. Barnard, P.A. Radcliffe, K.A. Mitrophanous, E.A. Elliot, and H.M. Sang. 2007. From the Cover: Oviduct-specific expression of two therapeutic proteins in transgenic hens. *Proc. Natl. Acad. Sci. U.S.A.* 104:1771–1776.

Liu, H., A.K. Rajasekaran, P. Moy, Y. Xia, S. Kim, V. Navarro, R. Rahmati, and N.H. Bander. 1998. Constitutive and antibody-induced internalization of prostate-specific membrane antigen. *Cancer Res.* 58:4055–4060.

Love, J., C. Gribbin, C. Mather, and H. Sang. 1994. Transgenic birds by DNA microinjection. *Bio/Technology* 12:60–63.

Ma, J.K., P.M. Drake, and P. Christou. 2003. The production of recombinant pharmaceutical proteins in plants. *Nature Rev. Genet.* 4:794–805.

McGrew, M.J., A. Sherman, F.M. Ellard, S.G. Lillico, H.J. Gilhooley, A.J. Kingsman, K.A. Mitrophanous, and H. Sang. 2004. Efficient production of germline transgenic chickens using lentiviral vectors. *EMBO Rep.* 5:728–733.

McKnight, G.S. 1978. The induction of ovalbumin and conalbumin mRNA by estrogen and progesterone in chick oviduct explant cultures. *Cell* 14:403–413.

McKnight, R.A., A. Shamay, L. Sankaran, R.J. Wall, and L. Hennighausen. 1992. Matrix-attachment regions can impart position-independent regulation of a tissue-specific gene in transgenic mice. *Proc. Natl. Acad. Sci. U.S.A.* 89:6943–6947.

Mizuarai, S., K. Ono, K. Yamaguchi, K. Nishijima, M. Kamihira, and S. Iijima. 2001. Production of transgenic quails with high frequency of germ-line transmission using VSV-G pseudotyped retroviral vector. *Biochem. Biophys. Res. Commun.* 286:456–463.

Mozdziak, P.E., S. Borwornpinyo, D.W. McCoy, and J.N. Petitte. 2003. Development of transgenic chickens expressing bacterial beta-galactosidase. *Dev. Dyn.* 226:439–445.

Muramatsu, T., T. Imai, H.M. Park, H. Watanabe, A. Nakamura, and J. Okumura. 1998. Gene gun-mediated in vivo analysis of tissue-specific repression of gene transcription driven by the chicken ovalbumin promoter in the liver and oviduct of laying hens. *Mol. Cell. Biochem.* 185:27–32.

Niwa, R., S. Hatanaka, E. Shoji-Hosaka, M. Sakurada, Y. Kobayashi, A. Uehara, H. Yokoi, K. Nakamura, and K. Shitara. 2004a. Enhancement of the antibody-dependent cellular cytotoxicity of low-fucose IgG1 is independent of FcgammaRIIIa functional polymorphism. *Clin. Cancer Res.* 10:6248–6255.

Niwa, R., E. Shoji-Hosaka, M. Sakurada, T. Shinkawa, K. Uchida, K. Nakamura, K. Matsushima, R. Ueda, N. Hanai, and K. Shitara. 2004b. Defucosylated chimeric anti-CC chemokine receptor 4 IgG1 with enhanced antibody-dependent cellular cytotoxicity shows potent therapeutic activity to T-cell leukemia and lymphoma. *Cancer Res.* 64:2127–2133.

Okazaki, A., E. Shoji-Hosaka, K. Nakamura, M. Wakitani, K. Uchida, S. Kakita, K. Tsumoto, I. Kumagai, and K. Shitara. 2004. Fucose depletion from human IgG1 oligosaccharide enhances binding enthalpy and association rate between IgG1 and FcgammaRIIIa. *J. Mol. Biol.* 336:1239–1249.

Pain, B., M.E. Clark, M. Shen, H. Nakazawa, M. Sakurai, J. Samarut, and R.J. Etches. 1996. Long-term in vitro culture and characterisation of avian embryonic stem cells with multiple morphogenetic potentialities. *Development* 122: 2339–2348.

Petitte, J.N., M.E. Clark, G. Liu, and R.J. Verrinder Gibbins. 1990. Production of somatic and germline chimeras in the chicken by transfer of early blastodermal cells. *Development* 108:185–189.

Petitte, J.N., G. Liu, and Z. Yang. 2004. Avian pluripotent stem cells. *Mech. Dev.* 121:1159–1168.

Petropoulos, C.J., W. Payne, D.W. Salter, and S.H. Hughes. 1992. Using avian retroviral vectors for gene transfer. *J. Virol.* 66:3391–3397.

Pollock, D.P., J.P. Kutzko, E. Birck-Wilson, J.L. Williams, Y. Echelard, and H.M. Meade. 1999. Transgenic milk as a method for the production of recombinant antibodies. *J. Immunol. Methods* 231:147–157.

Raju, T.S., J.B. Briggs, S.M. Borge, and A.J. Jones. 2000. Species-specific variation in glycosylation of IgG: Evidence for the species-specific sialylation and branch-specific galactosylation and importance for engineering recombinant glycoprotein therapeutics. *Glycobiology* 10:477–486.

Rapp, J.C., A.J. Harvey, G.L. Speksnijder, W. Hu, and R. Ivarie. 2003. Biologically active human interferon alpha-2b produced in the egg white of transgenic hens. *Transgenic Res.* 12:569–575.

Ripka, J., A. Adamany, and P. Stanley. 1986. Two Chinese hamster ovary glycosylation mutants affected in the conversion of GDP-mannose to GDP-fucose. *Arch. Biochem. Biophys.* 249:533–545.

Saba, J.A., X. Shen, J.C. Jamieson, and H. Perreault. 2001. Investigation of different combinations of derivatization, separation methods and electrospray ionization mass spectrometry for standard oligosaccharides and glycans from ovalbumin. *J. Mass Spectrom.* 36:563–574.

Sanders, M.M., and G.S. McKnight. 1985. Chicken egg white genes: Multihormonal regulation in a primary cell culture system. *Endocrinology* 116:398–405.

Sang, H., C. Gribbin, C. Mather, D. Morrice, and M. Perry. 1993. Transfection of chick embryos maintained under in vitro conditions. In: R.J. Etches and Ann M.V. Gibbins (eds), *Manipulation of the Avian Genome, Chapter 8*, Boca Raton, FL: CRC Press. pp. 121–133.

Sang, H., and M.M. Perry. 1989. Episomal replication of cloned DNA injected into the fertilised ovum of the hen, *Gallus domesticus. Mol. Reprod. Dev.* 1:98–106.

Sheridan, C. 2007. Pharma consolidates its grip on post-antibody landscape. *Nature Biotechnol.* 25:365–366.

Sherman, A., A. Dawson, C. Mather, H. Gilhooley, Y. Li, R. Mitchell, D. Finnegan, and H. Sang. 1998. Transposition of the *Drosophila* element mariner into the chicken germ line. *Nature Biotechnol.* 16:1050–1053 [see comments] [published erratum appears in *Nature Biotechnol.* 17, no. 1 (January 1999):81].

Shields, R.L., J. Lai, R. Keck, L.Y. O'Connell, K. Hong, Y.G. Meng, S.H. Weikert, and L.G. Presta. 2002. Lack of fucose on human IgG1 N-linked oligosaccharide improves binding to human Fcgamma RIII and antibody-dependent cellular toxicity. *J. Biol. Chem.* 277:26733–26740.

Shinkawa, T., K. Nakamura, N. Yamane, E. Shoji-Hosaka, Y. Kanda, M. Sakurada, K. Uchida, H. Anazawa, M. Satoh, M. Yamasaki, et al. 2003. The absence of fucose but not the presence of galactose or bisecting N-acetylglucosamine of human IgG1 complex-type oligosaccharides shows the critical role of enhancing antibody-dependent cellular cytotoxicity. *J. Biol. Chem.* 278:3466–3473.

Suzuki, N., K.H. Khoo, C.M. Chen, H.C. Chen, and Y.C. Lee. 2003. N-glycan structures of pigeon IgG: A major serum glycoprotein containing Galalpha1-4 Gal termini. *J. Biol. Chem.* 278:46293–46306.

Takahashi, N., K.H. Khoo, N. Suzuki, J.R. Johnson, and Y.C. Lee. 2001. N-glycan structures from the major glycoproteins of pigeon egg white: predominance of terminal Galalpha(1)Gal. *J. Biol. Chem.* 276:23230–23239.

Takahashi, N., T. Matsuda, K. Shikami, I. Shimada, Y. Arata, and R. Nakamura. 1993. A structural study of the asparagine-linked oligosaccharide moiety of duck ovomucoid. *Glycoconj. J.* 10:425–434.

Temin, H.M. 1989. Retrovirus vectors: Promise and reality. *Science* 2465:983.

van de Lavoir, M.C., J.H. Diamond, P.A. Leighton, C. Mather-Love, B.S. Heyer, R. Bradshaw, A. Kerchner, L.T. Hooi, T.M. Gessaro, S.E. Swanberg, et al. 2006a. Germline transmission of genetically modified primordial germ cells. *Nature* 441:766–769.

van de Lavoir, M.C., and C. Mather-Love. 2006. Avian embryonic stem cells. *Methods Enzymol.* 418:38–64.

van de Lavoir, M.C., C. Mather-Love, P. Leighton, J.H. Diamond, B.S. Heyer, R. Roberts, L. Zhu, P. Winters-Digiacinto, A. Kerchner, T. Gessaro, et al. 2006b. High-grade transgenic somatic chimeras from chicken embryonic stem cells. *Mech. Dev.* 123:31–41.

Vile, R.G., A. Tuszynski, and S. Castleden. 1996. Retroviral vectors. From laboratory tools to molecular medicine. *Mol. Biotechnol.* 5:139–158.

Wang, Z., P. Engler, A. Longacre, and U. Storb. 2001. An efficient method for high-fidelity BAC/PAC retrofitting with a selectable marker for mammalian cell transfection. *Genome Res.* 11:137–142.

Webster, J., G. Donofrio, R. Wallace, A.J. Clark, and C.B. Whitelaw. 1997. Intronic sequences modulate the sensitivity of beta-lactoglobulin transgenes to position effects. *Gene* 193:239–243.

Whitelaw, C.B., A.L. Archibald, S. Harris, M. McClenaghan, J.P. Simons, and A.J. Clark. 1991. Targeting expression to the mammary gland: Intronic sequences can enhance the efficiency of gene expression in transgenic mice. *Transgenic Res.* 1:3–13.

Yamane-Ohnuki, N., S. Kinoshita, M. Inoue-Urakubo, M. Kusunoki, S. Iida, R. Nakano, M. Wakitani, R. Niwa, M. Sakurada, K. Uchida, et al. 2004. Establishment of FUT8 knockout Chinese hamster ovary cells: An ideal host cell line for producing completely defucosylated antibodies with enhanced antibody-dependent cellular cytotoxicity. *Biotechnol. Bioeng.* 87:614–622.

Zhou, Q., J. Kyazike, Y. Echelard, H.M. Meade, E. Higgins, E.S. Cole, and T. Edmunds. 2005. Effect of genetic background on glycosylation heterogeneity in human antithrombin produced in the mammary gland of transgenic goats. *J. Biotechnol.* 117:57–72.

Zhu, L., M.C. van de Lavoir, J. Albanese, D.O. Beenhouwer, P.M. Cardarelli, S. Cuison, D.F. Deng, S. Deshpande, J.H. Diamond, L. Green, et al. 2005. Production of human monoclonal antibody in eggs of chimeric chickens. *Nature Biotechnol.* 23:1159–1169.

CHAPTER 29

Production of Antibodies in Plants

KEVIN M. COX, JEFFREY T. REGAN, JASON D. STERLING,
VINCENT P. M. WINGATE, and LYNN F. DICKEY

ABSTRACT

Plants have been demonstrated as a viable alternative to traditional cell culture systems for the production of recombinant proteins. To date, there have been numerous examples of successful expression of monoclonal antibodies in a broad range of plant species. Recent advances in the expression of recombinant antibodies have made plants an attractive alternative production system to meet the increasing demands of monoclonal antibody production. Plant expression systems offer several advantages over current mammalian systems, such as scalability, reduced production costs, fast timeline for construction of manufacturing facilities, and reduced safety concerns from viral or prion contamination. Plant-produced antibodies have reached a level of quality and quantity that has allowed scale-up to commercial manufacturing levels. Major breakthroughs in the area of *N*-glycan engineering have surpassed many of the traditional expression systems and have paved the way for large-scale production of mAbs with humanized *N*-glycans. These additional benefits have been

Therapeutic Monoclonal Antibodies: From Bench to Clinic. Edited by Zhiqiang An

realized with the removal of plant-specific sugars, leading to enhanced Fc functionality and resulting in a more efficacious and potent mAb.

29.1 INTRODUCTION

In recent years, significant progress has been made toward the development of plant-based recombinant protein expression systems. Plant systems have been identified as a viable alternative to traditional cell culture systems and utilized to express a broad range of pharmaceutical proteins with varying degrees of size and complexity. Hiatt, Cafferkey, and Bowdish (1989) first reported on the successful production of a correctly folded and assembled full length IgG1 antibody in transgenic tobacco plants. Since this ground-breaking work, there have been a multitude of reports on the expression of functional full length monoclonal antibodies (mAbs) and antibody fragments, including single chain Fv, bispecific Fvs, Fv-Fcs, and Fabs in plants (De Neve et al. 1993; Fiedler and Conrad 1995; Fischer et al. 1999; Frigerio et al. 2000; Perrin et al. 2000; Ko and Koprowski 2005; Van Droogenbroeck et al. 2007). This chapter will focus on the production of full length mAbs in plants.

To date, mAbs have been produced successfully in a broad range of plant species, including tobacco (*Nicotiana tabacum*; Hiatt, Cafferkey, and Bowdish 1989), *Arabidopsis thaliana* (Bouquin et al. 2002), moss (*Physcomitrella patens*; Nechansky et al. 2007), alfalfa (*Medicago sativa*; Bardor et al. 2003b), soybean (*Glycine max*; Zeitlin et al. 1998), lettuce (*Lactuca sativa*; Negrouk et al. 2005), cowpea (*Vigna unguiculata*; Alamillo et al. 2006), corn (*Zea mays*; Law et al. 2006), and duckweed (*Lemna minor*; Gasdaska, Spencer, and Dickey 2003). MAbs have been expressed in many tissue types, including plant callus (De Neve et al. 1993), plant cell cultures (Fujiyama et al. 2006); leaves (Khoudi et al. 1999); roots (Komarnytsky et al. 2006); and seeds (Fiedler and Conrad 1995).

29.2 PLANT DERIVED mAbs AS PHARMACEUTICAL AGENTS

A number of groups have reported on the production of mAbs in plants for use as therapeutic agents (Ma et al. 1998; Zeitlin et al. 1998; Ko et al. 2003, 2005). Plant-derived mAbs have been shown to be effective both in topical applications and when administered intravenously. In particular, topical application of plant-derived mAbs has been shown to be an effective treatment against infectious disease agents such as *Streptococcus mutans* and Herpes simplex virus (Ma et al. 1998; Zeitlin et al. 1998). For these applications, the potential immunogenicity from native plant-derived *N*-glycans is not a concern since these mAbs are not required to circulate in the bloodstream. Furthermore, these nonmammalian sugars on the mAbs do not affect their function *in vivo* and have caused no adverse effects in human patients (Ma et al. 1998; Zeitlin et al. 1998). In these two examples, the primary consideration is that the mAb recognize the appropriate antigen. CaroRx, the Guy's 13 mAb produced in tobacco plants, has been shown to prevent colonization of *S. mutans* and subsequent development of dental caries in human patients (Ma et al. 1998). This mAb is the first plant-derived mAb to be clinically tested and is currently in Phase II clinical trials in the United States, and expected to be on the market within three years. An anti-HSV-2 mAb produced in transgenic soybean was shown to provide protection against HSV-2 in a mouse vaginal transmission model (Zeitlin et al. 1998). This mAb demonstrated comparable levels of protection as the mammalian cell-derived versions of the anti-HSV-2 mAb.

Plant-derived mAbs have also been administered intravenously in animal models targeting diseases such as Rabies virus and colorectal cancer and have shown appropriate *in vivo* antibody activity (Ko et al. 2003, 2005). Additionally, an intravenously administered mAb produced in plants has been shown to neutralize human chorionic gonadotropin (HCG; Kathuria et al. 2002). Since the therapeutic mAb is to be administered intravenously to human patients, consideration must be given to the *N*-glycan profile. Ideally the *N*-glycans from plant-derived mAbs will be free of plant-specific sugar linkages and of complex order (i.e., not high-mannose) to deliver optimal pharmacokinetics. Removal of these sugar residues is desirable since their presence may be immunogenic to humans, and the presence of fucose may limit antibody-dependent cellular cytotoxicity (ADCC) and FcR binding. The SO57 mAb expressed in tobacco plants was shown to provide comparable postexposure prophylaxis to the rabies virus in hamsters as the mammalian-derived mAb (Ko et al. 2003). Additionally, the CO17-1A mAb produced in tobacco plants was shown to have comparable binding activity and tumor inhibition activity for human colorectal cancer cells in nude mice as its mammalian cell-derived counterpart (Ko et al. 2005). The PIPP antibody produced transiently in tobacco was shown to inhibit HCG-stimulated production of testosterone and delay uterine weight gain in a mouse model (Ma, Drake, and Christou 2003). Finally, a corn-derived mAb, Avicidin, administered in humans has shown anticancer activity in the treatment of colorectal cancer, confirming the *in vivo* functionality of a plant-derived mAb. Avicidin is the first example of a plant-derived mAb administered in humans and showed comparable activity to the equivalent mammalian-produced mAb (Ma, Drake, and Christou 2003).

29.3 PLANT TRANSFORMATION APPROACHES

The vast majority of stable transgenic plant expression systems producing therapeutic proteins have relied on the *Agrobacterium*-mediated method of nuclear transformation (Giddings et al. 2000) and to a lesser extent biolistic transformation (Larrick et al. 2001). Stably transformed expression systems that require sexual reproduction for subsequent generations can take a few years from the initial transformation event to obtaining gram quantities of purified antibody. This time frame can be reduced in certain crops with high seed number (e.g., tobacco) and with multiple plantings per year. In clonal systems such as *Lemna* and moss, tissue is maintained by vegetative rather than sexual reproduction, thus allowing for the perpetual propagation of transgenic lines. The time frame for production of gram quantities of antibody in *Lemna* has been reported to be six months from the initiation of plant transformation.

Alternative non-nuclear approaches for protein production include chloroplast transformation (Daniell, Streatfield, and Wycoff 2001) and transient expression. Transient expression systems rely primarily on either *Agrobacterium* infiltration (Vaquero et al. 2002) or viral vector systems (Verch, Yusibov, and Koprowski 1998). The major drawback of the initial transient expression approaches was the inability to produce high level co-expression of more than one recombinant protein. An improved transient expression system known as Magnifection has been developed based on *Agrobacterium* infection of whole plants with T-DNAs encoding viral replicons. This technology allows for the expression of mAbs at a concentration of 0.5 g/kg fresh weight within a few weeks in contained growth chambers (Giritch et al. 2006; Hiatt and Pauly 2006). The scalability of individual plant inoculation and product consistency in this system remains to be demonstrated.

Containment of field-grown transgenic plant material and segregation of the transgenic material from the food supply are challenges that must be overcome if these expression systems are to be successful. Since 1997, a number of transgene releases and/or contaminations of the food supply have been reported (Ledford 2007). These incidents have further heightened public awareness and concern, resulting in increased governmental oversight (Wolt et al. 2004; Elbehri 2005). The ability to outcross the transgene through pollen must be reduced through the use of self-pollinated crops or prevented by using male sterile plants (Huang 2004). Containment of the transgenic material can be ensured by growing the transgenic plants in greenhouses or controlled growth rooms. This has been accomplished with systems such as *Lemna* and moss, and through the use of plant root secretion and plant cell culture (Gasdaska, Spencer, and Dickey 2003; Decker and Reski 2004; Fujiyama et al. 2006; Komarnytsky et al. 2006).

29.4 BENEFITS OF PLANT-BASED ANTIBODY EXPRESSION

There are a number of benefits to using a plant-based expression system for the production of mAbs (Stoger et al. 2002; Twyman et al. 2003; Schillberg, Twyman, and Fischer 2005). Both the capital equipment and production costs of producing antibodies in plants are significantly less than production in stainless steel bioreactors currently used for mammalian cell-based systems. In general, plant systems can be economically scaled simply by increasing acreage, greenhouse or growth room space in comparison to the large capital investment required for increasing the footprint of a mammalian cell culture production facility. Due to the simplicity of the transgenic plant production format, the time to build a commercial facility is significantly reduced. As a result, the decision on when to build a commercial facility can be delayed until much later in clinical testing.

An additional advantage of plant expression systems over mammalian systems is that there is no risk of animal pathogenic contaminants, including viruses, bacterial toxins, endotoxins, oncogenic DNA and prions (Twyman et al. 2003). In mammalian cell production systems, safeguards must be included in both the production and purification processes to ensure that animal pathogenic contaminants do not compromise the final product.

29.5 PRODUCTION OF PLANT-MADE ANTIBODIES

There are several considerations when determining a suitable plant expression system for use in the production of mAbs. Criteria such as expression level, growth cycle, biomass yield, set-up and infrastructure required for scale-up (including land, labor, and processing facilities), and product consistency and containment all play a significant role in the choice of the production system and the total production costs. Additionally, total protein content, ease of extractability, and storage of biomass impact the downstream processing costs. Requirements for downstream processing techniques that comply with GMP regulations involve extensive quality control and assurance programs, including the establishment of master plant banking procedures (Schillberg, Twyman, and Fischer 2005).

29.5.1 Expression Level

Transgenic plant systems show large variations in the expression levels of mAbs. A comprehensive comparison of the individual expression systems is difficult because no single mAb has been produced in all available systems to provide comparable yield data. However, a significant amount of expression data exists in each of the systems. Expression levels are typically represented in three ways: grams of mAb per kilogram of fresh weight (g/kg FW), grams of mAb per kilogram of dry weight (g/kg DW), and percent of the total soluble protein (percent TSP). It is difficult to compare fresh weight and dry weight values due to variation in the total water content of a particular system. In some leafy tissues, the water content can be as high as 90 percent of the total weight. The use of percent TSP is useful when comparing expression systems, but differences in extraction techniques and total protein content of the expression system must be taken into account when using this comparison. IgG1 mAb production has been reported in transgenic tobacco via stable transformation with expression levels up to 1.3 percent TSP (Hiatt, Cafferkey, and Bowdish 1989), and a SIgA mAb was reported at 500 mg/kg (Ma et al. 1995). Transient expression of IgG1 mAbs have been reported at levels up to 40 mg/kg FW (Kathuria et al. 2002), while expression in roots has been reported at levels of 11.7 mg/kg FW (Drake et al. 2003) and in tobacco cell culture at levels up to 0.6 percent TSP (Yano, Maeda, and Takekoshi 2004). Using the Magnifection transient expression system in tobacco, expression levels of 500 mg/kg FW have also been reported (Giritch et al. 2006). MAbs have also been produced in alfalfa at 1 percent TSP (Khoudi et al. 1999), lettuce at 80 mg/kg FW (Negrouk et al. 2005), corn at 0.3 percent TSP (Hood, Woodard, and Horn 2002) and *Arabidopsis* with expression levels up to 1.3 percent TSP (De Wilde et al. 1998). Transgenic moss was capable of the expression of an IgG1 antibody up to 14.0 mg/L (Schuster et al. 2007), and an IgG4 antibody up to 6.5 percent TSP (Gorr and Jost 2005). MAb expression in *Lemna* has recently been reported at expression levels up to 6.0 percent TSP (Cox et al. 2006), equivalent to 1 g/kg FW and 10 g/kg DW. A comprehensive list of antibodies produced in plants can be found in Table 29.1.

29.5.2 Factors Affecting Antibody Yield

In addition to mAb expression level, another important factor affecting the total biomass yield is the length of the growth cycle for each production system. In general, green leafy plants such as tobacco, alfalfa, and *Lemna* have a higher biomass yield because they can be harvested multiple times per year and require little time to scale-up production. A potential downfall of recovery from leaf tissue is the requirement for immediate processing or frozen storage of the material upon harvest to preserve the mAb integrity (Schillberg, Twyman, and Fischer 2005). This requires significant infrastructure for the transport and processing of the harvested material, which impacts total production costs. Cereals and legumes, such as rice and wheat, offer the advantage of protein accumulation in seeds where long-term storage at ambient temperature does not impact protein quality and the recombinant protein is in high concentration in the seeds, potentially minimizing downstream manipulation of the extracted material (Khoudi et al. 1999; Petruccelli et al. 2006). The use of fruits and vegetables for mAb production allows for simpler extraction techniques, but the aqueous extraction also may increase the potential for proteolysis (Moloney 1995).

The overall yield of mAbs produced in plants must also take into account a comparison of the expression level and the *in planta* stability of the protein in a particular production system. In the majority of cases, the recombinant protein accumulation and stability *in planta* vary significantly during the growth cycle depending on the growth conditions. For example, the product quality and concentration of mAbs produced in transgenic tobacco has been shown to vary considerably with variations in light intensity and temperature as a result of tissue senescence and total protein content (Elbers et al. 2001). Additional studies have reported significant *in planta* proteolysis of recombinant IgG in transgenic tobacco through stable intermediates (Ma et al. 1994; Wongsamuth and Doran 1997; Sharp and Doran 2001) typically as a result of tissue senescence and plant stress when nutrients are remobilized to other plant tissues (Stevens et al. 2000; Elbers et al. 2001).

Product consistency between multiple transgenic lines and upon scale-up has been a limitation in many expression systems, frequently leading to batch losses due to heterogeneous final product. A thorough assessment of plant-produced mAbs requires analysis of posttranslational modifications and batch-to-batch variation to ensure product homogeneity, stability, and activity (Schillberg, Twyman, and Fischer 2005). Regulatory agencies are particularly concerned about batch-to-batch heterogeneity of mAb *N*-glycans, which has been a consistent problem in mammalian expression systems. Many plant expression systems also show heterogeneity in their mAb *N*-glycan profiles based on varied growth scales and conditions. One notable exception is seen with mAbs produced in *Lemna* as is discussed later in this chapter.

29.5.3 Storage and Extraction

The majority of the total production cost of mAbs (up to 60 percent depending on the expression system) occurs in downstream processing—extraction, purification, and analytical characterization of the mAb. As a result, several criteria must be evaluated when choosing an efficient plant production system. Total protein content, ease of extractability, and storage of biomass impact the overall production cost. Ease of handling, as well as high protein content, makes seed storage an attractive method for mAb production. Conversely, it is easier to extract protein from watery tissues than from dry material that requires significant downstream processing steps to extract protein (Schillberg, Twyman, and Fischer 2005). Protein extraction from plant cells requires physical disruption of the plant tissue and seeds using equipment that often requires large capital expenditures. Efficient cell breakage is required to maximize protein extraction and yield, while maintaining product quality. Downstream processing techniques for protein extraction include milling, aqueous extraction, vacuum infiltration, and ultrafiltration. Transgenic tobacco releases toxic alkaloids and polyphenols upon extraction that must be removed. Seed-based systems do not suffer the same toxin release problems, but often require more intense extraction procedures to extract protein. A series of clarification techniques, centrifugation and ultrafiltration, are often employed prior to traditional chromatography to

TABLE 29.1 Comparison of Monoclonal Antibody Expression Levels from Various Plant Production Systems

Production System	Antibody Type	Antigen	Expression Level	References
Tobacco (*Nicotiana tabacum*)	IgG1	Human creatine kinase	0.055% TSP	De Neve et al. (1993)
		Streptococcal surface antigen (I/II)	1.1% TSP	Vine et al. (2001)
		Hepatitis B virus surface antigen	25 mg/kg FW	Valdez et al. (2003a, 2003b)
		Rabies virus	0.07% TSP	Ko et al. (2003)
		Tumor-associated antigen EpCAM	0.02% TSP/0.9 mg/kg FW	Ko et al. (2005)
		Hepatitus B surface antigen	0.5% TSP	Ramirez et al. (2003)
		Phosphonate Ester (P3)	1.3% TSP	Hiatt, Cafferkey, and Bowdish (1989)
	SIgA	Streptococcal surface antigen (I/II)	500 mg/kg FW	Ma et al. (1995)
	IgM	NP (B 1-8)	Not reported	During et al. (1990)
Tobacco (*Nicotiana tabacum*) transient expression	IgG1	Human choriogonadotrophin	40 mg/kg FW	Kathuria et al. (2002)
		Human choriogonadotrophin	0.1% TSP	Sriraman et al. (2004)
		Human carcinoembryonic antigen	1 mg/kg FW	Vaquero et al. (1999)
		Epidermal growth factor receptor	Not reported	Rodriquez et al. (2005)
Tobacco (*Nicotiana tabacum*) root expression	IgG1	Streptococcal surface antigen (I/II)	11.7 mg/kg DW	Drake et al. (2003)
Tobacco (*Nicotiana tabacum*) cultured cells	IgG1	Hepatitis B virus surface antigen	0.6% TSP	Yano, Maeda, and Takekoshi (2004)
		RNaseA	Not reported	Fujiyama et al. (2006)
Tobacco (*Nicotiana benthamiana*) transient expression	IgG1	Protective antigen of *Bacillus anthracis*	Not reported	Hull et al. (2005)
Tobacco (*Nicotiana benthamiana*) viral expression	IgG1	Colon cancer antigen	Not reported	Verch, Yusibov, and Koprowski (1998)
		Human tumor specific (A5)	500 mg/kg	Giritch et al. (2006)

Moss (*Physcomitrella patens*)	IgG4	Not reported	6.5% TSP (5.8 g/kg DW)	Gorr and Jost (2005)
Moss (*Physcomitrella patens*) transient expression	IgG1	Tumor antigen-specific (IGN314)	14.0 mg/L	Schuster et al. (2007)
Soybean (*Glycine max*)	IgG1	Herpes simplex virus 2	Not reported	Zeitlin et al. (1998)
Alfalfa (*Medicago sativa*)	IgG1	Human IgG	1.0% TSP	Khoudi et al. (1999)
Lettuce (*Lactuca sativa*) transient expression	IgG1	Anti-tissue factor antibody (hOAT)	80 mg/kg FW	Negrouk el al. (2005)
Maize (*Zea mays*)	SIgA	Not reported	0.3% TSP	Hood, Woodard, and Horn (2002)
Duckweed (*Lemna minor*)	IgG1	CD-30	6.0% TSP	Cox et al. (2006)
		Not reported	2.8% TSP (5.6 g/kg DW)	Gasdaska, Spencer, and Dickey (2003)
Thale cress (*Arabidopsis thaliana*)	IgG1	Human creatine kinase	1.3% TSP	De Wilde et al. (1998)
		Rhesus D antigen	0.6% TSP	Bouquin et al. (2002)
		HIV	0.2% TSP	Schahs et al. (2007)
Winter cherry (*Physalis alkekengi*) transient expression	IgG1	Hepatitis B virus surface antigen	Not reported	Sriraman et al. (2004)

remove the tissue extract components and concentrate the feed stream. The amount of time the protein is exposed to the crude extract during clarification can also impact the final product yield due to proteolysis (Pujol et al. 2005). For example, proteolytic degradation in tissue extract has been shown to produce Fab and Fab′2-like proteins in transgenic tobacco that requires the addition of protease inhibitors and protein stabilizers to the tissue extract during processing (Sharp and Doran 2001).

29.5.4 Purification

Plant production systems are unique in that they require no viral inactivation steps because plant viruses are not known to infect humans or animals, and potentially pathogenic human and animal viruses are not capable of replicating in plant cells (Miele 1997). Several traditional affinity chromatography techniques exist for the purification of mAbs. Affinity purification using protein A or protein G resins takes advantage of an interaction between the Fc portion of mAbs and the resins through hydrophobic and hydrogen bonding interactions at neutral pH (Ghose et al. 2005). In addition, affinity purification using protein L resin takes advantage of an interaction with the resin and the light chain of mAbs and their fragments, including Fab and scFV fragments. This resin has been shown to have low binding capacity and requires the presence of a kappa light chain. In addition, the resin has proven problematic because of the ability to bind antibody fragments. Elution from all of the resins is accomplished with a simple pH adjustment to a low pH buffer that is sufficient to disrupt the resin interactions and elute the protein. Typical protein purity of greater than 90 percent can be achieved with a single chromatography step. Recombinant protein A Streamline technology has been developed as a single step purification strategy used to handle unclarified material to provide a combination of clarification, concentration, and purification in a single step. This resin is optimal because downstream processing can compromise the stability and recovery of the mAb. In many cases, the use of Streamline protein A resin has proven problematic with plant systems because of the complexity of the tissue extract, and an initial clarification step is often required prior to chromatography to remove plant particulates and avoid column blockage. In addition, green components in the tissue extract often interact with the resin and cause the eluted material to contain a yellow color while reducing resin lifetimes due to fouling (Valdes et al. 2003a). Concentration techniques such as ammonium sulfate precipitation and ultrafiltration are also used as a means to concentrate the material and remove some of the small molecular weight impurities prior to chromatography.

In addition to the classic binding site, some mAbs interact with protein A resin through an interaction with the heavy chain variable domain (Inganas 1981; Vidal and Conde 1985; Akerstrom et al. 1994; Starovasnik et al. 1999). Proteolytic degradation of mAbs has been shown to produce Fab and Fab′2-like proteins that can interact with protein A through a variable region interaction and must be removed from the intact material during purification (Sharp and Doran 2001). A resin designed by GE Healthcare, mAbSelect SuRe, to withstand alkaline conditions for resin regeneration procedures has been shown to eliminate that variable region interaction by removing several of the protein A-binding domains. In addition to the benefits of higher cleanability, the resin is efficient in mAb fragment removal because the fragments will not bind to the resin. A final benefit of the resin is the adjustment of the elution pH to a higher pH for some mAbs, minimizing problems associated with aggregation induced by low pH (Ghose et al. 2005).

29.6 PLANT SECRETORY PATHWAY

Plants and animals are highly similar with respect to protein synthesis and posttranslational modifications. The secretory pathways of both systems share the same mechanisms in the endoplasmic reticulum (ER) and Golgi apparatus for signal peptide recognition, protein folding, disulfide bond formation, *N*-glycan addition, and protein trafficking. In plant cells, signal peptides direct the light and heavy chains to the ER (Hiatt, Cafferkey, and Bowdish 1989; Hein et al. 1991), where the signal peptides are proteolytically removed and the mAb is assembled into the heterotetrameric

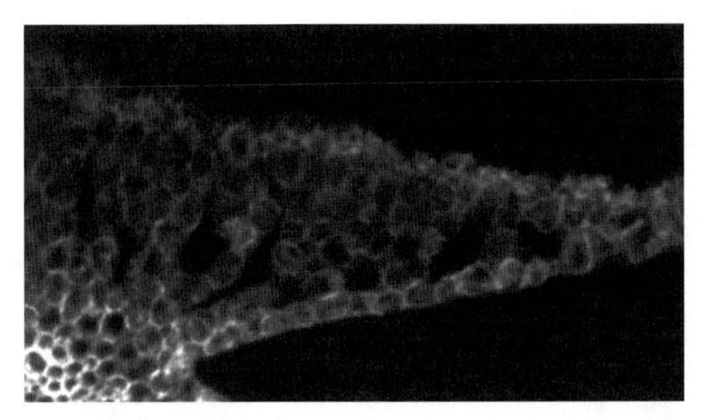

Figure 29.1 Immunolocalization of IgG1 mAb expressed in *Lemna* (cross section). IgG1 mAb expressed in a stable transgenic *Lemna* line; immunolocalized using a goat anti-human IgG1 followed by a donkey anti-goat flourescently labeled antibody. The recombinant antibody visualized in blue is localized to the apoplast of the *Lemna* frond and is expressed evenly throughout. (See color insert.)

polypeptide. Interestingly, signal peptides from either plant or animal origin are adequate to direct translation into the ER of plants. In the plant secretory pathway, chaperone proteins and other similar enzymes such as protein disulfide isomerase and the enzymes involved in the early stages of *N*-glycosylation are highly similar to their mammalian counterparts. It has been shown in plants, as in mammalian systems, that the expression of antibody molecules is associated with an increase in expression of the BiP chaperone protein and protein disulfide isomerase (Nuttall et al. 2002). In plant cells, following assembly and posttranslational modification in the ER and Golgi endomembrane system, vesicles fuse with the plasma membrane and empty the contents into the apoplastic space, defined as cell wall and intracellular space (Fig. 29.1).

29.6.1 *N*-Glycosylation of Plant Derived mAbs

Glycosylation plays a critical role in the structure and function of all mAb isotypes (IgG, IgD, IgA, IgM, and IgE; Rudd et al. 1991, 2001; Bjorklund, Karlsson, and Magnusson 1999; Arnold et al. 2004, 2005). To this end, correct glycosylation is critical to the success of recombinantly expressed mAbs. Arguably, one of the biggest concerns with the expression of therapeutic mAbs in plant expression systems is the difference between plant and human *N*-glycan structures. Not only is *N*-glycosylation considered to be the largest source of heterogeneity for recombinantly expressed mAbs (Jefferis 2005; Bardor et al. 2006), the *in vitro* and *in vivo* effector function of human and recombinant mAbs is highly dependent on their *N*-glycan profile (Malhotra et al. 1995; Niwa et al. 2004; Kaneko, Nimmerjahn, and Ravetch 2006).

Comparision of *N*-glycan profiles from numerous monocot and dicot species (Altmann 1998; Wilson and Altmann 1998; Rayon et al. 1999; Kimura and Matsuo 2000; Bakker et al. 2001; Wilson et al. 2001; Bardor et al. 2003a, 2003b; Misaki et al. 2003), ferns (Fitchette et al. 1999), and moss (Koprivova et al. 2003; Vietor et al. 2003) indicate that plant *N*-glycosylation pathways are highly conserved across all plant species. *N*-glycosylation in plants begins as it does for all eukaryotes with the synthesis of $Glc_3Man_9GlcNAc_2$ in the endoplasmic reticulum (Freeze and Aebi 2005). The oligosaccharide is transferred *en masse* to the asparagine (Asn) residue of an Asn-Xxx-Ser/Thr sequon (where Xxx is any amino acid except Pro or Asp) on a nascent polypeptide by oligosaccharyltransferase (Kelleher and Gilmore 2006). Following the removal of two Glc residues and folding of the glycoprotein via the action of the calnexin/calreticulin protein folding machinery (Moremen and Molinari 2006), the last Glc residue and one Man residue are removed and the glycoprotein is transported to the Golgi apparatus.

TABLE 29.2 Reference Table for Plant-Derived mAb *N*-Glycan Structures

Glycan	Symbol	Structure	Glycan	Symbol	Structure
GnM		Manα1↘6 Manβ1→4GlcNAcβ1→4GlcNAcβ GlcNAcβ1→2Manα1↗3	Man3		Manα1↘6 Manβ1→4GlcNAcβ1→4GlcNAcβ Manα1↗3
GnM_{iso}		GlcNAcβ1→2{Manα1↘6 / Manα1↗3} Manβ1→4GlcNAcβ1→4GlcNAcβ	MMX		Manα1↘6 Manβ1→4GlcNAcβ1→4GlcNAcβ Manα1↗3 2↑ Xylβ1
GnGnXF		GlcNAcβ1→2Manα1↘6 Manβ1→4GlcNAcβ1→4GlcNAcβ GlcNAcβ1→2Manα1↗3 2↑ Xylβ1 3↑ Fucα1	Man5		Manα1↘6 Manα1↘6 Manα1↗3 Manβ1→4GlcNAcβ1→4GlcNAcβ Manα1↗3
GnGn		GlcNAcβ1→2Manα1↘6 Manβ1→4GlcNAcβ1→4GlcNAcβ GlcNAcβ1→2Manα1↗3	Man5A		Manα1↘6 Manα1↘6 Manα1↗3 Manβ1→4GlcNAcβ1→4GlcNAcβ Galβ1→4GlcNAcβ1→2Manα1↗3
MMXF		Manα1↘6 Manβ1→4GlcNAcβ1→4GlcNAcβ Manα1↗3 2↑ Xylβ1 3↑ Fucα1	Man5Gn		Manα1↘6 Manα1↘6 Manα1↗3 Manβ1→4GlcNAcβ1→4GlcNAcβ GlcNAcβ1→2Manα1↗3
$GnM_{iso}XF$		GlcNAcβ1→2{Manα1↘6 / Manα1↗3} Manβ1→4GlcNAcβ1→4GlcNAcβ 2↑ Xylβ1 3↑ Fucα1	Man6		Manα1→2{Manα1↘6 Manα1 / Manα1↗3}↘6 Manβ1→4GlcNAcβ1→4GlcNAcβ Manα1↗3
$GnM_{iso}X$		GlcNAcβ1→2{Manα1↘6 / Manα1↗3} Manβ1→4GlcNAcβ1→4GlcNAcβ 2↑ Xylβ1	Man7		{Manα1↘6 Manα1 / Manα1↗3}↘6 Manβ1→4GlcNAcβ1→4GlcNAcβ {Manα1→2Manα1↗3}
GnGnX		GlcNAcβ1→2Manα1↘6 Manβ1→4GlcNAcβ1→4GlcNAcβ GlcNAcβ1→2Manα1↗3 2↑ Xylβ1	Man8		Manα1→2{Manα1↘6 Manα1 / Manα1↗3}↘6 Manβ1→4GlcNAcβ1→4GlcNAcβ Manα1→2Manα1→2Manα1↗3
AM_{iso}		Galβ1→3GlcNAcβ1→2{Manα1↘6 / Manα1↗3} Manβ1→4GlcNAcβ1→4GlcNAcβ			

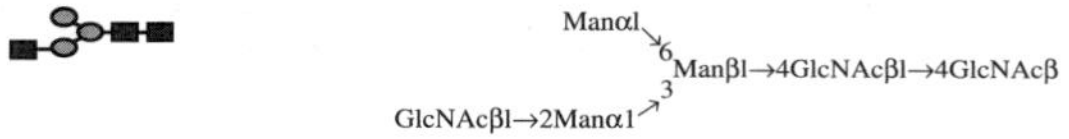

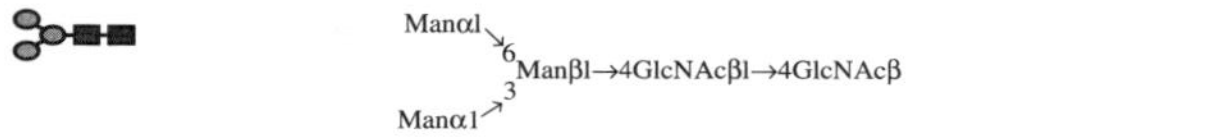

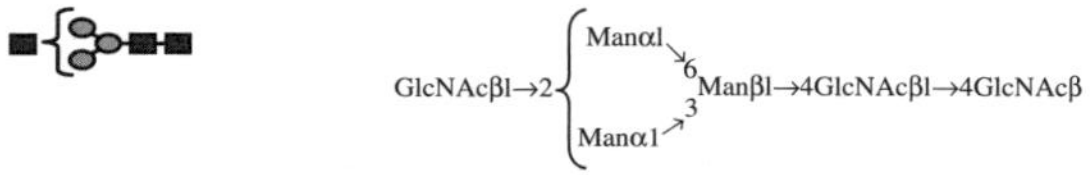

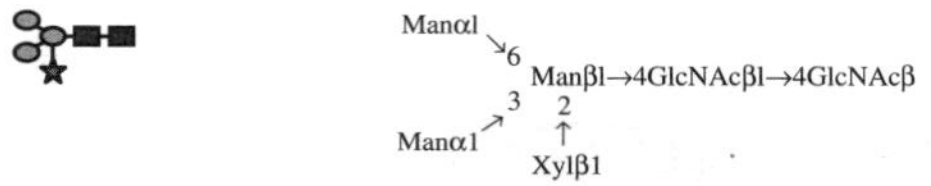

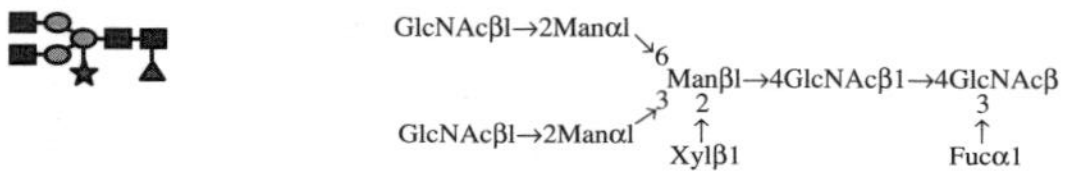

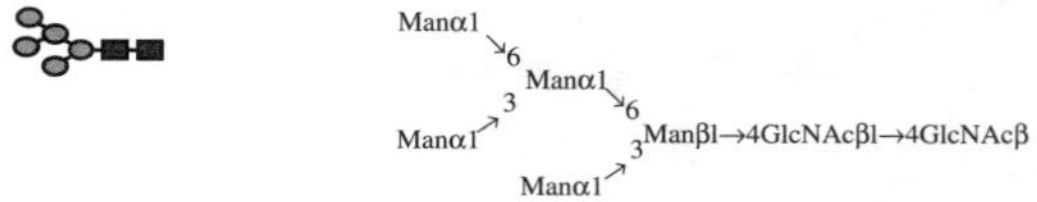

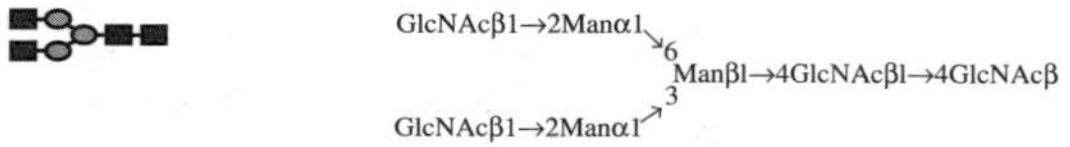

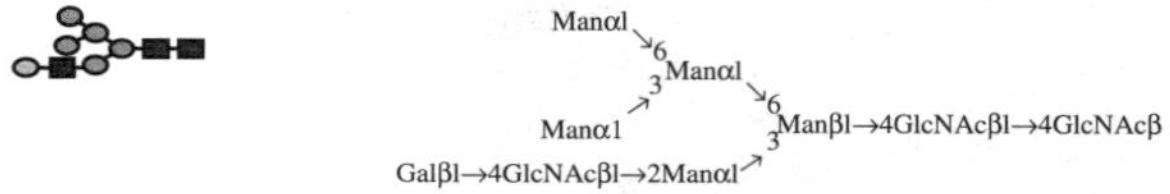

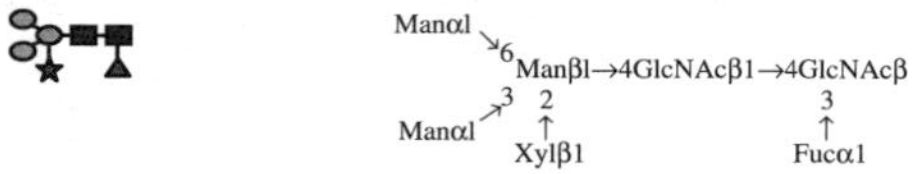

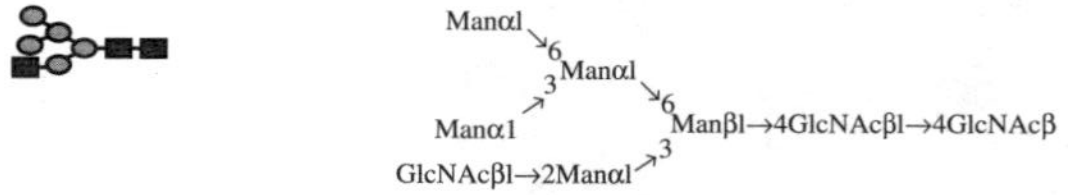

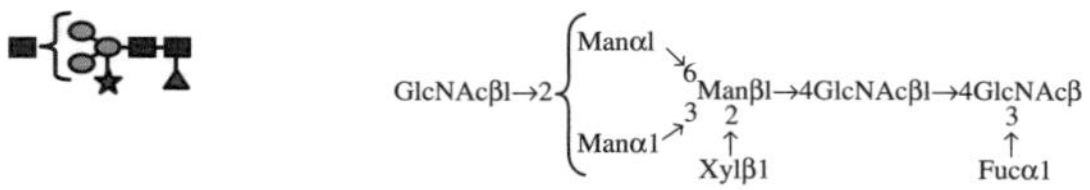

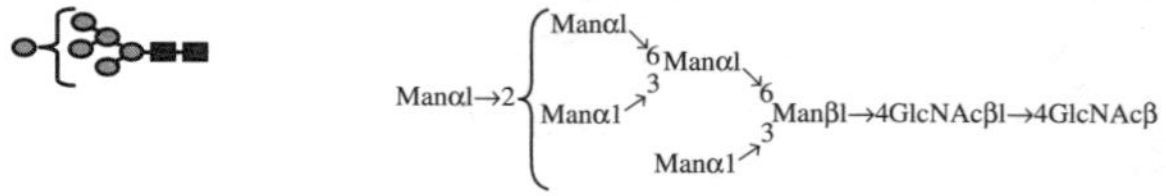

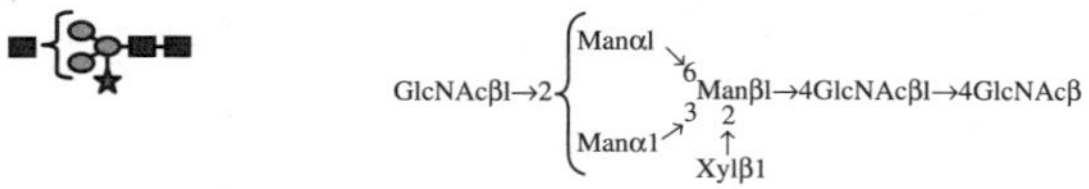

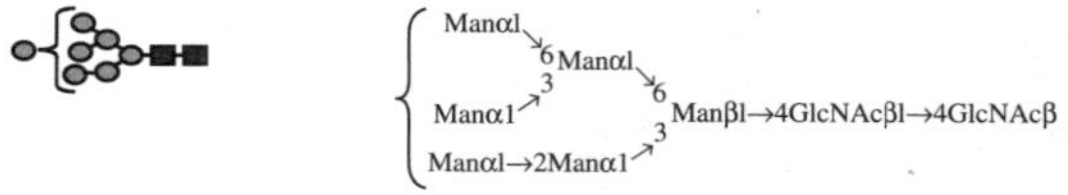

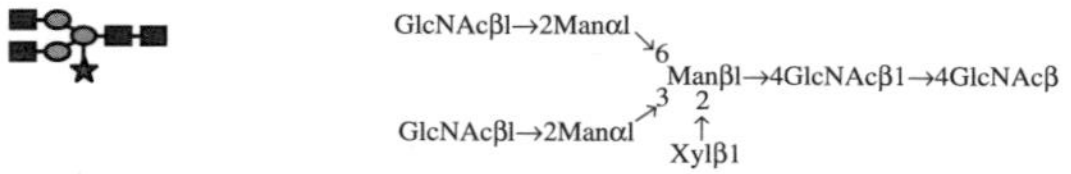

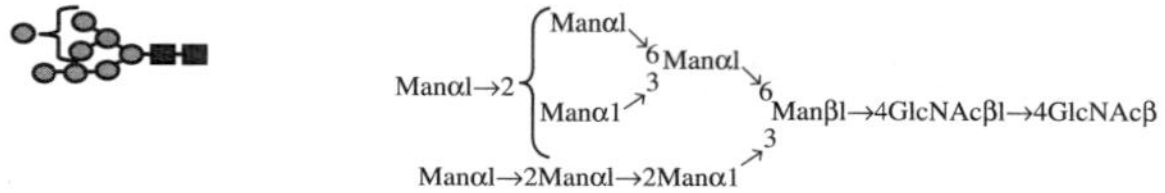

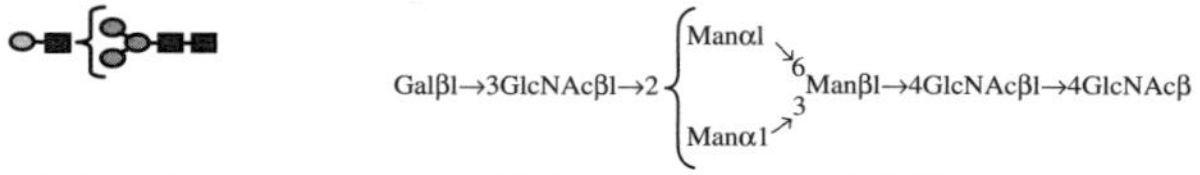

TABLE 29.3 ***N*-Glycan Composition of Plant-Derived mAbs: Comparison of Wild-Type and Glyco-Optimized Products**

Plant Species	mAb Type	mAb Name	No. of *N*-Glycan Structures (Wild-Type)	Glyco-Optimization Strategy	No. of *N*-Glycan Structures (after Glyco-Optimization)	References
Tobacco (*Nicotiana tabacum*)	IgG1	MGR48	7–17	—	—	Elbers et al. (2001)
Tobacco (*Nicotiana tabacum*)	IgG1	Guy's 13	8	—	—	Cabanes-Macheteau et al. (1999)
Tobacco—BY2 cells (*Nicotiana tabacum*)	IgG1	anti-RNaseA	7	—	—	Fujiyama et al. (2006)
Alfalfa (*Medicago sativa*)	IgG1	C5-1	6	—	—	Bardor et al. (2003b)
Duckweed (*Lemna minor*)	IgG1	MDX-060	3	XylT/FucT KO (RNAi)	1	Cox et al. (2006)
Tobacco (*Nicotiana tabacum*)	IgG1	SO57	—	KDEL	6	Ko et al. (2003)
Tobacco (*Nicotiana tabacum*)	IgG1	anti-Rabies	10	KDEL	6	Tekoah et al. (2004)
Tobacco (*Nicotiana tabacum*)	IgG2b	CB.Hep1	—	KDEL	9	Triguero et al. (2005)
Tobacco (*Nicotiana tabacum*)	IgG1	cPIPP	8	KDEL	4	Sriraman et al. (2004)
Tobacco (*Nicotiana tabacum*)	IgG1	MGR48	9	β-1,4-GalT	14	Bakker et al. (2001)
Tobacco (*Nicotiana tabacum*)	IgG1	MGR48	—	β-1,4-GalT	7	Bakker et al. (2006)
Moss (*Physcomitrella patens*)	IgG1	IGN311	—	XylT/FucT KO (HR)	3	Schuster et al. (2007)
Thale cress (*Arabidopsis thaliana*)	IgG1	2G12	5	XylT/FucT KO (Mut)	4	Schahs et al. (2007)

The divergence of plant *N*-glycosylation pathways from other eukaryotic systems begins in the Golgi apparatus with the concerted action of Golgi resident mannosidases (Moremen 2002) and *N*-acetylglucosyltransferases (Strasser et al. 1999a, 1999b), which convert Man_8 to GnM. GnM is then further modified by Golgi-resident glycosyltransferases to produce complex-type, diantennary *N*-glycans terminating in mannose (Man) or N-acetylglucosamine (GlcNAc) with xylose (Xyl) β-1,2-linked to the core β-linked Man residue and fucose (Fuc) α-1,3-linked to the Asn-linked GlcNAc residue of the *N*-glycan (Chen et al. 2005). This is in contrast to *N*-glycans found on human glycoproteins, which range from di- to tetraantennary structures and usually contain Fuc α-1,6-linked to Asn-linked GlcNAc residue and terminate in galactose (Gal) or sialic acid (NeuA) (Faye et al. 2005). A compilation of *N*-glycan structures commonly present on plant-derived antibodies is shown in Table 29.2.

The *N*-glycan profiles of recombinant plant mAbs have been elucidated in a small subset of plant species, including tobacco (Cabanes-Macheteau et al. 1999; Elbers et al. 2001; Fujiyama et al. 2006), alfalfa (Bardor et al. 2003b), *Lemna* (Cox et al. 2006), *Arabidopsis* (Schahs et al. 2007), and moss (Schuster et al. 2007). Results from these studies suggest that plants mainly attach complex or paucimannosidic *N*-glycans containing β-1,2-linked Xyl and/or α-1,3-linked Fuc, and terminating in GlcNAc. High-mannose structures are also found on plant mAbs while the presence of hybrid *N*-glycans is uncommon and has only been detected on mAbs expressed in tobacco BY2 cell cultures (Triguero et al. 2005). The exact complement (type and number) of *N*-glycan structures added onto each recombinant mAb is dependent on the individual plant expression system (Table 29.3) and partially on the spatial constraints imposed on the plant glycosyltransferases by the central cavity created between the two mAb heavy chains (Harris, Skaletsky, and McPherson 1998; Krapp et al. 2003; Jones, Krag, and Betenbaugh 2005; Bardor et al. 2006). MAbs produced under standard growth conditions and without glycoengineering typically contain from three to eight different major *N*-glycan structures. Conversely, the developmental stage and growth conditions of the plant may impact *N*-glycan profiles. In particular, a mouse IgG1 produced in tobacco and isolated from leaves grown under different conditions showed a range from 7 to 17 different *N*-glycan structures (Elbers et al. 2001).

29.7 *N*-GLYCAN OPTIMIZATION OF PLANT-DERIVED mAbs

N-glycans containing β-1,2-linked Xyl and α-1,3-linked Fuc in plants represent a structural difference between plant- and human-derived glycoproteins. These plant-specific sugars have proven immunogenic in rabbits injected with plant glycoproteins (Faye et al. 1993). Furthermore, a portion of the human population contains circulating IgEs that cross-react with these carbohydrate structures (van Ree et al. 2000; Bardor et al. 2003a; Bencurova et al. 2004). At present, the clinical relevance of these structural differences has yet to be elucidated as the immunization of mice with a plant-derived mAb containing native *N*-glycans did not result in an antiglycan immune reaction (Chargelegue et al. 2000; Jin et al. 2006) and plant-derived therapeutic mAbs containing these structures have been used successfully in human immunotherapeutic studies (Ma et al. 1998; Zeitlin et al. 1998). Nevertheless, several groups have made progress in the area of *N*-glycan humanization by eliminating the presence of α-1,3-Fuc and β-1,2-Xyl from plant-derived mAbs. Various strategies have been employed successfully, including fusion of a KDEL ER retention sequence to the mAb heavy and/or light chains, transformation of plants with human β-1,4-galactosyltransferase (β-1,4-GalT), and targeted silencing or knockout of the endogenous β-1,2-xylosyltransferase (β-1,2-XylT) and α-1,3-fucosyltransferase (α-1,3-FucT) proteins (Table 29.3).

The addition of an ER retention sequence to either the heavy and/or light chain of mAbs expressed in tobacco plants results primarily in high mannose *N*-glycan structures ranging from reduced to undetectable levels of Fuc and Xyl (Ko et al. 2003; Sriraman et al. 2004; Tekoah et al. 2004; Triguero et al. 2005). Sriraman et al. (2004) reported *N*-glycans completely devoid of Fuc and Xyl on mAbs containing an ER retention sequence while Ko et al. (2003) and Tekoah et al. (2004) reported that *N*-glycans contained Xyl, but no detectable levels of Fuc. Further complicating the issue, Triguero et al. (2005) reported the presence of *N*-glycans containing trace amounts of both Fuc and Xyl. These results could reflect differences in the efficiencies of the different KDEL ER retention signals or the differential preference of the tobacco *N*-glycosylation machinery for IgG1 versus IgG2b isoforms. However, a more likely explanation is the difference between analytical methodologies used in *N*-glycan detection. Triguero et al. labeled *N*-glycans with the fluorophore 2-aminobenzamide prior to analysis, a method that has been shown to greatly enhance the identification of low abundant *N*-glycans (Bigge et al. 1995). Differences could also be observed due to the method used to release *N*-glycans from plant glycoproteins. Hydrazinolysis is a commonly used method for the isolation of *N*-glycans from glycoproteins (Patel and Parekh 1994); however, this is not a viable method for the isolation of plant *N*-glycans as hydrazine treatment partially degrades *N*-glycans containing α-1,3-linked Fuc (Bardor et al. 2006). The endoglycanase PNGase F also cannot be used to isolate plant *N*-glycans as this enzyme is inhibited by the presence of α-1,3-linked Fuc (Tretter, Altmann, and Marz 1991).

The conversion of mAb *N*-glycans into predominantly high-mannose *N*-glycans is helpful from the standpoint of immunogenicity but not of great benefit for therapeutic mAbs in that these structures are substrates for circulating mannan-binding lectin (Arnold et al. 2006) and could be rapidly cleared from human serum. This phenomenon has already been shown to occur in BALB/c mice injected with the tobacco SO57 antibody containing high oligo-mannose *N*-glycans (Ko et al. 2003).

N-glycans terminating in β-1,4-linked Gal are common structures found on mAbs expressed in mammalian systems (Cabanes-Macheteau et al. 1999; Bakker et al. 2001; Bardor et al. 2003b; Ko et al. 2003; Jefferis 2005; Cox et al. 2006). *N*-glycans terminating in galactose are inefficient acceptors for plant *N*-glycan β-1,2-XylT and α-1,3-FucT activity (Leiter et al. 1999; Bencur et al. 2005). Therefore, the attachment of β-1,4-Gal residues to plant mAb *N*-glycans via the action of a co-expressed human β-1,4-GalT is a suitable glyco-optimization strategy, as it might inhibit the addition of Fuc and Xyl. Co-expression of native β-1,4-GalT and the mouse MGR48 mAb in tobacco plants resulted in the production of five different terminally galactosylated *N*-glycans representing approximately 30 percent of the total (Bakker et al. 2001). However, no significant change in the relative abundance of xylosylated and fucosylated *N*-glycans on the co-expressed MGR48 was observed. These results suggest that while human β-1,4-GalT was efficiently targeted to the Golgi apparatus of tobacco cells, the enzyme was not targeted early enough in the secretory pathway to cause the inhibition of Fuc and Xyl addition. In an attempt to target human β-1,4-GalT to earlier Golgi cisternae, a modified version of the human enzyme containing the cytoplasmic tail, transmembrane domain, and stem regions of the *Arabidopsis* β-1,2-XylT was co-expressed in tobacco with the MGR48 mAb (Bakker et al. 2006). Analysis of MGR48 glycopeptides indicated the presence of three hybrid-type *N*-glycans, Man5A, Man5Gn and AM_{iso}, as well as high-mannose structures ranging from Man5-9, for a total of seven different structures. Xylosylated and fucosylated *N*-glycans were significantly reduced with only 0.4 percent of structures containing Fuc and the complete absence of Xyl-containing *N*-glycans.

Similar efforts to inhibit the addition of Fuc and Xyl were made by co-expressing human GlcNAcTIII with a mAb in tobacco plants (Rouwendal et al. 2007). GlcNAcTIII catalyzes the transfer of GlcNAc in a β-1,4-linkage to the first Man of the core, forming a bisected *N*-glycan (Nishikawa et al. 1992). These structures are of relatively low abundance on human IgG's (Jefferis 2005); however, it has been observed in mammalian cells that these bisected *N*-glycans are poor substrates for fucosylation (Longmore and Schachter 1982; Ferrara et al. 2006). Analysis of the acceptor substrate specificity of recombinant plant β-1,2-XylT and α-1,3-FucT indicated that the plant enzymes are incapable of transferring Xyl or Fuc to bisected *N*-glycans (Rouwendal et al. 2007). Therefore, it was expected that the expression of GlcNAcTIII in plants will result in a reduction of Xyl- and Fuc-containing *N*-glycans on recombinant mAbs.

When the mAb MGR48 was co-expressed in a tobacco line transformed with human GlcNAcTIII (Rouwendal et al. 2007) no paucimannosidic *N*-glycans were detected and a relatively small amount of Xyl-containing *N*-glycans were detected. In contrast, a higher proportion of fucosylated *N*-glycans were detected, suggesting that GlcNAcTIII expression had a negative effect on tobacco XylT and hexosaminidase activity *in vivo*.

Other glyco-optimization strategies have been employed with the overall objective of maintaining *N*-glycan complexity with the concurrent elimination of Fuc and Xyl. Another strategy involves silencing of the genes responsible for addition of the plant-specific Xyl and Fuc. Direct gene disruption approaches were implemented in moss and *Arabidopsis* plants for the elimination of FucT and XylT expression using homologous recombination and mutagenesis, respectively. As expected, mAbs produced in these mutant plant lines contained *N*-glycans without detectable Fuc and Xyl (Schahs et al. 2007; Schuster et al. 2007). The moss-derived IGN314 mAb contained three major *N*-glycan structures, GnGn, GnM, and Man3, whereas *Arabidopsis*-derived mAbs contained four primary *N*-glycan structures, GnGn, GnM, Man7, and Man8.

An RNA interference (RNAi)-based approach was utilized with the *Lemna* expression system. The MDX-060 mAb was expressed in *Lemna* plants in conjunction with an RNA interference (RNAi) cassette designed to silence the endogenous α-1,3-FucT and β-1,2-XylT expression (Cox et al. 2006). This strategy resulted in the production of mAb with a single major (GnGn) *N*-glycan structure comprising greater than 95 percent of the total mAb *N*-glycans with undetectable Fuc and Xyl. This *N*-glycan homogeneity was further demonstrated in mAbs expressed from different *Lemna* plant lines grown at different scales.

29.8 ENHANCED Fc FUNCTIONALITY OF GLYCO-OPTIMIZED mAbs

Another reason for the increased interest in the area of mAb glyco-engineering is the potential for enhanced *in vivo* effector function activity of the mAbs with optimized glycosylation. Fc receptor-mediated effector cell function has been shown to be important for the *in vivo* activity of many therapeutic mAbs, and these interactions are highly impacted by the presence and type of *N*-glycans found on the mAb heavy chains. FcγRIIIa is the receptor expressed on NK (natural killer) cells and macrophages and is responsible for antibody-dependent cell-mediated cytotoxicity (ADCC). It has been shown that the absence of core Fuc residues leads to an increase in ADCC activity of mAbs (Shields et al. 2002).

MAbs with humanized *N*-glycans from moss (IGN314) and *Lemna* (MDX-060) were shown to have significantly enhanced ADCC activity irrespective of donor FcγRIIIa genotype (Cox et al. 2006; Schuster et al. 2007). This increase in ADCC activity was observed with homozygous FcγRIIIaPhe/Phe^{158} and heterozygous FcγRIIIaPhe/Val^{158} effector cell donors assayed with *Lemna*-derived MDX-060 and homozygous FcγRIIIaPhe/Phe^{158} and FcγRIIIaVal/Val^{158} cell donors assayed with moss-derived IGN314. ADCC activity was calculated to be increased approximately 20- to 40-fold depending on donor cell type. This effect is undoubtedly due to the removal of Fuc which has been shown previously to increase FcR binding and enhance ADCC function (Shields et al. 2002; Shinkawa et al. 2003; Niwa et al. 2004). These results demonstrate that removal of the α-1,3-linked Fuc from plant-derived mAbs has a similar effect on mAb function as the removal of α-1,6-linked Fuc from mammalian-derived mAbs. This enhanced ADCC activity is predicted to result in an increase in patients responding to treatments that require this Fc functionality. This increase in efficacy and potency may also result in a decrease of the required dosage of a given mAb therapy. Importantly, the mAbs also showed identical antigen binding between the plant-derived mAbs and their mammalian cell counterparts. In addition, the complement-dependent cytotoxicity (CDC) activities of the moss and *Lemna* glyco-optimized mAbs were characterized. The results show a significant reduction in activity, likely the result of the lack of terminal galactose residues on the mAbs (Schuster et al. 2007; Biolex unpublished).

The production of glyco-optimized mAbs in *Lemna*, moss, and *Arabidopsis* represents a significant step forward in the production of mAbs in plants. The implementation of similar *N*-glycan engineering strategies in mammalian systems has been attempted with less success, typically resulting in lethality (Furukawa and Sato 1999; Wang, Schachter, and Marth 2002). While plants require a functional *N*-glycosylation pathway to be viable (Burn et al. 2002; Gillmor et al. 2002; Lerouxel et al. 2005), there is no requirement for the presence of plant-specific *N*-glycan structures on glycoproteins as evidenced by the studies described in this chapter.

29.9 *O*-GLYCOSYLATION OF PLANT-DERIVED mAbs

O-glycosylation is another carbohydrate-based posttranslational modification that has been shown to occur to human immunoglobulins. This type of glycosylation has been shown to only occur on IgA and IgD molecules. Analysis of IgAs (Mattu et al. 1998) and IgDs (Arnold et al. 2004) from human serum reveals the presence of mucin-type *O*-glycans attached to the hinge region of the heavy chains of both mAb isotypes. A consensus sequence describing the precise amino acid residues targeted for *O*-glycosylation in mammals has yet to be defined (Julenius et al. 2005). The lack of a well-defined *O*-glycosylation consensus sequence is caused mainly by the considerable amount of site-specific heterogeneity that exists on *O*-glycoproteins, making the determination of the precise locations of *O*-glycan attachment difficult to ascertain (Renfrow et al. 2005).

Plants are also capable of producing proteins with *O*-glycans. All plant *O*-glycoproteins enter the secretory pathway and are localized to specialized areas of the plant cell, such as the cell wall and vacuole (Kishimoto et al. 1999; Kieliszewski 2001). The majority of plant *O*-glycoproteins are part of a super-family of hydroxyproline-rich glycoproteins (HRGP) found in the plant cell wall. Analysis of *O*-glycans from several *O*-glycoproteins has revealed three main types of structures attached to Ser/Thr and hydroxyproline (Hyp) residues. They are characterized by the attachment of Gal to Ser (solanaceous lectin-type; Lamport, Katona, and Roerig 1973; Yong-Pill and Chrispeels 1976), the attachment of Ara to Hyp (extensin-type) (Lamport 1967; Lamport and Miller 1971), and the attachment of type II arabinogalactans to Hyp [arabinogalactan protein (AGP) type; Tan, Leykam, and Kieliszewski 2003].

To date, only one group has attempted to express an *O*-glycan-containing mAb in plants. Karnoup et al. (2005) expressed two IgA molecules, HX8 and H6-3C4, in maize and the peptides encompassing the human *O*-glycosylation site on the hinge region (DVTVPCPVPSTPPTPSPSTPPTPSPSCCHPR) were isolated. Despite the ability of plants to attach mucin-type disaccharides onto Ser residues (Kishimoto et al. 1999), analysis of the hinge region peptides by MALDI-MS revealed a highly heterogeneous mixture of *O*-glycans. Further analysis revealed that both IgAs had roughly 80 to 94 percent of their hinge region Pro residues converted to Hyp or Hyp-Ara, and that they had identical levels of arabinosylation. The *O*-glycopeptides were highly heterogeneous with respect to the amount of Hyp and arabinosylation. Overall these results demonstrate that the hinge region of two human IgA1 mAbs expressed in maize were modified with extensin-type *O*-glycans.

29.10 SUMMARY

Over the past 20 years there has been significant advancement in the knowledge base for expression of recombinant proteins in plants. From the examples presented in this chapter it is clear that plants offer a viable alternative production format for therapeutic antibodies. Due to recent advances in *N*-glycan engineering, plants are now capable of generating final mAb product that is indistinguishable to that of their mammalian cell culture counterpart. Furthermore, the power of glyco-engineering in some of the plant systems offers the ability to modulate effector function producing more potent and efficacious antibodies. Upstream processing of plant biomass and subsequent purification via standard protein A chromatography methods allow for the simple extraction and purification of mAb from plant tissue to a high level of purity. Plant systems offer a more cost effective production system compared to the traditional mammalian cell culture systems and, in some cases, simpler and faster production of antibodies. With the successful demonstration *in vivo* of plant-derived antibodies as therapeutic agents, plant production systems have advanced beyond the proof-of-concept stage and are on a path toward commercial development.

REFERENCES

Akerstrom, B., B.H. Nilson, H.R. Hoogenboom, and L. Bjorck. 1994. On the interaction between single chain Fv antibodies and bacterial immunoglobulin-binding proteins. *J. Immunol. Methods* 177:151–163.

Alamillo, J.M., W. Monger, I. Sola, B. Garcia, Y. Perrin, M. Bestagno, O.R. Burrone, P. Sabella, J. Plana-Duran, L. Enjuanes, G.P. Lomonossoff, and J.A. Garcia. 2006. Use of virus vectors for the expression in plants of active full-length and single chain anti-coronavirus antibodies. *Biotechnol. J.* 1:1103–1111.

Altmann, F. 1998. Structures of the *N*-linked carbohydrate of ascorbic acid oxidase from zucchini. *Glycoconj. J.* 15:79–82.

Arnold, J.N., R.A. Dwek, P.M. Rudd, and R.B. Sim. 2006. Mannan binding lectin and its interaction with immunoglobulins in health and in disease. *Immunol. Lett.* 106:103–110.

Arnold, J.N., C.M. Radcliffe, M.R. Wormald, L. Royle, D.J. Harvey, M. Crispin, R.A. Dwek, R.B. Sim, and P.M. Rudd. 2004. The glycosylation of human serum IgD and IgE and the accessibility of identified oligomannose structures for interaction with mannan-binding lectin. *J. Immunol.* 173:6831–6840.

Arnold, J.N., M.R. Wormald, D.M. Suter, C.M. Radcliffe, D.J. Harvey, R.A. Dwek, P.M. Rudd, and R.B. Sim. 2005. Human Serum IgM Glycosylation: Identification of glycoforms that can bind to mannan-binding lectin. *J. Biol. Chem.* 280:29080–29087.

Bakker, H., M. Bardor, J. Molthoff, V. Gomord, I. Elbers, L. Stevens, W. Jordi, A. Lommen, L. Faye, P. Lerouge, and D. Bosch. 2001. Galactose-extended glycans of antibodies produced by transgenic plants. *Proc. Natl. Acad. Sci. U.S.A.* 98:2899–2904.

Bakker, H., G. Rouwendal, A. Kamoup, D. Florack, G. Stoopen, J. Helsper, R. van Ree, I. Die, and D. Bosch. 2006. An antibody produced in tobacco expressing a hybrid beta-1,4-galactosyltransferase is essentially devoid of plant carbohydrate epitopes. *Proc. Natl. Acad. Sci. U.S.A.* 103:7577–7582.

Bardor, M., G. Cabrera, P.M. Rudd, R.A. Dwek, J.A. Cremata, and P. Lerouge. 2006. Analytical strategies to investigate plant *N*-glycan profiles in the context of plant-made pharmaceuticals. *Curr. Opin. Struct. Biol. Carbohydr. Glycoconj. Biophys. Methods* 16:576–583.

Bardor, M., C. Faveeuw, A.-C. Fitchette, D. Gilbert, L. Galas, F. Trottein, L. Faye, and P. Lerouge. 2003a. Immunoreactivity in mammals of two typical plant glyco-epitopes, core {alpha}(1,3)-fucose and core xylose. *Glycobiology* 13:427–434.

Bardor, M., C. Loutelier-Bourhis, T. Paccalet, P. Cosette, A. Fitchette, L. Vézina, S. Trépanier, M. Dargis, R. Lemieux, C. Lange, L. Faye, and P. Lerouge. 2003b. Monoclonal C5-1 antibody produced in transgenic alfalfa plants exhibits a *N*-glycosylation that is homogenous and suitable for glyco-engineering into human-compatible structures. *Plant Biotechnol. J.* 1:451–462.

Bencur, P., H. Steinkellner, B. Svoboda, J. Mucha, R. Strasser, D. Kolarich, S. Hann, G. Kollensperger, J. Glossl, F. Altmann, and L. Mach. 2005. *Arabidopsis thaliana* alpha1,2-xylosyltransferase: An unusual glycosyltransferase with the potential to act at multiple stages of the plant *N*-glycosylation pathway. *Biochem. J.* 388:515–525.

Bencurova, M., W. Hemmer, M. Focke-Tejkl, I. Wilson, and F. Altmann. 2004. Specificity of IgG and IgE antibodies against plant and insect glycoprotein glycans determined with artificial glycoforms of human transferrin. *Glycobiology* 14:457–466.

Bigge, J., T. Patel, J. Bruce, P. Goulding, S. Charles, and R. Parekh. 1995. Nonselective and efficient fluorescent labeling of glycans using 2-amino benzamide and anthranilic acid. *Anal. Biochem.* 230:229–238.

Bjorklund, J.E.M., T. Karlsson, and C.G.M. Magnusson. 1999. *N*-glycosylation influences epitope expression and receptor binding structures in human IgE. *Mol. Immunol.* 36:213–221.

Bouquin, T., M. Thomsen, L.K. Nielsen, T.H. Green, J. Mundy, and M. Hanefeld Dziegiel. 2002. Human anti-rhesus D IgG1 antibody produced in transgenic plants. *Transgenic Res.* 11:115–122.

Burn, J.E., U.A. Hurley, R.J. Birch, T. Arioli, A. Cork, and R.E. Williamson. 2002. The cellulose-deficient Arabidopsis mutant rsw3 is defective in a gene encoding a putative glucosidase II, an enzyme processing *N*-glycans during ER quality control. *Plant J.* 32:949–960.

Cabanes-Macheteau, M., A. Fitchette-Laine, C. Loutelier-Bourhis, C. Lange, N. Vine, J. Ma, P. Lerouge, and L. Faye. 1999. *N*-Glycosylation of a mouse IgG expressed in transgenic tobacco plants. *Glycobiology* 9:365–372.

Chargelegue, D., N. Vine, C. van Dolleweerd, P. Drake, and J. Ma. 2000. A murine monoclonal antibody produced in transgenic plants with plant-specific glycans is not immunogenic in mice. *Transgenic Res.* 9:187–194.

Chen, M., X. Liu, Z. Wang, J. Song, Q. Qi, and P. Wang. 2005. Modification of plant *N*-glycans processing: The future of producing therapeutic protein by transgenic plants. *Med. Res. Rev.* 25:343–360.

Cox, K.M., J.D. Sterling, J.T. Regan, J.R. Gasdaska, K.K. Frantz, C.G. Peele, A. Black, D. Passmore, C. Moldovan-Loomis, M. Srinivasan, S. Cuison, P.M. Cardarelli, and L.F. Dickey. 2006. Glycan optimization of a human monoclonal antibody in the aquatic plant *Lemna minor. Nature Biotechnol.* 24:1591–1597.

Daniell, H., S.J. Streatfield, and K. Wycoff. 2001. Medical molecular farming: production of antibodies, biopharmaceuticals and edible vaccines in plants. *Trends Plant Sci.* 6:219–226.

Decker, E., and R. Reski. 2004. The moss bioreactor. *Curr. Opin. Plant Biol.* 7:166–170.

De Neve, M., M. De Loose, A. Jacobs, H. van Houdt, B. Kaluza, U. Weidle, M. van Montagu, and A. Depicker. 1993. Assembly of an antibody and its derived antibody fragment in Nicotiana and Arabidopsis. *Transgenic Res.* 2:227–237.

De Wilde, C., R. De Rycke, T. Beeckman, M. De Neve, M. van Montagu, G. Engler, and A. Depicker. 1998. Accumulation pattern of IgG antibodies and Fab fragments in transgenic *Arabidopsis thaliana* plants. *Plant Cell Physiol.* 39:639–646.

Drake, P.M., D.M. Chargelegue, N.D. Vine, C.J. van Dolleweerd, P. Obregon, and J.K. Ma. 2003. Rhizosecretion of a monoclonal antibody protein complex from transgenic tobacco roots. *Plant Mol. Biol.* 52:233–241.

During, K., S. Hippe, F. Kreuzaler, and J. Schell. 1990. Synthesis and self-assembly of a functional monoclonal antibody in transgenic *Nicotiana tabacum. Plant Mol. Biol.* 15:281–293.

Elbehri, A. 2005. Biopharming and the food system: Examining the Potential benefits and risks. *AgBioForum* 8:18–25.

Elbers, I., G. Stoopen, H. Bakker, L. Stevens, M. Bardor, J. Molthoff, W. Jordi, D. Bosch, and A. Lommen. 2001. Influence of growth conditions and developmental stage on *N*-glycan heterogeneity of transgenic immunoglobulin G and endogenous proteins in tobacco leaves. *Plant Physiol.* 126:1314–1322.

Faye, L., A. Boulaflous, M. Benchabane, V. Gomord, and D. Michaud. 2005. Protein modifications in the plant secratory pathway: Current status and practical implications in molecular pharming. *Vaccine* 23:1770–1778.

Faye, L., V. Gomord, A. Fitchette-Laine, and M. Chrispeels. 1993. Affinity purification of antibodies specific for Asn-linked glycans containing a-1,3-Fuc or b-1,2-Xyl. *Anal. Biochem.* 209:104–108.

Ferrara, C., P. Brunker, T. Suter, S. Moser, U. Puntener, and P. Umana. 2006. Modulation of therapeutic antibody effector functions by glycosylation engineering: Influence of Golgi enzyme localization domain and co-expression of heterologous beta1,4-*N*-acetylglucosaminyltransferase III and Golgi a-mannosidase II. *Biotechnol. Bioeng.* 93:851–861.

Fiedler, U., and U. Conrad. 1995. High-level production and long-term storage of engineered antibodies in transgenic tobacco seeds. *Biotechnology (NY)* 13:1090–1093.

Fischer, R., D. Schumann, S. Zimmermann, J. Drossard, M. Sack, and S. Schillberg. 1999. Expression and characterization of bispecific single-chain Fv fragments produced in transgenic plants. *Eur. J. Biochem.* 262:810–816.

Fitchette, A., M. Cabanes-Macheteau, L. Marvin, B. Martin, B. Satiat-Jeunemaitre, V. Gomord, K. Crooks, P. Lerouge, L. Faye, and C. Hawes. 1999. Biosynthesis and immunolocalization of Lewis a-containing *N*-glycans in the plant cell. *Plant Physiol.* 121:333–343.

Freeze, H.H., and M. Aebi. 2005. Altered glycan structures: The molecular basis of congenital disorders of glycosylation. *Curr. Opin. Struct. Biol. Carbohydr. Glycoconj. Biophys. Methods* 15:490–498.

Frigerio, L., N.D. Vine, E. Pedrazzini, M.B. Hein, F. Wang, J.K. Ma, and A. Vitale. 2000. Assembly, secretion, and vacuolar delivery of a hybrid immunoglobulin in plants. *Plant Physiol.* 123:1483–1494.

Fujiyama, K., R. Misaki, A. Katsura, T. Tanaka, A. Furukawa, T. Omasa, and T. Seki. 2006. *N*-linked glycan structures of a mouse monoclonal antibody produced from tobacco BY2 suspension-cultured cells. *J. Biosci. Bioeng.* 101:212–218.

Furukawa, K., and T. Sato. 1999. [beta]-1,4-Galactosylation of *N*-glycans is a complex process. *Biochim. Biophy. Acta General Subjects* 1473:54–66.

Gasdaska, J., D. Spencer, and L. Dickey. 2003. Advantages of therapeutic protein production in the aquatic plant *Lemna. Bioprocessing J.* Mar/Apr:50–56.

Ghose, S., M. Allen, B. Hubbard, C. Brooks, and S. Cramer. 2005. Antibody variable region interactions with protein A: Implications for the development of generic purification processes. *Biotechnol. Bioeng.* 92(6):665–673.

Giddings, G., G. Allison, D. Brooks, and A. Carter. 2000. Transgenic plants as factories for biopharmaceuticals. *Nature Biotechnol.* 18:1151–1155.

Gillmor, C.S., P. Poindexter, J. Lorieau, M.M. Palcic, and C. Somerville. 2002. {alpha}-Glucosidase I is required for cellulose biosynthesis and morphogenesis in Arabidopsi. *J. Cell Biol.* 156:1003–1013.

Giritch, A., S. Marillonnet, C. Engler, G. van Eldik, J. Botterman, V. Klimyuk, and Y. Gleba. 2006. Rapid high-yield expression of full-size IgG antibodies in plants coinfected with noncompeting viral vectors. *Proc. Natl. Acad. Sci. U.S.A.* 103:14701–14706.

Gorr, G., and W. Jost. 2005. Glycosylation design in transgenic moss for better product efficacy. *Bioprocessing J.* 4:26–30.

Harris, L., E. Skaletsky, and A. McPherson. 1998. Crystallographic structure of an intact IgG1 monoclonal antibody. *J. Mol. Biol.* 275:861–872.

Hein, M.B., Y. Tang, D.A. McLeod, K.D. Janda, and A. Hiatt. 1991. Evaluation of immunoglobulins from plant cells. *Biotechnol. Prog.* 7:455–461.

Hiatt, A., R. Cafferkey, and K. Bowdish. 1989. Production of antibodies in transgenic plants. *Nature* 342:76–78.

Hiatt, A., and M. Pauly. 2006. Monoclonal antibodies from plants: A new speed record. *Proc. Natl. Acad. Sci. U.S.A.* 103:14645–14646.

Hood, E., S. Woodard, and M. Horn. 2002. Monoclonal antibody manufacturing in transgenic plants: Myths and realities. *Curr. Opin. Biotechnol.* 13:630–635.

Huang, N. 2004. High-level protein expression system uses self-pollinating crops as hosts. *BioProcess Int.* 2:54–59.

Hull, A.K., C.J. Criscuolo, V. Mett, H. Groen, W. Steeman, H. Westra, G. Chapman, B. Legutki, L. Baillie, and V. Yusibov. 2005. Human-derived, plant-produced monoclonal antibody for the treatment of anthrax. *Vaccine* 23:2082–2086.

Inganas, M. 1981. Comparison of mechanisms of interaction between protein A from *Staphylococcus aureus* and human monoclonal IgG, IgA and IgM in relation to the classical FC gamma and the alternative F(ab′)2 epsilon protein A interactions. *Scand. J. Immunol.* 13:343–352.

Jefferis, R. 2005. Glycosylation of recombinant antibody therapeutics. *Biotechnol. Prog.* 21:11–16.

Jin, C., M. Bencurova, N. Borth, B. Ferko, E. Jensen-Jarolim, F. Altmann, and B. Hantusch. 2006. Immunoglobulin G specifically binding plant *N*-glycans with high affinity could be generated in rabbits but not in mice. *Glycobiology* 16:349–357.

Jones, J., S. Krag, and M. Betenbaugh. 2005. Controlling *N*-linked glycan site occupancy. *Biochim. Biophys. Acta* 1726:121–137.

Julenius, K., A. Molgaard, R. Gupta, and S. Brunak. 2005. Prediction, conservation analysis, and structural characterization of mammalian mucin-type *O*-glycosylation sites. *Glycobiology* 15:153–164.

Kaneko, Y., F. Nimmerjahn, and J.V. Ravetch. 2006. Anti-Inflammatory activity of immunoglobulin G resulting from Fc sialylation. *Science* 313:670–673.

Karnoup, A., V. Turkelson, and W. Anderson. 2005. O-linked glycosylation in maize-expressed human IgA1. *Glycobiology* 15:965–981.

Kathuria, S., R. Sriraman, R. Nath, M. Sack, R. Pal, O. Artsaenko, G.P. Talwar, R. Fischer, and R. Finnern. 2002. Efficacy of plant-produced recombinant antibodies against HCG. *Human Reproduction* 17:2054–2061.

Kelleher, D.J., and R. Gilmore. 2006. An evolving view of the eukaryotic oligosaccharyltransferase. *Glycobiology* 16:47R–62R.

Khoudi, H., S. Laberge, J. Ferullo, R. Bazin, A. Darveau, Y. Castonguay, G. Allard, R. Lemieux, and L. Vezina. 1999. Production of a diagnostic monoclonal antibody in perennial alfalfa plants. *Biotechnol. Bioeng.* 64:135–143.

Kieliszewski, M.J. 2001. The latest hype on Hyp-*O*-glycosylation codes. *Phytochemistry* 57:319–323.

Kimura, Y., and S. Matsuo. 2000. Changes in *N*-linked oligosaccharides during seed development of *Ginko biloba*. *Biosci. Biotechnol. Biochem.* 64:562–568.

Kishimoto, T., M. Watanabe, T. Mitsui, and H. Hori. 1999. Glutelin basic subunits have a mammalian mucin-type *O*-linked disaccharide side chain. *Arch. Biochem. Biophys.* 370:271–277.

Ko, K., and H. Koprowski. 2005. Plant biopharming of monoclonal antibodies. *Virus Res.* 111(1):93–100.

Ko, K., Z. Steplewski, M. Glogowska, and H. Koprowski. 2005. Inhibition of tumor growth by plant-derived mAb. *Proc. Natl. Acad. Sci. U.S.A.* 102:7026–7030.

Ko, K., Y. Tekoah, P. Rudd, D. Harvey, R. Dwek, S. Spitsin, C. Hanlon, C. Rupprecht, B. Dietzschold, M. Golovkin, and H. Koprowski. 2003. Function and glycosylation of plant-derived antiviral monoclonal antibody. *Proc. Natl. Acad. Sci. U.S.A.* 100:8013–8018.

Komarnytsky, S., N. Borisjuk, N. Yakoby, A. Garvey, and I. Raskin. 2006. Cosecretion of protease inhibitor stabilizes antibodies produced by plant roots. *Plant Physiol.* 141:1185–1193.

Koprivova, A., F. Altmann, G. Gorr, S. Kopriva, R. Reski, and E. Decker. 2003. *N*-glycosylation in the moss *Physcomitrella patens* is organized similarly to that in higher plants. *Plant Biol.* 5:582–591.

Krapp, S., Y. Mimura, R. Jefferis, R. Huber, and P. Sondermann. 2003. Structural analysis of human IgG-Fc glycoforms reveals a correlation between glycosylation and structural integrity. *J. Mol. Biol.* 325:979–989.

Lamport, D. 1967. Hydroxyproline-*O*-glycosidic linkage of the plant cell wall glycoprotein extensin. *Nature* 216:1322–1324

Lamport, D.T., L. Katona, and S. Roerig. 1973. Galactosylserine in extensin. *Biochem. J.* 133:125–132.

Lamport, D.T.A., and D.H. Miller. 1971. Hydroxyproline arabinosides in the plant kingdom. *Plant Physiol.* 48:454–456.

Larrick, J.W., L. Yu, C. Naftzger, S. Jaiswal, and K. Wycoff. 2001. Production of secretory IgA antibodies in plants. *Biomol. Eng.* 18:87–94.

Law, R.D., D.A. Russell, L.C. Thompson, S.C. Schroeder, C.M. Middle, M.T. Tremaine, T.P. Jury, X. Delannay, and S.C. Slater. 2006. Biochemical limitations to high-level expression of humanized monoclonal antibodies in transgenic maize seed endosperm. *Biochim. Biophys. Acta* 1760:1434–1444.

Ledford, H. 2007. Out of bounds. *Nature* 445:132–133.

Leiter, H., J. Mucha, E. Staudacher, R. Grimm, J. Glossl, and F. Altmann. 1999. Purification, cDNA cloning, and expression of GDP-L-Fuc:Asn-linked GlcNAc a-1,3-fucosyltransferase from mung beans. *J. Biol. Chem.* 274:21830–21839.

Lerouxel, O., G. Mouille, C. Andeme-Onzighi, M. Bruyant, M. Seveno, C. Loutelier-Bourhis, A. Driouich, H. Hofte, and P. Lerouge. 2005. Mutants in defective glycosylation, an Arabidopsis homolog of an oligosaccharyltransferase complex subunit, show protein underglycosylation and defects in cell differentiation and growth. *Plant J.* 42(4):455–468.

Longmore, G.D., and H. Schachter. 1982. Product-identification and substrate-specificity studies of the GDP--fucose: 2-acetamido-2-deoxy-[beta]--glucoside (fuc--> asn-linked GlcNAc) 6-[alpha]--fucosyltransferase in a golgi-rich fraction from porcine liver. *Carbohydr. Res.* 100:365–392.

Ma, J., P. Drake, D. Chargelegue, P. Obregon, and A. Prada. 2005. Antibody processing and engineering in plants, and new strategies for vaccine production. *Vaccine* 23:1814–1818.

Ma, J., P. Drake, and P. Christou. 2003. The production of recombinant pharmaceutical proteins in plants. *Nature Rev. Genet.* 4:794–805.

Ma, J., A. Hiatt, M. Hein, N.D. Vine, F. Wang, P. Stabila, C. van Dolleweerd, K. Mostov, and T. Lehner. 1995. Generation and assembly of secretory antibodies in plants. *Science* 268:716–719.

Ma, J., B. Hikmat, K. Wycoff, N. Vine, D. Chargelegue, L. Yu, M. Hein, and T. Lehner. 1998. Characterization of a recombinant plant monoclonal secretory antibody and preventative immunotherapy in humans. *Nature Med.* 4:601–606.

Ma, J., T. Lehner, P. Stabila, C.I. Fux, and A. Hiatt. 1994. Assembly of monoclonal antibodies with IgG1 and IgA heavy chain domains in transgenic tobacco plants. *Eur. J. Immunol.* 24:131–138.

Malhotra, R., M.R. Wormald, P.M. Rudd, P.B. Fischer, R.A. Dwek, and R.B. Sim. 1995. Glycosylation changes of IgG associated with rheumatoid arthritis can activate complement via the mannose-binding protein. *Nature Med.* 1:237–243.

Mattu, T., R. Pleass, A. Willis, M. Kilian, M. Wormald, A. Lellouch, P. Rudd, J. Woofi, and R. Dwek. 1998. The glycosylation and structure of human serum IgA1, Fab, and Fc regions and the role of *N*-glycosylation on Fca receptor interactions. *J. Biol. Chem.* 273:2260–2272.

Miele, L. 1997. Plants as bioreactors for biopharmaceuticals: Regulatory considerations. *Trends Biotechnol.* 15:45–50.

Misaki, R., Y. Kimura, N. Palacpac, S. Yoshida, K. Fujiyama, and T. Seki. 2003. Plant cultured cells expressing human beta-1,4-galactosyltransferase secrete glycoproteins with galactose-extended *N*-linked glycans. *Glycobiology* 13:199–205.

Moloney, M. 1995. "Molecular farming" in plants: Achievements and prospects. *Biotechnol. Biotechnological Equipment* 9:3–9.

Moremen, K. 2002. Golgi a-mannosidase II deficiency in vertebrate systems: Implications for asparagine-linked oligosaccharide processing in mammals. *Biochim. Biophys. Acta* 1573:225–235.

Moremen, K., and M. Molinari. 2006. *N*-linked glycan recognition and processing: The molecular basis of endoplasmic reticulum quality control. *Curr. Opin. Struct. Biol. Carbohydr. Glycoconj. Biophys. Methods* 16:592–599.

Nechansky, A., M. Schuster, W. Jost, P. Siegl, S. Wiederkum, G. Gorr, and R. Kircheis. 2007. Compensation of endogenous IgG mediated inhibition of antibody-dependent cellular cytotoxicity by glyco-engineering of therapeutic antibodies. *Mol. Immunol.* 44:1815–1817.

Negrouk, V., G. Eisner, H. Lee, K. Han, D. Taylor, and H. Wong. 2005. Highly efficient transient expression of functional recombinant antibodies in lettuce. *Plant Sci.* 169:433–438.

Nishikawa, A., Y. Ihara, M. Hatakeyama, K. Kangawa, and N. Taniguchi. 1992. Purification, cDNA cloning, and expression of UDP-*N*-acetylglucosamine: Beta-D-mannoside beta-1,4 *N*-acetylglucosaminyltransferase III from rat kidney. *J. Biol. Chem.* 267:18199–18204.

Niwa, R., S. Hatanaka, E. Shoji-Hosaka, M. Sakurada, Y. Kobayashi, A. Uehara, H. Yokoi, K. Nakamura, and K. Shitara. 2004. Enhancement of the antibody-dependent cellular cytotoxicity of low-fucose IgG1 is independent of FcgRIIIa functional polymorphism. *Clin. Cancer Res.* 10:6248–6255.

Nuttall, J., N. Vine, J.L. Hadlington, P. Drake, L. Frigerio, and J.K. Ma. 2002. ER-resident chaperone interactions with recombinant antibodies in transgenic plants. *Eur. J. Biochem.* 269:6042–6051.

Patel, T.P., and R.B. Parekh. 1994. Release of oligosaccharides from glycoproteins by hydrazinolysis. In *Guide to Techniques in Glycobiology*, ed. W.J.L.A.G.W. Hart, 57–66. *Methods Enzymology.* New York: Academic Press.

Perrin, Y., C. Vaquero, I. Gerrard, M. Sack, J. Drossard, E. Stoger, P. Christou, and R. Fischer. 2000. Transgenic pea seeds as bioreactors for the production of a single chain Fv fragment (scFV) antibody used in cancer diagnosis and therapy. *Mol. Breeding* 6:345–352.

Petruccelli, S., M.S. Otegui, F. Lareu, O. Tran Dinh, A.C. Fitchette, A. Circosta, M. Rumbo, M. Bardor, R. Carcamo, V. Gomord, and R.N. Beachy. 2006. A KDEL-tagged monoclonal antibody is efficiently retained in the endoplasmic reticulum in leaves, but is both partially secreted and sorted to protein storage vacuoles in seeds. *Plant Biotechnol. J.* 4:511–527.

Pujol, M., N. Ramirez, M. Ayala, J. Gavilondo, R. Valdes, M. Rodriguez, J. Brito, S. Padilla, L. Gomez, B. Reyes, R. Peral, M. Perez, J.L. Marcelo, L. Mila, R.F. Sanchez, R. Paez, J.A. Cremata, G. Enriquez, O. Mendoza, M. Ortega, and C. Borroto. 2005. An integral approach towards a practical application for a plant-made monoclonal antibody in vaccine purification. *Vaccine* 23:1833–1837.

Ramirez, N., M. Rodriguez, M. Ayala, J. Cremata, M. Perez, A. Martinez, M. Linares, Y. Hevia, R. Paez, R. Valdes, J.V. Gavilondo, and G. Selman-Housein. 2003. Expression and characterization of an anti-(hepatitis B surface antigen) glycosylated mouse antibody in transgenic tobacco (*Nicotiana tabacum*) plants and its use in the immunopurification of its target antigen. *Biotechnol. Appl. Biochem.* 38:223–230.

Rayon, C., M. Cabanes-Macheteau, C. Loutelier-Bourhis, I. Salliot-Maire, J. Lemoine, W. Reiter, P. Lerouge, and L. Faye. 1999. Characterization of *N*-glycans from *Arabidopsis*. Application to a fucose-deficient mutant. *Plant Physiol.* 119:725–734.

Renfrow, M., H.J. Cooper, M. Tomana, R. Kulhavy, Y. Hiki, K. Toma, M.R. Emmett, J. Mestecky, A.G. Marshall, and J. Novak. 2005. Determination of aberrant *O*-glycosylation in the IgA1 hinge region by electron capture dissociation Fourier transform-ion cyclotron resonance mass spectrometry. *J. Biol. Chem.* 280:19136–19145.

Rodriguez, M., N.I. Ramirez, M. Ayala, F. Freyre, L. Perez, A. Triguero, C. Mateo, G. Selman-Housein, J.V. Gavilondo, and M. Pujol. 2005. Transient expression in tobacco leaves of an aglycosylated recombinant antibody against the epidermal growth factor receptor. *Biotechnol. Bioeng.* 89:188–194.

Rouwendal, G.J.A., M. Wuhrer, D.E.A. Florack, C.A.M. Koeleman, A.M. Deelder, H. Bakker, G.M. Stoopen, I. van Die, J.P.F.G. Helsper, C.H. Hokke, and D. Bosch. 2007. Efficient introduction of a bisecting GlcNAc residue in tobacco *N*-glycans by expression of the gene encoding human *N*-acetylglucosaminyltransferase III. *Glycobiology* 17:334–344.

Rudd, P., T. Elliott, P. Cresswell, I. Wilson, and R. Dwek. 2001. Glycosylation and the immune system. *Science* 291:2370–2376.

Rudd, P., R. Leatherbarrow, T. Rademacher, and R. Dwek. 1991. Diversification of the IgG molecule by oligosaccharides. *Mol. Immunol.* 28:1369–1378.

Schahs, M., R. Strasser, J. Stadlmann, R. Kunert, T. Rademacher, and H. Steinkellner. 2007. Production of a monoclonal antibody in plants with a humanized *N*-glycosylation pattern. *Plant Biotechnol. J.* 5:657–663.

Schillberg, S., R.M. Twyman, and R. Fischer. 2005. Opportunities for recombinant antigen and antibody expression in transgenic plants: Technology assessment. *Vaccine* 23:1764–1769.

Schuster, M., W. Jost, G.C. Mudde, S. Wiederkum, C. Schwager, E. Janzek, F. Altmann, J. Stadlmann, C. Stemmer, and G. Gorr. 2007. In vivo glyco-engineered antibody with improved lytic potential produced by an innovative non-mammalian expression system. *Biotechnol. J.* 2:700–708.

Sharp, J., and P. Doran. 2001. Characterization of monoclonal antibody fragments produced by plant cells. *Biotechnol. Bioeng.* 73:338–346.

Shields, R., J. Lai, R. Keck, L. O'Connell, K. Hong, Y. Meng, S. Weikert, and L. Presta. 2002. Lack of fucose on human IgG1 *N*-linked oligosaccharide improves binding to human FcgRIII and antibody-dependent cellular toxicity. *J. Biol. Chem.* 277:26733–26740.

Shinkawa, T., K. Nakamura, N. Yamane, E. Shoji-Hosaka, Y. Kanda, M. Sakurada, K. Uchida, H. Anazawa, M. Satoh, M. Yamasaki, N. Hanai, and K. Shitara. 2003. The absence of fucose but not the presence of galactose or bisecting *N*-acetylglucosamine of human IgG1 complex-type oligosaccharides shows the critical role of enhancing antibody-dependent cellular cytotoxicity. *J. Biol. Chem.* 278:3466–3473.

Sriraman, R., M. Bardor, M. Sack, C. Vaquero, L. Faye, R. Fischer, R. Finnern, and P. Lerouge. 2004. Recombinant anti-hCG antibodies retained in the endoplasmic reticulum of transformed plants lack core-xylose and core-α(1-3)-fucose residues. *Plant Biotechnol. J.* 2:279–287.

Starovasnik, M., M. O'Connell, W. Fairbrother, and R. Kelley. 1999. Antibody variable region binding by Staphylococcal protein A: Thermodynamic analysis and location of the Fv binding site on E-domain. *Protein Sci.* 8:1423–1431.

Stevens, L.H., G.M. Stoopen, I.J. Elbers, J.W. Molthoff, H.A. Bakker, A. Lommen, D. Bosch, and W. Jordi. 2000. Effect of climate conditions and plant developmental stage on the stability of antibodies expressed in transgenic tobacco. *Plant Physiol.* 124:173–182.

Stoger, E., M. Sack, Y. Perrin, C. Vaquero, E. Torres, R.M. Twyman, P. Christou, and R. Fischer. 2002. Practical considerations for pharmaceutical antibody production in different crop systems. *Mol. Breeding* 9:149–158.

Strasser, R., J. Mucha, H. Schwihla, F. Altmann, J. Glossl, and H. Steinkellner. 1999a. Molecular cloning and characterization of cDNA coding for beta1, 2*N*-acetylglucosaminyltransferase I (GlcNAc-TI) from *Nicotiana tabacum*. *Glycobiology* 9:779–785.

Strasser, R., H. Steinkellner, M. Boren, F. Altmann, L. Mach, J. Glossl, and J. Mucha. 1999b. Molecular cloning of cDNA encoding *N*-acetylglucosaminyltransferase II from *Arabidopsis thaliana*. *Glycoconj. J.* 16:787–791.

Tan, L., J.F. Leykam, and M.J. Kieliszewski. 2003. Glycosylation motifs that direct arabinogalactan addition to arabinogalactan-proteins. *Plant Physiol.* 132:1362–1369.

Tekoah, Y., K. Ko, H. Koprowski, D. Harvey, M. Wormald, R. Dwek, and P. Rudd. 2004. Controlled glycosylation of therapeutic antibodies in plants. *Arch. Biochem. Biophys.* 426:266–278.

Tretter, V., F. Altmann, and L. Marz. 1991. Peptide-N^4-(*N*-acetyl-β-glucosaminyl)asparagine amidase F cannot release glycans with fucose attached $\alpha 1 \rightarrow 3$ to the asparagine-linked *N*-acetylglucosamine residue. *Eur. J. Biochem.* 199:647–652.

Triguero, A., G. Cabrera, J. Cremata, C. Yuen, J. Wheeler, and N. Ramírez. 2005. Plant-derived mouse IgG monoclonal antibody fused to KDEL endoplasmic reticulum-retention signal is *N*-glycosylated homogeneously throughout the plant with mostly high-mannose-type *N*-glycans. *Plant Biotechnol. J.* 3(4):449–457.

Twyman, R., E. Stoger, S. Schillberg, P. Christou, and R. Fischer. 2003. Molecular farming in plants: Host systems and expression technology. *Trends Biotechnol.* 21:570–578.

Umana, P., J. Jean-Mairet, R. Moudry, H. Amstutz, and J.E. Bailey. 1999. Engineered glycoforms of an anti-neuroblastoma IgG1 with optimized antibody-dependent cellular cytotoxic activity. *Nature Biotechnol.* 17:176–180.

Valdes, R., L. Gomez, S. Padilla, J. Brito, B. Reyes, T. Alvarez, O. Mendoza, O. Herrera, W. Ferro, M. Pujol, V. Leal, M. Linares, Y. Hevia, C. Garcia, L. Mila, O. Garcia, R. Sanchez, A. Acosta, D. Geada, R. Paez, J. Luis Vega, and C. Borroto. 2003a. Large-scale purification of an antibody directed against hepatitis B surface antigen from transgenic tobacco plants. *Biochem. Biophys. Res. Commun.* 308:94–100.

Valdes, R., B. Reyes, T. Alvarez, J. Garcia, J.A. Montero, A. Figueroa, L. Gomez, S. Padilla, D. Geada, M.C. Abrahantes, L. Dorta, D. Fernandez, O. Mendoza, N. Ramirez, M. Rodriguez, M. Pujol, C. Borroto, and J. Brito. 2003b. Hepatitis B surface antigen immunopurification using a plant-derived specific antibody produced in large scale. *Biochem. Biophys. Res. Commun.* 310:742–747.

van Droogenbroeck, B., J. Cao, J. Stadlmann, F. Altmann, S. Colanesi, S. Hillmer, D.G. Robinson, E. van Lerberge, N. Terryn, M. van Montagu, M. Liang, A. Depicker, and G. De Jaeger. 2007. Aberrant localization and underglycosylation of highly accumulating single-chain Fv-Fc antibodies in transgenic Arabidopsis seeds. *Proc. Natl. Acad. Sci. U.S.A.* 104:1430–1435.

van Ree, R., M. Cabanes-Macheteau, J. Akkerdaas, J. Milazzo, C. Loutelier-Bourhisi, C. Rayon, M. Villalba, S. Koppelman, R. Aalberse, R. Rodriguez, L. Faye, and P. Lerouge. 2000. b(1,2)-xylose and a(1,3)-fucose residues have a strong contribution in IgE binding to plant glycoallergens. *J. Biol. Chem.* 274:11451–11458.

Vaquero, C., M. Sack, J. Chandler, J. Drossard, F. Schuster, M. Monecke, S. Schillberg, and R. Fischer. 1999. Transient expression of a tumor-specific single-chain fragment and a chimeric antibody in tobacco leaves. *Proc. Natl. Acad. Sci. U.S.A.* 96:11128–11133.

Vaquero, C., M. Sack, F. Schuster, R. Finnern, J. Drossard, D. Schumann, A. Reimann, and R. Fischer. 2002. A carcinoembryonic antigen-specific diabody produced in tobacco. *FASEB J.* 16:408–410.

Verch, T., V. Yusibov, and H. Koprowski. 1998. Expression and assembly of a full-length monoclonal antibody in plants using a plant virus vector. *J. Immunol. Methods* 220:69–75.

Vidal, M., and F. Conde. 1985. Alternative mechanism of protein A-immunoglobulin interaction: The VH-associated reactivity of a monoclonal human IgM. *J. Immunol.* 135:1232–1238.

Vietor, R., C. Loutelier-Bourhis, A. Fitchette, P. Margerie, M. Gonneau, L. Faye, and P. Lerouge. 2003. Protein *N*-glycosylation is similar in the moss *Physcomitrella patens* and in higher plants. *Planta* 218:269–275.

Vine, N., P. Drake, A. Hiatt, and J. Ma. 2001. Assembly and plasma membrane targeting of recombinant immunoglobulin chains in plants with a murine immunoglobulin transmembrane sequence. *Plant Mol. Biol.* 45:159–167.

Wang, Y., H. Schachter, and J.D. Marth. 2002. Mice with a homozygous deletion of the Mgat2 gene encoding UDP-*N*-acetylglucosamine:[alpha]-6--mannoside [beta]1,2-*N*-acetylglucosaminyltransferase II: A model for congenital disorder of glycosylation type IIa. *Biochim. Biophys. Acta General Subjects* 1573:301–311.

Wilson, I., and F. Altmann. 1998. Structural analysis of *N*-glycans from allergenic grass, ragweed and tree pollens: Core a1,3-linked fucose and xylose present in all pollens examined. *Glycoconj. J.* 15:1055–1070.

Wilson, I., R. Zeleny, D. Kolarich, E. Staudacher, C. Stroop, J. Kamerling, and F. Altmann. 2001. Analysis of Asn-linked glycans from vegetable foodstuffs: Widespread occurrence of Lewis^a, core a-1,3-linked fucose and xylose substitutions. *Glycobiology* 11:261–274.

Wolt, J., Y. Shyy, P. Christensen, K. Dorman, and M. Misra. 2004. Quantitative exposure assessment for confinement of maize biogenic systems. *Environ. Biosafety Res.* 3:183–196.

Wongsamuth, R., and P. Doran. 1997. Production of monoclonal antibodies by tobacco hairy roots. *Biotechnol. Bioeng.* 54:401–415.

Yano, A., F. Maeda, and M. Takekoshi. 2004. Transgenic tobacco cells producing the human monoclonal antibody to hepatitis B virus surface antigen. *J. Med. Virol.* 73:208–215.

Yong-Pill, C., and M. Chrispeels. 1976. Serine-*O*-galactosyl linkages in glycopeptides from carrot cell walls. *Phytochemistry* 15:165–169.

Zeitlin, L., S. Olmsted, T. Moench, M. Co, B. Martinell, V. Paradkar, D. Russell, C. Queen, R. Cone, and K. Whaley. 1998. A humanized monoclonal antibody produced in transgenic plants for immunoprotection of the vagina against genital herpes. *Nature Biotechnol.* 16(13):1361–1364.

PART VIII

THERAPEUTIC ANTIBODIES

CHAPTER 30

The Formulation and Delivery of Monoclonal Antibodies

VIKAS K. SHARMA, HUNG-WEI CHIH, RANDALL J. MRSNY, and ANN L. DAUGHERTY

30.1 INTRODUCTION

The ability to generate monoclonal antibodies through hybridoma technology in somatic cells has enabled large-scale production of monoclonal antibodies (MAbs) for therapeutic use (Jones et al. 1986; Kohler and Milstein 1975; Peterson 2005). Therefore, antibodies have become one of the fastest growing classes of protein therapeutics in the last two decades, in large part because of advances made in recombinant DNA technology, and because of our enhanced understanding of the underlying mechanisms of various diseases. The U.S. Food and Drug Administration (FDA) approved the first murine monoclonal antibody in 1986, muronomab-CD3 (Orthoclone OKT3®), against a T-cell receptor protein complex, to protect patients from life-threatening tissue rejection following kidney transplantation (Chatenoud 2003). Since that time, several technological advances have been made to address the clinical and economic challenges that this class of therapeutic molecules faces, such as immunogenicity and production yields. The observed immune response against fully murine MAbs led to the development of chimeric, humanized antibodies, and more recently, fully human MAbs. Likewise, the clinical potential of these molecules was increased by further process developments that have resulted in significantly higher production yields of MAbs owing to advances in both cell culture methods and recovery/purification processes. The widespread therapeutic applications for

Therapeutic Monoclonal Antibodies: From Bench to Clinic. Edited by Zhiqiang An

MAbs and MAb-related molecules have led several biotechnology companies, as well as traditional pharmaceutical companies, to engage in the continued development of these molecules, creating a highly competitive market for treating many of the same indications. As of this writing, there are over 20 FDA-approved MAb-related therapeutics in the market and more than 2000 clinical trials utilizing MAb-based therapy are underway against a variety of cancers, immunological disorders, and other significant unmet medical needs (http://clinicaltrials.gov/ct2/results?term=antibody). In the year 2008, sales from engineered MAbs for clinical use were projected to be over a third (approximately $US 17 billion) of all biotech products (Pavlou and Belsey 2005). The dramatic growth in the production and development of MAbs for clinical use is primarily due to the ability to design MAbs that have exquisite specificity towards antigens involved in a particular disease state or pathology. This therapeutic approach potentially lowers the risks of side effects and provides opportunities for specific cellular or tissue targeting.

In addition to full length MAbs, a number of modified MAbs have also been developed and are currently marketed for therapeutic indications. These modifications include antibody fragments, such as Fabs, Fab′2, monovalent antibodies, and single chain Fv, and conjugated antibodies. The development of antibody-drug conjugates, which uses a small molecule cytotoxic agent linked to the antibody, capitalizes on the targeting of the antibody in order to deliver its potent cytotoxic payload to a specific cell that expresses the relevant receptor. This approach potentially limits the collateral damage to normal cells that express little or no targeted ligand. One such antibody-drug conjugate is Mylotarg®.This anti-CD33 antibody conjugated to calcheamicin, a potent cytotoxic agent that is extremely difficult to deliver as a single agent due to nonspecific effects on normal tissues (Zein et al. 1988), is currently being marketed by Wyeth for the treatment of acute myeloid leukemia in patients over 60 years of age who are not considered candidates for other cytotoxic chemotherapies. Another antibody-drug conjugate, composed of the anti-HER2 MAb, trastuzumab, conjugated to a potent anti-tubule agent, the maytansine derivative DM1, is currently being studied in clinical trials conducted by Genentech and Immunogen in patients with HER2-positive metastatic breast cancer (Burris et al. 2006). The conjugate is internalized following binding to HER2-positive cells and subsequently releases the DM1 molecule, which is responsible for the cytotoxic effects. Trastuzumab-DM1 is designed to deliver DM1 preferentially to HER2+ tumor cells, thereby increasing the therapeutic index of DM1 and avoiding excessive toxicity while maintaining the biologic activity of trastuzumab. In a trastuzumab-resistant xenograft model, trastuzumab-DM1 was able to virtually eliminate the HER2+ tumor, while nonconjugated trastuzumab was ineffective (Burris et al. 2006). Another type of antibody conjugates are the radioimmunoconjugates, which have been used for treatments of certain cancers and are being used as contrast agents for radioimaging (Orciuolo et al. 2007; Iagaru et al. 2007). Examples of radioimmunoconjugates include anti-CD 20 murine antibodies conjugated with the radioisotopes iodine-131 and yttrium-90, Zevalin® and Bexxar®, respectively, being used for the treatment of non-Hodgkin lymphoma.

Besides linkage to small molecules, antibody fragments have also been linked to other therapeutic proteins to form fusion proteins. Enbrel® (etanercept) is one such example that consists of an extracellular ligand-binding portion of the human 75 kDa tumor necrosis factor receptor (TNFR) linked to the Fc portion of a human IgG1 (Klareskog et al. 2006). Antibody fragments themselves are another class of modified MAbs. Lucentis® is a Fab fragment of a full length anti-VEGF molecule, and is currently marketed by Genentech for use in treating age-related wet macular degeneration (Mojica et al. 2008).

Antibodies are proteins and hence are complex molecules composed of numerous reactive chemical groups and possess a delicate multidomain three-dimensional structure. Preservation of the native tertiary structure and preventing the modification of reactive amino-side side chains is essential to retain the biological activity and efficacy of an MAb (Daugherty and Mrsny 2006; Wang et al. 2007; Wakankar et al. 2007a, 2007b). Furthermore, compared to some of the other protein therapeutics, MAbs are typically required to be administered at higher doses for oncology applications as well as for the treatment of certain immunological and allergic disorders, depending on the therapeutic molecule (Shire, Shahrokh, and Liu 2004). To attain such high doses, as well as administer subcutaneous (SC) doses with acceptable dose volumes, the typical concentration requirements for MAb products range from 5 mg/mL to higher than 100 mg/mL. These doses are significantly higher than those seen with other therapeutic proteins such as growth factors, for example, human growth hormone (Carroll et al. 1998), interferons/interleukins (Cacopardo et al. 1998; Recchia et al. 2007), or clotting factors such as factor VIII (Yoshioka et al. 2006), for all of which the doses are in the microgram per kilogram range. Therefore, formulation development of MAbs presents challenges that may be distinct from other protein formulations and such development is vital to the clinical and commercial success of a new MAb therapeutic.

Formulation development essentially involves optimizing the composition of the therapeutic entity, either as a solution or as a dry form such as a lyophilized powder, in order to minimize physical (denaturation, aggregation) and/or chemical (oxidation, deamidation, isomerization, hydrolysis, etc.) degradation so as to achieve a desired shelf-life (Arakawa et al. 1993; Cleland, Powell, and Shire 1993; Frokjaer and Otzen Daniel 2005). Following an involved production process of protein harvest, isolation and purification, pharmaceutical development work is aimed at ensuring that the final product delivered to patients retains its stability, efficacy, and safety during the product's shelf-life (B. Sharma 2007).

In this chapter, we present the considerations involved in the pharmaceutical development of an MAb therapeutic from a stability and formulation perspective for optimal delivery and clinical performance.

30.2 FORMULATION DEVELOPMENT OF MONOCLONAL ANTIBODIES

30.2.1 General Features of Monoclonal Antibodies and Mechanisms of Instability

Currently, all of the FDA-approved monoclonal antibodies are produced by recombinant DNA technology. Full length recombinant antibodies are produced in mammalian cell expression using Chinese hamster ovary (CHO) or murine lymphoid cell lines; however, antibody fragments have been produced in bacterial systems such as *Escherichia coli.* Furthermore, most of the currently marketed MAbs belong primarily to the IgG1 isotype; however, a few IgG2 isotypes (Orthoclone OKT3®, Bexxar®) and IgG4 isotypes (Mylotarg®, Tysabri®) have also been developed (Table 30.1). The details of the antibody structure and the variety of techniques and methodologies to obtain monoclonal antibodies for therapeutic use have been discussed in some of the previous chapters and hence will not be presented here.

30.2.1.1 Heterogeneity in Monoclonal Antibodies The development of MAbs intended for therapeutic use requires a comprehensive characterization of their structural integrity, purity, and stability. Molecular alterations can take place at every stage of manufacturing: cell culture, purification, formulation, and storage. Heterogeneity can be introduced through intracellular processes (Awdeh, Askonas, and Williamson 1967) or extracellularly in serum or cell culture medium (Compton et al. 1989). In addition, heterogeneity can be introduced by exposure to various buffers and solutions during the production process and during storage under different stress conditions, such as elevated temperature, exposure to intense light, or exposure to chemicals (Liu, Gaza-Bulseco, and Sun 2006; Lam, Yang, and Cleland 1997; Krotkiewski, Laskowska, and Krotkiewska 1995; Chu Grace et al. 2007). Heterogeneity in MAbs can be broadly categorized into three types of variants according to size, chemical/charge, and conformational alterations.

30.2.1.1.1 Size Heterogeneity Size heterogeneity arises primarily due to aggregation and fragmentation (Morrison 2005; Jefferis 2005; Hooker and James 1998; Perkins et al. 2000; Marino et al. 1997; Harris et al. 2001). Monoclonal antibodies, like all proteins, are prone to aggregation.

TABLE 30.1 FDA-Approved and Marketed Monoclonal Antibody Products

	Antibody Product (Generic name, Company, Year Approved)	IgG Isotype, Target	Delivery Route	MAb Concentration	Buffer Components, pH	Excipients
1.	Lucentis® (Ranibizumab; Genentech, 2006)	Humanized IgG1κ fragment, Anti-VEGF	Intravitreal	10 mg/mL	Histidine-HCl, pH 5.5	Trehalose dihydrate, polysorbate 20
2.	Vectibix® (Panitumumab; Amgen, 2006)	Fully human IgG2κ, anti-EGFR	IV	100 mg/mL	Na acetate, pH 5.6–6.0	NaCl
3.	Avastin ® (Bevacizumab; Genentech, 2004)	Humanized IgG1κ, anti-VEGF	IV	25 mg/mL	Na phosphate, pH 6.42	Trehalose dihydrate, polysorbate 20
4.	Erbitux® (Cetuximab; Imclone, BMS, 2004)	Chimeric IgG1κ, anti-EGFR	IV	2 mg/mL	Na phosphate, pH 7.2	NaCl
5.	Tysabri® (Natalizumab; Biogen-IDEC, 2004)	Humanized IgG4, Anti-4α-integrin	IV	60 mg/mL	Na phosphate, pH 6.1	NaCl, polysorbate 80
6.	Xolair®[a] (Omalizumab; Genentech, 2003)	Humanized IgG1aκ, Anti-IgE	SC	125 mg/mL	Histidine	Sucrose, polysorbate 20
7.	Bexxar® (Tositumomab; Corixa-Glaxo-Smithkline, 2003)	Murine IgG 2aλ–radiolabeled ^{131}I, anti-CD20	IV	14 mg/mL	Na phosphate, pH 7.2	NaCl, maltose, povidone, ascorbic acid
8.	Raptiva®[a] (Efalizumab; Genentech, 2003)	Humanized IgG1aκ, anti-CD11a	SC	100 mg/mL	Histidine, pH 6.2	Sucrose, polysorbate 20
9.	Humira® (Adalimumab; Abbott Laboratories, 2002)	Human IgG1Kκ, anti-TNFα	SC	50 mg/mL	Na phosphate, Na citrate, pH 5.2	Mannitol, polysorbate 80
10.	Zevalin®[b] (Ibritumomab-Tiuxetan; Biogen-IDEC, 2002)	Murine IgG1, anti-CD20	IV	3.2 mg/mL	Na phosphate, pH 7.1	NaCl

11.	Campath® (Alemtuzumab, Bayer-Genzyme, 2001)	Humanized IgG1κ, anti-CD52	IV	30 mg/mL	Na K-phosphate, pH 7	NaCl, KCl, Disodium edetate, polysorbate 80
12.	Mylotarg®[a] (Gemtuzumab ozogamicin; Wyeth, 2000)	Humanized IgG4κ, Anti-CD33	IV	4 mg/mL	Na phosphate	NaCl, sucrose, dextran 40
13.	Herceptin®[a] (Trastuzumab; Genentech, 1998)	Humanized IgG1κ, anti-HER2	IV	21 mg/mL	Histidine	Trehalose, polysorbate 20
14.	Remicade®[a] (Infliximab; Centocor, 1998)	Chimeric IgG1κ, anti-TNFα	IV	10 mg/mL	Na phosphate, pH 7.2	Sucrose, polysorbate 80
15.	Simulect®[a] (Basiliximab; Novartis, 1998)	Chimeric IgG, anti-CD25	IV	4 mg/mL	Na K-phosphate	Glycine, NaCl, mannitol, sucrose, polysorbate 80
16.	Synagis®[a] (Palivizumab; Medimmune, 1998)	Humanized IgG1, anti-RSV	IV	100 mg/mL	Histidine, pH 6	Glycine, mannitol
17.	Rituxan® (Rituximab; Biogen-IDEC, Genentech, 1997)	Chimeric mouse/human IgG1κ, anti-CD20	IV	10 mg/mL	Na citrate, pH 6.5	NaCl, polysorbate 80
18.	Zenapax® (Daclizumab; Roche, 1997)	Daclizumab, Humanized IgG1, anti-CD25	IV	25 mg/mL	Na phosphate, pH 6.9	NaCl, polysorbate 80
19.	ReoPro® (Abciximab; Centocor-Lilly, 1994)	chimeric IgG1 Fab, anti-GPIIb/IIIab	IV	2 mg/mL	Na phosphate, pH 7.2	NaCl, polysorbate 80
20.	Orthoclone® OKT3, (Muromomab; J&J, Ortho-Biotech, 1986)	Murine IgG2a, anti-CD3	IV	1 mg/mL	Na phosphate, pH 7	NaCl, polysorbate 80

Abbreviations: RSV = respiratory syncytial virus; TNFa = tissue necrosis factor alpha; EGFR = epidermal growth factor receptor; VEGF = vascular endothelial growth factor.
[a]Lyophilized product, concentrations and pH are following reconstitution.
[b]Kit where components are mixed prior to administration.

Furthermore, fragmentation can occur between different domains such as the Fab and the Fc. Either process can affect the efficacy and safety of the MAb and hence needs to be minimized during development.

30.2.1.1.2 Chemical/Charge Heterogeneity Chemical and/or charge heterogeneity involves a modification of the primary sequence of monoclonal antibodies. Common alterations may include changes to the disulfide bond, modifications in N-glycosylation, C-terminal lysine processing, glycation of Lys residues, deamidation, isomerization, oxidation, and hydrolysis/fragmentation (Yu et al. 2007; Harris 2005). Glycosylation and C-terminal Lys processing are briefly discussed in this section. Other modifications such as deamidation, oxidation, and glycation are discussed in more detail in the stability section later in the chapter. Incomplete C-terminal processing of the heavy chain is one of the other most common modifications of recombinant Mabs. It is often found in recombinant proteins, including those from mammalian cell culture, resulting in antibody heavy chains with zero or one C-terminal lysines per heavy chain. Although the presence of a C-terminal lysine residue on the heavy chain may not have any effect on the structure or stability of IgG molecules (Perkins et al. 2000; H. Liu, Bulseco, and Sun 2006; Lazar, Kloczewiak, and Mazsaroff 2004), this modification should be differentiated from other charge variants during stability studies.

30.2.1.1.3 Glycosylation Glycosylation is one of the common posttranslational modifications in a recombinant MAb produced using mammalian cell lines (Jefferis 2002). The most common site of glycosylation in recombinant MAbs produced using a CHO cell line is through N-linkage of the Asn 297 side chain on the CH2 domain of the heavy chain. Oligosaccharides present at the N-glycosylation site are known to affect biological and pharmacological properties of IgGs, such as antibody-dependent cellular cytotoxicity (ADCC) and complement-mediated cytotoxicity (CDC; Jefferis 2005; Harris 2005; Barbin et al. 2006; Mori et al. 2007; Scallon et al. 2007). For example, it has been demonstrated that removal of the fucose from the oligosaccharides (afucosylation) enhances the ADCC activity of MAbs (Mori et al. 2007; Imai-Nishiya et al. 2007). Therefore, for those MAbs that exhibit their therapeutic activity through ADCC, characterization of the N-glycosylation pattern during stability testing, as well as for release of the material, may be critical.

30.2.1.1.4 Conformational Heterogeneity Conformational heterogeneity relates to the distribution of the conformational states as defined by the intrinsic thermodynamic stability of an MAb under a given solution condition (James, Roversi, and Tawfik 2003). While the native conformation could be defined as the one that is prevalent under physiological conditions, altered conformations could exist under different solution conditions or may be formed during the cell culture process as a result of misfolding. The altered state may be less biologically active, as well as exhibit a different stability profile such as aggregation under given solution conditions. Although precise determination of the non-native states may be challenging, an overall estimation of the protein conformational stability can be assessed by spectroscopic techniques such as circular dichroism spectroscopy, fluorescence spectroscopy, and/or differential scanning calorimetry (Harn et al. 2007; Rothlisberger, Honegger, and Plückthun 2005).

30.2.1.2 Degradation Pathways Antibodies, similar to other recombinant proteins, are subjected to a variety of physical and chemical stresses during manufacturing, processing, and storage as a liquid, in a frozen state, or in a dried state (for example during and following lyophilization). Physical stresses include shear and agitation (stirring, shaking, and pumping), exposure to various surfaces/air-liquid interfaces, freezing, and dehydration. Most common routes of physical degradation include denaturation, aggregation, precipitation, and adsorption. Chemical stresses include alteration in solution pH, ionic strength, and chemical composition of the solution, and exposure to oxygen and light. Common routes of chemical degradation include deamidation, isomerization, oxidation, and fragmentation. Physical and chemical stresses may be interrelated, and are often affected by the temperature of the solution and the surroundings. Various degradation pathways relevant to formulation of MAbs are discussed in detail in the following section.

30.2.1.2.1 Physical Degradation

30.2.1.2.1.1 DENATURATION Denaturation of antibodies can occur under a variety of conditions, including exposure to heat, pH extremes, shear, the presence of a denaturing cosolute, and surface interaction. Denaturation of antibodies often leads to aggregation. A partly denatured antibody would also tend to have a more flexible structure than the native one. The flexibility in structure could further accelerate chemical degradation (Ma et al. 2001). Formulation excipients are designed to lessen if not obviate the denaturation that some of these factors may cause in proteins such as antibodies. Each of the factors mentioned imposes various outcomes that are predictable; only the extent of damage is unpredictable and variable. Stability studies under a variety of stress conditions are integral to formulation development.

Heat figures as a prominent factor in protein denaturation, particularly as most proteins have a melting temperature, T_m, below 70°C (W. Wang 1999). In the case of MAbs, the Fab region is typically more heat stable, exhibiting a T_m above 70°C, whereas at least one of the domains in the Fc region exhibits a T_m around 60°C. The heat denaturation of antibodies is dependent on solution conditions (S.Q. Li, Bomser, and Zhang 2005; Ionescu et al. 2008). For example, acidic pH tends to destabilize MAbs towards heat-induced denaturation (Szenczi et al. 2006; Chung et al. 2005). Therefore, exposure of MAbs to solution extremes during production and purification could lead to formation of structurally altered species, which could potentially impact its long-term stability (Ejima et al. 2007; Arakawa et al. 2004).

Production processes such as pumping liquid solution from cell culture and during purification can exert physical stress on MAbs. A variety of pumps, such as a peristaltic pump or a diaphragm pump, are commonly used throughout the industry and their diverse pumping mechanisms produces various types of stresses on the protein solutions being transferred, such as shear, agitation, and cavitation, each of which may lead to antibody denaturation and subsequent aggregation.

30.2.1.2.1.2 AGGREGATION Aggregation is one of the major challenges for formulation development of antibodies. Aggregation may alter not only the therapeutic, pharmacokinetic, and pharmacodynamic profile of a monoclonal antibody, but may also have a damaging impact on the safety profile, as aggregates are often assumed to be associated with immunogenicity, due to the presence of multiple epitopes and potential conformational changes (Braun et al. 1997; Koren, Zuckerman, and Mire-Sluis 2002; Hermeling et al. 2004). Depending on the mechanism of association, a variety of aggregates may be formed and further, such aggregates may or may not be reversible. Formation of reversible aggregates is primarily due to antibody self-association, presumably arising from Fab-Fab, Fab-Fc, and/or Fc-Fc interactions. Such aggregation is generally concentration dependent and the aggregates thus formed may dissociate on dilution. On the other hand, irreversible aggregates, once formed, do not reverse on dilution and could be covalent or noncovalent in nature. When an MAb is denatured by thermal, mechanical (Chi et al. 2003), and/or chemical stress, the hydrophobic domains, initially folded inside to minimize the interaction with water, are disrupted and can become exposed to the surrounding solvent. In order to reduce the overall free energy of the system, these hydrophobic domains tend to interact with each other and form noncovalent irreversible aggregates (Ramsland and Farrugia 2002).

Antibodies may also form covalent irreversible aggregates through intermolecular disulfide cross-links. In the majority of antibody structures there are typically 32 cysteine residues, which form 12 interchain and 4 intrachain disulfide bonds (Edelman et al. 1969). In CHO cells, disulfide bond formation takes place upon translocation of the nascent peptides to the lumen of the endoplasmic reticulum (ER; Frand, Cuozzo, and Kaiser 2000). Incomplete disulfide formation may occur at this point given that the production of antibodies involves transcription of two separate genes and therefore the assembly of the two heavy chains and two light chains is often the rate-limiting step (Strutzenberger et al. 1999). The cross-linking of free sulfhydryl groups from unpaired cysteine residues could then lead to formation of covalent irreversible aggregates. Since the thiolate ion is the initiator of the disulfide exchange, this reaction often proceeds faster when the solution pH is increased. For example, the percent of dimers increased of an antifebrin T2G1s Fab′ (IgG1 k) containing free –SH groups with increasing pH from 5.8 to 9.5 (Kamat, Tolman, and Brown 1996). Another mechanism through which covalent disulfide-linked aggregates can form is through disulfide exchange in solutions at a range of near-neutral to basic pH, whereby a reducing impurity at the elevated pH could reduce a native disulfide. The disulfide does not reform in its original positions, instead forming a disulfide bond with another cysteine on a different antibody molecule.

Antibody aggregation is often concentration dependent, since association between two or more antibodies is required, irrespective of the type of aggregate formed. This presents a formidable challenge for the development of high concentration formulations. In addition to the propensity towards aggregation, high concentration formulations often are associated with high viscosities that present difficulties during production, such as during the ultrafiltration/diafiltration step, as well as during dose administration, for example, larger needle size needed and longer dosing time required. Additional discussion of issues related to high concentration antibody formulations follow in a later section of this chapter.

Similar to denaturation, a variety of physical and chemical stresses can also induce aggregation. Unfolding of an antibody is an endothermic process and requires the absorption of heat. An unfolded antibody has a higher tendency to form aggregates and therefore, the thermal conditions to which the antibody is subjected are important to monitor and at times control when developing an antibody formulation. Aggregation can also be induced by low temperatures and, in fact, certain antibodies, termed cryoimmunoglobulins, can undergo denaturation and aggregation in the temperature range of 0°C to 10°C (Middaugh et al. 1978). A serum cyroimmunoglobin has been shown to form reversible aggregates below 37°C. It has also been reported that precipitation and a gel were formed when an IgG antibody was formulated above 18 mg/mL (Sukumar et al. 2004). Freezing may also lead to aggregation through several mechanisms. During the freezing process, one of the buffer components that has a lower solubility than its counterparts may crystallize, resulting in a drastic pH shift, inducing antibody aggregation. Freezing may also lead to cold denaturation and subsequent aggregation. Furthermore, on freezing, because of ice formation and therefore removal of water into ice crystals, antibody molecules could experience dehydration, leading to potential denaturation and subsequently to aggregation.

30.2.1.2.1.3 ADSORPTION Adsorption of an MAb to a variety of surfaces may lead to a reduction of solution concentration that likely only poses a problem at low concentrations (typically less than 1 mg/mL). Generally this loss of protein to contact surfaces is not significant for antibody formulations, as discussed earlier, since therapeutic concentrations are usually greater than those of other protein therapeutics. Nonetheless, there will be losses of protein to surfaces during manufacturing processes, from production through filling into the final configuration and potentially minor losses during dose administration, for example, losses to the surface of the primary packaging and the administration apparatus (syringe, IV bag, and infusion set), that really are only of significance for low dose concentrations, atypical of antibody therapeutics.

30.2.1.2.2 Chemical Degradation

30.2.1.2.2.1 DEAMIDATION Protein deamidation is a common degradation pathway that has been extensively studied and reviewed (Robinson and Robinson 2001, 2004; Robinson 2002; Li et al. 2006; Wakankar and Borchardt 2006). The most prominent site for protein deamidation is the asparagine residue side chain; however, the glutamine residue side chain can also undergo deamidation to a lesser degree (Cleland, Powell, and Shire; M. Xie and Schowen 1999). The initial step, also the rate-determining step of asparagine deamidation, involves cyclization and formation of a cyclic imide intermediate. The cyclic imide intermediate undergoes ring opening and results in either an aspartate or an isoaspartate residue (Wakankar and Borchardt 2006). Under neutral and basic pH conditions, the ratio of aspartate to isoaspartate is normally 1 : 3 to 4 (Li et al. 2003). Deamidation is a common occurrence in proteins in general and can be observed even after cell culture production and purification processes. A specific antibody example is the humanized monoclonal antibody against HER2, trastuzumab (Herceptin®), whose deamidated variants were detected using cation exchange chromatography, which on further analysis, demonstrated that both Asn30 on the light chain and Asn55 on the heavy chain could be deamidated (Harris et al. 2001). Interestingly, all the deamidated Asn30 resulted in aspartate, not isoaspartate. This suggests that deamidation of asparagine might go through an alternate mechanism in cell culture conditions.

The rate of deamidation is largely affected by pH. In the typical formulation pH range of 4 to 8 for MAb formulation, deamidation occurs faster at basic pH compared to acid pH (Patel and Borchardt 1990; Zheng and Janis 2006). Deamidation is base catalyzed and the rate increases as the concentration of hydroxide ion increases. At neutral and basic pH, amino acid residues on the carboxylate side of asparagine also play a role in determining the rate of asparagine deamidation. Asparagine residues in the Asp-Gly sequence are prone to deamidation as much as 70-fold faster than asparagine residues with a large side chain on the adjacent amino acid. This observation of differential deamidation rates depending on the adjacent amino acid sequence does not hold, however, at acidic pH (Tyler-Cross and Schirch 1991).

Therefore, based on the formulation conditions, deamidation of MAbs can occur during storage as well. Antibody deamidation is a major source of charge heterogeneity and often results in more acidic variants when antibody sequences contain more carboxylate groups. The deamidated antibody can be detected by charge-based analytical methods such as ion-exchange chromatography and isoelectric focusing (IEF). For example, in the case of an IgG2a antibody, OKT3, 90 percent of the H386 Asn deamidated upon storage for two months at 37°C in a PBS formulation containing polysorbate 80 (Kroon, Baldwin-Ferro, and Lalan 1992). Wakankar et al. (2007a) recently reported deamidation of Asn in the CDR of two closely related MAbs and discussed the impact of CDR conformation and neighboring residues on the different rate of deamidation. In this case, deamidation had an impact on the potency of the MAbs. In general, deamidation may lead to a loss of potency if the deamidated amino acid is located in the complementary determining region (CDR) and is involved in antigen recognition. Deamidation may also impact local tertiary structure, as the formation of isoaspartate adds one more carbon to the amino acid backbone.

30.2.1.2.2.2 ISOMERIZATION Aspartate isomerization is closely related to asparagine deamidation, as both reactions go through the same cyclic imide intermediate. Nucleophilic attack of the ionized backbone nitrogen on the carbonyl carbon of the aspartate side chain is the first and rate-determining step of isomerization and forms a cyclic imide intermediate. The second step is the same as seen during deamidation and involves cyclic imide ring opening at the carbon to form isoaspartate or to return to an aspartate residue. Similar to deamidation, the ratio of aspartate to isoaspartate is about 1 : 3 or 1 : 4.

Though isomerization and deamidation share a similar mechanism, the pH dependency of aspartate isomerization is quite different. A study conducted with the heptapeptide VYPDGA found that between pH 5 and 6, isomerization was acid catalyzed (Oliyai and Borchardt 1993). The rate of aspartate isormerization increases with decreasing pH, as when the pH is lowered, the equilibrium of the aspartate sidechain is shifted toward the carboxylic acid form. The formation of the cyclic imides, the first and rate-determining step of aspartame summarization, involves dehydrolysis, since the hydroxyl group (-OH) of the carboxylic acid is a better leaving group than the corresponding anion ($-O^-$) of the carboxlate anion. Thus, isomerization occurs faster at the lower pH. Increasing pH also promotes ionization of nitrogen on the peptide backbone; however, that leads to a higher rate of nucleophilic attack and thus an increased rate of cyclic imide formation. At a pH above 6 these opposing driving forces tend to offset each other and no pH dependency can be seen clearly.

Similar to asparagine deamidation, adjacent amino acid residues on the carboxylate side of aspartate also influence the isomerization rate. The greatest amount of isomerization is observed when Gly is the adjacent amino acid, while bulky side chains on the adjacent amino acid appear to hinder aspartate isomerization (Tyler-Cross and Schirch 1991).

Antibody activity may be severely impacted by isomerization when the isomerized aspartate is involved in the interaction between the antibody and its antigen. For example, the humanized monoclonal antibody anti-HER2 trastuzumab (Herceptin®) loses 80 to 90 percent of its activity when Asp120 on the heavy chain is isomerized to isoaspartate (Harris et al. 2001). A monoclonal antibody against IgE, omalizumab (Xolair®), similarly has a reduced activity of 42 percent compared to its wild type when Asp32 on one of the light chains is isomerized. When Asp32 on both light chains are isomerized, only 15 percent of the activity remains (Cacia et al. 1996).

30.2.1.2.2.3 OXIDATION Several amino acid residues, methionine, cysteine, tyrosine, tryptophan, phenylalanine, and histidine, are subject to oxidation (Griffiths 2000). Among them, methionine residues are the most susceptible and well documented (Harris 2005). Methionine primarily oxidizes to methionine sulfoxide. For example, in a study of humanized monoclonal antibody HER2 (Herceptin®), it was reported that Met255 on the Fc region of the heavy chain oxidized readily when treated with *t*-butyl hydrogen peroxide to force oxidation. Met431 also undergoes oxidation when subjected to more aggressive conditions (Shen, Keck, and Harris 1996). Oxidation may still occur during storage in the absence of strong oxidizing agent. For example, several methionine residues of OKT3®, an IgG2a monoclonal antibody formulated in a phosphate buffer at pH 7, were oxidized when stored at 5°C (Kroon, Baldwin-Ferro, and Lalan 1992). Oxidation of a monoclonal anti-CD20 antibody during storage has also been reported (Lam et al. 2000). Oxidation reactions can also be catalyzed by light. For example, exposure of the humanized monoclonal antibody anti-HER2 trastuzumab (Herceptin®) to light leads to increased oxidation of Met255 and Met432 on the heavy chain.

As mentioned earlier, although oxidation of other residues such as tyrosine, histidine, tryptophan, etc., have been reported for various proteins and peptides (Kerwin and Remmele 2007), literature reports on oxidation of these residues in MAbs are somewhat lacking. Nevertheless, if these residues are present in the CDR sequence, their susceptibility to oxidation and the subsequent possible impact on stability should be considered and explored.

30.2.1.2.2.4 FRAGMENTATION Hydrolysis of the peptide backbone results in antibody fragmentation. The amide bonds of Asp-Gly and Asp-Pro are most susceptible to hydrolytic cleavage. Other sequences, such as Asn-Ser have also been found to be subject to hydrolysis (Tyler-Cross and Schirch 1991). Hydrolysis often occurs at the hinge region, a region between the Fab and Fc domains, because this region is the most flexible domain of the antibody. A hydrolytic cleavage at this domain often leads to fragments that are readily dissociated and detectable by size-exclusion high performance liquid chromatography (SE-HPLC) or other analytical methods such as SDS-PAGE and CE-SDS. Sometimes antibody fragments are not readily dissociable as they are held together by disulfide bond. In this case, pretreatment of the antibody with reducing agents, such as DTT and TCEP, is needed to detect the cleaved fragments.

Antibody fragmentation can be induced by a variety of conditions, such as the acidic pH environment that the antibody would experience during elution from protein A columns. Cell culture media used to culture CHO cells is kept between 30°C and 37°C. The temperature at which the cell culture is performed may result in some antibody fragmentation. Other treatments, such as freezing and thawing of drug substance bulks (Martsev et al. 2004) and storage (Jiskoot et al. 1990), may also lead to the formation of antibody fragments. Compared to intact antibody, cleaved antibody is more prone to protease digestion and often has a shorter half-life. Therefore, fragmentation of therapeutic antibodies often leads to reduced activity as not only is its ability to bind antigen reduced, its pharmacokinetics properties may be significantly compromised.

30.2.1.2.2.5 DISULFIDE FORMATION/EXCHANGE As mentioned previously, MAbs typically contain about 16 inter- and intrachain disulfides involving 32 cysteines. Hence, MAbs are prone to disulfide exchange, resulting in intermolecule cross-linking and formation of covalent aggregates, a propensity that is even greater if the antibody contains a small fraction of free sulfhydryls. Since the thiolate ion is the initiator of the disulfide exchange, this reaction often proceeds faster as the solution pH is raised.

30.2.1.2.2.6 OTHER ROUTES OF DEGRADATION In addition to the above-mentioned routes of chemical degradation, other major degradation pathways that may affect the stability and activity of monoclonal antibodies are glycation (reaction of reducing sugars, e.g., glucose with lysine side chains through Maillard reaction), formation of stable basic succinimide intermediates, and formation of pyruGlu of the N-terminal Glu residue (Dick et al. 2007).

30.2.2 Strategies and Configurations of Monoclonal Antibody Formulations

30.2.2.1 General Formulation Development Issues Formulation development of MAbs essentially entails providing a dosage form, either in a liquid or a solid form, commonly administered either intravenously (IV) or subcutaneously (SC), but on occasion intramuscularly (IM), or by some other nonparenteral route of administration or delivery. While liquid formulations are either injected as provided or diluted with a diluent prior to administration, dry formulations need to be reconstituted with a prescribed vehicle before injection. The primary goal of formulation development of any biological pharmaceutical, including MAbs, is to provide a composition that will support its stability during final production, shipping, storage, and in use, to ensure the efficacy and safety of the therapeutic antibody for its desired shelf-life with minimized/controlled degradation, as well as to deliver the dose determined to achieve its full clinical efficacy.

A liquid formulation is preferable over a lyophilized formulation, as it provides several benefits: no reconstitution is required, the volume in the container is well defined, there is greater ease of use, and dose administration has greater accuracy. Furthermore, liquid formulations may have a lower manufacturing cost since no specialized equipment, for example, a lyophilizer, is required following filling into vials. In a lyophilized formulation, the net volume following reconstitution depends on the volume of the diluent and the volume occupied by dissolved solids and, therefore, precise addition and complete dissolution need to be assured before dose administration.

For certain antibodies, however, the enhanced rate of degradation under any given solution storage condition may preclude liquid formulation development and necessitate the development of a lyophilized formulation. It is commonly accepted and observed that the rate of degradation (physical and chemical) is significantly lower in a dry state compared to a solution. A lyophilized formulation is also preferred in certain instances where refrigerated shipping (commonly used for liquid MAb products) is not feasible, such as shipping to economically challenged locations that could lack sufficient refrigerated storage space. Once a lyophilized formulation is developed and marketed, further advances in the technology and science could provide an opportunity for a viable liquid formulation. Post approval formulation change to a liquid product brings fresh challenges since the new liquid product must demonstrate comparability, and should exhibit similar stability profile as well as efficacy and safety profile to that of the approved lyophilized product. This could warrant additional toxicity assessment studies and clinical trials and demonstration of the fact that the newly developed liquid formulation is equivalent to the approved lyophilized formulation in all aspects.

In the following sections, general considerations for the development of a liquid formulation are discussed first, which is applicable to a lyophilized formulation as well; as even in a lyophilized formulation, the antibody is in a liquid environment prior to lyophilization and after reconstitution. Additional considerations relevant to developing a lyophilized formulation follow. In a separate section, the unique challenges and approaches to formulate MAbs at relatively high concentrations (>100 mg/mL) are presented.

30.2.2.2 Liquid Formulations Liquid formulation development involves a combination of approaches, such as selection of an appropriate solution pH and addition of stabilizers to minimize and control degradation. The various critical factors that should be considered while developing a liquid MAb formulation are discussed below.

30.2.2.2.1 Formulation pH Since chemical degradation could be one of the major degradation pathways of MAbs and because chemical degradation pathways often exhibit a pH dependence, solution pH is one of the critical parameters to consider first when developing a liquid antibody formulation (Jiskoot et al. 1990; Lam et al. 2000; Paborji et al. 1994). Further, physical aggregation may also exhibit pH dependence, since pH has an impact on the tertiary structure conformation of the antibody as well as the net charge on the protein, thus affecting net protein-protein interactions (Ejima et al. 2007). Additionally, while solution environment of higher than pH 7.0 can promote deamidation of the asparagine residues, disulfide exchange, and aggregation, there may be added complications at neutral to basic pH such as solubility because of proximity to the MAb's isoelectric point (pI). On the other hand, lower pH values (pH 4.0 and below) can promote isomerization, hydrolysis, and fragmentation (Cleland, Powell, and Shire 1993; Zheng and Janis 2006; Cacia et al. 1996). A very low pH (3.0 and lower) will also affect the tertiary structure conformation of an MAb and in certain cases, lead to precipitation. While oxidation may often show a pH-independent behavior, in certain cases, an alteration in solution pH could significantly affect the rate of oxidation as well, especially if a change in solution pH perturbs the protein's native structure (Griffiths and Cooney 2002).

Since different antibodies will vary somewhat in primary sequence, especially in the CDR, as well as have different pI, the optimal solution pH will differ based on the balance between various degradation processes, despite an overall similar framework. In general, currently marketed antibodies have been formulated in the pH range of 5.0 to 7.2, with most of the antibodies being formulated at slightly acidic pH (Table 30.1).

30.2.2.2.2 Formulation Buffer The choice of formulation buffer species and molarity is of paramount importance for any solution formulation in order to ensure stable pH during the desired storage shelf-life. Typical formulation buffers used in antibody formulations include phosphate, acetate, histidine, citrate, and succinate. The choice of the buffer system and concentration is based on the buffering capacity needed to provide optimal protein stability under the conditions to which the formulation will be exposed. For example, freezing of a phosphate-buffered formulation (consisting of certain hydrate forms of dibasic sodium phosphate) may result in a pH shift (Gomez, Pikal, and Rodriguez-Hornedo 2001). Furthermore, a higher concentration of a buffer may induce antibody aggregation due to a higher ionic strength, as a result of a shielding of surface charges, leading to a decrease in the repulsive electrostatic interactions (Zheng and Janis 2006). In addition, a higher buffer concentration may also increase the rate of a certain chemical reaction through buffer catalysis (Zheng and Janis 2006).

30.2.2.2.3 Formulation Excipients The use of formulation excipients or stabilizers is an established strategy to stabilize antibodies/proteins against degradation in solution. A majority of excipients such as sugars (common examples being trehalose, sucrose, dextrose, and maltose), nonionic surfactants (e.g., polysorbate 20 or polysorbate 80), polyols (e.g., mannitol, sorbitol), amino acids (e.g., arginine), are used to prevent or minimize primarily physical degradation pathways such as aggregation or adsorption (Wang 1999, 2005). Certain excipients such as methionine, histidine, or EDTA can be included in a formulation to prevent oxidation of susceptible side chains (Lam, Yang, and Cleland 1997; Griffiths 2000; Hovorka and Schoneich 2001). This latter class of excipients acts primarily as a scavenger for reactive oxygen species (ROS) such as peroxide and free radicals, and the addition of such excipients minimizes the oxidation of susceptible amino acid side chains. EDTA is particularly effective in preventing metal-catalyzed oxidation through chelation of metal ions.

The mechanisms by which various excipients protect MAbs and other proteins against aggregation in solution have been investigated extensively. It is generally believed that sugars and polyols exert their protective effect through stabilization of the native state by being "preferentially excluded" from the protein surface (Xie and Timasheff 1997a, 1997b; Auton, Ferreon, and Bolen 2006; Street, Bolen, and Rose 2006). Since a denatured state has a higher solvent accessible surface area, the preferential exclusion effect is further exaggerated. The net result is a greater difference in the free energy difference between the native and the denatured state for a sugar/polyol-containing solution compared to a solution lacking these excipients. Therefore, the conformational stability is enhanced in sugar/polyol solutions, leading to preservation of the native structure and thus preventing formation of structurally perturbed states and possible subsequent aggregation. Sugars and polyols have been shown to also affect the colloidal stability of proteins in native state as measured by an increase in the second virial coefficient value (B) of proteins in sugar/polyol-containing solutions (Valente et al. 2005). An increase in B value indicates increased net repulsive interactions and could help prevent protein–protein interactions. As yet, there is no clear mechanism to explain the increased B values, or increased net repulsive forces for proteins formulated with carbohydrates.

Surfactants such as polysorbate-20 or polysorbate-80 are utilized primarily in liquid formulations to prevent adsorption of proteins on various interfaces (Mahler et al. 2005; Vidanovic et al. 2003). For low concentration protein solutions, adsorption to the liquid-solid interface can lead to a loss in the amount of protein delivered. Adsorption of proteins at the air-water interface is also particularly undesirable as proteins have a tendency to unfold at this interface, and for that matter, at any hydrophobic solid interface such as plastic or Teflon, and subsequently undergo aggregation. Upon agitation of the protein-containing solution, protein aggregation may occur due to repetitive processing (adsorption, unfolding, and aggregation) of the proteins at the interface, either by shaking or stirring. Surfactants prevent this adverse event through preferential adsorption at the interface due to their smaller size (faster diffusion to a new interface) and higher affinity for the interface. Therefore, surfactants are effective stabilizers to prevent adsorption-based losses as well as agitation-induced aggregation of MAbs.

One excipient that has been finding increased application in biopharmaceutical processes and formulation is arginine (Arakawa et al. 2004, 2007b). Arginine has been widely used to inhibit protein aggregation, to increase protein solubility, and recently, to reduce viscosity of protein solutions (Liu and Shire 2007). The mechanisms through which arginine modulates protein–protein interactions are not well understood; nevertheless, the utilization of arginine in protein formulations continues to grow. Only recently have reports been published on its purported mechanism of action (Arakawa et al. 2007a; Baynes, Wang, and Trout 2005). It has been postulated that arginine exhibits its effects by affecting solvent properties such as surface tension as well as through its weak interactions with amino acid side chains and peptide bonds.

Preservatives as a class of excipients are not broadly used in antibody formulations, are a necessity for multidose parenteral formulations. A number of preservatives are available, such as benzyl alcohol, phenol, *m*-cresol, methyl paraben, propyl paraben, chlorbutanol, thiomersal, etc., that have been used in various multidose injectable therapeutics (generally, SC and IM). A recent review has discussed the use of preservatives in injectable dosage forms in detail (Meyer et al. 2007). The choice of a preservative depends on a number of factors, including pH, presence of nonionic surfactants (polysorbates may inactivate certain preservatives), solubility, adsorption to containers, etc. Most importantly, in the case of proteins, protein-preservative compatibility must be assessed as well. For example, a model monoclonal antibody was shown not compatible with phenol and *m*-cresol and precipitated in solution, but was found to be compatible with benzyl alcohol in a concentration-dependent manner (Gupta and Kaisheva 2003). Similar examples also exist in the literature for proteins other than antibodies, where it has been shown that benzyl alcohol can affect protein structure and stability (Thirumangalathu et al. 2006; Tobler et al. 2004). In addition to the physicochemical activity, the preservative must also fulfill the regulatory requirement of preservative efficacy test (PET) to demonstrate antimicrobial efficacy. The requirements for PET differ considerably for the United States Pharmacopeia and the European Pharmacopeia, so the country in which the therapeutic will be marketed will also impact this aspect of formulation development and design (Meyer et al. 2007). Lastly, the maximum concentration of a preservative used is also limited by its tolerability and safety profile (Brown et al. 1982).

30.2.2.3 Lyophilized Formulations

30.2.2.3.1 General Considerations Often a faster rate of degradation in solution will preclude liquid formulation development of MAbs, in order to achieve stability for the desired shelf-life. For such rapidly degrading MAbs, lyophilization offers an alternate formulation strategy in order to achieve sufficient long-term stability. Several reviews in the literature discuss lyophilization in detail as applied to protein formulation development, lyophilization process development, and stabilization of proteins during and following lyophilization (Schwegman, Hardwick, and Akers 2005; Kett, McMahon, and Ward 2004; Wang 2000; Bhatnagar, Bogner, and Pikal 2007; Costantino and Pikal 2004). In this section, some of the critical factors relevant to the development of lyophilized formulations of MAbs are briefly presented.

Lyophilization is a process that primarily consists of three steps: a freezing step that involves lowering the temperature significantly below the freezing point of the solution, a primary drying step that removes ice through the process of sublimation under low pressure, and a secondary drying step that subsequently removes unfrozen and bound water at low pressure and somewhat elevated temperatures, ultimately leading to the formation of a dry formulation cake. Even though lyophilization is designed to prepare dry formulations to enhance long-term storage stability, the process of lyophilization itself generates several stresses on the protein during the multiple steps involved. These stresses include the concentration of solutes during the freezing step, exposure of proteins to an ice-water interface, potential for cold-induced denaturation of proteins, and the dehydration stress during freezing, primary drying, and secondary drying (Bhatnagar, Bogner, and Pikal 2007). Exposure to these stresses may compromise protein quality and therefore warrants development of a robust lyophilization cycle and formulation to minimize protein degradation during the process itself.

The stress of dehydration is particularly critical, as proteins typically tend to unfold or undergo structural changes upon the removal of water, as water is essential for proteins to maintain their native folded structure. A minimum amount of water is essential to maintain the tertiary structure of proteins (typically >20 percent w/w) and therefore excessive drying may lead to a loss in the folded structure of the protein (Gregory 1995). Therefore, stabilizers such as carbohydrates, for example, trehalose or sucrose, are often added to preserve protein structure during and following dehydration. Multiple mechanisms can play a role in the stabilization of proteins by carbohydrates during the process of lyophilization and following lyophilization. Carbohydrates often act as both cryoprotectants and lyoprotectants (Wang 2000; Allison et al. 1999, 2000; Carpenter, Izutsu, and Randolph 1999; Kendrick et al. 1997). The cryoprotectant effect of sugars protects proteins during the step of freezing against the stress of dehydration (removal of water as ice) and a potential low temperature-induced cold denaturation. This effect is attributed to the preferential exclusion of carbohydrates in the presence of proteins as described earlier in the text, which stabilizes the native form of the protein. The lyoprotectant effect relates to the stabilization of proteins by carbohydrates in the dried state and is attributed to the water-substitution properties of carbohydrates as well as their ability to vitrify the system through formation of amorphous glasses, thereby considerably slowing the mobility of the system (Allison et al. 1999; Crowe, Carpenter, and Crowe 1998).

Formulation components can have an affect on the lyophilization cycle that is developed, since the properties of the solid phase in the vial, for example, morphology of the ice crystals, specific surface area available for drying following partial or complete removal of ice, solids content, and the amorphous/crystalline nature of the solids, govern the heat and mass transfer during lyophilization as well as the product quality and appearance. Therefore, lyophilized formulation development involves consideration of formulation components not only from a stability and appearance point of view, but also for their suitability for developing a robust lyophilization cycle. With these points of consideration in mind, various excipients that are commonly utilized to develop a lyophilized antibody formulation are discussed below.

30.2.2.3.2 Stabilizers

30.2.2.3.2.1 SUGARS Sugars are the most widely used stabilizers in lyophilized protein formulations. They act as cryoprotectants and lyoprotectants to protect against freezing-induced and dehydration-induced changes in protein structure, respectively, and against protein aggregation (Carpenter and Crowe 1989; Chang et al. 2004; Ándya et al. 1996; Cleland et al. 2001). Of the various sugars, sucrose and trehalose remain the stabilizers of choice, primarily due to their nonreducing nature in the pH range used for formulating MAbs. They are also the stabilizers of choice for their ability to remain amorphous following lyophilization, and their higher relative glass transition temperatures (Pikal 1990; Duddu and Monte 1997).

The glass transition temperature, T_g, is a critical parameter that affects development of a lyophilization cycle for proteins as well as the stability of a lyophilized formulation. Two types of glass transition temperatures are defined for lyophilized formulation development. The first is the glass transition temperature of the freeze concentrate, designated as T_g'. This temperature reflects the transition of the noncrystalline freeze concentrate from a viscous liquid to a brittle glass, following freezing and ice/excipient crystallization. The second glass transition temperature is the commonly known T_g of the lyophilized dried amorphous cake formed following removal of water (Chang et al. 2005). The importance of knowing the T_g' is that it informs the lyophilization cycle development, as it is desirable to carry out primary drying below the T_g' of the product in order to prevent melt back or a collapse of the cake. While the consequence of cake collapse on product stability is still debatable, collapse can affect the formation of cake structure during primary drying, leading to a higher resistance for subsequent water loss, and can also affect the appearance of the final lyophilized cake as well its reconstitution time (Pikal 1985; Overcashier, Patapoff, and Hsu 1999).

The glass transition of the lyophilized cake, T_g, can impact the long-term storage stability of the dried product. It is desirable to maximize the T_g and therefore to store the lyophilized product several degrees below T_g to minimize mobility within the amorphous cake (Duddu and Monte 1997). Moisture has a tremendous impact on the T_g of an amorphous system and can lead to a decrease in the T_g. Therefore, moisture levels of the final product must be kept low and properly controlled to ensure adequate long-term storage stability of a lyophilized antibody formulation (Breen et al. 2001).

Both trehalose and sucrose are known to stabilize proteins during lyophilization and in the dried state through water substitution as well as through vitrification. The T_g' of a 1 : 1 w/w ratio of sugar : MAb for an IgG1a has been reported to be −26°C for sucrose and −25.3°C for trehalose (Chang et al. 2005). Therefore, either excipient should not complicate the development of the lyophilization cycle. On the other hand, the T_g of the dry lyophilized cake having less than 1 percent moisture of a 1 : 1 w/w ratio of sugar : Mab was reported to be 93°C for sucrose and 131°C for trehalose (Chang et al. 2005). Even though these temperatures are considerably higher than intended storage temperatures (25°C or 5°C), it seems that trehalose formulations would have a significant lower mobility compared to sucrose formulations at a given temperature below their respective T_g and may provide greater benefits for some formulations.

Another consideration regarding the addition of sugar as a stabilizer is the sugar : protein molar (w/w) ratio. From a water-substitution hypothesis point of view, a minimum amount of sugar may be necessary to provide sufficient H-bonding sites for preservation of protein native structure. On the other hand, excessive amounts of sugar will affect the T_g' as well as the T_g of the dried amorphous cake containing a protein. Addition of excessive amounts of sugar may decrease the T_g' of an antibody-containing formulation requiring a lower temperature at which primary drying needs to be conducted. This can significantly increase the freeze-drying run time as the sublimation rate of ice would decrease with a decrease in the primary drying temperature. An increase in the weight fraction of sucrose has also been shown to decrease the T_g of an antibody in the dried cake, which may have consequences for stability at elevated temperatures (Duddu and Monte 1997). Furthermore, a higher amount of solids could affect the cake density and thus impact the primary drying rate.

In a study utilizing a model antibody, anti HER2 [trastuzumab (Herceptin®)], Cleland et al. (2001) showed that a sugar-to-protein molar ratio of at least 360 : 1 was essential to prevent protein aggregation, deamidation, and oxidation in the lyophilized formulation in order to achieve the desirable storage stability with no significant change in protein quality. In fact, utilizing a higher sugar : antibody molar ration of 600 : 1, the antibody remained stable for more than two years at 40°C. It was also shown that while amorphous mannitol alone was ineffective in conferring similar stability, mannitol in combination with sucrose or trehalose proved to be as effective as sucrose or trehalose alone. Through FTIR studies, it was concluded that maintaining the native structure following lyophilization was essential to achieve long-term stability in the dried form.

In another report, Andya, Hsu, and Shire (2003) have shown that a sugar : antibody molar ratio of 500 : 1 was essential to provide stability for a lyophilized antibody formulation. The authors estimated the number of water-binding sites on the surface of the antibody and reported this number to be 550 sites per antibody. Considering that a sucrose/trehalose molecule may have multiple H-bonding sites on a protein surface in the absence of water, it seems only reasonable that sufficient stability could be attained for various antibodies at a sugar : antibody molar ratio of at least 360 : 1 and no greater than 550 : 1, unless additional sugar is required to achieve isotonicity.

The importance of maintaining antibody structure has also been reported by Sane, Wong, and Hsu (2004), utilizing Raman spectroscopy. These authors further confirmed that following drying, the native structure of a model antibody was preserved only at a sucrose : antibody ratio of 360 : 1 or higher. These authors further demonstrated the correlation of structural perturbation to aggregation upon long-term storage stability of a lyophilized formulation. The authors also noted that while the structural perturbation of the antibody in the absence of a stabilizer was reversible upon reconstitution, long-term stability was compromised, as the structurally perturbed antibody in the dried form showed aggregation.

While overall, sucrose and trehalose continue to be the choice of carbohydrate stabilizers for preservation of protein native structure during and following lyophilization and to confer long-term stability to lyophilized antibody formulations, some of the other carbohydrate/polyol-based stabilizers reported in the literature to provide stability to proteins in lyophilized formulations include sorbitol, dextrans, and cyclodextrins (Chang et al. 2004, 2005; Ressing et al. 1992).

30.2.2.3.2.2 BULKING AGENTS A bulking agent is typically added to a low protein concentration formulation that will be lyophilized in order to obtain an acceptable appearance of the lyophilized cake inside the vial. Mannitol and glycine are typical bulking agents because of their ability to crystallize and thereby provide bulk to the lyophilized cake (Liao, Krishnamurthy, and Suryanarayanan 2005; Izutsu et al. 2004). It should be noted, however, that a crystalline solute typically will not act as a stabilizer for the antibody. Various studies have supported this statement and it has been shown, especially with mannitol, that when a bulking agent is in an amorphous state, it can provide stability and preserve protein structure following lyophilization, whereas the crystalline form is ineffective as a stabilizer (Pyne, Chatterjee, and Suryanarayanan 2003; Tian et al. 2006; V.K. Sharma and Kalonia 2004). This is a factor to be considered in developing a lyophilization cycle. If the intention is to use a solute as a crystalline bulking agent then the freezing cycle must be developed in a way to allow for complete crystallization. Incomplete crystallization would render the solute partly amorphous, which could subsequently crystallize on storage and may even release trapped water, thus compromising protein stability (Lechuga-Ballesteros, Miller, and Duddu 2004).

As mentioned earlier, a bulking agent is required primarily for low concentration protein solutions that will be formulated as lyophilized products. Typically, antibodies that will be lyophilized are formulated at somewhat higher concentrations ($>\sim 5$ mg/mL), compared to other proteins such as clotting factors, growth factors, and cytokines (Carroll and Christ 1998; Ccopardo et al. 1998; Recchia et al. 2007; Yoshioka et al. 2006). Frequently, the amount of a stabilizer (e.g., sucrose or trehalose) required to achieve the desired molar ratio that will provide a stable lyophilized formulation will preclude the need to add a bulking agent; the protein and the stabilizer can provide sufficient bulk to the lyophilized formulation. Examples exist in currently marketed products where lyophilized formulations of antibodies do not contain a bulking agent (Table 30.1).

30.2.2.3.2.3 pH AND BUFFERS Attention must be paid to the choice of buffering agent used to control the pH of the Mab formulation prior to lyophilization and following reconstitution. This is primarily because after freezing and freeze-concentration, certain buffer solutes may crystallize out of a solution due to their solubility limitations, which could in turn adversely affect the pH of the freeze-concentrate. A common example of this phenomenon is the use of sodium phosphate buffers (Gomez, Pikal, and Rodriguez-Hornedo 2001; Pikal-Cleland et al. 2002). It has been shown that the dibasic sodium phosphate salt has a lower solubility than the monobasic sodium phosphate salt and hence tends to crystallize on freezing. Crystallization of the disodium salt in the freeze-concentrate could lead to a considerable decrease in the pH and this can adversely affect protein structure and stability in the frozen, and subsequently in the dried, state. Since this situation is specific to the freezing step, crystallization of one of the components of the buffering species also has implications for the frozen storage of the bulk antibody solution, commonly referred to as the drug substance, prior to lyophilization. The appearance of crystals of a given component in a frozen solution or in a dry state could further be limited by the nucleation kinetics and crystal growth kinetics. Therefore, crystallization tendency, that is, propensity to crystallize, of a solute under a variety of conditions should be investigated a priori, before selecting a buffer component or for that matter, any excipient, which may form a component of a lyophilized formulation. Another aspect of buffer selection for a lyophilized formulation involves volatility. Since a volatile buffer component could be lost during the drying process, particularly at low pressures, leading to a shift in the pH of the freeze concentrate, volatile buffer components such as acetic acid or hydrochloric acid, should not be used in a lyophilized formulation.

In addition, buffer species may have their own specific effect on protein stability such as protein aggregation. For example, histidine has been shown to stabilize antibodies during and following lyophilization upon storage against aggregation (Chen et al. 2003).

30.2.2.3.2.4 MOISTURE The level of moisture in the final lyophilized product is critical to the stability of the protein in the dried cake. Moisture can affect protein stability in two ways. First, moisture has a tremendous effect on the T_g of the amorphous cake and an increase in moisture will result in a lower T_g (Duddu and M.P.R.D. 1997; Chang et al. 2005). A lower T_g would increase the mobility of the protein, thus affecting the kinetics of degradation reactions. Although a limited mobility exists below T_g, it is generally recommended to store an amorphous material at least 50°C below its T_g (Hancock, Shamblin, and Zografi 1995). The second way that moisture can affect protein stability is that water itself participates and catalyzes several chemical reactions and, therefore, higher moisture content can affect the chemical stability of the protein (Hill, Shalaev, and Zografi 2005; Li et al. 2005).

Thus, it may seem only reasonable to target the moisture content of the lyophilized cake to be as low as possible to maximize T_g and minimize degradation and achieve long-term stability. The role of moisture on protein stability may be more complex than a simple linear effect, however, and often, over-drying may be detrimental to the stability of the protein. For example, in a study by Breen et al. (2001), it was shown that an intermediate level of moisture of 2 to 3 percent was essential for optimal stability. While higher moisture levels ($\sim$8 percent w/w) led to an increase in the chemical (isoaspartate formation) and physical (soluble aggregate formation) degradation of a model monoclonal antibody, aggregation was also observed for samples with less than 1 percent w/w moisture content. A plausible explanation of this phenomenon, based on earlier work performed with bovine serum albumin, is that at low moisture levels, the protein structure could be perturbed and the low amount of moisture may still provide enough mobility for aggregation to occur, while not enough to assist in the refolding of the protein (Costantino, Langer, and Klibanov 1994, 1995). At somewhat intermediate moisture levels, the protein may refold to its native state and minimize this aggregation. At elevated moisture levels, the rate of reaction is further enhanced due to a significantly higher mobility, as well as water facilitating some of the degradation reactions.

30.2.3 Formulation Concerns for Specialized Monoclonal Antibodies

30.2.3.1 High Concentration Formulations For long-term treatment of a majority of chronic diseases that may be treated by monoclonal antibodies, such as autoimmune diseases, asthma, and certain allergic conditions, an outpatient setting or even at-home treatment, possibly self-administration, would be preferred. Such indications may require doses of the relevant therapeutic antibody to be on the order of more than 1 mg/kg or as high as 100 mg/dose. Furthermore, since a subcutaneous route of administration by a single injection is preferred, a high concentration formulation of antibodies is required, given the low volume limit for subcutaneous injections (<1.5 mL; Shire, Shahrokh, and Liu 2004). This volume restriction and the need to dose large amounts of therapeutic antibody have led to the development of several high concentration antibody formulations that exceed 100 mg/mL for therapeutic use. Such high concentration formulations present additional challenges to develop, for example, elevated solution viscosities, increased propensity to undergo concentration-dependent aggregation, and precipitation due to exceeding the solubility limit of the antibody in a given solution condition.

Concentration-dependent increases in viscosity of antibody solutions could have impact on the purification/recovery process, as a high viscosity solution could present challenges during the tangential flow filtration step, commonly used to formulate and concentrate antibody solutions. From a delivery point of view, a high viscosity solution would be challenging to draw up and then inject through a small bore needle (about 25 gauge) preferred for SC injection (Shire, Shahrokh, and Liu 2004).

The increase in the viscosity of an antibody solution could simply arise from an increase in the number of molecules in solution, with the assumption that antibodies could be modeled as hard spheres, or enhanced viscosity could be observed for certain antibody solutions beyond the hard spheres contribution due to protein–protein interactions (Saluja et al. 2007). At least one example exists in the literature where a higher viscosity obtained for the solution of a monoclonal antibody was attributed to the existence of self-association behavior under specific solution conditions (Liu et al. 2005). In this case, the enhanced viscosity was attributed to the presence of electrostatic interactions, as the viscosity was reduced considerably in the presence of sodium chloride.

As previously mentioned, aggregation is a bimolecular reaction and hence its rate of formation would be directly affected by antibody concentration. Aggregation may in fact be the primary route of degradation for high concentration antibody solutions. The extent of covalent and noncovalent aggregation may increase at elevated concentrations and hence antibody liquid solutions must be formulated in a way to minimize this degradation pathway. Additionally, high concentration antibody solutions can be prone to precipitation due a lower solubility of the protein in a given solution. Both the aggregation and solubility of a protein may be affected by several factors, such as pH, ionic strength, temperature, and type and concentration of a cosolute such as salts, sugars, surfactants, amino acids, etc., in a formulation. The solubility of a protein is typically lowest near its isoelectric point, pI, and increases on either side of the pI. Hence, one may overcome the solubility issue by choosing appropriate solution conditions and excipients to achieve the final desired antibody concentration. Additionally and most importantly, the solubility of the antibody should also be evaluated under physiological conditions of pH, tonicity, and temperature, for example in phosphate-buffered saline at pH 7.0 to 7.4, especially for high concentration antibody solutions administered SC. While the antibody may be completely soluble in its formulation, precipitation could occur following injection and exposure to physiological conditions, leading to local inflammation and possibly even prompting an antigenic response.

Several formulation approaches could be adopted to formulate and deliver antibodies at high concentration for subcutaneous delivery. The first choice, obviously, would be a liquid formulation, an approach that may be limited by the possibility of high solution viscosity. Nevertheless, recent patents have described low viscosity high concentration antibody solutions that make use of a combination of excipients such as arginine, histidine, polysorbates, and sugars (Liu and Shire 2007; Matheus and Mahler 2007). The formulations thus obtained exhibited low viscosity (50 cs or less), acceptable stability, low turbidity, and provided isoosmotic solution conditions. Besides excipients, alteration in solution pH could help lower viscosity as typically protein-protein interactions leading to increased viscosity exhibit a pH dependence. The selection of the pH, however, should be balanced against other possible chemical and physical degradation routes.

Alternatively, a high concentration antibody solution could be obtained by reconstituting a lyophilized formulation at a lower volume compared to that used for the initial fill into the vial. This approach provides a reasonable and cost-effective approach for generating high concentration antibody solutions since lyophilization of low protein concentration solutions with a partial volume reconstitution is a well-established strategy (Andya et al. 2001). Furthermore, a lyophilized product should possess better stability on storage compared to a liquid formulation. For this approach, a loading volume of an antibody, V_L, of concentration C_L is lyophilized and then reconstituted prior to administration by a volume, V_R, where $V_R < V_L$. The final concentration of the antibody, C_F, in the reconstituted solution will be

$$C_F = C_L V_L / (V_R + V_S)$$

where V_S is the volume contribution from solids. While this approach is suitable to achieve antibody solutions with concentrations as high as 150 mg/mL following reconstitution (initial concentration of about 50 mg/mL), lyophilization of an antibody solution having initial concentrations up to

100 mg/ml has been reported recently, that makes use of a lyoprotectant such as sucrose and does not incorporate a bulk forming crystalline solute (Colandene et al. 2007).

30.2.3.2 Antibody-Drug Conjugates (ADCs) To access the therapeutic power of the cytotoxic drugs and minimize the associated nonspecific toxicity, a strategy of using a tumor targeting antibody has been used as a carrier to deliver certain cytotoxic small molecules specifically to the tumor cells (Zein et al. 1988; Doronina et al. 2003; Erickson et al. 2006). Arming antibodies with cytotoxic molecules/reagents thus provides targeted delivery of the small molecule to the tumor cell, increases the antitumor activities of the antibody (if the antibody has antitumor activity by itself) and/or reduces the minimum effective dose (MED), ultimately may lead to a wider therapeutic window and potentially a more effective clinical outcome.

Three types of cytotoxic reagents have been used in conjugation with antibodies in clinical studies. These are protein toxins (antibodies conjugated to toxins from plant source such as ricin and gleonin; Schrama, Reisfeld, and Becker 2006), radionuclides (yttrium and iodine; Iagaru et al. 2007), and cytotoxic small molecules (calcheamicin, maytansinoids, auristatins, etc.; Zein et al. 1988; Costantino, Langer, and Klibanov 1994, 1995). In particular, ADCs constituting the antibody conjugated to a small cytotoxic molecule have gained popularity in recent years as at least one conjugated molecule is currently being approved (Mylotarg, Table 30.1), and a few others have appeared in a number of clinical trials, including trastuzumab-DM1 for the treatment of advanced HER2+ breast cancer (Krop et al. 2007; G.Q. Li, Sliwkowski, and Phillips 2005; Chan et al. 2003). Therefore, while the concept of antibody-drug conjugates has been covered in detail in another chapter, we will specifically discuss in this chapter the formulation challenges that surround the product development of these ADCs, such as those formed by the linkage of the small cytotoxic molecules with the antibody.

ADCs are typically produced through conjugation of the cytotoxic molecule to the antibody through a linker. ADCs are designed in a way such that the linker-drug combination stays attached to the antibody in the circulation; however, once internalized, releases the active drug to achieve the desired cell killing effect (Alley et al. 2008; Kovtun and Goldmacher 2007; Ricart and Tolcher 2007). The development of different linkers has been described in detail in recent publications along with a variety of cytotoxic molecules (Temming et al. 2007; Polson et al. 2007). The conjugation of the linker-drug combination is typically achieved on the lysine on cysteine side chains. Although native cysteines have been utilized to achieved cysteine-based conjugation, a novel technique has also been developed that utilizes engineered cysteines on the antibody (Eigenbrot et al. 2005). From a formulation development point of view, it becomes critical that the ADC retains its desired chemical integrity during its shelf-life. This includes minimizing the degradation of the linker-drug molecule while still attached to the antibody and preventing the cleavage of the linker-drug/drug moiety from the antibody during final production/manufacturing, shipping, storage, and in use. The latter degradation is particularly undesirable, since any free linker-drug/drug released in the solution during storage could potentially have much higher toxicity than the ADC itself.

Formulation development of an ADC would typically constitute stabilization of the antibody itself with respect to various characteristics as described above, as well as stability of the conjugated linker-drug system. This is akin to the formulation development and stability of any given traditional therapeutically relevant small molecule in solution. As expected, the stability of the linker-drug itself, as well as its release from the antibody as free drug could be affected by solution pH, temperature, buffer solutes, and other cosolutes present in solution. Therefore, these various parameters need to be optimized to achieve maximal stability for the MAb itself while minimizing degradation of the linker-drug/drug in solution. As mentioned earlier, if stabilization of the antibody or of the linker-drug moiety is unachievable in a liquid formulation, lyophilization provides a suitable alternative to attain the desired shelf-life.

The current literature is lacking on the comprehensive formulation development of ADCs that would cover all of the above-mentioned aspects. With the elevated interest in recent years in ADCs and the increasing number of clinical trials, more reports describing the formulation aspects of ADCs can be expected to appear in the near future.

30.3 NOVEL DELIVERY OF MONOCLONAL ANTIBODIES

Antibodies are typically administered as liquid formulations by one of several routes, IV, SC, and IM. Delivery by these routes of administration is easily accomplished by injection using a standard syringe, prefilled syringe, needle-free syringes or catheters attached to IV bags, depending on dose volume and route of administration. While many full length antibodies may have long half-lives in circulation, in order to increase the duration of an antibody or to protect the antibody from degradation, various routes of administration and novel delivery modalities may be utilized, as is reviewed below.

30.3.1 Formulation Issues for Alternative Routes of Administration

Injectable routes of administration have dominated antibody delivery strategies. IV antibody infusions have a long history (Sgouris 1970) and infusion of bio-oncology antibodies has clear advantages for reaching inaccessible or diffuse tumor sites. As doses are on the order of one gram of protein, an IV infusion represents an effective way to deliver this high a dose. For many antibody-based therapies a critical safety concern involves the onset of acute or delayed systemic reactions following infusion (Cook-Bruns 2001; Crandall and Mackner 2003); acute events commonly occur within 10 minutes to 4 hours and delayed events occur 24 hours to 4 days post-dosing. While antibody formation to the administered antibody drug is a risk factor for infusion reaction events (Baert et al. 2003), infusion reactions can occur after the first exposure to the therapeutic antibody (Cheifetz and Mayer 2005) and such reactions do not appear to be due to IgE responses or mast cell activation (Cheifetz et al. 2003). Fortunately, treatment protocols provide the possibility of retreatment of patients after an infusion reaction event (Cheifetz and Mayer 2005) with concomitant administration of immunomodulation molecules (Baert et al. 2003). Infusion reactions and protocols for dealing with such reactions have been well described in the literature (Cheifetz and Mayer 2005); however, many questions remain unanswered about the mechanism and prevention of adverse infusion reaction events associated with IV infusion of antibody drugs. One option to mitigate the safety risks associated with IV infusion of antibodies is to explore IM and SC delivery whenever possible, although these delivery approaches may also have confounding aspects of tissue response and volume limitations, as well as having differing pharmacokinetic parameters. The limitations of SC dose volume may be reversed by the use of the enzyme hyaluronidase at the time of injection, a delivery option currently being explored in several clinical trials involving antibodies (Thomas et al. 2007).

The expanding numbers of antibody therapeutics in clinical trials and on the market, coupled with the increased exploration of non-life-threatening indications being treated by antibodies, is prompting the investigation of delivering these agents by alternative routes. These routes may be more effective, less invasive, have an improved safety profile, be easier to use, permit self-administration, and thereby be more convenient and may even have lower costs if less active agent can be used. Further support for examining an alternate route to IV administration is seen by results of studies that suggest that antibodies administered by IV infusion are more susceptible to proteolysis than by IM injection (Page et al. 1995) and sucrose-stabilized immunoglobulins have been associated with acute renal failure following IV infusion (Chapman et al. 2004). As more antibody drugs enter the market, there will be a trend toward antibody-based drugs being dosed SC, with formulations efforts being focused on addressing the challenges of higher concentrations and devices being designed to accommodate larger volumes to be administered SC over time. Based on the desired actions for a number of antibody-based drugs, however, routes of administration other than IV, IM, or SC may be desirable. These alternative routes can pose unique formulation and stability challenges.

30.3.1.1 Oral Delivery Antibodies have been showed to be taken up from the intestinal lumen through fetal Fc receptors (FcRn) expressed at the apical surface of enterocytes (Lencer and Blumberg 2005). This uptake mechanism for antibodies was initially identified in the gastrointestinal (GI) tract of neonatal rodents for passive immunity and the expression of the receptor was thought to diminish and disappear after closure of the intestine after weaning. The recent finding that the expression of this receptor continues in the adult intestine has prompted efforts to utilize this receptor system to administer antibodies orally. The obvious obstacle to the full utilization of this pathway is the impact of the hostile enzymatic environment and extreme pH conditions on antibody stability in the GI tract (Reilly, Domingo, and Sandhu 1997), and competition from endogenous antibodies can dramatically reduce antibody uptake after oral delivery. To date, no other such transcellular mechanism of efficient antibody transport across epithelial cells of the GI tract has been demonstrated. Nonetheless, topical application of antibodies has been shown to produce a desired clinical outcome, with the oral delivery of chicken-derived (IgY) antibodies for treating and preventing infectious diseases of the GI tract (Reilly, Domingo, and Sandhu 1997; Horie et al. 2004; Mine and Kovacs-Nolan 2002); IgY antibodies do not appear to bind to or be transported by FcRn (Lencer and Blumberg 2005).

30.3.1.2 Delivery by the Respiratory Tract Respiratory tract (intranasal and pulmonary) administration provides protein therapeutics another portal for entering the circulation and has shown more promising clinical outcomes than oral protein drug delivery. The large surface area and ease of accessibility of both the respiratory and gastrointestinal systems are an advantage; however, while the gastrointestinal system poses chemical challenges, the respiratory system poses physical challenges in the form of clearance mechanisms in place to remove foreign particles, which act to reduce residence time dramatically following administration. Thus, clearing mechanisms must be overcome or taken into account. The use of a cream emulsion and a polymer delivery system has been shown to significantly decrease the clearance rate for a monoclonal antibody when applied to the nasal mucosa (Walsh et al. 2004). While the topical application of a therapeutic antibody may provide some clinical benefit at the mucosal surface, an appreciable amount of an antibody is not likely to be absorbed into the systemic circulation from this site. Compared to the lung epithelial lining, the nasal mucosa affords only a small absorptive surface area and may provide very limited access for systemic absorption for monoclonal antibodies.

Aerosol delivery of antibodies to the lung has been performed with both liquid and dry powder formulations that produce particles in the respirable range, with a mass mean aerodynamic diameter small enough to permit deposition in the deep lung, but large and dense enough that they are not readily exhaled or just impact the back of the throat. The particle size range that provides this performance has been reported to be between 1 and 5 μm depending on the formulation used and in one case, far larger, 20 μm where the particles are dense ($<0.4\ g/cm^3$) and highly porous (Edwards et al. 1997). In addition to formulation considerations for this route of administration, there is a profound need for devices that can produce the desired particle sizes, can be efficient at delivering the material to the respiratory tract with minimal loss to the device, can administer the intended dose in a short duration, and may be portable, with the antibody to be delivered stable for short periods at ambient temperature. Several interesting devices now exist and novel device technologies are beginning to achieve success in clinical studies, indicating the promise and potential for this route of administration, despite some recent commercial set backs seen in this therapeutic space for other recombinant proteins such as insulin (Mack 2007).

Formulation studies focused on producing fine dry powder particles of an anti-IgE monoclonal antibody (E25) in the size range desired for delivery to the lung (Costantino et al. 1998) showed that sodium phosphate was capable of extending the usefulness of mannitol as a protective excipient in spray-dried antibody formulations. Further, mannitol was seen to play a role both in maintaining protein stability and producing a suitable aerosol preparation. Additional studies to assess carbohydrate excipients for spray-dried aerosol E25 formulations showed that the excipient to protein ratios were critical and that degradation issues involved not only protein aggregation but also protein glycation where covalent sugar addition can occur, typically at free amine sites on the protein (Andya et al. 1999). In addition to the uptake through the alveolar membrane and through the cell membrane of the underlying capillaries for uptake into the systemic circulation, delivery to the lung also provides the potential for systemic delivery of that protein via uptake through the FcRn expressed at this mucosal surface. The distribution and function, as well as the application, of this transport route for systemic deliver of an Fc-containing protein chimera has been demonstrated (Bitonti et al. 2004; Spiekermann et al. 2002). The potential for systemic delivery of antibody-based drugs via the respiratory tract looks promising with the advances made in device designs, molecular constructs that can utilize endogenous transport mechanisms, and an enhanced understanding of pulmonary physiology.

30.3.1.3 Miscellaneous Delivery Sites of the Body Antibody-polymer formulations have been designed for intravitreal injection (Mordenti et al. 1999), topical vaginal delivery (Kuo, Sherwood, and Saltzman 1998; Saltzman et al. 2000), for sustained brain delivery using hyaluronic acid hydrogels (Tian et al. 2005), and to prevent postsurgical infections (administered to the intraperitoneal cavity) using CMC gel containing antibodies (Poelstra et al. 2002). These methods focus on strategies to administer large amounts of an antibody at a specific site in order to provide a sustained level to improve efficacy and/or to minimize injection frequency. Other strategies have been described to deliver antibodies to miscellaneous sites throughout the body, such as an anti-transferrin antibody (OX-26) that is described as a carrier for delivery of covalently attached therapeutic molecules across the blood-brain barrier following IV injection (Granholm et al. 1998). An anti-tumor necrosis factor antibody has been adsorbed onto cardiovascular stent wires for local neutralization of this cytokine in order to limit restenosis events following stent placement (Javed et al. 2002).

30.3.1.3.1 Intracellular Targeting Many of the current antibody drug targets are intracellular. One promising strategy is to chemically couple an antibody to a pH-sensitive polymer that functions to disrupt membranes at acidic pH. The application of this type of platform approach is based on the finding that many antibodies are internalized through a receptor-mediated uptake process that leads to their delivery into early and then late endosomes. During endosomal maturation after receptor-antibody internalization, the pH of these vesicles will drop from near neutral to below pH 6. This approach was demonstrated in a format where the pH-responsive polymer poly(propylacrylic acid) was covalently attached to a monoclonal antibody to facilitate its delivery to the cytoplasm of target cells (Lackey et al. 2002).

30.3.1.3.2 Other Delivery Issues A wide range of novel drugs and drug conjugates are being developed using monoclonal antibodies and engineered antibody fragments (Wu and Senter 2005). For some of these novel therapeutic approaches to be successful, they require additional or nontraditional delivery strategies. A cursory discussion of some of these approaches as they relate to intracellular targeting, site placement, and engineered antibody fragments follows.

30.3.1.3.2.1 ANTIBODY FRAGMENTS An entire spectrum of antibody fragments has been identified through protein engineering studies, with many of these showing remarkable potential for unique and novel clinical opportunities (Holliger and Hudson 2005). These smaller recombinant antibody fragments, the monovalent antibody fragments, Fab and single chain Fv and the engineered variants, diabodies, triabodies, minibodies, and single-domain antibodies are an emerging form of antibody therapy that are less costly to manufacture while still retaining the targeting specificity of full length antibodies. In fact some may have greater efficacy and more applications than whole monoclonal antibodies. It may be anticipated that this new subset of antibodies is posing new formulation challenges; however, to date, publications detailing stability studies and formulation development for these molecules are rare in the scientific literature so formulation data on antibody fragments is scant. The single chain Fv domains have been found to be poorly soluble and have a tendency to aggregate (Holliger and Hudson 2005), but these problems are being overcome by the identification of mutants that minimize the hydrophobic interface and selected by phage display (Jespers et al. 2004). Six Fab molecules have been approved as of this writing, but many are in early clinical trials and most in preclinical development. One study has been described that focused on Fab fragments. In general, the binding of an antibody to an effector protein does not necessarily inactivate that target's function

(Kawade 1985). When an antibody is effectively inactivating, however, the Fab fragments in the absence of an Fc domain can frequently function as highly potent binding agents capable of masking binding sites or enzymatic elements on macromolecules. For example, Fab fragments prepared from sheep anti-snake venom antibodies (Digibind, from GSK) can be effective antivenom. Storage of liquid formulations of such antivenoms in subtropical countries where refrigeration may not be possible provides a significant formulation challenge. Although lyophilized products can perform well under these conditions, a pH 4.0 (acetate buffer) formulation was found to also be stable and potent for at least one year at room temperature (Al-Abdulla et al. 2003), presumably at room temperatures characteristic of the tropical and subtropical regions where the material would be used. Strikingly, this stable formulation did not require the presence of a protective carbohydrate excipient. To improve drug delivery, antibody fragments have been fused to many types of molecules, such as radionuclides, cytotoxic drugs, toxins, peptides and proteins, enzymes, and liposomes (Nakamura et al. 2004). An immunoliposome made with an MAb fragment is reported to deliver drugs to the brain as such a molecule is able to cross the formidable barrier of the blood-brain barrier (Schnyder and Huwyler 2005).

30.3.1.3.2.2 IMMUNOCONJUGATES As discussed earlier in this chapter, arming antibodies with materials for selective cytotoxicity is a promising approach to increase the utility of these proteins and provide a more selective chemotherapy regimen, as expertly reviewed by Wu and Senter (2005). Antibodies made into radiopharmaceuticals have been useful for site- or tissue-specific localization using gamma scintigraphy (Bogard et al. 1989; Tuncay et al. 2000) or for enhanced cell killing.

In a preceding section of this chapter, we discussed that the challenges of formulating immunoconjugates include maintaining both the drug and the antibody's potency and stability. While some antibodies are more therapeutically active after being armed with a toxin, drug, and radionuclide (Brannon-Peppas and Blanchette 2004), this modification can make the resulting product more difficult to prepare, stabilize, and deliver. The potential for this approach spawns interest in investigating other ways of giving this type of antibody preparation. Antibodies can be combined with a generic delivery system, such as a liposome, to even deliver a variety or even multiple agents simultaneously through antibody-directed targeting (Park, Benz, and Martin 2004). For example, studies have been described for an antibody directed against the HER2/*neu* antigen to delivery liposomes containing a chemotherapeutic (Park et al. 1995), antibodies used to deliver cationic liposomes for the administration of nucleic acid material for gene therapy (Stuart, Kao, and Allen 2000), and antibodies used in the targeted delivery of modified gelatin nanoparticles (Balthasar et al. 2005) that can be coupled through a disulfide linkage (Dinauer et al. 2005).

30.3.2 Polymer Delivery Systems

Solid phase antibody preparations, prepared by lyophilization or spray freeze-drying, may be added to a polymer matrix to provide an implantable dosage form that will release the antibody over time as the polymer erodes or if nonerodible, permits release of the incorporated therapeutic from the matrix over a predesigned duration. For example, the widely nused polymer polylactide-co-glycolide (PLGA) may be made into a single solid or gel-like implant or be used to form microspheres, generated through a solid-in-oil-in-water (S/O/W) encapsulation process (Wang, Chua, and Wang 2004). Excipients such as mannitol, sucrose, and trehalose may be used to stabilize the antibody during spray freeze-drying, as the double-emulsion solvent evaporation method used to load spray freeze-dried human IgG into PLGA microspheres has been shown to produce extensive antibody aggregation (Wang, Chua, and Wang 2004). This is likely due to thermodynamic stress experienced at the water/organic solvent interface, and a variety of process methods and excipients can be use to minimize this concern (Jones et al. 1997).

A point for consideration in the use of PLGA microparticle preparations is their capacity for immunorecognition and their ability to stimulate immunization against incorporated proteins (Gupta, Chang, and Siber 1998; Lavelle et al. 1999; O'Hagan et al. 1991). While this is a clear advantage for vaccines, there may be drawbacks for protein delivery. Additionally, mild to moderate local inflammation has been observed at SC injections sites after delivery of hydrophobic microparticles (Daugherty et al. 1997), a result that could be partially mediated through macrophage recruitment (Luzardo-Alvarez et al. 2005). Not only is the potential for antibody formation increased by inflammation, such an injection site reaction can be clinically unacceptable.

In spite of these risks, polymeric microspheres and delivery systems can be successful antibody formulations if administered to the appropriate sites, using certain types of polymers and preparing microspheres or polymeric formulations by methods that are more protective of the antibody. An example of a certain delivery site that may have a decreased tendency towards an inflammatory response is the vitreal chamber of the eye, an immune privileged site, where PLGA microspheres loaded with an anti-VEGF antibody were injected for the treatment of age related macular degeneration (AMD) (Mordenti et al. 1999). Resorbable polymers with a lower potential to incite inflammation, for example, that have smoother surface characteristics, might be used to deliver antibodies in a sustained manner. Ethylene-vinyl acetate copolymer (EVAc) has also been examined as a potential polymer-based antibody delivery system for the delivery of IgG antibodies (C.H. Wang, Sengothi, and Lee 1999). Similar to PLGA systems, EVAc microspheres can extend the time course of IgG release. The possibility for extended release makes polymers loaded with antibodies promising prospects not only for injectable sustained release delivery formats but also for topical applications at mucosal surfaces. An example of mucosal administration is the long-term vaginal delivery of an antibody achieved in a mouse model using polymer vaginal rings (Kuo, Sherwood, and Saltzman 1998; Saltzman et al. 2000). Formulations of a biomedical grade polyurethane hydrogel with an antibody coating have also been prepared and shown to release bioactive protein (Rojas, Slunt, and Grainger 2000). Many of these approaches, while exciting for their potential to deliver the desired amount of the antibody and for the desired duration, nonetheless require a polymer solubilization step using a solvent, such as methylene chloride or isopropanol, which can compromise protein drug stability. Sustained release formulation approaches such as microspheres prepared using water, for example Epic Therapeutics's process of making microspheres, PROMAXX, or crystallization of proteins (Yang et al. 2003) can bypass the use of solvents that are incompatible with proteins and antibodies.

Direct, local delivery of antibodies using hydrogels has been reported for both immediate and sustained release. An example of an immediate release gel is a carboxymethylcellulose aqueous gel used to locally deliver polyclonal antibodies as a postsurgical anti-infective strategy (Poelstra et al. 2002). Antibodies covalently coupled to a biodegradable hyaluronic acid hydrogel through a labile hydrazone linkage have been used for the sustained delivery of antibodies to sites within the central nervous system (Tian et al. 2005).

30.4 CONCLUSIONS: CHALLENGES AND OPPORTUNITIES

Antibody-based therapeutics hold tremendous promise in their potential to treat currently unmet medical needs. To achieve clinical success, these novel molecules require the development of stable formulations that can be used for specific delivery methods and sites of administration. Antibodies, being endogenous molecules, have physiological features that may challenge the formulation scientist seeking to design a stable formulation as normal physiological function would dictate the clearance of these proteins from the body (Robinson, McKerrow, and Cary 1970). As such, the ultimate stability of these products may be limited by inherent mechanisms of degradation predesigned for the rate of clearance optimal for their function and turnover in the body. Protein engineering efforts have resulted in safely bypassing some of these normal degradation pathways and additional advances will be made in the near future (reviewed in Holliger and Hudson 2005). Protein modifications can be engineered to improve stability and efficacy in ways that cannot be achieved solely, if at all, by formulation changes. Information about primary sequences, especially in the complementary determining region (CDR), can help identify potential hot spots for oxidation (Met, Trp, Tyr, His), deamidation (Asn, Gln), isomerization (Asn and Asp), and other chemical degradation routes and further help design appropriate formulation conditions. More research efforts must be invested in the relatively new area of engineered antibody fragments for clinical use (Holliger and Hudson 2005), as they have the potential for significant future pay-off in their diagnostic and therapeutic applications.

The physical stability of a formulation should be evaluated separately under different stress conditions, such as thermal stress, agitation, freeze-thaw stress in the intended storage container, as the effect of these stresses on an antibody is less predictable based on sequence. Further, aggregation and viscosity need to be addressed for high concentration formulations both for stability as well as for delivery considerations. In designing antibody formulations, several competing degradation pathways are weighed against each other to minimize their overall impact on the stability of the therapeutic molecule. The pH where the minimal amount of degradation events such as deamidation and oxidation will occur is determined and a buffer system selected that can stabilize the formulation at that pH without damaging the antibody. To reduce the rate of many of these chemical degradation events, a lyophilized or freeze-dried antibody preparation can be made if a liquid formulation is seen to be unstable, but even these can be compromised by aggregation issues due to thermodynamic stress that occurs during freezing, drying, and reconstitution. Protective excipients such as carbohydrates can decrease antibody aggregation during these stressful processes. Further, the formulation must be compatible with whatever delivery is chosen and is generally isotonic. The addition of a surface active agent, such as a nonionic detergent at a low concentration (Mahler et al. 2005) to help with liquid stability to avoid the formation of aggregates in response to agitation, as well as the possible addition of preservatives for multiuse systems (Gupta and Kaisheva 2003) may be required to optimize antibody formulation stability.

Until recently, monoclonal antibodies were administered almost exclusively by the IV route. With the expansion of clinical indications being treated by the ever-increasing number of therapeutic monoclonal antibodies becoming available, the use of alternative routes of administration and novel delivery modalities has become integral in product development and life-cycle management strategies. Selection of an optimal delivery technology should be based on several factors, including the molecule's stability and safety profiles, the desired pharmacokinetic parameters, dosing frequency, the disease state being treated, convenience to the patient if self-administered, reimbursement policies, and health care economics. The ideal delivery system for a given therapeutic will be a decision maximizing the potential to satisfy all of these considerations. Clearly this is not always a straightforward decision and as yet, there is a gaping need for more technical development for protein delivery systems. While delivery options for small molecules have benefited from almost 40 years of development efforts (with the inception of ALZA), delivery technologies for macromolecules are in their relative adolescence and are yet to reach their full potential. In the coming decade, we should see many exciting therapies using delivery technologies be made possible by the integration of advances in many separate domains, including pharmaceutics, materials and polymer science, bioengineering, miniaturization of electronics, and device development. Our growing understanding of the complex biology of pathological states will drive our rational approaches to deliver the right drug to the right place for the right duration.

Looking forward, more complex antibody-based therapeutic molecules will be developed to meet the needs of treating diseases of which we are currently gaining an understanding. Foremost in the development of a clinically and commercial acceptable antibody molecule is designing a formulation that addresses the critical aspects of stability for the protein or the materials associated with the protein. Antibody engineering will have a huge impact on antibody design and small changes in sequence can provide improved stability, but also cause equally impressive changes in efficacy, as even a single amino acid change in the Fv region of an antibody can result in dramatic differences in antigen-binding specificity (Rudikoff et al. 1982). Alterations of specific amino acids in the Fc domain can also greatly change functional aspects of an antibody. Any changes that affect function and safety must be considered most important in the search for a successful formulation. The ultimate goal of developing and delivering a stable antibody formulation is to provide a maximally potent and safe therapeutic with minimal physical and chemical stability losses to provide patients with a product that has the desired shelf life, stability during in-use conditions, and optimal *in vivo* performance.

ACKNOWLEDGMENTS

We gratefully acknowledge the review of this work by Jim Andya, Jamie Moore, and Tom Patapoff. The references cited in this chapter were not intended to be inclusive of all of the seminal publications on the formulation and stability of this rapidly growing class of therapeutic molecules as we intended to highlight specific aspects within the scope of this review. Many excellent publications could not be cited that have made important contributions to this field.

REFERENCES

Al-Abdulla, I., J.M. Garnvwa, S. Rawat, D.S. Smith, J. Landon, and A. Nasidi. 2003. Formulation of a liquid ovine Fab-based antivenom for the treatment of envenomation by the Nigerian carpet viper (*Echis ocellatus*). *Toxicon* 42:399–404.

Alley, S.C., D.R. Benjamin, S.C. Jeffrey, N.M. Okeley, D.L. Meyer, R.J. Sanderson, and P.D. Senter. 2008. Contribution of linker stability to the activities of anticancer immunoconjugates. *Bioconjug. Chem.* 19:759–765.

Allison, S.D., B. Chang, T.W. Randolph, and J.F. Carpenter. 1999. Hydrogen bonding between sugar and protein is responsible for inhibition of dehydration-induced protein unfolding. *Arch. Biochem. Biophys.* 365:289–298.

Allison, S.D., M.C. Manning, T.W. Randolph, K. Middleton, A. Davis, and J.F. Carpenter. 2000. Optimization of storage stability of lyophilized actin using combinations of disaccharides and dextran. *J. Pharm. Sci.* 89:199–214.

Andya, J., J.L. Cleland, C.C. Hsu, X.M. Lam, D.E. Overcashier, S.J. Shire, J. Yang, and S.S.-Y. Wu. 2001. Protein formulation. U.S. Patent 6,685,940.

Andya, J.D., C.C. Hsu, and S.J. Shire. 2003. Mechanisms of aggregate formation and carbohydrate excipient stabilization of lyophilized humanized monoclonal antibody formulations. *AAPS PharmSci.* 5, Article 10.

Andya, J.D., Y.F. Maa, H.R. Costantino, P.A. Nguyen, N. Dasovich, T.D. Sweeney, C.C. Hsu, and S.J. Shire. 1999. The effect of formulation excipients on protein stability and aerosol performance of spray-dried powders of a recombinant humanized anti-IgE monoclonal antibody. *Pharm. Res.* 16:350–358.

Andya, J.D., S. Wu, C.C. Hsu, and S.J. Shire. 1996. The effect of sugars and buffer excipients on the stabilization of a lyophilized formulation for an anti-IgE humanized monoclonal antibody. *Pharm. Res.* 13:S78.

Arakawa, T., D. Ejima, K. Tsumoto, N. Obeyama, Y. Tanaka, Y. Kita, and S.N. Timasheff. 2007a. Suppression of protein interactions by arginine: A proposed mechanism of the arginine effects. *Biophys. Chem.* 127:1–8.

Arakawa, T., J.S. Philo, K. Tsumoto, R. Yumioka, and D. Ejima. 2004. Elution of antibodies from a protein-A column by aqueous arginine solutions. *Protein Expr. Purif.* 36:244–248.

Arakawa, T., S.J. Prestrelski, W.C. Kenney, and J.F. Carpenter. 1993. Factors affecting short-term and long-term stabilities of proteins. *Adv. Drug Delivery Rev.* 10:1–28.

Arakawa, T., K. Tsumoto, Y. Kita, B. Chang, and D. Ejima. 2007b. Biotechnology applications of amino acids in protein purification and formulations. *Amino Acids* 33:587–605.

Auton, M., A.C.M. Ferreon, and D.W. Bolen. 2006. Metrics that differentiate the origins of osmolyte effects on protein stability: A test of the surface tension proposal. *J. Mol. Biol.* 361:983–992.

Awdeh, Z.L., B.A. Askonas, and A.R. Williamson. 1967. Homogeneous gamma-G-immunoglobulin produced by mouse plasmacytoma 5563 and its subsequent heterogeneity in serum. *Biochem. J.* 102:548–553.

Baert, F., M. Noman, S. Vermeire, G. Van Assche, D.H.G.A. Carbonez, and P. Rutgeerts. 2003. Influence of immunogenicity on the long-term efficacy of infliximab in Crohn's disease. *N. Engl. J. Med.* 348:601–608.

Balthasar, S., K. Michaelis, N. Dinauer, H. von Briesen, J. Kreuter, and K. Langer. 2005. Preparation and characterisation of antibody modified gelatin nanoparticles as drug carrier system for uptake in lymphocytes. *Biomaterials* 26:2723–2732.

Barbin, K., J. Stieglmaier, D. Saul, K. Stieglmaier, B. Stockmeyer, M. Pfeiffer, P. Lang, and G.H. Fey. 2006. Influence of variable N-glycosylation on the cytolytic potential of chimeric CD19 antibodies. *J. Immunother.* 29:122–133.

Baynes, B.M., D.I.C. Wang, and B.L. Trout. 2005. Role of arginine in the stabilization of proteins against aggregation. *Biochemistry (Mosc).* 44:4919–4925.

Bhatnagar, B.S., R.H. Bogner, and M.J. Pikal. 2007. Protein stability during freezing: Separation of stresses and mechanisms of protein stabilization. *Pharm. Dev. Technol.* 12:505–523.

Bitonti, A.J., J.A. Dumont, S.C. Low, R.T. Peters, K.E. Kropp, V.J. Palombella, J.M. Stattel, Y. Lu, C.A. Tan, J.J. Song, A.M. Garcia, N.E. Simister, G.M. Spiekermann, W.I. Lencer, and R.S. Blumberg. 2004. Pulmonary delivery of an erythropoietin Fc fusion protein in non-human primates through an immunoglobulin transport pathway. *Proc. Natl. Acad. Sci. U.S.A.* 101:9763–9768.

Bogard, W.C., Jr., R.T. Dean, Y. Deo, R. Fuchs, J.A. Mattis, A.A. McLean, and H.J. Berger. 1989. Practical considerations in the production, purification, and formulation of monoclonal antibodies for immunoscintigraphy and immunotherapy. *Semin. Nuclear Med.* 19:202–220.

Brannon-Peppas, L., and J.O. Blanchette. 2004. Nanoparticle and targeted systems for cancer therapy. *Adv. Drug Deliv. Rev.* 56:1649–1659.

Braun, A., L. Kwee, M.A. Labow, and J. Alsenz. 1997. Protein aggregates seem to play a key role among the parameters influencing the antigenicity of interferon alpha (IFN-alpha) in normal and transgenic mice. *Pharm. Res.* 14:1472–1478.

Breen, E.D., J.G. Curley, D.E. Overcashier, C.C. Hsu, and S.J. Shire. 2001. Effect of moisture on the stability of a lyophilized humanized monoclonal antibody formulation. *Pharm. Res.* 18:1345–1353.

Brown, W.J., N.R.M. Buist, H.T.C. Gipson, R.K. Huston, and N.G. Kennaway. 1982. Fatal benzyl alcohol poisoning in a neonatal intensive-care unit. *Lancet* 1:1250.

Burris, H.A., I.E. Krop, S. Modi, J. Rodon, S.G. Lutzker, S.N. Holden, S. Bhattacharya, M. Sliwkowski, and M. Beeram. 2006. A phase I study of a first-in-class HER2 antibody-drug conjugate in patients with HER2-overexpressing metastatic breast cancer. *Breast Cancer Res Treat.* 100:S104.

Cacia, J., R. Keck, L.G. Presta, and J. Frenz. 1996. Isomerization of an aspartic acid residue in the complementarity-determining regions of a recombinant antibody to human IgE: Identification and effect on binding affinity. *Biochemistry* 35:1897–1903.

Cacopardo, B., F. Benanti, G. Brancati, F. Romano, and A. Nunnari. 1998. Leucocyte interferon-alpha retreatment for chronic hepatitis C patients previously intolerant to other interferons. *J. Viral Hepat.* 5:333–339.

Carpenter, J.F., and J.H. Crowe. n.d. An infrared spectroscopic study of the interactions of carbohydrates with dried proteins. *Biochemistry (Mosc.)* 28:3916–3922.

Carpenter, J.F., K. Izutsu, and T.W. Randolph. 1999. Freezing- and drying-induced perturbations of protein structure and mechanisms of protein protection by stabilizing additives. *Freeze-Drying/Lyophilization of Pharmaceutical and Biological Products* 96:123–160.

Carroll, P.V., E.R. Christ, B.A. Bengtsson, L. Carlsson, J.S. Christiansen, D. Clemmons, R. Hintz, K. Ho, Z. Laron, P. Sizonenko, P.H. Sönksen, T. Tanaka, and M. Thorne. 1998. Growth hormone deficiency in adulthood and the effects of growth hormone replacement: A review. *J. Clin. Endocrinol. Metab.* 83:382–395.

Chan, S.Y., A.N. Gordon, R.E. Coleman, J.B. Hall, M.S. Berger, M.L. Sherman, C.B. Eten, and N.J. Finkler. 2003. A phase 2 study of the cytotoxic immunoconjugate CMB-401 (hCTM01-calicheamicin) in patients with platinum-sensitive recurrent epithelial ovarian carcinoma. *Cancer Immunol. Immunother.* 52:243–248.

Chang, L.L., D. Shepherd, J. Sun, X.C. Tang, and M.J. Pikal. 2004. Mechanism of protein stabilization by sugars during freeze drying and storage: Specific interaction or matrix effect? *J. Pharm. Sci.* 94:1427–1444.

Chang, L., D. Shepherd, J. Sun, X. Tang, and M.J. Pikal. 2005. Effect of sorbitol and residual moisture on the stability of lyophilized antibodies: Implications for the mechanism of protein stabilization in the solid state. *J. Pharm. Sci.* 94:1445–1455.

Chapman, S.A., K.L. Gilkerson, T.D. Davin, and M.R. Pritzker. 2004. Acute renal failure and intravenous immune globulin: Occurs with sucrose-stabilized, but not with D-sorbitol-stabilized, formulation. *Ann. Pharmacother.* 38:2059–2067.

Chatenoud, L. 2003. CD3-specific antibody-induced active tolerance: From bench to bedside. *Nature Rev. Immunol.* 3:123–132.

Cheifetz, A., and L. Mayer. 2005. Monoclonal antibodies, immunogenicity, and associated infusion reactions. *Mt. Sinai J. Med.* 72:250–256.

Cheifetz, A., M. Smedley, S. Martin, M. Reiter, G. Leone, L. Mayer, and S. Plevy. 2003. The incidence and management of infusion reactions to infliximab: A large center experience. *Am. J. Gastroenterol.* 98:1315–1324.

Chen, B., R. Bautista, K. Yu, G.A. Zapata, M.G. Mulkerrin, and S.M. Chamow. 2003. Influence of histidine on the stability and physical properties of a fully human antibody in aqueous and solid forms. *Pharm. Res.* 20:1952–1960.

Chi, E.Y., S. Krishnan, T.W. Randolph, and J.F. Carpenter. 2003. Physical stability of proteins in aqueous solution: Mechanism and driving forces in nonnative protein aggregation. *Pharm. Res.* 20:1325–1336.

Chu Grace, C., D. Chelius, G. Xiao, K. Khor Hui, S. Coulibaly, and V. Bondarenko Pavel. 2007. Accumulation of succinimide in a recombinant monoclonal antibody in mildly acidic buffers under elevated temperatures. *Pharm. Res.* 24:1145–1156.

Chung, C.M., J.D. Chiu, L.H. Connors, O. Gursky, A. Lim, A.B. Dykstra, J. Liepnieks, M.D. Benson, C.E. Costello, M. Skinner, and M.T. Walsh. 2005. Thermodynamic stability of a kappa I immunoglobulin light chain: Relevance to multiple myeloma. *Biophys. J.* 88:4232–4242.

Cleland, J.L., X. Lam, B. Kendrick, J. Yang, T.-H. Yang, D. Overcashier, D. Brooks, C. Hsu, and J.F. Carpenter. 2001. A specific molar ratio of stabilizer to protein is required for storage stability of a lyophilized monoclonal antibody. *J. Pharm. Sci.* 90:310–321.

Cleland, J.L., M.F. Powell, and S.J. Shire. 1993. The development of stable protein formulations: A close look at protein aggregation, deamidation, and oxidation. *Crit. Rev. Ther. Drug Carrier Syst.* 10:307–377.

Colandene, J.D., L.M. Maldonado, A.T. Creagh, J.S. Vrettos, K.G. Goad, and T.M. Spitznagel. 2007. Lyophilization cycle development for a high-concentration monoclonal antibody formulation lacking a crystalline bulking agent. *J. Pharm. Sci.* 96:1598–1608.

Compton, B.J., J.S. Gerald, D.A. Lowe, and R.P. Elander. 1989. Micro isoelectric point heterogeneity of a murine monoclonal antibody (L6) originating from cell cultivation conditions. *Biotechnol. Techniques* 3:349–354.

Cook-Bruns, N. 2001. Retrospective analysis of the safety of Herceptin immunotherapy in metastatic breast cancer. *Oncology* 61(Suppl 2):58–66.

Costantino, H.R., J.D. Andya, P.A. Nguyen, N. Dasovich, T.D. Sweeney, S.J. Shire, C.C. Hsu, and Y.F. Maa. 1998. Effect of mannitol crystallization on the stability and aerosol performance of a spray-dried pharmaceutical protein, recombinant humanized anti-IgE monoclonal antibody. *J. Pharm. Sci.* 87:1406–1411.

Costantino, H.R., R. Langer, and A.M. Klibanov. 1994. Moisture-induced aggregation of lyophilized insulin. *Pharm. Res.* 11:21–29.

Costantino, H.R., R. Langer, and A.M. Klibanov. 1995. Aggregation of a lyophilized pharmaceutical protein, recombinant human albumin: Effect of moisture and stabilization by excipients. *Bio-Technology.* 13:493–496.

Costantino, H.R., and M.J. Pikal (eds.). 2004. *Lyophilization of Biopharmaceuticals*. Arlington, VA: American Association of Pharmaceutical Scientists.

Crandall, W.V., and L.M. Mackner. 2003. Infusion reactions to infliximab in children and adolescents: Frequency, outcome and a predictive model. *Aliment. Pharmacol. Ther.* 17:75–84.

Crowe, J.H., J.F. Carpenter, and L.M. Crowe. 1998. The role of vitrification in anhydrobiosis. *Annu. Rev. Physiol.* 60:73–103.

Daugherty, A.L., J.L. Cleland, E.M. Duenas, and R.J. Mrsny. 1997. Pharmacological modulation of the tissue response to implanted polylactic-co-glycolic acid microspheres. *Eur. J. Pharmacol. Biopharm.* 44:89–102.

Daugherty, A.L., and R.J. Mrsny. 2006. Formulation and delivery issues for monoclonal antibody therapeutics. *Adv. Drug Delivery Rev.* 58:686–706.

Dick, L.W., C. Kim, D.F. Qiu, and K.C. Cheng. 2007. Determination of the origin of the N-terminal pyro-glutamate variation in monoclonal antibodies using model peptides. *Biotechnol. Bioeng.* 97:544–553.

Dinauer, N., S. Balthasar, C. Weber, J. Kreuter, K. Langer, and H. von Briesen. 2005. Selective targeting of antibody-conjugated nanoparticles to leukemic cells and primary T-lymphocytes. *Biomaterials* 26:5898–5906.

Doronina, S.O., B.E. Toki, M.Y. Torgov, B.A. Mendelsohn, C.G. Cerveny, D.F. Chace, R.L. DeBlanc, R.P. Gearing, T.D. Bovee, C.B. Siegall, J.A. Francisco, A.F. Wahl, D.L. Meyer, and P.D. Senter. 2003. Development of potent monoclonal antibody auristatin conjugates for cancer therapy. *Nature Biotechnol.* 21:778–784.

Duddu, S.P., and P.R.D. Monte. 1997. Effect of glass transition temperature on the stability of lyophilized formulations containing a chimeric therapeutic monoclonal antibody. *Pharm. Res.* 14:591–595.

Edelman, G.M., B.A. Cunningham, W.E. Gall, P.D. Gottlieb, U. Rutishauser, and M.J. Waxdal. 1969. The covalent structure of an entire gammaG immunoglobulin molecule. *Proc. Natl. Acad. Sci. U.S.A.* 63:78–85.

Edwards, D.A., J. Hanes, G. Caponetti, J. Hrkach, A. Ben-Jebria, M.L. Eskew, J. Mintzes, D. Deaver, N. Lotan, and R. Langer. 1997. Large porous particles for pulmonary drug delivery. *Science* 276:1868–1871.

Eigenbrot, C.W., J.R. Junutula, H. Lowman, H.E. Raab, and R. Vandlen. 2005. Cysteine engineered antibodies and conjugates. U.S. Patent Application 20070092940.

Ejima, D., K. Tsumoto, H. Fukada, R. Yumioka, K. Nagase, T. Arakawa, and J.S. Philo. 2007. Effects of acid exposure on the conformation, stability, and aggregation of monoclonal antibodies. *Proteins Struct. Function Bioinformat.* 66:954–962.

Erickson, H.K., P.U. Park, W.C. Widdison, Y.V. Kovtun, L.M. Garrett, K. Hoffman, R.J. Lutz, V.S. Goldmacher, and W.A. Blattler. 2006. Antibody-maytansinoid conjugates are activated in targeted cancer cells by lysosomal degradation and linker-dependent intracellular processing. *Cancer Res.* 66:4426–4433.

Frand, A.R., J.W. Cuozzo, and C.A. Kaiser. 2000. Pathways for protein disulphide bond formation. *Trends Cell Biol.* 10:203–210.

Frokjaer, S., and E. Otzen Daniel. 2005. Protein drug stability: A formulation challenge. *Nature Rev. Drug Discovery* 4:298–306.

Gomez, G., M.J. Pikal, and N. Rodriguez-Hornedo. 2001. Effect of initial buffer composition on pH changes during far-from-equilibrium freezing of sodium phosphate buffer solutions. *Pharm. Res.* 18:90–97.

Granholm, A.C., D. Albeck, C. Backman, M. Curtis, T. Ebendal, P. Friden, M. Henry, B. Hoffer, J. Kordower, G.M. Rose, S. Soderstrom, and R.T. Bartus. 1998. A non-invasive system for delivering neural growth factors across the blood-brain barrier: A review. *Rev. Neurosci.* 9:31–55.

Gregory, R.B. 1995. Protein hydration and glass transition behavior. In *Protein-Solvent Interactions*, ed. R.B. Gregory, 191–264. New York: Marcel Dekker.

Griffiths, H.R. 2000. Antioxidants and protein oxidation. *Free Radic. Res.* 33(Suppl):S47–S58.

Griffiths, S.W., and C.L. Cooney. 2002. Relationship between protein structure and methionine oxidation in recombinant human alpha 1-antitrypsin. *Biochemistry (Mosc).* 41:6245–6252.

Gupta, R.K., A.C. Chang, and G.R. Siber. 1998. Biodegradable polymer microspheres as vaccine adjuvants and delivery systems. *Dev. Biol. Stand.* 92:63–78.

Gupta, S., and E. Kaisheva. 2003. Development of a multidose formulation for a humanized monoclonal antibody using experimental design techniques. *AAPS J.* 5, Article 8.

Hancock, B.C., S.L. Shamblin, and G. Zografi. 1995. Molecular mobility of amorphous pharmaceutical solids below their glass transition temperatures. *Pharm. Res.* 12:799–806.

Harn, N., C. Allan, C. Oliver, and C.R. Middaugh. 2007. Highly concentrated monoclonal antibody solutions: Direct analysis of physical structure and thermal stability. *J. Pharm. Sci.* 96:532–546.

Harris, R.J. 2005. Heterogeneity of recombinant antibodies: Linking structure to function. *Dev. Biologicals* 122:117–127.

Harris, R.J., B. Kabakoff, F.D. Macchi, F.J. Shen, M. Kwong, J.D. Andya, S.J. Shire, N. Bjork, K. Totpal, and A.B. Chen. 2001. Identification of multiple sources of charge heterogeneity in a recombinant antibody. *J. Chromatogr. B Biomed. Sci. Appl.* 752:233–245.

Hermeling, S., D.J. Crommelin, H. Schellekens, and W. Jiskoot. 2004. Structure-immunogenicity relationships of therapeutic proteins. *Pharm. Res.* 21:897–903.

Hill, J.J., E.Y. Shalaev, and G. Zografi. 2005. Thermodynamic and dynamic factors involved in the stability of native protein structure in amorphous solids in relation to levels of hydration. *J. Pharm. Sci.* 94:1636–1667.

Holliger, P., and P.J. Hudson. 2005. Engineered antibody fragments and the rise of single domains. *Nature Biotechnol.* 23:1126–1136.

Hooker, A., and D. James. 1998. The glycosylation heterogeneity of recombinant human IFN-gamma. *J. Interferon Cytokine Res.* 18:287–295.

Horie, K., N. Horie, A.M. Abdou, J.O. Yang, S.S. Yun, H.N. Chun, C.K. Park, M. Kim, and H. Hatta. 2004. Suppressive effect of functional drinking yogurt containing specific egg yolk immunoglobulin on *Helicobacter pylori* in humans. *J. Dairy Sci.* 87:4073–4079.

Hovorka, S.W., and C. Schoneich. 2001. Oxidative degradation of pharmaceuticals: Theory, mechanisms and inhibition. *J. Pharm. Sci.* 90:253–269.

Iagaru, A., H. Zhu, C. Mari, S.J. Knox, K. Ganjoo, and M.L. Goris. 2007. Comparison of I-131-tositumomab (Bexxar®) and Y-90-ibritumomab (Zevalin®) therapy of refractory/relapsed non-Hodgkin lymphoma. *Eur. J. Nuclear Med. Mol. Imaging* 34:S168.

Imai-Nishiya, H., K. Mori, M. Inoue, M. Wakitani, S. Iida, K. Shitara, and M. Satoh. 2007. Double knockdown of alpha 1,6-fucosyltransferase (FUT8) and GDP-mannose 4,6-dehydratase (GMD) in antibody-producing cells: A new strategy for generating fully non-fucosylated therapeutic antibodies with enhanced ADCC. *BMC Biotechnol.* 7:84 DOI: 10.1186/1472-6750-7-84.

Ionescu, R.M., J. Vlasak, C. Price, and M. Kirchmeier. 2008. Contribution of variable domains to the stability of humanized IgG1 monoclonal antibodies. *J. Pharm. Sci.* 97:1414–1426.

Izutsu, K., S.O. Ocheda, N. Aoyagi, and S. Kojima. 2004. Effects of sodium tetraborate and boric acid on non-isothermal mannitol crystallization in frozen solutions and freeze-dried solids. *Int. J. Pharm.* 273:85–93.

James, L.C., P. Roversi, and D.S. Tawfik. 2003. Antibody multispecificity mediated by conformational diversity. *Science* 299:1362–1367.

Javed, Q., N. Swanson, H. Vohra, H. Thurston, and A.H. Gershlick. 2002. Tumor necrosis factor-alpha antibody eluting stents reduce vascular smooth muscle cell proliferation in saphenous vein organ culture. *Exp. Mol. Pathol.* 73:104–111.

Jefferis, R. 2002. Glycosylation of recombinant IgG antibodies and its relevance for therapeutic applications (a soupcon of sugar helps the medicine work). *Cell Eng.* 3:93–107.

Jefferis, R. 2005. Glycosylation of recombinant antibody therapeutics. *Biotechnol. Prog.* 21:11–16.

Jespers, L., O. Schon, L.C. James, D. Veprintsev, and G. Winter. 2004. Crystal structure of HEL4, a soluble, refoldable human V(H) single domain with a germ-line scaffold. *J. Mol. Biol.* 337:893–903.

Jiskoot, W., E.C. Beuvery, A.A. de Koning, J.N. Herron, and D.J. Crommelin. 1990. Analytical approaches to the study of monoclonal antibody stability. *Pharm. Res.* 7:1234–1241.

Jones, A.J., S. Putney, O.L. Johnson, and J.L. Cleland. 1997. Recombinant human growth hormone poly(lactic-co-glycolic acid) microsphere formulation development. *Adv. Drug Deliv. Rev.* 28:71–84.

Jones, P.T., P.H. Dear, J. Foote, M.S. Neuberger, and G. Winter. 1986. Replacing the complementarity-determining regions in a human antibody with those from a mouse. *Nature (London)* 321:522–525.

Kamat, M.S., G.L. Tolman, and J.M. Brown. 1996. Formulation development of an antifibrin monoclonal antibody radiopharmaceutical. *Pharm. Biotechnol.* 9:343–364.

Kawade, Y. 1985. Neutralization of activity of effector protein by monoclonal antibody: Formulation of antibody dose-dependence of neutralization for an equilibrium system of antibody, effector, and its cellular receptor. *Immunology* 56:497–504.

Kendrick, B.S., B.S. Chang, T. Arakawa, B. Peterson, T.W. Randolph, M.C. Manning, and J.F. Carpenter. 1997. Preferential exclusion of sucrose from recombinant interleukin-1 receptor antagonist: Role in restricted conformational mobility and compaction of native state. *Proc. Natl. Acad. Sci. U.S.A.* 94:11917–11922.

Kerwin, B.A., and J.R.L. Remmele. 2007. Protect from light: Photodegradation and protein biologics. *J. Pharm. Sci.* 96:1468–1479.

Kett, V.L., D. McMahon, and K. Ward. 2004. Freeze-drying of protein pharmaceuticals: The application of thermal analysis. *Cryoletters* 25:389–404.

Klareskog, L., M. Gaubitz, V. Rodriguez-Valverde, M. Malaise, M. Dougados, J. Wajdula, and Etanercept Study. 2006. A long-term, open-label trial of the safety and efficacy of etanercept (Enbrel) in patients with rheumatoid arthritis not treated with other disease-modifying antirheumatic drugs. *Ann. Rheum. Dis.* 65:1578–1584.

Kohler, G., and C. Milstein. 1975. Continuous culture of fused cells secreting antibody of predefined specificity. *Nature* 256:495–497.

Koren, E., L.A. Zuckerman, and A.R. Mire-Sluis. 2002. Immune responses to therapeutic proteins in human: Clinical significance, assessment and prediction. *Curr. Pharm. Biotechnol.* 3:349–360.

Kovtun, Y.V., and V.S. Goldmacher. 2007. Cell killing by antibody-drug conjugates. *Cancer Lett.* 255:232–240.

Kroon, D.J., A. Baldwin-Ferro, and P. Lalan. 1992. Identification of sites of degradation in a therapeutic monoclonal antibody by peptide mapping. *Pharm. Res.* 9:1386–1393.

Krop, I.E., M. Beeram, S. Modi, N. Rabbee, S. Girish, J. Tibbitts, S.N. Holden, S.G. Lutzker, and H.A. Burris. 2007. A phase I study of trastuzumab-DM1, a first-in-class HER2 antibody-drug conjugate, in patients with advanced HER2+ breast cancer. *Breast Cancer Res Treat.* 106:S33.

Krotkiewski, H., A. Laskowska, and B. Krotkiewska. 1995. Mouse monoclonal IgA antibodies lack interchain disulfide bonds. *Arch. Immunol. Ther. Exp. (Warsz).* 43:167–172.

Kuo, P.Y., J.K. Sherwood, and W.M. Saltzman. 1998. Topical antibody delivery systems produce sustained levels in mucosal tissue and blood. *Nature Biotechnol.* 16:163–167.

Lackey, C.A., O.W. Press, A.S. Hoffman, and P.S. Stayton. 2002. A biomimetic pH-responsive polymer directs endosomal release and intracellular delivery of an endocytosed antibody complex. *Bioconjug. Chem.* 13:996–1001.

Lam, X., J.Q. Oeswein, B. Ongpipattanakul, Z. Shahrokh, S.X. Wang, R.P. Weissburg, and R.L. Wong. 2000. Stabilized antibody formulation. U.S. Patent 6991790 B1.

Lam, X.M., J.Y. Yang, and J.L. Cleland. 1997. Antioxidants for prevention of methionine oxidation in recombinant monoclonal antibody HER2. *J. Pharm. Sci.* 86:1250–1255.

Lavelle, E.C., M.K. Yeh, A.G. Coombes, and S.S. Davis. 1999. The stability and immunogenicity of a protein antigen encapsulated in biodegradable microparticles based on blends of lactide polymers and polyethylene glycol. *Vaccine* 17:512–529.

Lazar, A.C., M.A. Kloczewiak, and I. Mazsaroff. 2004. Matrix-assisted laser desorption/ionization mass spectrometry for the evaluation of the C-terminal lysine distribution of a recombinant monoclonal antibody. *Rapid Commun. Mass. Spectrom.* 18:239–244.

Lechuga-Ballesteros, D., D.P. Miller, and S.P. Duddu. 2004. Thermal analysis of lyophilized pharmaceutical peptide and protein formulations. In *Lyophilization of Biopharmaceuticals*, ed. H.R. Costantino and M.J. Pikal, 271–336. Arlington, VA: American Association of Pharmaceutical Scientists.

Lencer, W.I., and R.S. Blumberg. 2005. A passionate kiss, then run: Exocytosis and recycling of IgG by FcRn. *Trends Cell Biol.* 15:5–9.

Li, B., R.T. Borchardt, E.M. Topp, D. VanderVelde, and R.L. Schowen. 2003. Racemization of an asparagine residue during peptide deamidation. *J. Am. Chem. Soc.* 125:11486–11487.

Li, B., M.H. O'Meara, J.W. Lubach, R.L. Schowen, E.M. Topp, E.J. Munson, and R.T. Borchardt. 2005. Effects of sucrose and mannitol on asparagine deamidation rates of model peptides in solution and in the solid state. *J. Pharm. Sci.* 94:1723–1735.

Li, B., R.L. Schowen, E.M. Topp, and R.T. Borchardt. 2006. Effect of N-1 and N-2 residues on peptide deamidation rate in solution and solid state. AAPS J. 8:E166–E173.

Li, G.Q., M.X. Sliwkowski, and G.D.L. Phillips. 2005. Biological activity of trastuzumab-DM1 immunoconjugate in human tumor cells. *Clin. Cancer Res.* 11:8983S–8983S.

Li, S.Q., J.A. Bomser, and Q.H. Zhang. 2005. Effects of pulsed electric fields and heat treatment on stability and secondary structure of bovine immunoglobulin G. *J. Agric. Food Chem.* 53:663–670.

Liao, X.M., R. Krishnamurthy, and R. Suryanarayanan. 2005. Influence of the active pharmaceutical ingredient concentration on the physical state of mannitol-implications in freeze-drying. *Pharm Res.* 22:1978–1985.

Liu, H., G.-G. Bulseco, and J. Sun. 2006. Effect of posttranslational modifications on the thermal stability of a recombinant monoclonal antibody. *Immunol. Lett.* 106:144–153.

Liu, H., G. Gaza-Bulseco, and J. Sun. 2006. Characterization of the stability of a fully human monoclonal IgG after prolonged incubation at elevated temperature. *J. Chromatogr. B Anal. Technol. Biomed. Life Sci.* 837:35–43.

Liu, J., M.D.H. Nguyen, J.D. Andya, and S.J. Shire. 2005. Reversible self-association increases the viscosity of a concentrated monoclonal antibody in aqueous solution. *J. Pharm. Sci.* 94:1928–1940.

Liu, J., and S.J. Shire. 2007. Reduced-viscosity concentrated protein formulations. U.S. Patent Application 20070116700.

Luzardo-Alvarez, A., N. Blarer, K. Peter, J.F. Romero, C. Reymond, G. Corradin, and B. Gander. 2005. Biodegradable microspheres alone do not stimulate murine macrophages in vitro, but prolong antigen presentation by macrophages in vitro and stimulate a solid immune response in mice. *J. Control. Release* 109:62–76.

Ma, X., D.Q. Wang, R. Bouffard, and A. MacKenzie. 2001. Characterization of murine monoclonal antibody to tumor necrosis factor (TNF-MAb) formulation for freeze-drying cycle development. *Pharm. Res.* 18:196–202.

Mack, G.S. 2007. Pfizer dumps Exubera. *Nature Biotechnol.* 25:1331–1332.

Mahler, H.C., R. Muller, W. Friess, A. Delille, and S. Matheus. 2005. Induction and analysis of aggregates in a liquid IgG1-antibody formulation. *Eur. J. Pharm. Biopharm.* 59:407–417.

Marino, M., A. Corti, A. Ippolito, G. Cassani, and G. Fassina. 1997. Effect of bench-scale culture conditions on murine IgG heterogeneity. *Biotechnol. Bioeng.* 54:17–25.

Martsev, S.P., Y.I. Tsybovsky, O.A. Stremovskiy, S.G. Odintsov, T.G. Balandin, P. Arosio, Z.I. Kravchuk, and S.M. Deyev. 2004. Fusion of the antiferritin antibody VL domain to barnase results in enhanced solubility and altered pH stability. *Protein Eng. Design Select.* 17:85–93.

Matheus, S., and H.C. Mahler. 2007. Highly concentrated, liquid formulations of anti-egfr antibodies. U.S. Patent Application 20070172475.

Meyer, B.K., A. Ni, B. Hu, and L. Shi. 2007. Antimicrobial preservative use in parenteral products: Past and present. *J. Pharm. Sci.* 96:3155–3167.

Middaugh, C.R., B. Gerber-Jenson, A. Hurvitz, A. Paluszek, C. Scheffel, and G.W. Litman. 1978. Physicochemical characterization of six monoclonal cryoimmunoglobulins: Possible basis for cold-dependent insolubility. *Proc. Natl. Acad. Sci. U.S.A.* 75:3440–3444.

Mine, Y., and J. Kovacs-Nolan. 2002. Chicken egg yolk antibodies as therapeutics in enteric infectious disease: A review. *J. Med. Food* 5:159–169.

Mojica, G., S.M. Hariprasad, R.D. Jager, and W.F. Mieler. 2008. Short-term intraocular pressure trends following intravitreal injections of ranibizumab (Lucentis) for the treatment of wet age-related macular degeneration. *Br. J. Ophthalmol.* 92:584–584.

Mordenti, J., K. Thomsen, V. Licko, L. Berleau, J.W. Kahn, R.A. Cuthbertson, E.T. Duenas, A.M. Ryan, C. Schofield, T.W. Berger, Y.G. Meng, and J. Cleland. 1999. Intraocular pharmacokinetics and safety of a humanized monoclonal antibody in rabbits after intravitreal administration of a solution or a PLGA microsphere formulation. *Toxicol. Sci.* 52:101–106.

Mori, K., S. Iida, N. Yamane-Ohnuki, Y. Kanda, R. Kuni-Kamochi, R. Nakano, H. Imai-Nishiya, A. Okazaki, T. Shinkawa, A. Natsume, R. Niwa, K. Shitara, and M. Satoh. 2007. Non-fucosylated therapeutic antibodies: The next generation of therapeutic antibodies. *Cytotechnology.* 55:109–114.

Morrison, S.L. 2005. The role of glycosylation in engineered antibodies. In *Handbook of Carbohydrate Engineering* 563–586.

Nakamura, T., K.W. Peng, S. Vongpunsawad, M. Harvey, H. Mizuguchi, T. Hayakawa, R. Cattaneo, and S.J. Russell. 2004. Antibody-targeted cell fusion. *Nature Biotechnol.* 22:331–336.

O'Hagan, D.T., D. Rahman, J.P. McGee, H. Jeffery, M.C. Davies, P. Williams, S.S. Davis, and S.J. Challacombe. 1991. Biodegradable microparticles as controlled release antigen delivery systems. *Immunology* 73:239–242.

Oliyai, C., and R.T. Borchardt. 1993. Chemical pathways of peptide degradation. IV. Pathways, kinetics, and mechanism of degradation of an aspartyl residue in a model hexapeptide. *Pharm. Res.* 10:95–102.

Orciuolo, E., G. Buda, S. Galimberti, G. Boni, N. Cecconi, and M. Petrini. 2007. Role of yttrium-90 ibritumomab tiuxetan (Zevalin®) in inducing and maintaining complete molecular response in B non-Hodgkin's lymphoma patients in clinical complete remission after chemotherapy regimen. *Blood* 110:196B.

Overcashier, D.E., T.W. Patapoff, and C.C. Hsu. 1999. Lyophilization of protein formulations in vials: Investigation of the relationship between resistance to vapor flow during primary drying and small-scale product collapse. *J. Pharm. Sci.* 88:688–695.

Paborji, M., N.L. Pochopin, W.P. Coppola, and J.B. Bogardus. 1994. Chemical and physical stability of chimeric L6, a mouse-human monoclonal-antibody. *Pharm. Res.* 11:764–771.

Page, M., C. Ling, P. Dilger, M. Bentley, T. Forsey, C. Longstaff, and R. Thorpe. 1995. Fragmentation of therapeutic human immunoglobulin preparations. *Vox Sang.* 69:183–194.

Park, J.W., C.C. Benz, and F.J. Martin. 2004. Future directions of liposome- and immunoliposome-based cancer therapeutics. *Semin. Oncol.* 31:196–205.

Park, J.W., K. Hong, P. Carter, H. Asgari, L.Y. Guo, G.A. Keller, C. Wirth, R. Shalaby, C. Kotts, W.I. Wood, et al. 1995. Development of anti-p185HER2 immunoliposomes for cancer therapy. *Proc. Natl. Acad. Sci. U.S.A.* 92:1327–1331.

Patel, K., and R.T. Borchardt. 1990. Chemical pathways of peptide degradation. III. Effect of primary sequence on the pathways of deamidation of asparaginyl residues in hexapeptides. *Pharm. Res.* 7:787–793.

Pavlou, A.K., and M.J. Belsey. 2005. The therapeutic antibodies market to 2008. *Eur. J. Pharm. Biopharm.* 59:389–396.

Perkins, M., R. Theiler, S. Lunte, and M. Jeschke. 2000. Determination of the origin of charge heterogeneity in a murine monoclonal antibody. *Pharm. Res.* 17:1110–1117.

Peterson, N.C. 2005. Advances in monoclonal antibody technology: Genetic engineering of mice, cells and immunoglobulins. *Ilar J.* 46:314–319.

Pikal, M.J. 1985. Use of laboratory data in freeze-drying process design: Heat and mass transfer coefficients and the computer simulation of freeze-drying. *J. Parenter. Sci. Technol.* 39:115–139.

Pikal, M.J. 1990. Freeze-drying of proteins. II. Formulation selection. *BioPharm.*, October, 26–30.

Pikal-Cleland, K.A., J.L. Cleland, T.J. Anchordoquy, and J.F. Carpenter. 2002. Effect of glycine on pH changes and protein stability during freeze-thawing in phosphate buffer systems. *J. Pharm. Sci.* 91:1969–1979.

Poelstra, K.A., N.A. Barekzi, A.M. Rediske, A.G. Felts, J.B. Slunt, and D.W. Grainger. 2002. Prophylactic treatment of gram-positive and gram-negative abdominal implant infections using locally delivered polyclonal antibodies. *J. Biomed. Mater. Res.* 60:206–215.

Polson, A.G., S.C. Lark, C. Du, D. Eaton, K. Elkins, R.N. Fuji, M.A. Go, A. Gray, T. Januario, H. Koeppen, J. Lau, W. Leach, E. McDorman, K. McKeever, S. Prabhu, S.J. Scales, D. Slaga, S. Spencer, M. Williams, D. Xie, S.F. Yu, B. Zheng, and A. Ebens. 2007. Antibody-drug conjugates with uncleavable linkers for the treatment of non-Hodgkin lymphoma. *Blood* 110:162A.

Pyne, A., K. Chatterjee, and R. Suryanarayanan. 2003. Solute crystallization in mannitol-glycine systems: Implications on protein stabilization in freeze-dried formulations. *J. Pharm. Sci.* 92:2272–2283.

Ramsland, P.A., and W. Farrugia. 2002. Crystal structures of human antibodies: A detailed and unfinished tapestry of immunoglobulin gene products. *J. Mol. Recognit.* 15:248–259.

Recchia, F., G. Saggio, A. Cesta, G. Candeloro, A. Di Blasio, G. Amiconi, M. Lombardo, A. Nuzzo, A. Lalli, E. Alesse, S. Necozione, and S. Rea. 2007. Phase II study of interleukin-2 and 13-cis-retinoic acid as maintenance therapy in metastatic colorectal cancer. *Cancer Immunol. Immunother.* 56:699–708.

Reilly, R.M., R. Domingo, and J. Sandhu. 1997. Oral delivery of antibodies. Future pharmacokinetic trends. *Clin. Pharmacokinet.* 32:313–323.

Ressing, M.E., W. Jiskoot, H. Talsma, C.W. Vaningen, E.C. Beuvery, and D.J.A. Crommelin. 1992. The influence of sucrose, dextran, and hydroxypropyl-beta-cyclodextrin as lyoprotectants for a freeze-dried mouse Igg2a monoclonal-antibody (Mn12). *Pharm. Res.* 9:266–270.

Ricart, A.D., and A.W. Tolcher. 2007. Technology Insight: Cytotoxic drug immunoconjugates for cancer therapy. *Nature Clin. Practice Oncol.* 4:245–255.

Robinson, A.B., J.H. McKerrow, and P. Cary. 1970. Controlled deamidation of peptides and proteins: An experimental hazard and a possible biological timer. *Proc. Natl. Acad. Sci. U.S.A.* 66:753–757.

Robinson, N.E. 2002. Protein deamidation. *Proc. Natl. Acad. Sci. U.S.A.* 99:5283–5288.

Robinson, N.E., and A.B. Robinson. 2001. Deamidation of human proteins. *Proc. Natl. Acad. Sci. U.S.A.* 98:12409–12413.

Robinson, N.E., and A.B. Robinson. 2004. Prediction of primary structure deamidation rates of asparaginyl and glutaminyl peptides through steric and catalytic effects. *J. Pept. Res.* 63:437–448.

Rojas, I.A., J.B. Slunt, and D.W. Grainger. 2000. Polyurethane coatings release bioactive antibodies to reduce bacterial adhesion. *J. Control. Release* 63:175–189.

Rothlisberger, D., A. Honegger, and A. Plückthun. 2005. Domain interactions in the Fab fragment: A comparative evaluation of the single-chain Fv and Fab format engineered with variable domains of different stability. *J. Mol. Biol.* 347:773–789.

Rudikoff, S., A.M. Giusti, W.D. Cook, and M.D. Scharff. 1982. Single amino acid substitution altering antigen-binding specificity. *Proc. Natl. Acad. Sci. U.S.A.* 79:1979–1983.

Saltzman, W.M., J.K. Sherwood, D.R. Adams, P. Castle, and P. Haller. 2000. Long-term vaginal antibody delivery: Delivery systems and biodistribution. *Biotechnol. Bioeng.* 67:253–264.

Saluja, A., A.V. Badkar, D.L. Zeng, and D.S. Kalonia. 2007. Ultrasonic rheology of a monoclonal antibody (IgG(2)) solution: Implications for physical stability of proteins in high concentration formulations. *J. Pharm. Sci.* 96:3181–3195.

Sane, S.U., R. Wong, and C.C. Hsu. 2004. Raman spectroscopic characterization of drying-induced structural changes in a therapeutic antibody: Correlating structural changes with long-term stability. *J. Pharm. Sci.* 93:1005–1018.

Scallon, B.J., S.H. Tam, S.G. McCarthy, A.N. Cal, and T.S. Raju. 2007. Higher levels of sialylated Fc glycans in immunoglobulin G molecules can adversely impact functionality. *Mol. Immunol.* 44:1524–1534.

Schnyder, A., and J. Huwyler. 2005. Drug transport to brain with targeted liposomes. *NeuroRx* 2:99–107.

Schrama, D., R.A. Reisfeld, and J.C. Becker. 2006. Antibody targeted drugs as cancer therapeutics. *Nature Rev. Drug Discovery* 5:147–159.

Schwegman, J.J., L.M. Hardwick, and M.J. Akers. 2005. Practical formulation and process development of freeze-dried products. *Pharm. Dev. Technol.* 10:151–173.

Sgouris, J.T. 1970. Studies on immune serum globulin (IgG) and its modification for intravenous administration. *Prog. Immunobiol. Stand.* 4:104–113.

Sharma, B. 2007. Immunogenicity of therapeutic proteins. II. Impact of container closures. *Biotechnol. Adv.* 25:318–324.

Sharma, V.K., and D.S. Kalonia. 2004. Effect of vacuum drying on protein-mannitol interactions: The physical state of mannitol and protein structure in the dried state. *AAPS PharmSciTech.* 5:Article 10.

Shen, M.Y.K.F.J., R.G. Keck, and R.J. Harris. 1996. The application of tert-butylhydroperoxide oxidation to study sites of potential methionine oxidation in a recombinant antibody. In *Techniques in Protein Chemistry*, ed. D.R. Marshak, 275–284. Vol. 7, New York: Academic Press.

Shire, S.J., Z. Shahrokh, and J. Liu. 2004. Challenges in the development of high protein concentration formulations. *J. Pharm. Sci.* 93:1390–1402.

Spiekermann, G.M., P.W. Finn, E.S. Ward, J. Dumont, B.L. Dickinson, R.S. Blumberg, and W.I. Lencer. 2002. Receptor-mediated immunoglobulin G transport across mucosal barriers in adult life: Functional expression of FcRn in the mammalian lung. *J. Exp. Med.* 196:303–310.

Street, T.O., D.W. Bolen, and G.D. Rose. 2006. A molecular mechanism for osmolyte-induced protein stability. *Proc. Natl. Acad. Sci. U.S.A.* 103:13997–14002.

Strutzenberger, K., N. Borth, R. Kunert, W. Steinfellner, and H. Katinger. 1999. Changes during subclone development and ageing of human antibody-producing recombinant CHO cells. *J. Biotechnol.* 69:215–226.

Stuart, D.D., G.Y. Kao, and T.M. Allen. 2000. A novel, long-circulating, and functional liposomal formulation of antisense oligodeoxynucleotides targeted against MDR1. *Cancer Gene Ther.* 7:466–475.

Sukumar, M., B.L. Doyle, J.L. Combs, and A.H. Pekar. 2004. Opalescent appearance of an IgG1 antibody at high concentrations and its relationship to noncovalent association. *Pharm. Res.* 21:1087–1093.

Szenczi, A., J. Kardos, G.A. Medgyesi, and P. Zavodszky. 2006. The effect of solvent environment on the conformation and stability of human polyclonal IgG in solution. *Biologicals* 34:5–14.

Temming, K., D.L. Meyer, R. Zabinski, P.D. Senter, K. Poelstra, G. Molema, and R.J. Kok. 2007. Improved efficacy alpha(v)beta(6)-targeted albumin conjugates by conjugation of a novel auristatin derivative. *Mol. Pharm.* 4:686–694.

Thirumangalathu, R., S. Krishnan, D.N. Brems, T.W. Randolph, and J.F. Carpenter. 2006. Effects of pH, temperature, and sucrose on benzyl alcohol-induced aggregation of recombinant human granulocyte colony stimulating factor. *J. Pharm. Sci.* 95:1480–1497.

Thomas, J.R., R.C. Yocum, M.F. Haller, and C.F. von Gunten. 2007. Assessing the role of human recombinant hyaluronidase in gravity-driven subcutaneous hydration: The INFUSE-LR study. *J. Palliative Med.* 10:1312–1320.

Tian, W., D. Wong, R. McLaughlin, and D. Pandya. 2006. Mannitol crystallisation in freeze dried dosage forms. *J. Pharm. Pharmacol.* 58:A23–A24.

Tian, W.M., C.L. Zhang, S.P. Hou, X. Yu, F.Z. Cui, Q.Y. Xu, S.L. Sheng, H. Cui, and H.D. Li. 2005. Hyaluronic acid hydrogel as Nogo-66 receptor antibody delivery system for the repairing of injured rat brain: in vitro. *J. Control. Release* 102:13–22.

Tobler, S.A., B.W. Holmes, M.E.M. Cromwell, and E.J. Fernandez. 2004. Benzyl alcohol-induced destabilization of interferon-gamma: A study by hydrogen-deuterium isotope exchange. *J. Pharm. Sci.* 93:1605–1617.

Tuncay, M., S. Calis, H.S. Kas, M.T. Ercan, I. Peksoy, and A.A. Hincal. 2000. In vitro and in vivo evaluation of diclofenac sodium loaded albumin microspheres. *J. Microencapsul.* 17:145–155.

Tyler-Cross, R., and V. Schirch. 1991. Effects of amino acid sequence, buffers, and ionic strength on the rate and mechanism of deamidation of asparagine residues in small peptides. *J. Biol. Chem.* 266:22549–22556.

Valente, J.J., K.S. Verma, M.C. Manning, W.W. Wilson, and C.S. Henry. 2005. Second virial coeffcient studies of cosolvent-induced protein self-interaction. *Biophys. J.* 89:4211–4218.

Vidanovic, D., J.M. Askrabic, M. Stankovic, and V. Poprzen. 2003. Effects of nonionic surfactants on the physical stability of immunoglobulin G in aqueous solution during mechanical agitation. *Pharmazie* 58:399–404.

Wakankar, A.A., and R.T. Borchardt. 2006. Formulation considerations for proteins susceptible to asparagine deamidation and aspartate isomerization. *J. Pharm. Sci.* 95:2321–2336.

Wakankar, A.A., R.T. Borchardt, C.E. Eigenbrot, S. Shia, Y.J. Wang, S.J. Shire, and J.L. Liu. 2007a. Aspartate isomerization in the complementarity-determining regions of two closely related monoclonal antibodies. *Biochemistry (Mosc).* 46:1534–1544.

Wakankar, A.A., J. Liu, D. Vandervelde, Y.J. Wang, S.J. Shirem, and R.T. Borchardt. 2007b. The effect of cosolutes on the isomerization of aspartic acid residues and conformational stability in a monoclonal antibody. *J. Pharm. Sci.* 96:1708–1718.

Walsh, S., J. Kokai-Kun, A. Shah, and J. Mond. 2004. Extended nasal residence time of lysostaphin and an anti-staphylococcal monoclonal antibody by delivery in semisolid or polymeric carriers. *Pharm. Res.* 21:1770–1775.

Wang, C.H., K. Sengothi, and T. Lee. 1999. Controlled release of human immunoglobulin G. I. Release kinetics studies. *J. Pharm. Sci.* 88:215–220.

Wang, J., K.M. Chua, and C.H. Wang. 2004. Stabilization and encapsulation of human immunoglobulin G into biodegradable microspheres. *J. Colloid Interface Sci.* 271:92–101.

Wang, W. 1999. Instability, stabilization, and formulation of liquid protein pharmaceuticals. *Int. J. Pharm.* 185:129–188.

Wang, W. 2000. Lyophilization and development of solid protein pharmaceuticals. *Int. J. Pharm.* 203:1–60.

Wang, W. 2005. Protein aggregation and its inhibition in biopharmaceutics. *Int. J. Pharm.* 289:1–30.

Wang, W., S. Singh, D.L. Zeng, K. King, and S. Nema. 2007. Antibody structure, instability, and formulation. *J. Pharm. Sci.* 96:1–26.

Wu, A.M., and P.D. Senter. 2005. Arming antibodies: Prospects and challenges for immunoconjugates. *Nature Biotechnol.* 23:1137–1146.

Xie, G.F., and S.N. Timasheff. 1997a. The thermodynamic mechanism of protein stabilization by trehalose. *Biophys. Chem.* 64:25–43.

Xie, G.F., and S.N. Timasheff. 1997b. Mechanism of the stabilization of ribonuclease A by sorbitol: Preferential hydration is greater for the denatured than for the native protein. *Protein Sci.* 6:211–221.

Xie, M., and R.L. Schowen. 1999. Secondary structure and protein deamidation. *J. Pharm. Sci.* 88:8–13.

Yang, M.X., B. Shenoy, M. Disttler, R. Patel, M. McGrath, S. Pechenov, and A.L. Margolin. 2003. Crystalline monoclonal antibodies for subcutaneous delivery. *Proc. Natl. Acad. Sci. U.S.A.* 100:6934–6939.

Yoshioka, A., K. Fukutake, J. Takamatsu, A. Shirahata, and S. Kogenate. 2006. Post-marketing clinical evaluation of recombinant factor VIII preparation (Kogenate) in previously treated patients with hemophilia A: Descriptive meta-analysis of post-marketing study data. *Int. J. Hematol.* 84:158–165.

Yu, B.L., A. Vizel, M. Young, A. Morando, and B. He. 2007. Impact of degradations on bioactivity: A reflection from a monoclonal antibody. Abstracts of Papers, 234th ACS National Meeting, Boston, MA, August 19–23, 2007:BIOT-136.

Zein, N., A.M. Sinha, W.J. McGahren, and G.A. Ellestad. 1988. Calicheamicin-gamma-1-I: An antitumor antibiotic that cleaves double-stranded DNA site specifically. *Science* 240:1198–1201.

Zheng, J.Y., and L.J. Janis. 2006. Influence of pH, buffer species, and storage temperature on physicochemical stability of a humanized monoclonal antibody LA298. *Int. J. Pharm.* 308:46–51.

CHAPTER 31

Therapeutic Antibodies in Clinical Use and Leading Clinical Candidates

NINGYAN ZHANG*, BRENT R. WILLIAMS*, PING LU, ZHIQIANG AN, and CHEN-NI CHIN

*N. Zhang and B. Williams contributed equally to this work

Therapeutic Monoclonal Antibodies: From Bench to Clinic. Edited by Zhiqiang An

ABSTRACT

Since the first therapeutic mAb OrthoClone OKT3 was approved in 1986 for clinical use for the treatment of organ transplant rejection (Thistlethwaite et al. 1987), remarkable growth has been seen in this field. As of May, 2009, at least 25 mAbs are in clinical use (Table 31.1) and the current pipeline of therapeutic mAbs contains more than 180 products (Datamonitor 2007). This chapter covers 24 therapeutic mAb products currently in clinical use and 17 therapeutic mAbs in late stage phase III clinical development, with an overview on the practice of discovery, development, and clinical application of therapeutic mAbs.

31.1 APPROVED THERAPEUTIC mAbs IN ONCOLOGY

The therapeutic area where mAbs has the most prominent presence is oncology. Those mAbs achieve their therapeutic effect through various mechanisms. They can have direct effects in producing programmed cell death; blocking growth factor receptors, arresting proliferation of tumor cells; or inducing antibody-mediated cell killing, that is, antibody-dependent cell-mediated cytotoxicity (ADCC) and complement-dependent cytotoxicity (CDC). Section 31.1 provides a review of nine mAbs currently marketed for the treatment of cancers.

31.1.1 Rituxan/Rituximab

Rituximab is a mouse-human chimeric mAb directed against the CD20 antigen found on normal and malignant human B-cells. It was discovered in the laboratories of IDEC Pharmaceuticals, where the original mouse anti-CD20 antibody, IDEC-2B8, was engineered to contain human IgG1 and kappa constant regions while retaining the original mouse variable regions (Reff et al. 1994). CD20 is a transmembrane protein with four predicted hydrophobic regions that cross the membrane and two extracellular loops. It is a B-cell marker that is expressed from the pre-B-cell stage until the plasma cell stage. CD20 is not expressed on stem cells or plasma cells. Although the exact function of CD20 is still not completely understood, data indicate that it is possibly a Ca^{2+} channel (Bubien et al. 1993) and may be involved in B-cell growth and activation (Tedder et al. 1985). CD20 is also expressed in more than 90 percent of B-cell non-Hodgkin's lymphomas and 10 to 15 percent of chronic lymphocytic leukemia B-cells (Almasri et al. 1992; Anderson et al. 1984). CD20 is close to an ideal target for

TABLE 31.1 mAb Therapeutics Approved for Clinical Use

Generic Name Trade Name Manufacturer	Launch Date	Therapy Area	Major Indication	Target	Protein Form/Isotype	Delivery	References
Muromonab Orthoclone/OKT3 Johnson & Johnson	1986	AIID	Transplant rejection	CD3	Murine IgG2a	IV	Cohen et al. (1989)
Abciximab ReoPro Eli Lilly	1995	CV	Cardiovascular disease	CD41	Chimeric Fab	IV	Faulds and Sorkin (1994)
Rituximab Rituxan/MabThera Genentech/Roche	1997	Oncology	Non-Hodgkin lymphoma	CD20	Chimeric IgG1	IV	Maloney et al. (1997)
Daclizumab Zenapax Roche	1997	AIID	Transplant rejection	CD25	Humanized IgG1	IV	Vincenti et al. (1998)
Basiliximab Simulect Novartis	1998	AIID	Transplant rejection	CD25	Chimeric IgG1	IV	Nashan et al. (1997)
Infliximab Remicade Centocor	1998	AIID	Rheumatoid arthritis	TNFα	Chimeric IgG1	IV	Onrust and Lamb (1998)
Palivizumab Synagis MedImmune	1998	ID	Respiratory syncytial virus	RSV F-protein	Humanized IgG1	IM	Storch (1998)
Trastuzumab Herceptin Genentech	1998	Oncology	Breast cancer	Her2	Humanized IgG1	IV	Albanell and Baselga (1999)
Gemtuzumab/ozogamicin Mylotarg Wyeth	2000	Oncology	Acute myelogenous leukemia	CD33	Humanized IgG4 conjugated with ozogamicin	IV	Sorokin (2000)

(Continued)

TABLE 31.1 *Continued*

Generic Name Trade Name Manufacturer	Launch Date	Therapy Area	Major Indication	Target	Protein Form/Isotype	Delivery	References
Alemtuzumab Campath Bayer-Schering	2001	Oncology	Chronic lymphocytic leukemia	CD52	Humanized IgG1	IV	Ferrajoli, O'Brien, and Keating (2001)
Ibritumomab tiuxetan Zevalin Biogen/Idec	2002	Oncology	Non-Hodgkin lymphoma	CD20	Murine IgG1 conjugated with Yttrium 90	IV	Krasner and Joyce (2001)
Omalizumab Xolair Genentech/Novartis	2003	Respiratory	Asthma	IgE	Humanized IgG1	SC	Davis (2004)
Efalizumab Raptiva Genentech	2003	AIID	Psoriasis	CD11A	Humanized IgG1	SC	Gauvreau et al. (2003)
Tositumomab Bexxar GSK	2003	Oncology	Non-Hodgkin lymphoma	CD20	Murine IgG2a conjugated with Iodine-131	IV	Davies (2004)
Adalimumab Humira Abbott	2003	AIID	Rheumatoid arthritis	TNFα	Human IgG1	SC	Weinblatt et al. (2003)
Cetuximab Erbitux ImClone/BMS	2003	Oncology	Colorectal cancer	EGFR	Chimeric IgG1	IV	Kies and Harari (2002)

I-131 ch-TNT Shanghai Medipharm Biotech Co.	2003	Oncology	Advanced lung cancer	Intracellular DNA in tumors	Chimeric IgG1 conjugated with I-131	IV	S. Chen et al. (2005)
Bevacizumab Avastin Genentech	2004	Oncology	Colorectal and non-small cell lung cancer	VEGF	Humanized IgG1	IV	Kerr (2004)
Natalizumab Tysabri Biogen IDEC/Elan	2004	CNS/AIID	Multiple sclerosis	VLA4	Humanized IgG1	IV	Rudick and Sandrock (2004)
Tocilizumab Actemra Roche/Chugai	2005	AIID	Castleman's disease	IL-6R	Humanized IgG1	IV	Paul-Pletzer (2006)
Ranibizumab Lucentis Genentech/Novartis	2006	Ophthalmology	Wet age-related macular degeneration	VEGF	Humanized mAb fragment of Avastin	Injection into the eye	Kenneth and Kertes (2006)
Panitumumab Vectibix Amgen	2006	Oncology	Colorectal cancer	EGFR	Human IgG2	IV	Cohenuram and Saif (2007)
Certolizumab pegol Cimzia UCB-Schwarz	2007	AIID	Rheumatoid arthritis	TNFα	PEGylated fragment	SC	Rutgeerts et al. (2007)
Eculizumab Soliris Alexion	2007	Hematology	PNH (chronic hemolysis)	C5a	Humanized IgG2/IgG4 hybrid	IV	Rother et al. (2007)
Golimumab Simponi Centocor	2009	AIID	Rheumatoid arthritis	TNFα	Human IgG1	IV	Kay et al. (2008); Keystone et al. (2009)

AIID: autoimmune and inflammatory disorders; ID: infectious disease; CNS: central nervous system; CV: cardiovascular.

an mAb such as rituximab, as it is expressed at high levels on B-cells, is relatively resistant to internalization, and is not shed, thereby allowing the antibody to persist on the cell surface (Einfeld et al. 1988; Johnson and Glennie 2003; Liu et al. 1987; O.W. Press et al. 1989).

Rituximab binds to an epitope on the large extracellular loop of CD20 (Binder et al. 2006; Du et al. 2007; Polyak and Deans 2002) on normal and malignant B-cells with an approximate affinity of 8 nM. Several different mechanisms of action have been proposed to explain its anticancer activity, including induction of apoptosis (Vega et al. 2005a, 2005b; Zhang et al. 2005) and lysis through complement-dependent cytotoxicity (CDC) and antibody-dependent cell-mediated cytotoxicity (ADCC) (Cvetkovic and Perry 2006; Reff et al. 1994). Rituximab binding may also sensitize the cells to the effects of chemotherapy (Demidem et al. 1997; Vega et al. 2005b). Most studies testing the clinical efficacy of Rituxan in NHL (non-Hodgkin's lymphoma) employed treatment regimens that included 375 mg/m^2 of rituximab by intravenous infusion (IV), either in combination with chemotherapy (Coiffier et al. 2002; Feugier et al. 2005) or as a monotherapy (Davis et al. 1999; McLaughlin et al. 1998). In most B-cell NHL patients, rituximab causes a swift and profound depletion of circulating B-cells which lasts for three to six months (Maloney et al. 1994; McLaughlin et al. 1998). In one pivotal trial demonstrating the safety and efficacy of rituximab for the treatment of relapsed or refractory NHL, patients receiving 375 mg/m^2 of rituximab IV weekly for four doses, an overall response rate (ORR) of 48 percent and a complete response rate (CR) of 6 percent were seen (McLaughlin et al. 1998). The median time to the onset of response was 50 days and the median duration of the response was 11.2 months. Clinical studies of rituximab in relapsed or refractory, CD-20 positive, B-cell NHL and diffuse large B-cell lymphoma have demonstrated that combination chemo/immunotherapy is superior to either treatment alone (Coiffier et al. 2002; Czuczman et al. 2004; Feugier et al. 2005; Pfreundschuh et al. 2006; Schulz et al. 2007). When pharmacokinetics were studied in a phase I clinical trial in patients with NHL given rituximab by IV infusion, drug serum levels and half-life were proportional to dose (Maloney et al. 1994). However, in further clinical trials, serum concentrations were inversely proportional to the level of circulating peripheral B-cells and measurement of tumor bulk at baseline (Berinstein et al. 1998). In these same trials, IV administration of rituximab in four weekly doses resulted in a mean serum half-life of 76.3 hours after the first dose and 205.8 hours after the fourth dose. Rituximab was still detectable in the serum of patients at three and six months post-treatment. A similar pharmacokinetic profile was seen when rituximab was either administered alone or as six infusions in combination with CHOP chemotherapy (Plosker and Figgitt 2003). Recently, the use of rituximab to deplete B-cells for the treatment of autoimmune diseases has become an area of study (see Arkfeld 2008 for a review). Early studies demonstrating the efficacy of rituximab in the treatment of rheumatoid arthritis (RA) led to further interest and larger clinical trials (De Vita et al. 2002; Edwards and Cambridge 2001; Leandro, Edwards, and Cambridge 2002; Protheroe et al. 1999).

Rituximab, the first antibody to be approved by the FDA for the treatment of cancer, was approved in 1997 for the treatment of NHL. The larger clinical trial studies of rituximab in RA led to its approval by the U.S. Food and Drug Administration (FDA) in 2006 for the treatment of patients with RA. Rituximab's current FDA approved indications include the treatment of patients with relapsed or refractory, low-grade or follicular, CD20-positive, B-cell NHL. It is also indicated for the first-line treatment of follicular or diffuse large B-cell, CD20 positive, non-Hodgkin's lymphoma in combination with CVP or CHOP (or other anthracycline-based) chemotherapy regimens, respectively. A third approved usage is for the treatment of low-grade, CD20 positive, B-cell NHL in patients with stable disease or who achieve a partial or complete response following first-line treatment with CVP therapy. Finally, rituximab is approved in a nononcology indication, for use in combination with methotrexate in patients with moderately to severely active rheumatoid arthritis who have had an inadequate response to TNF antagonist therapies. Rituximab has proven to be a well-tolerated drug in patients, having one of the longest records of safety and tolerability data of oncology-related mAbs. The most common adverse event is infusion reactions. In most cases, these are mild to moderate (fever and chills/rigor) and occur during the first infusion, with time to onset of 30 to 120 minutes. Other adverse events included infections (bacterial, most common), hematologic (lymphopenia, most common), and pulmonary events (cough, rhinitis, bronchospasm, dyspnea, and sinusitis, most common). Boxed warnings are included for fatal infusion reactions, tumor lysis syndrome, severe mucocutaneous reactions, and progressive multifocal leukoencephalopathy (see Biogen Idec, Inc. and Genentech, Inc. 2007).

31.1.2 Herceptin/Trastuzumab

Trastuzumab is a humanized mAb directed against the extracellular domain of human epidermal growth factor receptor 2, HER2. It was originally a mouse mAb, mumAb4D5, discovered at Genentech through immunization of mice with 3T3 cells transfected to express the HER2/neu gene (Fendly et al. 1990; Hudziak et al. 1989). It was later humanized by Genentech scientists to create an antibody with human IgG1 and kappa constant as well as human framework regions while retaining the complementarity-determining regions of mumAb4D5 (Carter et al. 1992). Its antigen, HER2, is a 185 kDa transmembrane receptor that is a member of the epidermal growth factor receptor family of receptor tyrosine kinases. While a ligand has not yet been identified for HER2, the receptor has been shown to heterodimerize with other family members where it mediates signaling to the others. HER2 normally regulates cell proliferation. Signaling from the receptor reduces expression of cyclin D and c-myc, ultimately leading to increased cellular proliferation (Lane et al. 2000; Neve et al. 2000). HER2 gene amplification and/or receptor overexpression occurs in approximately 20 to 30 percent of primary breast cancers where it is associated with reduced overall survival (M.F. Press et al. 1997; Slamon et al. 1987, 1989). Additionally, HER2 overexpression has been shown to be transforming in numerous studies (Di Fiore et al. 1987; Hudziak, Schlessinger, and Ullrich 1987; Muthuswamy et al. 1994, 2001; Woods Ignatoski et al. 2003), making this an attractive target for development of a cancer therapeutic.

Trastuzumab binds to an epitope in the juxtamembrane region of HER2 (Cho et al. 2003) on breast cancer cells with an approximate affinity of 5 nM (Genentech 06-Nov-2006). Several different mechanisms of action have been proposed to explain its antineoplastic activity, including inhibition of cancer cell proliferation and ADCC. Trastuzumab has also been shown to cause receptor internalization, degradation, decreased downstream signaling (Baselga et al. 2001; Lane et al. 2000; Sliwkowski et al. 1999), inhibition of metalloproteinase-induced cleavage of the extracellular domain (Baselga et al. 2001), apoptosis (Mohsin et al. 2005), and be antiangiogenic (Izumi et al. 2002). Early clinical studies of trastuzumab tested in patients with HER2-overexpressing metastatic breast cancer as monotherapy (Baselga et al. 1996; Cobleigh et al. 1999). In the larger part of these studies (Cobleigh et al. 1999), patients who were treated with an IV loading dose of 4 mg/kg followed by 2 mg/kg weekly maintenance doses had an objective response rate of 15 percent and a 9.1 month median duration of response. In later phase II and phase III trials in metastatic breast cancer patients, trastuzumab combined with chemotherapy showed better response rates, increased time to disease progression, and overall survival as compared to either chemotherapy or trastuzumab alone (Esteva et al. 2002; Marty et al. 2005; Slamon et al. 2001). Additionally, better response rates were seen in patients who had tumors with overexpression of HER2 as compared to those with normal expression (Seidman et al. 2001). The pharmacokinetics of trastuzumab was also studied in metastatic breast cancer patients. Studies using trastuzumab at 4 mg/kg loading followed by 2 mg/kg with weekly dosing, calculated a mean half-life of approximately six days (Cobleigh et al. 1999; Genentech 06-Nov-2006). In later studies, a longer interval between doses was used (8 mg/kg loading followed by 6 mg/kg every three weeks) and a longer mean half-life of 16 to 18 days was demonstrated (Baselga et al. 2005; Leyland-Jones et al. 2003).

Trastuzumab was approved in 1998 for the treatment of HER2-overexpressing breast cancer. Its current FDA-approved indications include adjuvant treatment of patients with HER2-overexpressing, node-positive breast cancer as part of a regimen containing doxorubicin, cyclophosphamide, and paclitaxel. It is also approved for use in combination with paclitaxel in metastatic breast cancer patients whose tumors overexpress HER2 and who have received prior chemotherapy. Finally, it is approved for use as a single agent in patients with metastatic breast cancer whose tumors overexpress HER2 and who have received one or more chemotherapy regimens. Most adverse events in patients receiving trastuzumab were infusion-related, usually fever and chills. Other more common adverse reactions reported have included nausea, vomiting, diarrhea, rash, neutropenia, anemia, and myalgia. The most serious toxicities in patients receiving trastuzumab have included cardiomyopathy, pulmonary toxicity, severe infusion reactions, and febrile neutropenia/exacerbation of chemotherapy-induced neutropenia. Due to the risk of cardiomyopathy (left ventricular dysfunction), it is recommended that potential patients undergo a thorough cardiac assessment before beginning therapy and that cardiac function should be monitored throughout therapy (see Genentech 06-Nov-2006). Boxed warnings are included on the label for cardiomyopathy and infusion reactions.

31.1.3 Mylotarg/Emtuzumab Ozogamicin

Gemtuzumab ozogamicin is a humanized antibody conjugated with a cytotoxic antitumor antibiotic, calicheamicin. The antibody moiety is directed against the CD33 antigen found on the surface of leukemic blasts and immature normal cells of the myelomonocytic lineage. The current humanized antibody is derived from a mouse IgG1 antibody, called P67.6, prepared by the scientists at Celltech Therapeutics (now part of UCB S.A.) (Hamann et al. 2002a). The mouse antibody was humanized to create the current antibody containing human IgG4 and kappa constant as well as human framework regions while retaining the complementarity-determining regions of the original mouse antibody (Hamann et al. 2002b). The cytotoxic antibiotic portion of the molecule, *N*-acetyl-gamma calicheamicin, is linked to the antibody via a bifunctional linker (Lee et al. 1987). Approximately 50 percent of the antibodies in the product are loaded with 4 to 6 moles of calcheamicin per mole of antibody while the remaining 50 percent of the antibodies are not linked (Wyeth Pharmaceuticals Inc., 23-Jan-2006). The antibody target, CD33 (also known as Siglec-3), is a type I transmembrane sialoglycoprotein that is a member of the immunoglobulin superfamily subset of sialic acid-binding immunoglobulin-like lectins (see Crocker, Paulson, and Varki 2007 for a review). CD33 is expressed on the surface of myeloid and early multilineage hematopoietic progenitor cells and monocytes. It is not expressed on normal pluripotent hematopoietic stem cells or on nonhematopoietic cells (Brendel and Neubauer 2000). The ligand and exact biologic function of CD33 are not yet completely known. Studies have pointed to a functional role for CD33 in cell-cell interactions and signaling in the immune system, including modulating leukocyte behavior such as inhibition of cellular proliferation and activation (Balaian, Zhong, and Ball 2003; Paul et al. 2000; Vitale et al. 1999) and induction of proinflammatory cytokine secretion (Lajaunias, Dayer, and Chizzoline 2005). CD33 is also expressed on the surface of leukemic blasts in approximately 90 percent of adult and pediatric acute myeloid leukemia (AML) patients (Dinndorf et al. 1986; Legrand et al. 2000), making it an attractive therapeutic target.

Gemtuzumab ozogamicin binds to cell surface-expressed CD33 on leukemic blasts with an approximate affinity of 0.08 nM (Carter 2006). The antibody functions as a vehicle, delivering the conjugated toxic payload to CD33-expressing cells. Binding of the antibody to CD33 results in endocytosis, cleavage of the link between the antibody and calicheamicin, and release of calicheamicin (van Der Velden et al. 2001). The released calicheamicin is reduced by glutathione forming a reactive intermediate that binds to DNA causing double-strand breaks and inducing apoptosis (Dedon, Salzberg, and Xu 1993; Wyeth Pharmaceuticals Inc., 23-Jan-2006; Zein et al. 1988). *In vitro* studies of gemtuzumab ozogamicin in CD33-positive human myeloid leukemia cell lines showed that three out four lines tested were sensitive to the treatment, with IC_{50} in the range of 2 to 6 ng/ml (Amico et al. 2003). In a preclinical human promyelocytic leukemia tumor xenograft study in nude mice, treatment with gemtuzumab ozogamicin resulted in long-term tumor-free survivors (Hamann et al. 2002b). The efficacy and safety of gemtuzumab ozogamicin was demonstrated in three pivotal, single arm phase II studies (Larson et al. 2005). In these studies, 277 patients with CD33-positive acute myeloid leukemia received monotherapy with gemtuzumab ozogamicin in two doses separated by two weeks. Of these patients, 26% achieved remission consisting of 13% of the patients with complete remission with platelet recovery (CR) and 13% of the patients with complete remission without platelet recovery (CRp). The median recurrence-free survival for patients who achieved CR was 6.4 months and 4.5 months for those who achieved CRp. In examination of pharmacokinetics, patients given a first dose of 9 mg/m^2 of gemtuzumab ozogamicin as a two hour infusion showed elimination half-lives of total and unconjugated calicheamicin of 41 and 143 hours, respectively (Wyeth Pharmaceuticals Inc., 23-Jan-2006). After patients were given the second 9 mg/m^2 dose, the half-life of total calicheamicin was increased to about 64 hours and the area under the concentration-time curve was increased 30 percent for the unconjugated calcheamicin.

The encouraging clinical results led to the FDA approval of gemtuzumab ozogamicin in 2000 for the treatment of patients with CD33-positive acute myeloid leukemia. Its current FDA approved indications include treatment of patients with CD33-positive acute myeloid leukemia in first relapse who are $\geq$60 years old and who are not considered candidates for other cytotoxic chemotherapy. Treatment with intravenous gemtuzumab ozogamicin is usually moderately tolerated. However, adverse events related to infusion reactions or drug toxicity are fairly common (Wyeth Pharmaceuticals Inc., 23-Jan-2006; Larson et al. 2005). Infusion reactions usually occurred with the first dose on the same day as infusion and fewer events were noted after the second dose. The most common (58 to 82 percent) infusion-related events included vomiting, chills, nausea, and fever. Other serious adverse events noted during treatment with gemtuzumab ozogamicin include myelosuppression, neutropenia, anemia, thrombocytopenia, infection, bleeding, mucositis, hepatotoxicity, and veno-occlusive disease. Boxed warinings are included on the label for hypersensitivity reactions and hepatotoxicity.

31.1.4 Campath/Alemtuzumab

Alemtuzumab is a humanized mAb directed against the CD52 antigen. It was originally a rat mAb, Cambridge Pathology-1G or CAMPATH-1G, which was discovered through immunization of rats with human lymphocytes and screening of hybridomas (Hale et al. 1983, 1987; Waldmann and Hale 2005). This was done as part of an original strategy to find anti-lymphocyte antibodies to control graft-versus-host disease. The investigators decided that in order to minimize the risk of immunogenicity that would potentially occur through use of a rat antibody in humans, they would humanize the CAMPATH-1G antibody. This led to the development of CAMPATH-1H (alemtuzumab) which contains human IgG1 and kappa as well as framework regions while retaining the complementarity-determining regions of CAMPATH-1G (Riechmann et al. 1988). Its antigen, CD52, is a small (21 to 28 kDa) glycosylphosphatidylinositol-anchored cell surface glycoprotein. CD52 is highly expressed ($\sim$500,000 molecules/cell on lymphocytes) on B- and T-cells, as well as in monocytes, macrophages, eosinophils, natural killer cells, dendritic cells, and epithelial cells of the male reproductive tract (Buggins et al. 2002; Genzyme Corporation and Bayer Healthcare Pharmaceuticals, 19-Dec-2006; Elsner et al. 1996; Ginaldi et al. 1998; Ratzinger et al. 2003). The exact function of CD52 is not yet known, but its proposed functions include mediation of signal transduction, promotion of cell-cell adhesion, and protection of the cell from environmental insult. A recent study suggests that CD52 may contribute to the induction of CD4-positive T-regulatory cells (Watanabe et al. 2006). The CD52 antigen is an interesting target for an oncology-related antibody therapeutic as it is expressed on tumor cells, including those from T-cell prolymphocytic leukemia, chronic lymphocytic leukemia, hairy cell leukemia, acute lymphoblastic leukemia, and non-Hodgkin's lymphoma (Cheson 2006; Ginaldi et al. 1998).

Alemtuzumab binds to CD52 with an approximate affinity of 10 to 32 nM at an epitope consisting of the C-terminal peptide and the GPI anchor (Carter 2006; Xia et al. 1993). Early clinical studies of the parent rat anti-CD52 antibody in patients with a variety of lymphoid malignancies showed that the antibody was capable of inducing a long-lasting depletion of lymphocytes from the blood and marrow and thus had potential application in lymphoid malignancies (Dyer et al. 1989). An early clinical study was also done using the humanized form of the antibody (alemtuzumab) in two patients with non-Hodgkin's lymphoma (Hale et al. 1988). Remission was induced in both patients with clearance of lymphoma cells from the blood and bone marrow and resolution of splenomegaly. The proposed mechanisms of action of alemtuzumab include ADCC, complement dependent cytolysis (Crowe et al. 1992), and induction of apoptosis (Rowan et al. 1998; Stanglmaier, Reis, and Hallek 2004). Alemtuzumab was evaluated in three pivotal multicenter clinical trials in 149 patients with B-cell chronic lymphocytic leukemia (B-CLL) who were previously treated with alkylating agents, fludarabine, or other chemotherapies (Genzyme Corporation and Bayer Healthcare Pharmaceuticals, 19-Dec-2006). Patients were treated with 30 mg of alemtuzumab, intravenously, three times per week for up to 12 weeks. The overall response rate in these trials ranged from 21 to 33 percent, with 0 to 2 percent of the patients having a complete response and 21 to 31 percent of the patients having a partial response. The progression-free survival achieved was four to seven months. A further clinical study of alemtuzumab was conducted in 297 previously untreated patients with B-CLL, Rai stage I-IV, with evidence of progressive disease requiring therapy (Genzyme Corporation and Bayer Healthcare Pharmaceuticals, 19-Sept-2007). Patients were treated with either chlorambucil, 40 mg/m^2 by mouth once daily for 28 days, for a maximum of 12 cycles or alemtzumab, 30 mg intravenously three times per week for up to 12 weeks. The overall response rate (OR) in the chlorambucil arm (CA) was 55 percent while the OR in the alemtuzumab arm (AA) was 83 percent, with complete response rates of 2 percent in the CA and 24 percent in the AA. The progression-free survival achieved in the CA was 11.7 months versus 14.6 months in the AA. Pharmacokinetics were studied in 30 alemtuzumab-naïve patients with B-CLL who had failed previous therapy with purine analogs (Genzyme Corporation and Bayer Healthcare Pharmaceuticals, 19-Dec-2006). It was administered at the recommended dosing schedule, starting at 3 mg and increasing to 30 mg three times per week for up to 12 weeks. Alemtuzumab displayed nonlinear elimination kinetics. The mean half-life was 11 hours after the first 30 mg does and 6 days after the last 30 mg dose.

The encouraging clinical results led to the FDA approval of alemtuzumab in 2001 for the treatment of B-CLL. Its current FDA approved indications include use in the treatment of B-CLL in patients who have been treated with alkylating agents and who have failed fludarabine therapy. It is also indicated for use in previously untreated B-CLL patients. Treatment of patients with alemtuzumab is usually moderately tolerated, with initial infusion reactions occurring in 70 to 80 percent of patients. Acute infusion reactions were most common during the first week of therapy. The most common (33 to 89 percent) infusion-related adverse events included rigors, fever, nausea, and vomiting. Other more serious adverse events noted during treatment with alemtuzumab include infections and hematologic events (cytopenias). Boxed warnings are included on the label for infusion reactions, hematologic toxicity, and infections.

31.1.5 Zevalin/Ibritumomab Tiuxetan

Ibritumomab tiuxetan is an Indium-111 (^{111}In) or Yttrium-90 (^{90}Y) radiolabeled antibody. The antibody moiety, Ibritumomab is an entirely murine mAb directed against the CD20 antigen found on human normal and malignant B-cells. The covalently attached linker-chelator part of the therapeutic, tiuxetan, provides a high affinity, conformationally restricted chelation site for labeling the antibody with the γ-emitter radionuclide ^{111}In for imaging or the β-emitter radionuclide ^{90}Y for therapy. The antibody portion is a mouse IgG1 kappa antibody that is the parent antibody of the mouse-human chimerized antibody rituximab. The discovery of the mouse antibody and the biology and rationale for targeting its antigen, CD20, are described in Section 31.1.1.

Ibritumomab tiuxetan binds to an epitope on the large extracellular loop of CD20 (Binder et al. 2006; Du et al. 2007; Polyak and Deans 2002) on normal and malignant B-cells with an approximate affinity between 14 and 18 nM (Biogen Idec, Inc., 13-Sept-2005). The antibody is expected to retain some of the same mechanisms of action of rituximab, including apoptosis (see Section 31.1.1 on rituximab). However, since it is an entirely mouse IgG1 antibody whereas rituximab has human IgG1 constant regions, it probably lacks the capability of mediating ADCC and CDC in human patients. Limited data examining these capabilities in radioimmunotherapy exist. The ^{90}Y-labeled antibody exerts its antitumor effects by binding to CD20+ cells and emitting β-radiation to the targeted cells. This radiation induces cellular damage by the formation of free radicals in the target as well as the neighboring cells (Chakrabarti et al. 1996; Biogen Idec, Inc., 13-Sept-2005). The β-emitter's long path length (100 to 250 cell diameters; 5 to 10 mm) allow it to kill neighboring cancer cells without the antibody actually binding to those neighboring cells, an effect known as the crossfire effect (Johnston, Bondly, and

Micallef 2006). It is hypothesized that this may enable the therapeutic to have an increased effect, as compared to a naked antibody, on bulky or poorly vascularized tumors. An early clinical study assessed the pharmacokinetics and biodistribution of radiolabeled antibody (Knox et al. 1996). The need for preadministration of unlabeled antibody was assessed by administering ^{111}In-labeled ibritumomab tiuxetan, used for imaging, either before or after unlabeled ibritumomab. When ^{111}In-labeled antibody was given before unlabeled antibody, only 18 percent of known disease sites were imaged, whereas if given after unlabeled antibody, 56 to 92 percent of known disease sites were imaged. Thus, unlabeled antibody was required to clear peripheral B-cells and facilitate tumor targeting by the labeled antibody. This led to the Zevalin regimen. The regimen is administered in two steps: In step 1, rituximab is given (250 mg/m^2) to clear peripheral B-cells and maximize tumor targeting. Following this, an imaging dose (5.0 mCi/1.6 mg total antibody dose) of ^{111}In-ibritumomab tiuxetan is given. Biodistribution is assessed by performing whole body scans 48 to 72 hours later. If biodistribution is acceptable, the next step is undertaken seven to nine days after step 1. In step 2, a second infusion of rituximab is first given (250 mg/m^2) followed by the therapeutic dose of ^{90}Y-ibritumomab tiuxetan (0.4 mCi/kg) (Biogen Idec, Inc., 13-Sept-2005). In patients receiving the regimen, a mean effective half-life for ^{90}Y activity in blood was 30 hours and a median of 7.2 percent of the injected activity was excreted in the urine over seven days (Biogen Idec, Inc., 13-Sept-2005). Additionally, the regimen caused a sustained depletion of B-cells, with a median number of zero circulating B-cells at 4 weeks. B-cell recovery began at approximately 12 weeks after treatment and levels were normal by 9 months. Two pivotal clinical trials tested the safety and efficacy of ibritumomab tiuxetan in patients with either relapsed follicular lymphoma that are refractory to rituximab treatment (study 1) or with relapsed or refractory low-grade or follicular NHL or transformed B-cell NHL (study 2). In both studies patients received the Zevalin therapeutic regimen described above. In study 1, an overall response rate (ORR) of 74 percent was observed, with a median time to disease progression (TTP) of 6.8 months (Witzig et al. 2002a). A complete response rate (CR) of 15 percent was seen. In study 2, patients received either the Zevalin therapeutic regimen or rituximab alone (Witzig et al. 2002b). The ORR seen in patients receiving the therapeutic regimen was 80 percent, significantly higher than the 56 percent ORR seen in the rituximab-treated patients. The TTP in both groups was not significantly different, with a TTP of 10.6 months in the therapeutic regimen group versus 10.1 months in the rituximab group. The CR was higher in the therapeutic regimen group, with 34 percent of the patients having a CR while 20 percent of the patients treated with rituximab alone had a CR.

The encouraging clinical results led to the FDA approval of ibritumomab tiuxetan in 2002 for treatment of NHL. It was the first radioimmunoconjugate to be approved by the FDA for the treatment of cancer. Its current FDA approved indications include the treatment, as part of the Zevalin therapeutic regimen, of patients with relapsed or refractory low-grade, follicular, or transformed B-cell non-Hodgkin's lymphoma, including patients with Rituximab refractory follicular non-Hodgkin's lymphoma. An analysis of safety data available from five clinical studies in which 349 patients were treated with the Zevalin therapeutic regimen showed the most common adverse events were neutropenia, thrombocytopenia, anemia, gastrointestinal symptoms (nausea, vomiting, abdominal pain, and diarrhea), increased cough, dyspnea, dizziness, arthralgia, anorexia, anxiety, and ecchymosis (Biogen Idec, Inc., 13-Sept-2005; Witzig et al. 2003). Hematologic toxicity was the most common adverse event observed in clinical trials and was often severe and prolonged. The most serious events associated with the Zevalin therapeutic regimen were prolonged and severe cytopenias, infections, hemorrhage while thrombocytopenic, and allergic reactions. Boxed warnings are included for infusion reactions (reaction usually occurs following the first infusion of rituximab), prolonged and severe cytopenias, and severe cutaneous and mucocutaneous reactions.

31.1.6 Bexxar/Tositumomab/Iodine-131 Tositumomab

Tositumomab is an entirely murine mAb directed against the CD20 antigen found on normal and malignant human B-cells. Iodine-131 (^{131}I) tositumomab is the murine antibody covalently linked to ^{131}I. The antibody, originally called anti-B1, is a murine IgG2a lambda antibody that was discovered through immunization of mice with tumor cells obtained from a patient with Burkitt's lymphoma (Stashenko et al. 1980). The anti-B1 antibody was discovered through screening of antibodies from the hybridomas created from the immunized mice for reactivity to the immunizing tumor. The biology and rationale for targeting its antigen, CD20, are described in Section 31.1.1.

Tositumomab binds to an epitope on the extracellular loop of CD20 on normal and malignant B–cells, with an approximate affinity of 1.4 nM (Cardarelli et al. 2002). The tositumomab epitope was shown to overlap with the rituximab epitope on CD20 (Cardarelli et al. 2002). Tositumomab has been shown to exert antitumor effects both *in vitro* and in xenograft studies through a number of different mechanisms, including ADCC and CDC (Buchsbaum et al. 1992b; Cardarelli et al. 2002) and induction of apoptosis (Cardarelli et al. 2002; Shan, Ledbetter, and Press 1998). Unlabeled tositumomab has been shown to have antitumor activity *in vivo* (Buchsbaum et al. 1992b; Davis et al. 2004). However, radiolabeling the antibody with ^{131}I was shown to significantly enhance all therapeutic outcome measures (Davis et al. 2004). Radiolabeled tositumomab exerts its antitumor effects through delivery of radiation to the CD20+ cells inducing their death. The radiation may also affect neighboring malignant cells at a distance of 20 to 90 cell diameters, causing their death in an effect known as the crossfire effect (O'Donoghue, Bardies, and Wheldon 1995). Early preclinical xenograft studies with the ^{131}I-tositumomab showed that predosing with unlabeled tositumomab prior to administration of radiolabeled antibody improves the delivery of the radiolabeled antibody to the tumor (Buchsbaum et al. 1992a). The same was shown to be true in clinical studies in NHL patients (Kaminski et al. 1993; Wahl, Kroll, and Zasadny 1998; Wahl et al. 1998), leading to the development of the Bexxar therapeutic regimen. The regimen is administered in two steps consisting of a dosimetric and therapeutic step (Corixa Corp. and GlaxoSmithKline, 22-Dec-2004), similar to that done for ibritumomab tiuxetan. In the dosimetric step (day 0), 450 mg of unlabeled tositumomab is administered IV to saturate nonspecific binding sites and specific binding sites in nonmalignant B–cells, thus improving the labeled antibody delivery. This is followed by infusion of 5.0 mCi ^{131}I-tositumomab to perform dosimetry and study biodistribution. Whole body dosimetry (gamma counting) is performed on day 0, then days 2, 3, or 4, followed by days 6 or 7. If biodistribution is determined to be acceptable, on days 7 to 14, the therapeutic step is undertaken. In the therapeutic step 450 mg of unlabeled tositumomab is administered IV followed by a prescribed mCi dose of ^{131}I-tositumomab to deliver 75 cGy total body dose calculated from the dosimetry results. In NHL patients receiving the Bexxar regimen in clinical studies, a median effective half-life, measured by total body gamma camera counts, was 67 hours, and 98 percent of the whole body clearance was accounted for in the urine (Corixa Corp. and GlaxoSmithKline, 22-Dec-2004). Additionally, the regimen caused a sustained depletion of circulating CD20+ cells. After seven weeks, the median number of circulating CD20+ cells was zero. Lymphocyte recovery began approximately 12 weeks after treatment. Two pivotal clinical trials tested the efficacy of the Bexxar therapeutic regimen in patients with low-grade, transformed low-grade, or follicular large-cell lymphoma. Both trials were single arm studies where the patients received the Bexxar therapeutic regimen described above. Study 1 included 40 patients whose disease had failed to respond to or progressed after at least four doses of rituximab therapy (Corixa Corp. and GlaxoSmithKline, 22-Dec-2004). The overall response rate (ORR) was 68 percent, with a median duration of response of 16 months. A complete response rate (CRR) of 33 percent was achieved, with a median duration of CR longer than 26 months. Study 2 included 60 chemorefractory patients who had a median number of four prior chemotherapy regimens (Corixa Corp. and GlaxoSmithKline, 22-Dec-2004; Kaminski et al. 2001). The ORR was 47 percent, with a median duration of response of 12 months. A CRR of 20 percent was achieved, with a median duration of CR longer than 47 months.

The encouraging clinical results led to the FDA approval of the Bexxar therapeutic regimen in 2003 for the treatment of NHL. It is currently only approved for a single course of treatment. Its current FDA approved indications include the treatment of patients with CD20 antigen-expressing relapsed or refractory, low-grade, follicular, or transformed NHL, including patients with Rituximab-refractory NHL. It is not indicated for the initial treatment of patients with CD20 positive NHL. The safety and tolerability of the Bexxar therapeutic regimen has been determined from data reported from 995 patients, including 230 NHL patients from five clinical trials and an additional 765 patients from an expanded access program to further characterize the delayed adverse events (Corixa Corp. and GlaxoSmithKline, 22-Dec-2004). The most common adverse event observed in clinical trials was hematologic toxicity (thrombocytopenia 53 percent, neutropenia 63 percent, and anemia 29 percent). The most common nonhematologic adverse events included asthenia, fever, nausea, infection, increased cough, pain, chills, rash, headache, vomiting, and abdominal pain. The most serious adverse reactions seen in clinical trials of the Bexxar therapeutic regimen included severe and prolonged cytopenias (including infections and hemorrhage due to the cytopenias), allergic reactions (bronchospasm and angioedema), secondary leukemia, and myelodysplasia. Boxed warnings are included for hypersensitivity reactions and prolonged and severe cytopenias.

31.1.7 Avastin/Bevacizumab

Bevacizumab is a humanized mAb directed against human vascular endothelial growth factor (VEGF). It was originally a mouse mAb, A.4.6.1, discovered in the laboratories at Genentech through immunization of mice with human VEGF165 (K.J. Kim et al. 1992). It was later humanized by Genentech scientists to create an antibody with human IgG1 and kappa constant as well as human framework regions while retaining the mouse complementarity-determining regions (and several key framework residues) of A.4.6.1 (Presta et al. 1997). Its antigen, VEGF, is a potent proangiogenic growth factor whose activity is mediated primarily through binding to two tyrosine kinase receptors, VEGFR-1 (Flt-1) and VEGFR-2 (Flk-1/KDR). These receptors are located on the surface of endothelial cells of blood vessels and lymphatics. Stimulation of the cells with ligand leads to cellular proliferation, inhibition of apoptosis, and new blood vessel formation (see N. Ferrara, Gerber, and LeCouter 2003 for a review). VEGFR-2 is the main receptor through which the mitogenic and angiogenic effects of VEGF occur. The VEGF superfamily consists of several members, including VEGFA (referred to as simply VEGF here), the main molecule of interest. There are also several different VEGF isoforms, including 121, 165, 189, and 206 amino acid isoforms formed by alternate splicing (N. Ferrara et al. 1991; Houck et al. 1991; Tischer et al. 1991). VEGF165 is the dominant isoform, but bevacizumab effectively binds to and neutralizes all isoforms (K.J. Kim et al. 1992). VEGF has been shown to play a key role in the regulation of both normal and abnormal angiogenesis (N. Ferrara, Gerber, and LeCouter 2003). VEGF is involved in neovascularization in a number of pathologic conditions, including tumors and intraocular neovascularization. VEGF production is elevated in a number of cancers, including colorectal, breast, lung, and others (see N. Ferrara 2005; N. Ferrara, Gerber, and LeCouter 2003 for a review). It is secreted by the malignant cells as well as by tumor-associated stroma (Fukumura et al. 1998; Gerber et al. 2000). Inhibition of many different tumor cell lines has been demonstrated by anti-VEGF antibodies as well as small molecule and antisense oligonucleotides (Drevs et al. 2000; N. Ferrara and Davis-Smyth 1997; K.J. Kim et al. 1992, 1993; Smolich et al. 2001; Wood et al. 2000), making VEGF an attractive target to use in the treatment of cancer.

Bevacizumab binds to VEGF at an epitope that is distinct from the receptor-binding site with an approximate affinity of 2.2 nM (Liang et al. 2006; Muller et al. 1997). Bevacizumab blocks VEGF binding to receptor by steric hindrance, as the antibody epitope is in close proximity to the receptor-binding site (Fuh et al. 2006; Muller et al. 1997). Bevacizumab was first found to have tumor inhibitory capabilities, similar to its parent murine antibody, in a human rhabdomyosarcoma and breast carcinoma xenograft model (Presta et al. 1997). It caused a 90 to 95 percent decrease in tumor weight in treated animals as compared to a control group. In further animal xenograft models, bevacizumab was shown to have a marked effect on tumor vasculature, including decreased vascular density and permeability (Salnikov et al. 2006; Yuan et al. 1996). Finally, bevacizumab was shown to act synergistically when combined with chemotherapy (see Blagosklonny 2005 for a review). Clinical studies of bevacizumab in combination with chemotherapy led to the conclusion that bevacizumab did not exacerbate chemotherapy-associated toxicities (Margolin et al. 2001) and the combination had the most promising results in colorectal and non-small cell lung cancer (D.H. Johnson et al. 2004; Kabbinavar et al. 2003). A pivotal phase III trial was conducted to evaluate the efficacy of bevacizumab in combination with bolus-IFL (irinotecan, 5-fluorouracil, and leucovorin) or 5-FU (5-fluorocuracil)/LV (leucovorin) chemotherapy in first-line treatment of metastatic carcinoma of the colon or rectum (Genentech 11-Oct-2006; Hurwitz et al. 2004). In the group receiving bevacizumab plus IFL the median duration of survival was 20.3 months versus 15.6 in the group given IFL alone. The median duration of response was 10.4 months in the bevacizumab-IFL group compared to 7.1 months in the IFL group. Only 110 patients were enrolled in the bevacizumab-5FU/LV arm as the data monitoring committee determined that the safety profile of bevacizumab-IFL was acceptable. However, the median overall survival among these 110 patients was 18.3 months, with a median duration of response of 8.5 months. A phase I pharmacokinetic study of bevacizumab in which patients received IV dosages of 0.1 to 10 mg/kg on days 0, 28, 35, and 42 revealed a linear pharmacokinetic profile and a 21 day half-life (Gordon et al. 2001).

Bevacizumab was approved by the FDA in 2004 for the first-line treatment of metastatic colorectal cancer. Its current FDA-approved indications include use in combination with IV 5-fluorouracil-based chemotherapy for first- or second-line treatment of patients with metastatic carcinoma of the colon or rectum. It is also indicated for use in combination with carboplatin and paclitaxel for first-line treatment of locally advanced, recurrent or metastatic non-squamous, non-small cell lung cancer. In clinical studies of bevacizumab, infusions were well tolerated. Infusion reactions with the first dose of bevacizumab were uncommon (<3 percent) and severe reactions occurred in 0.2 percent of patients (Genentech 11-Oct-2006). The most common adverse events seen in people receiving bevacizumab were asthenia, pain, headache, hypertension, diarrhea, nausea, vomiting, anorexia, stomatitis, constipation, upper respiratory infection, epistaxis, dyspnea, exfoliative dermatitis, and proteinura (Genentech 11-Oct-2006; Gordon et al. 2001). The most serious adverse reactions seen in patients receiving bevacizumab were gastrointestinal perforations, wound healing complications, hemorrhage, arterial thromboembolic events, hypertensive crises, reversible posterior leukoencephalopathy syndrome, neutropenia and infection, nephritic syndrome, and congestive heart failure (Genentech 11-Oct-2006). Boxed warnings are included on the label for gastrointestinal perforations, wound healing complications, and hemorrhage.

31.1.8 Erbitux/Cetuximab

Cetuximab is a mouse-human chimeric mAb directed against epidermal growth factor receptor (EGFR). It was originally a mouse mAb, IMC-225, discovered through immunization of mice with partially purified EGFR from A431 human epidermoid carcinoma cells (Kawamoto et al. 1983; J.D. Sato et al. 1983). An early phase I study of the mouse antibody was conducted in patients with squamous cell lung cancer where, although no toxicity was observed, it was noted that all of the patients produced anti-mouse antibodies (Divgi et al. 1991). The investigators decided to try to minimize the anti-mouse antibody response and therefore created a chimeric version of IMC-225, containing human IgG1 and kappa constant regions and the original mouse variable regions (Goldstein et al. 1995; Naramura et al. 1993). Cetuximab's antigen, EGFR, is a member of the ErbB type I receptor tyrosine kinase family, which also includes HER2, HER3, and HER4. It is a 170 kDa transmembrane receptor consisting of an extracellular, ligand-binding region, a transmembrane region, and an intracellular region containing the kinase domain. Ligands, such as epidermal growth factor (EGF) and transforming growth factor-alpha (TGF-α), bind to the receptor inducing homo- and heterodimerization with other members of the EGFR family ultimately leading to phosphorylation and signaling that regulate cell growth, invasion, adhesion, differentiation, and survival. The EGFR is expressed in many human normal epithelial tissues and is overexpressed in many epithelial cancers, including tumors of the head and neck, colon, lung, breast, ovary, and kidney (see Mendelsohn and Baselga 2006 for a review). Its expression and abnormal activation in cancer has been correlated with a poor prognosis due to faster disease progression, reduced response to treatment and increased resistance to chemotherapy, and decreased survival (Brabender et al. 2001; Galizia et al. 2006; Meyers et al. 1988; Wosikowski et al. 1997). Early studies of anti-EGFR mAbs demonstrated the ability of such therapeutics in the inhibition of human cancer cells, making EGFR an attractive anticancer target (Gill et al. 1984; Kawamoto et al. 1983; Masui et al. 1984; J.D. Sato et al. 1983).

Cetuximab binds to EGFR with an approximate affinity of 0.2 nM (Goldstein et al. 1995) at an epitope located in domain III of the extracellular region of the receptor (S. Li et al. 2005). Antibody binding at this site partially blocks the ligand-binding region of the receptor and prevents the receptor from adopting a conformation required for dimerization. Cetuximab binding to EGFR competitively blocks ligand binding, including EGF and TGF-α, thereby inhibiting receptor activation and downstream signaling (Goldstein et al. 1995; Masui et al. 1984). Cetuximab has been shown to downregulate the receptor, inhibit cancer cell growth, induce apoptosis, and decrease the production of angiogenic factors, *in vitro* and *in vivo* (Fan et al. 1994; Goldstein et al. 1995; ImClone Systems, Inc. and Bristol-Myers Squibb Co., 01Mar2006; Peng et al. 1996; Petit et al. 1997; X. Wu et al. 1995), as well as being able to mediate ADCC (ImClone Systems, Inc. and Bristol-Myers Squibb Co., 01Mar2006; Kimura et al. 2007). The antibody has shown antitumor activity against a number of different cancer cell lines (Goldstein et al. 1995; Oliveira et al. 2006; Speake, Holloway, and Costello 2005). Adding the antibody to chemotherapy (irinotecan or irinotecan plus 5-fluorouracil) or radiation therapy results in an increase in antitumor efficacy compared to either therapy alone (ImClone Systems, Inc. and Bristol-Myers Squibb Co., 01Mar2006). A pivotal clinical trial was conducted in 329 patients with metastatic colorectal cancer whose disease had progressed after receiving an irinotecan-containing regimen. The patients received either cetuximab monotherapy (400 mg/m^2 initial dose followed by 250 mg/m^2 weekly) or cetuximab plus irinotecan (Cunningham et al. 2004). An objective response rate (ORR) of 10.8 percent was seen in patients receiving cetuximab alone and an ORR of 22.9 percent was seen in patients receiving the combination therapy. The median duration of response was 4.2 months in the monotherapy group and 5.7 months in the combination group. Finally, the median overall survival in the monotherapy group was 6.9 months compared to 8.6 months in the combination group. Clincial trials comparing the use of cetuximab in combination with radiation therapy versus radiation therapy alone were also conducted in patients with squamous cell carcinoma of the head and neck (SCCHN; Bonner et al. 2006; Robert et al. 2001). In clinical studies examining pharmacokinetics, cetuximab administration exhibited nonlinear pharmacokinetics and was similar in patients with SSCHN and those with colorectal cancer (ImClone Systems, Inc. and Bristol-Myers Squibb Co., 01Mar2006). The mean half-life of cetuximab was approximately 112 hours.

The encouraging clinical results led to the FDA approval of cetuximab in 2004 for the treatment of colorectal cancer. Its current FDA-approved indications include use in combination with irinotecan for the treatment of EGFR-expressing, metastatic colorectal cancer in patients who are refractory to irinotecan-based chemotherapy. It is also indicated for use as a single agent for the treatment of EGFR-expressing, metastatic colorectal cancer in patients who are intolerant to irinotecan-based chemotherapy. It was further approved for use in combination with radiation therapy in patients with locally or regionally advanced SSCHN. It was approved as a single agent for the treatment of patients with recurrent or metastatic SSCHN for whom prior platinum-based therapy has failed. Treatment with cetuximab, either alone or in combination with chemo- or radiation therapy, is generally well tolerated. The most common adverse events seen in colorectal cancer patients receiving cetuximab plus irinotecan or cetuximab alone were acneform rash, asthenia/malaise, nausea, abdominal pain, and diarrhea (ImClone Systems, Inc. and Bristol-Myers Squibb Co., 01Mar2006). The most serious adverse reactions associated with cetuximab in colorectal cancer patients were infusion reactions, dermatologic toxicity, interstitial lung disease, fever, sepsis, kidney failure, pulmonary embolus, dehydration, and diarrhea. The most common adverse events seen in head and neck cancer patients receiving either monotherapy or combination with radiation therapy were similar, including acneform rash and asthenia, while the most serious events were infusion reaction, cardiopulmonary arrest, dermatologic toxicity, mucositis, radiation dermatitis, confusion, and diarrhea. Boxed warnings are included on the label for infusion reactions and cardiopulmonary arrest. Of note, the acneform rash was one of the most common adverse events reported in both colorectal and head and neck cancer patients, occurring in 76 to 90 percent of the patients receiving combination or monotherapy. Interestingly, skin reactions have been correlated with efficacy (Cunningham et al. 2004; Saltz et al. 2003). For example, in a study of colorectal cancer patients receiving either cetuximab monotherapy or cetuximab combined with irinotecan, a higher response rate was seen in those patients who had a skin reaction versus those who did not (Cunningham et al. 2004). In the monotherapy group a response rate of 13.0 percent was seen in the patients with skin reactions versus a rate of 0.0 percent seen in patients without a skin reaction; in the combination group a response rate of 25.8 percent was seen in those patients with skin reactions versus a 6.3 percent response rate in those patients without a skin reaction.

31.1.9 Vectibix/Panitumumab

Panitumumab is an entirely human mAb directed against the epidermal growth factor receptor (EGFR). It was generated by Abgenix scientists through immunization of a strain of mouse called XenoMouse-G2 (Mendez et al. 1997), that produces fully human IgG2κ antibodies with EGFR-expressing A431 human epidermoid carcinoma cells (Yang et al. 1999). Panitumumab, initially called clone E7.6.3, was discovered after screening more than 70 anti-EGFR hybridomas produced from these mice for specificity, ability to block ligand binding and phosphorylation of EGFR, and high affinity. It is a human antibody containing human IgG2 and kappa constant regions. An introduction to the biology and rationale for targeting its antigen, EGFR, is provided in Section 31.1.8.

Panitumumab binds to EGFR with an approximate affinity of 0.05 nM (Yang et al. 1999) at an epitope located in domain III of the extracellular region of the receptor (Jakobovits et al. 2007). Antibody binding to the receptor blocks ligand binding, including EGF and TGF-α, prevents receptor activation, induces receptor internalization and degradation, causes cell cycle arrest, and suppresses production of some angiogenic factors from tumor cells (Jakobovits et al. 2007). It is not expected to mediate ADCC since the antibody is of IgG2 isotype. Panitumumab was able to prevent or eradicate tumor growth in xenograft studies using a number of different EGFR-expressing human cancer cell lines (Yang et al. 1999, 2001). A pivotal phase III trial was conducted in people with previously treated metastatic colorectal cancer who had progression after fluoropyrimidine-, irinotecan-, and oxaliplatin-containing chemotherapy regimens (Van Cutsem et al. 2007). A total of 463 patients were randomized to receive either panitumumab (6 mg/kg every two weeks) in combination with best supportive care (BSC) or BSC alone. The panitumumab/BSC combination significantly increased mean progression-free survival time to 13.8 weeks while BSC alone had a mean progression-free surviva time of 8.5 weeks. An objective response rate (ORR) of 10 percent was seen in patients receiving the combination therapy versus an ORR of 0 percent in the patients receiving BSC alone. No difference in overall survival (OS) was seen between the groups. However, interpretation of the OS is problematic as 76 percent of the patients receiving BSC alone switched over to receive panitumumab at a median time of 8.4 weeks, and similar activity of panitumumab was seen in the crossover study. Examination of the pharmacokinetics of panitumumab as single agent showed nonlinear pharmacokinetics (Amgen Inc., 27-Sept-2006). The mean half-life of antibody at the recommended dose (6 mg/kg every other week) was approximately 7.5 days (Amgen Inc., 27-Sept-2006).

The clinical studies of panitumumab led to its approval by the FDA in 2006 for the treatment of EGFR-expressing colorectal cancer. Panitumumab's current FDA-approved indications include the treatment of patients with EGFR-expressing, metastatic colorectal cancer with disease progression on or following fluoropyrimidine-, oxaliplatin-, and irinotecan-containing chemotherapy regimens. An analysis of safety data available from 15 clinical trials in which 1293 patients received panitumumab monotherapy and 174 received panitumumab in combination with chemotherapy showed the most common adverse events were skin rash, hypomagnesemia, paronychia, fatigue, abdominal pain, nausea, and diarrhea (Amgen Inc., 27-Sept-2006). The most serious adverse events were pulmonary fibrosis, severe dermatologic toxicity complicated by infection and septic death, infusion reactions, abdominal pain, hypomagnesemia, nausea, vomiting, and constipation. Boxed warnings are included on the label for the dermatologic toxicity and infusion reactions.

31.2 APPROVED THERAPEUTIC mAbs FOR THE TREATMENT OF AUTOIMMUNE AND INFLAMMATORY DISORDERS (AIID)

AIID is the second largest category for therapeutic use of mAbs. In this category, disease modifiying antirheumatic drugs (DMARDs) such as Remicade and Humira (both are anti-TNFα products) are the major market driver because of the prevalence of rheumatoid arthritis and other autoimmune conditions. Ten AIID mAb products currently in clinical use are reviewed in Section 31.2.

31.2.1 OrthoClone OKT3/Muromonab

Muromonab, developed by Ortho Biotech, was a mouse IgG2a that recognizes the human CD3ε chain for the prevention and treatment of transplantation rejection. It was the first mAb approved for clinical use in humans (Smith 1996; Thistlethwaite et al. 1987). CD3 in complex with T-cell receptor (TCR) plays a vital role in T-cell activation upon recognition of antigen. Muromonab modulates the immunological status by suppressing, enhancing, or redirecting T-cell responses to antigens. Despite its effectiveness, over the last decades, the use of muromonab has been limited by two major factors: (1) human anti-murine IgG antibody (HAMA) responses which reduce the efficacy and (2) the possibility of cytokine release due to Fc receptor (FcR)-mediated T-cell mitogenic activity.

The toxicity profile of murmonab renders it unsuitable for many patients; therefore, pretreatment with methylprednisolone, a synthetic corticosteroid, is recommended to minimize symptoms of cytokine release syndrome. However, as the complexities of CD3 and T-cell responses become better understood, the second generation of anti-CD3 mAbs has been created by humanization and by reducing its binding to FcR with glycosylation mutations in the FC region (Bolt et al. 1993; Carpenter et al. 2000). Teplizumab [also known as MGA031 and hOKT3γ1(Ala-Ala), by MarcoGenic and Lilly] is a humanized version of OKT3 where the complementarity determining regions (CDRs) were grafted onto human IgG1 and the CH2 region has been mutated to eliminate FcR-binding activity (Brown 2006; Xu et al. 2000). Teplizumab is in phase II/III clinical trail (in 2007) for the treatment of type I diabetes. With the better safety features of humanized, non-FcR-binding anti-CD3 mAbs, the therapeutic potential of anti-CD3 antibodies has been expanded from the treatment of kidney allograft rejection to wider autoimmune disease indications. Other anti-CD3 mAbs in the current pharmaceutical pipeline include visilizumab (also called Nuvion, by PDL) for inflammatory bowel disease (IBD), otelixizumab (developed by GSK and TolerRx) for type I diabetes, and CHAglyCD3, also for type I diabetes (Chatenoud and Bluestone 2007).

31.2.2 Zenapax/Daclizumab

Developed by PDL and marketed by Roche, Zenapax (daclizumab) is an anti-IL-2Rα antibody used as an immunosuppressant in organ transplantations. By combining the CDRs of a murine antibody with human framework and constant regions, daclizumab is composed of 90 percent human and 10 percent murine antibody sequences with a binding affinity of 3 nM to IL-2Rα, about one-third that of its murine parental antibody (Queen et al. 1989). Daclizumab was approved by the FDA in 1997 as the first humanized therapeutic mAb.

31.2.3 Simulect/Basiliximab/SDZ-CHI-621

Simulect (basiliximab, SDZ-CHI-621) is a mouse-human chimeric mAb (IgG1κ) that binds to the IL-2Rα-chain, also known as CD25 antigen, on the surface of activated T-lymphocytes. It was developed by Novartis and approved by the FDA in 1998 for the prevention of rejection in organ transplantation, especially in kidney transplants. It is used as part of a standard immunosuppressive regimen comprised of other medications. High affinity binding (Ka = 10 nM) of basiliximab to IL-2Rα competitively inhibits IL-2-mediated activation and proliferation of T-cells and also prevents T-cells from activating B-cells, thus blocking a critical pathway in the cellular immune response involved in acute transplant rejection (Kovarik et al. 1996).

Both basiliximab and daclizumab have demonstrated the efficacy as induction therapy to reduce the rate of acute rejection after renal and liver transplantations with no increase in opportunistic infections or adverse effects, proven to be a class of effective and specific immunosuppressive agents (Adu et al. 2003; Niemeyer et al. 2002; Vincenti, Nashan, and Light 1998). The two antibodies bind to different epitopes on IL-2Rα. Daclizumab has a half-life of 20 days and basiliximab has a half-life of 7 days; the longer half-life of daclizumab is probably a result of higher human content in the antibody sequence.

The approved dosing for basiliximab is a two-dose regimen, that is 20 mg (without adjustment for body weight) given on the day of transplantation and repeated on day 4. The approved daclizumab dosing calls for five doses of 1 mg/kg each, given initially on the day of transplantation and repeated at 14-day intervals (Van Gelder, Warle, and Ter Meulen 2004). The difference in dosing regimens arises from the uses of the two drugs in their respective phase III trials that led to approval by the FDA, not from the differences in biochemical and pharmacokinetic properties. No studies have been published in head-to-head comparison of basiliximab and dacilizumab in renal transplantation. Daclizumab has also been developed as a therapy for multiple sclerosis (codeveloped by PDL and Biogen Idec). Improvements in clinical scores have been observed (Rose et al. 2007).

31.2.4 Remicade/Infliximab/cA2

Remicade (inflixmab), developed by Centocor (a subsidiary of Johnson and Johnson) and Schering-Plough) is a murine/human chimeric IgG. Infliximab was the first of the three mAbs targeted tumor necrosis factor-α (TNFα) currently on the market (the other two mAbs being Humira and Cimzia). Infliximab is indicated for the treatment of autoimmune diseases such as plaque psoriasis, rheumatoid arthritis, psoriatic arthritis, Crohn's disease, ulcerative colitis, and ankylosing spondylitis (Knight et al. 1993; Siegel et al. 1995). TNFα is a proinflammatory cytokine known as a mediator of chronic inflammatory diseases. Infliximab binds to TNFα with a high affinity (Ka = 1.8 nM) and strong avidity, thus it effectively prevents TNFα from binding to its receptors (Scallon et al. 1995). Different from Enbrel (a TNFα antagonist in the format of recombinant FC fusion protein), anti-TNFα mAbs such as Remicade and Humira are able to neutralize to all forms of TNFα (soluble, transmembrane, and receptor-bound forms), which renders the ability of Remicade and Humira in lysing cells involved in the inflammatory processes; however, the clinical significance of this activity is not clear.

Infliximab is administered by intravenous infusion. It has a half-life of 8 to 10 days. The approved dosing is 3 to 10 mg/kg at 0, 2, and 4 weeks and then every 8 weeks; it is often adjusted upward depending on clinical response (Geletka and St. Clair 2005). In patients with RA, the anti-inflammatory effects of infliximab were shown by the rapid reduction of levels of proinflammatory cytokines (e.g. IL6) and other inflammatory markers. Infliximab has a reasonable safety profile and infusions are generally well tolerated. Serious adverse events on the postmarketing label are increased risk of tuberculosis, opportunistic infections, and worsening of preexisting heart failure. Some patients developed hypersensitivity to the drug and efficacy reduced over time due to anti-drug antibody production. This immune response is alleviated by coadministration of methotrexate (MTX) or other immunosuppressive drugs.

31.2.5 Raptiva/Efalizumab/hu1124

Raptiva (efalizumab) is a humanized mAb developed by Genentech and was approved by the FDA in 2003 for the treatment of moderate to severe plaque psoriasis (Lebwohl et al. 2003). Efalizumab binds to CD11a, an α-subunit of the lymphocyte-function-associated antigen (LFA)-1, expressed on the cell surface of T-lymphocytes. Interactions between LFA-1 and intercellular adhesion molecule 1 (ICAM-1) on antigen-presenting cells mediate the pathogenesis of psoriasis, including the migration of T-lymphocytes from the circulation into dermal and epidermal tissues, with subsequent reactivation. The blockage of the interaction between LFA-1 and ICAM-1 was demonstrated by a murine anti-CD11 antibody MHM24 (Hildreth and August 1985). Efalizumab is a humanized version of MHM24 made by CDR grafting onto human VH3 and Vκ1 framework. Variants of a humanized version of MHM24 were expressed and selected in the format of Fab fragments. The resulting humanized antibody was then made into full length IgG1. The binding affinity of efalizumab to LFA-1 was comparable to MHM24 with Kd of 0.1 to 0.2 nM, measured by binding to peripheral blood mononuclear cells from human donors (Werther et al. 1996).

Efalzumab is indicated for the treatment of adult patients with chronic moderate to severe plaque psoriasis. It is administered once a week subcutaneously, and noticeable results usually can be seen in 12 weeks. It has a generally good safety profile for long term use, and, as many immunosuppressive agents, the potential risks associated with decreased activity of the immune system, such as infection, are noted.

31.2.6 Humira/Adalimumab/D2E7

Humira (adalimumab) from Abbott is the first fully human antibody approved for clinical use. It was first approved by the FDA in 2002 for the treatment of RA, and later approved for the treatments of other TNFα-mediated chronic inflammatory diseases, including polyarticular juvenile idiopathic arthritis, psoriatic arthritis, ankylosing spondylitis, Crohn's disease, and chronic plaque psoriasis. Adalimumab was derived from a murine antibody MAK195 using guided selection phage display from CAT (Osbourn, Groves, and Vaughan 2005). This technology involved in a series of selections of human HC and LC variable regions by pairing them with the variable regions of MAK195. The resulting variable domains were then fused with human IgG1κ constant domain. The fully human, affinity matured clone D2E7, produced in CHO cells, has a binding affinity of 100 pM to TNFα (Santora et al. 2001).

Adalimumab is administered by subcutaneous injection at 40 mg every one or two weeks for rheumatic diseases. The maximum serum concentration (*Cmax*) and the time to reach the maximum concentration (*Tmax*) were 4.7 ± 1.6 μg/mL and 131 ± 56 hours, respectively, following a single 40 mg subcutaneous administration of adalimumab to healthy adult subjects. The average absolute bioavailability of adalimumab estimated from three studies following a single 40 mg subcutaneous dose was 64 percent. The mean serum half-life ranged from 10 to 20 days when given intravenously (Voulgari and Drosos 2006). In patients with RA, a rapid decrease in levels of markers associated with inflammation, such as C-reactive protein, fibrinogen, and cytokines was observed. Despite its human origin, adalimumab still elicits immune response in about 5 percent of the RA patients (Humira prescription label). It is marketed both as preloaded syringes and as preloaded pen devices, which can typically be self-injected.

31.2.7 Tysabri/Natalizumab/AN100226

Tysabri (natalizumab) is a humanized anti-α4 integrin mAb (IgG4κ) marketed by Biogen Idec and Elan and it was approved by the FDA in 2004 for the treatment of patients with relapsing forms of multiple sclerosis (MS). It binds to $\alpha_4\beta$ integrins, blocking the entry of inflammatory lymphocyte into specific tissue, such as central nervous system or gut, during disease pathogenesis. $\alpha_4\beta_1$ integrin [also called very late antigen 4 (VLA4) or CD49D/CD29] is a key cell surface receptor expressed on leukocytes that binds to vascular cell adhesion molecule-1 (VCAM-1) and fibronectin, regulating cellular adhesion and migration of inflammatory cells into tissues. Natalizumab and the murine parent antibody AN100226 were shown to have similar binding affinity for integrin on human Jurkat cells with the Kd of 0.3 nM (Leger et al. 1997).

Natalizumab has been shown to slow MS disease progression by reducing inflammatory cell infiltration into the brain; it has been shown to reduce new lesions by 93 percent and MS relapses by 50 percent. It is dosed as 300 mg every four weeks. The half-life is 11 to 15 days. It is administered intravenously at monthly intervals (Hutchinson 2007). In 2005, two patients given natalizumab developed a rare brain disease called progressive multifocal leukoencephalophy (PML). It is thought to be triggered by the JC virus infection under conditions of immunosuppression. In 2006 natalizumab was reintroduced for the treatment of MS with a warning label stating the risks of PML, and patients who receive any other immunosuppressive therapy or with underlying immunosuppressive disorders should be precluded from the use of natalizumab. In 2008 it was reapproved by the FDA for the treatment of Crohn's disease (Bjorn 2008).

31.2.8 Actemra/Tocilizumab

Under development by Roche, tocilizumab is an anti-human interleukin-6 (IL-6) receptor antibody genetically engineered by grafting the CDRs from a mouse antibody into human IgG1 (K. Sato et al. 1993). Different from currently marketed immune response modifiers, tocilizumab targets interleukin-6 (IL-6), a cytokine that is over-produced in the joints of RA patients. Tocilizumab blocks the actions of IL-6 by competitive binding to the IL-6 receptor. Humanization of tocilizumab has resulted in decrease in immunogenicity and extended half-life in humans. It was first launched in 2005 in Japan by Chugai as a therapy for Castleman's disease, a rare condition that causes severe lymph node enlargement. Tocilizumab now has been developed as a treatment for RA, juvenile idiopathic arthritis (JIA), and Crohn's disease (Ito et al. 2004; Yokota et al. 2005).

As of 2008, four out of five phase III clinical trials designed to evaluate clinical findings of tocilizumab have shown clinical benefit. In March 2008, two trials demonstrated that tolcilixumab could be an effective therapy in patients with moderate to severe active RA and systemic JIA, particularly with patients who had a poor response to conventional disease modifying antirheumatic drugs (DMARDs; e.g., methotrexate; Smolen et al. 2008; Yokota et al. 2008). In the OPTION trials, patients who received tocilizumab 4 mg/kg or 8 mg/kg intravenously every four weeks with weekly methotrexate, effective reduction in RA symptoms was achieved. Tocilisumab is generally well tolerated. The most common adverse events reported in clinical studies were upper respiratory tract infection, nasopharyngitis, headache, and hypertension. Although no direct comparison data with TNF inhibitors is available, tocilizumab shows promise as a monotherapy for RA.

31.2.9 Cimzia/Certolizumab Pegol/CDP-870

Cimzia is a Fab′ fragment of a humanized murine antibody of TNF*a*, developed by UCB. It is PEGylated using Nektar's technology platform and manufactured using XOMA's bacterial cell expression technology. It was approved and subsequently launched in 2008 in Switzerland for the treatment of Crohn's disease. Compared with infliximab and adalimumab, certolizumab pegol has a higher binding affinity to TNF*a* (K_D of 89.3 pM) and has shown a higher potency *in vitro* with an IC_{50} of 0.35 ng/ml versus that of 6 and 5 ng/ml for infliximab and adalimumab, respectively (Nesbitt et al. 2007). Due to the absence of Fc portion, certolizumab pegol did not induce apoptosis of peripheral lymophocytes or monocytes as other anti-TNF*a* agents, nor it induce ADCC or CDC (Kaushik and Moots 2005).

Certolizumab pegol can be administered subcutaneously or intravenously. Site-specific peglyation has increased the half-life of the antibody to 11 to 14 days. The antibody fragments can be synthesized by fed-batch fermentation in *Esherichia coli*, lowering the manufacturing costs by 10- to 20-fold compared with the conventional antibody production in mammalian cell culture (Rose-John and Schooltink 2003). In January of 2008, the FDA accepted for filing and reviewing a biologics license application (BLA) for Cimzia submitted by UCB for the treatment of adult patients with active rheumatoid arthritis and it was approved by FDA in May, 2009.

31.2.10 Soliris/Eculizumab

Soliris is an mAb developed by Alexion for the treatment of paroxysmal nocturnal hemoglobinuria (PNH), a rare form of hemolytic anemia. PNH is an acquired genetic disease caused by clonal expansion of hematopoietic stem cells due to somatic mutations of the X-linked gene, phosphatidylinositol glycan-complementation class A (PIGA), a molecule involved in synthesis of GPI-anchored proteins (Bessler et al. 1994; Takeda et al. 1993). Due to the lack of GPI-anchored proteins, for example CD59, the PNH red blood cells are vulnerable to hemolysis by terminal complement complex (TCC). The primary symptoms include intravascular hemolytic anemia, thromboses in vessels, and bone marrow failure (Johnson and Hillmen 2002). Eculizumab binds to complement component C5 blocking the formation of C5a and C5b, thus preventing the generation of proinflammatory responses and cell-lytic TCC (i.e., C5b-9 complex). It was approved in 2007 as a first-in-class inhibitor targeting complement system (Rother et al. 2007).

Eculizumab was derived from a murine antibody engineered to reduce immunogenicity and to eliminate effector functions. It was humanized by CDR grafting and the constant region of IgG was a combination of human IgG2 (CH1 and the hinge region) and human IgG4 (CH2 and CH3) plus human kappa light chain, therefore it is unable to bind FC receptors or to activate complement cascade. It retained the similar binding affinity to human C5 ($K_D = 120$ pM) of the parental murine antibody (Rother et al. 2007). The half-life of the molecule is 272 ± 82 hours when administered intravenously.

In a double-blind, randomized, placebo-controlled, multicenter, phase III trial, half of the PNH patients treated with eculizumab showed reduction in hemolysis, stabilization of hemoglobin concentrations, and less dependency on blood transfusions (Hillmen et al. 2004). Eculizumab is generally safe and well tolerated, demonstrated in the three main clinical studies. Because it is a complement C5 inhibitor, patients treated with eculizumab are at a higher risk of meningococcal infection; therefore, patients must receive meningococcal vaccination prior to treatment with eculizumab (Zareba 2007).

31.2.11 Simponi/Golimumab/CNTO-148

Golimumab is the next generation of anti-TNFα mAb that was developed by Centocor. It was approved by the FDA in April, 2009, alone or in combination with methotrexate, for the treatment of moderate to severe rheumatoid arthritis, active psoriatic arthritis and active ankylosing spondylitis. Golimumab is a fully human IgG1 (VH3 and Vk3) generated using the Medarex's UltiMAb technology. Its variable domains are nearly 100% identical to the human germline sequences and its constant domains are identical to the corresponding regions of infliximab in amino acid sequences (Shealy et al. 2007; Zhou et al. 2007). It has been shown that golimumab bound to both membrane-bound and soluble TNFα with subpicomolar binding affinity and it effectively neutralized TNFα *in vitro*. The protein structure is stable and well-behaved when formulated at 100 mg/ml (Shealy et al. 2007). Golimumab has a serum half-life of 7 to 20 days with a C_{max} of 70.8 ± 18.9 mg/mL when administered at 3 mg/kg I.V. (Zhou et al. 2007; see review of Tracey et al. 2008).

Different from infliximab which is administered through an I.V., golimumab was approved for subcutaneous injection once monthly, which is a less frequent dose than two other leading biologic agents, etanercept and adalimumab, in the class of anti-TNFα therapy.

31.3 APPROVED THERAPEUTIC mAbs FOR INFECTIOUS DISEASES, RESPIRATORY AND CARDIOVASCULAR DISORDERS

In additional to oncology and AIID, other areas of clinical applications for mAbs include anti-infective, respiratory, ophthalmology, and hematology. Section 31.3 reviews two therapeutic mAbs for infectious disease, one mAb for the treatment of asthma, and one mAb for the prevention of cardiac ischemic complications.

31.3.1 ReoPro/Abciximab

ReoPro (abciximab) is an antibody against glycoprotein (GP) IIb/IIIa receptor on the platelet, first approved as an adjunct for prevention of cardiac ischemic complications in percutaneous coronary intervention (PCI). It blocks interaction of fibrinogen with the GP IIb/IIIa receptor and prevents platelet aggregates, which is the first stage of formation of an arterial thrombus (Gabriel and Oliveira 2006). Abciximab has been shown to be effective both in the prevention of thrombus formation in PCI and in the reduction of existing thrombus.

Abciximab is a 47 kDa Fab fragment of the human-murine chimeric mAb with the variable regions from the murine parental antibody and the constant region from IgG1 (Coller 1995). It is produced by continuous perfusion in mammalian cell culture and purified by a series of steps involving viral removal procedures, papain digestion, and column chromatography (Coller 1995). In an unbound state, abciximab has a very short plasma half-life. While a very small amount of abciximab is eliminated by the kidneys, the majority of the Fab fragment is bound to the receptor and is removed from the body when the platelets are cleared from the blood by the spleen. The longevity of platelets accounts for abciximab's longer half-life (de Belder and Sutton 1998; Faulds and Sorkin 1994). Blocking of GP IIb/IIIa receptor at a low level can last for more than 10 days following the abciximab infusion, and it falls to 10 percent at 15 days and the residual binding was observed as late as 21 days (Mascelli et al. 1998).

Abciximab was launched in 1995 by Centocor (a subsidiary of Johnson & Johnson) and Lilly and it is only marketed drug for hemostasis. In addition to PCI patients, ReoPro was approved in 1997 for use as adjunctive therapy to prevent cardiac ischemic complications in patients undergoing a broad range of angioplansty.

31.3.2 Synagis/Palivizumab/MEDI-493

Synagis is a humanized IgG1 mAb directed against the F protein of respiratory syncytial virus (RSV) and it is produced in mouse myeloma cells (NSO). The discovery and development of palivizumab is as follows. A total of 18 different murine mAbs directed to the RSV F protein were analyzed for their biological and immunological properties (Beeler and van Wyke Coelingh 1989). One of the mAbs (mAb 1129) directed to the antigenic site A on the RSV F protein, was selected for CDR-grafting based humanization, that is, the antigen-binding motifs (CDRs) of the murine antibody was grafted onto the human variable framework segments (S. Johnson et al. 1997). Kinetic analysis of binding by Biacore evaluation revealed that palivizumab had binding affinity of 3.46 nM to the RSV F protein. In a single dose PK study in cynomolgus monkeys, palivizumab had a $T_{1/2}\alpha$ that ranged from 8.8 to 12.6 hours and the $T_{1/2}\beta$ was 8.6 days; and palivizumab was shown to be safe and well tolerated by the animals at 10 mg/kg and 30 mg/kg. In humans, PK of palivizumab had a $T_{1/2}\beta$ value of 17 days and the drug was safe and well tolerated in single and repeated doses ranging from 1 to 15 mg/kg in adults (Groothuis et al. 1993; Pollack and Groothuis 2002).

In September 1998, MedImmune (now a subsidiary of AstraZeneca) launched Synagis in the United States for immunoprophylaxis of respiratory syncytial virus (RSV) infection. RSV is classified in the genus Pneumovirus, in the family Paramyxoviridae and it is the leading cause of serious respiratory tract disease in infants and children. Up to 70 percent of infants are infected in the first year of life, and virtually every baby has experienced RSV infection by his or her second birthday (Krilov 2002). Of the infected babies, about 4 million babies born in the United States are at risk of serious consequences. RSV infection causes about 4500 deaths and hospitalization of about 90,000 babies every year in the United States. Synagis has been shown to cut the chance of RSV high-risk infant hospitalization by 50 percent. The drug's label recommends monthly 15 mpk doses throughout the RSV season (November to April in the northern hemisphere). Synagis is the first mAb to treat an infectious disease. Synagis had $1.1 billion in 2001 sales and costs about $5000 for a typical full course of treatment. Due to the high cost associated with Synagis, physicians are taking the lead to reduce the costs by narrowing the population of premature infants who receive the drug. Synagis is distributed by Abbott Laboratories in Europe.

31.3.3 Xolair/Omalizumab

Xolair (omalizumab) is a recombinant humanized mAb that selectively binds to human immunoglobulin E (IgE). It was approved by the FDA in 2003 for the indication of severe allergy-related asthma that is inadequately controlled with inhaled corticosteroids alone. The antibody is an IgG1κ that contains 95 percent human and 5 percent murine sequences (Presta et al. 1993). IgE naturally occurs in the body in small amounts. It binds to specific receptors (FcεRI and FcεRII) on basophils and mast cells. In an allergic reaction, allergens bind to IgE and cross-link IgE-Fc receptors on the cell surface, thus triggering cells to release mediators such as histamine, prostaglandins, and leukotrienes that cause the swelling and inflammation of the airways. Omalizumab blocks the circulating IgE binding to the cells and decreases the release of inflammatory mediators (Chang 2000).

In clinical studies, the levels of unbound IgE in the serum were reduced in a dose-dependent manner within one hour following the first dose of omalizumab. Using recommended doses, the decrease of serum IgE can be greater than 96 percent (Milgrom et al. 1999; Strunk and Bloomberg 2006). Total IgE levels returned to pretreatment levels within one year after discontinuation of omalizumab. Following a single SC dose in adult and adolescent patients with asthma, omalizumab was absorbed slowly, reaching peak serum concentrations after an average of seven to eight days. In asthma patients, omalizumab serum elimination half-life averaged 26 days. Omalizumab is administered subcutaneously in one to three injections every two to four weeks. Xolair was developed jointly by Novartis, Genentech, and Tanox (acquired by Genentech) and co-marketed by Genentech and Novartis.

31.3.4 Lucentis/Ranibizumab/rhuFabV2

Ranibizumab binds to and inhibits the biological activity of human vascular endothelial growth factor A (VEGF-A). It is a recombinant Fab fragment designed for intraocular use. Developed by Genentech, it was approved by the FDA for the treatment of neovascular age-related macular degeneration (AMD) in 2006. It was a Fab fragment generated from the same murine parental antibody A4.6.1 that gave rise to Avastin (bevacizumab) via different engineering processes. The humanized and affinity matured lead antibody, rhuFabV2, was the result of a series of engineering and selection processes (Baca et al. 1997; Y. Chen et al. 1999; Muller et al. 1998). Ranibizumab is produced by an *E. coli* expression system.

Ranibizumab binds to active forms of VEGF-A and $VEGF_{110}$ (a cleaved form of VEGF-A) to an epitope that partially overlaps with the receptor-binding site, thus blocking receptor binding. VEGF-A promotes neovascularization and leakage in ocular angiogenesis and is known to contribute to the progression of the neovascular AMD. Blocking of VEGF-A binding to its receptors (VEGFR1 and VEGFR2) on endothelial cells can reduce cell proliferation, vascular leakage, and new blood vessel formation (N. Ferrara and Alitalo 1999). Ranibizumab is delivered intravitreally. The small size of ranibizumab allows penetration into the rentina and rapid clearance from the circulation (Gaudreault et al. 2005). Pharmacokinetic studies in cynomolgus monkeys or rabbits showed that the serum level of ranibizumab was <0.1 percent of the ocular level after one intravitreal injection (Gaudreault et al. 2005). In clinical studies, no patients developed anti-drug antibodies. In phase III trials of ranibizumab treatment for neovascular AMD, patients showed improved visual acuity compared to untreated patients at one and two years (Rosenfeld, Rich, and Lalwani 2006). The most common adverse event is intraocular inflammation, and no significant toxicity was observed.

31.4 THERAPEUTIC ANTIBODIES IN CLINICAL PHASE III TRIALS FOR ONCOLOGY

With the success of mAb drugs on the market, investment in research and development of antibody therapeutics has grown significantly in the past 10 years. Many antibodies have entered into clinical trials for various disease types. Antibodies for treatment of cancers account for almost half of the current antibody drugs in late clinical development. To date, there are at least 10 new mAbs in clinical phase III trials for treatment of various types of cancers and this number does not include those trials for new indications with existing mAb drugs, such as Rituxan, Herceptin, and others. Table 31.2 shows a partial list of new antibodies in clinical phase III trials for oncology indications. Proxinium, which targets EpCAM, is being developed as a conjugated antibody and catumaxomab as a trifunctional (EpCAM, Her2/neu, and macrophage) antibody for treatment of cancer. These two antibody candidates are not reviewed in this chapter and the following section will focus on nonconjugated (naked) antibodies in late stage clinical trials (phase III) for treatment of different types of cancers.

TABLE 31.2 Cancer Monoclonal Antibodies in Late Stage Clinical Programs

mAb Names	Target Name	IgG Type	Clinical Stage	Cancer Type	Sponsor Company
Pertuzumab	HER2	Humanized	Phase III	Her2+ breast cancer, ovarian cancer	Genentech
Ipilimumab	CTLA-4	Human mAb	Phase III	Metastatic melanoma	Medarex/ BMS
Ticilimumab	CTLA4, cytotoxic T-lymphocyte protein	Human mAb	Phase III	melanoma	Pfizer
Ofatumumab	CD20	Human mAb	Phase III	Chronic lymphocytic leukemia, NHL	Genmab/ GSK
Lumiliximab	CD-23	Chimeric, IgG1-κ	Phase III	CLL	Biogen IDEC
Rencarex/ Wx-G250	Carbonic anhydrase regulator	Chimeric, IgG1-κ	Phase III	Renal cell carcinoma	Centocor
Zalutumumab	EGFR	Human mAb	Phase III	Head/neck, NSCLC	Genmab
Zanolimumab	CD4	Human mAb, IgG1/k	Phase III	Cutaneous T-cell lymphoma	Genmab/ Merck-Serono
Galiximab	CD80			Follicular NHL	Biogen IDEC
Epratuzumab	CD22	Humanized	PhaseII/ III	NHL, CLL	UCB Schwarz
Proxinium	EpCAM	Conjugated mAb	Phase III	Head/neck, bladder cancer	Viventia Biotech
Catumaxomab	CD3/EpCAM	Trifunctional-Mab	Phase III	Ovarian, gastric	Trion Pharm

31.4.1 Pertuzumab

Her-2 (neu, ErbB-2) is a member of the receptor tyrosine kinase family. Overexpression of Her-2 was found in a high percentage of breast cancers (25 to 30 percent overexpression) and other cancer types, including ovarian cancer and lung cancer. Upon overexpression of Her-2, it forms homodimers, oligomers, and heterodimers with other members of the tyrosine kinase family, which activates the kinase signaling pathway and leads to abnormal regulation of cell growth/survival and has significant implication in cancer development. Pertuzumab (Omnitarg^{TM}, 2C4) is the second antibody therapy (Herceptin being the first anti-Her-2 mab) directed against Her-2-dependent tumors. This mAb is a humanized IgG1 that blocks HER-2-mediated signal transduction by sterically blocking dimerization between HER-2 and other members of the ErbB family (Adams et al. 2006; Franklin et al. 2004; Rabindran 2005). In a preclinical study by Agus et al. (2002), they found that pertuzumab inhibited the heregulin-dependent proliferation of breast cancer cells expressing either low or high levels of HER-2 *in vitro* and pertuzumab was as active as trastuzumab in tumor xenografts derived from breast cancer cells expressing high levels of HER-2 (Agus et al. 2002). Responses in breast and non-small cell lung cancer (NSCLC) xenografts are correlated with the formation of HER-2/HER-3 dimers, but not with expression levels of EGFR, HER-2, or HER-3 alone (Bossenmaier et al. 2004). In phase I clinical trials, pertuzumab was well tolerated at doses up to 15 mg/kg, administered once every three weeks. The pharmacokinetic profile was comparable to other humanized IgG antibodies, with a half-life of about 21 days. Steady-state trough levels exceed the concentration estimated for antitumor activity *in vitro*. The most frequently reported adverse effects (grade 1 or 2) were nausea, vomiting, fatigue, diarrhea, rash, abdominal pains, and anemia (Allison et al. 2004). Pertuzumab is currently in phase II trials for ovarian, lung, and prostate cancers. Phase III trials for treatment of Her2+ first-line metastatic breast cancer in combination with Herceptin were also initiated in 2007.

31.4.2 Ipilimumab/MDX-010

Cytotoxic T-lymphocyte antigen (CTLA-4/CD152) is expressed on activated T-cells and acts as a negative regulator of CD28/CD80 stimulatory signal. CD28 is a receptor constitutively expressed on T-cells. Upon initial interaction at the immune synapse between MHC-presented antigens (CD80, CD86) and CD28 on T-cells, CTLA-4 is mobilized from intracellular stores and relocalized to the cell surface within the immune synapse. Preferential bindings of CD80 and CD86 to CTLA-4 result in a reduction of IL-2 production and T-cell proliferation.

MDX-010 (Ipilimumab) is a fully human mAb against CTLA-4. The antibody was produced in Medarex's transgenic mice with human antibody genes (Langer, Clay, and Morse 2007). Currently MDX-010 is in phase III trial by Medarex and Bristol-Myers jointly for treatment of metastatic melanoma. During the preclinical development, a surrogate antibody raised in hamster against mouse CTLA-4 was used in a mouse model due to lacking cross-reactivity of MDX-010 to the mouse antigen (Leach, Krummel, and Allison 1996). Most preclinical studies were done in cynomolgus monkeys and *in vitro* human T-cells. Interaction of CTLA-4 and CD80 can be blocked efficiently at concentrations of 1 to 10 μg/mL *in vitro* even when the CTLA-4 is expressed at high levels. It was predicted that *in vivo* saturation of the target CTLA-4 would need lower concentrations (Phan et al. 2003). Phase II dose trial with 1 mg/kg intravenous infusion (i.v.) every three weeks found that a mean peak concentration (C_{max}) of the antibody reached 72 + 33 μg/mL, with a trough level of 12 + 7 μg/mL prior to the second dose. Modest accumulation was reported after repeated dosing. No correlation was found between antibody clearance, plasma concentration, and toxicity (Small et al. 2007). Half-life was >12.5 + 5.3 days in phase I studies in prostate cancer and melanoma. Considerable variability in pharmacokinetic parameters was observed among cohorts. There are several ongoing phase II trials with mono- and combined therapy protocols in many malignancies, including prostate cancer, urothelial carcinoma, non-small cell lung cancer, melanomas, and other cancer types. Common adverse effects were erythematous skin rashes and colitis/diarrhea. These toxicities are usually related to the immune mechanism of blocking CTLA-4 signaling. There was a strong correlation between toxicity and tumor response. One of the phase II studies showed that 5 out 14 patients who experienced grade III/IV toxicity showed tumor regression, whereas only 2 out of 42 patients experienced a response in the absence of toxicity (Attia et al. 2005). There are three registrational phase III trials in metastatic melanoma, one is the second line, single arm monotherapy and the other two are combination therapies with either dacarbazine in the first line setting or MDX-1379 (peptide vaccine) at the second line treatment. The results from phase III trials will further shed light on the importance of this class of antibodies in immune modulation in general and tumor therapy.

31.4.3 Ticilimumab/CP-675,206

Ticilimumab is also a fully human mAb raised against CTLA-4 and currently in phase III trial by Pfizer. The half-life of Ticilimumab was reported as 25.6 days, with clearance at 0.132 mL/h/kg and distribution at 81.2 mL/kg (Reuben et al. 2006). One of the phase I studies reported that there was little accumulation of ticilimumab over multiple infusions and no evidence of immunogenicity (Ribas et al. 2005). Multiple phase I and phase II studies are ongoing for patients with advanced melanoma (stage IIIc or IV) and other solid malignancies in monotherapy and combination. In a phase I trial in 39 patients with solid malignancies with a single i.v. dose of 0.01 to 15 mg/kg, four patients experienced grade 3 dermatitis or diarrhea. Other adverse effects observed include vitiligo, pruritis, and liver enzyme elevations such as lipase and amylase (Ribas et al. 2005). Currently, there is one phase III study reported ongoing for CP-675,206, with the objective to compare it with temozolomide and dacarbazine in patients naïve to therapy. The three arm study was designed to run with up to 630 melanoma patients randomized to receive either infusion of CP-675,206 or two other therapies; CP-675,206 is infused once every 90 days and up to a year of treatment, decarbazine infusion was conducted every three weeks for 36 weeks, and temozolomide once weekly for one month. No reports are yet available for this study.

31.4.4 HuMax-CD20/Ofatumumab

CD20 acts as a calcium ion channel and regulates early steps in B-lymphocyte activation. The molecule is not shed from the cell surface and is not internalized on antibody binding. CD20 is expressed on >90 percent of B-cell non-Hodgkin's lymphomas (NHL) and to a lesser degree on B-cell chronic lymphocytic leukemia (CLL) cells, but not expressed on hematopoietic stem cells. Differing from rituximab (anti-CD20 commercial mAb), ofatumumab targets a different epitope of CD20 and is a novel fully human antibody (Hagenbeek et al. 2008). It is in clinical development for the treatment of NHL and CLL in addition to rheumatoid arthritis (RA). Genmab and GSK have announced plans to expand the HuMax-CD20 program into diffuse large B-cell lymphoma (DLBCL) and remitting relapsing multiple sclerosis (RRMS).

Data from preclinical laboratory tests showed that HuMax-CD20 appeared to kill tumor cells that were resistant to rituximab. The antibody was highly effective in inducing CDC of B-cell tumors. Additional data showed that HuMax-CD20 was also effective in inducing natural killer cell-mediated cytotoxicity of B-cell tumors. HuMax-CD20 appears to induce more effective killing of cells expressing low levels of CD20 than rituximab in laboratory experiments where a panel of cell lines expressing varying amounts of CD20 molecules per cell (4,500 to 135,000 molecules) were tested with Humax-CD20 as compared to rituximab. HuMax-CD20 appeared to induce significant lysis of cells at the lowest CD20 expression level tested, whereas such cells seemed resistant to rituximab. In animal studies, HuMax-CD20 can deplete B-cells effectively. HuMax-CD20 is also effective at binding to the disease target and releases very slowly from the target over time.

A phase I/II trial by Coiffier et al. (2008) studied 33 patients with relapsed/refractory CLL treated with weekly i.v. infusions of ofatumumab for 4 weeks and followed for 12 months. About 46 percent of overall response rate (nPR + PR) at the highest doses was shown with a favorable toxicity profile. Another phase II study of HuMax-CD20 in combination with cyclophosphamide, doxorubicin, vincristine, and prednisone (CHOP) in patients with previously untreated follicular NHL was initiated in June 2007. The primary endpoint in the study is objective response from start of treatment until three months after last treatment assessed according to the standardized response criteria for NHL at 30 weeks (Cragg and Glennie 2004; Cragg et al. 2003; Teeling et al. 2004). A pivotal phase III study with HuMax-CD20 is currently in progress to treat follicular NHL patients who are refractory to rituxan in combination with chemotherapy or to rituxan mAb given as maintenance treatment. Disease status will be assessed every three months for two years. The objective of the pivotal study is to determine the efficacy and safety of HuMax-CD20. A phase III pivotal study with HuMax-CD20 to treat patients with refractory B-cell CLL is also ongoing. The study has been amended to include approximately 150 CLL patients and two different patient populations. The main patient populations to be examined in the study are those patients who are refractory to both fludarabine and alemtuzumab. Patients who are fludarabine refractory are considered inappropriate candidates for alemtuzumab due to bulky tumor in their lymph nodes. Each group will consist of approximately 66 patients and will be analyzed separately in this trial. Due to the highly unmet medical need among these patients, registration of HuMax-CD20 could be possible in each indication, depending on the data generated from this study.

HuMax-CD20 was well tolerated by patients in the study. No dose limiting toxicities were reported during the study and the maximum tolerated dose was not reached. The most frequently reported adverse events were rigors, fatigue, headache, and rash. Hematological toxicity was reported in 6 of 40 patients with grade 1 neutropenia and no patients reporting thrombocytopenia.

31.4.5 Lumiliximab

CD23 is a 43 kDa transmembrane glycoprotein and acts as a low-affinity cell surface receptor for IgE (Delespesse et al. 1991). CD23 exists as both membrane-bound (mCD23) and soluble form (sCD23). The mCD23 is continuously cleaved into soluble CD23 fragments, which binds and regulates IgE function. The sCD23 is believed to have important physiological functions in promoting B-cell and T-cell growth, proliferation and rescuing germinal center B-cells from apoptosis (Armant et al. 1994; Delespesse et al. 1991). Patients with CLL have increased levels of sCD23 compared with normal individuals (Sarfati and Delespesse 1988). Primary B-CLL cells express more mCD23 than normal B-cells (Sarfati et al. 1990). It is believed that CD23 plays an important role in the regulation of malignant B-cell viability and proliferation (Fournier et al. 1994; Schwarzmeier et al. 2002).

Lumiliximab is a chimeric macaque/human anti-CD23 IgG1-κ mAb currently in phase III trial by Biogen IDEC for treatment of CLL. Possible mechanism of action involves ADCC, CDC, and induction of apoptosis. The lumiliximab binds mCD23 ($K_D = 0.12$ nM) slightly better than sCD23 ($K_D = 0.3$ nM) (Reichert 2004) and epitope for the antibody was not reported.

A phase I study for relapsed and refractory CLL in 25 patients with lumiliximab using i.v. was reported. Multiple doses (125, 250, 375, and 500 mg/m^2) were administered in six different schedules and objective of the study was to assess safety, pharmacokinetics, and efficacy of lumiliximab (Byrd et al. 2007; Reichert 2004). It has been reported that lumiliximab treatment is generally safe and some indication of clinical response was observed in patients with heavily pretreated CLL. Lumiliximab exhibited low saturation pharmacokinetics, low clearance, and a low volume of distribution without an immunogenic response in the multicentered phase I study. The phase II study with combination of FCR (fludaribine, cyclophosphamide, and rituximab) showed response in 52 percent of patients when compared with FCR alone at 25 percent of response rate, but the data was not generated as a two arm study and the reported response rates came from two separate studies. In 2007, Biogen IDEC initiated a randomized, controlled, and registrational phase III trial to evaluate efficacy in combination with FCR or FCR alone for treatment of CLL that has relapsed or failed in initial therapy; no results are reported yet.

In phase I/II trials, the most common adverse events in lumiliximab treatment of CLL were fatigue, nausea, headache, cough, and increased sweating. Grade 4 neutropenia and headache were reported by two patients at the highest dose of 500 mg/m^2 among the 25 patients participating in the trial.

31.4.6 Rencarex/Wx-G250

The antigen for Rencarex/Wx-G250 is carbonic anhydrase IX (G250), which was discovered through study of the mouse antibody G250 (Kennett 2003; Sonpavde and Hutson 2007). The G250 antigen is expressed on greater than 90 percent of renal carcinoma cells (RCC), but it was not found in normal tissues and cells, except detection on some gastric mucosa and large bile duct epithelial cells (G. Li et al. 2001). Rencarex/Wx-G250 is a chimeric version of G250 mouse antibody with IgG1 human constant region and mouse variable region from G250. The possible mechanism of action of Wx-G250 in cancer treatment is by internalization and ADCC.

A multicenter phase II study of 36 patients was performed to evaluate Wx-G250 in metastatic RCC (Bleumer et al. 2004, 2006). Wx-G250 was given weekly by intravenous infusion for 12 weeks. Patients with stable disease or with response after the 12 weeks were eligible to receive additional treatment for 8 weeks. A total of 10 patients (~30 percent) had stable disease (SD) and received extended treatment. One patient showed complete response (CR) and a significant regression of tumor. There were five patients with progressive disease at study entry who were stable for more than 6 months after study entry and the median survival was 15 months. A combination of Wx-G250 with low-dose IL-2 was evaluated in a 35-patient phase II trial to improve ADCC activity of Wx-G250 (Bleumer et al. 2006). Patients received weekly infusions of Wx-G250 for 11 weeks combined with a daily IL-2 regimen. A durable clinical benefit was achieved in 23 percent of patients, including three patients with a partial response and five with stabilization at $\geq$24 weeks. Median survival was

22 months. A randomized and placebo-controlled phase III trial to study efficacy in nonmetastatic RCC is ongoing with Wx-G250.

PhaseI/II study of radioconjugate with G250 and Wx-G250 reached maximum tolerated dose at 90 mci/m^2. Immunogenicity was detected for G250 mouse antibody but not the humanized Wx-G250. Phase II study with Wx-G250 at 50 mg/week for 12 weeks in 36 patients did not show evidence of toxicity or allergy at or greater than grade 3. Among 36 patients in the trial, 2 exhibited grade 2 adverse effect, one event of gastritis and two events of vomiting. There were no serious toxicities reported (Kennett 2003).

31.4.7 Zalutumumab

Zalutumumab is a fully human, high-affinity antibody targeting epidermal growth factor receptor (EGFR). EGFR is a type I transmembrane tyrosine kinase receptor. Aberrant activation in mutated EGFR has been implicated in many cancer types including, non-small cell lung cancer (NSCLC). EGFR is one of the most targeted molecules for cancer therapeutics. Zalutumumab is currently in clinical development to treat non-small cell lung cancer and head and neck cancer patients in whom standard therapies have previously failed. In preclinical study, zalutumumab was shown to lock the EGF receptor in an inactive conformation that prevents receptor activation and the binding of growth factors. Zalutumumab was shown to inhibit EGF receptor signaling by preventing receptor dimerization, the pairing of two receptor molecules, which starts the signaling cascade. The mAb also showed ADCC activity in preclinical study (Bleeker et al. 2004).

A comprehensive review on progress in clinical trials of zalutumumab was published by Rivera et al. (2009). Open label phase II study of zalutumumab in combination with chemoradiation to treat NSCLC will include a maximum of 270 previously untreated patients with locally advanced NSCLC. Another phase I/II study in combination with chemoradiation as first-line treatment of head and neck cancer by Genmab included a total of 36 patients with advanced squamous cell carcinoma of the head and neck (SCCHN). The open label study consists of an initial dose-escalation and a subsequent parallel group design. Zalutumumab was also tested in combination with cisplatin and three different regimens of accelerated radiotherapy. Genmab is also conducting a phase III pivotal study to treat patients with head and neck cancer refractory in standard treatment. Design of the pivotal study includes a maximum of 273 patients with SCCHN who are refractory to or intolerant of standard plantinum-based chemotherapy (Rivera et al. 2009).

The objective of the phase III study is to evaluate the efficacy of zalutumumab in combination with best supportive care as compared to best supportive care alone in terms of overall survival. The second phase III study to treat approximately 600 previously untreated head and neck cancer patients was also initiated. Patients in the study are randomized to treatment with radiotherapy or antibody combination with radiotherapy. Clinical response evaluated by computerized tomography (CT scan) and FDG-PET indicated 2 of 19 patients achieved partial response (PR) and 9 patients had stable disease (SD) according to RECIST criteria. In the two highest-dose groups, 7 out of 10 patients obtained PR or SD (Rivera et al. 2009).

Results from the trial indicate that zalutumumab was well tolerated by head and neck cancer patients. No patients experienced dose limiting toxicity (DLT) when treated with the highest dose of 8 mg/kg. The most frequent adverse event was acneiform rash which has been considered a positive correlation with biological activity in 56 percent of the patients. The occurrence of rash increased with dose and 10 of 11 patients in the 4 and 8 mg/kg dose groups experienced rash. Other adverse events included rigors, fatigue, pyrexia, nausea, flushing, and increased sweating.

31.4.8 Zanolimumab/HuMax-CD4

CD4 is a 55 kDa glycoprotein that is expressed on ~60 percent of peripheral blood T-lymphocytes and activates T-lymphocytes by binding to the nonpolymorphic region of the MHC class II molecules. Zanolimumab is a fully human anti-CD4 mAb generated in transgenic mice. The antibody is specific for the CD4 expressed on most T-lymphocytes and to a lesser extent on macrophages. Zanolimumab prevents interaction between the CD4 receptor and the MHC class II antigens and interferes with T-cell activation. Studies *in vitro* showed that zanolimumab had ADCC activity, but not CDC (Assaf and Sterry 2007; Y.H. Kim et al. 2007; Villadsen et al. 2007).

Zanolimumab is currently in development for cutaneous T-cell lymphoma (CTCL) and non-cutaneous T-cell lymphoma (NCTCL) using i.v. administration. Two open-label, clinical phase II studies with zanolimumab were reported in patients with therapy-resistant CTCL of $CD4^+$ mycosis fungoides (MF) and Sezary syndrome (SS) subtypes. Marked efficacy in both early and advanced stages of MF was reported in the highest dose groups (56 percent overall response rate) (Chentouf et al. 2007). Phase II trials with patients at both early and late stages of CTCL showed a clinical response between 36 percent (4 of 11) for early stage MF patients and 22 percent (2 of 9) for the advanced stage MF patients with 280 mg treatment. The treatment doses included 280 mg, 560 mg, or 980 mg of zanolimumab once a week for 16 weeks. At the 560 mg dose level, 50 percent (7 of 14) of patients showed partial response and 75 percent (3 of 4) of advanced stage patients treated with 980 mg of the antibody obtained partial responses. The ongoing phase III pivotal study includes patients with profiles such as the most common form of CTCL and MF who are refractory to or intolerant of Targretin® and one other standard therapy. The trial will consist of two stages and the first stage of the trial is open-label and consists of three dose cohorts (280 mg, 560 mg, and 980 mg). The second stage of the trial is blinded and will include 70 patients randomized once weekly for 12 weeks. Preliminary results from the first stage of the pivotal study for the end point phase III trial were reported in December 2006 (Y.H. Kim et al. 2007). Clinical response was shown in 5 out 12 patients (42 percent) in the two highest dose groups. Patients treated at the 280 mg dose level had no responses observed.

Zanolimumab was considered well tolerated with no dose-related toxicity (Chentouf et al. 2007). However, out of 47 treated patients reported, five had grade three adverse events, including hypersensitivity, elevated liver enzymes, aggravated pruritus, and muscle fiber rupture.

31.4.9 Galiximab

CD80 (B7.1) is a membrane-bound costimulatory molecule and it is known for its role in regulating T-cell activity (Chambers and Allison 1999). Several studies have suggested that CD80 may play a role in the regulation of normal and malignant B-cells (Suvas et al. 2002). CD80 is transiently expressed on the surface of activated B-cells, antigen-presenting cells (APCs), and T-cells. However, it has been found that CD80 is constitutively expressed on a variety of non-Hodgkin lymphoma (NHLs), including follicular lymphoma, making it an attractive target for lymphoma therapy (Vyth-Dreese et al. 1998).

Galiximab is a chimeric (primate/human), anti-CD80 IgG1/lambda mAb with human constant regions and primate (cynomologous macaque) variable regions. Currently the antibody is in phase III trial by Biogen IDEC for treatment of relapsed and refractory follicular NHLs. Galiximab can bind to human CD80 on lymphoma cell lines and it has also been shown to bind CD80 on T-cells and block CD80-CD28 interaction without interfering with the interaction between CD80 and CTLA-4. ADCC has been reported as the main mechanism of action for galiximab and it is possible that other modes of action may also be responsible for its clinical activity. Galiximab is formulated for intravenous injection and a dose range of 125, 250, 375, and 500 mg/m^2 was studied in phase I trial. The mean half-life of galiximab was reported in a range of 13 to 24 days. Phase I safety study reported no severe adverse event (grades 3 and 4) up to the dose at 500 mg/m^2. No dose limited toxicity was observed. Multicenter, phase I/II clinical studies using galiximab alone or in combination with rituximab were studied to evaluate safety, efficacy, and pharmacokinetics (PK) in patients with relapsed or refractory follicular lymphoma (Czuczman et al. 2005; Leonard et al. 2007). In galiximab single agent treatment, overall response rate was 11 percent (4 out of 35) and 34 percent of patients maintained stable disease (Czuczman et al. 2005) and in combination regimen with rituximab, overall response rate (ORR) was 66 percent (Leonard et al. 2007). Phase III study is designed to be a randomized, double-blind study of galiximab for patients with lymphoma. The trial was designed to compare treatment with galiximab in combination with rituximab to the treatment with the rituximab in combination with placebo. Patients for the trial will be with follicular NHL that has relapsed or failed to respond to initial therapy. The study enrolled the first patient in November 2006 and will ultimately enroll approximately 700 patients worldwide at more than 150 centers.

Galiximab was generally well tolerated, with a good safety profile. The most common side effects reported included lymphopenia, fatigue, nausea, and headache. All but one of the events was at grade 1 or 2.

31.4.10 Epratuzumab

The CD22 antigen is a 135 kDa transmembrane glycoprotein and belongs to the immunoglobulin superfamily. Expression of CD22 is restricted to B-cells and possibly basophils (Han et al. 1999; Tedder et al. 1997). CD22 expression is highly regulated during B-cell development and its expression is lost after differentiation into plasma cells or after B-cell activation, but CD22 expression is retained in most B-cell malignancies, including B-cell chronic lymphocytic leukemia (B-CLL).

Epratuzumab is a humanized IgG1 mAb directed against CD22 (Yee and O'Brien 2006). On binding to B-lymphocytes, epratuzumab is rapidly internalized. Preclinical study showed that epratuzumab prolonged the survival in the xenograft mice models of NHL. Preclinical data in cell lines and xenograft mice models of NHL indicated that its main mechanism of action is via induction of ADCC (Carnahan et al. 2003). Internalization of the CD22/epratuzumab complex may also result in activation of nonreceptor tyrosine kinases (Leonard et al. 2005).

Phase I/II trials for evaluating epratuzumab monotherapy in patients with relapsed or refractory aggressive and indolent NHL have been completed. No patient developed human antihuman antibodies. Mean serum half-life was 23 days and overall response rate (ORR) was 18 percent, including 6 percent of CR and 12 percent of PR. Median duration of responses was 79.3 weeks (range: 11.1 to 143.3). Responses were observed only in patients with follicular lymphoma. Three phase II studies were conducted for evaluation of weekly therapy with epratuzumab 360 mg/m^2 i.v. followed by standard dose rituximab for four consecutive weeks in 137 patients with relapsed or refractory indolent or aggressive NHL. Median duration of response did not reach objectives (Leonard et al. 2005). A phase III multicenter trial of epratuzumab in patients with relapsed or refractory low-grade follicular B-cell NHL was initiated and subsequently suspended due to the expected lack of commercial viability of single-agent epratuzumab in this patient population. Currently, epratuzumab is in phase III trial for AIID indications. Epratuzumab was well tolerated in the trial and no dose limiting toxicities (DLTs) were reported.

31.5 THERAPEUTIC ANTIBODIES IN CLINICAL PHASE III TRIALS IN AIID, INFECTIOUS DISEASES, AND OTHER DISORDERS

Table 31.3 shows a list of monoclonal antibodies in late stage clinical trials for AIID, infectious diseases. and other disorders. Due to page limitations, only a few selected antibodies in late stage development are reviewed in this section.

31.5.1 Mycograb/Efungumab

Mycograb is an antifungal human mAb against fungal heat shock protein 90 (Hsp90) for the treatment of invasive *Candida* infection in combination with amphotericin B. Efungumab was developed by NeuTec Pharma (acquired by Novartis) and produced as a 28 kDa antibody fragment in an *E. coli.* system. Infection by *Candida* has a mortality rate of 35 percent despite advances in the antifungal drugs (McNeil et al. 2001). Mycograb acts synergistically with amphotericin B against a broad spectrum of clinically relevant species of *Candida*, including *C. albicans*, *C. krusei*, *C. tropicalis*,

TABLE 31.3 Monoclonal Antibodies in Late Stage Clinical Trials for AIID, Infectious Diseases, and Other Disorders

mAb Names	Target Name	IgG Type	Clinical Stage	Major Indication	Sponsor Company
Mycograb Efungumad	*Candida* HSP90	scFv fragment	Phase III	Fungal infection	Novartis
Mepolizumab Bosatria	IL-5	Humanized	Phase III	Hypereosinophilic syndrome	GSK
Inolimomab Leukotac	IL-2/CD-25	Murine mAb	Phase III	Acute graft-versus-host disease	EUSA Pharma
Motavizumab Numax MEDI-542	RSV F protein	Chimeric, IgG1	Phase III	RSV infection	Medimmune
Ustekinumab CNTO-1275	IL-12 and IL-23	Human mAb	Phase III	AIID such as psoriasis	Genmab
Denosumab AMG 162	RANKL	Human mAb, IgG2	Phase III	Bone loss such as osteoporosis	Amgen
Epratuzumab	CD22	Humanized	Phase III	Lupus	UCB Schwarz
Visilizumab Nuvion HuM291	CD3	Humanized	Phase III	Inflammatory bowel disease	PDL Biopharma
Teplizumab MGA031 hOKT3γ1 (Ala-Ala)	CD3	Humanized	Phase III	Type I diabetes	Lilly
Otelixizumab	CD3	Humanized	Phase III	Type I diabetes	GSK and Tolerx
ACZ885	IL-1	Human mAb	Phase III	CAPA	Novartis

C. parapsilosis, and *C. glabrata*. Mycograb increased the clearance of fluconazole-sensitive species of *Candida* (Burnie and Matthews 2004; Matthews et al. 2003).

The discovery of Mycograb was based on an early observation that in patients infected by invasive candidiasis, there was a strong correlation between recovery and the presence of protective antibodies to Hsp90 (Matthews et al. 1984). Hsp90 is a molecular chaperon involved in multiple cellular functions such as mitogenic signaling and protein folding in fungi. Mycograb was derived from a dominant sequence of Hsp90 antibody identified in survivors from infection. It binds to an immunodominant epitope on candida Hsp90 that is essential for the function of the molecule. In fungi, Hsp90 was induced during hyphal formation to facilitate the folding of cell wall kinases; therefore, inhibition of Hsp90 could weaken the cell wall. This may explain the synergy seen between Mycograb and other antifungal agents targeting cell wall (Matthews and Burnie 2005).

Mycograb has been granted orphan drug status in the United States and Europe. However, the European Medicines Agency has refused twice to grant marketing authorization for Mycograb due to product safety and quality issues. The concerns regarding the folding and aggregation of the product, the possible immune responses in some patients, and potential cases of cytokine release syndrome need to be addressed.

31.5.2 Bosatria/Mepolizumab

Mepolizumab is an anti-interleukin-5 (IL-5) humanized mAb (Hart et al. 2001). Initially, the drug was investigated in larger indications where eosinophils play an important role in the disease progression. Phase II trials in asthma and in atopic dermatitis were initiated in 1998 and 2002; however, mepolizumab showed little efficacy in studies for atopic dermatitis and it failed to sufficiently deplete eosinophils for asthma. It is currently in development for the treatment of the hypereosinophilic syndrome (HES) by GlaxoSmithKlein and has received orphan drug status granted by the FDA.

HES encompasses a wide range of clinical manifestations characterized by sustained eosinophilia (with a blood eosinophil count >1500 per microliter for more than six months) and eosinophil-related organ damage without other causes such as parasite infection, allergy, malignancy, and collagen vascular disease (Chusid et al. 1975). It is a rare condition, with an estimated 2000 to 5000 patients in the United States. Although the disease often can be managed by long term systemic corticosteroids, it can also lead to serious morbidity. IL-5 contributes to the pathogenesis of HES by playing a role in the development of eosinophil from hematopoietic progenitors, involving eosinophil maturation, differentiation, mobilization, and activation. Mepolizumab lowers blood eosinophil level by binding to free IL-5, which blocks the interaction between IL5 and its receptor on eosinophils and eosinophil progenitors.

Mepolizumab is an IgG1 with a half-life of 19 days. Results from a phase III study demonstrated that patients treated with mepolizumab were able to maintain control of the disease with a reduced dose of corticosteroids, compared with patients who received placebo (Rothenberg et al. 2008). In this study the drug was generally well tolerated. It may provide a new therapeutic option for physicians to treat HES patients while reducing corticosteroid use.

31.5.3 Leukotac/Inolimomab/BT563

Inolimomab is a murine IgG1 mAb against interleukin-2 (IL-2) receptor developed by OPi SA (acquired by EUSA Pharma) for the potential treatment of graft-versus-host disease (GvHD). Steroid-refractory acute GvHD and opportunistic infection after hematopoietic stem cell transplantation are the major contributors to a high mortality rate (Wingard et al. 1989). The first-line treatment of acute GvHD is usually a high dose of steroids; however, steroid resistance is observed in 40 percent of the GvHD patients. Acute GvHD is a condition induced by activation of donor T-cells against recipients' tissues or organs. One of the important cytokines for T-cell differentiation and proliferation is IL-2. Inolimomab is a murine antibody directed against the alpha chain of the IL-2 receptor (CD25) specifically expressed on the surface of activated T-cells (J.L. Ferrara and Deeg 1991).

The initial clinical trial of inolimomab in combination with corticosteroids showed that the drug is well tolerated (Cuthbert et al. 1992). Several clinical trials have been carried out in corticosteroid-resistant acute GvHD patients and positive responses have been observed (Herbelin et al. 1994; Hertenstein et al. 1994; Herve et al. 1990). Inolimomab has shown better efficacy when compared with other anti-IL2R antibodies, possibly due to patient selection in the trials, the pharmacokinetic profile, or the selective binding of inolimomab to cell-bound IL2R versus soluble IL2R (Bay et al. 2005). In 2001 orphan status was granted by the European Commission to inolimomab for the treatment of GvHD and clinical trials are ongoing.

31.5.4 Numax/Motavizumab/MEDI-524

Although Synagis demonstrated efficacy by reducing hospitalization due to the RSV infection by approximately 50 percent in high risk infants, it is conceivable that preventive therapy could be significantly improved. An enhanced anti-RSV mAb may potentially allow for superior viral neutralization activity in the upper and lower respiratory tracts, and result in improved efficacy for the prophylaxis of severe RSV illness in high risk children.

Numax (motavizumab) was derived by *in vitro* affinity maturation of the murine CDRs of the heavy chain and light chain of palivizumab (Synagis, MEDI-493). Motavizumab exhibited a 70-fold increased affinity for the RSV-F protein A antigen (50 pM to the RSV F protein) (4-fold increased k_{on}/17-fold decreased k_{off}), compared to Synagis (3.46 nM to the RSV-F protein). During the affinity maturation process, combinatorial libraries were screened and selected Fab fragments were introduced into the Synagis IgG1κ backbone. Motavizumab was selected for its increased *in vitro* bioactivity and its enhanced *in vivo* potency profile due to the improved pharmacokinetics and biodistribution. There are one, two, and three amino acid changes in HC CDR1, CDR2, and CDR3, respectively. There is one amino acid change in frame 4 of the heavy chain. There are five amino acid changes in LC CDR1 and there is one amino acid change in frame 4 of the light chain (H. Wu et al. 2005, 2007).

The potency of motavizumab was evaluated *in vitro* and was shown to be 18 times more potent than Synagis at neutralizing RSV infection in a susceptible cell line (Hep-2). Both preclinical and clinical studies showed motavizumab had a similar safety profile as that of Synagis. In the preclinical cotton rat animal model (1×10^5 to 2×10^7 pfu RSV was challenged 24 hours after the administration of motavizumab or palivizumab), motavizumab was 50 to 100 times more potent at reducing RSV titers than palivizumab at equivalent serum levels (approximately 20 μg/mL). Specifically, motavizumab prophylaxis resulted in 3 to 4 log and 2 log reductions in pulmonary and nasal RSV titers, respectively. PK studies in cynomolgus monkeys ($n = 6$) show that motavizumab had a $T_{1/2}$ of 8.6 days. In toxicity studies of cynomolgus monkeys, motavizumab was shown to be safe and well tolerated by the animals at 100 mpk.

As of 2008, a pivotal phase III study of motavizumab showed that it had a 26 percent reduction in RSV-related hospitalization compared to that of palivizumab. Motavizumab also showed 50 percent reduction in RSV-specific outpatient lower respiratory infection, which was not an end point for palivizumab. A recent phase III RSV study demonstrated the effectiveness of motavizumab in that it reduced hospitalizations due to RSV infection by 83 percent as compared to placebo, and also reduced physician-attended outpatient cases by 71 percent. This study involves 1410 full-term infants less than six months of age from Southwest Native American populations who are at high risk of RSV infection.

31.5.5 CNTO-1275/Ustekinumab

CNTO-1275 is a fully human mAb, isolated using Medarex's HuMAb technology. It is an IgGκ with high affinity binding to the p40 subunit of human interleukin-12 (IL-12) and interleukin-23 (IL-23) for the potential treatment of autoimmune diseases. It is currently in phase III clinical development for psoriasis and in phase II clinical development for CD, MS, and PA (Wittig 2007).

IL-12 and IL-23 are cytokines produced by inflammatory cells including marcophages, neutrophils, and dendritic cells that activate immune responses involved in the pathogenesis of psoriasis. IL-12 and IL-23 are heterodimeric cytokines with a common p40 subunit which was found to be overexpressed in psoriasis plaques (Toichi et al. 2006). The potential use of blocking the activities of IL-12 and IL-23 by anti-p40 antibodies has been demonstrated in several animal models for autoimmune diseases such as MS, arthritis, and colitis (Leonard, Waldburger, and Goldman 1995; Matthys et al. 1998). The inhibitory activity of CNTO-1275 on IL-12 induced expression of cutaneous lymphocyte antigen (CLA), IL-2α, IL12 receptor, and IFNγ were shown *in vitro* in peripheral blood mononuclear cells (PBMCs) (Reddy et al. 2007). The efficacy of CNTO-1275 *in vivo* in the prevention of experimental autoimmune encephalomyelitis has also been demonstrated (Brok et al. 2002).

In a single dose study of 17 psoriasis patients (0.3 to 3.0 mg/kg sc), CNTO-1275 was absorbed slowly into the circulation, with a *Cmax* value of 12 days and a $T_{1/2}$ of 20 days (Gottlieb et al. 2007). Early clinical studies did not show toxic effects, and a phase II trial of CNTO-1275 for the treatment of psoriasis showed that adverse events occurred in 79 percent of the patients in the CNTO-1275 group as compared with 72 percent in the placebo group (Krueger et al. 2007). Recent phase III data showed promising efficacy and safety profiles where CNTO-1275 sustained an overall 75 percent improvement in psoriasis over a year and overall incidence of adverse events similar to the placebo group. Known as STELARA™, ustekinumab has been approved by the European Commission and Canada for the treatment of moderate to severe plaque psoriasis.

31.5.6 AMG 162/Denosumab

Denosumab is a fully human mAb for the treatment of skeletal disorders associated with bone loss, including osteoporosis, treatment induced bone loss, bone metastasis, rheumatoid arthritis, and multiple myeloma (Hamdy 2008). It binds to the receptor activator of nuclear factor kappa B ligand

(RANKL) and blocks the effect of RANKL, inhibiting the process of bone resorption. Denosumab was developed using the Xenomouse technology with low picomolar binding affinity to RANKL. It was converted to IgG2 format to reduce the potential cytotoxic effects caused by binding of IgG1 to RANKL-producing cells such as stromal cells, osteoblasts, and T-cells (Schwarz and Ritchlin 2007).

RANKL belongs to the TNF superfamily and is produced by osteoblasts, bone marrow stromal cells, and activated T-lymphocytes. The RANKL-mediated signaling pathway is an essential in the dynamic process of bone remodeling. Binding of RANKL to its receptor RANK expressing on the surface of hematopoietic precursors triggers the formation, activation, and survival of osteoclasts for the breakdown of bone. RANKL inhibition therefore is a rational approach for suppression of bone loss due to excess bone resorption. The pathology of bone loss is often associated with aging, inflammatory disorders, and malignancy treatments.

A phase I single dose, dose-escalation study was conducted to address the effect of single subcutaneous dose of denosumab ranging from 0.01 mg/kg to 3 mg/kg in healthy postmenopausal women. A rapid and sustained suppression of bone resorption, up to 9 months at highest doses of 1 mg/kg and 3 mg/kg, was observed (Bekker et al. 2004). In this study, the $T_{1/2}$ of denosumab at the higher doses of 1 and 3 mg/kg was 32 days, and the drug was well tolerated. Data from a phase II study in postmenopausal female patients demonstrated that the substantial reduction in bone resorption induced by three- to six-monthly administration of denosumab is associated with a significant increase in bone mineral density (BMD) (McClung et al. 2006). Several other clinical studies are ongoing, addressing the efficacy, safety, and tolerability for the treatments of osteoporosis, rheumatoid arthritis, and prevention of bone loss and skeletal-related events in malignancy (Hamdy 2008). Most recently, a phase III study in women ($n = 332$) with menopause-related osteoporosis showed that twice-yearly subcutaneous injections of denosumab effectively increased BMD (Bone et al. 2008). A large, pivotal phase III study (the FREEDOM trial) in 2008 showed the risk of bone fracture was significantly reduced by the treatment of denosumab in women with postmenopausal osteoporosis. This study also showed that denosumab decreased the breakdown of bone and increased bone mineral density. Good safety profile and tolerability were observed (Cummings et al. 2008).

31.5.7 ACZ885

ACZ885 is a human mAb againt interleukin-1-beta (IL-1β) for the treatment of autoinflammatory diseases mediated by IL-1. It is under development by Novartis and was generated using Medarex's UltiMAb technology with a binding affinity of 30 pM to IL-1β measured by BIAcore. In April 2008, data from a phase III clinical trial indicated that ACZ885 could develop into a therapeutic for the treatment of cryopyrin-associated periodic syndromes (CAPA), a group of rare but life long diseases including Muckel Wells syndrome. The investigation of ACZ885 in the treatment of more common autoimmune diseases such as rheumatoid arthritis is underway.

31.6 MARKET OUTLOOK

Of the 24 mAb products in clinical use (Table 31.1), five (Avastin, Herceptin, Humira, Remicade, and Rituxan) of them accounted for 80 percent of sales in 2006. Small molecule drugs and mAbs accounted for 83.3 percent and 4.6 percent of total pharmaceutical revenue in 2006, respectively. mAbs together with therapeutic proteins were projected to account for 70 percent of the growth in prescription pharmaceutical annual sales between 2006 and 2012 (Datamonitor 2007). Revenues from antibody therapeutics are predicted to increase from US$19,573 million in 2006 to US$43,381 million in 2012. This is because mAbs have the advantage of primarily addressing high unmet medical need in areas such as oncology and AIID. Of the approximately 180 mAbs in clinical use and in clinical development, 100 of them are being developed for oncology. The second concentrated therapeutic category is in the AIID area, with 31 antibodies in clinical development. Infectious disease is becoming a major disease area for antibody therapeutics, with about 27 antibody molecules at various stages of clinical development. It is encouraging to see that a number of antibody drugs that are in clinical use or in clinical development fall in the areas of CNS, CV, bone loss, diabetes, endocrinology, hematology, ophthalmology, and respiratory diseases (Datamonitor 2007). It is also worth noting that about 20 of the 180 antibody therapeutics in various stages of clinical use and development are antibody : drug and antibody : radioisotope conjugates.

ACKNOWLEDGMENTS

The authors thank Triveni Bhatt and Thu Ho for their careful review of the manuscript.

REFERENCES

Adams, C.W., D.E. Allison, K. Flagella, L. Presta, J. Clarke, N. Dybdal, K. McKeever, and M.X. Sliwkowski. 2006. Humanization of a recombinant monoclonal antibody to produce a therapeutic HER dimerization inhibitor, pertuzumab. *Cancer Immunol. Immunother.* 55:717–727.

Adu, D., P. Cockwell, N.J. Ives, J. Shaw, and K. Wheatley. 2003. Interleukin-2 receptor monoclonal antibodies in renal transplantation: Meta-analysis of randomised trials. *Br. Med. J.* 326:789.

Agus, D.B., R.W. Akita, W.D. Fox, G.D. Lewis, B. Higgins, P.I. Pisacane, J.A. Lofgren, C. Tindell, D.P. Evans, K. Maiese, et al. 2002. Targeting ligand-activated ErbB2 signaling inhibits breast and prostate tumor growth. *Cancer Cell* 2:127–137.

Albanell, J., and J. Baselga. 1999. Trastuzumab, a humanized anti-HER2 monoclonal antibody, for the treatment of breast cancer. *Drugs Today (Barc)* 35:931–946.

Allison, D.E., M. Malik, F. Qureshi, D. Baker, S. Kelsey, and G. Fyfe. 2004. Pharmacokinetics of Her-2 targeted rhuMan 2C4 (pertuzumab) in patients with advanced solid malignancies: Phase 1a results. *Proc. Am. Assoc. Clin. Oncol.* 22:790.

Almasri, N.M., R.E. Duque, J. Iturraspe, E. Everett, and R.C. Braylan. 1992. Reduced expression of CD20 antigen as a characteristic marker for chronic lymphocytic leukemia. *Am. J. Hematol.* 40:259–263.

Amico, D., A.M. Barbui, E. Erba, A. Rambaldi, M. Introna, and J. Golay. 2003. Differential response of human acute myeloid leukemia cells to gemtuzumab ozogamicin in vitro: Role of Chk1 and Chk2 phosphorylation and caspase 3. *Blood* 101:4589–4597.

Anderson, K.C., M.P. Bates, B.L. Slaughenhoupt, G.S. Pinkus, S.F. Schlossman, and L.M. Nadler. 1984. Expression of human B cell-associated antigens on leukemias and lymphomas: A model of human B cell differentiation. *Blood* 63:1424–1433.

Arkfeld, D.G. 2008. The potential utility of B cell-directed biologic therapy in autoimmune diseases. *Rheumatol Int.* 28:205–215.

Armant, M., H. Ishihara, M. Rubio, G. Delespesse, and M. Sarfati. 1994. Regulation of cytokine production by soluble CD23: Costimulation of interferon gamma secretion and triggering of tumor necrosis factor alpha release. *J. Exp. Med.* 180:1005–1011.

Assaf, C., and W. Sterry. 2007. Drug evaluation: Zanolimumab, a human monoclonal antibody targeted against CD4. *Curr. Opin. Mol. Ther.* 9:197–203.

Attia, P., G.Q. Phan, A.V. Maker, M.R. Robinson, M.M. Quezado, J.C. Yang, R.M. Sherry, S.L. Topalian, U.S. Kammula, R.E. Royal, et al. 2005. Autoimmunity correlates with tumor regression in patients with metastatic melanoma treated with anti-cytotoxic T-lymphocyte antigen-4. *J. Clin. Oncol.* 23:6043–6053.

Baca, M., T.S. Scanlan, R.C. Stephenson, and J.A. Wells. 1997. Phage display of a catalytic antibody to optimize affinity for transition-state analog binding. *Proc. Natl. Acad. Sci. U.S.A.* 94:10063–10068.

Balaian, L., R.K. Zhong, and E.D. Ball. 2003. The inhibitory effect of anti-CD33 monoclonal antibodies on AML cell growth correlates with Syk and/or ZAP-70 expression. *Exp. Hematol.* 31:363–371.

Baselga, J., J. Albanell, M.A. Molina, and J. Arribas. 2001. Mechanism of action of trastuzumab and scientific update. *Semin. Oncol.* 28:4–11.

Baselga, J., X. Carbonell, N.J. Castaneda-Soto, M. Clemens, M. Green, V. Harvey, S. Morales, C. Barton, and P. Ghahramani. 2005. Phase II study of efficacy, safety, and pharmacokinetics of trastuzumab monotherapy administered on a 3-weekly schedule. *J. Clin. Oncol.* 23:2162–2171.

Baselga, J., D. Tripathy, J. Mendelsohn, S. Baughman, C.C. Benz, L. Dantis, N.T. Sklarin, A.D. Seidman, C.A. Hudis, J. Moore, et al. 1996. Phase II study of weekly intravenous recombinant humanized anti-p185HER2 monoclonal antibody in patients with HER2/neu-overexpressing metastatic breast cancer. *J. Clin. Oncol.* 14:737–744.

Bay, J.O., N. Dhedin, M. Goerner, J.P. Vannier, A. Marie-Cardine, A. Stamatoullas, J.P. Jouet, I. Yakoub-Agha, R. Tabrizi, C. Faucher, et al. 2005. Inolimomab in steroid-refractory acute graft-versus-host disease following allogeneic hematopoietic stem cell transplantation: Retrospective analysis and comparison with other interleukin-2 receptor antibodies. *Transplantation* 80:782–788.

Beeler, J.A., and K. van Wyke Coelingh. 1989. Neutralization epitopes of the F glycoprotein of respiratory syncytial virus: Effect of mutation upon fusion function. *J. Virol.* 63:2941–2950.

Bekker, P.J., D.L. Holloway, A.S. Rasmussen, R. Murphy, S.W. Martin, P.T. Leese, G.B. Holmes, C.R. Dunstan, and A.M. DePaoli. 2004. A single-dose placebo-controlled study of AMG 162, a fully human monoclonal antibody to RANKL, in postmenopausal women. *J. Bone Miner. Res.* 19:1059–1066.

Berinstein, N.L., A.J. Grillo-Lopez, C.A. White, I. Bence-Bruckler, D. Maloney, M. Czuczman, D. Green, J. Rosenberg, P. McLaughlin, and D. Shen. 1998. Association of serum Rituximab (IDEC-C2B8) concentration and anti-tumor response in the treatment of recurrent low-grade or follicular non-Hodgkin's lymphoma. *Ann. Oncol.* 9:995–1001.

Bessler, M., P.J. Mason, P. Hillmen, T. Miyata, N. Yamada, J. Takeda, L. Luzzatto, and T. Kinoshita. 1994. Paroxysmal nocturnal haemoglobinuria (PNH) is caused by somatic mutations in the PIG-A gene. *EMBO J.* 13:110–117.

Binder, M., F. Otto, R. Mertelsmann, H. Veelken, and M. Trepel. 2006. The epitope recognized by rituximab. *Blood* 108:1975–1978.

Biogen Idec, Inc. (13-Sept-2005). Zevalin (Ibritumomab Tiuxetan): Label and Prescribing Information.

Biogen Idec, Inc. and Genentech (21-Feb-2007). Rituxan (Rituximab): Label and Prescribing Information.

Bjorn, G. 2008. Despite potential side effects, two drugs make a comeback. *Nature Med.* 14:226.

Blagosklonny, M.V. 2005. How Avastin potentiates chemotherapeutic drugs: Action and reaction in antiangiogenic therapy. *Cancer Biol. Ther.* 4:1307–1310.

Bleeker, W.K., J.J. Lammerts van Bueren, H.H. van Ojik, A.F. Gerritsen, M. Pluyter, M. Houtkamp, E. Halk, J. Goldstein, J. Schuurman, M.A. van Dijk, et al. 2004. Dual mode of action of a human anti-epidermal growth factor receptor monoclonal antibody for cancer therapy. *J. Immunol.* 173:4699–4707.

Bleumer, I., A. Knuth, E. Oosterwijk, R. Hofmann, Z. Varga, C. Lamers, W. Kruit, S. Melchior, C. Mala, S. Ullrich, et al. 2004. A phase II trial of chimeric monoclonal antibody G250 for advanced renal cell carcinoma patients. *Br. J. Cancer* 90:985–990.

Bleumer, I., E. Oosterwijk, J.C. Oosterwijk-Wakka, M.C. Voller, S. Melchior, S.O. Warnaar, C. Mala, J. Beck, and P.F. Mulders. 2006. A clinical trial with chimeric monoclonal antibody WX-G250 and low dose interleukin-2 pulsing scheme for advanced renal cell carcinoma. *J. Urol.* 175:57–62.

Bolt, S., E. Routledge, I. Lloyd, L. Chatenoud, H. Pope, S.D. Gorman, M. Clark, and H. Waldmann. 1993. The generation of a humanized, non-mitogenic CD3 monoclonal antibody which retains in vitro immunosuppressive properties. *Eur. J. Immunol.* 23:403–411.

Bone, H.G., M.A. Bolognese, C.K. Yuen, D.L. Kendler, H. Wang, Y. Liu, and J. San Martin. 2008. Effects of denosumab on bone mineral density and bone turnover in postmenopausal women. *J. Clin. Endocrinol. Metab.* 93:2149–2157.

Bonner, J.A., P.M. Harari, J. Giralt, N. Azarnia, D.M. Shin, R.B. Cohen, C.U. Jones, R. Sur, D. Raben, J. Jassem, et al. 2006. Radiotherapy plus cetuximab for squamous-cell carcinoma of the head and neck. *N. Engl. J. Med.* 354:567–578.

Bossenmaier, B., M. Hasmann, H. Koll, H.H. Fiebig, R.W. Akita, M.X. Sliwkowski, and H.J. Muler. 2004. Presence of Her-2/Her-3 heterodimers predict antitumor effects of pertuzumab (Omnitarg) in different human xenograft models. *Proc. Am. Assoc. Cancer Res.* 45:5342.

Brabender, J., K.D. Danenberg, R. Metzger, P.M. Schneider, J. Park, D. Salonga, A.H. Holscher, and P.V. Danenberg. 2001. Epidermal growth factor receptor and HER2-neu mRNA expression in non-small cell lung cancer is correlated with survival. *Clin. Cancer Res.* 7:1850–1855.

Brendel, C., and A. Neubauer. 2000. Characteristics and analysis of normal and leukemic stem cells: Current concepts and future directions. *Leukemia* 14:1711–1717.

Brok, H.P., M. van Meurs, E. Blezer, A. Schantz, D. Peritt, G. Treacy, J.D. Laman, J. Bauer, and B.A. Hart. 2002. Prevention of experimental autoimmune encephalomyelitis in common marmosets using an anti-IL-12p40 monoclonal antibody. *J. Immunol.* 169:6554–6563.

Brown, W.M. 2006. Anti-CD3 antibody MacroGenics Inc. *Curr. Opin. Investig. Drugs* 7:381–388.

Bubien, J.K., L.J. Zhou, P.D. Bell, R.A. Frizzell, and T.F. Tedder. 1993. Transfection of the CD20 cell surface molecule into ectopic cell types generates a Ca^{2+} conductance found constitutively in B lymphocytes. *J. Cell Biol.* 121:1121–1132.

Buchsbaum, D.J., R.L. Wahl, S.D. Glenn, D.P. Normolle, and M.S. Kaminski. 1992a. Improved delivery of radiolabeled anti-B1 monoclonal antibody to Raji lymphoma xenografts by predosing with unlabeled anti-B1 monoclonal antibody. *Cancer Res.* 52:637–642.

Buchsbaum, D.J., R.L. Wahl, D.P. Normolle, and M.S. Kaminski. 1992b. Therapy with unlabeled and ^{131}I-labeled pan-B-cell monoclonal antibodies in nude mice bearing Raji Burkitt's lymphoma xenografts. *Cancer Res.* 52:6476–6481.

Buggins, A.G., G.J. Mufti, J. Salisbury, J. Codd, N. Westwood, M. Arno, K. Fishlock, A. Pagliuca, and S. Devereux. 2002. Peripheral blood but not tissue dendritic cells express CD52 and are depleted by treatment with alemtuzumab. *Blood* 100:1715–1720.

Burnie, J., and R. Matthews. 2004. Genetically recombinant antibodies: New therapeutics against candidiasis. *Expert Opin. Biol. Ther.* 4:233–241.

Byrd, J.C., S. O'Brien, I.W. Flinn, T.J. Kipps, M. Weiss, K. Rai, T.S. Lin, J. Woodworth, D. Wynne, J. Reid, et al. 2007. Phase 1 study of lumiliximab with detailed pharmacokinetic and pharmacodynamic measurements in patients with relapsed or refractory chronic lymphocytic leukemia. *Clin. Cancer Res.* 13:4448–4455.

Cardarelli, P.M., M. Quinn, D. Buckman, Y. Fang, D. Colcher, D.J. King, C. Bebbington, and G. Yarranton. 2002. Binding to CD20 by anti-B1 antibody or F(ab′)(2) is sufficient for induction of apoptosis in B-cell lines. *Cancer Immunol. Immunother.* 51:15–24.

Carnahan, J., P. Wang, R. Kendall, C. Chen, S. Hu, T. Boone, T. Juan, J. Talvenheimo, S. Montestruque, J. Sun, et al. 2003. Epratuzumab, a humanized monoclonal antibody targeting CD22: Characterization of in vitro properties. *Clin. Cancer Res.* 9:3982S–3990S.

Carpenter, P.A., S. Pavlovic, J.Y. Tso, O.W. Press, T. Gooley, X.Z. Yu, and C. Anasetti. 2000. Non-Fc receptor-binding humanized anti-CD3 antibodies induce apoptosis of activated human T cells. *J. Immunol.* 165:6205–6213.

Carter, P.J. 2006. Potent antibody therapeutics by design. *Nature Rev. Immunol.* 6:343–357.

Carter, P., L. Presta, C.M. Gorman, J.B. Ridgway, D. Henner, W.L. Wong, A.M. Rowland, C. Kotts, M.E. Carver, and H.M. Shepard. 1992. Humanization of an anti-p185HER2 antibody for human cancer therapy. *Proc. Natl. Acad. Sci. U.S.A.* 89:4285–4289.

Chakrabarti, M.C., N. Le, C.H. Paik, W.G. De Graff, and J.A. Carrasquillo. 1996. Prevention of radiolysis of monoclonal antibody during labeling. *J. Nuclear Med.* 37:1384–1388.

Chambers, C.A., and J.P. Allison. 1999. CTLA-4: The costimulatory molecule that doesn't—regulation of T-cell responses by inhibition. *Cold Spring Harb. Symp. Quant. Biol.* 64:303–312.

Chang, T.W. 2000. The pharmacological basis of anti-IgE therapy. *Nature Biotechnol.* 18:157–162.

Chatenoud, L., and J.A. Bluestone. 2007. CD3-specific antibodies: A portal to the treatment of autoimmunity. *Nature Rev. Immunol.* 7:622–632.

Chen, S., L. Yu, C. Jiang, Y. Zhao, D. Sun, S. Li, G. Liao, Y. Chen, Q. Fu, Q. Tao, et al. 2005. Pivotal study of iodine-131-labeled chimeric tumor necrosis treatment radioimmunotherapy in patients with advanced lung cancer. *J. Clin. Oncol.* 23:1538–1547.

Chen, Y., C. Wiesmann, G. Fuh, B. Li, H.W. Christinger, P. McKay, A.M. de Vos, and H.B. Lowman. 1999. Selection and analysis of an optimized anti-VEGF antibody: Crystal structure of an affinity-matured Fab in complex with antigen. *J. Mol. Biol.* 293:865–881.

Chentouf, M., S. Ghannam, C. Bes, S. Troadec, M. Cerutti, and T. Chardes. 2007. Recombinant anti-CD4 antibody 13B8.2 blocks membrane-proximal events by excluding the Zap70 molecule and downstream targets SLP-76, PLC gamma 1, and Vav-1 from the CD4-segregated Brij 98 detergent-resistant raft domains. *J. Immunol.* 179:409–420.

Cheson, B.D. 2006. Monoclonal antibody therapy of chronic lymphocytic leukemia. *Cancer Immunol. Immunother.* 55:188–196.

Cho, H.S., K. Mason, K.X. Ramyar, A.M. Stanley, S.B. Gabelli, D.W. Denney, Jr., and D.J. Leahy. 2003. Structure of the extracellular region of HER2 alone and in complex with the Herceptin Fab. *Nature* 421:756–760.

Chusid, M.J., D.C. Dale, B.C. West, and S.M. Wolff. 1975. The hypereosinophilic syndrome: Analysis of fourteen cases with review of the literature. *Medicine (Baltimore)* 54:1–27.

Cobleigh, M.A., C.L. Vogel, D. Tripathy, N.J. Robert, S. Scholl, L. Fehrenbacher, J.M. Wolter, V. Paton, S. Shak, G. Lieberman, and D.J. Slamon. 1999. Multinational study of the efficacy and safety of humanized anti-HER2 monoclonal antibody in women who have HER2-overexpressing metastatic breast cancer that has progressed after chemotherapy for metastatic disease. *J. Clin. Oncol.* 17:2639–2648.

Cohen, D.J., A.I. Benvenisty, J. Cianci, and M.A. Hardy. 1989. OKT3 prophylaxis in cadaveric kidney transplant recipients with delayed graft function. *Am. J. Kidney Dis.* 14:19–27.

Cohenuram, M., and M.W. Saif. 2007. Panitumumab the first fully human monoclonal antibody: From the bench to the clinic. *Anticancer Drugs* 18:7–15.

Coiffier, B., E. Lepage, J. Briere, R. Herbrecht, H. Tilly, R. Bouabdallah, P. Morel, E. Van Den Neste, G. Salles, P. Gaulard. et al. 2002. CHOP chemotherapy plus rituximab compared with CHOP alone in elderly patients with diffuse large-B-cell lymphoma. *N. Engl. J. Med.* 346:235–242.

Coiffier, B., S. Lepretre, L.M. Pedersen, O. Gadeberg, H. Fredriksen, M.H. van Oers, J. Wooldridge, J. Kloczko, J. Holowiecki, A. Hellmann, et al. 2008. Safety and efficacy of ofatumumab, a fully human monoclonal anti-CD20 antibody, in patients with relapsed or refractory B-cell chronic lymphocytic leukemia: A phase 1-2 study. *Blood* 111:1094–1100.

Coller, B.S. 1995. Blockade of platelet GPIIb/IIIa receptors as an antithrombotic strategy. *Circulation* 92:2373–2380.

Corixa Corp. and GlaxoSmithKline (22-Dec-2004). Bexxar (Tositumomab and Iodine 131 Tositumomab): Label and Prescribing Information.

Cragg, M.S., and M.J. Glennie. 2004. Antibody specificity controls in vivo effector mechanisms of anti-CD20 reagents. *Blood* 103:2738–2743.

Cragg, M.S., S.M. Morgan, H.T. Chan, B.P. Morgan, A.V. Filatov, P.W. Johnson, R.R. French, and M.J. Glennie. 2003. Complement-mediated lysis by anti-CD20 mAb correlates with segregation into lipid rafts. *Blood* 101:1045–1052.

Crocker, P.R., J.C. Paulson, and A. Varki. 2007. Siglecs and their roles in the immune system. *Nature Rev. Immunol.* 7:255–266.

Crowe, J.S., V.S. Hall, M.A. Smith, H.J. Cooper, and J.P. Tite. 1992. Humanized monoclonal antibody CAMPATH-1H: Myeloma cell expression of genomic constructs, nucleotide sequence of cDNA constructs and comparison of effector mechanisms of myeloma and Chinese hamster ovary cell-derived material. *Clin. Exp. Immunol.* 87:105–110.

Cunningham, D., Y. Humblet, S. Siena, D. Khayat, H. Bleiberg, A. Santoro, D. Bets, M. Mueser, A. Harstrick, C. Verslype, et al. 2004. Cetuximab monotherapy and cetuximab plus irinotecan in irinotecan-refractory metastatic colorectal cancer. *N. Engl. J. Med.* 351:337–345.

Cummings S.R., M.R. McClung, C. Christiansen, E. Siris, S. Adami, S. Kutilek, I.R. Reid, J.R. Zanchetta, J. San Martin, C. Libanati, S. Siddhanti, A. Wang, P.D Delmas. 2008. A phase III study of the effects of denosumab on vertebral, nonvertebral, and hip fracture in women with osteoporosis: Results from the FREEDOM trial. *J. Bone Miner. Res.* 2(Suppl):S80 (abstr 1286).

Cuthbert, R.J., G.L. Phillips, M.J. Barnett, S.H. Nantel, D.E. Reece, J.D. Shepherd, and H.G. Klingemann. 1992. Anti-interleukin-2 receptor monoclonal antibody (BT 563) in the treatment of severe acute GVHD refractory to systemic corticosteroid therapy. *Bone Marrow Transplant* 10:451–455.

Cvetkovic, R.S., and C.M. Perry. 2006. Rituximab: A review of its use in non-Hodgkin's lymphoma and chronic lymphocytic leukaemia. *Drugs* 66:791–820.

Czuczman, M.S., A. Thall, T.E. Witzig, J.M. Vose, A. Younes, C. Emmanouilides, T.P. Miller, J.O. Moore, J.P. Leonard, L.I. Gordon, et al. 2005. Phase I/II study of galiximab, an anti-CD80 antibody, for relapsed or refractory follicular lymphoma. *J. Clin. Oncol.* 23:4390–4398.

Czuczman, M.S., R. Weaver, B. Alkuzweny, J. Berlfein, and A.J. Grillo-Lopez. 2004. Prolonged clinical and molecular remission in patients with low-grade or follicular non-Hodgkin's lymphoma treated with rituximab plus CHOP chemotherapy: 9-year follow-up. *J. Clin. Oncol.* 22:4711–4716.

Datamonitor. 2007. Monoclonal Antibodies Report, Part I. New York: Datamonitor.

Davies, A.J. 2004. Tositumomab and iodine [131I] tositumomab in the management of follicular lymphoma. An oncologist's view. *Quart. J. Nuclear Med. Mol. Imaging* 48:305–316.

Davis, L.A. 2004. Omalizumab: A novel therapy for allergic asthma. *Ann. Pharmacother.* 38:1236–1242.

Davis, T.A., M.S. Kaminski, J.P. Leonard, F.J. Hsu, M. Wilkinson, A. Zelenetz, R.L. Wahl, S. Kroll, M. Coleman, M. Goris, et al. 2004. The radioisotope contributes significantly to the activity of radioimmunotherapy. *Clin. Cancer Res.* 10:7792–7798.

Davis, T.A., C.A. White, A.J. Grillo-Lopez, W.S. Velasquez, B. Link, D.G. Maloney, R.O. Dillman, M.E. Williams, A. Mohrbacher, R. Weaver, et al. 1999. Single-agent monoclonal antibody efficacy in bulky non-Hodgkin's lymphoma: Results of a phase II trial of rituximab. *J. Clin. Oncol.* 17:1851–1857.

de Belder, M.A., and A.G. Sutton. 1998. Abciximab (Reopro): A clinically effective glycoprotein IIb/IIIa receptor blocker. *Expert Opin. Investig. Drugs* 7:1701–1717.

Dedon, P.C., A.A. Salzberg, and J. Xu. 1993. Exclusive production of bistranded DNA damage by calicheamicin. *Biochemistry* 32:3617–3622.

Delespesse, G., U. Suter, D. Mossalayi, B. Bettler, M. Sarfati, H. Hofstetter, E. Kilcherr, P. Debre, and A. Dalloul. 1991. Expression, structure, and function of the CD23 antigen. *Adv. Immunol.* 49:149–191.

Demidem, A., T. Lam, S. Alas, K. Hariharan, N. Hanna, and B. Bonavida. 1997. Chimeric anti-CD20 (IDEC-C2B8) monoclonal antibody sensitizes a B cell lymphoma cell line to cell killing by cytotoxic drugs. *Cancer Biother. Radiopharm.* 12:177–186.

De Vita, S., F. Zaja, S. Sacco, A. De Candia, R. Fanin, and G. Ferraccioli. 2002. Efficacy of selective B cell blockade in the treatment of rheumatoid arthritis: Evidence for a pathogenetic role of B cells. *Arthritis Rheum.* 46:2029–2033.

Di Fiore, P.P., J.H. Pierce, M.H. Kraus, O. Segatto, C.R. King, and S.A. Aaronson. 1987. erbB-2 is a potent oncogene when overexpressed in NIH/3T3 cells. *Science* 237:178–182.

Dinndorf, P.A., R.G. Andrews, D. Benjamin, D. Ridgway, L. Wolff, and I.D. Bernstein. 1986. Expression of normal myeloid-associated antigens by acute leukemia cells. *Blood* 67:1048–1053.

Divgi, C.R., S. Welt, M. Kris, F.X. Real, S.D. Yeh, R. Gralla, B. Merchant, S. Schweighart, M. Unger, S.M. Larson, et al. 1991. Phase I and imaging trial of indium 111-labeled anti-epidermal growth factor receptor monoclonal antibody 225 in patients with squamous cell lung carcinoma. *J. Natl. Cancer Inst.* 83:97–104.

Drevs, J., I. Hofmann, H. Hugenschmidt, C. Wittig, H. Madjar, M. Muller, J. Wood, G. Martiny-Baron, C. Unger, and D. Marme. 2000. Effects of PTK787/ZK 222584, a specific inhibitor of vascular endothelial growth factor receptor tyrosine kinases, on primary tumor, metastasis, vessel density, and blood flow in a murine renal cell carcinoma model. *Cancer Res.* 60:4819–4824.

Du, J., H. Wang, C. Zhong, B. Peng, M. Zhang, B. Li, S. Huo, Y. Guo, and J. Ding. 2007. Structural basis for recognition of CD20 by therapeutic antibody Rituximab. *J. Biol. Chem.* 282:15073–15080.

Dyer, M.J., G. Hale, F.G. Hayhoe, and H. Waldmann. 1989. Effects of CAMPATH-1 antibodies in vivo in patients with lymphoid malignancies: Influence of antibody isotype. *Blood* 73:1431–1439.

Edwards, J.C., and G. Cambridge. 2001. Sustained improvement in rheumatoid arthritis following a protocol designed to deplete B lymphocytes. *Rheumatology (Oxford)* 40:205–211.

Einfeld, D.A., J.P. Brown, M.A. Valentine, E.A. Clark, and J.A. Ledbetter. 1988. Molecular cloning of the human B cell CD20 receptor predicts a hydrophobic protein with multiple transmembrane domains. *EMBO J.* 7:711–717.

Elsner, J., R. Hochstetter, K. Spiekermann, and A. Kapp. 1996. Surface and mRNA expression of the CD52 antigen by human eosinophils but not by neutrophils. *Blood* 88:4684–4693.

Esteva, F.J., V. Valero, D. Booser, L.T. Guerra, J.L. Murray, L. Pusztai, M. Cristofanilli, B. Arun, B. Esmaeli, H.A. Fritsche, et al. 2002. Phase II study of weekly docetaxel and trastuzumab for patients with HER-2-overexpressing metastatic breast cancer. *J. Clin. Oncol.* 20:1800–1808.

Fan, Z., Y. Lu, X. Wu, and J. Mendelsohn. 1994. Antibody-induced epidermal growth factor receptor dimerization mediates inhibition of autocrine proliferation of A431 squamous carcinoma cells. *J. Biol. Chem.* 269:27595–27602.

Faulds, D., and E.M. Sorkin. 1994. Abciximab (c7E3 Fab). A review of its pharmacology and therapeutic potential in ischaemic heart disease. *Drugs* 48:583–598.

Fendly, B.M., M. Winget, R.M. Hudziak, M.T. Lipari, M.A. Napier, and A. Ullrich. 1990. Characterization of murine monoclonal antibodies reactive to either the human epidermal growth factor receptor or HER2/neu gene product. *Cancer Res.* 50:1550–1558.

Ferrajoli, A., S. O'Brien, and M.J. Keating. 2001. Alemtuzumab: A novel monoclonal antibody. *Expert Opin. Biol. Ther.* 1:1059–1065.

Ferrara, J.L., and H.J. Deeg. 1991. Graft-versus-host disease. *N. Engl. J. Med.* 324:667–674.

Ferrara, N. 2005. VEGF as a therapeutic target in cancer. *Oncology* 69(Suppl 3):11–16.

Ferrara, N., and K. Alitalo. 1999. Clinical applications of angiogenic growth factors and their inhibitors. *Nature Med.* 5:1359–1364.

Ferrara, N., and T. Davis-Smyth. 1997. The biology of vascular endothelial growth factor. *Endocr. Rev.* 18:4–25.

Ferrara, N., H.P. Gerber, and J. LeCouter. 2003. The biology of VEGF and its receptors. *Nature Med.* 9:669–676.

Ferrara, N., K.A. Houck, L.B. Jakeman, J. Winer, and D.W. Leung. 1991. The vascular endothelial growth factor family of polypeptides. *J. Cell. Biochem.* 47:211–218.

Feugier, P., A. Van Hoof, C. Sebban, P. Solal-Celigny, R. Bouabdallah, C. Ferme, B. Christian, E. Lepage, H. Tilly, F. Morschhauser, et al. 2005. Long-term results of the R-CHOP study in the treatment of elderly patients with diffuse large B-cell lymphoma: A study by the Groupe d'Etude des Lymphomes de l'Adulte. *J. Clin. Oncol.* 23:4117–4126.

Fournier, S., M. Rubio, G. Delespesse, and M. Sarfati. 1994. Role for low-affinity receptor for IgE (CD23) in normal and leukemic B-cell proliferation. *Blood* 84:1881–1886.

Franklin, M.C., K.D. Carey, F.F. Vajdos, D.J. Leahy, A.M. de Vos, and M.X. Sliwkowski. 2004. Insights into ErbB signaling from the structure of the ErbB2-pertuzumab complex. *Cancer Cell* 5:317–328.

Fuh, G., P. Wu, W.C. Liang, M. Ultsch, C.V. Lee, B. Moffat, and C. Wiesmann. 2006. Structure-function studies of two synthetic anti-vascular endothelial growth factor Fabs and comparison with the Avastin Fab. *J Biol Chem* 281:6625–6631.

Fukumura, D., R. Xavier, T. Sugiura, Y. Chen, E.C. Park, N. Lu, M. Selig, G. Nielsen, T. Taksir, R.K. Jain, and B. Seed. 1998. Tumor induction of VEGF promoter activity in stromal cells. *Cell* 94:715–725.

Gabriel, H.M., and E.I. Oliveira. 2006. Role of abciximab in the treatment of coronary artery disease. *Expert Opin. Biol. Ther.* 6:935–942.

Galizia, G., E. Lieto, F. Ferraraccio, F. De Vita, P. Castellano, M. Orditura, V. Imperatore, A. La Mura, G. La Manna, M. Pinto, et al. 2006. Prognostic significance of epidermal growth factor receptor expression in colon cancer patients undergoing curative surgery. *Ann. Surg. Oncol.* 13:823–835.

Gaudreault, J., D. Fei, J. Rusit, P. Suboc, and V. Shiu. 2005. Preclinical pharmacokinetics of Ranibizumab (rhuFabV2) after a single intravitreal administration. *Invest. Ophthalmol. Vis. Sci.* 46:726–733.

Gauvreau, G.M., A.B. Becker, L.P. Boulet, J. Chakir, R.B. Fick, W.L. Greene, K.J. Killian, P.M. O'Byrne, J.K. Reid, and D.W. Cockcroft. 2003. The effects of an anti-CD11a mAb, efalizumab, on allergen-induced airway responses and airway inflammation in subjects with atopic asthma. *J. Allergy Clin. Immunol.* 112:331–338.

Geletka, R.C., and E.W. St. Clair. 2005. Infliximab for the treatment of early rheumatoid arthritis. *Expert Opin. Biol. Ther.* 5:405–417.

Genentech, Inc. (06-Nov-2006). Herception (Trastuzumab): Label and Prescribing Information.

Genentech, Inc. (11-Oct-2006). Avastin (Bevacizumab): Label and Prescribing Information.

Genzyme Corporation and Bayer HealthCare Pharmaceuticals (19-Dec-2006). Campath (Alemtuzumab): Label and Prescribing Information.

Genzyme Corporation and Bayer HealthCare Pharmaceuticals (19-Sept-2007). Campath (Alemtuzumab): Highlights of prescribing information.

Gerber, H.P., J. Kowalski, D. Sherman, D.A. Eberhard, and N. Ferrara. 2000. Complete inhibition of rhabdomyosarcoma xenograft growth and neovascularization requires blockade of both tumor and host vascular endothelial growth factor. *Cancer Res.* 60:6253–6258.

Gill, G.N., T. Kawamoto, C. Cochet, A. Le, J.D. Sato, H. Masui, C. McLeod, and J. Mendelsohn. 1984. Monoclonal anti-epidermal growth factor receptor antibodies which are inhibitors of epidermal growth factor binding and antagonists of epidermal growth factor-stimulated tyrosine protein kinase activity. *J. Biol. Chem.* 259:7755–7760.

Ginaldi, L., M. De Martinis, E. Matutes, N. Farahat, R. Morilla, M.J. Dyer, and D. Catovsky. 1998. Levels of expression of CD52 in normal and leukemic B and T cells: Correlation with in vivo therapeutic responses to Campath-1H. *Leuk. Res.* 22:185–191.

Goldstein, N.I., M. Prewett, K. Zuklys, P. Rockwell, and J. Mendelsohn. 1995. Biological efficacy of a chimeric antibody to the epidermal growth factor receptor in a human tumor xenograft model. *Clin. Cancer Res.* 1:1311–1318.

Gordon, M.S., K. Margolin, M. Talpaz, G.W. Sledge, Jr., E. Holmgren, R. Benjamin, S. Stalter, S. Shak, and D. Adelman. 2001. Phase I safety and pharmacokinetic study of recombinant human anti-vascular endothelial growth factor in patients with advanced cancer. *J. Clin. Oncol.* 19:843–850.

Gottlieb, A.B., K.D. Cooper, T.S. McCormick, E. Toichi, D.E. Everitt, B. Frederick, Y. Zhu, C.E. Pendley, M.A. Graham, and M.A. Mascelli. 2007. A phase 1, double-blind, placebo-controlled study evaluating single subcutaneous administrations of a human interleukin-12/23 monoclonal antibody in subjects with plaque psoriasis. *Curr. Med. Res. Opin.* 23:1081–1092.

Groothuis, J.R., E.A. Simoes, M.J. Levin, C.B. Hall, C.E. Long, W.J. Rodriguez, J. Arrobio, H.C. Meissner, D.R. Fulton, R.C. Welliver, et al. 1993. Prophylactic administration of respiratory syncytial virus immune globulin to high-risk infants and young children. The Respiratory Syncytial Virus Immune Globulin Study Group. *N. Engl. J. Med.* 329:1524–1530.

Hagenbeek, A., O. Gadeberg, P. Johnson, L.M. Pedersen, J. Walewski, A. Hellmann, B.K. Link, T. Robak, M. Wojtukiewicz, M. Pfreundschuh, et al. 2008. First clinical use of ofatumumab, a novel fully human anti-CD20 monoclonal antibody in relapsed or refractory follicular lymphoma: Results of a phase I/II trial. *Blood.* 111:5486–5495.

Hale, G., S.P. Cobbold, H. Waldmann, G. Easter, P. Matejtschuk, and R.R. Coombs. 1987. Isolation of low-frequency class-switch variants from rat hybrid myelomas. *J. Immunol. Methods* 103:59–67.

Hale, G., M.J. Dyer, M.R. Clark, J.M. Phillips, R. Marcus, L. Riechmann, G. Winter, and H. Waldmann. 1988. Remission induction in non-Hodgkin lymphoma with reshaped human monoclonal antibody CAMPATH-1H. *Lancet* 2:1394–1399.

Hale, G., T. Hoang, T. Prospero, S.M. Watt, and H. Waldmann. 1983. Removal of T cells from bone marrow for transplantation. Comparison of rat monoclonal anti-lymphocyte antibodies of different isotypes. *Mol. Biol. Med.* 1:305–319.

Hamann, P.R., L.M. Hinman, C.F. Beyer, D. Lindh, J. Upeslacis, D.A. Flowers, and I. Bernstein. 2002a. An anti-CD33 antibody-calicheamicin conjugate for treatment of acute myeloid leukemia. Choice of linker. *Bioconjug. Chem.* 13:40–46.

Hamann, P.R., L.M. Hinman, I. Hollander, C.F. Beyer, D. Lindh, R. Holcomb, W. Hallett, H.R. Tsou, J. Upeslacis, D. Shochat, et al. 2002b. Gemtuzumab ozogamicin, a potent and selective anti-CD33 antibody-calicheamicin conjugate for treatment of acute myeloid leukemia. *Bioconjug. Chem.* 13:47–58.

Hamdy, N.A. 2008. Denosumab: RANKL inhibition in the management of bone loss. *Drugs Today (Barc)* 44:7–21.

Han, K., Y. Kim, J. Lee, J. Lim, K.Y. Lee, C.S. Kang, W.I. Kim, B.K. Kim, S.I. Shim, and S.M. Kim. 1999. Human basophils express CD22 without expression of CD19. *Cytometry* 37:178–183.

Hart, T.K., R.M. Cook, P. Zia-Amirhosseini, E. Minthorn, T.S. Sellers, B.E. Maleeff, S. Eustis, L.W. Schwartz, P. Tsui, E.R. Appelbaum, et al. 2001. Preclinical efficacy and safety of mepolizumab (SB-240563), a humanized monoclonal antibody to IL-5, in cynomolgus monkeys. *J. Allergy Clin. Immunol.* 108:250–257.

Herbelin, C., J.L. Stephan, J. Donadieu, F. Le Deist, E. Racadot, J. Wijdenes, and A. Fischer. 1994. Treatment of steroid-resistant acute graft-versus-host disease with an anti-IL-2-receptor monoclonal antibody (BT 563) in children who received T cell-depleted, partially matched, related bone marrow transplants. *Bone Marrow Transplant* 13:563–569.

Hertenstein, B., M. Stefanic, M. Sandherr, D. Bunjes, T. Mertens, and R. Arnold. 1994. Treatment of steroid-resistant acute graft-vs-host disease after allogeneic marrow transplantation with anti-interleukin-2 receptor antibody (BT563). *Transplant Proc.* 26:3114–3116.

Herve, P., J. Wijdenes, J.P. Bergerat, P. Bordigoni, N. Milpied, J.Y. Cahn, C. Clement, R. Beliard, B. Morel-Fourrier, E. Racadot, et al. 1990. Treatment of corticosteroid resistant acute graft-versus-host disease by in vivo administration of anti-interleukin-2 receptor monoclonal antibody (B-B10). *Blood* 75:1017–1023.

Hildreth, J.E., and J.T. August. 1985. The human lymphocyte function-associated (HLFA) antigen and a related macrophage differentiation antigen (HMac-1): Functional effects of subunit-specific monoclonal antibodies. *J. Immunol.* 134:3272–3280.

Hillmen, P., C. Hall, J.C. Marsh, M. Elebute, M.P. Bombara, B.E. Petro, M.J. Cullen, S.J. Richards, S.A. Rollins, C.F. Mojcik, and R.P. Rother. 2004. Effect of eculizumab on hemolysis and transfusion requirements in patients with paroxysmal nocturnal hemoglobinuria. *N. Engl. J. Med.* 350:552–559.

Houck, K.A., N. Ferrara, J. Winer, G. Cachianes, B. Li, and D.W. Leung. 1991. The vascular endothelial growth factor family: Identification of a fourth molecular species and characterization of alternative splicing of RNA. *Mol. Endocrinol.* 5:1806–1814.

Hudziak, R.M., G.D. Lewis, M. Winget, B.M. Fendly, H.M. Shepard, and A. Ullrich. 1989. p185HER2 monoclonal antibody has antiproliferative effects in vitro and sensitizes human breast tumor cells to tumor necrosis factor. *Mol. Cell. Biol.* 9:1165–1172.

Hudziak, R.M., J. Schlessinger, and A. Ullrich. 1987. Increased expression of the putative growth factor receptor p185HER2 causes transformation and tumorigenesis of NIH 3T3 cells. *Proc. Natl. Acad. Sci. U.S.A.* 84:7159–7163.

Hurwitz, H., L. Fehrenbacher, W. Novotny, T. Cartwright, J. Hainsworth, W. Heim, J. Berlin, A. Baron, S. Griffing, E. Holmgren, et al. 2004. Bevacizumab plus irinotecan, fluorouracil, and leucovorin for metastatic colorectal cancer. *N. Engl. J. Med.* 350:2335–2342.

Hutchinson, M. 2007. Natalizumab: A new treatment for relapsing remitting multiple sclerosis. *Ther. Clin. Risk Manag.* 3:259–268.

ImClone Systems, Inc. and Bristol-Myers Squibb, Co. (01Mar2006). Erbitux (Cetuximab): Label and Prescribing Information.

Ito, H., M. Takazoe, Y. Fukuda, T. Hibi, K. Kusugami, A. Andoh, T. Matsumoto, T. Yamamura, J. Azuma, N. Nishimoto, et al. 2004. A pilot randomized trial of a human anti-interleukin-6 receptor monoclonal antibody in active Crohn's disease. *Gastroenterology* 126:989–996; discussion 947.

Izumi, Y., L. Xu, E. di Tomaso, D. Fukumura, and R.K. Jain. 2002. Tumour biology: Herceptin acts as an anti-angiogenic cocktail. *Nature* 416:279–280.

Jakobovits, A., R.G. Amado, X. Yang, L. Roskos, and G. Schwab. 2007. From XenoMouse technology to panitumumab, the first fully human antibody product from transgenic mice. *Nature Biotechnol.* 25:1134–1143.

Johnson, D.H., L. Fehrenbacher, W.F. Novotny, R.S. Herbst, J.J. Nemunaitis, D.M. Jablons, C.J. Langer, R.F. DeVore III, J. Gaudreault, L.A. Damico, et al. 2004. Randomized phase II trial comparing bevacizumab plus carboplatin and paclitaxel with carboplatin and paclitaxel alone in previously untreated locally advanced or metastatic non-small-cell lung cancer. *J. Clin. Oncol.* 22:2184–2191.

Johnson, P., and M. Glennie. 2003. The mechanisms of action of rituximab in the elimination of tumor cells. *Semin. Oncol.* 30:3–8.

Johnson, R.J., and P. Hillmen. 2002. Paroxysmal nocturnal haemoglobinuria: Nature's gene therapy? *Mol. Pathol.* 55:145–152.

Johnson, S., C. Oliver, G.A. Prince, V.G. Hemming, D.S. Pfarr, S.C. Wang, M. Dormitzer, J. O'Grady, S. Koenig, J.K. Tamura, et al. 1997. Development of a humanized monoclonal antibody (MEDI-493) with potent in vitro and in vivo activity against respiratory syncytial virus. *J. Infect. Dis.* 176:1215–1224.

Johnston, P.B., C. Bondly, and I.N. Micallef. 2006. Ibritumomab tiuxetan for non-Hodgkin's lymphoma. *Expert Rev. Anticancer Ther.* 6:861–869.

Kabbinavar, F., H.I. Hurwitz, L. Fehrenbacher, N.J. Meropol, W.F. Novotny, G. Lieberman, S. Griffing, and E. Bergsland. 2003. Phase II, randomized trial comparing bevacizumab plus fluorouracil (FU)/leucovorin (LV) with FU/LV alone in patients with metastatic colorectal cancer. *J. Clin. Oncol.* 21:60–65.

Kaminski, M.S., K.R. Zasadny, I.R. Francis, A.W. Milik, C.W. Ross, S.D. Moon, S.M. Crawford, J.M. Burgess, N.A. Petry, G.M. Butchko, et al. 1993. Radioimmunotherapy of B-cell lymphoma with [131I]anti-B1 (anti-CD20) antibody. *N. Engl. J. Med.* 329:459–465.

Kaminski, M.S., A.D. Zelenetz, O.W. Press, M. Saleh, J. Leonard, L. Fehrenbacher, T.A. Lister, R.J. Stagg, G.F. Tidmarsh, S. Kroll, et al. 2001. Pivotal study of iodine I 131 tositumomab for chemotherapy-refractory low-grade or transformed low-grade B-cell non-Hodgkin's lymphomas. *J. Clin. Oncol.* 19:3918–3928.

Kaushik, V.V., and R.J. Moots. 2005. CDP-870 (certolizumab) in rheumatoid arthritis. *Expert Opin. Biol. Ther.* 5:601–606.

Kawamoto, T., J.D. Sato, A. Le, J. Polikoff, G.H. Sato, and J. Mendelsohn. 1983. Growth stimulation of A431 cells by epidermal growth factor: Identification of high-affinity receptors for epidermal growth factor by an anti-receptor monoclonal antibody. *Proc. Natl. Acad. Sci. U.S.A.* 80:1337–1341.

Kay, J., E.L. Matteson, B. Dasgupta, P. Nash, P. Durez, S. Hall, E.C. Hsia, J. Han, C. Wagner, Z. Xu, et al. 2008. Golimumab in patients with active rheumatoid arthritis despite treatment with methotrexate: A randomized, double-blind, placebo-controlled, dose-ranging study. *Arthritis Rheum.* 58:964–975.

Kenneth, T.E., and P.J. Kertes. 2006. Ranibizumab in neovascular age-related macular degeneration. *Clin. Interv. Aging* 1:451–466.

Kennett, R.H. 2003. Technology evaluation: WX-G250, Wilex/Ludwig Institute for Cancer Research. *Curr. Opin. Mol. Ther.* 5:70–75.

Kerr, D.J. 2004. Targeting angiogenesis in cancer: Clinical development of bevacizumab. *Nature Clin. Pract. Oncol.* 1:39–43.

Keystone, E.C., M.C. Genovese, L. Klareskog, E.C. Hsia, S.T. Hall, P.C. Miranda, J. Pazdur, S-C. Bae, W. Palmer, J. Zrubek, M. Wiekowski, S. Viswanathan, Z. Wu, and M.U. Rahman. 2009. Golimumab, a human antibody to tumour necrosis factor as given by monthly subcutaneous injections, in active rheumatoid arthritis despite methotrexate therapy: The GO-FORWARD Study. *Ann. Rheum. Dis.* 68:789–796.

Kies, M.S., and P.M. Harari. 2002. Cetuximab (Imclone/Merck/Bristol-Myers Squibb). *Curr. Opin. Investig. Drugs* 3:1092–1100.

Kim, K.J., B. Li, K. Houck, J. Winer, and N. Ferrara. 1992. The vascular endothelial growth factor proteins: Identification of biologically relevant regions by neutralizing monoclonal antibodies. *Growth Factors* 7:53–64.

Kim, K.J., B. Li, J. Winer, M. Armanini, N. Gillett, H.S. Phillips, and N. Ferrara. 1993. Inhibition of vascular endothelial growth factor-induced angiogenesis suppresses tumour growth in vivo. *Nature* 362:841–844.

Kim, Y.H., M. Duvic, E. Obitz, R. Gniadecki, L. Iversen, A. Osterborg, S. Whittaker, T.M. Illidge, T. Schwarz, R. Kaufmann, et al. 2007. Clinical efficacy of zanolimumab (HuMax-CD4): Two phase 2 studies in refractory cutaneous T-cell lymphoma. *Blood* 109:4655–4662.

Kimura, H., K. Sakai, T. Arao, T. Shimoyama, T. Tamura, and K. Nishio. 2007. Antibody-dependent cellular cytotoxicity of cetuximab against tumor cells with wild-type or mutant epidermal growth factor receptor. *Cancer Sci.* 98:1275–1280.

Knight, D.M., H. Trinh, J. Le, S. Siegel, D. Shealy, M. McDonough, B. Scallon, M.A. Moore, J. Vilcek, P. Daddona, et al. 1993. Construction and initial characterization of a mouse-human chimeric anti-TNF antibody. *Mol. Immunol.* 30:1443–1453.

Knox, S.J., M.L. Goris, K. Trisler, R. Negrin, T. Davis, T.M. Liles, A. Grillo-Lopez, P. Chinn, C. Varns, S.C. Ning, et al. 1996. Yttrium-90-labeled anti-CD20 monoclonal antibody therapy of recurrent B-cell lymphoma. *Clin. Cancer Res.* 2:457–470.

Kovarik, J.M., E. Rawlings, P. Sweny, O. Fernando, R. Moore, P.J. Griffin, P. Fauchald, D. Albrechtsen, G. Sodal, K. Nordal, and P.L. Amlot. 1996. Pharmacokinetics and immunodynamics of chimeric IL-2 receptor monoclonal antibody SDZ CHI 621 in renal allograft recipients. *Transpl. Int.* 9(Suppl 1):S32–S33.

Krasner, C., and R.M. Joyce. 2001. Zevalin: 90-yttrium labeled anti-CD20 (ibritumomab tiuxetan), a new treatment for non-Hodgkin's lymphoma. *Curr. Pharm. Biotechnol.* 2:341–349.

Krilov, L.R. 2002. Palivizumab in the prevention of respiratory syncytial virus disease. *Expert Opin. Biol. Ther.* 2:763–769.

Krueger, G.G., R.G. Langley, C. Leonardi, N. Yeilding, C. Guzzo, Y. Wang, L.T. Dooley, and M. Lebwohl. 2007. A human interleukin-12/23 monoclonal antibody for the treatment of psoriasis. *N. Engl. J. Med.* 356:580–592.

Lajaunias, F., J.M. Dayer, and C. Chizzolini. 2005. Constitutive repressor activity of CD33 on human monocytes requires sialic acid recognition and phosphoinositide 3-kinase-mediated intracellular signaling. *Eur. J. Immunol.* 35:243–251.

Lane, H.A., I. Beuvink, A.B. Motoyama, J.M. Daly, R.M. Neve, and N.E. Hynes. 2000. ErbB2 potentiates breast tumor proliferation through modulation of p27(Kip1)-Cdk2 complex formation: Receptor overexpression does not determine growth dependency. *Mol. Cell. Biol.* 20:3210–3223.

Langer, L.F., T.M. Clay, and M.A. Morse. 2007. Update on anti-CTLA-4 antibodies in clinical trials. *Expert Opin. Biol. Ther.* 7:1245–1256.

Larson, R.A., E.L. Sievers, E.A. Stadtmauer, B. Lowenberg, E.H. Estey, H. Dombret, M. Theobald, D. Voliotis, J.M. Bennett, M. Richie, et al. 2005. Final report of the efficacy and safety of gemtuzumab ozogamicin (Mylotarg) in patients with CD33-positive acute myeloid leukemia in first recurrence. *Cancer* 104:1442–1452.

Leach, D.R., M.F. Krummel, and J.P. Allison. 1996. Enhancement of antitumor immunity by CTLA-4 blockade. *Science* 271:1734–1736.

Leandro, M.J., J.C. Edwards, and G. Cambridge. 2002. Clinical outcome in 22 patients with rheumatoid arthritis treated with B lymphocyte depletion. *Ann. Rheum. Dis.* 61:883–888.

Lebwohl, M., S.K. Tyring, T.K. Hamilton, D. Toth, S. Glazer, N.H. Tawfik, P. Walicke, W. Dummer, X. Wang, M.R. Garovoy, and D. Pariser. 2003. A novel targeted T-cell modulator, efalizumab, for plaque psoriasis. *N. Engl. J. Med.* 349:2004–2013.

Lee, M.D., T.M. Dunne, M.M. Siegel, C.C. Chang, G.O. Morton, and D.B. Borders. 1987. Calicheamicins, a novel family of antitumor antibiotics. 1. Chemistry and partial structure of calicheamicin $g_1{}^{I}$. *J. Am. Chem. Soc.* 109:3464–3466.

Leger, O.J., T.A. Yednock, L. Tanner, H.C. Horner, D.K. Hines, S. Keen, J. Saldanha, S.T. Jones, L.C. Fritz, and M.M. Bendig. 1997. Humanization of a mouse antibody against human alpha-4 integrin: A potential therapeutic for the treatment of multiple sclerosis. *Hum. Antibodies* 8:3–16.

Legrand, O., J.Y. Perrot, M. Baudard, A. Cordier, R. Lautier, G. Simonin, R. Zittoun, N. Casadevall, and J.P. Marie. 2000. The immunophenotype of 177 adults with acute myeloid leukemia: Proposal of a prognostic score. *Blood* 96:870–877.

Leonard, J.P., M. Coleman, J. Ketas, M. Ashe, J.M. Fiore, R.R. Furman, R. Niesvizky, T. Shore, A. Chadburn, H. Horne, et al. 2005. Combination antibody therapy with epratuzumab and rituximab in relapsed or refractory non-Hodgkin's lymphoma. *J. Clin. Oncol.* 23:5044–5051.

Leonard, J.P., J.W. Friedberg, A. Younes, D. Fisher, L.I. Gordon, J. Moore, M. Czuczman, T. Miller, P. Stiff, B.D. Cheson, et al. 2007. A phase I/II study of galiximab (an anti-CD80 monoclonal antibody) in combination with rituximab for relapsed or refractory, follicular lymphoma. *Ann. Oncol.* 18:1216–1223.

Leonard, J.P., K.E. Waldburger, and S.J. Goldman. 1995. Prevention of experimental autoimmune encephalomyelitis by antibodies against interleukin 12. *J. Exp. Med.* 181:381–386.

Leyland-Jones, B., K. Gelmon, J.P. Ayoub, A. Arnold, S. Verma, R. Dias, and P. Ghahramani. 2003. Pharmacokinetics, safety, and efficacy of trastuzumab administered every three weeks in combination with paclitaxel. *J. Clin. Oncol.* 21:3965–3971.

Li, G., K. Passebosc-Faure, C. Lambert, A. Gentil-Perret, F. Blanc, E. Oosterwijk, J.F. Mosnier, C. Genin, and J. Tostain. 2001. The expression of G250/mn/CA9 antigen by flow cytometry: Its possible implication for detection of micrometastatic renal cancer cells. *Clin. Cancer Res.* 7:89–92.

Li, S., K.R. Schmitz, P.D. Jeffrey, J.J. Wiltzius, P. Kussie, and K.M. Ferguson. 2005. Structural basis for inhibition of the epidermal growth factor receptor by cetuximab. *Cancer Cell* 7:301–311.

Liang, W.C., X. Wu, F.V. Peale, C.V. Lee, Y.G. Meng, J. Gutierrez, L. Fu, A.K. Malik, H.P. Gerber, N. Ferrara, and G. Fuh. 2006. Cross-species vascular endothelial growth factor (VEGF)-blocking antibodies completely inhibit the growth of human tumor xenografts and measure the contribution of stromal VEGF. *J. Biol. Chem.* 281:951–961.

Liu, A.Y., R.R. Robinson, E.D. Murray, Jr., J.A. Ledbetter, I. Hellstrom, and K.E. Hellstrom. 1987. Production of a mouse-human chimeric monoclonal antibody to CD20 with potent Fc-dependent biologic activity. *J. Immunol.* 139:3521–3526.

Maloney, D.G., A.J. Grillo-Lopez, C.A. White, D. Bodkin, R.J. Schilder, J.A. Neidhart, N. Janakiraman, K.A. Foon, T.M. Liles, B.K. Dallaire, et al. 1997. IDEC-C2B8 (Rituximab) anti-CD20 monoclonal antibody therapy in patients with relapsed low-grade non-Hodgkin's lymphoma. *Blood* 90:2188–2195.

Maloney, D.G., T.M. Liles, D.K. Czerwinski, C. Waldichuk, J. Rosenberg, A. Grillo-Lopez, and R. Levy. 1994. Phase I clinical trial using escalating single-dose infusion of chimeric anti-CD20 monoclonal antibody (IDEC-C2B8) in patients with recurrent B-cell lymphoma. *Blood* 84:2457–2466.

Margolin, K., M.S. Gordon, E. Holmgren, J. Gaudreault, W. Novotny, G. Fyfe, D. Adelman, S. Stalter, and J. Breed. 2001. Phase Ib trial of intravenous recombinant humanized monoclonal antibody to vascular endothelial growth factor in combination with chemotherapy in patients with advanced cancer: pharmacologic and long-term safety data. *J. Clin. Oncol.* 19:851–856.

Marty, M., F. Cognetti, D. Maraninchi, R. Snyder, L. Mauriac, M. Tubiana-Hulin, S. Chan, D. Grimes, A. Anton, A. Lluch, et al. 2005. Randomized phase II trial of the efficacy and safety of trastuzumab combined with docetaxel in patients with human epidermal growth factor receptor 2-positive metastatic breast cancer administered as first-line treatment: The M77001 study group. *J. Clin. Oncol.* 23:4265–4274.

Mascelli, M.A., E.T. Lance, L. Damaraju, C.L. Wagner, H.F. Weisman, and R.E. Jordan. 1998. Pharmacodynamic profile of short-term abciximab treatment demonstrates prolonged platelet inhibition with gradual recovery from GP IIb/IIIa receptor blockade. *Circulation* 97:1680–1688.

Masui, H., T. Kawamoto, J.D. Sato, B. Wolf, G. Sato, and J. Mendelsohn. 1984. Growth inhibition of human tumor cells in athymic mice by anti-epidermal growth factor receptor monoclonal antibodies. *Cancer Res.* 44:1002–1007.

Matthews, R.C., and J.P. Burnie. 2005. Human recombinant antibody to HSP90: A natural partner in combination therapy. *Curr. Mol. Med.* 5:403–411.

Matthews, R.C., J.P. Burnie, and S. Tabaqchali. 1984. Immunoblot analysis of the serological response in systemic candidosis. *Lancet* 2:1415–1418.

Matthews, R.C., G. Rigg, S. Hodgetts, T. Carter, C. Chapman, C. Gregory, C. Illidge, and J. Burnie. 2003. Preclinical assessment of the efficacy of mycograb, a human recombinant antibody against fungal HSP90. *Antimicrob. Agents Chemother.* 47:2208–2216.

Matthys, P., K. Vermeire, T. Mitera, H. Heremans, S. Huang, and A. Billiau. 1998. Anti-IL-12 antibody prevents the development and progression of collagen-induced arthritis in IFN-gamma receptor-deficient mice. *Eur. J. Immunol.* 28:2143–2151.

McClung, M.R., E.M. Lewiecki, S.B. Cohen, M.A. Bolognese, G.C. Woodson, A.H. Moffett, M. Peacock, P.D. Miller, S.N. Lederman, C.H. Chesnut, et al. 2006. Denosumab in postmenopausal women with low bone mineral density. *N. Engl. J. Med.* 354:821–831.

McLaughlin, P., A.J. Grillo-Lopez, B.K. Link, R. Levy, M.S. Czuczman, M.E. Williams, M.R. Heyman, I. Bence-Bruckler, C.A. White, F. Cabanillas, et al. 1998. Rituximab chimeric anti-CD20 monoclonal antibody therapy for relapsed indolent lymphoma: Half of patients respond to a four-dose treatment program. *J. Clin. Oncol.* 16:2825–2833.

McNeil, M.M., S.L. Nash, R.A. Hajjeh, M.A. Phelan, L.A. Conn, B.D. Plikaytis, and D.W. Warnock. 2001. Trends in mortality due to invasive mycotic diseases in the United States, 1980–1997. *Clin. Infect. Dis.* 33:641–647.

Mendelsohn, J., and J. Baselga. 2006. Epidermal growth factor receptor targeting in cancer. *Semin. Oncol.* 33:369–385.

Mendez, M.J., L.L. Green, J.R. Corvalan, X.C. Jia, C.E. Maynard-Currie, X.D. Yang, M.L. Gallo, D.M. Louie, D.V. Lee, K.L. Erickson, et al. 1997. Functional transplant of megabase human immunoglobulin loci recapitulates human antibody response in mice. *Nature Genet.* 15:146–156.

Meyers, M.B., W.P. Shen, B.A. Spengler, V. Ciccarone, J.P. O'Brien, D.B. Donner, M.E. Furth, and J.L. Biedler. 1988. Increased epidermal growth factor receptor in multidrug-resistant human neuroblastoma cells. *J. Cell. Biochem.* 38:87–97.

Milgrom, H., R.B. Fick, Jr., J.Q. Su, J.D. Reimann, R.K. Bush, M.L. Watrous, and W.J. Metzger. 1999. Treatment of allergic asthma with monoclonal anti-IgE antibody. rhuMAb-E25 Study Group. *N. Engl. J. Med.* 341:1966–1973.

Mohsin, S.K., H.L. Weiss, M.C. Gutierrez, G.C. Chamness, R. Schiff, M.P. Digiovanna, C.X. Wang, S.G. Hilsenbeck, C.K. Osborne, D.C. Allred, et al. 2005. Neoadjuvant trastuzumab induces apoptosis in primary breast cancers. *J. Clin. Oncol.* 23:2460–2468.

Muller, Y.A., Y. Chen, H.W. Christinger, B. Li, B.C. Cunningham, H.B. Lowman, and A.M. de Vos. 1998. VEGF and the Fab fragment of a humanized neutralizing antibody: Crystal structure of the complex at 2.4 A resolution and mutational analysis of the interface. *Structure* 6:1153–1167.

Muller, Y.A., B. Li, H.W. Christinger, J.A. Wells, B.C. Cunningham, and A.M. de Vos. 1997. Vascular endothelial growth factor: Crystal structure and functional mapping of the kinase domain receptor binding site. *Proc. Natl. Acad. Sci. U.S.A.* 94:7192–7197.

Muthuswamy, S.K., D. Li, S. Lelievre, M.J. Bissell, and J.S. Brugge. 2001. ErbB2, but not ErbB1, reinitiates proliferation and induces luminal repopulation in epithelial acini. *Nature Cell Biol.* 3:785–792.

Muthuswamy, S.K., P.M. Siegel, D.L. Dankort, M.A. Webster, and W.J. Muller. 1994. Mammary tumors expressing the neu proto-oncogene possess elevated c-Src tyrosine kinase activity. *Mol. Cell Biol.* 14:735–743.

Naramura, M., S.D. Gillies, J. Mendelsohn, R.A. Reisfeld, and B.M. Mueller. 1993. Therapeutic potential of chimeric and murine anti-(epidermal growth factor receptor) antibodies in a metastasis model for human melanoma. *Cancer Immunol. Immunother.* 37:343–349.

Nashan, B., R. Moore, P. Amlot, A.G. Schmidt, K. Abeywickrama, and J.P. Soulillou. 1997. Randomised trial of basiliximab versus placebo for control of acute cellular rejection in renal allograft recipients. CHIB 201 International Study Group. *Lancet* 350:1193–1198.

Nesbitt, A., G. Fossati, M. Bergin, P. Stephens, S. Stephens, R. Foulkes, D. Brown, M. Robinson, and T. Bourne. 2007. Mechanism of action of certolizumab pegol (CDP870): In vitro comparison with other anti-tumor necrosis factor alpha agents. *Inflamm. Bowel Dis.* 13:1323–1332.

Neve, R.M., H. Sutterluty, N. Pullen, H.A. Lane, J.M. Daly, W. Krek, and N.E. Hynes. 2000. Effects of oncogenic ErbB2 on G1 cell cycle regulators in breast tumour cells. *Oncogene* 19:1647–1656.

Niemeyer, G., M. Koch, S. Light, E.R. Kuse, and B. Nashan. 2002. Long-term safety, tolerability and efficacy of daclizumab (Zenapax) in a two-dose regimen in liver transplant recipients. *Am. J. Transplant.* 2:454–460.

O'Donoghue, J.A., M. Bardies, and T.E. Wheldon. 1995. Relationships between tumor size and curability for uniformly targeted therapy with beta-emitting radionuclides. *J. Nuclear Med.* 36:1902–1909.

Oliveira, S., P.M. van Bergen en Henegouwen, G. Storm, and R.M. Schiffelers. 2006. Molecular biology of epidermal growth factor receptor inhibition for cancer therapy. *Expert Opin. Biol. Ther.* 6:605–617.

Onrust, S.V., and H.M. Lamb. 1998. Infliximab: A review of its use in Crohn's disease and rheumatoid arthritis. *BioDrugs* 10:397–422.

Osbourn, J., M. Groves, and T. Vaughan. 2005. From rodent reagents to human therapeutics using antibody guided selection. *Methods* 36:61–68.

Paul, S.P., L.S. Taylor, E.K. Stansbury, and D.W. McVicar. 2000. Myeloid specific human CD33 is an inhibitory receptor with differential ITIM function in recruiting the phosphatases SHP-1 and SHP-2. *Blood* 96:483–490.

Paul-Pletzer, K. 2006. Tocilizumab: Blockade of interleukin-6 signaling pathway as a therapeutic strategy for inflammatory disorders. *Drugs Today (Barc)* 42:559–576.

Peng, D., Z. Fan, Y. Lu, T. DeBlasio, H. Scher, and J. Mendelsohn. 1996. Anti-epidermal growth factor receptor monoclonal antibody 225 up-regulates p27KIP1 and induces G1 arrest in prostatic cancer cell line DU145. *Cancer Res.* 56:3666–3669.

Petit, A.M., J. Rak, M.C. Hung, P. Rockwell, N. Goldstein, B. Fendly, and R.S. Kerbel. 1997. Neutralizing antibodies against epidermal growth factor and ErbB-2/neu receptor tyrosine kinases down-regulate vascular endothelial growth factor production by tumor cells in vitro and in vivo: angiogenic implications for signal transduction therapy of solid tumors. *Am. J. Pathol.* 151:1523–1530.

Pfreundschuh, M., L. Trumper, A. Osterborg, R. Pettengell, M. Trneny, K. Imrie, D. Ma, D. Gill, J. Walewski, P.L. Zinzani, et al. 2006. CHOP-like chemotherapy plus rituximab versus CHOP-like chemotherapy alone in young patients with good-prognosis diffuse large-B-cell lymphoma: A randomised controlled trial by the MabThera International Trial (MInT) Group. *Lancet Oncol.* 7:379–391.

Phan, G.Q., J.C. Yang, R.M. Sherry, P. Hwu, S.L. Topalian, D.J. Schwartzentruber, N.P. Restifo, L.R. Haworth, C.A. Seipp, L.J. Freezer, et al. 2003. Cancer regression and autoimmunity induced by cytotoxic T lymphocyte-associated antigen 4 blockade in patients with metastatic melanoma. *Proc. Natl. Acad. Sci. U.S.A.* 100:8372–8377.

Plosker, G.L., and D.P. Figgitt. 2003. Rituximab: Areview of its use in non-Hodgkin's lymphoma and chronic lymphocytic leukaemia. *Drugs* 63:803–843.

Pollack, P., and J.R. Groothuis. 2002. Development and use of palivizumab (Synagis): A passive immunoprophylactic agent for RSV. *J. Infect. Chemother.* 8:201–206.

Polyak, M.J., and J.P. Deans. 2002. Alanine-170 and proline-172 are critical determinants for extracellular CD20 epitopes; heterogeneity in the fine specificity of CD20 monoclonal antibodies is defined by additional requirements imposed by both amino acid sequence and quaternary structure. *Blood* 99:3256–3262.

Press, M.F., L. Bernstein, P.A. Thomas, L.F. Meisner, J.Y. Zhou, Y. Ma, G. Hung, R.A. Robinson, C. Harris, A. El-Naggar, et al. 1997. HER-2/neu gene amplification characterized by fluorescence in situ hybridization: Poor prognosis in node-negative breast carcinomas. *J. Clin. Oncol.* 15:2894–2904.

Press, O.W., A.G. Farr, K.I. Borroz, S.K. Anderson, and P.J. Martin. 1989. Endocytosis and degradation of monoclonal antibodies targeting human B-cell malignancies. *Cancer Res* 49:4906–4912.

Presta, L.G., H. Chen, S.J. O'Connor, V. Chisholm, Y.G. Meng, L. Krummen, M. Winkler, and N. Ferrara. 1997. Humanization of an anti-vascular endothelial growth factor monoclonal antibody for the therapy of solid tumors and other disorders. *Cancer Res.* 57:4593–4599.

Presta, L.G., S.J. Lahr, R.L. Shields, J.P. Porter, C.M. Gorman, B.M. Fendly, and P.M. Jardieu. 1993. Humanization of an antibody directed against IgE. *J. Immunol.* 151:2623–2632.

Protheroe, A., J.C. Edwards, A. Simmons, K. Maclennan, and P. Selby. 1999. Remission of inflammatory arthropathy in association with anti-CD20 therapy for non-Hodgkin's lymphoma. *Rheumatology (Oxford)* 38:1150–1152.

Queen, C., W.P. Schneider, H.E. Selick, P.W. Payne, N.F. Landolfi, J.F. Duncan, N.M. Avdalovic, M. Levitt, R.P. Junghans, and T.A. Waldmann. 1989. A humanized antibody that binds to the interleukin 2 receptor. *Proc. Natl. Acad. Sci. U.S.A.* 86:10029–10033.

Rabindran, S.K. 2005. Antitumor activity of HER-2 inhibitors. *Cancer Lett.* 227:9–23.

Ratzinger, G., J.L. Reagan, G. Heller, K.J. Busam, and J.W. Young. 2003. Differential CD52 expression by distinct myeloid dendritic cell subsets: Implications for alemtuzumab activity at the level of antigen presentation in allogeneic graft-host interactions in transplantation. *Blood* 101:1422–1429.

Reddy, M., C. Davis, J. Wong, P. Marsters, C. Pendley, and U. Prabhakar. 2007. Modulation of CLA, IL-12R, CD40L, and IL-2Ralpha expression and inhibition of IL-12- and IL-23-induced cytokine secretion by CNTO 1275. *Cell. Immunol.* 247:1–11.

Reff, M.E., K. Carner, K.S. Chambers, P.C. Chinn, J.E. Leonard, R. Raab, R.A. Newman, N. Hanna, and D.R. Anderson. 1994. Depletion of B cells in vivo by a chimeric mouse human monoclonal antibody to CD20. *Blood* 83:435–445.

Reichert, J.M. 2004. Technology evaluation: Lumiliximab, Biogen Idec. *Curr. Opin. Mol. Ther.* 6:675–683.

Reuben, J.M., B.N. Lee, C. Li, J. Gomez-Navarro, V.A. Bozon, C.A. Parker, I.M. Hernandez, C. Gutierrez, G. Lopez-Berestein, and L.H. Camacho. 2006. Biologic and immunomodulatory events after CTLA-4 blockade with ticilimumab in patients with advanced malignant melanoma. *Cancer* 106:2437–2444.

Ribas, A., L.H. Camacho, G. Lopez-Berestein, D. Pavlov, C.A. Bulanhagui, R. Millham, B. Comin-Anduix, J.M. Reuben, E. Seja, C.A. Parker, et al. 2005. Antitumor activity in melanoma and anti-self responses in a phase I trial with the anti-cytotoxic T lymphocyte-associated antigen 4 monoclonal antibody CP-675,206. *J. Clin. Oncol.* 23:8968–8977.

Riechmann, L., M. Clark, H. Waldmann, and G. Winter. 1988. Reshaping human antibodies for therapy. *Nature* 332:323–327.

Rivera, F., M. Salcedo, N. Vega, Y. Blanco, and C. Lopez. 2009. Current situation of zalutumumab. *Expert Opinion on Biological Therapy* 9:667–674.

Robert, F., M.P. Ezekiel, S.A. Spencer, R.F. Meredith, J.A. Bonner, M.B. Khazaeli, M.N. Saleh, D. Carey, A.F. LoBuglio, R.H. Wheeler, et al. 2001. Phase I study of anti-epidermal growth factor receptor antibody cetuximab in combination with radiation therapy in patients with advanced head and neck cancer. *J. Clin. Oncol.* 19:3234–3243.

Rose, J.W., J.B. Burns, J. Bjorklund, J. Klein, H.E. Watt, and N.G. Carlson. 2007. Daclizumab phase II trial in relapsing and remitting multiple sclerosis: MRI and clinical results. *Neurology* 69:785–789.

Rose-John, S., and H. Schooltink. 2003. CDP-870. Celltech/Pfizer. *Curr. Opin. Investig. Drugs* 4:588–592.

Rosenfeld, P.J., R.M. Rich, and G.A. Lalwani. 2006. Ranibizumab: Phase III clinical trial results. *Ophthalmol. Clin. North Am.* 19:361–372.

Rothenberg, M.E., A.D. Klion, F.E. Roufosse, J.E. Kahn, P.F. Weller, H.U. Simon, L.B. Schwartz, L.J. Rosenwasser, J. Ring, E.F. Griffin, et al. 2008. Treatment of patients with the hypereosinophilic syndrome with mepolizumab. *N. Engl. J. Med.* 358:1215–1228.

Rother, R.P., S.A. Rollins, C.F. Mojcik, R.A. Brodsky, and L. Bell. 2007. Discovery and development of the complement inhibitor eculizumab for the treatment of paroxysmal nocturnal hemoglobinuria. *Nature Biotechnol.* 25:1256–1264.

Rowan, W., J. Tite, P. Topley, and S.J. Brett. 1998. Cross-linking of the CAMPATH-1 antigen (CD52) mediates growth inhibition in human B- and T-lymphoma cell lines, and subsequent emergence of CD52-deficient cells. *Immunology* 95:427–436.

Rudick, R.A., and A. Sandrock. 2004. Natalizumab: Alpha 4-integrin antagonist selective adhesion molecule inhibitors for MS. *Expert Rev. Neurother.* 4:571–580.

Rutgeerts, P., S. Schreiber, B. Feagan, D.L. Keininger, L. O'Neil, and R.N. Fedorak. 2007. Certolizumab pegol, a monthly subcutaneously administered Fc-free anti-TNFalpha, improves health-related quality of life in patients with moderate to severe Crohn's disease. *Int. J. Colorectal Dis.* 23:289–296.

Salnikov, A.V., N.E. Heldin, L.B. Stuhr, H. Wiig, H. Gerber, R.K. Reed, and K. Rubin. 2006. Inhibition of carcinoma cell-derived VEGF reduces inflammatory characteristics in xenograft carcinoma. *Int. J. Cancer* 119:2795–2802.

Saltz, L., M. Kies, J.L. Abbruzzese, N. Azarnia, and M. Needle. 2003. The presence and intensity of the cetuximab-induced acne-like rash predicts increased survival in studies across multiple malignancies. *Proc. Am. Soc. Clin. Oncol.* 22:204 (abstr 817).

Santora, L.C., Z. Kaymakcalan, P. Sakorafas, I.S. Krull, and K. Grant. 2001. Characterization of noncovalent complexes of recombinant human monoclonal antibody and antigen using cation exchange, size exclusion chromatography, and BIAcore. *Anal. Biochem.* 299:119–129.

Sarfati, M., and G. Delespesse. 1988. Possible role of human lymphocyte receptor for IgE (CD23) or its soluble fragments in the in vitro synthesis of human IgE. *J. Immunol.* 141:2195–2199.

Sarfati, M., S. Fournier, M. Christoffersen, and G. Biron. 1990. Expression of CD23 antigen and its regulation by IL-4 in chronic lymphocytic leukemia. *Leuk. Res.* 14:47–55.

Sato, J.D., T. Kawamoto, A.D. Le, J. Mendelsohn, J. Polikoff, and G.H. Sato. 1983. Biological effects in vitro of monoclonal antibodies to human epidermal growth factor receptors. *Mol. Biol. Med.* 1:511–529.

Sato, K., M. Tsuchiya, J. Saldanha, Y. Koishihara, Y. Ohsugi, T. Kishimoto, and M.M. Bendig. 1993. Reshaping a human antibody to inhibit the interleukin 6-dependent tumor cell growth. *Cancer Res.* 53:851–856.

Scallon, B.J., M.A. Moore, H. Trinh, D.M. Knight, and J. Ghrayeb. 1995. Chimeric anti-TNF-alpha monoclonal antibody cA2 binds recombinant transmembrane TNF-alpha and activates immune effector functions. *Cytokine* 7:251–259.

Schulz, H., J. Bohlius, N. Skoetz, S. Trelle, T. Kober, M. Reiser, M. Dreyling, M. Herold, G. Schwarzer, M. Hallek, and A. Engert. 2007. Chemotherapy plus Rituximab versus chemotherapy alone for B-cell non-Hodgkin's lymphoma. *Cochrane Database Syst. Rev.*, CD003805.

Schwarz, E.M., and C.T. Ritchlin. 2007. Clinical development of anti-RANKL therapy. *Arthritis Res. Ther.* 9(Suppl 1):S7.

Schwarzmeier, J.D., M. Shehata, M. Hilgarth, I. Marschitz, N. Louda, R. Hubmann, and R. Greil. 2002. The role of soluble CD23 in distinguishing stable and progressive forms of B-chronic lymphocytic leukemia. *Leuk. Lymphoma* 43:549–554.

Seidman, A.D., M.N. Fornier, F.J. Esteva, L. Tan, S. Kaptain, A. Bach, K.S. Panageas, C. Arroyo, V. Valero, V. Currie, et al. 2001. Weekly trastuzumab and paclitaxel therapy for metastatic breast cancer with analysis of efficacy by HER2 immunophenotype and gene amplification. *J. Clin. Oncol.* 19:2587–2595.

Shan, D., J.A. Ledbetter, and O.W. Press. 1998. Apoptosis of malignant human B cells by ligation of CD20 with monoclonal antibodies. *Blood* 91:1644–1652.

Shealy D., A. Cai, E. Lacy, T. Nesspor, K. Staquet, L. Johns, A. Baker, M. Brigham-Burke, E. Emmell, P. Bugelski, P. Verzaal, L. Nagelkerken, B. Scallon, and J. Giles-Komar. 2007. Characterization of Golimumab (CNTO 148), a novel fully human monoclonal antibody specific for human TNFα. (EULAR abstract THU0088). *Ann. Rheum. Dis.* 66(Suppl II):151.

Siegel, S.A., D.J. Shealy, M.T. Nakada, J. Le, D.S. Woulfe, L. Probert, G. Kollias, J. Ghrayeb, J. Vilcek, and P.E. Daddona. 1995. The mouse/human chimeric monoclonal antibody cA2 neutralizes TNF in vitro and protects transgenic mice from cachexia and TNF lethality in vivo. *Cytokine* 7:15–25.

Slamon, D.J., G.M. Clark, S.G. Wong, W.J. Levin, A. Ullrich, and W.L. McGuire. 1987. Human breast cancer: Correlation of relapse and survival with amplification of the HER-2/neu oncogene. *Science* 235:177–182.

Slamon, D.J., W. Godolphin, L.A. Jones, J.A. Holt, S.G. Wong, D.E. Keith, W.J. Levin, S.G. Stuart, J. Udove, A. Ullrich, et al. 1989. Studies of the HER-2/neu proto-oncogene in human breast and ovarian cancer. *Science* 244:707–712.

Slamon, D.J., B. Leyland-Jones, S. Shak, H. Fuchs, V. Paton, A. Bajamonde, T. Fleming, W. Eiermann, J. Wolter, M. Pegram, et al. 2001. Use of chemotherapy plus a monoclonal antibody against HER2 for metastatic breast cancer that overexpresses HER2. *N. Engl. J. Med.* 344:783–792.

Sliwkowski, M.X., J.A. Lofgren, G.D. Lewis, T.E. Hotaling, B.M. Fendly, and J.A. Fox. 1999. Nonclinical studies addressing the mechanism of action of trastuzumab (Herceptin). *Semin. Oncol.* 26:60–70.

Small, E.J., N.S. Tchekmedyian, B.I. Rini, L. Fong, I. Lowy, and J.P. Allison. 2007. A pilot trial of CTLA-4 blockade with human anti-CTLA-4 in patients with hormone-refractory prostate cancer. *Clin. Cancer Res.* 13:1810–1815.

Smith, S.L. 1996. Ten years of Orthoclone OKT3 (muromonab-CD3): A review. *J. Transpl. Coord.* 6:109–119; quiz 120–121.

Smolen, J.S., A. Beaulieu, A. Rubbert-Roth, C. Ramos-Remus, J. Rovensky, E. Alecock, T. Woodworth, and R. Alten. 2008. Effect of interleukin-6 receptor inhibition with tocilizumab in patients with rheumatoid arthritis (OPTION study): A double-blind, placebo-controlled, randomised trial. *Lancet* 371:987–997.

Smolich, B.D., H.A. Yuen, K.A. West, F.J. Giles, M. Albitar, and J.M. Cherrington. 2001. The antiangiogenic protein kinase inhibitors SU5416 and SU6668 inhibit the SCF receptor (c-kit) in a human myeloid leukemia cell line and in acute myeloid leukemia blasts. *Blood* 97:1413–1421.

Sonpavde, G., and T.E. Hutson. 2007. Recent advances in the therapy of renal cancer. *Expert Opin. Biol. Ther.* 7:233–242.

Sorokin, P. 2000. Mylotarg approved for patients with CD33+ acute myeloid leukemia. *Clin. J. Oncol. Nurs.* 4:279–280.

Speake, G., B. Holloway, and G. Costello. 2005. Recent developments related to the EGFR as a target for cancer chemotherapy. *Curr. Opin. Pharmacol.* 5:343–349.

Stanglmaier, M., S. Reis, and M. Hallek. 2004. Rituximab and alemtuzumab induce a nonclassic, caspase-independent apoptotic pathway in B-lymphoid cell lines and in chronic lymphocytic leukemia cells. *Ann. Hematol.* 83:634–645.

Stashenko, P., L.M. Nadler, R. Hardy, and S.F. Schlossman. 1980. Characterization of a human B lymphocyte-specific antigen. *J. Immunol.* 125:1678–1685.

Storch, G.A. 1998. Humanized monoclonal antibody for prevention of respiratory syncytial virus infection. *Pediatrics* 102:648–651.

Strunk, R.C., and G.R. Bloomberg. 2006. Omalizumab for asthma. *N. Engl. J. Med.* 354:2689–2695.

Suvas, S., V. Singh, S. Sahdev, H. Vohra, and J.N. Agrewala. 2002. Distinct role of CD80 and CD86 in the regulation of the activation of B cell and B cell lymphoma. *J. Biol. Chem.* 277:7766–7775.

Takeda, J., T. Miyata, K. Kawagoe, Y. Iida, Y. Endo, T. Fujita, M. Takahashi, T. Kitani, and T. Kinoshita. 1993. Deficiency of the GPI anchor caused by a somatic mutation of the PIG-A gene in paroxysmal nocturnal hemoglobinuria. *Cell* 73:703–711.

Tedder, T.F., A.W. Boyd, A.S. Freedman, L.M. Nadler, and S.F. Schlossman. 1985. The B cell surface molecule B1 is functionally linked with B cell activation and differentiation. *J. Immunol.* 135:973–979.

Tedder, T.F., J. Tuscano, S. Sato, and J.H. Kehrl. 1997. CD22, a B lymphocyte-specific adhesion molecule that regulates antigen receptor signaling. *Annu. Rev. Immunol.* 15:481–504.

Teeling, J.L., R.R. French, M.S. Cragg, J. van den Brakel, M. Pluyter, H. Huang, C. Chan, P.W. Parren, C.E. Hack, M. Dechant, et al. 2004. Characterization of new human CD20 monoclonal antibodies with potent cytolytic activity against non-Hodgkin lymphomas. *Blood* 104:1793–1800.

Thistlethwaite, J.R., Jr., A.O. Gaber, B.W. Haag, A.J. Aronson, C.E. Broelsch, J.K. Stuart, and F.P. Stuart. 1987. OKT3 treatment of steroid-resistant renal allograft rejection. *Transplantation* 43:176–184.

Tischer, E., R. Mitchell, T. Hartman, M. Silva, D. Gospodarowicz, J.C. Fiddes, and J.A. Abraham. 1991. The human gene for vascular endothelial growth factor. Multiple protein forms are encoded through alternative exon splicing. *J. Biol. Chem.* 266:11947–11954.

Toichi, E., G. Torres, T.S. McCormick, T. Chang, M.A. Mascelli, C.L. Kauffman, N. Aria, A.B. Gottlieb, D.E. Everitt, B. Frederick, et al. 2006. An anti-IL-12p40 antibody down-regulates type 1 cytokines, chemokines, and IL-12/IL-23 in psoriasis. *J. Immunol.* 177:4917–4926.

Tracey, D., L. Klareskog, E.H. Sasso, J.G. Salfeld, and P.P. Tak. 2008. Tumor necrosis factor antagonist mechanisms of action: A comprehensive review. *Pharmacol. Ther.* 117:244–279.

Van Cutsem, E., M. Peeters, S. Siena, Y. Humblet, A. Hendlisz, B. Neyns, J.L. Canon, J.L. Van Laethem, J. Maurel, G. Richardson, et al. 2007. Open-label phase III trial of panitumumab plus best supportive care compared with best supportive care alone in patients with chemotherapy-refractory metastatic colorectal cancer. *J. Clin. Oncol.* 25:1658–1664.

van der Velden, V.H., J.G. te Marvelde, P.G. Hoogeveen, I.D. Bernstein, A.B. Houtsmuller, M.S. Berger, and J.J. van Dongen. 2001. Targeting of the CD33-calicheamicin immunoconjugate Mylotarg (CMA-676) in acute myeloid leukemia: In vivo and in vitro saturation and internalization by leukemic and normal myeloid cells. *Blood* 97:3197–3204.

Van Gelder, T., M. Warle, and R.G. Ter Meulen. 2004. Anti-interleukin-2 receptor antibodies in transplantation: What is the basis for choice? *Drugs* 64:1737–1741.

Vega, M.I., S. Huerta-Yepez, A.R. Jazirehi, H. Garban, and B. Bonavida. 2005a. Rituximab (chimeric anti-CD20) sensitizes B-NHL cell lines to Fas-induced apoptosis. *Oncogene* 24:8114–8127.

Vega, M.I., A.R. Jazirehi, S. Huerta-Yepez, and B. Bonavida. 2005b. Rituximab-induced inhibition of YY1 and Bcl-xL expression in Ramos non-Hodgkin's lymphoma cell line via inhibition of NF-kappa B activity: Role of YY1 and Bcl-xL in Fas resistance and chemoresistance, respectively. *J. Immunol.* 175:2174–2183.

Villadsen, L.S., L. Skov, T.N. Dam, F. Dagnaes-Hansen, J. Rygaard, J. Schuurman, P.W. Parren, J.G. van de Winkel, and O. Baadsgaard. 2007. In situ depletion of CD4+ T cells in human skin by Zanolimumab. *Arch. Dermatol. Res.* 298:449–455.

Vincenti, F., R. Kirkman, S. Light, G. Bumgardner, M. Pescovitz, P. Halloran, J. Neylan, A. Wilkinson, H. Ekberg, R. Gaston, et al. 1998. Interleukin-2-receptor blockade with daclizumab to prevent acute rejection in renal transplantation. Daclizumab Triple Therapy Study Group. *N. Engl. J. Med.* 338:161–165.

Vincenti, F., B. Nashan, and S. Light. 1998. Daclizumab: Outcome of phase III trials and mechanism of action. Double Therapy and the Triple Therapy Study Groups. *Transplant. Proc.* 30:2155–2158.

Vitale, C., C. Romagnani, M. Falco, M. Ponte, M. Vitale, A. Moretta, A. Bacigalupo, L. Moretta, and M.C. Mingari. 1999. Engagement of p75/AIRM1 or CD33 inhibits the proliferation of normal or leukemic myeloid cells. *Proc. Natl. Acad. Sci. U.S.A.* 96:15091–15096.

Voulgari, P.V., and A.A. Drosos. 2006. Adalimumab for rheumatoid arthritis. *Expert Opin. Biol. Ther.* 6:1349–1360.

Vyth-Dreese, F.A., H. Boot, T.A. Dellemijn, D.M. Majoor, L.C. Oomen, J.D. Laman, M. van Meurs, R.A. De Weger, and D. De Jong. 1998. Localization in situ of costimulatory molecules and cytokines in B-cell non-Hodgkin's lymphoma. *Immunology* 94:580–586.

Wahl, R.L., S. Kroll, and K.R. Zasadny. 1998. Patient-specific whole-body dosimetry: Principles and a simplified method for clinical implementation. *J. Nuclear Med.* 39:14S–20S.

Wahl, R.L., K.R. Zasadny, D. MacFarlane, I.R. Francis, C.W. Ross, J. Estes, S. Fisher, D. Regan, S. Kroll, and M.S. Kaminski. 1998. Iodine-131 anti-B1 antibody for B-cell lymphoma: An update on the Michigan Phase I experience. *J. Nuclear Med.* 39:21S–27S.

Waldmann, H., and G. Hale. 2005. CAMPATH: From concept to clinic. *Philos. Trans. R. Soc. London B Biol. Sci.* 360:1707–1711.

Watanabe, T., J. Masuyama, Y. Sohma, H. Inazawa, K. Horie, K. Kojima, Y. Uemura, Y. Aoki, S. Kaga, S. Minota, et al. 2006. CD52 is a novel costimulatory molecule for induction of CD4+ regulatory T cells. *Clin. Immunol.* 120:247–259.

Weinblatt, M.E., E.C. Keystone, D.E. Furst, L.W. Moreland, M.H. Weisman, C.A. Birbara, L.A. Teoh, S.A. Fischkoff, and E.K. Chartash. 2003. Adalimumab, a fully human anti-tumor necrosis factor alpha monoclonal antibody, for the treatment of rheumatoid arthritis in patients taking concomitant methotrexate: The ARMADA trial. *Arthritis Rheum.* 48:35–45.

Werther, W.A., T.N. Gonzalez, S.J. O'Connor, S. McCabe, B. Chan, T. Hotaling, M. Champe, J.A. Fox, P.M. Jardieu, P.W. Berman, and L.G. Presta. 1996. Humanization of an anti-lymphocyte function-associated antigen (LFA)-1 monoclonal antibody and reengineering of the humanized antibody for binding to rhesus LFA-1. *J. Immunol.* 157:4986–4995.

Wingard, J.R., S. Piantadosi, G.B. Vogelsang, E.R. Farmer, D.A. Jabs, L.S. Levin, W.E. Beschorner, R.A. Cahill, D.F. Miller, D. Harrison, et al. 1989. Predictors of death from chronic graft-versus-host disease after bone marrow transplantation. *Blood* 74:1428–1435.

Wittig, B.M. 2007. Drug evaluation: CNTO-1275, a mAb against IL-12/IL-23p40 for the potential treatment of inflammatory diseases. *Curr. Opin. Investig. Drugs* 8:947–954.

Witzig, T.E., I.W. Flinn, L.I. Gordon, C. Emmanouilides, M.S. Czuczman, M.N. Saleh, L. Cripe, G. Wiseman, T. Olejnik, P.S. Multani, and C.A. White. 2002a. Treatment with ibritumomab tiuxetan radioimmunotherapy in patients with rituximab-refractory follicular non-Hodgkin's lymphoma. *J. Clin. Oncol.* 20:3262–3269.

Witzig, T.E., L.I. Gordon, F. Cabanillas, M.S. Czuczman, C. Emmanouilides, R. Joyce, B.L. Pohlman, N.L. Bartlett, G.A. Wiseman, N. Padre, et al. 2002b. Randomized controlled trial of yttrium-90-labeled ibritumomab tiuxetan radioimmunotherapy versus rituximab immunotherapy for patients with relapsed or refractory low-grade, follicular, or transformed B-cell non-Hodgkin's lymphoma. *J. Clin. Oncol.* 20:2453–2463.

Witzig, T.E., C.A. White, L.I. Gordon, G.A. Wiseman, C. Emmanouilides, J.L. Murray, J. Lister, and P.S. Multani. 2003. Safety of yttrium-90 ibritumomab tiuxetan radioimmunotherapy for relapsed low-grade, follicular, or transformed non-Hodgkin's lymphoma. *J. Clin. Oncol.* 21:1263–1270.

Wood, J.M., G. Bold, E. Buchdunger, R. Cozens, S. Ferrari, J. Frei, F. Hofmann, J. Mestan, H. Mett, T. O'Reilly, et al. 2000. PTK787/ZK 222584, a novel and potent inhibitor of vascular endothelial growth factor receptor tyrosine kinases, impairs vascular endothelial growth factor-induced responses and tumor growth after oral administration. *Cancer Res.* 60:2178–2189.

Woods Ignatoski, K.M., N.K. Grewal, S. Markwart, D.L. Livant, and S.P. Ethier. 2003. p38MAPK induces cell surface alpha4 integrin downregulation to facilitate erbB-2-mediated invasion. *Neoplasia* 5:128–134.

Wosikowski, K., D. Schuurhuis, G.J. Kops, M. Saceda, and S.E. Bates. 1997. Altered gene expression in drug-resistant human breast cancer cells. *Clin. Cancer Res.* 3:2405–2414.

Wu, H., D.S. Pfarr, S. Johnson, Y.A. Brewah, R.M. Woods, N.K. Patel, W.I. White, J.F. Young, and P.A. Kiener. 2007. Development of motavizumab, an ultra-potent antibody for the prevention of respiratory syncytial virus infection in the upper and lower respiratory tract. *J. Mol. Biol.* 368:652–665.

Wu, H., D.S. Pfarr, Y. Tang, L.L. An, N.K. Patel, J.D. Watkins, W.D. Huse, P.A. Kiener, and J.F. Young. 2005. Ultra-potent antibodies against respiratory syncytial virus: Effects of binding kinetics and binding valence on viral neutralization. *J. Mol. Biol.* 350:126–144.

Wu, X., Z. Fan, H. Masui, N. Rosen, and J. Mendelsohn. 1995. Apoptosis induced by an anti-epidermal growth factor receptor monoclonal antibody in a human colorectal carcinoma cell line and its delay by insulin. *J. Clin. Invest.* 95:1897–1905.

Wyeth Pharmaceuticals, Inc. (23-Jan-2006). Mylotarg (Gemtuzumab Ozogamicin): Label and Prescribing Information.

Xia, M.Q., G. Hale, M.R. Lifely, M.A. Ferguson, D. Campbell, L. Packman, and H. Waldmann. 1993. Structure of the CAMPATH-1 antigen, a glycosylphosphatidylinositol-anchored glycoprotein which is an exceptionally good target for complement lysis. *Biochem. J.* 293(Pt 3):633–640.

Xu, D., M.L. Alegre, S.S. Varga, A.L. Rothermel, A.M. Collins, V.L. Pulito, L.S. Hanna, K.P. Dolan, P.W. Parren, J.A. Bluestone, et al. 2000. In vitro characterization of five humanized OKT3 effector function variant antibodies. *Cell. Immunol.* 200:16–26.

Yang, X.D., X.C. Jia, J.R. Corvalan, P. Wang, and C.G. Davis. 2001. Development of ABX-EGF, a fully human anti-EGF receptor monoclonal antibody, for cancer therapy. *Crit. Rev. Oncol. Hematol.* 38:17–23.

Yang, X.D., X.C. Jia, J.R. Corvalan, P. Wang, C.G. Davis, and A. Jakobovits. 1999. Eradication of established tumors by a fully human monoclonal antibody to the epidermal growth factor receptor without concomitant chemotherapy. *Cancer Res.* 59:1236–1243.

Yee, K.W., and S.M. O'Brien. 2006. Emerging drugs for chronic lymphocytic leukaemia. *Expert Opin. Emerg. Drugs* 11:167–189.

Yokota, S., T. Imagawa, M. Mori, T. Miyamae, Y. Aihara, S. Takei, N. Iwata, H. Umebayashi, T. Murata, M. Miyoshi, et al. 2008. Efficacy and safety of tocilizumab in patients with systemic-onset juvenile idiopathic arthritis: A randomised, double-blind, placebo-controlled, withdrawal phase III trial. *Lancet* 371:998–1006.

Yokota, S., T. Miyamae, T. Imagawa, N. Iwata, S. Katakura, M. Mori, P. Woo, N. Nishimoto, K. Yoshizaki, and T. Kishimoto. 2005. Therapeutic efficacy of humanized recombinant anti-interleukin-6 receptor antibody in children with systemic-onset juvenile idiopathic arthritis. *Arthritis Rheum.* 52:818–825.

Yuan, F., Y. Chen, M. Dellian, N. Safabakhsh, N. Ferrara, and R.K. Jain. 1996. Time-dependent vascular regression and permeability changes in established human tumor xenografts induced by an anti-vascular endothelial growth factor/vascular permeability factor antibody. *Proc. Natl. Acad. Sci. U.S.A.* 93:14765–14770.

Zareba, K.M. 2007. Eculizumab: A novel therapy for paroxysmal nocturnal hemoglobinuria. *Drugs Today (Barc)* 43:539–546.

Zein, N., A.M. Sinha, W.J. McGahren, and G.A. Ellestad. 1988. Calicheamicin gamma 1I: An antitumor antibiotic that cleaves double-stranded DNA site specifically. *Science* 240:1198–1201.

Zhang, N., L.A. Khawli, P. Hu, and A.L. Epstein. 2005. Generation of rituximab polymer may cause hyper-cross-linking-induced apoptosis in non-Hodgkin's lymphomas. *Clin. Cancer Res.* 11:5971–5980.

Zhou, H., H. Jang, R.M. Fleischmann, E. Bouman-Thio, Z. Xu, J.C. Marini, C. Pendley, Q. Jiao, G. Shankar, S.J. Marciniak, et al. 2007. Pharmacokinetics and safety of golimumab, a fully human anti-TNF-alpha monoclonal antibody, in subjects with rheumatoid arthritis. *J. Clin. Pharmacol.* 47:383–396.

CHAPTER 32

Follow-On Protein Products: What, Where, When, How?

BRENT R. WILLIAMS and WILLIAM R. STROHL

ABSTRACT

As patent protections have begun to expire for many of the original protein therapeutics on the market, a wide-ranging interest in and debate about the scientific and regulatory issues associated with the development of generic protein therapeutics has evolved. However, "generic" protein therapeutics is not a term used by the scientific community, the U.S. Food and Drug Adminstration (FDA), or the European Medicines Agency (EMEA) to describe such molecules. Rather, the molecules have been called follow-on protein products (FOPPs), or follow-on biologics by the FDA, and biosimilars by the EMEA, to reflect the difficulties in their characterization and comparison to a reference innovator product. While the U.S. regulatory agency and Congress have not yet decided on an approval route for FOPPs, the EMEA has passed guidelines for the abbreviated approval of FOPPs and even approved several small, simple (complexity) FOPPs. From the examples and guidelines that are available to date, it is clear that the various categories of protein therapeutics will be handled differently, as each presents different challenges of proving comparability, consistency, efficacy, and safety (especially immunogenicity) as compared to an already approved reference product. Currently, the development and approval of FOPP therapeutic antibodies, one of the largest and most molecularly complex categories, are an exceptional challenge. This chapter will introduce the concept of FOPPs, provide a summary of the current regulatory status and debate in Europe and the United States, present several examples of currently approved FOPPs, and conclude with a broader discussion and opinion on the future development of FOPPs.

Therapeutic Monoclonal Antibodies: From Bench to Clinic. Edited by Zhiqiang An

32.1 INTRODUCTION

In the United States, the Public Health Service Act (PHS Act) has defined a biological product as "a virus, therapeutic serum, toxin, antitoxin, vaccine, blood, blood component or derivative, allergenic product, or analogous product, or arsphenamine or derivative of arsphenamine (or any other trivalent organic arsenic compound), applicable to the prevention, treatment, or cure of a disease or condition of human beings." Some of the various uses of biological products as therapeutics include attack of cancerous cells or pathogenic organisms, neutralization of disease-related molecules, and supplementation or replacement of factors that are deficient due to a condition or unrelated treatment. Having an array of uses, these protein therapeutics fall into a wide variety of categories, including monoclonal antibodies, interferons, anti-hemophilic factors, hormones, insulin, vaccine products, and others. The variety and clinical success of protein therapeutics has resulted in their seizure of a large proportion of the U.S. drug market, growing at twice the rate of conventional drugs, with sales expected to exceed $60 billion by 2010 (Hofelich 2006). As patent protections have begun to expire for many of the innovative protein therapeutics on the market, a wide-ranging interest and debate about the scientific and regulatory issues associated with the development of generic (i.e., follow-on) protein therapeutics has evolved.

What does the term generic drug mean? According to the Office of Generic Drugs (OGD), which is part of the U.S. Food and Drug Administration's Center for Drug Evaluation and Research (CDER), "a generic drug is identical, or bioequivalent to a brand name drug in dosage form, safety, strength, route of administration, quality, performance characteristics and intended use" (FDA/Center for Drug Evaluation and Research 2008). Additionally, OGD states that for a generic version of an innovator drug to gain FDA approval, the generic must contain the same active ingredients; be identical in strength, dosage form, and route of administration; have the same use indications; be bioequivalent; meet the same batch requirements for identity, strength, purity, and quality; and be manufactured under the same strict standards of FDA's good manufacturing practice as the innovator drug. These definitions relate to conventional drugs, that is, non-protein small molecule, chemically synthesized drugs, that apply for market entry through an Abbreviated New Drug Application (ANDA) under an amendment to the Federal Food, Drug and Cosmetic Act (FDC Act) called the Drug Price Competition and Patent Term Restoration Act (commonly known as the Hatch-Waxman Act). They do not necessarily apply to generic protein therapeutics, which are fundamentally different than small molecule drugs. Protein therapeutics have molecular weights ranging from thousands to several hundred thousand daltons, while small molecule drugs are typically only several hundred daltons. Protein products are structurally complex, containing primary, secondary, tertiary, and quaternary structures and, in addition, may contain complex glycosylation patterns. While most small molecule drugs (with the exception of natural products, such as penicillin, erythromycin, doxorubicin, etc.) are manufactured utilizing a series of chemical syntheses and purifications, protein therapeutics are produced by complex biosynthetic manufacturing processes, utilizing living organisms, including mammalian cell lines, bacteria, or yeast as producers. As a result, the final protein product will usually be well defined yet heterogeneous, while a chemically synthesized product will be a pure and homogeneous molecule. Even small molecule natural products, which are typically produced by bacterial or fungal fermentations, or by harvesting plants, are purified to homogeneity and characterized based on chemical structure. The difference between natural products and chemically synthesized small molecules is that the process for production and purification of natural products often results in minor residual related components; characterization of these and development of manufacturing processes to maintain consistency of the minor residuals is part of the natural product drug development and licensing process. Proteins, on the other hand, add the additional complexity of higher order three-dimensional structure, and even the exact same protein therapeutic produced using different manufacturing processes (e.g., a different cell line) may differ in key biological properties (e.g., glycosylation, immunogenicity, microheterogeneity, including N- and C- terminal degradation, etc.). While it has been possible to accurately characterize generic small molecule drug composition as being pure, homogeneous, and containing the same active ingredient as the innovator drug, protein therapeutics cannot be characterized to the same degree using currently available science and analytical technology. As a result, the term "generic protein therapeutics" has not been adopted by the scientific community, the FDA, or the European Medicines Agency (EMEA) to describe such molecules. Instead, the molecules have been called "follow-on protein products" (FOPPs) or "follow-on biologics" by the FDA and "biosimilars" by the EMEA to reflect the difficulties in their characterization and comparison to a reference innovator product.

In the United States, an abbreviated approval pathway, analogous to an ANDA for conventional generic small molecule drugs through the Hatch-Waxman Act, is not available for licensure of follow-on versions of a protein drug product. Rather, protein therapeutics are regulated under the Public Health Service Act (PHS Act) and must apply for market entry through a Biologics License Application (BLA) under that act. The Hatch-Waxman Act does not apply to protein therapeutics marketed under a BLA and thus a barrier to entry of FOPPs in the U.S. market remains. Some older, less complex protein therapeutics (most were natural-source), such as insulin, glucagon, and human growth hormone, are historical exceptions. They applied for market entry under the FDC Act and are regulated as drugs. The regulatory path to market for FOPPs is currently under intense debate in the U.S. Congress. Bills have been introduced in Congress to amend the PHS Act and provide for the licensing of FOPPs, but have not yet reached a vote (see GovTrack.us 2007). While the U.S. regulatory agency and Congress have not yet decided on a route, the EMEA has passed guidelines for the abbreviated approval of simpler FOPPs. In fact, several small, simple (complexity) FOPPs have already been approved under the European system, including human growth hormone and epoetin FOPPs.

The primary drivers behind the interest in development of FOPPs include interest of the generic industry in tapping into the large protein therapeutic drug market and the ability to make a more affordable version of protein therapeutics available to all patients needing them. Protein therapeutics are typically very expensive compared with small molecule drugs; one study has stated that the average biologic is about 22 times more expensive than the average small molecule innovative drug (Shapiro, Singh, and Mukim 2008). A more affordable version of a protein therapeutic would result in the possibility of utilization and accessibility of the drugs to a larger population of patients as well as cost savings to those patients, insurance companies, and federal budgets. According to the Generic Pharmaceutical Association (2008), a conventional generic drug can cost from 30 to 80 percent less than its equivalent brand name drug. In a study analyzing potential savings from substituting conventional generic drugs for brand name drugs during a period from 1997 to 2000 in the United States, it was estimated that substitution of a generic for a brand-name drug whenever available would have saved approximately \$46 per year for adults under 65 years old and approximately \$78 per year for adults older than 65 (Haas et al. 2005). This would have equated to a national savings of approximately \$6 billion for adults under 65 years old and approximately \$3 billion for adults older than 65 during that time period. Due to their higher development and manufacturing costs, most agree that FOPPs will not be discounted at the same levels as seen for conventional drug generics. Moreover, interchangeability and substitutability are issues with FOPPs, further complicating the pricing debate. Most estimates of projected discounts for FOPPs in the United States range from 10 to 30 percent off the innovator product price (Marchant 2007; Datamonitor 2007; Grabowski et al. 2007). Much uncertainty remains, however, and answers to the scientific, regulatory, and marketing debates will come as experience is gained with FOPPs worldwide.

32.2 REGULATORY LANDSCAPE

Regulatory agencies in both the United States and Europe have realized that a pathway for marketing approval of FOPPs will require a different structure and guidelines as compared to conventional generics to ensure the quality, safety, and efficacy of FOPPs. However, similarities to the generic pathway will need to be built in, including balancing patient safety, making reasonably priced alternative protein therapeutics available, and protecting and promoting innovation. Due to the inability to define FOPPs as exact copies of the innovator product, marketing approval of an FOPP does not automatically grant interchangeability/substitutability with the innovator product. In fact, France and Spain have passed legislation stating that there cannot be automatic substitution of an FOPP for the innovator product (Moran 2008). The European Union is ahead of the United States in dealing with the regulatory issues associated with marketing approval of FOPPs, having already laid out a regulatory pathway for approval. The current and potential regulations and guidelines in Europe and the United States are discussed below.

32.2.1 Europe

Traditionally, marketing authorization for medicinal products could occur through two different routes under the European system. Under the first route, called the centralized procedure, application for market approval is submitted to the EMEA (similar to the FDA in the United States) where it is assessed. The EMEA issues an opinion (favorable or unfavorable) to the European Commission (EC) who then makes the final approval. Approval by the EC is valid in all EU member states. Under the second route, called the mutual recognition procedure, application for market approval is submitted to one or more EU member states. The reference member state then assesses the application and issues a report to all concerned member states (i.e., other states to which the identical application has been made). The report and comments issued by the reference and concerned member states are sent to the EMEA for an opinion and the final decision is made by the EC. The EC decision is then binding to all concerned member states as well as any member state that receives a market application for the same product.

The centralized procedure has been made mandatory for all products derived from biotechnology as well as certain other categories of drugs. EU regulations define and handle FOPPs distinctly and separately from conventional generic drugs. As defined in EU Directive 2004/27/EC which amends 2001/83/EC, FOPPs "similar to a reference medicinal product do not usually meet all the conditions to be considered as a generic medicinal product mainly due to manufacturing process characteristics, raw materials used, molecular characteristics and therapeutic modes of action. When a biological medicinal product does not meet all the conditions to be considered as a generic medicinal product, the results of appropriate tests should be provided in order to fulfill the requirements related to safety (pre-clinical tests) or to efficacy (clinical tests) or to both." Additionally, this directive refers one to Annex I of Directive 2001/83/EC and related detailed guidelines when deciding what type and quantity of data will be required. Essentially, Directive 2001/83/EC and updates 2003/63/EC and 2004/27/EC lay out a general pathway for approval of FOPPs, which is separate from that of conventional generic drugs. According to these directives, comparability studies will be required to establish the similarity between the FOPP and the reference protein therapeutic (which must already have marketing authorization in the EU) in terms of quality, safety, and efficacy. Since the reference protein therapeutic was already authorized and marketed in the EU, clinical data and experience with the product are available. As a result, the FOPP to the reference therapeutic will usually require less nonclinical and clinical data as it can rely on some of the reference product data if the products are comparable.

EU regulations establishing a regulatory pathway for approval of FOPPs have been purposely left rather vague as the requirements for approval of such products will be dealt with on a case-by-case basis. The amount of studies required will vary based on the level of complexity of the therapeutic, understanding of the mechanisms of action and biology behind the therapeutic, and the clinical experience acquired with the therapeutic. Companies developing FOPPs for approval in the EU have sought advice from the Committee for Medicinal Products for Human Use (CHMP) which is part of EMEA. As part of the required centralized approval scheme, companies submit their applications for marketing approval of the FOPP to the EMEA which then issues an opinion to the EC for the final approval decision. The EMEA, in order to better guide these applicants, has issued a "Guideline on Similar Biological Medicinal Products," which went into effect in October 2005 (Committee for Medicinal Products for Human Use 2005). This guideline introduces and outlines the principles to be applied to FOPPs and serves as a user guide for directing applicants in establishing similarity. This guideline does not apply nor define an approach for any specific category of FOPP, but rather broadly defines the requirements for FOPPs based on their complexity. Those FOPPs that cannot currently be fully characterized due to their complexity (structural, manufacturing, etc.), or lack acceptable analytical procedures and technology or clinical and regulatory experience, may not be acceptable using the "similar biological medicinal product" approach. Any quality data for the FOPP must, "fulfill all requirements for Module 3 as defined in Annex I to Directive 2001/83/EC and satisfy the technical requirements of the monographs of the European Pharmacopoeia and any additional requirements, such as defined in relevant CHMP and ICH guidelines." The EMEA issued, and continues to issue, further guidelines for demonstrating therapeutic equivalence of specific categories of FOPP products. The guidelines cover all of the current major biosimilar candidates, including recombinant erythropoietins (CHMP/945226/05), granulocyte-colony stimulating factor (CHMP/31329/05), somatropin (CHMP/4528/05), and insulin (CHMP/32775/05). A draft guideline has been issued for recombinant interferon alpha (CHMP/BMWP/102046/06). A detailed analysis of each of these specific guidelines is beyond the scope of this chapter. The reader is referred to the EMEA website for full access to all of the guidance documents (http://www.emea.europa.eu/htms/human/humanguidelines/multidiscipline.htm).

Comparability exercises, demonstrating similarity of highly purified and characterized FOPPs, are required under the EU legislation and guidelines described above. For these comparability exercises, the FOPP is compared to a chosen reference innovator protein therapeutic approved for the same indications for which the application is being made. This reference protein therapeutic must be approved in the EU and the marketing exclusivity must be expired. For most FOPPs, the comparability exercises proceed in a step-wise fashion. In the first step, a comparison of the quality of the FOPP and the reference protein therapeutic is made. Any differences or concerns in the quality between the FOPP and the reference product will affect the details and amount of comparative nonclinical and clinical bridging studies that will be required in further steps. Clinical comparability studies would usually consist of pharmacokinetic and pharmacodynamic studies in healthy volunteers, followed by comparative safety and efficacy trials in patients with the most sensitive and relevant indication (specific target study populations are recommended in the EMEA guidelines for each of the current categories of biosimilars). Importantly, the guidelines require applicants to provide comparative immunogenicity data as well as a pharmacovigilance plan, paying special attention to immunogenicity, in line with EU legislation and guidelines. Optimally, a positive opinion for approval would be granted if the FOPP is well characterized, is similar in quality, and has similar efficacy and immunogenicity as compared to the reference product.

32.2.2 The United States

Conventional, small molecule drugs are regulated under the FDC Act in the United States. Under this act, new innovator drugs must apply for marketing approval through a New Drug Application (NDA) under Section 505(b)(1). The Drug Price Competition and Patent Term Restoration Act (Hatch-Waxman Act), enacted in 1984, amended the FDC Act and laid out a pathway for approval of generic versions of innovator conventional drugs. Utilizing these amendments for approval of a generic drug, an ANDA is submitted to the FDA under Section 505(j). To do so, the generic version of an innovator drug must essentially be identical to the innovator drug in active ingredient, dosage form, strength, route of administration, labeling, quality, performance characteristics, and intended use, as described in the Introduction section. As a result, the generic applicant has decreased data requirements for approval, as it can rely on safety and efficacy data submitted by the innovator drug. Another ANDA route for generic drug approval also added to the FDC Act by the Hatch-Waxman Act, Section 505(b)(2), can be utilized by drugs that have been modified as compared to the previously approved innovator drug yet are comparable. Such an application contains full reports of safety and investigation, but some of the investigations "were not conducted by or for the applicant and for which the applicant has not obtained a right of reference or use from the person by or for whom the investigations were conducted" (21 U.S.C. 355(b)(2)). Drugs utilizing this approach can include new chemical or molecular entities and drugs that have been changed as compared to the previously approved version.

According to CDER draft guidance, "an applicant should file a 505(b)(2) application if it is seeking approval of a change to an approved drug that would not be permitted under Section 505(j), because approval will require the review of clinical data. However, Section 505(b)(2) applications should not be submitted for duplicates of approved products that are eligible for approval under 505(j) (see 21 CFR 314.101(d)(9))" (FDA/Center for Drug Evaluation and Research 1999). These abbreviated approaches were developed under the Hatch-Waxman Act under the principle that, "it is wasteful and unnecessary to carry out studies to demonstrate what is already known about a drug" (FDA/Center for Drug Evaluation and Research 1999).

The FDA handles biological products as a subset of drugs and both are regulated under the FDC Act. However, due to differences between biological products and conventional drugs, biologics must apply for marketing approval through a BLA under the PHS Act. The requirements for approval of biological products are different than those for conventional drugs as the final biological product is typically less well defined and characterized. Additionally, relatively minor variations in the manufacturing conditions of a biologic product can lead to large effects on the quality, safety, and efficacy of the final product that often cannot be detected by available analytical methods. For example, changes in the manufacturing of Eprex® (epoetin-α; approved in the United States and Europe), a protein therapeutic similar to Epogen®, were associated with an increase in the incidence of pure red-cell aplasia. A detailed examination discovered that leachables associated with the uncoated rubber stoppers of syringes exposed to polysorbate 80, used in a new formulation and packaging system (prefilled syringe), were present in the product. It has been hypothesized that these may have acted as an adjuvant, enhancing the immune response to both the therapeutic as well as endogenous erythropoietin (Boven et al. 2005; Louet 2003). It is for these reasons that the FDA requires not only that the product is safe, pure, and potent, but also that the manufacturing process and facilities meet the requirements for continued consistent production of a safe, pure, and potent product. Thus, the PHS Act "emphasizes the importance of appropriate manufacturing control for products" and "provides for a system of controls over all aspects of the manufacturing process" (FDA/Center for Drug Evaluation and Research 2006). Any changes in the manufacturing process or facilities may require additional studies. Thus, the FDA's final approval decision is based on the data obtained in preclinical and clinical studies showing the product's safety and efficacy as well as inspection of, and data from, the manufacturing process and facilities.

Currently, development of an FOPP version of an innovator protein therapeutic licensed under the PHS Act would require a full BLA for approval. An approval scheme, analogous to an ANDA under the Hatch-Waxman Act, cannot be applied to FOPP due to the inability to declare the follow-on version identical to the innovator product and described differences to conventional small molecule drugs. However, historically some simpler protein therapeutics have been approved under the FDC Act, including insulin, hyaluronidase, menotropins, and human growth hormones. Since these products have been approved under this act, it is possible to get follow-on versions of them approved through an ANDA process under 505(b)(2). In fact, several FOPPs have been approved in the United States under Section 505(b)(2) including Hylenex (hyaluronidase), Hydase™ (hyaluronidase), Amphadase™ (hyaluronidase), Fortical® (calcitonin), GlucaGen® (glucagon), and Omnitrope™ (somatropin). The appropriateness of a Section 505(b)(2) application, and the information required for each of the FOPP approvals, have been decided on a case-by-case basis by the FDA. However, for the majority of protein therapeutics approved under the PHS Act, an ANDA pathway is not applicable. As explained by Janet Woodcock of the FDA in testimony to the U.S. House of Representatives on March 26, 2007, due to "the variability and complexity of protein molecules, current limitations of analytical methods, and the difficulties in manufacturing a consistent product, it is unlikely that, for most proteins, a manufacturer of a follow-on protein product could demonstrate that its product is identical to an already approved product. Therefore, the Section 505(j) generic drug approval pathway, which is predicated on a finding of the same active ingredient, will not ordinarily be available for protein products" (Woodcock 2007). Additionally, according to Janet Woodcock in this same testimony, the FDA is in the process of developing guidance documents that are broadly applicable to development and approval of FOPPs in general.

The best process for the approval of an abbreviated biological product application in the United States is still being developed and debated. To gain a better understanding of what the FDA may require of companies developing FOPPs under such a regulatory path in the future, one can examine the several follow-on protein products it approved under Section 505(b)(2) of the FDC Act. Additionally, one can examine the several bills introduced in the U.S. Congress to amend the PHS Act to create such a pathway (GovTrack.us 2007) as well as an opinion paper written by members of the working group on follow-on proteins at the FDA entitled, "The FDA's Assessment of Follow-On Protein Products: A Historical Perspective" (Woodcock et al. 2007).

According to these authors some of the factors that have influenced the amount and type of data required by the FDA for marketing approval of FOPPs (under FDC Act Section 505(b)(2)) and second-generation products in the past include:

> "1) Evidence of integrity and consistency of the manufacturing process; 2) Conformance of manufacturing standards to existing regulations (if any); 3) Demonstrations of a product's consistency with appropriate reference standards or comparators (using relevant assays), including comparative pharmacokinetic and pharmacodynamic data; and 4) The extent to which the existing body of clinical data and experience with the approved product can be relied on."

The general data included for FDA approval of the FOPPs (glucagon, salmon calcitonin nasal spray, somatropin, erythropoietin-α, and interferon β1a) included data showing high structural similarity, highly similar pharmacokinetics and pharmacodynamics, and immunogenicity and safety data that were comparable to (or better than) the reference product. Despite approval of these FOPPs by the FDA, these products have not been declared as directly substitutable for the reference products. The FDA would require additional clinical data demonstrating that repeated switches between the follow-on and reference product would not have a negative effect on the safety or the effectiveness of the products.

An examination of one of the most recent bills introduced in 2007 to the 110th U.S. Congress, S. 1695: Biologics Price Competition and Innovation Act of 2007, gives an idea as to what a regulatory pathway for licensure of FOPPs might look like. This bill, if approved, would amend the PHS Act to allow a person to apply for marketing approval of an FOPP based on comparisons to an approved reference product. It would require licensure of the FOPP if it was biosimilar to, or interchangeable with, the reference product. An FOPP could be declared interchangeable if it were: "1) biosimilar to the reference product and can be expected to produce the same clinical result in any given patient; and 2) the risk in terms of safety or diminished efficacy of alternating or switching between the products is not greater than the risk of using the reference product without switching" (GovTrack.us 2007). Additionally, it would offer protections for promotion of innovation. For example, it would prohibit licensure of an FOPP until 12 years after the initial licensure of the reference product and would offer one year of market exclusivity to the first FOPP approved as interchangeable. The bill tries to balance the incentives and protection for development of innovative products with ability to approve and market follow-on versions of those products, similar to the balances incorporated into the Hatch-Waxman Act. The bill was not voted on by the 110th U.S. Congress, and future versions may change as debate and experience continues to develop.

32.3 CASE STUDIES: CURRENTLY MARKETED FOPPs

Several FOPPs have been approved in the EU under their biosimilars regulations and guidelines, including human growth hormone FOPPs Omnitrope™ and Valtropin®, and erythropoietin FOPPs Binocrit®, Epoetin alfa HEXAL®, Abseamed®, Silapo®, and Retacrit™. In the United States, several follow-on versions of protein therapeutics originally approved under the FDC Act have gained approval under Section 505(b)(2), including Omnitrope™. However, the approvals in the United

States are limited to those original products approved under the FDC Act, not the majority of protein therapeutics approved under the PHS Act.

As of September 2007, worldwide there were seven different marketed EPO FOPPs, seven marketed human growth hormone (HGH) FOPPs, six different marketed insulin FOPPs, eight marketed granulocyte colony-stimulating factor (G-CSF) FOPPs, and seven different interferon FOPPs (Shapiro, Singh, and Mukim 2008). A discussion of the approvals of both growth hormone and erythropoietin FOPPs in Europe and the United States is included below.

32.3.1 Human Growth Hormone

HGH is a pituitary-derived 191 amino acid single chain nonglycosylated polypeptide hormone with a molecular weight of 22 kDa. HGH has been used clinically as a replacement therapy in pediatric patients who have growth failure due to inadequate secretion of normal endogenous growth hormone. It has also been used as replacement therapy in adult patients with growth hormone deficiency and some other select indications, depending on the product. Early, original versions of HGH were obtained from the pituitary glands of human cadavers. More recently, recombinant versions of HGH, also known as somatropin, were developed and approved in both Europe and the United States. Branded HGH is available from a number of different manufacturers, including Pfizer, Eli Lilly, Novo Nordisk, Genentech, and Serono. With global sales of HGH in 2006 of $2.5 billion, new players continue to tap into the HGH market, including those who are developing follow-on versions of the hormone (Marchant 2007).

In the EU, two follow-on versions of HGH have been approved in 2006 using an abbreviated application process under Article 10(4) of Directive 2001/83/EC [amended by Directive 2004/27/EC (Similar Biological Medicinal Product)]. The products, Omnitrope (Sandoz GmbH) and Valtropin (BioPartners GmbH), were both granted European Union marketing approval in April 2006. The applications for both products were developed and submitted before adoption of guidelines addressing the development of follow-on versions of HGH, which came into effect in June 2006.

The development program of the FOPP Omnitrope™ in Europe used Genotropin®, a somatropin product produced by Pfizer originally approved in the EU in 1988, as the reference medicinal product. During the comparability exercises, Omnitrope™ was compared to EU-produced Genotropin® from a few different EU markets, using spectrometric, sequence, and physicochemical data. No significant differences were detected between the products using these methods. Nonclinical studies comparing the properties/activity of Omnitrope™ with Genotropin® included primary pharmacodynamic (rat weight gain assay), potency (rat tibia width assay), and repeat dose toxicity (including toxicokinetics) studies. The results from these studies showed that Omnitrope™ was comparable to the reference medicinal product, Genotropin®. The clinical efficacy of Omnitrope™ was compared to Genotropin® in 89 prepubertal children who were HGH deficient and treatment naïve in a pivotal 9 month phase III trial. The primary endpoints included the height velocity between months 0 and 9 and height at month 9. The velocity and increases in height were similar, with an increase of 10.7 cm/year observed with both medicines. Although there were patients receiving the Omnitrope™ treatment who did develop anti-HGH antibodies, these did not affect the growth rate seen with treatment. To address this immunogenicity, additional purification steps were added during the manufacture of Omnitrope™, resulting in no patients developing anti-HGH antibodies in a subsequent study. An initial application submitted in May 2001 received a positive review from the EMEA but was denied approval by the EC in November 2003 as a legal basis for FOPP approval did not exist. Additional directives, establishing an approval pathway for FOPPs, were eventually issued and Sandoz resubmitted its application (with additional data) using the newer directives. For the final application, the EMEA concluded that Omnitrope™ was shown to have a comparable quality, safety, and efficacy profile to Genotropin® and therefore recommended marketing authorization on January 26, 2006. The EC granted marketing authorization, valid throughout the EU, on April 12, 2006.

The development program of the FOPP Valtropin® in Europe used Humatrope®, a somatropin product produced by Eli Lilly originally approved in the EU in 1990, as the reference medicinal product.

During the comparability exercises, Valtropin® was compared to Humatrope®, using mass spectrometry, peptide mapping, N-terminal sequencing, C-terminal sequencing, amino acid composition analysis, and electrophoretic, chromatographic, and spectroscopic methods. No significant differences were detected between the products using these methods. Nonclinical studies comparing the properties/activity of Valtropin® with Humatrope® included primary pharmacodynamic (rat weight gain assay), potency (rat tibia width assay), pharmacokinetic (in rabbit) and repeat dose toxicity (including toxicokinetics) studies. The results from these studies showed that Valtropin® was comparable to the reference medicinal product, Humatrope®. The clinical efficacy and safety of Valtropin® was compared to Humatrope® in 149 prepubertal children who were HGH deficient in a pivotal 12 month phase III trial. The primary endpoint of the study was height velocity (cm/year) after the first 12 months of treatment. The increases in height velocity were similar between the two products, with mean increases from 3.4 cm/year at baseline to11.3 cm/year at month 12 for Valtropin® and from 3.2 cm/year at baseline to 10.5 cm/year with Humatrope®. An additional, supportive, uncontrolled 12 month trial was conducted in 30 treatment naïve girls, aged two to nine years, with short stature due to Turner syndrome. The height velocity was significantly increased from 3.75 cm/year at baseline to 9.73 cm/year at month 12. Although a firm conclusion regarding efficacy was not possible from this supportive study, the results indicated that treatment had the expected growth-promoting effect of a somatropin-containing product. There was no difference in the development of anti-HGH antibodies between the two products, with approximately 3 percent of the patients developing anti-HGH antibodies in the Valtropin® groups and approximately 2 percent of the patients developing anti-HGH antibodies in the Genotropin® group. The development of these antibodies did not affect growth. The EMEA concluded that Valtropin® was shown to have a comparable quality, safety, and efficacy profile to Humatrope® and therefore recommended marketing authorization on February 23, 2006. The EC granted marketing authorization, valid throughout the EU, on April 24, 2006.

Since approval of each of the FOPP somatropins, the EMEA has constructed and put into effect a guideline addressing the nonclinical and clinical issues associated with development of FOPPs containing somatropin (Committee for Medicinal Products for Human Use 2006a). The somatropin-specific guideline presents the recommendation of the CHMP for demonstrating the comparability of two recombinant human somatropin-containing products. The guideline consists of five main sections, which include: (1) Nonclinical studies that should be performed to detect differences in the pharmaco-toxicological response between the FOPP and reference product before initiating clinical studies. (2) Clinical studies, which should include a clinical efficacy study in a recommended target study population (for somatropins the recommended population is prepubertal, treatment naïve children with HGH deficiency). (3) Clinical safety data including 12 month comparative immunogenicity data of patients who participated in the efficacy trial(s). (4) A pharmacovigilance plan where an applicant will present a risk management program or pharmacovigilance plan that takes into account identified or potential risks and immunogenicity. (5) Extension of indication where demonstration of efficacy in the target population may allow extrapolation to other indications if justified by the applicant. The specifics of this guideline must be considered along with the more general guidelines adopted by the EMEA and the requirements laid down in EU Pharmaceutical legislation.

Omnitrope™ was also approved in the United States by the FDA using an abbreviated pathway under Section 505(b)(2) of the FDC Act. This abbreviated pathway was available for Omnitrope™ as it used Genotropin®, a somatropin product produced by Pfizer, originally approved in 1985 under Section 505(b)(1) of the FDC Act, as the reference product. In its Omnitrope™ application, Sandoz submitted data from physicochemical tests that established the structural similarity of Omnitrope™ to Genotropin®. It also submitted new nonclinical pharmacology and toxicology data specific to Omnitrope™ as well as pharmacokinetic, pharmacodynamic, and comparative bioavailability data establishing the similarity to Genotropin® based on these results. Clinical efficacy and safety data were submitted from two controlled trials that compared Omnitrope™ with Genotropin® in the treatment of 89 pediatric patients with growth failure due to HGH deficiency. Additional clinical data were also submitted including clinical data establishing the safety and effectiveness of a reformulated version of Omnitrope™ and safety data from a 24 month uncontrolled trial establishing longer-term safety and immunogenicity. The abbreviated application, filed in July 2003, was not immediately approved by the FDA despite a lack of any deficiencies in the application identified by the FDA. Eventually, after Sandoz sued the FDA in U.S. court, the FDA approved Omnitrope™ in May 2006. Importantly, Omnitrope™ has not been rated by the FDA as substitutable for any other approved HGH product.

32.3.2 Erythropoietin

Erythropoietin (EPO) is a 165 amino acid glycoprotein produced mainly in the kidney in humans. It has a molecular weight of 30 kDa of which 40 percent is carbohydrate (Mocini et al. 2007). EPO is used clinically to treat anemia in chronic renal failure patients, in cancer patients with anemia due to chemo- or radiotherapy, and to reduce the need for blood transfusion in adults with mild anemia or expected complications before major surgery. Recombinant EPO was developed and approved in both Europe and the United States. Branded EPO is available from a number of different manufacturers, including Amgen and Johnson & Johnson. EPO is the leading class of protein therapeutics in terms of sales, with global sales of EPO in 2006 of $12 billion (Marchant 2007). The large EPO market, coupled with patent expiration in Europe and pending patent expiration in the United States, make EPO the most likely protein therapeutic to be most immediately affected by FOPPs.

In the EU, five follow-on versions of EPO were approved in 2007 using an abbreviated application process under Article 10(4) of Directive 2001/83/EC [amended by Directive 2004/27/EC (Similar Biological Medicinal Product)]. Three of the FOPP EPO products, Binocrit® (Sandoz GmbH), Epoetin alfa HEXAL® (Hexal Biotech Forschungs GmbH), Abseamed® (Medice Arzneimittel Putter GmbH & Co.) are all the same recombinant epoetin alfa product produced by the same manufacturer, Rentschler Biotechnologie GmbH, and marketed separately by each of the respective companies. Since the product is sold by three different companies using three different names, three separate yet essentially identical applications for marketing authorization were submitted to the EMEA. All were granted European Union marketing approval on August 28, 2007. The other two FOPP EPO products, Silapro™ (Stada Arzneimittel AG) and Retacrit™ (Hospira Enterprises B.V.), are the same recombinant epoetin zeta product produced by the same manufacturer, Norbitec GmbH, and marketed separately by each of the respective companies. Two separate yet essentially identical applications for marketing authorization were submitted to the EMEA and both were granted European Union marketing approval on December 18, 2007. All of the companies, with the exception of Hospira Enterprises B.V. (Retacrit), developed and submitted their applications to the EMEA before the guideline addressing the development of follow-on versions of EPO came into effect [Guidance on Similar Medicinal Products Containing Recombinant Erythropoietins came into effect on July 1, 2006 (Committee for Medicinal Products for Human Use 2006b)].

The guideline addressing the nonclinical and clinical issues associated with development of FOPPs containing erythropoietins presents the recommendation of the CHMP for demonstrating the comparability of two recombinant human erythropoietin-containing products (Committee for Medicinal Products for Human Use 2006b). The guideline is similar in format to the somatropin-specific guideline and consists of five main sections, including: (1) Nonclinical studies that should be performed to detect differences in the pharmacotoxicological response between the FOPP and the reference product. These studies should include *in vitro* studies (comparative bioassays), *in vivo* studies (comparing erythrogenic activity), and toxicological studies (at least one repeat dose toxicity and local tolerance study). (2) Clincal studies, including pharmacokinetic and pharmacodynamic studies (can be performed in healthy volunteers), as well as clinical efficacy studies. The clinical efficacy studies should consist of at least two adequately powered, randomized, parallel group clinical trials in the recommended target study population (for erythropoietins the recommended population is patients with renal anemia). (3) Clinical safety data from the efficacy trials. This should include at least 12 month comparative immunogenicity data. (4) A pharmacovigilance plan where an applicant will present a risk management program or pharmacovigilance plan that takes into account identified or potential risks and immunogenicity. Particular mention is made of monitoring for rare serious adverse events, such as immune-mediated PRCA. (5) Extension of indication where demonstration of efficacy in the target population may allow extrapolation to other indications if justified by the applicant. The specifics of this guideline must be considered along with the more general guidelines adopted by the EMEA and the requirements laid down in EU Pharmaceutical legislation.

The development programs for the FOPPs Binocrit®, Epoetin alfa HEXAL®, and Abseamed® (all referred to as HX575) were identical, using the same data and reference product. The reference medicinal product used for the comparability exercises was EPREX/ERYPO® (epoetin alfa, Janssen-Cilag GmbH). For the application, the protein structure of HX575 was compared with the reference product through peptide mapping (showing correct formation of disulfide bonds and sequence confirmation), investigations of secondary and tertiary structure, structural analysis of *N*- and *O*-glycans, and site specific glycan analysis. No significant differences with the reference product were revealed. There were, however, some differences observed at the glycosylation level that did not impact the safety or efficacy of HX575. Nonclinical studies comparing the properties and activity of HX575 with Eprex/Erypo® included *in vitro* studies (ELISA, plasmon resonance spectroscopy) and *in vivo* studies in mice, dogs, and rabbits for pharmacokinetics/pharmacodynamics and toxicology. Clinical studies submitted included five pharmacology studies examining pharmacokinetics/pharmacodynamics after single and multiple dose intravenous and subcutaneous administration in healthy male volunteers. Clincal studies also included a phase III trial comparing the efficacy and safety of intravenous HX575 and Erypo® in 479 patients with renal anemia (all were previously taking Eprex/Erypo®, intravenously, for at least 8 weeks) and a phase III trial in 114 cancer patients with anemia due to chemotherapy. The results from the clinical studies supported a conclusion of therapeutic equivalence between HX575 and Eprex/Erypo®. Finally, clinical safety was studied in five phase I studies in healthy volunteers as well as in the two phase III trials mentioned above. The safety studies concluded that was no significant difference in adverse event incidence or type between groups treated with HX575 or Eprex/Erypo®. Importantly, no increase in immunogenicity was found for HX575 compared to Eprex/Erypo. Based on the application, the EMEA concluded that HX575 was shown to have a comparable quality, safety, and efficacy profile to Eprex/Erypo and recommended marketing authorization on June 21, 2007. The EC granted marketing authorization for Binocrit, Epoetin alfa HEXAL, and Abseamed (all HX575), valid throughout the EU, on August 28, 2007.

The development programs for the FOPPs Silapo™ and Retacrit™ (both referred to as SB309) were identical, using the same data and reference product. The reference medicinal product used for the comparability exercises was EPREX/ERYPO® (epoetin alfa, Janssen-Cilag GmbH). For the application, the protein structure of SB309 was compared with the reference product through peptide mapping (showing correct formation of disulfide bonds and sequence confirmation), investigations of secondary structure, and characterization of carbohydrate moieties. No significant differences with the reference product were revealed. Nonclinical studies comparing the properties/activity of SB309 with Eprex/Erypo® included *in vitro* assays (receptor binding and proliferation and second messenger activation in cultured cells), *in vivo* pharmacodynamic and repeat dose toxicity (with toxicokinetics) studies in rats and dogs, and local tolerance tests by a number of routes. From the nonclinical studies it was concluded that SB309 and the reference product could be distinguished from each other in certain *in vitro* pharmacology tests and in some aspects of the *in vivo* tests. For example, a slight increase in the erythropoietic effect of SB309 as compared to Eprex/Erypo® was noted in the rat and antibodies to erythropoietin occurred more frequently in dogs treated with SB309 than dogs treated with Eprex/Erypo®. It was concluded, however, that the differences were not toxicologically meaningful. Clinical studies submitted included two pharmacokinetic studies in healthy volunteers receiving single dose subcutaneous or intravenous administration, clinical efficacy studies in two phase III trials involving 922 patients with anemia due to chronic renal failure, and an uncontrolled safety trial in 261 cancer patients with chemotherapy-induced anemia. The clinical results showed that SB309 was as clinically effective as Eprex/Erypo®. Clinical safety data was also obtained from the phase III trials in 922 chronic renal failure patients with anemia and the trial in 261 cancer patients with chemotherapy-induced anemia. It was concluded that the overall safety profile of SB309 and the reference product are similar and SB309 was in line with the safety profile of other authorized epoetin-containing medicinal products. Based on the application, the EMEA concluded that SB309 was shown to have a comparable quality, safety, and efficacy profile to Eprex/Erypo® and recommended marketing authorization on October 18, 2007. The EC granted marketing authorization for Silapo™ and Retacrit™ (both SB309), valid throughout the EU, on December 18, 2007.

In the United States, the innovator version of recombinant epoetin alfa, Epogen/Procrit® (Amgen), was originally approved on June 1, 1989 through a BLA under the PHS Act. Thus, a regulatory pathway for the abbreviated approval of these EPO FOPPs is not yet available. Additionally, the U.S. patent on Epogen® has not yet expired (see Lin 1996). However, once a regulatory path forward is put into place and patent expiration occurs in the United States, EPO is expected to be the category of protein therapeutics most affected by the introduction of FOPP versions.

32.4 DISCUSSION/PROSPECTIVE

The evolution of guidelines and regulations directing the development of FOPPs is currently occurring, worldwide. The EU has been leading the way, with sets of guidelines and regulations already in place. Many countries, including China, India, and South Korea also have regulations in place to deal with FOPPs, and already have a number of licensed FOPPs approved under their current regulations (Joung et al. 2008). U.S. authorities continue to struggle with and debate the creation of guidelines and legislation to allow the approval of FOPPs. The U.S. 2009 budget proposal includes an initiative to pass legislation enabling FOPP approval that "protects patient safety, promotes innovation, and includes a financing structure to cover the costs of this activity through user fees" (Budget of the United States Government 2008). Health Canada, the Canadian regulatory authority, recently released a draft guidance document containing information and regulatory requirements for FOPPs in Canada (Health Products and Food Branch 2008). If the draft guidance were adopted, as is, an FOPP applicant would have to show that the FOPP is similar to a previously approved protein therapeutic, relying in part on publicly available safety and efficacy data. Additionally, Health Canada believes it could approve FOPPs using existing regulations while these regulations are amended to provide a more comprehensive framework for approval.

An examination of the proposed or existing regulations and systems leads one to realize that there are many commonalities among the different systems. One of the biggest shared messages has been that development and approval of FOPPs cannot be handled in the same manner as traditional generic drugs. The data requirements to gain approval of an FOPP are greater, more complex, and are examined on a case-by-case basis. It is clear that the various categories of protein therapeutics will be handled differently, as each presents different challenges of proving similarity, consistency, efficacy, and safety (especially immunogenicity) as compared with an already approved reference product. As a result, at present there is more uncertainty, cost, and risk associated with the development of an FOPP as compared to a conventional generic drug, thus limiting competition.

The challenges for FOPP development and approval grow larger as the complexity of the therapeutic is increased. The development and approval of FOPP therapeutic antibodies, one of the largest and most molecularly complex categories, are currently an exceptional challenge. Antibodies are large glycoprotein molecules, composed of four polypeptide chains (heavy and light chains) with an approximate molecular weight of 150 kDa (IgG). In comparing the size and molecular complexity of antibodies with currently approved FOPPS, such as the somatropin Omnitrope™ (approved in the United States and EU), one realizes that assessment of similarity of an antibody will be quite challenging. Additional challenges lie in the manufacturing of a highly complex FOPP, such as antibodies, as the production methods, characterization, and formulation are often protected by a complex array of patents or are trade secrets. There currently are no specific guidelines and/or regulations dealing with FOPP antibodies in either the EU or United States. In the United States, the FDA has highlighted the likely high regulatory hurdles that will be encountered by complex protein products, as although a demonstration of similarity "may be currently possible for some relatively simple protein products, technology is not yet sufficiently advanced to allow this type of comparison for more complex protein products" (Woodcock 2007). Additionally, according to the FDA, when the mechanism of action for the product is not well understood, as is the case for many currently approved antibodies, "even very extensive structural and functional comparisons between a follow-on and a comparable innovator product may not be sufficient to allow broad reliance on conclusions regarding a prior product" (Woodcock 2007). Thus, with a lack of guidance and experience in approving FOPP antibodies (or other more complex protein therapeutics), one can only speculate that preclinical and clinical requirements, cost, and risk will be higher for such FOPPs.

An Indian Company, Dr. Reddy's Laboratories, already has developed and received marketing approval for an FOPP version of the monoclonal antibody rituximab (Genentech, Inc. and Biogen Idec, Inc.), called Reditux™, in India in 2007 (Dr. Reddy's Laboratories Ltd. 2007). While the requirements for FOPP approval were likely much less than any requirements for even simple FOPPs in the United States or the EU, the data and experience gained post-approval can be used to identify potential issues associated with approval of such an FOPP. Although the degree of similarity, safety, and efficacy of Reditux™ compared to Rituxan® are not clear, a spokesperson for Dr. Reddy's claims the drug would act exactly the same way and had the same properties as Rituxan® (Reymond 2007). Reditux™, currently only available in India where it can be used for the treatment of non-Hodgkin lymphoma, has been priced 50 percent lower in India than the innovator product.

In addition to the complexities of FOPP development and approval, uncertainties also exist in the naming and pharmacovigilance of approved molecules. The International Nonproprietary Names (INN) program directed through the World Health Organization (WHO) has the purpose of assigning nonproprietary names to medicinal substances. These unique, public names allow each medicinal substance to be recognized globally and identification of active pharmaceutical ingredients. INNs are assigned to both chemical and protein therapeutic drugs. Assigning INNs to protein therapeutics is often a difficult process due to their complexity. The debate around assigning INNs to FOPPs centers around the inability to completely characterize the follow-on version as an exact copy of the reference product. Additionally, FOPPs will have varying degrees of similarity with the reference product. Assigning an INN to an FOPP can affect issues such as interchangeability, tracing, and pharmocovigilance. Should the same INN as the reference product be given to an FOPP? Currently, there are several marketed protein therapeutics that have the same INN despite slight differences between the products, including glycosylation. However, where significant differences exist, different names are assigned based on scientific considerations of molecular characteristics and pharmacological class. For example, where products have significant differences in glycosylation, such as with the epoetin products, different Greek letters are attached to their INN. The idea of an FOPP is based on regulations, while the system for assigning INNs needs to be based on scientific characterization of the FOPP and comparison with a reference product.

A meeting was held by the WHO in 2006 to debate the potential policies for INNs of FOPPs (Joung et al. 2008; World Health Organization 2006). The meeting included representatives from the regulatory authorities of many different countries, including the United Kingdom, Canada, Japan, Republic of Korea, Australia, Belgium, United States, and Poland. A consensus was reached at the meeting that INNs should be independent of regulatory processes, considerations of interchangeability, or use in pharmacovigilance. Rather, they concluded an INN policy for biosimilars should be based on scientific considerations. The proposed INN policy for FOPPs would be the same policy as that used for novel protein therapeutics, requiring no special consideration in naming and with no distinction built into the FOPP names (Joung et al. 2008). Of course, the naming policy for FOPPs by both the WHO and individual regulatory authorities, many of which have their own naming systems, will most likely change and adapt as experience is gained with marketed FOPPs.

Since modern protein therapeutics were first approved in the early 1980's, they have become an increasingly important part of health care. However, treatment with modern protein therapeutics can cost from $15,000 to $150,000 per year, an unbearable burden for all payers involved, including public and private insurances and patients themselves. As a result, many of these cutting-edge, breakthrough treatments are unavailable to the very patients who need them. It is important to maintain the proper perspective as to why these medicines were developed, the majority after many years of intense research: To enhance, extend, or save the lives of people who are sick. It is therefore imperative that the proper regulations, balancing the need for cheaper follow-on versions of protein therapeutics with the incentives and protections for development of new, novel protein therapeutics, are put into place worldwide (Manheim, Granahan, and Dow 2006). These should allow follow-on versions of these important medicines to be efficiently developed, rapidly introduced into the markets, and more easily and widely accessed by the sick people who need them.

REFERENCES

Boven, K., J. Knight, F. Bader, J. Rossert, K.U. Eckardt, and N. Casadevall. 2005. Epoetin-associated pure red cell aplasia in patients with chronic kidney disease: Solving the mystery. *Nephrol. Dial. Transplant.* 20 (Suppl 3): 33–40.

Budget of the United States Government, Fiscal Year 2009. 2008. Washington, D.C.: U.S. Government Printing Office.

Committee for Medicinal Products for Human Use. 2005. Guideline on Similar Biological Medicinal Products – CHMP/437/04. London: European Medicines Agency.

Committee for Medicinal Products for Human Use. 2006a. Annex to Guideline on Similar Biological Medicinal Products Containing Biotechnology-Derived Proteins as Active Substance: Non-Clinical and Clinical Issues—Guidance on Similar Medicinal Products Containing Somatropin – EMEA/CHMP/BMWP/94528/2005. London: European Medicines Agency.

Committee for Medicinal Products for Human Use. 2006b. Annex to Guideline on Similar Biological Medicinal Products Containing Biotechnology-Derived Proteins as Active Substance: Non-Clinical and Clinical Issues—Guidance on Similar Medicinal Products Containing Erythropoietins – EMEA/CHMP/BMWP/94526/2005 Corr. London: European Medicines Agency.

Datamonitor. 2007. Biosimilars Series: Strategic Issues; Potential Remains Uncertain. New York: Datamonitor USA.

Dr. Reddy's Laboratories Ltd. 2007. Dr. Reddy's Launches Reditux: Monoclonal Antibody Treatment for Non-Hodgkin's Lymphoma. Available at http://www.drreddys.com/newsroom/2007yrnews.htm# (accessed March 24, 2008).

FDA/Center for Drug Evaluation and Research. 1999. Guidance for Industry: Applications Covered by Section 505(b)(2) (Draft). Available at http://www.fda.gov/CDER/guidance/2853dft.htm#P109_2037 (accessed January 10, 2008).

FDA/Center for Drug Evaluation and Research. 2006. Frequently Asked Questions About Therapeutic Biological Products. Available at http://www.fda.gov/cder/biologics/qa.htm (Accessed January 10, 2008).

FDA/Center for Drug Evaluation and Research. 2008. What Are Generic Drugs. (available at http://www.fda.gov/cder/ogd (accessed January 10, 2008).

Generic Pharmaceutical Association. 2008. Generics: Lower Cost. Available at http://www.gphaonline.org/AM/Template.cfm?Section=FAQs&Template=/CM/HTMLDisplay.m&ContentID=2497 (accessed January 10, 2008).

GovTrack.us. 2007. S. 1695–110th Congress. Biologics Price Competition and Innovation Act of 2007. Available at http://www.govtrack.us/congress/bill.xpd?bill=s110-1695 (accessed January 10, 2008).

Grabowski, H., I. Cockburn, G. Long, R. Mortimer, and S. Johnson. 2007. The Effect on Federal Spending of Legislation Creating a Regulatory Framework for Follow-on Biologics: Key Issues and Assumptions. White Paper. Available at http://www.bio.org/healthcare/followonbkg/Federal_Spending_of_followonbkg200709.pdf (accessed January 10, 2008).

Haas, J.S., K.A. Phillips, E.P. Gerstenberger, and A.C. Seger. 2005. Potential savings from substituting generic drugs for brand-name drugs: Medical expenditure panel survey, 1997–2000. *Ann. Intern. Med.* 142:891–897.

Health Products and Food Branch. 2008. Draft Guidance for Sponsors: Information and Submission Requirements for Subsequent Entry Biologics. Ottawa: Health Canada.

Hofelich, A. 2006. GPhA Hails Introduction of Legislation to Bring Affordable Generic Biopharmaceutical Medicines to Consumers. Available at http://www.gphaonline.org/AM/PrinterTemplate.cfm?Section=Media&Template=/CM/HTMLDisplay.cfm&ContentID=2849 (accessed January 10, 2008).

Joung, J., J.S. Robertson, E. Griffiths, and I. Knezevic. 2008. WHO informal consultation on regulatory evaluation of therapeutic biological medicinal products held at WHO Headquarters, Geneva, April 19–20, 2007. *Biologicals* 36:269–276.

Lin, F.K., inventor; Kiren-Amgen, Inc., assignee. 1996. Production of erythropoietin. U.S. patent 5,547,933. August 20, 1996.

Louet, S. 2003. Lessons from Eprex for biogeneric firms. *Nature Biotechnol.* 21:956–957.

Manheim, B.S., Jr., P. Granahan, and K.J. Dow. 2006. Follow-on biologics: Ensuring continued innovation in the biotechnology industry. *Biotech Industry* March/April, 394–404.

Marchant, J. 2007. *The Future of Biosimilars. Key Opportunities and Emerging Therapies.* London: Business Insights Ltd.

Mocini, D., T. Leone, M. Tubaro, M. Santini, and M. Penco. 2007. Structure, production and function of erythropoietin: Implications for therapeutical use in cardiovascular disease. *Curr. Med. Chem.* 14:2278–2287.

Moran, N. 2008. Fractured European market undermines biosimilar launches. *Nature Biotechnol.* 26:5–6.

Public Health Service Act. 2006. 42 U.S.C. 262.

Reymond, E. 2007. Dr. Reddy's launches Rituxan biosimilar in India. Breaking News on Drug Discovery. Available at http://www.drugresearcher.com/news/ng.asp?n=76261-dr-reddy-s-roche-biosimilars-india (accessed March 24, 2008).

Shapiro, R.J., K. Singh, and M. Mukim. 2008. The Potential American Market for Generic Biological Treatments and the Associated Costs Savings. Available at: www.insmed.com/pdf/Biogeneric_Savings.pdf (accessed August 14, 2008).

Woodcock, J. 2007. Statement of Dr. Janet Woodcock before the U.S. House of Representatives, March 26, 2007, "Follow-on Protein Products". Available at http://www.fda.gov/ola/2007/protein32607.html (accessed January 10, 2008).

Woodcock, J., J. Griffin, R. Behrman, B. Cherney, T. Crescenzi, B. Fraser, D. Hixon, C. Joneckis, S. Kozlowski, A. Rosenberg, L. Schrager, E. Shacter, R. Temple, K. Webber, and H. Winkle. 2007. The FDA's assessment of follow-on protein products: A historical perspective. *Nature Rev. Drug Discov.* 6:437–442.

World Health Organization. 2006. WHO Informal Consultation on International Nonproprietary Names (INN) Policy for Biosimilar Products. Available at http://www.who.int/medicines/services/inn/BiosimilarsINN_ReportSept2006.pdf (accessed March 24, 2008).

CHAPTER 33

Monomeric Fc Fusion Molecules

JENNIFER A. DUMONT, SUSAN C. LOW, ROBERT T. PETERS, and ALAN J. BITONTI

ABSTRACT

Delivery of therapeutic proteins is traditionally achieved through injections since noninvasive routes of administration have proven challenging. Recently we have shown that a naturally occurring receptor, the neonatal Fc receptor (FcRn) can be utilized to deliver large therapeutic proteins fused to a portion of the Fc domain of IgG (the natural ligand for FcRn) through epithelial barriers such as the lung, resulting in delivery into the bloodstream. First-generation dimeric Fc fusion proteins were successfully transported by the pulmonary route and biological activity was demonstrated in both nonhuman primates and human volunteers. However, the bioavailability of dimeric Fc fusion proteins was not as robust as anticipated, so second-generation molecules were created in which a single effector molecule was fused to an Fc dimer. These molecules have been named Fc fusion monomers. Several different therapeutic protein Fc monomers have been created, and have demonstrated improved transport efficiency, achieving higher bioavailabilities for pulmonary delivery in nonhuman primates. While traditional Fc fusion dimers have enhanced half-lives compared to the unconjugated protein, monomers have been

Therapeutic Monoclonal Antibodies: From Bench to Clinic. Edited by Zhiqiang An

shown to result in even greater extension of the circulating half-life, which improves the pharmacokinetic parameters for the therapeutic proteins whether administered by the pulmonary route or by injection. In addition, many of the monomeric Fc fusion proteins have enhanced biological activity. Due to these advantages, the monomer configuration for Fc fusion proteins promises to be an enabling advance to achieve clinically relevant, noninvasive delivery with the potential for less frequent administration for a broad range of protein therapeutics. Several examples of novel Fc monomeric fusion proteins are described here.

33.1 INTRODUCTION

Immunoglobulins have long serum half-lives due to binding through their Fc domains to the neonatal Fc receptor (FcRn) the main functions of which include protection of immunoglobulins from degradation and transport of immunoglobulins across epithelial cells. In attempts to utilize the FcRn pathway to increase the circulating half-lives of effector molecules, numerous fusion proteins have been generated. These molecules are most commonly made as homodimers with an effector molecule attached to each chain of an Fc dimer. Our initial work with Fc fusion proteins focused on the potential to use these molecules for transepithelial delivery of therapeutics, with an emphasis on delivery into the circulation via inhalation. Prototype Fc fusion proteins were absorbed through the lung, but delivery was not as efficient as anticipated. We proposed that Fc monomers, which consist of a single effector molecule fused to an Fc dimer, would have superior pulmonary uptake and efficacy compared to the more common fusion dimer due to a reduction in size, charge, and steric hindrance.

33.2 NEONATAL Fc RECEPTOR

33.2.1 Background

Over 40 years ago, F.W.R. Brambell proposed a link between the transmission of passive immunity in the form of transfer of maternal antibodies to the fetus, and the protection of immunoglobulin G (IgG) catabolism (Brambell, Hemmings, and Morris 1964). He suggested that a single saturable receptor might mediate both processes and that the Fc domain of IgG was involved in the transport of immunoglobulins as well as the protection from degradation (Brambell 1966). This was based on studies that showed transport of maternal IgG to newborn rodents in the intestine (Halliday 1955a; Halliday 1955b) and the demonstration of increased IgG metabolism upon hyperimmunization (Fahey and Robinson 1963). Numerous studies have now proven this hypothesis. These include animal models in which FcRn has been genetically deleted, resulting in a reduced serum IgG half-life and low serum IgG concentrations (Roopenian et al. 2003; Ghetie et al. 1996; Israel et al. 1995, 1996; Junghans and Anderson 1996).

The neonatal Fc receptor (FcRn) was first isolated from the brush borders of intestinal epithelial cells from neonatal rats, the site at which maternal transfer of immunoglobulins takes place during the first few weeks of life (Simister and Rees 1985). The receptor is comprised of two subunits, one of which is a membrane-bound heavy chain that is structurally related to the major histocompatibility complex (MHC) class I molecules and the other subunit is soluble light chain β2-microglobulin (β2m: Simister and Mostov 1989). The regions on immunoglobulin G (IgG) that are critical for the control of IgG catabolism were localized to the Fc fragment, in the CH2 and CH3 domains (Kim et al. 1994). Site-directed mutagenesis identified three amino acids (Ile 253, His 310, and His 435) that are involved in the binding to FcRn and play a role in regulation of the serum half-life of IgG (Kim et al. 1999). Crystal structures of FcRn with Fc have been determined for both rat (Burmeister et al. 1994) and human receptors (West and Bjorkman 2000) and the interactions between the molecules have been more precisely identified. This interaction of IgG with FcRn is pH dependent as binding occurs with high affinity at acidic pH (<6.5), but not at neutral pH (7 to 7.4) (Rodewald 1976; Raghavan, Gastinel, and Bjorkman 1993; Vaughn and Bjorkman 1998).

33.2.2 FcRn Expression and Function

Newborn rats acquire passive immunity through the uptake of maternal IgG from ingested colostrum and milk. FcRn is expressed at relatively high levels in the proximal small intestine during the first few weeks of life and IgG absorption/transport from maternal milk reaches a maximum in rodents during this same period of time (Morris and Morris 1974; Jones and Waldmann 1972; Rodewald 1973). At the time of weaning, FcRn expression in rodents decreases and remains low throughout adult life. FcRn is also expressed in the yolk sac of rats where maternal transfer of immunoglobulins takes place during gestation (Roberts, Guenthert, and Rodewald 1990).

In humans, FcRn is expressed in the syncytiotrophoblast of the placenta where it mediates maternofetal transfer of immunoglobulins during gestation (Simister et al. 1996). In contrast to rodents, expression of FcRn in humans remains relatively high throughout life in the epithelial cells lining the intestine, and also in the lung (Israel et al. 1997; Dickinson et al. 1999; Spiekermann et al. 2002; Bitonti et al. 2004). Consistent with this, the human receptor is functional in adult life in epithelia, as modeled in a humanized rodent model (Yoshida et al. 2006). Further examination of expression in the lung showed that FcRn levels were most prominent in the upper and central airways with less receptor being found in the peripheral lung regions, that is, the alveoli in humans and nonhuman primates. Other tissues that express FcRn include the kidney, where the receptor is thought to play a role in the reabsorption of IgG (Haymann et al. 2000; Akilesh et al. 2008), and also in the liver, where the receptor may contribute to maintaining the levels of circulating IgG (Blumberg et al. 1995; Telleman and Junghans 2000).

FcRn is also expressed in the vascular endothelial cells, one of the sites at which IgG protection occurs, resulting in the maintenance and long half-life of IgG in the circulation (Ghetie et al. 1996; Borvak et al. 1998). More recently, it has been reported that bone marrow-derived cells from mice and humans show high expression of FcRn and also contribute to IgG homeostasis (Zhu et al. 2001; Akilesh et al. 2007). FcRn is located primarily in intracellular compartments of the cell in most cell types that express FcRn, with limited expression on the cell surface (Borvak et al. 1998; Antohe et al. 2001). It is here that FcRn acts as a protective receptor, through the binding of IgG in acidic endosomes. The IgG bound to FcRn in the acidic endosomes is then shuttled back to the plasma membrane and released into the circulation at neutral pH, whereas IgG that has not interacted with FcRn appears to be degraded in lysosomes (Ward et al. 2003). For a review of FcRn, please refer to Roopenian and Akilesh (2007).

33.3 FcRn FOR DELIVERY OF Fc-FUSION MONOMERS AND DIMERS

The FcRn pathway was explored as a means to deliver large therapeutic proteins noninvasively through the lung (Bitonti and Dumont 2006; Dumont et al. 2006; Low et al. 2005). Expression of FcRn was confirmed in the lungs of adult mice, monkeys, and humans (Spiekermann et al. 2002; Bitonti et al. 2004) and transcellular transport of IgG has also been demonstrated in human lung epithelial cells, as well as intestinal epithelial cells (Dickinson et al. 1999; Kim et al. 2004). In epithelial cells, pH-dependent binding to FcRn occurs in acidic intracellular compartments, but in contrast to the trafficking in endothelial cells, the IgG is transported across the cell and is released at neutral pH at the basolateral surface.

Fc fusion proteins are most commonly generated as homodimers containing two effector molecules fused to a dimer of Fc (Fig. 33.1). However, depending on the therapeutic protein fused to the Fc dimer, these molecules may be very large and because of this, steric hindrance between the two effector molecules or between the effector molecules and the Fc may result. In addition, therapeutic homodimers may be heavily charged molecules due to glycosylation on the therapeutic protein portion of the molecule. All of these properties present a possible challenge for uptake through the lung.

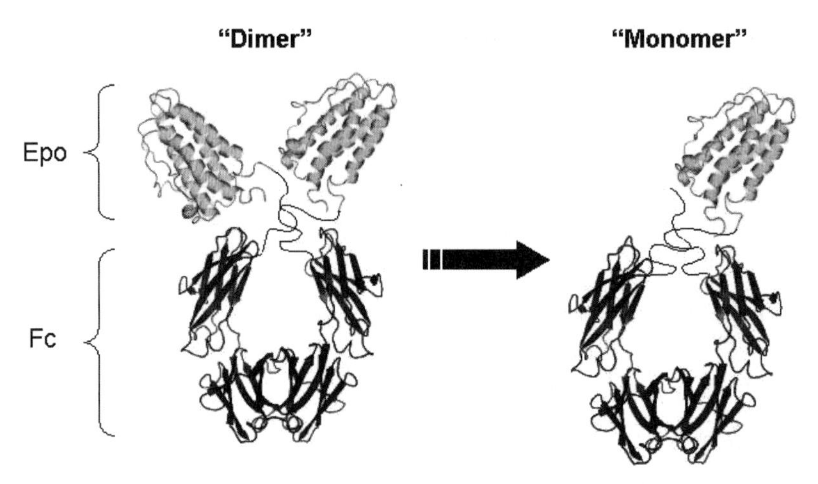

Figure 33.1 Schematic diagram for erythropoietin Fc (EpoFc) dimer and monomer. The dimer is comprised of two molecules of Epo joined to dimeric Fc and the monomer has a single molecule of Epo joined to dimeric Fc. (See color insert.)

Airway mucus provides a protective environment for the airways and can be a considerable barrier for the absorption of molecules through the lung (Rubin 1996; Knowles and Boucher 2002; Lillehoj and Kim 2002). It is known that size can affect the ability of molecules to penetrate the mucus layer. In fact, there is an inverse relationship between molecular weight and the diffusion of particles through the mucus, which is most apparent at molecular weights of >30 kDa (Desai, Mutlu, and Vadgama 1992). It has also been shown that charge can affect endocytosis in epithelial cells, which is an important step for uptake through the lung (Williams 1984). Thus, monomer Fc fusion proteins consisting of one effector molecule fused to a dimer of Fc (Fig. 33.1) would be reduced in size and charge compared to the traditional homodimer effector Fc fusion protein, due to the removal of one effector molecule. Synthesis of Fc monomers would also potentially reduce steric hindrance introduced by two effector molecules, which could adversely affect the binding to both effector protein receptors and Fc receptors.

33.4 EpoFc MONOMER AND DIMER PROTOTYPE MOLECULES

33.4.1 EpoFc Monomer and Dimer

An Fc-fusion molecule comprised of erythropoietin (Epo) linked to the Fc domain of IgG1 was chosen as the prototype molecule to test the concept of delivery of a large therapeutic molecule by noninvasive means through the lung. Epo was selected because it is a potent cytokine that elicits a readily measured biologic response (i.e., differentiation of hematopoietic precursors to reticulocytes; Graber and Krantz 1978; Lacombe and Mayeux 1998). Epo is a 30 kDa molecule used in the treatment of anemia that has three N-linked glycosylation sites and one O-linked sugar with the carbohydrate structures comprising approximately 50 percent of the mass of fully glycosylated Epo (Tsuda et al. 1990). The EpoFc dimer was created by recombinantly fusing human erythropoietin (Epo) to the Fc domain (hinge-CH2-CH3) of IgG1. EpoFc dimer was produced recombinantly in Chinese hamster ovary (CHO) cells by transfection of a single plasmid expressing EpoFc. In contrast, the EpoFc monomer was produced by cotransfecting CHO cells with an expression plasmid encoding EpoFc as well as with a separate plasmid expressing Fc alone (Bitonti et al. 2004). In this way, three proteins were produced, namely EpoFc dimer, Fc dimer, and EpoFc monomer. A purification process was established to isolate the EpoFc monomer. The molecular weights of the dimer and monomer were approximately 112 kDa and 82 kDa, respectively.

TABLE 33.1 Biological Activity of EpoFc Monomer and Dimer

		Receptor Binding		
Drug	TF-1 Proliferation[a] (EC_{50}, nM)	EpoR Binding[b] K_d, nM	FcRn[c] K_{d1}, nM	FcRn[c] K_{d2}, nM
EpoFc monomer	0.09 ± 0.02	0.30 ± 0.10	10 ± 1.0	808 ± 147
EpoFc dimer	0.07 ± 0.02	0.28 ± 0.13	23 ± 8	994 ± 545
Epo	0.03 ± 0.01	0.19 ± 0.06	NA	NA
hIgG1	NA[d]	NA	10 ± 1.2	1923 ± 775

[a]Growth of human erythroleukemia cells was measured for 48 h. Values represent the concentration of drug that stimulated growth by 50% of maximum (EC_{50}).
[b]Epo receptor binding was determined in TF-1 cells by competition of binding of ^{125}I-Epogen.
[c]Surface plasmon resonance analysis using a BiaCore 3000 instrument was utilized to determine the binding affinity to soluble human FcRn at pH 6.0.
[d]NA = not applicable.

33.4.2 EpoFc Monomer and Dimer Characterization *In Vitro*

The biologic activity and binding affinity to the Epo receptor and FcRn were compared for the EpoFc monomer and dimer. The *in vitro* biologic activity was assessed by measuring the proliferation of TF-1 human erythroleukemia cells which are stimulated to grow in the presence of Epo. There was no significant difference in the EC_{50} (concentration of drug that stimulated growth to 50 percent of maximum) determined for either EpoFc molecule toward TF-1 cell proliferation (Table 33.1). The EpoFc monomer and dimer both bound to the Epo receptor with similar affinity, which was comparable to Epo alone. However, the EpoFc monomer bound to FcRn with affinity that was approximately twofold greater than the dimer and with the same affinity as human IgG1. These data suggest that there may be some steric hindrance affecting the binding of the dimeric Fc portion of the fusion molecule to FcRn.

33.4.3 Pulmonary Delivery of EpoFc Monomer and Dimer in Nonhuman Primates

EpoFc monomer and dimer pulmonary delivery was targeted to the upper and central airways of nonhuman primates since FcRn expression was demonstrated to be highest in these regions of the human and

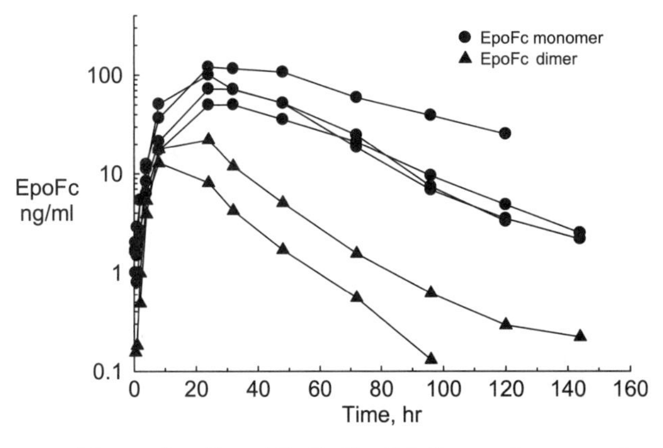

Figure 33.2 Pulmonary delivery of erythropoietin Fc (EpoFc) dimer and monomer in nonhuman primates. Equivalent doses (20 μg/kg) of EpoFc dimer or EpoFc monomer were aerosolized with an Aeroneb Pro® nebulizer and delivered directly into the lungs of intubated cynomolgus monkeys using shallow breathing maneuvers. Serum concentrations of the dimer and monomer were measured with an Epo-specific ELISA. The number of animals per group were $n = 2$ for EpoFc dimer and $n = 4$ for EpoFc monomer.

TABLE 33.2 Summary of Bioactivity and Pharmacokinetics for Dimeric and Monomeric Fc Fusion Proteins

		Pharmacokinetics					*In Vitro* Bioactivity	
Fusion Protein	Species and Route	Construct	Dose (μg/kg)	*Cmax* (ng/mL)	AUC (ng · hr · mL^{-1})	$t_{1/2}$ (hr)	Assay	Result
EpoFc	Monkey/pulmonary	Dimer	20	17	557	16	TF-1 proliferation assay[a] (EC_{50})	0.07 nM
		Monomer	20	86	5279	25		0.09 nM
IFN-β-Fc	Monkey/pulmonary	Dimer	20	3.8	123	11	CPE assay[b] (specific activity)	0.45×10^6 IU/nmol
		Monomer	20	22.3	987	27		1.22×10^6 IU/nmol
IFN-α-Fc[c]	Monkey/pulmonary	Dimer	15	107	18,760	107	CPE assay[b] (specific activity)	2.3×10^5 IU/nmol
		Monomer	15	102	17,325	87		5.2×10^5 IU/nmol
FIXFc	Neonatal rat/oral	Dimer	1370	590	9,570,00	14	Not determined	
		Monomer	1000	5830	121,250,000	19		
	Adult rat/IV	Dimer	5000	7500	109,000	22		
		Monomer	5000	33,000	509,000	35		
	Adult FIX-deficient mouse/IV	Dimer	5000	10,100	167,000	53		
		Monomer	5000	33,400	761,000	46		

[a]Growth of human erythroleukemia cells was measured for 48 hours. Values represent the concentration of drug that stimulated growth by 50% of maximum (EC_{50}).

[b]Antiviral activity was measured in a standard CPE assay using human lung A549 cells and encephalomyocarditis virus.

[c]Constructs with 15 amino acid linkers.

AUC = area under the serum or plasma concentration vs time curve; *Cmax* = maximum serum or plasma concentration; CPE = cytopathic effect; EC_{50} = concentration of drug that stimulates growth to 50% of maximum; EpoFc = erythropoietin-Fc; FIXFc = Factor IX-Fc; IFN = interferon; IU = international units; $t_{1/2}$ = terminal half-life in plasma or serum; TF-1 = human erythroleukemia cells.

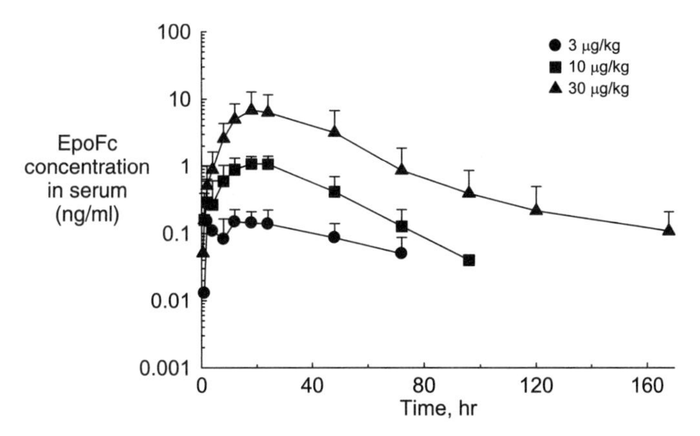

Figure 33.3 Pulmonary delivery of EpoFc dimer to human volunteers. Healthy volunteers were administered a single dose of EpoFc by inhalation using an Aeroneb Pro® nebulizer. The lung deposited doses were 3, 10, and 30 μg/kg. The number of subjects per group were 7, 6, and 8, respectively. Serum concentrations of the dimer and monomer were measured with an Epo-specific ELISA.

nonhuman primate lung, with very little expression in the alveolar regions (Spiekermann et al. 2002; Bitonti et al. 2004). This was accomplished by administering aerosolized protein with a particle size of 4 to 6 μm to intubated, anesthetized cynomolgus monkeys that were allowed to breathe spontaneously. Uptake of EpoFc monomer was markedly improved for EpoFc monomer compared to the dimer (Fig. 33.2, Table 33.2). The result was a 4- to 5-fold increase in the *Cmax*, a 6- to 10-fold increase in the area under the serum concentration-time curve (AUC) and an increase in the circulating half-life of the monomer (25 hours) compared to the dimer (16 hours). As a result, the bioavailability of EpoFc monomer was >30 percent (compared to intravenous administration), which is a significant improvement over the 5 to 10 percent bioavailability observed for EpoFc dimer. Therefore, reducing the size and charge of the molecule resulted in a striking increase in transport of the molecule via the lung epithelium. Moreover, this uptake was achieved almost completely via the upper and central airways since forced delivery of the EpoFc dimer to the deep lung resulted in less transport into the bloodstream. Similarly, the transport was proven to be FcRn mediated since mutation of three amino acids in the Fc domain required for FcRn binding (I253A, H310A, H435A) resulted in poor absorption when administered either to the upper airways or deep lung (Bitonti et al. 2004).

33.4.4 Pulmonary Delivery of EpoFc Dimer in Human Volunteers

EpoFc dimer was administered to human volunteers in order to demonstrate that the FcRn pathway functioned similarly in humans (Dumont et al. 2005). A single dose of aerosolized EpoFc dimer (3, 10, or 30 μg/kg) was administered by inhalation to healthy human volunteers who were advised to breath normally in order to favor deposition to the central airways. Uptake of the drug was dose dependent and serum concentration versus time curves were similar for the individuals within each dose group (Fig. 33.3). A biologic response, that is, an increase in circulating reticulocytes was observed for the volunteers treated with 30 μg/kg EpoFc, which demonstrated that not only was the molecule transported across the lung epithelium, but also it retained its activity. This study also demonstrated that the preclinical nonhuman primate model was predictive of the transport pathway in humans.

33.5 OTHER MONOMERIC Fc FUSION PROTEINS

Several other Fc fusion proteins were constructed as monomers with respect to the effector molecule and in each case, improvements were made in the bioactivity and/or the pharmacokinetic parameters (Dumont et al. 2006).

33.5.1 Interferon Alpha-Fc Monomer

Interferon-α (IFNα) is an 18 kDa cytokine that is used in the treatment of hepatitis C, hepatitis B, and several forms of cancer (Brassard, Grace, and Bordens 2002). IFNα has a relatively short half-life of 5 hours (Kozlowski and Harris 2001), necessitating a subcutaneous dosing regimen of three times per week. Pegylation of IFNα (Pegasys®, PEG-Intron®) extends the half-life of the molecule, allowing for once weekly dosing, but these improved molecules still require injection. We generated monomeric and dimeric Fc fusion proteins using IFNα in order to determine whether Fc extends the half-life of IFNα, to see if the molecules can be delivered by the pulmonary route and to examine whether monomers of IFNα have improved pharmacokinetic and efficacy properties compared to dimers.

IFNαFc monomers and dimers were constructed in a similar manner to that described for EpoFc. An *in vitro* assay commonly used to assess IFNα activity was used in which the protective effect of IFNα was measured in cells challenged with a cytopathic virus (Rubenstein, Famillenti, and Pestka 1981). In this case, the antiviral efficacy of purified IFNαFc monomers and dimers was tested in A549 lung carcinoma cells challenged with encephalomyocarditis virus (EMCV). With dimeric IFNαFc fusion proteins, antiviral activity was highly dependent on the length of the amino acid linker between the IFNα and the Fc domain (Table 33.3). Increasing the linker from 0 to 8 amino acids resulted in an approximate twofold increase in activity, while further lengthening to 15 amino acids resulted in an additional fivefold increase in activity indicating that there is likely steric hindrance between the two IFNα moieties in the dimeric IFNαFc.

When monomeric IFNαFc fusion proteins that contained only a single IFNα attached to the Fc domain were tested, there was less of an effect of linker length on antiviral activity of the molecules. However, in each case, the monomeric IFNαFc fusion proteins had significantly greater antiviral activity than the dimeric homolog. *In vitro* assays revealed a 10-fold increase in cytopathic activity when comparing the monomer to the dimer with no linker, and a similar 10-fold increase when comparing the eight amino acid linker constructs (Table 33.3). There was only an approximately twofold increase in activity when comparing the monomer to the dimer with the 15 amino acid linker, indicating that the majority of the steric hindrance was already relieved by the longer linker such that the monomer configuration had less of an effect.

Pulmonary administration of IFNαFc monomer and dimer (both with a 15 amino acid linker) to the central airways of cynomolgus monkeys showed significant and similar absorption of the two proteins (Fig. 33.4). Not only was the uptake much greater for both fusion proteins compared to unconjugated IFNα (Fig. 33.4), but also the serum half-life ($t_{1/2}$) was much longer (100 hours compared to 5 hours). The pharmacokinetic parameters of pulmonary uptake of IFNαFc monomer and dimer (i.e., $Cmax$, AUC, and $t_{1/2}$) were similar after inhalation of the same dose in nonhuman primates (Table 33.2).

Oligoadenylate synthetase (OAS) is a gene that is regulated by interferons (Wandinger et al. 2001). OAS activity was measured in serum from nonhuman primates treated with IFNαFc monomer and

TABLE 33.3 Antiviral Activity of IFNαFc Monomer and Dimer Constructs with Linkers of Varying Lengths[a]

	Dimeric IFNαFc Constructs			Monomeric IFNαFc Constructs		
Linker Length[b]	0	8	15	0	8	15
IU/mg ± SD	$2.5 \times 10^5 \pm 0.8 \times 10^5$	$4.9 \times 10^5 \pm 3.0 \times 10^5$	$2.5 \times 10^6 \pm 1.1 \times 10^6$	3.4×10^6	$6.3 \times 10^6 \pm 1.6 \times 10^6$	7.3×10^6
IU/nmol	2.2×10^4	4.5×10^4	2.3×10^5	2.4×10^5	4.5×10^5	5.2×10^5
n	3	10	13	2	3	2

[a]Antiviral activity was measured in a standard cytopathic effect assay using human lung A549 cells and encephalomyocarditis virus.
[b]Number of amino acids between the IFNα and Fc sequences.
IU = international units; n = number of assay replicates; SD = standard deviation.

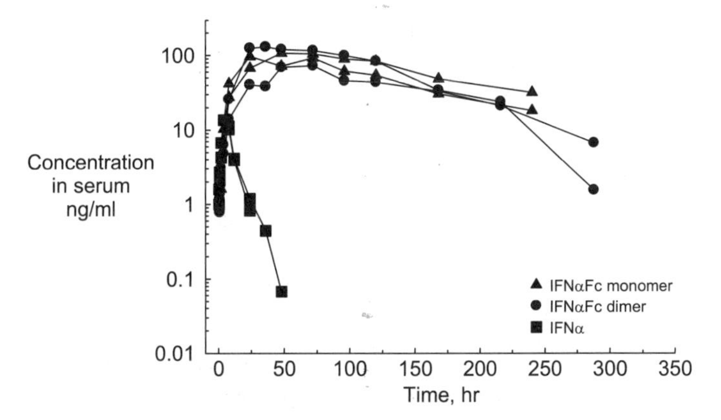

Figure 33.4 Pharmacokinetics of interferon α-Fc (IFNαFc) dimer, IFNαFc monomer, and IFNα after a single pulmonary dose in nonhuman primates. Equivalent doses (22.5 μg/kg) of IFNαFc dimer, IFNαFc monomer, or IFNα were aerosolized with an Aeroneb Pro® nebulizer and delivered directly into the lungs of intubated cynomolgus monkeys using shallow breathing maneuvers. Serum concentrations of the dimer and monomer were measured with an IFNα-specific ELISA. The number of animals per group were $n = 2$ for IFNαFc dimer, $n = 2$ for IFNαFc monomer, and $n = 2$ for IFNα.

dimer to monitor the biological activity of the fusion proteins and to confirm that the molecules were intact after pulmonary delivery. OAS activity increased in animals treated with both molecules but the peak OAS response was approximately twofold higher after treatment with the IFNαFc monomer compared to the dimer (Fig. 33.5). The twofold higher *in vivo* activity of IFNαFc monomer compared to the dimer is in good agreement with the *in vitro* cytopathic effect assays.

Therefore, the reduction in antiviral activity for both the monomer and dimer IFNαFc proteins appears to be due to steric hindrance of the Fc moiety with binding to the IFN receptor and increasing the distance of these two domains from one another via a linker seems to have relieved the constraints. In addition, the functionality of the IFNα moiety appears to be slightly better for the monomeric construct when compared on a molar basis to the dimeric construct. This *in vitro* activity was mirrored by OAS activity in nonhuman primates following pulmonary delivery of equivalent doses of IFNαFc

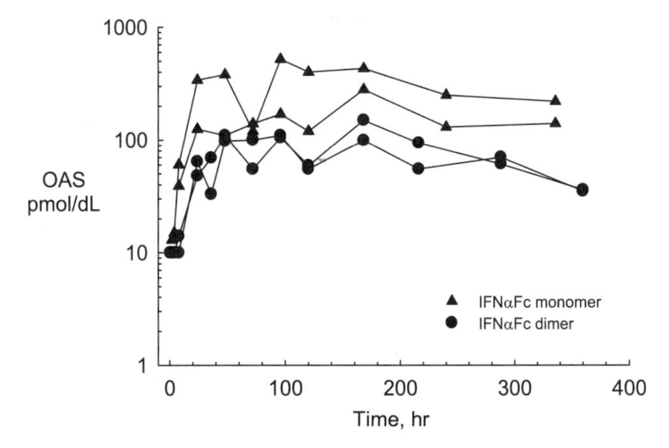

Figure 33.5 Bioactivity of IFNαFc dimer or IFNαFc monomer after a single pulmonary dose in nonhuman primates. Oligoadenylate synthetase (OAS) levels were quantified in serum samples from the nonhuman primates dosed with pulmonary IFNαFc dimer and IFNαFc monomer (22.5 μg/kg) as illustrated in Figure 33.4 using an OAS ELISA kit. The number of animals per group were $n = 2$ for IFNαFc dimer and $n = 2$ for IFNαFc monomer.

monomer and dimer. Unlike EpoFc in which the pulmonary uptake was significantly improved with the monomer, IFNαFc pulmonary uptake in primates was similar for both monomeric and dimeric molecules, which indicates that interaction with FcRn is equivalent for both molecules.

33.5.2 Interferon Beta-Fc Monomer

Interferon beta (IFNβ) is a 22.5 kDa glycoprotein used in the treatment of multiple sclerosis (MS; Noseworthy et al. 2000). Although the mechanisms by which IFNβ exerts its effects in MS are not well understood, IFNβ is known to regulate the expression of numerous genes (Wandinger et al. 2001) and some of these are commonly used as biomarkers of IFNβ efficacy, including oligoadenylate synthetase, neopterin and β2 m (Stürzebecher et al. 1999; Buchwalder et al. 2000). IFNβ is also known to protect cells from viral challenge, and an *in vitro* assay in which the protective effect of IFNβ was measured in cells challenged with a cytopathic virus (Rubenstein, Famillenti, and Pestka 1981) was used to characterize the specific activity of IFNβ. Similar to the assay used to measure *in vitro* activity of IFNαFc fusion proteins, the antiviral efficacy of purified IFNβFc monomers and dimers was tested in A549 lung carcinoma cells challenged with encephalomyocarditis virus (EMCV).

IFNβFc monomers and dimers were constructed in a similar manner to that described for EpoFc and IFNαFc. Unlike IFNαFc fusion molecules, linker length (0 to 15 amino acids) in IFNβFc dimeric proteins had little effect on the *in vitro* cytopathic effect assay (Table 33.4). However creating an IFNβFc monomer with an eight amino acid linker increased the *in vitro* activity of the molecule fourfold compared to the dimeric protein with the same linker. This suggests that having two molecules of IFNβ in the Fc fusion protein may cause some hindrance in the ability of both IFNβ molecules to bind to the IFN receptor, whereas a construct with only a single IFNβ molecule may have better access to the receptor.

Pulmonary uptake of equivalent doses of IFNβFc monomer and dimer were compared in nonhuman primates (Fig. 33.6). Pulmonary uptake was markedly improved for the monomer compared to the dimer, as indicated by a sixfold increase in the maximum serum concentration (*Cmax*) and an approximately twofold increase in the half-life resulting in an eightfold increase in the area under the serum concentration-time curve (AUC; Table 33.2). The improved pharmacokinetic parameters for IFNβFc monomer were mirrored by improved biological response compared to the IFNβFc dimer in the same nonhuman primates. Serum neopterin levels, a biomarker used to monitor activity of interferons (Stürzebecher et al. 1999; Buchwalder et al. 2000), were measured in each case and IFNβFc monomer was found to stimulate neopterin approximately twofold higher than IFNβFc dimer (Fig. 33.7).

Thus, in the case of IFNβFc, steric hindrance appears to be an issue only with dimeric Fc fusion proteins since the IFNβ antiviral activity is increased by making the molecule monomeric with respect to IFNβ. This *in vitro* activity was mirrored by neopterin activity in nonhuman primates following pulmonary delivery of equivalent doses of IFNβFc monomer and dimer (Bitonti and Dumont 2006).

TABLE 33.4 Antiviral Activity of Interferon β-Fc Monomer and Dimer Constructs with Varying Linker Lengths[a]

	Dimeric IFNβFc Constructs			Monomeric IFNβFc
Linker Length[b]	0	8	15	8
IU/mg ± SD	$4.1 \times 10^6 \pm 1.1 \times 10^6$	4.0×10^6	$4.7 \times 10^6 \pm 2.8 \times 10^6$	$17.2 \times 10^6 \pm 2.6 \times 10^6$
IU/nmol	3.7×10^5	3.7×10^5	4.4×10^5	15.9×10^5
n	3	2	3	4

[a]Antiviral activity was measured in a standard cytopathic effect assay using human lung A549 cells and encephalomyocarditis virus.
[b]Number of amino acids between the IFNβ and Fc sequences.
IU = international units; n = number of assay replicates; SD = standard deviation.

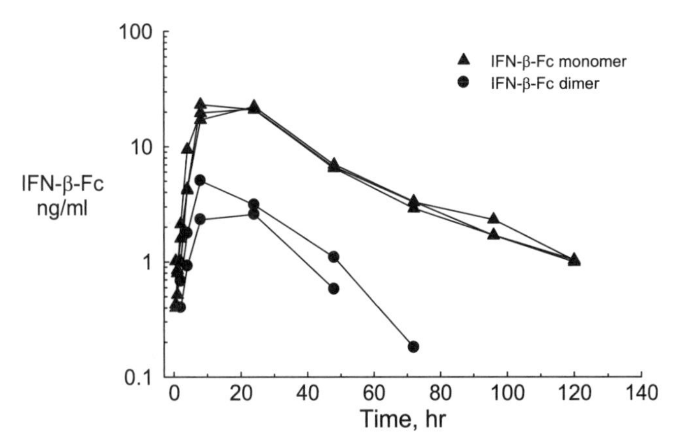

Figure 33.6 Pharmacokinetics of interferon β-Fc (IFNβFc) dimer or IFNβFc monomer after a single pulmonary dose in nonhuman primates. Equivalent doses (20 μg/kg) of IFNβFc dimer or IFNβFc monomer were aerosolized with an Aeroneb Pro® nebulizer and delivered directly into the lungs of intubated cynomolgus monkeys using shallow breathing maneuvers. Serum concentrations of the dimer and monomer were measured with an IFNβ-specific ELISA. The number of animals per group were $n = 2$ for IFNβFc dimer and $n = 3$ for IFNβFc monomer.

Unlike IFNαFc in which the pulmonary uptake was similar for both monomeric and dimeric molecules, IFNβFc pulmonary uptake in primates was significantly improved for IFNβFc monomer compared to IFNβFc dimer.

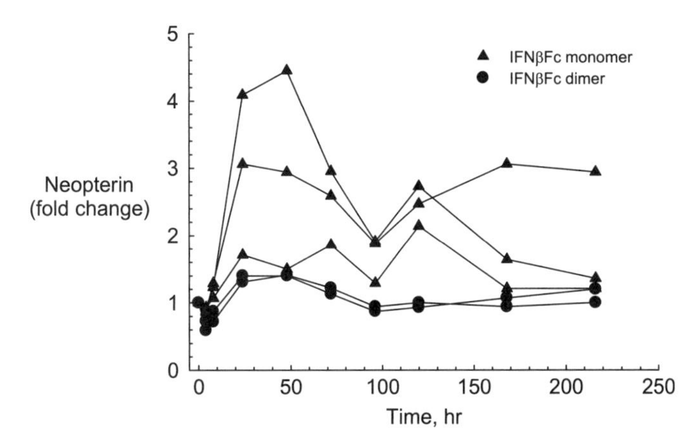

Figure 33.7 Bioactivity of IFNβFc dimer or IFNβFc monomer after a single pulmonary dose in nonhuman primates. Neopterin levels were quantified in serum samples from the nonhuman primates dosed with pulmonary IFNβFc dimer and IFNβFc monomer (20 μg/kg) shown in Figure 33.6, using a Neopterin ELISA kit. The number of animals per group were $n = 2$ for IFNβFc dimer and $n = 3$ for IFNβFc monomer.

33.5.3 Factor IX-Fc Monomer

Factor IX is a 55 kDa serine protease that functions as an essential component of the coagulation cascade (Limentani et al. 1993; Green, Naylor, and Giannelli 1995; Bolton-Maggs and Pasi 2003). A deficiency of factor IX results in an inherited disorder known as hemophilia B or Christmas disease (Giangrande 2005). Both plasma-derived and recombinant factor IX products are utilized for both the on-demand and prophylactic treatment of this disease. Fusion of factor IX (FIX) with Fc was pursued as

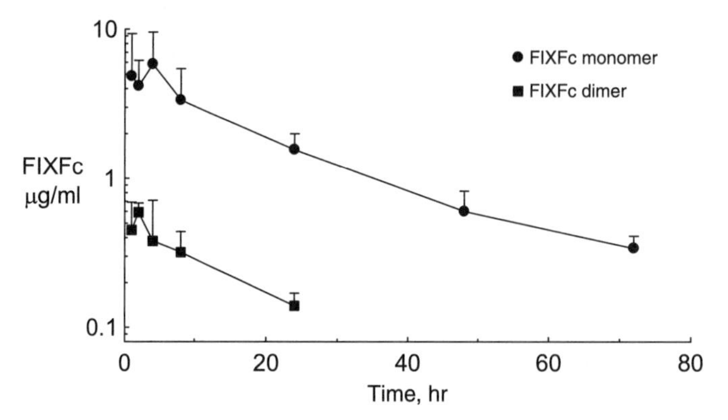

Figure 33.8 Comparison of the pharmacokinetics of factor IX (FIX)-Fc dimer with FIXFc monomer after a single oral dose in neonatal rats. Ten-day-old rats were administered equimolar doses ($\sim$10 nmol/kg) of FIXFc monomer or dimer. Blood was collected at various time points and the protein in the plasma was measured using a FIX:Fc sandwich ELISA ($n = 4$ animals per time point).

an approach to extend the half-life of this clotting factor and thereby reduce the frequency of injections for hemophilia B patients.

Factor IX-Fc (FIXFc) monomer and dimer were produced recombinantly in a manner similar to the other constructs described above. The pharmacokinetics of the two proteins were studied after a single oral dose in 10-day-old neonatal rats. At this age, the expression of FcRn and IgG transport in the intestine of rodents is relatively high, secretion of stomach acid low, and the level of digestive enzymes is low (Morris and Morris 1974; Simister and Rees 1985). As a result, formulation with trypsin inhibitor allows for the transport of intact immunoglobulins or Fc-fusion molecules across the intestinal epithelium and into the circulation. These studies demonstrated more efficient uptake of the FIXFc monomer into the circulation compared to the dimer after a similar molar dose, with a 10-fold increase in *Cmax* and greater than a 12-fold increase in AUC (Fig. 33.8; Table 33.2). In addition, the terminal half-life was somewhat longer for the monomer (19 hours) compared to the dimer (14 hours). The

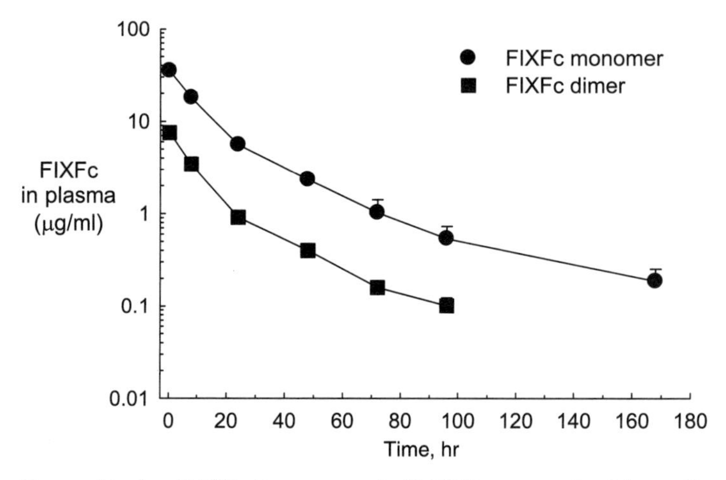

Figure 33.9 Pharmacokinetics of FIXFc dimer compared with FIXFc monomer in adult rats. Rats were given a single intravenous dose (5 mg/kg) of FIXFc dimer or FIXFc monomer. Blood was collected at various time points and the concentration of FIXFc was determined using a FIX:Fc sandwich ELISA ($n = 4$ animals per time point, curves represent the mean $\pm$ SD).

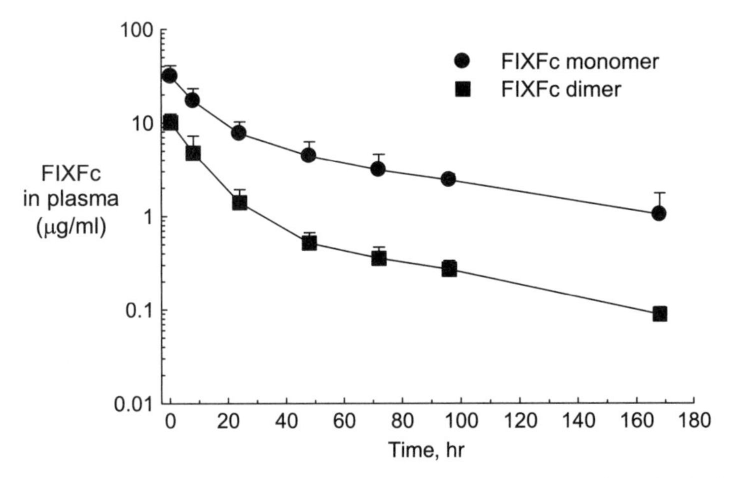

Figure 33.10 Pharmacokinetics of FIXFc dimer compared with FIXFc monomer in adult FIX-deficient mice. Mice were given a single intravenous dose (5 mg/kg) of FIXFc dimer or FIXFc monomer. Blood was collected at various time points and the concentration of FIXFc was determined using a FIX:Fc sandwich ELISA ($n = 4$ animals per time point, curves represent the mean $\pm$ SD).

pharmacokinetics of the monomer and dimer constructs were also studied in adult rats and in adult FIX-deficient mice after a single intravenous dose. The FIX-deficient mice are of a knockout strain that was produced using selective gene targeting (Lin et al. 1997) and as a result, these mice have no endogenous clotting activity, making them a good model for severe human hemophilia B. In both the rat and mouse studies, the *Cmax* and AUC for the monomer were considerably greater (three- to fourfold and fivefold, respectively) compared to that of the dimer, and the terminal half-life of the monomer was again greater in the adult rats compared to the dimer (Figs. 33.9 and 33.10, Table 33.2). However, in the FIX-deficient mice, the half-life for the two constructs was similar. These data show that the monomer construct has markedly improved pharmacokinetics after both oral and intravenous administration in rodents in comparison to the dimer construct. This may be due to a difference in recovery and also to an extended half-life of the monomer compared to the dimer.

33.6 SUMMARY

Previous work has shown that conjugation of molecules to the Fc domain of IgG results in an extension of half-life in the circulation. These Fc fusion proteins typically are comprised of two molecules of an effector molecule, each attached to one arm of dimeric Fc, thereby forming a homodimeric complex. We initially created Fc fusion dimers for the noninvasive delivery of large therapeutic proteins through the lung using an endogenous receptor called FcRn as the transporter. These proteins were successfully absorbed from the lung after administration of aerosolized protein; however, the bioavailability was not as robust as was anticipated. It was hypothesized that the uptake could possibly be improved by decreasing the size of the molecule, reducing the charge, and potentially decreasing any steric hindrance for either the effector portion of the molecule or Fc and binding to the respective receptors. Hence, a monomer of Fc and chimerized ligand was created as described herein through the removal of one effector molecule, which resulted in a reduction in size, a decrease in charge, and possibly a decrease in steric hindrance.

EpoFc was studied as both a monomer and a dimer for effects on biochemical properties, and pharmacokinetics following pulmonary delivery in nonhuman primates. Not only was the uptake of the monomer greatly enhanced compared to the dimer, but also the half-life was improved, resulting in bioavailability of approximately 30 percent, which is similar to that obtained for subcutaneously administered unconjugated Epo (Salmonsen, Danielson, and Wikström 1990).

Monomeric Fc fusion constructs were also produced for other proteins and in each instance, the monomer configuration demonstrated improved bioactivity and/or pharmacokinetics. Of these other fusion proteins, IFNαFc showed an increase in bioactivity as a monomer construct compared to the dimer, but there was no apparent difference in pharmacokinetic parameters. The transport properties of the IFNαFc dimer were already very good as a dimer, with approximately 50 percent pulmonary bioavailability compared to intravenous dosing in nonhuman primates and the half-life was 20 times longer compared to unconjugated IFNα. In contrast, IFNβFc showed improvement in both the bioactivity and pharmacokinetics as a monomer compared to the dimer. FIXFc was also studied as a monomer and dimer for effects on pharmacokinetics after oral or intravenous administration in rodents. Once again, the transport of the monomeric protein was notably enhanced after oral administration compared to the dimer and the half-life was extended. In addition, intravenous administration in adult rats and FIX-deficient mice both resulted in a much greater *Cmax* and AUC compared to the dimer. Therefore, the superior pharmacokinetic parameters of the monomer may not solely be due to better transport.

It is remarkable that the removal of one effector molecule from an Fc fusion construct has such a marked impact on the biological activity and/or pharmacokinetics of a molecule. The creation of these monomeric Fc fusion proteins represents an important advance in the area of Fc fusion molecules. As such, FcRn significantly enables mucosal delivery and/or half-life extension of therapeutic proteins with maintenance of the important biological properties of these macromolecules.

ACKNOWLEDGMENTS

We gratefully acknowledge Dr. Richard Blumberg (Brigham and Women's Hospital, Harvard Medical School, Boston, MA) for his expert review of this chapter. The authors are associated with Syntonix Pharmaceuticals, Inc., a subsidiary of Biogen Idec (Waltham, MA). The work with FIXFc was performed in partnership with Biovitrum (Stockholm, Sweden).

REFERENCES

Akilesh, S., G.J. Christianson, D.C. Roopenian, and A.S. Shaw. 2007. Neonatal FcR expression in bone marrow-derived cells to protect serum IgG from catabolism. *J. Immunol.* 179:4580–4588.

Akilesh, S., T.B. Huber, H. Wu, G. Wang, B. Hartleben, J.B. Kopp, J.H. Miner, D.C. Roopenian, E.R. Unanue, and A.S. Shaw. 2008. Podocytes use FcRn to clear IgG from the glomerular basement membrane. *Proc. Natl. Acad. Sci. U.S.A.* 105(3):967–972.

Antohe, F., L. Rădulescu, A. Gafencu, V. Ghetie, and M. Simionescu. 2001. Expression of functionally active FcRn and the differentiated bidirectional transport of IgG in human placental endothelial cells. *Human Immunol.* 62:93–105.

Bitonti, A.J., and J.A. Dumont. 2006. Pulmonary administration of therapeutic proteins using an immunoglobulin transport pathway. *Adv. Drug Delivery Rev.* 58:1106–1118.

Bitonti, A.J., J.A. Dumont, S.C. Low, R.T. Peters, K.E. Kropp, V.J. Palombella, J.M. Stattel, Y. Lu, C.A. Tan, J.J. Song, A.M. Garcia, N.E. Simister, G.M. Spiekermann, W.I. Lencer, and R.S. Blumberg. 2004. Pulmonary delivery of an erythropoietin Fc fusion protein in non-human primates through an immunoglobulin transport pathway. *Proc. Natl. Acad. Sci. U.S.A.* 101(26):9763–9768.

Blumberg, R.S., T. Koss, C.M. Story, D. Barisani, J. Polischuk, A. Lipin, L. Pable, R. Green, and N.E. Simister. 1995. A major histocompatibility complex class I-related Fc receptor for IgG on rat hepatocytes. *J. Clin. Invest.* 95:2397–2402.

Bolton-Maggs, P.H.B., and K.J. Pasi. 2003. Haemophilias A and B. *Lancet* 361:1801–1809.

Borvak, J., J. Richardson, C. Medesan, F. Antohe, C. Radu, M. Simionescu, V. Ghetie, and E.S. Ward. 1998. Functional expression of the MHC class I-related receptor, FcRn, in endothelial cells of mice. *Int. Immunol.* 10(9):1289–1298.

Brambell, F.W.R. 1966. The transmission of immunity from mother to young and the catabolism of immunoglobulins. *Lancet* 2(7473):1087–1093.

Brambell, F.W.R., W.A. Hemmings, and I.G. Morris. 1964. A theoretical model of γ-globulin catabolism. *Nature* 203:1352–1355.

Brassard, D.L., M.J. Grace, and R.W. Bordens. 2002. Interferon-alpha as an immunotherapeutic protein. *J. Leukoc. Biol.* 71(4):565–581.

Buchwalder, P.-A., T. Buclin, S. Trinchard, A. Munafo, and J. Biollaz. 2000. Pharmacokinetics and pharmacodynamics of IFNβ-1a in healthy volunteers. *J. Interferon Cytokine Res.* 20(10):857–866.

Burmeister, W.P., L.N. Gastinel, N.E. Simister, M.L. Blum, and P.J. Bjorkman. 1994. Crystal structure at 2.2 Å resolution of the MHC-related neonatal Fc receptor. *Nature* 372:336–343.

Desai, M.A., M. Mutlu, and P. Vadgama. 1992. A study of macromolecular diffusion through native porcine mucus. *Experientia* 48(1):22–26.

Dickinson, B.L., K. Badizadegan, Z. Wu, J.C. Ahouse, X. Zhu, N.E. Simister, R.S. Blumberg, and W.I. Lencer. 1999. Bidirectional FcRn-dependent IgG transport in a polarized human epithelial cell line. *J. Clin. Invest.* 104(7):903–911.

Dumont, J.A., A.J. Bitonti, D. Clark, S. Evans, M. Pickford, and S.P. Newman. 2005. Delivery of an erythropoietin-Fc fusion protein by inhalation in humans through an immunogloblulin transport pathway. *J. Aerosol Med.* 18(3):294–303.

Dumont, J.A., S.C. Low, R.T. Peters, and A.J. Bitonti. 2006. Monomeric Fc fusions: Impact on pharmacokinetic and biological activity of protein therapeutics. *BioDrugs* 20(3):151–160.

Fahey, J.L., and A.G. Robinson. 1963. Factors controlling serum γ-globulin concentration. *J. Exp. Med.* 118:845–868.

Ghetie, V., J.G. Hubbard, J.-K. Kim, M.-F. Tsen, Y. Lee, and E.S. Ward. 1996. Abnormally short serum half-lives of IgG in β2m-microglobulin-deficient mice. *Eur. J. Immunol.* 26:690–696.

Giangrande, P. 2005. Haemophilia B: Christmas disease. *Expert Opin. Pharmacother.* 6(9):1517–1524.

Graber, S.E., and S.B. Krantz. 1978. Erythropoietin and the control of red cell production. *Annu. Rev. Med.* 29:51–66.

Green, P.M., J.A. Naylor, and F. Giannelli. 1995. The hemophilias. *Adv. Genet.* 32:99–139.

Halliday, R. 1955a. The absorption of antibodies from immune sera by the gut of the young rat. *Proc. R. Soc. London B. Biol. Soc.* 143(912):408–413.

Halliday, R. 1955b. Prenatal and postnatal transmission of passive immunity to young rats. *Proc. R. Soc. London B.* 144(916):427–430.

Haymann, J.-P., J.-P. Levraud, S. Bouet, V. Kappes, J. Hagege, G. Nguyen, Y. Xu, E. Rondeau, and J.-D. Sraer. 2000. Characterization and localization of the neonatal Fc receptor in adult human kidney. *J. Am. Soc. Nephrol.* 11:632–639.

Israel, E.J., V.K. Patel, S.F. Taylor, A. Marshak-Rothstein, and N.E. Simister. 1995. Requirement for a β2m-microglobulin-associated Fc receptor for acquisition of maternal IgG by fetal and neonatal mice. *J. Immunol.* 154:6246–6251.

Israel, E.J., S. Taylor, Z. Wu, E. Mizoguchi, R.S. Blumberg, A. Bhan, and N.E. Simister. 1997. Expression of the neonatal Fc receptor, FcRn, on human intestinal epithelial cells. *Immunology* 92:69–74.

Israel, E.J., D.F. Wilsker, K.C. Hayes, D. Schoenfeld, and N.E. Simister. 1996. Increased clearance of IgG in mice that lack β2m-microglobulin: Possible protective role of FcRn. *Immunology* 89:573–578.

Jones, E.A., and T.A. Waldmann. 1972. The mechanism of intestinal uptake and transcellular transport of IgG in the neonatal rat. *J. Clin. Invest.* 51:2916–2927.

Junghans, R.P., and C.L. Anderson. 1996. The protection receptor for IgG catabolism is the β2m-microglobulin-containing neonatal intestinal transport receptor. *Proc. Natl. Acad. Sci.* 93:5512–5516.

Kim, J.K., M.-F. Tsen, V. Ghetie, and E.S. Ward. 1994. Localization of the site of the murine IgG1 molecule that is involved in binding to the murine intestinal Fc receptor. *Eur. J. Immunol.* 24:2429–2434.

Kim, J.K., M. Firan, C.G. Radu, C.-H. Kim, V. Ghetie, and E.S. Ward. 1999. Mapping the site on human IgG for binding of the MHC class I-related receptor, FcRn. *Eur. J. Immunol.* 29:2819–2825.

Kim, K.J., T.E. Fandy, V.H. Lee, D.K. Ann, A. Borok, and E.D. Crandall. 2004. Net absorption of IgG via FcRn-mediated transcytosis across rat alveolar epithelial cell monolayers. *Am. J. Physiol. Lung Cell Mol. Physiol.* 287(3):L616–L622.

Knowles, M.R., and R.C. Boucher. 2002. Mucus clearance as a primary innate defense mechanism for mammalian airways. *J. Clin. Invest.* 109(5):571–577.

Kozlowski, A., and M.J. Harris. 2001. Improvements in protein PEGylation: Pegylated interferons for the treatment of hepatitis C. *J. Control. Release* 72:217–224.

Lacombe, C., and P. Mayeux. 1998. Biology of erythropoietin. *Haemotologica* 83:724–732.

Lillehoj, E.R., and K.C. Kim. 2002. Airway mucus: Its components and function. *Arch. Pharm. Res.* 25(6):770–780.

Limentani, S.A., D.A. Roth, B.C. Furie, and B. Furie. 1993. Recombinant blood clotting proteins for hemophilia therapy. *Semin. Thromb. Hemostasis* 19:62–72.

Lin, H.-F., N. Maeda, O. Smithies, D.L. Straight, and D.W. Stafford. 1997. A coagulation factor IX-deficient mouse model for human hemophilia. *Blood* 90(10):3962–3966.

Low, S.C., S.L. Nunes, A.J. Bitonti, and J.A. Dumont. 2005. Oral and pulmonary delivery of FSH-Fc fusion proteins via neonatal receptor-mediated transcytosis. *Hum. Reprod.* 20(7):1805–1813.

Morris, B., and R. Morris. 1974. The absorption of ^{125}I-labelled immunoglobulin G by different regions of the gut in young rats. *J. Physiol.* 241:761–770.

Noseworthy, J.H., C. Lucchinetti, M. Rodriguez, and B.G. Weinshenker. 2000. Multiple sclerosis. *N. Engl. J. Med.* 343(13):938–952.

Raghavan, M., L.N. Gastinel, and P.J. Bjorkman. 1993. The class I major histocompatibility complex related Fc receptor shows pH-dependent stability differences correlating with immunoglobulin binding and release. *Biochemistry* 32:8654–8660.

Roberts, D.M., M. Guenthert, and R. Rodewald. 1990. Isolation and characterization of the Fc receptor from the fetal yolk sac of the rat. *J. Cell. Biol.* 111:1867–1876.

Rodewald, R. 1973. Intestinal transport of antibodies in the newborn intestine. *J. Cell. Biol.* 58:189–211.

Rodewald, R. 1976. pH-dependent binding of immunoglobulins to intestinal cells of the neonatal rat. *J. Cell. Biol.* 71:666–670.

Roopenian, D.C., and S. Akilesh. 2007. FcRn: The neonatal Fc receptor comes of age. *Nature Rev.* 7:715–725.

Roopenian, D.C., G.J. Christianson, T.J. Sproule, A.C. Brown, S. Akilesh, N. Jung, S. Petkova, L. Avanessian, E.Y. Choi, D.J. Shaffer, P.A. Eden, and C.L. Anderson. 2003. The MHC class I-like IgG receptor controls perinatal IgG transport, IgG homeostasis, and fate of IgG-Fc-coupled drugs. *J. Immunol.* 170:3528–3533.

Rubin, B.K. 1996. Therapeutic aerosols and airway secretions. *J. Aerosol Med.* 9(1):123–130.

Rubenstein, S., P.C. Famillenti, and S. Pestka. 1981. Convenient assay for interferons. *J. Virol.* 37(2):755–758.

Salmonsen, T., B.G. Danielson, and B. Wikström. 1990. The pharmacokinetics of recombinant human erythropoietin after intravenous and subcutaneous administration to healthy subjects. *Br. J. Clin. Pharmacol.* 29:709–713.

Simister, N.E., and Mostov, K.E. 1989. An Fc receptor structurally related to MHC class I antigens. *Nature* 337:184–187.

Simister, N.E., and A.R. Rees. 1985. Isolation and characterization of an Fc receptor from neonatal rat small intestine. *Eur. J. Immunol.* 15:733–738.

Simister, N.E., C.M. Story, H.L. Chen, and J.S. Hunt. 1996. An IgG-transporting Fc receptor expressed in the syncytiotrophoblast of human placenta. *Eur. J. Immunol.* 26:1527–1531.

Spiekermann, G.M., P.W. Finn, E.S. Ward, J. Dumont, B.L. Dickinson, R.S. Blumberg, and W.I. Lencer. 2002. Receptor-mediated immunoglobulin G transport across mucosal barriers in adult life: Functional expression of FcRn in the mammalian lung. *J. Exp. Med.* 196(3):303–310.

Stürzebecher, S., R. Maibauer, A. Heuner, K. Beckmann, and B. Aufdembrinke. 1999. Pharmacodynamic comparison of single doses of IFN-beta 1a and IFN-beta 1b in healthy volunteers. *J. Interferon Cytokine Res.* 19(11):1257–1264.

Telleman, P., and R.P. Junghans. 2000. The role of the Brambell receptor (FcRB) in liver: Protection of endocytosed immunoglobulin G (IgG) from catabolism in hepatocytes rather than transport of IgG to bile. *Immunology* 100(2):245–251.

Tsuda, E., G. Kawanishi, M. Ueda, S. Masuda, and R. Sasaki. 1990. The role of carbohydrate in recombinant human erythropoietin. *Eur. J. Biochem.* 188:405–411.

Vaughn, D.E., and P.J. Bjorkman. 1998. Structural basis of pH-dependent antibody binding by the neonatal Fc receptor. *Structure* 6:63–73.

Wandinger, K.P., C.S. Stürzebecher, B. Bielekova, G. Detore, A. Rosenwald, L.M. Staudt, H.F. McFarland, and R. Martin. 2001. The complex immunomodulatory effects of interferon-β in multiple sclerosis include the upregulation of T helper 1-associated marker genes. *Ann. Neurol.* 50(3):349–357.

Ward, E.S., J. Zhou, V. Ghetie, and R.J. Ober. 2003. Evidence to support the cellular mechanism involved in serum IgG homeostasis in humans. *Int. Immunol.* 15(2):187–195.

West, A.P., and P.J. Bjorkman. 2000. Crystal structure and immunoglobulin G binding properties of the human major histocompatibility complex-related Fc receptor. *Biochemistry* 39:9698–9708.

Williams, M.C. 1984. Endocytosis in alveolar type II cells: Effect of charge and size of tracers. *Proc. Natl. Acad. Sci. U.S.A.* 81(19):6054–6058.

Yoshida, M., A. Masuda, T.T. Kuo, K. Kobayashi, S.M. Claypool, T. Takagawa, H. Kutsumi, T. Azuma, W.I. Lencer, and R.S. Blumberg. 2006. IgG transport across mucosal barriers by neonatal Fc receptor for IgG and mucosal immunity. *Springer Semin. Immun.* 28:397–403.

Zhu, X., G. Meng, B.L. Dickinson, X. Li, E. Mizoguchi, L. Miao, Y. Wang, C. Robert, B. Wu, P.D. Smith, W.I. Lencer, and R.S. Blumberg. 2001. MHC class I-related neonatal Fc receptor for IgG is functionally expressed in monocytes, intestinal macrophages, and dendritic cells. *J. Immunol.* 166:3266–3276.

CHAPTER 34

Radioimmunotherapy: Current Status and Future Directions

NEETA PANDIT-TASKAR and CHAITANYA R. DIVGI

ABSTRACT

A number of new antibodies have been developed for cancer therapy in the past decade. Following approval by the U.S. Food and Drug Administration and the success of two radiolabeled antiCD20 antibodies against lymphoma, there has been renewed interest in radioimmunotherapy. Development of new radiotherapeutic agents has been most marked in the hematologic neoplasms for myeloablative or nonmyeloablative treatment strategies. However, radioimmunotherapy of solid tumors is still a challenge mostly due to tumor heterogeneity, the immunogenicity of murine proteins, and the relatively slow clearance of humanized intact immunoglobulins. Development of a variety of antigen-binding constructs though genetic engineering is being evaluated and may be more valuable due to variation

Therapeutic Monoclonal Antibodies: From Bench to Clinic. Edited by Zhiqiang An

in size and immunobiologic characteristics. A variety of therapeutic radionuclides are now available for use with antibodies through advances in radiochemistry enabling more optimal imaging for dosimetry. Such progress has generated pivotal studies that will establish the radiobiologic paradigms for successful radioimmunotherapy in solid tumor. This chapter describes studies that have guided a better understanding of radioimmunotherapy in solid tumors.

34.1 INTRODUCTION

Radioimmunotherapy (RIT) has emerged as one of many promising therapeutic strategies that are encompassed by the "magic bullet" concept. RIT utilizes a combination of an antibody that recognizes tumor-associated antigens and carries cytotoxic radionuclides to a target, thereby destroying cancer cells. A landmark achievement was the approval of two radiolabeled antibodies by the U.S. Food and Drug Administration (FDA) for the treatment of lymphoma, which renewed interest in and offered hope for the use of RIT in the management of cancer. Developing an effective RIT is challenging, especially in the treatment of solid tumors, where progress has been slow.

Initially proof-of-principle studies with radiolabeled antibodies utilized polyclonal antibodies that lacked uniformity and therefore as therapeutic agents. Monoclonal antibodies have uniform reactivity and immunobiologic characteristics. However, these are produced in murine systems and the invariable immunogenicity of murine proteins (in all but patients with B-cell lymphoma) necessitated the development of less immunogenic chimeric and humanized antibodies. Some of the chimeric and humanized monoclonal antibodies are now part of the therapeutic armamentarium for lymphoma and solid tumors. However, their slow serum clearance makes them suitable for immunotherapy but not ideal as carriers of cytotoxic radioisotopes.

RIT has the advantages of selective delivery of cytotoxic radiation to tumor cells following preferential uptake of radiolabeled antibody by tumor; high residence time in tumor; cross-fire effect by particle emissions and minimal deleterious effect on normal tissues and organs. Radioimmunotherapy supplements the cytotoxic immune mechanisms of the unconjugated antibody therapy with the cross-fire and bystander effects and by overcoming the innate resistance of tumor cells to direct antitumor mechanisms.

Currently, the only radiolabeled antibodies with FDA approval are for the treatment of B-cell non-Hodgkin lymphoma (NHL); they have demonstrated satisfactory response in relapsed and refractory disease, as well as potential for use with certain chemotherapeutic agents and autologous stem cell transplant (Witzig et al. 2003a; Horning 2003; Press 2003; Press et al. 2003; Kaminski et al. 2001). Success in the radioimmunotherapy of solid tumors has been hampered by the immunogenicity of murine proteins and the relatively slow clearance of humanized intact immunoglobulins. Response to RIT has been studied in various tumor types, including breast, ovarian, colorectal, renal, prostate, and brain with less encouraging results. Overall responses have not been adequate to warrant phase II studies or further development with nonmyeloablative radioimmunotherapy. However, much of the initial work had provided significant insight in understanding the variations in the physiology in solid tumors and shed considerable light on optimizing antigen-binding construct design, choice of radionuclides for tumors of various sizes, and design of treatment delivery systems that will result in novel approaches to targeted radiopharmaceutical therapy.

Designing optimal RIT for a solid tumor depends on many variables, most importantly the size of constructs. Other factors that are key in optimizing the design of therapeutic radiolabeled antibody include immunogenicity, heterogeneity of antigen expression, uptake by tumor secondary to vascular and tumor permeability, deleterious effect on normal tissues, and residence time in the body (O'Donaghue 1999; Schlom et al. 1990). Recombinant techniques, as well as the development of small antigen-binding constructs, are just some of the methods being utilized to improve the optimal targeting of a tumor. Understanding of the potential and limitations of radionuclides as well as antigen-binding constructs is necessary.

34.2 RADIONUCLIDES

Radionuclides form a very important component of RIT. Physical and chemical properties of the radionuclide, its fate after antibody metabolism *in vivo*, and the nature of the emitted radiation are key determinants of the suitability of a radionuclide for therapy. Cytotoxic radionuclides may be divided into three groups of radiochemicals: halogens (iodine, ^{211}At); metals (^{90}Y, ^{67}Cu, ^{213}Bi; ^{212}Bi); and transitional elements (^{188}Re). Radionuclides can also be categorized into four types based on the type of emissions: pure beta emitters (^{67}Cu and ^{90}Y), alpha emitters (^{213}Bi, ^{211}At), beta emitters that also emit gamma radiation (^{177}Lu, ^{186}Re, ^{188}Re, ^{131}I), and Auger electron emitting radionuclides that decay by internal conversion (^{125}I and ^{67}Ga) (Table 34.1).

^{131}I is the most widely used radionuclide for the treatment of cancer and is a mainstay in the treatment of differentiated thyroid cancer. The biodistribution has been well studied and high energy gamma emissions allow for imaging and localization of tumor to confirm targeting. It has a half-life of eight days and the labeling process is easy. The high energy gamma emissions also raise potential concerns regarding the safety of patients and exposure to those around; specific guidelines, however, do permit radioimmunotherapy with ^{131}I to be given as outpatient treatment. Hospitals in the United States may release patients only if the expected exposure to an individual due to proximity to the patient is no greater than 5 mSv (500 mrem; Zanzonico, Siegel, and St. Germain 2000).

Radioactive iodine (^{131}I) if conjugated to antibodies that are internalized undergoes dehalogenation of the complex and release of the radionuclide, thus limiting tumor dose. Other radionuclides therefore have been studied especially for antibodies that may be internalized. ^{90}Y, a pure beta emitting radiometal, has been explored as a high energy alternative to ^{131}I. Conjugation of ^{90}Y with internalized antibody-antigen complexes is stable as radiometal antibodies are infrequently degraded internally. The success of ^{90}Y for use in RIT is evident as the first RIT to be approved by the FDA is an anti-CD20 antibody conjugated to ^{90}Y [Zevalin®, Cell Therapeutics, Inc. (CTI), Seattle, Washington]. One limitation of yttrium (^{90}Y), however, is its affinity for bone, which could result in high levels of radiation to marrow when ^{90}Y is detached from the chelating agent. Both ^{186}Re and ^{188}Re have been under investigation for use in RIT and have been linked to a variety of antibodies. ^{188}Re is produced from a Tungsten-188 generator, an advantage if not having cyclotron. Its short half-life of only 17 hours renders it suitable for locoregional therapies or for therapies with small molecules that are cleared from the body rapidly. Because of its physical properties, including longer half-life, ^{186}Re has been more extensively studied. Rhenium nuclides have gamma emissions, which allow for detection of distribution and clearance by imaging. Early clinical trials used gamma emitting ^{186}Re to get dosimetric analyses of absorbed doses to tumors and normal organs and have demonstrated its safety for use in RIT for solid tumors (Postema et al. 2003a).

Recent developments in radiochemistry have led to increased interest in alpha-emitters for therapy. Alpha particles are high energy helium nuclei with high linear energy transfer (LET). In addition to having short half-lives, an alpha particle's range is only 50 to 80 μm, making it most useful for individual cells, such as micrometastases and circulating tumor cells. In fact, ^{213}Bi conjugated to a

TABLE 34.1 Radionuclides Used for Therapy

Radionuclide (Therapeutic Emission)	$T_{1/2}$	Average Energy
^{131}I (beta)	8 days	191 Kev
^{90}Y (beta)	2.67 days	934 Kev
^{177}Lu (beta)	6.7 day	150 Kev
^{186}Re (beta)	3.78 day	362 Kev
^{188}Re (beta)	17 h	795 Kev
^{211}At (alpha)	7 h	7.5 Mev (^{211}Po)
^{213}Bi (alpha)	46 min	8.4 Mev (^{213}Po)
^{225}Ac (alpha)	10 days	8.4 Mev

Source: http://ie.lbl.gov/toi/nucSearch.asp.

humanized anti-CD33 antibody is being evaluated in clinical trials for the treatment of refractory and relapsed acute myelogenous leukemia. In addition to ^{212}Bi and ^{213}Bi, obtained from ^{224}Ra and ^{225}Ac generators, cyclotron-produced ^{211}At is under investigation as therapy for several tumor types. Clinical trials have been reported using ^{211}At-labeled chimeric anti-tenascin antibody 81C6 for gliomas (Zalutsky et al. 2002).

In vivo generators of alpha particles are being investigated for their ability to deliver cytotoxic particles to sites of micrometastases. These generators overcome the limitations posed by the short half-lives of alpha particles. These nanogenerators have been studied in various tumor types in mice. One study conjugated a ^{225}Ac generator to an internalizing antibody that targeted Her-2/neu (Trastuzumab) to study its efficacy in treating disease in a mouse model of advanced ovarian cancer (Borchardt et al. 2003). This generator produces three alpha particle-emitting radioisotopes, including ^{213}Bi.

Auger emitters deposit high LET over extremely short distances and are therefore most effective when the decay occurs in the nucleus, and less so when the decay occurs in the cytoplasm (Adelstein et al. 2003). ^{125}I is the prototypical radionuclide but its long $T_{1/2}$ renders it less than optimal for therapy. Other similar radionuclides that have been studied, although not with antibodies, have included ^{111}In. In both cases, the amount of radioactivity necessary is economically prohibitive. A radionuclide that is gaining increasing attention in this category is ^{67}Ga. Improvements in chelation chemistry have resulted in stable radioimmunoconjugates with ^{67}Ga that are under investigation.

Alpha emitters deposit high energy, over shorter distances and therefore are extremely cytotoxic for small clusters of tumor cells ($<200\ \mu m$). Antibodies conjugated with alpha emitters have shown promise in leukemia and when instilled into the surgically created brain tumor resection cavity, and will be most useful when there is rapid targeting of radioimmunoconjugate to tumor cells. However, this high LET may preclude their use in solid tumor therapy (except perhaps in the adjuvant situation), where irradiation of normal tissue may preclude delivery of adequate cytotoxic radiation to the tumor.

34.3 RADIOIMMUNOTHERAPY OF LYMPHOMA

Non-Hodgkin lymphoma (NHL) is the fifth most common malignancy in the United States. The most common forms of NHL are large B-cell lymphoma and follicular lymphoma. Over 90 percent of B-cell lymphomas are CD20+ (Advani, Rosenberg, and Horning 2004). In the United States, two FDA-approved radioimmunotherapy agents are available for use in low-grade follicular or transformed lymphomas that are refractory to rituximab. These radiolabeled antibodies are anti-CD20 antibodies that tagged with a beta radiation emitter uses the added cross-fire effect of ionizing radiation to augment the antitumor effect. Other antigens, including CD19, CD21b, CD22, and HLA-DR have also been targeted against lymphomas, of which anti-CD22 antibodies have been studied extensively. Relatively slow clearance of humanized anti-CD22 antibody makes it an unlikely candidate for radioimmunotherapy (Postema et al. 2003b; Sharkey et al. 2003).

34.3.1 Anti-CD20 Antibodies

Rituximab, an unlabeled monoclonal antibody, directed against the B-lymphocyte-specific antigen CD20, has demonstrated significant single-agent activity in follicular lymphoma, with response rates of 50 to 70 percent and median response duration of approximately 12 months (Hainsworth et al. 2000). Rituximab combined with chemotherapy has shown improved response rates in indolent lymphoma that may be as high as 93 to 95 percent (Czuczman 1999, 2002).

Two radiolabeled anti-CD20 antibodies, ibritumomab tiuxetan (Zevalin®; Cell Therapeutics, Inc., a ^{90}yttrium-labeled agent), and tositumomab (Bexxar®, GlaxoSmithKline Inc., a ^{131}iodine-labeled agent), have been approved by the U.S. FDA. Both ^{90}Y-ibritumomab tiuxetan and ^{131}I tositumomab are approved for use in patients with relapsed or refractory low grade, follicular, or transformed B-cell NHL, including follicular lymphoma refractory to rituximab.

34.3.1.1 Ibritumomab Tiuxetan Ibritumomab, is the parent IgG_1 monoclonal antibody murine antibody of the chimeric murine-human monoclonal antibody rituximab (IDEC Pharmaceutical Corp. 2001). The linker-chelator tiuxetan (MX-DTPA) is bound via a covalent bond to ibritumomab and provides a high-affinity chelation site for radioisotopes such as indium-111 or yttrium-90. Rapid clearance of murine antibody helps in limiting the total-body irradiation as compared to chimeric antibody. The therapeutic effect of the agent appears to be secondary to a combination of apoptosis induction, antibody-dependent cellular toxicity, and complement-mediated cell lysis mediated by rituximab (Shan, Ledbetter, and Press 1998; Flieger et al. 2000), along with radiolysis caused by ^{90}Y. Data from four clinical trials in a total of 179 patients, demonstrated highest median radiation absorbed doses for ^{90}Y (administered at 0.4 mCi/kg, up to a maximum of 32 mCi) to spleen and liver (Wiseman et al. 2003).

Radioimmunotherapy with ibritumomab tiuxetan is carried out in two phases, an imaging phase and a therapeutic phase given as an outpatient procedure (patient condition permitting), as ^{90}Y is a pure beta emitter. The first phase consists of an infusion of rituximab (250 mg/m^2 administered following the same guidelines as for rituximab given alone) followed by an imaging dose of 5 mCi of ^{111}In-labeled ibritumomab tiuxetan. Administration of the cold antibody before administration of radiolabeled agent diminishes nonspecific binding to circulating CD20+ B-cells, improving biodistribution and tumor targeting, and theoretically saturates peripheral CD20 sites on bulky tumors, allowing greater penetration of the radioimmunoconjugate dose. Whole-body gamma camera imaging is done at 48 to 72 hours to ensure acceptable biodistribution of the radioimmunoconjugate. Unfavourable biodistribution that would prevent administration of therapeutic dose includes presence of significant radioactivity in the vasculature in delayed images or increased radioactivity in the kidneys or the bowel that would suggest the presence of an immune response to the murine protein. Visualization of targeting to tumor is not a prerequisite for therapy. Dosimetry is not required before therapy in the defined patient population; imaging is carried out to exclude altered biodistribution that may predict increased toxicity (Witzig et al. 2003a). If the biodistribution is found favorable, the dose for therapy is decided based on the platelet counts. Dose of 0.4 mCi/Kg is given if the platelet count is $>$150,000/μL; 0.3 mCi/Kg when the platelet count is between 100,000 and 149,000/μL, up to a maximum in either case of 32 mCi. Treatment is contraindicated in those with platelet counts $<$100,000/μL.

Figure 34.1 depicts anterior and posterior early and delayed whole body planar images of a patient at 48 and 96 hours after ^{111}In-ibritumomab tiuxetan. Early images show tracer activity in the blood pool and accumulation in the liver and spleen, with clearance over time. Accumulation in kidneys is mild; no significant bowel activity is seen. The therapy is given in the following week; again after infusion of 250 mg/m^2 rituximab followed by a therapeutic dose of ^{90}Y-labeled ibritumomab tiuxetan infused over 10 minutes.

The phase I/II study with ibritumomab tiuxetan established the maximum tolerated dose to be 0.4 mCi/kg in patients with platelet counts greater than 150,000/μL, and 0.3 mCi/kg in patients with platelet counts between 100,000 and 150,000/μL. Follow-up studies in patients with relapsed or refractory low grade or transformed NHL with mild thrombocytopenia (100 to 149 K) confirmed the effectiveness of the 0.3 mCi/kg dose regimen, with an overall response rate (ORR) of 83 percent [complete recovery (CR) = 37 percent; unconfirmed CR = 6.7 percent; partial response (PR) = 40 percent]. The median time to progression was 9.4 months (Witzig 2003).

34.3.1.1.1 Clinical Trials: Standard Use Two pivotal trials reported good responses. In a phase III randomized, controlled study of 143 patients with relapsed or refractory low grade, follicular, or CD20-positive transformed B-cell NHL, ibritumomab tiuxetan combined with rituximab had a higher ORR compared with rituximab alone (80 percent versus 56 percent). The CR rate with ibritumomab tiuxetan was also higher than that of rituximab alone (30 percent versus 16 percent; Witzig et al. 2002b).

In patients with follicular NHL who were refractory to rituximab, ORR of 74 percent, with 15 percent of patients achieving CRs was seen (Witzig et al. 2002a). Phase II trials in relapsed indolent

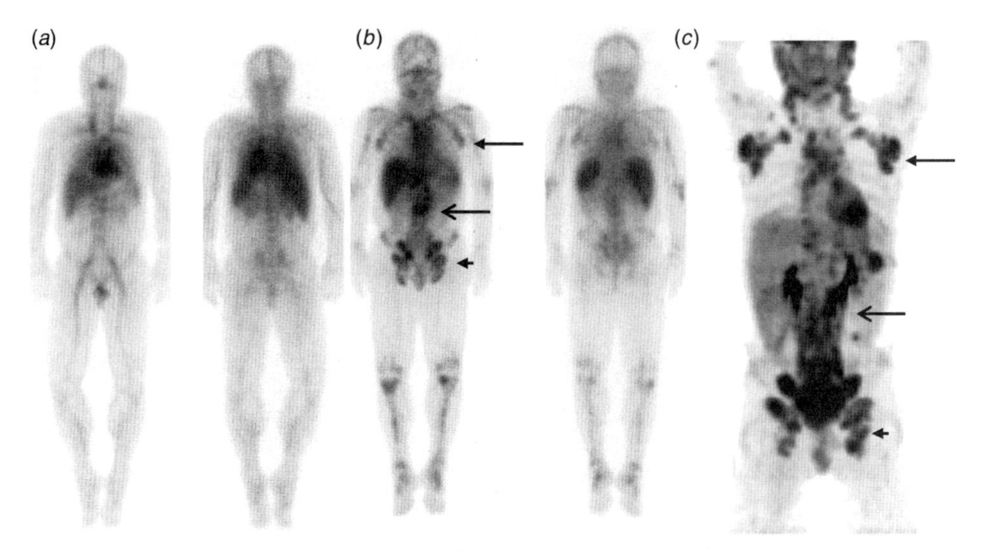

Figure 34.1 Patient with lymphoma injected with ^{111}In ibritumomab tuixetan. Anterior and posterior whole body images show tracer in heart, liver, and spleen and mild activity in kidneys (a). There is visualization of tumor in later images (b) in the axillary (upper arrow), abdominal (middle open arrow), pelvic and inguinal adenopathy (arrow head), similar to distribution seen on PET images (c).

lymphomas, response rates ranged from 67 to 83 percent (CR 15 to 37 percent; Witzig et al. 2002a; Wiseman et al. 2002).

For transformed B-cell NHL the data is limited. Reported overall response in a study was 53 percent (13 percent CR; 40 percent PR). The median time to progression was 8.5 months, and two patients continued to be in remission at 30 months (Bartlett et al. 2002).

34.3.1.1.2 Other Investigational Applications: Upfront Therapy Early results from a study in which standard CHOP-R or cyclophosphamide, vincristine, prednisone-R chemotherapy was given for three cycles followed by ^{90}Y ibritumomab tiuxetan showed encouraging results in patients with follicular lymphoma (Shipley et al. 2005). Responses were seen in all 39 patients who completed treatment, with CRs in 26 (67 percent). In 15 of 28 (54 percent) patients where PR was seen after chemotherapy, RIT led to CR.

A randomized phase III trial explored the use of ^{90}Y ibritumomab tiuxetan in patients with partial response or a CR. Following chemotherapy patients received either or no further treatment (Radford et al. 2003). Witzig and colleagues (2003b) studied the efficacy and safety of a second dose of ^{90}Y ibritumomab tiuxetan given 12 to 24 weeks after the first dose. The second dose was escalated from 0.2 mCi/kg to 0.3 mCi/kg. The OR rate was 73 percent (11/15), with all patients with a response remaining in remission at up to 27 months of follow-up. The maximum tolerated second dose of ^{90}Y ibritumomab tiuxetan was established at 0.2 mCi/kg. Further dose escalation with the use of prophylactic granulocyte colony-stimulating factor and interleukin-11 is being studied.

Safety of use of ^{90}Y ibritumomab tiuxetan in previously heavily treated patients with NHL who have relapsed after high-dose chemotherapy and autologous peripheral stem cell transplantation, using lower dose (up to 0.2 mCi/kg) has been established (Vose et al. 2003). Use of RIT as conditioning regimens before stem cell rescue has been investigated. Preliminary results show safety for use of ibritumoab tiuxetan, with no increase in transplantation-related toxicity. No engraftment delay is seen. It is possible that this treatment may help in increasing the rate and quality of remissions (Krishnan et al. 2005; Winter et al.2004; Nademanee et al. 2005).

34.3.1.1.3 Use in DLCBL and Mantle Cell Lymphoma A phase II trial of ^{90}Y ibritumomab tiuxetan after first relapse in patients older than 60 years who have relapsed or have refractory diffuse large-cell lymphoma and are not eligible for transplant is being conducted in Europe (Morschhauser et al. 2004). The OR rate in this high-risk population is 44 percent, indicating that ^{90}Y ibritumomab tiuxetan is also effective in relapsed, diffuse, large-cell lymphoma. The role of consolidation therapy with RIT after frontline CHOP-R in patients with diffuse large-cell lymphoma is also being evaluated (Hamlin et al. 2005). In a study, single-agent ^{90}Y ibritumomab tiuxetan produced a response in 8 (36 percent) of 22 patients with relapsed or refractory mantle-cell lymphoma, including complete responses in 5 (Younes et al. 2005).

34.3.1.1.4 Side Effects and Contraindications Primary toxicity associated with ibritumomab tiuxetan is hematologic which is generally transient and reversible. Thrombocytopenia is most commonly seen. The safety data on 349 patients in five studies most commonly showed grade 1 or 2 toxicities with no significant organ toxicities. No dose-limiting nonhematologic toxicity has been observed. Incidents of neutropenia, thrombocytopenia, and anemia occurred in 30, 10, and 3 percent of patients, respectively, in those taking 0.4 mCi/kg, compared with 35, 14, and 8 percent, respectively, in those taking 0.3 mCi/kg (Witzig et al. 2003a). In patients with limited marrow reserve, manifested primarily by thrombocytopenia, a reduced dose (0.3 mCi/kg or 11 MBq/kg ^{90}Y) of ibritumomab tiuxetan has been found to be safe and effective, with a toxicity profile comparable with that of patients with normal platelet counts. Patients with more than 25 percent marrow involvement by disease; a platelet count less than 100,000/μL; an absolute neutrophil count less than 1500/μL; prior bone marrow transplant (either autologous or peripheral stem cell); hypocellular marrow (less than 15 percent cellularity or a marked reduction in bone marrow precursors); or a history of failed stem cell collection are not eligible to get this therapy under current FDA-approved indications. Patients with a history of type 1 hypersensitivity or anaphylactic reactions to murine proteins should not get this treatment. The incidence of human anti-rituximab antibody (HACA) and human anti-mouse antibody (HAMA) is less than 2 percent (Gordon et al. 2001). The incidence of myelodysplasia and acute leukemia with ibritumomab is 1 percent, though the follow-up is short (Kaminski et al. 1993).

34.3.1.2 Tositumomab Tositumomab is an IgG_{2a} murine monoclonal antibody that binds only to CD20-positive cells and radiolabeled antibody has a cross-fire effect. Unconjugated murine tositumomab is administered prior to ^{131}I-labeled antibody, which is also murine. Indications are similar to ibritumomab, including patients who have less than 25 percent marrow involvement by lymphoma (preferably documented by bilateral iliac crest biopsy); patients with >15 percent normal marrow cellularity. Clearance of the antibody is affected by tumor size, splenomegaly, and the amount of bone marrow involvement. The treatment is based on a delivered total body dose of either 75 or 65 cGy for those with normal platelet counts and those having lower platelets. Using dosimetry with iodine-131-labeled antibodies enables physicians to measure the clearance rate directly in order to prospectively individualize the therapeutic dose. Individualizing treatment using dosimetric methodology can therefore be an important consideration in the choice of radioimmunotherapy (Figure 34.2).

The treatment schema includes infusion of 450 mg unlabeled tositumomab (typically given over 1 hour) followed by injection of 5 mCi of ^{131}I-labeled tositumomab for dosimetry. Serial quantitative imaging (carried out after administration and before voiding, between 2 and 4 days, and finally between 6 and 7 days after administration) allows measurement of residence time of radioactive antibody, permitting calculation of radiation absorbed dose to the body based on patient mass using a simple look-up table (Kaminski et al. 2001). The therapeutic dose is administered a week later. Again, 450 mg of cold tositumomab antibody is followed by ^{131}I-tositumomab, the amount of ^{131}I now determined by the patient-specific dosimetry. Patients with platelet counts of >150,000/μL are treated to a whole body dose no greater than 0.75 Gy; patients with platelet counts between 100,000 and 149,000/μL are treated to 0.65 Gy whole body radiation absorbed dose.

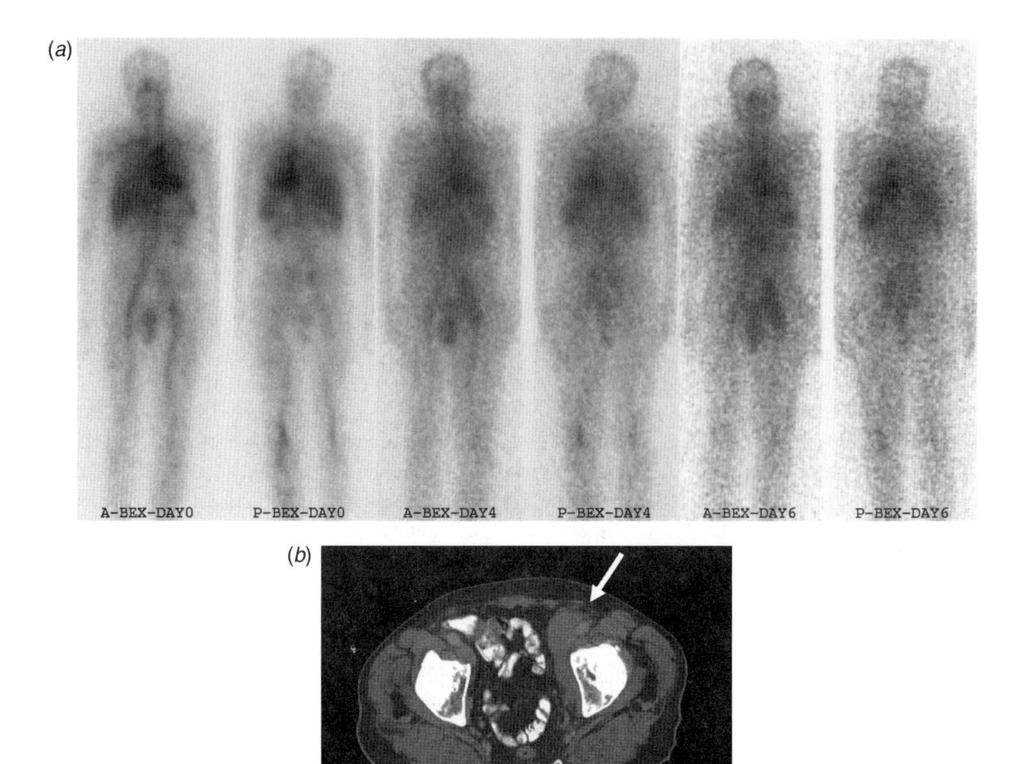

Figure 34.2 Anterior and posterior whole body scans in a patient injected with 5 mCi of ^{131}I-Bexxar (a). The scans are acquired at higher speed with intention of calculating clearance. Targeting of tumor is seen in left inguinal and pelvic region (arrow) on day 4 and day 6 scans corresponding to the adenopathy seen on CT images (b).

A multicenter phase II trial assessed the safety and efficacy of the nonmyeloablative dose of the tositumomab regimen specifically in relapsed or refractory, low grade, and transformed NHL (Flieger et al. 2000) with ORR and CR rates similar in patients with low grade and transformed NHL (60 and 50 percent, respectively). Phase III clinical trials of patients with refractory or transformed low grade B-cell NHL showed improved complete response rates and median response durations compared to their last prior chemotherapy (Zelenetz et al. 2002; Vose et al. 2000).

Myelotoxicity is dose limiting and infusion-related adverse reactions are minimal. Because of dehalogenation of iodine, thyroid blockage is required for therapy with ^{131}I-tositumomab. Subclinical hypothyroidism [rise in serum thyroid-stimulating hormone (TSH)] may occur in about 7 to 10 percent of patients. Clinical hypothyroidism has not been reported. The annualized incidence of myelodysplasia and leukemia appears to be no greater than in patients who have not received radioimmunotherapy.

^{131}I-tositumomab therapy has been evaluated for treatment of other lymphomas, including mantle-cell and DLCBCL. Preliminary results in previously untreated mantle-cell lymphoma, using initial therapy with ^{131}I-tositumomab followed by CHOP, showed CR in 9 (75 percent) of 12 patients (Zelenetz et al. 2003) Whereas low dose therapy offers palliation, high dose therapy may provide prolonged relapse-free survival and the potential for permanent cure. The myelotoxicity of high dose therapy is greater, requiring stem cell rescue, with more incidence of hypothyroidism. In a phase I trial myeloablative doses of single-agent ^{131}I-tositumomab followed by autologous stem cell transplantation produced complete responses in 79 percent of patients with relapsed NHL (S.Y. Liu et al. 1998; Gopal et al. 2002). As a first-line or consolidative therapy, the Southwest Oncology Group (SWOG)

S9911 study with CHOP followed by ^{131}I-tositumomab in patients with advanced-stage follicular lymphoma showed CRs in 27 (57 percent) of the 47 patients in whom there was a partial response or unconfirmed CR after CHOP with no significant increase in toxicity (Press et al. 2003) Similar adjuvant therapy with ^{131}I-tositumomab in combination with other chemotherapy like CVP (Link et al. 2004) and fludarabine have also been done (Leonard et al. 2005).

Alpha emitters have also been investigated for potential use in treatment of NHL. A preclinical study of rituximab labeled with ^{211}astatine appears to show promise for clinical trials (Aurlien et al. 2000), and clinical studies with shorter-acting ^{213}Bi are also being carried out.

There has been considerable progress and promise in use of RIT for refractory Hodgkin disease and leukemia. Preliminary data using ^{111}In- and ^{90}Y-labeled polyclonal antiferritin in refractory Hodgkin lymphoma showed that myelotoxicity was dose limiting (Vriesendorp et al. 1985). In a more recent study with 10 patients using a mean dose of 12 MBq/kg (0.32 mCi/kg), one complete response and six partial responses were observed (overall response rate 70 percent), with a median duration of response of 8 months (range: 7 to 12 months). Toxicity was mainly grade 1 or 2 hematological (Decaudin et al. 2007). Radiolabeled anti-CD33 antibody humanized M195 (IgG_1) has shown significant remission rates in myelogenous leukemias (Caron et al. 1994) and acute promyelocytic leukemia (Jurcic et al. 2000) through antibody-dependent cellular cytotoxicity (Scheinberg et al. 1991). Various cytotoxic radionuclides have been used, including ^{131}I, ^{90}Y, and alpha emitter bismuth (^{213}Bi).

Beta emitters ^{131}I and ^{90}Y used with HuM195 helped eliminate bulky disease (Scheinberg et al. 1991; Burke et al. 2003), but produced prolonged myelosuppression requiring hematopoietic stem cell transplantation at high doses. Alpha emissions have higher energy but short range and can enhance the potency of native HuM195, when conjugated for RIT. ^{213}Bi conjugated to HuM195 proved to be safe and feasible and was the first proof of concept for targeted systemic alpha immunotherapy in humans (Jurcic et al. 2002). The short half-life of bismuth-213 makes its use less practical. Its parent isotope Ac225 with longer half-life may be a more attractive isotope for alpha-based therapy of leukemia. A Phase II study to determine the efficacy (clinical and molecular) of ^{213}Bi-huM195 and 225 Ac-huM195 in patients with refractory CD33-positive leukemias is being conducted.

34.4 RADIOIMMUNOTHERAPY FOR SOLID TUMORS

34.4.1 Colon Cancer

Secreted antigen systems such as carcinoembryonic antigen (CEA) and TAG-72 (sialyl Tn); cell surface antigens including 17-1A, Lewis-y, and A33; and stromal antigens, notably fibroblast activation-protein-alpha, have been studied for effectiveness against colon cancer.

34.4.1.1 Carcinoembryonic Antigen Goldenberg and colleagues pioneered the use of antibodies using radio-iodinated polyclonal anti-CEA antibody in colon cancer detection and therapy (Juweid et al. 1996a, 1996b, 1998; Behr et al. 1997). The maximum tolerated dose for ^{131}I-labeled humanized anti-CEA Ab (hMN-14, Immunomedics, Inc., Morris Plains, New Jersey) is 2220 MBq/m^2 (60 mCi/m^2). A phase II trial of 21 patients treated with ^{131}I-labeled hMN-14, studied response to RIT in patients with small volume metastasis refractory to treatment, with an overall response rate of 58 percent, with a mean duration of response of 9 months (Behr et al. 2002). Of the nine patients receiving RIT in an adjuvant setting following surgery, seven remained disease free at 36 months post-treatment. Although these results appear promising, they underscore the suitability of current methods of RIT for small volume disease rather than bulky disease. Using the chimeric anti-CEA antibody, RIT studies with ^{90}Y-labeled antibody in a nonmyeloablative trial (Wong et al. 1999) demonstrated no responses. A subsequent myeloablative trial (Wong et al. 2000) was initiated.

Various antigen-binding constructs are being explored (Wu et al. 2000). ^{131}I-labeled antigen-binding constructs [using $F(ab)'_2$ fragments; cross-linked divalent {DFM} and trivalent {TFM} versions] showed selective targeting of iodinated sFv to tumor, the degree of uptake was perhaps not adequate to result in responses with ^{131}I (Begent et al. 1989). Development of a single chain Fv fragment of their anti-CEA antibody grown in *Escherichia coli* exhibits impressive targeting abilities (Begent et al. 1996).

The *in vivo* half-life of small therapeutic proteins/peptides can be prolonged by use of albumin fusion proteins, thus increasing their therapeutic efficacy. One such fusion protein of anti-CEA, T84.66 scFv-albumin fusion protein, demonstrated highly specific tumor uptake comparable to cognate recombinant antibody fragments (Yazaki et al. 2008). The ^{111}In-labeled H310A/H435Q protein exhibited intermediate serum half-life and tumor uptake, the highest liver uptake (23.5 percent ID/g at 24 h) and dose estimations reflected ^{131}I-labeled scFv-Fc H310A/H435Q as a promising candidate for radioimmunotherapy (Kenanova et al. 2007).

34.4.1.2 TAG-72 MAb CC49 is a murine IgG targeting the antigen (Tn/sialylTn) expressed on a tumor-associated mucin, TAG-72, that is expressed heterogeneously in most adenocarcinomas, but is rarely expressed in most normal tissues, which makes it a potential target for diagnosis and therapy of several cancers. CC49 IgG exhibits high reactivity against tumor cells in most adenocarcinomas from colorectum, ovary, breast, stomach, and pancreas, with very little reactivity against normal tissues (Molinolo et al. 1990). Phase I studies with ^{131}I-CC49 in colorectal cancer demonstrated low toxicity, a maximum tolerated dose (MTD) of 2775 Mbq/m^2 (75 mCi/m^2) with excellent tumor targeting (Divgi et al. 1994, 1995). Most patients developed human anti-mouse antibodies (HAMA). While tumor doses of 0.19 to 6.67 Gy (19 to 667 rads) are reached, objective tumor responses are not seen (Divgi et al. 1995; Murray et al. 1994).

Initial clinical trials with humanized antibody (Forero et al. 2003) have shown that clearance and targeting characteristics are comparable to murine counterpart with no immunogenicity. Retention of the CH_3 domain retains the immunobiologic functions of the antibody, and therefore a Phase 1 trial to assess both the immunobiologic properties of CH_2-deleted huCC49 and the safety of ^{131}I-labeled antibody was assessed (Glaser et al. 2005). Such CH2 deleted antibody (Delta CH2) was also used intraperitoneally for therapy. Using a ^{131}I-labeled system, the maximum tolerated dose was more than three times that of intact HuCC49. In animals, ^{177}Lu-labeled antibody showed an increase in median survival compared to controls (67.5 ± 7.5 days versus controls of 32 ± 3.3). Intraperitoneal radioimmunotherapy using radiolabeled HuCC49DeltaCH2 should allow higher radiation doses to be administered with less toxicity and potentially improved efficacy (Rogers et al. 2005).

This antibody has been studied in other solid tumors too, notably breast and prostate cancers (Macey et al. 1997; Meredith et al. 1996, 1999; Mulligan et al. 1995). A phase I/II trial of intraperitoneal ^{177}Lu-CC49 Ab in 27 patients with recurrent ovarian cancer showed the maximum tolerated dose (MTD) was 1665 MBq/m^2 (45 mCi/m^2) with bone marrow toxicity dose limiting (Alvarez et al. 1997). Prolonged disease-free survival was observed only in patients with microscopic disease while most of the patients with gross disease progress after therapy. Intraperitoneal RIT with iodine isotope ^{131}I-CC49 and ^{90}Y-CC49 has also been carried out in ovarian cancer (Meredith et al. 2001; Alvarez et al.2002).

The limitations in the clinical utility of radiolabeled CC49 IgG are primarily due to normal tissue toxicity, immunogenicity, and relatively poor penetration into tumor, which may be responsible for lack of significant responses. Some of these limitations can be overcome by using genetically engineered single chain antibody fragments (sFvs). Initial work has shown that sFvs had accelerated clearance from the vasculature, excellent penetration into the tumor from the vasculature, reduced immunogenicity, and higher tumor to background ratios than corresponding IgG, $F(ab')_2$, or Fab′ fragments in animal models (Colcher et al. 1990; Milenic et al. 1991; Adams et al. 1993). However, due to their small size and monovalency, sFvs clear the body too rapidly to allow for sufficient tumor uptake and retention for therapeutic applications. Moreover, early sFvs were generated in bacterial systems and may not have been too stable *in vivo*.

34.4.1.3 A33 A33 is a more promising target for RIT as it is found to be expressed to normal colon and distal small-bowel epithelium. The A33 antigen, a transmembrane glycoprotein, is not secreted and is highly expressed by cells of colonic origin. In addition, mAb A33 is rapidly internalized into a cell after binding to the A33 antigen on the cell membrane. All of these factors promote rapid tumor localization as well as high uptake by tumors. Phase I/II studies of ^{131}I-mAb A33 RIT were carried out in patients with colon cancer (Welt et al. 1994a). MTD was determined to be 2775 MBq/m^2 (75 mCi/m^2) of ^{131}I. Although the isotope showed variable uptake in the normal bowel, gastrointestinal symptoms were mild or absent. A more recent study in advanced colorectal cancer patients conducted dose escalation in 15 patients with treatment at 20 to 50 mCi/m^2. The maximum tolerated dose was determined to be 40 mCi/m^2. The serum $T_{1/2}\beta$ of ^{131}I-huA33 was 135.2 $\pm$ 46.9 hours. The mean absorbed tumor dose was 6.49 $\pm$ 2.47 Gy/GBq. At restaging, 4 patients had stable disease, whereas 11 patients had progressive disease (Chong et al. 2005).

^{125}I-murine A33, as the radioiodinated antibody, does not appear to undergo catabolism following internalization, which could provide a therapeutic advantage by transporting low-energy electron emitting isotopes close to the tumor cell nucleus using internalizing antibodies (Daghighian et al. 1996). A phase I/II study of ^{125}I-labeled mAb A33 demonstrated no MTD reached, up to a dose of 350 mCi/m^2, and low rates of toxicity (Welt et al. 1996). Of the 21 patients, one demonstrated a mixed response based on imaging studies and two patients with stable disease demonstrated decreased serum CEA levels. Higher levels of response were seen in patients who received further chemotherapy following RIT, suggesting a potential role of combined RIT and chemotherapy. Humanized A33 is less immunogenic, with a similar biodistribution profile to murine protein. Studies comparing the efficacy of combined radiosensitizing chemotherapy (capecitabine) and radioimmunotherapy with ^{131}I-huA33 are currently underway.

34.4.1.4 Other Antigens Radioimmunotherapy targets include not only cell surface antigens but also other components of the tumor, including stroma and vasculature. A novel targeting approach to colon cancer therapy can be achieved by utilixzingan antibody, F19, against fibroblast activation protein-alpha (FAP-α). FAP-α is highly expressed by activated fibroblasts abundant in most solid tumors, including more than 95 percent of primary and metastatic colorectal carcinomas, but is not expressed by normal tissue (other than healing scar).

The toxicity, imaging, and biodistribution characteristics of iodine ^{131}I-mAb F19 have been studied. There is highly selective localization in the tumor, with minimal uptake in the normal tissues. Lesions as small as 1 cm can be visualized by scintigraphy (Welt et al. 1994b). Humanized antibody sibrotuzumab (BIBH 1) has been developed (Scott et al. 2003), and RIT studies are being carried out.

Another antigen that has been explored for RIT is MX35, which is expressed homogeneously in 90 percent of epithelial ovarian cancers and is also expressed in epithelial cells of the lung, sweat glands, kidney, collecting ducts, thyroid, fallopian tube, cervix, and uterus (Rubin et al. 1993). Several studies utilizing ^{131}I- and ^{125}I-labeled Ab MX35 showed excellent localization to sites of ovarian tumors, including micrometastases in the peritoneal wall (Findtad et al. 1997). Using alpha emitters like astatine (At), intraperitoneal disease can be targeted. Feasibility of intraperitoneal therapy was established in mice (Elgqvist et al. 2005, 2006a). Fractionated dosing and use of Fab′ fragments leads to lesser toxicity; increased therapeutic efficacy was also seen with labeled Fab′ fragments (Elgqvist et al. 2006b, 2006c).

Ovarian cancer is limited to the peritoneal cavity for most of its natural progression, and therefore newer intraperitoneal therapies would be extremely attractive. ^{131}I chimeric MOv18 Ab that also targets the membrane folate receptor was studied in three patients and showed moderate visualization of tumor sites, tumor absorbed doses of 600 to 3800 cGy, and stable disease for 2 to $>$6 months (Van Zanten-Przybysz et al. 2000). No human anti-chimeric antibody (HACA) was seen. Intraperitoneal RIT using ^{90}Y-labeled murine anti-HMFG$_1$ (analogous to the muc-1 antigen) as an adjuvant to chemotherapy compared to chemotherapy alone showed increased overall survival in patients who received the RIT compared to those who received chemotherapy alone (Nicholson et al. 1998).

Systemic radioimmunotherapy studies with anti-muc1 antibodies have been carried out largely in breast cancer using ^{90}Y isotope. A novel humanized anti-muc1 antibody, BrE3, labeled with

^{90}Y (DeNardo et al. 1997) has been shown to target very well using ^{111}In as a surrogate (Kramer et al. 1998).

Another antigen that has been used to target cancer is the Lewis Y (Le^{y}) antigen that is expressed in a high proportion of epithelial cancers, including breast, colon, ovarian, and lung cancer. In a phase II study murine anti-Le^{y} antibody B3 was labeled either with ^{111}In to study biodistribution or ^{90}Y for therapy (Richardson et al. 1990). The demonstrated good tumor localization and the MTD was found to be 740 MBq of ^{90}Y. Development of humanized mAb B3 and evaluation of the use of higher doses of ^{90}Y-mAb B3 with autologous stem cell support is underway. Efficacy in patients with advanced GI malignancies is being studied. The Lewis-y antigen is abundantly expressed in both gastric and ovarian cancers, which frequently spread to peritoneum. An intraperitoneal RIT phase 1 study was carried out using ^{90}Y labeled to another humanized anti-Le^{y} antibody, hu3S193, in patients with minimal residual disease ovarian cancer. Alpha therapy appears attractive in treating minimal disease. 213Bi-hu3S193 was assessed in animals; resulted in >90 percent cytotoxicity *in vitro* and led to significant retardation of tumor growth. The efficacy appears enhanced by paclitaxel (Kelly et al. 2006, 2007). Studies with ^{111}In huS3193 in lung cancer showed good targeting to lesions (Krug et al. 2007).

34.4.2 Renal Cell Carcinoma: Carbonic Anhydrase IX

Carbonic anhydrase IX is expressed in over 80 percent of primary renal cell carcinomas and therefore is an excellent therapeutic target. Antibody G250, developed by Oosterwijk et al. (1993) recognizes carbonic anhydrase-IX with normal tissue cross-reactivity limited only to biliary epithelium. An early study of mouse monoclonal Ab G250 demonstrated excellent targeting to renal tumors, but treatment was limited by the development of a host immune response (Divgi et al. 1998) precluding the use of multiple administrations for critical treatment of solid tumors. The chimeric monoclonal antibody cG250 recognizes the CAIX/MN antigen and induces antibody-dependent cellular cytotoxicity (ADCC) responses *in vitro*. ^{131}I chimeric Ab G250 is less immunogenic, and its effect can be enhanced by use of IL-2 (Davis et al. 2007).

Fractionated cG250 using <1110 MBq (30 mCi) ^{131}I-cG250 doses in a whole body radiation absorbed dose-based schema, demonstrated differential distribution of radioactivity following each treatment dose with MTD of 0.75 Gy ^{131}I limited by hematological toxicity (Divgi et al. 2004). Measurements of whole-body and serum clearance accurately predicted the clearance of subsequent administrations of cG250, enabling rational treatment planning and real-time pharmacokinetic monitoring to reduce toxic side effects. Dose-limiting whole body radiation absorbed dose is not different in "rapidly" fractionated and single large dose treatment schema; no benefit was seen towards hematopoietic toxicity (Divgi et al. 2004). PET imaging with ^{124}I-cG250 has been used in preoperative identification of tumor type that could have important implications for the choice of treatment for renal cancers, particularly identifying clear-cell renal carcinomas that over-expressed carbonic anhydrase-IX. ^{124}I cG250 PET imaging in 26 patients showed accurate prediction of clear-cell renal carcinoma. Surgery was performed 1 week after a single intravenous infusion of 185 MBq/10 mg of ^{124}I-cG250 over 20 minutes. Fifteen of sixteen clear-cell carcinomas were identified accurately by antibody PET, and all nine non-clear-cell renal masses were negative for the tracer. The sensitivity of ^{124}I-cG250 PET for clear-cell kidney carcinoma was 94 percent with negative predictive value of 90 percent and specificity and positive predictive accuracy of 100 percent each. Stratification of patients with renal masses by ^{124}I-cG250 PET can identify aggressive tumors and help decide treatment (Divgi et al. 2007). Figure 34.3 shows the coronal and transaxial images of a patient with clear renal cell carcinoma following an injection of with ^{124}I-cG250.

Biliary cancers are sensitive to radiation. Since the G250 antibody binds to epithelial cells of the large bile duct and gall bladder, RIT with WX-G250RIT may have clinical benefits in the treatment of biliary cancer patients. A ^{131}I-labeled antibody WX-G250 RIT is designed to identify the tumor cells and deliver tumor-destroying doses of radiation directly to the tumor cells.

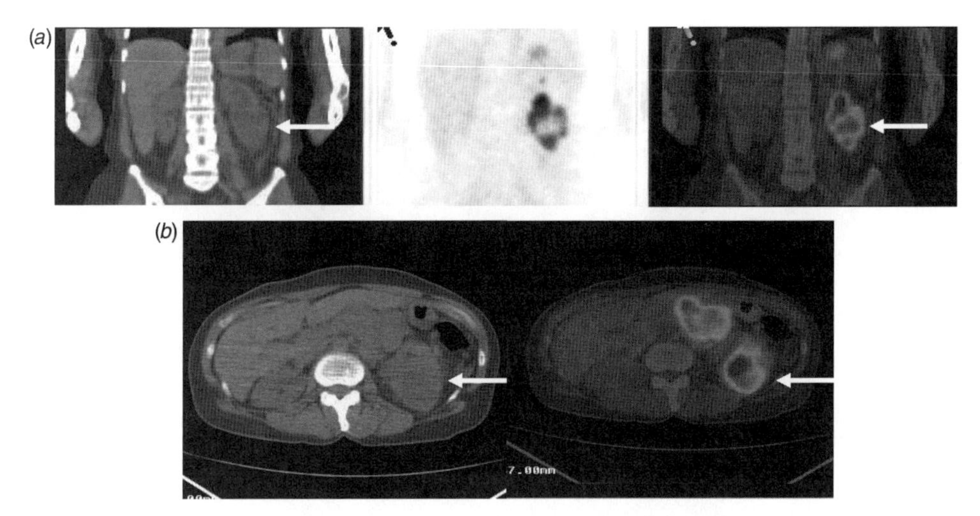

Figure 34.3 Coronal (a) and transaxial (b) images of a patient with clear-cell renal cell carcinoma following an injection with ^{124}I cG250 antibody. There is excellent targeting of antibody to tumor (white arrow). (See color insert.)

34.4.3 Prostate Cancer: Prostate Specific Membrane Antigen

Prostate specific membrane antigen (PSMA) is a 100 kDa type 2 transmembrane glycoprotein on prostate epithelial cells with a large external domain ($PSMA_X$) comprised of over 700 amino acids responsible for its enzymatic action as a hydrolase. PSMA is expressed in almost all prostate cancer cells, from primary to metastatic disease, and may be maximal following androgen withdrawal. Antibodies against PSMA labeled with radionuclides are therefore attractive both as diagnostic and therapeutic agents. Overall success of RIT for the disease has, however, lagged behind.

Prostascint® (capromab pendetide) consists of an intact murine monoclonal antibody, 7E11-C5.3 to which a linker chelator (GYK-DTPA-HCL) is bound; the chelator permits labeling with indium-111 (^{111}In). The antibody targets the intracellular epitope of prostate specific membrane antigen (PSMA), a transmembrane molecule (which perhaps limits its detection to membrane-permeable cells). Scintigraphic (gamma camera and SPECT) imaging with ^{111}In-capromab pendetide is approved in detection tumor of prostatic origin. Its use in the therapy, however, has been limited, likely due to the fact that it likely binds more to necrotic or dead tissue due to targeting of intracellular epitope. Studies of 7E11/CY356 labeled with ^{90}Y demonstrated no therapeutic efficacy. A modified anti-PSMA monoclonal antibody, J591 (Millennium Pharmaceuticals, Cambridge, Massachusetts), is a monoclonal IgG1 antibody that targets the external domain of PSMA and exhibits immune effector function by inducing ADCC Bander et al. 2003a). The antibody can be conjugated to a variety of radiometals (Bander et al. 2003a; Smith-Jones et al. 2003).

Pilot studies show that J591 is safe to administer and localized well to bone and soft tissue metastatic sites (Bander et al. 2003b, 2005; Morris et al. 2005, 2007; Vallabhajosula et al. 2005). The feasibility of radioimmunotherapy using J591 has been explored in phase I studies of ^{90}Y- and ^{177}Lu-labeled humanized J591 Ab that showed potential to induce biochemical and objective responses. A phase I dose-escalation radioimmunotherapy study using ^{177}Lu-J591 (370 to 2590 MBq/ m^2) showed similar biodistribution and kinetics as ^{111}In J591 (Vallabhajosula et al. 2005). Bone marrow was the dose-limiting organ followed by liver (Morris et al. 2005, 2007; Vallabhajosula et al. 2005; Bander et al. 2005). Figure 34.4 shows anterior and posterior whole body images of a patient imaged at 72 hours following an injection with 5 mCi of ^{111}In J591. Images show targeting of antibody in multiple metastatic lesions in spine, pelvis, and femur (arrow). PSMA-directed therapy to include the neovasculature of non-prostate solid tumors has shown safety of administration and encouraging results in targeting

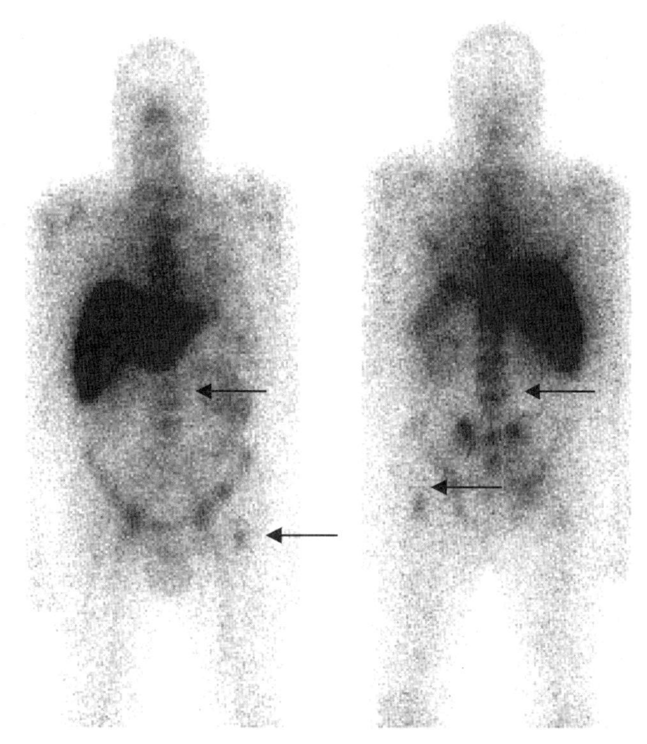

Figure 34.4 Anterior and posterior whole body images of a patient imaged at 72 hours following an injection with 5 mCi of ^{111}In J591. Images show targeting of antibody in multiple metastatic lesions in spine, pelvis, and femur (arrow).

lesions for therapy (Vallabhajosula et al. 2005; H. Liu et al. 1997). Studies in solid tumors using ^{90}Y- or ^{177}Lu-labeled J591 are being performed.

34.5 PRETARGETING STRATEGIES FOR RIT

Multistep targeting helps to minimize the radiation to normal tissue caused by prolonged residence time of radiolabeled antibody in the body. Various approaches are being explored, of which one utilizes the interaction of avidin and biotin. The targeting antibody is biotinylated prior to injection. After injection, avidin is injected in order to bind to the antibody in the tumor. Finally, radiolabeled biotin is injected and attaches to the avidin in the tumor. The use of a clearing agent may further reduce the amount of antibody in normal tissues. It is anticipated that dose-limiting toxicity with most multistep techniques will not be hematological but probably renal. This poses additional issues, as nephropathy is usually not evident until several years after renal insult.

Promising results using the pretargeting approach have been seen in gliomas. Studies of biotinylated anti-tenascin monoclonal antibody with ^{90}Y-DOTA biotin as the radionuclide have demonstrated response as sole therapy for recurrent disease as well as in the adjuvant setting (Breitz et al. 2000). Bispecific constructs that react against a radiometal-chelate complex, on the other hand, are unlikely to be immunogenic and therefore hold promise as therapeutic agents. Studies are now being planned that would utilize these novel systems to deliver higher amounts of relative tumor radiation absorbed dose than would be possible with single-step radiolabeled antibody methods (Breitz et al. 1998; Paganelli et al. 2001).

34.6 FUTURE STRATEGIES

The radioimmunotherapy studies above demonstrate that the early lack of success in solid tumor RIT has been an impetus for the development of a plethora of constructs linked to radionuclides with varying radiobiologic characteristics. Patient-based dosing, fractionated therapy, and multistep targeting are just some of the methods being investigated to optimize treatment. A new generation of intelligently designed clinical trials are either being initiated or in planning stages. These trials share the following characteristics.

34.6.1 Advances in Radiochemistry

These have led to successful trials of labeling auger emitters and alpha emitters to conjugates and pharmaceuticals resulting in stable compounds that can be used for imaging and therapy. Also the development of positron emitting isotopes of copper, yttrium, and iodine has the potential to allow PET-based dosimetry that will permit calculation of tumor radiation absorbed doses allowing accurate dose delivery. While ^{131}I, ^{186}Re, and ^{188}Re are ideal radionuclides for external scintigraphy, surrogate gamma emitters must be used to evaluate the distribution and clearance of pure beta emitters. ^{111}In has been considered to be an appropriate surrogate for ^{90}Y. Their half-lives are almost identical and both are readily incorporated into the same metal chelating agents. A recent study using PET imaging to compare ^{86}Y and ^{111}In as surrogates for ^{90}Y showed that, although ^{111}In and ^{86}Y have similar biodistribution, ^{86}Y remained in organs such as bone for a longer period of time (Meredith et al. 1996). Given yttrium's slower clearance kinetics and bone-seeking properties, ^{86}Y is a more suitable surrogate for ^{90}Y. This feature has been used to obtain more accurate dosimetry in bone-seeking radiopharmaceuticals (Mulligan et al. 1995) and with radiolabeled peptides (Alvarez et al. 1997).

34.6.2 Genetic Engineering

The development of nonimmunogenic antibody constructs as well as pretargeting strategies, such as an affinity-enhancement system, appears to enhance tumor to nontumor ratios and may increase radiation doses to tumor more selectively than directly labeled antibodies. The pretargeting studies described above demonstrate that, although this strategy is still early in its development, it holds great potential. Approaches to genetic modification of antigen-binding constructs to make them suitable for radioimmunotherapy have focused on decreasing size—this has usually entailed a change to univalency, as with single chain constructs. Moreover, production in bacterial systems results in deglycosylated proteins that may thus have suboptimal affinity/avidity characteristics. Making bivalent diabodies has resulted in molecules that are retained to a very great degree in the kidney (Meredith et al. 2001), thus raising the specter of unacceptable nephrotoxicity. Several groups have attempted to retain bivalency and minimize renal accumulation by adding other moieties, notably the CH_3 domain (Alvarez et al. 2002), or adding cytotoxic agents, such as TNF dimers (Colcher et al. 1990).

Interest in use of small antibody constructs for RIT is increasing as it can likely help improve tumor penetration and increase clearance from normal tissues. Fab′ and $F(ab)'_2$ are two such constructs that are being investigated in mouse models and humans with good results. Single-chain antigen-binding proteins (sFv) are linear constructs of light and heavy Fv fragments that clear rapidly from the blood and may have lower renal retention compared to Fab′ fragments. Other constructs with rapid clearance are minibodies, which consist of two sFv fragments linked by a component of the heavy chain region (e.g. CH_3), and diabodies, which comprise of two sFv fragments joined chemically by disulfide bonds or by genetic engineering of the sequences.

34.6.3 Understanding Radiobiology

Utilizing radionuclides with physical characteristics that are tailored to the individual disease condition will in turn improve our ability to treat tumors appropriately. Creating smart systemic targeted radiotherapy depends not only on the careful selection of appropriate antigen targets and antibody constructs, but also depends on choosing radionuclides appropriate for the extent and type of disease. Choosing alpha emitters for microscopic disease or energetic beta-minus emitters for bulky disease with beta-minus emitters of strengths that are proportional to disease bulk will maximize efficacy while limiting side effects to normal tissue whenever possible. A prime concern with these multiple agents will be toxicity. It will therefore be important to combine nuclides with differing pharmacodynamic properties (e.g. bone-seeking beta-minus emitting yttrium-90 with gallium-67 which has Auger emission and is cleared largely through the GI tract), for combination therapies; the concept of sequential therapies with different nuclides based on tumor burden and other characteristics will also play a role in nuclide selection. There appears to be little doubt, however, that combination radioimmunotherapy will be as important to successful therapy as combination chemotherapy has been.

34.6.4 Combination Multimodality Therapy

Using chemotherapy and external beam radiotherapy in conjunction with RIT earlier in the treatment of solid tumors has the potential to maximize the potential of RIT. Complementary modalities have great potential—a chemotherapeutic agent such as paclitaxel may not only have independent cytotoxicity but also act as a radiosensitizer and enhance the efficacy of RIT. We are investigating the potential of small molecule inhibitors such as Iressa® (Astra Zeneca) that can cause downstream metabolic effects that will change tumor uptake of, as well as the susceptibility of tumor cells to, RIT. Agents that change the hypoxic fraction of tumors may also enhance RIT efficacy. Use of multiple modalities can cause changes in the permeability and vascularity of tumors that will permit radioimmunoconjugates access to tumor regions that may otherwise not be reached. The use of anti-angiogenic agents in particular is of great interest as it appears clear that these agents will actually decrease tumor uptake of radioimmunoconjugate, and conversely decrease egress of radioimmunoconjugate if the anti-angiogenic agent is administered at an as yet undetermined time after RIT. These exciting studies are currently being designed in preclinical models and will soon be applied in the clinic.

34.6.5 Administration Schedule

Use of fractionated multidose RIT instead of single larger dose RIT may result in a slower rate of cell repopulation. Theoretical models have compared the effects of large single dose administration and rapid fractionation (Adams et al. 1996). While a large single dose may have a large rate of cell killing, fractionated therapy offers the advantages of lower toxicity and prolonged tumor response. In addition, similar to the rationale behind multimodality therapy, preceding doses may cause architectural changes in the tumor that may allow subsequent doses to target previously inaccessible regions. However, our studies with fractionated and single large dose RIT have shown no advantage in safety or total tumor radiation absorbed dose. We are currently analyzing the data to determine whether there has been any change in intratumoral distribution of radioactivity (Kenanova et al. 2007). Nevertheless, fractionation may have promise when combined in a multimodality therapeutic strategy.

These strategies will pave the way toward tailored RIT for solid tumors that is both safe and effective. After the success of RIT for hematologic cancers, there is little doubt that RIT will find its niche in cancer therapy, both in the adjuvant situation and in bulky disease. Since cancer therapy must target both isolated cancer cells as well as bulky tumors, combination radioimmunotherapy, either given in sequence or simultaneously, will soon be a crucial player in the era of molecular therapeutics.

34.7 CONCLUSION

At a time when molecular medicine is becoming a reality, radioimmunotherapy has already proven its role in the treatment of non-Hodgkin lymphomas and is poised to find a role in the treatment of solid tumors. Humanized and other nonimmunogenic antigen-binding constructs by themselves are finding their place in cancer therapy, as are small molecules that modulate a variety of processes, from cell surface receptor expression to intracellular enzyme activity. The toxicity inherent in radioactivity will dictate that the role of radioimmunotherapy will eventually have to be weighed against the benefits and side-effects of a multitude of targeted molecular therapies. Nevertheless, this may well result in treatments that exploit multiple cellular processes sequentially, each one rendering the tumor more amenable to treatment with the next, with radioimmunotherapy finding a place in the vast new armamentarium that awaits the future of targeted therapy.

REFERENCES

Adams, G.P., J.E. McCartney, M.S. Tai, et al. 1993. Highly specific in vivo tumor targeting by monovalent and divalent forms of 741F8 anti-c-erbB-2 single-chain Fv. *Cancer Res.* 53:4026–4034.

Adelstein, S.J., A.I. Kassis, L. Bodei, and G. Mariani. 2003. Radiotoxicity of iodine-125 and other auger-electron-emitting radionuclides: Background to therapy. *Cancer Biother. Radiopharm.* 18:301–316.

Advani, R., S.A. Rosenberg, and S.J. Horning. 2004. Stage I and II follicular non-Hodgkin's lymphoma: Long-term follow-up of no initial therapy. *J. Clin. Oncol.* 22(8):1454–1459.

Alvarez, R.D., W.K. Huh, M.B. Khazaeli, et al. 2002. A phase I study of combined modality (90)Yttrium-CC49 intraperitoneal radioimmunotherapy for ovarian cancer. *Clin. Cancer Res.* 8:2806–2811.

Alvarez, R.D., E.E. Partridge, M.B. Khazaeli, et al. 1997. Intraperitoneal radioimmunotherapy of ovarian cancer with 177Lu-CC49: A phase I/II study. *Gynecol. Oncol.* 65:94–101.

Aurlien, E., R.H. Larsen, G. Kvalheim, and O.S. Bruland. 2000. Demonstration of highly specific toxicity of the alpha emitting radioimmunoconjugate 211At-rituximab against non Hodgkin lymphoma cells. *Br. J. Cancer* 83:1375–1379.

Bander, N.H., M.I. Milowsky, D.M. Nanus, L. Kostakoglu, S. Vallabhajosula, and S.J. Goldsmith. 2005. Phase I trial of ^{177}lutetium-labeled J591, a monoclonal antibody to prostate-specific membrane antigen, in patients with androgen-independent prostate cancer. *J. Clin. Oncol.* 23(21):4591–4601.

Bander, N.H., D.M. Nanus, M.I. Milowsky, L. Kostakoglu, S. Vallabahajosula, and S.J. Goldsmith. 2003a. Targeted systemic therapy of prostate cancer with a monoclonal antibody to prostate-specific membrane antigen. *Semin. Oncol.* 30:667–676.

Bander, N.H., E.J. Trabulsi, L. Kostakoglu, D. Yao, S. Vallabhajosula, P. Smith-Jones, M.A. Joyce, M. Milowsky, D.M. Nanus, and S.J. Goldsmith. 2003b. Targeting metastatic prostate cancer with radiolabeled monoclonal antibody J591 to the extracellular domain of prostate specific membrane antigen. *J. Urol.* 170(5):1717–1721.

Bartlett, N.L., T.E. Witzig, L. Murray, et al. 2002. ^{90}Y-ibritumomab tiuxetan (Zevalin) immunotherapy for transformed B cell non Hodgkin's lymphoma (NHL) patients. Abstract. *Proc. ASCO* 21:51.

Begent, R.H., J.A. Ledermann, A.J. Green et al. 1989. Antibody distribution and dosimetry in patients receiving radiolabeled antibody therapy for colorectal cancer. *Br. J. Cancer* 60:406–412.

Begent, R.H., M.J. Verhaar, K.A. Chester, et al. 1996. Clinical evidence of efficient tumor targeting based on single-chain Fv antibody selected from a combinatorial library. *Nature Med.* 2:979–984.

Behr, T.M., T. Liersch, L. Greiner-Bechert, et al. 2002. Radioimmunotherapy of small-volume disease of metastatic colorectal cancer. *Cancer* 94 (Suppl):1373–1381.

Behr, T.M., R.M. Sharkey, M.E. Juweid, et al. 1997. Phase I/II clinical radioimmunotherapy with an iodine-131-labeled anti-carcinoembryonic antigen murine monoclonal antibody IgG. *J. Nuclear Med.* 38:858–870.

Borchardt, P.E., R.R. Yuan, M. Miederer, M.R. McDevitt, and D.A. Scheinberg. 2003. Targeted actinium-225 in vivo generators for therapy of ovarian cancer. *Cancer Res.* 63:5084–5090.

Breitz, H., S. Knox, P. Widen, et al. 1998. Pretargeted radioimmunotherapy with antibody-streptavidin and Y-90 DOTA-biotin (Avicidin): Result of a dose escalation study. *J. Nuclear Med.* 39:71P.

Breitz, H.B., P.L. Weiden, P.L. Beaumier, et al. 2000. Clinical optimization of pretargeted radioimmunotherapy with antibody-streptavidin conjugate and ^{90}Y-DOTA-biotin. *J. Nuclear Med.* 41:131–140.

Burke, J.M., P.C. Caron, E.B. Papadopoulos, C.R. Divgi, G. Sgouros, K.S. Panageas, R.D. Finn, S.M. Larson, R.J. O'Reilly, D.A. Scheinberg, and J.G. Jurcic. 2003. Cytoreduction with iodine-131-anti-CD33 antibodies before bone marrow transplantation for advanced myeloid leukemias. *Bone Marrow Transplant* 32:549–556.

Caron, P.C., J.G. Jurcic, A.M. Scott, R.D. Finn, C.R. Divgi, M.C. Graham, I.M. Jureidini, G. Sgouros, D. Tyson, L.J. Old, et al. 1994. A phase 1B trial of humanized monoclonal antibody M195 (anti-CD33) in myeloid leukemia: Specific targeting without immunogenicity. *Blood* 83:1760–1768.

Chong, G., F.T. Lee, W. Hopkins, N. Tebbutt, J.S. Cebon, A.J. Mountain, B. Chappell, A. Papenfuss, P. Schleyer, U P, R. Murphy, V. Wirth, F.E. Smyth, N. Potasz, A. Poon, I.D. Davis, T. Saunder, G.J. O'Keefe, A.W. Burgess, E.W. Hoffman, L.J. Old, and A.M. Scott. 2005. Phase I trial of ^{131}I-huA33 in patients with advanced colorectal carcinoma. *Clin. Cancer Res.* 11(13):4818–4826.

Colcher, D., R. Bird, M. Roselli, et al. 1990. In vivo tumor targeting of a recombinant single-chain antigen-binding protein. *J. Natl. Cancer Inst.* 82:1191–1197.

Czuczman, M.S. 1999. CHOP plus rituximab chemoimmunotherapy of indolent B-cell lymphoma. *Semin. Oncol.* 26 (Suppl 14):88–96.

Czuczman, M.S. 2002. Immunochemotherapy in indolent non-Hodgkin's lymphoma. *Semin. Oncol.* 29 (Suppl 6):11–17.

Daghighian, F., E. Barendswaard, S. Welt, et al. 1996. Enhancement of radiation dose to the nucleus by vesicular internalization of iodine-125-labeled A33 monoclonal antibody. *J. Nuclear Med.* 37:1052–1057.

Davis, I.D., Z. Liu, W. Saunders, F.T. Lee, V. Spirkoska, W. Hopkins, F.E. Smyth, G. Chong, A.T. Papenfuss, B. Chappell, A. Poon, T.H. Saunder, E.W. Hoffman, L.J. Old, and A.M. Scott. 2007. A pilot study of monoclonal antibody cG250 and low dose subcutaneous IL-2 in patients with advanced renal cell carcinoma. *Cancer Immun.* 7:14.

Decaudin, D., R. Levy, F. Lokiec, F. Morschhauser, M. Djeridane, J. Kadouche, and A. Pecking. 2007. Radioimmunotherapy of refractory or relapsed Hodgkin's lymphoma with 90Y-labelled antiferritin antibody. *Anticancer Drugs* 18(6):725–731.

DeNardo, S.J., E.L. Kramer, R.T. O'Donnell, et al. 1997. Radioimmunotherapy for breast cancer using indium-111/yttrium-90 BrE-3: Results of a phase I clinical trial. *J. Nuclear Med.* 38:1180–1185.

Divgi, C.R., N.H. Bander, A.M. Scott, et al. 1998. Phase I/II radioimmunotherapy trial with iodine-131-labeled monoclonal antibody G250 in metastatic renal cell carcinoma. *Clin. Cancer Res.* 4:2729–2739.

Divgi, C.R., J.A. O'Donoghue, S. Welt, et al. 2004. Phase I clinical trial with fractionated radioimmunotherapy using iodine-131 labeled chimeric G250 (^{131}I-cG250) in metastatic renal cancer. *J. Nuclear Med.* 45(8):1412–1421.

Divgi, C.R., N. Pandit-Taskar, A.A. Jungbluth, V.E. Reuter, M. Gönen, S. Ruan, C. Pierre, A. Nagel, D.A. Pryma, J. Humm, S.M. Larson, L.J. Old, and P. Russo. 2007. Preoperative characterisation of clear-cell renal carcinoma using iodine-124-labelled antibody chimeric G250 (124I-cG250) and PET in patients with renal masses: A phase I trial. *Lancet Oncol.* 8(4):304–310.

Divgi, C.R., A.M. Scott, L. Dantis, et al. 1995. Phase I radioimmunotherapy trial with iodine-131-CC49 in metastatic colon carcinoma. *J. Nuclear Med.* 36:586–592.

Divgi, C.R., A.M. Scott, K. McDermott, et al. 1994. Clinical comparison of radiolocalization of two monoclonal antibodies (mAbs) against the TAG-72 antigen. *Nuclear Med. Biol.* 21:9–15.

Elgqvist, J., H. Andersson, T. Bäck, I. Claesson, R. Hultborn, H. Jensen, B.R. Johansson, S. Lindegren, M. Olsson, S. Palm, E. Warnhammar, and L. Jacobsson. 2006a. Alpha-radioimmunotherapy of intraperitoneally growing OVCAR-3 tumors of variable dimensions: Outcome related to measured tumor size and mean absorbed dose. *J. Nuclear Med.* 47(8):1342–1350.

Elgqvist, J., H. Andersson, T. Bäck, I. Claesson, R. Hultborn, H. Jensen, S. Lindegren, M. Olsson, S. Palm, E. Warnhammar, and L. Jacobsson. 2006b. Fractionated radioimmunotherapy of intraperitoneally growing ovarian cancer in nude mice with 211At-MX35 F(ab′)2: Therapeutic efficacy and myelotoxicity. *Nuclear Med. Biol.* 33(8):1065–1072.

Elgqvist, J., H. Andersson, T. Bäck, R. Hultborn, H. Jensen, B. Karlsson, S. Lindegren, S. Palm, E. Warnhammar, and L. Jacobsson. 2005. Therapeutic efficacy and tumor dose estimations in radioimmunotherapy of intraperitoneally growing OVCAR-3 cells in nude mice with (211)At-labeled monoclonal antibody MX35. *J. Nuclear Med.* 46(11):1907–1915.

Elgqvist, J., H. Andersson, P. Bernhardt, T. Bäck, I. Claesson, R. Hultborn, H. Jensen, B.R. Johansson, S. Lindegren, M. Olsson, S. Palm, E. Warnhammar, and L. Jacobsson. 2006c. Administered activity and metastatic cure probability during radioimmunotherapy of ovarian cancer in nude mice with 211At-MX35 F(ab′)2. *Int. J. Radiat. Oncol. Biol. Phys.* 66(4):1228–1237.

Finstad, C.L., K.O. Lloyd, M.G. Federici, et al. 1997. Distribution of radiolabeled monoclonal antibody MX35 $F(ab')_2$ in tissue samples by storage phosphor screen image analysis: Evaluation of antibody localization to micrometastatic disease in epithelial ovarian cancer. *Clin. Cancer Res.* (8):1433–1442.

Flieger, D., S. Renoth, I. Beier, et al. 2000. Mechanism of cytotoxicity induced by chimeric mouse human monoclonal antibody IDEC-C2B8 in CD20-expressing lymphoma cell lines. *Cell. Immunol.* 204:55–63.

Forero, A., R.F. Meredith, M.B. Khazaeli, D.M. Carpenter, S. Shen, J. Thornton, J. Schlom, and A.F. LoBuglio. 2003. A novel monoclonal antibody design for radioimmunotherapy. *Cancer Biother. Radiopharm.* 18:751–759.

Glaser, S.M., I.E. Hughes, J.R. Hopp, K. Hathaway, D. Perret, and M.E. Reff. 2005. Novel antibody hinge regions for efficient production of CH2 domain-deleted antibodies. *J. Biol. Chem.* 280(50):41494–41503.

Gopal, A.K., J.G. Rajendran, S.H. Petersdorf, et al. 2002. High-dose chemo-radioimmunotherapy with autologous stem cell support for relapsed mantle cell lymphoma. *Blood* 99:3158–3162.

Gordon, L.I., C.A. White, J.P. Leonard, et al. 2001. Zevalin radioimmunotherapy is associated with a low incidence of human anti-mouse antibody (HAMA) and human anti Rituxan® antibody (HACA) response. Abstract. *Blood* 98:22b.

Hainsworth, J.D., H.A. Burris III, L.H. Morrissey, et al. 2000. Rituximab monoclonal antibody as initial systemic therapy for patients with low-grade non-Hodgkin lymphoma. *Blood* 95:3052–3056.

Hamlin, P.A., C.H. Moskowitz, B.C. Wegner, et al. 2005. Early safety and efficacy analysis of a phase II study of sequential R-CHOP and yttrium-90 ibritumomab tiuxetan (Zevalin) for elderly high risk patients with untreated DLBCL. Abstract. *Blood* 106:272a.

Horning, S.J. 2003. Future directions in radioimmunotherapy for B-cell lymphoma. *Semin. Oncol.* 30(Suppl): 29–34.

IDEC Pharmaceutical Corp. 2001 (December 21). Ibritumomab Tiuxetan (Zevalin®). Package insert. San Diego, CA:IDEC.

Jurcic, J.G., T. DeBlasio, L. Dumont, T.J. Yao, and D.A. Scheinberg. 2000. Molecular remission induction with retinoic acid and anti-CD33 monoclonal antibody HuM195 in acute promyelocytic leukemia. *Clin. Cancer Res.* 6:372–380.

Jurcic, J.G., S.M. Larson, G. Sgouros, M.R. McDevitt, R.D. Finn, C.R. Divgi, A.M. Ballangrud, K.A. Hamacher, D. Ma, J.L. Humm, M.W. Brechbiel, R. Molinet, and D.A. Scheinberg. 2002. Targeted alpha particle immunotherapy for myeloid leukemia. *Blood* 100:1233–1239.

Juweid, M.E., R.M. Sharkey, T. Behr, et al. 1996a. Radioimmunotherapy of patients with small-volume tumors using iodine-131-labeled anti-CEA monoclonal antibody NP-4 F(ab′)2. *J. Nuclear Med.* 37:1504–1510.

Juweid, M., R.M. Sharkey, T.M. Behr, et al. 1996b. Clinical evaluation of tumor targeting with the anticarcinoembryonic antigen murine monoclonal antibody fragment, MN-14 F(ab′)2. *Cancer* 78:157–168.

Juweid, M., R.M. Sharkey, L.C. Swayne, G.L. Griffiths, R. Dunn, and D.M. Goldenberg. 1998. Pharmacokinetics, dosimetry and toxicity of rhenium-188-labeled anti-carcinoembryonic antigen monoclonal antibody, MN-14, in gastrointestinal cancer. *J. Nuclear Med.* 39:34–42.

Kaminski, M.S., K.R. Zasadny, I.R. Francis, et al. 1993. Radioimmunotherapy of B cell lymphoma with [iodine 131] anti-B1(anti CD 20) antibody. *N. Engl. J. Med.* 329:459–465.

Kaminski, M.S., A.D. Zelenetz, O.W. Press, et al. 2001. Pivotal study of iodine I 131 tositumomab for chemotherapy-refractory low-grade or transformed low-grade B-cell non-Hodgkin's lymphomas. *J. Clin. Oncol.* 19:3918–3928.

Kelly, M.P., F.T. Lee, F.E. Smyth, M.W. Brechbiel, and A.M. Scott. 2006. Enhanced efficacy of ^{90}Y-radiolabeled anti-Lewis Y humanized monoclonal antibody hu3S193 and paclitaxel combined-modality radioimmunotherapy in a breast cancer model. *J. Nuclear Med.* 47(4):716–725.

Kelly, M.P., F.T. Lee, K. Tahtis, F.E. Smyth, M.W. Brechbiel, and A.M. Scott. 2007. Radioimmunotherapy with alpha-particle emitting ^{213}Bi-C-functionalized trans-cyclohexyl-diethylenetriaminepentaacetic acid-humanized 3S193 is enhanced by combination with paclitaxel chemotherapy. *Clin. Cancer Res.* 13(18 Pt 2): 5604s–5612s.

Kenanova, V., T. Olafsen, L.E. Williams, N.H. Ruel, J. Longmate, P.J. Yazaki, J.E. Shively, D. Colcher, A.A. Raubitschek, and A.M. Wu. 2007. Radioiodinated versus radiometal-labeled anti-carcinoembryonic antigen single-chain Fv-Fc antibody fragments: Optimal pharmacokinetics for therapy. *Cancer Res.* 67(2):718–726.

Kramer, E.L., L. Liebes, C. Wasserheit, et al. 1998. Initial clinical evaluation of radiolabeled MX-DTPA humanized BrE-3 antibody in patients with advanced breast cancer. *Clin. Cancer Res.* 4:1679–1688.

Krishnan, A., A. Nademanee, S.J. Forman, et al. 2005. The outcome of ZBEAM, a regimen combining yttrium 90 (^{90}Y) ibritumomab tiuxetan with high-dose chemotherapy, in elderly patients with non-Hodgkin's lymphoma (NHL). Abstract. *J. Clin. Oncol.* 23(Suppl):576s.

Krug, L.M., D.T. Milton, A.A. Jungbluth, L.C. Chen, E. Quaia, N. Pandit-Taskar, A. Nagel, J. Jones, M.G. Kris, R. Finn, P. Smith-Jones, A.M. Scott, L. Old, and C. Divgi. 2007. Targeting Lewis Y (Le(y)) in small cell lung cancer with a humanized monoclonal antibody, hu3S193: A pilot trial testing two dose levels. *J. Thorac. Oncol.* 2(10):947–952.

Leonard, J.P., M. Coleman, L. Kostakoglu, et al. 2005. Abbreviated chemotherapy with fludarabine followed by tositumomab and iodine I 131 tositumomab for untreated follicular lymphoma. *J. Clin. Oncol.* 23:5696–5704.

Link, B., M.S. Kaminski, M. Coleman, et al. 2004. Phase II study of CVP followed by tositumomab and iodine I 131 tositumomab (Bexxar therapeutic regimen) in patients with untreated follicular non-Hodgkin's lymphoma (NHL). Abstract. *Proc. Am. Soc. Clin. Oncol.* 23:560.

Liu, H., P. Moy, S. Kim, Y. Xia, A. Rajasekaran, V. Navarro, B. Knudsen, and N.H. Bander. 1997. Monoclonal antibodies to the extracellular domain of prostate-specific membrane antigen also react with tumor vascular endothelium. *Cancer Res.* 57(17):3629–3634.

Liu, S.Y., J.F. Eary, S.H. Petersdorf, et al. 1998. Follow-up of relapsed B-cell lymphoma patients treated with iodine-131-labeled anti-CD20 antibody and autologous stem-cell rescue. *J. Clin. Oncol.* 16:3270–3278.

Macey, D.J., E.J. Grant, L. Kasi, et al. 1997. Effect of recombinant alpha-interferon on pharmacokinetics, biodistribution, toxicity, and efficacy of ^{131}I-labeled monoclonal antibody CC49 in breast cancer: A phase II trial. *Clin. Cancer Res.* 3:1547–1555.

Meredith, R.F., R.D. Alvarez, E.E. Partridge, et al. 2001. Intraperitoneal radioimmunochemotherapy of ovarian cancer: A phase I study. *Cancer Biother. Radiopharm.* 16:305–315.

Meredith, R.F., M.B. Khazaeli, D.J. Macey, et al. 1999. Phase II study of interferon-enhanced ^{131}I-labeled high affinity CC49 monoclonal antibody therapy in patients with metastatic prostate cancer. *Clin. Cancer Res.* 5(Suppl):3254s–3258s.

Meredith, R.F., M.B. Khazaeli, W.E. Plott, et al. 1996. Phase II study of dual ^{131}I-labeled monoclonal antibody therapy with interferon in patients with metastatic colorectal cancer. *Clin. Cancer Res.* 2:1811–1818.

Milenic, D.E., T. Yokota, D.R. Filpula, et al. 1991. Construction, binding properties, metabolism, and tumor targeting of a single-chain Fv derived from the pan carcinoma monoclonal antibody CC49. *Cancer Res.* 51:6363–6371.

Molinolo, A., J.F. Simpson, A. Thor, and J. Schlom. 1990. Enhanced tumor binding using immunohistochemical analyses by second generation anti-tumor-associated glycoprotein 72 monoclonal antibodies versus monoclonal antibody B72.3 in human tissue. *Cancer Res.* 50:1291–1298.

Morris, M.J., C.R. Divgi, N. Pandit-Taskar, et al. 2005. Pilot trial of unlabeled and indium-111-labeled anti-prostate-specific membrane antigen antibody J591 for castrate metastatic prostate cancer. *Clin. Cancer Res.* 11:7454–7461.

Morris, M.J., N. Pandit-Taskar, C.R. Divgi, S. Bender, J.A. O'Donoghue, A. Nacca, P. Smith-Jones, L. Schwartz, S. Slovin, R. Finn, S. Larson, and H.I. Scher. 2007. Phase I evaluation of J591 as a vascular targeting agent in progressive solid tumors. *Clin. Cancer Res.* 13(9):2707.

Morschhauser, F., D. Huglo, G. Martinelli, et al. 2004. Yttrium-90 ibritumomab tiuxetan (Zevalin) for patients with relapsed/refractory diffuse large B-cell lymphoma not appropriate for autologous stem cell transplantation: Results of an open-label phase II trial. Abstract. *Blood* 104:41a.

Mulligan, T., J.A. Carrasquillo, Y. Chung, et al. 1995. Phase I study of intravenous Lu-labeled CC49 murine monoclonal antibody in patients with advanced adenocarcinoma. *Clin. Cancer Res.* 1:1447–1454.

Murray, J.L., D.J. Macey, L.P. Kasi, et al. 1994. Phase II radioimmunotherapy trial with ^{131}I-CC49 in colorectal cancer. *Cancer* 73(Suppl):1057–1066.

Nademanee, A., S. Forman, A. Molina, et al. 2005. A phase 1/2 trial of high-dose yttrium-90-ibritumomab tiuxetan in combination with high-dose etoposide and cyclophosphamide followed by autologous stem cell transplantation in patients with poor-risk or relapsed non-Hodgkin lymphoma. *Blood* 106:2896–2902.

Nicholson, S., C.S. Gooden, V. Hird, et al. 1998. Radioimmunotherapy after chemotherapy compared to chemotherapy alone in the treatment of advanced ovarian cancer: A matched analysis. *Oncol. Rep.* 5:223–226.

O'Donoghue, J.A. 1999. The implications of non-uniform tumor doses for radioimmunotherapy. *J. Nucl. Med.* 40:1337–1341.

Oosterwijk, E., N.H. Bander, C.R. Divgi, et al. 1993. Antibody localization in human renal cell carcinoma: A phase I study of monoclonal antibody G250. *J. Clin. Oncol.* 11:738–750.

Paganelli, G., M. Bartolomei, M. Ferrari, et al. 2001. Pre-targeted locoregional radioimmunotherapy with ^{90}Y-biotin in glioma patients: Phase I study and preliminary therapeutic results. *Cancer Biother. Radiopharm.* 16:227–235.

Postema, E.J., P.K. Borjesson, W.C. Buijs, et al. 2003a. Dosimetric analysis of radioimmunotherapy with ^{186}Re-labeled bivatuzumab in patients with head and neck cancer. *J. Nuclear Med.* 44:1690–1699.

Postema, E.J., J.M. Raemaekers, W.J. Oyen, O.C. Boerman, C.M. Mandigers, D.M. Goldenberg, G.A. van Dongen, and F.H. Corstens. 2003b. Final results of a phase I radio immunotherapy trial using (186) Re-epratuzumab for the treatment of patients with non-Hodgkin's lymphoma. *Clin. Cancer Res.* 9 (10 Pt 2):3995S–4002S.

Press, O.W. 2003. Radioimmunotherapy for non-Hodgkin's lymphomas: A historical perspective. *Semin. Oncol. Suppl.* 10–21.

Press, O.W., J.M. Unger, R.M. Braziel, et al. 2003. A phase 2 trial of CHOP chemotherapy followed by tositumomab/iodine I 131 tositumomab for previously untreated follicular non-Hodgkin lymphoma: Southwest Oncology Group Protocol S9911. *Blood* 102:1606–1612.

Radford, J.A., N. Ketterer, C. Sebban, et al. 2003. Ibritumomab tiuxetan (Zevalin^{TM}) therapy is feasible and safe for the treatment of patients with advanced B-cell follicular NHL in first remission: Interim analysis for safety of a multicenter, phase III clinical trial. Abstract. *Blood* 102:408–409a.

Richardson, R.B., J.T. Kemshead, A.G. Davies, et al. 1990. Dosimetry of intrathecal iodine-131 monoclonal antibody in cases of neoplastic meningitis. *Eur. J. Nucl. Med.* 17:42–48.

Rogers, B.E., P.L. Roberson, S. Shen, M.B. Khazaeli, M. Carpenter, S. Yokoyama, M.W. Brechbiel, A.F. LoBuglio, and D.J. Buchsbaum. 2005. Intraperitoneal radioimmunotherapy with a humanized anti-TAG-72 (CC49) antibody with a deleted CH2 region. *Cancer Biother. Radiopharm.* 20(5):502–513.

Rubin, S.C., L. Kostakoglu, C. Divgi, et al. 1993. Biodistribution and intraoperative evaluation of radiolabeled monoclonal antibody MX35 in patients with epithelial ovarian cancer. *Gynecol. Oncol.* 51:61–66.

Scheinberg, D.A., D. Lovett, C.R. Divgi, M.C. Graham, E. Berman, K. Pentlow, N. Feirt, R.D. Finn, B.D. Clarkson, T.S. Gee, et al. 1991. A phase I trial of monoclonal antibody M195 in acute myelogenous leukemia: Specific bone marrow targeting and internalization of radionuclide. *J. Clin. Oncol.* 9:478–490.

Schlom, J., A. Molinolo, J.F. Simpson, et al. 1990. Advantage of dose fractionation in monoclonal antibody-targeted radioimmunotherapy. *J. Natl. Cancer Inst.* 82:763–771.

Scott, A.M., G. Wiseman, S. Welt, et al. 2003. A phase I dose-escalation study of sibrotuzumab in patients with advanced or metastatic fibroblast activation protein-positive cancer. *Clin. Cancer Res.* 9:1639–1647.

Shan, D., J.A. Ledbetter, and O.W. Press. 1998. Apoptosis of malignant human B cells by ligation of CD20 with monoclonal antibodies. *Blood* 91:1644–1652.

Sharkey, R.M., A. Brenner, J. Burton, G. Hajjar, S.P. Toder, A. Alavi, A. Matthies, D.E. Tsai, S.J. Schuster, E.A. Stadtmauer, M.S. Czuczman, D. Lamonica, F. Kraeber-Bodere, B. Mahe, J.F. Chatal, A. Rogatko, G. Mardirrosian, and D.M. Goldenberg. 2003. Radioimmunotherapy of non-Hodgkin's lymphoma with 90Y-DOTA humanized anti-CD22 IgG (90Y-Epratuzumab): Do tumor targeting and dosimetry predict therapeutic response? *J. Nuclear Med.* 44:2000–2018.

Shipley, D.L., F.A. Greco, D.R. Spigel, et al. 2005. Rituximab with short duration chemotherapy followed by ^{90}Y ibritumomab tiuxetan as first-line treatment for patients with follicular lymphoma: Update of a Minnie Pearl Cancer Research Network phase II trial. Abstract. *J. Clin. Oncol.* 23 (Suppl):579s.

Smith-Jones, P.M., S. Vallabhajosula, V. Navarro, D. Bastidas, S.J. Goldsmith, and N.H. Bander. 2003. Radiolabeled monoclonal antibodies specific to the extracellular domain of prostate-specific membrane antigen: Preclinical studies in nude mice bearing LNCaP human prostate tumor. *J. Nuclear Med.* 44(4):610–617.

Vallabhajosula, S., I. Kuji, K.A. Hamacher, S. Konishi, L. Kostakoglu, P.A. Kothari, M.I. Milowski, D.M. Nanus, N.H. Bander, and S.J. Goldsmith. 2005. Pharmacokinetics and biodistribution of ^{111}In- and ^{177}Lu-labeled J591 antibody specific for prostate-specific membrane antigen: Prediction of ^{90}Y-J591 radiation dosimetry based on ^{111}In or ^{177}Lu? *J. Nuclear Med.* 46(4):634–641.

Van Zanten-Przybysz, I., C.F. Molthoff, J.C. Roos, et al. 2000. Radioimmunotherapy with intravenously administered ^{131}I-labeled chimeric monoclonal antibody MOv18 in patients with ovarian cancer. *J. Nuclear Med.* 41:1168–1176.

Vose, J.M., P.J. Bierman, J.C. Lynch, et al. 2003. Phase I clinical trial of Zevalin (90Y-ibritumomab) in patients with B-cell non-Hodgkin's lymphoma (NHL) with relapsed disease following high-dose chemotherapy and autologous stem cell transplantation (ASCT). Abstract. *Blood* 102:30a.

Vose, J.M., R.L. Wahl, M. Saleh, et al. 2000. Multicenter phase II study of iodine-131 tositumomab for chemotherapy-relapsed/refractory low-grade and transformed low-grade B-cell non-Hodgkin's lymphomas. *J. Clin. Oncol.* 18:1316–1323.

Vriesendorp, H.M., J.M. Herpst, P.K. Leischner, et al. 1985. Polyclonal Y-90 labeled antiferritin for refractory HD. *J. Clin. Oncol.* 3:1296.

Welt, S., C.R. Divgi, N. Kemeny, et al. 1994a. Phase I/II study of iodine 131-labeled monoclonal antibody A33 in patients with advanced colon cancer. *J. Clin. Oncol.* 12:1561–1571.

Welt, S., C.R. Divgi, A.M. Scott, et al. 1994b. Antibody targeting in metastatic colon cancer: A phase I study of monoclonal antibody F19 against a cell-surface protein of reactive tumor stromal fibroblasts. *J. Clin. Oncol.* 12:1193–1203.

Welt, S., A.M. Scott, C.R. Divgi, et al. 1996. Phase I/II study of iodine 125-labeled monoclonal antibody A33 in patients with advanced colon cancer. *J. Clin. Oncol.* 14:1787–1797.

Winter, J.N., D.J. Inwards, S. Spies. et al. 2004. ^{90}Y ibritumomab tiuxetan (Zevalin; 90YZ) doses higher than 4 mCi/kg may be safely combined with high-dose BEAM and autotransplant: The role for dosimetry. Abstract. *Blood* 104:329a.

Wiseman, G.A., L.I. Gordon, P.S. Multani, T.E. Witzig, S. Spies, N.L. Bartlett, R.J. Schilder, J.L. Murray, M. Saleh, R.S. Allen, A.J. Grillo-López, and C.A. White. 2002. Ibritumomab tiuxetan radioimmunotherapy for patients with relapsed or refractory non-Hodgkin lymphoma and mild thrombocytopenia: a phase II multicenter trial. *Blood* 99(12):4336–4342.

Wiseman, G.A., E. Kornmehl, B. Leigh, W.D. Erwin, D.A. Podoloff, S. Spies, R.B. Sparks, M.G. Stabin, T. Witzig, and C.A. White. 2003. Radiation dosimetry results and safety correlations from ^{90}Y-ibritumomab tiuxetan radioimmunotherapy for relapsed or refractory non-Hodgkin's lymphoma: Combined data from 4 clinical trials. *J. Nuclear Med.* 44:465–474.

Witzig, T.E. 2003. Efficacy and safety of ^{90}Y ibritumomab tiuxetan (Zevalin) radioimmunotherapy for non-Hodgkin's lymphoma. *Semin. Oncol.* 30 (6 Suppl 17):11–16.

Witzig, T.E., I.W. Flinn, L.I. Gordon, C. Emmanouilides, M.S. Czuczman, M.N. Saleh, L. Cripe, G. Wiseman, T. Olejnik, P.S. Multani, and C.A. White. 2002a. Treatment with ibritumomab tiuxetan radioimmunotherapy in patients with rituximab-refractory follicular non-Hodgkin's lymphoma. *J. Clin. Oncol.* 20(15):3262–3269.

Witzig, T.E., L.I. Gordon, F. Cabanillas, et al. 2002b. Randomized controlled trial of yttrium-90-labeled ibritumomab tiuxetan radioimmunotherapy for patients with relapsed or refractory low grade, follicular or transformed B cell non Hodgkin's lymphoma. *J. Clin. Oncol.* 20:2453–2463.

Witzig, T.E., C.A. White, L.I. Gordon, et al. 2003a. Safety of yttrium-90 ibritumomab tiuxetan radioimmunotherapy for relapsed low-grade, follicular, or transformed non-Hodgkin's lymphoma. *J. Clin. Oncol.* 21:1263–1270.

Witzig, T.E., G.A. Wiseman, S.M. Geyer, et al. 2003b. A phase I trial of two sequential doses of Zevalin radioimmunotherapy for relapsed low-grade B-cell non-Hodgkin's lymphoma. Abstract. *Blood* 102:406a.

Wong, J.Y.C., D.Z. Chu, D.M. Yamauchi, et al. 2000. A phase I radioimmunotherapy trial evaluating 90-yttrium-labeled anti-carcinoembryonic antigen (CEA) chimeric T84.66 in patients with metastatic CEA-producing malignancies. *Clin. Cancer Res.* 6:3855–3863.

Wong, J.Y., G. Somlo, T. Odom-Maryon, et al. 1999. Initial clinical experience evaluating yttrium-90-chimeric T84.66 anticarcinoembryonic antigen antibody and autologous hematopoietic stem cell support in patients with carcinoembryonic antigen-producing metastatic breast cancer. *Clin. Cancer Res.* 5 (Suppl):3224s–3231s.

Wu, A.M., P.J. Yazaki, S. Tsai, et al. 2000. High-resolution microPET imaging of carcinoembryonic antigen-positive xenografts by using a copper-64-labeled engineered antibody fragment. *Proc. Natl. Acad. Sci. U.S.A.* 97:8495–8500.

Yazaki, P.J., T. Kassa, C.W. Cheung, D.M. Crow, M.A. Sherman, J.R. Bading, A.L. Anderson, D. Colcher, and A. Raubitschek. 2008. Biodistribution and tumor imaging of an anti-CEA single-chain antibody-albumin fusion protein. *Nuclear Med. Biol.* 35(2):151–158.

Younes, A., B. Pro, M.A. Rodriguez, et al. 2005. Activity of yttrium 90 (^{90}Y) ibritumomab tiuxetan (Zevalin) in 22 patients with relapsed and refractory mantle cell lymphoma (MCL). Abstract. *Blood* 106:689a–690a.

Zalutsky, M., D. Reardon, G. Akabani, et al. 2002. Astatine-211 labeled human/mouse chimeric anti-tenascin monoclonal antibody via surgically created resection cavities for patients with recurrent gliomas: Phase I study. Abstract. *Neuro-Oncology* 4:S103.

Zanzonico, P.B., J.A. Siegel, and J.St. Germain. 2000. A generalized algorithm for determining the time of release and the duration of post-release radiation precautions following radionuclide therapy. *Health Phys* 78:648–659.

Zelenetz, A.D., G. Donnelly, J. Halaas, et al. 2003. Initial treatment of mantle cell lymphoma with sequential radioimmunotherapy with tositumomab/iodine I131 I-tositumomab followed by CHOP chemotherapy results in a high complete remission rate. Abstract. *Blood* 102:406a.

Zelenetz, A., J.P. Leonard, J.M. Bennett, et al. 2002. Long term follow up of patients with low grade and transformed low grade NHL treated with Bexxar therapy. Abstract. *Proc ASCO* 21:1128.

CHAPTER 35

Antibody-Drug Conjugate Therapy

STEPHEN C. ALLEY, DENNIS BENJAMIN, and CHE-LEUNG LAW

ABSTRACT

Antibodies have become a major source of therapeutics for treatment of a wide spectrum of human diseases. Many strategies are being developed to improve the potency of therapeutic antibodies. One of such is to exploit the specificity of antibodies to delivery potent small molecule drugs to the target tissues in the forms of antibody-drug conjugates (ADCs). This chapter summarizes the cell

Therapeutic Monoclonal Antibodies: From Bench to Clinic. Edited by Zhiqiang An

biology pathways involved in ADC targeting, current efforts in optimizing the ADC technology, and the major challenges faced in the development of ADCs.

35.1 INTRODUCTION

To date, more than 20 antibody-based drugs are marketed for the treatment of diseases ranging from viral infection, macular degeneration, hemolysis, ischemic stroke, autoimmunity and inflammatory reactions, to cancers. These drugs utilize a plethora of mechanisms to deliver their therapeutic effects. As summarized in Table 35.1, these include interference with T-cell functions, inhibition of cell adhesion, antibody-dependent cellular cytotoxicity (ADCC) and complement-dependent cytotoxicity (CDC), virus neutralization, ligand neutralization and receptor antagonism, ligand antagonism and receptor binding, proliferation inhibition or apoptosis induction, sensitization toward chemotherapeutics, inhibition of complement protein degradation, and delivery of radioactive or cytotoxic payloads.

One area of intense research has been in discovering ways to improve potency and efficacy of therapeutic antibodies. Boosting antibody effector function, prolonging antibody serum half-life, and radioactive therapeutic antibodies are subjects of other chapters. Delivery of cytotoxic payload is another approach to augment the potency of therapeutic antibodies. The concept of immunoconjugates has been validated clinically, as four of the currently marketed immunotherapeutics are different forms of immunoconjugates (Tables 35.1 and 35.2). Payload delivery is a particularly attractive strategy for antibodies that inherently do not possess strong pharmacodynamic effects against their target cells.

TABLE 35.1 FDA-Approved Antibody-Based Therapeutics[a]

Therapeutic Antibody, Format	Target Antigen	Mechanism(s) of Action	Approved Indications
Muromonab-CD3, mouse IgG2a	CD3 subunit of the T-cell receptor complex	Interference with T-cell functions	Acute organ transplant rejection
Abciximab, chimeric Fab′	Glycoprotein IIb/IIIa	Inhibition of platelet aggregation	Cardiac ischemic complications
Edrecolomab, mouse IgG2a	EpCAM	Antibody-dependent cellular cytotoxicity (ADCC), complement-dependent cytotoxicity (CDC), and the induction of an anti-idiotypic network	Colorectal cancer (Germany only)
Daclizumab, chimeric human IgG1	CD25/IL-2 receptor (IL-2R) α subunit	Receptor binding, inhibits IL-2 binding to IL-2R	Transplant rejection
Rituximab, chimeric human IgG1	CD20	ADCC, CDC, apoptosis induction, chemo-sensitization	Non-Hodgkin lymphoma (NHL); rheumatoid arthritis (RA)
Basiliximab, chimeric IgG1	CD25	Receptor binding, inhibits IL-2 binding to IL-2R	Transplant rejection
Infliximab, chimeric human IgG1	TNFα	Ligand neutralization, receptor antagonism, CDC against cell expressing transmembrane TNFα	RA; psoriatic arthritis (PsA); Crohn's disease (CD)

(*Continued*)

TABLE 35.1 *Continued*

Therapeutic Antibody, Format	Target Antigen	Mechanism(s) of Action	Approved Indications
Palivizumab, humanized IgG1	F protein of respiratory synctial virus (RSV)	RSV neutralization	RSV infection
Trastuzuma, humanized IgG1	HER2	Proliferation inhibition of HER2-expressing cells, ADCC, chemo-sensitization	Metastatic breast cancer
Gemtuzumab ozogamicin, humanized IgG4	CD33	Delivery of the cytotoxic drug calicheamicin	Relapse acute myelogenous leukemia
Alemtuzumab, humanized IgG1	CD52	ADCC and CDC	Chronic lymphocytic leukemia
Adalimumab, human IgG1	TNFα	Ligand neutralization, receptor antagonism, CDC against cell expressing transmembrane TNFα	RA; PsA; ankylosing spondylitis
Efalizumab, humanized IgG1	CD11a	Inhibition of CD11a-mediated leukocyte adhesion	Chronic moderate to severe plaque psoriasis
Ibritumomab tiuxetan, mouse IgG1	CD20	Radiation-mediated cell death, apoptosis induction	Refractive NHL
^{131}I-Tositumomab, mouse IgG2a	CD20	Radiation-mediated cell death, ADCC, CDC	Refractive NHL
Bevacizumab, chimeric IgG1	Vascular endothelial growth factor (VEGF)	Ligand neutralization, receptor antagonism, inhibits angiogenesis	Metastatic colorectal cancer; recurrent or metastatic non-squamous, non-small cell lung carcinoma; metastatic breast cancer
Cetuximab, chimeric IgG1	Epidermal growth factor receptor (EGFR)	Receptor binding, blockade of ligand binding	Metastatic colorectal cancer
Omalizumab, humanized IgG1	Immunoglobulin E (IgE)	Inhibition of IgE binding to the high affinity IgE receptor	Moderate to severe persistent asthma
Natalizumab, humanized IgG1	α4 subunit of α4β1 and α4β7 integrins	Inhibition of α4-mediated leukocyte adhesion	Relapsing multiple sclerosis, CD
Panitumumab, human IgG1	EGFR	Receptor binding, blockade of ligand binding	Metastatic colorectal cancer
Ranibizumab, humanized Fab	VEGF	Ligand neutralization, receptor antagonism, inhibits angiogenesis	Neovascular, age-related macular degeneration
Eculizumab, humanized IgG1	Complement system protein C5	Inhibition of C5 cleavage to C5a and C5b	Paroxysmal nocturnal hemoglobinuria

[a]Information on antibody format, target antigen, mechanism(s) of action, and approved indications are obtained from the corresponding drug package inserts (labels).

TABLE 35.2 FDA-Approved Protein Therapeutics for Targeted Delivery of Cytotoxic Molecules

Generic Name Product Brand Name; (Company)	Target Antigen	Vehicle Format	Payload	Proposed Mechanisms of Action	Approved Indication
Gemtuzumab ozogamicin Mylotarg; (Wyeth)	CD33	Humanized IgG4 kappa	Calicheamicin	Calicheamicin-mediated double-stranded DNA breaks and cytotoxicity	CD33-expressing AML
Ibritumomab tiuxetan Zevalin; (Biogen Idec Inc.)	CD20	Murine IgG1 kappa	^{90}Y	Radiation-mediated cytotoxicity	NHL
^{131}I-tositumomab Bexxar; (GlaxoSmithKline)	CD20	Murine IgG2a lambda	^{131}I	Radiation-mediated cytotoxicity; ADCC and CDC	NHL
Denileukin diftitox Ontak; (Seragen Inc.)	IL-2 receptor complex	Recombinant human IL-2	Diphtheria toxin fragments A & B	Inhibition of protein synthesis and induction of cytotoxicity	IL-2 receptor-expressing CTCL

ADCC, antibody-dependent cellular cytotoxicity; AML, acute myelogenous leukemia; CDC, complement-dependent cytotoxicity; CTCL, cutaneous T-cell lymphoma; NHL, non-Hodgkin lymphoma.

The basic principle underlying antibody-drug conjugates (ADCs) is very simple. By attaching cytotoxic drugs onto targeting antibodies, ADCs increase the specificity, improve the pharmacokinetic properties, and decrease the systemic toxicity of the delivered small molecule drug. The three key components are the targeting antibody, the linker for payload coupling, and the cytotoxic payload, as illustrated in Figure 35.1. Studies in the 1980s have provided experimental evidence that cytotoxic drugs such as antifolates (Manabe et al. 1984; Shen and Ryser 1984; Uadia et al. 1984, 1985), vinca alkaloids (Bumol et al. 1988; Johnson and Laguzza 1987; Laguzza et al. 1989), and anthracyclines (Iwahashi et al. 1989; Pimm et al. 1988) could be successfully carried by antibodies to target and kill transformed cells. Besides the approval of four immunoconjugates by the U.S. Food and Drug Administration (FDA), ADCs specific against a variety of receptors expressed on transformed cells are in different stages of clinical and preclinical investigations, as summarized in Tables 35.3 and 35.4. We will discuss the current understanding in the cell biology of drug delivery using gemtuzumab ozogamicin as an example, properties of tumor-associated antigens suitable for ADC targeting, and ongoing efforts to optimize drug-linker chemistries to improve ADC potency and efficacy. The main focus of this chapter is on ADCs for treatment of cancers, but it is noteworthy that the ADC approach is not restricted to cancer indications. In fact, small molecule cytotoxic and immunosuppressive agents have targeted to specific lymphocyte subsets by antibodies (Jelinkova et al. 1998; Rihová et al. 1992; St'astny et al. 1997), offering the rationale for developing ADCs for the treatment of autoimmune and inflammatory conditions.

Figure 35.1 A schematic representation of an antibody drug conjugate. A small molecule cytotoxic drug is usually chemically conjugated onto the targeting monoclonal antibody (mAb) by a chemical linker.

TABLE 35.3 Investigational Antibody Drug Conjugates

Target Antigen	Targeting Molecule	Antibody Vehicle Format	Payload	Proposed Mechanism of Action	Clinical Trial Indication(s)	Status	References
CD22	Inotuzumab ozogamicin; CMC-544	Humanized IgG4	Calicheamicin	Double-stranded DNA breaks and cytotoxicity	NHL	Phase III	DiJoseph et al. (2007a, 2007b)
MUC-1	huC242-DM4	Humanized	Maytansine DM4	Microtubule network disruption, cell cycle arrest, apoptosis induction	Unresectable solid tumors	Phase II	Erickson et al. (2006)
CD56	huN901-DM1	Humanized	Maytansine DM1	Same as above	Refractory multiple myeloma	Phase I, II	Ishitsuka et al. (2008); Tassone et al. (2004b)
Her2	Trastuzumab-DM1	Humanized IgG1	Maytansine DM1	Same as above	Her2-expressing metastatic breast carcinoma	Phase II	Beeram et al. (2007)
CD19	SAR3419	Humanized IgG	Maytansine DM4	Same as above	Refractory B-cell NHL	Phase I	Aboukameel et al. (2007)
CD33	AVE9633	Humanized IgG1	Maytansine DM4	Same as above	CD33-expressing AML	Phase I	Giles et al. (2006); Legrand et al. (2007)
CD30	SGN-35, cAC10-vcMMAE	Chimeric IgG1	Monomethyl-auristatin E (MMAE)	Microtubule network disruption, cell cycle arrest, apoptosis induction	CD30-expressing hematologic malignancies	Phase I	Francisco et al. (2003); Hamblett et al. (2004); Younes et al. (2007)
GPNMB	CR011-vcMMAE	Humanized IgG2	MMAE	Same as above	Unresectable stage III or IV melanoma	Phase II	Pollack et al. (2007); Sznol et al. (2007); Tse et al. (2006)

(*Continued*)

TABLE 35.3 *Continued*

Target Antigen	Targeting Molecule	Antibody Vehicle Format	Payload	Proposed Mechanism of Action	Clinical Trial Indication(s)	Status	References
Lewis Y antigen	SGN-15, cBR96-Dox	Chimeric IgG1	Doxorubicin	Topoisomerase II inhibitor, DNA intercalating agent, inhibit DNA synthesis	Hormone refractory prostate carcinoma, non-small cell lung carcinoma, metastatic or recurrent breast carcinoma, ovarian carcinoma	Discontinued	H.J. Ross et al. (2006)
MUC-1	CMB-401	Humanized IgG4	Calicheamicin	Double-stranded DNA breaks and cytotoxicity	Epithelial ovarian carcinoma	Discontinued	Chan et al. (2003); Gillespie et al. (2000)
MUC-1	Cantuzumab mertansine; huC242-DM1	Humanized IgG1	Maytansine DM1	Microtubule network disruption, cell cycle arrest, apoptosis induction	Colorectal, gastric, pancreatic, and lung carcinoma	Discontinued	Helft et al. (2004); Tolcher et al. (2003); Xie et al. (2004)
CD44v6	Bivatuzumab mertansine	Humanized IgG1	Maytansine DM1	Same as above	Squamous cell carcinoma of head and neck	Discontinued	A. Sauter et al. (2007); Tijink et al. (2006)
PSMA	MLN2704	Humanized IgG1	Maytansine DM1	Same as above	Androgen-independent prostate carcinoma	Discontinued	Henry et al. (2004)
gp40	LY256787, LY203728	Murine IgG2a	Vinblastine	Same as above	Colorectal, pancreatic, ovarian, lung carcinoma	Discontinued	Petersen et al. (1991)

TABLE 35.4 Experimental Antibody Drug Conjugates

Target Antigen	Targeting Molecule	Antibody Vehicle Format	Payload	Proposed Mechanism of Action	Preclinical *In Vivo* Antitumor Activity	References
Ganglioside GD2	14G2a	Murine IgG2a	Calicheamicin	Double-stranded DNA breaks and cytotoxicity	Syngeneic NXS2 metastatic murine neuroblastoma	Lode et al. (1998)
γ-glutamyl-transferase	138H11	Murine IgG1	Calicheamicin	Same as above	Caki-1 renal cell carcinoma xenografts	Knoll et al. (2000)
Lewis Y antigen	hu3S193	Humanized IgG1	Calicheamicin	Same as above	N87 gastric, LNCaP prostatic, and LOVO colon carcinoma xenografts	Boghaert et al. (2004)
MUC1	hCTM01	Humanized IgG4	Calicheamicin	Same as above	OVCAR-3 ovarian and MX-1 breast carcinoma xenografts	Hamann et al. (2005a, 2005b)
CD64	H22	Humanized IgG1	Calicheamicin	Same as above	*In vitro* cytotoxicity against CD64-expressing cell lines and synovial macrophages from rheumatoid arthritis patients	van Roon et al. (2005)
CD20	Rituximab	Chimeric IgG1	Calicheamicin	Same as above	Ramos Burkitt's lymphoma xenografts	DiJoseph et al. (2007b)
PSCA	8D11	Murine IgG	Maytansine DM1	Microtubule network disruption, cell cycle arrest, apoptosis induction	PC3 prostate carcinoma xenografts overexpressing PSCA (PC3.gd.PSCA); SW780 bladder carcinoma xenografts	Ross et al. (2002)
CD138	B-B4	Murine IgG1	Maytansine DM1	Same as above	OPM1 and OPM2 multiple myeloma xenografts	Tassone et al. (2004a)
CD44v integrins	CNTO 95	Fully human IgG1	Maytansine DM1 & DM4	Same as above, anti-angiogenesis	A549 lung carcinoma, HT-29 colon carcinoma, and A375.S2 melanoma xenografts	Q. Chen et al. (2007a); Trikha et al. (2004)
CD79b	2F2	Chimeric IgG1	Maytansine DM1	Microtubule network disruption, cell cycle arrest, apoptosis induction	BJAB Burkitt's, Granta-519 mantle cell, and DoHH2 follicular lymphoma xenografts	Polson et al. (2007)

(*Continued*)

TABLE 35.4 *Continued*

Target Antigen	Targeting Molecule	Antibody Vehicle Format	Payload	Proposed Mechanism of Action	Preclinical *In Vivo* Antitumor Activity	References
CD19	B496	Murine IgG1	Maytansine DM1	Same as above	*In vitro* cytotoxicity against CD19-expressing cell lines; efficacy studies in xenograft models not available	Ingle et al. (2008)
E-selectin	5D11	Murine IgG1	MMAE	Microtubule network disruption, cell cycle arrest, apoptosis induction	HCT116 colon carcinoma xenografts	Bhaskar et al. (2003)
Lewis^Y antigen	cBR96	Chimeric IgG1	MMAE	Same as above	L2987 lung carcinoma xenografts	Doronina et al. (2003)
PSMA	PSMA mAb	Fully human IgG1	MMAE	Same as above	C4-2 androgen-independent prostate carcinoma xenograft	Ma et al. (2006)
CD20	1F5	Murine IgG2a	MMAE	Same as above	Ramos Burkitt's lymphoma xenografts	Law et al. (2004)
CD20	rituximab	Chimeric IgG1	MMAE	Same as above	Ramos Burkitt's lymphoma xenografts	Law et al. (2004)
EphB2	2H9	Murine IgG1	MMAE	Same as above	HT1080 fibrosarcoma xenograft overexpressing EphB2 (HT1080-ephB2)	Mao et al. (2004)
CD70	1F6	Chimeric IgG1	Momethyl-auristatin F (MMAF) & auristatin phenylalanine phenylene-diamine (AFP)	Same as above	Caki-1 and 786-O renal cell carcinoma xenografts	Law et al. (2006)

Melanotrans-ferrin/p97	L49	Murine IgG1	MMAF	Same as above	*In vitro* cytotoxicity against p97-expressing cell lines; efficacy studies in xenograft models not available	Smith et al. (2006)
MUC16	3A5	Chimeric human IgG1	MMAE & MMAF	Same as above	OVCAR-3 ovarian carcinoma xenografts	Y. Chen et al. (2007b)
CD79b	2F2	Chimeric IgG1	MMAF	Same as above	BJAB Burkitt's, Granta-519 mantle cell, and DoHH2 follicular lymphoma xenografts	Polson et al. (2007)
BCMA	SG1	Rat IgG	MMAF	Same as above	*In vitro* cytotoxicity against BCMA-expressing cell lines; efficacy studies in xenograft models not available	Ryan et al. (2007)
CD74	LL1	Murine IgG1	Doxorubicin	Topoisomerase II inhibitor, DNA intercalating agent, inhibit DNA synthesis	Raji Burkitt's lymphoma xenografts	Griffiths et al. (2003)
PSMA	MLN591/huJ591	Humanized IgG1	^{90}Y or ^{131}I	Radiation-mediated cytotoxicity	LNCaP prostate cancinoma xenografts	Smith-Jones (2004); Vallabhajosula et al. (2004)

35.2 THE ROLE OF RECEPTOR-MEDIATED ENDOCYTOSIS IN DRUG DELIVERY BY IMMUNOCONJUGATES

Harnessing cellular receptor-mediated endocytic pathways for drug delivery is a common feature shared by immunoconjugates. Instead of passive diffusion through the plasma membrane, small molecule drugs access subcellular vesicles via various endocytic pathways, depending on the nature of the receptors recognized by the antibodies (Fig. 35.2). Interestingly, small molecules delivered by antibodies and released into the cytosol from these vesicular compartments appear to be less sensitive to efflux catalyzed by P-glycoprotein (Guillemard and Uri Saragovi 2004), leading to the possibility that the ADC approach may potentially circumvent the multiple drug resistance (MDR) phenotype.

Endocytosis through clathrin-coated vesicles is the most commonly used receptor-mediated endocytosis (RME) pathway by the majority of cell membrane proteins subsequent to ligand binding. This process initiates with invaginations in the plasma membrane that are coated with the fibrous triskelion protein clathrin on the cytoplasmic face of the membrane. This is followed by rapid budding, for example, within minutes, of intracellular vesicles. In association with clathrin in the coated vesicles are several heterotetrameric adaptor protein (AP) complexes (AP-1, AP-2, AP-3, and AP-4) (Kirchhausen 1999) and accessory factors (e.g., GGA, Dab2, epsin, eps15; Slepnev and De Camilli 2000) that facilitate the polymerization of clathrin subunits into a polyhedral lattice scaffold.

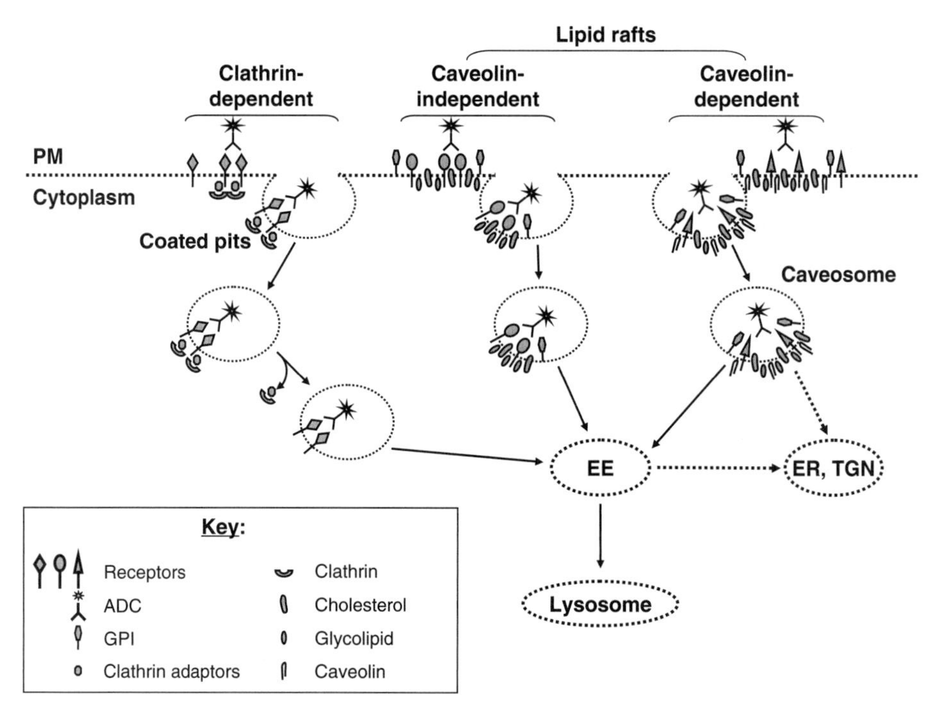

Figure 35.2 Major receptor-mediated endocytosis pathways available for internalization of antibody drug conjugates. Three pathways: clathrin-dependent endocytosis, clathrin-independent non-caveolar endocytosis, and clathrin-independent caveolar endocytosis have been described to mediate internalization of antibody drug conjugates. All three pathways can converge at the early endosomes and eventually end at the lysosomes. The dotted (green in color insert) arrows indicate retrograde trafficking from the early endosomes or caveosome to the endoplasmic reticulum or trans-Golgi network bypassing the lysosomal compartment. Abbreviations: PM, plasma membrane; EE, early endosome; ER, endoplasmic reticulum; TGN, trans-Golgi network; ADC, antibody drug conjugate; GPI, glycophosphatidylinostol-linked proteins. (See color insert.)

A small cytosolic GTPase, dynamin, then assembles around the neck of the coated vesicle to facilitate scission and release of the vesicles to the cytosol. Once detached from the plasma membrane, internalized vesicles rapidly shed their clathrin coat with the aid of the heat shock protein hsp70 and then traffic along the cytoskeleton to fuse with early or sorting endosomes (Fig. 35.2).

A number of amino acid motifs within the cytosolic domains of membrane proteins are responsible for interaction with adaptor proteins during RME and formation of the coated vesicles. Two of these are tyrosine-based motifs conforming to the NPXY or YXXØ (the Ø position represents residues with bulky hydrophobic side chains) consensus sequence (Bonifacino and Traub 2003). Other consensus sequences are dileucine-based motifs of [DE]XXXL[LI] or DXXLL (Bonifacino and Traub 2003). Besides physically interacting with the adaptor proteins in the clathrin-coated vesicles, these motifs also provide the signals for vesicular sorting and trafficking. For example, the tyrosine motifs YXXØ found in many signaling receptors become phosphorylated upon ligand binding. This is followed by ubiquitination of the cytoplasmic tail through the recruitment of ubiquitin ligases, a process that allows sorting of internalized membrane proteins to the lysosomal vesicels for proteolytic degradation (d'Azzo, Bongiovanni, and Nastasi 2005; Hicke and Dunn 2003; Marmor and Yarden 2004). Not surprisingly, most of the ADC targets presented in Tables 35.2 to 35.4 have been reported to undergo internalization via clathrin-coated vesicles. This includes CD19 (Ingle et al. 2008), CD22 (Tateno et al. 2007), CD30 (Sutherland et al. 2006), CD74 (Hofmann et al. 1999; Odorizzi et al. 1994), CD79b (Cassard et al. 1998), MUC1 (Altschuler et al. 2000; Kinlough et al. 2004), HER2 (Gilboa et al. 1995), E-selectin (Kluger et al. 2002; Setiadi and McEver 2008), and PSMA (Rajasekaran et al. 2003). The presence of either the tyrosine-based (Altschuler et al. 2000; Cassard et al. 1998; Gilboa et al. 1995; Ingle et al. 2008; Kinlough et al. 2004; Tateno et al. 2007), or dileucine-based (Hofmann et al. 1999; Kluger et al. 2002; Odorizzi et al. 1994; Rajasekaran et al. 2003) motifs in these ADC targets is consistent with their use of clathrin-coated vesicles as their port of intracellular entry.

Instead of clathrin-coated vesicles, other membrane proteins internalize through unique membrane areas termed lipid rafts, characterized as areas that are insoluble in the nonionic detergent Triton X-100 in low temperatures. They are usually rich in saturated lipid, cholesterol, and glycosphingolipids (Laude and Prior 2004). Lipid rafts are also enriched with glycophosphatidylinositol (GPI)-linked proteins and acylated (palmitoylated and myristoylated) proteins on the extracellular and cytoplasmic layers, respectively, of the plasma membrane. Many of the acylated proteins on the cytoplasmic phase of lipid rafts are signaling intermediates like Src-family kinases and GTP-binding proteins. Hence, lipid rafts may play an important role in concentrating and sequestering signaling proteins within a specific region of the plasma membrane to increase signaling intensity, proximity, and specificity (Simons and Ikonen 1997; Simons and Toomre 2000). Some lipid rafts associate with the protein caveolin, whose N- and C-termini are located in the cytoplasm linked together by a hydrophobic hairpin that is inserted into the membrane (Fig. 35.2). One function of caveolin is to organize the plasma membrane into small invaginations called caveolae (latin for small caves). Internalization of membrane receptors has been demonstrated to take place through both caveolin-dependent and caveolin-independent vesicles (Lajoie and Nabi 2007; Pelkmans and Helenius 2002; Tarrago-Trani and Storrie 2007; Fig. 35.2). The receptors for folic acid, albumin, and oxidized low density lipoprotein are examples of membrane proteins that internalize through the caveolae pathway (Kirkham and Parton 2005; Pelkmans and Helenius 2002). On the other hand, the Shiga toxin or Shiga-like toxin receptor glyosphingolipid Gb3 appears to internalize exclusively through the lipid rafts (Tarrago-Trani and Storrie 2007). The receptor for cholera toxin can utilize all three internalization pathways illustrated in Figure 35.2 (Tarrago-Trani and Storrie 2007). For the immunoconjugate targets, melanotransferrin (p97; Smith et al. 2006) and the high affinity interleukin-2 receptor (IL-2R; Grassart et al. 2008; Lamaze et al. 2001) have been reported to internalize via the caveolae pathway, whereas CD44 internalizes through a clathrin- and caveolin-independent pathway (Tammi et al. 2001), and therefore CD44v6 may be an example of an ADC target that internalizes with lipid rafts.

Three of the changes that occur in vesicles along the endolysosomal pathway have been exploited for releasing the cytotoxic drugs delivered by antibodies: (1) the gradual drop in pH from the physiological range on the cell surface of 7.2–7.4 to 5.5–6.0 within the endosomes and finally to a relatively acidic environment of pH ~5.0 in the lysosomes (Mellman, Fuchs, and Helenius 1986); (2) the increase in redox potential due to the higher concentrations of reduced glutathione and cysteine that can reduce disulfide bonds (Saito, Swanson, and Lee 2003); and (3) the presence of hydrolases that catalyze breakdown of disulfide, ester, sugar, or peptide bonds (Pillay, Elliott, and Dennison 2002; Saito et al. 2003; Winchester 2005). Cytotoxic drugs have been conjugated onto antibody vehicles through acid-labile hydrazone linkers, sterically hindered disulfide linkers, glucuronide sugar linkers, as well as peptide linkers to take advantage of the environment in the endolysosomal pathway for drug release (Garnett 2001; Jeffrey et al. 2006; Wu and Senter 2005). Interestingly, in some situations complete degradation of the targeting antibody in the lysosomes may be sufficient to release an active cytotoxic agent that is coupled onto the antibody by a chemically and enzymatically inert linker (Doronina et al. 2006; Polson et al. 2007). Most investigations in ADCs have focused on the endosomal-lysosomal pathway for drug release. Intracellular vesicles also traffic through retrograde pathways from early endosomes or caveosomes to the endoplasmic reticulum and trans-Golgi network bypassing the hydrolytic environment of the endolysosomal pathway (Fig. 35.2). These are possibilities of alternate routes for ADC trafficking. Studies to exploit the unique physicochemical and enzymatic environments of these pathways for drug delivery and release will add flexibility to the ADC technology (Tarrago-Trani and Storrie 2007).

35.3 CHARACTERISTICS OF AN FDA-APPROVED IMMUNOCONJUGATE

35.3.1 Gemtuzumab Ozogamicin (Mylotarg®)

The first FDA-approved ADC is gemtuzumab ozogamicin (GO; Bross et al. 2001; Stasi et al. 2008). The target for GO is CD33, also known as Siglec-3 as it belongs to the sialic acid binding Ig-like lectin (Siglec) family (Crocker, Paulson, and Varki 2007; Varki and Angata 2006). CD33 expression is highly specific to the hematopoietic system and is restricted to myeloid progenitor cells, monocytes, macrophages, and granulocytes, with no known expression in nonhematopoietic tissues. CD33 contains tyrosine-based inhibitory motifs (ITIMs) in its cytoplasmic domain and may regulate myeloid cell biology through interaction with sialic-acid containing ligands and intracellular signaling involving intermediates such as SHP-1 and c-Cbl (Taylor et al. 1999; Walter et al. 2008a, 2008b). Nevertheless, the real biologic role of CD33 remains obscure. While no specific ligand has been identified for CD33, it is also unclear how CD33-mediated signaling can regulate functions of myeloid cells. Despite its functional obscurity, CD33 has proven to be a suitable ADC target. About 90 percent of acute myeloid leukemia (AML) cases are CD33 positive (Jilani et al. 2002). In addition, CD33 is expressed to varying levels in myelodysplastic syndromes, chronic myeloid leukemia, and myeloproliferative disorders (Jilani et al. 2002). The lack of CD33 expression on pluripotent hematopoietic stem cells ensures that therapies targeting CD33 will allow recovery of the hematopoietic system.

GO is a 1 : 1 mixture of the unconjugated anti-CD33 monoclonal antibody (mAb), humanized (h) P67.6 (IgG4), and hP67.6 conjugated to 4–6 molar equivalents of *N*-acetyl-γ calicheamicin through a dimethyl hydrazide linker (Bross et al. 2001; Stasi et al. 2008). GO is rapidly internalized into CD33-expressing leukemia cells, followed by the release of calicheamicin in the acidic compartments of the endolysosomal pathway (Golay et al. 2005; van Der Velden et al. 2001). Internalization of GO-bound CD33 is dependent on the tyrosine phosphorylation of its ITIM and ubiquitination of CD33 (Walter et al. 2008a, 2008b). The sensitivity of cells to GO is proportional to the number of surface-expressed CD33 molecules (Walter et al. 2005). Conversely, resistance to GO treatment is correlated to the presence of the multi-drug resistance (MDR) phenotype (Linenberger et al. 2001). The *in vitro* cytotoxicity of GO has been correlated to antitumor efects in xenograft models of CD33-expressing leukemia (Golay et al. 2005; Hamann et al. 2002).

In a phase I clinical trial in AML, GO induced a reduction in marrow leukemic blasts in eight patients (20 percent). Three of the eight patients achieved a complete remission while two achieved complete remission with thrombocytopenia (Sievers et al. 1999). Three pivotal phase II trials in multiple centers were carried out. Collectively, about 26 percent of the patients achieved complete remission (13 percent) or complete remission with thrombocytopenia (13 percent) (Larson et al. 2002, 2005; Sievers et al. 2001). Based on these clinical data GO is currently approved for the treatment of patients with CD33-positive AML in first relapse who are 60 years of age or older and who are not candidates for cytotoxic chemotherapy.

35.3.2 Lessons from Gemtuzumab Ozogamicin

Cytotoxic payload delivery by antibodies is clearly a validated approach for clinical practice. There is certainly room for improvement. The choice of a target antigen and the mAb vehicle remains a critical challenge to produce ADCs that yield the best antitumor activity while at the same time impart the least on- and off-target toxicities to normal tissues. In the case of GO, myelosuppression due to targeting of $CD33^+$ normal myeloid progenitor cells by GO, resistance to the released calicheamicin resulting from the MDR phenotype in a subset of the patients, and drug linker stability are key barriers limiting its efficacy in AML patients. Hence, potency of the payloads and their metabolic routes for elimination, availability of appropriate conjugation chemistries to append payloads onto mAbs, as well as the stability of drug linkers in circulation are all important factors in determining the ultimate therapeutic indices of ADCs.

35.4 TUMOR CELL SURFACE TARGETS FOR ADCs

35.4.1 Target Identification

Since the ultimate goal is to localize as precisely as possible an ADC to its targeted tumor, an ADC target should be expressed homogeneously and at a high level on the surface of the majority of neoplastic cells within the tumor. Other highly desirable properties include high frequency of tumor expression of the target antigen in patients as well as expression in multiple cancer indications. Although surface expression on tumor cells is a prerequisite, determining normal tissue expression profile is an essential step in target validation. This is particularly important when evaluating ADCs, as the payload may contribute to on-target toxicity in any normal tissues that express the target antigen. In the ideal situation, target antigens should be absent from normal tissues. However, more often than not targets do demonstrate varying extents of normal tissue expression. To lessen potential toxicity of ADCs careful expression profiling to compare tumor versus normal tissues is key to ensure that only receptors with minimal expression in vital tissues are considered as targets for ADCs.

A variety of genomic and proteomic techniques have been developed in the past 20 years to identify surface receptors overexpressed on neoplastic cells relative to normal tissues (Carter, Smith, and Ryan 2004). Immunohistochemistry (IHC) analysis remains the gold standard to confirm and validate target expression in both tumor and normal tissues. The throughput and economics of IHC analysis has radically increased with the advent of tissue microarray (TMA) technologies (Kallioniemi et al. 2001; Kononen et al. 1998; G. Sauter, Simon, and Hillan 2003). Cylindrical biopsy cores from fixed or frozen tumor and normal tissues are arrayed at high density into recipient blocks (Kallioniemi et al. 2001), sometimes with density as high as 1000 cores per recipient block. Serial sections from these blocks are prepared and mounted onto microscopic slides. A distinct advantage of TMA for target validation is the ability to interrogate DNA, mRNA, and protein expression in serial sections prepared from the same set of specimens (Kallioniemi et al. 2001).

35.4.2 Leukocyte Differentiation Antigens as ADC Targets

Cell surface receptors restricted to the hematopoietic system have been popular choices as targets for ADCs. Of the 26 receptors in preclinical evaluation for ADC targeting listed in Tables 35.2 to 35.4, 13 of them are leukocyte differentiation antigens (CD19, CD20, CD22, CD30, CD33, CD56,

CD64, CD70, CD74, CD79b, CD138, IL-2R, and BCMA). More importantly, all four approved immunoconjugates and radiolabeled conjugates target classical leukocyte differentiation surface antigens: CD33, CD20, and IL-2R (Table 35.1). The success in targeting leukocyte differentiation antigens can be partially attributed to the fact that most of them are not expressed to any detectable level in normal nonhematopoietic tissues. Consequently, very little, if any, on-target toxicity in nonhematopoietic tissues is expected from ADCs targeting leukocyte differentiation antigens, potentially sparing many vital organs and tissues. This is supported by the observation that the only reported on-target toxicity for GO is its effect on normal $CD33^+$ myeloid progenitors as manifected as myelosuppression and neutropenia (Turturro 2007).

A subset of leukocyte differentiation antigens, exemplified by CD25, CD30, and CD70, is even more attractive as targets for ADCs. They are the so-called lymphocyte activation antigens whose expression is induced only transiently during lymphocyte activation. The number of cells expressing these activation markers, even within the hematopoietic system, is extremely low in normal individuals who are not undergoing any immune responses. Interestingly, aberrant expression of CD30 and CD70 has been reported in a number of solid tumors derived from nonhematopoietic cells. Thus, CD30 has been detected in nasopharyngeal carcinoma (Kneile et al. 2006), germ cell tumors (Durkop et al. 2000), and mesothelioma (Dunphy, Gardner, and Bee 2000; Durkop et al. 2000) and CD70 in renal cell carcinoma (Law et al. 2006), nasopharyngeal carcinoma (Agathanggelou et al. 1995), glioblastoma (Chahlavi et al. 2005; Wischhusen et al. 2002), and thymic carcinoma (Hishima et al. 2000). ADCs targeting CD30 or CD70 for treating the above solid tumors may elicit minimal or no on-target toxicity to normal tissues. Further transcript and protein expression profiling in solid tumor indications to examine expression of leukocyte differentiation antigens and activation markers may yield additional receptors suitable as ADC targets on solid tumors.

35.4.3 Characteristics of Target Antigen that Affect ADC Potency

A multitude of factors, including copy number of the targeted receptor, rate of internalization of ADC-target complexes, rate of ADC-target trafficking to the appropriate subcellular compartment(s) for drug release, rate of ADC-target recycling back to the cell surface, and potency of the payload (to be discussed later) can affect the potency of any given ADC.

The presence of either tyrosine- or dileucine-based internalization motifs in the cytoplasmic tail of a membrane protein may predict rapid internalization by RME (Bonifacino and Traub 2003). Many ADC targets listed in Tables 35.2 to 35.4 contain these motifs and have been shown to undergo rapid RME. In general, it appears that the potency of an ADC is dependent on the cell surface copy number of the antigen that it targets. For example, the sensitivity of cells toward GO has been correlated to the number of CD33 molecules they express on their cell surface (Walter et al. 2005). In the case of melanotransferrin/p97, a threshold may be required to confer sensitivity toward an auristatin-based ADC. Most cell lines expressing less than 80,000 copies of p97 are refractive to the anti-p97 ADC. This resistance appears to result from a failure of the internalized ADC to traffic from the caveolae to the lysosomes (Smith et al. 2006). Furthermore, the rate of internalization can be affected by co-receptors that interact with the ADC targets. For example, some CD19 molecules expressed on B-cell surface form a molecular complex with the surface Ig and CD21 to regulate signaling by the B-cell receptor (Tedder, Inaoki, and Sato 1997). CD21 is a relatively slow internalizing antigen, and it has recently been reported that as the CD21 copy number on B-cell surface increases the rate of ADC-induced internalization of CD19 declines. Thus, overexpression of CD21 on a lymphoma B-cell may render it completely refractory to an anti-CD19 ADC (Ingle et al. 2008). Finally, the anatomical localization of surface antigens may affect physical access by targeting ADCs. The mucin MUC1 primarily localizes to the apical membrane of normal epithelia lining the bronchus, breast, salivary gland, pancreas, prostate, and uterus, away from the underlining vascularized stromal tissues (Ho et al. 1993). Interestingly, MUC1 becomes more diffusely distributed along the cellular membrane or concentrated on the stromal side of breast, gallbladder, and colon carcinoma cells (Kawamoto et al. 2004; Rakha et al. 2005; Suzuki et al. 2004). Such relocalization may enable ADCs targeting MUC1 to gain tumor access from the vascular space.

35.4.4 Other Potential ADC Targets

Fast internalizing surface antigens with an acceptable tumor to normal tissue expression differential probably constitute a small fraction of the human proteome. Additional strategies are therefore needed to identify other classes of antigens that may be suitable for ADC targeting. A number of avenues can potentially be explored toward this goal.

Evidence is available that drug conjugation may significantly alter the biological activity of certain antibodies. The B-cell associated antigen CD20 is not known to be a fast internalizing antigen, even when it is bound by anti-CD20 mAb (Law et al. 2004). Interestingly, anti-CD20 ADCs consisting of either a tubulin-binding drug, an auristatin (Law et al. 2004), or the DNA-binding drug, calicheamicin (DiJoseph et al. 2007b), can efficiently mediate internalization of CD20 molecules from B-cell surface and induce cytotoxicity indicative of intracellular ADC processing and drug release. Thus, it is important to examine empirically the ability of any given ADC to induce internalization of its cognate antigen even though the antigen recognized by the ADC has been defined as a noninternalizing surface protein.

The tumor microenvironment may provide conditions in favor of drug release from ADCs. It is known that the tumor extracellular milieu is more acidic than normal tissues, potentially favoring hydrolysis of acid-labile drug linkers, for example, hydrazone linkers. In cases of tumors with high metastatic potential their microenvironment may be enriched with a variety of proteases, such as matrix metalloproteinases (Deryugina and Quigley 2006), urokinase (Duffy 2004), and cysteine cathepsins (Mohamed and Sloane 2006; Palermo and Joyce 2008). It is therefore conceivable that ADCs with acid-labile or peptide drug linkers that are differentially susceptible to degradation in the tumor microenvironment can release their cytotoxic payload extracellularly. As a result, even if these ADCs are directed against truly noninternalizing tumor-associated antigens, the released cytotoxic drugs should still be able to exert their pharmacologic effects given they have the ability to diffuse across the plasma membrane.

Approaches to target antigens that are normally not expressed on the cell surface have also been reported. One such antigen is the LY6D protein, a GPI-linked molecule involved in keratinocyte cell-cell adhesion. Treatment of colorectal carcinoma cell lines with irinotecan (CPT-11) induces surface expression of LY6D on these otherwise LY6D-negative cell lines (Rubinfeld et al. 2006). More interestingly, a combination of CPT-11 and an ADC targeting LY6D demonstrate potent antitumor effects in a xenograft model using one of these colorectal cell lines, whereas either CPT-11 or the ADC alone is much less effective, confirming that CPT-11 does indeed induce surface expression of LY6D *in vivo*. A second example is the La antigen, a molecular chaperone for RNA polymerase III often overexpressed in neoplastic cells. Upon cell death induced by chemotherapeutic drugs and loss of membrane integrity, intracellular La becomes accessible to antibody binding (Al-Ejeh et al. 2007). Tumor targeting *in vivo* by an anti-LA mAb has been confirmed subsequent to treatment of tumor-bearing mice with chemotherapeutics (Al-Ejeh, Darby, and Brown 2007).

Finally, the recent FDA approvals of bevacizumab, sorafenib, and sunitinib, which interfere with angiogenesis, have validated the clinical application of the anti-angiogenic approach to treatment of solid tumors. Ligand-directed vascular targeting molecules against vascular endothelial growth factor receptor 2 (VEGFR-2), integrin $\alpha v\beta 4$, CD13 (aminopeptidase N), or CD105 (endoglin) have shown encouraging preclinical activities (Gerber, Grewal, and Oflazoglu 2009). It is therefore possible that ADCs against one or more of these receptors involved in tumor-associated angiogenesis may be efficacious through selective destruction of blood vessels within tumors, thereby restricting oxygen and nutrient supply to tumor cells.

35.5 DRUG DELIVERY VEHICLES

35.5.1 IgG and Antibody Fragments

In addition to conferring targeting specificity, the antibody components are key determinants for the pharmacokinetic properties of the ADCs. The more than 40 immunoconjugates described in Tables 35.2 to 35.4 all consist of intact IgG of different isotypes. Human IgG have relatively long serum half-lives, ranging from approximately 1 week for IgG3 to more than 21 days for some IgG1. The neonatal Fc receptor (FcRn) is the key receptor expressed by cells in the reticuloendothelial system (RES), for example, monocytes, macrophages, Kupffer cells, and tissue histocytes, which regulates homeostasis of serum IgG (Ghetie and Ward 1997, 2002). FcRn salvages IgG from the lysosomal degradation pathway and prevents its clearance from serum. Certain amino acid substitutions in the Fc regions of IgG can increase their affinity to FcRn at pH 6 in the early endosomal compartment, but not at the physiological pH of 7.2 on the cell surface, thereby favoring IgG recycling and significantly prolonging IgG serum half-lives (Ghetie et al. 1997; Shields et al. 2001). Understanding how serum half-life and clearance contribute to the efficacy and toxicity are among the key considerations in ADC development. ADCs with stronger binding to FcRn will have lower rates of degradation, longer serum half-lives, and therefore may reduce exposure of the RES to the released cytotoxic payloads. For ADCs with stable drug linkers in circulation, reduced clearance from serum will increase drug exposure at sites of the targeted tumor. However, it should be noted that on-target toxicity due to expression of the tumor-associated antigens in normal tissues is also a function of ADC serum half-lives.

It is well established that Fc regions of IgG mediate antibody effector functions. With respect to ADC biology, the relevant questions are how drug conjugation would affect antibody effector functions, how much antibody effector function would be desired for ADCs that are already carrying potent cytotoxic payloads, and how antibody effector functions would potentially contribute to the toxicity of ADCs. The best defined Fc-dependent effector functions are ADCC, phagocytosis, and CDC. ADCC and cellular phagocytosis are both dependent on interactions between Fc domains of IgG and Fcγ receptors (FcγRs) expressed on immune effector cells (Nimmerjahn and Ravetch 2007; Sigal 2007). The activating and inhibitory, high and low affinity FcγRs are broadly expressed on hematopoietic cells, including macrophages, monocytes, polymorphonuclear cells, esoinophils, natural killer (NK) cells, B- and T-cells (Sigal 2007). In nonhematopoietic cell, FcγRs have also been detected on liver sinusoidal endothelial cells (Mousavi et al. 2007; Muro et al. 1993), peripheral nerve cells (Vedeler et al. 1991) and microglial cells (Ulvestad et al. 1994). One cell type can often express multiple kinds of FcγRs on its surface. It is instinctive to assume that ADC binding to FcγR-bearing cells may contribute to toxicity. As the binding affinity of human IgG to FcγRs is in the general order of IgG3 $\geq$ IgG1 $>$ IgG2 $\geq$IgG4, one hypothesis is that IgG2 and IgG4 may be better antibody backbones for ADCs with respect to limiting toxicity elicited by binding to FcγR-expressing normal cells. In fact, the vehicles of gemtuzumab ozogamicin and ADCs targeting CD22, GPNMB, and MUC1 are either IgG4 or IgG2 (Tables 35.2 to 35.4). However, systematic analysis to compare the toxicity profiles of ADCs consisting of identical targeting specificity and payload but differing only in the IgG isotypes of the vehicles has not been conducted to test this hypothesis. In contrast, antibody effector functions may contribute to the overall potency of ADCs. A chimeric form of an anti-CD70 mAb of IgG1 isotype has recently been reported to mediate a full complement of antibody effector functions against CD70-expressing tumor cells (McEarchern et al. 2007). ADCs consisting of this anti-CD70 mAb conjugated with auristatin-based cytotoxic drugs elicit potent CD70-dependent cytotoxicity against renal cell carcinoma cells (Law et al. 2006). Drug conjugation, however, has minimal impact on the effector functions of the parent mAb (unpublished results). It would therefore be interesting to use this set of anti-CD70 ADCs to further elucidate the role of antibody effector functions in the potency of ADCs.

Similar to intact IgG, antibody fragments provide the specificity needed for cytotoxic payload delivery. Antibody fragments of varying molecular sizes, for example, Fab (~25 kDa), scFv (~25 kDa), diabodies (~55 kDa), minibodies (~80 kDa), and nanobodies (VHH of camelid IgG, ~15 kDa), have been engineered and are being evaluated for clinical applications (Harmsen and De Haard 2007; Jain, Kamal, and Batra 2007). Varying degrees of success have been reported in using antibody fragments in preclinical and clinical settings to deliver radionuclides for tumor imaging and radioimmunotherapy (Wu and Senter 2005) or to carry protein toxins to tumor cells expressing receptors like CD19, CD22, CD25, CD30, the Lewis Y antigen, or erbB2 (Kreitman and Pastan 2006; Pastan et al. 2007). The smaller sizes of antibody fragments and the lack of any Fc domains in Fab, scFv, diabodies, and nanobodies will certainly yield ADCs with profound differences in pharmacokinetic profiles compared to those with intact IgG. Antibody fragments of less than ~50 kDa are below the renal threshold and are efficiently excreted, resulting in drastically reduced serum half-lives (Filpula 2007; Mahmood and Green 2005). The lack of Fc domains also means that Fab, scFv, diabodies, and nanobodies will not be recycled by the FcRn salvage pathway, further accelerating their clearance from plasma. One advantage offered by the smaller sizes of antibody fragments may be better tumor penetration and more homogeneous tissue distribution within tumors (Adams et al. 2001). This combined with shorter serum residence time seems to be desirable for tumor imaging, as the use of radiolabeled immunoconjugates consisting of scFv give higher tumor to normal tissue ratios of radioactivity and better contrast in the images generated (Wu and Senter 2005). Whether ADCs consisting of antibody fragments will give more favorable tumor to normal tissue ratios for the cytotoxic payloads and whether sufficient quantities of cytotoxic drugs will be delivered to induce tumor cell killing in the face of reduced serum half-lives remain to be decided. Another perceived benefit for antibody fragments that do not contain Fc domains is less Fc-dependent toxicity. If FcγR interaction is proven to be a major culprit behind off-target toxicity for certain ADCs made up of intact IgG, a combination of antibody fragments and approaches for enhancing half-lives, for example, pegylation (Yang et al. 2003) or complex formation with a protein with longer plasma half-life like albumin (Breton et al. 1995), may be a strategy to generate ADCs with better therapeutic indices.

35.5.2 Alternative Vehicles for Drug Delivery

As specificity and selectivity are the principal parameters for choosing vehicles for drug delivery, there is not an *a priori* reason to restrict vehicles only to antibody or antibody fragments. There are many examples of high affinity ligands that can serve as delivery vehicles. For example, recombinant toxin molecules made up of cytokines have been generated to target tumor cells expressing receptors for VEFG (Akiyama et al. 2005; Mohamedali et al. 2005), Blys (Lyu et al. 2007), IL-4 (Shimamura, Husain, and Puri 2006), IL-13 (Stish et al. 2007), or EGF (Stish et al. 2007). Chemical conjugates have also been made based on small molecule microtubule disrupting drugs carried by folate and RDG peptides to target the folate receptor (Ladino et al. 1997; Lu et al. 2004; Reddy et al. 2007) and the αvβ3 integrin (Temming et al. 2006, 2007), respectively; membrane lytic peptide carried by a 15 amino acid segment of chorionic gonadotropin β chain to target the chorionic gonadotropin receptor (Leuschner and Hansel 2005); or a derivative of doxobicin carried by somatostatin to target its receptors (Nagy and Schally 2005). Beyond naturally occurring ligands, artificial binding proteins have become available in recent years due to the advent of display technologies. These proteins are usually composed of a compact and structurally rigid core scaffolding that presents surface loops of varying sequence and length of amino acids. Diverse libraries can be generated through random mutagenesis of the displayed loops. Binders of desired specificity and affinity can then be selected from these libraries by standard binding techniques. Protein scaffolds evaluated for pharmaceutical applications include lipocalin, thioredoxin, ankryin repeat, protein A, C-type lectin, bovine trypsin inhibitor, and γ-crystalline (Hey et al. 2005; Skerra 2000, 2007). In principle, any one of these ligands could potentially be a suitable vehicle for cytotoxic payload delivery. Nevertheless, challenges facing the development of ADCs such as target specificity and selectivity, pharmacokinetic properties, on- and off-target side effects will remain the same for non-antibody-based drug carriers. In addition, potential immunogenicity of these artificial proteins in humans will need to be addressed as well.

35.5.3 Bispecific Vehicles

ADC technology appears to be disproportionately applied to hematological malignancies despite the obvious and more serious unmet medical needs in many solid tumor indications. As discussed earlier, lineage-specific differentiation antigens expressed by hematologic cells and their malignant counterparts is probably one of the reasons behind this discrepancy. The absence of truly tumor-specific antigens expressed only on transformed but not normal epithelial cells has been a formidable obstacle toward the identification of more antigens suitable for ADC targeting in solid tumor indications.

Bispecific antibodies have been used to bridge cancer cells and immune effector cells like NK and T-cells, for example, bispecific antibodies recognizing CD3 and CD19 or EphA2, CD16 and CD19, Her2, ErbB2, or CD30 (Lum, Davol, and Lee 2006; Sanz, Blanco, and Alvarez-Vallina 2004; Wolf et al. 2005). More recently bispecific antibodies that modify cytotoxic signaling to lymphoma B-cells (Qu et al. 2008) or simultaneously neutralizing two proinflammatory cytokines (Wu et al. 2007) have been reported. Another potential application of this technology is to increase target cell selectivity. In this case, the two arms to the bispecific antibody each recognizes a distinct antigen expressed on the same target tumor cells. Simultaneous engagement of both arms with antigens provides high avidity stable binding, whereas interaction of a single arm to its target antigen will not give sufficiently high affinity to sustain binding. Such an approach has been applied to bispecific antibodies targeting Fas and CD19, CD20 or CD40 to focus Fas-induced apoptosis to $Fas^{+}CD19^{+}$, $Fas^{+}CD20^{+}$ or $Fas^{+}CD40^{+}$ lymphoma cells (Jung et al. 2001) and $Fas^{+}CD40^{+}$neuroblastoma cells (Wischhusen et al. 2005). This approach should be applicable to refine the selectivity of ADCs by targeting two antigens on the surface of solid tumors as has recently been shown by the bispecifc antibody targeting MUC1 and TAG-72 (Natsume et al. 2006) and the bispecific recombinant cytokine-toxin conjugate containing IL-13 and EGF (Stish et al. 2007). Although attractive in theory, careful molecular design and fine tuning of affinity for each arm of such bispecific vehicles to ensure low affinity monovalent interaction with fast off-rates but high avidity and stable bivalent binding are required for success in this approach.

35.6 THE CHOICE OF DRUG

Many classes of drugs have been tested as ADCs, but the number that has shown substantial activity is quite low. The first ADCs were made with drugs already approved by the FDA as single agents, while current ADCs use drugs that were typically found to be too toxic on their own but of substantially higher potency. These high potency drugs have been a key improvement in the development of ADC-based drug delivery, similar to the improvements offered by chimeric and humanized antibodies over murine antibodies. Figure 35.3 shows the structures of the five classes of drugs that have been incorporated into ADCs listed in Tables 35.2 to 35.4: the anthracycline doxorubicin, the vinca alkaloid vinblastine, the DNA double-strand-cleaving agent calicheamicin, and the tubulin-binding agents maytansine and auristatin. Multiple chemical approaches have been employed to conjugate these drugs onto antibodies; each takes advantage of unique cellular mechanisms for targeted drug release.

35.6.1 Potency

35.6.1.1 Early ADCs The first ADCs were made from drugs that were already approved by the FDA for cancer treatment. This offered two perceived advantages: familiarity with the pharmacology and toxicology of these drugs, as well as an abundant commercial supply of these drugs. One early ADC, chimeric BR96-Dox (cBR96-Dox; also known as BMS-182248 and later SGN-15), was composed of a chimeric anti-Lewis Y mAb conjugated to doxorubicin, a topoisomerase II inhibitor and DNA-binding agent obtained from cultures of a chemically mutated strain of the bacterium *Streptomyces peucetius* (Arcamone et al. 1969) and approved for a wide variety of cancers, including leukemias and carcinomas. Lewis Y is expressed on many carcinomas, providing the rationale for the

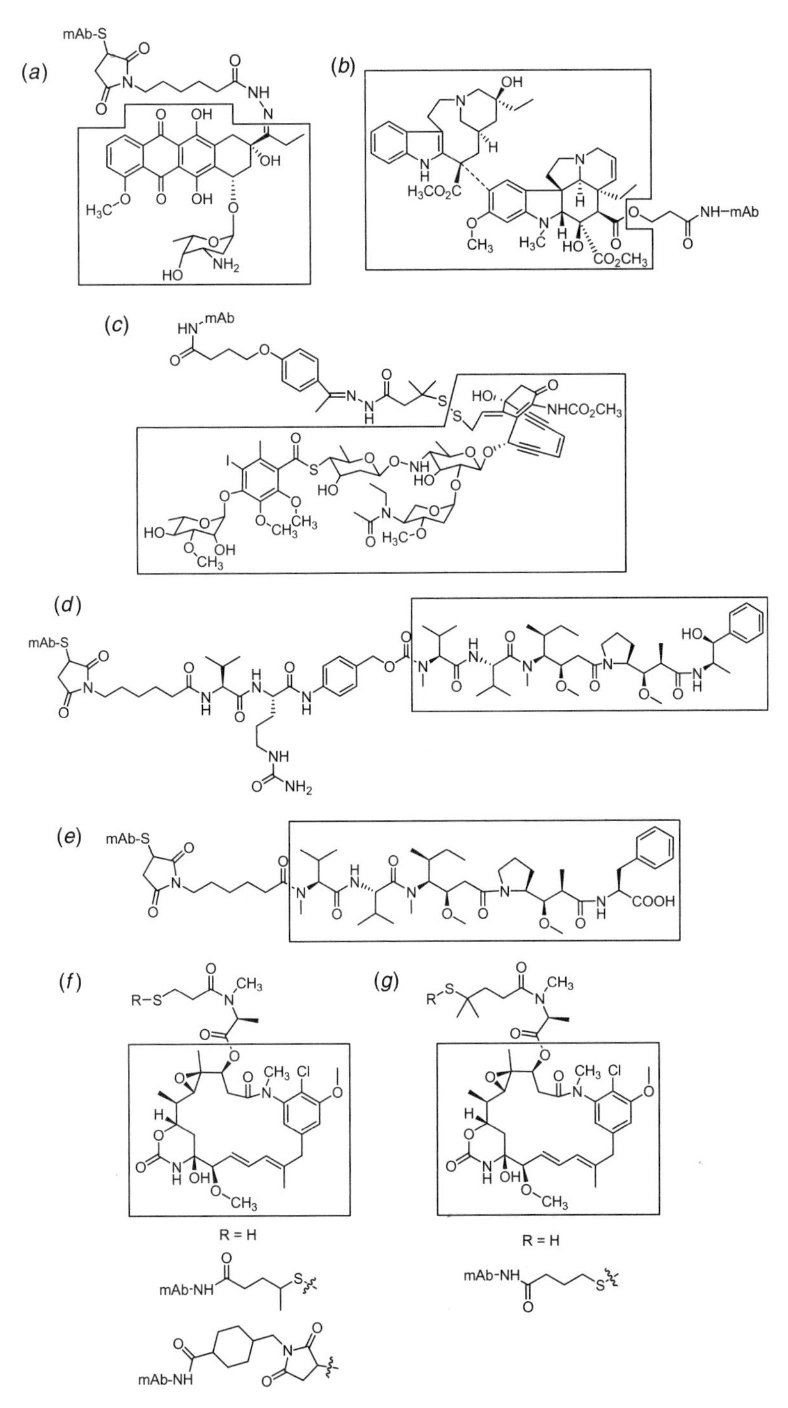

Figure 35.3 Structures of drug and linker systems described in this chapter. Parent drug molecules are shown in boxes. (a) Doxorubicin and the drug linker used in cBR96-Dox. (b) 4-desacetylvinblastine and the drug linker used in LY256787. (c) Calicheamicin and the drug linker used in Mylotarg®. (d) Monomethyl auristatin E and the drug linker vcMMAE. (e) Monomethyl auristatin F and the drug linker mcMMAF. (f) Maytansinol, and DM1, SPDP-DM1, and SMCC-DM1. (g) Maytansinol, and DM4 and SPDB-DM4.

cBR96-Dox ADC. This conjugate has a modest IC_{50} value of 2.7 μM for antigen positive cells *in vitro* and could yield tumor regressions in mouse xenograft models using the relatively high dose of 5 to 10 mg/kg when given in multiple administrations (Trail et al. 1993). Chimeric BR96-Dox has been tested in a variety of phase I and II clinical trials, but its toxicity limits its clinical utility. Relatively high doses of up to 350 mg/m^2 or about 9 mg/kg for a 65 kg adult have been administered to patients (H.J. Ross et al. 2006). Interestingly, a patient dosed with unconjugated cBR96 showed the same type and severity of toxicity, suggesting that the antibody rather than the drug might be responsible for the observed toxicity (Saleh et al. 2000). Accordingly, a more potent drug might have enabled better clinical activity to be observed before the onset of mAb toxicity. As cBR96 has been demonstrated to have strong Fc-FcγR-mediated antibody effector functions (Schreiber, Hellstrom, and Hellstrom 1992), antibody engineering to suppress its FcγR binding activity could eliminate one potential route of toxicity.

Another early ADC, LY256787, is composed of the murine mAb KS1/4 conjugated to 4-desacetylvinblastine. Vinblastine is an antitubulin alkaloid extracted from the periwinkle *Vinca rosea* approved for the treatment of lymphomas and carcinomas. KS1/4 binds to an antigen found on a variety of carcinomas, again providing rationale for the choice of conjugated drug. Patients received as much as 550 mg of this ADC in a single dose, or about 8 mg/kg for a 65 kg adult, which again is a rather high dose. There were strong and rapid anti-mouse responses, which ultimately limited the clinical utility of this agent (Petersen et al. 1991). Chimerization or humanization of KS1/4 would probably have mitigated the immune response against the murine antibody. A more potent drug, on the other hand, should have also enabled lower doses of the ADC.

35.6.1.2 New ADCs More recently, the trend has moved away from using FDA-approved drugs in ADCs towards using drugs that are much more. The first and to date only FDA-approved ADC GO carries such cytotoxic payload calicheamicin (Bross et al. 2001). After its discovery (Lee 1987a, 1987b), it was demonstrated that calicheamicin has a unique DNA sequence-specific strand cleaving mechanism (Zein et al. 1988). Because of its high potency, calicheamicin first found its way into the clinic as an ADC rather than a small molecule agent. GO is substantially more potent than cBR96-Dox, demonstrating an IC_{50} for antigen-positive cells of about 0.3 pM (Hamann et al. 2002), an improvement of about 10^7-fold. The recommended clinical dosage of GO is 9 mg/m^2, or about 0.22 mg/kg for a 65 kg adult. This dose is nearly 100-fold lower than was used for doxorubicin and vinblastine conjugates. While causing profound neutropenia and thrombocytopenia, GO achieved a complete remission rate of 26 percent (Larson et al. 2002, 2005; Sievers et al. 2001). CD33 is found on myeloid progenitor cells, which likely explains the observed toxicity. With the use of highly potent drugs, targeted as well as nontargeted toxicities due to the drug are likely to limit the therapeutic dose, as opposed to the antibody-related limiting toxicity observed for cBR96-Dox and LY256787.

Maytansinoids represent another highly potent drug class that has been used in the clinic. Maytansine was first isolated from the Ethiopian shrub *Maytenus ovatus* (Kupchan 1972), although it is likely produced by or in conjunction with associated microorganisms (Cassady et al. 2004) as it was later found to be produced by isolated bacteria (Higashide et al. 1977). It is a macrolide with a 19-member ring that binds to tubulin at the same site as the vinca alkaloids (Cassady et al. 2004). Clinical trials were conducted with maytansine as a single small molecule agent in the 1970s, and toxicity was encountered before any significant evidence for clinical benefit was observed. Two different maytansine derivatives, DM1 and DM4, have been conjugated with multiple linkers. An early DM1 anti-CanAg ADC demonstrated potent cytotoxicity *in vitro*, achieving an IC_{50} of 32 pM on antigen positive cells (Liu et al. 1996). Clinical trials with DM1 ADCs against CD56 and Her2 and a DM4 ADC against CanAg have been dosed up to 4.8 mg/kg and have shown evidence of clinical response at tolerated doses (Beeram et al. 2007; Ishitsuka et al. 2008; Tolcher et al. 2003).

The auristatins, derived from dolastatin 10, represent a third class of highly potent drugs. A wide variety of highly cytotoxic peptides have been isolated from the Indian Ocean sea hare *Dolabella auricularia*, with the tenth in the series being the precursor for the auristatins. The first isolation required 1000 kg of sea hares to yield 28.7 mg of dolastatin 10 (Pettit 1987). Dolastatin 10 has been found to be produced by a marine cyanobacterium (Luesch et al. 2001). So similar to the maytansines, a microorganism associated with *Dolabella auricularia* may be the actual organism producing dolastatin 10. Viable chemical routes have been found for the total synthesis of dolastatin 10 and many of its analogs (Pettit 1989). Substituting norephedrine (Pettit 1997) or phenylalanine (Doronina et al. 2006) for the dolaphenine unit yields auristatin E and F, respectively. Dolastatin 10 and its analogs bind to the same site on tubulin as maytansine and vinblastine (Bai, Pettit, and Hamel 1990), although dolastatin 10 is a noncompetitive inhibitor of tubulin polymerization while maytansine and vinblastine are competitive inhibitors (Bai, Pettit, and Hamel 1990). Many dolastatins and synthetic analogs have been tested in clinical trials, but none have progressed past phase II due to dose-limiting toxicity. The analog monomethylauristatin E (MMAE) conjugated to anti-CD30 (Doronina et al. 2003; Francisco et al. 2003; Hamblett et al. 2004) and anti-LewisY (Doronina et al. 2003; Francisco et al. 2003; Hamblett et al. 2004) mAbs had IC_{50} values against hematologic and carcinoma cell lines between 0.24 and 4.8 nM. Two MMAE ADCs, SGN-35 targeting CD30 (Francisco et al. 2003; Hamblett et al. 2004; Younes et al. 2007) and CRO-11 targeting GPNMB (Pollack et al. 2007; Sznol et al. 2007; Tse et al. 2006), are currently being tested in clinical trials, each showing evidence of clinical benefit at doses less than 2 mg/kg without encountering dose-limiting toxicity.

These new ADCs have been more successful in the clinic due to a number of features, with the use of chimeric or humanized mAbs playing a significant role. The much reduced immune response to ADCs utilizing chimerized or humanized mAbs has enabled multiple rounds of dosing and a more sustained level of treatment. Increasing the potency of the drug is also a substantial benefit. Because there are a finite number of antigens that can internalize the ADC into the target cells, once saturated the amount of ADC internalized cannot increase unless a secondary nonspecific mechanism operates. A more potent drug will therefore increase the odds of achieving an intracellular concentration of the released drug higher than the minimum concentration necessary for a therapeutic effect.

35.6.2 Safety

Only a small fraction of the ADC can be expected to localize to the tumor: in humans, perhaps 0.01 percent of the injected dose. The remainder of the ADC will be eliminated somewhere else in the body. Several studies have established the sites of antibody metabolism as including liver, muscle, and skin. As the stability of the linkage between drug and antibody has increased and the systemic release of free drug has decreased, drug release by normal antibody metabolism becomes the dominant mechanism by which the body is exposed to the free drug. The toxicity profile of an ADC could be markedly different than that of the free drug, and special attention to sensitivity to drug at known sites of drug release may be key in minimizing ADC toxicity. For instance, doxorubicin use is ultimately limited by cardiotoxicity (Ng and Green 2007). In clinical trials of the doxorubicin ADC cBR96-Dox, cardiotoxicity was not observed. In one phase II study that contained both doxorubicin and cBR96-Dox treatment arms, the observed toxicities of the two agents were different (Tolcher et al. 1999). A lack of Lewis Y expression on normal heart tissues might have helped to minimize targeted delivery of doxorubicin to the heart, and the systemically released doxorubicin from cBR96-Dox could have exposed the heart to much less doxorubicin than a comparable dose of the free drug. In mice treated with BR96-doxorubicin and an equivalent dose of doxorubicin, much less doxorubicin was observed in heart tissues for BR96-doxorubicin treatment than free doxorubicin treatment, resulting in a fourfold lower exposure (Mosure et al. 1997).

Previous reviews of ADCs have identified minimization of systemic toxicity as a goal of targeted therapies. One of the consequences of ADCs, however, is an alteration of the biodistribution and the rate of elimination of the drug. The body is still exposed to the free drug following degradation of the ADC, but the concentrations and kinetics of drug in various organs has been substantially altered. An effective targeted drug may be one whose mechanism of toxicity is the most compatible with the mechanism of drug release and elimination. Drugs that are tolerated by the hepatobiliary system, gastrointestinal tract, muscle, and skin should make better ADCs.

35.6.3 Stability and Trafficking of Released Drug

As stated earlier, drug release from ADCs occurs after internalization to lysosomes with an acidic and reducing environment and an abundance of hydrolases. None of the drugs commonly used in ADCs has its cellular target in the lysosome: doxorubicin and calicheamicin need to travel to the nucleus to target DNA, while vinblastine, maytansine, and auristatin need to travel to the cytosol to target tubulin. The environment within lysosomes may not be entirely compatible with the chemical structures of the released drugs, and a drug that is unstable to low pH or proteases would be a poor choice for an ADC. Drugs that are sequestered in lysosomes and are not able to reach their cellular targets would also make poor choices. It has been observed that cells with the multidrug resistance phenotype can accumulate doxorubicin in lysosomes and keep it out of the nucleus (Rajagopal and Simon 2003). Similarly, a conjugate of the protein toxin gelonin to the antibody HuM195 also accumulated in lysosomes of multidrug resistant cells, and the sequestered conjugate was also degraded in the lysosomes (McGrath et al. 2003).

35.7 BYSTANDER EFFECT

Tumors are most typically heterogeneous in composition, with multiple types of cells as well as different levels of antigen expression within the tumor. For maximal therapeutic effects, most, if not all, of the neoplastic cells within the tumor should be killed. This poses a problem if the targeted agent can only kill highly antigen-positive cells or if the distribution of the ADC within the tumor is heterogeneous (Thurber, Schmidt, and Wittrup 2008). Attempts have been made to demonstrate that ADCs could also kill neighboring antigen-negative cells. GO has shown no clear correlation between clinical benefit and CD33 expression, leading to studies that showed non-CD33-mediated uptake via endocytosis is responsible for some activity (Jedema et al. 2004). Ideally, this non-antigen-mediated uptake would only occur in the tumor, or an increase in toxicity would be observed. Another mechanism for killing neighboring cells would be through a bystander effect, where antigen-mediated uptake results in a locally high concentration of the released free drug that can diffuse into neighboring tumor cells. The concentration gradient outside the tumor would be very steep, because the surrounding antigen-negative normal cells would ideally not produce free drug by an antigen-mediated process. ADCs made with an anti-CanAg mAb and either DM1 or a CC1065 analog DC1 were sometimes able to kill both antigen-positive and antigen-negative cells in mixed population *in vitro* and *in vivo*. The observation of this bystander effect occurred with a disulfide linkage but not a thioether linkage, presumably because the former linkage releases a cell permeable drug while the latter does not (Kovtun et al. 2006).

35.8 LINKER CHOICE

As mentioned in Section 35.3, the unique lysosomal environment with respect to pH, redox potential, and wealth of hydrolytic enzymes has been exploited for the release of drugs. LY256787 is linked through an ester, which would need to hydrolyze to release the conjugated 4-desacetylvinblastine (DALVB; Fig. 35.3). After injecting LY256787 into animals, in early time points in plasma the hemisuccinate was observed, while at later time points DALVB and the N-oxide of DALVB were observed (Spearman et al. 1987). The observation of the hemisuccinate is consistent with proteolysis of the amide linkage to the antibody lysine, with subsequent ester hydrolysis releasing DALVB.

Chimeric BR96-Dox was conjugated via an acid-sensitive hydrazone bond. At neutral pH, hydrazones are relatively stable, but upon acidification they become more unstable. Following dosing of cBR96-DOX, doxorubicin was observed in tumors in mouse xenograft models indicating linker hydrolysis and drug release (Mosure et al. 1997). GO has both a hydrazone and a hindered disulfide bonds in its linker. In patients, calicheamicin has been observed in circulation using a competition

ELISA with an anti-calicheamicin antibody (Dowell et al. 2001). While this method does not allow for the identity of the released drug to be identified, it does demonstrate that a calicheamicin-derived material is being released from the ADC.

DM1 and DM4 have been linked by disulfides as well as noncleavable thioesters. The C-3 alcohol maytansinol (Fig. 35.3) can be further elucidated to yield the *N*-methyl *N*-(thiopropanoyl)-L-alanine C-3 amide to form DM1 (Chari et al. 1992) or the *N*-methyl *N*-(4-thio-4,4-dimethylbutanoyl)-L-alanine C-3 amide to form DM4. A variety of cross-linking agents can link the thiol to lysines of antibodies: SPDP-DM1 and SPDB-DM4 are disulfide linked ADCs, and SMCC-DM1 is a thioether linked ADC (Fig. 35.3). Although the disulfide-linked ADCs were envisioned as being reduced in the lysosome to release DM1 and DM4, recent evidence suggests that the lysosomal compartment may not provide the redox potential sufficient to reduce the disulfide bonds (Austin et al. 2005). Indeed, a study investigating cellular degradation of a disulfide-linked SPDB-DM4 ADC showed a variety of products including lysine-N^{ε}-SPDB-DM4, which would result from antibody degradation without disulfide reduction. Additional products included DM4, *S*-methyl-DM4 and *S*-cysteinyl-DM4, which require disulfide reduction. The exclusive product from a thioether-linked SMCC-DM1 ADC was lysine-N^{ε}-SMCC-DM1, the product of antibody degradation in the strongly proteolytic environment of the lysosome (Erickson et al. 2006).

Finally, linkers have been designed to take advantage of the proteolytic environment of the lysosome to release unmodified forms of linked drugs. The auristatin MMAE (Fig. 35.3) has been linked to antibodies with the cathepsin B-labile linker maleimidyl-valine-citrulline-PABA (Doronina et al. 2003). The ubiquitous lysosomal protease cathepsin B can cleave the dipeptide (Dubowchik and Firestone 1998; Dubowchik et al. 1998), while the self-immolative PABA subsequently fragments to release MMAE in a tracerless fashion. This sort of linkage is extremely stable in circulation, because of the general absence of proteases in the blood, while efficiently releasing drug upon internalization of ADC. In cells, inhibition of cysteine proteases, which include cathepsin B, causes a decrease in the cytotoxicity of these dipeptide-linked ADCs (Sutherland et al. 2006). Auristatins have also been linked through thioether linkers: the drug linker mcMMAF (Fig. 35.3) does not contain the valine-citrulline-PABA group, but was still shown to be highly active as an ADC. The product of ADC degradation is cysteine-mcMMAF, with the cysteine coming from the antibody due to the attachment of the linker-drug moiety to a reduced antibody disulfide (Doronina et al. 2006).

35.8.1 Drug Linker Stability

Conjugation of a highly cytotoxic drug to an antibody can be considered a pro-drug strategy, where the ADC is the inactive pro-drug that needs to be converted into the released free drug to restore activity. The stability of the linkage between the antibody and drug is important. It should be stable enough to prevent systemic release of the active drug, at the same time allowing optimal exposure of the tumor to the released drug. Several studies have been used to establish the stability of the linkage between the antibody and drug. The doxorubicin component of cBR96-Dox was shown to have a half-life of about 43 hours in a phase I trial, compared to about 300 hours for the antibody component (Saleh et al. 2000). The hydrazone used to conjugate the drug to the antibody was significantly less stable than the half-life of the antibody, and could be improved upon. GO has also been investigated for its drug linker stability in mice. A surface plasmon resonance assay was developed to assay for the antibody and calicheamicin components of the ADC (Boghaert et al. 2008). The calicheamicin component has a shorter half-life than the antibody component, 47 and 109 hours, respectively, with a decrease in the average number of drugs per antibody observed over time. Both of these hydrazone linker-containing ADCs use a pH-dependent chemical cleavage mechanism, with each decrease in one pH unit causing a 10-fold decrease in stability. The pH difference between the blood and lysosomes is about 3 units, so a 1000-fold difference in stability is the maximum expectation. The stability of hydrazones can be varied by chemical modification, allowing more fine tuning to take place to achieve the most efficient drug release in the lysosome with the minimum loss in circulation, but this type of release mechanism will still have a significant release rate in circulation.

The stability of a disulfide linked DM1 ADC has also been investigated. In mice, the drug and antibody components had half-lives of 24 and 100 hours, respectively (Xie et al. 2004), while in humans these values were 41 and 230 hours with dosing every three weeks (Tolcher et al. 2003). In another study with dosing three times per week, the drug-linker had a similar half-life of two days (Rodon et al. 2008). In the mouse experiment, a plot of DM1 per antibody showed a linear decrease over time, suggesting a first order process. Drug release from this conjugate will occur upon disulfide reduction, with the maximum stability equal to the difference in reducing potential between the blood and the lysosome or cytoplasm. The one- to two-day half-lives of the hydrazone and disulfide drug linkers may limit the *in vivo* antitumor activity of these ADCs. Replacing DM1-SPDP ADCs with the more stable SMCC-DM1 and SPDB-DM4 ADCs may yield improved antitumor activity.

In contrast to hydrazones and disulfides, protease sensitive linkers offer an increase in stability due to the enzyme mediated cleavage mechanism. An MMAE ADC with a valine-citrulline linker was shown to have a half-life in mice of 6 days and in cynomolgus monkeys of roughly 10 days (Sanderson et al. 2005). In this study, ADCs with a fixed number of drugs per antibody were used after purification by hydrophobic interaction chromatography, rather than mixtures of variably loaded ADCs. In this way, any differential clearance of more heavily loaded ADCs was removed from the complexity of this measurement. The increase in stability from one to two days to six to nine days was possible because of the use of proteases in the lysosome to release the drug, which are either absent or inactive in circulation. Interestingly, the source of the loss of drug in circulation from this type of ADC does not appear to be due to cleavage of the valine-citrulline dipeptide in the linker. ADCs made with mcMMAF that do not have the dipeptide also have a half-life in circulation of about seven days. A drug linker that substituted a bromoacetamide for the maleimide was completely stable over two weeks, suggesting that the source of drug loss may be due to the maleimide group rather than the valine-citrulline (Alley et al. 2008).

35.9 DRUG CONJUGATION SITE

Several strategies have been employed to conjugate drugs to antibodies. Lysine residues are commonly found in antibodies and can be easily modified with *N*-hydroxysuccinimide (NHS) esters and peptide bond forming agents. For instance, LY256787 was conjugated using the NHS ester of DAVLB succinate (Spearman et al. 1987; Spearman, Goodwin, and Kau 1987). GO is similarly conjugated by treating the antibody with an NHS ester of calicheamicin (Hamann et al. 2002). To form disulfide-linked DM1 ADCs, the antibody was first treated with the NHS ester of a protected thiopropionic acid (Chari et al. 1992). This reagent reacts with lysines and installs a disulfide that can be further treated with a DM1 thiol. The final product is a lysine-linked disulfide. Conjugation through lysines will yield ADCs that are highly heterogeneous, since there are many potential conjugation sites. In practice, some lysines are more reactive than others (Wang et al. 2005), reducing but not eliminating the heterogeneity.

An alternative to lysine conjugation is conjugation through cysteines. There are four interchain disulfides in human IgG1 antibodies, and these can be selectively reduced to yield eight reactive cysteines. Chimeric BR96-Dox was conjugated in this way, yielding a completely homogeneous ADC (Trail et al. 1993). The auristatins were conjugated with four drugs on cysteines. This required either partial reduction followed by conjugation or full reduction followed by partial reoxidation and conjugation (Sun et al. 2005). The species formed by either of these methods was not random, demonstrating a preference for certain isomers over others. Interestingly, the two routes produced unique distributions of isomers, but the biological properties were similar.

A final method for conjugation to cysteines is through the use of genetically engineered mutants. Auristatins were conjugated to a series of mutants in which antibody interchain disulfides were substituted by serine residues, allowing homogeneous populations of two, four, or six drugs per antibody depending on the specific mutant (McDonagh et al. 2006). Again, the biological properties of these ADCs were similar to chemically produced mixtures. Alternatively, additional cysteines were genetically engineered into antibodies (Eigenbrot 2007; Junutula et al. 2008). This final method yielded a homogeneous population with two drugs per antibody. It will be interesting to see if such site-specific methods of conjugation lead to improved biological properties.

35.10 OPTIMIZING THE NUMBER OF DRUGS PER ANTIBODY

While the location of drugs in ADCs has yet to show a strong effect on biological properties, the absolute number of drugs per antibody has a significant effect. ADCs made with vcMMAE were chromatographically fractionated into samples with two, four, six, or eight drugs per antibody and their pharmacokinetic properties investigated. It was found that the higher the drug loading, the more rapid the clearance of the ADC in mice. Consequently, the exposure to antibody was about three times greater for four- than eight-loaded ADCs, which probably translated into higher drug exposure at the tumor for the four-loaded ADC despite injecting half as much drug (Hamblett et al. 2004). As expected, when the differentially loaded ADCs were tested for safety the maximum tolerated dose correlated with the absolute amount of delivered MMAE. In contrast, the *in vivo* antitumor activity was dependent on the stoichiometry of drug loading; a four-loaded ADC was more active than an eight-loaded ADC delivering the molar equivalent of MMAE (Hamblett et al. 2004). Hence, understanding the effects of chemical modification on the pharmacokinetic properties of antibodies and the optimal stoichiometry of drug loading would help the design of ADCs with enhanced therapeutic indices.

35.11 SUMMARY

The approval of gemtuzumab ozogamicin in 2000 has clearly validated ADCs as viable approaches to create novel and clinically active therapeutics. Significant advance have been made on multiple fronts since then. Expression profiling studies on normal and neoplastic cells at both the genomic and proteomic levels have revealed a variety of surface antigens that can be potentially exploited as targets to internalize ADCs for drug delivery. The endocytic pathways for ADC trafficking as well as the chemical and enzymatic mechanisms that promote drug release from ADCs are also being better defined. With respect to the payload, highly potent small molecule drugs, most notably microtubule-disrupting agents, including the maytansines and auristatins, and new chemistries for antibody conjugation have been examined in a multitude of experimental systems. Evolving preclinical and early clinical data on this new generation of ADCs have provided some very encouraging and exciting new data. In parallel with the progress in antibody engineering and site-directed drug conjugation novel ADCs with fine tuned target specificity, drug linker stability, and drug release mechanisms, with the ultimate goal of widened therapeutic indices, are currently on the horizon. It is likely that therapeutic application of ADCs will be expanded to multiple cancer indications including carcinoma in which significant unmet medical needs exist. More importantly, in addition to delivering cytotoxic drugs for cancer cell killing, one could also speculate on the possibility that future ADCs will be equipped to carry small molecule drugs capable of interfering with specific metabolic or signaling pathways in target pathogenic cells for use in diseases beyond oncology.

REFERENCES

Aboukameel, A., A.-S. Goustin, R. Mohammad, et al. 2007. Superior anti-tumor activity of the CD19-directed immunotoxin, SAR3419 to rituximab in non-Hodgkins xenograft animal models: Preclinical evaluation. Abstract No. 2339. *Blood* 110(11).

Adams, G.P., R. Schier, A.M. McCall, et al. 2001. High affinity restricts the localization and tumor penetration of single-chain fv antibody molecules. *Cancer Res.* 61:4750–4755.

Agathanggelou, A., G. Niedobitek, R. Chen, et al. 1995. Expression of immune regulatory molecules in Epstein-Barr virus-associated nasopharyngeal carcinomas with prominent lymphoid stroma. Evidence for a functional interaction between epithelial tumor cells and infiltrating lymphoid cells. *Am. J. Pathol.* 147:1152–1160.

Akiyama, H., K.A. Mohamedali, R.L. E Silva, et al. 2005. Vascular targeting of ocular neovascularization with a vascular endothelial growth factor121/gelonin chimeric protein. *Mol. Pharmacol.* 68:1543–1550.

Al-Ejeh, F., J.M. Darby, and M.P. Brown. 2007. The La autoantigen is a malignancy-associated cell death target that is induced by DNA-damaging drugs. *Clin. Cancer Res.* 13:5509s–5518s.

Al-Ejeh, F., J.M. Darby, K. Pensa, et al. 2007. In vivo targeting of dead tumor cells in a murine tumor model using a monoclonal antibody specific for the La autoantigen. *Clin. Cancer Res.* 13:5519s–5527s.

Alley, S.C., D.R. Benjamin, S.C. Jeffrey, et al. 2008. Contribution of linker stability to the activities of anticancer immunoconjugates. *Bioconjug. Chem.* 19:759–765.

Altschuler, Y., C.L. Kinlough, P.A. Poland, et al. 2000. Clathrin-mediated endocytosis of MUC1 is modulated by its glycosylation state. *Mol. Biol. Cell* 11:819–831.

Arcamone, F., G. Cassinelli, G. Fantini, et al. 1969. Adriamycin, 14-hydroxydaunomycin, a new antitumor antibiotic from *S. peucetius* var. caesius. *Biotechnol. Bioeng.* 11:1101–1110.

Austin, C.D., X. Wen, L. Gazzard, et al. 2005. Oxidizing potential of endosomes and lysosomes limits intracellular cleavage of disulfide-based antibody-drug conjugates. *Proc. Natl. Acad. Sci. U.S.A.* 102:17987–17992.

Bai, R.L., G.R. Pettit, and E. Hamel. 1990. Binding of dolastatin 10 to tubulin at a distinct site for peptide antimitotic agents near the exchangeable nucleotide and vinca alkaloid sites. *J. Biol. Chem.* 265:17141–17149.

Beeram, M., I. Krop, S. Modi, et al. 2007. A phase I study of trastuzumab-MCC-DM1 (T-DM1), a first-in-class HER2 antibody-drug conjugate (ADC), in patients (pts) with HER2+ metastatic breast cancer (BC). Abstract No. 1042. *J. Clin. Oncol.* 25(18S).

Bhaskar, V., D.A. Law, E. Ibsen, et al. 2003. E-selectin up-regulation allows for targeted drug delivery in prostate cancer. *Cancer Res.* 63:6387–6394.

Boghaert, E.R., K.M. Khandke, L. Sridharan, et al. 2008. Determination of pharmacokinetic values of calicheamicin-antibody conjugates in mice by plasmon resonance analysis of small (5 mul) blood samples. *Cancer Chemother. Pharmacol.* 61:1027–1035.

Boghaert, E.R., L. Sridharan, D.C. Armellino, et al. 2004. Antibody-targeted chemotherapy with the calicheamicin conjugate hu3S193-N-acetyl gamma calicheamicin dimethyl hydrazide targets Lewisy and eliminates Lewisy-positive human carcinoma cells and xenografts. *Clin. Cancer Res.* 10:4538–4549.

Bonifacino, J.S., and L.M. Traub. 2003. Signals for sorting of transmembrane proteins to endosomes and lysosomes. *Annu. Rev. Biochem.* 72:395–447.

Breton, J., N. Pezzi, A. Molinari, et al. 1995. Prolonged half-life in the circulation of a chemical conjugate between a pro-urokinase derivative and human serum albumin. *Eur. J. Biochem.* 231:563–569.

Bross, P.F., J. Beitz, G. Chen, et al. 2001. Approval summary: Gemtuzumab ozogamicin in relapsed acute myeloid leukemia. *Clin. Cancer Res.* 7:1490–1496.

Bumol, T.F., A.L. Baker, E.L. Andrews, et al. 1988. KS1/4-DAVLB, a monoclonal antibody-vinca alkaloid conjugate for site-directed therapy of epithelial malignancies. *Targeted Diagn. Ther.* 1:55–79.

Carter, P., L. Smith, and M. Ryan. 2004. Identification and validation of cell surface antigens for antibody targeting in oncology. *Endocr. Relat. Cancer* 11:659–687.

Cassady, J.M., K.K. Chan, H.G. Floss, et al. 2004. Recent developments in the maytansinoid antitumor agents. *Chem. Pharm. Bull. (Tokyo)* 52:1–26.

Cassard, S., J. Salamero, D. Hanau, et al. 1998. A tyrosine-based signal present in Ig alpha mediates B cell receptor constitutive internalization. *J. Immunol.* 160:1767–1773.

Chahlavi, A., P. Rayman, A.L. Richmond, et al. 2005. Glioblastomas induce T-lymphocyte death by two distinct pathways involving gangliosides and CD70. *Cancer Res.* 65:5428–5438.

Chan, S.Y., A.N. Gordon, R.E. Coleman, et al. 2003. A phase 2 study of the cytotoxic immunoconjugate CMB-401 (hCTM01-calicheamicin) in patients with platinum-sensitive recurrent epithelial ovarian carcinoma. *Cancer Immunol. Immunother.* 52:243–248.

Chari, R.V., B.A. Martell, J.L. Gross, et al. 1992. Immunoconjugates containing novel maytansinoids: Promising anticancer drugs. *Cancer Res.* 52:127–131.

Chen, Q., H.J. Millar, F.L. McCabe, et al. 2007a. Alphav integrin-targeted immunoconjugates regress established human tumors in xenograft models. *Clin. Cancer Res.* 13:3689–3695.

Chen, Y., S. Clark, T. Wong, et al. 2007b. Armed antibodies targeting the mucin repeats of the ovarian cancer antigen, MUC16, are highly efficacious in animal tumor models. *Cancer Res.* 67:4924–4932.

Crocker, P.R., J.C. Paulson, and A. Varki. 2007. Siglecs and their roles in the immune system. *Nature Rev. Immunol.* 7:255–266.

d'Azzo, A., A. Bongiovanni, and T. Nastasi. 2005. E3 ubiquitin ligases as regulators of membrane protein trafficking and degradation. *Traffic* 6:429–441.

Deryugina, E.I., and J.P. Quigley. 2006. Matrix metalloproteinases and tumor metastasis. *Cancer Metastasis Rev.* 25:9–34.

DiJoseph, J.F., M.M. Dougher, D.C. Armellino, et al. 2007a. Therapeutic potential of CD22-specific antibody-targeted chemotherapy using inotuzumab ozogamicin (CMC-544) for the treatment of acute lymphoblastic leukemia. *Leukemia* 21:2240–2245.

DiJoseph, J.F., M.M. Dougher, D.C. Armellino, et al. 2007b. CD20-specific antibody-targeted chemotherapy of non-Hodgkin's B-cell lymphoma using calicheamicin-conjugated rituximab. *Cancer Immunol. Immunother.* 56:1107–1117.

Doronina, S.O., B.A. Mendelsohn, T.D. Bovee, et al. 2006. Enhanced activity of monomethylauristatin F through monoclonal antibody delivery: Effects of linker technology on efficacy and toxicity. *Bioconjug. Chem.* 17:114–124.

Doronina, S.O., B.E. Toki, M.Y. Torgov, et al. 2003. Development of potent monoclonal antibody auristatin conjugates for cancer therapy. *Nature Biotechnol.* 21:778–784.

Dowell, J.A., J. Korth-Bradley, H. Liu, et al. 2001. Pharmacokinetics of gemtuzumab ozogamicin, an antibody-targeted chemotherapy agent for the treatment of patients with acute myeloid leukemia in first relapse. *J. Clin. Pharmacol.* 41:1206–1214.

Dubowchik, G.M., and R.A. Firestone. 1998. Cathepsin B-sensitive dipeptide prodrugs. I. A model study of structural requirements for efficient release of doxorubicin. *Bioorg. Med. Chem. Lett.* 8:3341–3346.

Dubowchik, G.M., K. Mosure, J.O. Knipe, et al. 1998. Cathepsin B-sensitive dipeptide prodrugs. II. Models of anticancer drugs paclitaxel (Taxol), mitomycin C and doxorubicin. *Bioorg. Med. Chem. Lett.* 8:3347–3352.

Duffy, M.J. 2004. The urokinase plasminogen activator system: Role in malignancy. *Curr. Pharm. Design* 10:39–49.

Dunphy, C.H., L.J. Gardner, and C.S. Bee. 2000. Malignant mesothelioma with CD30-positivity. A case report and review of the literature. *Arch. Pathol. Lab. Med.* 124:1077–1079.

Durkop, H., H.D. Foss, F. Eitelbach, et al. 2000. Expression of the CD30 antigen in non-lymphoid tissues and cells. *J. Pathol.* 190:613–618.

Eigenbrot, C.W., inventor; Genentech, assignee. 2007. Cysteine engineered antibodies and conjugates. U.S. Patent 20070092940.

Erickson, H.K., P.U. Park, W.C. Widdison, et al. 2006. Antibody-maytansinoid conjugates are activated in targeted cancer cells by lysosomal degradation and linker-dependent intracellular processing. *Cancer Res.* 66:4426–4433.

Filpula, D. 2007. Antibody engineering and modification technologies. *Biomol. Eng.* 24:201–215.

Francisco, J.A., C.G. Cerveny, D.L. Meyer, et al. 2003. cAC10-vcMMAE, an anti-CD30-monomethyl auristatin E conjugate with potent and selective antitumor activity. *Blood* 102:1458–1465.

Garnett, M.C. 2001. Targeted drug conjugates: Principles and progress. *Adv. Drug Deliv. Rev.* 53:171–216.

Gerber, H.P., I.S. Grewal, and E. Oflazoglu. 2009. The role of vascular targeting agents in the treatment of solid tumors: Current and future developments. In *Emerging Protein Biotherapeutics*, ed. I.S. Grewal.

Ghetie, V., S. Popov, J. Borvak, et al. 1997. Increasing the serum persistence of an IgG fragment by random mutagenesis. *Nature Biotechnol.* 15:637–640.

Ghetie, V., and E.S. Ward. 1997. FcRn: The MHC class I-related receptor that is more than an IgG transporter. *Immunol. Today* 18:592–598.

Ghetie, V., and E.S. Ward. 2002. Transcytosis and catabolism of antibody. *Immunol. Res.* 25:97–113.

Gilboa, L., R. Ben-Levy, Y. Yarden, et al. 1995. Roles for a cytoplasmic tyrosine and tyrosine kinase activity in the interactions of Neu receptors with coated pits. *J. Biol. Chem.* 270:7061–7067.

Giles, F., R. Morariu-Zamfir, J. Lambert, et al. 2006. Phase I study of AVE9633, an anti-CD33-maytansinoid immunoconjugate, administered as an intravenous infusion in patients with refractory/relapsed CD33-positive acute myeloid leukemia (AML). Abstract No. 4548. *Blood* 108.

Gillespie, A.M., T.J. Broadhead, S.Y. Chan, et al. 2000. Phase I open study of the effects of ascending doses of the cytotoxic immunoconjugate CMB-401 (hCTMO1-calicheamicin) in patients with epithelial ovarian cancer. *Ann. Oncol.* 11:735–741.

Golay, J., N. Di Gaetano, D. Amico, et al. 2005. Gemtuzumab ozogamicin (Mylotarg) has therapeutic activity against CD33 acute lymphoblastic leukaemias in vitro and in vivo. *Br. J. Haematol.* 128:310–317.

Grassart, A., A. Dujeancourt, P.B. Lazarow, et al. 2008. Clathrin-independent endocytosis used by the IL-2 receptor is regulated by Rac1, Pak1 and Pak2. *EMBO Rep.* 9:356–362.

Griffiths, G.L., M.J. Mattes, R. Stein, et al. 2003. Cure of SCID mice bearing human B-lymphoma xenografts by an anti-CD74 antibody-anthracycline drug conjugate. *Clin. Cancer Res.* 9:6567–6571.

Guillemard, V., and H. Uri Saragovi. 2004. Prodrug chemotherapeutics bypass p-glycoprotein resistance and kill tumors in vivo with high efficacy and target-dependent selectivity. *Oncogene* 23:3613–3621.

Hamann, P.R., L.M. Hinman, C.F. Beyer, et al. 2005a. An anti-MUC1 antibody-calicheamicin conjugate for treatment of solid tumors. Choice of linker and overcoming drug resistance. *Bioconjug. Chem.* 16:346–353.

Hamann, P.R., L.M. Hinman, C.F. Beyer, et al. 2005b. A calicheamicin conjugate with a fully humanized anti-MUC1 antibody shows potent antitumor effects in breast and ovarian tumor xenografts. *Bioconjug. Chem.* 16:354–360.

Hamann, P.R., L.M. Hinman, I. Hollander, et al. 2002. Gemtuzumab ozogamicin, a potent and selective anti-CD33 antibody-calicheamicin conjugate for treatment of acute myeloid leukemia. *Bioconjug. Chem.* 13:47–58.

Hamblett, K.J., P.D. Senter, D.F. Chace, et al. 2004. Effects of drug loading on the antitumor activity of a monoclonal antibody drug conjugate. *Clin. Cancer Res.* 10:7063–7070.

Harmsen, M.M., and H.J. De Haard. 2007. Properties, production, and applications of camelid single-domain antibody fragments. *Appl. Microbiol. Biotechnol.* 77:13–22.

Helft, P.R., R.L. Schilsky, F.J. Hoke, et al. 2004. A phase I study of cantuzumab mertansine administered as a single intravenous infusion once weekly in patients with advanced solid tumors. *Clin. Cancer Res.* 10:4363–4368.

Henry, M.D., S. Wen, M.D. Silva, et al. 2004. A prostate-specific membrane antigen-targeted monoclonal antibody-chemotherapeutic conjugate designed for the treatment of prostate cancer. *Cancer Res.* 64:7995–8001.

Hey, T., E. Fiedler, R. Rudolph, et al. 2005. Artificial, non-antibody binding proteins for pharmaceutical and industrial applications. *Trends Biotechnol.* 23:514–522.

Hicke, L., and R. Dunn. 2003. Regulation of membrane protein transport by ubiquitin and ubiquitin-binding proteins. *Annu. Rev. Cell Dev. Biol.* 19:141–172.

Higashide, E., M. Asai, K. Ootsu, et al. 1977. Ansamitocin, a group of novel maytansinoid antibiotics with antitumour properties from Nocardia. *Nature* 270:721–722.

Hishima, T., M. Fukayama, Y. Hayashi, et al. 2000. CD70 expression in thymic carcinoma. *Am. J. Surg. Pathol.* 24:742–746.

Ho, S.B., G.A. Niehans, C. Lyftogt, et al. 1993. Heterogeneity of mucin gene expression in normal and neoplastic tissues. *Cancer Res.* 53:641–651.

Hofmann, M.W., S. Honing, D. Rodionov, et al. 1999. The leucine-based sorting motifs in the cytoplasmic domain of the invariant chain are recognized by the clathrin adaptors AP1 and AP2 and their medium chains. *J. Biol. Chem.* 274:36153–36158.

Ingle, G.S., P. Chan, J.M. Elliott, et al. 2008. High CD21 expression inhibits internalization of anti-CD19 antibodies and cytotoxicity of an anti-CD19-drug conjugate. *Br. J. Haematol.* 140:46–58.

Ishitsuka, K., S. Jimi, V.S. Goldmacher, et al. 2008. Targeting CD56 by the maytansinoid immunoconjugate IMGN901 (huN901-DM1): A potential therapeutic modality implication against natural killer/T cell malignancy. *Br. J. Haematol.* 141:129–131.

Iwahashi, T., Y. Tone, J. Usui, et al. 1989. Selective killing of carcinoembryonic-antigen (CEA)-producing cells in vitro by the immunoconjugate cytorhodin-S and CEA-reactive cytorhodin-S antibody CA208. *Cancer Immunol. Immunother.* 30:239–246.

Jain, M., N. Kamal, and S.K. Batra. 2007. Engineering antibodies for clinical applications. *Trends Biotechnol.* 25:307–316.

Jedema, I., R.M. Barge, V.H. van der Velden, et al. 2004. Internalization and cell cycle-dependent killing of leukemic cells by gemtuzumab ozogamicin: Rationale for efficacy in CD33-negative malignancies with endocytic capacity. *Leukemia* 18:316–325.

Jeffrey, S.C., J.B. Andreyka, S.X. Bernhardt, et al. 2006. Development and properties of beta-glucuronide linkers for monoclonal antibody-drug conjugates. *Bioconjug. Chem.* 17:831–840.

Jelinkova, M., J. Strohalm, D. Plocova, et al. 1998. Targeting of human and mouse T-lymphocytes by monoclonal antibody-HPMA copolymer-doxorubicin conjugates directed against different T-cell surface antigens. *J. Control. Release* 52:253–270.

Jilani, I., E. Estey, Y. Huh, et al. 2002. Differences in CD33 intensity between various myeloid neoplasms. *Am. J. Clin. Pathol.* 118:560–566.

Johnson, D.A., and B.C. Laguzza. 1987. Antitumor xenograft activity with a conjugate of a vinca derivative and the squamous carcinoma-reactive monoclonal antibody PF1/D. *Cancer Res.* 47:3118–3122.

Jung, G., L. Grosse-Hovest, P.H. Krammer, et al. 2001. Target cell-restricted triggering of the CD95 (APO-1/Fas) death receptor with bispecific antibody fragments. *Cancer Res.* 61:1846–1848.

Junutula, J.R., S. Bhakta, H. Raab, et al. 2008. Rapid identification of reactive cysteine residues for site-specific labeling of antibody-Fabs. *J. Immunol. Methods* 332:41–52.

Kallioniemi, O.P., U. Wagner, J. Kononen, et al. 2001. Tissue microarray technology for high-throughput molecular profiling of cancer. *Hum. Mol. Genet.* 10:657–662.

Kawamoto, T., J. Shoda, N. Miyahara, et al. 2004. Expression of MUC1 recognized by a monoclonal antibody MY.1E12 is a useful biomarker for tumor aggressiveness of carcinoma of the gallbladder. *Clin. Exp. Metastasis* 21:353–362.

Kinlough, C.L., P.A. Poland, J.B. Bruns, et al. 2004. MUC1 membrane trafficking is modulated by multiple interactions. *J. Biol. Chem.* 279:53071–53077.

Kirchhausen, T. 1999. Adaptors for clathrin-mediated traffic. *Annu. Rev. Cell Dev. Biol.* 15:705–732.

Kirkham, M., and R.G. Parton. 2005. Clathrin-independent endocytosis: New insights into caveolae and non-caveolar lipid raft carriers. *Biochim. Biophys. Acta* 1746:349–363.

Kluger, M.S., S.L. Shiao, A.L. Bothwell, et al. 2002. Cutting Edge. Internalization of transduced E-selectin by cultured human endothelial cells: Comparison of dermal microvascular and umbilical vein cells and identification of a phosphoserine-type di-leucine motif. *J. Immunol.* 168:2091–2095.

Kneile, J.R., G. Tan, S. Suster, et al. 2006. Expression of CD30 (Ber-H2) in nasopharyngeal carcinoma, undifferentiated type and lymphoepithelioma-like carcinoma. A comparison study with anaplastic large cell lymphoma. *Histopathology* 48:855–861.

Knoll, K., W. Wrasidlo, J.E. Scherberich, et al. 2000. Targeted therapy of experimental renal cell carcinoma with a novel conjugate of monoclonal antibody 138H11 and calicheamicin thetaI1. *Cancer Res.* 60:6089–6094.

Kononen, J., L. Bubendorf, A. Kallioniemi, et al. 1998. Tissue microarrays for high-throughput molecular profiling of tumor specimens. *Nature Med.* 4:844–847.

Kovtun, Y.V., C.A. Audette, Y. Ye, et al. 2006. Antibody-drug conjugates designed to eradicate tumors with homogeneous and heterogeneous expression of the target antigen. *Cancer Res.* 66:3214–3221.

Kreitman, R.J., and I. Pastan. 2006. Immunotoxins in the treatment of hematologic malignancies. *Curr. Drug Targets* 7:1301–1311.

Kupchan, S. 1972. Maytansine. *J. Am. Chem. Soc.* 94:1354–1356.

Ladino, C.A., R.V. Chari, L.A. Bourret, et al. 1997. Folate-maytansinoids: Target-selective drugs of low molecular weight. *Int. J. Cancer* 73:859–864.

Laguzza, B.C., C.L. Nichols, S.L. Briggs, et al. 1989. New antitumor monoclonal antibody-vinca conjugates LY203725 and related compounds: Design, preparation, and representative in vivo activity. *J. Med. Chem.* 32:548–555.

Lajoie, P., and I.R. Nabi. 2007. Regulation of raft-dependent endocytosis. *J. Cell. Mol. Med.* 11:644–653.

Lamaze, C., A. Dujeancourt, T. Baba, et al. 2001. Interleukin 2 receptors and detergent-resistant membrane domains define a clathrin-independent endocytic pathway. *Mol. Cell* 7:661–671.

Larson, R.A., M. Boogaerts, E. Estey, et al. 2002. Antibody-targeted chemotherapy of older patients with acute myeloid leukemia in first relapse using Mylotarg (gemtuzumab ozogamicin). *Leukemia* 16:1627–1636.

Larson, R.A., E.L. Sievers, E.A. Stadtmauer, et al. 2005. Final report of the efficacy and safety of gemtuzumab ozogamicin (Mylotarg) in patients with CD33-positive acute myeloid leukemia in first recurrence. *Cancer* 104:1442–1452.

Laude, A.J., and I.A. Prior. 2004. Plasma membrane microdomains: Organization, function and trafficking. *Mol. Membr. Biol.* 21:193–205.

Law, C.L., C.G. Cerveny, K.A. Gordon, et al. 2004. Efficient elimination of B-lineage lymphomas by anti-CD20-auristatin conjugates. *Clin. Cancer Res.* 10:7842–7851.

Law, C.L., K.A. Gordon, B.E. Toki, et al. 2006. Lymphocyte activation antigen CD70 expressed by renal cell carcinoma is a potential therapeutic target for anti-CD70 antibody-drug conjugates. *Cancer Res.* 66:2328–2337.

Lee, M. 1987a. Calicheamicins. I. *J. Am. Chem. Soc.* 109:3464–3466.

Lee, M. 1987b. Calicheamicins. II. *J. Am. Chem. Soc.* 109:3466–3468.

Legrand, O., M.B. Vidriales, X. Thomas, et al. 2007. An open label, dose escalation study of AVE9633 administered as a single agent by intravenous (IV) infusion weekly for 2 weeks in 4-week cycle to patients with relapsed or refractory CD33-positive acute myeloid leukemia (AML). Abstract No. 1850. *Blood* 110(11).

Leuschner, C., and W. Hansel. 2005. Targeting breast and prostate cancers through their hormone receptors. *Biol. Reprod.* 73:860–865.

Linenberger, M.L., T. Hong, D. Flowers, et al. 2001. Multidrug-resistance phenotype and clinical responses to gemtuzumab ozogamicin. *Blood* 98:988–994.

Liu, C., B.M. Tadayoni, L.A. Bourret, et al. 1996. Eradication of large colon tumor xenografts by targeted delivery of maytansinoids. *Proc. Natl. Acad. Sci. U.S.A.* 93:8618–8623.

Lode, H.N., R.A. Reisfeld, R. Handgretinger, et al. 1998. Targeted therapy with a novel enediyene antibiotic calicheamicin theta(I)1 effectively suppresses growth and dissemination of liver metastases in a syngeneic model of murine neuroblastoma. *Cancer Res.* 58:2925–2928.

Lu, Y., E. Sega, C.P. Leamon, et al. 2004. Folate receptor-targeted immunotherapy of cancer: Mechanism and therapeutic potential. *Adv. Drug Deliv. Rev.* 56:1161–1176.

Luesch, H., R.E. Moore, V.J. Paul, et al. 2001. Isolation of dolastatin 10 from the marine cyanobacterium Symploca species VP642 and total stereochemistry and biological evaluation of its analogue symplostatin 1. *J. Nat. Prod.* 64:907–910.

Lum, L.G., P.A. Davol, and R.J. Lee. 2006. The new face of bispecific antibodies: Targeting cancer and much more. *Exp. Hematol.* 34:1–6.

Lyu, M.A., L.H. Cheung, W.N. Hittelman, et al. 2007. The rGel/BLyS fusion toxin specifically targets malignant B cells expressing the BLyS receptors BAFF-R, TACI, and BCMA. *Mol. Cancer Ther.* 6:460–470.

Ma, D., C.E. Hopf, A.D. Malewicz, et al. 2006. Potent antitumor activity of an auristatin-conjugated, fully human monoclonal antibody to prostate-specific membrane antigen. *Clin. Cancer Res.* 12:2591–2596.

Mahmood, I., and M.D. Green. 2005. Pharmacokinetic and pharmacodynamic considerations in the development of therapeutic proteins. *Clin. Pharmacokinet.* 44:331–347.

Manabe, Y., T. Tsubota, Y. Haruta, et al. 1984. Production of a monoclonal antibody-methotrexate conjugate utilizing dextran T-40 and its biologic activity. *J. Lab. Clin. Med.* 104:445–454.

Mao, W., E. Luis, S. Ross, et al. 2004. EphB2 as a therapeutic antibody drug target for the treatment of colorectal cancer. *Cancer Res.* 64:781–788.

Marmor, M.D., and Y. Yarden. 2004. Role of protein ubiquitylation in regulating endocytosis of receptor tyrosine kinases. *Oncogene* 23:2057–2070.

McDonagh, C.F., E. Turcott, L. Westendorf, et al. 2006. Engineered antibody-drug conjugates with defined sites and stoichiometries of drug attachment. *Protein Eng. Design Select.* 19:299–307.

McEarchern, J.A., E. Oflazoglu, L. Francisco, et al. 2007. Engineered anti-CD70 antibody with multiple effector functions exhibits in vitro and in vivo antitumor activities. *Blood* 109:1185–1192.

McGrath, M.S., M.G. Rosenblum, M.R. Philips, et al. 2003. Immunotoxin resistance in multidrug resistant cells. *Cancer Res.* 63:72–79.

Mellman, I., R. Fuchs, and A. Helenius. 1986. Acidification of the endocytic and exocytic pathways. *Annu. Rev. Biochem.* 55:663–700.

Mohamed, M.M., and B.F. Sloane. 2006. Cysteine cathepsins: Multifunctional enzymes in cancer. *Nature Rev. Cancer* 6:764–775.

Mohamedali, K.A., D. Kedar, P. Sweeney, et al. 2005. The vascular-targeting fusion toxin VEGF121/rGel inhibits the growth of orthotopic human bladder carcinoma tumors. *Neoplasia* 7:912–920.

Mosure, K.W., A.J. Henderson, L.J. Klunk, et al. 1997. Disposition of conjugate-bound and free doxorubicin in tumor-bearing mice following administration of a BR96-doxorubicin immunoconjugate (BMS 182248). *Cancer Chemother. Pharmacol.* 40:251–258.

Mousavi, S.A., M. Sporstol, C. Fladeby, et al. 2007. Receptor-mediated endocytosis of immune complexes in rat liver sinusoidal endothelial cells is mediated by FcgammaRIIb2. *Hepatology* 46:871–884.

Muro, H., H. Shirasawa, I. Kosugi, et al. 1993. Defect of Fc receptors and phenotypical changes in sinusoidal endothelial cells in human liver cirrhosis. *Am. J. Pathol.* 143:105–120.

Nagy, A., and A.V. Schally. 2005. Targeting cytotoxic conjugates of somatostatin, luteinizing hormone-releasing hormone and bombesin to cancers expressing their receptors: A "smarter" chemotherapy. *Curr. Pharm. Design* 11:1167–1180.

Natsume, A., M. Wakitani, N. Yamane-Ohnuki, et al. 2006. Fucose removal from complex-type oligosaccharide enhances the antibody-dependent cellular cytotoxicity of single-gene-encoded bispecific antibody comprising of two single-chain antibodies linked to the antibody constant region. *J. Biochem.* 140:359–368.

Ng, R., and M.D. Green. 2007. Managing cardiotoxicity in anthracycline-treated breast cancers. *Expert Opin. Drug Saf.* 6:315–321.

Nimmerjahn, F., and J.V. Ravetch. 2007. Antibodies, Fc receptors and cancer. *Curr. Opin. Immunol.* 19:239–245.

Odorizzi, C.G., I.S. Trowbridge, L. Xue, et al. 1994. Sorting signals in the MHC class II invariant chain cytoplasmic tail and transmembrane region determine trafficking to an endocytic processing compartment. *J. Cell Biol.* 126:317–330.

Palermo, C., and J.A. Joyce. 2008. Cysteine cathepsin proteases as pharmacological targets in cancer. *Trends Pharmacol. Sci.* 29:22–28.

Pastan, I., R. Hassan, D.J. FitzGerald, et al. 2007. Immunotoxin treatment of cancer. *Annu. Rev. Med.* 58:221–237.

Pelkmans, L., and A. Helenius. 2002. Endocytosis via caveolae. *Traffic* 3:311–320.

Petersen, B.H., S.V. DeHerdt, D.W. Schneck, et al. 1991. The human immune response to KS1/4-desacetylvinblastine (LY256787) and KS1/4-desacetylvinblastine hydrazide (LY203728) in single and multiple dose clinical studies. *Cancer Res.* 51:2286–2290.

Pettit, G. 1987. The isolation and structure of a remarkable marine animal antineoplastic constituent: Dolastatin 10. *J. Am. Chem. Soc.* 109:6883–6885.

Pettit, G. 1989. The absolute configuration and synthesis of natural (-)-dolastatin 10. *J. Am. Chem. Soc.* 111:5463–5465.

Pettit, G.R.B., and J. Barkoczy, inventors; University of Arizona, assignee. 1997. Tumor inhibiting tetrapeptide bearing modified phenethyl amines. U.S. Patent 5635483.

Pillay, C.S., E. Elliott, and C. Dennison. 2002. Endolysosomal proteolysis and its regulation. *Biochem. J.* 363:417–429.

Pimm, M.V., M.A. Paul, Y. Ogumuyiwa, et al. 1988. Biodistribution and tumour localisation of a daunomycin-monoclonal antibody conjugate in nude mice with human tumour xenografts. *Cancer Immunol. Immunother.* 27:267–271.

Pollack, V.A., E. Alvarez, K.F. Tse, et al. 2007. Treatment parameters modulating regression of human melanoma xenografts by an antibody-drug conjugate (CR011-vcMMAE) targeting GPNMB. *Cancer Chemother. Pharmacol.* 60:423–435.

Polson, A.G., S.F. Yu, K. Elkins, et al. 2007. Antibody-drug conjugates targeted to CD79 for the treatment of non-Hodgkin lymphoma. *Blood* 110:616–623.

Qu, Z., D.M. Goldenberg, T.M. Cardillo, et al. 2008. Bispecific anti-CD20/22 antibodies inhibit B-cell lymphoma proliferation by a unique mechanism of action. *Blood* 111:2211–2219.

Rajagopal, A., and S.M. Simon. 2003. Subcellular localization and activity of multidrug resistance proteins. *Mol. Biol. Cell* 14:3389–3399.

Rajasekaran, S.A., G. Anilkumar, E. Oshima, et al. 2003. A novel cytoplasmic tail MXXXL motif mediates the internalization of prostate-specific membrane antigen. *Mol. Biol. Cell* 14:4835–4845.

Rakha, E.A., R.W. Boyce, D. Abd El-Rehim, et al. 2005. Expression of mucins (MUC1, MUC2, MUC3, MUC4, MUC5AC and MUC6) and their prognostic significance in human breast cancer. *Mod. Pathol.* 18:1295–1304.

Reddy, J.A., E. Westrick, H.K. Santhapuram, et al. 2007. Folate receptor-specific antitumor activity of EC131, a folate-maytansinoid conjugate. *Cancer Res.* 67:6376–6382.

Rihová, B., A. Jegorov, J. Strohalm, et al. 1992. Antibody-targeted cyclosporin A. *J. Control. Release* 19:25–39.

Rodon, J., M. Garrison, L.A. Hammond, et al. 2008. Cantuzumab mertansine in a three-times a week schedule: A phase I and pharmacokinetic study. *Cancer Chemother. Pharmacol.* 62:911–919.

Ross, H.J., L.L. Hart, P.M. Swanson, et al. 2006. A randomized, multicenter study to determine the safety and efficacy of the immunoconjugate SGN-15 plus docetaxel for the treatment of non-small cell lung carcinoma. *Lung Cancer* 54:69–77.

Ross, S., S.D. Spencer, I. Holcomb, et al. 2002. Prostate stem cell antigen as therapy target: Tissue expression and in vivo efficacy of an immunoconjugate. *Cancer Res.* 62:2546–2553.

Rubinfeld, B., A. Upadhyay, S.L. Clark, et al. 2006. Identification and immunotherapeutic targeting of antigens induced by chemotherapy. *Nature Biotechnol.* 24:205–209.

Ryan, M.C., M. Hering, D. Peckham, et al. 2007. Antibody targeting of B-cell maturation antigen on malignant plasma cells. *Mol. Cancer Ther.* 6:3009–3018.

Saito, G., J.A. Swanson, and K.D. Lee. 2003. Drug delivery strategy utilizing conjugation via reversible disulfide linkages: Role and site of cellular reducing activities. *Adv. Drug Deliv. Rev.* 55:199–215.

Saleh, M.N., S. Sugarman, J. Murray, et al. 2000. Phase I trial of the anti-Lewis Y drug immunoconjugate BR96-doxorubicin in patients with lewis Y-expressing epithelial tumors. *J. Clin. Oncol.* 18:2282–2292.

Sanderson, R.J., M.A. Hering, S.F. James, et al. 2005. In vivo drug-linker stability of an anti-CD30 dipeptide-linked auristatin immunoconjugate. *Clin. Cancer Res.* 11:843–852.

Sanz, L., B. Blanco, and L. Alvarez-Vallina. 2004. Antibodies and gene therapy: Teaching old "magic bullets" new tricks. *Trends Immunol.* 25:85–91.

Sauter, A., C. Kloft, S. Gronau, et al. 2007. Pharmacokinetics, immunogenicity and safety of bivatuzumab mertansine, a novel CD44v6-targeting immunoconjugate, in patients with squamous cell carcinoma of the head and neck. *Int. J. Oncol.* 30:927–935.

Sauter, G., R. Simon, and K. Hillan. 2003. Tissue microarrays in drug discovery. *Nature Rev. Drug Discov.* 2:962–972.

Schreiber, G.J., K.E. Hellstrom, and I. Hellstrom. 1992. An unmodified anticarcinoma antibody, BR96, localizes to and inhibits the outgrowth of human tumors in nude mice. *Cancer Res.* 52:3262–3266.

Setiadi, H., and R.P. McEver. 2008. Clustering endothelial E-selectin in clathrin-coated pits and lipid rafts enhances leukocyte adhesion under flow. *Blood* 111:1989–1998.

Shen, W.C., and H.J. Ryser. 1984. Selective killing of Fc-receptor-bearing tumor cells through endocytosis of a drug-carrying immune complex. *Proc. Natl. Acad. Sci. U.S.A.* 81:1445–1447.

Shields, R.L., A.K. Namenuk, K. Hong, et al. 2001. High resolution mapping of the binding site on human IgG1 for Fc gamma RI, Fc gamma RII, Fc gamma RIII, and FcRn and design of IgG1 variants with improved binding to the Fc gamma R. *J. Biol. Chem.* 276:6591–6604.

Shimamura, T., S.R. Husain, and R.K. Puri. 2006. The IL-4 and IL-13 pseudomonas exotoxins: New hope for brain tumor therapy. *Neurosurg. Focus* 20:E11.

Sievers, E.L., F.R. Appelbaum, R.T. Spielberger, et al. 1999. Selective ablation of acute myeloid leukemia using antibody-targeted chemotherapy: A phase I study of an anti-CD33 calicheamicin immunoconjugate. *Blood* 93:3678–3684.

Sievers, E.L., R.A. Larson, E.A. Stadtmauer, et al. 2001. Efficacy and safety of gemtuzumab ozogamicin in patients with CD33-positive acute myeloid leukemia in first relapse. *J. Clin. Oncol.* 19:3244–3254.

Sigal, L.H. 2007. Basic science for the clinician 46: Fcgamma receptors. *J. Clin. Rheumatol.* 13:355–358.

Simons, K., and E. Ikonen. 1997. Functional rafts in cell membranes. *Nature* 387:569–572.

Simons, K., and D. Toomre. 2000. Lipid rafts and signal transduction. *Nature Rev. Mol. Cell. Biol.* 1:31–39.

Skerra, A. 2000. Engineered protein scaffolds for molecular recognition. *J. Mol. Recognit.* 13:167–187.

Skerra, A. 2007. Alternative non-antibody scaffolds for molecular recognition. *Curr. Opin. Biotechnol.* 18:295–304.

Slepnev, V.I., and P. De Camilli. 2000. Accessory factors in clathrin-dependent synaptic vesicle endocytosis. *Nature Rev. Neurosci.* 1:161–172.

Smith-Jones, P.M. 2004. Radioimmunotherapy of prostate cancer. *Quart. J. Nuclear Med. Mol. Imaging* 48:297–304.

Smith, L.M., A. Nesterova, S.C. Alley, et al. 2006. Potent cytotoxicity of an auristatin-containing antibody-drug conjugate targeting melanoma cells expressing melanotransferrin/p97. *Mol. Cancer Ther.* 5:1474–1482.

Spearman, M.E., R.M. Goodwin, L.D. Apelgren, et al. 1987. Disposition of the monoclonal antibody-vinca alkaloid conjugate KS1/4-DAVLB (LY256787) and free 4-desacetylvinblastine in tumor-bearing nude mice. *J. Pharmacol. Exp. Ther.* 241:695–703.

Spearman, M.E., R.M. Goodwin, and D. Kau. 1987. Disposition of the monoclonal antibody-vinca alkaloid conjugate, KS1/4-DAVLB (LY256787), in Fischer 344 rats and rhesus monkeys. *Drug Metab. Dispos.* 15:640–647.

St'astny, M., K. Ulbrich, J. Strohalm, et al. 1997. Abnormal differentiation of thymocytes induced by free cyclosporine is avoided when cyclosporine bound to N-(2-hydroxypropyl)methacrylamide copolymer carrier is used. *Transplantation* 63:1818–1827.

Stasi, R., M.L. Evangelista, F. Buccisano, et al. 2008. Gemtuzumab ozogamicin in the treatment of acute myeloid leukemia. *Cancer Treat. Rev.* 34:49–60.

Stish, B.J., H. Chen, Y. Shu, et al. 2007. A bispecific recombinant cytotoxin (DTEGF13) targeting human interleukin-13 and epidermal growth factor receptors in a mouse xenograft model of prostate cancer. *Clin. Cancer Res.* 13:6486–6493.

Sun, M.M., K.S. Beam, C.G. Cerveny, et al. 2005. Reduction-alkylation strategies for the modification of specific monoclonal antibody disulfides. *Bioconjug. Chem.* 16:1282–1290.

Sutherland, M.S., R.J. Sanderson, K.A. Gordon, et al. 2006. Lysosomal trafficking and cysteine protease metabolism confer target-specific cytotoxicity by peptide-linked anti-CD30-auristatin conjugates. *J. Biol. Chem.* 281:10540–10547.

Suzuki, H., J. Shoda, T. Kawamoto, et al. 2004. Expression of MUC1 recognized by monoclonal antibody MY.1E12 is a useful biomarker for tumor aggressiveness of advanced colon carcinoma. *Clin. Exp. Metastasis* 21:321–329.

Sznol, M., D. Sanders, H. Kluger, et al. 2007. A phase I and pharmacokinetic study of CR011-vcMMAE, an antibody toxin conjugate drug, in patients with unresectable stage III /IV melanoma. 19th EORTC-NCI-AACR Symposium: Molecular Targets and Cancer Therapeutics: Abstract # B47.

Tammi, R., K. Rilla, J.P. Pienimaki, et al. 2001. Hyaluronan enters keratinocytes by a novel endocytic route for catabolism. *J. Biol. Chem.* 276:35111–35122.

Tarrago-Trani, M.T., and B. Storrie. 2007. Alternate routes for drug delivery to the cell interior: Pathways to the Golgi apparatus and endoplasmic reticulum. *Adv. Drug Deliv. Rev.* 59:782–797.

Tassone, P., V.S. Goldmacher, P. Neri, et al. 2004a. Cytotoxic activity of the maytansinoid immunoconjugate B-B4-DM1 against CD138+ multiple myeloma cells. *Blood* 104:3688–3696.

Tassone, P., A. Gozzini, V. Goldmacher, et al. 2004b. In vitro and in vivo activity of the maytansinoid immunoconjugate huN901-N2′-deacetyl-N2′-(3-mercapto-1-oxopropyl)-maytansine against CD56+ multiple myeloma cells. *Cancer Res.* 64:4629–4636.

Tateno, H., H. Li, M.J. Schur, et al. 2007. Distinct endocytic mechanisms of CD22 (Siglec-2) and Siglec-F reflect roles in cell signaling and innate immunity. *Mol. Cell. Biol.* 27:5699–5710.

Taylor, V.C., C.D. Buckley, M. Douglas, et al. 1999. The myeloid-specific sialic acid-binding receptor, CD33, associates with the protein-tyrosine phosphatases, SHP-1 and SHP-2. *J. Biol. Chem.* 274:11505–11512.

Tedder, T.F., M. Inaoki, and S. Sato. 1997. The CD19-CD21 complex regulates signal transduction thresholds governing humoral immunity and autoimmunity. *Immunity* 6:107–118.

Temming, K., D.L. Meyer, R. Zabinski, et al. 2006. Evaluation of RGD-targeted albumin carriers for specific delivery of auristatin E to tumor blood vessels. *Bioconjug. Chem.* 17:1385–1394.

Temming, K., D.L. Meyer, R. Zabinski, et al. 2007. Improved efficacy of alphavbeta3-targeted albumin conjugates by conjugation of a novel auristatin derivative. *Mol. Pharm.* 4:686–694.

Thurber, G.M., M.M. Schmidt, and K.D. Wittrup. 2008. Factors determining antibody distribution in tumors. *Trends Pharmacol. Sci.* 29:57–61.

Tijink, B.M., J. Buter, R. de Bree, et al. 2006. A phase I dose escalation study with anti-CD44v6 bivatuzumab mertansine in patients with incurable squamous cell carcinoma of the head and neck or esophagus. *Clin. Cancer Res.* 12:6064–6072.

Tolcher, A.W., L. Ochoa, L.A. Hammond, et al. 2003. Cantuzumab mertansine, a maytansinoid immunoconjugate directed to the CanAg antigen: A phase I, pharmacokinetic, and biologic correlative study. *J. Clin. Oncol.* 21:211–222.

Tolcher, A.W., S. Sugarman, K.A. Gelmon, et al. 1999. Randomized phase II study of BR96-doxorubicin conjugate in patients with metastatic breast cancer. *J. Clin. Oncol.* 17:478–484.

Trail, P.A., D. Willner, S.J. Lasch, et al. 1993. Cure of xenografted human carcinomas by BR96-doxorubicin immunoconjugates. *Science* 261:212–215.

Trikha, M., Z. Zhou, J.A. Nemeth, et al. 2004. CNTO 95, a fully human monoclonal antibody that inhibits alphav integrins, has antitumor and antiangiogenic activity in vivo. *Int. J. Cancer* 110:326–335.

Tse, K.F., M. Jeffers, V.A. Pollack, et al. 2006. CR011, a fully human monoclonal antibody-auristatin E conjugate, for the treatment of melanoma. *Clin. Cancer Res.* 12:1373–1382.

Turturro, F. 2007. Denileukin diftitox: A biotherapeutic paradigm shift in the treatment of lymphoid-derived disorders. *Expert Rev. Anticancer Ther.* 7:11–17.

Uadia, P., A.H. Blair, and T. Ghose. 1984. Tumor and tissue distribution of a methotrexate-anti-EL4 immunoglobulin conjugate in EL4 lymphoma-bearing mice. *Cancer Res.* 44:4263–4266.

Uadia, P., A.H. Blair, T. Ghose, et al. 1985. Uptake of methotrexate linked to polyclonal and monoclonal antimelanoma antibodies by a human melanoma cell line. *J. Natl. Cancer Inst.* 74:29–35.

Ulvestad, E., K. Williams, C. Vedeler, et al. 1994. Reactive microglia in multiple sclerosis lesions have an increased expression of receptors for the Fc part of IgG. *J. Neurol. Sci.* 121:125–131.

Vallabhajosula, S., P.M. Smith-Jones, V. Navarro, et al. 2004. Radioimmunotherapy of prostate cancer in human xenografts using monoclonal antibodies specific to prostate specific membrane antigen (PSMA): Studies in nude mice. *Prostate* 58:145–155.

van Der Velden, V.H., J.G. te Marvelde, P.G. Hoogeveen, et al. 2001. Targeting of the CD33-calicheamicin immunoconjugate Mylotarg (CMA-676) in acute myeloid leukemia: In vivo and in vitro saturation and internalization by leukemic and normal myeloid cells. *Blood* 97:3197–3204.

van Roon, J.A., J.W. Bijlsma, J.G. van de Winkel, et al. 2005. Depletion of synovial macrophages in rheumatoid arthritis by an anti-FcgammaRI-calicheamicin immunoconjugate. *Ann. Rheum. Dis.* 64:865–870.

Varki, A., and T. Angata. 2006. Siglecs: The major subfamily of I-type lectins. *Glycobiology* 16:1R–27R.

Vedeler, C.A., R. Matre, E.K. Kristoffersen, et al. 1991. IgG Fc receptor heterogeneity in human peripheral nerves. *Acta Neurol. Scand.* 84:177–180.

Walter, R.B., P. Hausermann, B.W. Raden, et al. 2008a. Phosphorylated ITIMs enable ubiquitylation of an inhibitory cell surface receptor. *Traffic* 9:267–279.

Walter, R.B., B.W. Raden, D.M. Kamikura, et al. 2005. Influence of CD33 expression levels and ITIM-dependent internalization on gemtuzumab ozogamicin-induced cytotoxicity. *Blood* 105:1295–1302.

Walter, R.B., B.W. Raden, R. Zeng, et al. 2008b. ITIM-dependent endocytosis of CD33-related Siglecs: Role of intracellular domain, tyrosine phosphorylation, and the tyrosine phosphatases, Shp1 and Shp2. *J. Leukoc. Biol.* 83:200–211.

Wang, L., G. Amphlett, W.A. Blattler, et al. 2005. Structural characterization of the maytansinoid-monoclonal antibody immunoconjugate, huN901-DM1, by mass spectrometry. *Protein Sci.* 14:2436–2446.

Winchester, B. 2005. Lysosomal metabolism of glycoproteins. *Glycobiology* 15:1R–15R.

Wischhusen, J., G. Jung, I. Radovanovic, et al. 2002. Identification of CD70-mediated apoptosis of immune effector cells as a novel immune escape pathway of human glioblastoma. *Cancer Res.* 62:2592–2599.

Wischhusen, J., D. Schneider, M. Mittelbronn, et al. 2005. Death receptor-mediated apoptosis in human malignant glioma cells: Modulation by the CD40/CD40L system. *J. Neuroimmunol.* 162:28–42.

Wolf, E., R. Hofmeister, P. Kufer, et al. 2005. BiTEs: Bispecific antibody constructs with unique anti-tumor activity. *Drug Discov. Today* 10:1237–1244.

Wu, A.M., and P.D. Senter. 2005. Arming antibodies: Prospects and challenges for immunoconjugates. *Nature Biotechnol.* 23:1137–1146.

Wu, C., H. Ying, C. Grinnell, et al. 2007. Simultaneous targeting of multiple disease mediators by a dual-variable-domain immunoglobulin. *Nature Biotechnol.* 25:1290–1297.

Xie, H., C. Audette, M. Hoffee, et al. 2004. Pharmacokinetics and biodistribution of the antitumor immunoconjugate, cantuzumab mertansine (huC242-DM1), and its two components in mice. *J. Pharmacol. Exp. Ther.* 308:1073–1082.

Yang, K., A. Basu, M. Wang, et al. 2003. Tailoring structure-function and pharmacokinetic properties of single-chain Fv proteins by site-specific PEGylation. *Protein Eng.* 16:761–770.

Younes, A., A. Forero-Torres, N. Bartlett, et al. 2007. A novel antibody-drug conjugate, SGN-35 (anti-CD30-auristatin), induces objective responses in patients with relapsed or refractory Hodgkin lymphoma: Preliminary results of a phase I tolerability study. *Haematologica* 92(s5):63.

Zein, N., A.M. Sinha, W.J. McGahren, et al. 1988. Calicheamicin gamma 1I: An antitumor antibiotic that cleaves double-stranded DNA site specifically. *Science* 240:1198–1201.

ABBREVIATIONS

ABP	Albumin-binding protein
ADA	Anti-drug antibody
ADC	Antibody-drug conjugate
ADCC	Antibody-dependent cellular cytotoxicity
ADEPT	Antibody-directed enzyme prodrug therapy
ADHP	10-acetyl-3,7-dihydroxyphenoxazine
AE	Adverse event
AMD	Age-related macular degeneration
ANDA	Abbreviated new drug application
AOX1	Alcohol oxidase I gene
APC	Antigen presenting cell
APEx	Anchored periplasmic expression
API	Active pharmaceutical ingredient
AS-MS	Affinity selection/mass spectrometry
ASA	Accessible surface area
AUC	Area under the curve
BA	Bioavailability
BBB	Blood-brain barrier
BCL2	B-cell leukemia/lymphoma 2
B-CLL	B-cell chronic lymphocytic leukemia
BCR	B-cell receptor
BDNF	Brain-derived neurotrophic factor
BE	Bioequivalence
bFGF	Basic fibroblast growth factor
BHK	Baby hamster kidney
BIC	Best in class
BLA	Biologics license application
BMD	Bone mineral density
BMI	Body mass index
BoNT/A	Botulinum neurotoxin type A
CAPA	Cryopyrin-associated periodic syndrome
CD	Circular dichroism
CDC	Complement dependent cytotoxicity
CDER	Center for drug evaluation and research
CDG	Congenital disorders of glycosylation
CDR	Complementarity determining regions
cDSC	Capillary differential scanning calorimetry
CE	Capillary electrophoresis
CEA	Carcinoembryonic antigen
cES	Chicken embryonic stem cell

CEX	Cation exchange chromatography
cGMP	Current good manufacturing procedures
CHMP	Committee for medicinal products for human use
CHO	Chinese hamster ovary
CHOP	Cyclophosphamide, doxorubicin, Oncovin (vincristine), and prednisone
CIC	Circulating immune complexes
cIEF	Capillary isoelectric focusing
CLA	Cutaneous lymphocyte antigen
CLL	Chronic lymphocytic leukemia
CMC	Complement-mediated cell lysis
CMC	Chemical manufacturing & control
CMCase	Carboxymethylcellulase
CMV	Human cytomegalovirus
CNS	Central nervous system
COGs	Cost of goods
CR1	Complement receptor 1
CRP	Complement regulatory protein
CRR	Complete response rate
CSF	Cerebrospinal fluid
CT	Computed tomography
CTCL	Cutaneous T-cell lymphoma
CTLA-4	Cytotoxic T-lymphocyte antigen
CTR1	Copper transporter 1 receptor
CVO	Circumventricular organ
dAb	Single domain antibody
DAF	Decay-accelerating factor
DALVB	4-desacetylvinblastine
DART	Developmental and reproductive toxicity
DC	Dendritic cell
DD	Drug delivery
DDI	Drug-drug interactions
DFM	Di-Fab maleimide
DHB	Dihydroxybenzoic acid
DHFR	Dihydrofolate reductase
DLBCL	Diffuse large B-cell lymphoma
DLS	Dynamic light scattering
DLT	Dose limiting toxicity
DM	Drug metabolism
DP	Drug product
DS	Drug substance
DSC	Differential scanning calorimetry
EBA	Expanded bed absorption
EBV	Epstein-Barr virus
EC	European commission
EGF	Epidermal growth factor
EGFR	Epidermal growth factor receptor
EIAV	Equine infectious anemia virus
ELISA	Enzyme-linked immunosorbent assay
EM	Electron microscope
EMCV	Encephalomyocarditis virus

EMEA	European medicines agency
EPO	Erythropoietin
EpoFc	Erythropoietin Fc
ER	Endoplasmic reticulum
ERE	Estrogen-responsive enhancer element
ES	Embryonic stem cell
ESI	Electrospray ionization
ESI-TOF	Electrospray ionization-time-of-flight mass spectrometry
EVAc	Ethylene-vinyl acetate copolymer
Fab	Fragment antigen binding
FACS	Fluorescence activated cell sorting
FAP-α	Fibroblast activation protein-alpha
FBS	Fetal bovine serum
Fc	Fragment-crystallizable
FCR	Fludarabine, cyclophosphamide, and rituximab
FcRn	Neonatal Fc receptor
FcγR	Fc gamma receptor
FDA	Food and drug administration
FDC	Fixed dose combination
FDG-[^{18}F]	18-fluoro-2-deoxy-D-glucose
FFF	Field-flow fractionation
FIM	First in man
FOPP	Follow-on protein product
FPV	Fowlpox virus
FR	Framework region
FT-IR	Fourier transform infrared
FVT1	Follicular lymphoma variant translocation 1
G1F	Monogalactosylated glycoform
GAL	Galactose-inducible promoter
GALT	Gut-associated lymphoid tissues
GAPDH	Glyceraldehyde 3-phosphate dehydrogenase
GAS	Group A streptococcus
GCS	Group C streptococcus
G-CSF	Granulocyte colony-stimulating factor
GFAP	Glial fibrillary acidic protein
GFP	Green fluorescent protein
GGS	Group G streptococcus
GI	Gastrointestinal
GLP	Good laboratory practices
GMP	Good manufacturing practices
GO	Gemtuzumab ozogamicin
GP	Glycoprotein
GPCR	G-protein coupled receptor
GPI	Glycosyl phosphatidyl inositol
GS	Glutamine synthetase
GVHD	Graft-versus-host disease
HA	Hemagglutinin
HACA	Human anti-chimeric antibody
HAHA	Human anti-human antibody

HAIA	Human anti-animal Ig antibody
HAMA	Human anti-mouse antibody
HB-EGF	Heparin-binding epidermal growth factor-like growth factor
HCC	Hepatocellular carcinoma
HCV	Hepatitus C virus
HD	Huntington's disease
HEK	Human embryonic kidney
HER2	Human epidermal growth factor receptor 2
HES	Hypereosinophilic syndrome
HGH	Human growth hormone
HIC	Hydrophobic interaction chromatography
HPAE-PAD	High performance anion exchange chromatography with pulsed amperometric detection
HPLC	High performance liquid chromatography
HPSEC	High performance size exclusion chromatography
HRF	Homologous restriction factor
HRGP	Hydroxyproline-rich glycoprotein
HSC	Human string content
HSG	Histamine-succinyl-glycine
HSP	Heat shock protein
HSV	Herpes simplex virus
HT	High throughput
HTS	High throughput screening
IBD	Inflammatory bowel disease
IBP	Immunoglobulin-binding protein
ICAM-1	Intercellular adhesion molecule 1
ICV	Intracerebroventricular
ID	Infectious disease
IEC	Ion exchange chromatography
IEF	Isoelectric focusing
IFNα	Interferon α
IFNβ	Interferon β
IFNγ	Interferon γ
IFOM	Imaging figure of merit
IG	Immunoglobulin
Ig-Ag	Ig-antigen complex
IgE	Immunoglobulin E
IGF-II	Insulin-like growth factor II
IHC	Immunohistochemistry
IL	Interleukin
IM	Intramuscular
IMAC	Immobilized metal affinity chromatography
IND	Investigational new drug
INN	International nonproprietary name
INP	Ice nucleation protein
IP	Intellectual property
ITAM	Immunoreceptor tyrosine-based activation motif
ITIM	Tyrosine-based inhibitory motif
ITP	Idiopathic thrombocytopenic purpura
IV	Intravenous
IVIG	Intravenous Ig

JIA	Juvenile idiopathic arthritis
K_D	Equilibrium dissociation constant
KinExA	Kinetics exclusion assay
K_{off}	Dissociation constant
LC	Liquid chromatography
LCM	Ligand concentration multiplier
LC-MS	Liquid chromatography-mass spectrometry
LET	Linear energy transfer
LFA	Lymphocyte-function-associated antigen
LRP	Low density lipoprotein receptor
M6P	Mannose 6-phosphate
MAC	Membrane attack complex
MACS	Magnetic-activated cell sorting
MAR	Matrix attachment region
MASP	Mannan-binding lectin associated serine protease
MBL	Mannose-binding lectin
MED	Minimum effective dose
MF	Mycosis fungoides
MHC	Major histocompatibility complex
MLST	Multi-locus sequence typing
MMAE	Monomethyl-auristatin E
MoMLV	Moloney murine leukemia virus
MRI	Magnetic resonance imaging
MRSA	Methicillin-resistant *Staphylococcus aureus*
MS	Multiple sclerosis
MS	Mass spectrometry
MSCRAMM	Microbial surface components recognizing adhesive matrix molecules
MSCV	Mouse stem cell virus
MTD	Maximum tolerated dose
MTX	Methotrexate
MX-DTPA	*p*-isothiocyanatobenzyl-diethylenetriamine-pentaacetic acid
NCE	New chemical entity
NCTCL	Noncutaneous T-cell lymphoma
NDA	New drug application
NGF	Nerve growth factor
NHL	Non-Hodgkin lymphoma
NHS	*N*-hydroxysuccinimide
NIH	National institutes of health
NIR	Near infra-red spectroscopy
NK	Natural killer cell
NME	New molecular entity
NMR	Nuclear magnetic resonance
NRE	Negative-response element
NSCLC	Non-small cell lung cancer
OAS	Oligoadenylate synthetase
OGD	Office of generic drugs
OMP	Outer membrane protein

OmpA	Outer membrane protein A
OmpF	Outer membrane protein F
ORR	Overall response rate
Ov	Ovalbumin
PBMC	Peripheral blood mononuclear cell
PCI	Percutaneous coronary intervention
PCR	Polymerase chain reaction
PD	Pharmacodynamics
PDB	Protein data bank
PDI	Protein disulfide isomerase
PECS	Periplasmic expression with cytometric screening
PEG	Polyethylene glycol
PEI	Polyethylenimines
PelB	Pectate lyase B
PET	Positron emission tomography
Pfu-Pol	Pyrococcus furiosus polymerase
PGC	Primordial germ cell
P-gp	P-glycoprotein
PHA	Polyhydroxyalkanoate
PHS	Public health service
pI	Isoelectric point
PIGA	Phosphatidylinositol glycan-complementation class A
pIgR	Polymeric Ig receptor
PK	Pharmacokinetics
PML	Progressive multifocal leukoencephalophy
PMMA	Polymethyl methacrylate
PNA	Peptide nucleic acid
PNGase F	Peptide-*N*-glycosidase F
PNH	Paroxysmal nocturnal hemoglobinuria
POC	Proof of concept
POS	Probability of success
PPIase	Peptidyl-prolyl *cis-trans* isomerase
PSMA	Prostate specific membrane antigen
PTM	Posttranslational modification
PVDF	Polyvinylidene fluoride
QA	Quality assurance
RA	Rheumatoid arthritis
RabMAb	Rabbit monoclonal antibody
RANKL	Receptor activator of nuclear factor kappa B ligand
RAP	Receptor-associated protein
RCC	Renal carcinoma cell
rEPO	Recombinant erythropoietin
RES	Reticuloendothelial system
RIA	Radioactive ligand immunoassay
RIGS	Radioimmunoguided surgery
RIT	Radioimmunotherapy
RME	Receptor-mediated endocytosis
RNAi	Ribonucleic acid interference

RP-HPLC Reversed-phase high performance liquid chromatography
RRMS Remitting relapsing multiple sclerosis
RSV Respiratory syncytial virus
RT Reverse transcription
RU Resonance units

SA Safety assessment
SACI *S. aureus* strain Cowan I
SAR Structure-activity relationships
SC Subcutaneous
scFv Single chain fragment variable
SCRS Systemic cytokine release syndrome
scTCR Single-chain T-cell receptor
SDR Specificity-determining residue
SDRE Steroid-dependent response element
SDS-PAGE Sodium dodecyl sulfate polyacrylamide gel electrophoresis
SEC Size exclusion chromatography
SEC-HPLC Size-exclusion high-performance liquid chromatography
shRNA Short hairpin RNA
SIGLEC Sialic acid binding Ig-like lectin
siRNA Small interfering RNA
SLE Systemic lupus erythematosus
SOD Superoxide dismutase
SOP Standard operating procedure
SPECT Single-photon emission computed tomography
SPR Surface plasmon resonance
SS Sézary syndrome
SUV Standardized uptake value

TAG Tumor-associated glycoprotein
TAM Tumor-associated macrophage
Taq-Pol Thermus aquaticus polymerase
TCA cycle Tricarboxylic acid cycle
TfR Transferrin receptor
TGEV Transmissible gastroenteritis coronavirus
TGF-α Transforming growth factor-alpha
TH Tyrosine hydroxylase
TLR Toll-like receptor
T*m* Melting temperature
TMA Tissue microarray
TNFR Tumor necrosis factor receptor
TNF-α Tumor necrosis factor-alpha
TOF Time-of-flight
TR T cell receptor
Treg Regulatory T-cell
TRX Thioredoxin
TSH Thyroid-stimulating hormone
TSP Tail specific protease
TTP Time to disease progression
TVCV Turnip vein-clearing virus

UTR Untranslated region

VCAM-1	Vascular cell adhesion molecule-1
VEE virus	Venezuelan equine encephalitis virus
VEGF	Vascular endothelial growth factor
Vmax	Maximum elimination rate
VPS4B	Vacuolar sorting protein 4b
vWF	Von Willebrand factor
WHIP	Werner helicase-interacting protein
WHO	World health organization
WM	Waldenström's macroglobulinemia
WNV	West nile virus
YAC	Yeast artificial chromosome

INDEX

Therapeutic Monoclonal Antibodies: From Bench to Clinic. Edited by Zhiqiang An